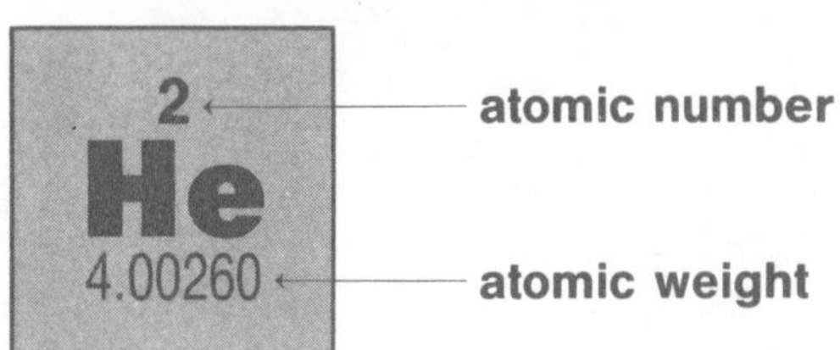

ELEMENTS

8B	9B	10B	11B	12B	3	4	5	6	7	8
										2 **He** 4.00260
					5 **B** 10.81	6 **C** 12.011	7 **N** 14.0067	8 **O** 15.9994	9 **F** 18.998403	10 **Ne** 20.179
					13 **Al** 26.98154	14 **Si** 28.0855	15 **P** 30.97376	16 **S** 32.06	17 **Cl** 35.453	18 **Ar** 39.948
26 **Fe** 55.847	27 **Co** 58.9332	28 **Ni** 58.69	29 **Cu** 63.546	30 **Zn** 65.38	31 **Ga** 69.72	32 **Ge** 72.59	33 **As** 74.9216	34 **Se** 78.96	35 **Br** 79.904	36 **Kr** 83.80
44 **Ru** 101.07	45 **Rh** 102.9055	46 **Pd** 106.42	47 **Ag** 107.8682	48 **Cd** 112.41	49 **In** 114.82	50 **Sn** 118.69	51 **Sb** 121.75	52 **Te** 127.60	53 **I** 126.9045	54 **Xe** 131.29
76 **Os** 190.2	77 **Ir** 192.22	78 **Pt** 195.08	79 **Au** 196.9665	80 **Hg** 200.59	81 **Tl** 204.383	82 **Pb** 207.2	83 **Bi** 208.9804	84 **Po** (209)	85 **At** (210)	86 **Rn** (222)
	109 **Une**									

The heavy line approximately separates the metallic elements (left of the line) from the nonmetallic elements.

61 **Pm** (145)	62 **Sm** 150.36	63 **Eu** 151.96	64 **Gd** 157.25	65 **Tb** 158.9254	66 **Dy** 162.50	67 **Ho** 164.9304	68 **Er** 167.26	69 **Tm** 168.9342	70 **Yb** 173.04	71 **Lu** 174.967
93 **Np** 237.0482	94 **Pu** 244	95 **Am** (243)	96 **Cm** (247)	97 **Bk** (247)	98 **Cf** (251)	99 **Es** (252)	100 **Fm** (257)	101 **Md** (258)	102 **No** (259)	103 **Lr** (260)

GENERAL CHEMISTRY

FIFTH EDITION

GENERAL CHEMISTRY

FIFTH EDITION

Frank Brescia
late of The City College of the City University of New York

John Arents
The City College of the City University of New York

Herbert Meislich
The City College of the City University of New York

Amos Turk
The City College of the City University of New York

Harcourt Brace Jovanovich, Publishers
and its subsidiary, Academic Press
San Diego New York Chicago Austin Washington, D.C.
London Sydney Tokyo Toronto

Cover: Photo by Steve Thompson/© Uniphoto Picture Agency

ISBN: 0-15-529538-1

Library of Congress Catalog Card Number: 86-70498

Printed in the United States of America

The first four editions were entitled *Fundamentals of Chemistry*.

Illustration Credits

Page 6: National Bureau of Standards, United States Department of Commerce; **33:** Cavendish Laboratory, University of Cambridge; **36:** Physics Department, University of Manchester; **62:** Cavendish Laboratory, University of Cambridge; **128:** Courtesy of *PSSC Physics*, D. C. Heath, Lexington, MA; **129:** Courtesy of *PSSC Physics*, D. C. Heath, Lexington, MA; **146:** Science Museum, London; **200:** Courtesy, Ward's Natural Science Establishment, Rochester, NY; **285:** Grant Heilman Photography; **326:** British Crown Copyright, Science Museum, London; **379:** Courtesy, Eastman Kodak Laboratories, Rochester, NY; **387:** Courtesy, P. R. Mallory & Company; **409:** E. J. Pasierb, RCA David Sarnoff Research Center, Princeton, NJ; **463:** Courtesy, Link Belt/FMC; **630:** Museo Nazionale della Scienze e della Techica, Milan; **658:** Courtesy of Markson Science, Inc.; **689:** Science Museum, London; **729:** Bettmann Newsphotos; **876:** Courtesy, Luis Alvarez; **886:** Photographed by P. I. Dee and C. W. Gilbert, *Proceedings of the Royal Society* (1936); **889:** Courtesy, Dr. Rosalyn S. Yalow; **901:** Deutsches Museum, Munich; **907:** Courtesy of the Institute of Human Origins; **976:** *J. Ultrastructure Research 2*, 8 (1958). Courtesy of Dr. R. W. G. Wyckoff; **986:** Courtesy of Harvard Apparatus.

We dedicate this book to Frank Brescia (1908–1987), our beloved and respected friend, colleague, and coauthor. Professor Brescia was an outstanding teacher and scientist who inspired several generations of students and who set an example of professional commitment and inquiry for all of us. He was active in the preparation of this edition almost to the very end of its production, and we deeply regret that he will not see it. It is fitting that it be a memorial to him.

J. A.
H. M.
A. T.

PREFACE

This Fifth Edition of *General Chemistry* is a major revision. We have retained the strengths and the integrity of the four previous editions, but we have changed the text to conform to changes in chemical education. As with previous editions, our objective has been to make the book understandable, attractive, and interesting without sacrificing rigor. In particular, we have developed principles from experimental foundations and have made every effort to motivate the student by presenting material so that each topic leads naturally to the next. At every step we have considered the student's background and the student's needs.

What are those needs? For many students, the need they feel most strongly during their general chemistry course is to do well—to understand the material, to be able to answer the questions, and to get a good grade. Doing well in any task, apart from other goals, is internally satisfying and a powerful motivation for continued excellence. The prime purpose of this book is therefore to help the student in every possible way toward the realization of that objective. But students also realize that chemistry is much more than a college course that ends with a grade, and introductory textbooks generally point out the many ways in which chemistry affects our daily lives. We feel that even more is needed, however. Is it possible for a textbook to offer everything that an introductory course requires and, at the same time, to integrate glimpses of the real world of chemistry and the activities of working chemists? We have aimed to do just that, and so to convey to students the message that chemistry does not end in the classroom and that it contributes to their liberal as well as technical education.

In addition to the rewriting that occurs throughout the book, the most dramatic changes in this Fifth Edition are in three categories: a significant increase in the amount of descriptive chemistry covered, the reorganization of certain topics, and the integration of topics that were once separate chapters. Let us elaborate on these changes.

Descriptive Chemistry. We introduce the descriptive chemistry of the elements fairly early (in Chapter 6). Shortly thereafter, after electronic structure and chemical bonding have been treated, we present a completely new chapter of descriptive chemistry—Chapter 9, "Some Types of Chemical Reactions." Descriptive chemistry cannot be separated from chemical principles; therefore, it appears in all chapters of this text. The main body of descriptive chemistry, in addition to Chapter 9, is incorporated in the last six chapters: "Representative Elements," "Transition Elements," "Metals and Metallurgy," "Nuclear Chemistry," "Organic Chemistry," and "Polymers and Biochemicals." These chapters are last not because they are least, but rather because descriptive chemistry can best be learned, understood, and appreciated when the student has a good grasp of chemical principles.

Chemical Bonding. We have grappled with the organization of the classical ("ball and stick") and the modern ("quantum mechanical") discussions of chemical bonding. In this edition, the treatments are set in sequence early, in Chapters 7 and 8. This way, the concept of atomic orbitals (developed in Chapter 5) is still fresh when it is applied to chemical bonding, and contributing (resonance) structures of compounds can be better understood in terms of delocalized π-bonding. However, the instructor may elect to postpone the material of Chapter 8 until later—for example, after Chapter 19. One of the most significant recent advances in chemistry has been the application of molecular orbital theory to the concepts of orbital symmetry and frontier orbitals. Therefore, we believe that the general chemistry course should provide an introduction to the fundamental ideas of MO theory. Whatever the treatment, we consider it important to emphasize the experimental basis for the various classifications of types of chemical bonding. We have carefully selected examples that are related to the student's previous knowledge and experience.

Thermodynamics. Chapter 4 treats the elementary ideas of thermochemistry: calorimetry, enthalpy of reaction and of formation, Hess's law, and bond energy. However, the first law itself and details like pressure-volume work are left for Chapter 18, most of which deals with the second law. Our approach to chemical thermodynamics conforms to the logic of our text development. We present material related to the second law *after* our coverage of electrochemistry. This is because we believe it is pedagogically valuable for the concrete to precede the abstract. The best chemical examples of second-law concepts—reversibility, free energy, entropy—are found in the study of galvanic cells. These concepts, difficult at best, are quite beyond the understanding of most students unless they are related to actual physical processes, such as the operation of the common automobile battery. Free energy is given priority over entropy because of its closer relationship to familiar measurements and its greater practical importance in chemistry.

Integration of Topics. Certain topics, such as "environmental chemistry" and "radiation and matter," were presented as complete chapters in the previous edition. This subject matter retains its importance in our chemical world, and it has not disappeared from the book. Although we have eliminated those chapters, we have integrated the important aspects of each of them into appropriate sections of this new text.

Polymers. Whether or not polymers should be part of the general chemistry curriculum is an important issue. Professional chemists support it, because polymer chemistry, after all, is what the majority of chemists do. Since the subject involves new concepts, it is interesting and challenging in itself. For these reasons, we chose to make our Chapter 25 (the last one) a more comprehensive introduction to polymer chemistry than is common in other texts. It is a natural extension of the preceding chapter on organic chemistry.

LEARNING AIDS

A number of learning aids will help the student better understand *General Chemistry*, Fifth Edition:

- Chapter introductions set the stage and tell the student what to expect.
- Sections are designed to proceed logically from the concrete and experimental to the abstract and theoretical.
- Numerous worked-out examples in the text clarify important concepts.
- In-text problems immediately follow most of the worked-out examples to provide a quick drill and check of comprehension.
- Important terms are printed in boldface type and defined where they are first used.
- Analogies appear throughout the text to provide students with concrete understanding of abstract concepts.
- Illustrations and tables have been completely redesigned for this edition.
- Boxed essays expand in various ways upon the basic material presented in the text.
- A generous selection of Additional Problems at the end of each chapter ranges from simple drills to thought-provoking problems.
- Answers to numerical problems are given at the back of the book.
- A Self-Test covers the essential material in each chapter. Answers to the Self-Test questions are at the back of the book.
- Chapter summaries restate the important concepts developed in each chapter.
- A glossary of important terms includes terms that appear boldfaced in the text.
- Appendixes provide useful reference tables, reviews of physical concepts, and important mathematical tools.

ANCILLARY MATERIALS

A complete set of helpful supplements supports *General Chemistry*, Fifth Edition:

- *Experiments in General Chemistry*, Fifth Edition, by Eugene Weiner of the University of Denver
- *Instructor's Manual for Experiments in General Chemistry*, also by Eugene Weiner
- *Study Guide to accompany General Chemistry*, Fifth Edition, by Jo Allen Beran of Texas A&I University
- *Instructor's Solutions Manual for General Chemistry*, Fifth Edition, by Brescia et al.
- *Transparencies to accompany General Chemistry*, Fifth Edition, by Brescia et al.
- *Test Bank to accompany General Chemistry*, Fifth Edition, by Brescia et al.

ACKNOWLEDGMENTS

The manuscript for this revised edition was extensively reviewed during its preparation. For their extremely valuable comments and suggestions, we extend our sincere thanks to the reviewers who have helped this book grow successfully through its five editions. It would be impossible to enumerate all of the people who have influenced us over our many years of teaching and writing. Without the help of countless reviewers, colleagues, and especially students, this text could never have been written. We would, however, like to specifically thank reviewers of the Fifth Edition for their helpful advice, encouragement, and dedication to this project. We are especially grateful to the following people: Russell G. Baughman, *Northeast Missouri State University*; William H. Breazeale, *Francis Marion College*; Donald Campbell, *University of Wisconsin—Eau Claire*; Carolyn Collins, *Clark County Community College*; Dan Covey, *Los Angeles Pierce College*; Terry L. Eyrich, *Merced College*; Larry Epstein, *University of Pittsburgh*; James Fanning, *Clemson University*; John H. Forsberg, *St. Louis University*; Izzy Goodman, *Los Angeles Pierce College*; Ray G. Garvey, *North Dakota State University*; Harold R. Hunt, Jr., *Georgia Institute of Technology*; Paul W. W. Hunter, *Michigan State University*; Earl S. Huyser, *University of Kansas*; Loretta Jones, *University of Illinois*; O. Lloyd Jones, *United States Naval Academy*; Terry L. Morris, *Southwest Virginia Community College*; Patricia L. Samuel, *Boston University*; John D. Scott, *University of Montana*; Duane Sell, *William Rainey Harper College*; Donald H. Stedman, *University of Denver*; James L. Stewart, *Cypress College*; Pelham Wilder, Jr., *Duke University*. While we believe we have synthesized the best ideas, responsibility for the final product is nonetheless our own. We welcome suggestions for further improvements.

John Arents
Herbert Meislich
Amos Turk

TO THE STUDENT

Although there are many ways to study a textbook, we have incorporated some special features into *General Chemistry*, Fifth Edition, that can help you. You should take advantage of them. We suggest that you go over each chapter twice—first for a quick orientation, and second for more serious study, as follows:

First Round. Before your instructor's lecture on the chapter material, read the short chapter introduction. It sets the stage and tells you what to expect. Then scan the chapter attentively but quickly, so that you can·anticipate what you will be hearing in the lecture. Look at the worked-out Examples; they will prepare you for the types of problems you will be assigned. You will find it a great help to do all this *before* you attend your instructor's lecture—the lecture will then be much clearer to you.

Second Round. Now you are ready for a serious study of the chapter. As you read the chapter text again, pay special attention to the Examples. After most Examples there is at least one Problem for you to do on your own. It is closely related to the preceding Example so that you can test your understanding of what you have just read. If you have difficulty, return to the Example and go over the solution carefully, making sure you understand each step.

When you have finished studying the chapter, do the Additional Problems at the end of the chapter that your instructor has assigned. Check your answers with the answers at the back of the book. At some point in the process—whenever you feel ready—do the Self-Test at the end of the chapter, check your answers against the answers at the back of the book, and determine your score.

CONTENTS

GENERAL CHEMISTRY

FIFTH EDITION

1

SOME FUNDAMENTAL TOOLS OF CHEMISTRY

This chapter introduces the methods and language of chemistry. Chemical units, significant figures, conversion factors, and the distinction between precision and accuracy are explained. Finally, the meaning of chemical purity and the importance of pure substances to the study of chemistry are discussed.

Refer to the appendixes for reviews of some physical concepts and mathematical procedures, explanations of units of measurement, and tables of data.

1.1 SCIENTIFIC METHOD

Many of the great architectural works of the ancient world still exist—the pyramids in Egypt, the Great Wall of China, the Greek temples, the Roman aqueducts—and pottery, jewelry, and other artifacts of ancient peoples may be seen in museums. Yet the selection of materials available to ancient builders and artisans was extremely limited. Only six metals are mentioned in the Bible: gold, silver, copper, iron, lead, and tin. Copper and tin were melted together to make bronze, after which an entire age of ancient history has been named. Mercury, too, was known, but not other metals—not aluminum, not tungsten, not chromium, not platinum. Most other materials that are familiar to us today were also unknown to the ancients; for example, there were no disinfectants to sterilize drinking water, no anesthetics, few medicinals.

The abundance of materials now available to us has come about mainly through a type of human activity we refer to as the "scientific method." In this book we are concerned with **chemistry**, the science that deals with the properties and transformations of materials.

People are interested in chemistry for two reasons. The first reason, as old as the human species itself, is that chemical changes are part of life and of the living environment. For example, growth, which is the transformation of nutrients such as sugar, starch, fats, and protein into body tissue, is a chemical change. Civilization began when humans learned to control chemical changes in a systematic way, and this began with agriculture. Other crafts developed in ancient times included metallurgy, cheese making, leather tanning, pottery making, brewing, dyeing, and the manufacture of soap and glass. The second reason people are interested in chemistry is related to basic human curiosity, the desire to understand, which flowered early among ancient philosophers in both the West and the East.

What, then, is the modern scientific method, and how do chemists use it? If you were a chemist starting to work in a laboratory, you would not be "starting from scratch" in your pursuit of scientific knowledge. Even Sir Isaac Newton, not known for modesty, acknowledged his debt to others: "If I have seen farther, it is by standing on the shoulders of giants." Therefore, do not ask whether scientists begin with facts or with theories; they start with an ample supply of both. If you as a chemist had a specific objective, such as synthesizing a new compound, you would start with an idea that you thought would be successful. Such an idea is called a **hypothesis**. This idea would not be just a wild guess; most wild guesses are wrong. The hypothesis would not conflict with generally accepted chemical concepts. For example, you would not hypothesize that you can synthesize your new compound out of nothing, or with insufficient starting material. Your hypothesis would be consistent with a set of ideas that chemists accept. If, after effort and practice, you concluded that your attempted synthesis was unsuccessful, you would abandon or modify your hypothesis and start anew on the basis of another idea, founded in part on what you learned from your failure.

Imagine now that your hypothesis is fruitful and your experiment works, and furthermore, that the hypothesis suggests a large variety of other experiments and leads to correct predictions of their outcome. Your idea would then become more precious to you, and your desire to hold onto it, stronger. Its name, too, would be elevated; you would begin to call it a **theory**. Another possibility is that you might discover something that is a fully reliable predictor of how matter will behave under given conditions and that can be expressed in a concise statement or a mathematical equation. Such an expression is called a **law**. Scientific laws do not necessarily originate from a basic understanding of the phenomenon in question. A noteworthy example is Newton's law of universal gravitation, which came about from a unification of many empirical facts, not from a knowledge of why bodies attract each other.

Hypothesis, *theory*, and *law* thus convey different implications about guesswork, understanding, and certainty, but their usage is not always consistent. In unscientific language, "theory" sometimes means an unproven or even wild guess, while in science it refers to a more fundamental, solidly based idea, or even to an entire conceptual system for explaining a large body of facts. In many instances, concepts are referred to as hypotheses, theories, or laws for historical reasons. Thus we refer to Avogadro's "hypothesis" about molecules in a gas (Chapter 10), which no one doubts any longer, to Dalton's atomic "theory" (Chapter 2), which was in fact strongly doubted for many years after it was proposed, and to Faraday's "laws" of electrochemistry (Chapter 18), which are indeed very accurate and can be expressed by mathematical equations.

It is, of course, correct to say that any scientific concept, by whatever name, must stand the test of being consistent with the experimental facts. If not, the facts remain and the hypothesis or theory must be modified or rejected. Alas, the reality of life in the laboratory is not so simple. For example, if you mixed some chemicals together in a closed container and allowed them to react and then found that the final weight of the products was greater than the weight of the starting materials, would you conclude that matter had been created out of nothing or that you had made an experimental error? If you made the latter choice, you would be favoring theory over the "facts." Of course, if all scientists always behaved that way, there would be no progress. Nonetheless, if all well-accepted theories were continually being challenged, the demands on scientists' time would be overwhelming.

What is a scientist to do then? Unfortunately, there is no definite rule. Perhaps an example will be helpful. By the last decades of the nineteenth century, chemists assumed that the investigations into the composition of the Earth's atmosphere were complete, that there was nothing new to be found in the air. After all, it had been studied by scientific methods since the seventeenth century. An interesting experiment conducted by Henry Cavendish in 1785, however, had been ignored and then forgotten. Cavendish had attempted to remove all the gases, one by one, from a sample of air until nothing was left. He used such methods as electric sparks to make atmospheric nitrogen and oxygen combine with each other to form acids that could be washed out with appropriate chemicals. After he did all he could, he found that about 1/120 of his original sample remained. His conclusion was, in effect, a simple, conservative, and perhaps even timid statement of his results. Cavendish wrote, ". . . if there is any part of the . . . air . . . which differs from the rest, we may safely conclude that it is not more than 1/120 part of the whole." No one paid any attention.

In 1882, Lord Rayleigh (born Robert John Strutt) started a new study of the atmosphere for the purpose of making accurate measurements of the densities of its gases. He found, to his puzzlement, that the nitrogen isolated from air was slightly denser than the nitrogen prepared by chemical methods from a nitrogen-containing substance such as ammonia. The discrepancy was about five parts in a thousand. When Rayleigh and his collaborator William Ramsay tried to remove all the nitrogen from a sample of air by chemical methods, they obtained an unreacted residue amounting to 1/80 of the original volume. (Recall Cavendish's "1/120 part of the whole.") This time the results were not ignored; Rayleigh and Ramsay showed that they had discovered a new element, argon. Further study of the atmosphere revealed the presence of a previously unknown group of gaseous elements chemically related to argon: helium, neon, krypton, and xenon.

We must not conclude from this story that Rayleigh and Ramsay were more clever or more "scientific" than Cavendish. By the end of the nineteenth century the accuracy of measurements had advanced to the point where a difference in gas volume of one part in a hundred or a discrepancy in density of five parts in a thousand was no longer assumed to be an experimental error. What's more, Rayleigh and Ramsay had a new instrument, the spectroscope, that could identify the presence of a new element. Thus when Ramsay saw something new in the spectrum of argon, he "was puzzled, but began to smell a rat," as he wrote later. The "rat" turned out to be helium.

Is there a lesson here about the scientific method? Perhaps all one should say, once again, is that progress in concepts and in experimental methods go hand in hand; it is idle to speculate about which precedes and which follows. The scientist must always exercise judgment in deciding which unexpected results are worth pursuing and which are likely to be fruitless. Good judgments and careful experimentation in such matters are the keys to success. In addition, the scientist must have a keen awareness of the reliability and the limits of measurements. It is to these matters that we turn our attention in the following sections.

1.2 THE INTERNATIONAL SYSTEM OF UNITS

For information to be valuable, it must be possible to preserve it and, when necessary, to retrieve it. In this way an interesting experiment or a useful process can be duplicated at different times and in different places. Such information must be expressed in **quantitative**, or measured, terms. The expressions of measurement must convey the same information to everyone. This requirement is satisfied by the establishment of uniform standards of measurement.

Charles Augustin Coulomb (1736–1806) is noted mainly for his studies on electrical attraction and repulsion.

Introduced as an innovation of the French Revolution in 1799, the metric system was established by an international treaty at the Metric Convention in Paris in 1875. Gradually it was adopted by most of the world—first in the scientific community, then in commerce and everyday life, and last by the English-speaking countries, where its acceptance is still incomplete. The present-day descendant of the metric system is called the International System of Units (Système International d'Unités). Its international abbreviation is SI.

The SI contains base units, derived units, and prefixes. The seven base units are carefully defined and are regarded as being independent of each other. They are the meter, the kilogram, the second, the ampere, the kelvin, the mole, and the candela.

Each of the derived units is formed from one or more base units by arithmetical relations such as multiplication or division. Thus, meter per second is the derived unit for speed. Some derived units have special names, most of which honor scientists that have made significant contributions. For example, the derived unit for electric charge is the ampere-second and its name is the coulomb. André Ampère and Charles Coulomb are honored in these cases.

It would be quite awkward to have nothing but SI base units and derived units for expressing all quantities. Many numbers would be very large, such as those for astronomical distances, and others would be very small, such as those for the masses of dust particles. The SI eases the problem by providing prefixes for denoting multiples and fractions of the SI base units and derived units.

Finally, the base units, derived units, and prefixes can be expressed by symbols. This allows all SI quantities to be written in an internationally recognized shorthand.

USING THE SI BASE UNITS AND PREFIXES

The base units and their symbols are shown in Table 1.1. The SI rules specify that the symbols are not to be followed by periods or changed in the plural. Thus we write, "The river is 10 m wide" (not "10 m. wide" or "10 ms wide"). The SI symbol for second, s, is often replaced by the unofficial sec, since an expression like 10 s is easily misread as "tens." This book uses sec for second.

The SI prefixes are shown in Table 1.2. Those from *kilo-*, 10^3, to *pico-*, 10^{-12}, are most commonly used by chemists. Conversions involving SI prefixes require the use of exponential notation, both positive and negative. You will use such notation throughout your studies in science, so you must be able to use it well. If you need refreshing, refer to Appendix B.

TABLE 1.1
BASE UNITS OF THE INTERNATIONAL SYSTEM

QUANTITY	UNIT	SYMBOL
Length	meter	m
Mass	kilogram	kg
Time	second	s
Temperature	kelvin	K
Electric current	ampere	A
Amount of substance	mole	mol
Luminous intensity	candela†	cd

† The candela is not used in this book; it is included here only to complete the set.

TABLE 1.2
SI PREFIXES

MULTIPLE OR FRACTION	PREFIX	SYMBOL
10^{12}	*tera-*	T
10^{9}	*giga-*	G
10^{6}	*mega-*	M
10^{3}	*kilo-*	k
10^{-1}	*deci-*	d
10^{-2}	*centi-*	c
10^{-3}	*milli-*	m
10^{-6}	*micro-*	μ†
10^{-9}	*nano-*	n
10^{-12}	*pico-*	p
10^{-15}	*femto-*	f
10^{-18}	*atto-*	a

† μ is the Greek letter *mu*.

EXAMPLE 1.1 (a) What is the symbol for picogram? For milligram? (b) What is the value of each of these units in grams? (c) How many picograms are there in a milligram? (Do this one "in your head," without pencil, paper, or calculator.)

ANSWER

(a) The symbol for *pico-* is p and for gram is g, so a picogram is pg. Similarly, milligram is mg.

(b) *Pico-* is the prefix for 10^{-12}, so a pg is 10^{-12} g. Similarly, *milli-* is 10^{-3}, so a mg is 10^{-3} g.

(c) A picogram is much smaller than a milligram, so there will be many picograms in a milligram, just as there are many seconds in a year. The ratio of mg to pg is $10^{-3}/10^{-12}$, or 10^{9}, so there are 10^{9} or one billion picograms in a milligram. ■

A more general method for such calculations will be described later in this chapter, in section 1.5. Also, see Appendix B.3 for more hints on mental arithmetic.

PROBLEM 1.1 How many (a) micrometers are in a centimeter? (b) Nanograms in a kilogram? □

The base units for length, mass, time, and temperature are discussed in the following paragraphs.

LENGTH

The SI base unit for linear measurement is the **meter**, m. The standard is based on the speed of light. The official definition of the meter is "the distance traveled by light in a vacuum during 1/299,792,458th of a second." Don't memorize this. A kitchen counter or laboratory benchtop is almost a meter high; an average kindergartener is about one meter tall.

MASS

The SI base unit of mass is the **kilogram**, kg. The mass of a body is the quantity of matter it contains and it is the same everywhere—on Earth, on the Moon, and

FIGURE 1.1 Prototype Kilogram Number 20, the national standard of mass for the United States. It is a platinum-iridium cylinder 39 millimeters in diameter and 39 millimeters high.

in outer space. The mass of our 1-meter-tall kindergarten child is about 20 kg. The **weight** of a body, however, is the force of gravity exerted on it; therefore, weight is not an invariant property of a body. A body weighs less on Mt. Everest, where gravity is less, than it does at sea level. It weighs still less on the Moon and is weightless in outer space, but its mass is the same everywhere. Two objects that have the same weight at the same location on Earth (or on any other planet) must also have the same mass. An instrument that matches an unknown weight against standard weights is called a **balance**, and the matching procedure is called **weighing**. It is for this reason that ''weight'' is often used informally to mean the same as mass. Strictly speaking, such usage is incorrect, but in common practice it is usually harmless.

The SI standard of mass is a piece of platinum-iridium alloy known as the Prototype Kilogram Number 1 that is kept at the International Bureau of Weights and Measures in Sèvres, France (see Figure 1.1).

TIME

The SI base unit of time is the second, which was originally defined as 1/86,400 of the mean solar day. It is now defined as the duration of 92,192,631,770 cycles of a specific radiation in the emission spectrum of the cesium-133 atom—another number you need not memorize.

TEMPERATURE

The difference between heat and temperature was confusing to scientists for many years. Heat will be discussed in various later sections of this book. For now it is sufficient for you to recognize that heat is a form of energy and that it can be transferred from one body to another. Heat always flows spontaneously from a body at a higher temperature to one at a lower temperature, never the other way. Temperature, on the other hand, is the property of a body that determines the direction of heat flow. The higher a body's temperature, the greater is the tendency for heat to flow away from it.

Many properties of substances change with changes of temperature. Two examples are color and electrical conductivity. A set of values in which temperature is related to some measured property is called a **temperature scale**. Most liquids expand as their temperatures increase. The temperature scale that makes use of this phenomenon is the familiar liquid-in-glass thermometer (Figure 1.2).

The triple point of water is discussed in Chapter 10.

The SI unit of temperature is the **kelvin**, K. There is only one temperature at which liquid water, ice, and water vapor can exist together in the absence of air or any other substance. This temperature, called the **triple point of water**, serves as the SI reference point, and it is assigned the exact value of 273.16 K. The lowest temperature on the kelvin scale is 0 K. No substance can be cooled below 0 K; therefore, this temperature is called **absolute zero**. The kelvin is then defined as exactly 1/273.16 of the triple point of water, which is the same as 1/273.16 of the difference between absolute zero and the triple point of water.

After Anders Celsius (1701–1744), who proposed the scale in 1742.

The SI also accepts **Celsius** (formerly called centigrade) temperature, expressed in degrees Celsius, °C. On this scale the triple point of water is 0.01°C. The ''degree Celsius'' has the same magnitude as the kelvin. Let

t = Celsius temperature, °C

T = Kelvin temperature, K

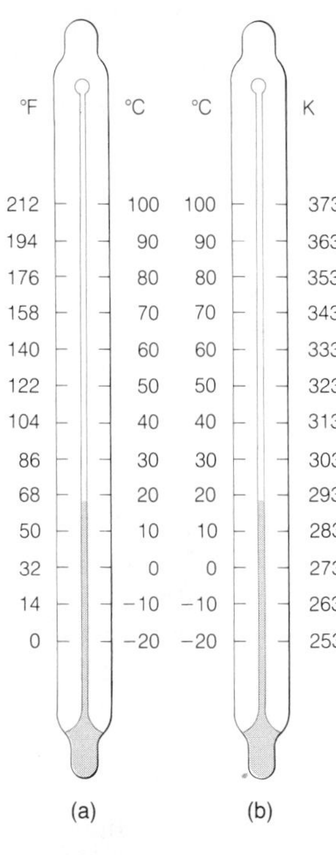

FIGURE 1.2
(a) °F, °C thermometer.
(b) °C, K thermometer.

The difference between the temperatures expressed on these two scales is 273.16 − 0.01, or 273.15. Therefore, we may write $T - t = 273.15$ and

$$\mathbf{T = t + 273.15} \tag{1}$$

In words, kelvin temperature is numerically 273.15 degrees *higher* than the equivalent Celsius temperature. When the highest precision is not needed, this number is approximated as 273. Remember it.

A difference between two temperatures may be expressed either in kelvins or in degrees Celsius. Thus when the temperature rises by 10°C, it also rises by 10 K.

At this point you may be wondering what happened to those old familiar landmarks, 0°C for the freezing point of water and 100°C for the boiling point. The answer is that, although they are not so accurately reproducible as the triple point of water, they still are widely used as practical reference points for calibrating thermometers. Some of the more important practical reference temperatures are:

REFERENCE (At standard atmospheric pressure)	TEMPERATURE, °C
Boiling point of oxygen	−182.97
Freezing point of mercury	−38.87
Ice point (melting point of ice)	0.00
Triple point of water	0.01
Steam point (boiling point of water)	100.00
Boiling point of sulfur	444.60
Freezing point of silver	960.8

The **Fahrenheit** scale is still used in commerce and industry in the United States. It designates the ice point as 32°F and the steam point as 212°F. Its relationship to the Celsius scale is derived from the fact that the temperature difference between steam and ice, 100°C − 0°C, is equal to 212°F − 32°F on the Fahrenheit scale. The ratio is therefore 100°C/180°F, which reduces to 5°C/9°F. The 32° difference in the ice points must also be taken into account. The relationship is

$$\mathbf{t_C = 5/9\,(t_F - 32)} \tag{2}$$

EXAMPLE 1.2 Express (a) room temperature, 20°C, in kelvins and in °F; (b) normal body temperature, 98.6°F, in °C and in kelvins.

ANSWER

(a) From equation 1,

$$T = t + 273 \text{ K}$$
$$= 20 + 273 \text{ K} = 293 \text{ K}$$

Solving equation 2 for t_F gives:

$$t_F = (9/5)20 + 32 = 68°\text{F}$$

(b) From equation 2,

$$t_C = (5/9)(98.6 - 32) = 37.0°\text{C}$$

and from equation 1,

$$T = 37.0 + 273.15 = 310.2 \text{ K}$$ ■

PROBLEM 1.2 Express (a) the freezing point of mercury (−38.87°C) in °F and in kelvins; (b) 500°F in °C and in kelvins. □

SI DERIVED UNITS

Various derived units, such as those that express pressure, energy, and concentration, will be taken up in those sections of this book where they are needed. Four other derived units—frequency, area, volume, and density—will be discussed here.

FREQUENCY Think of frequency as how often (how frequently) something happens. Anything. How often your clock ticks, how often waves break against a shore, how often geese fly south. These are events that repeat themselves; they are said to occur in cycles. Frequency is therefore expressed as repetitions, or cycles, of something per unit time. Since the SI unit of time is the second, the unit of frequency is simply its reciprocal, 1/sec, or sec^{-1}. The name of this unit is the hertz, the symbol for which is Hz.

Named after Heinrich Hertz (1857–1894), who detected radio waves produced by an alternating electric current.

AREA Area is derived by multiplying length times length, and its unit is therefore the square meter (m^2). Of course, there is no rule against using units such as square centimeter, cm^2, or square kilometer, km^2.

VOLUME Volume is expressed by the cubic meter (m^3). The SI also recognizes the **liter**, L, which is 10^{-3} m^3, or 1 dm^3. The liter was formerly defined as the volume of 1 kg of pure water at 3.98°C and at normal atmospheric pressure at sea level. Although this relationship is quite accurate, it is no longer exact (section 1.6). Because the liter is accepted for use with the SI, so are its multiples and fractions, of which the most common is the milliliter, mL.

DENSITY Density is mass divided by volume:

Density = mass/volume (3)

Volume is expressed in cubic meters. The derived SI unit of density is therefore kg/m^3. However, hardly anyone uses this unit. Since everyone is very familiar with water, it is a more convenient standard for density. A cubic meter of water weighs just about 1000 kg (sometimes called a metric ton), so the density of water is 1000 kg/m^3. Chemists prefer an expression in which the density of water is 1 unit, not 1000 units. The mass of ice-cold water needed to fill a 1.000-liter container is 1.000 kg. Therefore, if kg/L is selected as the derived unit of density, then the density of water is 1.000 kg/L. This value can also be expressed as 1.000 g/mL or 1.000 g/cm^3. Of course, all of these values are the same: 1000 kg/m^3 = 1.000 kg/L = 1.000 g/mL = 1.000 g/cm^3. The most common units for solids and liquids are g/mL and g/cm^3. For gases, g/L is most common.

Think of density only in terms of mass and volume. Density is not determined by hardness. (Lead is much denser than diamond, though lead is soft and diamond is hard.) Nor is it determined by physical state. (Mercury, a liquid, is denser than most solids. A gas, too, can be very dense if it is compressed enough.) Just remember, mass over volume.

1.3 UNITS OUTSIDE THE SI

The SI has perfect internal consistency. The Director of the International Bureau of Weights and Measures has characterized the SI as "the universal language of scientists," and has stated that it is also becoming "the universal language of measurement in all industrial, technical, and commercial areas; in all branches of human activity; and in all countries." Perhaps this will be the case sometime in the future, but it is not at present. Nor is it true that all scientists everywhere have accepted the SI in its entirety. Some have objected that the demands for consistency have led to the imposition of awkward, strange, and inconvenient units and to the abandonment of older units that are handy, familiar, and cherished.

Even the SI itself accepts some units that are different from its official ones because they are so widely used. Other units are accepted only "temporarily." Some units, however, are not given even temporary acceptance; the International Bureau prefers to avoid them. Hour, day, and liter are examples of accepted units; the angstrom (a measure of length) is temporarily accepted; and the calorie is one that is to be avoided. Not everyone listens to the International Bureau, however. In fact, the older **British system** of units is still in use in many areas of commerce and industry in English-speaking countries, especially in the United States, and the transition to the SI is slow and erratic. For this reason, it is convenient to have a table of conversion factors for units of different kinds. You will find such a table inside the back cover of this book.

1.4 SIGNIFICANT FIGURES

Suppose you weigh an object four times on a balance that provides readings to the hundredth of a gram and you get these values:

5.14 g
5.13 g
5.12 g
5.13 g

You assume that the mass is actually constant; why, then, did you get different values? Perhaps the balance was influenced by uneven air currents, or maybe you didn't eye the scale the same way each time. Those variations could indeed change the reading over a range of one or two hundredths of a gram, so it would be reasonable to state that the mass of the object is between 5.12 g and 5.14 g, or 5.13 ± 0.01 g. More commonly, the value would be written simply as 5.13 g, with the understanding that there may be some uncertainty in the information provided by the last figure, the 3.

Now suppose that the same object is weighed on a more sensitive balance, with the following results:

5.12904 g
5.12903 g
5.12904 g

The weight would then be expressed as 5.12904 ± 0.00001 g, or simply as 5.12904 g, again with the understanding that there may be some uncertainty in the last figure. With the more sensitive balance you would be sure of the 5.1290, but the last place would still be uncertain.

In the value of 5.13 g there are three digits that provide significant information; in the value 5.12904 g there are six such digits. A **significant figure**, therefore, is defined as a digit that is believed to be correct or nearly so (Figure 1.3). The value 5.13 g has three significant figures, and 5.12904 g has six. Note that even

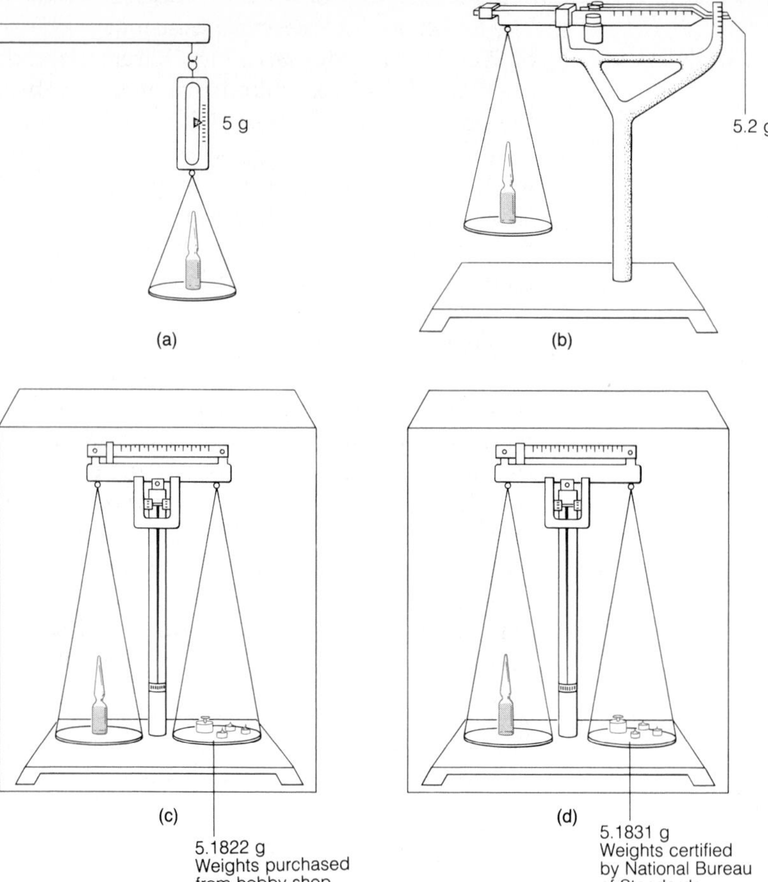

FIGURE 1.3
The same object is weighed on (a) a spring scale—one significant figure, (b) a trip scale—two significant figures, (c) a two-arm balance—five significant figures, and (d) a two-arm balance—five significant figures. The value in (d) is more accurate than that in (c).

though the last number is uncertain, it is nonetheless counted as a significant figure since its uncertainty is limited, usually to ± 1 or 2.

Decimal points have nothing to do with significant figures. If the value of 5.12904 g is expressed in mg, it becomes 5129.04 mg, which still has six significant figures.

The concept of significant figures does not apply to all kinds of numbers; it applies only to numbers that express measurements or computations derived from measurements. Five types of numbers are listed here; significant figures generally apply to only the last two types.

1. *Counted numbers,* or tallies, are exact. A fly has exactly six legs, not approximately six legs. Significant figures do not apply.

2. *Defined or designated numbers* are not measurements. There are 100 centimeters in a meter, exactly. Many relationships between units of different systems of measurement are also defined exactly, such as 1 inch equals exactly 2.54 cm. Other conversions that may not be exact have been measured so accurately that they afford more significant figures than are usually needed.

3. *Mathematical numbers* such as π (3.14 . . .) or e (2.71 . . .), the base of natural logarithms, do not impose any limit on the number of significant figures in a calculation.

BOX 1.1

DENSITY AND SPECIFIC GRAVITY

A traditional method of expressing the density of a substance uses a reference substance as a standard. The ratio between the two densities is called **specific gravity**.

$$\text{specific gravity} = \frac{\text{density of a given substance}}{\text{density of a standard substance}}$$

The temperatures of both substances must be specified; they are frequently, but not always, the same. For solids and liquids the standard substance is water; for gases it is usually air, and other times hydrogen.

Thus, the specific gravity value for carbon tetrachloride, $1.594^{20°/4°}$, expresses that the density of this liquid at 20°C is 1.594 times the density of water at 4°C. Since the density of water at 4°C is practically 1 g/cm^3, the density of carbon tetrachloride at 20°C is 1.594 g/cm^3. Thus, the specific gravity $^{x°/4°}$ is numerically equal to the density in g/cm^3 (or g/mL, which is the same) at x°C. In such case, the only difference between density and specific gravity is that density has units and specific gravity, being a ratio, does not.

4. *Measured numbers* do involve significant figures.
5. *Computed numbers* derived from measurements also involve significant figures.

The following rules govern the use of significant figures:

RULE 1 To count the significant figures in a measured number, read the number from left to right starting with *the first digit that is not zero.* Count that first digit and all the digits that follow, including all the later zeros. The position of the decimal point, if any, is irrelevant. Thus, the value 0.10 mg has two significant figures. If the same mass is expressed as 0.00010 g, it still has two significant figures.

There are instances, however, where this rule is not followed and common sense will allow you to recognize such exceptions. For example, if you read a statement that Mt. Everest is "about 9000 meters high," you would probably not conclude that the statement means 9000 $\pm$ 1 m, implying four significant figures. First, you may doubt that the measurement could be that precise; second, you may think it is unlikely that the height would be a whole number of thousands of meters; and third, the word *about* denotes an approximation. The three zeros are in fact spacers, or place-holders, to define the magnitude of the first digit. The ambiguity could be avoided by using exponential notation, such as 9×10^3 m, which shows only one significant figure, or 9.0×10^3 m, which shows two significant figures, as explained in rule 2 on the next page.

EXAMPLE 1.3 How many significant figures are there in each of the following measured quantities? (a) 0.00406 nm; (b) 31.020 L; (c) 0.020 sec; (d) 6.00×10^8 kg; (e) 50,000,000 years.

ANSWER (a) 3; (b) 5; (c) 2; (d) 3; (e) ambiguous, but probably only 1. Note that if a fossil is "about 50 million" years old, then a thousand years later it will still be "about 50 million" years old. ■

PROBLEM 1.3 What type of number is expressed by each of the following values? How many significant figures are there in values to which the concept applies? (a) There are 26 letters in the English alphabet. (b) There are 12 inches in 1 foot. (c) The density of mercury at room temperature is 13.546 g/mL. □

Exponential notation is reviewed in Appendix B.1.

RULE 2 If the number of digits needed to express the magnitude of a measurement exceeds the number of significant figures in the measured number, exponential notation should be used. For example, assume that a length is measured to be 5.2 meters, a value containing two significant figures. How would that length be expressed in mm? Multiplying 5.2 mm by 1000 gives 5200 mm, but that appears to have four significant figures, which is too many. The answer should therefore be expressed as 5.2×10^3 mm, which expresses the correct magnitude and retains the number of significant figures.

RULE 3 In addition and subtraction, the value with the *fewest decimal places* determines how many significant figures are used in the answer:

	308.7810 g	(4 decimal places; 7 significant figures)
	0.00034 g	(5 decimal places; 2 significant figures)
	10.31 g	(2 decimal places, the fewest; 4 significant figures)
Sum:	319.09 g	(2 decimal places; 5 significant figures)

RULE 4 In multiplication and division, the value with the *fewest significant figures* determines the number of significant figures the answer should have:

$$\frac{3.0 \times 4297}{0.0721} = 1.8 \times 10^5$$

Note that the number with the fewest, two, significant figures is 3.0; thus the answer must also have two significant figures.

RULE 5 When a number is "rounded off" (nonsignificant figures discarded), the last significant figure is increased by 1 if the discarded figure is 5 or greater, and is unchanged if the discarded figure is less than 5:

4.6349, rounded off to four significant figures ⟶ 4.635

4.6349, rounded off to three significant figures ⟶ 4.63

2.815, rounded off to three significant figures ⟶ 2.82

OTHER RULES Some special rules that apply to significant figures in logarithms and antilogarithms are found in Appendix B.1.

EXAMPLE 1.4 Evaluate the expression

$$V = 4.4 \times \frac{311.8}{273.1} \times \frac{760}{784 - 3.1}\ \text{L}$$

ANSWER First, note the term 784 − 3.1. By rule 3, its value is 781. The entire expression can then be calculated as

$$V = 4.4 \times \frac{311.8}{273.1} \times \frac{760}{781} = 4.8884328\ \text{L}$$

Let your pocket calculator give you all the figures it can during the sequence of multiplications and divisions. You will need to round off only at the end. The 4.4 factor limits the answer to two significant figures, and

$$V = 4.9\ \text{L}$$

■

EXAMPLE 1.5 Round off the number 64,629 to one, two, three, and four significant figures, using exponential notation in each case.

ANSWER

6×10^4	(1 significant figure)
6.5×10^4	(2 significant figures)
6.46×10^4	(3 significant figures)
6.463×10^4	(4 significant figures)

■

PROBLEM 1.4 (a) Evaluate the expression

$$\frac{3.46 \times 10^4 \times 0.003547}{(373 - 75) \times 2}$$

assuming that all of the numbers are measured. (b) Repeat the calculation, assuming that the 2 is an exact number. □

1.5 CONVERSION FACTORS

Often it is necessary to express a particular quantity in a different unit of measurement. Sometimes the task is so simple that it can be done by mental arithmetic. However, when the units are unfamiliar it is easy to make careless errors. This section describes a procedure that should help to ensure correct conversions from one unit to another.

To start with very familiar units, we can ask, "How many seconds are in 1.5 minutes?" You know immediately that the answer is 90 seconds, but let us examine the reasoning process. We start with the statement of equivalence:

$$60 \text{ sec} = 1 \text{ min}$$

Not all conversion factors are exact. The table inside the back cover provides three or four significant figures for the inexact factors, and handbooks of chemistry and physics provide more. For example, there are 28.316847 liters per cubic foot. It is the availability of such information that supports the earlier statement that conversion factors afford more significant figures than are usually needed.

The fraction 60 sec/1 min is called a **conversion factor** because it can be used to "convert" minutes to seconds. How do you decide which factor will give the correct answer to the problem? Trying both will show which factor is right and which is wrong:

$$1.5 \cancel{\text{min}} \times \frac{60 \text{ sec}}{1 \cancel{\text{min}}} = 90 \text{ sec} \qquad \text{(correct answer)}$$

$$1.5 \text{ min} \times \frac{1 \text{ min}}{60 \text{ sec}} = 0.025 \frac{\text{min}^2}{\text{sec}} \qquad \text{(incorrect answer)}$$

The second answer cannot be correct because the units are wrong. *It is impossible for an incorrect use of a conversion factor to give a correct answer.* In general,

$$\textbf{given quantity and unit} \times \textbf{conversion factor} = \textbf{desired quantity and unit}$$

The conversion factor takes the form of a ratio in which *the desired unit is in the numerator and the given unit is in the denominator.* The question of significant figures in the conversion factor 60 sec/1 min does not arise since the numbers are exact. Note also that we cancel out dimensional units just as we ordinarily cancel out numerical or algebraic factors. For this reason this method is known as **dimensional analysis**.

EXAMPLE 1.6 (a) How many mg are in 2.50 g? (b) There are 1.61 km in 1 mile. How many miles are in 7.65 km?

ANSWER[†]

(a) The conversion factor is 10^3 mg/1 g because mg is the desired unit.

$$2.50\,\cancel{g} \times \frac{10^3\text{ mg}}{1\,\cancel{g}} = 2.50 \times 10^3\text{ mg}$$

(b) The conversion factor is 1 mi/1.61 km because mi is the desired unit.

$$7.65\,\cancel{km} \times \frac{1\text{ mi}}{1.61\,\cancel{km}} = 4.75\text{ mi}$$

■

PROBLEM 1.5 Express the United States highway speed limit of 55 mi/h in km/h, in cm/sec, and in m/min. □

Such a yield indicates a profitable mine.

We can also apply such dimensional analysis to chemical conversions. For example, imagine you discover a gold mine that yields 75 mg of gold per kilogram of ore. You may write

$$1\text{ kg ore} \xrightarrow{\text{yields}} 75\text{ mg gold}$$

This expression is not an equation, and the conversion factor 75 mg gold/1 kg ore is not equal to 1, because 1 kg ore does not *equal* 75 mg gold. However, since "yield" is what you are interested in, the factor does "convert" ore into gold numerically. We will therefore use such factors in the same way we use the conversion factors just described.

EXAMPLE 1.7 How many grams of gold can you get from 51 kg of ore in your mine?

ANSWER Here you need two conversion factors, one to "convert" ore into gold and the other to change mg to g. You may use as many conversion factors in sequence as needed for a given calculation.

The conversion factors are 75 mg gold/1 kg ore and 1 g/10^3 mg. Then

$$51\,\cancel{\text{kg ore}} \times \frac{75\,\cancel{\text{mg}}\text{ gold}}{1\,\cancel{\text{kg ore}}} \times \frac{1\text{ g}}{10^3\,\cancel{\text{mg}}} = 3.8\text{ g gold}$$

■

PROBLEM 1.6 The density of copper at room temperature is 8.9 g/cm^3. Express this value in g/mL, in mg/mL, and in kg/m^3. □

1.6 PRECISION AND ACCURACY

The fact that a measured value is expressed by a limited number of significant figures reminds us that measurements are not absolutely accurate; they are subject to error. If a measurement is repeated over and over, one usually finds that the results vary to a limited degree and that an occasional individual result may be "way off." The fact that such variations are the rule makes it important for us to make more than one measurement of any quantity. The value that is reported

[†] Differences of 1 or 2 units in the last significant figure of the answer to a problem, such as 103 g and 104 g, are often obtained by different calculations. Such a discrepancy does not imply that one of the values is "wrong." Instead, the discrepancy may be due to the number of digits in the conversion factors used.

as a "final" answer is obtained by assuming, first, that a very widely deviating (or "way off") result may be discarded because it is probably due to a large error that is not likely to recur. The remaining results are then subjected to one of a number of standard statistical procedures, the simplest of which is described here.

The **precision** of a measurement is an expression of the mutual agreement of repeated determinations; it is a measure of the reproducibility of an experiment. The arithmetical average of the values is usually taken as the "best" value. The simplest measure of precision is the **average deviation**, calculated by first determining the average of the series of measurements. Then the difference of each individual measurement from the average is calculated. These differences, each treated as a *positive* quantity, are the deviations. Finally, the deviations are averaged. The reported value is often given as the average value plus or minus (±) the average deviation, as shown in the following example.

EXAMPLE 1.8 Student A determined the diameter of a steel rod with a micrometer caliper; student B used a plastic ruler (Figure 1.4). The results of four experiments are

STUDENT A	STUDENT B
28.246 mm	27.9 mm
28.244	28.0
28.246	27.8
28.248	28.1

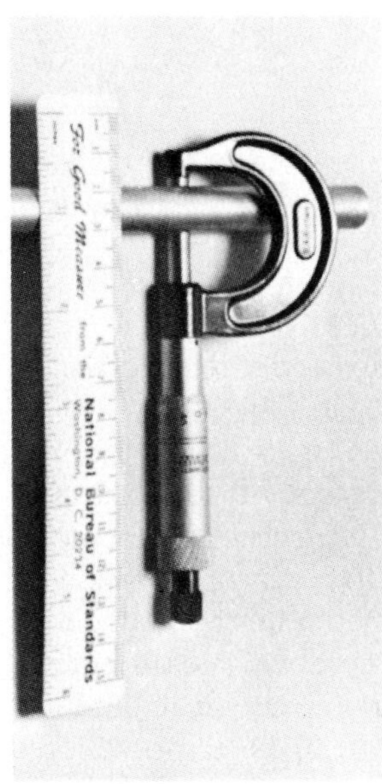

FIGURE 1.4
Steel rod with micrometer caliper and a plastic ruler.

For each series, calculate the average ("best") value and the average deviation.

ANSWER

STUDENT A		STUDENT B	
Measurements	Deviations from the average	Measurements	Deviations from the average
28.246	0.000	27.9	0.1
28.244	0.002	28.0	0.0
28.246	0.000	27.8	0.2
28.248	0.002	28.1	0.1
4 ⌊112.984	4 ⌊0.004	4 ⌊111.8	4 ⌊0.4
28.246	0.001	27.95	0.1
Av. 28.246	Av. dev. 0.001	Av. 28.0	Av. dev. 0.1

Student A would report these results as (28.246 ± 0.001) mm and student B, as (28.0 ± 0.1) mm. ■

Precise measurements, however, are not necessarily accurate. **Accuracy** is the agreement of a measurement with the accepted value of the quantity. Accuracy is expressed in terms of the error, which is the experimentally determined value minus the accepted value.

Who determines what is accepted, however, and does "accepted" mean "true"? In the case of a measurement of mass, the "truth" originates in Sèvres, France, where the Prototype Kilogram Number 1 is kept. The United States Standard is Prototype Kilogram Number 20 (Figure 1.1), and the most careful experiments that science can provide indicate that it is equivalent in mass to the Number 1 Kilogram in France. The accuracy of the U.S. kilogram is therefore certified, or said to be true, or simply "accepted." The U.S. Bureau of Standards certifies masses of other values, such as 100 g, 50 g, 1 g, 100 mg, and so on, to be accurate by comparison with its Kilogram. The balance in your laboratory was certified by its manufacturer against such certified masses of the Bureau of Standards or against another balance that had been so certified. These kinds of procedures and assumptions underlie the concepts of accuracy and error. Thus,

$$\textbf{Error} = \textbf{determined value} - \textbf{accepted value} \tag{4}$$

and

$$\textbf{Percent error} = \frac{\textbf{error}}{\textbf{accepted value}} \times 100\% \tag{5}$$

EXAMPLE 1.9 The steel rod of Example 1.8 has a certified diameter of 28.054 mm. Calculate the error and percent error for each student's determination of diameter. Whose determination is more precise? Whose is more accurate?

ANSWER

STUDENT A		STUDENT B
28.246 ←	determined value	→ 28.0
−28.054 ←	accepted value	→ −28.1
+0.192 ←	error	→ −0.1

Student A's results are more precise, but Student B's results are more accurate.

■

In Examples 1.8 and 1.9 the more precise value has the lower accuracy. Intuitively, you may feel that this case is exceptional—that precise measurements tend to be more accurate. Maybe so, but there is no *logical* reason for such an association. Precision and accuracy are different aspects of measurement and they do not necessarily go hand in hand. A precise but inaccurate set of measurements suffers from a *systematic error*; that is, an error that occurs repeatedly in the same direction with about the same magnitude. (Think about it: if the errors differed widely, the measurements wouldn't be precise.) In these examples, the micrometer caliper may have been defective or improperly calibrated or used.

More accurate thermometers are based on the change in the electrical resistance of platinum wire with changes in temperature.

PROBLEM 1.7 Two students are asked to determine the freezing point of an unknown liquid in a container that is open to the air. They are not told that the liquid is pure water. Student A uses an alcohol-in-glass thermometer. Student B uses a more accurate thermometer. Their results are

STUDENT A	STUDENT B
−0.3°C	273.13 K
0.2°C	273.17 K
0.0°C	273.15 K
−0.3°C	273.19 K

(a) For each student's series, calculate the average value and the average deviation of the individual measurements. (b) Calculate the error for each student's average value. Whose average value is more precise? Whose is more accurate? □

1.7 CLASSIFICATIONS OF MATTER

Every discipline has its own language, and chemistry is no exception. Since chemistry deals with the properties and transformations of materials, chemists describe and classify matter in various ways. Most of this usage will emerge as this book unfolds, but it will be helpful, as a start, to outline some of the language of chemistry.

It is easy to recognize that cutting a wooden chair causes a change in the chair different from the change brought about by burning it, because the properties we associate with wood *as a material*—such as its color, density, and hardness—are changed by burning but not by cutting. Properties that depend on the size of a sample of matter are called **extensive properties**; those that are independent of the size of the sample are **intensive properties**. Mass and volume are examples of extensive properties—they are characteristic of a particular object. Color, hardness, density, melting point, and freezing point are examples of intensive properties—they are characteristic of a particular **substance** or **material**, in other words, a particular *kind* of matter. Handbooks of chemistry, physics, and engineering list the intensive properties of materials. Catalogs and marketing brochures for automobiles, furniture, and computers specify extensive properties of those items in their descriptions.

Extensive properties are also called "accidental properties," and intensive ones are called "specific properties."

Matter may be classified according to its purity, its homogeneity, and its physical state, as well as by many other attributes. Physical states—such as gas, liquid, and solid—are the subjects of Chapters 10 and 11. Here we will introduce the concepts of purity and homogeneity.

PURITY

A material or substance is characterized by its intensive properties. Sugar (table sugar, also called sucrose) is a sweet solid at room temperature; water is a liquid with little taste. Suppose you taste a clear, colorless liquid and find it to be sweet. You know that it cannot be sugar, for it is liquid, and that it cannot be water, since it is not tasteless. Imagine further that you heat the liquid to boiling and the escaping vapor, when cooled, condenses to form water. When the vaporization is complete, sugar remains behind. It would be reasonable to conclude from these results that the original sweet liquid was a **mixture** of sugar and water, because the component substances separated out of the mixture and because some of their intensive properties, such as sweetness, persisted in the mixture.

If you have had even a brief introduction to chemistry or to general science before reading this book, you may consider the preceding discussion to be obvious and unnecessary. Of course sugar water is a mixture, you may know, since it contains water molecules and sugar molecules, which are different. Water consists only of water molecules, and sugar only of sugar molecules, so these are pure substances. These statements are correct, and a pure substance may therefore be defined as one that contains only molecules or atoms of a single kind. If all of the atoms are alike, the substance is an **element**. If there is more than one kind of atom but all of the molecules are alike, the substance is a **compound**. This classification can be pictured:

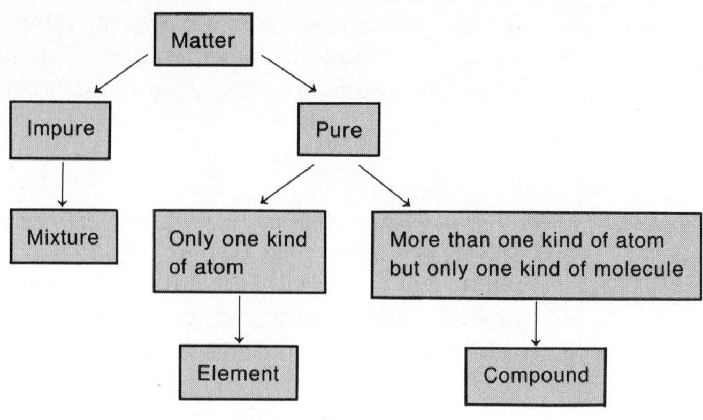

If you don't know the atomic or molecular composition of a substance, however, how do you know whether or not it is pure? The answer to this is not obvious. The question can be resolved only by experiments of the type described in the previous paragraph about sugar water. The intensive properties of an impure substance, or mixture, can change progressively as various processes are applied to it. Sugar water gradually becomes sweeter when water evaporates from it and the remaining sugar becomes more concentrated. Is gasoline a pure substance? Let a gallon of it sit outdoors in an open dish (away from flames) until it has evaporated to half its original volume. If you test the octane rating (resistance to engine knock) of the remaining half-gallon, you will find that it is different from that of the original, unevaporated gallon. From this fact alone, and without knowing anything about gasoline's molecular composition, you can be sure that it is an impure substance.

Gasoline is in fact a highly complex mixture of many different kinds of molecules.

A pure substance has a fixed set of intensive properties. Because of this constancy, pure substances are more suitable than mixtures for experiments that are designed to elucidate chemical principles. The tasks of tracing the pathways of chemical transformations, learning how to synthesize new substances, or finding how to prevent the formation of unwanted ones are difficult enough when one deals with pure substances. Therefore, chemists generally prefer to start with pure materials and apply the principles that are learned from them to mixtures.

HOMOGENEITY

If you divided a liter of gasoline into one hundred 10-mL portions and measured the properties of each portion, you would find that the properties of each portion are identical. The same identities would hold if you divided one of the 10-mL portions into droplets. In fact, all samples, no matter how small, that could be manipulated by laboratory equipment or observed with the most powerful microscope would be found to be identical. Such a uniform substance is said to be **homogeneous**. A substance whose properties differ from sample to sample is **heterogeneous**.

In a sample of matter, any homogeneous portion that has uniform properties and is separated from other parts of the sample by a definite surface or boundary is called a **phase**. Thus, a mixture of ice and water is a two-phase system (Figure 1.5a). If sand and cottonseed oil are added to that mixture, the sand sinks to the bottom and the oil and ice float to the top, making it a four-phase system of ice, water, sand, and oil (Figure 1.5b). A phase need not be all in one piece. Each ice cube shown in Figure 1.5a is not a separate phase.

Homogeneous matter is not necessarily pure. An impure substance, or mixture, can be homogeneous if the molecules or atoms of its various components are

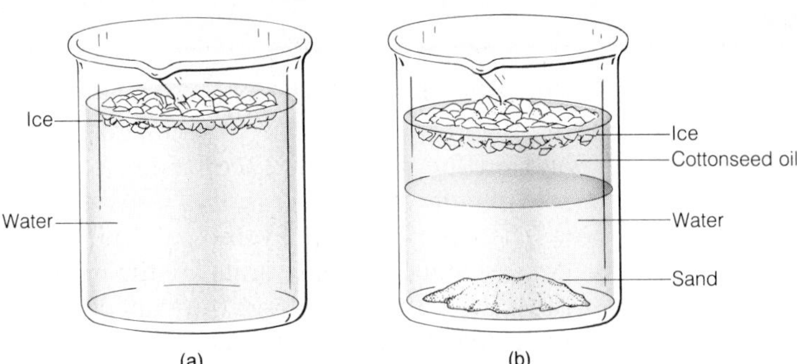

FIGURE 1.5
(a) Ice and water—a two-phase system. (b) Ice, water, sand, and cottonseed oil—a four-phase system.

uniformly mixed. The reason is that the smallest sample of a substance that can be handled or manipulated contains very many atoms or molecules, and if the atoms or molecules are uniformly mixed, all samples will be identical. Such a uniform mixture is called a **solution**. A solution is therefore a single phase. Gasoline is a solution. So are air, vinegar, olive oil, and the carbonated mixture kept under pressure in a bottle or can of soft drink. When solids or gases are dissolved in a liquid, the solid or gas is called the **solute** and the liquid is called the **solvent**. Thus, sugar or ammonia (the solute) dissolves in water (the solvent). When one liquid dissolves in another liquid, the more abundant substance is usually considered to be the solvent, but when the quantities are nearly equal, it doesn't matter which is designated the solvent and which the solute.

Similarly, heterogeneous matter is not necessarily impure. A mixture of ice and water is heterogeneous since it consists of more than one phase, but it is a single substance, because each phase consists only of water molecules. Accordingly, matter can be classified according to its homogeneity by this scheme:

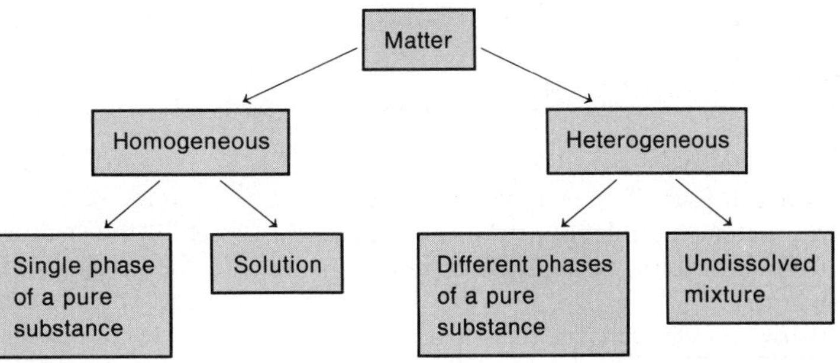

Words are not always used in everyday language as they are defined in books. "Pure water" often means water that contains small natural concentrations of harmless and perhaps tasty mineral matter but which is free from noxious or smelly contaminants. Similarly, "pure air" consists of more than one kind of molecule; it is a mixture of various gases. It is said to be "pure" if it is not polluted with unwanted substances. Is "homogenized" milk homogeneous? The answer is yes if you compare different samples of it teaspoon by teaspoon. But if it is examined by more refined optical methods, it is seen to contain droplets of milk fat dispersed in a "skim-milk" medium; the answer, then, is that it is heterogeneous.

Remember that the meaning of a word is often in its context.

SUMMARY

The scientific method involves a complex interplay between, on the one hand, the findings obtained by experiment or by observations of natural phenomena, and on the other hand, concepts variously termed **hypotheses**, **theories**, and scientific **laws**.

The **International System of Units (SI)** consists of base units, prefixes, and derived units. The base units are the meter, the kilogram, the second, the ampere, the kelvin, the mole, and the candela. All can be expressed by symbols. Some units outside the SI are accepted ''temporarily,'' and others, though not accepted, are nevertheless used in some areas of commerce and industry.

A **significant figure** is a digit that is believed to be correct or nearly so. This concept applies only to numbers that express measurements or computations derived from measurements. The use of significant figures is governed by a set of rules designed to assure that only the information provided by the measurements is preserved in the reported values.

A quantity expressed in one unit can be expressed in another unit by the use of a **conversion factor**.

The **precision** of a measurement is an expression of the mutual agreement of repeated determinations. The **accuracy** expresses the agreement of the measurement with the accepted value. The **error** of a measurement is the difference between the determined value and the accepted value.

Extensive properties of a sample of matter depend on the size of the sample; **intensive** ones do not. A **pure substance** is one that contains only molecules or atoms of a single kind. If all of the atoms in a substance are alike, the substance is an **element**. If there is more than one kind of atom in a substance but all the molecules are alike, the substance is a **compound**. A uniform substance, which has identical properties throughout, is said to be **homogeneous**; a nonuniform one is **heterogeneous**. A homogeneous portion of matter that is separated from other parts of the sample by a surface or boundary is called a **phase**. A uniform mixture of atoms or molecules of different kinds is a **solution**.

1.8 ADDITIONAL PROBLEMS

UNITS OF MEASUREMENT

1.8 Which of the four SI base units discussed in this chapter—the meter, the kilogram, the second, or the kelvin—is defined in terms of an artificial rather than a natural standard?

1.9 How many (a) milligrams are there in a kilogram; (b) microliters in a milliliter; (c) microliters in a cubic centimeter; (d) nanometers in a centimeter; (e) picoseconds in a millisecond?

1.10 An angstrom (Å) is 10^{-10} m. How many angstroms are there in (a) a centimeter; (b) a millimeter; (c) a nanometer?

1.11 A particular medicine dropper delivers 25 drops of water to make 1.0 mL. (a) What is the volume of one drop in cubic centimeters; in microliters? (b) If the drop is spherical, what is its diameter in millimeters? (Volume of a sphere = $4/3\ \pi r^3$.)

DENSITY

1.12 Mercury, poured into a glass of water, sinks to the bottom, while gasoline poured into the same glass floats on the surface of the water. A piece of paraffin, dropped into the mixture, comes to rest between the water and the gasoline, while a piece of iron comes to rest between the water and the mercury. List these five substances in order of increasing density (least dense first).

1.13 A 1.00-carat diamond weighs 200 mg and occupies a volume of 57.0 mm^3 at 20°C. What is the density of diamond in mg/mm^3 at this temperature? Is this the same value as the density in g/cm^3?

1.14 A piece of gold weighing 31.2 g is dropped into 5.00 mL of water. The water level rises until the total volume (water + gold) is 6.62 mL. Calculate the density of gold in g/mL at the temperature of the experiment.

1.15 The density of uranium at room temperature is 18.9 g/cm^3. What is the volume in cm^3 of a 10.0-kg mass of uranium?

1.16 The density of mercury at 20°C is 13.6 g/mL. What is the mass in kg of 750 mL of mercury at this temperature?

TEMPERATURE SCALES

1.17 Which represents a larger temperature interval, (a) a degree Celsius or a degree Fahrenheit; (b) a kelvin or a degree Fahrenheit?

1.18 Express (a) 373°C in K; (b) 10.15 K in °C; (c) −32.0°C in °F; (d) −32.0°F in °C.

1.19 At what temperature are the Celsius and Fahrenheit values identical? (*Hint*: Let $t_C = t_F = x$ in equation 2 and solve for x.)

1.20 On the Réaumur scale, which is no longer used, water freezes at 0°R and boils at 80°R. Derive an equation that relates this to the Celsius scale.

1.21 In 1701 Isaac Newton devised a temperature scale based on the melting point of ice, labeled 0°N, and the temperature of the armpit of a healthy Englishman (98.5°F), labeled 12°N. What is the boiling point of water on Newton's scale?

SIGNIFICANT FIGURES AND CONVERSION FACTORS

1.22 To which of the quantities appearing in the following statements would the concept of significant figures apply? Where it would apply, count the number of significant figures. (a) The density of platinum at 20°C is 21.45 g/cm^3. (b) Wilbur Shaw won the Indianapolis 500 race in 1940 with an average speed of 114.277 mi/h. (c) A child was found to have a mass of 22.0 kg. (d) A mile is defined as 1760 yards. (e) The International Committee for Weights and Measures "accepts that the curie be . . . retained as a unit of radioactivity, with the value 3.7×10^{10} sec^{-1}." (This resolution was passed in 1964.)

1.23 If you ran a mile in 4.00 minutes, what would be your average speed in km/h; in cm/sec?

1.24 In the United States gasoline is priced in units of dollars per U.S. gallon. By what two conversion factors would you multiply this price to convert it to cents per liter to the nearest cent? (One U.S. gallon = 3.785 L.)

1.25 The price of gold on the London market one day was $302.10 per ounce (troy weight). If you had the following quantities of gold on that day, what would the total mass have been worth? Look up the current value and calculate what the same mass is worth today. (One oz (troy) = 31.1035 g.)
(a) 0.62 mg
(b) 5.80 g
(c) 13.33 oz (troy)
(d) 0.04214 oz (troy)

1.26 Water flows through a pipe at a rate of 104 L/min. The cross-sectional area of the pipe is 41.56 cm^2. Calculate the speed of the flowing water in cm/sec and in km/h. The formula is

$$\text{speed}\left(\frac{\text{distance}}{\text{time}}\right) = \frac{\text{flow rate}\left(\frac{\text{volume}}{\text{time}}\right)}{\text{cross-sectional area of pipe}}$$

1.27 An advertisement in a chemical journal by a manufacturer of measuring equipment asks, "Is a discovery hiding behind your last significant figure?" Discuss the implications of this question.

PRECISION AND ACCURACY

1.28 A student is asked to determine the density at 20°C of a cylindrical aluminum bar of accurately known mass. The student measures the volume of the bar by determining the volume of the water it displaces (method I) and by measuring its length and diameter with a micrometer caliper (method II), and then calculates the density (mass/volume), obtaining these results:

METHOD I	METHOD II
2.2 g/mL	2.703 g/mL
2.3	2.701
2.7	2.705
2.4	5.811

The accepted value is 2.702 g/mL. (a) For each method, calculate the average density and the average deviation of the individual values. Should all of the values be included in your calculations? If not, justify any omissions. (b) Calculate the error for each method's average value. Which method's average value is more precise? Which method is more accurate?

CLASSIFICATIONS OF MATTER

1.29 Which of the following properties of a sample of matter are extensive; which are intensive: (a) density; (b) melting point; (c) volume; (d) mass; (e) electrical conductivity?

1.30 Classify each of the following substances as a mixture or a pure substance. (a) A piece of "dry ice," commonly used as a coolant, sublimes (transforms itself directly from solid to gas) at −78.5°C under normal atmospheric pressure. As the dry ice continues to sublime, it maintains a steady −78.5°C temperature. (b) An alcoholic liquid labeled "100 proof neutral spirit" is divided into two equal portions, each of which is placed in an open dish. When a lighted match is applied to the first dish, the liquid burns with a pale blue flame. The liquid in the second dish is allowed to stand undisturbed until much of it has evaporated. The residue cannot be ignited with a match. (c) The label on a bottle of "hydrogen peroxide" includes the statement, "Contains 3% hydrogen peroxide. Inert ingredients 97%." (d) The mercury in a thermometer consists entirely of mercury atoms.

1.31 Classify each of the following samples of matter as (1) a pure, homogeneous substance, (2) a solution, (3) a pure heterogeneous substance, or (4) an impure heterogeneous substance: (a) air; (b) ammonia water (a clear liquid); (c) topsoil; (d) a single crystal of the element sulfur; (e) a mixture of liquid and crystalline sulfur.

SELF-TEST

1.32 (15 points) How many (a) picograms are there in a kilogram; (b) nanoseconds in a microsecond; (c) centimeters in a kilometer; (d) milliliters in a cubic meter?
1.33 (15 points) Express (a) 100°F (hot summer day) in °C; (b) 20°C (room temperature) in °F; (c) −40°C in K.
1.34 (15 points) How many significant figures are there in each of the following measured quantities: (a) 5.0045 g; (b) 0.008700 L; (c) 5.5×10^{22} molecules; (d) 3.0040 kg; (e) 0.001 mL?
1.35 (20 points) (a) The volume of an object is 1.2 m^3. Express this in L and in mL. (b) The mass of this object is 3.8×10^3 kg. What is its density, expressed in g/cm^3; in g/mL?
1.36 (20 points) Two students are asked to determine the melting point of aspirin (accepted value 135°C). Student X obtained the values 134°C, 136°C, 133°C, and 138°C. Student Y got 138°C, 137°C, 138°C, and 138°C. (a) For each student, calculate the average value, the average deviation of the individual measurements, and the error. (b) Which student's average value is more precise? Which student's is more accurate?
1.37 (15 points) Classify each of the following substances as a pure homogeneous substance, a solution, a pure heterogeneous substance, or an impure heterogeneous substance: (a) a single crystal of the compound ascorbic acid (vitamin C); (b) a mixture of water and steam; (c) a chocolate-chip cookie; (d) a clear liquid, with no internal boundaries, consisting of corn oil and olive oil; (e) liquid copper (an element).

2

ATOMS, MOLECULES, AND IONS

Chemists study matter and how it changes under controlled conditions. In Chapter 1, you learned that a pure substance can be recognized by its specific properties. For instance, oxygen boils at −183°C, water boils at 100°C, table salt melts at 801°C, and diamond melts at 3800°C. Why do these differences exist? A major goal of chemistry is to develop theories or models (mental pictures) to explain these and other observable properties of matter. We must therefore think in terms of the internal structures of the many varieties of matter in our environment. The concept that matter is atomic in nature is very old; such thoughts were expressed even by ancient philosophers. These ideas have since been greatly refined, and no scientist today doubts that matter is composed of many individual particles. This chapter is devoted to a study of these tiny chemical building blocks—atoms, molecules, and ions.

After memorizing a small amount of information, you will be able to predict the chemical formulas of a large number of simple compounds. You will also learn how to name these compounds.

2.1 HISTORICAL PERSPECTIVE

A cupful of mercury, a bar of gold, and a rod of iron all appear to fill completely the spaces they occupy. If we cut the gold bar in half, the two resulting pieces are still solid gold, not something else. Let us imagine that we could continue cutting indefinitely. Would we *always* produce smaller and smaller pieces of gold? The answer to this question is generally credited to the fifth-century B.C. Greek philosophers Leucippos and Democritos. They could not accept the idea that subdivision of matter could go on forever. Ultimately, they thought, such a process *must* end when a particle is obtained that can no longer be cut. This concept led them to conclude that all matter is composed of individual particles called **atoms**. In "Concerning the Nature of Things," the Roman poet Lucretius beautifully summarized the Greek theory of the atom:

From the Greek *a tomos*, "not cut."

. . . the nature of the universe
Consists then, in its essence, of two things;
For there are atoms and there is void.

The void is the empty space between atoms. Lucretius concluded that substances vary because they are composed of atoms of different size, shape, mass, and motion. The ancients' belief in the atom could not be fully appreciated until the

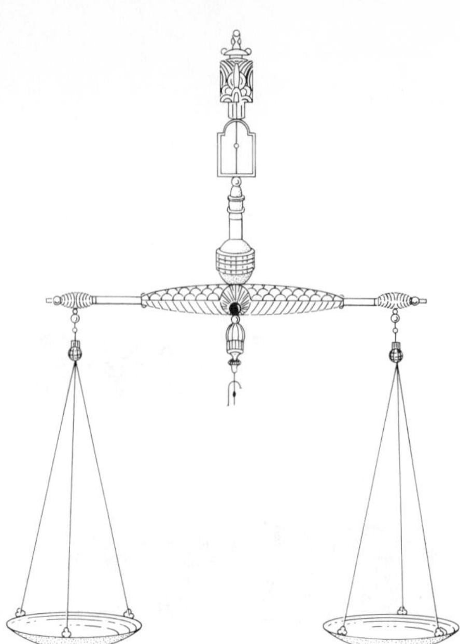

FIGURE 2.1
Balance used by Lomonosov in his experiments on the conservation of matter. (*Source*: Naum M. Raskin, *Khimicheskaia Laboratoriia M.V. Lomonosova*, Moscow: Academy of Sciences Press, 1962, p. 68)

experimental studies of the eighteenth and nineteenth centuries gave rise to modern chemistry. In 1803 John Dalton used the theory of the atom to explain the laws of chemical combination that were then known: the laws of conservation of matter and of definite composition. Let us look at each of these laws.

2.2 LAW OF CONSERVATION OF MATTER

Democritos stated, "Nothing is created out of nothing or is destroyed into nothing." The proof of this theory came in 1756 when Mikhail Lomonosov burned metals in tightly sealed vessels so that no matter could enter or escape. For example, he showed that when a reaction between tin and oxygen (from air), forming tin oxide, is carried out in a sealed container, the total mass measured after the reaction is the same as before the reaction. The balance he used for making his measurements is illustrated in Figure 2.1. Such measurements prove that, within experimental error, the mass of substances within a sealed container does not change during a chemical change. Figure 2.2 shows some types of apparatus used by chemists to detect changes in mass in chemical reactions. The results of many such experiments are summarized in the **law of conservation of matter**: **The mass of a chemically reacting system remains constant.**

Precision of best available balance: 6 parts per 110,000,000 parts by mass.

Detectable changes in mass occur in nuclear reactions (Chapter 23).

2.3 LAW OF DEFINITE COMPOSITION

In 1799 Joseph Proust showed that copper carbonate prepared in the laboratory has the same composition as naturally occurring copper carbonate. In both cases the percent composition by mass of the constituent elements is 51.4% Cu (copper), 9.7% C (carbon), and 38.9% O (oxygen). He concluded that "a compound is a substance to which nature assigns a fixed ratio."

When elements form a given compound, they always combine in the same mass ratio. For example, the compound silicon dioxide (quartz) always contains the elements silicon, Si, and oxygen, O, in the mass ratio of 0.876 Si to 1.00 O, or 46.7% Si and 53.3% O. This percent composition is always the same regardless of the source or the method of formation. Here are some examples:

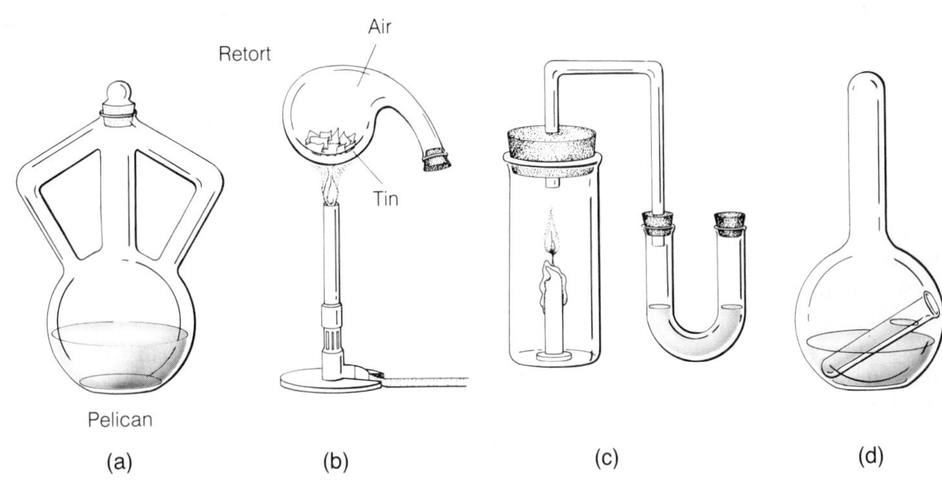

FIGURE 2.2
Apparatus used in the quantitative studies of matter conservation. (a) Slowly heated, water continuously evaporates and condenses and also dissolves the glass, forming a solid dispersed in the water. (b) Metal combines with oxygen. (c) Sodium hydroxide U-tube to absorb gaseous products of burning paraffin (candle). (d) Test tube (containing barium chloride solution) within a sealed flask (containing sodium sulfate solution). Turn apparatus upside down and the chemicals react.

SOURCE OF SILICON DIOXIDE	MASS PERCENT Si
Sand	46.7
Granite	46.7
Flint	46.7
Lava	46.7

This knowledge is summarized in the **law of definite composition**: *The mass composition of a given compound is constant.*

2.4 THE ATOMIC THEORY

John Dalton is generally recognized as the originator of the **atomic theory**, although he was anticipated by other scientists. Dalton postulated (1803) that:

1. The elements are composed of indivisible particles called atoms. Today we express this view by saying that matter is ''quantized''; that is, it consists of separated individual particles in empty space. *Quantum,* Latin for ''how much'' or ''how many,'' denotes a quantity that can exist only in whole-number amounts, not in fractions. (See Figure 2.3.)

We now know that the atoms of some elements do not have uniform masses (section 2.6). However, the average mass of the atoms of an element is constant (section 3.3).

2. All the atoms of a given element possess identical properties. For example, all atoms of a given element are assumed to have the same mass.
3. The atoms of different elements have different properties.
4. Atoms are the units of chemical changes; chemical changes involve the combination, separation, or rearrangement of atoms; atoms are not destroyed, created, or changed. An **atom** is the smallest particle of an element that can enter into chemical reactions. The chemical symbol for an element is also the symbol for the atom of the element.

FIGURE 2.3
Half-dollar coins are quantized. Any amount of half-dollars must be a whole number of such coins.

Jöns Berzelius originated the present system of letter symbols for the elements (1813). Dalton, preferring circles and dots, commented, "Berzelius's symbols are horrifying."

Cu is the symbol for the element copper as well as for an atom of copper.

5. When atoms combine, they combine in fixed ratios of whole numbers, forming particles known as molecules. A **molecule** is a tightly bound combination of atoms that acts as a single particle. Molecules can be combinations of atoms of the same element, such as H_2, O_2, or P_4. However, most molecules are combinations of atoms of different elements forming compounds, such as water, H_2O, and carbon monoxide, CO. In any compound, the kinds of atoms and their relative numbers are fixed. *Since atoms are indivisible, a fraction of an atom cannot combine.*

This theory advanced the understanding of matter in that it attributed definite properties to atoms—particularly definite masses—and limited the number of kinds of atoms to the number of known elements.

The atomic theory also offers acceptable explanations for the laws of chemical change:

By volume, 0.1% carbon monoxide produces unconsciousness in 1 hour and death in 4 hours. It forms explosive mixtures with air in the range of 12.5 to 74% CO by volume.

- *The conservation of matter.* During chemical reactions, atoms may combine with each other or separate from each other in various patterns, but they do not split apart, disappear, or change into atoms of other elements. Therefore, the number of atoms of each element in the **products** of a reaction is the same as the number of atoms of each element in the **reactants** or starting materials of the reaction. Moreover, we assume that the mass of an atom does not change. It would follow, then, that the mass of a chemically reacting system remains constant. Let us illustrate this conclusion with carbon monoxide, CO, a colorless, odorless, and extremely poisonous gas produced when fuels burn incompletely. Imagine 10 atoms of carbon, C, and 8 atoms of oxygen, O, (see Figure 2.4): 3 C atoms combine with 3 O atoms to form 3 molecules of carbon monoxide, CO. We now have 3 C atoms and 3 O atoms combined as CO molecules plus 7 uncombined C atoms and 5 uncombined O atoms. These add up to the original number of atoms. Because the mass of an atom does not change during a chemical change, the total mass remains constant.

- *The law of definite composition.* (a) A pure compound contains only one kind of molecule. For example, carbon monoxide consists only of CO molecules. (b) A molecule always contains the same whole number of atoms of each element. Each CO molecule contains one C atom and one O atom. (c) Each kind of atom has a fixed mass. A

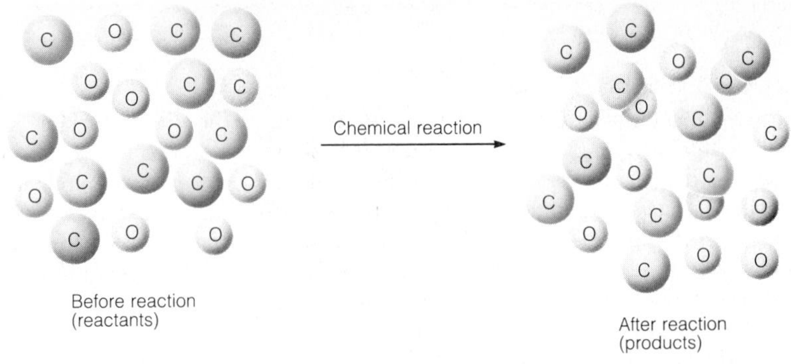

FIGURE 2.4
Illustration of the atomic theory explanation of the conservation of matter. (It is not unusual for some atoms or molecules to remain unreacted after a reaction occurs.)

C atom is like every other C atom; it is *assumed* that each has the same mass, combined or uncombined. Similarly, an O atom is *assumed* to be like every other O atom so that each has the same mass. It then follows that the mass ratio of the elements in a given compound must be constant. For example, 3 C atoms combine with 3 O atoms to form 3 CO molecules. Each molecule contains the same number of C and O atoms. The mass of each of these atoms is also constant; therefore, the mass ratio of carbon to oxygen must be constant in any given number of carbon monoxide molecules.

The atomic theory also explains the fixed composition of compounds made up of molecules containing more than one atom of a particular element. Water, for example, consists of molecules composed of hydrogen and oxygen atoms (Chapter 3). One molecule of water, H_2O, contains 2 H atoms and 1 O atom, and 12 molecules of H_2O contain 24 H atoms and 12 O atoms—the same mass ratio as in one molecule of H_2O (Figure 2.5):

$$\frac{\text{24 H atoms}}{\text{12 O atoms}} = \frac{\text{2 H atoms}}{\text{1 O atom}}$$

The mass ratio is the same for any number of H_2O molecules; for instance, 10^9 H_2O molecules contain 2×10^9 H atoms and 10^9 O atoms. Hence, any sample of water has the same mass composition (Box 2.1).

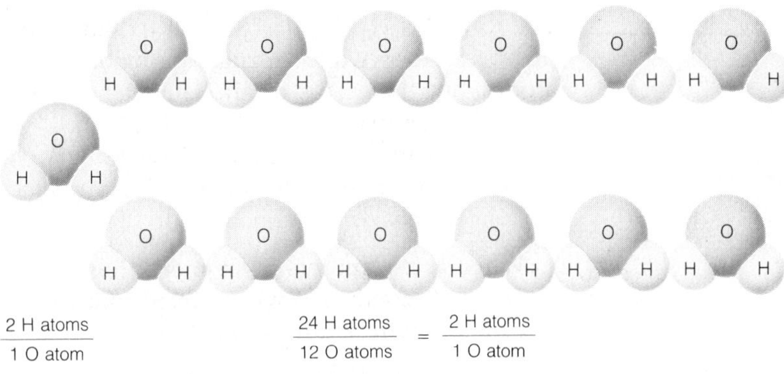

FIGURE 2.5
One H_2O molecule and 12 H_2O molecules.

BOX 2.1
BATTLE OF THE BALANCE

Claude Berthollet challenged the law of definite composition in 1803 because his studies of the composition of glasses and solutions had convinced him that the composition of compounds could vary. Berthollet's objections were swept aside when Dalton successfully applied the atomic theory; Dalton's theory rejected variable composition in the formation of compounds.

Nevertheless, Berthollet was not wrong. Today, glasses and solutions are recognized as homogeneous mixtures of variable composition. Although compounds generally obey the law of definite composition, some of them do not. Iron is typical of metals that form compounds in which some of the atoms may be missing. For example, depending upon the method of preparation, iron sulfide covers the range of $Fe_{0.858}S$ to FeS. Such deviations are essential for the preparation of transistors (section 22.8). Berthollet may have lost the "battle of the balance" in the 1800s, but today every transistor radio proclaims that victory is finally his.

LAW OF MULTIPLE PROPORTIONS

The laws of conservation of matter and of definite composition formed the basis of Dalton's atomic theory. Specifically, Dalton's concept of the atom meant that an atom of a given element does not appear, disappear, or change in mass during a chemical reaction. This is *all* that the laws imply. They do not state that an atom cannot change the number of linkages it makes with other atoms. On this basis, Dalton made a bold prediction. He reasoned that similar atoms can combine in more than one way. For example, if hydrogen and oxygen form more than one compound, a relationship must exist among the masses of hydrogen and oxygen in these compounds. Two such compounds are known:

COMPOUND 1	COMPOUND 2
Water	Hydrogen peroxide (pale blue liquid, corrosive and explosive)
H O H	H O O H

The formulas H_2O and H_2O_2 tell us that water has 1 oxygen atom for every 2 hydrogen atoms while hydrogen peroxide has 2 oxygen atoms for every 2 hydrogen atoms. But the atoms of an element have a fixed mass. Thus, the 2 hydrogen atoms in each compound have a fixed mass, while the mass of 1 O atom in H_2O is not the same as the mass of 2 O atoms in H_2O_2. However, the masses of the oxygen atoms in the two compounds are related as whole numbers, 1 to 2. Dalton postulated that, in general,

> **When two elements form more than one compound, the masses of one element, combined with a fixed mass of the other element, are in ratios of whole numbers.**

This statement is known as the **law of multiple proportions**. Many other examples, such as SO_2 and SO_3, CO and CO_2, and CuCl and $CuCl_2$, are known. Verification of the prediction stimulated much research and generated much confidence in the assumption that matter is quantized.

PROBLEM 2.1 Show that (a) the uranium borides, UB_2, UB_4, and UB_{12}, and (b) the solid compounds $AgClO_3$ and $AgClO_4$, illustrate the law of multiple proportions. □

EXAMPLE 2.1 Phosphorus forms two compounds with chlorine. In one, 7.75 g of phosphorus combines with 26.60 g of chlorine. In the other, 3.10 g of phosphorus combines with 17.73 g of chlorine. Show that these compounds confirm the law of multiple proportions.

ANSWER First, calculate the mass of one element, for example, chlorine, that combines with a fixed mass, for example, 1.00 g, of the other element. For the first compound:

Note that one insignificant figure is carried, to be discarded later.

$$1.00\ \text{g}\ \cancel{\text{P}} \times \frac{26.60\ \text{g Cl}}{7.75\ \text{g}\ \cancel{\text{P}}} = 3.432\ \text{g Cl}$$

For the second compound:

$$1.00\ \text{g}\ \cancel{\text{P}} \times \frac{17.73\ \text{g Cl}}{3.10\ \text{g}\ \cancel{\text{P}}} = 5.719\ \text{g Cl}$$

Now find the ratio of chlorine in the two compounds:

$$\frac{3.432\ \text{g}\ \cancel{\text{Cl}}}{5.719\ \text{g}\ \cancel{\text{Cl}}} = 0.600$$

The 0.600 is the same as 3/5, a ratio of whole numbers consistent with the law of multiple proportions. (Of course, this example could have been answered by calculating the mass of phosphorus that combines with 1.00 g Cl.) ■

PROBLEM 2.2 Show that the data for two compounds of nitrogen and oxygen are consistent with the law of multiple proportions:

	COMPOUND 1	COMPOUND 2
Grams nitrogen	3.500	4.000
Grams oxygen ⟷	4.000 ⟷	2.286

□

2.5 THE ELECTRICAL NATURE OF MATTER

Dalton's contribution to atomic theory was inspired, but it was also limited. His inspiration was his concept that atoms of any particular element have a particular mass, and that the mass relationships in chemical reactions result from the combination of atoms. But that was as far as he could go. Dalton's atoms had no structure. Consequently, his theory could not explain why various groups of elements, such as lithium, sodium, and potassium, or chlorine, bromine, and iodine, have chemical properties in common. If we try to imagine why elements resemble each other, Dalton's structureless and indivisible atoms do not help us at all. Nor can we guess what might account for differences in chemical properties—why does one fluorine atom combine with one hydrogen atom, whereas one nitrogen atom combines with three hydrogen atoms?

John Dalton, 1766–1844
In Dalton's day, "electric fluid" was the term used for electric charge.

The first clues that atoms might possess an internal structure resulted from experiments that began in the nineteenth century. Those experiments established the electrical nature of matter and included the discovery of X rays. However, radioactivity, the key that unlocked the door to a structured atom, was not discovered until 1896, fifty-two years after Dalton's death. Radioactivity (Chapter 23) is the spontaneous disintegration of atoms of certain elements into atoms of other elements with the emission of smaller particles. For example, a radium atom naturally changes to a radon atom with the emission of a high-speed electrically charged helium atom (section 23.2). This behavior contradicts Dalton's hypothesis that atoms are unalterable. In turn, these discoveries gave rise to a more complete theory of the structure of atoms and reaffirmed the atom as the unit of chemical change. We have learned from the atomic theory that matter is quantized. For example, solid iron, liquid water, and gaseous hydrogen are composed of atoms or molecules in otherwise empty space. Electricity in Dalton's time, however, was regarded as a continuous fluid. This view could not survive when confronted by the discoveries of the late nineteenth century.

ELECTRONS

From *electron*, Greek for "amber"; amber is *electrified* when rubbed with wool.

ELECTRICITY AND CHEMICAL CHANGE Compounds can be decomposed by electricity. For example, electricity decomposes copper chloride to copper and chlorine. Michael Faraday studied (1833) the relationship between the quantity of electricity that produces a certain chemical change and the quantity of product of the chemical change (Chapter 17).

Faraday's experimental results led George Johnstone Stoney to conclude in 1874 that electricity, like matter, consists of particles (section 17.2 and Problem 2.26). These particles, he implied, were associated with atoms. He named a single particle of electricity the **electron**. Further, he estimated that the charge on an electron is 10^{-20} coulomb (C), later corrected to 1.6×10^{-19} C. An electric current is thus a stream of electrons, each electron carrying a charge of 1.6×10^{-19} C.

The coulomb is the SI unit of electric charge. It is equal to the quantity of charge passing a point in a circuit when an electric current of 1 ampere flows for 1 second. It is the quantity of charge that will deposit 1.1180×10^{-3} g of silver from a water solution of silver nitrate, $AgNO_3$. Stoney introduced the term *electron* for the "atom of electricity" in 1891.

ELECTRIC DISCHARGE TUBES Michael Faraday, William Crookes, and many others studied the effects of passing an electric current through a gas. The results of these studies are best explained in terms of the electron particle suggested by Stoney.

An electric discharge tube (Figure 2.6a) is a partially evacuated glass tube with a metal rod (an electrode) sealed into each end. When this apparatus is attached to a high-voltage source, the gas in the tube acts as a conductor and current flows through the gas from one electrode to the other. At the same time, the gas emits light as in a neon sign. When more gas is removed, emission of light by the gas ceases. However, the current continues to flow between the electrodes, and the glass at the positive end of the tube glows. It thus appears that a radiation capable of causing the glow is emitted from the cathode (negative end). Because of its origin, this radiation is named the **cathode ray**. A television picture tube is a cathode-ray tube. The television screen, coated with phosphorescent salts, forms a picture when hit by electrons.

These rays behave like particles in motion. They possess mass, move in straight lines, and are capable of penetrating very thin metallic sheets. They are also deflected in electric and magnetic fields in a manner characteristic of negatively charged particles (Figure 2.6b).

An electric charge produces a surrounding *electric field*; a magnet produces a surrounding *magnetic field*.

Such experiments, especially those performed by Joseph J. Thomson, established the particle character of the cathode rays. Thomson concluded (1897) that the ray consists of a beam of very fast moving particles with a negative charge.

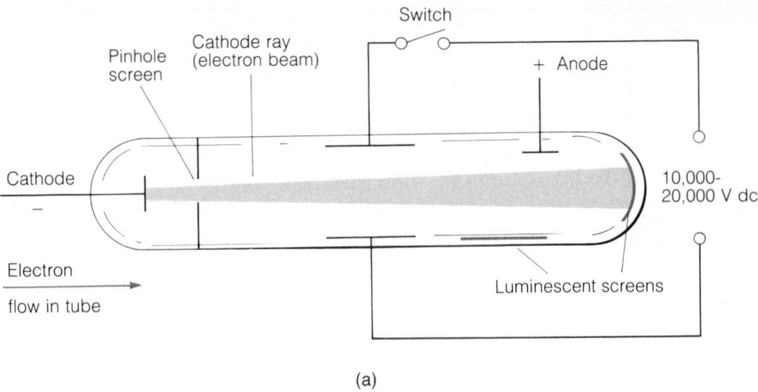

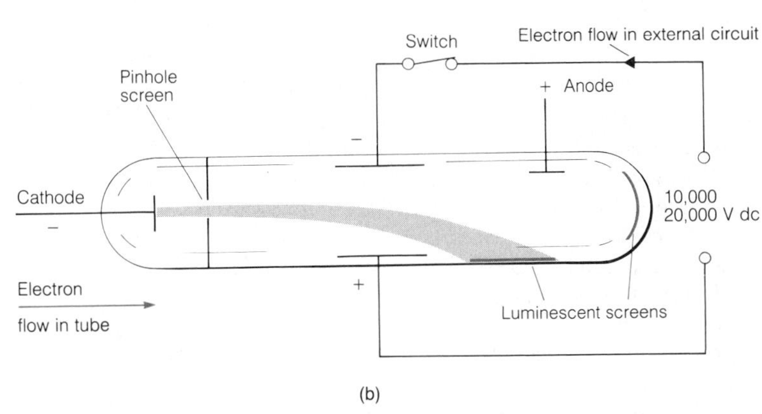

FIGURE 2.6
Electric discharge tube experiment, showing that the cathode ray possesses a negative charge.
(a) Cathode ray travels in a straight line in the absence of an electric field. (b) Position of cathode ray in an electric field. The ray is repelled by the negative plate and attracted to the positive plate. A cathode ray is obtained only when the gas pressure in the tube is very low. Electrons flow from cathode to anode.

Thomson wrote, "However we twist and deflect the cathode rays, the negative electricity follows the same path as the cathode ray." He preferred the name *corpuscles* for the negatively charged particles.

Moreover, he showed the particles to be *electrons* by measuring the ratio of the charge of the particle, e, to its mass, m, e/m (see Box 2.2):

$$e/m = 1.76 \times 10^8 \text{ C/g}$$

The mass of the electron is known from other measurements to be 9.11×10^{-28} g (Problem 2.28). Then, knowing the value of m and Thomson's value of e/m, the charge of the particle e is calculated to be 1.60×10^{-19} C:

$$e = m \times 1.76 \times 10^8 \frac{\text{C}}{\text{g}} = 9.11 \times 10^{-28}\,\cancel{\text{g}} \times 1.76 \times 10^8 \frac{\text{C}}{\cancel{\text{g}}}$$

$$= 1.60 \times 10^{-19} \text{ C}$$

This value agrees with the calculated electronic charge based on Faraday's discoveries. It was reconfirmed by Robert Millikan in 1911; by direct experimentation, he proved that the electron always carries the charge of 1.6×10^{-19} C (Problem 2.27).

The ratio e/m is independent of the kind of gas in the discharge tube and the materials of which the discharge tube is composed. Also, electrons from other sources, such as particles liberated from metals by light (light-meters) and from filaments by heat, have the same charge and mass. Furthermore, the mass of the electron is about 2×10^3 times smaller than the mass of the hydrogen atom, the lightest atom known. These facts indicate that the electron is a universal constituent of matter. The electron charge is therefore used as a unit of charge, abbreviated as e. The more modern name for cathode rays is **electron beams**.

BOX 2.2

THE DETERMINATION OF e/m

Charged particles, unlike light rays, are deflected by electrically charged plates (Figure 2.6) or by magnets (Figure 1). It is interesting to note that the early attempts (1883) of Heinrich Hertz, discoverer of electromagnetic waves, to deflect a beam of cathode rays were unsuccessful. His failures were due to the fact that high-vacuum technology had not yet been developed. Consequently, the gas pressure in the discharge tube was too high, leading him to the conclusion that cathode rays were not charged. Clearly, this incident illustrates the necessary marriage between experimental science and technology.

Imagine that the magnetic field in Figure 1 lies in the *plane* of the page. Then, the deflection is upward; the ray enters the field *across* the page and exits *above* the page. In Figure 2.6, the negative plate is above while the positive plate is below the plane of the paper. The deflection is then downward; the ray is repelled from the negative metal plate and attracted toward the positively charged plate. The ray enters the field *across* the page and exits *below* the page. The important feature to observe is the opposing effect of the two fields: when the electric and magnetic fields are placed perpendicular as in Figure 2, the actions of the fields oppose each other. Consequently, the strengths of the fields can be adjusted so that the path of the ray is upward, downward, or straight across the plane of the paper. By setting up the cathode-ray tube shown in Figure 2, and by adjusting the electric field exactly to cancel the measured deflection produced by a magnetic field of fixed strength, Thomson (Figure 3) determined the ratio of the charge to the mass for an electron. The result, $1.2 \times 10^8 \frac{\text{C}}{\text{g}}$, later corrected to $1.76 \times 10^8 \frac{\text{C}}{\text{g}}$, is a large ratio.

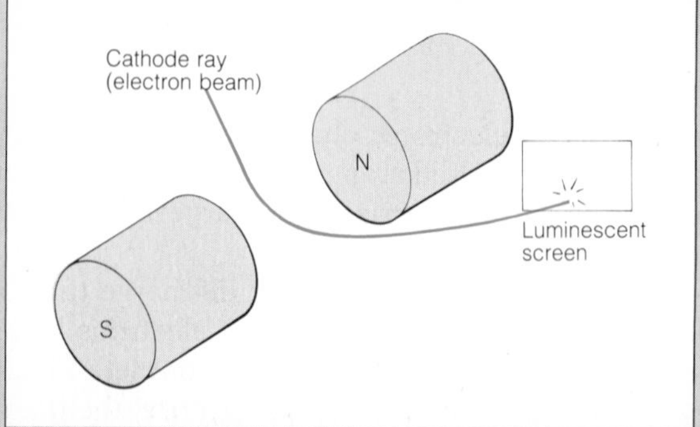

FIGURE 1 Path of a cathode ray upon passing through a magnetic field.

Although neither e nor m can be separately determined from these experiments, the ratio does show that an electron has either a very small mass or a very large charge. Use of Stoney's electronic charge based on Faraday's work permits an estimation of the electron mass:

$$m = \frac{10^{-20}\ \text{C}}{1.76 \times 10^8 \dfrac{\text{C}}{\text{g}}} = 10^{-28} \frac{\text{g}}{\text{electron}}$$

Wilhelm Wien confirmed (1898) that positively charged atoms are also formed in discharge tubes (Figure 2.8) and, in fact, made the first e/M_H measurement of a charged hydrogen atom. Assuming that e of the charged atom (Problem 2.28) is the same as the charge of an electron, Thomson calculated the relative masses of the charged H atoms and the electron:

$$\frac{\dfrac{e}{m_e}}{\dfrac{e}{\text{M}_\text{H}}} = \frac{1.76 \times 10^8 \dfrac{\not{\text{C}}}{\not{\text{g}}}}{10^5 \dfrac{\not{\text{C}}}{\not{\text{g}}}} = \frac{\text{M}_\text{H}}{m_e} = 2 \times 10^3$$

The mass of the electron must therefore be about 2×10^3 times smaller than that of the lightest known atom, H.

Thomson then made the first direct measurement of the electronic charge (1899) by measuring the total charge, in coulombs, of a cloud, $\frac{\text{C}}{\text{cloud}}$, of water droplets. It was known that electric charges act as condensation nuclei for water vapor. The number of droplets in the cloud, $\frac{\text{N droplets}}{\text{cloud}}$, was obtained from the mass of water condensed from the cloud, $\text{M}\frac{\text{g}}{\text{cloud}}$, divided by the average mass of a droplet, $m\frac{\text{g}}{\text{droplet}}$:

$$\frac{\text{M}\dfrac{\not{\text{g}}}{\text{cloud}}}{m\dfrac{\not{\text{g}}}{\text{droplet}}} = \text{N}\frac{\text{droplets}}{\text{cloud}}$$

The mass of the droplets was obtained from their rate of fall under the effect of gravity. Then, the charge on an average droplet is found from the total charge carried by the cloud:

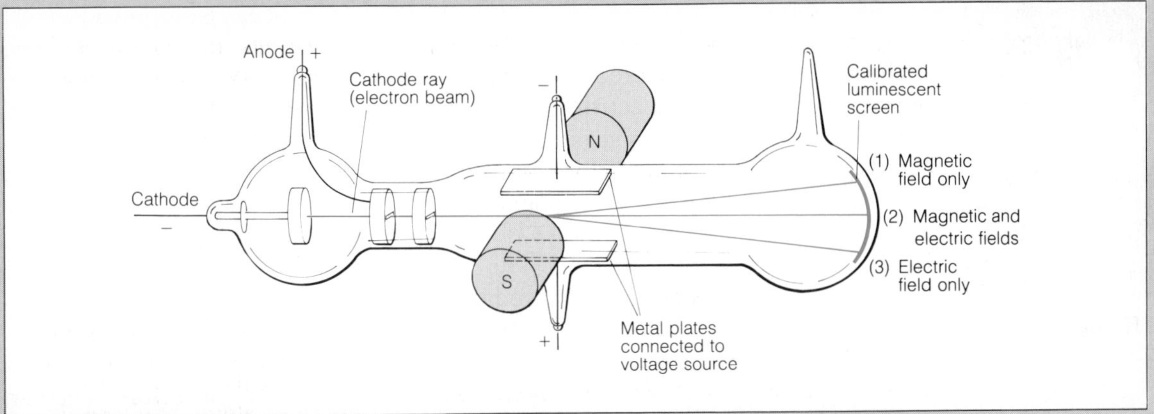

FIGURE 2 Thomson's cathode-ray tube with electric and magnetic fields for measuring e/m. (1) Position of cathode ray after deflection by magnetic field. (2) Position of cathode ray when magnetic field deflection is balanced by the electric field. (3) Position of cathode ray after deflection by electric field.

$$\frac{\frac{\text{C}}{\cancel{\text{cloud}}}}{\text{N}\frac{\text{droplets}}{\cancel{\text{cloud}}}} = 10^{-19}\frac{\text{C}}{\text{droplet}}$$

On the assumption that each droplet condenses on a single particle carrying one electron charge, e is obtained:

$$10^{-19}\frac{\text{C}}{\text{droplet}} = 10^{-19}\frac{\text{C}}{\text{electron}} = e$$

The method, refined by Robert Millikan to observe individual oil drops, yielded precise values (1910) for the electronic charge (Problem 2.27). The smallest charge ever observed on any droplet is 1.60×10^{-19} C. $\frac{e}{\text{M}_\text{H}}$ measurements $\left(9.58 \times 10^4 \frac{\text{C}}{\text{g}}\right)$ made in discharge tubes or measurements made during the passage of electricity through acid solutions and the known mass of a proton (1.67×10^{-24} g) permit a calculation of the proton charge (1.60×10^{-19} C).

As well as identifying the cathode ray with the electrons based on Faraday's work (section 17.2), Thomson's experiments also clearly proved the existence of a particle much smaller than the atom. Since these particles—electrons—come out of metals (the cathode in an electric circuit) and radioactive elements (Chapter 23), the atom must have some kind of structure.

FIGURE 3 J. J. Thomson giving a lecture demonstration of an e/m measurement.

PROBLEM 2.3 How many electrons per second pass through an appliance rated at 100 watts at 110 volts, corresponding to $0.909\,\frac{\text{C}}{\text{sec}}$? □

Wilhelm Röntgen, 1845–1923

In 1895, while performing experiments on light produced by electron beams, Wilhelm Röntgen discovered a radiation that penetrates metal plates that are opaque to light and to electrons. This radiation, named the **X ray**, is produced when high-speed electrons strike an object (Figure 2.7). X rays possess the characteristic properties of light; for example, they darken a photographic plate and are not deflected by electric or magnetic fields.

IONS

Atoms and molecules become charged when they gain or lose electrons. The conversion of atoms or molecules to charged particles is called **ionization**, and a charged particle is called an **ion**. When an electron is gained, the particle acquires a negative charge. Such a particle is called a "negative ion" or an **anion**. When an electron is lost, the particle acquires a positive charge. Such a particle is called a "positive ion" or a **cation**. Charges are measured in electronic units, +1, +2, . . . ; −1, −2, ...

$\text{O atom} \xrightarrow{-e^-} \text{O}^+$ (positive ion, cation)

$\text{O atom} \xrightarrow{+e^-} \text{O}^-$ (negative ion, anion)

Ions may have more than one unit charge.

In a discharge tube, atoms or molecules of the enclosed gas lose electrons, forming positive ions; typical is the ionization of the neon, Ne, atom (Figure 2.8):

$$\text{Ne}(atom) + \text{energy} \longrightarrow \text{Ne}^+(ion) + e^-(electron)$$

The positive ions move toward the negative electrode (the cathode). If a perforated cathode is used, some positive ions will pass through and emerge behind the cathode (Figure 2.9). Eugen Goldstein performed such an experiment and detected the positive ions (1886). Unlike electrons, positive ions have properties that *are characteristic* of the gas in the tube.

The positive ion and the electron produced by ionization of an atom are by no means equal in mass. In fact, the relative mass of the electron is so small that the mass of the positive ion is almost equal to that of the atom from which it was formed. Of all the positive ions whose masses have been determined, the *lightest ion found* is obtained from the hydrogen atom, H. This ion, H^+, is named the

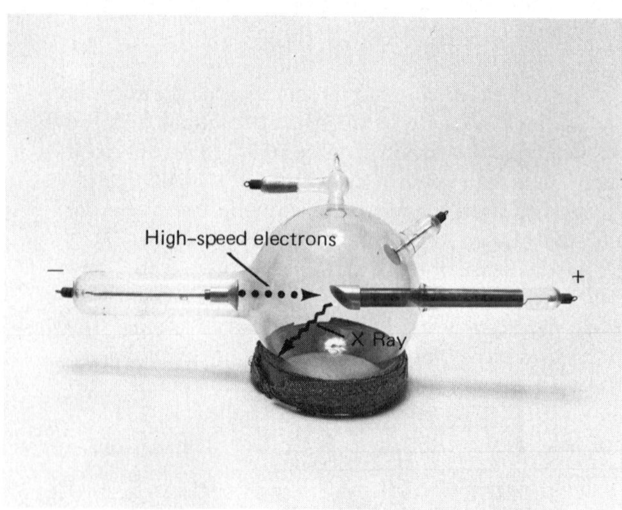

FIGURE 2.7
A Coolidge X-ray tube (1913). When high-speed electrons strike the heavy metal plate (anode), X rays are produced.

FIGURE 2.8 In a discharge tube, the negative electrode (cathode) is the source of electrons that flow across the tube to the positive electrode (anode). An electron is shown colliding with a neon atom and then continuing its journey to the anode. The neon atom is thus ionized, forming Ne^+, which moves toward the cathode, and an electron that moves toward the anode.

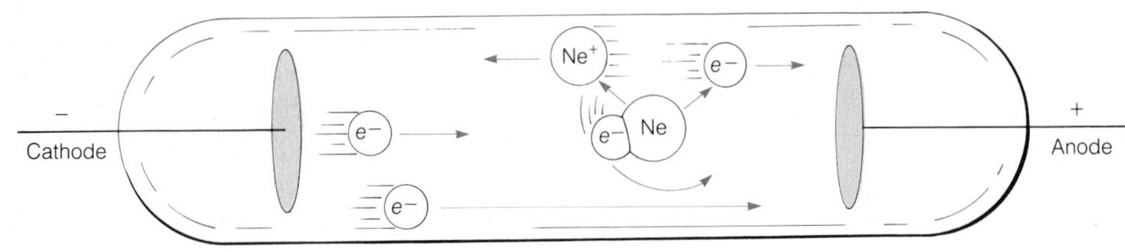

From *protos*, Greek for "first."

proton. The proton has a mass nearly equal to that of the H atom and it carries a charge equal to that of an electron but opposite in sign (Box 2.2). Similarly, the removal of two electrons from a helium atom, He, produces an **alpha (α) particle**, He^{2+}. This is also the particle that is emitted when a radium atom disintegrates and becomes a radon atom (section 23.2). This particle was used by Rutherford in his famous experiments, which will be discussed in the next section.

The experiments discussed in this section establish the electrical character of matter. Electrons are constituents of all atoms. Atoms and molecules, however, are electrically neutral particles. Thus each atom must somehow contain a positively charged portion to balance its negatively charged electrons. Where is the positive part of the atom? What is its size and shape? In the next section, we will see how scientists began to solve this puzzle early in the twentieth century.

2.6 THE RUTHERFORD THEORY OF THE ATOM

The first attempt to account for the presence of electrons in atoms was made by Joseph J. Thomson in 1904. He proposed that an atom is composed of a uniform, positively charged sphere in which are embedded a number of electrons equal to the positive charge (Figure 2.10). In other words, if the total positive charge is 10, there would be 10 electrons dispersed throughout the atom. Since the number of positive charges equals the number of negative charges (electrons), the atom has no net charge—it is electrically neutral. But, as we have explained, the mass of an electron is exceedingly small compared to the mass of an atom. This means that *nearly all of the mass of an atom is associated with the positive charge*. Thus Thomson predicted that fast-moving alpha particles, He^{2+}, should

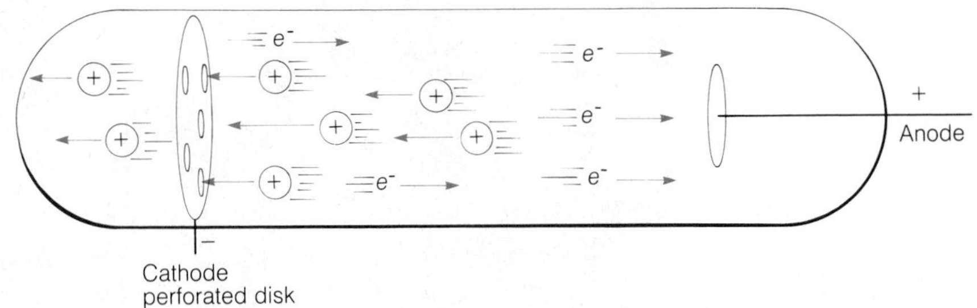

FIGURE 2.9 Discharge tube for producing positive ions.

Positive charge of uniform density

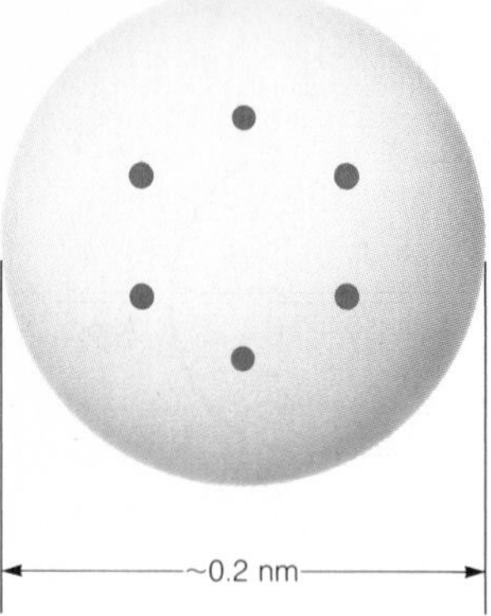

FIGURE 2.10
The Thomson atom, a sphere of positive electricity, +6, in which six electrons (*black dots*) are embedded so that the atom is electrically neutral (uncharged); 10^7 nm = 1 cm.

pass straight through a metal film. He reasoned that the alpha particle should be repelled equally from all sides by the positive charge, assumed to be evenly spread out through the entire atom.

To test Thomson's model, one of the most important experiments of all time was performed in 1909. Ernest Rutherford (Figure 2.11), with Hans Geiger and Ernest Marsden, measured the extent of the deflections of alpha particles aimed at a metal film (Figure 2.12). In their experiments, they used high-speed alpha particles from a radioactive source. As predicted, practically all (about 99.9%) of the particles passed through the film with almost no deviation from the original path. However, much to their amazement, some particles were deflected through large angles, and a few almost completely reversed their direction. Rutherford is quoted to have said, "It was almost as incredible as if you fired a fifteen-inch shell at a piece of tissue paper and it came back and hit you."

To explain these results, Rutherford proposed in 1911 that the atom must be almost completely empty space. This would account for the passage of most of the alpha particles straight through the film. To account for the repulsive force required to produce the observed large deflections, he assumed that the positive charge of the atom must be concentrated in a very small volume. Then, because of the small mass of electrons, nearly all the mass of the atom is in this small volume. Rutherford called this concentrated region of mass and positive charge the **nucleus** of the atom. When an alpha particle approaches a nucleus, it is repelled; the closer the approach to the nucleus, the greater is the deflection (Figure 2.13).

Because an atom is electrically neutral, the number of electrons around the nucleus must be equal to the positive charge of the nucleus. The space the electrons occupy determines the volume of the atom. It is this space through which alpha particles pass easily.

FIGURE 2.11
Ernest Rutherford (*right*) with Hans Geiger in their laboratory at The University, Manchester, England. (Courtesy of the Physics Department, The University, Manchester.)

FIGURE 2.12
Rutherford's alpha-particle deflection experiment. Positively charged alpha particles are directed at a metal film, and the effect on the particles is observed with detecting devices placed at various angles. The angle (ϕ) is a measure of the deflection. The speed of the alpha particle is about 1.9×10^4 km/sec.

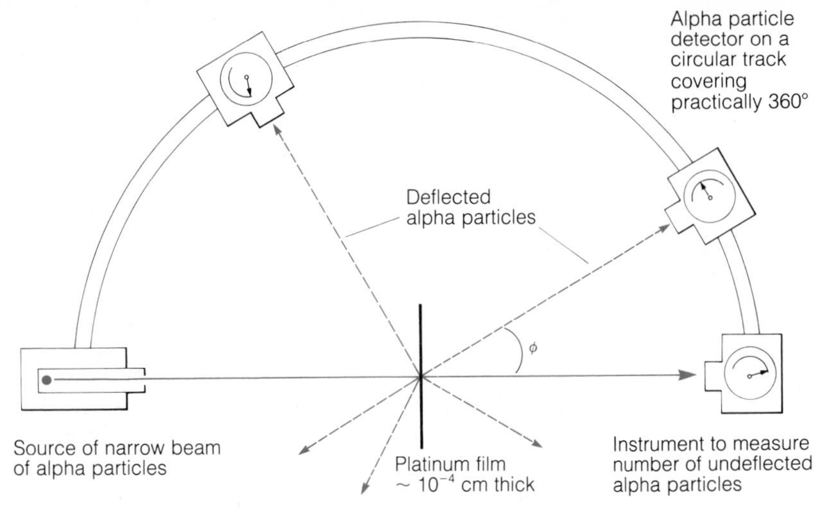

Rutherford also calculated a rough value for the diameter of the nucleus—about 10^{-6} nm. To give you an idea of the relative sizes of atoms and nuclei, imagine the following: If the diameter of a nucleus is represented by a pinhead 2 mm wide, the diameter of an atom would be roughly 200 m—about twice the length of a football field.

On the basis of his experiments with the deflection of alpha particles, Rutherford found that the greater the nuclear charge, the larger the number of alpha particles deflected at a given angle. With this charge he was able to calculate the numbers of unit positive charges on the nuclei of aluminum, copper, and gold as Al, 13; Cu, 29; and Au, 79. The *number of unit positive charges on a nucleus* is called the **atomic number** of the element. Since the proton has a charge of +1, the atomic number is equal to the number of protons in the nucleus of the atom:

The atomic number is also known as the *proton number*.

$$\textbf{number of protons in nucleus} = \textbf{atomic number of the element}$$

An element is characterized by its atomic number. For example, chlorine, Cl, has an atomic number of 17. The nucleus of every Cl atom has 17 protons, and every neutral Cl atom has 17 electrons balancing the charge of the 17 protons. (See table inside the back cover.) The atomic number also determines the position of an element in the periodic table (section 2.7).

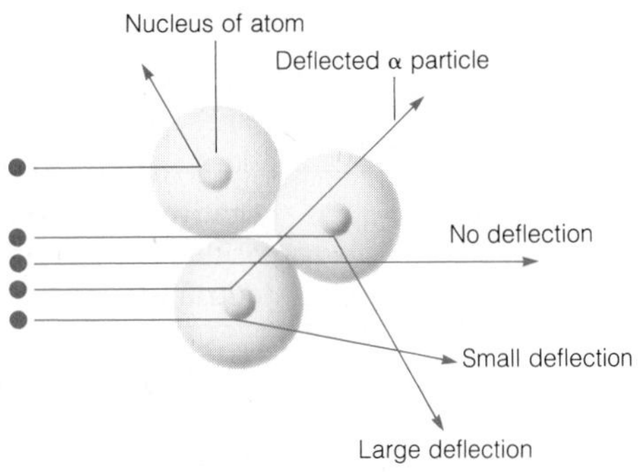

FIGURE 2.13
Representation of the deflection of alpha particles, He^{2+}, by a metal film, as predicted by the Rutherford nuclear model of an atom.

COMPOSITION OF THE NUCLEUS

Except for hydrogen, H, the number of protons in a nucleus accounts for only part of the mass of the atom, usually less than half of it. The remainder of the mass must be made up by other particles in the nucleus. These particles are called **neutrons**. The neutron, an electrically neutral particle with a mass nearly the same as the proton mass (Table 2.1), was predicted by Rutherford in 1927. It was discovered in 1932 by James Chadwick. The number of neutrons in a nucleus together with the number of protons in the nucleus make up the **mass number** of the atom:

The mass number is also known as the *nucleon number*.

mass number = number of neutrons + number of protons

Fluorine, F, for example, has an atomic number of 9 and a mass number of 19. Each F nucleus therefore has 9 protons and 10 neutrons.

Most elements are composed of a mixture of atoms with different mass numbers. Atoms that have the same atomic number but different mass numbers are called **isotopes**.† Thus the isotopes of a given element all have the same atomic number but different masses.

The three isotopes of neon, Ne, have the same atomic number, 10, but different mass numbers: 20, 21, and 22. This means that every Ne nucleus has 10 protons. However, a nucleus of the isotope of mass number 20 has 10 neutrons, the one of mass number 21 has 11 neutrons, and the one of mass number 22 has 12 neutrons. Symbolically, these nuclei are represented as $^{20}_{10}Ne$, $^{21}_{10}Ne$, and $^{22}_{10}Ne$. In these symbols the superscripts 20, 21, and 22 are the mass numbers, and the subscript 10 is the atomic number. The symbols $^{20}_{10}Ne$ and ^{20}Ne are both read as "neon-20."

Hydrogen is the only element whose isotopes have special names. They are protium for $^{1}_{1}H$, deuterium, D, for $^{2}_{1}H$, and tritium, T, for $^{3}_{1}H$. They have special names and symbols because the ratios between their masses are so far from 1.

Protons and neutrons are called **nucleons** when there is no need to distinguish between them. The mass number thus represents the total number of nucleons in the nucleus.

A **nuclide** refers to a particular atomic species characterized by its atomic number and mass number, such as $^{12}_{6}C$ or $^{20}_{10}Ne$. Different nuclides having the same atomic number (for example, $^{20}_{10}Ne$, $^{21}_{10}Ne$, and $^{22}_{10}Ne$) are isotopes.

Rutherford was also correct when he said that the space the electrons occupy determines the volume of the atom. In neon, for instance, the volume is determined by the 10 electrons moving about the nucleus.

EXAMPLE 2.2 Copper has two isotopes, $^{63}_{29}Cu$ and $^{65}_{29}Cu$. How many (a) protons, (b) electrons, and (c) neutrons make up the atoms of these isotopes?

TABLE 2.1
CONSTITUENT PARTICLES OF ATOMS

PARTICLE	CHARGE	MASS	MASS NUMBER
Electron	negative (−1), 1.60×10^{-19} C	9.110×10^{-28} g	0
Proton	positive (+1), 1.60×10^{-19} C	1.673×10^{-24} g	1
Neutron	neutral (0)	1.675×10^{-24} g	1

† From the Greek word meaning "same place." The name was chosen by Frederick Soddy (1911) to indicate that these atoms occupy the same position in the periodic system of elements (section 2.7).

ANSWER

(a) The number of protons is equal to the atomic number, 29, which is the same for each kind of Cu atom.

(b) Atoms are electrically neutral. The number of electrons is therefore equal to the atomic number 29 and is the same for each kind of Cu atom.

(c) The mass number of a nucleus (63 or 65) equals the sum of the number of protons and neutrons. The number of neutrons is therefore the mass number minus the atomic number (29):

$^{63}_{29}Cu$: 63 nucleons − 29 protons = 34 neutrons

$^{65}_{29}Cu$: 65 nucleons − 29 protons = 36 neutrons ■

PROBLEM 2.4 What is the nuclear composition of the isotopes of uranium (atomic number 92) with mass numbers 233, 234, and 235? □

PROBLEM 2.5 How many protons, neutrons, and electrons are in $^{40}_{19}K$ and in $^{40}_{19}K^+$? □

2.7 ATOMIC NUMBER AND THE PERIODIC TABLE

There are 108 known elements. Their names are derived from the names of planets, ancient gods, modern scientists, countries, cities, continents, and various Greek and Latin words (Table 2.2). Each name is also denoted by a symbol of one or two letters, usually an abbreviation of the English name, but sometimes of the Latin one. The complete list appears inside the back cover.

Chapter 6 is devoted to the periodic table.

A brief introduction to the *periodic table* (inside the front cover) is presented here because the table organizes a great mass of information about the properties of elements. It also serves as a powerful aid to remembering some of this information.

Symbols of some common elements:

hydrogen	H
nitrogen	N
fluorine	F
magnesium	Mg
silicon	Si
chlorine	Cl
carbon	C
oxygen	O
sodium	Na
aluminum	Al
sulfur	S
iron	Fe

Starting with $_1H$, the elements are placed in the table in the order of increasing atomic numbers from left to right. The atomic number is the number printed above the symbol of the element. In this way, the table groups the elements into rows and columns so that elements with similar chemical properties fall into the same column, called a *group*. The so-called *main groups*, also known as *representative elements*, are identified by numbers. Starting at the left with 1 for the ^{1}H–^{87}Fr group and skipping over the central groups of elements (labeled 3B–12B), the numbering is continued with 3 for the ^{5}B–^{81}Tl group and ends with 8 for the ^{2}He–^{86}Rn group. The elements in a given group have similar chemical properties, while elements of different groups often bear little resemblance to each other. For example, in chemical combinations, the elements in Group 1 form positive ions, of which Na^+ is typical. At the other end, chlorine forms a negative ion, Cl^-, typical of Group 7. The elements of Group 8, called the *noble gases*, were once believed to be chemically unreactive; however, in recent years several compounds of a few noble gases, particularly krypton, Kr, and Xenon, Xe, have been synthesized. The central groups of elements (3B–12B) are known as *transition elements*. The zigzag line separates the metals (on the left side of the line) from the nonmetals (on the right side).

2.8 WRITING CHEMICAL FORMULAS

Substances and their molecules are represented by chemical formulas. **Chemical formulas** are combinations of atomic symbols that show the composition of substances. The chemical formula for cane (table) sugar (sucrose) is $C_{12}H_{22}O_{11}$. The

TABLE 2.2
SOURCES OF NAMES OF SOME ELEMENTS

ELEMENT	SOURCE OF NAME
Uranium	The planet Uranus, the outermost known planet of the Solar System when uranium was the end of the series of known elements. (More distant planets and heavier elements have since been discovered.)
Mercury	Mercury, the messenger god, noted for his quickness. The element is a shiny metal that is liquid at room temperature. It flows quickly and looks alive—hence its other name, *quicksilver.*
Curium; einsteinium; fermium	Marie Curie; Albert Einstein; Enrico Fermi
Francium; germanium; polonium	France; Germany; Poland
Berkelium; yttrium; ytterbium; erbium; terbium	Berkeley, California; all the others are derived from Ytterby, Sweden
Europium	Europe
Xenon	Greek *xenos,* ''a stranger''
Hydrogen	Greek *hydōr,* ''water,'' and *genēs,* ''producing''

Read C–12–H–22–O–11.

The subscript 1 is omitted; CH_4 means C_1H_4.

subscripts, 12, 22, and 11, tell us the number of atoms of each element in a molecule of sucrose. Thus $C_{12}H_{22}O_{11}$ means that the sugar molecule consists of 12 atoms of carbon, 22 atoms of hydrogen, and 11 atoms of oxygen. The formula for methane (natural gas), CH_4, indicates that the methane molecule consists of 1 atom of carbon and 4 atoms of hydrogen. Cane sugar, methane, and water, H_2O, are typical **molecular compounds**; they are composed of molecules containing more than one kind of atom. The formulas of molecular substances are known as molecular formulas (Chapter 3). Molecular compounds are generally formed when the nonmetals (right side of the heavy line in the periodic table) combine with each other.

Elements can be represented by their formulas as well as by their symbols. The molecular formula for an element indicates the number of atoms in one molecule of that element:

ATOM	MOLECULAR FORMULA
Oxygen, O	oxygen, O_2, or ozone, O_3
Chlorine, Cl	chlorine, Cl_2
Phosphorus, P	phosphorus, P_4
Helium, He	helium, He

This table tells us that oxygen consists of **diatomic** molecules; that is, each molecule of oxygen consists of two oxygen atoms. Fluorine, chlorine, bromine, and iodine (Group 7) all consist of diatomic molecules, F_2, Cl_2, Br_2, and I_2. The ozone molecule is **triatomic**; it consists of three oxygen atoms. Phosphorus occurs as **tetratomic** molecules. Helium is **monatomic**; each helium molecule consists of just one helium atom (Figure 2.14).

Atoms and molecules are electrically neutral (uncharged) particles. Ions are electrically charged particles, atoms, or molecules with an excess or deficiency of electrons. A fluorine atom, $^{19}_{9}F$, has 9 protons and 9 electrons; a fluoride ion, F^-, has 9 protons and 10 electrons. A hydrogen atom, H, has 1 proton and 1 elec-

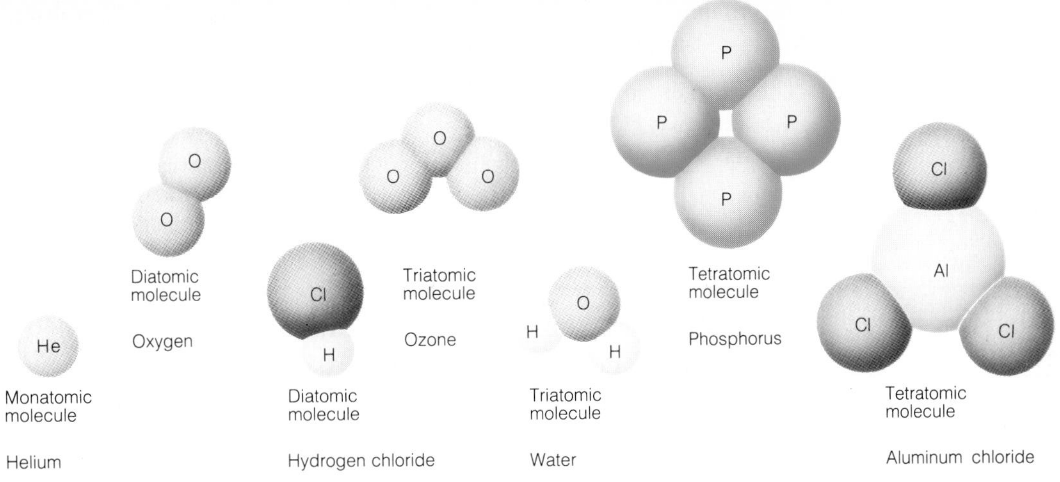

FIGURE 2.14
Some common molecules.

tron; removal of the electron leaves the hydrogen ion, H^+, a proton; addition of an electron yields the hydride ion, H^-, with 1 proton and 2 electrons. When an aluminum atom Al, gives up 3 electrons, it becomes the Al^{3+} ion:

An illustration of the law of conservation of charges: The sum of the charges of products $(+3 - 3 = 0)$ must equal the sum of the charges of reactants (0).

$$Al \longrightarrow Al^{3+} + 3e^-$$

$$\textit{charges:} \quad 0 \longrightarrow 3+ \quad +3-$$

The common ions are given in Tables 2.3 and 2.4. An ion formed from one atom (for example, Na^+, O^{2-}, or Al^{3+}) is called a **monatomic ion**. An ion formed from more than one atom ($SO_4{}^{2-}$, $O_2{}^{2-}$, $PO_4{}^{3-}$, or $NH_4{}^+$, is called a **polyatomic ion.**

Ionic compounds are solid compounds composed of positive and negative ions; they do not contain molecules. Each Na^+ ion is surrounded by six Cl^- ions; and in turn, each Cl^- ion is surrounded by six Na^+ ions (Figure 2.15). An ionic compound is generally formed when a metal (left side of the line in the periodic table) combines with a nonmetal, except a noble gas (right side of the line). Any compound, however, must be electrically neutral. Therefore,

We can write the formula for an ionic compound by combining a number of positive and negative ions so as to balance the charges.

Common practice dictates the use of formulas *without* charges for ionic compounds: NaCl in place of Na^+Cl^-.

For example, we may write NaCl (ordinary table salt) for sodium chloride, a typical ionic compound. The $+1$ charge of the sodium ion balances the -1 charge of the chloride ion. By convention, the symbol of the positive ion is written first, followed by the negative ion.

EXAMPLE 2.3 Write the formulas for (a) calcium sulfate, (b) potassium phosphate, (c) magnesium hydroxide, and (d) zinc phosphate from the ions of which each is composed. (Use Tables 2.3 and 2.4.)

ANSWER

(a) The calcium ion is Ca^{2+}. The sulfate ion is $SO_4{}^{2-}$. One Ca^{2+} balances the negative charge of an $SO_4{}^{2-}$ ion. Therefore, the formula of the neutral compound is $CaSO_4$. (Note that this is not written $Ca_1(SO_4)_1$.)

(b) The potassium ion is K^+. The phosphate ion is $PO_4{}^{3-}$. Three K^+ are needed to balance the negative charge of a $PO_4{}^{3-}$ ion. Therefore, the formula of the neutral compound is K_3PO_4.

(c) The magnesium ion is Mg^{2+}. The hydroxide ion is OH^-. Therefore, the formula of the neutral compound is $Mg(OH)_2$.

(d) The charges on the ions Zn^{2+} and PO_4^{3-} require that three Zn^{2+} ($3 \times +2 = +6$) and two PO_4^{3-} ($2 \times -3 = -6$) are needed to form a neutral compound, written as $Zn_3(PO_4)_2$. ■

TABLE 2.3 COMMON MONATOMIC IONS

POSITIVE IONS	NEGATIVE IONS
Copper(I) (cuprous), Cu^+	Bromide, Br^-
Copper(II) (cupric), Cu^{2+}	Chloride, Cl^-
Potassium, K^+	Fluoride, F^-
Silver, Ag^+	Hydride, H^-
Sodium, Na^+	Iodide, I^-
Barium, Ba^{2+}	Nitride, N^{3-}
Cadmium, Cd^{2+}	Oxide, O^{2-}
Calcium, Ca^{2+}	Sulfide, S^{2-}
Cobalt(II) (cobaltous), Co^{2+}	
Cobalt(III) (cobaltic), Co^{3+}	
Iron(II) (ferrous), Fe^{2+}	
Iron(III) (ferric), Fe^{3+}	
Lead(II), Pb^{2+}	
Magnesium, Mg^{2+}	
Mercury(II) (mercuric), Hg^{2+}	
Nickel, Ni^{2+}	
Strontium, Sr^{2+}	
Tin(II) (stannous), Sn^{2+}	
Tin(IV) (stannic), Sn^{4+}	
Zinc, Zn^{2+}	
Aluminum, Al^{3+}	
Chromium, Cr^{3+}	

TABLE 2.4 COMMON POLYATOMIC IONS

POSITIVE IONS	
Ammonium, NH_4^+	Mercury(I) (mercurous), Hg_2^{2+}
NEGATIVE IONS	
Acetate, $C_2H_3O_2^-$	Carbonate, CO_3^{2-}
Bromate, BrO_3^-	Hydrogen carbonate (bicarbonate), HCO_3^-
Hypochlorite, ClO^-	Chromate, CrO_4^{2-}
Chlorite, ClO_2^-	Dichromate, $Cr_2O_7^{2-}$
Chlorate, ClO_3^-	Peroxide, O_2^{2-}
Perchlorate, ClO_4^-	Sulfite, SO_3^{2-}
Cyanide, CN^-	Sulfate, SO_4^{2-}
Hydroxide, OH^-	Hydrogen sulfate (bisulfate), HSO_4^-
Nitrite, NO_2^-	Phosphate, PO_4^{3-}
Nitrate, NO_3^-	Hydrogen phosphate, HPO_4^{2-}
Permanganate, MnO_4^-	Dihydrogen phosphate, $H_2PO_4^-$

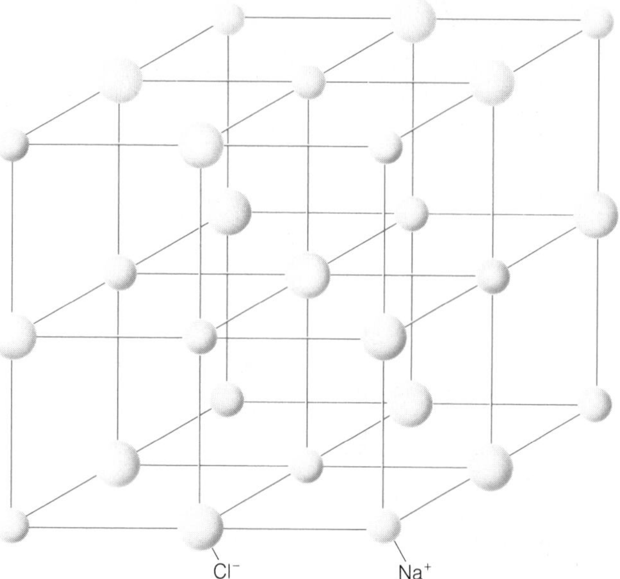

FIGURE 2.15
A representation of solid ionic NaCl. Experimentally, NaCl molecules are not detected in solid ionic NaCl.

PROBLEM 2.6 Write the formulas for (a) magnesium sulfate, (b) sodium carbonate, and (c) aluminum sulfate. □

Tables 2.3 and 2.4 may lead one to conclude that all compounds are ionic. Using Table 2.3, one could correctly write $CrBr_3$ for chromium bromide; but that would not necessarily mean that the compound is ionic. Further, although neither antimony nor arsenic ions, Sb^{3+} and As^{3+}, have been detected, many molecular compounds like SbI_3 and AsF_3 do exist. Thus, many molecular compounds have the molecular formulas that would be predicted for them if they were ionic. This fact makes Tables 2.3 and 2.4 useful for predicting the chemical formulas of ionic *and* molecular compounds, although it cannot be used to predict their ionic or molecular character. However, the system of naming compounds affords another method of writing the formulas of both ionic and molecular compounds, the subject of the next section.

In the next chapter we will study how formulas are derived from experimental data, and in Chapter 7 we will explore the tendency of atoms to form either positive or negative ions.

2.9 NOMENCLATURE OF COMPOUNDS—NAMING COMPOUNDS

Communication is impossible without a rational system of naming substances. Each discipline finds it necessary to develop its own systematic language. We cannot derive much comfort or knowledge from such old names as ''gas inflammable'' (hydrogen), ''lunar caustic'' (silver nitrate), and ''liver of sulphur'' (potassium sulfide). Ancient symbols show no relationship to the chemical composition of the substances they represent (Figure 2.16). However, modern rules of nomenclature (naming substances) issued by the International Union of Pure and Applied Chemistry (IUPAC) make it possible to name millions of compounds in a systematic way with a minimum of memorization.

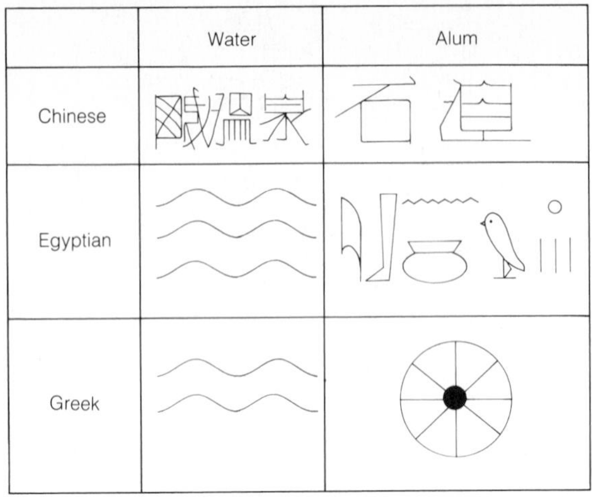

(a)

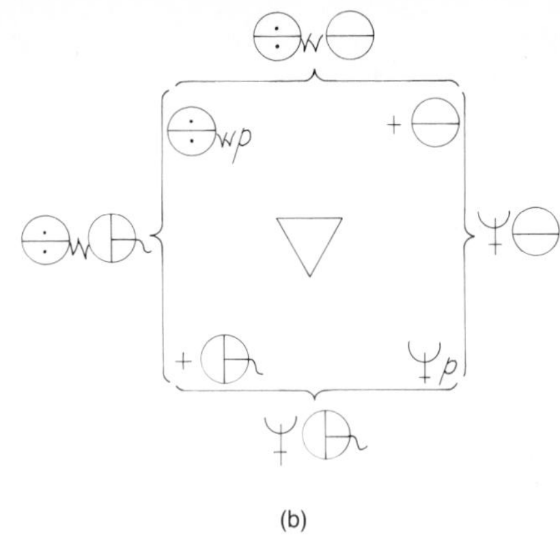

FIGURE 2.16
(a) The symbols used for water (H_2O) and alum ($K_2Al_2(SO_4)_4(H_2O)_{24}$) in three ancient civilizations. (b) Bergman's representation (1775) of the reaction between calcium chloride and potassium sulfate: $CaCl_2 + K_2SO_4 \rightarrow CaSO_4 + 2KCl$. (*Source*: D. H. Rouvray, *Endeavour*, *1*, 1977, p. 23)

BINARY COMPOUNDS

The nomenclature of **binary compounds**, compounds consisting of two elements, makes it possible to write the formulas of these compounds from their names. Binary compounds may be ionic or molecular. Typical binary compounds include HCl, NaCl, NH_3, Al_2O_3, $AlCl_3$, and PCl_5.

Binary ionic compounds consist of a metal cation (such as Na^+ or Al^{3+}) and a nonmetal anion (such as Cl^- or O^{2-}). The name is derived from the names of the elements (see inside back cover). The element that forms a positive ion is written first, and the second element is given the suffix *-ide.* The metals of the representative elements generally form *only one positive ion*; sodium ion, Na^+, magnesium ion, Mg^{2+}, and aluminum ion, Al^{3+} are typical. Zinc is another element that forms only one positive ion, Zn^{2+}. However, the transition elements generally form *more than one positive ion*; Fe^{2+} and Fe^{3+} are typical. When an element forms more than one positive ion, its charge is shown as a Roman numeral; for example, Fe^{2+}, iron(II) and Fe^{3+}, iron(III). In the traditional system (not recommended by IUPAC) these elements are distinguished by the suffix *-ic* for the higher charge and the suffix *-ous* for the lower charge. Examples are given in Table 2.5.

Alfred Stock, famous for his work on the chemistry of boron and silicon hydrides (1910–1936), introduced the system of Roman numerals—the "Stock system."

Binary molecular compounds frequently consist of two nonmetals; H_2O, NH_3, and SF_6 are typical. When naming binary molecular compounds, the use of a Greek prefix to denote the number of atoms of each element is preferable (see marginal note). The prefix *mono-* is generally omitted unless it is needed to distinguish among different binary compounds of the same element. Examples are given in Table 2.6.

1 = *mono-*, 2 = *di-*,
3 = *tri-*, 4 = *tetra-*,
5 = *penta-*, 6 = *hexa-*,
7 = *hepta-*.

Traditional names are used for well-known substances such as water (H_2O), ammonia (NH_3), and ozone (O_3). Bottles of chemicals are still labeled with traditional names such as "ferric chloride" for $FeCl_3$ and "stannous chloride" for $SnCl_2$. See Tables 2.5 and 2.6.

According to the rule, ammonia should be written H_3N, but NH_3 is well established.

In going from the name to the formula, convert each given prefix into a numerical subscript after the element symbol. The Roman numerals give you the charge of the metal ion from which the formula of the compound may be written.

TABLE 2.5
NAMES OF TYPICAL BINARY IONIC COMPOUNDS

COMPOUND	IUPAC NAME	TRADITIONAL NAME
$NaCl$	sodium chloride	sodium chloride
BaF_2	barium fluoride	barium fluoride
MgO	magnesium oxide	magnesium oxide
Li_3N	lithium nitride	lithium nitride
$CuCl$	copper(I) chloride	cuprous chloride
$CuCl_2$	copper(II) chloride	cupric chloride
$SnCl_2$	tin(II) chloride	stannous chloride
$SnCl_4$	tin(IV) chloride	stannic chloride
$FeCl_2$	iron(II) chloride	ferrous chloride
$FeCl_3$	iron(III) chloride	ferric chloride
MnO_2	manganese(IV) oxide	manganese dioxide

TABLE 2.6
NAMES OF BINARY MOLECULAR COMPOUNDS AND ELEMENTS

SUBSTANCE	IUPAC NAME†
N_2O	dinitrogen oxide
NO	nitrogen oxide
N_2O_3	dinitrogen trioxide
NO_2	nitrogen dioxide
N_2O_4	dinitrogen tetroxide
CO	carbon monoxide
CO_2	carbon dioxide
AsF_5	arsenic pentafluoride
H	monohydrogen
H_2	dihydrogen
O_3	trioxygen

† When ambiguity will not result, vowels are frequently condensed, as in phosphorus pentoxide and carbon monoxide, but never when it might be confusing as in diiron trioxide, Fe_2O_3.

For example:

hydrogen chloride, HCl
calcium chloride, $CaCl_2$
manganese(III) oxide, Mn_2O_3
arsenic tribromide, $AsBr_3$
tetraphosphorus decaoxide, P_4O_{10}
nitrogen trichloride, NCl_3
sulfur hexafluoride, SF_6
magnesium nitride, Mg_3N_2
cobalt(III) chloride, $CoCl_3$
chromium(II) iodide, CrI_2

EXAMPLE 2.4 Write the formulas for the compounds diphosphorus tetraiodide, dinitrogen pentoxide, iron(III) oxide, manganese(III) fluoride, and copper(I) oxide.

ANSWER P_2I_4, N_2O_5, Fe_2O_3, MnF_3, Cu_2O. ■

PROBLEM 2.7 Write a formula for each space of this table:

	MONATOMIC	DIATOMIC	TRIATOMIC	TETRATOMIC
Element				
Binary Compound	×			

□

COMPOUNDS CONTAINING POLYATOMIC IONS

Polyatomic ions (Table 2.4) generally behave as discrete units and remain unchanged in most chemical reactions. Compounds containing them are therefore named in the same way as binary compounds: the positive ion is named first. A few polyatomic ions have names ending in *-ide*; for example, hydroxide for OH^-, cyanide for CN^-, and peroxide for O_2^{2-}.

However, many polyatomic ions contain oxygen and another element, such as NO_2^-, MnO_4^-, ClO_3^-. These polyatomic ions are called **oxoanions**. When elements such as N, S, and Cl form more than one oxoanion, the following rules apply:

- Two oxoanions of the same element form: The name of the oxoanion with fewer oxygen atoms ends in *-ite*; the name of the one with more oxygen atoms ends in *-ate*. For example,

  SO_3^{2-} sulf*ite* NO_2^- nitr*ite*
  SO_4^{2-} sulf*ate* NO_3^- nitr*ate*

- Three or four oxoanions of the same element form: The prefixes *hypo-* and *per-* are used. *Hypo-* is assigned to the oxoanion with fewer oxygen atoms than the *-ite*, while *per-* indicates more oxygen atoms than the *-ate*. For example,

  ClO^- (fewer O atoms than chlor*ite*) *hypo*chlor*ite*
  ClO_2^- chlor*ite*
  ClO_3^- chlor*ate*
  ClO_4^- (more O atoms than chlor*ate*) *per*chlor*ate*

In chemistry, *hypo-* generally denotes the compound with the smallest number of oxygen atoms per anion or molecule in a series of compounds, while *per-* denotes the one with the greatest number of oxygen atoms in the series.

Some examples are

$Mg(ClO_4)_2$	magnesium perchlorate
$(NH_4)_3PO_4$	ammonium phosphate
$Sn(CrO_4)_2$	tin(IV) chromate
$Fe_2(SO_4)_3$	iron(III) sulfate
$Ba(OH)_2$	barium hydroxide
K_2CO_3	potassium carbonate
$Fe(NO_3)_2$	iron(II) nitrate
$Cu(CN)_2$	copper(II) cyanide

Notice that, as for the K^+ ion, the charge of the ammonium ion, NH_4^+, a very common positive polyatomic ion, is not shown as a Roman numeral.

If you learn the names and charges of 20 positive and 20 negative ions, you can name and write the formulas for 400 compounds.

PROBLEM 2.8 (a) Name the following: CuI_2, Na_2HPO_4, $Cr_2O_7^{2-}$, $CoCl_2$, $Co(OH)_3$, and $(NH_4)_2CO_3$. (b) Write the formulas for chromium(III) sulfide, mercury(II) chlorate, zinc perchlorate, and ammonium permanganate. □

The nomenclature of other compounds (acids, organic compounds, and coordination compounds) is considered in later chapters.

PERIODICITY OF MONATOMIC IONS

The periodic table is very useful in learning the charges of the more common monatomic ions; see Table 2.7. Notice that the ions of the elements of Groups 1, 2, and 3 have charges equal to the group number, G. On the other hand, the charges for the elements of Groups 5, 6, and 7 are equal to the group number

TABLE 2.7
PERIODICITY OF SOME COMMON MONATOMIC IONS

	1						
1	$H^{+\dagger}$	2	3	4	5	6	7
2	Li^+	$Be^{2+\dagger}$	$B^{3+\dagger}$	C^{4-}	N^{3-}	O^{2-}	F^-
3	Na^+	Mg^{2+}	Al^{3+}	$Si^{4+\dagger}$	P^{3-}	S^{2-}	Cl^-

† These elements do not exist as ions but combine in the same atomic ratios they would have if they were charged. The H^- ion does exist in hydrides, but H^+ exists only bound to other molecules, such as H^+ bound to H_2O, forming H_3O^+.

minus eight, G − 8, and are therefore negative. The Group-4 elements do not usually exist as 4+ or 4− ions. The ions of the transition elements are characterized by variable charges.

PROBLEM 2.9 What is the charge of the strontium ion (Group 2) and of the tellurium ion (Group 6)? Write the formula for strontium telluride. □

The generalizations offered in this section do not always correctly predict which compounds exist and which do not. However, nearly all of the elements form many compounds whose formulas correspond to these generalizations. While N_2O_5 and PCl_5 do exist, NCl_5 does not; and the formulas for compounds such as propane, C_4H_8, benzene, C_6H_6, and sucrose, $C_{12}H_{22}O_{11}$, cannot be written by these methods.

2.10 CHEMICAL EQUATIONS

The greatest application of chemical formulas is in writing chemical equations. A **chemical equation** is a shorthand method of describing a chemical change. In chemical changes, atoms in molecules are separated and then rearranged, forming different molecules. An equation thus shows the substances that react and the substances that are produced in a chemical change.

The products of a reaction are determined by experimentation. The physical states of the substances may also be indicated in the equation according to the following usage:

National Bureau of Standards (U.S.A.) publications of chemical data use (cr) to designate crystalline solids. Some textbooks use (s) to designate both crystalline and noncrystalline solids.

SYMBOL	MEANING
(g)	gas
(ℓ)	liquid
(c)	crystalline solid
(amorph)	amorphous solid
(aq)	aqueous solution, relatively large amount of water present
(sol)	solution other than aqueous

For example, in an experiment calcium(c) is burned in oxygen(g), forming calcium oxide(c), which is then treated with water, forming calcium hydroxide(c). These reactions are abbreviated in chemical equations as shown:

$$Ca(c) + O_2(g) \longrightarrow CaO(c) \qquad \text{(not balanced)}$$

$$CaO(c) + H_2O \longrightarrow Ca(OH)_2(c) \qquad \text{(balanced)}$$

Here the plus sign means "reacts with" and the arrow means "yields" or "produces." The reacting substances, Ca and O_2 or CaO and H_2O, are called *reactants*. The substances formed, CaO and $Ca(OH)_2$, are called *products*. A plus sign is also used on the right side of the arrow when more than one product is obtained.

EXAMPLE 2.5 When potassium nitrate(c) is heated, it decomposes to potassium nitrite(c) and oxygen(g). Write the unbalanced equation for the reaction.

ANSWER

$$KNO_3(c) \longrightarrow KNO_2(c) + O_2(g) \qquad \text{(not balanced)}$$ ■

PROBLEM 2.10 Air (oxygen) is passed over a mixture of cobalt(III) sulfide(c) and calcium oxide(c) at high temperature, forming cobalt(c) and calcium sulfite(c). Write the unbalanced equation for the reaction. □

These unbalanced equations, however, are not correct as written, because they violate the law of conservation of matter. Thus, in the presence of a limited amount of oxygen, carbon reacts to form carbon monoxide:

$$C(amorph) + O_2(g) \longrightarrow CO(g) \qquad \text{(not balanced)}$$

We show 2 atoms of oxygen reacting to form a product containing only 1 atom of carbon. Since atoms are not created or destroyed in chemical changes but merely rearranged, we **balance the equation** by making the number of atoms of each element in the reactants the same as the number of them in the products. *This is accomplished by placing the required number before each formula.* Thus the 2 atoms in O_2 must form 2 molecules of CO:

$$C(amorph) + O_2(g) \longrightarrow 2CO(g) \qquad \text{(not balanced)}$$

But 2 molecules of CO require 2 atoms of carbon. Therefore, the balanced equation is

$$2C(amorph) + O_2(g) \longrightarrow 2CO(g) \qquad \text{(balanced)}$$

The coefficient 1 is omitted; O_2 denotes one molecure of O_2.

A number before a formula, such as the 2 before the C or the CO, is known as its **coefficient**. A coefficient is a multiplier for the entire formula, never for only a part of it. Thus, changing 2CO to 4CO makes it 4 C atoms and 4 O atoms; $4H_2O$ denotes 8 H atoms and 4 O atoms; $3CaCl_2(H_2O)_6$ includes 3 Ca atoms, 6 Cl atoms, 36 H atoms, and 18 O atoms. It should be stressed that

Subscripts in formulas must never be altered.

This error would change the substances involved. For instance, the equation

$$C + O_2 \longrightarrow CO \qquad \text{(not balanced)}$$

cannot be balanced correctly by changing CO to CO_2 or by changing O_2 to O. CO_2 is the formula for carbon dioxide, a substance different from carbon monoxide, CO. The formula O represents atomic oxygen, not molecular oxygen, O_2. Thus, changing a *subscript* in a formula changes the *identity* of the substance; changing a *coefficient* changes the *quantity* of the substance.

PROBLEM 2.11 Balance

$$H_2 + O_2 \longrightarrow H_2O$$ □

$C_3H_8(g)$ is easily liquefied and transported in tanks for use as a fuel—"liquid petroleum gas."

Another example: *hydrocarbons* (carbon-hydrogen compounds) burn in the presence of sufficient oxygen to form carbon dioxide and water. Propane, C_3H_8, is typical. Knowing the formulas for these substances, we may write

$$C_3H_8(g) + O_2(g) \longrightarrow CO_2(g) + H_2O(g) \qquad \text{(not balanced)}$$

$$3\text{ C, 8 H, 2 O atoms} \longrightarrow 1\text{ C, 2 H, 3 O atoms} \qquad \text{(not balanced)}$$

Thus the 3 atoms of carbon must form 3 molecules of CO_2, and the 8 atoms of hydrogen in C_3H_8 must form 4 molecules of H_2O. But the 3 molecules of CO_2 and the 4 molecules of H_2O require 10 atoms or 5 molecules of oxygen. The balanced equation is therefore

$$C_3H_8(g) + 5O_2(g) \longrightarrow 3CO_2(g) + 4H_2O(g) \qquad \text{(balanced)}$$

$$3\text{ C, 8 H, 10 O atoms} \longrightarrow 3\text{ C, 8 H, 10 O atoms} \qquad \text{(balanced)}$$

Always verify that the number of atoms of each element in the reactants is the same as the number in the products.

Note that in balancing the equation for this typical combustion reaction, the sequence is to *first balance the C atoms, then the H atoms,* and finally, the O atoms. Why? In these kinds of reactions, it is easier to start by balancing atoms that appear in only one reactant and one product. For example, C appears only in C_3H_8 (reactant) and CO_2 (product). Also, H appears only in C_3H_8 (reactant) and H_2O (product). Oxygen, however, appears in both products, CO_2 and H_2O. Therefore, save the O atoms for last.

EXAMPLE 2.6 Balance the equation for the combustion of glucose (corn sugar):

$$C_6H_{12}O_6(c) + O_2(g) \longrightarrow CO_2(g) + H_2O(\ell) \qquad \text{(not balanced)}$$

Since this is a trial-and-error process, there are several acceptable methods.

ANSWER Start with carbon. There are 6 C atoms in the reactants and only 1 C atom in the products. Thus the CO_2 molecule is multiplied by 6:

$$C_6H_{12}O_6 + O_2 \longrightarrow 6CO_2 + H_2O \qquad \text{(not balanced)}$$

There are 12 H atoms in the reactants and only 2 in the products. Thus the H_2O molecule is multiplied by 6:

$$C_6H_{12}O_6 + O_2 \longrightarrow 6CO_2 + 6H_2O \qquad \text{(not balanced)}$$

There are now 18 O atoms in the products (6×2 in $6CO_2$ + 6 in $6H_2O$) but only 8 in the reactants. The coefficient of $C_6H_{12}O_6$, however, cannot be changed without upsetting the carbon and hydrogen balance already achieved. We therefore must multiply the O_2 by 6. This balances the equation:

$$C_6H_{12}O_6 + 6O_2 \longrightarrow 6CO_2 + 6H_2O \qquad \text{(balanced)}$$

Verify: 6 C, 12 H, 18 O atoms $\longrightarrow$ 6 C, 12 H, 18 O atoms ■

EXAMPLE 2.7 Balance the equation for the thermal decomposition of potassium chlorate, a common source of small quantities of oxygen[†]

$$KClO_3(c) \longrightarrow KCl(c) + O_2(g) \qquad \text{(not balanced)}$$

[†]**SAFETY NOTE:** The preparation of O_2 from $KClO_3$ must be performed carefully. The presence of organic or combustible material, such as rubber, cork, paper, or sulfur, can lead to disastrous explosions.

Recall that the least common multiple of several whole numbers is the smallest number that is a multiple of all of them: 6 is the least common multiple of 2 and 3, since 2 × 3 = 6 and 3 × 2 = 6.

ANSWER We see that there are equal numbers of K and Cl atoms on each side of the equation. The left side, however, has 3 O atoms and the right side has 2 O atoms. Therefore, find the least common multiple of 3 and 2, which is 6. Thus multiplying $KClO_3$ by 2 and O_2 by 3 balances the O atoms:

$$2KClO_3 \longrightarrow KCl + 3O_2 \qquad \text{(not balanced)}$$

Then multiplying KCl by 2 balances the equation:

$$2KClO_3 \longrightarrow 2KCl + 3O_2$$

Verify: 2 K, 2 Cl, 6 O atoms ⟶ 2 K, 2 Cl, 6 O atoms ■

EXAMPLE 2.8 Balance the equation

$$Al(c) + HCl(aq) \longrightarrow AlCl_3(aq) + H_2(g)$$

for preparing hydrogen from a metal and hydrochloric acid.

ANSWER Note that Al is balanced but that Cl is not because of the subscript 3 in $AlCl_3$. Therefore, place a coefficient of 3 before HCl:

$$Al + 3HCl \longrightarrow AlCl_3 + H_2 \qquad \text{(not balanced)}$$

But the 3 H atoms in 3HCl does not balance the 2 H atoms in the product. The least common multiple of 3 and 2 is 6; so multiply the 3HCl by 2 and the H_2 by 3:

$$Al + 6HCl \longrightarrow AlCl_3 + 3H_2 \qquad \text{(not balanced)}$$

Now, multiplying $AlCl_3$ by 2 balances the Cl atoms:

$$Al + 6HCl \longrightarrow 2AlCl_3 + 3H_2 \qquad \text{(not balanced)}$$

Finally, multiplying Al by 2 balances the equation:

$$2Al + 6HCl \longrightarrow 2AlCl_3 + 3H_2 \qquad \text{(balanced)}$$

Verify: 2 Al, 6 H, 6 Cl atoms ⟶ 2 Al, 6 Cl, 6 H atoms ■

ALTERNATE ROUTE: Remember, at this point, balancing equations is a matter of trial and error. So we could start by balancing the hydrogen atoms:

$$Al + 2HCl \longrightarrow AlCl_3 + H_2 \qquad \text{(not balanced)}$$

This balances the Al and H atoms. But you cannot get 3 atoms of Cl (in $AlCl_3$) from 2 Cl atoms (in 2HCl). Therefore, use the least common multiple of 2 and 3, namely, 6. Thus,

$$Al + 6HCl \longrightarrow 2AlCl_3 + H_2 \qquad \text{(not balanced)}$$

This balances the Cl atoms but unbalances the Al and H atoms. So, multiplying Al by 2 and H_2 by 3 balances the equation:

$$2Al + 6HCl \longrightarrow 2AlCl_3 + 3H_2 \qquad \text{(balanced)}$$

■

In the next example, an equation containing polyatomic ions is balanced.

EXAMPLE 2.9 *Phosphate rock*, $Ca_3(PO_4)_2$, may be converted to phosphoric acid, H_3PO_4, by the reaction

$$Ca_3(PO_4)_2(c) + H_2SO_4(aq) \longrightarrow CaSO_4(c) + H_3PO_4(aq)$$

Balance the equation.

ANSWER Since the composition of polyatomic ions is not changed in going from reactants to products, they may be treated as single units and counted as if they were atoms.

There are 3 Ca atoms and 2 (PO_4) units in the reactants. Thus multiply $CaSO_4$ by 3 and H_3PO_4 by 2:

$$Ca_3(PO_4)_2 + H_2SO_4 \longrightarrow 3CaSO_4 + 2H_3PO_4 \qquad \text{(not balanced)}$$

There are now 6 H atoms and 3 (SO_4) units in the products. Therefore, multiplying H_2SO_4 by 3 balances the equation:

$$Ca_3(PO_4)_2 + 3H_2SO_4 \longrightarrow 3CaSO_4 + 2H_3PO_4 \qquad \text{(balanced)}$$

Verify: 3 Ca, 2 (PO_4), 6 H, 3 (SO_4) atoms and units $\longrightarrow$ 3 Ca, 3 (SO_4), 6 H, 2 (PO_4) atoms and units ■

The balanced equation summarizes the observed chemical change. It does *not*, however, show the actual processes by which reactants are converted to products.

PROBLEM 2.12 Balance each of these equations:

(a) $KClO_4 \longrightarrow KCl + O_2$
(b) $KO_3 \longrightarrow KO_2 + O_2$
(c) $Zn + H_3PO_4 \longrightarrow Zn_3(PO_4)_2 + H_2$
(d) $Al + Cr_2O_3 \longrightarrow Cr + Al_2O_3$
(e) $C_2H_4 + O_2 \longrightarrow CO_2 + H_2O$ □

In the next chapter, you will learn that a balanced chemical equation is more than just a description of a chemical reaction. It is a quantitative statement from which much practical information can be extracted.

SUMMARY

Modern chemistry began with the **atomic theory** that explains the **laws of conservation of matter** and **definite composition** and predicted the **law of multiple proportions**. The later discoveries of the **electron**, the **cathode ray**, and the **alpha particle** led to the discovery of **X rays**. The Rutherford **nuclear model** of the atom explains the deflection of alpha particles by metal foils.

Atoms are the fundamental units of matter. The **atom** is the smallest unit of an element that can combine with other atoms. A **molecule** is a combination of atoms chemically bonded to each other. An atom consists of a nucleus made up of **protons** and **neutrons**. A proton has a unit positive charge; a neutron has no charge. A proton or a neutron is a **nucleon**. The **mass** (nucleon) **number** of an atom is the number of its nucleons. The nuclear charge, the **atomic** (proton) **number**, is the number of protons in the nucleus of an atom. Electrons swarming outside the nucleus make the atom electrically neutral. The electron is the unit particle of electricity carrying a negative charge of 1.60×10^{-19} C. **Isotopes** are atoms of the same atomic number but different mass numbers. **Ions** are electrically charged atoms (**monatomic ions**) or a group of atoms (**polyatomic ions**) formed from neutral atoms or molecules by **ionization**. Ions may be positively charged (electrons are lost) or negatively charged (electrons are gained). By organizing the elements into a table according to atomic numbers, the charge of monatomic ions can be predicted.

The **formula** of a compound can be written from the charges of its constituent ions or from its IUPAC name. A **chemical equation** describes a chemical change by showing the formulas of the **reactants** and **products**. In accord with the law of conservation of matter, a **balanced equation** has equal numbers of atoms of each element in the reactants and in the products. Chemical equations are balanced by placing an appropriate **coefficient** before each formula. The coefficient is a multiplier for the entire formula.

2.11 ADDITIONAL PROBLEMS

DEFINITE COMPOSITION

2.13 An 8.99-g sample of sulfur trioxide, SO_3, contains 3.60 g S. Find the mass of oxygen in the sample.

2.14 Ethanol (grain alcohol) consists of 52.2% C, 13.0% H, and 34.8% O by mass. What is the mass in grams of C, H, and O in 23.0 g of ethanol?

ATOMIC THEORY

2.15 The relative masses of nitrogen and oxygen in five substances are given as O/N: 1.60/1.40, 2.00/3.50, 6.40/2.80, 1.20/0.700, and 10.0/3.50. (a) For each substance calculate the mass of oxygen combined with 7.00 g nitrogen. Are these five different compounds? Explain. (b) Is there something worth noting about the masses of O that combine with the same mass of N?

2.16 For a given set of conditions, how should the number of alpha particles deflected through a given large angle (a) vary (increase or decrease) with kinetic energy of the particle and (b) be related (directly or inversely) to the number of unit positive charges on the nucleus of the atom? (c) Should the angle of deflection increase or decrease as the alpha particle approaches the nucleus more closely?

2.17 For an atom of $^{64}_{30}Zn$, find the density in g/cm^3 (a) of the nucleus: nuclear radius = 4.8×10^{-6} nm, mass of ^{64}Zn atom = 1.06×10^{-22} g; (b) of the space occupied by the electrons: atomic radius = 0.125 nm, electronic mass = 9.11×10^{-28} g. What statement can you make regarding the density of the various parts of an atom? $V = 4/3\ \pi r^3$. The density of osmium, Os, the densest known element, is 22.5 g/mL. How many times denser than Os is the Zn nucleus?

NUCLEAR COMPOSITION

2.18 What is the nuclear composition of each of the following isotopes of platinum, $_{78}Pt$? mass numbers 192, 194, 195, 196, 198. How many electrons are there in each of these atoms?

2.19 How many neutrons, protons, and electrons are in each of the following atoms or ions? $^{209}_{83}Bi^{3+}$, $^{193}_{77}Ir$, $^{51}_{23}V^{5+}$, $^{81}_{35}Br^-$, $^{98}_{42}Mo^{4+}$, $^{32}_{16}S^{2-}$.

2.20 Lithium consists of two isotopes: ^6_3Li and ^7_3Li. For each isotope, (a) how many protons and neutrons are in the nucleus? (b) How many electrons are in each atom? (c) How many protons, neutrons, and electrons are in each of the following? $^6Li^{1+}$, $^7Li^{2+}$, $^6Li^{3+}$, $^7Li^{3+}$.

2.21 How many electrons, protons, and neutrons are in ^{105}Pd, $^{80}Se^{2-}$, $^{114}Cd^+$? (See inside back cover.)

2.22 Complete this table by substituting a numerical value where *w*, *x*, *y*, or *z* appears and the symbol for an element where E appears. (See periodic table, inside front cover.)

ISOTOPE	MASS NUMBER	ATOMIC NUMBER	NEUTRONS	ELECTRONS
^x_yAu	197	*y*	*z*	*w*
$^x_yCo^{2+}$	*x*	*y*	32	*w*
^x_yE	*x*	30	36	*w*
$^x_yE^{2-}$	80	*y*	46	*w*
^x_yKr	84	*y*	*z*	*w*
$^x_yF^-$	*x*	*y*	10	*w*
$^{18}_yO^{2-}$	*x*	*y*	*z*	*w*
^x_yZn	*x*	*y*	34	*w*

2.23 Which of these pairs of atoms contains the same number of (a) protons; (b) neutrons? (1) ^{238}U and ^{238}Pu; (2) ^{78}Br and ^{80}Br; (3) ^{55}Mn and ^{58}Ni; (4) ^{15}N and ^{16}O; (5) ^{209}Bi and ^{209}Po.

2.24 Which of these isotopes has 38 protons and 52 neutrons? (a) $^{87}_{38}E$, (b) $^{90}_{52}E$, (c) $^{90}_{38}E$, (d) $^{52}_{38}E$.

ELECTRONIC CHARGE AND MASS

2.25 Using the table inside the back cover, determine the charge on each of the following species: (a) Ca with 18 electrons; (b) Cu with 28 electrons; (c) Cu with 26 electrons; (d) Pt with 74 electrons; (e) F with 10 electrons; (f) C with 6 electrons; (g) C with 7 electrons; (h) C with 5 electrons; (i) CO with 13 electrons.

2.26 Stoney calculated the magnitude of the electronic charge from the following ideas: during the passage of electricity, a silver ion, Ag^+, picks up one electron, becoming a silver atom. Let *e* be the charge associated with one electron. Then the quantity of electricity required to liberate 6.0×10^{23} Ag atoms is $6.0 \times 10^{23} \times e$. This quantity of electricity is easily measurable: 9.65×10^4 C. (a) Calculate "the natural unit of electricity." (b) Calculate the electron charge using Stoney's original data (1874) obtained at room conditions: (1) 100 coulombs of electricity liberates 10^{-3} g H; (2) $\dfrac{10^{24}\ \text{H atoms}}{1\ \text{L}}$; (3) $\dfrac{0.1\ \text{g H}}{1\ \text{L}}$; (4) $H^+ + e \longrightarrow H$.

2.27 *Millikan oil drop experiment.* The velocity of rise of a positively charged oil drop in an electric field is measured in an apparatus (see diagram). The oil drop's velocity of fall under the action of gravity is measured with the electric field turned off. The charge on the drop may be calculated from these data. Some typical results (1917) are

OIL DROP NUMBER	MEASURED CHARGE ON DROP, coulombs
9	1.59×10^{-19}
14	11.1×10^{-19}
12	9.54×10^{-19}
3	15.9×10^{-19}
15	6.36×10^{-19}

What is (a) the electronic charge and (b) the charge on each drop in units of the electronic charge? (c) What charge on a drop would have to be measured to compel a change in the accepted value of the electronic charge? (d) Use your answers for (b) to find the electronic charge obtained for each drop. Calculate the average deviation and the error for the determination of the electronic charge. The accepted value is 1.60×10^{-19} C. What statement can you make about the precision and the accuracy of these measurements?

2.28 (a) After e/m_e was measured by Thomson, e/m_H was calculated from measurements made during the passage of electricity through acid solutions. The ratio of these two measurements is $m_H/m_e = 1845$. The accepted mass of an H atom was 1.68×10^{-24} g. Find the mass of an electron based on these measurements. (b) More recent measurements show that the ratio of proton mass to electron mass is 1836.1527. The proton mass is $1.6726230 \times 10^{-24}$ g. Calculate the mass of an electron.

FORMULAS

2.29 Write the formula of each of the following compounds: (a) calcium chlorate, (b) potassium oxide, (c) zinc fluoride, (d) silver(I) carbonate, (e) ammonium sulfate, (f) barium nitrate, (g) aluminum chloride, (h) tin(II) fluoride, (i) lead(II) sulfide, (j) magnesium sulfide, (k) sodium phosphate, (l) copper(II) hydroxide.

2.30 Write the formula of each of the following: (a) barium phosphate, (b) aluminum phosphate, (c) mercury(II) chloride, (d) tetraaluminum tricarbide, (e) iron(II) phosphate, (f) chromium(III) sulfate.

2.31 Write the formula of the compound that each of the following elements forms with a sulfate ion: (a) Mg, (b) Al, (c) Ba, (d) Sn(II), (e) Sn(IV), (f) Zn, (g) Cr(III).

2.32 Correct the following formulas. Some of them are incorrect; others are written in unconventional ways: (a) $BaOH_2$, (b) $(Na)_2(CO_3)$, (c) $Mg(CO_3)$, (d) $K(ClO_3)$, (e) $KCl(O)_3$, (f) $Zn(CO_3)_2$, (g) $(NH_4{}^2)SO_4$.

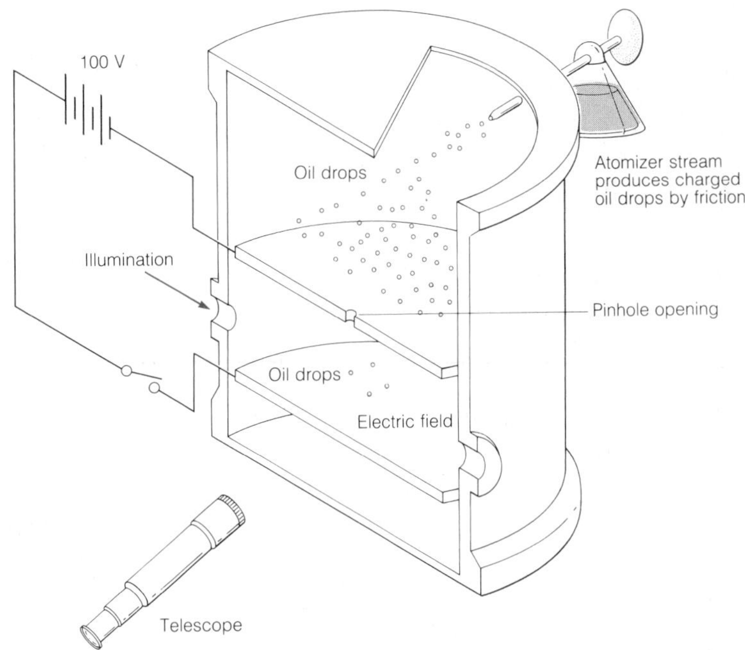

The Millikan apparatus for determining the electronic charge. Oil drops pass through the pinhole into the electric field. A telescope containing a length scale is used to measure the velocity of rise and fall. Oil drops also become charged by picking up ionized air molecules.

2.33 Using this chart,

1	2	3	4	5	6	7

predict and write in charges of the monatomic ions of the representative elements in period 5 of the periodic table, starting with Rb. (a) Write the formulas of the fluorides of the Group-1 through Group-4 elements in period 5. (b) Can you be certain of the existence of these fluoride compounds? (c) Can you be certain of the existence of these ions?

2.34 What is the charge of the cation in each of these substances (assume they are ionic): (a) V_2O_5, (b) ScF_3, (c) MnO_2, (d) $CoCl_3$, (e) SnO_2, (f) $La_2(SO_4)_3$.

NOMENCLATURE

2.35 Name these substances: (a) O, O_2, O_3, (b) Bi_2O_5, (c) $Mg(OH)_2$, (d) $Sn(SO_4)_2$, (e) $ZnSO_4$, (f) NH_4ClO_3, (g) Na_2SO_4, (h) CaF_2, (i) K_2S, (j) Li_2CO_3, (k) $FeBr_2$.

2.36 Name the following: (a) $Pb(C_2H_3O_2)_2$, (b) Ag_2O, (c) $Mg(HCO_3)_2$, (d) BaI_2, (e) KCN, (f) $Cu(NO_3)_2$, (g) $Cd(NO_3)_2$, (h) $Mg(ClO_3)_2$, (i) Li_2O_2, (j) $CaHPO_4$, (k) MnO_4^-, (l) $H_2PO_4^-$, (m) $Cr_2O_7^{2-}$.

2.37 In each box of this chart write the formula of the compound formed by the ions in that column and row. Name each compound.

	F^-	OH^-	SO_3^{2-}	PO_4^{3-}
NH_4^+				
K^+				
Mg^{2+}				
Cu^{2+}				
Fe^{3+}				

2.38 Name these compounds: (a) Cl_2O, (b) ICl, (c) CF_4, (d) P_4O_6, (e) XeO_4, (f) Li_3N, (g) Al_2S_3, (h) $Al_2(SO_3)_3$, (i) $Al_2(SO_4)_3$, (j) UCl_3, (k) UCl_6.

2.39 Write a name for each formula and a formula for each name:

(a) MgF_2 (b) cadmium dichromate
(c) $(NH_4)_2Cr_2O_7$ (d) ammonia
(e) PbS (f) iron(III) sulfate
(g) Na_3N (h) tin(IV) oxide
(i) BrF_5 (j) nitrogen trichloride
(k) $Al(OH)_3$ (l) cobalt(III) hydride

2.40 Write a chemical formula for each of the following: (a) hydrogen peroxide, (b) diphosphorus pentasulfide, (c) dichlorine heptoxide, (d) tetrasulfur tetranitride, (e) tetraphosphorus decaoxide, (f) sulfur monochloride, (g) disulfur dichloride, (h) chromium(VI) oxide, (i) nickel(III) oxide, (j) vanadium(V) chloride.

CHEMICAL EQUATIONS

2.41 Balance these equations:

(a) $H_2O_2 \longrightarrow H_2O + O_2$
(b) $Zn + HCl \longrightarrow ZnCl_2 + H_2$
(c) $Ca + H_3PO_4 \longrightarrow Ca_3(PO_4)_2 + H_2$
(d) $NaClO_4 \longrightarrow NaCl + O_2$
(e) $N_2O_5 + H_2O \longrightarrow HNO_3$
(f) $Fe + H_2O \longrightarrow Fe_3O_4 + H_2$
(g) $C_4H_{10} + O_2 \longrightarrow H_2O + CO_2$
(h) $P_4O_{10} + H_2O \longrightarrow H_3PO_4$
(i) $Ba(OH)_2 + H_3AsO_4 \longrightarrow Ba_3(AsO_4)_2 + H_2O$
(j) $Mn_3Ga_2S_6 \longrightarrow MnS + Ga_2S + S_2$
(k) $TiO_2 + C + Cl_2 \longrightarrow TiCl_4 + CO$

2.42 Balance these equations:

(a) $K_2CO_3 + HCl \longrightarrow KCl + H_2O + CO_2$
(b) $XeF_6 + H_2O \longrightarrow XeO_3 + HF$
(c) $PuO_2 + O_2F_2 \longrightarrow PuF_6 + O_2$
(d) $MnO_2 + HCl \longrightarrow MnCl_2 + Cl_2 + H_2O$
(e) $CO + H_2 \longrightarrow CH_4 + H_2O$
(f) $S_2Cl_2 + C \longrightarrow CCl_4 + S$
(g) $NH_3 + O_2 \longrightarrow N_2 + H_2O$
(h) $Na_3N + H_2O \longrightarrow NH_3 + NaOH$
(i) $Mg_3N_2 + HCl \longrightarrow NH_4Cl + MgCl_2$
(j) $Ca(OH)_2 + H_3PO_3 \longrightarrow CaHPO_3 + H_2O$
(k) $FeCl_3 + NH_3 + H_2O \longrightarrow Fe(OH)_3 + NH_4Cl$

2.43 Write the formula of each of these compounds and balance each equation:

(a) barium carbonate + hydrogen bromide ⟶ barium bromide + carbon dioxide + water
(b) sodium sulfide + water ⟶ sodium hydroxide + hydrogen sulfide
(c) potassium hydroxide + potassium dihydrogen phosphate ⟶ potassium phosphate + water
(d) mercury + ammonium iodide ⟶ mercury(II) iodide + hydrogen + ammonia

SELF-TEST

2.44 (8 points) A 2.50-mg sample of magnesium powder is ignited with 2.00 mg oxygen in a sealed container. All of the magnesium is consumed, and 4.15 mg of a white solid, magnesium oxide, is formed. Find the mass of the unreacted oxygen.

2.45 (10 points) The mass ratios of O/Mg determined for different masses of magnesium in an oxide of magnesium are $\frac{1.60}{2.43}$, $\frac{0.658}{1.00}$, and $\frac{2.29}{3.48}$. Do these ratios confirm the law of definite composition?

2.46 (10 points) (a) A negatively charged drop carries a charge of 3.68×10^{-18} C. How many excess electrons are on the drop? (b) How many electrons have been removed from a drop carrying a charge of 2.24×10^{-18} C?
2.47 (12 points) What is the nuclear composition of and the number of electrons in each of these atoms or ions? ^{233}U, $^{235}U^{2+}$, $^{69}Ga^{3+}$, $^{12}C^{4-}$, $^{239}Pu^{+}$, $^{14}N^{3-}$.

2.48 (10 points) Use the periodic table inside the front cover to complete this table:

SYMBOL	NUMBER OF PROTONS	NUMBER OF NEUTRONS	ION CHARGE	NUMBER OF ELECTRONS
$^{19}_{9}F^-$				
	12	12	0	
	38	52	2+	
	16	16		18

2.49 (8 points) Find the charge of the ion formed in each of the following ionization reactions: (a) Fe → Fe ion + $2e^-$; (b) SO_2 → SO_2 ion + $3e^-$; (c) P + $3e^-$ → P ion.
2.50 (10 points) Write a formula for each of these compounds: magnesium phosphate, gallium sulfate, cobalt(III) carbonate, ammonium sulfide, chromium(III) carbonate.
2.51 (12 points) Name the given compounds and ions: $PbCl_2$, $PbCl_4$, $(NH_4)_2SO_3$, $Ba(NO_3)_2$, $Cu(C_2H_3O_2)_2$, CuBr, $NaNO_2$, ClO_3^-, O_2^{2-}.
2.52 (10 points) Balance these equations:
(a) $SiO_2 + C \longrightarrow Si + CO$
(b) $PBr_3 + H_2O \longrightarrow H_3PO_3 + HBr$
(c) $C_{12}H_{22}O_{11} + O_2 \longrightarrow H_2O + CO_2$
(d) $NO_2 + H_2O \longrightarrow HNO_3 + NO$
(e) $(NH_4)_2Cr_2O_7 \longrightarrow Cr_2O_3 + H_2O + N_2$
2.53 (10 points) Write the formula of each of the substances and balance the equations:
(a) aluminum + iron(III) oxide ⟶ aluminum oxide + iron
(b) potassium hydroxide + zinc chlorate ⟶ zinc hydroxide + potassium chlorate
(c) silver(I) nitrate + hydrogen sulfide ⟶ silver(I) sulfide + hydrogen nitrate
(d) sodium carbonate + hydrogen chloride ⟶ sodium chloride + carbon dioxide + water
(e) ammonium sulfate + sodium hydroxide ⟶ ammonia + sodium sulfate + water

3

ATOMIC AND MOLECULAR WEIGHTS, CHEMICAL ARITHMETIC

Our discussion of the atomic theory in Chapter 2 established that mass is a fixed property of an atom. The mass of an atom remains constant during chemical changes, reflecting the laws of conservation of matter and definite composition. Different kinds of atoms differ in mass because they contain different numbers of protons and neutrons. In this chapter, the concepts of Chapter 2 will be studied quantitatively. We will learn how chemical reactions studied in a laboratory yield information about the mass composition of substances. Measuring the mass composition is the first step in finding the kind and number of atoms in molecules.

Early attempts to determine how much heavier one kind of atom is than another met with enormous confusion, ambiguity, and frustration. This important problem plagued chemists and hindered progress through half a century following Dalton's applications of the atomic theory. In fact, the idea that laboratory measurements could provide a method for determining how much heavier one kind of molecule is than another was not widely accepted until 1858. Scientists still use a relative mass scale because it is convenient and because relative masses are more accurately measured than are the actual masses of atoms and molecules.

Here, we will present only the modern method of measuring relative masses of particles. Then we will see how formulas are obtained from this information and how these formulas are used in chemistry, particularly for determining mass relationships in chemical reactions. Determination of the quantity of a substance is referred to as **quantitative analysis**. This chapter introduces the general method of **volumetric analysis** in which most of the measurements are of liquid volumes.

3.1 INTRODUCTION

Of all the ideas that chemists have contributed to human thought, certainly the most valuable and productive is the concept of atoms and molecules. In fact, when the atomic theory finally emerged, chemists found it so useful that many of them accepted it before they really believed in it. (See Box 3.1.)

In the eighteenth and nineteenth centuries, when chemists began to use that marvelous device, the balance, to weigh the reactants and products of chemical changes, the atomic theory proved to be extremely useful for systematizing their

BOX 3.1
DID YOUR GREAT-GRANDPARENTS BELIEVE IN ATOMS?[†]

"The doctrine of atoms is still alive although it came into being about one hundred years ago. It has proved to be illogical just as the ether that fills all space has been shown to be incapable of existence. Properties must be ascribed to the atom that it cannot possess and the same is true of the ether. What are we to do? Throw over the atom and the ether? Although both have been convicted of being illogical, I do not think it would be logical to give them up, for they are helpful in spite of their shortcomings, and in some way they suggest great truths. They are symbolic."

[†] From Ira Remsen's presidential address to the American Chemical Society, 1902.

experimental results. Although it would have been more satisfactory to be able to weigh atoms one at a time, atoms were still imaginary concepts; they could not be detected, let alone placed on a balance pan. However, chemists realized that it was just as useful to know the relative masses of different atoms.

3.2 THE MOLE

When we talk of *relative masses* of atoms, we mean "*how many times heavier*" one atom is than another. The relative masses of different atoms are called **atomic weights.**[†] To determine atomic weights, we must be able to weigh the *same* large number of atoms of each element. The relative masses of these large numbers are also the relative masses of the individual atoms. For example, one helium atom, ^{4}He, is four times heavier than one hydrogen atom, ^{1}H; 100 ^{4}He atoms are four times heavier than 100 ^{1}H atoms, and 10^{23} ^{4}He atoms are four times heavier than 10^{23} ^{1}H atoms. Methods for determining relative masses were in fact developed. This idea is important, and the following analogy might be helpful.

Imagine that you want to determine the mass of a dime relative to the mass of a nickel. This is analogous to a chemist wanting to determine the mass of an oxygen atom relative to the mass of a carbon atom. Although it is possible to detect a single atom, it is not yet possible to weigh one. Therefore, for illustrative purposes, assume that your laboratory scale can weigh nothing smaller than a kilogram; it is too insensitive to weigh a small number of coins. However, you have a large bag of quarters and a change-making machine that gives out 2 dimes and 1 nickel for each quarter (Figure 3.1). Empty the bag of coins into the machine. You need not know how many quarters are in the bag; just call the number N. Then

$$1 \text{ quarter} \longrightarrow 2 \text{ dimes} + 1 \text{ nickel}$$

and

$$N \text{ quarters} \longrightarrow 2N \text{ dimes} + N \text{ nickels}$$

[†] Atomic weights are dimensionless numbers because they are ratios of masses. A new quarter has a mass of 5.00 g and a new penny has a mass of 3.15 g. Thus a quarter is 1.59 times heavier than a penny; it is not 1.59 g heavier. Compared to one penny, its relative mass is 1.59:

$$\frac{\text{mass of 1 quarter}}{\text{mass of 1 penny}} = \frac{1.59 \times \cancel{\text{mass of one penny}}}{1 \times \cancel{\text{mass of one penny}}}$$

The concept of atomic weight originated with John Dalton.

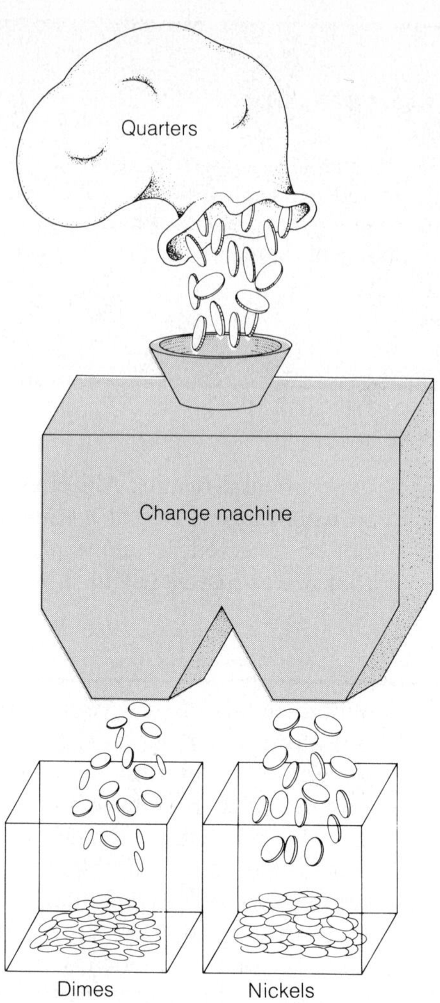

FIGURE 3.1
A change machine gives 2 dimes and 1 nickel for each quarter.

Now, weigh all the dimes, and then all the nickels. Since you have twice as many dimes as nickels, divide the mass of the dimes by 2 and get the relative mass ("weight") of 1 dime and 1 nickel. You now have

$$\frac{\frac{1}{2}(\text{mass of } 2\ N \text{ dimes})}{\text{mass of } N \text{ nickels}} = \frac{\text{mass of } N \text{ dimes}}{\text{mass of } N \text{ nickels}} = \frac{\text{mass of 1 dime}}{\text{mass of 1 nickel}}$$

Any one dime or nickel may have a mass slightly different from other dimes or nickels, but the average mass of a large number of dimes or nickels is a constant. The result is the same for the masses of 1000 dimes and 1000 nickels or for any equal number of dimes and nickels. The task has been accomplished. You have determined the relative mass of a dime to a nickel without bothering to count out quarters, dimes, or nickels.

Actually, they are *relative atomic masses*, but the expression "atomic weight" is too well established to be changed.

When such procedures, using different but analogous methods, are carried out with atoms, the results obtained are the *atomic weights*. However, in place of ratios, such as that of the mass of 1 C atom to the mass of 1 H atom, it is much more convenient to determine atomic weights relative to some fixed mass of one kind of atom. The chosen mass contains a definite number of atoms and it becomes the standard. All other atomic weights are then determined relative to this stan-

By a majority vote of chemists taken in 1905, exactly 16 g of natural oxygen was chosen to avoid using a number smaller than 1 for the atomic weight of hydrogen. In 1960 this was changed to exactly 12 g of ^{12}C. This produced only small changes in atomic weights; for example, the atomic weight of oxygen changed from 16 (exactly) to 15.9994.

dard. Through history, several fixed masses of hydrogen and oxygen have been selected. The standard now in use is exactly 12 g of the most common isotope of carbon, carbon-12, ^{12}C. This exact mass of ^{12}C contains an exact whole number of atoms. However, that number is not known exactly because it has not been possible to count the ^{12}C atoms in 12 g of ^{12}C as one would tally the number of apples in a box or the number of books on a shelf. Nevertheless, it is possible to obtain the number of ^{12}C atoms in exactly 12 g ^{12}C by several independent methods to six significant figures; that number is 6.02214×10^{23}. In this way we assign a *relative mass* of 12 (exactly) to the ^{12}C atom. This is the same as saying

$$\frac{\text{mass of } 6.02214 \times 10^{23}\ {}^{1}\text{H atoms}}{\text{mass of } 6.02214 \times 10^{23}\ {}^{12}\text{C atoms}} = \frac{\text{mass of 1 } {}^{1}\text{H atom}}{\text{mass of 1 } {}^{12}\text{C atom}}$$

and it is analogous to

$$\frac{\text{mass of } N \text{ dimes}}{\text{mass of } N \text{ nickels}} = \frac{\text{mass of 1 dime}}{\text{mass of 1 nickel}}$$

When this ratio is measured, the mass of one ^{1}H atom *relative* to the mass of one ^{12}C atom, the atomic weight of ^{1}H, is determined.

The number of ^{12}C atoms in exactly 12 g ^{12}C, 6.02214×10^{23}, is called the **Avogadro number**, symbolized N_A.[†] The *amount of any substance that contains the Avogadro number of particles* is called a **mole**.

The word mole, from the Latin *moles* for "mass" or "bulk," was introduced by Wilhelm Ostwald in 1896 for describing a collection of a fixed number of similar atoms or molecules.

The given data in many of the exercises in this book limit the number of significant figures to three. In those cases it is convenient to use the value:

1 mol = 6.02×10^{23} particles

When the word *mole* is used, the particles must be specified (or at least understood). They may be atoms, molecules, electrons, ions, and so on. It is therefore perfectly correct to speak of a "mole of electrons," which means as many electrons as there are carbon atoms in 12 g of carbon-12. When a chemist speaks of a "mole of water," it is understood to mean a mole of water molecules, since water consists of molecules. Therefore the complete meaning of a mole of water is "an amount of water that contains as many molecules as there are atoms in 12 g of carbon-12." The symbol for mole is *mol.* A mole thus contains 6.02×10^{23} particles, written as 6.02×10^{23} particles/mol. We can also say that 1 mole of ^{12}C has a mass of 12 g or that the **molar mass** of ^{12}C is 12 g/mol, written as 12 g ^{12}C/mol. Thus, the molar mass of a substance is the mass in grams of one mole of the substance and its unit is g/mol. We can also say that the mass of an Avogadro number of particles of any species is a molar mass of that species: 1 mol of ^{12}C contains 6.02×10^{23} atoms of ^{12}C and has a mass of 12 g (exactly).

As used in chemistry, *species* refers to any kind of particle—atom, molecule, ion, etc.

The molar mass of ^{12}C is 12 g ^{12}C/mol.

1 mole of He contains 6.02×10^{23} atoms of He and has a mass of 4.00 g.

The molar mass of He is 4.00 g He/mol.

The molar mass of several common elements:

aluminum	26.9815 g/mol
copper	63.546 g/mol
iron	55.847 g/mol
lead	207.2 g/mol

You should recognize that the Avogadro number, 6.02×10^{23} particles/mol, is a special name for a standard (if very large) number, analogous to 12 (12 eggs/dozen), 144 (144 nails/gross), or 500 sheets (500 sheets/ream).

[†] In honor of Amedeo Avogadro. You will read more about him in Chapter 10 and learn one way to determine the number named after him.

EXAMPLE 3.1 You place this order with a general store:

12 = dozen
144 = gross
500 sheets = ream
6.02×10^{23} particles = Avogardro number

2 dozen eggs
2 reams of letter paper
2.00 moles of helium atoms, He
2.00 moles of hydrogen molecules, H_2

How many (a) eggs, (b) sheets of paper, (c) atoms of He, (d) molecules of H_2 have you ordered?

ANSWER To answer these questions, you must know how many eggs there are in a dozen, how many sheets in a ream, and how many particles in a mole.

(a) There are 12 objects in 1 dozen. The conversion factor is therefore 12 eggs/doz:

$$2\ \cancel{\text{doz}} \times \frac{12\ \text{eggs}}{1\ \cancel{\text{doz}}} = 24\ \text{eggs}$$

(b) There are 500 sheets in 1 ream. The conversion factor is 500 sheets/ream:

$$2\ \cancel{\text{reams}} \times \frac{500\ \text{sheets}}{1\ \cancel{\text{ream}}} = 1000\ \text{sheets}$$

(c) There are 6.02×10^{23} atoms in 1 mole of atoms. The conversion factor is $\frac{6.02 \times 10^{23}\ \text{He atoms}}{1\ \text{mol He}}$:

$$2.00\ \cancel{\text{mol He}} \times \frac{6.02 \times 10^{23}\ \text{He atoms}}{1\ \cancel{\text{mol He}}} = 1.20 \times 10^{24}\ \text{atoms He}$$

(d) There are 6.02×10^{23} molecules in 1 mole of molecules. The conversion factor is $\frac{6.02 \times 10^{23}\ H_2\ \text{molecules}}{1\ \text{mol}\ H_2}$:

$$2.00\ \cancel{\text{mol}\ H_2} \times \frac{6.02 \times 10^{23}\ H_2\ \text{molecules}}{1\ \cancel{\text{mol}\ H_2}} = 1.20 \times 10^{24}\ \text{molecules}\ H_2$$ ■

PROBLEM 3.1 (a) How many molecules are in 0.45 mol H_2O? (b) You are given 1 lb each of navy beans and rice kernels. Do you have an equal number of beans and kernels? If you count and weigh an equal number of each, will you have equal weights of beans and kernels? Will you have determined their relative weights? □

3.3 ATOMIC WEIGHTS AND THE MASS SPECTROMETER

Stanislao Cannizzaro deduced the first set of accepted atomic weights from the molecular weights of substances (section 3.4). The Cannizzaro method (1858), though clever and intellectually satisfying, is now only of historical interest (Problem 3.31). Today, we can obtain very precise atomic weights using the instrument known as a **mass spectrometer** (Figure 3.2 and Box 3.2). When we examine atoms of nitrogen in a mass spectrometer, two kinds of nitrogen atoms, which differ in mass, are detected; one atom of nitrogen is 1.0712 times heavier than the other atom. Nitrogen thus has two isotopes. Oxygen and carbon each have three isotopes. Although a few elements, such as beryllium and phosphorus, consist of only one kind of atom, most elements exist naturally as mixtures of isotopes. However, most of these mixtures have constant composition. (The few exceptions are not significant for our purposes.)

^{14}N, ^{15}N
^{12}C, ^{13}C, ^{14}C
^{16}O, ^{17}O, ^{18}O

In the magnetic field the ions travel in a circular path.

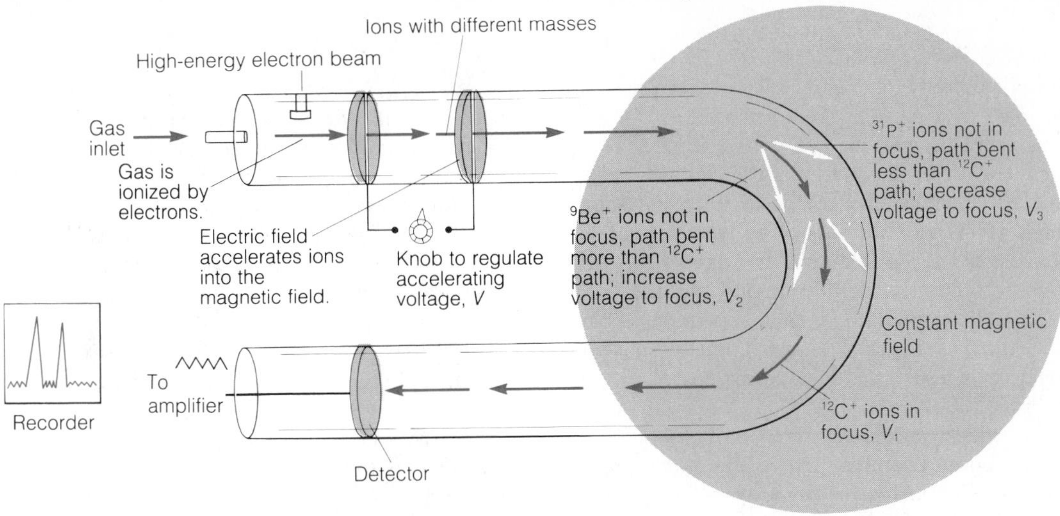
Ions with different masses
High-energy electron beam
Gas inlet
Gas is ionized by electrons.
Electric field accelerates ions into the magnetic field.
Knob to regulate accelerating voltage, V
⁹Be⁺ ions not in focus, path bent more than ¹²C⁺ path; increase voltage to focus, V₂
³¹P⁺ ions not in focus, path bent less than ¹²C⁺ path; decrease voltage to focus, V₃
Constant magnetic field
¹²C⁺ ions in focus, V₁
To amplifier
Recorder
Detector

FIGURE 3.2
Schematic diagram of a mass spectrometer. Heavier ions, more difficult to deflect in the magnetic field, are focused by decreasing the voltage, which decreases their velocity The apparatus is vacuum sealed. Mass spectrometers are now also used to identify complex molecules such as pesticides and to measure the quantities of cocaine in coca plants.

As previously stated, the masses of atoms are measured relative to the mass of the most common isotope of carbon, carbon-12, $^{12}_{6}C$. As an example, let us determine the atomic weight of beryllium, Be. The mass spectrometer shows that the mass of the beryllium atom is slightly more than 3/4, or 0.751015, of the mass of the ^{12}C atom:

$$\frac{\text{mass of Be atom}}{\text{mass of } ^{12}\text{C atom}} = 0.751015$$

or

$$\text{mass of one Be atom} = \text{mass of one } ^{12}\text{C atom} \times 0.751015$$

and

12 is an exact number by definition.

$$\text{atomic weight of Be} = 12 \times 0.751015 = 9.01218$$

The atomic weight of beryllium is thus recorded as 9.01218 (see inside back cover).

Provided that we have the *same number* of Be and ^{12}C atoms, the mass ratio for the atoms is the same as the mass ratio of one atom of Be to one atom of ^{12}C. Thus, since 1 mole of ^{12}C atoms has a mass of 12 g (exactly), 1 mole of Be atoms has a mass of 9.01218 g, or the molar mass of Be is 9.01218 g/mol. We can also simply write 9.01218 g Be/mol (Table 3.1 and Box 3.3).

In this way, we find that the atomic weight of the most common isotope of oxygen is 15.9949. This isotope is called oxygen-16, written as ^{16}O. The atomic weights of the other isotopes of oxygen are 16.9991 and 17.9992. They are called oxygen-17 and oxygen-18 and written as ^{17}O and ^{18}O. The atomic weights of the isotopes of nitrogen, ^{14}N and ^{15}N, are 14.003 and 15.000. We now see that the mass number of an isotope is the whole number nearest its atomic weight. Also note that the *isotopic weights*, the atomic weights of isotopes, are nearly whole numbers.

BOX 3.2

THE MASS SPECTROMETER

The mass spectrometer (Figure 3.2) measures the ratio of the mass of a particle, M, to its charge, Q: M/Q. The charge of the particle is a whole number of unit charges. However, the energy of the electron beam is adjusted to get singly charged positive ions (+1 ions). For high precision, the mass of an electron is added to the mass of the ion to obtain the mass of the atom, but this correction is significant only for the lighter atoms (see Problem 3.56). It thus appears possible to measure the mass in grams of an isotope, but it is more practical to measure the relative masses of different isotopes.

In the mass spectrometer, a beam of gaseous atoms enters a vacuum chamber, where the atoms are ionized (atom $\longrightarrow$ ion$^+$ + e^-) by an electron beam. The ions are accelerated by a voltage into a magnetic field of fixed strength. The magnetic field then deflects the ions into a circular path. For a magnetic field of given strength and a given accelerating voltage, V, the ion path of the heavier +1 ions is bent less than the path of the lighter +1 ions. Thus the circular path traced out by each ion depends on the mass of the ion and the accelerating voltage. A device at the end of the apparatus detects the ions. However, the ions must circulate at a fixed radius to reach the detector (Figure 3.2). By changing the voltage, ions of different mass are focused at the detector. Decreasing the voltage decreases the radius of the circular path of an ion. Increasing the voltage increases the radius of the path. It is thus possible to scan a range of ion masses. The voltage scale is easily converted to a mass scale, since at constant magnetic field, the ion mass is inversely related to the voltage:

$$\frac{\text{mass of atom}^+}{\text{mass of }^{12}\text{C}^+} = \frac{V_1}{V_2} \qquad \frac{\text{mass of }^{9}\text{Be}}{\text{mass of }^{12}\text{C}} = 0.751015$$

The measurement of V_1 (for ^{12}C) and V_2 (for the other atom) yields the mass of other atoms relative to ^{12}C. If all of the atoms of an element have the same mass, as in Be or He, all of them reach the detector at the same voltage. If they differ in mass, more than one voltage is required to detect the isotopes. As we mentioned before, radioactivity provided the first clue to the existence of isotopes, but their separation and mass determination were first achieved in a mass spectrograph. A mass spectrograph is the same as a mass spectrometer except that a photographic plate is used as the ion detector instead of a metal plate (ion collector) and signal amplifier.

The original mass spectrograph set up by Francis Aston in the Cavendish Laboratory, Cambridge University, in 1919. Pioneers in the field of mass spectroscopy (about 1910) include Joseph J. Thomson.

TABLE 3.1
MASS EXPRESSIONS FOR BERYLLIUM

MASS NUMBER	9 or ${}^{9}_{4}$Be or Beryllium-9, from 4 protons plus 5 neutrons
ATOMIC WEIGHT	9.01218, from $\frac{\text{mass of Be atom}}{\text{mass of }^{12}\text{C atom}} = 0.751015$ and $12 \times 0.751015 = 9.01218$ (ratios, no units)
MOLAR MASS	9.01218 g/mol, from mass of 1 mole of ^{12}C = 12 g (exactly, by definition) and $\frac{12\text{ g }^{12}\text{C}}{1\text{ mol}} \times \frac{0.751015\text{ g Be}}{1\text{ g }^{12}\text{C}} = 9.01218\text{ g/mol}$
MASS OF 1 ATOM	$\frac{9.01218\text{ g/mol}}{N_A} = \frac{9.01218\text{ g/mol}}{6.022137 \times 10^{23}\text{ atoms/mol}} = 1.496509 \times 10^{-23}\text{ g/atom}$
ATOMIC MASS UNITS (amu) OR DALTONS PER ATOM	$9.01218\,\frac{\text{amu}}{\text{atom}}$ or $9.01218\,\frac{\text{daltons}}{\text{atom}}$, from amu $= \frac{1}{N_A}$ (by definition) $= 1.6605402 \times 10^{-24}\,\frac{\text{g}}{\text{amu}}$ and 1.496509×10^{-23} g/atom, thus $\frac{1.496509 \times 10^{-23}\text{ g/atom}}{1.6605402 \times 10^{-24}\text{ g/amu}} = 9.01218\,\frac{\text{amu}}{\text{atom}}$

Note: This book will use only atomic (molecular) weight and molar mass.

The relative number is generally referred to as the *relative abundance*.

A spectrum (Latin for "image") is a separated group of components arranged in some sequence.

Reminder: the term *atomic weight* is firmly entrenched, but its use refers to the masses of atoms of elements relative to the ^{12}C atom.

To convert the atomic weights of the isotopes of an element to the atomic weight of the element as found in nature, one must also know the relative number of atoms of each isotope that are present. In mass spectroscopy, the relative abundance of isotopes is obtained from the relative heights of the peaks plotted by the recorder as the ions of the isotopes reach the detector (Figure 3.2). Figure 3.3 illustrates the **mass spectrum** of chlorine; the graph shows the relative abundance plotted against mass number. The relative abundances of the naturally occurring isotopes of chlorine are 75.770% ^{35}Cl, isotopic weight 34.969, and 24.23% ^{37}Cl, isotopic weight 36.966. The atomic weight of chlorine is obtained by calculating the average atomic weight from the isotopic weights and the relative abundance of each isotope:

$$\textbf{atomic weight of chlorine} = (0.75770 \times 34.969) + (0.2423 \times 36.966) = 35.453^{\dagger}$$

EXAMPLE 3.2 Calculate the atomic weight of oxygen found in air from the isotopic weights 15.9949, 16.9991, and 17.9992 and their relative abundances: 99.7587% ^{16}O, 0.0374% ^{17}O, and 0.2039% ^{18}O. Find the molar mass of O.

† The dimensional analysis is shown:

$$0.75770\,\frac{\cancel{\text{mol }^{35}\text{Cl}}}{\text{mol Cl}} \times 34.969\,\frac{\text{g }^{35}\text{Cl}}{\cancel{\text{mol }^{35}\text{Cl}}} + 0.2423\,\frac{\cancel{\text{mol }^{37}\text{Cl}}}{\text{mol Cl}} \times 36.966\,\frac{\text{g }^{37}\text{Cl}}{\cancel{\text{mol }^{37}\text{Cl}}} = 35.453\,\frac{\text{g Cl}}{\text{mol Cl}}$$

However, units are usually omitted in calculating molar masses (atomic weights) from spectroscopic data.

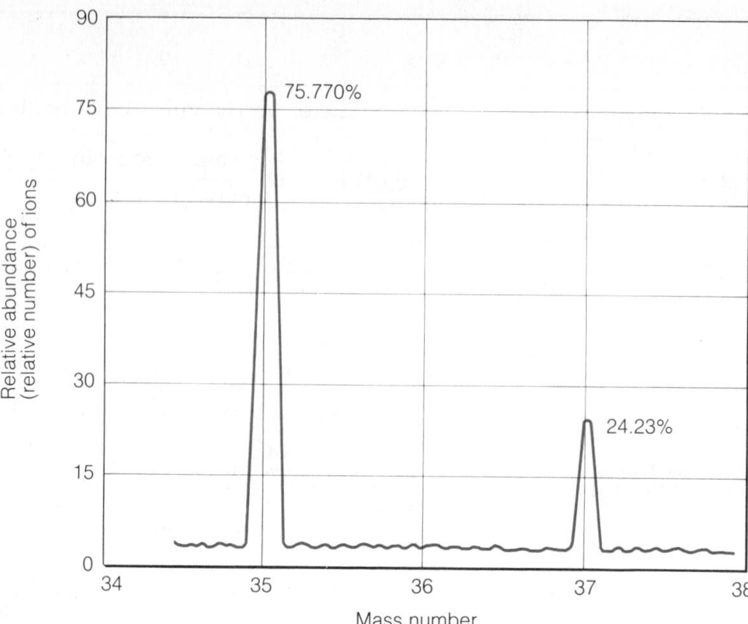

FIGURE 3.3
The mass spectrum of chlorine. The plot shows the relative abundances (peaks) of the two naturally occurring isotopes of chlorine.

The Greek letter for *s*, Σ (sigma), is commonly used to abbreviate "the sum of," so that average atomic weight = Σ (relative abundances) × (isotopic weights)

ANSWER The atomic weight of an element is obtained by calculating the average atomic weight from the isotopic weight and the relative abundance of each isotope. Thus the atomic weight of atmospheric oxygen is

$$15.9949 \times 0.997587 + 16.9991 \times 0.000374 + 17.9992 \times 0.002039 = 15.9994$$

The molar mass of O is 15.9994 g/mol. ■

The calculation of the atomic weights of elements is analogous to the calculation of the average weight of a collection of pennies and dimes. Imagine that you have a collection consisting of 3 pennies and 7 dimes; the relative abundance of pennies would be 0.30 and the relative abundance of dimes would be 0.70. Suppose each penny weighs 3.15 g and each dime weighs 2.40 g. Then

$$\begin{array}{r|l} & 3.15 \text{ g} \\ & 3.15 \\ & 3.15 \\ & 2.40 \\ & 2.40 \\ & 2.40 \\ & 2.40 \\ & 2.40 \\ & 2.40 \\ & 2.40 \\ \hline 10 & 26.25 \text{ g} \\ \hline & 2.63 \text{ g} = \text{average coin weight} \end{array}$$

However, you would not live long enough to do this calculation with one mole of atoms! The shortcut calculation is

BOX 3.3

THE ATOMIC MASS UNIT (THE DALTON)

The mass unit known as an **atomic mass unit** (**amu**, or **dalton**) is defined as exactly 1/12 of the mass of a ^{12}C atom. The mass of a ^{12}C atom is therefore 12 amu by definition, 12 amu/atom ^{12}C. Expressed in grams, 1 amu = $1.6605402 \times 10^{-24}$ g, as shown:

$$\underbrace{\frac{12\text{ g }^{12}\text{C}}{1\ \cancel{\text{mol}}} \times \frac{1\ \cancel{\text{mol}}}{6.0221367 \times 10^{23}\ \cancel{\text{atoms}}\ ^{12}\text{C}}}_{\text{mass of one }^{12}\text{C atom}} \times \underbrace{\frac{1\ \cancel{\text{atom}}}{12\text{ amu}}}_{\text{by definition}} = \frac{1}{N_A} = 1.6605402 \times 10^{-24}\ \frac{\text{g}}{\text{amu}}$$

Thus, the mass in grams of an amu is a very small number; in fact, it is the reciprocal of the Avogadro number, $1/N_A$, expressed in grams.

The number of amu's per atom of any element is then obtained as follows, using Be as an example:

$$\frac{9.01218\ \cancel{\text{g}}\text{ Be}}{1\ \cancel{\text{mol Be}}} \times \frac{1\ \cancel{\text{mol Be}}}{6.0221367 \times 10^{23}\text{ atoms Be}} \times \frac{1\text{ amu}}{1.6605402 \times 10^{-24}\ \cancel{\text{g}}} = 9.01218\ \frac{\text{amu}}{\text{atom}}$$

Thus, the number of amu's per atom of an element is numerically the same as the atomic weight of the element. The SI, however, does not recognize the amu because, unlike the kilogram, it is based on an experimental value, namely, Avogadro's number.

It would take the entire world population, working full time and counting one atom per sec, a billion years to count the atoms in one mole.

$$\frac{3.15\ \dfrac{\text{g}}{\cancel{\text{penny}}} \times 3\ \dfrac{\cancel{\text{pennies}}}{\cancel{\text{collection}}} + 2.40\ \dfrac{\text{g}}{\cancel{\text{dime}}} \times 7\ \dfrac{\cancel{\text{dimes}}}{\cancel{\text{collection}}}}{(3+7)\ \dfrac{\text{coins}}{\cancel{\text{collection}}}} = 2.63\ \frac{\text{g}}{\text{coin}}$$

or

$$\frac{3.15\text{ g} \times 0.30 + 2.40\text{ g} \times 0.70}{0.30 + 0.70} = 2.63\text{ g}$$

This is the same as the sum of the atomic weight of each isotope multiplied by its relative abundance.

PROBLEM 3.2 Calculate the atomic weight of nitrogen from the isotopic weights, 14.003 and 15.000, and their relative abundances: ^{14}N, 99.634%, and ^{15}N, 0.366%. What is the molar mass of N? □

In a mixture of a mole of Cl atoms and a mole of N atoms, the mass ratio of any one atom of Cl to any one atom of N may vary. These elements consist of several isotopes so that, for example, the mass ratios

See note d, Table of Atomic Weights, inside back cover.

$$\frac{\text{mass of 1 atom }^{35}\text{Cl}}{\text{mass of 1 atom }^{14}\text{N}} \quad \text{and} \quad \frac{\text{mass of 1 atom }^{37}\text{Cl}}{\text{mass of 1 atom }^{15}\text{N}}$$

differ. But the isotopic composition of most elements is constant, so that the average mass ratio of a large number of atoms such as

$$\frac{\text{mass of }6 \times 10^{23}\text{ atoms Cl}}{\text{mass of }6 \times 10^{23}\text{ atoms N}}$$

is constant. For this reason, the experimentally determined mass composition of compounds is constant, as predicted by Dalton's assumption that all of the atoms of an element are identical. Even though this assumption is now known to be incorrect, it leads to right answers as long as we are dealing with large numbers of atoms.

The International Commission on Atomic Weights was established in 1900 by the International Union of Pure and Applied Chemistry (IUPAC). The Commission was given the duty of periodically issuing a table of atomic weights after considering all papers dealing with the subject. The values chosen by the Commission are the **accepted atomic weights**. Inside the back cover of this book you will find the currently accepted weights.

3.4 MOLECULAR WEIGHTS

Molecular weight is analogous to atomic weight. It is *not* the mass of one molecule; rather, it is the mass of a molecule relative to the mass of a ^{12}C atom. When chemists say, "The molar mass of water is 18 g/mol and the molar mass of carbon dioxide is 44 g/mol," they mean that there are as many molecules in 18 g of water and in 44 g of carbon dioxide as there are atoms in 12 g of ^{12}C.

Like *atomic weight*, the term *molecular weight* is outmoded but still in use.

The **molecular weight** of a molecule is the sum of the atomic weights of its constituent atoms. For example, the molecular weight of water, H_2O, is

$$\begin{aligned}&2 \text{ H atoms} \times \text{atomic weight of hydrogen}\\&\quad + 1 \text{ O atom} \times \text{atomic weight of oxygen} = 2 \times 1.01 + 1 \times 16.0 = 18.0\end{aligned}$$

Molecular weights obtained from mass spectroscopy are for substances in the gaseous state. Molecular weights are also obtained from the properties of gases (Chapter 10) and solutions (Chapter 12).

Molecular weights are obtained from mass spectroscopy in the same way that atomic weights are obtained.

We already know that most elements consist naturally of several isotopes. It then follows that the molecules of a given substance may also have different molecular weights. Hydrogen fluoride, HF, for example, consists of $^{1}H^{19}F$ and $^{2}H^{19}F$ molecules, while nitrogen, N_2, consists of $^{14}N^{14}N$, $^{14}N^{15}N$, and $^{15}N^{15}N$ molecules. However, *the relative abundances of the heavy isotopes of hydrogen, carbon, oxygen, and nitrogen are sufficiently small to be ignored for most purposes.* This permits us to find molecular weights with an *accuracy of three significant figures* quite easily.

The answer is in 3 significant figures because molecules containing the heavier isotopes of H, C, O, or N are discarded.

Thus, analogous to the calculation of an atomic weight, the molecular weight of carbon monoxide is obtained by finding the average molecular weight from the molecular weight of each of its molecules detected in a mass spectrometer and their relative abundances (Figure 3.4): the molecular weights and relative abundances are 27.995 (98.5%), 29.00 (1.14%), and 30.00 (0.39%). The molecular weight of carbon monoxide is

$$\mathbf{0.985 \times 27.995 + 0.0114 \times 29.00 + 0.0039 \times 30.00 = 28.02}$$

However, ignoring the $^{13}C^{16}O$ and $^{12}C^{18}O$ molecules, the calculation simplifies to

molecular weight of carbon monoxide = 100% × 27.995 = 28.0

See Problem 3.55 for the determination of highly accurate molecular weights by mass spectroscopy. Fragmentation (break-up) of molecules is minimized by use of Ar^+ or Xe^+ to ionize molecules.

Like carbon monoxide, many other compounds, each consisting mainly of only one kind of molecule, are known. Some examples are acetylene (26.0), benzene (78.0), and tetraphosphorus decaoxide (284). Their molecular weights, 26.0, 78.0, and 284, accurate to three significant figures, suffice for determining their molecular formulas (section 3.8).

EXAMPLE 3.3 The molecular weights and relative abundances of the molecules of hydrogen perbromate are 144 (50.7%) and 146 (49.3%). Find the molecular weight of hydrogen perbromate.

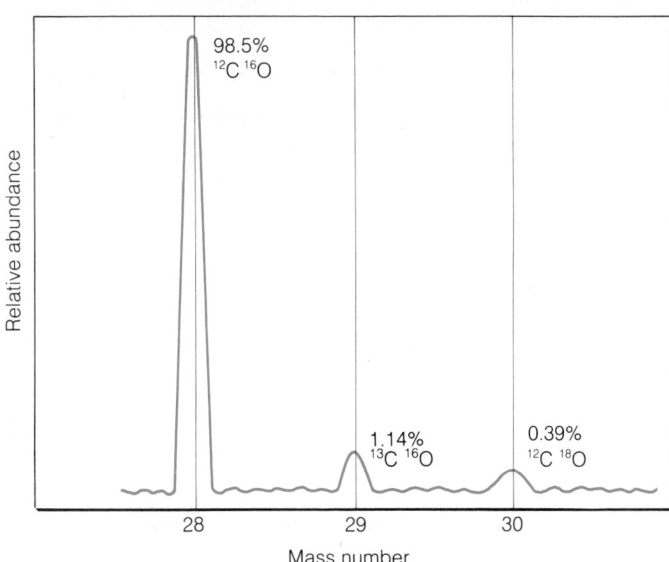

FIGURE 3.4
The mass spectrum of carbon monoxide.

ANSWER The molecular weight of a substance is found in the same way that the atomic weight of an element is obtained (Example 3.2); it is the average molecular weight calculated from the relative abundances of its molecules:

$$144 \times 0.507 + 146 \times 0.493 = 145$$

The molecular weight of hydrogen perbromate is 145. ■

PROBLEM 3.3 Find the molecular weight of a chlorofluoromethane from the relative abundances of its molecules and their molecular weights, 68.0 (75.8%) and 70.0 (24.2%). □

3.5 MOLE RELATIONSHIPS: SAME QUANTITY, DIFFERENT UNITS

Depending upon the nature of the discussion, the same quantity of a substance, hydrogen for example, may be expressed in units of mass, amount of substance (mol), or number of particles. The conversion factor is then

6.02×10^{23} particle/1 mol, or the molar mass, g/1 mol

Always be careful to specify the nature of the particle, atom (H), molecule (H_2), or ion (H^+). Use 1.0 g H/1 mol H, not 1.0 g/1 mol hydrogen, and 2.0 g H_2/1 mol H_2, not 2.0 g/1 mol hydrogen.

EXAMPLE 3.4 A test minirocket engine is loaded with 66.0 g liquid hydrazine, N_2H_4, 32.0 g/mol. Find (a) the number of moles of N_2H_4, (b) the number of N_2H_4 molecules, and (c) the number of H atoms.

ANSWER In solving the problem, first write the quantities given and the quantities sought. (Don't forget the units! *Writing quantities without units will be disastrous.*)

(a) Quantity given: 66.0 g N_2H_4

Quantity sought: moles of N_2H_4

We want to convert grams of N_2H_4 to moles of N_2H_4. Hence, we need a relationship between grams and moles. The needed conversion factor is therefore the molar mass, in this case 32.0 g/mol. However, the unit we want, mol N_2H_4, must be in the numerator; hence we use

$$\frac{1 \text{ mol } N_2H_4}{32.0 \text{ g } N_2H_4}$$

Make sure that the unit to be canceled is in the denominator (section 1.5). Now convert grams of N_2H_4 to moles of N_2H_4:

$$\text{STEP 1.}\quad 66.0\,\cancel{\text{g } N_2H_4} \times \frac{1 \text{ mol } N_2H_4}{32.0\,\cancel{\text{g } N_2H_4}} = 2.06 \text{ mol } N_2H_4$$

The correct unit in the answer tells us that the equation is set up properly (section 1.5).

(b) Quantity given: 66.0 g N_2H_4

Quantity sought: number of N_2H_4 molecules

We want to find the number of molecules and we know the number of moles from (a). To go from moles to molecules, we need a relationship between moles and molecules, namely, the Avogadro number. Thus, the needed conversion factor is

$$\frac{6.02 \times 10^{23} \text{ molecules}}{1 \text{ mol}}$$

Again, the wanted unit is in the numerator and the unit to be canceled is in the denominator. Now go from moles of N_2H_4 to molecules of N_2H_4:

$$\text{STEP 2.}\quad 2.06\,\cancel{\text{mol } N_2H_4} \times \frac{6.02 \times 10^{23} \text{ } N_2H_4 \text{ molecules}}{1\,\cancel{\text{mol } N_2H_4}}$$

$$= 1.24 \times 10^{24} \text{ } N_2H_4 \text{ molecules}$$

Steps 1 and 2 may be combined into one step:

$$66.0\,\cancel{\text{g } N_2H_4} \times \frac{1\,\cancel{\text{mol } N_2H_4}}{32.0\,\cancel{\text{g } N_2H_4}} \times \frac{6.02 \times 10^{23} \text{ } N_2H_4 \text{ molecules}}{1\,\cancel{\text{mol } N_2H_4}}$$

$$= 1.24 \times 10^{24} \text{ } N_2H_4 \text{ molecules}$$

$$\textbf{grams} \xrightarrow[\text{mass}]{\text{molar}} \textbf{mol} \xrightarrow[\text{number}]{\text{Avogadro}} \textbf{molecules}$$

(c) Quantity given: 66.0 g N_2H_4

Quantity sought: number of H atoms

We want to find the number of H atoms. The molecular formula, N_2H_4, shows how many atoms of N and H are in each molecule. Each N_2H_4 molecule has 4 H atoms and we know the number of molecules from part b. Thus the conversion factor is

$$\frac{4 \text{ H atoms}}{1 \text{ } N_2H_4 \text{ molecule}}$$

Again, notice that the wanted unit is in the numerator. We now convert molecules of N_2H_4 to atoms of H:

$$\text{STEP 3.}\quad 1.24 \times 10^{24}\ \cancel{N_2H_4\ \text{molecules}} \times \frac{4\ \text{H atoms}}{1\ \cancel{N_2H_4\ \text{molecule}}} = 4.96 \times 10^{24}\ \text{H atoms}$$

Steps 1, 2, and 3 may be combined into one step:

$$66.0\ \cancel{g\ N_2H_4} \times \frac{1\ \cancel{mol\ N_2H_4}}{32.0\ \cancel{g\ N_2H_4}} \times \frac{6.02 \times 10^{23}\ \cancel{N_2H_4\ \text{molecules}}}{1\ \cancel{mol\ N_2H_4}} \times \frac{4\ \text{H atoms}}{1\ \cancel{N_2H_4\ \text{molecule}}} = 4.96 \times 10^{24}\ \text{H atoms}$$

$$\textbf{grams} \xrightarrow[\text{mass}]{\text{molar}} \textbf{mol} \xrightarrow[\text{number}]{\text{Avogadro}} \textbf{molecules} \xrightarrow{\text{formula}} \textbf{atoms}$$ ■

Remember that for each conversion in a one-step setup, the unit to be canceled is in the denominator. This use of dimensional analysis illustrates an approach that is useful in solving many types of numerical problems: set up the conversion factors in a sequence that cancels an unwanted unit of a preceding conversion factor, leaving only the desired unit in the answer (section 1.5). *Also remember* that

- the conversion between grams and moles of a substance requires its molar mass;
- the conversion between moles of a substance and the number of particles requires the Avogadro number;
- the conversion between the number of molecules of a substance and the number of atoms of a constituent element requires the formula of the substance.

PROBLEM 3.4 How many grams of ozone, O_3, are in 0.155 mol O_3? □

PROBLEM 3.5 Find (a) the number of moles, (b) the number of molecules, and (c) the number of O atoms in 72.0 g SO_3. □

PROBLEM 3.6 Find the mass in grams of 3.95×10^{25} H_2SO_4 molecules. □

3.6 MOLECULAR AND EMPIRICAL FORMULAS

Consider a chemical symbol such as C. The symbol C is used to represent the element C as well as an atom of C; it also stands for 1 mole of carbon atoms and for 12.0 g of carbon. C_6H_6 is the molecular formula as well as the symbol for benzene. It tells you that there are 6 carbon atoms and 6 hydrogen atoms in 1 molecule of benzene. Furthermore, it tells you that there are 6 moles of carbon atoms and 6 moles of hydrogen atoms in 1 mole of benzene. Thus a **molecular formula** expresses an *actual* molecular composition; it gives us the number of each kind of atom in one molecule. This value is the same as the number of moles of each kind of atom in one mole of substance. The atom ratio is the same as the mole ratio.

Since a molecule is composed of atoms, molecular weight is the sum of the atomic weights of a molecule's constituent atoms (section 3.4). The molecular formula C_6H_6 also represents a definite quantity of the substance, 1 mole, or 78.113 g of benzene ($6 \times 12.011 + 6 \times 1.0079$).

The mass composition expresses the relative masses of the constituent elements in a particular compound. For example, the molecular formulas of benzene, C_6H_6, and acetylene, C_2H_2, tell us that their mass compositions are the same:

$$C_6H_6: \frac{6 \text{ mol C}}{6 \text{ mol H}} = \frac{6 \times 12.0 \text{ g C}}{6 \times 1.01 \text{ g H}} = \frac{12.0 \text{ g C}}{1.01 \text{ g H}}$$

$$C_2H_2: \frac{2 \text{ mol C}}{2 \text{ mol H}} = \frac{2 \times 12.0 \text{ g C}}{2 \times 1.01 \text{ g H}} = \frac{12.0 \text{ g C}}{1.01 \text{ g H}}$$

An empirical formula is also called the *simplest* formula.

Thus the mass composition or atom ratios of C_6H_6 and C_2H_2 can be represented by CH. CH is a typical **empirical formula**, a formula that gives the simplest whole number ratios in which atoms combine to form a compound. These ratios are expressed as whole numbers to conform to the fact that fractions of atoms cannot combine. However, the empirical and molecular formulas of a substance may differ. For example, the empirical formula for the antifreeze ethylene glycol, $C_2H_6O_2$, is CH_3O, whereas both the molecular and the empirical formula of methane is CH_4. The molecular formula is a whole number, n, times the empirical formula:

$$\textbf{(empirical formula)}_n = \textbf{molecular formula}$$

For benzene, $n = 6$; for acetylene and ethylene glycol, $n = 2$; and for methane, $n = 1$.

PROBLEM 3.7 Find the empirical formula of (a) an oxide of phosphorus, P_4O_{10}; (b) ascorbic acid (vitamin C), $C_6H_8O_6$; (c) butene, C_4H_8; and (d) toluene, C_7H_8. □

Empirical formulas are calculated from the known mass composition of compounds. This will be discussed in the next section.

3.7 DETERMINATION OF MASS COMPOSITION

The mass composition of compounds is determined by experimentation. The first step is to determine what elements are present in the given compound (*qualitative analysis*). Then the mass percentages of the elements are determined (*quantitative analysis*). A typical traditional method of determining mass composition is illustrated in Figure 3.5.

EXAMPLE 3.5 A sample of an iron oxide weighing 0.550 g is heated in a furnace in a stream of hydrogen (Figure 3.5a). The oxygen in the oxide reacts with hydrogen to form water vapor, which is allowed to escape with any unreacted hydrogen. The remaining pure iron weighs 0.384 g. Find the mass percentage of Fe and O in the iron oxide.

ANSWER The term *percentage* is derived from the Latin word *percentum* meaning "for each hundred." It is defined as

$$\frac{\text{part}}{\cancel{\text{whole}}} \times 100\% \ \cancel{\text{whole}} = x\% \text{ part}$$

or simply,

$$\frac{\text{part}}{\text{whole}} \times 100\% = x\% \text{ part}$$

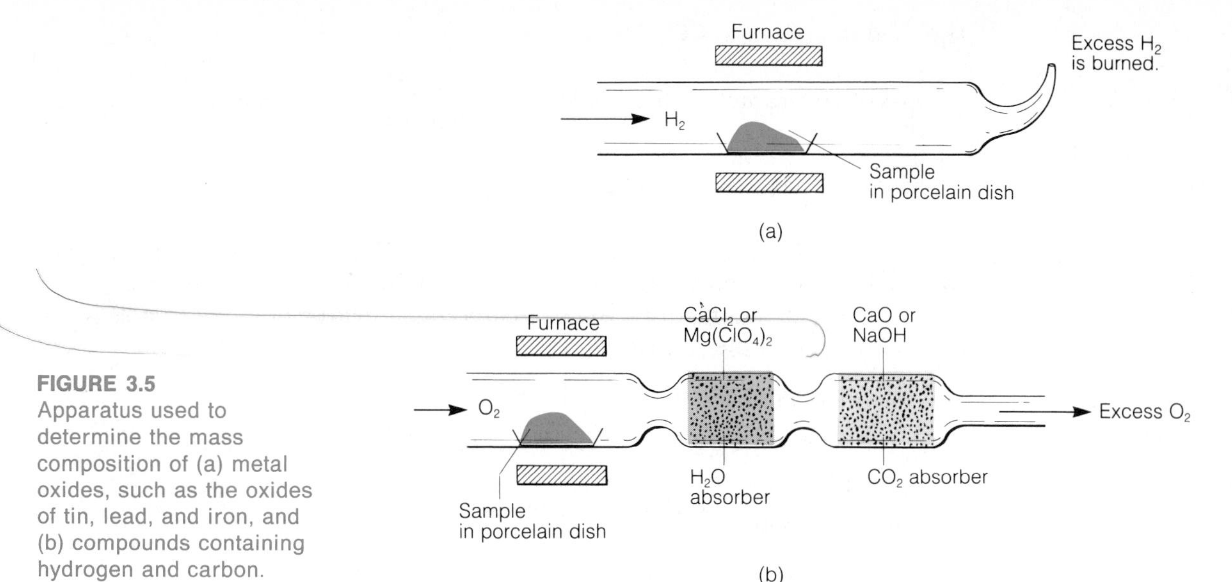

FIGURE 3.5 Apparatus used to determine the mass composition of (a) metal oxides, such as the oxides of tin, lead, and iron, and (b) compounds containing hydrogen and carbon.

Then,

$$\text{mass percentage of iron} = \frac{0.384\text{ g Fe}}{0.550\text{ g iron oxide}} \times 100\% = 69.8\%\text{ Fe}$$

and

$$\text{mass percentage of oxygen} = 100\%\text{ iron oxide} - 69.8\%\text{ iron} = 30.2\%\text{ oxygen}$$

■

About 6 million compounds are known.

Most of the known compounds contain carbon and hydrogen, and many compounds contain only carbon, hydrogen, and oxygen. These compounds are analyzed by burning them in excess oxygen. The combustion reaction converts the hydrogen to water vapor and the carbon to carbon dioxide, which are trapped in separate tubes (Figure 3.5b). The increase in the mass of each absorption tube gives us the quantities of water and carbon dioxide that are produced from a weighed quantity of a purified sample. For example, the products of photosynthesis, *starch*, the mainstay of most diets, and *cellulose*, the major component of wood, paper, and cotton, have the same empirical formula. Analyses show that they both contain carbon, hydrogen, and oxygen. When 0.700 g of starch is completely burned in oxygen, 1.14 g CO_2 and 0.389 g H_2O are collected. So what is the mass composition of starch? From other analyses, the mass percentage of hydrogen in water, 11.2% H (11.2 g H/100 g H_2O), and of carbon in carbon dioxide, 27.3% C (27.3 g C/100 g CO_2), are known. How is the mass composition of starch obtained from these data?

Our first task is to find the masses of carbon, hydrogen, and oxygen in the 0.700-g sample. The sample is the only source of carbon and hydrogen. The masses of carbon and hydrogen may then be calculated from the weighed quantities of trapped CO_2 and H_2O and the known mass percentages of hydrogen in

H_2O and of carbon in CO_2:

$$\text{mass of H in sample} = 0.389 \cancel{\text{g H}_2\text{O}} \times \frac{11.2 \text{ g H}}{100 \cancel{\text{g H}_2\text{O}}} = 0.0436 \text{ g H}$$

$$\text{mass of C in sample} = 1.14 \cancel{\text{g CO}_2} \times \frac{27.3 \text{ g C}}{100 \cancel{\text{g CO}_2}} = 0.311 \text{ g C}$$

Next, we find the mass of oxygen in the sample. The compound consists only of carbon, hydrogen, and oxygen. Therefore, the difference between the mass of the sample, 0.700 g, and the total mass of carbon and hydrogen in the sample is the mass of oxygen in the sample:

$$0.700 \text{ g compound} - 0.0436 \text{ g H} - 0.311 \text{ g C} = 0.345 \text{ g O}$$

Finally, we calculate the mass percentage of each element in starch:

$$\frac{0.0436 \text{ g H}}{0.700 \text{ g compound}} \times 100\% = 6.23\% \text{ H}$$

$$\frac{0.311 \text{ g C}}{0.700 \text{ g compound}} \times 100\% = 44.4\% \text{ C}$$

$$\frac{0.345 \text{ g O}}{0.700 \text{ g compound}} \times 100\% = 49.3\% \text{ O}$$

Or, the percentage of oxygen may be found by subtracting the percentage of C and H from 100%:

If 4 significant figures are carried, the answers will agree to 3 significant figures:

100% compound − 6.229% H − 44.43% C = 49.3% C

$$100\% \text{ compound} - 6.23\% \text{ H} - 44.4\% \text{ C} = 49.4\% \text{ O}$$

The difference between 49.4% O and 49.3% O is within the experimental error.

PROBLEM 3.8 Combustion analysis of 1.00 mg of methane yields 2.75 mg CO_2. Find the mass percentage of C and H in methane, a compound containing only carbon and hydrogen. □

3.8 FORMULA FROM MASS COMPOSITION

In the previous section you learned how the mass composition of compounds is determined experimentally. Now you will learn how this information is used to calculate empirical formulas. The method has three steps:

STEP 1. Convert the mass of each element in a given sample of a substance to moles.

STEP 2. Write the simplest mole ratio calculated in Step 1.

STEP 3. Convert the ratio to the simplest whole number of atoms.

For example, a sample of magnesium bromide contains 3.30 g Mg (24.3 g/mol) and 21.7 g Br (79.9 g/mol). Its empirical formula is found by first converting the given masses to moles:

$$21.7 \cancel{\text{g Br}} \times \frac{1 \text{ mol Br}}{79.9 \cancel{\text{g Br}}} = 0.272 \text{ mol Br}$$

$$3.30 \cancel{\text{g Mg}} \times \frac{1 \text{ mol Mg}}{24.3 \cancel{\text{g Mg}}} = 0.136 \text{ mol Mg}$$

The empirical formula might then be written as $Mg_{0.136}Br_{0.272}$. The mole ratio is correct, but this seems to deny the indivisibility of atoms. We therefore divide all subscripts by the smallest subscript, 0.136 in this case, to find the simplest ratio of atoms that can be expressed in whole numbers: $MgBr_2$, the empirical formula of magnesium bromide.

EXAMPLE 3.6 Sodium acid pyrophosphate (SAP) is added to frankfurters and other sausages to accelerate development of the desired red color. Its mass composition is 20.7% sodium, 0.910% hydrogen, 27.9% phosphorus, and 50.5% oxygen. Find its empirical formula.

ANSWER First, assume a 100-g sample so that the given percentages yield 20.7 g Na, 0.910 g H, 27.9 g P, and 50.5 g O. Next, convert the mass of each element to moles of atoms of the element:

$$\text{sodium: } 20.7\ \cancel{\text{g Na}} \times \frac{1\ \text{mol Na}}{23.0\ \cancel{\text{g Na}}} = 0.900\ \text{mol Na}$$

$$\text{hydrogen: } 0.910\ \cancel{\text{g H}} \times \frac{1\ \text{mol H}}{1.01\ \cancel{\text{g H}}} = 0.901\ \text{mol H}$$

$$\text{phosphorus: } 27.9\ \cancel{\text{g P}} \times \frac{1\ \text{mol P}}{31.0\ \cancel{\text{g P}}} = 0.900\ \text{mol P}$$

$$\text{oxygen: } 50.5\ \cancel{\text{g O}} \times \frac{1\ \text{mol O}}{16.0\ \cancel{\text{g O}}} = 3.16\ \text{mol O}$$

This calculation allows us to write the formula $Na_{0.900}H_{0.901}P_{0.900}O_{3.16}$. Finally, write the simplest formula in terms of whole numbers by dividing all subscripts by the smallest subscript, in this case 0.900: $Na_1H_1P_1O_{3.5}$. Such a division often yields whole numbers, but in this case multiplication by 2 is required, yielding the empirical formula of the compound, $Na_2H_2P_2O_7$. In summary:

	Na	H	P	O
g of element per 100 g of compound	20.7 g	0.910 g	27.9 g	50.5 g
Relative number of moles of atoms	0.900 mol	0.901 mol	0.900 mol	3.16 mol
Divide by 0.900	1 mol	1 mol	1 mol	3.5 mol
Scale to whole numbers (multiply by 2)	2 mol	2 mol	2 mol	7 mol

This compound is named disodium dihydrogen diphosphate (old name, sodium acid pyrophosphate). ■

PROBLEM 3.9 The mass composition of starch (section 3.7) is 6.23% H, 44.4% C, and 49.3% O. Find its empirical formula. □

The **formula weight** is the sum of the atomic weights of all the atoms in a formula *as it is written.* This is regardless of whether or not the written formula corresponds to the actual molecular composition. Some formulas, once thought to represent molecules, have since been shown to be correct only as empirical formulas; they are incorrect as molecular formulas. An example of such an empirical formula is P_2O_5. This substance is still often called "phosphorus pentoxide,"

The molecular weight of tetraphosphorus decaoxide is 284 (section 3.4).

although it actually consists of P_4O_{10} molecules. Therefore we may write:

	EMPIRICAL FORMULA P_2O_5	MOLECULAR FORMULA P_4O_{10}
Formula weight	142	284
Molecular weight	—	284

Hydrogen peroxide provides another illustration. Its molecular formula is H_2O_2, while its empirical formula is HO. The formula weights are 34 for H_2O_2 ($2 \times 1.0 + 2 \times 16$) and 17 for HO; the molecular weight of H_2O_2 is 34. However, we cannot call the formula weight of 17 its molecular weight because the molecular formula of hydrogen peroxide is *known* to be H_2O_2.

The fact that *a molecular formula must be a whole-number multiple of the empirical formula makes it possible to find the molecular formula* (section 3.6). The molecular formula can be obtained from the empirical formula *only* when the molecular weight of the substance is known.

EXAMPLE 3.7 The empirical formula of the insecticide hexachlorocyclohexane is CHCl. Its molecular weight is 291. Find its molecular formula.

ANSWER The molecular weight of a substance is a whole-number multiple of the formula weight calculated from its empirical formula (section 3.6):

$$\text{(empirical formula)}_n = \text{molecular formula}$$

or

$$\text{formula weight} \times n = \text{molecular weight}$$

The whole number is obtained by dividing the molecular weight by the formula weight of CHCl:

$$\text{formula weight of CHCl} = 12.0 + 1.0 + 35.5 = 48.5$$

$$\frac{\text{molecular weight}}{\text{formula weight}} = \frac{291}{48.5} = 6$$

The molecular formula is therefore $(CHCl)_6 = C_6H_6Cl_6$. ■

EXAMPLE 3.8 Insects metabolize DDT, $C_{14}H_9Cl_5$, to DDD, the mass composition of which is 52.5% C, 3.15% H, and 44.4% Cl. (a) Calculate the empirical formula of DDD. (b) The molecular weight of DDD is 320. Find its molecular formula.

ANSWER

(a) First, convert relative masses to relative numbers of moles,

$$52.5\ \text{g}\cancel{\text{C}} \times \frac{1\ \text{mol C}}{12.0\ \text{g}\cancel{\text{C}}} = 4.38\ \text{mol C}$$

$$3.15\ \text{g}\cancel{\text{H}} \times \frac{1\ \text{mol H}}{1.01\ \text{g}\cancel{\text{H}}} = 3.12\ \text{mol H}$$

$$44.4\ \text{g}\cancel{\text{Cl}} \times \frac{1\ \text{mol Cl}}{35.5\ \text{g}\cancel{\text{Cl}}} = 1.25\ \text{mol Cl}$$

obtaining $C_{4.38}H_{3.12}Cl_{1.25}$. Next, to find the simplest relative whole numbers, divide by the smallest subscript, 1.25:

$$C_{3.50}H_{2.50}Cl_{1.00}$$

and, finally, multiply by 2 to get whole numbers:

$$C_7H_5Cl_2$$

(b) As in Example 3.7:

$$\text{formula weight} = 7 \times \text{C} + 5 \times \text{H} + 2 \times \text{Cl}$$
$$= 7 \times 12.0 + 5 \times 1.01 + 2 \times 35.5 = 160$$

$$\frac{\text{molecular weight}}{\text{formula weight}} = \frac{320}{160} = 2$$

The molecular formula is therefore $(C_7H_5Cl_2)_2 = C_{14}H_{10}Cl_4$. ■

PROBLEM 3.10 The mass composition of a gas with a molecular weight of 92.0 is 30.4% nitrogen and 69.6% oxygen. (a) Find the empirical formula of the gas. (b) Write its molecular formula. □

The determination of molecular formulas from approximate molecular weights permits the calculation of more accurate molecular weights.

EXAMPLE 3.9 Find the accurate molecular weight of (a) DDD, $C_{14}H_{10}Cl_4$, and (b) diethyl ether (an anesthetic), $C_2H_5OC_2H_5$.

ANSWER

(a) The molecular weight of $C_{14}H_{10}Cl_4$ is the sum of the atomic weights of its constituent atoms, obtained from the table of atomic weights (inside back cover): 14 C atoms × atomic weight of carbon + 10 H atoms × atomic weight of hydrogen + 4 Cl atoms × atomic weight of chlorine = 14 × 12.011 + 10 × 1.0079 + 4 × 35.453 = 320.045.

(b) The molecular weight of $C_2H_5OC_2H_5$, which can be written as $C_4H_{10}O$ = 4 C atoms × atomic weight of carbon + 10 H atoms × atomic weight of hydrogen + 1 O atom × atomic weight of oxygen = 4 × 12.011 + 10 × 1.00794 + 1 × 15.9994 = 74.123. ■

PROBLEM 3.11 Calculate the accurate molecular weight of dimethyl ether, CH_3OCH_3. □

IONIC SOLIDS

Formulas derived from molecular weights of substances in the gaseous state are valid only for the gaseous state. We cannot assume that the particles of liquids and solids are identical to those of the corresponding gaseous state. For example, the molecular weight of sodium chloride vapor is about 59, determined at 1970°C. This indicates that the vapor consists mostly of NaCl molecules; NaCl is thus the molecular formula for sodium chloride in the vapor state at about 2000°C. But solid sodium chloride is a typical ionic compound; it does not exist as molecules in the solid state (Figure 2.15). It is impossible to write a molecular formula for an ionic solid.[†] Since NaCl is the simplest formula that can be written for sodium chloride, its empirical formula is also NaCl. The formula weight of NaCl is $23.0 + 35.5 = 58.5$.

By general usage, the formula weight calculated from an empirical formula for a substance that does not exist in the molecular form is called a molecular weight. Thus we can say that the molecular weight of solid NaCl is 58.5, or that its molar mass is 58.5 g/mol, even though molecules do not exist in solid sodium chloride.

EXAMPLE 3.10 Find the molecular weight and molar mass of calcium phosphate, $Ca_3(PO_4)_2$.

ANSWER The subscript 2 outside the parentheses doubles all atoms within the parentheses. Thus there are 3 Ca atoms, 2 P atoms, and 8 O atoms in $Ca_3(PO_4)_2$. The molecular weight is the sum of (3 × atomic weight of Ca) + (2 × atomic weight of P) + (8 × atomic weight of O):

$$3 \times 40.08 + 2 \times 30.974 + 8 \times 16.00 = 310.19$$

The molar mass of $Ca_3(PO_4)_2$ is 310.19 g/mol. ■

PROBLEM 3.12 Find the molar mass of $Al_2(SO_4)_3$. □

3.9 MASS COMPOSITION AND MOLARITY

In the following sections you will learn that formulas can be put to many uses. Here we consider only two uses.

MASS COMPOSITION

A formula allows us to calculate the percent composition by mass for each constituent element. We do this by dividing the mass of each element in one mole of the compound by the molar mass and then multiplying by 100%:

Do the molecular and empirical formulas of a compound give the same mass percentages?

$$\textbf{\% element} = \frac{\textbf{mass of element in 1 mol compound}}{\textbf{molar mass of compound}} \times \textbf{100\%}$$

EXAMPLE 3.11 The formula of monosodium glutamate (MSG), a naturally occurring salt used as a flavor enhancer, is $C_5H_8O_2Na$. Find (a) the mass percentage of carbon, C, in the MSG and (b) the mass in mg of hydrogen, H, in 10.0 mg of MSG.

Monosodium glutamate acts by increasing the sensitivity of nerves.

ANSWER

(a) The molar mass of $C_5H_8O_2Na$ is $5 \times 12.0 + 8 \times 1.0 + 2 \times 16.0 + 1 \times 23.0 =$ 123 g/mol, and the molar mass of C is 12.0 g/mol. The formula shows 5 moles of C in 1 mole of $C_5H_8O_2Na$. Therefore,

$$\% \text{ C} = \frac{\text{mass of C in 1 mol } C_5H_8O_2Na}{\text{mass of 1 mol } C_5H_8O_2Na} \times 100\%$$

$$= \frac{5 \times 12.0 \text{ g C}}{123 \text{ g } C_5H_8O_2Na} \times 100\% = 48.8\% \text{ C}$$

[†] Ionic solids might be more clearly represented by ionic formulas; for example, Na^+Cl^-. However, common practice (section 2.8) dictates the use of empirical formulas, NaCl in this case.

(b) Here we need the relationship between the mass of H and the mass of $C_5H_8O_2Na$. The conversion factor is

$$\frac{8 \times 1.01 \text{ g H}}{123 \text{ g } C_5H_8O_2Na} \quad \text{or} \quad \frac{8 \times 1.01 \text{ mg H}}{123 \text{ mg } C_5H_8O_2Na}$$

The mass of H in the given quantity is then given by

$$10.0 \cancel{\text{mg } C_5H_8O_2Na} \times \frac{8 \times 1.01 \text{ mg H}}{123 \cancel{\text{mg } C_5H_8O_2Na}} = 0.657 \text{ mg H}$$ ■

PROBLEM 3.13 There is concern about excess sodium in diets. (a) Find the mass of Na in 100 mg of $C_5H_8O_2Na$ (MSG). (b) How many milligrams of NaCl have the same mass of Na? □

Solutions are covered more extensively in Chapter 12.

SOLUTION MOLARITY

You may already be aware of the importance of specifying solution concentrations because of your laboratory experience. There are many ways in which the quantity of solute in a given quantity of solvent or solution may be expressed. One of the most useful methods of expressing the composition of a solution is by specifying the molarity or the concentration.[†] The **molarity, M,** of a dissolved substance is given by the number of moles of solute, *n*, divided by the volume of the solution, *V*, in liters:

IUPAC defines concentration of solute B as moles of solute B divided by the volume in liters of the solution, symbol c_B. Other ways of expressing solution composition are given in Chapter 12.

$$\textbf{molarity, M} = \frac{\textbf{moles of solute}}{\textbf{solution volume in liters}} = \frac{n}{V}$$

Thus, 49.0 g H_3PO_4, a 0.500 mol of phosphoric acid (98.0 g/mol), dissolved in 1.00 L of solution is a 0.500 M solution of H_3PO_4, or more briefly, 0.500 M H_3PO_4:

$$\frac{0.500 \text{ mol } H_3PO_4}{1.00 \text{ L solution}} = 0.500 \frac{\text{mol } H_3PO_4}{\text{L}} = 0.500 \text{ M } H_3PO_4$$

500 mL (0.500 L) of another solution containing 9.80 g H_3PO_4 (0.100 mol H_3PO_4) is 0.200 M H_3PO_4:

$$\frac{0.100 \text{ mol } H_3PO_4}{0.500 \text{ L solution}} = 0.200 \frac{\text{mol } H_3PO_4}{\text{L}} = 0.200 \text{ M } H_3PO_4$$

This solution may be described in several equally correct ways: the solution is 0.200 M H_3PO_4; the molarity (concentration) of H_3PO_4 is 0.200 M; the molarity of the solution with respect to H_3PO_4 is 0.200; or, the H_3PO_4 molarity is 0.200 mol/L. Notice that the *formula* of the solute is essential in the description of solution molarity or concentration.

EXAMPLE 3.12 Find the molarity of a solution prepared by adding water to 25.5 g of sodium carbonate, Na_2CO_3, to obtain 300 mL of solution.

ANSWER The solution contains 25.5 g Na_2CO_3 in 300 mL of solution, or

$$\frac{25.5 \text{ g } Na_2CO_3}{300 \text{ mL}}$$

[†] The word *concentration* is sometimes used for "quantity per unit volume" in other units—such as grams per liter or molecules per mL. However, moles per liter is so common that "concentration" is usually understood as a synonym for "molarity" unless otherwise specified.

But, by definition, molarity is moles per liter of solution. Therefore, convert 25.5 g Na_2CO_3 to moles of Na_2CO_3 and 300 mL of solution to liters of solution:

$$25.5 \cancel{\text{g Na}_2\text{CO}_3} \times \frac{1 \text{ mol Na}_2\text{CO}_3}{106 \cancel{\text{g Na}_2\text{CO}_3}} = 0.241 \text{ mol Na}_2\text{CO}_3 = n$$

1000 mL = 1 L

$$300 \cancel{\text{mL solution}} \times \frac{1 \text{ L solution}}{1000 \cancel{\text{mL solution}}} = 0.300 \text{ L solution} = V$$

and the molarity is

$$\text{M} = \frac{n}{V} = \frac{0.241 \text{ mol}}{0.300 \text{ L solution}} = 0.803 \frac{\text{mol Na}_2\text{CO}_3}{\text{L}}$$

Or in one step:

$$\text{M} = \frac{25.5 \cancel{\text{g Na}_2\text{CO}_3}}{300 \cancel{\text{mL}}} \times \frac{1 \text{ mol Na}_2\text{CO}_3}{106 \cancel{\text{g Na}_2\text{CO}_3}} \times \frac{1000 \cancel{\text{mL}}}{\text{L}} = 0.803 \frac{\text{mol Na}_2\text{CO}_3}{\text{L}}$$

grams $\xrightarrow[\text{mass}]{\text{molar}}$ **mol** ↘ $\frac{\textbf{mol}}{\textbf{L}}$

mL $\xrightarrow{\text{definition of M}}$ **L** ↗

The advantage of expressing solution composition in molarity is that it enables us to deliver a definite amount of solute by measuring a definite volume of solution, a rather easy measurement. ■

EXAMPLE 3.13 (a) How many moles of nitric acid, HNO_3, are in 2.5 L of 0.20 M HNO_3? (b) What volume of 0.10 M HNO_3 is needed to supply 0.30 mol HNO_3?

ANSWER

(a) Quantity given: 2.5 L of 0.20 M HNO_3
Quantity sought: moles of HNO_3

To go from liters of solution of known molarity to moles of solute, the definition of molarity is used; 0.20 M HNO_3 is the same as $\frac{0.20 \text{ mol}}{1 \text{ L}}$. This is the conversion factor for going from liters of the solution to moles of solute:

$$2.5 \cancel{\text{L HNO}_3} \times \frac{0.20 \text{ mol HNO}_3}{1 \cancel{\text{L HNO}_3}} = 0.50 \text{ mol HNO}_3$$

(b) Quantity given: 0.30 mol HNO_3
Quantity sought: liters of 0.10 M HNO_3

To go from moles of solute to liters of a solution of known molarity, the conversion factor is $\frac{1 \text{ L}}{0.10 \text{ mol}}$ and the required volume is

$$0.30 \cancel{\text{mol HNO}_3} \times \frac{1 \text{ L HNO}_3}{0.10 \cancel{\text{mol HNO}_3}} = 3.0 \text{ L HNO}_3$$

■

EXAMPLE 3.14 How many grams of potassium permanganate, $KMnO_4$, are in 25.0 mL of 0.100 M $KMnO_4$?

ANSWER

Quantity given: 25.0 mL of 0.100 M $KMnO_4$

Quantity sought: grams of $KMnO_4$

You want to go from molarity to grams of solute, but the number of grams is related to number of moles, not to molarity. The first problem, then, is to find the number of moles in the 25.0 mL of solution. So the key to this problem is the definition of molarity (Example 3.13):

$$0.100 \text{ M means } \frac{0.100 \text{ mol}}{1 \text{ L}} \quad \text{or} \quad \frac{0.100 \text{ mol}}{1000 \text{ mL}}$$

This is the conversion factor for going from mL of solution of known molarity to moles of solute:

$$25.0 \cancel{\text{mL solution}} \times \frac{0.100 \text{ mol KMnO}_4}{1000 \cancel{\text{mL solution}}} = 0.00250 \text{ mol KMnO}_4$$

Now, go from moles to grams using the molar mass ($KMnO_4$, 158 g/mol) as the conversion factor:

$$0.00250 \cancel{\text{mol KMnO}_4} \times \frac{158 \text{ g KMnO}_4}{1 \cancel{\text{mol KMnO}_4}} = 3.95 \text{ g KMnO}_4$$

Or in one step:

$$25.0 \cancel{\text{mL}} \times \frac{0.100 \cancel{\text{mol KMnO}_4}}{1000 \cancel{\text{mL}}} \times \frac{158 \text{ g KMnO}_4}{1 \cancel{\text{mol KMnO}_4}} = 3.95 \text{ g KMnO}_4$$

$$\boldsymbol{V} \xrightarrow{\text{M}} \textbf{mol} \xrightarrow[\text{mass}]{\text{molar}} \textbf{grams}$$

If we were asked, ''How many grams of $KMnO_4$ are needed to prepare 25.0 mL of 0.100 M solution?'' the answer would be exactly the same as this (Box 3.4). ■

EXAMPLE 3.15 Find the volume in mL of 0.175 M H_3PO_4 needed to obtain 0.326 g H_3PO_4.

ANSWER First, convert 0.326 g H_3PO_4 to moles:

$$0.326 \cancel{\text{g H}_3\text{PO}_4} \times \frac{1 \text{ mol H}_3\text{PO}_4}{98.0 \cancel{\text{g H}_3\text{PO}_4}} = 0.00333 \text{ mol H}_3\text{PO}_4$$

The given solution contains 0.175 mol H_3PO_4 in 1 L (1000 mL) of solution. The conversion factor is therefore 1000 mL/0.175 mol H_3PO_4, and the required volume is given by

$$0.00333 \cancel{\text{mol H}_3\text{PO}_4} \times \frac{1000 \text{ mL}}{0.175 \cancel{\text{mol H}_3\text{PO}_4}} = 19.0 \text{ mL}$$

Or in one step:

$$0.326 \cancel{\text{g H}_3\text{PO}_4} \times \frac{1 \cancel{\text{mol H}_3\text{PO}_4}}{98.0 \cancel{\text{g H}_3\text{PO}_4}} \times \frac{1000 \text{ mL}}{0.175 \cancel{\text{mol H}_3\text{PO}_4}} = 19.0 \text{ mL}$$

$$\textbf{grams} \xrightarrow[\text{mass}]{\text{molar}} \textbf{mol} \xrightarrow{\text{M}} \boldsymbol{V}$$

■

BOX 3.4

THE VOLUMETRIC FLASK

With the aid of a **volumetric flask** we can prepare a solution of known molarity. An accurately measured mass of solute is dissolved in the solvent, and the solution is transferred completely to the flask. Solvent is then added to slightly below the mark on the neck. When the solution is well mixed, solvent is added carefully to the mark, and the solution is mixed again. We now have a known volume of solution containing a known number of moles of solute and can easily calculate the molarity.

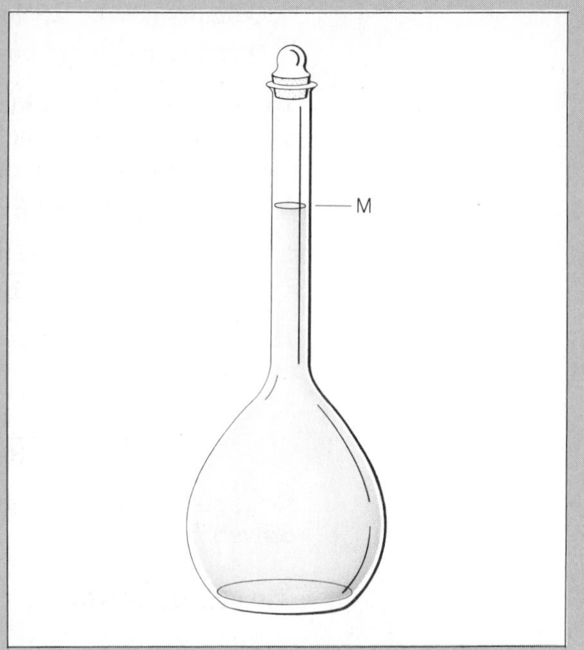

Volumetric flask. The capacity, up to the mark *M*, is accurately known.

PROBLEM 3.14 How many grams of $BaCl_2$ are needed to prepare 100 mL of 0.150 M solution? □

PROBLEM 3.15 What volume in mL of 0.195 M HCl is needed to obtain 0.375 g HCl? □

3.10 QUANTITATIVE INFORMATION FROM CHEMICAL EQUATIONS

A balanced chemical equation (section 2.10) reveals much more than the observed chemical change. It is also a statement of the *relative numbers* of moles of reactants and products involved in a chemical change. It is thus concerned with *quantities* of reactants and products.

For example, the *coefficients* in a balanced equation give the relative number of molecules of each reactant and the relative number of molecules of each product:

$$C_3H_8(g) + 5O_2(g) \longrightarrow 3CO_2(g) + 4H_2O(g)$$

In other words, for each molecule of propane, 5 oxygen molecules react, producing 3 carbon dioxide molecules and 4 water molecules. Or, in larger units, for each mole of propane consumed, 5 moles of oxygen are consumed, producing 3 moles of carbon dioxide and 4 moles of water.

These statements may be abbreviated:

$C_3H_8(g)$	+	$5O_2(g)$	$\longrightarrow$	$3CO_2(g)$	+	$4H_2O(g)$
1 mol C_3H_8		5 mol O_2		3 mol CO_2		4 mol H_2O
6.02×10^{23} C_3H_8 molecules		$5 \times 6.02 \times 10^{23}$ O_2 molecules		$3 \times 6.02 \times 10^{23}$ CO_2 molecules		$4 \times 6.02 \times 10^{23}$ H_2O molecules
44.1 g C_3H_8		5×32.0 g O_2		3×44.0 g CO_2		4×18.0 g H_2O

The quantities *under each formula* represent the same amount of substance. For example, 1 mol C_3H_8 is the same quantity as 6.02×10^{23} molecules and 44.1 g of C_3H_8. Thus we can choose any molar quantity to express the amounts of substances undergoing chemical change. From the coefficients in the balanced equation,

$$3H_2(g) + Fe_2O_3(c) \longrightarrow 2Fe(c) + 3H_2O(g)$$

it follows that

$$3\text{ mol } H_2 + 1\text{ mol } Fe_2O_3 \longrightarrow 2\text{ mol Fe} + 3\text{ mol } H_2O$$

or

Note that each quantity includes a unit.

$$3 \times 2.0\text{ g } H_2 + 1\text{ mol } Fe_2O_3 \longrightarrow 2 \times 6.02 \times 10^{23}\text{ atoms Fe} + 3 \times 18\text{ g } H_2O$$

Balanced equations with fractional coefficients, such as

$$H_2(g) + \tfrac{1}{2}O_2(g) \longrightarrow H_2O(\ell)$$

are frequently used. This equation is read as "1 mole of H_2 combines with $\frac{1}{2}$ mole of O_2 to form 1 mole H_2O," *not* as "1 molecule of H_2 combines with $\frac{1}{2}$ molecule of O_2." Unless oxygen atoms are actually used in the experiment, it is incorrect to rewrite the equation as $H_2(g) + O(g) \longrightarrow H_2O(\ell)$. That equation describes a different reaction and is read as "1 mole of H_2 combines with 1 mole of O (oxygen atoms)." The substance O is different from O_2. But $\frac{1}{2}O_2$ and $3O_2$ are not substances different from O_2; they are merely specific amounts of O_2. We cannot write O for $\frac{1}{2}O_2$, or $\frac{1}{2}O_2$ for O, or O_6 for $3O_2$.

Chemical equations make it possible to calculate quantities of substances when given a definite quantity of another substance. We can illustrate this with typical examples. For instance, how many moles of O_2 are required to burn 3.50 mol of liquefied petroleum gas, propane, C_3H_8? The balanced equation is

$$C_3H_8(g) + 5O_2(g) \longrightarrow 3CO_2(g) + 4H_2O(g)$$

In solving the problem, the procedure set up for Example 3.4 may be followed:

Quantity given: 3.50 mol C_3H_8

Quantity sought: moles of O_2

The key is to find a relationship between moles of C_3H_8 and moles of O_2. This relationship is given in the balanced equation. It tells us that 5 moles of O_2 react with 1 mole of C_3H_8. To go from moles of C_3H_8 to moles of O_2, the conversion factor is therefore

$$\frac{5\text{ mol } O_2}{1\text{ mol } C_3H_8}$$

and the number of moles of O_2 needed to burn 3.50 mol C_3H_8 is

$$3.50\,\cancel{\text{mol } C_3H_8} \times \frac{5\text{ mol } O_2}{1\,\cancel{\text{mol } C_3H_8}} = 17.5\text{ mol } O_2$$

EXAMPLE 3.16 The reaction of hydrazine, N_2H_4, with dinitrogen tetroxide, N_2O_4, is used to operate a rocket motor:

$$2N_2H_4(\ell) + N_2O_4(\ell) \qquad 3N_2(g) + 4H_2O(g)$$

Find how many grams of N_2 are produced when 0.0300 mol N_2H_4 is burned in a rocket motor test.

ANSWER

Quantity given: 0.0300 mol N_2H_4
Quantity sought: grams of N_2

Again, concentrate on the balanced equation to find the relationship between the moles of N_2H_4 and the moles of N_2. The equation tells us that 2 mol N_2H_4 yields 3 mol N_2. This relationship will give us the number of moles of N_2 produced from the given 0.0300 mol N_2H_4. We must then convert the moles of N_2 to grams of N_2. The calculation is thus carried out in two steps:

STEP 1. Convert moles of N_2H_4 to moles of N_2 (from the coefficients in the balanced equation).

STEP 2. Convert moles of N_2 to grams of N_2 (from the molar mass of N_2, 28.0 g/mol).

These two steps are combined into one step:

See Appendix B.1 for a suggested method of solving this equation.

$$0.0300\,\cancel{\text{mol N}_2\text{H}_4} \times \frac{3\,\cancel{\text{mol N}_2}}{2\,\cancel{\text{mol N}_2\text{H}_4}} \times \frac{28.0\text{ g N}_2}{1\,\cancel{\text{mol N}_2}} = 1.26\text{ g N}_2$$

$$\textbf{mol N}_2\textbf{H}_4 \xrightarrow[\text{equation}]{\text{balanced}} \textbf{mol N}_2 \xrightarrow[\text{mass}]{\text{molar}} \textbf{grams N}_2$$ ■

EXAMPLE 3.17 Hydrogen is produced in large quantities by electrolysis of water for use as a rocket fuel:

$$2H_2O(\ell) \longrightarrow 2H_2(g) + O_2(g)$$

Find the mass in grams of hydrogen produced from 50.0 g H_2O.

ANSWER

Quantity given: 50.0 g H_2O
Quantity sought: grams of H_2

This kind of calculation is more realistic because balances measure mass in kilogram, gram, or milligram units. This example is answered in three steps:

STEP 1. Convert 50.0 g H_2O to moles of H_2O (from the molar mass of H_2O, 18.0 g/mol).

STEP 2. Convert moles of H_2O to moles of H_2 (from the coefficients in the balanced equation).

STEP 3. Convert moles of H_2 to grams of H_2 (from the molar mass of H_2, 2.02 g/mol).

These three steps are combined into one step:

$$50.0\ \cancel{g\ H_2O} \times \frac{1\ \cancel{mol\ H_2O}}{18.0\ \cancel{g\ H_2O}} \times \frac{1\ \cancel{mol\ H_2}}{1\ \cancel{mol\ H_2O}} \times \frac{2.02\ g\ H_2}{1\ \cancel{mol\ H_2}} = 5.61\ g\ H_2$$

$$\textbf{grams } H_2O \xrightarrow[\text{mass}]{\text{molar}} \textbf{mol } H_2O \xrightarrow[\text{equation}]{\text{balanced}} \textbf{mol } H_2 \xrightarrow[\text{mass}]{\text{molar}} \textbf{grams } H_2$$ ■

Remember that converting moles of one substance to moles of another substance requires a molar ratio from a balanced equation or a formula. The procedure or "road map" we used to solve these problems can be summarized:

- If the given quantity is not in moles, convert to moles.
- Convert moles of the given substance to moles of the substance sought.
- If necessary, convert the moles of substance to the unit wanted.

EXAMPLE 3.18 The fertilizer ammonia, NH_3, is produced from nitrogen (from air) and hydrogen (from natural gas):

$$N_2(g) + 3H_2(g) \longrightarrow 2NH_3(g)$$

Find the mass in grams of NH_3 formed for every 10.0 grams of H_2 that react.

Natural gas, mainly CH_4, is the major industrial source of H_2:

$$CH_4(g) + 2H_2O(g) \longrightarrow CO_2(g) + 4H_2(g)$$

ANSWER

Quantity given: 10.0 g H_2
Quantity sought: grams of NH_3

We follow the road map given in Example 3.17:

STEP 1. Convert 10.0 g H_2 to moles of H_2 (from the molar mass of H_2, 2.02 g/mol).

STEP 2. Convert moles of H_2 to moles of NH_3 (from the coefficients in the balanced equation).

STEP 3. Convert moles of NH_3 to grams of NH_3 (from the molar mass of NH_3, 17.0 g/mol).

Or, in one step:

$$10.0\ \cancel{g\ H_2} \times \frac{1\ \cancel{mol\ H_2}}{2.02\ \cancel{g\ H_2}} \times \frac{2\ \cancel{mol\ NH_3}}{3\ \cancel{mol\ H_2}} \times \frac{17.0\ g\ NH_3}{1\ \cancel{mol\ NH_3}} = 56.1\ g\ NH_3$$

$$\textbf{grams } H_2 \xrightarrow[\text{mass}]{\text{molar}} \textbf{mol } H_2 \xrightarrow[\text{equation}]{\text{balanced}} \textbf{mol } NH_3 \xrightarrow[\text{mass}]{\text{molar}} \textbf{grams } NH_3$$ ■

EXAMPLE 3.19 Find the number of grams of iron, Fe, in 2.55 g iron ore, Fe_2O_3 (160 g/mol).

ANSWER

Quantity given: 2.55 g Fe_2O_3
Quantity sought: grams of Fe

STEP 1. Convert 2.55 g Fe_2O_3 to moles of Fe_2O_3 (from the molar mass, 160 g/mol).

STEP 2. Convert moles of Fe_2O_3 to moles of Fe (from the subscripts in the formula, Fe_2O_3).

STEP 3. Convert moles of Fe to grams of Fe (from the molar mass of Fe, 55.8 g/mol.)

In one step, the calculation is:

$$2.55\ \text{g}\ \cancel{\text{Fe}_2\text{O}_3} \times \frac{1\ \cancel{\text{mol Fe}_2\text{O}_3}}{160\ \cancel{\text{g Fe}_2\text{O}_3}} \times \frac{2\ \cancel{\text{mol Fe}}}{1\ \cancel{\text{mol Fe}_2\text{O}_3}} \times \frac{55.8\ \text{g Fe}}{1\ \cancel{\text{mol Fe}}} = 1.78\ \text{g Fe}$$

$$\textbf{g Fe}_2\textbf{O}_3 \xrightarrow[\text{mass}]{\text{molar}} \textbf{mol Fe}_2\textbf{O}_3 \xrightarrow{\text{formula}} \textbf{mol Fe} \xrightarrow[\text{mass}]{\text{molar}} \textbf{g Fe}$$ ■

PROBLEM 3.16 In the production of iron, iron(III) oxide is converted to iron in a blast furnace:

$$Fe_2O_3(c) + 3CO(g) \longrightarrow 3CO_2(g) + 2Fe(\ell)$$

How many moles of CO must be consumed to produce 25.0 grams of Fe? □

PROBLEM 3.17 How many grams of H_2O are produced when 25.0 grams of methanol, CH_3OH, are burned?

$$2CH_3OH(\ell) + 3O_2(g) \longrightarrow 2CO_2(g) + 4H_2O(\ell)$$ □

PROBLEM 3.18 Find the mass in grams of oxygen, O, in 15.0 g Al_2O_3. □

3.11 VOLUMETRIC ANALYSIS AND TITRATION

By definition, molarity (concentration) relates the volume of a solution and the number of moles of a solute (section 3.9). Application of this definition makes it possible to analyze for the quantity of a solute in solution. For example, the waste sulfuric acid in an electroplating solution is converted to harmless carbon dioxide and *gypsum*, $CaSO_4(H_2O)_2$, by reaction with calcium carbonate:

$$H_2SO_4(aq) + CaCO_3(c) + H_2O \longrightarrow CaSO_4(H_2O)_2(c) + CO_2(g)$$

However, it is first necessary to find the quantity of H_2SO_4 per liter of waste solution in order to fix the amount of $CaCO_3$ needed to react with the acid.

In **volumetric analysis**, the approach is to determine what volume of a solution of known concentration reacts with an unknown quantity of some other substance. If you know the concentration of the solution, its volume tells you how many moles are used. Then from the chemical equation, you can figure out how many moles it reacts with.

Suppose our task is to determine how many moles of sulfuric acid are in a given volume of the waste solution. The given volume is **titrated** with a solution of sodium hydroxide whose concentration (molarity) is known:

$$H_2SO_4(aq) + 2NaOH(aq) \longrightarrow Na_2SO_4(aq) + 2H_2O$$

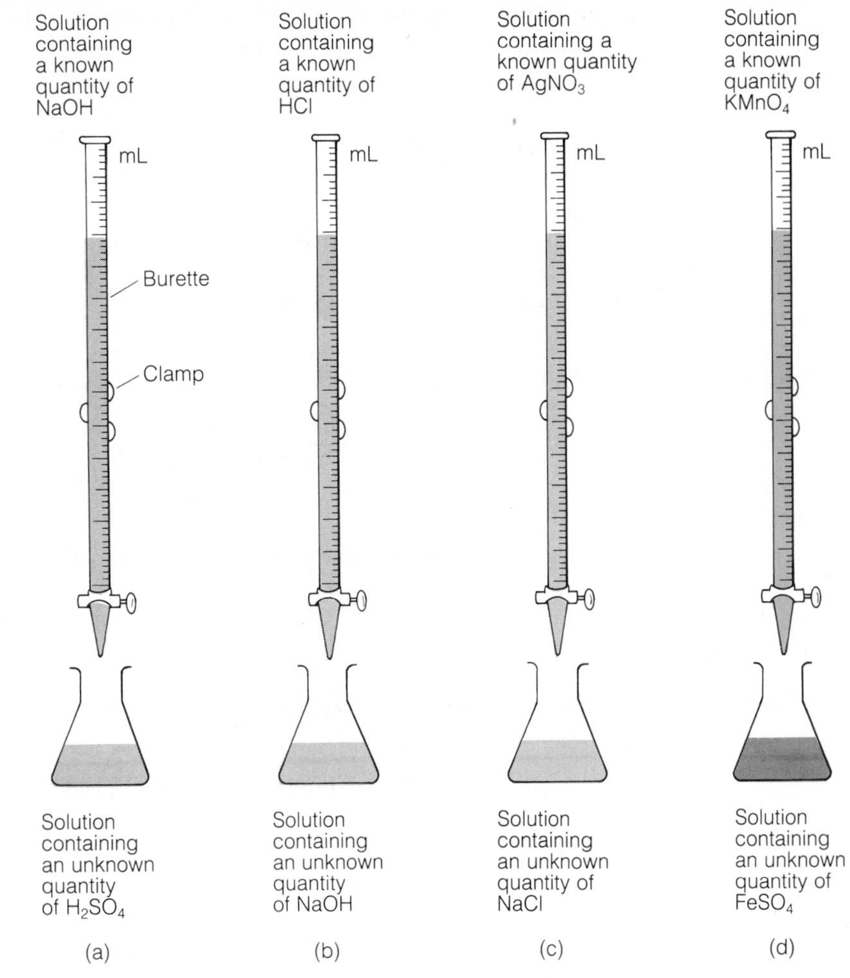

FIGURE 3.6
Setups for some typical titrations. The reactions are (a) $H_2SO_4(aq) + 2NaOH(aq) \rightarrow Na_2SO_4(aq) + 2H_2O$; (b) $HCl(aq) + NaOH(aq) \rightarrow NaCl(aq) + H_2O$; (c) $AgNO_3(aq) + NaCl(aq) \rightarrow AgCl(c) + NaNO_3(aq)$; (d) $2KMnO_4(aq) + 10FeSO_4 + 8H_2SO_4 \rightarrow 5Fe_2(SO_4)_3(aq) + K_2SO_4(aq) + 2MnSO_4(aq) + 8H_2O$.

Detailed instructions are found in laboratory manuals.

We measure the solution of NaOH from a **burette**, a long tube with volume markings and a stopcock (valve) at the bottom (Figure 3.6). Then we allow the NaOH solution to drain into the sample until we have added just enough NaOH to react with the H_2SO_4. This is the **equivalence point**, so we stop adding NaOH solution and record the volume that we have added. The equivalence point is reached when the *molar ratio* of the known (NaOH) added to the unknown (H_2SO_4) conforms to the molar ratio in the chemical equation. In titrating H_2SO_4 with NaOH, 2 moles of NaOH must be added for every mole of H_2SO_4.

Indicators are usually dyes that produce the desired color change.

How do we know when just enough solution has been added? There must be some signal that announces that the equivalence point has been reached. Usually, the signal is a color change. In most cases, an *indicator* (sections 9.1 and 15.4) is added to produce the color change (Color plate 11). The stage in a titration when a color change occurs and the addition of solution should stop is referred to as the **end point**. The end point is practically the same as the equivalence point.

EXAMPLE 3.20 45.50 mL of 0.1000 M NaOH reacts with 50.01 mL of waste H_2SO_4 solution. Find the molarity of the waste H_2SO_4 solution.

$$H_2SO_4 + 2NaOH \longrightarrow Na_2SO_4 + 2H_2O$$

ANSWER What are we given? Recognize that, as in Example 3.14, we are given a number of moles of NaOH: a known volume (45.50 mL) of a solution of known molarity (0.1000 M):

$$45.50 \text{ mL NaOH} \times \frac{0.1000 \text{ mol NaOH}}{1000 \text{ mL NaOH}} = 4.550 \times 10^{-3} \text{ mol NaOH}$$

But the chemical equation tells us that the reaction is 1 mol H_2SO_4 to 2 mol NaOH. Therefore, 4.550×10^{-3} mol NaOH reacts with 2.275×10^{-3} mol H_2SO_4:

$$4.550 \times 10^{-3} \text{ mol NaOH} \times \frac{1 \text{ mol } H_2SO_4}{2 \text{ mol NaOH}} = 2.275 \times 10^{-3} \text{ mol } H_2SO_4$$

Or, in one step:

$$45.50 \text{ mL NaOH} \times \frac{0.1000 \text{ mol NaOH}}{1000 \text{ mL NaOH}} \times \frac{1 \text{ mol } H_2SO_4}{2 \text{ mol NaOH}}$$

$$= 2.275 \times 10^{-3} \text{ mol } H_2SO_4$$

The molarity of the waste H_2SO_4 solution is therefore

$$\frac{2.275 \times 10^{-3} \text{ mol } H_2SO_4}{50.01 \text{ mL } H_2SO_4} \times \frac{1000 \text{ mL}}{1 \text{ L}} = 0.04550 \text{ M}_{H_2SO_4}$$

■

EXAMPLE 3.21 Find the mass of $CaCO_3$ that reacts with 10.0 L of waste solution 0.0455 M with respect to H_2SO_4.

$$H_2SO_4 + CaCO_3(c) + H_2O \longrightarrow CaSO_4(H_2O)_2(c) + CO_2(g)$$

ANSWER As in previous examples,

$$10.0 \text{ L } H_2SO_4 \times \frac{0.0455 \text{ mol } H_2SO_4}{1 \text{ L } H_2SO_4} \times \frac{1 \text{ mol } CaCO_3}{1 \text{ mol } H_2SO_4} \times \frac{100 \text{ g } CaCO_3}{1 \text{ mol } CaCO_3}$$

$$= 45.5 \text{ g } CaCO_3$$

■

EXAMPLE 3.22 What volume of 0.1005 M HCl is needed to react with 30.06 mL of 0.2509 M NaOH? The reaction is

$$HCl(aq) + NaOH(aq) \longrightarrow NaCl(aq) + H_2O$$

ANSWER First find the number of moles of HCl that react with the given number of moles of NaOH:

$$30.06 \text{ mL NaOH} \times \frac{0.2509 \text{ mol NaOH}}{1000 \text{ mL NaOH}} \times \frac{1 \text{ mol HCl}}{1 \text{ mol NaOH}} = 7.542 \times 10^{-3} \text{ mol HCl}$$

We now find the volume of 0.1005 M HCl that contains 7.542×10^{-3} mol HCl. As in Example 3.15, the conversion factor is

$$\frac{1000 \text{ mL HCl}}{0.1005 \text{ mol HCl}}$$

so,

$$7.542 \times 10^{-3} \text{ mol HCl} \times \frac{1000 \text{ mL HCl}}{0.1005 \text{ mol HCl}} = 75.05 \text{ mL HCl}$$

Or, in one step:

$$30.06 \text{ mL NaOH} \times \frac{0.2509 \text{ mol NaOH}}{1000 \text{ mL NaOH}} \times \frac{1 \text{ mol HCl}}{1 \text{ mol NaOH}} \times \frac{1000 \text{ mL HCl}}{0.1005 \text{ mol HCl}} = 75.05 \text{ mL HCl}$$

Thus, 75.05 mL of 0.1005 M HCl reacts with 30.06 mL of 0.2509 M NaOH.

PROBLEM 3.19 What volume of 0.125 M HI reacts with 35.0 mL of 0.175 M KOH? The reaction is $HI(aq) + KOH(aq) \longrightarrow KI(aq) + H_2O$. □

3.12 LIMITING REACTANT

We seldom add reactants in the exact relative amounts required by the balanced equation. Often, only one reactant is entirely consumed. Its quantity fixes the quantity of product(s) obtained. The other reactant(s) are present *in excess.*

This situation is analogous to baking cookies. A recipe calls for $3\frac{1}{2}$ cups of flour and 2 well-beaten eggs (in addition to other ingredients) to make 60 cookies. The baker has 8.75 cups of flour and 9 well-beaten eggs. What is the maximum number of cookies that can be prepared? Which of these two ingredients is in excess, and by how much?

First find the number of cookies that can be prepared from 8.75 cups of flour and from 9 eggs:

$$8.75 \text{ cup flour} \times \frac{60 \text{ cookies}}{3.5 \text{ cup flour}} = 150 \text{ cookies}$$

$$9 \text{ eggs} \times \frac{60 \text{ cookies}}{2 \text{ eggs}} = 270 \text{ cookies}$$

The maximum number of cookies that can be prepared is 150. Evidently, it is *impossible* to prepare 270 of the desired cookies with just 8.75 cups of flour. The eggs are in excess. How much in excess? Calculate the number of eggs needed to prepare 150 cookies:

$$150 \text{ cookies} \times \frac{2 \text{ eggs}}{60 \text{ cookies}} = 5 \text{ eggs}$$

The baker will have 4 eggs left over, but the flour will be completely consumed. The flour is the *limiting ingredient.*

The manufacture of ethyl alcohol from ethylene and water, $C_2H_4(g) + H_2O(g) \longrightarrow C_2H_5OH(\ell)$, offers a chemical example: the C_2H_4 and H_2O react in equimolar quantities, but they are *not mixed* in equimolar quantities. Actually, 1.2 moles of H_2O are added to 1 mole of C_2H_4. The C_2H_4 is entirely consumed, combining with 1.0 mole of H_2O and producing 1 mole of C_2H_5OH. Thus the water is in *excess* by 0.2 mol. The C_2H_4, which is entirely consumed, is called the **limiting reactant** because it limits—fixes—the amount of ethyl alcohol that is produced.

The advantage of using an excess of a reactant is the subject of Chapter 13.

Excess $H_2O = 1.2$ mol H_2O (added) − 1.0 mol H_2O (consumed) = 0.2 mol H_2O

PROBLEM 3.20 An automobile assembler requires 1 car body and 4 wheels for each car. If 100 car bodies and 360 wheels are available, how many cars can be assembled? What is the limiting item? What item is in excess and by how many? □

EXAMPLE 3.23 Nitrogen oxide, NO, introduced into the atmosphere is converted to NO_2:

$$2NO(g) + O_2(g) \longrightarrow 2NO_2(g)$$

How many moles of NO_2 can be obtained when 0.10 mol NO and 0.40 mol O_2 are mixed? Which reactant is present in excess and by how many moles?

ANSWER First find the maximum amount of NO_2 that could be formed from each reactant:

$$0.10 \cancel{\text{mol NO}} \times \frac{2 \text{ mol } NO_2}{2 \cancel{\text{mol NO}}} = 0.10 \text{ mol } NO_2$$

$$0.40 \cancel{\text{mol } O_2} \times \frac{2 \text{ mol } NO_2}{1 \cancel{\text{mol } O_2}} = 0.80 \text{ mol } NO_2$$

Note that 0.10 mol NO_2 obtained from 0.10 mol NO is less than 0.80 mol NO_2 obtained from 0.40 mol O_2. It is *impossible* to obtain 0.80 mol NO_2 from only 0.10 mol NO. There are simply not enough N atoms in 0.10 mol NO to make that much NO_2. Thus, the quantity of NO_2 formed is 0.10 mol. We also see that not all of the O_2 is consumed. Therefore, the O_2 is present in excess, and NO, which is completely consumed, is the limiting reactant.

We find the amount of the excess O_2 by calculating the quantity of O_2 needed to prepare 0.10 mol NO_2:

$$0.10 \cancel{\text{mol } NO_2} \times \frac{1 \text{ mol } O_2}{2 \cancel{\text{mol } NO_2}} = 0.050 \text{ mol } O_2$$

Thus,

$$0.40 \text{ mol } \textit{given} - 0.050 \text{ mol } \textit{consumed} = 0.35 \text{ mol } O_2 \textit{ excess}$$

■

EXAMPLE 3.24 4.00 g of lead(II) nitrate, $Pb(NO_3)_2$, is added to an aqueous solution containing 1.00 g of potassium iodide, KI, and insoluble lead iodide forms:

$$Pb(NO_3)_2(aq) + 2KI(aq) \longrightarrow PbI_2(c) + 2KNO_3(aq)$$

Find the mass of PbI_2 formed. Which reactant is present in excess and by how many grams?

ANSWER First find how many moles of PbI_2 could be formed from each reactant. For $Pb(NO_3)_2$ (331 g/mol), as in previous examples:

$$4.00 \cancel{\text{g } Pb(NO_3)_2} \times \frac{1 \cancel{\text{mol } Pb(NO_3)_2}}{331 \cancel{\text{g } Pb(NO_3)_2}} \times \frac{1 \text{ mol } PbI_2}{1 \cancel{\text{mol } Pb(NO_3)_2}} = 0.0121 \text{ mol } PbI_2$$

Thus 4.00 g $Pb(NO_3)_2$ could form 1.21×10^{-2} mol PbI_2. Do the same kind of calculations for KI (166 g/mol):

$$1.00 \cancel{\text{g KI}} \times \frac{1 \cancel{\text{mol KI}}}{166 \cancel{\text{g KI}}} \times \frac{1 \text{ mol } PbI_2}{2 \cancel{\text{mol KI}}} = 0.00301 \text{ mol } PbI_2$$

Thus 1.00 g KI could form 0.00301 mol PbI_2.

Now observe that 0.00301 mol PbI_2 calculated from 1.00 g KI is less than the 0.0121 mol PbI_2 calculated from 4.00 g $Pb(NO_3)_2$. You *cannot get* 0.0121 mol PbI_2 from only 1.00 g KI. Therefore, the quantity of PbI_2 obtained is the smaller amount,

0.00301 mol PbI_2, and the quantity in grams is

$$3.01 \times 10^{-3}\ \cancel{\text{mol PbI}_2} \times \frac{461\ \text{g PbI}_2}{1\ \cancel{\text{mol PbI}_2}} = 1.39\ \text{g PbI}_2$$

Not all of the $Pb(NO_3)_2$ can be consumed and thus *it* is present in excess. To find the quantity of excess $Pb(NO_3)_2$, find the mass of $Pb(NO_3)_2$ required to produce 3.01×10^{-3} mol PbI_2:

Quantities before and after reaction:

$Pb(NO_3)_2$ +	$2KI \longrightarrow$	PbI_2
4.00 g	1.00 g	0.00 g
3.00 g	0.00 g	1.39 g

$$3.01 \times 10^{-3}\ \cancel{\text{mol PbI}_2} \times \frac{1\ \cancel{\text{mol Pb(NO}_3)_2}}{1\ \cancel{\text{mol PbI}_2}} \times \frac{331\ \text{g Pb(NO}_3)_2}{1\ \cancel{\text{mol Pb(NO}_3)_2}} = 0.996\ \text{Pb(NO}_3)_2$$

Then the excess $Pb(NO_3)_2$ is

$$4.00\ \text{g Pb(NO}_3)_2\ \textit{given} - 0.996\ \text{g Pb(NO}_3)_2\ \textit{consumed} = 3.00\ \text{g Pb(NO}_3)_2\ \textit{excess}$$

■

This reaction is used to obtain marketable sulfur from sulfur-containing coals in the pollution-free coal gasification electric generating plant operated by Southern California Edison Co. near Barstow, California.

PROBLEM 3.21 (a) Volcanoes emit hydrogen sulfide, H_2S, and sulfur dioxide, SO_2, which accounts for many of the sulfur deposits found in nature:

$$SO_2(g) + 2H_2S(g) \longrightarrow 3S(c) + 2H_2O(g)$$

However, an excess of either reactant, H_2S or SO_2, is a pollutant. In an eruption, a volcano emits 2.00×10^6 moles of SO_2 and 4.03×10^6 moles of H_2S. Which reactant is present in excess and by how many moles? (b) In an experiment, 6.60 g SO_2 and 5.00 g H_2S are mixed. How many grams of S are formed? Name the reactant present in excess and express the amount of the excess in grams. □

3.13 PERCENT YIELD

The conversion of reactants to products is frequently incomplete (Chapter 13).

For a variety of reasons, reactants often yield quantities of products that are less than those calculated from the balanced chemical equation. For example, the reactants may not react completely, or some of the desired product may be lost, or some other products may be formed. The quantity of product calculated from the chemical equation is called the **theoretical yield**. The **percent yield** is given by

$$\textbf{percent yield} = \frac{\textbf{actual yield}}{\textbf{theoretical yield}} \times \mathbf{100\%}$$

EXAMPLE 3.25 Ethylene oxide, C_2H_4O, is manufactured from ethylene and air:

Recall section 3.10 regarding the use of $\frac{1}{2}O_2$.

$$C_2H_4 + \tfrac{1}{2}O_2 \longrightarrow C_2H_4O$$

If 60 g of C_2H_4O is obtained from 42 g of C_2H_4, what is the percent yield?

Undesirable possible outcomes of this reaction include the failure of some ethylene to react; the conversion of some C_2H_4 to formaldehyde (H_2CO), CO, and CO_2; and the decomposition of some C_2H_4 to carbon (smoke).

ANSWER The theoretical yield, the maximum quantity of C_2H_4O that could be obtained from 42 g of C_2H_4, is found as in previous examples:

$$42\ \cancel{\text{g C}_2\text{H}_4} \times \frac{1\ \cancel{\text{mol C}_2\text{H}_4}}{28\ \cancel{\text{g C}_2\text{H}_4}} \times \frac{1\ \cancel{\text{mol C}_2\text{H}_4\text{O}}}{1\ \cancel{\text{mol C}_2\text{H}_4}} \times \frac{44\ \text{g C}_2\text{H}_4\text{O}}{1\ \cancel{\text{mol C}_2\text{H}_4\text{O}}} = 66\ \text{g C}_2\text{H}_4\text{O}$$

The actual yield obtained, however, is 60 g C_2H_4O. The percent yield is therefore

$$\frac{60\ \cancel{\text{g C}_2\text{H}_4\text{O}}}{66\ \cancel{\text{g C}_2\text{H}_4\text{O}}} \times 100\% = 91\%$$

■

PROBLEM 3.22 Carborundum, SiC, is manufactured by the reaction

$$SiO_2(c) + 3C(coke) \longrightarrow SiC(c) + 2CO(g)$$

A mixture containing 150 g SiO_2 yields 30.0 g SiC. What is the percent yield of SiC? □

In many chemical reactions, none of the reactants is completely consumed. This is illustrated in the next example.

EXAMPLE 3.26 In a laboratory synthesis of ammonia, NH_3, by the reaction $N_2(g) + 3H_2(g) \longrightarrow 2NH_3(g)$, only 55% of the nitrogen reacts. Starting with 4.00 mol N_2 and 13.0 mol H_2, how many moles of NH_3, N_2, and H_2 are in the final mixture when the production of NH_3 stops?

ANSWER First, find how many moles of N_2 react. We are given 4.00 mol N_2, but only 55% of it reacts. The conversion factor is

$$\frac{55.0 \text{ mol } N_2 \text{ reacts}}{100 \text{ mol } N_2 \text{ given}}$$

Thus,

$$4.00 \cancel{\text{mol } N_2 \text{ given}} \times \frac{55.0 \text{ mol } N_2 \text{ reacts}}{100 \cancel{\text{mol } N_2 \text{ given}}} = 2.20 \text{ mol } N_2 \text{ reacts}$$

Now find how many moles of NH_3 are formed, as in previous examples:

$$2.20 \cancel{\text{mol } N_2} \times \frac{2 \text{ mol } NH_3}{1 \cancel{\text{mol } N_2}} = 4.40 \text{ mol } NH_3 \text{ formed}$$

and how many moles of H_2 are consumed:

$$2.20 \cancel{\text{mol } N_2} \times \frac{3 \text{ mol } H_2}{1 \cancel{\text{mol } N_2}} = 6.60 \text{ mol } H_2 \text{ reacts}$$

Finally, from the given initial amounts and the amounts that react or form, the number of moles of NH_3, N_2, and H_2 in the final mixture are calculated:

	N_2	+	$3H_2$	$\longrightarrow$	$2NH_3$
Initial amount	4.00 mol		13.0 mol		0
Change	−2.20 mol		−6.60 mol		+4.40 mol
Final amount	1.80 mol		6.4 mol		4.40 mol

■

PROBLEM 3.23 In a laboratory test of the reaction

$$2SO_2(g) + O_2(g) \longrightarrow 2SO_3(g)$$

2.5 moles of SO_2 are mixed with 3.0 moles of O_2. Under the test conditions, 40% of the oxygen reacts. Find the number of moles of SO_3, SO_2, and O_2 in the mixture when the production of SO_3 stops. □

The **atomic weight** of an atom or the **molecular weight** of a molecule is the mass of the atom or molecule *relative* to the mass of an atom of ^{12}C, the major isotope of carbon to which a relative mass of 12 is assigned. These relative masses are experimentally determined in a *mass spectrometer*. The atomic weight of an element is obtained from the atomic weights and natural abundances of its isotopes. The molecular weight of a molecule is the sum of the atomic weights of all of the atoms in the molecule. Molecular weights may be determined in a mass spectrometer.

The assignment of a relative mass of 12 to the ^{12}C isotope fixes the **Avogadro number**: the number of ^{12}C atoms in exactly 12 g of ^{12}C. This number to three significant figures is 6.02×10^{23} particles. The **mole** is the amount of substance that contains the Avogadro number of particles, independent of the kind of particle. The mass of one mole is known as a **molar mass** and has the units of g/mol, specified for a particular particle, such as a molecule, atom, ion, proton, and so on. Thus we can write

$$\frac{6.02 \times 10^{23} \text{ H}_2 \text{ molecule}}{1 \text{ mol H}_2 \text{ molecule}} \quad \text{or} \quad \frac{2.02 \text{ g H}_2 \text{ molecule}}{1 \text{ mol H}_2 \text{ molecule}}$$

and,

$$\frac{6.02 \times 10^{23} \text{ H}_2\text{O molecule}}{1 \text{ mol H}_2\text{O molecule}} \quad \text{or} \quad \frac{18.0 \text{ g H}_2\text{O molecule}}{1 \text{ mol H}_2\text{O molecule}}$$

Formula weights calculated from empirical formulas for substances like NaCl that do not exist as individual molecules in the solid state are nevertheless called molecular weights.

The **simplest (empirical) formula** of a substance is determined from its mass percentage composition and the atomic weights of its constituent elements. The **molecular formula** of a substance is determined from its empirical formula and its molecular weight.

The concept of the balanced equation, the mole, and molar mass makes it convenient to relate quantities of reactants and products in any chemical change. For example, the following "road map" is useful when converting a given quantity of a substance to the quantity of another substance:

- If the given quantity is not in moles, convert to moles (grams $\longrightarrow$ moles of a substance, use its molar mass; number of particles $\longrightarrow$ moles, use the Avogadro number).
- Convert moles of the given substance to moles of the substance sought (use molar ratio—coefficients—from the balanced equation).
- If necessary, convert the moles of the substance sought to the unit wanted (moles $\longrightarrow$ grams of a substance, use its molar mass; moles $\longrightarrow$ number of particles, use the Avogadro number).

A common method of expressing the composition of a solution is called the **concentration** or the **molarity** (M) of the solution:

$$\text{M} = \frac{\text{moles of solute}}{\text{solution volume in liters}} = \frac{n}{V}$$

This diagram summarizes the conversions of quantities of substances:

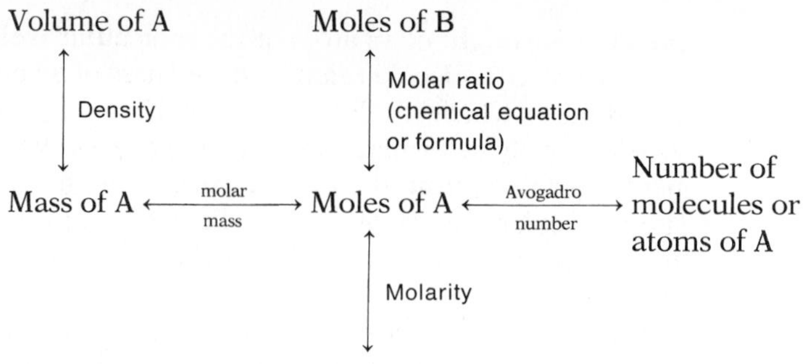

A **titration** in a **volumetric analysis** involves the addition of a solution of known concentration (molarity) from a **burette** to a sample. The known solution contains a substance that reacts completely with the substance being determined. The titration is stopped when the **end point**, recognized by a color change, is reached. Ideally, the end point is the same as the **equivalence point**, the point at which just enough solution has been added to react with the sample. An indicator is added to signal the end point.

In most chemical reactions, an excess of one substance, the *excess reactant*, remains while another reactant is entirely consumed. The reactant that is entirely consumed is called the **limiting reactant**. Also, in many reactions, unwanted products are obtained in addition to the products given in the balanced chemical equation. These undesirable reactions reduce the yield—the quantity of the desired product that is obtained. The quantity calculated from the chemical equation is called the **theoretical yield**. The **percent yield** is the ratio of the actual yield to the theoretical yield multiplied by 100%.

3.14 ADDITIONAL PROBLEMS

ATOMIC WEIGHT

3.24 What is the mass ratio (4 significant figures) of 1 atom of P to 1 atom of F?

3.25 A steel ball has a mass of 120 g and a wooden ball has a mass of 12.0 g. How many times heavier is the steel ball? Show that 100 of these steel balls would be the same number of times heavier than 100 of these wooden balls.

3.26 Magnesium consists of 78.99% ^{24}Mg, 23.9850; 10.00% ^{25}Mg, 24.9858; and 11.01% ^{26}Mg, 25.9826. Find the atomic weight of magnesium.

3.27 Chromium consists of 4.35% ^{50}Cr, 49.9460; 83.79% ^{52}Cr, 51.9405; 9.50% ^{53}Cr, 52.9407; 2.36% ^{54}Cr, 53.9389. Calculate the atomic weight of chromium.

3.28 (a) Boron consists of two isotopes: 19.8% ^{10}B, 10.01, and ^{11}B. Find the isotopic weight of ^{11}B. (b) Copper also consists of two isotopes: ^{63}Cu, 62.930, and ^{65}Cu, 64.928. What are the relative abundances of the two isotopes?

3.29 4.05 g of magnesium combines with 6.33 g of fluorine, forming magnesium fluoride, MgF_2. Find the relative masses of the atoms of magnesium and fluorine. Check your answer using the table of atomic weights (inside back cover). If the formula were not known, could you still answer this question?

3.30 The mass spectrometer shows that a Be atom is 0.751 times as heavy as a ^{12}C atom. Show that 0.751 g Be contains the same number of atoms as 1.00 g ^{12}C. Be consists of only one isotope.

TABLE 3.2
ATOMIC WEIGHTS BY THE CANNIZZARO METHOD

SUBSTANCE	MOLAR MASS, g/mol	MASS PERCENT COMPOSITION				MASS COMPOSITION, g/mol of compound			
		C	H	O	Cl	C	H	O	Cl
Carbon monoxide	28.0	42.9		57.1		12.0		16.0	
Hydrogen chloride	36.5		2.74		97.3		1.00		35.5
Methanol	32.0	37.5	12.2	50.0		12.0	4.00	16.0	
Formaldehyde	30.0	40.0	6.67	53.3		12.0	2.00	16.0	
Acetic acid	60.0	40.0	6.67	53.3		24.0	4.00	32.0	
Oxygen	32.0			100				32.0	
Hydrogen	2.00		100				2.00		
Water	18.0		11.1	88.9			2.00	16.0	
Oxide of carbon	68.0	52.9		47.1		36.0		32.0	

3.31 In 1858 Cannizzaro deduced the first universally accepted atomic weights and molecular formulas from the molecular weights of gases. The mass of a molecule is the sum of the masses of all the atoms in the molecule. Therefore the mass of an element found in 1 mole of its compounds is the atomic weight of that element, or is the atomic weight multiplied by a whole number. Data on compounds of carbon, hydrogen, oxygen, and chlorine are given in Table 3.2. (a) Using Table 3.2, assign atomic weights to carbon, hydrogen, and oxygen. (b) The atomic weight of Cl is 35.5. Assign a molecular formula to each of these substances. (c) What data would you have to discover to compel a change in the atomic weight of oxygen? Would you undertake such a research project?

STANDARDS

3.32 (a) The atomic weights of B and H are 10.81 and 1.0079, respectively. What is the relative mass of one atom of B to one atom of H? (b) In the *Tables Atomiques* by Le Bon L. J. Thenard (1827), the atomic weight of H is given as 6.2175. On this standard, what would be the atomic weight of B? Show that this shift does not change the relative mass of 1 atom of B to 1 atom of H. (c) Is the Avogadro number changed? If so, in what direction, upward or downward?

3.33 Assume the mole is redefined as the amount of substance that has the same number of particles as there are atoms in exactly 6 grams of the major isotope of carbon. Make any necessary changes in the following symbols and values: carbon-12, ^{12}C, $N_A = 6.02 \times 10^{23}$ particles, $^{16}O = 15.9949$, $O = 15.9994$, $H = 1.00794$, $^1_1H^+$ (proton) $= 1.00728$, ^1_0n (neutron) $= 1.00867$, $N = 14.0067$. Is the ratio of the mass of one atom of ^{12}C to the mass of one atom of ^1H changed? *Define the mass* (nucleon) *number* as the number of nucleons in the nucleus of an atom.

3.34 Standards may come and go, but the atomic mass unit (amu) is always defined so that there are as many amu in one atom of the chosen standard isotope as there are grams in one mole of the standard. For example, 1 mole of ^{12}C has a mass of 12.0 g (exactly), and the mass of one ^{12}C atom is 12 amu. Show that one gram is equal to an Avogadro number (N_A) of amu independent of the standard used. Is the value of N_A independent of the standard chosen?

MOLECULAR WEIGHT

3.35 Find the molecular weight of each of the following compounds from the molecular weights and relative abundances of their molecules: (a) ammonia, 17.0 (100%); (b) nitrogen trichloride, 119 (43.6 %), 121 (41.7%), 123 (13.2%), 125 (1.4%); (c) bromofluoroethane 126 (50.7%), 128 (49.3%); (d) cesium sulfide, 298 (95.0%), 299 (0.75%), 300 (4.20%).

3.36 (a) Find the accurate molecular weight of each of the following substances: (1) chlorine, Cl_2; (2) water, H_2O; (3) saccharin, $C_7H_5NSO_3$; (4) a PCB (polychlorinated biphenyl), $C_{12}H_6Cl_4$; (5) the sulfa drug sulfanilamide, $C_6H_4SO_2(NH_2)_2$; (6) $(NH_4)_4CdCl_6$; (7) $(UO_2)_3(PO_4)_2$. (b) Can you calculate the molecular weight of a substance from atomic weights without its molecular formula? Can you calculate the molecular formula of a substance without its molecular weight?

3.37 The formula of a molecule is X_2Y_7. X is 3.00 times heavier than ^{12}C and Y is 1.33 times heavier than ^{12}C. Find the molecular weight of X_2Y_7.

MOLE

3.38 The masses of 3 coins are given: 5.00 g/quarter, 3.15 g/penny, 2.20 g/dime. Given 400-g sets of each kind of these coins, (a) which set has the most coins and which has the fewest coins? (b) Which set has the largest and which has the smallest monetary value?

3.39 The mass of one round-head wood screw is 3.71 g. Find the mass of one dozen, one gross, and one mole of these wood screws. Compare the molar mass with the mass of the Earth, 5.98×10^{24} kg.

3.40 Sulfur molecules exist as S_8, S_6, S_4, S_2, and S, depending upon the temperature. Is the mass of sulfur in one mole of each molecule the same? Is the number of molecules in one mole of each molecule the same?
3.41 How many grams of (a) cholesterol, $C_{27}H_{45}OH$, are in 0.0285 mol cholesterol? (b) silver in 0.0285 mol Ag? (c) ammonia in 0.200 mol NH_3?
3.42 What mass in grams should be weighed for an experiment that calls for 0.135 mol $(NH_4)_2HPO_4$?
3.43 How many moles are in 75.0 g of each of the following? (a) S, (b) Cl, (c) Cl_2, (d) TiO_2, (e) C_4H_{10}, (f) a soap, $C_{17}H_{35}COONa$, (g) HCl, (h) $PbCr_2O_7$, (i) nickel, (j) gold.
3.44 How many molecules are in 28.0 g of each of the following substances? (a) CO; (b) N_2; (c) P_4; (d) P_2. Is the quantity of phosphorus in grams the same in (c) and (d)?
3.45 (a) How many moles and grams of P and Na are in 0.450 mole $Na_4P_2O_7$? (b) How many grams of nitrogen are in 0.20 mol $(NH_4)_2Cr_2O_7$?
3.46 How many atoms of C, H, and O are in (a) 2.50×10^{19} glucose ($C_6H_{12}O_6$) molecules; (b) 0.300 g glucose?
3.47 Express the following quantities of bromine in moles of Br_2 molecules: (a) 79.9×10^{23} Br_2 molecules, (b) 79.9 g bromine, (c) 79.9×10^{23} Br atoms, (d) 79.9 mol Br.
3.48 Express the following quantities of bromine in moles of Br atoms: (a) 79.9×10^{23} Br_2 molecules, (b) 79.9 g bromine, (c) 79.9×10^{23} Br atoms, (d) 79.9 mol Br_2.
3.49 Find the number of moles of Ag needed to form (a) 0.263 mol Ag_2S, (b) 0.263 mol Ag_2O, (c) 0.263 g Ag_2S, (d) 2.63×10^{20} Ag_2S molecules.
3.50 What is the *maximum* number of moles of CO_2 that can be obtained from (a) 1.00 mol $Fe_2(CO_3)_3$; (b) 2.00 mol $CaCO_3$; (c) 1.00 mol $Ni(CN)_4$? Does it matter how the conversions are made?
3.51 Find the smallest and the largest mass of silver among the following quantities: 0.0100 mol Ag, 0.0100 mol Ag_2O, 200 mg Ag, 3.01×10^{21} Ag atoms, 3.01×10^{21} Ag_2 molecules.
3.52 Complete this table:

MOLES OF COMPOUND	MOLES OF CATIONS	MOLES OF ANIONS
1 mol NaCl		
2 mol Na_2SO_4		
0.1 mol $Ca(NO_3)_2$		
0.2 mol $La_2(SO_4)_3$		
	0.75 mol NH_4^+	0.25 mol PO_4^{3-}

3.53 1/8 mol P_4O_{10} reacts with water, forming 1/2 mol H_3PO_4. Write the balanced equation for the reaction.
3.54 What volume of glycerine, $C_3H_8O_3$, density = 1.26 g/ml, should be taken to obtain 2.50 mol $C_3H_8O_3$?

MASS SPECTROMETER

3.55 Two molecules, H_2CO and N_2H_2, have the same molecular weight to three significant figures. A mass spectrometer shows two molecules: 30.01057 and 30.02180. Identify each molecule. Isotopic weights: $^{14}N = 14.003074$, $^{16}O = 15.994915$, $^{12}C = 12$ exactly, $^{1}H = 1.007825$.
3.56 Mass spectrometry determines the atomic weights of ions but not of atoms. Atomic weights are given: ^1H (1.007825), ^2H (2.01410), ^6Li (6.01512), ^{25}Mg (24.98584), electron (5.486×10^{-4}). Find the atomic weights of the ions $^1H^+$, $^2H^+$, $^6Li^+$, and $^{25}Mg^+$. What is the percentage error for each isotope if the ion mass is not corrected?

EMPIRICAL FORMULA

3.57 Write the empirical formula of each of these substances: (a) C_2H_2, (b) C_7H_7, (c) $C_{10}H_{10}$, (d) $C_{20}H_{20}$, (e) H_2, (f) O_2, (g) O_3, (h) Xe, (i) N_2O_4, (j) $K_2S_2O_8$, (k) Cr_2O_3, (l) C_3H_6, (m) Ag_2CO_3, (n) $C_2H_6O_2$, (o) Al_2Cl_6.
3.58 The mass composition of baking soda is 27.37% Na, 57.14% O, 14.30% C, and 1.200% H. Calculate its empirical formula.
3.59 Calculate the empirical formula of these compounds: (a) a bleaching agent, 30.9% Na, 21.5% O, and 47.7% Cl by mass; (b) a rocket fuel, 40.0% C, 13.4% H, and 46.6% N; (c) cocaine, 67.30% C, 6.930% H, 21.15% O, and 4.62% N.
3.60 A sample of a compound (triazanium) having a composition of 0.600 g C, 0.150 mol H, and 3.01×10^{22} N atoms is being studied as a solid rocket fuel. Calculate its empirical formula.
3.61 A 2.31-g sample of an oxide of iron, heated in a stream of H_2, produces 0.720 g H_2O. Find the empirical formula of the oxide.
3.62 Complicated chemical reactions occur at hot springs on the ocean floor. One compound involved in producing various ores consists of Mg, Si, H, and O. From a 334-mg sample, Mg is recovered as 0.115 g MgO; H is recovered as 25.7 mg H_2O; and Si as 0.172 g SiO_2. Find the empirical formula and molar mass of this ionic compound.
3.63 Combustion of 0.5707 mg of a hydrocarbon (C—H compound) produces 1.759 mg CO_2. Find the empirical formula of the hydrocarbon.
3.64 Analysis of a 20.0-mg sample of aminochlorobromoethane for H, N, and C yields 1.99 mg H_2O, 1.25 mg NH_3, and 6.47 mg CO_2. The Cl and Br are recovered as a mixture of AgCl and AgBr with a mass of 48.8 mg. Finally, when all the AgBr is converted to AgCl, the mass of AgCl formed from AgBr plus the mass of AgCl originally present in the mixture is 42.3 mg. Calculate the empirical formula of aminochlorobromoethane.

MOLECULAR FORMULA

3.65 If you are given the mass composition of a substance, do you have enough data to calculate a molecular formula? If not, what else do you need?
3.66 Three molecules of phosphorus exist with molecular weights of 62.0, 31.0, and 124.0. Write the formula of each molecule.
3.67 Evaporation of solid sodium chloride yields molecules with molecular weights of 58 and 116. Write the formulas of these molecules. Find the accurate molar mass of solid sodium chloride, an ionic compound.
3.68 The β-blocker drug, timolol, is expected to reduce the need for heart by-pass operations. Its composition by mass is 47.2% C, 6.55% H, 13.0% N, 25.9% O, and 7.43% S. 0.0100 mol timolol has a mass of 4.32 g. Calculate its molecular formula. What is its empirical formula?
3.69 0.250 mole of a hydrocarbon is burned in excess chlorine, forming 0.500 mol CCl_4 and 1.50 mol HCl. Write (a) the molecular formula of the hydrocarbon and (b) the balanced equation for the reaction.

MASS COMPOSITION

3.70 Find the mass percentage of H in (a) C_3H_4, (b) C_6H_8, (c) C_2H_5OH, and (d) CaH_2, and of As in (e) $NiAs_3$ and (f) AsH_3.
3.71 Calculate the mass percentage of K in KCl and potassium hydrogen tartrate, $KHC_4H_4O_6$, ingredients of No Salt, a table salt substitute.
3.72 How many grams of oxygen and atoms of O are in (a) 9.00 H_2O and (b) 19.0 g $Fe_2(HPO_4)_3$?

MOLARITY (CONCENTRATION)

3.73 What is the molarity of a solution containing 0.155 mol H_3PO_4 in 200 mL solution?
3.74 A solution contains 0.100 mole per liter of each of the following acids: HCl, H_2SO_4, H_3PO_4. (a) Is the molarity the same for each acid? (b) Is the number of molecules per liter the same for each acid? (c) Is the mass per liter the same for each acid?
3.75 A solution contains 1.05 g of a rubbing alcohol, $(CH_3)_2CHOH$, in 100 mL solution. Find (a) the molarity of the solution and (b) the number of moles in 1.00 mL of solution.
3.76 (a) Calculate the molarity of caffeine in a 12-oz cola drink containing 50 mg caffeine, $C_8H_{10}N_4O_2$. (b) Cola drinks are usually 5.06×10^{-3} M with respect to H_3PO_4. How much of this acid is in a 250-mL drink (1 oz = 29.6 mL)?
3.77 How many grams of the cleansing agent Na_3PO_4 (a) are needed to prepare 200 mL of 0.25 M solution; (b) are in 200 mL of 0.25 M solution?
3.78 How many kg of ethylene glycol, $C_2H_6O_2$, are needed to prepare a 9.00 M solution to protect a 15.0-liter car radiator against freezing? What is the mass of $C_2H_6O_2$ in 15.0 L of 9.00 M solution?
3.79 (a) You need 2.75 g H_2SO_4. What volume of 0.175 M H_2SO_4 should you use? (b) What volume of 0.175 M H_2SO_4 contains 2.75 g H_2SO_4?
3.80 How many grams of washing soda, $Na_2CO_3(H_2O)_{10}$, are needed to prepare 250 mL of a 5.05 M solution of Na_2CO_3?

CHEMICAL ARITHMETIC

3.81 Nickel is produced by the reaction

$$Ni(NH_3)_2SO_4(aq) + H_2(g) \longrightarrow Ni(c) + (NH_4)_2SO_4(aq)$$

What is the number of (a) moles of H_2 consumed to produce 2.50 mol Ni; (b) moles of $Ni(NH_3)_2SO_4$ consumed when 4.50 mol $(NH_4)_2SO_4$, a fertilizer, is obtained; (c) moles of $(NH_4)_2SO_4$ obtained when 3.45 mol Ni forms; (d) grams of $(NH_4)_2SO_4$ obtained when 25.0 g Ni forms?
3.82 Find the mass in grams of Cl_2 formed when 25.5 g HCl reacts as shown:
(a) $2HCl(aq) + O_2(g) \longrightarrow Cl_2(g) + H_2O_2(aq)$
(b) $4HCl(g) + O_2(g) \longrightarrow 2Cl_2(g) + 2H_2O(g)$
3.83 How many grams of P_4O_{10} must react to form 15.8 g $Ca_3(PO_4)_2$?

$$6CaO(c) + P_4O_{10}(c) \longrightarrow 2Ca_3(PO_4)_2(c)$$

3.84 Hypochlorites, used as disinfecting agents, react with ammonia, forming nitrogen trichloride, an eye irritant:

$$NH_3 + HOCl \longrightarrow NH_2Cl + H_2O$$
$$NH_2Cl + HOCl \longrightarrow NHCl_2 + H_2O$$
$$NHCl_2 + HOCl \longrightarrow NCl_3 + H_2O$$

Starting with 1.69 mol NH_3 and assuming that all of the NH_3 goes to NH_2Cl, all of the NH_2Cl goes to $NHCl_2$, and all of the $NHCl_2$ goes to NCl_3, (a) how many grams of HOCl are consumed and (b) how many grams of NCl_3 are formed?
3.85 The first lighter-than-air balloon to make a flight (1783) was loaded with the most flammable gas known prepared from "1000 pounds of iron filings" (454 g/lb) and sulfuric acid:

$$Fe(c) + H_2SO_4(aq) \longrightarrow FeSO_4(aq) + H_2(g)$$

What is the maximum mass in pounds of H_2 obtained? If the density of the gas is 0.083 g/L, what is the volume of the gas?
3.86 How many grams of $Cr_2(SO_3)_3$ are consumed when all of the acid in 100 mL of 3.00 M H_2SO_4 reacts?

$$Cr_2(SO_3)_3(aq) + 3H_2SO_4(aq) \longrightarrow Cr_2(SO_4)_3(aq) + 3SO_2(g) + 3H_2O$$

3.87 During volcanic action S_8 is converted to S, which is then converted to H_2S; in turn, the H_2S reacts with Fe, forming FeS_2. In water containing O_2, the FeS_2 goes to "mine acid," H_2SO_4. Find the maximum mass in grams of H_2SO_4 that can be formed from 0.366 mol S_8. (*Hint*: Do not waste time writing chemical equations; they are not needed.)

3.88 A 3.000-g sample of element E forms 3.615 g E_2O. Find the molar mass of E and name it.

VOLUMETRIC ANALYSIS

3.89 What volume of 0.100 M $AgNO_3$ reacts with 20.0 mL of 0.250 M NaCl?

$$NaCl(aq) + AgNO_3(aq) \longrightarrow AgCl(c) + NaNO_3(aq)$$

3.90 A 0.200-g sample of NaCl reacts with 30.1 mL of $AgNO_3$ solution. Find the molarity of the $AgNO_3$ solution. See previous problem for the chemical equation.

3.91 A tablet of the antacid milk of magnesia contains 0.450 g magnesium hydroxide, $Mg(OH)_2$. Find the volume of 0.100 M HCl (about the HCl concentration of stomach fluid) that reacts with one tablet.

$$Mg(OH)_2(c) + 2HCl(aq) \longrightarrow MgCl_2(aq) + 2H_2O$$

EXCESS REACTANT

3.92 Aluminum reacts with hydroxides, forming hydrogen. 3.00 g Al and 7.20 g KOH are mixed. Find the mass of potassium aluminate, $KAl(OH)_4$, that is formed. Which reactant is in excess and by how many grams, and which reactant is the limiting one?

$$2Al(c) + 2KOH(aq) + 6H_2O \longrightarrow 3H_2(g) + 2KAl(OH)_4(aq)$$

3.93 When 10.0 g Fe_2O_3 and 6.00 g CO are heated, producing iron and carbon dioxide, one reactant is completely consumed. Write the balanced chemical equation. Which reactant is in excess and by how many grams? How much Fe is produced?

YIELD

3.94 Zinc is produced from ZnS ore. In a test run, 0.637 g Zn is obtained from 1.00 g ZnS. What is the percent yield?

3.95 P_4O_{10} serves as the only source of P in the production of a nerve gas, $PO_2FC_4H_{10}$; 0.467 g $PO_2FC_4H_{10}$ is obtained from 1.10 g P_4O_{10}. Find the percent yield of nerve gas.

3.96 In the production of a drug, 2.00 moles of the limiting reactant (265 g/mol) forms only 0.250 mole of the desired product (200 g/mol) for an actual yield of 25.0%. (a) If the yield were 100%, what mass of product would be produced for the consumption of 2 mol limiting reactant? (b) What is the minimum mass of limiting reactant that must be used to obtain 125 g product?

SELF-TEST

3.97 (15 points) (a) Silver consists of 51.8392% ^{107}Ag, 106.9051, and 48.1608% ^{109}Ag, 108.9048. Find its atomic weight. (b) Under certain conditions, gaseous lithium consists of four molecules: 21.048 (79.150%), 20.047 (19.25%), 19.046 (1.56%), and 18.045 (0.04%). Find its molecular formula.

3.98 (10 points) How many moles, grams, and atoms of Pb are in 85.0 g red lead, Pb_3O_4?

3.99 (10 points) One mole of Cl_2 reacts with ammonia, NH_3, forming two compounds, samples of which have these compositions: (a) 5.60 g N, 14.2 g Cl, and 0.800 mol H; (b) 26.2% N, 7.5% H, and 66.4% Cl. Write the balanced equation for the reaction, showing the formation of these two compounds.

3.100 (10 points) A compound extracted from a recently rediscovered Mexican herb is 1000 times sweeter than sucrose and is composed of 7.23×10^{24} H atoms, 1 mol O, and 90.0 g C. Mass spectroscopic analysis shows the presence of three molecules: 236 (98.50%), 237 (1.14%), and 238 (0.39%). Find the compound's empirical formula, molecular formula, and accurate molecular weight.

3.101 (10 points) A 11.7-g sample of a nickel sulfide is converted to 8.60 g NiO. (a) What is the mass composition of the sulfide? (b) Calculate its empirical formula.

3.102 (10 points) Air bags in automobiles are inflated by decomposition of sodium azide, which decomposes as shown:

$$2NaN_3(c) \longrightarrow 2Na(\ell) + 3N_2(g)$$

How many grams of NaN_3 must decompose to produce 168 g N_2?

3.103 (15 points) A 20.0-mL sample of sulfuric acid from a battery reacts with 52.2 mL of 1.50 M NaOH. What is the concentration of the battery acid?

$$H_2SO_4(aq) + 2NaOH(aq) \longrightarrow Na_2SO_4(aq) + 2H_2O$$

3.104 (10 points) Highly toxic diborane, B_2H_6, used in the preparation of C—H—B compounds, is synthesized as shown:

$$3LiAlH_4 + 4BF_3(g) \longrightarrow 2B_2H_6(g) + 3LiAlF_4$$

A mixture contains 7.68 g BF_3 and 5.00 g $LiAlH_4$. Find the quantity of B_2H_6 formed and the quantity of reactant left in excess.

3.105 (10 points) When 2.00 mol CH_4 and 3.00 mol steam are mixed, 60.0% of the CH_4, reacts as shown:

$$CH_4 + 2H_2O \longrightarrow 4H_2 + CO_2$$

The remainder of the CH_4 does not undergo reaction. Find the number of moles of CH_4, H_2O, CO_2, and H_2 present when the H_2 production stops.

4

ENERGY AND THERMOCHEMISTRY

Chemists are concerned with energy as well as matter. In Chapters 2 and 3 you learned how to use formulas and equations to represent chemical changes. The story is not complete, however, until we explore the energy changes that accompany chemical changes. For example, the combustion of fossil fuels, such as coal, oil, and natural gas, is our major source of energy. All combustion reactions emit energy. Another type of chemical change is photosynthesis, the conversion of carbon dioxide and water to oxygen, sugars, and other plant products. This reaction absorbs energy from the sun. In this chapter, we will study how these energy changes are measured and how they are expressed in balanced chemical equations. Then you will be able to predict the energy changes of reactions. Finally, we will relate these energy changes to molecular composition.

4.1 ENERGY, WHAT IS IT?

A living organism evolves heat, does work, and requires energy. An interesting sentence, but what do "work," "energy," and "heat" mean? We encounter several familiar forms of energy every day: **mechanical energy** (drives our cars), **electrical energy** (lights our lamps), **heat energy** (cooks our food), **solar** (light) **energy** (needed to grow our food), and **chemical energy** (batteries that start our cars, fuels that push our cars, trucks, trains, and spaceships, and explosives used in constructing our roads). Yet heat, work, and energy, unlike matter, cannot be put into an apparatus, chopped up, and analyzed. They are abstract concepts. Nonetheless, they are useful because they help us to interpret many common experiences. Moreover, they are definite, measurable quantities.

Force is discussed in Appendix A.3.

Let us start with the term work, which is used in the sciences in a restricted sense: **work** is done only when a force is exerted on an object so that it moves. Thus, "mental work" is not considered work since no motion is involved. For the same reason, the effort of holding a 5-kg weight in your hand is not considered work. The work, w, done by an object (solid, liquid, gas) is defined and measured by *the product of the force, F,* and *the distance, ℓ,* through which *the object moves while the force is acting on it.* When you push or pull an object, you are exerting a force. Thus, work is done in lifting a weight. Work is also done when a spring is stretched, a gas is compressed in a container, or positively charged ions are separated from negatively charged ions. A frictionless piston moving against a

BOX 4.1
THOMAS YOUNG AND $\frac{1}{2}mu^2$

In 1805 Thomas Young suggested substituting the word *energy* (Greek *energos*, "work") for "living force" or "vis viva." He used the term for what we now call "kinetic" energy. Young said, "The term energy may be applied, with great propriety, to the product of the mass or weight of a body, into the square of the number expressing its velocity." The term *kinetic energy* was introduced by William Thomson (Lord Kelvin) in 1856, being derived from *kinema*, Greek for "motion."

Thomas Young, 1773–1829

vacuum (opposing pressure is zero) does no work, however, because it is not opposed by a force.

Energy is defined as the *capacity to do work* on an object, as in a piston engine, or to *transfer heat*, as in a refrigerator. Any moving object can do work by colliding with another object. This *energy of motion* is called **kinetic energy** (Box 4.1).

In many cases an object may possess energy because of some property other than motion. A compressed spring, for example, may do work as it expands, a suspended weight as it falls, a liter of gasoline as it burns, a stick of dynamite as it explodes, a kilogram of uranium as it undergoes fission, a quietly standing mule as it kicks, and water trapped behind a dam as it spills over. The energy of such stationary objects is said to be "potential" (Latin *potens*, "being able") in the sense that they can do work, but they may not do work right away, or ever (Box 4.2). The energy stored in an object because of its position or chemical composition is called **potential energy** (Table 4.1). Coal, oil, and gasoline possess energy—potential energy—because combustion liberates energy that can be used to do work. A suspended weight has potential energy because of its position. The heavier the weight or the greater the height, the greater is the potential energy. When released, it drops to the ground because it is attracted by gravity. Its energy is not lost. Instead it appears as kinetic energy because of its motion, and as heat because of air friction and impact.

Conversion of energy to matter may occur in nuclear reactions (Chapter 23).

The gain in kinetic energy of a freely falling object (in a vacuum, no friction) is exactly equal to the loss of potential energy. This is a statement of **the law of conservation of energy:** *Energy may be transformed from one form into another, but it cannot be created or destroyed.* Consequently, *the total amount of energy in the universe never changes.*

PROBLEM 4.1 A billiard ball with a kinetic energy of 100 J hits a stationary billiard ball in outer space. After the collision, the kinetic energy of one ball is 75 J. What is the kinetic energy of the other ball? □

TABLE 4.1
EXAMPLES OF POTENTIAL ENERGIES OF OBJECTS

POSITION	CHEMICAL COMPOSITION
Suspended weight	gasoline
Water behind a dam	coal
Compressed spring	dynamite
Stationary mule	fissionable uranium

BOX 4.2
INVENTIONS AND POTENTIAL ENERGY

The ignition temperature of anthracite (hard coal) is so high that it was formerly thought of as rock. Before the discovery of nuclear fission and fusion (Chapter 23), uranium and lithium hydride ($^{7}Li^{1}H$) were considered valueless as fuels. Thus, how much potential energy an object has depends on what kind of process has been developed to make it available as work.

Energy, work, and heat are discussed in more detail in Chapter 18.

Heat is the name of the energy that is transferred from one object to another because of a temperature difference (section 1.2). Heat always flows from a warmer to a neighboring cooler object. If a hot stone is dropped into cold water, the water warms up while the stone cools; this heat transfer stops when the stone and water temperatures become identical. The reverse process—the flow of heat from a cooler to a warmer object—never occurs spontaneously. Heat is the most common form of energy absorbed or produced in chemical reactions.

The first (jet) steam toy engine (an engine converts heat to work) was invented by Hero of Alexandria, a Greek who lived about 2100 years ago.

Heat can be transformed into *work*. For instance, the combustion of gasoline—the change of gasoline and oxygen to carbon dioxide and water—provides the push that propels a car. The combustion of coal or oil liberates heat that is used to do work—for example, to turn a turbine that drives an electric generator. When a given object does work, its energy is decreased; a storage battery that is not recharged eventually will not have enough energy to start a car.

Conversely, work can be converted to heat. For example, a paddle wheel rotating in water (doing work on the water) raises the temperature of the water. The effect is the same as if a flame were applied to the water. We can start a fire by rapidly rubbing sticks together (doing work) or by using a match. Thus, we can transfer energy to an object either by doing work on it or by heating it.

The concept of temperature, heat, and work at the molecular level is discussed in Chapters 10 and 18.

Energy concerns us because chemical reactions are accompanied by energy changes—the absorption or emission of energy. All life processes, for example, can be discussed in these terms, a subject referred to as bioenergetics.

4.2 ENERGY CHANGES AND CHEMICAL CHANGES

Why do energy changes accompany chemical changes? The answer involves a force that holds atoms together in molecules. The force is called a **chemical bond**, or simply a bond. A dash is used to represent a bond. Thus, H_2 may be written as H—H. We may then say that the rearrangement of atoms involves breaking chemical bonds in reactant molecules and forming new bonds in product molecules. The atoms themselves are not created or destroyed; they are simply rearranged.

For the reaction $H_2(g) + Cl_2(g) \longrightarrow 2HCl(g)$ to occur, the H—H and Cl—Cl bonds must be broken and two H—Cl bonds must form. *In separating atoms, energy is needed to overcome the forces that hold the atoms together.* In short, *energy must be absorbed to break a bond.* Conversely, *when a bond forms, energy is evolved.* The energy needed to break a bond and the energy given off when a bond forms are definite and characteristic for each bond (section 4.7). For example, energies needed to break the H—H, Cl—Cl, and H—Cl bonds are all different. The bond energies of the product molecules may therefore be greater or smaller than the bond energies of the reactant molecules. Thus, chemical changes are accompanied by energy changes that occur as bonds break and other bonds form as shown:

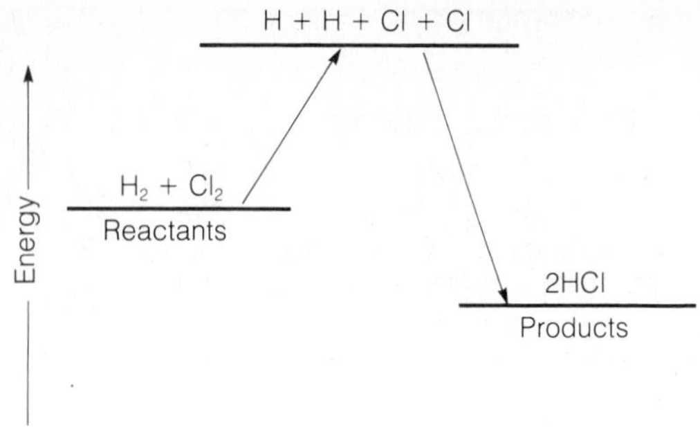

The net energy changes resulting from breaking and making chemical bonds are known as **heats of reaction**. They are called the *heats* of reaction because heat is the most common form of energy that is absorbed or given out in chemical reactions. **Thermochemistry** deals with the measurement and calculation (sections 4.6 and 4.7) of heats of reaction. The heat of a reaction is a characteristic property of the chemical reaction and a major factor in determining whether it will occur.

4.3 MEASURING HEAT

Heat measurements always involve the flow of heat. Devices known as **calorimeters** (Latin *calor,* "heat," and *metrum,* "measure") use the change in the temperature of a known mass of water to measure the quantity of heat that is transferred.

A joule (Appendix A.5) is the kinetic energy ($\frac{1}{2}mu^2$) of an object with a mass of 2 kg (exactly) moving at a speed of exactly one m/sec. One kilojoule, kJ, equals 10^3 J. The units of energy, work, and heat are the same.

The SI unit of energy is the **joule**, J. By definition, one **calorie**, another unit that is still in use, is equal to 4.184 J (exactly).

1 cal = 4.184 J

One calorie (cal) raises the temperature of 1 g of water 1°C.† The **specific heat** of a substance is the quantity of heat required to warm 1 g of the substance 1°C. The same quantity of heat is removed to cool 1 g of the substance 1°C. Thus, the specific heat of water is 1 cal/(g °C) or 4.184 J/(g °C). The specific heat of copper is 0.091 cal/(g °C) or 0.38 J/(g °C). This means it takes 0.38 J (0.091 cal) to raise the temperature of 1 g of copper 1°C, much less than for water.

The quantity of heat transferred (lost or gained), q, is the product of the temperature change, the mass, m, of the substance, and the specific heat of the substance:

$$q = \text{specific heat} \times \text{mass} \times \Delta t$$

Δt is the temperature change; that is, the final temperature minus the initial temperature ($\Delta t = t_{\text{final}} - t_{\text{initial}}$). The Greek letter delta, Δ, denotes a difference in a quantity: Δ = *final quantity* – *initial quantity.* When an object is warmed (absorbs heat), its temperature increases; Δt is therefore positive and the sign of q is plus. When an object is cooled (gives up heat), its temperature decreases; Δt is therefore negative and q takes a minus sign. In heat measurements, the heat lost by one object is gained by another. Since water is the liquid most commonly used in heat measurements, it follows that

$$q_{\text{object}} = -q_{\text{water}}$$

† Historically, the calorie has been defined in various ways that are almost identical. There is no need to study these definitions, but if you are curious about them, refer to Appendix A.5.

For example, assume that the object is a warm piece of nickel and that the water absorbs 50 J. Then q water = 50 J and

$$q_{nickel} = -q_{water} = -50 \text{ J}$$

The minus sign tells us that the *nickel evolved* 50 J. Next, assume that a cold piece of silver *cools* the water by 100 J. Then, $q_{water} = -100$ J and

$$q_{silver} = -q_{water} = -(-100 \text{ J}) = +100 \text{ J}$$

The positive sign indicates that the silver absorbed 100 J.

EXAMPLE 4.1 An 18.5-g sample of a metal absorbs 1170 J as its temperature increases from 25.0 to 92.5°C. What is the specific heat of the metal?

ANSWER The sample absorbs heat so that $q = 1170$ J.

$$q = \text{specific heat} \times \text{mass} \times \Delta t$$

$$\text{specific heat} = \frac{q}{m \times \Delta t} \quad \text{and} \quad \Delta t = t_{final} - t_{initial}$$

$$= \frac{1170 \text{ J}}{18.5 \text{ g} \times (92.5°\text{C} - 25.0°\text{C})}$$

$$= 0.937 \text{ J/(g °C)}$$ ■

EXAMPLE 4.2 When 25.0 g of a metal at a temperature of 90.00°C is added to 50.0 g of water at 25.00°C, the water temperature rises to 29.80°C. The specific heat of water is 4.18 J/(g °C). Find the specific heat of the metal.

ANSWER The heat gained by the water equals the heat lost by the metal. We therefore find the heat gained by the water:

$$q = \text{specific heat} \times \text{mass} \times \Delta t$$

$$q_{water} = \frac{4.18 \text{ J}}{\cancel{\text{g}}\ \cancel{°\text{C}}} \times 50.0\ \cancel{\text{g}} \times (29.80\cancel{°\text{C}} - 25.00\cancel{°\text{C}}) = +1003 \text{ J}$$

Retaining 4 significant figures in these calculations to obtain an answer with the proper number of significant figures is acceptable.

We can now find the specific heat of the metal. The metal gave up the same quantity of heat that the water gained, so $q_{metal} = -1003$ J:

$$q_{metal} = -q_{water} = -1003 \text{ J}$$

The final temperature of the metal is the same as that of the water so that Δt = 29.80°C − 90.00°C. Then

Note here the use of dimensional analysis: the specific heat is multiplied by the mass and by Δt, eliminating the units g and °C and leaving the unit J for the answer. In the second part, the quantity of heat is divided by the mass and Δt, giving the units for specific heat for the answer.

$$\text{specific heat} = \frac{-1003 \text{ J}}{25.0 \text{ g} \times (29.80°\text{C} - 90.00°\text{C})} = 0.666 \frac{\text{J}}{\text{g °C}}$$ ■

Specific heats vary markedly from substance to substance and, to some extent, with temperature, for any given substance. Some approximate values at ordinary room conditions are given in Table 4.2.

PROBLEM 4.2 (a) How many kilojoules are needed to raise the temperature of 30.5 g Al, specific heat 0.895 J/(g °C), from 18.6 to 48.2°C? (b) How many joules are required to raise the temperature of 1.00 mole of atoms of each of the following

TABLE 4.2
SPECIFIC† HEATS OF SOME SUBSTANCES AT ORDINARY ROOM CONDITIONS, J/(g °C)

Air	1.0	Ethyl alcohol	2.2
Water (0°C)	4.218	Copper	0.38
Water (17°C)	4.184	Iron	0.46
Water (25°C)	4.180	Mercury	0.14
Water (100°C)	4.216	Silver	0.23
Ice	2.1		

† *specific* means "per unit mass."

solids 1°C (specific heats given)? Sn (0.23 J/(g °C)), Fe (0.46), Pt (0.131), U (0.12), Li (3.9), I_2 (0.22). What can you say about the quantity of heat required to raise the temperature of the same number of atoms of solid elements 1°C? (Alexis Petit and Pierre Dulong said it in 1819.) (c) Is the same quantity of heat required to raise the temperature of equal quantities, in grams, of different metals from 10 to 20°C? □

PROBLEM 4.3 14.6 g metal at 52.62°C is stirred with 25.0 g water at 20.00°C. The final temperature is 20.80°C. Find the specific heat of the metal. □

Specific heat is also known as *specific heat capacity*.

The quantity of heat required to raise the temperature of a given object 1°C is called its **heat capacity**:

$$\textbf{heat capacity} = \frac{q}{\Delta t}$$

The units of heat capacity are J/°C. When the heat capacity is given for 1 g of a substance, it is called the specific heat, J/(g °C). We find the heat capacity of an object by putting in a known amount of energy and measuring the temperature rise. When a given object requires a comparatively large quantity of heat to raise its temperature 1°C, we say that it has a large heat capacity.

EXAMPLE 4.3

Joules = volts × coulombs.

The temperature of a calorimeter (next section) is raised 2.2052°C by passing a current through an electric heater immersed in the calorimeter. The current delivers 3820.5 J. Find the heat capacity in J/°C of the calorimeter.

ANSWER The heat capacity is the quantity of heat required to raise the temperature of the calorimeter 1°C. The calorimeter absorbed 3820.5 J while its temperature increased by 2.2052°C. Thus

$$q = 3820.5\ \text{J}$$

$$\Delta t = 2.2052°\text{C}$$

$$\text{heat capacity} = \frac{q}{\Delta t}$$

$$\text{heat capacity of calorimeter} = \frac{3820.5\ \text{J}}{2.2052°\text{C}} = 1732.5\ \frac{\text{J}}{°\text{C}}$$ ■

4.4 MEASURING HEATS OF REACTION: THERMOCHEMICAL EQUATIONS

Heats of reaction are measured by mixing known amounts of reactants in a calorimeter (Figure 4.1) at a known temperature (for example, 25.02°C) and letting them react. The heat evolved or absorbed by the reaction is equal to the heat that is absorbed or evolved by the calorimeter. In an **exothermic reaction**, the chemical reaction *gives off* heat to the surroundings, raising the temperature

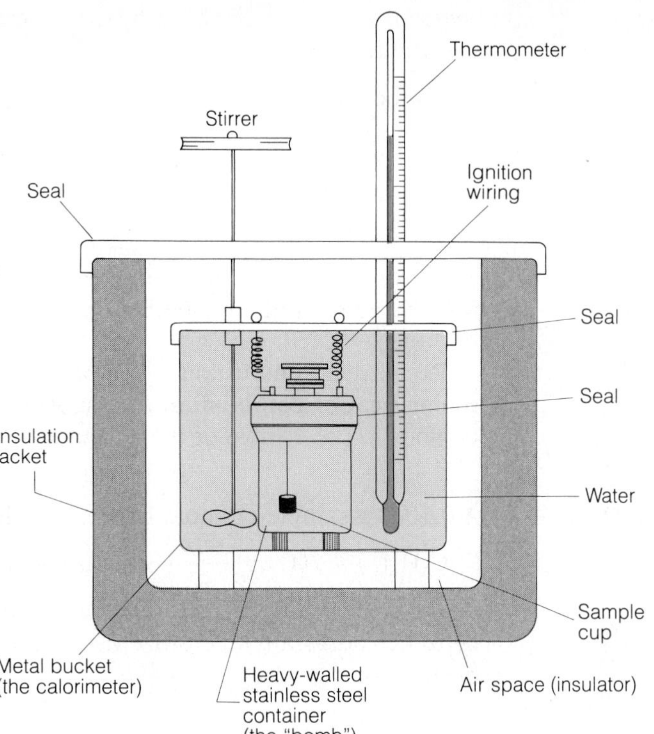

FIGURE 4.1
A bomb calorimeter. The stirrer maintains uniform temperature. The reaction is set off by a wire heater in the chamber and occurs with explosive speed. We determine the heat produced by the reaction by measuring the rise in the temperature of the water in the calorimeter and the heat capacity of the calorimeter.

of the reaction products above the temperature of the calorimeter. The hot products cool by *transferring heat* to the calorimeter; as it absorbs this heat, the calorimeter becomes warmer. The transfer stops when the temperatures of the products and the calorimeter are equal (for example, 26.07°C). The calorimeter thus measures the quantity of heat that must now be *removed* from the calorimeter and the products (26.07°C) to restore them to the original temperature (25.02°C).

Marcellin Berthelot, a founder of thermochemistry, introduced (1879) the terms exothermic and endothermic from *exo* (Greek, ''outside''), *endon* (''inside''), and *therme* (''heat'').

In an **endothermic reaction**, the chemical reaction *absorbs* heat from the surroundings, lowering the temperature of the calorimeter and the reaction products (for example, to 23.98°C) below the original temperature (25.02°C). The calorimeter thus measures the quantity of heat that must be *added* to the calorimeter and the products (23.98°C) to restore them to the original temperature (25.02°C). Thus, the general idea is to determine the rise or fall in the temperature of the calorimeter.

A calorimeter consists of an insulated chamber containing a known quantity of water, a metal bucket to hold the water, a metal reaction chamber containing known quantities of reactants, a stirrer, and a thermometer. The heat absorbed or evolved by a calorimeter depends on its heat capacity and the temperature change:

$$q = \textbf{heat capacity} \times \Delta t$$

The measured change in the temperature of the calorimeter and its known heat capacity allow us to calculate heats of reaction.

EXAMPLE 4.4 A fuel is burned in the calorimeter whose heat capacity, 1732.5 J/°C, was determined in Example 4.3. Its temperature rises from 22.3102 to 23.9863°C. Calculate the heat absorbed by the calorimeter and evolved by the fuel.

ANSWER The heat absorbed by the calorimeter depends on both the heat capacity and the temperature change:

$$q = \text{heat capacity} \times \Delta t$$

$$q = 1732.5 \frac{\text{J}}{°\cancel{\text{C}}} \times (23.9863°\cancel{\text{C}} - 22.3102°\cancel{\text{C}})$$

$$= 2903.8 \text{ J}$$

The heat evolved by the fuel equals the heat absorbed by the calorimeter. The fuel therefore emitted 2903.8 J. ■

Recall that combustion is the process in which a substance burns in air. The **molar heat of combustion** of a substance is the amount of heat evolved in the combustion of 1 mole of the substance; its units are kJ/mol.

EXAMPLE 4.5 A 0.1000-g sample of liquid benzene, C_6H_6, is burned

$$C_6H_6(\ell) + 7\tfrac{1}{2}O_2(g) \longrightarrow 3H_2O(\ell) + 6CO_2(g)$$

in a closed calorimeter whose heat capacity is 1602 J/°C. The combustion of benzene causes a temperature rise of 2.609°C. Calculate the molar heat of combustion for benzene.

ANSWER First, find the quantity of heat evolved:

$$q = \text{heat capacity} \times \Delta t$$

$$q = 1602 \frac{\text{J}}{°\cancel{\text{C}}} \times 2.609°\cancel{\text{C}} = 4.180 \times 10^3 \text{ J}$$

Thus 4180 J are evolved during the combustion of 0.1000 g of C_6H_6.

We want to calculate the heat evolved by the combustion of 1 mole of C_6H_6 (78.11 g/mol). The conversion factor is

$$\frac{4.180 \times 10^3 \text{ J}}{0.1000 \text{ g } C_6H_6}$$

Then the quantity of heat evolved by 1 mol C_6H_6 is given by

$$\frac{78.11 \cancel{\text{g } C_6H_6}}{1 \text{ mol } C_6H_6} \times \frac{4.180 \times 10^3 \cancel{\text{J}}}{0.1000 \cancel{\text{g } C_6H_6}} \times \frac{1 \text{ kJ}}{10^3 \cancel{\text{J}}} = \frac{3265 \text{ kJ}}{\text{mol } C_6H_6}$$

The molar heat of combustion of $C_6H_6(\ell)$ is 3265 kJ/mol. ■

The combustion of C_6H_6 is a typical exothermic reaction. Combustion of compounds containing C, H, and O always produces CO_2 and H_2O.

Calorimetric measurements of reactions involving gases are made in containers of fixed volume so that the chemical reactions occur at *constant volume.* Because a volume cannot change under this condition, expansion or contraction is impossible during the reaction. However, reactions are frequently carried out in open vessels at constant barometric pressure. When 1 mole of $C_6H_6(\ell)$ is burned at constant pressure, the heat evolved is 3269 kJ, not 3265 kJ (as measured at constant volume). The difference is small but real and it may be calculated (Chapter 18).

If expansion occurs, work is done in pushing back the atmosphere and less heat is emitted. If contraction occurs, work is done by the atmosphere on the products and more heat is emitted.

The data for heats of reaction are generally recorded for the quantity of heat evolved or absorbed at *constant pressure*; this quantity is given the symbol ΔH. That is, $\Delta H = q_p$, measured at constant pressure (volume may vary). The symbol ΔH is called the change in the **enthalpy**† of a chemical reaction. By convention, differences represented by Δ are always the product quantity minus the reactant quantity. So ΔH, the **enthalpy change** accompanying a chemical reaction, is equal to the enthalpy of the products, H_{products}, minus the enthalpy of the reactants, $H_{\text{reactants}}$:

Δ = final quantity − initial quantity here becomes Δ = product quantity − reactant quantity

$$\Delta H = H_{\text{products}} - H_{\text{reactants}}$$

For an *exothermic reaction* (heat is evolved), *the enthalpy of the products, H_{products}, must be less than the enthalpy of the reactants, $H_{\text{reactants}}$*. This makes ΔH negative. When 1 mole of liquid benzene reacts with $7\frac{1}{2}$ moles of gaseous oxygen, forming 3 moles of liquid water and 6 moles of gaseous carbon dioxide at constant pressure, 3269 kJ are evolved. Therefore the enthalpy of the 3 moles of $H_2O(\ell)$ and 6 moles of $CO_2(g)$ is 3269 kJ less than the enthalpy of 1 mole of $C_6H_6(\ell)$ and $7\frac{1}{2}$ moles of $O_2(g)$, as illustrated:

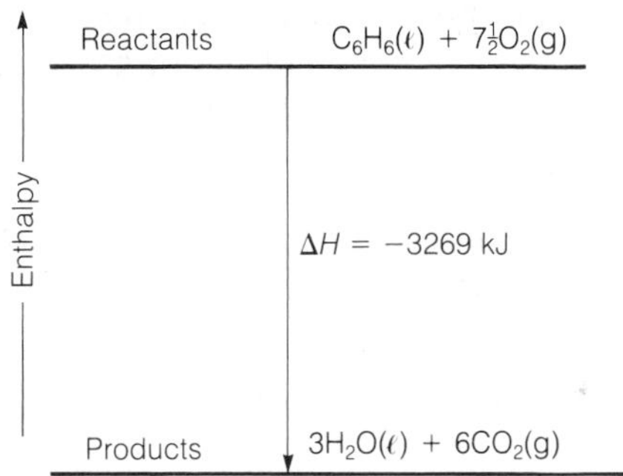

Since $\Delta H = H_{\text{products}} - H_{\text{reactants}}$, ΔH is negative for the combustion of $C_6H_6(\ell)$: $\Delta H = -3269$ kJ. No attempt is made to define the separate values of H_{products} and $H_{\text{reactants}}$ because these values cannot be determined experimentally. We concern ourselves only with the difference, ΔH, because this is the quantity measured.

A chemical equation that describes both a chemical reaction and its associated heat change (ΔH) is called a **thermochemical equation**. For example, the thermochemical equation for the combustion of liquid C_6H_6 is

The energy given out is sometimes shown as a reaction product:

$$C_6H_6(\ell) + 7\tfrac{1}{2}O_2(g) \longrightarrow 3H_2O(\ell) + 6CO_2(g) + 3269 \text{ kJ}$$

$$C_6H_6(\ell) + 7\tfrac{1}{2}O_2(g) \longrightarrow 3H_2O(\ell) + 6CO_2(g)$$

$$\Delta H = -3269 \text{ kJ} \qquad \text{(exothermic reaction)}$$

This means that when 1 mole of liquid benzene reacts with $7\frac{1}{2}$ moles of gaseous oxygen, forming 3 moles of liquid water and 6 moles of gaseous carbon dioxide at constant pressure, 3269 kJ are evolved.

† Derived from the Greek word *enthalpo*, meaning "warming up." The more descriptive but misleading term, *heat content*, is sometimes used synonymously with enthalpy. This obsolete term accounts for the symbol *H*.

The quantity of heat liberated or absorbed in a given reaction depends on the mass or amount (moles) of reactants or products. Thus, if 2 moles of $C_6H_6(\ell)$ are burned instead of 1 mole, ΔH is doubled and 2×3269 kJ is given out; $\frac{1}{2}$ mole of $C_6H_6(\ell)$ gives out $\frac{3269}{2}$ kJ; if 1 mole of $H_2O(\ell)$ is produced in the reaction, $\frac{3269}{3}$ kJ is given out.

For practical purposes, the change in enthalpy (ΔH) may be used as the change in energy (ΔE), the heat emitted at constant temperature and constant volume. For the combustion of benzene, the difference is 0.12%.

We must also indicate the state of each reactant and product because ΔH is different for different states; for example,

$$H_2(g) + \tfrac{1}{2}O_2(g) \longrightarrow H_2O(\ell) \qquad \Delta H = -285.8 \text{ kJ}$$

$$H_2(g) + \tfrac{1}{2}O_2(g) \longrightarrow H_2O(g) \qquad \Delta H = -241.8 \text{ kJ}$$

Less heat is given out in forming 1 mole of $H_2O(g)$ because it takes 44.0 kJ to vaporize 1 mole of liquid H_2O to 1 mole of water vapor at constant pressure:

$$H_2O(\ell) \longrightarrow H_2O(g) \qquad \Delta H = +44.0 \text{ kJ}$$

Thus, if we specify $H_2O(g)$ as the product instead of $H_2O(\ell)$ for the combustion of $C_6H_6(\ell)$, the heat given out is less by 3×44.0 kJ for 3 moles of $H_2O(g)$:

$$C_6H_6(\ell) + 7\tfrac{1}{2}O_2(g) \longrightarrow 6CO_2(g) + 3H_2O(g) \qquad \Delta H = -3269 \text{ kJ} + 3 \times 44.0 \text{ kJ} = -3137 \text{ kJ}$$

Increasing the temperature from 25°C to 100°C increases ΔH for $H_2(g) + \frac{1}{2}O_2(g) \longrightarrow H_2O(\ell)$ by less than 1%.

When any substance is heated, it absorbs heat and its enthalpy must increase. But heat capacities of substances change with temperature and pressure. Thus, the magnitude of ΔH of a reaction depends on the temperature and pressure. All data are given at 25°C and 1 atmosphere (atm) pressure unless otherwise specified.

EXAMPLE 4.6

CAUTION: Do not perform this experiment without proper guidance and equipment; mercury vapor is very toxic.

In his discovery of oxygen (1774), Joseph Priestley used solar energy to decompose red mercury(II) oxide. The decomposition of 0.2000 g HgO(c) in an open vessel absorbs 0.08387 kJ.

$$HgO(c) \longrightarrow Hg(\ell) + \tfrac{1}{2}O_2(g)$$

Calculate the heat absorbed in decomposing 1.000 mole of HgO(c).

ANSWER

Quantity given: 1.000 mol HgO (216.59 g/mol)

Quantity sought: kJ absorbed

From the given data, the conversion factor is

$$\frac{0.08387 \text{ kJ}}{0.2000 \text{ g HgO}}$$

and the quantity of heat absorbed by 1 mol HgO is given by

$$1.000 \cancel{\text{mol HgO}} \times \frac{216.59 \cancel{\text{g HgO}}}{1 \cancel{\text{mol HgO}}} \times \frac{0.08387 \text{ kJ}}{0.2000 \cancel{\text{g HgO}}} = 90.83 \text{ kJ}$$

■

The decomposition of HgO(c) is a typical endothermic reaction. The substances absorb heat as the reaction proceeds. Since heat is absorbed, the enthalpy of the products, H_{products}, of an endothermic reaction must be greater than that of the reactants, $H_{\text{reactants}}$. This makes ΔH positive. For example, the enthalpy of 1 mole of $Hg(\ell)$ and $\frac{1}{2}$ mole of $O_2(g)$ is 90.83 kJ more than the enthalpy of HgO(c). Then

$$\Delta H = H_{\text{products}} - H_{\text{reactants}} = +90.83 \text{ kJ}$$

Endothermic reactions:

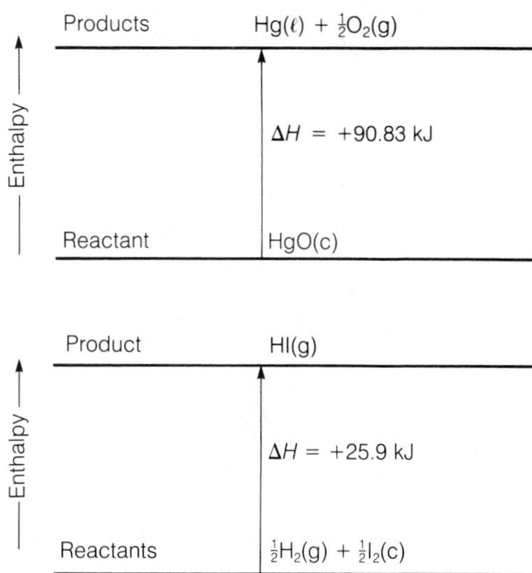

is positive. The thermochemical equation for the decomposition reaction at constant pressure is

$$HgO(c) \longrightarrow Hg(\ell) + \tfrac{1}{2}O_2(g) \qquad \Delta H = +90.83 \text{ kJ} \qquad \text{(endothermic reaction)}$$

The thermochemical equation for the formation of hydrogen iodide is

$$\tfrac{1}{2}H_2(g) + \tfrac{1}{2}I_2(c) \longrightarrow HI(g) \qquad \Delta H = +25.9 \text{ kJ}$$

This means that when $\frac{1}{2}$ mole of gaseous hydrogen reacts with $\frac{1}{2}$ mole of solid iodine, forming 1 mole of gaseous hydrogen iodide at constant pressure, 25.9 kJ is absorbed from the surroundings. The enthalpy of the products is larger than the enthalpy of reactants. Therefore, ΔH for the reaction is positive. See diagrams above.

In summary:

EXOTHERMIC REACTION	ENDOTHERMIC REACTION
Heat is given out.	Heat is taken in.
$H_{reactants}$ is greater than $H_{products}$.	$H_{products}$ is greater than $H_{reactants}$.
ΔH is negative.	ΔH is positive.

EXAMPLE 4.7 The thermochemical equation for the combustion of methanol at constant pressure is

A "new fuel" that has a high octane rating and burns cleanly.

$$CH_3OH(g) + 1\tfrac{1}{2}O_2(g) \longrightarrow CO_2(g) + 2H_2O(\ell) \qquad \Delta H = -76.2 \text{ kJ}$$

(a) Write the thermochemical equation for this reaction in which the products are written $\frac{1}{2}CO_2(g)$ and $H_2O(\ell)$. (b) How many grams and kilograms of methanol must be burned to obtain 3.57×10^5 kJ?

ANSWER

(a) ΔH is directly proportional to the number of moles of reactants or products.

Dividing the chemical equation by 2 divides ΔH by 2, so the thermochemical equation is written:

$$\tfrac{1}{2}CH_3OH(g) + \tfrac{3}{4}O_2(g) \longrightarrow \tfrac{1}{2}CO_2(g) + H_2O(\ell)$$

$$\Delta H = \frac{-76.2 \text{ kJ}}{2} = -38.1 \text{ kJ}$$

(b) Quantity given: 3.57×10^5 kJ

Quantity sought: grams of CH_3OH

From the thermochemical equation, combustion of 1 mol CH_3OH (32.0 g/mol) emits 76.2 kJ. The conversion factor is

$$\frac{32.0 \text{ g } CH_3OH}{76.2 \text{ kJ}}$$

and

$$3.57 \times 10^5 \cancel{\text{kJ}} \times \frac{32.0 \text{ g } CH_3OH}{76.2 \cancel{\text{kJ}}} = 1.50 \times 10^5 \text{ g } CH_3OH$$

$$= 150 \text{ kg } CH_3OH$$ ■

PROBLEM 4.4 The combustion of 0.614 g $C_{10}H_{22}$(c) at constant pressure raises the temperature of a calorimeter 1.80°C. The heat capacity of the calorimeter is 4.060 kJ/°C. (a) Find ΔH for the combustion of 1 mol $C_{10}H_{22}$ and write the thermochemical equation. (b) How many kilograms are burned when the heat output is 1.20×10^6 kJ? □

EXAMPLE 4.8 Find the number of kilojoules given out when 0.360 g Al(c) forms Al_2O_3(c). The molar heat of combustion of Al(c) is 838 kJ.

ANSWER From the given molar heat of combustion, the thermochemical equation is

$$Al(c) + \tfrac{3}{4}O_2(g) \longrightarrow \tfrac{1}{2}Al_2O_3(c) \qquad \Delta H = -838 \text{ kJ}$$

Quantity given: 0.360 g Al

Quantity sought: number of kJ

From the thermochemical equation, we see that the combustion of 1 mol Al emits 838 kJ. The molar mass of Al is 27.0 g/mol. The conversion factor is

$$\frac{838 \text{ kJ}}{27.0 \text{ g Al}}$$

Then, the number of kJ emitted by the combustion of 0.360 g Al is

$$0.360 \cancel{\text{g Al}} \times \frac{838 \text{ kJ}}{27.0 \cancel{\text{g Al}}} = 11.2 \text{ kJ}$$ ■

PROBLEM 4.5 The molar heat of combustion of $C_2H_5OH(\ell)$ is 1922 kJ. Find the heat evolved when 25.5 g C_2H_5OH burns. □

4.5 HESS'S LAW

Chemical reactions can be carried out in one step or in several steps. Either way, the net chemical change is the same. For example, we can burn carbon directly to carbon dioxide:

$$\text{C(graphite)} + O_2(g) \longrightarrow CO_2(g) \qquad \Delta H = -393.509 \text{ kJ} \qquad (1)$$

Or we can burn carbon to carbon monoxide and then burn the carbon monoxide to carbon dioxide:

$$\begin{array}{lll} \text{C(graphite)} + \frac{1}{2}O_2(g) \longrightarrow \cancel{CO(g)} & \Delta H = -110.524 \text{ kJ} & (2) \\ \cancel{CO(g)} + \frac{1}{2}O_2(g) \longrightarrow CO_2(g) & \Delta H = -282.985 \text{ kJ} & (3) \\ \hline \text{C(graphite)} + O_2(g) \longrightarrow CO_2(g) & \Delta H = -393.509 \text{ kJ} & (1) \end{array}$$

Discovered by Germain Hess in 1840.

The sum of these two steps, reactions 2 and 3, gives reaction 1. Also notice that the sum of the ΔH values of reactions 2 and 3 is exactly the same as the ΔH value of reaction 1. This fact is expressed by **Hess's law: We cannot alter the value of ΔH of a given chemical reaction by changing the method of carrying out the reaction. The ΔH of the reaction is equal to the sum of the enthalpy changes for the individual steps.** The heat of a reaction (ΔH) is independent of the number of steps involved in going from the given reactants to the given products (Figure 4.2). Consequently, Hess's law allows us to predict heats of reactions. It is therefore frequently used to calculate ΔH for reactions whose heats are unknown or difficult to measure (Box 4.3).

In applying Hess's law, keep in mind two properties of thermochemical equations:

1. *They may be reversed.* When we reverse an equation, the value of ΔH remains the same but its sign is also reversed. For example, when we reverse

$$H_2(g) + \frac{1}{2}O_2(g) \longrightarrow H_2O(\ell) \qquad \Delta H = -286 \text{ kJ} \qquad \text{(exothermic)} \qquad (4)$$

we also reverse the sign of ΔH:

$$H_2O(\ell) \longrightarrow H_2(g) + \frac{1}{2}O_2(g) \qquad \Delta H = +286 \text{ kJ} \qquad \text{(endothermic)} \qquad (5)$$

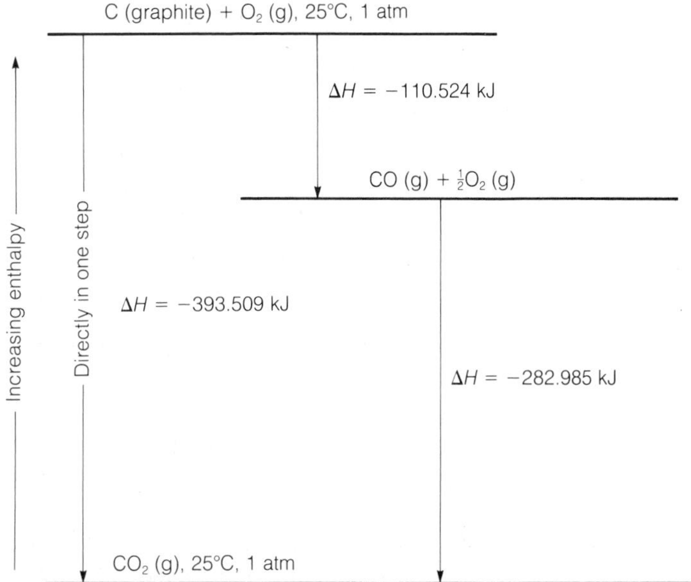

FIGURE 4.2
The ΔH for the heat of combustion of graphite is the same whether the reaction occurs in one step or in two steps.

BOX 4.3

HESS'S LAW AND PERPETUAL MOTION

What if Hess's law were not true? Then, in the direct combustion of graphite to CO_2, the ΔH would not be the same as the sum of the steps $C + \frac{1}{2}O_2 \longrightarrow CO$ and $CO + \frac{1}{2}O_2 \longrightarrow CO_2$. If it were not the same, it would have to be larger or smaller.

Let us imagine that the direct combustion is more exothermic, such as −394.509 kJ. Then

$$\begin{array}{rlr} C + O_2 \longrightarrow & CO_2 & \Delta H = -394.509 \text{ kJ} \\ CO_2 \longrightarrow & CO + \frac{1}{2}O_2 & \Delta H = +282.985 \text{ kJ} \\ CO \longrightarrow & C + \frac{1}{2}O_2 & \Delta H = +110.524 \text{ kJ} \\ \hline \text{Net: } C + O_2 \xrightarrow[\text{reaction}]{\text{no}} & C + O_2 & \Delta H = -1.000 \text{ kJ} \end{array}$$

This means that we could design a process that burns C to CO_2 in one step and then decomposes the CO_2 back to $C + O_2$ in two steps while giving out 1 kJ of heat per mole of C. So, at the end of the cycle, we would still have our initial $C + O_2$ *plus* 1 kJ of heat available for our use. Such a process would be nothing less than a perpetual motion machine, a violation of the law of energy conservation.

If Hess's law were wrong in the other direction (the ΔH for the direct combustion of C to CO_2 would be less exothermic, such as −392.509 kJ), another perpetual motion machine could be designed: burn C in two steps to CO_2 ($\Delta H = -393.509$ kJ) and then decompose CO_2 in one step back to $C + O_2$ ($\Delta H = +392.509$ kJ), leaving 1 kJ of heat for our use, also a violation of the conservation law.

Therefore, Hess's law cannot be wrong in *either* direction. And if it cannot be wrong in either direction, it must be right.

This means that the heat evolved in the formation of 1 mole of liquid water (reaction 4) is equal to the heat that would be required to decompose 1 mole of liquid water (reaction 5).

2. *They may be treated as algebraic equations.* However, when doing so, remember that
   - **(a)** If you multiply or divide an equation by a number, you must also multiply or divide the ΔH (section 4.4).
   - **(b)** If you add an equation as if it were an ordinary algebraic equation, also add the ΔH values. For example, addition of the two thermochemical equations

$$C_2H_4(g) + H_2(g) + \cancel{3\tfrac{1}{2}O_2(g)} \longrightarrow \cancel{2CO_2(g)} + \cancel{3H_2O(\ell)} \qquad \Delta H = -1696.6 \text{ kJ}$$

$$\cancel{2CO_2(g)} + \cancel{3H_2O(\ell)} \longrightarrow C_2H_6(g) + \cancel{3\tfrac{1}{2}O_2(g)} \qquad \Delta H = +1559.8 \text{ kJ}$$

predicts that the heat of reaction of H_2 and ethylene, C_2H_4, to form ethane, C_2H_6, is

$$C_2H_4(g) + H_2(g) \longrightarrow C_2H_6(g) \qquad \Delta H = -1696.6 + 1559.8 = -136.8 \text{ kJ}$$

The experimentally determined heat of reaction is

$$C_2H_4(g) + H_2(g) \longrightarrow C_2H_6(g) \qquad \Delta H = -136.4 \pm 0.4 \text{ kJ}$$

   - **(c)** When adding equations, a formula appearing on both sides of the arrows can be canceled. The substance will not appear in the final equation if it is produced and consumed in equal quantities.

However, we must be careful to *cancel a formula only when the state* (g, ℓ, c) *is the same on both sides of the arrow.* For instance, the enthalpy of $H_2O(g)$ is greater than that of $H_2O(\ell)$ (section 4.4) by 44.0 kJ/mol:

$$H_2O(\ell) \longrightarrow H_2O(g) \qquad \Delta H = +44.0 \text{ kJ}$$

Hess's law thus shows that ΔH depends only on the final products and initial reactants, and is independent of the path or the number of steps by which the reaction is carried out.

EXAMPLE 4.9 From the reactions

$$C_3H_8(g) + 5O_2(g) \longrightarrow 3CO_2(g) + 4H_2O(\ell) \qquad \Delta H = -2218.8 \text{ kJ} \qquad (6)$$

$$C(\text{graphite}) + O_2(g) \longrightarrow CO_2(g) \qquad \Delta H = -393.5 \text{ kJ} \qquad (7)$$

$$H_2O(\ell) \longrightarrow H_2(g) + \tfrac{1}{2}O_2(g) \qquad \Delta H = +285.8 \text{ kJ} \qquad (8)$$

predict ΔH for the reaction

$$3C(\text{graphite}) + 4H_2(g) \longrightarrow C_3H_8(g) \qquad \Delta H = ? \qquad (9)$$

which cannot be measured directly.

ANSWER We want ΔH for equation 9. So concentrate on its reactants and product. The object is to manipulate the given equations 6, 7, and 8 so that after addition, all substances except those in equation 9 cancel. But we want 1 mole of product C_3H_8, so reverse equation 6 and change the sign of ΔH:

$$3CO_2(g) + 4H_2O(\ell) \longrightarrow C_3H_8(g) + 5O_2(g) \qquad \Delta H = +2218.8 \text{ kJ} \qquad (6')$$

Since 3 moles of the reactant C(graphite) are needed, multiply equation 7 by 3; this also multiplies ΔH by 3:

$$3C(\text{graphite}) + 3O_2(g) \longrightarrow 3CO_2(g)$$
$$\Delta H = 3 \times -393.5 \text{ kJ} = -1180.5 \text{ kJ} \qquad (7')$$

We also need 4 moles of reactant H_2; therefore, we reverse and multiply equation 8 by 4; ΔH is multiplied by 4 and its sign is changed:

$$4H_2(g) + 2O_2(g) \longrightarrow 4H_2O(\ell) \qquad \Delta H = 4 \times -285.8 \text{ kJ} = -1143.2 \text{ kJ} \qquad (8')$$

Adding equations 6′, 7′, and 8′ cancels the $3CO_2$, $4H_2O$, and $5O_2$ in the equations, leaving equation 9:

$$3C(\text{graphite}) + 4H_2(g) \longrightarrow C_4H_8(g)$$
$$\Delta H = +2218.8 \text{ kJ} - 1180.5 \text{ kJ} - 1143.2 \text{ kJ} = -104.9 \text{ kJ} \qquad (9)$$

■

PROBLEM 4.6 Use the thermochemical equations

$$4PCl_5(c) \longrightarrow P_4(c) + 10Cl_2(g) \qquad \Delta H = +1774.0 \text{ kJ}$$

$$4PCl_3(\ell) \longrightarrow P_4(c) + 6Cl_2(g) \qquad \Delta H = +1278.8 \text{ kJ}$$

to find ΔH for the reaction

$$PCl_5(c) \longrightarrow PCl_3(\ell) + Cl_2(g)$$

□

4.6 STANDARD HEATS OF FORMATION

It would be impractical to list all the chemical reactions for which ΔH values have been determined. Hess's law, however, allows us to summarize these known enthalpy changes in terms of a limited number of ΔH's called **standard heats of formation**. It is possible to calculate from these data the enthalpy changes of many reactions, including reactions for which the ΔH's have not been measured.

Before defining the standard heat of formation, however, we define the term "standard state." The **standard state** of any pure substance is its stable form—solid, liquid, or gas—at a *pressure of one atmosphere* and a specified temperature, usually 25°C. Then, **the standard heat of formation, ΔH_f^0, of a substance is the change in enthalpy when 1 mole of substance in its standard state is formed from its elements in their standard states.** The superscript zero merely tells us that the reaction was carried out at the standard condition of one atmosphere pressure. For example, at 25°C, the standard heats of formation of ammonia, NH_3, and sodium chlorate, $NaClO_3$, are the enthalpy changes for these reactions:

The U.S. National Bureau of Standards publishes a comprehensive collection of standard heats of formation at 25°C.

$$\tfrac{1}{2}N_2(g) + 1\tfrac{1}{2}H_2(g) \longrightarrow NH_3(g) \qquad \Delta H_f^0 = -46.11 \text{ kJ}$$

$$Na(c) + \tfrac{1}{2}Cl_2(g) + 1\tfrac{1}{2}O_2(g) \longrightarrow NaClO_3(c) \qquad \Delta H_f^0 = -361.1 \text{ kJ}$$

From the definition of standard heat of formation, it follows that the standard heat of formation of any element in its stable form at 25°C and 1 atm pressure is zero. This is another way of saying that there is no enthalpy change in the reaction of forming the element from itself, because there is actually no reaction:

$$O_2(25°C, 1 \text{ atm}) \longrightarrow O_2(25°C, 1 \text{ atm}) \qquad \Delta H_f^0 = 0$$

$$He(25°C, 1 \text{ atm}) \longrightarrow He(25°C, 1 \text{ atm}) \qquad \Delta H_f^0 = 0$$

The stable form of oxygen at 25°C and 1 atm is O_2, not O or O_3 (ozone). The addition of energy is required to convert O_2 to O or O_3:

$$1\tfrac{1}{2}O_2(g) \text{ (stable)} \longrightarrow O_3(g) \text{ (unstable)} \qquad \Delta H_f^0 = +142.7 \text{ kJ}$$

$$O_2(g) \text{ (stable)} \longrightarrow 2O(g) \text{ (unstable)} \qquad \Delta H_f^0 = +498 \text{ kJ}$$

Similarly, the stable form of carbon at 25°C and 1 atm pressure is C(graphite), not C(diamond).

Diamond, unstable with respect to graphite, is another form of carbon.

Let us first study how standard heats of formation are obtained from measured heats of combustion. Then, we shall study how these heats of formation are used to calculate ΔH^0 for reactions.

The ΔH_f^0's of most organic compounds (carbon-containing compounds) cannot be measured directly from the one-step reaction of the elements. Acetylene, C_2H_2, is typical; it is not prepared from the reaction

$$2C(\text{graphite}) + H_2(g) \longrightarrow C_2H_2(g) \qquad \Delta H_f^0 = ? \tag{10}$$

Therefore, its ΔH_f^0 cannot be measured in such a direct manner. Instead, ΔH_f^0's are generally obtained from heats of combustion, which can be measured with high accuracy.

The heat of combustion of acetylene, C_2H_2, is

$$C_2H_2(g) + 2\tfrac{1}{2}O_2(g) \longrightarrow 2CO_2(g) + H_2O(\ell) \qquad \Delta H^0 = -1299.6 \text{ kJ} \tag{11}$$

The ΔH_f^0's of the combustion products have also been measured:

$$C(\text{graphite}) + O_2(g) \longrightarrow CO_2(g) \qquad \Delta H_f^0 = -393.5 \text{ kJ} \tag{12}$$

$$H_2(g) + \tfrac{1}{2}O_2(g) \longrightarrow H_2O(\ell) \qquad \Delta H_f^0 = -285.9 \text{ kJ} \tag{13}$$

Hess's law now permits us to calculate the ΔH_f^0 of $C_2H_2(g)$. Reverse equation 11, making C_2H_2 a product, and change the sign of its ΔH^0. Then multiply equation 12 and its ΔH_f^0 by 2:

$$\cancel{2CO_2(g)} + \cancel{H_2O(\ell)} \longrightarrow C_2H_2(g) + \cancel{2\tfrac{1}{2}O_2(g)} \qquad \Delta H^0 = +1299.6 \text{ kJ} \qquad (11')$$

$$2C(\text{graphite}) + \cancel{2O_2(g)} \longrightarrow \cancel{2CO_2(g)} \qquad \Delta H^0 = 2 \times -393.5 \text{ kJ} \qquad (12')$$

$$H_2(g) + \cancel{\tfrac{1}{2}O_2(g)} \longrightarrow \cancel{H_2O(\ell)} \qquad \Delta H_f^0 = -285.9 \text{ kJ} \qquad (13)$$

Addition of equations 11′, 12′, and 13 cancels the $2CO_2$, H_2O, and the $2\frac{1}{2}O_2$ on both sides of the equations, yielding the standard heat of formation of C_2H_2:

$$2C(\text{graphite}) + H_2(g) \longrightarrow C_2H_2(g)$$
$$\Delta H_f^0 = +1299.6 \text{ kJ} - 787.0 \text{ kJ} - 285.9 \text{ kJ} = +226.7 \text{ kJ} \qquad (10)$$

Acetylene may explode, but it is safely stored as a solution in acetone, $(CH_3)_2O$.

An examination of the table of standard heats of formation, Appendix C, shows that most of the heats of formation are negative. This tells us that most compounds require a heat input for decomposition to the elements. Acetylene, C_2H_2, on the other hand, has a positive heat of formation and decomposes with a large output of heat.

Let us now calculate the heat of combustion of carbon disulfide, CS_2, an excellent solvent for waxes, greases, and cellulose,

$$CS_2(\ell) + 3O_2(g) \longrightarrow 2SO_2(g) + CO_2(g) \qquad \Delta H^0 = ? \qquad (14)$$

from the following ΔH_f^0's:

$$C(\text{graphite}) + 2S(\text{rhombic}) \longrightarrow CS_2(\ell) \qquad \Delta H_f^0 = +89.7 \text{ kJ} \qquad (15)$$

$$S(\text{rhombic}) + O_2(g) \longrightarrow SO_2(g) \qquad \Delta H_f^0 = -296.8 \text{ kJ} \qquad (16)$$

$$C(\text{graphite}) + O_2(g) \longrightarrow CO_2(g) \qquad \Delta H_f^0 = -393.5 \text{ kJ} \qquad (17)$$

To obtain the heat of combustion of $CS_2(\ell)$, reverse equation 15 to make CS_2 a reactant and change the sign of its ΔH_f^0; multiply equation 16 and its ΔH_f^0 by 2 to get 2 moles of SO_2; then add the three equations:

$$CS_2(\ell) \longrightarrow \cancel{C(\text{graphite})} + \cancel{2S(\text{rhombic})} \qquad \Delta H = -\Delta H_f^0 = -89.7 \text{ kJ} \qquad (15')$$

$$\cancel{2S(\text{rhombic})} + 2O_2(g) \longrightarrow 2SO_2(g) \qquad \Delta H = 2 \times \Delta H_f^0 = 2 \times -296.8 \text{ kJ} \qquad (16')$$

$$\cancel{C(\text{graphite})} + O_2(g) \longrightarrow CO_2(g) \qquad \Delta H = \Delta H_f^0 = -393.5 \text{ kJ} \qquad (17)$$

$$CS_2(\ell) + 3O_2(g) \longrightarrow 2SO_2(g) + CO_2(g) \qquad \Delta H^0 = -1076.8 \text{ kJ} \qquad (14)$$

Let us examine what we have done:

- We added the heats of formation of the products, CO_2 and SO_2, after adjusting for the number of moles written in the given chemical equation.

- But notice that the heat of formation of the reactant, CS_2, was *not* added. By reversing and changing the sign of the ΔH_f^0 of CS_2, we subtracted the heats of formation of the reactants.

This procedure may be summarized as follows: The **enthalpy change of a reaction, ΔH^0**, equals the sum of the heats of formation of the products minus the sum of the heats of formation of the reactants, each multiplied by its coefficients in the chemical equation, abbreviated as

May also be written as

$$\Delta H^0_{\text{reaction}} = \Sigma n \Delta H^0_{\text{f (products)}} - \Sigma n \Delta H^0_{\text{f (reactants)}}$$

in which n refers to the coefficients in the chemical equation as written.

$$\boldsymbol{\Delta H^0_{\text{reaction}} = \text{total } \Delta H^0_{\text{f products}} - \text{total } \Delta H^0_{\text{f reactants}}}$$

$$\begin{aligned}\Delta H^0_{\text{reaction}} &= [2 \times \Delta H^0_f\ SO_2(g) + \Delta H^0_f\ CO_2(g)] \\ &\quad - [\Delta H^0_f\ CS_2(\ell) + 3 \times \Delta H^0_f\ O_2(g)] \\ &= [(2 \times -296.8\ \text{kJ}) - 393.5\ \text{kJ}] - [+89.7\ \text{kJ} + 0\ \text{kJ}] \\ &= -1076.8\ \text{kJ, the same as above}\end{aligned}$$

EXAMPLE 4.10 The heat of combustion of methane is

$$CH_4(g) + 2O_2(g) \longrightarrow CO_2(g) + 2H_2O(\ell) \qquad \Delta H^0 = -890.41\ \text{kJ}$$

Use the heats of formation of CO_2 and H_2O given in Appendix C to check the value for the heat of formation of methane, CH_4, given in that table.

ANSWER

$$\Delta H^0_{\text{reaction}} = \text{total } \Delta H^0_{\text{f products}} - \text{total } \Delta H^0_{\text{f reactants}}$$

$$\begin{aligned}\Delta H^0_{\text{reaction}} &= [\Delta H^0_f\ CO_2(g) + (2 \times \Delta H^0_f\ H_2O(\ell))] \\ &\quad - [\Delta H^0_f\ CH_4(g) + (2 \times \Delta H^0_f\ O_2(g))]\end{aligned}$$

From Appendix C, $\Delta H^0_f\ CO_2(g) = -393.475$ kJ, $\Delta H^0_f\ O_2(g) = 0$ kJ, and $\Delta H^0_f\ H_2O(\ell) = -285.83$ kJ:

$$-890.41\ \text{kJ} = -393.475\ \text{kJ} + (2 \times -285.83\ \text{kJ}) - \Delta H^0_f\ CH_4(g)$$

$$\Delta H^0_f\ CH_4(g) = -74.73\ \text{kJ}$$

■

EXAMPLE 4.11 Predict ΔH^0 for these reactions from standard heats of formation given in Appendix C: (a) $3C_2H_2(g) \longrightarrow C_6H_6(g)$; (b) $C_2H_4(g) + 6Cl_2(g) \longrightarrow 2CCl_4(\ell) + 4HCl(g)$.

ANSWER

$$\Delta H^0_{\text{reaction}} = \text{total } \Delta H^0_{\text{f products}} - \text{total } \Delta H^0_{\text{f reactants}}$$

(a) $\Delta H^0_{\text{reaction}} = \Delta H^0_f\ C_6H_6(g) - 3 \times \Delta H^0_f\ C_2H_2(g)$

From Appendix C, $\Delta H^0_f\ C_2H_2(g) = +226.7$ kJ and $\Delta H^0_f\ C_6H_6(g) = -83.68$ kJ.

$$\Delta H^0_{\text{reaction}} = -83.68\ \text{kJ} - (3 \times 226.7\ \text{kJ}) = -763.8\ \text{kJ}$$

(b) $\Delta H^0_{\text{reaction}} = [2 \times \Delta H^0_f\ CCl_4(\ell) + 4 \times \Delta H^0_f\ HCl(g)] - [\Delta H^0_f\ C_2H_4(g) + 6 \times \Delta H^0_f\ Cl_2(g)]$

From Appendix C, $\Delta H^0_f\ C_2H_4(g) = +52.26$ kJ, $\Delta H^0_f\ CCl_4(\ell) = -135.4$ kJ, and $\Delta H^0_f\ HCl(g) = -92.307$ kJ.

$$\begin{aligned}\Delta H^0_{\text{reaction}} &= [(2 \times -135.4\ \text{kJ}) + (4 \times -92.307\ \text{kJ})] - [52.26\ \text{kJ} + 0] \\ &= -692.3\ \text{kJ}\end{aligned}$$

■

PROBLEM 4.7 Calculate ΔH^0 for the reaction

$$C_2H_4(g) + 4Cl_2(g) \longrightarrow C_2Cl_4(\ell) + 4HCl(g)$$

from standard heats of formation given in Appendix C. □

4.7 BOND ENERGY

Common experience teaches us that an input of energy is required to break any object. Strong materials like chromium and stainless steel require considerable energy to pull them apart. In 1927, the hydrogen welding torch (Figure 4.3) was invented. It operates in this way: molecular hydrogen, H_2, is passed through an electric arc and is dissociated (broken) into atomic hydrogen, H, with an energy input of 436 kJ/mol H_2. As the atomic hydrogen emerges it recombines, forming molecular hydrogen with a heat output that raises the flame temperature above 3000°C, sufficient to melt even tungsten (m.p. 3370°C). An *input of energy* is required to break or pull apart any molecule because *chemical bonds* (section 4.2) between its atoms must be *broken.*

Strictly speaking, the term *molar bond enthalpy* should be used. *Bond energy*, however, is in common use. The gaseous state is specified because the energy of particles in liquids and solids is greatly influenced by intermolecular forces (Chapter 10).

Bond energy is the energy associated with breaking a chemical bond between atoms in a gaseous molecule. The dissociation (bond breaking) of 1 mole of HI(g) into gaseous atoms absorbs 298.3 kJ, while the formation of 1 mole of HI(g) from gaseous atoms evolves 298.3 kJ.

The higher a given weight (the longer the distance from the Earth), the larger is the potential energy. Similarly, the longer the distance of an H atom from an I atom, the larger is the potential energy. Thus, when a molecule forms there is a reduction in potential energy. In general, a molecule always has a lower energy than the atoms from which it is formed. This is another way of saying that energy is required to separate a molecule into its atoms.

The quantity of energy (ΔH) required to break one mole of a bond in one mole of molecules in the gaseous state, forming gaseous products, is called the **bond dissociation energy**, or simply the **bond energy**, abbreviated BE. **Bond breaking in gaseous molecules is always endothermic; thus, ΔH for a bond energy is always positive. On the other hand, bond formation from gaseous atoms is always exothermic; and that ΔH is negative.** For example, the bond energy for H—I(g) is 298.3 kJ. This means that in the reaction

$$HI(g) \longrightarrow H(g) + I(g) \qquad \text{(endothermic)} \qquad \Delta H = BE = +298.3 \text{ kJ}$$

heat is absorbed, while in the reaction

$$H(g) + I(g) \longrightarrow HI(g) \qquad \text{(exothermic)} \qquad \Delta H = -BE = -298.3 \text{ kJ}$$

the same quantity of heat is evolved:

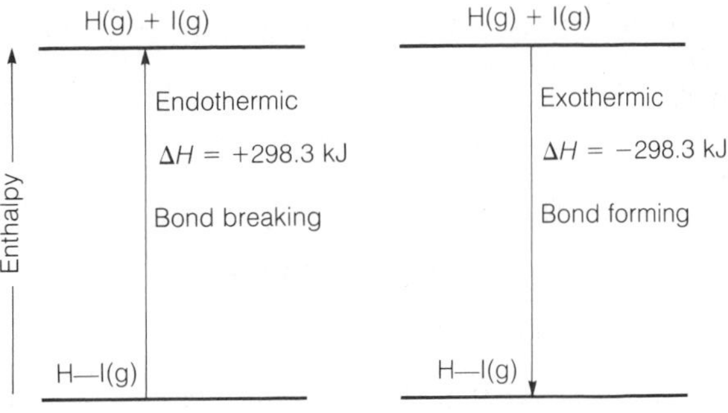

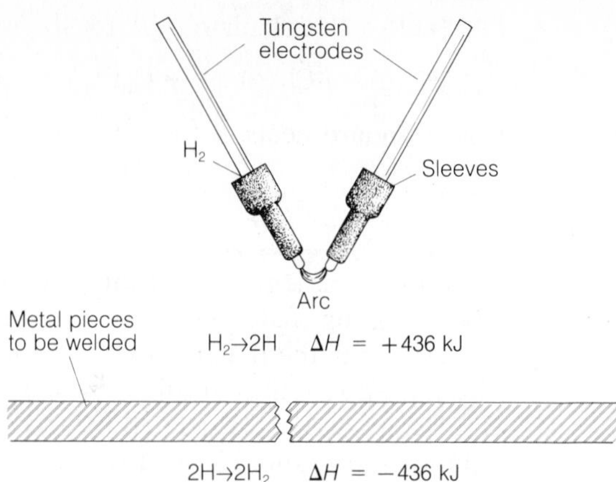

FIGURE 4.3
The hydrogen welding torch

For molecules with two or more identical atoms attached to another atom, such as H_2O, NH_3, and CH_4, it is possible to *assign* a bond energy to each of the bonds. To do so, first find the total quantity of energy needed to break all the bonds in the molecule. Then, divide that total by the total number of bonds in the molecule. This gives an *average bond energy*, as illustrated in Example 4.12. An average bond energy is also simply called a bond energy (Table 4.3).

EXAMPLE 4.12 Assign an average bond energy to the C—H bond from the given heat of reaction:

$$\begin{array}{c} \text{H} \\ | \\ \text{H—C—H(g)} \\ | \\ \text{H} \end{array} \longrightarrow \text{C(g)} + 4\text{H(g)} \qquad \Delta H = +1663.29 \text{ kJ} \qquad (18)$$

ANSWER In all calculations involving bond energies, recall the three basic rules:

1. Energy is always required to break chemical bonds in gaseous molecules.
2. Energy is always given out when gaseous atoms combine.
3. Note the kind and number of bonds broken or formed.

Here, notice that the only reaction occurring is the breakage of four moles of C—H bonds. Thus, 1663.29 kJ is the total quantity of energy that is absorbed in breaking four moles of C—H bonds. Division by four thus assigns an average bond energy to the C—H bonds:

It is called an average bond energy because the energy needed to remove any one H atom from CH_4 differs from that needed to remove an H atom from CH_3, CH_2, or CH.

$$\text{C—H(g)} \longrightarrow \text{C(g)} + \text{H(g)} \qquad \Delta H = \text{BE} = \frac{+1663.29 \text{ kJ}}{1 \cancel{\text{mol CH}_4}} \times \frac{1 \cancel{\text{mol CH}_4}}{4 \text{ mol (C—H)}}$$

$$= +416 \frac{\text{kJ}}{\text{mol (C—H)}}$$

This means that for the reaction

$$\text{C—H(g)} \longrightarrow \text{C(g)} + \text{H(g)} \qquad \text{(bond breaking)} \qquad \Delta H = \text{BE} = +416 \text{ kJ}$$

and for the reaction

$$\text{C(g)} + \text{H(g)} \longrightarrow \text{C—H(g)} \qquad \text{(bond formation)} \qquad \Delta H = -\text{BE} = -416 \text{ kJ}$$

■

TABLE 4.3
BOND ENERGY (ΔH) VALUES FOR ONE MOLE OF BONDS, kJ/mole[†]

C—H	416	C—I	213	N_2	945.6
C—C	339	H—F	569	O_2	498
C—O	336	H—Cl	431	H_2	435.9
C—N	285	H—Br	366	F_2	158
C—F	490	H—I	298	Cl_2	244
C—Cl	326	O—H	464	Br_2	193
C—Br	285	N—H	391	I_2	151

† Both reactants and products are gaseous.

Reactions like reaction 18, in which a molecule is dissociated into atoms, can seldom be carried out directly. Bond energies usually have to be calculated by Hess's law. In particular, bond energies can be obtained from known heats of formation. Typical is the calculation of the bond energy for the N—H bond from ΔH_f^0 NH_3.

Bond energies involve gaseous atoms as either reactants or products:

Note that the ΔH subscript numbers are the same as the equation numbers.

$$N(g) + 3H(g) \longrightarrow NH_3(g) \qquad \Delta H_{19} = ? \tag{19}$$

However, the thermochemical equation for ΔH_f^0 NH_3 (Appendix C) involves only molecules:

$$\tfrac{1}{2}N_2(g) + 1\tfrac{1}{2}H_2(g) \longrightarrow NH_3(g) \qquad \Delta H^0_{f,20} = -46.11\text{ kJ} \tag{20}$$

Therefore, first convert $N_2(g)$ and $H_2(g)$ to N(g) and H(g); from Table 4.3:

$$N_2(g) \longrightarrow 2N(g) \quad \text{(bond breaking, endothermic)} \quad \Delta H_{21} = +945.6\text{ kJ} \tag{21}$$

$$H_2(g) \longrightarrow 2H(g) \quad \text{(bond breaking, endothermic)} \quad \Delta H_{22} = +435.9\text{ kJ} \tag{22}$$

We need ΔH_{19} for equation 19. Therefore, reverse equations 21 and 22 and multiply equation 21 by $\frac{1}{2}$ and equation 22 by $1\frac{1}{2}$:

$$N(g) \longrightarrow \tfrac{1}{2}N_2(g) \quad \text{(bond formation, exothermic)} \quad \Delta H_{21'} = \tfrac{1}{2} \times -945.6\text{ kJ} = -472.8\text{ kJ} \tag{21'}$$

$$3H(g) \longrightarrow 1\tfrac{1}{2}H_2(g) \quad \text{(bond formation, exothermic)} \quad \Delta H_{22'} = 1\tfrac{1}{2} \times -435.9\text{ kJ} = -653.9\text{ kJ} \tag{22'}$$

Then adding equations 20, 21′, and 22′ gives the desired equation 19:

$$\cancel{\tfrac{1}{2}N_2(g)} + \cancel{1\tfrac{1}{2}H_2(g)} \longrightarrow NH_3(g) \qquad \Delta H^0_{f,20} = -46.11\text{ kJ} \tag{20}$$

$$N(g) \longrightarrow \cancel{\tfrac{1}{2}N_2(g)} \qquad \Delta H_{21'} = -472.8\text{ kJ} \tag{21'}$$

$$3H(g) \longrightarrow \cancel{1\tfrac{1}{2}H_2(g)} \qquad \Delta H_{22'} = -653.9\text{ kJ} \tag{22'}$$

$$N(g) + 3H(g) \longrightarrow NH_3(g) \qquad \Delta H_{19} = -1172.8\text{ kJ} \tag{19}$$

But by definition, bond energy is the quantity of energy needed to dissociate (break) a mole of a bond. Therefore, reverse equation 19:

$$\begin{array}{l} \text{H} \\ | \\ \text{N—H} \longrightarrow \text{N(g)} + 3\text{H(g)} \qquad \Delta H_{19'} = +1172.8 \text{ kJ/mol NH}_3 \\ | \\ \text{H} \end{array} \tag{19'}$$

However, 1172.8 kJ is the total quantity of energy that is absorbed in breaking three moles of N—H bonds. Division by three assigns an average bond energy to the N—H bond:

$$\text{N—H(g)} \longrightarrow \text{N(g)} + \text{H(g)} \qquad \Delta H = \text{BE} = \frac{+1172.8 \text{ kJ}}{1 \cancel{\text{mol NH}_3}} \times \frac{1 \cancel{\text{mol NH}_3}}{3 \text{ mol (N—H)}}$$

$$= +391 \frac{\text{kJ}}{\text{mol (N—H)}}$$

This result agrees with the value in Table 4.3.

The greater the bond energy is, the greater also are the energy released in forming the bond, the energy needed to break the bond, and the **bond stability**. Thus H—H (436 kJ/mol) is a more stable bond than F—F (158 kJ/mol): more energy is needed to break the H—H bond. Bond stability should not be confused with *unreactivity*, the tendency not to react. Under given conditions, an unreactive substance has little tendency to undergo chemical changes *with itself or other substances*. High bond stability does not necessarily mean that the molecule is unreactive. For instance, oxygen, O_2, has a relatively high bond energy of 498 kJ/mol, but it is very reactive. It reacts with other elements to form oxides.

Some bond energies are given in Table 4.3. Those energies may be used for any molecule in which the particular bonds appear, because of the *assumption* that the bond for a given pair of atoms is independent of the molecule in which it resides. However, average bond energies vary somewhat in different compounds—such as the C—H bond in CH_4, C_2H_4, and C_2H_2—but not enough to prevent them from being useful. Consequently, bond energies are used to estimate the ΔH for reactions in the gas phase whose heats of reaction are unknown. Failure to keep track of signs and coefficients must lead to the wrong answer. *Remember:* bond breaking is always *endothermic* and its ΔH is *positive*, as given in Table 4.3. Bond formation, on the other hand, is always *exothermic*, and its ΔH is *negative*. If the total energy input (bond breaking) exceeds the total output (bond formation), the reaction is endothermic (ΔH is positive); if the total output exceeds the total input, the reaction is exothermic (ΔH is negative). Some examples follow.

EXAMPLE 4.13 Use the data in Table 4.3 to calculate the heat of the reaction:

$$\begin{array}{ccc} \text{H} & & \text{Cl} \\ | & & | \\ \text{H—C—H(g)} + 4\text{Cl}_2\text{(g)} & \longrightarrow & \text{Cl—C—Cl(g)} + 4\text{HCl(g)} \\ | & & | \\ \text{H} & & \text{Cl} \end{array}$$

ANSWER List the kind and number of moles of bonds broken (endothermic) and formed (exothermic) with their bond energies and ΔH values as shown.

Bond Broken	Moles of Bond Broken	BE, kJ/mol	REACTION	ΔH
C—H	4	416	$CH_4 \longrightarrow C + 4H$	$\Delta H = +1664$ kJ
Cl—Cl	4	244	$4Cl_2 \longrightarrow 8Cl$	$\Delta H = +976$ kJ
Bond Formed	**Moles of Bond Formed**			
C—Cl	4	326	$C + 4Cl \longrightarrow CCl_4$	$\Delta H = -1304$ kJ
H—Cl	4	431	$4H + 4Cl \longrightarrow 4HCl$	$\Delta H = -1724$ kJ

This can be illustrated:

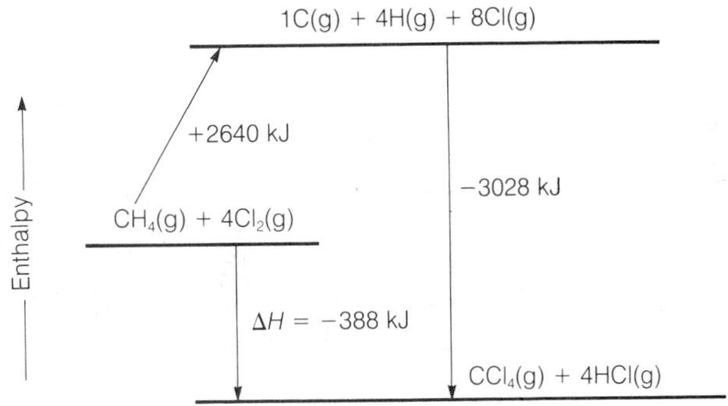

Addition of the ΔH's yields:

$$\Delta H_{\text{reaction}} = +1664 \text{ kJ} + 976 \text{ kJ} - 1304 \text{ kJ} - 1724 \text{ kJ} = -388 \text{ kJ}$$

The experimental value is -395 kJ/mol CH_4. ■

PROBLEM 4.8 Calculate ΔH for the reaction $CH_4(g) + 4F_2(g) \longrightarrow CF_4(g) + 4HF(g)$. The experimental value is -1937 kJ/(mol CH_4). □

Although agreement between the calculated ΔH and the experimental ΔH is usually good, significant differences occur for many reactions. These discrepancies compel us to reexamine our assumption that the properties of a bond are independent of its molecular environment. The reexamination, in Chapter 8, will lead us to a better understanding of the nature of bonds holding atoms together in molecules.

The term "bond energy" may be misleading because it may imply a bond full of vim and vigor, perhaps ready to explode. Such a notion is incorrect. Bond energy is rather the energy that the bonded atoms *do not have,* and which they will *not* get until some outside source of energy comes along and breaks the bond. *Thus, a high bond energy means that the bond requires a high energy input for separation of the atoms.* An explosive compound like TNT or nitroglycerine can do its work because a reaction converts (weaker) bonds of lower bond energy (high potential energy) to (stronger) bonds of higher bond energy (lower potential energy) (Figure 4.4).

TNT, trinitrotoluene, $C_7H_5(NO_2)_3$, is a yellow solid. Nitroglycerine, $C_3H_5(NO_3)_3$, is a clear, colorless, syrupy liquid.

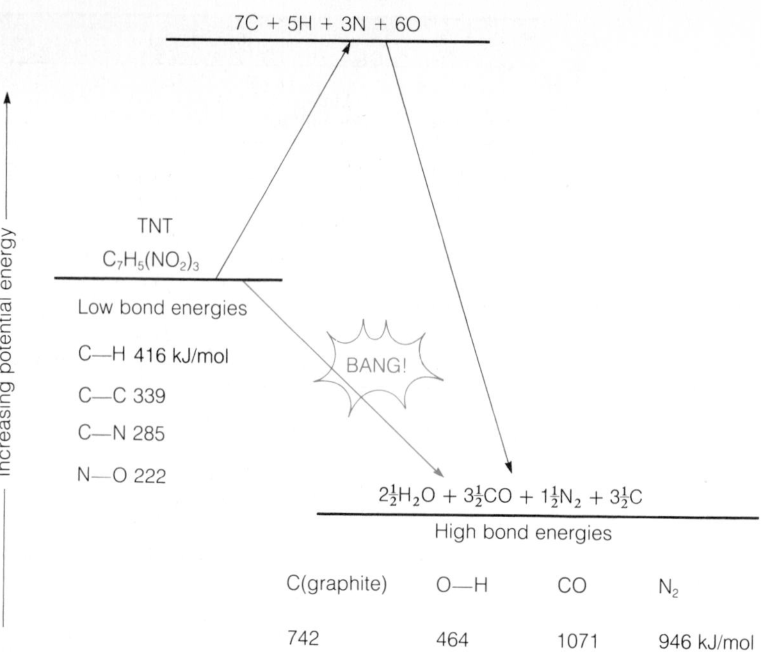

FIGURE 4.4 Potential energy decreases as stronger bonds are formed from weaker bonds.

PROBLEM 4.9 (a) Can the exhaust products of combustion of gasoline (a mixture of C—H compounds such as C_8H_{18}) be used to drive a car? (b) Do the products or the reactants have the higher bond energies? □

SUMMARY

Energy is the capacity to do work or to transfer heat. Familiar forms of energy include *mechanical energy, light energy, heat energy,* and *electrical energy.* Energy may be classified as either kinetic or potential. A body in motion possesses **kinetic energy**. A stationary body possesses **potential energy** because of its position or composition. Energy can be converted from one form to another or transferred from one body to another, but it is neither destroyed nor created (law of conservation of energy).

Chemical reactions are accompanied by energy changes, mainly in the form of heat emitted or absorbed. Most chemical reactions are **exothermic**—energy is *released* to the surroundings; others are **endothermic**—energy is *absorbed* from the surroundings.

The SI unit of energy is the **joule**, J. The older and still much used unit is the **calorie**, cal (1 cal = 4.184 J exactly). **Heat capacity** is the quantity of heat needed to warm or cool a given object by one degree Celsius, J/°C. The heat capacity per gram is known as the **specific heat**, J/(g °C). Thus the quantity of heat lost or gained, q, is given by

$$q = \text{specific heat} \times \text{mass} \times (t_{\text{final}} - t_{\text{initial}})$$

or

$$q = \text{heat capacity} \times (t_{\text{final}} - t_{\text{initial}})$$

Heat is the energy that is transferred between two neighboring objects because of a temperature difference; the transfer is always from the warmer object to the cooler object.

Thermochemistry is the study of heat absorbed or emitted by chemical reactions. These heats of reaction are measured with a calorimeter. They are calculated from the temperature change of the calorimeter of known heat capacity and the known mass of the reactant. Heats of reaction are generally reported for the reaction of 1 mole of a specified reactant. The **molar heat of combustion** is the heat released when one mole of a substance burns at constant pressure (1 atm) at 25°C.

When the heat of reaction is measured at constant temperature and constant pressure (the calorimeter is open to the atmosphere), it is called the **enthalpy change** of the reaction, ΔH. The Δ always represents the product quantity minus the reactant quantity, so that $\Delta H = H_{\text{products}} - H_{\text{reactants}}$. These separate quantities are not defined because they cannot be measured experimentally; open calorimeters measure ΔH values. In an exothermic reaction heat is released to the calorimeter, and the enthalpy change (ΔH) is negative; in an endothermic reaction heat is absorbed from the calorimeter, and the enthalpy change (ΔH) is positive.

The heat of reaction, ΔE, measured at constant temperature and constant volume (the calorimeter is closed) may differ from ΔH, but the difference is not large.

When the ΔH of a reaction is included in the chemical equation, the equation is called a **thermochemical equation**:

$$CH_3OH(g) + 1\tfrac{1}{2}O_2(g) \longrightarrow CO_2(g) + 2H_2O(\ell)$$

$$\Delta H = -76.2 \text{ kJ} \qquad \text{(exothermic)}$$

ΔH depends on the number of moles of reactants or products and the physical state (g, ℓ, c, aq) of each reactant and product.

Hess's law states that the enthalpy change of a reaction does not depend on how the reaction is carried out. The law is used to calculate the ΔH of reactions by proper algebraic manipulation of thermochemical equations. Enthalpy changes for reactions can also be calculated from **standard heats of formation, ΔH^0_f**:

$$\Delta H^0_{\text{reaction}} = \begin{matrix}\text{sum of heats of}\\ \text{formation of products}\end{matrix} - \begin{matrix}\text{sum of heats of}\\ \text{formation of reactants}\end{matrix}$$

The superscript zero indicates that the reaction was carried out at the standard pressure of one atmosphere.

Hess's law is also used to calculate **bond energies** from known thermochemical equations by breaking bonds in gaseous reactants to gaseous atoms and forming bonds in gaseous products from gaseous atoms. In turn, these bond energies are used to estimate enthalpy changes of reactions. Bond breaking is *always* endothermic; bond formation is *always* exothermic. The larger the energy needed to break a bond, the greater is the **bond stability**.

4.8 ADDITIONAL PROBLEMS

ENERGY

4.10 What kind of energy (kinetic or potential) does each of the following possess? (a) a moving locomotive, (b) a stationary locomotive, (c) an oil pool 1500 m underground, (d) diesel oil, (e) a moving electron.

4.11 How much heat is needed to raise the temperature of (a) 100 g H_2O, specific heat 4.18 J/(g °C), from 25.0°C to 50.0°C, (b) 100 g gold, 0.0135 J/(g °C), from 25°C to 200°C? Which heated sample, the gold or the water, is at a higher temperature? Which heated sample will give up more heat in cooling to room temperature?

4.12 A bullet at 20°C with a mass (m) of 20.0 g traveling at a velocity (u) of 600 m/sec in a vacuum becomes embedded in a 1.25-kg block of wood at 20°C. What is the final temperature of the metal and the wood? Specific heat of the wood = 1.8 J/(g °C), and of the metal = 0.136 J/(g °C). Kinetic energy = $\frac{1}{2}mu^2$ and 1 J = 1 kg m^2/sec^2.

THERMOCHEMISTRY

4.13 Potassium iodide, KI, is added to water in a test tube. As KI dissolves, the test tube and its contents become cold. (a) Is the reaction $KI(c) + H_2O \longrightarrow KI(aq)$ endothermic or exothermic? (b) On holding the test tube, your hand "feels cold." Is heat flowing to your hand or from your hand?

4.14 You have two substances, A and G. Their molar masses and molar heats of combustion are the same, but the density of A is greater than that of G. A sample of A has the same volume as a sample of G. Which sample gives out the greater quantity of heat upon complete combustion? Explain.

SPECIFIC HEAT

4.15 The specific heat of Al is 0.895 J/(g °C). How much heat in joules is required to heat a 15.8-g sample from 27.0°C to 41.0°C?

4.16 Find the number of joules given out when 46.2 g Hg, 0.139 J/(g °C), is cooled from 42.2°C to 11.2°C.

4.17 A 20.0-g sample of Ca, 0.628 J/(g °C), initially at 26.1°C, absorbs 905 J. What is the final temperature of the calcium?

4.18 When 2.60 g powdered Ni at 47.044°C is rapidly stirred in 15.0 g H_2O at 22.200°C, the water temperature rises to 22.644°C. (a) Find the specific heat of Ni. (b) Find the number of joules required to raise the temperature of 1 mol Ni 1.00°C.

4.19 10.0 g Au, 0.0135 J/(g °C), at 90.0°C is stirred into 15.0 g of water at 5.000°C. Find the final water temperature. Use 4.20 for the specific heat of water.

4.20 0.20 g Pt, 0.136 J/(g °C), at 100°C is plunged into 12.0 kg H_2O at 20.0°C. Find the heat in joules transferred from the Pt to the water. Use 4.20 for the specific heat of water. (*Hint*: What reasonable assumption may you make about the temperature rise of the water? Verify your assumption after you have your answer.)

ΔH

4.21 For each of these reactions, (a) does the enthalpy increase or decrease; (b) is $H_{reactant} > H_{product}$ or is $H_{product} > H_{reactant}$; (c) is ΔH positive or negative?
(1) $Al_2O_3(c) \longrightarrow 2Al(c) + 1\frac{1}{2}O_2(g)$ (endothermic)
(2) $Sn(c) + Cl_2(g) \longrightarrow SnCl_2(c)$ (exothermic)

4.22 (a) The combustion of 0.0222 g octane vapor, C_8H_{18}, at constant pressure raises the temperature of a calorimeter 0.400°C. The heat capacity of the calorimeter is 2.48 kJ/°C. Find the molar heat of combustion of C_8H_{18}:

$$C_8H_{18}(g) + 12\tfrac{1}{2}O_2(g) \longrightarrow 8CO_2(g) + 9H_2O(\ell) \qquad \Delta H = ?$$

(b) How many grams of $C_8H_{18}(g)$ must be burned to obtain 105 kJ?

4.23 Methanol, CH_3OH, is an efficient fuel with a high octane rating that can be produced from coal and hydrogen:

$$CH_3OH(g) + 1\tfrac{1}{2}O_2(g) \longrightarrow CO_2(g) + 2H_2O(\ell) \qquad \Delta H = -76.2 \text{ kJ}$$

Find (a) the heat evolved when 30.0 g $CH_3OH(g)$ burns; (b) the mass of O_2 consumed when 950 kJ is given out.

4.24 Methylhydrazine is burned with dinitrogen tetroxide in the altitude-control engines of the space shuttle:

$$CH_6N_2(\ell) + 1\tfrac{1}{4}N_2O_4(\ell) \longrightarrow CO_2(g) + 3H_2O(\ell) + 2\tfrac{1}{4}N_2(g) \qquad \Delta H = ?$$

The two substances ignite instantly on contact, producing a flame temperature of 3000 K. The energy liberated per 0.100 g CH_6N_2 at constant atmospheric pressure after the products are cooled back to 25°C is 750 J. (a) Find ΔH for the reaction as written. (b) How many kilojoules are liberated when (1) 32.0 g N_2 is produced, (2) 13.5 g NO_2 is converted to N_2O_4 and consumed?

4.25 How many kilojoules are absorbed or evolved when (a) 18.0 g Al reacts, forming Al_2O_3; (b) 36.0 g Al is formed from Al_2O_3? See Appendix C for necessary data.

4.26 Find the mass of mercury that can be obtained from excess HgO by the absorption of 100 kJ while decomposing. See Appendix C for required data.

4.27 Coal is converted to coke, which in turn is converted to a mixture of H_2 and CO known as "water gas":

$$C(\text{coke}) + H_2O(g) \xrightarrow{600°C} CO(g) + H_2(g)$$

Assume the coke is pure carbon (graphite). (a) Calculate the heat obtainable from the combustion of 1 mole of coke and from the combustion of the products obtained from 1 mole of coke at constant pressure at 25°C. (b) Does the combustion of the products produce more or less energy than the combustion of the reactant coke? (c) Does your answer disprove the law of conservation of energy?

BODY ENERGY†

4.28 The energy output of major body fuels are carbohydrates, 4.18 kcal/g; protein, 4.32 kcal/g; and triglycerides (fats), 9.46 kcal/g. An individual runs at 10 mph for $1\frac{1}{2}$ h at an energy output of 19 kcal/min. The runner then decides to enjoy a nice snack consisting of a steak (190 g protein, 2.2 g carbohydrate, 28 g triglycerides), a malted

† Max Rubner demonstrated in 1894 by direct calorimeter measurements on animals that the law of conservation of energy applies to living organisms. The heat of reaction of food with oxygen is the same whether carried out in a body (flameless) or a bomb calorimeter (explosive).

milk shake (17 g protein, 67 g carbohydrate, 17 g triglycerides, 100 g H_2O), and a piece of apple pie (64 g protein, 23 g carbohydrate, 1.2 g triglycerides). Does this meal balance the energy output? If not, will the runner lose or gain weight? (To maintain body weight, fuel intake should balance energy output.)

4.29 ATP (adenosine triphosphate), the fuel used by the body to power muscular work, is produced from stored carbohydrate (glycogen) or fat (triglycerides). Average energy output per minute for various activities are given: sitting (1.7 kcal); walking, level, 3.5 mph (5.5 kcal); cycling, level, 13 mph (10 kcal); swimming (8.4 kcal); running 10 mph (19 kcal). Approximate energy values of some common foods are also given: large apple (100 kcal); 8-oz cola drink (105 kcal); malted milk (8 oz milk) shake (500 kcal); $\frac{3}{4}$ cup pasta with tomato sauce and cheese (195 kcal); hamburger on bun with burger sauce (350 kcal); 10-oz sirloin steak, including fat (1000 kcal). To maintain body weight, fuel intake should balance energy output. Prepare a table showing (a) each given food and (b) its fuel value, and (c) the minutes of each activity that would balance the kcal of each food.

4.30 The average North American consumes 47 kg of sucrose, $C_{12}H_{22}O_{11}$, 5648 kJ/mol, per year (365 days/yr). A recent search of early (1573) Mexican literature uncovered a reference by Francisco Hernandez to a "sweet herb." A compound ($C_{15}H_{24}O_2$) isolated from that herb, and named *hernandulcin* in Hernandez's honor, has been found to be 1000 times sweeter than sucrose. (a) What should be the heat of combustion in kJ/g to effect an energy input reduction of 500 kcal/day? (b) If the heat of combustion of hernandulcin is nearly the same as it is for sucrose, say, 6000 kJ/mol, what would be the reduction in energy input per day?

HESS'S LAW

4.31 Find ΔH for the reaction

$$2HCl(g) + F_2(g) \longrightarrow 2HF(\ell) + Cl_2(g)$$

from

$$4HCl(g) + O_2(g) \longrightarrow 2H_2O(\ell) + 2Cl_2(g) \qquad \Delta H = -148.4 \text{ kJ}$$

$$HF(\ell) \longrightarrow \tfrac{1}{2}H_2(g) + \tfrac{1}{2}F_2(g) \qquad \Delta H = +600.0 \text{ kJ}$$

$$H_2(g) + \tfrac{1}{2}O_2(g) \longrightarrow H_2O(\ell) \qquad \Delta H = -285.8 \text{ kJ}$$

4.32 Calculate ΔH for

$$Ca^{2+}(aq) + 2OH^-(aq) + CO_2(g) \longrightarrow CaCO_3(c) + H_2O(\ell)$$

from

$$CaCO_3(c) \longrightarrow CaO(c) + CO_2(g) \qquad \Delta H = +178.1 \text{ kJ}$$

$$CaO(c) + H_2O(\ell) \longrightarrow Ca(OH)_2(c) \qquad \Delta H = -64.8 \text{ kJ}$$

$$Ca(OH)_2(c) \longrightarrow Ca^{2+}(aq) + 2OH^-(aq) \qquad \Delta H = +11.7 \text{ kJ}$$

4.33 Write the thermochemical equations you would need to find ΔH for the reaction

$$C_3H_6(g) + 3\tfrac{1}{2}O_2(g) \longrightarrow 2CO(g) + CO_2(g) + 3H_2O(\ell)$$

ΔH_f^0

4.34 True or false: The heat of formation of $H_2(g)$ is zero; therefore, the heat of formation of H(g) is also zero. Defend your answer.

4.35 Which of these reactions corresponds to the heat of formation of KOH(c)?

(a) $K(c) + NaOH(c) \longrightarrow KOH(c) + Na(c)$
(b) $KNO_3(aq) + LiOH(aq) \longrightarrow KOH(c) + LiNO_3(aq)$
(c) $K(c) + \frac{1}{2}O_2(g) + \frac{1}{2}H_2(g) \longrightarrow KOH(c)$
(d) $K(g) + OH(g) \longrightarrow KOH(c)$
(e) $K(g) + O(g) + H(g) \longrightarrow KOH(c)$

4.36 Use Appendix C and ΔH_f^0 HF(ℓ) = −300.0 kJ to calculate ΔH for each of these reactions:

(a) $Al_2O_3(c) + 6HF(\ell) \longrightarrow 2AlF_3(c) + 3H_2O(g)$
(b) $2NaClO_3(c) \longrightarrow 2NaCl(c) + 3O_2(g)$
(c) $PbO(c) + H_2(g) \longrightarrow Pb(c) + H_2O(g)$
(d) $H_2S(g) + 1\frac{1}{2}O_2(g) \longrightarrow SO_2(g) + H_2O(g)$
(e) $H_2S(g) + 2O_2(g) \longrightarrow H_2SO_4(\ell)$
(f) $BaCl_2(c) + H_2SO_4(\ell) \longrightarrow BaSO_4(c) + 2HCl(g)$
(g) C(diamond) $\longrightarrow$ C(g, not graphite)
(h) $C_2H_6(g) + 5Cl_2(g) \longrightarrow C_2Cl_4(\ell) + 6HCl(g)$
(i) $CH_4(g) + 4Cl_2(g) \longrightarrow CCl_4(\ell) + 4HCl(g)$

4.37 Find the heat of formation of hydrogen peroxide, $H_2O_2(\ell)$, from $2H_2O_2(\ell) \longrightarrow 2H_2O(\ell) + O_2(g)$; $\Delta H = -196.0$ kJ and ΔH_f^0 $H_2O(\ell) = -285.8$ kJ.

4.38 The heat of combustion of morphine, $C_{17}H_{19}O_8N$, is given:

$$2C_{17}H_{19}O_8N(c) + 37\tfrac{1}{2}O_2(g) \longrightarrow 34CO_2(g) + 19H_2O(\ell) + 2NO_2(g) \qquad \Delta H = 2 \times -8980.1 \text{ kJ}$$

Find the heat of formation of morphine. Use Appendix C.

4.39 From the heat of formation of Fe_2O_3(c) (Appendix C), calculate the heat evolved or absorbed (a) when 0.589 g Fe_2O_3 is decomposed to iron and oxygen and (b) when 0.602 g Fe is converted to Fe_2O_3(c). (c) Find the mass in grams of Fe consumed when 806 kJ are evolved.

BOND ENERGY

4.40 How many bonds are in H_2O? Assign a bond energy to the O—H bond from the following data and check your answer with Table 4.3.

$$H_2(g) + \tfrac{1}{2}O_2(g) \longrightarrow H_2O(\ell) \qquad \Delta H = -285.83 \text{ kJ}$$
$$H_2O(\ell) \longrightarrow H_2O(g) \qquad \Delta H = +44.01 \text{ kJ}$$
$$O_2(g) \longrightarrow 2O(g) \qquad \Delta H = +498 \text{ kJ}$$
$$H_2(g) \longrightarrow 2H(g) \qquad \Delta H = +435.9 \text{ kJ}$$

4.41 Use Table 4.3 to calculate ΔH for each of the following reactions in the gas phase:
(a) $2HBr + O_2 \longrightarrow Br_2 + H_2O$

(b)
$$\begin{array}{c} \text{Cl} \\ | \\ \text{Cl—C—H} \\ | \\ \text{F} \end{array} + 2F_2 \longrightarrow CF_4 + HF + Cl_2$$

(c)
$$\begin{array}{c} \text{H}\ \ \text{H} \\ |\ \ \ | \\ \text{H—C—C—O—H} \\ |\ \ \ | \\ \text{H}\ \ \text{H} \end{array} + 1\tfrac{1}{2}Cl_2 \longrightarrow 2C + O + 3H + 3HCl$$

(d) $F + H_2 \longrightarrow HF + H$
(e) $2HI + F_2 \longrightarrow 2HF + I_2$

4.42 Given:

$$2C(\text{graphite}) + 2H_2(g) + Cl_2 \longrightarrow \begin{array}{c} \text{H}\ \ \text{H} \\ |\ \ \ | \\ \text{H—C—C—Cl(g)} \\ |\ \ \ | \\ \text{H}\ \ \text{Cl} \end{array} \qquad \Delta H = -106 \text{ kJ}$$

Use Table 4.3 and Appendix C to assign a bond energy to the C—Cl bond. Use Table 4.3 to check your answer.

4.43 Is this true? $\Delta H^0_{\text{reaction}}$ = total $\Delta H^0_{\text{f products}}$ − total $\Delta H^0_{\text{f reactants}}$. Therefore, $\Delta H_{\text{reaction}}$ = bond energies of products − bond energies of reactants. Explain.

CALORIMETRY

4.44 (a) The heat capacity of a calorimeter is generally determined by passing an electric current through a heater inside the calorimeter. When a current of 2.121 amperes (amp) is passed through a heater with a resistance of 625.0 ohm for 2.495 sec, the temperature of the calorimeter, including all parts, is raised 1.255°C. Calculate the heat capacity of the calorimeter in kJ/deg. Energy in joules = current^2 (amp^2) × resistance (ohm) × time (sec). (b) How many kilograms of water (4.184 J/(g °C)) would be raised 1.255°C by the same energy input?

4.45 All the heat from the combustion of 1.000×10^{-3} mol $C_{10}H_{22}(\ell)$, decane, is transferred to 2.158 kg water at an initial temperature of 25.010°C. Find the temperature rise in the water. Molar heat of combustion of $C_{10}H_{22}(\ell)$, $\Delta H = -6737$ kJ. Use 4.180 for the specific heat of water.

SOLAR ENERGY

4.46 (a) The Earth's surface receives solar energy at the rate of about 2.1 J/(cm^2 min). Solar One, a solar power plant at Barstow, California (Color plate 2), consists of 1818 heliostats (movable silver mirror reflectors), each with a reflective area of 39.3 m^2. Sunlight is reflected to a boiler on top of a tower connected to an electric generator at ground level. Calculate the number of kJ received by the boiler during 9.0 h of sunlight, assuming 90% efficiency. (b) An underground thermal storage tank connected to the Solar One boiler, for use during periods of darkness, is filled with 6.4×10^6 kg of rock, 0.84 J/(g °C), and 908 m^3 of oil 1.0 J/(g °C), density = 0.90 g/mL. What is the maximum rise in the temperature of the tank, assuming all of the energy received by the boiler (your answer to part a) is transferred to the tank?

4.47 (a) The daily energy requirement for the average household in the U.S. is estimated at 3.0×10^5 kJ/day. A solar collector is exposed to 2.1 J/(cm^2 min) for an average of 7.0 h. Find the required solar collector area in m^2. (b) The solar collector contains a solution of a salt that decomposes when exposed to sunlight (heat); for example,

$$[Co(H_2O)_6]^{2+}(aq) + 4Cl^-(aq) \longrightarrow [CoCl_4]^{2-}(aq) + 6H_2O \qquad \text{(endothermic)}$$

and reacts with water on cooling

$$[CoCl_4]^{2-}(aq) + 6H_2O \longrightarrow [Co(H_2O)_6]^{2+}(aq) + 4Cl^-(aq) \qquad \text{(exothermic)}$$

The heat absorbed or evolved by the reaction is 94 kcal/L. The specific heat of the solution is 3.15 J/(mL °C). Find the volume in liters per day of the heat storage solution required. Assume that the temperature of the cooled solution is 18°C and the temperature of the heated solution is 68°C; the overall efficiency is 70%.

ENERGY CRISIS

4.48 A catalytic coal gasification process is summarized in these reactions:

$$\begin{array}{lll} 2C + 2H_2O \longrightarrow 2CO + 2H_2 & & \Delta H = +268 \text{ kJ} \\ CO + H_2O \longrightarrow CO_2 + H_2 & & \Delta H = -34 \text{ kJ} \\ CO + 3H_2 \longrightarrow CH_4 + H_2O & & \Delta H = -226 \text{ kJ} \end{array}$$

(a) What is ΔH for the net reaction? (b) Taking the coal as C(graphite), what is the energy gain or loss in converting coal to CH_4, a clean fuel? Take a guess: is the actual gain or loss more than, less than, or the same as calculated?

4.49 The methane production rates in the bacterial conversion of plant biomass are too low for it to be a commercial source of methane. The methane fermentation of glucose can be represented as

$$C_6H_{12}O_6(aq) \longrightarrow 3CH_4(g) + 3CO_2(g)$$

Find the energy gain or loss per mole of reactant by comparing the heat given out by the combustion of the methane with the heat of combustion of glucose. The molar heat of combustion of glucose is $\Delta H = -2803$ kJ. Use data in Appendix C.

THERMAL POLLUTION

4.50 Hurricanes have been called the "Mack trucks of rainmakers." Hurricane Allen (1980) averaged 2.8 inches of rainfall over an area of 1.55×10^5 mi^2. A thunderstorm complex (1978) covered 5.75×10^5 mi^2 and dumped 5.8 inches of rainfall when "adjusted to an area as that affected by Hurricane Allen." Calculate the heat absorbed from or liberated to the atmosphere in kJ by each storm. Required data are in this chapter.

SELF-TEST

4.51 (10 points) During rainstorms, water vapor condenses to liquid water. Is this process exothermic or endothermic? Should this process cause an increase or a decrease in the surrounding air temperature? On the other hand, the wind accelerates the evaporation of raindrops. Is that process exothermic or endothermic? Should that cool or warm the surrounding air?

4.52 (10 points) How many joules are needed to heat 25.5 g Au, specific heat 0.0135 J/(g °C), from 25.0°C to 36.5°C?

4.53 (10 points) A mixture of 25.0 g Fe at 38.1°C and 25.0 g H_2O at 24.00°C are stirred rapidly. The water temperature increases to 25.40°C. (a) Find the specific heat of iron. (b) Find the heat in joules needed to raise the temperature of 12.0 g, 144 g, and 1 mole of Fe 1.00°C. Use 4.180 for the specific heat of water.

4.54 (15 points) (a) When 0.167 g C_2N_2, cyanogen, is burned at constant pressure in oxygen in a calorimeter whose heat capacity is 3.00 kJ/°C, the calorimeter temperature increases by 0.350°C. Calculate ΔH for the reaction

$$C_2N_2(g) + O_2(g) \longrightarrow 2CO(g) + N_2(g) \qquad \Delta H = ?$$

(b) How many kilojoules are evolved when (1) 2.78 g $C_2N_2(g)$ is consumed; (2) 12.5 g O_2 is used up? (c) How many grams of $C_2N_2(g)$ must be burned to get 24.5 kJ?

4.55 (10 points) How many grams of $H_2S(g)$ would have to be burned to raise the temperature of 2.855 kg of water by 0.325°C? The molar heat of combustion of $H_2S(g)$ is $\Delta H = -562.0$ kJ. Use 4.180 for the specific heat of water. (Assume no loss of heat.)

4.56 (10 points) Predict ΔH for the reaction used in the production of chloroform

$$CH_4(g) + 3Cl_2(g) \rightarrow 3HCl(g) + CHCl_3(\ell)$$

from

$$CH_4(g) + 2O_2(g) \rightarrow 2H_2O(\ell) + CO_2(g) \quad \Delta H = -890.4 \text{ kJ}$$

$$HCl(g) \rightarrow \tfrac{1}{2}H_2(g) + \tfrac{1}{2}Cl_2(g) \quad \Delta H = +92.3 \text{ kJ}$$

$$C(\text{graphite}) + O_2(g) \rightarrow CO_2(g) \quad \Delta H = -393.5 \text{ kJ}$$

$$H_2(g) + \tfrac{1}{2}O_2(g) \rightarrow H_2O(\ell) \quad \Delta H = -285.8 \text{ kJ}$$

and ΔH_f^0 $CHCl_3(\ell) = -134.5$ kJ.

4.57 (10 points) Use the heats of formation given in Appendix C to find ΔH for these reactions:
(a) $2Na(g) + 2HCl(g) \longrightarrow H_2(g) + 2NaCl(c)$
(b) $NH_4Cl(c) \longrightarrow NH_3(g) + HCl(g)$
(c) $2C_6H_6(g) + 15O_2(g) \longrightarrow 6H_2O(\ell) + 12CO_2(g)$

4.58 (15 points) Find the heat of formation of $NH_3(\ell)$ using the data in Table 4.3 and $NH_3(\ell) \longrightarrow NH_3(g)$, $\Delta H = +20$ kJ.

4.59 (10 points) (a) The accurately known molar heat of the combustion of naphthalene, $C_{10}H_8(c)$, $\Delta H = -5156.8$ kJ, is used to calibrate calorimeters. The complete combustion of 0.01520 g $C_{10}H_8$ at constant pressure raises the temperature of a calorimeter by 0.212°C. Find the heat capacity of the calorimeter. (b) The initial temperature of the calorimeter (part a) is 22.102°C. 0.1040 g $C_8H_{18}(\ell)$, octane, molar heat of combustion $\Delta H = -1303$ kJ, is completely burned in the calorimeter. Find the final temperature of the calorimeter.

5

THE ELECTRONIC STRUCTURE OF ATOMS

In previous chapters, the successes and limitations of the atomic theory of matter were examined. We saw how Dalton's structureless atom gave way to Rutherford's concept that the atom consists of a nucleus, composed of protons and neutrons, surrounded by electrons. Since atoms are neutral particles, the number of electrons outside the nucleus must equal the number of protons in the nucleus. In this and the following chapters, you will learn that the arrangement of electrons in atoms—the electronic structure of atoms—accounts for the positioning of the elements in a logical classification scheme (Chapter 6) and also provides an acceptable explanation of chemical combination (Chapter 7). You will see how the chemistry of an element is related to the electronic structure of its atoms. We will begin our exploration of electronic structure in the same way that the pioneering scientists did, by studying the nature of light.

5.1 THE NATURE OF LIGHT

Red light shines from the glowing embers of a wood fire. If you sprinkle some crystals of copper sulfate on the embers, a blue light will appear. Substitute barium salts and you will see a green light. Compounds of potassium will contribute a violet hue. These colored lights are *emitted* from the hot materials, the color of the light being determined by the chemical composition of the heated substance. Fireworks are another entertaining application of this (Color plate 3). Because the properties of substances depend on the structures of their atoms, it is reasonable to infer that the color of the emitted light is related to the atomic structure of the light source. This knowledge, however, has evolved through centuries of laboratory and theoretical studies of the properties of light. This section will introduce you to the characteristic properties and concepts of light.

For many years scientists attempted to determine whether light travels in waves or as a stream of particles. Since light can be transmitted in a vacuum, scientists knew that its "waves" are not *ordinary* ones (like sound waves, which require a medium such as air or water, or like water waves, which cannot exist without water). In addition, since light does not transport matter, any "particles" it has cannot be ordinary particles. Let us first consider the evidence for wave motion.

Waves can also be transmitted through other materials—guitar strings, for example.

We can visualize wave motion by thinking of a familiar example, water waves. The highest points of the wave (Figure 5.1) are called **crests** and the lowest, **troughs**. The distance between successive crests or troughs is the length of an

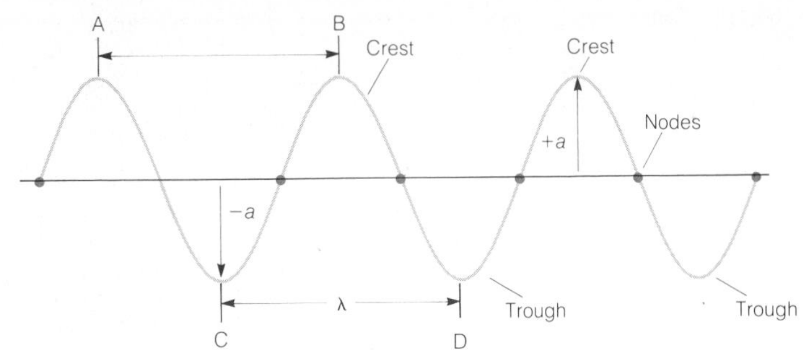

FIGURE 5.1
A wavelength, λ, is the distance between two consecutive crests (AB) or two consecutive troughs (CD). The amplitude, $+a$ or $-a$, of the wave is shown. The points of zero amplitude are the nodes.

individual wave. This length is called the **wavelength**, designated by the lowercase Greek letter *lambda*, λ. The **frequency**, designated by the lowercase Greek letter *nu*, ν, is the number of crests or complete waves that pass a given point per unit of time. The unit of frequency is therefore waves per second or simply "per second" or $\sec^{-1}$, which is called a *hertz*, Hz (section 1.2). The speed of a wave is the product of these two properties:

$$\begin{aligned} \text{speed} &= \text{wavelength} \times \text{frequency} \\ &= \lambda \times \nu \\ &= \text{m} \times \frac{1}{\text{sec}} = \text{m/sec or m sec}^{-1} \end{aligned}$$

The intensity is proportional to the square of the amplitude.

The height of the crest, $+a$, or the depth of the trough, $-a$, is called the *amplitude* of the wave (Figure 5.1). The loudness of a sound or the brightness of a light source is related to the amplitude of the wave. For a given frequency, the larger the amplitude, the greater is the intensity of the sound (loudness) or the intensity of the light (brightness). The points of zero amplitude (Figure 5.1) are called *nodes* (Box 5.1).

We know that sound waves diffuse around obstacles and spread out in passing through a narrow opening, just as water waves spread out when a pebble is dropped into water. This property of sound waves is called *diffraction.*

BOX 5.1
MUSIC AND STANDING WAVES

Anyone who loves music appreciates the waveforms known as *standing* (*stationary*) *waves.* Described as "standing" because they stay in place, these waves are the source of every musical instrument's sound. They also play an important role in the theoretical (quantum mechanical) description of electrons in atoms (section 5.4) and molecules (Chapter 8).

When a guitar string is properly plucked, standing waves are set up in it. See Figure 1. The string is divided into a whole number of loops, or vibrating segments, of equal length. Some points on the string, however, remain stationary; they do not vibrate or move in any way. These stationary points are the nodes.

Figure 2 shows four standing waves being set up in a rubber tube. A standing wave can be divided only into a whole number of loops, 1, 2, 3, 4, . . . , as seen in Figure 2. We can therefore say that standing waves are quantized. Figure 3 illustrates that standing waves can form unusual patterns.

BOX 5.1 (*continued*)

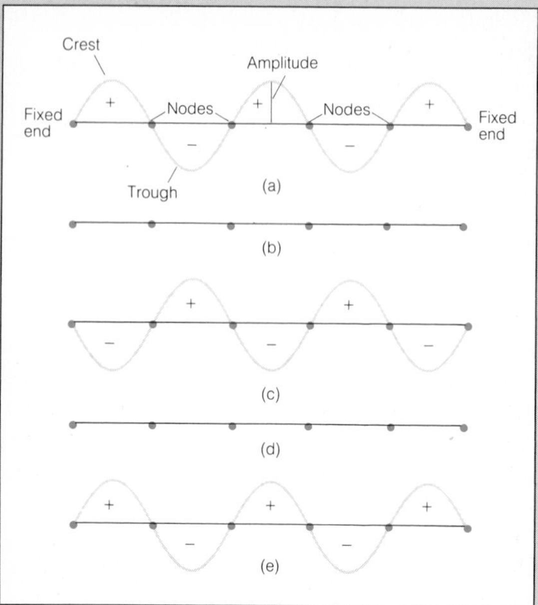

FIGURE 1 The motion of a standing wave as illustrated by a vibrating guitar string. When the string vibrates, the crests become troughs, the troughs become crests, and the nodes remain stationary (compare waves a, c, and e). The wave does not move along the string. A plus sign refers to the wave displacement above the reference line (crest), and a minus sign to the displacement below the reference line (trough). The distance from the reference line is referred to as the amplitude of the wave.

FIGURE 2 Four standing waves of different wavelengths set up in a tight rubber tube whose ends were fixed. The larger the number of loops, the shorter the wavelengths. (Courtesy of PSSC *Physics*, D. C. Heath, Lexington, Mass.)

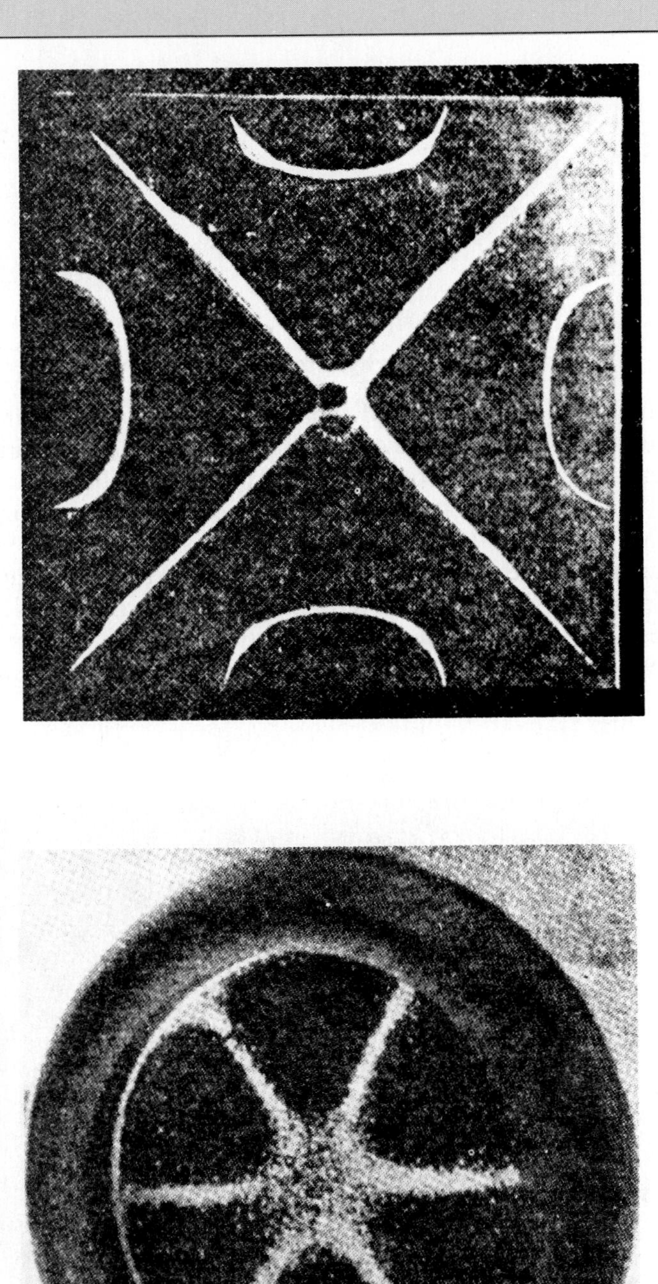

FIGURE 3 Standing wave patterns set up on a drum (*left*) and on the diaphragm (*below*) of a telephone receiver, illustrating that standing waves can form unusual patterns.

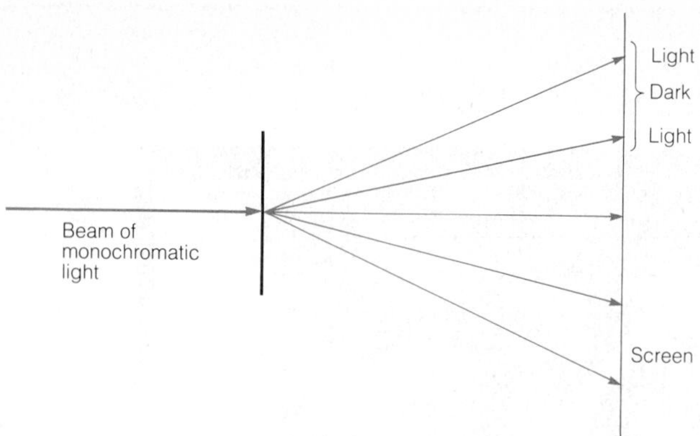

FIGURE 5.2
Diffraction of a light beam. The original beam is split upon passing through a pinhole. An actual pattern seen on the screen is illustrated in Figure 5.3.

A beam of monochromatic light consists of light of a single wavelength.

The speed of sound differs from the speed of light. The speeds of light through water, air, and a vacuum differ.

In 1801 Thomas Young carried out a crucial experiment that established the wave nature of light. Young showed that a beam of *monochromatic* light shining through a pinhole spreads out in a series of dark and bright rings. This proved that light, like sound, is diffracted. **Diffraction** is typical of all wave motions (Figures 5.2 and 5.3). We now know that visible light is only a small portion of a broad range of wavelengths called **electromagnetic radiation**, or radiant energy, which also includes X rays, γ rays, infrared and ultraviolet radiation, microwaves (radar), and radio waves (Color plate 4). All of these radiations travel at the same speed in a vacuum: 2.998×10^8 m/sec, designated c. Then for electromagnetic radiation,

$$c = \lambda \times \nu \tag{1}$$

FIGURE 5.3
The diffraction pattern obtained when light shines through a pinhole. The pinhole has about the same diameter as the wavelength of the light, about 500 nm.

EXAMPLE 5.1 The wavelength of violet light from a source is 420 nm. Find its frequency (10^9 nm = 1 m).

ANSWER The frequency and wavelength of light are related by $\lambda \times \nu = c$, so

$$\nu = \frac{c}{\lambda} = \frac{3.00 \times 10^8 \dfrac{\cancel{m}}{\text{sec}}}{420 \cancel{nm}} \times \frac{10^9 \cancel{nm}}{1 \cancel{m}} = 7.14 \times 10^{14}\ \text{sec}^{-1}$$ ■

PROBLEM 5.1 The frequency of a radio transmitter is $1.00 \times 10^6\ \text{sec}^{-1}$. Find the wavelength. □

Two properties of electromagnetic radiation are difficult to visualize as wave motion and are best interpreted in terms of particles. One of these properties is the distribution of frequencies of radiation from a hot, black surface; the other is observed when radiation strikes a metal surface. We begin our study of these properties by giving some historical perspective.

Throughout the late nineteenth century, scientists studied the distribution of the frequencies and the energies of the light waves emitted by hot bodies, such as a hot block of carbon or a hot tungsten filament. Scientists of that time recognized that both matter and electricity are quantized—matter consisting of particles called *atoms*, and electricity consisting of particles called *electrons*. Because matter is quantized, masses of a given element do not add up in a continuous range; they have definite, discrete values. Because electric charge is quantized, the quantity of *any* charge is a whole-number multiple of *e*.

PROBLEM 5.2 Is the quantization of electrical energy or matter evident in your laboratory work? Answer yes or no, and verify your answer by calculating the loss in weight of 4.35784 g H_2O after removing 10^{15} H_2O molecules with a "high-tech" instrument. □

Max Planck, 1858–1947

By a complicated mathematical derivation, Planck showed that his assumption leads to the correct frequency distribution.

Wave theory assumed that the energy of a radiation may have *any* value—from infinitely small to infinitely large. However, no arguments based on this assumption accounted for the frequency distribution found experimentally. A plot of the energy of the radiation emitted by the sun against the frequency of the radiation is shown in Figure 5.4. Notice that the energy of the emitted light drops sharply after passing through a maximum. The wave theory, with its assumption that radiation is not quantized, predicts incorrectly that the energy should continue to increase (the dashed line in Figure 5.4). In fact, the difference between the experimental results and the wave theory predictions was so great that scientists referred to the discrepancy as the "ultraviolet catastrophe." How was the puzzle solved?

Max Planck proposed a solution in 1900: the **quantum theory**. According to the quantum theory, radiation, like matter and electricity, *is quantized*, consisting of particles named **photons**. Thus, radiation is emitted, transmitted, and absorbed only in whole numbers of *photons*. It was Planck who named *the smallest quantity of any radiation*—the "bundle" of radiant energy, a **quantum**. Just as any mass of ^{1}H can be only a whole-number multiple of the mass of one ^{1}H atom, or as any charge can be only the charge of a whole number of electrons, a light beam can be only a whole number of photons.

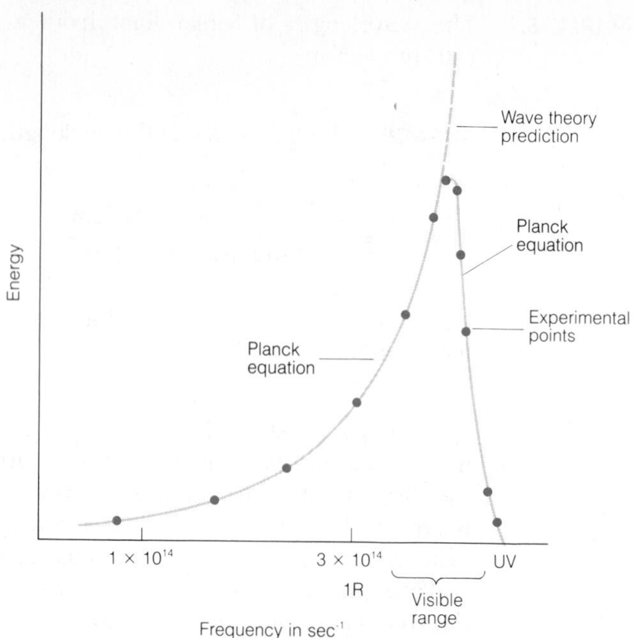

FIGURE 5.4
The experimental distribution of the frequencies and energies per unit time and per unit surface area of the light emitted by the surface of the sun (5700°C). (*IR* = infrared, *UV* = ultraviolet)

The name *photon* was suggested in 1926 by the chemist Gilbert N. Lewis (Chapter 7).

Matter has mass, energy (potential or kinetic), and de Broglie wave properties (next section). A photon has energy (radiant energy), travels only at the speed of light, and has wave properties ($c = \lambda v$). The energy of a photon, E, is given by

$$\boldsymbol{E = h \times v} \qquad (2)$$

where v is the frequency of the radiation and h is a constant, 6.6261×10^{-34} J sec/particle, known as **Planck's constant.**

THE PHOTOELECTRIC EFFECT

The second nineteenth-century experiment that led to the demise of the original wave theory involved the discovery that light can eject electrons from metallic surfaces. This emission of electrons is known as the **photoelectric effect** (Figure 5.5). Experiments show that the maximum kinetic energy of the ejected electrons depends on the *frequency,* but not on the brightness, of the monochromatic light. The higher the frequency, the greater is the maximum kinetic energy of the ejected electron (Figure 5.6). Increasing the brightness of the light increases only the number of ejected electrons and *not the maximum energy* of each electron.

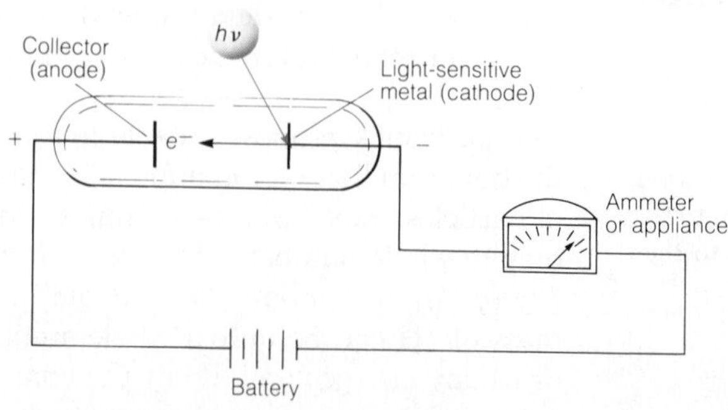

FIGURE 5.5
The photoelectric effect. The light-sensitive electrode of a photoelectric cell in an electric circuit emits electrons and causes a current to flow when light strikes the light-sensitive electrode. (Electrodes are enclosed in a vacuum or inert gas.)

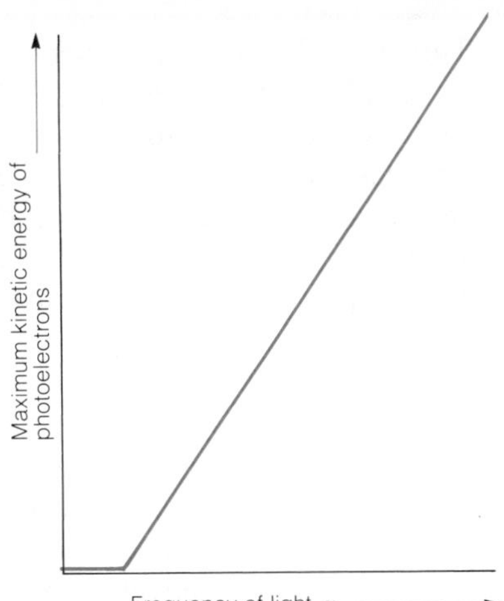

FIGURE 5.6
A plot of the measured maximum kinetic energy of photoelectrons against the frequency of the incident light. The slope of the line is Planck's constant.

Planck's photon was the key to Albert Einstein's explanation of these results (1905): The energies of all photons of a given frequency, ν, are identical. Each photon has the energy $h\nu$. A photon strikes an atom in the metal, delivers all of its energy to one electron, and so ejects that electron. Part of the photon energy ($h\nu$) is used to remove the electron from the metal (a constant, characteristic property of the metal), and the remainder appears as the maximum kinetic energy ($\frac{1}{2}mu^2$) of the ejected electron. Then, by the law of conservation of energy,

The frequency below which electrons are not emitted is a characteristic property of each metal and determines the value of the constant. More energy is needed to knock an electron out of the metal interior than out of the metal surface. Therefore, not all electrons escape with the maximum energy.

$$h\nu = \text{constant} + \tfrac{1}{2}mu^2$$

where m is the mass of an electron and u is its maximum speed. This equation predicts that as the frequency of the light is increased, $h\nu$ increases and the kinetic energy of the photoelectrons should increase. Experimental results agree with this prediction (Figure 5.6).

The explanation of the photoelectric effect and the energy distribution of light emitted by a hot object was decisive in convincing scientists that Planck's proposed photons are real (Box 5.2). The photoelectric cell has many applications, such as in automatic door-openers, light-exposure meters, television cameras, and motion picture sound projectors.

BOX 5.2

PLANCK, EINSTEIN, AND THE PHOTOELECTRIC EFFECT

It is interesting to note that the quantum concept seemed so revolutionary that physicists were quite reluctant to accept it. In fact, they searched diligently but unsuccessfully among the accepted concepts for possible alternatives. Planck assumed that the atom could lose or gain energy only in discrete amounts. However, it was still believed that the energy of a light beam was spread continuously throughout the beam. It remained for Einstein to show that the quantum idea is valid not only for the emission and absorption, but also for the transmission of all electromagnetic radiation through his simple explanation of what seemed to be a complicated phenomenon, the photoelectric effect. The concept of quantization of energy represented a radical departure from the older, classical physics.

5.2 MATTER WAVES

Louis de Broglie, 1875–1960

Thus far it has been sufficient for us to accept the electron as a particle, a bit of matter. However, we have just learned that light has particle properties. In 1924 Louis de Broglie reasoned that if light can be a particle as well as a wave, perhaps an electron can be a wave as well as a particle. An electron, a proton, an atom, a molecule, any piece of matter, he predicted, should have a wavelength associated with it. The *wave associated with a piece of matter* is called a **matter wave**. De Broglie predicted that the wavelength of a matter wave is

$$\lambda = \frac{h}{m \times u} \qquad (3)$$

A matter wave is also called a de Broglie wave to distinguish it from radiation.

where m is the mass of a particle and u is its speed. The equation predicts that an electron beam, like light, should produce a diffraction pattern, and this prediction has been verified by numerous experiments (Figure 5.7). However, matter waves are *not* electromagnetic radiation. They are *not* radiated into space or emitted by the particle; they *never leave the particle*. The particle does *not* have a wavelike motion. The speed of a matter wave is not the same as the speed of light or the speed of the particle; nor is it a constant. When we speak of a matter wave, we are simply referring to wave properties associated with the non-wave motion of particles.

The diffraction pattern in Figure 5.7b was obtained with electrons (9.11×10^{-31} kg/electron) accelerated to a speed of 1.3×10^8 m/sec and corresponding to the wavelength calculated as shown:

$$1\,\text{J} = \frac{\text{kg m}^2}{\text{sec}^2}$$

$$\lambda = \frac{h}{mu}$$

$$= \frac{6.63 \times 10^{-34}\,\frac{\cancel{\text{J sec}}}{\cancel{\text{electron}}} \times \frac{\cancel{\text{kg m}^2}}{\cancel{\text{J sec}^2}} \times 10^9\,\frac{\text{nm}}{\cancel{\text{m}}}}{9.11 \times 10^{-31}\,\frac{\cancel{\text{kg}}}{\cancel{\text{electron}}} \times 1.3 \times 10^8\,\frac{\cancel{\text{m}}}{\cancel{\text{sec}}}}$$

$$= 5.6 \times 10^{-3}\ \text{nm}$$

A similar diffraction pattern is produced by a radiation of the same wavelength. Accelerated atoms (He) and molecules (H_2) also show diffraction effects. The short wavelengths attainable with accelerated electrons reveal detailed features of viruses and molecules. The electron microscope, now a common laboratory tool, is an application of the de Broglie concept (Figure 25.7).

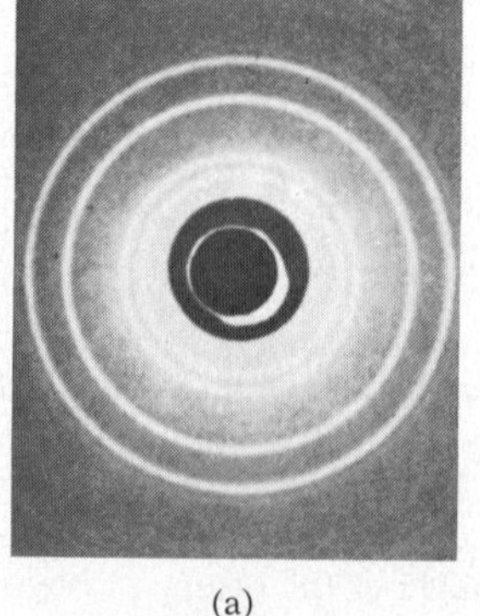

(a)

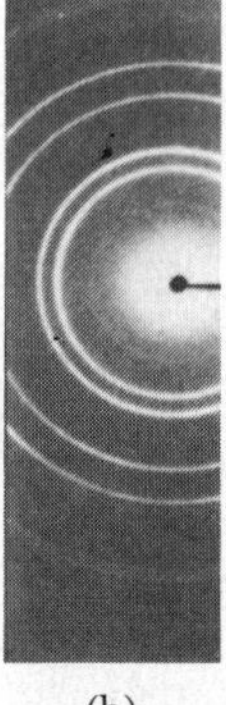

(b)

FIGURE 5.7 Comparison of X-ray and electron diffraction. (a) Diffraction of X rays by gold foil. (b) Diffraction of an electron beam by gold foil.

PROBLEM 5.3 (a) Calculate the wavelength (in nm) associated with a 100-g ball moving at 30 m/sec (67 mi/h). (b) How fast must the ball travel to have the wavelength of 5.6×10^{-3} nm? (c) Would you consider doing research on the wave properties of baseballs? □

5.3 SPECTRA OF ELEMENTS

A light beam may consist of one wavelength (such as a laser beam, Color plate 5), several different wavelengths (such as an H or He light beam, Color plate 6), or all wavelengths (the sun, Color plate 4).

When a narrow beam of white light—sunlight or light from a hot tungsten light bulb—passes through a triangular glass prism, it is separated into a ''rainbow'' of colors (Color plate 7). Each color of this ''rainbow'' has a characteristic wavelength. This separation of light into its component wavelengths is the basis of the **spectroscope**, an instrument that allows each wavelength to be seen on a screen (as in Color plate 7), or detected by an electronic device. This pattern of wavelengths is called a **spectrum**.

EMISSION AND ABSORPTION BY GASES AT LOW PRESSURE

When the light *emitted* from a gas—such as the light emitted by a mercury vapor lamp—is examined in a spectroscope, the spectrum does not include all wavelengths. Rather it consists of only several narrow, separated colored bands of wavelengths called **spectral lines** on a black background. Such a spectrum is therefore called an **emission line spectrum** (Color plate 6). This kind of spectrum is obtained from gases in discharge tubes or from gases at low pressure heated to high temperatures. If sodium, sodium chloride, or sodium bromide is sprayed into a flame, the same distinctive spectrum of sodium is obtained. *The emission line spectrum is a characteristic property of an element regardless of the element's source.* Therefore, just as fingerprints are used to identify persons, emission line spectra can be used to identify elements.

Some characteristic emission spectral lines of some elements, in nm (1 nm = 10 Å):

Li	670.8
Na	589.6
	589.0
Ca	616.2
	487.8
Hg	546.1
	435.8

All gases also *absorb* light. Examination of the spectrum obtained after white light from some source has passed through a gas shows that absorption occurs only at certain wavelengths. Thus, just like emission, absorption is selective, and an absorption spectrum consists of all colors interspersed with black lines. It is therefore called an **absorption line spectrum**. For example, the emission spectrum of sodium vapor consists of two yellow lines on a black background (Color plate 8a). However, when white light is passed through sodium vapor, the two sodium lines show up as two black lines on a rainbowlike background (Color plate 8b). Since absorption lines have the same wavelengths as emission lines, they, also, serve as ''fingerprints'' for the elements. In fact, analyzing absorption lines in the emission spectra of the sun and stars gives us information about their chemical composition.

EMISSION BY SOLIDS, LIQUIDS, AND DENSE GASES

Solids, liquids, and dense gases glow at high temperatures: a hot tungsten filament, a hot coal, and the sun all emit light. Examined in a spectroscope, the emitted light consists of a *continuous* band of colors, called a **continuous spectrum**. Unlike a line spectrum, a continuous spectrum consists of all wavelengths (Color plates 4 and 8b and Figure 5.4).

Since each element has a characteristic line spectrum, we conclude that it is the atoms of the elements that emit the line spectra. Therefore, studying the spectra of elements helps us to determine the structures of atoms.

Each spectral line is produced by photons possessing a definite energy. Emission and absorption spectra are obtainable over the entire radiation range (Color plate 4).

EXAMPLE 5.2 What is the energy of a photon associated with the violet line, 410.2 nm, in the emission spectrum of hydrogen?

ANSWER The energy of the photon is given by

$$E = h \times \nu = \frac{h \times c}{\lambda}$$

$$= \frac{6.626 \times 10^{-34} \dfrac{\text{J } \cancel{\text{sec}}}{\text{photon}} \times 2.998 \times 10^{8} \dfrac{\cancel{\text{m}}}{\cancel{\text{sec}}} \times 10^{9} \dfrac{\cancel{\text{nm}}}{\cancel{\text{m}}}}{410.2 \ \cancel{\text{nm}}}$$

$$= 4.843 \times 10^{-19} \frac{\text{J}}{\text{photon}}$$ ■

PROBLEM 5.4 (a) What is the energy of a photon corresponding to a wavelength of 30.38 nm? (b) Which photons, gamma photons (10^{-12} nm) or infrared photons (10^{-5} nm), have the greater energy per photon, and how many times greater is the energy of the more energetic photon? □

At first glance atomic spectra (Color plate 6) may appear to be disorderly arrays of colored lines. However, atomic spectra are actually highly ordered. The frequencies of the lines in the hydrogen atom spectrum, for example, can be represented by a relationship *involving dimensionless whole numbers, a and b,*

$$\nu = 3.2886 \times 10^{15} \text{ sec}^{-1}\left(\frac{1}{a^2} - \frac{1}{b^2}\right) \qquad (4)$$

Named for Johannes Robert Rydberg, who discovered this empirical relationship (1890) after considering much data and an equation derived (1885) by Johann J. Balmer (a high school teacher) for the wavelengths of the lines in the visible region of the hydrogen spectrum.

where ν is the frequency of the line in sec^{-1}, and $3.2886 \times 10^{15}\ \text{sec}^{-1}$ is a constant for hydrogen, known as the *Rydberg constant*; b is greater than a. Thus, when $a = 1$, b may be any whole number greater than 1, and each value of b yields a frequency, ν, corresponding to a line in the spectrum of hydrogen. When $a = 2$, b may be any whole number greater than 2, each value yielding a frequency corresponding to a line in the hydrogen spectrum. The relationship for atoms more complex than hydrogen is more complicated, but each frequency can still be represented as a difference of two frequency terms. The atoms therefore emit or absorb only photons of definite, characteristic frequencies or energies, depending upon the values of whole numbers a and b. This restriction means that *the energy absorbed or emitted by atoms is quantized.* This should not surprise us; after all, quantization appears to be a universal property: matter is quantized, electricity is quantized, and radiation is quantized. Spectra can be explained only by quantum mechanics, as we will learn in the next section.

By multiplying equation 4 by Planck's constant, you obtain an up-to-date version of the equation,

$$h\nu = E = 2.1790 \times 10^{-18} \frac{\text{J}}{\text{photon}}\left(\frac{1}{a^2} - \frac{1}{b^2}\right) \qquad (4')$$

where E is the energy absorbed or emitted by an H atom.

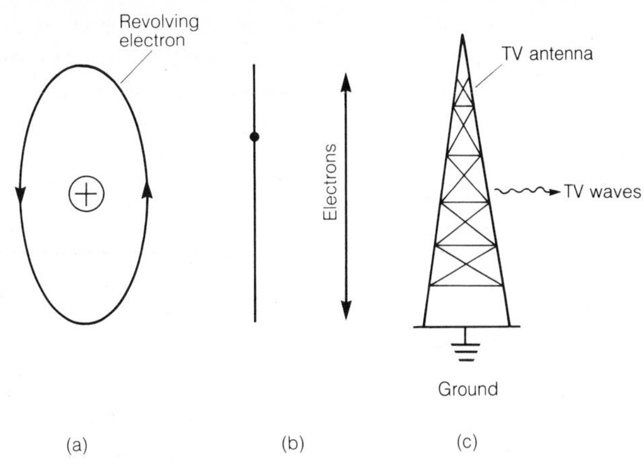

FIGURE 5.8
Emission of radiation by periodic electron motion. (a) A revolving electron. (b) Side view of the revolving electron. (a) Television antenna. The top of the antenna is connected to an oscillator that alternately attracts electrons from the ground and repels them to the ground.†

5.4 QUANTIZATION OF THE ENERGY OF AN ELECTRON IN AN ATOM

THE RUTHERFORD ATOM

Rutherford's work (section 2.6) provided a general model of the internal structure of the atom. However, his model did not explain how the electrons are situated and why atoms are stable. For example, why don't the electrons fall into the nucleus or escape into space? To answer these questions, let us consider the simplest atom, the hydrogen atom. If the negative electron were stationary, the electrical attractive force would cause it to fall into the positive nucleus. Therefore Rutherford suggested that the electron revolves about the proton (nucleus) just as the Moon revolves about the Earth. The electron does not escape into space because the attractive force acting on the electron prevents it from moving away from its orbit.

$F \propto \frac{q_1 q_2}{r^2}$ (Appendix A.6)

Further consideration of Rutherford's model presents us with a more serious dilemma. We know that the hydrogen atom consists of charged particles (an electron and a proton), but the wave (electromagnetic) theory of light predicts that the revolving electron will emit radiation. The periodic motion of a revolving electron is similar to the periodic electron motion in a television antenna that is sending out television waves (Figure 5.8). The revolving electron therefore should lose energy and spiral toward the proton. In doing so, it should emit a "rainbow" of colors and finally fall into the proton. Rutherford's theory thus incorrectly predicts a *continuous spectrum and a collapsing atom.*

Niels Bohr, 1885–1962

THE BOHR ATOM

In 1913 Niels Bohr avoided this problem when he proposed that the electron revolves with a definite fixed energy in a fixed path, *without emitting or absorbing energy*. The marvelous success of the Bohr theory in explaining the spectrum of the H atom created an emotional impact among scientists that is impossible to recapture today. His calculated wavelengths agreed perfectly with the experimental wavelengths. Such matching of theory with experimental numbers is successful science.

Nevertheless, Bohr's concept of a fixed path was eventually abandoned when a fundamental principle of quantum mechanics (1927) made it unacceptable. That principle is that we cannot precisely know or measure both the location and the

A version of the Heisenberg (Werner Heisenberg, 1927) uncertainty principle.

† Here is a simple experiment that illustrates periodic motion. While looking in a mirror, extend one arm out to your side and rotate it clockwise or counterclockwise. You can observe an up-and-down periodic motion of your fingers that is similar to the motion of an electron in an atom.

velocity (energy) of an object like an electron, even with perfect instruments. The more precisely we measure the energy of the electron in the hydrogen atom, the less we can know about its position. This is because measurement affects the object that is being measured; all measurements involve some use of energy, even if it is only the light that illuminates what we are measuring. If the object is an electron, then any exchange of energy will shift it to another position. It will then not be in the same place it was when the "measurement" was made. The original position of the electron thus remains a mystery, and each succeeding measurement will fail in the same way. Yes, the electron is elusive. No one has devised an experiment capable of locating the electron—and if it were possible to do so, we would know almost nothing about its energy. In short, it is impossible to assign at the same time both a definite energy and a definite path to the electron. Since experiments give us the precise energy of the electron in a hydrogen atom (seven to eight significant figures), we satisfy ourselves with this knowledge and must forgo knowing the exact position of the electron. For this reason, we do not assign a definite path to an electron in an atom. However, quantum mechanical theory does permit us to talk about the *probability* of finding the electron at every point in space about the nucleus of the H atom.

A photon strikes an isolated electron at rest. The electron acquires a velocity (kinetic energy) and moves away while the photon is scattered with a lower frequency, ν'.

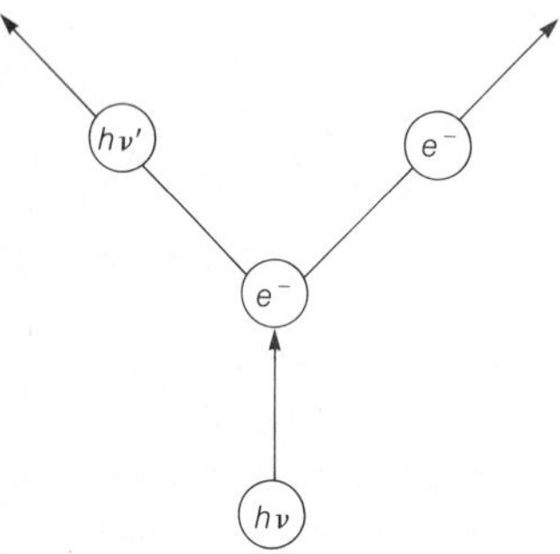

THE MODERN ATOM

It was taken for granted that the equations of motion developed by the great physicists of past centuries were applicable to all bodies, regardless of size. Quantum mechanics, of necessity, contradicts this concept; it *does not depict the structure of the atom in terms that are familiar to us from the behavior of visible bodies.* Nevertheless, quantum mechanics is a part of the evolution of our knowledge. In 1926, Erwin Schrödinger conceived an equation that fuses the wave and particle properties of matter. Schrödinger developed his equation by fitting de Broglie's equation into the equations already known for the propagation of water waves, sound waves, and light waves. The resulting **Schrödinger equation** is the fundamental assumption of the quantum mechanical theory. It offers a mathematical approach to understanding the quantum behavior of small particles. To do this, the electron is treated mathematically as if it were an electron wave, and the electron wave is treated as a standing wave. The Schrödinger equation provides the standing wave patterns of the electron.

Developed by Augustin Fresnel (1788–1827), Jean Fourier (1768–1830), and William Hamilton (1805–1860).

Erwin Schrödinger 1887–1961

The form of the Schrödinger equation need not concern us. For our purposes, the main points can be learned without the knowledge of the mathematics. The equation relates the energy, E, of an electron in an atom to the distribution of the electron is treated mathematically as if it were an electron wave, and the equation tells us that the *energy of the electron* in the H atom (also referred to as the energy of the H atom) is *quantized.* Only certain fixed energies are possible. These energy values are referred to as the **energy levels** of the electron in the H atom or as the energy levels of the atom. Each energy level has a characteristic electron wave pattern and has an energy value fixed by one to four dimensionless numbers, depending upon the environment of the H atom (section 5.6). These dimensionless numbers, called **quantum numbers**, arise naturally when the Schrödinger equation is solved for electron wave patterns and their energies; that is, they arise without putting any assumptions regarding quantum numbers into the equation.

The simplest case means that the H atom is not in an external magnetic field and its spectrum is examined in a spectroscope of low precision.

In the *simplest case* of an isolated H atom, its energy levels are determined almost entirely by one quantum number represented by n. Because n makes the largest contribution to the energy of the H atom, n is known as the *principal quantum number*. The number n can be any positive whole number. The energy of an electron increases when work is done on it to force it away from the nucleus. An electron that absorbs energy and is thus forced away from the nucleus *has* this additional energy. This process is analogous to increasing the potential energy of a weight by lifting it from, say, 5 m to 10 m above the ground. But these electronic energies are quantized, and the electron is said to be at a higher energy level when it is farther from the nucleus. Just as a and b in equation 4′, the quantity n cannot be 0, negative, or a fractional number such as 1.5.

ENERGY-LEVEL DIAGRAMS

The energy of the electron in the H atom is related to n^2 as $E = -k/n^2$. The proportionality constant, k, has the value of 2.179×10^{-18} J for the H atom. The energy of the electron in the H atom can take only the values given by

$$E_n = \frac{-2.179 \times 10^{-18}\ \text{J}}{n^2} \tag{5}$$

It is *not necessary to memorize* the number -2.179×10^{-18} J, since it is good only for $_1H$. What is important is to recognize that the energy of the energy levels of one-electron atoms and ions—for example, $_1H$, $_2He^+$, $_3Li^{2+}$—is inversely proportional to the square of the principal quantum number, n^2. The higher-charged nuclei of these hydrogenlike ions attract the single electron more strongly, but their spectra, as predicted by equation 5, are similar to the H spectrum (Problems 5.30 and 5.31).

An atom or ion with only one electron is described as "hydrogenlike."

What does the minus sign in equation 5 mean? Experimental measurements of energy (ΔH and ΔE, Chapter 4) give the energy difference between two states. This is true for any energy measurement, including the measurement of the energy of the electron. If we wish to assign an energy value to the electron as in equation 5, we must start with some arbitrarily assigned value of energy—a ruler mark—from which *energy differences* can be measured. The most simply defined state is the state in which the electron and the proton (nucleus) are separated to the point where there is no attraction between them. In this state the H atom is ionized. Under these conditions, we can assign any arbitrary value of energy to the electron, *but zero is the most convenient choice.* We followed the same procedure in assigning $\Delta H_f^0 = 0$ to elements (section 4.6) in calculating $\Delta H^0{}_{\text{reaction}}$.

When the electron approaches the proton, energy is *given out* and the energy of the electron decreases. Since we designated the energy as zero when the electron was far from the nucleus, it now becomes less than zero. A minus sign indicates that energy is lost as the electron approaches the proton. It also tells us that energy must be *added* to move the electron farther away from the region of the proton. Thus, the minus sign means that the electron is bound to the proton; it does not have the energy to escape from the atom. When $n = 1$, the energy of the electron in the H atom is -2.179×10^{-18} J and it must be increased by 2.179×10^{-18} J to ionize the atom (see diagram below). We can also say that when $n = 1$ and the H atom absorbs 2.179×10^{-18} J, then -2.179×10^{-18} J +

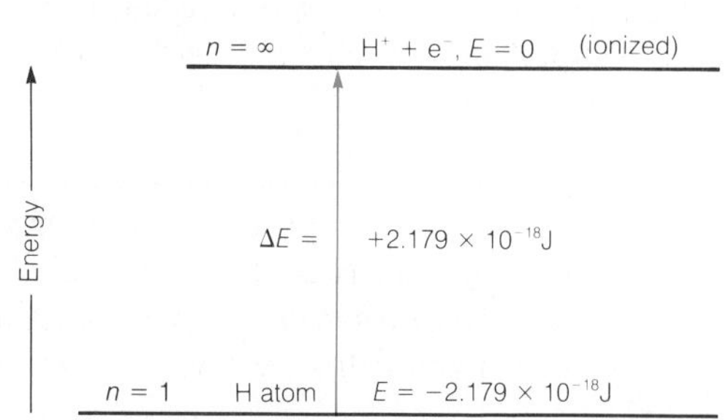

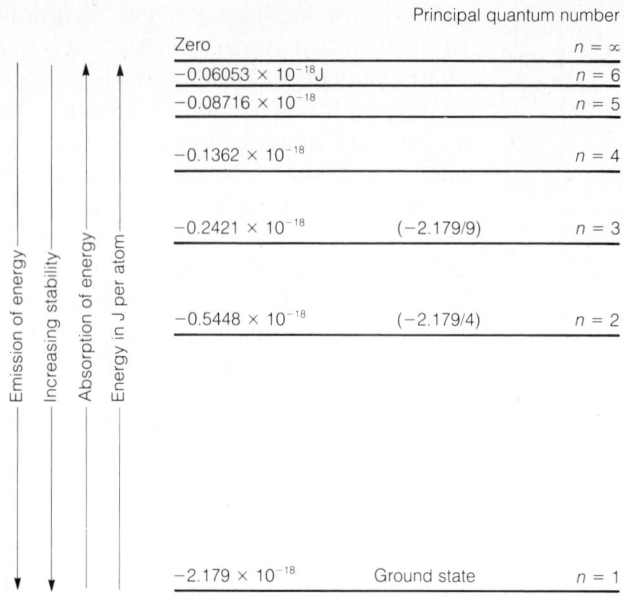

Principal quantum number

Zero $n = \infty$

-0.06053×10^{-18} J $n = 6$

-0.08716×10^{-18} $n = 5$

-0.1362×10^{-18} $n = 4$

-0.2421×10^{-18} $(-2.179/9)$ $n = 3$

-0.5448×10^{-18} $(-2.179/4)$ $n = 2$

-2.179×10^{-18} Ground state $n = 1$

Emission of energy

Increasing stability

Absorption of energy

Energy in J per atom

FIGURE 5.9
Energy-level diagram of the hydrogen atom (not to scale).

2.179×10^{-18} J $= 0$, or the atom is ionized. When $n = 10$, the energy of the electron in the H atom is -2.179×10^{-20} J:

$$E_n = \frac{-2.179 \times 10^{-18}\text{ J}}{n^2} = \frac{-2.179 \times 10^{-18}\text{ J}}{10^2} = -2.179 \times 10^{-20}\text{ J}$$

When this H atom absorbs 2.179×10^{-20} J, then $-2.179 \times 10^{-20}\text{ J} + 2.179 \times 10^{-20}\text{ J} = 0$, and the atom is ionized.

Note that as the value of n increases, E_n becomes *less negative* (closer to zero) and the energy of the electron increases. When $n = 10$, the H atom is at a higher energy level (-2.18×10^{-20} J) than the H atom at $n = 1$ (-2.18×10^{-18} J). Consequently, less energy is required to ionize the H atom when $n = 10$. Thus equation 5 tells us the energy that is needed to ionize an H atom whose energy is fixed by the value of n.

PROBLEM 5.5 Calculate the energy in J/atom and kJ/mole needed to ionize the H atom in the state $n = 4$.

The lowest energy level, the level at which $n = 1$, corresponds to the **ground state** of the H atom. It is the most stable state of the H atom. All the higher energy levels are known as **excited states**. A convenient method of representing the ground and excited states of the hydrogen atom (Figure 5.9) shows the energy levels, calculated from equation 5, as lines. This figure is called an **energy-level diagram**.

5.5 ORIGIN OF SPECTRAL LINES

We are now prepared to understand the observed spectrum of the hydrogen atom. In 1913 Niels Bohr explained the emission and absorption of photons: *The emission of a photon corresponds to an electron transition between two energy levels, from a higher level to a lower level.* The electron "drops" from a higher

PLATE 1
Sulfuric acid titrated with sodium hydroxide. Bromthymol blue is the indicator. Upper photo: yellow in H_2SO_4 solutions. Lower photo: blue on addition of an excess of one drop of NaOH solution. *(Small photos by E.R. Degginger; large photo by Runk/Schoenberger from Grant Heilman)*

PLATE 2
Solar One Project. Sunlight is reflected to the boiler at the top of the tower. The steam generates electricity for 6000 homes. *(Courtesy of Southern California Edison Co.)*

PLATE 3
Fireworks display. Chlorates and nitrates are basic ingredients of fireworks. Colors are produced by the metal constituent of the salt: Sr (scarlet), Ba (green), Na (yellow), Cu (blue); Al, Mg, and Fe powders produce white. *(Fireworks by Grucci, photograph by Manis)*

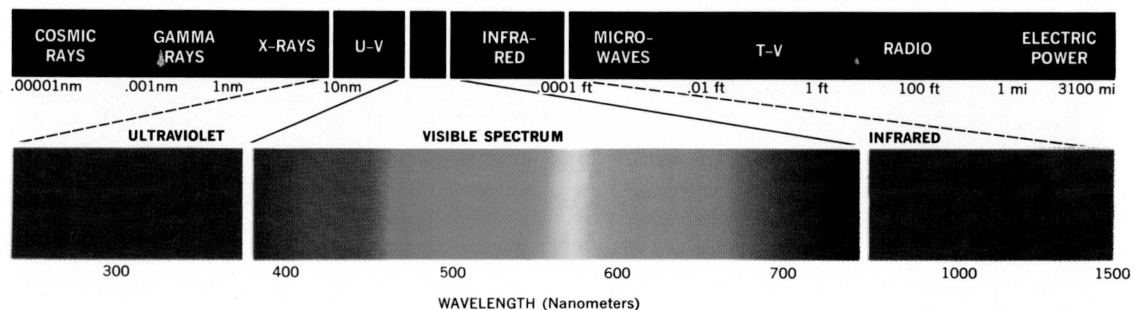

PLATE 4

The spectrum of electromagnetic radiation. Visible light is a very small portion of the spectrum. *(Reproduced with permission from* Light and Color, © *General Electric Company, Large Lamp Department, Nela Park, Cleveland, Ohio)*

Gamma rays: radiation from atomic nuclei.

X rays: radiation from electrons striking a target.

Ultraviolet rays: radiation from arcs and electric discharges in gases.

Visible light: radiation from stars, hot objects, hot gases, electric discharges.

Infrared rays: radiation from warm objects.

Hertzian waves: radiation from alternating electric currents—radio waves, television, microwaves, radar.

10 Å = 1 nanometer (nm)

10^9 nm = 1 m

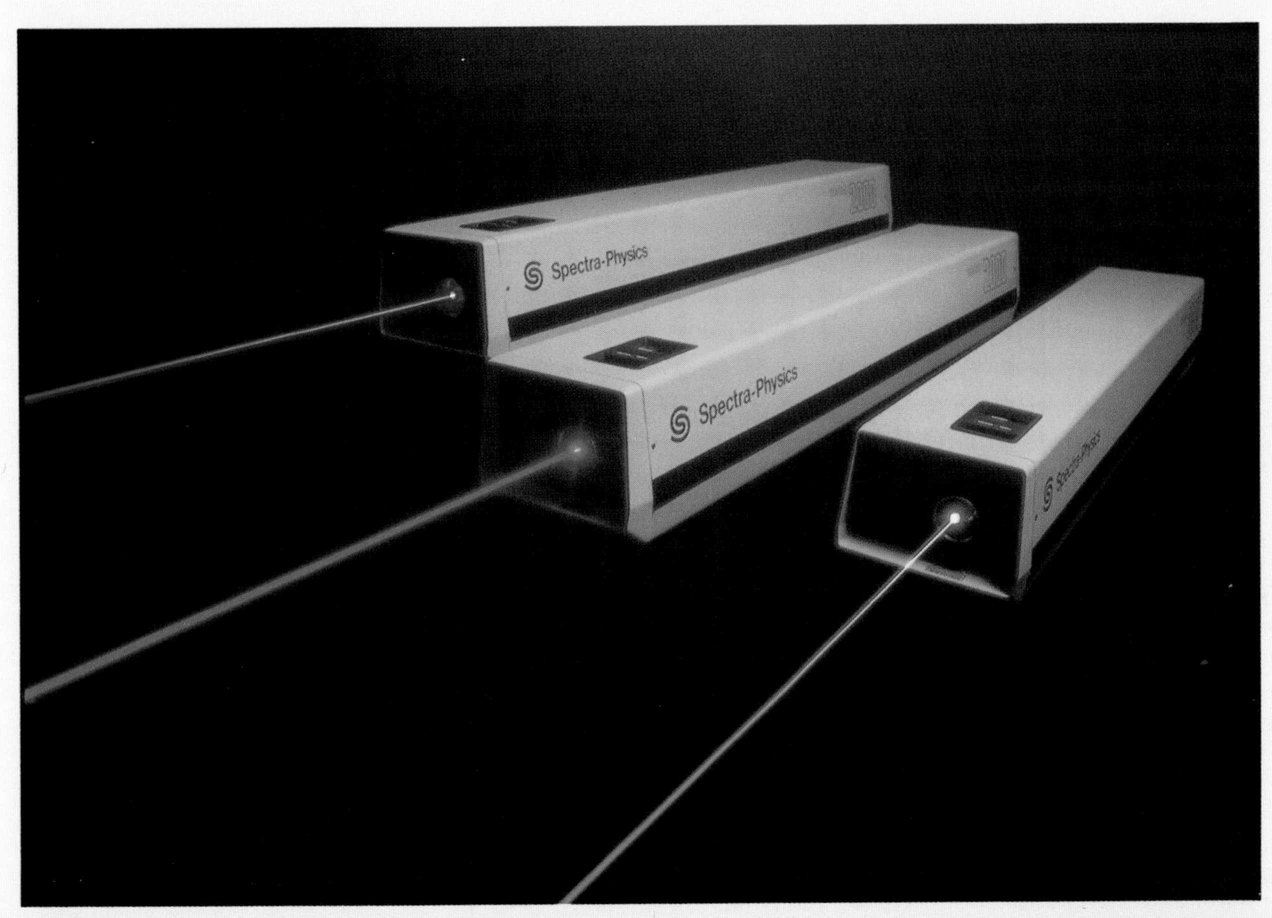

PLATE 5
One wavelength (monochromatic light) emitted from each of three lasers: $\lambda = 457.9$ mm (blue), $\lambda = 661.8$ mm (red), and $\lambda = 514.5$ mm (green).
(Courtesy of Laser Products Division, Spectra-Physics)

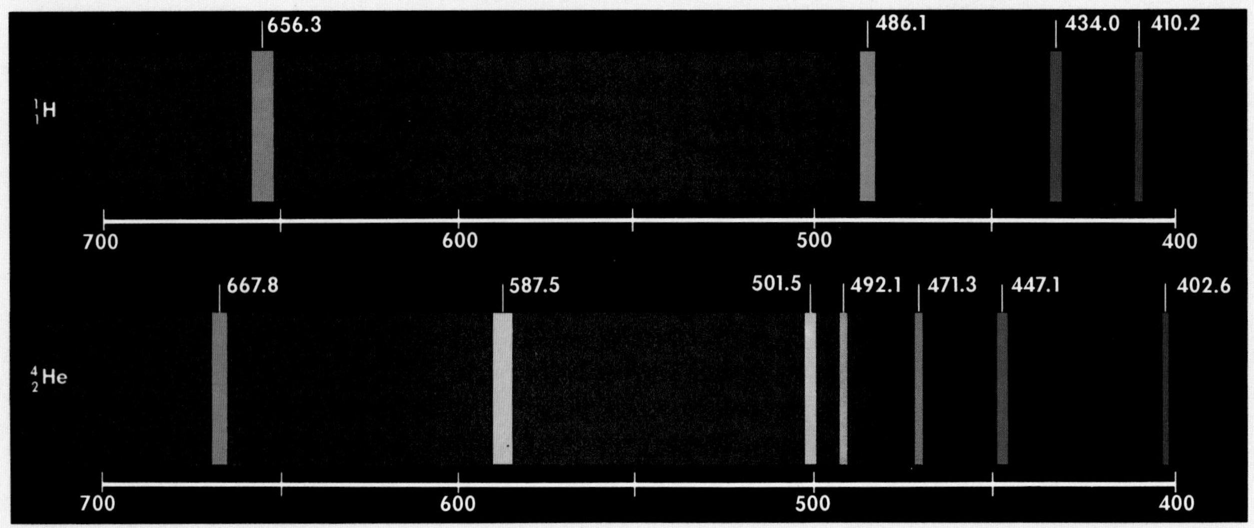

PLATE 6
The emission line spectra of hydrogen and helium in the visible range. The wavelengths are given in nanometers. The visible spectrum of the H atom is called the Balmer series in honor of Johann J. Balmer.

PLATE 7
Representation of a prism bending rays of white light. The effect of the prism is to bend shorter wavelengths more than longer wavelengths, separating them into distinctly identifiable lines of color. *(Reproduced with permission from* Light and Color, *© General Electric Company, Large Lamp Department, Nela Park, Cleveland, Ohio)*

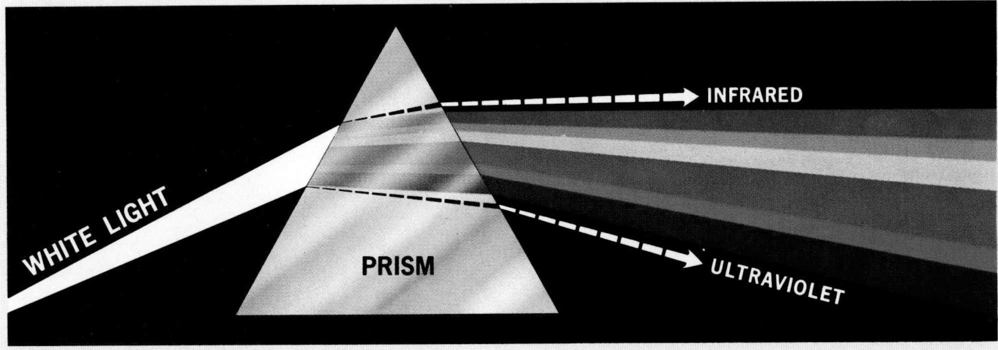

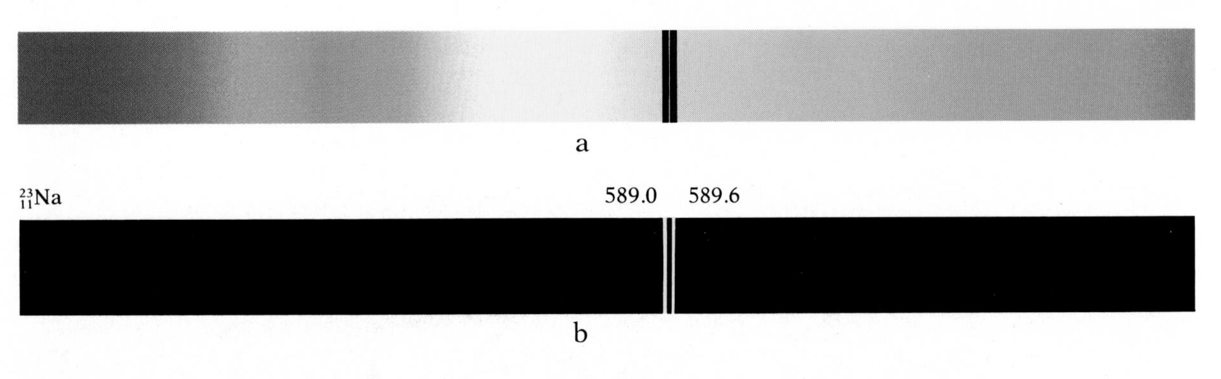

PLATE 8
(a) The absorption line spectrum of sodium.
(b) The emission line spectrum of sodium.

$n = 3$

Photon

$n = 2$

$n = 2$

Photon

$n = 1$

to a lower energy level. The emitted energy appears as a photon. The energy of the photon can only be the difference in energy of the two levels between which the electron transition occurs. Now for the reverse case: The absorption of a photon corresponds to an electron transition between two energy levels, from a *lower* level to a *higher* level. An electron "jumps" from a lower to a higher level, and the energy of the electron increases. The energy is gained as a photon disappears. The gain can only be the difference in energy of the two levels between which the electron transition occurs. The energy of the photon absorbed or emitted must exactly equal the gain or loss in energy experienced by the atom (conservation of energy). This gain or loss is represented by an arrow between two energy levels.

Since the energy of the electron is quantized, *the hydrogen atom (or any other atom) can emit or absorb only photons whose energy is equal to the difference between two energy levels.* A spectral line thus originates when the electron "jumps" or "drops" from one energy level to another:

Note: We use the convention established in Chapter 4, $\Delta E = E_{\text{final}} - E_{\text{initial}}$.

$$\underset{\text{(energy of photon emitted or absorbed)}}{\Delta E} = E_{\text{final level}} - E_{\text{initial level}} = h\nu \qquad (6)$$

A positive ΔE indicates photon absorption and electron excitation. A negative ΔE indicates photon emission and loss of electronic energy.

EXAMPLE 5.3 The electron of an H atom is excited from its first ($n = 1$) to its sixth energy level ($n = 6$). The electron then "drops" to its second energy level ($n = 2$). What is the energy of the photon (a) absorbed and (b) emitted?

ANSWER

(a) The energy of the photon absorbed is the difference between the $n = 6$ and the $n = 1$ energy levels. From Figure 5.9 or equation 5:

$$E_{\text{final level}} = E_6 = -0.06053 \times 10^{-18}\,\frac{\text{J}}{\text{atom}}$$

$$E_{\text{initial level}} = E_1 = -2.179 \times 10^{-18}\,\frac{\text{J}}{\text{atom}}$$

and

$$\Delta E = -0.06053 \times 10^{-18}\,\frac{\text{J}}{\text{atom}} - \left(-2.179 \times 10^{-18}\,\frac{\text{J}}{\text{atom}}\right)$$

$$= +2.118 \times 10^{-18}\,\frac{\text{J}}{\text{atom}}$$

The energy of the absorbed photon is therefore 2.118×10^{-18} J.

(b) The energy of the emitted photon is the difference between the $n = 2$ and the $n = 6$ energy levels. From Figure 5.9 or equation 5,

$$E_{\text{final level}} = E_2 = -0.5448 \times 10^{-18}\,\frac{\text{J}}{\text{atom}}$$

$$E_{\text{initial level}} = E_6 = -0.06053 \times 10^{-18}\,\frac{\text{J}}{\text{atom}}$$

and

$$\Delta E = -0.5448 \times 10^{-18} \frac{\text{J}}{\text{atom}} - \left(-0.06053 \times 10^{-18} \frac{\text{J}}{\text{atom}}\right)$$

$$= -0.4843 \times 10^{-18} \frac{\text{J}}{\text{atom}}$$

The energy of the emitted photon is 4.843×10^{-19} J. The answer agrees with experimental equation 4′ when $a = 2$ and $b = 6$. ■

Notice that while ΔE may be positive (photon absorption) or negative (photon emission), the energy of the photon (absorbed or emitted) is always positive.

PROBLEM 5.6 A photon excites the electron of an H atom from $n = 2$ to $n = 5$. What is the energy of the absorbed photon in J/atom? What is the energy of the photon emitted in the transition $n = 5$ to $n = 2$ in J/photon and kJ/mol? □

See Example 5.2.

The energy of the emitted photon in Example 5.3b (4.843×10^{-19} J) corresponds to the wavelength 410.2 nm, which is in perfect agreement with one of the lines in the spectrum of hydrogen (Color plate 6).

The lines in the ultraviolet range given in Figure 5.10, as well as other series of lines, were discovered *after* Bohr predicted their existence.

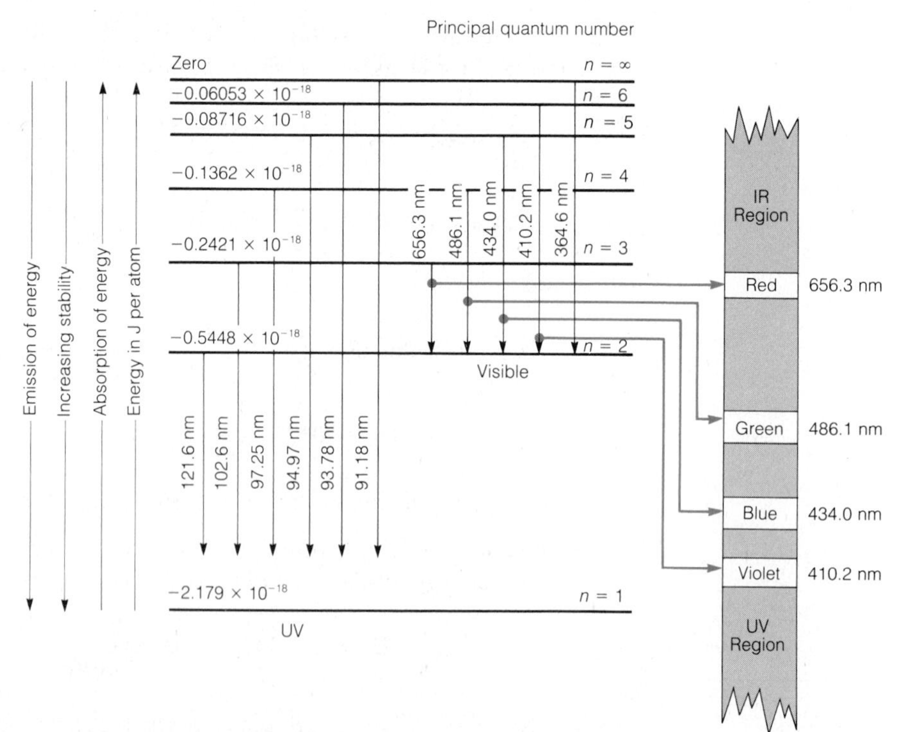

FIGURE 5.10 Energy-level diagram (not to scale) used to predict the spectrum of the hydrogen atom. The energies of the photons emitted or absorbed are given by the energy difference between two levels. The greater the difference, the greater the energy and the frequency of the photon, and the shorter its wavelength. A comparison is made between the predicted and measured wavelengths for four lines in the visible region of the hydrogen spectrum. (See Color plate 6 for colors.)

EXAMPLE 5.4 The energy of the photon emitted for the transition $n = 6 \longrightarrow n = 1$ is 2.118×10^{-18} J (see Example 5.3a). Predict the wavelength of the spectral line corresponding to this energy.

ANSWER

$$E = h \times \nu = \frac{h \times c}{\lambda}$$

$$\lambda = \frac{h \times c}{E} = \frac{6.626 \times 10^{-34} \dfrac{\cancel{\text{J}}\ \cancel{\text{sec}}}{\cancel{\text{photon}}} \times 2.998 \times 10^{8} \dfrac{\cancel{\text{m}}}{\cancel{\text{sec}}} \times 10^{9} \dfrac{\text{nm}}{\cancel{\text{m}}}}{2.118 \times 10^{-18} \dfrac{\cancel{\text{J}}}{\cancel{\text{photon}}}} = 93.79 \text{ nm}$$

The measured value, shown in Figure 5.10, is 93.78 nm. ■

PROBLEM 5.7 Find the wavelength of the spectral line for the $n = 5 \longrightarrow n = 2$ transition. □

5.6 THE ELUSIVE ELECTRON AND ATOMIC ORBITALS

The total energy of the electron in a given energy level (or energy state) is the sum of its potential and kinetic energies. For example, as an electron moves away from the positive nucleus, its potential energy increases but its kinetic energy decreases so that the total energy does not change for a given energy state.

Earlier in this chapter we said that the exact position and the exact energy of electrons in atoms cannot be measured at the same time. This means that an electron in a *definite energy level* can move anywhere relative to the nucleus *without changing its total energy*. However, we can talk about finding the electron in a specified region of space about the nucleus. The probability that an electron will be in a particular region of space about the nucleus is commonly referred to as the *electron density*. The higher the electron density in a region of space, the higher is the probability of finding the electron in that region.

For each energy level, the Schrödinger equation makes it possible to calculate the amplitude, represented by the Greek letter psi, ψ, of the electron standing wave in an atom. ψ describes the shapes of the electron stationary waves and so affords a picture of the behavior of an electron at a particular energy level. The square of ψ, ψ^2, has a more direct physical interpretation. It gives the probability distribution, that is, the electron density, in an atom or molecule (Box 5.3). Applied to the H atom, ψ^2 specifies the probability of finding the electron in a unit of volume at a given distance from the nucleus (Figure 5.11 and Problem 5.58).

If we draw a diagram of a hydrogen atom in the ground state by making darker shadings where there is a greater chance of finding the electron and lighter shadings where there is less chance, the result looks like a spherical cloud of

BOX 5.3 MAX BORN AND PROBABILITY

The accepted interpretation of ψ and ψ^2 is credited to Max Born, who suggested in 1926 that the probability of finding the electron in a small volume of space is proportional to the intensity (ψ^2) of the electron wave at that point. He related his idea to the theory of light: on the wave model, the intensity of light is proportional to the *square* of the amplitude of the wave (section 5.1). On the photon model, the intensity is proportional to the photon density, the number of photons per unit volume of the beam. Photon density is simply the probability of finding a photon in a small volume of space.

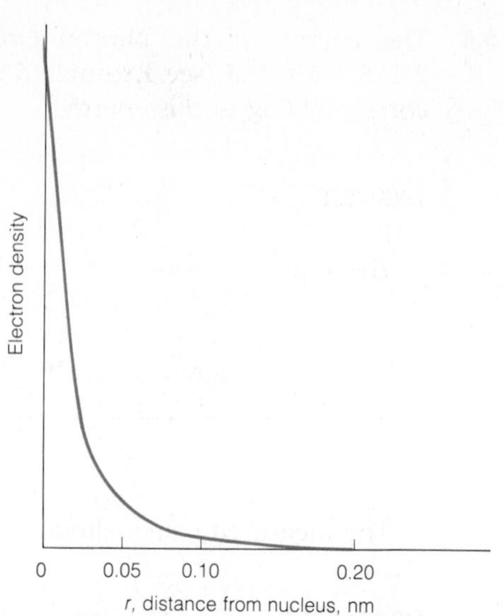

FIGURE 5.11
A plot of the electron density in the H atom in the ground state against distance from the nucleus. The probability of finding the electron in the ground state is the same in all directions but varies with the distance from the nucleus.

FIGURE 5.12
A crude representation of the electron cloud of the H atom in the ground state.

negative charge around the nucleus (Figure 5.12). In fact, such a picture is called an **electron cloud diagram**, and the space in which an electron is likely to be found is called its **atomic orbital**. An atom, then, is but a speck of a dense positively charged core submerged in a cloud of electrons.

ORBITALS AND QUANTUM NUMBERS

The larger the value of n, the higher the energy of the electron. n can be any positive whole number, 1, 2, 3, 4, ... Orbitals having a given value of n are said to belong to a given **shell**. The letters K, L, M, N, . . . are sometimes used to represent successive n values 1, 2, 3, 4, ...

Shells and subshells are also known as *levels* and *sublevels*.

Experimental results show that a principal energy level (except for $n = 1$) is split into a set of closely spaced energy levels. Each closely spaced energy level corresponds to a **subshell. The number of subshells in a given shell equals *n*, the principal quantum number of that shell.** The subshells are represented by the letters *s, p, d, f, g,* and so on. Thus, the $n = 1$ shell has one subshell, designated 1*s*; the $n = 2$ shell has two subshells, designated 2*s* and 2*p*; the $n = 3$ shell has three subshells, designated 3*s*, 3*p*, and 3*d*. The *s, p, d,* . . . orbitals in a given shell have slightly different energies. In the H atom, these differences are small compared to the differences between principal energy levels.

Each subshell consists of a set of orbitals. All the orbitals in the same subshell have the same energy. The number of orbitals in a given subshell varies in a systematic way as shown (a dash represents an orbital):

Before the advent of quantum mechanics, these letters were used to classify spectral lines into various series: *s* (sharp), *p* (principal), *d* (diffuse), *f* (fundamental).

SUBSHELL DESIGNATION	NUMBER OF ORBITALS
s	one *s* orbital —
p	three *p* orbitals — — —
d	five *d* orbitals — — — — —
f	seven *f* orbitals — — — — — — —

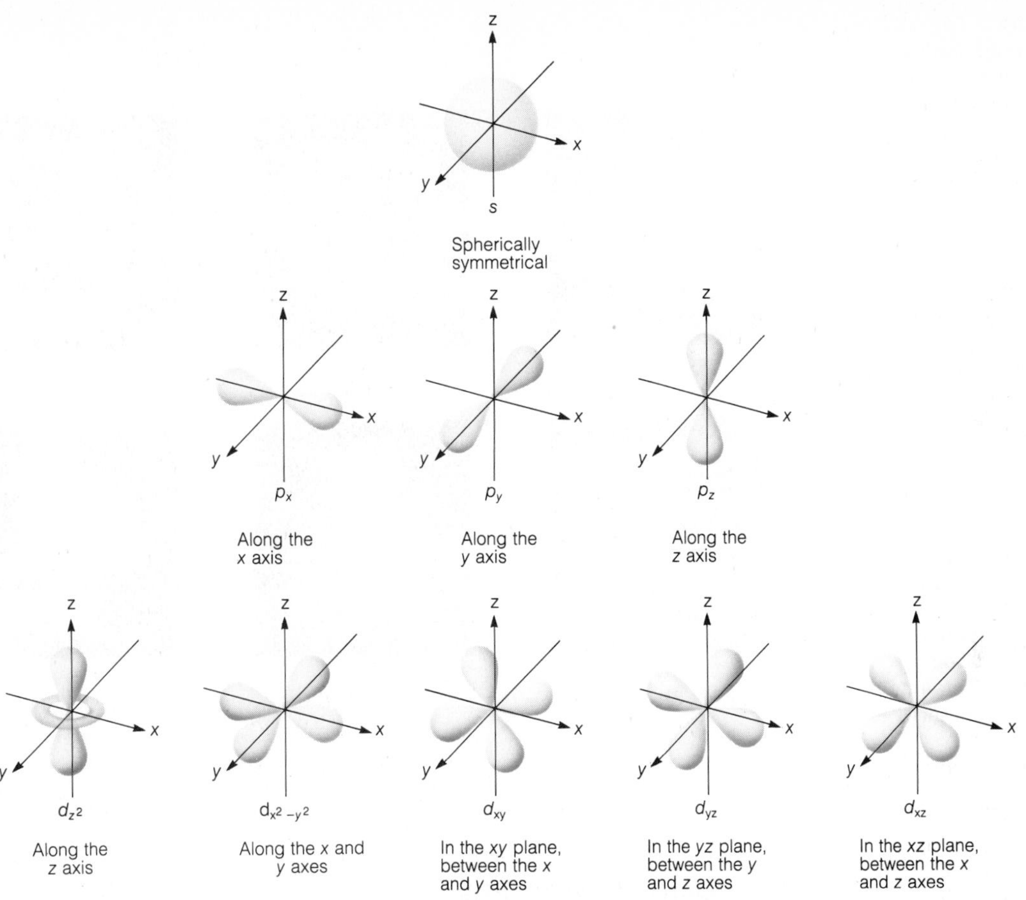

FIGURE 5.13
Schematic representations of the *s*, the three *p*, and the five *d* orbitals (electrons). The nucleus is at the origin of the axes. These representations may be viewed as electron standing wave patterns.

The diagrams in Figure 5.13 show the shapes of the regions that encompass about 90% to 95% of the electron density.

Different orbitals have different sizes, shapes, and spatial orientations. Schematic representations of the *s*, *p*, and *d* orbitals are given in Figure 5.13. Only the *s* orbital is spherical. Note that the three *p* orbitals differ. One is oriented along the *x* axis (p_x), a second is oriented along the *y* axis (p_y), and the third is oriented along the *z* axis (p_z).

The shape of one of the five *d* orbitals (d_{z^2}) looks like that of a *p* orbital in a "doughnut." The four other *d* orbitals resemble four-leaf clovers. Visualize one "clover" ($d_{x^2-y^2}$) along the *x* and *y* axes; rotate the clover 45° without tilting and you will produce the d_{xy} orbital. Imagine another clover (d_{yz}) upright between the *y* and *z* axes. Rotate this clover 90° so that it faces you now; you are looking at the d_{xz} orbital. The shapes of orbitals (electron cloud diagrams) are *pictorial representations* of the mathematical solutions of the Schrödinger equation. They are *not* pictures of electric charges or of electrons (Figures 5.13 and 5.14).

Figure 5.14 represents the standing wave patterns of the electron in several states of the hydrogen atom. Like the standing waves of a struck guitar string or a snapped rubber tube (Figures 2 and 3 in Box 5.1), electron standing waves may have nodes. Note particularly in Figure 5.14 that the 2*s* orbital has one node whereas the 3*s* orbital has two nodes.

Table 5.1 summarizes the shells, subshells, and orbitals within the first three shells.

FIGURE 5.14
Electron cloud diagrams of several states of the hydrogen atom according to the predictions of quantum mechanics for several values of n, ℓ, and m_ℓ. These should be visualized as three-dimensional clouds surrounding the nucleus of the hydrogen atom. The lightness of the shading (the cloud) is proportional to ψ^2. The lighter the cloud, the higher the chances of finding the electron. (H. E. White, *Phys. Rev.* **37** (1931), 1416.) *Note*: These are photographs not of electrons, but of mechanical models whose rotations correspond to the electron standing wave patterns obtained from the mathematical solutions of the Schrödinger equation.

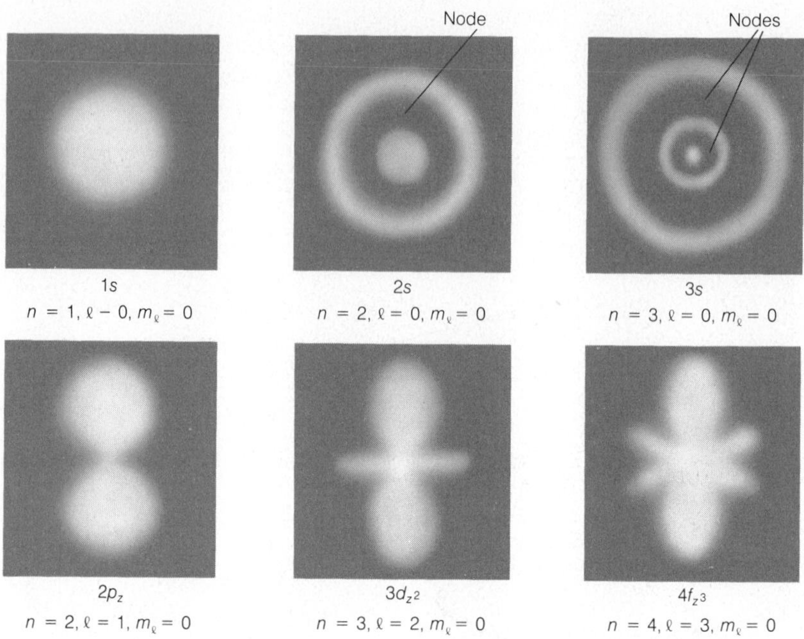

MORE QUANTUM NUMBERS

The quantum mechanical treatment of the H atom yields the three quantum numbers n, ℓ, m_ℓ, and their relationships.

Recall that a shell is labeled by n, the principal quantum number. However, only one spectral line can originate between two given energy levels. For example, in the transition $n = 2 \longrightarrow n = 1$ in the H atom, only one line can be radiated (Figure 5.10). But if a spectroscope of high precision is used, *two slightly separated lines* are observed. If the light source—the atoms emitting radiation—is placed in a strong magnetic field, each spectral line will be further split. For example, the two yellow sodium lines are split into eight very closely spaced lines. These additional lines are explained by splitting each principal energy level into a number of sublevels. Increasing the number of energy levels increases the number of spectral lines that may be radiated. These sublevels are designated by three additional quantum numbers: ℓ (identifying the subshells), m_ℓ (identifying the orbitals in the subshell), and m_s (distinguishing the electrons in an orbital). The ℓ quantum number is related to n, and the m_ℓ quantum number is related to ℓ, as shown in the following paragraphs.

TABLE 5.1
SUBSHELLS AND ORBITALS WITHIN THE FIRST THREE SHELLS

Energy ↑	SHELL	
	$n = 3$	$\overline{3s}$ $\overline{3p_x}\,\overline{3p_y}\,\overline{3p_z}$ $\overline{3d_{z^2}}\,\overline{3d_{x^2-y^2}}\,\overline{3d_{xy}}\,\overline{3d_{yz}}\,3d_{xz}$
	$n = 2$	$\overline{2s}$ $\overline{2p_x}\,\overline{2p_y}\,\overline{2p_z}$
	$n = 1$	$\overline{1s}$

THE QUANTUM NUMBER ℓ The ℓ values are whole numbers starting with 0. For any given n, the ℓ values go up to $n - 1$. For example, when $n = 1$, ℓ can be only 0. When $n = 4$, ℓ can be 0, 1, 2, and 3. *Notice that there are* n *different values of* ℓ. This number of ℓ values equals the number of subshells:

n	ℓ	NUMBER OF SUBSHELLS	DESIGNATION OF SUBSHELLS
1	0	1	1*s*
2	0, 1	2	2*s*, 2*p*
3	0, 1, 2	3	3*s*, 3*p*, 3*d*
4	0, 1, 2, 3	4	4*s*, 4*p*, 4*d*, 4*f*

The letters *s*, *p*, *d*, *f*, and so on, correspond, respectively, to ℓ values of 0, 1, 2, 3, 4, and so on.

$\ell = 0$	1	2	3	4 . . .
s	*p*	*d*	*f*	*g*

THE MAGNETIC QUANTUM NUMBER, m_ℓ The m_ℓ quantum number labels the orbitals that make up a subshell. m_ℓ can be a whole number having the value of $+\ell$ and descending to zero and then to $-\ell$:

$$m_\ell = +\ell, \ell - 1, \ell - 2, \ldots, 0, -1, -2, \ldots, -\ell$$

For example, when $\ell = 0$, m_ℓ can only be 0. When $\ell = 1$, m_ℓ can have values of 1, 0, -1; when $\ell = 2$, m_ℓ takes the values of 2, 1, 0, -1, -2. *Notice that there are* $(2\ell + 1)$ *values of* m_ℓ. The number of m_ℓ values equals the number of orbitals in a subshell, as shown in this table:

n	ℓ	m_ℓ	DESIGNATION OF SUBSHELL	NUMBER OF ORBITALS IN SUBSHELL	NUMBER OF ORBITALS IN SHELL n^2
1	0	0	1*s*	1	1
2	0	0	2*s*	1	
	1	$+1, 0, -1$	2*p*	3	4
3	0	0	3*s*	1	
	1	$+1, 0, -1$	3*p*	3	
	2	$+2, +1, 0, -1, -2$	3*d*	5	9
4	0	0	4*s*	1	
	1	$+1, 0, -1$	4*p*	3	
	2	$+2, +1, 0, -1, -2$	4*d*	5	
	3	$+3, +2, +1, 0, -1, -2, -3$	4*f*	7	16

An atomic orbital is thus defined by the three quantum numbers n, ℓ, and m_ℓ.

Recall that orbitals may differ from one another in size, shape, and spatial orientation. Their differences are related to the quantum numbers in these ways:

n determines the size of an orbital;

ℓ determines the shape of an orbital; and

m_ℓ determines the spatial orientation of an orbital.

See Figures 5.13 and 5.14.

An orbital or an electron is designated by a symbol that gives its n and ℓ values. For example, a $3d$ orbital ($n = 3$, $\ell = 2$) refers to an orbital in the $\ell = 2$ subshell in the $n = 3$ shell; a $2p$ orbital ($n = 2$, $\ell = 1$) is in the $\ell = 1$ subshell in the $n = 2$ shell.

PROBLEM 5.8 (a) List the n, ℓ, and m_ℓ values for orbitals in the $3d$, $4d$, $6d$, $7p$, and $5f$ subshells. (b) How many orbitals are in the $4f$ and $6f$ subshells? □

PROBLEM 5.9 List the possible values of m_ℓ for all the orbitals in the $\ell = 4$ subshell in the $n = 5$ shell. □

THE SPIN QUANTUM NUMBER, m_s This quantum number takes only two values: $+\frac{1}{2}$ and $-\frac{1}{2}$. The trajectory of the electron is not described in quantum mechanics. Nevertheless, it is useful to associate the m_s quantum number with some kind of definite movement of the electron in an atom. It is assumed that the electron spins like a top. The values $+\frac{1}{2}$ and $-\frac{1}{2}$ are interpreted to mean that the spin is oriented in one of two possible ways, clockwise or counterclockwise.†

It is assumed that a single spinning electron *acts as if it were a small bar magnet.* We conclude that the electron is the "atom of magnetism," the basic unit of magnetism, the permanent "micromagnet" that generates the magnetic fields observed in experiments. The magnetism of an isolated atom resulting from the spin of the electron around its axis is illustrated for the hydrogen atom in Figure 5.15. The electron-magnetic-bar analogy in Figure 5.16 may help you to visualize the magnetic property of an electron. However, *only two spin orientations* (relative to the magnetic field) are permitted for an electron, corresponding to the m_s quantum numbers $+\frac{1}{2}$ and $-\frac{1}{2}$ (Figure 5.15). These orientations are arbitrarily "pictured" as

$$\uparrow \text{ and } \downarrow \text{ or } \overset{\text{N}}{\underset{\text{S}}{\uparrow}} \text{ and } \overset{\text{S}}{\underset{\text{N}}{\downarrow}}$$

The four quantum numbers provide sufficient energy levels to account for all spectral lines that appear in the spectrum of an element in strong magnetic fields. In addition, spin quantum numbers play a major role in determining the number of electrons an orbital may hold (next section).

5.7 DISTRIBUTION OF ELECTRONS IN ATOMS

Any atom other than a hydrogen atom has more than one electron. The presence of several electrons creates complex mathematical problems because electrons repel each other. Even the most modern computers are not capable of supplying exact answers to the Schrödinger equation when it is applied to many-electron atoms. However, approximate results show that the energy of an electron in a many-electron atom has a greater dependence on ℓ than the energy of an electron in an H atom. Fortunately, the orbitals of an H atom and a many-electron atom are sufficiently similar so that *we may use orbitals like those found in the H atom to describe the electron configuration of atoms with more than one electron.*

† The "electron-spin" hypothesis was formulated in 1925 by George Uhlenbeck and Samuel Goudsmit to account for the splitting of spectral lines in a magnetic field. In 1928 Paul Dirac made the "electron spin" quantum numbers, $+\frac{1}{2}$ and $-\frac{1}{2}$, come out of the theory naturally by fitting the theory of relativity into the Schrödinger equation.

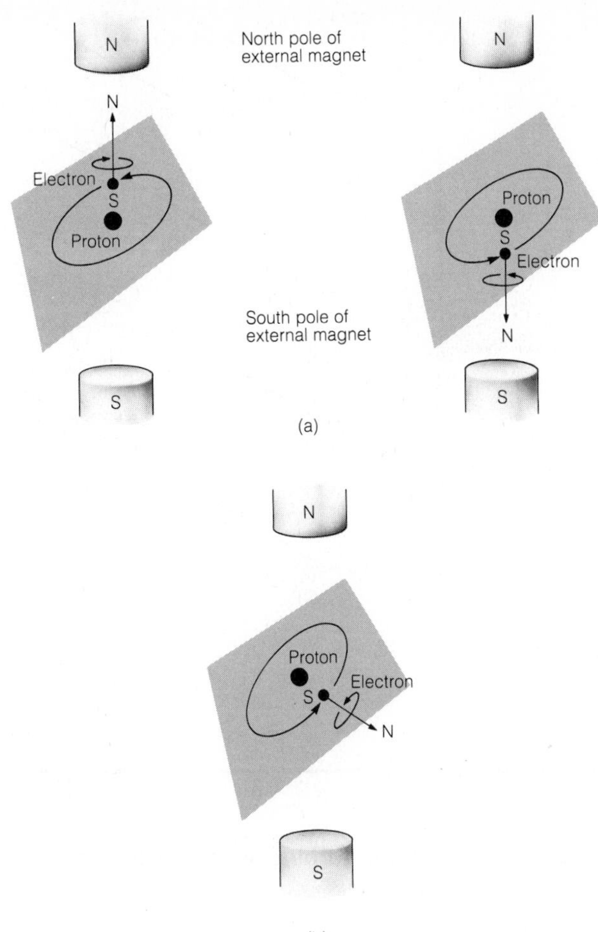

FIGURE 5.15
(a) The two spin orientations of the electron, represented as
N S
↑ ↓, may occur in a
S N
magnetic field. Any other position, such as the one illustrated in (b), is impossible. (*N* = north, *S* = south)

The $(n + \ell)$ rule, applicable in the absence of an external magnetic field.

All the orbitals in a given subshell have the same energy because they have the same n and ℓ values. The three $4p$ orbitals have the same n and ℓ values, $n = 4$ and $\ell = 1$, and therefore have the same energy. In neutral isolated atoms, *the orbital with the lowest energy is the one for which the sum of n and ℓ is lowest.* When two orbitals have the same $(n + \ell)$ value, the orbital with the lower n value has the lower energy. The orders of the energy and stability of the orbitals are given in Figure 5.17.

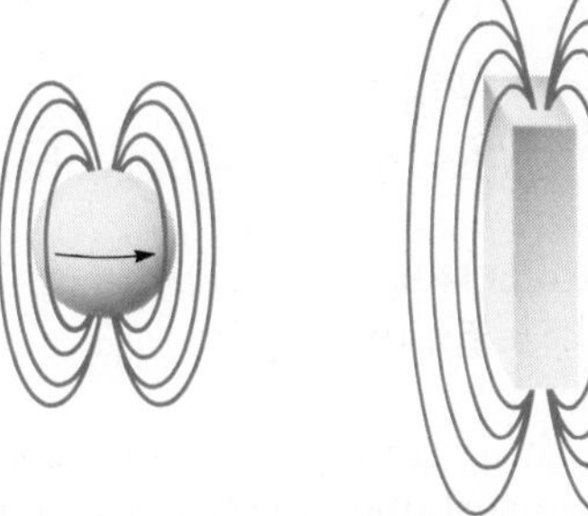

FIGURE 5.16
The electron in motion sets up a magnetic field whose properties are identical to the field set up by a bar magnet. (The magnetic moments, of course, differ.)

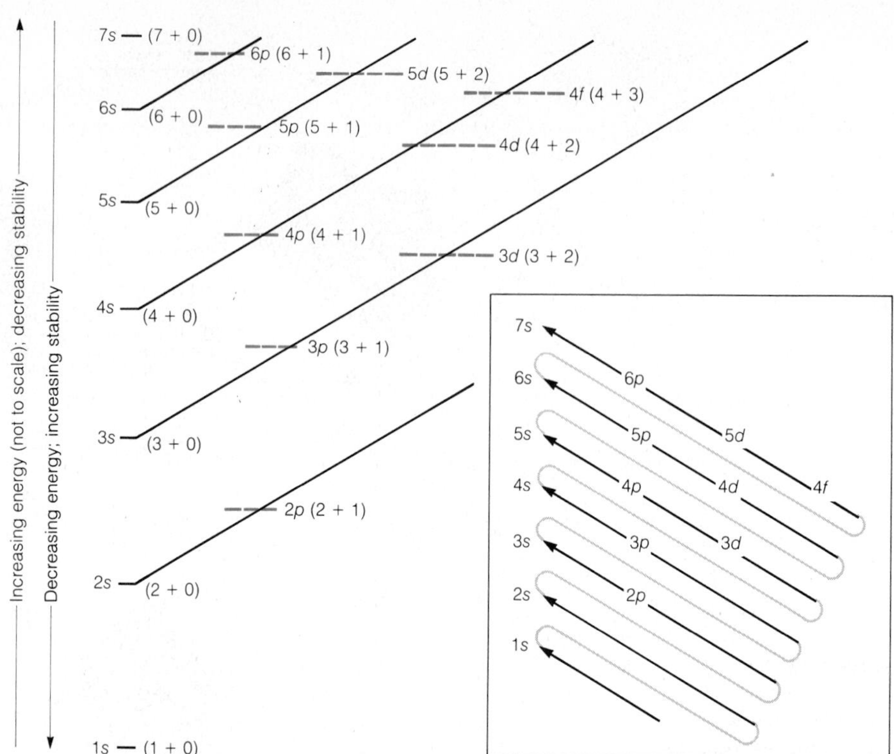

FIGURE 5.17
$(n + \ell)$ values. The order of the energies and stabilities of atomic orbitals in neutral isolated atoms; not applicable to positive ions. The faint lines show the spread of the orbitals with the same n value. The filling order is summarized in the inset; follow the arrows, starting at the bottom.

EXAMPLE 5.5 Select the orbital with the lower energy in each pair: (a) $3d$, $4s$; (b) $2p$, $3s$; (c) $3d$, $4p$.

ANSWER

(a) For any orbital in the $3d$ subshell: $n = 3$, $\ell = 2$; $(n + \ell) = 3 + 2 = 5$.

$4s$: $n = 4$, $\ell = 0$; $(n + \ell) = 4 + 0 = 4$.

The $4s$ orbital has a lower energy than the $3d$.

(b) $2p$: $n = 2$, $\ell = 1$; $(n + \ell) = 2 + 1 = 3$

$3s$: $n = 3$, $\ell = 0$; $(n + \ell) = 3 + 0 = 3$

The sums are the same, but n makes a greater contribution to the energy; therefore the $2p$ has a lower energy than the $3s$.

(c) $3d$: $n = 3$, $\ell = 2$; $(n + \ell) = 5$

$4p$: $n = 4$, $\ell = 1$; $(n + \ell) = 5$

$3d$ has a lower energy than $4p$. ■

PROBLEM 5.10 Select the orbital with the lower energy in each of these pairs: (a) $5p$, $6s$; (b) $4d$, $5s$. □

The order of atomic orbitals shown in Figure 5.17 is very useful in determining the distribution of electrons among the available atomic orbitals. This distribution is known as the **electron configuration** or the **electronic structure** of a gaseous atom. The electron configuration with the *lowest* energy is the *ground state*; other configurations are *excited states.*

The $n = 1$ shell consists of one $1s$ orbital. The $n = 2$ shell contains one $2s$ orbital and three $2p$ orbitals; the three $2p$ orbitals have *equal energies.* The $n = 3$ shell contains one $3s$ orbital, three $3p$ orbitals of *equal energy*, and five $3d$ orbitals of *equal energy*.

Wolfgang Pauli, 1900–1958

Before we can assign electrons to the various orbitals, we must know how many electrons can be accommodated in each orbital. From his study of atomic spectra and the periodic table (Chapter 6), Wolfgang Pauli discovered (1925) a fundamental law, referred to as the **Pauli exclusion principle**: *No two electrons in an atom can have the same four quantum numbers.* Recall that an atomic orbital is defined by the three quantum numbers, n, ℓ, and m_ℓ. However, the complete description of an electron requires the fourth quantum number, m_s, which can have only two values ($+\frac{1}{2}$ and $-\frac{1}{2}$). Therefore, *an atomic orbital can accommodate a maximum of 2 electrons.* A $3s$ orbital, for example, can accommodate only 2 electrons. Each electron in this orbital is described by four quantum numbers:

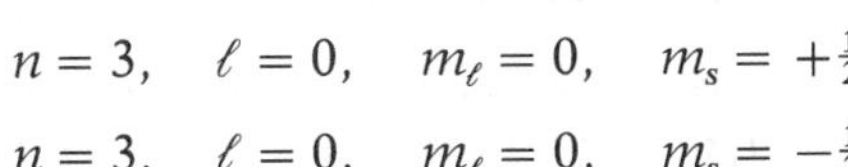

$$n = 3, \quad \ell = 0, \quad m_\ell = 0, \quad m_s = +\tfrac{1}{2}$$

$$n = 3, \quad \ell = 0, \quad m_\ell = 0, \quad m_s = -\tfrac{1}{2}$$

It follows that:

One s orbital can accommodate 2 electrons.

Three p orbitals can accommodate a total of 6 electrons.

Five d orbitals can accommodate a total of 10 electrons.

Seven f orbitals can accommodate a total of 14 electrons.

In summary:

SHELL	MAXIMUM NUMBER OF ELECTRONS IN SUBSHELLS				MAXIMUM NUMBER OF ELECTRONS IN SHELL
n	s	p	d	f	$2n^2$
1	2				2
2	2	6			8
3	2	6	10		18
4	2	6	10	14	32

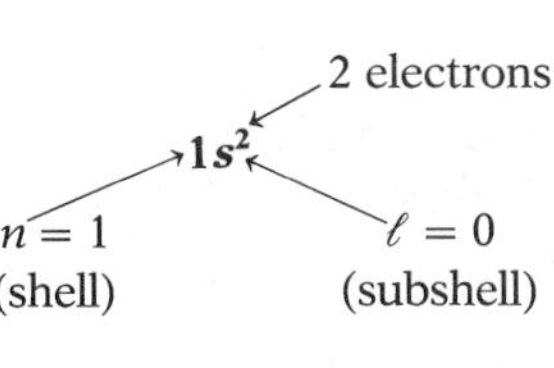

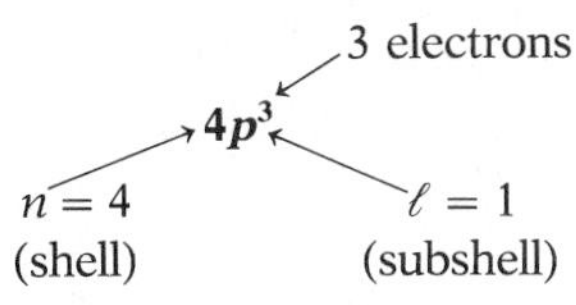

The symbol $1s^2$ represents 2 electrons for which $n = 1$ and $\ell = 0$. The symbol $4p^3$ represents 3 electrons for which $n = 4$ and $\ell = 1$. The symbol $3d^4$ represents 4 electrons for which $n = 3$ and $\ell = 2$.

EXAMPLE 5.6 Select the impossible cases from the following symbols: $6s^2$, $6s^3$, $3p^5$, $3p^7$, $3d^9$, $3d^{11}$.

ANSWER The impossible cases are $6s^3$, $3p^7$, $3d^{11}$. Independent of the n value, an s orbital can house only 2 electrons, three p orbitals can house only 6 electrons, and five d orbitals can house only 10 electrons. ■

PROBLEM 5.11 Which of the following symbols show an impossible number of electrons per subshell? $2p^6$, $4p^8$, $1s^1$, $1s^2$, $2s^3$, $5d^{10}$, $6d^{12}$, $3d^0$. □

WRITING ELECTRON CONFIGURATIONS

We are now ready to distribute the electrons in the ground states of the atoms. The basis of the distribution is that

Known as the *aufbau* (German, "building up") principle.

(1) the number of electrons in an atom is equal to the atomic number;
(2) added electrons enter the orbitals in the order of increasing energy (Figure 5.17); and
(3) an orbital cannot hold more than 2 electrons.

- $_1$H The electron occupies the $1s$ orbital; the electron configuration of hydrogen is therefore $1s^1$.

- $_2$He with 2 electrons Both can go into the $1s$ orbital; the configuration of helium is thus $1s^2$. Since the $1s$ orbital can hold only 2 electrons, the $1s$ subshell is now filled or "closed."

- $_3$Li with 3 electrons Since the $1s$ orbital can hold only 2 electrons, we put 2 electrons in $1s$ and the third one in the next available orbital, namely $2s$. The configuration of lithium is therefore $1s^22s^1$.

- $_4$Be with 4 electrons Since the $2s$ orbital can take 2 electrons, we put the fourth electron into it, yielding $1s^22s^2$ as the configuration of beryllium. This closes the $2s$ subshell.

- $_5$B with 5 electrons The fifth is put into the next available subshell, $2p$. The configuration of boron is therefore $1s^22s^22p^1$.

- $_6$C with 6 electrons Since the $2p$ subshell can take 6 electrons, we put the sixth electron into $2p$ and the configuration of carbon is $1s^22s^22p^2$. The same reasoning gives the following configurations:

  $_7$N $1s^22s^22p^3$ $_8$O $1s^22s^22p^4$
  $_9$F $1s^22s^22p^5$ $_{10}$Ne $1s^22s^22p^6$

  With $_{10}$Ne, the $2p$ subshell is closed.

- $_{11}$Na with 11 electrons Since the $2p$ subshell is closed, the eleventh electron is put into the next available subshell, $3s$, yielding $1s^22s^22p^63s^1$ for sodium.

  $_{13}$Al $1s^22s^22p^63s^23p^1$
  $_{18}$Ar $1s^22s^22p^63s^23p^6$ closes the $3p$ subshell.

EXAMPLE 5.7 Write the electron configurations of $_{19}$K and $_{20}$Ca.

ANSWER $_{18}$Ar with 18 electrons, $1s^22s^22p^63s^23p^6$, closes all orbitals through the $3p$ subshell. The next available subshell is $4s$. (Recall that the $4s$ orbital has a lower energy than the $3d$.) The configurations are therefore

$_{19}$K $1s^22s^22p^63s^23p^64s^1$ $_{20}$Ca $1s^22s^22p^63s^23p^64s^2$ ■

EXAMPLE 5.8 Write the electron configuration of $_{21}$Sc, $_{25}$Mn, and $_{30}$Zn.

ANSWER $_{20}$Ca with 20 electrons closes the $4s^2$ subshell. The next available subshell is the $3d$, which can take 10 electrons. The configurations are therefore

$_{21}$Sc $1s^22s^22p^63s^23p^64s^23d^1$

$_{25}$Mn $1s^22s^22p^63s^23p^64s^23d^5$

$_{30}$Zn $1s^22s^22p^63s^23p^64s^23d^{10}$ ■

A *shorthand method* of writing electron configurations of atoms uses [He core] or [He] to represent the $1s^2$ configuration of $_2$He; [Ne core] or [Ne] to represent the $1s^22s^22p^6$ configuration of $_{10}$Ne; [Ar core] or [Ar] to represent the $1s^22s^22p^63s^23p^6$ configuration of $_{18}$Ar; and so on. Thus we may write

$_{12}$Mg [Ne]$3s^2$ $_{21}$Sc [Ar]$4s^23d^1$ $_{30}$Zn [Ar]$4s^23d^{10}$

Notice that in the shorthand form, the atomic number of the core plus the sum of the superscripts should equal the atomic number of the element. Also notice that each core represents a noble gas at the end of a period in the periodic table.

PROBLEM 5.12 Write the electron configurations of $_{15}$P and $_{22}$Ti and then rewrite them in the shorthand form. □

Electron configurations may be assigned to the remaining elements by similar reasoning (Table 5.2).

DISTRIBUTION OF ELECTRONS IN A SUBSHELL

Thus far, we have asked only how many electrons are in each subshell. The next questions we must ask are, How are the electrons distributed among the orbitals in the subshell? and How are the spins of the two electrons in an orbital shown?

In an *orbital diagram,* a dash is used to represent an orbital and an arrow over the dash represents an electron. One arrow points up (↑) to represent one spin orientation. A second arrow points down (↓) to represent the other spin orientation. The electron configuration of beryllium is represented as $_4\text{Be}\ \underset{1s}{\underline{\uparrow\downarrow}}\ \underset{2s}{\underline{\uparrow\downarrow}}$. There is only one way in which the electrons can be assigned to the orbitals in $_4$Be. Two electrons having opposite spins in the same orbital (↑↓) are said to be **paired**. The next electron in $_5$B $1s^22s^22p^1$ enters a p orbital. Since all three orientations of the p orbitals are equivalent, it makes no difference which p orbital is used in $_5$B. With carbon, however, the order of occupancy is significant: do the $2p^2$ electrons occupy one or two orbitals?

Named in honor of Friedrich Hund for his work with atomic and molecular spectra.

The question is answered by **Hund's rule**, which applies to atoms, ions, and molecules: *Each of the orbitals of the same energy must house one electron before pairing occurs.* Then, the order of occupancy of three $2p$ orbitals is

	1s	2s	2p		1s	2s	2p
$_5$B	↑↓	↑↓	↑ _ _	$_6$C	↑↓	↑↓	↑ ↑ _
$_7$N	↑↓	↑↓	↑ ↑ ↑	$_8$O	↑↓	↑↓	↑↓ ↑ ↑
$_9$F	↑↓	↑↓	↑↓ ↑↓ ↑	$_{10}$Ne	↑↓	↑↓	↑↓ ↑↓ ↑↓

TABLE 5.2 GROUND-STATE ELECTRONIC STRUCTURES OF GASEOUS ATOMS OF ELEMENTS

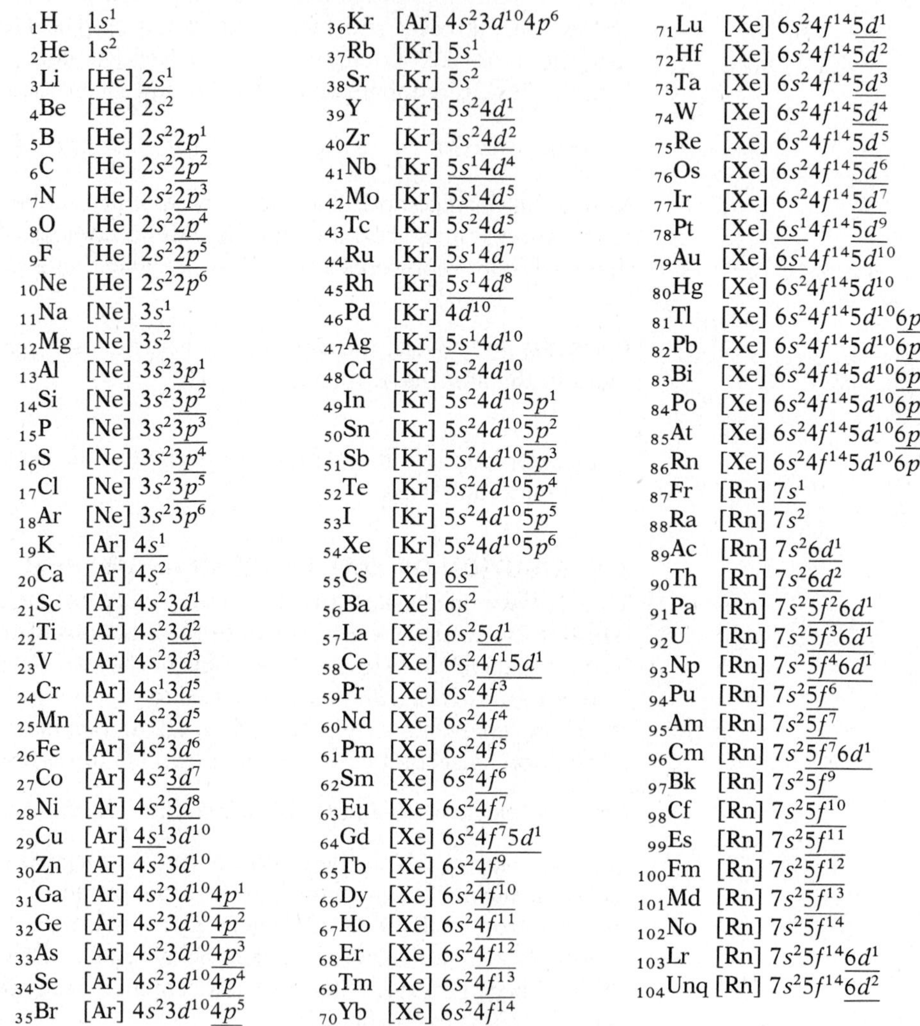

Abbreviations:
[He] = [He core] = $1s^2$
[Ne] = [Ne core] = [He] $2s^22p^6$
[Ar] = [Ar core] = [Ne] $3s^23p^6$
[Kr] = [Kr core] = [Ar] $4s^23d^{10}4p^6$
[Xe] = [Xe core] = [Kr] $5s^24d^{10}5p^6$
[Rn] = [Rn core] = [Xe] $6s^24f^{14}5d^{10}6p^6$

Note: Underlining indicates partly filled subshells.

Element	Configuration	Element	Configuration	Element	Configuration
$_{1}$H	$\underline{1s^1}$	$_{36}$Kr	[Ar] $4s^23d^{10}4p^6$	$_{71}$Lu	[Xe] $6s^24f^{14}\underline{5d^1}$
$_{2}$He	$1s^2$	$_{37}$Rb	[Kr] $\underline{5s^1}$	$_{72}$Hf	[Xe] $6s^24f^{14}\underline{5d^2}$
$_{3}$Li	[He] $\underline{2s^1}$	$_{38}$Sr	[Kr] $5s^2$	$_{73}$Ta	[Xe] $6s^24f^{14}\underline{5d^3}$
$_{4}$Be	[He] $2s^2$	$_{39}$Y	[Kr] $5s^2\underline{4d^1}$	$_{74}$W	[Xe] $6s^24f^{14}\underline{5d^4}$
$_{5}$B	[He] $2s^2\underline{2p^1}$	$_{40}$Zr	[Kr] $5s^2\underline{4d^2}$	$_{75}$Re	[Xe] $6s^24f^{14}\underline{5d^5}$
$_{6}$C	[He] $2s^2\underline{2p^2}$	$_{41}$Nb	[Kr] $\underline{5s^14d^4}$	$_{76}$Os	[Xe] $6s^24f^{14}\underline{5d^6}$
$_{7}$N	[He] $2s^2\underline{2p^3}$	$_{42}$Mo	[Kr] $\underline{5s^14d^5}$	$_{77}$Ir	[Xe] $6s^24f^{14}\underline{5d^7}$
$_{8}$O	[He] $2s^2\underline{2p^4}$	$_{43}$Tc	[Kr] $5s^2\underline{4d^5}$	$_{78}$Pt	[Xe] $\underline{6s^1}4f^{14}\underline{5d^9}$
$_{9}$F	[He] $2s^2\underline{2p^5}$	$_{44}$Ru	[Kr] $\underline{5s^14d^7}$	$_{79}$Au	[Xe] $\underline{6s^1}4f^{14}5d^{10}$
$_{10}$Ne	[He] $2s^22p^6$	$_{45}$Rh	[Kr] $\underline{5s^14d^8}$	$_{80}$Hg	[Xe] $6s^24f^{14}5d^{10}$
$_{11}$Na	[Ne] $\underline{3s^1}$	$_{46}$Pd	[Kr] $4d^{10}$	$_{81}$Tl	[Xe] $6s^24f^{14}5d^{10}\underline{6p^1}$
$_{12}$Mg	[Ne] $3s^2$	$_{47}$Ag	[Kr] $\underline{5s^1}4d^{10}$	$_{82}$Pb	[Xe] $6s^24f^{14}5d^{10}\underline{6p^2}$
$_{13}$Al	[Ne] $3s^2\underline{3p^1}$	$_{48}$Cd	[Kr] $5s^24d^{10}$	$_{83}$Bi	[Xe] $6s^24f^{14}5d^{10}\underline{6p^3}$
$_{14}$Si	[Ne] $3s^2\underline{3p^2}$	$_{49}$In	[Kr] $5s^24d^{10}\underline{5p^1}$	$_{84}$Po	[Xe] $6s^24f^{14}5d^{10}\underline{6p^4}$
$_{15}$P	[Ne] $3s^2\underline{3p^3}$	$_{50}$Sn	[Kr] $5s^24d^{10}\underline{5p^2}$	$_{85}$At	[Xe] $6s^24f^{14}5d^{10}\underline{6p^5}$
$_{16}$S	[Ne] $3s^2\underline{3p^4}$	$_{51}$Sb	[Kr] $5s^24d^{10}\underline{5p^3}$	$_{86}$Rn	[Xe] $6s^24f^{14}5d^{10}6p^6$
$_{17}$Cl	[Ne] $3s^2\underline{3p^5}$	$_{52}$Te	[Kr] $5s^24d^{10}\underline{5p^4}$	$_{87}$Fr	[Rn] $\underline{7s^1}$
$_{18}$Ar	[Ne] $3s^23p^6$	$_{53}$I	[Kr] $5s^24d^{10}\underline{5p^5}$	$_{88}$Ra	[Rn] $7s^2$
$_{19}$K	[Ar] $\underline{4s^1}$	$_{54}$Xe	[Kr] $5s^24d^{10}5p^6$	$_{89}$Ac	[Rn] $7s^2\underline{6d^1}$
$_{20}$Ca	[Ar] $4s^2$	$_{55}$Cs	[Xe] $\underline{6s^1}$	$_{90}$Th	[Rn] $7s^2\underline{6d^2}$
$_{21}$Sc	[Ar] $4s^2\underline{3d^1}$	$_{56}$Ba	[Xe] $6s^2$	$_{91}$Pa	[Rn] $7s^2\underline{5f^26d^1}$
$_{22}$Ti	[Ar] $4s^2\underline{3d^2}$	$_{57}$La	[Xe] $6s^2\underline{5d^1}$	$_{92}$U	[Rn] $7s^2\underline{5f^36d^1}$
$_{23}$V	[Ar] $4s^2\underline{3d^3}$	$_{58}$Ce	[Xe] $6s^2\underline{4f^15d^1}$	$_{93}$Np	[Rn] $7s^2\underline{5f^46d^1}$
$_{24}$Cr	[Ar] $\underline{4s^13d^5}$	$_{59}$Pr	[Xe] $6s^2\underline{4f^3}$	$_{94}$Pu	[Rn] $7s^2\underline{5f^6}$
$_{25}$Mn	[Ar] $4s^2\underline{3d^5}$	$_{60}$Nd	[Xe] $6s^2\underline{4f^4}$	$_{95}$Am	[Rn] $7s^2\underline{5f^7}$
$_{26}$Fe	[Ar] $4s^2\underline{3d^6}$	$_{61}$Pm	[Xe] $6s^2\underline{4f^5}$	$_{96}$Cm	[Rn] $7s^2\underline{5f^76d^1}$
$_{27}$Co	[Ar] $4s^2\underline{3d^7}$	$_{62}$Sm	[Xe] $6s^2\underline{4f^6}$	$_{97}$Bk	[Rn] $7s^2\underline{5f^9}$
$_{28}$Ni	[Ar] $4s^2\underline{3d^8}$	$_{63}$Eu	[Xe] $6s^2\underline{4f^7}$	$_{98}$Cf	[Rn] $7s^2\underline{5f^{10}}$
$_{29}$Cu	[Ar] $\underline{4s^1}3d^{10}$	$_{64}$Gd	[Xe] $6s^2\underline{4f^75d^1}$	$_{99}$Es	[Rn] $7s^2\underline{5f^{11}}$
$_{30}$Zn	[Ar] $4s^23d^{10}$	$_{65}$Tb	[Xe] $6s^2\underline{4f^9}$	$_{100}$Fm	[Rn] $7s^2\underline{5f^{12}}$
$_{31}$Ga	[Ar] $4s^23d^{10}\underline{4p^1}$	$_{66}$Dy	[Xe] $6s^2\underline{4f^{10}}$	$_{101}$Md	[Rn] $7s^2\underline{5f^{13}}$
$_{32}$Ge	[Ar] $4s^23d^{10}\underline{4p^2}$	$_{67}$Ho	[Xe] $6s^2\underline{4f^{11}}$	$_{102}$No	[Rn] $7s^25f^{14}$
$_{33}$As	[Ar] $4s^23d^{10}\underline{4p^3}$	$_{68}$Er	[Xe] $6s^2\underline{4f^{12}}$	$_{103}$Lr	[Rn] $7s^25f^{14}\underline{6d^1}$
$_{34}$Se	[Ar] $4s^23d^{10}\underline{4p^4}$	$_{69}$Tm	[Xe] $6s^2\underline{4f^{13}}$	$_{104}$Unq	[Rn] $7s^25f^{14}\underline{6d^2}$
$_{35}$Br	[Ar] $4s^23d^{10}\underline{4p^5}$	$_{70}$Yb	[Xe] $6s^24f^{14}$		

For $_{25}$Mn, the order of occupancy is

$_{25}$Mn [Ne]	↑↓	↑↓ ↑↓ ↑↓	↑ ↑ ↑ ↑ ↑	↑↓
	$3s$	$3p$	$3d$	$4s$

A lone electron in an orbital, ↑, is referred to as an **unpaired electron**. For example, $_{25}$Mn has five unpaired electrons. $_{10}$Ne has no unpaired electrons; all electrons are paired. Never write two electrons with the same spin, ↑↑ or ↓↓, in the same orbital. Such representations would violate the Pauli exclusion principle.

The four quantum numbers would be the same for $\frac{\uparrow\uparrow}{1s}$; 1, 0, 0, $+\frac{1}{2}$. What would the quantum numbers be for the incorrect representation, $\frac{\downarrow\downarrow}{2s}$?

PROBLEM 5.13 How many unpaired electrons are in each atom: (a) $_{15}P$, (b) $_{10}Ne$, (c) $_{16}S$, (d) $_{23}V$? □

Hund's rule is most likely associated with repulsion effects between electrons. In carbon atoms, for example, repulsion is greater when the $2p$ electrons are in the same orbital than when they are in separate orbitals. Consequently, a half-filled subshell (each orbital containing one electron) is a comparatively stable structure. The fourth electron of a p^4 set is removed more easily than an electron of a p^3 set.

Although orbital energies depend on the atomic number, the order given in Figure 5.17 closely parallels the order of stability obtained from experimentally determined properties of the elements. However, some of the predicted electron configurations do not agree with the experimentally assigned configurations. For example, the predicted electron configurations for the transition elements $_{24}Cr$ [Ar]$4s^2 3d^4$ and $_{29}Cu$ [Ar]$4s^2 3d^9$ do not agree with the experimentally assigned configurations, $_{24}Cr$ [Ar]$4s^1 3d^5$ and $_{29}Cu$ [Ar]$4s^1 3d^{10}$. These irregularities are associated with the greater stability of half-filled (d^5) and completely filled (d^{10}) subshells. In fact, the atoms of the noble gases are characterized by completely filled subshells. A unique filling order valid for all elements does not exist.

It is not necessary to memorize exceptions to the filling order (Figure 5.17).

Similar properties of elements occur periodically with increasing atomic number (section 2.7). In the next chapter we will learn that this periodic variation is best understood in terms of the variation of the electron configuration of atoms with increasing atomic number.

It is also customary to write the electron configurations in order of increasing n values, such as $_{30}Zn$ [Ar]$3d^{10}4s^2$.

BOX 5.4

THE 4s VERSUS 3d BATTLE OF ELECTROSTATICS

It is not surprising that many students ask, "If the $4s$ orbital is lower in energy than the $3d$, why do the atoms of the transition elements lose the $4s$ electrons more easily?" The circumstances of this are complex.

Look at the situation this way: in writing the electron configurations of many-electron atoms, we are adding electrons and protons to the hydrogen atom. And we assume that the added electron does not interact with the electrons already present. This assumption leads to the correct configuration (in most cases), but each added electron *does interact* with the nucleus and all other electrons. It just becomes another member of the electron mob, indistinguishable from any other electron. Nature does not label electrons; they are labeled "$3d$" or "$4s$" only for our bookkeeping purposes.

However, ionization converts a *neutral atom* into an *ion, changing the interactions* among the remaining electrons and nucleus. Therefore, the question is not "Which electron is lost," but rather, "What is the configuration of the resulting ion?" Examination of the isoelectronic (same number of electrons) species,

$_{20}Ca$	[Ar]$4s^2$
$_{21}Sc^+$	[Ar]$4s^1 3d^1$
$_{22}Ti^{2+}$	[Ar]$3d^2$

shows that protons are added to the nucleus while the number of electrons remains constant. Consequently, *nuclear attraction* becomes more important in comparison to electron repulsion. This is the same as saying that the orbitals in positively charged ions become more *hydrogenlike* where nuclear attraction is the only force and n alone determines the filling order (the n rule).

Theory, however, cannot always be simple *and exact* at the same time, and the n rule is not always reliable. The ionization of $_{39}Y$ [Kr]$5s^2 4d^1$ yields $_{39}Y^+$ [Kr]$5s^2$.

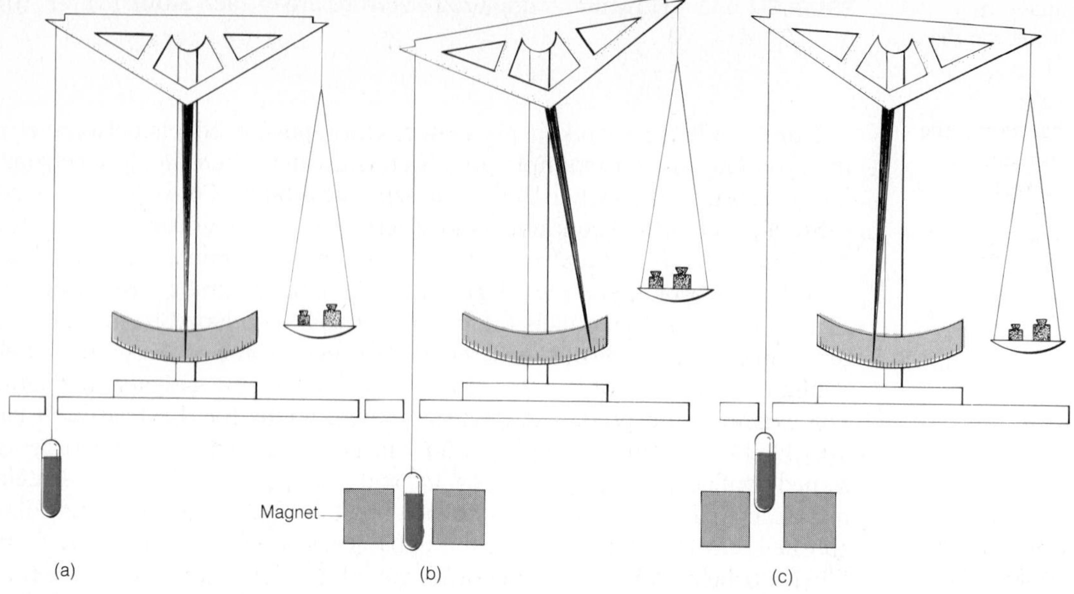

FIGURE 5.18
A weighed sample (a) in the absence of a magnetic field is (b) attracted into the magnetic field and appears to have gained weight if it is paramagnetic, and (c) repelled and appears to have lost weight if it is diamagnetic.

FORMATION OF POSITIVE IONS

In the *ionization process*—by which positive ions are formed when electrons are removed from atoms—*the electrons with the highest value of n are removed first* (Box 5.4). It is important to remember this when writing electron configurations of ions of elements in which *d* orbitals are occupied. The electron configurations of S, S^{2+}, Ti, and Ti^{2+} are illustrative:

$$_{16}S \quad 1s^2 2s^2 2p^6 3s^2 3p^4 \qquad _{16}S^{2+} \quad 1s^2 2s^2 2p^6 3s^2 3p^2$$
$$_{22}Ti \quad 1s^2 2s^2 2p^6 3s^2 3p^6 4s^2 3d^2 \qquad _{22}Ti^{2+} \quad 1s^2 2s^2 2p^6 3s^2 3p^6 4s^0 3d^2$$

PROBLEM 5.14 How many unpaired electrons are in (a) $_{26}Fe$, (b) $_{26}Fe^{2+}$, (c) $_{26}Fe^{5+}$, (d) $_{21}Sc^{2+}$? □

5.8 PARAMAGNETISM

Metallic substances, like iron, that are very strongly attracted to magnets and that tend to remain magnetized are *ferromagnetic.* Ferromagnetism is about a million times stronger than paramagnetism.

Atomic sodium, atomic hydrogen, molecular oxygen, and chromium are attracted into magnetic fields (Figure 5.18b). They are typical **paramagnetic** substances. *Paramagnetism is associated with unpaired electrons in atoms, ions, and molecules.* These electrons possess permanent magnetic moments and behave like microscopic magnets. They are attracted into a magnetic field, just like the magnetic needle of a compass is. Experimentally, the attraction of isolated atoms into the field is related to the number of unpaired electrons obtained from the orbital diagrams of atoms.†

† The contribution of the motion of an unpaired electron about the nucleus to the magnetic moment can usually be ignored. The contribution of nuclei with unpaired nucleons (Chapter 23) is insignificantly small, but the magnetic moment of nuclei are of paramount importance in the nuclear magnetic imaging technique used in diagnostic medicine (Chapter 23).

Diamagnetism is not associated with electron spin. All substances (except H atoms) are slightly repelled because the motion of *paired electrons* in atoms or molecules generates a magnetic field that opposes the applied external field. Unpaired electrons easily overcome the repulsion effect of paired electrons.

Substances like helium, without unpaired electrons, are **diamagnetic**. They are weakly repelled by a magnetic field (Figure 5.18c). The observation that paired electrons are not paramagnetic is explained as follows: when two electrons occupy the same orbital, they spin in opposite directions so that the paramagnetism of one electron, ↑, is canceled by the paramagnetism of the other electron, ↓. Diamagnetism is associated with paired electrons.

PROBLEM 5.15 (a) Which of these particles are paramagnetic? (1) ${}_{18}Ar$, (2) ${}_{9}F$, (3) ${}_{4}Be$, (4) ${}_{23}V$, (5) ${}_{30}Zn^{2+}$. (b) If the electron spins for ${}_{2}He$ were written as $\frac{\uparrow\uparrow}{1s^2}$, would this representation agree with experimental results? □

The paramagnetic properties of O_2 are used by anesthesiologists to monitor the concentration of oxygen in blood.

Magnetic recording tape is an interesting application of the magnetic properties of atoms. The tape contains very small, needle-shaped particles of CrO_2, which are very strongly attracted to magnets. When a tape is passed through a magnetic field generated by an electric current, the particles, like midget compasses, are oriented. The orientation depends on the strength of the magnetic field, which is determined by the variable electric current corresponding to the original sound.

Later chapters discuss the role of unpaired electrons in determining the magnetic properties of elements and the kind of chemical bonds in their compounds.

Finally, although most scientists regard quantum mechanics as a beautifully finished product from the *practical* point of view, many scientists find it unsatisfactory from a philosophical point of view. "This confusion between what things are in themselves and what we can know of them is quite an elementary confusion."†

SUMMARY

Visible light is a very small portion of a broad range of frequencies or radiation that includes X rays, visible light, and radio waves. The speed (c) of any radiation is the product of its **wavelength** (λ) and **frequency** (ν):

$$c = \lambda \times \nu = \text{meter (m)} \times \text{hertz}\left(\frac{1}{\text{sec}}\right) = \text{m/sec}$$

Radiation, like matter and electricity, is quantized; it consists of a stream of particles called **photons**. The energy of a photon is given by

$$E = h \times \nu$$

where h is *Planck's constant*, 6.6261×10^{-34} J sec/particle, and ν is the frequency of the radiation. The ejection of electrons from the surface of a metal by the absorption of light (photons) is known as the **photoelectric effect**.

The **spectrum** of the radiation from the sun is **continuous**; it consists of a continuous range of frequencies, as does a rainbow. The spectrum of the light from an atomic vapor such as a sodium vapor lamp or a neon sign is an **emission line spectrum**; it consists of only a relatively small number of colored lines, specific

† J. M. Jauch, *Are Quanta Real? A Galilean Dialogue*, Simplicio, Indiana University Press, Bloomington, Indiana, 1973, p. 54.

and characteristic frequencies on a black background. Atoms not only *emit* characteristic frequencies of light when excited in a discharge tube; they also *absorb* light of the same characteristic frequencies. When white light is passed through an atomic vapor, an **absorption line spectrum** is seen as black lines in a "rainbow" background.

Line spectra indicate that only photons of definite energies are emitted or absorbed by atoms. This means that the energy of the electron in atoms is quantized; the electron is *restricted* to certain energies referred to as **energy levels** in an atom. These energy values are fixed by one to four *quantum numbers* (dimensionless numbers), depending on the environment of the atom and the precision of the instrumentation. The major contribution is made by the **principal quantum number**, n, so that the energy of the electron in the H atom can take only the values given by

$$E_n = \frac{-2.179 \times 10^{-18}\ \text{J}}{n^2}$$

These values correspond to the principal energy levels (shells) in the H atom; n can only be an integer—1, 2, 3, 4, ...

The *emission* of a photon corresponds to an electron transition from a *higher energy level* to a *lower energy level*. The *absorption* of a photon corresponds to an electron transition from a *lower energy level* to a *higher energy level*.

$$\underset{\text{(energy of photon emitted or absorbed)}}{\Delta E} = E_{\text{final level}} - E_{\text{initial level}}$$

The **ground state** of an atom corresponds to the lowest energy level or the most stable state. All higher energy levels are called **excited states**.

The mathematical model of atoms also leads to the concept that electrons are distributed in the space about the positive nucleus. The same model predicts that if you pinpoint the position of an electron you will lose track of its energy, and if you fix its energy you will blur its position. Since the energy of electrons in atoms can be measured very accurately, no attempt is made to assign definite paths to electrons in atoms. For a definite energy, however, the model does give the density of electronic charge, an **electron cloud density**, or the probability of finding an electron at a given distance from the nucleus. The cloud is most dense when the probability of finding the electron is high and fades out when the probability is low. The point at which the electron density is zero is called a *node*. Three-dimensional plots of these probabilities have definite shapes called **atomic orbitals**, which are determined by quantum numbers. An orbital is described by the principal quantum number, n, and a letter, s, p, d, f, . . . ; n not only makes the largest contribution to the energy of an orbital, it also determines the size of the orbitals.

The *quantum number*, ℓ, determines the shape of an orbital. The letters s, p, d, f, . . . correspond to $\ell = 0, 1, 2, 3, \ldots$.

The *magnetic quantum number*, m_ℓ, determines the spatial orientation of the orbital. The possible values of m_ℓ are fixed by the value of ℓ.

Atomic orbitals having a given value of n constitute a **shell**. Each shell consists of **subshells** whose energies are closely spaced. The number of subshells in a shell is fixed by the value of n, and the energy of a subshell is fixed by the sum of $(n + \ell)$. The number of orbitals in a subshell is fixed by the number of m_ℓ values. An orbital may have no more than 2 electrons:

n	ℓ (n values of ℓ; 0 to $n-1$)	m_ℓ ($+\ell$–0–$-\ell$)	SUBSHELL DESIGNATION	ORBITALS IN SUBSHELL ($2\ell + 1$)	MAXIMUM ELECTRONS IN SUBSHELL	ORBITAL ENERGY ($n + \ell$)
2	0	0	$2s$	1	2	$(2 + 0) = 2$
	1	$+1, 0, -1$	$2p$	3	6	$(2 + 1) = 3$

The *spin quantum number*, m_s, takes only two values, $+\frac{1}{2}$ and $-\frac{1}{2}$. Substances with unpaired electrons are drawn into a magnetic field; they are **paramagnetic**. Substances with paired electrons are repelled from a magnetic field; they are **diamagnetic**.

The ground state **electron configuration** of atoms having more than one electron can be built up by using orbitals similar to the H atom orbitals as follows:

- The atomic number fixes the number of electrons in an atom.
- Each added electron goes into the orbitals in the order of increasing energy, usually the order of increasing $(n + \ell)$.
- An orbital cannot accommodate more than 2 electrons (**Pauli exclusion principle**).
- Each orbital in a given subshell houses a single electron before electrons are paired (**Hund's rule**). All electrons in singly occupied orbitals have their spins in the same direction.

5.9 ADDITIONAL PROBLEMS

λ AND ν

5.16 Microwaves are reflected by metals and pass through glass, paper, and plastic, but they are absorbed by foods. This absorption causes water molecules to rotate and so produce the heat that cooks foods in microwave ovens. A microwave has a wavelength of 10 cm. Find its frequency.

5.17 *Voyager* detected a transition of carbon monoxide, CO, at 115.271 GHz (gigahertz) on Titan, a satellite of Saturn. Find the wavelength of this transition.

5.18 (a) The SI unit of length is now defined as the distance traveled by light in a vacuum in (1/299,792,458) sec. The velocity of light in a vacuum is 299,792,458 m/sec. Find the SI unit of length and the frequency of a wave whose length is equal to the SI unit of length. (b) Find ΔE in kJ/mol for the electron transition that emits such a spectral line.

QUANTUM THEORY

5.19 The human eye receives a 2.500×10^{-14} J-signal consisting of photons of blue light, $\lambda = 470$ nm. How many photons reach the eye?

5.20 A Ne atom in the ground state requires 2090 kJ/mol for ionization; $Ne(g) \longrightarrow Ne^+(g) + e^-$. Find the wavelength of the photon that will ionize a Ne atom.

5.21 Use Figure 5.9 to find the work in joules per H atom and kilojoules per mole of H atoms needed to remove the electron in the H atom to infinity from the excited state $n = 5$.

SPECTRUM

5.22 Find the energy of the photons corresponding to the red line, 657.3 nm, in the spectrum of the Ca atom.

5.23 Hydrogen atoms absorb energy so that the electrons are excited to the energy level $n = 7$. Electrons then undergo these transitions: (1) $n = 7 \longrightarrow n = 1$, (2) $n = 7 \longrightarrow n = 6$, (3) $n = 2 \longrightarrow n = 1$. Which transition will produce the photon with (a) the smallest energy; (b) the highest frequency; (c) the shortest wavelength? (d) What is the frequency of a photon resulting from the transition $n = 6 \longrightarrow n = 1$?

5.24 Calculate the shortest-wavelength photon that an H atom can emit.

5.25 Predict the minimum kinetic energy that an electron must possess so that, upon collision with a mercury atom, it causes the appearance of the intense ultraviolet line (dangerous to the eye) in the mercury spectrum, $\lambda = 253.7$ nm.[†]

5.26 A laser[§] puts out 6.0×10^6 J/sec. The specific heat of solid and liquid iron averages 0.50 J/(g °C). It melts at 1530°C and boils at 3000°C. How many seconds are required for the laser to raise the temperature of 100 g Fe to its boiling point from 25°C? 28 kJ are needed to melt the iron. 80% of the laser output is absorbed by the sample.

5.27 Five energy levels of the He atom are given in J/atom: (1) 6.000×10^{-19}, (2) 8.812×10^{-19}, (3) 9.381×10^{-19}, (4) 10.443×10^{-19}, (5) 10.934×10^{-19}. Construct an energy-level diagram for He and find the energy of the photon (a) absorbed for the electron transition from level 1 to level 4 and (b) emitted for the electron transition from level 5 to level 2.

5.28 The deep red line, 670.8 nm, in the Li spectrum, the basis of its flame test, results from a transition between the ground state and the first excited state. (a) Does this line represent the minimum energy that a Li atom in the ground state can absorb? (b) Find the energy of the wallop a Li atom must receive from another particle in the flame for this line to show up. (c) If the particle is a LiCl(g) molecule, find the minimum velocity of the particle required to yield a positive Li flame test.

5.29 Hydrogen atoms in the ground state are subjected to collisions with electrons having a maximum kinetic energy of 2.029×10^{-18} J per electron. What is the maximum number of lines in the spectrum emitted by the H atoms? Use Figure 5.9.

5.30 A theoretical analysis of the Rydberg constant shows that it is dependent upon the mass of a nucleus even for the same atomic number. This, in fact, was the difference Harold Urey used in 1931 to identify and so discover deuterium, (D, ^{2}H). Calculate the energy and frequency of the spectral line resulting from the electron transition $n = 5 \longrightarrow n = 2$ for the ^{1}H and D atoms. Use

$$E_H = \frac{-2.17872 \times 10^{-18}\ \text{J}}{n^2}, \quad E_D = \frac{-2.17931 \times 10^{-18}\ \text{J}}{n^2},$$

$$\text{and } h = 6.626076 \times 10^{-34}\ \frac{\text{J sec}}{\text{particle}}$$

What difference in wavelength, in nm, was Urey looking for in his spectroscope? Use $c = 2.997925 \times 10^8$ m/sec.

5.31 The energy of a hydrogenlike atom or ion such as $^2_1$H, $^4_1$H, $^4_2$He^{1+}, $_3$Li^{2+}, or $^{16}_8$O^{7+} is given by

$$E_n = \frac{-RZ^2}{n^2}$$

where R is the Rydberg constant expressed in units of energy, 2.179×10^{-18} J, and Z is the atomic number of the species. (a) Calculate the energy and wavelength of the spectral line of $_4$Be^{3+} for the $n = 2$ to $n = 1$ transition and compare them with those of the same transition in the ^{1}H atom (Figure 5.10). (b) Compare the energy required to remove the last electron of $_6$C^{5+} in the ground state with the energy needed to ionize a ^{1}H atom in the ground state.

QUANTUM NUMBERS

5.32 What is the maximum number of electrons that can be housed in subshells *s*, *p*, *d*, and *f*?

5.33 Write the quantum numbers of each electron in an atom of $^{12}_6$C and $^{13}_6$C.

5.34 (a) List the quantum numbers of each electron in (1) a $_9$F atom, (2) a $_{15}$P atom, (3) a $_{25}$Mn atom. (b) How many unpaired electrons are in each of these atoms?

5.35 (a) Write the possible values of ℓ when $n = 5$. (b) Write the allowed number of orbitals (1) with the quantum numbers $n = 4$, $\ell = 3$; (2) with the quantum number $n = 4$; (3) with the quantum numbers $n = 7$, $\ell = 6$, $m_\ell = 6$; (4) with quantum numbers $n = 6$, $\ell = 5$.

5.36 Write the total number of electrons that can have the given quantum number(s) or designation in an atom (some answers may be zero): (a) $n = 4$; (b) 1*p*; (c) 2*p*; (d) 6*d*; (e) 5*d*; (f) 3*d*; (g) $n = 5$, $\ell = 2$; (h) $n = 6$, $\ell = 0$; (i) $n = 7$, $\ell = 2$, $m_\ell = -1$; (j) $n = 7$, $\ell = 2$, $m_\ell = 0$; (k) $n = 6$, $\ell = 4$, $m_\ell = 0$, $m_s = -\frac{1}{2}$; (l) $n = 0$; (m) $n = 7$, $\ell = 5$, $m_\ell = 6$.

5.37 The following incorrect sets of quantum numbers in the order n, ℓ, m_ℓ, m_s are written for paired electrons or for one electron in an orbital. Correct them, assuming n values are correct. (a) 1, 0, 0, $+\frac{1}{2}$, $+\frac{1}{2}$; (b) 2, 2, 1, $\pm\frac{1}{2}$; (c) 3, 2, 3, $\pm\frac{1}{2}$; (d) 3, 1, 2, $+\frac{1}{2}$; (e) 2, 1, -1, 0; (f) 3, 0, -1, $-\frac{1}{2}$.

† This prediction was verified in 1913 by the experiment of James Franck and Gustav Hertz in which electrons of precisely known kinetic energy were passed through mercury vapor at low pressure.

§ In a *laser* (*l*ight *a*mplification by *s*timulated *e*mission of *r*adiation, the theory for which was discovered in 1955 by Charles Townes), one and only one excited state is highly populated, whereas the number of atoms in the ground state is relatively small. Then, *all* of the excited atoms are made to *emit photons* together at the same time. The resulting photon beam concentrates a large number of photons (energy) in a pinpoint beam. A laser beam may be used for surgery on individual organs of a single living cell, or it can carry a "wallop" sufficient to kill a horse or bore a hole in steel.

5.38 Complete this table:

SHELL	SUB-SHELL	ORBITALS PER SUBSHELL	MAX. e^-'s PER SUBSHELL	MAX. e^-'s PER SHELL
n				$2n^2$
1	$1s$	1	2	2
2				
3				
4				

ELECTRON CONFIGURATION

5.39 (a) Write the electron configuration of each atom or ion: H, Al, P, Ca, Ni, Zn, Rb, Fe, Mn^{4+}, Co^{2+}. (b) How many unpaired electrons are in each?

5.40 Write the electron configuration of (a) ${}^1_1H$ and (b) the heavy isotopes ${}^2_1H$ and ${}^5_1H$.

5.41 Write the electron configurations of the Group-1 elements Li, Na, and K (see inside front cover). What similarity do you observe?

5.42 Write the electron configurations of the following: ${}_8O^{2-}$, ${}_9F^-$, ${}_{10}Ne$, ${}_{11}Na^+$, ${}_{12}Mg^{2+}$, ${}_{13}Al^{3+}$. What do they have in common?

5.43 Identify these atoms: (a) $1s^22s^22p^6$, (b) $[Ne]3s^23p^6$, (c) $[Ne]3s^23p^64s^23d^3$, (d) $[Ne]3s^23p^64s^23d^5$, (e) $[Kr]4d^55s^1$, (f) $[Xe]4f^{12}6s^2$.

5.44 (a) Write the electron configuration of He as usual in the ground state. (b) How would you write the configuration of the excited state of an He atom with one electron (1) in the $2s$ subshell? (2) in the $3p$ subshell?

5.45 Write the electron configuration of ${}_{26}Fe$ in an excited state for which the magnetic moment corresponds to 8 unpaired electrons.

5.46 Find the total number of s, p, and d electrons in ${}_{14}Si$, ${}_{18}Ar$, ${}_{28}Ni$, ${}_{30}Zn$, and ${}_{37}Rb$.

5.47 Use the $(n + \ell)$ rule to predict the electron configuration of ${}_{57}La$ and ${}_{71}Lu$. Use Table 5.2 to check your answers.

PARAMAGNETISM

5.48 How many unpaired electrons are in atoms of Na, Ne, B, Be, Se, and Ti?

5.49 How many unpaired electrons are in gaseous Ni; in Ni^{2+}?

5.50 (a) Which ion has 2 unpaired electrons: ${}_{20}Ca^{2+}$, ${}_{14}Si^{2+}$, ${}_{29}Cu^{2+}$, ${}_{21}Sc^{2+}$, ${}_8O^{2+}$. (b) Which of these ions would be attracted into a magnetic field?

ORBITALS

5.51 What are the chances of finding an electron in an area or volume for which the calculated electron density is zero?

5.52 (a) Is the electron density for the H atom in the $1s$ state zero at 0.4 nm from the nucleus? Explain your answer. (b) Is the electron density of a $2p_z$ electron at the point on the x axis 0.05 nm from the nucleus high, low, or zero, compared to the same distance on the z axis?

5.53 How do the spatial distributions of the orbitals in each given set differ? (a) $1s$ and $2p_x$; (b) $2p_x$ and $2p_y$; (c) $2p_y$ and $2p_z$.

QUANTUM MECHANICS

5.54 (a) Has anyone proved that particles of atomic and subatomic dimensions obey Newton's laws of motion? (b) Has anyone proved that an electron lacks structure? (c) Has anyone ever observed (1) light waves, (2) photons with or without the aid of instruments? (d) Complete this statement correctly: Orbital shapes (patterns of electron distribution) are (1) established by experiment, (2) calculated from theory. (e) In studying the trajectory of an ordinary moving pendulum, is it possible to measure the position and the speed of the oscillation accurately at any instant? (f) Do electrons travel around the nucleus in (1) definite circular orbits, (2) elliptical orbits like that of our Earth around the Sun?

5.55 Which of these are observable? (a) Position of electron in H atom; (b) frequency of radiation emitted by H atoms; (c) path of electron in H atom; (d) wave motion of electrons; (e) diffraction patterns produced by electrons; (f) diffraction patterns produced by light; (g) energy required to remove electrons from H atoms; (h) an atom; (i) a molecule; (j) a water wave.

5.56 (a) What is the de Broglie wavelength of a proton moving at a speed of 3.00×10^7 m/sec? The proton mass is 1.67×10^{-24} g. (b) What is the de Broglie wavelength of a stone with a mass of 30.0 g moving at 2.00×10^3 m/h (≈ 100 mi/h)?

5.57 An H atom absorbs 8.72×10^{-18} J, more energy than the H atom requires for the electron to escape from it. Find the de Broglie wavelength of the emitted electron. Electron mass $= 9.11 \times 10^{-28}$ g.

5.58 Use these results for the hydrogen atom in the $2s$ state to plot ψ^2_{2s} (y axis) versus r (x axis):

r, cm	ψ_{2s}	ψ^2_{2s}, cm^{-3}
0.0	5.20×10^{11}	27.04×10^{22}
0.5×10^{-8}	1.70×10^{11}	2.89×10^{22}
1.0×10^{-8}	0.10×10^{11}	0.010×10^{22}
1.06×10^{-8}	0.00	0.00
1.5×10^{-8}	-0.52×10^{11}	0.27×10^{22}
2.0×10^{-8}	-0.65×10^{11}	0.42×10^{22}
3.0×10^{-8}	-0.58×10^{11}	0.34×10^{22}
5.0×10^{-8}	-0.18×10^{11}	0.032×10^{22}
10.0×10^{-8}	-0.0037×10^{11}	0.00001×10^{22}
∞	0.0	0.00

(a) Correlate your plot with the electron density diagram for the $2s$ electron (Figure 5.14). (b) Indicate the position of the node in your diagram.

5.59 In principle, is it possible to determine (a) the energy of an electron in the H atom with high precision and accuracy; (b) the position of a high-speed electron with high precision and accuracy; (c) at the same time, the energy and the position of a high-speed electron with high precision and accuracy?

5.60 Briefly explain the misconception in this question: "The energy of an electron in an H atom in its ground state is -2.179×10^{-18} J. We are also told that the electron does not revolve about the nucleus at a constant distance. But is not the energy of the electron greater when it is farther from the nucleus and less when it is closer to the nucleus? How can its energy be constant?"

5.61 The uncertainty (Heisenberg) principle may be written as

$$\Delta(mu) \times \Delta x \geq \frac{h}{4\pi} = 5.3 \times 10^{-35} \frac{\text{kg m}^2}{\text{sec}}$$

(Recall that 1 J = 1 kg m^2/sec^2.) Here $\Delta(mu)$ is the uncertainty in the momentum (mass × velocity) of a particle and Δx is the uncertainty in its position. (a) The position of a 1.00-g ball is known to within 10^{-2} mm. What is the uncertainty in its velocity? (b) The position of an electron, 9.1×10^{-28} g, is known to within 1.0×10^{-8} cm. What is the uncertainty in its velocity? (c) How much larger is the uncertainty in the velocity of the electron?

SELF-TEST

5.62 (10 points) Exposure to high dosages of microwaves can cause painful burns. Estimate how many photons, $\lambda = 12$ cm, must be absorbed by an eye to raise its temperature by 3.0°C. Assume the mass of the eye is 10 g and its specific heat is 4.0 J/(g °C).

5.63 (10 points) (a) Is it possible to detect a line in the spectrum of atomic hydrogen corresponding to a transition $n = 3.5 \longrightarrow n = 2.5$? (b) Calculate the energy in joules per photon of the hydrogen spectral line corresponding to the transition $n = 3 \longrightarrow n = 2$. Check your answer against Figure 5.10.

5.64 (10 points) The energy difference, ΔE, between two energy states of the H atom is 1.457×10^2 kJ/mol. Find the wavelength of this line in the H spectrum.

5.65 (10 points) A line in the H spectrum is found at 102.6 nm. Calculate the quantum number of the upper energy level if the lower level is $n = 1$.

5.66 (10 points) How many orbitals in any atom can have the given quantum number or designation? (a) $3p$, (b) $4p$, (c) $4p_x$, (d) $n = 5$, (e) $6d$, (f) $5d$, (g) $5f$, (h) $7s$.

5.67 (10 points) Write the electron configurations of $_{28}Ni^{3+}$, $_{23}V$, V^{3+}, V^{4+}, V^{5+}. How many unpaired electrons are in each?

5.68 (10 points) The electron configuration of Ca is [Ne] $3s^2 3p^6 4s^2$. Write the (a) chemical symbol and (b) electron configuration of the species formed when the nucleus of $_{20}Ca$ absorbs 3 protons without gaining or losing electrons.

5.69 (10 points) (a) Which of these orbital designations are impossible? $1s$, $1p$, $2s$, $2d$, $3f$, $3d$, $8s$, $4g$. (b) Find the total number of s, p, d, and f electrons in $_{40}Zr$.

5.70 (10 points) Which of these ions and atoms possess paramagnetic properties? F^-, Na^+, S, S^{2-}, Mn^{2+}, Co, Co^{3+}, Ar, Ar^-, Ar^+, and Zn.

5.71 (10 points) (a) Imagine that the nucleus of an H atom is located at the origin—the zero point—of an x, y, z graph. Assume that at the distance d, the probability of finding the $1s$ electron around $x = \text{d}$ is 1.0×10^{-4}. What is the probability of finding the electron at $y = \text{d}$? Is the probability of finding the electron at $z = \text{b}$, a distance less than d, greater, the same, or smaller? (b) The probability of finding a $2p_x$ electron at $x = \text{d}$ is 1×10^{-3}. What is the probability of finding the electron at $y = \text{d}$?

6

CHEMICAL PERIODICITY AND SURVEY OF THE ELEMENTS

Chemists have always been fascinated by the properties of the elements, because understandings about their natures lead to interpretations of the chemical behavior of all substances. Early chemists found it particularly interesting that elements could be grouped conveniently into categories by their chemical and physical properties. However, classifications based on quantitative measurements were not possible until the nineteenth century, when methods for determining valences and atomic weights became available.

This chapter describes the arrangement of elements into periodic tables. It also relates periodic classifications to the electronic structures of atoms and to their physical and chemical properties.

6.1 FROM DÖBEREINER TO MENDELEEV

The atomic weight of bromine is close to the average of the atomic weights of chlorine and iodine.

When Johann Wolfgang Döbereiner made the first significant observation of chemical periodicity in 1829, it seemed little more than a curiosity. Döbereiner noticed that the chemically similar elements chlorine, bromine, and iodine form a linear sequence of increasing atomic weights. He added that the same could be said of other groups of three elements: sulfur–selenium–tellurium, and lithium–sodium–potassium (Figure 6.1). These observations encouraged other chemists to look further into the possibility of discovering some orderly patterns among the properties of the elements.

In 1864 John Newlands noted that when the elements were arranged in the order of increasing atomic weight, similar properties recurred at periodic intervals. Thus, in the order of increasing atomic weights, lithium ⟶ beryllium ⟶ boron ⟶ carbon ⟶ nitrogen ⟶ oxygen ⟶ fluorine, the properties of the elements change considerably from one to the next. But then comes sodium, whose properties are much like those of lithium. (Actually, neon comes next, but it had not yet been discovered.) As we progress on through the next several elements to chlorine, we come to elements whose properties seem

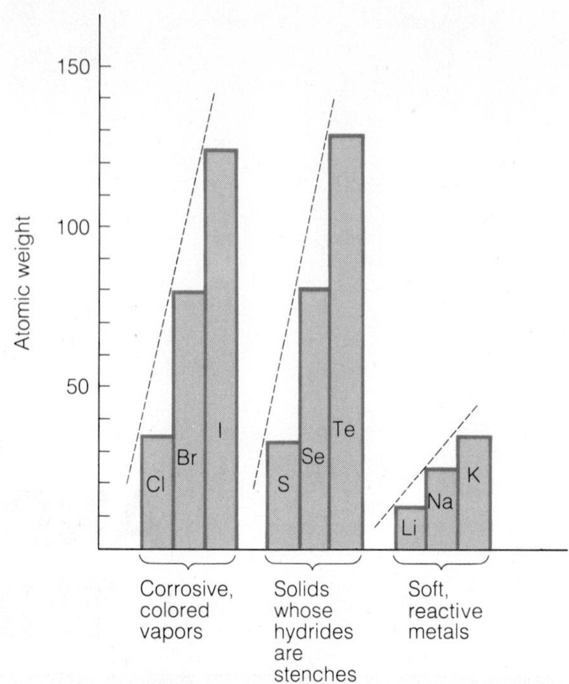

FIGURE 6.1
Döbereiner's triads.

to repeat those of the first sequence. For example, both Na and Li displace hydrogen from water and they form similar compounds: NaCl, LiCl, Na_2SO_4, Li_2SO_4. When Newlands suggested an analogy to the musical scale by calling these relationships the **law of octaves**, he was ridiculed by his contemporaries. In later years, however, his idea came to be recognized as important, and in 1887 he was awarded the Davy Medal.

Named in honor of Humphry Davy, discoverer of the alkali metals and the alkaline earth metals.

Attempts to classify the elements finally achieved the status of a serious and extremely valuable study in 1869, when Dmitri Ivanovich Mendeleev and Julius Lothar Meyer published independent versions of a periodic system of the elements based on the order of increasing atomic weight. Table 6.1 is an adaptation of

TABLE 6.1
PERIODIC TABLE OF THE ELEMENTS ADAPTED FROM MENDELEEV'S TABLE (1871)

GROUP:	1		2		3		4		5		6		7		8
SUBGROUP:	A	B	A	B	A	B	A	B	A	B	A	B	A	B	
Period 1		H													
2	Li		Be		B		C		N		O		F		
3		Na		Mg		Al		Si		P		S		Cl	
4	K		Ca		—		Ti		V		Cr		Mn		Fe, Co, Ni, (Cu)
5		(Cu)		Zn		—		—		As		Se		Br	
6	Rb		Sr		Yt?		Zr		Nb		Mo		—		Ru, Rh, Pd, (Ag)
7		(Ag)		Cd		In		Sn		Sb		Te		I	
8	Cs		Ba		Di?		Ce?		—		—		—		— — — —
9		—		—		—		—		—		—		—	
10	—		—		Er?		La?		Ta		W		—		Os, Ir, Pt, (Au)
11		(Au)		Hg		Tl		Pb		Bi					
12	—		—		—		Th		—		U		—		— — — —

Dmitri Ivanovich Mendeleev, 1834–1907.

Eka (Greek) means "first after" or "first beyond."

Mendeleev's periodic table of 1871. Note that despite its many gaps and uncertainties, and despite the fact that it classifies only the 60 or so elements then known, it has a recognizably modern form.

Mendeleev avoided what would otherwise be additional chemical irregularities by boldly leaving gaps in his table. He predicted that elements would be discovered to fill the gaps, and even described the properties of the undiscovered elements. For example, in 1871 the known element after zinc (Zn) was arsenic (As). Mendeleev recognized that arsenic does not belong in the same chemical group as either aluminum (Al) or silicon (Si). (See Table 6.1, Groups 3 and 4, period 5.) Since arsenic is chemically like phosphorus (P), he placed it in Group 5. Succeeding elements (selenium, bromine, and so on) then fell into reasonable locations. This showed that two elements, one similar to aluminum and the other similar to silicon, were missing.

Table 6.2 shows the degree to which Mendeleev's predictions were successful for "*eka*-silicon" (now called germanium). "*Eka*-aluminum" (gallium) and "*eka*-boron" (scandium) are other examples of Mendeleev's successful predictions.

Mendeleev recognized that when the atomic weight order produced a chemically unreasonable sequence of elements in the table, the atomic weight order must be sacrificed. For example, if iodine (determined atomic weight in 1869: 127) is placed before tellurium (atomic weight in 1869: 128), iodine appears in the group with sulfur and selenium, and tellurium finds itself in the company of chlorine and

TABLE 6.2
SUMMARY OF THE SUCCESS OF MENDELEEV'S PREDICTIONS ON GERMANIUM

PROPERTY	"EKA-SILICON" PREDICTED IN 1871 BY MENDELEEV	GERMANIUM (Ge)	
		Reported in 1886 by Winkler†	Currently Accepted
Atomic weight	72	72.32	72.59
Density, g/mL	5.5	5.47	5.35
Melting point	high	—	947°C
Specific heat, cal/g-deg	0.073	0.076	0.074
Molar volume§	13	13.22	13.6
Color	dark gray	grayish white	grayish white
Valence	4	4	4
Reaction with acids and bases	Es will be slightly attacked by such acids as HCl, but will resist attack by NaOH.	Ge is dissolved by neither HCl nor dilute NaOH, but is dissolved by concentrated NaOH.	Ge is dissolved by neither HCl nor dilute NaOH, but is dissolved by concentrated NaOH.
Boiling point of the tetraethyl derivative	160°C	160°C	185°–187°C
Density of the dioxide, g/mL	4.7	4.703	4.228
Density of the tetrachloride, g/mL	1.9	1.887	1.8443
Boiling point of the tetrachloride	100°C	86°C	84°C

† Clemens Alexander Winkler, who named the element after his country, Germany.
§ The volume occupied by 1 mole of atoms in the solid state, cm^3/mol.

bromine. However, iodine resembles chlorine and bromine and does not resemble sulfur and selenium. Tellurium is similar to sulfur and selenium, and not similar to chlorine and bromine. This chemical mismatching indicated to Mendeleev that either the atomic weights were wrong or atomic weight is not the fundamental basis of chemical periodicity. Mendeleev imagined that the weights were in error; actually, they were not far from today's accepted values. As we shall see, atomic weights are not the basis of chemical periodicity, although they yield the correct results most of the time.

6.2 PERIODIC TABLES DERIVED FROM THE ELECTRONIC STRUCTURES OF ATOMS

Mendeleev published his periodic table in 1869, 28 years before J. J. Thomson characterized the electron. Therefore, Mendeleev could not have known about atomic structure.

We presented a brief introduction to the periodic table in an earlier chapter (section 2.7) in order to show the periodicity of some monatomic ions. Recall that the vertical columns of elements are called **groups**, and the horizontal rows, **periods**. Now, after a study of atomic structure, we can construct a periodic table based on electronic configurations of atoms. The arrangement that emerges from this effort appears in Table 6.3. The basic assumption of this arrangement is that elements whose atoms are electronically similar have similar chemical properties, so that when such atoms are arranged in groups, chemical periodicity is evident. For example, H, Li, and Na each have 1 electron in their highest subshell (H, $1s^1$; Li, $2s^1$; Na, $3s^1$), and they form similar compounds, such as HCl, LiCl, and NaCl. As another example, F, Cl, and Br each have 7 electrons in their highest subshell (F, $2s^2$, $2p^5$; Cl, $3s^2$, $3p^5$; Br, $4s^2 4p^5$), and they form similar compounds, such as NaF, NaCl, and NaBr and CaF_2, $CaCl_2$, and $CaBr_2$.

Recall from section 5.7 that the order of filled subshells is $1s^2$, $2s^2$, $2p^6$, $3s^2$, $3p^6$, $4s^2$, $3d^{10}$, $4p^6$, and so on. If the symbols of the elements, taken in the order of increasing atomic number, are arranged so that recurring configurations of the highest *s*, *p*, and *d* subshells fall into vertical columns, the arrangement shown in Table 6.3 results. The elements appear in four sections:

1. The first section, composed only of hydrogen and helium, corresponds to the filling of the first shell, which consists only of the $1s$ orbital.
2. The second section consists of eight columns. The first two columns correspond to the filling of the *s* subshells from $2s$ on, and the next six columns correspond to the 6 electrons needed to fill a *p* subshell.
3. The third section consists of ten columns, corresponding to the 10 electrons needed to fill a *d* subshell.
4. The fourth section is fourteen elements wide, which corresponds to the 14 electrons needed to fill an *f* subshell.

This arrangement shows us the correct number of elements in each period, the number obtained by adding the elements in all of the sections for each period:

PERIOD	NUMBER OF ELEMENTS
1	2 = 2
2, 3	2 + 6 = 8
4, 5	2 + 6 + 10 = 18
6, 7†	2 + 6 + 10 + 14 = 32

† incomplete

TABLE 6.3
PERIODIC TABLE, SEPARATED FORM, BASED ON ELECTRONIC STRUCTURES

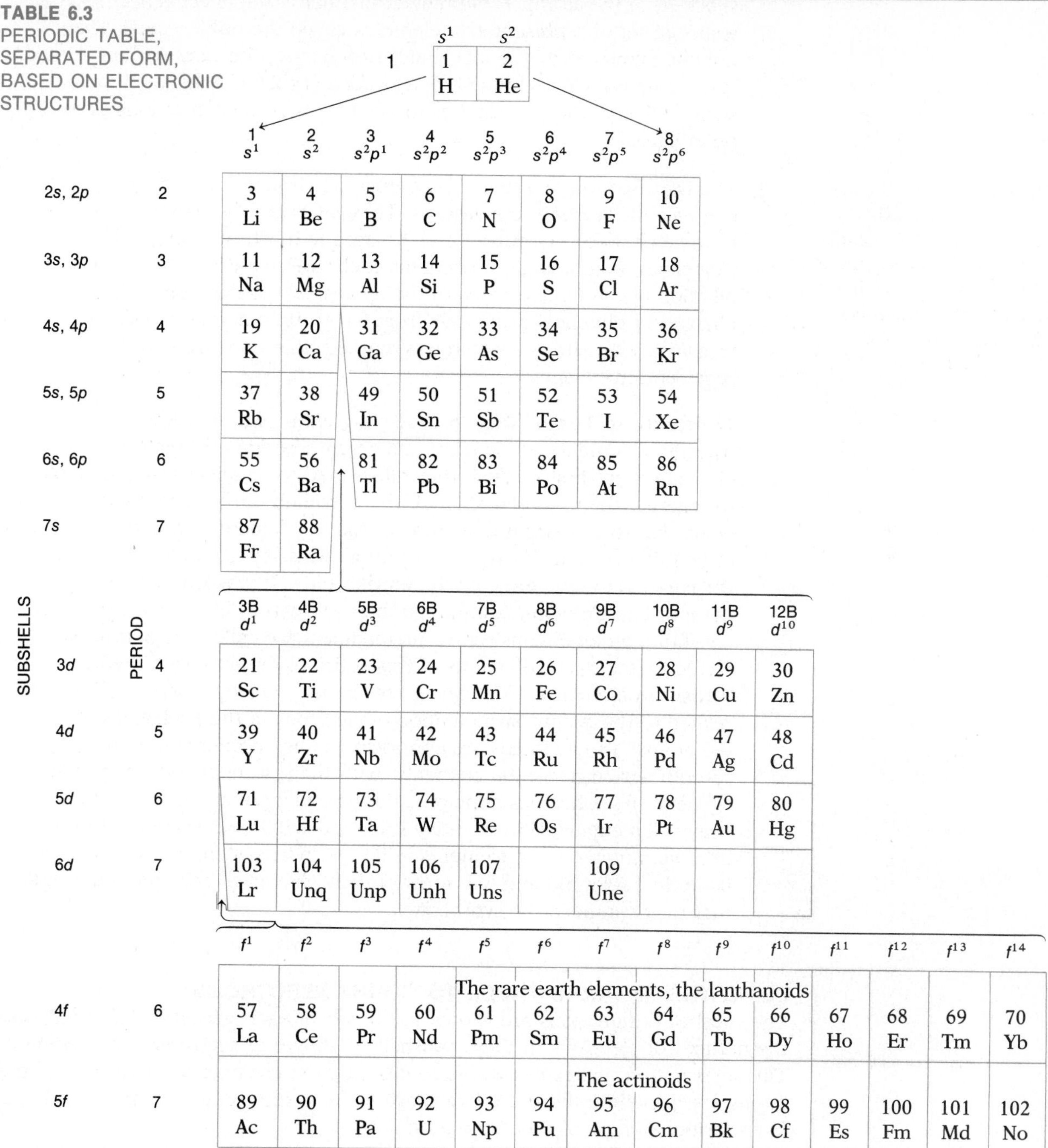

Subshells	Period	s^1	s^2
	1	1 H	2 He

Subshells	Period	1 s^1	2 s^2	3 s^2p^1	4 s^2p^2	5 s^2p^3	6 s^2p^4	7 s^2p^5	8 s^2p^6
2s, 2p	2	3 Li	4 Be	5 B	6 C	7 N	8 O	9 F	10 Ne
3s, 3p	3	11 Na	12 Mg	13 Al	14 Si	15 P	16 S	17 Cl	18 Ar
4s, 4p	4	19 K	20 Ca	31 Ga	32 Ge	33 As	34 Se	35 Br	36 Kr
5s, 5p	5	37 Rb	38 Sr	49 In	50 Sn	51 Sb	52 Te	53 I	54 Xe
6s, 6p	6	55 Cs	56 Ba	81 Tl	82 Pb	83 Bi	84 Po	85 At	86 Rn
7s	7	87 Fr	88 Ra						

Subshells	Period	3B d^1	4B d^2	5B d^3	6B d^4	7B d^5	8B d^6	9B d^7	10B d^8	11B d^9	12B d^{10}
3d	4	21 Sc	22 Ti	23 V	24 Cr	25 Mn	26 Fe	27 Co	28 Ni	29 Cu	30 Zn
4d	5	39 Y	40 Zr	41 Nb	42 Mo	43 Tc	44 Ru	45 Rh	46 Pd	47 Ag	48 Cd
5d	6	71 Lu	72 Hf	73 Ta	74 W	75 Re	76 Os	77 Ir	78 Pt	79 Au	80 Hg
6d	7	103 Lr	104 Unq	105 Unp	106 Unh	107 Uns		109 Une			

Subshells	Period	f^1	f^2	f^3	f^4	f^5	f^6	f^7	f^8	f^9	f^{10}	f^{11}	f^{12}	f^{13}	f^{14}
		The rare earth elements, the lanthanoids													
4f	6	57 La	58 Ce	59 Pr	60 Nd	61 Pm	62 Sm	63 Eu	64 Gd	65 Tb	66 Dy	67 Ho	68 Er	69 Tm	70 Yb
		The actinoids													
5f	7	89 Ac	90 Th	91 Pa	92 U	93 Np	94 Pu	95 Am	96 Cm	97 Bk	98 Cf	99 Es	100 Fm	101 Md	102 No

Table 6.3 has two important features: it highlights a convenient classification of the elements into three types, and it affords a quick count of the number of electrons in the various subshells. We will discuss each of these features.

THE THREE TYPES OF ELEMENTS

Also called *main-group* elements.

1. The elements in the first two sections of Table 6.3, comprising Groups 1 through 8, are called **representative elements**. This category is characterized by the stepwise addition of electrons into the *s* and *p*

subshells of the atoms. Helium and the other Group-8 elements make up a special set of representative elements called the **noble gases**. They are characterized by the fact that, in each case, the *next* element starts a new shell. Helium ($1s^2$) and neon ($1s^2 2s^2 2p^6$) have full outer shells. All the other Group-8 elements have s^2p^6 configurations in their outer shells.

The International Union of Pure and Applied Chemistry (IUPAC) defines transition elements as those whose atoms or ions have an incomplete *d* subshell. This definition excludes Zn, Cd, and Hg.

2. The third section of Table 6.3, comprising Groups 3B through 12B, contains the **transition elements**. They interrupt the representative elements between Groups 2 and 3 from the fourth period on. These sequences, containing ten elements each, are related to the stepwise addition of the ten electrons to the *d* subshells of the atoms. The changes of chemical properties from left to right across a period of transition elements are much less marked than they are with the representative elements.

3. The fourth section of Table 6.3, starting with $_{57}$La, precedes the transition elements of the sixth and seventh periods. When these elements were first studied, they all seemed to be alike; it was difficult to separate them, to characterize them, to find out how many there were, and to be certain that some of these "elements" would not prove to be mixtures after further purification. This series came to be called the **rare earth elements**, or the **lanthanoids**. It was learned later that the fourteen elements (atomic numbers 89 to 102) in the seventh period constitute an analogous series. Its members are called the **actinoids**. Taken together, the elements of this entire block are called the **inner transition elements**. The concept of "inner transition elements" is related to the progressive addition of electrons in the *f* subshells of the atoms. The actinoids are considered to be an interruption in the seventh-period transition elements, with the resumption starting with element 103. Chemists anticipate the possibility of synthesizing a few heavy elements with moderately stable nuclei. The expected positions of as-yet-undiscovered elements in the periodic system appear as empty boxes in Tables 6.3 and 6.4. Most of them, however, may be so unstable that they can never be synthesized.

USING THE PERIODIC TABLE TO COUNT ELECTRONS

The subshell designations are shown to the left of each period in Table 6.3, and the number of electrons in these subshells is shown at the top of each column. This information makes it possible to describe the electron configuration of the atoms except where there are irregularities from the usual aufbau rules. These irregularities were explained in section 5.7.

The electrons that play a dominant role in determining the chemical properties of the elements are called the **valence electrons**. For the representative elements, the valence electrons are those in the highest (outermost) shell. For this reason, the outermost shell is also called the **valence shell**. The electrons in the subshells of the valence or outermost shell can be counted directly from Table 6.3. For example, all Group-5 elements have the same valence shell configuration: s^2p^3.

EXAMPLE 6.1 How many (a) 4*s* and 4*p* electrons are in $_{20}$Ca; (b) 5*s* and 5*p* electrons are in $_{50}$Sn; and (c) 6*s* and 6*p* electrons are in $_{84}$Po?

ANSWER

(a) $4s$, $4p$ is shown to the left of $_{20}$Ca in Table 6.3, and s^2 is shown at the top of the Group-2 column that includes $_{20}$Ca. Therefore, $_{20}$Ca has $4s^24p^0$.

(b) The designations to the left of $_{50}$Sn show $5s$, $5p$, and the Group-5 column heading shows s^2p^2; therefore, $_{50}$Sn has $5s^25p^2$.

(c) Similarly, $6s^26p^4$ for $_{84}$Po. ■

PROBLEM 6.1 Find the number of (a) $5s$ and $5p$ electrons in $_{53}$I; (b) $2s$ and $2p$ electrons in $_9$F. □

In the case of the transition elements, both the outermost and the next-to-the-outermost shells play important roles in determining chemical properties. The electrons in the required subshells can also be counted using Table 6.3.

EXAMPLE 6.2 What are the electron configurations in the two highest shells of (a) $_{22}$Ti and (b) $_{74}$W?

ANSWER

(a) The designation to the left of $_{22}$Ti's period is $3d$, and on the top of its column is d^2, so the configuration of the valence electrons in its next-to-highest shell is $3d^2$. Now note that the entire $3d$ period, from atomic numbers 21 to 30, which includes $_{22}$Ti, belongs in the wedge right after $_{20}$Ca, whose outermost shell is $4s^2$. Therefore, $_{22}$Ti has $3d^24s^2$.

(b) The designation to the left of $_{74}$W is $5d$, and at the top of its column is d^4, so it has $5d^4$. Also, it belongs in the sixth period after $_{56}$Ba, which is $6s^2$. Therefore $_{74}$W has $6s^25d^4$. ■

PROBLEM 6.2 Use the periodic table to determine the electron configurations in the valence shells of (a) $_{40}$Zr and (b) $_{30}$Zn. □

THE LONG-FORM PERIODIC TABLE

The separated sections of Table 6.3 offer the advantage that representative, transition, and inner transition elements appear in uninterrupted sections. However, the separations impose discontinuities in the sequences of atomic numbers, such as those between Groups 2 and 3 of the representative elements. These separations can be overcome by inserting the elements that belong in the spaces shown by the "inserts" in Table 6.3. If the insertions include both the transition and inner transition elements, the table becomes too large to fit on a single page, or even on a two-page spread (see Table 6.4A). The usual solution to this problem is a compromise in which the transition elements are inserted into the table but the inner transition elements are left outside. The result is the long-form table shown as Table 6.4B and inside the front cover. This is the most common form used in textbooks of chemistry. For convenience, this table shows the atomic weights of the elements below their symbols and the atomic numbers above them.

The valence electrons can be counted from this table as well as from the separated form of Table 6.3. Thus, repeating Example 6.2a, $_{22}$Ti is in the fourth period, where the valence shells are $4s$, $3d$, $4p$. The heading above $_{22}$Ti shows d^2, and the heading just before the transition series is s^2, so $_{22}$Ti is $3d^24s^2$.

TABLE 6.4 PERIODIC TABLE, TWO LONG FORMS.

A. TABLE WITHOUT SEPARATIONS

REPRESENTATIVE, OR MAIN-GROUP, ELEMENTS

INNER TRANSITION ELEMENTS / TRANSITION ELEMENTS

PERIOD	1	2															3B	4B	5B	6B	7B	8B	9B	10B	11B	12B	3	4	5	6	7	8
1	1																															2
2	3	4																									5	6	7	8	9	10
3	11	12																									13	14	15	16	17	18
4	19	20															21	22	23	24	25	26	27	28	29	30	31	32	33	34	35	36
5	37	38															39	40	41	42	43	44	45	46	47	48	49	50	51	52	53	54
6	55	56	57	58	59	60	61	62	63	64	65	66	67	68	69	70	71	72	73	74	75	76	77	78	79	80	81	82	83	84	85	86
7	87	88	89	90	91	92	93	94	95	96	97	98	99	100	101	102	103	104	105	106	107		109									

B. USUAL FORM

REPRESENTATIVE, OR MAIN-GROUP, ELEMENTS

TRANSITION ELEMENTS

	PERIOD	1 s^1	2 s^2	3B d^1	4B d^2	5B d^3	6B d^4	7B d^5	8B d^6	9B d^7	10B d^8	11B d^9	12B d^{10}	3 p^1	4 p^2	5 p^3	6 p^4	7 p^5	8 p^6
1s	1	1 H 1.00794																	2 He 4.00260
2s, 2p	2	3 Li 6.941	4 Be 9.01218											5 B 10.81	6 C 12.011	7 N 14.0067	8 O 15.9994	9 F 18.998403	10 Ne 20.179
3s, 3p	3	11 Na 22.9897	12 Mg 24.305											13 Al 26.98154	14 Si 28.0855	15 P 30.97376	16 S 32.06	17 Cl 35.453	18 Ar 39.948
4s, 3d, 4p	4	19 K 39.0983	20 Ca 40.08	21 Sc 44.9559	22 Ti 47.88	23 V 50.9415	24 Cr 51.996	25 Mn 54.9380	26 Fe 55.847	27 Co 58.9332	28 Ni 58.69	29 Cu 63.546	30 Zn 65.38	31 Ga 69.72	32 Ge 72.59	33 As 74.9216	34 Se 78.96	35 Br 79.904	36 Kr 83.80
5s, 4d, 5p	5	37 Rb 85.4678	38 Sr 87.62	39 Y 88.9059	40 Zr 91.22	41 Nb 92.9064	42 Mo 95.94	43 Tc (98)	44 Ru 101.07	45 Rh 102.9055	46 Pd 106.42	47 Ag 107.8682	48 Cd 112.41	49 In 114.82	50 Sn 118.69	51 Sb 121.75	52 Te 127.60	53 I 126.9045	54 Xe 131.29
6s, 5d, 6p	6	55 Cs 132.9054	56 Ba 137.33	71 Lu 174.967	72 Hf 178.49	73 Ta 180.9479	74 W 183.85	75 Re 186.207	76 Os 190.2	77 Ir 192.22	78 Pt 195.08	79 Au 196.9665	80 Hg 200.59	81 Tl 204.383	82 Pb 207.2	83 Bi 208.9864	84 Po (209)	85 At (210)	86 Rn (222)
7s, 6d	7	87 Fr (223)	88 Ra 226.0254	103 Lr (260)	104 Unq (261)	105 Unp (262)	106 Unh (263)	107 Uns (262)		109 Une (266)									

		f^1	f^2	f^3	f^4	f^5	f^6	f^7	f^8	f^9	f^{10}	f^{11}	f^{12}	f^{13}	f^{14}
4f	Lanthanoids	57 La 138.9055	58 Ce 140.12	59 Pr 140.9077	60 Nd 144.24	61 Pm (145)	62 Sm 150.36	63 Eu 151.96	64 Gd 157.25	65 Tb 158.9254	66 Dy 162.50	67 Ho 164.9304	68 Er 167.26	69 Tm 168.9342	70 Yb 173.04
5f	Actinoids	89 Ac 227.0278	90 Th 232.0381	91 Pa 231.0359	92 U 238.0289	93 Np 237.0482	94 Pu (244)	95 Am (243)	96 Cm (247)	97 Bk (247)	98 Cf (251)	99 Es (252)	100 Fm (257)	101 Md (258)	102 No (259)

6.3 THE PERIODICITY OF VALENCE

Chemists use the word *valence* in various ways; they use it to refer to a particular kind of chemical bonding, to ionic charge, and to various aspects of chemical combinations. For the purpose of establishing a relationship between valence and the arrangement of the periodic table, however, we use the word here in its original, and simplest, sense to refer to *combining capacity*, without regard to the nature of the chemical bonds involved or to the charge, if any, on the bonded atoms. This concept of valence emerges when we consider the formulas of various binary compounds, particularly compounds of hydrogen. For example,

HCl, hydrogen chloride
NaH, sodium hydride
H_2O, water
CaH_2, calcium hydride
NH_3, ammonia
CH_4, methane

Note that in each of these compounds *an atom of some element combines with one or more hydrogen atoms.* Thus, the chemical combining capacities, or valences, of the non-hydrogen elements are equal to or greater than that of hydrogen.

Since no element has a valence *less* than that of hydrogen, it is convenient to assign a valence of 1 to hydrogen. Then the valence of an element is defined as the number of hydrogen atoms that combine with one atom of it. Thus the valences of the other elements in the formulas listed are

Cl and Na	1
O and Ca	2
N	3
C	4

Defined in this way, valence has nothing to do with plus or minus signs or with metallic, nonmetallic, or any other properties of the elements. Valence, here, relates only to the formula of a binary compound composed of hydrogen and another element. Since chlorine also has a valence of 1, it, too, can be used as a reference. Thus, the definition of valence can be extended to *the number of H or Cl atoms that combine with 1 atom of a given element.*

This chemical combining capacity is what *valence* meant to Mendeleev and his contemporaries. (They could hardly have thought of electron configurations before the electron was discovered.) As more was learned about the elements, it became apparent that the most striking and significant periodic variation among the elements is that of valence. Valence periodicity is exhibited most consistently among the representative elements.

The relationship between the typical valences of the representative elements and their positions in the periodic table (see section 2.7) can be expressed by two simple equations:

Valence of elements in Groups 1 to 4 = Group number (G) (1)

Valence of elements in Groups 5 to 8 = 8 − G (2)

Table 6.5 illustrates these relationships with typical binary compounds.

Equations 1 and 2 do not always predict correctly which compounds exist and which do not. You will learn about other bases for prediction through the discussions of chemical bonding, which begin with Chapters 7 and 8. However, all of

TABLE 6.5
PERIODICITY OF THE COMMON VALENCES OF SOME REPRESENTATIVE ELEMENTS

GROUP (G) 1	2	3	4	5	6	7	8
H HCl							He —
Li LiCl	Be $BeCl_2$	B BF_3	C CH_4	N NH_3	O H_2O	F HF	Ne —
Na NaF	Mg MgF_2	Al $AlCl_3$	Si $SiCl_4$	P PH_3	S H_2S	Cl HCl	Ar —
Valence = G				Valence = 8 − G			

the representative elements form many compounds whose formulas correspond to these relationships. Note also that the (8 − G) term of equation 2 predicts zero valences for the elements of Group 8, which is consistent with their general inertness.

EXAMPLE 6.3 Predict the formulas of (a) germanium iodide and (b) gallium oxide.

ANSWER

(a) Germanium is a representative element in Group 4, so a typical valence is 4. Iodine is in Group 7, so its typical valence is $8 - 7 = 1$, and the formula is GeI_4.

(b) Gallium, in Group 3, has a typical valence of 3, and oxygen, in Group 6, has a valence of $8 - 6 = 2$, so the formula is Ga_2O_3. ■

PROBLEM 6.3 Predict the formulas of (a) hydrogen selenide and (b) aluminum carbide. □

Using the periodic table to predict the valences of the transition elements is less direct, but some generalizations can be made. First, note that the transition elements are considered to be squeezed into their own periods in Group 2 of the representative elements. Therefore they all should show a valence of 2. Actually, the transition elements typically exhibit several different valences, the lowest one often being 2. The highest is usually 2 plus the position of the element in its period, reading from left to right, up to a maximum of 7, or occasionally 8. After the maximum is reached, the valences of the transition elements decrease to the right, so that 2 is the maximum valence of Zn, Cd, and Hg.

EXAMPLE 6.4 Predict the maximum and minimum valences of (a) manganese, Mn, and (b) tungsten, W.

ANSWER

(a) For Mn, as for any transition element, the expected typical minimum valence is 2. Mn is the fifth transition element in its period, so its expected maximum valence is $2 + 5 = 7$.

(b) The expected minimum is 2. Since W is the fourth transition element in its period, the expected maximum valence is $2 + 4 = 6$. ■

Such predictions hold true in many, but not all cases. The predictions do work for Mn and W, which is why they were chosen for Example 6.4. Examples of Mn compounds with minimum and maximum valences are $MnCl_2$ and Mn_2O_7. Parallel examples of W compounds are WCl_2 and WF_6. Also, transition elements generally exhibit valences between the extremes, as exemplified for Mn by MnF_3, $MnCl_4$, and MnO_3 and for W by WCl_4 and WCl_5.

PROBLEM 6.4 Predict the minimum and maximum valences of (a) Cr and (b) V. □

Since you know more about electronic configurations than Mendeleev even dreamed of, you can recognize that these relationships express the participation of the *ns* and $(n-1)d$ electrons in chemical bonding, where *n* is the highest principal quantum number, representing the outermost, or valence shell. The 2 electrons in the *ns* subshell account for the minimum valence of 2, and the electrons in the $(n-1)d$ subshell account for the additional valence up to the maximum.

Many of the inner transition elements exhibit multiple valences, a typical one being 3. More specific generalizations about them are complex and uncertain.

6.4 SIZES OF ATOMS AND IONS

How do we talk about the "size" of an atom? First, since the electron cloud completely surrounds the nucleus, the atom is regarded as a sphere. Therefore, the most convenient dimension to consider is the atom's radius.

Second, we do not deal with atoms one at a time, but rather in aggregates as they occur in elements or compounds. Therefore the size of an atom is not a constant; it varies with the chemical bonding of the atom, which in turn influences the size of the electron cloud. In this regard, atoms are best divided into two categories: (1) those that bear an electric charge, as do those in ionic compounds, and (2) those that are electrically neutral, as are those in elements and nonionic compounds. The radius of a neutral atom is called an **atomic radius**, and that of an ion is called an **ionic radius**. In the latter case, the charge on the ion must be specified, because the radius varies with the charge.

Finally, we must deal with the fact that the electron clouds surrounding nuclei do not have sharp, definite boundaries. So where does the atom "end"? Unfortunately, there is no precise answer to this question. The distance between the nuclei of two bonded atoms, however, can be measured, although bonded atoms do not remain still—they continuously vibrate as if they were attached to the ends of a spring being stretched and compressed. In Figure 6.2 a vibrating spring with

X-ray and electron diffraction methods are used.

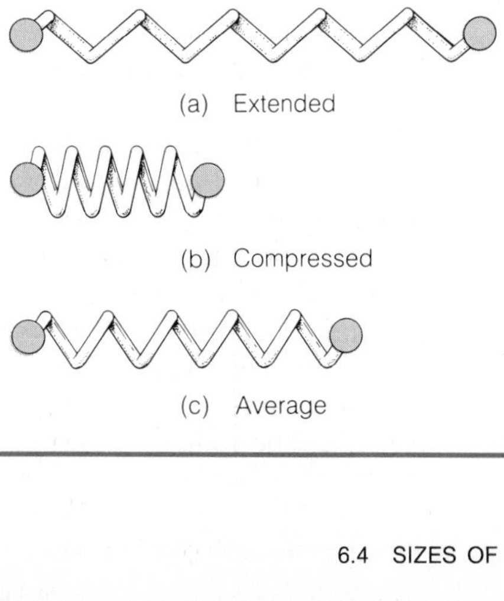

(a) Extended

(b) Compressed

(c) Average

FIGURE 6.2
A vibrating spring with a weight at each end.
(a) Extended state.
(b) Compressed state.
(c) Average length.

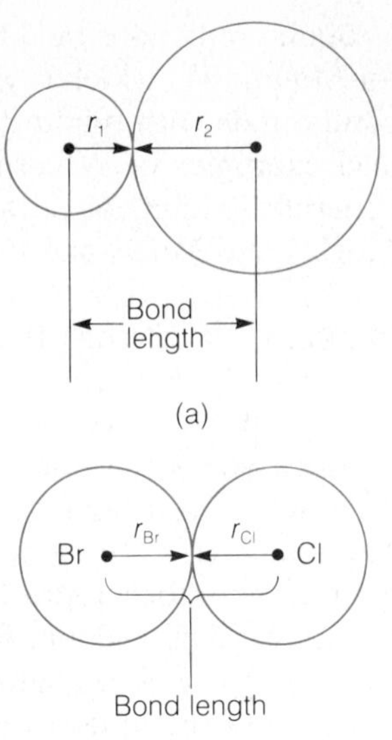

FIGURE 6.3
(a) Bond length $= r_1 + r_2$.
(b) Br—Cl bond length: r_{Br} = atomic radius of Br = 0.114 nm; r_{Cl} = atomic radius of Cl = 0.099 nm; $r_{Br} + r_{Cl}$ = 0.114 nm + 0.099 nm = 0.213 nm.

---

a weight at each end is shown in its extended and compressed states (Figure 6.2a and b). We can think of the weights as the nuclei of two bonded atoms. If the spring never stops vibrating, how can we talk about the distance between the two weights? One way is to regard the distance between the weights as the *average* between the two extremes (Figure 6.2c). In the same way, a chemical bond length is taken to be the distance separating the nuclei of the atoms in the *average condition.* This distance is experimentally determined and serves as the basis for assigning atomic radii to individual atoms (Table 6.6). Then the sum of the radii of two bonded atoms or ions is the **bond length**. This internuclear distance can then be apportioned between the two atoms or ions. That is how atomic and ionic radii are assigned (Figure 6.3).

Also called bond *distance*.

EXAMPLE 6.5 From the atomic radii given in Table 6.6, calculate the N—F bond length in NF_3.

ANSWER The atomic radii are 0.073 nm for N and 0.071 nm for F, and the calculated bond length is the sum of these radii, or 0.144 nm. The experimental value is 0.136 nm for the bond length, which shows that these calculations give approximate values, not necessarily accurate ones. ■

PROBLEM 6.5 Using the data in Table 6.6, calculate the Si—Cl bond length in $SiCl_4$. The experimental value is 0.203 nm. □

Unfortunately, such assignments of bond length are not always as accurate as we would like—particularly when atomic radii are involved, since the radius of a neutral atom depends to some extent on its molecular environment. An extreme example is the hydrogen atom, which must be assigned one radius when it combines with itself, a different radius when it combines with elements of the second period, yet another radius when it combines with elements of the third period, and so on. The radii of most atoms are not so drastically affected by their chemical environment, so that it is reasonable to assign a typical atomic radius to each element. The radii of most ions are less variable than those of the neutral atoms.

TABLE 6.6
ATOMIC (COVALENT)† AND IONIC RADII, in nanometers§

Period	Group 1	Group 2	Group 3	Group 4	Group 5	Group 6	Group 7
1	H .028 to .038 H^- .208						
2	Li .133 Li^+ .060	Be .090 Be^{2+} .031	B .080 B^{3+} .023	C .077 C^{4+} .016 C^{4-} .260	N .073 N^{3-} .171	O .074 O^{2-} .140	F .071 F^- .136
3	Na .154 Na^+ .095	Mg .136 Mg^{2+} .065	Al .125 Al^{3+} .050	Si .115 Si^{4+} .042	P .11 P^{3-} .212	S .102 S^{2-} .184	Cl .099 Cl^- .181
4	K .196 K^+ .133	Ca .174 Ca^{2+} .099	Ga .126 Ga^{3+} .062	Ge .122 Ge^{4+} .053	As .119 As^{3-} .222	Se .116 Se^{2-} .198	Br .114 Br^- .195
5	Rb .216 Rb^+ .148	Sr .192 Sr^{2+} .113	In .144 In^{3+} .081	Sn .141 Sn^{4+} .071	Sb .138 Sb^{3-} .245	Te .135 Te^{2-} .221	I .133 I^- .216
6	Cs .235 Cs^+ .169	Ba .198 Ba^{2+} .135	Tl .148	Pb .147	Bi .146 Bi^{3+} .096	Po	At

Period	3B	4B	5B	6B	7B	8B	9B	10B	11B	12B
4	Sc .144 Sc^{3+} .081	Ti .132 Ti^{2+} .090 Ti^{4+} .068	V .122 V^{3+} .074 V^{5+} .059	Cr .118 Cr^{3+} .065 Cr^{6+} .052	Mn .117 Mn^{2+} .080 Mn^{7+} .046	Fe .117 Fe^{2+} .075 Fe^{3+} .067	Co .116 Co^{2+} .082 Co^{3+} .029	Ni .115 Ni^{2+} .078 Ni^{3+} .035	Cu .117 Cu^+ .096 Cu^{2+} .072	Zn .125 Zn^{2+} .074
5	Y .162 Y^{3+} .093	Zr .145 Zr^{4+} .080	Nb .134	Mo .130	Tc .127	Ru .125	Rh .125	Pd .128	Ag .144 Ag^+ .126	Cd .148 Cd^{2+} .097
6	Lu .156	Hf .144	Ta .134	W .130	Re .128	Os .126	Ir .127	Pt .130	Au .144 Au^+ .137	Hg .149 Hg^{2+} .110

† Covalent radii (taken mostly from *Table of Interatomic Distances*, London, Chem. Soc., Special Publ., 1958, and from the publications of R. T. Sanderson) are applicable only to single-bonded atoms in mainly covalent molecules (Chapter 7). Ionic radii are from the publications of Linus Pauling.

§ Some crystallographers prefer to use angstroms for atomic radii and bond lengths because the numbers then range around 1 to 2. However, nanometers are SI units and are commonly used to express wavelengths in the visible and UV ranges. The use of the same units for atomic radii and electromagnetic radiation makes comparisons more convenient.

ATOMIC SIZE

Data for the noble gases are omitted because they are derived from interatomic distances in their solid (frozen) states, which are not the same as chemical bond distances.

The variation in the size of atoms and ions within a period and within a group is illustrated for various elements in Figure 6.4 and Table 6.6. Within a given period (going from left to right across the table), the atomic radius decreases with increasing atomic number. This observation is explained as follows: With each unit increase in atomic number there is an addition of one proton (+1) to the nucleus and one electron (−1) to the surrounding cloud. These additions have opposing effects on atomic size. The increased nuclear charge exerts a greater attraction on the electron cloud and tends to shrink it. At the same time, however, the added electron repels its neighboring electrons in the cloud and tends to expand it. Within the same period, new electrons enter the same valence shell. In this case, the contracting effect of the increased nuclear charge exceeds the expanding effect of the increased electron repulsions. Thus, within a given period, the atomic radius decreases as the atomic number increases.

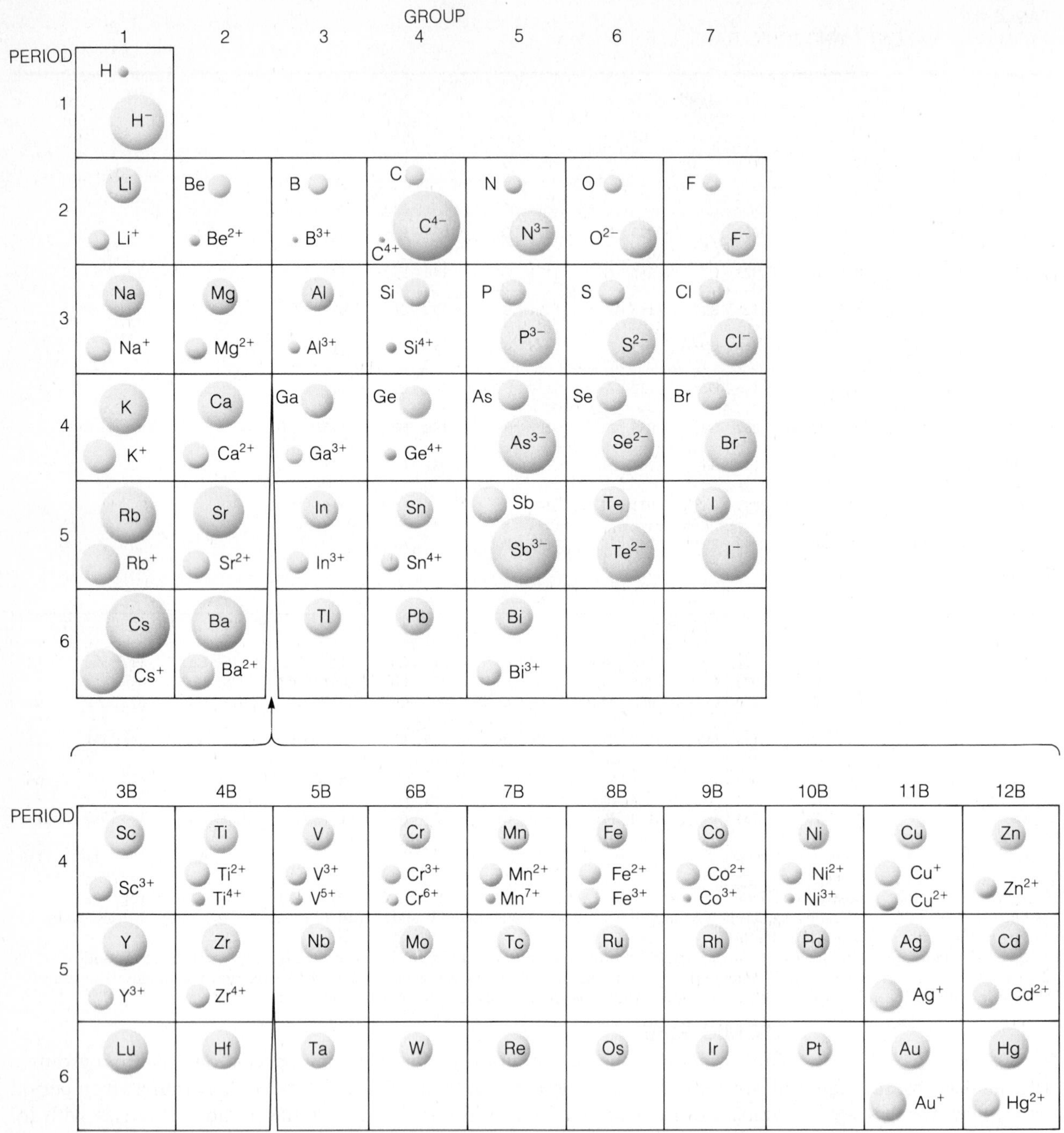

FIGURE 6.4
Relative sizes of atoms and ions.

Note the pattern for the second period, as shown in Figure 6.4 and the data in Table 6.6. The marked drop from Li to Be is associated with an increase in the nuclear charge when an electron is added to the same 2*s* subshell. As electrons are added to the 2*p* subshell (which extends further from the nucleus but is still in the same shell), the shrinking continues but by lesser amounts. Note that nitrogen has a half-filled subshell, $2p^3$, so that the next electron (in oxygen, $2p^4$) must enter an occupied orbital. One *p* orbital in oxygen then has a pair of electrons.

The resulting repulsion between these paired electrons more than offsets the attraction of the increased nuclear charge, and thus the atomic radius of oxygen is slightly greater than that of nitrogen. This effect constitutes a minor exception to the previous generalization that the atomic radius decreases with increasing atomic number within a given period.

Within a group, the atomic radius generally increases with atomic number. The reason for this is that elements within a given group are in different periods, and in going from one period to the next, electrons are added to *higher* shells. For example, the radius of Mg is larger than that of Be, as shown in Figure 6.4 and the data of Table 6.6. As we progress downward in Group 2, each step corresponds to an increase (of +8, +18, or +32) in the nuclear charge and to an addition of the corresponding number of electrons in the surrounding cloud. Again, there are the same opposing effects on atomic size: the increased nuclear charge tends to shrink the electron cloud, whereas the added electron repulsions tend to swell it. But, going down the group, the added electrons are entering higher shells ($n = 2 \longrightarrow n = 3 \longrightarrow n = 4$). As a result, the electron clouds extend a greater distance from the nucleus. Thus in Groups 1 and 2 the effect of electronic repulsion in larger electron clouds exceeds the effect of the increased nuclear attractions. These trends may be visualized as:

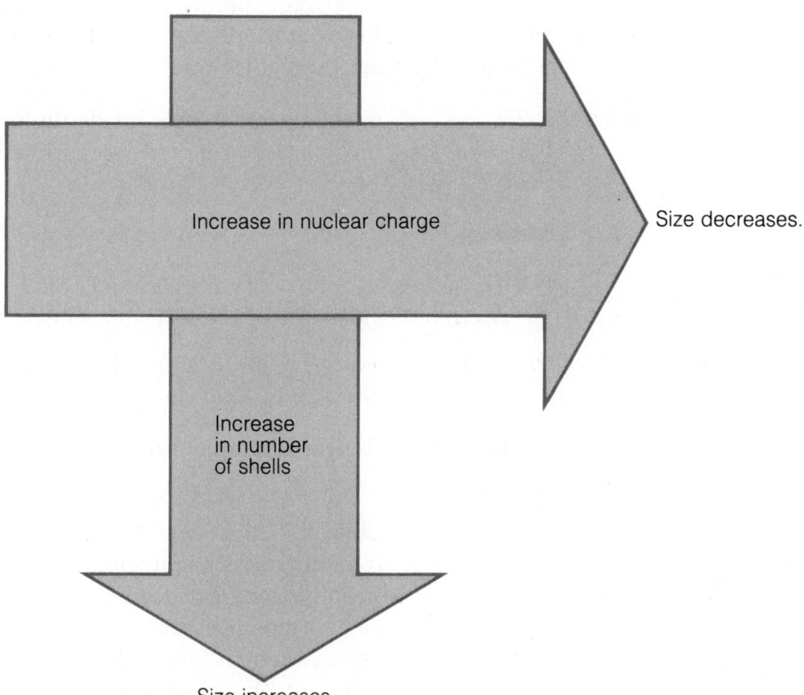

When we come to Group 3, however, a new trend arises. Consider aluminum (number 13) and gallium (number 31). The intervening elements include the ten transition elements from scandium (number 21) to zinc (number 30). In this transition the increased nuclear charge leads to a gradual contraction in atomic size (all electrons having been added to the *d* subshell). This contraction practically nullifies the expanding effect of adding electrons to higher shells, so that the two atoms have almost equal radii—Al, 0.125 nm, and Ga, 0.126 nm.

BOX 6.1

"SHIELDING"

When two atomic nuclei (bare nuclei, no electrons) approach each other, the repulsion between their positive charges increases. When two atoms approach, however, their electrons are attracted to both nuclei, and these attractions offset the nuclear repulsions. It is often said that the electrons "shield" the nuclei from their mutual repulsions.

Do not get the impression, however, that the repulsive force between two nuclei is somehow interrupted by the electron clouds between them. The repulsive force between nuclei depends on the nuclear charges and on the distance between the nuclei, and on nothing else. The *net* force between the nuclei is the resultant of all the attractive and repulsive forces, and the electronic effects can be characterized as "shielding" to the extent that they help to reduce this resultant.

In Group 3, the effect of the *d* subshell may be summarized as follows:

$_5B$ $1s^22s^22p^1$
$_{13}Al$ $1s^22s^22p^63s^23p^1$ } No *d* subshell added; 8 more electrons from $n = 2$ to $n = 3$; large increase in size

$_{31}Ga$ $1s^22s^22p^63s^23p^63d^{10}4s^24p^1$ } *d* subshell added; 18 more electrons from $n = 3$ to $n = 4$; very little change in size

The comparison between zirconium and hafnium (look for them among the transition elements) is even more dramatic; the heavier element, Hf, 0.144 nm, is just a bit smaller than the lighter one, Zr, 0.145 nm.

EXAMPLE 6.6 Referring only to the periodic table inside the front cover, list the elements in the following sets in the order of increasing atomic radii: (a) Na, Cl, Al, P; (b) Pb, C, Sn, Ge, Si.

ANSWER

(a) These are all representative elements of the third period, so the order is the reverse of the order of their atomic numbers: Cl < P < Al < Na.

(b) These elements are all in Group 4, so the order is the same as that of their atomic numbers: C < Si < Ge < Sn < Pb. ■

PROBLEM 6.6 Repeat Example 6.6 for (a) B, Li, F; (b) Sr, Ba, Be, Mg. □

IONIC SIZE

Let us now consider the effect on the size of a given atom of adding or removing an electron without changing the nuclear charge. Recall that the addition of electrons creates a negative ion (for example, $F + e^- \longrightarrow F^-$), whereas the removal of electrons creates a positive ion ($Na - e^- \longrightarrow Na^+$). In a negative ion, the added repulsions between the new electron(s) and those already present cause the cloud to spread out. As a result, the negative ion is considerably larger than the neutral atom. (Look at F^-, O^{2-}, and N^{3-} in Figure 6.4, for example.) In the case of a positive ion, the loss of electrons (often of the whole outer shell) and the decrease in total electron repulsion causes *considerable* shrinkage. (Look at Na^+, Ca^{2+}, and Al^{3+}, for example.) An extreme example is the contrast between the formation of a C^{4+} and a C^{4-} ion, as shown in Figure 6.4.

We can observe the same effect on atomic size in a series in which the number of electrons is the same but the nuclear charge varies. To do this, we must compare atoms and ions of *different elements* that have the same electronic configurations. Such ions are said to be **isoelectronic**. An example is $_3Li^+$ and $_4Be^{2+}$ (both

$1s^2$). The Be^{2+} ion has a greater nuclear charge with the same number of electrons, and is therefore much smaller. (Look up the data in Table 6.6.) An even more extreme example is the isoelectronic series ${}_{7}N^{3-}$, ${}_{8}O^{2-}$, ${}_{9}F^{-}$, ${}_{11}Na^{+}$, ${}_{12}Mg^{2+}$, and ${}_{13}Al^{3+}$ (all $1s^22s^22p^6$). Note the sharp decreases in size shown in Table 6.6.

6.5 IONIZATION ENERGY

The chemical behavior of atoms is related to their ability to gain or lose electrons, as you will learn in later chapters. It is therefore important that we consider the energies involved in such exchanges. We will also relate these energies to the positions of the elements in the periodic table. This section deals with the *loss* of electrons.

Also called ionization *potential.*

The minimum energy required to remove an electron (e^-) from a gaseous atom in its ground state is called the first **ionization energy**; the removal of a second electron from the positive ion requires the second ionization energy. For example,

IONIZATION PROCESS	IONIZATION ENERGY (always endothermic)
$He(g) \longrightarrow He^+(g) + e^-$	+2372 kJ/mol (first)
$He^+(g) \longrightarrow He^{2+}(g) + e^-$	+5247 kJ/mol (second)

A "gaseous" atom is isolated; it is not bonded in any way to other atoms. If it were bonded, some of the energy expended in the ionization process could be absorbed in breaking a bond. Table 6.7 presents the first ionization energies for the representative elements and for the transition elements of the fourth period. The following trends are illustrated by the data in the table:

1. Among the representative elements, the first ionization energy increases irregularly across a period. In the second and third periods, which are not interrupted by the transition elements, the first irregularity occurs between Groups 2 and 3. For example, note the decrease from Be to B, which coincides with the completion of an s subshell (Be, $2s^2$) and the start of a p subshell (B, $2p^1$). Recall that the $2p$ subshell is higher in energy than the $2s$. Because of this difference, less energy is required to remove the $2p$ electron from boron than to remove the $2s$ electron from beryllium.

   The second irregularity in these periods is the decrease in ionization energy between Groups 5 and 6, as exemplified by the change from N to O. This drop coincides with the half-filling of a subshell (N, $2p^3$) and the necessity for the next electron to enter an occupied orbital (O, $2p^4$). The resulting repulsion makes it easier to remove an electron from the oxygen.

BOX 6.2
MEASURING IONIZATION ENERGY

The ionization energy of a gaseous atom is measured in a discharge tube (section 2.5) containing the gas. Initially, the current flowing through the tube is practically zero. The voltage is gradually increased until ionization occurs. Ionization is signaled by a sudden, very large increase in the current. The voltage at which the sharp increase occurs is proportional to the ionization energy. With hydrogen atoms, for example, the sharp increase in current takes place at 13.60 V; the ionization energy of hydrogen is then said to be 13.60 eV/atom. One eV/atom = 96.5 kJ/mol, or 23.1 kcal/mol (see Appendix A.5).

TABLE 6.7
FIRST IONIZATION ENERGIES, kJ/mol

	1	2	3	4	5	6	7	8
1	H 1312							He 2373
2	Li 521	Be 897	B 801	C 1090	N 1399	O 1312	F 1679	Ne 2084
3	Na 492	Mg 733	Al 579	Si 781	P 1013	S 1003	Cl 1254	Ar 1524
4	K 415	Ca 589	Ga 579	Ge 762	As 946	Se 946	Br 1138	Kr 1351
5	Rb 405	Sr 550	In 560	Sn 704	Sb 830	Te 868	I 1013	Xe 1167
6	Cs 376	Ba 502	Tl 589	Pb 714	Bi 704	Po 810	At 917	Rn 1032

4	Sc 627	Ti 656	V 646	Cr 656	Mn 714	Fe 762	Co 762	Ni 733	Cu 743	Zn 907

2. Within a group, the first ionization energy decreases downward, with some exceptions.

The transition elements do not conform fully to these two patterns, but we can generalize that, among the representative elements, *the first ionization energy increases across a period and decreases downward in a group.* Visualize this as:

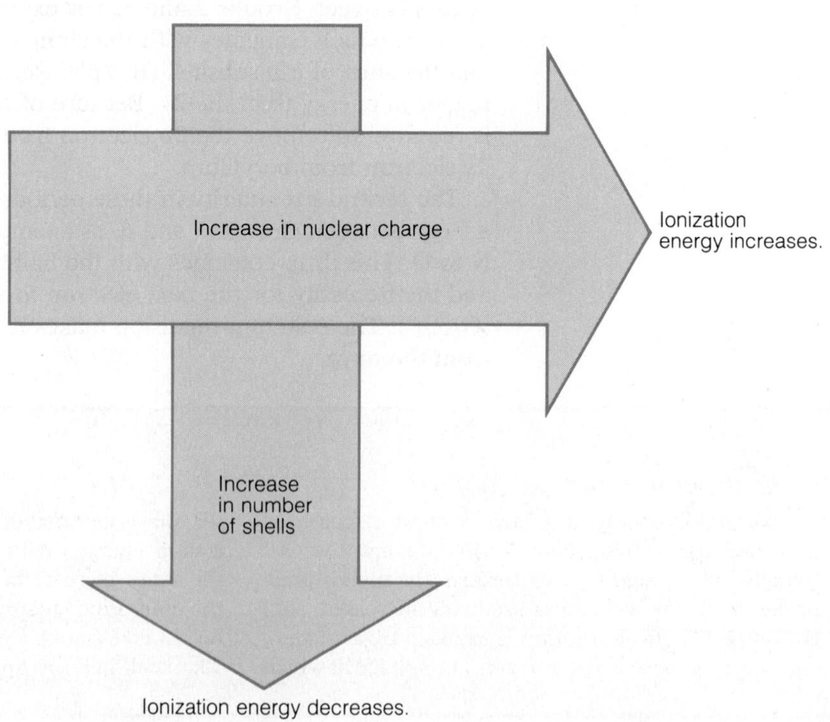

3. When ionization energies (Table 6.7) are compared with atomic radii (Table 6.6), the trends in ionization energies and atomic radii are usually found to run opposite to each other. The reason for this is that the outer, or valence, electrons in smaller atoms, being closer to the nucleus and less "shielded" by intervening shells, are more strongly attracted to the nucleus. They are, therefore, less easily removed. In larger atoms, the valence electrons are farther from the nucleus and shielded by intervening shells—and therefore more easily removed.

6.6 ELECTRON AFFINITY

The energy change that occurs when an electron is added to a gaseous atom in its ground state is called the **electron affinity**. Consider the process of moving a distant electron toward a neutral atom. To a first approximation, the atom is a neutral unit in space; it is exerting no force of attraction or repulsion on the approaching electron. However, when the electron comes close enough to form the −1 ion, its attraction to the nucleus and repulsion from the other electrons must be taken into account. If the attraction exceeds the repulsions, the energy of the −1 ion is lower than that of the separated atom and electron. Energy is then released. For example,

All the formulas in this section refer to the atoms or ions in their isolated, or gaseous, states. The symbol (g) should therefore be understood after each formula.

$$O + e^- \longrightarrow O^- \qquad \Delta H = -142 \text{ kJ/mol}$$

If the repulsions exceed the attraction, the energy of the ion is greater than that of the separated atom and electron, and energy is absorbed:

$$He + e^- \longrightarrow He^- \qquad \Delta H = +21 \text{ kJ/mol}$$

Electron affinities of most of the representative elements are displayed in Table 6.8. Electron affinities are more difficult to measure than ionization energies or atomic radii. As a result, data are unavailable for many elements and conflicting data have been reported for some elements whose electron affinities have been measured. Nevertheless, two contrasting sets of values are evident in Table 6.8—a set of exothermic values and a set of endothermic values.

TABLE 6.8
ELECTRON AFFINITIES, kJ/mol†

1	2	3	4	5	6	7	8
H −72							He 21
Li −60	Be 241	B −23	C −123	N 0	O −142	F −322	Ne 29
Na −53	Mg 231	Al −44	Si −120	P −74	S −201	Cl −348	Ar 35
K −48	Ca 156	Ga −36	Ge −116	As −77	Se −195	Br −324	Kr 39
Rb −47	Sr 120	In −34	Sn −121	Sb −101	Te −190	I −295	Xe 41
Cs −45	Ba 52	Tl −48	Pb −101	Bi −101	Po −174		

† The − sign convention in this table is consistent with our previous usage for ΔH values; thus the capture of an electron by a chlorine atom is exothermic. However, the opposite convention, in which the energy released is given a + sign, is used in some books.

- The most exothermic processes (negative ΔH) occur when the capture of an electron produces an ion that is isoelectronic with the atom of a noble gas. Such instances all involve Group-7 elements (the halogens) where the ΔH values are in the range of about -300 to -350 kJ/mol. For example,

$$\text{F}, 2s^22p^5 + e^- \longrightarrow \text{F}^-, 2s^22p^6 \qquad \Delta H = -322 \text{ kJ/mol}$$

  These atoms are the smallest in their respective periods, and have the highest nuclear charges. Note also in Table 6.7 that they have the highest ionization energies. So, it is difficult to get an electron out of these atoms, but easy to get one in.

- The most endothermic processes (positive ΔH) occur when a captured electron must enter a new shell or subshell. The first instance, an electron entering a shell, is exemplified by the noble gases (Group-8 elements). When a captured electron enters a new shell, it is further from the nucleus that attracts it and is repelled by the intervening electron shells. For example,

$$\text{Ar}, 3s^23p^6 + e^- \longrightarrow \text{Ar}^-, 3s^23p^64s^1 \qquad \Delta H = +35 \text{ kJ/mol}$$

  The entry of an electron to a subshell is represented by the elements of Group 2, which have the ns^2 configuration. A captured electron must enter a p subshell where, again, the attraction by the more distant nucleus cannot compensate for the repulsion by the intervening electron shells and subshells. For example,

$$\text{Be}, 2s^2 + e^- \longrightarrow \text{Be}^-, 2s^22p^1 \qquad \Delta H = +241 \text{ kJ/mol}$$

The variations within a group relate in complicated ways to various competing factors, and besides, the data are not all reliable, so it is well not to try to interpret them. Note, however, that the electron affinities of most elements are negative (exothermic processes).

The electron affinities of ions differ from those of neutral atoms in expected ways:

- A negative ion repels an electron, so work must be done to force another electron onto it. Thus,

$$\text{Neutral atom: } \text{O} + e^- \longrightarrow \text{O}^- \qquad \Delta H = -142 \text{ kJ/mol} \qquad \text{(exothermic)}$$

$$\text{Anion: } \text{O}^- + e^- \longrightarrow \text{O}^{2-} \qquad \Delta H = +879 \text{ kJ/mol} \qquad \text{(endothermic)}$$

- A positive ion attracts an electron, so energy is released when the electron is accepted:

$$\text{Na}^+ + e^- \longrightarrow \text{Na} \qquad \Delta H = -492 \text{ kJ/mol}$$

Of course, this is just the reverse of the process of ionization of Na, so the ΔH for electron affinity of a $1+$ ion is minus the ionization energy of the neutral atom. (Note the value of $+492$ kJ/mol in Table 6.7.)

Some of the data on ionization energy and electron affinity may seem puzzling in their apparent conflict with what we know about common chemical reactions. For instance, metallic sodium, Na(c), becomes a positive ion, Na^+, in its familiar chemical transformations. Yet the $\text{Na} \longrightarrow \text{Na}^+$ change requires energy ($\Delta H = +502$ kJ/mol), while the $\text{Na} \longrightarrow \text{Na}^-$ process releases energy ($\Delta H = -53$ kJ/mol). As another apparent anomaly, look back at the stepwise capture of

BOX 6.3
APPLICATIONS OF ELECTRON CAPTURE

A halogen atom, such as F(g), has an affinity for electrons, while its anion, F^-(g), does not. Nonionic compounds containing halogen atoms also tend to capture electrons, and this property can help us to detect and analyze such compounds. An unknown gas, passed through a kind of minature cathode-ray tube (section 2.5), will capture electrons and decrease the current if the gas contains compounds of the halogens, sulfur, or oxygen. (Why not nitrogen?) SF_6 is a gas that captures electrons very well. It is used around electrical contacts to prevent them from sparking. The SF_6 captures stray electrons before they go where they don't belong.

two electrons by an oxygen atom: the first step releases energy, but work must be done to add the second electron. Yet the typical ion that oxygen forms in its compounds is O^{2-}, not O^-. However, the ionization energy and electron affinity reactions refer to *isolated atoms,* not to ordinary substances in typical chemical environments. The chapter on bonding, which follows this one, will explain the roles of ionization energies and electron affinities, as well as those of other factors, in the energetics of chemical reactions.

Box 6.3 describes some interesting applications of electron capture.

6.7 THE ELEMENTS

The study of chemical principles cannot be separated from the study of the properties of chemical substances. Thus it is reasonable to begin any study of chemistry by studying the chemistry of the elements, because they are fundamental to all other materials.

For many years, elements were defined as substances that could not be decomposed by heat, chemical action, or electric current. It was known that silver, for example, can be heated, or can conduct an electric current, without being broken down into any simpler substances. Silver can be converted to another substance in the presence of, say, fluorine, but the new substance weighs more than the silver from which it was produced and is therefore not a product of decomposition. No chemical or electrical process to decompose silver has been found. Thus, silver is an element. On the other hand, when sugar is heated, steam is driven off and carbon remains behind; clearly, sugar is not an element. Neither is water, since it can be decomposed into hydrogen and oxygen gases by the passage of an electric current through it.

These criteria became inadequate with the discovery of radioactivity and the elucidation of atomic structures. Radium is known to spontaneously decompose into two gases, helium and radon, yet it is still classified as an element. Also, isotopes are considered to be different kinds of atoms of the same element. These differences can be accommodated by the definition of an element as a *substance consisting of atoms, all of which have the same atomic number.* With this definition in mind, let us classify the elements in various ways.

NATURAL ABUNDANCES

The naturally occurring element with the highest atomic number is uranium, number 92. However, four lighter elements are not found on Earth in more than trace quantities. Thus we can say that 88 elements occur naturally on Earth. Various elements have been synthesized in nuclear reactors and in high-energy nuclear accelerators. These include the four that occur naturally in only trace amounts—

technetium (43), promethium (61), astatine (85), and francium (87)—as well as various elements beyond atomic weight 92, so that the total number of known elements, natural and synthetic, is about 108.

About 80% of the naturally occurring elements are found as mixtures of two or more isotopes. The other 20% occur naturally only as single isotopes, some examples being aluminum, arsenic, beryllium, and phosphorus. Many additional isotopes of natural elements have been synthesized, however.

The most abundant element in the universe is hydrogen; it makes up about 79% by mass of all existing matter. Helium is next with close to 21%. The extremely small remaining proportion is made up of all the other elements, which are therefore only cosmic traces compared with hydrogen and helium. For the most part, this universal abundance reflects the compositions of stars and intergalactic matter, not the planets.

The planet Earth consists mostly of iron (39.9% by mass), oxygen (28.5%), silicon (14.3%), and magnesium (13.2%), but these elements are by no means uniformly distributed. Most of the iron is in the Earth's core, and therefore not available to us. The most abundant elements in the Earth's crust, the relatively thin layer that serves as the environment for living organisms, are oxygen (46.6% by mass), silicon (27.7%), aluminum (8.1%), iron (5.0%), and calcium (3.6%). Ninety-nine percent of the mass of the human body, on the other hand, is made up of oxygen (65%), carbon (18%), hydrogen (10%), nitrogen (3%), calcium (2%), and phosphorus (1%). These elemental abundances are shown graphically in Figure 6.5.

STABILITY

Ozone is an unstable substance, because it readily decomposes to dioxygen (usually called just "oxygen"):

$$2O_3 \longrightarrow 3O_2$$

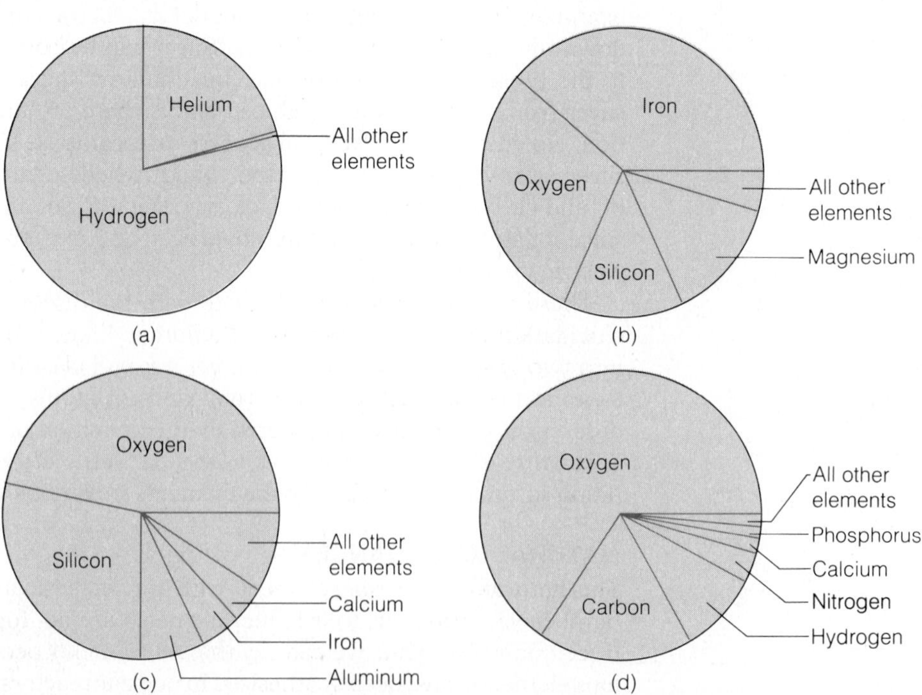

FIGURE 6.5
Elemental abundances by mass in (a) the universe, (b) the Earth, (c) the Earth's crust, and (d) the human body.

However, the nuclei of the abundant isotope in both ozone and oxygen are ^{16}O nuclei, which are stable. It is in this sense that we say that oxygen is a stable element. Since the age of the Earth is measured in billions of years, one might conclude that any naturally occurring elements would have to be very stable to have survived. However, some elements are continuously being formed on Earth by the radioactive decomposition of other elements; this process is taken up in Chapter 23, Nuclear Chemistry. All synthetic elements and synthetic isotopes of natural elements are unstable.

PHYSICAL AND CHEMICAL PROPERTIES

There is such an exhibit at the Museum of Science and Industry in Chicago. If you get the chance, visit it.

About 85 of the elements are metals (Chapter 22), 19 or 20 are nonmetals, and a few are something in between. The heavy line that crosses Groups 3 to 7 in Table 6.3 zigzags among these borderline elements, serving as an approximate boundary between the metals, on the left, and the nonmetals, on the right. Eleven elements are gases under ordinary conditions, and two are liquids (bromine and mercury). On a warm summer day, the metals cesium and gallium would join the company of the liquids. The radioactive metal francium would be yet another liquid if there were ever enough of it in one place to be anything. Figure 6.6 illus-

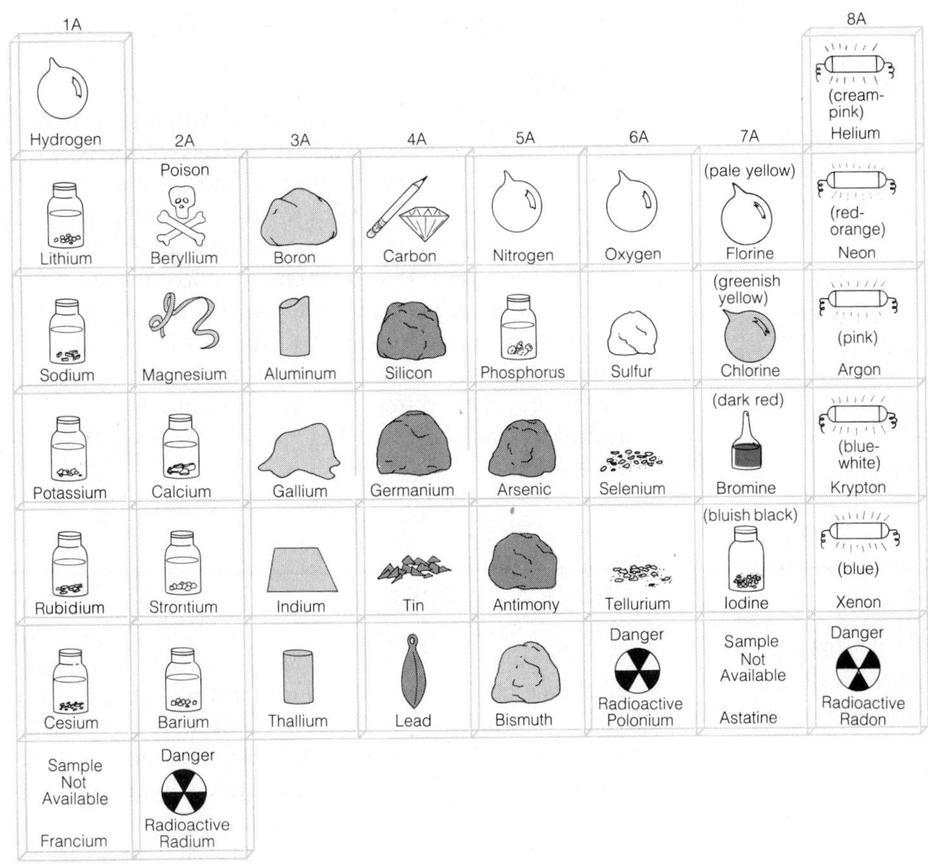

FIGURE 6.6
A sketch of what a collection of the representative elements arranged in the form of a periodic table might look like. The elements that are gases under ordinary conditions are displayed in glass bulbs. (The fluorine must be entirely free of any HF impurity, which attacks glass.) The noble gases (Group 8) are all colorless, but at low pressure in a discharge tube their emission spectra exhibit characteristic colors, as in "neon lights." The metals of Group 1 and calcium, strontium, and barium in Group 2 are so reactive to moist air that they are stored under an inert liquid such as mineral oil. White phosphorus is stored under water. Liquid bromine has a high vapor pressure and is so toxic that it is best kept in a sealed ampule. Iodine, too, has a high vapor pressure even though it is a solid, so it is stored in a closed container. Carbon is shown as diamond and as graphite (the "lead" of a pencil). The solid elements could have any shapes; the gallium is shown melting, as it would on a hot day.

TABLE 6.9
THE ALLOTROPE REGION OF THE PERIODIC TABLE
The shaded elements exist in allotropic forms.

	1	2	3	4	5	6	7	8
1								
2			B 5	C 6	N 7	O 8		
3			Al 13	Si 14	P 15	S 16		
4			Ga 31	Ge 32	As 33	Se 34		
5			In 49	Sn 50	Sb 51	Te 52		
6			Tl 81	Pb 82	Bi 83	Po 84		

trates what a collection of samples of the representative elements might look like if they were placed in bins arranged like a periodic table.

ALLOTROPY

Since diamond and graphite have different properties, they are different substances. Yet they each consist entirely of carbon atoms, and therefore, by the definition of an element, both diamond and graphite are the element carbon. Allotropes are sometimes defined as different forms of the same element. However, different physical states (solid, liquid, and gas) of an element are not considered allotropes. For example, gaseous oxygen, O_2, and gaseous ozone, O_3, are allotropes, whereas gaseous O_2 and liquid O_2 are not. A more rigorous definition is that **allotropes** are different forms of the same element in the same physical state. Allotropes may have different molecular structures, as oxygen and ozone do, or they may differ only in crystalline form. Allotropy is by no means a rare occurrence among elements; Table 6.9 shows the allotrope region of the periodic table.

6.8 SURVEY OF THE REPRESENTATIVE ELEMENTS BY PERIODIC GROUPS

GROUP 1, HYDROGEN

Although hydrogen appears in Group 1 of the periodic table, it is really in a class by itself, since it does not closely resemble any other element. The ground-state electronic configuration of the H atom is simply $1s^1$, and the loss of its electron would therefore leave only a bare proton, H^+. As will be explained in later chapters, bare protons do not exist in ordinary chemical systems, and when the symbol H^+ is written in the context of a chemical reaction, it is generally understood that the proton is attached to something. However, a hydrogen atom can gain an electron and become a negative ion, $H:^-$, and there are a few ionic compounds of hydrogen. These include hydrides of the Groups-1 and -2 elements, such as sodium hydride, NaH, and calcium hydride, CaH_2.

Under ordinary conditions, the stable form of hydrogen is the colorless, odorless, tasteless diatomic gas, H_2. Hydrogen gas is less dense than any other known substance; it is about 1/15 as dense as air. An undisturbed mixture of hydrogen and oxygen is stable. If a flame or even the smallest electrical spark is introduced, however, a violent explosion occurs as the two gases combine to form water vapor:

A flammable gas can explode in air only if its concentration falls within a particular range characteristic of the gas. For hydrogen, the minimum of this range (called the lower explosive limit) is 4% by volume and the maximum is 94%.

$$2H_2(g) + O_2(g) \longrightarrow 2H_2O(g) \qquad \Delta H^0 = -572 \text{ kJ}$$

Hydrogen is not toxic, and since it burns to form water, its combustion is not polluting. Therefore, if hydrogen were cheaply available, it would be an ideal fuel (the reaction would be controlled, not explosive). What is needed is an economic means of using solar energy directly to reverse the hydrogen and oxygen reaction and thus produce hydrogen by the decomposition of water.

Hydrogen combines with many other elements, exhibiting a valence of 1, as in HCl, H_2S, NH_3, and CH_4. Some of these compounds are described later in this section, as well as in many other parts of the book.

GROUP 1, THE ALKALI METALS

Li [He]$2s^1$
Na [Ne]$3s^1$
K [Ar]$4s^1$
Rb [Kr]$5s^1$
Cs [Xe]$6s^1$
Fr [Rn]$7s^1$

The ground-state electronic configurations of the other Group-1 elements, the alkali metals, all have a noble gas core plus one electron in the next-higher shell. The first ionization energies of these elements are low (see Table 6.7). Since the M^+ ions (M stands for metal) are isoelectronic with the noble gases, we would expect that the second ionization energies are very high—and in fact, they are about 10 times as high as the first. The alkali metals therefore typically exist as M^+ ions in their compounds. Lithium, having the smallest atomic radius, has the highest ionization energy and is somewhat less reactive than the others.

The "lithium" that is used as an antidepressant drug is a compound such as lithium carbonate or lithium citrate, not the metallic element.

The metals of Group 1 are soft and lustrous and so reactive that they must be kept from air and moisture in the laboratory. They are usually stored in a bottle of an inert liquid such as mineral oil. If you wanted to experience the softness and luster of sodium (the most abundant of the alkali metals), you could slice off a piece with a knife. Sodium cuts like hard cheese, and a freshly exposed surface is as bright as silver. However, if sodium is dropped into water, it reacts vigorously to produce hydrogen, which may be ignited by the heat of the reaction.

$$2Na(c) + H_2O(\ell) \longrightarrow H_2(g) + 2NaOH(aq)$$

Alkali metals are dangerous to handle; gloves and goggles required.

Lithium reacts with water more slowly than sodium; the other alkali metals react faster. They all form oxides of formula M_2O, and the larger alkali metal atoms (see Table 6.6) can combine with more oxygens:

$$4Li(c) + O_2(g) \longrightarrow 2Li_2O(c) \qquad \text{(lithium oxide)}$$

$$2Na(c) + O_2(g) \longrightarrow Na_2O_2(c) \qquad \text{(sodium peroxide)}$$

$$K(c) + O_2(g) \longrightarrow KO_2(c) \qquad \text{(potassium superoxide)}$$

KO_2 is *extremely* dangerous because it can explode spontaneously:

$$4KO_2 \longrightarrow 2K_2O + 3O_2$$

Therefore, potassium metal must never be stored where it can be exposed to air and form KO_2. Francium occurs only from the radioactive disintegration of actinium (atomic number 89). Francium is also radioactive, and at any one time there is probably less than an ounce of it on Earth.

The violent reaction between sodium and water may lead you to conclude that these substances should always be kept far away from each other. Nonetheless, hot liquid sodium is used in some nuclear reactors ("breeders"; see Chapter 23) as the medium for transferring heat from the nuclear core to the water that is converted to steam to drive the turbines. Of course, the liquid sodium is circulated between the core and the water in sealed pipes; there is no physical contact between the two chemicals. Liquid sodium offers several advantages as a means of heat transfer: (1) it is liquid over a wide temperature range, from its melting point at 98°C to its boiling point at 890°C at normal atmospheric pressure; (2) sodium is an excellent conductor of heat, better even than most other metals; (3) the liquid is easy to pump; and (4) sodium is not prohibitively expensive because it is obtained from salt, NaCl, which is abundant.

GROUP 2

Be	$[He]2s^2$
Mg	$[Ne]3s^2$
Ca	$[Ar]4s^2$
Sr	$[Kr]5s^2$
Ba	$[Xe]6s^2$
Ra	$[Rn]7s^2$

The elements of Group 2 have a noble gas core plus two electrons in the valence shell. These elements resemble the alkali metals but are much less reactive. Their typical valence, corresponding to their ns^2 electronic structure, is 2, as in $BeCl_2$ and MgO.

Beryllium is the least active metal in Group 2 (note the parallel with Cl in Group 1). A dark gray metal, it is less dense than aluminum. These properties make it useful as a structural material for high-speed aircraft and spacecraft, as well as for missiles. Beryllium and its compounds are very poisonous.

The other elements of Group 2 are known as the **alkaline earth metals**. The first of these, magnesium, is a light, lustrous, fairly tough metal. It does not react readily with water at ordinary temperatures. However, it does react with steam:

$$Mg(c) + 2H_2O(g) \longrightarrow Mg(OH)_2(c) + H_2(g)$$

Magnesium burns in air with a dazzling flame to produce a mixture of the oxide and nitride:

$$2Mg(c) + O_2(g) \longrightarrow 2MgO(c) \qquad \text{(magnesium oxide)}$$

$$3Mg(c) + N_2(g) \longrightarrow Mg_3N_2(c) \qquad \text{(magnesium nitride)}$$

Calcium is a fairly hard, silvery metal. It reacts with water, though much less rapidly than the alkali metals do.

PROBLEM 6.7 Using the reaction between magnesium and steam as a guide, write the equation for the reaction between calcium and liquid water, showing the physical states of all substances involved. Calcium hydroxide is slightly soluble in water, so you may assume that it is dissolved. □

Radium was discovered in 1898 by Marie and Pierre Curie, probably the most famous married couple in science. Marie Curie actually gave her life to her work. The dangers of radioactivity were not known in her day, and she died of leukemia as a result of exposure.

Strontium and barium are soft, reactive metals. Like the Group-1 metals, they are stored in mineral oil to prevent contact with air or moisture. A radioactive isotope of strontium, ^{90}Sr, occurs in fallout from nuclear explosions. Its environmental hazard lies in its chemical similarity to calcium, which it accompanies into the bones of animals and the milk of mammals. Radium is radioactive and extremely dangerous to handle.

GROUP 3

B	$[He]2s^22p^1$
Al	$[Ne]3s^23p^1$
Ga	$[Ar]4s^24p^1$
In	$[Kr]5s^25p^1$
Tl	$[Xe]6s^26p^1$

The electronic configurations of the Group-3 elements are shown in the margin.

Boron is usually available as an impure brownish powder, but it may also be obtained in the form of pure yellow crystals. It is the only nonmetallic Group-3 element. As expected from its electronic structure, boron typically exhibits a valence of 3, as in BCl_3 and B_2O_3.

Aluminum is well known as a light, shiny, easily worked, noncorroding metal with many household and industrial applications. That it resists corrosion does not mean it is unreactive, however. Instead, aluminum reacts rapidly with oxygen to form a very thin but impervious layer of the oxide, Al_2O_3, which protects the underlying metal from further reaction with oxygen. The fact that aluminum compounds are abundant in the Earth's crust and that aluminum metal is seen everywhere, from kitchen pots to roadside litter, may lead one to think that the metal is cheap to produce. In fact, a great deal of energy is required to produce aluminum. The conversion of naturally occurring Al_2O_3 to the metal,

$$2Al_2O_3 \longrightarrow 4Al + 3O_2 \qquad \Delta H = +3352 \text{ kJ}$$

represents one of the largest industrial consumptions of electrical energy. The recycling of aluminum beverage cans is therefore important for energy conservation.

Gallium, indium, and thallium are lustrous, moderately reactive metals. They are all rather rare elements, but each has special uses. Gallium has interesting properties: though it melts at only 30°C, it has a high boiling point (2403°C) and so is liquid over an exceptionally wide temperature range. Like water, it expands on freezing. Gallium arsenide, GaAs, is used in solar cells to convert solar energy to electricity. If the solar energy program expands, gallium reserves will be critical. Indium is used as an alloying material with various metals to impart resistance to corrosion. (An **alloy** is a mixture, compound, or solid solution of two or more metals.) Thallium compounds are very toxic; their use in depilatories (hair removers) has been discontinued.

GROUP 4

The elements of Group 4 all have the s^2p^2 electron configuration in their valence shells. They show a progression of nonmetallic to metallic character with increasing atomic weight from carbon (nonmetallic) to silicon and germanium (some metallic properties) to tin and lead (metals).

C $[He]2s^22p^2$, nonmetal
Si $[Ne]3s^23p^2$, some metallic properties
Ge $[Ar]4s^24p^2$, some metallic properties
Sn $[Kr]5s^25p^2$, metal
Pb $[Xe]6s^26p^2$, metal

Graphite and diamond, the predominant allotropes of carbon, are structurally different nonmetals. Their differences in bonding, which are shown in Chapters 7 and 24, make for very different properties. Whereas diamond is the hardest substance known, graphite, the so-called "lead" in pencils, is soft. Graphite conducts electricity; diamond does not. Both allotropes are stable in oxygen but can be made to burn. Diamonds are said to be "forever," but if you place one in an atmosphere of pure oxygen and bring it to a white heat, it will burn with a brilliant glow, but without a flame, until it has all been converted to carbon dioxide.

$$\text{C(diamond)} + O_2(g) \longrightarrow CO_2(g)$$

You will never get your diamond back.

Surely the most noteworthy chemical feature of carbon is that its known compounds outnumber those of all other elements. The subject of organic chemistry (Chapter 24) is devoted to these compounds.

Silicon and germanium fall between the nonmetal (carbon) and the metals (tin and lead) in Group 4 and they are said to be semimetallic. Their function as semiconductors has made them important in the electronics industry, as you will read in Chapter 22. Silicon occurs predominantly in the form of silica, SiO_2, the major component of sand and gravel, and in many silicate minerals. These substances are used in large quantities in the construction of highways, buildings, and dams. Annual world expenditures for these materials are very high, even though their cost per unit mass is low in comparison with other minerals.

Quartz is almost pure SiO_2.

Ordinary tin is a white metal with a slightly bluish tinge. It takes a high polish and is relatively unreactive with air and water. For these reasons it is used as a coating on other metals. The so-called "tin" can is actually steel with tin plating. When a sheet of tin is bent, it emits a high-pitched, crackling sound called the "tin cry." Though known since ancient times, tin is a relatively rare metal; its abundance in the Earth's crust is only 0.004%. It is a constituent of many valuable alloys such as bronze and pewter. A typical bronze is about 90% copper and 10% tin by mass. Tin also exists as a powdery nonmetallic allotrope known as "gray tin," to which metallic tin slowly changes below 13°C. This transformation, known as "tin disease," was first observed as blistery outbreaks on the surface of tin objects such as organ pipes in cold cathedrals. When the conversion is complete, the entire object collapses into a sandy powder.

The difference in energy between metallic and nonmetallic allotropes is generally quite small, often less than 15 kJ/mole.

Lead is a dull gray, soft, dense metal that is fairly resistant to corrosion and that has a rather low melting point (327°C). Moreover, most of its compounds, such as the sulfate, the carbonate, and the halides, are insoluble in water. As a result, the action of acids on lead, which converts the metal to these compounds, leaves an impervious layer that enables the lead to resist further attack. This unusual combination of properties offers many practical applications. Lead is widely used as an "acid-proof" construction material, especially for pipes. When alloyed with about half of its mass of tin, its melting point is reduced to 275°C and it becomes plumber's solder. Lead's softness makes it a good medium for absorbing sound and other vibrations. Lead is the cheapest dense solid (more than 11 times as dense as water) and is therefore cast into weights of various kinds, including sinkers for fishing lines and nets and plumbs for plumb lines. Lead also provides an excellent shield against radiation and so is used around X-ray equipment and nuclear reactors.

The words plumber and plumb, and the symbol Pb all derive from *plumbum,* Latin for lead.

Since compounds of lead are now known to be toxic, health codes require that lead pipes no longer be installed for carrying drinking water. For the same reason, lead pigments are no longer used in interior paints. The peelings from old painted surfaces are attractive to some children, possibly because they are sweet. The lead in leaded gasoline, which is discharged to the atmosphere in the form of lead chloride and bromide, finds its way into the dust on city streets and country roads. Fortunately, the use of lead compounds in gasoline is also being phased out. Children, especially, must be protected from all such exposures to lead.

Lead acetate, $Pb(C_2H_3O_2)_2$, is also known as "sugar of lead" in recognition of its dangerously sweet taste. (DO NOT try to confirm this!)

Isotopes of lead are the end products of the three series of radioactive decompositions that occur naturally on Earth: the uranium series, the actinium series, and the thorium series (Chapter 23).

GROUP 5

N [He]$2s^22p^3$, nonmetal
P [Ne]$3s^23p^3$, nonmetal
As [Ar]$4s^24p^3$, some metallic properties
Sb [Kr]$5s^25p^3$, metal
Bi [Xe]$6s^26p^3$, metal

The elements of Group 5 have the ns^2np^3 electron configuration. Like the Group-4 elements, they show a marked progression of nonmetallic to metallic character with increasing atomic number.

Nitrogen, as N_2, constitutes about 78 mole % of the atmosphere, and it is an essential element in all living organisms, plant and animal. Its atmospheric abundance, however, does not make it readily available to the organisms that need it, since N_2 is a very stable gas. It does combine with oxygen to produce oxides, but only slowly at flame temperatures (as in boilers and gasoline engines) or in lightning flashes.

Mole percent is the correct expression for what is often called "percent by volume" (Chapter 10).

$$N_2(g) + O_2(g) \longrightarrow 2NO(g) \quad \text{(nitrogen oxide or nitric oxide)}$$

$$2NO(g) + O_2(g) \longrightarrow 2NO_2(g) \quad \text{(nitrogen dioxide)}$$

There are various chemical pathways between the atmospheric reservoir of gaseous N_2 and living organisms. The industrial route for nitrogen fixation is the combination of nitrogen with hydrogen to form ammonia, which is used as a fertilizer:

The conversion of atmospheric nitrogen to a compound that is available as a nutrient for plants is called **nitrogen fixation**.

$$2N_2(g) + 3H_2(g) \longrightarrow 2NH_3(g)$$

Phosphorus is a soft solid that occurs in white and red allotropic forms. The white form, dangerous to handle, yields a vapor whose formula is P_4. The vapor ignites spontaneously in air or oxygen to form tetraphosphorus decoxide:

$$P_4(g) + 5O_2(g) \longrightarrow P_4O_{10}(c)$$

White phosphorus is stored under water. The red form is fairly stable and not so dangerous to handle. Both forms are quite toxic when they are ingested, inhaled, or absorbed through the skin.

Arsenic exists in three allotropic forms: a black metallic form, a gray metallic form, and a waxy, yellow, nonmetallic form that is too unstable to be kept at room temperature. Various arsenic compounds have been used as poisons for so many centuries, both in fiction and in real life, that the element has come to symbolize toxicity. The hydride arsine, AsH_3, is an extraordinarily toxic gas.

The search for chemical agents to fight disease is a search for chemicals that are more toxic to pathogens than to the affected hosts. Compounds of arsenic were among the early agents used in such chemotherapy. The search for a cure for syphilis led Paul Ehrlich through 605 compounds before he synthesized the 606th in 1910. Number 606 was his "magic bullet," arsphenamine, an organic arsenic compound. Some of the pesticides used in the early 1900s, especially those to protect fruit from insects, were arsenic compounds. Modern antibiotics and synthetic organic pesticides have displaced these arsenical agents.

Antimony, like arsenic, also exists in a gray metallic form and an unstable, nonmetallic yellow form. Bismuth, however, occurs only as a lustrous, brittle, and rather soft metal.

GROUP 6

O	$[He]2s^2 2p^4$, nonmetal
S	$[Ne]3s^2 3p^4$, nonmetal
Se	$[Ar]4s^2 4p^4$, nonmetal
Te	$[Kr]5s^2 5p^4$, nonmetal
Po	$[Xe]6s^2 6p^4$, metal

All of them having the ns^2np^4 electronic configuration, the Group-6 elements are nonmetals except for the highest member, polonium.

Oxygen exists in two allotropic forms: O_2 (dioxygen, although the *di-* is generally omitted), and O_3 (trioxygen, or ozone). Oxygen is a reactive element that combines with all of the other elements except some of the noble gases. As a gas it is odorless and colorless, but the liquid form is a beautiful pale blue. Its boiling point at normal atmospheric pressure is −183°C.

Ozone is a pale blue gas. Its odor is noticeable around electrical sparks, which convert some oxygen to ozone. It is much more reactive than oxygen and, in its pure form, is explosive as it decomposes to oxygen:

Ozone was once thought to be the odor of electricity. Hence, its name is derived from the Greek word for smell.

$$O_3(g) \longrightarrow 1\tfrac{1}{2}O_2(g) \qquad \Delta H = -143 \text{ kJ}$$

Ozone has sometimes been associated with pure air, and devices that produce ozone were once sold as air "fresheners" for homes and hospitals. However, ozone is toxic, and home appliances that generate it pollute the air rather than purify it.

Sulfur is a pale yellow solid obtained from underground beds of the element. The sulfur is melted and forced upward from the beds with a mixture of steam, superheated water, and compressed air (the **Frasch process**). That sulfur is readily obtained by the Frasch process is a detriment to the environment in one sense. If this cheap source of sulfur were not available, we would be forced to recover sulfur from coal and petroleum. Our coal and petroleum would then be freer of sulfur, and their combustion would be less polluting to the atmosphere, resulting in less acid rain (Chapter 20). The most common oxides of sulfur are the dioxide, SO_2, and the trioxide, SO_3. Sulfur combines with practically all metals to form sulfides. Its hydride is the poisonous, foul-smelling gas hydrogen sulfide, H_2S. Most sulfur is converted to sulfuric acid, H_2SO_4, which is widely used in the manufacture of fertilizers, explosives, dyestuffs, and many other products.

Selenium and tellurium both exist in various allotropic forms, some of which possess semimetallic properties. Their hydrides, H_2Se and H_2Te, have even more repulsive odors than hydrogen sulfide. In fact, their stench and toxicity have discouraged the investigation of their properties.

Polonium, the last element in Group 6, is a radioactive metal. It is found in tobacco leaves and therefore may be a contributor to the carcinogenicity of tobacco smoke.

GROUP 7: THE HALOGENS

F [He]$2s^22p^5$, nonmetal
Cl [Ne]$3s^23p^5$, nonmetal
Br [Ar]$4s^24p^5$, nonmetal
I [Kr]$5s^25p^5$, nonmetal
At [Xe]$6s^26p^5$, nonmetal

The electronic configurations of the Group-7 elements, the halogens, are shown in the margin.

In many compounds, the elements of Group 7 exist in the form of a singly charged negative ion, such as F^- and Cl^-. The strong resemblances among these elements have led chemists to designate them, collectively, by the symbol X. Thus, the formula NaX refers to any sodium halide, such as NaF, NaCl, and so on.

Fluorine, F_2, is a pale yellow, corrosive gas that reacts with practically everything. It is stored in containers made of metals such as iron or nickel, with which it forms a fluoride coating that protects it against further reaction.

Chlorine, Cl_2, is a greenish yellow gas that has the unhappy distinction of being the first poisonous gas used in warfare (World War I, in 1915). Chlorine is used as a disinfectant for water (section 12.18) and as a bleaching agent. Chlorine and fluorine are incorporated into many widely used organic compounds such as plastics, refrigerants, and pesticides.

Bromine, Br_2, is a dense, dark red liquid that evaporates to a reddish brown vapor. It, too, is acutely toxic and must be handled with great care. The liquid form is especially corrosive to the skin, even on very brief contact.

A "tincture" is a solution in ethanol.

Iodine, I_2, is a bluish black, lustrous solid that evaporates to a purple vapor. It dissolves in ethanol (ethyl alcohol) to form a brown solution, once widely used as an antiseptic called "tincture of iodine."

Astatine, like the Group-1 element francium, is a radioactive element that occurs in such minute quantities on Earth that little is known about its chemistry. However, if a trace amount of astatine, detectible by its radioactivity, is mixed with iodine, and the iodine is subjected to various physical and chemical transformations, the astatine accompanies the iodine. This is the type of evidence from which we conclude that astatine should be classified as a halogen. Extrapolation of the properties of the Group-7 elements, however, leads to the possibility that astatine may also have some metallic characteristics.

The hydrides of the halogens are well-known compounds: HF (hydrogen fluoride), HCl (hydrogen chloride), HBr (hydrogen bromide), and HI (hydrogen iodide). They are gases that dissolve in water to form the solutions hydrofluoric acid, hydrochloric acid, hydrobromic acid, and hydroiodic acid, respectively.

GROUP 8: THE NOBLE GASES

Nobles don't associate with the common folk, just as these gases don't combine much with other elements.

He $1s^2$
Ne $[He]2s^22p^6$
Ar $[Ne]3s^23p^6$
Kr $[Ar]3d^{10}4s^24p^6$
Xe $[Kr]4d^{10}5s^25p^6$
Rn $[Xe]4f^{14}5d^{10}6s^26p^6$

Referred to as the noble gases, the elements of Group 8 have the ns^2np^6 configuration. They are all colorless, monatomic gases and they exhibit little reactivity. (The limited chemistry of these elements is discussed in Chapter 20.) However, their visible emission spectra show characteristic colors, as in "neon lights." Radon is a radioactive gas emitted by many minerals, masonry materials, and soils. The pollution of indoor air by radon is discussed in Chapter 23.

The most abundant of these gases in the atmosphere is argon (about 1 mole %; much more than carbon dioxide, which is only about 0.3 mole %). Helium, however, is the most useful noble gas, because of its unique properties. The second-lightest gas (hydrogen is first), helium is truly chemically inert; it does not react with anything, nor can it be altered even by neutron bombardment. Recall (from section 2.5) that the nucleus of the helium-4 atom is an alpha particle, which is *extremely* stable. It is the product of nuclear fusion in the Sun (Chapter 23). Helium has the lowest boiling point of any known substance, $-269°C$. When it does condense to a liquid, it remains liquid down to absolute zero.

Because of these properties helium has some important applications. Helium gas is ideal for lifting lighter-than-air craft, such as balloons and blimps, because it cannot burn. It is also used to create an absolutely inert atmosphere and to test for gas leaks in anything. Liquid helium is the ultimate cooling agent; it is a liquid in the region of 4 K and is therefore the medium of choice for studying superconductivity at very low temperatures. Helium, unlike hydrogen, poses no danger of fire or explosion in any of these applications.

If helium doesn't leak from a pipe or container, nothing will.

Helium is found in the atmosphere (0.0005 mole %) but is about 800 times as concentrated (about 0.4 mole %) in some pockets of natural gas, its commercial source. However, billions of cubic feet of natural gas are burned each year without first extracting the helium, which then enters the atmosphere. Unfortunately, helium is so light that the Earth's gravity does not hold it strongly enough, and eventually, in the fullness of geological time, this valuable gas diffuses out into space.

6.9 TRANSITION AND INNER TRANSITION ELEMENTS

THE TRANSITION ELEMENTS

The transition elements are all recognizably metallic—dense, lustrous, good conductors of heat and electricity, and in most cases, hard. Some, like platinum and gold, are also rather expensive. These two elements are used in jewelry because they are unreactive in air and do not tarnish. Copper and gold are the two colored metals; their colors are so well known that they are named after the elements. Mercury, the one liquid among the transition elements, is so dense (13.5 g/mL) that lead, a very dense solid (11 g/mL), will float on it. A liter of mercury weighs 13.6 kg—too heavy for a child to lift. The densest metal of all is osmium (22.6 g/mL); it would sink in mercury almost like lead in water. A sphere of osmium the size of a grapefruit (about 15 cm diameter) would surprise anyone who tried to pick it up, for it would weigh about 40 kg—close to 90 pounds.

THE INNER TRANSITION ELEMENTS

The lanthanoids, or rare earths (from $_{57}La$ to $_{70}Yb$), are all metals, famous (or infamous, in the minds of past generations of frustrated chemists) for the similarity of their chemical and physical properties. The actinoids (from $_{89}Ac$ to $_{102}No$) are all radioactive. All of the elements with atomic numbers beyond uranium are produced artificially by nuclear reactions. (See Box 6.4.)

Chapter 21 is devoted to the chemistry of the transition elements.

BOX 6.4

**THE "EARTH" AND METALLIC OXIDES—
A VOCABULARY LESSON**

The word *earth* formerly referred to any of various stable metallic oxides and, by extension, to their metals. The expression *alkaline earth metals* (Group-2 elements) is still in common use. *Rare earths* refers to the metals of a group of oxides once thought to be quite uncommon. In fact, however, they are rather widely distributed. These metals later came to be called *lanthanides* and *actinides*, after the first element of each series. The modern suffix is *-oid* rather than *-ide*.

6.10 GROUP 12B: THE ZINC SUBGROUP

Zinc, cadmium, and mercury make up the last group in the "transition elements" section of the periodic table. They are often called, collectively, the "zinc subgroup." From the viewpoint of electronic structure, an element may be classified as transitional if its neutral atom or any of its common ions has a partially filled *d* subshell. By this criterion, the elements of the zinc subgroup are not transition elements. Rather, their ns^2 configurations invite a comparison with the representative elements of Group 2:

ZINC-SUBGROUP ELEMENT	GROUP-2 ELEMENT
$_{30}Zn$ $[Ar]3d^{10}4s^2$	$_{20}Ca$ $[Ar]4s^2$
$_{48}Cd$ $[Kr]4d^{10}5s^2$	$_{38}Sr$ $[Kr]5s^2$
$_{80}Hg$ $[Xe]4f^{14}5d^{10}6s^2$	$_{56}Ba$ $[Xe]6s^2$

Atomic radii, nm:
Ca, 0.174
Sr, 0.192
Ba, 0.198
Zn, 0.125
Cd, 0.148
Hg, 0.149

Note that Zn and Ca, for example, both have $4s^2$ valence-shell configurations. However, zinc is higher by 10 in atomic number and has a full $3d^{10}$ subshell. As explained in section 6.4, the net effect of higher nuclear charge and electrons added to *an existing shell* (the $n = 3$ shell) is smaller atomic size. Thus, a zinc atom is smaller than a calcium atom. The valence electrons of a Zn atom are therefore closer to the nucleus and held more tightly by it. As a result, Zn has a greater ionization energy than Ca and it is chemically less reactive.

Similar comparisons may be made between cadmium and strontium, and between mercury and barium. Thus, unlike Ca, Sr, and Ba, the elements of the zinc subgroup do not react with cold water. Cadmium is so resistant to oxidation that it is used as a protective coating for nails and bolts. Mercury is so unreactive that it occurs in some natural deposits as the metallic element rather than a mercury compound.

The elements of the zinc subgroup also undergo different biochemical reactions than those of Group 2. Zinc is an essential element in the human body. In contrast, the compounds formed by cadmium and mercury are extremely toxic, and they have been responsible for serious pollution of waters. The metallic forms are not particularly dangerous to handle, but mercury vapor and cadmium dust are both toxic.

SUMMARY

Mendeleev's periodic table classified a broad range of chemical and physical properties and included predictions of elements to be discovered. Modern periodic tables are based on the electronic configurations of atoms. The vertical columns (**groups**) are numbered according to the number of electrons in the valence shells of the atoms. The horizontal rows (**periods**) are numbered according to the number of shells and subshells of the atoms. These relationships allow us to classify the elements into three types:

TYPE OF ELEMENT	CHARACTERIZED BY STEPWISE ADDITION OF ELECTRONS INTO:
Representative (main-group)	the *s* and *p* subshells
Transition	the *d* subshells
Inner transition	the *f* subshells

The noble gases are a special set of representative elements characterized by the fact that, in each case, the next element starts a new shell.

If **valence** is defined in the simple sense of chemical combining capacity, without regard to plus or minus signs, then the typical valence of a representative element can be expressed in terms of the element's periodic table group number. Similar, but less reliable, predictions can be made for the transition elements.

The radius of a neutral atom is its **atomic radius**, and that of a charged atom—an ion—is its **ionic radius**. Atomic radii generally decrease from left to right within a given period and increase downward in a group. Ionic radii depend strongly on the ionic charge: the more negative the charge, the larger the ion; and the more positive the charge, the smaller the ion.

The minimum energy needed to remove an electron from an atom is its first **ionization energy**. Among the representative elements, first ionization energies generally increase from left to right across a period and decrease downward in a group. The trends in ionization energies run counter to those in atomic radii. The **electron affinity** is the energy change that occurs when an electron is added to a gaseous atom in its ground state. The most exothermic processes all involve halogens (Group-7 elements), and the most endothermic ones occur when a captured electron enters a new shell or subshell.

An element is a substance consisting of atoms that all have the same atomic number. The most abundant element in the Universe is hydrogen. The most abundant element on Earth is iron. Oxygen, however, is the most abundant element in the Earth's crust and in the human body. The nuclei of most naturally occurring elements are stable; a few naturally occurring isotopes and all synthetic ones are unstable, that is, radioactive. Some elements exist as **allotropes**, elements with the same atomic number and physical state but different physical properties.

Hydrogen bears no close resemblance to any other element. The properties of the other representative elements may be briefly summarized as follows:

Group 1. soft, reactive metals that typically exist as M^+ ions in their compounds

Group 2. less reactive metals than those of Group 1; typically form M^{2+} ions

Group 3. metals with typical valence of 3

Group 4. progression of nonmetallic to metallic character with increasing atomic number; typical valence of 4

Group 5. progression of nonmetallic to metallic character with increasing atomic number; typical valence of 3

Group 6. nonmetallic except for the highest member, polonium; typical valence of 2

Group 7. the halogens: reactive nonmetals with typical valence of 1

Group 8. the noble gases: colorless monatomic gases with little reactivity

The transition and inner transition elements are all metals. The zinc subgroup (Zn, Cd, and Hg) does not have partially filled *d* subshells either as neutral atoms or as ions. Their atoms are smaller than those of their counterparts in Group 2; they have higher ionization energies and they are less reactive.

6.11 ADDITIONAL PROBLEMS

THE PERIODIC TABLE

6.8 On the periodic table outlines provided, label the sections containing the representative elements, the transition elements, and the inner transition elements. Add the group numbers and period numbers. Which group contains the noble gases?

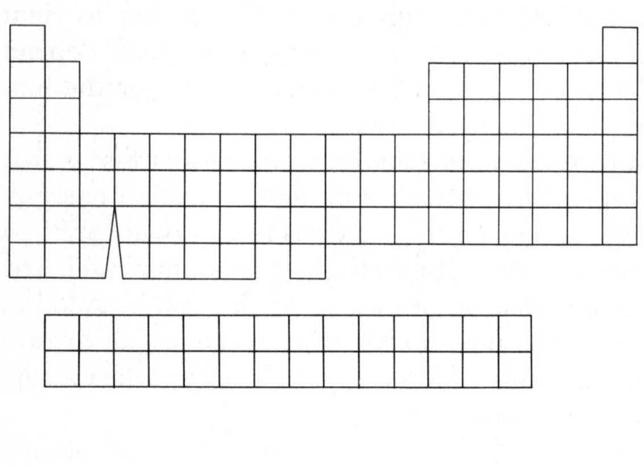

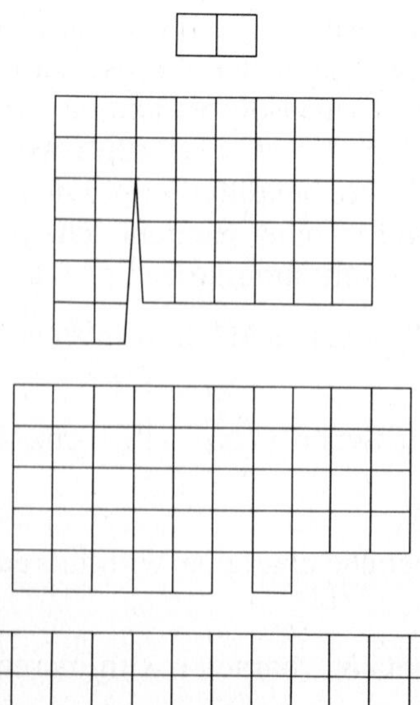

6.9 Use Table 6.3 to predict the electron configurations of (a) $_{50}$Sn; (b) $_{56}$Ba; (c) $_{35}$Br; (d) $_{40}$Zr; (e) $_{77}$Ir; (f) $_{103}$Lr. Explain how the same predictions can be made with the aid of Table 6.4B. Compare your predictions with the values given in Table 5.2 (section 5.7).

6.10 Chemists have postulated that a hitherto undiscovered element, *eka*-platinum, may be stable enough to exist in minute traces on Earth. (a) Predict the atomic number and electron configuration of *eka*-platinum. (b) In minerals of what elements would you look for it?

6.11 State whether each of the following characteristics is more likely to describe the representative elements in a period or a group of the periodic table: (a) all of the elements have the same valence; (b) the elements show sharp differences in chemical properties; (c) the elements differ from each other in the number of electron shells; (d) the atomic numbers of the elements lie between those of successive noble gases.

6.12 Classify each of the following groups of elements as representative, transition, or inner transition. (If it is a group of representative elements, state whether or not they are noble gases.) (a) The highest shells have an s^2p^6 electron configuration; (b) the three highest shells are partially filled; (c) the electron configuration consists of a noble gas core plus a partially filled highest shell; (d) the electron configuration consists of a noble gas core plus two partially filled shells.

6.13 Where in Table 6.3 would elements 110 through 121 be located? Where would they be located in Table 6.4B? (*Hint*: Use Table 6.4A as a guide to answering this question.)

THE PERIODICITY OF VALENCE

Note: Refer to the periodic table inside the front cover.

6.14 Predict the formulas of (a) strontium bromide, (b) cesium sulfide, (c) boron fluoride, (d) gallium arsenide, and (e) magnesium nitride.

6.15 For each of the following formulas, state whether the valence of the underlined representative element corresponds to equation 1, equation 2, or neither. For valences of elements not underlined, assume that H, F, Cl,

Br, I, $(NO_3) = 1$; O, S, Se, Te, $(SO_4) = 2$; N, P, $(PO_4) = 3$. (a) $\underline{Al}PO_4$; (b) $\underline{Ga}_2S_3$; (c) $\underline{Ge}H_4$; (d) $\underline{Ge}Cl_2$; (e) $\underline{P}_2I_4$; (f) $\underline{Pb}Se$; (g) $\underline{B}_2H_6$; (h) $\underline{I}_2O_5$; (i) $\underline{Fr}F$; (j) $\underline{Sb}_2Te_3$; (k) $\underline{SiC}$; (l) $\underline{Rb}_2S$; (m) $\underline{Al}N$; (n) $\underline{Be}_3(PO_4)_2$; (o) $\underline{B}_4H_{10}$; (p) $NaH\underline{S}$; (q) $\underline{Mg}S$; (r) $\underline{As}F_5$; (s) $\underline{Pb}_3O_4$; (t) $\underline{Rb}NO_3$; (u) $\underline{Tl}NO_3$.

ATOMIC AND IONIC RADII

6.16 Using the atomic radii given in Table 6.6, calculate the lengths of these bonds: (a) B—Br in BBr_3; (b) Si—F and Si—Cl in $SiClF_3$. (The experimental values are (a) 0.197 nm and (b) 0.156 nm and 0.199 nm.)

6.17 Using the ionic radii given in Table 6.6, calculate the K^+—Br^- distance in a potassium bromide crystal, and the Mg^{2+}—O^{2-} distance in a magnesium oxide crystal. (The experimental values are 0.3293 and 0.2104 nm, respectively.)

6.18 Explain these data: the decrease in the radius for the elements $_{11}Na$ to $_{17}Cl$ is 0.055 nm, whereas the decrease for $_{21}Sc$ to $_{30}Zn$ is only 0.019 nm.

6.19 Note the atomic and ionic radii of the atoms from $_{14}Si$ to $_{21}Sc$ given in Table 6.6. (a) With what neutral atom are all of these ions isoelectronic? Explain (b) the variation in atomic size and (c) the relationship between the size of each ion and its parent atom.

6.20 Account for the very large difference between the ionic radii of Sb and Bi shown in Table 6.6.

ISOELECTRONIC SPECIES

6.21 What electrical charge, if any, must be assigned to each of these atoms to make it isoelectronic with Cl^-? (a) P; (b) Ca; (c) S; (d) Ar; (e) K.

6.22 Identify the species in each set that is *not* isoelectronic with the other three: (a) O^-, F^-, Mg^{2+}, Ne; (b) Br, Se^-, As^{2-}, Rb^+; (c) Xe^+, I^-, Te^-, Cs^{2+}.

6.23 How many electrons would have to be removed from a K atom so that the resulting ion would be isoelectronic with Na^+? What is the charge on this hypothetical K ion? Would you expect this ion to be larger or smaller than a Na^+ ion; than a Li^+ ion? Explain.

IONIZATION ENERGY

6.24 (a) Choose the correct words in the parentheses: In general, ionization energies of the representative elements (*increase, decrease*) from left to right across a period and (*increase, decrease*) down a group. These trends are (*parallel, opposite*) to those shown by the atomic radii. (b) Explain these trends.

6.25 Identify the regions of the periodic table in which (a) elements differing by only 1 in atomic number show the largest differences in ionization energy; (b) elements with successively greater atomic numbers show irregular trends in ionization energy. Account for these effects.

6.26 Using the data in Table 6.7, plot the first ionization energy (y axis) versus atomic number (x axis) for the third period and for the transition elements Sc to Cu on the same scale. What conclusion(s) could you draw from these graphs?

6.27 Refer to Figure 5.9 (section 5.4) and note the energies in J/atom given for several values of n from 1 to ∞. The ionization energy of hydrogen corresponds to the transfer of an electron from $n = ?$ to $n = ?$ Verify your answer by calculating the ionization energy in kJ/mol and comparing it with the value given in Table 6.7.

6.28 Would the ionization energy of a hydrogen atom in an excited state be greater or less than the value given in Table 6.7?

6.29 Predict a value for the ionization energy of francium.

6.30 Helium atoms can be promoted to an excited state whose energy is 19.6 eV/atom. This excited helium can ionize other gases by transferring its excess energy to them. Which element(s) shown in Table 6.7 *cannot* be ionized by excited helium?

6.31 True or false: The second ionization energy of an atom is always greater than the first ionization energy. Defend your answer.

6.32 All first ionization energies are positive, whereas some elements' electron affinities are positive and others' are negative. Explain why electron affinities do not show the same consistency of charge sign.

ELECTRON AFFINITY

6.33 True or false: The electron affinity of a negative ion is always more positive (endothermic) than that of the neutral atom. Defend your answer.

6.34 Note in Table 6.8 that all of the elements in Groups 2 and 8 have positive electron affinities. (a) Is the capture of an electron by one of these atoms exothermic or endothermic? (b) Suggest explanations for the positive values of the Group-2 elements and (c) the Group-8 elements.

6.35 Which of the following substances could be readily detected by their ability to capture electrons? For which of them would detection depend on the electron affinity of their atoms? (a) magnesium nitride, Mg_3N_2; (b) chlorofluoromethanes, such as CCl_2F_2 and CCl_3F; (c) sulfur oxybromide, $SOBr_2$; (d) krypton, Kr.

DESCRIPTIVE CHEMISTRY

6.36 Which of these pairs of substances are (a) allotropes, (b) isotopes, (c) different physical states of the same element, and (d) different elements? (1) radon-215 and radium-215; (2) rhombic sulfur (prismlike crystals) and monoclinic sulfur (needlelike crystals); (3) ^2H and ^3H; (4) liquid bromine and gaseous bromine, both Br_2.

6.37 Point out the deficiencies in each of these definitions of an element: An element is a substance that (a) cannot be decomposed; (b) consists of atoms that are all alike.

6.38 (a) What is the most abundant metal in the Earth; in the Earth's crust; in the human body? (b) What nonmetal is most abundant in all three of these places?

6.39 Iron pyrite, FeS_2, a yellow solid with a density of 5 g/cm^3, is sometimes called "fool's gold." How could such a nickname have originated? If you thought you had discovered gold, what quick test could you conduct to make certain your find was not fool's gold?

6.40 Wax can be hammered into a sheet; silk or nylon can be formed into a wiry filament; molten salt conducts electricity; cinnabar, HgS, is very dense (8 g/cm^3). Yet none of these substances is metallic. Explain in each case why the property described does not demonstrate that the substance is a metal.

6.41 Referring to the periodic table, identify the representative elements that (a) are gaseous at 1 atm pressure and 20°C; (b) are liquid at these conditions; (c) are colored; (d) do not occur naturally on Earth in any significant quantities.

6.42 Aluminum is a reactive metal, yet it resists corrosion in air. Explain why.

6.43 A gas may be odorous and toxic, nonodorous and toxic, or nonodorous and nontoxic. Write the formulas for the following gases and classify them according to those categories: (a) carbon monoxide; (b) dinitrogen; (c) ozone; (d) chlorine; (e) arsine; (f) dioxygen.

6.44 Give two important industrial applications of compounds of (a) silicon, (b) sulfur, and (c) lead.

6.45 Explain why the elements of the zinc subgroup are much less reactive than those of Group 2.

SELF-TEST

Note: You may refer to the periodic table inside the front cover.

6.46 (10 points) Study the periodic table and sketch a blank outline of the first four periods. Then, without referring to the table, write the symbols and names of elements 1 through 20 in your outline.

6.47 (8 points) K, Mg, Si, and Te each form a hydride in which the valence of the element equals its periodic table group number. Write the formulas for these hydrides.

6.48 (8 points) Identify each of the following pairs as (1) allotropes, (2) isoelectronic species, (3) noble gases, or (4) transition elements: (a) Rb^+ and Br^-; (b) gray, crystalline selenium and red, noncrystalline selenium; (c) Ti and element 105; (d) Xe and Rn.

6.49 (8 points) (a) Select from the following list the atom with (1) the largest radius and (2) the smallest radius: $_{14}Si$, $_{17}Cl$, $_{20}Ca$, $_{12}Mg$. (b) From this list, select the ion with (1) the largest radius and (2) the smallest radius: Mg^{2+}, P^{3-}, Si^{4+}, Al^{3+}.

6.50 (8 points) From each of the following sets, choose the atom or ion with (1) the highest ionization energy and (2) the lowest ionization energy: (a) $_3Li$, $_6C$, $_9F$; (b) $_{11}Na$, $_{19}K$, $_{37}Rb$; (c) $_4Be$, $_7N$, $_{10}Ne$; (d) Ag, Ag^+, Ag^{2+}.

6.51 (8 points) From each of the following sets, select the atom with the greatest tendency to gain an electron (that is, the element that evolves the largest quantity of heat from the gain of an electron): (a) $_3Li$, $_6C$, $_9F$; (b) $_{14}Si$, $_{17}Cl$, $_{13}Al$.

6.52 (24 points) Write balanced chemical equations for these 24 transformations:

- reaction of sodium with water, followed by spontaneous ignition of the hydrogen (two equations)
- combustion of Mg in air (two equations—one showing reaction with oxygen, the other with nitrogen)
- reaction of strontium with water
- decomposition of Ag_2O to silver and oxygen
- burning of diamond in oxygen
- burning of graphite to yield carbon monoxide
- oxidation of nitrogen to NO_2 (two steps)
- combination of N_2 with hydrogen to produce ammonia
- decomposition of arsine (the hydride of arsenic) to its elements
- combustion of phosphorus vapor in air to form P_4O_{10}
- conversion of oxygen to ozone
- oxidation of sulfur vapor, S_8, to the dioxide
- oxidation of sulfur dioxide to the trioxide
- oxidation of hydrogen selenide to water and selenium dioxide
- reaction of hydrogen sulfide with sulfur dioxide to produce sulfur and water
- combination of fluorine with sodium, with magnesium, and with aluminum (separate equations)
- reaction of chlorine, bromine, and iodine with hydrogen gas to form their respective hydrides (separate equations)

6.53 (26 points) From the choices in parentheses, select the correct description(s) of each element or group of elements. (Score 2 points for each answer.)

- hydrogen (colorless, shiny, denser than ozone, less dense than air)
- alkali metals (waterproof, lustrous, soft, very dense, resistant to oxidation, rapidly reactive with water)
- Group-2 elements (very hard, very soft, semimetallic, waterproof, reactive with oxygen when heated, divalent)
- Group-3 elements (all metallic except boron, all shiny except boron, moderately reactive, inert)
- Group-4 elements (metallic, nonmetallic, range from nonmetallic to metallic, include some semimetals)
- Group-5 elements (metallic, nonmetallic, range from nonmetallic to metallic, include elements that exhibit allotropy)

- Group-6 elements (metallic, nonmetallic, range from nonmetallic to metallic, exhibit allotropy, include some elements whose hydrides stink)
- Group-7 elements (include elements in three different physical states at normal temperatures, the *only* group in the periodic system that includes elements in all three states, all are colored, all are nontoxic)
- noble gases (colored, reactive, colorless, unreactive)
- transition elements (mostly metals, all metals, all less dense than granite rock, include the densest substances known)
- zinc subgroup (typical transition elements, typical representative elements, form some toxic compounds)
- lanthanoids (mostly metals, all metals, similar to each other in many properties)
- actinoids (mostly metals, all metals, mostly radioactive, all radioactive)

7

TYPES OF CHEMICAL BONDS

Molecules consist of combined atoms. Even most elements normally exist as molecules of more than one atom. For example, oxygen, nitrogen, hydrogen, and the halogens are made up of diatomic molecules. Yellow sulfur and white phosphorus exist as molecules whose formulas are S_8 and P_4. The "molecules" of diamond and graphite (both forms of carbon) and of red phosphorus consist of trillions of atoms. Metallic elements, too, such as copper and potassium, are composed of bonded atoms, generally in a crystalline form.

How do atoms combine and what are the forces that bind them? These questions are fundamental to the study of chemistry since chemical change is essentially an alteration of chemical bonds. Of the various forces that exist in nature, only electrical attractions between atoms can account for the observed bond energies. This chapter discusses the bonding of atoms in terms of their electronic structure (Chapter 5) and their periodicity (Chapter 6).

7.1 BONDING TYPES

To illustrate the different types of bonds, we will describe some tests with four familiar substances: a piece of copper (Cu), a large crystal of table salt (sodium chloride, NaCl), a chunk of ice (H_2O), and some pure, white sand (quartz, silicon dioxide, SiO_2; Figure 7.1). These particular substances have been chosen because they represent, as you will see, four distinct types of chemical bonding. These substances are tested with the electrical conductivity apparatus shown in Figure 7.2. Electrical conduction requires the movement of charged particles. Hence, we are testing the four substances for the presence of movable charged particles.

FIGURE 7.1
Quartz crystals. Pure sand is pulverized quartz. (Courtesy of Ward's Natural Science Establishment, Rochester, NY.)

When the rods touch the copper slab, the bulb glows brightly, as in Figure 7.2a. Copper, a typical metal, is a good conductor of electricity. No matter how long the current flows, the copper wire is unchanged. This ability to conduct electricity without undergoing chemical change is typical of metals. In metals the movable charged particles are electrons. This is discussed in more detail in section 22.2.

When the rods touch the NaCl crystal, the bulb does not light (Figure 7.2b). If the NaCl crystal in a porcelain dish is melted (at about 800°C) and then touched by the rods, the bulb glows. Thus, whereas solid NaCl does not conduct an electric current, liquid NaCl does. There is one major difference, however, between the liquid salt and a metal. In contrast to metallic copper, the salt is chemically changed by the current. In the absence of air the salt is decomposed into chlorine (Cl_2) gas, which appears at one rod, and molten sodium metal (Na), which collects

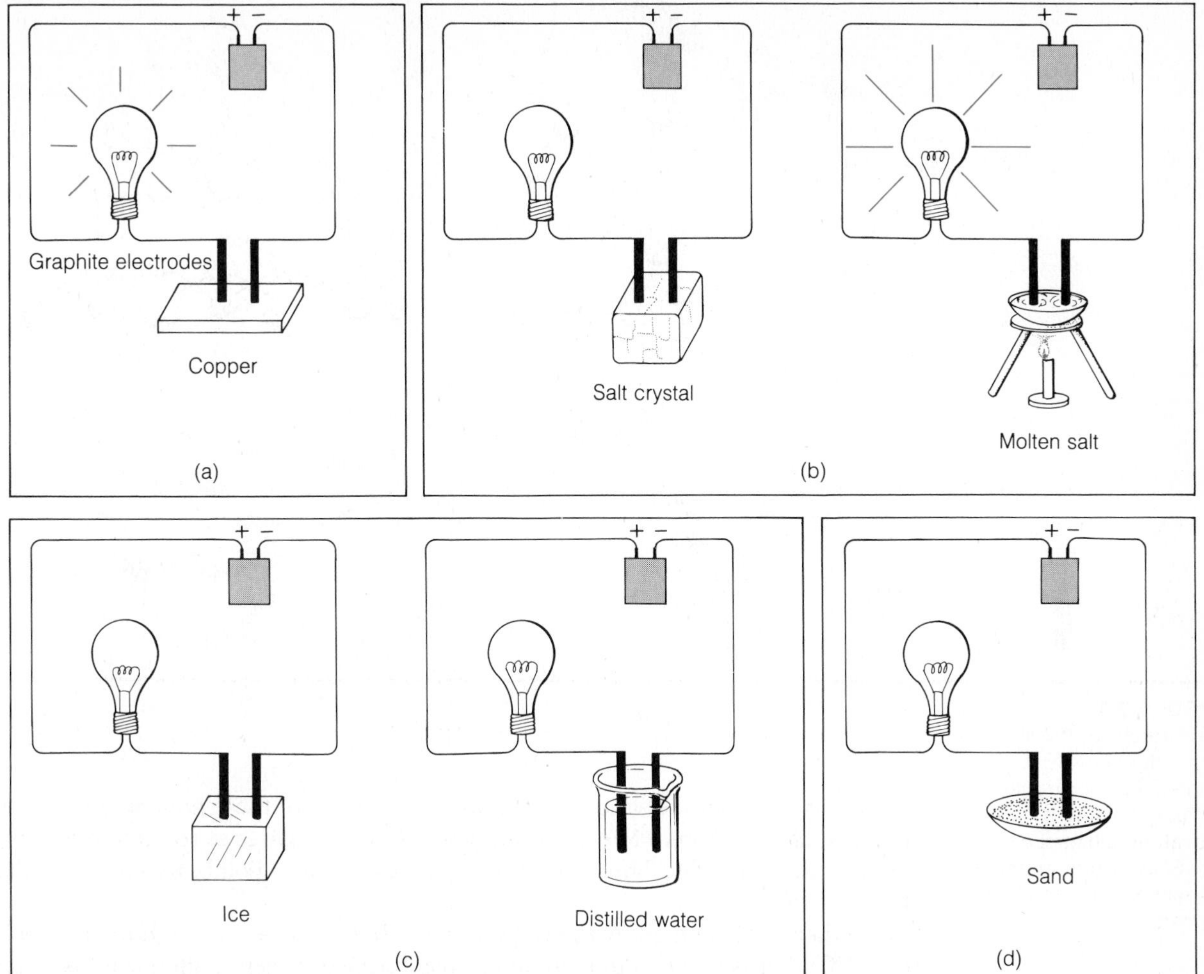

FIGURE 7.2
Conductivity apparatus. Bulb lights up when (a) copper and (b) molten sodium chloride are tested. Bulb does not light up when (b) a salt crystal, (c) ice or liquid water, or (d) quartz is tested.

Safety Note: Cl_2 is toxic; this experiment is done in a hood.

at the other rod:

$$2NaCl(c) \xrightarrow[\text{current}]{\text{electric}} 2Na(\ell) + Cl_2(g)$$

In liquid NaCl, the movable charged particles are the ions, Na^+ and Cl^-. NaCl is a typical *ionic compound,* also called a **salt**. Salts are composed of ions in the solid state as well as in the liquid state. However, the solid is not a conductor because the positive and negative ions in the crystal attract each other so strongly that they do not move about. In the liquid state the ions of salts are free to migrate and conduct an electric current because the increase in kinetic energy of the ions partially overcomes the attraction between the ions in the solid state. Much energy, in the form of heat, is needed to overcome the attractive forces between the oppositely charged ions in the solid state. For this reason salts have high melting and boiling points.

Tap water may conduct because of dissolved salts.

When ice is tested for electrical conductivity, the bulb does not glow (Figure 7.2c). Even when the ice is melted, testing the resulting liquid water does not cause the bulb to glow. Pure water is typical of substances that do not conduct

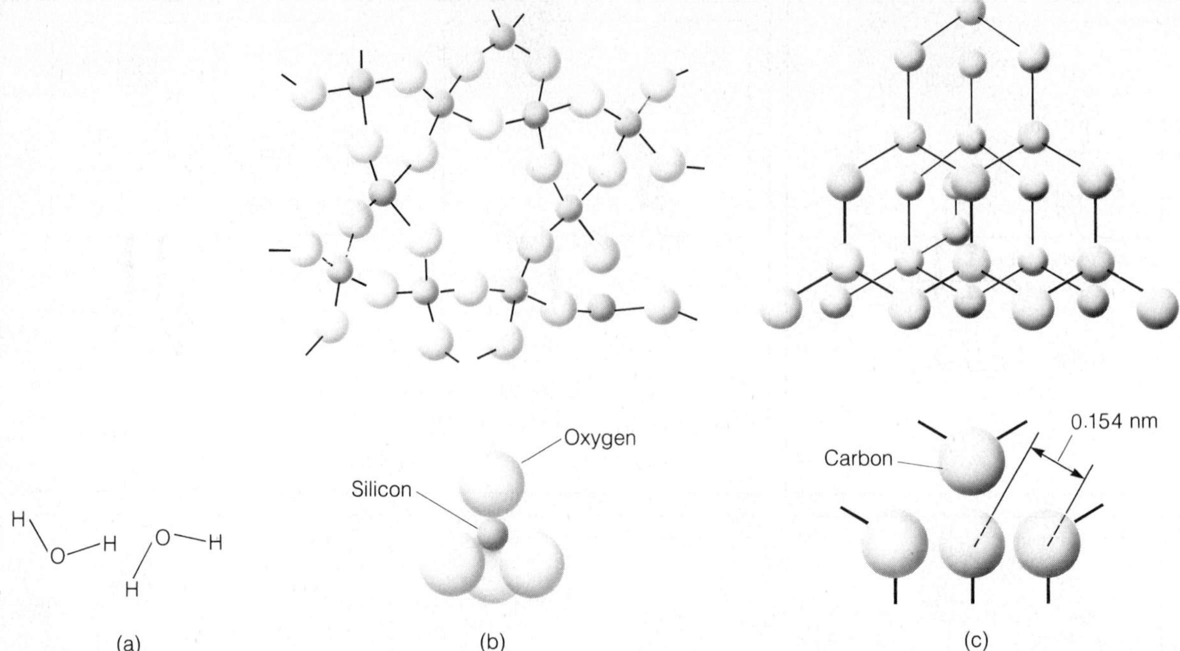

FIGURE 7.3
(a) Structure of water, an individual covalent molecule. (b) Structure of silicon dioxide, a network covalent substance. (c) Structure of diamond, a network covalent substance.

electricity in either the solid or the liquid state. Some of these nonconducting substances, like ammonia, NH_3, and methane, CH_4, are gases at room temperature. Others are solids and liquids with melting and boiling points much lower than those of salts.

Neither does sand allow the bulb to glow, but it has a very high melting point (1710°C) that is reached only in an electric furnace. When molten sand is tested, the bulb still does not glow. Sand is typical of substances that do not conduct electricity in any state, but which, unlike H_2O, have very high melting and boiling points.

These four typical substances have strikingly different physical properties, which indicate their bonding differences. The atoms in copper, a typical metal, are held together by **metallic bonds**. The bond in NaCl is an **ionic bond**. Since water and sand are nonconductors, they have no movable charged particles; they have neither ionic nor metallic bonds. Their atoms are said to be joined by **covalent bonds**. However, the very large differences in melting and boiling points between water (m.p. 0°C, b.p. 100°C) and sand (m.p. 1710°C, b.p. 2650°C) indicate some other difference in bonding. Melting and boiling require the separation of the particles composing the solid and liquid states, respectively. These separations occur more easily in H_2O than in SiO_2. The difference is attributed to the nature of the unit particles that have to be separated. In H_2O, each H atom is covalently bonded to the O atom to give a discrete molecule, H_2O (Figure 7.3a). H_2O is a **covalent molecule**; its molecules are the particles that must be separated from each other. No covalent bonds between atoms are broken when H_2O melts or boils.

SiO_2 is quite different from water. The Si atom is not bonded to two O atoms to give an individual SiO_2 molecule. Instead, each Si atom is bonded to four O

TABLE 7.1
PROPERTIES OF SUBSTANCES CLASSIFIED BY BONDING TYPE

TYPE OF BOND	ELECTRICAL CONDUCTIVITY Solid	ELECTRICAL CONDUCTIVITY Liquid	BOILING POINT RANGE	UNIT PARTICLE	EXAMPLES
Ionic	no[‖]	yes[†]	high (700°C to 3600°C)	ion	NaCl, KF
Covalent Molecule	no	no	low (−253°C to 600°C)	molecule	H_2O, NH_3, CO_2
Covalent Network	no	no	very high (2000°C to 6000°C)	atom	SiO_2, diamond
Metallic	yes[§]	yes[§]	high to very high (375°C to 6000°C)	atom	Cu, Na, Hg

[†] Chemical change occurs.
[§] No chemical change occurs.
[‖] Some exceptions exist.

atoms, and each O atom is bonded to two Si atoms, as shown in Figure 7.3b. The bonding between Si and O goes on and on in three dimensions; *there are no individual molecules of SiO_2*. The entire piece of solid can be considered a giant molecule, a **network covalent** substance. The formula SiO_2 is really the empirical formula—not the molecular formula—of the network substance. The particle in the solid that must be separated is the bonded atom. Therefore SiO_2 melts and boils only when covalent bonds between atoms are broken. *It takes much more energy to break covalent bonds than to separate molecules from each other.* Therefore, network covalent substances have much higher melting and boiling points than molecular covalent substances. Table 7.1 summarizes the differences in properties of the four types of substances. This chapter examines the formation of ionic and covalent bonds. Metallic bonding will be discussed in Chapter 22.

7.2 LEWIS THEORY OF BONDING, THE OCTET RULE

Bonding between atoms may be described in terms of a model that involves interaction between the nuclei and the electrons of the atoms. With the representative elements, the interacting electrons are in the highest occupied shell. These electrons are called **valence** or **outer-shell electrons**. Bonding of transition elements may involve electrons in the next-to-highest as well as the highest shell. Our discussion in this chapter is restricted to bonding of the representative elements; bonding of transition elements will be discussed in Chapter 21.

Symbols suggested by Gilbert N. Lewis in 1916 are used to represent the representative elements. A **Lewis symbol** consists of the symbol for the particular element and a dot for each valence electron it has. The number of valence electrons in an unbonded atom is the same as the atom's periodic group number. Lewis symbols are adapted to represent ions by adding a dot for every negative charge on the ion and removing a dot for every positive charge on the ion. Hence, an ion and its respective unbonded atom have a different number of valence electrons.

Lewis symbols for the first ten elements and their common ions are shown.

1	2	3	4	5	6	7	8
H· H^+ $H:^-$							He:
Li· Li^+	Be: Be^{2+}	·B:	·Ċ:	·N̈· $:\ddot{N}:^{3-}$	:Ö· $:\ddot{O}:^{2-}$	:F̈· $:\ddot{F}:^{-}$	:N̈e:

Note that $:\ddot{N}:^{3-}$, $:\ddot{O}:^{2-}$, and $:\ddot{F}:^-$ have the same valence shell electronic arrangement as :N̈e: (s^2p^6). These ions are **isoelectronic** with :N̈e:. The ions $H:^-$, Li^+, and Be^{2+} are isoelectronic with He: because they have the $1s^2$ electron arrangement.

·Ċ: would indicate the arrangement in the atomic orbitals.

Lewis symbols need not necessarily show how electrons are paired in atomic orbitals. The symbol for carbon, for example, could be written as ·Ċ·, ·Ċ:, or :C:.

EXAMPLE 7.1 Write the Lewis symbols for Al, Al^{3+}, Ba, Ba^{2+}, Se, Se^{2-}, P, and P^{3-}.

ANSWER These are all atoms or ions of representative elements. For the uncharged atoms, the number of valence electrons is the same as the group number. Therefore we have ·Al: (Group 3), Ba: (Group 2), :Ṡe: (Group 6), and :Ṗ· (Group 5).

Positive ions result when the atom loses electrons; negative ions, when the atom gains electrons. Thus when Ba: loses 2 electrons it becomes Ba^{2+}, and when :Ȧl loses 3 electrons it becomes Al^{3+}. (No valence electrons remain, so no dots are shown.) For the negative ions, :Ṡe: becomes $:\ddot{Se}:^{2-}$ on gaining 2 electrons, and :Ṗ· becomes $:\ddot{P}:^{3-}$ on gaining 3 electrons. ■

Irving Langmuir, 1881–1957

PROBLEM 7.1 (a) Write the Lewis symbols for the third-period (Na to Ar) elements. (b) Write the Lewis symbols for ions of these elements that are isoelectronic with Ar. (c) Write the Lewis symbols for ions of these elements that are isoelectronic with Ne. (Avoid forming ions with charges of more than +3 or −3.) □

Compounds of the noble gas Xe were first discovered in 1962. Compounds of Kr have also been made.

An important clue to the understanding of chemical bonding was the discovery, in the late nineteenth century, of the noble gases and their apparent resistance to chemical change. Gilbert Lewis proposed the relationship between the bonding of atoms and the electronic configuration of noble gases in 1916, and Irving Langmuir extended it in 1919. Lewis and Langmuir reasoned as follows: Electrons in atoms are involved in chemical bonding. Atoms bond with one another so as to acquire new and more stable electron configurations. Since no compounds of the noble gases were known, the arrangements of their electrons must already be stable. Therefore, *all other atoms undergo bonding by gaining, losing, or sharing electrons so as to acquire the electronic arrangement of the nearest noble gas.* With the exception of :He, which has a $1s^2$ electron arrangement, each noble gas has 8 electrons with an s^2p^6 distribution in its highest shell (Table 6.3):

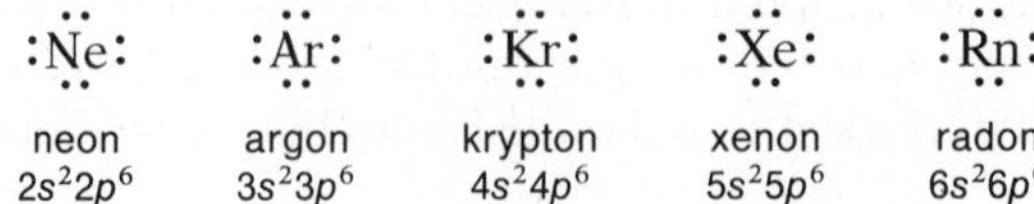

The stable ions of N, O, and F, for example, also have octets of electrons:

$:\ddot{\underset{..}{N}}:^{3-}$	$:\ddot{\underset{..}{O}}:^{2-}$	$:\ddot{\underset{..}{F}}:^{-}$
nitride ion	oxide ion	fluoride ion

The need for 8 electrons gives the name **octet rule** to this concept. The elements hydrogen, lithium (Li), and beryllium (Be) are close to He in the periodic table and therefore react to acquire a **duet** of electrons instead of an octet.

7.3 IONIC BOND

How do the ions of a salt form from the atoms? We use potassium bromide, KBr, a typical ionic salt, to examine how this occurs on an atomic level. A potassium atom, K· (atomic number 19), loses 1 electron to become K^+, an ion isoelectronic with argon (atomic number 18), a noble gas. A bromine atom, $:\ddot{\underset{..}{Br}}\cdot$ (atomic number 35), gains 1 electron to become $:\ddot{\underset{..}{Br}}:^-$, an ion isoelectronic with krypton (atomic number 36), another noble gas. Such *transfer of electrons* results in the formation of ions that strongly attract each other to form the **ionic bond**. Lewis symbols are used here to show the electron transfer to form KBr.

A half-headed arrow (⇀) indicates movement of a single electron.

transfer of an electron

$$\mathbf{K\cdot\ [Ar]4s^1 + :\ddot{\underset{..}{Br}}\cdot\ [Ar]3d^{10}4s^24p^5 \longrightarrow K^+[Ar] + :\ddot{\underset{..}{Br}}:^-[Kr]}$$

argon configuration **krypton configuration**

or

$$K\cdot + :\ddot{\underset{..}{Br}}\cdot \longrightarrow K^+:\ddot{\underset{..}{Br}}:^- \quad \text{(KBr, an ionic salt)}$$

lewis symbols a lewis formula

Diagrams that use Lewis symbols to show the structure of compounds are called Lewis structural formulas, or simply **Lewis structures** or **electron-dot** structures. When atoms react by electron transfer, the number of electrons *gained and lost must be equal* because the resulting ionic salt is neutral. We illustrate this idea by using Lewis symbols to show the formation of ionic aluminum fluoride, AlF_3, from aluminum (Group 3) and fluorine (Group 7).

$$\dot{Al}\cdot + 3\,\cdot\ddot{\underset{..}{F}}: \longrightarrow Al^{3+} + 3:\ddot{\underset{..}{F}}:^-$$

$$Al + 3F \longrightarrow AlF_3 \quad (AlF_3\text{, aluminum fluoride})$$

A chemical formula, such as AlF_3, is written without the ionic charges.

$\dot{\underset{.}{Al}}\cdot$ must lose 3 electrons to form a stable ion, but $\cdot\ddot{\underset{..}{F}}:$ can gain only 1 electron. Therefore 3 $\cdot\ddot{\underset{..}{F}}:$ atoms are needed to accept the 3 electrons lost by 1 $\dot{\underset{.}{Al}}\cdot$ atom.

EXAMPLE 7.2 Use Lewis symbols to depict the formation of these ionic compounds: (a) calcium hydride; (b) lithium oxide; (c) magnesium nitride.

ANSWER

(a) The Lewis symbols are H· and Ca: (Group 2). Two H· atoms are needed to accept the 2 electrons from one Ca: atom,

$$Ca\!: + 2H\cdot \longrightarrow Ca^{2+} + 2H\!:^{-} \qquad (CaH_2\text{, calcium hydride})$$

(b) Two Li· (Group 1) atoms are needed to furnish the 2 electrons needed by the single :Ö: (Group 6) atom.

$$2Li\cdot + :\ddot{O}: \longrightarrow 2Li^{+} + :\ddot{O}:^{2-} \qquad (Li_2O\text{, lithium oxide})$$

(c) Mg: (Group 2) must lose 2 electrons, but ·N· (Group 5) needs 3 electrons. Since 2 and 3 are both factors of 6, a total of 6 electrons must be transferred:

$$3Mg\!: \;+\; 2\cdot\dot{N}\cdot \longrightarrow 3Mg^{2+} + 2\!:\ddot{N}\!:^{3-} \qquad (Mg_3N_2\text{, magnesium nitride})$$

loses 6 electrons; gains 6 electrons ■

PROBLEM 7.2 Use Lewis symbols to depict the formation of these ionic compounds: (a) cesium fluoride, (b) calcium oxide, (c) aluminum sulfide. □

Atoms that lose electrons to form positive ions are generally the metals. Atoms that gain electrons to form negative ions are generally the nonmetals. In general, metallic and nonmetallic elements react to form salts. The loss of electrons is called **oxidation**, and the gain of electrons is called **reduction**.[†] By these definitions, the formation of an ionic bond from *elements* must involve an **oxidation–reduction** (or **redox** for short) reaction. The metallic element is oxidized, and the nonmetallic element is reduced.

7.4 FORMATION OF IONIC SOLIDS

We can introduce this topic by asking the question, what factors most favor ionic bond formation between atoms? For example, it is observed that whereas Al reacts with F_2 to form AlF_3, an ionic compound, it reacts with Cl_2 to form Al_2Cl_6, a covalent molecule. The likelihood of an ionic compound being formed is related to the stability of the salt that is formed. One factor that determines that stability is the heat of formation (ΔH_f^0) of the salt, as shown by the general equation:

M and N must be in their standard states (section 4.6).

$$\underset{\text{metal}}{M} + \underset{\text{nonmetal}}{N} \longrightarrow \underset{\text{salt}}{MN(c)} \qquad (\Delta H_f^0\text{, the heat of formation})$$

If ΔH_f^0 is negative, the reaction is exothermic, which means the salt is stable and likely to be formed. This section is therefore concerned with the thermochemistry of ionic bonding.

In the previous section the formation of KBr was described as a reaction between an individual potassium atom, K, and an individual bromine atom, Br. However, combining K and Br atoms is not a practical way to make KBr. Potassium is a soft, silvery *solid*, K(c). Bromine is a red-brown *liquid* consisting of molecules of Br_2. Since these are the standard states of K and Br_2, the thermo-

K is usually coated with a layer of dull oxide.

[†] Originally, oxidation referred to the addition of oxygen to elements, and reduction meant a return to the elementary metallic state. The broader concept of oxidation as a loss of electrons and reduction as a gain was first suggested by Wilhelm Ostwald in 1903.

chemical equation for the enthalpy of formation, ΔH_f^0, of KBr (section 4.6) is

Safety Note: This reaction should be done in a hood.

$$K(c) + \tfrac{1}{2}Br_2(\ell) \longrightarrow KBr(c) \qquad \Delta H_f^0 = -392.0 \text{ kJ} \tag{1}$$

not $K(g) + Br(g) \longrightarrow KBr(c)$. The large negative value for the ΔH_f^0 of KBr indicates an exothermic reaction. This exothermicity becomes apparent when a piece of potassium is dropped into liquid Br_2 in a flask (Color plate 11). The reaction makes noise and emits sparks of light as the flask becomes hot. The negative enthalpy of formation of KBr formation is typical of the formation of ionic compounds.

See Appendix C for ΔH_f^0 values of several common ionic compounds.

Why is the formation of an ionic compound, such as KBr, from its elements exothermic? Does the exothermicity arise from the transfer of electrons from K(g) atoms to Br(g) atoms (reaction 2)?

$$K\cdot(g) + :\ddot{\underset{\cdot\cdot}{Br}}\cdot(g) \longrightarrow K^+(g) + :\ddot{\underset{\cdot\cdot}{Br}}:^-(g) \tag{2}$$

Since separated atoms and ions are considered to be in the gaseous state in thermochemical equations, they are shown with the symbol (g). Although the ΔH of reaction 2 cannot be measured, it can be calculated using Hess's law (section 4.5) and shown to have a value of +418 kJ. Clearly, reaction 2 is endothermic, but we have not told the complete story. The ΔH of reaction 2 is not the ΔH_f of KBr. In reaction 2 the reactants are gaseous atoms and the products are gaseous ions. However, the reaction is actually between *solid* potassium and *liquid* Br_2 and gives *solid* KBr rather than separated ions. Somehow the solid potassium and the liquid Br_2 are broken up, and crystalline KBr salt is produced. We can postulate how this might occur by suggesting some reasonable steps with measurable ΔH values. The sequence of these steps is immaterial because ΔH is independent of the pathway. Then, according to Hess's law, the sum of these individual ΔH values must equal the heat of formation of the KBr (reaction 1). This application of Hess's law of thermochemical analysis is called the **Born–Haber cycle**. The steps of the cycle are illustrated for KBr(c) in Figure 7.4 and explained briefly here:

STEP 1. Vaporization, the change of liquid Br_2 to gaseous Br_2 (*endothermic*). Since $\tfrac{1}{2}$ mole of Br_2 is used, half of the molar heat of vaporization is used; $\tfrac{1}{2}\Delta H_{vap} = +15.0$ kJ.

STEP 2. Bond energy of Br_2 (*endothermic*). Again, since $\tfrac{1}{2}$ mol Br_2 is dissociated, half of the bond energy is used. $\tfrac{1}{2}\Delta H_{BE} = +96.6$ kJ.

STEP 3. Electron affinity of Br (*exothermic*); $\Delta H_{EA} = -341.4$ kJ.

STEP 4. Sublimation, the change of solid K(c) to gaseous K(g) (*endothermic*); $\Delta H_{sub} = 89.9$ kJ.

STEP 5. Ionization energy of K; $\Delta H_{IE} = 418.4$ kJ (*endothermic*).

STEP 6. Combination of the gaseous ions to form the solid salt (*very exothermic*). This process is the reverse of the conversion of the solid into gaseous ions, the ΔH for which is called the **lattice energy**, ΔH_{LE}.

The lattice energy is much greater than the energy needed to melt or vaporize the salt. Salt vapors contain molecules.

$$KBr(c) \longrightarrow K^+(g) + Br^-(g) \qquad \Delta H_{LE} = +668.4 \text{ kJ}$$

This process is endothermic because, to get free ions, the electrostatic attraction of the oppositely charged ions must be overcome. The sign of ΔH for step 6 is, therefore, the reverse of ΔH_{LE}; ΔH for step 6 = -668.4 kJ. This step is highly exothermic because the ions arrange themselves in the crystal in a pattern that

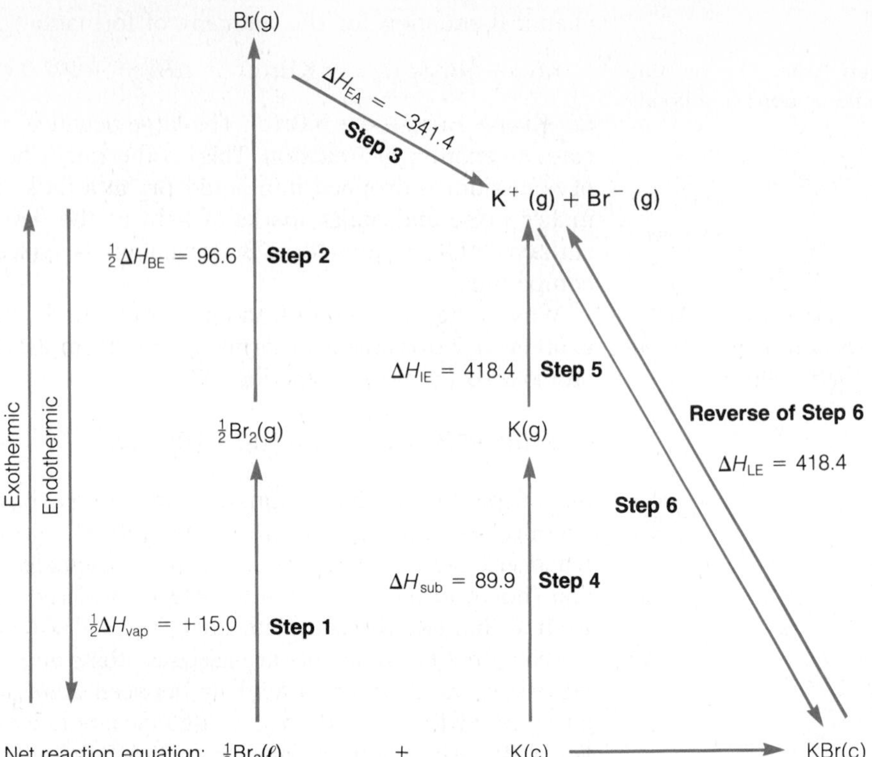

FIGURE 7.4
Born–Haber cycle for KBr(c). All ΔH values are in kJ.

Fritz Haber, 1868–1934

maximizes the electrostatic attraction between oppositely charged ions while minimizing the repulsion between ions with like charges (section 10.5).

The addition of these six steps gives reaction 1 with a calculated value of -389.8 kJ—close to the experimental value of -392.0 kJ for the heat of formation (ΔH_f^0) of KBr.

An important aspect of the energetics of this process is revealed when the first five steps are added. The net equation thus obtained is for the formation of gaseous K^+ and Br^- from their elements.

$$\text{Sum of first five steps} = K(c) + \tfrac{1}{2}Br_2(\ell) \longrightarrow K^+(g) + Br^-(g) \qquad \Delta H = +278.5 \text{ kJ}$$

This reaction is endothermic. Therefore, it is the very exothermic step (step 6) that accounts for the large release of energy when KBr(c) is formed from K(c) and $Br_2(\ell)$. The Born–Haber discussion of all ionic solids shows the same behavior; the lattice energy largely accounts for the high stability of ionic solids. Remember that, as defined, lattice energy is always positive because the process of separating ions is endothermic. Therefore, the larger the lattice energy, the more stable is the salt.

This is like bond energies; the more positive the ΔH_{BE}, the more stable is the bond.

Except for the lattice energies, all of the ΔH values used in the Born–Haber cycle are obtainable experimentally. Lattice energies are most often obtained by using the cycle as indicated in Example 7.3.

EXAMPLE 7.3 Calculate the lattice energy of LiF(c) from the following data in kJ/mol. ΔH_f^0 for LiF(c) $= -594.1$, ΔH_{sub} for Li(c) $= +155.2$, ΔH_{IE} for Li(g) $= +520$, ΔH_{BE} for F_2(g) $= +150.6$, ΔH_{EA} for F(g) $= -333$.

BOX 7.1

MORE ON $K(c) + Br_2(\ell)$

You may wonder why this reaction ever gets started in the first place. After all, how do the K atoms in the solid state and the Br_2 molecules in the liquid state "know" that the energy invested in their conversion into gaseous ions will be more than repaid by the lattice energy? The answer, of course, is that K and Br_2 don't know anything at all, and therefore the reaction probably does not proceed by the steps of the Born–Haber cycle. Some other sequence may occur. The beauty of measuring ΔH values is that the differences between the enthalpies of any two states are *independent of the reaction pathways.* Even if the chosen pathway is wrong, the calculated ΔH value must come out correct if the ΔH value for each step is correct. However, do not think that chemists do not care about the details of reaction pathways. Their concern is the subject of Chapter 19.

ANSWER The steps, equations, and ΔH values are

STEP	EQUATION	ΔH (kJ)
Sublimation of Li(c)	$Li(c) \longrightarrow Li(g)$	155.2
Ionization energy of Li(g)	$Li(g) \longrightarrow Li^+(g) + e^-$	520.0
Dissociation of $\frac{1}{2}$ mol $F_2(g)$	$\frac{1}{2}F_2(g) \longrightarrow F(g)$	75.3
Electron affinity of F(g)	$F(g) + e^- \longrightarrow F^-(g)$	-333.0
Reverse of lattice energy	$Li^+(g) + F^-(g) \longrightarrow LiF(c)$	$-\Delta H_{LE}$
Net reaction:	$Li(c) + \frac{1}{2}F_2(g) \longrightarrow LiF(c)$	$\Delta H_f^0 = -594.1$

Adding the equations and ΔH values gives the net reaction and ΔH_f^0 for LiF(c).

$$155.2 + 520.0 + 75.3 - 333.0 + (-\Delta H_{LE}) = -594.1 \text{ kJ}$$

$$(-\Delta H_{LE}) = -1011.6 \text{ kJ}$$

Since this value is for an equation that is the reverse of the one for the lattice energy, the lattice energy is $+1011.6$ kJ/mol. ■

PROBLEM 7.3 Calculate the lattice energy for NaCl(c) given these ΔH values in kJ/mol: ΔH_f^0 for NaCl(c) $= -411$, ΔH_{sub} for Na(c) $= 108$, ΔH_{IE} for Na(g) $= 496$, ΔH_{BE} for $Cl_2(g) = +243$, ΔH_{EA} for Cl(g) $= -348$. □

The steps of the Born–Haber cycle can be used to answer the question asked at the beginning of this section, what factors most favor ionic bond formation between two atoms? The ΔH_f^0 is significantly influenced by three ΔH values in the Born–Haber cycle:

1. Ionization energy (ΔH_{IE}) of the metal, which is always positive. As we saw in section 6.5, the greater the charge on the cation, the more positive is ΔH_{IE} and the more difficult it is to form the cation. On going downward in a group of the periodic table, ΔH_{IE} becomes less positive and formation of a cation becomes more likely.
2. Electron affinity, ΔH_{EA}, of the nonmetal, which may be negative or positive. When one electron is added to a neutral nonmetallic atom, ΔH_{EA} is negative; when more than one electron is added, ΔH_{EA} is positive.

The greater the charge on the anion, the more positive is ΔH_{EA} and the more difficult it is to form the anion. On going down a group, with few exceptions, ΔH_{EA} becomes less negative (Table 6.8, section 6.6) and formation of an anion becomes less likely.

The sizes of the ions also influence ΔH_{LE}; the smaller the ions, the closer they are and the greater is ΔH_{LE}.

3. Lattice energy, ΔH_{LE}, of the salt, which is always positive. The more positive the ΔH_{LE} is, the more likely is the formation of the ionic bond. The greater the charges on the ions, the greater is ΔH_{LE}. Therefore, the difficulties of forming highly charged cations and anions may be overcome by a large lattice energy.

You are not asked to make predictions; that is too difficult.

The following examples show how this knowledge can help us understand trends in ionic bonding.

EXAMPLE 7.4 Explain why Li tends to form ionic bonds, whereas B (boron) does not (it forms covalent molecules).

ANSWER Li loses one electron to give Li^+ and therefore has a smaller positive ΔH_{IE} than B, which would have to lose 3 electrons to give B^{3+}. ■

EXAMPLE 7.5 Explain why ionic fluoride compounds outnumber ionic iodide compounds. (Consult Table 6.8.)

ANSWER F and I are in the same group. F has a more negative ΔH_{EA} and is thus more likely to form an ion. Also, F^- is smaller and closer to the positive ion than I^- so that fluorides have larger ΔH_{LE} values than iodides. ■

EXAMPLE 7.6 Is the ΔH_{EA} of O, the ΔH_{IE} of Al, or the ΔH_{LE} most responsible for Al_2O_3's being an ionic compound?

ANSWER Since the O atom gains 2 electrons to give O^{2-} ($O + 2e^- \longrightarrow O^{2-}$), we would correctly expect O to have a net positive ΔH_{EA}. Since the Al loses 3 electrons to become Al^{3+} ($Al - 3e^- \longrightarrow Al^{3+}$), we would correctly expect Al to have a large positive ΔH_{IE}. These two positive values would tend to discourage ionic bonding. However, the lattice energy between a +3 ion (Al^{3+}) and a −2 ion (O^{2-}) should be very large. ΔH_{LE} is therefore the overriding factor. ■

PROBLEM 7.4 Explain (a) why, with rare exceptions, carbon does not form ionic compounds having the C^{4-} ion; and (b) why aluminum carbide, Al_4C_3, is an ionic compound. □

7.5 COVALENT BOND; COVALENCE

We have just seen that electron transfer leading to an ionic bond occurs when a nonmetallic atom gains electrons and a metallic atom loses electrons. In this way ions are produced that are usually isoelectronic with noble gases. But what happens when two nonmetallic atoms, both of which *require* electrons to become isoelectronic with noble gases, react? The simplest example of such a situation is the combination of two H· atoms to form an H_2 molecule. Each H· atom needs one electron to become isoelectronic with :He, the nearest noble gas. A transfer of electrons cannot satisfy the requirements of both, and therefore does not happen. Instead, the H· atoms *mutually share their electrons.* The shared pair is said to "belong" to both. Each H· atom can be considered to have gained an

TABLE 7.2
COVALENT BONDS FORMED BY H AND SECOND-PERIOD ELEMENTS IN GROUPS 4, 5, 6, 7, AND 8

GROUP	H	4	5	6	7	8
Lewis symbol	H·	·Ċ· (4 dots)	·N̈·	·Ö·	·F̤̈:	:N̤̈e:
Electrons needed for octet	1[†]	4	3	2	1	0
Number of covalent bonds commonly formed (covalence)	1	4	3	2	1	0
Potential bonds and unshared pairs of electrons in a neutral molecule[§]	H—	—C— (with bonds up and down)	—N̈— (with bond down)	—Ö— (with two unshared pairs)	—F: (with three unshared pairs)	no bonding
Compounds with H atoms[‖]	H—H	H—C—H with H above and below (methane)	H—N̈—H with H below (ammonia)	H—Ö—H (water)	H—F: (hydrogen fluoride)	

[†] Needed for a duet.
[§] A dash to a single atom represents a single electron. It is a potential bond.
[‖] A dash between atoms represents a bond consisting of two shared electrons.

electron and to have acquired the heliumlike electron structure:

$$\text{H}\cdot + \text{H}\cdot \longrightarrow \text{H:H} \qquad \Delta H = -435.1 \text{ kJ}$$

each H has a duet of electrons (both are satisfied).

Such a sharing of a *pair* of electrons results in a **covalent bond**. In a Lewis structure, a covalent bond is usually depicted by a dash, H—H, or a pair of dots, H:H.

Table 7.2 lists H and second-period elements in Groups 4 through 8 that form covalent bonds.

An important relationship should be noted from Table 7.2. *The number of covalent bonds usually formed (the covalence) equals the number of electrons needed by a nonmetallic atom to acquire an octet.* Note also that among elements in Groups 5, 6, 7, and 8, some of the valence electrons are present as **unshared pairs**. We shall see in Chapter 14 that these unshared pairs of electrons significantly affect the chemical behavior of molecules.

Unshared pairs are also called lone pairs or nonbonded pairs.

The covalent bonding indicated in Table 7.2 also applies to elements in the higher periods of these groups. For example, the usual covalences observed for the third-period elements are Si(4), P(3), S(2), and Cl(1). However, elements in the third and higher periods show many exceptions to these usual values, as will be discussed in section 7.6. Representative elements in Groups 1, 2, and 3 typically form ionic bonds. However, Be and B, the smallest elements in Groups 2 and 3, respectively, form only covalent bonds. Be: has two valence electrons and in its molecules forms two covalent bonds. ·B: has three valence electrons and forms three covalent bonds in its molecules. Some elements, such as Al, form both kinds of bonds; AlF_3 is ionic and Al_2Br_6 has covalent bonds. The largest elements in Groups 4 (Sn and Pb) and 5 (Bi) tend to be metallic and form ionic bonds, although some covalent bonding also occurs (see Chapter 21).

Heavy metallic elements may form ions that are not isoelectronic with noble gases; for example, Bi^{3+}, Tl^{+}.

MULTIPLE COVALENT BONDS

Two atoms may share more than one pair of electrons with each other, forming **multiple bonds**. Two pairs of electrons are shared in a **double bond**, and three pairs are shared in a **triple bond**. All electrons in multiple bonds belong to each bonded atom, as illustrated by CO_2:

$$:\ddot{O}::C::\ddot{O}: \quad \text{or} \quad :\ddot{O}{=}C{=}\ddot{O}:$$

carbon dioxide

$$:\ddot{O}\boxed{::C::}\ddot{O}: \qquad (:\ddot{O}::)C(::\ddot{O}:)$$

emphasizing 8 e⁻ around C — emphasizing 8 e⁻ around each O

Phosgene, nitrogen, and acetylene molecules provide further examples:

$$:\ddot{C}l:\overset{:O:}{\overset{::}{C}}:\ddot{C}l: \quad \text{or} \quad :\ddot{C}l{-}\overset{:O:}{\overset{\|}{C}}{-}\ddot{C}l: \qquad :N:::N: \quad \text{or} \quad :N{\equiv}N:$$

phosgene — nitrogen

$$H:C:::C:H \quad \text{or} \quad H{-}C{\equiv}C{-}H$$

acetylene

S, a third-period element, occasionally forms a double bond; for example, CS_2, $:\ddot{S}{=}C{=}\ddot{S}:$.

When seeking an *octet* of electrons, it is mainly the second-period elements C, N, and O that form multiple bonds as well as single bonds, as illustrated in Table 7.3.

COORDINATE COVALENT BONDS

In most cases each atom contributes one electron to make up the shared pair of electrons in the covalent bond:

$$A\cdot + \cdot B \longrightarrow A:B \qquad \text{(a typical covalent bond)}$$

However, sometimes both electrons are furnished by the same atom, the **donor atom**. The other atom, the **acceptor atom**, brings no electrons for bonding.

A bond formed this way is called a **coordinate covalent bond** to indicate the manner in which it was formed.

TABLE 7.3
COVALENT BONDING FOR C, N, AND O WITH USUAL COVALENCES

BONDING FOR C				BONDING FOR N			BONDING FOR O	
$-\overset{\vert}{\underset{\vert}{C}}-$	$-\overset{\vert}{C}{=}$	$=C=$	$-C{\equiv}$	$-\ddot{N}-$ (with third bond below)	$-\ddot{N}{=}$	$:N{\equiv}$	$-\ddot{\underset{..}{O}}-$	$\ddot{\underset{..}{O}}{=}$
4 single	2 single 1 double	2 double	1 single 1 triple	3 single	1 single 1 double	1 triple	2 single	1 double
CH_4: H–C(H)(H)–H	H₂C=CH₂	O=C=O	H–C≡C–H	$H{-}\ddot{N}{-}H$, H	$H{-}\ddot{\underset{..}{O}}{-}\ddot{N}{=}\ddot{O}:$	$:N{\equiv}C{-}H$	$H{-}\ddot{\underset{..}{O}}{-}H$	$\ddot{\underset{..}{O}}{=}CH_2$
methane	ethene (ethylene)	carbon dioxide	ethyne (acetylene)	ammonia	nitrous acid	hydrogen cyanide	water	formaldehyde

$$A \quad + \quad :B \longrightarrow A:B \quad \text{(a typical coordinate covalent bond)}$$

accepts electrons (acceptor) — donates both electrons (donor)

Two examples are given:

$$H^+ + :\overset{\displaystyle H}{\underset{\displaystyle H}{N}}-H \longrightarrow \left[H-\overset{\displaystyle H}{\underset{\displaystyle H}{N}}-H\right]^+ \qquad \text{and} \qquad :\ddot{\underset{..}{O}} + :\overset{\displaystyle Cl}{\underset{\displaystyle Cl}{\ddot{S}}}: \longrightarrow :\ddot{\underset{..}{O}}:\ddot{S}\begin{matrix}Cl\\ \\ Cl\end{matrix}$$

ammonium ion — thionyl chloride

Once formed, the coordinate covalent bond is *no different* from the covalent bond formed by each atom contributing one electron. Thus, the four N—H bonds in NH_4^+ are identical. The two examples shown illustrate an important consequence of forming a coordinate covalent bond: *when atoms bond this way they often do not have their usual number of covalent bonds.* The N atom in NH_4^+ has four covalent bonds, rather than three. In $SOCl_2$, the O has one rather than two bonds and the S has three rather than two bonds.

NH_4^+ is a *polyatomic* ion (section 2.9). All polyatomic ions have covalently bonded atoms. There is usually a **central atom**, to which most of the other atoms are bonded. In NH_4^+ the central atom is N.

COVALENT BONDING AND ATOMIC ORBITALS

The nature of the covalent bond has been a subject of much scientific thought. A more complete presentation of modern theories is found in Chapter 8. In Chapter 8 we will see that covalent bonding is an interplay of attractive and repulsive forces between the bonding electrons and nuclei of the bonding atoms. Since we believe that electrons move about in atomic orbitals (section 5.6), we ascribe a role to atomic orbitals in bonding. The discussions of molecules in this book are limited to the involvement of the *s*, *p*, and *d* atomic orbitals. The important point to make now is that *for each covalent bond and unshared pair of electrons, an atom must provide an atomic orbital.* The octet rule can be interpreted in terms of this statement. Eight electrons, distributed in four pairs, must be housed in four atomic orbitals. Invariably for representative elements, these four orbitals are the one *s* and the three *p* orbitals of the outermost shell.

7.6 SOME EXCEPTIONS TO THE OCTET RULE

Although the octet rule is a useful generalization, there are exceptions to it. Some bonded atoms have less than the octet of valence electrons; they are said to have **incomplete octets**. Others have more than eight valence electrons and they are said to have **expanded valence shells**.

INCOMPLETE OCTETS

Atoms with incomplete octets are those with fewer than 4 valence electrons to share when forming covalent bonds. Boron, ·B:, the first member of Group 3, and beryllium, Be:, the first member of Group 2, are typical examples. Boron and fluorine, for instance, form the covalent compound boron trifluoride, BF_3 (b.p. = −101°C):

$$\dot{B}: + 3:\ddot{\underset{..}{F}}\cdot \longrightarrow :\ddot{\underset{..}{F}}:\overset{\displaystyle :\ddot{F}:}{\ddot{B}}:\ddot{\underset{..}{F}}:$$

The B atom in BF_3 is surrounded by only 6 valence electrons, not by 8.

Solid beryllium chloride is a network covalent substance; individual gas molecules exist at about 750°C.

In gaseous $BeCl_2$, Be has only 4 valence electrons:

$$\text{Be}\!: + 2\,\text{:}\ddot{\underset{..}{\text{Cl}}}\cdot \longrightarrow \text{:}\ddot{\underset{..}{\text{Cl}}}\text{:Be:}\ddot{\underset{..}{\text{Cl}}}\text{:}$$

FREE RADICALS

The few radicals with two unpaired electrons have even numbers.

A unique group of molecules and ions have at least one atom with an **unpaired** electron. Such species are called **free radicals**, or more simply, **radicals**. With few exceptions, the atom with the unpaired electron has seven valence electrons—it has an incomplete octet. Free-radical molecules can be recognized by adding the number of valence electrons of the individual atoms; an *odd* number indicates a free radical. Two examples are shown:

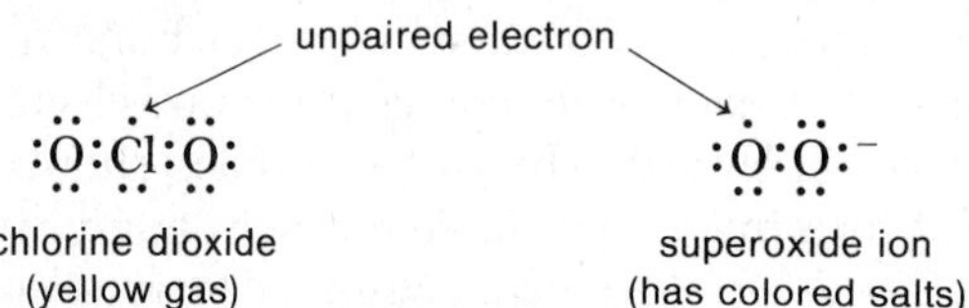

Most free radicals are unstable. ClO_2 is explosive. NO, O_2, and NO_2 are exceptions.

Unpaired electrons make radicals paramagnetic (section 5.8), as well as colored in most cases.

PROBLEM 7.5 Which of these oxides of nitrogen, NO, NO_2, N_2O, N_2O_3, and N_2O_5, are free radicals? □

EXPANDED VALENCE SHELLS

Some molecules and polyatomic ions have atoms with more than eight valence electrons. Examples are

$[SiF_6]^{2-}$ (Si bonded to six F)	PCl_5 (P bonded to five Cl)	SF_6 (S bonded to six F)	ICl_3 (I bonded to three Cl, with two lone pairs)
SiF_6^{2-} silicon hexafluoride anion	PCl_5 phosphorus pentachloride	SF_6 sulfur hexafluoride	ICl_3 iodine trichloride

In these examples, P and I are surrounded by 10 valence electrons, and Si and S are surrounded by 12 valence electrons. The valence electrons of Si, S, and P are in the third shell ($n = 3$), and the valence electrons of I are in the fifth shell ($n = 5$). The third and higher shells have a number of *d* orbitals as well as one *s* and three *p* orbitals available for bonding. Therefore, elements with these shells are not limited by the octet rule. Third-period elements can house up to twelve valence electrons, although they often have just an octet. Second-period elements have no *d* orbitals in the valence shell and therefore cannot acquire more than eight electrons when they form bonds. Here the octet rule applies; no stable molecule is known in which Li, Be, B, C, N, O, or F has more than eight valence electrons.

Elements in the fourth and higher shells infrequently acquire more than 12 e^-'s; for example, IF_7, TeF_8^{2-}.

7.7 POLAR COVALENT BONDS; ELECTRONEGATIVITY

Equal sharing of a pair of electrons occurs in diatomic molecules of identical atoms, as in H:H and Cl:Cl. This means that the shared electrons are most likely to be midway between the atoms. When two bonded atoms are dissimilar, as in gaseous HCl (hydrogen chloride), the sharing is unequal and the shared electrons are not midway between the atoms. One atom is likely to attract electrons more strongly than the other atom. This difference in electron attraction for different atoms is not unexpected, because each atom has a different electron arrangement and nuclear charge. The atom that attracts electrons more strongly develops some negative charge; the other atom develops some positive charge. These charges have *less* than unit $+1$ or -1 values and are therefore called **partial charges,** δ^+ and δ^-. For example, a Cl atom is more electron-attracting than an H atom; hydrogen chloride is therefore depicted as

$$\overset{\delta^+}{\mathrm{H}}-\overset{\delta^-}{\ddot{\underset{\cdot\cdot}{\mathrm{Cl}}}}:$$

Such covalent bonds are said to be **polar**. Bonds such as Cl—Cl and H—H, in which the sharing of electrons is equal, are called **nonpolar**. The most extreme case of unequal sharing of a pair of electrons is the ionic bond in a compound such as CsF or CaF_2. According to our simplified picture of ionic bonding, one atom takes the entire pair of electrons and acquires a full negative charge. The other atom is completely deprived of these electrons and therefore acquires a full positive charge. Thus, the *nonpolar covalent bond and the ionic bond may be considered extremes in the way a pair of electrons can be distributed between two nuclei.* Between these extremes are the polar covalent bonds that have varying amounts of partial charges. In summary (see Figure 7.5):

$Cs^+ :\ddot{\underset{\cdot\cdot}{F}}:^-$	$\overset{\delta^+}{:\ddot{\underset{\cdot\cdot}{Cl}}}\,\overset{\delta^-}{:\ddot{\underset{\cdot\cdot}{F}}:}$	$:\ddot{\underset{\cdot\cdot}{F}}:\ddot{\underset{\cdot\cdot}{F}}:$
ionic bond	**polar covalent bond**	**nonpolar covalent bond**

The net tendency of a bonded atom in a molecule to attract electrons is called **electronegativity**. This term does not imply that the atom has a full negative

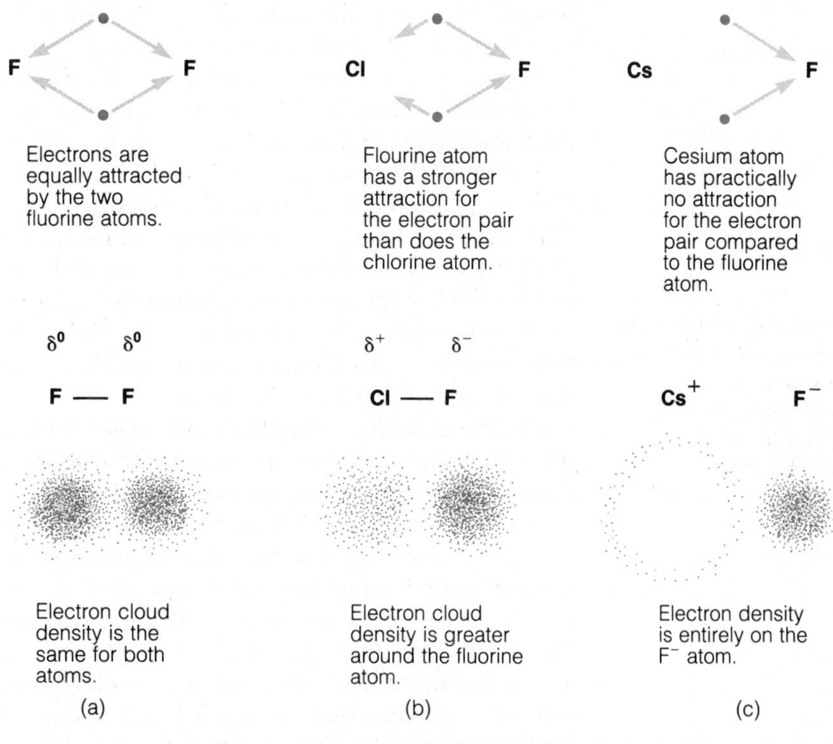
F
F
Electrons are equally attracted by the two fluorine atoms.
Cl
F
Flourine atom has a stronger attraction for the electron pair than does the chlorine atom.
Cs
F
Cesium atom has practically no attraction for the electron pair compared to the fluorine atom.
δ0 δ0
F — F
δ+ δ−
Cl — F
Cs+ F−
Electron cloud density is the same for both atoms.
(a)
Electron cloud density is greater around the fluorine atom.
(b)
Electron density is entirely on the F− atom.
(c)

FIGURE 7.5
Attraction of atoms in molecules for the electron pair. (a) Equal attraction, nonpolar bond. (b) Unequal attraction, polar bond. (c) Extremely unequal attraction, ionic bond.

TABLE 7.4
ELECTRONEGATIVITIES OF REPRESENTATIVE ELEMENTS† (PAULING VALUES)

$_{1}$H 2.1							
$_{3}$Li 1.0	$_{4}$Be 1.5	$_{5}$B 2.0	$_{6}$C 2.5	$_{7}$N 3.0	$_{8}$O 3.5	$_{9}$F 4.0	
$_{11}$Na 0.9	$_{12}$Mg 1.2	$_{13}$Al 1.5	$_{14}$Si 1.9	$_{15}$P 2.1	$_{16}$S 2.5	$_{17}$Cl 3.0	
$_{19}$K 0.8	$_{20}$Ca 1.0	$_{31}$Ga 1.8	$_{32}$Ge 2.0	$_{33}$As 2.0	$_{34}$Se 2.4	$_{35}$Br 2.8	
$_{37}$Rb 0.8	$_{38}$Sr 1.0	$_{49}$In 1.5	$_{50}$Sn 1.7	$_{51}$Sb 1.9	$_{52}$Te 2.1	$_{53}$I 2.4	
$_{55}$Cs 0.7	$_{56}$Ba 0.9	$_{81}$Tl 1.4	$_{82}$Pb 1.6	$_{83}$Bi 1.9	$_{84}$Po 1.9	$_{85}$At 2.1	
$_{87}$Fr 0.8	$_{88}$Ra 1.1						

† Values for noble gases are uncertain and therefore not included.

R. S. Mulliken suggested (1934) that electronegativity is proportional to the average of the ionization energy and the electron affinity (disregarding signs).

charge. It just refers to the atom's *tendency* to acquire some negative charge when it is bonded in a covalent molecule. Thus, while F is highly electronegative, F^-, which already has its extra electron, is not. Electronegativity is *not* synonymous with electron affinity. Electronegativity is the net effect of electron affinity and ionization energy. An analogy can be made between electronegativity and the acquisition of wealth. Your wealth (electronegativity) increases with the amount of money you earn (electron affinity) as well as with your reluctance to spend or give away money (ionization energy).

Periodic Table

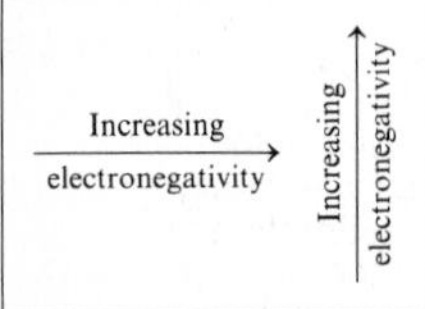

Approximate relative electronegativities of elements suggested by Linus Pauling are given in Table 7.4. The order of decreasing electronegativities for the most electronegative elements is F > O > N, Cl > Br > S, I, C.

The periodic trends are noteworthy. *Electronegativities increase from left to right in a period* and, with few exceptions, *from the bottom to the top of a group.* Thus the most highly electronegative elements, such as fluorine (4.0) and oxygen (3.5), are found in the upper right corner of the periodic table (ignoring the noble gases). The least electronegative elements, such as Cs (0.7) and Rb (0.8), are found in the lower left corner of the periodic table. Elements with low values tend to form positive ions—these are the metals. Elements with high values tend to form negative ions—these are the nonmetals.

Elements with electronegativity values of 1.0 or less are said to be electropositive because they tend to be electron-donating.

Electronegativity values are not precise. They are *approximations* of the *relative* tendencies of atoms to attract electrons in a covalent bond to themselves. Furthermore, the values are not constant, since they depend on the kind and number of atoms bonded to the atom under consideration. Electronegativity values, therefore, must be regarded as approximate. Nevertheless, they are useful in rationalizing a fair amount of chemical behavior. For example, they can be used to make *rough* predictions of the type of bonding found in a compound. The greater the difference in electronegativity values (ΔEn) of two bonding elements, the more likely they are to form ionic bonds. Also, the electronegativity difference between two covalently bonded atoms determines the polarity of their bond: *the larger the difference in electronegativity, the more polar is the covalent bond,* with the more electronegative element bearing the partial negative charge,

δ^-. In summary:

ΔEn	BOND TYPE
Zero or small (<0.4, approximately)	covalent
Intermediate	polar covalent
Large (>1.7, approximately)	ionic

EXAMPLE 7.7 Which element would be the most electronegative, and which would be the least electronegative? An element with (a) high ionization energy (*IE*) and low electron affinity (*EA*); (b) low *IE* and high *EA*; (c) low *IE* and low *EA*; (d) high *IE* and high *EA*.

High EA means "very negative"; low EA means "close to zero."

ANSWER The most electronegative element is the one with the greatest affinity for electrons (high *EA*) and the least tendency to lose electrons (high *IE*). Element d fits this description. The converse is true for the least electronegative element—the lowest affinity for electrons (low *IE*) and the greatest tendency to lose an electron (low *IE*). Element c fits this description. ■

EXAMPLE 7.8 (a) Which two pairs of elements form ionic bonds? Na and O, Al and Cl, Al and F, O and F, P and H. (b) Which of the three remaining pairs forms the least polar covalent bond? Use Table 7.4 for electronegativity values.

ANSWER

(a) The electronegativity values of the elements are (Na) 0.9, (O) 3.5, (Al) 1.5, (Cl) 3.0, (F) 4.0, (P) 2.1, and (H) 2.1. The two pairs with the largest differences in electronegativity (ΔEn) are most likely to form ionic bonds. These are Na and O ($\Delta En = 3.5 - 0.9 = 2.6$) and Al and F ($\Delta En = 4.0 - 1.5 = 2.5$). Notice that both of these ΔEn values are greater than 1.7.

(b) ΔEn values for the other pairs are (Al and Cl) $\Delta En = 3.0 - 1.5 = 1.5$, (O and F) $\Delta En = 4.0 - 3.5 = 0.5$, and (P and H) $\Delta En = 2.1 - 2.1 = 0$. The least polar covalent bond is P—H. It has the smallest ΔEn value. ■

PROBLEM 7.6 Indicate the partial charges, if any, on the atoms in each of these bonds: (a) F—O, (b) N—Cl, (c) C—H, (d) S—H, (e) As—H. Refer to Table 7.4. □

7.8 BOND POLARITY, BOND ENERGY, AND BOND ANGLE

We have already discussed two important properties of covalent bonds, bond length (section 6.4) and bond energy (section 4.7). Now we will examine how these properties relate to each other and how they are influenced by bond polarity. As a beginning, the following generalization can be made: *the shorter the distance between a given pair of covalently bonded atoms, the greater is the bond energy and the stronger is the bond.* This generalization is supported by the observed distances and energies of the bond between carbon atoms in ethane, ethylene, and acetylene:

	$H_3C—CH_3$	$H_2C{=}CH_2$	$HC{\equiv}CH$
Bond length, nm	0.154	0.133	0.120
Bond energy, kJ/mol	333.9	589.9	811.7

TABLE 7.5
SINGLE- AND MULTIPLE-BOND COVALENT RADII (in nanometers, nm)

ELEMENT	SINGLE-BOND RADIUS	DOUBLE-BOND RADIUS	TRIPLE-BOND RADIUS
C	.077	.0665	.060
N	.073	.060	.055
O	.074	.055	
S	.102	.094	

These values also show that *the distance between two given atoms decreases when the number of bonds between the atoms increases.* Multiple bonding results in a greater electron density between the bonded nuclei. The nuclei are more strongly attracted by this greater electron density and are brought closer together. Multiple-bond covalent radii are shown in Table 7.5 for C, N, O, and S.

Bond polarity may cause contraction of bond length. For example, the calculated length of the very polar Si—F bond, obtained from Table 6.6 by adding r_{Si} and r_F, is 0.186 nm. However, the measured bond length in SiF_4 is only 0.154 nm. On the other hand, when we compare the calculated and observed lengths of bonds of low polarity, the correspondence is very good. Thus in bromine chloride, Br—Cl, the calculated bond length is 0.213 nm and the observed bond length is 0.214 nm. Along with contraction of a bond length goes an increase in bond energy. Hence, *an increase in polarity of a covalent bond strengthens the covalent bond.*

EXAMPLE 7.9 The observed P—H bond length is 0.144 nm. Use the data for atomic radii in Table 6.6 to get the calculated bond length (use the average radius for H). Is P—H a polar or a practically nonpolar bond?

ANSWER The listed radii (in nm) are P = 0.11 and H (average) = 0.35. The P—H bond length is 0.145, the sum of these radii. The observed and calculated values are very close to each other; therefore the bond is practically nonpolar. ■

PROBLEM 7.7 Given the following observed bond lengths (in nm): N—O = 0.124, As—Br = 0.233, and S—F = 0.159. Compare these values with the bond lengths calculated from data in Table 6.6 in order to determine the most polar and least polar bonds. □

BOND ANGLE

When an atom has two or more atoms bonded to it, another property must be considered, the **bond angle**. A bond angle can be assigned for any sequence of three atoms in a molecule. For example,

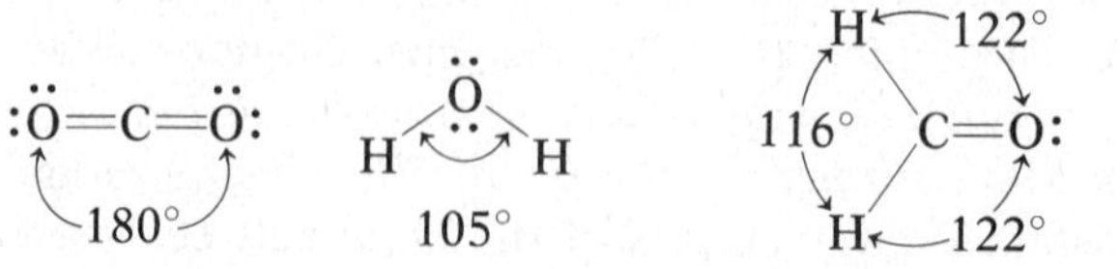

:Ö=C=Ö: (180°) — CO_2 is linear.

H–Ö–H (105°) — H_2O is bent.

H_2C=Ö: (H–C–H 116°, H–C=O 122°, 122°) — formaldehyde, H_2C=O, is triangular.

Molecular vibrations alternately compress and expand the bond angles (Figure 7.6). Therefore, the experimentally determined angles, like bond lengths, are average values.

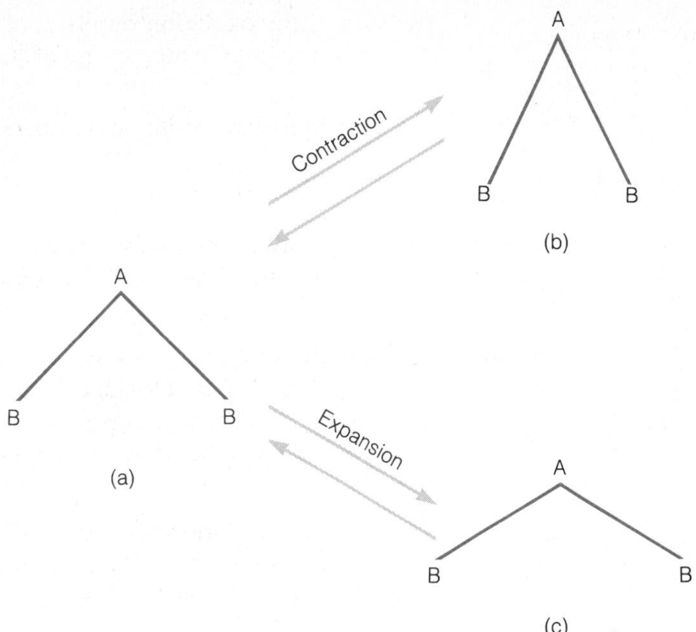

FIGURE 7.6
Vibrational changes in a bond angle. (a) Average bond angle. (b) Smaller bond angle resulting from contraction. (c) Larger bond angle resulting from expansion.

7.9 OXIDATION NUMBER AND FORMAL CHARGE

OXIDATION NUMBER

A question arises when covalent bonds in molecules or polyatomic ions change in chemical reactions. The question is how to decide if an oxidation–reduction (redox) reaction has occurred. Redox reactions are best recognized when a transfer of electrons is apparent, as in the formation of an ionic bond. However, is the change of CH_4 to CH_3Cl or of $CrO_4{}^{2-}$ to $Cr_2O_7{}^{2-}$ an oxidation step, a reduction step, or neither? The concept of **oxidation number** (also called **oxidation state**) can be used to help recognize such redox reactions. Oxidation number can be defined as the "charge" on an atom if the electrons in each covalent bond are arbitrarily assigned to the more electronegative element. However, *the oxidation number is not a real charge.* This assignment of bonding electrons overemphasizes the bond polarity by treating polar bonds as if they were ionic.

Proper assignment of electrons requires a knowledge of writing Lewis structures, a subject discussed in more detail in section 7.10. However, it is not always easy to write Lewis structures, especially for compounds with transition elements. Fortunately, using some simple empirical rules allows us to obtain oxidation numbers directly from molecular or empirical formulas.

1. The oxidation number of any atom in its elementary state is zero, as it is for the atoms in N_2, Cl_2, C, S_8, P_4, O_2, and O_3.

2. The oxidation number of any monoatomic ion, such as Na^+, Cu^{2+}, Fe^{3+}, Br^-, or H^-, is the charge on the ion. Alkali metals of Group 1 are always +1, and Group-2 elements are always +2.

O has an oxidation number of −1 when bonded to itself and other atoms as in $\overset{+1}{H}-\overset{-1}{O}-\overset{-1}{O}-\overset{+1}{H}$ and +2 in OF_2.

3. The usual oxidation numbers of some common atoms are oxygen, −2; covalently bonded H, +1; and the halogens, when bonded to less electronegative atoms, −1. The oxidation number of F is always −1 when it is bonded to another element, because it is the most electronegative element.

4. The sum of the oxidation numbers of all atoms in a molecule is zero. In an ion, the sum is the charge of the ion.

EXAMPLE 7.10 Use these rules to assign oxidation numbers to (a) Na_2S, (b) CH_4, and (c) CH_3Cl.

ANSWER

(a) Na, a Group-1 alkali metal, forms ions. Na_2S is a salt, $(Na^+)_2S^{2-}$. The charges on the monoatomic ions are the oxidation numbers; $+1$ for Na and -2 for S.

(b) The oxidation number of each H is $+1$, as C is more electronegative than H (Table 7.4). The sum of all oxidation numbers of this molecule is zero. Therefore, since the oxidation numbers of the four H atoms add up to $+4$, the C atom must have an oxidation number of -4 ($0 = 4H + C$, or $0 = 4 + C$, and $C = -4$).

(c) Again, H has an oxidation number of $+1$. Cl has an oxidation number of -1. Since CH_3Cl is a molecule, the sum of all oxidation numbers ($Cl + 3H + C$) is zero. Therefore, $0 = -1 + 3 + C$; $C = -2$. ■

EXAMPLE 7.11 Calculate the oxidation number of each underlined atom: (a) $Na_2\underline{Cr}O_4$; (b) $K_2\underline{Cr}_2O_7$; (c) $\underline{S}_2O_3^{2-}$.

ANSWER

(a) Na, a Group-1 element, has an oxidation number of $+1$. Each O atom has an oxidation number of -2. Let ON equal the oxidation number of Cr. Since the sum of all oxidation numbers equals zero, we write

$$2Na + Cr + 4O = 0$$

$$2(+1) + ON + 4(-2) = 0$$

from which $ON = +6$.

(b) A similar equation is set up for $K_2Cr_2O_7$:

$$2K + 2Cr + 7O = 0$$

$$2(+1) + 2(ON) + 7(-2) = 0$$

from which $ON = +6$.

(c)

$$2S + 3O = -2 \quad \text{(the ion charge)}$$

$$2(ON) + 3(-2) = -2$$

then

$$ON = +2$$

■

PROBLEM 7.8 Assign an oxidation number to each underlined atom: (a) $\underline{C}OCl_2$ (phosgene); (b) $K_3\underline{V}_5O_{14}$ (potassium vanadate); (c) $H\underline{Cl}O_4$ (perchloric acid). □

The maximum oxidation number allowed for a representative element is the same as the number of the element's group in the periodic table; for example:

Group Number: 4 5 6 7

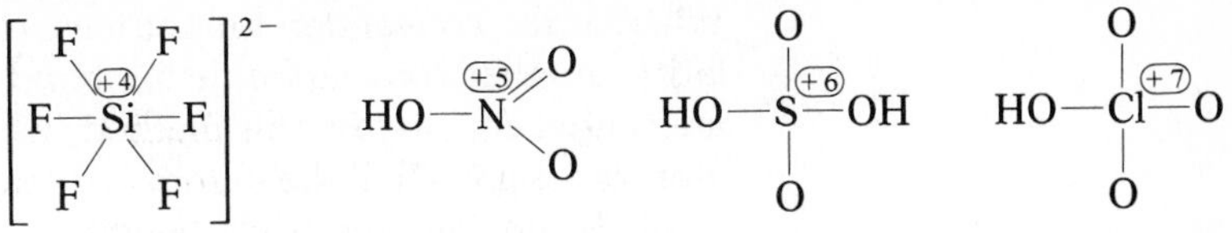

$$\left[\begin{matrix} F & & F \\ & \overset{(+4)}{Si} & \\ F & & F \end{matrix}\right]^{2-} \text{ (with F—Si—F)} \qquad HO—\overset{(+5)}{N}(=O)O \qquad HO—\overset{(+6)}{S}(O)(O)—OH \qquad HO—\overset{(+7)}{Cl}(O)(O)—O$$

C compounds	Ox. no. of C
CO_2	+4
$H_2C_2O_4$	+3
$HCHO_2$	+2
$C_2H_2O_2$	+1
CH_2O	0
C_2H_2	−1
C_2H_4	−2
C_2H_6	−3
CH_4	−4

REDOX REACTIONS AND OXIDATION NUMBERS

We now see how oxidation numbers are used to recognize redox reactions. *A decrease in oxidation number is a reduction, and an increase in oxidation number is an oxidation.* Let us apply this definition to the two chemical changes mentioned at the beginning of this section. The oxidation number of C in CH_4 is -4; in CH_3Cl it is -2 (see Example 7.10b and c). This is an *increase in oxidation number* of Cl, and the change is an *oxidation.* The reduction that must go along with every oxidation is revealed by examining the complete reaction:

$$CH_4 + Cl_2 \longrightarrow CH_3Cl + HCl$$

The oxidation number of Cl in Cl_2 is 0; in CH_3Cl and in HCl it is -1 (see Example 7.10c). This is a *decrease in oxidation number* of Cl, and the change is a *reduction.* CH_4 is oxidized, Cl_2 is reduced. However, the change

$$2CrO_4^{2-} + 2H^+ \longrightarrow Cr_2O_7^{2-} + H_2O$$

is not a redox reaction because the oxidation numbers of Cr and O are $+6$ and -2, respectively, in CrO_4^{2-} and $Cr_2O_7^{2-}$ (see Example 7.11a and b).

We can summarize the two ways of recognizing oxidations and reductions with illustrations:

CHANGE	ILLUSTRATION
Oxidation	
Loss of electrons	$Na - e^-$, giving Na^+
Increase in oxidation number	CH_4 going to CH_3Cl
Reduction	
Gain of electrons	$Cl + e^-$, giving Cl^-
Decrease in oxidation number	Cl_2 going to HCl

EXAMPLE 7.12 Does the reaction $Cl_2 + H_2O \longrightarrow HCl + HOCl$ involve an oxidation–reduction?

ANSWER First assign oxidation numbers to each atom of the reactants and the products. These are shown with each atom

Recall that O is more electronegative than Cl.

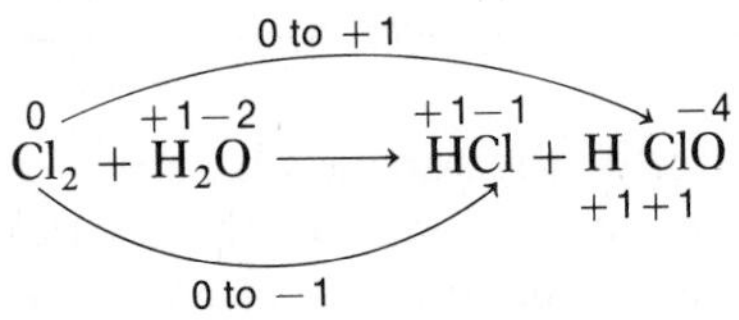

$$\overset{0}{Cl_2} + \overset{+1-2}{H_2O} \longrightarrow \overset{+1-1}{HCl} + \underset{+1+1}{H\,Cl}\overset{-4}{O}$$

0 to +1 (Cl in Cl_2 to Cl in HClO); 0 to −1 (Cl in Cl_2 to Cl in HCl)

The changes in the oxidation number of Cl indicate a redox reaction. ■

PROBLEM 7.9 Is the reaction $PBr_3 + 3H_2O \longrightarrow H_3PO_3 + 3HBr$ a redox reaction? □

FORMAL CHARGE

Whereas the convention for getting oxidation numbers overemphasizes bond polarity, another convention denies bond polarity by assuming all bonded electrons are shared equally. By this method, when two atoms form a covalent bond, each atom is assigned half the number of shared electrons. The assignment of electrons in single, double, and triple bonds is shown:

A:B single bond A::B double bond A:::B triple bond

The numbers calculated from this method of assigning electrons are called **formal charges**. Formal charge is defined as

$$\text{formal charge (FC)} = \frac{\text{number of valence electrons}}{\text{in free atom (G)}} - \frac{\text{total number of}}{\text{assigned electrons}}$$

(Group number)

The assigned number of shared electrons equals the number of covalent bonds. This number (C) plus the number of unshared electrons (U) equals the total number of assigned electrons. This method is summarized as

$$\textbf{formal charge (FC)} = \begin{matrix}\textbf{number of}\\ \textbf{valence electrons}\\ \textbf{in lone atom (G)}\end{matrix} - \left[\begin{matrix}\textbf{unshared}\\ \textbf{electrons (U)}\end{matrix} + \begin{matrix}\textbf{number of}\\ \textbf{covalent}\\ \textbf{bonds (C)}\end{matrix}\right]$$

Although these calculated numbers are called charges, they are *not* actual charges like those on an electron, a proton, or an ion such as Na^+ or Cl^-. Nevertheless, this convention helps us to understand various aspects of the covalent bond. In particular, the *formal charge helps us to write proper Lewis structures,* which will be discussed in section 7.10. Formal charges are differentiated from actual charges by encircling the plus and minus signs: ⊕ and ⊖. We illustrate the convention in tabular form with boron trifluoride hydrate:

```
H       :F:
 \     /
  :O—B——:F:
 /     \
H       :F:
```

ATOM	G	UNSHARED ELECTRONS(U)	BONDS(C)	G − (U + C) = FC
Each F	7	6	1	7 − (6 + 1) = 0
O	6	2	3	6 − (2 + 3) = +1
B	3	0	4	3 − (0 + 4) = −1
Each H	1	0	1	1 − (0 + 1) = 0

Notice that the sum of the formal charges is zero, and indeed $H_2\overset{\oplus}{O}\overset{\ominus}{B}F_3$ is a neutral molecule. For ions, the sum of the formal charges is the charge on the ion. If the sum is positive, the ion is a cation; if it is negative, the ion is an anion.

COORDINATE COVALENT BONDING AND FORMAL CHARGE

$H_2\overset{\oplus}{O}\overset{\ominus}{B}F_3$ is formed as shown:

$$H_2\ddot{O}\!:(\ell) + BF_3(g) \longrightarrow H_2\overset{\oplus}{\ddot{O}}\!-\!\overset{\ominus}{B}F_3(aq)$$

Since the O atom contributes *both* electrons, the O—B bond is an example of a coordinate covalent bond. When any covalent bond is formed this way, the atom contributing the pair of electrons (in this case the O) gets a positive formal charge. The atom that brings no electrons to the bonding "marriage" (in this case the B) gets a negative formal charge. Thus, the presence of formal charge on adjacent atoms tells us about the formation of the bond. Also notice that the coordinate covalently bonded atoms, in this case the O and B atoms, do not have their normal covalence. However, when an H^+ (or any atom with a positive charge) accepts the pair of electrons, it gains negative charge and ends up with zero formal charge in the product. For example,

$$H^+ + :NH_3 \longrightarrow [H_3N^{\oplus}\!-\!H]^+$$

acceptor donor

The N atom donates the pair of electrons for the bond and, as expected, develops a positive formal charge. The absence of formal charge is typical of *all* covalently bonded H atoms.

We now examine the situation in which the donor atom has a negative charge and the acceptor atom has no charge, as shown:

$$:\ddot{F}:^- + BF_3 \longrightarrow [:\ddot{F}-{}^{\ominus}BF_3]^-$$

donates pair of electrons (donor) accepts pair of electrons (acceptor)

Since the B atom starts with no charge and accepts the pair of electrons to form the bond, it acquires a negative formal charge. The F^- loses its negative charge as a result of donating the electron pair and ends up with zero formal charge in the product, BF_4^-. In general, *atoms that do not have more than eight valence electrons will have formal charges whenever they do not have their normal covalence.*

PROBLEM 7.10 Calculate the formal charges of the atoms of thionyl chloride:

$$:\ddot{Cl}-\ddot{S}(-\ddot{O}:)-\ddot{Cl}:$$

(Note that S and O do not have their normal covalence.) □

7.10 WRITING LEWIS STRUCTURES

It is very useful to be able to translate a molecular formula into a Lewis structure. Knowing the arrangement of the atoms and the presence or absence of unshared electrons and multiple bonds in a species helps us to understand its chemical and physical properties. There is no foolproof recipe that will ensure success with all molecular formulas, but we can give you some guidance.

Writing the Skeleton. Write the **skeleton** of the molecule, showing the bonding arrangement of the atoms. The skeleton of a molecule with more than two atoms will have at least one central atom and terminal atoms. The central atom must be able to form at least two covalent bonds (**multicovalent**). Atoms, such as H and F, that form only one covalent bond (**unicovalent**) must be terminal. The other halogen atoms, Cl, Br, and I, are also terminal when they form one covalent bond. However, unlike F, they may be central atoms because they can also be multicovalent. Multicovalent atoms, such as O and to a lesser extent N, may also be terminal when they form multiple bonds, as for example in $HC{\equiv}N$ and $H_2C{=}O$.

Except for H, terminal atoms are often more electronegative than central atoms.

When there is more than one multicovalent atom in the molecule, *bond the multicovalent atoms to each other to get the skeleton.* When you achieve the skeleton, then bond the unicovalent terminal atoms so as to satisfy the usual covalences of the central atom or atoms. If the number of univalent atoms is insufficient for this purpose, multiple bonding is necessary. If, after satisfying these covalences, unicovalent atoms are left over, exceed the usual covalences. *Never exceed the octet of second-period elements.* However, the valence shells of elements in the third and higher periods, such as S, P, Si, Cl, Br, and I, may have more than an octet of electrons.

Counting and Placing the Valence Electrons. Count the number of valence electrons of each atom in the molecule. With ions, add an electron for each negative charge and subtract an electron for each positive charge. Place this total number of valence electrons, regardless of where they come from, onto the skeleton, distributed two per bond and as unshared pairs. The number of unshared electrons equals the total number of valence electrons minus two electrons for each bond. Unshared electrons are then used to complete octets and exceed octets when appropriate. Remember that Be and B may have less than an octet.†

Free radicals are exceptions—they have at least one unpaired, unshared electron.

When organizing the skeleton and distributing the electrons within the confines of the octet, try to avoid formal charges. There will, however, be Lewis structures where formal charges must be present. (See Problem 7.59 for useful hints about writing Lewis structures with formal charges.)

The following examples illustrate the procedure.

EXAMPLE 7.13 Write the Lewis structures for (a) chloromethane, $CHCl_3$; (b) methanol, CH_4O; (c) formaldehyde, CH_2O; and (d) sulfur tetrafluoride, SF_4.

ANSWER

(a) The skeleton of chloromethane has the multicovalent C as the central atom bonded to the four terminal atoms—the three H's and the Cl—as shown:

$$\begin{array}{ccccc} & & H & & \\ & & | & & \\ H & - & C & - & Cl \\ & & | & & \\ & & H & & \end{array}$$

† The problems found in this book can be answered correctly by properly applying this procedure. However, the procedure is not foolproof because the structure of a molecule does not necessarily follow the rules. For example, the procedure would predict that $SOCl_2$, thionyl chloride, has the Lewis structure, $:\ddot{\underset{..}{Cl}}-\ddot{\underset{..}{O}}-\ddot{\underset{..}{S}}-\ddot{\underset{..}{Cl}}:$, which happens to be *incorrect*. Usually the central atom is the less electronegative one. In $SOCl_2$, S is less electronegative than O and is the central atom in the molecule; the correct structure is

$$\begin{array}{ccccc} & & {}^{\ominus}:\ddot{O}: & & \\ & & {}^{\oplus}| & & \\ :\ddot{\underset{..}{Cl}} & - & \underset{..}{S} & - & \ddot{\underset{..}{Cl}}: \end{array}$$

The number of valence electrons is 3H + C + Cl

$(3 \times 1) + 4 + 7 = 14$

The four bonds shown in the skeleton account for 8 electrons, which leaves 6 electrons (14 − 8) to be placed on the Cl as three unshared pairs to complete its octet. The Lewis structure is

```
    H
    |      ..
 H—C—Cl:
    |      ..
    H
```

(b) To get the skeleton of methanol, the multicovalent C and O atoms are first bonded to each other. Then the four terminal H atoms are distributed so that C and O have their usual covalences of 4 and 2, respectively, as shown:

```
    H
    |
 H—C—O—H
    |
    H
```

(Convince yourself that placing two H's on both the C and O would lead to an incorrect Lewis structure with formal charges.)

The number of valence electrons is 4H + C + O

$(4 \times 1) + 4 + 6 = 14$

The five bonds in the skeleton account for 10 electrons, which leaves 4 electrons (14 − 10) to be placed as two unshared pairs on the O atom. The Lewis formula is

```
    H
    |   ..
 H—C—O—H
    |   ..
    H
```

(c) In the skeleton of formaldehyde, the multicovalent C and O atoms are bonded to each other. The 2 H's are insufficient to satisfy the covalences of C and O, so a multiple bond is needed. A double bond is placed between the C and O, C=O, and the 2 H's are bonded to the C to give the skeleton

```
 H—C=O
    |
    H
```

that satisfies all the usual covalencies. The total number of valence electrons is

2H + C + O

$(2 \times 1) + 4 + 6 = 12$

There are 8 electrons in the four bonds, leaving 4 unshared electrons placed on O as two pairs. The octet of O is now complete, and the Lewis structure is

```
       ..
 H—C=O:
    |
    H
```

(Show that putting an H on both O and C leads to an incorrect structure with formal charges.)

(d) The S in sulfur tetrafluoride is the central atom bonded to the four terminal F atoms, to give the skeleton

```
     F
     |
  F—S—F
     |
     F
```

The number of valence electrons is 4F + S

$$(4 \times 7) + 6 = 34$$

The skeleton has 8 electrons in four bonds. 24 electrons are used to supply each F with three unshared electron pairs, thereby completing the octet of the F's. Up to this point we have used 32 (24 + 8) electrons. The remaining 2 electrons are placed as an unshared pair on S because S can acquire more than an octet. The Lewis structure is

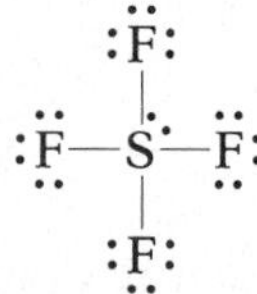

Notice that unless we had correctly counted the number of valence electrons, we might have overlooked the unshared pair on the S atom. (Convince yourself that these 2 electrons could not have been used to form a double bond between S and F.) ■

EXAMPLE 7.14 Write Lewis structures for (a) NH_4^+ and (b) BF_4^-. Locate all formal charges.

ANSWER

(a) The multicovalent N is the central atom bonded to the four H's to give the skeleton

```
⎡    H    ⎤+
⎢    |    ⎥
⎢ H—N—H ⎥
⎢    |    ⎥
⎣    H    ⎦
```

The number of valence electrons is 4H + N − 1 (because of + charge on ion)

$$(4 \times 1) + 5 - 1 = 8$$

These 8 electrons are in the four covalent bonds of the skeleton; the skeleton is also the Lewis structure. The N has a formal charge of +1.

(b) B is the central atom bonded to the four terminal F atoms to give the skeleton

```
⎡    F    ⎤−
⎢    |    ⎥
⎢ F—B—F ⎥
⎢    |    ⎥
⎣    F    ⎦
```

The number of valence electrons is 4F + B + 1 (because of − charge on ion)

$$(4 \times 7) + 3 + 1 = 32$$

Eight electrons are in the four covalent bonds, and the remaining 24 electrons are placed so that each F gets an octet. This requires that each F gets three unshared pairs of electrons. The Lewis structure is

```
 [      ..      ]-
 [     :F:      ]
 [  ..  |  ..   ]
 [ :F — B — F:  ]
 [  ..  |  ..   ]
 [     :F:      ]
 [      ..      ]
```

The B has a negative formal charge. ■

CH$_2$Cl$_2$ is used to decaffeinate coffee.

PROBLEM 7.11 Write Lewis structures for (a) methylene chloride, CH_2Cl_2; (b) hydrogen peroxide, H_2O_2; (c) ethylene (ethene), C_2H_4; (d) bromine trifluoride, BrF_3. □

PROBLEM 7.12 Write Lewis structures for (a) H_3O^+, (b) AlF_6^-, and (c) HCO_3^-. Locate all formal charges. □

LEWIS STRUCTURES AND ISOMERS

Our discussion of Lewis structures and isomers begins with an example.

EXAMPLE 7.15 Write a Lewis structure for C_2H_6O.

ANSWER The three multicovalent atoms can be bonded to give two different "backbones" as shown:

C—C—O and C—O—C

If the H's are placed so that C and O acquire their usual covalences of 4 and 2, respectively, two skeletons are obtained:

```
     H  H              H       H
     |  |              |       |
  H—C—C—O—H        H—C—O—C—H
     |  |              |       |
     H  H              H       H
```

The number of valence electrons is 6H + 2C + O

$(6 \times 1) + (2 \times 4) + 6 = 20$

Sixteen electrons are used for the eight bonds of the skeleton, and the remaining 4 electrons are placed as two unshared pairs on O to give two Lewis structures:

Dimethyl ether, b.p. −24°C; ethanol, b.p. 78.5°C.

```
     H       H                 H  H
     |   ..  |                 |  |  ..
  H—C—O—C—H      and      H—C—C—O—H
     |   ..  |                 |  |  ..
     H       H                 H  H
   dimethyl ether              ethanol
```

These are acceptable Lewis structures for two different compounds. Therefore both answers are correct. ■

This example reveals an important fact: *there can be more than one compound with the same molecular formula.* Such compounds are called **isomers**. Other examples of isomers will be discussed in Chapters 21 and 24.

For convenience, Lewis structures are often **condensed**; that is, not all bonds are shown, and the atoms (or groups of atoms) bonded to a multicovalent atom are grouped together. For example,

$$CH_3\ddot{\underset{\cdot\cdot}{O}}CH_3 \qquad CH_3CH_2\ddot{\underset{\cdot\cdot}{O}}H$$

dimethyl ether   ethanol

PROBLEM 7.13 Write Lewis structures for two isomers with the molecular formula C_2H_7N. □

LEWIS STRUCTURES OF OXO ACIDS

Oxo acids have the general formula H_mXO_n. Their chemistry is discussed in Chapter 14. Here, we are concerned only with writing their Lewis structures. In most cases the O atoms (or atom) are bonded to atom X, the central atom, and the H atoms are bonded to the O atoms. For example, the skeleton for nitric acid, HNO_3, is

Sometimes an H may be bonded to X.

```
    O
    |
O—N—O—H
```

PROBLEM 7.14 Write the Lewis structures of (a) hypochlorous acid, HClO, and (b) the other oxo acids of Cl: chlorous acid, $HClO_2$; chloric acid, $HClO_3$; and perchloric acid, $HClO_4$. Show the formal charges. □

Different students may have written different structures for Problem 7.14b. For example, for $HClO_2$, some might have written

$$H-\ddot{\underset{\cdot\cdot}{O}}-\overset{\oplus}{\ddot{\underset{\cdot\cdot}{Cl}}}-\overset{\ominus}{\ddot{\underset{\cdot\cdot}{O}}}:$$

Others, knowing that Cl can expand its valence shell beyond the octet, might have written

$$H-\ddot{\underset{\cdot\cdot}{O}}-\ddot{\underset{\cdot\cdot}{Cl}}=\underset{\cdot\cdot}{O}:$$

In this way each O acquires its usual covalence. Which Lewis structure is correct? This issue will be addressed in section 8.5. This condition often prevails among the oxo acids of elements in the third and higher periods. For now, either structure is acceptable.

7.11 SHAPES OF MOLECULES AND VSEPR THEORY

The Lewis structures drawn so far in this chapter tell nothing about the bond angles in the molecules. For example, is H_2O a linear molecule with a bond angle of 180°, $H:\ddot{\underset{\cdot\cdot}{O}}:H$, or is it a bent molecule with an angle less than 180°, $\overset{H\quad H}{\diagdown\ddot{\underset{\cdot\cdot}{O}}\diagup}$? As you can see, the bond angle determines the **molecular shape**. The structures drawn so far also imply flat molecular shapes because all of the atomic nuclei are shown lying in the plane of the paper. Yet, with molecules of four or more atoms, this is not often the case. Many such molecules have three-dimensional molecular shapes, several of which are described later in this section. Molecular shape greatly influences the chemical and physical properties of a substance. Molecular shape is especially important in biochemical reactions. The reactivities of large, complex molecules such as enzymes and proteins are determined by their ability to acquire a definite shape. This section first discusses simple molecules (and polyatomic ions) of representative elements of the XY_n-type such as CH_4, H_2O, BF_3, PCl_5, and SF_6. *The molecular shape refers only to the position of the atomic nuclei.* The instruments used to determine the shapes of molecules usually do not detect the presence of unshared pairs of electrons, nor do they distinguish between single and multiple bonds. However, the presence of unshared pairs of electrons greatly influences the shape.

A Nobel Prize was awarded in 1962 to Max F. Perutz and John C. Kendrew for determining the shape of hemoglobin.

In an XY_n-type molecule or polyatomic ion, the positions of the Y atoms about the central X atom are given in terms of the Y—X—Y bond angle. This angle can vary from about 90° to 180°. One general theory that has been suggested to account for the observed bond angles is based on the simple idea that pairs of electrons repel each other. The formal name of the theory is the **valence shell electron pair repulsion (VSEPR) theory**, and its basic assumption is that an electron pair is attracted to a nucleus but repelled by other electron pairs surrounding the central atom. According to VSEPR theory, the major factor influencing molecular shape is *the tendency of the electron pairs surrounding the central atom to minimize their mutual repulsion.* This means that the electron pairs repel each other so as to attain *maximum distances* of separation. When the electron pairs are farther apart, the repulsive force is smaller. The concept applies both to bonded and to unshared electron pairs. To use this theory, you must know the Lewis structure of the molecule.

Reminder: The valence shell is the highest-energy shell.

MOLECULES WITHOUT UNSHARED PAIRS ON THE CENTRAL ATOM

We will discuss the five XY_n-types of molecules in this subsection. The types are listed here, and we give a typical example of each.

TYPE				
XY_2	XY_3	XY_4	XY_5	XY_6
:C̈l—Be—C̈l:	B bonded to three :F̈:	H—C—H with H above and below	P bonded to five :C̈l:	S bonded to six :F̈:
$BeCl_2$	BF_3	CH_4	PCl_5	SF_6

The central atoms in each type are bonded, respectively, to 2, 3, 4, 5, and 6 atoms and they have no unshared pairs of electrons.

The maximum angle of separation of two electron pairs is 180°. Therefore, $BeCl_2$ (XY_2-type) is predicted to be a **linear** molecule. See Table 7.6a.

The B atom in BF_3 (XY_3-type) is surrounded by three bonding electron pairs. See Table 7.6b. The maximum angle of separation of three pairs is 120°. Therefore, we assume that the B atom sits in the center of an equilateral triangle with

TABLE 7.6
MOLECULES AND POLYATOMIC IONS WHOSE CENTRAL ATOMS HAVE NO UNSHARED ELECTRON PAIRS†

TYPE	EXAMPLES	A SPECIFIC STRUCTURE	A GENERAL STRUCTURE	SHAPE
(a) XY_2	$BeCl_2$, CO_2, NO_2^+	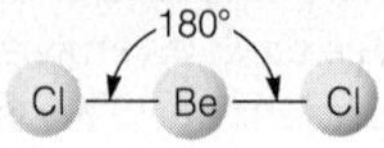 Cl—Be—Cl, 180°	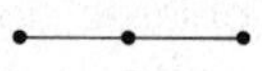	Linear
(b) XY_3	BF_3, NO_3^-, CO_3^{2-}	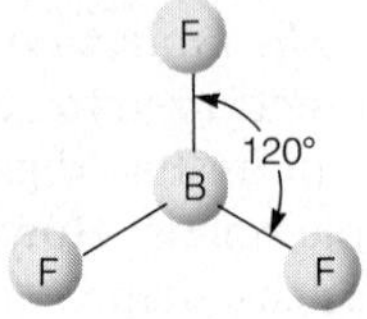 BF_3, 120°	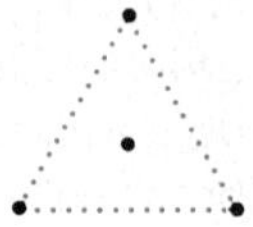	Trigonal-planar
(c) XY_4	CH_4, PO_4^{3-}, NH_4^+	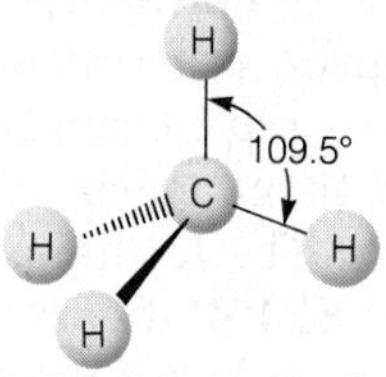 CH_4, 109.5°	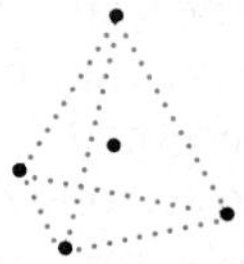	Tetrahedral§
(d) AB_4	$PtCl_4^{2-}$	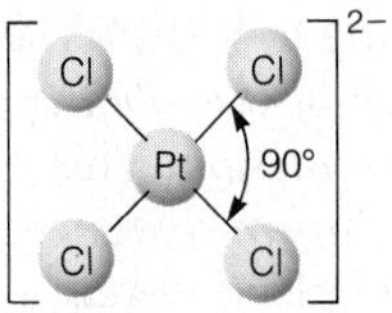 $[PtCl_4]^{2-}$, 90°	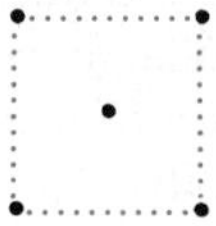	Square-planar‖
(e) XY_5	PCl_5, $SnCl_5^-$	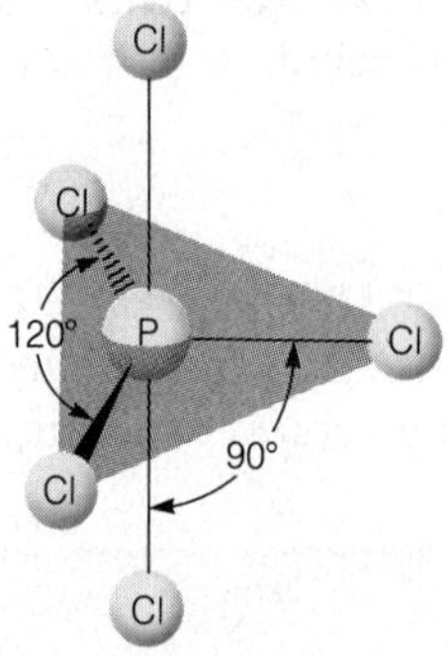 PCl_5, 120°, 90°	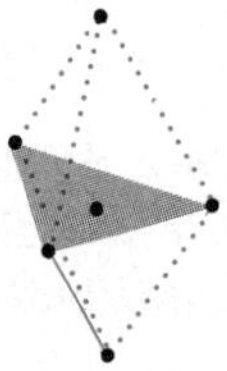	Trigonal-bipyramidal
(f) XY_6	SF_6, AlF_6^{3-}	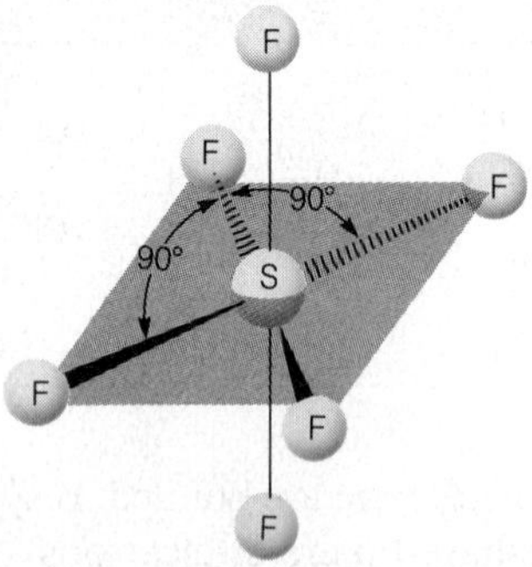 SF_6, 90°, 90°	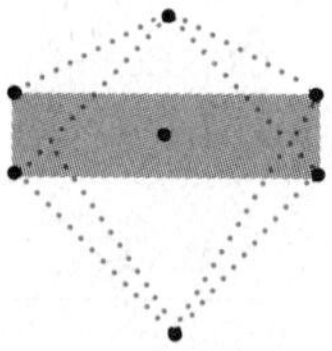	Octahedral

† Some general remarks about Table 7.6: (1) Unshared pairs on terminal atoms are not shown. (2) Dotted lines (· · · ·) are not bonds; they are tielines to indicate shape. (3) Solid wedges (▶) represent bonds that project toward the viewer. Broken wedges (ııı·) represent bonds that project away from the viewer. Solid lines represent bonds in the plane of the paper.

§ Always observed for a representative central atom.

‖ Occasionally observed for some transition elements.

an F atom at each corner. Since the F atoms form a triangle and all four atoms lie in a plane, the shape of BF_3 is **trigonal-planar**. The three F\B/F bond angles are observed to be the same, namely 120°.

As a typical example the XY_4-type of molecule, CH_4, has been drawn with the C and four H atoms lying in the plane of the page, thereby indicating a planar (flat) molecular structure. It would appear from this drawing that the shape is a square with C at the center and an H at each corner. This molecular shape, described as **square-planar**, is shown in Table 7.6d for a molecule of $PtCl_4^{2-}$, indicating the typical 90° bond angles. However, CH_4 is *not* a square-planar molecule, but rather has a **tetrahedral** shape. A tetrahedron is a three-dimensional structure with four triangular sides; the C atom is in the center and an H atom is at each of the four corners. Table 7.6c emphasizes the tetrahedral shape of CH_4. The bond angles in the tetrahedron are 109°, rather than 90° as in the square-planar structure. This larger bond angle places the electron pairs farther apart than they would be with a 90° angle. For this reason, XY_4-type molecules and ions are usually tetrahedral; this is always the case when X is a representative element. When X is bonded to the same four atoms or groups, the four Y—X—Y bond angles are the same, 109°.

The tielines in the drawing are not bonds.

In SF_6 (XY_6-type), the S atom is surrounded by six pairs of valence electrons in six bonds. The maximum angle of separation among the six pairs is 90°. The predicted and observed three-dimensional shape having these angles is **octahedral** (Figure 7.6f). Visualize this octahedron: the S atom sits in the center of a square (shaded area in Table 7.6f) with an F atom at each corner. One of the two remaining F atoms is above, and the other is below the S atom. All six S—F bond lengths are equal. Each F—S(—F) bond angle between nearest F atoms is 90°. Any four of the six F atoms can be visualized as lying at the corners of a square.

The P atom in an individual PCl_5 (XY_5-type) molecule is surrounded by five bonding pairs of valence electrons. The observed and predicted arrangement of these five pairs differs in one important way from the four molecular shapes just discussed. These four molecular shapes each have a uniform bond angle between nearest members, and all X—Y bonds are alike. However, the shape for PCl_5 has two different sets of bond angles. The three-dimensional shape of PCl_5 (Table 7.6e) can be visualized by thinking of the P atom sitting at the center of an equilateral triangle (shaded area in Figure 7.6e) with a Cl atom at each corner, which gives a Cl to P to Cl bond angle of 120°. Of the two remaining Cl atoms, one is above and the other is below the P atom. The bond angle from one of these Cl atoms to a Cl atom forming the triangle is 90°. This shape is **trigonal-bipyramidal**. The shapes of all of these molecules, predicted from the VSEPR theory, agree with the observed shapes.

These three Cl atoms are described as "radial" or "equatorial" atoms.

These two Cl atoms are described as "axial" atoms.

For purposes of predicting molecular shape, multiple bonds are treated as if they were single bonds. What is significant is the number of atoms that are bonded to the central atom, not whether the atoms are singly or multiply bonded. Thus, carbon dioxide, $:\ddot{O}{=}C{=}\ddot{O}:$, is considered to be an XY_2-type molecule, and indeed it is linear.

EXAMPLE 7.16 Give the shapes of these covalent species: BF_4^-, $AlBr_3$, NH_4^+, $HgCl_2$, SO_4^{2-}, NO_3^-, and H_2CO.

ANSWER First write the Lewis structures:

$$\left[\begin{array}{c} :\ddot{F}: \\ | \\ :\ddot{F}-B-\ddot{F}: \\ | \\ :\ddot{F}: \end{array}\right]^- \qquad :\ddot{Br}:-Al(-:\ddot{Br}:)_2 \qquad \left[\begin{array}{c} H \\ | \\ H-N-H \\ | \\ H \end{array}\right]^+ \qquad :\ddot{Cl}-Hg-\ddot{Cl}:$$

$$\left[\begin{array}{c} :\ddot{O}: \\ | \\ :\ddot{O}-S-\ddot{O}: \\ | \\ :\ddot{O}: \end{array}\right]^{2-} \qquad H_2C=\ddot{O}: \qquad \left[:\ddot{O}-N(=\ddot{O})-\ddot{O}:\right]^-$$

$$\left[\begin{array}{c} :O: \\ \| \\ :\ddot{O}-S-\ddot{O}: \\ \| \\ :O: \end{array}\right]^{2-}$$

Gives same shape as structure with single bonds.

The presence of multiple bonds in NO_3^- and H_2CO has no appreciable effect on their shapes. Those structures in which a central atom is bonded to four atoms are tetrahedral; they are BF_4^-, NH_4^+, and SO_4^{2-}. Those in which the central atom is bonded to three atoms are trigonal-planar; they are $AlBr_3$, NO_3^-, and H_2CO (formaldehyde). The only molecule with just two bonds, $HgCl_2$, is linear. ■

PROBLEM 7.15 Write the Lewis structures and give the shapes of these covalent species: SiF_6^{2-}, CCl_4, $POCl_3$, NO_2^+, $AsCl_5$, and CO_3^{2-}. □

MOLECULES WITH UNSHARED ELECTRON PAIRS ON THE CENTRAL ATOM

We first consider molecules in which the central atom does *not* have more than eight valence electrons. Typical molecules in this category are $:NH_3$, $H_2\ddot{O}:$, and $H:\ddot{F}:$. In each case, the four electron pairs (unshared and in bonds) are oriented in space toward the corners of a tetrahedron, as in Table 7.7c. If an electron pair could be located experimentally, $:NH_3$, $H_2\ddot{O}:$, and $H:\ddot{F}:$ would each have a tetrahedral shape like that of CH_4. But unshared pairs *cannot* be "seen." Therefore, we visualize $:NH_3$ as a tetrahedral molecule from which a corner atom has been removed. The remaining atoms then give the $:NH_3$ molecule the shape described as *trigonal-pyramidal* (Table 7.7a). The NH_3 pyramid and the CH_4 pyramid differ in that the C is at the center of its pyramid, while the N is at a corner of its pyramid.

Water is a *bent* (*V-shaped* or *angular*) molecule; see Table 7.7b. Hydrogen fluoride, like any diatomic molecule, must necessarily be linear; see Table 7.7c.

Sulfur dioxide, $:\ddot{O}^{\ominus}-\ddot{S}^{\oplus}=\ddot{O}:$, is an interesting case. The double bond is counted as if it were a single bond with one bonding pair of electrons. Therefore, $:SO_2$ should have the shape expected of an $:XY_2$-type molecule. The double bond, single bond, and unshared pair of electrons should make for a trigonal-planar arrangement. Since the unshared pair cannot be "seen," the molecule is bent (Table 7.7d).

These examples make clear that although unshared pairs of electrons on the central atom are "invisible," they nevertheless have profound effects on molecular shape. Thus, were it not for the unshared pairs of electrons on the O atom, $H_2\ddot{O}:$ would be linear like $BeCl_2$ instead of bent.

TABLE 7.7 MOLECULES OR IONS WITH UNSHARED ELECTRON PAIRS ON THE CENTRAL ATOM

TYPE†	NUMBER OF BONDED ATOMS	NUMBER OF UNSHARED e^- PAIRS	EXAMPLES	SPECIFIC STRUCTURE§	GENERAL STRUCTURE	SHAPE
With Eight Valence e^-'s Based on a Tetrahedral Electronic Arrangement						
(a) $:XY_3$	3	1	$:NH_3$, $:SO_3{}^{2-}$, $H_3\ddot{O}^+$	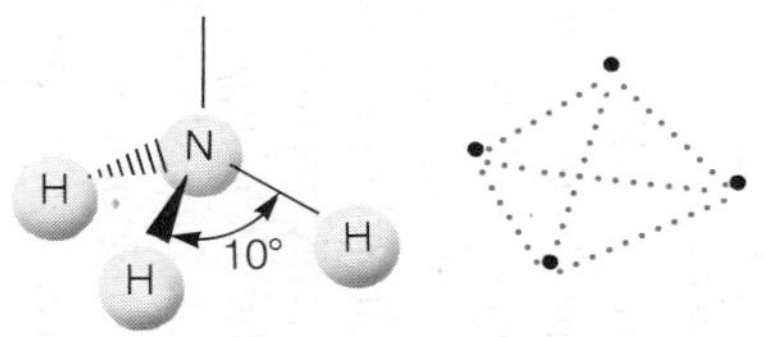	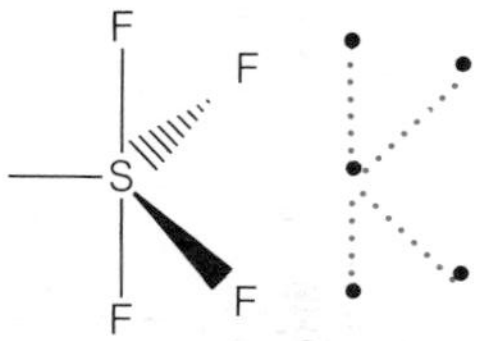	Pyramidal
(b) $:\ddot{X}Y_2$	2	2	$H_2\ddot{O}:$			Bent or V-shaped
(c) $:\ddot{\underset{..}{X}}Y$	1	3	$H\ddot{\underset{..}{F}}:$			Linear
Based on Trigonal-Planar Electronic Arrangement						
(d) $:XY_2$	2	1	$:SO_2$, $:NO_2{}^-$, $\ddot{S}nCl_2$	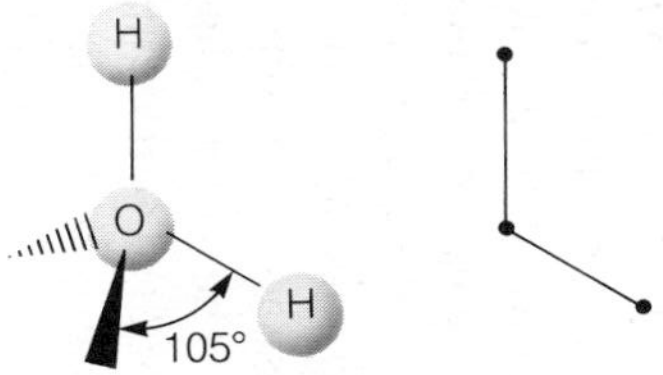	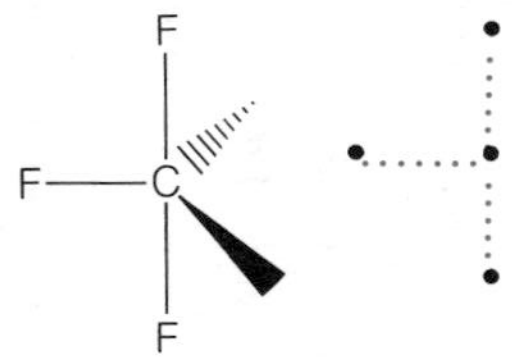	Bent or V-shaped
Based on Six-Pair Octahedral Arrangement (Table 7.6f)†						
(e) $:XY_5$	5	1	$:BrF_5$	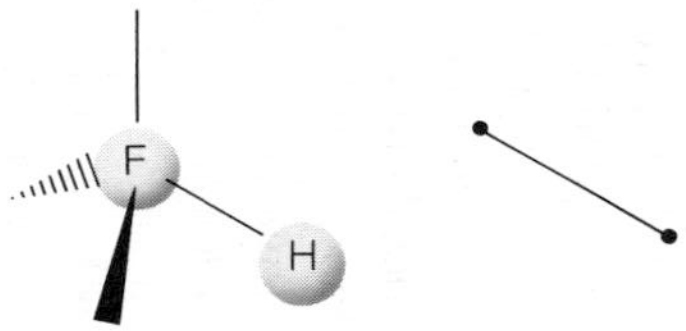	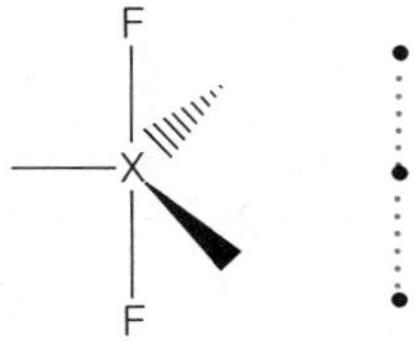	Square-pyramidal
(f) $:\ddot{X}Y_4$	4	2	$:\ddot{X}eF_4$			Square-planar
Based on Five-Pair Trigonal-Bipyramidal Electronic Arrangement (Table 7.6e)						
(g) $:XY_4$	4	1	$:SF_4$	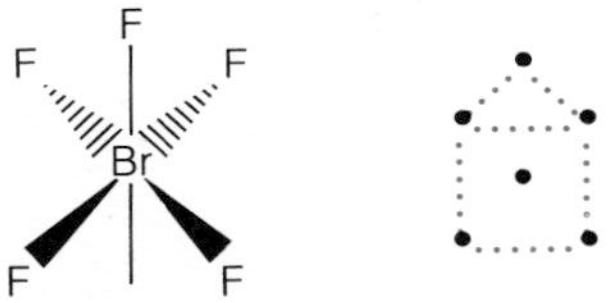	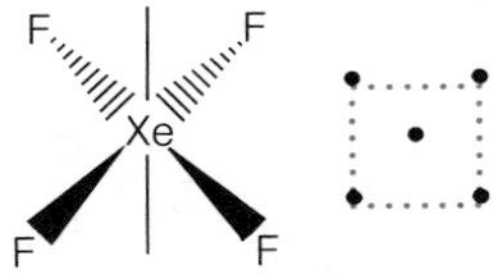	Distorted-tetrahedral
(h) $:\ddot{X}Y_3$	3	2	$:\ddot{C}lF_3$			T-shaped
(i) $:\ddot{\underset{..}{X}}Y_2$	2	3	$:\ddot{\underset{..}{X}}eF_2$			Linear

† Missing types, such as $:\ddot{\underset{..}{X}}Y_3$, have no known examples.

§ In the specific structure diagrams, a dash or wedge from the central atom represents a pair of electrons. If nothing else is attached to the dash or wedge, the electrons are unshared.

We note from the figures in Table 7.7a and b that the bond angles in $:NH_3$ and $H_2\ddot{O}:$ are 107° and 105°, respectively. The contraction from the characteristic tetrahedral angle of 109.5° can be explained by the secondary effects of unshared pairs: *Unshared pairs repel adjacent electron pairs more strongly than do bonding pairs.* As a rule, the repulsive forces decrease in this order:

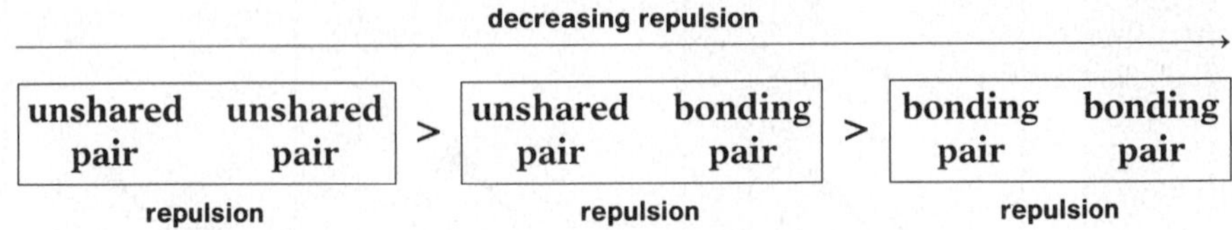

Consequently, the greater repulsion of unshared pairs contracts the angles made by bonding pairs. The two unshared electron pairs on the O atom of a water molecule cause greater contraction (to 105°) than the contraction caused by the single unshared pair on the N atom of $:NH_3$ (to 107°).

We now consider molecules in which the central atom has an expanded valence shell containing at least one unshared electron pair. In one group of such molecules, a central atom has six pairs of valence electrons, each pointing to the corner of an octahedron. In a typical octahedron (XY_6), all bonds are equivalent, and therefore, it makes no difference which bond is replaced by the unshared electron pair. However, once an unshared pair replaces a shared pair, the observed shapes are no longer octahedral. The actual shape of the $:XY_5$-type, as illustrated by $:BrF_5$ in Table 7.7e, is designated **square-pyramidal**; four of the F atoms form the square base, and the fifth F is at the apex above (or below) the Br. Replacing any of the six bonds with the unshared pair gives the same shape. With the $:\ddot{X}Y_4$-type, as illustrated with $:\ddot{X}eF_4$ in Table 7.7f, the unshared pairs are 180° away from each other, which minimizes their strong repulsive force. As a result, $:\ddot{X}eF_4$, the $:\ddot{X}Y_4$-type, is square-planar.

$:XY_5$ and $:\ddot{X}Y_4$ are the common types of general formulas.

In the incorrect structure, the unshared pairs are closer to each other, 90° apart.

The next group has five pairs of electrons on the central atom. Recall that the five bonds in the trigonal-bipyramidal shape of XY_5 are not equivalent: the three at the corners of the triangle (radial) are different from the other two (axial). Any unshared electron pairs always replace the bonds at the corners of the triangle. In this way the repulsion exerted by the lone pair is minimized. Hence, $:SF_4$, the $:XY_4$-type, is an **irregular tetrahedron** (Table 7.7g); $:\ddot{C}lF_3$, the $:\ddot{X}Y_3$-type, is **T-shaped** (Table 7.7h); and $:\ddot{\underset{..}{X}}eF_2$, the $:\ddot{\underset{..}{X}}Y_2$-type, is linear (Table 7.7i).

EXAMPLE 7.17 Predict the shapes of these molecules and ions: (a) I_3^-, (b) AsF_4^-, (c) ClF_3, (d) ClF_4^-, and (e) ClF_5.

ANSWER First draw the Lewis structures and determine the general types. Check for the correct number of valence electrons in each Lewis structure so that you will have the correct number of unshared pairs of electrons. This number is the key to predicting the shapes.

(a) The skeleton of I_3^- is $[I—I—I]^-$. The number of valence electrons is 3I + 1, or $(3 \times 7) + 1 = 22$. Four electrons are in the two bonds shown in the skeleton, which leaves 18 e^-'s. Eight pairs of electrons (16 e^-'s) are used to complete the octet of each I. The two remaining electrons are placed on the central I, rather than on a terminal I in order to minimize formal charges. The correct Lewis structure is $[:\ddot{\underset{..}{I}}—\ddot{\underset{..}{I}}—\ddot{\underset{..}{I}}:]^-$, an $:\ddot{\underset{..}{X}}Y_2$-type. It has a linear shape.

(b) The skeleton of AsF_4^- is

$$\left[\begin{array}{ccc} & F & \\ & | & \\ F- & As & -F \\ & | & \\ & F & \end{array}\right]^-$$

The number of valence electrons is As + 4F + 1, or 5 + 4(7) + 1 = 34. Distributing these electrons gives

$$\left[\begin{array}{ccc} :\ddot{\underset{..}{F}} & & \ddot{\underset{..}{F}}: \\ & \ddot{As} & \\ :\ddot{\underset{..}{F}} & & \ddot{\underset{..}{F}}: \end{array}\right]^-$$

an $:XY_4$-type having an irregular tetrahedral shape.

(c) The skeleton of ClF_3 is

$$\begin{array}{cc} F & \\ | & \\ Cl- & F \\ | & \\ F & \end{array}$$

Placing the 28 (7 × 4) valence electrons gives

$$\begin{array}{cc} :\ddot{F}: & \\ | & \\ \dot{\underset{\cdot}{:Cl}}- & \ddot{\underset{..}{F}}: \\ | & \\ :\underset{..}{F}: & \end{array}$$

an $:\ddot{X}Y_3$-type that is T-shaped.

(d) The skeleton of ClF_4^- is

$$\left[\begin{array}{ccc} & F & \\ & | & \\ F- & Cl & -F \\ & | & \\ & F & \end{array}\right]^-$$

Placing the 36 valence electrons, Cl + 4F + 1 or (5 × 7) + 1 = 36, gives the Lewis structure

$$\left[\begin{array}{ccc} & :\ddot{F}: & \\ & | & \\ :\ddot{\underset{..}{F}}- & \dot{\underset{\cdot}{Cl}}\dot{} & -\ddot{\underset{..}{F}}: \\ & | & \\ & :\underset{..}{F}: & \end{array}\right]^-$$

an $:\ddot{X}Y_4$-type with a square-planar shape.

(e) The skeleton of ClF_5 is

$$\begin{array}{ccc} F & & F \\ & Cl & \\ F & | & F \\ & F & \end{array}$$

Placing 42 valence electrons, Cl + 5F or 6 × 7 = 42, gives the Lewis structure

$$\begin{array}{c} :\ddot{F}: \quad \ddot{F}: \\ \ddot{Cl} \\ :\ddot{F}: \; | \; \ddot{F}: \\ :\ddot{F}: \end{array}$$

an $:XY_5$-type with a square-pyramidal shape. ■

Notice that in Example 7.17b–e the univalent F had to be the terminal atom. Furthermore, notice how important it is to get the correct count and distribution of the valence electrons. You must know the number of unshared electron pairs on the central atom in order to predict the observed molecular shapes.

PROBLEM 7.16 Predict the shapes of (a) ClF_4^+, (b) ClF_2^-, (c) XeF_3^+, and (d) $XeOF_4$. □

7.12 DIPOLE MOMENTS

In this section we discuss a property of molecules that you can use experimentally to help you decide between alternative molecular shapes. Earlier we showed that certain covalent bonds are polar. Because of the polarity of individual bonds, a molecule *may* have separated centers of positive and negative partial charge. Such a molecule, said to be a **polar molecule**, constitutes a **dipole**. An example of a polar molecule is H—F.

$$\overset{\delta+}{H}—\overset{\delta-}{F} \quad \text{or} \quad \overset{+\!\!\longrightarrow}{H—F}$$

Chemists symbolize the dipole by $+\!\!\!\longrightarrow$, with the arrowhead pointing toward the center of negative charge, δ^-, and the + end of the arrow pointing toward the center of positive charge, δ^+. Polar molecules possess a **dipole moment**, μ, which is the product of the magnitude of *partial charge, q*, and the distance, *d*, between the centers of opposite charge.

$$\textbf{dipole moment}(\mu) = \begin{array}{c}\textbf{amount of}\\ \textbf{partial charge}(q)\end{array} \times \begin{array}{c}\textbf{distance}(d)\\ \textbf{between charges}\end{array}$$

The unit for μ is the *debye* (D), named in honor of Peter Debye, who pioneered in this field of study.

As a consequence of the dipole moment, polar molecules tend to be oriented in an electric field with their positive ends directed toward the negative electric plate and their negative ends toward the positive plate (Figure 7.7). These orientations are far from exact because of the randomness imposed on them by their kinetic energies. **Nonpolar** molecules have a zero dipole moment and, therefore, are *not* oriented in an electric field.

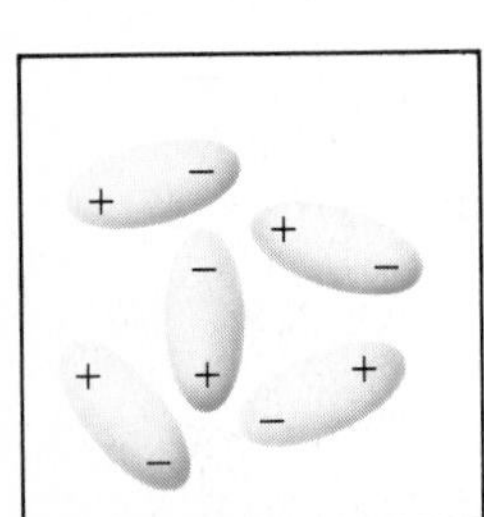

(a)

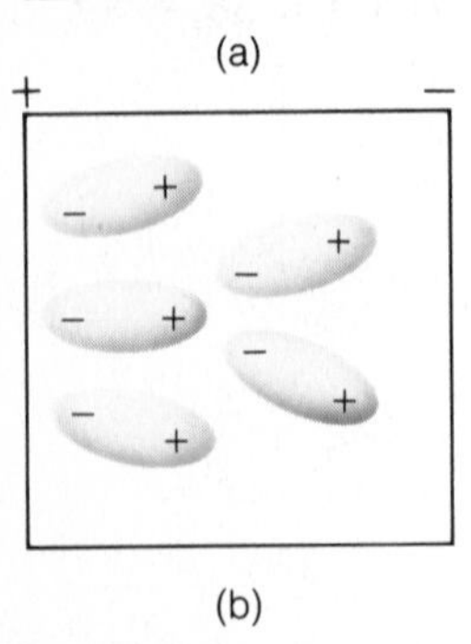

(b)

FIGURE 7.7
Orientation of polar molecules in an electric field. (a) Field off; (b) field on.

When considering the dipole moment of a molecule, it is useful to imagine that the molecule is separated into pairs of its bonded atoms. A dipole moment, called a **bond moment**, may then be assigned to each pair. For example, the molecule of $BeCl_2$

$$Cl\overset{\longleftarrow\!\!+}{—}Be\overset{+\!\!\longrightarrow}{—}Cl$$

is considered to consist of two Be—Cl bonds, each with its own bond moment. An unshared pair of electrons on an atom can also be considered as having the equivalent of a bond moment pointing toward the unshared pair. Thus $:NH_3$ would

Peter Debye, 1884–1966

have three N—H bond moments and an unshared pair bond moment:

[Diagram: NH_3 with bond moment arrows on each N—H bond and on the unshared pair]

The dipole moment of a molecule is a resultant of the individual bond moments and the moments of any unshared pairs. The bond and unshared pair moments are considered as vectors (Appendix B.5), the sum of which gives the dipole moment of the molecule.

DIPOLE MOMENTS AND MOLECULAR SHAPE

Now let us look at some simple cases of how experimentally obtained values for dipole moments are used for predicting the shapes of molecules. A molecule with only nonpolar bonds is nonpolar. On the other hand, polar bonds may or may not make a molecule polar; it depends on the shape of the molecule. For this reason, the presence or absence of dipole moments in molecules with polar bonds is a clue to the shape of the molecule. For some shapes the bond moments cancel each other. For example, in CO_2 each C to O bond is polar because the O atom is more electronegative than the C atom. Furthermore, each O has moments from unshared electron pairs. Yet CO_2 is observed to be a nonpolar molecule. The observed nonpolarity of CO_2 confirms the linear shape predicted for the molecule using the VSEPR theory.

$$\overset{\longleftarrow\!\!+}{:\ddot{O}}=C=\overset{+\!\!\longrightarrow}{\ddot{O}:}$$

(Bond moments and unshared electron pair moments cancel; molecule is nonpolar, $\mu = 0$.)

The bond polarities are equal and opposite in direction, so they cancel each other. Overall, therefore, the molecule is nonpolar.

Unlike CO_2, water, H_2O, has a large net dipole moment (Table 7.8). This fact suggests that the individual $\overset{+\!\!\longrightarrow}{H—O}$ bond moments do not cancel. Indeed, the water molecule has a bent shape, as shown in Table 7.7b. The dipole moment of water is illustrated in Figure 7.8a.

With the linear, trigonal-planar, tetrahedral, octahedral, and trigonal-bipyramidal shapes, the sum of the individual bond moment vectors is zero if the central atom is bonded to only one kind of atom or group. This fact is illustrated for the linear, trigonal-planar, and tetrahedral structures in Figure 7.9. The terminal atom is usually more electronegative than the central atom, as indicated in the figure. The fact that compounds such as $BeCl_2$, BF_3, and CCl_4 have zero dipole moments supports (but does not prove) the assigned structures. If the bonded atoms are

TABLE 7.8
DIPOLE MOMENTS OF SOME SIMPLE MOLECULES

MOLECULE	DIPOLE MOMENT, DEBYES	MOLECULE	DIPOLE MOMENT, DEBYES
HF	1.92	H_2S	1.10
HCl	1.08	NH_3	1.46
HBr	0.78	NF_3	0.24
HI	0.38	SO_2	1.60
H_2O	1.87	CO_2	0.00

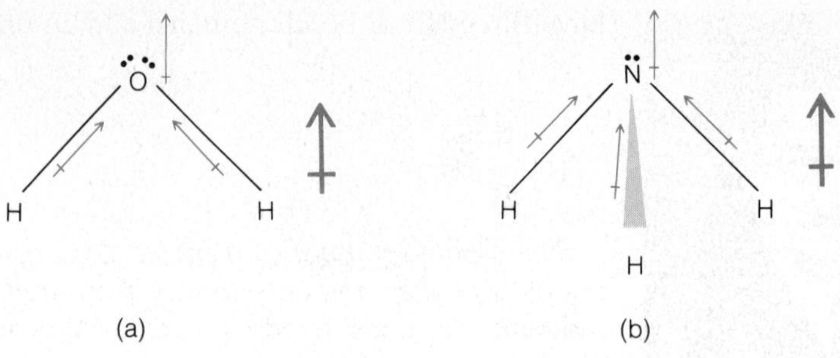

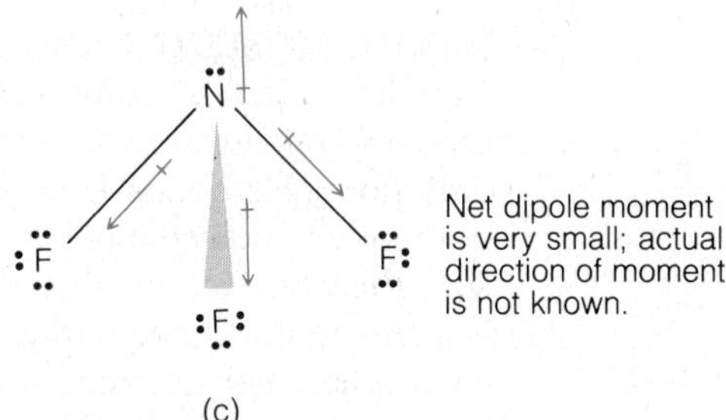

FIGURE 7.8
Polar molecules (a) $H_2\ddot{O}:$, (b) $:NH_3$, and (c) $:NF_3$ ⟼ indicates individual bond moments. A heavy arrow indicates the dipole moment of the molecule.

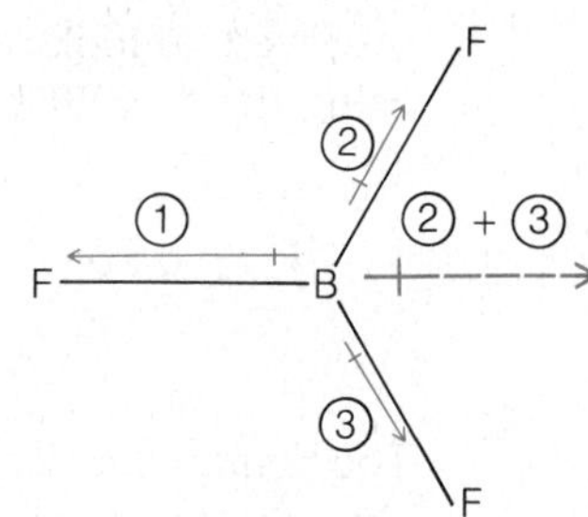

① ②

Cl ——— Be ——— Cl

Vector ① + vector ② = dipole moment

(①) + (②) = 0

(a)

Vector ① + resultant vector of ② + ③ = dipole moment

(①) + (② + ③) = 0

(b)

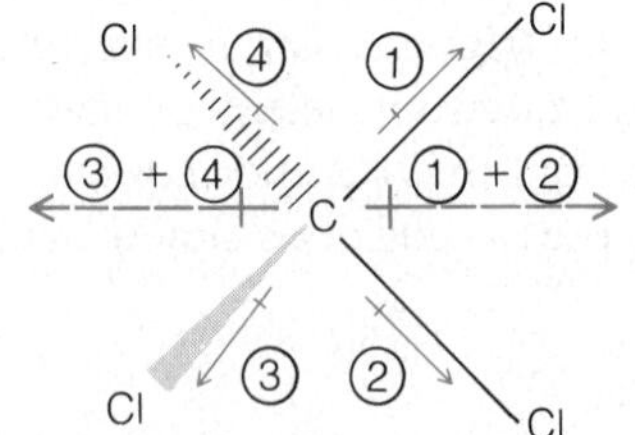

Resultant vector of ① + ② + resultant vector of ③ + ④ = dipole moment

(③ + ④) + (① + ②) = 0

(c)

FIGURE 7.9
Vector representation of zero dipole moments for (a) linear, $BeCl_2$; (b) trigonal-planar, BF_3; and (c) tetrahedral, CCl_4, molecules. The dashed arrows are resultant vectors.

BOX 7.2

HEATING IN MICROWAVE OVENS

The dipole moments of water molecules are utilized in microwave heating in a simple way. Generated by a device called a magnetron, microwaves are released into the oven chamber, where they are absorbed by the food. When water molecules in the food are exposed to microwaves they tend to align their dipole moments with the field of the microwave radiation. However, the direction of the radiation's electric field changes about 10^9 times/sec. Just as water molecules become partially aligned with the field, the field direction reverses and the molecules must realign. The aligning and realigning water molecules transmit some of their kinetic energy to other molecules in the food, thereby heating the food.

H_2O molecules in ice do not absorb significant microwave radiation because hydrogen bonding (section 10.3) prevents them from rotating; they cannot readily change their orientation. For this reason, frozen foods must be defrosted before they can be warmed in a microwave oven. The microwave heating process takes much less time and consumes much less energy than traditional heating methods.

not the same, the bond moments do not cancel. For example, whereas carbon dioxide, $:\ddot{O}{=}C{=}\ddot{O}:$, has a zero dipole moment, carbon oxysulfide, $:\ddot{O}{=}C{=}\ddot{S}:$, has a significant dipole moment. Table 7.8 lists dipole moments for some simple molecules.

EXAMPLE 7.18 Explain why $:NH_3$ has a much larger dipole moment than $:NF_3$ has (see Table 7.8).

ANSWER In $:NH_3$, the dipole moments of the three N—H bonds are directed toward the N atom; see Figure 7.8b. The unshared pair of electrons on the N atom strengthens this effect—that is, it piles more negative charge on the N atom. As a result, the unshared pair increases the dipole moment of the NH_3 molecule. The dipole moments of the three N—F bonds in $:NF_3$ (Figure 7.8c) are directed toward the F atoms, in *opposition* to the effect of the unshared pair. ■

PROBLEM 7.17 In terms of individual bond moments, would CCl_4 be a polar molecule if it had a square-planar shape? □

SUMMARY

Chemical bonds hold atoms or ions together. **Ionic bonding** between atoms results from the transfer of electrons from one atom to another atom, forming a cation and an anion, respectively. Electron loss is called **oxidation**; electron gain is called **reduction**. Ionic compounds are typically high-melting solids that conduct an electric current in the molten state. The relatively high energy needed to separate closely packed ions from the solid state into individual ions in the gaseous state is called the **lattice energy**. The **Born–Haber cycle**, an application of Hess's law, shows that the exothermicity of ionic bonding comes mainly from the strong electrostatic attraction between the oppositely charged ions aggregated in the solid state.

Covalent bonds hold atoms together in molecules and polyatomic ions by sharing electron pairs. In typical covalent bonding, each atom contributes one electron to the shared pair. In **coordinate covalent** bonding, the **donor** atom contributes both electrons for the bond to the **acceptor** atom. Although the force of attraction between the atoms in covalently bonded molecules is strong, the attraction between the *molecules* is weak. Molecules separate from each other easily and therefore have relatively low boiling and melting points. Since they are not composed of ions, covalent molecules are nonconductors of electricity. **Network covalent** substances are not composed of individual molecules but rather have individual atoms covalently bonded to each other throughout the entire piece of solid. Because much energy is required to break covalent bonds, such substances have extremely high melting and boiling points.

Atoms are designated by **Lewis symbols**, which include the symbol of the element with a dot for each **valence (outer-shell) electron**. The **octet rule**, based on knowledge about the stability of the electron configuration of the noble gases, states that atoms undergo bonding so as to acquire eight electrons in their valence shells. Exceptions are H and Li, which because they are close to He, tend to acquire a duet of electrons. The bond between a metallic element and a nonmetallic element is typically ionic. The bond between nonmetals is always covalent. The number of electrons a nonmetallic element needs for its octet equals the number of covalent bonds it usually forms, called its **covalence**. The covalence of a representative element is eight minus the element's periodic-group number. Atoms participating in coordinate covalent bonding do not have their usual covalence. Atoms may share more than one pair of electrons to form a **multiple bond**. Two shared pairs of electrons constitute a **double bond**; three shared pairs, a **triple bond**. A **structural formula** shows the arrangement of the bonded atoms in a molecule. A **Lewis structure** uses Lewis symbols to show bonded atoms and dashes to represent shared pairs of electrons. Different compounds with the same molecular formula (CH_3OCH_3, CH_3CH_2OH) are called **isomers**. For each shared and unshared pair of electrons, an atom must provide an atomic orbital; for our purposes, these are the *s*, *p*, and *d* orbitals. A combination of the *s* and the three *p* orbitals holding two electrons each constitutes the octet of electrons.

Exceptions to the octet rule include molecules or ions whose atoms have either an expanded valence shell or an incomplete octet. Only atoms in the third or higher periods have the necessary *d* orbitals to accommodate more than eight valence electrons. Species with at least one unshared electron, **free radicals**, are also exceptions to the octet rule. In most cases they have an odd number of electrons.

A **polar covalent** bond results from the unequal sharing of an electron pair. The atom that attracts the shared pair more strongly has a **negative partial charge** (δ^-); the other atom has a **positive partial charge** (δ^+). In a **nonpolar covalent** bond there is equal sharing of electrons. The separation of opposite charges in a polar bond forms a **dipole** whose **dipole moment** is the product of the partial charge and the distance between the charge centers. The polarity of the entire molecule is a resultant of the individual bond moments and the moments of unshared pairs. A **polar molecule** has a dipole moment because the individual bond moments do not cancel each other. A **nonpolar molecule** has no dipole moment, either because it has no polar bonds, or it has polar bonds whose moments cancel.

The net tendency of an atom to attract shared pairs of electrons is known as **electronegativity**. Electronegativities generally increase from left to right in a period and from bottom to top in a periodic group. The larger the difference in electronegativity, the more polar is the covalent bond. The shorter the bond distance and the greater the bond polarity, the stronger is the bond. Bond energies also increase as more electron pairs are shared between two atoms. A **bond angle** appears whenever three atoms are bonded.

An atom's **oxidation number (oxidation state)** is determined by overemphasizing the ionic quality of a covalent bond and assigning all the bonding electrons to the more electronegative atom. An increase in oxidation number indicates an oxidation; a decrease, a reduction. **Formal charge** is determined by overemphasizing the covalent quality of covalent bonds and assigning half of the shared electrons to each bonded atom. Coordinate covalent bonding results in the appearance of formal charge.

Lewis structures are written by first arranging the atoms in a **skeleton** composed of a backbone of bonded multicovalent **central** atoms and then attaching terminal atoms that are often unicovalent. If there are not enough univalent atoms to achieve the usual covalences of the central atoms, multiple bonds are needed. If there are too many univalent atoms, the usual covalences are exceeded. The total of valence electrons minus the electrons in bonds gives the number of unshared electrons.

The **shape** of a molecule, which refers only to the positions of the atomic nuclei and not to unshared pairs of electrons, is determined mainly by its bond angles. The **valence shell electron pair repulsion (VSEPR) theory** states that molecular shape is influenced by the tendency of the electron pairs (bonded and unshared) surrounding the central atom to minimize their mutual repulsion by attaining the maximum distance of separation.

7.13 ADDITIONAL PROBLEMS

LEWIS SYMBOLS AND STRUCTURES

7.18 (a) Write Lewis symbols for tellurium, germanium, magnesium, krypton, cesium, gallium, bromine, arsenic. (b) Which of these elements are likely to be found in salts as positive ions?

7.19 (a) Write Lewis symbols for the positive and negative ions in these salts: $SrBr_2$, Li_2S, Ca_3P_2, PbF_2, CaTe, Bi_2O_3. (b) Which ions do not have a noble gas configuration?

7.20 Find the total number of valence electrons for (a) $SnCl_4$, (b) NH_2^-, (c) CH_5O^+, (d) XeO_3, (e) CN_2H_2.

7.21 Write Lewis structures for the covalent molecules (a) hydrogen sulfide, (b) phosphorus trichloride, (c) boron trichloride, (d) sulfur tetrafluoride, (e) silicon tetrahydride (silane), (f) iodine monochloride.

7.22 Write Lewis structures for (a) CH_4S, (b) CH_2ClBr, (c) S_2Cl_2, (d) NH_2Cl, (e) CH_5P, (f) C_2H_6S (two structural possibilities), (g) H_3NO, (h) H_4N_2, (i) Si_2H_5F.

7.23 Write Lewis structures without formal charges for these multiply bonded compounds: (a) CS_2, (b) ClNO, (c) C_2N_2, (d) C_2H_2O (2 possibilities), (e) C_3H_4 (2 possibilities), (f) C_3O_2.

7.24 (a) Write Lewis structures for the anions (negative ions) in these salts: BaO_2, barium peroxide; RbO_2, rubidium superoxide; KO_3, potassium ozonide. (b) Give the Lewis structure of O_2^+. (c) Which oxygen-containing ions in a and b are paramagnetic?

7.25 Write Lewis structures for each of these species, and indicate the cases in which atoms have expanded valence shells: (a) NO_2Cl, (b) H_2NCN, (c) I_3^-, (d) ICl_4^-, (e) $COCl_2$, (f) AsF_5.

7.26 Write Lewis structures for these noble gas compounds: XeF_2, XeF_4, $XeOF_4$, XeO_3, Na_4XeO_6 (a salt).

THERMOCHEMISTRY AND BONDING

7.27 What thermodynamic factor accounts for the fact that ionic compounds of Bi^{3+} but not of N^{3+} are known?

7.28 Calculate the enthalpy of formation for RbBr; $Rb(c) + \frac{1}{2}Br_2(\ell) \rightarrow RbBr(c)$, given these ΔH values in kJ/mol: ΔH_{sub} for Rb(c) = 75.8; ΔH_{vap} for $Br_2(\ell)$ = 30.9; ΔH_{BE} of $Br_2(g)$ = 192; ΔH_{IE} of Rb(g) = 403; ΔH_{EA} of Br(g) = −324; and ΔH_{LE} of RbBr(c) = −670.

7.29 Calculate the ΔH_{EA} of Cl from these ΔH values, in kJ/mol: ΔH_f for $CaCl_2(c)$ = −799.1; the first and second ΔH_{IE} values for Ca = 589.5 and 1144.7, respectively; ΔH_{BE} for Cl_2 = 241.8; ΔH_{sub} for Ca = 138.1; ΔH_{LE} for $CaCl_2(c)$ = −2217.1.

7.30 Calculate the ΔH_{LE} of KI from these ΔH values, in kJ/mol: ΔH_{sub} of K = 89; ΔH_{IE} of K = 419; ΔH_{sub} of I_2 = 62; ΔH_{EA} of I = −295; ΔH_f of KI(c) = −328; ΔH_{BE} of I_2 = 151.

BOND TYPE

7.31 Use Lewis symbols to show the formation of ionic compounds from these pairs of elements: (a) Rb and I, (b) Ba and O, (c) Na and H, (d) Ga and F, (e) Al and O, (f) Ca and N, (g) Ra and Br.

7.32 Use Lewis symbols to show the formation of covalent molecules from these pairs of elements (avoid valence shell expansion): (a) P and H, (b) Se and Br, (c) C and F, (d) Si and H, (e) O and F, (f) As and Cl, (g) I and Cl.

7.33 With what type of element does an H atom form (a) an ionic bond; (b) a covalent bond? To what ion is the H atom always changed when it forms an ionic bond?

7.34 The elements X, Y, and Z are in the same period of the periodic table. Their group numbers are 2, 6, and 7, respectively. (a) Write the Lewis formula for the compound most likely to be formed between X and Z. Will this compound most likely be ionic or covalent? (b) Write the Lewis structure for the most probable compound of Y and Z. Is this compound ionic or covalent?

7.35 State whether the structure of each substance is molecular covalent, network covalent, or ionic, and give your reason for each answer: (a) HBr, m.p. $-85.5°C$, b.p. $-67°C$; (b) $MgCl_2$, m.p. 708°C, b.p. 1412°C (the molten substance conducts electricity); (c) $AsCl_3$, m.p. $-8.5°C$, b.p. 130.2°C; (d) CS_2, has double bonds; (e) diamond, m.p. 3800°C, b.p. 4830°C (a nonconductor of electricity).

7.36 Given these facts, $BeCl_2$ and BF_3 are covalent compounds, whereas $MgCl_2$ and AlF_3 are ionic compounds, suggest a useful generalization that relates the position of an element in its periodic group to the type of bonding it exhibits.

ELECTRONEGATIVITY AND DIPOLE MOMENTS

(Consult Table 7.4.)

7.37 (a) Which *two* of these pairs of elements are most likely to form ionic bonds: Te and H, C and F, Ba and F, N and F, K and O? (b) Of the remaining three pairs, which one forms the least, and which the most polar covalent bond?

7.38 (a) List five reasonable nonpolar covalent bonds between dissimilar atoms. (b) List three ionic bonds that should exhibit extreme ionic character.

7.39 Why do the electronegativities of elements decrease downward in a group in the periodic table?

7.40 True or false? (a) A molecule with polar bonds must have a dipole moment. Justify your answer with the help of two illustrations. (b) BrF is a more polar molecule than BrCl. Explain.

7.41 How would you expect the polarity of a bond to vary with an increase in the oxidation number of the less electronegative element? Compare the polarity of the S—F bond in SF_2 and SF_6.

7.42 Write the Lewis structures, with appropriate shapes, of the following covalent molecules and indicate which of them have appreciable dipole moments. Represent the dipole moment by ($\mapsto$). (a) H_2O, (b) $BeCl_2$, (c) $CHCl_3$, (d) BF_3, (e) $BClF_2$, (f) H_2CO, (g) HCN.

7.43 Account for the fact that O=C=S has a dipole moment, whereas S=C=S and O=C=O do not.

7.44 In view of the fact that Si—F and P—F are both polar bonds, explain why SiF_4 has zero dipole moment, whereas for PF_3 $\mu = 1.02$ D.

OXIDATION NUMBER AND REDOX

7.45 What is the relationship, if any, between the periodic group of an element and the highest oxidation state the element can possess?

7.46 Evaluate the oxidation number of N in NO, NO_2, N_2O_3, N_2O_5, N_2H_4, and H_2NOH.

7.47 Evaluate the oxidation number of (a) C in CH_4, CH_3OH, $H_2C{=}O$, CO, and CO_2; (b) Cl in Cl_2, HClO, $HClO_2$, $HClO_3$, and $HClO_4$.

7.48 Evaluate the oxidation number of the underlined atom in (a) $\underline{C}O_2$, (b) $\underline{S}OCl_2$, (c) $Na_2\underline{S}$, (d) $K_3\underline{P}O_4$, (e) $K_2\underline{P}HO_3$, (f) $K\underline{P}H_2O_2$, (g) $K_4\underline{P}_2O_7$, (h) $\underline{P}OCl_3$, (i) $K_2\underline{S}_2O_7$, (j) $K_2\underline{Mn}O_4$, (k) $H\underline{C}O_2H$, (l) $\underline{N}_2H_4$. Assume that the other atoms have their common oxidation numbers.

7.49 These facts are given: $GeCl_2$ (unstable), $GeCl_4$ (stable), $SnCl_2$ and $SnCl_4$ (both stable), $PbCl_2$ (stable), and $PbCl_4$ (unstable). Suggest a useful generalization for how the stability of compounds of elements in different oxidation states is related to the elements' placement in their periodic table groups.

7.50 Which of these equations represent oxidation–reduction reactions?

(a) $CH_4 + 2O_2 \rightarrow CO_2 + 2H_2O$ (combustion of natural gas)
(b) $2KI + Pb(NO_3)_2 \rightarrow PbI_2 + 2KNO_3$
(c) $CH_3Br + NaOH \rightarrow CH_3OH + NaBr$
(d) $PCl_3 + 3H_2O \rightarrow H_3PO_3 + 3HCl$
(e) $Cl_2 + 2NaBr \rightarrow Br_2 + 2NaCl$
(f) $2AsH_3 \rightarrow 2As + 3H_2$
(g) $NI_3 + 3H_2O \rightarrow 3HOI + NH_3$
(h) $2NF_2H + 2KF \rightarrow 2KHF_2 + N_2F_2$
(i) $Cl_2 + H_2 \rightarrow 2HCl$
(j) $3O_2 \rightarrow 2O_3$

7.51 Which of these transformations of carbon-containing (organic) molecules represent an oxidation or reduction? (*Hint*: Focus attention on the oxidation state of carbon.)

(a) $H_2C{=}O \rightarrow H_3COH$; (b) $H_2C{=}O \rightarrow H{-}\overset{\overset{OH}{|}}{C}{=}O$;
(c) $HCO_2H \rightarrow CO_2$; (d) $CH_3CH_2OH \rightarrow H_2C{=}CH_2$;
(e) $CH_3CH_2Cl \rightarrow H_2C{=}CH_2$;
(f) $H_2C{=}CH_2 \rightarrow CH_3CH_3$;
(g) $6CO_2 + 6H_2O \rightarrow C_6H_{12}O_6 + 6O_2$ (photosynthesis).

BOND PROPERTIES AND FORMAL CHARGES

7.52 Calculate (a) the double bond length in O_2 and (b) the triple bond length in HC≡N from the following multiple bond lengths: C=O is 0.121 nm, C=C is 0.133 nm, C≡C is 0.060 nm, and N≡N is 0.055 nm.

7.53 Discuss the relationship of covalent bond strength and (a) bond length; (b) bond polarity.

7.54 Compare the bond lengths and predict the relative bond stabilities of (a) —C≡C—, N≡N; (b) $>C{=}O$, $>C{=}S$; (c) —N=O, —N=C—. Use Table 7.5.

7.55 Account for the fact that the C—C, C—H, H—H, H—Cl, C—O, C—N, and C—Cl bonds are much stronger than Cl—Cl, O—Cl, N—N, O—O, and N—Cl bonds. (*Hint*: Think about the presence of unshared pairs of electrons on adjacent atoms.)

7.56 (a) Find the formal charge on the N and B atoms in NH_3, BF_3, NH_4^+, BF_4^-, NH_2^-, and $H_3B—NH_3$. (b) For cases in which the valence shell has more than eight electrons, generalize about the presence of formal charge and normal covalency.

7.57 Find the formal charge of each atom, other than H and halogen atoms, in (a) $Cl_3P—O$, (b) 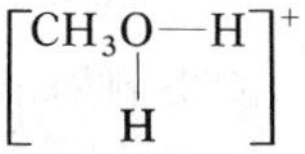 $[CH_3O(H)—H]^+$, and (c) 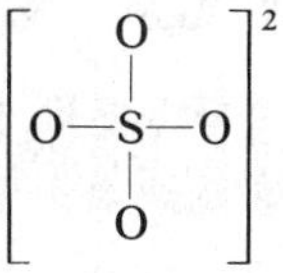 $[SO_4]^{2-}$ (O—S—O with O above and below S).

7.58 Show with the aid of two examples that the formation of a coordinate covalent bond creates formal charge.

7.59 With the aid of formal charges, explain which Lewis structure is more likely to be correct for each given molecule or ion. (a) For Cl_2O, :C̈l—Ö—C̈l: *or* :C̈l—C̈l—Ö:? (b) For SO_2, :Ö—S̈=Ö: *or* :S̈—Ö=Ö:? (*Hint*: If a Lewis structure must have formal charges, it is better to have the negative formal charge on the more electronegative atom.) (c) For N_2O, :N̈=O=N̈: *or* :N≡N—Ö:? (*Hint*: Minimize the amount of formal charge on each atom.)

7.60 Write Lewis structures for three isomers with the molecular formula HCNO. Indicate all formal charges that may exist. Predict which isomer is likely to be the least stable and justify your selection.

SHAPES OF COVALENT SPECIES

7.61 What two shapes can a triatomic species have?

7.62 Give the shapes of these covalent molecules: CCl_4, $CdCl_2$, $AlCl_3$, H_2S, PCl_3, and $AsCl_5$.

7.63 Give the shapes of these polyatomic ions: BO_3^{3-}, PO_4^{3-}, SO_3^{2-}, NO_2^-, H_3O^+ (hydronium ion), GeF_3^-, ClF_6^{3-}, $IO_2F_2^-$.

7.64 Draw Lewis structures and predict the shapes of (a) ICl_4^-; (b) $TeCl_4$; (c) XeO_3; (d) nitrosyl bromide, BrNO; (e) nitryl chloride, whose skeleton is Cl—N(O)(O); and (f) thionyl chloride, whose skeleton is Cl—S(O)—Cl.

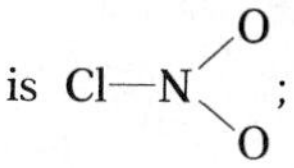

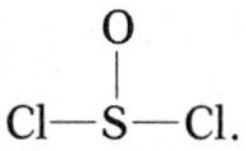

7.65 Give the Lewis structures and predict the shapes of these very reactive carbon-containing species: H_3C^+ (a carbocation), $H_3C:^-$ (a carbanion), and $:CH_2$ (a carbene whose unshared electrons are paired).

7.66 The methyl free radical $\cdot CH_3$ has bond angles of about 120°, whereas the methyl carbon ion $:CH_3^-$ has bond angles of about 109°. What can you infer from these facts about the repulsive force exerted by an unpaired, unshared electron as compared to that exerted by an unshared pair of electrons?

7.67 Two isomeric Lewis structures can be written for the square-planar molecule $PtCl_2Br_2$:

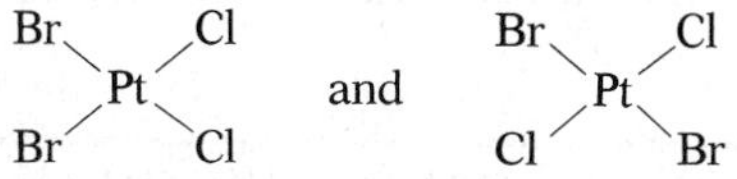

(Br, Br on left; Cl, Cl on right around Pt) and (Br, Cl on top; Cl, Br on bottom around Pt)

Show how a difference in dipole moments can distinguish between these two isomers.

SELF-TEST

(Consult any table you need.)

7.68 (20 points) (a) Write Lewis symbols for strontium, chlorine, and silicon. (b) Use the appropriate pairs of these atoms to show formation of (1) an ionic compound and (2) a covalent molecule. (c) Indicate the polarity of the bond in the covalent molecule.

7.69 (15 points) (a) Write Lewis structures for (1) CH_3Br, (2) BrF_2^+, and (3) ClCN. (b) Predict the shape of each molecule and ion.

7.70 (10 points) Write Lewis structures consistent with the octet rule for (a) HN_3 (skeleton H—N—N—N) and (b) carbon monoxide, CO. Locate all formal charges.

7.71 (15 points) (a) Find the oxidation number of the underlined atoms in (1) $\underline{S}O_2Cl_2$ and (2) $K_2\underline{V}_4O_9$. (b) Does either of these reactions involve an oxidation–reduction?
(1) $Cl_2 + H_2O \rightarrow HCl + HOCl$
(2) $PBr_3 + 3H_2O \rightarrow H_3PO_3 + 3HBr$

7.72 (5 points) (a) Which element has the greatest tendency to form (1) a cation; (2) an anion: As, Se, Sr, Bi, F, I? (b) Account for the fact that barium has a greater tendency than magnesium to form an ionic bond.

7.73 (35 points) (a) Give a three-dimensional Lewis structure for PF_2Cl_3 that (1) has a dipole moment; (2) has a zero dipole moment. (b) Account for the fact that the calculated bond distance for P—F is about 0.15 nm, whereas the observed value is 0.130 nm. (c) In the compound BN each atom has a covalency of three, and there are no multiple bonds. Classify BN as to bonding type. (d) Write Lewis structures for the isomers that have the molecular formula $C_2H_4Cl_2$. (e) Explain why HBr has a dipole moment and Br_2 does not. (f) In terms of bond lengths, account for this decreasing order of bond energies (for > read "is greater than"): C—F > C—Cl > C—Br > C—I. (g) Arrange as many as possible of these species into isoelectronic groups: CO_2, CN^-, CO, N_3, SiO_2, I_2, SO_2, C_2^{2-}, H_2S.

8

CONCEPTS OF COVALENT BONDING

In Chapter 7 we mentioned that we assume covalent bonding involves atomic orbitals. We will discuss this orbital involvement in more detail in this chapter by applying two concepts of covalent bonding derived from quantum mechanics. These are the valence bond concept and the molecular orbital concept. We will also address the question that arose in section 7.10 when we found that in some cases there was more than one way electrons could be placed on a Lewis structure skeleton for a given compound.

8.1 THE VALENCE BOND (VB) CONCEPT

ORBITAL OVERLAP

The valence bond concept assumes that the formation of a covalent bond requires the interaction of an atomic orbital (AO) of one atom with an atomic orbital of its bonding mate. When the appropriate bond length is attained, the atomic orbitals of the bonding atoms *occupy some of the same space.* We say that atomic orbitals overlap to form the covalent bond. This idea of **orbital overlap** is illustrated in Figure 8.1 for the formation of H:H from two H· atoms.

Figure 8.1a shows two separated H· atoms, each with an electron in a 1*s* orbital. The separated atoms approach each other, going through a stage where they are close but not yet bonded (Figure 8.1b). As the approach continues, overlap occurs so that the two orbitals share a region between the atoms (Figure 8.1c).

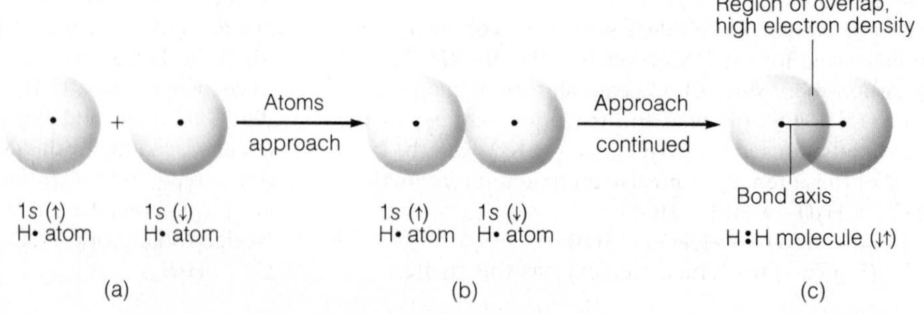

FIGURE 8.1 Overlap of atomic orbitals to form a covalent bond. (a) Two separated H· atoms, each with an electron of opposite spin. (b) Two H· atoms just prior to becoming bonded. (c) Orbitals overlap to form a covalent bond in H_2 molecule.

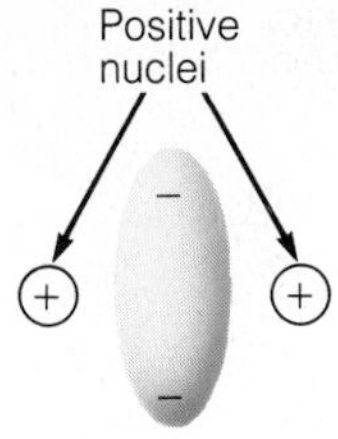

FIGURE 8.2
The nature of electrostatic attraction in a covalent bond.

The shared electrons in this region of orbital overlap constitute the covalent bond. The overlap results in a region of high electron density between the atoms.

We can now better picture the electrical nature of the covalent bond. The bond is made up of two positive atomic nuclei attracted by a high density of negative charge, as shown in Figure 8.2. The shared electrons are separated from each other, thereby minimizing their mutual repulsion. These attractive forces exceed the repulsive forces between the two positive nuclei and between the two electrons. Otherwise, the molecule would not exist; it would be less stable than the unbonded atoms. The optimum bond distance is such as to maximize the *net* attractive force, thereby giving the strongest possible bond. *Although the force of attraction between bonded atoms is very strong, the force of attraction between molecules is very weak.*

Once the bond forms, the electrons in the shared pair are indistinguishable. That is, an electron no longer belongs only to the atom that brought it to the bond. The shared pair belongs to both atoms.

SIGMA (σ) BOND

The bond in Figure 8.3a is labeled σ_s because it results from the overlap of two s atomic orbitals.

The bond just described in H_2 is a single bond. Characteristic of single bonds, its electron density is generally distributed symmetrically around an imaginary straight line between the nuclei. This line is called the **bond axis** (Figure 8.3a). Such bonds are called **sigma (σ) bonds.**

This one is labeled $\sigma_{s,p}$.

Another way of forming a sigma bond is illustrated in Figure 8.3b. There you see an H atom and an F atom combine to form an HF molecule. The $1s$ atomic orbital of the H atom overlaps with one lobe of a $2p$ atomic orbital of the F atom in a *head-on* fashion.

This one is labeled σ_p.

The bonding of two F atoms to form an F_2 molecule illustrates a third way a sigma bond is formed. As shown in Figure 8.3c, one lobe of a p orbital of one F atom overlaps head-on, along the bond axis, with one lobe of a p orbital of the other F atom to form a σ bond.

Note that *each σ bond* illustrated in Figure 8.3 is *formed by a head-on overlap of atomic orbitals* regardless of the atomic orbital used. Single covalent bonds are typically σ bonds.

Because of the symmetry of the electron cloud about the bond axis, a σ-bonded atom can be rotated about its bond axis without affecting the orbital overlap and, therefore, without destroying the bond. "Rotation of an atom" includes rotation of the bonding orbital, any atoms attached to the atom being rotated, and any unshared pairs on the atom. Thus, if we rotate the C in $H_2N—CH_3$, we rotate the entire CH_3 group with respect to the H_2N group.

In the cases just described, each atom forming the σ bond has one electron in the atomic orbital used for bonding. This is not the case with coordinate covalent bonding when the donor atom has both electrons in its atomic orbital (AO) and the acceptor atom has none. In the reaction $H^+ + :\ddot{F}:^- \longrightarrow H:\ddot{F}:$, the empty s orbital of the H^+ overlaps with any one of the filled p AO's of F^-:

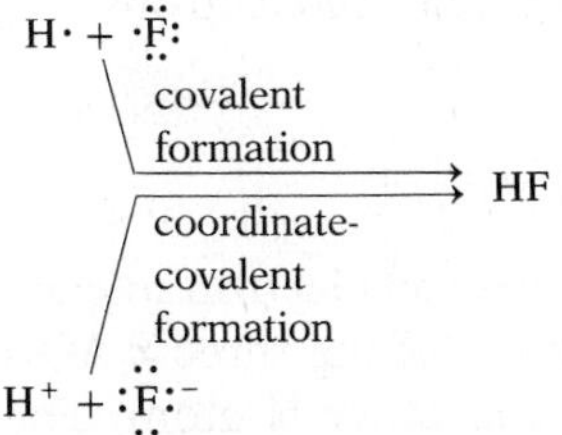

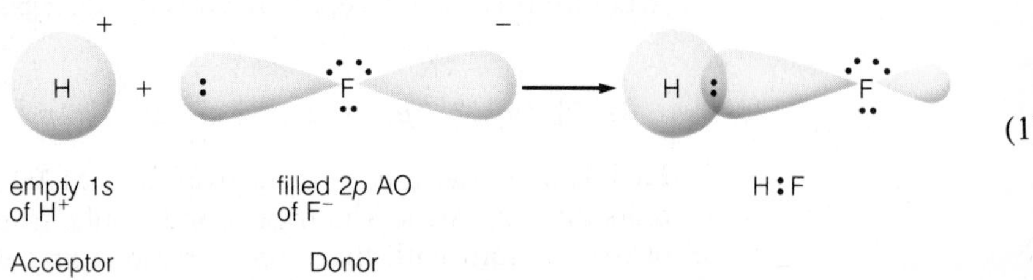

(1)

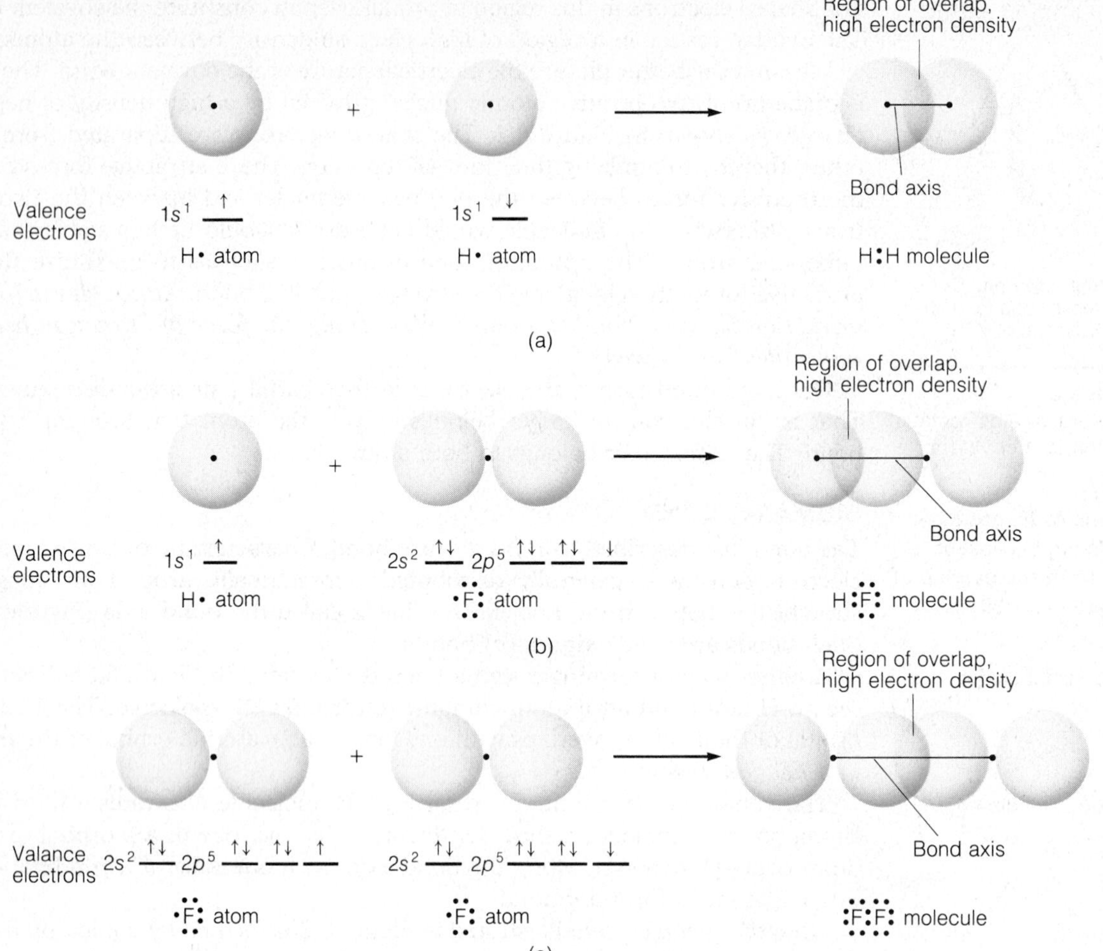

FIGURE 8.3
Formation of sigma (σ) bonds by sharing electrons in overlapping AO's: (a) $s + s$, (b) $s + p$, and (c) $p + p$. (The dots represent the nuclei.) The region of overlap designates the σ bond.

The HF molecule formed this way is *identical* to the HF molecule formed when the H and F atoms each contribute an electron (Figure 8.3b); there is only one kind of HF. The electrons have no "memory" of where they came from, so that a coordinate covalent bond once formed is no different from any other covalent bond.

PI (π) BONDS

We now consider the formation of multiple bonds. A simple example is the combination of two :N· atoms to form N_2, a molecule with a triple bond. The Lewis structure is shown here, along with the distribution of the valence electrons:

$$
\begin{array}{ccc}
:\dot{\underset{\cdot}{N}}\cdot & + & \cdot\dot{N}: \longrightarrow :N:::N: \\
2s^2 \ \underline{\uparrow\downarrow}\ 2p_x{}^1\ \underline{\uparrow}\ 2p_y{}^1\ \underline{\uparrow}\ 2p_z{}^1\ \underline{\uparrow} & & 2s^2\ \underline{\uparrow\downarrow}\ 2p_x{}^1\ \underline{\downarrow}\ 2p_y{}^1\ \underline{\downarrow}\ 2p_z{}^1\ \underline{\downarrow}
\end{array}
$$

Each N atom has a pair of electrons in a $2s$ AO and an unpaired electron in each of its three $2p$ AO's. The triple bond results from the overlap of the three p AO's of one N atom with the three corresponding p AO's of the other N atom. The

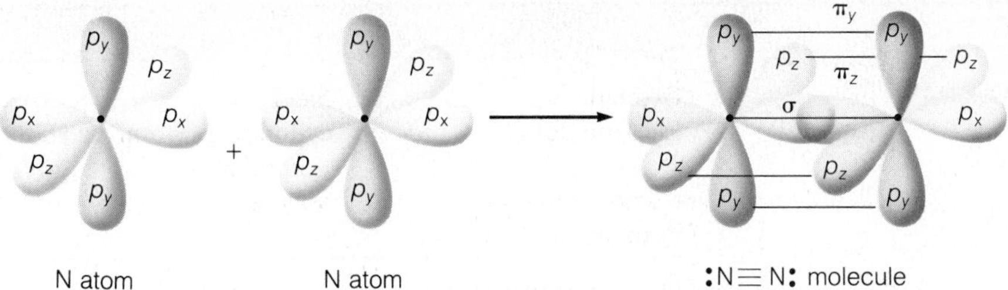

FIGURE 8.4
Formation of two π bonds and one σ bond in N_2.

A σ_p type.

direction of approach is arbitrarily chosen as the x axis. See Figure 8.4. The $2s$ orbital of each N atom has been omitted from the illustration in order to simplify this figure. The electron pairs in these $2s$ AO's are unshared and so do not contribute to the formation of the triple bond. With an approach along the x axis, the p_x AO's overlap head-on to form a σ bond. The two p_y AO's *cannot* overlap head-on also. Instead, they overlap *side-to-side* (laterally). Side-to-side overlap also occurs for the two p_z AO's. The bond resulting from side-to-side overlap of AO's is called a **pi (π) bond**. An N_2 molecule thus has two π bonds, one from overlap of the p_y AO's and one from overlap of the p_z AO's. A triple bond therefore consists of one σ bond and two π bonds. Generally, between representative elements *in any multiple bond, one bond is a σ bond, and the other bonds are π bonds.* A double bond consists of one σ bond and one π bond.

Let us examine π bond formation by looking at just one pair of overlapping p orbitals, say the p_y AO of each atom (Figure 8.5a). The upper lobe of one p_y orbital overlaps with the upper lobe of the other p_y orbital. Likewise the lower lobes overlap. But these overlappings result in only *one π bond.* Do not be misled into thinking that two π bonds result when both the upper p_y lobes and the lower

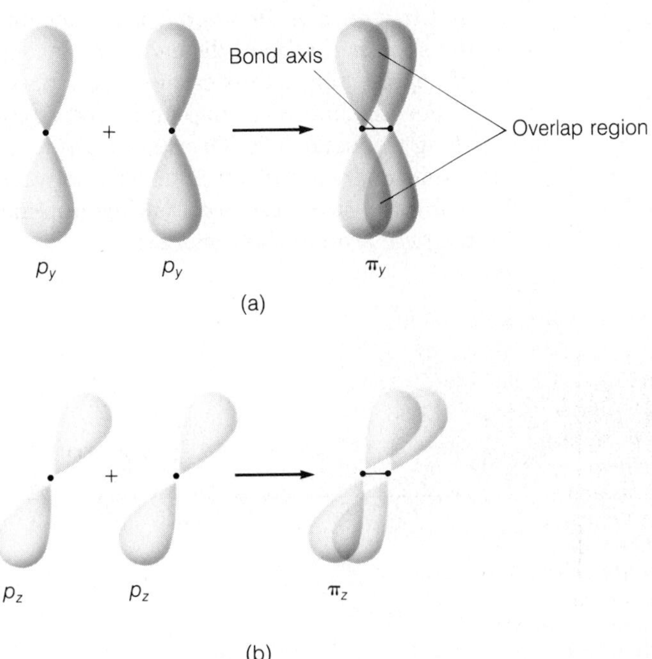

FIGURE 8.5
Schematic representation of π bond formation (a) π_y from overlap of two p_y AO's. (b) π_z from overlap of two p_z AO's.

TABLE 8.1
DIFFERENCES BETWEEN σ AND π BONDS

	σ BOND	π BOND
Overlap	head-on	side-to-side
Electron density	cylindrical symmetry about bond axis	maximum in the plane of overlapping orbitals
Number of bonds	only one	one or two†
Rotation about bond	yes	no

† For representative elements.

p_y lobes overlap. The electron density of a π bond is concentrated *above and below the bond axis.* Unlike a σ bond, a π bond has no uniform electron density about the bond axis. A strong π bond can result only when the individual p orbitals are parallel. Thus a π bond can be formed from overlap of two p_y or two p_z orbitals, but not from overlap of a p_y and a p_z. When two p_y AO's overlap, the π bond is designated π_y (Figure 8.5a); two p_z AO's give a π_z bond (Figure 8.5b).

Earlier it was shown that rotation about the σ bond axis can occur without destroying the bond. Let us see if this is true for the π bond. If one of the π-bonded atoms is held stationary while its mate is rotated, the overlap and, hence, the π bond is destroyed, because after the rotation the two p orbitals are no longer side by side. For this reason free rotation about a π bond is prohibited.

Ultraviolet radiation may supply the energy to induce rotation.

Table 8.1 summarizes the differences between σ and π bonds.

ATOMIC ORBITAL OVERLAP AND BOND STRENGTH

The strength of a bond between two atoms is due to the electron density between the atomic nuclei. The greater the overlap of the AO's, the greater is the electron density between the nuclei. Consequently, *the closer the atoms approach each other, the more the atomic orbitals can overlap, and the higher is the electron density between the nuclei.* This concept is consistent with the observed fact that the shorter the bond distance, the stronger the bond. When the atomic orbitals of approaching atoms cannot overlap, no bonding occurs.

For maximum overlap and bond strength, σ-bonding should occur by overlap along the bond axis. Overlap of a p and s AO along the bond axis (Figure 8.6a) results in greater electron density between the nuclei than off-axis overlap (Figure 8.6b); the strongest bond is thus obtained. This explains why σ bonds are symmetrical around the bond axis.

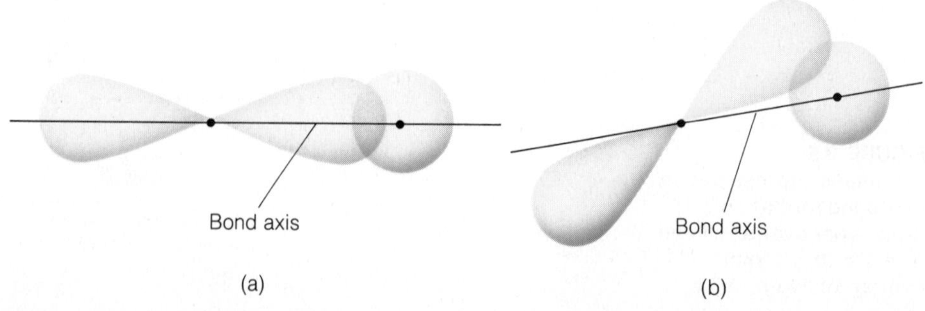

FIGURE 8.6
Overlap of two p AO's. (a) Along the bond axis. Electron density is concentrated between the nuclei. (b) Off the bond axis. Region of concentrated electron density is not directly between the nuclei. There is also less orbital overlap.

8.2 HYBRID ATOMIC ORBITALS (HO's)

Thus far the concept of overlapping atomic orbitals has been applied to diatomic molecules. With AX_n-type molecules, where n is more than one and where A may have some number of unshared pairs, complications arise. Any proposed model for bonding must be consistent with the observed properties of the molecule. The model must predict the observed bond angle and molecular shape (section 7.12). It must also predict whether all of the A—X bonds are identical; that is, if they have the same bond lengths and the same chemical reactivities.

Our discussion of these predictions begins with the example of beryllium dichloride, $BeCl_2$. In order to form two covalent bonds, Be needs two AO's to overlap with a p AO of each Cl atom. One might be the $2s$ and the other one could be one of the $2p$ AO's. Since two different AO's are used, we would then expect the two Be—Cl bonds to be different; they should have different energies and different lengths. However, experiment shows that *the two Be—Cl bonds are identical.* Thus, this proposed model for bonding in $BeCl_2$ must be wrong. Any proposed model must be consistent with the observed facts; the facts dictate the model.

We might also have suggested that Be uses two of its $2p$ AO's to get two identical bonds. But any model of chemical bonding must also be consistent with the observed molecular shape and bond angles. For example, $BeCl_2$ is a linear molecule with bond angles of 180°. Are these bond angles expected from the overlap of the two p AO's of Be with AO's of the Cl atoms? Sigma bond angles are nearly always determined by the axes of the overlapping AO's of the central atom, since orbital overlap is maximal with a head-on approach (Figure 8.6). Since the axes of the p_x, p_y, and p_z AO's are 90° apart, the angles of the bonds formed from these AO's would be predicted to be 90°, as shown in Figure 8.7 for overlap with an s AO. Since the prediction (90°) does not correspond to the observed value (180°), this model is also wrong.

sp HYBRID ATOMIC ORBITALS

To be valid, the VB concept of atomic orbital overlap must account for the fact that the two Be—Cl bonds in $BeCl_2$ are identical and have a 180° bond angle. To get two identical bonds with a 180° bond angle, it is necessary for Be to use two identical AO's, but not two p AO's. The VB method assumes that these two identical AO's come from a "blending" of the $2s$ and any one of $2p$ AO's of beryllium. This blending is a mathematical process called **hybridization**, and the "blended" atomic orbitals are called **hybrid orbitals** (HO). HO's are labeled by showing the number of each atomic orbital in the blend. Since Be uses an s and *one* of the p's, its HO is called *sp*. The Be atom still retains two empty p orbitals.

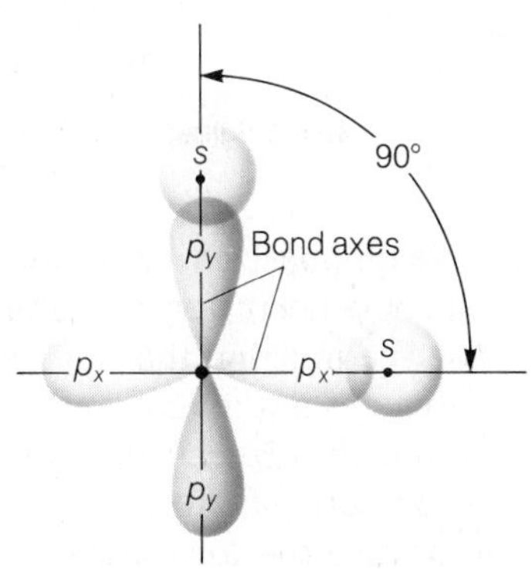

FIGURE 8.7
Representation of two σ bonds formed from the overlap of two p orbitals showing the 90° bond angle.

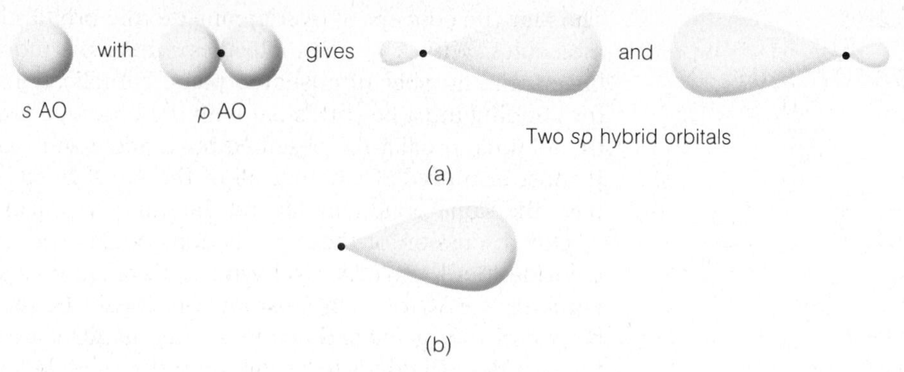

FIGURE 8.8
(a) The hybridization of an *s* AO with a *p* AO to give two new hybrid orbitals known as *sp* HO's. Note the shape of the *sp* HO with a large "head" and small "tail." This is the shape of all hybrid orbitals made up from *s* and *p* orbitals, the so-called *s-p*-type hybrid orbitals.
(b) The *sp* HO with "tail" omitted.

Hybridization involves a redistribution of the ground-state valence AO's; it *does not result in any change in the number of orbitals.* It is helpful to imagine bonding as occurring in two steps:

1. The *s* and one of the *p* ground-state valence AO's of Be are first reorganized (hybridized) to two identical HO's. For each atomic orbital we hybridize, we get a hybrid atomic orbital. *There is always a conservation of orbitals.* Each HO has one electron (Hund's rule, section 5.7) as shown:

Valence-shell orbitals of beryllium:

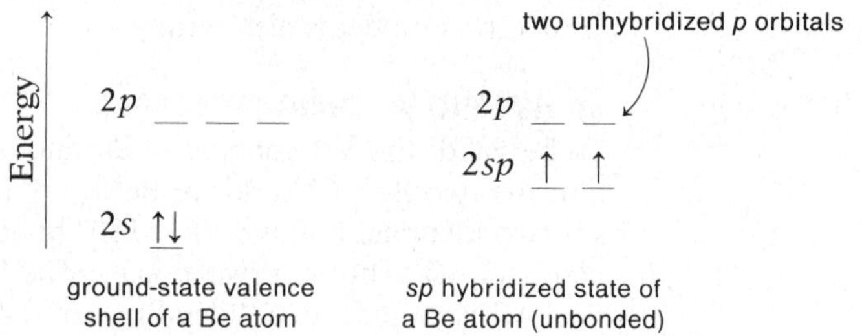

2. Then the two HO's of Be overlap with the *S* AO's of two *H* atoms. These overlappings give the bonded state:

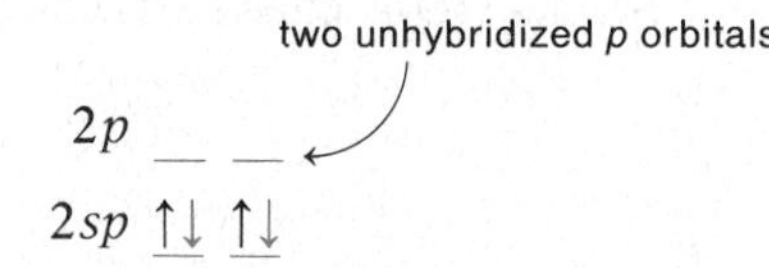

sp hybridized state of Be after bonding (Each colored arrow represents an e^- from Cl.)

Even isolated ground-state Be atoms are hard to come by.

We emphasize that although the hybridized state of an atom is a useful mathematical concept for explaining the bonding and structure of molecules, it is *not real.* We imagine that an isolated Be atom has hybridized orbitals, but such an entity does not actually exist. In summary, we merely assume that somehow the two unhybridized AO's of Be are converted to two identical HO's. One thing we can be sure of, however, is that the electronic conditions attained by bonded atoms give the lowest-energy molecule. There is an investment of energy in the

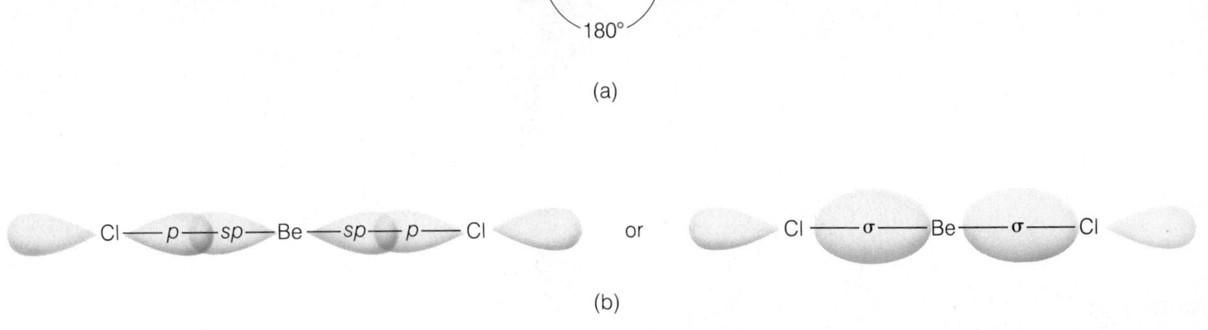

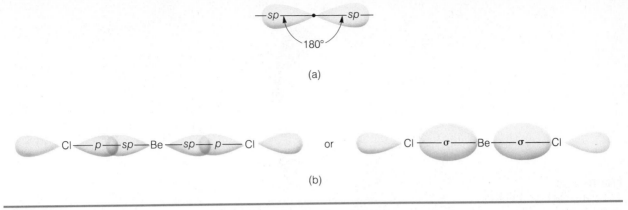

FIGURE 8.9
(a) Linear orientation of the two *sp* hybrid orbitals. (The small lobes of HO's are omitted.) (b) Orbital representation of $BeCl_2$ showing formation of two Be—Cl σ bonds.

hybridization process, but the return after bonding occurs is in excess of the investment. The molecule formed from hybrid orbitals would be at a lower energy than if ground-state orbitals had been used.

Figure 8.8 is an illustration of the net hybridization. It shows that each *sp* hybrid orbital is shaped like a *p* orbital except that its two lobes are of unequal size. Bonding overlap occurs with the larger lobe (the head); the small lobe (the tail) plays no role. For this reason and in order to simplify illustrations, the small lobes are often omitted from diagrams of hybrid orbitals in molecules. In general, *bonds formed from HO's fabricated from s and p hybrid orbitals overlap head-on to give sigma bonds.*

The mathematical procedures are beyond the scope of this book.

When hybridization is treated mathematically, two *sp* HO's are predicted to be oriented at angles of 180° (Figure 8.9a), in agreement with the observed linear bond angle (Table 7.7a, section 7.11). Each *sp* HO of the central atom overlaps with some orbital of the terminal atom; for example, with a 3*p* AO of a Cl atom (Figure 8.9b). The success of this prediction supports the concept of hybridization. It also agrees with the prediction of the VSEPR concept (section 7.12).

sp^2 HYBRID ORBITALS

The concept of hybridization is also used to account for the three identical B—F bonds observed in boron trifluoride; all three bonds have the same bond length (0.129 nm) and bond energy. BF_3 is a trigonal-planar molecule; each F—B—F bond angle is 120° (Table 7.7b).

The p_x and p_y AO's were arbitrarily chosen to form the sp^2 hybrid AO's.

Three similar bonds require three similar orbitals, which are deemed to result from a blending of *one* 2*s* and *two* of the 2*p* AO's of boron. Therefore, the new hybrid orbitals are called sp^2. The B atom is left with one empty *p* orbital as shown:

sp^2 state of B when bonded:

2*p* —

2sp^2 ↑↓ ↑↓ ↑↓

Valence-shell orbitals of boron:

Energy ↑	ground-state valence shell of boron	sp^2 hybridized state of a boron atom (unbonded)
	2*p* ↑ — —	2*p* — (one unhybridized *p* orbital)
		2sp^2 ↑ ↑ ↑
	2*s* ↑↓	

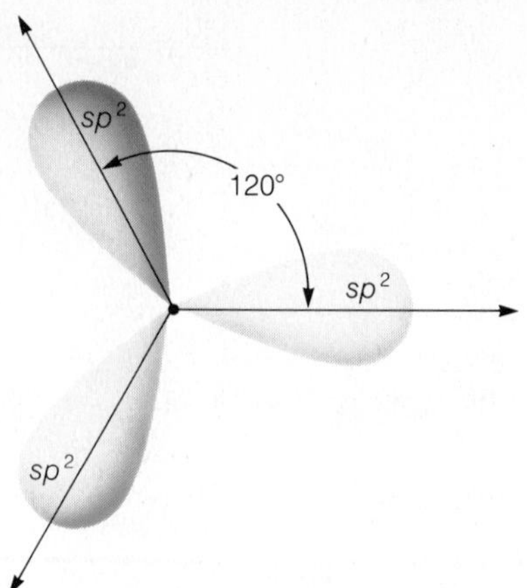

FIGURE 8.10
Trigonal-planar orientation of sp^2 hybrid orbitals.

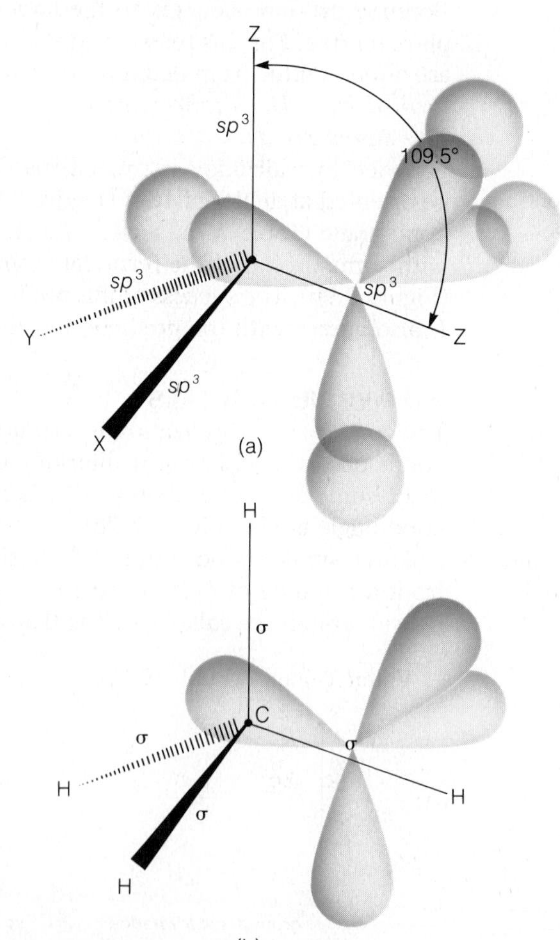

FIGURE 8.11
(a) Orientation of tetrahedral sp^3 hybrid orbitals. (The small lobes have been omitted.) Solid-wedge bond to X projects toward the viewer. Broken-wedge bond to Y projects away from viewer. Solid-line bond to each Z is in the plane of the paper. (b) Orbital representation of CH_4 showing formation of four C—H σ bonds.

The mathematically predicted orientation of the hybrid orbitals agrees with the observed geometry (Table 7.7b) of the molecule; the sp^2 HO's are planar and separated by 120° (Figure 8.10). Each sp^2 HO of the central atom overlaps with an orbital of the terminal atom to form three σ bonds. This is also the arrangement that permits the maximum bond angle (VSEPR theory). The sp^2 HO's always have this trigonal-planar orientation.

sp^3 HYBRID ORBITALS

The concept of hybridization also accounts for the four identical C—H bonds in CH_4; they all have the same bond length (0.1093 nm) and the same bond energy (416 kJ). The concept also predicts the observed tetrahedral shape of CH_4. In order to get four similar orbitals for these four identical bonds, we hybridize the *one* 2*s* and all *three* 2*p* AO's. The HO's formed this way are called sp^3.

The 1*s* AO of C is not involved in bonding.

Valence-shell atomic orbitals of carbon:

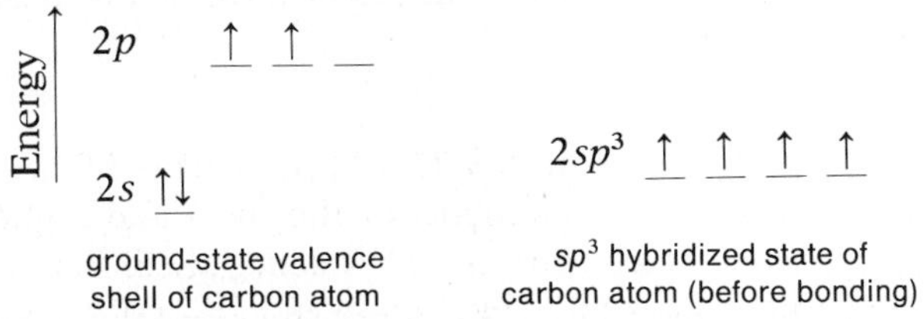

Energy

2*p* ↑ ↑ _

2*s* ↑↓

ground-state valence shell of carbon atom

$2sp^3$ ↑ ↑ ↑ ↑

sp^3 hybridized state of carbon atom (before bonding)

Again, when hybridization is treated mathematically, the four sp^3 HO's are predicted to be oriented at angles of 109.5° (Figure 8.11a). Each sp^3 HO overlaps with an orbital of the bonded atom to give four σ bonds, as shown in Figure 8.11b. This agrees with the observed tetrahedral bond angles in CH_4 (Table 7.7c, section 7.11) and with the prediction of the VSEPR concept (section 7.12).

RELATIVE ENERGIES OF *s-p*-TYPE HO's

The sequence of relative energy levels of the *s-p*-type orbital is

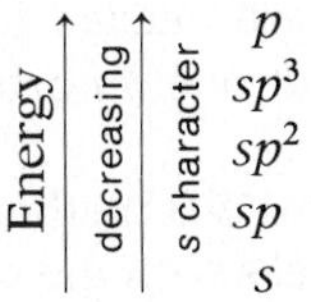

To find the *s* character, divide 1 (for the single *s* AO) by the total number of hybridized AO's. For pure *s* and *p* AO's, the values are 1 and 0, respectively. See Table 8.2 for values in percentages.

Energy ↑ decreasing *s* character ↑

p
sp^3
sp^2
sp
s

We usually say that, for a given atom, *the more* s *character* (or less *p* character) *in an orbital, the lower is the energy of the electrons in the orbital and the closer*

TABLE 8.2
PROPERTIES OF σ BONDS FORMED FROM *s*, *p*, AND *s-p*-TYPE ATOMIC ORBITALS

NUMBER OF HYBRID ORBITALS	TYPE	PERCENT *s* CHARACTER	PERCENT *p* CHARACTER	BOND ANGLE	GEOMETRIC ARRANGEMENT
0	*s*†	100	0	—	—
2	*sp*	50	50	180°	linear
3	sp^2	33	67	120°	trigonal-planar
4	sp^3	25	75	109.5°	tetrahedral
0	*p*	0	100	90°	right angle

† Atom using an *s* AO must be terminal; there is no bond angle.

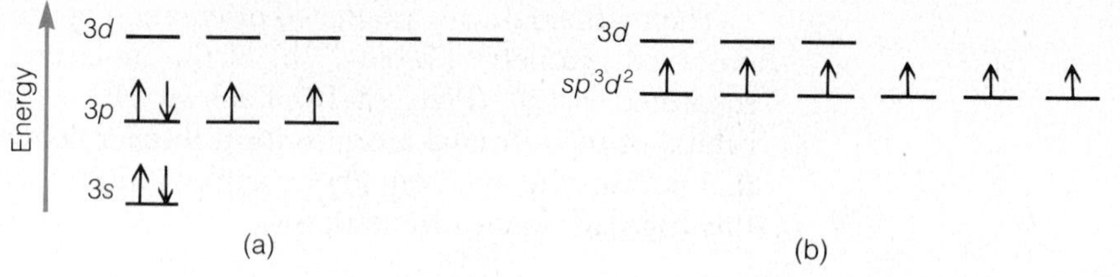

FIGURE 8.12
Electronic structure of the valence shell of sulfur.
(a) Ground state.
(b) Hybridized sp^3d^2 state.

are the electrons to the nucleus. If the electrons are unshared, the closer they are to the nucleus, the less available they are for coordinate covalent–type bonding to another atom.

Table 8.2 summarizes the properties of the hybrid AO's formed from *s* and *p* orbitals, the so-called *s-p*-type hybrids. Note that when a bonded atom has 8 or fewer valence electrons, hybridization involves the *s* orbital and one, two, or three *p* orbitals.

HYBRIDIZATION WITH *s*, *p*, AND *d* AO's

Atoms of the elements of the third and higher periods may have more than an octet of electrons in their valence shell. Examples include P in PCl_5, S in SF_6 and $:SF_4$, and Cl in $:\ddot{Cl}F_3$. Since the *s* and three *p* AO's can accommodate only 8 electrons, valence-shell *d* orbitals are used to accommodate the extra electrons. This subsection describes the hybridization of *s*, *p*, and *d* orbitals of representative elements surrounded by five or six pairs of electrons.

Transition elements are discussed in Chapter 21.

In a molecule of SF_6, the six S—F bonds are identical. Hence, S must have six identical orbitals to accommodate the six pairs of electrons. We get these six orbitals by hybridizing the one 3*s*, three 3*p*'s, and two of the five 3*d* orbitals (Figure 8.12) to give six equivalent sp^3d^2 HO's.

The molecular shape predicted by the VB method from the use of sp^3d^2 HO's is octahedral. This is the shape observed for SF_6 (Table 7.7f) and predicted by the VSEPR theory.

In the PCl_5 molecule five orbitals are needed for the five pairs of electrons. We hybridize the one 3*s*, three 3*p*, and one 3*d* orbitals to form five sp^3d hybrid orbitals (Figure 8.13).

The molecular shape predicted by the VB method from the use of sp^3d hybrid HO's is *trigonal-bipyramidal.* This agrees with the shape observed for PCl_5 (Table 7.7e) and predicted by the VSEPR theory.

Table 8.3 summarizes the HO's and their spatial orientations.

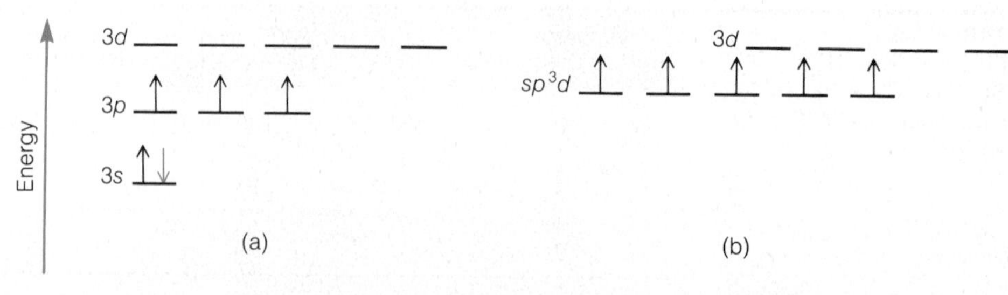

FIGURE 8.13
Electronic structure of the valence shell of phosphorus. (a) Ground state. (b) Hybridized sp^3d state.

TABLE 8.3
HYBRID ORBITALS AND THEIR SPATIAL ORIENTATION

AO's of CENTRAL ATOM	HYBRID STATE	NUMBER OF HO's	SPATIAL ORIENTATION	EXAMPLES
s and 1 *p*	*sp*	2	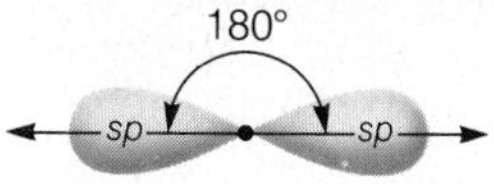 180°	BeH_2, $HgCl_2$
s and 2 *p*'s	sp^2	3	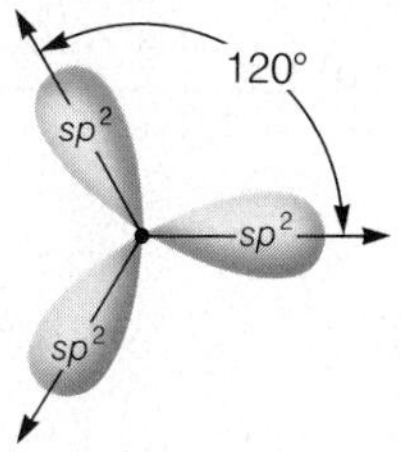 120°	BF_3, AlH_3
s and 3 *p*'s	sp^3	4	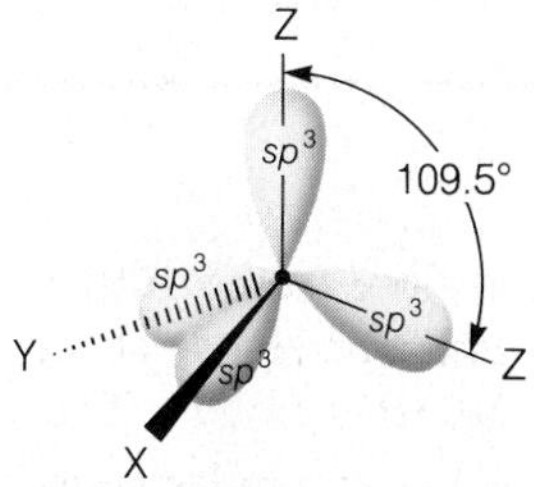 109.5°	CH_4, NH_4^+, SO_4^{2-}
s, 3 *p*'s, and *d*	sp^3d	5	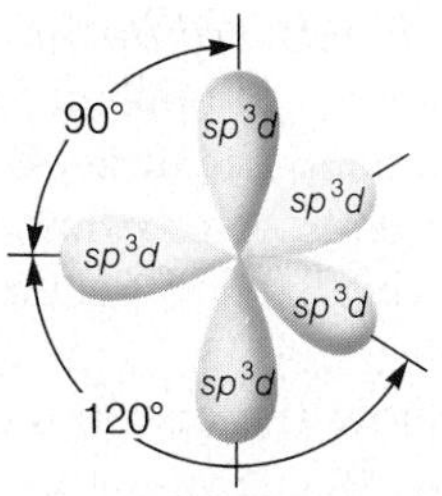 90°, 120°	
s, 3 *p*'s, and 2 *d*'s	sp^3d^2	6	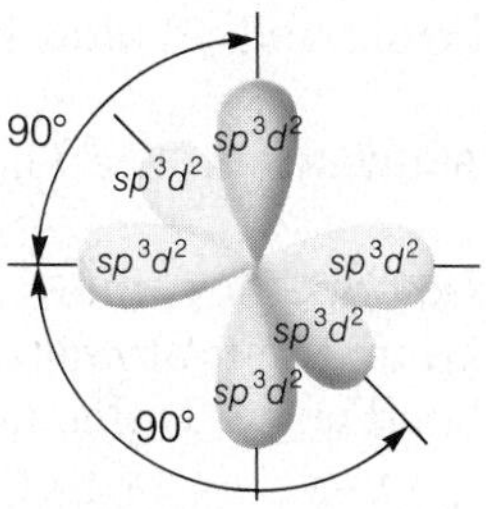 90°	SF_6, $AlCl_6^{3-}$

8.3 PREDICTING HYBRIDIZATION

In the previous section the hybridization of an atom was predicted from the number of its σ bonds. In this section we see how the prediction of the hybridization is affected by the presence of unshared pairs of electrons and π bonds.

UNSHARED PAIRS AND HYBRIDIZATION

Bond angle shrinkage was explained in section 7.12.

The best clue to the type of orbitals used by a central atom is the observed bond angle. For example, in ammonia, $:NH_3$, the H–N–H bond angle is 107°, which

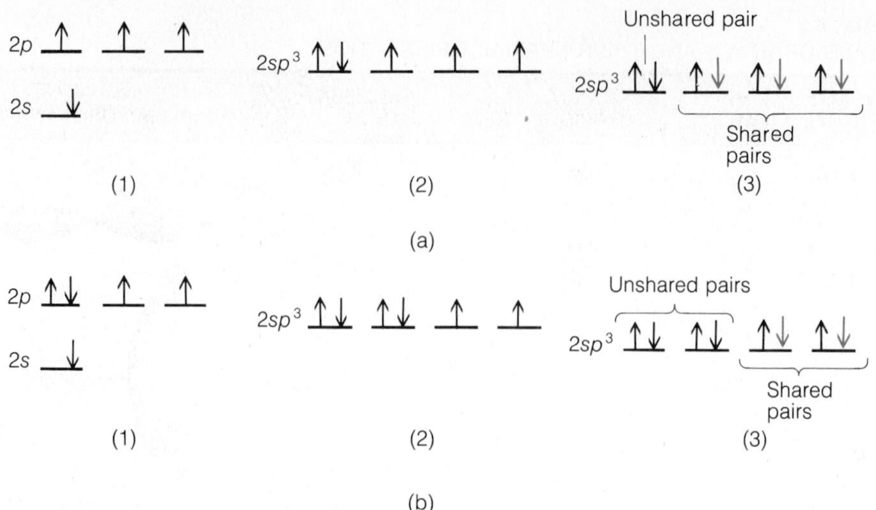

FIGURE 8.14 The distribution of valence electrons of (a) N in $:\ddot{N}H_3$ and (b) O in $H_2\ddot{O}:$ in (1) the ground state, (2) the hybridized state before bonding, and (3) the hybridized state after bonding. The colored arrows in (3) represent the electrons from the H atoms.

is close to the 109.5° tetrahedral angle. Therefore, it is assumed that the N atom uses sp^3 HO's. Three of these HO's overlap the H's 1*s* orbitals to form the three N—H σ bonds. The fourth sp^3 HO houses the unshared pair. This, along with other examples such as $H_2\ddot{O}:$ (bond angle = 105°), lead to the generalization that *σ bonds and unshared pairs of electrons require hybrid orbitals.* This generalization is not applied to terminal halogen atoms, F, Cl, Br, and I, which we *assume* to be in the unhybridized ground state when they use a *p* orbital to form a single σ bond. Note that terminal halogen atoms have no bond angle to indicate if hybridization occurs.

See Problem 8.19 for evidence that Cl might use *sp* HO's in HCl.

EXAMPLE 8.1 Show the distribution of the valence electrons for (a) the N atom in $:NH_3$ and (b) the O atom in $H_2\ddot{O}:$ in (1) the ground state and (2) the hybridized state before and (3) after bonding.

ANSWER (a) See Figure 8.14a. (b) See Figure 8.14b. ■

EXAMPLE 8.2 Describe how the N atom in $:NH_3$ can form three identical N—H bonds from ground-state atomic orbitals. On the basis of that prediction, predict the H—N—H bond angle. Is this model correct?

ANSWER The ground-state valence shell of N is $2s\,\underline{\uparrow\downarrow}\;2p\,\underline{\uparrow}\;\underline{\uparrow}\;\underline{\uparrow}$. The unshared pair of $:NH_3$ could be in the 2*s* atomic orbital and each *p* orbital could overlap with the *s* orbital of an H atom. Since the three *p* orbitals have the same energy, the three N—H bonds would be identical. Such a bonding situation would lead to H—N—H bond angles of 90° (see Figure 8.7). This model is incorrect because the observed angles are 107°. ■

PROBLEM 8.1 Show that the O atom in $H_2\ddot{O}:$ can form two identical O—H bonds from ground-state atomic orbitals. What would be the H—O—H bond angle? □

The answers to Example 8.2 and Problem 8.1 lead us to ask, "Why is bonding involving N and O atoms best described in terms of hybrid orbitals even though

BOX 8.1
INTERMEDIATE HYBRIDIZATION

The F—N—F bond angle in NF_3 is 102°. This means the N atom can use either sp^3 HO's (bond angle 109.5°) or *p* AO's (bond angle 90°), because 102° is between those values. Since the angle is closer to 109.5° than to 90°, however, it is claimed that N uses sp^3 hybrid AO's. In rationalizing the difference in bond angle between 102° and 109.5°, however, it is further claimed that the hybrid orbitals used are not typically sp^3 but rather have "some more *p* character."

This example reveals an important aspect of the concept of hybridization. It is not an either-or choice of *p*, *sp*, sp^2, or sp^3. Rather, intermediate degrees of hybridization are possible. The hybridization is such as to provide a molecule with the lowest possible energy. The central atoms do not "know" about the pure hybrid states that we imagine. Nature resists being "pigeonholed." In the same way, most bonds are neither purely ionic nor purely covalent, but are something in between.

these atoms could use ground-state atomic orbitals?" There are two reasons why hybrid orbitals give stronger σ bonds than *p* orbitals give:

1. The resulting bond angles are greater—about 109° for sp^3 hybrids and about 90° when unhybridized *p* orbitals are used. The larger the angle, the less repulsion there is between pairs of electrons.
2. The bonds are stronger. Only one lobe of a *p* orbital is used in σ bond formation. The unused lobe is "wasted." In an *s-p*-type hybrid orbital there is greater electron density in one, "the head" lobe, and it is the head that overlaps with an orbital of the other atom when a bond is formed. Thus the *s-p*-type of HO has a greater electron density in its bonding region than does a *p* orbital (Figure 8.15). For this reason the σ bond from an sp^3 hybrid AO is stronger than the σ bond from a *p* orbital.

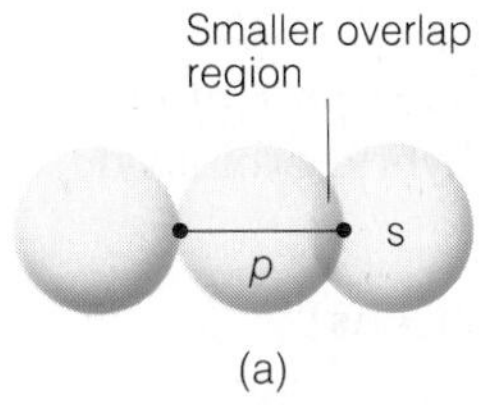

(a)

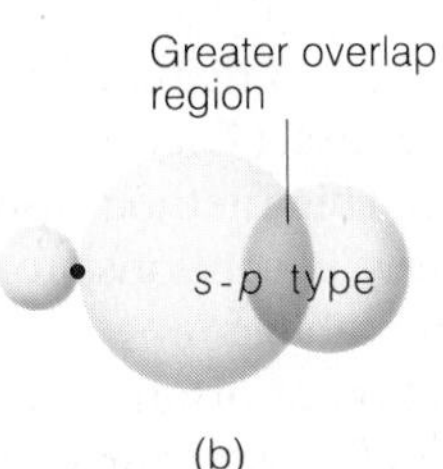

(b)

FIGURE 8.15
Comparison of the orbital overlap between (a) *p* and *s* AO's and (b) an *s-p*-type HO and an *s* AO. Note the greater overlap in illustration b.

MULTIPLE BONDS AND HYBRIDIZATION

We have seen that multiple bonds consist of σ and π bonds and that *no more than one bond between two atoms can be a σ bond*; thus, *all bonds in excess of one are π bonds:*

single bond: 1 σ bond
double bond: 1 σ bond + 1 π bond
triple bond: 1 σ bond + 2 π bonds

Since π bonds are formed by side-to-side overlap of *p* orbitals, *they do not require hybrid orbitals*. This is consistent with observed bond angles. For example, the bond angles in formaldehyde, $H_2C{=}O$, are approximately 120°, a value associated with sp^2 hybrid orbitals. We assume the C atom uses sp^2 HO's to form the three σ bonds.

H
σ
116° C $\overset{\pi}{\underset{\sigma}{=}}$ Ö:
σ
H

TABLE 8.4
RELATIONSHIP OF HYBRID ORBITAL NUMBER TO HYBRID STATE

CALCULATED HYBRID ORBITAL NUMBER (HON)	PREDICTED HYBRID STATE
2	*sp*
3	sp^2
4	sp^3
5	sp^3d
6	sp^3d^2

TABLE 8.5
ORBITAL HYBRIDIZATION, BONDING, AND SHAPE

HYBRID AO's		sp^2			sp^3	
Bonding†	$—E$ (tetrahedral)	$\ddot{E}$	$\ddot{E}$ (two lone pairs)	E	$E=$	$\ddot{E}=$
Shape	tetrahedral	trigonal-pyramidal	bent	trigonal planar	trigonal planar	bent
Examples‖	CH_4, NH_4^+	$\ddot{N}H_3$	$H_2\ddot{O}:$	BF_3	$Cl_2C=\ddot{O}:$	$[O—\ddot{N}=O]^-$

† E stands for any element.
§ Shape is determined by the element attached to E, which is terminal in these examples.
‖ Bonding type refers to the colored atom in each example.

HYBRID ORBITAL NUMBER (HON) RULE

The odd electron of a free radical rarely requires an HO.

By using the generalization that *each unshared and σ-bonded pair of electrons needs a hybrid orbital but π bonds do not,* the number of hybrid orbitals needed by the central atom can be obtained as indicated:

no. of HO's = no. of (σ bonds + unshared pairs of electrons)

The hybridized state of the atom can then be predicted from the calculated *h*ybrid *o*rbital *n*umber (HON) as summarized in Table 8.4.

One important advantage of this method is that it allows us to deduce hybrid conditions from Lewis structures without knowing the bond angles. (This is another reason for learning how to write correct Lewis structures.) This method applies to Be, B, C, N, and O in the second period and, with a few exceptions, to the multicovalent elements in the higher periods of the periodic table.†

Table 8.5 shows relationships between *s-p*-type hybridization of orbitals, bonding, and shape. The table should be used as a reference. When E is a terminal atom in a molecule, as in $:\ddot{E}=$ and $\equiv E:$, it has no bond angle to help us predict the kind of orbitals that are used. In fact, such an atom may use hybridized or ground-state orbitals. For the sake of consistency, we arbitrarily assume that these atoms use hybrid orbitals.

EXAMPLE 8.3 Determine the type of hybridized AO's used by the central atom and show the orbital distribution of valence electrons of the central atom before and after bonding in forming identical σ bonds in these species:

```
    H          Cl       ┌   F   ┐⁻
    |          |        |   |   |
H—Si—H,    B—Cl,        | F—B—F | ,  and  H—Be—H
    |          |        |   |   |
    H          Cl       └   F   ┘
```

ANSWER None of the central atoms have any unshared pairs; therefore, you should count only the σ bonds.

† Exceptions are the hydrides, such as PH_3, H_2S, AsH_3, H_2Se, and H_2Te, whose bond angles are close to 90°. Here the central atoms use p orbitals.

		sp		
:Ë=	—E≡	=E=	—E—	≡E:
§	linear	linear	linear	§
:Ö=N̈—ÖH	H—C≡N:	:Ö=C=Ö:	Cl—Be—Cl	H—C≡N:

(SiH_4) Since all four bonds are identical, Si uses sp^3 HO's:

sp^3 ↑ ↑ ↑ ↑ (before bonding) sp^3 ↑↓ ↑↓ ↑↓ ↑↓ (after bonding)

(BCl_3) B has three identical covalent bonds and needs three HO's. It uses sp^2 HO's:

before bonding: p __ (empty); sp^2 ↑ ↑ ↑

after bonding: p __ (empty); sp^2 ↑↓ ↑↓ ↑↓

(BF_4^-) B now has four identical covalent bonds and needs four sp^3 HO's:

sp^3 ↑ ↑ ↑ __ (before bonding) sp^3 ↑↓ ↑↓ ↑↓ ↑↓ (after bonding)

On forming BF_4^- ($BF_3 + F^- \longrightarrow BF_4^-$), the B atom undergoes a change in hybridization ($sp^2 \longrightarrow sp^3$). ($BeH_2$) Be has two identical covalent bonds and uses two sp HO's:

before bonding: p __ __ (both empty); sp ↑ ↓

after bonding: p __ __ (both empty); sp ↑↓ ↑↓

■

PROBLEM 8.2 State the type of orbitals used by the central atom and show the orbital distribution of valence electrons before and after bonding when forming identical bonds in (a) AlH_3, (b) NH_4^+, and (c) H_3O^+. □

EXAMPLE 8.4 Use the hybrid orbital number method to determine the kind of hybrid orbital used by the central atom in (a) ClO_4^-, (b) ClO_3^-, (c) ClO_2^-, (d) ClF_3, and (e) O=PCl_3. Tell the shape of each ion or molecule.

ANSWER Since the Cl atoms in (a), (b), (c), and (d) are multicovalent, they are assumed to use hybrid orbitals. It is first necessary to write the Lewis structures and in particular to be able to locate any unshared pairs of electrons on the central atoms. At this time you should review the procedure for writing Lewis structures; look back to section 7.10.

(a) $\left[\begin{array}{c} :\ddot{O}: \\ | \\ :\ddot{O}-Cl-\ddot{O}: \\ | \\ :\underset{..}{O}: \end{array}\right]^-$

The Cl has 4 σ bonds and no unshared pairs. The HON value is 4 and Cl uses sp^3 hybrid orbitals. The shape is tetrahedral.

(b) $\left[\begin{array}{c} :\ddot{O}-\ddot{Cl}-\ddot{O}: \\ | \\ :\underset{..}{O}: \end{array}\right]^-$

The Cl has 3 σ bonds and one shared pair. The HON value is again 4, 3 + 1, and Cl uses sp^3 hybrid orbitals. The ion is trigonal-pyramidal.

(c) $[:\ddot{O}-\ddot{Cl}-\ddot{O}:]^-$

The Cl has two σ bonds and two unshared pairs. Again the HON value is 4, (2 + 2), and Cl uses sp^3 HO's. The ion is bent.

(d) $:\ddot{Cl}(-\ddot{F}:)_3$

The Cl has three σ bonds and two unshared pairs. Now the HON value is 5 and Cl uses sp^3d HO's. The molecule is T-shaped.

(e) $:\ddot{O}=P(-\ddot{Cl}:)_3$

P uses an empty *d* AO to form the π bond with a *p* AO of O (called $p\pi \rightarrow d\pi$ bonding).

The P has four σ bonds and no unshared pairs. The π bond does not count in determining the hybrid state because hybrid orbitals form only σ bonds. The HON value is 4 and P uses sp^3 HO's. The molecule is tetrahedral. ■

PROBLEM 8.3 Use the HON method to determine the hybrid state of the central atom in (a) SO_2, (b) SO_3^{2-}, (c) NO_3^-, (d) SCl_2, and (e) XeF_4. □

8.4 HYBRID ORBITALS OF CARBON

As you will learn in Chapters 24 and 25, the study of carbon-containing compounds makes up an important field of chemistry. For this reason we will now discuss examples of *sp*, sp^2, and sp^3 bonding observed for carbon.

ETHANE, $\begin{array}{c} \quad H \quad H \\ \quad | \quad\ | \\ H-C-C-H \\ \quad | \quad\ | \\ \quad H \quad H \end{array}$

The four bonds on each C in ethane are observed to have a tetrahedral arrangement. Therefore each C is assumed to use sp^3 HO's. We could have predicted this using the HON rule without knowing the shape. Each C has four σ bonds, HON values of 4, and they are sp^3 hybridized. Three HO's of each C overlap with

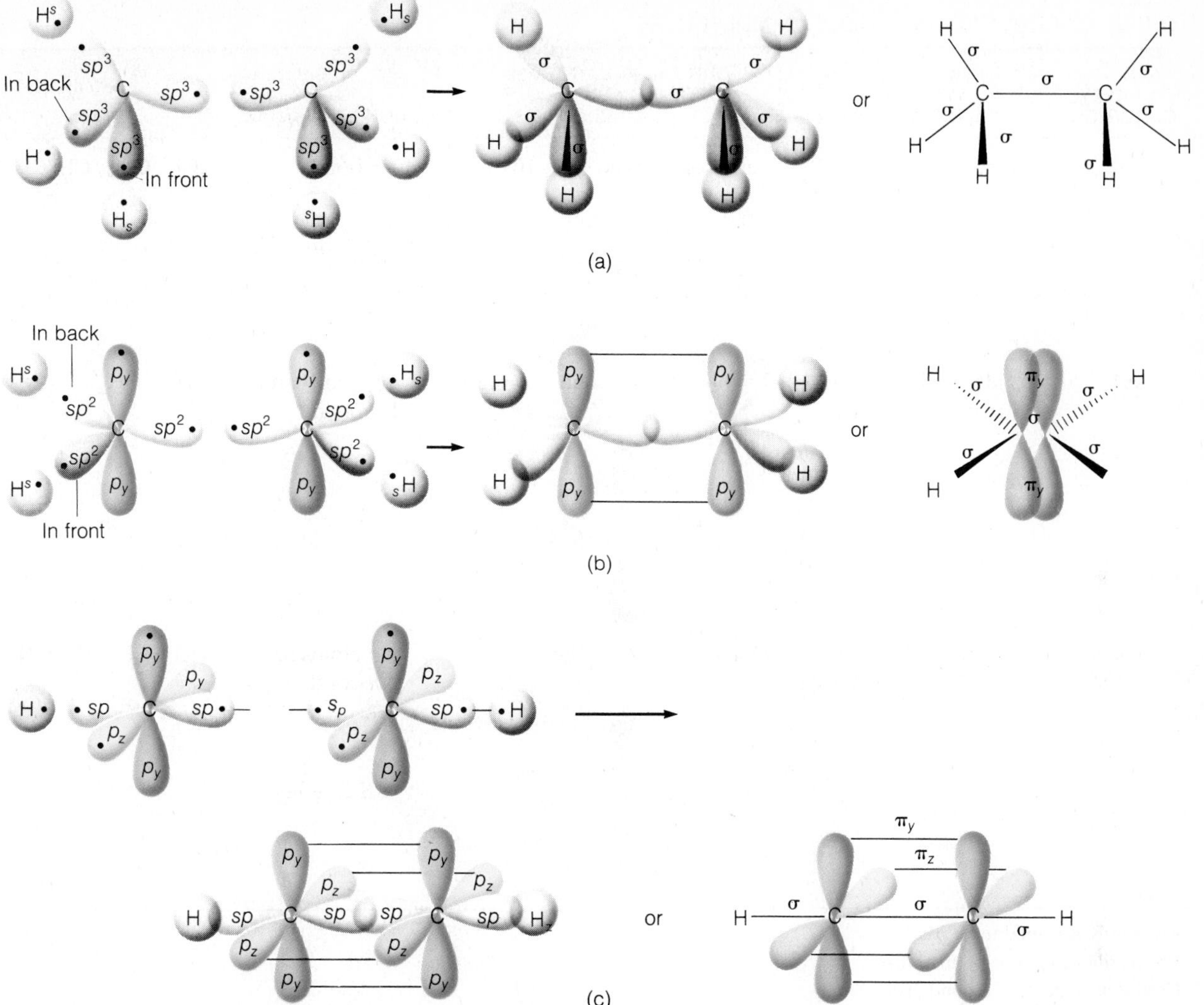

FIGURE 8.16
Assembly and orbital representation of (a) ethane, CH_3CH_3; (b) ethene (ethylene), $CH_2{=}CH_2$; and (c) ethyne (acetylene), $HC{\equiv}CH$. (Black dots represent electrons.)

the *s* atomic orbital of an H atom to form the six C—H bonds. The fourth sp^3 HO of each C overlap with each other to form the C—C bond. Each C atom is tetrahedral and each tetrahedron is joined at a corner. The bonding arrangement is shown in Figure 8.16a.

ETHENE (Ethylene), $H_2C{=}CH_2$

Let us predict the hybrid condition of each C in ethene and the shape of the molecule. Each C has three σ bonds and no unshared pairs. The π bond does not count in determining the HON value. The HON value is 3 and the C atoms are predicted to use sp^2 HO's. Two of these sp^2 HO's of each C overlap with the 1*s* atomic orbital of an H atom to form the four C—H bonds. The third sp^2 HO of each C overlaps with each other to give the σ bond of the double bond. The remaining *p* orbitals on the two adjacent C atoms have one electron each. These *p*

TABLE 8.6
HYBRIDIZED CONDITION OF CARBON IN MOLECULES

HYBRIDIZED CONDITION	ORBITALS USED BY C	ORBITAL ARRANGEMENT SHOWING e^-'s BEFORE BONDING	BONDS FORMED	EXAMPLES
sp^3	4 identical HO's	Tetrahedral bond ∡ = 109.5° 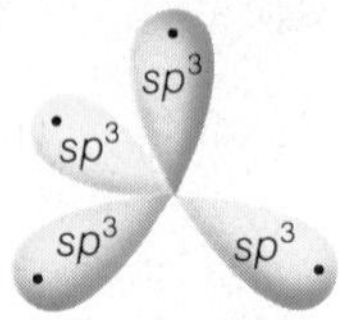	4 σ bonds 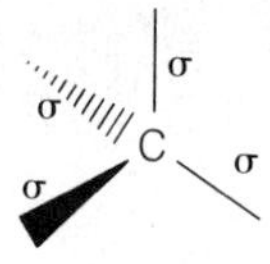	CH_4, CCl_4, CH_3CH_3‖
sp^2	3 identical HO's and 1 unchanged p AO†	Trigonal-planar bond ∡ = 120° 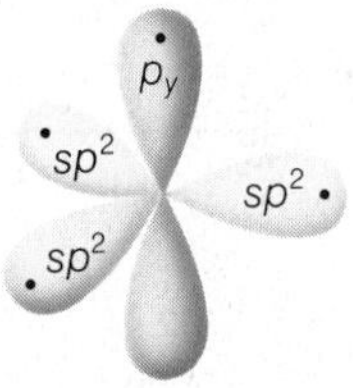	3 σ bonds and 1 π bond 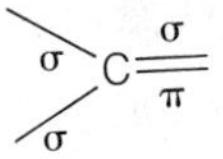	$CH_2{=}CH_2$§, $Cl_2C{=}O$
sp	2 identical HO's and 2 unchanged p AO's§	Linear bond ∡ = 180° 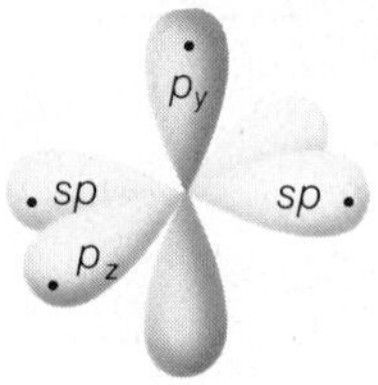	2 σ bonds and 2 π bonds 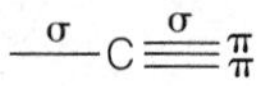	$HC{\equiv}CH$‖, $HC{\equiv}N$

† Arbitrarily assumed to be p_y.
§ Arbitrarily assume to be p_y and p_z.
‖ Orbital arrangement around each C atom.

The same bonding picture is obtained by using any unhybridized p AO to form the π bond.

orbitals, arbitrarily assumed to be p_y, overlap laterally with each other to give the π bond of the double bond. The bonding arrangement is shown in Figure 8.16b. The σ bonds of each sp^2 hybridized C atom in ethylene should have a triangular-planar arrangement with bond angles about 120°. Furthermore, in order for the p_y AO's to overlap laterally, the σ bonds on each C must lie in the same plane. This bonding model is consistent with reality; ethene is indeed a planar molecule with the predicted bond angles:

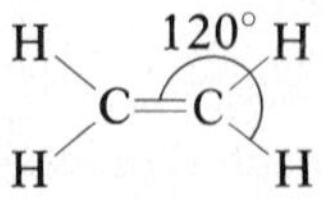

$$\mathrm{H_2C{=}CH_2}\quad (120^\circ)$$

ETHYNE (Acetylene), H—C≡C—H

Again we predict the hybrid condition of each C and the shape of the acetylene molecule. Each C has two σ bonds and no unshared pairs. The HON value is 2, and the C atoms should use sp HO's. It is arbitrarily assumed that it is the $2p_x$ AO that hybridizes with the $2s$ AO. One sp HO of each C overlaps with the $1s$

orbital of an H atom to give the C—H bonds, and the second ones overlap with each other to give the σ bond of the triple bond. The p_y orbitals on each adjacent C overlap laterally to form a π_y bond in the xy plane. The p_z orbitals overlap to form a π_z bond in the xz plane. The planes of the two π bonds are at right angles to each other and to the C—C σ bond axis. This bonding arrangement is shown in Figure 8.16c. The two σ bonds of each C are derived from sp HO's and therefore have bond angles of 180°. Since one σ bond is common to each C, acetylene is correctly predicted to be a linear molecule, H—C≡C—H. Table 8.6 summarizes the bonding nature of carbon.

It is believed that the two π bonds merge into a cylindrical shape.

EXAMPLE 8.5 Describe the bonding in formaldehyde, $H_2C{=}\ddot{O}$:, in terms of the orbitals used. Predict the shape. (Assume that O uses hybrid orbitals.)

ANSWER The C atom has three σ bonds and no unshared pairs of electrons (type >C=). It needs three sp^2 HO's. The O atom has one σ bond and two unshared pairs (type :$\ddot{E}$=) and needs three sp^2 HO's. The sp^2 HO's of C form three σ bonds, one with each H and one with an sp^2 hybrid orbital of the O atom. The two unshared pairs of electrons on the O atom are in its two other sp^2 hybrid AO's. The remaining p orbital on C and the p orbital on O overlap to form the π bond. ■

PROBLEM 8.4 Describe the bonding in terms of the orbitals in (a) cyanide ion, :C≡N:$^-$; (b) methylene imine, $H_2C{=}\ddot{N}H$; and (c) carbon dioxide, :$\ddot{O}$=C=$\ddot{O}$:. □

8.5 THE RESONANCE CONCEPT AND DELOCALIZED π BONDING

Our discussion of the resonance concept and delocalized π bonding starts with writing a Lewis structure for dinitrogen oxide, N_2O (also called nitrous oxide). Its *skeleton* is N—N—O and it has 16 valence electrons. One possible Lewis structure is (a) :$\overset{\ominus}{\ddot{N}}$=$\overset{\oplus}{N}$=$\ddot{O}$:, shown with formal charges. The structure (b) :N≡$\overset{\oplus}{N}$—$\overset{\ominus}{\underset{..}{\ddot{O}}}$: is another possibility. Lewis structures a and b differ only in the arrangement of multiple bonds and unshared electrons; that is, they have different **electronic structures**. But which Lewis structure, *if any*, is correct? This question can be answered only by doing experiments to differentiate between the predicted possible structures.

One such experiment is the measurement of the N to N and the N to O bond distances. The measured bond distances are then compared with those calculated using the appropriate atomic radii in Table 6.6 (section 6.4) and Table 7.5 (section 7.8). Table 8.7 summarizes the calculated results and shows the observed results for our N_2O molecule.

The results are revealing; neither set of predicted bond lengths matches the observed values. The actual bond lengths are intermediate between the values

TABLE 8.7
BOND LENGTHS (nm) IN DINITROGEN OXIDE, N_2O

	CALCULATED BOND LENGTH		
	(a) :$\overset{\ominus}{\ddot{N}}$=$\overset{\oplus}{N}$=$\ddot{O}$:	(b) :N≡$\overset{\oplus}{N}$—$\overset{\ominus}{\underset{..}{\ddot{O}}}$:	OBSERVED BOND LENGTH
N to N bond	0.120	0.110	0.112
N to O bond	0.115	0.147	0.119

predicted from Lewis structures a and b. We conclude that *neither Lewis structure by itself is correct.* But then what *is* the correct Lewis structure? We turn for help to the concept of **resonance**. According to this concept, the actual structure of a molecule such as N_2O *cannot be accurately depicted by a single Lewis structure.* Instead, the molecule is depicted by *two or more Lewis structures,* such as the two that we predicted for N_2O. Taken together, these serve as a better description than any single one. Thus, whereas either a or b, taken by itself, predicts an incorrect bond length, the two structures taken together suggest that the bond is something between a double and a triple bond. While not precise, this is less incorrect. Thus, by writing plausible but inaccurate Lewis structures, we can *imagine* the correct structure for which we cannot write a single Lewis structure.

Any one of the contributing structures can usually be used to predict the molecular shape.

These inaccurate Lewis structures are called **contributing** or **resonance structures**. The actual molecule is called a **resonance hybrid**. The resonance hybrid exists because it has less energy and is more stable than any contributing structure.

Do not be misled by the term *hybrid.* In biology a hybrid is a species that is a "cross" between two real species. For example, a mule, a real animal, is a hybrid of two other real animals, a horse and a donkey. In resonance theory, the resonance hybrid is a "cross" between two *fictitious* Lewis structures. A fitting analogy is the description of a rhinoceros offered by a child after a first visit to the zoo as "an animal that looks like a dragon and a unicorn." The dragon and unicorn are mythical, as are contributing (resonance) structures; the rhinoceros, like the resonance hybrid, is real.

Traditionally, a double-headed arrow is placed between the contributing structures to indicate resonance:

$$:\overset{\ominus}{\underset{\cdot\cdot}{N}}=\overset{\oplus}{N}=\overset{\cdot\cdot}{O}: \longleftrightarrow :N\equiv\overset{\oplus}{N}-\overset{\ominus}{\underset{\cdot\cdot}{\overset{\cdot\cdot}{O}}}:$$

It is unfortunate that the word *resonance* was chosen for this concept. If you should mistakenly apply the meaning of resonance you learned in physics class to the problem of chemical bonding in a molecule, misconceptions may result. These misconceptions must be dispelled. For example, the contributing structures do *not* oscillate or resonate back and forth. *They do not exist at all.* In our rhinoceros analogy, the actual "structure," the rhino, is not a unicorn now and a dragon later; it is always a rhino. Neither is a group of rhinos a mixture of unicorns and dragons. The group consists only of rhinos.

When we examine the contributing structures of N_2O we notice that they essentially differ in two interrelated ways:

- the number of bonds between pairs of bonded atoms—whether the bonds are single, double, or triple bonds—and
- the number of unshared electrons on the atoms.

These differences in *electronic structure* prevail for contributing structures of any molecule to which the resonance concept is applied.

There are some important restrictions on permissible contributing structures:

1. Atomic nuclei must have the same positions in all structures; only electrons can have different positions.
2. They must have the correct number of valence electrons.
3. With second-period elements, do not exceed the octet.

An exception to formal charge minimizing is

$>C{=}\ddot{O}{:} \longleftrightarrow >\overset{\oplus}{C}{-}\overset{\ominus}{\ddot{\underset{..}{O}}}{:}$;

they are both acceptable.

4. Although there may be different numbers of bonds and unshared pairs, the number of electron pairs must be the same in all structures.

5. Don't have less than an octet (except B and Be). Maximize normal covalences and minimize formal charges.

EXAMPLE 8.6 Which of these pairs are permissible contributing structures?

(a) $:\ddot{\underset{..}{O}}{-}\dot{N}{=}\ddot{O}{:}$ and $:\ddot{\underset{..}{O}}{-}\ddot{N}{=}\ddot{O}{:}$, (b) $H{-}\ddot{N}{=}C{=}\ddot{O}{:}$ and $:N{\equiv}C{-}\ddot{\underset{..}{O}}{-}H$,

(c) $:\ddot{O}{::}\ddot{O}{:}$ and $:\ddot{\underset{.}{O}}{:}\ddot{\underset{.}{O}}{:}$, (d) $H_2C{=}\underset{..}{O}{:}$ and $H_2\overset{\ominus}{C}{\equiv}\overset{\oplus}{O}{:}$,

(e) $H_2C{=}CH_2$ and $H_2\overset{\ominus}{\underset{..}{C}}{-}\overset{\oplus}{C}H_2$, (f) $:\overset{\ominus}{\underset{..}{N}}{=}\overset{\oplus}{N}{=}\underset{..}{O}{:}$ and $:\overset{2-}{\ddot{\underset{..}{N}}}{-}\overset{\oplus}{N}{\equiv}\overset{\oplus}{O}{:}$

ANSWER

(a) These are permissible structures; they have the same skeleton (ONO), the same number of pairs of electrons (8 pairs), and the same number of unpaired electrons (1).

(b) These are *not* permissible structures because they have different skeletons. They are Lewis structures for two different compounds; they are isomers.

(c) These are not permissible structures since they have different numbers of paired and unpaired electrons. The first Lewis structure has 6 pairs of electrons. The second Lewis structure has 5 pairs of electrons and 2 unpaired electrons.

(d) $H_2\overset{\ominus}{C}{\equiv}\overset{\oplus}{O}{:}$ is not permissible because C, a second-period element, has 10 valence electrons.

(e) $H_2\overset{\ominus}{\underset{..}{C}}{\equiv}\overset{\oplus}{C}H_2$ has a very high energy because it has formal charge, and its C's do not have normal covalences. It is not permissible.

(f) N_2O is an example of a molecule whose contributing structures must have formal charge. But since $:\overset{2-}{\ddot{\underset{..}{N}}}{-}\overset{\oplus}{N}{\equiv}\overset{\oplus}{O}{:}$ has too much formal charge, it is not permissible; it has a very high energy. ■

PROBLEM 8.5 Which pairs of Lewis structures are valid contributing resonance structures?

(a) $H_2C{=}\overset{\oplus}{N}{=}\overset{\ominus}{\underset{..}{N}}{:}$, $H_2C{=}\underset{..}{N}{-}\underset{..}{N}{:}$

(b) $H{-}C(={\underset{..}{O}}{:}){-}\ddot{\underset{..}{O}}{-}H$, $H{-}C(-{:}\underset{..}{O}{:}^{\ominus}){=}\overset{\oplus}{\ddot{O}}{-}H$

(c) $H\ddot{\underset{..}{O}}{-}\ddot{P}(-{:}\underset{..}{O}H){-}\ddot{O}H$, $:\overset{\ominus}{\ddot{\underset{..}{O}}}{-}\overset{\oplus}{P}(H)(-{:}\underset{..}{O}H){-}\ddot{\underset{..}{O}}H$ □

EXAMPLE 8.7 Write two contributing structures with formal charges for diazomethane (H_2CN_2), whose skeleton is

$$\begin{matrix} H \\ & C{-}N{-}N \\ H \end{matrix}$$

ANSWER First, determine the total number of valence electrons that must show in the structure. This number is 2 e^-'s (2 H's) + 4 e^-'s (1 C) + 10 e^-'s (2 N's) = 16 e^-'s. Therefore, 8 e^-'s must be added, as bonded or unshared, to the 8 bonded pairs shown in the skeleton. For N and C, getting an octet has priority over having the normal covalence. The contributing structures are

$$H_2C{=}\overset{\oplus}{N}{=}\overset{\ominus}{\underset{\cdot\cdot}{N}}\!: \longleftrightarrow H_2\overset{\ominus}{\underset{\cdot\cdot}{C}}{-}\overset{\oplus}{N}{\equiv}N:$$

(Why would $H_2\overset{\oplus}{C}{-}\overset{\cdot\cdot}{N}{=}\overset{\ominus}{\underset{\cdot\cdot}{N}}\!:$ *not* be a valid contributing structure?) ■

PROBLEM 8.6 (a) Write two contributing structures with formal charges for hydrazoic acid, HN_3, whose skeleton is H—N—N—N. (b) Write two electronic structures, with the correct skeletons and showing formal charges, that would not be valid contributing structures. Explain your choices. □

EXAMPLE 8.8 There are three valid contributing structures for the nitrate ion, NO_3^-, one of which is

$$\left[\begin{matrix} :\ddot{O} \\ & \overset{\oplus}{N}{-}\overset{\ominus}{\underset{\cdot\cdot}{\ddot{O}}}: \\ \overset{\ominus}{:\underset{\cdot\cdot}{\ddot{O}}}: \end{matrix}\right]^-$$

(with a double bond from N to the upper O)

Give the other two structures.

ANSWER The double bond is placed in each of three equally possible N to O positions, giving two more contributing structures, for a total of three:

$$\left[\begin{matrix} \overset{\ominus}{:\ddot{O}:} \\ & \overset{\oplus}{N}{-}\overset{\ominus}{\underset{\cdot\cdot}{\ddot{O}}}: \\ :\underset{\cdot\cdot}{O} \end{matrix} \longleftrightarrow \begin{matrix} \overset{\ominus}{:\ddot{O}:} \\ & \overset{\oplus}{N}{=}\underset{\cdot\cdot}{O}: \\ \overset{\ominus}{:\underset{\cdot\cdot}{\ddot{O}}}: \end{matrix}\right]^-$$

■

The three contributing structures for NO_3^- look alike. In each structure the N atom has a single bond to each of two O atoms, and a double bond to a third O atom. Nevertheless, all three contributing structures must be written to help us visualize the real structure. The real structure is different from any of these contributing structures. Any one structure by itself implies, *incorrectly*, two N—O bond distances of the same length and one shorter N=O bond distance. The three structures taken together imply something different. They imply that each of the three N to O bonds has *some* double-bond character and *some* single-bond character. No one O atom can be said to be doubly bonded to the N atom. The resonance concept predicts that all three N to O bond lengths are the same. This prediction is experimentally verified.

PROBLEM 8.7 Write the contributing structures for ozone, O_3, showing formal charges and discuss the nature of the two O to O bonds. □

When the resonance concept describes the real structure as being "something in between" unreal structures, we are left unsatisfied. We would be much happier with a single structure. The atomic orbital overlap concept comes to our rescue by providing a more adequate description and a reasonable notation for species for which single Lewis structures cannot be written. The nitrite anion, NO_2^-, shown here by two contributing structures, will serve as an example.

$$\left[{}^{\alpha}\!:\!\ddot{\underset{\cdot\cdot}{O}}\!:\!-\ddot{N}\!=\!\underset{\cdot\cdot}{O}\!:^{\beta} \longleftrightarrow {}^{\alpha}\!:\!\underset{\cdot\cdot}{O}\!=\!\ddot{N}\!-\!\ddot{\underset{\cdot\cdot}{O}}\!:^{\beta} \right]^-$$

a b

Nearly all molecules whose structures need to be explained by the resonance concept have a common feature: the electrons that are assigned to different positions in the contributing structures are either unshared pairs or in π bonds. The positions of electrons in sigma bonds are left unchanged in all contributing structures. Therefore, when applying the idea of orbital overlap, we are concerned only with π bonds. If contributing structure a of NO_2^- were correct, there would be a π bond between the N and O^{β} atoms formed by side-by-side overlap of p orbitals, one on each atom. Contributing structure b tells us that a π bond is also between the N and O^{α} atoms, likewise formed by side-by-side overlap of p orbitals. For these contributing structures to exist, the same p orbital on the N atom must overlap with either a p orbital on O^{α} or a p orbital on O^{β}.

At this point, we must introduce a new idea: nothing prohibits the p orbital of the N atom from overlapping with the p orbital of O^{α} and with the p orbital of O^{β} *at the same time* (Figure 8.17). Therefore, we say that the correct structure has *simultaneous* π bonding between N and each O atom. The π bonding is extended over three atoms—the N and the two O atoms; it is *not* localized between the N and just one O atom as pictured by contributing structures a and b.

A π bond involving only two atoms is said to be "localized."

Because the π bonding extends over more than two atoms—in this case three atoms—the π-bonding electrons are said to be **delocalized**. This type of bonding gives an extended or **delocalized π system**. The π electrons in the delocalized π system are more spread out and therefore suffer from less repulsion than do the

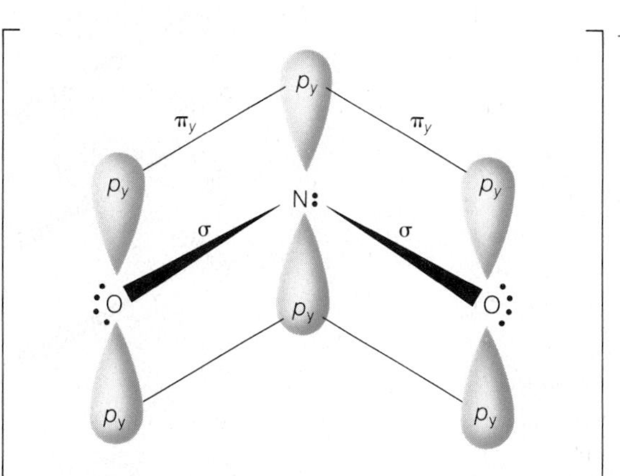

FIGURE 8.17
Delocalized π bonding in the nitrite ion, NO_2^-

localized electrons in simple π bonds. This is the main reason that the real structure, with delocalized electrons, has less energy and greater stability than an unreal contributing structure with localized electrons.

The real molecule or ion is depicted by a single structural formula in which dashed lines are used to represent delocalized π bonds, as shown for NO_2^-:

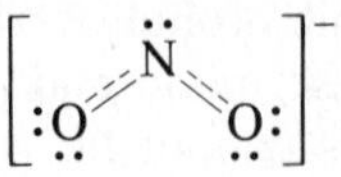

There are 4 electrons in this delocalized π system.

dashed-line representation of delocalized π bonding (the resonance hybrid)

We can say each O atom has one-half the negative charge.

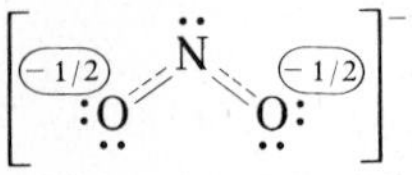

The unit negative charge on the ion is shared equally by each O atom. Each O atom bears some formal negative charge, the charge not being localized on just one O atom.

EXAMPLE 8.9 Three contributing structures describe the nitrate ion, NO_3^-, in terms of delocalized π bonding. Write a delocalized π structure.

ANSWER Each contributing structure shows a double bond between N and one O atom. Each double bond would be formed by the side-to-side overlap of, let us say, a p_y orbital of the N atom with a p_y orbital of an O atom. But the p_y orbital on N can have simultaneous side-to-side overlap with a p_y orbital of each O atom (Figure 8.18a). The delocalized π bonding generated from these four p AO's extends over the four atoms. Figure 8.18b represents the delocalized π structure for NO_3^-. The solid lines are for the N—O σ bonds. Since there is no fixed double

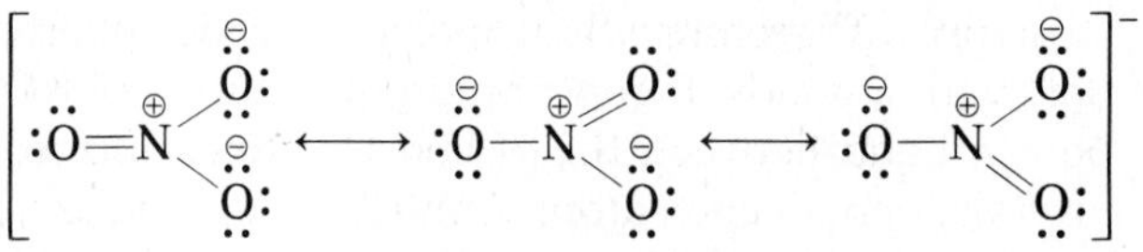

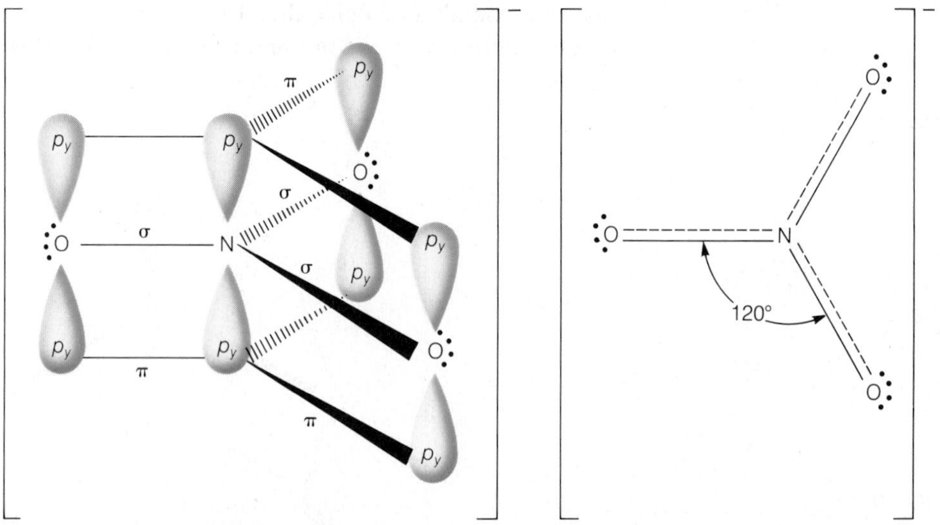

We could have used the p_x or p_z AO's just as well.

FIGURE 8.18
(a) Extended p orbital overlap of the four atoms of the nitrate ion, NO_3^-
(b) Structure showing delocalized π bonding.

(a)

(b)

Here one could say that each O bears a $-\frac{2}{3}$ formal charge.

bond between the N and any given O, dashed lines are used for all of the π bonds. Since the smallest number of unshared electrons on each O atom in any of the contributing structures is four, that is the number shown on each O atom in the delocalized structure. The formal charge on N in Figure 8.18b is +1 since that charge is present on N in all three contributing structures. ■

PROBLEM 8.8 Describe the carbonate anion, CO_3^{2-}, in terms of delocalized π bonding. Draw the delocalized π structure. □

8.6 THE MOLECULAR ORBITAL (MO) CONCEPT

Although the valence bond concept explains most covalent bonding situations, it fails in some cases. For example, how do we describe the electronic structure of H_2^-, a very unstable but detectable ion? H_2^- has three electrons, but has only one σ bond. If there can be only a pair of electrons in a σ bond and each H can acquire only 2 electrons, where is the third electron?

Since O_2 has an even number of e^-'s, it cannot have just one unshared e^-.

The structure of O_2 also challenges the VB concept. Judging from bond length measurements, O_2 should have a double bond. A reasonable Lewis structure would be $:\ddot{O}{=}\ddot{O}:$, with the O atoms bonded by a σ bond and a π bond. Yet O_2 is paramagnetic, having two unpaired electrons. Based upon this information, $:\dot{\ddot{O}}{-}\dot{\ddot{O}}:$ could be the Lewis structure. Neither formula can adequately account for all the structural facts. Furthermore, the resonance concept is of no help. These formulas are *not* contributing structures because they do not have the same number of paired electrons. A second quantum mechanical model for covalent bonding, called the **molecular orbital (MO) concept**, can explain the structure of H_2^- and O_2. The MO concept was developed about 1932 by Freidrich Hund, Robert Mulliken, Erich Hückel, and John Lennard-Jones.

In Chapter 5 we saw that electrons in isolated atoms occupy certain fixed (quantized) energy states called *atomic orbitals.* The molecular orbital concept is also extended to molecules. *Electrons in molecules are in fixed energy states called molecular orbitals.* Scientists determine the shapes and energies of molecular orbitals mathematically by combining atomic orbitals. But, two atomic orbitals can be combined in two different ways to give two molecular orbitals. An

The AO's can be added or subtracted.

The waves in Figure 8.19a and b are called *fundamental* waves.

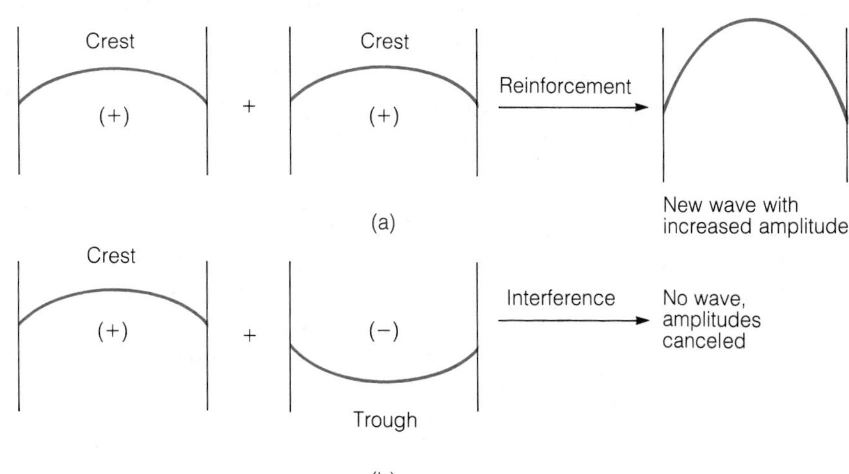

FIGURE 8.19
Interaction of two standing waves. (a) Reinforcement (+ with +). (b) Interference (+ with −).

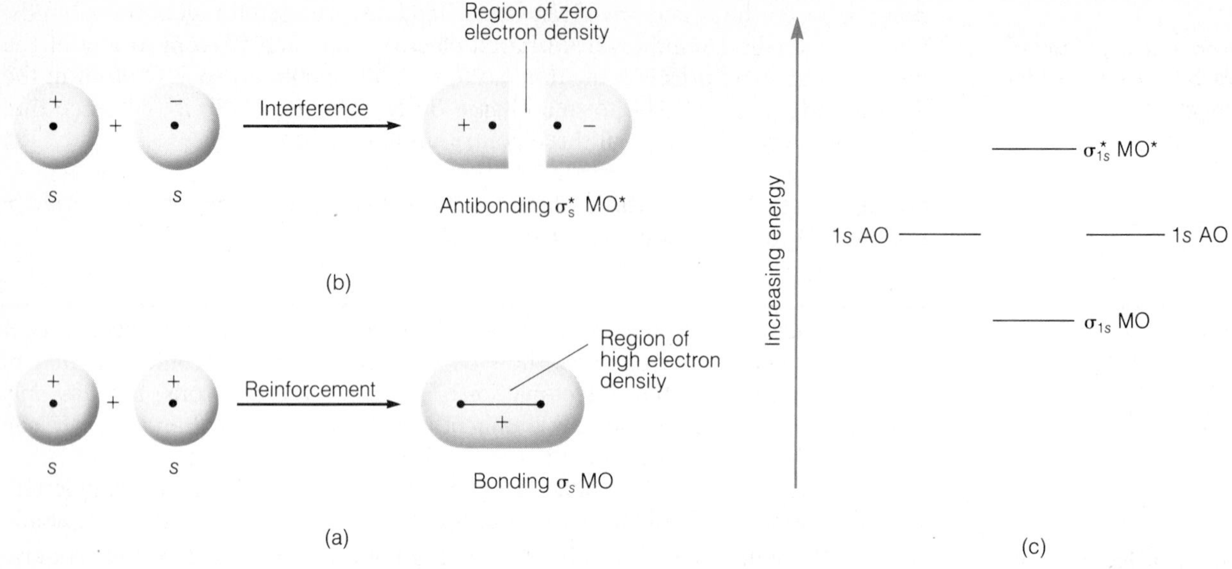

FIGURE 8.20
Overlap of two 1*s* AO's (H atoms in this case) to form (a) a low-energy bonding MO (of H_2) by reinforcement and (b) a high-energy antibonding MO* (also of H_2) by interference. (c) An energy-level diagram showing the combination of two AO's to form two MO's, one bonding MO and one antibonding MO*. In the actual mathematical treatment, the wave functions of the AO's are subtracted.

oversimplified but useful demonstration of these two ways is based on the standing wave properties of electrons. Standing waves can be combined in two ways. Two waves (section 5.1) can be combined so that their crests (designated positive with a plus sign) coincide. When they combine this way they reinforce each other because their amplitudes are added together. Such reinforcement, shown in Figure 8.19a, leads to a new wave with a larger amplitude. This reinforcement is analogous to the combination of two 1*s* AO's of like signs, as shown in Figure 8.20a for formation of hydrogen. The plus and minus signs in Figure 8.20 indicate the amplitude (section 5.1) of the wave functions of the atomic orbitals—*do not confuse them with electrical charges.* The combination of the two 1*s* AO's with plus signs (Figure 8.20a) is like the interaction of the crests of two stationary waves (Figure 8.19a).

Reinforcement leads to a concentration of electron density between the nuclei of the bonding atoms. Because the electrons are closer to the two nuclei, a molecular orbital formed this way has a *lower energy and is more stable than the atomic orbitals from which it is formed*—it is called a **bonding molecular orbital**, designated as **MO** (Figure 8.20c).

Standing waves can also combine crest-to-trough, as illustrated in Figure 8.19b, thereby canceling each other because their amplitudes are subtracted. Such an interference interaction produces no wave. Interference is like the combination of a "plus" 1*s* AO with a "minus" 1*s* AO (Figure 8.20b). This way of combining AO's leads to less electron density between the nuclei of the bonding atoms than would exist between the close but unbonded atoms. The molecular orbital formed this way has a *higher energy and is less stable than the atomic orbitals from which it is formed.* It is called an **antibonding molecular orbital** and is designated **MO*** (Figure 8.20c). The asterisk denotes antibonding.

Reinforcement is also called *in-phase interaction,* and interference, *out-of-phase interaction.*

The symbols for MO's and MO*'s include subscripts that denote the AO's used to generate the molecular orbitals. Thus, the molecular orbitals shown in Figure 8.20 are designated σ_{1s} and σ_{1s}^* since they are formed from two 1*s* atomic orbitals.

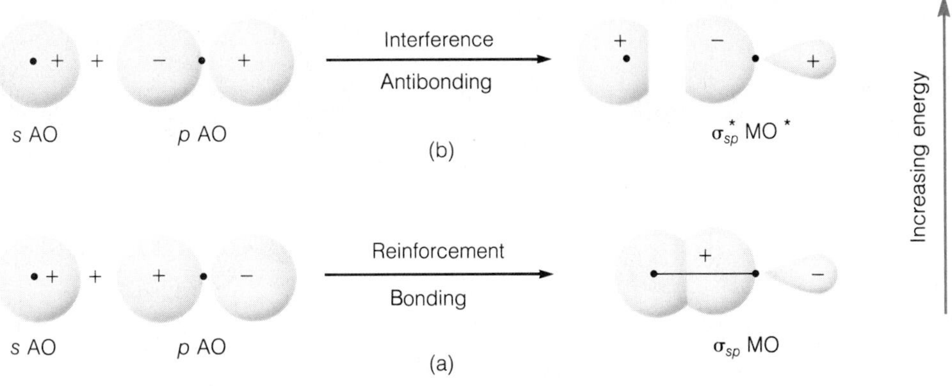

FIGURE 8.21
Sigma molecular orbitals formed from head-on overlap of an *s* and *p* AO. (a) Formation of a bonding σ_{sp} MO. (b) Formation of an antibonding σ_{sp} MO*.

The bonding and antibonding molecular orbitals are regions of space about the entire molecule. If the electrons in a molecular orbital were visible, at any given instant they could be found anywhere within the orbital. Molecular orbitals, like atomic orbitals, are subject to the Pauli exclusion principle—they can house a maximum of two electrons each, provided the electrons have opposite spins.

We saw in Figure 8.20 that the interaction of two AO's gives two molecular orbitals. Yet most bonding atoms have more than one AO. Then how do we know how many molecular orbitals to expect? The rule of the **conservation of orbitals** gives the answer: *The number of molecular orbitals formed must equal the number of atomic orbitals combined*:

***n* atomic orbitals ⟶ *n* molecular orbitals**

Do not get the impression from this discussion that antibonding molecular orbitals do not exist—they do. They represent quantized energy levels and are regions of space capable of housing a pair of electrons subject to the Pauli exclusion principle.

TYPES OF MOLECULAR ORBITALS

Recall from our earlier discussion that atomic orbitals can overlap head-on or side-to-side to give σ and π bonds, respectively. The σ and π designations are extended to molecular orbitals. A head-on combination generates σ-bonding MO's and σ^*-antibonding MO*'s. The combination of two *s* orbitals to give σ_s and σ^*_s molecular orbitals is shown in Figure 8.20a and b, respectively.

We use σ_{1s}, σ_{2s}, etc., to show the shell of the *s* orbitals used.

A head-on interaction of *p* AO's also affords sigma bonds. Since the *p* orbital has a node (section 5.1) it can be thought of as being like a stationary wave with a node. Therefore, a plus sign is assigned to one lobe (the crest) of *p* AO and a minus sign to the other lobe (the trough) as shown in Figure 8.21. (Figure 8.23 shows these assignments of signs to the stationary wave.) A head-on combination of a "plus" *s* AO with the "plus" lobe of a *p* AO gives a bonding σ_{sp} MO (Figure 8.21a). Combination of a "plus" *s* AO with the "minus" lobe of a *p* AO gives an antibonding σ^*_{sp} MO* (Figure 8.21b). When the lobes of *p* AO's with the same sign interact, a bonding σ_p MO results (Figure 8.22a). Interaction of lobes of unlike signs gives an antibonding σ^*_p MO* (Figure 8.22b).

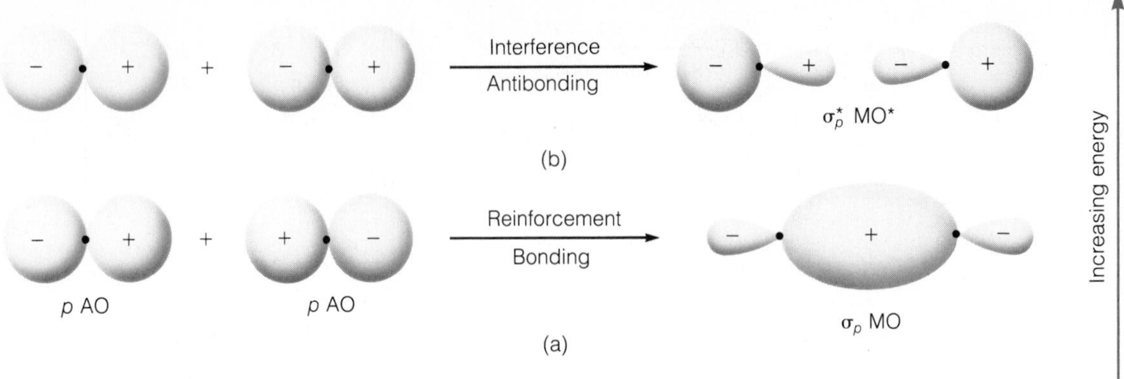

FIGURE 8.22
Sigma molecular orbitals formed from head-on overlap of two p AO's. (a) Formation of bonding σ_p MO. (b) Formation of antibonding σ_p MO*.

Two p AO's can combine side-to-side to give a π-bonding MO and a π^*-antibonding MO*. For a standing wave with a node, when crests (+) coincide and troughs (−) coincide there is reinforcement (Figure 8.23a). Similarly, when two p AO's are combined side-by-side so that the "plus" lobes overlap each other as do the "minus" lobes, a bonding π MO results (Figure 8.23c). When a crest coincides with a trough there is interference (Figure 8.23b). In terms of side-by-side overlap of two p AO's, an antibonding π MO* results when the "plus" lobes overlap with the "minus" lobes (Figure 8.23c). The π MO has a lower energy and the π^* MO* a higher energy than the individual p AO's (Figure 8.23d).

8.7 MOLECULAR ORBITAL ENERGY LEVELS

We described the electron configuration of an atom in Chapter 5 by listing the AO's with the number of electrons in each AO. For the ground state of an atom, the electrons are placed in AO's of the lowest possible energies. The electron configuration of a *molecule* is described in the same way by using *molecular* orbitals. We first summarize the procedure.

- Set up a series of molecular orbitals in order of increasing energies, as was done for combining two s AO's in Figure 8.20. Remember that the number of molecular orbitals equals the number of AO's combined.

- The number of electrons to be inserted is the sum of all the electrons of the bonded atoms. For ions, to this total add an electron for each negative charge and subtract an electron for each positive charge.

- Fill the molecular orbitals with electrons, remembering that (1) each molecular orbital can house up to two electrons (Pauli exclusion principle), and that (2) Hund's rule applies for molecular orbitals of equal energy; that is, each molecular orbital takes one electron, with parallel spin, before any pairing occurs.

Following this procedure gives the electron arrangement called the **ground state** of the molecule.

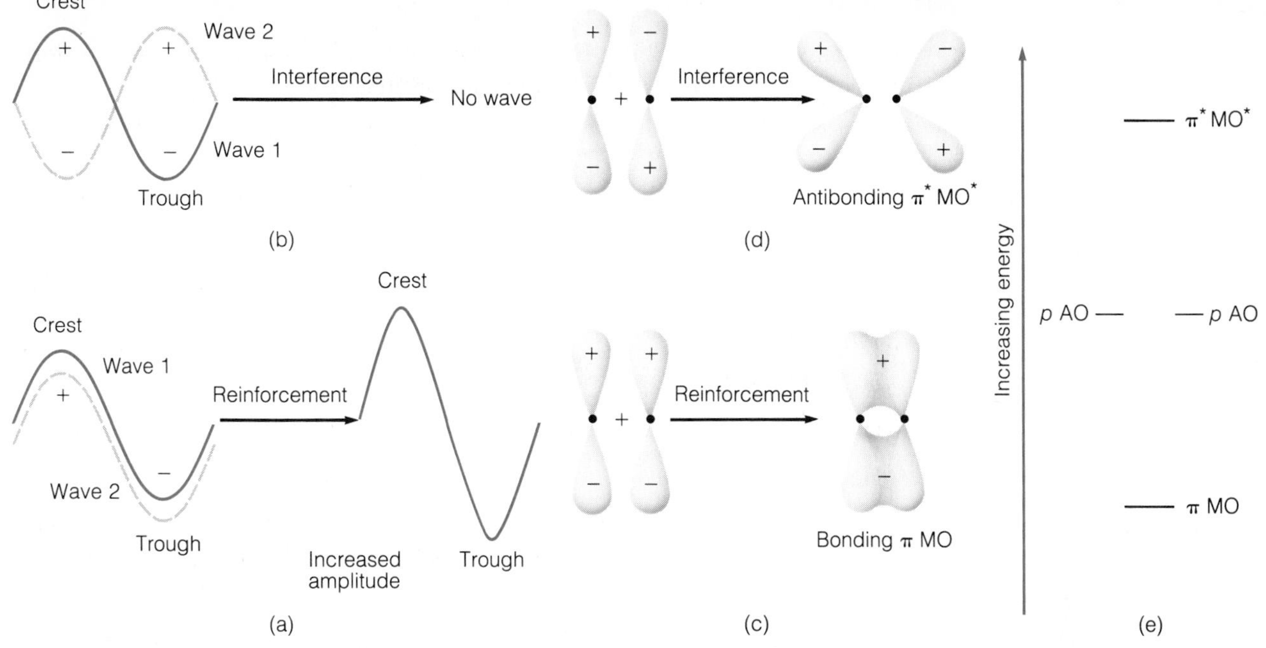

FIGURE 8.23
Combination of two standing waves, each having a node, (a) by reinforcement and (b) by interference. Related combination of two p AO's (c) by reinforcement to give a bonding π MO and (d) by interference to give an antibonding π^* MO*. (e) Relative energy levels of p AO's and π and π^* molecular orbital. (*Note*: The p AO's can be drawn with their "plus" lobes up or down.)

How can this procedure be used to predict the stability of molecules (charged and uncharged)? At the outset, we must recognize two important factors:

- *In order for atoms to be bonded, there must be more electrons in bonding MO's than there are in antibonding MO*'s.*
- *The more electrons there are in bonding MO's than in antibonding MO*'s, the stronger the bond is.* This is true because electrons in MO*'s are at higher energies, which reduces the bond stabilization effect of electrons in lower-energy MO's.

Stabilities of molecules can be qualitatively related to the **bond order**, which is defined as:

$$\textbf{bond order} = \frac{\textbf{number of } e^-\textbf{'s in MO's} - \textbf{number of } e^-\textbf{'s in MO*'s}}{\textbf{2}}$$

Fractional parts of bond orders exist in many resonance hybrids.

We apply this definition just to the valence electrons. Usually the bond order is the number of σ and π bonds between two atoms. Hence, bond orders are typically 1 for a single bond, 2 for a double bond, and 3 for a triple bond. The greater the bond order, the shorter is the bond distance and the greater is the bond energy (see Table 8.8). A 0 bond order means the bond has no stability—the two atoms are not bonded. In Example 8.10, these ideas are applied to diatomic molecules and ions of H and He.

EXAMPLE 8.10 (a) Use Figure 8.20c to show the electron distribution in H_2^+, H_2, H_2^-, and He_2. (b) Determine the bond orders of the species in each case. (c) Predict the relative stabilities of the four species.

ANSWER

(a) H_2^+ has 1 e^-, H_2 has 2 e^-, H_2^- has 3 e^-, and He_2 has 4 e^-. The distribution of electrons into the σ and σ^* molecular orbitals is

σ^*_{1s}	__	__	↑	↑↓
σ_{1s}	↑	↑↓	↑↓	↑↓
Species:	H_2^+	H_2	H_2^-	He_2

(b) The bond orders are

H_2^+ $(1 - 0)/2 = 0.5$

H_2 $(2 - 0)/2 = 1$

H_2^- $(2 - 1)/2 = 0.5$

He_2 $(2 - 2)/2 = 0$

(c) The order of bond stability is $H_2 > H_2^+ \approx H_2^- > He_2$. Since He_2 has a zero bond order, it does not exist. He_2 is actually less stable than two He atoms. ■

Note that H_2^+ and H_2^- (both free radicals, section 7.6) have odd numbers of electrons and so, characteristically, have fractional bond orders.

PROBLEM 8.9 Answer the questions in Example 8.10 for (a) He_2^+ and (b) HHe. □

The symbolism for representing the **MO electronic structure** of H_2^-, for example, is $(\sigma_{1s})^2(\sigma^*_{1s})^1$, and of He_2 is $(\sigma_{1s})^2(\sigma^*_{1s})^2$. This symbolism is similar to that used for AO's of atoms. Note that the MO concept predicts that both H_2^+ and H_2^- can exist, and indeed they both have been detected. Typical Lewis structures with shared pairs of electrons cannot be drawn for these ions. H_2^+ would need a one-electron bond—and where would you put the third electron of H_2^-?

We will now combine two *like atoms* of the second period of the periodic table to get *homonuclear diatomic* molecules (A_2) or ions (A_2^{n+} or A_2^{n-}). Since each atom has five atomic orbitals, $1s$, $2s$, $2p_x$, $2p_y$, and $2p_z$, we get ten molecular orbitals—again, the number of molecular orbitals is equal to the number of atomic orbitals. Five of the molecular orbitals are bonding and five are antibonding. Since the σ_{1s} and σ^*_{1s} molecular orbitals are created from inner-shell electrons, they have no effect on bonding and we do not bother with them. We are concerned only with the eight molecular orbitals formed by combining the valence-shell AO's, the $2s$ and the three $2p$'s. Figure 8.24 is an energy-level diagram showing the relative energies of these eight valence-shell molecular orbitals of homonuclear diatomic species. The AO's, listed by their relative energies, are combined to give a lower-energy MO and a higher-energy MO*. The energy levels illustrated in Figure 8.24a are used for diatomic molecules and ions from Li_2 to C_2. The levels in Figure 8.24b are used for N_2, O_2, F_2, and hypothetical Ne_2. There is only one difference. In Figure 8.24a, the π_y and π_z MO's have a lower energy than the σ_{2p} MO. These relative energy levels are switched in Figure 8.24b. Only the valence electrons are placed in the molecular orbitals shown in Figure 8.24a and b. We are not concerned with the $1s$ inner-shell electrons.

A small energy difference between the s and p orbitals leads to the arrangement shown in Figure 8.24a; a large difference leads to the arrangement in Figure 8.24b.

The π_y MO arising from the overlap of the $2p_y$ atomic orbitals has the same energy as the π_z MO resulting from the overlap of the $2p_z$ orbitals. Although

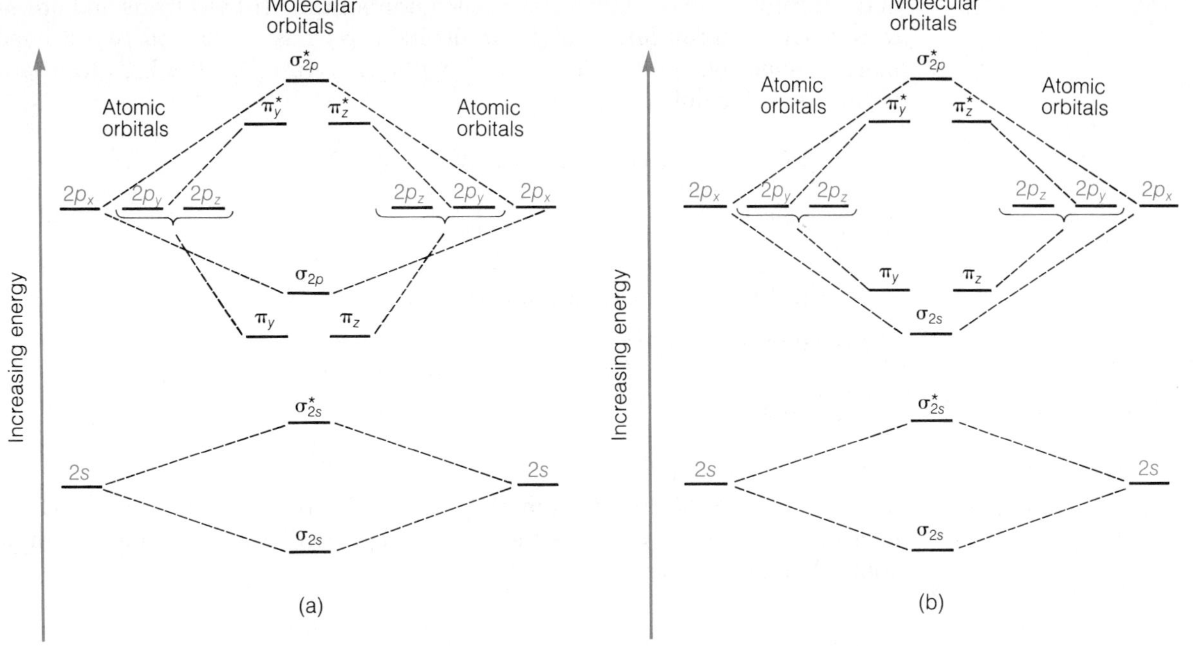

FIGURE 8.24 Energy-level diagrams showing energy levels of valence-shell molecular orbitals of diatomic molecules and ions arising by overlap of the valence-shell AO's of the second-period elements. (a) For Li, Be, B, N, and C; (b) for O, F, and (hypothetically) Ne.

these two π MO's have the same energy, they are not identical because they have different spatial orientations. The maximum electron density of the π_y orbital is in the xy plane, and that of the π_z orbital is in the xz plane. Likewise, the antibonding π_y^* and π_z^* molecular orbitals are equal in energy. Note that an antibonding MO* such as σ_{2s}^* formed from a lower-energy AO (2s) has *less* energy than a bonding MO such as σ_{2p} formed from a higher-energy AO (2p).

The method is now applied to Li_2 and N_2 by using Figure 8.24a and b.

Li_2

The number of valence electrons contributed by each Li· equals (atomic number − 2).

We first find the total number of valence electrons in Li_2. Each Li· has one 2s electron; therefore, Li_2 has 2 valence electrons (2 × 1) that must be assigned to the molecular orbital energy levels in Figure 8.24a. The 2 valence electrons are placed in the lowest-energy molecular orbital shown in Figure 8.24a. This is the σ_{2s} bonding MO, giving $\underline{\uparrow\downarrow}\ \sigma_{2s}$. Li_2 has a bond order of 1 and should be stable. The molecule has been detected in the vapor state of lithium. This MO-valence electronic configuration of Li_2 may be represented as $(\sigma_{2s})^2$.

Although MO theory accurately predicts that Li_2 can exist, it does not predict that Li_2 is the most stable form of lithium under ordinary conditions. In fact, lithium is most stable in the metallic state. (This will be discussed in section 22.3.)

N_2

$\pi_y\pi_z$ $\underline{\uparrow\downarrow}$ $\underline{\uparrow\downarrow}$ } bonding
σ_{2p} $\underline{\uparrow\downarrow}$
σ_{2s}^* $\underline{\uparrow\downarrow}$ } nonbonding
σ_{2s} $\underline{\uparrow\downarrow}$

Since each N has 5 valence electrons, 10 electrons must be assigned to the valence-MO energy levels to give the molecular orbital valence-electron configuration:

$$(\sigma_{2s})^2(\sigma_{2s}^*)^2(\sigma_{2p})^2(\pi_y)^2(\pi_z)^2$$

All 10 valence electrons appear in the Lewis structure. The $(\sigma_{2s})^2$ and $(\sigma_{2s}^*)^2$ cancel and make no net contribution to the bond energy. They are the unshared pairs in the formula. Unshared pairs are called **nonbonding (n) electrons** and are said to populate **nonbonding molecular orbitals**. N_2 has one σ and two π bonds; more specifically, the bonds are $(\pi_y)^2(\pi_z)^2(\sigma_{2p})^2$. We relate the MO designation to the Lewis formula as shown:

$(\sigma_{2s})^2(\sigma_{2s}^*)^2$ (nonbonding electrons)

:N≡N:

$(\sigma_{2p})^2(\pi_y)^2(\pi_z)^2$ (bonding electrons)

The bond order of N_2 is

$$\frac{8-2}{2} = 3$$

EXAMPLE 8.11 (a) Use Figure 8.24b to give the MO-valence electron configuration for O_2. (b) Find the bond order. (c) Does the MO concept predict the paramagnetism and double-bond character of O_2?

ANSWER

(a) The 12, (6 × 2), valence electrons are distributed as shown:

$\pi_y^*\pi_z^*$ antibonding
$\pi_y\pi_z$, σ_{2p} bonding
σ_{2s}^*, σ_{2s} nonbonding

$$(\sigma_{2s})^2(\sigma_{2s}^*)^2(\sigma_{2p})^2(\pi_y)^2(\pi_z)^2(\pi_y^*)^1(\pi_z^*)^1$$

In accordance with Hund's rule, the last two electrons are distributed, one each, into the equal-energy π_y^* and π_z^* MO*'s.

(b) $$\text{Bond order} = \frac{8-4}{2} = 2$$

(c) The MO concept predicts both the paramagnetism and the double-bond character of O_2. O_2 has a bond order of 2. In terms of the Lewis structure, this would be like having a double bond between the O atoms. The observed bond length in an O_2 molecule is indeed the value expected for a double bond and not for a single bond. The MO concept also predicts that the π_y^* and π_z^* each have a single electron. These unpaired electrons have the same spin. For this reason, O_2 is predicted to be paramagnetic—which it is. From this analysis it is evident that neither (a) :Ö=Ö: nor (b) :Ö̇—Ö̇: is an adequate formula for O_2. Structure a emphasizes the double-bond character, but overlooks the radical properties. Structure b emphasizes the radical character, but denies the double-bond character. ■

A molecule with 2 unpaired e^-'s is a diradical.

PROBLEM 8.10 (a) Use Figure 8.24a to get the correct MO valence electron configuration and bond order of B_2. (b) Use Figure 8.24b to get the incorrect MO valence electron configuration and bond order of B_2. (c) How could the determination of whether or not B_2 is paramagnetic permit you to decide which electron configuration is correct? □

Table 8.8 relates bond properties to bond orders of homonuclear diatomic molecules of the first- and second-period elements. Figure 8.24 can also be used to get

TABLE 8.8
BOND PROPERTIES OF SOME HOMONUCLEAR DIATOMIC MOLECULES AND IONS

MOLECULE	BOND ORDER	BOND LENGTH, nm	BOND ENERGY, kJ/mol
H_2	1	0.074	435
H_2^+	0.5	0.106	256
He_2	0	nonexistent	0
Li_2	1	0.267	104.6
Be_2	0	nonexistent	0
B_2	1	0.159	288.7
C_2	2	0.131	627.6
N_2	3	0.110	941.4
O_2	2	0.121	498.7
F_2	1	0.142	150.6
Ne_2	0	nonexistent	0

electron distributions and bond orders of diatomic molecules or ions of different (heteronuclear; AB, AB^+, or AB^-) second-period atoms. (The relative energy levels of the molecular orbitals are slightly different.)

The *l*owest-energy *u*noccupied *m*olecular *o*rbital is referred to as the **LUMO**.

Recall that when the first ionization energy of an atom is measured, the electron is lost from the atomic orbital with the highest energy. The same holds for molecules—the electron is lost from the *h*ighest-energy *o*ccupied *m*olecular *o*rbital. The (italicized) first letters give the name **HOMO** to this molecular orbital. We find that the first ionization energies of different molecules are related to the relative energies of their respective HOMO's. Generally, it takes less energy to lose an electron from a higher-energy HOMO than from a lower-energy HOMO. Thus, it takes less energy to remove an electron from an antibonding HOMO than from a bonding HOMO when both are in the same valence shell.

EXAMPLE 8.12 In terms of the MO valence electron configurations in Figure 8.24b, explain why the ionization energy for NO going to NO^+ is lower than that of N_2 going to N_2^+.

ANSWER For N_2, $(\sigma_{2s})^2(\sigma_{2s}^*)^2(\sigma_{2p})^2(\pi_y)^2(\pi_z)^2$, the HOMO from which the electron is lost, is π_y or π_z, a bonding MO. NO has one more electron than N_2, which is in (π_y^*). Consequently, in NO the electron is lost from π_y^*, which is a higher-energy antibonding MO*. We use the MO concept to explain the fact that NO^+ forms more easily than N_2^+ by suggesting that the electron lost by NO comes from a higher-energy antibonding MO*, whereas the electron lost by N_2 comes from a lower-energy bonding MO. ■

PROBLEM 8.11 Does CN or O_2 have the higher ionization energy? □

For diatomic molecules the MO concept is compatible with the writing of Lewis structures, as we saw for N_2 (page 276). However, a problem arises with molecules of more than two atoms. After all, molecular orbitals encompass the entire molecule, and electrons in molecular orbitals can move about the entire molecule. Yet when we draw a bond in a Lewis structure, we say that that bond shows where the electrons are. Fortunately, the problem is solvable. Even though the electrons can roam about the molecule, they are most likely to be found between pairs of atoms. This means that electrons in typical sigma and pi bonds are assumed to be **localized** in the bond as drawn in the Lewis structure. Exceptions are π bonds in species with delocalized π bonding (section 8.4).

SUMMARY

The **valence bond** (VB) concept assumes that a covalent bond results from overlap of an atomic orbital of each of the bonding atoms. Head-on **orbital overlap** of *p* or *s* AO's provides a **sigma (σ) bond**, in which the electron density is distributed symmetrically about the bond axis. This enables either atom with all of its attached groups and unshared electrons to rotate about the **bond axis**. Side-to-side overlap of *p* (or *d*) AO's provides a **pi (π) bond**, in which the electron density is mainly in the plane of the *p* AO's used, thus preventing rotation about the bond axis. The smaller the bond length, the greater is the orbital overlap and the stronger is the bond.

The concept of **hybridization** accounts for the formation of identical bonds to the central atom in molecules with at least three atoms and for the observed bond angles. **Hybrid orbitals (HO's)** result from a blending of the energies of some set of AO's. Some HO types are sp^3 (from an *s* and three *p*'s), sp^2 (from an *s* and two *p*'s), *sp* (from an *s* and a *p*), sp^3d^2 (from an *s*, three *p*'s, and two *d*'s), and sp^3d (from an *s*, three *p*'s, and one *d*). For *s* and *p* AO's and *s-p*-type HO's (those without *d*'s), the more *s* character in the orbital, the lower is the energy of the electrons in the orbital and the closer are the electrons to the nucleus. The number of HO's needed by a central atom, the **hybrid orbital number** (HON), equals the number of its σ bonds plus the number of its unshared pairs of electrons. All bonds between two atoms in excess of one are π bonds formed from unhybridized *p* AO's. HO's are often used instead of pure *p* AO's, because stronger bonds with larger bond angles (less repulsion) are formed.

Some species have two or more plausible Lewis structures that differ only in the position of multiple bonds and unshared pairs of electrons. These structures do not represent isomers because they have the same arrangement of atoms. In such cases the **resonance concept** holds that none of these **contributing (resonance) structures** is correct in itself. However, taken together, they help describe the correct structure (**resonance hybrid**). Bonding in resonance hybrids can be pictured in terms of π bonding extending over more than two atoms. The **delocalization** of the electrons in an **extended π system** stabilizes the resonance hybrid, causing it to have less energy and more stability than any unreal contributing structure in which the π electrons are localized.

The **molecular orbital (MO) concept** assumes that AO's are combined to give **molecular orbitals** that extend over the entire molecule. We combine AO's of bonding atoms with each other in one of two ways. Two AO's with like signs (+ and + or − and −) interact to give a **bonding MO** lower in energy than the AO's. Two AO's with unlike signs (+ and −) interact to give an **antibonding MO*** higher in energy than the AO's. There is a **conservation of orbitals**; that is, the number of molecular orbitals formed equals the number of combined AO's. Molecular orbitals are also designated σ and π. The molecular orbitals of molecules are filled in order of increasing energy following the Pauli exclusion principle and Hund's rule. For bonding to occur, there must be more electrons in MO's than there are in MO*'s. For diatomic molecules, the greater the excess of electrons in MO's over those in MO*'s, as indicated by the **bond order**, the stronger is the bond.

HYBRIDIZATION AND SHAPE

8.12 Explain what is meant by orbital overlap.

8.13 What form of hybridization is associated with these shapes: trigonal-planar, linear, tetrahedral, octahedral, trigonal-bipyramidal?

8.14 Compare the shape of a *p* AO with that of an *sp* HO. Which one has a more directionally concentrated electron density? Which would you expect to form the stronger bond?

8.15 Justify these statements: (a) A bond formed from an *sp*-type hybrid AO is stronger than one formed from a *p* AO. (b) The presence of unshared pairs of electrons on a central atom reduces the bond angle to the central atom. (c) Except for the noble gases, unbonded atoms do not naturally exist in the ground state. (d) All atoms using hybridized AO's are in a bonded state. (e) Within a given group of the periodic table, the tendency to use pure *p* AO's rather than hybridized AO's increases as the size of the atom increases. (*Hint*: Think in terms of the effect of bond length on electron pair repulsion.) (f) The idea of delocalized π bonding is applied when we draw the structure of a species for which a single Lewis formula cannot be drawn.

8.16 (a) What kind of HO's are used by the central atom in each of these covalent species? (1) $CHCl_3$, (2) BeH_2, (3) NCl_3, (4) ClO_3^-, (5) IF_6^+, (6) SiF_6^{2-}. (b) Give the shape of each species.

8.17 What kind of hybrid orbitals are used by the underlined atoms in $\underline{C}_2Cl_4$, $\underline{C}_2Cl_2$, $\underline{N}_2F_2$, and $H_2\underline{N}CO$?

8.18 (a) Describe the hybridization of N in NO_2^+ and NO_2^-. (b) Predict the bond angle in each case.

8.19 After comparing experimental and calculated dipole moments, Charles A. Coulson suggested that the Cl atom in HCl is *sp* hybridized. (a) Give the orbital electronic structure for an *sp*-hybridized Cl atom. (b) Which HCl molecule would have a larger dipole moment—one in which the chlorine uses pure *p* orbitals for bonding with the H atom or one in which *sp*-hybrid orbitals are used?

8.20 Describe the orbitals (*s*, *p*, sp^2, and so on) of the central atom used for bond orbitals and unshared pairs in (a) H_2S (bond angle 91°); (b) CH_3OCH_3 (C—O—C angle 110°); (c) S_8 (S—S—S angle 105°); (d) $(CH_3)_2Sn^{2+}$ (C—Sn—C angle 180°); (e) $(CH_3)_3Sn^+$ (planar with respect to the Sn and 3 C atoms); (f) $SnCl_2$ (angle 95°).

8.21 A water solution of cadmium bromide, $CdBr_2$, contains not only Cd^{2+} and Br^-, but also $CdBr^+$, $CdBr_2$, $CdBr_3^-$, and $CdBr_4^{2-}$. Give the type of hybrid orbital used by Cd in each polyatomic species and describe the shape of the species.

8.22 In their crystalline states, PCl_5 exists as $(PCl_4)^+(PCl_6)^-$ and PBr_5 exists as $PBr_4^+Br^-$. (a) Predict the shapes of all the polyatomic ions. (b) Indicate the electronic orbital structure for the P atom in each of its different types of ions.

RESONANCE AND DELOCALIZED π BONDING

8.23 With the aid of an example, explain the difference between a contributing (resonance) structure and a resonance hybrid.

8.24 Write acceptable contributing structures with formal charges for nitryl chloride, $ClNO_2$.

8.25 State whether each of these contributing structures is reasonable, impossible, or possible but with a very high energy. Explain your choices. (a) $(:\underset{\cdot\cdot}{\ddot{O}}—\ddot{N}=\ddot{O}:)^-$ for NO_2^-; (b) $(:\ddot{O}=\ddot{N}=\ddot{O}:)^-$ for NO_2^-; (c) $:\ddot{O}—\ddot{N}=\ddot{O}:$ for NO_2; (d) $:\underset{\cdot}{\ddot{O}}—\ddot{N}=\ddot{O}:$ for NO_2; (e) $:\underset{\cdot\cdot}{O}=\underset{\cdot}{N}$ for $:\underset{\cdot\cdot}{O}=\underset{\cdot}{N}:$; (f) $:C=\ddot{O}:$ for CO; (g) $:\ddot{O}=C—\underset{\cdot\cdot}{\ddot{O}}:$ for CO_2.

8.26 For each of these pairs explain why the second contributing structure is not as permissible as the first one.

(a) $H_2C=CH_2$, $H_2\overset{\ominus}{\ddot{C}}—\overset{\oplus}{C}H_2$; (b) $:\overset{\ominus}{C}\equiv\overset{\oplus}{O}:$, $:C=\ddot{O}:$;

(c) $H—\ddot{N}=\overset{\oplus}{N}=\overset{\ominus}{\ddot{N}}:$, $H—\overset{\oplus}{N}\equiv\overset{\oplus}{N}—\overset{2-}{\underset{\cdot\cdot}{\ddot{N}}}:$; (d) $(H—C\equiv\overset{\oplus}{O}:)^+$, $(H—\overset{\oplus}{C}=\ddot{O}:)^+$.

8.27 In terms of resonance, account for (a) these differences in C—F bond length: 0.138 nm in $H_3C—F$ and 0.127 nm in F_3C^+; and for (b) why BF_3 is stable, whereas BH_3 is not.

8.28 Write a contributing structure for sulfur trioxide, SO_3 (S is the central atom bonded to each of three terminal O atoms), in which the S atom has a formal charge of (a) 0, (b) +1, (c) +2, and (d) +3. Give the formal charge on each O atom in each of the contributing structures. Which structure would not be permissible (make insignificant contribution)? Explain your answer.

8.29 In terms of delocalized π bonding, explain why the carbocation $H_2C=CH\overset{\oplus}{C}H_2$ is more stable than $CH_3CH_2\overset{\oplus}{C}H_2$.

8.30 Show that butadiene, $\overset{H}{\underset{H}{}}\!\!>C=C<\!\!\overset{H}{\underset{C=C}{}}$ — structure: H(H)C=CH—CH=C(H)H — is a molecule with delocalized π bonding.

MO THEORY AND DIATOMIC SPECIES

8.31 In terms of overlapping AO's describe the three ways a σ MO can be formed.

8.32 Explain why (a) electrons in filled shells have no effect on bonding strengths; (b) there is relatively free rotation about the C—C bond of $H_3C—CH_3$ but not about the C=C bond of $H_2C=CH_2$.

8.33 Compare and illustrate the differences between (a) atomic orbitals and molecular orbitals, (b) bonding and antibonding molecular orbitals, (c) σ bonds and π bonds, and (d) localized and delocalized molecular orbitals.

8.34 Draw an MO energy-level diagram for Cl_2 using only the $3s$ and $3p$ AO's and the valence electrons.

8.35 (a) Give the MO designations for O_2, O_2^-, O_2^{2-}, O_2^+, and O_2^{2+}. (b) Match these species with these observed bond lengths, in nm: 0.104, 0.112, 0.121, 0.133, and 0.149. (c) Give the bond order in each case.

8.36 (a) Give the MO designation for (1) Be_2, (2) F_2, (3) HeH^+, (4) OF, (5) Ne_2. (b) Which of these species is/are unlikely to exist? Explain.

8.37 Which of these species would you expect to be paramagnetic? (a) He_2^+, (b) NO, (c) NO^+, (d) N_2^{2+}, (e) CO^{2-}, (f) F_2^+.

8.38 Explain in terms of MO theory (Figure 8.24) why the first ionization energy of nitric oxide (NO) is less than that of CO.

8.39 Explain why the ionization energy of H_2 is greater than that of H, whereas the ionization energy of O_2 is less than that of O.

8.40 To increase the strength of the bond in BO, would you add or subtract an electron? Explain your answer with the aid of an MO electron structure.

8.41 Which of the diatomic molecules of the first ten elements of the periodic table, H to Ne, are (a) bonded only by π bonds with no σ bond, (b) paramagnetic?

8.42 In terms of the MO theory, choose the species in each pair that has the greater electron affinity. (Consult Figure 8.24.) (a) OF, NF; (b) O_2^{2+}, N_2^{2+}; (c) CN, NO. If you cannot choose, state why you cannot.

8.43 Give the distribution of electrons in the molecular orbitals of BN if π_y, π_z, and σ_{2p} have the same energy.

8.44 When considering the first *three* ionization energies of an O_2 molecule, where would you expect to find the largest jump in energy?

8.45 An excited electron can go from an occupied lower-energy MO (or MO*) to an unoccupied higher-energy molecular orbit, often an MO*. Draw an MO energy diagram for excited He_2. Explain why He_2 does not exist whereas excited He_2 has been detected.

8.46 Figure 8.18 shows the delocalized π system of NO_3^- with four p AO's. According to the MO concept, the combination of four p AO's results in four π-type molecular orbitals. By using plus and minus signs for the lobes of the p AO's, show the origin of these π-type molecular orbitals by finding the four ways of getting interactions between lobes of like signs and opposite signs. (b) List the four π-type molecular orbitals in order of increasing energy and indicate which two are bonding and which two are antibonding. (*Hint*: When the molecular orbital has an excess of overlaps between lobes of like signs, it is bonding. Furthermore, the more like-sign overlaps there are, the lower is the energy of the molecular orbital, and the more unlike-sign overlaps there are, the higher is the energy of the molecular orbital.) (c) Which of your four π-type molecular orbitals is the one represented in Figure 8.18?

SELF-TEST

8.47 (10 points) True or false? (a) Electrons are never found in an antibonding MO*. (b) All antibonding MO*'s in a molecule have a higher energy than the bonding MO's. (c) σ bonds are usually formed by head-on overlap of atomic orbitals. (d) Free rotation can occur about an occupied bonding π molecular orbital. (e) It always takes less energy to remove an electron from an antibonding MO* than from a bonding MO of the same shell.

8.48 (10 points) (a) Give the symbols and spatial orientations of the three s-p-type HO's. (b) Give their relative energies with respect to pure s and pure p AO's. (c) Give two reasons for an atom's using hybrid rather than pure p AO's. (d) Give the symbols and spatial orientations of two HO's involving d orbitals.

8.49 (12 points) (a) Which HO's are used by N in each of these species: (1) $:NH_3$, (2) NH_4^+, (3) $H\ddot{N}{=}\ddot{N}H$, (4) $HC{\equiv}N:$. (b) Give an orbital description for each species, specifying the location of any unshared pairs and the orbitals used for the multiple bonds.

8.50 (18 points) Draw the Lewis structures and predict the orbital types and the shapes of these polyatomic ions and covalent molecules: (a) $HgCl_2$, (b) BF_3, (c) BF_4^-, (d) SeF_2, (e) $AsCl_5$, (f) SbF_6^-.

8.51 (10 points) Write four valid contributing structures for sulfuric acid, $(HO)_2SO_2$ (condensed structural formula). Remember that S can expand its valence shell. Show all formal charges that may exist.

8.52 (12 points) (a) What is the hybridized state of each C in these molecules? (1) $Cl_2C{=}O$; (2) $HC{\equiv}N$; (3) CH_3CH_3; (4) ketene, $H_2C{=}C{=}O$. (b) Describe the shape of each molecule.

8.53 (12 points) For each of the species (1) H_2, (2) He_2^+, (3) He_2, and (4) He_2^-, (a) write the MO electron structure and (b) indicate the bond order. (c) Give the relative stabilities of these species.

8.54 (8 points) (a) Using Figure 8.24a, give the molecular orbital structure for C_2. (b) Should C_2 be paramagnetic? (c) What is the bond order?

8.55 (8 points) Describe the structures of (a) $\left[H{-}C\begin{matrix}{\overset{..}{O}}: \\ \ddot{O}:\end{matrix}\right]^-$ (formate ion) and (b) SO_2 in terms of overlapping p AO's. Draw a structural formula indicating the delocalized π bonding.

9

SOME TYPES OF CHEMICAL REACTIONS

In previous chapters we have described atoms, ions, and molecules and explained how they are bonded to each other in ionic and covalent compounds. This chapter deals mainly with compounds that dissolve in water to form solutions that contain ions. We will examine how these ions participate in chemical changes. The chapter will also introduce you to some types of chemical reactions that do not occur in solution, as well as to a scheme that can be used for classifying many reactions.

Many of the solution reactions studied in this chapter provide the basis for analyzing substances by volumetric analysis (section 3.11) and gravimetric analysis, which is done entirely by weighing.

9.1 ACIDS AND BASES

Acids and bases are as common in daily life as they are in chemistry laboratories. We encounter acids in car batteries (sulfuric acid), vinegar (acetic acid), vitamin C (ascorbic acid), aspirin (acetylsalicylic acid), citrus fruits (citric acid), grapes (tartaric acid), and even in the human stomach (hydrochloric acid). We encounter bases in household drain and oven cleaners (sodium hydroxide), window cleaners (ammonia water, ammonium hydroxide), and milk of magnesia (magnesium hydroxide).

Recall that the proton was discovered in hydrogen gas at very low pressures in electrical discharge tubes (Chapter 2).

Two ions are especially important because their presence in water makes the resulting solutions either "acids" or "bases." One of these is the hydrogen ion. When it is present in water in sufficient concentration it forms an *acidic solution.* Strictly speaking, a hydrogen ion, H^+, being a hydrogen atom without the electron, is nothing more than a proton. However, because a proton is very small in comparison with any atom, its positive charge is highly concentrated in its tiny volume. As a result, protons do not exist in chemical environments under ordinary conditions. Instead, they attach themselves to other molecules. In water, a proton attaches itself to a water molecule, forming $[H(H_2O)]^+$, more simply written as H_3O^+ and called a **hydronium ion**.

Most protons in water are believed to be associated with four water molecules, $[H(H_2O)_4]^+$ or $H_9O_4{}^+$, a formula too cumbersome for everyday use. In fact, all ions are hydrated in water solutions.

The other important ion is the **hydroxide ion**, OH^-. A *basic solution* is produced when a substance dissolved in water yields OH^- ions.

Reactions of an acid with a base in water constitute a large class of reactions involving ions and molecules. However, before we can discuss these reactions, we must first learn to classify compounds as either acids or bases and, furthermore, as either weak or strong electrolytes or nonelectrolytes.

EARLY CONCEPTS OF ACIDS AND BASES

Seventeenth-century chemists did not know about the structure of molecules. Instead, they classified substances according to their chemical and physical properties. Their early definitions of acidic and basic substances or their water solutions illustrate this. The early chemists knew that vinegar and lemon juice have a sour taste. They also knew that these sour substances change the colors of certain natural dyes. You, too, may have noticed that lemon juice changes the color of tea from brown to yellow. Vinegar added to tea does the same thing. The colored substance (tannic acid) in tea that changes color is an **indicator**. A common indicator is litmus, the reddish blue dye of "red" cabbage that turns red in vinegar. When a metal such as zinc or magnesium is added to vinegar, hydrogen is released:

Juices of such fruits as cherries, purple grapes, and blueberries are indicators. Try them with vinegar (acid) and soapy water (base).

$$\underset{\text{acetic acid (active ingredient in vinegar)}}{2HC_2H_3O_2(aq)} + Mg(c) \longrightarrow Mg(C_2H_3O_2)_2(aq) + H_2(g)$$

Early chemists described solutions with these properties—sour taste, turn blue litmus red, react with some metals to release H_2—as **acidic**.

Pure (99.5%) acetic acid is called *glacial* acetic acid (freezing point 16.0°C), because it freezes in cold laboratories.

Some other solutions, such as soapy water, feel slippery, taste bitter, and change red litmus to blue. Solutions of washing soda (sodium carbonate) and lye (sodium hydroxide) also have these properties. Such solutions were described as **basic**. Acidic and basic solutions have long been known to react with each other to give a salt and water:

Safety Note: Lye is a deadly poison. Since household drain and oven cleaners contain lye, such products should be kept in locked cabinets that are absolutely unreachable by children.

$$\underset{\substack{\text{acetic acid}\\ \text{(an acid)}}}{HC_2H_3O_2(aq)} + \underset{\substack{\text{sodium hydroxide}\\ \text{(a base)}}}{NaOH(aq)} \longrightarrow \underset{\substack{\text{sodium acetate}\\ \text{(a salt)}}}{NaC_2H_3O_2(aq)} + \underset{\text{(water)}}{H_2O}$$

ELECTROLYTES AND NONELECTROLYTES

In the late 1800s, chemists began to wonder about the structural features that account for the observed behavior. They realized that all acids have at least *one H atom per molecule.* In 1887, Svante Arrhenius, the father of the ionization theory, suggested how these H atoms are related to acidity. The clue for Arrhenius came from conductivity measurements, which can be demonstrated by using the kind of apparatus shown in Figure 9.1. When the metal rods are dipped in a *benzene* solution of hydrogen chloride (HCl) or acetic acid ($HC_2H_3O_2$), the light bulb does *not* glow. These solutions are not conductors of electricity. When dilute *water* solutions of (about 0.01 M to 1.0 M) HCl and $HC_2H_3O_2$ are tested, the bulb does glow. A substance whose aqueous solution conducts an electric current is called an **electrolyte**. For example, a 0.5 M solution of HCl causes the bulb to glow brightly; thus HCl is a **strong electrolyte**. NaCl, KF, H_2SO_4, and HNO_3 are typical strong electrolytes. A 0.5 M solution of $HC_2H_3O_2$ causes the bulb to glow only dimly; acetic acid is a **weak electrolyte**. $HC_2H_3O_2$, HCN, HF, and NH_3 are typical weak electrolytes. However, most molecular compounds are **nonelectrolytes**, compounds that have no measurable tendency to form ions in water. Ethyl alcohol (ethanol), C_2H_5OH, and the sugars glucose, $C_6H_{12}O_6$, and sucrose, $C_{12}H_{22}O_{11}$, are typical nonelectrolytes (Table 9.1).

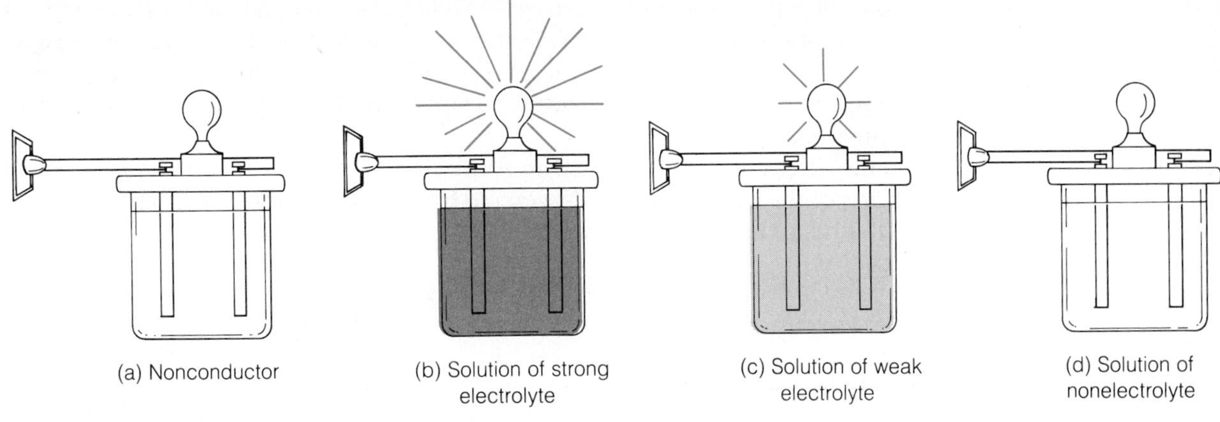

FIGURE 9.1
Testing solutions for conductivity. (a) The light bulb does not glow when the rods are dipped in hydrogen chloride (HCl) or ammonia (NH_3) dissolved in benzene. (b) The bulb glows brightly when the rods are dipped in a 0.5 M aqueous solution of hydrochloric acid (a strong electrolyte). (c) The bulb glows dimly when the rods are dipped in a 0.5 M aqueous solution of acetic acid, $HC_2H_3O_2$ (a weak electrolyte). (d) The bulb does not glow when the rods are dipped in 0.5 M aqueous solution of glucose, $C_6H_{12}O_6$ (nonelectrolyte). The molarities are immaterial, provided they are the same for each solution.

ELECTROLYTES AND STRENGTH OF ACIDS AND BASES

First, *note that conductivities are compared by using aqueous solutions with the same molarity*. Using this knowledge, Arrhenius reasoned as follows: hydrogen chloride, HCl, is a molecule that remains unchanged when it is dissolved in benzene. This solution does not conduct electricity because there are no ions present to conduct the electric current. To explain the conductivity of the water solution, Arrhenius assumed that the HCl molecules ionize to give H^+ and Cl^- ions; these ions conduct the electric current. Furthermore, he said that *the H^+ ions impart the familiar acidic properties to the solution.* This water solution of HCl is called *hydrochloric acid*. Any substance that furnishes H^+ ions when dissolved in water is an **acid**.

Strong electrolytes are almost completely ionized and produce many more ions than do weak electrolytes of the same molarity. This difference is illustrated by comparing the ionization equations for HCl and $HC_2H_3O_2$:

$$HCl \rightleftharpoons H^+ + Cl^- \quad \text{(strong electrolyte, almost completely ionized, strong acid)}$$

$$HC_2H_3O_2 \rightleftharpoons H^+ + C_2H_3O_2^- \quad \text{(weak electrolyte, little ionized, weak acid)}$$

TABLE 9.1
ELECTROLYTIC PROPERTIES OF TYPICAL AQUEOUS SOLUTIONS

STRONG ELECTROLYTES			
Ionic Compound	Molecular Compound	WEAK ELECTROLYTES	NONELECTROLYTES
NaCl	HCl	NH_3 (ammonia)	C_2H_5OH (ethanol)
MgF_2	HBr	$HC_2H_3O_2$ (acetic acid)	$C_{12}H_{22}O_{11}$ (sucrose)
KOH	HI	H_2CO_3 (carbonic acid)[†]	$CO(NH_2)_2$ (urea)
$Al_2(SO_4)_3$	HNO_3	HNO_2 (nitrous acid)	CH_2O (formaldehyde)
$Cu(NO_3)_2$	H_2SO_4	H_2SO_3 (sulfurous acid)[§]	CH_3OH (methanol)
$Ca(OH)_2$	$HClO_4$	H_2S (hydrogen sulfide)	$CO(CH_3)_2$ (acetone)

[†] Solution of CO_2 in water.
[§] Solution of SO_2 in water.

Arrows of different lengths are used to indicate strong and weak electrolytes. A longer left-to-right arrow ($\rightleftharpoons$) signifies a strong electrolyte, while a longer right-to-left arrow ($\rightleftharpoons$) signifies a weak electrolyte. In fact, for strong electrolytes, the tendency to reform molecules is so small that we may omit the short arrow for the right-to-left reaction.

Since H^+ combines with H_2O to form H_3O^+, the ionization reaction of acids consists of the transfer of a hydrogen ion from the acid molecule to H_2O, forming H_3O^+ and an anion (a negative ion):

$$\underset{\text{hydrochloric acid}}{HCl(g)} + H_2O(\ell) \longrightarrow H_3O^+(aq) + Cl^-(aq)$$

(*strong electrolyte*, almost completely ionized, little HCl remains, *high* H_3O^+ concentration, *strong acid*)

$$\underset{\text{acetic acid}}{HC_2H_3O_2(aq)} + H_2O(\ell) \rightleftharpoons H_3O^+(aq) + C_2H_3O_2^-(aq)$$

(*weak electrolyte*, little ionized, most of $HC_2H_3O_2$ remains, *low* H_3O^+ concentration, *weak acid*)

The symbol $H^+(aq)$ or H^+ is commonly used to represent H_3O^+ when the role of water is not emphasized.

Since the acidity of a solution is determined by its concentration of H^+, the strength of an acid depends on how much H^+ is formed from the molecules. Thus HCl is a typical *strong acid* because it is a *strong electrolyte*; its molecules ionize almost completely. $HC_2H_3O_2$ is a typical *weak acid* because it is a *weak electrolyte*; relatively few of its molecules are ionized.

The acidity of carbonated water is due to the reaction of carbon dioxide with water forming carbonic acid, a weak acid:

$$CO_2(g) + H_2O \rightleftharpoons H_2CO_3(aq) \rightleftharpoons H^+(aq) + HCO_3^-(aq)$$

Such solutions are known by a variety of common names, including seltzer, soda water, carbonated water, and carbonic acid.

During Arrhenius's time, all bases were known to possess the hydroxyl group (OH) and to conduct a current when dissolved in water. For example, aqueous solutions of NaOH and KOH are good conductors of electricity. The water solution of ammonia, then written as NH_4OH, is a poor conductor of electricity. He therefore assumed that a **base** is a substance that ionizes in water to produce OH^- ions and positive ions:

$$NaOH(aq) \longrightarrow Na^+(aq) + OH^-(aq)$$

The OH^- ions impart the basic properties to the solution. The difference between NaOH, a strong electrolyte, and NH_4OH, a weak electrolyte, results from the difference in the extent of ionization:

$$NaOH(aq) \longrightarrow Na^+(aq) + OH^-(aq)$$

(*strong electrolyte*, almost completely ionized, *high* OH^- concentration, *strong base*)

$$NH_4OH(aq) \rightleftharpoons NH_4^+(aq) + OH^-(aq)$$

(*weak electrolyte*, little ionized, most of NH_4OH remains, *low* OH^- concentration, *weak base*)

Thus strong electrolytes like HCl and NaOH are designated strong acids and strong bases. Weak electrolytes like $HC_2H_3O_2$ and NH_4OH are called weak acids and weak bases.

FIGURE 9.2
A tank of liquid ammonia.

A water solution of ammonia is better represented by

$$NH_3(aq) + H_2O \rightleftharpoons \underset{\text{ammonium hydroxide}}{NH_4^+(aq) + OH^-(aq)}$$

(weak electrolyte, little ionized, most of NH_3 remains, *low* OH^- concentration, weak base)

A commercial use of NH_3 is illustrated in Figure 9.2. See also Table 9.2.

While strong electrolytes such as HCl(g) are molecular compounds, strong electrolytes such as NaOH(c) and NaCl(c) are ionic compounds. They consist of positive and negative ions held together by strong electrostatic forces (Chapter 7). Thus, the ions are not free to move in the solid state. However, when the crystals are dissolved in water, the ions are liberated so that they are free to move throughout the solution.

The process of solution is examined in more detail in Chapter 12.

TABLE 9.2
U.S. PRODUCTION OF IMPORTANT INDUSTRIAL CHEMICALS, billions of pounds

H_2SO_4	73.6	HNO_3	13.1
N_2	48.6	CO_2(c and ℓ)	8.5
CaO (lime)	30.3	HCl	6.0
O_2	33.0	K_2CO_3 (potash)	2.6
NH_3	28.0	C_2H_4	32.8
NaOH (lye)	22.0	CH_3OH	7.3
Cl_2	21.0	$C_2H_6O_2$ (antifreeze)	4.8
H_3PO_4	18.4	$HC_2H_3O_2$	2.9
NH_4NO_3, $(NH_4)_2SO_4$	15.6	C_2H_5OH	1.1
Na_2CO_3 (soda ash)	17.2		

Source: *Chemical and Engineering News*, 13 April 1987, p. 21.

PROBLEM 9.1 Label each substance as either a strong or weak electrolyte, a nonelectrolyte, a strong or weak acid, or a strong or weak base in water.

HI (strong acid/______________)
$HClO_2$ (weak electrolyte/______________)
RbOH (strong base/______________)
CsOH (good electrical conductor/______________)
HNO_2 (weak acid/______________)
$C_{12}H_{22}O_{11}$ (nonconductor/______________)
$Ba(OH)_2$ (strong electrolyte/______________)
CH_3NH_2 (weak base/______________)
HNO_3 (strong electrolyte/______________)
H_2S (poor conductor/______________) □

Broader and more general concepts of acids and bases are discussed in Chapter 14.

ACID–BASE NEUTRALIZATION REACTION

When an acidic solution like HCl(aq) is added to a basic solution like NaOH(aq) in the right proportions, the properties of each disappear, are "neutralized." The new solution, after mixing, is no longer sour (acidic) or bitter (basic). Instead the new solution, NaCl(aq), has a salty taste.

The tastes of many salts are known because early chemists tasted everything they made and recorded the results. Many of those chemists poisoned themselves that way. *Do not imitate them.*

The reaction between an acid and a base that forms the solvent and a salt is called **neutralization**. In water, the neutralization reaction is the formation of the solvent water and a salt as shown:

$$\underset{\text{acid}}{HCl(aq)} + \underset{\text{base}}{NaOH(aq)} \longrightarrow \underset{\text{solvent}}{H_2O} + \underset{\text{salt}}{NaCl(aq)}$$

$$\underset{\text{acid}}{HC_2H_3O_2} + \underset{\text{base}}{KOH(aq)} \longrightarrow \underset{\text{solvent}}{H_2O} + \underset{\text{salt}}{KC_2H_3O_2(aq)}$$

Since NaCl(c) and $KC_2H_3O_2$(c) dissolve easily in water, these solid salts do not form unless the solutions are evaporated. The neutralization reaction is, in fact, a very important method of preparing salts.

EXAMPLE 9.1 *Potash*[†], K_2CO_3, is used in the preparation of fertilizers and shaving soaps. What substances would you use to prepare K_2CO_3 by a neutralization reaction?

ANSWER You would use potassium hydroxide, KOH, to furnish the potassium ion and carbonic acid, H_2CO_3, to furnish the carbonate ion. But carbonic acid is a solution of CO_2 in water:

$$CO_2 + H_2O \rightleftharpoons H_2CO_3$$

$$H_2CO_3 + 2KOH \longrightarrow K_2CO_3 + 2H_2O$$

[†] Bases are also known as **alkalis**, from the Arabic word *alqili*, "plant ash." The ashes of burnt vegetation were early sources of bases. It was common practice to extract the alkali from wood ashes in iron pots. Thus, the two common alkalis produced by this method were called *potash* (potassium carbonate) and *caustic potash* (potassium hydroxide). These were used to make soap from animal fat (Chapter 24).

Compounds in which a salt and water are combined in a definite molar proportion, such as magnesium sulfate-water (1/7), are called *hydrates* (section 12.2).

Or, bubble CO_2 into a solution of KOH:

$$CO_2 + 2KOH \longrightarrow K_2CO_3 + H_2O$$ ■

PROBLEM 9.2 Magnesium sulfate, $MgSO_4$, is usually sold as Epsom salt, $MgSO_4(H_2O)_7$, which is used as a laxative and in the manufacture of insulation and polishing materials. What acid and base would you use to prepare $MgSO_4$? Write the balanced equation for the reaction. □

9.2 NET IONIC EQUATIONS

Recall (Table 2.4) that O_2^{2-} is the formula of the peroxide ion. H_2SO_4 gives up two protons per molecule in two steps:

$$H_2SO_4 \longrightarrow H^+ + HSO_4^-$$

$$HSO_4^- \longrightarrow H^+ + SO_4^{2-}$$

A more general treatment of the driving force of chemical reactions appears in Chapter 18.

The fact that all reactions between acids and bases that take place in water *produce* water leads us to an important idea. *It is the tendency to form water that makes acids and bases react with each other.* But why do they tend to form water? Water is a weak electrolyte, almost a nonelectrolyte. We can generalize this concept by postulating that *strong electrolytes tend to react with each other to form weak electrolytes.* This idea makes sense because it is simply another expression of the tendency of opposite charges to attract each other and form a weak electrolyte. If this generalization is correct, it should also apply to other reactions in which *ions are removed* or in which *weak electrolytes are formed.* It does. For example, sodium peroxide, Na_2O_2, and H_2SO_4 are strong electrolytes that react to form Na_2SO_4, a strong electrolyte, and hydrogen peroxide, H_2O_2, a weak electrolyte. It is the removal of H^+ and O_2^{2-} ions, the formation of the weak electrolyte H_2O_2, that pulls this reaction forward (from left to right):

$$\underbrace{2Na^+(aq) + O_2^{2-}(aq)}_{\text{strong electrolyte}} + \underbrace{2H^+(aq) + SO_4^{2-}(aq)}_{\text{strong electrolyte}} \longrightarrow$$

$$\underbrace{2Na^{2+}(aq) + SO_4^{2-}(aq)}_{\text{strong electrolyte}} + \underbrace{H_2O_2(aq)}_{\text{weak electrolyte}}$$

WRITING IONIC CHEMICAL EQUATIONS

Since we recognize that ions exist in solution as more or less independent chemical species, it becomes natural to write chemical equations involving electrolytes in terms of ions. Therefore, in this section, we will first present the scheme for writing chemical equations in ionic form. Then we will apply that method to reactions in which weak electrolytes are formed and to reactions in which ions are removed as gases, insoluble substances, and coordination ions. The advantage of first focusing our attention on writing ionic equations is that it will show you the common essential features of these reactions. With that knowledge, you will be able to predict these kinds of reactions in aqueous solutions (next section).

In order to decide when and how to write an equation in ionic form, we must first know which substances are present as ions in solution (strong electrolytes) and which are present mostly as molecules (weak electrolytes). The rules for classifying acids, bases, and salts as strong or weak electrolytes are

Some *weakly ionized salts*—$CdCl_2$, $HgCl_2$, $Pb(C_2H_3O_2)_2$—exist in solution mostly as neutral molecules.

1. Almost all salts are composed of ions and therefore are strong electrolytes.
2. The common strong acids HCl, HBr, HI, HNO_3, H_2SO_4, and $HClO_4$ are almost completely ionized in water. All other common acids are weak electrolytes and therefore weak acids; they exist mostly as molecules in water and are written in molecular form.
3. All metal hydroxides are strong electrolytes and therefore strong bases.

4. NH_3 and related compounds such as CH_3NH_2 are weak electrolytes and therefore weak bases; they exist mostly as molecules in water and are represented by molecular formulas.

The general scheme for writing ionic chemical equations can be summarized as shown:

COMPOUND	REPRESENTED BY	EXAMPLES
gas	molecular formula	CO_2, H_2S
liquid	molecular formula	H_2O, CCl_4
insoluble solid†	"molecular" (actually empirical) formula	$AgCl(c)$, $BaSO_4(c)$, $Fe_2O_3(c)$, $Fe(OH)_3(c)$
soluble ionic solid in water	ions	$Na^+ + OH^-$ for $NaOH(aq)$ $Na^+ + Cl^-$§ for $NaCl(aq)$ $3K^+ + PO_4^{3-}$ for $K_3PO_4(aq)$ $2Al^{3+} + 3SO_4^{2-}$ for $Al_2(SO_4)_3$

† Insoluble salts and hydroxides such as $BaSO_4$ and $Mg(OH)_2$ are strong electrolytes because they are ionic, but their solutions are poor electrical conductors since so few of their ions are in solution.
§ We could write $Na^+(aq) + Cl^-(aq)$. You may assume that all ions are "(aq)" unless otherwise specified.

To decide whether to write the formula of a salt or metal hydroxide in molecular or ionic form, you must determine whether the compound is soluble or insoluble.

Some salts dissolve easily in water; others are nearly (but never completely) insoluble. There is no sharp line between "soluble" and "insoluble" salts. However, as with other qualities that occur as gradations (such as tall/short, strong/

TABLE 9.3
RULES FOR IDENTIFYING THE SOLUBILITY OF COMMON SALTS IN WATER

RULE 1: MOST SALTS OF THESE ANIONS ARE SOLUBLE:	Exceptions	Borderline
NO_3^-		
$C_2H_3O_2^-$		$AgC_2H_3O_2$
ClO_3^-		
ClO_4^-		$KClO_4$
Cl^-	$AgCl$, Hg_2Cl_2, $PbCl_2$	
Br^-	$AgBr$, Hg_2Br_2, $HgBr_2$, $PbBr_2$	
I^-	AgI, Hg_2I_2, HgI_2, PbI_2	
SO_4^{2-}	$CaSO_4$, $SrSO_4$, $BaSO_4$, Ag_2SO_4, Hg_2SO_4, $PbSO_4$	

RULE 2: SALTS OF THESE ANIONS ARE SOLUBLE:	Borderline
Hydroxides (OH^-) of Group-1 cations, NH_4^+, Sr^{2+}, Ba^{2+}	$Ca(OH)_2$
Fluorides (F^-) of Na^+, K^+, Rb^+, Cs^+, NH_4^+, Al^{3+}	
Sulfides (S^{2-})† and sulfites (SO_3^{2-}) of Group-1 cations, NH_4^+, Ba^{2+}	
Carbonates (CO_3^{2-}), bicarbonates (HCO_3^-), chromates (CrO_4^{2-}), and phosphates (PO_4^{3-}) of Group-1 cations, NH_4^+	

† CaS and Al_2S_3 dissolve, but the hydroxides then precipitate:

$$Al_2S_3(c) + 6H_2O \longrightarrow 3H_2S(g) + 2Al(OH)_3(c)$$

weak), many cases are much closer to one end than the other, and it is helpful to be able to recognize them. Table 9.3 gives some rules for identifying soluble and "insoluble" salts. Only the most common ions are included. For our purposes here, a soluble salt is one that will not precipitate when approximately 0.1 M solutions containing the two ions are mixed; for example,

$$Na^+ + Cl^- \longrightarrow \text{no reaction}$$

so NaCl is "soluble." In the reaction

$$Ca^{2+} + CO_3^{2-} \longrightarrow CaCO_3(c)$$

$CaCO_3$ precipitates and is thus an "insoluble" salt.

The *key point* here is to represent substances in the form (molecules or ions) in which they largely exist when they are added to water.

These rules can be applied to writing ionic equations for a variety of reactions.

FORMATION OF A WEAK ELECTROLYTE

When solutions of hydrochloric acid and sodium hydroxide are mixed, water and sodium chloride are formed:

$$HCl(aq) + NaOH(aq) \longrightarrow NaCl(aq) + H_2O \qquad \text{(molecular form)}$$

Many such reactions are of the type known as **double displacement**; the cations and anions of the two reactants "exchange" partners to form two products. Other names for this type of reaction are double replacement, metathesis (Greek for "transpose"), and exchange reaction. The formulas of the strong acid (HCl), the strong base (NaOH), and the salt are better written in ionic form:

$$H^+ + Cl^- + Na^+ + OH^- \longrightarrow Na^+ + Cl^- + H_2O \qquad \text{(ionic form)} \qquad (1)$$

Inspection of equation 1 shows that Na^+ and Cl^- appear on both sides. This means they were present initially and are not changed by the reaction. *Any species remaining unchanged can be omitted from a chemical equation.* Equation 1 is therefore simplified to

$$H^+ + OH^- \longrightarrow H_2O \qquad \text{(net ionic equation)} \qquad (2)$$

But H_2O is never written as HHO.

This form of the equation emphasizes that the actual chemical change is a combination of H^+ and OH^- ions. It is for this reason that H_2O is sometimes written as HOH. Equation 2 is called a **net ionic equation** because nonreacting species have been omitted, leaving only those that actually participate in the reaction. We know that positive ions must have been present with OH^- and negative ions with H^+, but the equation does not tell us what they are, nor do we necessarily care. The reaction is essentially the same when any strong acid reacts with a strong base. Thus if we use HI and KOH, the net ionic reaction is the same as it is for HCl and NaOH, equation 2. Ions that are present but do not participate are called **spectator ions**. Na^+ and Cl^- are spectator ions in equation 1; so are K^+ and I^- in the reaction between HI and KOH.

Remember that all reactions in this chapter, unless otherwise specified, are in aqueous solutions, making the (aq) notation unnecessary.

Three examples will further illustrate how to write net ionic equations for reactions in water:

1. $H_2SO_4 + 2LiOH \longrightarrow Li_2SO_4 + 2H_2O \qquad \text{(molecular form)}$

   We represent the acid (a strong acid), base (a strong base), and salt in solution as ions and the solvent by its molecular formula:

   $$2H^+ + SO_4^{2-} + 2Li^+ + 2OH^- \longrightarrow 2Li^+ + SO_4^{2-} + 2H_2O \qquad \text{(ionic form)}$$

The spectator ions are those that appear on both sides, Li^+ and SO_4^{2-}. We omit them to get the net ionic equation:

$$2H^+ + 2OH^- \longrightarrow 2H_2O \quad \text{or} \quad H^+ + OH^- \longrightarrow H_2O$$

(net ionic equation)

Note that net ionic equations are balanced with respect to both number of atoms *and* number of charges:

$$+2 - 2 = 0 \quad \text{or} \quad +1 - 1 = 0$$

2. $HCHO_2 + NaOH \longrightarrow NaCHO_2 + H_2O$

formic acid — Sodium formate

The Lewis structure for formic acid is

```
 :O:
  ||  ..
H—C—O—H
      ..
```

The H atom bonded to oxygen is the one that ionizes.

In this chemical equation, $HCHO_2$, a weak acid, is represented by its molecular formula:

$$HCHO_2 + Na^+ + OH^- \longrightarrow Na^+ + CHO_2^- + H_2O$$

The net ionic equation is

$$HCHO_2 + OH^- \longrightarrow CHO_2^- + H_2O$$

3. $NaC_2H_3O_2 + HBr \longrightarrow NaBr + HC_2H_3O_2$

Here, acetic acid, a weak acid, is written in molecular form:

$$Na^+ + C_2H_3O_2^- + H^+ + Br^- \longrightarrow Na^+ + Br^- + HC_2H_3O_2$$

The net ionic equation is

$$C_2H_3O_2^- + H^+ \longrightarrow HC_2H_3O_2$$

The drive is so strong to form H_2O from an acid, weak or strong (HCN, HCl), and a base, weak or strong (NH_3, NaOH), that an acid will even react with a solid metal oxide forming a salt and water; for example,

$$CaO(c) + 2HCl \longrightarrow CaCl_2 + H_2O$$

In ionic form

$$CaO(c) + 2H^+ + 2Cl^- \longrightarrow Ca^{2+} + 2Cl^- + H_2O$$

The net ionic equation is

$$CaO(c) + 2H^+ \longrightarrow Ca^{2+} + H_2O$$

Note that the charges balance: $+2 = +2$. The net charge on each side need not be zero.

PROBLEM 9.3 Write the balanced net ionic equation for the reaction

$$2HNO_3 + Ba(OH)_2 \longrightarrow Ba(NO_3)_2 + 2H_2O$$

□

TABLE 9.4 SOME COMMON GASES FORMED FROM IONS IN WATER	
	H_2S
	HCN
	NH_3
	CO_2 (Unstable carbonic acid, H_2CO_3, decomposes to CO_2 and H_2O.)
	SO_2 (Unstable sulfurous acid, H_2SO_3, decomposes to SO_2 and H_2O.)

FORMATION OF A GAS

Table 9.4 lists some common gases formed from ions in water. Let us examine the reaction that may occur when solutions of hydrochloric acid and sodium sulfide, Na_2S, are mixed. In molecular form, we write

$$HCl + Na_2S \longrightarrow ?$$

and in the ionic form, we write

$$H^+ + Cl^- + 2Na^+ + S^{2-} \longrightarrow ?$$

Referring to Table 9.4, we see that H_2S is a possible product:

$$H^+ + Cl^- + 2Na^+ + S^{2-} \longrightarrow H_2S(g) + 2Na^+ + Cl^-$$

Balancing requires $2H^+$ and, therefore, $2Cl^-$:

$$2H^+ + 2Cl^- + 2Na^+ + S^{2-} \longrightarrow H_2S(g) + 2Na^{2+} + 2Cl^-$$

Omitting the spectator ions (Na^+ and Cl^-) gives us the net ionic equation:

$$2H^+ + S^{2-} \longrightarrow H_2S(g)$$

Thus, any strong acid added to any soluble sulfide will form $H_2S(g)$ (Box 9.1). In this case, as in the formation of a weak electrolyte, it is the removal of H^+ and S^{2-} ions, *the formation of a gas*, that *pulls the reaction forward*. In general, carbonates, sulfides, sulfites, and cyanides react with acids to form, respectively, the gases CO_2, H_2S, SO_2, and HCN.

PROBLEM 9.4 Write the balanced net ionic equation for the reaction

$$Sr(CN)_2 + 2HBr \longrightarrow SrBr_2 + 2HCN(g)$$ □

FORMATION OF AN INSOLUBLE SOLID

When solutions of the soluble salts KCl and $AgNO_3$ are mixed, a white precipitate of silver chloride, AgCl, is formed:

$$KCl + AgNO_3 \longrightarrow AgCl(c) + KNO_3$$

We write the formulas of the soluble salts in ionic form, but silver chloride, which is present almost entirely as the solid, is represented as AgCl(c):

$$K^+ + Cl^- + Ag^+ + NO_3^- \longrightarrow AgCl(c) + K^+ + NO_3^- \qquad (3)$$

BOX 9.1
SAFETY NOTE

HCN and H_2S are both deadly gases. The important difference between them is not in their toxicity, but in their odor. HCN has a pleasant, almondlike scent; H_2S smells like rotten eggs. H_2S, therefore, may give an earlier warning of its presence. However, exposure to moderate concentrations of H_2S anesthetizes the sense of smell and thus destroys a person's ability to detect greater concentrations. One or two inhalations of a concentrated H_2S mixture causes respiratory failure and death within seconds.

CN^- and S^{2-} ions in solution do not evaporate, but when the solutions are acidified, the gases form. Therefore, do not add acid to CN^- or S^{2-} solutions. SO_2 is not so deadly, but it is poisonous enough, corrosive, and very irritating. Thus, it is a good idea not to acidify sulfites.

Equation 3 shows K^+ and NO_3^- as spectator ions. The net ionic equation is therefore written as

$$Cl^- + Ag^+ \longrightarrow AgCl(c)$$

In this case, it is the removal of Ag^+ and Cl^- ions, the *formation of a precipitate*, an insoluble solid, that *pulls the reaction forward.* Any soluble chloride added to any soluble silver salt will form AgCl(c). Thus, the formation of white insoluble AgCl may be used to test for soluble chlorides or soluble silver salts.

You will learn later how to predict precipitation in a more quantitative way (section 16.1).

EXAMPLE 9.2 Write the net ionic equations for these reactions:

(a) $Pb(NO_3)_2 + (NH_4)_2SO_4 \longrightarrow PbSO_4(c) + 2NH_4NO_3$ (molecular form)

(b) $CO_2(g) + Ca(OH)_2 \longrightarrow CaCO_3(c) + H_2O$ (molecular form)

ANSWER

(a) The dissolved salts are represented as ions. The insoluble salt, $PbSO_4$ (Rule 1 in Table 9.3), is represented by its molecular formula:

$$Pb^{2+} + 2NO_3^- + 2NH_4^+ + SO_4^{2-} \longrightarrow PbSO_4(c) + 2NH_4^+ + 2NO_3^- \quad \text{(ionic form)}$$

The spectator ions, NH_4^+ and NO_3^-, are omitted to give the net ionic equation:

$$Pb^{2+} + SO_4^{2-} \longrightarrow PbSO_4(c) \quad \text{(net ionic equation)}$$

(b) $Ca(OH)_2$ (Rule 2 in Table 9.3) is a dissolved ionic base, written as ions. The gas CO_2, the solid $CaCO_3$, and the liquid H_2O are written in molecular form. The equation in ionic form is

$$CO_2(g) + Ca^{2+} + 2OH^- \longrightarrow CaCO_3(c) + H_2O \quad \text{(net ionic equation)}$$

In this case, there are no spectator ions—nothing drops out of the equation—so the net ionic equation is the same as the ionic form. ■

PROBLEM 9.5 Write the balanced net ionic equation for the reaction

$$AlCl_3 + K_3PO_4 \longrightarrow 3KCl + AlPO_4(c)$$ □

FORMATION OF A COORDINATION ION

A **coordination ion** is formed in solution when a positive metal ion, usually a transition metal, combines with a number of uncharged molecules or negative ions. A typical example is the formation of the coordination ion $[Ag(NH_3)_2]^+$ in the reaction between Ag^+ and NH_3:

$$Ag^+ + 2NH_3 \longrightarrow [Ag(NH_3)_2]^+$$

The Group-3 and -4 representative elements often form coordination ions; for example, Al^{3+} forms $[AlF_6]^{3-}$ (Chapter 21). Coordination ions are also known as *complex* ions.

The charge on the coordination ion is the algebraic sum of the charges of its parts. The charge may come out to be zero; to allow for this possibility, the term coordination *compound* is used. Some other examples of coordination ions are $[Ag(S_2O_3)_2]^{3-}$, $[Cu(NH_3)_4]^{2+}$, $[Fe(CN)_6]^{3-}$, and $[Zn(OH)_4]^{2-}$.

The metal atom with a zero or positive oxidation number is called the *central atom.* The number of atoms directly bonded to the central atom is called the

coordination number. For example:

COORDINATION ION	CENTRAL ATOM	NUMBER OF ATOMS ATTACHED TO CENTRAL ATOM	COORDINATION NUMBER	CHARGE ON COORDINATION ION
$[FeCl_6]^{3-}$	Fe^{3+}	$6[:\ddot{\underset{..}{Cl}}:]^-$	6	$+3+6(-1)=-3$
$[Co(CN)_6]^{3-}$	Co^{3+}	$6[:C{\equiv}N:]^-$	6	$+3+6(-1)=-3$
$[Ag(NH_3)_2]^+$	Ag^+	$2[:NH_3]^0$ (H—N—H with H, lone pair on N)	2	$+1+2(0)=+1$
$[Ag(S_2O_3)_2]^{3-}$	Ag^+	$2[S_2O_3]^{2-}$ (:S: bonded to S, with :O—S—O: and :O: below)	2	$+1+2(-2)=-3$

The factors that determine the coordination number cannot be properly evaluated to allow reliable predictions. Therefore, we will restrict our discussion to transition-metal cations having coordination numbers that are twice their charges. Some typical values are 2 for Ag^+, 4 for Cu^{2+}, 6 for Co^{3+}, 4 for Ni^{2+}, and 6 for Fe^{3+}.

EXAMPLE 9.3 Write the net ionic equation for the reaction

$$FeCl_3 + 6KCN \longrightarrow K_3[Fe(CN)_6] + 3KCl$$

ANSWER The soluble salts are written as ions:

$$Fe^{3+} + 3Cl^- + 6K^+ + 6CN^- \longrightarrow 3K^+ + [Fe(CN)_6]^{3-} + 3K^+ + 3Cl^-$$

The spectator ions (K^+ and Cl^-) are omitted to get the net ionic equation:

$$Fe^{3+} + 6CN^- \longrightarrow [Fe(CN)_6]^{3-}$$ ■

The great stability of coordination ions is indicated by enclosing the coordination ion in brackets. The formula $K_3[Fe(CN)_6]$ informs us that $[Fe(CN)_6]^{3-}$ is a coordination ion whose identity is not lost in a solution. Just as NaCl(c) dissolves to Na^+(aq) and Cl^-(aq), $K_3[Fe(CN)_6]$(c) dissolves to $3K^+$(aq) and $[Fe(CN)_6]^{3-}$(aq). Coordination ions are discussed in more detail in Chapter 21.

When no other anion or molecule is present, transition-metal cations in aqueous solution are combined with water molecules; for example, Fe^{3+}(aq) represents $[Fe(H_2O)_6]^{3+}$.

PROBLEM 9.6 Write the net ionic equation for the reaction

$$AgNO_3 + 2Na_2S_2O_3 \longrightarrow Na_3[Ag(S_2O_3)_2] + NaNO_3$$ □

Notice that the *removal of ions is the driving force for the formation of a coordination ion*, just as it is in the formation of a weak electrolyte or a gas or a precipitate. The fact that the formation of $[Ag(NH_3)_2]^+$ involves a neutral molecule does not alter the basic concept. The removal of Ag^+ from solution as $[Ag(NH_3)_2]^+$ is treated in a more quantitative way in section 16.3.

9.3 PREDICTING REACTIONS IN WATER SOLUTIONS

Now you are ready to learn how to predict the products formed, if any, when aqueous solutions are mixed. In each example **?** represents the product, if any, of the reaction. The product may be (1) a weak electrolyte, (2) a gas, (3) an insoluble solid, or (4) a coordination ion. If you need to review the guidelines, refer to these pages:

- for generalizations about weak and strong electrolytes, page 287
- for generalizations about the removal of ions as the driving force for the

  formation of a weak electrolyte, page 287
  formation of a gas, page 291 and Table 9.4
  formation of a precipitate, page 292 and Table 9.3
  formation of a coordination ion, page 293

EXAMPLE 9.4 $NaOH + FeCl_3 \longrightarrow ?$

ANSWER First write the given reactants as ions:

$$Na^+ + OH^- + Fe^{3+} + 3Cl^- \longrightarrow ?$$

Then, look for possible products. $Fe(OH)_3$ and NaCl can be formed by an exchange of ions (double displacement). Are these weak electrolytes? From our generalizations (page 287), the answer is no. Are these gases? No (Table 9.4). Is a precipitate possible? NaCl is soluble by Rule 1 in Table 9.3 and by universal experience in the kitchen. Most hydroxides are insoluble (Rule 2, Table 9.3). $Fe(OH)_3$ is not on the list of soluble hydroxides; therefore it is insoluble:

$$Na^+ + OH^- + Fe^{3+} + 3Cl^- \longrightarrow Fe(OH)_3(c) + Na^+ + 3Cl^-$$

Omit the spectator ions (Na^+ and Cl^-)

$$OH^- + Fe^{3+} \longrightarrow Fe(OH)_3(c)$$

and then balance to give

$$3OH^- + Fe^{3+} \longrightarrow Fe(OH)_3(c)$$

This result tells us that any soluble hydroxide added to any soluble Fe^{3+} salt will form $Fe(OH)_3(c)$. ■

EXAMPLE 9.5 $BaCl_2 + K_2SO_4 \longrightarrow ?$

ANSWER In ionic form, the reactants are

$$Ba^{2+} + 2Cl^- + 2K^+ + SO_4^{2-} \longrightarrow ?$$

The possible products are KCl and $BaSO_4$. They are not weak electrolytes, nor are they gases. KCl is soluble but $BaSO_4$ is insoluble:

$$Ba^{2+} + 2Cl^- + 2K^+ + SO_4^{2-} \longrightarrow BaSO_4(c) + 2K^+ + 2Cl^-$$

The net ionic equation is

$$Ba^{2+} + SO_4^{2-} \longrightarrow BaSO_4(c)$$

(Can $BaSO_4$ be made from a different pair of compounds?) ■

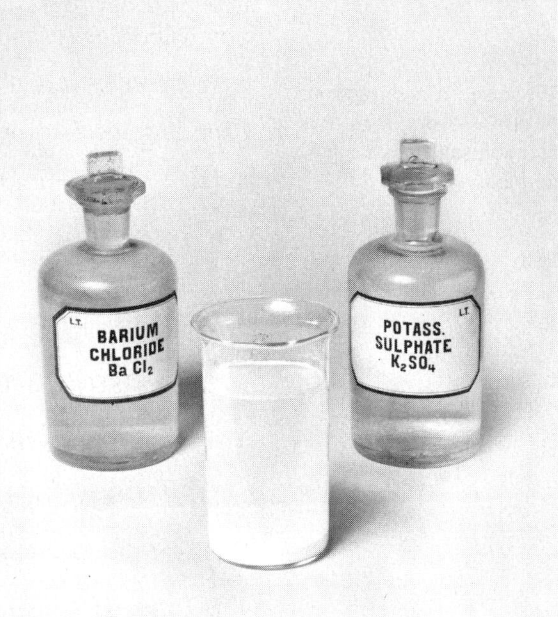

$BaCl_2(aq) + K_2SO_4(aq) \longrightarrow BaSO_4(c) + 2KCl(aq)$

EXAMPLE 9.6 $NaBr + NH_4Cl \longrightarrow ?$

ANSWER In ionic form

$$Na^+ + Br^- + NH_4^+ + Cl^- \longrightarrow ?$$

The possible products are NaCl and NH_4Br. Neither of them is a weak electrolyte, and both are soluble. $NH_3(g)$ does not form because there is no OH^- to remove an H^+ from NH_4^+. None of the reactants is a transition element. We therefore assume that coordination ions are not formed. Thus all four salts are in solution and should be written in ionic form:

$$Na^+ + Br^- + NH_4^+ + Cl^- \longrightarrow Na^+ + Cl^- + NH_4^+ + Br^- \qquad \text{(no reaction)}$$

When the ions appearing on both sides are canceled, nothing is left. *There is no reaction* when dissolved salts are mixed and the products are soluble or are not weak electrolytes, gases, or coordination ions. ■

EXAMPLE 9.7 $LiBr + HCN \longrightarrow ?$

ANSWER LiBr is a soluble ionic salt. HCN, a weak acid, is written in molecular form:

$$Li^+ + Br^- + HCN \longrightarrow ?$$

The possible products, HBr and LiCN, are soluble, strong electrolytes and they are not gases or coordination ions. Thus a reaction

$$Li^+ + Br^- + HCN \longrightarrow H^+ + Br^- + Li^+ + CN^- \longrightarrow Li^+ + Br^- + HCN \qquad \text{(no reaction)}$$

will not occur. In fact, if HBr and LiCN are mixed, the *reverse* (right-to-left) *reaction* will occur, forming the weak electrolyte HCN from two strong electrolytes:

Warning: it is dangerous to add a strong acid to a cyanide salt because HCN is *deadly*!

$$H^+ + Br^- + Li^+ + CN^- \longrightarrow Li^+ + Br^- + HCN$$

Canceling spectator ions gives the reaction

$$H^+ + CN^- \longrightarrow HCN(g)$$

■

EXAMPLE 9.8 $Na_2SO_3 + HI \longrightarrow ?$

ANSWER The ionic form of the reactants is

$$2Na^+ + SO_3^{2-} + H^+ + I^- \longrightarrow ?$$

The possible products are NaI and H_2SO_3. Recognize H_2SO_3 as a weak electrolyte:

$$2Na^+ + SO_3^{2-} + H^+ + I^- \longrightarrow 2Na^+ + I^- + H_2SO_3$$

Omitting the spectator ions (Na^+ and I^-),

$$SO_3^{2-} + H^+ \longrightarrow H_2SO_3$$

H_2SO_3 decomposes to $SO_2(g)$ and H_2O, and balancing yields the net ionic equation:

$$SO_3^{2-} + 2H^+ \longrightarrow H_2SO_3 \longrightarrow SO_2(g) + H_2O$$

■

EXAMPLE 9.9 $KOH + NH_4Cl \longrightarrow ?$

ANSWER The ionic form of the reactants is

$$K^+ + OH^- + NH_4^+ + Cl^- \longrightarrow ?$$

Among the possible products, we recognize that NH_3 can be formed because OH^- is a reactant that can pull an H^+ from NH_4^+ forming H_2O:

$$K^+ + OH^- + NH_4^+ + Cl^- \longrightarrow K^+ + Cl^- + NH_3(g) + H_2O$$

The net ionic equation is

$$OH^- + NH_4^+ \longrightarrow NH_3(g) + H_2O$$

Indeed the formation of $NH_3(g)$ upon the addition of any strong base serves as a test for the presence of ammonium salts. ■

EXAMPLE 9.10 $NiCl_2 + NaCN \longrightarrow ?$

ANSWER The presence of a transition element suggests that a coordination ion may form. According to the rule (page 293), the coordination number of Ni^{2+} is 4:

$$Ni^{2+} + 2Cl^- + Na^+ + CN^- \longrightarrow [Ni(CN)_4]^{2-} + Na^+ + 2Cl^-$$

Omitting the spectator ions and balancing,

$$Ni^{2+} + 4CN^- \longrightarrow [Ni(CN)_4]^{2-}$$

■

It is evident that the ion removai rules provide an acceptable framework for correctly predicting an enormous number of reactions with a minimum of memorization. Of course, no classification system is always correct.†

PROBLEM 9.7 Write the balanced net ionic equation for these reactions. If no reaction occurs, write "no reaction."

(a) $(NH_4)_2SO_4 + NaOH \longrightarrow ?$	(e) $KOH + HC_3H_5O_2 \quad ?$
(b) $Cu(NO_3)_2 + NH_3 \longrightarrow ?$	(f) $BaCl_2 + Na_2CO_3 \quad ?$
(c) $KOH + H_2SO_4 \longrightarrow ?$	(g) $HCl + Na_2CO_3 \longrightarrow ?$
(d) $NaCl + RbNO_3 \longrightarrow ?$	

□

9.4 OXIDATION–REDUCTION REACTIONS

When a copper wire is dipped into an aqueous solution of silver nitrate, the copper becomes coated with silver (silver-plated) and the solution changes to the blue color characteristic of $Cu^{2+}(aq)$. The equation for the reaction is

$$Cu(c) + 2AgNO_3(aq) \longrightarrow Cu(NO_3)_2 + 2Ag(c)$$

or, in net ionic form,

$$Cu(c) + 2Ag^+(aq) \longrightarrow Cu^{2+}(aq) + 2Ag(c)$$

The traditional name for this type of reaction is **displacement**; "Cu displaces Ag from $AgNO_3$."

Evidently, a Cu atom is changed to a Cu^{2+} ion. For this to happen, the Cu atom must lose two electrons: Cu is *oxidized* to Cu^{2+} (section 7.9). At the same time, an Ag^+ ion is changed to an Ag atom. For this to happen, the Ag^+ ion must gain an electron: Ag^+ is *reduced* to Ag.

Cu is a **reducing agent** in this redox reaction since it *gives up electrons* to Ag^+, reducing it to Ag. Ag^+ (or $AgNO_3$) is an **oxidizing agent** because it *accepts electrons* from Cu, oxidizing it to Cu^{2+}. Thus, the substance that is being reduced (Ag^+) is the one that oxidizes another substance (Cu). The substance that is being oxidized (Cu) is the one that reduces another substance (Ag^+). In summary:

Cu *loses* electrons.	**Ag^+ *gains* electrons.**
Cu is *oxidized* to Cu^{2+}.	**Ag^+ is *reduced* to Ag.**
Cu is the *reducing* agent.	**Ag^+ is the *oxidizing* agent.**
Cu *reduces* Ag^+.	**Ag^+ *oxidizes* the Cu.**

PROBLEM 9.8 (a) What is oxidized and what is reduced in the following reaction? (b) What is the oxidizing agent and what is the reducing agent?

$$Zn(c) + 2H^+(aq) \longrightarrow Zn^{2+}(aq) + H_2(g)$$

□

---

† For a mixture of HCl and K_2CrO_4, you might predict that H_2CrO_4 is the only product:

$$2H^+ + 2Cl^- + \cancel{2K^+} + CrO_4{}^{2-} \longrightarrow H_2CrO_4 + \cancel{2K^+} + \cancel{2Cl^-}$$

However, the actual products are Cr^{3+}, Cl_2, and H_2O. (You will be able to predict the products of such mixtures after you have studied Chapter 17.) And if H_2SO_4 is substituted for HCl, the yellow chromate ion is changed to the orange-red dichromate ion:

$$2CrO_4{}^{2-} + 2H^+ \qquad Cr_2O_7{}^{2-} + H_2O$$

All redox reactions involve changes in charge—transfers of electrons—or changes in oxidation numbers (section 7.9). In this respect, they differ from the ion removal reactions we just studied.

Redox reactions occur practically everywhere. They are present in the power sources of batteries and the oxidation of foods and fuels, and in the processes of photography, bleaching, and metallurgy. However, our discussion of redox reactions in this section is restricted to displacement reactions that involve metals and metal ions; that is, reactions in which a metal displaces another metal in a compound.

ACTIVITY SERIES OF THE METALS

We can now learn how to predict the products of redox reactions involving metals and metal ions. Table 9.5 is a list of metals arranged in the order of their tendency to be oxidized.

Lithium, at the top, has the strongest tendency to be oxidized (Li $\quad$ Li^+). Li is therefore the strongest reducing agent in the series; it is the most active metal. Actually, the table lists the metals and their corresponding ions. A metal higher in the series will reduce a lower ion. For example, Cu is above Ag in the table. This means that Cu will reduce Ag^+ and thus displace Ag from any silver salt. Since Pb is above Cu in the table, Cu will not reduce Pb^{2+}:

$$Pb^{2+}(aq) + Cu(c) \longrightarrow \text{no reaction}$$

TABLE 9.5
ACTIVITY SERIES OF THE METALS

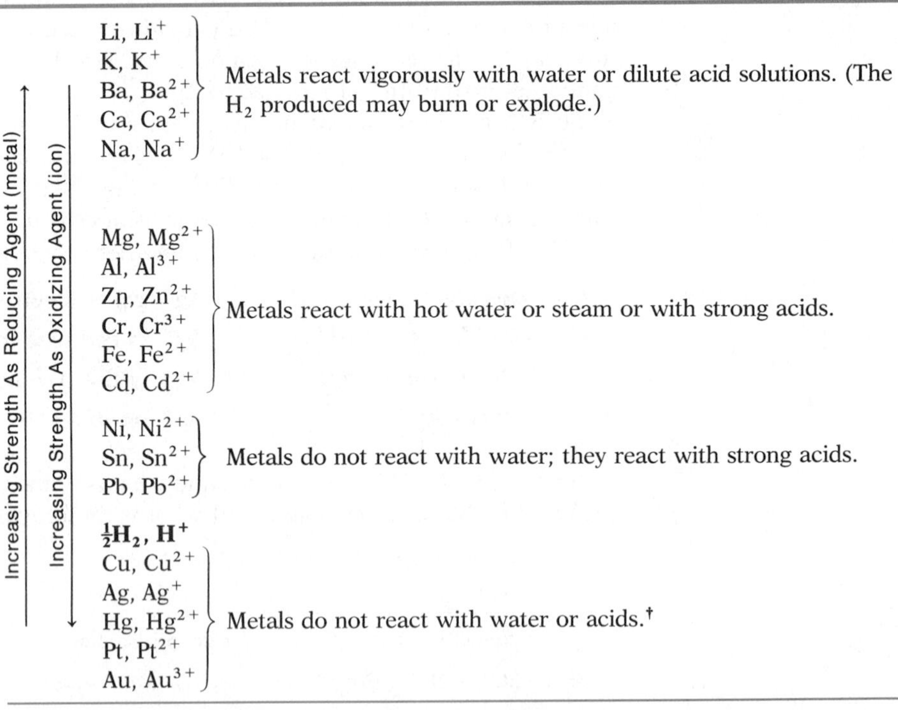

Increasing Strength As Reducing Agent (metal) ↑ (upward)
Increasing Strength As Oxidizing Agent (ion) ↓ (downward)

Li, Li^+ K, K^+ Ba, Ba^{2+} Ca, Ca^{2+} Na, Na^+	Metals react vigorously with water or dilute acid solutions. (The H_2 produced may burn or explode.)
Mg, Mg^{2+} Al, Al^{3+} Zn, Zn^{2+} Cr, Cr^{3+} Fe, Fe^{2+} Cd, Cd^{2+}	Metals react with hot water or steam or with strong acids.
Ni, Ni^{2+} Sn, Sn^{2+} Pb, Pb^{2+}	Metals do not react with water; they react with strong acids.
$\frac{1}{2}H_2$, H^+	
Cu, Cu^{2+} Ag, Ag^+ Hg, Hg^{2+} Pt, Pt^{2+} Au, Au^{3+}	Metals do not react with water or acids.†

† These metals react with concentrated HNO_3 or with a mixture of concentrated HCl and HNO_3 (*aqua regia*, 3:1 by volume), but H_2 is not produced. For example,

$$4Cu(c) + 11HNO_3 \longrightarrow 4Cu(NO_3)_2 + HNO_2 + 2NO(g) + 5H_2O$$

and

$$Au(c) + 4H^+ + 4Cl^- + NO_3^- \longrightarrow [AuCl_4]^- + NO(g) + 2H_2O$$

Instead, Pb will reduce Cu^{2+} and so displace Cu from any copper salt:

$$Pb(c) + Cu^{2+}(aq) \longrightarrow Pb^{2+}(aq) + Cu(c)$$

EXAMPLE 9.11 Will this reaction occur?

$$Al(c) + Fe_2O_3(c) \longrightarrow ?$$

Table 9.5 applies to reactions involving ions in aqueous solutions. However, it predicts correct answers for most reactions of solid compounds also.

If your answer is yes, write the balanced equation for the reaction.

ANSWER The reaction will occur because Al is above Fe in Table 9.5. The reaction† is

$$Al(c) + Fe_2O_3(c) \longrightarrow Fe(\ell) + Al_2O_3(c)$$

and balancing,

Mixture is ignited with a magnesium fuse.

$$2Al(c) + Fe_2O_3(c) \longrightarrow 2Fe(\ell) + Al_2O_3(c)$$ ■

In a mixture of Mg and Al_2O_3, however, Mg will reduce Al^{3+} (displace Al) since Mg is above Al in Table 9.5. That reaction is

$$3Mg(c) + Al_2O_3(c) \longrightarrow 2Al(\ell) + 3MgO(c)$$

PROBLEM 9.9 When Henri Deville prepared aluminum in 1854, the metal was sufficiently expensive to be considered as a material for "the manufacture of jewelry and other articles of luxury." Select the reactants he used (a or b) and write the balanced equation for the reaction.

(a) $Zn(c) + AlCl_3(c) \longrightarrow ?$
(b) $Na(c) + AlCl_3(c) \longrightarrow ?$ □

PROBLEM 9.10 Describe what you would expect to observe if you placed a copper rod in an aqueous solution of $AuCl_3$. Write a net ionic equation for any reaction that may occur. □

Safety Note: Never handle alkali metals with bare fingers, and never add water to them. Dangerous quantities of heat are generated rapidly by these exothermic reactions. Alkali metals are usually stored in vacuum or under mineral oil to shield the metal from air and moisture.

Hydrogen, a nonmetal, is included in Table 9.5 because of its great natural abundance as part of water molecules. It is also the standard by which the relative reducing power of metals is measured (Chapter 17). The metals above hydrogen in the table displace it from acids and water. Natural waters contain CO_2 ($CO_2 + H_2O \longrightarrow H^+ + HCO_3^-$) and therefore are slightly acidic. Generally, the farther above hydrogen a metal is, the more rapid is the displacement. The alkali (Group-1) and alkaline earth (Group-2) metals react very rapidly with water:

$$Na(c) + H_2O(\ell) \longrightarrow \tfrac{1}{2}H_2(g) + NaOH(aq) \qquad \Delta H = -184.3 \text{ kJ}$$

On the other hand, Mg and Zn displace H_2 from water only at elevated temperatures but easily displace it from acidic solutions. The metals below hydrogen in the table do not react with water or acids.

† This reaction is known as the *Goldschmidt reaction* or the *thermite reaction*, discovered by Hans Goldschmidt in 1905. It is used in the welding of iron or steel. In the repair of railroad tracks, it provides an on-site source of molten iron to pour into a crack or gap to fuse the broken parts together. Sufficient heat is emitted to melt the iron.

PROBLEM 9.11 Write the net ionic equation for the reaction

$$Zn(c) + HCl(aq) \longrightarrow ?$$ □

Later, in Chapter 17, you will learn how Table 9.5 is obtained experimentally. In the meantime, it will be useful for predicting hundreds of reactions.

9.5 BALANCING OXIDATION–REDUCTION EQUATIONS

Many volumetric analytical procedures (section 3.11) depend on oxidation–reduction reactions. However, we cannot always balance equations for redox reactions by inspection. Rather frequently we must use a systematic method. The task may seem tedious at first, but with a little practice and careful attention to details, you will be successful.

In this section we describe the **ion-electron method** of balancing equations for redox reactions. The method is especially appropriate for reactions involving ions in solution. The reaction is resolved into two parts, each called a **partial equation**. One partial equation represents a loss of electrons (oxidation); the other partial equation represents a gain of electrons (reduction). In reality, most oxidation–reduction reactions probably do not proceed by the steps shown in the partial equations. However, the partial equations can correspond to physical reality in a galvanic or electrolytic cell (Chapter 17). Oxidation and reduction occur in two separate places in these cells. The two processes can be described by the partial equations that are obtained by this method.

The ion-electron method is essentially a bookkeeping device that enables us to deal separately with the oxidation and the reduction and then to combine the results into a balanced equation. Although this artificial procedure gives the correct equation for the net reaction, *do not interpret the procedure as an explanation or a pathway of the reaction.*

BALANCING IN ACIDIC SOLUTIONS

We will illustrate the ion-electron method with the reaction

$$Cu(c) + HNO_3(aq) \longrightarrow Cu(NO_3)_2(aq) + H_2O + NO(g)$$

which occurs when copper is added to dilute nitric acid.

STEP 1 The first step is to write the equation in ionic form. HNO_3 and $Cu(NO_3)_2$ are strong electrolytes and are represented as ions:

$$Cu + H^+ + NO_3^- \longrightarrow Cu^{2+} + 2NO_3^- + H_2O + NO$$

STEP 2 Next, observe which species are oxidized or reduced (although it is not yet necessary to decide which process is which). Look for an atom that appears on both sides of the equation, but with different oxidation numbers (section 7.9). In this reaction, H and O have their usual oxidation numbers of +1 and −2, respectively, on both sides of the equation. However, copper goes from 0 (in Cu) to +2 (in Cu^{2+}). At least one other element must also change its oxidation number, and since we have excluded H and O, the only possibility is nitrogen. It would not be difficult to calculate its oxidation numbers in NO_3^- and NO, but you do not need to do that.

STEP 3 Write "skeleton equations" for oxidation and for reduction. Here, Cu becomes Cu^{2+}

$$Cu \longrightarrow Cu^{2+} \qquad \text{(unbalanced skeleton)} \qquad (4)$$

and NO_3^- somehow becomes NO:

$$NO_3^- \longrightarrow NO \qquad \text{(unbalanced skeleton)} \tag{5}$$

STEP 4 Balance each partial equation separately. Equation 4 is almost balanced, but not quite. Not only must an equation have the same number of atoms of each kind on each side; it must also have the *same total electric charge on each side.* Charge imbalance is corrected by adding the appropriate number of electrons (e^-) to each side—hence the name, "ion-electron method." Equation 4 has a charge of 0 on the left and +2 on the right. This situation is corrected by adding 2 electrons to the right-hand side:

$$Cu \longrightarrow Cu^{2+} + 2e^- \qquad \text{(oxidation)} \tag{6}$$

This is a balanced partial equation for the oxidation of Cu.

Equation 5 is unbalanced both electrically and atomically. We first balance it atomically. There are 3 O atoms on the left and 1 on the right. Where did the other two go? They did not become O_2 gas; O_2 is not a product of the reaction. The only other fate possible for the O atoms in an acidic solution is to become part of the water molecules:

Basic solutions are discussed later in this section.

$$NO_3^- \longrightarrow NO + H_2O$$

H now appears on the right side, and it must also appear on the left. The source of the needed H must be the H^+ ion:

$$NO_3^- + H^+ \longrightarrow NO + H_2O$$

The next step is to balance the equation with respect to numbers of atoms. Balancing O requires 2 H_2O, and balancing H then requires 4 H^+:

$$NO_3^- + 4H^+ \longrightarrow NO + 2H_2O$$

For an acidic solution, the general procedure is:

- **To the side deficient in O, add 1 H_2O for each O atom needed.**
- **To the side deficient in H, add 1 H^+ for each H atom needed.**

There are now equal numbers of atoms of each kind on each side of the equation. However, the charge is not balanced. The net charge is $-1 + 4 = +3$ on the left and 0 on the right. Add 3 electrons to the left to balance the charge:

$$NO_3^- + 4H^+ + 3e^- \longrightarrow NO + 2H_2O \qquad \text{(reduction)} \tag{7}$$

STEP 5 Add the balanced partial equations to obtain a balanced equation for the net reaction. Simply adding them may not give you the answer, however. They must be multiplied by coefficients that insure that *the number of electrons given out in the oxidation equals the number of electrons taken up in the reduction.* Inspection of equations 6 and 7 shows you that each Cu atom loses 2 electrons and each NO_3^- ion gains 3 electrons. For every 3 Cu atoms oxidized, 2 NO_3^- ions must be reduced, so that 6 electrons are lost and 6 gained. Multiply equation 6 by 3 and equation 7 by 2, and then add them:

$$\begin{array}{rcl} 3Cu & \longrightarrow & 3Cu^{2+} + 6e^- \\ 2NO_3^- + 8H^+ + 6e^- & \longrightarrow & 2NO + 4H_2O \\ \hline 3Cu + 2NO_3^- + 8H^+ & \longrightarrow & 3Cu^{2+} + 2NO + 4H_2O \end{array} \qquad \text{(ionic equation)}$$

In checking the balancing of an equation like this, be certain to verify that the net charge is the same on both sides—as well as to count the atoms, of course.

	LEFT SIDE	RIGHT SIDE
Net charge	6+	6+
Cu atoms	3	3
N atoms	2	2
O atoms	6	6
H atoms	8	8

Another good method of balancing equations uses the changes in oxidation numbers instead of partial equations.

Observe that the number of electrons appearing in the partial equation is the same as the change in oxidation number of the atom being oxidized or reduced. In the oxidation, $Cu \longrightarrow Cu^{2+} + 2e^-$, the oxidation number of Cu changes from 0 to +2, corresponding to a *loss* of 2 electrons. In the reduction, the oxidation number of N changes from +5 (in NO_3^-) to +2 (in NO), a change of −3; accordingly, the partial equation shows that NO_3^- *gains* 3 electrons. Oxidation numbers are thus useful as a check on the correctness of the partial equations.

When the equation to be balanced is given, H_2O and H^+ are often omitted. If the reaction occurs in an acidic solution (understood to be aqueous), then H_2O is available as a source of O and H^+ as a source of H. For instance, the problem might have been to balance the equation $Cu + NO_3^- \longrightarrow Cu^{2+} + NO$ in acidic solution. We would then have been justified in adding H_2O and H^+ to either side as needed to balance the partial equations.

Molecular is not quite the right word, since only empirical formulas can be written for ionic compounds.

An ionic equation is the best representation of a reaction involving ionic compounds in solution. However, there cannot be just one kind of ion in a solution. What is actually used in a reaction is a compound made from ions of both signs. Therefore, we sometimes want to rewrite the final equation in a form that shows the actual compounds that went into the solution and that are obtained when the water is evaporated. Such an equation is said to be in "molecular" form. To convert an ionic equation to molecular form, decide what ions are present in the solution, add as many of them to each side as needed to make the net charge zero, and combine the ions into compounds. In the reaction just described, the solution contains reactants 8 H^+, 2 NO_3^- and product 3 Cu^{2+}. Therefore, add 6 NO_3^- to each side to cancel the 6+ charges:

$$3Cu + 8NO_3^- + 8H^+ \longrightarrow 3Cu^{2+} + 6NO_3^- + 2NO + 4H_2O$$

Now the positive and negative ions can be combined:

$$3Cu + 8HNO_3 \longrightarrow 3Cu(NO_3)_2 + 2NO + 4H_2O \qquad \text{(molecular form)}$$

EXAMPLE 9.12 Balance the equation

$$I^- + H_2O_2 \longrightarrow I_2 + H_2O \qquad \text{(acidic solution)}$$

ANSWER First write unbalanced partial equations:

$$I^- \longrightarrow I_2 \qquad H_2O_2 \longrightarrow H_2O$$

Then balance with respect to atoms:

$$2I^- \longrightarrow I_2 \qquad H_2O_2 \longrightarrow 2H_2O \qquad \text{(balanced for O)}$$

$$H_2O_2 + 2H^+ \longrightarrow 2H_2O \qquad \text{(balanced for H and O)}$$

Balance the charge:

$$2I^- \longrightarrow I_2 + 2e^- \qquad H_2O_2 + 2H^+ + 2e^- \longrightarrow 2H_2O$$

(oxidation) (reduction)

(In this case, the number of electrons lost happens to equal the number gained.) Finally, add the partial equations:

$$H_2O_2 + 2H^+ + 2I^- \longrightarrow I_2 + 2H_2O$$ ■

PROBLEM 9.12 Balance the equation

$$Cr_2O_7^{2-} + H_3AsO_3 \longrightarrow Cr^{3+} + H_3AsO_4 \qquad \text{(acidic solution)}$$ □

BALANCING IN BASIC SOLUTIONS

If a solution is basic, it contains very little H^+ but has a more or less high concentration of OH^-. We should balance the equation for a reaction in such a solution by using H_2O as the source of hydrogen and OH^- as the source of oxygen, as follows:

To Balance O Atoms: To the side deficient in O atoms add 2 OH^- for each 1 O atom needed. To the other side add 1 H_2O for each 2 OH^-.

$$:\ddot{O}:H^- + :\ddot{O}:H^- \longrightarrow H:\ddot{O}:H$$

To Balance H Atoms: To the side deficient in H atoms add 1 H_2O for each 1 H atom needed. To the other side add 1 OH^- for each 1 H_2O.

$$H\,:\ddot{O}:H \longrightarrow :\ddot{O}:H^-$$

Consider the reaction:

$$MnO_4^- + SO_3^{2-} \longrightarrow MnO_2(c) + SO_4^{2-} \qquad \text{(basic solution)}$$

The unbalanced partial equations are

$$MnO_4^- \longrightarrow MnO_2 \qquad (8)$$

and

$$SO_3^{2-} \longrightarrow SO_4^{2-} \qquad (9)$$

Equation 8 is short by 2 O on the right; therefore, add 4 OH^- to the right side:

$$MnO_4^- \longrightarrow MnO_2 + 4OH^-$$

Balancing the numbers of atoms requires 2 H_2O on the left side:

$$MnO_4^- + 2H_2O \longrightarrow MnO_2 + 4OH^-$$

The net charge is -1 on the left and -4 on the right. Add 3 e^- to the left:

$$MnO_4^- + 2H_2O + 3e^- \longrightarrow MnO_2 + 4OH^- \qquad \text{(reduction)} \qquad (10)$$

Equation 9 is balanced similarly. Add 2 OH^- on the left, which leaves H_2O on the right:

$$SO_3^{2-} + 2OH^- \longrightarrow SO_4^{2-} + H_2O$$

For electrical balancing, add 2 e^- to the right:

$$SO_3^{2-} + 2OH^- \longrightarrow SO_4^{2-} + H_2O + 2e^- \quad \text{(oxidation)} \qquad (11)$$

Before adding the equations, multiply equation 10 by 2 and equation 11 by 3:

$$\begin{array}{rcl} 2MnO_4^- + 4H_2O + 6e^- & \longrightarrow & 2MnO_2 + 8OH^- \\ 3SO_3^{2-} + 6OH^- & \longrightarrow & 3SO_4^{2-} + 3H_2O + 6e^- \\ \hline 2MnO_4^- + 3SO_3^{2-} + 4H_2O + 6OH^- & \longrightarrow & 2MnO_2 + 3SO_4^{2-} + 3H_2O + 8OH^- \end{array} \qquad (12)$$

Equation 12 can be simplified. H_2O and OH^- appear on both sides and should be canceled in part:

$$2MnO_4^- + 3SO_3^{2-} + H_2O \longrightarrow 2MnO_2 + 3SO_4^{2-} + 2OH^- \qquad (13)$$

EXAMPLE 9.13 Balance the equation $CrO_4^{2-} + I^- \longrightarrow Cr(OH)_4^- + IO_3^-$ in basic solution.

ANSWER

$CrO_4^{2-} \longrightarrow Cr(OH)_4^-$ (unbalanced)

Short 4 H on the left. Add 4 H_2O to the left and 4 OH^- to the right:

$$CrO_4^{2-} + 4H_2O \longrightarrow Cr(OH)_4^- + 4OH^-$$

$$CrO_4^{2-} + 4H_2O + 3e^- \longrightarrow Cr(OH)_4^- + 4OH^-$$

$$2CrO_4^{2-} + 8H_2O + 6e^- \longrightarrow 2Cr(OH)_4^- + 8OH^-$$

$I^- \longrightarrow IO_3^-$ (unbalanced)

Short 3 O on the left. Add 6 OH^- to the left and 3 H_2O to the right:

$$I^- + 6OH^- \longrightarrow IO_3^- + 3H_2O \quad \text{(balance atomically)}$$

$$I^- + 6OH^- \longrightarrow IO_3^- + 3H_2O + 6e^- \quad \text{(balance charges)}$$

$$I^- + 6OH^- \longrightarrow IO_3^- + 3H_2O + 6e^- \quad \text{(electron gain = electron loss)}$$

$$2CrO_4^{2-} + 8H_2O + I^- + 6OH^- \longrightarrow 2Cr(OH)_4^- + 8OH^- + IO_3^- + 3H_2O \quad \text{(add partial equations)}$$

$$2CrO_4^{2-} + 5H_2O + I^- \longrightarrow 2Cr(OH)_4^- + 2OH^- + IO_3^- \quad \text{(simplify)}$$ ■

PROBLEM 9.13 Balance the equation (in basic solution)

$$Cu(OH)_2(c) + HSnO_2^- \longrightarrow Cu(c) + Sn(OH)_6^{2-}$$ □

9.6 CHEMICAL ANALYSIS

OXIDATION–REDUCTION TITRATION REACTIONS

The reagent most often used in redox titrations is potassium permanganate, $KMnO_4$ (section 3.11). It is a strong oxidizing agent, which means that it will oxidize most reducing agents. It has another special virtue. MnO_4^- is very deep purple. A reaction such as

$$5Fe^{2+} + MnO_4^- + 8H^+ \longrightarrow 5Fe^{3+} + Mn^{2+} + 4H_2O$$

makes the purple color disappear as fast as $KMnO_4$ solution is added to a solution containing Fe^{2+}. When the last trace of Fe^{2+} has been consumed, the next drop of $KMnO_4$ solution has nothing to react with and it turns the whole solution faintly purple (usually called "pink"). Thus, no indicator is needed; $KMnO_4$ serves as its own indicator (Color plate 11).

EXAMPLE 9.14 In an acidic solution, 25.00 mL of 0.2100 M $KMnO_4$ is used to oxidize Fe^{2+} to Fe^{3+}. (a) Calculate the mass of Fe oxidized in the reaction

$$5Fe^{2+} + MnO_4^- + 8H^+ \longrightarrow Mn^{2+} + 5Fe^{3+} + 4H_2O$$

(b) If the mass of the Fe sample is 15.50 g, calculate the percentage of Fe in the sample.

ANSWER

(a) The number of moles of MnO_4^- is

$$25.00\ \text{mL KMnO}_4 \times \frac{0.2100\ \text{mol KMnO}_4}{1000\ \text{mL KMnO}_4} = 5.250 \times 10^{-3}\ \text{mol KMnO}_4$$

$$= 5.250 \times 10^{-3}\ \text{mol MnO}_4^-$$

From the balanced equation, 5 mol of Fe^{2+} reacts with 1 mol of MnO_4^- so that the mass of Fe is

$$5.250 \times 10^{-3}\ \text{mol MnO}_4 \times \frac{5\ \text{mol Fe}^{2+}}{1\ \text{mol KMnO}_4} \times \frac{55.85\ \text{g Fe}^{2+}}{1\ \text{mol Fe}^{2+}} = 1.466\ \text{g Fe}$$

(b) The percentage is (part/whole) × 100%:

$$\frac{1.466\ \text{g Fe}}{15.50\ \text{g sample}} \times 100\% = 9.458\%\ \text{Fe}$$

■

PROBLEM 9.14 Another method of determining iron is by titration of Fe^{2+} with potassium dichromate, $K_2Cr_2O_7$:

$$6Fe^{2+} + Cr_2O_7^{2-} + 14H^+ \longrightarrow 6Fe^{3+} + 2Cr^{3+} + 7H_2O$$

A 2.682-g sample is dissolved in H_2SO_4 solution; any Fe^{3+} is reduced to Fe^{2+}, and the solution is titrated with 39.62 mL of 0.02169 M $K_2Cr_2O_7$ solution. Calculate the percentage of Fe in the sample. □

PRECIPITATION REACTIONS

In addition to acid-base and redox reactions, the third category of titration reactions consists of those in which an insoluble salt is precipitated. The following example involves a familiar reaction: precipitation of white silver chloride; $Ag^+ + Cl^- \longrightarrow AgCl(c)$.

EXAMPLE 9.15 39.12 mL of 0.1102 M $AgNO_3$ is required to react with a 0.7515-g sample of a commercial mixture used for the preparation of club soda. The mixture contains NaCl. Calculate the percentage of NaCl in the sample.

ANSWER The number of moles of Ag^+ is

$$38.12\ \text{mL AgNO}_3 \times \frac{0.1102\ \text{mol AgNO}_3}{1000\ \text{mL AgNO}_3} = 4.311 \times 10^{-3}\ \text{mol AgNO}_3$$

$$= 4.311 \times 10^{-3}\ \text{mol Ag}^+$$

From the reaction, $Ag^+ + Cl^- \longrightarrow AgCl(c)$, 1 mol of Ag^+ reacts with 1 mol of NaCl (58.44 g). Therefore, the mass of NaCl in the sample is

$$4.311 \times 10^{-3}\ \text{mol Ag}^+ \times \frac{1\ \text{mol NaCl}}{1\ \text{mol Ag}^+} \times \frac{58.44\ \text{g NaCl}}{1\ \text{mol NaCl}} = 0.2519\ \text{g NaCl}$$

and the percentage of NaCl in the sample is

$$\frac{0.2519\ \text{g NaCl}}{0.7515\ \text{g sample}} \times 100\% = 33.52\%\ \text{NaCl}$$ ■

How can the end point in Example 9.15 be recognized? The answer is that chromate ion, CrO_4^{2-}, is used as an indicator. Silver chromate is more soluble than silver chloride. Addition of Ag^+ therefore first removes Cl^- as AgCl(c). After the Cl^- ions are removed, the next drop of $AgNO_3$ solution precipitates brick-red silver chromate, Ag_2CrO_4(c), thus signaling the end point.

PROBLEM 9.15 Oxalate ($C_2O_4^{2-}$) can be determined by titration with $Pb(NO_3)_2$ solution:

$$Pb^{2+} + C_2O_4^{2-} \qquad PbC_2O_4(\text{c, white})$$

Potassium iodide is used as an indicator. At the end point, PbI_2 precipitates:

$$Pb^{2+} + 2I^- \longrightarrow PbI_2(\text{c, yellow})$$

A 1.974-g sample is titrated with 23.68 mL of a 0.08724 M $Pb(NO_3)_2$ solution. What is the percentage of $Na_2C_2O_4$ in the sample? □

Torbern Bergman is regarded as the father of quantitative analysis; his method depended on the formation of a precipitate of known composition.

Precipitation reactions also form the basis of **gravimetric analysis**. In a gravimetric analysis, a known mass of sample is dissolved and a reagent is added to precipitate the substance whose quantity is being determined. The precipitate is then separated, dried, and weighed. Figure 9.3 outlines the general procedure.

EXAMPLE 9.16 A sample weighing 0.4572 g contains chlorine in the form of Cl^- ions. The Cl^- is precipitated as AgCl by the addition of excess $AgNO_3$. The mass of AgCl obtained is 0.4598 g. Calculate the percentage of Cl in the sample.

ANSWER First find the mass of Cl in 0.4598 g AgCl:

$$0.4598\ \text{g AgCl} \times \frac{1\ \text{mol AgCl}}{143.3\ \text{g AgCl}} \times \frac{1\ \text{mol Cl}}{1\ \text{mol AgCl}} \times \frac{35.45\ \text{g Cl}}{1\ \text{mol Cl}} = 0.1137\ \text{g Cl}$$

Then the percentage of Cl is

$$\frac{0.1137\ \text{g Cl}}{0.4572\ \text{g sample}} \times 100\% = 24.87\%\ \text{Cl}$$ ■

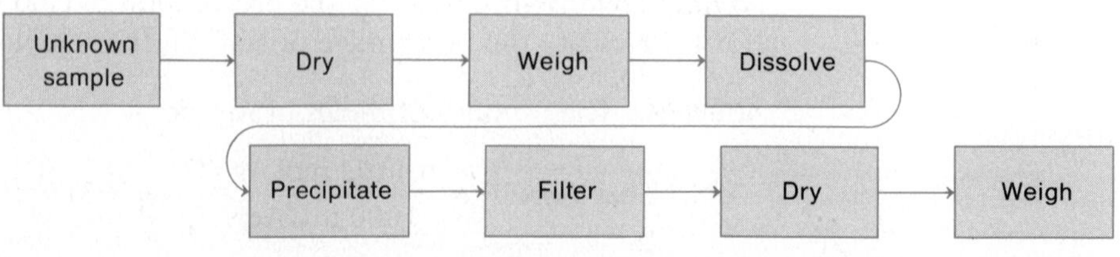

FIGURE 9.3
A representation of the steps in a typical gravimetric analysis.

PROBLEM 9.16 3.550 g of stainless steel is analyzed for nickel. Addition of dimethylglyoxime, $C_4H_8N_2O_2$, in a slightly basic solution, yields 1.470 g of red nickel dimethylglyoximate, $Ni(C_4H_7N_2O_2)_2$ (288.9 g/mol). Calculate the percentage of Ni in the steel. □

9.7 DECOMPOSITION REACTIONS

These reactions do not refer to reactions in aqueous solutions.

The breakdown of a compound into two or more substances is called a **decomposition reaction**. Examples of typical decomposition reactions are

1. The decomposition of water by passing an electric current through it (electrolysis, section 17.2):

$$2H_2O(\ell) \longrightarrow 2H_2(g) + O_2(g) \qquad \Delta H = +571.6 \text{ kJ}$$

2. The decomposition of ammonium dichromate (the "volcano reaction") (Box 9.2):

$$(NH_4)_2Cr_2O_7(c) \longrightarrow Cr_2O_3(c) + 4H_2O(g) + N_2(g) \qquad \Delta H = -300.2 \text{ kJ}$$

3. The decomposition of high molecular–weight hydrocarbons at high temperatures into smaller hydrocarbon molecules. This process is known as *cracking*. It is used by the petroleum industry in the manufacture of gasoline:

$$C_{10}H_{22}(g) \longrightarrow C_8H_{18}(g) + C_2H_4(g) \qquad \Delta H = +93.3 \text{ kJ}$$

4. The thermal decomposition (decomposition by heating) of limestone, calcium carbonate, $CaCO_3$ (section 20.3). Limestone and *dolomite* ($CaCO_3MgCO_3$) mountains are very common. Limestone is used in the manufacture of lime, CaO (Table 9.2):

$$CaCO_3(c) \longrightarrow CaO(c) + CO_2(g) \qquad \Delta H = +178.3 \text{ kJ}$$

This section deals with the *thermal decomposition* of some common salts—the carbonates (CO_3^{2-}), sulfates (SO_4^{2-}), sulfites (SO_3^{2-}), chlorates (ClO_3^-), perchlorates (ClO_4^-), and peroxides (O_2^{2-}). Three generalizations permit us to predict the products of thermal decomposition:

1. A sulfate, carbonate, or sulfite decomposes to the corresponding metal oxide and the nonmetal oxide without changes in oxidation numbers. For example:

$$\underset{+3\quad +6\,-2}{Fe_2(SO_4)_3(c)} \longrightarrow \underset{+3\quad -2}{Fe_2O_3(c)} + \underset{+6\,-2}{3SO_3(g)}$$

> **BOX 9.2**
> **SAFETY NOTE**
>
> While $(NH_4)_2Cr_2O_7$ (the "volcano") reacts at a controllable speed, many ammonium salts decompose at explosive speeds. Reactions involving salts with the N atom in its lowest oxidation state (−3) and another atom in its higher oxidation state may occur explosively. For example, ammonium nitrate, a fertilizer, is produced in large quantities (Table 9.2). Yet the explosion of a ship loaded with NH_4NO_3 demolished Texas City, Texas, in 1947. The reader is therefore advised to recognize the danger in heating ammonium salts.

DANGER: Chlorates and perchlorates can react explosively with organic matter.

2. A chlorate or perchlorate decomposes to the corresponding metal chloride and oxygen with changes in the oxidation numbers of the nonmetallic elements. For example:

$$\underset{+1\ +5\ -2}{KClO_3(c)} \longrightarrow \underset{+1\ -1}{KCl(c)} + 1\tfrac{1}{2}\underset{0}{O_2(g)}$$

$$\underset{+2\ \ +7-2}{Mg(ClO_4)_2(c)} \longrightarrow \underset{+2\ -1}{MgCl_2(c)} + 4\underset{0}{O_2(g)}$$

Safety Note: Heating peroxides is dangerous because many peroxides are explosive.

3. A peroxide decomposes to the corresponding metal oxide and oxygen with changes in the oxidation number of oxygen. For example:

$$\underset{+1\ -1}{Na_2O_2(c)} \longrightarrow \underset{+1\ -2}{Na_2O(c)} + \tfrac{1}{2}\underset{0}{O_2(g)}$$

EXAMPLE 9.17 With the aid of the three generalizations, predict the products and write a balanced equation for the thermal decomposition of each of these compounds: (a) scandium(III) sulfate, (b) lithium chlorate, (c) rubidium carbonate, (d) strontium peroxide, (e) barium perchlorate, and (f) lead(II) sulfite.

ANSWER

(a) The generalization for sulfates is

$$\text{sulfate} \longrightarrow \text{metal oxide} + \text{oxide of S}$$

without changes in oxidation numbers. Thus, we first find the oxidation numbers of the metal and the sulfur in SO_4^{2-}:

$$\underset{+3\ +6\ -2}{Sc_2(SO_4)_3}$$

The products are $\overset{+3\ -2}{Sc_2O_3}$ and $\overset{+6\ -2}{SO_3}$. The balanced equation is

$$Sc_2(SO_4)_3(c) \longrightarrow Sc_2O_3(c) + 3SO_3(g)$$

(b) The formula of lithium chlorate is $LiClO_3(c)$, and from the generalizations,

$$\text{chlorate} \longrightarrow \text{metal chloride} + O_2$$

$$\underset{+1+5-2}{LiClO_3(c)} \longrightarrow \underset{+1-1}{LiCl(c)} + 1\tfrac{1}{2}\underset{0}{O_2(g)}$$

(c) The formula of rubidium carbonate is $Rb_2CO_3(c)$, and from the generalizations,

$$\text{carbonate} \longrightarrow \text{metal oxide} + \text{oxide of C}$$

without changes in oxidation numbers. Find the oxidation numbers:

$$\underset{+1\ +4\ -2}{Rb_2CO_3}$$

The oxides are $\overset{+1\ -2}{Rb_2O}$ and $\overset{+4\ -2}{CO_2}$, and the balanced equation is

$$Rb_2CO_3(c) \longrightarrow Rb_2O(c) + CO_2(g)$$

(d) The formula of strontium peroxide is SrO_2(c), and from the generalizations,

$$\text{peroxides} \longrightarrow \text{metal oxide} + O_2$$

with changes in the oxidation number of O in O_2^{2-}:

$$\underset{+2\,-1}{SrO_2} \longrightarrow \underset{+2-2}{SrO} + \underset{0}{O_2} \qquad \text{(not balanced)}$$

Balancing:

$$SrO_2(c) \longrightarrow SrO(c) + \tfrac{1}{2}O_2(g)$$

(e) The formula of barium perchlorate is $Ba(ClO_4)_2$(c), and the generalization for perchlorates is

$$\text{perchlorate} \longrightarrow \text{metal chloride} + O_2$$

$$\underset{+2\ +7\ -2}{Ba(ClO_4)_2} \longrightarrow \underset{+2\ -1}{BaCl_2} + \underset{0}{O_2} \qquad \text{(not balanced)}$$

Balancing,

$$Ba(ClO_4)_2(c) \longrightarrow BaCl_2(c) + 4O_2$$

(f) The formula of lead(II) sulfite is $PbSO_3$(c); the generalization is

$$\text{sulfites} \longrightarrow \text{metal oxide} + \text{oxide of S}$$

without changes in oxidation numbers. Find the oxidation numbers:

$$\underset{+2+4-2}{PbSO_3}$$

The oxides are PbO and SO_2, and the equation is

$$PbSO_3(c) \longrightarrow \overset{+2\ -2}{PbO}(c) + \overset{+4\ -2}{SO_2}(g)$$

■

PROBLEM 9.17 Write a balanced equation for the thermal decomposition of each of these compounds: (a) BaO_2, (b) $MgCO_3$, (c) Na_2SO_3, (d) $Ba(ClO_3)_2$, (e) $NaClO_4$. (See Box 9.3.) □

STABILITIES OF METAL OXIDES

The activity series (Table 9.5) correctly predicts the thermal stabilities of metal oxides, as well as the ability of hydrogen to reduce the metal oxides to the metals:

Safety Note: This reaction, involved in the discovery of oxygen (Example 4.6), is frequently used in college chemistry laboratories for preparing small amounts of oxygen. Read the cautionary note on Example 4.6 before heating HgO.

- The oxides of the last four metals (Hg, Ag, Pt, and Au) decompose easily when they are heated; the decomposition of mercury(II) oxide is typical:

$$HgO(c) \longrightarrow Hg(\ell) + \tfrac{1}{2}O_2(g)$$

- The oxides of the metals above silver up to iron (Fe, Fe^{2+}) in Table 9.5 are not decomposed when they are heated alone at flame temperatures. However, they can be reduced to the metal when heated in the presence of hydrogen. Fe_2O_3 is typical:

$$Fe_2O_3(c) + 3H_2(g) \longrightarrow 3H_2O(g) + 2Fe(c)$$

BOX 9.3
DECOMPOSITION AND THERMAL ANALYSIS

Decomposition reactions are sufficiently predictable so that substances can be identified by **thermal analysis**: the substance to be identified is heated in the absence of oxygen and the changes in its mass are automatically recorded as a function of temperature. The thermal decomposition of calcium oxalate-water (1/1), $CaC_2O_4 \cdot H_2O$, also called calcium oxalate hydrate, is typical.

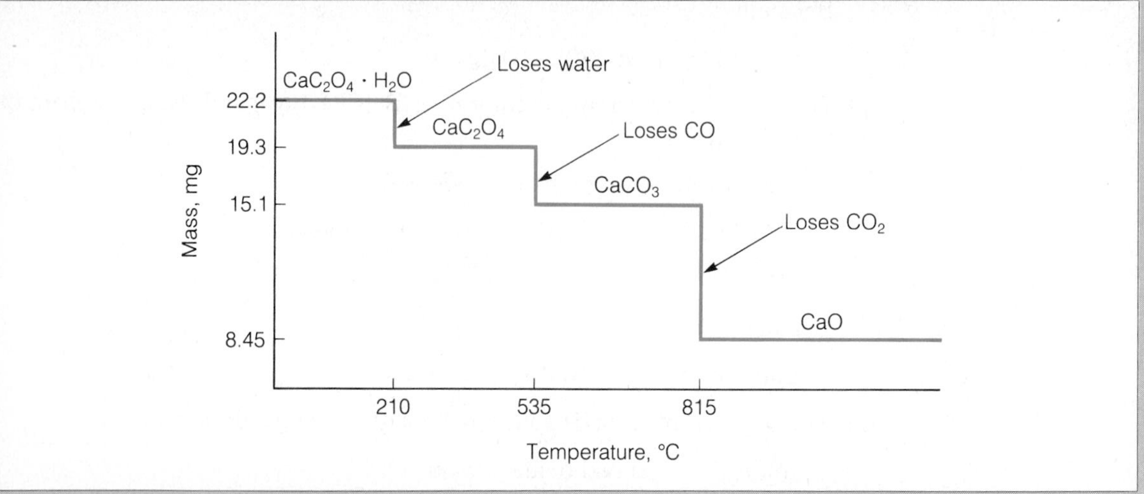

The thermal decomposition of $CaC_2O_4 \cdot H_2O$, a typical thermal analysis.

Li, Li^+ — Li_2O

Increasingly difficult to reduce ↑ — Increasing stability of oxides ↑

Au, Au^{3+} — Au_2O_3

■ The oxides of the metals above iron in the table are very stable. Magnesium oxide (*magnesia*, MgO) and calcium oxide (CaO), for example, are not decomposed at the highest temperatures obtained in electric furnaces. In fact, they are used for lining furnaces. Furthermore, it is impractical to reduce these metal oxides by treating them with hydrogen at high temperatures.

PROBLEM 9.18 Write a balanced equation for each reaction. If no reaction occurs, write "no reaction." *Heat* refers to heating with an ordinary Bunsen burner.

(a) $Na_2O + \text{heat} \longrightarrow ?$
(b) $Ag_2O + \text{heat} \longrightarrow ?$
(c) $Al_2O_3 + H_2 + \text{heat} \longrightarrow ?$
(d) $PbO + H_2 + \text{heat} \longrightarrow ?$ □

Used as early as 3000 years ago, iron became the principal metal about 1200 years ago. Can you explain the order of utilization?

These stability rules predict correctly that only the metals below hydrogen (in Table 9.5) are found uncombined in nature. They do not reduce H^+ in naturally acidic waters and they are not oxidized in air (oxygen). Those above hydrogen, if they existed uncombined in nature, would displace hydrogen (reduce H^+) from natural acid solutions. It is only in exceptional cases, such as iron in meteorites, that metals above hydrogen are found in nature. Historically, the use of the metals by humans follows the order from the bottom of Table 9.5 upward: gold and silver, 10,000–20,000 years ago; copper, 6000–7000 years ago; bronze (Cu-Sn), 2500 years ago; iron, 1200 years ago; aluminum, 130 years ago.

9.8 COMBINATION REACTIONS

Reactions in which two or more substances combine to form another substance are called **combination reactions**. These reactions may involve changes in oxidation numbers. Elements and compounds can combine in three ways:

1. *Two elements may combine to form a compound.* The reaction of the elements of Groups 1, 2, 3, 4, and 5 with those of Groups 6 and 7 are typical of this kind of combination reaction. For example, when phosphorus reacts with a limited supply of chlorine the product formed is PCl_3, phosphorus trichloride:

   $$2P(c) + 3Cl_2(g) \longrightarrow 2PCl_3(\ell)$$

   With an excess of chlorine, however, the product is PCl_5, phosphorus pentachloride:

   $$2P(c) + 5Cl_2(g) \longrightarrow 2PCl_5(c)$$

   Most of the metals also combine with sulfur to form sulfides, of which Na_2S, Al_2S_3, FeS, and Fe_2S_3 are typical.

2. *An element and a compound may combine to form another compound.* An example is the manufacture of methanol ("wood alcohol"), CH_3OH (Table 9.2), from carbon monoxide and hydrogen at high temperature and high pressure:

   $$CO(g) + 2H_2(g) \longrightarrow CH_3OH(\ell)$$

3. *Two compounds may combine to form another compound.* The preparation of sulfuric acid from SO_3 (section 20.2) and the preparation of phosphoric acid from tetraphosphorus decaoxide are typical:

   $$SO_3(g) + H_2O(\ell) \longrightarrow H_2SO_4(aq)$$

   $$P_4O_{10}(c) + 6H_2O(\ell) \longrightarrow 4H_3PO_4(aq)$$

   Dilute phosphoric acid is used in sodas, jams, and jellies (Table 9.2).

We are ignoring the unusual cases in which oxidation numbers change when two compounds combine.

Notice that *changes in oxidation numbers generally occur in combination reactions only when elements are involved.* The oxidation number of an element is always zero. In a compound, an element only rarely has an oxidation number of zero.

EXAMPLE 9.18 Several gaseous oxides of nitrogen exist: N_2O, NO, NO_2, N_2O_3, N_2O_4, and N_2O_5. Which oxide will react with water to form HNO_3?

ANSWER Assume that no change in oxidation number occurs when two compounds combine to form a third compound. The oxidation number of N in HNO_3 is +5. Therefore, the N in the oxide has the same oxidation number, +5; the oxide is N_2O_5:

$$N_2O_5(g) + H_2O \longrightarrow 2HNO_3(aq)$$ ■

EXAMPLE 9.19 Predict the formula of the hydroxide formed by the combination of CaO and H_2O and write the balanced equation.

ANSWER Assume that when two compounds combine, the oxidation numbers remain constant. The oxidation number of Ca is +2. The formula of the hydroxide is therefore $Ca(OH)_2$, and the balanced equation is

$$CaO(c) + H_2O \longrightarrow Ca(OH)_2(c)$$ ■

When calcium oxide (''quicklime'') reacts with water, the product, slightly soluble calcium hydroxide, is called *slaked lime* (Danish *slukke,* ''to quench thirst''). Mortar, used in the construction industry, is made by mixing slaked lime with sand and sufficient water to form a pasty mass. Plaster used for coating walls and ceilings is a mortar to which a fibrous binding material is added to hold it in place. Calcium hydroxide is the cheapest of the hydroxides and is therefore widely used in the chemical industry to neutralize acids.

PROBLEM 9.19 Write the formula of the compound formed by the combination of ammonia(g) and nitric acid(g). Do oxidation numbers change in this reaction? □

SUMMARY

Knowledge of the properties of substances in water solution enables us to predict certain kinds of solution reactions. **Electrolytes** are substances that form solutions in water that conduct an electric current. Electrolytes vary in strength. **Strong electrolytes** exist almost completely as ions in water. **Weak electrolytes** exist mainly as molecules along with a small percentage of ions in water. **Nonelectrolytes** exist completely as molecules in water, yielding solutions that do not conduct an electric current. According to the **Arrhenius definition**, an **acid** is a compound that produces hydrogen ions, H^+, or **hydronium ions**, H_3O^+, in water. **Strong acids** exist almost completely as ions in water. A **weak acid** is only slightly ionized in water. According to the Arrhenius definition, a **base** is a compound that produces OH^- ions in water. A **strong base**, also known as an **alkali**, exists almost completely as ions in water. A **weak base** exists mainly as molecules with a small percentage of ions in water. **Salts**, products of the reactions between acids and bases (**neutralization**), are strong electrolytes. A **net ionic equation** is obtained by writing strong electrolytes as ions in a chemical equation and then canceling the **spectator ions**, identical ions that are on both the left and right sides of the equation.

A reaction in water solution can be predicted by recognizing the formation of either a weak electrolyte, a **precipitate**, an insoluble solid, an insoluble gas, or a **coordination ion**. A coordination ion is generally a transition metal ion that forms covalent bonds with anions or molecules.

Oxidation–reduction (redox) reactions are reactions in which some atoms change their oxidation numbers or gain or lose electrons. Oxidation and reduction always occur simultaneously in a reaction. The substance that is oxidized is called the **reducing agent**; it is an *electron donor* and reduces the other substance. The substance that is reduced is called the **oxidizing agent**; it is an *electron acceptor* and oxidizes the other substance.

A listing of the metals in the order of their tendency to be oxidized—the **activity series** of the metals—enables us to predict the redox reactions between metals and metal ions: a metal higher in the series will reduce a lower ion. The activity series also predicts that the metals above hydrogen will displace it from water or acidic solutions. A reaction in which one element displaces another from a compound is called a **displacement reaction**.

We can balance equations for oxidation–reduction reactions by first writing separate **partial equations**, one for oxidation and one for reduction. If the solution is acidic, we balance each partial equation by adding H_2O to the side deficient in O and H^+ to the side deficient in H. If the solution is basic, we balance each partial equation by adding OH^- to the side deficient in O and H_2O to the side deficient in H. Then we add electrons to each partial equation to balance the charges, and multiply each partial equation by a coefficient. This equalizes the gain and loss of electrons. Finally, we add the partial equations.

In addition to acid–base reactions, titrations make use of oxidation–reduction reactions and precipitation reactions. Precipitation reactions also form the basis of **gravimetric analysis**, done entirely by weighing.

Decomposition reactions involve the breakdown of a substance into two or more substances. A sulfate, carbonate, or sulfite decomposes to the corresponding metal oxide and nonmetal oxide without any changes in oxidation numbers. A chlorate or perchlorate decomposes to the corresponding metal chloride and oxygen with changes in the oxidation numbers of the nonmetallic elements. A peroxide decomposes to the corresponding metal oxide and oxygen with changes in the oxidation number of oxygen. The activity series predicts the thermal stabilities of metal oxides and the ability of hydrogen to reduce a metal oxide to the metal.

Combination reactions are reactions in which two or more substances combine to form another substance. There is a change in oxidation numbers only when an element is involved.

9.9 ADDITIONAL PROBLEMS

ACIDS, BASES, SALTS

9.20 List at least three characteristic properties of acids and of bases.

9.21 Summarize the electrical properties of strong electrolytes, weak electrolytes, and nonelectrolytes.

9.22 Write the formulas of two soluble and two insoluble chlorides, sulfates, and hydroxides.

9.23 Describe an experiment for classifying each of these compounds as a strong electrolyte, a weak electrolyte, or a nonelectrolyte: K_2CO_3, HCN, C_2H_5COOH, CH_3OH, H_2S, H_2SO_4, H_2CO, NH_3.

9.24 Which of these are acids? HBr, NH_3, H_2SeO_4, BF_3, H_3SbO_4, $Al(OH)_3$, H_2S, C_6H_6, CsOH, H_3BO_3, and HCN.

9.25 Which of these are bases? NaOH, H_2Se, BaO, BCl_3, NH_3.

9.26 Classify each substance as either an electrolyte or a nonelectrolyte: NH_4Cl, HI, C_6H_6, RaF_2, $Zn(C_2H_3O_2)_2$, $Cu(NO_3)_2$, $HC_2H_3O_2$, $C_{12}H_{22}O_{11}$, LiOH, $KHCO_3$, CCl_4, $La_2(SO_4)_3$, I_2.

9.27 Classify each substance as either a strong or weak electrolyte, and then list (a) the strong acids, (b) the strong bases, (c) the weak acids, (d) the weak bases: NaCl, $MgSO_4$, HCl, $H_2C_2O_4$, $Ba(NO_3)_2$, H_3PO_4, $Sr(OH)_2$, HNO_3, HI, $Ba(OH)_2$, LiOH, $HC_3H_5O_2$, NH_3("NH_4OH"), CH_3NH_2, KOH, $MgMoO_4$, HCN, $HClO_4$.

9.28 Which is the stronger acid in each pair? (a) H_2SO_4, H_3PO_4; (b) HCN, HCl; (c) HNO_2, HNO_3; (d) $HClO_4$, HClO.

9.29 Write a balanced equation for the preparation of each salt by a neutralization reaction: $Ca(NO_3)_2$, SrC_2O_4, $Zr(SO_4)_2$, $ZnSO_3$, $(NH_4)_2CO_3$.

9.30 Classify each salt as either soluble or insoluble in water: CuS, $CuSO_4$, $RaSO_4$, $NiCO_3$, $Fe(OH)_3$.

NET IONIC EQUATIONS

9.31 Write net ionic equations for the following reactions in water.

(a) $Pb(C_2H_3O_2)_2 + Cs_2SO_4 \longrightarrow PbSO_4 + CsC_2H_3O_2$
(b) $CO_2 + Ba(OH)_2 \longrightarrow BaCO_3 + H_2O$
(c) $Na_3PO_4 + CaCl_2 \longrightarrow Ca_3(PO_4)_2 + NaCl$
(d) $HCHO_2$(formic acid) $+ Mg(OH)_2 \longrightarrow Mg(CHO_2)_2 + H_2O$
(e) $FeS + H_2SO_4 \longrightarrow FeSO_4 + H_2S$
(f) $Cu(OH)_2 + NH_3 \longrightarrow Cu(NH_3)_4(OH)_2$

9.32 Complete and balance each equation in molecular form. Then rewrite the equation in its ionic form and in net ionic form. (All reactions are in aqueous solutions.)

(a) $Zn(OH)_2 + HCl \longrightarrow$
(b) $Al(OH)_3 + H_2SO_4 \longrightarrow$
(c) $Li_2CO_3 + HF \longrightarrow$
(d) $Mg(CN)_2 + HI \longrightarrow$
(e) $NaCl + AgNO_3 \longrightarrow$
(f) $BaCl_2$ r $K_2SO_4 \longrightarrow$
(g) $Ca(NO_3)_2 + Li_3PO_4 \longrightarrow$
(h) $CoCl_3 + NaCN \longrightarrow$
(i) $(NH_4)_2SO_4 + NaOH \longrightarrow$

9.33 Predict the product, if any, and write the net ionic equation for the reaction that occurs when the following aqueous solutions are mixed:

(a) $BaCl_2 + (NH_4)_2CrO_4 \longrightarrow$
(b) $Mg(ClO_3)_2 + K_2CO_3 \longrightarrow$

(c) $Al(OH)_3(c) + H_3PO_4 \longrightarrow$
(d) $KNO_3 + (NH_4)_2SO_4 \longrightarrow$
(e) $Cr(NO_3)_3 + NaCN \longrightarrow$
(f) $Cu(NO_3)_2 + NaOH \longrightarrow$
(g) $ZnSO_4 + (NH_4)_2S \longrightarrow$
(h) $Ba(OH)_2 + HC_2H_3O_2 \longrightarrow$

9.34 Write net ionic equations for these reactions in water:
(a) potassium sulfite + hydrochloric acid $\longrightarrow$
(b) cesium hydroxide + ammonium sulfate $\longrightarrow$
(c) nickel(II) nitrate + potassium cyanide $\longrightarrow$
(d) sulfuric acid + barium hydroxide $\longrightarrow$
(e) silver nitrate + lithium chloride $\longrightarrow$
(f) zinc chloride + hydrogen sulfide $\longrightarrow$
(g) potassium hydroxide + hydrogen cyanide $\longrightarrow$

9.35 Write the formulas of any two substances that can be mixed in water to obtain each of the following: $CuCO_3(c)$, $Fe(CN)_6{}^{3-}$, $Ag_2CrO_4(c)$, $Hg_3(PO_4)_2(c)$, $NH_3(g)$, $Ni(OH)_2(c)$.

ACTIVITY SERIES

9.36 Define and illustrate the following by using the concept of electron transfer and changes in oxidation numbers: (a) oxidation, (b) reduction, (c) oxidizing agent, (d) reducing agent.

9.37 What is oxidized, what is reduced, what is the oxidizing agent, and what is the reducing agent in each reaction?
(a) $Mg(c) + Sn^{2+}(aq) \longrightarrow Sn(c) + Mg^{2+}(aq)$
(b) $3Zn(c) + 2CoCl_3(aq) \longrightarrow 3ZnCl_2(aq) + 2Co(c)$
(c) $3H_2SO_3(aq) + HIO_3(aq) \longrightarrow 3H_2SO_4(aq) + HI(aq)$
(d) $CH_4(g) + 4Cl_2(g) \longrightarrow CCl_4(\ell) + 4HCl(g)$

9.38 Name two common metals that (a) *do not* displace hydrogen (1) from water, (2) from water or acid solutions; (b) *do* displace hydrogen (1) from water, (2) from acid solutions but not from water. Write net ionic equations for the reactions that occur.

9.39 Predict the products of each mixture. If a reaction occurs, write the net ionic equation. If no reaction occurs, write "no reaction."
(a) $Ba(c) + Al_2O_3(c) \longrightarrow$
(b) $Sn(c) + Al_2O_3(c) \longrightarrow$
(c) $Cd^{2+}(aq) + Al(c) \longrightarrow$
(d) $Ca(c) + H_2O \longrightarrow$
(e) $Ni(c) + H_2O \longrightarrow$
(f) $Hg(\ell) + H_2O \longrightarrow$
(g) $Hg(\ell) + HCl(aq) \longrightarrow$
(h) $Ni(c) + H_2SO_4(aq) \longrightarrow$
(i) $Fe(c) + H_2SO_4(aq) \longrightarrow$

9.40 Which is the stronger oxidizing agent in each pair? (a) K^+, Ca^{2+}; (b) Ca, Ca^{2+}; (c) Li, Au^{3+}; (d) Na^+, Pt^{2+}.

9.41 On the basis of (first) ionization energy, which metal in each pair is more easily oxidized? (a) Na, K; (b) Na, Al; (c) Li, Na; (d) Mg, Al; (e) K, Ca; (f) Ca, Ba. Do your answers agree with Table 9.5?

9.42 The least reactive metals are located in which part of the periodic table (inside front cover)?

9.43 What mass of Zn is needed to displace 12.5 g Cr from $[Cr(NH_3)_6]Cl_3$?

BALANCING OXIDATION–REDUCTION EQUATIONS

9.44 Balance these equations for reactions in acidic solutions by the ion-electron method. Show the balanced partial equations for oxidation and reduction and the ionic equation for the net reaction.
(a) $MnO_4{}^- + H_2C_2O_4 \longrightarrow Mn^{2+} + CO_2$
(b) $IO_3{}^- + Cl^- + N_2H_4 \longrightarrow ICl_2{}^- + N_2$
(c) $Zn + NO_3{}^- \longrightarrow Zn^{2+} + NH_4{}^+$
(d) $I_2 + S_2O_3{}^{2-} \longrightarrow S_4O_6{}^{2-} + I^-$
(e) $NO_2{}^- + I^- \longrightarrow I_2 + NO(g)$
(f) $Ag^+ + AsH_3 \longrightarrow H_3AsO_4 + Ag(c)$

9.45 Balance these equations for reactions in basic solutions by the ion-electron method. Show the balanced partial equations for oxidation and reduction and the ionic equation for the net reaction.
(a) $MnO_4{}^- + IO_3{}^- \longrightarrow IO_4{}^- + MnO_2$
(b) $SO_3{}^{2-} + MnO_4{}^- \longrightarrow SO_4{}^{2-} + MnO_4{}^{2-}$
(c) $Cl_2 \longrightarrow Cl^- + ClO_3{}^-$

9.46 Chemists had long wanted to synthesize "squaric acid,"

```
HO         OH
  \       /
   C === C
   |     |
   C --- C
  //      \\
 O          O
```

both as an interesting compound and as a starting point for useful products. Finally it was synthesized in 1972 by electrochemical reduction of CO under pressure in a nonaqueous solvent containing solvated H^+ ions. Write a partial equation for this reduction.

CHEMICAL ANALYSIS

9.47 A 25.00-mL solution of $Ba(OH)_2$ is titrated with 34.26 mL of 0.1163 M HCl

$$Ba(OH)_2 + 2HCl \longrightarrow BaCl_2 + 2H_2O$$

Find the molarity of the $Ba(OH)_2$ solution.

9.48 A solution of $KMnO_4$ contains 95.5 g $KMnO_4$ in 500 mL. Find the molarity with respect to (a) $KMnO_4$ and (b) $MnO_4{}^-$.

9.49 What mass of Fe is needed to react with 36.9 mL of 0.135 M $KMnO_4$? The Fe is first dissolved in acid to give Fe^{2+}.

$$5Fe^{2+} + MnO_4{}^- + 8H^+ \longrightarrow 5Fe^{3+} + Mn^{2+} + 4H_2O$$

9.50 The iron in a sample containing some Fe_2O_3 is reduced to Fe^{2+}. The Fe^{2+} is titrated with 12.02 mL of 0.1167 M $K_2Cr_2O_7$ in an acid solution.

$$6Fe^{2+} + Cr_2O_7{}^{2-} + 14H^+ \longrightarrow 6Fe^{3+} + 2Cr^{3+} + 7H_2O$$

Find (a) the mass of Fe and (b) the percentage of Fe in a 5.675-g sample.

9.51 Copper(II) ion, Cu^{2+}, can be determined by the net reaction:

$$2Cu^{2+} + 2I^- + 2S_2O_3^{2-} \longrightarrow 2CuI(c) + S_4O_6^{2-}$$

A 2.075-g sample containing $CuSO_4$ and excess KI is titrated with 41.75 mL of 0.1214 M solution of $Na_2S_2O_3$. What is the percentage of $CuSO_4$ (159.6 g/mol) in the sample?

9.52 Find the volume of 0.150 M HI solution required to titrate
(a) 25.0 mL of 0.100 M NaOH
(b) 5.03 g of $AgNO_3$ ($Ag^+ + I^- \longrightarrow AgI(c)$)
(c) 0.621 g $CuSO_4$ ($2Cu^{2+} + 4I^- \longrightarrow 2CuI(c) + I_2(c)$

9.53 An excess of $AgNO_3$ added to a 0.1555-g sample of groundwater containing $MgCl_2$ precipitates 0.01655 g AgCl. Find the percentage of $MgCl_2$ in the groundwater.

9.54 A 0.9346-g sample of brass, dissolved and analyzed for zinc, precipitates $ZnNH_4PO_4$ on addition of $(NH_4)_2HPO_4$. After the precipitate is dried and heated at about 900°C, 0.3502 g $Zn_2P_2O_7$ (304.7 g/mol) is obtained. Find the percentage of Zn in the brass.

DECOMPOSITION

9.55 Predict the products and write a balanced equation for the thermal decomposition of each substance: (a) $NaClO_3$, (b) $Cr_2(SO_4)_3$, (c) $BaCO_3$, (d) MgO_2, (e) $Ca(ClO_4)_2$, (f) $FeSO_3$.

9.56 At about 500°C, potassium chlorate goes to potassium perchlorate and potassium chloride and then the potassium perchlorate decomposes. Write balanced equations for these two reactions.

9.57 Name two metal oxides that are (a) decomposed easily when heated; (b) not decomposed at high temperatures; (c) reduced to the corresponding metal only when heated in an atmosphere of hydrogen. Write a balanced equation for each reaction.

9.58 Complete and balance these decomposition reactions. The given formulas and coefficients should not be changed. Substitute a formula (with coefficient) for each question mark.
(a) $2CO_2 \longrightarrow O_2 + ?$
(b) $C_2H_5OH \longrightarrow H_2O + ?$
(c) $CuSO_4(H_2O)_5 \longrightarrow 2H_2O + ?$
(d) $2HIO_3 \longrightarrow H_2O + ?$
(e) $N_2O_4 \longrightarrow ?$
(f) $FeCl_3 \longrightarrow FeCl_2 + ?$
(g) $(PbO)_2PbCO_3 \longrightarrow CO_2 + ?$
(h) $3PbO_2 \longrightarrow O_2 + ?$

9.59 Write the balanced equation for the thermal decomposition of solid silver carbonate.

COMBINATION

9.60 Define and illustrate three types of combination reactions.

9.61 Write the *empirical* formula (with coefficient) of the product obtained from each set of reactants by a combination reaction. Indicate any changes in oxidation numbers.
(a) $Ca + H_2 \longrightarrow$
(b) $2Fe + 3Cl_2 \longrightarrow$
(c) $K_2O + \frac{1}{2}O_2 \longrightarrow$
(d) $2Cu + \frac{1}{2}O_2 \longrightarrow$
(e) $2Cu + O_2 \longrightarrow$
(f) $Cl_2O_5 + H_2O \longrightarrow$
(g) $SeO_3 + K_2O \longrightarrow$
(h) $8PH_3 + P_4 \longrightarrow$
(i) $Na_4P_2O_7 + H_2O \longrightarrow$
(j) $P_4O_{10} + 6Na_2O \longrightarrow$
(k) $4Fe_3O_4 + O_2 \longrightarrow$
(l) $3H_3PO_4 + PH_3 \longrightarrow$

9.62 Write the formula of the oxide you would add to water to produce each compound: H_3PO_4, $Ba(OH)_2$, $HClO_4$, HNO_2, KOH.

SELF-TEST

9.63 (10 points) Which of the following salts are insoluble in water? silver chlorate, silver chloride, lead sulfide, lead chloride, lead nitrate, calcium sulfate, calcium hydrogen carbonate, sodium chromate, barium chromate, ammonium phosphate, mercury(II) nitrate, mercury(I) chloride, mercury(II) chloride, mercury(I) sulfate, bismuth phosphate.

9.64 (10 points) What dissolved salt (aq) would you add to an aqueous solution containing (a) Cl^-, NO_3^-, ClO_3^-, and $C_2H_3O_2^-$ to separate Cl^- from the other anions; (b) SO_4^{2-}, Cl^-, $C_2H_3O_2^-$, and Br^- to separate SO_4^{2-} from the other anions; (c) Ag^+, NH_4^+, Ba^{2+}, and Na^+ to separate Ag^+ from the other cations; (d) Pb^{2+}, Zn^{2+}, Co^{3+}, and K^+ to separate Pb^{2+} from the other cations?

9.65 (10 points) Write net ionic equations only for the reactions that occur in water:
(a) $Rb_2SO_4 + HCl \longrightarrow$
(b) $Hg_2(NO_3)_2 + MgCl_2 \longrightarrow$
(c) $Ba(ClO_3)_2 + NaNO_3 \longrightarrow$
(d) $NaClO_4 + HCN \longrightarrow$
(e) $HClO_4 + NaCN \longrightarrow$

9.66 (10 points) Predict the products formed in each of the following mixtures. If a reaction occurs, write the net ionic equation and write the formula of the oxidizing agent. If no reaction occurs, write "no reaction."
(a) $Al(c) + Cr_2O_3(c) \longrightarrow$
(b) $Cu(c) + H_2SO_4(aq) \longrightarrow$
(c) $Al(c) + H_2SO_4(aq) \longrightarrow$
(d) $Cr^{3+}(aq) + Zn(c) \longrightarrow$
(e) $Ni(c) + Zn^{2+}(aq) \longrightarrow$

9.67 (8 points) Which is the stronger reducing agent in each pair? (a) Ca, Ca^{2+}; (b) K, Ca; (c) Li^+, Zn; (d) Sn, Zn; (e) H_2, Hg.

9.68 (8 points) Write a balanced equation for the thermal decomposition of each of these solids: (a) $ZnCO_3$, (b) $TlClO_3$, (c) $Cr_2(SO_3)_3$.

9.69 (5 points) Which metal oxide in each pair is more easily reduced to the metal? (a) ZnO, CdO; (b) Na_2O, ZnO; (c) CuO, CdO; (d) SnO, PtO.

9.70 (10 points) Write a balanced equation for the combination reaction by which each of these compounds is prepared from two reactants: $CaCl_2$, NH_3, Cr_2S_3, SO_2, P_4O_6, As_2O_5, $CaSiO_3$, $BaCO_3$, $Mg(OH)_2$, H_3PO_3, LiOH, H_3AsO_4, H_2SO_4.

9.71 (10 points) Balance these equations by the ion-electron method:

(a) $Cr_2O_7^{2-} + H_2S \longrightarrow Cr^{3+} + S(c)$ (acidic solution)

(b) $ClO_3^- + N_2H_4 \longrightarrow NO_3^- + Cl^-$ (basic solution)

9.72 (10 points) (a) A solution of sodium thiosulfate, $Na_2S_2O_3$, is 0.1455 M. 25.00 mL of this solution reacts with 26.36 mL of I_2 solution. What is the molarity of the I_2 solution?

$$2Na_2S_2O_3 + I_2 \longrightarrow Na_2S_4O_6 + 2NaI$$

(b) 25.32 mL of the I_2 solution is required to titrate a sample containing As_2O_3. Calculate the mass of As_2O_3 (197.8 g/mol) in the sample.

$$As_2O_3 + 5H_2O + 2I_2 \longrightarrow 2H_3AsO_4 + 4HI$$

9.73 (9 points) A 0.5483-g sample containing aluminum is dissolved, precipitated as $Al(OH)_3$, and dehydrated to Al_2O_3; 0.1702 g Al_2O_3 is obtained. Find the percentage of Al in the sample.

10 GASES

The word *gas* was coined by Jan van Helmont in 1624—it is derived from the Greek *chaos* for "empty space" or "confusion." Gases have always been regarded as somewhat more mysterious than solids or liquids, because gases are usually invisible, although many of them can be heard and felt (such as a howling wind) or smelled. In fact, the ancients barely considered gases to be ordinary matter. Such modern day expressions as "spirits of ammonia" still convey the connotation that gases are somewhat unreal. It was only in the seventeenth century, when techniques became available for confining gases at elevated or reduced pressures, that their properties could be studied in a quantitative way.

The scientific study of gases led to an understanding of molecules and to methods of measuring their relative masses, "molecular weights." Experiments with gases, especially at low pressures, have continued to be important to the present time, because such research enables chemists to study the structures and properties of molecules when they exist as individual, separated particles. Many lasers are based on purely gas-phase processes.

This chapter, then, describes and interprets the behavior of gases.

10.1 THE PROPERTIES OF GASES

Air is a gas, and so is the "natural gas" (mostly methane, CH_4) that we use for heating our homes and cooking. What you smell from a flower or from a rotten egg is also a gas. Let us take a look at the distinguishing properties of gases:

1. They are transparent. Most of the common gases are colorless, but there are some well-known colored ones, such as fluorine and chlorine (both greenish yellow) and nitrogen dioxide (reddish brown).
2. They show little resistance to flow compared with liquids or solids; they spread out rapidly in space and flow through very small openings.
3. They can be mixed with each other in all proportions (assuming they do not react chemically); such mixtures are therefore true solutions. Once mixed, gases in closed containers never separate from each other spontaneously, nor can they be separated by filters. For example, carbon monoxide gas is found in all cigarette smoke; no filter can remove it.
4. Under ordinary conditions, they are much less dense than liquids or solids and they can be compressed much more easily.

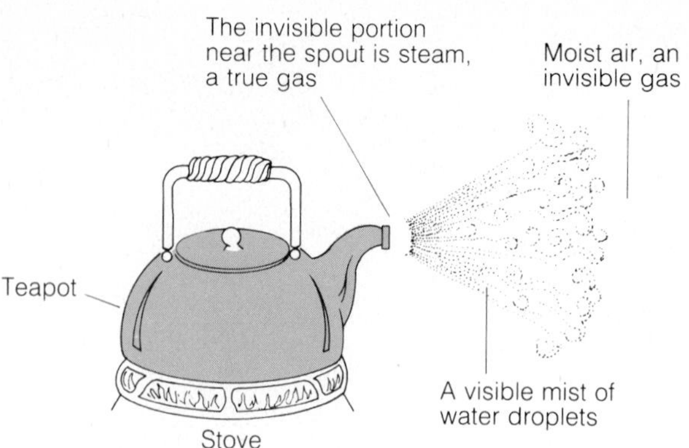

FIGURE 10.1
Mist is visible; steam is invisible.

5. They expand when they are heated and contract when they are cooled to a much greater extent than liquids or solids do.

The properties of gases are best understood in terms of the motions of their molecules. A gas may be characterized as that *state of matter in which the molecules have no orderly pattern of arrangement* and in which the *molecules' attractions for each other are too weak to establish a definite boundary*. For this reason *a gas assumes the shape and volume of its container.*

Smokes and mists, which are not transparent, are not gases; they consist of solid particles or liquid droplets that are much larger than typical gas molecules. **Steam** is the gaseous state of water, but the word is often misused. What you see coming out of a teakettle of boiling water is not steam, but the **mist** of water droplets that forms after steam cools; the invisible matter between the spout and the mist (Figure 10.1) is the steam. When you get a chance, look carefully at a smokestack that is discharging a white smoke. If the smoke rises sharply before it starts to drift with the wind, it is probably quite hot, and if there is an invisible gap between the top of the stack and the beginning of the white smoke, the material in the gap may well be steam and the "smoke" may just be a water mist, with or without some other matter.

If all the air in a 20-cubic-meter room (the size of a typical office) at ordinary atmospheric conditions were liquefied, it would occupy just slightly more than 20 liters (a large paint bucket). If one assumes that the molecules in a liquid are packed closely together, then it follows that the molecules in air occupy only about 0.1 percent of the total space; the other 99.9 percent is empty.

10.2 THE PRESSURE OF A GAS

We often use the word *pressure* in ordinary conversation, as when we say, "The pressure in my front tire is low," or "The nurse took my blood pressure." You may also be familiar with the fact that the pressure of water on a diver increases with the depth of the dive but is not affected by the surface area of the water. Thus, the pressure at a depth of 10 meters in Lake Superior is the same as the pressure 10 meters below the surface of a pond or a narrow tank of water.

The pascal is named for Blaise Pascal, who recognized in 1654 that pressure is a force per unit area.

Pressure is force per unit area. The SI unit of force is the **newton**, N, which is 1 kg m/sec^2. The SI unit of pressure is the **pascal**, Pa, which is one newton per square meter. A pascal represents a rather small pressure. For example, one pascal is the pressure of the atmosphere at an elevation of about 97 km (60 mi), far above the stratosphere. Or as another example, the weight of a U.S. penny, lying flat in the palm of your hand, exerts a pressure of about 100 Pa. A more convenient

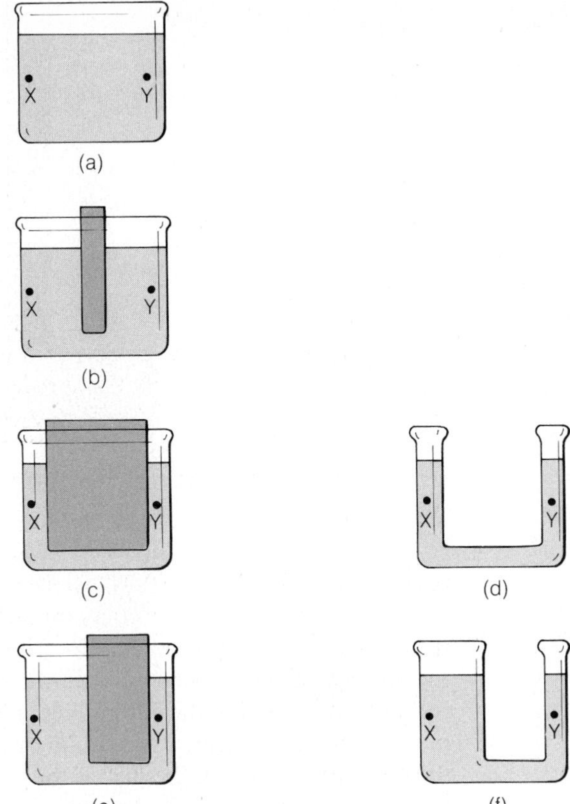

FIGURE 10.2
Points X and Y are at the same levels and therefore at equal pressures. Inserting a barrier or changing the shape of the container does not change this equality.

unit is the kilopascal, kPa, which is 10^3 Pa, the pressure of water at a depth of about 10 cm. Another more widely adopted SI unit is the **bar**, which is close to atmospheric pressure at sea level.

$$1 \text{ bar} = 10^5 \text{ Pa}$$

Pressures have long been measured and expressed in other standard units, and some of these older units persist in common use. One set of units is related to the height (or depth) of a column of liquid. The most convenient standard liquid is mercury. Two principles govern the use of a column of liquid to measure pressure:

1. The pressures in a given liquid are equal at equal levels, as implied by the earlier statement about the pressure on a diver. Points X and Y in Figure 10.2a are at equal levels and therefore at equal pressures. This equality is not changed by the intrusion of a partial barrier between the two points, as shown in Figure 10.2b–f.

2. The pressure at any point in a liquid is exerted equally in all directions.

THE BAROMETER

Longer than about the height of a dinner table.

In 1643 Evangelista Torricelli discovered that when a long tube filled with mercury is inverted in a dish of mercury, some of the mercury in the tube flows down into the dish. This lowering of the mercury level in the tube leaves a space in the upper

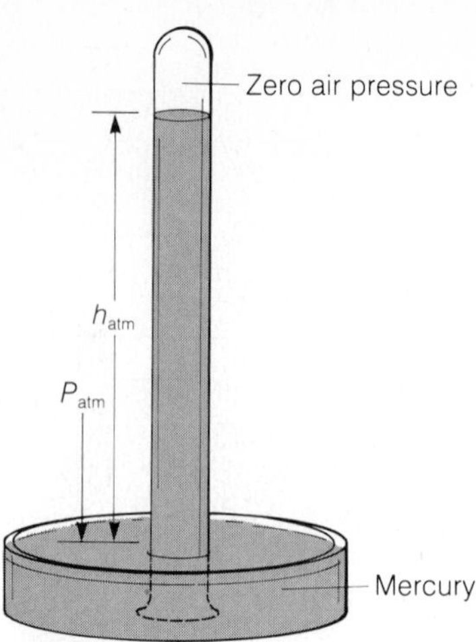

FIGURE 10.3
A mercury barometer.

part of the tube (Figure 10.3). There is no air in this space because there was no air to begin with (only mercury), and there is no way for air to get in. The space is therefore empty; it is a vacuum. This means that no air is exerting pressure on the surface of the mercury inside the tube. But why doesn't *all* of the mercury run out? What holds the remaining column of mercury in the tube? The answer must be that the mercury is held up by some pressure. The pressure at the level of the mercury surface in the dish is the pressure exerted by the Earth's atmosphere, and it is therefore called **atmospheric pressure**. Thus, the pressure exerted by the mercury column of height h_{atm} equals the pressure of the atmosphere, P_{atm}. If it is a nice day and the atmospheric pressure rises, h_{atm} will increase. If a hurricane is approaching and the atmospheric pressure is falling, h_{atm} will decrease. Torricelli's clever device is called a **barometer**.

Some mercury does evaporate into this space, but the amount is negligible.

Mercury is convenient because it is very dense and has a very low vapor pressure. If, say, olive oil were used, the column would have to be about 11 m (about 36 ft) high to exert a pressure of 1 atm.

Under normal atmospheric conditions at sea level, h_{atm} is about 760 mm or 76 cm of a column of mercury. Knowing these conditions allows us to define some convenient units of pressure:

- 1 **standard atmosphere** (atm) is the pressure exerted by a column of mercury 760 mm high at 0°C at standard gravity (sea level). The temperature of the mercury, 0°C, must be specified because the density of mercury varies with the temperature.

BOX 10.1
TIRE GAUGES

Tire gauges indicate tire pressure in pounds, lb, which means "pounds per square inch of pressure." Newer gauges show pressure in kilopascals, kPa, as well (Figure 10.4). Both of these scales refer to the pressure in the tire *in excess* of atmospheric pressure. (A flat tire is not a vacuum; it still contains some air at 1 atm pressure.) One atmosphere is about 14.7 lb/in^2, or 101 kPa. A tire gauge test of a tire containing 7.25 atm gas pressure would respond to the 6.25 atm *above* normal pressure, and would therefore show 6.25 × 14.7 lb/in^2, or 91.9 "pounds" of pressure. The kilopascal scale would show 6.25 × 101 kPa, or 631 kPa.

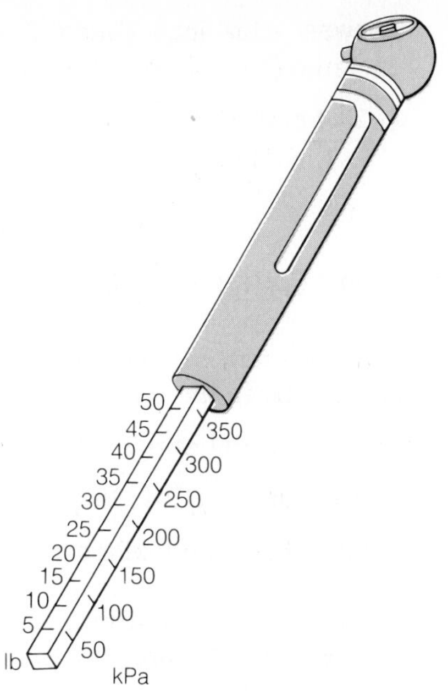

FIGURE 10.4
A tire gauge. The "lb" scale measures air pressure in pounds per square inch (psi).

- 1 **torr** = 1/760 atm = the pressure exerted by a column of mercury 1 mm high at 0°C and at standard gravity. The older name for torr is mm Hg.

- **Standard pressure** means 1 standard atmosphere, or 760 torr.

The units of pressure commonly used in science, medicine, commerce, and industry in English-speaking countries include kilopascals, bars, atmospheres, torrs, mm of mercury, inches of mercury, inches of water, and pounds per square inch (psi). For using this book, you need to be familiar with bars, atmospheres, and torrs. The relationship between bars and atmospheres is explained in Appendix A.4, and some handy conversion factors are given in Table 10.1.

EXAMPLE 10.1 A barometer reading is reported as "28.2 inches of mercury." Calculate the barometric pressure in torrs, atmospheres, bars, and kilopascals.

TABLE 10.1
PRESSURE CONVERSIONS

FROM	TO	MULTIPLY BY
atmosphere	torr	760 torr/atm (exactly)
atmosphere	lb/in^2	14.6960 $lb/(in^2$ atm)
atmosphere	kilopascal	101.325 kPa/atm
atmosphere	bar	1.01325 bar/atm
bar	pascal	10^5 Pa/bar (exactly)
mm of mercury	torr	1 torr/mm mercury (exactly)
pound(force)/in^2	pascal	6894.73 Pa in^2/lb

ANSWER One inch equals 2.54 cm or 25.4 mm. Therefore 1 in (Hg) equals 25.4 mm (Hg), so the conversion factor is

$$\frac{25.4 \text{ mm (Hg)}}{1 \text{ in (Hg)}}$$

Then,

$$28.2 \cancel{\text{in (Hg)}} \times \frac{25.4 \text{ mm (Hg)}}{1 \cancel{\text{in (Hg)}}} = 716 \text{ mm (Hg)}$$

Since a torr is defined in terms of mm of mercury in a barometer, the pressure is 716 torr. In the other units, the pressure is

$$716 \text{ torr} \times (1 \text{ atm}/760 \text{ torr}) = 0.942 \text{ atm}$$

$$0.942 \text{ atm} \times (1.01325 \text{ bar}/1 \text{ atm}) = 0.954 \text{ bar}$$

$$0.954 \text{ bar} \times (100 \text{ kPa/bar}) = 95.4 \text{ kPa}$$ ■

The summit of Mt. Everest is about 8.9 km above sea level, and the Dead Sea about 395 m below.

PROBLEM 10.1 The average barometric pressure at an altitude of 10 km is 210 torr. Express this pressure in atmospheres, bars, and kilopascals. □

10.3 THE GAS LAWS: A PREVIEW

Think of some particular samples of gas, such as the hydrogen in a balloon, the air in a bicycle tire, and the helium-oxygen mixture in a diver's breathing tank. Each of these samples is characterized by four interdependent, variable properties; that is, if any one of the variables is changed, there *must* be a change in at least one of the other three. The four variables are

the volume of the gas,
the pressure of the gas,
the temperature of the gas, and
the number of molecules of the gas.

Consider, for example, what might happen if the air in the bicycle tire is heated: the pressure may increase; the tire could swell up so that the volume of the gas expands; perhaps some air would leak out so that the number of molecules of air in the tire decreases, or some combination of these effects would occur. No matter what, something would have to happen to at least one of the other three variables. It would be convenient if we could express all of these possibilities in one simple equation. Such an equation does in fact exist, and it is not a difficult one, but it was not discovered or derived all at once. Instead, various scientists studied these variables—two at a time—and discovered a set of simple gas laws, which, when combined, give us the useful general equation. The next few sections will trace this development and lead to the general equation in section 10.7.

10.4 BOYLE'S LAW

Robert Boyle (1627–1691) was first to discover a law relating two of the variable properties of gases; his law relates pressure and volume. To establish the relationship, he had to keep the temperature and the number of molecules constant. Actually, Boyle didn't take elaborate precautions to maintain a constant temperature; he carried out his experiments indoors, where the temperature did not vary too much. What's more, at that time scientists did not have a clear conception of molecules. However, Boyle used a simple and clever method for trapping the gas he was working with so that nothing could leak in or out. As a result, whether he knew it or not, the number of molecules in his experiments did not change.

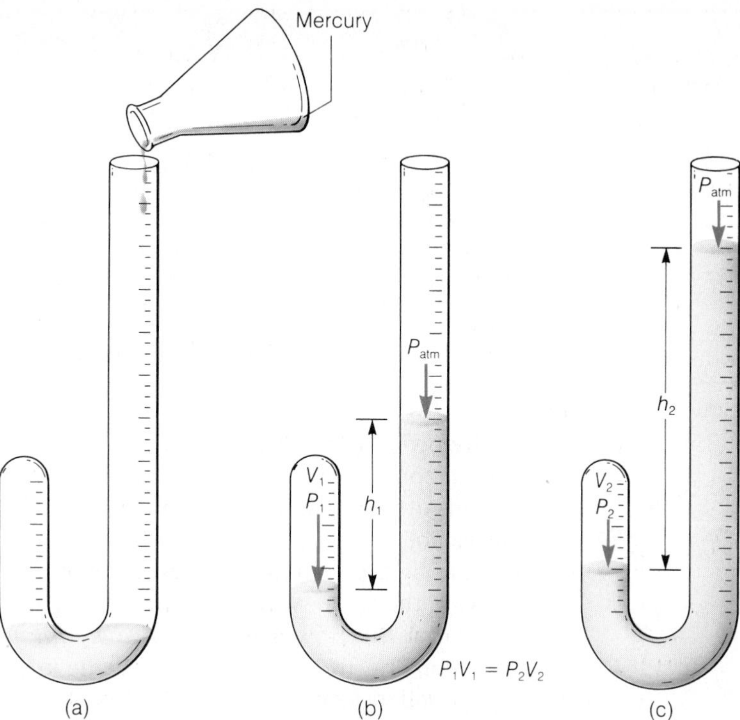

FIGURE 10.5 Boyle's apparatus. The liquid in the tubes is mercury; pressure is measured in terms of the height of the mercury in the column. The addition of the barometer reading takes into account the atmospheric pressure on the right arms of the J-tubes.

Boyle poured successive quantities of mercury into the open arm of a J-shaped tube (Figure 10.5). As soon as the mercury filled the bottom portion of the tube (Figure 10.5a), the gas in the closed end was trapped—none of it could get out through the mercury. This trapping satisfied the condition that the number of molecules in the sample of gas be held constant. Now look at Figure 10.5b, which represents the experiment after some more mercury has been poured in. Consider how the volume and pressure of the trapped gas, V_1 and P_1, are measured. The volume of the gas is easy—it is the same as the volume of the inside of the tube above the mercury. The pressure, however, requires two measurements. The pressure of the trapped gas, P_1, is the same as the pressure in the other arm of the tube at the same level (remember, equal pressures at equal levels). But the pressure P_1 in the open arm is exerted by the weight of the mercury column of height h_1 and the atmospheric pressure above that column, P_{atm}. As Torricelli had shown, P_{atm} is measured by the height of mercury in his barometer. Therefore, adding the two measurements, we have

$$P_1 = P_{h_1} + P_{atm}$$

EXAMPLE 10.2 If h_1 in Figure 10.5b is 520 mm and the barometric reading is 759 mm, what is the pressure of the trapped gas?

ANSWER Since the liquid is mercury, P_{h_1} is 520 torr and P_{atm} is 759 torr. Therefore, P_1 is the sum of these pressures, (520 + 759) torr, or 1279 torr. ■

Now that we know how to measure the volume and the pressure of the trapped gas, we can, in our imagination, continue watching the experiment. Boyle pours some more mercury into the open arm of the tube, and notices that the volume of

TABLE 10.2
BOYLE'S DATA (1660), SHOWING THAT
PV = constant

PRESSURE, $P^{\dagger}$	VOLUME, $V^{\S}$	$P \times V$
$0 + 29\frac{1}{8} = 29\frac{1}{8}$	4.8	140
$6\frac{3}{16} + 29\frac{1}{8} = 35\frac{5}{16}$	4.0	141
$21\frac{3}{16} + 29\frac{1}{8} = 50\frac{5}{16}$	2.8	141
$45 + 29\frac{1}{8} = 74\frac{1}{8}$	1.9	141
$63\frac{15}{16} + 29\frac{1}{8} = 93\frac{1}{16}$	1.5	140

† As measured by the height of the mercury column in inches plus the height of mercury in the barometer ($29\frac{1}{8}$ in).
§ Volume of air in the closed leg of the J-tube, measured in arbitrary units.

the trapped gas decreases while the mercury column, h_2, lengthens (Figure 10.5c). Boyle records his measurements, and continues the experiment, using additional portions of mercury. When he has all of his results, he does something with them that establishes his place in the history of chemistry forever: he multiplies the pressure by the volume obtained in each experiment and notes that the product is always the same, within the limits of error of his work (Table 10.2). His observation that *the pressure multiplied by the volume of a fixed mass of gas at a fixed temperature is a constant* became known as **Boyle's law.**

In equation form it is stated in one of two ways:

$PV = constant$ (temperature and mass fixed), or

$$V = \frac{constant}{P}$$

The second form of the equation leads to another expression of Boyle's law: *the volume of a fixed mass of gas at a fixed temperature is inversely proportional to the pressure.*

Problems dealing with *P-V* relationships among gases can be solved as follows: Let P_1, V_1 refer to one set of conditions and P_2, V_2 to another set of conditions of the same mass of the same gas at the same temperature. Then both P_1V_1 and P_2V_2 equal the same constant and therefore are equal to each other:

$$\boldsymbol{P_1V_1 = P_2V_2} \tag{1}$$

The effect of a change in pressure on gas volume can be calculated by substituting values in equation 1.

Refer to Appendix B.4 for a review of graphs and proportionality.

Alternatively, we can convert the old volume (V_1) to the new volume (V_2) by multiplying by the ratio of the pressures:

$$V_2 = V_1 \times \text{correction for pressure change}$$

The form of this correction factor is established by our "common sense" knowledge of gas behavior. Thus an increase in pressure will decrease the volume, and the pressure correction must be less than 1. The larger of the two pressures then becomes the denominator. A decrease in pressure increases the volume; therefore the larger of the two pressures becomes the numerator.

EXAMPLE 10.3 A sample of gas occupies 10.0 L at 760 torr. The volume is to be changed so as to reduce the pressure to 700 torr at constant temperature and without any loss of gas. What must the new volume be?

ANSWER

Quantities given: $V_1 = 10.0$ L
$P_1 = 760$ torr
$P_2 = 700$ torr

Quantity sought: V_2

FROM EQUATION 1	COMMON-SENSE METHOD
$P_1V_1 = P_2V_2$	New volume (V_2) = original volume (V_1) × pressure correction
$V_2 = V_1 \times \frac{P_1}{P_2}$	(The gas will expand because of pressure reduction, so the correction factor is greater than one and the larger pressure becomes the numerator.)
$= 10.0 \text{ L} \times \frac{760 \text{ torr}}{700 \text{ torr}}$	$= 10.0 \text{ L} \times \frac{760 \text{ torr}}{700 \text{ torr}}$

Therefore, $V_2 = 10.9$ L. ■

EXAMPLE 10.4 A cylinder (Figure 10.6) contains 580 mL of gas at 0.200 atm pressure. What will the new pressure be after the piston is moved so as to reduce the volume of the gas to 100 mL? Assume that the temperature and mass of the sample are constant.

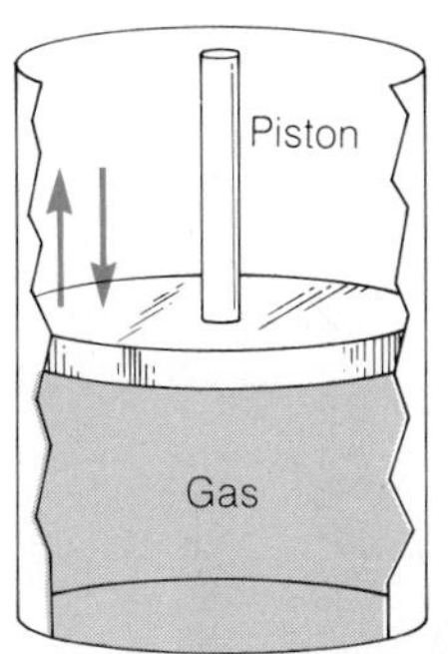

FIGURE 10.6
Gas and piston in a cylinder.

ANSWER

Quantities given: $P_1 = 0.200$ atm
$V_1 = 580$ mL
$V_2 = 100$ mL

Quantity sought: P_2

FROM EQUATION 1	COMMON-SENSE METHOD
$P_2 = P_1 \times \frac{V_1}{V_2}$	New pressure (P_2) = original pressure × volume correction
$= 0.200 \text{ atm} \times \frac{580 \text{ mL}}{100 \text{ mL}}$	(The pressure must be increased to compress the gas, so the correction factor is greater than one and the larger volume becomes the numerator.)
$= 1.16$ atm	$= 0.200 \text{ atm} \times \frac{580 \text{ mL}}{100 \text{ mL}} = 1.16$ atm

■

PROBLEM 10.2 850 mL of helium in a sealed balloon at constant temperature is reduced to 550 mL when the balloon is squeezed until the helium pressure reaches 1.44 atm. What was the original pressure of the helium? □

10.5 CHARLES'S LAW; KELVIN TEMPERATURE

Thermometers were first developed in Europe in the early 1600s (Figure 10.7). Earliest Western concepts of gas expansion by heating probably originated with the mechanical toys of the Hellenistic period, such as Hero's engine, a pinwheel powered by steam jets.

The two variable properties of gases that were next found to be related by a simple law were temperature and volume. Jacques Charles carried out experiments that established a relationship between them around 1787, but his data are not available to us today. His work was confirmed in 1802 by Joseph Louis Gay-Lussac.

One method of showing the relationship between the temperature and the volume of a fixed sample of gas at constant pressure is shown in Figure 10.8. The volume of the gas increases as the temperature increases. The same rise in temperature of a given mass of gas always produces the same added volume. Thus, the volume of a fixed mass of gas at constant pressure increases *linearly* with increasing temperature. If any quantity of gas is warmed by 1°C, it expands by 1/273 of the volume it occupied at 0°C. For example, if 100 mL of gas at 0°C is warmed to 1°C, it expands by 1/273 × 100 mL, or 0.366 mL. If 273 mL of gas at 0°C is warmed to 1°C, it expands by 1/273 × 273 mL, or 1 mL; the final volume is thus 273 mL + 1 mL, or 274 mL. This *same sample* of gas—which is then 274 mL—would continue to expand by 1 mL with any further rise in temperature of 1 degree Celsius (or 1 kelvin). For example, the 274-mL sample of gas at 1°C would expand to 275 mL if the temperature were raised to 2°C. And it would contract by 1 mL for any 1-degree drop in temperature. A graph of this relationship is a straight-line plot of gas volume versus temperature (Figure 10.9).

The scale is named after Lord Kelvin (1824–1907), who explained its fundamental significance.

No substance can be cooled below zero kelvin. The precise Celsius equivalent of 0 kelvin is −273.15°C. This temperature is taken as the zero point of the **Kelvin scale** of absolute temperature and it is called **absolute zero**. A gas that obeys Charles's law would theoretically shrink down to zero volume at this temperature. Of course, real gases do not disappear; they liquefy first. The left portions of the lines of Figure 10.9 are dashed, not solid, to show that the reduction to zero volume is theoretical, not actually measured.

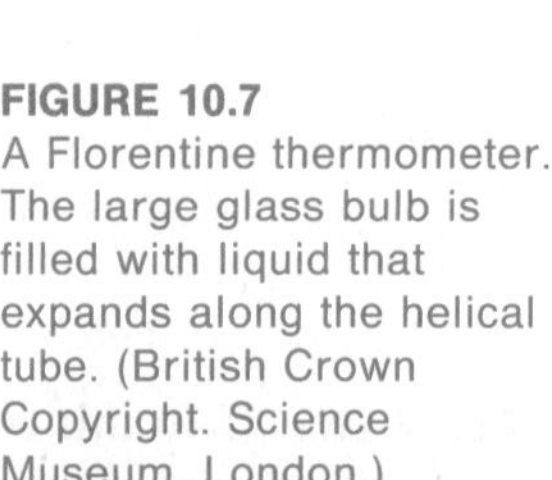

FIGURE 10.7
A Florentine thermometer. The large glass bulb is filled with liquid that expands along the helical tube. (British Crown Copyright. Science Museum, London.)

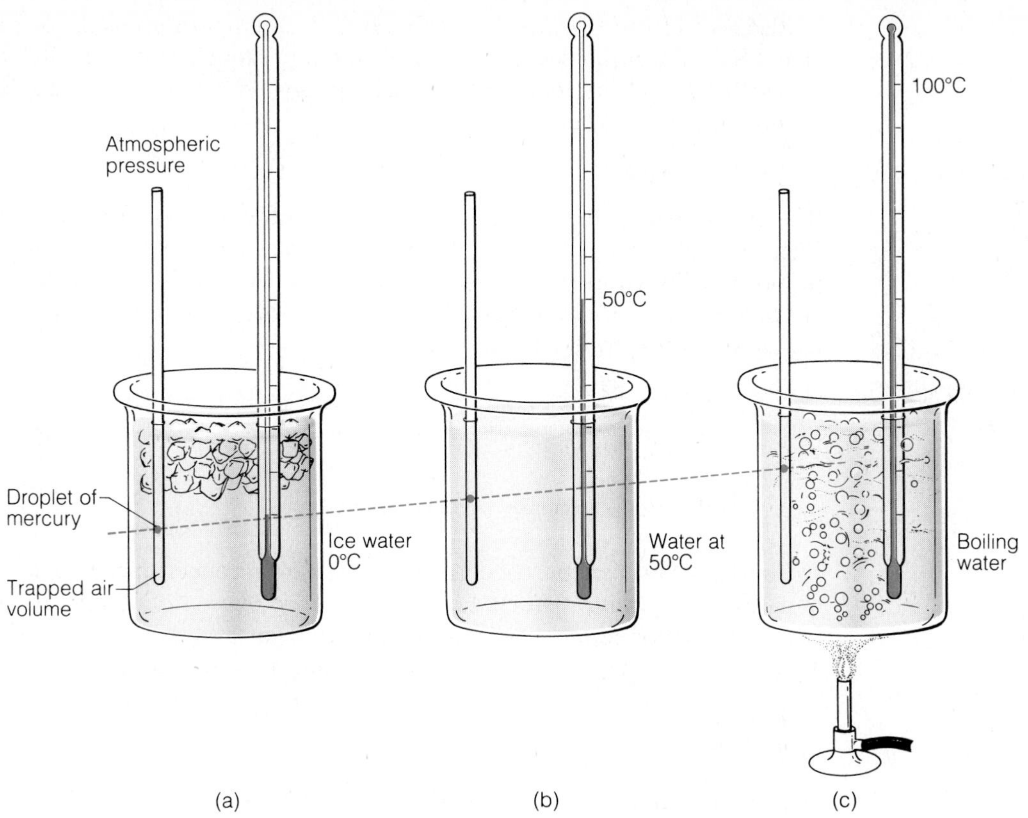

FIGURE 10.8
An illustration of Charles's law. (a) A droplet of mercury traps a sample of air below it at 0°C and a fixed pressure. (b) The same air sample at the same pressure, at 50°C and (c) at 100°C.

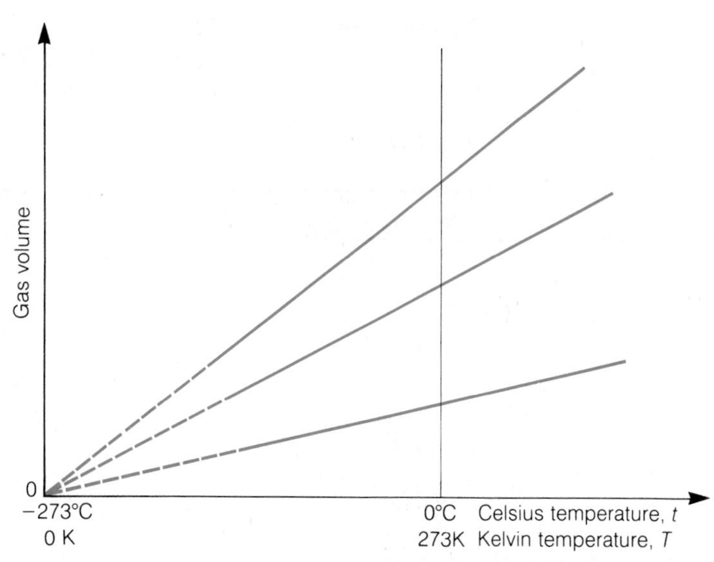

FIGURE 10.9
Volume-temperature relationships for three different quantities of gas, obtained from experiments at constant low pressure. Charles's law is illustrated by the straight lines.

Also recall from Chapter 1 that the SI unit of temperature is the kelvin, abbreviated K (without the degree sign). Kelvin temperatures, if unspecified, are represented by T. Celsius temperatures are designated by °C, or if unspecified, by lowercase letter t. Then

Remember 273 for calculations limited to three significant figures.

$$T = t + 273.15 \quad (2)$$

If temperatures are expressed in kelvins, then 0 K corresponds to zero gas volume (Figure 10.9). Such a linear relationship that passes through the origin is a direct proportionality (Appendix B.4). Figure 10.9 thus expresses **Charles's law**: *the volume of a fixed mass of gas at constant pressure is directly proportional to the absolute temperature.* In equation form,

The symbol $\propto$ means "is proportional to."

$$V \propto T \text{ or}$$

$$V = constant \times T \text{ or}$$

$$\frac{V}{T} = constant \qquad \text{(pressure and mass fixed)}$$

Charles's law can be used to solve problems concerning V–T relationships of gases at constant pressure. Let V_1, T_1 refer to one set of conditions and V_2, T_2 to another set of conditions on a quantity of a gas at a particular pressure. Then V_1/T_1 and V_2/T_2 equal the same constant and therefore are equal:

$$\frac{V_1}{T_1} = \frac{V_2}{T_2} \quad (3)$$

The "common sense" application of correction factors also can be used.

The term **standard temperature and pressure** (STP), or **standard conditions**, as used in calculations involving gases, means 0°C and 760 torr.

EXAMPLE 10.5 10.0 mL of a fixed mass of gas at 20.0°C is cooled at constant pressure to 0°C. What is the new volume?

ANSWER

Quantities given: $V_1 = 10.0$ mL
$t_1 = 20°C$; therefore $T_1 = (20 + 273)$ K
$t_2 = 0°C$; therefore $T_2 = (0 + 273)$ K

Quantity sought: V_2

FROM EQUATION 3	COMMON-SENSE METHOD
$\frac{V_1}{T_1} = \frac{V_2}{T_2}$	New volume (V_2) = original volume (V_1) × temperature correction (In this case, cooling reduces gas volume, so the correction factor must be less than one.)
$V_2 = V_1 \times \frac{T_2}{T_1}$ $= 10.0 \text{ mL} \times \frac{273 \text{ K}}{293 \text{ K}}$ $= 9.32$ mL	$= 10.0 \text{ mL} \times \frac{273 \text{ K}}{293 \text{ K}} = 9.32$ mL

■

PROBLEM 10.3 To what temperature, °C, must a 25.5-mL sample of oxygen at 90°C be cooled for its volume to shrink to 21.5 mL? Assume that the pressure and mass of the gas are fixed. □

10.6 THE LAW OF COMBINING VOLUMES AND AVOGADRO'S LAW

Boyle's law and Charles's law are both concerned with volumes of gases. It is important to remember that these laws apply *only* to gases, not to solids or liquids. Furthermore, they apply to *fixed* masses and therefore fixed numbers of molecules of gases; this means that they apply to gases that are not undergoing chemical reactions. If a chemical reaction did occur, the number of molecules in the gas might change.

The study of the relationships between volumes of gases that *are* undergoing chemical reactions fascinated many scientists, among them Gay-Lussac, who made some remarkable observations. He found that the ratios of volumes involved in gas reactions at fixed temperatures and pressures could be expressed by small whole numbers. Here are some examples that illustrate his law:

1. hydrogen gas + oxygen gas ⟶ water vapor (steam)
   2 liters + 1 liter ⟶ 2 liters
2. hydrogen gas + nitrogen gas ⟶ ammonia gas
   3 liters + 1 liter ⟶ 2 liters
3. hydrogen gas + chlorine gas ⟶ hydrogen chloride gas
   1 liter + 1 liter ⟶ 2 liters

Note that the *volumes* of gases in chemical reactions are not always conserved; that is, the volumes reacting may or may not equal the volumes produced. More important is the fact that the volumes combine in small whole-number ratios. This is stated in Gay-Lussac's **law of combining volumes** (1808): *When gases react, the volumes consumed and produced, measured at the same temperature and pressure, are in ratios of small whole numbers.*

What is to be made of such an observation? One thought is that gas volumes must have something to do with molecules, because reacting molecules are also consumed and produced in ratios of small whole numbers. The simplest assumption is that the volumes of the different gases are in the same ratios as the numbers of molecules in them. In 1811 Amedeo Avogadro made the first clear statement of this relationship. In modern terms, *Equal volumes of all gases at the same temperature and pressure contain the same number of molecules.* Expressed as an equation, where N is the number of molecules, Avogadro's law is

$$N \propto V \quad \text{or} \quad \frac{N}{V} = \mathit{constant} \qquad \text{(temperature and pressure fixed)}$$

For two different samples of a gas at the same temperature and pressure, the relationship is

$$\frac{\mathbf{N_1}}{\mathbf{V_1}} = \frac{\mathbf{N_2}}{\mathbf{V_2}} \tag{4}$$

This means that, say, 10 mL of *any* gas, whether pure or a mixture—hydrogen, nitrogen, ammonia, radon, air, natural gas, and so on—at the same temperature and pressure contains the same number of molecules. The only way to double the volume at a fixed temperature and pressure is to double the number of molecules.

EXAMPLE 10.6 One gram of radium emits alpha particles (He^{2+} ions) at a rate of 1.16×10^{18} particles/year. Each alpha particle becomes a molecule of helium gas:

$$He^{2+} + 2e^- \longrightarrow He$$

The total volume of the 1.16×10^{18} molecules is 0.0430 mL at STP. How many molecules are there in 1.00 liter of helium at STP; in 1.00 liter of oxygen; in 1.00 liter of any other gas at STP?

ANSWER

Quantities given: $N_1 = 1.16 \times 10^{18}$ molecules of He
$V_1 = 0.0430$ mL of He at STP
$V_2 = 1.00$ L, or 1.00×10^3 mL

Quantity sought: N_2 for He, O_2, and any other gas

From equation 4,

$$N_2 = N_1\left(\frac{V_2}{V_1}\right)$$

$$= 1.16 \times 10^{18} \text{ molecules} \times \frac{1.00 \times 10^3 \text{ mL}}{0.0430 \text{ mL}}$$

$$= 2.70 \times 10^{22} \text{ molecules}$$

Avogadro's law tells us that the number of molecules of a gas in a given volume is a constant at a fixed temperature and pressure, regardless of the nature of the gas. Therefore the answer applies to helium, oxygen, or any other gas. ■

PROBLEM 10.4 The air in a tire pump occupies 1.5 L and contains 1.5×10^{23} molecules. If air is forced out of the pump and there is no change in the temperature or pressure, how many molecules will remain in the pump when the volume has been reduced to 0.85 L? □

A very useful relationship is now available to us. A liter of oxygen gas has sufficient mass that it can be weighed accurately. The same is true of a liter of helium gas. Since the two volumes (at the same temperature and pressure) are equal, they have the same number of molecules. We don't have to know how many; just call the number N. Then,

$$\frac{\text{mass of 1 L of oxygen}}{\text{mass of 1 L of helium}} = \frac{\text{mass of } N \text{ molecules of oxygen}}{\text{mass of } N \text{ molecules of helium}}$$

$$= \frac{\text{mass of 1 molecule of oxygen}}{\text{mass of 1 molecule of helium}}$$

Thus, we can now measure the relative masses of molecules of oxygen and helium gases, because Avogadro's law tells us how to select equal numbers of them in quantities large enough to weigh. Since the mass of one liter of gas is its density, the relationship just given may be expressed as a general equation for any two gases, X and Y:

$$\frac{\text{density of gas X}}{\text{density of gas Y}} = \frac{\text{mass of 1 molecule of gas X}}{\text{mass of 1 molecule of gas Y}} \qquad (5)$$

EXAMPLE 10.7 1.0 liter of chlorine at STP weighs 3.16 g, and 1.0 liter of neon at STP weighs 0.90 g. What are the relative masses of a molecule of chlorine and a molecule of neon?

ANSWER

Quantities given: density of chlorine = 3.16 g/L
density of neon = 0.90 g/L

Quantity sought: $\dfrac{\text{mass per molecule of chlorine}}{\text{mass per molecule of neon}}$

From equation 5,

$$\frac{\text{density of chlorine}}{\text{density of neon}} = \frac{3.16\ \text{g/L}}{0.90\ \text{g/L}} = \frac{\text{mass of 1 molecule of chlorine}}{\text{mass of 1 molecule of neon}} = 3.5$$

The mass of a molecule of chlorine is 3.5 times the mass of a molecule of neon. (Check this answer with the aid of the table of atomic weights.) ■

Actually, these values refer to the average mass of one molecule, since not all molecules of a given element have the same isotope composition or the same mass.

PROBLEM 10.5 The mass of a molecule of oxygen is 8.00 times that of a molecule of helium. A balloon containing 0.031 kg of oxygen gas under pressure is opened and then squeezed completely flat. Helium gas is then pumped into it until the original pressure, volume, and temperature are restored. What is the mass of helium in the balloon? □

Recall that a mole is the amount of any substance that contains as many particles as there are carbon atoms in 12 g of carbon-12. Since the particles of gases are molecules, a mole of gas contains as many *molecules* as there are carbon atoms in 12 g of carbon-12. Since equal numbers of molecules of all gases occupy the same volume (at constant temperature and pressure), the volume of 1 mole of any gas must be a standard quantity. This volume, at 0°C and 1 atm, has been determined to be 22.4138 liters. The number of molecules in a mole, designated N_A, is called **Avogadro's number** (Chapter 3).

Remember, 3 significant figures: 22.4 L/mol.

The volume of a mole of *any* substance—gas, liquid, or solid—is its **molar volume**. However, the molar volume at a given temperature and pressure is *not* the same for all liquids and solids; the value of 22.4 L at STP applies only to gases.

EXAMPLE 10.8 Using the data of Example 10.6, calculate the Avogadro number.

ANSWER

Quantity given: $N/V = 2.70 \times 10^{22}$ molecules/L at STP (from Example 10.6)

Quantity sought: the Avogadro number

Avogadro's number is the number of molecules in 1 mole of a gas (22.4 liters at STP). From Example 10.6, this is

$$2.70 \times 10^{22}\,\frac{\text{molecules}}{\cancel{\text{L}}} \times 22.4\,\frac{\cancel{\text{L}}}{\text{mol}} = 6.05 \times 10^{23}\ \text{molecules/mol}$$ ■

Avogadro's number has been determined by several methods in addition to the one described by Example 10.6, and the currently accepted value is 6.0221367×10^{23} molecules/mol. Expressed in three significant figures, 6.02×10^{23} molecules (1 mole) of any gas at STP will occupy a volume of 22.4 liters.

10.7 THE IDEAL GAS LAW

We promised in section 10.3 that a combination of simple gas laws dealing with only two variables at a time can give us a general equation that includes all four variables. Now it is time to fulfill that promise.

Boyle's, Charles's, and Avogadro's laws each relate the volume of a gas to one other variable:

Boyle	$V = k/P$	(at constant T and N)
Charles	$V = k' \times T$	(at constant P and N)
Avogadro	$V = k'' \times N$	(at constant T and P)

Combining these relationships yields the equation

$V = constant \times NT/P$, or

$PV = constant \times NT$

This combined equation merely restates that PV is constant for a fixed number of molecules and fixed temperature; that V is directly proportional to T for a fixed mass and a fixed pressure; and that N is fixed when P, V, and T are fixed.

The number of molecules, N, is proportional to the number of moles, n. Therefore we may write

$PV = constant \times nT$

This relationship is called the **ideal gas law**. The constant is the same for all gases and is designated by R, which is called the **gas constant**.

$PV = RnT$

or, as it is usually written,

$$\boldsymbol{PV = nRT} \quad (6)$$

We evaluate R as follows: Suppose that we have 1 mole of gas at standard conditions (STP). Then

$n = 1$ mole

$$\left.\begin{array}{l} P = 1 \text{ atm} \\ T = 273.15 \text{ K} \end{array}\right\} \text{STP}$$

$V = 22.4138$ L

We may now calculate the value of R from the equation $PV = nRT$. Rearranging to solve for R, we obtain

$$R = \frac{PV}{nT} = \frac{1 \text{ atm} \times 22.4138 \text{ L}}{1 \text{ mol} \times 273.15 \text{ K}}$$

$$= 0.082057 \text{ L atm/(mol K)}$$

Most of the experimental data cited in this chapter are expressed with no more than three significant figures; therefore it will usually be sufficient to use 0.0821 L atm/(mol K), 273 K, and 22.4 L for R and the standard values of T and V, respectively.

We can now do calculations on a broader basis than is permitted by the two-variable laws (Boyle's, Charles's, Avogadro's). There are two different approaches:

1. Use $PV = nRT$ when only one set of conditions is given. The conditions *must* be expressed in units consistent with those used for R.

This method is appropriate when the problem furnishes the values for three of the four variables, P, V, n, and T. Solve for the fourth variable. Make sure that the units of R and of the variables are consistent with each other.

2. Refer to two sets of conditions:

$$P_1V_1 = n_1RT_1, \quad \text{and} \quad P_2V_2 = n_2RT_2$$

Combining the equations and canceling the R's gives

$$\frac{P_1V_1}{n_1T_1} = \frac{P_2V_2}{n_2T_2}$$

Use this method, or a common-sense approach, when two sets of conditions are given. The problem may refer to a given sample of gas, so that the number of moles is constant and the n factors, as well as the R's, cancel. The equation then becomes

$$\boldsymbol{\frac{P_1V_1}{T_1} = \frac{P_2V_2}{T_2}} \qquad (7)$$

EXAMPLE 10.9 A fixed mass of gas occupies 10.0 L at STP. What volume will it occupy at 20°C and 700 torr?

ANSWER Two sets of conditions are given, so method 2 may be used.

$V_1 = 10.0$ L

$P_1 = 760$ torr (standard pressure)	$P_2 = 700$ torr
$T_1 = 273$ K (standard temperature)	$T_2 = 293$ K
$V_1 = 10.0$ L	$V_2 = ?$

Solving equation 7 for V_2,

$$V_2 = V_1 \times \frac{T_2}{T_1} \times \frac{P_1}{P_2}$$

$$V_2 = 10.0\text{ L} \times \frac{293\text{ K}}{273\text{ K}} \times \frac{760\text{ torr}}{700\text{ torr}} = 11.7\text{ L}$$

Note that the temperature correction is greater than 1 because the temperature goes up, and the pressure correction is also greater than 1 because the pressure goes down. ■

EXAMPLE 10.10 Assume that a bicycle tire containing 0.406 mol of air will burst if its internal pressure reaches 7.25 atm, at which time its internal volume would be 1.52 liters. To what temperature, in °C, would the air in the tube need to be heated to cause a blowout?

ANSWER

Quantities given: $n = 0.406$ mol
$P = 7.25$ atm
$V = 1.52$ L

Quantity sought: t

Since only one set of conditions is given, we must solve $PV = nRT$ for T and then convert to t.

$$T = \frac{PV}{nR} = \frac{7.25 \text{ atm} \times 1.52 \text{ L}}{0.406 \text{ mol} \times 0.0821 \text{ L atm/(mol K)}} = 331 \text{ K}$$

$$t = T - 273 = 331 - 273 = 58°\text{C}$$

■

EXAMPLE 10.11 The temperature of the atmosphere on Mars can be as high as 27°C at the equator at noon, and the atmospheric pressure is about 8 torr. If a spacecraft could collect 10 m^3 of this atmosphere, compress it to a small volume, and send it back to Earth, how many moles would the sample contain?

BOX 10.2

USE OF THE SI FOR GAS CALCULATIONS

As pointed out in section 10.2, the SI unit of pressure is the *pascal*, not the atmosphere or the torr. However, using the pascal—less than a hundredth of a torr—as the standard pressure for gases leads to awkward and unfamiliar expressions, and chemists have rejected its adoption. Metric countries have been using the *kilopascal*, kPa, but that unit, too, is numerically far from the older, more familiar ones. The Commission on Thermodynamics of the IUPAC has recommended the adoption of the *bar* as the defined pressure for tabulating standard data. The bar is close to 1 atmosphere, so when it is used values of R and the molar gas volume are close to the old values.

It has also been suggested that the "standard temperature" associated with the gas laws be changed from the ice point (0°C, 273.15 K) to standard ambient temperature (25°C, 298 K), to be consistent with the standard temperature used for tabulating thermodynamic data. However, this change has not yet been made. The consequences of these two changes, and the values of R in various other units, are shown here.

If STP were 1 bar and 273.15 K:
The molar volume could be calculated from Boyle's law, where

$$P_1 = 1 \text{ atm} \qquad P_2 = 1 \text{ bar}$$

$$V_1 = 22.4138 \text{ L} \qquad V_2 = ?$$

and from the conversion:

$$1 \text{ atm} = 1.01325 \text{ bar} \qquad \text{(Appendix A.4)}$$

Then,

$$V_2 = V_1 \times (P_1/P_2)$$

$$= 22.4138 \text{ L} \times \frac{1 \text{ atm}}{1 \text{ bar}} \times \frac{1.01325 \text{ bar}}{1 \text{ atm}}$$

$$= 22.7108 \text{ L}$$

The value of R would be:

$$R = \frac{PV}{nT} = \frac{1 \text{ bar} \times 22.7108 \text{ L}}{1 \text{ mol} \times 273.15 \text{ K}}$$

$$= 0.083144 \text{ L bar/(mol K)}$$

If STP were 1 bar and 298.15 K:
The molar volume could be calculated from the preceding value of 22.7108 L and the Charles's law correction for temperature. Let the new volume be V_3. Then,

$$V_3 = V_2 \text{ (298.15 K/273.15 K)}$$

$$= 22.7108 \text{ L (298.15 K/273.15 K)}$$

$$= 24.7894 \text{ L}$$

Pressure applied through a volume, like force over a distance, is work. (Think of pumping up a bicycle tire.) Therefore PV has the units of energy (Appendix A.5), and other energy units may be substituted for "liter atmosphere." The following conversions are used:

$$1 \text{ L atm} = 101.325 \text{ J} = 24.2173 \text{ cal}$$

Then,

$$R = \frac{PV}{nT} = \frac{1 \text{ atm} \times 22.4138 \text{ L} \times 101.325 \text{ J/(L atm)}}{1 \text{ mol} \times 273.15 \text{ K}}$$

$$= 8.3144 \text{ J/(mol K)}$$

$$= \frac{1 \text{ atm} \times 22.4138 \text{ L} \times 24.2173 \text{ cal/(L atm)}}{1 \text{ mol} \times 273.15 \text{ K}}$$

$$= 1.9872 \text{ cal/(mol K)}$$

ANSWER

Quantities given: $\left.\begin{aligned} P &= 8 \text{ torr} \\ V &= 10 \text{ m}^3 \\ T &= 27°\text{C} \end{aligned}\right\}$ You must change these to units consistent with $R = 0.0821$ L atm/(mol K).

Quantity sought: n

The number of moles of any gas is determined by the pressure, volume, and temperature of the original sample. Assuming that no chemical reaction occurs, that number is not affected by any subsequent changes in these conditions. Therefore, we solve $PV = nRT$ for n. However, if we use the value of 0.0821 L atm/(mol K) for R, then P, V, and T must be expressed in consistent units. The conversions are

$$P = 8 \cancel{\text{torr}} \times \frac{1 \text{ atm}}{760 \cancel{\text{torr}}} = 0.011 \text{ atm}$$

$$V = 10 \cancel{\text{m}^3} \times \frac{10^3 \text{ L}}{\cancel{\text{m}^3}} = 1.0 \times 10^4 \text{ L}$$

$$T = (27°\text{C} + 273) = 300 \text{ K}$$

Then

$$n = \frac{PV}{RT} = \frac{(0.011 \cancel{\text{atm}})(1.0 \times 10^4 \cancel{\text{L}})}{0.0821 \cancel{\text{L}} \cancel{\text{atm}}/(\text{mol} \cancel{\text{K}}) \times 300 \cancel{\text{K}}} = 4 \text{ mol}$$ ■

PROBLEM 10.6 A fixed mass of gas occupies 18.6 liters at 24°C and 765 torr. What will its volume be at 240°C and 76.5 torr? □

PROBLEM 10.7 A chemist wishes to collect 5.00 moles of gas being discharged from a chimney to the outside atmosphere. The gas is hot (180°C). Atmospheric pressure is 770 torr. What volume must the chemist collect? □

10.8 MOLECULAR WEIGHTS OF GASES

Recall that molecular weight, without units, has the same numerical value as the molar mass in g/mol.

The molar mass or molecular weight (M) of a gas can be calculated from its volume, temperature, and pressure by using the ideal gas equation. Molecular weights were obtained in this way long before the use of mass spectroscopy. The older method is less precise, but it has the advantage of needing only inexpensive equipment.

The number of moles, n, equals the mass of gas in grams, m, divided by the molar mass, M, in grams/mole:

$$n = m/M$$

Substituting this value in the ideal gas equation gives

$$PV = \left(\frac{m}{M}\right)RT$$

and

$$\boldsymbol{M = \frac{mRT}{PV}} \tag{8}$$

But m/V is the density, d, of the gas in grams per liter at temperature T. Then, substituting d for m/V, equation 8 becomes,

$$\boldsymbol{M = \frac{dRT}{P}} \tag{9}$$

EXAMPLE 10.12 When 0.482 g of pentane is injected into a 204-mL container at 102°C, it exerts a pressure of 767 torr. Calculate its molar mass.

ANSWER First use the given data (expressing all quantities in the proper units) to obtain n:

$$R = 0.0821 \text{ L atm/(mol K)}$$

$$T = 102°\text{C} + 273 = 375 \text{ K}$$

$$P = 767 \cancel{\text{torr}} \times \frac{1 \text{ atm}}{760 \cancel{\text{torr}}} = 1.01 \text{ atm}$$

$$V = 204 \cancel{\text{mL}} \times \frac{1 \text{ L}}{10^3 \cancel{\text{mL}}} = 0.204 \text{ L}$$

From equation 8,

$$M = \frac{0.482 \text{ g} \times 0.0821 \cancel{\text{L}} \cancel{\text{atm}}/(\text{mol} \cancel{\text{K}}) \times 375 \cancel{\text{K}}}{1.01 \cancel{\text{atm}} \times 0.204 \cancel{\text{L}}}$$

$$= 72.0 \text{ g/mol}$$ ■

EXAMPLE 10.13 The density of methyl fluoride is 0.259 g/L at 400 K and 190 torr. Find its molar mass.

ANSWER First, change 190 torr to 0.250 atm to make it consistent with the units of R. Then, from equation 9,

$$M = \frac{0.259 \text{ g}/\cancel{\text{L}} \times 0.0821 \cancel{\text{L}} \cancel{\text{atm}}/(\text{mol} \cancel{\text{K}}) \times 400 \cancel{\text{K}}}{0.250 \cancel{\text{atm}}}$$

$$= 34.0 \text{ g/mol}$$ ■

PROBLEM 10.8 Find the molar mass of Freon-11 (a chlorofluoromethane). 8.29 L of vapor at 200°C and 790 torr has a mass of 30.5 g. □

BOX 10.3

THE BUOYANCY OF A GAS

The buoyancy (lifting ability) of a gas in air is proportional to the difference between the densities of the gas and air. The density of an ideal gas (equation 9) is given by

$$density = PM/RT$$

where M is the molecular weight, or molar mass, of the gas.

Hot-air balloonists use heated air to lift their balloons. Suppose a balloonist asked, "How hot would I have to heat air to make it as buoyant as helium at 25°C? Would it be practical?"

The molecular weights of air and helium are 29 and 4, respectively. To answer the balloonist's question, note that for the two buoyancies to be equal, the densities of the two gases would also have to be equal, and

$$PM_{air}/RT_{air} = PM_{He}/RT_{He}$$

Since R is a constant, and assuming that the pressure inside the balloon is the same as outside, the R and P terms cancel each other. Solving for T_{air},

$$T_{air} = T_{He}\,(M_{air}/M_{He})$$

$$= (25 + 273) \text{ K } (29/4)$$

$$= 2160 \text{ K, or } 1887°\text{C}$$ (an impractical temperature for a balloon)

PROBLEM 10.9 Calculate the density of methane gas, CH_4, at 25°C and standard atmospheric pressure. □

MOLECULAR WEIGHT OF A MIXTURE OF GASES

We usually think of a molecular weight as the property of a pure substance such as water or cane sugar. However, a molecular weight can also be assigned to a mixture by averaging the molecular weights of its components. The procedure is strictly analogous to the calculation of the average atomic weight of a mixture of isotopes, in which each isotope is counted in proportion to its relative abundance. The concept of an "average molecular weight" of a mixture is particularly meaningful for gases, because the molecular weight of a gas, whether pure or a mixture, is directly proportional to its density, as shown by equation 9. (Why is this not true for liquids or solids?)

PROBLEM 10.10 The density of dry air at STP is 1.29 g/L. Calculate its average molecular weight. □

10.9 CHEMICAL REACTIONS INVOLVING GASES

The gas laws can be applied to calculations on chemical reactions involving gases. Recall from Chapter 3 that the information given by a balanced chemical equation can be interpreted to apply to molar quantities of substances. When the substance is a gas, the balanced chemical equation can also refer to its volume. Remember that the molar volume is the same for all ideal gases (but *not* for liquids or solids) at the same temperature and pressure. At STP, the value for gases is 22.4 L/mol. For example,

$$\underset{\substack{\text{2 moles}\\ 2\times 22.4\text{ L}\\ \text{(STP)}}}{2H_2S(g)} + \underset{\substack{\text{1 mole}\\ 22.4\text{ L}\\ \text{(STP)}}}{SO_2(g)} \longrightarrow \underbrace{\underset{\text{2 moles}}{2H_2O(\ell)} + \underset{\text{3 moles}}{3S(c)}}_{\substack{\text{not gases, so molar gas}\\ \text{volumes do not apply}}}$$

Now, what if the gases are *not* at STP? In that event, the gas laws are used to make the necessary conversion.

EXAMPLE 10.14 Referring to the preceding reaction, calculate the mass of sulfur that can be produced by the reaction of 107 L of H_2S, measured at 745 torr and 25°C. Assume that the SO_2 is present in excess.

ANSWER

Quantities given: 107 liters of H_2S at 748 torr and 25°C

Quantity sought: grams of S

The chemical equation tells us that 2 moles of H_2S yield 3 moles of S. This "conversion" is the key to the calculation, which is carried out in three steps, as previously described in Chapter 3.

STEP 1. Liters of $H_2S \longrightarrow$ moles of H_2S (from $PV = nRT$).

$$n_{H_2S} = \frac{PV}{RT} = \frac{(748\ \cancel{\text{torr}} \times 1\ \cancel{\text{atm}}/760\ \cancel{\text{torr}}) \times 107\ \cancel{\text{L}}}{0.0821\ \cancel{\text{L}}\ \cancel{\text{atm}}/(\text{mol}\ \cancel{\text{K}}) \times (25 + 273)\ \cancel{\text{K}}}$$

$$= 4.30 \text{ mol } H_2S$$

STEP 2. Moles of $H_2S \longrightarrow$ moles of S (from the chemical equation).

$$n_S = n_{H_2S} \times \frac{3 \text{ mol S}}{2 \text{ mol } H_2S} = 4.30 \cancel{\text{mol } H_2S} \times \frac{3 \text{ mol S}}{2 \cancel{\text{mol } H_2S}}$$

$$= 6.45 \text{ mol S}$$

STEP 3. Moles of S $\longrightarrow$ grams of S (from the molar mass of S).

$$\text{Mass of S} = 6.45 \text{ mol S} \times \text{molar mass of S} \quad \text{(from atomic weight table)}$$

$$= 6.45 \cancel{\text{mol S}} \times \frac{32.06 \text{ g S}}{1 \cancel{\text{mol S}}}$$

$$= 207 \text{ g S}$$

■

EXAMPLE 10.15 Referring again to the H_2S–SO_2 reaction, what volume of SO_2, measured at 810 torr and 35°C, is needed to produce 28.3 g of sulfur, assuming an excess of available H_2S?

ANSWER

Quantity given: 28.3 grams of S

Quantity sought: Liters of SO_2 at 810 torr and 35°C

STEP 1. Grams of S $\longrightarrow$ moles of S (from the molar mass of S).

STEP 2. Moles of S $\longrightarrow$ moles of SO_2 (from the chemical equation).

STEP 3. Moles of SO_2 to liters of SO_2 (from $PV = nRT$).

Thus

$$\text{STEP 1. Moles of S} = 28.3 \cancel{\text{g S}} \times \frac{1 \text{ mol S}}{32.06 \cancel{\text{g S}}}$$

$$= 0.883 \text{ mol S}$$

$$\text{STEP 2. } n_{SO_2} = n_S \times \frac{1 \text{ mol } SO_2}{3 \text{ mol S}} = 0.883 \cancel{\text{mol S}} \times \frac{1 \text{ mol } SO_2}{3 \cancel{\text{mol S}}}$$

$$= 0.294 \text{ mol } SO_2$$

$$\text{STEP 3. } V_{SO_2} = \frac{nRT}{P} = \frac{0.294 \cancel{\text{mol}} \times 0.0821 \text{ L} \cancel{\text{atm}}/(\cancel{\text{mol K}}) \times (35 + 273) \cancel{\text{K}}}{810 \cancel{\text{torr}} \times 1 \cancel{\text{atm}}/760 \cancel{\text{torr}}}$$

$$= 6.98 \text{ L}$$

■

EXAMPLE 10.16 SO_2, used in the manufacture of sulfuric acid, is obtainable from sulfide ores:

$$4FeS_2(c) + 11O_2(g) \longrightarrow 2Fe_2O_3(c) + 8SO_2(g)$$

Find the mass of oxygen in grams that react when 75.0 liters of SO_2 are produced at 100°C and 1.04 atm.

ANSWER

Quantity given: 75.0 liters of SO_2 at 100°C and 1.04 atm

Quantity sought: grams of O_2

We follow the usual procedure.

STEP 1. Liters of $SO_2 \longrightarrow$ moles of SO_2 (from $PV = nRT$).

$$n_{SO_2} = \frac{1.04\ \cancel{atm} \times 75.0\ \cancel{L}}{0.0821\ \cancel{L}\ \cancel{atm}(\text{mol}\ \cancel{K})(100 + 273)\ \cancel{K}}$$

$$= 2.55 \text{ mol } SO_2$$

STEP 2. Moles of $SO_2 \longrightarrow$ moles of O_2 (from the chemical equation).

$$n_{O_2} = n_{SO_2} \times \frac{11 \text{ mol } O_2}{8 \text{ mol } SO_2} = 2.55\ \cancel{\text{mol } SO_2} \times \frac{11 \text{ mol } O_2}{8\ \cancel{\text{mol } SO_2}}$$

$$= 3.51 \text{ mol } O_2$$

STEP 3. Moles of $O_2 \longrightarrow$ g O_2 (from the molar mass of O_2).

$$\text{mass } O_2 = 3.51\ \cancel{\text{mol } O_2} \times \frac{32.0 \text{ g } O_2}{1\ \cancel{\text{mol } O_2}}$$

$$= 112 \text{ g } O_2$$ ■

PROBLEM 10.11 Oxygen masks use canisters containing potassium superoxide:

$$4KO_2(c) + 2CO_2(g) \longrightarrow 2K_2CO_3(c) + 3O_2(g)$$

Calculate the mass in grams of KO_2 used up when 8.90 L CO_2 at 22.0°C and 767 torr react. □

PROBLEM 10.12 Dichlorodifluoromethane (Freon-12) is made by the reaction

$$CCl_4(\ell) + 2HF(g) \longrightarrow CCl_2F_2(g) + 2HCl(g)$$

What volume of CCl_2F_2 at 50.0°C and 780 torr can be made from 56.5 g HF? □

10.10 HOW CHEMISTS ARRIVE AT MOLECULAR FORMULAS

In the early 1800s, the chemical formulas of even simple substances like hydrogen, chlorine, and water were not known with certainty (Figure 10.10). Avogadro's law provided a way out of this confusion. Consider, for example, the relationships of gas volumes in the reaction between hydrogen gas and chlorine gas to produce hydrogen chloride gas. Experiment shows that, at a fixed temperature and pressure, these two gases combine with each other in equal volumes, producing two volumes of hydrogen chloride. Thus, for example, 1 liter of hydrogen reacts with 1 liter of chlorine to produce 2 liters of hydrogen chloride. From Avogadro's law we know that the ratios of volumes are also the ratios of numbers of molecules:

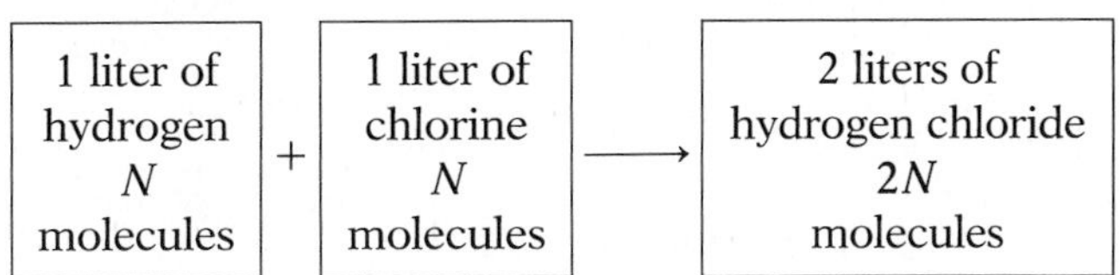

This relationship also means that

1 molecule of hydrogen gas + 1 molecule of chlorine gas ⟶ 2 molecules of hydrogen chloride gas

Since the atoms are indivisible, the simplest possible formula for hydrogen chloride is HCl. If this formula is correct, 1 *molecule* of hydrogen chloride would contain 1 *atom* each of hydrogen and chlorine. Likewise, 2 *molecules* of hydrogen chloride would contain 2 *atoms* of hydrogen and 2 *atoms* of chlorine. But note from the diagram above that the 2 atoms of hydrogen come from only 1 molecule of hydrogen.

Mass spectroscopy confirms that H_2 and Cl_2 are the correct formulas.

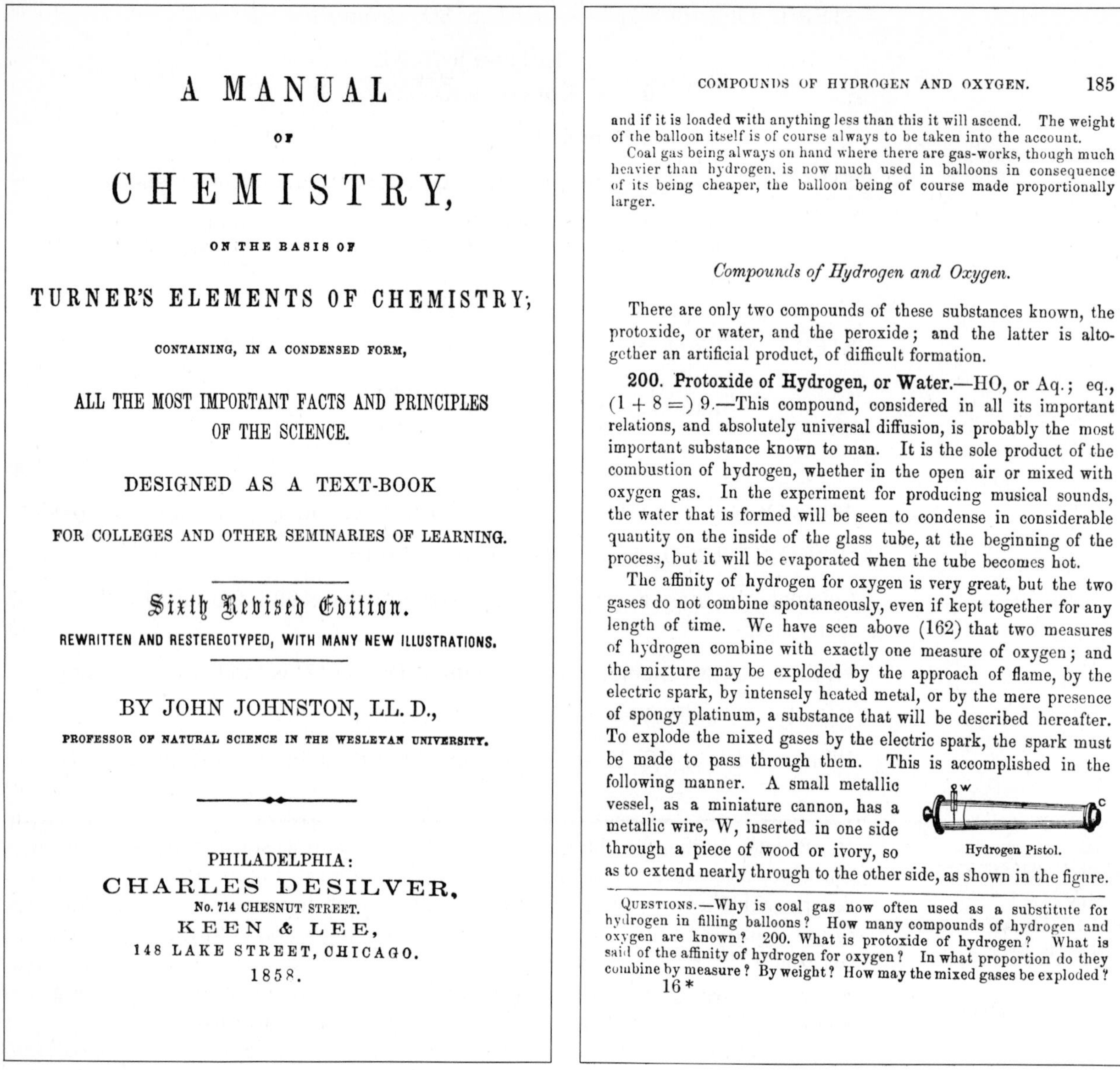

A MANUAL

OF

CHEMISTRY,

ON THE BASIS OF

TURNER'S ELEMENTS OF CHEMISTRY;

CONTAINING, IN A CONDENSED FORM,

ALL THE MOST IMPORTANT FACTS AND PRINCIPLES
OF THE SCIENCE.

DESIGNED AS A TEXT-BOOK

FOR COLLEGES AND OTHER SEMINARIES OF LEARNING.

Sixth Revised Edition.

REWRITTEN AND RESTEREOTYPED, WITH MANY NEW ILLUSTRATIONS.

BY JOHN JOHNSTON, LL. D.,

PROFESSOR OF NATURAL SCIENCE IN THE WESLEYAN UNIVERSITY.

PHILADELPHIA:
CHARLES DESILVER,
No. 714 CHESNUT STREET.
KEEN & LEE,
148 LAKE STREET, CHICAGO.
1858.

COMPOUNDS OF HYDROGEN AND OXYGEN. 185

and if it is loaded with anything less than this it will ascend. The weight of the balloon itself is of course always to be taken into the account.

Coal gas being always on hand where there are gas-works, though much heavier than hydrogen, is now much used in balloons in consequence of its being cheaper, the balloon being of course made proportionally larger.

Compounds of Hydrogen and Oxygen.

There are only two compounds of these substances known, the protoxide, or water, and the peroxide; and the latter is altogether an artificial product, of difficult formation.

200. Protoxide of Hydrogen, or Water.—HO, or Aq.; eq., (1 + 8 =) 9.—This compound, considered in all its important relations, and absolutely universal diffusion, is probably the most important substance known to man. It is the sole product of the combustion of hydrogen, whether in the open air or mixed with oxygen gas. In the experiment for producing musical sounds, the water that is formed will be seen to condense in considerable quantity on the inside of the glass tube, at the beginning of the process, but it will be evaporated when the tube becomes hot.

The affinity of hydrogen for oxygen is very great, but the two gases do not combine spontaneously, even if kept together for any length of time. We have seen above (162) that two measures of hydrogen combine with exactly one measure of oxygen; and the mixture may be exploded by the approach of flame, by the electric spark, by intensely heated metal, or by the mere presence of spongy platinum, a substance that will be described hereafter. To explode the mixed gases by the electric spark, the spark must be made to pass through them. This is accomplished in the following manner. A small metallic vessel, as a miniature cannon, has a metallic wire, W, inserted in one side through a piece of wood or ivory, so as to extend nearly through to the other side, as shown in the figure.

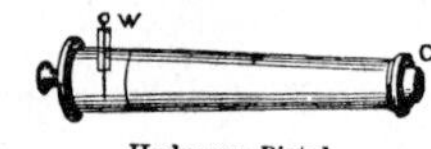

Hydrogen Pistol.

QUESTIONS.—Why is coal gas now often used as a substitute for hydrogen in filling balloons? How many compounds of hydrogen and oxygen are known? 200. What is protoxide of hydrogen? What is said of the affinity of hydrogen for oxygen? In what proportion do they combine by measure? By weight? How may the mixed gases be exploded?

16*

FIGURE 10.10
A chemistry book published in 1858. The formula for water was still written as HO at that time.

Thus we may write

$$1 \text{ molecule of hydrogen} \longrightarrow 2 \text{ atoms of hydrogen}$$

Of course, the only way that a molecule of hydrogen can yield 2 atoms of hydrogen is for the molecule to *contain* 2 atoms. Thus the molecular formula for hydrogen is H_2. Again, note that we need not know the value of N to arrive at this conclusion. The same reasoning leads to the molecular formula Cl_2 for chlorine.

EXAMPLE 10.17 What would the formulas for hydrogen and chlorine be if hydrogen chloride were H_2Cl? H_xCl_y?

ANSWER Regardless of the formula for hydrogen chloride, it would still be true that

1 molecule hydrogen + 1 molecule chlorine ⟶
2 molecules hydrogen chloride

If the formula were H_2Cl, then 2 molecules of H_2Cl would contain 4 H atoms, and the formula for hydrogen would be H_4. Chlorine would still be Cl_2. (Why?) In general, the volume relationships tell us that

$$H_{2x} + Cl_{2y} \longrightarrow 2H_xCl_y$$

So the formulas for hydrogen and chlorine must contain even numbers of atoms. ■

PROBLEM 10.13 The volume relationships in the combination of oxygen gas with fluorine gas, at a fixed temperature and pressure, are

1 volume of oxygen + 2 volumes of fluorine ⟶
2 volumes of fluorine oxide

From this information, what are the possible formulas for oxygen gas? What is the simplest possible formula for oxygen gas? □

10.11 DALTON'S LAW OF PARTIAL PRESSURES

A novice diver asks, "How full is this oxygen tank?"

The old-timer taps thoughtfully up and down the tank, stops about halfway, and answers jokingly, "It's down to here!"

The novice's question was foolish because a tank that contains only gas is *always* full of gas. If the volume of the tank is 20 L, then there are *always* 20 L of oxygen in the tank, since a gas occupies all of the volume available to it. If the gas in the tank is a mixture of oxygen and helium, then the volume of the oxygen is 20 L and the volume of the helium is also 20 L; each gas occupies the entire volume of the tank, even when mixed. Of course, if the tank had previously been used by another diver, there would be fewer molecules of gas left in it—fewer moles, less mass—and a lower *pressure*. The novice should have asked, "What's the pressure in this tank?" It is essential to keep this point in mind when thinking about the pressure of each gas in a mixture of gases.

Now let's consider two gases: chlorine, Cl_2, which is greenish yellow, and neon, Ne, which is colorless. Suppose that we pump chlorine into a 2-L container at 25°C (298 K) until the pressure reaches 250 torr, and we pump neon into another 2-L container until it, too, reaches a pressure of 250 torr. Then we transfer all of the neon into the container in which we have pumped the chlorine. (No chemical reaction takes place.) You would find that the pressure of the gas mixture is 500 torr — the sum of the two pressures before the gases were mixed (Figure 10.11). Note that each gas would occupy the same volume (2 liters) before and after mixing, and hence exert the same pressure in the mixture that it exerted when it was alone.

The *partial pressure, p,* of a particular gas in a mixture of gases is defined as the pressure the gas would exert if it occupied the container alone. In our chlorine-neon mixture, the partial pressure of each gas is 250 torr, just as when each gas was alone in its own container. From experiments of this type, John Dalton (of atomic theory fame) found that when gases are mixed, the pressure of the mixture

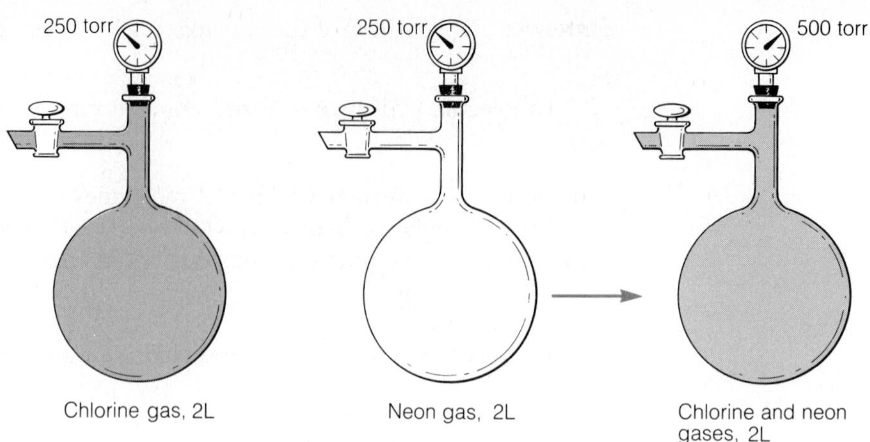

FIGURE 10.11
Dalton's law of partial pressures.

is the sum of the pressures each gas exerts by itself in the same container. Dalton's law (1803) may be stated: *In a mixture of gases, the total pressure equals the sum of the partial pressures.* In equation form,

$$P_{tot} = p_1 + p_2 + \ldots \tag{10}$$

The observation applies only to gases that do not react with each other.

Gas mixtures are important in many contexts, and you will often see references to "composition by volume" of the atmosphere, fuel gases, and so on. Since each component of a gaseous mixture occupies the entire volume, not just a portion of it, the expression is misleading. What is really meant is "composition by number of molecules or moles," which is molar composition. Thus, when someone says that 21% of the air is oxygen, he or she means that 21% of the molecules in air are O_2 molecules, or that 21% of the moles in air are moles of O_2.

From $PV = nRT$ it follows that, in a container of fixed volume at constant temperature, the pressure is proportional to the number of moles, $P = \left(\frac{RT}{V}\right)n$. This relationship must be equally true for the partial pressure, p_1, of any one gas in a gaseous mixture and for the total pressure of the mixture, P_{tot}.

$$p_1 = constant \times n_1 \quad \text{and} \quad P_{tot} = constant \times n_{tot}$$

Therefore,

$$\frac{p_1}{P_{tot}} = \frac{n_1}{n_{tot}} \quad \text{or}$$

$$p_1 = P_{tot}\frac{n_1}{n_{tot}} \tag{11}$$

In words, equation 11 says that the partial pressure of any gas equals its mole fraction times the total pressure. The **mole fraction** of any substance in a mixture is the number of moles of that substance divided by the total number of moles in the mixture.

EXAMPLE 10.18 The molar composition of dry air is 78.1% nitrogen, 20.9% oxygen, and 1.0% other gases. Calculate the partial pressures, in atmospheres, in a tank of dry air compressed to 10.0 atm.

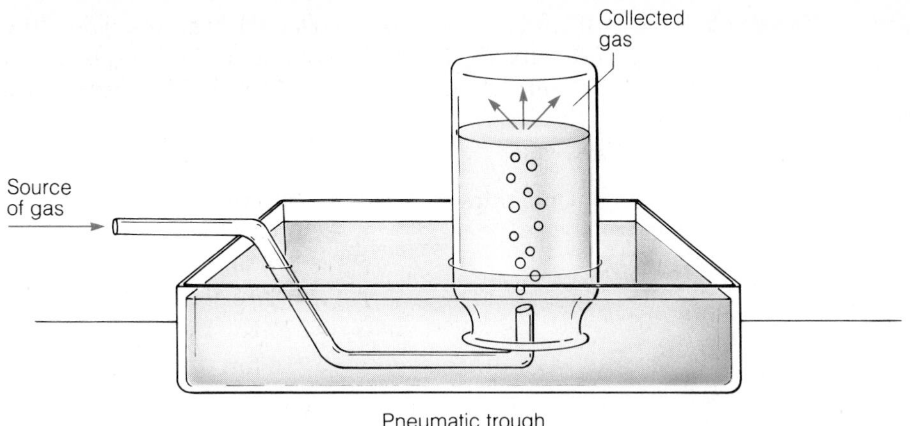

FIGURE 10.12 Collection of a gas over water. A bottle is filled with water and inverted into a trough of water. As the gas bubbles up into the bottle, an equal volume of water is displaced into the trough. The collected gas is said to be "wet" because it contains water vapor at a partial pressure equal to the vapor pressure of water at the temperature of the system.

ANSWER From equation 11,

$$p_{\text{nitrogen}} = \frac{78.1}{100} \times 10.0 \text{ atm} = 7.81 \text{ atm}$$

$$p_{\text{oxygen}} = \frac{20.9}{100} \times 10.0 \text{ atm} = 2.09 \text{ atm}$$

$$p_{\text{other gases}} = \frac{1.00}{100} \times 10.0 \text{ atm} = 0.100 \text{ atm}$$

$$\text{Total} = 10.0 \text{ atm}$$

By Dalton's law, the sum of the partial pressures is the total pressure:

$$P_{\text{tot}} = p_{N_2} + p_{O_2} + p_{\text{other gases}} = 10.0 \text{ atm}$$ ■

PROBLEM 10.14 A cylinder of compressed gas is labeled, "Composition: 4.5% H_2S, 3.0% CO_2, balance N_2." The pressure gauge attached to the cylinder reads 46 atm. Calculate the partial pressure of each gas in atmospheres. □

Vapor pressures of water at different temperatures are given in Appendix E.

Gases used in medicine and industry are usually stored in steel cylinders. In the laboratory, however, it is sometimes convenient to collect and confine a gas over a liquid such as mercury (as Boyle did) or water. Whereas the evaporation of mercury is usually negligible, the evaporation of water adds enough molecules to the gas to make up a significant part of the total pressure. Fortunately, the pressure contribution from such evaporation at constant temperature is a definite value called the **vapor pressure** of the liquid.

A gas that is collected "over water" (Figure 10.12) must be essentially insoluble in water; otherwise it will dissolve and the effort to collect it will be frustrated. Ammonia, NH_3, and sulfur dioxide, SO_2, are gases that dissolve in water and therefore cannot be collected over it. Oxygen, hydrogen, and nitrogen are nearly insoluble and therefore can be collected over water. A gas that is collected over water and is thus saturated with water vapor is said to be "wet," even though there are no droplets of water in it and the gas does not feel wet to the touch. A gas that contains no water vapor is said to be "dry." In accordance with Dalton's law, the pressure of the wet gas is the total pressure of the gas mixture, including the water vapor.

$$P_{\text{tot}} = p_{\text{wet gas}} = p_{\text{dry gas}} + p_{H_2O \text{ vapor}}$$

EXAMPLE 10.19 10.0 mL of oxygen is collected over water at 20°C and at a total pressure of 770.0 torr. What is the volume of the dry gas at standard conditions? The vapor pressure of water, p_{water}, at 20°C is 17.5 torr (Appendix E).

ANSWER

Quantities given: $V_1 = 10.0$ mL
$t_1 = 20°C$
$P_{tot} = 770.0$ torr
$p_{water} = 17.5$ torr at 20°C
$\left.\begin{array}{l} T_2 = 273 \text{ K} \\ p_2 = 760 \text{ torr} \end{array}\right\}$ STP

Quantity sought: V_2 (of dry oxygen)

To find p_1, use Dalton's law, $P_{tot} = p_{oxygen} + p_{water}$.

$$p_1 = p_{oxygen} = P_{tot} - p_{water} = 770.0 \text{ torr} - 17.5 \text{ torr}$$
$$= 752.5 \text{ torr}$$

Then

$$V_2 = V_1 \times \frac{T_2}{T_1} \times \frac{p_1}{p_2}$$
$$= 10.0 \text{ mL} \times \frac{273 \not{\text{K}}}{293 \not{\text{K}}} \times \frac{752.5 \not{\text{torr}}}{760 \not{\text{torr}}}$$
$$= 9.23 \text{ mL}$$

■

PROBLEM 10.15 A sample of nitrogen collected over water at 25°C and 752 torr occupies 53.5 mL. What is the volume of the dry gas at standard conditions? □

PROBLEM 10.16 100 mL of hydrogen gas saturated with water vapor at 20°C exerts a pressure of 760.0 torr. If the gas is dried and kept at a volume of 100 mL, what pressure will it exert? To what volume must the dry gas be adjusted to exert a pressure of 760.0 torr? Assume no change in temperature. □

10.12 IDEAL GASES; THE KINETIC MOLECULAR THEORY

Recall the properties of gases described at the beginning of this chapter: gases are transparent; they can spread out readily in space; they can flow through very small openings; they can be compressed much more easily than liquids or solids. In 1738 Daniel Bernoulli proposed a novel concept to account for these properties. We must question what Bernoulli could have known in 1738. Boyle had done his experiments in 1662, of course, but Dalton's atomic theory was not to come until 1803. The laws of Charles, Gay-Lussac, and Avogadro were also not to be formulated until early in the nineteenth century. The idea that matter consisted of small fundamental particles was fairly well appreciated at that time, but the distinction between molecules and atoms was not clear.

Bernoulli based his theory of gas behavior on the concept that molecules are in *motion.* This idea was quite advanced for its day, and it was not accepted until it was revived about a century later and gradually won out over rival theories. Today, Bernoulli's concept is called the **kinetic molecular theory of gases**, and the model on which it is based is called the **ideal gas**. It will be apparent from this description of the ideal gas that it exists only in our imagination.

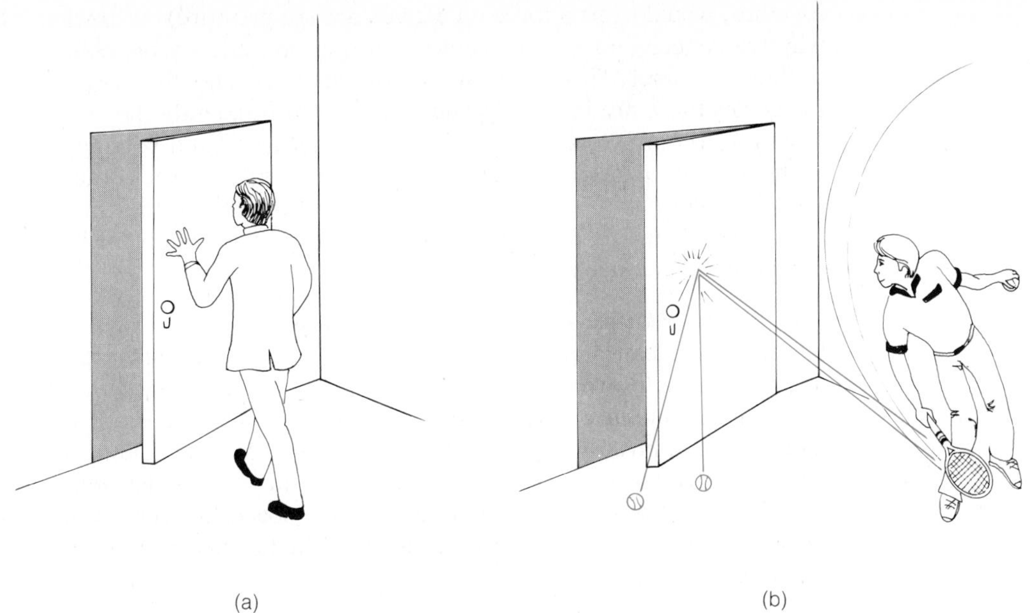

FIGURE 10.13
Pressure exerted by a steady push (a) and by a series of collisions (b).

The ideal gas consists of molecules that

- are points in space (in other words, have no volume);
- have mass and velocity;
- do not attract or repel each other or the walls of their container;
- collide with each other or with the walls of the container with no loss of total kinetic energy.

One molecule may lose energy during a collision, but this energy will be transferred to other molecules. Of course, if the molecules were really volumeless points, they would not collide. Real molecules, however, do occupy a small volume and collide with each other frequently.

Thus, the theory is not an experimental description of a real gas. Rather, it is a concept, or a **model**, of an ideal gas. The model is pretty good, but not perfect. It explains the properties of many gases with reasonable accuracy ($\pm$ about 1% at ordinary temperatures and pressures). The following discussions will interpret the gas laws in terms of the ideal model. In section 10.14 we will examine the defects of the model.

THE IDEAL MODEL AND BOYLE'S LAW

To start with, the model offers a very satisfactory explanation of why gases can be compressed much more easily than liquids or solids. The reason is that only *empty space* is reduced when molecules are crowded more closely together; the molecules themselves are not squeezed into smaller sizes.

Let us now consider how a gas being compressed would exert pressure on the walls of its container. Pressure is defined as force per unit area, and we think of force as a push—as when you force a mass to move. If the mass were a door, you could push it by exerting force on it with your body (Figure 10.13a). But you could also push it by hitting it with tennis balls (Figure 10.13b), which would bounce back at you as the door gradually swung shut. The bouncing balls, taken together, would exert a force on a given area (a pressure) on the door. This is the way we conceive of gas molecules exerting pressure by bouncing off the walls of their container. This visualization now suggests some further points. The more molecules there are in a given space, the more frequently they will hit the walls. Therefore the pressure of a gas in a container at a fixed volume is proportional to the number of gas molecules present. If the volume of the container is reduced, the molecules have less distance to travel to the walls, so they hit the walls more frequently. Thus, as the volume is decreased, the pressure increases. This is a qualitative statement of Boyle's law.

THE IDEAL MODEL AND CHARLES'S LAW

Faster molecules (larger u) will hit the walls of a container harder and more frequently than slower ones. It follows, then, that an increase in molecular speed results in an increase in gas pressure, or if the container is not rigid, the faster molecules can push the walls back and thus increase the volume of the gas. At a given speed, a more massive molecule will hit the wall harder than a less massive one. The effect of mass and the two effects of speed (harder collisions, more frequent collisions) correspond to the effect of kinetic energy, E_k, of a moving body. The relationship is

$$E_k = \tfrac{1}{2}mu^2$$

The u in the expression $\frac{1}{2}mu^2$ for a gas is the velocity of a molecule that possesses average kinetic energy.

We know from experiment that these effects are produced by raising the temperature. Evidently, the temperature of a gas is related to the kinetic energy of its molecules. Thus, a rise in temperature corresponds to an increase in the kinetic energy of the gas molecule and results in an increase in the volume of the gas at constant pressure—in agreement with Charles's law—or an increase in pressure at a fixed volume.

THE IDEAL MODEL AND AVOGADRO'S AND DALTON'S LAWS

The ideal model is consistent with the belief that the pressure of a gas in a rigid container is determined only by the temperature and the number of molecules. It does not matter what *kind* of molecules the gas consists of. For example, consider a quantity of helium in a rigid container at a given temperature. Imagine that the helium atoms (4 g/mol) are replaced by an equal number of neon atoms (20 g/mol). The neon atoms bring a greater mass but a lower average speed and therefore collide with the walls with less frequency. A detailed mathematical treatment shows that the opposing effects just compensate each other, and the pressure of the neon is the same as that exerted by the helium at the same temperature.

Thus, in a rigid container,

- kinetic molecular theory says that P depends on $\begin{cases}\text{number of molecules,}\\ \text{kinetic energy of}\\ \text{molecules}\end{cases}$
- experiment ($P = nRT/V$) says that P depends on $\begin{cases}\text{number of molecules,}\\ \text{absolute temperature}\end{cases}$

A reasonable conclusion is that a given absolute temperature corresponds to a given kinetic energy for *any* kind of molecule. If we fix the temperature and pressure, then the volume of the gas must depend only on the *number* of molecules—a statement of Avogadro's law. It follows that "any kind of molecule" also includes mixtures of different kinds, so Dalton's law also may take a bow.

A more mathematical approach to this qualitative discussion is given in section 10.16.

EXAMPLE 10.20 Two equal-sized tanks contain compressed gas at the same temperature. One tank contains 2×10^{25} molecules of O_2; the other contains 2×10^{25} molecules of He. (a) Are the pressures in the two tanks the same or different? If they are different, which tank has the greater pressure? (b) Are the average kinetic energies of the molecules in the two tanks the same or different? If they are different, which kinetic energy is greater? (c) Are the average speeds of the molecules in the two tanks the same or different? If they are different, which molecules are faster?

ANSWER

(a) $P = nRT/V$. Since n, R, T, and V are the same in both tanks, the two pressures must be the same.

(b) Because the two temperatures are the same, the average kinetic energies of the molecules in the two tanks must be the same.

(c) Because the two energies are the same, the average values of $\frac{1}{2}mu^2$ are the same. However, the masses of the two kinds of molecules are different, and therefore the speeds of the molecules must also be different. The lighter molecules are the speedier ones, as we explain in the next section. ■

10.13 GRAHAM'S LAW

Imagine an astronaut on a "space walk," supplied from a spaceship with a breathing gas consisting of oxygen and helium. The astronaut's suit, though well constructed, is not perfect—there are some tiny pores in it through which molecules of the hydrogen and helium gas can escape to the vacuum outside. The flow of a gas through a small hole or a porous material is called **effusion**. Will one of the gases effuse more rapidly than the other and, if so, which molecules will effuse faster, oxygen or helium? An He molecule (molecular weight 4) has one-eighth the mass of an O_2 molecule (molecular weight 32), but at the same temperature the average kinetic energies of helium and oxygen molecules are equal. Therefore the helium molecules must be speedier. Speedier molecules will effuse more rapidly because they will reach the tiny holes in less time. This idea can be confirmed, and quantified, by experiment as well as by theory.

Early experiments on effusion were carried out by Thomas Graham in 1846. Graham found that *the rates of effusion of gases are inversely proportional to the square roots of their densities.* The density of a gas depends on its temperature and pressure. It is therefore meaningful to compare the densities of different gases *only* at the same temperature and pressure. We can understand this relationship by recalling from the preceding section that the temperature of a gas is related to the average kinetic energy of its molecules. Therefore, for two gases at the same temperature,

$$T_1 = T_2$$

$$E_{k_1} = E_{k_2}$$

and

$$\tfrac{1}{2}m_1{u_1}^2 = \tfrac{1}{2}m_2{u_2}^2$$

Then,

$$\frac{{u_1}^2}{{u_2}^2} = \frac{m_2}{m_1}$$

and, taking square roots of both sides,

$$\frac{u_1}{u_2} = \sqrt{\frac{m_2}{m_1}}$$

The molecular speed, u, is proportional to the rate of effusion because the faster the molecule, the sooner it will reach an opening in the wall. At a given temperature and pressure, the mass of the molecule, m, is proportional to the density of the gas because density equals mass/volume. Thus we may write

$$\frac{\textbf{rate}_1}{\textbf{rate}_2} = \sqrt{\frac{\textbf{density}_2}{\textbf{density}_1}} \tag{12}$$

which is the mathematical statement of Graham's law.

The molar mass (molecular weight) of a gas is proportional to the masses of its molecules. Therefore we may also write

$$\frac{\textbf{rate}_1}{\textbf{rate}_2} = \sqrt{\frac{(\textbf{molar mass})_2}{(\textbf{molar mass})_1}} \tag{13}$$

EXAMPLE 10.21 Again, consider the astronaut that is being supplied with helium and oxygen. The density of helium at STP is 0.179 g/L and of oxygen is 1.43 g/L. Which gas effuses faster? How much faster?

ANSWER From equation 12

$$\frac{\text{rate of effusion of helium}}{\text{rate of effusion of oxygen}} = \sqrt{\frac{1.43 \text{ g/L}}{0.179 \text{ g/L}}} = 2.83$$

The helium effuses 2.83 times as fast as the oxygen. ■

EXAMPLE 10.22 One of the methods of separating uranium-235 (a nuclear fuel) from uranium-238 (nonfuel) is by the effusion of their gaseous hexafluorides, UF_6. How much faster will $^{235}UF_6$ effuse than $^{238}UF_6$? The molar masses are 235.0439 g/mol for ^{235}U, 238.0508 g/mol for ^{238}U, and 18.99840 g/mol for F.

ANSWER Since the given data are molar masses, equation 13 is applicable.

Graham's law also applies, less strictly, to **diffusion**, the process by which gases disperse in space by the random motion of their molecules. The application to diffusion is satisfactory when the molecules of the different gases are of equal size.

$$\text{molar mass of } ^{235}UF_6 = 235.0439 + 6(18.99840) = 349.034 \text{ g/mol}$$

$$\text{molar mass of } ^{238}UF_6 = 238.0508 + 6(18.99840) = 352.041 \text{ g/mol}$$

$$\frac{\text{rate for } ^{235}U_6}{\text{rate for } ^{238}UF_6} = \sqrt{\frac{352.041}{349.034}} = 1.00430$$

$^{235}UF_6$ effuses 1.00430 times faster than $^{238}UF_6$ (not much!). ■

PROBLEM 10.17 Argon is ten times as dense as helium at the same temperature and pressure. Which gas effuses faster? How much faster? □

10.14 THE DISTRIBUTION OF MOLECULAR SPEEDS

We pointed out in section 10.11 that real molecules occupy volume and therefore collide with each other frequently. After some collisions, one molecule speeds up and the other slows down. An individual molecule in a gas therefore has different speeds at different times, even if the temperature of the gas is constant. However, a measurable sample of gas has so many molecules that at any one time the fractions of molecules having speeds close to particular values are constant. A

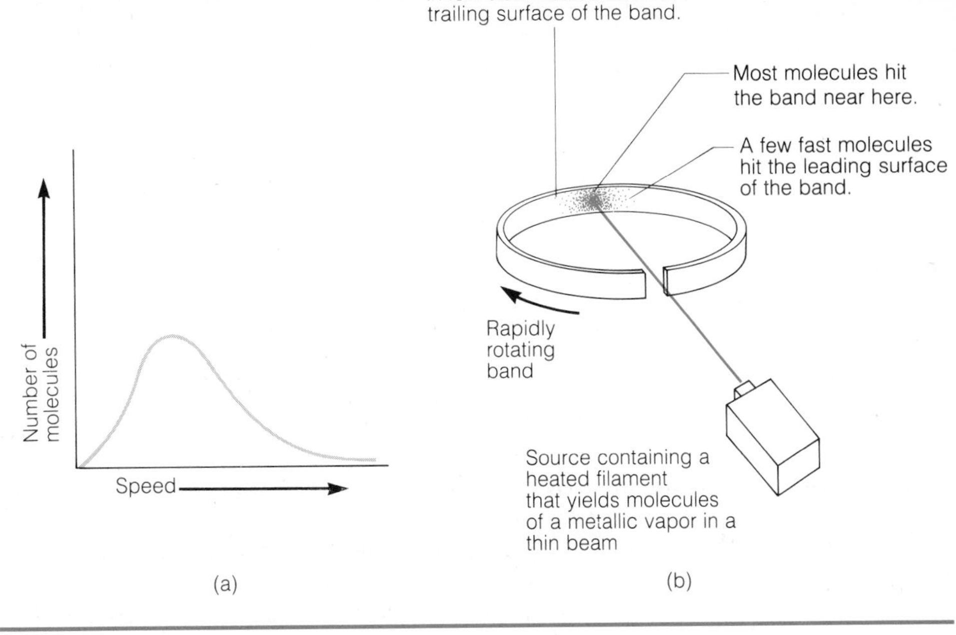
A few slow molecules hit the trailing surface of the band.
Most molecules hit the band near here.
A few fast molecules hit the leading surface of the band.
Number of molecules
Speed
Rapidly rotating band
Source containing a heated filament that yields molecules of a metallic vapor in a thin beam
(a)
(b)

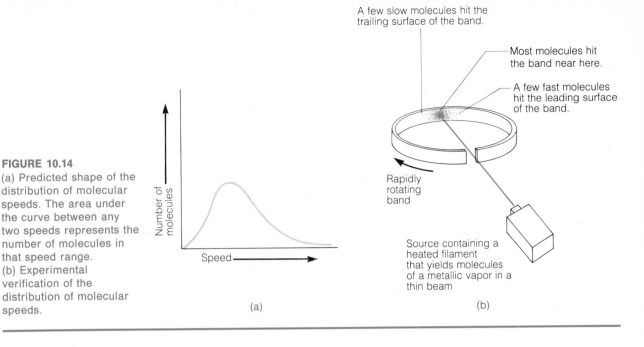

FIGURE 10.14
(a) Predicted shape of the distribution of molecular speeds. The area under the curve between any two speeds represents the number of molecules in that speed range.
(b) Experimental verification of the distribution of molecular speeds.

very small fraction of the molecules have speeds close to zero; another small fraction have relatively high speeds; most of the molecules have intermediate speeds.

The predicted shape of this distribution is shown in Figure 10.14a, which plots the relative number of molecules against their speeds. The *area* under the curve represents *the total number of molecules in the sample.* The curve is not symmetrical because it is limited by zero on the left end but is not limited on the right. The flat "tail" on the right represents a small proportion of very speedy molecules. The theory predicts that there will be a diminishing number of molecules at increasingly higher speeds, as shown by the graph.

A typical average molecular speed is about 400 m/sec for nitrogen at 0°C.

The experimental evidence for the distribution of molecular speeds in a gas supports the theoretical prediction. See Figure 10.14b. In an experiment, a narrow beam of gas molecules is aimed at a rapidly rotating solid band that is split in one position. Once every rotation some molecules enter the slit, hit the inside surface of the band opposite the beam of gas, and stick there. Not all of them hit the same spot, however, because they are traveling at different speeds. The fast ones arrive first and strike the leading surface of the band. Then, the molecules traveling at average speeds arrive, hitting the band to the left of the fast molecules. Finally, the slow molecules arrive, hitting farther to the left, on the trailing surface of the band. Where any given molecule hits the band is thus related to its speed, and a plot of the number of "hits" per unit area therefore shows the distribution of molecular speeds.

If a gas sample is heated from T_1 to a higher temperature T_2, the average speed of the molecules must increase, but of course, the number of molecules (represented by the area under the curve) remains constant, and the lowest possible speed is still zero. The result is that the curve of molecular speeds at T_2 is flatter and more skewed to the right than the curve at T_1 (Figure 10.15). The important consequence is not so much that the average speed has increased, but that the proportion of really speedy molecules is considerably larger. This shift to the right with increasing temperature is a very significant factor in determining the speed of chemical reactions. This effect is discussed in Chapter 19.

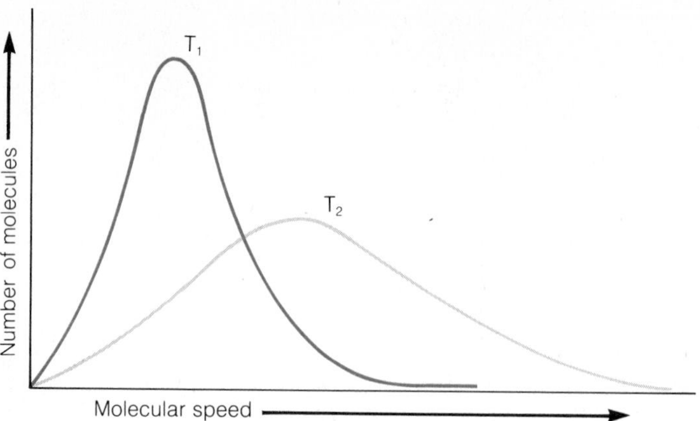

FIGURE 10.15
Distribution of molecular speeds at T_1 (lower temperature) and T_2 (higher temperature).

10.15 DEVIATIONS FROM IDEAL BEHAVIOR

The ideal gas law is not an exact expression of the behavior of real gases. It is a reasonably good approximation at ordinary temperatures and pressures, but under some conditions the deviations become larger. The important characteristics of real gases overlooked by the ideal gas model are that (1) real gas molecules attract each other, and (2) they occupy volume. Let us consider the consequences of each of these differences between real and ideal gases.

The attraction between molecules. When two molecules collide, they may actually remain together for a short time (a sticky collision). During that short time, what were formerly two molecules are then one molecule with a larger mass (Figure 10.16). Remember that in a given container at a given temperature, the pressure of a gas depends on the *number of molecules,* not on their masses ($P \propto NT/V$). Such "doubling" generally lasts for only a tiny fraction of a second and then the molecules become unstuck, but in a large sample of gas there is always a certain proportion of these doubled molecules. The gas pressure is reduced as a result, and the pressure of a real gas is thus less than the pressure of the ideal gas. Table 10.3 illustrates this effect for methane at 0°C and pressures from 50 to 300 atmospheres.

These doubled molecules have been detected and studied.

The volume occupied by molecules. There are two ways we can think about how the volumes of molecules influence the properties of gases.

1. Consider an "ideal" molecule (a point; see Figure 10.17a) in a container. The molecule must travel a distance x to hit the wall. Now consider a real molecule in the same circumstances (Figure 10.17b).

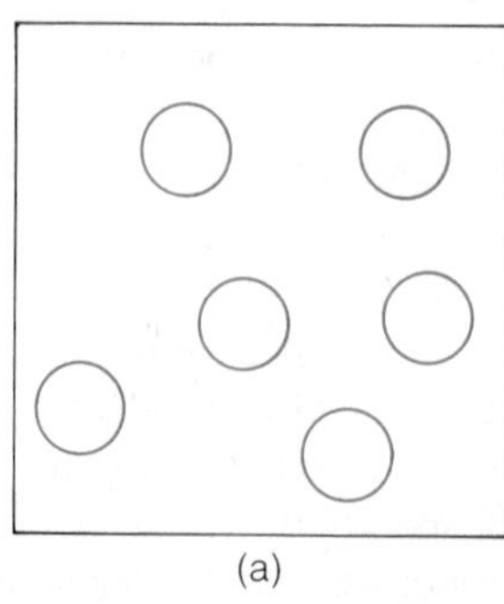

(a)

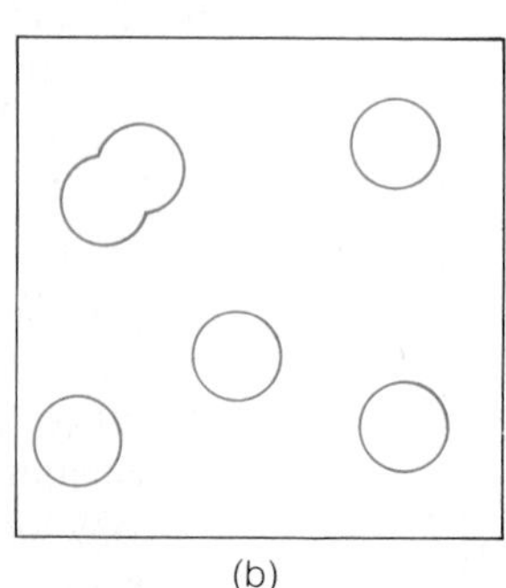

(b)

FIGURE 10.16
(a) Six molecules of gas in a container. (b) After a sticky collision, there are only five molecules, even though the mass in the container is the same. Since there are fewer molecules, the pressure exerted by the gas is less.

TABLE 10.3
PV VERSUS *P* FOR 1 MOLE OF METHANE AT 0°C

P_{real}	×	V	=	PV	P_{ideal}†	CONDITION OF GAS
0.100 atm		224 L		22.4 L atm	0.100 atm	Ideal
1.00		22.4		22.4	1.00	
50		0.374		18.7	60	Real pressures are less than ideal. Attractive forces predominate.
100		0.169		16.9	133	
200		0.0885		17.7	253	
300		0.0683		20.5	328	
400		0.0608		24.3	368	Real pressures are greater than ideal. Effect of molecular volume predominates.
600		0.0420		25.2	533	

† If the gas were ideal, *PV* would be constant at 22.4 L atm, or $P_{ideal} \times V = 22.4$ L atm. Therefore, $P_{ideal} = 22.4$ L atm/V. For example, 22.4 L atm/0.374 L = 60 atm.

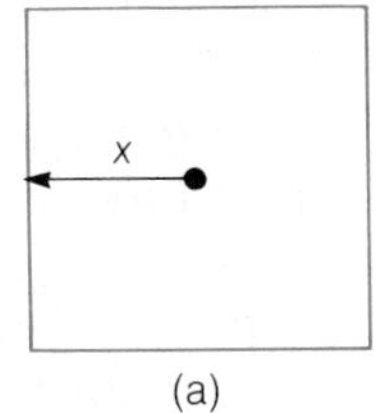

(a)

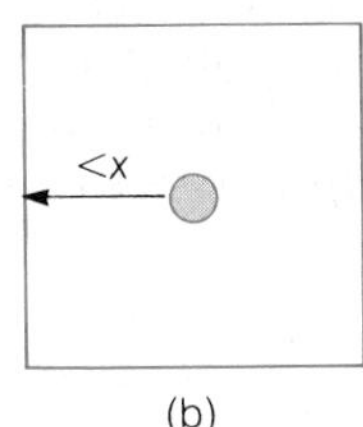

(b)

FIGURE 10.17
(a) An ideal molecule (a point) in a container must travel a distance *x* to hit the wall. (b) A real molecule, which occupies volume, need not travel quite so far to hit the wall.

Since the real molecule occupies volume, it need not travel quite so far until its *outer edge* hits the wall. Since it travels less distance, it hits the wall more frequently, and hence it exerts more pressure than the ideal molecule.

2. A second explanation leads to the same conclusion. In a real gas, the space in which a molecule can move is *less* than the volume of the container because some of the space is occupied by the molecules. Therefore the molecules act as if they were *in a smaller container*, and they hit the wall more frequently. More frequent collisions result in a greater pressure. According to the ideal gas law ($P = nRT/V$), if the result of the molecular size is that the effective volume of the container is smaller, then the pressure must be greater. At higher pressures the molecules are more crowded. The molecules themselves are not compressed; only the empty space between them is reduced. The effect of the reduction of the free space in which the molecules can move about becomes intensified, and the molecules hit the wall much more frequently than would be predicted for an ideal gas (Figure 10.18). Consequently, the measured (real) pressure is much greater than that calculated for an ideal gas. This effect can be illustrated by the data for methane at 0°C and at pressures of 400 atmospheres and above (Table 10.3).

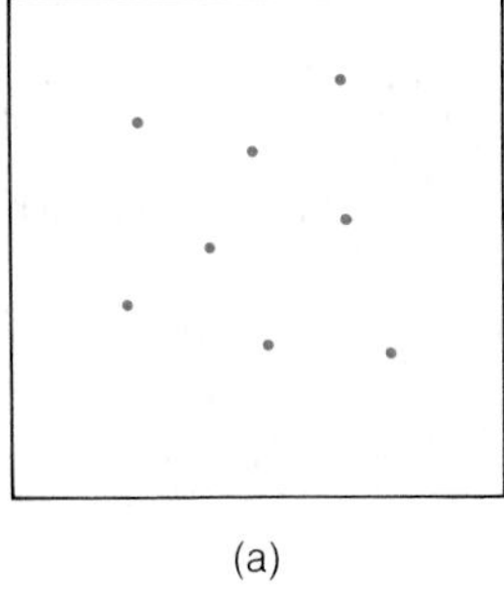
(a)

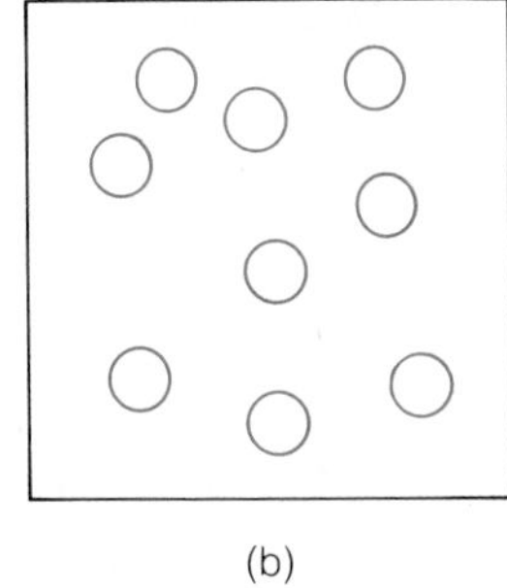
(b)

FIGURE 10.18
(a) Ideal gas. Molecules are points, occupying no volume. The space available to molecules equals the volume of the container. (b) Real gas. Molecules occupy volume. Space available to molecules equals the volume of the container *minus* the volume of the molecules.

The lowest boiling point of all, −269°C or 4 K, belongs to helium.

Finally we return to Charles's law (section 10.5) and its puzzling implication that a gas cooled to absolute zero will shrink to zero volume. On the contrary, it is the *space between* the molecules that will shrink to nothing, not the molecules themselves. Therefore, even if a gas did not liquefy (they all do), it would not disappear at absolute zero.

In any real gas, both of the effects we have just described operate at the same time, but they oppose each other. The experimental conditions will determine which effect predominates. For an ideal gas, $PV = nRT$ and $PV/nRT = 1$. The observed deviations from ideal behavior are shown in Figure 10.19. For hydrogen at 0°C, the molecular attractive forces are low. As a result, the effect of the volume of the molecules overwhelms the effect of molecular attraction, so that the PV/nRT values for H_2 exceed the ideal value of 1 at all pressures. This is shown in Figure 10.19 as a positive deviation from ideality.

For nitrogen, the intermolecular attractive forces are great enough to yield negative deviations ($PV/nRT < 1$) up to about 150 atm. Intermolecular attractions in carbon dioxide, even at 40°C, are still more important than those in nitrogen at 0°C. For both nitrogen and carbon dioxide a pressure exists (150 or 600 atm, respectively) at which the two effects cancel each other and the PV/nRT value is ideal; that is, it equals 1. However, at higher pressures, the effect of molecular volume overwhelms the molecular attraction effect and the PV/nRT values become greater than 1.

At low pressures, the volume of a given quantity of gas is large, and the fraction of the volume occupied by the molecules is relatively small. Because the molecules are more widely separated at low pressures, their mutual attractions and tendencies to stick are lower. As the temperature increases, the molecules gain greater kinetic energy and their tendencies to stick lessen. Thus *high temperatures and low pressures both favor ideal gas behavior.* At room temperature and 1 atm pressure, deviations from ideality are of the order of 1% for most gases.

Chemists have devised several equations relating P, V, n, and T that give better approximations of the behavior of real gases than the ideal gas law provides.

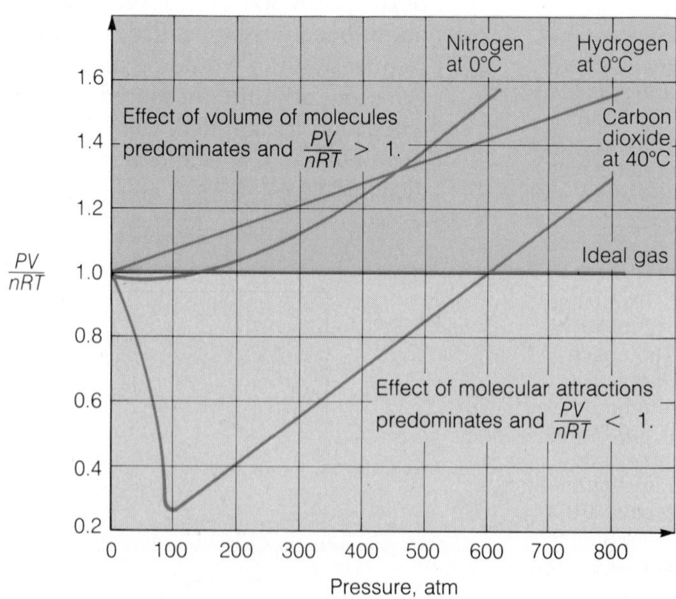

FIGURE 10.19
Deviations from ideal gas behavior. $PV/nRT = 1$ for an ideal gas.

Empirical constants are constants evaluated from experimental data.

These equations are more complicated than $PV = nRT$ because they include several empirical constants to account for the nonideal effects we have described. The first of these equations was introduced in 1873 by Johannes van der Waals, in whose honor intermolecular attractive forces are called **van der Waals forces**, and the doubled molecules are called **van der Waals molecules**. Van der Waals's equation is explained in the next section.

10.16 SPECIAL TOPIC: MATHEMATICAL DERIVATION OF THE IDEAL GAS LAW AND OF THE VAN DER WAALS EQUATION FOR NONIDEAL GASES

THE IDEAL GAS LAW

Referring to Figure 10.20, let us assume that a molecule of mass m approaches the wall at velocity $-u$ (one direction) and rebounds at equal speed with velocity $+u$ (reverse direction). The change in velocity, Δu, is the final velocity minus the initial velocity, or $u - (-u) = 2u$. The force exerted by this collision with the wall of the container is given by the product of the mass of the molecule and its acceleration:

$$\frac{\text{force}}{\text{collision}} = \text{mass} \times \text{acceleration}$$

The velocity of a body is a vector that gives its *speed* and *direction*. A reversal of direction is indicated by a minus sign.

But the acceleration of a body is the rate at which its velocity changes with time:

$$\frac{\text{force}}{\text{collision}} = \text{mass} \times \frac{\text{change in velocity}}{\text{time between collisions}} = m \times \frac{2u}{\text{time}} = \frac{2mu}{\text{time}} \tag{14}$$

We assume that collisions occur so frequently that the pressure remains practically constant. Then

$$\text{pressure } (P) = \frac{\text{force}}{\text{collision}} \times \frac{\text{collisions}}{\text{area}}$$

Substituting from equation 14, we have

$$P = \frac{2mu}{\text{time}} \times \frac{\text{collisions}}{\text{area}} = 2mu \times \frac{\text{collisions}}{\text{time} \times \text{area}}$$

The factor, collisions/(time × area), or "collisions per unit time per unit area," is the key to this derivation. Let this factor be designated by Z. The value of Z depends on the number of molecules, their speed, and the volume of the container. Then

$$P = 2muZ \tag{15}$$

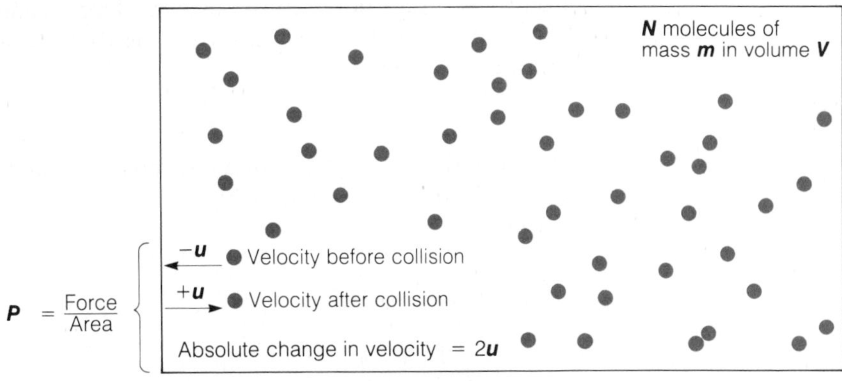

FIGURE 10.20
Gas molecules confined in a container.

We can obtain an expression for Z by making the following assumptions:

- Z is proportional to the number of molecules per unit volume, N/V. The more molecules in a unit of volume, the more frequent will be the collisions with the walls.
- Z is proportional to the molecular speed, u. The faster the molecules move, the more frequent will be the collisions with the walls.

Thus

$$Z \propto \frac{N}{V} \times u$$

or

$$Z = constant \times \frac{Nu}{V}$$

Assume that the container is a cube. Since a cube has six sides, only 1/6 of the molecules approach any given wall. The constant in the equation, which is 1/6, is related to this fact. A rigorous proof (which we do not present here) confirms that the constant 1/6 applies to containers of any shape. Then

$$Z = \frac{1}{6}\frac{Nu}{V}$$

Substituting this value for Z in equation 15 we have,

$$P = 2mu \times \frac{Nu}{6V} = \frac{1}{3}\frac{Nmu^2}{V}$$

or

$$PV = \tfrac{1}{3}Nmu^2 \tag{16}$$

Since the kinetic energy of a moving molecule is $\frac{1}{2}mu^2$, it is helpful to rewrite equation 16 as

$$PV = \tfrac{2}{3}N(\tfrac{1}{2}mu^2)$$

However, not all molecules of a given sample of gas have the same speed at any one time. We therefore interpret $\frac{1}{2}mu^2$ to be the *average* kinetic energy of a molecule, written as $\bar{E}_k$. Substituting $\bar{E}_k$ for $\frac{1}{2}mu^2$ gives

$$PV = (\tfrac{2}{3}N)\bar{E}_k \tag{17}$$

Equation 17 shows that for any given number of molecules N of a gas, the PV product is proportional to the average kinetic energy of all the molecules, $N\bar{E}_k$. But Boyle's and Charles's laws tell us that PV is proportional to the absolute temperature, T:

$$PV = (nR)T \tag{18}$$

Therefore the temperature of a gas must be related to the average kinetic energy of its molecules. The relationship can be obtained by combining equations 17 and 18 to give

$$\left(\frac{2}{3}N\right)\bar{E}_k = (nR)T, \text{ or}$$

$$\bar{E}_k = \frac{3}{2}\left(\frac{n}{N}\right)RT \tag{19}$$

The number of moles (n) equals the number of molecules (N) divided by the Avogadro number (N_A), or

$$n = N/N_A \quad \text{and} \quad \frac{n}{N} = \frac{1}{N_A}$$

Substituting this value in equation 19 gives

$$\bar{E}_k = \frac{3}{2}(RT/N_A), \text{ and}$$

$$N_A\bar{E}_k = \frac{3}{2}RT \tag{20}$$

where $\bar{E}_k$ is the average kinetic energy per molecule. $N_A\bar{E}_k$ is then the kinetic energy of 1 mole of any gas, averaged over all of its molecules. Equation 20 therefore tells us that *the average kinetic energy per mole of a gas is fixed by its temperature.*

The ratio R/N_A is called *Boltzmann's constant, k,* and has the value 1.38×10^{-23} J/(K molecule). Then $\bar{E}_k = \frac{3}{2}kT$.

THE VAN DER WAALS EQUATION

Van der Waals realized that the real volume of a gas is the "ideal" volume (the empty space between the molecules) plus the volume occupied by the molecules (Figure 10.21). The volume of the molecules is proportional to their number, and therefore is proportional to the number of moles of gas, n. The volume also depends upon the size of the molecules, which may be expressed as a constant that is characteristic of the particular gas. This constant is designated b, and the correction for the volume occupied by the molecules is therefore nb. Then

$$V_{\text{real}} = V_{\text{ideal}} + nb$$

and

$$V_{\text{ideal}} = V_{\text{real}} - nb \tag{21}$$

Van der Waals also realized that real molecules attract each other and that the resulting sticky collisions produce doubled, and hence fewer, molecules. Fewer molecules produce fewer collisions with the walls. The magnitude of the effect is derived as follows: the decrease in the total number of molecules equals the number of double molecules present at any moment. For example, if 3 sticky collisions occurred among 1000 molecules, there would be 994 single molecules

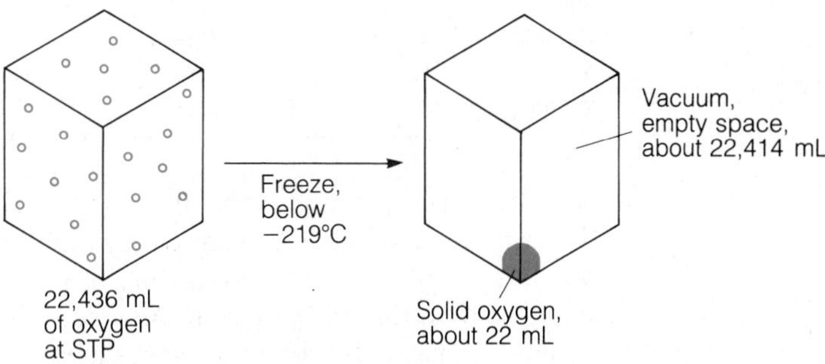

FIGURE 10.21
$V_{\text{total}} = V_{\text{space}} + V_{\text{molecules}}$
22,436 mL = 22,414 mL + 22 mL

remaining plus 3 double molecules, for a total of 997 molecules. The reduction is then 1000 − 997, or 3 molecules.

The reduction in pressure is proportional to the reduction in number of molecules, and thus proportional to the number of double molecules. When you study chemical equilibrium later (Chapter 13), you will learn that the number of double molecules per unit volume is proportional to the *square* of the number of single molecules per unit volume. Thus,

$$\text{reduction in pressure} \propto \left(\frac{n}{V}\right)^2$$

The proportionality constant is designated as a. Then,

$$P_{real} = P_{ideal} - \text{reduction of pressure}$$

$$= P_{ideal} - \left(\frac{n}{V}\right)^2 a$$

and

$$P_{ideal} = P_{real} + \frac{n^2 a}{V^2} \qquad (22)$$

We now substitute the corrected pressure (equation 22) and volume (equation 21) in the ideal gas equation

$$P_{ideal} V_{ideal} = nRT$$

and obtain

$$\left(P_{real} + \frac{n^2 a}{V^2}\right)(V_{real} - nb) = nRT$$

P_{real} and V_{real} are the measured pressures and volumes of real gases, represented simply by P and V.

$$\left(P + \frac{n^2 a}{V^2}\right)(V - nb) = nRT$$

This is the van der Waals equation for a real gas. Some typical values of the constants are as follows:

GAS	a, L^2 atm/mol^2	b, L/mol	BOILING POINT, K
Hydrogen, H_2	0.24	0.027	20
Nitrogen, N_2	1.39	0.039	77
Carbon dioxide, CO_2	3.59	0.043	195
Ammonia, NH_3	4.17	0.037	240
Water, H_2O	5.46	0.030	373

Note that a, a measure of the intermolecular attractive forces, is greater for gases that are easier to liquefy, as shown by their higher boiling points. Water, for example, has a higher boiling point than ammonia, and the boiling point of nitrogen is higher than that of hydrogen. Also note that the constant b, which is a measure of molecular size, is greater for gases whose molecules are larger and occupy more volume.

The van der Waals equation is not an exact expression of the behavior of gases under all conditions; it is only the next-best approximation after the ideal gas law. Other, more complex, equations with more empirical constants have been derived that fit the experimental facts over wider pressure ranges.

PROBLEM 10.18 Assume that $PV/nRT = 0.9$ for a certain gas at 700 atm pressure. Predict the direction in which PV/nRT will change (a) as the gas is compressed at constant temperature; (b) as more gas is added with no change in volume; (c) as the temperature is raised at constant pressure. □

10.17 SPECIAL TOPIC: THE EARTH'S ATMOSPHERE

The predominantly gaseous blanket that surrounds the Earth is its **atmosphere**, and the stuff of which it consists is **air**. The nongaseous components of air include liquids such as the water droplets in clouds and fogs, as well as solid particles such as airborne soil granules, volcanic dust, salts from the evaporation of sea spray, and spores. Since this chapter deals with gases, this section will be devoted to the gaseous components of air, not to its mists and dusts.

Atmospheric moisture is quite variable; the fraction of water vapor ranges from negligibly small in a cold desert to about 5% (by moles) in a steaming jungle. The other gaseous components of the atmosphere, however, are quite uniform if they are not altered by human activities or by natural disasters such as volcanic eruptions. If samples of unpolluted air anywhere near the Earth's surface were filtered and dried, their compositions would be very close to the values shown in Table 10.4.

Nitrogen, the most abundant atmospheric gas, is chemically rather inert and therefore not directly accessible to most organisms. Nitrogen is essential to life, however, and various chemical pathways make atmospheric nitrogen available to organisms. The reactions that lead to oxides of nitrogen are discussed in Chapter 20. Some of these oxides react with water to form acids that contribute to acid rain, a subject also discussed in that chapter.

TABLE 10.4
GASEOUS COMPOSITION OF NATURAL DRY AIR

GAS	MOLAR COMPOSITION, ppm†	PERCENT
Nitrogen, N_2	780,900	78.09
Oxygen, O_2	209,400	20.94
Argon, 0.93%, and other noble gases, 0.0025%	9,325	0.93
Carbon dioxide, CO_2 (variable)	350	0.03
Methane, CH_4 (variable)	1.5	
Hydrogen, H_2	0.5	
Oxides of nitrogen, mostly N_2O produced by biological action (variable)	0.3	
Carbon monoxide, CO, from oxidation of methane and other natural sources (variable)	0.2	
Ozone, O_3, produced by solar radiation and by lightning (variable)	0.02	

† Molar composition is equivalent to composition by number of molecules, and is the same as what is usually called "composition by volume."

Oxygen is essential for respiration by animal life and is replenished by photosynthesis. It has been suggested that the remarkable constancy of the oxygen concentration in the atmosphere over geological time (close to the present 21 mole %) has been maintained by the collective action of the biological community itself. One implication of this idea is that any serious threat to plant life, such as might result from the poisoning of the blue-green algae of the oceans by pollutants, or from the destruction of the rain forests around the world for commercial development, might thus also threaten the O_2 level of the atmosphere. We do not have reliable information about the resistance of the plant community of the Earth to the stresses imposed by human activities. However, the total quantity of oxygen in the atmosphere is about 10^{21} grams, whereas the quantity of oxygen globally recycled by the processes of photosynthesis, respiration, and decay is only about 10^{13} or 10^{14} grams per year. Therefore, even if photosynthesis were to stop altogether, the atmospheric store of oxygen would last for thousands of years.

Argon and other noble gases are unreactive in the atmosphere. The most useful of these gases is helium, for reasons that were discussed in section 6.8.

Carbon dioxide is released to the atmosphere by the respiration of animals and by the combustion of coal, oil, and natural gas. It is removed from the atmosphere by photosynthesis. It is also exchanged between air and water, because the solubility of CO_2 depends on the temperature and on the acidity of the water. The atmospheric abundance of CO_2 has been slowly rising, presumably as a result of the increased rate of combustion of fossil fuels and the destruction of forests. There is concern among climatologists that this increase in atmospheric CO_2 may cause a general warming of the Earth's surface, because CO_2 absorbs some of the heat radiation from the Earth that would normally go directly to outer space. This phenomenon is known as the **greenhouse effect**, a reference to the fact that the air trapped in greenhouses is generally warmer than the outside atmosphere.

Pollutants are introduced into the atmosphere when air is used as an oxidizing agent (to support combustion), as a coolant, as a carrier, or as a medium in which to dispose of unwanted gaseous wastes. When air is used as a carrier gas (especially in drying operations in which hot air carries away moisture) or as a cooling agent (as in cooling towers), any pollutants that can evaporate or be otherwise introduced into the airstream are carried out into the atmosphere. These pollutants include particulate matter as well as gases and vapors. Since air is an oxidizing medium, many pollutant gases are oxides, some of which will be discussed in Chapter 20.

SUMMARY

Gases are transparent, flow readily, mix with each other in all proportions, are much less dense and can be compressed much more readily than solids and liquids, and expand and contract with changes in temperature much more than solids and liquids do. The properties of gases are best understood in terms of the random motions of their molecules.

Pressure is force per unit area. The common SI units of pressure are the pascal, kilopascal, and bar (10^5 Pa). Older units of pressure, which are related to the height of a column of mercury, are the **standard atmosphere** and the **torr**. Atmospheric pressure is conveniently measured by the height of a column of mercury in a barometer.

Four variable properties of a sample of gas that are dependent on one another are its volume, pressure, temperature, and the number of molecules it contains. The relationships between them can be expressed by simple equations.

Boyle's law states that the pressure multiplied by the volume of a fixed mass of gas at a fixed temperature is a constant.

Charles's law states that the volume of a fixed mass of gas at constant pressure is directly proportional to its absolute temperature. Standard temperature and pressure (STP) are 0°C (273.15 K) and 1 atm (760 torr), respectively.

Gay-Lussac's law of combining volumes states that when gases react, the volumes consumed and produced, measured at the same temperature and pressure, are in ratios of small whole numbers. This finding led to **Avogadro's law**, which is that equal volumes of all gases at the same temperature and pressure contain the same number of molecules. The molar volume of a gas at STP is 22.4 L.

The **ideal gas law**, which relates the four variable properties of a sample of gas to each other, is expressed by $PV = nRT$, where the symbols denote pressure, volume, number of moles, a constant, and absolute temperature, respectively. The constant R is usually expressed as 0.0821 L atm/(mol K), but can also be expressed in other units.

The ideal gas equation can be used to calculate the molecular weights or molar masses of gases, knowing that the number of moles, n, equals the mass in grams, m, divided by the molar mass, M, in grams per mole. This substitution leads to the relationship $M = mRT/(PV)$. Since m/V is density, the molar masses of gases can be calculated directly from their densities.

The gas laws can be applied to chemical reactions involving gases by converting between gas volumes and moles of gas. Numbers of moles can then be converted to mass as the problem may require.

The partial pressure of a particular gas in a mixture of gases is the pressure the gas would exert if it occupied the container alone. **Dalton's law of partial pressures** states that in a mixture of gases, the total pressure equals the sum of the partial pressures. A gas that is collected over water is saturated with water vapor, and the total pressure of the "wet" gas therefore equals the sum of the partial pressure of the collected gas and the partial pressure of the water vapor (its vapor pressure).

The **kinetic molecular theory of gases** is based on a model called the **ideal gas**, whose molecules are points in space that have mass and velocity, do not attract or repel each other, and collide with the walls of the container and with each other without a net loss of kinetic energy. The model relates gas pressure to the collisions of molecules with the walls of the container, and relates gas temperature to the average kinetic energy of the molecules.

Graham's law states that a gas's rate of effusion is inversely proportional to the square root of its density. The relationship is derived from the expression for the kinetic energy of a moving molecule and the direct proportionality between the density of a gas and the masses of its molecules.

In any sample of gas at constant temperature, the speeds of the molecules at any instant are distributed so that there is a small proportion of slow molecules, a large proportion with speeds close to the average value, and a small proportion of very speedy molecules. The plot of the distribution is a distorted bell-shaped curve that flattens out in the direction of increased speed—to the right. Increasing the temperature of the gas has the effect of distorting the curve toward the right, increasing the proportion of speedy molecules very significantly.

Real gases differ from the ideal model because their molecules attract each other and occupy volume. The effect of the molecular attractions is to decrease the gas pressure, making PV/nRT less than 1. The effect of the volume occupied by the molecules is to increase the gas pressure, making PV/nRT greater than 1.

The molecular attraction effect predominates at low temperatures, and the effect of molecular volume predominates at high pressures. Therefore, gases are closest to the ideal model at high temperatures and low pressures.

10.18 ADDITIONAL PROBLEMS

PROPERTIES OF GASES

10.19 State whether each property is characteristic of all gases, some gases, or no gas: (a) transparent to light, (b) colorless, (c) unable to pass through filter paper, (d) more difficult to compress than water, (e) odorless, (f) settles on standing.

10.20 Suppose you were asked to supply a particular mass of a specified gas in a container of fixed volume at a specified pressure and temperature. Is it likely that you could fulfill the request? Explain.

10.21 State whether each of the following samples of matter is a gas. If the information is insufficient for you to decide, write "insufficient information." (a) A material is in a steel tank at 100 atm pressure. When the tank is opened to the atmosphere, the material suddenly expands, increasing its volume by 10%. (b) A material, on being emitted from an industrial smokestack, rises about 10 m into the air. Viewed against a clear sky, it has a dense, white appearance. (c) 1.0 mL of material weighs 8.2 g. (d) When a material is released from a point 30 ft below the level of a lake at sea level (equivalent in pressure to about 76 cm of mercury), it rises rapidly to the surface, at the same time doubling its volume. (e) A material is transparent and pale green in color. (f) One cubic meter of a material contains as many molecules as 1 m^3 of air at the same temperature and pressure.

GAS PRESSURE

10.22 With the aid of the conversion factors given in Table 10.1, complete this table:

	atm	torr	Pa	kPa	bar
Standard atmosphere	1				
Partial pressure of nitrogen in the atmosphere		593			
A tank of compressed hydrogen					133
Atmospheric pressure at the summit of Mt. Everest				33.7	

10.23 Consider a container of mercury open to the atmosphere. Calculate the total pressure, in torrs and atmospheres, within the mercury at depths of (a) 100 mm and (b) 5.04 cm. The barometric pressure is 758 torr.

10.24 The densities of mercury and corn oil are 13.5 g/mL and 0.92 g/mL, respectively. If corn oil were used in a barometer, what would be the height of the column in meters at standard atmospheric pressure? (The vapor pressure of the oil is negligible.)

BOYLE'S LAW

10.25 A sample of krypton gas occupies 40.0 mL at 0.400 atm. If the temperature remained constant, what volume would the krypton occupy at (a) 4.00 atm, (b) 0.00400 atm, (c) 765 torr, (d) 4.00 torr, and (e) 3.5×10^{-2} torr?

10.26 A 50-liter sample of gas of the upper atmosphere at a pressure of 6.5 torr is compressed into a 150-mL container at the same temperature. (a) What is the new pressure, in atmospheres? (b) To what volume would the original sample have had to be compressed to reach a pressure of 10.0 atm?

10.27 A cylinder containing 44 L of helium gas at a pressure of 170 atm is to be used to fill toy balloons to a pressure of 1.1 atm. Each inflated balloon has a volume of 2.0 L. What is the maximum number of balloons that can be inflated? (Remember that 44 L of helium at 1.1 atm will remain in the "exhausted" cylinder.)

10.28 Assume that the value of k in $PV = k$ is 12. (a) Plot the graph of P (x axis) versus V (y axis) for the values $V = 1, 2, 3, 4, 6,$ and 12. What is the shape of the curve? (b) Plot the graph of $1/P$ (x axis) versus V (y axis) for the same values of V. What plot is obtained?

CHARLES'S LAW AND ABSOLUTE TEMPERATURE

10.29 Complete this table by making the required temperature conversions. Pay attention to significant figures.

	TEMPERATURE	
	K	°C
Normal boiling point of water		100
Standard for thermodynamic data	298.15	
Dry ice becomes a gas at atmospheric pressure		−78.5
The center of the sun (more or less)	1.5×10^7	

10.30 A gas occupies a volume of 30.3 L at 17.0°C. If the gas temperature rises to 34.0°C at constant pressure, (a) would you expect the volume to double to 60.6 L? Explain. Calculate the new volume (b) at 34.0°C, (c) at 400 K, (d) at −34.0°C.

10.31 Imagine that you live in a cabin with an interior volume of 150 m^3. On a cold morning your indoor air temperature is 10°C, but by the afternoon the sun has warmed the cabin air to 18°C. The cabin is not sealed; therefore the pressure inside is the same as it is outdoors. Assume that the pressure remains constant during the day. How many cubic meters of air would have been forced out of the cabin by the sun's warming? How many liters?

10.32 Gases at 250°C are being discharged from a chimney to the outside atmosphere at a rate of 200 m^3/min. To what temperature would these gases have to be cooled to reduce the emission rate to 175 m^3/min?

10.33 The device shown in Figure 10.22 is a **gas thermometer**. (a) At the ice point, the gas volume is 1.400 liters. What would be the new volume if the gas temperature was raised from the ice point to 6.0°C? (b) Assume the cross-sectional area of the graduated arm is 1.0 cm^2. What would be the difference in height if the gas temperature went up from 0°C to 6.0°C? (c) What modifications could be made to increase the sensitivity of the thermometer?

10.34 One gram of dry air occupies 1060 mL at 100°C and 776 mL at 0.00°C. Using graph paper, plot these values on coordinates as in Figure 10.9. Draw a straight line between the two points and extend it to obtain the value of "absolute zero" in °C. (If you prefer, you may answer this question by using mathematical extrapolation in place of a graph.)

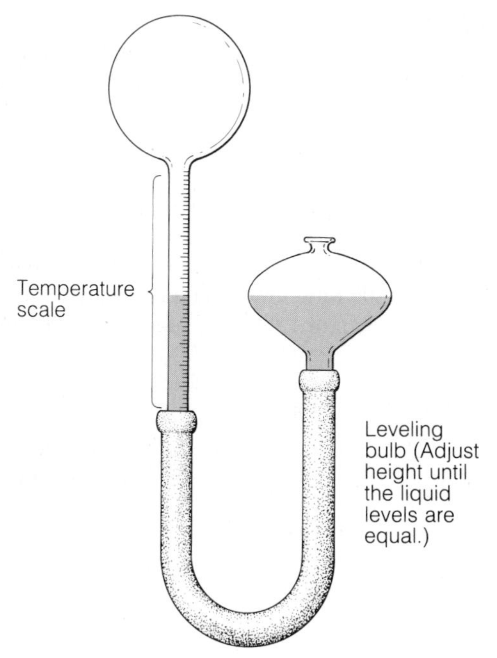

FIGURE 10.22
Gas thermometer.

LAW OF COMBINING VOLUMES AND AVOGADRO'S LAW

10.35 When a radium atom emits an alpha particle, it is transformed to an atom of radon, a monatomic gas. 1.5×10^{19} molecules of radon occupy 0.56 mL at STP. The molar gas volume is 22.4 L/mol. From this information, calculate the Avogadro number.

10.36 If the formulas for ammonia, nitrogen, and hydrogen gases were not known, the volume relationships in the decomposition of ammonia could be expressed in equation form as follows:

$$2N_?H_? \longrightarrow N_? + 3H_?$$

(a) How many liters of hydrogen are produced by the decomposition of 1 liter of ammonia? How many liters of nitrogen? (b) Balance the equation by substituting numbers for the question marks. Since the coefficients represent the experimentally determined ratios of volumes, they may not be changed. Is there more than one solution to the problem? Does the evidence rule out the possibility that nitrogen gas is monatomic? How about hydrogen gas?

10.37 As shown in Figure 10.10, the formula for water in early chemistry books was HO. However, the volume relationships for its formation were known at the time and they can be represented, using the old formula for water, as

$$2H_? + O_? \longrightarrow 2HO$$

Balance this equation by substituting numbers for the question marks. Do not change anything else. If the formula for water really was HO, what would be the simplest formula for oxygen, and for hydrogen?

10.38 The four tires (all of equal volume) of an automobile are filled with four different gases—air, neon, helium, and an unknown gas (Figure 10.23), each to a total pressure of 3.0 atm at 25°C. The masses are air, 116 g; neon, 80.7 g; helium, 16.0 g; unknown gas, 160 g. (a) Are all four pressures the same? If not, which gas is at the highest pressure? (b) Do all four tires contain the same number of gas molecules? If not, which one has the most? (c) How many times heavier is a molecule of the unknown gas than a molecule of helium? (d) Use the same procedure to calculate how many times heavier a "molecule of air" is than a molecule of helium. Since air is a mixture of gases, does your answer refer to the mass of the heaviest molecules, the lightest ones, the most numerous ones, an average molecule—or is it a meaningless number? Explain your answer.

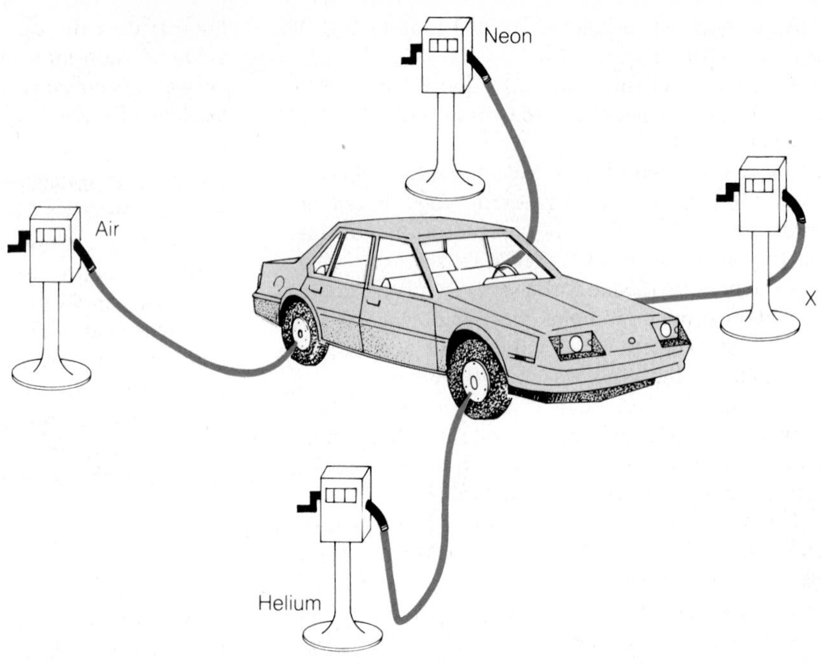

FIGURE 10.23
Four automobile tires are filled with four different gases. See Problem 10.38.

THE IDEAL GAS LAW

10.39 A sample of gas occupies 18.4 liters at 24°C and 460 torr. What will its volume be at 240°C and 46.0 torr?

10.40 A cylinder of fixed volume contains 81.2 mol of nitrogen gas at a pressure of 110 atm and a temperature of 28°C. Nitrogen leaks out of the cylinder until the pressure drops to 6.81 atm. At the same time, the temperature falls to 19°C. How many moles of nitrogen remain in the cylinder?

10.41 A toy balloon is filled with helium gas to a gauge pressure of 22 torr at 25°C. The volume of the gas is 300 mL, and the barometric pressure is 755 torr. How many moles of helium are in the balloon? (Remember, gauge pressure = total pressure − barometric pressure.)

10.42 A chemist is preparing to carry out a reaction at high pressure that requires 45 moles of hydrogen gas. The chemist pumps the hydrogen into a 10.5-L rigid steel vessel at 25°C. To what pressure (in atmospheres) must the hydrogen be pumped?

10.43 What volume, in liters, is occupied by 22.4 moles of argon gas at 1.12×10^{-2} torr and 100°C?

10.44 A 350-mL flask contains 0.0131 moles of neon gas at a pressure of 744 torr. Are these data sufficient to allow you to calculate the temperature of the gas? If not, what is missing? If so, what is the temperature in °C?

MOLECULAR WEIGHTS OF GASES

10.45 Find the molar mass of Freon-12 (a chlorofluoromethane): 8.29 L of vapor at 200°C and 790 torr has a mass of 22.1 g.

10.46 Ethylene dibromide (EDB, formerly used as a fumigant for fruits and grains, but now banned because of its potential carcinogenicity) is a liquid that boils at 109°C. Its molar mass is 188 g/mol. Calculate the density of its vapor at 200°C and 0.50 atm.

10.47 The density of ozone is 2.194 g/L at 850 torr and 298 K. Find its molecular weight.

CHEMICAL REACTIONS INVOLVING GASES

10.48 "Air" bags for automobiles are inflated during a collision by the explosion of sodium azide, NaN_3. The equation for the decomposition is

$$2NaN_3 \longrightarrow 2Na + 3N_2$$

What mass of sodium azide would be needed to inflate a 25.0-L bag to a pressure of 1.3 atm at 25°C?

10.49 Calculate the volume of methane, CH_4, measured at 300 K and 800 torr, that can be produced by the bacterial breakdown of 1.00 kg of sugar according to the reaction

$$C_6H_{12}O_6 \longrightarrow 3CH_4 + 3CO_2$$

10.50 Oxygen in a 20.0-L iron tank at 250°C and 50.0 atm reacts with the iron according to the equation

$$2O_2 + 3Fe \longrightarrow Fe_3O_4$$

If all of the oxygen reacts, what mass of iron is consumed?

DALTON'S LAW

10.51 A sample of hydrogen, collected over water at 20°C and 765 torr, occupies 28.8 mL. What volume would the dry gas occupy at STP?

10.52 A sample of dry nitrogen occupies 331 mL at STP. What would its volume be if it were collected over water at 26°C and 740 torr?

10.53 A study of climbers who reached the summit of Mt. Everest without supplemental oxygen revealed that the partial pressures of oxygen and CO_2 in their lungs were 35 torr and 7.5 torr, respectively. The barometric pressure at the summit was 253 torr. Assume that the lung gases are saturated with moisture at a body temperature of 37°C. Calculate the partial pressure of inert gas (mostly nitrogen) in the climbers' lungs.

10.54 Apples last longer in storage when they are held in a "controlled atmosphere" rather than in ordinary air. A typical composition of such an atmosphere, at 5°C, is oxygen (O_2), 4.00% "by volume"; carbon dioxide (CO_2), 3.50% "by volume"; "inert" gases, including nitrogen (N_2); and water vapor at 90% relative humidity. (This means that the partial pressure of H_2O is 90% of its vapor pressure at 5°C.) Given that the total pressure in such a storage room is 770.0 torr, calculate the partial pressures of O_2, CO_2, H_2O, and "inert" gases. Could you live in such an atmosphere?

KINETIC MOLECULAR THEORY

10.55 Explain in your own words how the kinetic molecular theory accounts for (a) Boyle's law, (b) Charles's law, (c) Avogadro's law, and (d) Graham's law.

GRAHAM'S LAW

10.56 Referring to Problem 10.38, which describes an automobile whose tires each contain a different gas, let the volume of each tire be x liters. (a) Express the density of each gas in terms of x. (b) Given that all four tires are equally porous, calculate the following ratios of effusion rates: neon/air, helium/air, unknown gas/air.

10.57 The average speed of hydrogen molecules at 25°C is about 31 km/min. Pentane vapor is about 36 times as dense as hydrogen at the same temperature and pressure. What is the average speed of the pentane molecules at 25°C?

DEVIATIONS FROM IDEAL BEHAVIOR

10.58 (a) Does the effect of intermolecular attraction on the properties of a gas become more significant or less significant if (1) the gas is compressed to a smaller volume at constant temperature; (2) more gas is forced into the same volume at constant temperature; (3) the temperature of the gas is raised at constant pressure? (b) With regard to each of the circumstances described in part a, does the effect of molecular volume on the properties of a gas become more significant or less significant?

MATHEMATICAL TREATMENT OF THE GAS LAWS

10.59 Copy the van der Waals equation, multiply out the left side, and solve for PV/nRT. Refer to the equation in this form to answer the following questions: Does the behavior of a gas become more ideal ($PV/nRT \rightarrow 1$) or less ideal if (a) the gas is compressed to a smaller volume at constant temperature; (b) more gas is forced into the same volume at constant temperature; (c) the temperature of the gas is raised at constant volume?

SELF-TEST

10.60 (12 points) State each law in words and as an equation: (a) Boyle's law, (b) Charles's law, (c) Avogadro's law, (d) Dalton's law, (e) the ideal gas law, (f) Graham's law.

10.61 (12 points) A sample of neon gas occupies 318 mL at −5°C and 300 torr. What volume will it occupy (a) at 300°C and 100 torr and (b) at STP?

10.62 (12 points) A 22.0-L tank contains 6.17 mol of hydrogen gas at 22°C. Calculate the pressure in the tank, in atmospheres and in bars.

10.63 (10 points) 61.1 mL of oxygen are collected over water at 745 torr and 22.4°C. Calculate the volume of the dry gas at STP.

10.64 (10 points) The density of liquefied petroleum gas (LPG) at 50°C and 0.110 atm is 0.183 g/L. Calculate its molar mass.

10.65 (12 points) What volume of oxygen, measured at 20°C and 740 torr, can be formed by the decomposition of 5.76 g of silver oxide, Ag_2O?

$$2Ag_2O \longrightarrow 4Ag + O_2$$

10.66 (12 points) What mass of ammonium pentasulfide, $(NH_4)_2S_5$, can be produced by the reaction of 115 L of hydrogen sulfide gas, measured at 115 torr and 30°C, according to this equation?

$$5H_2S + 2NH_3 + 2O_2 \longrightarrow (NH_4)_2S_5 + 4H_2O$$

10.67 (10 points) The molecular weights of xenon and argon are 131 and 39.9, respectively. Which gas effuses faster through a narrow opening? How much faster?

10.68 (10 points) (a) Referring to Figure 10.19, compare qualitatively the intermolecular attractive forces among nitrogen, hydrogen, and carbon dioxide at 100 atm pressure. (b) At 600 atm and 0°C, the PV product for nitrogen exceeds that for hydrogen. Does this mean that the intermolecular forces in hydrogen are the greater at this pressure? Why or why not? (*Hint:* Nitrogen molecules are much larger than hydrogen molecules.)

11

INTERMOLECULAR FORCES AND CONDENSED STATES OF MATTER

In Chapter 10 we dealt with gas molecules as though they were completely independent of each other, treating the forces of attraction among the molecules as deviations from an ideal condition. When we discuss liquids and solids, however, we must deal with groups of molecules, because the forces of attraction among them determine many of the properties of the substances they make up. Since these intermolecular forces are generally much weaker than the forces that bind atoms together within a molecule, they are not classified as ordinary chemical bonds.

This chapter opens with a classification of the states of matter and descriptions of melting, boiling, and other changes of state. A discussion of intermolecular forces then serves as a basis for a more detailed treatment of the condensed states of matter. We give special attention to crystalline solids.

Later in the chapter we return to the subject of changes of state and ask some fundamental questions about when and why such changes occur. In that investigation, some concepts of thermodynamics arise that we will explore in greater depth in Chapter 18.

11.1 THE STATES OF MATTER

In Chapter 1, a homogeneous substance was characterized as one that is uniform throughout. This section deals with the aspects of molecular arrangements in a homogeneous substance that determine its physical form. Two factors are most significant:

- the degree to which molecules are orderly in their positions in space, and
- the degree to which they stick together (cohere).

We can obtain a simple but adequate classification scheme by defining two degrees of orderliness and three degrees of cohesiveness:

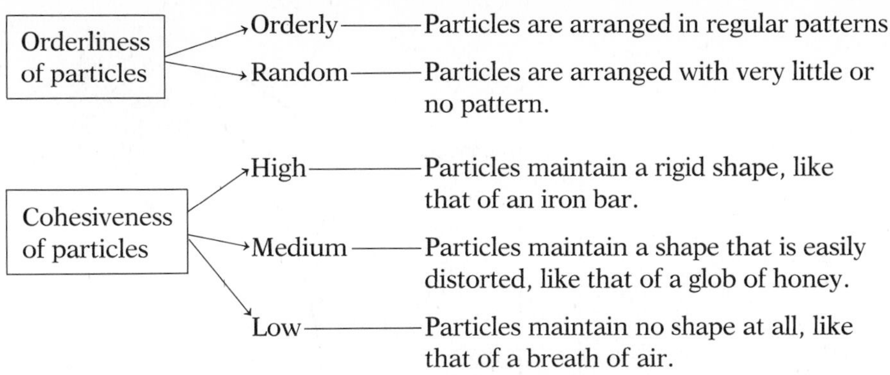

This classification would give us six states of matter, as shown in Table 11.1. The more usual classification is based only on the degrees of cohesiveness among molecules and gives three states: gas, liquid, and solid. That familiar system groups crystalline and noncrystalline solids (states 3 and 6 of Table 11.1) together, and does not account for "liquid crystals" (state 5). No system, however, adequately classifies all forms of matter. Thus, many substances usually regarded as noncrystalline or amorphous—such as rubber, plastics, and textiles—are composed of large molecules whose arrangements are not completely random. Consequently, such materials have definite and measurable degrees of crystallinity.

Amorphous: from Greek *amorphos,* "without shape."

In everyday language, the word *fluid* often means "liquid." A fluid substance, however, is one that flows readily. Therefore, in the scientific sense, fluids include both liquids and gases.

The various states of matter are generally interconvertible. It will be helpful to you through the remainder of this course to learn the vocabulary of changes of state now. They are illustrated in Figure 11.1. Some of the words in the illustration are used more frequently than others. For now, become familiar with these:

Sometimes the term *condensed state* is used to describe anything that is not a gas.

- **Condensation** is the change from gas to liquid.
- **Liquefaction** is the formation of a liquid from any other state.
- **Evaporation** or **vaporization** is the change from the liquid to the gaseous state.

TABLE 11.1
THE STATES OF MATTER

DEGREE OF COHESIVENESS	ARRANGEMENT OF MOLECULES	
	Random	Orderly
Low	(1) gas	(4) exists only at extremely low temperatures (for example, solid helium)
Medium	(2) liquid	(5) "liquid crystals," such as those used in the digital displays of pocket calculators
High	(3) glass; that is, various kinds of noncrystalline or amorphous solids; sometimes regarded as extremely viscous liquids	(6) crystalline solid

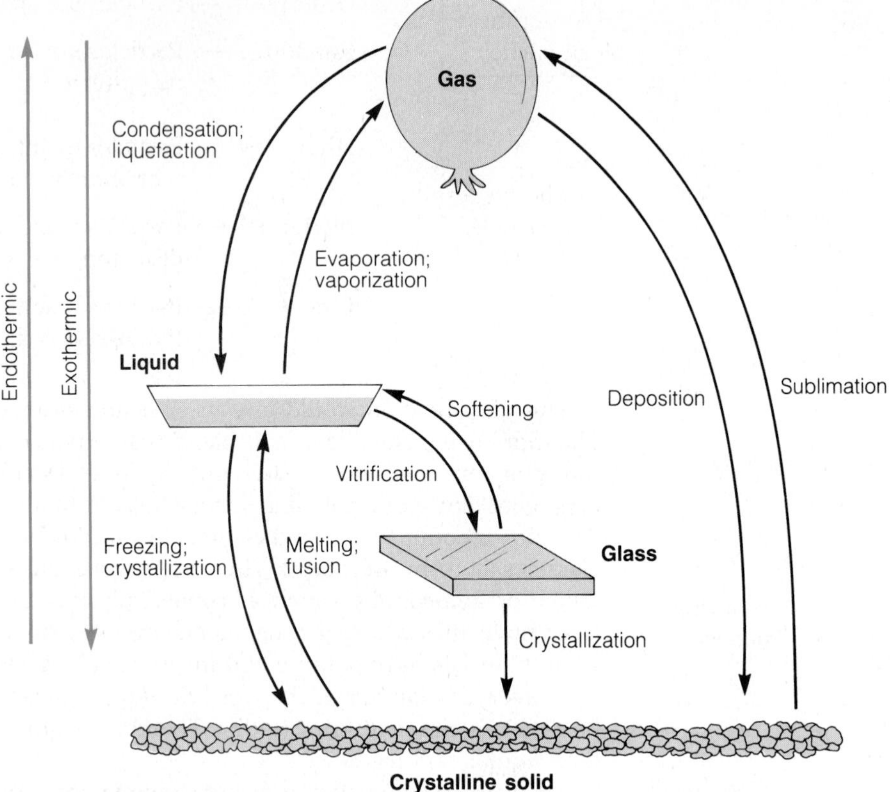

FIGURE 11.1
Vocabulary of changes of state. The enthalpies associated with changes of state are called "heat of fusion" (solid $\rightleftharpoons$ liquid), "heat of vaporization" (liquid $\rightleftharpoons$ gas), and "heat of sublimation" (solid $\rightleftharpoons$ gas).

- **Freezing** and **melting** are the familiar terms that refer to changes from liquid to solid and solid to liquid, respectively.

- **Fusion** is another word for melting, although in common usage it also connotes "sticking together"; it may help you to think of *fused* as meaning "united by melting."

- **Sublimation** can mean either (1) the direct transfer from the solid to the gaseous state, or (2) the two-step process of going from solid to vapor and back to solid. For example, if iodine crystals in the bottom of a test tube are gently heated, they are converted to iodine vapor, which rises to the upper, cooler portion of the tube where it deposits as crystals again. The statement, "The iodine sublimes," may then refer only to the vaporization of the solid, or to the entire solid $\longrightarrow$ vapor $\longrightarrow$ solid transformation.

- **Crystallization** refers to the formation of the crystalline state from the liquid, gas, or noncrystalline solid state.

Note especially that all the changes in the direction of greater disorder (solid $\longrightarrow$ liquid $\longrightarrow$ gas) absorb energy and are endothermic. All those in the opposite direction (gas $\longrightarrow$ liquid $\longrightarrow$ solid) release energy and are exothermic.

In Chapter 7 you learned that covalent bonds link atoms to form molecules, and as with quartz (silicon dioxide), such bonded atoms may extend indefinitely in three dimensions to form a solid network. Ions, too, can arrange themselves into a pattern of alternating positive and negative centers of electrical charge to form such familiar solids as sodium chloride and copper sulfate. But liquids? Think of it this way: A drop of benzene, on being slightly warmed, evaporates. The vaporized material is a gas, in which state its molecules can move independently. But the benzene molecules in the liquid phase must attract each other, or else they would not have to be urged, by means of heat, to enter the gaseous state. Furthermore, if liquid benzene is left outdoors on a cold winter day, it crystallizes. The crystals will gradually disappear (like mothballs) as they sublime. The crystalline benzene, therefore, is different from salt or quartz, which would remain indefinitely in the solid state, winter or summer.

Benzene, C_6H_6, boils at 80°C and is therefore liquid at room temperature. It is also quite toxic, and contact with either the liquid or the vapor should be avoided.

The molecules of benzene remain the same during these changes from one physical state to another—the covalent bonds between the atoms of any benzene molecule are not altered. In the case of liquid or crystalline benzene, therefore, we must be dealing with forces of attraction *between molecules*—forces that are much weaker than the chemical bonds discussed in Chapter 7. We dealt with these forces briefly in the discussion of deviations from ideal gas behavior (section 10.15); now it is time for us to consider them in more detail.

The temperature at which a solid melts, called the **melting point**, provides clues about the relative strengths of the intermolecular forces in the solid and liquid states. In most cases, the liquid occupies a bit more volume than the solid. This is because the liquid is somewhat less dense than the solid that produced it; the molecules are slightly farther apart and more loosely held. (Water is an important exception to this.)

When a liquid boils, it enters the gaseous state, in which the intermolecular forces have been almost completely overcome. The temperature at which this change occurs, called the **boiling point**, provides clues about the relative strengths of the molecular forces in the liquid and gaseous states. The higher the boiling point, the greater the kinetic energy required to overcome the intermolecular forces in the liquid state, and therefore the greater these forces must be. Boiling points therefore teach us much more than melting points can about intermolecular forces. It is for this reason that boiling point data are used as indicators of the strengths of intermolecular forces.

Melting point and freezing point refer to the same temperature. The difference is only in the direction from which the transition is approached. The terms will be defined rigorously in section 11.12.

11.2 INTERMOLECULAR FORCES

Chapter 10 pointed out that molecules of a gas attract each other. In fact, the only significant forces that operate in chemical systems are electrostatic. But molecules of a gas do not bear any net charge. The electrostatic forces between them must therefore involve more complexities than the forces between oppositely charged ions. To understand those complexities, you need to know some things about the boiling points of various gases.

The explanation begins with the fact that all gases can be liquefied, including those with nonpolar molecules such as O_2, N_2, and F_2, and even the monatomic noble gases like Ne and He. Boiling points of different gases, however, vary widely. Helium boils at -269°C (4 K), whereas some gases such as butane (b.p. -0.5°C) liquefy on a cold day. In general, a low boiling point means that the intermolecular forces of the substance are relatively weak, because the molecules can hold together only when their kinetic energies are greatly lowered. Conversely, a high boiling point implies stronger intermolecular forces, because the molecules can hold together even when they have greater kinetic energy. We will first consider the weakest of the intermolecular forces and then go on to the stronger ones.

The boiling point of a gas is also its *condensation* point, although no such term is used. However, the temperature at which the transition occurs is the same in either direction, just as with the melting/freezing point.

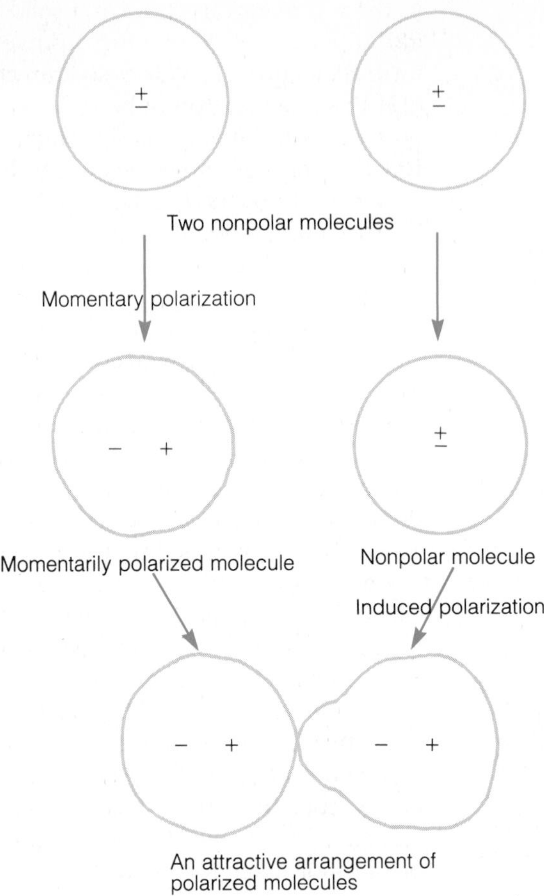

FIGURE 11.2
A momentarily polar molecule induces a dipole moment in a nonpolar molecule.

LONDON FORCES

A theoretical explanation of the weak intermolecular forces among gases with low boiling points was developed in 1928 by Fritz London.

The electrons in an atom or molecule are in constant motion. The average distribution of electronic charge in the He atom, for example, is spherically symmetrical about the nucleus. At any instant, however, an atom or nonpolar molecule may be momentarily distorted into a shape that has an unbalanced charge distribution. The result is a short-lived dipole. Any such momentarily polarized molecule induces transient dipole moments in neighboring molecules (Figure 11.2). The weak attractions between such momentarily polarized molecules are called **London forces**. They exist between all molecules, but are generally overshadowed when stronger forces are present.

Recall that a dipole is a species with separated centers of positive and negative charge.

London forces are also known as dispersion forces, van der Waals forces, and induced dipole–induced dipole interactions.

The data in Figure 11.3 show a correlation between the boiling point and the number of electrons for some noble gases and other nonpolar substances. The boiling point increases as the number of electrons increases, which suggests that the London forces increase in strength as the number of electrons per molecule increases. This relationship results from the fact that a larger electron cloud can be polarized more easily than a smaller one, assuming that the molecular structures are comparable. The number of electrons is the same as the sum of the

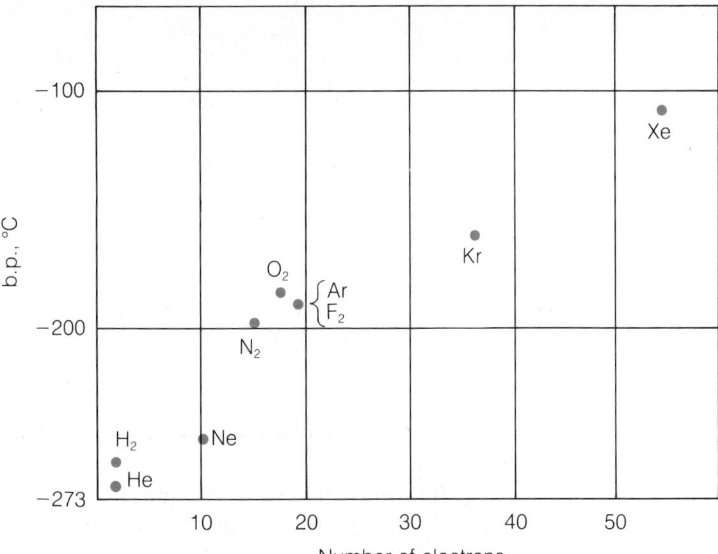

FIGURE 11.3
Boiling points of nonpolar substances plotted against number of electrons per molecule.

atomic numbers, which is also the trend with molar mass, or molecular weight. Therefore, a similar graph would result from plotting boiling point against molecular weight. Theoretically, the number of electrons is responsible for the observed trend, even though the trend is usually expressed in terms of molecular weight. Thus, for any given class of nonpolar substances, the greater the molecular weight, the higher the boiling point.

The boiling point is also considerably influenced by the shape of the molecule. A comparison of the boiling points of *n*-pentane and neopentane, which have the same molecular formula, C_5H_{12}, reveals this dependence. Both molecules have the same kind and number of atoms and therefore the same number of electrons, yet the boiling points differ by 27°C. A molecule of *n*-pentane may be regarded as a zigzag chain (Figure 11.4a), and a molecule of neopentane as almost a sphere (Figure 11.4b). For *n*-pentane, the approach between the two molecules can occur

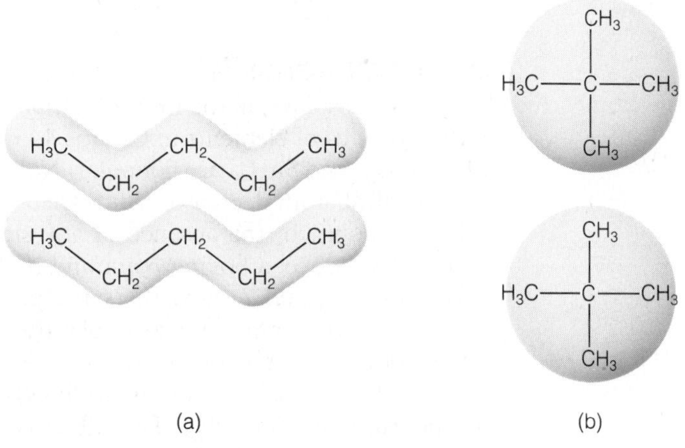

(a) (b)

FIGURE 11.4
Molecular shape of (a) *n*-pentane (b.p. 36.2°C) and (b) neopentane (b.p. 9.5°C).

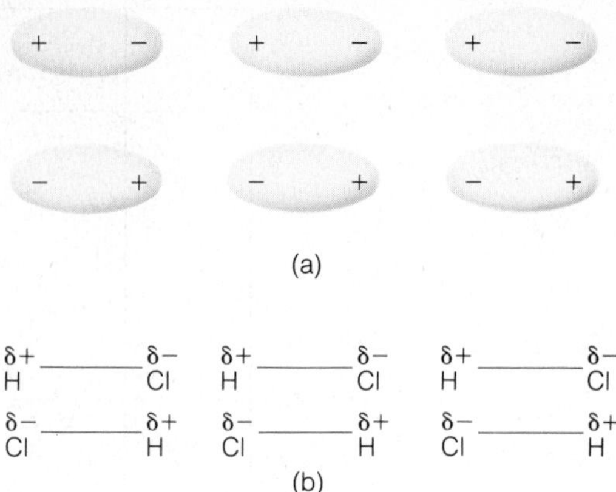

FIGURE 11.5
(a) Dipole-dipole attraction.
(b) Dipole-dipole attraction of HCl molecules.

over the entire length of the chain, whereas for neopentane the approach can occur in only a limited area. The side-to-side approach of two *n*-pentane molecules involves more contact, which increases the London forces; therefore the boiling point of *n*-pentane is higher.

DIPOLE-DIPOLE INTERACTIONS

When polar molecules attract each other, they orient themselves so that the positive pole of one molecule is close to the negative pole of another molecule (Figure 11.5). The strength of this dipole-dipole interaction is one factor that determines the boiling points of polar substances. With other factors such as molar mass and molecular shape being roughly equal, a substance with zero dipole moment will have a lower boiling point than a polar substance. Thus the nonpolar substances N_2 and O_2 have boiling points of $-196°C$ and $-183°C$, respectively, whereas the slightly polar NO ($\mu = 0.070$ D) boils at a higher temperature, $-151°C$.

There can also be interactions among local dipoles, even when the molecule itself has a zero dipole moment. An example is the linear molecule carbon dioxide, O═C═O, in which the individual C═O bond moments cancel each other out. CO_2, molecular weight 44, sublimes at $-78.5°C$. By contrast, the substances fluorine, argon, krypton, and xenon, with molecular weights ranging from 38 to 131 but with no bond moments, all have boiling points in the $-200°C$ to $-100°C$ range (Figure 11.3).

HYDROGEN BONDING

Figure 11.6 displays, in the form of a fragmented periodic table, the boiling points of the hydrides of elements in Groups 4, 5, 6, and 7. The Group-4 hydrides, CH_4, SiH_4, GeH_4, and SnH_4, are nonpolar. Note in the figure that, as the atomic weight of the central atom decreases (Sn > Ge > Si > C), there is a progressive decrease in boiling point. In the absence of any significant dipole-dipole interaction, this relationship between boiling point and molecular weight is typical. The hydrides of Groups 5, 6, and 7 show the same pattern, but with one striking reversal in each group. In Group 5, for example, the boiling points of the hydrides decrease as the atomic weight of the central atom decreases (Sb > As > P), but the sequence is reversed at nitrogen, whose hydride, NH_3, has a higher boiling point than does phosphorus's hydride, PH_3. This effect is even more marked in Group 6, where water—the hydride of oxygen, the lightest element in the group—has a *much* higher boiling point than the others. HF stands out similarly in Group 7.

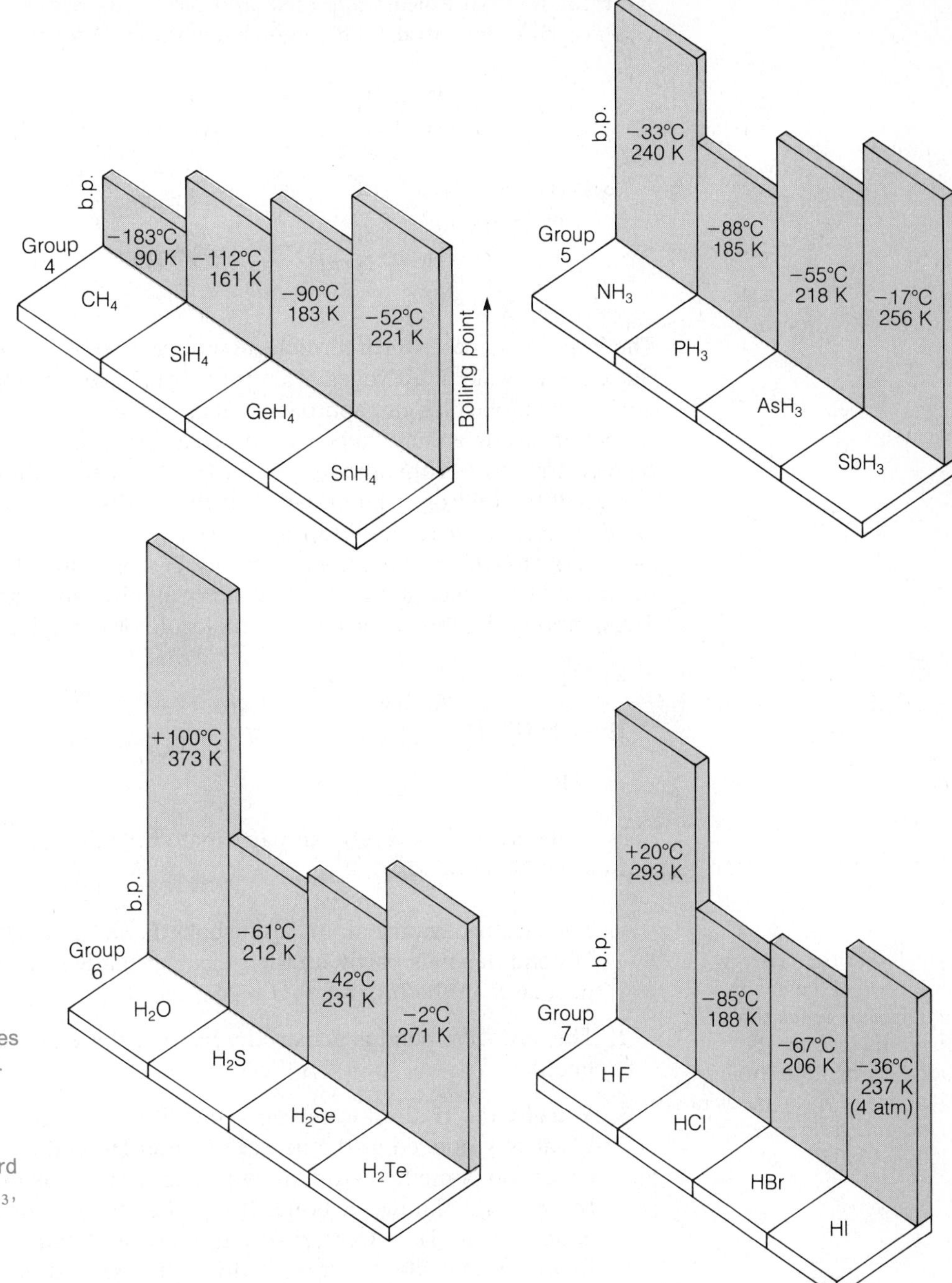

FIGURE 11.6
Boiling points of hydrides of Groups 4, 5, 6, and 7. Note that the boiling points decrease as the molecular weight decreases (going upward in a group), but that NH_3, H_2O, and HF do not fit the pattern. These hydrides are strongly hydrogen-bonded.

Most of the hydrides of Groups 5, 6, and 7 are polar. NH_3, H_2O, and HF are the most polar ones in their respective groups, because N, O, and F are the most electronegative elements. Nevertheless, the effects shown in Figure 11.6 are too great to be accounted for by dipole-dipole interactions alone. There must be something more.

Note that H_2O, HF, and NH_3 have a structural feature in common: Each has at least one hydrogen atom covalently bonded to a highly electronegative atom

(O, F, or N) with at least one unshared pair of electrons. The covalently bonded O, F, or N atoms attract the somewhat positive H atom of another molecule:

$$\overset{\delta-}{:\ddot{O}}-\overset{\delta+}{H} + \overset{\delta-}{:\ddot{O}}-\overset{\delta+}{H} \longrightarrow :\ddot{O}-H\cdots\cdots\ddot{O}-H$$

covalent bonds — hydrogen bonds

$$\overset{\delta-}{:\ddot{O}}-\overset{\delta+}{H} + \overset{\delta-}{:N}-\overset{\delta+}{H} \longrightarrow :\ddot{O}-H\cdots\cdots N-H$$

The H atom increases the intermolecular attraction by bridging the two molecules. Since the very small hydrogen atom is the bridge between the two electronegative atoms, the molecules can approach sufficiently close to each other to produce an attraction strong enough to be considered a *bond*, rather than just another dipole-dipole attraction. This unique type of bond is called the **hydrogen bond**. The energy of the hydrogen bond, which is much less than that of a typical covalent bond, ranges from 13 to 40 kJ/mol.

Hydrogen-bonding does *not* occur with H atoms linked to carbon. (Refer again to Figure 11.6, which shows that CH_4 is not atypical among the Group-4 hydrides.) Thus, among the four H atoms in a molecule of methyl alcohol,

```
     H
     |
  H—C—O—H
     |
     H
```

only the H linked to oxygen can participate in hydrogen bonding.

In a typical H bond,

Recall that the "energy" of a bond is the energy required to *break* the bond; the greater the bond energy, the stronger the bond. Typical covalent bond energies range from 200 to 1000 kJ/mol.

1. The electronegative atoms are usually F, O, or N. The energy of the H bond depends partly on their electronegativities and usually decreases in the order F > O > N.
2. The strongest bond is achieved when the three atoms are in a straight line.
3. Usually, the H atom is nearer to one atom than the other. It is covalently bonded to the nearer one and H-bonded to the other. An exception is the hydrogen difluoride ion, HF_2^- or FHF^-, which has the strongest known H bond. It may be regarded as two F^- ions held equally by an H^+. Another strong hydrogen bond is observed in hydrogen fluoride, the formula for which should be written $(HF)_n$, because it exists as open-chain or cyclic aggregates (Figure 11.7a and b).

If it is true that the hydrogen bond in $(HF)_n$ is stronger than the hydrogen bond in water, how can we account for the observation that water (b.p. 100°C, molecular weight 18) has a much higher boiling point than hydrogen fluoride (b.p. 19.4°C, molecular weight 20)? To answer this we must consider the geometry of the two hydrogen-bonded systems. In the case of hydrogen fluoride, any one F atom is surrounded by only two hydrogen atoms, and so can participate in only one hydrogen bond. In both ice and liquid water, each oxygen atom is surrounded

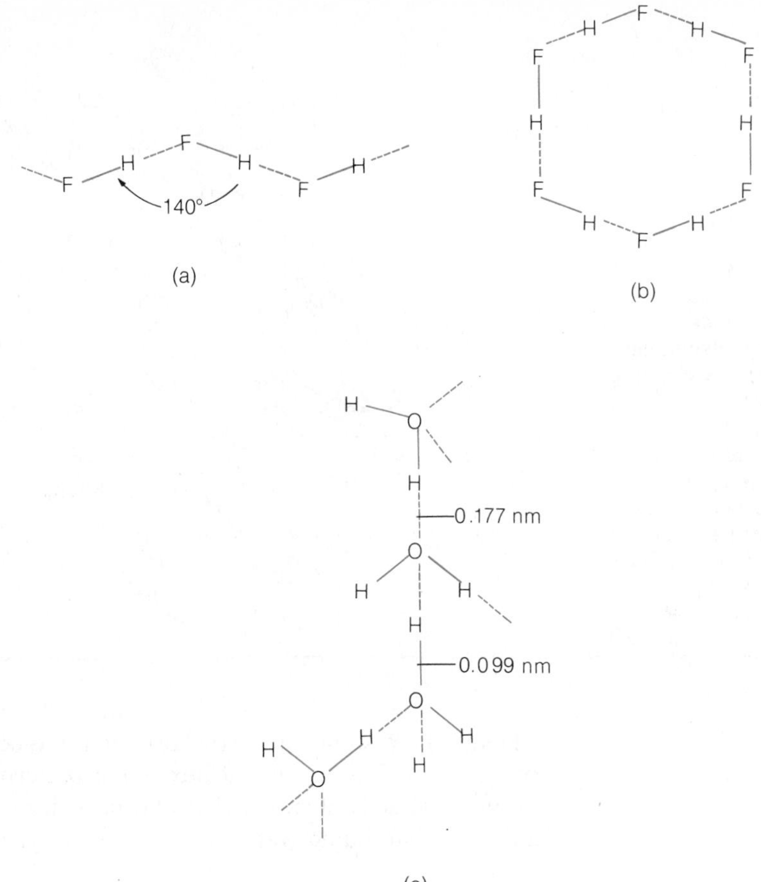

FIGURE 11.7
H-bonded structures of (a) $(HF)_n$, open-chain; (b) H_6F_6, cyclic; and (c) ice, cross-linked. Dashed lines indicate H bonds.

by four hydrogen atoms. Two of these are covalently bonded to the O atom to form the molecule of H_2O. These covalent bonds are the shorter, stronger ones shown in Figure 11.7c. The other two H atoms participate in the longer, weaker, hydrogen bonds (also shown in Figure 11.7c). The presence of two hydrogen bonds per $H_2\ddot{O}$: molecule makes it more difficult to separate any one water molecule from the others. Anything with this kind of network structure holds together firmly because a molecule can get free only if several bonds break at the same time.

The unusually high melting point of 0°C for a molecule with a molecular weight of only 18 is also accounted for by the cross-linked nature of hydrogen bonds. Another consequence of the cross-linking is that as ice melts, some H bonds break, partially collapsing the open network structure. Liquid water is therefore denser than ice. Up to 4°C, the continuing breakdown of the structure brings the H_2O molecules closer together. The density of water therefore increases from 0°C to a maximum at 4°C. With increasing temperature above 4°C, the increase in kinetic energy of the molecules is sufficient to cause the molecules to begin to disperse, and the density steadily decreases. In the vapor phase, water exists as individual H_2O molecules.

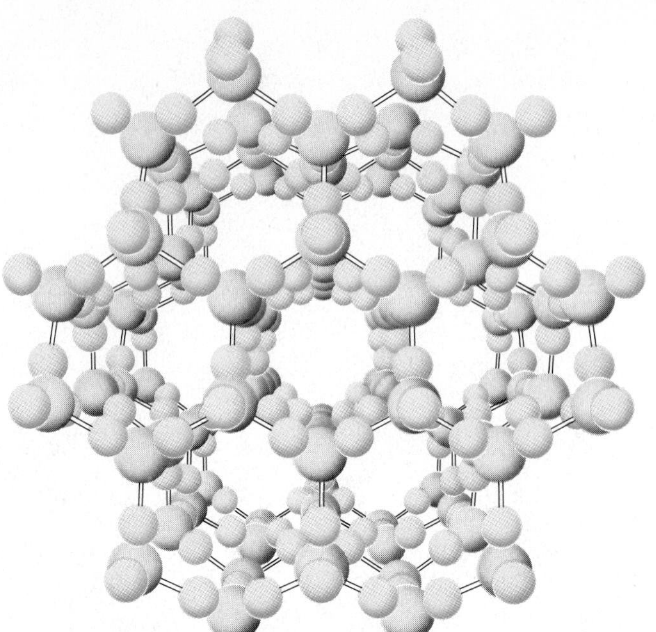

FIGURE 11.8 Hydrogen-bonded H_2O molecules in ice or snow. Each H_2O molecule is represented by one large sphere (the O atom) closely attached to two smaller ones (the H atoms). The hydrogen bonds are represented by the rods between the O atoms of the molecule and the H atoms of an adjacent one.

Figure 11.8 shows an array of water molecules held together by hydrogen bonds, producing an overall hexagonal pattern. This bonding accounts for the hexagonal shapes of snowflakes and provides the empty channels that make ice less dense than liquid water.

EXAMPLE 11.1 In terms of the intermolecular forces involved, account for the fact that the sequence of boiling points of these three substances runs counter to the sequence of their molecular weights.

SUBSTANCE	MOLECULAR WEIGHT	BOILING POINT, °C
Krypton, Kr	84	−153
Hydrogen sulfide, H_2S	34	−60
Ammonia, NH_3	17	−33

ANSWER The increasing order of boiling points, $-153°C < -60°C < -33°C$, parallels the increasing order of intermolecular forces: London forces (Kr) < dipole-dipole attractions (H_2S) < H-bonding (NH_3). This sequence is more important than the opposing decrease in molecular weights. ■

PROBLEM 11.1 State the substance in each pair that you think has the higher boiling point: (a) O_2, oxygen, or H_2S, hydrogen sulfide; (b) Ar, argon, or Xe, xenon; (c) CH_3CH_2OH, ethanol, or CH_3OCH_3, dimethyl ether; (d) $CH_3CH_2CH_2CH_2OH$, *n*-butyl alcohol, or $(CH_3)_3COH$, *t*-butyl alcohol; (e) $(CH_3)_3N$, trimethylamine, or $CH_3CH_2CH_2NH_2$, *n*-propyl amine. Explain your choices. Check your results with a handbook. □

11.3 LIQUIDS AND GLASSES

Note: This is only a thought experiment. Chlorine is a deadly gas.

Imagine two 1-liter containers, one filled with chlorine gas and the other with water. You are going to try to pour the contents of each into another 1-liter container. The water pours out readily, and its total volume in the new container is still 1 liter. The chlorine, too, flows readily, but it disperses into space and overflows its container (Figure 11.9).

Typically, the solid is about 10% denser.

Thus, liquids resemble gases in their ability to flow. Liquids differ from gases, however, in that liquids cannot easily be compressed, nor are their densities affected much by changes in temperature. Under ordinary conditions, in fact, the density of a liquid is fairly close to that of the solid that it forms on freezing.

In 1827 Robert Brown, a botanist, made another observation. He noted that small pollen grains (about 1 μm in diameter) suspended in a liquid appeared to have a quivering motion. The quivers were random in direction, were not brought about by stirring, and never slowed down. This phenomenon is called **Brownian motion**.

These observations give important insights into the properties of molecules in the liquid state. Pollen grains suspended in a liquid move because they are being pushed by something. Since they are in contact only with the liquid, the source of the push can only be the molecules of the liquid. If the molecules of liquid were stationary, they could not push the pollen grains. Thus, Brownian motion indicates that the molecules of a liquid, like those of a gas, are in constant thermal motion. We cannot see the movement of molecules of water, even with a microscope, because they are too small. But the water molecules can transmit their motion to a 1-μm pollen grain, which is small enough to be moved by these collisions and yet large enough to be observed.

The fact that liquids are incompressible indicates that the distances between molecules are small in spite of their thermal motions. A molecule of a liquid need not travel very far (less than its own radius, in fact) to hit another molecule. Furthermore, such closeness is associated with substantial intermolecular forces. Why, then, do liquids flow so easily—more like gases than like solids? Why, in fact, aren't liquids *solid*? To approach this question, we must consider the degree of orderliness of molecules in the liquid state. For identical objects to be packed densely into a space (any objects—chairs, pencils, teacups, baseballs, or molecules) they must be arranged in an orderly pattern; jumbles always occupy more volume.

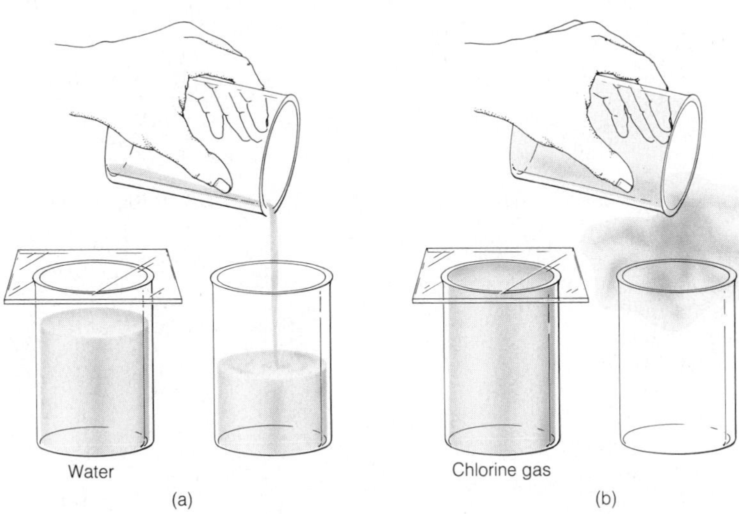

FIGURE 11.9
Liquids cohere; gases disperse.

The arrangements of molecules in liquids therefore must be much more orderly than those of molecules in gases, which are random. However, orderly patterns give rise to properties that are not the same in all directions, and an important characteristic of liquids is that their properties *are* the same in all directions. The speed of light or the electrical conductivity through a liquid, for example, is the same in any direction. Therefore, liquids cannot be perfectly orderly in their molecular arrangements.

If liquids can be neither random nor orderly, they must be something in between. X-ray examinations provide some clues. They indicate order when considered over short ranges (up to 2 or 3 molecular diameters) and disorder over longer ranges.

To visualize the effects of such circumstances, try to arrange circular "molecules" (pennies perhaps) so that any given "molecule" touches five others (Figure 11.10). It will be easy enough to get started, but impossible for you to continue without creating defects in the pattern. Gaps, or "holes," will appear. What you will have is a two-dimensional model of the liquid state. Note the short-range order: any one molecule is part of a more or less orderly array with the molecules nearest to it. Note also the *long-range disorder*: there is no regular pattern that repeats itself throughout the entire model. On a large scale, the disorder will look the same from any direction. Now look at the effect of the "holes." If you push (exert pressure on) the array, it will not decrease much in volume, because the molecules are, after all, rather close to each other. But it is not at all difficult to push a molecule partway into one of the "holes," thus creating a new hole in a space formerly occupied. Since the holes are scattered throughout the liquid, these displacements can spread out in every direction. It is this continual movement of *real* molecules that permits liquids to flow.

Even at ordinary temperatures, glass does flow, although quite slowly. Old windows, as in cathedrals, are somewhat thicker toward the bottom due to downward flow of the glass.

When a liquid is cooled, the kinetic energies of its molecules are decreased, and flow becomes slower. In some cases, the flow can slow down to such an extent that the substance becomes almost rigid. If, at the same time, the long-range disorder of the liquid is preserved, the material is called a **glass** (see Table 11.1). When a glass is heated, fluid properties increase gradually, since no sudden breakup of an orderly structure is required. There is no sharp melting point.

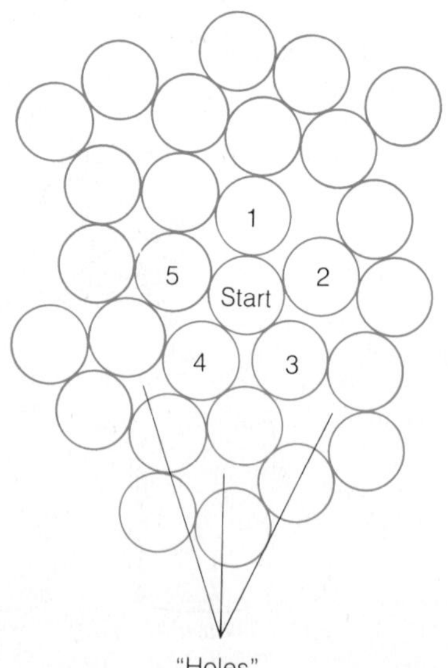

FIGURE 11.10
An attempt to make a continuous pattern in which five circles touch a given circle starts out well but cannot be maintained. This provides a two-dimensional model of the long-range disorder of the liquid state.

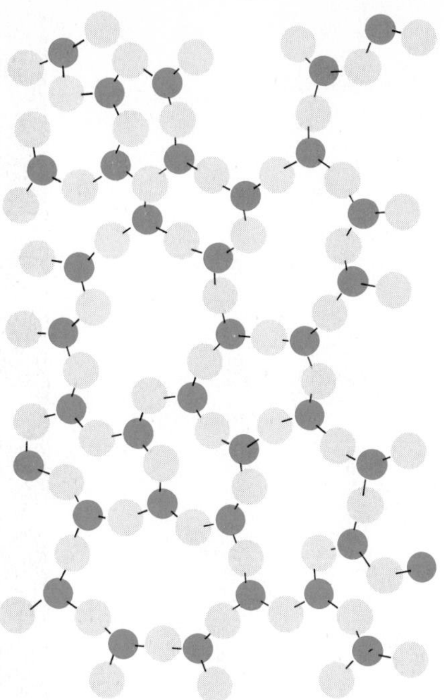

FIGURE 11.11
"Order-disorder" molecular arrangement in a glass.

Examples of liquids that, on cooling, are likely to produce glasses rather than crystalline solids include the following:

1. Melted sugar mixtures frequently fail to crystallize on cooling. Ordinary "hard candy" is a glassy solid obtained from such melts. "Rock candy," in contrast, is crystalline sugar.
2. A molten mixture of sodium and calcium silicates cools to give the material whose common name is "glass" (Figure 11.11).
3. Molten tar, on cooling, forms a glassy product. If you have ever broken a piece of cold solid tar, you have seen the glassiness of the fresh surface.

11.4 CRYSTALLINE SOLIDS

Many people admire the beauty of snowflakes (Figure 11.12) and sugar crystals, not to mention the more durable symmetries of diamonds and emeralds. On the other hand, few people get excited over the physical appearance of a bottle of milk, a glob of tar, or a drop of rainwater (Figure 11.13). Order is usually more appealing to the eye than disorder. As will be elaborated on in section 11.14, this observation conceals unexpected subtleties. Crystals are characterized by the orderly, cohesive arrangement of their atoms, ions, or molecules. As a result of this internal orderliness, crystals assume recognizable external shapes. The angles at which a crystal's surfaces (faces) meet each other are a characteristic and reproducible property of the particular crystal. These geometrical patterns have fascinated people since earliest civilization.

The *internal* structures of crystals can be studied more effectively by measuring the degree to which they reflect or scatter radiation. Since the distances between atoms in crystals are within the range of X-ray wavelengths, X-ray diffraction patterns give useful information about the arrangement of atoms in crystals (Figure

FIGURE 11.12
Photograph of a snowflake crystal.

11.14). This technique was introduced by Max von Laue in 1912. In 1913 William Henry Bragg and William Lawrence Bragg (father and son) calculated the spacing between layers of atoms by measuring the intensities of X rays reflected from crystals at different angles. X-ray analysis is a very powerful, if complicated, method of examining orderly arrangements in crystals. For example, the structure of DNA, the genetic material of living organisms, was confirmed by X-ray analysis.

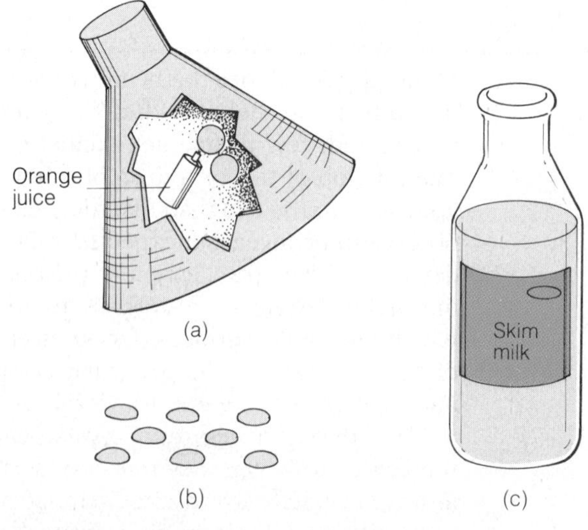

FIGURE 11.13
Liquids occur in various shapes: (a) not falling—spherical blobs; (b) falling—drops; (c) fallen—in the shape of the container.

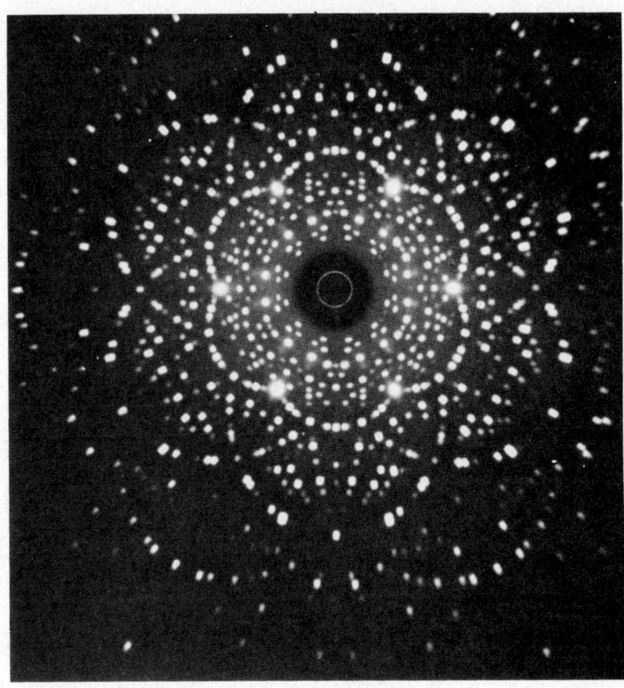

FIGURE 11.14
X-ray diffraction pattern of beryl, $Be_3Al_2Si_6O_{18}$. (Courtesy of Eastman Kodak Laboratories)

11.5 THE CRYSTAL LATTICE

The internal structure of a crystal is characterized by regularity in three dimensions. Consider the crystal of an element, in which all the atoms are alike. If we represent the center of each atom by a point corresponding to its position in space, the arrangement of these points is called a **crystal lattice**. An illustration of a crystal lattice is shown in Figure 11.15. Note that the straight lines connecting the lattice points outline a block of unit shapes referred to as *cells*. All such cells

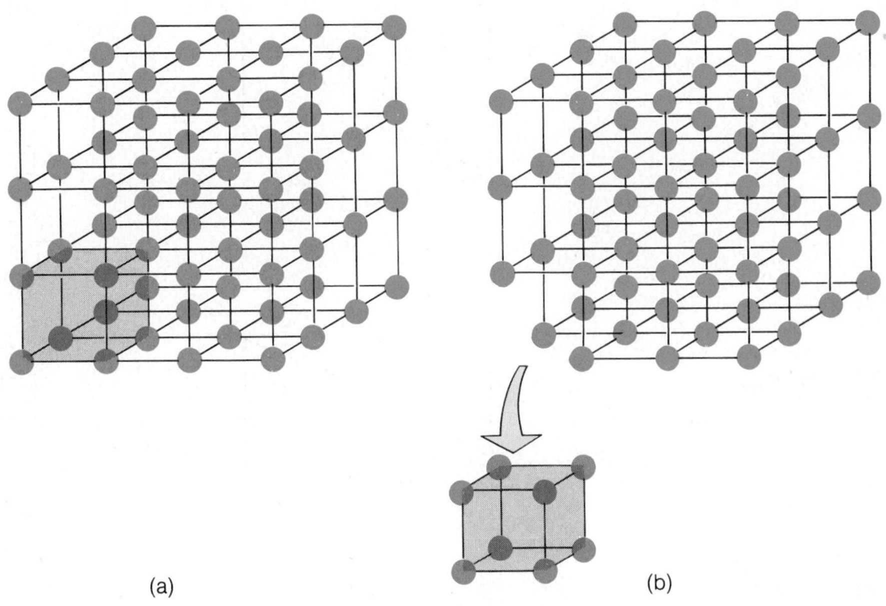

FIGURE 11.15
(a) Crystal lattice. Each circle represents an atom. The shaded portion is a unit cell. (b) A unit cell separated from the crystal lattice. Note that only 1/8 of each corner atom lies within a given unit cell.

are identical, and the smallest one that represents the overall crystal geometry is the **unit cell**. The crystal lattice may be considered to consist of an indefinite number of unit cells, all similarly oriented in space and in contact with each other.

The simplest unit cells are cubic. The atoms in a crystal whose lattice consists of cubic unit cells can be assigned to specific locations on or within these cells. These locations include corners, edges, faces, and body centers (Figure 11.16a). Note that a cube has 8 corners, 12 edges, 6 faces, and 1 body center.

Since the atoms vibrate in the crystal because of their thermal energy, we really mean an average position, or a "center of vibration."

An atom has volume; it is not a point in space. Therefore when we speak of the "location of an atom," we refer to the *position of its center*. With these conventions in mind, let us consider what portion of an atom at a given location actually lies *within* the cube.

A **corner atom** (Figure 11.16b) is shared equally by all the unit cells that touch the same point. In two dimensions, four squares can touch a given point. In three dimensions, eight cubes can touch a given point. Therefore only one-eighth of a corner atom belongs to any one unit cell.

An **edge atom** (Figure 11.16c) is shared equally among the four unit cells that have a common edge. Therefore one-fourth of an edge atom is assigned to each of the four unit cells.

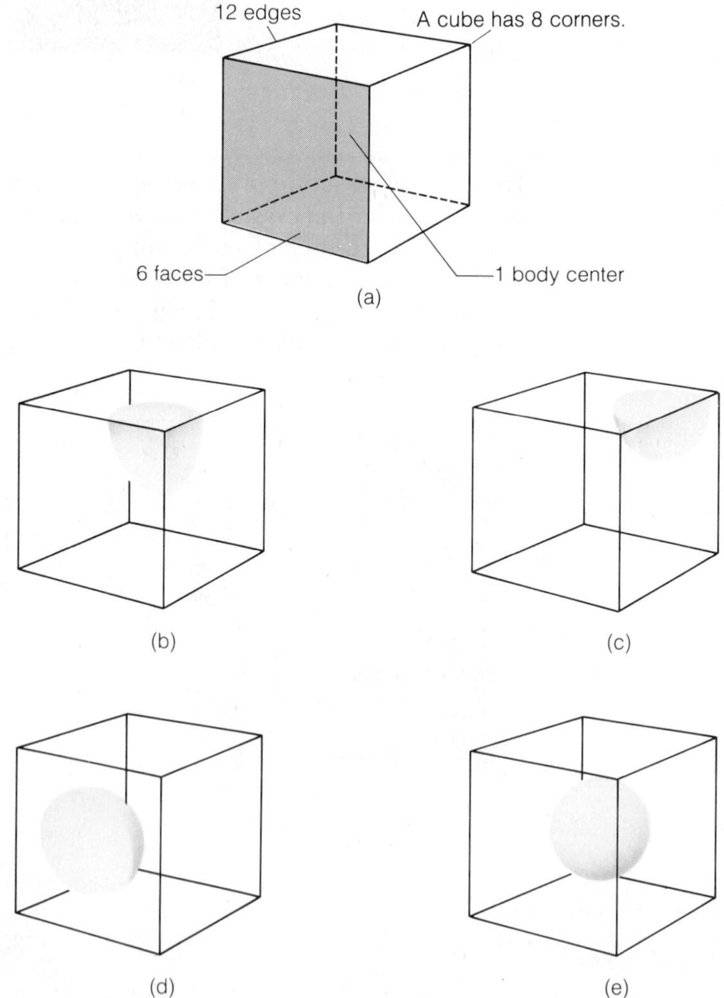

FIGURE 11.16
(a) Parts of a cube. (b) Corner atom; 1/8 of the atom is inside the cube. (c) Edge atom; 1/4 is inside cube. (d) Face atom; 1/2 is inside cube. (e) Body-centered atom is entirely inside the cube.

A **face atom** (Figure 11.16d) is shared equally between the two cells that face each other. Therefore one-half of a face atom is assigned to one unit cell.

A **body-centered atom** belongs entirely to its unit cell (Figure 11.16e).

In summary:

LOCATION OF ATOM	PORTION WITHIN A CUBIC UNIT CELL
Corner	$\frac{1}{8}$
Edge	$\frac{1}{4}$
Face	$\frac{1}{2}$
Center	1

With these pictures in mind, we may now visualize three cubic lattices (Figure 11.17):

1. In a **simple cubic lattice**, atoms would be located only at the eight corners of each unit cell. How many atoms, then, are in each cell?

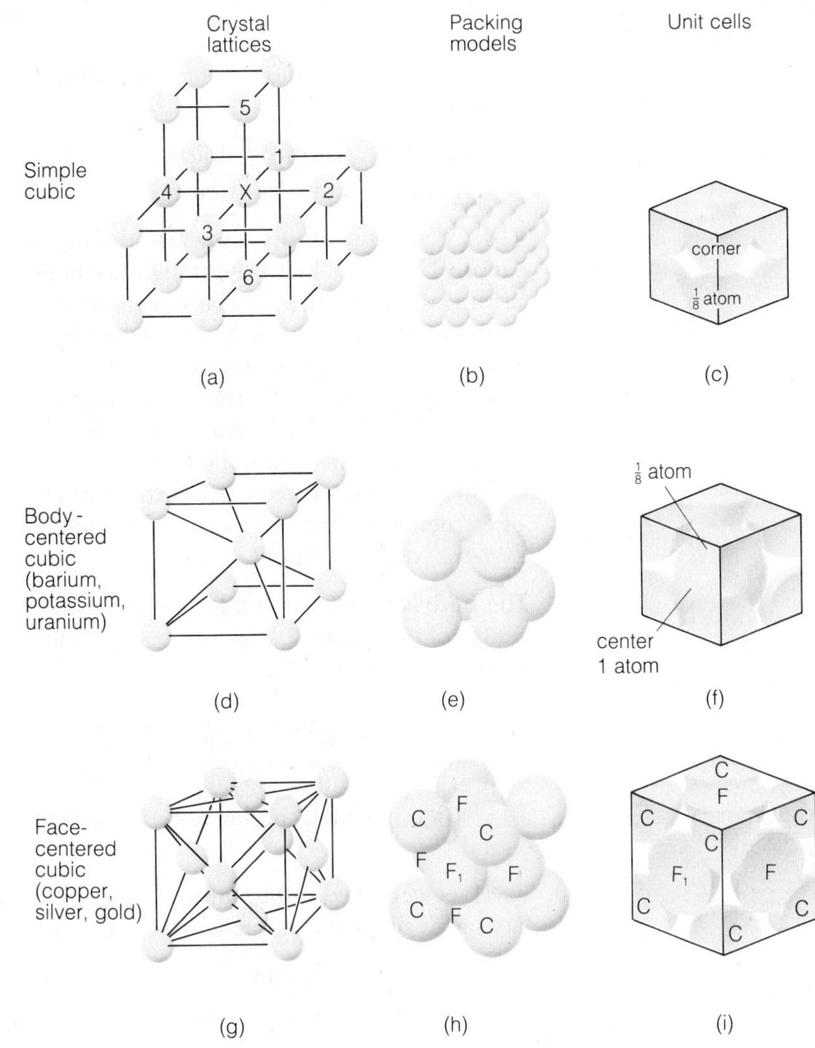

FIGURE 11.17
Cubic crystal systems. C = corner atom, F = face atom.

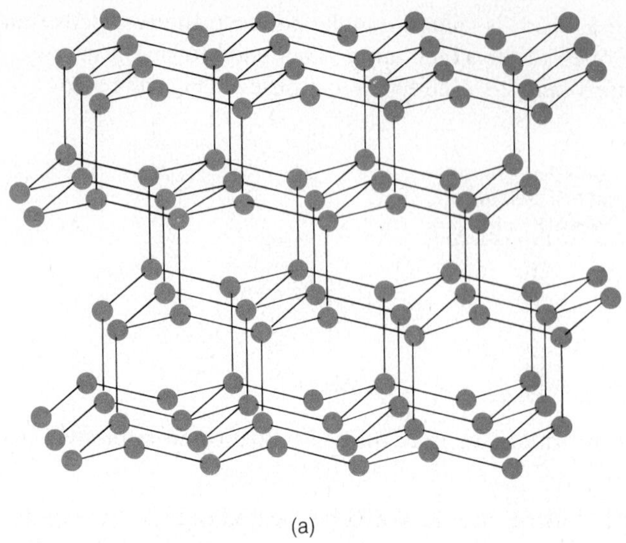

(a)

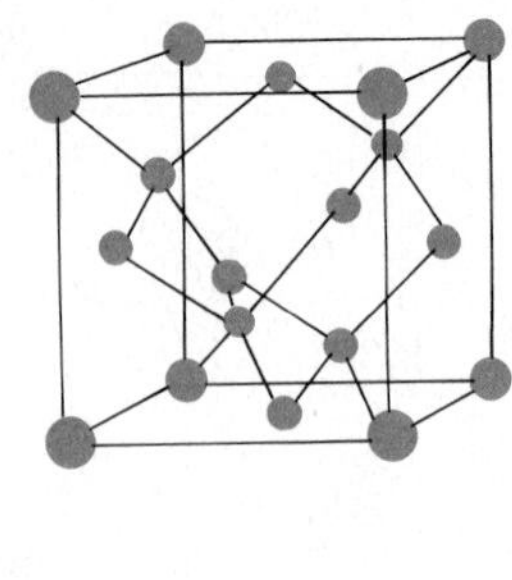

(b)

FIGURE 11.18
(a) The crystal lattice of diamond. (b) The unit cell of diamond.

Although there are no whole atoms in the cell, we can add up the parts as follows:

$$\frac{8\ \cancel{\text{corners}}}{\text{unit cell}} \times \frac{\frac{1}{8}\ \text{atom}}{\cancel{\text{corner}}} = \frac{1\ \text{atom}}{\text{unit cell}}$$

These relationships are shown in the top row of Figure 11.17. Figure 11.17a includes the imaginary lines that illustrate the cubic arrangement. Figure 11.17b is a more realistic picture, because it presents the atoms as spheres packed in contact with each other. Finally, Figure 11.17c shows the unit cell itself, with fractional atoms at its eight corners. Of course, this cell is a geometrical concept, a model, not a real object. The result of stacking them all together would be the packing model of Figure 11.17b.

There are no simple cubic lattices of elements. The typical lattices of elements are the body-centered cubic and face-centered cubic systems, described next.

2. The **body-centered cubic lattice** contains one atom in the body center of each unit cell as well as one atom at each corner. The number of atoms per unit cell is

$$\left(\frac{8\ \cancel{\text{corners}}}{\text{unit cell}} \times \frac{\frac{1}{8}\ \text{atom}}{\cancel{\text{corner}}}\right) + \left(\frac{1\ \cancel{\text{body center}}}{\text{unit cell}} \times \frac{1\ \text{atom}}{\cancel{\text{body center}}}\right) = \frac{2\ \text{atoms}}{\text{unit cell}}$$

The lattice, packing model, and unit cell are shown in Figure 11.17d, e, and f.

3. The **face-centered cubic lattice** contains one atom in each of the six faces as well as one at each corner. The number of atoms per unit cell is

$$\left(\frac{8\ \cancel{\text{corners}}}{\text{unit cell}} \times \frac{\frac{1}{8}\ \text{atom}}{\cancel{\text{corner}}}\right) + \left(\frac{6\ \cancel{\text{faces}}}{\text{unit cell}} \times \frac{\frac{1}{2}\ \text{atom}}{\cancel{\text{face}}}\right) = \frac{4\ \text{atoms}}{\text{unit cell}}$$

The lattice, packing model, and unit cell are shown in Figure 11.17g, h, and i.

PROBLEM 11.2 The diamond structure (Figure 11.18) is based on a cubic unit cell that has four carbon atoms within the cube and carbon atoms at all corners and faces. How many carbon atoms are assigned to this unit cell? □

11.6 COORDINATION NUMBER AND THE UNIT CELL

If you examine the packing models shown in Figure 11.17b it appears that some of the atoms touch each other. The concept of "touching" has little meaning on an atomic scale, however, because the electron clouds surrounding the nucleus do not have definite boundaries. We can avoid this problem by considering how many atoms are the *nearest neighbors* of a given atom, without worrying about whether they really "touch" or not. This number of nearest neighbors is called the **coordination number**.

The coordination number of atoms in the **simple cubic system** can readily be visualized from Figure 11.17a. The nearest neighbors to the atom labeled X are six atoms, labeled 1 through 6. If you were to construct a model of the simple cubic lattice from sticky spheres such as gumballs, you would find that each ball touches six others.

From Figure 11.17e or f, you can readily see that the coordination number of the **body-centered cubic system** is 8.

The coordination number of atoms in the **face-centered cubic system** is more difficult to visualize. Refer to Figure 11.17h or i and note that the face atom labeled F_1 touches four adjacent corner atoms and the four face atoms behind it. In addition, atom F_1 would touch another four atoms (not shown in the figure) that would be in front of it—four of the face atoms in the unit cell that contains the other half of the face atom F_1 we are considering. This arrangement, which gives a coordination number of 12, is called **cubic closest packing**. It is impossible to pack more uniform spheres into a given space in any other way. In summary:

There is another, equally closely packed arrangement based on a hexagonal crystal type called *hexagonal closest packing.*

LATTICE OF UNIFORM SPHERES	COORDINATION NUMBER	ATOMS PER UNIT CELL
Simple cubic	6	1
Body-centered cubic	8	2
Face-centered cubic	12	4

UNIT CELL CALCULATIONS

The key to the problems that follow is in the properties of the unit cell—specifically, its volume and the number of atoms it contains. Multiplying the mass of one atom by the number of atoms per unit cell gives the mass of the unit cell. The relationships, expressed in convenient units, are

Unit cell edges are typically in the range of 3 to 5 Å, or 0.3 to 0.5 nm. We use angstroms here because they lead to fewer errors in cubing, especially without a calculator. Thus, $(3\ \text{Å})^3 = 27\ \text{Å}^3$, while $(0.3\ \text{nm})^3 = 0.027\ \text{nm}^3$. Crystallographers also use angstroms.

$$\text{volume of unit cell, cm}^3 = (\text{edge, }\cancel{\text{Å}})^3 \times \left(\frac{1\ \text{cm}}{10^8\ \cancel{\text{Å}}}\right)^3 = (\text{edge, }\cancel{\text{Å}})^3 \times 10^{-24}\ \frac{\text{cm}^3}{\cancel{\text{Å}^3}}$$

$$\begin{aligned}\text{atoms per unit cell} &= 4 \text{ for face-centered cube}\\ &= 2 \text{ for body-centered cube}\\ &= 1 \text{ for simple cube}\end{aligned}$$

$$\begin{aligned}\text{mass of unit cell} &= \text{mass of 1 atom} \times \text{atoms/unit cell}\\ &= \frac{\text{molar mass (g/}\cancel{\text{mol}})}{6.02 \times 10^{23}\ \cancel{\text{atoms/mol}}} \times \cancel{\text{atoms}}\text{/unit cell}\\ &= \frac{\text{g}}{\text{unit cell}}\end{aligned}$$

Since the unit cell is a representative sample of the entire crystal, the density of the unit cell must be the same as the density of the crystal. Then,

$$\text{density of crystal} = \frac{\text{mass of crystal}}{\text{volume of crystal}}$$

and,

$$\textbf{density of crystal} = \frac{\textbf{mass of unit cell}}{\textbf{volume of unit cell}} \qquad (1)$$

EXAMPLE 11.2 Lithium (molar mass 6.94 g/mol) crystallizes in a body-centered cubic lattice whose unit cell edge is 3.51 Å at 20°C. Calculate the density of lithium at 20°C.

ANSWER Using equation 1, the quantities needed are the mass of a unit cell and the volume of a unit cell.

$$\text{mass of unit cell} = \frac{6.94 \text{ g/}\cancel{\text{mol}}}{6.022 \times 10^{23} \cancel{\text{atoms/mol}}} \times \frac{2 \cancel{\text{atoms}}}{\text{unit cell}}$$

$$= 2.304 \times 10^{-23} \text{ g/unit cell}$$

$$\text{volume of unit cell} = \frac{(3.51 \cancel{\text{Å}})^3}{\text{unit cell}} \times 10^{-24} \frac{\text{cm}^3}{\cancel{\text{Å}^3}}$$

$$= 4.324 \times 10^{-23} \text{ cm}^3\text{/unit cell}$$

$$\text{density of Li} = \frac{2.304 \times 10^{-23} \text{ g/}\cancel{\text{unit cell}}}{4.324 \times 10^{-23} \text{ cm}^3/\cancel{\text{unit cell}}}$$

$$= 0.533 \text{ g/cm}^3$$

■

PROBLEM 11.3 Given that the density of lithium is 0.533 g/cm^3 at 20°C and that its unit cell edge is 3.51 Å, show that its cubic unit cell must be body-centered. □

11.7 CRYSTAL LATTICES OF IONIC COMPOUNDS

The crystal structures of ionic solids must satisfy several requirements that do not apply to elements:

1. The total positive charge of all the cations must equal the total negative charge of all the anions, so that the crystal itself is electrically neutral. This is the same requirement that must be satisfied by the empirical formula, such as NaCl, $CaCl_2$, and so on.
2. The crystal structure must accommodate ions of different sizes. (Recall that anions are larger than cations. For data, refer to Table 6.6, section 6.4.)
3. Since like charges repel each other and unlike charges attract, oppositely charged ions must alternate with each other in the spatial array of an ionic crystal. In this way, unlike charges are closer to each other, like charges are farther apart, and the crystal is stable.

With these requirements in mind, consider first the crystal lattice structures of the alkali metal halides, which have the general formula MX: The ionic radii are

Li^+, 0.60 Å F^-, 1.36 Å
Na^+, 0.95 Å Cl^-, 1.81 Å
K^+, 1.33 Å Br^-, 1.95 Å
Rb^+, 1.48 Å I^-, 2.16 Å
Cs^+, 1.69 Å

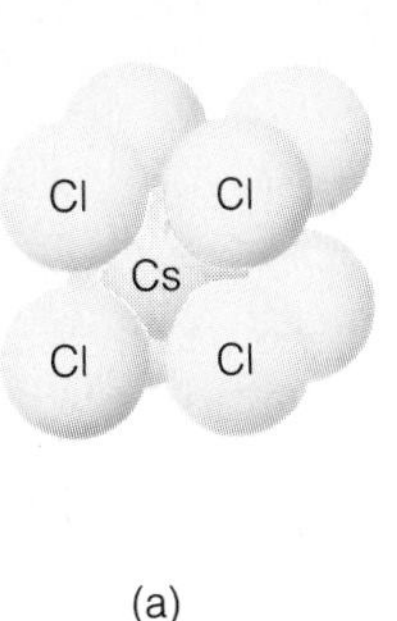

(a)

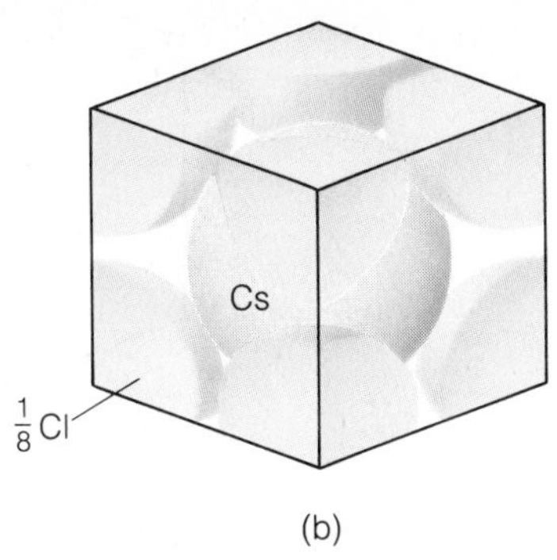

(b)

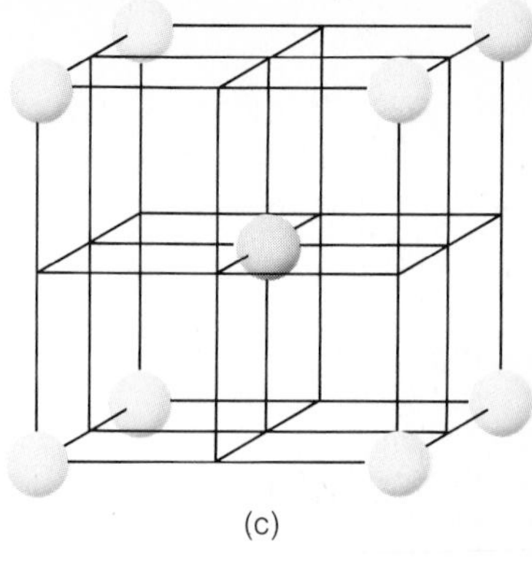
(c)

FIGURE 11.19
Ionic crystal structure of cesium chloride. (a) Packing model. (b) Unit cell. (c) Crystal lattice.

The halides in which the cationic and anionic radii are most nearly equal are KF, RbF, and CsCl. For these compounds, it is possible to achieve a crystal structure in which each ion is the nearest neighbor to eight ions of the opposite charge, while ions of like charges do not "touch" each other. Such an arrangement is shown in Figure 11.19a for cesium chloride, CsCl, with the Cs^+ ion in the center and eight Cl^- ions at the corners. Since only 1/8 of each Cl^- ion is inside the cubic unit cell (Figure 11.19b) the composition inside the cell is CsCl, which corresponds to the formula. Alternatively, if we picture an extended CsCl lattice (Figure 11.19c), it can be seen that each Cl^- ion is the nearest neighbor to eight Cs^+ ions, so everything comes out right. Note also that if all the Cs^+ ions were removed from the lattice (an imaginary process only) the remaining Cl^- ions would make up a simple cubic system. (What if the Cl^- ions were removed, leaving only the Cs^+ ions?)

When the ionic radii are very different, as in LiCl or NaCl, the type of crystal structure used by cesium chloride is no longer the most stable one. This is because an Li^+ or Na^+ ion is so much smaller than a Cl^- ion that it could not "touch" eight of them no matter how close the Cl^- ions were to each other (Figure 11.20). A more stable arrangement is a structure of the type shown in Figure 11.21 for sodium chloride, in which the Na^+ ions are large enough to prevent the Cl^- ions from touching each other. This structure can be visualized as a face-centered cubic

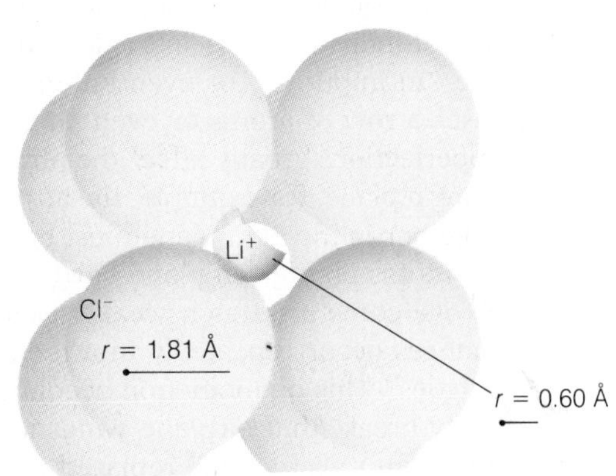

FIGURE 11.20
A lattice like that of CsCl would not be the best one for LiCl, because the Li^+ ion is too small to prevent the Cl^- ions from "touching" each other.

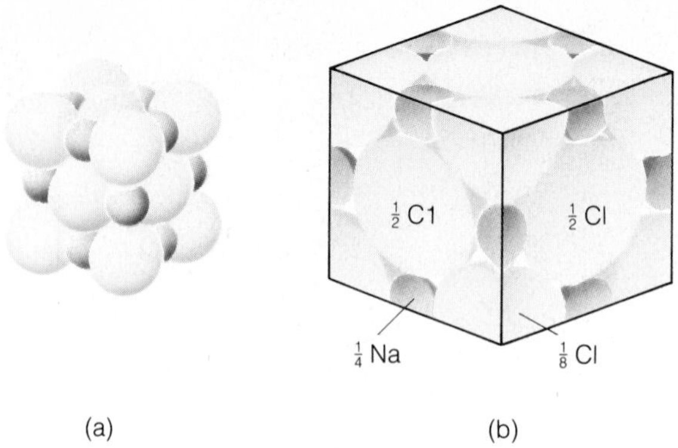

$\frac{1}{2}$ C1 $\frac{1}{2}$ Cl

$\frac{1}{4}$ Na $\frac{1}{8}$ Cl

(a) (b)

FIGURE 11.21
Crystal structure of sodium chloride.
(a) Packing model.
(b) Unit cell.

FIGURE 11.22
Rutile (TiO_2) structure. Each Ti atom "touches" four O atoms, while each O atom "touches" two Ti atoms.

lattice of Cl^- ions with the smaller cations tucked into the spaces between them. Each ion "touches" six ions of the opposite charge. Thus, for LiCl and NaCl, this type of crystal lattice is more stable than a CsCl-type structure would be because it brings unlike charges closer together.

PROBLEM 11.4 How many Cl^- ions are in the unit cell shown in Figure 11.21? How many Na^+ ions? □

Some crystal structures are made up of ions that have unequal charges. The anionic and cationic coordination numbers cannot be equal if the empirical formula, or the condition of electrical neutrality, is to be satisfied. These crystal structures are therefore necessarily more complex than those of the simple MX type. An example is titanium dioxide, TiO_2. X-ray diffraction studies of the TiO_2 crystal, known as **rutile**, show the structure to be that of the model shown in Figure 11.22, where the coordination numbers of Ti and O are 4 and 2, respectively.

11.8 REAL CRYSTALS

Like different buildings constructed of identical bricks, real crystals may not resemble the crystal lattices or unit cells of which they are composed. The shape of real crystals is determined in large part by their relative rates of growth in different directions. For example, a face-centered cubic structure may grow to form a cube, an octahedron, or a cube with its corners cut off (Figure 11.23).

The shape in which a crystal usually grows is called its *habit*.

The perfect order described in the previous sections is the *ideal* picture. Real crystals, on the other hand, are very likely to be imperfect. The faults may occur only at individual sites in the lattice—for example, a misplaced atom, or a foreign one (an impurity), or even an empty space where an atom should be. In other cases a row of atoms or even an entire plane may be altered or displaced. Such imperfections greatly affect the physical properties of a crystal. To break a crystal along a plane, for example, the attraction of each atom in the plane for its nearest neighbor in the next plane must be overcome. If the crystal has a defect in which the atoms on a given plane have been partially displaced, the attractive forces to be overcome are much weaker, and the crystal breaks more easily. When such changes occur repeatedly in a metal over a period of time, the result is "metal fatigue." This phenomenon accounts for the facts that automobile springs eventually break, that airplane wings must be periodically inspected for cracks, and that train rails must be replaced from time to time.

Imperfections in a crystal lattice also allow solids to diffuse into each other. For example, gold that is plated on lead will gradually diffuse into and actually dissolve

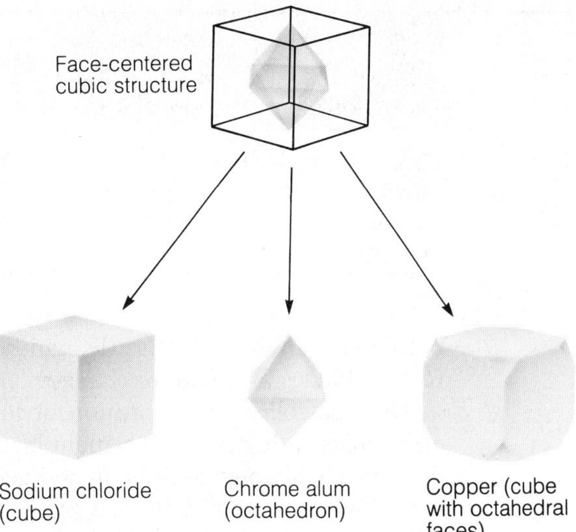

FIGURE 11.23
Crystal habits derived from a face-centered cubic lattice. The figure formed by connecting the centers of all adjoining faces with straight lines is an octahedron.

in the lead. The atoms in a crystal are not static; they vibrate, and at any one time some vibrations are more energetic than others. Since atoms of lead and gold are almost the same size, a vigorously vibrating atom of either kind can move and fit into an adjacent lattice if there is a vacancy to receive it. This action is solid-state diffusion. The phenomenon can also occur with unequally sized atoms if the smaller atoms can slip in among the larger ones (for example, carbon in iron), like a cyclist in stalled truck traffic.

In the absence of imperfections, crystals possess unusual properties. This was first noted in 1952 as a result of investigations of tiny crystal growths that produced short circuits in electronic equipment. These growths, called "whiskers," were found to be substantially perfect crystals of unusually high strength relative to their cross-sectional areas (Figure 11.24 and Table 11.2).

Crystal whiskers of silicon carbide, SiC, can be used to strengthen ceramic materials. The strengthened ceramics, which are resistant to heat and corrosion

FIGURE 11.24
Crystal whiskers of aluminum oxide.

TABLE 11.2
STRENGTHS OF CRYSTALS

SUBSTANCE	STRENGTH, pressure in bars needed to crush crystal: Ordinary Crystal	Whiskers
Iron	275	130,000
Gold	15	16,000
Graphite	2950	200,000
Copper	350	40,000

but would otherwise be brittle, can be used in the manufacture of such items as turbine blades and heat exchangers.

The controlled incorporation of impurities into a crystal is practiced by the electronics industry in the manufacture of transistors and other "solid-state" devices. Highly purified silicon and germanium have low electrical conductivities, but when very small amounts of specific types of impurities, such as arsenic or boron, are added to the silicon or germanium, electrical properties of the type needed by transistors are created. These phenomena are discussed in more detail in Chapter 22.

11.9 CHANGES OF STATE

As the energy of a chemical system is increased, its particles move more vigorously and their arrangement becomes more disordered (more random). Adding heat to ice at 0°C changes it to liquid water (melts it); the liquid has a less orderly arrangement of molecules than the crystal had. Further heating changes liquid water to water vapor—an example of that supremely disordered state, a gas. Conversely, as the energy of a system is decreased, the probability of an orderly arrangement increases. The reason is that as the energy of motion decreases, the cohesive forces between molecules are better able to hold them together. Thus we can change the physical state of a substance by putting in or taking out energy, usually in the form of heat.

When there is a change in physical state, the temperature of the substance remains constant as long as both phases are present. For example, when ice has been warmed to 0°C, it begins to melt. As heat flows in from the surroundings, the temperature stays at 0°C until all of the ice has melted. The **heat of fusion**, ΔH_{fus}, of a substance is the amount of heat necessary to melt 1 gram or 1 mole (we must specify which it is) of the solid to a liquid *at the same temperature* (the melting point) and at constant pressure (1 atm). The heat of fusion is absorbed by the substance from the environment. The opposite process (freezing) involves the same quantity of heat per gram (or mole) of substance, but the heat is released by the substance to the environment. Therefore, fusion is endothermic, and freezing is exothermic.

Similarly, the **heat of vaporization**, ΔH_{vap}, is the heat required to vaporize a unit quantity of liquid, and the **heat of sublimation**, ΔH_{sub}, is the heat required to sublime a unit quantity of a solid. These conversions, too, occur at a fixed temperature and at 1 atm pressure. Recall that sublimation is the conversion of a solid to a gas without passing through the liquid state. Therefore, vaporization and sublimation are endothermic, and condensation is exothermic. Note that the various ΔH values express the amounts of energy at 1 atm needed for the transition when the substance is at the transition temperature. If it is not at the transition temperature, the energy needed to reach that temperature must also be calculated. This calculation was described in section 4.3.

TABLE 11.3
HEATS OF FUSION AND VAPORIZATION

SUBSTANCE	MELTING POINT, °C	BOILING POINT, °C	HEAT, J/g[†]	
			Fusion	Vaporization
Aluminum	658	2057	393	1.0×10^4
Beeswax	62	[§]	177	—
Carbon tetrachloride	−24	76.8	18	194
Ethanol	−114	78.5	104	854
Mercury	−39	357	12	295
TNT	79	[‖]	93	—
Water[#]	0	100	333	2258

[†] The heat of fusion is given at the melting point; the heat of vaporization is given at the boiling point. They are somewhat different at other temperatures.

[§] Decomposes before reaching boiling point.

[‖] Explodes before reaching boiling point.

[#] The heats of fusion and vaporization of water are frequently used values and are often given in calories. The conversions are

$$\text{heat of fusion of } H_2O(c) = 333 \frac{\cancel{J}}{g} \times \frac{1 \text{ cal}}{4.184 \cancel{J}} = 79.6 \frac{\text{cal}}{g}$$

$$\text{heat of vaporization of } H_2O(\ell) = 2258 \frac{\cancel{J}}{g} \times \frac{1 \text{ cal}}{4.184 \cancel{J}} = 539.7 \frac{\text{cal}}{g}$$

Any process that proceeds in the direction solid ⟶ liquid ⟶ gas produces a twofold change in the affected matter:

1. The energy of the matter is increased because the molecules are moving against their attractive forces and are farther apart. (The melting of ice seems like an exception—the molecules are closer together in the liquid—but some of the hydrogen bonds between them have been broken.)

2. The orderliness of the molecular arrangement is decreased.

Conversely, in any process that proceeds in the direction gas ⟶ liquid ⟶ solid, the energy is decreased and the order is increased.

Table 11.3 presents data on heats of fusion and vaporization for selected substances. Note that the highest ΔH values among the compound substances belong to those with H-bonding (water and ethanol).

EXAMPLE 11.3 (a) How much heat, in joules, is needed to melt 2.00 kg of ice? (b) What mass of ethanol could be vaporized at its boiling point by this amount of heat?

ANSWER

(a) The heat of fusion of ice (from Table 11.3) is 333 J/g. Do not forget to convert kilograms to grams:

$$2.00 \cancel{\text{kg}} \times \frac{10^3 \text{ g}}{\cancel{\text{kg}}} \times \frac{333 \text{ J}}{\cancel{g}} = 6.66 \times 10^5 \text{ J}$$

(b) The heat of vaporization of ethanol (Table 11.3) is 854 J/g.

$$6.66 \times 10^5 \cancel{J} \times \frac{1 \text{ g}}{854 \cancel{J}} = 780 \text{ g}$$

■

EXAMPLE 11.4 How much heat in kilojoules is needed to convert 75.0 g of water at 20°C to steam at 100°C?

ANSWER Two processes are involved here. The heat required for each must be calculated separately, and the two heats added. The first process is the warming of the water from 20°C to its boiling point, 100°C. Using 4.20 J/(g °C) for the specific heat of water, we have (section 4.3)

$$\text{heat of warming} = \text{mass} \times \text{specific heat} \times \text{temperature change}$$

$$= \underbrace{75.0\ \cancel{g} \times 4.20\ \frac{\cancel{J}}{\cancel{g}\,°\cancel{C}} \times (100 - 20)°\cancel{C}}_{\text{heat in joules}} \quad \underbrace{\times \frac{1\ \text{kJ}}{1000\ \cancel{J}}}_{\text{conversion to kJ}}$$

$$= 25.2\ \text{kJ}$$

The second process is vaporization at 100°C. From Table 11.3, the heat of vaporization of water is 2258 J/g:

$$\text{heat required for vaporization} = 75.0\ \cancel{g} \times 2258\ \frac{\cancel{J}}{\cancel{g}} \times \frac{1\ \text{kJ}}{1000\ \cancel{J}} = 169\ \text{kJ}$$

The total heat is then 25.2 kJ + 169 kJ = 194 kJ. ■

PROBLEM 11.5 How much heat in kilojoules must be extracted from 50 g of water at 90°C to convert it to ice at 0°C? □

Hess's law (section 4.5) may be used to calculate the heat of sublimation from the heats of fusion and vaporization:

$$\begin{array}{ll} \text{solid} \longrightarrow \cancel{\text{liquid}} & \Delta H_{fus} \\ \cancel{\text{liquid}} \longrightarrow \text{gas} & \Delta H_{vap} \\ \hline \text{solid} \longrightarrow \text{gas} & \Delta H_{sub} = \Delta H_{fus} + \Delta H_{vap} \end{array}$$

EXAMPLE 11.5 The heat of vaporization of liquid water at 0°C is 2501 J/g. Using this datum and Table 11.3, calculate the heat of sublimation of ice at 0°C.

ANSWER

$$\Delta H_{sub} = \Delta H_{fus} + \Delta H_{vap}$$
$$= 333\ \text{J/g} + 2501\ \text{J/g} = 2834\ \text{J/g}$$

The heat of fusion given in the table applies to a process at 0°C and can therefore be used directly. However, the table gives the heat of vaporization at 100°C, which should not be used in this calculation because the sublimation occurs at 0°C. ■

PROBLEM 11.6 At −39°C, the heat of vaporization of liquid mercury is 297 J/g. Calculate the heat of sublimation in J/g of solid mercury at this temperature. □

11.10 VAPOR PRESSURE

A wet towel, rolled up into a ball and sealed in a plastic bag, will never dry out. Suppose you want it to dry. Everyone knows what to do: spread it out in the sun, preferably on a breezy day, or tumble it in a clothes dryer. Vaporization (evaporation) is favored by (1) high temperature—to give the molecules more kinetic energy; (2) small attractive forces in the liquid—so that less energy is required to separate the molecules; (3) a large surface area—to offer more molecules the opportunity to escape; (4) low atmospheric pressure above the liquid—to decrease

BOX 11.1
WHAT IS THE DIFFERENCE BETWEEN A GAS AND A VAPOR?

A vapor *is* a gas. The distinction is that a vapor is a gas that is—or could be—formed by the vaporization of a liquid or a solid. Such a transformation is possible if a condensed phase could exist in equilibrium with the gas phase; that is, if the substance is below its critical temperature. Thus, all gaseous H_2O below 374.1°C (Figure 11.29) may be called water "vapor."

This usage is not universal, however. Sometimes a gas is called a vapor only when it is at or below the normal boiling point of the liquid and we visualize it as resulting from the vaporization of the liquid at normal pressure. When we cannot see the liquid, even though it is present, we tend to avoid the word *vapor*. Thus, a steel cylinder containing chlorine in both liquid and gas phases is said to contain chlorine gas, whereas a glass bottle half full of liquid bromine is said to be in equilibrium with bromine vapor. Furthermore, no one refers to a tank of liquid and gaseous cooking fuel (usually propane) as "bottled vapor," even though most of the contents are in the liquid phase, vaporizing only as the gas above it is withdrawn.

Despite these inconsistencies, two usages are clearly incorrect. A gas that is above its critical temperature should never be called a vapor, and a mist—such as the condensate from your exhaled breath that is visible on a cold day—is neither a vapor nor a gas, but a collection of liquid droplets.

the number of collisions with other molecules and thereby minimize return of molecules to the liquid; and (5) a motion of the atmosphere (breeze) above the liquid—to carry away vaporized molecules and minimize reentry.

Items 1 and 2 deserve a few comments. Molecules are held together in the liquid by attractive forces. The only molecules that can escape from the liquid are those with especially high kinetic energy—just as a rocket that is to escape from the Earth's gravity must be given a large kinetic energy. A rise in temperature drastically increases the proportion of molecules with enough kinetic energy to escape (Figure 10.15, section 10.14) and thus speeds up vaporization. How about item 2? *Can* we decrease the energy needed to remove a molecule from the liquid? The only way to do this is to use a different liquid. If the towel is wet with gasoline instead of water, it will dry much faster because gasoline has weaker intermolecular forces than water. Gasoline molecules and water molecules have the same kinetic energy at the same temperature, but more gasoline molecules will escape in a given time because they need less energy to do so.

The preceding discussion hints at a concept that is very important in chemistry—**rate**. The rate of a process is its speed: the number of molecules doing something (reacting, escaping, colliding) per unit time. The rates we are studying now are the rates of vaporization and condensation. The *rate of vaporization* is the number of molecules leaving the liquid per second. The *rate of condensation* is the number of molecules entering the liquid per second. Of course, the rates can be in other units, such as moles per second or grams per minute.

But let's get back to the wet towel in the sealed plastic bag. Why won't it dry out? Water molecules have no memory. They cannot be expected to know that there is no sense in leaving the liquid because they will just encounter an impenetrable barrier anyway. The fact is that they *are* escaping from the liquid to the vapor, just as fast as if the towel were out of the bag at the same temperature. However, something else is happening: molecules of vapor are returning to the liquid. Because the bag traps the escaping vapor, the partial pressure of the vapor builds up and, if the temperature is held constant, a condition will be reached in which molecules are returning just as fast as they are leaving, or

rate of vaporization = rate of condensation

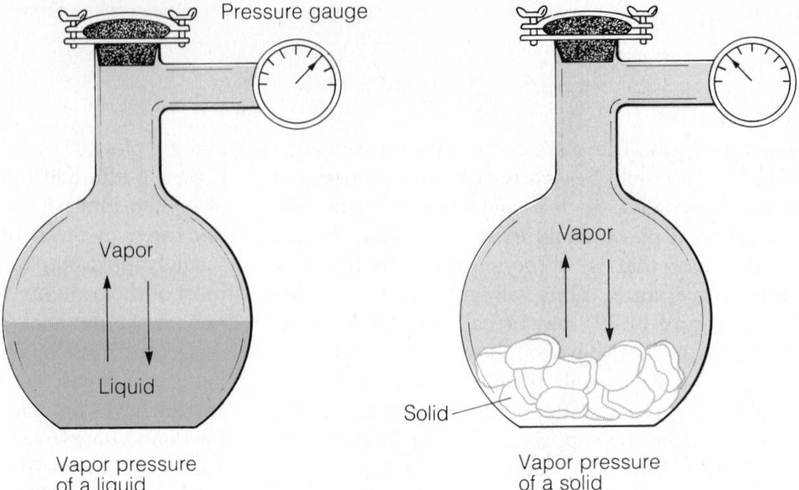

FIGURE 11.25
Vapor pressures of liquids are generally higher than those of solids at the same temperature.

The system is said to be in **equilibrium**. Under these conditions, the amount of water in the towel and the amount of water in the bag remain constant.

A system at equilibrium is defined as one in which two opposing processes are occurring in the same place and at the same rate. The effect is no net change.

A **saturated vapor** is a vapor that is in equilibrium with the liquid at a definite temperature. The **vapor pressure** of a liquid is the pressure exerted by the saturated vapor. The vapor pressure depends on the temperature and on the nature of the liquid; in general, it is independent of the surface area of the liquid and of the volume available to the vapor, as well as of the presence of other gases. Solids, too, exert vapor pressures, but substances that are solids at a given temperature usually have lower vapor pressures than those that are liquids at the same temperature (Figure 11.25). Figure 11.26 illustrates the relationships between vapor pressure and temperature for several liquids.

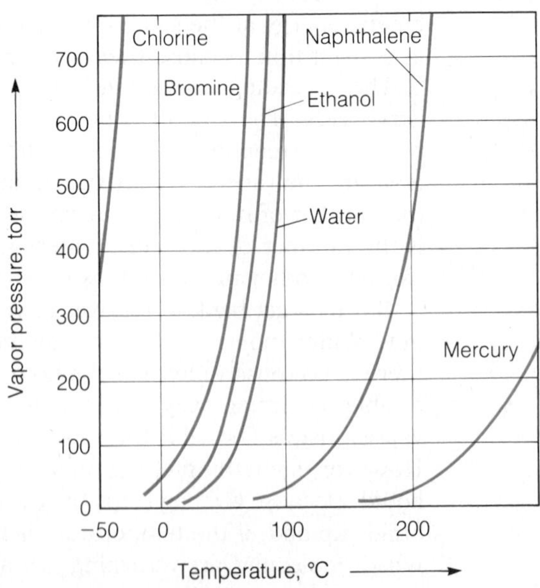

FIGURE 11.26
Vapor pressures of some liquids.

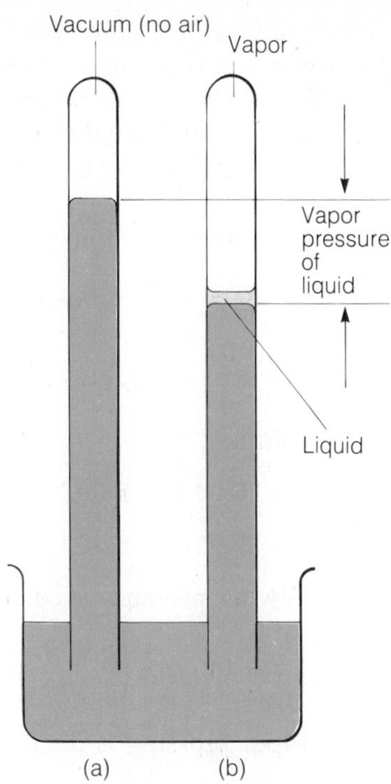

FIGURE 11.27
An apparatus for measuring the vapor pressure of a liquid.

To make approximate measurements of vapor pressure, we will set up the apparatus shown in Figure 11.27. We first prepare two mercury barometers (section 10.2) with the mercury in each tube falling to a height corresponding to the barometric pressure (about 760 mm at sea level). Above the mercury in each tube is a vacuum (almost—the vapor pressure of mercury is small). Now we introduce a little liquid into tube b. It rises to the top of the mercury column (most liquids are less dense than mercury) and some of it evaporates. The pressure of the vapor pushes the mercury down; how *far* down tells us what the pressure is. As long as we use enough liquid so that not all of it evaporates, liquid and vapor are in equilibrium. The measured pressure then gives us the vapor pressure of the liquid.

As illustrated by the tubes in Figure 11.27, the vapor pressure is measured in a sealed system containing only the liquid or solid and its vapor. No other substance, like air, is present. Under these conditions, liquid (or solid) and vapor can coexist only when the pressure has the correct value—the vapor pressure—for the given temperature. Suppose that below the mercury in Figure 11.27b was a piston that we could move up and down. If we tried to increase the pressure by pushing the mercury up, all we would do is condense some of the vapor. The pressure remains constant as long as both liquid and vapor are present. If we tried to decrease the pressure by pulling the mercury down, the pressure would remain constant until all of the liquid had evaporated. When air is present, the vapor mixes with air; equilibrium is reached when the *partial* pressure of the vapor equals the equilibrium vapor pressure of the liquid. This partial pressure—but not the total pressure—remains constant as the volume changes, and is the same as it would be without the air. All these equilibria are established in *closed* containers. If the container is open, the vapor diffuses away through the air until all of the liquid or solid has evaporated. There is no equilibrium in that case.

EXAMPLE 11.6 A 10.00-L cylinder contains 5.00 L of a liquid at 25°C. The remaining volume contains air and vapor at a total pressure of 765 torr. The vapor pressure of the liquid at 25°C is 625 torr. A piston is then pushed down (without leakage) until the total volume is 7.50 L. Find the new pressure in the container. Assume that the liquid volume remains 5.00 L and the temperature is constant.

ANSWER From Dalton's law,

$$765 \text{ torr} = p_{\text{air},1} + p_{\text{vapor}}$$
$$= p_{\text{air},1} + 625 \text{ torr}$$
$$P_{\text{air},1} = 765 \text{ torr} - 625 \text{ torr} = 140 \text{ torr}$$

Gas volumes:

$$V_1 = 10.00 \text{ L} - 5.00 \text{ L} = 5.00 \text{ L}$$
$$V_2 = 7.50 \text{ L} - 5.00 \text{ L} = 2.50 \text{ L}$$

The new air pressure is calculated from Boyle's law:

$$p_{\text{air},2} = 140 \text{ torr}\left(\frac{5.00 \text{ L}}{2.50 \text{ L}}\right) = 280 \text{ torr}$$

The vapor pressure is unchanged and remains at 625 torr. Then,

$$P_{\text{total},2} = 280 \text{ torr} + 625 \text{ torr} = 905 \text{ torr}$$ ■

PROBLEM 11.7 Referring to the previous example, find the new pressure in the container if the piston is pulled up until the volume is 12.50 L. □

11.11 BOILING AND CONDENSATION

As long as the temperature of a liquid remains below its boiling point, evaporation occurs only from its surface. As the liquid is heated, the temperature rises until a condition is reached in which bubbles of vapor appear within the body of the liquid and rise to the surface. At this time the temperature becomes constant, even though heat is still being added. This constant-temperature process is what we call **boiling**. Boiling occurs when the vapor pressure of the liquid is equal to the pressure of the atmosphere in contact with the liquid. The *boiling point* of a liquid is *the temperature at which the vapor pressure of the liquid is equal to the pressure of the surrounding atmosphere.* The boiling point of a liquid therefore varies with changes in the atmospheric pressure. The **normal boiling point** is measured at standard atmospheric pressure (1 atm or 760 torr). At lower atmospheric pressures, such as at the top of a mountain, the boiling point is lower than normal. Below sea level, such as at Death Valley (California) or the Dead Sea (Israel), the atmospheric pressure is higher, and so is the boiling point.

Boiling usually occurs when the container is open or the vapor is being pumped away; thus, no equilibrium is attained. Whether the liquid boils is determined not by the partial pressure of its vapor, but by the *total* gas pressure (the atmosphere) above the liquid. As long as the vapor pressure is less than this external pressure, no vapor bubbles form in the liquid. Evaporation *from the surface* occurs at any temperature and air pressure, even well below the boiling point. What happens at the boiling point is that the vapor attains enough pressure to form bubbles *inside* the liquid. Evaporation then occurs throughout the boiling liquid, not only at its surface. As a result, the *rate* of evaporation increases greatly when boiling begins.

When you have observed a gently boiling liquid, you may have noticed that bubbles seem to keep coming from the same points on the wall of the container. You may have thought that there must be something at those points that helps bubbles to form, and that is indeed true. The "something" is usually a tiny bubble adhering to a rough spot on the wall. A bubble, like a balloon, is hard to begin blowing up, but easier to expand once you get past the initial effort. The most likely place for bubbles to appear is at one of these ready-made bubbles. The trapped bubble serves as a *nucleus*, that is, a beginning from which growth can occur. As liquid evaporates into it, the bubble expands and breaks off, leaving a bit of itself behind to help the next bubble get started.

Suppose that these natural places for bubble formation are absent. A pure liquid, heated carefully in a smooth, clean vessel, can reach a temperature considerably above its boiling point without bubbling. This phenomenon is called **superheating**. In the absence of nuclei, bubbles do not form, and superheating occurs instead. Eventually, the temperature becomes high enough so that a bubble forms without a preexisting nucleus, and then grows rapidly—almost explosively. This often dangerous behavior is called "bumping." To insure smooth boiling, nuclei are introduced deliberately; for example, a piece of porous clay (a "boiling chip") holds air in its pores and thus provides ready-made bubbles.

On a large scale in industrial vats the effect can actually be explosive.

The same reluctant initiation of a phase change is observed in condensation. When a vapor is cooled to the temperature where the vapor pressure of the liquid (or solid) equals the actual pressure of the vapor, we expect condensation to begin. However, a vapor can often be **undercooled** (or less logically, "supercooled") below this temperature without any condensation. The reason is the same as for superheating: the vapor needs a nucleus on which it can condense. An important example of undercooling is found in the upper atmosphere. It often contains undercooled water vapor that farmers desperately need as rain. Cloud-seeding is the process of introducing nuclei—commonly dry ice (solid CO_2) and silver iodide—in an attempt to start the growth of raindrops or snowflakes.

11.12 MELTING AND FREEZING

Liquid water freezes at 0°C. This means that when heat is removed from liquid water at 1 atm pressure, it cools until it reaches 0°C, at which temperature ice begins to form. The temperature then remains constant at 0°C as the heat of fusion is removed. Finally, when all the water has crystallized, the temperature again begins to fall. Conversely, when ice is warmed, it begins to melt at 0°C and the temperature remains constant until all of it has melted.

The **melting point** and the **freezing point** of a substance are the same; they are the temperature at which liquid and solid are in equilibrium at standard atmospheric pressure. The melting point is reached by warming the solid; the freezing point, by cooling the liquid. The effect of a change in atmospheric pressure is so small that it usually need not be considered.

This is in marked contrast to the boiling point, which is significantly affected by changes in external pressure.

If heat is slowly and continuously removed from any pure liquid substance, the liquid will cool, like water, until it starts to freeze; then the liquid will freeze at constant temperature until it is all solid, and finally the solid will start to cool. The graphical representation of this sequence, called a **cooling curve**, is shown in Figure 11.28. Note that the liquid may cool *below* its freezing point before crystallization occurs and solid–liquid equilibrium is established. This is the same undercooling effect that is observed in the vapor. It is even more common when the expected new phase is crystalline. Undercooling can be avoided by introducing nuclei, preferably crystals of the same substance, on which crystals can grow.

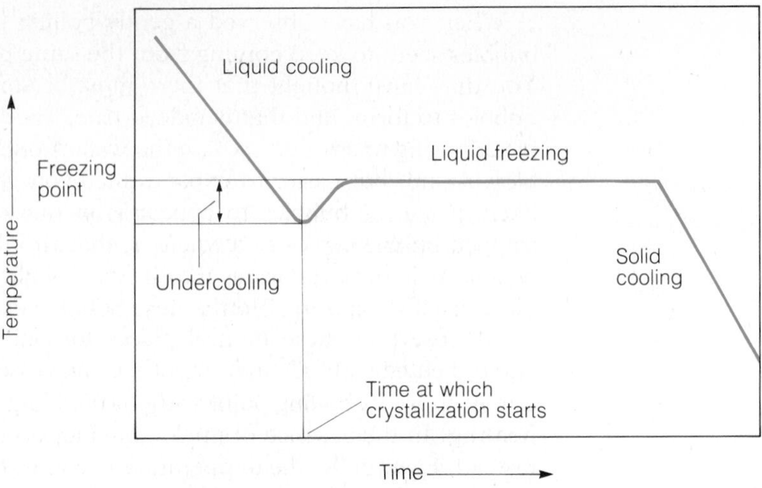

FIGURE 11.28
Cooling curve of a pure liquid, with undercooling.

Cooling curves obtained in practice do not show such sharp changes of direction because it is difficult to keep the temperature of the entire sample uniform, especially when much of it is solid.

The time direction of this curve may be reversed by warming the solid. Such a graph is called a **warming curve**. Superheating does not occur during warming of a crystal.

11.13 THE PHASE DIAGRAM

Only water is present; there is no air or other substance that could mix with the vapor.

Recall from section 1.7 that any homogeneous portion of matter that is separated from other parts of the sample by a definite surface or boundary is called a **phase**. Different phases need not be in different physical states. For example, a layer of oil floating on water is a two-phase system, both phases being liquid.

In Figure 11.26, the vapor pressures of various liquids, including water, are plotted against temperature. The plot for water alone also appears in Figure 11.29; it is the curve marked "liquid–vapor equilibrium." (Figure 11.29 is not drawn to scale, however; it is distorted to make everything large enough.) Each point on this curve represents a pressure-temperature combination at which liquid water and water vapor (steam) can coexist at equilibrium. An example shown in the diagram is 760 torr and 100°C. For a given temperature, if the pressure is higher than what this curve specifies, only liquid is stable; any vapor would be squeezed down to liquid. If the pressure is lower than what this curve reads, only vapor is stable; any liquid would evaporate. This liquid–vapor curve thus represents a two-phase equilibrium; off the curve, there can be only one phase.

EXAMPLE 11.7 Steam at 99°C and 500 torr is suddenly compressed to 800 torr at the same temperature. What type of change would occur?

ANSWER The point on the graph corresponding to 99°C and 800 torr falls in the area where only liquid water is stable. Therefore, the steam would condense and form liquid water. ■

PROBLEM 11.8 Liquid water at 95°C and 1 atm is suddenly decompressed to 5 torr. What type of change would occur? □

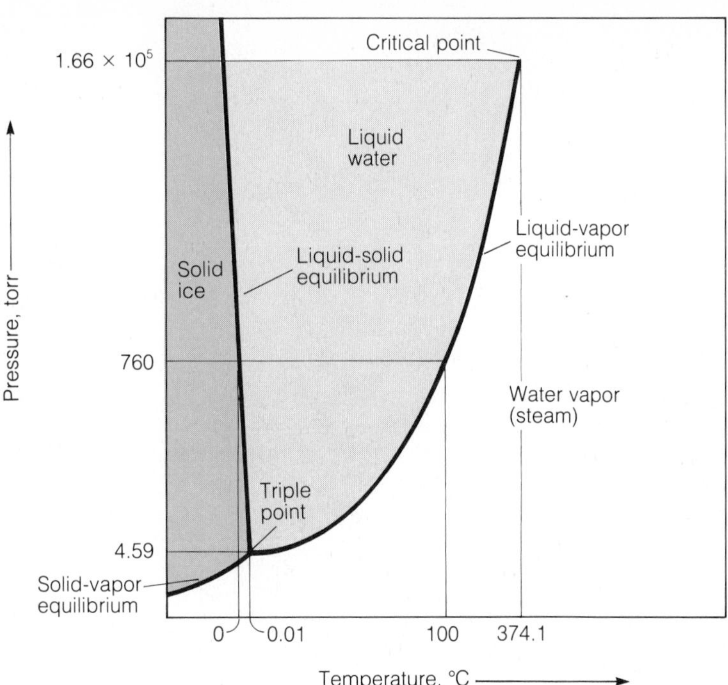

FIGURE 11.29
Phase diagram for water (not to scale).

In the same diagram (Figure 11.29) the vapor pressure of ice is plotted against temperature. (This curve is labeled "solid–vapor equilibrium.") This curve also represents a two-phase equilibrium. At the point where the solid–vapor and liquid–vapor curves meet, ice and liquid water have the same vapor pressure.

In addition to equilibria between liquid and vapor and between solid and vapor, there is a third possibility, equilibrium between solid and liquid. This curve is best understood as a plot of melting point against pressure. Since pressure has only a small effect on melting point, the curve goes up very steeply. The fact that the curve for water leans to the left (its slope is negative) is unusual; water is one of the few substances whose melting point decreases as the pressure increases.

The intersection of the liquid–vapor, solid–vapor, and solid–liquid curves is called the **triple point**, because it corresponds to an equilibrium among three phases. At this unique set of conditions (0.01°C, 4.59 torr), ice, liquid water, and water vapor can all coexist at equilibrium.

The regions between the curves—the whole diagram except for the curves—are one-phase regions. When the temperature and pressure correspond to a point in one of these regions, only one phase can exist: solid in the upper left region (high pressure, low temperature), liquid in the upper middle region (high pressure, intermediate temperature), and vapor everywhere else. Figure 11.29 is known as a **phase diagram** because it tells us what phase or phases can be present for any given pressure-temperature combination.

The "critical point" shown in Figure 11.29 will be explained shortly.

The meaning of the points, curves, and regions in a phase diagram can be summarized thus:

At a point where three lines meet, three phases coexist in equilibrium.
On a curve, two phases coexist in equilibrium.
In a region, one phase exists.

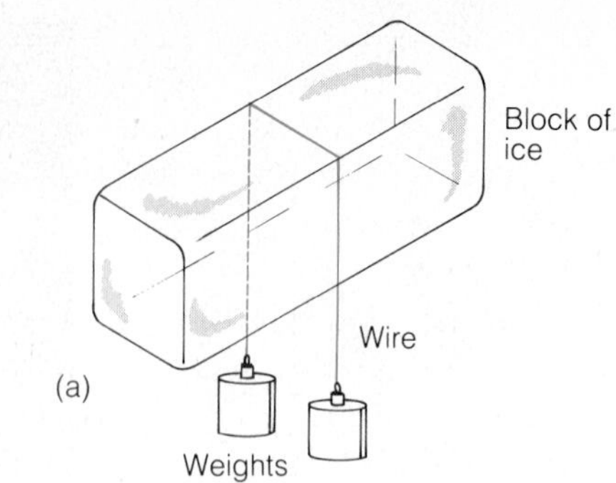

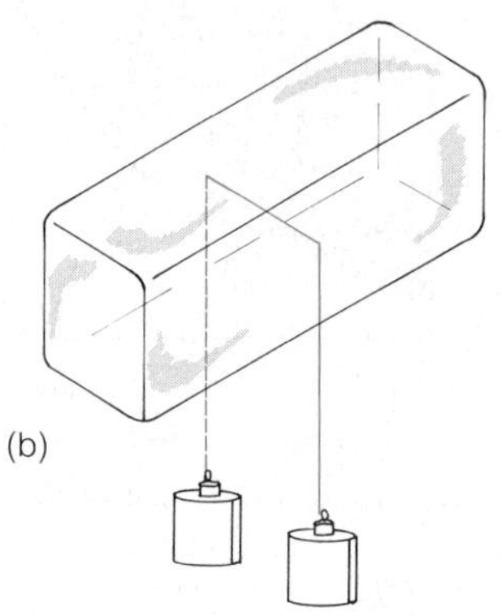

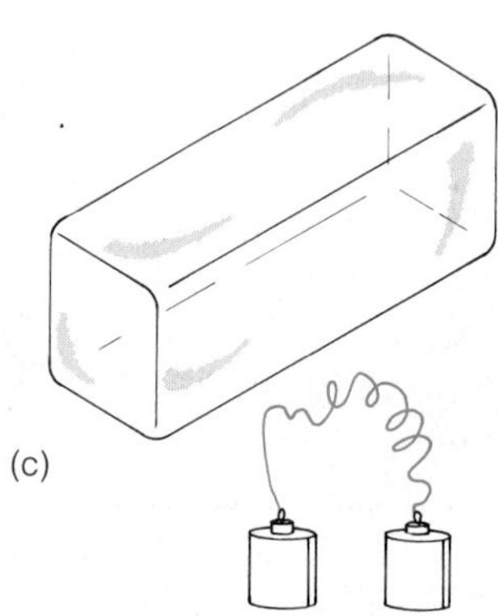

FIGURE 11.30
Ice melts under pressure. (a) A fine wire is pulled over a block of ice by a weight suspended on each side. (b) The pressure exerted by the wire melts the ice, allowing the wire to be pulled down by the weights. When the pressure above the wire is released, the water refreezes. (c) After the wire has been pulled entirely through the ice, the block is still in one piece.

Ice (less dense)
↓ Applied Pressure
Liquid water (more dense)

The effect of pressure on the melting point is related to the relative densities of liquid and solid. Liquid water is denser than ice, which is why ice floats on water. An increase in the external pressure tends to push the molecules closer together, thereby increasing the density. This is what happens when pressure applied to ice results in melting (Figure 11.30); the less dense solid (larger volume) changes to the denser liquid (smaller volume). The more typical behavior of other materials is just the reverse. When the solid phase is denser than the liquid (the usual situation), an increase in pressure favors freezing, and the solid–liquid equilibrium line leans to the right (has a positive slope). Figure 11.31 is a phase diagram for carbon dioxide, in which the liquid–solid line leans in the usual direction. Note also that solid–vapor equilibrium occurs for carbon dioxide at pressures up to 5.2 atm (about 3950 torr). That is why dry ice does not melt but sublimes at ordinary atmospheric pressure. Liquid CO_2 exists only under high pressure, as in a fire extinguisher.

In water, the solid–vapor equilibrium is possible only at pressures below 4.59 torr. In an open system, however, such as exists when a substance is exposed to the atmosphere, equilibrium cannot be attained, and ice does sublime at ordinary pressures.

EXAMPLE 11.8 What phase or phases can be present in each of the following systems? Refer to Figures 11.29 and 11.31. (a) Water at 10°C and 760 torr. (b) CO_2 at -57°C and 5.2 atm. (c) CO_2 at -60°C and 10 atm. (Remember: No air or other foreign substance is present.)

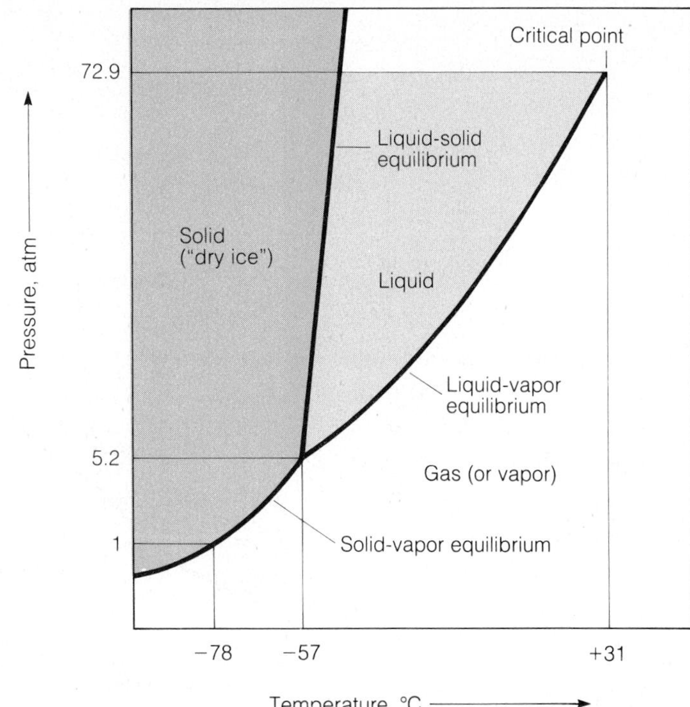

FIGURE 11.31 Phase diagram for carbon dioxide (not to scale).

ANSWER

(a) This point is in the ''liquid'' region. Only the liquid is present.

(b) These conditions correspond to the triple point. Solid, liquid, and gaseous CO_2 can all be present in equilibrium.

(c) This point is to the left of −57°C and above 5.2 atm—in the ''solid'' region. Only the solid is present. ■

EXAMPLE 11.9 Copy the phase diagram in Figure 11.31 for CO_2. Draw an arrow on it to represent each of the following processes, and describe in words the phase changes that occur. (a) Dry ice is warmed at a constant pressure of 10 atm from −78 to 31°C. (b) Gaseous CO_2 at 1 atm is compressed to 50 atm at a constant temperature of −57°C.

ANSWER Refer to Figure 11.32.

(a) The solid melts a little above −57°C. The liquid vaporizes at the temperature where the arrow crosses the curve, somewhere between −57°C and +31°C (actually −40°C). Above −40°C there is only vapor.

(b) The gas condenses to a mixture of gas, solid, and liquid at the triple point, −57°C and 5.2 atm. When the pressure is increased above 5.2 atm, all gas condenses and the liquid freezes because the freezing point increases with increasing pressure. ■

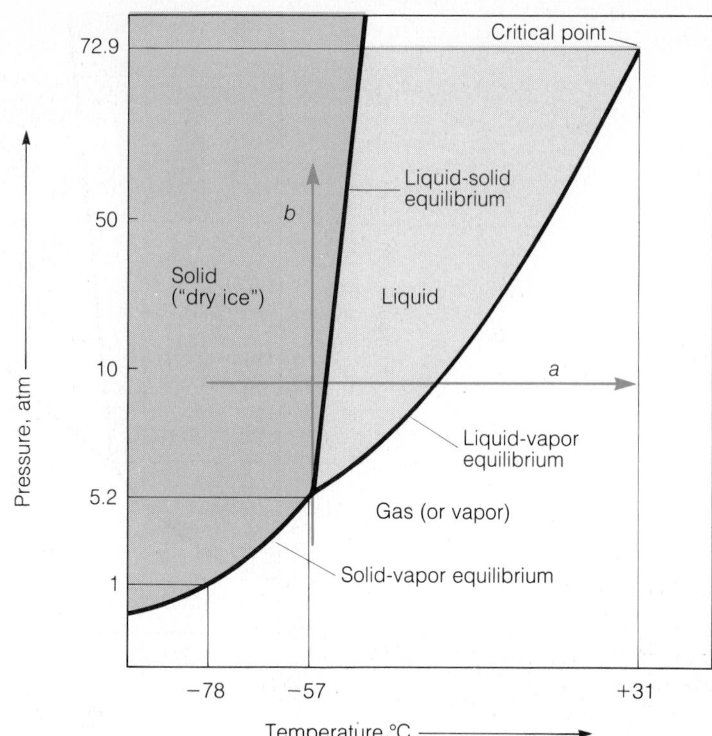

FIGURE 11.32
Diagram for Example 11.9.

PROBLEM 11.9 Figure 11.33 is a duplicate of the graph in Figure 11.29. The straight lines labeled A through E represent the five processes described here. Match lines A through E with processes 1 through 5.

1. Converting ice to steam on a mountain
2. Converting ice to steam at sea level

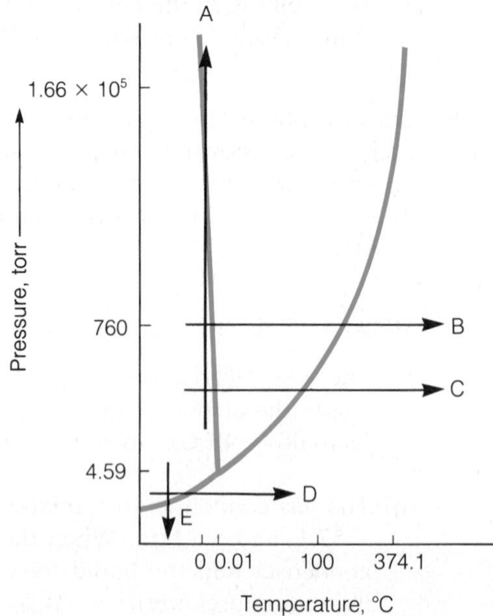

FIGURE 11.33
Diagram for Problem 11.9.

3. Sublimation of ice at constant pressure
4. Freeze-drying (converting ice to water vapor at constant temperature)
5. Melting ice by pressure □

The physical states of a substance are by no means limited to three phases. The coexistence of more than one solid phase is quite common (for example, the diamond and graphite forms of carbon, or the rhombic and monoclinic forms of sulfur; see Problem 11.40). Any three-phase equilibrium, such as two solid phases plus one liquid phase, may be represented by a triple point.

The solid–liquid equilibrium curve is different; it continues to the highest pressures at which measurements have been made.

When a phase diagram like Figure 11.29 or 11.31 is extended to higher temperatures, we find something interesting: the liquid–vapor equilibrium curve comes to an end. Why? Think what happens to a liquid in equilibrium with its vapor in a sealed tube as its temperature is raised. It expands, becoming less and less dense—more like a gas. Meanwhile, what is happening to the saturated vapor? As the temperature increases, the vapor pressure increases, there are more molecules per unit volume, and the density of the vapor increases—making it more like a liquid. Figure 11.34 illustrates these changes: the higher (liquid) density decreasing and the lower (vapor) density increasing. At some temperature, the two halves of the curve must meet. The liquid and the vapor then have the same density—the **critical density**—and indeed all of their properties become identical. The temperature at which the liquid and vapor properties become identical is called the **critical temperature** of the substance. The vapor pressure at (more precisely, just below) the critical temperature is the **critical pressure**. This pressure–temperature combination is the **critical point**. Above the critical temperature, the liquid–gas distinction is absent. There are never two fluid phases in equilibrium above the critical temperature. The substance can be compressed as much as our equipment allows—even though we call it a "gas," it may be made as dense as a liquid or even denser—but it never liquefies.

The critical temperature of a substance, like its boiling point, is related to the magnitude of its intermolecular forces. If the forces are great, liquefaction can be achieved even at elevated temperatures. The critical temperature is therefore high. If the forces are small, liquefaction is more difficult, and the critical temperature is low. For example, the values given for water in Table 11.4 indicate that

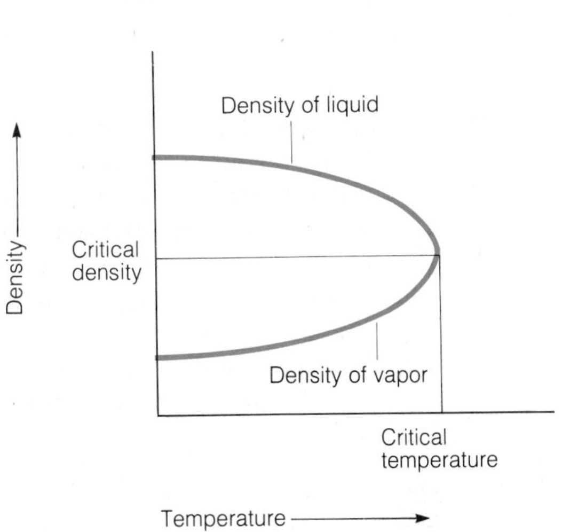

FIGURE 11.34 Densities of liquid and saturated vapor plotted against temperature.

TABLE 11.4
CRITICAL CONSTANTS

SUBSTANCE	TEMPERATURE, °C	PRESSURE, atm
Carbon dioxide	31.0	72.9
Ethanol	243	63.0
Helium	−267.96	2.26
Hydrogen	−239.9	12.8
Mercury	900	180
Nitrogen	−147	33.5
Oxygen	−118.57	49.7
Water	374.2	218.3

water can exist as a liquid up to 374.2°C if the pressure at that temperature is 218.3 atm. Water cannot exist as a liquid above that temperature, regardless of the pressure.

11.14 SPONTANEOUS PHASE CHANGES; ENTROPY

Two factors determine whether a substance will change its state under a given set of conditions:

1. *Energy* Heat naturally flows from a warmer body to a cooler body. Therefore, if a body's environment is warmer than the body, heat will flow into the body, and the process that absorbs heat will be "favored."

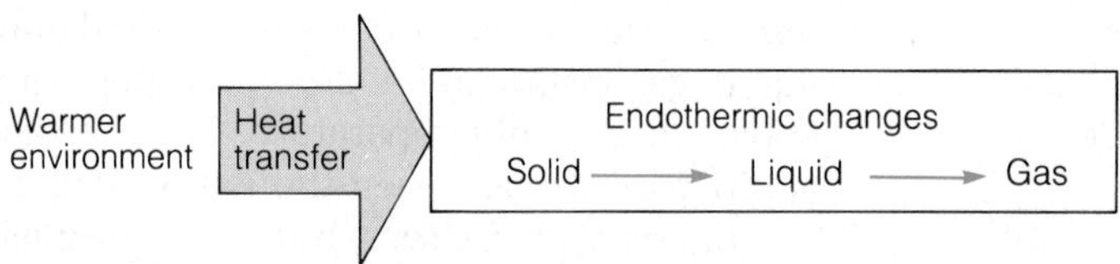

If the surroundings are cooler than the body, heat will flow out of the body, and the process that releases heat will be favored.

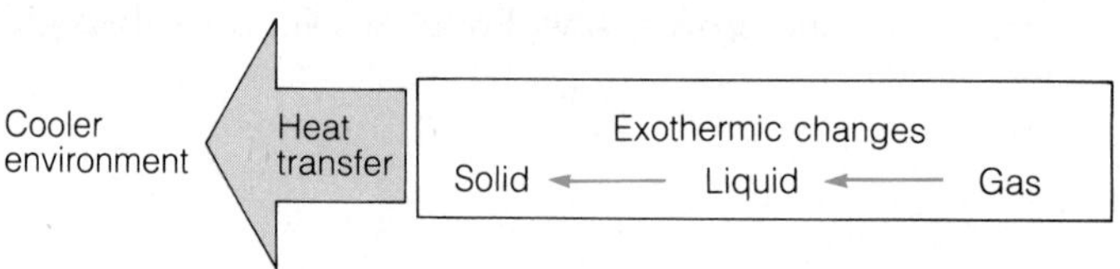

2. *Entropy* There is a natural tendency toward disorder. This factor always favors the solid ⟶ liquid ⟶ gas transformations, regardless of the direction of heat flow. Thus it is not necessary to place solids or liquids in a warm environment for them to evaporate; evaporation will occur even when they are cooling. For example, snow sublimes even on a cold winter day, and moth-repellent crystals sublime in a closet, even though the temperature may be low. **Entropy** is a measure of the degree of disorder or randomness in a substance. *The greater the disorder, the higher the entropy.*

The concept of entropy was introduced by Rudolf Clausius (1850). The order–disorder interpretation of entropy was proposed by Ludwig Boltzmann in 1866. Entropy is defined more rigorously in Chapter 18.

Disordered conditions are more probable than ordered ones because there are many more of the former. To understand this, we must first think about what we mean by *order* and *disorder*. If we walked into a library and saw books on the floor, tables, chairs, and shelves, many of the books open (but not to the same page number), and the books oriented in various directions in space, we would describe the arrangement as disorderly. However, if all the books were on the shelves, in the sequence of their call numbers as given in the catalog file, we would consider the books orderly. Why? Because *order is characterized by repetition.* The books on the shelves occur *repeatedly* in the same orientation (vertical and parallel); their arrangement by number *repeats* the numbers in the catalog file. In fact, the catalog arrangement itself is orderly because its sequence *repeats* the alphabet. The books lying around in various ways are disorderly because *there is very little repetition.* The page numbers to which the books lie open are all different, their orientations in space are random, and the sequence of books in any direction does not match the sequence of their call numbers or any other familiar sequence. It would take a long time to find a specific book, because the disorderly arrangement provides no information that can be used to locate it.

BOX 11.2

STATES OF MATTER, ENTROPY, AND THE RECYCLING OF BEVERAGE CONTAINERS

A century ago, milk, beer, cooking oil, and other liquids were shipped to general stores in barrels. Customers then filled their own reusable jars. Later, the beverage industry added a new approach—it supplied its own glass containers in the form of deposit bottles. Deposit values of two to five cents were assigned to the bottles as an incentive for the customer to return them. The system worked so well at first that a deposit bottle averaged about thirty round trips. Some, of course, were broken, lost, neglected, or used as flower vases and therefore never returned. Gradually, consumers became so indifferent to the deposit that by 1960 the average number of round trips had declined to four.

This decline initiated the shift to no-deposit, no-return containers. These bottles, too, can be recycled, but the consumer must take them to a recycling center, getting little or no money in return. As a result, only a small fraction of the containers find their way back to the manufacturer. The consequent littering in city and country has led to the enactment of laws in several states mandating the reestablishment of deposits on beverage containers.

We may visualize the beverage containers as existing in three different states of aggregation, analogous to the physical states of matter discussed in this chapter. A highly ordered state, with low entropy, exists at the store, where the containers are stacked neatly on shelves. A less-ordered state exists in the home, where the containers may be in the refrigerator, on the shelf or table, or in the wastebasket. The least-ordered state is in the outdoor environment, where they may be almost anywhere. The natural tendency is for the containers to disperse, because there are more places for them to be outdoors than there are in the home or store. The situation can be reversed only by doing work on the system (picking up containers and transporting them back to the store), and in our culture, work is more often done when it is rewarded by money. Hence, deposit bottles do not find their way into the outdoor environment as fast as no-deposit bottles do.

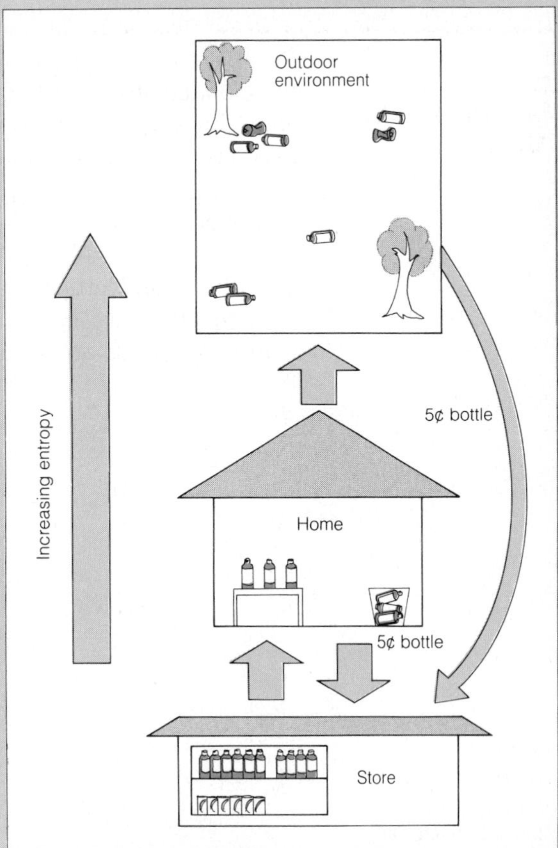

Entropy of beverage containers.

If we started with an orderly college library and permitted the students to use it but did not permit the library staff to reshelve the books, we would soon notice a trend to disorder—and the disorder would increase day by day. This tendency is natural, because there are *many* more ways to replace books randomly than to arrange them in any orderly pattern. When a book is replaced on a shelf, *only one* position will maintain the order. Every other position will increase the disorder of the collection. Thus, disorder is favored or, in other words, *entropy tends to increase.* Furthermore, work must be done (in this case by the library staff) to restore order; that is, to decrease the entropy.

Now, let us turn from library books to solids, liquids, and gases. There are many more ways for atoms or molecules to be partially disordered in a liquid than there are in the highly ordered arrangement of a crystal, and there are still more ways to be disordered in the completely disordered arrangement of a gas. For this reason there is always a tendency toward the most disordered, most highly entropic state.

The tendency toward high entropy is frustrated at low temperatures because states of high entropy, as exist in gases, have high energies. Therefore, the most probable state of a substance at low temperature is a low-energy, low-entropy state, like that of a crystal. As the temperature of a substance increases, the higher-energy, higher-entropy states become more probable. The quantitative expressions of the criteria that determine the direction of spontaneous changes are dealt with by the study of thermodynamics, the subject of Chapter 18.

11.15 SPECIAL TOPIC: ATOMIC RADII AND THE UNIT CELL

Recall from Chapters 6 and 7 that a bond length is the sum of the radii of the two bonded atoms. When the substance is an element, all of the atoms are alike and therefore have the same radius. The bond length is then divided equally between the two atoms, and

$$\text{atomic radius} = \frac{\text{bond length}}{2}$$

In the crystal lattice of an element, the atoms that are considered to be bonded to each other are the nearest neighbors, the atoms that are "touching." The number of bonds an atom has is the same as its coordination number.

FIGURE 11.35 In a simple cubic crystal lattice, the edge of the unit cell equals two atomic radii.

With these principles in mind, let us consider how to calculate the atomic radius from the dimensions of the unit cell for the three types of cubic crystal lattices shown in Figure 11.17.

THE SIMPLE CUBIC (sc) LATTICE

Figure 11.35 shows a plane of the simple cubic lattice and the face of one unit cell. It is clear that the atoms "touch" only along the edge of the unit cell; the atoms that are diagonal to each other do not touch. Therefore the edge of the cell is the length of two atomic radii.

$$\text{edge of sc cell} = 2r, \text{ and}$$

$$r = \frac{\text{edge of sc cell}}{2}$$

THE FACE-CENTERED CUBIC (fcc) LATTICE

Note from Figure 11.17i that the atoms touch each other along the face diagonal of the fcc unit cell. Then (Figure 11.36),

$$\text{face diagonal} = 4r$$

The square of the hypotenuse equals the sum of the squares of the arms.

The face diagonal is the hypotenuse of the right triangle whose arms are the edges of the fcc unit cell (Figure 11.37c), and

$$\text{face diagonal} = \sqrt{\text{edge}^2 + \text{edge}^2} = \sqrt{2\ \text{edge}^2} = \text{edge}\sqrt{2}$$

where each edge is the edge of the fcc cell. Then, from the previous equation,

$$4r = \text{edge}\ \sqrt{2}, \text{ and}$$

$$\mathbf{r = \frac{fcc\ edge\ \sqrt{2}}{4}} \qquad (2)$$

EXAMPLE 11.10 Nickel crystallizes in a face-centered cubic lattice whose unit cell edge is 3.52 Å. Calculate the atomic radius of nickel in nm.

ANSWER

$$r = \frac{3.52\ \text{Å}\sqrt{2}}{4} = 1.24\ \text{Å} \times \frac{1\ \text{nm}}{10\ \text{Å}} = 0.124\ \text{nm}$$

■

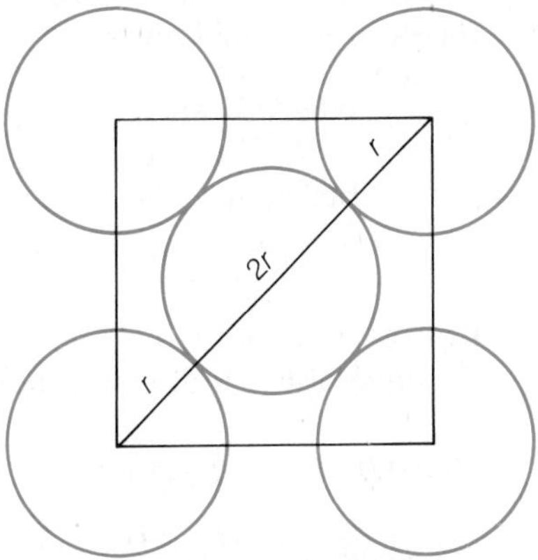

FIGURE 11.36
The diagonal of a face-centered cubic unit cell equals four radii.

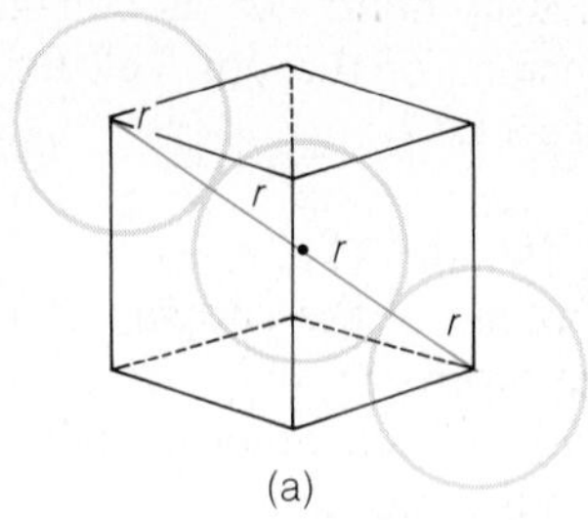

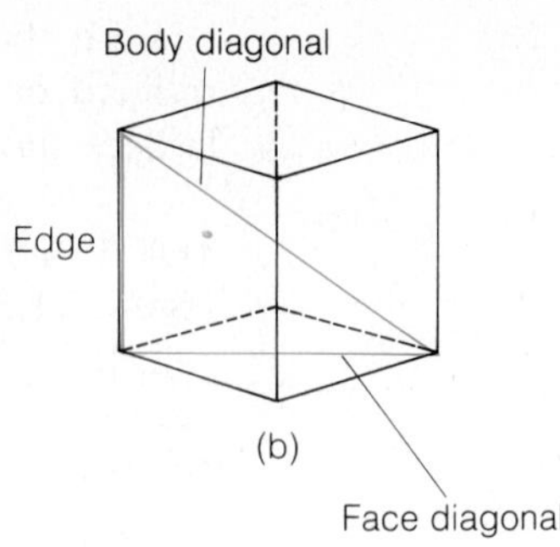

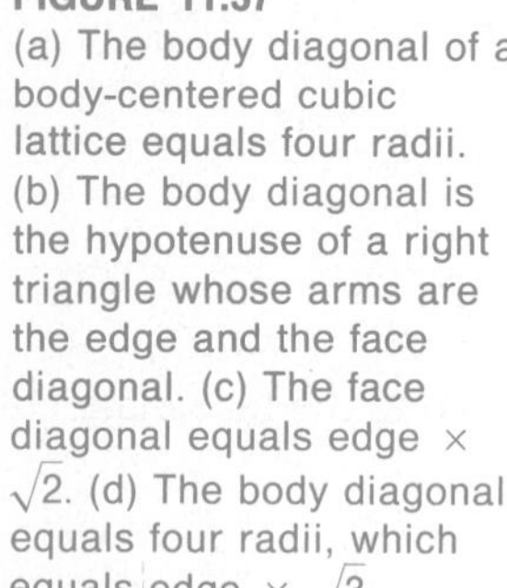

FIGURE 11.37
(a) The body diagonal of a body-centered cubic lattice equals four radii. (b) The body diagonal is the hypotenuse of a right triangle whose arms are the edge and the face diagonal. (c) The face diagonal equals edge × $\sqrt{2}$. (d) The body diagonal equals four radii, which equals edge × $\sqrt{3}$.

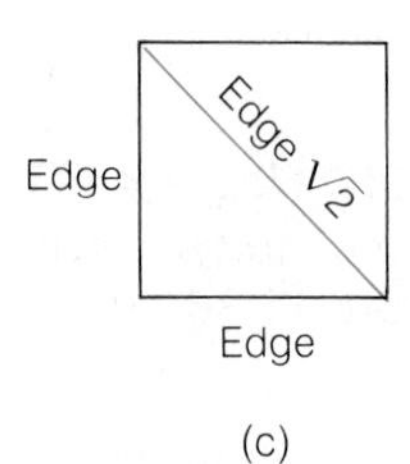

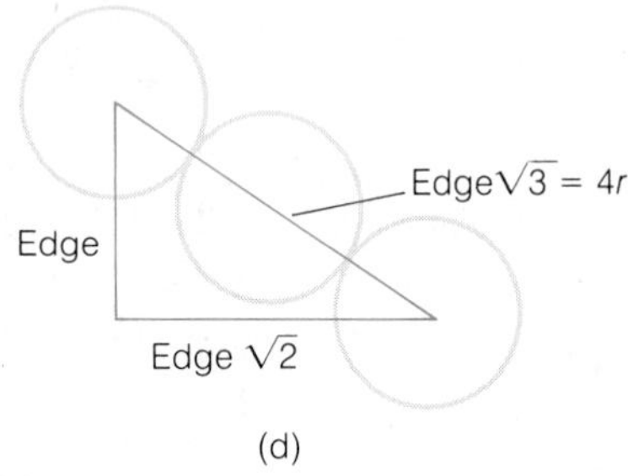

THE BODY-CENTERED CUBIC (bcc) LATTICE

Note from Figure 11.17f that none of the atoms at the corners of the unit cell "touches" another corner atom. Instead, it is the body-centered atom that "touches" all the corner atoms. The straight line that goes from one corner of the cube through the body-centered atom to the opposite corner is the "body diagonal" of the cube. Figure 11.37 shows that the length of this line equals 4 atomic radii:

$$\text{body diagonal} = 4r \tag{3}$$

The length of the body diagonal can be calculated from the right triangle shown in Figure 11.37b. This triangle slices through the center of the cube; one arm is the edge, the other arm is a diagonal along the bottom face of the unit cell, and the body diagonal is the hypotenuse. From equation 2, the face diagonal = edge $\sqrt{2}$. Then, from Figure 11.37c and d,

$$\text{body diagonal} = \sqrt{\text{edge}^2 + (\text{edge}\sqrt{2})^2} = \sqrt{3\ \text{edge}^2} = \text{edge}\sqrt{3} \tag{4}$$

where each edge is the edge of the bcc cell. Combining equations 3 and 4 for the body diagonal gives

$$4r = \text{edge}\sqrt{3}, \text{ and}$$

$$\mathbf{r = \frac{bcc\ edge\sqrt{3}}{4}} \tag{5}$$

EXAMPLE 11.11 Iron, atomic radius 1.24 Å, crystallizes in a body-centered cubic lattice. Calculate the edge of the unit cell in nm.

ANSWER From equation 5,

$$\text{bcc edge} = \frac{4r}{\sqrt{3}} = \frac{4 \times 1.24\ \text{Å}}{\sqrt{3}} = 2.86\ \text{Å} \times \frac{1\ \text{nm}}{10\ \text{Å}} = 0.286\ \text{nm}$$ ■

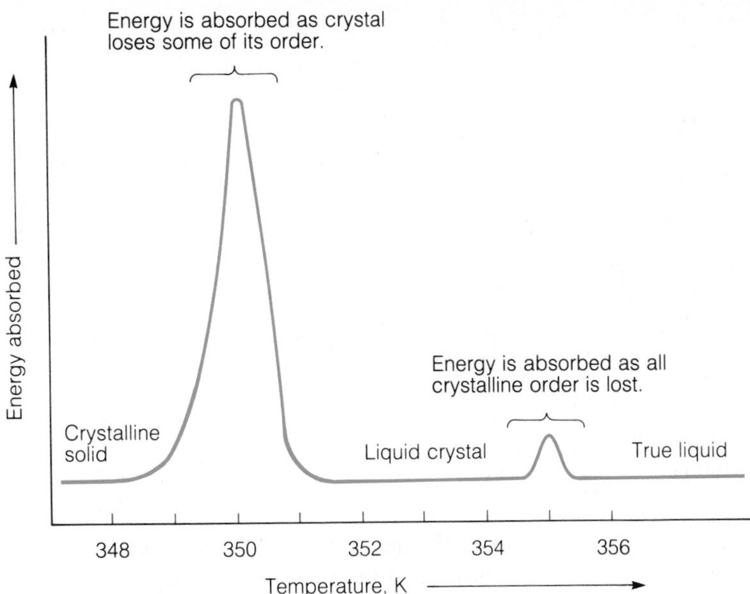
Energy is absorbed as crystal loses some of its order.
Energy absorbed
Energy is absorbed as all crystalline order is lost.
Crystalline solid
Liquid crystal
True liquid
348
350
352
354
356
Temperature, K

FIGURE 11.38
Energy changes during the heating of cholesteryl palmitate.

11.16 SPECIAL TOPIC: LIQUID CRYSTALS

The three commonly cited states of matter (solid, liquid, and gas) or the six states described in Table 11.1 are, to some extent, arbitrary classifications. When water boils and becomes steam, the change occurs at a definite set of conditions of temperature and pressure, and a definite transfer of energy is associated with the change. The same is true when ice melts and becomes liquid water. Therefore, the classification of ice, water, and steam as distinct states of matter is consistent with the fact that there are distinct differences of energy between them. To cite a contrary example, the classification of chicken eggs into "small," "medium," "large," "extra large," and "jumbo" is arbitrary and does not reflect any discontinuities in size. The distinction between liquid and glass is also arbitrary. If a piece of ordinary glass is heated, it gradually softens until, if heated enough, it can be poured from one container to another; it is *then* said to be a liquid. However, the transition is not sharp; it occurs continuously over a broad temperature range. Thus the distinction between liquid and glass is entirely different from the distinction between water and ice, which is a true crystalline solid.

Palmitic acid is an acid obtained from palm oil.

Cholesteryl palmitate is the compound formed when cholesterol and palmitic acid react. When cholesteryl palmitate is heated, it becomes liquid at 355 K, with absorption of heat. However, a careful measurement of energy transfer shows another distinct absorption of energy at 351 K, four kelvins below the melting point (Figure 11.38). What's more, the energy absorbed at 351 K is much greater than the amount absorbed at 355 K. If distinct energy changes are associated with changes of state, there must be another state of matter for cholesteryl palmitate between 351 K and 355 K. Indeed, its properties in this temperature range are different from those of a true crystal or a true liquid. Such intermediate transitions between crystalline solids and true liquids are not uncommon, and the states that exist in these ranges are called, collectively, **liquid crystals**.

Since the molecules of liquid crystals are typically long and rather rigid, they tend to lie parallel to each other. These features are exemplified by the structure of cholesteryl palmitate,

Palmitic acid portion　　　　Cholesterol portion

where the rings in the cholesterol portion give the molecule rigidity, and the palmitic acid portion adds greatly to the length of the molecule.

As we would expect, a liquid crystal is more orderly than a true liquid and less orderly than a solid crystal. (Figure 11.39). The conversions between liquid crystals and other states are thus true phase transitions, similar to those that occur when solids melt and liquids boil. For substances that have a liquid crystal phase, the progression from solid crystal to vapor can be represented as

$$\text{solid crystal} \rightleftharpoons \text{liquid crystal} \rightleftharpoons \text{liquid} \rightleftharpoons \text{gas}$$

Liquid crystals display some of the properties of crystalline solids and some that resemble those of ordinary liquids. They are generally opaque or translucent.

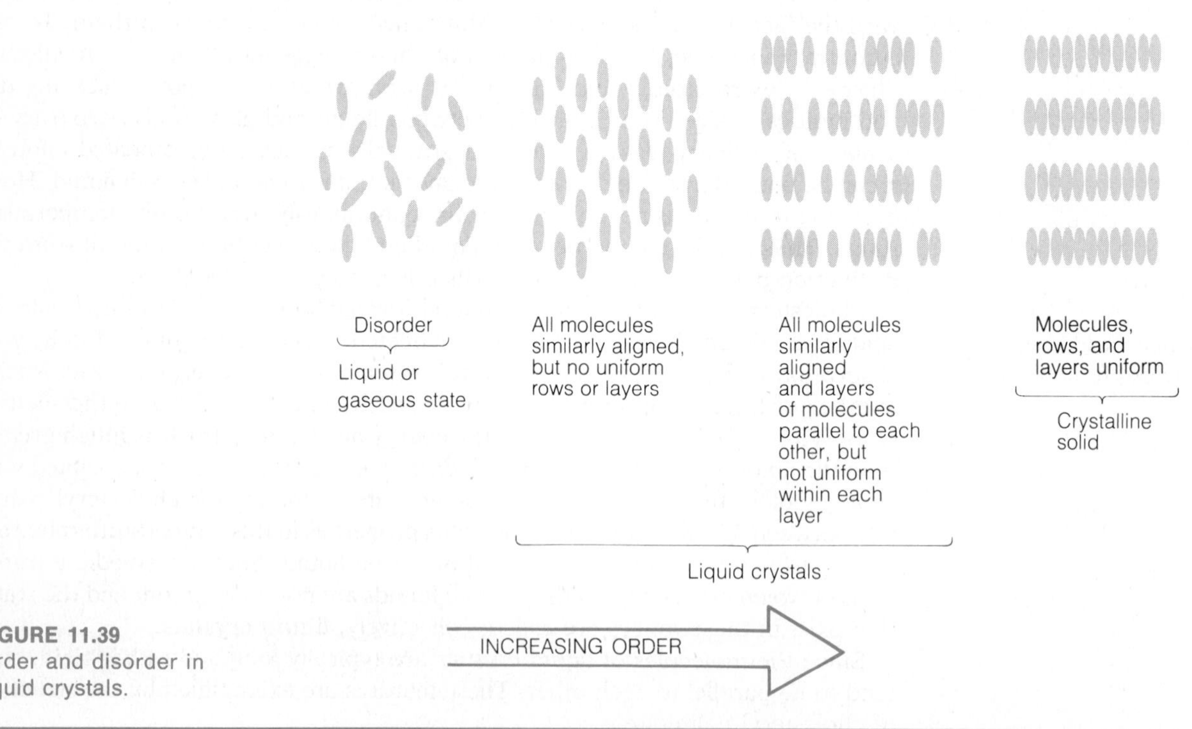

FIGURE 11.39
Order and disorder in liquid crystals.

Some liquid crystals can maintain a definite shape, but even they can be made to flow with only very small inputs of mechanical or electrical energy. One practical application of liquid crystals takes advantage of the fact that light reflected from a very thin layer of liquid—such as an oil slick only a few molecular layers deep floating on water—is colored. Some wavelengths of light are reflected from the upper surface and other wavelengths from the lower one. The particular wavelengths that reinforce each other determine the color of the reflected light. Since the thickness of the layer determines which wavelengths reinforce each other, the color reflected depends on the thickness. In some liquid crystals, the distance between the layers is very sensitive to temperature. This property is utilized in "color-mapping" portions of the human body. When a section of skin is coated with an appropriate liquid crystal, the warm areas over blood vessels and certain organs take on a color different from that of the neighboring cooler areas, and their precise locations are thus identified.

The popular liquid crystal displays on pocket calculators and wristwatches are another application. Here the liquid crystal materials exist as a very thin layer between thin sheets of transparent glass or plastic in which electrical conductors are embedded. These electrodes are arranged in patterns that can represent letters or numerals. When voltage is applied in a particular pattern, the molecular orientation of adjacent liquid crystals is changed—resulting in an opaque readout. (Recall that transitions between liquid crystal and other states involve only very small inputs of energy.) In this way a desired set of letters or numbers becomes visible (Figure 11.40).

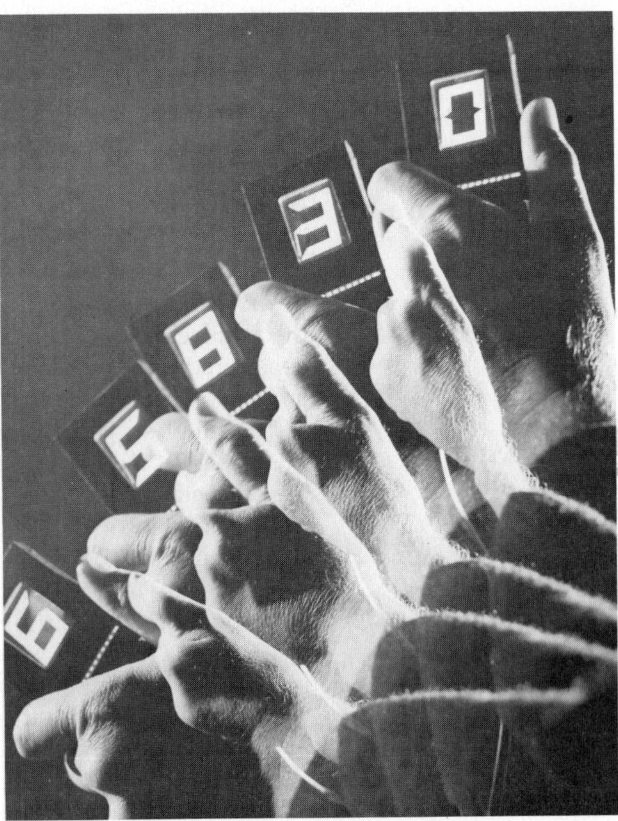

FIGURE 11.40
Composite photograph of a single-digit liquid crystal display. (Courtesy of E. J. Pasierb, RCA David Sarnoff Research Center, Princeton, N.J.)

PROBLEM 11.10 In cholesteryl palmitate, which transition do you think is accompanied by the greater degree of disordering, the first one at 351 K or the second one at 355 K? Explain. □

SUMMARY

Matter can be classified according to the orderliness of the atomic or molecular arrangements and the strengths of the forces that hold the atoms or molecules together. The states of matter derived from such classification include **solid**, **liquid**, **gas**, **liquid crystal**, and **glass**. Transformations in the direction of greater disorder (solid ⟶ liquid ⟶ gas) are **endothermic**, and those in the opposite direction are **exothermic**. The higher the boiling point of a liquid, the greater are its intermolecular forces.

Intermolecular forces may be classified into three types, in the order of increasing strength: **London forces** < **dipole-dipole interactions** < **hydrogen bonding**. London forces result from the mutual attractions of momentarily polarized molecules, and their magnitudes depend on the number of electrons per molecule, which is approximately related to the molecular weight. Dipole-dipole attractions result from the mutual attractions of polar molecules. Hydrogen-bonding is the formation of a weak chemical bond, represented by the dashed line in the formulation :A—H --- A or :A—H --- B, where A and B are highly electronegative atoms (usually F, O, or N) with at least one unshared pair of electrons.

Liquids resemble gases in their ability to flow, but are more like solids in their density and incompressibility. The fact that the properties of liquids are the same in all directions shows that their molecular arrangements cannot be perfectly orderly. The liquid state is best visualized as one in which the molecular arrangements are orderly over ranges of only two or three molecular diameters. The easy displacement from one partially disordered arrangement to another accounts for the fluidity of liquids. When a liquid is cooled to the extent that flow is practically stopped and it becomes rigid but does not crystallize, the resulting solid is classified as a glass.

Crystalline solids are characterized by the orderly arrangements of their atoms or molecules. If the center of each atom in a crystal is represented by a point, the arrangement of such points in space is the **crystal lattice**. The lattice may be considered a collection of unit shapes, all identical, similarly oriented in space, and in contact with each other. Each such shape is called a **unit cell**.

The simplest types of unit cells of crystals of the elements are cubic. An atom that is part of a cubic unit cell may be located at a corner ($\frac{1}{8}$ of the atom is inside the cell), an edge ($\frac{1}{4}$ is inside), a face ($\frac{1}{2}$ is inside), or in the body center (the entire atom is inside).

In the **simple cubic** system, atoms are located only at the corners ⟶ 1 atom per unit cell. The **body-centered cubic** system contains atoms at the corners plus one atom at the body center ⟶ 2 atoms per unit cell. The **face-centered cubic** system contains atoms at the corners and faces ⟶ 4 atoms per unit cell. The number of nearest neighbors (coordination number) is 6 for the simple cubic, 8 for the body-centered cubic, and 12 (the theoretical maximum for an element) for the face-centered cubic system.

Calculations related to the dimensions of the unit cell are based on the fact that the unit cell represents the entire crystal. Therefore the density of the crystal equals the mass of the unit cell divided by the volume of the unit cell.

The crystal structure of an ionic compound must satisfy the requirement of electrical neutrality, accommodate ions that may differ in size, and be arranged so that unlike charges are closer to each other and like charges are farther apart in order to achieve stability.

Real crystals are generally imperfect; some atoms or groups of atoms are displaced or missing from where they would be in a perfect lattice. Such imperfections greatly weaken a crystal.

The **heat of fusion** is the amount of heat required to melt a unit quantity of a substance. The **heat of vaporization** is the heat needed to vaporize a unit quantity of a substance. The **heat of sublimation** is the heat required to sublime a unit quantity of a substance. All of these conversions occur at a fixed temperature and pressure, and these processes are all endothermic.

A liquid that evaporates into a confined, closed space reaches a condition of equilibrium with its vapor in which the rate of vaporization equals the rate of condensation. The vapor is then said to be **saturated**. The **vapor pressure** of a liquid is the pressure exerted by the saturated vapor.

The **boiling point** of a liquid is the temperature at which its vapor pressure equals the pressure of the surrounding atmosphere. Therefore, the higher the surrounding pressure, the higher the boiling point. Boiling is characterized by the formation of bubbles within the body of the liquid. The **melting point** of a solid or the **freezing point** of a liquid is the temperature at which liquid and solid are in equilibrium at standard atmospheric pressure. The effect of a change in atmospheric pressure is usually insignificant.

Any homogeneous portion of matter that is separated by a boundary from other matter in a sample is called a **phase**. A graph that shows the pressure–temperature conditions of equilibrium between different phases is called a **phase diagram**. The coexistence of three phases is represented by a point at which three lines meet; the coexistence of two phases is represented by a line. The regions between lines represent the existence of a single phase. When a liquid in equilibrium with its vapor is heated in a sealed tube, a **critical temperature** is reached at which the two phases become identical. The pressure at the critical temperature is called the **critical pressure**. The stronger the intermolecular forces, the higher the critical temperature.

The tendency for a substance to reach the temperature of its environment favors endothermic changes (solid $\longrightarrow$ liquid $\longrightarrow$ gas) when the surroundings are warm, and exothermic changes (gas $\longrightarrow$ liquid $\longrightarrow$ solid) when the surroundings are cool. However, the natural tendency for increased disorder always favors the solid $\longrightarrow$ liquid $\longrightarrow$ gas transformations, regardless of the direction of heat flow. **Entropy** is a measure of the degree of disorder in a substance. Disordered states are more probable than ordered ones.

11.17 ADDITIONAL PROBLEMS

STATES OF MATTER

11.11 A simpler classification of the states of matter than the one in Table 11.1 characterizes *gases* as substances that do not cohere and have no definite shape, *liquids* as substances that do cohere and have no definite shape, and *solids* as substances that cohere and maintain a definite shape. How would the six states of matter in Table 11.1 be classified under this simpler system? Which of the six states would be difficult to classify? Explain the difficulties.

11.12 A zinc metal rod and a glass rod are heated in the absence of air to a temperature somewhat above 419°C. The zinc rod remains rigid, but liquid zinc begins to drip from it. The glass rod sags, but does not drip. Account for these phenomena.

11.13 Referring to Figure 11.1, give the correct names for these changes in state: (a) Crystals of *para*-dichlorobenzene, used as a moth repellent, gradually become vapor without passing through the liquid phase. (b) As you enter a warm room from the outdoors on a cold winter day, your eyeglasses become fogged with a film of moisture. (c) On the same winter day, a pan of water is left outdoors. Some of the water turns to vapor, the rest to ice. (d) Molten lava flows down the sides of an erupting volcano. As the lava cools, it flows more slowly; finally it stops flowing altogether and becomes hard, but it does not crystallize.

11.14 The normal boiling point of trichlorofluoromethane, CCl_3F, is 24°C, and its freezing point is −111°C. Complete these sentences by supplying the proper terms. (a) At standard temperature and pressure, CCl_3F is a ________. (b) In an arctic winter at −40°C and 1 atm pressure, CCl_3F is a ________. If it is cooled further to −196°C and the molecules arrange themselves in an orderly lattice, the CCl_3F will ________ and become a ________. However, if the molecules become fixed in a disorderly pattern, the CCl_3F will ________ and become a ________. (c) If crystalline CCl_3F is held at a temperature of −120°C while a stream of helium gas is blown over it, the crystals will gradually disappear by the process of ________. If liquid CCl_3F is boiled at atmospheric pressure, it is converted to a ________ at a temperature of ________.

INTERMOLECULAR FORCES

11.15 (a) Explain in your own words why boiling points provide more information about intermolecular forces than melting points do. (b) What choice would you make between melting points and sublimation points (temperatures at which solids vaporize) for studying intermolecular forces? Justify your answer.

11.16 These dipole moments are observed for three hydrogen halides: HF, 1.9 D; HI, 0.38 D; HCl, 1.03 D. (a) Predict a value for the dipole moment of HBr. (b) Account for the observed order of dipole moments and justify your prediction.

11.17 Select the substance in each pair that has the higher boiling point. For each choice, state whether the reason is stronger dipole interactions, stronger London forces, or stronger hydrogen bonding.

(a) Br_2 or ICl (They have about the same molecular weight.)

(b) neon (20 g/mol) or krypton (84 g/mol)

(c) $CH_3—CH_2OH$ (ethanol) or $H_2C—CH_2$ bridged by O (ethylene oxide)

(d) $CH_3—CH_2—CH_2—CH_2—CH_2—CH_2—CH_3$ (n-heptane)

or $CH_3—C(CH_3)_2—CH(CH_3)—CH_3$ (2,2,3-trimethylbutane, "triptane")

(e) $CH_2—CH_2—CH_2—CH_2—CH_2—N(H)$ (piperidine) or $H_2C—CH_2—CH_2—CH_2—N(CH_3)$ (N-methylpyrrolidine)

11.18 Hydrogen bonding can occur between molecules of the following pairs of substances. For each pair, draw structural formulas to show which atoms are hydrogen-bonded to each other. (a) HF and HBr; (b) CH_3OH and HF (two possibilities); (c) H_2O and $CH_3—O—CH_3$; (d) $(CH_3)_2NH$ and H_2O (two possibilities); (e) $CH_3—O—CH_3$ and NH_3.

11.19 Account for these facts: (a) Although ethyl alcohol, C_2H_5OH (b.p. 80°C), has a higher molecular weight than water (b.p. 100°C), it has a lower boiling point. (b) Salts of the HCl_2^- anion are known. (c) Mixing 50 mL each of water and ethyl alcohol gives a solution whose volume is less than 100 mL.

11.20 Acetic acid, $CH_3—C(=O)—O—H$, has structural features that enable a pair of molecules to form *two* hydrogen bonds with each other. (a) Identify these features and draw a likely structural formula of a doubled molecule (dimer) of acetic acid. (b) Propyl alcohol, $CH_3—CH_2—CH_2OH$, has the same molecular weight (60) as acetic acid. Which compound do you think has the higher boiling point? (c) At the same temperature, which vapor do you think is denser? (Assume that some acetic acid dimers survive in the vapor, and remember that vapor density is directly proportional to molecular weight.)

LIQUIDS AND GLASSES

11.21 In which of these properties do liquids resemble gases; solids; neither gases nor solids? (a) density, (b) incompressibility, (c) ability to flow, (d) appearing the same in all directions, (e) Brownian motion, (f) X-ray diffraction patterns that show short-range order but long-range disorder.

11.22 In which of these properties do glasses resemble liquids; crystalline solids? (a) rigidity, (b) appearing the same in all directions, (c) disordered molecular arrangements.

CRYSTAL LATTICE AND THE UNIT CELL

11.23 Outline a two-dimensional unit cell in each of these patterns:

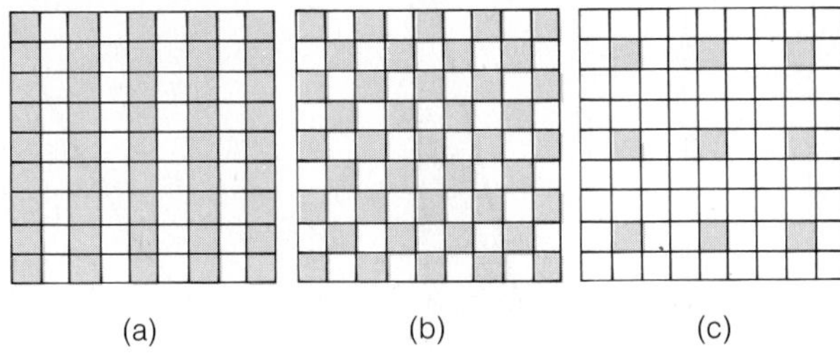

11.24 What fraction of a circle is located within a two-dimensional square unit cell in each of these locations? (a) corner, (b) side, (c) center.

11.25 Two identical empty swimming pools are filled with uniform spheres of ice packed as closely as possible. The spheres in the first pool are the size of grains of sand; those in the second pool are the size of oranges. The ice in both pools melts. In which pool, if either, will the water level be higher? (Neglect any differences in filling space at the planes next to the walls and the bottom.)

11.26 Refer to Figure 11.19, the ionic crystal structure of cesium chloride. (a) What is the coordination number of the cesium; of the chlorine? (b) How many Cs^+ ions are assigned to a unit cell? How many Cl^- ions?

11.27 Gold (197.0 g/mol) crystallizes in a face-centered cubic lattice whose unit cell edge is 4.10 Å. Predict the density of gold.

11.28 Tungsten (183.9 g/mol) has a density of 19.3 g/cm^3 and crystallizes in a cubic lattice whose unit cell edge is 3.16 Å. Which type of cubic lattice is it?

11.29 A Group-4 element with a density of 11.35 g/cm^3 crystallizes in a face-centered cubic lattice whose unit cell edge is 4.95 Å. Calculate its atomic weight. What is the element?

11.30 Chromium (atomic weight 52.0 and density 7.2 g/cm^3) crystallizes in a body-centered cubic lattice whose unit cell edge is 2.88 Å. Calculate the Avogadro number.

CHANGES OF STATE

11.31 The specific heat of silver is 0.237 J/(g °C). Its melting point is 961°C. Its heat of fusion is 11 J/g. How much heat is needed to change 5.00 g of silver from solid at 25°C to liquid at 961°C?

11.32 The heat of fusion of thallium is 21 J/g, and its heat of vaporization is 795 J/g. The melting and boiling points are 304°C and 1457°C. The specific heat of liquid thallium is 0.13 J/(g °C). How much heat is needed to change 1.00 g of solid thallium at 304°C to vapor at 1457°C and 1 atm?

11.33 The specific heat of liquid mercury is 0.137 J/(g °C). Calculate the amount of heat required to warm 100 g of mercury from 25°C to its boiling point and then to vaporize it. Refer to Table 11.3.

11.34 Aluminum is produced as a liquid. How much heat must be removed to cool 1 metric ton (10^6 g) of liquid aluminum at 1000°C to solid aluminum at 25°C? Assume that both Al(c) and Al(ℓ) have specific heat 0.92 J/(g °C). Refer to Table 11.3.

VAPOR PRESSURE

11.35 Figure 11.41 is a plot of vapor pressure versus temperature for dichlorodifluoromethane, CCl_2F_2. The heat of vaporization of CCl_2F_2 is 165 kJ/kg, and the specific heat of the liquid is about 1.0 kJ/(kg °C). (a) What is the normal boiling point of CCl_2F_2? (b) A steel cylinder containing 25 kg of CCl_2F_2 in the form of liquid and vapor is set outdoors on a mild (25°C) day. What is the pressure of the vapor in the cylinder (approximately)? (c) The valve is opened, and CCl_2F_2 vapor blows out of the cylinder in a rapid flow. Soon, however, the flow becomes quite slow, and the outside of the cylinder becomes coated with ice frost. When the valve is closed and the cylinder is reweighed, it is found that 20 kg of CCl_2F_2 is still inside. Why is the flow very rapid at first? Why does it slow down long before the cylinder is empty? Why does the outside become icy? (d) Which of the following procedures would be effective in emptying the cylinder rapidly? Explain your answer. (1) Turn the cylinder upside down and open the valve. (2) Cool the cylinder with dry ice and open the

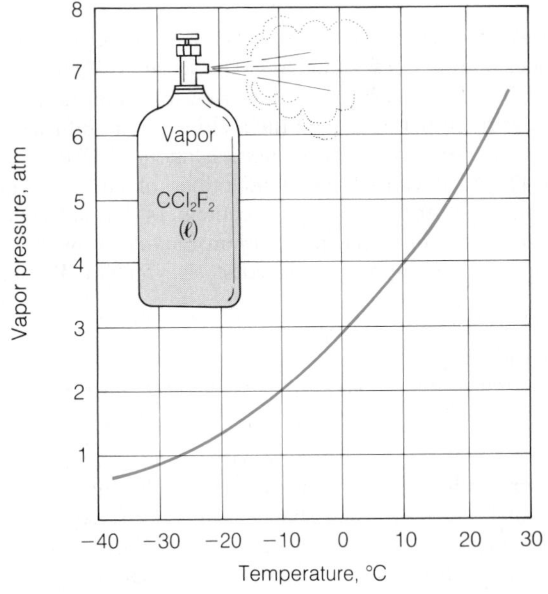

FIGURE 11.41
Vapor pressure of dichlorodifluoromethane, CCl_2F_2.

valve. (3) Knock off the top of the cylinder, valve and all, with a sledge hammer.

11.36 The vapor pressure of mercury is 1.85×10^{-4} torr at 0°C and 0.2729 torr at 100°C. The Hg molecule is monatomic. Calculate the density (in g/L) of Hg vapor at each of these temperatures.

11.37 A student comes to you with this problem: "I looked up the vapor pressure of water in a table; it is 26.7 torr at 300 K and 92,826 torr at 600 K. That means that when the absolute temperature doubles, the vapor pressure is multiplied by 3477. But I thought the pressure was proportional to the absolute temperature, $P = nRT/V$. Why doesn't the pressure just double?" How would you help the student?

COOLING

11.38 Meteorologists have found water droplets at −35°C in high cirrus clouds. Since the freezing point of water is 0°C, how is such a temperature possible?

11.39 Draw a complete time (x axis)–temperature (y axis) cooling curve for water at a pressure of 1 atm, showing: the cooling of steam; steam's condensation at 100°C; the cooling, undercooling, and freezing of water; and the cooling of ice. The specific heats of steam and ice are both about 2 J/(g °C).

PHASE DIAGRAM

11.40 Figure 11.42 is the phase diagram for sulfur, S_8. There are two crystalline forms, rhombic and monoclinic. Name the phases in equilibrium at each of the triple points.

11.41 Sketch a phase diagram for ethanol, C_2H_5OH. Label points, curves, and regions. Refer to Tables 11.3 and 11.4 for data. Assume that the triple point is slightly lower in temperature than the melting point and that the vapor pressure at the triple point is about 10^{-5} torr.

11.42 At steel mills large quantities of liquid oxygen are transferred to and kept in containers that are vented to the atmosphere. Liquid carbon dioxide, however, is handled and shipped in closed steel cylinders. Explain.

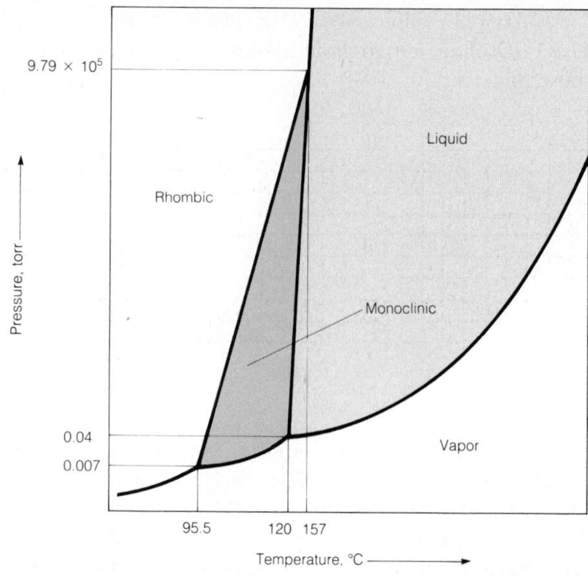

FIGURE 11.42
Phase diagram for sulfur (not to scale).

ENTROPY

11.43 Does the entropy of the system increase or decrease when (a) magnesium is extracted from the sea and stored in a warehouse in the form of bars of the metal; (b) gold bars are removed from a vault, cast into coins, and distributed in commerce; (c) mold grows on cheese; (d) honey crystallizes; (e) wedding guests throw rice at the bride and groom?

ATOMIC RADII AND THE UNIT CELL

11.44 Calculate the atomic radius of gold from the data given in Problem 11.27.

11.45 Calculate the atomic radius of tungsten from the data given in Problem 11.28.

11.46 Calculate the atomic radius of chromium from the data given in Problem 11.30.

SELF-TEST

11.47 (10 points) Identify the states of matter of these substances: (a) A dark purple substance occupies the lower tenth of a long test tube. The substance consists of small particles, each shaped like a flat scale or plate with the same interfacial angles. (b) Above this substance is a transparent purple material that occupies the remainder of the tube. It is light in color near the particles on the bottom, is increasingly lighter toward the upper parts of the tube, and is practically invisible near the top. (c) The bottom of a second test tube is filled with a blue substance whose upper boundary is sharp and level. When the tube is tilted, the substance shifts so that its upper boundary is still level and the bottom part of the tube is still filled. (d) A rigid vertical bar gradually thickens toward the bottom after many years. (e) X rays reflected from a substance at a certain angle show sharp patterns of reinforcement and interference.

11.48 (10 points) The normal boiling point of ammonia, NH_3, is −33°C, and its freezing point is −78°C. Fill in the blanks. (a) At STP (0°C, 1 atm pressure), NH_3 is a ________. (b) If the temperature drops to −40°C, the ammonia will ________ and become a ________.

(c) If the temperature drops further to $-80°C$ and the molecules arrange themselves in an orderly pattern, the ammonia will __________ and become a __________. However, if the molecules become fixed in a random pattern, the ammonia will become a __________. (d) If crystals of ammonia are left on the planet Mars at a temperature of $-100°C$, they will gradually disappear by the process of __________ and become a __________.

11.49 (10 points) In which combinations of substances will significant hydrogen bonding occur between the molecules? (a) NH_3 and H_2O, (b) CH_4 and H_2O, (c) HBr and HCl, (d) HF and HCl, (e) H_2S and H_2O, (f) HBr and HI, (g) NH_3 and CH_4.

11.50 (10 points) (a) List the following forces in the order of increasing ability to bind atoms or molecules to each other: (1) hydrogen bonding, (2) gravitation, (3) London forces, (4) dipole-dipole attraction, (5) attraction between NH_4^+ and NO_3^- ions. (b) Which of these forces is/are electrical?

11.51 (14 points) Tantalum (181 g/mole, density 16.7 g/cm^3) crystallizes in a cubic lattice whose unit cell edge is 3.32 Å. How many atoms are in one unit cell? What is the type of cubic crystal lattice?

11.52 (10 points) True or false? (a) The rate of vaporization of a liquid is independent of its surface area. (b) The vapor pressure of a liquid is doubled when the absolute temperature is doubled. (c) At the normal boiling point of a liquid, its vapor pressure is one atm. (d) Superheating can be avoided by heating a very pure liquid in a clean vessel. (e) At the triple point, the solid and the liquid have the same vapor pressure. (f) In order for a solid to sublime, the pressure of its vapor must be lower than the vapor pressure at the triple point. (g) A curve in a phase diagram corresponds to pressure–temperature combinations at which only one phase can exist. (h) Above the critical temperature, only one fluid (nonsolid) phase can exist. (i) States of higher entropy are more probable because they usually have lower energy. (j) States of higher entropy are more probable because there are more of them.

11.53 (12 points) Phenanthrene, $C_{14}H_{10}$, melts at 101°C and boils at 340°C. The liquid is less dense than the solid. Select the correct word or phrase. (a) At the normal boiling point of water, phenanthrene is a (*solid, liquid, gas*). (b) At the boiling point of water on the top of a mountain, phenanthrene is a (*solid, liquid, gas*). (c) The triple point of phenanthrene is (*the same as, slightly lower than, slightly higher than, much higher than*) its melting point. (d) Phenanthrene can be sublimed only when (*the temperature is above 340°C, the pressure is above 1 atm, the temperature is below the triple point*).

11.54 (12 points) Butane, C_4H_{10}, has its freezing point at $-138°C$ and its normal boiling point at 0°C. The triple point is slightly lower in temperature than the freezing point. The vapor pressure at the triple point is 3×10^{-5} atm. The critical temperature is 152°C; the critical pressure is 38 atm. (a) On the axes provided, sketch the phase diagram of butane. Label each region to show the phase present when the system is represented by a point in that region. (b) Butane at 1 atm and 140°C is compressed to 40 atm. Are two phases present at any time during this process? (c) Butane at 1 atm and 160°C is compressed to 40 atm. Are two phases present at any time during this process?

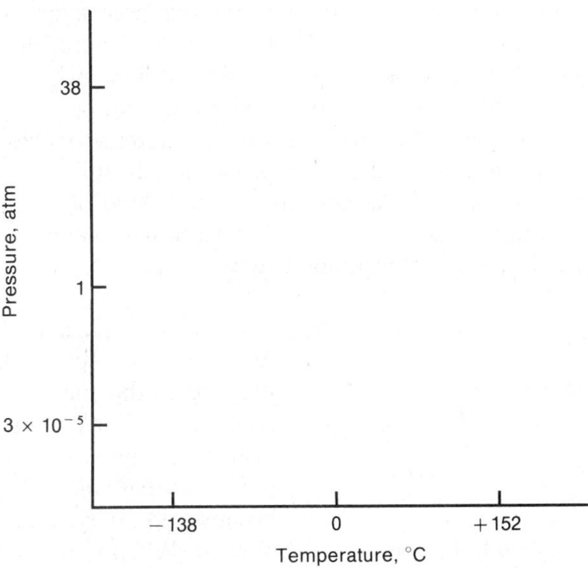

11.55 (12 points) The heat of fusion of mercury is 11.3 J/g and its heat of vaporization is 276 J/g. Its melting point is $-39°C$ and its normal boiling point is 357°C. The specific heat of the liquid is 0.138 J/(g °C). How much heat is required to change 20.0 g of solid mercury at its freezing point to mercury vapor at its normal boiling point?

12
SOLUTIONS

In Chapter 11 we discussed some important ideas about matter and its transformations. However, we dealt only with pure substances, and pure, or nearly pure, substances are rare in this world; they seldom occur naturally and they can be very expensive to prepare. Furthermore, mixtures often have more desirable properties. In this chapter we will discuss an important kind of mixture: solutions. You will learn how to keep track of their composition, why some combinations of substances form solutions and others do not, and how the properties of solutions are related to the properties of their components. You will see why water is such a marvelous solvent (but not for grease), why cars need antifreeze in winter, and why seawater is not good to drink. You will also learn about colloids, which are not quite solutions, and how dirty water and seawater can be made fit to drink.

12.1 WHAT IS A SOLUTION?

A solution is a special kind of mixture. Mixtures can be classified according to the size of the particles of one component that are dispersed through another component. For example, if you mix some sand and water, the sand grains will disperse in the water and you will be able to see the grains with the naked eye; this mixture is called a **suspension**. After a while the sand will settle to the bottom because of gravity. Imagine doing this several times with progressively finer grains of sand. When you reached the point where the grains are very small, no bigger than dust particles, you would find that they do not sink to the bottom, no matter how long you wait. Then you would have a **colloidal dispersion** (section 12.17). Although the individual grains would be invisible, the mixture would appear cloudy in a strong beam of light.

It would feel like talcum powder, not sand.

Now imagine that you stir some sugar into a glass of water. The grains disappear, and you have a clear liquid that looks just like pure water. Although solid sugar is no longer present, the liquid has a distinctive property of sugar, the sweet taste. You could correctly conclude that sugar molecules have dispersed among the water molecules and that sugar and water thus form a true solution.

A solution is a mixture of two or more substances dispersed as molecules, atoms, or ions

rather than as larger aggregates. A solution may be a gas, a liquid, or a solid. Table 12.1 lists a few common examples of each. You can supply many other examples, such as tea, soda, windshield-washing liquid—almost everything we encounter is made of or contains solutions.

TABLE 12.1
EXAMPLES OF SOLUTIONS

SOLUTION	COMPONENTS
GASES	
Air	N_2, O_2, other gases
Water gas (fuel)	H_2, CO
LIQUIDS	
Seawater	H_2O, NaCl, $MgCl_2$, other salts
Gasoline	C_7H_{16}, C_8H_{18}, etc.
SOLIDS	
Brass	Cu, Zn
Soda-lime glass	Ca_2SiO_4, $Ca_2Si_2O_6$, Na_4SiO_4, $Na_4Si_2O_6$, etc.

The **solvent** is the component of a solution that is visualized as dissolving another component, called a **solute** (section 1.6). Usually, the solvent is the component that is present in larger quantity. (The exceptions are liquid solutions containing more than 50% of a solid or gaseous solute; the liquid is still considered the solvent. An example is a liquid solution prepared from 60% $NH_4I(c)$, the solute, and 40% $H_2O(\ell)$, the solvent.) A **dilute solution** is one that contains only a small quantity of solute (or solutes) relative to the quantity of solvent. A **concentrated solution** contains a large proportion of solute. These terms are no more precise than "large" and "small." Many a student has been surprised to learn that the "dilute" hydrochloric acid being used in a laboratory exercise is 20 or 30 times more concentrated than the acid the experiment calls for.

12.2 PERCENTAGE AND MOLARITY

When discussing solutions, we must be able to specify their compositions, that is, the relative amounts of the components. Composition is expressed in a number of ways. All methods of specifying composition use the ratio of the quantity of one component to the quantity of the other or to the total quantity of solution. "Quantity" is expressed in one of three ways, as mass, number of moles, or volume. The ratio usually takes the form

$$\frac{\text{quantity of solute}}{\text{quantity of solution}} \quad \text{or} \quad \frac{\text{quantity of solute}}{\text{quantity of solvent}}$$

We will consider only the measures of composition that are in most common use. Two more of them will be defined in sections 12.9 and 12.12.

Let us assume that we have a solution of two components, A (solvent) and B (solute). We will use the following notation to express the various "quantities":

w_A, w_B = mass ("weight") in grams of A and B in the solution

n_A, n_B = number of moles of A and B

V = total volume of the solution in liters

MASS PERCENTAGE

The ratio *mass of solute/mass of solution* is usually expressed in the form of **mass percentage**, also called "weight percentage" or "percent by weight." Mass percentage was discussed in section 3.7 in connection with the composition of compounds; it is used even more for solutions. The mass percentage of the solute B is

$$\frac{\textbf{mass of solute (B)}}{\textbf{mass of solution}} \times 100\% = \frac{w_B}{w_A + w_B} \times 100\% \qquad (1)$$

Likewise, the mass percentage of the solvent A is

$$\frac{\text{mass of solvent (A)}}{\text{mass of solution}} \times 100\% = \frac{w_A}{w_A + w_B} \times 100\% \qquad (2)$$

If there is more than one solute (B, C, . . .), the denominator is $w_A + w_B + w_C + \cdots$. It is usually understood that *percentage* means "mass percentage" unless otherwise specified.

EXAMPLE 12.1 A solution contains 15 grams of sodium chloride, NaCl, in 100 grams of water. What is the percentage by mass of NaCl in the solution?

Do not confuse the percentage composition of a solution with the percentage composition (in terms of elements) of any of its components. NaCl is 39.34% Na and 60.66% Cl by mass. This information has no relationship to the percentage of NaCl in a solution.

ANSWER The total mass of the solution is

$$15 \text{ g NaCl} + 100 \text{ g } H_2O = 115 \text{ g solution}$$

Then the percentage of NaCl is

$$\frac{15 \text{ g NaCl}}{115 \text{ g solution}} \times 100\% = 13\% \text{ NaCl}$$

Note that the solution is *not* 15% NaCl; that would mean that the 15 g of NaCl are dissolved in 85 g water to form 100 g solution. ■

EXAMPLE 12.2 A sample of vinegar is 5.0% acetic acid, $HC_2H_3O_2$, and 95.0% water. How much of that vinegar would you need in order to have 80 grams of acetic acid?

ANSWER A solution that is 5.0% acetic acid contains 5.0 g acetic acid in every 100 g solution. The percentage provides the conversion factor for going from grams of acetic acid to grams of solution (vinegar):

$$80 \cancel{\text{ g acetic acid}} \times \frac{100 \text{ g vinegar}}{5.0 \cancel{\text{ g acetic acid}}} = 1600 \text{ g vinegar} = 1.6 \text{ kg vinegar}$$ ■

PROBLEM 12.1 A solution is prepared by dissolving 15.0 g magnesium sulfate, $MgSO_4$, in 75.0 g H_2O. (a) What is the percentage of $MgSO_4$ in this solution? (b) If you need 1.00 kg of $MgSO_4$, what mass of solution should you take? □

MOLARITY

Recall from section 3.9 that the **molarity** or **concentration** of a solute B (also called the molarity of the solution) is

$$[\mathbf{B}] = \frac{\textbf{number of moles of B}}{\textbf{volume of solution in liters}} = \frac{n_B}{V} \qquad (3)$$

Molarity is denoted by enclosing the formula of the solute in square brackets; for example, $[H_2SO_4]$. When [B] = 0.1 mol/L, we say the solution is 0.1 molar, or 0.1 M.

EXAMPLE 12.3 A solution is prepared by dissolving 25.00 g biphenyl, $C_{12}H_{10}$, in enough benzene, C_6H_6, to bring the volume of the solution up to 500.0 mL. What is the molarity of the solution?

ANSWER The molar mass of biphenyl is

$$(12 \times 12.01) + (10 \times 1.008) = 154.20 \text{ g/mol}$$

Then the number of moles dissolved is

$$25.00 \not{g} \times \frac{1 \text{ mol}}{154.20 \not{g}} = 0.1621 \text{ mol}$$

The molarity is the number of moles of solute divided by the volume of the solution in liters:

$$\frac{0.1621 \text{ mol}}{500.0 \not{\text{mL}}} \times \frac{1000 \not{\text{mL}}}{1 \text{ L}} = 0.3242 \text{ mol/L or } 0.3242 \text{ M}$$

(Don't forget to convert milliliters to liters.) We do not need to know how much benzene is used or even what the solvent *is*. Only the final volume of the solution is needed. ■

PROBLEM 12.2 15.0 g $MgSO_4$ is dissolved in water and water is added to make the total volume 100.0 mL. What is the molarity of the solution? □

The millimolarity of a solution is the number of millimoles per liter. The micromolarity is the number of micromoles per liter (1 μmol = 10^{-6} mol). These are common measures of composition for very dilute solutions.

In the laboratory, volumes are measured in milliliters more often than in liters. It is therefore convenient to express molarity in such units that the denominator is in milliliters rather than liters. The way to do this is to express the amount of substance in **millimoles** and the mass in milligrams. A millimole (mmol) is 10^{-3} mol; a milligram is 10^{-3} g; a milliliter is 10^{-3} L. The molar mass is the same number in mg/mmol as in g/mol. For example, the solute has molar mass 50 g/mol and the solution is 0.1 M:

$$\frac{50 \not{g}}{1 \not{\text{mol}}} \times \frac{10^3 \text{ mg}}{1 \not{g}} \times \frac{1 \not{\text{mol}}}{10^3 \text{ mmol}} = \frac{50 \text{ mg}}{1 \text{ mmol}}$$

The molarity is the same number in mmol/mL as in mol/L:

$$\frac{0.1 \not{\text{mol}}}{1 \not{L}} \times \frac{10^3 \text{ mmol}}{1 \not{\text{mol}}} \times \frac{1 \not{L}}{10^3 \text{ mL}} = \frac{0.1 \text{ mmol}}{1 \text{ mL}}$$

To avoid confusion, convert *everything* or *nothing* to milliquantities.

Percentage by volume is seldom used in scientific work (except for gases; section 10.11), but is commercially important. The idea is to compare the volume of one pure component (before mixing) with the volume of the solution:

$$\% \text{ by volume of B} = \frac{\text{volume of pure B}}{\text{volume of solution}} \times 100\%$$

In the United States, the "proof number" of an alcoholic beverage is twice the percentage by volume of ethanol, measured at 60°F; for example, 100 proof means 50% by volume and 86 proof means 43%. Other countries have different systems for specifying the strength of liquor. The word *proof* is used in the sense of "test" or "trial"; "proof spirit" (100 proof) is a standard of strength with which other drinks are compared. The original test was that proof spirit is sufficiently concentrated to burn and ignite gunpowder.

It is noteworthy that the volume of a solution is not, in general, equal to the sum of the volumes of its components (Problem 12.23). Thus, if the percentages by volume of both ethanol and water in a solution are calculated, they will add up to more than 100% because the volume of the solution is, in this case, less than the total volume of the two pure components.

In any line of study or work, it is necessary to be alert to usages peculiar to that specialty. In the health professions, for example, mass of solute per 100 mL of solution is often called "percentage." If the solution is aqueous and dilute, this "weight/volume percentage" is close to the mass/mass percentage because 100 mL of solution has a mass of about 100 g.

HYDRATES Suppose that an inexperienced chemist is asked to prepare 1.00 L of a 0.100 M solution of sodium sulfate, Na_2SO_4. He or she calculates the molar mass, 142.0 g/mol, and weighs out 0.100 mol, or 14.20 g. However, when the solution is used, it turns out to be too dilute, only 0.044 M. The lesson the chemist would learn from this is that many salts exist as **hydrates**, with water molecules (*water of hydration*) as essential parts of the crystal structure. The most common form of sodium sulfate is a hydrate, sodium sulfate-water (1/10), $Na_2SO_4(H_2O)_{10}$—also written $Na_2SO_4 \cdot 10H_2O$. The formula $Na_2SO_4(H_2O)_{10}$ means that for every two Na^+ ions in the crystal, there are one $SO_4{}^{2-}$ ion and ten H_2O molecules. What our budding chemist thought was Na_2SO_4 was actually $Na_2SO_4(H_2O)_{10}$, and therefore more than half of the mass was water.

Older name: sodium sulfate decahydrate. Still-older names: Glauber's salt and *sal mirabilis* ("miraculous salt," from its effectiveness as a laxative).

EXAMPLE 12.4 A solution is prepared by dissolving 15.0 g of $Na_2SO_4(H_2O)_{10}$ in water to make 500.0 mL of solution. Calculate the molarity of the solution.

ANSWER In calculating the number of moles of a hydrated salt, the molar mass to be used must include the water of hydration, for that is part of the weighed crystal:

$$\underset{(Na)}{2 \times 22.99} + \underset{(S)}{32.06} + \underset{(O)}{4 \times 16.00} + \underset{(H)}{20 \times 1.008} + \underset{(O)}{10 \times 16.00} = 322.2 \text{ g/mol}$$

Then the number of moles of hydrated salt is

$$15.0 \cancel{g} \times \frac{1 \text{ mol}}{322.2 \cancel{g}} = 0.0466 \text{ mol } Na_2SO_4(H_2O)_{10}$$

The molarity is

$$\frac{0.0466 \text{ mol}}{500.0 \cancel{mL}} \times \frac{1000 \cancel{mL}}{1 \text{ L}} = 0.0932 \text{ mol/L}$$

■

EXAMPLE 12.5 A solution is prepared by dissolving 10.0 g $Na_2SO_4(H_2O)_{10}$ in 100.0 g H_2O. What is the percentage by mass of the anhydrous salt (salt without water), Na_2SO_4, in the solution?

ANSWER First, calculate what mass of the solute is Na_2SO_4. The molar masses are 142.0 g/mol for Na_2SO_4 and 322.2 g/mol for $Na_2SO_4(H_2O)_{10}$. In 10.0 g $Na_2SO_4(H_2O)_{10}$ there is

$$10.0 \cancel{g\ Na_2SO_4(H_2O)_{10}} \times \frac{142.0 \text{ g } Na_2SO_4}{322.2 \cancel{g\ Na_2SO_4(H_2O)_{10}}} = 4.41 \text{ g } Na_2SO_4$$

Now we must find the total mass of the solution:

$$100.0 \text{ g } H_2O + 10.0 \text{ g hydrate} = 110.0 \text{ g}$$

The percentage of Na_2SO_4 is thus

$$\frac{4.41 \text{ g } Na_2SO_4}{110.0 \text{ g solution}} \times 100\% = 4.01\% \ Na_2SO_4$$ ■

Note that the *molarity* is the same whether the hydrate or the anhydrous salt is considered the solute: 1 mole of hydrate contains 1 mole of anhydrous salt. However, the *percentage* is different because the mass of hydrate is greater than the mass of anhydrous salt.

PROBLEM 12.3 A solution is prepared by dissolving 17.83 g of cobalt(II) chloride-water (1/6), $CoCl_2(H_2O)_6$, in water to make 1000.0 mL of solution. The total mass of this solution is 1007 g. (a) What is the molarity of the solution? (b) What is the percentage of anhydrous $CoCl_2$ by mass? □

Molarity and percentage by mass are the two most common ways of expressing the composition of a solution. Molarity has *moles* of solute, but percentage has *mass* of solute, in the numerator. Molarity has *volume* of solution in the denominator; percentage has *mass* of solution in the denominator. To calculate one of these from the other, you need to know the density (g/mL) of the solution as well as the molecular weight of the solute. To make life difficult, the density of a solution depends on its composition and temperature; it must be measured for every solution or looked up in a reference book. Common handbooks contain density tables for solutions in water at 20°C or 25°C. Densities of nonaqueous solutions, as well as aqueous solutions at other temperatures, are harder to find. However, the density of a *dilute* solution is close to the density of the solvent, 1.0 g/mL for water. This two-significant-figure approximation can be used for solutions more dilute than 0.1 M, and sometimes even up to 0.5 M. An approximate answer can thus be obtained when better information is not available.

EXAMPLE 12.6 A concentrated solution of HCl contains 37.8% HCl by mass and has a density of 1.19 g/mL at 20°C. What is its molarity?

ANSWER This problem is an ordinary unit conversion, but since it is a little complicated, we will break it down into steps. The given density tells us that 1 mL of solution has a mass of 1.19 g. We want mol HCl/L solution. It is therefore necessary to convert (1) 1.19 g of solution to grams of HCl using the 37.8% HCl,

$$1.19 \text{ g } \cancel{\text{solution}} \times \frac{37.8 \text{ g HCl}}{100 \text{ g } \cancel{\text{solution}}} = 0.450 \text{ g HCl}$$

(2) grams of HCl to moles of HCl using the molar mass, 36.5 g/mol,

$$0.450 \text{ g } \cancel{\text{HCl}} \times \frac{1 \text{ mol HCl}}{36.5 \text{ g } \cancel{\text{HCl}}} = 0.0123 \text{ mol HCl} = n$$

and then (3) 1 mL to L,

$$1 \ \cancel{\text{mL solution}} \times \frac{1 \text{ L solution}}{1000 \ \cancel{\text{mL solution}}} = 10^{-3} \text{ L solution} = V$$

$$\text{M} = \frac{n}{V} = \frac{0.0123 \text{ mol HCl}}{10^{-3} \text{ L}} = 12.3 \ \frac{\text{mol HCl}}{\text{L}}$$ ■

EXAMPLE 12.7 The density of a 1.000 M solution of sodium nitrate, $NaNO_3$, is 1.053 g/mL at 20°C. What is the mass percentage of $NaNO_3$ in this solution?

ANSWER The percentage of $NaNO_3$ is

$$\frac{\text{mass of NaNO}_3}{\text{mass of solution}} \times 100\%$$

The molar mass of $NaNO_3$, 84.99 g/mol, is needed for converting moles to mass. The mass of solute per liter of solution is

$$\frac{1.000\ \cancel{\text{mol}}}{1\text{ L}} \times \frac{84.99\text{ g NaNO}_3}{1\ \cancel{\text{mol}}} = 84.99\text{ g NaNO}_3\text{/L}$$

The next calculation is the mass of 1 L of solution. From the given density,

$$\frac{1.053\text{ g solution}}{1\ \cancel{\text{mL}}} \times \frac{10^3\ \cancel{\text{mL}}}{1\text{ L}} = 1.053 \times 10^3\text{ g solution/L}$$

Then the percentage is

$$\frac{84.99\text{ g NaNO}_3}{1\ \cancel{\text{L}}} \times \frac{1\ \cancel{\text{L}}}{1.053 \times 10^3\text{ g solution}} \times 100\% = 8.071\%\text{ NaNO}_3$$ ■

Do not confuse g *solute*/mL solution, g *solvent*/mL solution, and g *solution*/mL solution. The last of these is the density: total mass over total volume.

PROBLEM 12.4 A 10.0% solution of lactic acid ($HC_3H_5O_3$, 90.1 g/mol) has the density 1.020 g/mL. What is its molarity? □

DILUTION PROBLEMS

A common laboratory procedure is to convert a solution of known molarity to a more dilute solution by adding solvent. Adding solvent increases the volume of the solution *but does not change the quantity of solute.* Thus, the number of moles of solute in the solution is the same before and after dilution:

$$(\text{moles of solute})_{\text{concentrated solution}} = (\text{moles of solute})_{\text{dilute solution}} \quad (4)$$

where the subscripts refer to the original (concentrated) solution and the final (dilute) solution. Now

$$\text{moles of solute} = \text{volume of solution in liters} \times \text{molarity in mol/L} = VM$$

and equation 4 becomes

$$\begin{array}{c}\text{volume of}\\ \text{concentrated solution } (V_c)\end{array} \times \begin{array}{c}\text{molarity of}\\ \text{concentrated solution } (\text{M}_c)\end{array} = \begin{array}{c}\text{volume of}\\ \text{dilute solution } (V_d)\end{array} \times \begin{array}{c}\text{molarity of}\\ \text{dilute solution } (\text{M}_d)\end{array}$$

$$\boldsymbol{V_c\text{M}_c = V_d\text{M}_d} \quad (5)$$

The volume can be in any units, as long as the units are the same on the two sides of the equation. The same goes for the concentration: we could use molecules/mL, g/m^3, lb/gal, or any other units expressing mass per unit volume of solution or number of molecules per unit volume of solution:

$$\left(\frac{\text{quantity}}{\text{volume}}\right)_c \times (\text{volume})_c = \left(\frac{\text{quantity}}{\text{volume}}\right)_d \times (\text{volume})_d$$

EXAMPLE 12.8 You are given a 6.00 M aqueous (water) solution of HCl and asked to prepare 500 mL of a 1.00 M solution. What volume of the 6.00 M HCl should you use?

ANSWER You have V_c = ?, M_c = 6.00 mol/L, V_d = 500 mL, and M_d = 1.00 mol/L.

$$V_c = V_d \times \frac{M_d}{M_c}$$

$$= 500\ \text{mL} \times \frac{1.00\ \cancel{\text{mol/L}}}{6.00\ \cancel{\text{mol/L}}}$$

$$= 83.3\ \text{mL}$$

Whenever you calculate something, ask yourself, "Is the answer reasonable, or is it obviously absurd?" (Appendix B.3)

You must dilute 83.3 mL of the concentrated (6.00 M) acid to a *total* volume of 500 mL. (This does *not* mean adding 500 mL of water.) If you made a mistake and wrote 6.00/1.00, the answer would be 3000 mL. This is obviously wrong because the volume of the concentrated solution must be less than the volume (500 mL) of the dilute solution. ■

EXAMPLE 12.9 A 2.00 M aqueous solution of $AgNO_3$ is available. You dilute 20.0 mL of this solution with water to the mark in a 1000-mL volumetric flask (section 3.9). What is the molarity of the solution thus prepared?

ANSWER

$$V_c = 20.0\ \text{mL},\ M_c = 2.00\ \text{mol/L},\ V_d = 1000\ \text{mL},\ M_d = ?$$

An error would be obvious because the answer must be less than 2.00 mol/L. You cannot make a solution more concentrated by diluting it.

$$M_d = M_c \times \frac{V_c}{V_d}$$

$$= 2.00\ \text{mol/L} \times \frac{20.0\ \cancel{\text{mL}}}{1000\ \cancel{\text{mL}}}$$

$$= 0.0400\ \text{mol/L}$$

■

PROBLEM 12.5 A 3.00 M solution of NH_3 is on the shelf. (a) How much of this solution should you use to make 1000 mL of a 0.600 M solution? (b) If 10.0 mL of the 3.00 M solution is diluted to a total volume of 250 mL, what is the molarity of the resulting solution? □

12.3 SOLUBILITY

When a solute is added to a liquid solvent at a given temperature, all of the solute may eventually dissolve, but it may not. When dissolving stops in spite of the presence of some undissolved solute, the solution is said to be **saturated**. *Saturated* usually means "filled to capacity," as when we say a sponge is *saturated* with water. To test a sponge for saturation, you see whether it can soak up a puddle of water. Likewise, to test whether a solution is saturated, you add solute. If some of the added solute remains undissolved, then you know the solution is saturated. The **solubility** of a solute is a number that tells how much of it can be dissolved. It is specified by the composition of its saturated solution, which may be expressed in mass percentage, molarity, or other measures of composition.

Why does a solution become saturated? The question is really the same one asked in section 11.10: why does a liquid come to equilibrium with its "saturated" vapor? As in all cases of equilibrium, it is because two opposing processes occur at the same rate. Consider a gas dissolving in a liquid, with the entire gas-liquid system enclosed. The gas molecules, moving about in all directions, frequently strike the surface of the liquid; some merely bounce off, but others enter the liquid. Conversely, a molecule already in the liquid may happen to reach the surface with enough kinetic energy to escape and become part of the gas again. At a fixed temperature (and thus fixed average kinetic energy), the rate (section 11.10) at which gas molecules enter the liquid depends on the number of collisions with the liquid surface, and thus on the pressure of the gas (section 12.4). The rate at which gas molecules leave the liquid depends on the number already dissolved. As more molecules dissolve, the rate of escape of gas from the solution increases. When the rate of escape equals the rate at which molecules are dissolving, the composition of the solution remains constant:

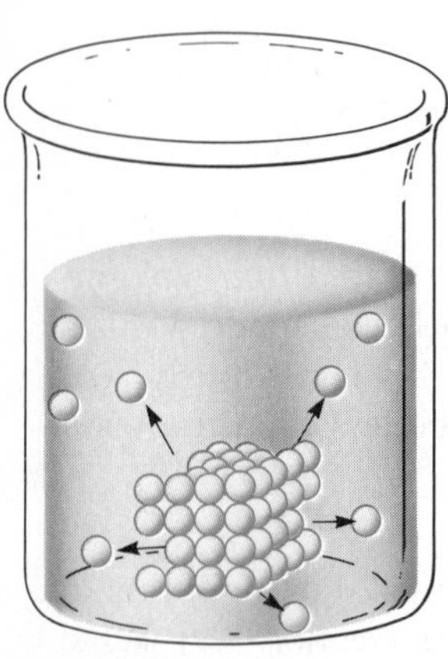

FIGURE 12.1
A crystal being dissolved. The size of the solute molecules is greatly exaggerated.

$$\frac{\text{number of molecules escaping}}{\text{per unit time}} = \frac{\text{number of molecules dissolving}}{\text{per unit time}}$$

The amount of dissolved gas no longer changes; equilibrium has been reached. The processes of capture and escape are still taking place, but their rates are equal, and there is no further net change in the pressure of the gas or the composition of the solution. The solution is saturated.

A solid dissolves in a liquid in much the same way as a gas dissolves in a liquid, but we do not picture the solute molecules as colliding with the solvent molecules. Rather, they are *attracted by* solvent molecules. As a result, the solute molecules become surrounded by (and to some extent, attached to) the solvent molecules, and drift away from the surface of the solid (Figure 12.1). After some of the solid has gone into solution, the remaining solid can recapture solute molecules from the solution. The rate of this redeposition increases as the solution becomes more concentrated. If enough solute is present, the solution reaches a composition at which there is a molecule being deposited from the solution for every molecule that goes into solution. When this equilibrium is attained, the solution is saturated.

Imagine that we double the surface area of the solid by cutting it into pieces. The rate of dissolving is then also doubled, as is the rate of deposition. The two rates will remain equal, and the solution will still be saturated, because the solubility of a solid is independent of its surface area.

An *extremely* small particle has above-normal solubility (section 12.5).

Note the distinction between *solubility* and *rate of dissolving*. A finely divided solid, with a large surface area, dissolves much more rapidly than a large lump—the solution becomes saturated sooner—but this does not mean that the total amount that can be dissolved (the solubility) is increased if a solid is cut up. Similarly, stirring accelerates the dissolving process, but once the solution is saturated, further stirring will not cause another milligram to dissolve.

Solubility is commonly expressed in several ways besides mass percentage and molarity:

1. grams of solute per 100 grams of solvent
2. grams of solute per 100 mL of solvent
3. grams of solute per 100 mL of solution

Number 1 is often confused with mass percentage; they are *not* the same. Mass percentage is grams of solute per 100 grams of *solution*. Number 2 is nearly the same as number 1 when the solvent is water, since the density of liquid water is close to 1 g/mL. Number 3 is the "weight/volume percentage" mentioned in section 12.2.

EXAMPLE 12.10 The solubility of potassium chloride, KCl, is 34.7 g KCl per 100 mL H_2O at 20°C. The density of H_2O at 20°C is 0.9982 g/mL. Calculate the mass percentage of KCl in the saturated solution.

ANSWER The mass of 100 mL H_2O is

$$100\ \text{mL} \times 0.9982\ \text{g/mL} = 99.8\ \text{g}$$

The percentage of KCl is

$$\frac{34.7\ \text{g KCl}}{99.8\ \text{g H}_2\text{O} + 34.7\ \text{g KCl}} \times 100\% = 25.8\%$$

The same answer (to 3 significant figures) is obtained if the density of water is assumed to be 1.00 g/mL instead of 0.9982 g/mL. ■

EXAMPLE 12.11 The solubility of copper(II) sulfate-water (1/5), $CuSO_4(H_2O)_5$, is 31.6 g per 100 g H_2O at 0°C. Calculate the mass percentage of $CuSO_4$ in the saturated solution.

Older name: copper(II) sulfate pentahydrate. Still-older name: blue vitriol.

ANSWER The percentage is the mass of $CuSO_4$ (without the 5 H_2O) per 100 g solution. The first step is to calculate how much $CuSO_4$ is present in 31.6 g of the hydrate. The molar masses are 159.6 g/mol for $CuSO_4$ and 249.7 for $CuSO_4(H_2O)_5$.

$$31.6\ \text{g CuSO}_4\text{(H}_2\text{O)}_5 \times \frac{159.6\ \text{g CuSO}_4}{249.7\ \text{g CuSO}_4\text{(H}_2\text{O)}_5} = 20.2\ \text{g CuSO}_4$$

The total mass of the solution equals 100 g H_2O plus 31.6 g $CuSO_4(H_2O)_5$, or 131.6 g, so

$$\frac{20.2\ \text{g CuSO}_4}{131.6\ \text{g solution}} \times 100\% = 15.3\%\ \text{CuSO}_4$$ ■

PROBLEM 12.6 A saturated solution of lithium metaborate, $LiBO_2$, at 80°C contains 11.83 g $LiBO_2$ per 100 mL H_2O. The density of water at 80°C is 0.9718 g/mL. What is the solubility of $LiBO_2$ as mass percentage? □

When solubility is very small, one substance is commonly said to be "insoluble" in the other. Barium sulfate ($BaSO_4$), for instance, is said to be "insoluble" in water (section 9.2). However, a solubility is never exactly zero; for example, 0.2 mg $BaSO_4$ dissolves in 100 g H_2O at 25°C. Therefore, *insoluble* should be interpreted as "very slightly soluble—not soluble enough to be important." When the solubility is small but enough to be concerned about, the solute is said to be *slightly soluble*. "Soluble" usually means that it is possible to make a solution with enough solute in it to be useful, troublesome, or at least not overlooked.

When two liquids can dissolve in one another in all proportions, they are said to be completely **miscible**. Examples of completely miscible liquids include water and acetic acid, water and glycerol, and benzene and toluene. In such cases there is no saturated solution. Other pairs of liquids, however, are only partially miscible; that is, each dissolves in the other to some extent to give two saturated solutions. An example of appreciable partial miscibility is provided by water and diethyl ether, $(C_2H_5)_2O$. When these two liquids are shaken together, two saturated layers are obtained, with both water and ether in each layer (Figure 12.2). The composition of each layer is independent of the relative masses of the layers, just as the composition of any saturated solution is independent of the relative masses of the solution and undissolved solute.

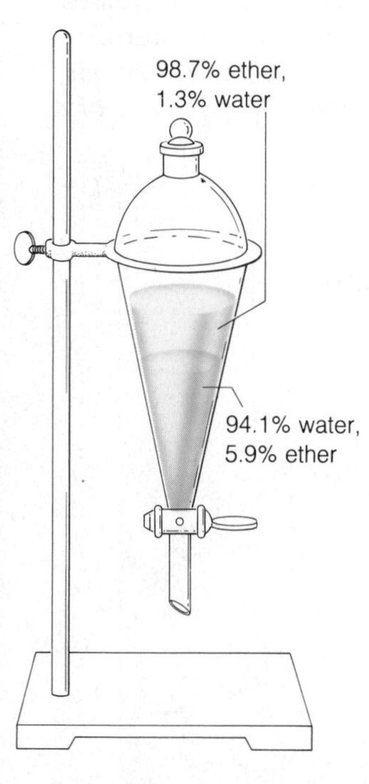

FIGURE 12.2
Two solutions are obtained by shaking water and diethyl ether together at 25°C. The glass apparatus is a separatory funnel.

Although we devote this chapter entirely to liquid solutions, solid solutions also are very common and important. Alloys, minerals, and the "doped" semiconductors used in electronics (section 22.4) are examples of solid solutions. Solids are usually less soluble in each other than they are in liquids because of the difficulty of accommodating "foreign" atoms, molecules, or ions in regular crystal structures. However, the concepts of solubility apply to solid solutions as well as to liquid solutions; it is just that the numbers and usually the time scales are different. Whereas equilibrium may be established rapidly with liquids (if they are stirred), it can take millions or billions of years with solids.

12.4 SOLUBILITY AND PRESSURE

A carbonated ("sparkling") beverage contains CO_2 dissolved in water. It is bottled (or canned) under a pressure of 3 to 4 atm. When the container is opened, the pressure of CO_2 in the bottle drops to about 1 atm and the beverage fizzes; that is, bubbles of CO_2 form. Why? The solubility of CO_2 in water must have decreased as a result of the decrease in pressure. The rule is that *the solubility of a gas in a liquid increases as its pressure increases, and decreases as its pressure decreases.* The kinetic molecular theory (section 10.12) can help to explain this. When the pressure of a gas in contact with a liquid is increased, the number of collisions with the surface of the liquid is increased (Figure 12.3). When the number of collisions per unit time increases, the rate of the capture of gas molecules by the liquid also increases. Molecules in solution sometimes reach the surface with enough energy to escape. The greater the concentration, the more molecules escape per unit time. Thus, as the pressure increases, the quantity of dissolved gas increases; the rate of escape therefore increases until it equals the new rate of capture.

After William Henry (1774–1836).

For gases that are only moderately soluble, the effect of pressure on solubility is given by a simple approximate relationship, **Henry's law**: *The molarity of a*

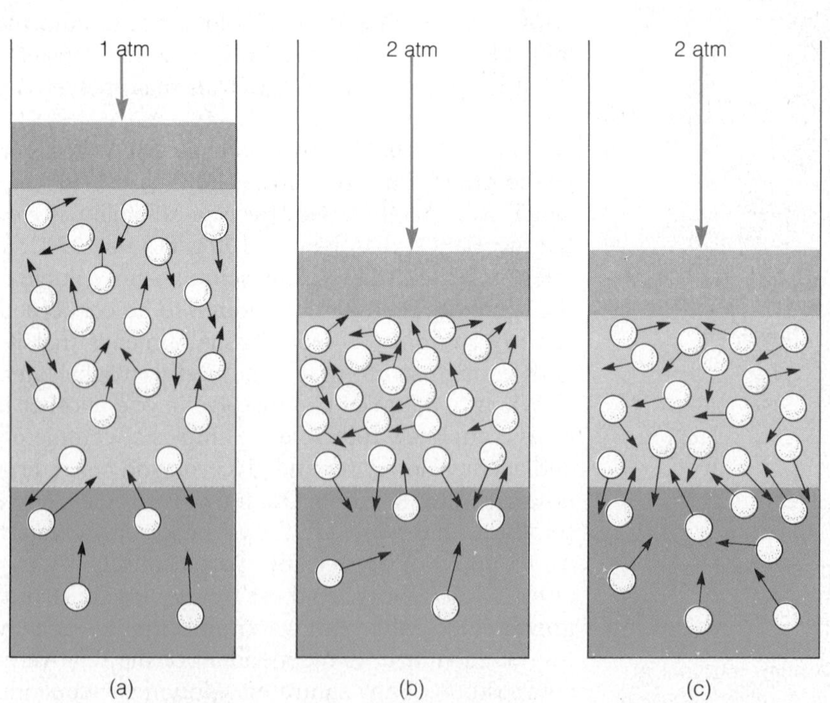

FIGURE 12.3
A simple illustration of Henry's law. (a) Gaseous solute B is in equilibrium with the solution. Equal numbers of B molecules (2 in this example) enter and leave the solution per unit time. (b) The pressure of B has been doubled. Now 4 molecules enter the solution per unit time, and only 2 leave. (c) More B molecules are now dissolved. The number escaping (4) equals the number entering. Equilibrium has been restored.

TABLE 12.2
SOLUBILITIES OF GASES IN WATER[†]

GAS	AT 0°C	AT 25°C	AT 50°C
N_2	0.00294	0.00175	0.00122
O_2	0.00694	0.00393	0.00266
H_2	0.000192	0.000154	0.000129
CO_2	0.335	0.145	0.076
SO_2	22.83	9.41	

[†] In g gas/100 g H_2O, total pressure 1 atm (including water-vapor pressure).

gaseous solute B in a saturated solution is approximately proportional to the pressure of the gas:

$$[\mathbf{B}] \approx k_\mathbf{B} P_\mathbf{B} \tag{6}$$

[B] is the concentration (mol/L) of the solute B in the solution. k_B is a constant; it depends on the temperature as well as on what the solute and solvent are. P_B is the pressure of B in the gas phase. If B is in a mixture of gases, its solubility is determined by its partial pressure, p_B.

Tension formerly meant "pressure"; this old usage persists only for blood gases.

Henry's law can be turned around to read $P_B \approx [B]/k_B$ or, for a gas in a mixture, $p_B \approx [B]/k_B$. This version says that the partial pressure of B in equilibrium with a solution is approximately proportional to the molarity [B]. The concentrations of gases, especially O_2 and CO_2, in blood are commonly expressed by giving the partial pressure of the gas in equilibrium with blood. These pressures are referred to as *gas tensions*.

Table 12.2 gives the solubilities of some common gases in water.

EXAMPLE 12.12 From the data in Table 12.2, calculate the Henry's law constant (k_{O_2}) for O_2 in water at 25°C. Assume that 100 mL of solution contains 100 g of water.

ANSWER The pressure to be used in Henry's law is the *partial* pressure of O_2. In the table, the *total* pressure is 1 atm. Part of this pressure is the vapor pressure of water (Appendix E); the remainder is the pressure of O_2. Using the formula $P_{\text{total}} = p_1 + p_2 + \cdots$ (Dalton's law of partial pressure; section 10.11), we have

$$P_{\text{total}} = p_{O_2} + p_{H_2O}$$

$$p_{O_2} = P_{\text{total}} - p_{H_2O}$$
$$= 760 \text{ torr} - 24 \text{ torr} = 736 \text{ torr}$$

We assume that the vapor pressure of this dilute solution is the same as the vapor pressure of pure water. To calculate the molarity of O_2, convert grams of O_2 to moles:

$$0.00393 \not{g}\ O_2 \times \frac{1 \text{ mol}}{32.0 \not{g}} = 1.23 \times 10^{-4} \text{ mol } O_2$$

This quantity of O_2 is dissolved in 100 g H_2O. We are assuming that the volume of solution containing 100 g H_2O is 100 mL. Then

$$[O_2] = \frac{1.23 \times 10^{-4} \text{ mol } O_2}{100 \cancel{\text{mL}} \text{ solution}} \times 1000 \frac{\cancel{\text{mL}}}{\text{L}}$$
$$= 1.23 \times 10^{-3} \text{ mol } O_2/\text{L solution}$$

Finally,

$$k_{O_2} = \frac{[O_2]}{p_{O_2}} = \frac{1.23 \times 10^{-3} \text{ mol } O_2/\text{L solution}}{736 \text{ torr}}$$

$$= 1.67 \times 10^{-6} \text{ mol/(L torr)}$$ ■

PROBLEM 12.7 The Henry's law constant for Cl_2 in H_2O at 25°C is 0.062 mol/(L atm). What mass of Cl_2 can be dissolved in 100 g H_2O when the total pressure is 800 torr? Make reasonable approximations. □

"Hyperbaric" is from Greek *hyper*, "above," and *baros*, "weight"; therefore, high pressure.

The pressure dependence of solubility has some interesting medical applications. In a *hyperbaric chamber*, a patient breathes air under a pressure of (usually) 3 to 6 atm. The concentration of O_2 in the blood plasma is thus three to six times normal. This increase is beneficial in treating gangrene, tetanus, and other conditions caused by anaerobic bacteria (those that cannot survive in oxygen). It also helps infants with defective hearts. Deep-sea divers and underwater construction workers must work under a high total pressure to keep water out of their helmets or working space, and therefore they breathe air at high pressure. Nitrogen dissolves in the blood, and even more in fatty tissues, where it is more soluble than in water. If the pressure is released suddenly, the solubility decreases and bubbles of nitrogen appear in the blood and other tissues, with painful, sometimes fatal results ("decompression sickness" or "the bends"). Decompression must be carried out slowly to allow the nitrogen time to escape via the lungs. (Oxygen is less of a problem because it is consumed by normal metabolism.) Professional divers use "breathing air"—a mixture of helium, oxygen, and sometimes nitrogen—for high-pressure work because He is much less soluble than N_2 and effuses faster in the lungs (Graham's law; section 10.13).

Gas solubility is sensitive to pressure because of the large decrease in volume when a gas dissolves in a liquid. Indeed, it is a general rule that an increase in pressure favors a change to smaller volume (section 13.6). In contrast, *the solubilities of solids and liquids are affected very little by changes of pressure*, because the changes in volume when they dissolve are so small. For example, the solubility of sodium chloride, NaCl, in water is increased by only 2.5% when the pressure is increased from 1 to 1000 atm.

12.5 SOLUBILITY AND TEMPERATURE

When water is heated, bubbles appear at temperatures far below the boiling point. They are calm, nonviolent bubbles. They are bubbles of dissolved gases—O_2, N_2, CO_2—coming out of solution. This observation, and more precise scientific ones, show that *increasing the temperature usually decreases the solubility of a gas in a liquid.* Although higher temperature results in more frequent collisions of gas molecules with the liquid surface, the principal effect of raising the temperature is that a larger fraction of the dissolved molecules have the kinetic energy needed in order to escape from the liquid (section 11.10).

The solubility of a *solid* in a liquid, however, may change with temperature in either direction, as illustrated by these examples:

	SOLUBILITY, g solute/100 g H_2O				CHANGE IN SOLUBILITY WITH INCREASING TEMPERATURE
SOLUTE	20°C	40°C	60°C	80°C	
Sucrose ($C_{12}H_{22}O_{11}$)	204	238	287	362	increase
Lithium carbonate (Li_2CO_3)	1.33	1.17	1.01	0.85	decrease

The typical behavior is that of sucrose; *the solubility of a solid or a liquid in a liquid usually increases with increasing temperature.*

Dissolving a crystalline solid in a liquid requires that the crystal structure be broken down, just as melting the solid requires. Thus, we might expect the dissolving of a solid, like melting, to be endothermic—and it is in most cases. However, the process of dissolving a solid is a little more complicated than melting.

It might help you to understand the process if you visualize that it occurs in three steps. In reality, however, the three steps all occur at the same time. The three steps, then, are as follows:

Similarly, melting and vaporization require an input of energy to pull molecules away from each other.

1. Solute molecules are separated from each other, which requires an input of energy. You may think of this step as vaporization of the solute.
2. Solute molecules become associated with solvent molecules, giving out energy.
3. Solvent molecules are pushed away from each other by the intrusion of solute molecules between them. This requires an input of energy (usually less than the energy given out in step 2).

These energy changes are illustrated in Figure 12.4. ΔH for the net dissolving process—the sum of the ΔH's for the three steps—is the **heat of solution**. Steps 2 and 3 together are referred to as **solvation** or, when the solvent is water, **hydration**. For most solutions of *solids* in liquids, the energy required to separate the molecules is greater than the energy released when the molecules are solvated.

We can predict the change of solubility with temperature if we know whether energy is given out or absorbed when the solution is formed. When 1 mole of Li_2CO_3 is dissolved in water, 13 kJ is given out, and the solubility decreases with increasing temperature. However, when 1 mole of sucrose is dissolved in water, 5.5 kJ of heat is absorbed. In this case, which is more typical, the solubility increases with increasing temperature.

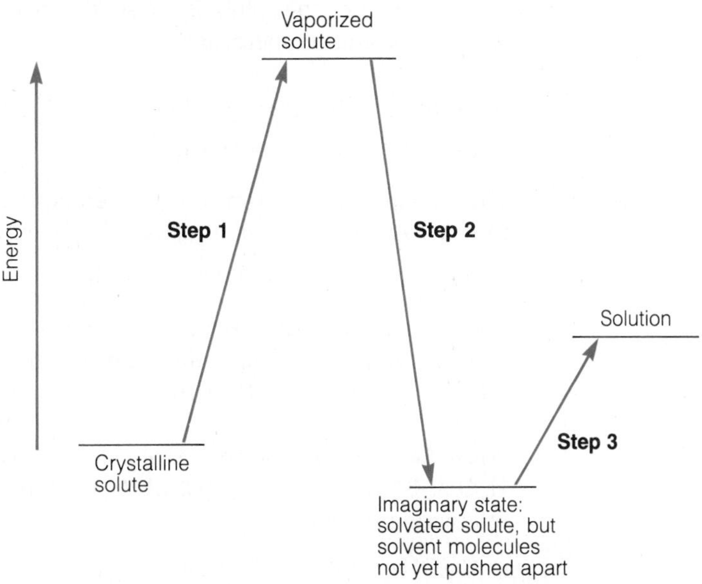

FIGURE 12.4
Energy changes in the formation of a solution.

How solubility changes with temperature is determined by the ΔH for the addition of solute to a nearly saturated solution. There are instances in which heat is emitted on forming a dilute solution and absorbed on adding solute to a concentrated solution (for example, NaOH in H_2O), or conversely. It is also necessary to specify what the solid phase is. For example, the solubility of $Na_2SO_4(H_2O)_{10}$ increases with temperature, but the solubility of Na_2SO_4 decreases with temperature.

Dissolving a gas in a liquid does, however, require that *solvent* molecules be separated from each other (Step 3). There are instances in which this effect predominates and the process is endothermic.

When a *gas* dissolves in a liquid, there is no need to separate the solute (gas) molecules because they are already separated. Therefore, only effects 2 and 3 are present, and heat is usually given out. The solubility of SO_2 in water, like the solubility of all gases in water, decreases as temperature is increased: when 1 mole of SO_2 is dissolved in water, 36 kJ of heat is given out.

The rule is summarized in this table:

HEAT TRANSFER DURING FORMATION OF SOLUTION	SIGN OF ΔH FOR DISSOLVING	CHANGE IN SOLUBILITY WITH INCREASING TEMPERATURE	SOLUTES
Heat given out (exothermic)	−	decrease	most gases; some solids and liquids
Heat absorbed (endothermic)	+	increase	most solids and liquids

EXAMPLE 12.13 The following enthalpy changes at 25°C are given:

$$HgCl_2(c) \longrightarrow HgCl_2(g) \qquad \Delta H = +78.7 \text{ kJ} \qquad \text{(heat of sublimation)}$$

separating $HgCl_2$ molecules from each other

$$HgCl_2(g) + aq \longrightarrow HgCl_2(aq) \qquad \Delta H = -64.9 \text{ kJ} \qquad \text{(heat of hydration)}$$

hydrating $HgCl_2$ molecules

Reminder: "aq" represents a large quantity of water used as a solvent.

Predict whether the solubility of solid mercury(II) chloride decreases or increases with increasing temperature.

ANSWER For the process of dissolving $HgCl_2$, the equation is

$$HgCl_2(c) + aq \longrightarrow HgCl_2(aq)$$

This equation is the sum of the given equations. Hess's law (section 4.5) tells us that ΔH for the dissolving process is obtained by adding the given ΔH's:

$$\Delta H = (+78.7 - 64.9) \text{ kJ} = +13.8 \text{ kJ}$$

Positive ΔH corresponds to absorption of heat. Therefore the solubility should increase with increasing temperature. Experimentally, the solubility increases by a factor of 7 from 20°C to 100°C. ■

PROBLEM 12.8 The solubility of calcium hydroxide, $Ca(OH)_2$, is 0.185 g/100 mL H_2O at 0°C and 0.077 g/100 mL H_2O at 100°C. Is ΔH positive or negative in the following process?

$$Ca(OH)_2(c) + aq \longrightarrow Ca(OH)_2(aq)$$ □

FIGURE 12.5
(a) A supersaturated solution. (b) The solution just after a seed is added. (c) A few minutes later, the excess solute has crystallized on the seed. The remaining solution is saturated.

(a)

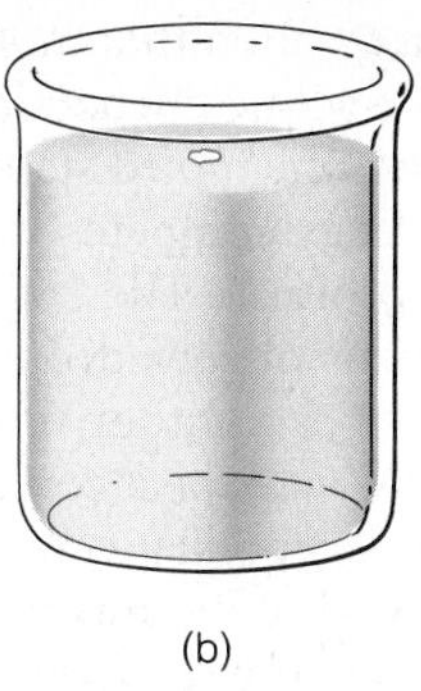
(b)

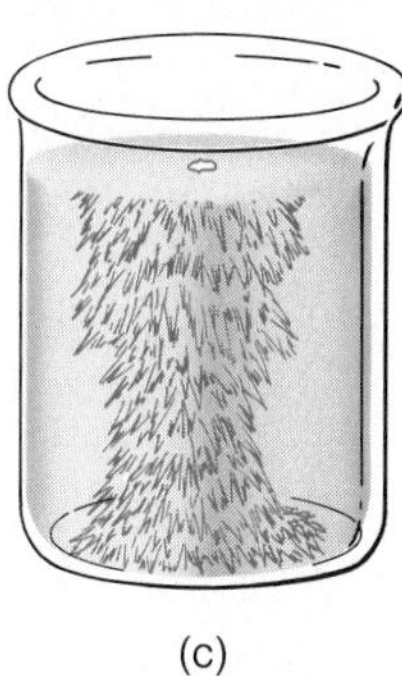
(c)

Suppose that we prepare an unsaturated solution of a solid in a liquid at a high temperature, and then cool the solution. If the solubility decreases on cooling, the solution may, at some temperature, become saturated. If we expect a crystal to appear at the temperature of saturation, we may be disappointed. Sometimes it is possible to cool the solution far below this temperature without the appearance of solid. That is because a large crystal must begin as a small crystal. An extremely small crystal—a few molecules—is more soluble than a larger crystal. From the small crystal's point of view, the solution is unsaturated, and it may therefore dissolve instead of growing. A solution more concentrated than a saturated solution at the same temperature is said to be **supersaturated**.

Supersaturation is the same phenomenon as undercooling (section 11.11). An undercooled liquid, an undercooled vapor, and a supersaturated solution are all analogous to a book standing on end. Each may remain that way indefinitely, but each could exist in a more stable condition. The book can be toppled into its more stable condition (lying flat) if it is given a little nudge. The supersaturated solution can bypass the small-crystal stage if solute molecules are deposited on a dust particle, the container wall, or any other solid present. The best crystallization nucleus is a fragment of the solute itself. When such a "seed" is introduced, crystals usually form very rapidly, leaving a saturated solution (Figure 12.5).

In August 1986, about 2000 people were suffocated in Cameroon, Africa, when CO_2 was suddenly released from a supersaturated lake.

Supersaturated solutions of gases in liquids are also common. (Soda water retains its gas for some time after the pressure is released.) These solutions, like superheated liquids (section 11.11), need nuclei in order to form bubbles.

12.6 SOLUBILITY AND MOLECULAR STRUCTURE

In section 11.14 we discussed one of the central ideas in chemistry and physics: there is a universal tendency toward disorder. A disordered state is more probable than an ordered state because the overwhelming majority of possible arrangements give disordered states. A solution is more disordered than the separated components in the same way that a shuffled deck of 52 cards is more disordered than a pile of 26 red cards plus a pile of 26 black cards. We are thus led to the incorrect prediction that when two substances are brought together, they should always mix to give a more disordered, therefore more probable, state: a solution. What is wrong?

A more detailed discussion of the maximum-disorder principle (section 18.12) leads to the conclusion that, at any given temperature, the tendency toward maximum disorder (entropy) competes with a tendency toward minimum energy. As noted in the last section, dissolving a solid in a liquid (as well as a liquid in a liquid) is more often than not an endothermic process—because the solvent-solvent and solute-solute attractions are stronger than the solvent-solute attraction. When this is true, the mixed state has a higher energy than the separated state. This increase in energy competes with the increase in entropy, so that the mixture arrives at a compromise. The compromise is a saturated solution—some, but not all, of the solute is dissolved. In the less common case where solvent and solute molecules attract each other more strongly than solvent attracts solvent or solute attracts solute, we expect complete miscibility; that is, we expect that when they are mixed the energy will go down, the entropy will go up, and the two effects will cooperate in favoring the mixed state.

Thus, the stronger the attractive forces *between solute and solvent molecules*, the greater is the solubility. The stronger the forces *between molecules of the solute*, the *less* is its solubility. Strong forces between solute molecules inhibit solubility unless the molecules of the solvent can exert a comparable attraction on the solute molecules. Similarly, strong forces between solvent molecules prevent solute molecules from elbowing in, unless there is strong solute-solvent attraction.

It is a time-honored, though ambiguous, rule among chemists that "like dissolves like." This tendency is not surprising. When the solute and solvent molecules are similar, solute molecules can easily replace solvent molecules. The reason is that forces between *similar* molecules are comparable to forces between *identical* molecules. In section 11.2 we classified the forces between molecules into several kinds, including London forces, dipole-dipole forces, and hydrogen bonds. Of these, hydrogen bonding is by far the strongest force between uncharged molecules. Liquids and solids can be classified, in a rough way, according to the kind of force primarily responsible for holding their molecules together. Two substances usually mix more easily if their intermolecular attractions fall into the same category. The following table gives some examples. "Strong" refers to hydrogen bonding; "weak," to anything else, such as dipole-dipole and London forces.

Ionic compounds will be discussed in section 12.7.

INTERMOLECULAR FORCES			EXAMPLES		
Solvent-solvent	Solute-solute	Solvent-solute	Solvent	Solute	Result
Weak	weak	weak	hexane, C_6H_{14}	carbon tetrachloride, CCl_4	completely miscible
Strong	strong	strong	H_2O	methanol, CH_3OH	completely miscible
Strong	weak	weak	H_2O	C_6H_{14}	almost immiscible
Strong	weak	strong	H_2O	acetone, $(CH_3)_2CO$	completely miscible
Weak	strong	weak	C_6H_{14}	sucrose, $C_{12}H_{22}O_{11}$	almost insoluble

Water is the most common and important hydrogen-bonding substance. Water molecules hang firmly together (section 11.2). If another molecule is to break into this stable arrangement, it must be able to participate in hydrogen bonding. That is, it should have hydrogen atoms bonded to highly electronegative atoms (especially N, O, F); these bonds are strongly polar, with the H atoms bearing positive partial charges. At least the other molecule should have electronegative atoms with unshared pairs of electrons that can attract the electron-deficient H atoms of H_2O. Molecules containing —OH and —NH_2 groups are most likely to be soluble in water. These molecules form hydrogen bonds with water, and with each other, like the hydrogen bonds between water molecules:

$$\begin{array}{lcl} \mathrm{H{-}\ddot{\underset{|}{O}}{:}\cdots H{-}\ddot{\underset{|}{O}}{:}} \\ \quad\ \mathrm{H} \qquad\quad\ \ \mathrm{H} \end{array} \qquad \begin{array}{l} \mathrm{H_3C{-}\ddot{\underset{|}{O}}{:}\cdots H{-}\ddot{\underset{|}{O}}{:}} \\ \qquad\quad \mathrm{H} \qquad\quad\ \ \mathrm{H} \end{array}$$

water and water — methanol and water

$$\begin{array}{l} \mathrm{H_3C{-}\ddot{\underset{|}{O}}{:}\cdots H{-}\ddot{\underset{|}{O}}{:}} \\ \qquad\quad \mathrm{H} \qquad\quad\ \ \mathrm{CH_3} \end{array} \qquad \begin{array}{l} \qquad \mathrm{H} \\ \qquad | \\ \mathrm{H{-}\underset{|}{N}{:}\cdots H{-}\ddot{\underset{|}{O}}{:}} \\ \qquad \mathrm{H} \qquad\quad\ \ \mathrm{H} \end{array}$$

methanol and methanol — ammonia and water

A molecule with no electron-deficient hydrogen atom may still form hydrogen bonds with water if it has unshared electron pairs on N, O, or F:

$$\begin{array}{l} \mathrm{H_3C} \\ \qquad \mathrm{C{=}\ddot{O}{:}\cdots H{-}\ddot{\underset{|}{O}}{:}} \\ \mathrm{H_3C} \qquad\qquad\qquad\quad \mathrm{H} \end{array} \qquad \begin{array}{l} \qquad\quad \mathrm{CH_3} \\ \qquad\quad | \\ \mathrm{H_3C{-}\underset{|}{N}{:}\cdots H{-}\ddot{\underset{|}{O}}{:}} \\ \qquad\quad \mathrm{CH_3} \qquad\quad \mathrm{H} \end{array}$$

acetone and water — trimethylamine and water

In contrast to hydrogen-bonding molecules like methanol, hydrocarbons (C and H only), such as hexane, have no atoms that can form hydrogen bonds:

$$\begin{array}{ccccccccccccc} & & \mathrm{H} & & \mathrm{H} & & \mathrm{H} & & \mathrm{H} & & \mathrm{H} & & \mathrm{H} \\ & & | & & | & & | & & | & & | & & | \\ \mathrm{H} & - & \mathrm{C} & - & \mathrm{C} & - & \mathrm{C} & - & \mathrm{C} & - & \mathrm{C} & - & \mathrm{C}-\mathrm{H} \\ & & | & & | & & | & & | & & | & & | \\ & & \mathrm{H} & & \mathrm{H} & & \mathrm{H} & & \mathrm{H} & & \mathrm{H} & & \mathrm{H} \end{array}$$

hexane

H atoms in the almost nonpolar C—H bonds do not form hydrogen bonds. It should not surprise us that water and hexane are almost immiscible. Any hydrocarbon molecule, or hydrocarbon-like portion of a molecule, makes a greasy blob with little attraction for water molecules. The water molecules will therefore not allow the hydrocarbon molecules to intrude between them.

As a rough rule, *hydrogen-bonding substances are soluble in each other; non-hydrogen–bonding substances are soluble in each other; hydrogen-bonding and non-hydrogen–bonding substances are not very soluble in each other.* This, like all rules about solubility, should be used with caution because solubility is the net result of opposing tendencies. A small change in one of these tendencies can make a large change in the solubility.

The more hydrogen-bonding groups a molecule has, the more opportunity there is for hydrogen bonding with water and the more soluble in water it is likely to be. The best examples are the sugars, which have at least one hydrogen-bonding atom (O in this case) on every carbon atom:

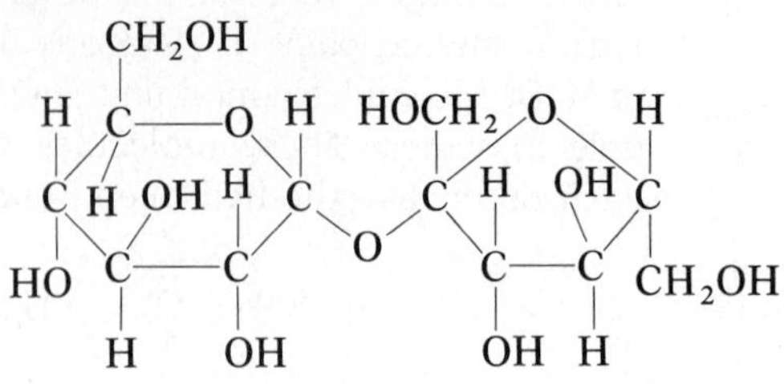

```
          CH2OH
           |
      H   C------O  H   HOCH2  O      H
        /          \ |      |  /  \    |
      C  H  OH  H   C       C   H  OH  C
       |  \     |  /    \  /  \ |     \ |
   HO   C------C         O      C------C   CH2OH
        |      |                |      |
        H      OH               OH     H
```

sucrose, $C_{12}H_{22}O_{11}$ (cane or beet sugar)

Cetyl alcohol (Latin *cetus*, "whale") is made from spermaceti, a wax obtained from whales and formerly used in making candles and ointments.

The solubility of sugars in water is well known, and it is essential for their biological function. Going in the other direction, the larger the hydrocarbon-like portion of the molecule, the less soluble the molecule is in water. Like CH_3OH, C_2H_5OH and C_3H_7OH are miscible with water in all proportions. C_4H_9OH is only partially miscible with water, and the solubility decreases as more C and H atoms are added. Cetyl alcohol, $C_{16}H_{33}OH$, is almost insoluble in water.

EXAMPLE 12.14 In each pair, select the solute that is more soluble in water. Explain your choices.

(a) $CH_3CH_2\underset{\underset{O}{\|}}{C}CH_2CH_3$ or $CH_3CH_2CH_2\underset{\underset{O}{\|}}{C}CH_2CH_2CH_3$

(b) $CH_3CH_2CH_2CH_3PH_2$ or $CH_3CH_2CH_2CH_2NH_2$

ANSWER

(a) $(CH_3CH_2)_2CO$. This molecule has the smaller hydrocarbon portions and thus needs to push fewer H_2O molecules apart.

(b) $CH_3(CH_2)_3NH_2$. The NH_2 group is much more effective in forming hydrogen bonds than the PH_2 group because N is more electronegative than P. ■

PROBLEM 12.9 In each pair, select the solute that is more soluble in water. Explain your choices.

(a) C_6H_6 or C_6H_5OH

(b) $CH_3CH_2CH_2\underset{\underset{N(CH_3)_2}{|}}{CH_2}$ or $\underset{\underset{OH}{|}}{CH_2}CH_2CH_2\underset{\underset{N(CH_3)_2}{|}}{CH_2}$

(c) $CH_3\underset{\underset{NH_2}{|}}{C}HCH_2CH_3$ or $CH_3\underset{\underset{NH_2}{|}}{C}HCH_2CH_2CH_2CH_3$ □

12.7 SOLUBILITY OF IONIC CRYSTALS

In section 9.2 we encountered the most remarkable example of solubility: ionic crystals are often soluble in water. When Svante Arrhenius proposed (1887) that some solutes exist in solution as ions, most chemists were unimpressed. The principal objection was that forces between charged particles are very strong, and it seemed unlikely that, in solution, these particles would be separated from each other and free to move independently. An explanation is suggested by the fact

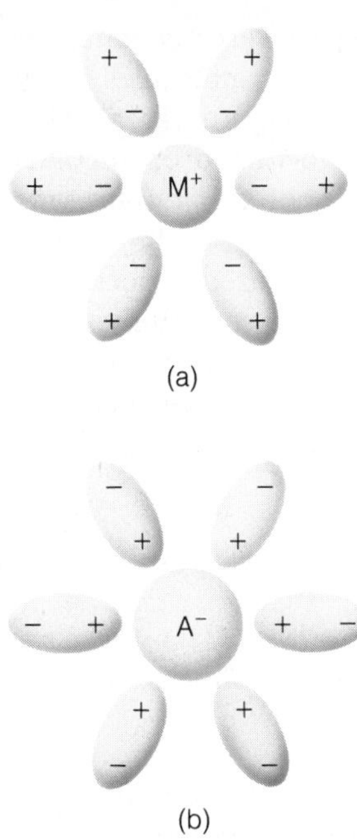

FIGURE 12.6
Ion-dipole attraction to (a) a positive ion and (b) a negative ion.

that the solvents in which ionic crystals dissolve consist of molecules with large dipole moments (section 7.12). An ion attracts the oppositely charged ends of such solvent molecules almost as strongly as it attracts another ion. The ions become *solvated* (or, when the solvent is water, *hydrated*). Ionic solutions are not exactly instances of "like dissolves like"—water is not ionic—but they come fairly close. A polar molecule like water has partial (fractional) charges—positive on H, negative on O—and thus resembles ions with their full charges.

What happens when two oppositely charged ions are immersed in a polar solvent? Think first of the positive ion. The solvent molecules are predominantly aligned with their negative ends adjacent to the positive ion (Figure 12.6a). These negative charges around the positive ion cancel part of its charge, decreasing its ability to attract negative ions. Similarly, solvent molecules are aligned with their positive ends adjacent to a negative ion (Figure 12.6b), and the negative ion's ability to attract positive ions is thus also decreased. The attraction between positive and negative ions is therefore smaller than it would be if the ions were at the same distance in a vacuum.

The liquids with the most polar molecules (the largest dipole moments) should then be the most effective in decreasing the forces between separated ions. Indeed, they are the best solvents for ionic crystals. In addition to H_2O (the world champion), ionizing solvents include liquid NH_3 (boiling point −33°C), HF (b.p. 20°C), HCN (b.p. 26°C), CH_3OH (b.p. 65°C), and SO_2 (b.p. −10°C). However, nonpolar (and slightly polar) molecules—for example, C_6H_6 (benzene), CCl_4 (carbon tetrachloride), CS_2 (carbon disulfide), and C_8H_{18} (octane)—have no tendency to orient themselves around an ion. Ions in these solvents would not be solvated. Ionic salts are almost insoluble in these liquids.

The number of water molecules closely associated with a given ion ranges from 4 to 6, increasing with the size and charge of the ion. We usually write (for instance) Na^+ and Cl^-, rather than $Na(H_2O)_x{}^+$ and $Cl(H_2O)_y{}^-$, for simplicity and because x and y are not accurately known or even precisely defined. However, when there is strong bonding between the ion and water (as with many of the transition metals), it is reasonable to write a definite formula showing the water molecules; for example, $Fe(H_2O)_6{}^{3+}$.

BOX 12.1
SOLUBILITY AND DIELECTRIC CONSTANT

The dielectric constant D (Appendix A.6) of a substance gives the decrease in force between charged bodies immersed in that substance. In a vacuum, D equals 1; in air, D is close to 1. The force is less than it would be in a vacuum by the factor $1/D$. (This relationship applies only roughly to particles as small as ions.) The table shows D for some good and poor solvents for ionic compounds.

GOOD SOLVENTS

Liquid	D	Temperature, °C
HCN	115	20
HF	84	0
H_2O	78.5	25
NH_3	22.4	−33
SO_2	17.6	−20

POOR SOLVENTS

Liquid	D	Temperature, °C
SO_3	3.11	18
$SnCl_4$	2.87	20
CS_2	2.64	20
C_6H_6	2.27	25
C_5H_{10}	2.10	20

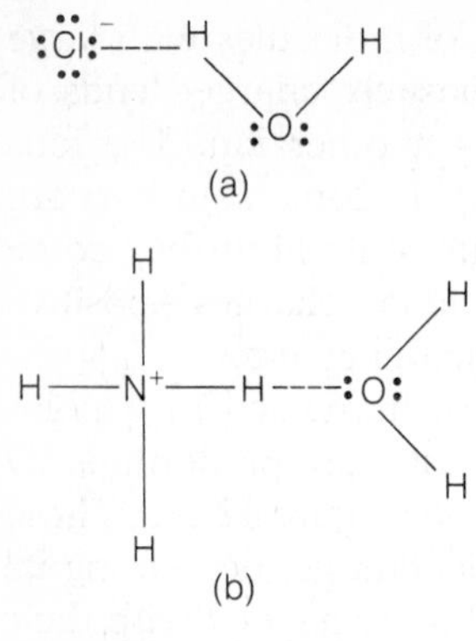

FIGURE 12.7
Hydrogen bonding between ions and water.
(a) Chloride ion.
(b) Ammonium ion.

A negative ion has unshared electron pairs. Its negative charge greatly enhances the ability of these electrons to form hydrogen bonds to water molecules, even when the best hydrogen bonders (N, O, F) are not present (Figure 12.7a). One common positive ion, the ammonium ion (NH_4^+), can also form hydrogen bonds (Figure 12.7b); this is the reason for the very large solubility of ammonium salts in water.

The process of dissolving an ionic crystal in water can be visualized as occurring in two steps, each of which would be catastrophic if it occurred separately. They are essentially the same as step 1 and steps 2 and 3 together in section 12.5, but the energy changes are much larger with ions because of the strong forces between charged particles. However, the large energy changes in these steps mostly cancel out, leaving only a small net change.

1. Destroy the crystal by separating the ions to a great distance from each other. The result would be ions in the gaseous state:

$$KCl(c) \longrightarrow K^+(g) + Cl^-(g) \qquad \Delta H_{LE} = \text{lattice energy} = 690 \text{ kJ/mol} \tag{7}$$

As in other instances, what is commonly called "energy" is really calculated as enthalpy.

An ionic crystal represents a state of low energy because the oppositely charged ions are close together (section 7.4). Therefore, separating the ions from each other against their electrostatic forces of attraction would require a large energy input, the lattice energy, as shown by the positive ΔH in equation 7.

2. We now imagine that the gaseous ions are inserted into water, forming a dilute solution. Each ion attracts water molecules and becomes hydrated. The energy of the hydrated ions is much lower than the energy of the free ions because energy is released when the ions attract the solvent molecules. The difference is the **hydration energy**:

$$K^+(g) + Cl^-(g) + aq \longrightarrow K^+(aq) + Cl^-(aq)$$
$$\Delta H_h = \text{hydration energy} = -674 \text{ kJ} \tag{8}$$

The process of dissolving the free ions is always very exothermic.

The actual process of dissolving a crystal is the sum of these two imaginary steps, converting the crystal to gaseous ions and dissolving the separated ions. By Hess's law (section 4.5), adding the reactions means adding the ΔH's. Their sum is the heat of solution (section 12.5):

$\boldsymbol{\Delta H_s}$ **= heat of solution**

$$\boldsymbol{\Delta H_s = \Delta H_{LE} + \Delta H_h} \tag{9}$$

$KCl(c) \longrightarrow$	$K^+(g) + Cl^-(g)$	$\Delta H_{LE} = +690$ kJ/mol
$K^+(g) + Cl^-(g) + aq \longrightarrow$	$K^+(aq) + Cl^-(aq)$	$\Delta H_h = -674$ kJ/mol
$KCl(c) + aq$	$K^+(aq) + Cl^-(aq)$	$\Delta H_s = 16$ kJ/mol

The heat of solution may be either positive (as with KCl and $AgNO_3$) or negative (as with Li_2CO_3 and $Ce_2(SO_4)_3$), depending upon the relative magnitudes of the lattice energy and hydration energy. When two numbers are large and close together, it is difficult to predict even the sign of their difference.

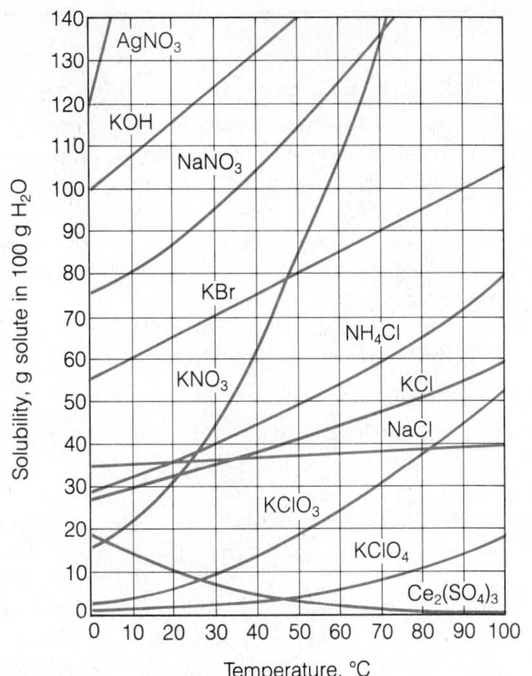

FIGURE 12.8
Curves showing the solubilities of some salts in water plotted against temperature.

EXAMPLE 12.15 The lattice energy of $BaCl_2$ is 2056 kJ/mol. Its heat of solution is -13.4 kJ/mol. What is its hydration energy?

ANSWER The hydration energy is ΔH for the process

$$Ba^{2+}(g) + 2Cl^-(g) + aq \qquad Ba^{2+}(aq) + 2Cl^-(aq)$$

This is the sum of the processes

$$Ba^{2+}(g) + 2Cl^-(g) \longrightarrow BaCl_2(c) \qquad \Delta H = -\Delta H_{LE} = -2056 \text{ kJ/mol}$$

$$BaCl_2(c) + aq \longrightarrow Ba^{2+}(aq) + 2Cl^-(aq) \qquad \Delta H = \Delta H_s = -13.4 \text{ kJ/mol}$$

Then by Hess's law,

$$\Delta H_h = \Delta H_s - \Delta H_{LE} = -13.4 - 2056 = -2069 \text{ kJ/mol}$$ ■

PROBLEM 12.10 For $LaCl_3$, the heat of solution is -137.7 kJ/mol and the hydration energy is -4380 kJ/mol. Calculate the lattice energy. □

The solubility of a salt changes with temperature in a way determined by the heat of solution (section 12.5). When dissolving is endothermic, the solubility increases with temperature. When dissolving is exothermic, the solubility decreases with temperature. Figure 12.8 shows that the solubilities of most salts in water increase with increasing temperature. This tells us that the process of dissolving a salt in water is usually endothermic. Many salts have solubilities too low to be shown on this graph. In section 9.3 we explained how reactions between ions can be predicted when we know which salts are nearly insoluble.

BOX 12.2

PORTABLE HOT AND COLD PACKS

The absorption or emission of heat when a solution is formed provides a convenient, portable way of producing high or low temperature. Instant hot and cold packs are sold for treating minor injuries. An outer plastic bag contains a pouch of water and a dry salt. When the bag is struck, the pouch breaks and the pack becomes warm if the dissolving process is exothermic, or cold if it is endothermic. The salts commonly used are $MgSO_4$ and $CaCl_2$ for hot packs, and NH_4NO_3 for cold packs:

$$MgSO_4(c) + aq \longrightarrow Mg^{2+}(aq) + SO_4^{2-}(aq) \qquad \Delta H = -91 \text{ kJ/mol}$$

$$CaCl_2(c) + aq \longrightarrow Ca^{2+}(aq) + 2Cl^-(aq) \qquad \Delta H = -81 \text{ kJ/mol}$$

$$NH_4NO_3(c) + aq \longrightarrow NH_4^+(aq) + NO_3^-(aq) \qquad \Delta H = +26 \text{ kJ/mol}$$

How warm or cold will the pack become? That is a difficult question, best answered by experiment. However, a rough estimate can be made. Suppose we have 35 g $MgSO_4$ and 100 g H_2O. The given ΔH's are for forming a very dilute solution but are still useful estimates when the solution is concentrated. Assume that all of the heat is used to raise the temperature of the water [specific heat = 4.18 J/(g °C)]. Then the heat liberated is

$$35 \text{ g } MgSO_4 \times \frac{1 \cancel{\text{mol}}}{120.4 \text{ g}} \times \frac{91 \text{ kJ}}{1 \cancel{\text{mol}}} = 26 \text{ kJ}$$

$$= 2.6 \times 10^4 \text{ J}$$

and the temperature increase is

$$\Delta t = \frac{2.6 \times 10^4 \cancel{\text{J}}}{100 \cancel{\text{g}} \times 4.18 \cancel{\text{J}}/(\cancel{\text{g}}\ °\text{C})} = 62°\text{C}$$

If the initial temperature is 20°C (68°F), the temperature just after the pouch is broken should be 82°C (180°F). The actual final temperature will be lower, because the $MgSO_4$ and the bag must also be heated and because heat starts to flow out as soon as dissolving begins.

12.8 VAPOR PRESSURES OF SOLUTIONS

We saw in section 11.10 that a solid or liquid in an enclosed space evaporates to some extent. The pressure of the vapor increases until equilibrium is attained. Recall that the pressure of the vapor at equilibrium is referred to as the vapor pressure of the substance.

Volatile means "having a vapor pressure considerably different from zero."

A solution also has a vapor pressure, and *each* of its components makes a contribution to it. In many solutions, however, the solute has a vapor pressure so low that we may assume it is zero. Such a solute is said to be "nonvolatile." A solution of any ionic salt in water is an example. A simple apparatus for measuring the vapor pressure of a solution of a nonvolatile solute is shown in Figure 12.9. It is like the apparatus in Figure 11.27 except that it has a third tube. A vacuum is above the mercury in tube a. Pure solvent is on top of the mercury in tube b. Most of the solvent vaporizes, and the vapor pressure is measured by how far down the mercury is pushed below its level in tube a. To insure that the vapor is saturated, enough solvent must be present so that a little of it remains after vaporization. Tube c is the same as tube b except that it contains a *solution* instead of pure solvent. The depression of the mercury below the level in (a) is a measure of the vapor pressure of the solution. The results of such measurements are always as illustrated in this figure:

The vapor pressure of the solvent in a solution is less than the vapor pressure of the pure solvent.

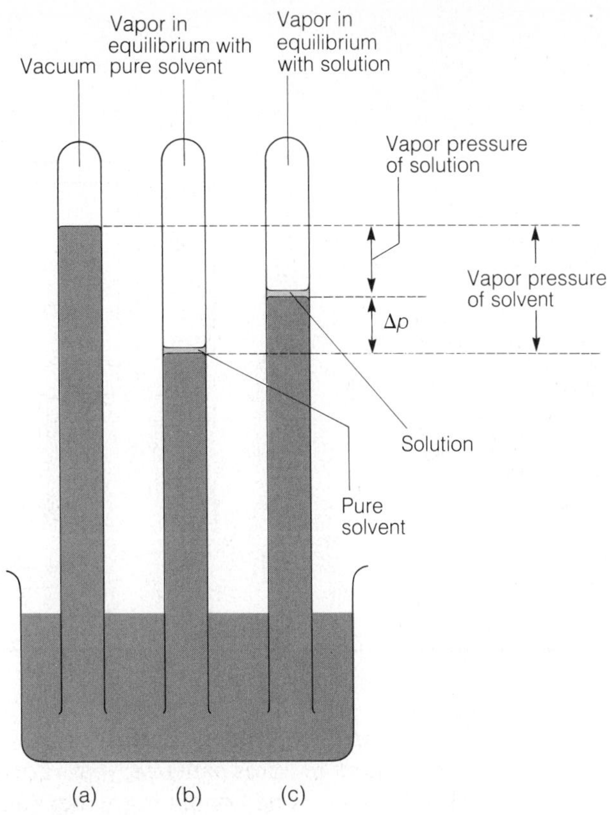

FIGURE 12.9
An apparatus for measuring the vapor pressures of a pure solvent and a solution in the same solvent. Δp indicates the vapor-pressure depression, the decrease caused by the presence of the solute.

Now consider a solution in which both solute and solvent contribute to the vapor pressure. Solutions of ethanol in water and acetic acid in water (vinegar) are familiar examples: our sense of smell is sufficient to tell us that these solutes are volatile. When both solvent and solute are volatile, we must distinguish among (1) the partial vapor pressure of the solvent (really the contribution of the solvent to the vapor pressure of the solution), (2) the partial vapor pressure of the solute (the contribution of the solute to the vapor pressure of the solution), and (3) the vapor pressure of the solution (the sum of the contributions of solvent and solute):

The word *partial* is often omitted.

$$P_{\text{solution}} = p_{\text{solvent}} + p_{\text{solute}} \tag{10}$$

(Dalton's law; section 10.11). This case is more troublesome experimentally. The apparatus in Figure 12.9 still gives the *total* vapor pressure of the solution, but measuring the separate contributions requires a second step: analyzing a sample of the vapor to determine how much of it is solvent and how much is solute. The partial pressures of the two components are thus obtained. For example, the total vapor pressure at 20°C of a 50% (by mass) solution of ethanol, C_2H_5OH, in water at 20°C is 38.0 torr. Analysis of the vapor shows that the partial pressure of H_2O in equilibrium with this solution is 14.5 torr and the partial pressure of ethanol in equilibrium with this solution is 23.5 torr. These partial pressures may be compared with the vapor pressure at 20°C of pure water, 17.5 torr, and of pure ethanol, 43.6 torr. This solution illustrates a general fact: the contribution of *each* component to the vapor pressure is *less* than the vapor pressure of the pure component (or equal if the solution is saturated, as the next page explains).

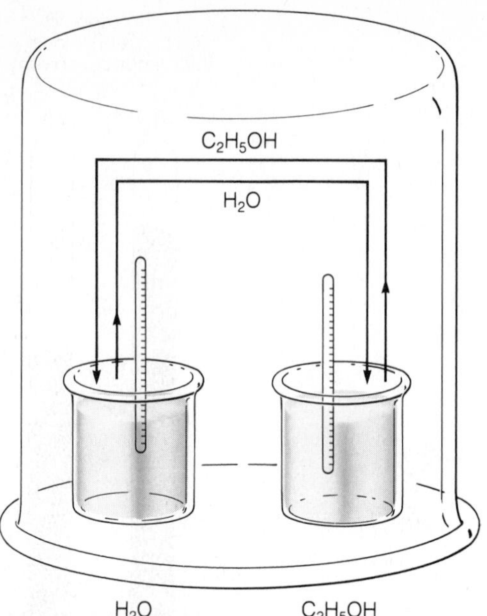

FIGURE 12.10
Transfer of vapor between liquid solutions. Each solution contains a hydrometer to show its density, and thus its composition.

A simple experiment illustrates an important aspect of vapor pressure. (The experiment requires patience; years are required to complete it.) See Figure 12.10. Two beakers, one containing water (on the left) and the other, ethanol (on the right), are in an enclosed space held at constant temperature. Water molecules escape from the left beaker into the space above. Some of these molecules collide with the ethanol surface in the right beaker and are captured. The water vapor is then partly depleted, and more water evaporates in an attempt to restore the equilibrium pressure. Water is therefore being transferred from the left beaker to the right beaker. Similarly, ethanol is being transferred from the right beaker to the left beaker. This process continues until the solutions become identical in composition. The partial vapor pressure of each component is then the same in the two solutions.

Put a hydrometer in each beaker to show the density of the liquid. You can thus follow the changes in composition.

The general rule this experiment illustrates is that *the possible direction of transfer of a substance is from the place where its vapor pressure is higher to the place where its vapor pressure is lower.* When the vapor pressure of a component in a solution is lower than the vapor pressure of the component as a pure substance, it can transfer itself to the solution—that is, dissolve. When the vapor pressure of every component is the same in one phase as in another, the two phases are in equilibrium. Thus, the vapor pressure of the solute in a *saturated* solution is the same as the vapor pressure of the pure solute, for a saturated solution, by definition, is in equilibrium with the pure solute.

12.9 VAPOR PRESSURE AND COMPOSITION

MOLE FRACTION

In section 12.2 we described several ways of expressing the composition of solutions. Now, in order to consider the vapor pressure of solutions, we need to introduce another composition measure, perhaps the most fundamental of all. The **mole fraction** of a component is its fraction expressed by the *number of molecules or number of moles* instead of by mass (section 10.11). The mole fraction of A is represented by x_A:

$$x_A = \frac{\text{number of molecules of A}}{\text{total number of molecules}} = \frac{\text{number of moles of A}}{\text{total number of moles}} \tag{11}$$

Recall that the number of moles is the number of molecules divided by Avogadro's number (N_A). In a solution of n_A mol A and n_B mol B, the total number of moles is $n_A + n_B$:

$$\mathbf{x_A = \frac{n_A}{n_A + n_B}} \tag{12}$$

It follows from the definition that $x_A + x_B = 1$ or, for more than two components, $x_A + x_B + x_C + \ldots = 1$.

EXAMPLE 12.16 A solution contains 10.0 g of carbon tetrachloride, CCl_4, and 20.0 g of benzene, C_6H_6. Calculate the mole fraction of each component.

ANSWER The molecular weights are 153.8 for CCl_4 and 78.1 for C_6H_6. Calculate the number of moles of each component:

$$n_{CCl_4} = 10.0 \text{ g} \times \frac{1 \text{ mol}}{153.8 \text{ g}} = 0.0650 \text{ mol } CCl_4$$

$$n_{C_6H_6} = 20.0 \text{ g} \times \frac{1 \text{ mol}}{78.1 \text{ g}} = 0.256 \text{ mol } C_6H_6$$

$$0.0650 \text{ mol } CCl_4 + 0.256 \text{ mol } C_6H_6 = 0.321 \text{ mol total}$$

Then divide the number of moles of each component by the total number of moles to find the mole fraction of each component:

$$x_{CCl_4} = \frac{0.0650 \text{ mol } CCl_4}{0.321 \text{ mol total}} = 0.202$$

The sum is 0.202 + 0.798 = 1.000, a useful check.

$$x_{C_6H_6} = \frac{0.256 \text{ mol } C_6H_6}{0.321 \text{ mol total}} = 0.798$$

■

Mole percentage is mole fraction multiplied by 100%; the solution in the preceding example is 20.2 mol % CCl_4 and 79.8 mol % C_6H_6.

PROBLEM 12.11 Calculate the mole fraction of each component in a solution that is 50.0% H_2O and 50.0% C_2H_5OH by mass. □

RAOULT'S LAW

Now, how does vapor pressure depend on the composition? The answer is determined by the components of the solution. Let us do an experiment with the two liquids in Example 12.16. We would expect them to show simple behavior—nonpolar, no hydrogen bonding, vapor pressures high enough to be measured.

A series of solutions of C_6H_6 and CCl_4 is prepared, ranging from pure C_6H_6 to pure CCl_4. Then the vapor pressures of C_6H_6 and CCl_4 over each solution are measured. Figure 12.11 is a graph that plots the partial vapor pressure of CCl_4 against the mole fraction of CCl_4 and the partial vapor pressure of C_6H_6 against the mole fraction of C_6H_6. When the mole fraction of CCl_4 is zero, its vapor

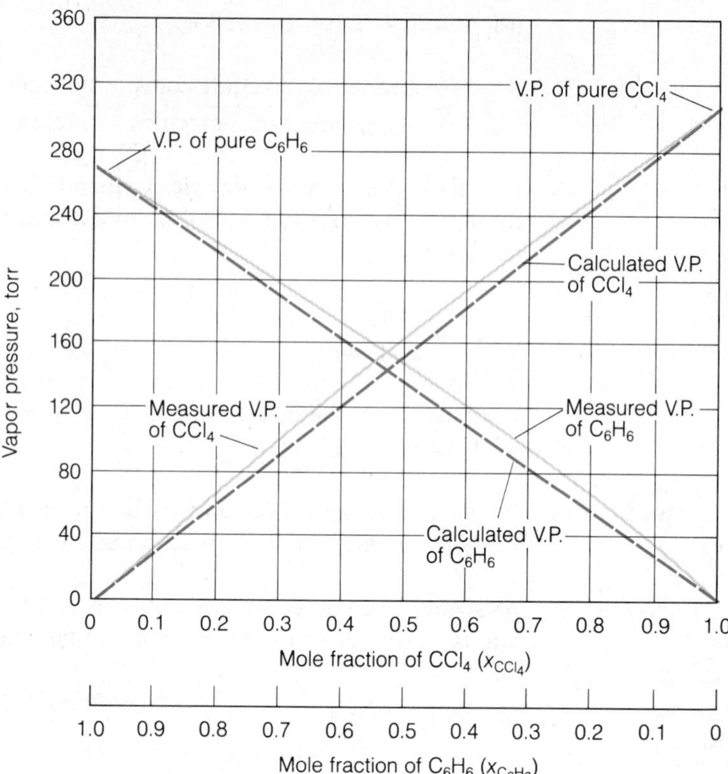

FIGURE 12.11
Vapor pressures (V.P.) of benzene (C_6H_6)–carbon tetrachloride (CCl_4) solutions at 49.99°C. The solid curves represent the experimental vapor pressures; the dashed lines represent the products $x_{C_6H_6}p^0_{C_6H_6}$ and $x_{CCl_4}p^0_{CCl_4}$.

pressure is, of course, zero; when the mole fraction is 1, its vapor pressure (p_{CCl_4}) is that of the pure substance ($p^0_{CCl_4}$). In between, the vapor pressure of either component increases as its mole fraction increases.

Figure 12.11 also shows, as a dashed line, the product of the mole fraction of CCl_4 (x_{CCl_4}) and the vapor pressure of *pure* CCl_4, $x_{CCl_4}p^0_{CCl_4}$. This dashed line shows the value that p_{CCl_4} would have if it were directly proportional at all compositions to the mole fraction (x_{CCl_4}) of CCl_4. The solid curve represents the actual values of p_{CCl_4} measured over the various solutions. Note that the solid and dashed curves are, in this case, quite close together. Similarly, the solid (experimental) curve for C_6H_6 is close to the dashed line for C_6H_6; this dashed line is a plot of $x_{C_6H_6}p^0_{C_6H_6}$.

There are many liquid solutions in which *the partial vapor pressure of each component is nearly equal to the mole fraction of that component multiplied by the vapor pressure of the pure component*:

x_A = mole fraction of A

p^0_A = vapor pressure of pure A

p_A = contribution of A to the vapor pressure of the solution = partial vapor pressure of A

$$\mathbf{p_A = x_A p^0_A} \tag{13}$$

where x_A is the mole fraction of A, p^0_A is the vapor pressure of pure A, and p_A is the contribution of A to the vapor pressure of the solution (or the partial vapor pressure of A). This relationship was first pointed out by François-Marie Raoult in 1886 and it is called **Raoult's law**. It is a good approximation for solutions in which the different molecules are very similar in size and polarity. A solution whose components have vapor pressures as given by Raoult's law is called an **ideal solution**. (An "ideal" solution has nothing to do with an "ideal" gas, except that each is described by a simple law.) Probably there are no exactly ideal solutions, but many solutions, such as CCl_4 and C_6H_6, are nearly ideal.

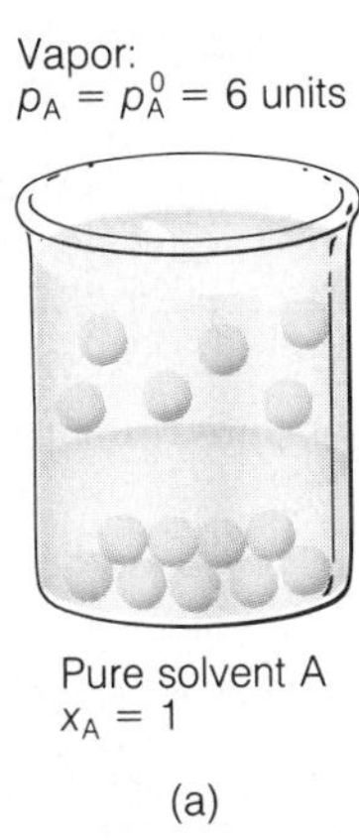

FIGURE 12.12 Simplified illustration of the change of vapor pressure with mole fraction of solvent (Raoult's law). Light spheres represent solvent (A) molecules; dark spheres represent solute (B) molecules.

Figure 12.12 shows a simple model for explaining Raoult's law. Imagine a liquid A that is in equilibrium with its vapor. We assume that there are 15 molecules of A in the container (Figure 12.12a). In a unit time interval, each molecule in the liquid has a certain probability of leaving the liquid, and each gas molecule has a certain probability of returning to the liquid. All (100%) of the molecules in the liquid are A molecules. Then 2 B (solute) molecules are added to the 15 A molecules in the container (Figure 12.12b). Now there is a liquid consisting of 10 A molecules and 2 B molecules. Only $10/12 = 5/6 = 83\%$ of the molecules in the liquid are A molecules. The mole fraction of A in the liquid is $x_A = 5/6$. The probability that, in unit time, an A molecule will escape from the solution has been decreased. We can suppose that the probability that some A molecule will escape from the liquid per unit time is reduced by the factor 5/6. This reasoning is based on the assumption that each individual A molecule has the same chance of escaping whether it is surrounded by all A's or by some A's and some B's, which is true only if A and B are very much alike. However, the rate of return of A molecules from gas to liquid is unaffected by the presence of B. When the liquid and vapor are in equilibrium, the rate of escape and the rate of return must be equal. To restore equality between the two rates, the number of A molecules in the vapor must decrease by the same factor, 5/6. The vapor pressure will thus be only 5/6 of what it was before, in agreement with Raoult's law: the partial vapor pressure of A is proportional to the mole fraction of A in the liquid.

NONIDEAL SOLUTIONS

When the attraction between different molecules is weaker than the attraction between identical molecules, each kind of molecule can escape from the solution more easily than it could from the pure substance. The vapor pressures of such solutions are therefore greater than we would predict from Raoult's law. These **nonideal solutions** are said to exhibit **positive deviation** from Raoult's law. CCl_4 and C_6H_6 illustrate slight positive deviations; their solutions are not quite ideal. Solutions of ethanol and water show strong positive deviation (Figure 12.13). Water molecules and the hydrocarbon (C_2H_5) parts of ethanol molecules are "uncomfortable" together. Not surprisingly, the forces between C_2H_5OH and H_2O, though strong enough for complete miscibility, are weaker than the forces between water molecules or between ethanol molecules.

Paradoxically, there is a loss of volume when C_2H_5OH and H_2O are mixed, because the open structure of H_2O is disrupted.

When the attraction is *stronger* between different molecules than between identical molecules—a less common situation—the molecules are held more firmly

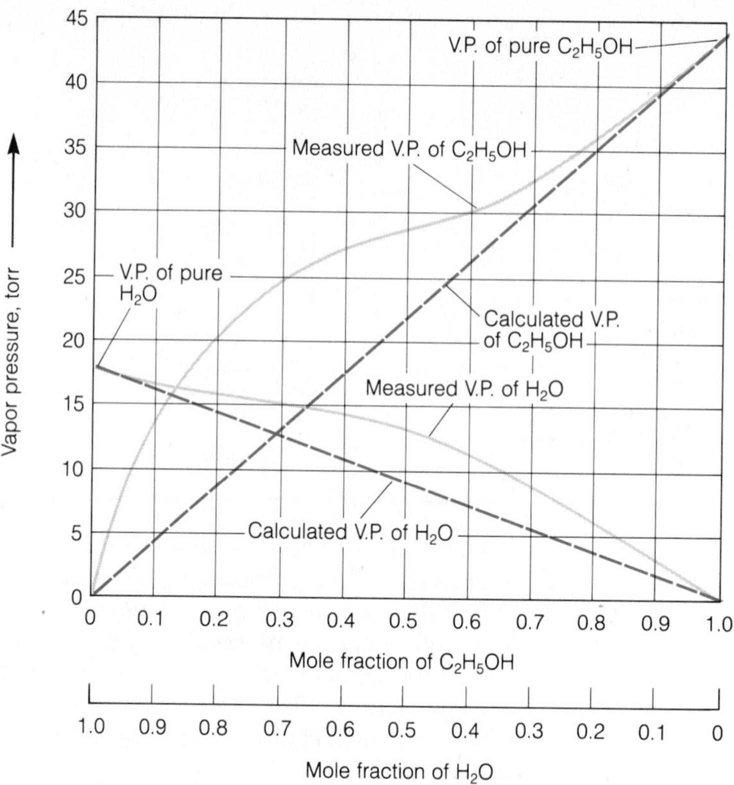

FIGURE 12.13 Vapor pressures (V.P.) of water-ethanol solutions at 20°C. The solid curves represent the experimental vapor pressures; the dashed lines represent the vapor pressures predicted by Raoult's law.

in the solution than in the pure substances, and the vapor pressures are lower than predicted. Such nonideal solutions exhibit **negative deviation** from Raoult's law.

EXAMPLE 12.17 Use Raoult's law to calculate the vapor pressure of each component and the total vapor pressure in a 50.0% (by mass) solution of ethanol, C_2H_5OH, in H_2O at 20°C. The mole fractions (calculated in Problem 12.11) are $x_{H_2O} = 0.719$ and $x_{C_2H_5OH} = 0.281$. The vapor pressures in solution (section 12.8) are $p_{H_2O} = 14.5$ torr, $p_{C_2H_5OH} = 23.5$ torr. Compare the calculated and actual vapor pressures.

ANSWER Each mole fraction is multiplied by the vapor pressure of the pure component:

$$p_{H_2O} = x_{H_2O}p^0_{H_2O} = 0.719 \times 17.5 \text{ torr} = 12.6 \text{ torr}$$

$$p_{C_2H_5OH} = x_{C_2H_5OH}p^0_{C_2H_5OH} = 0.281 \times 43.6 \text{ torr} = 12.3 \text{ torr}$$

$$P_{\text{total}} = 12.6 \text{ torr} + 12.3 \text{ torr} = 24.9 \text{ torr}$$

The measured vapor pressures are

$$p_{H_2O} = 14.5 \text{ torr}$$

$$p_{C_2H_5OH} = 23.5 \text{ torr}$$

$$P_{\text{total}} = 38.0 \text{ torr}$$

The deviation is positive. ■

PROBLEM 12.12 (a) A solution is 42.32 mol % chloroform, $CHCl_3$, and 57.68 mol % acetone, $(CH_3)_2CO$. The vapor pressures of the pure liquids at 35.17°C are 293.1 torr for chloroform and 344.5 torr for acetone. Calculate the total vapor pressure of the solution by Raoult's law. (b) The total vapor pressure of this solution is found experimentally to be 263.2 torr. Is the deviation from Raoult's law positive or negative? □

Raoult's law applies to a component (call it "A") in a series of solutions, provided that an A molecule has about the same environment in all the solutions. As noted, one way to make the environment about the same is to choose another component ("B") with molecules similar to A molecules. There is also another way: consider only very dilute solutions of the solute B, mostly A and not much B. Then an A molecule is surrounded mostly by other A molecules. Experimentally, *the vapor pressure of the solvent in any dilute solution is given approximately by Raoult's law*. That is, the small difference between the vapor pressure of the solvent in a dilute solution and the vapor pressure of the pure solvent can be calculated from Raoult's law. How dilute the solution must be depends on the identities of the solvent and solute. This generalization is illustrated by the curve for water in Figure 12.13. For the solutions containing mostly water and only a little ethanol (up to about $x_{C_2H_5OH} = 0.1$), the vapor pressure of water is close to the Raoult's law line.

12.10 COLLIGATIVE PROPERTIES OF SOLUTIONS

The vapor pressure of each substance in a solution is less than the vapor pressure of the pure substance. For the *solvent* in a *dilute* solution, there is a general relationship, Raoult's law, for how much less the vapor pressure is. The remarkable thing about Raoult's law is that it says nothing about what the solute is. All that appear in the equation are the mole fraction of the solvent and the vapor pressure of the pure solvent. Since $x_{solute} = 1 - x_{solvent}$, the lowering of the vapor pressure depends on the mole fraction, not the identity, of the solute. Molecule for molecule, one solute is therefore as good as another for lowering the vapor pressure of a given solvent. Along with the lowering of vapor pressure, several other things happen to a solvent when something is dissolved in it; its boiling point is raised, its freezing point is depressed, and its osmotic pressure (section 12.15) is increased. These four properties—vapor-pressure depression, freezing-point depression, boiling-point elevation, and osmotic pressure—are known as **colligative properties**. Their common feature is that they depend on the mole fraction of the solute, not on its identity—that is, on the relative *number* of solute molecules, not on the *kind* of molecules. We will now study how large these effects are and how we can use them to obtain useful information.

Literally, *colligative* means "bound together," because the four properties are related to each other.

12.11 VAPOR-PRESSURE DEPRESSION

Raoult's law says that, in any dilute solution, the vapor pressure of the solvent A is approximately equal to the mole fraction of A multiplied by the vapor pressure of pure A:

$$p_{solvent} \approx x_{solvent}p^0_{solvent}, \text{ or } p_A \approx x_A p^0_A$$

Let us assume that there is only one solute, B. Then

$$x_{solvent} = 1 - x_{solute}$$

$$x_A = 1 - x_B$$

and after substitution, Raoult's law may be written

$$p_A \approx (1 - x_B)p_A^0$$

$$p_A \approx p_A^0 - x_B p_A^0$$

or, by rearrangement,

$$p_A^0 - p_A \approx x_B p_A^0 \tag{14}$$

This equation illustrates how Raoult's law can be used to calculate the small difference between the vapor pressure of a solvent and that of a dilute solution (section 12.9). The quantity $p_A^0 - p_A$ is the **vapor-pressure depression** caused by the addition of the solute to the solvent. The vapor-pressure depression is often represented by Δp:

$$\Delta p = p_A^0 - p_A$$

$$\Delta p \approx x_B p_A^0 \tag{15}$$

This definition is chosen to make Δp positive. Δp depends on the identity of the solvent (which, together with the temperature, determines p_A^0) and the mole fraction of the solute (the relative number of solute molecules), but it does not depend on the identity of the solute.

12.12 MOLALITY

You need to become acquainted with another measure of composition before we go further into colligative properties. The **molality**, m, of a solution is

$$m = \frac{\text{moles of solute}}{\text{kilograms of solvent}} \tag{16}$$

m is also known as the molality of the solute or the molality of the solution with respect to the solute. When there is more than one solute (B, C, . . .), we must identify the molality with respect to each (m_B, m_C, . . .).

EXAMPLE 12.18 Calculate the molality of a 2.00% (by mass) solution of ethylene glycol, $C_2H_6O_2$, in H_2O.

ANSWER The molecular weight of $C_2H_6O_2$ is 62.1. When percentages are given, it is convenient to consider a 100.00-g sample of solution. In this sample there are 2.00 g of $C_2H_6O_2$ and 98.00 g of H_2O. The number of moles of solute is

$$2.00\text{ g} \times \frac{1\text{ mol}}{62.1\text{ g}} = 0.0322\text{ mol}$$

The mass of the solvent is

$$98.00\text{ g} \times \frac{1\text{ kg}}{1000\text{ g}} = 0.09800\text{ kg}$$

Then the molality is

$$m = \frac{0.0322\text{ mol solute}}{0.09800\text{ kg solvent}} = 0.329\,\frac{\text{mol solute}}{\text{kg solvent}}$$

This solution is 0.329 molal, or 0.329 m. ■

PROBLEM 12.13 Calculate the molality of a 4.00% (by mass) solution of methanol, CH_3OH, in H_2O. □

EXAMPLE 12.19 A solution of potassium nitrate, KNO_3, is 0.520 *m* (molal). Its density is 1.0298 g/mL. Calculate its molarity (M).

ANSWER As usual, you can use any specific amount of solution for your calculation. Since the molality is given, it is natural to consider the quantity of solution containing 1 kg of solvent (assumed to be water, though its identity does not matter). The given molality says that this amount of solution contains 0.520 mol KNO_3. The question is, how many moles of KNO_3 are in one *liter* of solution? The given density relates the volume of the solution to its mass. First, however, the total mass of the solution must be calculated. The mass of the KNO_3 in 1 kg H_2O is

$$0.520\ \cancel{\text{mol}} \times \frac{101.1\ \text{g}}{1\ \cancel{\text{mol}}} = 52.6\ \text{g KNO}_3$$

The mass of the solution is 52.6 g KNO_3 + 1000.0 g H_2O = 1052.6 g. Its volume is obtained from its mass and density:

$$1052.6\ \cancel{\text{g}} \times \frac{1\ \cancel{\text{mL}}}{1.0298\ \cancel{\text{g}}} \times \frac{1\ \text{L}}{1000\ \cancel{\text{mL}}} = 1.0221\ \text{L}$$

The molarity is

$$\frac{0.520\ \text{mol KNO}_3}{1.0221\ \text{L solution}} = 0.509\ \text{mol/L or } 0.509\ \text{M}$$ ■

PROBLEM 12.14 Find the molarity of a 0.261 *m* solution of $CuSO_4$. Its density is 1.0403 g/mL. □

The preceding example and problem illustrate a useful approximation. For dilute aqueous solutions, molality and molarity are roughly equal, because one liter of solution contains about one kilogram of water.

BOX 12.3

MOLALITY AND MOLARITY

Molality (*m*) is sometimes confused with molarity (M)—and their names do not help—but they are not the same. Molarity has *liters* of *solution* in the denominator; molality has *kilograms* of *solvent* in the denominator. The advantage of *molality* is that it is determined only by the masses of the components, whereas the *molarity* of a solution changes slightly with temperature because the volume changes.

For dilute solutions, mass percentage, molarity, mole fraction, and molality of the solute are all approximately proportional to each other (Problem 12.70). If you have a dilute solution and double the percentage of the solute, you just about double its molarity, molality, and mole fraction. This *proportionality* is true for solutions in any solvent. However, the approximate *equality* between molarity and molality exists only with aqueous solutions.

12.13 FREEZING-POINT DEPRESSION AND BOILING-POINT ELEVATION

A familiar fact about solutions is that they usually freeze at lower temperatures than the pure solvents. For example, whereas pure water freezes at 0°C, seawater freezes at about −2°C. Antifreeze is added to automobile radiators to prevent the water in them from freezing. Freezing-point depression is the most easily measured of the colligative properties. The boiling-point elevation is less important, but the two properties are described by similar equations and are therefore discussed together. It is customary to express both of these properties in terms of the molality of the solution rather than its molarity or mole fraction.

The **freezing-point depression** is the difference between the freezing points of the pure solvent and the solution:

The definition is chosen to make Δt_f positive.

$$\Delta t_f = t_f - t_f'$$

where t_f is the freezing point of the pure solvent and t_f' is the freezing point of the solution.

Methanol is no longer used in automobile radiators because it is very volatile, but it is used in windshield-washing solutions.

Ethylene glycol, $(CH_2OH)_2$, and methanol, CH_3OH, are both used as antifreezes. Let us prepare three aqueous solutions of each, 2.00, 4.00, and 6.00% by mass, and measure their freezing points (t_f'). Since the solvent (water) freezes at 0°C, the freezing-point depression is $\Delta t_f = 0°C - t_f' = -t_f'$. For each solution, we calculate the molality and the ratio of the freezing-point depression to the molality. The results are shown in Table 12.3. Not only is the ratio approximately constant for each solute; it is about the same for both solutes. Indeed, the same ratio is obtained for *any* solute in water (except electrolytes, which we will discuss very shortly). This observation is expressed by the equation

$$\frac{\Delta t_f}{m} \approx K_f$$

or

$$\boldsymbol{\Delta t_f \approx K_f m} \tag{17}$$

K_f is a constant characteristic of the solvent, called the **freezing-point depression constant**, and m is the molality of the solution. For water, the best value of K_f is 1.86°C kg/mol. Table 12.4 lists the values of K_f for some common solvents.

A solution of a nonvolatile solute boils at a *higher* temperature than the pure solvent. Since the vapor pressure of the solvent at each temperature is lower than the vapor pressure of the pure solvent, a higher temperature is needed to get the

TABLE 12.3
FREEZING-POINT DEPRESSIONS OF TWO AQUEOUS SOLUTIONS

MASS PERCENTAGE	m, mol/kg	Δt_f, °C	$\frac{\Delta t_f}{m}$, °C kg/mol
$C_2H_6O_2$			
2.00	0.329	0.60	1.8
4.00	0.671	1.24	1.85
6.00	1.028	1.92	1.87
CH_3OH			
2.00	0.637	1.14	1.79
4.00	1.300	2.37	1.82
6.00	1.992	3.71	1.86

TABLE 12.4
FREEZING-POINT DEPRESSION AND BOILING-POINT ELEVATION CONSTANTS

SOLVENT	FREEZING POINT, °C	K_f, °C kg/mol	BOILING POINT AT 1 atm, °C	K_b, °C kg/mol
Acetic acid, CH_3COOH	16.6	3.90	118.1	3.07
Benzene, C_6H_6	5.51	4.90	80.1	2.53
Biphenyl, $C_{12}H_{10}$	70	8.00	—	—
Camphor, $C_{10}H_{16}O$	178.4	37.7	205	5.41
Carbon tetrachloride, CCl_4	−22.8	31.8	76.8	5.03
Ethanol, C_2H_5OH	−117.3	1.99	78.5	1.22
Water	0.000	1.86	100.000	0.512

vapor pressure up to 1 atm. The **boiling-point elevation** is

Δt_b, like Δt_f, is defined so that it is positive.

$$\Delta t_b = t_b' - t_b$$

where t_b is the boiling point of the pure solvent and t_b' is the boiling point of the solution. The elevation is also approximately proportional to the molality of the solution:

$$\Delta t_b \approx K_b m \tag{18}$$

K_b is the **boiling-point elevation constant**. Values of K_b are also given in Table 12.4.

EXAMPLE 12.20 Find the freezing and boiling points of a solution containing 2.00 g sucrose, $C_{12}H_{22}O_{11}$, dissolved in 100 g of water.

ANSWER Both freezing-point depression and boiling-point elevation are calculated from the molality of the solution. Calculation of the molality requires converting grams to moles. The molecular weight of sucrose is 342.3. The number of moles of sucrose is

$$2.00 \cancel{g} \times \frac{1 \text{ mol}}{342.3 \cancel{g}} = 0.00584 \text{ mol}$$

This amount of sucrose is dissolved in 100 g = 0.100 kg H_2O. The molality is

$$m = \frac{0.00584 \text{ mol}}{0.100 \text{ kg}} = 0.0584 \frac{\text{mol}}{\text{kg}}$$

K_f and K_b for water are found in Table 12.4. Then the freezing-point depression is

$$\Delta t_f \approx K_f m = 1.86 \frac{°\text{C} \cancel{\text{kg}}}{\cancel{\text{mol}}} \times 0.0584 \frac{\cancel{\text{mol}}}{\cancel{\text{kg}}} = 0.109°\text{C}$$

and the freezing point is

$$0.000°\text{C} - 0.109°\text{C} = -0.109°\text{C}$$

The boiling-point elevation is

$$\Delta t_b \approx K_b m = 0.512 \frac{°\text{C} \cancel{\text{kg}}}{\cancel{\text{mol}}} \times 0.0584 \frac{\cancel{\text{mol}}}{\cancel{\text{kg}}} = 0.0299°\text{C}$$

and the boiling point is

$$100.000°\text{C} + 0.0299°\text{C} = 100.030°\text{C}$$

■

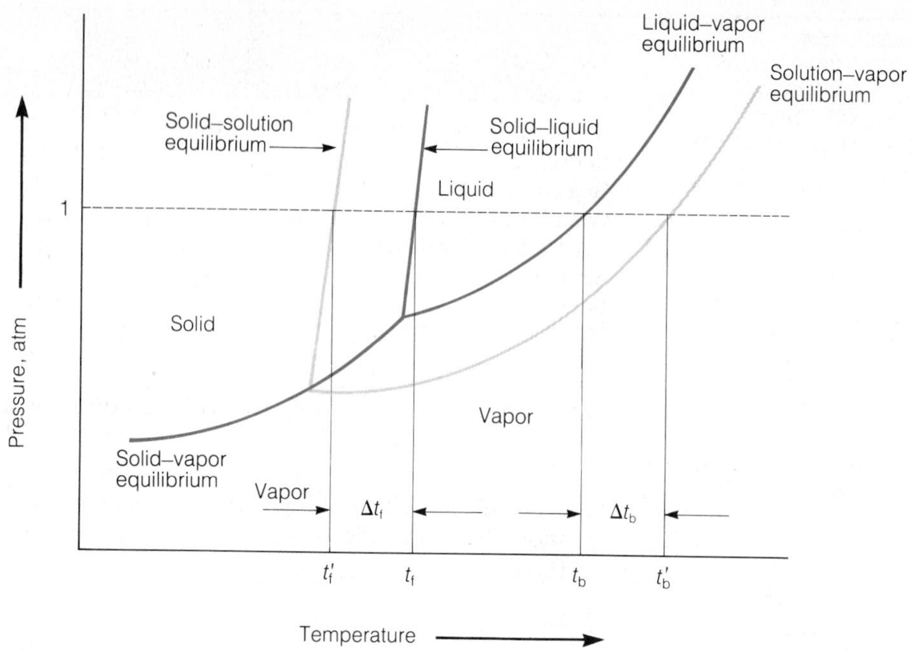

FIGURE 12.14
Phase diagram for pure solvent and solution. t_f and t_b are the freezing and boiling points of pure solvent; t_f' and t_b' are the freezing and boiling points of the solution. Δt_f is the freezing-point depression; Δt_b is the boiling-point elevation.

PROBLEM 12.15 A solution is prepared by mixing 0.384 g cetyl alcohol, $C_{16}H_{33}OH$, with 12.06 g of camphor. Calculate the freezing-point depression and the freezing point of the solution. □

Figure 12.14 shows the usual positive slope for the solid–liquid curve.

Freezing-point depression and boiling-point elevation are consequences of vapor-pressure depression. To understand why, we look first at the pressure–temperature phase diagram of a typical substance, as shown in Figure 11.31 (section 11.13). The black curves in Figure 12.14 are essentially that same phase diagram. They give the vapor pressures of the solid (frozen) solvent and of the pure liquid solvent plotted against temperature, and the melting point of the solid plotted against pressure. The freezing point (t_f) and the boiling point (t_b) of the pure solvent are the temperatures at which the horizontal dashed line (1 atm pressure) crosses these curves. Now dissolve a solute that has a negligible vapor pressure. The whole vapor-pressure curve for the liquid is lowered by the presence of this solute. The colored curve gives the vapor pressure of the *solution* plotted against temperature. This curve crosses the 1 atm line at a *higher* temperature than the black curve crosses it (t_b' instead of t_b), and the boiling point is thus raised by an amount Δt_b.

To understand freezing-point depression, remember that the solid and the liquid are in equilibrium when their vapor pressures are equal. The vapor pressure of the liquid is lowered by having solute dissolved in it; the vapor pressure of the solid in equilibrium with it must also decrease, and the only way to decrease the vapor pressure of the solid—still pure—is to lower the temperature. This is why the solid–liquid equilibrium curve moves to the left, to the position of the colored curve. The freezing point is thus *lowered* from t_f to t_f'. The solute has no effect on the vapor pressure of the solid solvent.

Vapor pressure is a measure of what is often called "escaping tendency." The depression of the vapor pressure means that the solvent is "happier," and therefore less likely to escape, when it is in solution than when it is a pure liquid; it is less likely to leave the liquid state and become a solid or gas. The addition of solute extends the range of liquidity at both ends; to freeze the solvent you must go to a lower temperature than for the pure liquid, and to boil it you must go to a higher temperature.

Equations 17 and 18 are applicable only under certain conditions:

1. They are approximations good only for dilute solutions.
2. Equation 17 assumes that the solid that appears at the freezing point is pure *solvent*—it is not the solute, a compound of solvent and solute, or a solid solution.
3. Equation 18 applies only to a *nonvolatile solute*—one whose vapor pressure is nearly equal to zero. The boiling points in Figure 12.14 are the temperatures at which the *solvent* vapor pressure becomes 1 atm. If the solute is also volatile, the total vapor pressure of the *solution* is 1 atm at some lower temperature, which is, by definition, the boiling point of the solution.

When the solution contains two or more solutes, the molality to be used in calculating the colligative properties is the *total molality of all solutes*:

$$m_{\text{total}} = m_{\text{B}} + m_{\text{C}} + \ldots \qquad (19)$$

The molalities are simply added because the identity of the solutes is irrelevant; a mole of one solute has the same effect as a mole of another.

12.14 COLLIGATIVE PROPERTIES OF IONIC SOLUTIONS

Let us repeat the experiment in Table 12.3 using solutions of sodium chloride, NaCl, and sodium sulfate, Na_2SO_4. We again measure the freezing-point depressions, and calculate the freezing-point depressions by the equation $\Delta t_f = 1.86\ m$. The results (Table 12.5) are puzzling at first. The actual freezing-point depressions are much larger than the values calculated from equation 17. The last column shows the ratio of the measured to the calculated depression. For NaCl, Δt_f is more than 1.8 times what it "should" be. For Na_2SO_4, the deviation of $\Delta t_{f,\text{measured}}$ from $\Delta t_{f,\text{calculated}}$ is even larger, approaching 3 in the more dilute solutions.

This peculiarity is characteristic of solutions of electrolytes (section 9.1). It was discovered in 1884 by François-Marie Raoult, of vapor-pressure fame, and was studied in detail by Jacobus Henricus van't Hoff in 1887. The enlargement of the

TABLE 12.5
FREEZING-POINT DEPRESSIONS OF SOME IONIC SOLUTIONS

MASS PERCENTAGE	m, mol/kg	$\Delta t_{f,\text{measured}}$, °C	$\Delta t_{f,\text{calculated}}$, °C	$\dfrac{\Delta t_{f,\text{measured}}}{\Delta t_{f,\text{calculated}}}$
NaCl				
0.00600	0.00103	0.00378	0.00192	1.97
0.0700	0.0120	0.0433	0.0223	1.94
0.500	0.0860	0.299	0.160	1.87
1.00	0.173	0.593	0.322	1.84
2.00	0.349	1.186	0.649	1.83
Na_2SO_4				
0.0700	0.00493	0.0257	0.00917	2.80
0.500	0.0354	0.165	0.0658	2.51
1.00	0.0711	0.320	0.132	2.42
2.00	0.144	0.606	0.268	2.26

freezing-point depression implies that the solution contains more solute particles—greater molality—than what we calculate from the masses and molecular weights of the solutes. It was also in 1887 that Svante Arrhenius provided the explanation: electrolytes exist in solution as ions. Thus, a 0.100 m solution of NaCl really contains two solutes, 0.100 m Na^+ and 0.100 m Cl^-. What we *should* use to calculate the freezing-point depression is the *total* molality:

$$m_{\text{total}} = m_{\text{Na}^+} + m_{\text{Cl}^-} = (0.100 + 0.100)\ \text{mol/kg} = 0.200\ \text{mol/kg}$$

A 0.100 m solution of Na_2SO_4 is 0.200 m with respect to Na^+ and 0.100 m with respect to SO_4^{2-}:

$$\underset{0.1\text{ mol}}{\text{Na}_2\text{SO}_4(\text{c})} \longrightarrow \underset{0.2\text{ mol}}{2\text{Na}^+} + \underset{0.1\text{ mol}}{\text{SO}_4^{2-}}$$

$$m_{\text{total}} = m_{\text{Na}^+} + m_{\text{SO}_4^{2-}} = (0.200 + 0.100)\ \text{mol/kg} = 0.300\ \text{mol/kg}$$

Thus, the apparent molality must be multiplied by the number of ions in the formula: 2 for NaCl, 3 for Na_2SO_4, 4 for $LaCl_3$, 5 for $Al_2(SO_4)_3$, and so on.

If ionization were ignored, the freezing-point depression would be calculated by equation 17 from the *apparent* molality (m) of the solution. The apparent molality is what you get from the mass and molecular weight of the compound—0.100 mol/kg for both the NaCl and the Na_2SO_4 solutions just mentioned. The freezing point calculated in this way is $\Delta t_{\text{f, calculated}} = K_f m$. *The ratio of the observed value of the freezing-point depression to the value calculated from the apparent molality* is called the **van't Hoff factor**, and it is represented by i:

$$i = \frac{\Delta t_{\text{f, measured}}}{\Delta t_{\text{f, calculated}}} = \frac{\Delta t_{\text{f, measured}}}{K_f m} \tag{20}$$

The numbers in the last column of Table 12.5 are van't Hoff factors. The van't Hoff factor is approximately the same for all colligative properties; vapor-pressure depressions, boiling-point elevations, and osmotic pressures (section 12.15) are all larger for electrolytes than for nonelectrolytes at the same molality.

The van't Hoff factor approaches a whole number (2, 3, and so on) only in very dilute solutions. In more concentrated solutions, the ions behave as if they were not all there. In 2% solutions, judging by the freezing-point depression, it seems that we have only 1.83 (not 2) moles of ions for each mole of NaCl, and only 2.26 (not 3) moles of ions for each mole of Na_2SO_4. This behavior is typical of ions. It results from the strong attraction and repulsion between ions. On the average, a positive ion has more negative ions than positive ions in its vicinity, and a negative ion has more positive ions than negative ions in its vicinity. The net result is about the same as if some of the positive and negative ions were paired; the colligative properties are lower than what we would calculate from the total molality of all the ions. Indeed, in more concentrated solutions, and especially in solvents less polar than water, most of the ions can be associated into *ion pairs* and even larger clusters.

EXAMPLE 12.21 A 0.00200 molal aqueous solution of $Co(NH_3)_4(NO_2)_2Cl$ freezes at -0.00732°C. How many moles of ions does one mole of this salt give on being dissolved?

ANSWER First calculate the freezing-point depression from the given molality:

$$\Delta t_f = K_f m = 1.86\ \frac{°\text{C}\ \cancel{\text{kg}}}{\cancel{\text{mol}}} \times 0.00200\ \frac{\cancel{\text{mol}}}{\cancel{\text{kg}}} = 3.72 \times 10^{-3}\ °\text{C}$$

Now compare this calculated freezing-point depression with the measured freezing-point depression, 7.32×10^{-3} °C. The van't Hoff factor is

$$i = \frac{7.32 \times 10^{-3}\ °C}{3.72 \times 10^{-3}\ °C} = 1.97 \approx 2$$

It appears that 1 mole of salt gives 2 moles of ions—actually, $Co(NH_3)_4(NO_2)_2^+$ and Cl^-. ■

PROBLEM 12.16 A 2.00% (by mass) aqueous solution of novocainium chloride ($C_{13}H_{21}ClN_2O_2$, 272.8 g/mol) freezes at -0.237°C. Calculate the van't Hoff factor. How many moles of ions are in solution per mole of compound? □

The usual symbol for osmolality is capital O. We use m_{os} instead to avoid confusion with the symbol for oxygen or with zero. Osmolality is expressed in "osmoles per kilogram." Brainteaser: Compose a definition of an *osmole*.

The **osmolality**, m_{os}, of a solution is its molality multiplied by the van't Hoff factor:

$$m_{os} = im = \frac{\Delta t_{f,\,measured}}{K_f}$$

For a very dilute solution, the osmolality is the same as the total molality of all ionic species. Although the word *osmolality* suggests osmotic pressure, it is defined in terms of the freezing-point depression. Osmolality is often used in describing the composition of body fluids and solutions used in treating patients, such as the solutions for injections and tube feeding.

EXAMPLE 12.22 From the data in Table 12.5, find the osmolality of a 2.00% solution of Na_2SO_4 in H_2O.

ANSWER The osmolality can be calculated from the measured freezing-point depression in the table:

$$m_{os} = \frac{\Delta t_f}{K_f} = \frac{0.606\ \cancel{°C}}{1.86\ \cancel{°C}\ \text{kg/mol}} = 0.326\ \text{mol/kg}$$ ■

PROBLEM 12.17 Calculate the osmolality of a 0.173 m aqueous solution of NaCl from the data in Table 12.5. □

12.15 OSMOTIC PRESSURE

Life could not exist were it not for a remarkable property of cell membranes: they allow some molecules and ions, but not others, to pass through them. This property is called **semipermeability**. It is also exhibited by cellophane and parchment paper, for example. Semipermeable membranes provide another benefit, somewhat less important than sustaining life; they enable us to measure a fourth colligative property of solutions—osmotic pressure.

Suppose we have a membrane that is permeable to the solvent but not to the solute in a certain solution. For example, cellophane is permeable to water but not to sugar. Suppose that we place a piece of cellophane between a sugar solution and pure water (Figure 12.15). Water can flow through the membrane in either direction. However, more water flows from the pure water to the solution. There is thus a slow net flow of water to the solution:

$$\text{pure water} \xrightarrow{\text{water flow}} \text{solution}$$

Why does this flow occur? Water molecules are continually colliding with the membrane from both sides, and some of them pass through. However, the

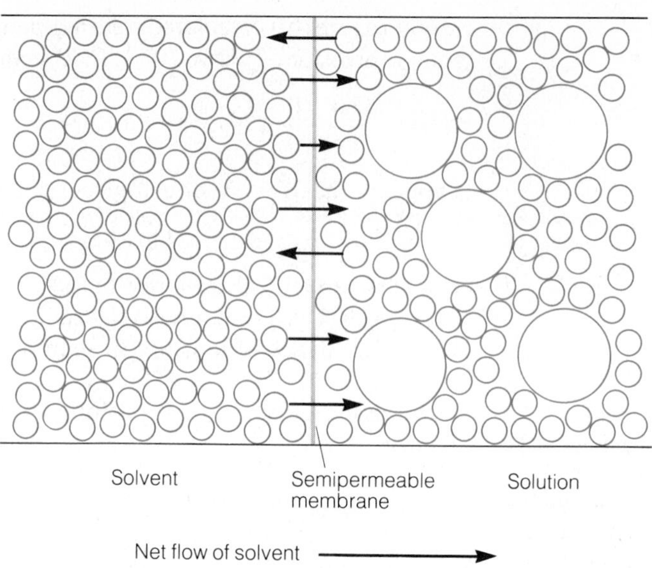

FIGURE 12.15
Osmotic flow from pure solvent to solution.

number of collisions per second is somewhat less on the solution side (*right*) than on the pure-water side (*left*). Since there are more collisions from the left than from the right, more water molecules pass through from left to right than from right to left. This net flow from solvent to solution, or from a dilute solution to a more concentrated solution, through a semipermeable membrane is known as **osmosis** (Greek, "push"). Sugar molecules also are colliding with the membrane from the right, but they are just butting their heads against a wall.

Water can also be made to flow through a membrane by applying a greater pressure to one side of it than to the other. Figure 12.16 illustrates an apparatus that can be used to do this. The application of a greater pressure (P_2) on the right piston than on the left (P_1) pushes water from the solution on the right into the solvent on the left. Thus the left-to-right osmotic flow can be stopped by just the correct excess pressure on the right piston. The excess pressure required to stop the flow and keep the solution in equilibrium with the pure solvent is the **osmotic pressure** (π) of the solution: $\pi = P_2 - P_1$. For dilute solutions, π (in atm) is given by the equation

$$\pi \approx [\mathbf{B}]\boldsymbol{RT} \tag{21}$$

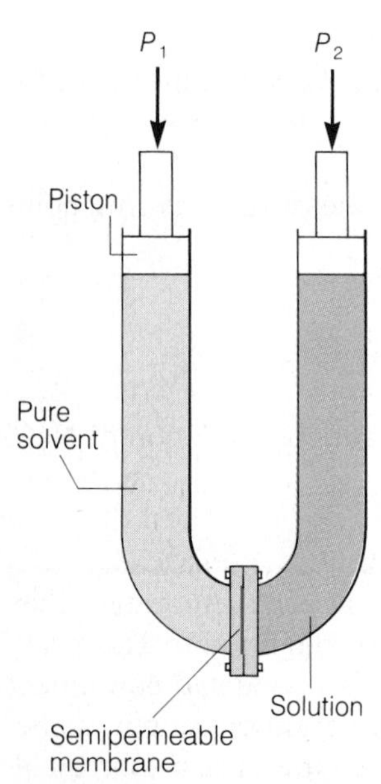

FIGURE 12.16
An apparatus for measuring osmotic pressure. To maintain equilibrium between solvent and solution, the pressure on the right piston, P_2, must be greater than the pressure on the left piston, P_1. The difference, $\pi = P_2 - P_1$, is the osmotic pressure of the solution.

where [B] represents the molarity of the solute (moles per liter); R is the gas constant, 0.08206 L atm/(mol K); and T is the absolute temperature. Since π is approximately proportional to the concentration of the solute but independent of the identity of the solute, it is a colligative property.

The higher the osmotic pressure, the lower is the concentration of the *solvent*. Therefore, solvent tends to flow from a solution with a lower osmotic pressure to a solution with a higher osmotic pressure. Two solutions with the same osmotic pressure are **isotonic** with each other. Human blood is isotonic with a 0.16 M NaCl solution. When a person ingests a solution, such as seawater or Epsom salts ($MgSO_4$), with a higher osmotic pressure than the blood, water in the body flows from the blood to the intestines—opposite to the effect when fresh water is ingested. The body thus loses water, which is all right for occasional treatment of constipation but fatal if prolonged.

When the excess pressure on the solution, $P_2 - P_1$ (Figure 12.16), is made *greater* than the osmotic pressure, there is a flow of solvent from solution to pure

BOX 12.4
A GASEOUS MEMBRANE

The apparatus in Figure 12.10 (section 12.8) illustrates osmotic flow, although this process may not be obvious. Suppose that the beaker on the left contains pure H_2O and the beaker on the right contains an aqueous solution of NaCl. Water evaporates from the pure water, where its vapor pressure is higher, and condenses in the solution, where its vapor pressure is lower. The air space is a semipermeable "membrane"; it is permeable to water, which can vaporize, but not to NaCl, which has practically zero vapor pressure. Spontaneous flow, whether through air or through a membrane, is from higher to lower vapor pressure.

solvent. This **reverse osmosis** can be used to purify polluted or salty water (section 12.18).

Calculating the osmotic pressure of an ionic solution requires multiplying the molarity by the van't Hoff factor i, the same as for the freezing-point depression:

$$\pi \approx i[\mathrm{B}]RT$$

where [B] is the apparent molarity, calculated without considering ionization.

Osmolarity, $\mathrm{M_{os}}$, is defined in analogy to osmolality (section 12.14):

$$\mathrm{M_{os}} = i\mathrm{M} = i[\mathrm{B}]$$

Then the osmotic pressure is

$$\pi = \mathrm{M_{os}}RT$$

For a very dilute solution, the osmolarity is simply the molarity multiplied by the number of ions in the formula.

EXAMPLE 12.23 Calculate the osmotic pressure of a 0.0120 M solution of NaCl in water at 0°C. Use the data in Table 12.5.

ANSWER Molarity and molality are about the same for dilute aqueous solutions. Assume that $i = 1.94$, as for the 0.0120 m solution in Table 12.5. Then

$$\pi = i[\mathrm{NaCl}]RT$$

$$= 1.94 \times 0.0120\,\frac{\cancel{\mathrm{mol}}}{\cancel{\mathrm{L}}} \times 0.08206\,\frac{\cancel{\mathrm{L}}\ \mathrm{atm}}{\cancel{\mathrm{mol}}\ \mathrm{K}} \times 273.2\ \mathrm{K}$$

$$= 0.522\ \mathrm{atm}$$

■

BOX 12.5
OSMOTIC PRESSURE AND SURVIVAL

The osmotic pressure of the environment is vitally important to living cells. A solution with higher osmotic pressure than the interior of a cell is *hypertonic* relative to the cell; a solution with lower osmotic pressure is *hypotonic.*

Suppose that a cell is placed in a hypertonic solution. Water flows toward the solution with the greater osmotic pressure (toward the more concentrated solution). As water flows out through the cell membrane into the hypertonic solution, the cell shrivels and perhaps dies. A cucumber placed in concentrated brine (NaCl solution) shrinks as it becomes a pickle. A similar fate awaits the foolhardy bacterium that invades salted meat or sugared fruit (jelly, jam, candied fruit). Conversely, a cell in a hypotonic solution—pure water being the extreme case—will absorb water, swell, and perhaps burst. The shriveling of red blood cells is called *crenation*, and the bursting is referred to as *hemolysis.*

A solution to be used for injection must be isotonic with blood. A 0.16 M NaCl solution, with the same osmotic pressure as blood, is a *physiological* (or *normal*) *saline solution.*

PROBLEM 12.18 Estimate the osmotic pressure of human blood at 37°C. It is isotonic with a 0.16 M NaCl solution. Assume that the van't Hoff factor of NaCl is 1.9. □

12.16 DETERMINING MOLECULAR WEIGHTS

One reason we are interested in colligative properties is that they provide us with a means of measuring molecular weights (molar masses). To calculate the mass per mole, we need two numbers, the mass and the number of moles in a given sample. Mass is measured with a balance, but how can moles be measured? One method is to measure a colligative property. This tells us the molality, molarity, or mole fraction of the solute, and thus the number of moles of solute in a unit quantity of the solution. Then the molar mass (in g/mol) is simply the mass (in grams) divided by the number of moles. (Make certain that the mass and the moles are adjusted to the same quantity of solution.) Freezing-point depression is the easiest of the properties to measure experimentally, except with solutes of very high molecular weight—such as proteins and starches (Chapter 25), for which osmotic pressure is more useful.

EXAMPLE 12.24 The freezing point of pure biphenyl was found to be 70.03°C. With the same biphenyl and equipment, the freezing point of a solution containing 0.100 g naphthalene in 10.0 g biphenyl was found to be 69.40°C. Find the molar mass of naphthalene.

ANSWER The freezing-point depression is

$$\Delta t_f = t_f - t_f' = 70.03° - 69.40° = 0.63°\text{C}$$

The freezing-point depression constant is

$$K_f = 8.00\,\frac{°\text{C kg}}{\text{mol}} \qquad \text{(Table 12.4)}$$

From the equation $\Delta t_f = K_f m$, we calculate the molality of naphthalene:

$$m = \frac{\Delta t_f}{K_f} = \frac{0.63°\cancel{\text{C}}}{8.00°\cancel{\text{C}}\text{ kg/mol}} = 0.079\,\frac{\text{mol naphthalene}}{\text{kg biphenyl}}$$

The mass of naphthalene is given for 10.0 g (not 1 kg) biphenyl. We must find either the number of moles of naphthalene in 10.0 g biphenyl or the mass of naphthalene per kilogram of biphenyl. We will take the second route:

$$w = \frac{0.100\text{ g naphthalene}}{10.0\text{ }\cancel{\text{g}}\text{ biphenyl}} \times 1000\,\frac{\cancel{\text{g}}}{\text{kg}} = 10.0\,\frac{\text{g naphthalene}}{\text{kg biphenyl}}$$

The molar mass is therefore

$$\frac{w}{m} = \frac{10.0\text{ g naphthalene}}{1\text{ }\cancel{\text{kg biphenyl}}} \times \frac{1\text{ }\cancel{\text{kg biphenyl}}}{0.079\text{ mol naphthalene}}$$

$$= 127\,\frac{\text{g}}{\text{mol}} \qquad \left(\text{actual value, }128\,\frac{\text{g}}{\text{mol}}\right)$$

This calculation assumes (correctly) that naphthalene exists in solution as molecules, not ions. ■

EXAMPLE 12.25 An aqueous solution containing 1.00 g bovine insulin (a protein, not ionized) per liter has an osmotic pressure of 3.1 torr at 25°C. Find the molar mass of bovine insulin.

ANSWER The osmotic pressure is related to the concentration by equation 21. First calculate the concentration of the insulin [B]:

$$[B] = \frac{\pi}{RT}$$

$$= 3.1 \cancel{\text{torr}} \times \frac{1 \cancel{\text{atm}}}{760 \cancel{\text{torr}}} \times \frac{1}{(0.08206 \text{ L} \cancel{\text{atm}}/\text{mol} \cancel{\text{K}})(298 \cancel{\text{K}})}$$

$$= 1.7 \times 10^{-4} \text{ mol/L}$$

We are told that there is 1.00 g insulin per liter. Then the molar mass is

$$\frac{1.00 \text{ g}}{1 \cancel{\text{L}}} \times \frac{1 \cancel{\text{L}}}{1.7 \times 10^{-4} \text{ mol}} = 6.0 \times 10^3 \frac{\text{g}}{\text{mol}} \qquad \left(\text{actual value, } 5733 \frac{\text{g}}{\text{mol}}\right)$$ ■

PROBLEM 12.19 (a) A solution is 4.00% (by mass) maltose and 96.00% H_2O. It freezes at $-0.229°C$. Calculate the molecular weight of maltose (not ionized). (b) The density of the solution is 1.014 g/mL. Calculate the osmotic pressure at 0°C. □

Mass spectrometry (section 3.3) is now the most common method of measuring molecular weights.

Solutes of high molecular weight, like insulin, give osmotic pressures that are easily measured, but they give imperceptibly small freezing-point depressions since the molalities are so small. Solutes of low molecular weight, like naphthalene, give convenient freezing-point depressions but very large osmotic pressures, and semipermeable membranes impermeable to these solutes can seldom be found. The two methods thus complement each other nicely. We use freezing-point depression when the molecular weight is low and osmotic pressure when it is high (Problems 12.85 and 12.86).

12.17 COLLOIDS

Imagine the strong beam of light from the projector in a darkened movie theater. Look at the beam from the side. If the air is very clean, you cannot see the beam at all. The air molecules—O_2, N_2, CO_2, H_2O—are much too small to scatter a detectable amount of light to your eyes. Now suppose that someone lights a cigarette. As the smoke drifts into the beam of light, the position of the beam becomes obvious. You cannot see the individual smoke particles, but they are large enough to scatter light at all angles. This observation is not peculiar to particles dispersed in gases. A beam of light passing through a transparent liquid or clear solid (like high-quality glass) is also invisible from the side. In a solution, the molecules do not scatter visible light. However, if the liquid contains particles with diameters of at least 5 nm (roughly), they will scatter the light as smoke particles in the air do. A dispersion of particles in this size range is a **colloidal dispersion**. The phase in which they are dispersed is the **dispersion medium** (analogous to the solvent in a solution). The ability of a colloidal dispersion to scatter light is called the **Tyndall effect**. Figure 12.17 illustrates an experiment in which the Tyndall effect provides a way to distinguish a colloidal dispersion from a true (molecular, atomic, or ionic) solution.

Discovered by Michael Faraday in 1857, and then studied in more detail by John Tyndall.

At the other end of the scale, what distinguishes a colloidal dispersion from a coarse suspension, like sand in water? The difference is that the particles in a coarse suspension settle to the bottom (or top) in a short time—seconds to hours—while colloidal particles settle very slowly or not at all. Figure 12.18 shows the particle sizes of some colloids and their approximate settling rates in air and water. Although the upper size limit of colloidal particles is often taken to be 200 nm in diameter, larger particles are sometimes included.

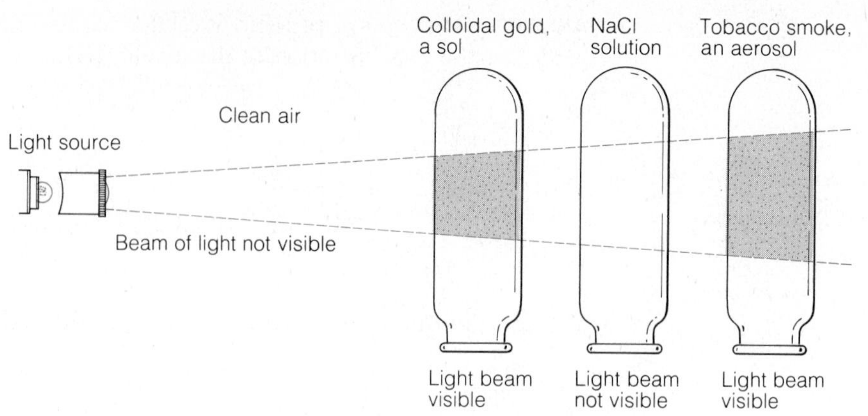

FIGURE 12.17
An experiment illustrating the Tyndall effect. Light is scattered by the colloids but not by the NaCl solution.

Some of the noticeable differences among solutions, colloidal dispersions, and coarse suspensions may be summarized as follows:

PROPERTY	SOLUTION	COLLOIDAL DISPERSION	COARSE SUSPENSION
Appearance in ordinary light	transparent	transparent	cloudy or opaque
Scattering of light from a strong beam	none	visible	visible (unless opaque)
Settling rate	zero	zero or slow	fast

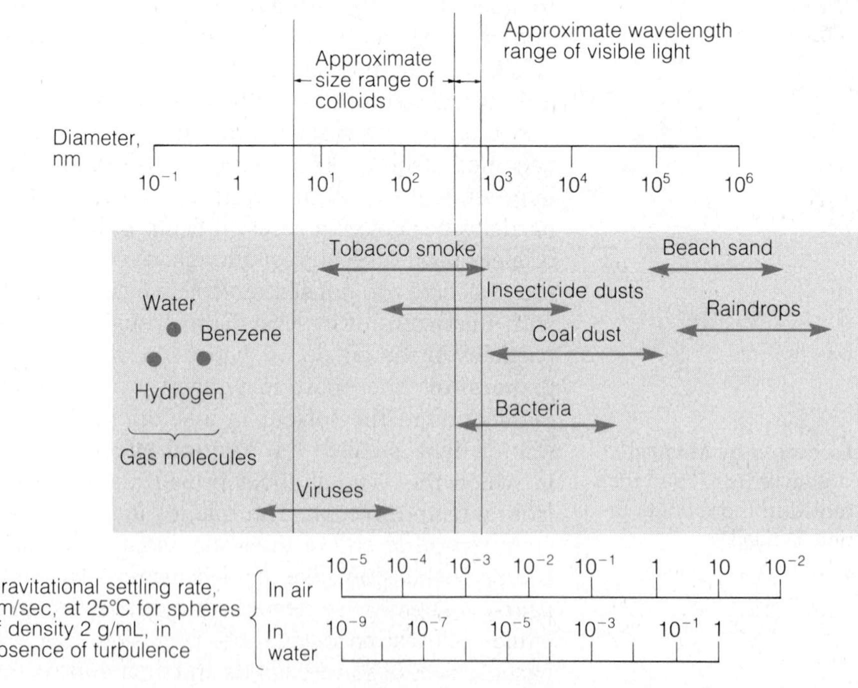

FIGURE 12.18
Sizes and settling rates of small particles.

BOX 12.6

DOES IT SETTLE OR DOESN'T IT?

Very small particles will not settle at all after they attain a certain equilibrium distribution. The equilibrium distribution of particles in a gravitational field is a condition in which the concentration of particles (the number of particles per unit volume) varies with height in a definite way. The numerical details of the distribution, but not its general appearance, depend on the mass of the particles, the buoyant effect of the medium, the temperature, and the local gravity. The same principles apply to gases, solutions, and colloids. With ordinary molecules in Earth's gravity, the equilibrium distribution is practically uniform—which is why air pressure does not vary much from the floor to the ceiling of a room. Over greater vertical distances, however, there is a considerable decrease in the concentration of molecules with increased altitude—this is why airplane cabins must be pressurized. The equilibrium distribution for colloidal particles involves much more settling because of their greater mass, but they are still kicked around by random collisions and therefore do not all stay at the bottom of the container.

An experiment that did much to convince people of the reality of molecules and the validity of the kinetic theory of gases was performed in 1908 by Jean Perrin. He prepared a dispersion in water of particles about 600 nm in diameter obtained from gutta-percha (a natural product similar to rubber). With the aid of a microscope, he counted the particles at each of several levels. The distribution he found was just what theory predicted for "molecules" the size of the rubber particles. He commented that the rarefaction of the particles in going up 10 μm was the same as the rarefaction of the atmosphere—where the particles have much smaller mass—in going up 6 km. He calculated Avogadro's number (known only roughly in 1908) from his observations and obtained 7×10^{23} particles/mol.

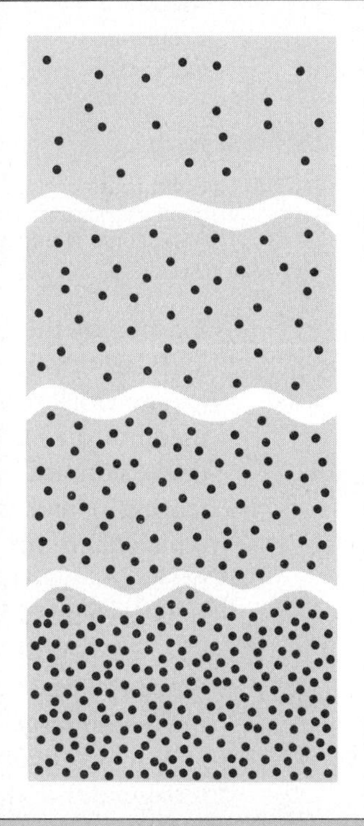

Distribution of particles as seen through a microscope focused at four different levels. The microscope is chosen to have a small depth of field, so that particles at other levels are out of focus. The levels are 10 μm apart. (After J. Perrin, *Annales de chimie et de physique*, **18**, 57 (1909).)

Some molecules, such as proteins, starch, and artificial polymers (Chapter 25), are so large that their "solutions" are colloidal. (*Colloid* means either "colloidal dispersion" or "the dispersed phase in a colloidal dispersion.") However, most colloidal particles consist of many molecules clumped together as they would be in an ordinary solid or liquid. Another type of colloidal particle is a tiny gas bubble. The dispersed phase and the dispersion medium can belong to any physical states (except both gaseous—gases always form solutions with each other). Table 12.6 lists the types of colloids and their names.

A colloid is notable for the very large surface area between the dispersed phase and the dispersion medium. When two immiscible liquids (like oil and water) or a liquid and an insoluble solid (like wax and water) are in contact, the area of contact counts as liquid surface. Thus the surface area of water is increased when droplets of oil are suspended in it. Work must be done to create this additional surface, just as work must be done by the lungs to blow up (increase the area of) a soap bubble. Work is necessary because molecules that were formerly in the

TABLE 12.6
TYPES OF COLLOIDAL DISPERSIONS

NAME	DISPERSED PHASE (particles)	DISPERSION MEDIUM	EXAMPLES
Aerosol	solid	gas	smoke
Sol	solid	liquid	AgCl, Au, etc., in H_2O; paint
	solid	solid	glass colored with dispersed metals
Aerosol	liquid	gas	fog, mist, clouds
Emulsion	liquid	liquid	homogenized milk
Gel	liquid	solid	jelly†
—	gas	gas	nonexistent
Foam	gas	liquid	whipped cream, shaving cream
Solid foam	gas	solid	pumice stone

† The liquid is trapped in long fibers of the solid.

interior of the liquid must be raised to the higher-energy state of being in the surface.

The work required depends on the degree of attraction between the dispersion medium and the molecules on the surfaces of the colloidal particles. Colloidal dispersions in water are commonly classified as **hydrophobic** (Greek, ''water-fearing'') and **hydrophilic** (''water-loving''). A hydrophobic colloid is one in which there are only weak attractive forces between the water and the surface of the colloidal particles. Examples are dispersions of metals and of nearly insoluble salts in water. When salts like AgCl and CuS precipitate (for example, $Ag^+ + Cl^- \longrightarrow AgCl$ or $Cu^{2+} + S^{2-} \longrightarrow CuS$), the result is often a colloidal dispersion. Since the reaction occurs too rapidly for ions to gather from long distances and make one large crystal, the ions can only form small particles.

Why do hydrophobic colloids exist? Why do the particles not come together (coagulate) and form bigger particles? The answer seems to be that the colloid particles carry electric charges of like sign. An AgCl particle, for example, will adsorb Ag^+ ions if they are present in substantial concentration, for there is a strong attraction between Ag^+ ions and the Cl^- ions in the surface of the particle. (The low solubility of AgCl is evidence of this attraction.) The particle thus becomes positively charged. In a solution containing a large concentration of Cl^-, an AgCl particle adsorbs Cl^- ions and becomes negatively charged. Of course, there must

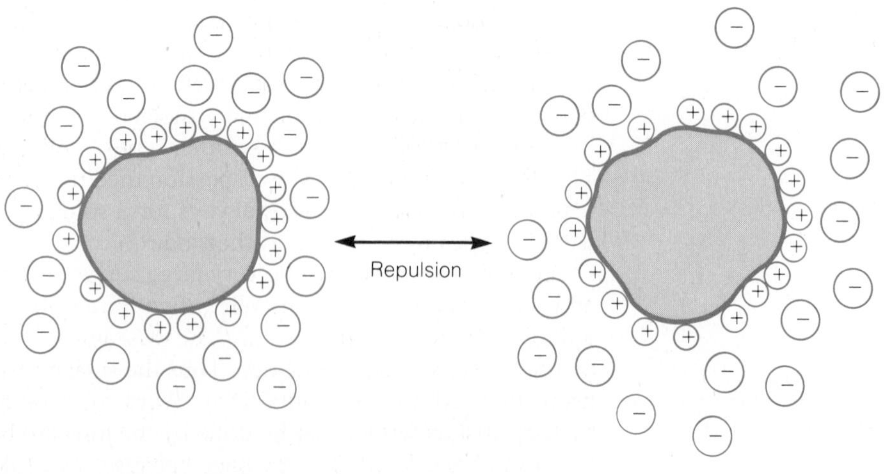

FIGURE 12.19
A hydrophobic colloid stabilized by positive ions adsorbed on each particle and an atmosphere of negative ions surrounding the particles.

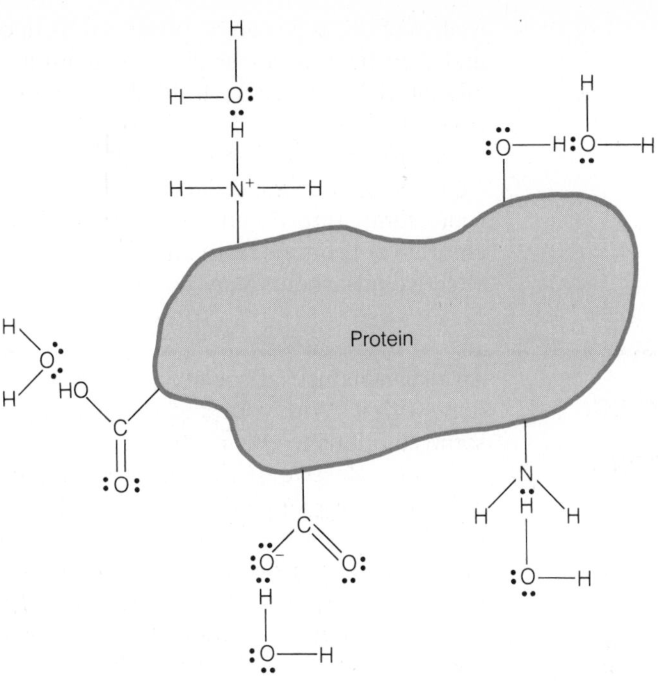

FIGURE 12.20
A hydrophilic colloidal particle stabilized by hydrogen bonding to water.

also be oppositely charged ions present in the solution—like NO_3^- with Ag^+ or Na^+ with Cl^-—and each colloid particle surrounds itself with a layer of these ions (Figure 12.19). The colloidal particles are thus prevented from coming into contact. If other ions are added that attach themselves directly to the colloidal particles and neutralize their charges, the colloid will coagulate. Coagulation may also occur slowly when the colloid is allowed to stand. The large particles grow and the small ones dissolve, because extremely small particles have above-normal solubility.

The particles in a hydrophilic colloid are often charged, but the charges are not necessary to the stability of the colloid.

Hydrophilic colloids are those that are strongly attracted to water molecules. They have groups like $-\ddot{\underset{\cdot\cdot}{O}}H$ and $-\ddot{N}H_2$ on their surfaces. These groups attach themselves to water molecules by hydrogen bonding, thus stabilizing the colloid (Figure 12.20). Proteins and starch are important examples, and homogenized milk is the most familiar example.

An **adsorbent** is a solid that can collect (*adsorb*) a substantial amount of another substance on its surface. A good adsorbent must have a relatively large surface area for its bulk volume. Colloids, because of their small particle size, are excellent adsorbents. Even large bodies, however, may have large surface-to-volume ratios if a network of pores within them provides extensive surface area (Figure 12.21).

(a)

(b)

(c)

FIGURE 12.21
Surface areas of solids. (a) Bulk solid has small surface area. (b) Small particles create large surface area. (c) Network of pores creates large surface area.

Matter of this type can be produced from coconut shells, for example, by charring and then heating the shells in a steam atmosphere to about 1500°C to drive out all material that can be gasified at that temperature. This process is called **activation**. The matter left behind, consisting mainly of carbon and retaining the many pores created by the activation process, is called **activated carbon**. The adsorbing capacity of such material is startlingly high. For example, activated carbon at room temperature can adsorb more than half its own weight of carbon tetrachloride or bromine from air that is saturated with the corresponding vapor. Other effective adsorbents may be made from metallic oxides, silicates, and metals.

12.18 SPECIAL TOPIC: WATER PURIFICATION

An urban industrial society pollutes enormous amounts of water. What can be done with it? With your knowledge of solutions, you are now prepared to consider some solutions to this problem.

A very effective method of purifying water is distillation. It is used all the time in laboratories, and indeed all of our fresh water is purified by natural distillation; heat from the Sun evaporates seawater and the vapor condenses as rain or snow. The trouble with distillation is that it is very expensive per unit mass of water purified. It requires 314 J to heat 1 g of water from 25° to 100°C, and another 2258 J to vaporize it—more than for any other substance because of the need to break hydrogen bonds. The only places where distillation has become practical for water purification on a large scale are those with plenty of petroleum and seawater but no fresh water, like Arabia. Therefore, different methods, operating at ordinary temperature and not requiring so much energy, are used in other places.

We will first outline the procedure for purifying a common category of polluted water, domestic sewage (Figure 12.22). The first stage, called (appropriately enough) **primary treatment**, is mechanical. The sewage is filtered through coarse

FIGURE 12.22 Sewage plant schematic, showing facilities for primary and secondary treatment. (From *The Living Waters*, U.S. Public Health Service Publication No. 382.)

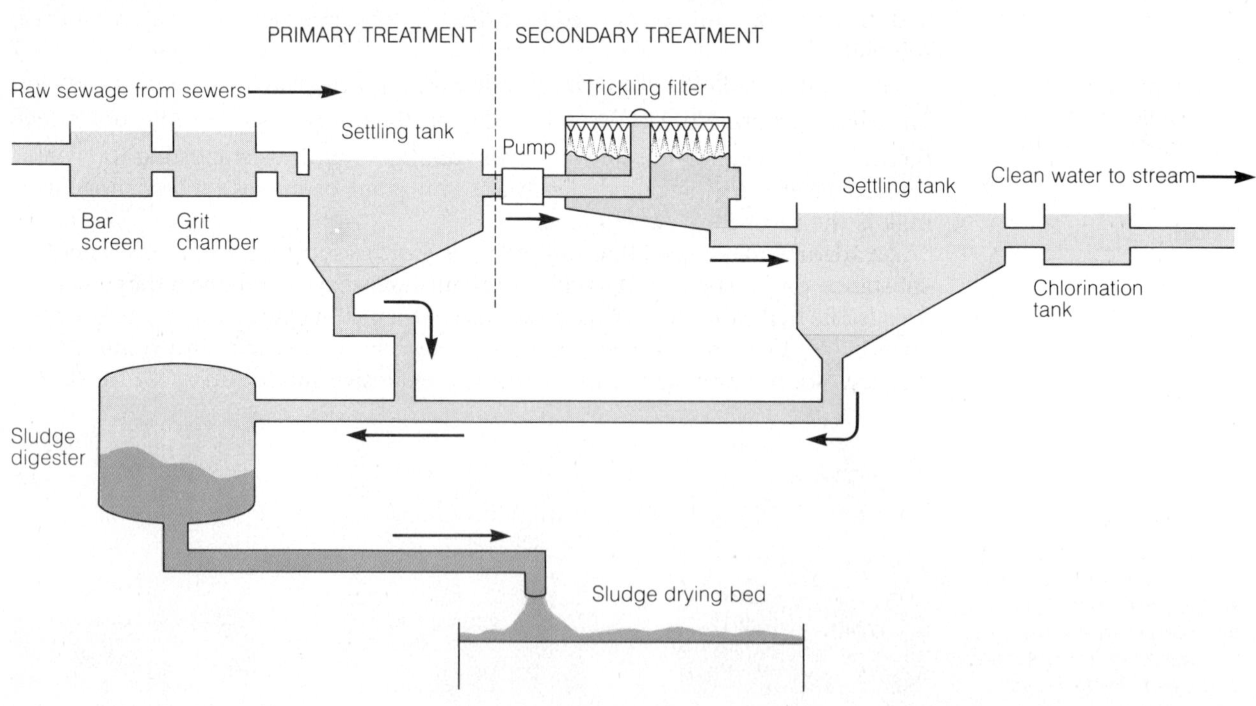

screens to remove large objects and then held in a pond or tank to remove dense particles, such as sand or grit, which settle out within an hour or so. These steps are not very costly and they are necessary preparation for further treatment, but they accomplish little. Though the resulting water may look cleaner, it is still heavily polluted with microorganisms and colloidal and dissolved matter.

The method of choice for purifying such water is *biologically induced oxidation*. Respiration is the oxidation of organic matter by living organisms. Of course, the organisms used for this purpose have not evolved simply to serve our needs by converting nutrients to carbon dioxide; respiration provides energy for their life processes, including growth and reproduction. These actions may be represented by this simple formulation:

Organic means "carbon-containing" (Chapter 24).

$$\text{organic nutrients} + O_2 \xrightarrow{\text{living organisms}} \begin{cases} CO_2 + H_2O \\ \text{increased body mass of organisms} \end{cases}$$

The important aspect of this conversion is that dissolved and colloidal organic matter is converted to CO_2 (which escapes as a gas) and to the bodies of organisms, which eventually coalesce into a sticky sludge that can be removed mechanically. This stage, which is called **secondary** or **biological treatment**, therefore requires a plentiful supply of oxygen (or air) and a large and varied population of microorganisms. These conditions may be satisfied by trickling the polluted water over beds of stones (Figure 12.23) on which many varieties of microbes and larger organisms are living. Alternatively, air may be pumped through a tank containing a mixture of polluted water and bacteria-laden ("activated") sludge.

The raw sludge is a watery, slimy, malodorous mixture of cellular protoplasm and other offensive (to us) residues. One environmentally sound way to dispose of sludge is to sterilize it and then utilize it as agricultural fertilizer. However, sludge may contain toxic metals, such as Pb and Cd, in which case it should not be used as fertilizer. Some treatment plants now digest the sludge anaerobically (without air), producing methane (CH_4), a valuable fuel.

FIGURE 12.23
A trickling filter with a section removed to show construction details. (Photo courtesy of Link-Belt/FMC.)

The water from which the sludge has been removed contains much less organic matter than before, but it is still laden with bacteria. The final step is therefore a disinfection process, usually chlorination, using Cl_2, or ozonization, using O_3. Both of these gases are strong oxidizing agents, fatal to living organisms from bacteria to human beings. Chlorine is produced from NaCl in large quantities and shipped as a liquid under pressure in tank cars, trucks, and barges. Ozone is prepared as needed by passing an electric discharge through air ($3\ O_2 \longrightarrow 2\ O_3$). The resulting mixture of O_3, O_2, N_2, etc., can be used as it is. The O_3 molecule is extremely reactive, readily giving up an O atom or combining with another molecule.

Although considerable purification is accomplished by the time wastewaters have passed through the primary and secondary stages, these treatments are still inadequate to deal with some complex aspects of water pollution. First, inorganic ions, such as nitrates and phosphates, remain in the treated waters. If chlorination is incomplete, microorganisms will remain. Chlorine can also react with dissolved organic matter; for example:

$$C_7H_{16} + Cl_2 \longrightarrow C_7H_{15}Cl + HCl$$

The chlorinated organic matter may impair the taste of the water and may even be poisonous. An advantage of ozonization is that these harmful compounds are not produced. It is also more effective than Cl_2 against viruses. A disadvantage of ozonization is that O_3 decomposes rapidly ($2\ O_3 \longrightarrow 3\ O_2$), while Cl_2 remains in the water to continue its work of sterilization.

Another problem is that many pollutants from such sources as factories, mines, and agricultural runoffs cannot be handled by municipal sewage treatment plants at all. Some synthetic organic chemicals from industrial wastes are not readily decomposed by living organisms; they are said to be **nonbiodegradable**. Such chemicals not only resist the bacteria of the purification system, but they may poison them and thereby prevent the biological oxidation that the bacteria would otherwise provide. There are also inorganic pollutants, including acids and metallic salts, that can poison the bacteria. Fine particles from roadways, construction sites, or irrigation runoff are troublesome before they settle, because they reduce the penetration of sunlight, and afterward, because they fill reservoirs, harbors, and stream channels with silt.

The treatment methods available for coping with these troublesome wastes are necessarily specific to the type of pollutant to be removed, and they are generally expensive. We will describe a few of these techniques, which collectively are called **tertiary treatment**.

1. *Coagulation and sedimentation.* As mentioned earlier in the discussion of biological treatment, it is advantageous to change little particles into big ones, because big ones settle faster. So it is also with inorganic pollutants. Various inorganic colloidal particles are water-loving (hydrophilic) and therefore rather adhesive. In their stickiness they sweep together many other colloidal particles that would otherwise fail to settle out in a reasonable time. This process is called **flocculation**. Salts of highly charged cations, such as the chlorides and sulfates of Fe(III) and Al(III), are among the best flocculating agents because they react with water to produce colloidal hydroxides (Problem 12.92).
2. *Adsorption.* The process of adsorption (section 12.17) is by no means restricted to gases; it also takes place in liquids. As in air, the agent of choice is activated carbon, which is particularly effective in removing

chemicals that produce offensive tastes and odors. These include the biologically resistant chlorinated hydrocarbons.

3. *Oxidizing agents.* In addition to their use as disinfectants, Cl_2 and O_3 can oxidize waterborne wastes that resist oxidation by air. Potassium permanganate, $KMnO_4$, is also used. Ozone has the important advantage that it produces no harmful by-products.

4. *Reverse osmosis.* Osmosis (section 12.15) is the process by which water passes through a membrane that is impermeable to dissolved ions. In the normal course of osmosis (Figure 12.24a), the system tends toward an equilibrium in which the concentrations on both sides of the membrane are equal. This means that the water flows from the pure side to the concentrated, "polluted" side. This is just what we do not want, for it increases the quantity of polluted water. However, if excess pressure is applied on the concentrated side (Figure 12.24b), the process can be reversed, and the pure water is squeezed through the membrane and thus freed of its dissolved ionic or other soluble pollutants.

5. *Deionization.* It is often necessary in the laboratory, home, or hospital to prepare modest quantities of **deionized water**—water that is free of ions. Deionized water may be needed, for example, as a solvent for chemical analysis, to replace evaporative loss from an automobile

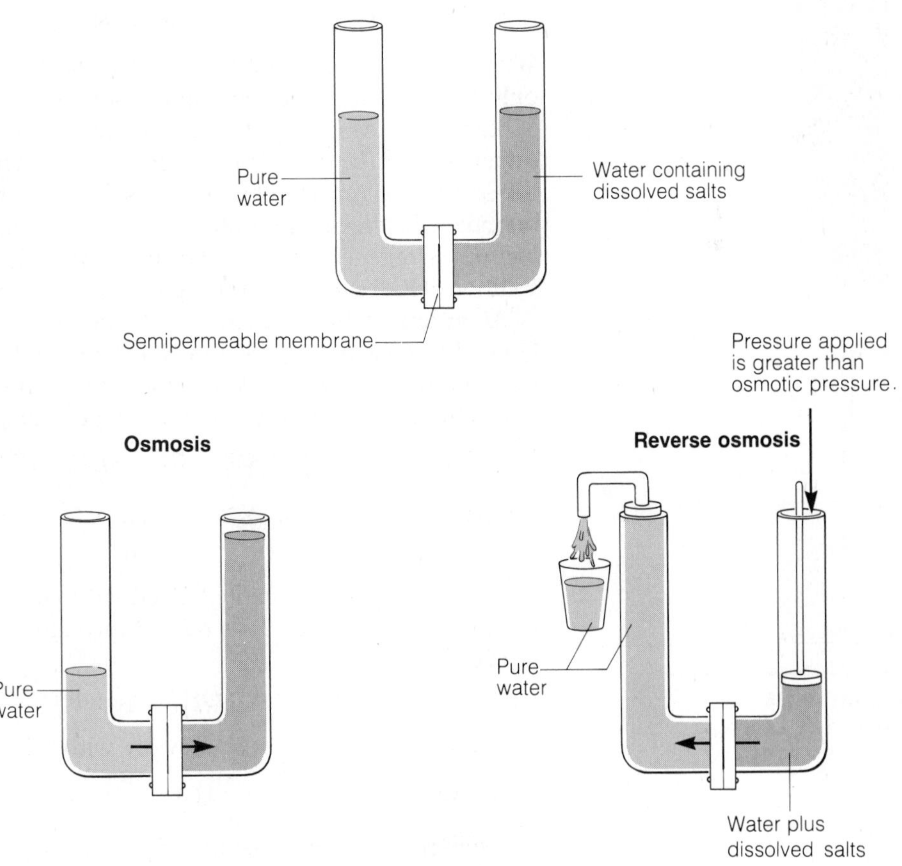

FIGURE 12.24
Osmosis and reverse osmosis.

battery, or to fill a steam iron. Apparatus for distillation or reverse osmosis is usually much too unwieldy for such applications. The objective can be accomplished by the use of **ion exchange**. The original ion exchangers are minerals called **zeolites**. They have been largely replaced by synthetic **ion-exchange resins**. These are polymers (giant molecules; Chapter 25) containing electrically charged sites of one sign (positive or negative) and mobile ions of the other sign. In a **cation** (positive ion) **exchanger** the immobile molecular network is negatively charged; the mobile ions are cations. A cation-exchange resin usually bears $—SO_3^-$ groups:

$$—CH_2—CH—CH_2—CH—CH_2—CH—CH_2—$$ (boxed group = R^-)

(each CH bears a benzene ring carrying $—SO_3^-$; mobile ions: H^+ H^+ H^+)

The hexagon represents a benzene ring (section 24.7).

Suppose the mobile positive ion is H^+ (short for H_3O^+; section 15.1). A solution containing other ions, say Na^+ and Cl^-, is passed through a column containing small beads of this resin (Figure 12.25). The negatively charged resin does not care what positive ions are in the vicinity, as long as there are enough to keep the net charge equal to zero. Ions can be transferred between the solution and the resin in either direction:

$$Na^+(aq) + H^+R^- \rightleftharpoons H^+(aq) + Na^+R^-$$

When the solution is first poured into the column, this reaction can go only from left to right because the solution contains Na^+ and the resin contains H^+. As the solution percolates down, most of the H^+ in the upper part of the resin is replaced by Na^+, while most of the Na^+ in the solution is replaced by H^+. As solution passes through, the boundary between Na^+-rich and H^+-rich resin moves down. The column is regenerated periodically by passing a concentrated H^+ solution (HCl or H_2SO_4) through it.

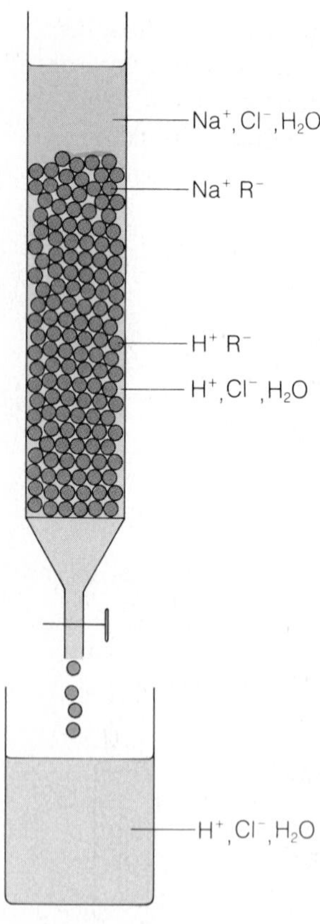

FIGURE 12.25
Column containing a cation-exchange resin.

An **anion-exchange resin** usually bears nitrogen atoms, each bonded to 4 carbon atoms (an NH_4^+ ion with all 4 H atoms replaced by C atoms). These positively charged sites are immobile; the negative ions (anions) attracted to them can be exchanged.

$$—CH_2—CH—CH_2—CH—CH_2—CH—CH_2—$$ (boxed group = R^+)

(each CH bears $(CH_3)_3N^+$ with OH^-: $(CH_3)_3N^+$ OH^- $(CH_3)_3N^+$ OH^- $(CH_3)_3N^+$ OH^-)

Suppose the resin is initially loaded with OH^- and the HCl solution produced in Figure 12.25 is passed through it. The Cl^- in solution is replaced by OH^-:

$$Cl^-(aq) + R^+OH^- \rightleftharpoons OH^-(aq) + R^+Cl^-$$

The H^+ in the solution reacts avidly with OH^-:

$$H^+(aq) + OH^-(aq) \longrightarrow H_2O$$

The sum of the three reactions is

$$Na^+(aq) + Cl^-(aq) + H^+R^- + R^+OH^- \longrightarrow Na^+R^- + R^+Cl^- + H_2O$$

Thus, the combined action of the cation- and anion-exchange resins removes ionic impurities from the water. A *deionizer* is made with both cation- and anion-exchange columns. The water passes through them in series and emerges as deionized (ion-free) water. Since most of the impurities in water are ionic, deionized water is as good for many purposes as distilled water.

SUMMARY

A **solution** is a mixture of two or more substances dispersed as atoms, molecules, or ions. The **solvent** (usually the majority component) dissolves the **solute**. A solution is **dilute** or **concentrated** when the solute/solvent ratio is small or large, respectively.

Solution compositions are commonly expressed in two ways:

$$\textbf{mass percentage} = \frac{\text{mass of solute}}{\text{mass of solution}} \times 100\%$$

$$\textbf{molarity} \text{ or } \textbf{concentration} \text{ of solute (B)} = [\text{B}] = \text{M}$$

$$= \frac{\text{number of moles of solute}}{\text{volume of solution in liters}} = \frac{\text{number of millimoles of solute}}{\text{volume of solution in milliliters}}$$

A **hydrate** is a salt that includes water in its crystal structure. Molarity and percentage can be interconverted if the density of the solution and the molar mass of the solute are known. When a solution is diluted, the molarities and volumes of the concentrated and dilute solutions are related by $M_c V_c = M_d V_d$.

A **saturated solution** is in equilibrium with undissolved solute. Its composition (as percentage of solute, molarity, etc.) is the **solubility** of the solute. Some liquids are completely **miscible** with each other; others are partially miscible and form two saturated solutions; some are nearly immiscible.

The solubility of a gas increases as the pressure increases. **Henry's law** says that the molarity of the saturated solution is approximately proportional to the pressure of the gas: $[\text{B}] \approx k_B p_B$.

The association of solute and solvent molecules is **solvation** (**hydration** if water is the solvent), which is usually exothermic. When the solute is a solid or liquid, solute molecules must be separated from each other in an endothermic process. The net dissolving process is usually exothermic for gases; it is endothermic, with many exceptions, for solid solutes. Solubility decreases with temperature when dissolving is exothermic and increases with temperature when dissolving is endothermic. A **supersaturated solution** is more concentrated than a saturated solution at the same temperature. It can be prepared by changing the temperature so that the solubility decreases and an unsaturated solution becomes supersaturated.

Solubility is favored by attractive forces between solvent and solute molecules, disfavored by solvent-solvent and solute-solute attraction. Substances are usually soluble in each other when their intermolecular attractions are similar: both strong (hydrogen bonding) or both weak (dipole-dipole and London forces). The more hydrogen-bonding sites a molecule has, the more soluble it is in water; the larger its nonpolar portion (as C and H), the less soluble it is in water.

Ionic crystals are often soluble in water because the water molecules, being polar, are attracted to the separated ions. The partial charges near the ions decrease their mutual attraction. The dissolving of an ionic crystal can be resolved into two steps: (1) Separate the ions to a great distance; ΔH_{LE} = **lattice energy**, positive. (2) Add the separated ions to water; ΔH_h = **hydration energy**, negative. For the net process, the **heat of solution**, $\Delta H_s = \Delta H_{LE} + \Delta H_h$, may be positive or negative.

The vapor pressure of a component in an unsaturated solution is less than the vapor pressure of the pure component at the same temperature. The direction of spontaneous transfer of a substance is to the place where its vapor pressure is lower. Two phases are in equilibrium when each component has the same vapor pressure in the two phases.

The **mole fraction** x_A of component A is the fraction of A by number of molecules or moles (n): $x_A = n_A/(n_A + n_B)$. The vapor pressure p_A of component A in a solution is often given to a useful approximation by **Raoult's law**: $p_A = x_A p_A^0$, where p_A^0 is the vapor pressure of pure A at the same temperature. This equation is applicable when the components are similar or when A is the solvent in a dilute solution. A solution described exactly by Raoult's law is an **ideal solution**. If the actual vapor pressure is greater than that predicted by Raoult's law, the deviation is positive—indicating stronger attraction between identical molecules than between different molecules. If the actual vapor pressure is less than predicted by Raoult's law, the deviation is negative—indicating stronger attraction between different than between identical molecules.

Colligative properties are vapor-pressure depression, freezing-point depression, boiling-point elevation, and osmotic pressure. They depend on the relative number, not the kind, of solute molecules.

In a dilute solution, the **vapor-pressure depression** of the solvent is $p_A^0 - p_A \approx x_B p_A^0$.

The **molality** (m) of a solution is the number of moles of solute per kilogram of solvent. For a dilute aqueous solution, it is approximately equal to the molarity.

The **freezing-point depression** is given by $\Delta t_f \approx K_f m$, where K_f is the **freezing-point depression constant** of the solvent. The **boiling-point elevation** is given by $\Delta t_b \approx K_b m$, where K_b is the **boiling-point elevation constant**. These changes are consequences of vapor-pressure depression, which extends the range of temperature in which the liquid is the stable phase at a fixed pressure.

For ionic solutions, the ratio of the measured colligative properties to the calculated colligative properties is the **van't Hoff factor**, i. For very dilute solutions, i is the number of ions in the formula. For more concentrated solutions, i is less because of interaction among the ions. The **osmolality** is im.

Osmosis is the transfer of solvent from the pure solvent to a solution through a **semipermeable membrane**. This membrane allows only solvent molecules to pass through. The **osmotic pressure** is the excess pressure that must be applied to the solution to stop osmosis. It is given by $\pi \approx [B]RT$, where [B] is the molarity, R is the gas constant, and T is the absolute temperature. When the excess pressure on the solution exceeds the osmotic pressure, **reverse osmosis** occurs. When the solute is ionic, the measured osmotic pressure is greater than calculated; their ratio is the van't Hoff factor.

Measurement of a colligative property gives the number of moles of solute. From this information and the mass of the solute, its molecular weight can be calculated. Freezing-point depression is most useful for solutes of low molecular weight, osmotic pressure for solutes of high molecular weight.

A **colloidal dispersion** contains particles larger than ordinary molecules but too small to be resolved with a visible-light microscope (about 5 to 200 nm in diameter). They are dispersed in a **dispersion medium**. They scatter light in all directions (the **Tyndall effect**). When the dispersion medium is water, a **hydro-**

phobic colloid has weak attraction between the dispersed particles and water; a **hydrophilic** colloid has strong attraction. A hydrophobic colloid is stabilized by charges resulting from ions adsorbed on the particles and oppositely charged ions nearby. A hydrophilic colloid has hydrogen-bonding groups on the surface of the particles. Colloidal particles have a very large surface area relative to their mass and are therefore excellent **adsorbents**.

12.19 ADDITIONAL PROBLEMS

PERCENTAGE AND MOLARITY

12.20 A solution is 8.00% H_2SO_4 by mass. (a) If you want 2.00 g of H_2SO_4, what mass of the solution do you need? (b) If you want 0.200 mol H_2SO_4, what mass of the solution do you need?

12.21 Calculate the molarity of each solution: (a) 15.0 g of NaCl in 250 mL of solution; (b) 75.0 g of C_2H_5OH in 500 mL of solution; (c) 17.0 g of $CuSO_4(H_2O)_5$ in 1000 mL solution.

12.22 At 20°C, the density of cyclopentane, C_5H_{10}, is 0.746 g/mL and the density of carbon tetrachloride, CCl_4, is 1.595 g/mL. Assuming no volume change on mixing, find the density of a solution prepared by mixing 40.0 g cyclopentane and 60.0 g carbon tetrachloride.

12.23 At 20°C, the density of ethanol, C_2H_5OH, is 0.7893 g/mL, the density of water is 0.9982 g/mL, and the density of a 50% (by mass) solution of ethanol in water is 0.9131 g/mL. A solution is prepared by mixing 50.0 g ethanol and 50.0 g water at 20°C. (a) What is the volume of the ethanol before mixing? (b) What is the volume of the water? (c) What is the volume of the solution? (d) Is the volume of the solution more or less than the sum of the volumes of the pure components? How much more or less?

12.24 An aqueous solution of potassium chromate is 20.00% K_2CrO_4 by mass. Its density is 1.172 g/mL at 20°C. Calculate the molarity of the solution.

12.25 The coefficient of expansion of H_2O near 20°C is 2.1×10^{-4}/°C; this means that the volume of a sample of H_2O is multiplied by 1.00021 for each 1°C that the temperature is raised. A certain aqueous solution is 0.1000 M at 20°C. Find its molarity at 25°C, assuming that the coefficient of expansion of the solution is the same as that of pure water.

12.26 A 0.210 M solution of KI in water has a density of 1.250 g/mL at 20°C. What is the percentage of KI in the solution?

12.27 A 0.600 M solution of sodium carbonate-water (1/10), $Na_2CO_3(H_2O)_{10}$, has a density of 1.061 g/mL at 20°C. Calculate the percentage of (a) Na_2CO_3 and (b) $Na_2CO_3(H_2O)_{10}$ in the solution.

12.28 A solution is prepared by dissolving 10.0 g lithium chloride, LiCl, in 50.0 g H_2O. (a) What is the percentage of LiCl in the solution? (b) Analysis of the solution shows that it is 4.31 M. Calculate its density.

DILUTION

12.29 A solution is prepared by diluting 50.0 mL of a 5.00 M solution of ammonia to a total volume of 1000 mL. What is the molarity of the resulting solution?

12.30 A 0.100 M solution of nitric acid is to be prepared by diluting a 3.00 M solution to 500 mL. What volume of the original solution should be used?

12.31 To make 1.00 L of a 0.025 M $KMnO_4$ solution, what volume of a 2.00 M solution must be diluted?

12.32 A 0.10 M solution of sodium thiosulfate, $Na_2S_2O_3$, is to be diluted to make 2.00 L of a 0.020 M solution. What volume of the 0.10 M solution should be used?

12.33 A bottle of concentrated ammonia solution ("ammonium hydroxide") has this on the label: "Assay 58% NH_4OH. Sp. gr. 0.90." The assay is the percentage by mass, and the specific gravity is numerically equal to the density in grams per milliliter. Find (a) the percentage by mass of NH_3 in the solution (note that NH_4OH is $NH_3 + H_2O$), (b) the molarity of NH_3, and (c) the volume of the solution that must be used to prepare 100 mL of 2.0 M solution by dilution with water.

12.34 A 25.00-mL sample of a Na_2CO_3 solution is diluted to 500.0 mL. Analysis of the resulting solution shows that 50.00 mL of it contains 0.2697 g of Na_2CO_3. What is the molarity of the original solution?

SOLUBILITY

12.35 The solubility of KCl in water at 20°C is 23.8 g KCl per 100 g H_2O. The density of the saturated solution is 1.133 g/mL. Calculate the solubility in (a) mass percentage and (b) molarity.

12.36 (a) Use the data in Table 12.2 to calculate the Henry's law constant for H_2 in water at 25°C. Remember to subtract the vapor pressure of water. Make reasonable approximations when necessary. (b) Calculate the molarity of a saturated solution of H_2 in water at 25°C in equilibrium with H_2 at a total pressure of 10.0 atm.

12.37 Estimate the solubility of O_2 in H_2O at 25°C under a total pressure of 10.0 atm in (a) grams O_2 per 100 g H_2O and in (b) volume in mL of O_2 (measured at 10.0 atm) per mL of H_2O. (Refer to Table 12.2.)

12.38 Dry air is about 3×10^{-4} mol % CO_2. Using Table 12.2, estimate the molarity of CO_2 in water in equilibrium with air when the total pressure is 1.00 atm and the temperature is 25°C.

12.39 A saturated solution of methane (CH_4; natural gas) in water is 1.35×10^{-5} M at 25°C when the pressure of methane is 1.00 atm. What is the solubility of methane in water at the same temperature and a pressure of 1000 atm, in (a) molarity and (b) mass percentage? Assume that the density of the solution is 1.00 g/mL.

12.40 Underline the correct words. When solid solute is added to an unsaturated solution, it (*will, will not*) dissolve. When 1 g of solute is added to a supersaturated solution, the amount of solid present at equilibrium will be (*more, less*) than 1 g. The solubility of K_2CO_3 in water increases with temperature, but the solubility of Li_2CO_3 in water decreases with temperature. To prepare a supersaturated solution of K_2CO_3, you must (*warm, cool*) an unsaturated solution. To prepare a supersaturated solution of Li_2CO_3, you must (*warm, cool*) an unsaturated solution. If you want a supersaturated solution, dust should be excluded because dust particles will (*make the solution impure, provide nuclei for crystal growth, prevent the solution from reaching equilibrium*).

12.41 For the process

$$KClO_3 + aq \longrightarrow KClO_3(aq)$$

$\Delta H = +41$ kJ. Does the solubility of $KClO_3$ increase or decrease as the temperature increases?

12.42 The solubility of sodium acetate-water (1/3), $NaC_2H_3O_2(H_2O)_3$, in water increases with increasing temperature. (a) Is the dissolving process exothermic or endothermic? (b) A crystal of this compound is added to a supersaturated aqueous solution. How does the temperature of this solution change as crystallization occurs?

SOLUBILITY AND MOLECULAR STRUCTURE

12.43 Draw the structural formula for a hydrogen-bonded combination of H_2O with (a) CH_3NH_2, (b) C_2H_5OH, and (c) $(CH_3)_2C{=}O$. The H atoms attached to C do not participate in hydrogen bonding.

12.44 (a) Would you expect N_2 to be more soluble in water or in oil (hydrocarbons)? Why? (b) How are your prediction and the fact that fatty tissue is less well supplied with blood vessels than other tissue related to decompression sickness?

12.45 In each pair, select the compound you would expect to be more soluble in water; explain.

(a) C_2H_5F and C_2H_5I
(b) C_3H_7OH and C_3H_7SH
(c) NH_3 and AsH_3
(d) $(CH_3)_3P$ and $(CH_3)_3N$
(e) $(C_3H_7)_3N$ and $(CH_3)_3N$
(f) benzene (ring of six CH: HC, CH, CH, CH, HC, CH) and pyridine (ring of HC, CH, CH, N, HC, CH)
(g) 1,4-dioxane (ring: O, CH_2, CH_2, O, CH_2, CH_2) and cyclohexane (ring of six CH_2)

12.46 In sickle-cell anemia, the protein hemoglobin has one abnormal amino acid portion: valine

$-N(H)-C(H)(CH(CH_3)_2)-C(=O)-$ instead of glutamic acid $-N(H)-C(H)(CH_2CH_2C(=O)OH)-C(=O)-$

Which form of hemoglobin, normal (glutamic acid) or sickle-cell (valine), should be more soluble in water?

12.47 In winter, an automobile can be disabled by ice in the gasoline line. To prevent this misfortune, something should be added to the gasoline that will mix with it and increase the solubility of water in it. Which of these compounds would you recommend?

(a) glycerol, $H_2C(OH)-CH(OH)-CH_2(OH)$
(b) methanol, CH_3OH
(c) dibromoethane, $H_2C(Br)-CH_2(Br)$

SOLUBILITY OF IONIC CRYSTALS

12.48 Refer to Figure 12.8. (a) A saturated solution of KBr containing 100 g of H_2O is prepared at 90°C. How many grams of KBr are dissolved? (b) This solution is cooled to 10°C. Assume that there is no supersaturation and that no water evaporates. What mass of KBr remains in solution? What mass of solid KBr precipitates? (c) What is the percentage by mass of KBr in each of the two solutions (hot and cold)?

12.49 The solubility of ammonium formate, NH_4CHO_2, in water is 102 g per 100 g of H_2O at 0°C and 546 g per 100 g of H_2O at 80°C. A solution is prepared by dissolving NH_4CHO_2 in 200 g H_2O until no more will dissolve at 80°C, and the solution is then cooled to 0°C. How many grams of NH_4CHO_2 precipitate? Assume that no water evaporates and that there is no supersaturation.

12.50 Saturated solutions of Na_2S in water contain 13.3% Na_2S (by mass) at 10°C and 37.2% Na_2S at 80°C. (a) A solution is prepared from 50.0 g Na_2S and 100.0 g H_2O at 80°C. Is this solution saturated? (b) The solution in part a is cooled from 80 to 10°C. What mass of Na_2S precipitates? Make the same assumptions as in the preceding problem.

12.51 The lattice energy of $CaCl_2$ is 2258 kJ/mol. Its heat of solution is -81.3 kJ/mol. Calculate the hydration energy of $Ca^{2+}(g) + 2Cl^-(g)$.

MOLE FRACTION

12.52 A solution contains 12.4 g NH_3 and 100.0 g H_2O. Find (a) the mass percentage and (b) the mole percentage of NH_3 in the solution. (c) Explain why the answers to (a) and (b) are nearly the same.

12.53 We want to prepare 100 g of a solution of ethylene glycol, $C_2H_6O_2$, and water in which the mole fraction of ethylene glycol will be 0.300. What mass of each component is needed?

12.54 Propellant (fuel and oxidizer) tanks in space vehicles are pressurized with various gases. These gases dissolve to some extent in the propellant, which may cause troublesome frothing when the pressure is decreased. The solubility of N_2 in liquid N_2O_4 at 20°C and 1 atm (N_2 pressure) is $x_{N_2} = 6.5 \times 10^{-4}$. Calculate (a) the mole fraction of N_2 in a saturated solution at 20°C and 100 atm (a reasonable tank pressure) and (b) the mass percentage of N_2 in the saturated solution. (c) What are you assuming about the relationship between solubility and pressure?

VAPOR PRESSURE

12.55 The vapor pressure of pure acetone, $(CH_3)_2CO$, at 25°C is 229 torr. (a) Calculate the vapor pressure of acetone in a 42.0 mol % solution of acetone in water, by Raoult's law. (b) The vapor pressure of acetone in this solution is 164 torr at 25°C. Is the deviation from Raoult's law positive or negative?

12.56 At 25°C, the vapor pressure of liquid benzene, C_6H_6, is 95 torr; the vapor pressure of liquid toluene, C_7H_8, is 28 torr. Assume that solutions of these two compounds are ideal. Find the total vapor pressure at 25°C of a liquid solution containing 60% benzene and 40% toluene by mass.

12.57 The vapor pressures of carbon disulfide, CS_2, and ethyl acetate, $CH_3COOC_2H_5$, at 27°C are 390 torr and 100 torr, respectively. Find the total vapor pressure at 27°C of a solution prepared by mixing equal masses of CS_2 and $CH_3COOC_2H_5$. Assume that Raoult's law is applicable.

12.58 Cigars are best stored at 18°C and 55% relative humidity. This means that the pressure of water vapor should be 55% of the vapor pressure of pure water at the same temperature. The proper humidity can be maintained by placing in the humidor a solution of glycerol, $C_3H_8O_3$, and water. The vapor pressure of glycerol is practically zero. Calculate the percentage by mass of glycerol that will lower the vapor pressure of water to the desired value. Assume that Raoult's law is applicable.

12.59 The vapor pressure of pure mercury at 322°C is 392 torr. An amalgam (alloy) is prepared by mixing equal numbers of moles of mercury and another metal. When the other metal is Cd, the vapor pressure of Hg in equilibrium with this amalgam is 133 torr at 322°C. When the other metal is Pb, the vapor pressure of Hg is 272 torr. For each amalgam, is the deviation from Raoult's law positive or negative?

12.60 A solution of mannitol, $C_6H_{14}O_6$, in water is 0.100 molal. (a) What is the mole fraction of mannitol in this solution? (b) What is the vapor-pressure depression at 25°C?

MOLALITY

12.61 Calculate the molality of each solution: (a) 40.0 g NaI in 1500 g H_2O; (b) 40.0 g NaI in 1500 g solution; (c) 30.0 g CH_3OH in 500 g H_2O.

12.62 A solution consists of 10.0 g glycine, H_2NCH_2COOH, and 40.0 g H_2O. (a) Calculate the molality of the solution with respect to glycine. (b) Calculate the mole fraction of glycine in the solution.

12.63 A solution of potassium sulfate, K_2SO_4, is 0.0500 molal. Calculate the mass percentage of K_2SO_4.

12.64 Formalin, used for preserving biological specimens and embalming the dead, is 37% formaldehyde (CH_2O) by mass and 63% water. Calculate (a) the mole fraction and (b) the molality of formaldehyde in the solution.

12.65 A 1.00 M solution of potassium chromate, K_2CrO_4, has a density of 1.143 g/mL at 20°C. Calculate its molality.

12.66 A solution is prepared by dissolving 15.0 g copper(II) nitrate-water (1/6), $Cu(NO_3)_2(H_2O)_6$, in 100 g H_2O. Calculate (a) the mass percentage and (b) the molality of $Cu(NO_3)_2$ in the solution. (Note that the mass of water in this solution is more than 100 g.)

12.67 A solution of sodium thiosulfate-water (1/5), $Na_2S_2O_3(H_2O)_5$, is 0.857 *m*. This means 0.857 mol $Na_2S_2O_3$ per kg *total* water. Its density is 1.100 g/mL. Calculate (a) the percentage by mass of $Na_2S_2O_3$, (b) the percentage by mass of $Na_2S_2O_3(H_2O)_5$, and (c) the molarity.

12.68 An unleaded gasoline contains 1.00×10^{-3}% by mass of the antiknock agent methylcyclopentadienyl manganese tricarbonyl, $C_9H_7O_3Mn$. The density of the gasoline is 0.78 g/mL. Calculate (a) the molarity and (b) the molality of the solution. (c) Are molarity and molality approximately equal in this case? Why or why not?

12.69 "Concentrated sulfuric acid" is 96% H_2SO_4 and 4% H_2O by mass. Its density is 1.84 g/mL. Treating H_2SO_4 as the solute, calculate (a) the molarity and (b) the molality of the solution. (c) What happens to the molality of a solution as all solvent is removed and it becomes pure solute?

12.70 Let B be the solute and A the solvent in a solution. (a) Express the molality, m_B, in terms of the mole fraction [$x_B = n_B/(n_A + n_B)$] and the molecular weights M_A and M_B. (b) Express the molarity [B] in terms of x_B, M_A, M_B, and d, the density of the solution. (*Suggestion*: First express m_B and [B] in terms of n_A, n_B, M_A, M_B, and if necessary, d.) (c) Show that m_B and [B] are both proportional to x_B in solutions that are sufficiently dilute (x_B much less than 1, and d nearly equal to the density, d_0, of pure A).

12.71 Two aqueous solutions of acetic acid, CH_3COOH, have mole fractions of acetic acid 0.0100 and 0.0200.

Their densities are 1.0029 and 1.0074 g/mL, respectively. Using the results of the preceding problem, find (a) the molalities and (b) the molarities of acetic acid in these two solutions. Find (c) the ratio of the molality in the second solution to the molality in the first, and (d) the corresponding ratio of molarities. Compare these ratios with the ratio (2:1) of the mole fractions.

FREEZING-POINT DEPRESSION

12.72 An engine coolant is prepared by mixing 1.00 kg of water and 1.11 kg of ethylene glycol ($C_2H_6O_2$). (a) What is the molality of the coolant, considering water the solvent? (b) Estimate its freezing point. Why is your answer only a rough estimate?

12.73 The two most commonly used antifreezes are methanol, CH_3OH, and ethylene glycol, $(CH_2OH)_2$. (a) Estimate the mass of each of these solutes that must be added to 1 kg of water to prevent the water from freezing at −10°C. Why are your answers only estimates? (b) A certain solution of water and methanol has a freezing point of −10°C. This means that pure ice deposits when the solution is cooled to −10°C. An automobile radiator is filled with this solution, tightly capped, and kept at −11°C for an indefinitely long time. Will the liquid in the radiator eventually freeze solid? Explain.

12.74 A solution is prepared by melting together 0.961 g of anthracene ($C_{14}H_{10}$) and 10.0 g of paradichlorobenzene ($C_6H_4Cl_2$; melting point 53.1°C). The solution freezes at 49.3°C. Calculate the freezing-point depression constant of paradichlorobenzene.

COLLIGATIVE PROPERTIES OF IONIC SOLUTIONS

12.75 Estimate the freezing point of a 0.0100 molal aqueous solution of (a) K_2CO_3; (b) $La(NO_3)_3$.

12.76 A 0.00200 molal aqueous solution of $Co(NH_3)_5NO_2Cl_2$ freezes at −0.0105°C. How many moles of ions are present in a dilute solution containing 1 mole of this compound?

12.77 A 1.00% (by mass) solution of tetracainium chloride ($C_{15}H_{25}ClN_2O_2$, 300.8 g/mol) in water freezes at −0.110°C. Calculate the van't Hoff factor. How many moles of ions are produced when 1 mole of this compound is dissolved to make a dilute solution?

12.78 A 2.00% solution of H_2SO_4 in water freezes at −0.796°C. (a) Calculate the van't Hoff factor for the solution. (b) Which of the following formulas best represents sulfuric acid in a dilute aqueous solution? H_2SO_4, $H^+ + HSO_4^-$, $2H^+ + SO_4^{2-}$.

OSMOTIC PRESSURE

12.79 A solution contains 3.00% phenylalanine, $C_9H_{11}NO_2$, and 97.00% water by mass. Assume that phenylalanine is nonionic and nonvolatile, that the solid that separates on freezing is pure ice, and that the density of the solution is 1.00 g/mL. Find (a) the vapor-pressure depression of the solution at 25°C, (b) the freezing point of the solution, (c) the boiling point of the solution at 1 atm pressure, and (d) the osmotic pressure of the solution at 25°C, with a membrane permeable to water but not to phenylalanine.

12.80 A protozoan (single-celled animal) that normally lives in the ocean is placed in fresh water. Will it shrivel or burst?

12.81 (a) A tree is 10 m tall. What must be the total molarity of solutes in its sap if the sap rises to the top of the tree by osmotic pressure at 20°C? Assume that the groundwater outside the tree is pure H_2O and that the density of the sap is 1.0 g/mL. (1 torr = 1 mm Hg = 13.6 mm H_2O.) (b) If the only solute in the sap is sucrose, $C_{12}H_{22}O_{11}$, what is its percentage by mass?

12.82 Concentrations of the principal ions in seawater:

ION	MOLARITY
Na^+	0.470
Mg^{2+}	0.054
Ca^{2+}	0.010
K^+	0.010
Cl^-	0.548
SO_4^{2-}	0.028
HCO_3^-	0.002
Br^-	0.001

(a) Calculate the freezing point of seawater. Make reasonable approximations. (b) Calculate the osmotic pressure of seawater at 25°C. What minimum pressure is needed to purify seawater by reverse osmosis?

DETERMINATION OF MOLECULAR WEIGHTS

12.83 When 0.163 g sulfur is finely ground and melted with 4.38 g camphor, the freezing point of the camphor is lowered by 5.47°C. What is the molecular weight of sulfur? What is its molecular formula?

12.84 An aqueous solution contains 0.180 g of an unknown solute and 50.0 g H_2O. It freezes at −0.040°C. Find the molecular weight of the solute.

12.85 An aqueous solution containing 10.0 g starch per liter gave an osmotic pressure of 3.8 torr at 25°C. (a) Find the molecular weight of the starch. (Because not all the starch molecules are identical, the result will be an average.) (b) What is the freezing point of the solution? Would it be easy to find the molecular weight of the starch by measuring the freezing-point depression? Assume that molality and molarity are equal for this solution.

12.86 (a) An aqueous solution containing 10.0 g thyroglobulin (a protein) per liter shows an osmotic pressure of 3.9 mm H_2O (0.29 torr) at 27°C. Find the molecular weight of the protein. (b) Estimate the freezing-point depression of the solution. (c) Which method, osmotic pressure or freezing-point depression, would you recommend for determining molecular weights of proteins?

12.87 A solution of 5.00 g acetic acid in 100 g benzene freezes at 3.38°C. A solution of 5.00 g acetic acid in 100 g water freezes at −1.49°C. Find the molecular weight of acetic acid from each of these data. What can you conclude about the state of the acetic acid molecules dissolved in each of these solvents?

COLLOIDS

12.88 Classify each of the following colloidal dispersions by identifying the state of matter of the dispersed phase (the particles) and the dispersion medium, as given in Table 12.6: (a) water polluted with viruses, (b) the hazy blue air that rises above hot oil in a frying pan, (c) blue diamonds. (Consult an encyclopedia to learn why some diamonds are blue.)

12.89 When solutions of $BaCl_2$ and Na_2SO_4 are mixed, a cloudy liquid is produced. After a few days, a white solid is on the bottom and a clear liquid is above it. (a) Write the equation for the reaction that occurs. (b) Why is the liquid cloudy at first? (c) What happens during the few days of waiting?

12.90 It is observed that when air containing gasoline vapor passes through a bed of activated carbon, the gasoline vapor is removed from the air and the temperature of the carbon bed rises. Account for these phenomena.

12.91 The dispersed phase in a certain colloidal dispersion consists of spheres of diameter 100 nm. (a) What is the volume and the surface area of each sphere? ($V = (4/3)\pi r^3$, $A = 4\pi r^2$) (b) How many spheres have a total volume of 1 cm^3? What is the total surface area of these spheres in square meters?

WATER PURIFICATION

12.92 Aluminum sulfate, $Al_2(SO_4)_3$, is often added to impure water. It reacts with H_2O to form solid $Al(OH)_3$, which collects colloidal impurities. Write the net ionic equation (section 9.2) for this reaction.

12.93 The following are present in a polluted stream: beer cans, algae, bacteria, sand, colloidal clay, $(NH_4)_3PO_4$ (from fertilizer), and 2,4-dichlorophenoxyacetic acid ("2,4-D," $C_8H_6Cl_2O_3$, a herbicide). In which stage of water treatment (primary, secondary, or tertiary) would each of these contaminants be removed? What specific kinds of tertiary treatment would you recommend?

12.94 It is possible to remove Ca^{2+} or Mg^{2+} ions from water and replace them with Na^+ ions (a process called *water softening*) by means of an ion-exchange resin. (a) What kind of ion-exchange resin (cation-exchange or anion-exchange) must be used? Write the equation for the reaction. (b) When the resin is exhausted, how can it be regenerated? Write the equation for the regeneration.

12.95 In a certain city, the water contains 6.0 mmol Ca^{2+} and 3.0 mmol Mg^{2+} per liter. The water is softened by ion exchange, as in the preceding problem. (a) How many millimoles of Na^+ per liter are needed to replace all of the Ca^{2+} and Mg^{2+}? (b) A household uses 2.0×10^4 L of water per month. What mass of NaCl per month is needed to soften the water?

SELF-TEST

12.96 (14 points) (a) A solution is prepared by dissolving 10.00 g sulfamic acid (HSO_3NH_2, 97.09 g/mol) in water to make 500.0 mL. What is the molarity of this solution? (b) A 50.00-mL portion of the solution prepared in (a) is diluted with water to a total volume of 250.0 mL. What is the molarity of the diluted solution?

12.97 (8 points) Calculate the molarity of a 50.0% (by mass) solution of methanol (CH_3OH) in water (considered the solvent). The density of the solution is 0.9156 g/mL.

12.98 (12 points) The solubility of gaseous dinitrogen oxide (N_2O) in water is 0.058 mol/L at 0°C and 1.00 atm. (a) What is the solubility in mol/L at 0°C and 5.00 atm? (b) What is the solubility at 0°C and 1.00 atm, expressed as liters of N_2O (measured at these conditions) per liter of solution? Pressures are N_2O partial pressures.

12.99 (8 points) In each pair, select the compound that is more soluble in water: (a) $(C_2H_5)_2S$, $(C_2H_5)_2O$; (b) $(C_2H_5)_2CO$, $(C_6H_{11})_2CO$.

12.100 (14 points) (a) A solution is prepared by mixing 10.0 g bromobenzene (C_6H_5Br, 157.01 g/mol) and 20.0 g iodobenzene (C_6H_5I, 204.01 g/mol). Calculate the mole fraction of each component in the solution. (b) The vapor pressures of pure bromobenzene and iodobenzene at 30°C are 5.68 torr and 1.46 torr, respectively. What is the total vapor pressure of the solution prepared in (a)? Assume that Raoult's law is applicable.

12.101 (12 points) A solution is prepared by dissolving 1.50 g of nicotinonitrile ($C_6H_4N_2$, 104.11 g/mol) in 100 g H_2O. (a) What is the molality of this solution? (b) What is its freezing point? (The solute is not ionized.)

12.102 (12 points) A 0.500% (by mass) aqueous solution of potassium hexacyanoferrate(III), $K_3Fe(CN)_6$ (329.3 g/mol) freezes at −0.093°C. (a) Calculate the molality of the solution. (b) What is the van't Hoff factor? (c) How many moles of ions are produced when 1 mole of this compound dissolves? Consider both the formula and the result of part b. What are the ions?

12.103 (12 points) A 1.00% (by mass) solution of creatinine in water freezes at −0.150°C. (a) Calculate the molecular weight of creatinine (not ionized). (b) What is the osmotic pressure of this solution at 0°C? Assume that 1 L of solution contains 1 kg H_2O.

12.104 (8 points) Starch contains C—C, C—H, C—O, and O—H bonds. Hydrocarbons contain only C—C and C—H bonds. Both starch and hydrocarbon oils can form colloidal dispersions in water. (a) Which dispersion is classified as hydrophobic? Which is hydrophilic? (b) Which dispersion requires more work for its formation?

13 CHEMICAL EQUILIBRIUM

Back in Chapters 3, 4, and 9, you studied chemical reactions. You learned how to calculate, from a balanced chemical equation, the mass of product that can be obtained. However, having a balanced equation on a slip of paper is no assurance that the reaction will actually occur. In those earlier chapters we just assumed that reactions proceed as the equations indicate. Now we are ready to explore that central question about a chemical reaction: will it occur or won't it? Or will it occur in the reverse direction? The answer is more complicated than "Yes, the reaction occurs," or "No, it doesn't." There are all degrees in between. Some reactions proceed very slowly; others, rapidly. Even if you waited forever, you would find that most reactions do not go completely in one direction or the other; they appear to stop with only some of the reactants converted to products. This chapter will show why a reaction *appears* to stop—it really does not—and how to predict the final composition of a reacting mixture.

13.1 THE APPROACH TO EQUILIBRIUM

Thus far in our study of chemical reactions, we have imagined that every reaction (if it occurs at all) goes to completion; that is, we have assumed that when A and B react to give C and D

$$A + B \longrightarrow C + D$$

the reaction continues until either A or B has been entirely consumed. Section 3.13 hints that this is not usually true. You have, indeed, studied some processes that arrive at a state of *equilibrium* with both starting materials and products present. A liquid in a closed container evaporates

$$H_2O(\ell) \longrightarrow H_2O(g)$$

but it may not *all* evaporate. If the container is not large enough, the vapor builds up to a certain pressure (the vapor pressure of the liquid) and the process appears to stop, because the reverse process

$$H_2O(g) \longrightarrow H_2O(\ell)$$

is occurring equally fast (section 11.10). Likewise, when a solution becomes saturated, a process like $NaCl(c) \longrightarrow NaCl(aq)$ is balanced by the reverse process, $NaCl(aq) \longrightarrow NaCl(c)$.

A chemical reaction can also reach a state of equilibrium whenever products are not allowed to escape from a container. Thus a gaseous mixture of hydrogen and iodine vapor, in a closed vessel, reacts to form hydrogen iodide:

$$H_2(g) + I_2(g) \longrightarrow 2HI(g) \tag{1}$$

If we had started with pure HI at the same temperature, it would have decomposed to give hydrogen and iodine:

$$2HI(g) \longrightarrow H_2(g) + I_2(g) \tag{2}$$

Now, the HI that is formed in reaction 1 acts like any other HI. As soon as some of it has been formed, the decomposition (reaction 2) begins. The H_2 and I_2 formed in reaction 2 will react to give HI, as in reaction 1. Thus, in any mixture of the three gases H_2, I_2, and HI, reactions 1 and 2 will *both* take place. This fact is expressed by writing the equation with a double arrow:

$$H_2 + I_2 \rightleftharpoons 2HI \tag{3}$$

or

$$2HI \rightleftharpoons H_2 + I_2 \tag{4}$$

Whatever appears on the left side of an equation is called a "reactant"; whatever appears on the right side is called a "product." This choice of words stems from the convention of reading from left to right, even though the double arrow indicates that the reaction can go either way. We say that a reaction goes "forward" or "as written" when it goes from left to right ($\longrightarrow$), and "backward" or "in reverse" when it goes from right to left ($\longleftarrow$). For reaction 3, $\Delta H = -10.4$ kJ; this means that -10.4 kJ is the enthalpy change in the *left-to-right* reaction. If we choose to write the reaction in the other direction, as in reaction 4, ΔH becomes $+10.4$ kJ, again for the reaction read from left to right.

REACTION RATES AND EQUILIBRIUM

The concept of **rate** is central in chemistry, as indeed in all of life. Rate is *something* per unit of time: speed in kilometers per hour, a current of water in liters per minute, or a child's growth in kilograms per month. We discussed rate in section 11.10. Recall that the rate of evaporation is the number of moles evaporating per second, and the rate of condensation is the number of moles returning to the liquid (or solid) per second. The units can, of course, be molecules or grams instead of moles, and minutes or hours instead of seconds. Equilibrium is attained when the two opposing rates—evaporation and condensation, forward and backward, making and unmaking—become equal. Once equilibrium has been reached, we can forget about rates because the system will remain unchanged forever, or until some condition is changed.

The rates of the forward and reverse reactions depend upon the concentrations of the reactants, of the products, and sometimes of other substances. They also depend on the temperature. The subject of rates is so interesting and important that Chapter 19 is entirely devoted to it. Here, it will suffice to note one property of reaction rates: they usually increase as the concentrations of the reactants increase, and decrease as the concentrations decrease.

Figure 13.1 shows the results of some experiments on the reaction

$$H_2(g) + I_2(g) \rightleftharpoons 2HI(g)$$

The horizontal axis represents the time elapsed since the beginning of the reaction. The vertical axis represents the composition of the reaction mixture, from $H_2 + I_2$ at the bottom to pure HI at the top. One series of experiments (the lower curve)

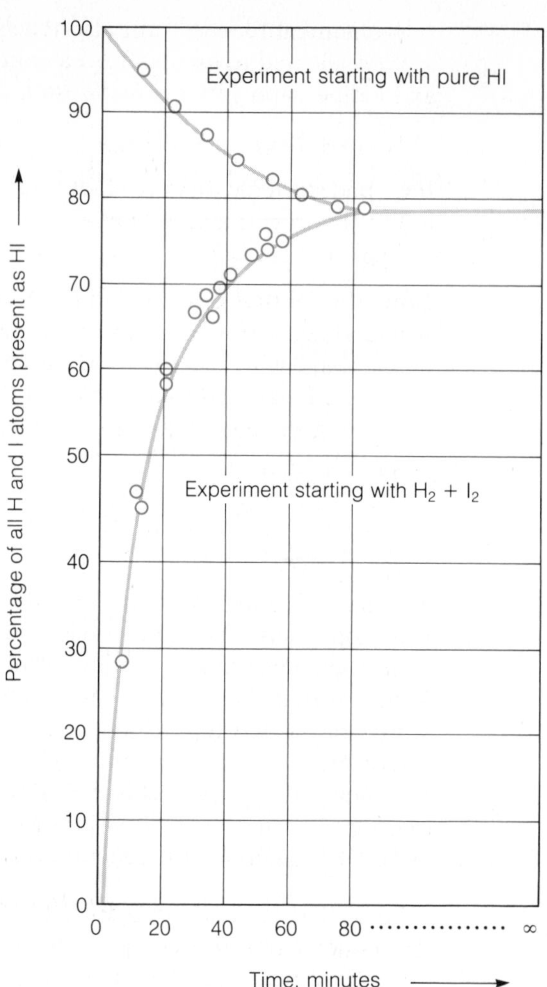

FIGURE 13.1
Experimental data on the reaction $H_2(g) + I_2(g) \rightleftharpoons 2HI(g)$ at 445°C. (After M. Bodenstein, *Z. Phys. Chem*, *13*, (1894) 111.)

starts with $H_2 + I_2$ (the same number of moles of one as of the other), but no HI. What happens as time passes? As the "forward" reaction, $H_2 + I_2 \longrightarrow 2HI$, occurs, the concentrations of H_2 and I_2 decrease. The rate of the forward reaction, r_f, therefore decreases. Meanwhile, the concentration of HI is increasing. The rate of the "reverse" reaction, $2HI \longrightarrow H_2 + I_2$, was initially zero because there was no HI, but it increases as [HI] increases. As the forward rate decreases and the reverse rate increases, the net rate of production of HI decreases. This decrease is apparent in Figure 13.1 in the leveling off of the lower curve. Eventually, the rates of the forward and reverse reactions become equal. When the rates are equal, there is no longer any *net* change in the concentrations. The reactions have not stopped; no one has told the molecules not to react when they meet each other. The forward and reverse reactions are still going on, but they are occurring at equal rates. The mixture is now said to be in *equilibrium.* The condition of equilibrium is recognized experimentally when concentrations that initially were changing have ceased to change. As long as concentrations are changing with time, a mixture has not reached equilibrium.

Another series of experiments (the upper curve in Figure 13.1) starts with pure HI. In this reaction, the rate of decomposition of HI ($2HI \longrightarrow H_2 + I_2$) decreases while the rate of production of HI ($H_2 + I_2 \longrightarrow 2HI$) increases. The curve levels off at the same equilibrium composition as in the experiments starting from $H_2 + I_2$.

EXAMPLE 13.1 Refer to Figure 13.1. (a) An experiment is started with 1.00 mole of HI at 445°C. How many moles of HI, H_2, and I_2 will be present after 20 minutes; 2 hours; 3 hours? (b) An experiment is started with 0.50 mole of H_2 and 0.50 mole of I_2. How many moles of HI, H_2, and I_2 will be present after 20 minutes; 2 hours; 3 hours?

ANSWER

(a) The upper curve shows that after 20 min, 91% of the H and I atoms will still be in the form of HI:

$$1.00 \text{ mol H} \times 91\% = 0.91 \text{ mol H (in HI)} = 0.91 \text{ mol HI}$$

The 0.09 mol HI that will have been consumed will have produced 0.045 mol H_2 and 0.045 mol I_2:

$$2HI \longrightarrow H_2 + I_2$$

$$0.09 \, \cancel{\text{mol HI}} \times \frac{1 \text{ mol } H_2}{2 \, \cancel{\text{mol HI}}} = 0.045 \text{ mol } H_2$$

and the same for I_2. After 2 hr, 3 hr, or some longer time, the reaction will have come to equilibrium, with the upper and lower curves merged at 78% HI:

$$1.00 \text{ mol H} \times 78\% = 0.78 \text{ mol H (in HI)} = 0.78 \text{ mol HI}$$

The mixture will then contain 0.78 mol HI, 0.11 mol H_2, and 0.11 mol I_2.

(b) The total number of H and I atoms is the same as in part a: 1.00 mol H and 1.00 mol I. The lower curve shows that after 20 min, approximately 56% will have become HI:

$$1.00 \text{ mol H} \times 56\% = 0.56 \text{ mol H (in HI)} = 0.56 \text{ mol HI}$$

The remaining 0.44 mol H will be present as 0.22 mol H_2; likewise, there will be 0.22 mol I_2. After 2 or 3 hr, the mixture will have reached equilibrium, with 0.78 mol HI, 0.11 mol H_2, and 0.11 mol I_2, the same as in part a. ■

In the reaction $H_2 + I_2 \rightleftharpoons 2HI$ at 400 to 500°C, the mixture reaches equilibrium in a few hours. In other reactions, it may take a tiny fraction of a second, and there are many reactions that would be nowhere close to equilibrium after a million years. Some reactions settle down to equilibrium after only minute amounts of the products have been formed. Other reactions go all the way to products, with only a minute amount of the limiting reactant remaining unreacted at equilibrium. Sometimes, as in the HI reaction, appreciable quantities of both reactants and products are present in the equilibrium mixture. Note that the reactants and products are *not* present in *equal* quantities; 78% of the mixture, well over half, is in the form of HI. "Equilibrium" refers to equality between forward and reverse rates, not to equality between quantities of reactants and products.

13.2 THE LAW OF CHEMICAL EQUILIBRIUM

It is of great practical importance to be able to predict or, still better, to control the equilibrium composition of a mixture. We would like to obtain the highest possible yield of the desired product at the least cost for reactants. If a reaction has not come to equilibrium, some catalyst (section 19.10) may be found that will accelerate it, or we can simply give it more time. However, once the reaction reaches equilibrium, patience is no longer a virtue. If we are not satisfied with the yield, we must change some of the conditions that determine the composition of the equilibrium mixture.

In 1888 Henry-Louis Le Chatelier (we will meet him again later) poked some fun at the British steel industry:

> It is known that in the blast furnace the reduction of iron oxide is produced by carbon monoxide, according to the reaction $Fe_2O_3 + 3CO \longrightarrow 2Fe + 3CO_2$, but the gas leaving the chimney contains a considerable proportion of carbon monoxide. . . . Because this incomplete reaction was thought to be due to an insufficiently prolonged contact between carbon monoxide and the iron ore, the dimensions of the furnaces have been increased. In England they have been made as high as thirty meters. But the proportion of carbon monoxide escaping has not diminished, thus demonstrating, by an experiment costing several hundred thousand francs, that the reduction of iron oxide by carbon monoxide is a limited reaction. Acquaintance with the laws of chemical equilibrium would have permitted the same conclusion to be reached more rapidly and far more economically.

A "limited" reaction is one that does not go to completion but reaches equilibrium with reactants still present.

In the century since then, most captains of industry have learned Le Chatelier's lesson that "these investigations of a rather theoretical sort are capable of much more immediate practical application than one would be inclined to believe. Indeed the phenomena of chemical equilibrium play a capital role in all operations of industrial chemistry."

The problem to be solved is this: there is a mixture in which a chemical reaction may occur. For the present let us assume it is a mixture of gases. Which way will the reaction go? How far will it go—that is, what will be the concentrations of the gases at equilibrium? If we do something to the system, like changing its volume or adding more of one substance, how will the equilibrium composition be affected?

When a mixture has come to equilibrium, the concentrations of the reactants and products satisfy an equation known as the **equilibrium condition** for the reaction. It will take a little time to describe the procedure for writing this equation. Remember that we are limiting our attention to mixtures of gases for now.

The equations in this section do not necessarily apply to reactions involving solids or liquids (sections 13.7, 13.8).

First we define the **reaction quotient**, Q. For the reaction

$$N_2O_4(g) \rightleftharpoons 2NO_2(g)$$

the reaction quotient is

$$Q = \frac{[NO_2]^2}{[N_2O_4]}$$

The concentration (moles per liter) of NO_2, $[NO_2]$, is written in the numerator because NO_2 is a product (on the right side of the equation). $[NO_2]$ has the exponent 2 because the coefficient of NO_2 in the chemical equation is 2. The concentration of N_2O_4 is in the denominator because N_2O_4 is a reactant (on the left side of the equation). The exponent of N_2O_4 is 1 (not written) because the coefficient of N_2O_4 in the equation is 1 (also not written).

Let's look at some other examples. For the reaction

$$H_2(g) + I_2(g) \rightleftharpoons 2HI(g)$$

the reaction quotient is

$$Q = \frac{[HI]^2}{[H_2][I_2]}$$

The concentrations of the two reactants, H_2 and I_2, appear in the denominator and are multiplied. For the reaction

$$2SO_2(g) + O_2(g) \rightleftharpoons 2SO_3(g) \qquad (5)$$

the reaction quotient is

$$Q = \frac{[SO_3]^2}{[SO_2]^2[O_2]}$$

For the reaction

$$2Cl_2(g) + 2H_2O(g) \rightleftharpoons 4HCl(g) + O_2(g)$$

$$Q = \frac{[HCl]^4[O_2]}{[Cl_2]^2[H_2O]^2}$$

For the general reaction

$$aA(g) + bB(g) \rightleftharpoons cC(g) + dD(g)$$

with coefficients a, b, c, and d, the reaction quotient is

$$\mathbf{Q = \frac{[C]^c[D]^d}{[A]^a[B]^b}} \qquad (6)$$

Remember these three rules for writing reaction quotients:

1. *Write the products over the reactants.*
2. *Each concentration is raised to a power equal to the corresponding coefficient in the chemical equation.*
3. *Concentrations are multiplied, never added.*

As the reaction proceeds, the concentrations change, and therefore Q changes. When equilibrium is reached, the concentrations and Q have settled down to being constant. The **law of chemical equilibrium** says that when the reaction has reached equilibrium, Q has a specific value, the **equilibrium constant** K:

We are not proving equation 7. That is left for more advanced books.

$$\mathbf{Q = K} \qquad (7)$$

The final (equilibrium) value of Q is the equilibrium constant K. It depends only on the temperature, not on the initial concentrations. This is why K is a ''constant.''

When the law of chemical equilibrium, $Q = K$, is applied to a specific reaction, the resulting equation is the *equilibrium condition* for the reaction. For the five reactions used as examples, the equilibrium conditions are

The K's are numbered as a reminder that different reactions have different equilibrium constants.

$$\frac{[NO_2]^2}{[N_2O_4]} = K_1 \qquad \frac{[HI]^2}{[H_2][I_2]} = K_2 \qquad \frac{[SO_3]^2}{[SO_2]^2[O_2]} = K_3$$

$$\frac{[HCl]^4[O_2]}{[Cl_2]^2[H_2O]^2} = K_4 \qquad \frac{[C]^c[D]^d}{[A]^a[B]^b} = K_5$$

If the mixture is not initially in equilibrium, *the concentrations will adjust themselves until the equilibrium condition is satisfied.*

Reaction 5 has been studied extensively, and data on it provide good examples of chemical equilibrium. This reaction is the critical step in the usual process for making sulfuric acid, H_2SO_4. The reaction is extremely slow under ordinary conditions, but proceeds quite rapidly above 500°C in the presence of a catalyst (section 19.10) such as divanadium pentoxide (V_2O_5) or platinum. Table 13.1 contains

TABLE 13.1
DETERMINATION OF THE EQUILIBRIUM CONSTANT OF $2SO_2(g) + O_2(g) \rightleftharpoons 2SO_3(g)$ AT 1000 K

EXPERIMENT	INITIAL CONCENTRATIONS, mol/L $[SO_2]$	$[O_2]$	CONCENTRATIONS AT EQUILIBRIUM, mol/L $[SO_2]$	$[O_2]$	$[SO_3]$	$Q = \frac{[SO_3]^2}{[SO_2]^2[O_2]}$
1	0.00675	0.00544	0.00377	0.00430	0.00412	278
2	0.00847	0.00372	0.00557	0.00219	0.00444	290
3	0.00864	0.00354	0.00590	0.00201	0.00435	270
4	0.00939	0.00280	0.00681	0.00123	0.00403	285
5	0.00433	0.00165	0.00302	0.000975	0.00156	274
					Average	279
					Relative average deviation†	2.3%

† Average deviation = 6.4 (section 1.6); relative average deviation = (6.4/279) × 100% = 2.3%.

data from some experiments in which the equilibrium constant of this reaction was measured. A mixture of SO_2 and O_2, with the concentrations shown, was passed through a tube containing platinum heated to 1000 K. The outcoming gas was analyzed for SO_2, SO_3, and O_2. The reaction quotient

$$Q = \frac{[SO_3]^2}{[SO_2]^2[O_2]}$$

was calculated for each mixture. The table shows that Q is approximately constant, as it should be at equilibrium. We can therefore assume that the gas mixtures have all reached equilibrium. If they have, the measured Q of the outcoming gas is the equilibrium constant, K.

It may seem that equilibrium constants should have units, like mol/L or L^2/mol^2. However, it is customary to omit them.

PREDICTING THE DIRECTION OF A REACTION

We will use the $SO_2 + O_2$ reaction to illustrate an important application of the law of chemical equilibrium. When the concentration of the product (SO_3) is smaller than it would be at equilibrium, and the concentrations of the reactants (SO_2 and O_2) are larger, then Q is less than K ($Q < K$). The net reaction (if any) must then be in the *forward* direction, $2SO_2 + O_2 \longrightarrow 2SO_3$, so that Q gets closer to K. Conversely, when the concentration of SO_3 is greater than it would be at equilibrium, and the concentrations of SO_2 and O_2 are smaller, then Q is greater than K ($Q > K$). The net reaction must then be in the *reverse* direction, $2SO_2 + O_2 \longleftarrow 2SO_3$, and again Q approaches K. In summary,

WHEN $Q < K$	WHEN $Q > K$
Numerator (product concentrations) is too small. OR Denominator (reactant concentrations) is too large. Q is too small. Reaction can proceed only $\longrightarrow$ (reactants to products).	Numerator (product concentrations) is too large. OR Denominator (reactant concentrations) is too small. Q is too large. Reaction can proceed only $\longleftarrow$ (products to reactants).

The reaction can proceed only in the direction that makes Q *more nearly* equal to K. However, the equilibrium constant tells us nothing about the rate of the reaction. The rate may be so close to zero that no reaction can be detected, even over geologic time spans. We may be sure, however, that the reaction will *not* go in the other direction unless some condition, like temperature or pressure, is changed (section 13.6).

EXAMPLE 13.2 A mixture of SO_2, O_2, and SO_3 at 1000 K contains the gases at the following concentrations: $[SO_2] = 5.0 \times 10^{-3}$ mol/L, $[O_2] = 1.9 \times 10^{-3}$ mol/L, and $[SO_3] = 6.9 \times 10^{-3}$ mol/L. Which way can the reaction $2SO_2 + O_2 \rightleftharpoons 2SO_3$ go to reach equilibrium? From Table 13.1, $K = 279$.

ANSWER Calculate the reaction quotient and then compare it with K:

$$Q = \frac{[SO_3]^2}{[SO_2]^2[O_2]}$$

$$= \frac{(6.9 \times 10^{-3})^2}{(5.0 \times 10^{-3})^2(1.9 \times 10^{-3})}$$

$$= 1.0 \times 10^3$$

Since Q is greater than K ($10^3 > 279$), the concentration of the product (SO_3) is too large relative to the concentrations of the reactants (SO_2 and O_2). To reach equilibrium, the reaction must go from right to left:

$$2SO_2 + O_2 \longleftarrow 2SO_3$$

■

EXAMPLE 13.3 For the reaction

$$I_2(g) \rightleftharpoons 2I(g)$$

at 500 K, $K = 5.6 \times 10^{-12}$. A mixture kept at 500 K contains I_2 at a concentration of 0.020 mol/L and I at a concentration of 2.0×10^{-8} mol/L. Which way must the reaction go to reach equilibrium?

ANSWER We first calculate the reaction quotient for the given reaction and the given mixture:

$$Q = \frac{[I]^2}{[I_2]} = \frac{(2.0 \times 10^{-8})^2}{0.020} = 2.0 \times 10^{-14}$$

Q is less than K ($2.0 \times 10^{-14} < 5.6 \times 10^{-12}$). Therefore the concentration of I is too small for equilibrium and the concentration of I_2 is too large. The reaction can go from left to right:

$$I_2 \longrightarrow 2I$$

■

PROBLEM 13.1 The equilibrium constant for the reaction

$$2NOCl(g) \rightleftharpoons 2NO(g) + Cl_2(g)$$

is 3.9×10^{-3} at 300°C. A mixture contains the gases at the following concentrations: $[NOCl] = 5.0 \times 10^{-3}$ mol/L, $[NO] = 2.5 \times 10^{-3}$ mol/L, and $[Cl_2] = 2.0 \times 10^{-3}$ mol/L. (a) Calculate the reaction quotient Q. (b) Which way can the reaction go at 300°C? □

Equilibrium constants come in all sizes, from nearly zero (like 10^{-50}) to enormous (like 10^{50}). When K is much larger than 1, the concentrations of the products at equilibrium are large in comparison to the concentrations of the reactants. We express this situation by saying that the reaction "goes (nearly) to completion," or "goes (mostly) to the right," or "goes forward," or "the equilibrium favors the products." Conversely, when K is much less than 1, the concentrations of the products are relatively small. The reaction then "goes (mostly) to the left," or "goes in reverse," or "goes backward," or "the equilibrium favors the reactants," or "the reaction does not occur as written." The small equilibrium constant (5.6×10^{-12}) for the reaction $I_2(g) \rightleftharpoons 2I(g)$ indicates that this reaction goes mostly to the left; that is, at 500 K, relatively few of the I_2 molecules are split up into I atoms.

Values of K like 10^{50} and 10^{-50} cannot be obtained by measuring concentrations of reactants and products, because some of the concentrations are too small to be detected. Instead, they are calculated from other kinds of measurements, as will be explained in Chapters 17 and 18. But when the value of K is so high that the equilibrium concentration of a reactant is undetectable, the chemist says that the reaction "goes to completion"; when K is so small that no product can be detected, the reaction "cannot go at all." In these extreme cases, the actual value of K is often unimportant; knowing that it is very large or very small may be sufficient.

EXAMPLE 13.4 An equilibrium mixture at 1500 K contains $N_2O_4(g)$ and $NO_2(g)$ at the following concentrations: $[N_2O_4] = 3.2 \times 10^{-10}$ mol/L, and $[NO_2] = 5.0 \times 10^{-3}$ mol/L. Calculate K for the reaction $N_2O_4(g) \rightleftharpoons 2NO_2(g)$.

ANSWER At equilibrium

$$\frac{[NO_2]^2}{[N_2O_4]} = K$$

Substitute the given equilibrium concentrations:

$$\frac{(5.0 \times 10^{-3})^2}{3.2 \times 10^{-10}} = 7.8 \times 10^4 = K$$

■

This large equilibrium constant is consistent with something that can be seen from the given concentrations: $[NO_2]$ is much larger than $[N_2O_4]$. Thus, the equilibrium favors the product, NO_2 in this case. Most N_2O_4 molecules are split up into NO_2 molecules at 1500 K.

PROBLEM 13.2 A mixture of the three gases H_2, I_2, and HI has come to equilibrium at 427°C. The following concentrations are present: $[H_2] = 0.0034$ mol/L, $[I_2] = 0.0072$ mol/L, and $[HI] = 0.036$ mol/L. Calculate K for the reaction $H_2(g) + I_2(g) \rightleftharpoons 2HI(g)$ at 427°C. □

THERMODYNAMIC AND KINETIC STABILITY

A substance (or mixture) that would be mostly converted into something else at equilibrium is said to be **thermodynamically unstable**. Thus N_2O_4 is thermodynamically unstable at 1500 K because when it reaches equilibrium, it is mostly split up into NO_2 molecules. A substance (or mixture) that is mostly *not* converted into anything else when it reaches equilibrium, such as I_2 at 500 K, is said to be

thermodynamically stable. Conversely, I *atoms* are thermodynamically unstable at 500 K because most of them pair off to form I_2 molecules.

There is another kind of stability that is quite different from thermodynamic stability. At 25°C, the reaction

$$2H_2(g) + O_2(g) \rightleftharpoons 2H_2O(\ell)$$

has $K = 7 \times 10^{38}$. This very large equilibrium constant indicates that a mixture of H_2 and O_2 is thermodynamically unstable. At equilibrium, nearly all of it would have reacted to form H_2O (except for the reactant present in excess). However, even though a mixture of H_2 and O_2 is not in equilibrium, it will keep indefinitely at room temperature—no change in composition can be detected—if no flames, sparks, or catalysts are present. The mixture is thus **kinetically stable**; it never comes to equilibrium. It reacts extremely slowly, or not at all. The rate constant for the forward reaction is practically zero. However, a small disturbance, like a spark, may greatly increase the rate of the reaction in a small region. The heat, or some reactive molecules, thus generated can increase the rate in a larger region, and so on. Some thermodynamically unstable but kinetically stable systems, including $H_2 + O_2$ and other fuel-air mixtures, are in a precarious condition, ready to explode on any provocation.

H_2 or O_2 alone is thermodynamically *stable.* A *mixture* of them is unstable.

Chapter 19 treats kinetic stability and instability in more detail.

13.3 EQUILIBRIUM CONDITIONS IN TERMS OF PRESSURES

The composition of a gas mixture can be represented by giving either the partial pressure, p_A, or the molar concentration, [A], of each component (A, B, . . .). Thus far in this chapter, equilibrium conditions have been expressed in terms of concentrations. However, for gases, partial pressures are more commonly used than concentrations; vapor pressure (sections 11.10 and 12.8) is used all the time, vapor concentration much less. You should therefore learn to deal with chemical equilibria either way, with pressures or with concentrations.

Pressure and concentration of a gas are related by the ideal gas law (section 10.7), $PV = nRT$, or $P = nRT/V$. But the ratio n/V is the concentration in moles per liter. Thus for any gas A, the partial pressure is

$$p_A = \frac{n_A}{V}RT = [A]RT \qquad (8)$$

Conversely, $[A] = p_A/RT$. Equilibrium conditions for reactions involving gases can therefore be expressed equally well in terms of the concentrations or the partial pressures of the gases: just substitute from equation 8. However, the numerical value of the equilibrium constant may be different in the two cases because of the RT factors. We distinguish by writing K_c when the composition is expressed in concentrations, and K_p when the composition is expressed in partial pressures.

The relationship between K_c and K_p depends on the reaction and how it is written. Specifically, it depends on the number of moles of *gaseous* reactants and products. The equation is

$$K_p = K_c(RT)^{\Delta n} \qquad (9)$$

where Δn is the number of moles of gaseous products minus the number of moles of gaseous reactants:

$$\Delta n = n_{\text{gaseous products}} - n_{\text{gaseous reactants}}$$

Δn may be positive, negative, or zero, depending on the reaction. If pressures are in atmospheres and concentrations are in moles per liter, the value of R to use is

0.08206 L atm/(mol K). We will not prove equation 9, but in Problem 13.26 you can verify it for a specific reaction.

EXAMPLE 13.5 Find Δn in equation 9 for these reactions:

(a) $PCl_3(g) + Cl_2(g) \rightleftharpoons PCl_5(g)$

(b) $2Cl_2(g) + 2H_2O(g) \rightleftharpoons 4HCl(g) + O_2(g)$

ANSWER

(a) There are 2 moles of reactants ($PCl_3 + Cl_2$) and 1 mole of product (PCl_5), all gaseous. Therefore

$$\Delta n = 1 \text{ mol product} - 2 \text{ mol reactant} = -1 \text{ mol}$$

(b) There are 4 moles of gaseous reactants ($2Cl_2 + 2H_2O$) and 5 moles of gaseous products ($4HCl + O_2$). Therefore

$$\Delta n = 5 \text{ mol product} - 4 \text{ mol reactant} = 1 \text{ mol}$$

■

EXAMPLE 13.6 Calculate K_p for the reaction

$$2SO_2(g) + O_2(g) \rightleftharpoons 2SO_2(g)$$

at 1000 K from the data in Table 13.1.

ANSWER

$$\Delta n = 2 \text{ mol product} - 3 \text{ mol reactant} = -1 \text{ mol}$$

The table gives $K_c = 279$. Then,

$$K_p = K_c(RT)^{\Delta n} = K_c(RT)^{-1}$$
$$= 279 \times (0.08206 \times 1000)^{-1}$$
$$= \frac{279}{82.06} = 3.40$$

■

PROBLEM 13.3 $K_p = 0.039$ at 250°C for the reaction

$$2NOCl(g) \rightleftharpoons 2NO(g) + Cl_2(g)$$

Calculate K_c.

□

PROBLEM 13.4 In Problem 13.2, you calculated $K_c = 53$ for the reaction

$$H_2(g) + I_2(g) \rightleftharpoons 2HI(g)$$

at 427°C. What is K_p?

□

13.4 DOES IT MATTER HOW YOU WRITE THE EQUATION?

The properties of a reacting mixture certainly do not depend on how we choose to write the equation for the reaction. For example, each of these equations

$$2NO_2 \rightleftharpoons N_2O_4$$
$$4NO_2 \rightleftharpoons 2N_2O_4$$
$$N_2O_4 \rightleftharpoons 2NO_2$$

is a correct way of describing the same reaction (except that the last version interchanges reactant and product). A hasty (and incorrect) conclusion might be that the equilibrium constant is the same in all of these ways of writing the equation. Note, however, that K_p or K_c is not the direct result of a measurement, but is *calculated* from the concentrations or partial pressures measured at equilibrium. These pressures or concentrations are not affected by how the equation is written—the chemicals cannot read our minds or books—but the rules for writing the reaction quotient Q and calculating K do depend on how the equation is written. Thus, the reaction

$$N_2O_4 \rightleftharpoons 2NO_2 \tag{10}$$

has the equilibrium condition

$$\frac{[NO_2]^2}{[N_2O_4]} = K_{c1} \tag{11}$$

However, the reaction may also be written

$$2N_2O_4 \rightleftharpoons 4NO_2$$

and the equilibrium condition is then

$$\frac{[NO_2]^4}{[N_2O_4]^2} = K_{c2} \tag{12}$$

The concentrations are the same in equations 11 and 12, but the K_c's cannot be the same. To get the left side of equation 12, the left side of equation 11 must be squared; then the right side must also be squared:

$$K_{c2} = K_{c1}{}^2$$

Reversing equation 10 interchanges reactant and product:

$$2NO_2 \rightleftharpoons N_2O_4$$

In keeping with the "products over reactants" rule, the equilibrium condition is now written with $[N_2O_4]$ as the numerator and $[NO_2]^2$ as the denominator:

$$\frac{[N_2O_4]}{[NO_2]^2} = K_{c3} = \frac{1}{K_{c1}} \tag{13}$$

For equations 11 and 13 to be consistent, the equilibrium constant must be replaced by its reciprocal when the equation is written in the opposite direction.

EXAMPLE 13.7 For each of the following reactions, write the equilibrium condition in terms of concentrations. For reactions b, c, and d, express the equilibrium constant K_c in terms of K_c for reaction a.

(a) $2SO_2 + O_2 \rightleftharpoons 2SO_3$

(b) $4SO_2 + 2O_2 \rightleftharpoons 4SO_3$

(c) $SO_2 + \frac{1}{2}O_2 \rightleftharpoons SO_3$

(d) $2SO_3 \rightleftharpoons 2SO_2 + O_2$

ANSWER

(a) $$\frac{[SO_3]^2}{[SO_2]^2[O_2]} = K_a$$

The subscripts refer to the parts of the example.

(b) $$\frac{[SO_3]^4}{[SO_2]^4[O_2]^2} = K_b$$

This reaction quotient is the square of the quotient in reaction a:

$$\left(\frac{[SO_3]^2}{[SO_2]^2[O_2]}\right)^2 = K_a^2$$

Then $K_b = K_a^2$.

(c) $$\frac{[SO_3]}{[SO_2][O_2]^{1/2}} = K_c$$

Then $K_c = K_a^{1/2} = \sqrt{K_a}$.

(d) $$\frac{[SO_2]^2[O_2]}{[SO_3]^2} = K_d$$

Then $K_d = K_a^{-1} = 1/K_a$. ■

In summary,

IF THE COEFFICIENTS IN THE CHEMICAL EQUATION ARE	THEN K_c OR K_p IS
doubled	squared
halved	replaced by its square root (the $\frac{1}{2}$ power)
reversed in sign (left and right sides interchanged)	inverted (the -1 power)
multiplied by any constant n	raised to the nth power

Thus, to speak of "the equilibrium constant for the formation of SO_3 from SO_2 and O_2" is ambiguous; before we know what the equilibrium constant means, the chemical equation must be specified and we must be told whether the constant is K_c or K_p. The normal assumption would be $2SO_2 + O_2 \rightleftharpoons 2SO_3$ (where the coefficients are the smallest possible integers), but it should be made explicit.

EXAMPLE 13.8 The equilibrium constant K_p for the reaction

$$CO_2(g) \rightleftharpoons CO(g) + \tfrac{1}{2}O_2(g) \tag{14}$$

at 1000 K is 6.03×10^{-11}. Calculate K_p for the reaction

$$2CO(g) + O_2(g) \rightleftharpoons 2CO_2(g) \tag{15}$$

ANSWER The given reaction (14) has CO_2 as a reactant and CO and O_2 as products. You are asked to find K_p for a reaction (15) that has CO_2 as a *product* and CO and O_2 as *reactants*. Therefore the equation must be reversed and K_p inverted. But that is not all; reaction 15 has coefficients twice as big as those in reaction 14. Therefore, the given equilibrium constant must be both inverted and squared:

$$K_p = \left(\frac{1}{6.03 \times 10^{-11}}\right)^2$$

$$= 2.75 \times 10^{20} \text{ for reaction 15}$$ ■

PROBLEM 13.5 The equilibrium constant for the reaction

$$H_2(g) + Cl_2(g) \rightleftharpoons 2HCl(g)$$

at 500 K is $K_p = 4.8 \times 10^{10}$. Calculate K_p for:

(a) $2H_2(g) + 2Cl_2(g) \rightleftharpoons 4HCl(g)$

(b) $HCl(g) \rightleftharpoons \frac{1}{2}H_2(g) + \frac{1}{2}Cl_2(g)$ □

13.5 COMBINATION OF EQUILIBRIA

An equilibrium constant is not always easy to find by experimenting or consulting reference books. Fortunately, it is possible to calculate the equilibrium constant of one reaction from the equilibrium constants of other reactions. We encountered the same problem with enthalpy changes (Chapter 4): they can be measured for some reactions but not for others. The key to the problem is Hess's law: When you add equations, you can add their ΔH's. With equilibrium constants, the idea is similar; find other reactions that can be added (or subtracted) to give the reaction you want.

We may be interested, for example, in the reaction

$$SO_2(g) + CO_2(g) \rightleftharpoons SO_3(g) + CO(g) \qquad (16)$$

but may find that data on it are not readily available. However, the equilibrium constants for the reactions

$$SO_2(g) + \tfrac{1}{2}O_2(g) \rightleftharpoons SO_3(g) \qquad (17)$$

and

$$CO_2(g) \rightleftharpoons CO(g) + \tfrac{1}{2}O_2(g) \qquad (18)$$

are well known. These two reactions add up to the desired reaction 16:

$$\begin{array}{rcll} SO_2(g) + \tfrac{1}{2}O_2(g) & \rightleftharpoons & SO_3(g) & (17) \\ CO_2(g) & \rightleftharpoons & CO(g) + \tfrac{1}{2}O_2(g) & (18) \\ \hline SO_2(g) + CO_2(g) & \rightleftharpoons & SO_3(g) + CO(g) & (16) \end{array}$$

We can write the equilibrium conditions

The constants are labeled by the same numbers as the equations.

$$\frac{[SO_3]}{[SO_2][O_2]^{1/2}} = K_{17}$$

$$\frac{[CO][O_2]^{1/2}}{[CO_2]} = K_{18}$$

In a mixture of all five substances—SO_2, SO_3, CO, CO_2, and O_2—both of these conditions must be satisfied at equilibrium. If we multiply one equation by the other, we obtain the equation

$$\frac{[SO_3]}{[SO_2]\cancel{[O_2]^{1/2}}} \times \frac{[CO]\cancel{[O_2]^{1/2}}}{[CO_2]} = K_{17} \times K_{18}$$

or

$$\frac{[SO_3][CO]}{[SO_2][CO_2]} = K_{17}K_{18} \qquad (19)$$

The equilibrium condition for equation 16 is

$$\frac{[SO_3][CO]}{[SO_2][CO_2]} = K_{16} \qquad (20)$$

On comparing equations 19 and 20, we see that

$$K_{16} = K_{17}K_{18}$$

Of course, K_{16} is the same whether or not O_2 is present.

This calculation illustrates the rule: *When two reactions are* **added** *to obtain a third reaction, their equilibrium constants are* **multiplied** *to give the equilibrium constant of the third reaction.* Generally, when any number of reactions are added, their equilibrium constants are *multiplied* together to give the equilibrium constant of the net reaction. The rule applies to either K_c's or K_p's, but we must not mix the two. The difference from Hess's law is that equilibrium constants are multiplied, whereas heats of reaction are added.

EXAMPLE 13.9 Use the following data to calculate the equilibrium constant K_p at 1000 K for the reaction

$$SO_2(g) + CO_2(g) \rightleftharpoons SO_3(g) + CO(g)$$

From Example 13.8, $K_p = 6.03 \times 10^{-11}$ at 1000 K for $CO_2(g) \rightleftharpoons CO(g) + \frac{1}{2}O_2(g)$. From Example 13.6, $K_p = 3.40$ at 1000 K for $2SO_2(g) + O_2(g) \rightleftharpoons 2SO_3(g)$.

ANSWER As noted, reaction 16 results from adding reactions 17 and 18. Reaction 17 is one-half the reaction in Example 13.6; therefore $K_{17} = 3.40^{1/2} = 1.84$. Then

$$\begin{array}{rll} SO_2(g) + \frac{1}{2}O_2(g) \rightleftharpoons & SO_3(g) & K_{17} = 1.84 \\ CO_2(g) \rightleftharpoons & CO(g) + \frac{1}{2}O_2(g) & K_{18} = 6.03 \times 10^{-11} \\ \hline SO_2(g) + CO_2(g) \rightleftharpoons & CO(g) + SO_3(g) & K_{16} = K_{17}K_{18} \end{array}$$

$$K_{16} = 1.84 \times 6.03 \times 10^{-11} = 1.11 \times 10^{-10}$$ ■

PROBLEM 13.6 The following equilibrium constants at 500 K are given:

$$H_2(g) + Br_2(g) \rightleftharpoons 2HBr(g) \qquad K_p = 7.9 \times 10^{11}$$

$$H_2(g) \rightleftharpoons 2H(g) \qquad K_p = 4.8 \times 10^{-41}$$

$$Br_2(g) \rightleftharpoons 2Br(g) \qquad K_p = 2.2 \times 10^{-15}$$

Calculate K_p at 500 K for the reaction

$$H(g) + Br(g) \rightleftharpoons HBr(g)$$ □

13.6 LE CHATELIER'S PRINCIPLE

Imagine an amusement arcade where people are playing with pinball machines, video games, fortune-telling machines, and other devices of various reputes. The population is approximately constant at 200, not because the same people stay all day but because people enter and leave at about the same rate; on the average, for each person that leaves, one enters. We might say that the system (arcade and people) is in equilibrium. Now a chartered bus arrives and 50 eager players get off. The population of the arcade suddenly rises to 250. It is no longer in equilibrium. The place is not big enough for 250 people, and the new arrivals make everyone uncomfortable; therefore, some change is to be expected.

Customers now leave at a greater rate than new ones arrive. When the population settles down again to being constant, it will probably be larger than the original 200, but smaller than 250. The equilibrium is disturbed by the arrival of fifty people at one time, but the system is self-regulating. It responds in a way that partially counteracts the effect of the disturbance, keeping the population from changing too much. It works the other way too. When the 50 players all leave at once to get on the bus, the arcade seems relatively empty, there is no waiting time for games, and more people come in, partially compensating for the sudden decrease in population.

A chemical system *in equilibrium* behaves in much the same way. You add a reactant or product to it and it responds by using up some of what you added. You squeeze it so as to increase its pressure and it reacts in a way that decreases the pressure. You try to raise its temperature and it reacts in a way that tends to keep it cool. That is what **Le Chatelier's principle** says:

> **When a system in equilibrium is disturbed, the equilibrium shifts in the direction that decreases the effect of the disturbance.**

This principle is more easily understood when it is applied to specific kinds of disturbance.

Suppose that a mixture is in equilibrium. Three common kinds of disturbance can be applied to it: (1) a change in the quantity of a reactant or product, (2) a change in its volume, and (3) a change in its temperature.

CHANGE IN QUANTITY

Assume that we have 1.00 liter of an equilibrium mixture of SO_3, SO_2, and O_2 at 1000 K, with $[SO_3] = 0.00412$ mol/L, $[SO_2] = 0.00377$ mol/L, and $[O_2] = 0.00430$ mol/L (the first line of Table 13.1). Then for the reaction

$$2SO_2(g) + O_2(g) \rightleftharpoons 2SO_3(g)$$

the reaction quotient Q (short for Q_c) is

$$Q = \frac{[SO_3]^2}{[SO_2]^2[O_2]} = \frac{(0.00412)^2}{(0.00377)^2 \times 0.00430} = 278 = K_c$$

Since the mixture is in equilibrium, the rate of the forward reaction ($2SO_2 + O_2 \longrightarrow 2SO_3$) is the same as the rate of the reverse reaction ($2SO_3 \longrightarrow 2SO_2 + O_2$). Let us assume that the reaction is slow enough so that we can disturb the mixture and then study it before it has changed toward its new equilibrium. We pump an additional 0.00500 mole of O_2 into the 1.00-L container, making the new concentration

$$[O_2] = 0.00430 \text{ mol/L} + 0.00500 \text{ mol/L} = 0.00930 \text{ mol/L}$$

The increased concentration of O_2 increases the number of collisions between O_2 and SO_2 molecules. Therefore, the "forward" reaction, $2SO_2 + O_2 \longrightarrow 2SO_3$, becomes faster than the "reverse" reaction, $2SO_3 \longrightarrow 2SO_2 + O_2$.

To restore equilibrium, the net process $2SO_2 + O_2 \longrightarrow 2SO_3$ must occur—a process described by saying that "the equilibrium shifts to the right." That the mixture is no longer in equilibrium can also be seen by calculating Q immediately after the O_2 is introduced:

$$Q = \frac{[SO_3]^2}{[SO_2]^2[O_2]} = \frac{(0.00412)^2}{(0.00377)^2 \times 0.00930} = 128$$

K_c is still 278, and since $Q < K_c$ (128 < 278), this mixture is not in equilibrium. The concentrations must now change in order to make Q again equal to K_c. Q must become larger. This is the same conclusion we would draw from the change of

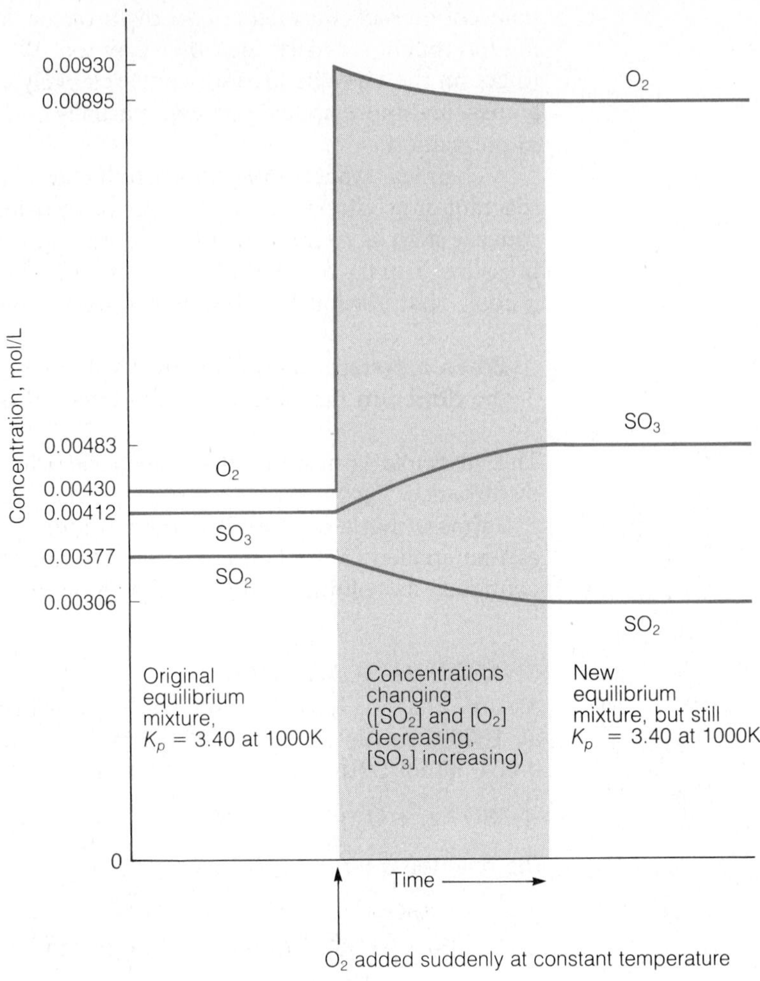

FIGURE 13.2
An illustration of how concentrations change when a mixture in equilibrium ($2SO_2 + O_2 \rightleftharpoons 2SO_3$) is disturbed by adding a reactant.

rates: $[SO_2]$ and $[O_2]$ (the denominator) must decrease, and $[SO_3]$ (the numerator) must increase. (SO_3 can be produced only by consuming SO_2 and O_2.)

Figure 13.2 shows how the concentrations change after O_2 is added. When SO_3 is produced at the expense of SO_2 and O_2, the reaction $2SO_2 + O_2 \rightleftharpoons 2SO_3$ is said to *shift to the right.* When this process has gone far enough, Q will have increased to 278 and the mixture will be in equilibrium again. The actual concentrations at equilibrium are $[SO_3] = 0.00483$ mol/L, $[SO_2] = 0.00306$ mol/L, and $[O_2] = 0.00895$ mol/L. Note that there is more O_2 in the container now than there was originally—naturally, O_2 was pumped in—but since the equilibrium has shifted to the right, not all of the added O_2 is still there. There is less SO_2 than before, because some of it was used up in reacting with O_2. Finally, there is more SO_3 than before because some of the added O_2 reacted with SO_2 to make SO_3.

You will learn how to do these calculations in section 13.10.

Similarly, adding SO_2 causes a shift to the right, and adding SO_3 causes a shift to the left (formation of more SO_2 and O_2). The general rule is that

Increasing the concentration (or partial pressure) of one substance in an equilibrium mixture displaces the equilibrium in the direction that consumes the added substance.

Conversely, decreasing the concentration of a substance causes the production of more of that substance.

CHANGE IN VOLUME OR TOTAL PRESSURE

If the volume of the container is changed, the pressures or concentrations of gases are changed, and an equilibrium mixture is thus disturbed. Concentration is moles *per liter*. When the volume is changed and the number of moles is not, the concentration must change, and so must the partial pressure ($p = nRT/V$).

Consider 1 liter of the same mixture we just discussed. Suppose we suddenly double the volume of the container to 2.00 liters at constant temperature. Each concentration is thereby divided by 2:

OLD CONCENTRATION	NEW CONCENTRATION
$[SO_3] = \frac{0.00412 \text{ mol}}{1.00 \text{ L}} = 0.00412 \text{ mol/L}$	$[SO_3] = \frac{0.00412 \text{ mol}}{2.00 \text{ L}} = 0.00206 \text{ mol/L}$
$[SO_2] = \frac{0.00377 \text{ mol}}{1.00 \text{ L}} = 0.00377 \text{ mol/L}$	$[SO_2] = \frac{0.00377 \text{ mol}}{2.00 \text{ L}} = 0.00189 \text{ mol/L}$
$[O_2] = \frac{0.00430 \text{ mol}}{1.00 \text{ L}} = 0.00430 \text{ mol/L}$	$[O_2] = \frac{0.00430 \text{ mol}}{2.00 \text{ L}} = 0.00215 \text{ mol/L}$

The new reaction quotient is

$$Q = \frac{(0.00206)^2}{(0.00189)^2 \times 0.00215} = 553$$

Since $Q > K_c$ ($553 > 278$), the mixture is no longer in equilibrium. To restore equilibrium, Q must become smaller. To make Q smaller, more SO_2 and O_2 must form at the expense of SO_3. The rule is that

The new equilibrium concentrations are actually $[SO_3] = 0.00175$ mol/L, $[SO_2] = 0.00219$ mol/L, and $[O_2] =$ 0.00230 mol/L.

Increasing the volume of the container shifts the equilibrium in the direction that produces more moles of gas.

Since $2SO_2 + O_2$ is 3 moles of gas and $2SO_3$ is 2 moles of gas, the equilibrium $2SO_2 + O_2 \rightleftharpoons 2SO_3$ is shifted to the *left* when the volume increases. Conversely, decreasing the volume results in the formation of more SO_3, because decreasing the volume favors that side of the equation showing fewer moles of gas, the *right* in this case. In Problem 13.35, you can work out these rules from the change in Q when the volume (and thus all the concentrations or partial pressures) is changed.

Here are some other examples of the effect on equilibrium of a change in volume:

		INCREASE IN VOLUME	DECREASE IN VOLUME
$CO(g) + Cl_2(g) \rightleftharpoons$	$COCl_2(g)$	shifts ⟵	shifts ⟶
2 moles of gas	1 mole of gas		
$N_2(g) + O_2(g) \rightleftharpoons$	$2NO(g)$	no effect	no effect
2 moles of gas	2 moles of gas		
$H_2S(g)$	$\rightleftharpoons H_2(g) + \frac{1}{2}S_2(g)$	shifts ⟶	shifts ⟵
1 mole of gas	$1\frac{1}{2}$ moles of gas		

Note that when there is no change in the number of moles of gas, a volume change does not shift the equilibrium.

A change in volume causes a change in the total pressure of the gas mixture. When the volume is decreased, the total pressure is increased; when the volume is increased, the pressure is decreased. *Increasing* the total pressure shifts the equilibrium toward *fewer moles of gas*; *decreasing* the total pressure shifts the equilibrium toward *more moles of gas*. The pressure version has one source of confusion. If the total pressure is increased by adding a gas that is not involved in the reaction, the equilibrium is *not* shifted. If helium, for example, is added to an equilibrium mixture of SO_2, O_2, and SO_3 (with the volume kept constant), these three kinds of molecules collide with, but otherwise ignore, the He atoms. Their collisions with each other are unaffected. The addition of helium increases the total pressure but has no effect on the partial pressures of the reacting gases. These partial pressures, not the total pressure of all the gases, determine whether the mixture is in equilibrium.

CHANGE IN TEMPERATURE

Recall from section 12.5 that the solubility of a solute is affected by a change in temperature. The direction of the change in solubility—decrease or increase—depends on whether dissolving is exothermic or endothermic. Raising the temperature increases the solubility when dissolving is endothermic; that is, raising the temperature favors an endothermic process. This rule is just a special case of Le Chatelier's principle.

Increasing the temperature causes a reaction in the direction that results in absorption of heat; decreasing the temperature causes a reaction in the direction that results in emission of heat.

In the formation of SO_3, heat is given out:

$$2SO_2 + O_2 \longrightarrow 2SO_3 \qquad \Delta H = -197 \text{ kJ}$$

Thus raising the temperature shifts the equilibrium to the left, and lowering the temperature shifts the equilibrium to the right.

One way to understand the effect of temperature is to imagine that heat is a substance, a fluid—as it was believed to be in the eighteenth century. In an exothermic reaction, heat is a "product":

$$2SO_2 + O_2 \rightleftharpoons 2SO_3 + 197 \text{ kJ of heat given out}$$

To raise the temperature, you add heat. This shifts the equilibrium to the left because you are adding the "product" heat, just as would happen if you added the product SO_3. To lower the temperature, you remove heat. That shifts the equilibrium to the right, just as removing SO_3 would do. On the other hand, in an endothermic reaction, heat is a "reactant":

$$N_2 + O_2 + 179 \text{ kJ} \rightleftharpoons 2NO$$

This time, raising the temperature (adding heat) shifts the equilibrium to the right, and lowering the temperature (removing heat) shifts the equilibrium to the left.

When we changed the quantity of a reactant or product, or changed the volume of the container, we still used the same equilibrium constant for determining when the system was in equilibrium. However, an equilibrium "constant" is constant only as long as the temperature does not change. A change in temperature results

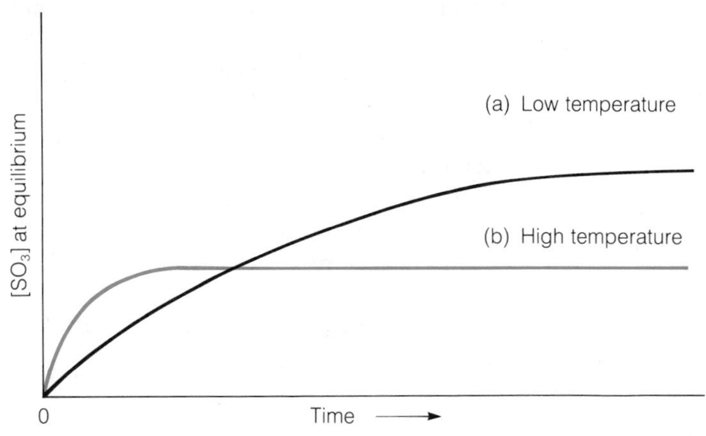

FIGURE 13.3
An illustration of the distinction between rate and equilibrium for an exothermic reaction such as $2SO_2 + O_2 \longrightarrow 2SO_3$. (a) At a low temperature (which may be far above room temperature), the reaction is slow, but a high concentration of product (SO_3) is present when the reaction eventually comes to equilibrium. (b) At a higher temperature, the reaction comes to equilibrium more quickly. However, less product is present at equilibrium.

in a change, usually large, in the equilibrium constant. When the equilibrium shifts to the right, the concentrations (or partial pressures) of products increase and the concentrations of reactants decrease. The change in K must be an increase. When the equilibrium shifts to the left, K decreases. Thus the direction in which K changes with temperature can be predicted if we know whether the reaction is exothermic or endothermic and remember the rule about which way the equilibrium shifts. The actual value of ΔH is not needed—only its sign.

There is no general relationship between the *rate* of a reaction and how far the reaction proceeds to reach equilibrium. The rate of attainment of equilibrium is *increased by raising the temperature,* regardless of the effect of temperature on the final state of equilibrium. An increase in temperature speeds up both the formation and the decomposition of SO_3, but speeds up the decomposition more than the formation. Thus less SO_3 is present when the two rates become equal. Figure 13.3 illustrates this distinction.

The rules given in this section are summarized by Le Chatelier's principle: *When a system in equilibrium is disturbed, the equilibrium shifts in the direction that decreases the effect of the disturbance.* Adding a reactant results in consumption of a fraction of the additional reactant. Raising the temperature requires an input of energy. The endothermic reaction that occurs uses up some of the added energy and makes the rise in temperature less than it would be if the same quantity of energy were added and no reaction occurred. The volume change can be fitted into this pattern if we consider the effect of a volume change on the pressure. When the volume of the container is *decreased,* the total pressure is *increased.* The shift in equilibrium then decreases the number of moles of gas, thereby *decreasing* the pressure. As a result, the net increase of pressure is *less* than it would be if the equilibrium had not shifted. It is the pressure increase, not the volume decrease, that is partly undone. Thus, when the volume is halved, the pressure at equilibrium is not doubled (as Boyle's law would predict); the pressure might be multiplied by only 1.8 or 1.9—somewhere between 1 and 2. Table 13.2 summarizes this.

A change that does not cause a shift in equilibrium is the addition or removal of a catalyst (section 19.10). A **catalyst** *is a substance that affects the rate of a reaction but is not consumed in the reaction.* It affects the rates of the forward and reverse reactions in the same way. Therefore it changes the rate of approach

TABLE 13.2 EFFECTS OF DISTURBANCES ON EQUILIBRIUM AND K

DISTURBANCE	CHANGE AS MIXTURE RETURNS TO EQUILIBRIUM	EFFECT ON EQUILIBRIUM	EFFECT ON K
Addition of a reactant	Some of added reactant is consumed.	shift to the right	no change
Addition of a product	Some of added product is consumed.	shift to the left	no change
Decrease in volume, increase in pressure	Pressure decreases.	shift toward fewer gas molecules	no change
Increase in volume, decrease in pressure	Pressure increases.	shift toward more gas molecules	no change
Rise in temperature	Heat is consumed.	shift in the endothermic direction	change
Drop in temperature	Heat is generated.	shift in the exothermic direction	change

to equilibrium, but the composition of the equilibrium mixture is unchanged by its presence.

EXAMPLE 13.10 For the reaction

$$PCl_3(g) + Cl_2(g) \rightleftharpoons PCl_5(g)$$

$\Delta H = -93$ kJ. An equilibrium mixture of PCl_3, Cl_2, and PCl_5 is in a container.

(a) Cl_2 is added to the container. What is the effect on the quantity of PCl_3 in the container? What is the effect on the quantity of Cl_2? What is the effect on K_p?

(b) The volume of the container is decreased. What is the effect on the quantity of PCl_3; of PCl_5?

(c) The temperature is increased. What is the effect on the quantity of Cl_2; of PCl_3; of PCl_5? What is the effect on the equilibrium constant?

ANSWER

(a) Adding Cl_2 shifts the equilibrium to the right: $PCl_3 + Cl_2 \longrightarrow PCl_5$. The quantity of PCl_3 is decreased. However, the quantity of Cl_2 is increased, because the reaction consumes only a part of the added Cl_2. K_p is unchanged.

(b) Decreasing the volume shifts the equilibrium to the right: there are 2 moles of gas on the left, only 1 on the right. The quantity of PCl_3 is decreased; the quantity of PCl_5 is increased.

(c) ΔH is negative; the left-to-right reaction is exothermic. Raising the temperature shifts the equilibrium to the left. Therefore the quantities of Cl_2 and of PCl_3 are increased, and the quantity of PCl_5 is decreased. The numerator decreases, and both factors in the denominator increase. K_p therefore decreases. ■

PROBLEM 13.7 A mixture of the three gases NOCl, NO, and Cl_2 is in equilibrium:

$$2NOCl(g) \rightleftharpoons 2NO(g) + Cl_2(g)$$

For this reaction, $\Delta H = 35$ kJ. How will the mass of NO be affected by each of the following disturbances? (a) Cl_2 is added. (b) NO is added. (c) The volume of the container is increased. (d) Argon is added. (e) The temperature is decreased. □

13.7 HETEROGENEOUS EQUILIBRIUM

When chemical equilibrium is established in just one phase—usually a mixture of gases or a liquid solution—it is described as a **homogeneous equilibrium**. This is the only kind of equilibrium we have considered thus far in this chapter. An equilibrium among substances present in more than one phase—gas and solid, for example, or liquid and solid—is said to be **heterogeneous**. A saturated solution in equilibrium with solute is a familiar instance of heterogeneous equilibrium.

We have studied the formation of HI from H_2 and I_2 at a temperature at which I_2 is a gas. At room temperature, however, I_2 is a solid. The composition of the equilibrium mixture of H_2 and HI in the reaction

$$H_2(g) + I_2(c) \rightleftharpoons 2HI(g) \tag{21}$$

is independent of the amount of solid I_2 present, as long as some of it is present. The equilibrium condition for reaction 21 is

$$\frac{[HI]^2}{[H_2]} = K_c \quad \text{or} \quad \frac{p_{HI}^2}{p_{H_2}} = K_p \tag{22}$$

Only the gases, HI and H_2, appear; *the solid,* $I_2(c)$, *is omitted from the reaction quotient.* Why? The reaction actually occurs in the gas phase. The partial pressure of $I_2(g)$, in equilibrium with $I_2(c)$ at 25°C, is the vapor pressure of I_2, 0.307 torr. This is a constant at each temperature, independent of the volume of the container and of the quantity of solid. The concentration of I_2 in the gas phase is also constant as long as *some* solid I_2 is present to replenish the vapor as it is consumed. The concentration of $I_2(g)$ is related to the vapor pressure of $I_2(c)$:

$$[I_2(g)] = \frac{p_{I_2}}{RT} = 1.65 \times 10^{-5} \, \frac{\text{mol}}{\text{L}} \text{ at } 25°\text{C}$$

For the all-gas reaction

$$H_2(g) + I_2(g) \rightleftharpoons 2HI(g)$$

the equilibrium condition is

$$\frac{[HI]^2}{[H_2][I_2]} = K_c' \quad \text{or} \quad \frac{p_{HI}^2}{p_{H_2}p_{I_2}} = K_p'$$

Because the concentration and pressure of $I_2(g)$ are constant when $I_2(c)$ is present, these equations can be written

$$\frac{[HI]^2}{[H_2]} = [I_2]K_c' = K_c \quad \text{or} \quad \frac{p_{HI}^2}{p_{H_2}} = p_{I_2}K_p' = K_p$$

Thus, K_c and K_p in equation 22 are constants, but only as long as $I_2(c)$ is present in the equilibrium mixture. Otherwise, the pressure of $I_2(g)$ would not necessarily be equal to the vapor pressure of $I_2(c)$.

A pure liquid is omitted from the equilibrium condition for the same reason. For the reaction

$$2Hg(\ell) + O_2(g) \rightleftharpoons 2HgO(c) \tag{23}$$

the equilibrium condition at 25°C is

$$\frac{1}{[O_2]} = K_c = 7.9 \times 10^{21} \quad \text{or} \quad \frac{1}{p_{O_2}} = K_p = 3.2 \times 10^{20} \tag{24}$$

Pure solids and pure liquids are omitted from the equilibrium condition.

EXAMPLE 13.11 Write the equilibrium condition in terms of partial pressures for these reactions:

(a) $CuSO_4(H_2O)_5(c) \rightleftharpoons CuSO_4(H_2O)_3(c) + 2H_2O(g)$
(b) $NH_3(g) + HCl(g) \rightleftharpoons NH_4Cl(c)$

ANSWER

(a) The solids are omitted. The only gas is H_2O:

$$p^2_{H_2O} = K_p$$

(b) There are two gases, both reactants:

$$\frac{1}{p_{NH_3}p_{HCl}} = K_p$$

■

PROBLEM 13.8 Write the equilibrium condition in terms of concentrations for each of these reactions:
(a) $CuO(c) + H_2(g) \rightleftharpoons Cu(c) + H_2O(g)$
(b) $C(graphite) + CO_2(g) \rightleftharpoons 2CO(g)$ □

Le Chatelier's principle also applies to heterogeneous equilibria, but with a few special features:

1. Changing the *quantity* of a pure solid or liquid does nothing to its equilibrium vapor pressure. Therefore, adding or subtracting a solid or liquid has no effect, as long as some of each substance involved in the reaction is still present.

2. Changing the pressure or concentration of a gas shifts the equilibrium in the direction that counteracts the change, just as in an all-gas reaction.

3. To decide what the effect of changing the volume of the container is, count only the moles of *gas*. Thus the equilibrium

$$\underset{\text{1 mole of gas}}{H_2(g) + I_2(c)} \rightleftharpoons \underset{\text{2 moles of gas}}{2HI(g)} \tag{21}$$

is shifted to the left when the volume of the container is decreased and to the right when it is increased, since there are fewer moles of gas on the left than on the right. The solid I_2 is ignored. The equilibrium

$$\underset{\text{1 mole of gas}}{2Hg(\ell) + O_2(g)} \rightleftharpoons \underset{\text{0 moles of gas}}{2HgO(c)} \tag{23}$$

is shifted to the right when the volume is decreased. The equilibrium

$$\underset{\text{4 moles of gas}}{4H_2O(g) + 3Fe(c)} \rightleftharpoons \underset{\text{4 moles of gas}}{4H_2(g) + Fe_3O_4(c)}$$

is unaffected by a change of volume. (The temperature is high enough so that no liquid water is present.)

The same ideas apply to the relationship between K_c and K_p for a heterogeneous reaction. In equation 9 (section 13.3), Δn counts only *gaseous* reactants and products. Solids and liquids are again ignored.

EXAMPLE 13.12 The equilibrium

$$CO_2(g) + C(graphite) \rightleftharpoons 2CO(g)$$

is established at high temperature, as in a blast furnace for making iron. For this reaction, $\Delta H = +172$ kJ.

(a) How is the quantity of CO in the equilibrium mixture affected when CO_2 is added; when graphite is added; when CO is added? (b) How is the quantity of CO affected when the volume is decreased? (c) How is the quantity of CO_2 affected when the temperature is increased?

ANSWER

(a) Adding CO_2 shifts the equilibrium to the right, increasing the quantity of CO. Adding graphite (a solid) has no effect on the equilibrium composition. Adding CO increases the quantity of CO, although some of the added CO is consumed because the equilibrium is shifted to the left.

(b) Decreasing the volume shifts the equilibrium to the left; there is 1 mole of gas (CO_2) on the left and there are 2 moles on the right. Therefore the quantity of CO is decreased.

(c) The left-to-right reaction is endothermic (positive ΔH). Therefore, increasing the temperature shifts the equilibrium to the right. The quantity of CO_2 is decreased. ■

PROBLEM 13.9 For the blast-furnace reaction

$$Fe_2O_3(c) + 3CO(g) \rightleftharpoons 2Fe(\ell) + 3CO_2(g)$$

$\Delta H = -26$ kJ. When these four substances are in equilibrium, how is the mass of Fe affected by each of the following disturbances? (a) Fe_2O_3 is added. (b) CO is added at constant volume. (c) The volume of the container is decreased. (d) The temperature is increased. □

EXAMPLE 13.13 For the reaction

$$H_2(g) + I_2(c) \rightleftharpoons 2HI(g)$$

$K_p = 0.501$ at 25°C. Calculate K_c.

ANSWER There are 2 moles of gas on the right and 1 mole (H_2) on the left:

$$\Delta n = 2 \text{ mol} - 1 \text{ mol} = 1 \text{ mol}$$

Solving equation 9 for K_c gives

$$K_c = K_p/(RT)^{\Delta n}$$
$$= 0.501/(0.08206 \times 298) = 2.05 \times 10^{-2}$$ ■

PROBLEM 13.10 For the reaction

$$2Hg(\ell) + O_2(g) \rightleftharpoons 2HgO(c)$$

$K_c = 7.9 \times 10^{21}$ at 25°C. Calculate K_p. Compare with the value (3.2×10^{20}) given in equation 24. □

13.8 EQUILIBRIUM IN SOLUTIONS

Chapters 15 and 16 are devoted entirely to equilibria in solutions.

Thus far, we have considered only equilibria in gases. In this section we will turn our attention to equilibria in liquid solutions. This is the most familiar kind of equilibrium; most reactions in living organisms and in the general chemistry laboratory occur in liquid solutions, usually with water as the solvent. We will consider homogeneous equilibria, in which all reactants and products are in solution, and heterogeneous equilibria, in which solids or gases may be reactants or products. We will confine our discussion to dilute solutions, in which one component—the solvent—is much more abundant than all the other components combined. For a dilute solution, the equilibrium condition is written in terms of the concentrations (molarities, mol/L) of the reactants and products. This is the same way equilibrium conditions for gas reactions are written when K_c is used.

For example, suppose that the reaction

$$\underset{\text{ethanol (an alcohol)}}{C_2H_5OH(aq)} + \underset{\text{acetic acid}}{CH_3COOH(aq)} \rightleftharpoons \underset{\text{ethyl acetate (an ester)}}{CH_3COOC_2H_5(aq)} + H_2O(\ell)$$

occurs in a dilute aqueous solution (mostly water). The equilibrium condition for the reaction is

$$\frac{[CH_3COOC_2H_5]}{[C_2H_5OH][CH_3COOH]} = K_c$$

The solvent (H_2O in this case) *is omitted from the equilibrium condition*, just as pure liquids and solids are omitted. As long as the solution is dilute, the solvent has about the same properties—concentration, mole fraction, vapor pressure—as if it were a pure substance. Since equilibrium conditions for solutions are always written in terms of concentrations, the equilibrium constant is usually represented simply by K, which is understood to mean K_c.

Whether or not the solvent appears in the chemical equation, the equilibrium constant for a reaction in solution *depends on the identity of the solvent.* For example, two different equilibrium constants are measured for the reaction $N_2O_4 \rightleftharpoons 2NO_2$ at 20°C, depending on whether the solvent is carbon disulfide (CS_2) or benzene (C_6H_6); K equals 1.4×10^{-4} in CS_2 and 2.3×10^{-5} in C_6H_6. The identity of the solvent is especially important in ionic reactions because of solvation effects (section 12.7). For the reaction

$$2AgCN(c) + Br^- \rightleftharpoons Ag(CN)_2^- + AgBr(c)$$

some values of K (at 25°C) are 8.6 in H_2O, 42.6 in $C_3H_5(OH)_3$ (glycerol), and too large to measure in CH_3OH (methanol).

Undissolved solids and immiscible liquids are omitted from the equilibrium condition (as in gas–solid and gas–liquid equilibria). Gases, however, must be included because compressing or expanding a gas changes its solubility (Henry's law, section 12.4). For a reaction involving a liquid solution and a gas phase, either concentrations or pressures of gases may be used. The equilibrium condition for the reaction

$$AgCl(c) \rightleftharpoons Ag^+(aq) + Cl^-(aq)$$

is

$$[Ag^+][Cl^-] = K$$

For the reaction

$$2H_2O_2(aq) \rightleftharpoons 2H_2O(\ell) + O_2(g)$$

the equilibrium condition is

$$\frac{[O_2]}{[H_2O_2]^2} = K_c \quad \text{or} \quad \frac{p_{O_2}}{[H_2O_2]^2} = K_p$$

H_2O is omitted because it is the solvent. For *solutes*, concentrations are always used. The subscripts in K_c and K_p tell whether the factors for *gases* appear as concentrations (c) or as pressures (p).

Le Chatelier's principle is applied to solution reactions the same way it is to gas reactions, but one comment is in order. Diluting a solution (adding solvent) separates the solute molecules from each other and is thus analogous to allowing a gaseous mixture to expand. Just as increasing the volume of a gas mixture shifts the equilibrium toward more moles of gas, increasing the volume of a solution favors the side of the equation that has more moles of *solute*. Thus, the reaction

$$\underset{\text{benzoic acid}}{C_6H_5COOH} + \underset{\text{hexene}}{C_6H_{12}} \rightleftharpoons \underset{\text{hexyl benzoate}}{C_6H_5COOC_6H_{13}}$$

occurs in the solvent benzene, C_6H_6. The equilibrium shifts to the left when the solution is diluted by adding benzene—there are 2 moles of solute on the left and 1 mole on the right. To bring this case within the Le Chatelier viewpoint, think of diluting the solution as decreasing the *total* concentration of all solutes. Then the equilibrium shifts in the direction that counteracts this change—the direction in which the total number of solute molecules is increased.

PROBLEM 13.11 For the reaction $C_6H_5COOH + C_6H_{12} \rightleftharpoons C_6H_5COOC_6H_{13}$, which of the following operations would change the equilibrium constant? (a) Adding more hexene (C_6H_{12}) to the solution. (b) Adding more solvent (benzene, C_6H_6) to the solution. (c) Raising the temperature. (d) Changing the solvent from C_6H_6 to xylene, $C_6H_4(CH_3)_2$. □

13.9 CALCULATING THE EQUILIBRIUM CONSTANT

When you know the values *at equilibrium* of all the concentrations or partial pressures, calculating the equilibrium constant simply involves substituting the numbers into the reaction quotient. However, be certain that the numbers you substitute are for a system that has indeed come to equilibrium.

EXAMPLE 13.14 A mixture of SO_2, O_2, and SO_3 is allowed to reach equilibrium at 852 K. The equilibrium concentrations are $[SO_2] = 3.61 \times 10^{-3}$ mol/L, $[O_2] = 6.11 \times 10^{-4}$ mol/L, and $[SO_3] = 1.01 \times 10^{-2}$ mol/L. Calculate the equilibrium constant (K_c) for the reaction $2SO_2(g) + O_2(g) \rightleftharpoons 2SO_3(g)$ at 852 K.

ANSWER First, write the equilibrium condition in terms of concentrations:

$$K_c = \frac{[SO_3]^2}{[SO_2]^2[O_2]}$$

Then substitute the equilibrium concentrations:

$$K_c = \frac{(1.01 \times 10^{-2})^2}{(3.61 \times 10^{-3})^2 \times 6.11 \times 10^{-4}}$$

$$= 1.28 \times 10^4$$

■

More commonly, we are given the initial quantity of each reactant or product and the final quantity of only one substance. All the quantities needed—concentrations or partial pressures—must then be calculated from that limited information. The remainder of this section explains these calculations.

In the reaction

$$2SO_2 + O_2 \rightleftharpoons 2SO_3$$

Recall limiting-reactant problems (section 3.12).

the equation tells us that, for every mole of O_2 that is consumed, 2 moles of SO_2 must also be consumed and 2 moles of SO_3 must be produced. Conversely, when the reaction goes from right to left, 2 moles of SO_2 and 1 mole of O_2 must be produced for every 2 moles of SO_3 consumed. Contrary to a common misconception, the equation does *not* require that SO_2 and O_2 must be mixed in the ratio of 2 moles SO_2 for each 1 mole O_2. You can start with any amounts. For example, put 1.00 mole SO_2 and 1.00 mole O_2, and no SO_3, into a 1.00-liter container. We cannot tell from this information how much of the SO_2 and O_2 will have reacted on reaching equilibrium, or how much SO_3 will then have been formed. However, some conclusions can be drawn from the chemical equation alone. Let y be the number of moles of O_2 consumed at equilibrium. The equation tells us that 2 moles of SO_2 are consumed for each 1 mole of O_2 consumed. Therefore, when y moles of O_2 are consumed, $2y$ moles of SO_2 must be consumed at the same time. Since 2 moles of SO_3 are formed for each 1 mole of O_2 consumed, $2y$ moles of SO_3 are formed when y moles of O_2 are consumed. The key idea is that all of the *changes* can be expressed in terms of a single unknown and the known coefficients in the equation. These numbers can be displayed in the form of an *equilibrium table*:

	$2SO_2$	+	O_2	$\rightleftharpoons$	$2SO_3$
Initial moles	1.00		1.00		0
Change	$-2y$		$-y$		$+2y$
Moles at equilibrium	$1.00 - 2y$		$1.00 - y$		$2y$

We now need to measure experimentally the final quantity of any one of the three gases. This will determine y, and thus the final quantities of the other two gases. From these numbers, we will be able to calculate the equilibrium constant.

The procedure used for setting up equilibrium tables throughout this book may be summarized as follows:

1. Write or rewrite the equation for the reaction, leaving plenty of space between formulas.
2. Below each formula, record the initial number of moles of each substance. As we will see in later examples, "number of moles" may be replaced by "concentration" or "partial pressure."
3. Record the change in number of moles (or concentration or partial pressure) of each substance in terms of a single unknown and the coefficients in the equation, with minus signs on one side (usually the left) and plus signs on the other.
4. Add the results of steps 2 and 3 to get the number of moles (or concentration or partial pressure) of each substance at equilibrium in terms of the one unknown.

EXAMPLE 13.15 1.00 mole of SO_2 and 1.00 mole of O_2 are confined at 1000 K in a 1.00-liter container. At equilibrium, 0.925 mole of SO_3 has been formed. Calculate K_c for the reaction $2SO_2(g) + O_2(g) \rightleftharpoons 2SO_3(g)$ at 1000 K.

ANSWER Refer to the equilibrium table written before. The number of moles of SO_3 at equilibrium is $2y = 0.925$ mol. Then $y = 0.463$ mol. The number of moles of SO_2 at equilibrium is

$$1.00 - 2y = 1.00 - 0.925 = 0.075 \text{ mol}$$

and the number of moles of O_2 is

$$1.00 - y = 1.00 - 0.463 = 0.537 \text{ mol}$$

For this problem, the equilibrium concentrations are the same as the numbers of moles, since the volume is 1.00 L: $[SO_3] = 0.925$ mol/L, $[SO_2] = 0.075$ mol/L, and $[O_2] = 0.537$ mol/L. Then,

$$K_c = \frac{[SO_3]^2}{[SO_2]^2[O_2]} = \frac{(0.925)^2}{(0.075)^2 \times 0.537} = 2.8 \times 10^2$$

■

EXAMPLE 13.16 3.00 moles of pure SO_3 are introduced into an 8.00-liter container at 1105 K. At equilibrium, 0.58 mole of O_2 has been formed. Calculate K_c for the reaction

$$2SO_3(g) \rightleftharpoons 2SO_2(g) + O_2(g)$$

at 1105 K.

ANSWER Let y be the number of moles of O_2 formed. Then,

	$2SO_3$	$\rightleftharpoons$	$2SO_2$	+	O_2
Initial moles	3.00		0		0
Change	$-2y$		$+2y$		$+y$
Moles at equilibrium	$3.00 - 2y$		$2y$		y

At equilibrium, there is 0.58 mol O_2:

$$y = 0.58 \text{ mol} \qquad 2y = 1.16 \text{ mol}$$

$$3.00 - 2y = 1.84 \text{ mol}$$

The concentrations are obtained by dividing the numbers of moles by 8.00 liters. Everything in the equilibrium table is now known:

	$2SO_3$	$\rightleftharpoons$	$2SO_2$	+	O_2
Moles at equilibrium	1.84		1.16		0.58
Concentration at equilibrium	1.84 mol/8.00 L = 0.073 mol/L		1.16 mol/8.00 L = 0.145 mol/L		0.58 mol/8.00 L = 0.230 mol/L

Then, substitution into the equilibrium condition gives

$$K_c = \frac{[SO_2]^2[O_2]}{[SO_3]^2} = \frac{(0.145)^2 \times 0.073}{(0.230)^2} = 2.9 \times 10^{-2}$$

■

PROBLEM 13.12 A mixture of 9.838×10^{-4} mole of H_2 and 1.377×10^{-3} mole of I_2 is sealed in a quartz tube and kept at 350°C for a week. By this time, the reaction

$$H_2(g) + I_2(g) \rightleftharpoons 2HI(g)$$

has come to equilibrium. The tube is broken and 4.725×10^{-4} mole of I_2 is found. (a) Calculate the number of moles of H_2 and of HI present at equilibrium. (b) Assume that the volume of the tube is 10.0 mL. Calculate K_c for the given reaction. (c) Is your answer different if the volume of the tube is 20.0 mL? □

Concentration and partial pressure are both proportional to the number of moles:

$$[SO_2] = \frac{n_{SO_2}}{V}$$

$$p_{SO_2} = \frac{n_{SO_2}RT}{V}$$

If temperature and volume are kept constant as the mixture comes to equilibrium (the only case we consider), the same kind of table can be set up in terms of concentrations or partial pressures rather than numbers of moles. In Example 13.16, the initial concentration of SO_3 is 3.00 mol/8.00 L = 0.375 mol/L and the final concentration of O_2 is 0.073 mol/L. Given this information, the equilibrium table can be written as follows:

	$2SO_3$	$\rightleftharpoons$	$2SO_2$	+	O_2
Initial concentration	0.375		0		0 mol/L
Change	$-2z$		$+2z$		$+z$ mol/L
Concentration at equilibrium	$0.375 - 2z$		$2z$		z mol/L

$z = 0.073$ mol/L (from Example 13.16). You can verify that the same answer is obtained for K_c.

Another way to convey the same information is to specify that the initial partial pressure of SO_2 is 34.0 atm and the equilibrium partial pressure of O_2 is 6.6 atm. If the data are presented in this way, the natural thing to do is to calculate K_p:

You can calculate these pressures by the ideal gas law from the concentrations (0.375 mol/L, 0.073 mol/L) and the temperature (1105 K).

	$2SO_3$	$\rightleftharpoons$	$2SO_2$	+	O_2
Initial pressure	34.0		0		0 atm
Change	$-2u$		$+2u$		$+u$ atm
Pressure at equilibrium	$34.0 - 2u$		$2u$		u atm

$$u = p_{O_2,\text{equil}} = 6.6 \text{ atm}$$

$$p_{SO_2} = 2u = 13.2 \text{ atm}$$

$$p_{SO_3} = 34.0 - 2u = 20.8 \text{ atm}$$

$$K_p = \frac{p_{SO_2}^2 p_{O_2}}{p_{SO_3}^2} = \frac{(13.2)^2 \times 6.6}{(20.8)^2} = 2.7$$

PROBLEM 13.13 A mixture of CO and Cl_2 has these initial partial pressures at 600 K: $p_{CO} = 0.500$ atm, $p_{Cl_2} = 0.300$ atm. When the reaction

$$CO(g) + Cl_2(g) \rightleftharpoons COCl_2(g)$$

has come to equilibrium at 600 K, the partial pressure of Cl_2 is 0.148 atm. (a) Calculate the partial pressures of CO and $COCl_2$ at equilibrium. (b) Calculate K_p. □

In a calculation involving dilute solutions, the equilibrium table has no entry for the solvent. It is present in such overwhelming abundance that the reaction has practically no effect on its concentration. Pure solids and pure liquids are also omitted from the equilibrium table for either a gas or a solution reaction.

EXAMPLE 13.17 1.00 mole of ethanol and 1.00 mole of acetic acid are dissolved in water and kept at 100°C. The volume of the solution is 250 mL. At equilibrium, 0.25 mole of acetic acid has been consumed in producing ethyl acetate and water. Calculate the equilibrium constant at 100°C for the reaction

$$C_2H_5OH(aq) + CH_3COOH(aq) \rightleftharpoons CH_3COOC_2H_5(aq) + H_2O(\ell)$$

ANSWER At equilibrium, 0.25 mole of the acid has been consumed; therefore 0.25 mole of ethanol must also have been consumed, and 0.25 mole of ethyl acetate has been formed:

	C_2H_5OH +	CH_3COOH $\rightleftharpoons$	$CH_3COOC_2H_5 + H_2O$
Initial moles	1.00	1.00	0
Change	−0.25	−0.25	+0.25
Moles at equilibrium	0.75	0.75	0.25
Concentration at equilibrium	$\frac{0.75\text{ mol}}{0.250\text{ L}}$ = 3.0 mol/L	$\frac{0.75\text{ mol}}{0.250\text{ L}}$ = 3.0 mol/L	$\frac{0.25\text{ mol}}{0.250\text{ L}}$ = 1.0 mol/L

$$K = K_c = \frac{[CH_3COOC_2H_5]}{[C_2H_5OH][CH_3COOH]}$$

$$= \frac{1.0}{(3.0)(3.0)} = 0.11$$

■

PROBLEM 13.14 A solution is prepared by dissolving 0.050 mole of diiodocyclohexane, $C_6H_{10}I_2$, per liter of solution in the solvent CCl_4. When the reaction

$$C_6H_{10}I_2 \rightleftharpoons C_6H_{10} + I_2$$

has come to equilibrium at 35°C, the concentration of I_2 is 0.035 mol/L. (a) What are the concentrations of cyclohexene (C_6H_{10}) and $C_6H_{10}I_2$ at equilibrium? (b) Calculate the equilibrium constant. □

13.10 CALCULATIONS FROM THE EQUILIBRIUM CONSTANT

Another type of problem is more common than those in the last section. We are given the equilibrium constant and the initial numbers of moles, concentrations, or partial pressures and asked to find the quantities present at equilibrium. Problems of this kind also call for setting up an equilibrium table in which one unknown (y) appears. The concentrations or partial pressures, expressed in terms of y, are then substituted into the equilibrium condition. The resulting equation may be quadratic, cubic, or worse. However, in many cases it can be simplified by shrewd use of approximations.

EXAMPLE 13.18 The reaction

$$N_2(g) + O_2(g) \rightleftharpoons 2NO(g)$$

contributes to air pollution whenever a fuel is burned in air at a high temperature, as in a gasoline engine. At 1500 K, $K_c = 1.0 \times 10^{-5}$. A sample of air is compressed and heated to 1500 K. Before any reaction occurs, $[N_2] = 0.80$ mol/L and $[O_2] = 0.20$ mol/L. Calculate the concentration of NO at equilibrium, assuming that no other reaction occurs.

ANSWER

Make a quick check to see that the coefficients of y are the same as the coefficients in the equation; − on the left and + on the right.

	N_2	+	O_2	$\rightleftharpoons$	2NO
Initial concentration	0.80		0.20		0 mol/L
Change	$-y$		$-y$		$+2y$
Concentration at equilibrium	$0.80 - y$		$0.20 - y$		$2y$

$$\frac{[NO]^2}{[N_2][O_2]} = K_c$$

$$\frac{(2y)^2}{(0.80 - y)(0.20 - y)} = 1.0 \times 10^{-5}$$

This is a quadratic equation, which can be rearranged to standard form:

$$(4 \times 10^5 - 1)y^2 + y - 0.16 = 0$$

Solving by the quadratic formula gives two roots:

$$y = 6.3 \times 10^{-4} \quad \text{or} \quad -6.3 \times 10^{-4}$$

The negative root is rejected because it makes [NO] negative at equilibrium. This moderately laborious calculation can be bypassed by noting that K_c is small. This means that $[NO]^2$ is much smaller than $[N_2][O_2]$. Thus very little of the N_2 and O_2 will have been converted to NO at equilibrium. To simplify the calculation, we try to approximate (Appendix B.2) by neglecting y in comparison to 0.80 and 0.20: Assume $y \ll$ ("is much less than") 0.20. Then $0.80 - y \approx 0.80$, $0.20 - y \approx 0.20$.

$$\frac{(2y)^2}{0.80 \times 0.20} = 1.0 \times 10^{-5}$$

$$y = 6.3 \times 10^{-4}$$

The answer is the same as that obtained by solving the quadratic equation.

Since you would normally do the simplified calculation without having solved the quadratic equation, the validity of the approximation needs to be checked. y is indeed much smaller than 0.20 (and therefore much smaller than 0.80):

$$\begin{array}{r} 0.20 \\ -0.00063 \\ \hline 0.20 \end{array}$$

The approximation is therefore justified for the number of significant figures in this problem.

$$[NO] = 2y = 1.3 \times 10^{-3} \text{ mol/L}$$

■

In most equilibrium calculations, a quantity may be neglected if it is less than 5% of the smallest quantity initially present. In the preceding example, $6.3 \times 10^{-4}/0.20 = 0.32\%$.

This approximate method appears at first to be circular reasoning—assuming what you want to prove—but it really is not. If the simplified equation has a solution that is small enough, then it follows, by sound logic, that this solution also satisfies the original equation within the desired precision.

EXAMPLE 13.19 At 1600 K, the equilibrium constant K_c of the reaction

$$Br_2(g) \rightleftharpoons 2Br(g)$$

is 1.94×10^{-3}. In a 1.0-liter container, 0.10 mole of Br_2 is confined and heated to 1600 K. Calculate (a) the concentration of Br atoms present at equilibrium and (b) the fraction of the initial Br_2 that is dissociated into atoms.

ANSWER The initial concentration of Br_2 (before dissociation) is

$$[Br_2] = \frac{0.10 \text{ mol}}{1.0 \text{ L}} = 0.10 \frac{\text{mol}}{\text{L}}$$

	Br_2	$\rightleftharpoons$	2Br
Initial concentration	0.10		0 mol/L
Change	$-y$		$+2y$
Concentration at equilibrium	$0.10 - y$		$2y$

$$\frac{[Br]^2}{[Br_2]} = K_c$$

$$\frac{(2y)^2}{0.10 - y} = 1.94 \times 10^{-3}$$

$$\frac{y^2}{0.10 - y} = 4.85 \times 10^{-4}$$

Let us try the simplification used in Example 13.18. Assume that y is much less than 0.10, so that $0.10 - y \approx 0.10$. Then,

$$y^2 = 0.10 \times 4.85 \times 10^{-4}$$
$$= 4.85 \times 10^{-5}$$
$$y = 7.0 \times 10^{-3}$$

However, this answer for y is not negligible by our standards; y is 7% of the initial concentration. The answer is therefore inaccurate. We must solve the quadratic equation:

$$\frac{y^2}{0.10 - y} = 4.85 \times 10^{-4}$$

$$y^2 + 4.85 \times 10^{-4}y - 4.85 \times 10^{-5} = 0$$

The quadratic formula (Appendix B.2) gives

$$y = \frac{-4.85 \times 10^{-4} \pm \sqrt{(4.85 \times 10^{-4})^2 + 4 \times 4.85 \times 10^{-5}}}{2}$$

$$= 6.7 \times 10^{-3}$$

(The negative root would make [Br] negative and is therefore rejected.)

(a) At equilibrium

$$[\text{Br}] = 2y = 1.3 \times 10^{-2} \text{ mol/L}$$

(b) To find the fraction dissociated, divide the quantity dissociated by the original quantity:

$$\frac{y}{0.10} = \frac{6.7 \times 10^{-3}}{0.10} = 0.067, \text{ or } 6.7\%$$

■

A root is rejected, as in the preceding example, when it makes *any* of the equilibrium concentrations (or pressures) negative. In this example, y itself must be positive, but in many others, a negative y is possible. Thus, if the equilibrium concentrations are $0.20 - y$ and $0.10 + 2y$, an acceptable value of y can be anywhere between $+0.20$ and -0.05.

EXAMPLE 13.20 For the reaction

$$2H_2S(g) \rightleftharpoons 2H_2(g) + S_2(g)$$

at 800 K, $K_p = 3.0 \times 10^{-7}$. A tank initially contains H_2S at 10.0 atm. The temperature is constant at 800 K. Calculate the partial pressure of $S_2(g)$ at equilibrium.

ANSWER

	$2H_2S$	$\rightleftharpoons$ $2H_2$	+ S_2
Initial pressure	10.0	0	0 atm
Change	$-2y$	$+2y$	$+y$ atm
Pressure at equilibrium	$10.0 - 2y$	$2y$	y atm

$$\frac{p_{H_2}^2 p_{S_2}}{p_{H_2S}^2} = K_p$$

$$\frac{(2y)^2 y}{(10.0 - 2y)^2} = 3.0 \times 10^{-7}$$

$$\frac{4y^3}{(10.0 - 2y)^2} = 3.0 \times 10^{-7}$$

This is a cubic equation, but it can be simplified. Because the right side of this equation (K_p) is small, y must also be small. If we assume that $2y \ll 10$, then $10.0 - 2y \approx 10.0$, and the equation becomes

$$\frac{4y^3}{(10.0)^2} = 3.0 \times 10^{-7}$$

This is still a cubic equation, but it is easily solved.

$$y^3 = \frac{3.0 \times 10^{-7} \times 100}{4}$$

$$= 7.5 \times 10^{-6}$$

$$y = \sqrt[3]{7.5 \times 10^{-6}} = \sqrt[3]{7.5} \times 10^{-2}$$
$$= 2.0 \times 10^{-2} \text{ atm} = p_{S_2}$$

We see that our assumption is justified; $2y$ is indeed much less than 10.0 atm:

$$\frac{2y}{10.0} = \frac{4.0 \times 10^{-2}}{10.0} = 4.0 \times 10^{-3} = 0.4\%$$ ■

PROBLEM 13.15 Graphite and carbon dioxide are kept at constant volume at 1000 K until the reaction

$$C(graphite) + CO_2(g) \rightleftharpoons 2CO(g)$$

has come to equilibrium. At this temperature, $K_p = 1.7$. The initial pressure of CO_2 is 1.00 atm. Calculate the equilibrium pressure of CO. □

PROBLEM 13.16 At 1000 K, $K_c = 33$ for the reaction

$$H_2(g) + I_2(g) \rightleftharpoons 2HI(g)$$

H_2 and I_2 are initially present at equal concentrations, 6.00×10^{-3} mol/L for each. Find the concentrations of H_2, I_2, and HI at equilibrium. □

You should have made one or the other of two observations in doing Problem 13.16. You may have noticed that the reaction quotient is a perfect square. Thus, taking the square root of each side of the equilibrium condition gives a simple linear equation. If you missed this trick, you got stuck with solving a quadratic equation and obtained two roots, both positive. One of them has to be rejected because it is greater than 6.00×10^{-3} and therefore makes the equilibrium concentrations of H_2 and I_2 negative.

PROBLEM 13.17 The equilibrium constant for the reaction

$$Pb(\ell) + H_2O(g) \rightleftharpoons PbO(c) + H_2(g)$$

is 1.3×10^{-4} at 1000 K. A mixture of gases initially contains H_2O at 1.0 atm and H_2 at 1.0×10^{-3} atm at 1000 K. It is kept in contact with Pb(ℓ) and PbO(c) at 1000 K until the reaction has come to equilibrium. What are the partial pressures of H_2O and H_2 at equilibrium? □

If you followed the usual procedure in doing this problem, you obtained a negative answer for y; $y = -8.7 \times 10^{-4}$ atm. This simply means that, with the given initial pressures, the reaction goes from right to left to reach equilibrium. A little more H_2O is formed (not enough to change 1.0 atm), and 87% of the H_2 is consumed.

The methods explained in this section enable us to calculate the composition of an equilibrium mixture after it has been disturbed. This is a step beyond Le Chatelier's principle, which merely predicts the direction of the change. The original equilibrium concentrations (or pressures), modified by addition or removal of a substance or by a change in the volume, are the initial values to be used in the calculation.

EXAMPLE 13.21 The equilibrium mixture of N_2, O_2, and NO in Example 13.18 is in a 2.00-L container. Based on the results of that example, the concentrations are $[NO] = 1.3 \times 10^{-3}$ mol/L, $[N_2] = 0.80 - y = 0.80$ mol/L, and $[O_2] = 0.20 - y = 0.20$ mol/L. $K_c = 1.0 \times 10^{-5}$ at 1500 K for $N_2(g) + O_2(g) \rightleftharpoons 2NO(g)$. The mixture is disturbed by adding 0.30 mole of O_2. What is the concentration of NO in the new equilibrium mixture?

ANSWER The first step is to calculate the new concentration of O_2 immediately after it is added. The additional O_2 is 0.30 mol in a 2.00-L container. The new concentration of O_2, before any reaction occurs, is

$$[O_2] = 0.20 \text{ mol/L} + \frac{0.30 \text{ mol}}{2.00 \text{ L}} = 0.35 \text{ mol/L}$$

The initial concentrations of the other gases are as before: $[N_2] = 0.80$ mol/L and $[NO] = 1.3 \times 10^{-3}$ mol/L. The second step is to calculate the concentrations at equilibrium. This calls for the usual equilibrium table:

	N_2	+	O_2	$\rightleftharpoons$	2NO
Initial concentration	0.80		0.35		1.3×10^{-3} mol/L
Change	$-z$		$-z$		$+2z$
Concentration at equilibrium	$0.80 - z$		$0.35 - z$		$1.3 \times 10^{-3} + 2z$

The equilibrium condition is

$$\frac{[NO]^2}{[N_2][O_2]} = K_c$$

$$\frac{(1.3 \times 10^{-3} + 2z)^2}{(0.80 - z)(0.35 - z)} = 1.0 \times 10^{-5}$$

Note that $2z$ is *not* negligible in comparison to the small initial quantity 1.3×10^{-3}.

As in Example 13.18, the approximation $z \ll 0.35$ is appropriate:

$$\frac{(1.3 \times 10^{-3} + 2z)^2}{0.80 \times 0.35} = 1.0 \times 10^{-5}$$

$$1.3 \times 10^{-3} + 2z = 1.7 \times 10^{-3} \text{ mol/L} = [NO]$$

$z = 2 \times 10^{-4}$, which is negligible in comparison to 0.80 and 0.35. Adding O_2 has increased the concentration of NO from 1.3×10^{-3} to 1.7×10^{-3} mol/L, in the direction that Le Chatelier's principle predicts. ■

PROBLEM 13.18 (a) A cylinder contains C(graphite), CO(g), and $CO_2(g)$ at 1000 K. The partial pressures of the gases are 1.10 atm for CO and 0.70 atm for CO_2. The mixture is in equilibrium. Calculate K_p for the reaction

$$2CO(g) \rightleftharpoons C(\text{graphite}) + CO_2(g)$$

(b) Additional CO is pumped into the cylinder until its partial pressure, before any reaction has occurred, is 2.00 atm. Calculate the partial pressure of each gas when the mixture has found its new equilibrium composition at 1000 K. (c) A similar cylinder, containing 1.10 atm CO and 0.70 atm CO_2, is disturbed by pushing down a piston until the gas volume has been cut in half. The temperature is again constant at 1000 K. Calculate the partial pressure of each gas and the total pressure before any reaction occurs. (d) Calculate the partial pressure of each gas and the total pressure when this second cylinder has come to equilibrium. How does the total pressure compare with what you calculated from Boyle's law in part c? □

The following example and problem illustrate typical calculations on equilibria in solutions.

EXAMPLE 13.22 A solution is made by dissolving 0.10 mole of $C_6H_{10}O$ in CH_3OH to make 0.250 liter of solution. For the reaction

$$\underset{\text{cyclohexanone}}{C_6H_{10}O} + \underset{\text{methanol}}{2CH_3OH} \rightleftharpoons \underset{\text{1,1-dimethoxycyclohexane}}{C_6H_{10}(OCH_3)_2} + H_2O$$

at 25°C in the solvent CH_3OH, $K = 3.7$. Calculate the concentration of H_2O in the equilibrium solution.

ANSWER The initial concentration of $C_6H_{10}O$ is

$$\frac{0.10\text{ mol}}{0.250\text{ L}} = 0.40\,\frac{\text{mol}}{\text{L}}$$

Since CH_3OH is the solvent, it does not appear in the equilibrium table or condition. Note that H_2O is a *solute* in this solution.

	$2CH_3OH$ + $C_6H_{10}O$	$\rightleftharpoons$ $C_6H_{10}(OCH_3)_2$	+ H_2O
Initial concentration	0.40	0	0 mol/L
Change	$-y$	$+y$	$+y$
Concentration at equilibrium	$0.40 - y$	y	y

$$\frac{[C_6H_{10}(OCH_3)_2][H_2O]}{[C_6H_{10}O]} = K$$

$$\frac{y^2}{0.40 - y} = 3.7$$

Use the quadratic formula.

$$y^2 + 3.7y - 1.48 = 0$$

$$[H_2O] = y = 0.36\text{ mol/L}$$

■

PROBLEM 13.19 Pentene and acetic acid, dissolved in carbon tetrachloride (CCl_4), react to produce pentyl acetate:

$$C_5H_{10} + CH_3COOH \rightleftharpoons CH_3COOC_5H_{11}$$

The equilibrium constant for this reaction is 103 at 25°C. A solution is initially 0.050 M with respect to C_5H_{10} and 0.020 M with respect to CH_3COOH. Calculate the concentrations of C_5H_{10}, CH_3COOH, and $CH_3COOC_5H_{11}$ at equilibrium. □

In Problem 13.19, the larger root of the quadratic equation must be rejected because it would make $[C_5H_{10}]$ and $[CH_3COOH]$ negative.

BOX 13.1
IS THERE LIFE AFTER EQUILIBRIUM?

The condition of equilibrium may be regarded as a kind of death. The analogy is hardly far-fetched. When the chemical reactions in a battery (Chapter 17) have attained equilibrium, we call it a "dead" battery. Indeed, if a living organism is held in a closed, insulated container in the dark, it dies, literally, as the system approaches equilibrium. The various chemical reactions that interest us—the processes of life, the natural cycles of materials on the Earth, the transformations of matter in the stars, the manufacture of useful substances—all occur under *nonequilibrium* conditions. The chemist is therefore very much interested in how far a system is from equilibrium, with what intensity it is pushing toward equilibrium, in how its drive toward equilibrium can be made to yield work, in how the system can be driven away from equilibrium (if desired), and in the rates at which these processes occur. These topics will be explored in later chapters. However, in the absence of external disturbances, the equilibrium condition is the final condition to which any system, if left to itself, tends to go, sooner or later.

SUMMARY

Reactions can reach **equilibrium** when the reactants and products are kept in a closed container at constant temperature. Reactions then occur equally in both directions, with no net change. By convention, the "forward" direction is left to right as written and the "reverse" direction is right to left.

The **rate** of a reaction is the change in concentration of a specified reactant or product per unit time, most often expressed in mol/(L sec). As reactants are consumed, the rate of the forward reaction usually decreases; as the concentrations of products increase, the rate of the reverse reaction usually increases. When these rates become equal, the reaction is in equilibrium.

The **reaction quotient** Q is a function of the concentrations of reactants and products. For the general reaction $aA + bB + \cdots \rightleftharpoons cC + dD + \cdots$,

$$Q = \frac{[C]^c[D]^d \cdots}{[A]^a[B]^b \cdots}$$

The **law of chemical equilibrium** requires that the **equilibrium condition**, $Q = K$, be satisfied at equilibrium. K is the **equilibrium constant** for the reaction. When the mixture is not in equilibrium, the only possible net reaction is in the direction that makes Q closer to K—left to right when $Q < K$ and right to left when $Q > K$. When K is very large, the reaction goes almost to completion; when K is very small, the reaction almost does not occur. A substance or mixture that would, at equilibrium, be converted into something else is **thermodynamically unstable**; if it is at or near equilibrium, it is **thermodynamically stable**. A thermodynamically unstable system may be **kinetically stable**, meaning that its approach to equilibrium is undetectably slow.

An equilibrium condition involving gases can be expressed in terms of partial pressures instead of concentrations. The reaction quotient is then

$$Q_p = \frac{p_C^c p_D^d \cdots}{p_A^a p_B^b \cdots}$$

The equilibrium condition is $Q_p = K_p$ in terms of pressures, or $Q_c = K_c$ in terms of concentrations. K_p and K_c are related by $K_p = K_c(RT)^{\Delta n}$, where Δn is the number of moles of gaseous products minus the number of moles of gaseous reactants.

The equilibrium constant for a reaction is raised to the nth power when all the coefficients in the chemical equation are multiplied by n: $K_{new} = K_{old}^n$. In particular, if the equation is reversed, $K_{new} = 1/K_{old}$. When two or more reactions are added to give a net reaction, the equilibrium constant of the net reaction is the product of the equilibrium constants of the reactions added.

Le Chatelier's principle says that when a system in equilibrium is disturbed, the equilibrium shifts in the direction that decreases the effect of the disturbance. Specifically, (1) When a reactant or product is added, the reaction that consumes part of the added substance is favored. (2) When the volume of the container is increased (the pressure is decreased), a reaction occurs in the direction that increases the number of moles of gas (increases the pressure). (3) When the temperature is increased, a reaction occurs in the endothermic direction. Inverse statements apply when a reactant or product is subtracted, the volume is decreased, or the temperature is decreased. Adding or removing a catalyst does not change the composition of an equilibrium mixture; it only changes the rate of attainment of equilibrium.

A **homogeneous equilibrium** is established in a one-phase mixture; a **heterogeneous equilibrium** in a mixture of two or more phases. The equilibrium condition contains no factor for a pure solid or liquid. Only moles of gas are counted in determining the effect of a volume change and in relating K_c to K_p.

The equilibrium condition for a reaction in a dilute solution is expressed in terms of the concentrations of the solutes. There is no factor for the solvent or for undissolved solids or liquids. However, K depends on the identity of the solvent. Diluting the solution shifts the equilibrium toward more total moles of solutes.

When quantities are known for all the components in the initial mixture and for one component at equilibrium, the other quantities present at equilibrium can be calculated. The "quantities" can be the numbers of moles, the concentrations, or the partial pressures. An equilibrium table, expressed in terms of an unknown (y), is used for the calculation:

	aA	+	bB	+ . . . ⇌	cC	+	dD	+ . . .
Initial quantity	n_A		n_B		n_C		n_D	
Change	$-ay$		$-by$		$+cy$		$+dy$	
Quantity at equilibrium	$n_A - ay$		$n_B - by$		$n_C + cy$		$n_D + dy$	

When y has been found from the known quantity, the equilibrium constant can be calculated.

When the equilibrium constant and the initial quantities are given, the same equilibrium table is used. The equilibrium condition is solved for y. A root that makes any of the quantities negative at equilibrium is rejected. After an equilibrium mixture is disturbed, the original equilibrium quantities, modified by addition or removal of a substance or a change in volume, are used as initial quantities.

13.11 ADDITIONAL PROBLEMS

LAW OF CHEMICAL EQUILIBRIUM

13.20 Write the equilibrium condition for each reaction. For gases, use either pressures or concentrations.

(a) $CO(g) + \frac{1}{2}O_2(g) \rightleftharpoons CO_2(g)$
(b) $C(c) + CO_2(g) \rightleftharpoons 2CO(g)$
(c) $\underset{\text{solute}}{C_6H_5COOH} + \underset{\text{solvent}}{C_2H_5OH} \rightleftharpoons \underset{\text{solute}}{C_6H_5COOC_2H_5} + \underset{\text{solute}}{H_2O}$
(d) $H_2C_4H_2O_4(aq) + H_2O(\ell) \rightleftharpoons H_2C_4H_4O_5(aq)$
(e) $Ni(c) + 4CO(g) \rightleftharpoons Ni(CO)_4(g)$

(f) $N_2(g) + 3H_2(g) \rightleftharpoons 2NH_3(g)$
(g) $(NH_4)_2CO_3(c) \rightleftharpoons 2NH_3(g) + CO_2(g) + H_2O(g)$
(h) $\underset{\text{solvent}}{Li_2CO_3(c)} + \underset{\text{solute}}{CO_2(\ell)} + H_2O \rightleftharpoons \underset{\text{solute}}{2LiHCO_3}$ (under high pressure)
(i) $P_4O_{10}(g) + 6PCl_5(g) \rightleftharpoons 10POCl_3(g)$

13.21 A sealed 25-mL tube containing H_2 and I_2 is kept at 448°C until the reaction

$$H_2(g) + I_2(g) \rightleftharpoons 2HI(g)$$

has come to equilibrium. The numbers of moles present at equilibrium in three experiments are

H_2	I_2	HI
(1) 3.45×10^{-4} mol	7.09×10^{-5} mol	1.15×10^{-3} mol
(2) 2.39×10^{-4}	6.92×10^{-5}	9.24×10^{-4}
(3) 7.72×10^{-5}	4.21×10^{-4}	1.25×10^{-3}

(a) Calculate the equilibrium constant for the given reaction from the results of each experiment. Calculate the average and the average deviation. (b) The volume of the tube is given, but do you really need to know it? (c) Do you need to distinguish between K_c and K_p?

13.22 $K_c = 0.21$ at 350°C for the reaction

$$2NO(g) + Br_2(g) \rightleftharpoons 2BrNO(g)$$

A 20.0-mL tube contains 1.0×10^{-4} mol NO, 2.0×10^{-4} mol Br_2, and 2.0×10^{-4} mol BrNO at 350°C. (a) Is the mixture in equilibrium? If it is not, which way can the reaction go? (Do not confuse number of moles and concentration.) (b) At equilibrium, the tube contains 2.8×10^{-4} mol NO and 1.6×10^{-5} mol BrNO. How many moles of Br_2 does it contain? (c) Do you need to know the volume of the tube to do this problem? (Compare this with Problem 13.21.)

13.23 (a) For each of the five equilibrium mixtures in Table 13.1, calculate the expression

$$\frac{[SO_3]}{[SO_2][O_2]}$$

Calculate the average deviation and the relative average deviation. Is this expression as nearly constant as the correct reaction quotient? (b) Which of the following expressions should be constant at equilibrium?

(1) $\dfrac{[SO_3]}{[SO_2][O_2]^{1/2}}$ (2) $\dfrac{[SO_3]^{1/2}}{[SO_2]^{1/2}[O_2]}$

(3) $\dfrac{[SO_2]^2[O_2]}{[SO_3]}$

13.24 Mixtures are thermodynamically and kinetically stable or unstable. There are four possibilities:

KINETICALLY	THERMODYNAMICALLY Stable	THERMODYNAMICALLY Unstable
Stable	Mixture 1	Mixture 2
Unstable	Mixture 3	Mixture 4

Match each mixture (1–4) with one of these descriptions:
(a) The reaction goes to completion when a catalyst is present, but without the catalyst no reaction can be detected.
(b) The reaction goes rapidly to completion.
(c) The reaction does not occur at all.
(d) The reaction reaches equilibrium rapidly, but very little of the products are present at equilibrium.

K_p AND K_c

13.25 For reactions a and b of Problem 13.20, express K_p in terms of K_c, R, and T.

13.26 (a) For the reaction $COCl_2(g) \rightleftharpoons CO(g) + Cl_2(g)$, write the equilibrium condition in terms of partial pressures. (b) Using the equation $p = cRT$, substitute the appropriate concentration expression for each partial pressure in the equilibrium condition. (c) Combine the RT factors and show that $K_p = K_cRT$ for this reaction.

13.27 A mixture of CO, H_2, CH_4, and H_2O is kept at 1133 K until the reaction

$$CO(g) + 3H_2(g) \rightleftharpoons CH_4(g) + H_2O(g)$$

has come to equilibrium. The volume of the container is 0.100 L. The equilibrium mixture contains 1.21×10^{-4} mol CO, 2.47×10^{-4} mol H_2, 1.21×10^{-4} mol CH_4, and 5.63×10^{-8} mol H_2O. Calculate K_p for this reaction at 1133 K.

FORM OF EQUATION

13.28 Given the reaction

$$H_2(g) + Br_2(g) \rightleftharpoons 2HBr(g)$$

(a) At 500 K, $K_c = 7.9 \times 10^{11}$. What is K_p?
(b) $\frac{1}{2}H_2(g) + \frac{1}{2}Br_2(g) \rightleftharpoons HBr(g)$ $K_c = ?$
(c) $2HBr(g) \rightleftharpoons H_2(g) + Br_2(g)$ $K_c = ?$
(d) $4HBr(g) \rightleftharpoons 2H_2(g) + 2Br_2(g)$ $K_c = ?$

COMBINATION OF EQUILIBRIA

13.29 These equilibrium constants at 700°C are given:

$$H_2(g) + CO_2(g) \rightleftharpoons H_2O(g) + CO(g) \qquad K_c = 0.62$$
$$FeO(g) + CO(g) \rightleftharpoons Fe(c) + CO_2(g) \qquad K_c = 0.68$$

Calculate K_c for the reaction

$$Fe(c) + H_2O(g) \rightleftharpoons FeO(c) + H_2(g)$$

13.30 K_c is given for these reactions at 1000 K:

$$Fe_3O_4(c) + 4CO(g) \rightleftharpoons 3Fe(c) + 4CO_2(g) \quad K_c = 1.67 \times 10^{-2}$$
$$C(\text{graphite}) + CO_2(g) \rightleftharpoons 2CO(g) \quad K_c = 2.03 \times 10^{-2}$$

Calculate (a) K_c and (b) K_p for the reaction

$$Fe_3O_4(c) + 2C(graphite) \rightleftharpoons 3Fe(c) + 2CO_2(g)$$

13.31 Equilibrium constants at 500 K are

$$H_2(g) \rightleftharpoons 2H(g) \qquad K_p = 4.8 \times 10^{-41}$$
$$Br_2(g) \rightleftharpoons 2Br(g) \qquad K_p = 2.2 \times 10^{-15}$$

Use these data and the answer to Problem 13.28a to calculate K_p for the reaction

$$Br(g) + H_2(g) \rightleftharpoons HBr(g) + H(g)$$

13.32 These equilibrium constants at 750°C are given:

$$SnO_2(c) + 2H_2(g) \rightleftharpoons Sn(c) + 2H_2O(g) \quad K_p = 8.12$$
$$H_2(g) + CO_2(g) \rightleftharpoons CO(g) + H_2O(g) \quad K_p = 0.771$$

Calculate K_p for the reaction

$$SnO_2(c) + 2CO(g) \rightleftharpoons Sn(c) + 2CO_2(g)$$

LE CHATELIER'S PRINCIPLE

13.33 Refer to reactions a and b in Problem 13.20. (a) When the volume of the container is decreased at constant temperature, which way is the equilibrium shifted in reaction a; in reaction b? (b) After O_2 is added to the mixture in reaction a and equilibrium is reestablished, which concentrations have become larger, and which have become smaller? (c) The equilibrium in reaction b shifts to the right when the temperature is raised. Is the reaction exothermic or endothermic? Is ΔH positive or negative?

13.34 For the reaction

$$Fe_2O_3(c) + 3CO(g) \rightleftharpoons 2Fe(c) + 3CO_2(g)$$

$K_c = 24.2$ at 1000 K and $\Delta H = -26$ kJ. (a) What is K_p for this reaction at 1000 K? (b) Calculate the ratio $[CO]/[CO_2]$ in the equilibrium mixture at 1000 K. (c) If you want this ratio to be smaller in a blast furnace, should the temperature be higher or lower than 1000 K? (See Figure 22.14, section 22.10.)

13.35 Consider the general gas reaction $a\text{A} \rightleftharpoons b\text{B}$. (a) Write the equilibrium condition in terms of concentrations. (b) The volume of the container is multiplied by f (which may be more or less than 1). Express the new concentrations ([A]′, [B]′) in terms of f and the initial concentrations ([A], [B]). (c) Express the new reaction quotient (Q') in terms of the initial reaction quotient (Q), f, a, and b. (d) Make a table showing which way the equilibrium shifts for each of the six combinations of f greater or less than 1 and a greater than, equal to, or less than b. Are your results compatible with Le Chatelier's principle?

HETEROGENEOUS EQUILIBRIUM

13.36 At −10°C, the solid compound $Cl_2(H_2O)_8$ is in equilibrium with gaseous chlorine, water vapor, and ice. The partial pressures of the two gases in equilibrium with a mixture of $Cl_2(H_2O)_8$ and ice are 0.20 atm for Cl_2 and 0.00262 atm for water vapor. Find the equilibrium constant K_p for each of these reactions:
(a) $Cl_2(H_2O)_8(c) \rightleftharpoons Cl_2(g) + 8H_2O(g)$
(b) $Cl_2(H_2O)_8(c) \rightleftharpoons Cl_2(g) + 8H_2O(c)$
Why are your two answers so different?

13.37 At −23.5°C, the total gas pressure in equilibrium with $HCl(H_2O)_2$ is 194 torr. The reaction is

$$HCl(H_2O)_2(c) \rightleftharpoons HCl(g) + 2H_2O(c)$$

The vapor pressure of ice at this temperature is 0.555 torr. (a) What is K_p for this reaction (with pressures in atm)? (b) What is K_p for the reaction $HCl(H_2O)_2(c) \rightleftharpoons HCl(g) + 2H_2O(g)$?

13.38 A flask contains $NH_4Cl(c)$ in equilibrium with its decomposition products:

$$NH_4Cl(c) \rightleftharpoons NH_3(g) + HCl(g)$$

For this reaction, $\Delta H = 176$ kJ. How is the mass of NH_3 in the flask affected by each of the following disturbances? (a) The temperature is increased. (b) NH_3 is added. (c) HCl is added. (d) NH_4Cl is added, with no appreciable change in the gas volume. (e) A large amount of NH_4Cl is added, decreasing the volume available to the gases.

13.39 The equilibrium constant for the reaction

$$H_2(g) + Br_2(\ell) \rightleftharpoons 2HBr(g)$$

is $K_p = 4.5 \times 10^{18}$ at 25°C. The vapor pressure of liquid Br_2 at this temperature is 0.28 atm. (a) Find K_p at 25°C for the reaction

$$H_2(g) + Br_2(g) \rightleftharpoons 2HBr(g)$$

(b) How will the equilibrium in part a be shifted by a decrease in the volume of the container if (1) liquid Br_2 is absent; (2) liquid Br_2 is present? Explain why the effect is different in these two cases.

13.40 A gaseous mixture of H_2 and H_2O is kept in contact with a solid mixture of Fe and Fe_3O_4 at 1150°C until the reaction

$$3Fe(c) + 4H_2O(g) \rightleftharpoons Fe_3O_4(c) + 4H_2(g)$$

has come to equilibrium. The following pressures are measured at equilibrium in three experiments:

	TOTAL PRESSURE torr	PARTIAL PRESSURE OF H_2O, torr
(1)	25.7	11.9
(2)	37.3	17.4
(3)	107.5	49.3

From the results of each experiment, find the equilibrium constant for the reaction. Calculate the average and the average deviation.

EQUILIBRIUM IN SOLUTIONS

13.41 This equilibrium is established in water:

$$\underset{\text{propionic acid}}{C_2H_5COOH(aq)} + \underset{\text{methanol}}{CH_3OH(aq)} \rightleftharpoons \underset{\text{methyl propionate}}{C_2H_5COOCH_3(aq)} + H_2O$$

How is the mass of methanol in the solution affected by each of these disturbances? (a) C_2H_5COOH is added. (b) H_2O is added. (c) $C_2H_5COOCH_3$ is added.

13.42 Hemoglobin (Hb) can form a complex with either O_2 or CO. For the reaction

$$O_2Hb(aq) + CO(g) \rightleftharpoons COHb(aq) + O_2(g)$$

at body temperature, $K \approx 200$. If the ratio [COHb]/[O_2Hb] comes close to 1, death is probable. What partial pressure of CO in the air is likely to be fatal? Assume that the partial pressure of O_2 is 0.2 atm. (Do you have to be told whether K is K_c or K_p?)

CALCULATING THE EQUILIBRIUM CONSTANT

13.43 NO(g) and O_2(g) are mixed in a container of fixed volume kept at 1000 K. Their initial concentrations are 0.0200 mol/L and 0.0300 mol/L, respectively. When the reaction

$$2NO(g) + O_2(g) \rightleftharpoons 2NO_2(g)$$

has come to equilibrium, the concentration of NO_2 is 2.2×10^{-3} mol/L. Calculate (a) the concentration of NO at equilibrium; (b) the concentration of O_2 at equilibrium; (c) the equilibrium constant K_c for the reaction.

13.44 A sealed tube initially contains 9.84×10^{-4} mol H_2 and 1.38×10^{-3} mol I_2. It is kept at 350°C until the reaction

$$H_2(g) + I_2(g) \rightleftharpoons 2HI(g)$$

comes to equilibrium. At equilibrium, 4.73×10^{-4} mol I_2 is present. Calculate (a) the numbers of moles of H_2 and HI present at equilibrium; (b) the equilibrium constant of the reaction.

13.45 CO_2 and H_2 are admitted to a container of constant volume kept at 959 K. Their initial partial pressures, before any reaction, are 1.50 atm and 3.00 atm, respectively. The reaction

$$CO_2(g) + H_2(g) \rightleftharpoons CO(g) + H_2O(g)$$

then occurs. At equilibrium, the partial pressure of H_2O is 0.86 atm. Calculate the partial pressures of CO_2, H_2, and CO at equilibrium, and K_p for the reaction.

13.46 0.0100 mole of NH_4Cl and 0.0100 mole of NH_3 are placed in a closed 2.00-liter container and heated to 603 K. At this temperature, all the NH_4Cl vaporizes. When the reaction

$$NH_4Cl(g) \rightleftharpoons NH_3(g) + HCl(g)$$

has come to equilibrium, 5.8×10^{-3} mole of HCl is present. Calculate (a) K_c and (b) K_p for this reaction at 603 K.

13.47 CO_2 is passed over graphite at 500 K. The emerging gas stream contains 4.0×10^{-3} mole percent CO. The total pressure is 1.00 atm. Assume that equilibrium is attained. Find K_p for the reaction

$$C(\text{graphite}) + CO_2(g) \rightleftharpoons 2CO(g)$$

13.48 The following equilibrium is established in the presence of water and a particular enzyme (catalyst):

$$\underset{\text{fumaric acid}}{H_2C_4H_2O_4(aq)} + H_2O \rightleftharpoons \underset{\text{malic acid}}{H_2C_4H_4O_5(aq)}$$

1.00 liter of solution was prepared from water and 0.300 mole of pure malic acid. At equilibrium, the solution contained 0.067 mole of fumaric acid, in addition to some unchanged malic acid. Find the equilibrium constant for the reaction.

CALCULATIONS FROM THE EQUILIBRIUM CONSTANT: GAS EQUILIBRIA

13.49 "Synthetic natural gas," CH_4, can be made from a mixture of CO and H_2 by the reaction

$$CO(g) + 3H_2(g) \rightleftharpoons CH_4(g) + H_2O(g)$$

At 1133 K, $K_p = 4.3 \times 10^{-3}$. An equilibrium mixture contains CO at 10 atm and H_2 at 20 atm. The partial pressures of CH_4 and H_2O are equal in this mixture. What is the partial pressure of CH_4 in the equilibrium mixture?

13.50 For the reaction

$$N_2(g) + O_2(g) \rightleftharpoons 2NO(g)$$

$K_p = 3.24 \times 10^{-3}$ at 2675 K. A mixture at this temperature initially contains N_2 at 0.50 atm and O_2 at 0.50 atm. Calculate the partial pressure of NO at equilibrium. (*Suggestion*: The equation to be solved can be simplified by taking the square root.)

13.51 For the reaction

$$2IF(g) \rightleftharpoons F_2(g) + I_2(g)$$

$K_p = 2.9 \times 10^{-14}$ at 1000 K. A sealed tube kept at 1000 K initially contains IF(g) at 0.80 atm. Nothing else is put into the tube. What is the partial pressure of F_2 in the tube at equilibrium?

13.52 At 25°C, the equilibrium constant for the reaction

$$N_2O_4(g) \rightleftharpoons 2NO_2(g)$$

is $K_c = 5.85 \times 10^{-3}$. 20.0 grams of N_2O_4 is confined in a 5.00-liter flask at 25°C. Calculate (a) the number of moles of NO_2 present at equilibrium; (b) the percentage of the original N_2O_4 that is dissociated.

13.53 The equilibrium constant K_c of the reaction

$$H_2(g) + Br_2(g) \rightleftharpoons 2HBr(g)$$

is 1.6×10^5 at 1297 K and 3.5×10^4 at 1495 K. (a) Is ΔH for this reaction positive or negative? (b) Find K_c for the reaction

$$\tfrac{1}{2}H_2(g) + \tfrac{1}{2}Br_2(g) \rightleftharpoons HBr(g)$$

at 1297 K. (c) Pure HBr is placed in a container of constant volume and heated to 1297 K. What percentage of the HBr is decomposed to H_2 and Br_2 at equilibrium? Is any superfluous information given in part c?

13.54 1.00 mole of H_2O is placed in a 100-liter container and heated to 1727°C. The equilibrium constant for the reaction

$$2H_2O(g) \rightleftharpoons 2H_2(g) + O_2(g)$$

is $K_c = 5.29 \times 10^{-10}$. (a) Find the number of moles of O_2 present at equilibrium. Make a reasonable approximation to avoid having to solve a cubic equation. (b) Calculate K_p for the reaction.

13.55 For the reaction

$$Cl_2(g) \rightleftharpoons 2Cl(g)$$

$K_p = 2.48 \times 10^{-5}$ at 1200 K. (a) In Cl_2 at 1.00 atm and 1200 K, what is the partial pressure of Cl? (b) The question in part a is ambiguous. It does not specify whether 1.00 atm is (1) the initial pressure of Cl_2, (2) the partial pressure of Cl_2 at equilibrium, or (3) the total pressure of Cl_2 and Cl. Write an equation that you would solve to obtain p_{Cl} for each of these interpretations. Are the results appreciably different? Why or why not?

13.56 When acetic acid is a gas, some of it exists as dimers (double molecules):

$$2CH_3COOH(g) \rightleftharpoons (CH_3COOH)_2(g)$$

For this reaction, $K_c = 3.18 \times 10^4$ at 25°C. (a) The vapor pressure of acetic acid at 25°C is 15.4 torr. Calculate the total concentration of acetic acid vapor, including both monomers (single molecules) and dimers, in equilibrium with the liquid at 25°C. (b) Let y be the concentration of monomer. What is the concentration of dimer in terms of y? (c) Calculate the concentrations of monomer and dimer and the percentage of the vapor molecules that are dimers.

13.57 For the reaction

$$PCl_3(g) + Cl_2(g) \rightleftharpoons PCl_5(g)$$

$K_c = 45$ at 500 K. (a) A mixture is in equilibrium, with $[PCl_3] = 2.0 \times 10^{-3}$ mol/L and $[PCl_5] = 1.0 \times 10^{-3}$ mol/L. What is the concentration of Cl_2? (b) The mixture in part a is disturbed by adding 2.5×10^{-3} mole of PCl_5 per liter. Calculate the new concentrations of the three gases when equilibrium has been restored.

13.58 For the reaction

$$COBr_2(g) \rightleftharpoons CO(g) + Br_2(g)$$

$K_p = 5.41$ at 346 K. Two identical flasks are prepared, each containing $COBr_2$ at 346 K, initially at 1.00 atm. (a) Calculate the partial pressure of CO in each flask when the reaction has come to equilibrium. (b) Flask 1 is disturbed by pumping in enough Br_2 to make its partial pressure 1.362 atm before the equilibrium has time to shift. Calculate the partial pressures of CO and of Br_2 when the mixture has reached its new equilibrium. (c) Flask 2 is disturbed by decreasing the volume available to the gases to one-half the initial volume. Calculate the partial pressure of CO when equilibrium is restored.

CALCULATIONS FROM THE EQUILIBRIUM CONSTANT: HETEROGENEOUS EQUILIBRIA

13.59 For the reaction

$$2CO(g) \rightleftharpoons C(graphite) + CO_2(g)$$

$K_p = 6.0 \times 10^{-4}$ at 1473 K. CO(g) is initially at 1.00 atm. It is kept in contact with graphite at 1473 K until the reaction has come to equilibrium. What is the partial pressure of CO_2 at equilibrium?

13.60 A stream of gas containing H_2 at an initial partial pressure of 0.200 atm is passed through a tube in which CuO is kept at 500 K. The reaction

$$CuO(c) + H_2(g) \rightleftharpoons Cu(c) + H_2O(g)$$

comes to equilibrium. For this reaction, $K_p = 1.6 \times 10^9$. What is the partial pressure of H_2 in the gas leaving the tube? Assume that the total pressure of the stream is unchanged.

13.61 In the distant future, when hydrogen may be cheaper than coal, steel mills may make iron by the reaction

$$Fe_2O_3(c) + 3H_2(g) \rightleftharpoons 2Fe(c) + 3H_2O(g)$$

For this reaction, $\Delta H = 96$ kJ and $K_c = 8.11$ at 1000 K. (a) What percentage of the H_2 remains unreacted after the reaction has come to equilibrium at 1000 K? (b) Is this percentage greater or less if the temperature is increased above 1000 K?

13.62 At 21.8°C, the equilibrium constant K_c for the reaction

$$NH_4HS(c) \rightleftharpoons NH_3(g) + H_2S(g)$$

is 1.2×10^{-4}. (a) Solid NH_4HS is introduced into an empty container and allowed to come to equilibrium at 21.8°C. Find the total concentration of the gas mixture at equilibrium. (b) Find the equilibrium concentration of H_2S when NH_4HS is introduced into a container that also contains NH_3 at an initial concentration of 0.010 mol/L. (c) A container contains $NH_4HS(c)$, $NH_3(g)$ at 0.030 mol/L, and $H_2S(g)$ at 0.010 mol/L. In which direction, forward or reverse, can the given reaction go at 21.8°C?

CALCULATIONS FROM THE EQUILIBRIUM CONSTANT: SOLUTION EQUILIBRIA

13.63 A solution contains 0.75 mole of benzoic acid, C_6H_5COOH, 1.00 mole of methanol, CH_3OH, and water in 1.00 liter of solution. The solution is kept at 200°C (under pressure) until the reaction

$$C_6H_5COOH(aq) + CH_3OH(aq) \rightleftharpoons C_6H_5COOCH_3(aq) + H_2O(\ell)$$

has come to equilibrium. At equilibrium, 5.6% of the benzoic acid has reacted. (a) Calculate the equilibrium constant for the reaction. (b) Calculate the percentage of the acid that reacts when 1.00 mole of benzoic acid and 2.00 moles of methanol are dissolved in water to make 1.00 liter and allowed to come to equilibrium at 200°C.

13.64 A solution is prepared by dissolving 2.70 grams of N_2O_4 in chlorobenzene, C_6H_5Cl, at 21°C to make 100 mL of solution. The equilibrium constant for the reaction $N_2O_4 \rightleftharpoons 2NO_2$ is 6.29×10^{-5} under these conditions. (a) What is the initial concentration of N_2O_4 in mol/L? (b) Calculate the concentrations of NO_2 and N_2O_4 in the solution at equilibrium.

13.65 When benzoic acid, C_6H_5COOH, is dissolved in benzene, C_6H_6, it forms a dimer (double molecule):

$$2C_6H_5COOH \rightleftharpoons (C_6H_5COOH)_2$$

The equilibrium constant for this reaction is 2.7×10^2 at 43.9°C. When 1.00 mol C_6H_5COOH is dissolved in C_6H_6 to make 1.00 L of solution, what percentage of the benzoic acid becomes dimers?

13.66 (a) A solution is prepared by dissolving 3.00×10^{-3} mol acetone, $(CH_3)_2CO$, in the solvent methanol, CH_3OH, to make 25 mL. When the reaction

$$(CH_3)_2CO + 2CH_3OH \rightleftharpoons (CH_3)_2C(OCH_3)_2 + H_2O$$

has come to equilibrium at 25°C, 67.3% of the acetone is still present. Calculate K for this reaction. (Note that H_2O is a solute.) (b) Another solution is prepared with the initial concentration of the acetone equal to 0.025 mol/L. What percentage of the acetone remains unreacted at equilibrium?

SELF-TEST

13.67 (12 points) Carbon tetrachloride, CCl_4, was formerly used in fire extinguishers, but at high temperature the reaction

$$CCl_4(g) + \tfrac{1}{2}O_2(g) \rightleftharpoons COCl_2(g) + Cl_2(g)$$

produces two poisonous gases, phosgene and chlorine. For this reaction at 1500 K, $K_c = 3.3 \times 10^7$. Calculate K_c and K_p for the reaction

$$2CCl_4(g) + O_2(g) \rightleftharpoons 2COCl_2(g) + 2Cl_2(g)$$

13.68 (14 points) The following equilibrium constants at 1362 K are given:

$$H_2(g) + \tfrac{1}{2}S_2(g) \rightleftharpoons H_2S(g) \qquad K_p = 0.80$$

$$H_2(g) + \tfrac{1}{2}O_2(g) \rightleftharpoons H_2O(g) \qquad K_p = 4.1 \times 10^6$$

$$\tfrac{1}{2}S_2(g) + O_2(g) \rightleftharpoons SO_2(g) \qquad K_p = 1.2 \times 10^{10}$$

Find K_p at 1362 K for the reaction

$$2H_2S(g) + 3O_2(g) \rightleftharpoons 2H_2O(g) + 2SO_2(g)$$

13.69 (11 points) Consider the reaction

$$2Cl_2(g) + 2H_2O(g) \rightleftharpoons 4HCl(g) + O_2(g) \qquad \Delta H = +113 \text{ kJ}$$

The four gases—Cl_2, H_2O, HCl, and O_2—are mixed and the reaction is allowed to come to equilibrium. State and explain the effect (*increase, decrease, no change*) of each operation in the left column of the table on the equilibrium value of the quantity in the right column. Consider each operation separately. Temperature and volume are constant except when stated otherwise.

OPERATION	QUANTITY
(a) Increasing the volume of the container	moles of H_2O
(b) Adding O_2	moles of H_2O
(c) Adding O_2	moles of O_2
(d) Adding O_2	moles of HCl
(e) Decreasing the volume of the container	moles of Cl_2
(f) Decreasing the volume of the container	partial pressure of Cl_2
(g) Decreasing the volume of the container	K_c
(h) Raising the temperature	K_c
(i) Raising the temperature	concentration of HCl
(j) Adding He	moles of HCl
(k) Adding a catalyst	moles of HCl

13.70 (15 points) A mixture of N_2 (initial pressure 50 atm) and H_2 (initial pressure 100 atm) is kept in contact with a catalyst until the reaction

$$N_2(g) + 3H_2(g) \rightleftharpoons 2NH_3(g)$$

comes to equilibrium at 773 K. The final partial pressure of NH_3 is 16 atm. (a) Calculate the equilibrium constant K_p. (b) Calculate K_c.

13.71 (16 points) At 2000°C, $K_p = 4.12 \times 10^{-6}$ for the reaction

$$2H_2O(g) \rightleftharpoons 2H_2(g) + O_2(g)$$

If H_2O is initially at 1.00 atm, what percentage of it will be dissociated to H_2 and O_2?

13.72 (16 points) For the reaction

$$\tfrac{1}{8}S_8(c, \text{rhombic}) + 2HI(g) \rightleftharpoons I_2(c) + H_2S(g)$$

$K_c = 3.13 \times 10^7$ at 300 K. H_2S is admitted to a rigid container until its concentration is 5.00×10^{-3} mol/L. Solid S_8 and solid I_2 are then introduced and the system is allowed to come to equilibrium at 300 K. Calculate the concentration of HI at equilibrium.

13.73 (16 points) A solution is prepared by dissolving 2.67 g N_2O_4 in dichloroethane, $C_2H_4Cl_2$, to make 100 mL of solution. At 14.2°C, 0.27% of the N_2O_4 is dissociated into NO_2. (a) Calculate the equilibrium constant for the reaction $N_2O_4 \rightleftharpoons 2NO_2$. (b) Another solution initially contains 4.00 g N_2O_4 in 100 mL solution. What percentage of the N_2O_4 is dissociated at equilibrium?

14

ACIDS AND BASES

Among the compounds that are used most widely and manufactured in the largest quantities are the acids sulfuric acid (H_2SO_4), hydrochloric acid (HCl), and nitric acid (HNO_3) and the bases sodium hydroxide ($NaOH$), calcium hydroxide ($Ca(OH)_2$), and ammonia (NH_3). Acids and bases are used in industry, agriculture, the home, and the laboratory. They are also essential to life processes. For example, stomach juices rich in hydrochloric acid are needed for digestion. In section 9.1, we defined acids and bases in terms of the electrolytic dissociation theory of Svante Arrhenius, which is that in water solution, acids furnish hydrogen ions, whereas bases supply hydroxide ions. In this chapter we discuss broader definitions of acids and bases and take a closer look at how they react. The ideas of chemical equilibrium that we introduced in Chapter 13 are qualitatively applied to this discussion. (Acidity and basicity will be treated quantitatively in Chapter 15.) In this chapter we will also explain how acidity and basicity are related to molecular structure.

14.1 BRÖNSTED–LOWRY ACID–BASE CONCEPT

Arrhenius's theory was an important advance when it was introduced in 1884, but during the next forty years certain limitations became apparent. The concept explained acid-base reactions in water reasonably well, but it did not explain the reactions in the absence of water. For example, Arrhenius would suggest that in water, HCl and NH_3 exist as ionic solutions of hydrochloric acid and ammonium hydroxide (NH_4OH), respectively. The products of their reaction are the salt ammonium chloride (NH_4Cl) and water:

$$\underbrace{H^+ + Cl^-}_{\text{hydrochloric acid}} + \underbrace{NH_4{}^+ + OH^-}_{\text{ammonium hydroxide}} \longrightarrow H_2O + \underbrace{NH_4{}^+ + Cl^-}_{\text{ammonium chloride}} \qquad \text{(reaction in water)}$$

But HCl and NH_3, as dry gases or when dissolved in a nonpolar solvent such as benzene, also react to give NH_4Cl:

$$HCl(g) + NH_3(g) \longrightarrow NH_4Cl(c) \qquad \text{(reaction of dry gases)} \qquad (1)$$

The formation of NH_4Cl in this reaction cannot be interpreted by the Arrhenius concept. NH_3 gas cannot furnish OH^-, and HCl gas does not dissociate into H^+ and Cl^-. Hence, in the absence of water, H^+ and OH^- are not present to react as they are in a water solution.

This limitation of the Arrhenius acid-base concept was resolved in 1923 when a more general theory of acids and bases that could explain both the aqueous and gaseous reactions of HCl and NH_3 was proposed independently by both Johannes N. Brönsted and Thomas M. Lowry. The **Brönsted–Lowry theory** states that an *acid-base reaction involves a proton transfer; the acid is the proton donor and the base is the proton acceptor.* This theory emphasizes the interdependence of an acid and a base; one is defined in terms of the other.

Proton transfer is nearly instantaneous.

We can now see how the Brönsted–Lowry theory extended the Arrhenius theory. HCl does not simply form H^+ in water:

$$HCl \longrightarrow H^+ + Cl^- \qquad \text{(Arrhenius's theory)}$$

Instead, it gives up the proton to water, the base, to form H_3O^+, the hydronium ion and Cl^-:

Curved arrows indicate the transfer of atoms, molecules, or ions.

$$H_2\ddot{O}: + H:\ddot{Cl}: \longrightarrow [H_2\ddot{O}{-}H]^+ + :\ddot{Cl}:^- \qquad \text{(Brönsted–Lowry theory)}$$

base acid hydronium ion ($H_3\ddot{O}^+$)

Arrhenius was correct when he said that all substances that furnish OH^- are bases. The hydroxide ion is indeed a base:

$$H:\ddot{Cl}: + :\ddot{O}H^- \longrightarrow H:\ddot{O}:H + :\ddot{Cl}:^-$$

acid base

But according to Brönsted and Lowry, OH^- is by no means the only base known. For example, in equation 1 gaseous NH_3 (ammonia) itself acts as the base toward gaseous HCl, the acid.

Bottles of concentrated solutions of HCl and NH_3 that are stored close together are soon coated with films of NH_4Cl.

$$:NH_3(g) + H:\ddot{Cl}:(g) \longrightarrow [NH_4]^+ :\ddot{Cl}:^-(c) \qquad \text{(written as } NH_4Cl)$$

base acid ammonium chloride

The ammonium chloride appears as a white cloud consisting of small crystals. This reaction is one of many that show the Brönsted–Lowry concept is much broader than the Arrhenius concept. It is not limited to water solutions, nor is it restricted to OH^- as the only base.

Since $:NH_3$ is a base, we can now see how ammonium hydroxide is formed when NH_3 is added to water. In this case water acts as an acid:

$$H:\ddot{O}:H + :NH_3 \rightleftharpoons [H{-}NH_3]^+ + :\ddot{O}H^-$$

acid base ammonium hydroxide

Notice that molecules of NH_4OH do not exist in this reaction. For this reason water solutions of NH_3 should be called "aqueous ammonia" and not "ammonium hydroxide."

EXAMPLE 14.1 Give the products in the following acid-base reactions. Indicate the acid and the base.

(a) $HNO_3 + H_2O$
(b) $H_3O^+ + F^-$
(c) $HSO_4^- + NO_2^-$

ANSWER

(a) HNO_3 is the acid and H_2O is the base; the products are NO_3^- and H_3O^+.
(b) H_3O^+ is the acid and F^- is the base; the products are H_2O and HF.
(c) HSO_4^- is the reactant with an H and therefore must be the acid; NO_2^- is the base. The products are SO_4^{2-} and HNO_2. ■

PROBLEM 14.1 Give the products in the following acid-base reactions. Indicate the acid and the base. (Recall that H_2O can be a base or an acid.)

(a) $HClO_4 + H_2O$
(b) $NH_4^+ + H_2O$
(c) $HCO_3^- + OH^-$
(d) $NH_2^- + H_2O$ □

14.2 CONJUGATE ACID–BASE PAIRS

According to the Brönsted–Lowry concept, in an aqueous solution, an acid (HA) transfers its proton to the base (H_2O), forming H_3O^+ and leaving $:A^-$ as shown:

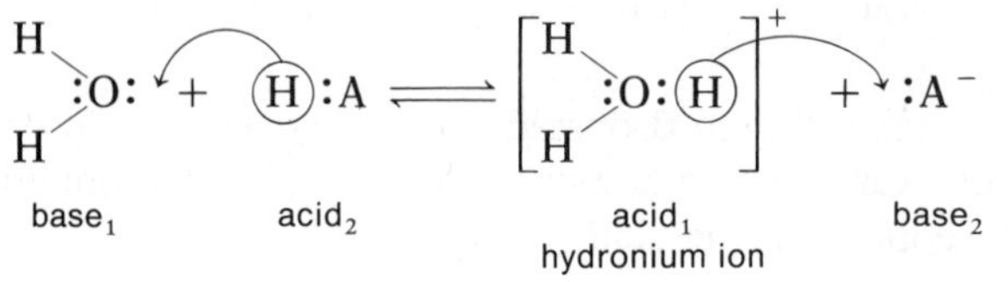

$$\underset{\text{base}_1}{H_2\ddot{O}:} + \underset{\text{acid}_2}{H:A} \rightleftharpoons \underset{\substack{\text{acid}_1\\ \text{hydronium ion}}}{\left[H_2\ddot{O}:H\right]^+} + \underset{\text{base}_2}{:A^-} \quad (2)$$

Once formed, H_3O^+ can act as an acid by transferring a proton back to A^-, which acts as a base. Acid-base reactions like equation 2 are reversible, and equilibrium is quickly established. In general, *the reaction of an acid and a base always leads to the formation of a new acid and a new base.* The product acid (H_3O^+) arises when the reactant base (H_2O) accepts a proton. H_3O^+ and H_2O are known as a **conjugate acid-base pair**. Similarly, the product base (A^-) results when the reactant acid (HA) loses a proton; HA and A^- also make up a conjugate acid-base pair. We can summarize this conjugate acid-base relationship with the aid of some typical examples:

■ *An acid loses a proton, leaving behind its conjugate base:*

ACID	−	H^+	⟶	CONJUGATE BASE
HBr	−	H^+	⟶	Br^-
H_3O^+	−	H^+	⟶	H_2O
HCO_3^-	−	H^+	⟶	CO_3^{2-}

Notice that because the acid loses an H^+, its conjugate base has one less H and one less positive charge (or one more negative charge) than the acid. From these examples it can be seen that acids can be molecules (HBr), positive ions (H_3O^+), or negative ions (HCO_3^-).

■ *A base gains a proton, forming its conjugate acid:*

BASE	+	H^+	$\longrightarrow$	CONJUGATE ACID
H_2O	+	H^+	$\longrightarrow$	H_3O^+
F^-	+	H^+	$\longrightarrow$	HF
SO_4^{2-}	+	H^+	$\longrightarrow$	HSO_4^-
$H_3NNH_2^+$	+	H^+	$\longrightarrow$	$H_3NNH_3^{2+}$

Notice that because the base gains an H^+, its conjugate acid has one more H and one more positive charge (or one less negative charge) than the base. From these examples it can be seen that bases can also be molecules (H_2O), negative ions (F^-, SO_4^{2-}), or positive ions ($H_3NNH_2^+$). One conjugate acid-base pair in the reaction is designated by the subscript 1 and the other pair by the subscript 2, as illustrated in equations 2 and 3.

$$\underset{\text{acid}_1}{H_2\ddot{O}:} + \underset{\text{base}_2}{:NH_3} \rightleftharpoons \underset{\text{base}_1}{H\ddot{O}:^-} + \underset{\text{acid}_2}{[NH_4]^+} \quad (3)$$

In the reaction shown in equation 3, NH_3 is the base and H_2O is the acid. The ammonium ion, NH_4^+, is the conjugate acid of NH_3, and OH^- is the conjugate base of H_2O. When gaseous NH_3 and HCl react, HCl (acid) donates a proton to NH_3 (base). The products are Cl^-, the conjugate base of HCl, and NH_4^+, the conjugate acid of NH_3, which form a salt (NH_4Cl):

$$\underset{\text{base}_1}{H_3N:(g)} + \underset{\text{acid}_2}{H:\ddot{Cl}:(g)} \longrightarrow \underset{\text{acid}_1}{[NH_4]^+} \underset{\text{base}_2}{:\ddot{Cl}:^-(c)}$$

Structural formulas have been used in the preceding equations to reveal an important structural feature of all bases: *all bases have an unshared pair of electrons that can form a covalent bond with the transferred proton.* We say that the atom with the unshared pair of electrons acts as the **basic site** for accepting the H^+.

EXAMPLE 14.2 (a) Give the conjugate acid of each of these bases: $\ddot{N}H_3$, $:\ddot{O}H^-$, $:\ddot{O}:^{2-}$, and $:H^-$. (b) Give the conjugate base of each of these acids: NH_4^+, HNO_3, HSO_4^-, and $H_2PO_4^-$.

ANSWER (a) To find the acid, add H^+. (b) To find the base, subtract H^+.

BASE	+	H^+	$\longrightarrow$	CONJUGATE ACID
$:NH_3$	+	H^+	$\longrightarrow$	NH_4^+
$:\ddot{O}H^-$	+	H^+	$\longrightarrow$	$H_2\ddot{O}:$
$:\ddot{O}:^{2-}$	+	H^+	$\longrightarrow$	$:\ddot{O}H^-$
$H:^-$	+	H^+	$\longrightarrow$	H_2

ACID	−	H^+	$\longrightarrow$	CONJUGATE BASE
NH_4^+	−	H^+	$\longrightarrow$	$:NH_3$
HNO_3	−	H^+	$\longrightarrow$	NO_3^-
HSO_4^-	−	H^+	$\longrightarrow$	SO_4^{2-}
$H_2PO_4^-$	−	H^+	$\longrightarrow$	HPO_4^{2-}

■

EXAMPLE 14.3 How many basic sites are present in hydroxylamine, $H_2\ddot{N}\ddot{O}H$? Write the structure for the conjugate acid formed by reaction at each site.

ANSWER The N and O each have unshared electrons, and therefore each is a basic site. If N picks up the proton, we get $H_3N\ddot{O}H^+$. If O picks up the proton, we get $H_2\ddot{N}\ddot{O}H_2{}^+$. ■

PROBLEM 14.2 (a) Give the conjugate acid for each of the bases $:\ddot{N}H_2{}^-$, $:\ddot{S}:^{2-}$, and $CO_3{}^{2-}$. (b) Give the conjugate base for each of the acids HI, $HPO_4{}^{2-}$, and H_2S. □

PROBLEM 14.3 Fill in the missing formula. $HSO_4{}^-$ is the conjugate acid of ________ and the conjugate base of ________. (Note the dual role of $HSO_4{}^-$.) □

Since an acid is a proton donor, it is reasonable to suggest that the *greater an acid's tendency to donate a proton, the stronger acid it is.* Thus a molecule of a strong acid tends to readily give up its proton, whereas a molecule of a weak acid tends to hold on to its proton. Likewise, since a base is the proton acceptor, *the greater is a base's tendency to accept a proton, the stronger base it is.* A molecule of a strong base readily accepts a proton, while a molecule of a weak base has little tendency to be a proton acceptor. Moreover, *once a strong base captures a proton, it tends to hold on to it.* These ideas of acid and base strengths lead to an important interdependent relationship between the strengths in a given conjugate acid-base pair. Saying that a strong acid, HA, readily donates a proton also means that the conjugate base, A^-, does not readily hold on to the proton. This behavior leads to the generalization that *the stronger an acid, the weaker is its conjugate base; the stronger a base, the weaker is its conjugate acid.* Conversely, *the weaker an acid, the stronger is its conjugate base; the weaker a base, the stronger is its conjugate acid.*

Table 14.1 lists some conjugate acid-base pairs and shows the interdependent relationship of their strengths. The H-containing substances are grouped as strong acids, weak acids, and nonacids, with H_3O^+ and H_2O serving as reference points;

strong acid: stronger than H_3O^+

weak acid: weaker than H_3O^+ but stronger than H_2O

nonacid: weaker than H_2O

Because of the broad range of weak acids, they are roughly subclassified:

Do not try to memorize Table 14.1.

STRENGTH	LOCATION IN TABLE 14.1	EXAMPLES
Moderately weak	top of B section	H_2SO_3, H_3PO_4
Weak	middle B section	$HC_2H_3O_2$, HF
Very weak	bottom B section	HCN, H_2O_2

The statement that "the weaker the acid, the stronger its conjugate base" serves only for making comparisons among two or more acids and their conjugate bases. It is not meant to express absolute values. Therefore, it does not follow that a weak acid necessarily has a strong conjugate base. For example, whereas acetic acid is a weak acid, acetate ion, its conjugate base, is a weak rather than a strong base.

TABLE 14.1
COMMON CONJUGATE ACID–BASE PAIRS

	ACID			CONJUGATE BASE	
	Name	Formula	Condensed Structure	Formula	Name
A	Perchloric acid	$HClO_4$	$HOClO_3$	ClO_4^-	Perchlorate ion
A	Sulfuric acid	H_2SO_4	$(HO)_2SO_2$	HSO_4^-	Hydrogen sulfate ion
A	Hydriodic acid	HI		I^-	Iodide ion
A	Hydrobromic acid	HBr		Br^-	Bromide ion
A	Hydrochloric acid	HCl		Cl^-	Chloride ion
A	Nitric acid	HNO_3	$HONO_2$	NO_3^-	Nitrate ion
	Hydronium ion	H_3O^+		H_2O	Water
B	Sulfurous acid	H_2SO_3	$(HO)_2SO$	HSO_3^-	Hydrogen sulfite ion
B	Hydrogen sulfate ion	HSO_4^-	$HOSO_3^-$	SO_4^{2-}	Sulfate ion
B	Phosphoric acid	H_3PO_4	$(HO)_3PO$	$H_2PO_4^-$	Dihydrogen phosphate ion
B	Phosphorous acid	H_2PHO_3	$(HO)_2PHO$	$HPHO_3^-$	Hydrogen phosphite ion
B	Nitrous acid	HNO_2	HONO	NO_2^-	Nitrite ion
B	Hydrofluoric acid	HF		F^-	Fluoride ion
B	Acetic acid	$HC_2H_3O_2$	CH_3COOH	$C_2H_3O_2^-$	Acetate ion
B	Carbonic acid	$CO_2 + H_2O(H_2CO_3)$	$(HO)_2CO$	HCO_3^-	Hydrogen carbonate ion
B	Hydrogen sulfide	H_2S		HS^-	Hydrosulfide ion
B	Hydrogen sulfite ion	HSO_3^-	$HOSO_2^-$	SO_3^{2-}	Sulfite ion
B	Dihydrogen phosphate ion	$H_2PO_4^-$	$(HO)_2PO_2^-$	HPO_4^{2-}	Hydrogen phosphate ion
B	Ammonium ion	NH_4^+		NH_3	Ammonia
B	Hydrocyanic acid	HCN		CN^-	Cyanide ion
B	Hydrogen carbonate ion	HCO_3^-	$HOCO_2^-$	CO_3^{2-}	Carbonate ion
B	Hydrogen peroxide	H_2O_2	HOOH	HO_2^-	Hydroperoxide ion
B	Hydrogen phosphate ion	HPO_4^{2-}	$HOPO_3^{2-}$	PO_4^{3-}	Phosphate ion
B	Hydrosulfide ion	HS^-		S^{2-}	Sulfide ion
	Water	H_2O	HOH	OH^-	Hydroxide ion
C	Ammonia	NH_3		NH_2^-	Amide ion
C	Hydrogen	H_2		H^-	Hydride ion†
C	Methane	CH_4		CH_3^-	Methide ion†

Increasing strength as an acid ↑ (acids, top to bottom: decreasing); Increasing strength as a base ↓

GROUP OF ACIDS	ARBITRARY CLASSIFICATION	RANGE OF K_a§ VALUES
A	strong	greater than 10^2
B	weak	10^{-1} to 10^{-14}
C	nonacidic	less than 10^{-14}

† These bases are not formed directly from H_2 and CH_4, respectively.
§ Discussed later in this section.

What the statement does permit us to say is that since hydrocyanic acid, HCN, is a weaker acid than acetic acid, its conjugate base, cyanide ion, CN^-, is a stronger base than acetate ion, the conjugate base of acetic acid.

For a molecule to be a Brönsted acid, it must possess an H atom. However, possession of an H is not enough to make the molecule an acid. To be called an acid in water solutions, the molecule should be able to transfer a proton to water to some measurable extent; by showing electrical conductivity, for example. Not all bonded H's are transferable to water. Molecules that cannot transfer their protons to water are the nonacids in Table 14.1. Nonacids show little activity toward OH^-; $AH + OH^- \rightleftharpoons A^- + H_2O$. With few exceptions (such as HCN) acidic H atoms are attached to electronegative elements such as O, F, Cl, Br, I, and S. In $-\ddot{N}-$H-type compounds, the basicity of the nitrogen site overshadows the acidity of the H. An exception is hydrazoic acid, HN_3.

You may wonder why the nonacids NH_3, H_2, and CH_4 were included in Table 14.1. They were included because they have conjugate bases formed by indirect methods rather than by direct proton transfer. For example, the methide ion, $:CH_3^-$, exists as the anion of the salt potassium methide, $K^+CH_3^-$. When this salt is placed in water, the methide ion reacts violently to give methane, CH_4:

$(CH_3)_2Hg + 2K$ (in Na) $\longrightarrow 2KCH_3 + Hg$ (in Na)

$$\underset{\text{base}_1}{:CH_3^-} + \underset{\text{acid}_2}{H\!:\!\ddot{\underset{..}{O}}\!:\!H} \longrightarrow \underset{\text{acid}_1}{CH_4} + \underset{\text{base}_2}{:\ddot{\underset{..}{O}}H^-}$$

By definition CH_4 is the conjugate acid of $:CH_3^-$, even though it shows absolutely no acidic properties—it does not react with any known base. Similarly, amide ion, NH_2^-, the anion of the salt sodium amide, $NaNH_2$, is the conjugate base of ammonia, NH_3. Although NH_3 is typically a base, it can give up a proton by reaction; for example, with $:CH_3^-$, a much stronger base:

$$\underset{\text{acid}_1}{:NH_3} + \underset{\text{base}_2}{:CH_3^-} \longrightarrow \underset{\text{base}_1}{:\ddot{N}H_2^-} + \underset{\text{acid}_2}{CH_4}$$

Notice in Table 14.1 that the molecular formula of acetic acid is written as $HC_2H_3O_2$ and the condensed structure is CH_3COOH. Only one of the four H atoms is acidic, the one bonded to the O atom. Typically, acidic H atoms are written first in the molecular formula, and nonacid H atoms are written after the central atom.

Also note from Table 14.1 that the hydroxide ion, OH^-, is a very strong base. Hence soluble ionic hydroxides, such as potassium hydroxide (KOH) and sodium hydroxide (NaOH), a much stronger base:

EQUILIBRIUM AND ACIDITY–BASICITY

We saw earlier that acid-base reactions are reversible, as shown for the general acid, HA, in water:

$$HA + H_2O \rightleftharpoons H_3O^+ + A^- \qquad \text{(acid dissociation)}$$

The equilibrium condition for this generalized acid dissociation (ionization) is

$$K_a = \frac{[H_3O^+][A^-]}{[HA]}$$

H_2O, the solvent, is omitted from the equation for K_a.

The magnitude of K_a is a measure of acidity or acid strength in water—the larger the value of K_a, the stronger is the acid. The acids in Table 14.1 are listed in order of decreasing K_a values. For example, the weak hydrofluoric acid (HF) has a $K_a = 6.4 \times 10^{-4}$, while the strong hydrochloric acid (HCl) has a very large K_a ($K_a \approx 10^7$).

0.100 M HF ionizes 8%; 0.100 M HCl ionizes practically 100%.

The difference in acid strengths can be explained in terms of the position of these equilibria for the dissociation of acids in water:

$$\underset{\substack{\text{acid}_1 \\ \text{(stronger} \\ \text{than } H_3O^+)}}{HCl} + \underset{\substack{\text{base}_2 \\ \text{(stronger} \\ \text{than } Cl^-)}}{H_2O} \rightleftharpoons \underset{\substack{\text{acid}_2 \\ \text{(weaker} \\ \text{than HCl)}}}{H_3O^+} + \underset{\substack{\text{base}_1 \\ \text{(weaker} \\ \text{than } H_2O)}}{Cl^-} \qquad K_a \approx 10^7 \qquad (4)$$

About 0.4% of the molecules in 2 M $HC_2H_3O_2$ ionize.

$$\underset{\substack{\text{acid}_1 \\ \text{(weaker} \\ \text{than } H_3O^+)}}{HC_2H_3O_2} + \underset{\substack{\text{base}_2 \\ \text{(weaker} \\ \text{than } C_2H_3O_2^-)}}{H_2O} \rightleftharpoons \underset{\substack{\text{acid}_2 \\ \text{(stronger} \\ \text{than } HC_2H_3O_2)}}{H_3O^+} + \underset{\substack{\text{base}_1 \\ \text{(stronger} \\ \text{than } H_2O)}}{C_2H_3O_2^-} \qquad K_a = 1.8 \times 10^{-5} \qquad (5)$$

With HCl, the equilibrium lies almost entirely to the right (very large K_a), as shown in equation 4. With $HC_2H_3O_2$, the equilibrium lies mainly to the left (small $K_a = 1.8 \times 10^{-5}$), as shown in equation 5. These trends are indicated by the direction of the longer arrow in each equation.

We summarize the difference in the behavior of dilute aqueous solutions of strong and weak acids:

$$\underset{\text{no molecules remain}}{HA_{\text{strong}}} + H_2O \longrightarrow \underset{\text{only species present}}{\underbrace{H_3O^+ + A^-}} \qquad \text{(100\% dissociated)}$$

$$\underset{\text{most molecules remain}}{HA_{\text{weak}}} + H_2O \rightleftharpoons \underset{\substack{\text{small number of} \\ \text{ions are present}}}{H_3O^+ + A^-} \qquad \text{(little dissociation)}$$

In general, Brönsted–Lowry acid-base equilibria *favor the side with the weaker acid and weaker base.* This means that the stronger acid and stronger base react to give their weaker conjugate base and acid, respectively:

$$\underset{\textbf{stronger acid}}{\mathbf{HA}} + \underset{\textbf{stronger base}}{\mathbf{B}} \longrightarrow \underset{\textbf{weaker base}}{\mathbf{A^-}} + \underset{\textbf{weaker acid}}{\mathbf{BH^+}}$$

Specific examples of this relationship are given in equations 4 and 5.

Base strength is also discussed in terms of chemical equilibrium:

$$\underset{\text{base}_1}{B{:}} + \underset{\text{acid}_2}{H{-}\ddot{\underset{..}{O}}{-}H} \rightleftharpoons \underset{\text{acid}_1}{B{:}H^+} + \underset{\text{base}_2}{{:}\ddot{\underset{..}{O}}H^-}$$

Ionization constants for bases, K_b, can also be calculated:

$$K_b = \frac{[BH^+][OH^-]}{[B]}$$

The magnitude of K_b is a measure of base strength—the larger the value of K_b, the stronger is the base. As usual the equilibrium favors the side with the weaker base and weaker acid.

All molecular (uncharged) bases are weaker than OH^-. Hence the equilibrium for the reaction of molecular bases favors the molecular base rather than the formation of OH^-. The weak ionization of the molecular base NH_3 illustrates this point:

$$\underset{\substack{\text{base}_1 \\ \text{(weaker} \\ \text{than } OH^-)}}{{:}NH_3} + \underset{\substack{\text{acid}_2 \\ \text{(weaker} \\ \text{than } NH_4^+)}}{H{-}\ddot{\underset{..}{O}}{-}H} \rightleftharpoons \underset{\substack{\text{acid}_1 \\ \text{(stronger} \\ \text{than } H_2O)}}{NH_4^+} + \underset{\substack{\text{base}_2 \\ \text{(stronger} \\ \text{than } NH_3)}}{{:}\ddot{\underset{..}{O}}H^-}$$

The equilibrium condition is

$$\frac{[NH_4^+][OH^-]}{[NH_3]} = K_b = 1.8 \times 10^{-5}$$

The small value of K_b indicates that the equilibrium in this reaction favors the reactants.

EXAMPLE 14.4 Identify the bases and use Table 14.1 to predict the favored direction of the equilibrium in these reactions.

(a) $HCN + OH^- \rightleftharpoons CN^- + H_2O$

(b) $HPO_4^{2-} + H_2O \rightleftharpoons H_3O^+ + PO_4^{3-}$

ANSWER

(a) The bases are OH^- and CN^-. According to Table 14.1, OH^- is a stronger base than CN^-, and HCN is a stronger acid than H_2O. Therefore the equilibrium lies mainly to the right where the weaker base, CN^-, and the weaker acid, H_2O, are found.

(b) The bases are H_2O and PO_4^{3-}. From Table 14.1, H_3O^+ is a stronger acid than HPO_4^{2-}, and PO_4^{3-} is a stronger base than H_2O. Therefore this equilibrium lies mainly to the left. ■

PROBLEM 14.4 Identify the bases and use Table 14.1 to predict the favored direction of the equilibrium for these reactions:

(a) $HS^- + NH_4^+ \rightleftharpoons NH_3 + H_2S$

(b) $H_2PO_4^- + HO_2^- \rightleftharpoons HPO_4^{2-} + H_2O_2$ □

NEUTRALIZATION

Section 9.1 called the reaction of an acid and a base to give a salt and water a *neutralization*. Arrhenius interpreted this reaction in terms of H^+ and OH^-.

$$H^+Cl^- + Na^+OH^- \longrightarrow Na^+Cl^- + H_2O \qquad \text{(Arrhenius theory)}$$

The major modification made by the Brönsted–Lowry theory replaces H^+ with H_3O^+:

$$H_3O^+ + Cl^- + Na^+ + OH^- \longrightarrow Na^+ + Cl^- + 2H_2O \qquad \text{(Brönsted–Lowry theory)}$$

Eliminating the spectator ions (Na^+ and Cl^-) gives the net ionic equation (section 9.2).

$$H_3O^+ + OH^- \longrightarrow 2H_2O \qquad \text{(net ionic equation for neutralization in water)} \qquad (6)$$

Equation 6 reveals a unique relationship among the participating species. H_3O^+ is the conjugate acid of water and OH^- is the conjugate base of water. Thus, in terms of the Brönsted–Lowry theory the general definition of **neutralization** is *the reaction between the conjugate acid and the conjugate base of the same substance.*

When NaOH is added to a weak acid such as acetic acid, $HC_2H_3O_2$, the net ionic equation is for the reaction between OH^- with $HC_2H_3O_2$ rather than with H_3O^+:

$$HC_2H_3O_2 + OH^- \longrightarrow H_2O + C_2H_3O_2^-$$

This reaction of OH^- and an acid is also loosely called a neutralization. The reaction between a weak base and a strong acid, such as that between NH_3 and HCl, is similarly loosely called a neutralization.

$NH_3(g)$ is condensed to $NH_3(\ell)$ at −33°C.

$$NH_3 + H_3O^+ \longrightarrow NH_4^+ + H_2O$$

The Brönsted–Lowry definition is not restricted to aqueous reactions in which H_2O is the product. For example, neutralizations also occur in liquid ammonia in the absence of water. The product of the neutralization is then NH_3 formed by the reaction of its conjugate acid (NH_4^+) and its conjugate base (NH_2^-);

molecular equation: $$NH_4Cl + NaNH_2 \longrightarrow NaCl + 2NH_3$$

ionic equation: $$\underset{\text{conjugate acid of } NH_3}{NH_4^+} + \underset{\text{conjugate base of } NH_3}{NH_2^-} \longrightarrow 2NH_3 \quad \text{(neutralization in liquid } NH_3) \qquad (7)$$

14.3 RELATIVE STRENGTHS OF ACIDS

RELATIVE NATURE OF ACIDITY

The intrinsic strength of an acid is determined by its tendency to donate a proton regardless of what is present to acquire the proton. Recently, such intrinsic acidities have been measured using low concentrations of gaseous molecules of the acid. Under these conditions the gaseous acid molecules behave independently of the other molecules. However, since acids are used most often in basic solvents, such as water, their intrinsic gas phase acidities are of little practical importance. Hence, the strength of an acid is determined by its ability to donate a proton to a base. Such *relative* acid strengths (acidities) can only be compared by reference to a common base. The base of choice is water since we live on a watery planet. Water is our most common solvent and reaction medium.

If chemists existed on Jupiter, their choice might be liquid ammonia.

A qualitative method for detecting differences in relative acid strengths is the reaction between acids and zinc (Figure 14.1).

$$Zn + 2H_3O^+ \longrightarrow H_2(g) + Zn^{2+} + 2H_2O$$

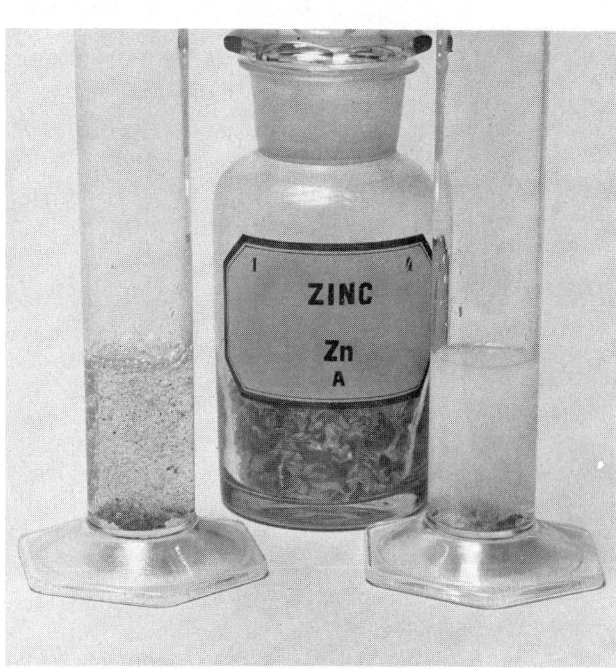

FIGURE 14.1
Reaction of zinc with solutions of hydrochloric acid, a strong acid (*left*), and acetic acid, a weak acid (*right*). (H_2O is not a strong enough acid to react with Zn.)

Why must the solutions of HCl and $HC_2H_3O_2$ have the same molarity in this test?

The stronger the acid is, the higher is the $[H_3O^+]$ and the more rapid is the evolution of hydrogen bubbles. Figure 14.1 shows that the evolution of H_2 from hydrochloric acid, a strong acid, is much faster than from acetic acid, a weak acid. Section 9.1 mentioned that electrical conductivities can be used to classify strong and weak acids. With good equipment, these experiments can be used quantitatively to measure acid strengths in terms of K_a values.

ACID STRENGTHS OF STRONG ACIDS—THE LEVELING EFFECT

It is easy to compare the acidities of a strong acid and a weak acid or of two weak acids. However, a problem arises when we compare the acid strengths of HCl and hydrobromic acid, HBr. Their K_a values in water are very large, approaching infinity. HCl and HBr are strong acids that are practically completely ionized in dilute aqueous solutions.

$$\left.\begin{array}{l} HCl + H_2O \longrightarrow H_3O^+ + Cl^- \\ HBr + H_2O \longrightarrow H_3O^+ + Br^- \end{array}\right\} \text{go practically to completion}$$

strong acids of different strength; all stronger than H_3O^+ — only acid present

Consequently, the only acid present is H_3O^+; *H_3O^+ is the strongest acid that can exist in a dilute aqueous solution.*

So how can we determine if HBr and HCl are equally strong acids on an absolute basis? Water is too strong a base for differences in their acid strength to be detected. This behavior is called the **leveling effect**: H_2O "levels out" the acidity, thus making strong acids like HBr and HCl appear to have equal acid strength. This is analogous to comparing the strengths of two persons by asking them each to lift a one-gram weight. They would appear equally strong because both of them could lift the weight easily.

Glacial acetic acid is called a *differentiating* solvent.

The acid strengths of strong acids must be compared by using a base weaker than water. One such substance is pure, anhydrous acetic acid—called "glacial" acetic acid. Although acetic acid is weakly acidic in water, in the absence of water it can accept a proton from strong acids. Thus in anhydrous solutions with strong acids, acetic acid is a weak base. Since anhydrous acetic acid is a weaker base than water, the strong acids do not dissociate completely. They are not leveled. Furthermore, strong acids dissociate to a different extent. Thus, anhydrous acetic acid serves as a medium for using electrical conductivities to measure the strengths of acids that are "leveled" in water.

$$HBr(g) + CH_3\overset{\overset{\displaystyle O}{\|}}{C}{-}OH(\ell) \rightleftharpoons \left[CH_3C\begin{matrix} {=}OH \\ {-}OH \end{matrix}\right]^+ + Br^-$$

$$HCl(g) + CH_3\overset{\overset{\displaystyle O}{\|}}{C}{-}OH(\ell) \rightleftharpoons \left[CH_3C\begin{matrix} {=}OH \\ {-}OH \end{matrix}\right]^+ + Cl^-$$

acid_1 base_2 acid_2 base_1

In this way, it is shown that HBr is a stronger acid than HCl. This method is analogous to comparing the strengths of two persons by measuring the height to which each can lift a heavy weight.

EXAMPLE 14.5 Which acids listed in Table 14.1 are leveled in H_2O?

ANSWER The acids that are stronger than H_3O^+ are leveled in H_2O. These are the strong acids listed in group A of Table 14.1: $HClO_4$, H_2SO_4, HI, HBr, HCl, and HNO_3. (You should memorize these.) ■

EXAMPLE 14.6 Discuss the validity of this statement: Strong acids leveled by water have conjugate bases weaker than water.

ANSWER The statement is true. Strong acids are leveled because all of their molecules donate protons to water. The conjugate bases of these strong acids do not take protons from H_3O^+, the only remaining acid in solution. Since H_2O competes better for the proton than do the conjugate bases, H_2O is a stronger base than the conjugate bases of these strong acids. Or we can simply say that because acid-base reactions favor the formation of the weak acid and the weak base, the conjugate bases of strong acids are weaker than water. ■

EXAMPLE 14.7 Explain this statement: A substance with an H atom, HA, would be called an acid if it reacted with OH^- to give an equilibrium state favoring the products $H_2O + A^-$.

ANSWER HA must be more acidic than H_2O in order to be called an acid in Table 14.1. The conjugate bases of all acids stronger than H_2O are weaker bases than OH^-, the conjugate base of H_2O. HA must therefore react with OH^-, with the equilibrium favoring H_2O and A^-, the weaker acid and weaker base:

$$\underbrace{\underset{\text{acid}_1}{HA} + \underset{\text{base}_2}{OH^-}}_{\text{stronger}} \rightleftharpoons \underbrace{\underset{\text{acid}_2}{H_2O} + \underset{\text{base}_1}{A^-}}_{\text{weaker}}$$

■

EXAMPLE 14.8 Explain why 100% perchloric acid ($HClO_4$) acts as an acid even when it is dissolved in 100% sulfuric acid (H_2SO_4). Write an equation for the reaction, indicating the acids and bases.

ANSWER H_2SO_4 possesses O atoms with unshared pairs of electrons capable of accepting protons. This potential basic property is almost always completely overshadowed by the very strong acidic property of H_2SO_4. The only time H_2SO_4 can be a base is when it reacts with an acid stronger than itself. The only such acid in Table 14.1 is $HClO_4$. Since sulfuric acid is an extremely weak base, $HClO_4$ dissociates only to a very slight extent.

The extent of the dissociation of $HClO_4$ in H_2SO_4 is determined by freezing-point depression techniques (section 12.13).

$$\underbrace{\underset{\text{acid}_1}{HClO_4} + \underset{\text{base}_2}{H_2SO_4}}_{\text{stronger}} \rightleftharpoons \underbrace{\underset{\text{base}_1}{ClO_4^-} + \underset{\text{acid}_2}{H_3SO_4^+}}_{\text{weaker}}$$

■

The facts in Example 14.8 support the importance of thinking in terms of the relative nature of acidity. Here we find that the very strong acid, H_2SO_4, can nevertheless also behave as a base. Furthermore, $HClO_4$, the strongest acid listed in Table 14.1, in this case acts as a weak acid.

PROBLEM 14.5 Use the listing in Table 14.1 to explain why 100% HNO_3 is a base when dissolved in 100% H_2SO_4. Write an equation and indicate the acids and bases. □

CONJUGATE ACIDS OF MOLECULAR BASES

The conjugate acids of molecular bases are weak acids in water solution. An example is the reaction of NH_4^+, the conjugate acid of NH_3:

$$\underbrace{\underset{\text{acid}_1}{NH_4^+} + \underset{\text{base}_2}{H_2O}}_{\text{weaker}} \rightleftharpoons \underbrace{\underset{\text{acid}_2}{H_3O^+} + \underset{\text{base}_1}{NH_3}}_{\text{stronger}} \tag{8}$$

The equilibrium condition for this reaction is

$$\frac{[H_3O^+][NH_3]}{[NH_4^+]} = K_a = 5.64 \times 10^{-10}$$

14.4 RELATIVE STRENGTHS OF BASES

The same principles are used to compare the strengths of bases. Relative base strengths are compared by using the same reference acid, usually H_2O. As we mentioned earlier, all molecular bases are weak.

BASIC ANIONS

CN^- is slightly more basic than NH_3.

Anions that are conjugate bases of weak acids act as bases toward water. Like uncharged bases, these anions are all weaker than OH^-. Hence, equilibrium favors the reactants, as in the reaction of CN^- in H_2O:

$$\underbrace{\underset{\text{base}_1}{:CN^-} + \underset{\text{acid}_2}{H\ddot{\underset{\cdot\cdot}{O}}H}}_{\text{weaker}} \rightleftharpoons \underbrace{\underset{\text{acid}_1}{HCN} + \underset{\text{base}_2}{:\ddot{\underset{\cdot\cdot}{O}}H^-}}_{\text{stronger}} \tag{9}$$

Reactions 8 and 9 are each sometimes called *hydrolysis*. In such a case, K, written K_h, is called the *hydrolysis constant*.

The equilibrium condition for this reaction is

$$\frac{[HCN][OH^-]}{[CN^-]} = K_b = 2.03 \times 10^{-5}$$

The weaker the acid, the more basic is the conjugate anion and the greater is K_b.

It used to be said that a water solution of NaCN was basic (turns red litmus blue) because of the presence of OH^-. In an Arrhenius sense this is true, but this statement overlooks the fact that CN^- is itself a base by the Brönsted–Lowry definition. It can act as a base in the absence of water and therefore without the formation of OH^-:

$$\underset{\text{base}_1}{Na^+CN^-(c)} + \underset{\text{acid}_2}{HC_2H_3O_2(\ell)} \longrightarrow \underset{\text{acid}_1}{HCN(g)} + \underset{\text{base}_2}{Na^+C_2H_3O_2^-(c)}$$

Anions that are conjugate bases of strong acids do not act as bases toward water. Neither do they react as bases toward acids weaker than H_3O^+. However, in the absence of water they can react as bases toward strong acids, especially those acids with weaker conjugate bases.

$$NaCl(c) + H_2SO_4(\ell) \longrightarrow HCl(g) + NaHSO_4(c) \tag{10}$$

The reaction shown in equation 10 is used to make gaseous hydrogen chloride.

EXAMPLE 14.9 (a) Select from among the following anions those that might be basic and those that are not basic toward water: HO_2^-, Br^-, S^{2-}, and ClO_4^-. Which one is the strongest base? (b) Write a chemical equation for each basic anion. (c) Which anion that does not react with H_2O can react with a strong acid? Use Table 14.1 to obtain your answers.

ANSWER

(a) Only those anions whose conjugate acids are weaker than H_3O^+ can act as bases toward H_2O. These are HO_2^-, whose conjugate acid is H_2O_2, and S^{2-}, whose conjugate acid is HS^-. The stronger basic anion is the one with the weaker conjugate acid. Since HS^- is a weaker acid than H_2O_2, the answer is S^{2-}.

(b) $HO_2^- + H_2O \longrightarrow H_2O_2 + OH^-$; $\quad S^{2-} + H_2O \longrightarrow HS^- + OH^-$.

(c) Br^- can react with any acid stronger than HBr, its conjugate acid. ClO_4^- is the weakest base and its conjugate acid, $HClO_4$, is the strongest acid shown in Table 14.1. Therefore, none of the acids shown in the table can react with ClO_4^-. ■

PROBLEM 14.6 (a) Select from Table 14.1 any two anions that potentially can act as a base toward H_2O. Write a chemical equation for each reaction and the equilibrium condition for one of these reactions. Which reaction has the larger K_b value? (b) Select two anions from the table that do not react as bases toward H_2O. Write a chemical equation for the reaction of one of these anions with $HClO_4$. Do not select any of the anions used in Example 14.9. □

The conjugate bases of the nonacidic hydrides (group C, Table 14.1) are extremely strong bases—much stronger than OH^-. Their base strengths cannot be compared in H_2O, because they react completely to give OH^-.

$$:\ddot{N}H_2^- + H_2O \longrightarrow NH_3 + OH^- \qquad \text{(as from sodium amide, } NaNH_2\text{)}$$

$$:H^- + H_2O \longrightarrow H_2 + OH^- \qquad \text{(as from calcium hydride, } CaH_2\text{)}$$

$$H_3C:^- + H_2O \longrightarrow CH_4 + OH^- \qquad \text{(as from potassium methide, } KCH_3\text{)}$$

base$_1$ (bases stronger than OH^-) — acid$_2$ — acid$_1$ — base$_2$ (only base present)

This is another example of the leveling effect.

CONJUGATE BASES OF POLYPROTIC ACIDS

Acids that donate more than one proton are called **polyprotic acids**. Examples are sulfuric acid, H_2SO_4, which is **diprotic**, and phosphoric acid, H_3PO_4, which is **triprotic**. One mole of a polyprotic acid can react with one or more moles of OH^- to give a series of salts, as shown for H_3PO_4:

$$H_3PO_4 + NaOH \longrightarrow NaH_2PO_4 + H_2O$$

$$H_3PO_4 + 2NaOH \longrightarrow Na_2HPO_4 + 2H_2O$$

$$H_3PO_4 + 3NaOH \longrightarrow Na_3PO_4 + 3H_2O$$

(NaH_2PO_4 and Na_2HPO_4 have acidic H atoms and are called acid salts.)

Polyprotic acids release their protons one at a time. For example, the diprotic acid H_2S ionizes in two steps:

STEP 1. $H_2S + H_2O \rightleftharpoons H_3O^+ + HS^- \qquad K_{a_1} = 6.3 \times 10^{-8}$

STEP 2. $HS^- + H_2O \rightleftharpoons H_3O^+ + S^{2-} \qquad K_{a_2} = 1 \times 10^{-14}$

The conjugate base in the first step (HS^-) becomes the acid in the second step. The values for the ionization constants reveal that HS^- is a much weaker acid than H_2S. This is understandable since the hydrosulfide ion, HS^-, has a negative charge and so is more likely to resist losing another proton to form the sulfide ion, S^{2-}. Conversely, the sulfide ion has a double negative charge and is therefore more likely to attract a proton than is the singly charged base, HS^-; S^{2-} is a much stronger base than HS^-. The base obtained by removing more than one proton from a polyprotic acid is a **polyprotic base** because it can accept more than one proton, as shown here for S^{2-}:

STEP 1. $S^{2-} + H_2O \rightleftharpoons HS^- + OH^-$ $\quad K_b = 1$

STEP 2. $HS^- + H_2O \rightleftharpoons H_2S + OH^-$ $\quad K_b = 1.6 \times 10^{-7}$

The difference in K_b values verifies that S^{2-} with a double negative charge is a much stronger base than HS^- with a single negative charge.

EXAMPLE 14.10 (a) Use ionic equations to show the stepwise conversion of $PO_4{}^{3-}$, a triprotic base, to H_3PO_4. (b) Is H_3PO_4 the conjugate acid of $PO_4{}^{3-}$?

ANSWER

(a) $PO_4{}^{3-}$ picks up three protons in three steps.

STEP 1. $PO_4{}^{3-} + H_2O \rightleftharpoons HPO_4{}^{2-} + OH^-$

STEP 2. $HPO_4{}^{2-} + H_2O \rightleftharpoons H_2PO_4{}^- + OH^-$

STEP 3. $H_2PO_4{}^- + H_2O \rightleftharpoons H_3PO_4 + OH^-$

$\quad\quad$ base$_1$ $\quad$ acid$_2$ $\quad$ acid$_1$ $\quad$ base$_2$

(b) No, H_3PO_4 is not the conjugate acid of $PO_4{}^{3-}$. The conjugate acid of $PO_4{}^{3-}$ is $HPO_4{}^{2-}$, the ion formed when *one* proton is accepted. ■

Note that the reverse arrows become longer relative to the forward arrows in each successive step because the bases are successively weaker as their negative charges decrease. Note the position of these species in Table 14.1.

PROBLEM 14.7 Use ionic equations for the stepwise conversion of carbonate ion, $CO_3{}^{2-}$, to carbonic acid, H_2CO_3 (or $CO_2 + H_2O$). □

EXAMPLE 14.11 Use ionic equations to show that the hydrogen sulfite anion, $HSO_3{}^-$, can act as an acid or as a base. This anion is the conjugate base of H_2SO_3, a weak diprotic acid that decomposes to SO_2 and H_2O.

ANSWER As an acid,

$HSO_3{}^- + H_2O \rightleftharpoons SO_3{}^{2-} + H_3O^+$

acid$_1$ $\quad$ base$_2$ $\quad$ base$_1$ $\quad$ acid$_2$

As a base,

$HSO_3{}^- + H_2O \rightleftharpoons H_2SO_3 + OH^-$

base$_1$ $\quad$ acid$_2$ $\quad$ acid$_1$ $\quad$ base$_2$ ■

PROBLEM 14.8 As in Example 14.11, write equations to illustrate the acid-base property of dihydrogen phosphate ion, $H_2PO_4{}^-$. □

14.5 SELF-IONIZATION

Water can be either a Brönsted base (equation 2) or a Brönsted acid (equation 3), depending upon the nature of the other reacting substance:

$$H_2\ddot{O}: \xrightarrow[\text{(as an acid)}]{-H^+} H\!:\!\ddot{O}\!:^- \text{ (conjugate base)}$$

$$H_2\ddot{O}: \xrightarrow[\text{(as a base)}]{+H^+} H_3\ddot{O}^+ \text{ (conjugate acid)}$$

A substance that can behave as an acid or a base is said to be **amphoteric**. Since H_2O is amphoteric, it might be expected to react with itself as shown:

$$\underset{\text{acid}_1}{H_2\ddot{O}} + \underset{\text{base}_2}{:\ddot{O}H_2} \underset{\text{neutralization}}{\overset{\text{self-ionization}}{\rightleftharpoons}} \underset{\text{base}_1}{[H\ddot{O}:]^-} + \underset{\text{acid}_2}{[H_3O:]^+} \quad (11)$$

This reaction, called a **self-ionization**, leads to the formation of the conjugate acid (H_3O^+) and the conjugate base (OH^-) of water (equation 11). The reverse of self-ionization is neutralization (equation 6). The equilibrium of equation 11 greatly favors neutralization, the reverse reaction, rather than self-ionization, the forward reaction.

The structural features that permit a molecule to self-ionize are an acidic H atom and a strongly electronegative atom with at least one unshared pair of electrons, which acts as the basic site. The atoms that meet the criterion of high electronegativity are the O atom as in H_2O, the F atom as in HF, the N atom as in NH_3, and to a lesser extent, the Cl atom as in HCl. Liquid NH_3 is an example.

$$\underset{\text{acid}_1}{H_2\ddot{N}H(\ell)} + \underset{\text{base}_2}{:NH_3(\ell)} \underset{\text{neutralization}}{\overset{\text{self-ionization}}{\rightleftharpoons}} \underset{\substack{\text{base}_1\\\text{conjugate}\\\text{base of } NH_3\\\text{(amide ion)}}}{H_2\ddot{N}:^-(\text{sol})} + \underset{\substack{\text{acid}_2\\\text{conjugate}\\\text{acid of } NH_3\\\text{(ammonium ion)}}}{NH_4^+(\text{sol})} \quad (12)$$

or simply, $2NH_3 \longrightarrow NH_2^- + NH_4^+$.

Again, the reverse of the self-ionization of NH_3 (equation 12) is a neutralization in liquid ammonia (equation 7).

EXAMPLE 14.12 Which of the following molecules are capable of self-ionizing? C_4H_{10}, H_2SO_4, PH_3, CH_3NH_2, CH_3OH.

ANSWER H_2SO_4 and CH_3OH, because of the O atom, and CH_3NH_2, because of the N atom. ■

EXAMPLE 14.13 Use Lewis structures to write an equation for the self-ionization of HF.

ANSWER

$$\mathrm{H}:\ddot{\underset{\cdot\cdot}{\mathrm{F}}}: + \mathrm{H}:\ddot{\underset{\cdot\cdot}{\mathrm{F}}}: \rightleftharpoons [\mathrm{H}:\ddot{\underset{\cdot\cdot}{\mathrm{F}}}:\mathrm{H}]^+ + \ :\ddot{\underset{\cdot\cdot}{\mathrm{F}}}:^-$$

conjugate acid of HF — conjugate base of HF

■

PROBLEM 14.9 Write equations for (a) a neutralization in liquid hydrogen fluoride, HF, and (b) a self-ionization of 100% H_2SO_4. □

14.6 pH

When pure water is tested in the apparatus shown in Figure 9.1 (section 9.1), the light bulb does not light. Water would be classified as a nonelectrolyte. However, more sensitive instruments show that even the purest water has some ability to conduct an electric current. This conductivity is attributed to the presence of H^+ and OH^- ions produced by the self-ionization reaction:

$$2H_2O \rightleftharpoons H_3O^+ + OH^-$$

or more simply, H^+ is used instead of H_3O^+:

$$H_2O \rightleftharpoons H^+ + OH^-$$

Recall that the solvent (H_2O in this case) is omitted from an equilibrium condition (section 14.3).

The equilibrium constant of this reaction is known as the *ion product of water* and is represented by K_w:

$$K_w = [H^+][OH^-] = 1.00 \times 10^{-14} \text{ at } 25°C \qquad (13)$$

EXAMPLE 14.14 The value of K_w increases with temperature. At 50° $K_w = 5.35 \times 10^{-14}$. Knowing this, complete the following statement by inserting *the same, more,* or *less* in each blank.

At higher temperatures (a) the acidity of H_2O is ________, (b) the basicity of H_2O is ________, (c) the acidity of H_3O^+ is ________, and (d) the basicity of OH^- is ________.

ANSWER As the temperature increases the equilibrium shifts slightly to the right, although it still lies mainly to the left. Because of this, you should have answered (a) more, (b) more, (c) less, and (d) less. ■

D, deuterium, is an isotope of H.

PROBLEM 14.10 (a) Write the chemical equation for the self-ionization of D_2O. The K for this reaction at 25° is 1.0×10^{-15}. (b) Compare the acidities and basicities of H_2O and D_2O. (c) Compare the acidities of H_3O^+ and D_3O^+. (d) Compare the basicities of OH^- and OD^-. Assume H_2O and D_2O have the same solvent effects. □

The concentrations of H^+ and OH^- must be equal in pure water for two reasons:

1. In the reaction $H_2O \rightleftharpoons H^+ + OH^-$, one H^+ is formed for every OH^- ion that is formed.
2. Any solution must be electrically neutral, which means that the total positive charge equals the total negative charge. In pure H_2O, this requires that $[H^+] = [OH^-]$.

The equilibrium concentrations of H^+ and OH^- in pure water at 25°C are found by letting $y = [H^+] = [OH^-]$ and substituting it into the equilibrium condition:

$$[H^+][OH^-] = K_w$$

$$y^2 = 1.00 \times 10^{-14}$$

$$y = 1.00 \times 10^{-7}\ \text{mol/L} = [H^+] = [OH^-]$$

If some other source of H^+ or OH^- is added to the water, we cannot expect that $[H^+]$ and $[OH^-]$ will remain equal. Of course, K_w, being an equilibrium constant, is unchanged. Thus, equation 13 must hold true for any aqueous solution at 25° regardless of the source of the H^+ and OH^-. For example, we might make $[H^+] = 0.10$ mol/L by adding a suitable quantity of acid. Then, at 25°C,

$$0.10 \times [OH^-] = 1.00 \times 10^{-14}$$

$$[OH^-] = \frac{K_w}{[H^+]} = \frac{1.00 \times 10^{-14}}{0.10} = 1.0 \times 10^{-13}\ \text{mol/L}$$

BOX 14.1

STOMACH ACIDITY

As soon as food reaches the stomach, acidic gastric juices are released by glands in the mucous lining of the stomach. The high acidity (pH about 1.5) is due to dissolved hydrochloric acid and is needed for the enzyme pepsin to catalyze the digestion of proteins in food. The H^+ is produced in the blood plasma by the ionization of $CO_2(aq)$. It is transported through the stomach lining along with Cl^-, which provides charge balance. Since the stomach wall contains protein, it is reasonable to wonder why the stomach would not digest itself. In fact, it sometimes does, producing a hole—an ulcer. Most often, however, the stomach wall resists the assault of the H^+, thanks to a protective layer of mucus-producing cells. These cells prevent the H^+ and Cl^- from diffusing back to the blood plasma. Ordinarily, these cells are continuously being sloughed off and replaced.

An overactive stomach may produce too much acid, overwhelming the protective layer. Heartburn is a frequent symptom of excess acidity, and it can be alleviated by ingesting antacids. The reaction of $Mg(OH)_2$ (milk of magnesia) is typical:

$$Mg(OH)_2(c) + 2H_3O^+ \longrightarrow Mg^{2+} + 4H_2O$$

The stomach wall can also be damaged by the action of aspirin. Aspirin, acetylsalicylic acid, is a weak carboxylic acid (section 14.7).

$C_6H_4(OCOCH_3)(COOH) \rightleftharpoons C_6H_4(OCOCH_3)(COO^-) + H^+$

acetylsalicylic acid (aspirin)

acetylsalicylate

At the pH of the stomach most of the conjugate base, acetylsalicylate, reacts to leave the acid largely unionized. The equilibrium in the equation shifts to the left. In this form it is able to penetrate the stomach wall. Once inside the wall, it is free of the high acidity and is able to ionize to produce H^+. The accumulation of H^+ in the wall causes bleeding, but the amount of blood lost per aspirin tablet is not generally harmful. However, when several tablets are taken at one time, bleeding may be more severe. Alcohol promotes the passage of aspirin through the membrane and augments the bleeding.

TABLE 14.2
APPROXIMATE pH VALUES OF COMMON MATERIALS AT 25°C

Gastric juice	1–3	Urine	5–8
Soft drinks	2.0–4.0	Water saturated with CO_2 from air	6
Lemons	2.2–2.4	Cow's milk	6.3–6.6
Vinegar	2.4–3.4	Saliva	6.5–7.5
Wine	2.8–3.8	Blood	7.3–7.5
Apples	2.9–3.3	Seawater	8–9
Oranges	3–4	Milk of magnesia	10.5
Beer	4–5	Household ammonia	12

Introduced by Sören P. L. Sörensen in 1909.

Many aqueous solutions have very small concentrations of H^+ and OH^-. By this time you are accustomed to "scientific notation," but numbers like 5.3×10^{-10}, cumbersome in any case, are bewildering to many persons who have little scientific background. To provide simpler expressions of acidity and basicity, the symbol **pH** was introduced as an abbreviation for "power of hydrogen." Its definition is

If $[H^+] = 1$ mol/L, $pH = 0$

If $[H^+] = 2$ mol/L, $pH = -\log 2 = -0.3$

If $[H^+] = 10^{-15}$ mol/L, $pH = 15$

$$pH = -\log [H^+] \quad \text{or} \quad pH = \log \frac{1}{[H^+]}$$

or

$$[H^+] = \text{antilog}\,(-pH) \quad \text{or} \quad [H^+] = 10^{-pH}$$

For now we are interested only in the use of pH; in Chapter 15 we will do calculations with pH. Values of pH are usually between 0 and 14, but they can also be zero, negative, and greater than 14. The relationships among acidity, basicity, $[H_3O^+]$, $[OH^-]$, and pH at 25°C are shown in this table:

Neutral has two meanings, (a) neither acidic nor basic (as used here) and (b) neither positively nor negatively charged.

	pH	SOLUTION IS
$[H_3O^+] = [OH^-]$	7.00	neutral
$[H_3O] > [OH^-]$	less than 7.00	acidic
$[H_3O^+] < [OH^-]$	greater than 7.00	basic (or alkaline)

A solution with a pH of 7 is called **neutral**, indicating that it is neither acidic or basic. For pH values less than 7, the smaller the value, the more acidic is the solution. For pH values greater than 7, the greater the value, the more basic is the solution. Table 14.2 gives approximate pH values of common materials.

14.7 NAMING ACIDS AND THEIR ANIONS

Acids can be classified into two main types. Compounds of hydrogen and one other element are called **binary hydrides**. Binary hydrides whose water solutions are acidic constitute the **binary acids**. Their general formula is H_nX, with X representing one of the more electronegative nonmetals. In the general formula of the acidic hydrides, H precedes X; examples are HCl (hydrogen chloride) and H_2S (hydrogen sulfide). The nonacidic hydrides are written with H after X; examples are CH_4(methane) and NH_3(ammonia).

Oxo acids consist of H, O, and a central element, which most often is nonmetallic (H_2SO_4, for example) but which may be metallic (as in the case of $HMnO_4$). They have the general formula H_xXO_y and possess at least one OH (hydroxyl) group attached to the central atom. Oxo acids may also have some number of lone O atoms attached to the central atom, $(HO)_nXO_m$, as does sulfuric acid, $(HO)_2SO_2$. Some may also have at least one of the H atoms attached to the central atom; for example, H_2PHO_3.

H—O—X—O
O ← lone O's

BINARY ACIDS AND THEIR ANIONS

A binary acid is named by "bracketing" the stem name of the particular atom X with the prefix *hydro-* and the suffix *-ic* and following it with the word *acid.* For example,

hydro(stem name)*ic acid*

Some examples, each with the stem name underlined, are given here with the names of the corresponding anions:

HCN is named as a binary acid.

Hydr- is used when the stem name starts with a vowel. *Azo,* rather than *nitro,* is the stem name for N in HN_3.

FORMULA	COMPOUND	ACID	ANION
HCl	hydrogen chloride	hydrochloric acid	chloride (Cl^-)
H_2S	hydrogen sulfide	hydrosulfuric acid	sulfide (S^{2-})
HCN	hydrogen cyanide	hydrocyanic acid	cyanide (CN^-)
HN_3	hydrogen azide	hydrazoic acid	azide (N_3^-)

OXO ACIDS AND ANIONS

Naming oxo acids is more complicated because the central atom may exhibit more than one oxidation state, with an acid corresponding to each state. An increase in oxidation state usually occurs as the number of lone O atoms bonded to the central atom increases. This relationship will become evident in the examples given later. One acid is invariably named by placing the suffix *-ic* after the stem name and following it with *acid.* The corresponding anion is named by dropping the word *acid* and replacing *-ic* with *-ate.* Some common *-ic* acids and their anions are shown in Table 14.3.

The *-ic* acids of boron and carbon exhibit only a single oxidation state.

In general, the word *acid* is dropped when naming anions. You should learn the formulas and names of these *-ic* acids because the names of the other oxo acids and anions are derived from them.

The scheme used to name the acids and anions of an element with different oxidation states depends on the change in the number of O atoms per atom of the central element. The number of bonded O atoms is compared to the reference number in the *-ic* acid (Section 2.9).

General scheme for naming oxo acids (H_mXO_n) *and anions*:

RELATIVE NUMBER OF O ATOMS PER X ATOMS	ACID	ANION
2 less	hypo______ous acid	hypo______ite
1 less	______ous acid	______ite
Reference number (learned)	______ic acid	______ate
1 more	per______ic acid	per______ate

TABLE 14.3
SOME COMMON *-ic* ACIDS AND THEIR ANIONS

OXO ACID			ANION	
Molecular Formula	Oxidation Number	Name	Formula	Name
H_3BO_3	+3	boric acid	BO_3^{3-}	borate
H_2CO_3	+4	carbonic acid	CO_3^{2-}	carbonate
HNO_3	+5	nitric acid	NO_3^-	nitrate
H_3PO_4	+5	phosphoric acid	PO_4^{3-}	phosphate
H_2SO_4	+6	sulfuric acid	SO_4^{2-}	sulfate
$HClO_3$	+5	chloric acid	ClO_3^-	chlorate

Notice that in general when naming the anion from the acid, *-ous* becomes *-ite* and *-ic* becomes *-ate*. The prefixes *per-* and *hypo-* are unchanged. Some typical examples of each type are given.

One less O atom than the -ic acid (see Table 14.3):

ACID			ANION	
Molecular Formula	Oxidation Number	Name	Formula	Name
HNO_2	+3	nitrous acid	NO_2^-	nitrite
H_2PHO_3	+3	phosphorous acid	PHO_3^{2-}	phosphite
H_2SO_3	+4	sulfurous acid	SO_3^{2-}	sulfite
$HClO_2$	+3	chlorous acid	ClO_2^-	chlorite

H
|
HO—P—OH
|
O

Notice that the central atom in the *-ous* acid has a lower oxidation number than the one in the *-ic* acid. (Compare with Table 14.3.)

Two less O atoms than the -ic acid:

ACID			ANION	
Molecular Formula	Oxidation Number	Name	Formula	Name
HPH_2O_2	+1	hypophosphorous acid	$PH_2O_2^-$	hypophosphite
HClO (HOCl)	+1	hypochlorous acid	ClO^-	hypochlorite

H
|
HO—P—O
|
H

One more O atom than the -ic acid:

ACID			ANION	
Molecular Formula	Oxidation Number	Name	Formula	Name
$HClO_4$	+7	perchloric acid	ClO_4^-	perchlorate

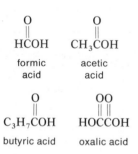

An important group of oxo acids are the carboxylic acids, RCOH (with =O on the C). (R stands for a group containing C's and H's.) Many of them have common names. Some typical ones are formic acid ($HCHO_2$), acetic acid ($HC_2H_3O_2$), butyric acid ($HC_4H_7O_2$), and oxalic acid ($H_2C_2O_4$).

HCOH (C=O) — formic acid
CH_3COH (C=O) — acetic acid
C_3H_7COH (C=O) — butyric acid
HOCCOH (both C=O) — oxalic acid

Polyprotic acids can give more than one anion differing by the number of H atoms present. These anions are differentiated by indicating the number of hydrogen atoms. For example, $H_2PO_4^-$ is named *di*hydrogen phosphate and HPO_4^{2-} is named *mono*hydrogen phosphate, although in practice the prefix *mono-* is frequently omitted. When an oxo anion has one H attached to an O, the prefix *bi-* may be used; HCO_3^- is commonly called *bi*carbonate.

EXAMPLE 14.15 Manganic acid is H_2MnO_4. Give the name for $HMnO_4$.

ANSWER The number of O atoms is the same; therefore, you must compare the oxidation numbers. The oxidation number of Mn in $HMnO_4$ is +7, which is higher than the value of +6 in the *-ic* acid, H_2MnO_4. The name for $HMnO_4$ is *permanganic* acid. ■

14.8 ACIDITY, BASICITY, AND STRUCTURE

We have already seen that some molecules possessing an H atom are very strong acids (such as $HClO_4$) and some are nonacids (such as CH_4). Likewise, any ion or molecule with an unshared pair of electrons may be a base. Yet some are potent bases (like CH_3^-) and others are practically nonbasic (like ClO_4^-). Why is there this disparity in behavior? To answer this question, we must become concerned with one of the most important aspects of modern chemistry—the relationship between structure and chemical properties. We will show you how basicity and acidity are related to the position in the periodic table of the central atom in the molecule and then explain the relationships.

BASICITY AND ANIONIC CHARGE

The simplest and most obvious factor affecting basicity is electric charge. For a series of oxo bases with the *same* basic-site atom, *basicity increases with an increase in negative charge.* The central atom does not have to be the same when making the comparisons.

Same central atom:

——— increasing basicities ——→

$$H_2PO_4^- < HPO_4^{2-} < PO_4^{3-}$$

— increasing negative charge —→

Different central atoms:

——— increasing basicities ——→

$$ClO_4^- < SO_4^{2-} < PO_4^{3-} < SiO_4^{4-}$$

——— increasing negative charge ——→

Comparisons cannot be made when the basic-site atoms are different. For example, even though the sulfate ion, SO_4^{2-} has a charge of -2, it is nevertheless a weaker base than F^-.

PERIODICITY AND ACIDITY OF BINARY ACIDS

Based on their Brönsted acid-base behavior, binary hydrides (H_nX) fall into four categories depending on the position of atom X in the periodic table. Only the binary hydrides of the representative elements are considered here. These compounds are discussed in more detail in Chapter 20.

CATEGORY	GROUP NUMBER OF X	EXAMPLES
Hydride salt (X^+H^-)	1 and 2	NaH, CaH_2
Nonacidic	3, 4, and 5	AlH_3, CH_4, PH_3
Acidic	6 and 7	H_2S, HCl
Basic	5 (only N)	NH_3

The hydride salts are basic because of the $:H^-$ anion. This discussion is concerned only with the covalent hydrides and not with hydride salts.

From the periodic listing in Table 14.4 we see that the *acidity of the covalent binary hydrides increases as we proceed to the right across a period and downward in a group.*

TABLE 14.4
PERIODICITY AND ACIDITY OF COVALENT BINARY HYDRIDES

Increasing acidity → (across groups 4 to 7); Increasing acidity ↓ (down each group)

4	5	6	7
CH_4 $K_a \approx 0^{\dagger}$	NH_3 $K_a = 10^{-32\dagger}$	H_2O $K_a = 10^{-14}$	HF $K_a = 6.4 \times 10^{-4}$
	PH_3 $K_a = 10^{-29\dagger}$	H_2S $K_a = 6.3 \times 10^{-8}$	HCl $K_a \approx 10^{7\dagger}$
		H_2Se $K_a = 10^{-4}$	HBr $K_a \approx 10^{9\dagger}$
		H_2Te $K_a = 10^{-3}$	HI $K_a \approx 10^{10\dagger}$

† These K_a values are *not* determined in aqueous solutions.

The hydrogen halides, HF, HCl, HBr, and HI, give distinctly acidic aqueous solutions, but HF is only weakly acidic:

$$HF + H_2O \rightleftharpoons H_3O^+ + F^- \qquad K_a = 6.4 \times 10^{-4}$$

The other three acids are strong; they are leveled in H_2O (section 14.3).

Of all the covalent binary hydrides in Table 14.4, ammonia, $:NH_3$, is unique; its *water* solution is distinctly basic. This basicity practically disappears on proceeding down Group 5. For example, the K_b of $:NH_3 = 1.8 \times 10^{-5}$, whereas the K_b of $:PH_3$ is about 1×10^{-25}.

PERIODICITY AND ACIDITY OF OXO ACIDS

The structural unit common to all oxo compounds is XOH; there may be more OH's, individual (lone) O's, or H's bonded to X. Based on their acid-base behavior, compounds with this structural unit fall into three categories depending on the position of atom X in the periodic table. Again, only the oxo compounds of the representative elements are considered.

CATEGORY	GROUP NUMBER OF X	EXAMPLES
Ionic hydroxides (X^+OH^-)	1 and 2 (all)	NaOH, $Ca(OH)_2$
Basic and acidic (amphoteric)	2 (Be), 3 (all except B), 4 (Sn and Pb), 5 (As(III) and Sb(III))	$Al(OH)_3$, $Sn(OH)_4$, $Pb(OH)_2$, $Sb(OH)_3$
Acidic	4, 5, 6, and 7	H_2CO_3, HNO_3, H_2SO_4, $HClO_4$

Caustic means destructive of living tissue and also means strongly basic.

The hydroxides of Groups 1 and 2 are sources of OH^-, but there are significant differences. The Group-1 hydroxides are soluble in water, and the resulting solutions are caustic because a mole of hydroxide gives a mole of OH^-.

$$MOH(c) \longrightarrow M^+(aq) + OH^-(aq)$$

The Group-2 hydroxides are much less soluble, and their saturated solutions are much less dangerous to handle. That is why $Mg(OH)_2$ is used medicinally as an antacid. It reduces excess acidity of the stomach without damaging the stomach wall as NaOH would. (See Box 14.1.)

$Al(OH)_3$ is a simplified formula; the formula is actually $Al(OH)_3(H_2O)_3$.

Recall that an amphoteric compound acts as both an acid and a base, as shown for $Al(OH)_3$:

TABLE 14.5
PEROIDICITY AND ACIDITY OF OXO ACIDS

Increasing acidity → (across); Decreasing acidity ↓ (down)

4	5	6	7
H_2CO_3 carbonic acid	HNO_3 nitric acid	—	—
H_4SiO_4 silicic acid	H_3PO_4 phosphoric acid	H_2SO_4 sulfuric acid	$HClO_4$ perchloric acid
	H_3AsO_4 arsenic acid	H_2SeO_4 selenic acid	$HBrO_4$ perbromic acid
	H_3SbO_4 antimonic acid	H_6TeO_6 telluric acid	H_5IO_6 periodic acid

$$Al(OH)_3(c) + 3H^+ \longrightarrow Al^{3+} + 3H_2O \qquad \text{(as a base)}$$
$$Al(OH)_3(c) + 3OH^- \longrightarrow \underset{\text{aluminate ion}}{AlO_3{}^{3-}} + 3H_2O \qquad \text{(as an acid)}$$

Table 14.5 lists the acidic oxo compounds according to the position of X in the periodic table. The acids listed are those in which the central element is in its highest oxidation state. We take our usual trip through the periodic table, first going across a period and then downward in a group. In addition, the acidity of oxo acids is also correlated with the number of lone O atoms on the central atom. The following trends are observed in Table 14.5:

1. *Acidity increases on proceeding from left to right across a period.*
2. *Acidity decreases on proceeding down a group.* Note that the trend down a group is opposite to that just observed for binary acids.
3. *Acidity increases with an increase in the number of lone O atoms bonded to the central atom.* This trend is illustrated with the oxo acids of chlorine using structural formulas to show the number of individual bonded O atoms.

ACID	STRUCTURAL FORMULA	NUMBER OF LONE O ATOMS	K_a
Hypochlorous	H—O—Cl	0	5×10^{-8}
Chlorous	H—O—Cl—O	1	1×10^{-2}
Chloric	H—O—Cl(—O)—O (one O above Cl)	2	1×10^{4}
Perchloric	H—O—Cl(—O)(—O)—O (O above and below Cl)	3	1×10^{11}

Except for $HClP_4$, these acids are known only in dilute water solutions. Pure $HClO_x$ is unstable.

TABLE 14.6
SCHEME FOR PREDICTING K_a VALUES OF OXO ACIDS, $(HO)_nXO_m$, BASED ON VALUE OF m

FORMULA	NUMBER OF LONE O ATOMS (m)	APPROXIMATE K_a	CLASS	EXAMPLES
HOX	0	10^{-8}	very weak	HOCl, $Te(OH)_6$
$(HO)_nXO$	1	10^{-2}	weak	HOClO, HON=O, $(HO)_2SO$, $(HO)_3PO$
$(HO)_nXO_2$	2	10^2	strong	$HOClO_2$, $(HO)_2SO_2$
$HOXO_3$	3	10^7	very strong	$HOClO_3$, $HOMnO_3$

A scheme for predicting approximate K_a values for oxo acids is based on the number of lone bonded O atoms. The scheme is shown in Table 14.6 with the aid of general structural formulas. The number of OH groups is immaterial. The lone O atoms can be singly or doubly bonded.

EXAMPLE 14.16 The molecular formulas (written without regard to ionizable H's) and K_a values for arsenious acid and phosphorous acid are AsH_3O_3 ($K_a = 6 \times 10^{-10}$) and PH_3O_3 ($K_a = 1.0 \times 10^{-2}$), respectively. Use Table 14.6 to predict the Lewis structures of these acids.

ANSWER From the K_a values we know that AsH_3O_3 has no lone O atoms, whereas PH_3O_3 has one lone O atom. The three O's in AsH_3O_3 must be part of OH groups. The Lewis structure is H—Ö—As—Ö—H (with :Ö—H bonded below As). Since PH_3O_3 has one lone O atom, the other two O's are part of two OH groups. The third H must be bonded to the P atom. The Lewis structure is H—Ö—P—Ö—H (with :Ö: bonded above P and H bonded below P). ■

PROBLEM 14.11 The molecular formula for hypophosphorous acid is PH_3O_2 ($K_a = 7 \times 10^{-2}$). Use Table 14.6 to predict the Lewis structure. □

14.9 RELATIVE BASICITY AND CHARGE DISPERSAL

Brönsted acidity depends on the ability of an acid to transfer a proton; the greater this tendency the stronger is the acid. However, the acidity of a substance also depends on the stability of its conjugate base. For example, the conjugate base of HCl is Cl^-, which is stable and has little tendency to accept a proton. Therefore, Cl^- is a very weak base and HCl, its conjugate acid, is a very strong acid. A careful analysis of acid-base strengths reveals that structural changes affect the stabilities of the conjugate bases more than the acids. Hence the following discussion of structure and strength focuses on the bases.

We have just seen that acidity is related to the position of the central atom in the periodic table. The acidity of oxo acids is also influenced by changes in the number of lone O atoms attached to the central atom. Predictions about relative strengths of a closely related group of acids or bases can be made with confidence provided that only *one* of these factors varies. Examples of closely related acids differing by only one factor are HF and HI (same group, different period), H_2O and HF (same period, different group), and H_2SO_4 and H_2SO_3 (same central atom, different number of lone O atoms).

Our generalizations about structure and acidity in the previous section are based on measured K_a values. However, the *theoretical explanations* for the observed facts are conjectural because the question is complex. Nevertheless, this discussion is simplified by a hypothesis that, fortunately, explains the observed facts. This hypothesis uses the idea that the ability of a base to accept an H^+ depends on the electron density at the basic site. We assume that *the more dispersed the electron density is, the less likely is the base to accept an* H^+ *and the weaker is the base.* The converse of this: the more localized the electron density is, the more likely is the base to accept an H^+ and the stronger is the base. Thus, any structural change that spreads out the electron density weakens the basicity. Since many bases are anions whose electron density is associated with negative charge, this generalization is called the **principle of charge dispersal** (or **delocalization**).

This principle is especially useful in the study of organic chemistry.

EXTENDED π BONDING IN OXO ANIONS

The most useful application of the principle of charge dispersal is for rationalizing the observed basicities of oxo anions. For oxo anions, *delocalization of charge results from extended* π *bonding* (section 8.5).

For example, in the nitrate ion (NO_3^-), a typical oxo anion, the charge is dispersed to all three oxygen atoms. *It is not just on the oxygen atom from which the proton was removed.* This charge delocalization greatly stabilizes the nitrate ion, causing the ion to be an extremely weak base and, hence, nitric acid to be a very strong acid. We suggest that *the more lone oxygen atoms bonded to a central atom, the more extended is the* π *bonding, the more delocalized is the charge, and the more stable and less basic should be the anion.* This suggestion is compatible with the observed facts. For example, the nitrate ion, with three oxygen atoms participating in extended π bonding (Example 8.9, section 8.5), is less basic than the nitrite ion, (NO_2^-), which has only two oxygen atoms participating in extended π bonding. (Elements in the third and higher periods use empty *d* orbitals to overlap with the filled *p* orbitals of the lone O atoms to form *p-d* π bonds, as shown in Figure 14.2.)

Therefore, nitric acid is a stronger acid than nitrous acid.

A second factor influencing the effectiveness of extended π bonding in all oxo anions is the X to lone O (X—O) bond length. *The shorter the X—O bond is, the more effective is the* π *overlap, the more delocalized is the charge, and the weaker should be the base.* The larger central atom always has the longer X—O bond. This second factor accounts for the fact that basicities of oxo anions *increase on*

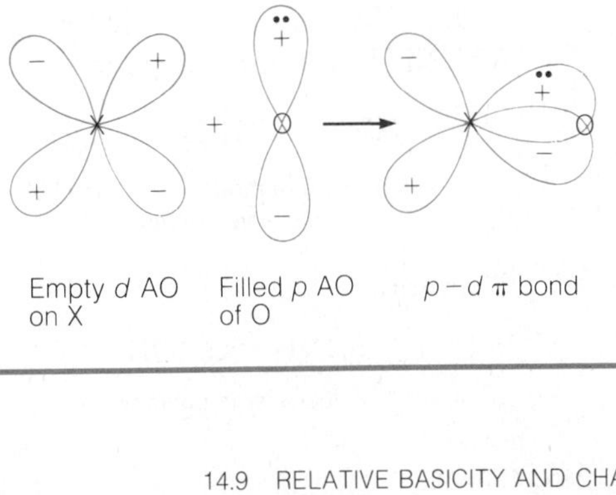

Empty *d* AO on X — Filled *p* AO of O — $p-d$ π bond

FIGURE 14.2
Formation of *p-d* π bond between S and O as in SO_4^{2-}.

going down a group, which is apparent when we compare $HSeO_4^-$ and HSO_4^-:

$HSeO_4^-$	HSO_4^-
Se is the larger atom.	S is the smaller atom.
↓	↓
Se—O bond is longer.	S—O bond is shorter.
↓	↓
Less effective extended π bonding	More effective extended π bonding
↓	↓
Less charge delocalization	More charge delocalization
↓	↓
Greater tendency to accept an H^+	Less tendency to accept an H^+
↓	↓
More basic	Less basic
↓	↓
H_2SeO_4 is the weaker acid.	H_2SO_4 is the stronger acid.

SIZE OF THE BASIC-SITE ATOM

The principle of charge dispersal also helps explain the relative basicities of the conjugate bases of binary acids of atoms in the *same* group. *The larger the basic-site atom is, the more dispersed (spread out) is the electron density (or charge), and the less basic is the species:*

$$F^- < Cl^- < Br^- < I^-$$

—— increasing size ⟶
— increasing charge dispersal ⟶

Therefore,

$$F^- > Cl^- > Br^- > I^-$$

— decreasing basicity ⟶

Consequently,

$$HF < HCl < HBr < HI$$

—— increasing acidity ⟶

Every oxo anion has an O atom as its basic site and, therefore, the size of the entire ion does not influence its basicity. However, as we have already seen, the size of the central atom is important because it influences the extended π bonding.

EFFECT OF NUMBER OF UNSHARED PAIRS OF ELECTRONS

On going across a period from left to right, the basic-site atom acquires more unshared electron pairs. Thus, $:\ddot{\underset{..}{F}}:^-$ has four unshared pairs whereas $:CH_3^-$ has only one. This trend is related to basicity by assuming that *the more unshared electron pairs there are, the more the charge is dispersed, and the less basic is the anion.* This idea holds only for basic-site atoms in the *same* period, as illustrated with the bases of the second period:

$$:CH_3^- < :\ddot{N}H_2^- < :\ddot{\underset{..}{O}}H^- < :\ddot{\underset{..}{F}}:^-$$

— increasing number of unshared electron pairs ⟶
———— increasing charge dispersal ————⟶

Therefore,

$$:CH_3^- > :NH_2^- > :OH^- > :F:^-$$

———— decreasing basicity ————⟶

Consequently,

$$CH_4 < NH_3 < H_2O < HF$$
$$\xrightarrow{\text{increasing acidity}}$$

Sizes of the basic-site atoms are not important for comparing these conjugate bases, because within the same period the sizes are about the same.

EXAMPLE 14.17 Explain why (a) NH_3 is a stronger base than PH_3 and (b) HS^- is a stronger base than Cl^-.

ANSWER

(a) Since N and P are in the same group, size difference is considered. The N atom is smaller than the P atom and, therefore, NH_3 is more basic.

(b) Since S and Cl are in the same period, we must consider the number of electron pairs. $H\ddot{\underset{..}{S}}\colon^-$ has 3 unshared pairs, which is less than the 4 pairs of $\colon\ddot{\underset{..}{Cl}}\colon^-$; therefore $HS\colon^-$ is more basic. ■

PROBLEM 14.12 Account for the difference in acidities of the Group-6 binary hydrides by focusing on their conjugate bases. □

RELATIVE ELECTRONEGATIVITIES; THE INDUCTIVE EFFECT

The basicity of a basic-site atom can be affected by the electronegativity of an atom bonded to it. This effect is illustrated by comparing the basicities of $\colon NH_3$ ($K_b = 1.8 \times 10^{-5}$) and hydroxylamine, $\colon NH_2OH$ ($K_b = 6.6 \times 10^{-9}$). To rationalize this difference in basicities, it is suggested that the electronegative O of the OH group in hydroxylamine draws electron density away from the basic site N atom. This has the effect of dispersing some of the electron density from the N to the O, making this N atom a weaker basic site than the N in NH_3. Hence, hydroxylamine is a weaker base than ammonia. This electron shift through σ bonds, shown by an arrow, $\colon N \rightarrow OH$, is an example of an **inductive effect**.

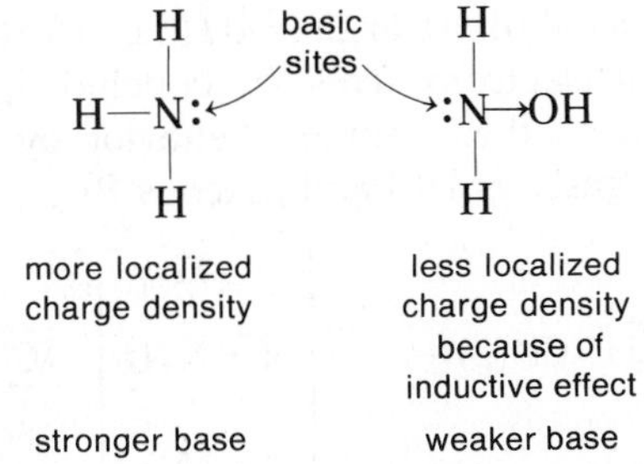

Example 14.18 shows that the inductive effect can proceed through more than one σ bond. However, the effect diminishes with an increase in the number of σ bonds through which the electron shift occurs.

EXAMPLE 14.18 In terms of the inductive effect, explain why fluorosulfuric acid, FSOH (with two O atoms bonded to S, one above and one below), is a stronger acid than sulfuric acid, HOSOH (with two O atoms bonded to S, one above and one below).

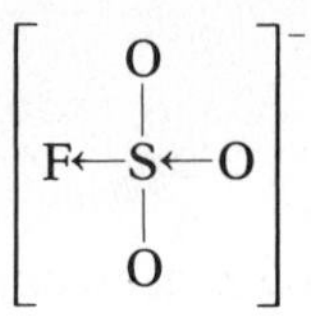

$$\left[\begin{array}{c} O \\ | \\ F\leftarrow S\leftarrow O \\ | \\ O \end{array}\right]^-$$

ANSWER The conjugate bases are hydrogen sulfate, $HOSO_3^-$, and fluorosulfate, FSO_3^-. The ions have the same number of lone O atoms to help delocalize the negative charge (making them both very weak bases). The only structural difference is the F—S bond in fluorosulfate instead of the HO—S bond in hydrogen sulfate. Since F is more electronegative than O, it exerts the greater electron-withdrawing inductive effect, making FSO_3^- the weaker base. ■

PROBLEM 14.13 (a) In terms of the inductive effect, predict which is the stronger base, ClO^- or BrO^-. (b) What does this tell you about the relative acidities of HOCl and HOBr? (c) Show how extended π bonding could also be used to compare the basicities of ClO^- and BrO^-. □

14.10 LEWIS ACID–BASE CONCEPT

As seen in the earlier sections, the Brönsted–Lowry concept helps us to unify our understanding of many acid-base reactions. Nevertheless, the concept leaves certain questions unanswered. Some of those questions arise from reactions catalyzed by acids. For example, the rates of many reactions are increased by the addition of acid catalysts such as concentrated H_2SO_4 and gaseous HCl. However, the gas boron trifluoride (BF_3) or anhydrous $AlCl_3$ can catalyze the same reactions. Obviously, there are no H atoms in BF_3 or $AlCl_3$. How, then, can these substances serve the same function as Brönsted acids?

This dilemma was resolved by G. N. Lewis (1923). Lewis recognized that both BF_3 and $AlCl_3$ are molecules having central atoms with less than an octet of valence electrons.

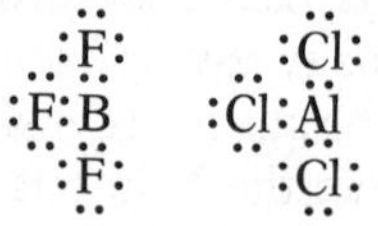

$$\begin{array}{c} :\ddot{F}: \\ :\ddot{F}:B \\ :\ddot{F}: \end{array} \qquad \begin{array}{c} :\ddot{Cl}: \\ :\ddot{Cl}:Al \\ :\ddot{Cl}: \end{array} \qquad \text{(Both B and Al have 6 valence electrons.)}$$

He also recognized that *the proton* (H^+) transferred from a typical Brönsted acid like HCl also *lacks electrons*; it does not have a duet of electrons. He emphasized that either a BF_3 molecule or a transferred H^+ can form a bond with an atom having an unshared pair of electrons. This, he concluded, accounts for their catalytic behavior. We illustrate this common behavior by comparing the reactions of gaseous $:NH_3$ with gaseous HCl and gaseous BF_3.

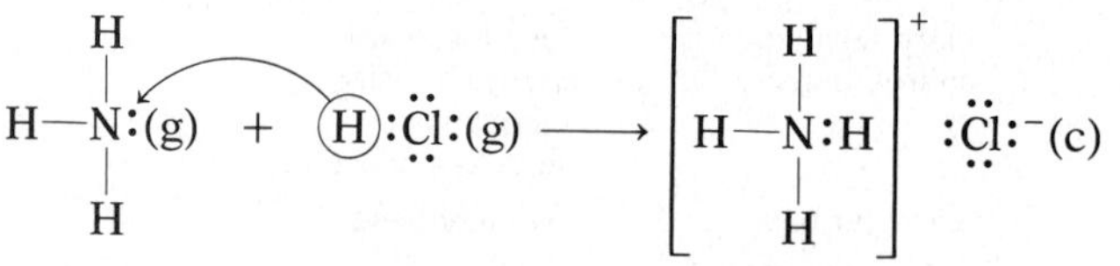

$$\underset{\text{Brönsted base}}{H-\overset{H}{\underset{H}{|}}{N}:(g)} + \underset{\text{Brönsted acid}}{(H):\ddot{Cl}:(g)} \longrightarrow \underset{\text{ammonium chloride}}{\left[H-\overset{H}{\underset{H}{N}}:H\right]^+ :\ddot{Cl}:^-(c)} \qquad (NH_4Cl(c))$$

$$H-\overset{H}{\underset{H}{N}}:(g) + \left(B\overset{F}{\underset{F}{-}}F\right)(g) \longrightarrow \underset{\text{ammonia boron trifluoride}}{H-\overset{H}{\underset{H}{N}}:\overset{F}{\underset{F}{B}}-F}$$

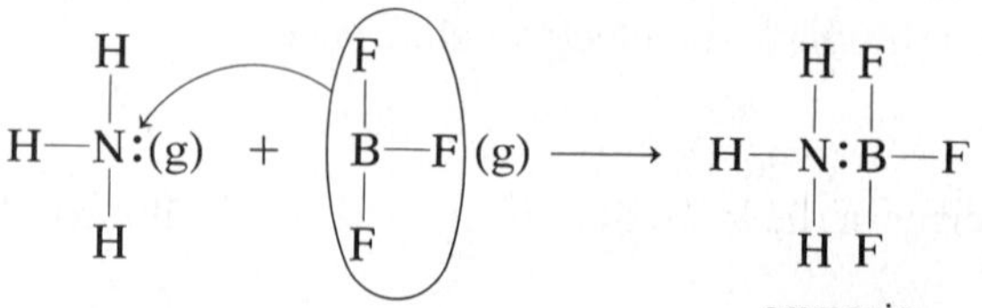

There is one major difference between HCl and BF_3 in these reactions. With HCl, only the H^+ is transferred; the Cl atom is left behind as the Cl^- anion. With

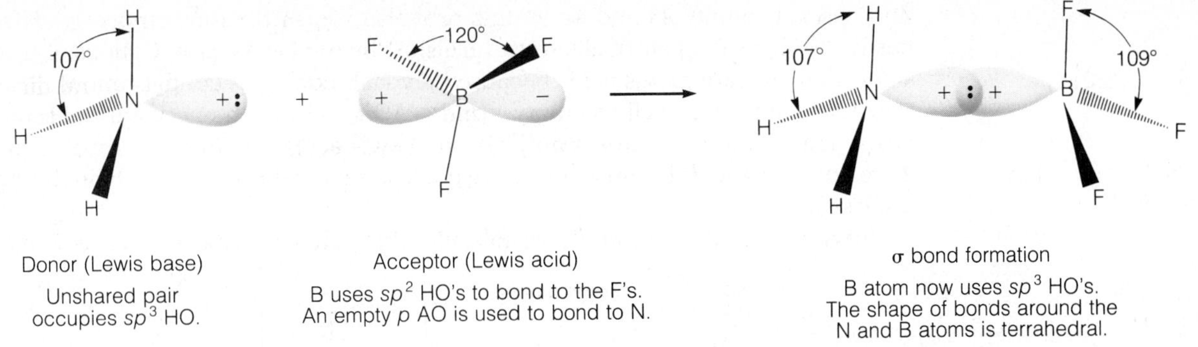

Donor (Lewis base)
Unshared pair occupies sp^3 HO.

Acceptor (Lewis acid)
B uses sp^2 HO's to bond to the F's. An empty p AO is used to bond to N.

σ bond formation
B atom now uses sp^3 HO's. The shape of bonds around the N and B atoms is terrahedral.

FIGURE 14.3 Orbital overlap in donor-acceptor (Lewis acid-base) reactions ($NH_3 + BF_3$). A coordinate-covalent type of bond formation.

BF_3, the B atom that forms the bond takes along the 3 F atoms so that BF_3 acts as a unit.

Lewis defined an acid-base reaction in terms of the sharing of an electron pair—*a Lewis base provides an electron pair for covalent bonding and a Lewis acid accepts the pair.* (The electron pair is *not* transferred from the base to the acid; it is shared.) The bond formed between the Lewis acid and base is coordinate covalent (section 7.5). The Lewis base is the donor and the Lewis acid is the acceptor. A filled orbital of the donor atom (the basic site) overlaps with an empty orbital of the electron-deficient, acceptor atom of the acid, as shown in Figure 14.3 for NH_3 and BF_3. Because Lewis acids seek to form bonds with substances having unshared electron pairs, they are also called **electrophiles** and are said to be **electrophilic** (electron-loving). Since Lewis bases seek to share pairs of electrons with nuclei of electron-deficient atoms in Lewis acids, they are also called **nucleophiles** and are said to be **nucleophilic** (nucleus-loving).

A base that accepts a proton is called a Brönsted base. A base that reacts with a Lewis acid, rather than with a Brönsted acid, *is called a Lewis base.* There is no structural difference between a Lewis base and a Brönsted base: they are both species with *unshared pairs of electrons.* They can be molecules, such as $:NH_3$, or anions, such as $:CN^-$. However, there may be sharp differences in their reactivities. For example, $:\ddot{\underset{\cdot\cdot}{Cl}}:^-$, the conjugate base of HCl, is an extremely weak proton-acceptor (Brönsted base). Yet it is quite reactive to some Lewis acids, as seen in equation 14. Because of such differences in reactivity, the distinction in name is retained.

The Lewis theory focuses attention on the electron pair rather than on the proton, and in so doing broadens the concept of acidity. The transferred proton of a Brönsted acid is a special case of a Lewis acid. Lewis acids appear in many guises, but they are all capable of sharing an electron pair when forming a covalent bond. All elementary cations are potential Lewis acids, but only the following kinds show this acidity to a significant extent:

Lewis acid   **Lewis base**

$$Al^{3+} + 4:\ddot{\underset{\cdot\cdot}{F}}:^- \longrightarrow AlF_4^- \qquad (14)$$

has a high charge (3+)

$$Zn^{2+}[Ar]3d^{10} + 4:\ddot{\underset{\cdot\cdot}{O}}H^- \longrightarrow \left[\begin{array}{c} OH \\ | \\ HO—Zn—OH \\ | \\ OH \end{array}\right]^-$$

lacks a noble gas configuration

Zn^{2+} uses its empty 4*s* and 4*p* atomic orbitals to form the four empty sp^3 HO's needed to share the pair of electrons furnished by the Lewis base. Cations of transition elements are Lewis acids because they lack noble gas configurations. They may use empty *d* as well as empty *s* and *p* AO's to form the necessary acceptor HO's. Transition-metal atoms may also be Lewis acids. Cations of Groups 1 and 2 are not Lewis acids because they have noble gas configurations and their charges are less than 3+.

Exception: Be^{2+} is a Lewis acid.

There are mainly three kinds of molecules that behave as Lewis acids, as shown:

Lewis acid **Lewis base**

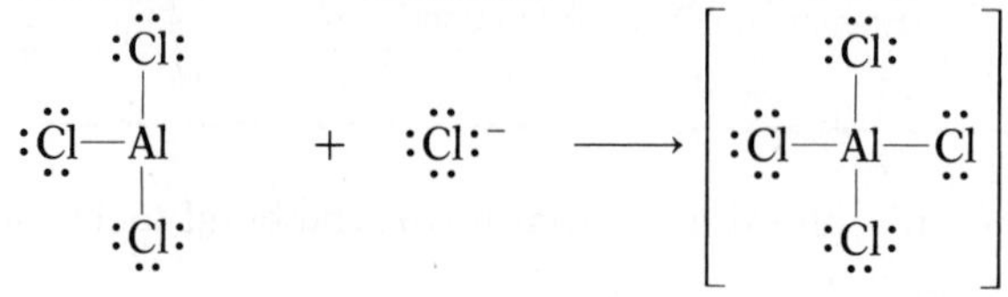

$AlCl_3 + :\ddot{Cl}:^- \longrightarrow [AlCl_4]^-$

Central atom (Al) has an incomplete octet.

$SnCl_4 + 2:\ddot{Cl}:^- \longrightarrow SnCl_6^{2-}$

Central atom (Sn) can expand its valence shell. (8 valence e^-'s in sp^3 HO's)

(12 valence e^-'s in sp^3d^2 HO's)

Elements in third and higher periods use *d* orbitals.

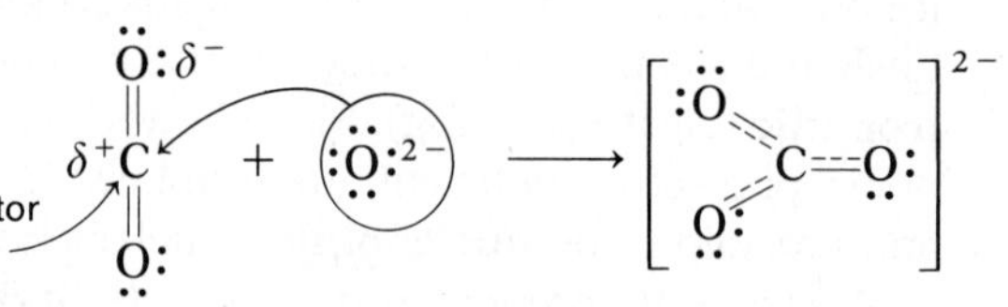

$O{:}\delta^- = \delta^+C = O: + :\ddot{O}:^{2-} \longrightarrow [CO_3]^{2-}$

acceptor site

has π-bonded atoms of dissimilar electronegativities

carbonate ion

A typical reaction is

$$CO_2 + CaO \longrightarrow CaCO_3$$

calcium oxide calcium carbonate

The less electronegative atom is electron-deficient and acts as the Lewis acid site. The contributing structures (section 8.5) for CO_2 reveal the electron deficiency on the less electronegative π-bonded atom.

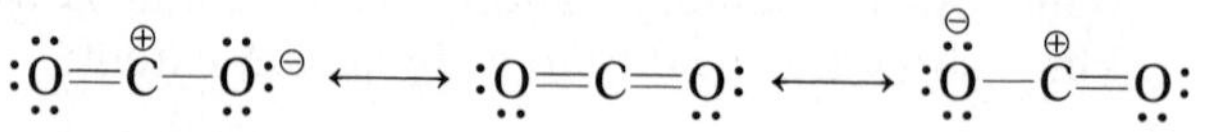

$$:\ddot{O}=\overset{\oplus}{C}-\ddot{O}:^{\ominus} \longleftrightarrow :\ddot{O}=C=\ddot{O}: \longleftrightarrow :\overset{\ominus}{\ddot{O}}-\overset{\oplus}{C}=\ddot{O}: \quad \text{(contributing structures)}$$

EXAMPLE 14.19 Sulfur dioxide, SO_2, reacts like CO_2 toward $:\ddot{O}:^{2-}$. Show the formation of SO_3^{2-} as a Lewis acid-base reaction. A typical reaction is $MgO + SO_2 \longrightarrow MgSO_3$

ANSWER The delocalized formula for SO_2 is $:O \cdots \ddot{S} \cdots O:$. However, because O is more electronegative than S, the O atom tends to attract the π electrons in the double bond, making the S atom an electron-deficient site.

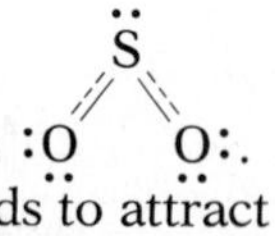

The unshared pair of e^-'s on S also make $:SO_2$ a Lewis base.

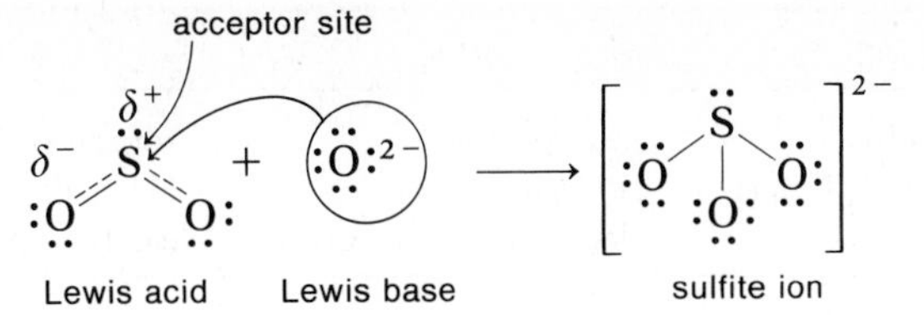

acceptor site

$$\overset{\delta-}{:\ddot{O}}{=}\overset{\delta+}{\ddot{S}}{=}\ddot{O}: \;+\; [:\ddot{O}:^{2-}] \longrightarrow \left[\begin{array}{c} \ddot{S} \\ :\ddot{O} \;|\; \ddot{O}: \\ :\ddot{O}: \end{array}\right]^{2-}$$

Lewis acid Lewis base sulfite ion

EXAMPLE 14.20 Select the Lewis acids from among the following species: (a) Na^+, (b) Co^{2+}, (c) $H_2C{=}\ddot{O}$, (d) CH_4, (e) $BeCl_2$ (a covalent molecule in gaseous state), (f) SiF_4.

ANSWER

(a) Na^+ is not a Lewis acid. It is univalent and has a noble-gas configuration. Thus it does not form covalent bonds with Lewis bases.

(b) Co^{2+} is a Lewis acid. It is a transition metal ion.

(c) $H_2C{=}O$ is a Lewis acid. There is a double bond between atoms of dissimilar electronegativity. The C atom is electron-deficient.

(d) CH_4 is not a Lewis acid.

(e) $BeCl_2$ is a Lewis acid. The Lewis structure, from $\cdot Be \cdot + 2{:}\ddot{Cl}\cdot$, is $:\ddot{Cl}{-}Be{-}\ddot{Cl}:$. The Be atom does not have an octet of valence electrons.

(f) SiF_4 is a Lewis acid. The Si atom has 8 valence electrons, but it is a third-period element capable of using *d* orbitals to acquire up to 12 electrons. ■

PROBLEM 14.14 Indicate whether or not the following substances are Lewis acids and explain your choices: (a) Cs^+, (b) CCl_4, (c) Fe^{3+}, (d) N_2, (e) $AlBr_3$. □

PROBLEM 14.15 Select the Lewis acids from among these species: (a) Ba^{2+}, (b) GeF_4, (c) Ni^{2+}, (d) AlH_3. □

PROBLEM 14.16 Complete the following reactions, specifying the Lewis acids and the Lewis bases. Use Lewis structures for reactants and products.

(a) $AlH_3 + H^- \longrightarrow$
(b) $Cd^{2+} + 4Cl^- \longrightarrow$
(c) $S^{2-} + SO_3 \longrightarrow$
(d) $GeF_4 + 2F^- \longrightarrow$ □

LEWIS ACID CATIONS AS BRÖNSTED ACIDS

A Lewis acid cation, such as Al^{3+}, can react with H_2O, acting as a Lewis base, to give a hydrated ion:

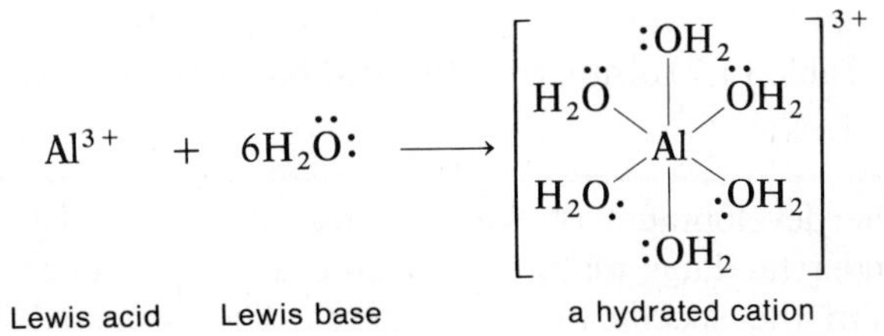

$$Al^{3+} + 6H_2\ddot{O}: \longrightarrow \left[\begin{array}{c} :OH_2 \\ H_2\ddot{O} \quad \ddot{O}H_2 \\ Al \\ H_2\ddot{O} \quad \ddot{O}H_2 \\ :OH_2 \end{array}\right]^{3+}$$

Lewis acid Lewis base a hydrated cation

TABLE 14.7
ACID–BASE CONCEPTS

CONCEPT	ACID	BASE
Arrhenius	source of H^+	source of OH^-
Brönsted–Lowry	H^+ donor	H^+ acceptor
Lewis	electron pair acceptor	electron pair donor

This highly charged hydrated cation is an oxo acid. $Al(H_2O)_6^{3+}$ is actually a triprotic acid that ionizes in water in three steps until the uncharged, insoluble $Al(OH)_3(H_2O)_3$ is formed.

STEP 1. $[Al(H_2O)_6]^{3+}$ (acid$_1$) $+ H_2O$ (base$_2$) $\rightleftharpoons$ $[Al(OH)(H_2O)_5]^{2+}$ (base$_1$) $+ [H_3O]^+$

STEP 2. $[Al(OH)(H_2O)_5]^{2+} + H_2O \longrightarrow [Al(OH)_2(H_2O)_4]^+ + H_3O^+$

STEP 3. $[Al(OH)_2(H_2O)_4]^+ + H_2O \longrightarrow [Al(OH)_3(H_2O)_3](c) + H_3O^+$

When a bound H_2O loses an H^+ it becomes a bound OH^-. As a result, the hydrated ion loses one positive charge for each H^+ lost. The number of H atoms that can dissociate to give the uncharged hydroxide equals the positive charge on the hydrated cation. The end product of the ionization is the insoluble, gelatinous precipitate of hydrated aluminum hydroxide. This precipitate has good adsorptive properties, and therefore Al(III) salts are used for water purification.

EXAMPLE 14.21 Write equations for each step in the ionization of the oxo acid $Zn(H_2O)_4^{2+}$ in water.

ANSWER This hydrated cation is a diprotic oxo acid because it has a charge of 2+.

STEP 1. $Zn(H_2O)_4^{2+} + H_2O \longrightarrow Zn(OH)(H_2O)_3^+ + H_3O^+$

STEP 2. $Zn(OH)(H_2O)_3^+ + H_2O \longrightarrow Zn(OH)_2(H_2O)_2(c) + H_3O^+$ ■

PROBLEM 14.17 What kind of polyprotic acid is $Cr(H_2O)_6^{3+}$? Write ionic equations for the dissociation of this acidic hydrated cation. □

Table 14.7 summarizes the acid-base concepts we have discussed in this chapter.

14.11 EPILOGUE

The development of the concepts of acids and bases illustrates how chemists' understandings mature with time and experience. The growth of knowledge is from the specific to the general. The Brönsted–Lowry concept included the

Arrhenius acids and bases. The Lewis concept incorporated the Brönsted acids and bases. Each concept offers something useful, although interest in Arrhenius's concept is mainly historical. For this reason we retain all of the concepts.

It is generally accepted that the broadening of a concept is an advance. However, there is a trade-off: much of the specificity of the older ideas is lost. If we admit BF_3 and $AlCl_3$ into the company of acids, then vinegar and lemon juice lose some of their uniqueness. Though we end our coverage of this topic here, keep in mind that chemists are continuing to reevaluate the definitions of acids and bases in order to form broader, more theoretical concepts.

SUMMARY

According to Brönsted and Lowry, an **acid** is a proton donor and a **base** is a proton acceptor. When an acid loses its proton, it leaves behind its **conjugate base**. A base gains a proton to form its **conjugate acid**. The newly formed conjugate acid and conjugate base can revert to the original acid and base, and the established equilibrium favors the presence of the weaker acid and base. For an acid and its conjugate base or a base and its conjugate acid (a conjugate acid-base pair), the stronger the acid, the weaker is the conjugate base. Conversely, the weaker the acid, the stronger is the conjugate base. **Neutralization** is the term for the reaction between the conjugate acid and the conjugate base of the same substance.

Relative acidities are measured with reference to a common base, water. Weak acids are only slightly ionized in water as measured by the value of K_a, the equilibrium constant for the ionization of the acid in water. Relative basicities are also measured with reference to water—water acting as an *acid*. Molecular bases are weak because they are slightly ionized in water as measured by K_b, the equilibrium constant for their ionization. Anionic conjugate bases of weak acids are basic toward water; cationic conjugate acids of molecular bases are weak acids. **Polyprotic acids** release their protons stepwise. The conjugate base with an H formed in one step acts as the acid in the ensuing step. The stepwise release of H^+'s occurs with increasing difficulty; the K_a values get smaller.

Strong acids are completely ionized in dilute aqueous solutions so that the only acid present is H_3O^+, the strongest acid that can exist in water. Because of this **leveling effect**, all strong acids appear to have the same acidity. Relative acidities of strong acids are obtained by observing the degree of dissociation in solvents such as 100% acetic acid that are weaker bases than water. Bases stronger than OH^- are leveled to OH^-, the strongest base that can exist in water.

Substances like water, which act as both acids and bases, are described as **amphoteric**. Water molecules act as acids or bases toward each other. In this way pure water undergoes **self-ionization** to furnish H_3O^+ and OH^- in equal concentrations. Self-ionization is the reverse of neutralization. Any molecule with an H atom attached to a strongly electronegative element (F, O, N, Cl) having at least one unshared pair of electrons can undergo self-ionization. The equilibrium constant for the self-ionization of H_2O, K_w, is called the **ion product of water**. Acidity is often expressed in units of **pH** ($pH = -\log[H^+]$). A **neutral** (neither acidic or basic) solution has a pH of 7 at 25°C. The pH of an acidic solution is less than 7; the pH of a basic solution is more than 7.

Acidity and basicity are influenced by the structure of the molecule or ion. When the basic site is the same atom, basicity increases with an increase in negative charge. When going across a period from left to right or downward in a group, **binary acids**, H_nX, become more acidic. **Oxo acids**, $(HO)_nX_m$, increase in acidity as more lone O atoms are bonded to the central atom. When the number of lone O atoms is the same, acidity decreases on proceeding down a periodic group.

The **principle of charge dispersal** correlates the basicities and structures of bases and, indirectly, the acidities of the conjugate acids. According to this principle, any structural change that increases the dispersal of negative charge (in ions) or electron density (in molecules) from the basic-site atom weakens the basicity of the ion or molecule. In oxo anions, charge dispersal increases as more lone O atoms are found on the central atom, X, and as the X—O bond gets shorter. For like-charged conjugate bases of binary acids, charge dispersal increases as the number of unshared pairs of electrons on the basic-site atom increases and, within the same periodic group, as the size of the basic-site atom increases. A strongly electronegative atom in the base can disperse electron density toward itself from the basic-site atom. This dispersal of electron density through σ bonds is called an **inductive effect.**

A broader view of acid and base behavior was suggested by G. N. Lewis, who said that acids and bases react to form a coordinate covalent type of bond. The base (a **nucleophile**) donates, and the acid (an **electrophile**) accepts the electrons. Any species with at least one unshared pair of electrons is a potential **Lewis base. Lewis acids** can be cations with high positive charge (3+) or cations lacking a noble gas electron configuration (includes the transferred H^+ of Brönsted acids). Lewis acid sites can also be atoms with less than an octet of electrons, atoms that can acquire more than an octet of electrons, or atoms that acquire partial charge when π-bonded to a more electronegative element.

14.12 ADDITIONAL PROBLEMS

BRÖNSTED ACID–BASE REACTIONS

14.18 In terms of Brönsted theory, state the differences between (a) a strong and a weak base and (b) a strong and a weak acid.

14.19 Illustrate with appropriate equations that these species are bases in water: NH_3, HS^-, $C_2H_3O_2^-$, and O^{2-}.

14.20 Illustrate with appropriate equations that the following species are acids in water: HBr, HNO_2, $HBrO_4$, and PH_4^+.

14.21 Give the products in the following acid-base reactions. Identify the conjugate acid-base pairs.
(a) $NH_4^+ + CN^-$
(b) $HS^- + HSO_4^-$
(c) $HClO_4 + [H_2NNH_3]^+$
(d) $H^- + H_2O$

14.22 List the conjugate acids of H_2O, OH^-, Cl^-, HCl, AsO_4^{3-}, NH_2^-, HPO_4^{2-}, and SO_4^{2-}.

14.23 List the conjugate bases of H_2O, HS^-, HCl, PH_4^+, and $HOCH_3$.

14.24 Identify the Brönsted acids and bases in these reactions and group them into conjugate acid-base pairs: (a) $NH_3 + HBr \rightleftharpoons NH_4^+ + Br^-$; (b) $NH_4^+ + OH^- \rightleftharpoons NH_3 + H_2O$; (c) $H_3O^+ + PO_4^{3-} \rightleftharpoons HPO_4^{2-} + H_2O$; (d) $HSO_3^- + CN^- \rightleftharpoons HCN + SO_3^{2-}$.

14.25 Referring to Table 14.1, state which of the following pairs react with each other; for those that do react, complete the equations. (a) HS^-, OH^-; (b) H^-, NH_3; (c) HCO_3^-, CN^-; (d) HCO_3^-, SO_3^{2-}; (e) $H_2PO_4^-$, NH_4^+; (f) NO_3^-, $HClO_4(\ell)$; (g) $HC_2H_3O_2$, HS^-; (h) HCN, $H_2PO_4^-$; (i) HSO_4^-, $C_2H_3O_2^-$; (j) CH_4, H^-.

14.26 Write equations and designate conjugate pairs for the stepwise reactions in water of (a) sulfuric acid, H_2SO_4; (b) phosphorous acid, H_2PHO_3; (c) H_3AsO_4; (d) ethylene diammonium ion, $(H_3NCH_2CH_2NH_3)^{2+}$.

14.27 Give the conjugate bases of $Al(H_2O)_6^{3+}$, $Al(OH)_3(H_2O)_3$, $Al(OH)_2(H_2O)_4^+$, $Al(OH)_4(H_2O)_2^-$, and $Al(OH)_6^{3-}$.

14.28 Use ionic equations, where appropriate, to describe the reaction (if any) expected when a strong aqueous acid such as HCl is added to (a) an active metal such as Mg; (b) a solid metal oxide such as CaO; (c) a solution of $Ba(OH)_2$; (d) barium peroxide, $BaO_2(c)$; (e) sodium iodide (c); (f) $Fe_2S_3(c)$.

14.29 Complete these equations by writing the formulas of the omitted compounds:
(a) $Ba(OH)_2 + ? \longrightarrow BaSO_4(c) + H_2O$
(b) $FeO(c) + ? \longrightarrow Fe(NO_3)_2(aq) + H_2O$
(c) $HCl(aq) + ? \longrightarrow AlCl_3(aq) + ?$
(d) $Na_2O + ? \longrightarrow 2NaOH(aq)$
(e) $NaOH + ? \longrightarrow Na_2HPO_4(aq) + ?$
(two possible answers)

14.30 Write the equation for the equilibrium established when (a) $HC_2H_3O_2$ is dissolved in D_2O; (b) $DC_2D_3O_2$ is dissolved in H_2O. (D, deuterium, is an isotope of H.)

14.31 (a) If the acid-base reaction $HA(aq) + B^-(aq) \longrightarrow HB(aq) + A^-(aq)$ has a $K = 10^{-4}$, which of the following statements are true? (1) HB is a stronger acid than HA; (2) HA is a stronger acid than HB; (3) HA and HB have the same acidity; (4) B^- is a stronger base than A^-; (5) A^- is a stronger base than B^-; (6) B^- and HB are a conjugate acid-base pair; (7) the acid and base strengths cannot be compared. (b) How would your answers change if (1) $K = 10^4$; (2) $K = 1$?

RELATIVE ACIDITY AND THE LEVELING EFFECT

14.32 (a) Which of the following acids are leveled in water? $HClO_4$, HBr, H_3PO_4, HF, HNO_3, HCN. (b) Which of the following bases are leveled in water? OH^-, NH_3, NH_2^-, H^-, S^{2-}, PO_4^{3-} (see Table 14.1).

14.33 (a) Elaborate on the statement, "The strongest base and acid that can exist in liquid NH_3 are NH_2^- and NH_4^+, respectively." (b) What are the strongest base and acid, respectively, that can exist in pure CH_3CH_2OH (ethyl alcohol)? (c) Generalize about the strongest acid and base that can exist in any self-ionizing solvent.

14.34 Write equations and account for the facts that (a) HCN is a strong acid in liquid NH_3; (b) HCl is a weak acid in pure acetic acid, $HC_2H_3O_2$; (c) sulfuric acid is a base in pure perchloric acid; (d) formaldehyde, $H_2C{=}O$, is a base in sulfuric acid.

14.35 To prepare sodium methoxide, $Na^+CH_3O^-$, sodium is added to pure methanol (methyl alcohol), CH_3OH, resulting in the generation of H_2. (a) Write an equation for the reaction. (b) Suggest a reason why sodium methoxide cannot be prepared by adding NaOH to CH_3OH.

14.36 If there were chemists on the planet Jupiter, which is rich in NH_3, they would probably classify both acetic and sulfuric acids as strong acids. Explain with the aid of equations.

SELF-IONIZATION AND NEUTRALIZATION

14.37 What structural features must a compound have in order to undergo self-ionization?

14.38 How many moles of NaOH can be neutralized by (a) 1 mole of H_3PO_4 and (b) 1 mole of H_2PHO_3? Write ionic equations for the reactions.

14.39 Write equations for the self-ionization of (a) $HOCH_3$, (b) HCl, and (c) $HC_2H_3O_2$.

14.40 Write ionic equations, using Lewis structures, for neutralization reactions in (a) formic acid, HCOOH, and (b) ethanol, CH_3CH_2OH.

14.41 Self-ionization occurs where an ion other than an H^+ is transferred, as exemplified by the transfer of a Cl^- ion from one PCl_5 molecule to another. Write an equation for this reaction. What are the shapes of the two ions that are formed?

LEWIS ACID–BASE REACTIONS

14.42 (a) List four types of Lewis acids and give an example of each. (b) What do all Lewis bases have in common?

14.43 Iodine, I_2, is much more soluble in a water solution of potassium iodide, KI, than it is in H_2O. The anion found in the solution is I_3^-. (a) Write an equation for this reaction, indicating the Lewis acid and the Lewis base. (b) Would you expect F_2 and F^- to react in the same way? Explain your answer.

14.44 For each reaction write the structures for the reactants and the products. Specify the Lewis acid and the Lewis base.

(a) $Mg^{2+}O^{2-} + CO_2 \longrightarrow$
(b) $Cu^{2+} + 4NH_3 \longrightarrow$
(c) $BF_3 + CH_3OH \longrightarrow$
(d) $BeF_2 + 2F^- \longrightarrow$
(e) $S + S^{2-} \longrightarrow$

14.45 A group of very strong acids are the fluoroacids, H_mXF_n. Two such acids are formed by Lewis acid-base reactions. (a) Identify the Lewis acid and the Lewis base:

$HF + SbF_5 \longrightarrow H(SbF_6)$ (called a "super" acid, hexafluoroantimonic acid)

$HF + BF_3 \longrightarrow H(BF_4)$ (tetrafluoroboric acid)

(b) To which atom is the H of the product bonded? How is the H bonded? (c) Explain why $HBCl_4$ is not known.

14.46 Boric acid is a weak acid. The ions formed in a solution of H_3BO_3 in water are $B(OH)_4^-$ and H_3O^+. Show how the formation of these ions first involves a Lewis acid-base reaction and then a Brönsted acid-base reaction.

```
     OH
     |
HO—B—OH
```

boric acid

STRUCTURE AND ACIDITY AND BASICITY

14.47 (a) Which is the stronger acid of each pair? (1) NH_4^+, NH_3; (2) H_2O, H_3O^+; (3) HS^-, H_2S. (b) How are acidity and charge related?

14.48 Arrange the members of each group in order of decreasing acidity: (a) H_2O, H_2Se, H_2S; (b) HI, HCl, HF, HBr; (c) H_2S, S^{2-}, HS^-; (d) SiH_4, HCl, PH_3, H_2S.

14.49 Volatile acids such as nitric acid, HNO_3, and acetic acid, $HC_2H_3O_2$, can be prepared by adding concentrated H_2SO_4 to salts of the acids. (a) Write chemical equations for the reaction of H_2SO_4 with (1) sodium acetate and (2) sodium nitrate (called chile saltpeter.) (b) Why can't a dilute aqueous solution of H_2SO_4 be used?

14.50 (a) Based on the inductive effect, which would you expect to be the stronger base, $:CF_3^-$ or $:CCl_3^-$? (b) $:CCl_3^-$ is actually the weaker base. Explain in terms of extended π bonding.

14.51 In terms of extended π bonding, account for the fact that $H_2\ddot{N}{-}N{\overset{\displaystyle O}{\underset{\displaystyle O}{\lessgtr}}}$ (nitramide) is a weaker base than $:NH_3$. Is this consistent with the expected inductive effect of the (NO_2) group?

14.52 In terms of the Lewis structures, explain why periodic acid, H_5IO_6, is a weak acid whereas perbromic acid, $HBrO_4$, is a very strong acid.
14.53 Focus on the conjugate bases and use the inductive effect to explain (a) why hydrogen peroxide, H_2O_2, is a stronger acid than water; (b) this order of acidities, $ClCH_2COOH > BrCH_2COOH > CH_3COOH$.
14.54 Account for the fact that many oxygen-containing compounds such as $CH_3CH_2CH_2CH_2CH_2OH$,

$CH_3CH_2OCH_2CH_3$, and $CH_3CH_2CH_2CH_2C(=O)H$ dissolve in cold, concentrated H_2SO_4 but not in a water solution of H_2SO_4. The compounds come out of solution when the solution in concentrated H_2SO_4 is added to H_2O.

APPLICATIONS OF ACID–BASE REACTIONS

14.55 Write ionic equations for these acid-base reactions:

(a) $NaHCO_3(aq) + H_2SO_4(aq)$
("wet" fire extinguisher)

(b) $NaHCO_3(aq) + K_2Al_2(SO_4)_4(H_2O)_{24}(aq)$
(baking powder reaction)

(c) calcium hydride(c) + water
(filling weather balloons)

(d) calcium acetylide(c) (Ca^{2+} $:C{\equiv}C:^{2-}$) + water
(preparation of acetylene)

(e) $Na_3PO_4(aq) + Ca(HCO_3)_2(aq)$
(softening of carbonate hard water)

14.56 Laundry detergents may contain any of the following salts: Na_3PO_4, Na_2CO_3, $Na_2B_4O_7$ (borax), and Na_4SiO_4 (all shown without water of hydration). What purpose do these salts have in common?
14.57 Limestone, $CaCO_3$, is a water-insoluble material, whereas $Ca(HCO_3)_2$ is soluble. Caves are formed when rainwater containing dissolved CO_2 passes over limestone for long periods of time. Write a chemical equation for the acid-base reaction.
14.58 Water containing Ca^{2+} and bicarbonate ion, HCO_3^-, is called "carbonate hard water." On being heated, the Ca^{2+}, which causes the hardness, precipitates as $CaCO_3$. (a) Write an ionic equation for the formation of CO_3^{2-} resulting from a self-ionization of HCO_3^-. (b) Ascribe a role to the heat. (c) Ammonia and sodium phosphate, Na_3PO_4, are used in household cleansers to soften carbonate hard water by causing precipitation of $CaCO_3$. Write equations for the acid-base reactions resulting in the formation of CO_3^{2-} from HCO_3^-. Indicate the conjugate acid-base pairs.
14.59 (a) The Solvay process for the manufacture of sodium bicarbonate, $NaHCO_3$, (baking soda) utilizes CO_2, H_2O, and NH_3 to generate the bicarbonate ion, HCO_3^-. (1) Write the equation for the formation of carbonic acid from carbon dioxide and water. (2) Write an ionic equation for the formation of HCO_3^- from carbonic acid and ammonia, indicating the conjugate acid-base pairs. (3) In terms of chemical equilibrium, explain why CO_2, alone, is not used to furnish the required concentration of HCO_3^-. (4) (a) Why cannot a base such as OH^- be used instead of NH_3 to produce the HCO_3^-? (b) Addition of NaCl to the concentrated solution of NH_4HCO_3 causes the precipitation of $NaHCO_3$, leaving ammonium chloride in solution. NH_3 is regenerated by heating the $NH_4Cl(aq)$ with calcium oxide, CaO. (The decomposition of $CaCO_3$ serves as the source of the CaO and CO_2.) (1) Write an ionic equation for the regeneration of the NH_3 from NH_4Cl and CaO solution. (2) What substance is the net by-product of the entire Solvay process? (Add up all equations.) (c) Sodium carbonate, Na_2CO_3 (washing soda is the hydrate $Na_2CO_3(H_2O)_{10}$), is made by heating $NaHCO_3$. The other products are H_2O and CO_2. Write an ionic equation for the reaction and indicate the type of Brönsted acid-base reaction it is.

MISCELLANEOUS

14.60 Classify each of the hydrides LiH, BeH_2, BH_3, CH_4, NH_3, H_2O, and HF as a Brönsted base, Brönsted acid, amphoteric, Lewis acid, or none of these.
14.61 In general, Brönsted acid-base reactions do not involve oxidation–reductions; an exception is the reaction of the base, H^-. Show that this is true by comparing these reactions:

$$CH_3OH + NH_2^- \longrightarrow CH_3O^- + NH_3$$

$$CH_3OH + H^- \longrightarrow CH_3O^- + H_2$$

14.62 When $:C{\equiv}N:^-$ acts as a base, why does the H^+ or Lewis acid bond to the C atom rather than to the N atom?
14.63 For the reaction at 25°C of NaOH and HCl, or KOH and HNO_3, $\Delta H = -57.82$ kJ/mol. For the reaction of $HC_2H_3O_2$ and NaOH, it is -56.57 kJ/mol. Is the dissociation of $HC_2H_3O_2$ exothermic or endothermic? Explain.
14.64 The reaction

$$H_2C{=}CH_2 + HBr \longrightarrow H_3CCH_2Br$$

can be considered to occur in two steps. The first step is a Brönsted acid-base reaction. The second step is a Lewis acid-base reaction involving the products of the first step. Write equations for the two steps and indicate the Brönsted and Lewis acids and bases.
14.65 A 0.1 M solution of copper(II) chloride, $CuCl_2$, causes the light bulb in Figure 9.1 (section 9.1) to glow brightly. When hydrogen sulfide, H_2S, a very weak acid ($K_a = 6.3 \times 10^{-8}$), is added to the solution, a black precipitate of copper(II) sulfide, CuS, forms and the bulb still glows brightly. The experiment is repeated with a 0.1 M solution of copper(II) acetate, $Cu(C_2H_3O_2)_2$, which also

causes the bulb to glow brightly. Again, CuS(c) forms, but this time the bulb glows dimly. With the aid of ionic equations, explain the difference in behavior of the $CuCl_2$ and $Cu(C_2H_3O_2)_2$ solutions.

14.66 Referring again to Figure 9.1, explain the following results of a conductivity experiment (use ionic equations): (a) Individual solutions of NaOH and HCl cause the bulb to glow brightly. When the solutions are mixed, the bulb still glows brightly. (b) Individual solutions of NH_3 and $HC_2H_3O_2$ cause the bulb to glow dimly. When the solutions are mixed, the bulb glows brightly.

ACID–BASE REACTIONS IN THE BODY

14.67 Some of the acid formed in tissues is excreted through the kidneys. One of the bases removing the acid is HPO_4^{2-}. Write the equation for the reaction. Could Cl^- serve this function?

14.68 Buffered aspirin has an added solid base. What is the rationale for adding the base? (See Box 14.1.)

14.69 Acid is also removed from tissues during respiration. In the lungs, hemoglobin (Hb for short) is converted to oxyhemoglobin, $(Hb)O_2^-$, which is transported in the blood of the arteries to the capillaries of the tissues. Here it comes in contact with H_3O^+ formed, along with HCO_3^-, by the ionization of $H_2CO_3(CO_2 + H_2O)$, which is a product of the metabolism of fats and sugars. (a) Write the equations, indicating the conjugate acid-base pairs, for the formation of the H_3O^+ and for the reaction of $(Hb)O_2^-$ to form $H(Hb)O_2$, which gives up O_2 to the tissues leaving H(Hb). (b) The HCO_3^- and H(Hb) are transported in the venous blood to the lungs, where H(Hb) combines with O_2 to give the acidic $H(Hb)O_2$. Write the equation for the reaction of this acid with HCO_3^-, which releases CO_2. Indicate the conjugate acid-base pairs.

SELF-TEST

14.70 (8 points) (a) Write equations for the reactions (1) $HCO_3^- + H_3O^+$ and (2) $HCO_3^- + OH^-$ and indicate the conjugate acid-base pairs in each case. (b) A substance such as HCO_3^- that reacts with both H_3O^+ and OH^- is said to be ________. (Fill in the missing word.)

14.71 (14 points) (a) List the conjugate bases of (1) H_3PO_4, (2) NH_4^+, and (3) OH^- and the conjugate acids of (4) HSO_4^-, (5) PH_3, and (6) PO_4^{3-}. (b) Given that NO_2^- is a stronger base than NO_3^-, which is the stronger acid, nitric acid, HNO_3, or nitrous acid, HNO_2?

14.72 (8 points) (a) Write the equation for the self-ionization of hydrogen cyanide, HCN(ℓ). (b) Write the equation for a neutralization in liquid HCN. (c) How are the self-ionization of HCN and the neutralization in HCN related?

14.73 (4 points) To determine the relative basicities of bases (a) stronger than OH^-, use an acid that is ________ (*stronger than, the same strength as,* or *weaker than*) H_2O; (b) weaker than H_2O, use an acid that is ________ (*stronger than, the same strength as,* or *weaker than*) H_3O^+.

14.74 (6 points) Write net ionic equations for the reactions of the amphoteric hydroxide $Sn(OH)_2$ with (a) HCl and (b) NaOH.

14.75 (8 points) Arrange the members of each group in order of increasing basicity: (a) CH_3^-, OH^-, F^-, NH_2^-; (b) H_2S, S^{2-}, HS^-; (c) HSO_4^-, $HTeO_4^-$, $HSeO_4^-$; (d) HSO_4^-, ClO_4^-, $H_2PO_4^-$, $H_3SiO_4^-$.

14.76 (5 points) The pH of 0.100 M HA is 3, whereas 0.100 M HB has a pH of 5. Which is the stronger acid? Which acid has the larger K_a value? Can acids of different molarities be compared in this way?

14.77 (4 points) Name the following iodine-containing oxo acids and their corresponding anions: H_5IO_6, HIO, HIO_2, and HIO_3. (HIO_2 does not exist.)

14.78 (13 points) Classify the following species as Lewis acids, Lewis bases, both, or neither. (A bonded Cl is not a good Lewis base site.) All molecules are covalent. (a) anhydrous $CdCl_2$, (b) $BeCl_2$, (c) CH_4, (d) I^-, (e) $GeCl_4$, (f) SO_2, (g) $H_2C{=}O$, (h) H_2, (i) AlH_3, (j) $(CH_3)_3C^+$, (k) $H_3C{:}^-$, (l) Ba^{2+}, (m) Zn^{2+}.

14.79 (15 points) For each reaction write the structural formulas for reactants and products. Specify the Lewis acids and Lewis bases and explain the behavior of the acids.

(a) $BH_3 + H^- \longrightarrow$
(b) $S^{2-} + SO_3 \longrightarrow$
(c) $Au^+ + 2CN^- \longrightarrow$
(d) $H_2C{=}O + H^- \longrightarrow$
(e) $GeF_4 + 2F^- \longrightarrow$

14.80 (15 points) Use the principle of charge dispersal to select the weaker base in each of the following pairs. Which aspect of charge dispersal do you use to get each answer?

(a) HS^-, HSe^-; (b) HSe^-, Br^-; (c) $HC(=O)O^-$, H_3CO^-; (d) PO_4^{3-}, AsO_4^{3-}; (e) BrO_3^-, BrO_4^-.

15

IONIC EQUILIBRIUM I: ACIDS AND BASES

Because of the great abundance of water and its remarkable effectiveness as a solvent, many chemical reactions occur in aqueous solutions. Among these are most of the reactions in living organisms and most of the reactions seen in the introductory chemistry laboratory. Most reactions in aqueous solutions involve ions, often accompanied by dissolved molecules or undissolved solids. Ionic reactions usually come to equilibrium rapidly, and the properties of the solution depend on the concentrations of the species present at equilibrium. For all of these reasons, ionic equilibria deserve careful study. The largest class of ionic reactions consists of the acid-base reactions discussed in Chapter 14. In this chapter we will apply the general principles of chemical equilibrium to acid-base reactions.

15.1 pH, pOH, AND p*K*

In this chapter, we will usually write H^+ instead of H_3O^+.

Section 14.6 acquainted you with the notation **pH**:

$$\mathbf{pH} = -\mathbf{log}\,[\mathbf{H_3O^+}]$$
$$= -\mathbf{log}\,[\mathbf{H^+}] \qquad (1)$$

or

$$[H^+] = \text{antilog}\,(-\text{pH}) = 10^{-\text{pH}} \qquad (2)$$

pH will be used a lot in this chapter, so our first order of business is to get some practice in using equations 1 and 2.

EXAMPLE 15.1 Find the pH of a solution in which $[H^+] = 6.38 \times 10^{-6}$ mol/L.

ANSWER

$$\begin{aligned} \text{pH} &= -\log\,[H^+] \\ &= -\log\,(6.38 \times 10^{-6}) \\ &= -\log 6.38 - \log 10^{-6} \\ &= -\log 6.38 - (-6) \\ &= -0.805 + 6 = 5.195 \end{aligned}$$

(See Appendix B.1.) ■

Note the fact, surprising at first, that the pH in this case is between 5 and 6, not between 6 and 7. The reason is that 6.38×10^{-6} is between 10^{-5} and 10^{-6}, not between 10^{-6} and 10^{-7}.

EXAMPLE 15.2 Calculate $[H^+]$ for a solution of pH 8.37.

ANSWER

$$\begin{aligned} pH &= -\log [H^+] \\ 8.37 &= -\log [H^+] \\ -8.37 &= \log [H^+] \\ [H^+] &= \text{antilog}\,(-8.37) = 10^{-8.37} = 10^{0.63} \times 10^{-9} \\ &= 4.3 \times 10^{-9} \text{ mol/L} \end{aligned}$$

■

Keep in mind that the *more acidic* the solution is, the *smaller* is the pH.

Calculations such as these are best done with a pocket calculator having logarithms and exponentials. However, you should also be able to use logarithm tables so that you will not be helpless when a calculator is not available. A calculator may give you many insignificant figures. See Appendix B.1 for the rules on significant figures in logarithms and antilogarithms.

PROBLEM 15.1 (a) In a certain solution, $[H^+] = 3.4 \times 10^{-10}$ mol/L. What is its pH? (b) Another solution has pH 3.62. Find $[H^+]$. □

Another calculation you will do many times uses the ion product of water (section 14.6) to relate $[H^+]$ and $[OH^-]$. Recall that in any aqueous solution,

$$\mathbf{[H^+][OH^-] = K_w} \qquad (3)$$
$$= 1.00 \times 10^{-14} \text{ at } 25°\text{C}$$

EXAMPLE 15.3 A solution has $[H^+] = 1.4 \times 10^{-9}$ mol/L at 25°C. Calculate $[OH^-]$.

ANSWER

$$[H^+][OH^-] = K_w$$

$$[OH^-] = \frac{K_w}{[H^+]}$$

$$= \frac{1.00 \times 10^{-14}}{1.4 \times 10^{-9}} = 7.1 \times 10^{-6} \text{ mol/L}$$

■

PROBLEM 15.2 What are $[H^+]$ and $[OH^-]$ in a solution with pH 6.23 at 25°C? □

The ion product of water has that easily remembered value 1.00×10^{-14} *only* at 25°C. Like other equilibrium constants, it varies with temperature. Therefore, the pH of pure water also varies with temperature. Table 15.1 gives K_w at several temperatures. K_w increases with increasing temperature; therefore, the reaction

TABLE 15.1
THE ION PRODUCT OF WATER

TEMPERATURE, °C	K_w
0	1.14×10^{-15}
25	1.00×10^{-14}
35	2.06×10^{-14}
50	5.35×10^{-14}

$H_2O \longrightarrow H^+ + OH^-$ is endothermic (ΔH is positive). To convince yourself that this is true, carry out the reverse reaction, $H^+ + OH^- \longrightarrow H_2O$, by mixing an acid and a base (not too concentrated, and wear goggles!) and observe that the beaker gets hot—showing that it is an exothermic reaction.

EXAMPLE 15.4 What is the pH of pure water at (a) 25°C and (b) 50°C?

ANSWER In pure water, $[H^+] = [OH^-]$. Let $y = [H^+] = [OH^-]$. Then

$$[H^+][OH^-] = K_w$$

$$y^2 = K_w$$

(a) At 25°C,

$$K_w = 1.00 \times 10^{-14} \qquad \text{(Table 15.1)}$$

$$y^2 = 1.00 \times 10^{-14}$$

$$y = 1.00 \times 10^{-7}$$

$$\text{pH} = -\log y = 7.000$$

(b) At 50°C,

$$K_w = 5.35 \times 10^{-14}$$

$$y^2 = 5.35 \times 10^{-14}$$

$$y = 2.31 \times 10^{-7}$$

$$\text{pH} = -\log y = 6.636$$

■

The pH of pure water, or of any neutral solution, serves as the dividing line between acidic and basic solutions (section 14.6). We commonly think of it as 7, but as Example 15.4 shows, this is correct only at 25°C.

EXAMPLE 15.5 A certain solution has pH 3.89 at 0°C. Find $[OH^-]$.

ANSWER From the pH, we can calculate $[H^+]$. Then equation 1 is used to find $[OH^-]$.

$$[H^+] = 10^{-\text{pH}} = 10^{-3.89} = 10^{0.11} \times 10^{-4} = 1.3 \times 10^{-4} \text{ mol/L}$$

$$[H^+][OH^-] = K_w = 1.14 \times 10^{-15} \qquad \text{(Table 15.1)}$$

$$[OH^-] = \frac{K_w}{[H^+]} = \frac{1.14 \times 10^{-15}}{1.3 \times 10^{-4}} = 8.8 \times 10^{-12} \text{ mol/L}$$

■

PROBLEM 15.3 The pH of a solution at 0°C is 7.00. What are its $[H^+]$ and $[OH^-]$?

□

The "p" notation has been extended to other concentrations; pOH = $-\log [OH^-]$, pNa = $-\log [Na^+]$, and so on. Note that the charge of the ion is omitted: pOH and pNa, not pOH^- and pNa^+. The same notation is used for the negative of the logarithm of an equilibrium constant:

$$pK = -\log K \tag{4}$$

$$K = \text{antilog}(-pK) = 10^{-pK}$$

EXAMPLE 15.6 The acidic ionization constant of phenol, C_6H_5OH, is $K_a = 1.3 \times 10^{-10}$. What is pK_a for phenol?

ANSWER

$$\begin{aligned} pK_a &= -\log K_a \\ &= -\log(1.3 \times 10^{-10}) \\ &= 9.89 \end{aligned}$$

■

EXAMPLE 15.7 For the weak base strychnine, $C_{21}H_{22}N_2O_2$, $pK_b = 5.74$. Calculate K_b.

ANSWER

$$\begin{aligned} K_b &= \text{antilog}(-pK_b) = 10^{-pK_b} \\ &= \text{antilog}(-5.74) = 10^{-5.74} \\ &= 1.8 \times 10^{-6} \end{aligned}$$

■

PROBLEM 15.4 For hydrocyanic acid, HCN, $K_a = 4.93 \times 10^{-10}$. What is pK_a for HCN? □

The advantage of p*K* is the same as that of pH: the numbers are more compact—easier to read, write, and especially, type. However, p*K* is afflicted by the paradox that the *stronger* the acid (or base) is, the *smaller* is its pK_a (or pK_b). The same goes for pH, pOH, and so on. A larger number means a smaller p-value because of the minus sign in the definition.

The equation

$$[H^+][OH^-] = K_w$$

is conveniently written in terms of pH, pOH, and pK_w. Take logarithms of both sides of the equation:

$$\log[H^+] + \log[OH^-] = \log K_w$$

Changing the signs (multiplying by -1) gives

$$-\log[H^+] - \log[OH^-] = -\log K_w$$

and, from the definitions,

$$\mathbf{pH + pOH = p}\boldsymbol{K_w} \qquad (5)$$

EXAMPLE 15.8 (a) What is pK_w at 25°C? (b) A solution has pOH 6.50 at 25°C. What is its pH?

ANSWER

(a) $K_w = 1.00 \times 10^{-14}$

$$pK_w = -\log(1.00 \times 10^{-14}) = 14.000$$

(b) Rearrange equation 5 and use the pK_w you got in part a.

$$\begin{aligned} pH &= pK_w - pOH \\ &= 14.000 - 6.50 \\ &= 7.50 \end{aligned}$$

■

PROBLEM 15.5 (a) What is pK_w at 35°C? (b) A solution has pH 4.00 at 35°C. What is its pOH? □

15.2 STRONG ACIDS AND BASES

Recall from Table 14.1 that there are three very common strong acids—HCl, H_2SO_4, HNO_3—and three that are a bit less common but still familiar—HBr, HI, and $HClO_4$. Their ionization reactions in dilute aqueous solutions go practically to completion:

$$HCl \longrightarrow H^+ + Cl^-$$

$$H_2SO_4 \longrightarrow H^+ + HSO_4^-$$

$$HNO_3 \longrightarrow H^+ + NO_3^-$$

$HSO_4^- \longrightarrow H^+ + SO_4^{2-}$ does not go to completion.

The common strong bases (section 14.2) are the soluble ionic hydroxides of the alkali and alkaline earth metals (Groups 1 and 2 in the periodic table). Just like other ionic compounds, they exist in dilute solution as ions:

$$NaOH(c) \longrightarrow Na^+ + OH^-$$

$$Ba(OH)_2(c) \longrightarrow Ba^{2+} + 2OH^-$$

For these strong acids and strong bases, $[H^+]$ or $[OH^-]$ is equal to the concentration of the solute multiplied by a whole number. The whole number is the number of moles of H^+ or OH^- produced by dissolving 1 mole of the solute:

$$HCl(g) \longrightarrow H^+ + Cl^- \qquad 1 \text{ mol HCl} \longrightarrow 1 \text{ mol } H^+$$

$$NaOH(c) \longrightarrow Na^+ + OH^- \qquad 1 \text{ mol NaOH} \longrightarrow 1 \text{ mol } OH^-$$

$$Ba(OH)_2(c) \longrightarrow Ba^{2+} + 2OH^- \qquad 1 \text{ mol } Ba(OH)_2 \longrightarrow 2 \text{ mol } OH^-$$

EXAMPLE 15.9 Calculate the pH of (a) 1.0×10^{-2} M HCl and (b) 5.0×10^{-3} M $Ba(OH)_2$ at 25°C.

ANSWER

(a) The reaction is

$$HCl + H_2O \longrightarrow H_3O^+ + Cl^-$$

or simply,

$$HCl \longrightarrow H^+ + Cl^-$$

$$[H^+] = 1.0 \times 10^{-2} \text{ mol/L}$$

$$pH = -\log(1.0 \times 10^{-2}) = -\log 1.0 + 2 = 2.00$$

(b) Each mole of $Ba(OH)_2$ that dissolves gives 2 moles OH^- in solution. Then, for 5.0×10^{-3} mol $Ba(OH)_2$/L,

$$[OH^-] = 2 \times 5.0 \times 10^{-3} \text{ mol/L} = 1.0 \times 10^{-2} \text{ mol/L}$$

To find the pH, we need $[H^+]$, which is related to $[OH^-]$ by equation 3:

$$[H^+][OH^-] = K_w$$

Then

$$[H^+] = \frac{K_w}{[OH^-]} = \frac{1.00 \times 10^{-14}}{1.0 \times 10^{-2}} = 1.0 \times 10^{-12} \text{ mol/L}$$

$$pH = -\log 1.0 + 12 = 12.00$$

■

PROBLEM 15.6 What is the pH at 25°C of (a) 5.0×10^{-4} M HNO_3 and (b) 3.0×10^{-3} M KOH? □

PROBLEM 15.7 A solution of NaOH has pH 12.50 at 25°C. What is the molarity of the solution? □

$10^{-7} + 10^{-10} = 10^{-7}$

If a solution is extremely dilute, however, this simple method fails. Suppose there is a 1.0×10^{-10} M solution of HCl. You might say that $[H^+] = 10^{-10}$ and pH = 10. But that is obviously wrong because pH 10 corresponds to a basic solution, and this solution is slightly acidic. We have overlooked the fact that the self-ionization of water also contributes something to $[H^+]$; in this case, it contributes far more than the meager 10^{-10} mol/L from the HCl. Since the water contributes most of the H^+ in this instance, the pH is almost the same as in pure water: 7.0. We will not consider intermediate cases in which the solute and the water make comparable contributions to $[H^+]$ or $[OH^-]$.

15.3 WEAK ACIDS AND BASES

Most acids and bases are weak electrolytes—only a small fraction is ionized at equilibrium. It is often necessary to find the concentration of H^+ or OH^-, or the pH, in a solution of a weak acid or base. When the acid or base is weak, it produces a concentration of H^+ or OH^- much less than its own concentration. Thus, whereas a 0.1 M solution of HCl (a strong acid) has $[H^+] = 0.1$ mol/L., a 0.1 M solution of HF (a weak acid) does not have $[H^+]$ anywhere near 0.1 mol/L.

It is necessary to do an equilibrium calculation like those in section 13.10. We write the equation for the ionization, set up an equilibrium table, and solve for the unknown. The ionization constant, K_a or K_b, must be given or previously calculated. Table 15.2 lists the values of K_a and K_b for a number of common acids and bases at 25°C. The equilibrium conditions are

In this chapter, you may assume that the temperature is 25°C unless otherwise stated.

WEAK ACID	WEAK BASE
$HA \rightleftharpoons H^+ + A^-$	$B + H_2O \rightleftharpoons BH^+ + OH^-$
$\frac{[H^+][A^-]}{[HA]} = K_a \quad (6)$	$\frac{[BH^+][OH^-]}{[B]} = K_b \quad (7)$

Where $[H^+]$ appears for acids, $[OH^-]$ appears for bases. Ignoring this difference is a frequent source of error.

Sometimes $[OH^-]$ is calculated and $[H^+]$ is wanted (or conversely). Equation 3 or 5 can then be used:

$$[H^+][OH^-] = K_w \quad \text{or} \quad pH + pOH = pK_w$$

EXAMPLE 15.10 Using Table 15.2, find the pH of a 0.20 M solution of formic acid.

Latin *formica,* "ant." Formic acid was first prepared by heating ants. Ants return the favor by injecting formic acid when they sting.

ANSWER Construct an equilibrium table as in section 13.9:

	$HCHO_2$	$\rightleftharpoons$ H^+	+ CHO_2^-
Initial concentration	0.20	0	0
Change	$-y$	$+y$	$+y$
Concentration at equilibrium	$0.20 - y$	y	y mol/L

TABLE 15.2 IONIZATION CONSTANTS OF ACIDS AND BASES IN WATER AT 25°C

MONOPROTIC ACID	FORMULA	K_a
Acetic acid	$HC_2H_3O_2$, CH_3COOH	1.75×10^{-5}
Benzoic acid	$HC_7H_5O_2$, C_6H_5COOH	6.46×10^{-5}
Bromoacetic acid	$HC_2H_2O_2Br$, $CH_2BrCOOH$	2.05×10^{-3}
Chloroacetic acid	$HC_2H_2O_2Cl$, $CH_2ClCOOH$	1.40×10^{-3}
Dichloroacetic acid	$HC_2HO_2Cl_2$, $CHCl_2COOH$	3.32×10^{-2}
Formic acid	$HCHO_2$, $HCOOH$	1.76×10^{-4}
Hydrocyanic acid	HCN	4.93×10^{-10}
Hydrofluoric acid	HF	3.53×10^{-4}
Nitrous acid	HNO_2	4.5×10^{-4}
Phenol (carbolic acid)	HC_6H_5O, C_6H_5OH	1.3×10^{-10}
Propionic acid	$HC_3H_5O_2$, CH_3CH_2COOH	1.34×10^{-5}
Trichloroacetic acid	$HC_2O_2Cl_3$, CCl_3COOH	2×10^{-1}

POLYPROTIC ACID	FORMULA	K_1	K_2	K_3
Carbonic acid[†]	H_2CO_3	4.30×10^{-7}	5.62×10^{-11}	—
Hydrogen sulfide	H_2S	6.3×10^{-8}	10^{-14}[§]	—
Maleic acid	$H_2C_4H_2O_4$, $C_2H_2(COOH)_2$	1.42×10^{-2}	8.57×10^{-7}	—
Malonic acid	$H_2C_3H_2O_4$, $CH_2(COOH)_2$	1.49×10^{-3}	2.03×10^{-6}	—
Oxalic acid	$H_2C_2O_4$, $(COOH)_2$	5.90×10^{-2}	6.40×10^{-5}	—
Phosphoric acid	H_3PO_4	7.52×10^{-3}	6.23×10^{-8}	4.8×10^{-13}
Phthalic acid	$H_2C_8H_4O_4$, $C_6H_4(COOH)_2$	1.26×10^{-3}	3.9×10^{-6}	—
Succinic acid	$H_2C_4H_4O_4$, $(CH_2COOH)_2$	6.89×10^{-5}	2.47×10^{-6}	—
Sulfuric acid	H_2SO_4	$\sim 10^{+11}$	1.2×10^{-2}	—
Sulfurous acid	H_2SO_3	1.7×10^{-2}	6.24×10^{-8}	—
Tartaric acid	$H_2C_4H_4O_6$, $(CHOHCOOH)_2$	1.04×10^{-3}	4.55×10^{-5}	—

ACIDIC CATION	FORMULA	K_1
Aluminum ion	$Al(H_2O)_6^{3+}$	1.2×10^{-5}
Ammonium ion	NH_4^+	5.64×10^{-10}
Beryllium ion	$Be(H_2O)_4^{2+}$	1.0×10^{-5}
Copper(II) ion	$Cu(H_2O)_4^{2+}$	1.0×10^{-8}
Iron(II) ion	$Fe(H_2O)_6^{2+}$	3.0×10^{-10}
Iron(III) ion	$Fe(H_2O)_6^{3+}$	4.0×10^{-3}
Zinc ion	$Zn(H_2O)_4^{2+}$	2.5×10^{-10}

MONOPROTIC BASE	FORMULA	K_b
Acetate ion	$C_2H_3O_2^-$, CH_3COO^-	5.71×10^{-10}
Ammonia	NH_3	1.77×10^{-5}
Aniline	$C_6H_5NH_2$	4.27×10^{-10}
Benzoate ion	$C_7H_5O_2^-$, $C_6H_5COO^-$	1.55×10^{-10}
Cyanide ion	CN^-	2.03×10^{-5}
Dimethylamine	$(CH_3)_2NH$	5.41×10^{-4}
Ethylamine	$CH_3CH_2NH_2$	4.71×10^{-4}
Hydroxylamine	$HONH_2$	1.1×10^{-8}
Methylamine	CH_3NH_2	3.70×10^{-4}
Pyridine	C_5H_5N	1.78×10^{-9}
Trimethylamine	$(CH_3)_3N$	6.45×10^{-5}

DIPROTIC BASE	FORMULA	K_1	K_2
Carbonate ion	CO_3^{2-}	1.78×10^{-4}	2.33×10^{-8}
Hydrazine	H_2NNH_2	9.1×10^{-7}	10^{-15}

[†] When CO_2 dissolves in water, a small fraction—less than 1%—reacts to form H_2CO_3: $CO_2 + H_2O \rightleftharpoons H_2CO_3$. Most of the dissolved CO_2 is present as CO_2. However, it is common practice to represent dissolved CO_2 by the formula H_2CO_3. $[H_2CO_3]$ therefore means the total concentration of CO_2 and H_2CO_3.

[§] The value of K_2 for H_2S has been a topic of controversy for decades. Recent measurements suggest that it may be as small as 10^{-19}. (R. J. Myers, *Journal of Chemical Education*, **63**, 687 (1986).) We will use 10^{-14} with the understanding that it is subject to revision.

y represents the number of moles of acid per liter that ionize. From Table 15.2, $K_a = 1.76 \times 10^{-4}$. The equilibrium condition, equation 6, is

$$\frac{[H^+][CHO_2^-]}{[HCHO_2]} = K_a$$

$$\frac{y^2}{0.20 - y} = 1.76 \times 10^{-4} \qquad (8)$$

This quadratic equation can be solved without undue labor, but the simplifying approximation described in section 13.10 is often applicable. The acid is weak, meaning that only a small fraction of it ionizes. We thus expect that y, representing the number of moles of acid that have ionized per liter, will be much less than the initial concentration of acid, $y \ll 0.20$, and therefore $0.20 - y \approx 0.20$. If this approximation is valid, $0.20 - y$ can be replaced by 0.20 without serious error, and equation 8 becomes

$$\frac{y^2}{0.20} = 1.76 \times 10^{-4} \qquad (9)$$

$$y^2 = 35.2 \times 10^{-6}$$

$$y = 5.9 \times 10^{-3} \text{ mol/L}$$

$$\text{pH} = -\log y = 3 - \log 5.9 = 2.23$$

It is necessary to check whether the assumption that only a negligible amount of acid ionizes ($0.20 - y \approx 0.20$) is valid. The answer for y makes $0.20 - y = 0.20 - 6 \times 10^{-3} = 0.19$. For most calculations involving ionic equilibria, the discrepancy between 0.20 and 0.19 is not serious. In that case, the exact equation 8 and the simplified equation 9 are essentially the same equation and therefore give the same answer. (More accurate solution of equation 8 with the quadratic formula gives $y = 5.8 \times 10^{-3}$, pH = 2.24.) ■

In this chapter, an error of ±5% in a concentration or of ±0.02 in a pH is quite acceptable. In the preceding example $y/0.20 = 5.9 \times 10^{-3}/0.20 = 3.0\%$, under the 5% limit.

PROBLEM 15.8 Find $[H^+]$ and pH in a 0.40 M solution of HF. □

The approximate method fails when K_a is too large or when the initial concentration is too small (the solution is too dilute). The following example illustrates a case in which $[H^+]$ cannot be neglected in comparison to the initial concentration because the ''weak'' acid is not weak enough. A quadratic equation must therefore be solved.

EXAMPLE 15.11 Find $[H^+]$ in a 0.50 M solution of trichloroacetic acid, $HC_2O_2Cl_3$.

ANSWER From Table 15.2, $K_a = 0.2$ for the reaction.

	$HC_2O_2Cl_3$ ⇌	H^+ +	$C_2O_2Cl_3^-$
Initial concentration	0.50	0	0
Change	$-y$	$+y$	$+y$
Concentration at equilibrium	$0.50 - y$	y	y mol/L

$$\frac{[H^+][C_2O_2Cl_3^-]}{[HC_2O_2Cl_3]} = 0.2$$

$$\frac{y^2}{0.50 - y} = 0.2 \qquad (10)$$

Forging ahead blindly and neglecting y gives $y^2/0.50 = 0.2$, $y = 0.32$ mol/L. However, 0.32 is far from negligible in comparison to 0.50; 0.32/0.50 = 64%. This tells us that the approximate method is inadequate and we must solve the quadratic equation. When equation 10 is written in the standard quadratic form, it becomes

$$y^2 + 0.2y - 0.1 = 0$$

Solving for y, we obtain

$$y = +0.23 \quad \text{or} \quad -0.43$$

A concentration cannot be negative, and the answer -0.43 must therefore be rejected; then

Why is only one significant figure retained?

$$[H^+] = y = 0.2 \text{ mol/L}$$

■

PROBLEM 15.9 Find $[H^+]$ and pH in a 0.00400 M solution of HF. □

It is interesting that in Problem 15.8, a sufficiently accurate answer could be obtained without solving a quadratic equation, while in Problem 15.9, the simplified procedure gives 1.19×10^{-3} mol/L and pH = 2.925, answers that are far from accurate. Therefore, the quadratic equation must be solved. The only difference is that the solution in Problem 15.9 is much less concentrated. In this more dilute solution, a larger fraction of the acid is ionized, spoiling the approximation that the acid concentration is practically unchanged at equilibrium. It is possible to write and memorize a rule for deciding when the approximation method is inadequate (Problem 15.46). However, it is easier to find the approximate solution and see whether it is negligible (less than 5%) in comparison to the initial concentration.

EXAMPLE 15.12 Find the pH and the percent ionized in a 0.100 M solution of ammonia, NH_3.

ANSWER

	NH_3	+ H_2O ⇌	NH_4^+	+ OH^-
Initial concentration	0.100		0	0
Change	$-y$		$+y$	$+y$
Concentration at equilibrium	$0.100 - y$		y	y mol/L

Note that y now represents $[OH^-]$ instead of $[H^+]$. The equilibrium condition is

$$\frac{[NH_4^+][OH^-]}{[NH_3]} = K_b = 1.77 \times 10^{-5} \qquad \text{(Table 15.2)}$$

$$y = [NH_4^+] = [OH^-]$$

Then

$$\frac{y^2}{0.100 - y} = 1.77 \times 10^{-5}$$

If $y \ll 0.100$,

$$\frac{y^2}{0.100} = 1.77 \times 10^{-5}$$

$$y^2 = 1.77 \times 10^{-6}$$

$$y = 1.33 \times 10^{-3} \text{ mol/L} = [OH^-]$$

y is less than 5% of 0.100 (5×10^{-3}), and thus meets our requirement for being negligible. By equation 3,

$$[H^+] = \frac{K_w}{[OH^-]}$$

$$= \frac{1.00 \times 10^{-14}}{1.33 \times 10^{-3}} = 7.5 \times 10^{-12}$$

$$pH = -\log(7.5 \times 10^{-12}) = -\log 7.5 + 12 = 11.12$$

The percent ionized is the concentration of NH_4^+ or OH^- divided by the initial NH_3 concentration, times 100%:

$$\frac{1.33 \times 10^{-3} \text{ mol/L}}{0.100 \text{ mol/L}} \times 100\% = 1.33\%$$ ■

The percentage of a weak acid or base that is ionized increases as the concentration decreases. This is in agreement with Le Chatelier's principle. Adding water increases the volume of the solution, and thus favors the side of the equation showing more solute particles (section 13.8). The reaction

$$HA \rightleftharpoons H^+ + A^- \quad \text{or} \quad HA + H_2O \rightleftharpoons H_3O^+ + A^-$$

has 2 moles of solute on the right and 1 mole on the left, and is thus shifted to the right when water is added. The same is true of the basic ionization $B + H_2O \rightleftharpoons BH^+ + OH^-$. However, the concentration of H^+ or of OH^- *decreases* as the solution is diluted. The following numbers for a solution of NH_3 illustrate the distinction:

H_2O, the solvent, does not count.

INITIAL $[NH_3]$	$[OH^-]$ AT EQUILIBRIUM	pH	PERCENT IONIZED
1.0 mol/L	4.2×10^{-3} mol/L	11.62	0.42%
0.10	1.3×10^{-3}	11.11	1.3
0.010	0.41×10^{-3}	10.61	4.1
0.0010	0.12×10^{-3}	10.08	12
	↓ Decreasing concentration	↓ Less basic	↓ Larger % ionized

As the solution becomes more dilute, the concentration of OH^- decreases, but less sharply than the concentration of NH_3 decreases. The ratio of $[OH^-]$ to $[NH]_3$, and thus the percent ionized, therefore increases.

PROBLEM 15.10 Find the pH and the percent ionized in a 0.030 M solution of propionic acid, $HC_3H_5O_2$. □

CHARGED ACIDS AND BASES

As we mentioned in sections 14.3 and 14.4, a salt may make a solution either acidic or basic, because one of its ions is an acid or a base. The following example illustrates a calculation in which the base is a negative ion.

EXAMPLE 15.13 Find the pH of a 0.200 M solution of sodium acetate, $NaC_2H_3O_2$.

ANSWER When a problem involves a salt, first ask yourself: What ions are present? What are their concentrations? Here, the ions are Na^+ and $C_2H_3O_2^-$, each of them 0.200 M. Na^+ is a spectator ion (section 9.2). The base is acetate ion:

$$C_2H_3O_2^- + H_2O \rightleftharpoons HC_2H_3O_2 + OH^-$$

$0.200 - y$ $\quad y \quad y$ mol/L at equilibrium

$$\frac{[HC_2H_3O_2][OH^-]}{[C_2H_3O_2^-]} = K_b$$

$$\frac{y^2}{0.200 - y} = 5.71 \times 10^{-10} \qquad \text{(Table 15.2)}$$

If $y \ll 0.200$, then

$$\frac{y^2}{0.200} = 5.71 \times 10^{-10}$$

$$y^2 = 1.14 \times 10^{-10}$$

$$y = 1.07 \times 10^{-5} \text{ mol/L} = [OH^-]$$

$$[H^+] = \frac{K_w}{[OH^-]} = \frac{1.00 \times 10^{-14}}{1.07 \times 10^{-5}} = 9.35 \times 10^{-10} \text{ mol/L}$$

$$pH = -\log(9.35 \times 10^{-10}) = 9.029$$

■

PROBLEM 15.11 Find the pH of a 0.150 M solution of potassium benzoate, $KC_7H_5O_2$. □

15.4 ACID–BASE INDICATORS

While calculating the pH's of various solutions, you must have wondered how pH is measured. The simplest method uses a special kind of acid or base. An **acid-base indicator** is a weak acid (or base) whose color is different from that of its conjugate base (or acid) (section 14.2). Let us represent the acid form of the indicator by HIn and its conjugate base form by In^-. An example is methyl orange, for which HIn is red and In^- is yellow:

$$HIn(aq) \rightleftharpoons H^+ + In^-$$

red $\quad$ yellow

If HIn and In^- are present in roughly equal concentrations, the solution is orange. The addition of an acid shifts the equilibrium to the left, and the solution turns from orange to red. Addition of a base removes H^+, shifting the equilibrium to the right, and the solution turns from orange to yellow. The equilibrium condition for the reaction is

$$\frac{[H^+][In^-]}{[HIn]} = K \quad \text{or} \quad \frac{[HIn]}{[In^-]} = \frac{[H^+]}{K}$$

or in logarithmic form,

$$\log\frac{[HIn]}{[In^-]} = \log[H^+] - \log K_a$$

$$= pK_a - pH \qquad (11)$$

TABLE 15.3
ACID–BASE INDICATORS

INDICATOR	COLOR: Acid	COLOR: Base	pH RANGE
Malachite green	yellow	blue-green	0.2–1.8
Thymol blue	red	yellow	1.2–2.8
Methyl orange	red	yellow	3.2–4.4
Bromcresol green	yellow	blue	3.8–5.4
Methyl red	red	yellow	4.8–6.0
Bromcresol purple	yellow	purple	5.2–6.8
Bromthymol blue	yellow	blue	6.0–7.6
Neutral red	red	yellow	6.8–8.0
Cresol red	yellow	red	7.0–8.8
p-Naphtholbenzene	orange	blue	8.2–10.0
Phenolphthalein	colorless	red	8.2–10.0
Thymolphthalein	colorless	blue	9.4–10.6
Alizarin yellow	yellow	violet	10.1–12.0
1,3,5-Trinitrobenzene	colorless	red	12.0–14.0

Acid-base indicators are used in small quantities so that practically all of the $[H^+]$ comes from other acids in the solution. It is therefore necessary that their colors be very intense. The ratio $[HIn]/[In^-]$, which determines the color, depends on the pH of the solution, as equation 11 shows. With methyl orange, the solution is red if $[HIn] \gg [In^-]$, yellow if $[In^-] \gg [HIn]$, and a shade of orange when $[HIn]$ and $[In^-]$ are about the same. Table 15.3 lists some common acid-base indicators, their colors, and the pH ranges in which they show perceptible gradations of color (Color plate 10). These indicators are good for rough estimation of pH or for detecting large changes in pH. The accuracy can be improved by matching colors in a special colorimeter. However, accurate measurements of pH are now made electrically rather than optically. The device used is called, logically, a **pH meter** (section 17.9).

A change of color is associated with a profound change in the electronic structure of a molecule or ion. All acid-base indicators are complicated organic molecules with the property that the addition or removal of a proton causes an extensive reorganization of the bonds. Figure 15.1 shows how the structure of methyl orange changes when it gains or loses a proton.

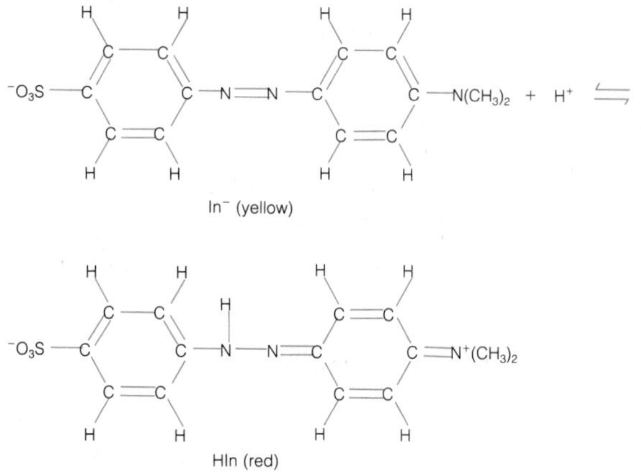

In^- (yellow)

HIn (red)

FIGURE 15.1
The basic and acidic forms of methyl orange. Note that the charges in HIn add up to zero.

EXAMPLE 15.14 Select from Table 15.3 an indicator suitable for estimating pH's near 8.

ANSWER Cresol red has pH 8 in about the middle of its range (7.0 to 8.8). It would show gradations of yellow-orange to red-orange around 8. Phenolphthalein also has a range that includes 8, but it changes from colorless to red, and the only gradation would therefore be in the depth of the pink color. ■

An indicator like phenolphthalein or thymolphthalein, in which one form is colorless, is useful for detecting large changes of pH. This is, indeed, one of the most important applications of indicators (section 15.10). However, two different colors are needed if the pH is to be estimated from the hue.

EXAMPLE 15.15 Assume that the pK_a of bromcresol green is 4.6 (the middle of its range). What is the ratio $[HIn]/[In^-]$ and the color when the pH is (a) 3.0, (b) 4.0, and (c) 5.0?

ANSWER

(a) Equation 11 gives

$$\log \frac{[HIn]}{[In^-]} = 4.6 - 3.0 = 1.6$$

$$\frac{[HIn]}{[In^-]} = 10^{1.6} = 40$$

For bromcresol green, the acid form [HIn] is yellow (Table 15.3). When the acid form is 40 times as abundant as the base form $[In^-]$, the solution is yellow.

(b)
$$\log \frac{[HIn]}{[In^-]} = 4.6 - 4.0 = 0.6$$

$$\frac{[HIn]}{[In^-]} = 10^{0.6} = 4$$

The solution is still yellow, but with $[In^-]$ as large as 1/4 of [HIn], a greenish tint would be noticeable. (The only way to find out what an indicator really looks like is to try it in the laboratory with solutions of known pH.)

(c)
$$\log \frac{[HIn]}{[In^-]} = 4.6 - 5.0 = -0.4$$

$$\frac{[HIn]}{[In^-]} = 10^{-0.4} = 0.4$$

This solution would be on the bluish side of green, with more In^- (blue) than HIn (yellow). ■

PROBLEM 15.12 When the pH of a solution is 10.00, the concentration ratio of the acid form to the base form of an indicator is 2.0. What are pK_a and K_a for this indicator? Use equation 11. □

15.5 CONJUGATE ACID–BASE PAIRS

Conjugate pairs were defined in section 14.2. HA and A^- are a *conjugate acid-base pair*:

$$\underset{\text{acid}}{HA} + H_2O \rightleftharpoons H_3O^+ + \underset{\text{base}}{A^-}$$

or, in the simplified form,

$$\underset{\text{acid}}{HA} \rightleftharpoons H^+ + \underset{\text{base}}{A^-}$$

There is a simple relationship between the ionization constants K_a of the acid HA and K_b of its conjugate base A^-. We write the equilibrium conditions for the acidic ionization of HA and the basic ionization of A^-, both in aqueous solution:

$$HA \rightleftharpoons H^+ + A^- \qquad A^- + H_2O \rightleftharpoons HA + OH^-$$

$$\frac{[H^+][A^-]}{[HA]} = K_a \qquad \frac{[HA][OH^-]}{[A^-]} = K_b$$

Now multiply one of these equations by the other:

$$\frac{[H^+][\cancel{A^-}]}{[\cancel{HA}]} \times \frac{[\cancel{HA}][OH^-]}{[\cancel{A^-}]} = K_aK_b$$

$$[H^+][OH^-] = K_aK_b$$

Recall that at equilibrium, $[H^+][OH^-]$ equals K_w, the ion product of water. K_w is independent of our choice of the acid-base pair. For *any* conjugate pair in aqueous solution,

$$\boldsymbol{K_aK_b = K_w} \tag{12}$$

Equation 12 saves paper by shrinking tables of ionization constants. It allows us to calculate K_a from K_b and K_b from K_a; thus, a table needs to give only one or the other. Table 15.2 includes both K_a and K_b for some important conjugate pairs (like $HC_2H_3O_2$ and $C_2H_3O_2^-$), but not for all of them.

EXAMPLE 15.16 Find K_b and pK_b for the formate ion, CHO_2^-.

ANSWER The conjugate acid of CHO_2^- is formic acid, $HCHO_2$. The ionization constant of $HCHO_2$ is $K_a = 1.76 \times 10^{-4}$; $K_w = 1.00 \times 10^{-14}$. Then

$$K_aK_b = K_w$$

$$K_b = \frac{K_w}{K_a}$$

$$= \frac{1.00 \times 10^{-14}}{1.76 \times 10^{-4}} = 5.68 \times 10^{-11}$$

$$pK_b = -\log(5.68 \times 10^{-11}) = 10.246$$

■

PROBLEM 15.13 What are K_a and pK_a of the ethylammonium ion, $C_2H_5NH_3^+$? (Look for its conjugate base in Table 15.2.) □

EXAMPLE 15.17 Find the pH of a 0.100 M solution of pyridinium chloride, $(C_5H_5NH)^+Cl^-$, a strong electrolyte.

ANSWER The pH of this solution is determined by the acidic ionization of $C_5H_5NH^+$: $C_5H_5NH^+ \rightleftharpoons H^+ + C_5H_5N$. The first step is to calculate the acidic ionization constant of $C_5H_5NH^+$, the conjugate acid of the base pyridine, C_5H_5N.

This constant is calculated from K_b for pyridine (Table 15.2):

$$K_a = \frac{K_w}{K_b}$$

$$= \frac{1.00 \times 10^{-14}}{1.78 \times 10^{-9}} = 5.62 \times 10^{-6}$$

Most uncharged weak bases are like NH_3 in that the unshared electron pair is on N.

The calculation of the pH follows the same pattern as with uncharged acids. The equilibrium condition is

$$\frac{[H^+][C_5H_5N]}{[C_5H_5NH^+]} = K_a = 5.62 \times 10^{-6}$$

$$\underset{0.100 - y}{C_5H_5NH^+} \rightleftharpoons \underset{y}{H^+} + \underset{y \text{ mol/L at equilibrium}}{C_5H_5N}$$

Then

$$\frac{y^2}{0.100 - y} = 5.62 \times 10^{-6}$$

If $y \ll 0.100$,

$$\frac{y^2}{0.100} = 5.62 \times 10^{-6}$$

$$y^2 = 0.562 \times 10^{-6}$$

$$y = 7.50 \times 10^{-4} \text{ mol/L} = [H^+]$$

$$pH = -\log (7.50 \times 10^{-4}) = 3.125$$

■

EXAMPLE 15.18 Find the pH of a 0.30 M solution of NaF.

ANSWER The pH is determined by the weak base F^-:

$$F^- + H_2O \rightleftharpoons HF + OH^-$$

$$\frac{[HF][OH^-]}{[F^-]} = K_b = \frac{K_w}{K_a}$$

$$= \frac{1.00 \times 10^{-14}}{3.53 \times 10^{-4}} = 2.83 \times 10^{-11}$$

Let $y = [OH^-] = [HF]$. Then

$$\frac{y^2}{0.30 - y} = 2.83 \times 10^{-11}$$

On the assumption that y is much less than 0.30,

$$y^2 = 0.30 \times 2.83 \times 10^{-11} = 8.5 \times 10^{-12}$$

$$y = 2.92 \times 10^{-6} \text{ mol/L} = [OH^-]$$

$$[H^+] = \frac{K_w}{[OH^-]} = \frac{1.00 \times 10^{-14}}{2.92 \times 10^{-6}} = 3.42 \times 10^{-9} \text{ mol/L}$$

$$pH = -\log (3.42 \times 10^{-9}) = 8.466$$

■

PROBLEM 15.14 Find the pH of a 0.50 M solution of hydroxylammonium nitrate, $HONH_3{}^+NO_3{}^-$. The hydroxylammonium ion is the conjugate acid of hydroxylamine (Table 15.2). □

PROBLEM 15.15 Find the pH of a 0.35 M solution of sodium cyanide, NaCN. □

15.6 POLYPROTIC ACIDS AND BASES

Recall that a *polyprotic* acid has more than one acidic hydrogen atom (section 14.4). Likewise, a polyprotic base can accept more than one proton. The specific terms *diprotic* and *triprotic* were introduced for the cases of two and three protons. The equilibrium constants for the two reactions

$$H_2A \rightleftharpoons H^+ + HA^- \qquad (13)$$

$$HA^- \rightleftharpoons H^+ + A^{2-} \qquad (14)$$

are represented by K_1 and K_2, respectively, and are called the *first* and *second ionization constants* of H_2A:

$$\frac{[H^+][HA^-]}{[H_2A]} = K_1 \quad \text{and} \quad \frac{[H^+][A^{2-}]}{[HA^-]} = K_2$$

For a triprotic acid, there is also a third constant, K_3. The ionization constants always decrease in the order $K_1 > K_2 > K_3$ (section 14.4). Note that HA^- can act either as an acid, in reaction 14, or as a base, in the reaction

$$HA^- + H_2O \rightleftharpoons H_2A + OH^- \qquad (15)$$

It is amphoteric (section 14.5). Likewise, A^{2-} is a base:

$$A^{2-} + H_2O \rightleftharpoons HA^- + OH^- \qquad (16)$$

The equilibrium constants for reactions 15 and 16 are obtained from K_1 and K_2 by the relationship (equation 12) between the ionization constants of an acid and its conjugate base:

$$HA^- + H_2O \rightleftharpoons H_2A + OH^- \qquad K_{b(HA^-)} = \frac{K_w}{K_{a(H_2A)}} = \frac{K_w}{K_1}$$

$$A^{2-} + H_2O \rightleftharpoons HA^- + OH^- \qquad K_{b(A^{2-})} = \frac{K_w}{K_{a(HA^-)}} = \frac{K_w}{K_2}$$

Be careful to select the correct conjugate acid for a given base and, conversely, the correct conjugate base for a given acid. For example, the conjugate acid of A^{2-} is HA^-, not H_2A. Likewise, for a diprotic base B, there are two ionization constants:

$$B + H_2O \rightleftharpoons BH^+ + OH^- \qquad (K_1)$$

$$BH^+ + H_2O \rightleftharpoons BH_2{}^{2+} + OH^- \qquad (K_2)$$

and, correspondingly, one can calculate from these the acidic ionization constants of BH^+ and $BH_2{}^{2+}$. Note that A^{2-} is a diprotic base and $BH_2{}^{2+}$ is a diprotic acid.

EXAMPLE 15.19 Calculate the basic ionization constant of hydrogen oxalate ion, $HC_2O_4{}^-$.

ANSWER Before using equation 12, you must decide what the conjugate acid of $HC_2O_4{}^-$ is. Add H^+ and get $H_2C_2O_4$, oxalic acid. For this acid, $K_a = K_1 = 5.90 \times 10^{-2}$ (Table 15.2).

$$K_{b(HC_2O_4^-)} = \frac{K_w}{K_{a(H_2C_2O_4)}} = \frac{1.00 \times 10^{-14}}{5.90 \times 10^{-2}} = 1.69 \times 10^{-13}$$

A common error is to do this calculation with K_a for $HC_2O_4^-$ rather than $H_2C_2O_4$. That would give you the answer to a different problem—K_b for $C_2O_4^{2-}$. ■

PROBLEM 15.16 Calculate the acidic ionization constant of hydrazinium ion, $H_2NNH_3^+$. (See "Diprotic Bases" in Table 15.2.) □

This approximation is good only for *weak* polyprotic acids.

Calculations involving the ionization of polyprotic acids and bases are not especially difficult, provided that we make a simplifying assumption. Since the second ionization constant is usually much smaller than the first, we can assume that *the first ionization accounts for most of the* H^+. Very little additional H^+ is produced by the second ionization. Calculations on polyprotic acids can then be performed in steps. First, calculate $[H^+]$ and $[HA^-]$ as if the acid were monoprotic. Then use the results to find $[A^{2-}]$. The same reasoning applies to the ionization of polyprotic bases.

EXAMPLE 15.20 Find $[H^+]$, $[HC_3H_2O_4^-]$, and $[C_3H_2O_4^{2-}]$ in a 0.700 M solution of malonic acid, $H_2C_3H_2O_4$.

Many life processes use compounds related to malonic acid.

ANSWER Let the acid ($H_2C_3H_2O_4$) and its ions, $HC_3H_2O_4^-$ and $C_3H_2O_4^{2-}$, be represented by H_2A, HA^-, and A^{2-}, respectively. $[H^+]$ and $[HA^-]$ can be calculated by considering only the first ionization:

$$\underset{0.700-y}{H_2A} \rightleftharpoons \underset{y}{H^+} + \underset{y \text{ mol/L at equilibrium}}{HA^-}$$

The equilibrium condition is

$$\frac{[H^+][HA^-]}{[H_2A]} = K_1 = 1.49 \times 10^{-3} \qquad \text{(Table 15.2)}$$

or

$$\frac{y^2}{0.700 - y} = 1.49 \times 10^{-3}$$

If $y \ll 0.700$,

$$\frac{y^2}{0.700} = 1.49 \times 10^{-3}$$

$y/0.70 = 4.6\%$, barely acceptable. Solving the quadratic equation gives $y = 3.16 \times 10^{-2}$.

$$y = 3.23 \times 10^{-2} \text{ mol/L} = [H^+] = [HA^-]$$

We have found $[H^+]$ and $[HA^-]$ produced by the first ionization. The equilibrium condition for the second ionization must now be solved to calculate $[A^{2-}]$ and the changes in $[H^+]$ and $[HA^-]$. We set up an equilibrium table, using the results of the previous calculation as initial concentrations:

	HA^-	$\rightleftharpoons$	H^+	+ A^{2-}
Initial concentration	3.23×10^{-2}		3.23×10^{-2}	0
Change	$-z$		$+z$	$+z$
Concentration at equilibrium	$3.23 \times 10^{-2} - z$		$3.23 \times 10^{-2} + z$	z mol/L

But HA^- is an even weaker acid than H_2A. We can therefore be confident that very little of it ionizes. Besides, H^+ is already present from the first ionization, and Le Chatelier's principle (section 13.6) tells us that the equilibrium is therefore shifted even more to the left. We can thus assume that z is negligible in comparison to 3.23×10^{-2}:

$$\frac{[H^+][A^{2-}]}{[HA^-]} = K_2 = 2.03 \times 10^{-6} \qquad \text{(Table 15.2)}$$

$$\frac{\cancel{3.23 \times 10^{-2}}[A^{2-}]}{\cancel{3.23 \times 10^{-2}}} = 2.03 \times 10^{-6}$$

$$[A^{2-}] = 2.03 \times 10^{-6} \text{ mol/L}$$

■

The result $[A^{2-}] = K_2$ will be true whenever HA^- is a much weaker acid than H_2A and the solution contains no source of a significant amount of H^+ or OH^- other than H_2A.

We should point out one source of confusion here. Each ionization step contributes something to $[H^+]$—more for the first ionization, less for the second. This observation sometimes leads to the misconception that one concentration of H^+ is produced by the first ionization and another concentration of H^+ is produced by the second ionization. This is *not* true. A given ion or molecule has only one concentration in a given solution. *The solution has only one value of* $[H^+]$, *to be used in all the equations in which* $[H^+]$ *appears.* After all, an ion has no memory—it cannot recall by what reaction it was produced.

PROBLEM 15.17 Find the pH and the concentration of maleate ion, $C_4H_2O_4{}^{2-}$, in a 0.40 M solution of maleic acid, $H_2C_4H_2O_4$. □

EXAMPLE 15.21 Find the pH and the concentration of H_2CO_3 in a 0.25 M solution of Na_2CO_3.

ANSWER The solution contains the diprotic base $CO_3{}^{2-}$:

$$CO_3{}^{2-} + H_2O \rightleftharpoons HCO_3{}^- + OH^-$$

$$\frac{[HCO_3{}^-][OH^-]}{[CO_3{}^{2-}]} = K_{b(CO_3{}^{2-})} = K_1 \text{ for } CO_3{}^{2-} = 1.78 \times 10^{-4} \tag{17}$$

$$HCO_3{}^- + H_2O \rightleftharpoons H_2CO_3 + OH^-$$

$$\frac{[H_2CO_3][OH^-]}{[HCO_3{}^-]} = K_{b(HCO_3{}^-)} = K_2 \text{ for } CO_3{}^{2-} = 2.33 \times 10^{-8} \tag{18}$$

We get these basic ionization constants from Table 15.2. Just as the second ionization constant of a diprotic acid is much smaller than the first, so also for the diprotic base $CO_3{}^{2-}$, $K_2 \ll K_1$. Therefore, $[H_2CO_3]$ will be much smaller than $[HCO_3{}^-]$. As for the diprotic acid in Example 15.20, we can confine our attention for now to the first ionization of the diprotic base $CO_3{}^{2-}$:

$$CO_3{}^{2-} + H_2O \rightleftharpoons HCO_3{}^- + OH^-$$

$0.25 - y$ $\qquad$ y $\qquad$ y mol/L at equilibrium

Equation 17 becomes

$$\frac{y^2}{0.25 - y} = 1.78 \times 10^{-4}$$

If $y \ll 0.25$, then

$$\frac{y^2}{0.25} = 1.78 \times 10^{-4}$$

$y/0.25 = 2.7\%$, less than 5%.

$$y = 6.7 \times 10^{-3} \text{ mol/L} = [OH^-]$$

$$[H^+] = \frac{K_w}{[OH^-]} = \frac{1.00 \times 10^{-14}}{6.7 \times 10^{-3}} = 1.49 \times 10^{-12}$$

$$pH = -\log(1.49 \times 10^{-12}) = 11.83$$

H_2CO_3 is produced by the basic ionization of HCO_3^-. The initial concentrations are those already calculated from the ionization of CO_3^{2-}:

$$HCO_3^- + H_2O \rightleftharpoons H_2CO_3 + OH^-$$

$6.7 \times 10^{-3} - z$ | z | $6.7 \times 10^{-3} + z$ mol/L at equilibrium

From equation 18,

$$\frac{z(6.7 \times 10^{-3} + z)}{6.7 \times 10^{-3} - z} = 2.33 \times 10^{-8}$$

On the assumption that z is much less than 6.7×10^{-3}, this equation gives

$$z = 2.33 \times 10^{-8} \text{ mol/L} = [H_2CO_3]$$

This calculation is analogous to the calculation of $[A^{2-}]$ in a solution of H_2A. ■

PROBLEM 15.18 (a) Calculate the basic ionization constants of S^{2-} and HS^-. (b) What are the pH and $[H_2S]$ in a 0.10 M solution of Na_2S? □

15.7 ACIDITY OF METAL IONS

Positive ions (cations) with three or more positive charges, as well as some with two, are acids (section 14.10). Thus, $Al(H_2O)_6^{3+}$ has 12 protons, but the ionization constants decrease as protons are removed. In water, $Al(H_2O)_6^{3+}$ is a triprotic acid:

$$Al(H_2O)_6^{3+} \rightleftharpoons Al(H_2O)_3(OH)_3(c) + 3H^+$$

Table 15.2 lists the first ionization constants (K_1) for some common cations. These are the equilibrium constants for reactions such as

$$Al(H_2O)_6^{3+} \rightleftharpoons H^+ + Al(H_2O)_5(OH)^{2+}$$

$$Be(H_2O)_4^{2+} \rightleftharpoons H^+ + Be(H_2O)_3(OH)^+$$

They are sufficient for calculating the pH of a solution containing only the acidic cation and anions (like Cl^- and NO_3^-) that are not appreciably basic—the conjugate bases of strong acids. The second, third, and subsequent ionizations can be ignored.

EXAMPLE 15.22 What is the pH of a 0.20 M solution of $Fe(NO_3)_3$? Fe^{3+} exists in solution as $Fe(H_2O)_6^{3+}$.

ANSWER Construct the usual table and write the equilibrium condition:

$$Fe(H_2O)_6^{3+} \rightleftharpoons H^+ + Fe(H_2O)_5(OH)^{2+}$$

$0.20 - y$ | y | y mol/L at equilibrium

$$\frac{[H^+][Fe(H_2O)_5(OH)^{2+}]}{[Fe(H_2O)_6{}^{3+}]} = K_1$$

$$\frac{y^2}{0.20 - y} = 4.0 \times 10^{-3} \qquad \text{(Table 15.2)}$$

The approximation $0.20 - y \approx 0.20$ gives $y = 2.8 \times 10^{-2}$, which is 14% of 0.20. Therefore, the approximation method is inadequate here. A quadratic equation must be solved:

$$y^2 + 4.0 \times 10^{-3}\, y - 8.0 \times 10^{-4} = 0$$

$$y = 2.6 \times 10^{-2} \text{ mol/L} = [H^+]$$

$$\text{pH} = -\log(2.6 \times 10^{-2}) = 1.59$$ ■

It is interesting that $Fe(H_2O)_6{}^{3+}$ is a stronger acid than acetic acid.

PROBLEM 15.19 Find the pH of a 0.050 M solution of $CuSO_4$. □

A use for the acidic property of Al^{3+} and Fe^{3+} was mentioned in section 12.18. When water is treated with $Al_2(SO_4)_3$ or $Fe_2(SO_4)_3$ and lime (CaO), a colloidal precipitate of the hydroxide is produced:

$$CaO + H_2O \longrightarrow Ca^{2+} + 2OH^-$$

$$Al(H_2O)_6{}^{3+} + 3OH^- \longrightarrow Al(H_2O)_3(OH)_3 + 3H_2O$$

$$Fe(H_2O)_6{}^{3+} + 3OH^- \longrightarrow Fe(H_2O)_3(OH)_3 + 3H_2O$$

The hydroxides form sticky particles that gather up other solid impurities, including colloids, from the water as they clump together. The precipitate is then coarse enough to be filtered out.

15.8 THE COMMON ION EFFECT

Le Chatelier's principle (section 13.6) predicts that an equilibrium is shifted to the left when the concentration of a product of the reaction is increased. Thus the weak acid equilibrium

$$HA \rightleftharpoons H^+ + A^-$$

The ionization is said to be "repressed."

is shifted to the left when H^+ (in the form of a strong acid, such as HCl) or A^- (in the form of a completely ionized salt, such as NaA) is added to the solution. This effect is referred to as the **common ion effect** because it occurs when an electrolyte having an ion in common with HA is added. If HA is acetic acid, then the shift can be produced by any strong acid (the common ion is H^+) or by sodium acetate (the common ion is A^-), but not by NaCl (no ion in common with the ions yielded by HA in solution). A shift to the left means that less HA ionizes. The effects are as shown:

ADD	$[H^+]$	$[A^-]$	SHIFT IN EQUILIBRIUM	SOLUTION BECOMES
H^+	increases	decreases	left	more acidic
A^-	decreases	increases	left	less acidic

When a weak acid is present in a solution that also contains a strong acid, the concentration of H^+ is determined almost entirely by the concentration of the strong acid. The same applies to the concentration of OH^- in a solution containing a soluble hydroxide and a weak base.

EXAMPLE 15.23 Find the concentration of $C_2H_3O_2^-$ in a solution 0.10 M with respect to HCl and 0.20 M with respect to $HC_2H_3O_2$.

ANSWER The solution contains 0.10 mol H^+/L from the strong acid HCl:

	$HC_2H_3O_2$ $\rightleftharpoons$	H^+	+ $C_2H_3O_2^-$
Initial concentration	0.20	0.10	0
Change	$-y$	$+y$	$+y$
Concentration at equilibrium	$0.20 - y$	$0.10 + y$	y mol/L

Since HCl is a much stronger acid than $HC_2H_3O_2$, practically all of the H^+ is contributed by the HCl, and y must be very small. Therefore, $[H^+] \approx 0.10$ mol/L.

$$\frac{[H^+][C_2H_3O_2^-]}{[HC_2H_3O_2]} = 1.75 \times 10^{-5} \qquad \text{(Table 15.2)}$$

Also, $0.20 - y \approx 0.20$, but it is already a simple linear equation.

$$\frac{0.10y}{0.20 - y} = 1.75 \times 10^{-5}$$

$$y = 3.5 \times 10^{-5} \text{ mol/L} = [C_2H_3O_2^-]$$

$$[H^+] = 0.10 + y \approx 0.10 \text{ mol/L, as assumed 8 on page 563,}$$ ■

Note that the acetate concentration is much decreased by the presence of the strong acid. If the HCl were absent, $[C_2H_3O_2^-]$ would be 1.9×10^{-3} instead of 3.5×10^{-5} mol/L.

The familiar relationship illustrated by equation 8 on page 563,

$$\frac{y^2}{0.20 - y} = K_a \qquad (8)$$

does not apply in the case of Example 15.23. This equation assumes that H^+ and $C_2H_3O_2^-$ are produced only by the ionization of $HC_2H_3O_2$, and thus that $[H^+] = [C_2H_3O_2^-] = y$. In the example, however, they are *not* equal because there is another source of H^+. Indeed, the H^+ contributed by HCl is much more abundant than the H^+ contributed by $HC_2H_3O_2$. Thus, one factor of y in equation 8, representing $[A^-]$, remains, while the other y, representing $[H^+]$, is replaced by 0.10.

PROBLEM 15.20 Calculate the concentration of propionate ion, $C_3H_5O_2^-$, in a solution 0.25 M with respect to propionic acid and 0.50 M with respect to HNO_3. □

EXAMPLE 15.24 Find the concentration of S^{2-} in a solution 0.10 M with respect to H_2S and 0.30 M with respect to HCl.

ANSWER For the reactions

$$H_2S \rightleftharpoons H^+ + HS^-$$

$$HS^- \rightleftharpoons H^+ + S^{2-}$$

we have the equilibrium conditions

$$\frac{[H^+][HS^-]}{[H_2S]} = K_1 = 6.3 \times 10^{-8}$$

$$\frac{[H^+][S^{2-}]}{[HS^-]} = K_2 = 10^{-14}$$

This calculation is sometimes misinterpreted as implying that $[H^+] = 2[S^{2-}]$—not true.

Since $[H^+]$ is known and $[HS^-]$ does not interest us in this problem, we can eliminate $[HS^-]$ by multiplying the two equilibrium conditions together:

$$\frac{[H^+]^2[S^{2-}]}{[H_2S]} = K_1K_2$$

$[H^+] = 0.30$ mol/L, because the H^+ contributed by the ionization of H_2S is insignificant compared to the H^+ from HCl.

$$[S^{2-}] = \frac{K_1K_2[H_2S]}{[H^+]^2}$$

$$= \frac{6.3 \times 10^{-8} \times 10^{-14} \times 0.10}{(0.30)^2} \qquad \text{(Table 15.2)}$$

$$= 7 \times 10^{-22} \text{ mol/L}$$

■

Note that $[S^{2-}]$ *is not equal to* K_2. That result (section 15.6) applied only to the case in which the diprotic acid was the only source of H^+ or OH^-. Here, the solution also contains HCl. If the HCl were absent, then $[S^{2-}]$ would indeed be $K_2 = 10^{-14}$ mol/L. The common ion (H^+) has decreased the concentration of S^{2-} by a factor of 7×10^{-8}.

PROBLEM 15.21 For the solution in Example 15.24, with $[H_2S] = 0.10$ mol/L and $[H^+] = 0.30$ mol/L, what is $[HS^-]$? How many times greater than $[S^{2-}]$ is $[HS^-]$? □

EXAMPLE 15.25 A solution contains 0.25 mole of Na_2CO_3 and 0.40 mole of NaOH per liter. What is the concentration of H_2CO_3 in the solution?

ANSWER This example is like the preceding one, except that the solution contains a diprotic base (CO_3^{2-}) instead of an acid (H_2S). The reactions we must consider are the first and second basic ionizations of CO_3^{2-}:

$$CO_3^{2-} + H_2O \rightleftharpoons HCO_3^- + OH^- \qquad K_1 = 1.78 \times 10^{-4} \qquad \text{(Table 15.2)}$$

$$HCO_3^- + H_2O \rightleftharpoons H_2CO_3 + OH^- \qquad K_2 = 2.33 \times 10^{-8}$$

As in the preceding example, the two equilibrium conditions can be multiplied to give

$$\frac{[OH^-]^2[H_2CO_3]}{[CO_3^{2-}]} = K_1K_2 = 4.15 \times 10^{-12}$$

Then

$$[H_2CO_3] = \frac{4.15 \times 10^{-12} \times [CO_3^{2-}]}{[OH^-]^2}$$

$$= \frac{4.15 \times 10^{-12} \times 0.25}{(0.40)^2}$$

$$= 6.5 \times 10^{-12} \text{ mol/L}$$

This answer may be compared with the one in Example 15.21, where the NaOH was absent and $[H_2CO_3] = K_2 = 2.33 \times 10^{-8}$ mol/L. ■

PROBLEM 15.22 (a) What are the basic ionization constants of SO_3^{2-} and HSO_3^-? (b) A solution contains 0.30 mole of Na_2SO_3 and 0.30 mole of NaOH per liter. What is the concentration of H_2SO_3? □

In Example 15.25 and Problem 15.22, the common ion is OH^-, which is produced by the ionization of a base such as CO_3^{2-} or SO_3^{2-}. The next example illustrates a case in which the added common ion is not H^+ or OH^- but the conjugate base of the acid.

EXAMPLE 15.26 Calculate the pH of a solution 0.20 M with respect to formic acid, $HCHO_2$, and 0.10 M with respect to sodium formate, $NaCHO_2$.

ANSWER

$$HCHO_2 \rightleftharpoons H^+ + CHO_2^-$$

$0.20 - y$ $\quad y \quad$ $0.10 + y$ mol/L at equilibrium

The acid is weak, and its ionization is further repressed by the presence of added CHO_2^- ion. We expect, therefore, that y will be very small. Assume that $[HCHO_2] = 0.20$ and $[CHO_2^-] = 0.10$ mol/L. The equilibrium condition is then

$$\frac{[H^+][CHO_2^-]}{[HCHO_2]} = K_a$$

$$\frac{y(0.10)}{0.20} = 1.76 \times 10^{-4} \qquad \text{(Table 15.2)}$$

$y/0.10 = 0.35\%$;
$y/0.20 = 0.18\%$

$$y = 3.52 \times 10^{-4} \text{ mol/L} = [H^+]$$

$$\text{pH} = \log (3.52 \times 10^{-4}) = 3.45$$ ■

Comparing this result with that of Example 15.10 (section 15.3) shows the effect of the common ion (formate) on the pH. The presence of sodium formate has increased the pH from 2.24 to 3.45. Although the solution is still acidic, it is less so than the solution of formic acid alone. Another way of looking at it is that a base, CHO_2^-, has been added to an acidic solution, which naturally makes the solution less acidic.

PROBLEM 15.23 A solution is prepared by dissolving 0.25 mol calcium acetate, $Ca(C_2H_3O_2)_2$, and 0.15 mol acetic acid in water to make 1.00 L. What is the pH of the solution? □

15.9 BUFFER SOLUTIONS

The pH of human blood is normally 7.40. In a state of good health, it deviates very little from this value. If the pH were to fall below 7.0 or rise above 7.8, death would be likely, for the rates of enzyme-catalyzed reactions are very sensitive to pH. How is the pH kept so constant, for example, when a person ingests baking soda (base) after eating sauerkraut (acid)?

A number of reactions help to keep the pH constant. The first line of defense, however, is for the blood to contain an acid that will react with any added base and a base that will react with any added acid. It would not do to have a strong acid (like HCl) and a strong base (like NaOH); they would simply react with each other, leaving the one that was in excess—$H^+ + OH^- \longrightarrow H_2O$. A strong base and a weak acid, or a strong acid and a weak base, would also neutralize each other—$HC_2H_3O_2 + OH^- \longrightarrow C_2H_3O_2^- + H_2O$.

Most practical buffers have pK_a's between 2 and 12.

So what is needed is a weak acid and a weak base. Specifically, the solution must contain a moderately weak acid and its conjugate base; for example, H_2CO_3 and HCO_3^-, or $H_2PO_4^-$ and HPO_4^{2-}, or NH_4^+ and NH_3. The first two of these pairs are present in blood. A solution made in this way is called a **buffer solution.** When a solution is a buffer, the addition of an acid or base, or water, causes only a relatively small change in pH.

The pH of pure water, or of a solution that is not a buffer, is very sensitive to additions of small quantities of acid or base. Suppose that we want a solution with pH 5.00. We can prepare such a solution by diluting a solution of a strong acid, such as HCl, with water until the concentration is down to 1.0×10^{-5} M. Since the acid is strong, this is also the H^+ concentration.

$$[H^+] = 1.00 \times 10^{-5} \text{ mol/L}$$

$$\text{pH} = -\log(1.00 \times 10^{-5}) = 5.00$$

Now a concentrated HCl solution is spilled into this dilute solution until 0.10 mole of HCl has been added per liter. The new $[H^+]$ is

$$[H^+] = 0.10 + 1.00 \times 10^{-5}$$

$$= 0.10 \text{ mol/L} \qquad \text{(The original 0.00001 is negligible.)}$$

$$\text{pH} = -\log(0.10) = 1.00$$

The pH is changed drastically.

Conversely, let us add 0.10 mole of NaOH per liter. Of the OH^- that is added, 1.00×10^{-5} mol/L is neutralized by the H^+ in the solution; in 1 liter,

$$10^{-5} \text{ mol } H^+ + 10^{-5} \text{ mol } OH^- \longrightarrow 10^{-5} \text{ mol } H_2O$$

The excess OH^- is $[OH^-] = 0.10 - 1.00 \times 10^{-5} = 0.10$ mol/L. Then

$$[H^+] = \frac{K_w}{[OH^-]} = 1.0 \times 10^{-13} \text{ mol/L}$$

$$\text{pH} = 13.00$$

$[H^+]$ has been changed by a factor of 100,000,000.

Again there has been a large change in pH, from 5 to 13.

However, when a strong acid (H^+) is added to a *buffer* solution, it reacts with the base (A^-) present in the solution to form the conjugate acid (HA):

$$H^+ + A^- \longrightarrow HA$$

Very little of the added H^+ remains. Any added base (OH^-) reacts almost completely with the acid (HA) in the solution, to form the conjugate base (A^-):

$$OH^- + HA \longrightarrow A^- + H_2O$$

The solution contains a *reservoir* of base that consumes the added acid (converting it to the weak acid HA), and a *reservoir* of acid to consume the added base (converting it to the weak base A^-). In each case, the acid/base ratio is changed, *but with only a moderate effect on the pH of the solution.*

THE HENDERSON–HASSELBALCH EQUATION

The pH of a buffer solution depends on the ratio of the concentrations of the conjugate acid and base. This dependence is especially clear if the equilibrium condition

$$\frac{[H^+][A^-]}{[HA]} = K_a$$

is written in logarithmic form. Take logarithms of both sides:

$$\log\left(\frac{[H^+][A^-]}{[HA]}\right) = \log K_a$$

From the properties of logarithms,

$$\log [H^+] + \log\frac{[A^-]}{[HA]} = \log K_a$$

On making a sign change and using the definitions of pH and pK, we have

$$-\log [H^+] - \log\frac{[A^-]}{[HA]} = -\log K_a$$

$$\text{pH} - \log\frac{[A^-]}{[HA]} = \text{p}K_a$$

$$\text{pH} = \text{p}K_a + \log\frac{[A^-]}{[HA]}$$

or

$$\mathbf{pH = p}\boldsymbol{K}_\mathbf{a} \mathbf{+ \log\frac{[base]}{[acid]}} \qquad (19)$$

Lawrence Henderson (United States, work published 1908) and Karl Hasselbalch (Denmark, 1916). Henderson discovered buffer action in his research on regulation in living systems.

This form of the equilibrium condition is known as the **Henderson–Hasselbalch equation**. It is convenient for calculations involving buffer solutions and is much used by biologists as well as chemists.

Equation 19 is essentially the same as equation 11 for indicators (section 15.4). The difference is that an indicator is used in small concentration, so that the other components of the solution determine the pH and thus the ratio [HIn]/[In$^-$]. In a buffer solution, the acid and base are used in large concentrations and the pH depends on their ratio.

EXAMPLE 15.27 What is the pH of a buffer solution that is 0.20 M with respect to ammonia, NH_3, and 0.20 M with respect to ammonium chloride, NH_4Cl?

ANSWER The equation for the basic ionization of NH_3 is

The calculation may also use the acidic ionization of NH_4^+. The answer must be the same.

	NH_3	$+ H_2O \rightleftharpoons$	NH_4^+	$+ OH^-$
Initial concentration	0.20		0.20	0
Change	$-y$		$+y$	$+y$
Concentration at equilibrium	$0.20 - y \approx 0.20$		$0.20 + y \approx 0.20$	y mol/L

Assume, as usual, that $[NH_3]$ and $[NH_4^+]$ are not changed by the ionization. Then

$$\frac{[NH_4^+][OH^-]}{[NH_3]} = K_b$$

$$\frac{0.20y}{0.20} = 1.77 \times 10^{-5}$$

One insignificant figure is retained in $[OH^-]$ and $[H^+]$.

$$y = 1.77 \times 10^{-5} \text{ mol/L} = [OH^-]$$

$$[H^+] = \frac{K_w}{[OH^-]} = \frac{1.00 \times 10^{-14}}{1.77 \times 10^{-5}} = 5.65 \times 10^{-10} \text{ mol/L}$$

$$pH = 9.25$$

$y/0.20 = 9 \times 10^{-3}\%$; y is negligible, as assumed.

The problem can also be solved by using the Henderson–Hasselbalch equation:

$$pH = pK_a + \log \frac{[NH_3]}{[NH_4^+]}$$

Here, pK_a refers to the NH_4^+ ion (the acid in this conjugate pair). From Table 15.2,

$$pK_a = -\log K_a = -\log (5.64 \times 10^{-10}) = 9.249$$

Then

$$pH = 9.249 + \log \frac{0.20}{0.20} = 9.25$$

■

A buffer solution of a given pH can be prepared by dissolving an appropriate acid and its conjugate base in the proper ratio of concentrations. This ratio is easily calculated when the ionization constant is known.

EXAMPLE 15.28 A buffer solution of pH 5.00 is to be prepared from acetic acid and sodium acetate. (a) What should be the ratio $[C_2H_3O_2^-]/[HC_2H_3O_2]$ in the solution? (b) A 0.50 M solution of acetic acid is available. How many moles of $NaC_2H_3O_2$ should be added to 1.00 liter of it to prepare this buffer solution?

ANSWER

(a) For the reaction

$$HC_2H_3O_2 \rightleftharpoons H^+ + C_2H_3O_2^-$$

the equilibrium condition is

$$\frac{[H^+][C_2H_3O_2^-]}{[HC_2H_3O_2]} = K_a$$

or in Henderson–Hasselbalch form,

$$pH = pK_a + \log \frac{[C_2H_3O_2^-]}{[HC_2H_3O_2]}$$

The desired ratio is obtained by rearranging this equation:

$$\log \frac{[C_2H_3O_2^-]}{[HC_2H_3O_2]} = pH - pK_a$$

From Table 15.2, $K_a = 1.75 \times 10^{-5}$ and $pK_a = 4.757$. Then

$$\log \frac{[C_2H_3O_2^-]}{[HC_2H_3O_2]} = 5.00 - 4.757 = 0.243$$

$$\frac{[C_2H_3O_2^-]}{[HC_2H_3O_2]} = 1.75$$

(b) The given solution has $[HC_2H_3O_2] = 0.50$ mol/L. Then

$$[C_2H_3O_2^-] = 1.75 \times [HC_2H_3O_2] = 1.75 \times 0.50 \text{ mol/L} = 0.88 \text{ mol/L}$$

It is necessary to add 0.88 mole of $NaC_2H_3O_2$ per liter. ■

PROBLEM 15.24 A solution is 0.50 M with respect to bromoacetic acid, $HC_2H_2BrO_2$, and 0.75 M with respect to potassium bromoacetate, $K^+C_2H_2BrO_2^-$. (a) Is the solution a buffer solution? (b) What is its pH? □

PROBLEM 15.25 A buffer solution of pH 8.50 is to be prepared from NH_3 and NH_4Cl. What should be the ratio $[NH_4^+]/[NH_3]$? □

To justify the claim that the solutions prepared in Examples 15.27 and 15.28, and Problems 15.24 and 15.25, are buffer solutions, it is necessary to calculate their pH's after some amount of strong acid or base has been added to them. For example, add 0.10 mole H^+ per liter to the solution of pH 5.00 prepared in Example 15.28. In this solution, $[HC_2H_3O_2] = 0.50$ mol/L and $[C_2H_3O_2^-] = 0.88$ mol/L. As is often done in chemistry, we think about a complicated process by resolving it into simple steps that add up to give the same result (the philosophy of Hess's law). Here, we imagine that the reaction of the added H^+ with the buffer occurs in three steps:

1. 0.10 mole of H^+/L is added to the solution, with no reaction occurring.
2. H^+ and $C_2H_3O_2^-$ react, and the reaction goes to completion (nearly true because the equilibrium constant for $H^+ + C_2H_3O_2^- \rightleftharpoons HC_2H_3O_2$ is a large number, $1/K_a$). These two steps can be summarized:

	$C_2H_3O_2^-$ +	H^+	$\longrightarrow$ $HC_2H_3O_2$
Initial concentration	0.88	small	0.50
Step 1	0.88	0.10	0.50
Step 2	0.88 − 0.10 = 0.78	0.10 − 0.10 ≈ 0	0.50 + 0.10 = 0.60 mol/L

Note the simplicity of this calculation—just increase the concentration of the weak acid by the moles of strong acid added per liter, and decrease the concentration of the weak base by the same amount. However, the solution has not yet come to equilibrium. That happens in the third step.

3. Use the concentrations after step 2 as the *initial* concentrations for the ionization of $HC_2H_3O_2$:

	$HC_2H_3O_2$ $\rightleftharpoons$	H^+ +	$C_2H_3O_2^-$
Initial concentration	0.60	0	0.78
Change	$-z$	$+z$	$+z$
Concentration at equilibrium	$0.60 - z$	z	$0.78 + z$ mol/L

The net result of the three steps is that most, but not quite all, of the added H^+ has been consumed by reacting with $C_2H_3O_2^-$. The solution thus becomes slightly more acidic—but only slightly. The remaining calculation can be simplified by a familiar approximation; since z is very small, $0.60 - z \approx 0.60$ and $0.78 + z \approx 0.78$. The equilibrium condition for the ionization is

$$\frac{[H^+][C_2H_3O_2^-]}{[HC_2H_3O_2]} = 1.75 \times 10^{-5}$$

or

$$pH = -\log(1.75 \times 10^{-5}) + \log\frac{[C_2H_3O_2^-]}{[HC_2H_3O_2]}$$

$$= 4.757 + \log\frac{0.78}{0.60}$$

$$= 4.87$$

The decrease in pH (from 5.00 to 4.87) is only 0.13. Compare this to the decrease of 4.0 for the unbuffered 10^{-5} M HCl solution we discussed early in this section.

EXAMPLE 15.29 The buffer solution of pH 5.00 in Example 15.28 has $[HC_2H_3O_2] = 0.50$ mol/L and $[C_2H_3O_2^-] = 0.88$ mol/L. To 1.00 L of this solution, 0.10 mol NaOH is added. What is the new pH of the solution?

ANSWER Assume that OH^- reacts completely with $HC_2H_3O_2$. $[HC_2H_3O_2]$ goes down from 0.50 to 0.40 mol/L; $[C_2H_3O_2^-]$ goes up from 0.88 to 0.98 mol/L:

	$HC_2H_3O_2$	+	OH^-	$\longrightarrow$	$C_2H_3O_2^-$	+	H_2O
Initial concentration	0.50		small		0.88 mol/L		
OH^- is added.	0.50		0.10		0.88		
Reaction occurs.	$0.50 - 0.10 = 0.40$		$0.10 - 0.10 \approx 0$		$0.88 + 0.10 = 0.98$		

This solution cannot be distinguished from a solution prepared by dissolving 0.40 mole $HC_2H_3O_2$ and 0.98 mole $NaC_2H_3O_2$ in 1 liter of solution. The Henderson–Hasselbalch equation is now

$$pH = 4.757 + \log\frac{0.98}{0.40}$$

$$= 5.15$$

Again, the change in pH is small (0.15). ■

PROBLEM 15.26 The buffer solution in Example 15.27 has $[NH_3] = 0.20$ mol/L and $[NH_4^+] = 0.20$ mol/L. (a) What is its pH after 0.050 mole of H^+ per liter is added to it? (b) What is the pH after 0.050 mole of OH^- per liter is added to another portion of it? □

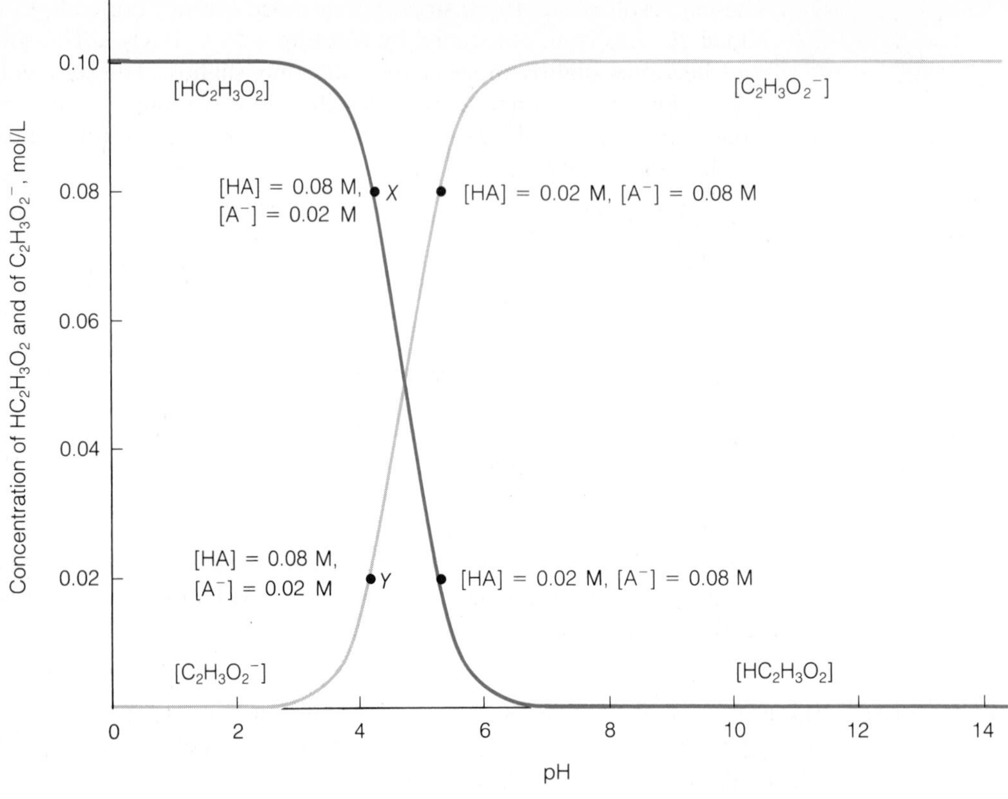

FIGURE 15.2
The concentrations of acetic acid (HA) and acetate ion (A^-) plotted against the pH of the solution. The total concentration of HA and A^- is $[HA] + [A^-] = 0.100$ mol/L. Two points at the same pH (for example, *X* and *Y*) represent the same solution; one curve (point *X*) shows [HA], the other curve (point *Y*) shows $[A^-]$.

A buffer solution retains its buffering action as long as the quantity of acid or base added is much less than the quantities of the weak acid and its conjugate base in the buffer solution. The more acid present in a buffer solution, the more strong base can be added to it without changing its pH drastically. If the number of moles of H^+ added exceeds the number of moles of A^- in the solution, all the A^- is converted to HA. The more base present, the more acid can be added without changing the pH drastically. If the number of moles of OH^- added exceeds the number of moles of HA in the solution, all of the HA is converted to A^-. In either case, the buffer is destroyed and there is a large change in pH.

Figure 15.2 shows how the concentrations of a typical weak acid—acetic acid—and its conjugate base depend on the pH of the solution—or, looking at the figure sideways, how the pH depends on these concentrations. The solution contains a total of 0.100 mole of acetate ($HC_2H_3O_2 + C_2H_3O_2^-$) per liter. When the pH is low (an acidic solution), most of the acetate is in the form of acetic acid. When the pH is high (a basic solution), most of the acetate is in the form of acetate ion. There is a range of pH about 2 units wide, centered at $pK_a = 4.76$, in which both $HC_2H_3O_2$ and $C_2H_3O_2^-$ are present in substantial concentration. This is the range of pH where an acetic acid–acetate ion buffer is useful.

The best buffer (with most nearly constant pH) is obtained when the concentrations of the acid and base are initially equal. In this case, $[H^+] = K_a$, or $pH = pK_a$, as can be seen from the Henderson–Hasselbalch equation. Similarly, with a base B and its conjugate acid BH^+, the equation

$$\frac{[BH^+][OH^-]}{[B]} = K_b$$

shows that when $[BH^+] = [B]$, then $[OH^-] = K_b$ or $pOH = pK_b$. Therefore, if you are told to prepare a buffer solution of given pH, you should seek an acid with $pK_a \approx pH$, or a base with $pK_b \approx pOH = pK_w - pH$. Since you cannot expect to find an available acid for which pK_a equals pH exactly, use the one that is closest and adjust the acid/base ratio to make the pH come out right.

$pH + pOH = pK_w$

EXAMPLE 15.30 (a) Choose from Table 15.2 a conjugate acid-base pair suitable for making a buffer solution of pH 7.40 (the pH of blood). (b) Calculate the acid/base ratio required in this buffer.

ANSWER

(a) We wish to have $pK_a \approx 7.4$, or $K_a \approx 4 \times 10^{-8}$. The nearest is K_2 for H_3PO_4 (K_a for $H_2PO_4^-$), corresponding to the pair $H_2PO_4^-$ (acid) + HPO_4^{2-} (base) (actually present in blood). Other possibilities are H_2CO_3 (also present in blood), H_2S (toxic), and HSO_3^- (toxic).

(b) $$H_2PO_4^- \rightleftharpoons H^+ + HPO_4^{2-}$$

$$\frac{[H^+][HPO_4^{2-}]}{[H_2PO_4^-]} = K_a \text{ for } H_2PO_4^- = K_2 \text{ for } H_3PO_4 = 6.22 \times 10^{-8}$$

$$\frac{[\text{acid}]}{[\text{base}]} = \frac{[H_2PO_4^-]}{[HPO_4^{2-}]} = \frac{[H^+]}{K_a}$$

For $pH = 7.40$, $[H^+] = 10^{-7.40} = 4.0 \times 10^{-8}$ mol/L. Then,

$$\frac{[H_2PO_4^-]}{[HPO_4^{2-}]} = \frac{[H^+]}{K_a}$$

$$= \frac{4.0 \times 10^{-8}}{6.22 \times 10^{-8}} = 0.64$$

■

The preceding example shows that the phosphate buffer in blood contains more base than acid and is thus more effective against acid than against base. It is probably not coincidental that many metabolic processes produce acids.

To prepare a buffer solution, it is not necessary to have both the acid and the base on hand. It can be prepared by partially neutralizing the acid with a strong base, or by partially neutralizing the base with a strong acid.

EXAMPLE 15.31 A solution is prepared by dissolving 1.00 mole propionic acid and 0.40 mole sodium hydroxide in water to make 1.00 liter. (a) What reaction occurs between these two solutes? Write the net ionic equation. (b) How many moles of the acid and how many moles of its conjugate base are present after the reaction? (c) Calculate the pH of the solution.

ANSWER

(a) Propionic acid reacts with OH^- ion:

$$HC_3H_5O_2 + OH^- \longrightarrow H_2O + C_3H_5O_2^-$$

(b) Of the 1.00 mole of $HC_3H_5O_2$ initially present, 0.40 mole reacts with OH^- and 0.60 mole remains unreacted:

	$HC_3H_5O_2$	+ OH^-	$\longrightarrow$ H_2O +	$C_3H_5O_2^-$
Initial concentration	1.00	≈ 0		≈ 0
OH^- is added.	1.00	0.40		≈ 0
Reaction occurs.	$1.00 - 0.40 = 0.60$	≈ 0		0.40 mol/L

Then (since the volume is 1.00 L), $[HC_3H_5O_2] = 0.60$ mol/L and $[C_3H_5O_2^-] = 0.40$ mol/L.

(c) Using the Henderson–Hasselbalch equation,

$$pH = pK_a + \log \frac{[C_3H_5O_2^-]}{[HC_3H_5O_2]}$$

$$= -\log(1.34 \times 10^{-5}) + \log \frac{0.40}{0.60}$$

$$= 4.873 - 0.176 = 4.70$$

■

PROBLEM 15.27 Calculate the pH of a buffer solution prepared by dissolving 0.30 mole of NH_3 and 0.10 mole of HCl in water to make 1.00 liter of solution.

□

15.10 ACID–BASE TITRATIONS

Section 3.11 described the process of *titration*, a common method of analyzing a sample. A *standard solution*—a solution with accurately known concentration—is added to the sample until a reaction is complete, as signaled most often by a color change. The most common kind of titration is an **acid-base titration**, a titration in which the standard solution is an acid and the sample being analyzed contains a base, or conversely. A typical example is the analysis of a sample of sulfuric acid, H_2SO_4, by titration with a standard solution of sodium hydroxide, NaOH. The reaction is

$$H_2SO_4 + 2NaOH \longrightarrow Na_2SO_4 + 2H_2O$$

or, in ionic form with spectator ions omitted,

$$2H^+ + 2OH^- \longrightarrow 2H_2O$$

It is also possible to titrate a weak acid with NaOH

$$HC_2H_3O_2 + OH^- \longrightarrow C_2H_3O_2^- + H_2O$$

or a weak base, like NH_3, with a strong acid

$$NH_3 + H^+ \longrightarrow NH_4^+$$

However, titrating a weak base with a weak acid does not work well, as we will see in the next section.

In a successful titration, the quantity of standard solution that has been added at the end point is *chemically equivalent to the quantity of the unknown.* This means that the *molar ratio* of the known added to the unknown must conform to the molar ratio in the chemical equation. In titrating H_2SO_4 with NaOH

($H_2SO_4 + 2NaOH \longrightarrow Na_2SO_4 + 2H_2O$), 2 moles of NaOH must be added for every mole of H_2SO_4. Adding an excess or an insufficient quantity of the known solution results in an incorrect analysis. In titrations of acids and bases, advantage is taken of the observation that *the pH of the solution being titrated changes rapidly as the equivalence point is approached.* It is because of this rapid change that an indicator changes color when just one drop of the standard solution is added after the equivalence point has been reached.

When an acid and a base have been mixed in equivalent quantities, we say they have *neutralized* each other. This word is misleading, because the resulting solution may actually be neutral, acidic, or basic. At the equivalence point, the solution contains only the product of neutralization of the acid and the base, and practically none of either the acid or the base used in the titration. Thus, if we mix 1 mmol of HCl and 1 mmol of NaOH, the reaction

$$HCl + NaOH \longrightarrow H_2O + NaCl$$

or, in ionic form,

The equilibrium constant for this reaction is $1/K_w = 10^{14}$ at 25°C.

$$H^+ + OH^- \longrightarrow H_2O$$

goes practically to completion. The solution contains only water and NaCl, and is therefore neutral (pH 7 at 25°C). Even when the acid or the base (but not both) is moderately weak, the reaction still goes nearly to completion. If 1 mmol of acetic acid and 1 mmol of NaOH are mixed, the reaction

$$HC_2H_3O_2 + NaOH \longrightarrow H_2O + NaC_2H_3O_2$$

or

K for this reaction is 1.75×10^9. It is the reciprocal of K_b for acetate ion, 5.71×10^{-10}.

$$HC_2H_3O_2 + OH^- \longrightarrow H_2O + C_2H_3O_2^-$$

goes practically to completion. The only product is an aqueous solution of sodium acetate, the same as if sodium acetate had been added to water. The solution is moderately basic, since acetate ion is a weak base:

$$C_2H_3O_2^- + H_2O \rightleftharpoons HC_2H_3O_2 + OH^-$$

The indicator used in an acid-base titration must be chosen with some care. It must be an indicator that changes color during the rapid change of pH at the equivalence point, not one that completes its color change before the equivalence point or that begins to change color after the equivalence point. *The pH of the solution at the equivalence point determines what indicator should be used.* Follow these steps when selecting an acid-base indicator:

1. Determine what salt is present at the equivalence point.
2. Estimate the salt's concentration at the equivalence point.
3. Calculate the pH of the solution at the equivalence point.
4. Select an indicator (from Table 15.3) that changes at about that pH.

EXAMPLE 15.32 2.500 mmol benzoic acid, $HC_7H_5O_2$, is titrated with 2.500 mmol NaOH. The volume of the solution at the equivalence point is 125 mL. (a) What solute is present in the solution at the equivalence point? (b) What is the concentration of the solute? (c) Calculate the pH at the equivalence point and select an indicator suitable for this titration.

ANSWER

(a) The reaction between benzoic acid and sodium hydroxide produces sodium benzoate:

$$HC_7H_5O_2 + NaOH \longrightarrow NaC_7H_5O_2 + H_2O$$

or, in ionic form,

$$HC_7H_5O_2 + OH^- \longrightarrow C_7H_5O_2^- + H_2O$$

(b) The equations in part a show that when 2.500 mmol of $HC_7H_5O_2$ reacts (almost completely) with OH^-, the product is 2.500 mmol $C_7H_5O_2^-$. Since the final volume of solution is 125 mL, the concentration is

Reminder: 1 mmol/mL = 1 mol/L.

$$[C_7H_5O_2^-] = \frac{2.500 \text{ mmol}}{125 \text{ mL}} = 2.00 \times 10^{-2} \text{ mol/L}$$

(c) We expect the solution to be somewhat basic because the benzoate ion is the conjugate base of a weak acid. The basic ionization constant of $C_7H_5O_2^-$ is calculated from K_a for benzoic acid (Table 15.2) as in section 15.5:

$$K_b = \frac{K_w}{K_a} = \frac{1.00 \times 10^{-14}}{6.46 \times 10^{-5}} = 1.55 \times 10^{-10}$$

This constant corresponds to the reaction

$$C_7H_5O_2^- + H_2O \rightleftharpoons HC_7H_5O_2 + OH^-:$$

$$\frac{[HC_7H_5O_2][OH^-]}{[C_7H_5O_2^-]} = K_b = 1.55 \times 10^{-10}$$

Since $HC_7H_5O_2$ and OH^- are produced in equal concentrations, let

$$[HC_7H_5O_2] = [OH^-] = y$$

Then

$$\frac{y^2}{0.0200 - y} = \frac{y^2}{0.0200} = 1.55 \times 10^{-10} \qquad (y \text{ is small compared to } 0.02.)$$

$$y = [OH^-] = 1.76 \times 10^{-6} \text{ mol/L}$$

$$[H^+] = \frac{K_w}{[OH^-]} = \frac{1.00 \times 10^{-14}}{1.76 \times 10^{-6}} = 5.68 \times 10^{-9} \text{ mol/L}$$

$$\text{pH} = 8.25$$

The indicator should change color at about pH 8.3. Table 15.3 shows that cresol red is suitable. Phenolphthalein could also be used but is less desirable because pH 8.3 is on the edge of its range, where its red color has just begun to appear. ■

EXAMPLE 15.33 In a titration of NH_3 with HCl, 3.550 mmol NH_3 is titrated with 3.550 mmol HCl. The volume of the solution at the equivalence point is 145 mL. (a) What solute is present at the equivalence point? (b) What is the pH of the solution? (c) Select an indicator for use in this titration.

ANSWER

(a) The reaction is $NH_3 + H^+ \longrightarrow NH_4^+$. Cl^- is a spectator ion (it does not react) in this instance. The solute at the equivalence point is NH_4Cl.

(b) For the reaction

$$NH_4^+ \rightleftharpoons H^+ + NH_3$$

the equilibrium condition is

$$\frac{[H^+][NH_3]}{[NH_4^+]} = 5.64 \times 10^{-10} \qquad \text{(Table 15.2)}$$

$$[NH_4^+] = \frac{3.550 \text{ mmol}}{145 \text{ mL}} = 2.45 \times 10^{-2} \text{ M}$$

$$[H^+] = [NH_3] = y$$

$$\frac{y^2}{0.0245 - y} = \frac{y^2}{0.0245} = 5.64 \times 10^{-10} \qquad (y \text{ is small compared to } 0.0245.)$$

$$y = [H^+] = 3.72 \times 10^{-6} \text{ M}$$

$$pH = -\log [H^+] = 5.43$$

(c) We need an indicator that changes at about pH 5.4. According to Table 15.3, methyl red would be a good choice. Bromcresol green or bromcresol purple could be used in a pinch, but it is better to have the desired pH near the middle of the indicator range. ■

PROBLEM 15.28 Dimethylamine is titrated with 8.00 mmol HCl:

$$(CH_3)_2NH + H^+ \longrightarrow (CH_3)_2NH_2^+$$

At the equivalence point, the volume of the solution in the flask is 100 mL. See Table 15.2 for the basic ionization constant of dimethylamine. Calculate (a) the acidic ionization constant of $(CH_3)_2NH_2^+$ and (b) the pH of the solution. (c) Select a suitable indicator. □

15.11 TITRATION CURVES

As an acid-base titration proceeds, the pH of the solution in the flask changes. The pH at any stage of the titration can be calculated if a few pieces of information are available: the identities and ionization constants of the acid and the base, the initial amount (millimoles) of the substance being titrated, the initial volume of the solution containing it, the molarity of the solution in the burette, and the volume of solution added up to the point of calculation.

STRONG ACID AND STRONG BASE

When a strong acid is titrated with a strong base, or vice versa, the net reaction is

$$H^+ + OH^- \longrightarrow H_2O$$

The pH can be calculated from the number of moles of acid remaining unreacted and the volume of the solution.

A flask contains 10.000 mmol HCl dissolved in 100 mL of solution. It is titrated at 25°C with a 0.2500 M solution of NaOH. The initial solution has

$$[H^+] = \frac{10.000 \text{ mmol}}{100 \text{ mL}} = 0.100 \text{ mol/L}$$

Its pH is $-\log(0.100) = 1.000$. When 30.00 mL NaOH solution has been added, then

$$30.00\ \cancel{\text{mL}} \times \frac{0.2500\ \text{mmol}}{1\ \cancel{\text{mL}}} = 7.500\ \text{mmol NaOH}$$

has been added. It reacts with 7.500 mmol HCl. The unreacted HCl is

$$10.000\ \text{mmol (initial)} - 7.500\ \text{mmol (reacted)} = 2.500\ \text{mmol HCl (unreacted)}$$

The volume of the solution is

$$100\ \text{mL (initial)} + 30.00\ \text{mL (added)} = 130\ \text{mL}$$

The concentration of H^+ is

$$[H^+] = \frac{2.500\ \text{mmol HCl}}{130\ \text{mL}} = 0.0192\ \text{mol HCl/L}$$

and the pH is 1.716. The pH has increased moderately.

Now suppose that 50.00 mL of NaOH solution is added. In this volume of solution, there is

$$50.00\ \cancel{\text{mL}} \times \frac{0.2500\ \text{mmol}}{1\ \cancel{\text{mL}}} = 12.50\ \text{mmol NaOH}$$

Since only 10.000 mmol HCl was present initially, there is

$$\begin{aligned} &12.50\ \text{mmol } OH^-\ \text{(added)} - 10.000\ \text{mmol } OH^-\ \text{(reacted)} \\ &\qquad = 2.50\ \text{mmol } OH^-\ \text{(unreacted)} \end{aligned}$$

present in 100 mL + 50 mL = 150 mL of solution:

$$[OH^-] = \frac{2.50\ \text{mmol}}{150\ \text{mL}} = 0.167\ \text{mol/L}$$

$$[H^+] = \frac{K_w}{[OH^-]} = \frac{1.00 \times 10^{-14}}{0.0167} = 6.00 \times 10^{-13}\ \text{mol/L}$$

$$\text{pH} = 12.22$$

The solution has become strongly basic, since the OH^- added is more than enough to react with all of the HCl. Similar calculations can be done for any volume of NaOH added to the HCl solution.

EXAMPLE 15.34 For the titration just described, calculate the pH when the volume of NaOH solution added is (a) 39.99 mL and (b) 40.00 mL.

ANSWER

(a) The volume of the solution is 100 mL + 39.99 mL = 140 mL. The number of mmol of NaOH added is

$$39.99\ \cancel{\text{mL}} \times \frac{0.2500\ \text{mmol}}{1\ \cancel{\text{mL}}} = 9.9975\ \text{mmol}$$

which reacts with 9.9975 mmol HCl. Of the original 10.000 mmol H^+, only $10.000 - 9.9975 = 2.5 \times 10^{-3}$ mmol remains unreacted. (Some insignificant

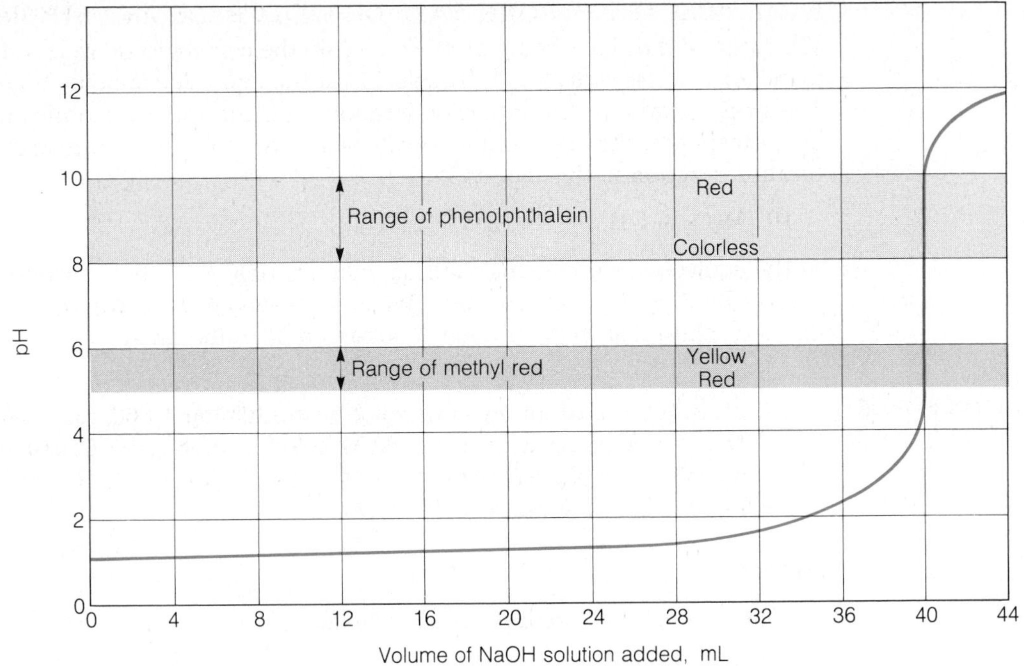

FIGURE 15.3
Titration curve for the titration of HCl, a strong acid, with NaOH, a strong base. The pH is plotted against the volume of 0.2500 M NaOH solution added. The solution in the flask initially contained 10.00 mmol acid in 100 mL.

figures are kept in this intermediate result.) Then

$$[H^+] = \frac{2.5 \times 10^{-3}\ \text{mmol}}{140\ \text{mL}} = 2 \times 10^{-5}\ \text{mol/L}$$

$$\text{pH} = 5$$

(b) The NaOH added is now 10.00 mmol, exactly enough to react with the HCl. The solution contains only H_2O and NaCl and is neutral; $[H^+] = [OH^-]$ and pH = 7.0. ■

PROBLEM 15.29 Calculate the pH when 40.01 mL of NaOH solution has been added in the titration of Example 15.34. □

The calculations of Example 15.34 and Problem 15.29 can be repeated for any volume of NaOH solution. When the pH is plotted against the volume added, the result is a **titration curve** (Figure 15.3). This curve shows why the titration of a strong acid with a strong base, or the converse, is an easy and usually successful experiment. As the volume of NaOH solution added goes from 0 to 39 mL, the pH changes gradually from 1 to 3. With another 0.99 mL, it rises to 5. Then a mere 0.02 mL of solution (about half a drop) changes the pH by 4 units—from 5 to 9. This sudden change in pH is what makes it possible for us to recognize the equivalence point so that we can stop adding solution at the proper time. The solution is indeed neutral at the equivalence point in this titration. Note that the total volume in the flask need not be known accurately; if we had started with 120 mL instead of 100 mL, the answers and the curve would be only slightly different.

WEAK ACID AND STRONG BASE OR WEAK BASE AND STRONG ACID

When the acid or base being titrated is weak, the calculations are less simple, the pH change is less sharp, and the solution at the equivalence point is not neutral. The next example will consider a titration with the same quantities as the last one, except that the flask contains acetic acid—a weak acid—instead of HCl. The titration reaction in this case is

$$HC_2H_3O_2 + OH^- \longrightarrow H_2O + C_2H_3O_2^-$$

At the equivalence point, the solution contains only Na^+ (the spectator ion) and acetate ion, $C_2H_3O_2^-$, a weak base. We expect—and will confirm by calculation—that the solution at the equivalence point is moderately basic.

EXAMPLE 15.35 In a flask is 100 mL of an aqueous solution containing 10.00 mmol acetic acid, $HC_2H_3O_2$. It is titrated with a 0.2500 M solution of NaOH. Calculate the pH when the volume of NaOH solution added to the flask is (a) 0 mL, (b) 20.00 mL, (c) 39.90 mL, (d) 40.00 mL, (e) 40.10 mL.

ANSWER

(a) This is a 0.100 M solution of acetic acid:

$$\frac{10.00 \text{ mmol}}{100 \text{ mL}} = 0.100 \text{ mol/L}$$

Its pH is obtained by solving the equation (section 15.3)

$$\frac{y^2}{0.100 - y} = K_a = 1.75 \times 10^{-5} \qquad \text{(Table 15.2)}$$

$$[H^+] = y = 1.31 \times 10^{-3} \text{ mol/L}$$

$$pH = 2.88$$

(b) As in the previous example, it is necessary to find the volume of the solution and the number of millimoles of acetic acid remaining unreacted. The number of mmol of NaOH added is

$$20.00 \cancel{\text{mL}} \times \frac{0.2500 \text{ mmol}}{1 \cancel{\text{mL}}} = 5.000 \text{ mmol}$$

Of the 10.00 mmol of acetic acid, half (5.00 mmol) has been converted to acetate ion by the reaction

$$HC_2H_3O_2 + OH^- \longrightarrow C_2H_3O_2^- + H_2O$$

leaving 5.00 mmol of acetic acid. The total volume is 100 mL + 20.00 mL = 120 mL. A buffer solution (section 15.9) has been prepared, with

$$[HC_2H_3O_2] = [C_2H_3O_2^-] = \frac{5.00 \text{ mmol}}{120 \text{ mL}} = 4.17 \times 10^{-2} \text{ mol/L}$$

Then the Henderson–Hasselbalch equation gives us

$$pH = pK_a + \log\left(\frac{[C_2H_3O_2^-]}{[HC_2H_3O_2]}\right)$$

$$= -\log(1.75 \times 10^{-5}) + \log\left(\frac{4.17 \times 10^{-2} \text{ mol/L}}{4.17 \times 10^{-2} \text{ mol/L}}\right) = 4.76$$

(c) The volume is 140 mL. The NaOH added is

$$39.90\ \cancel{\text{mL}} \times \frac{0.2500\ \text{mmol}}{1\ \cancel{\text{mL}}} = 9.975\ \text{mmol}$$

which produces an equal number of mmol of $C_2H_3O_2^-$, while 10.00 – 9.975 = 2.5×10^{-2} mmol of $HC_2H_3O_2$ remains unreacted. Then

$$[C_2H_3O_2^-] = \frac{9.975\ \text{mmol}}{140\ \text{mL}} = 7.13 \times 10^{-2}\ \text{mol/L}$$

$$[HC_2H_3O_2] = \frac{2.5 \times 10^{-2}\ \text{mmol}}{140\ \text{mL}} = 2 \times 10^{-4}\ \text{mol/L}$$

$$\text{pH} = \text{p}K_a + \log\left(\frac{[C_2H_3O_2^-]}{[HC_2H_3O_2]}\right)$$

$$= 4.76 + \log\left(\frac{7.13 \times 10^{-2}\ \text{mol/L}}{2 \times 10^{-4}\ \text{mol/L}}\right) = 7.3$$

(d) With 40.00 mL NaOH solution, the amount of NaOH added is

$$40.00\ \cancel{\text{mL}} \times \frac{0.2500\ \text{mmol}}{1\ \cancel{\text{mL}}} = 10.00\ \text{mmol}$$

This is just enough to react with the 10.00 mmol of acetic acid. At this point—the equivalence point—all of the acetic acid has reacted with OH^-:

$$HC_2H_3O_2 + OH^- \longrightarrow C_2H_3O_2^- + H_2O$$

The solution in the flask contains 10.00 mmol $NaC_2H_3O_2$. The volume is 100 mL + 40 mL = 140 mL, and the concentration is

$$[C_2H_3O_2^-] = \frac{10.00\ \text{mmol}}{140\ \text{mL}} = 0.0714\ \text{mol/L}$$

Since acetate ion is a weak base, the pH depends on its concentration. We calculated the pH of a solution like this in Example 15.13 (section 15.3):

$$\frac{[HC_2H_3O_2][OH^-]}{[C_2H_3O_2^-]} = K_b$$

$$\frac{y^2}{0.0714 - y} = 5.71 \times 10^{-10}$$

$$y = [OH^-] = 6.39 \times 10^{-6}\ \text{mol/L}$$

$$[H^+] = \frac{K_w}{[OH^-]} = 1.56 \times 10^{-9}\ \text{mol/L}$$

$$\text{pH} = 8.81$$

Although the acid and base have ''neutralized'' each other at the equivalence point, the solution is basic, not neutral.

(e) The NaOH added is now

$$40.10\ \cancel{\text{mL}} \times \frac{0.2500\ \text{mmol}}{1\ \cancel{\text{mL}}} = 10.025\ \text{mmol}$$

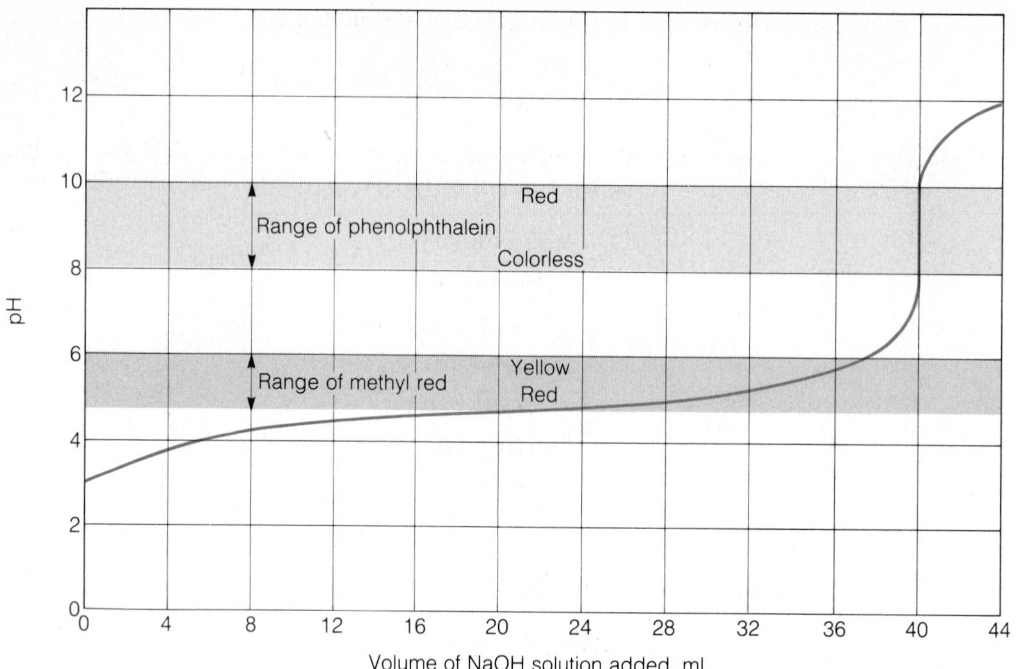

FIGURE 15.4
Titration curve for the titration of $HC_2H_3O_2$, a weak acid, with NaOH, a strong base. The pH is plotted against the volume of 0.2500 M NaOH solution added. The solution initially contains 10.00 mmol acid in 100 mL.

Of this amount, 10.00 mmol reacts with acetic acid to produce 10.00 mmol of $C_2H_3O_2^-$, while 2.5×10^{-2} mmol OH^- remains unreacted. This excess of OH^- determines the pH.

$$[OH^-] = \frac{2.5 \times 10^{-2} \text{ mmol}}{140 \text{ mL}} = 2 \times 10^{-4} \text{ mol/L}$$

$$[H^+] = \frac{K_w}{[OH^-]} = \frac{1.00 \times 10^{-14}}{2 \times 10^{-4}} = 5 \times 10^{-11} \text{ mol/L}$$

$$\text{pH} = 10.3$$

■

PROBLEM 15.30 Refer to Example 15.35. Calculate the pH when the volume of NaOH solution added is (a) 30.00 mL and (b) 39.00 mL. □

Figure 15.4 shows a titration curve plotted from the data in Example 15.35 and Problem 15.30. Inspection of the curve or the data shows that the change in pH around the equivalence point is less sharp than for the strong acid and strong base in Example 15.34. In Example 15.35, a pH change of 3 units (7 to 10) requires 0.2 mL, instead of the 0.02 mL that gives a change of 4 pH units in the strong acid–strong base titration. These volumes and pH's from Examples 15.34 and 15.35 and Problem 15.29 illustrate the difference.

For the weak acid, calculations at 39.99 mL and 40.01 mL would be difficult.

Strong acid and strong base:	39.99 mL	40.00 mL	40.01 mL
pH:	5	7	9
Weak acid and strong base:	39.90 mL	40.00 mL	40.10 mL
pH:	7.3	8.8	10.3

Still, NaOH and acetic acid are good titration partners.

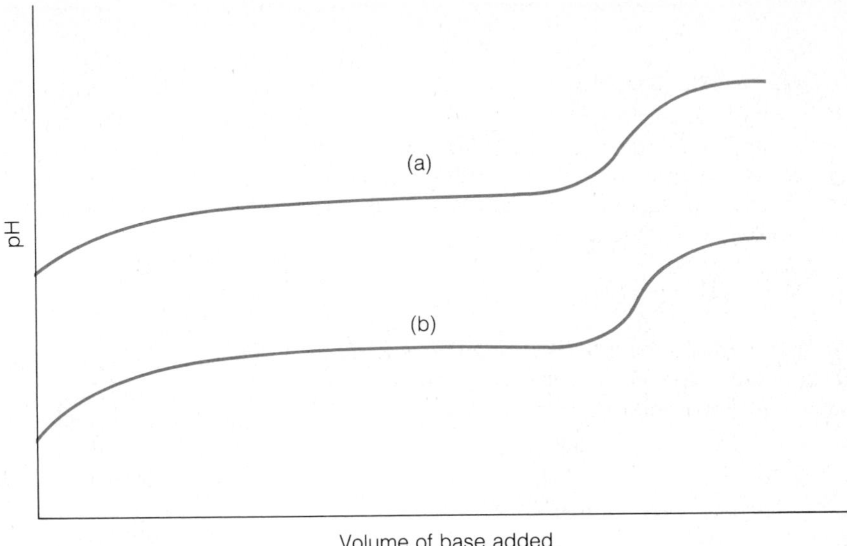

FIGURE 15.5
Titration curves for two unsuccessful titrations. (a) Extremely weak acid with strong base; for example, NH_4^+ with OH^-. (b) Moderately weak acid with moderately weak base; for example, $HC_2H_3O_2$ with NH_3.

The titration curves in Figures 15.3 and 15.4 both show that as the base is added to the acid, the pH rises slowly at first, but *the pH curve becomes almost vertical when equivalent quantities of acid and base are present.* All successful titrations have this feature. To provide the signal for the chemist to stop adding solution, the indicator chosen must be one that changes color somewhere within the pH range spanned by the steeply rising portion of the curve. The **end point**, signaled by a sudden change in the color of the solution, indicates that the titration is completed and the burette reading should be taken. Ideally, the end point and the equivalence point are the same, but this ideal is seldom attained. In both of these titrations, the difference between the end point and the equivalence point is not significant. Figure 15.3 shows that either phenolphthalein or methyl red is suitable for the HCl–NaOH titration. Figure 15.4 shows that phenolphthalein, but not methyl red, is suitable for the acetic acid–NaOH titration. In a titration involving a strong *acid* and a weak *base*, the indicator would have to change color on the acid side, because the product at the equivalence point makes the solution acidic.

PROBLEM 15.31 A sample containing NH_3 is titrated with a 0.1000 M HCl solution. Initially there are 3.000 mmol of NH_3 in 200 mL of solution. See Table 15.2 for the acidic ionization constant of NH_4^+. Calculate the pH when (a) 29.90 mL, (b) 30.00 mL, and (c) 30.10 mL of HCl have been added. □

If an acid is too weak, it cannot be titrated, even with a strong base. Figure 15.5a shows the general appearance of a titration curve for the titration of an acid whose K_a is around 10^{-10}—for example, phenol or NH_4^+—with a strong base. The buffer solution formed during the titration is only slightly less basic than the solution (after the end point) containing excess OH^-. There is no steeply rising portion of the curve, and therefore the end point cannot be identified precisely. The same curve, inverted, would be obtained in an attempt to titrate an extremely weak base, such as acetate ion, with a strong acid.

Another unsuccessful titration is illustrated in Figure 15.5b. In this case a weak acid—about the strength of acetic acid—is titrated with a weak base—about the strength of ammonia. They could be titrated with a strong base or a strong acid,

BOX 15.1
TITRATION OF CARBONATE

Titration of carbonate with strong acid is a common procedure, but doing it right takes a little practice. The graph shows a titration curve for Na_2CO_3 titrated with strong acid. The first break in the curve corresponds to completion of the reaction

$$CO_3^{2-} + H^+ \longrightarrow HCO_3^-$$

The pH of the resulting $NaHCO_3$ solution is about 8.3. The second break in the curve corresponds to completion of the reaction

$$HCO_3^- + H^+ \longrightarrow H_2CO_3$$

The pH of the solution is then about 4. However, neither end point is very sharp. A common procedure is to stop titrating shortly before the second end point and boil the solution to remove CO_2:

$$H_2CO_3 \longrightarrow CO_2(g) + H_2O$$

The pH is thus sent back up to about 8, as shown by the dashed line in the graph. The second half of the titration is now repeated, but with only a few drops of acid, since only a little HCO_3^- remains in solution. The curve is squeezed into the narrow space between the dashed line and the vertical line corresponding to the equivalence point. A sharp end point is thus obtained.

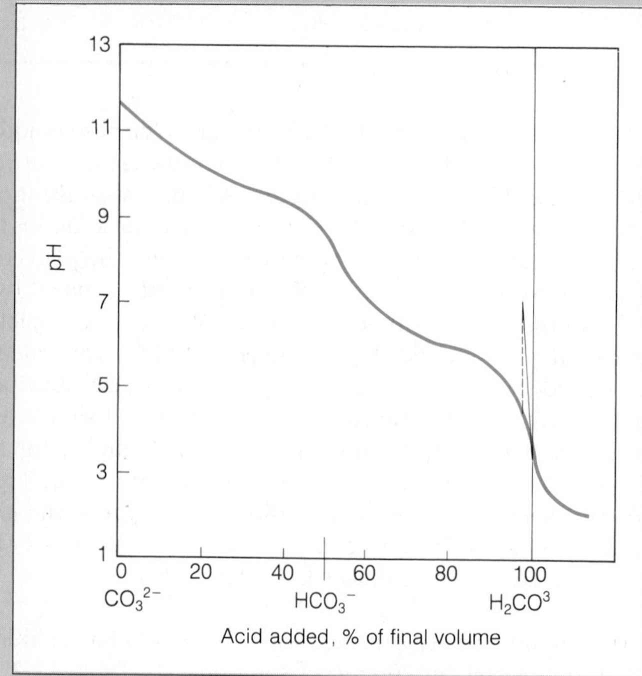

Titration curve for the titration of CO_3^{2-} with the HCl. The dashed line shows the increase in pH when the solution is boiled. The steeply descending curve represents the titration of the small amount of HCO_3^- that remains after boiling.

respectively, but they cannot be titrated with each other. Here, the solution containing excess weak base is not sufficiently different in pH from the buffer solution present just before the end point.

SUMMARY

$$pH = -\log [H^+] \quad \text{and} \quad [H^+] = \text{antilog}(-pH) = 10^{-pH}$$

The concentrations of H^+ and OH^- are related by $[H^+][OH^-] = K_w$, the ion product of water. At 25°C, $K_w = 1.00 \times 10^{-14}$. The pH of a neutral solution is 7.00 at 25°C. Other concentrations and equilibrium constants are represented by a notation analogous to pH:

$$pOH = -\log [OH^-]$$

$$pK = -\log K$$

$$pH + pOH = pK_w$$

Since a strong acid HA ionizes completely to $H^+ + A^-$, $[H^+] = [HA]$. A soluble strong base (ionic hydroxide) dissolves to give $[OH^-]$ equal to a whole number (1 for NaOH, 2 for $Ba(OH)_2$, and so on) times the concentration of the base. However, if the solution is extremely dilute ($\ll 10^{-7}$ M), it is practically neutral.

In a solution of a weak acid HA or weak base B, the equilibrium $HA \rightleftharpoons H^+ + A^-$ or $B + H_2O \rightleftharpoons BH^+ + OH^-$ is established. HA and B may be charged or uncharged. The corresponding equilibrium condition is

$$\frac{[H^+][A^-]}{[HA]} = K_a \quad \text{or} \quad \frac{[BH^+][OH^-]}{[B]} = K_b$$

The usual procedure for equilibrium calculations gives the equation

$$\frac{y^2}{c - y} = K_a \text{ (or } K_b\text{)}$$

where c is the initial concentration of HA (or B) and $y = [H^+]$ for acids (or $[OH^-]$ for bases). If the acid (or base) is weak enough, this equation simplifies to $y^2/c = K_a$ (or K_b). The percentage of the acid or base ionized at equilibrium, $y/c \times 100\%$, increases as the concentration decreases.

An **acid-base indicator** is a weak acid (or base) that differs in color from its conjugate base (or acid). The color is determined by the acid/base ratio, which depends on the pH. When the ratio is near 1, the pH can be estimated from the color of the indicator.

For a conjugate acid-base pair such as HA and A^-, the acidic ionization constant K_a of HA and the basic ionization constant K_b of A^- are related by $K_aK_b = K_w$. The pH of a solution of a salt such as Na^+A^- or BH^+Cl^- (NH_4Cl, for example) can be calculated when K_a for HA or K_b for B is known.

A polyprotic acid has more than one acidic H atom. A polyprotic base can accept more than one proton. Ionization occurs in steps:

$$\underset{K_1}{H_2A \rightleftharpoons H^+ + HA^-}, \qquad \underset{K_2}{HA^- \rightleftharpoons H^+ + A^{2-}}$$

and for bases,

$$\underset{K_1}{A^{2-} + H_2O \rightleftharpoons HA^- + OH^-}, \qquad \underset{K_2}{HA^- + H_2O \rightleftharpoons H_2A + OH^-}$$

For both acids and bases, $K_1 > K_2$. Calculations are simplified by calculating $[H^+]$ and $[HA^-]$ (or $[OH^-]$ and $[HA^-]$) as if only the first ionization occurred. These concentrations are used as the initial concentrations for the second ionization.

Solutions of positive metal ions are often acidic. Their pH's can be calculated from the first ionization constant of the hydrated ion; for example, $Al(H_2O)_6{}^{3+} \rightleftharpoons H^+ + Al(H_2O)_5(OH)^{2+}$.

An equilibrium such as $HA \rightleftharpoons H^+ + A^-$ is shifted to the left by adding a strong acid (increasing $[H^+]$) or a salt such as NaA (increasing $[A^-]$). This is known as the **common ion effect**. When a strong acid is added, nearly all of the

H^+ is provided by the strong acid. When A^- (a weak base) is added, the solution becomes less acidic. Corresponding statements apply to bases.

A **buffer solution** contains a weak acid HA and its weak conjugate base A^- (or a weak base B and its weak conjugate acid BH^+). It resists changes in pH because the acid or base present reacts with any added base or acid. The pH is determined by the acid/base ratio:

$$[H^+] = K_a\,[HA]/[A^-] \quad \text{or} \quad pH = pK_a + \log([A^-]/[HA])$$

(the **Henderson–Hasselbalch equation**). When a strong acid is added to a buffer solution, [HA] is increased and $[A^-]$ is decreased by the number of moles of acid added per liter. When a strong base is added, $[A^-]$ is increased and [HA] is decreased by the number of moles of base added per liter. The pH changes least on addition of acid or base when $[HA] = [A^-]$.

In **acid-base titrations**, an indicator is added to signal the end point. An indicator is suitable if the pH at the equivalence point is within the pH range where the indicator changes color.

A **titration curve** is a graph of pH plotted against the volume of solution added. The pH changes slowly except near the equivalence point, where the curve must become almost vertical if the titration is to be successful.

15.12 ADDITIONAL PROBLEMS

Note: When the temperature is not given in these problems, assume that it is the temperature for which the necessary data are tabulated. Refer to Tables 15.1, 15.2, and 15.3 as needed.

pH, pOH, AND pK

15.32 Calculate (a) $[H^+]$ and (b) pH of pure water at 35°C.

15.33 Fill in the blanks in this table for given solutions a, b, c, and d.

		CONCENTRATION, mol/L		
SOL.	TEMP., °C	$[H^+]$	$[OH^-]$	pH
(a)	25	1.00×10^{-6}	—	—
(b)	0	—	—	5.69
(c)	50	—	—	7.00
(d)	25	—	2.50×10^{-9}	—

15.34 The equilibrium constant of the reaction

$$2D_2O \rightleftharpoons D_3O^+ + OD^-$$

(where D is deuterium, ^2H) is 1.35×10^{-15} at 25°C. Calculate the pD of pure deuterium oxide (heavy water) at 25°C. What is the relationship between $[D_3O^+]$ and $[OD^-]$ in pure D_2O? Is pure D_2O acidic, basic, or neutral?

15.35 In liquid ammonia at −50°C, the equilibrium constant for the reaction

$$2NH_3(\ell) \rightleftharpoons NH_4^+ + NH_2^-$$

is $K = 1.0 \times 10^{-33}$. (a) What is the concentration of NH_4^+ in pure liquid ammonia? (b) What is the concentration of NH_4^+ in a 1.0×10^{-3} M solution of $NaNH_2$ ($Na^+ + NH_2^-$) in liquid ammonia?

15.36 Calculate the pH of a 1.0×10^{-4} M solution of $HClO_4$, a strong acid, at 25°C.

15.37 Calculate the pH of a 2.0×10^{-4} M solution of NaOH at 25°C.

15.38 What is the pH of a 1.0×10^{-11} M solution of the strong acid HCl at 25°C?

15.39 A solution of HNO_3 has pH 3.20. What is the molarity of the solution?

WEAK ACIDS AND BASES

15.40 Find the pH of a 0.20 M solution of phenol, C_6H_5OH.

15.41 Calculate $[H^+]$ in a 0.060 M solution of propionic acid, $HC_3H_5O_2$.

15.42 Find the pH of a 0.30 M solution of HCN.

15.43 Calculate the pH of a 0.020 M solution of dichloroacetic acid, $HC_2HCl_2O_2$.

15.44 Calculate the pH of a 0.50 M solution of ethylamine, $C_2H_5NH_2$, in H_2O.

15.45 Calculate $[H^+]$ in a 0.0100 M solution of hydrofluoric acid by (a) making the assumption that avoids a quadratic equation and by (b) solving the quadratic equation. Compare the two results.

15.46 (a) Solve the equation $y^2/(c - y) = K$ by the approximation method, assuming that y is much less than c. (b) Use your solution to write and simplify the ratio y/c. (c) If y/c is to be less than 0.05, what condition must be imposed on the ratio K/c?

15.47 The solubility of CO_2 in water at 25°C and 1 atm is given in Table 12.2 (section 12.4). (a) Convert this solubility to molarity. Assume that 100 mL solution contains 100 g H_2O. (b) Assume that the partial pressure of CO_2 in air is 3.3×10^{-4} atm. What is the solubility (in mol/L) of CO_2 at this pressure? Assume that Henry's law (section 12.4) is applicable. (c) Calculate the pH of water in equilibrium with "pure" air that is free of acidic pollutants such as SO_2 and NO_2.

15.48 A 5.00×10^{-3} M solution of butyric acid, C_3H_7COOH, has pH 3.58. Calculate K_a and pK_a for this acid.

15.49 A solution of acetic acid in water is to have a pH of 3.00. What should be the concentration of acetic acid?

15.50 You are told to add a 0.1000 M solution of acetic acid to 1.000 liter of water at 25°C until the pH is 3.80. (a) What should be the molarity of acetic acid in the diluted solution? (b) What volume of the acetic acid solution would you add?

ACID–BASE INDICATORS

15.51 Select from Table 15.3 an indicator suitable for estimating the pH's of solutions with pH in the range (a) 1.5 to 2.5; (b) 10.5 to 11.5. (c) Two indicators in the table have the same pH range, 8.2 to 10.0. Which would you select for estimating pH's in that range? Why?

15.52 Assume that the pK_a of alizarin yellow is 10.5. What is the [HIn]/[In^-] ratio and the approximate color of the indicator in a solution of pH (a) 10.0; (b) 10.5; (c) 11.0?

15.53 (a) Methyl orange and bromcresol purple are added to separate portions of an aqueous solution. In each case the solution is yellow. What can you conclude about the pH of the solution? (b) Select another indicator that would enable you to determine the pH more accurately.

CONJUGATE ACID–BASE PAIRS

15.54 Find the pH of a 0.15 M solution of sodium acetate, $NaC_2H_3O_2$.

15.55 Calculate (a) the pK_a of bromoacetic acid, $HC_2H_2O_2Br$, and (b) K_b and pK_b for the bromoacetate ion, $C_2H_2O_2Br^-$.

15.56 The weak base brucine, $C_{23}H_{26}N_2O_4$, has pK_b = 5.72 at 25°C. Calculate (a) K_b for brucine and (b) K_a for the brucinium ion, $C_{23}H_{27}N_2O_4^+$, the conjugate acid of brucine.

15.57 Calculate (a) the acidic ionization constant of the anilinium ion, $C_6H_5NH_3^+$, and (b) the pH of a 0.10 M solution of anilinium chloride, $(C_6H_5NH_3)^+Cl^-$.

15.58 Calculate the pH of a 0.30 M solution of sodium formate, $NaCHO_2$.

15.59 Saccharin ($HC_7H_4NO_3S$) is a weak acid with K_a = 2.1×10^{-12} at 25°C. It is used in the form of sodium saccharide, $NaC_7H_4NO_3S$. What is the pH of a 1.0×10^{-5} M solution of sodium saccharide? *Suggestion*: Let $z = 1.0 \times 10^{-5} - [OH^-]$. Write the equation to be solved in terms of z.

15.60 (a) For a certain weak acid, HA, $K_a = 10^{-6}$ at 25°C. What is K_b for its conjugate base A^-? Is A^- a strong base or a weak base? Is it true (as is sometimes said) that the conjugate base of a weak acid is a strong base? (b) What is the value of K_a such that K_a for an acid is equal to K_b for its conjugate base at 25°C?

15.61 (a) Consider an amphoteric species HA^-. It is a weaker acid than H_2A; $K_{a(HA^-)} < K_{a(H_2A)}$. Use this inequality and the equation $K_{a(H_2A)}K_{b(HA^-)} = K_w$ to show that

$$K_{a(HA^-)}K_{b(HA^-)} < K_w$$

(b) Obtain the same conclusion by using the fact that HA^- is a weaker base than A^{2-}. (c) Is it possible for the same species to be both (1) a strong acid and a strong base in the same solvent; (2) a strong acid and a weak base; (3) a weak acid and a strong base?

POLYPROTIC ACIDS AND BASES

15.62 Find the basic ionization constant of the phthalate ion, $C_8H_4O_4^{2-}$.

15.63 Calculate (a) the pH and (b) the concentration of SO_3^{2-} in a 0.20 M solution of H_2SO_3.

15.64 Vitamin C (ascorbic acid, $H_2C_6H_6O_6$, 176.1 g/mol) is a diprotic acid with $K_1 = 7.9 \times 10^{-5}$ and $K_2 = 1.6 \times 10^{-12}$. What is the pH of a solution prepared by dissolving 500 mg of vitamin C in 250 mL of water?

15.65 Calculate (a) [H^+], (b) [$H_2PO_4^-$], (c) [HPO_4^{2-}], and (d) [PO_4^{3-}] in a 0.10 M solution of phosphoric acid, H_3PO_4.

ACIDITY OF METAL IONS

15.66 Find the pH of a 0.10 M solution of $Be(NO_3)_2$.

15.67 The pH of a 0.010 M solution of $CoCl_3$ is 2.15. Calculate the acidic ionization constant (K_1) of $Co(H_2O)_6^{3+}$.

THE COMMON ION EFFECT

15.68 A solution is 0.40 M with respect to HCl and 1.00 M with respect to formic acid, $HCHO_2$. Calculate the concentration of CHO_2^- in the solution.

15.69 Find the concentration of (a) hydrogen succinate ion and (b) succinate ion in a solution 0.50 M with respect to $HClO_4$ and 0.050 M with respect to succinic acid, $H_2C_4H_4O_4$.

15.70 A solution is 0.15 M with respect to potassium malonate, $K_2C_3H_2O_4$, and 0.25 M with respect to KOH. What is the concentration of malonic acid in the solution?

15.71 In Problem 15.47, you calculated that water in equilibrium with air is 1.1×10^{-5} M with respect to H_2CO_3 and has pH 5.66. These are the properties of "pure" rain. A sample of "acid rain" has pH 4.00. (a) Assume that the low pH of acid rain results from the presence of a strong acid, such as H_2SO_4 or HNO_3. What is the concentration of this strong acid? (b) What is the concentration of HCO_3^- in the acid rain if the solubility of CO_2 is the same as it is in pure rain? (c) What is the total concentration of strong acid and H_2CO_3 in the acid-rain solution? (d) How many times greater is that than the concentration of H_2CO_3 in pure rain?

BUFFER SOLUTIONS

15.72 (a) Find the pH of a solution 0.50 M with respect to formic acid and 0.40 M with respect to sodium formate. (b) Find the pH of the solution after 0.05 mol HCl/L has been added to it.

15.73 (a) Find the pH of a solution that is 1.00 M with respect to NH_3 and 0.80 M with respect to NH_4Cl. (b) What would be the pH of the solution after 0.10 mol HCl/L had been added to it? (c) A solution of pH 9.34 is prepared by adding NaOH to pure water. What would be the pH of this solution after 0.10 mol HCl/L had been added to it?

15.74 A solution is 0.50 M with respect to sodium hydrogen tartrate, $NaHC_4H_4O_6$, and 0.60 M with respect to sodium tartrate, $Na_2C_4H_4O_6$. (a) Is this a buffer solution? If it is, what is the conjugate acid-base pair? (b) Calculate the pH of the solution. (c) Calculate the pH of the solution after 0.20 mole of HCl per liter has been added to it.

15.75 A buffer solution of pH 5.30 is to be prepared from propionic acid and sodium propionate. The concentration of sodium propionate must be 0.50 mol/L. What should be the concentration of the acid?

15.76 We need a buffer with pH 9.00. It can be prepared from NH_3 and NH_4Cl. What should be the $[NH_4^+]/[NH_3]$ ratio?

15.77 Calculate the concentrations of $HC_2H_3O_2$ and $C_2H_3O_2^-$ in the solution in Figure 15.2 (section 15.9) when the pH is 4.500.

15.78 A solution contains bromoacetic acid and sodium bromoacetate with a total concentration of 0.20 mol/L. If the pH is 3.00, what are the concentrations of the acid and the salt?

15.79 (a) A blood sample (cooled to 25°C) contains 0.025 mole H_2CO_3 per liter. This concentration includes both H_2CO_3 and CO_2. The pH of the sample is 7.44. What is the concentration of HCO_3^-? (b) Is the $H_2CO_3 + HCO_3^-$ buffer in blood more effective in maintaining the pH against added acid or against added base?

15.80 Assume that all of the monoprotic acids listed in Table 15.2 and their sodium salts are available. (a) Which acid-salt pair would you choose as a buffer solution of pH 2.90? (b) What should be the molar ratio of acid to salt in this buffer solution?

15.81 Calculate (a) the pH of a 0.20 M solution of propionic acid and (b) the pH of a 0.10 M solution of sodium propionate. (c) Equal volumes of the solutions in parts a and b are mixed. What is the pH of the resulting solution?

15.82 A buffer solution has $[HC_2H_3O_2] = 0.50$ mol/L and $[C_2H_3O_2^-] = 0.88$ mol/L (as in Example 15.28). Add 1.00 mol HCl to 1.00 L of this solution. What is the pH of the resulting solution? Has the solution served effectively as a buffer?

15.83 A solution contains 0.0200 mole of a weak acid HA. To this solution is added 0.0100 mole NaOH. The pH of the solution is then 4.50. (a) What is the $[HA]/[A^-]$ ratio in the solution after the addition of NaOH? (b) What is the ionization constant K_a of the acid?

15.84 The buffer solution with pH 5.00 in Example 15.28 (section 15.9) is diluted by adding water until the volume is doubled. What is the pH of the diluted solution?

15.85 (a) When enough strong acid is added to a buffer solution so that the acid/base ratio is multiplied by 2.0, by how much does the pH change? (b) By how much does the pH change if enough strong base is added to multiply the base/acid ratio by 2.0?

15.86 Classify each of these solutions as a good buffer solution, a poor buffer solution, or not a buffer solution at all: (a) 10^{-5} M $HC_2H_3O_2 + 10^{-5}$ M $NaC_2H_3O_2$, (b) 1.0 M HCl + 1.0 M NaCl, (c) 0.5 M $HC_2H_3O_2$ + 0.7 M $NaC_2H_3O_2$, (d) 0.10 M NH_3 + 0.10 M NH_4Cl, (e) 0.20 M $HC_2H_3O_2$ + 0.00020 M $NaC_2H_3O_2$.

15.87 What volumes of 0.1000 M acetic acid and 0.1000 M NaOH solutions should be mixed to prepare 1.000 liter of a buffer solution of pH 4.50 at 25°C?

15.88 A buffer solution is prepared by dissolving 0.50 mol tartaric acid and 0.20 mol NaOH in water to make 1.00 L. The resulting solution contains tartaric acid and its conjugate base, $HC_4H_4O_6^-$. What is the pH of this solution?

ACID–BASE TITRATIONS

15.89 Select an indicator suitable for determining the pH of solutions with pH around 7.

15.90 20.00 mL of 0.2000 M propionic acid ($HC_3H_5O_2$) is titrated with 20.00 mL of 0.2000 M KOH at 25°C. Calculate the pH at the equivalence point. Select a suitable indicator.

15.91 Calculate the pH at the equivalence point for the titration of a solution containing 150.0 mg of ethylamine, $C_2H_5NH_2$, with 0.1000 M HCl solution. The volume of the solution at the equivalence point is 200 mL. Select a suitable indicator.

TITRATION CURVES

15.92 A solution contains 2.000 mmol HF dissolved in 100 mL. This solution is titrated at 25°C with a 0.1000 M solution of NaOH. Calculate the pH when these volumes of the NaOH solution have been added: (a) 0 mL, (b) 10.00 mL, (c) 19.90 mL, (d) 20.00 mL, (e) 20.10 mL. (f) Plot these points on graph paper and sketch the titration curve. (g) Select a suitable indicator.

15.93 A solution contains an unknown weak monoprotic acid HA. It takes 46.24 mL of NaOH solution to titrate 50.00 mL of the HA solution. To another 50.00-mL sample of the same HA solution, 23.12 mL of the same NaOH solution is added. The pH of the resulting solution in the second experiment is 5.14. What are pK_a and K_a of HA?

15.94 (10 points) A solution has $[OH^-] = 2.0 \times 10^{-9}$ mol/L at 25°C. Calculate $[H^+]$ and pH.

15.95 (30 points) Calculate the pH at 25°C of these solutions: (a) 0.0050 M $Ca(OH)_2$; (b) 0.20 M chloroacetic acid, $HC_2H_2ClO_2$; (c) 0.040 M pyridine, C_5H_5N (a monoprotic base); (d) 0.15 M sodium phenoxide, NaC_6H_5O (the conjugate acid is phenol); (e) 0.50 M methylammonium iodide, CH_3NH_3I (the conjugate base is methylamine).

15.96 (8 points) Select from Table 15.3 two indicators suitable for estimating the pH's of solutions with pH between 5 and 6.

15.97 (14 points) A solution of succinic acid, $H_2C_4H_4O_4$, is 0.20 M. (a) Calculate the pH. (b) What is the concentration of the succinate ion, $C_4H_4O_4^{2-}$?

15.98 (6 points) What is the pH of a 0.50 M solution of zinc bromide, $ZnBr_2$?

15.99 (6 points) A solution is 0.050 M with respect to HCl and 0.025 M with respect to benzoic acid, $HC_7H_5O_2$. What is the concentration of benzoate ion, $C_7H_5O_2^-$?

15.100 (12 points) A buffer solution is prepared by dissolving 0.50 mol oxalic acid, $H_2C_2O_4$, and 0.75 mol sodium hydrogen oxalate, $NaHC_2O_4$, in water to make 1.00 L. (a) What is the pH of the solution? (b) What is the new pH after 0.25 mol NaOH has been added to 1.00 L of the solution?

15.101 (6 points) A buffer solution of pH 10.00 is to be prepared from trimethylamine, $(CH_3)_3N$, and trimethylammonium chloride, $(CH_3)_3NHCl$. What should be the ratio of the concentration of trimethylamine to the concentration of trimethylammonium chloride?

15.102 (8 points) Calculate the pH at the equivalence point for the titration of chloroacetic acid, $HC_2H_2O_2Cl$, with NaOH at 25°C. Assume that the concentration of the reaction product at the equivalence point is 0.100 mol/L. For the chloroacetate ion, $K_b = 7.1 \times 10^{-12}$. Select a suitable indicator from Table 15.3.

16

IONIC EQUILIBRIUM II: SLIGHTLY SOLUBLE SALTS AND COORDINATION IONS

In the last chapter we applied the principles of chemical equilibrium to the reactions of acids and bases. Many of the other reactions of ions in aqueous solutions involve slightly soluble salts—loosely called "insoluble" salts—or coordination ions, which are combinations of simpler ions with each other or with molecules. In this chapter you will learn how to do calculations for those reactions.

16.1 SLIGHTLY SOLUBLE SALTS

In section 9.2 we noted that some salts are "soluble" and others are "insoluble" in water. Actually, "insoluble" salts are slightly soluble, since nothing is completely insoluble in anything. It is just a question of what the concentration is when the solution becomes saturated. A solid salt in equilibrium with a saturated solution forms a typical equilibrium system: the concentrations of the ions in the saturated solution are determined by the law of chemical equilibrium. In this first section we will apply the law of chemical equilibrium to the dissolving process.

The dissolving of silver chloride (only a little of it) in water is represented by the equation

$$AgCl(c) \rightleftharpoons Ag^+ + Cl^-$$

As usual, we assume that any salt that dissolves is present in solution as ions. For this reaction to be in equilibrium, solid AgCl must be in contact with a saturated solution. The equilibrium condition is

$$[Ag^+][Cl^-] = K \qquad (1)$$

Since the solid is omitted (section 13.8), the reaction "quotient" is merely a product of concentrations. K is called the **solubility product** (or *solubility product constant*) of AgCl. It is also represented by K_{sp}, especially when it must be distinguished from other equilibrium constants. When this equilibrium is established, the solution is saturated, and the solubility product thus gives the product of the Ag^+ concentration multiplied by the Cl^- concentration in a saturated solution.

The individual concentrations may be changed—by adding NaCl or $AgNO_3$, for instance—but K, *at any one temperature, remains constant.* Solubility products are used only for slightly soluble salts, because the law of chemical equilibrium is accurate only for very dilute solutions. (Box 16.2 explains this.)

When the salt dissolves to give unequal numbers of positive and negative ions, each concentration must be raised to the power equal to the coefficient of that ion in the balanced chemical equation. This is the same rule that is used in writing any equilibrium condition. For example,

SALT	REACTION	EQUILIBRIUM CONDITION
CaF_2	$CaF_2(c) \rightleftharpoons Ca^{2+} + 2F^-$	$[Ca^{2+}][F^-]^2 = K$
Hg_2Cl_2	$Hg_2Cl_2(c) \rightleftharpoons Hg_2^{2+} + 2Cl^-$ (not $2Hg^+$)	$[Hg_2^{2+}][Cl^-]^2 = K$
$Al(OH)_3$	$Al(OH)_3(c) \rightleftharpoons Al^{3+} + 3OH^-$	$[Al^{3+}][OH^-]^3 = K$
$Ca_3(PO_4)_2$	$Ca_3(PO_4)_2(c) \rightleftharpoons 3Ca^{2+} + 2PO_4^{3-}$	$[Ca^{2+}]^3[PO_4^{3-}]^2 = K$

Solubility products of some slightly soluble salts are given in Table 16.1. Section 17.10 will explain one way to determine these constants experimentally.

Solubility and *solubility product* must not be confused with one another. The solubility of a salt is the quantity present in a unit amount of a saturated solution, expressed in moles per liter, grams per 100 mL, or other units. The *solubility product* is an equilibrium constant. The solubility depends on what else is in the solution, whereas the solubility product, for a given solute and solvent, depends only on temperature. There is a connection, of course, between the solubility and the solubility product; if one is given, the other can be calculated.

EXAMPLE 16.1 From the solubility product of AgCl, find its solubility in mol/L at 25°C in (a) pure water and (b) a 0.010 M solution of NaCl.

ANSWER

(a) The equation for the solubility equilibrium should first be written. In order to do this, you must know what ions are formed when the salt dissolves (usually, but not always, obvious). Then write the usual equilibrium table.

TABLE 16.1
SOLUBILITY PRODUCTS IN WATER (mostly at 25°C)

SALT	K	SALT	K
AgBr	5.32×10^{-13}	CuS	1.3×10^{-36}
AgCl	1.76×10^{-10}	$Fe(OH)_3$	3×10^{-39}
Ag_2CrO_4	1.2×10^{-12}	FeS	1.6×10^{-19}
AgI	8.49×10^{-17}	Hg_2Cl_2†	1.5×10^{-18}
Ag_2S	6.6×10^{-50}	HgI_2	2.9×10^{-29}
$Al(OH)_3$	10^{-32}	HgS	6×10^{-53}
$BaCO_3$	2.60×10^{-9}	MgC_2O_4(18°C)	8.57×10^{-5}
$BaCrO_4$	2.2×10^{-10}	$Mg(OH)_2$	5.7×10^{-12}
BaF_2(25.8°C)	1.73×10^{-6}	MnS	4.6×10^{-14}
$BaSO_4$	1.05×10^{-10}	NiS	1.1×10^{-21}
$CaCO_3$	4.95×10^{-9}	$PbCrO_4$	2.0×10^{-16}
$CaC_2O_4(H_2O)$	1.96×10^{-8}	PbI_2	8.5×10^{-9}
CaF_2	1.61×10^{-10}	PbS	8.8×10^{-29}
CdS	1.4×10^{-29}	$PbSO_4$	1.8×10^{-8}
CoS	9.7×10^{-21}	ZnS	2.9×10^{-25}

† $Hg_2Cl_2(c) \rightleftharpoons Hg_2^{2+} + 2Cl$

	$AgCl(c) \rightleftharpoons$	Ag^+	+	Cl^-
Initial concentration		0		0
Change		$+s$		$+s$
Concentration at equilibrium		s		s mol/L

The concentration of either of these ions (s) is equal to the solubility of AgCl in mol/L. The equilibrium condition is

$$[Ag^+][Cl^-] = K$$

$$s \times s = 1.76 \times 10^{-10} \qquad \text{(Table 16.1)}$$

$$s = 1.33 \times 10^{-5}\ \text{mol/L}$$

Note the difference between the *solubility* (1.33×10^{-5} mol/L) and the *solubility product* (1.76×10^{-10}).

(b) In this solution, $[Ag^+]$ and $[Cl^-]$ are *not* equal. By adding 0.010 mol/L of NaCl to the water, we have added 0.010 mol Cl^-/L. It is like any equilibrium in which one of the products is already present:

	$AgCl(c) \rightleftharpoons$	Ag^+	+	Cl^-
Initial concentration (before AgCl dissolves)		0		0.010 (from NaCl)
Change		$+y$		$+y$
Concentration at equilibrium		y		$0.010 + y$ mol/L

The solubility is the number of moles of AgCl that dissolve per liter. Each mole of AgCl that dissolves puts 1 mole of Ag^+ and 1 mole of Cl^- into solution. Thus, the increase (y) in each concentration is the solubility of AgCl. The solubility is the same as $[Ag^+]$ because Ag^+ comes only from the dissolving of AgCl. The solubility is *not* equal to $[Cl^-]$, because Cl^- is provided by NaCl as well as AgCl. Then, substituting into the equilibrium condition, we obtain

$$[Ag^+][Cl^-] = K$$

$$y(0.010 + y) = K$$

Since K is small, it is reasonable to expect that $y \ll 0.01$, which means that $0.010 + y = 0.010$. In other words, practically all of the Cl^- is contributed by the NaCl, making $[Cl^-] = 0.010$ mol/L. Then,

$$y(0.010) = K = 1.76 \times 10^{-10}$$

$$y = 1.76 \times 10^{-8}\ \text{mol/L}$$

This answer shows that $y \ll 0.01$ is indeed a good approximation. ■

Observe that the solubility (y) of AgCl is much less in the NaCl solution than it is in pure water (s). This result could have been predicted from Le Chatelier's principle: the equilibrium

$$AgCl(c) \rightleftharpoons Ag^+ + Cl^-$$

is shifted to the left by addition of Cl^- from the NaCl; that is, the solubility of AgCl is decreased. This effect—like the effect of a strong acid or a salt of the weak

BOX 16.1
COLLOIDAL PRECIPITATES

Colloidal precipitates (section 12.17) are usually undesirable because they cannot be separated by filtration or centrifugation. Most slightly soluble salts form hydrophobic colloids. They can often be coagulated simply by adding harmless ions to the solution—NH_4NO_3 is commonly used for this purpose—so that the colloidal particles attract whatever ions are opposite in charge to themselves and they thus become electrically neutral.

A familiar laboratory example of the formation and coagulation of a colloid is the precipitation of AgCl from a solution of NaCl by adding $AgNO_3$ solution. As precipitation begins, the solution contains Na^+ and Cl^- ions, as well as colloidal AgCl particles (part a of the figure). These AgCl particles are stabilized by adsorption of Cl^- ions on their surfaces, which gives them a negative charge. (The low solubility of AgCl tells us that Ag^+ and Cl^- ions attract each other strongly.) Na^+ ions surround each AgCl particle, providing a protective layer. As more Ag^+ is added, more AgCl precipitates and the concentration of Cl^- decreases. Meanwhile, more and more Ag^+ ions find their way to the surfaces of the AgCl colloidal particles. The net charges of the particles approach zero as the Cl^- ions in the solution are used up; the Na^+ ions no longer provide a protective layer, and the colloid coagulates when nearly enough Ag^+ has been added to react with all of the Cl^- (part b of the figure).

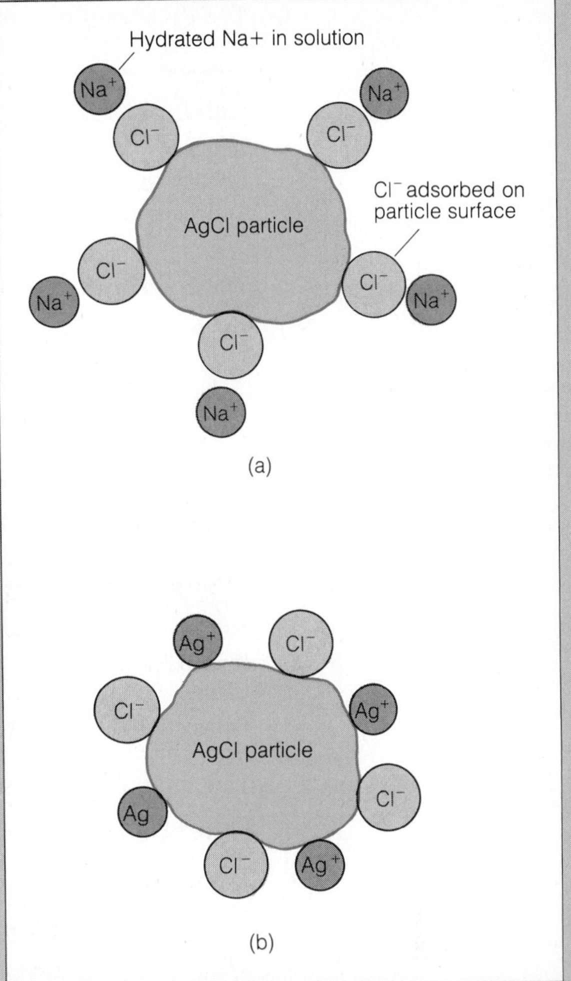

A colloidal particle of AgCl. (a) The solution contains excess NaCl. The particle is negatively charged because Cl^- ions outnumber Ag^+ ions on its surface. The particle is surrounded by Na^+ ions. (b) Most of the Cl^- in solution has been removed by reaction with Ag^+. Some Ag^+ ions have attached themselves to the colloidal particle, neutralizing its charge. The Na^+ ions no longer protect it from collision and combination with other AgCl particles.

acid in repressing the ionization of a weak acid (section 15.8)—is known as the **common ion effect**, because the decrease in solubility is caused by the addition of a salt that has an ion in common with the slightly soluble salt. The solubility would likewise be decreased if Ag^+ were added to the solution by adding, for example, $AgNO_3$.

When the salt has a more complicated formula than AgCl, the calculation of solubility is also a little more complicated.

EXAMPLE 16.2 Find the solubility (in mol/L) of PbI_2 at 25°C in (a) pure water and (b) a 0.50 M solution of $Pb(NO_3)_2$.

ANSWER

(a) The dissolving process is represented by

	$PbI_2(c) \rightleftharpoons$	Pb^{2+}	+ $2I^-$
Initial concentration		0	0
Change		$+s$	$+2s$
Concentration at equilibrium		s	$2s$ mol/L

The equilibrium condition is

$$[Pb^{2+}][I^-]^2 = K$$

$$s(2s)^2 = K$$

The calculation is easy if your calculator has a y^x button: $x = 0.3333$, $y = 8.5 \times 10^{-9}/4$.

$$4s^3 = K = 8.5 \times 10^{-9} \quad \text{(Table 16.1)}$$

$$s = (K/4)^{1/3} = 1.3 \times 10^{-3} \text{ mol/L}$$

(b) An equilibrium table is written as usual:

	$PbI_2(c) \rightleftharpoons$	Pb^{2+}	+ $2I^-$
Initial concentration		0.50	0
Change		$+y$	$+2y$
Concentration at equilibrium		$0.50 + y$	$2y$ mol/L

This time, the equilibrium condition is

$$(0.50 + y)(2y)^2 = 8.5 \times 10^{-9}$$

The approximation $0.50 + y \approx 0.50$ spares us from solving a cubic equation:

$$0.50 \times 4y^2 = 8.5 \times 10^{-9}$$

$$y = 6.5 \times 10^{-5} \text{ mol/L}$$

The approximation must be checked: 6.5×10^{-5} is indeed negligible (much less than 5%) in comparison to 0.50. y is the solubility, the number of moles of PbI_2 going into solution per liter. The solubility is much decreased by the presence of Pb^{2+} (the common ion effect). ■

Problems like the preceding example sometimes lead to such questions as, "Why do you double the concentration of I^-?" and, "Aren't you counting things *twice* when you double it *and* square it?" The concentration of I^- is *not* doubled. In part a, $[I^-]$ is calculated by doubling $[Pb^{2+}]$ because the balanced equation shows that 2 moles of I^- appear in solution for each mole of Pb^{2+}. In part b, for the same reason, $[I^-]$ is calculated by doubling the *change* in $[Pb^{2+}]$ that results from the dissolving of a little PbI_2. The power 2 in the equilibrium condition also comes from the coefficient 2 in the chemical equation.

PROBLEM 16.1 Calculate the solubility (in mol/L) of CaF_2 in (a) pure water and (b) a 0.20 M solution of NaF. □

EXAMPLE 16.3 The solubility of calcium phosphate, $Ca_3(PO_4)_2$, is 3.5×10^{-6} g/100 mL. (a) Convert this solubility to molarity. (b) Calculate the solubility product.

ANSWER

(a) The molar mass of $Ca_3(PO_4)_2$ is 310.2 g/mol. Then the solubility in mol/L is

$$\frac{3.5 \times 10^{-6} \text{ g}}{100 \text{ mL}} \times \frac{1 \text{ mol}}{310.2 \text{ g}} \times \frac{1000 \text{ mL}}{1 \text{ L}} = 1.1 \times 10^{-7} \text{ mol/L}$$

(b) When y mol $Ca_3(PO_4)_2$ dissolves, $3y$ mol Ca^{2+} and $2y$ mol $PO_4{}^{3-}$ are formed:

$$Ca_3(PO_4)_2(c) \rightleftharpoons 3Ca^{2+} + 2PO_4{}^{3-}$$

3y   2y mol/L at equilibrium

The solubility calculated in part a is y. Then the solubility product is

$$K = [Ca^{2+}]^3[PO_4{}^{3-}] = (3y)^3(2y)^2 = 108y^5$$
$$= 108 \times (1.1 \times 10^{-7})^5 = 1.7 \times 10^{-33}$$

■

PROBLEM 16.2 Calculate the solubility in mol/L of $Ca_3(PO_4)_2$ in a 0.010 M solution of $CaCl_2$. Use the result of Example 16.3. □

PRECIPITATION PROBLEMS

The rules for deciding which way a reaction can go (section 13.2) are the same for solubility equilibria as for any equilibrium. The reaction quotient Q—for example, $Q = [Ag^+][Cl^-]$—is compared with K. It is only in a saturated solution that equilibrium is established and $Q = K$:

With a few exceptions (notably MgC_2O_4), when a solution of a slightly soluble salt is supersaturated, precipitation occurs promptly.

$Q < K$	Solution is unsaturated. Solute can dissolve. Solute will not precipitate.
$Q = K$	Equilibrium. Solution is saturated.
$Q > K$	Solution is supersaturated. Solute can precipitate.

One use of the solubility product is to determine whether or not precipitation is possible when the ions have specified concentrations. We can also find what concentrations are necessary to cause or to prevent precipitation—important when two solutes are to be separated by precipitating one of them.

Remember that when two solutions are mixed, each is diluted; the new concentrations are less than in the initial solutions. These new concentrations are the ones to use in calculating Q. With dilute solutions, there is no significant error in calculating the volume of the mixed solution by adding the volumes of the original solutions.

EXAMPLE 16.4 A solution is prepared by mixing 50 mL of 0.0050 M $CaCl_2$ solution and 80 mL of 0.0060 M ZnF_2 solution at 25°C. Is it possible for CaF_2 to precipitate?

ANSWER First calculate the concentrations of Ca^{2+} and F^- ions in the mixed solution, *assuming that no precipitation occurs.* The total volume is 50 mL + 80 mL = 130 mL = 0.130 L. The number of moles of Ca^{2+} used is

$$50 \text{ mL} \times \frac{0.0050 \text{ mol}}{1000 \text{ mL}} = 2.5 \times 10^{-4} \text{ mol}$$

Its concentration in the mixed solution is

$$[Ca^{2+}] = \frac{2.5 \times 10^{-4} \text{ mol}}{0.130 \text{ L}} = 1.9 \times 10^{-3} \text{ mol/L}$$

This calculation can be simplified as in section 12.2:

$$[Ca^{2+}]_{final} = [Ca^{2+}]_{initial} \times \frac{\text{initial volume}}{\text{final volume}}$$

$$= 0.0050 \frac{mol}{L} \times \frac{50 \not{mL}}{130 \not{mL}}$$

$$= 1.9 \times 10^{-3} \text{ mol/L}$$

Since 1 mole of ZnF_2 yields 2 moles of F^- when dissolved, the quantity of F^- in the solution is

$$80 \not{mL} \times \frac{0.0060 \not{mol\ ZnF_2}}{1000 \not{mL}} \times \frac{2 \text{ mol } F^-}{1 \not{mol\ ZnF_2}} = 9.6 \times 10^{-4} \text{ mol } F^-$$

Its concentration is

$$[F^-] = \frac{9.6 \times 10^{-4} \text{ mol}}{0.130 \text{ L}} = 7.4 \times 10^{-3} \frac{mol}{L}$$

Then

$$Q = [Ca^{2+}][F^-]^2$$
$$= (1.9 \times 10^{-3})(7.4 \times 10^{-3})^2$$
$$= 1.0 \times 10^{-7}$$

$K = 1.61 \times 10^{-10}$ (Table 16.1). Since $Q > K$, the solution is supersaturated and it is possible for CaF_2 to precipitate. ■

PROBLEM 16.3 A solution is prepared by mixing 40 mL of 0.0080 M $Pb(NO_3)_2$ solution and 100 mL of 0.0010 M K_2SO_4 solution at 25°C. Calculate the reaction quotient (Q) for the dissolving of $PbSO_4$. Is it possible for $PbSO_4$ to precipitate? □

EXAMPLE 16.5 These solutions are mixed: 100 mL 0.0200 M $BaCl_2$ and 50.0 mL 0.0300 M Na_2SO_4. (a) Will $BaSO_4$ precipitate? (b) What are the final concentrations of Ba^{2+} and SO_4^{2-}? (c) If $BaSO_4$ precipitates, how many moles of it precipitate?

ANSWER

(a) The total volume is 150 mL. The concentrations of interest to us, after mixing and before precipitation, are

$$[Ba^{2+}] = 0.0200 \frac{mol}{L} \times \frac{100 \not{mL}}{150 \not{mL}} = 0.0133 \text{ mol/L}$$

$$[SO_4^{2-}] = 0.0300 \frac{mol}{L} \times \frac{50.0 \not{mL}}{150 \not{mL}} = 0.0100 \text{ mol/L}$$

Then $Q = [Ba^{2+}][SO_4^{2-}] = 0.0133 \times 0.0100 = 1.33 \times 10^{-4}$. Since $Q > K = 1.05 \times 10^{-10}$, precipitation is possible.

(b) When the precipitation reaction occurs, nearly all the SO_4^{2-} is used up:

SO_4^{2-} is the limiting reactant.

	Ba^{2+}	+	SO_4^{2-}	$\longrightarrow$	$BaSO_4(c)$
Initial concentration	0.0133		0.0100		
Change	−0.0100		−0.0100		
Concentration at equilibrium	0.0033		0 mol/L		

Of course, the concentration of SO_4^{2-} is not really zero, but it is close to it. Calculating its concentration is the usual equilibrium problem with a common ion:

$$BaSO_4(c) \rightleftharpoons \underset{0.0033 + y}{Ba^{2+}} + \underset{y \text{ mol/L}}{SO_4^{2-}}$$

In the saturated solution,

$$[Ba^{2+}][SO_4^{2-}] = K$$

$$(0.0033 + y)y = 1.05 \times 10^{-10}$$

The familiar assumption that $0.0033 + y \approx 0.0033$ is reasonable. Then,

$$y = 3.2 \times 10^{-8} \text{ mol/L} = [SO_4^{2-}]$$

$$[Ba^{2+}] = 3.3 \times 10^{-3} \text{ mol/L}$$

(c) Since $[SO_4^{2-}]$ decreases from 0.0100 mol/L to almost zero, 0.0100 mole of $BaSO_4$ precipitates per liter. The amount of $BaSO_4$ that precipitates is

$$\frac{0.0100 \text{ mol}}{1000 \cancel{\text{mL}}} \times 150 \cancel{\text{mL}} = 1.50 \times 10^{-3} \text{ mol}$$ ■

PROBLEM 16.4 A solution is prepared by mixing 50 mL 0.00500 M $Hg(NO_3)_2$ and 50 mL 0.00500 M NaI. The reaction $Hg^{2+} + 2I^- \longrightarrow HgI_2(c)$ occurs. (a) What are the final concentrations of Hg^{2+} and I^-? (b) How many moles of HgI_2 precipitate? □

EXAMPLE 16.6 What is the minimum concentration of S^{2-} that can cause precipitation of CdS from a 0.10 M solution of Cd^{2+} at 25°C?

ANSWER The minimum concentration is that corresponding to a saturated solution. In a saturated solution,

$$CdS(c) \rightleftharpoons Cd^{2+} + S^{2-}$$

$$[Cd^{2+}][S^{2-}] = 1.4 \times 10^{-29} \qquad \text{(Table 16.1)}$$

Since $[Cd^{2+}] = 0.10$ mol/L,

$$[S^{2-}] = \frac{1.4 \times 10^{-29}}{0.10} = 1.4 \times 10^{-28} \text{ mol/L}$$ ■

PROBLEM 16.5 What is the minimum concentration of I^- that can cause precipitation of PbI_2 from a 0.050 M solution of $Pb(NO_3)_2$? □

Both 10^{-5} mol/L and 10^{-10} mol/L are small concentrations, but they are very different.

It is often possible to separate two ions by precipitating first a salt that contains one of them, and then a salt that contains the other. The requirement is that the two salts have different solubilities, even though both are "slightly soluble." Precipitation of one may then be nearly complete before the other begins to precipitate.

EXAMPLE 16.7 A solution is 0.010 M with respect to NaCl and 0.0010 M with respect to K_2CrO_4. A solution of $AgNO_3$ is added slowly to the solution without changing its volume appreciably. (a) What is $[Ag^+]$ when AgCl begins to precipitate? (b) What is $[Ag^+]$ when Ag_2CrO_4 begins to precipitate? (c) What is $[Cl^-]$ when Ag_2CrO_4 begins to precipitate? What percentage is this of the original $[Cl^-]$?

BOX 16.2
ACTIVITY COEFFICIENTS

The law of chemical equilibrium is based on the assumption that the reacting particles are independent—that except for colliding and reacting, they exert no forces on each other. This is a good assumption for molecules in a gas at low pressure. It is also a good assumption for the solute molecules in a dilute solution, where the solute molecules are usually far apart and the forces exerted on the solute molecules by the solvent molecules can be regarded as a kind of constant background, independent of solute concentration. Ionic solutions are more complicated, however. The electrostatic forces between charged bodies are very strong and do not drop off sharply with increasing distance. Therefore, the law of chemical equilibrium is a rather poor approximation for all but extremely dilute ionic solutions.

This difficulty can be overcome—in theory if not always in practice—by writing equilibrium conditions in terms of **activities** instead of concentrations. For our purposes, two features of activity will suffice to define it:

1. The activity and the concentration become equal for a very dilute solution.
2. The law of chemical equilibrium is exact when expressed in terms of activities.

The ratio of activity (a) to concentration (molarity, c) is the **activity coefficient**, f:

$$f = a/c \quad \text{or} \quad a = fc \qquad (i)$$

f is the correction factor that changes concentration to activity.

This discussion of activities will, for brevity, be confined to solubility equilibria. However, the same principles apply to all equilibria involving ions. Consider, as a simple example, the process of dissolving AgCl in water:

$$AgCl(c) \rightleftharpoons Ag^+ + Cl^-$$

The equilibrium condition in terms of concentrations is

$$[Ag^+][Cl^-] = K$$

This is all right as long as $[Ag^+]$, $[Cl^-]$, and any other ionic concentrations in the solution are extremely small, like 10^{-6} M. In so dilute a solution, the ions are far enough apart so that the forces between them can be ignored. For larger concentrations, an accurate equilibrium condition is obtained only in terms of activities:

$$a_{Ag^+} \times a_{Cl^-} = K$$

Thus, the familiar equations containing concentrations, which are strictly correct only for very dilute solutions, can still be used for more concentrated solutions by replacing each concentration with the corresponding activity. Each activity can be expressed, by equation i, as the product of an activity coefficient and a concentration:

$$a_{Ag^+} = f_{Ag^+}[Ag^+] \quad \text{and} \quad a_{Cl^-} = f_{Cl^-}[Cl^-]$$

Then the equilibrium condition is

$$f_{Ag^+}[Ag^+]f_{Cl^-}[Cl^-] = K$$

or

$$[Ag^+][Cl^-] = \frac{K}{f_{Ag^+}f_{Cl^-}} \qquad (ii)$$

The activity coefficients f_{Ag^+} and f_{Cl^-} become 1 in very dilute solutions. In more concentrated solutions, they make equation ii exactly correct. They depend on the concentrations of Ag^+, Cl^-, and any other ions present.

Activity coefficients of ions are less than 1, and decrease as concentrations increase. Look at equation ii to see what this means. $K/f_{Ag^+}f_{Cl^-}$ is *greater* than K because the denominator is less than 1. The product of concentrations, $[Ag^+][Cl^-]$, in a saturated solution is therefore larger than it would be if the activity coefficients were 1. *The smaller the activity coefficients are, the more* AgCl *can be dissolved.* A decrease in activity coefficients indicates that the ions are more "contented"; that is, less "eager" to get out of solution and into the solid phase.

This table shows some activity coefficients calculated from the measured electromotive force of galvanic cells (section 17.8). Products of activity coefficients are given because the activity coefficients of individual ions cannot be obtained from experimental data. Even for ions with single charges, like K^+ and Cl^-, the assumption that activities can be replaced by concentrations—that activity coefficients are equal to 1—is not very good. When the charges are larger, as in $Zn^{2+}SO_4^{2-}$, the forces between the ions are stronger and the activity coefficients are far smaller. These data show that ignoring activity coefficients—pretending they are equal to 1—leads to answers that are, at best, only roughly correct.

CONCENTRATION, mol/L	KCl $f_{K^+}f_{Cl^-}$	$ZnSO_4$ $f_{Zn^{2+}}f_{SO_4^{2-}}$
0.005	0.86	0.23
0.01	0.81	0.15
0.02	0.76	0.089
0.05	0.67	0.041
0.10	0.60	0.022
0.20	0.53	0.0112
0.50	0.44	0.0045
1.00	0.38	0.0026

ANSWER

(a) When AgCl *begins* to precipitate, $[Cl^-]$ is still 0.010 mol/L. In this saturated solution,

$$[Ag^+][Cl^-] = K_{AgCl} = 1.76 \times 10^{-10}$$

$$[Ag^+] = \frac{1.76 \times 10^{-10}}{0.010} = 1.8 \times 10^{-8} \text{ mol/L}$$

(b) When Ag_2CrO_4 begins to precipitate, $[CrO_4{}^{2-}]$ is still 0.0010 mol/L. Then

$$[Ag^+]^2[CrO_4{}^{2-}] = K_{Ag_2CrO_4} = 1.2 \times 10^{-12}$$

$$[Ag^+]^2 = \frac{1.2 \times 10^{-12}}{0.0010} = 1.2 \times 10^{-9}$$

$$[Ag^+] = 3.5 \times 10^{-5} \text{ mol/L}$$

(c) The concentration of Ag^+ when Ag_2CrO_4 begins to precipitate was calculated in part b. Then

$$[Ag^+][Cl^-] = 1.76 \times 10^{-10}$$

$$[Cl^-] = \frac{1.76 \times 10^{-10}}{3.5 \times 10^{-5}} = 5.0 \times 10^{-6} \text{ mol/L}$$

The percentage of the original Cl^- still in solution is

$$\frac{5.0 \times 10^{-6} \text{ mol/L}}{0.010 \text{ mol/L}} \times 100\% = 0.050\%$$

Thus, if the AgCl is filtered out just before Ag_2CrO_4 begins to precipitate, 99.95% of the Cl^- will be separated from the $CrO_4{}^{2-}$. ■

Example 16.7 illustrates the central concept of inorganic qualitative analysis: the separation of elements and compounds by precipitation.

PROBLEM 16.6 A solution is 0.010 M with respect to $BaCl_2$ and 0.010 M with respect to $CaCl_2$. A solution of KF is added slowly without changing the volume appreciably. (a) What is $[F^-]$ when CaF_2 begins to precipitate? (b) What is $[F^-]$ when BaF_2 begins to precipitate? (c) What is $[Ca^{2+}]$ when BaF_2 begins to precipitate? What percentage of the Ca^{2+} is still in solution? (Ignore the difference between solubilities at 25° and 25.8°C.) □

16.2 SOLUBILITY AND pH

The cation of a salt may be acidic; the anion may be basic or, less often, acidic (sections 14.4 and 15.7). An acidic ion will react with a base added to the solution, decreasing its concentration and bringing more of the salt into solution to maintain equilibrium. Thus, adding a base to the solution increases the solubility of a salt that contains an acidic ion. Likewise, a basic ion will react with added acid, so that adding an *acid* increases the solubility of a salt that contains a *basic* ion. The most common cases of this kind are those in which the anion is a base. It is a base when it is the conjugate base of a weak acid. Thus, $NO_3{}^-$, $SO_4{}^{2-}$, and Cl^- are not bases of appreciable strength, but S^{2-}, F^-, and $PO_4{}^{3-}$ are; they are the conjugate bases of the weak acids HS^-, HF, and $HPO_4{}^{2-}$.

When the negative ion is a base, its concentration is decreased by adding H^+. For example, the reaction

$$HS^- \rightleftharpoons H^+ + S^{2-}$$

$$ZnS \rightleftharpoons Zn^{2+} + S^{2-}$$
$$+ H^+ \rightleftharpoons HS^-$$
$$+ H^+ \rightleftharpoons H_2S$$

is shifted to the left by an increase in $[H^+]$. In turn, the solubility of a metallic sulfide is affected by the concentration of S^{2-}. When acid is added to a saturated solution of ZnS, the concentration of S^{2-} is decreased because it is converted to HS^- and, to a smaller extent, to H_2S. The reaction

$$ZnS(c) \rightleftharpoons Zn^{2+} + S^{2-}$$

is thus shifted to the right, meaning that the solubility of ZnS increases. In general, *the solubility of a salt containing the conjugate base of a weak acid is increased by addition of a stronger acid to the solution.*

SOLUBILITY OF SULFIDES

The effect of acid on solubility is especially pronounced with sulfides, because S^{2-} is the conjugate base of the very weak acid HS^-. The sensitivity of solubility to pH makes sulfides valuable in separating metal ions, which is the major part of qualitative analysis.

If any two of the three concentrations $[H^+]$, $[S^{2-}]$, and $[H_2S]$ are given, the third can be calculated from the equation (Example 15.24, section 15.8) for 25°C,

H_2S is produced in solution by decomposition of thioacetamide:

$$CH_3C(=S)-NH_2 + 2H_2O \longrightarrow CH_3COO^- + NH_4^+ + H_2S.$$

$$\frac{[H^+]^2[S^{2-}]}{[H_2S]} = K_1K_2 = 6.3 \times 10^{-8} \times 10^{-14} = 6 \times 10^{-22} \qquad (2)$$

In a solution of H_2S, saturated at 1 atm pressure and 25°C, $[H_2S] = 0.10$ mol/L. This is the usual concentration of H_2S in the solutions used in qualitative analysis.

EXAMPLE 16.8 Find $[Zn^{2+}]$ in a saturated solution of ZnS in which $[H^+] = 0.20$ mol/L and $[H_2S] = 0.10$ mol/L at 25°C.

ANSWER In a saturated solution of ZnS, the equilibrium

$$ZnS(c) \rightleftharpoons Zn^{2+} + S^{2-}$$

is established. The equilibrium condition is

$$[Zn^{2+}][S^{2-}] = 2.9 \times 10^{-25} \qquad \text{(Table 16.1)}$$

$[Zn^{2+}]$ can be calculated if $[S^{2-}]$ is known. But $[S^{2-}]$ can be calculated from equation 2, since $[H^+]$ and $[H_2S]$ are given:

$$\frac{[H^+]^2[S^{2-}]}{[H_2S]} = 6 \times 10^{-22}$$

$$[S^{2-}] = \frac{6 \times 10^{-22}\,[H_2S]}{[H^+]^2} = \frac{6 \times 10^{-22} \times 0.10}{(0.20)^2} = 1.5 \times 10^{-21} \text{ mol/L}$$

Then

$$[Zn^{2+}][S^{2-}] = 2.9 \times 10^{-25} \qquad (3)$$

$$[Zn^{2+}] = \frac{2.9 \times 10^{-25}}{1.5 \times 10^{-21}} = 2 \times 10^{-4} \text{ mol/L}$$

■

We should point out again that in a given solution, there is only one concentration of any given molecule or ion, for example S^{2-}. The value of $[S^{2-}]$ must therefore be the same in the H_2S and the ZnS equilibrium conditions.

EXAMPLE 16.9 What is the maximum concentration of H^+ in a solution from which PbS will precipitate, when $[Pb^{2+}] = 0.0050$ and $[H_2S] = 0.10$ mol/L?

BOX 16.3

HOW NOT TO WORRY ABOUT K_2 FOR H_2S

A footnote to Table 15.2 pointed out that K_2 for H_2S is still uncertain, and may be much smaller than 10^{-14}. However, if K_2 is changed, the solubility products of the metallic sulfides are also changed. The net result is that the concentrations of metal ions, like $[Zn^{2+}]$, come out to be the same. The issue can be evaded by combining equations 4 and 5 so that $[S^{2-}]$ drops out:

$$[Zn^{2+}][S^{2-}] \times \frac{[H_2S]}{[H^+]^2[S^{2-}]} = \frac{[Zn^{2+}][H_2S]}{[H^+]^2}$$

$$= \frac{2.9 \times 10^{-25}}{6 \times 10^{-22}}$$

$$= 5 \times 10^{-4}$$

This is the equilibrium condition for the reaction

$$ZnS + 2H^+ \rightleftharpoons Zn^{2+} + H_2S$$

which is really what happens when ZnS dissolves in acid, regardless of how much or how little S^{2-} is present.

ANSWER In a saturated solution of PbS, the reaction

$$PbS(c) \rightleftharpoons Pb^{2+} + S^{2-}$$

is at equilibrium. Its equilibrium condition is

$$[Pb^{2+}][S^{2-}] = 8.8 \times 10^{-29}$$

Since $[Pb^{2+}]$ is given, $[S^{2-}]$ can be calculated:

$$[S^{2-}] = \frac{8.8 \times 10^{-29}}{0.0050} = 1.8 \times 10^{-26} \text{ mol/L}$$

Precipitation is possible if $[S^{2-}]$ is larger than this value. The concentration of H^+ needed to establish this S^{2-} concentration is calculated from equation 2:

$$[H^+]^2 = \frac{6 \times 10^{-22}\,[H_2S]}{[S^{2-}]} = \frac{6 \times 10^{-22} \times 0.10}{1.8 \times 10^{-26}} = 3 \times 10^3$$

$$[H^+] = 60 \text{ mol/L}$$

Preventing precipitation of PbS requires making $[S^{2-}]$ equal to or less than 1.8×10^{-26} mol/L. Smaller $[S^{2-}]$ means larger $[H^+]$; if $[S^{2-}]$ is less than 1.8×10^{-26} mol/L, $[H^+]$ must be greater than 60 mol/L. This is an absurdly high concentration—a saturated solution of HCl in water is only 12 M. Thus, PbS will precipitate (or remain undissolved) even in very concentrated strong acids. There is, of course, a *trace* of Pb^{2+} in solution, and its concentration increases as $[H^+]$ increases, but it remains only a trace. ■

PROBLEM 16.7 What is the minimum pH at which CoS can precipitate when $[Co^{2+}] = 0.010$ mol/L and $[H_2S] = 0.10$ mol/L? □

The following example illustrates an important step in qualitative analysis, the separation of metals according to the solubilities of their sulfides.

EXAMPLE 16.10 In a certain solution, $[H^+] = 0.30$, $[H_2S] = 0.10$, $[Cd^{2+}] = 0.010$, $[Fe^{2+}] = 0.010$, $[Hg^{2+}] = 0.010$, and $[Mn^{2+}] = 0.010$ mol/L. Which of the four sulfides, CdS, HgS, MnS, and ZnS, can precipitate? (Assume that $[H^+]$ and $[H_2S]$ are kept constant.)

ANSWER Whether a sulfide can precipitate from this solution depends on the concentration of S^{2-}. Since $[H^+]$ and $[H_2S]$ are given, $[S^{2-}]$ can be calculated from equation 2.

$$[S^{2-}] = \frac{6 \times 10^{-22} \times [H_2S]}{[H^+]^2} = \frac{6 \times 10^{-22} \times 0.10}{(0.30)^2} = 7 \times 10^{-22} \text{ mol/L}$$

For each of the metal ions (M^{2+}),

$$[M^{2+}] = 0.010 \text{ mol/L}$$

The reaction quotient for the reaction

$$MS(c) \rightleftharpoons M^{2+} + S^{2-}$$

is

$$Q = [M^{2+}][S^{2-}] = (0.010)(7 \times 10^{-22}) = 7 \times 10^{-24}$$

If, for a given sulfide, $7 \times 10^{-24} < K$, that sulfide will not precipitate; if $7 \times 10^{-24} > K$, there can be a precipitate. Inspection of Table 16.1 shows that CdS and HgS may precipitate, while FeS and MnS will not. ■

PROBLEM 16.8 In a solution of pH 9.0, $[H_2S] = 0.10$, $[Ag^+] = 0.0010$, and $[Ni^{2+}] = 0.020$ mol/L. Can Ag_2S precipitate? Can NiS precipitate? □

SOLUBILITY OF HYDROXIDES

Another application of acid-base equilibrium in qualitative analysis is the separation of metals by the different solubilities of their hydroxides. Here, more obviously than with sulfides, solubility is sensitive to pH: $[OH^-] = K_w/[H^+]$ and $pOH = pK_w - pH$. Adding H^+—decreasing the pH—makes any hydroxide more soluble by decreasing $[OH^-]$ and shifting an equilibrium like $Mg(OH)_2 \rightleftharpoons Mg^{2+} + 2OH^-$ to the right.

EXAMPLE 16.11 (a) A solution is 0.010 M with respect to Al^{3+}, 0.010 M with respect to Mg^{2+}, and 1.0 M with respect to NH_3. Is it possible for $Al(OH)_3$ to precipitate? Is it possible for $Mg(OH)_2$ to precipitate? (b) NH_4Cl is added to the solution in part a until it is 1.0 M with respect to NH_4^+. Is it now possible for $Al(OH)_3$ to precipitate; for $Mg(OH)_2$ to precipitate?

ANSWER

(a) To decide whether $Al(OH)_3$ or $Mg(OH)_2$ can precipitate, we need to know the concentration of Al^{3+} or Mg^{2+}, respectively (both are given), and of OH^-. $[OH^-]$ is calculated from the basic ionization constant of NH_3, as in section 15.3:

$$NH_3 + H_2O \rightleftharpoons NH_4^+ + OH^-$$

$1.0 - y$ y y mol/L at equilibrium

$$\frac{y^2}{1.0 - y} = K_b = 1.77 \times 10^{-5} \qquad \text{(Table 15.2)}$$

Assume $y \ll 1.0$. Then

$$\frac{y^2}{1.0} = 1.77 \times 10^{-5}$$

$$y = 4.2 \times 10^{-3} \text{ mol/L} = [OH^-]$$

For $Al(OH)_3$,

$$Q = [Al^{3+}][OH^-]^3$$
$$= (0.010)(4.2 \times 10^{-3})^3 = 7.4 \times 10^{-10}$$

$K_{sp} = 10^{-32}$ and $Q > K_{sp}$. $Al(OH)_3$ can precipitate. For $Mg(OH)_2$,

$$Q = [Mg^{2+}][OH^-]^2$$
$$= (0.010)(4.2 \times 10^{-3})^2 = 1.8 \times 10^{-7}$$

$K_{sp} = 5.7 \times 10^{-12}$ and $Q > K_{sp}$. $Mg(OH)_2$ can also precipitate. We have not succeeded in separating Al^{3+} from Mg^{2+}.

(b) When NH_4^+ is added to NH_3, the result is a buffer solution. To calculate $[OH^-]$ in this solution, proceed as in section 15.9. Let z be the mol/L of NH_3 that ionize:

$$\underset{1.0 - z \approx 1.0}{NH_3} + H_2O \rightleftharpoons \underset{1.0 + z \approx 1.0}{NH_4^+} + \underset{z \text{ mol/L at equilibrium}}{OH^-}$$

$$\frac{1.0z}{1.0} = 1.77 \times 10^{-5}$$

$$z = 1.77 \times 10^{-5} \text{ mol/L} = [OH^-]$$

For $Al(OH)_3$,

$$Q = [Al^{3+}][OH^-]^3$$
$$= (0.010)(1.77 \times 10^{-5})^3 = 5.5 \times 10^{-17}$$

Again, $Q > K_{sp}$ and precipitation is possible. For $Mg(OH)_2$,

$$Q = [Mg^{2+}][OH^-]^2$$
$$= (0.010)(1.77 \times 10^{-5})^2 = 3.1 \times 10^{-12}$$

This time $Q < K_{sp} = 5.7 \times 10^{-12}$, and precipitation is impossible. The addition of NH_4^+ decreases $[OH^-]$ sufficiently so that Al^{3+} and Mg^{2+} can be separated by precipitating only $Al(OH)_3$. ■

PROBLEM 16.9 A solution is 0.010 M with respect to Fe^{3+}. What is the lowest pH at which $Fe(OH)_3$ will precipitate from it? □

Some metal hydroxides are amphoteric (section 14.5). They not only dissolve in acids (thus behaving as bases), but they also dissolve in bases (thus behaving as acids). Examples are $Al(OH)_3$, $Cr(OH)_3$, $Pb(OH)_2$, $Sn(OH)_2$, and $Zn(OH)_2$.

EXAMPLE 16.12 The equilibrium constant of the reaction

$$Al(OH)_3 + OH^- \rightleftharpoons Al(OH)_4^-$$

is 40 at 25°C. What pH is necessary to dissolve 0.010 mole of $Al(OH)_3$ per liter as $Al(OH)_4^-$? Is this the maximum or the minimum pH?

ANSWER The equilibrium condition is

$$\frac{[Al(OH)_4^-]}{[OH^-]} = K$$

If 0.010 mole $Al(OH)_3$ is dissolved per liter, then $[Al(OH)_4^-] = 0.010$ mol/L, and

$$[OH^-] = \frac{[Al(OH)_4^-]}{K} = \frac{0.010}{40} = 2.5 \times 10^{-4} \text{ mol/L}$$

$$[H^+] = \frac{K_w}{[OH^-]} = \frac{1.00 \times 10^{-14}}{2.5 \times 10^{-4}} = 4.0 \times 10^{-11} \text{ mol/L}$$

$$pH = 10.40$$

This is the minimum pH. Increasing the pH means increasing $[OH^-]$, which favors the dissolving process. ■

PROBLEM 16.10 The equilibrium constant of the reaction

$$Pb(OH)_2 + OH^- \rightleftharpoons Pb(OH)_3^-$$

is 8×10^{-2} at 25°C. What concentration of OH^- is needed to dissolve 0.010 mole of $Pb(OH)_2$ per liter? □

Some sulfides are described as "amphoteric" because they dissolve in basic (as well as acidic) solutions. For example, antimony sulfide, Sb_2S_3, dissolves in basic solutions by several reactions, such as

$$Sb_2S_3 + 8OH^- \longrightarrow 2Sb(OH)_4^- + 3S^{2-}$$

$$Sb_2S_3 + 3S^{2-} \longrightarrow 2SbS_3^{3-}$$

In the second reaction, Sb_2S_3 is behaving as a Lewis acid in reacting with the base S^{2-}. As_2S_3 is also amphoteric. Although bismuth is just below arsenic and antimony in the periodic table, Bi_2S_3 does not dissolve in OH^- solutions and dissolves only slightly in S^{2-} solutions.

16.3 COORDINATION IONS

We introduced coordination ions (also known as "complex" ions) in section 9.2. Recall that a coordination ion consists of a metal atom or positive ion (usually a transition metal) and a number of **ligands**, uncharged molecules or negative ions. The metal ion is a Lewis acid (section 14.10); the ligands are bases. Theories of the structure of coordination ions are discussed in Chapter 21.

The dissociation of the coordination ion $Ag(NH_3)_2^+$ may be represented as

$$Ag(NH_3)_2^+ \rightleftharpoons Ag^+ + 2NH_3 \qquad (4)$$

The equilibrium constant for this reaction is called the **dissociation constant** of $Ag(NH_3)_2^+$,

$$\frac{[Ag^+][NH_3]^2}{[Ag(NH_3)_2^+]} = K \text{ or } K_{diss} \qquad (5)$$

The official (IUPAC) way to write the formulas is $[Ag(NH_3)_2]^+$, etc., but in this chapter we reserve [ ] for concentrations.

Table 16.2 gives dissociation constants for a number of coordination ions. A small dissociation constant, like 10^{-34} for $Co(NH_3)_6^{3+}$, belongs to a very stable coordination ion. A not-so-small dissociation constant, like 10^{-5} for $Co(NH_3)_6^{2+}$, identifies a not-so-stable ion.

Suppose that we have a 0.10 M solution of $AgNO_3$, and that we add NH_3 to make $[NH_3] = 4.0$ mol/L. The reaction

$$Ag^+ + 2NH_3 \rightleftharpoons Ag(NH_3)_2^+$$

is the reverse of the dissociation reaction; its equilibrium constant is therefore

K_{diss} is copied from Table 16.2.

$$K = \frac{1}{K_{diss}} = \frac{1}{6.15 \times 10^{-8}} = 1.63 \times 10^{7}$$

With K so large, the reaction goes practically to completion. In an approach that is familiar to you by now, we resolve the reaction conceptually into two steps:

STEP 1. Ag^+ and NH_3 react completely to produce the coordination ion. To see what the final concentrations are, write an equilibrium table:

	Ag^+	+	$2NH_3$	$\rightleftharpoons$	$Ag(NH_3)_2^+$
Initial concentration	0.10		4.0		0
Change	$-y$		$-2y$		$+y$
Concentration after step 1	$0.10 - y$		$4.0 - 2y$		y mol/L

The limiting reactant is Ag^+. Since it is nearly all used up, $0.10 - y \approx 0$; then $y \approx 0.10$ mol/L $= [Ag(NH_3)_2{}^+]$, and $4.0 - 2y \approx 3.8$ mol/L $= [NH_3]$. We thus know, with almost no calculation, the concentrations of NH_3 (the excess reactant) and $Ag(NH_3)_2{}^+$. However, not quite all of the Ag^+ reacts. The remaining small concentration of Ag^+ is the quantity that we will calculate in step 2.

TABLE 16.2
DISSOCIATION CONSTANTS OF COORDINATION IONS IN WATER AT 25°C†

COORDINATION ION	K
$Ag(CN)_2^-$	6.42×10^{-21}
$Ag(NH_3)_2^+$	6.15×10^{-8}
$Ag(S_2O_3)_2^{3-}$	5.47×10^{-14}
$AuBr_4^-$	2.2×10^{-33}
$AuCl_4^-$	7.0×10^{-26}
$Cd(CN)_4^{2-}$	5.0×10^{-19}
CdI_4^{2-}	3×10^{-7}
$Cd(NH_3)_4^{2+}$	2.57×10^{-7}
$Co(NH_3)_6^{2+}$	10^{-5}
$Co(NH_3)_6^{3+}$	10^{-34}
$Cu(CN)_4^{3-}$	3×10^{-31}
$Cu(NH_3)_4^{2+}$	4.4×10^{-13}
$Fe(CN)_6^{4-}$	10^{-35}
$Fe(CN)_6^{3-}$	10^{-42}
$HgBr_4^{2-}$	2.0×10^{-22}
$HgCl_4^{2-}$	1.1×10^{-16}
$Hg(CN)_4^{2-}$	4.9×10^{-42}
HgI_4^{2-}	7.0×10^{-31}
$Ni(NH_3)_4^{2+}$	1.12×10^{-8}
$Zn(NH_3)_4^{2+}$	3.47×10^{-10}
$Zn(OH)_4^{2-}$	4.5×10^{-16}

† K is for dissociation of all ligands; for example, $Ag(CN)_2^- \rightleftharpoons Ag^+ + 2CN^-$.

BOX 16.4
STEPWISE DISSOCIATION OF COORDINATION IONS

The dissociation of a coordination ion, like the ionization of a polyprotic acid, occurs in steps:

$$Ag(NH_3)_2^+ \rightleftharpoons AgNH_3^+ + NH_3 \qquad K_1 = 1.2 \times 10^{-4}$$

$$AgNH_3^+ \rightleftharpoons Ag^+ + NH_3 \qquad K_2 = 5.1 \times 10^{-4}$$

The net reaction is equation 4, so that $K_{diss} = K_1K_2$. You will recall (from section 15.6) that for a polyprotic acid, K_1 is greater than K_2 because the charge becomes more negative as protons are removed. The same reasoning applies to a coordination ion with charged ligands:

$$Ag(S_2O_3)_2^{3-} \rightleftharpoons AgS_2O_3^- + S_2O_3^{2-} \qquad K_1 = 4.4 \times 10^{-6}$$

$$AgS_2O_3^- \rightleftharpoons Ag^+ + S_2O_3^{2-} \qquad K_2 = 1.2 \times 10^{-8}$$

The first thiosulfate ion should be lost more easily because it is leaving behind an ion with a charge of −1: once the covalent bond is broken, $S_2O_3^{2-}$ is repelled by this negative ion. The second $S_2O_3^{2-}$ leaves behind an ion with a charge of +1 and is therefore attracted even without a covalent bond. Thus, $K_1 > K_2$. However, when the ligand is uncharged, as in $Ag(NH_3)_2^+$, there is no reason to assume any particular relationship between K_1 and K_2.

We will confine our attention to problems in which the solution contains an excess of the ligand. Then the equilibrium condition for the net reaction, like equation 5, will suffice. $[NH_3]$ is known—it is the excess—and $[Ag(NH_3)_2^+]$ is known because most of the Ag^+ is converted to $Ag(NH_3)_2^+$. Then $[Ag^+]$ is correctly calculated from equation 5. Some $AgNH_3^+$ is also present—indeed, more than Ag^+—but $[AgNH_3^+]$ is much less than $[Ag(NH_3)_2^+]$, because the excess NH_3 shifts both dissociation steps to the left. We simply do not calculate $[AgNH_3^+]$.

STEP 2. The coordination ion dissociates to a small extent, producing a little Ag^+ and a little extra NH_3. Think of $[NH_3] = 3.8$ mol/L and $[Ag(NH_3)_2^+] = 0.10$ mol/L as the initial concentrations. Then, for the dissociation reaction, the usual equilibrium table can be written:

If z cannot be neglected, stepwise dissociation (Box 16.4) must be taken into account. We will not concern ourselves with those cases.

	$Ag(NH_3)_2^+$ $\rightleftharpoons$	Ag^+ +	$2NH_3$
Initial concentration	0.10	0	3.8
Change in step 2	$-z$	$+z$	$+2z$
Concentration at equilibrium	$0.10 - z$ ≈ 0.10	z	$3.8 + 2z$ ≈ 3.8 mol/L

The equilibrium condition is

$$\frac{[Ag^+][NH_3]^2}{[Ag(NH_3)_2^+]} = 6.15 \times 10^{-8}$$

$$\frac{z(3.8)^2}{0.10} = 6.15 \times 10^{-8}$$

$$z = 4.3 \times 10^{-10}\ \text{mol/L} = [Ag^+]$$

PROBLEM 16.11 A solution is prepared by dissolving 0.500 mol $CdSO_4$ and 2.10 mol NaCN in water to make 1.00 L of solution. (a) Assuming that the reaction $Cd^{2+} + 4CN^- \longrightarrow Cd(CN)_4^{2-}$ goes to completion, what are the concentrations of $Cd(CN)_4^{2-}$ and CN^- in the solution? (b) Calculate $[Cd^{2+}]$ at equilibrium. □

The last two calculations illustrate the effect of adding a ligand, like NH_3 or CN^-, that reacts with a metal ion in solution to form a stable coordination ion. The

Ions written without ligands are actually hydrated. The water is customarily omitted from the formula.

concentration of the uncomplexed metal ion—Ag^+ or Cd^{2+}—is decreased to a very low value by the formation of the coordination ion. A result is that slightly soluble salts of this metal become more soluble:

$$\begin{array}{ccc} AgCl(c) \rightleftharpoons & Ag^+ & + Cl^- \\ & + & \\ & 2NH_3 & \\ & \updownarrow & \\ & Ag(NH_3)_2{}^+ & \end{array}$$

As Ag^+ is removed by reaction with NH_3, more AgCl goes into solution, according to Le Chatelier's principle.

Dissolving precipitates (or preventing them from forming) is a common application of coordination ion formation. We discussed one category of examples in the last section—using OH^- as the ligand to dissolve amphoteric hydroxides. Since many metals form stable coordination ions with NH_3, precipitates can often be dissolved in an ammonia solution. For example,

$$AgCl(c) + 2NH_3(aq) \longrightarrow Ag(NH_3)_2{}^+ + Cl^-$$

$$Zn(OH)_2(c) + 4NH_3(aq) \longrightarrow Zn(NH_3)_4{}^{2+} + 2OH^-$$

AgI is too insoluble to dissolve in NH_3 solution, but it will dissolve in a solution containing cyanide ion:

$$AgI(c) + 2CN^- \longrightarrow Ag(CN)_2{}^- + I^-$$

Formerly called *sodium hyposulfite*, hence its nickname "hypo."

The fixing solution used in photographic processing contains sodium thiosulfate, $Na_2S_2O_3$, which dissolves unreacted AgBr:

$$AgBr(c) + 2S_2O_3{}^{2-} \longrightarrow Ag(S_2O_3)_2{}^{3-} + Br^-$$

The negative ions that react with metal ions to form precipitates may also react with the precipitates to form coordination ions. The amphoteric hydroxides are good examples:

$$Al^{3+} + 3OH^- \rightleftharpoons Al(OH)_3 \qquad (6)$$

$$Al(OH)_3 + OH^- \rightleftharpoons Al(OH)_4{}^- \qquad (7)$$

If the goal is to precipitate nearly all of the Al^{3+} as $Al(OH)_3$, a delicate problem is confronted. A small excess of OH^- should be used to drive reaction 6 to the right and decrease $[Al^{3+}]$, but a large excess will drive reaction 7 to the right and increase $[Al(OH)_4{}^-]$. The problem is solved in this instance by a buffer, usually $NH_3 + NH_4{}^+$, that makes the solution only weakly basic.

SUMMARY

In a saturated solution of a slightly soluble salt such as AgCl or CaF_2, an equilibrium condition ($Q = K$) such as $[Ag^+][Cl^-] = K$ or $[Ca^{2+}][F^-]^2 = K$ is satisfied. K is the **solubility product** of the salt. The solubility of the salt in mol/L can be calculated from the solubility product by the same procedure used for other equilibrium calculations. If a soluble salt with an ion in common with the slightly soluble salt is present in solution, the solubility is decreased (the **common ion effect**). $Q = K$ only in a saturated solution; when $Q < K$, the solution is unsaturated and when $Q > K$, the solution is supersaturated. When solutions are mixed, Q can be calculated from the given concentrations. Precipitation is possible only if $Q > K$; dissolving is possible if $Q < K$. When two slightly soluble salts with an ion in common (AB, AC) differ greatly in solubility, one of them can be precipitated by adding A and removed before the other begins to precipitate. B and C can thus be separated.

When the negative ion in a slightly soluble salt is the conjugate base of a weak acid, its concentration is decreased by adding a stronger acid to the solution. The salt therefore becomes more soluble as the pH decreases. This effect is especially important for many sulfides (containing S^{2-}) and hydroxides (containing OH^-). By adjusting the pH, the solubilities of sulfides and hydroxides can be controlled and the compounds can be separated. Some hydroxides and sulfides are amphoteric, dissolving in base (OH^- or S^{2-}) as well as acid.

A **coordination ion** is a combination of a positive metal ion with several uncharged or negative **ligands**. The **dissociation constant** is K for a reaction such as $Ag(NH_3)_2{}^+ \rightleftharpoons Ag^+ + 2NH_3$. If an excess of ligand is added to a solution, nearly all the metal ion is converted to the complex ion and the ligand concentration is equal to the excess. The small concentration of the metal ion remaining uncomplexed can then be calculated from the equilibrium condition. Adding a ligand decreases the concentration of the metal ion and thus increases the solubility of slightly soluble salts.

16.4 ADDITIONAL PROBLEMS

Consult Tables 16.1, 16.2, and 15.2 (section 15.3) as needed.

SLIGHTLY SOLUBLE SALTS

16.12 Calculate the concentration of Ba^{2+} in a saturated solution of BaF_2 in water.

16.13 Calculate the concentration of Ba^{2+} in a solution saturated with BaF_2 and containing 0.20 mole of KF per liter.

16.14 Calculate the concentration of Pb^{2+} in a solution saturated with $PbSO_4$ and (a) containing no other solute; (b) 0.10 M with respect to $MgSO_4$.

16.15 The solubility of cesium permanganate, $CsMnO_4$, in water is 0.097 g/100 mL solution at 1°C. Calculate its solubility product. Does this salt conform to the solubility rules (Table 9.3, section 9.2)?

16.16 The solubility in water of calcium stearate, $Ca(C_{18}H_{35}O_2)_2$ (607 g/mol), is 4×10^{-3} g/100 mL at 15°C. Calculate (a) its solubility product and (b) its solubility, in grams per 100 mL solution, in 0.050 M $NaC_{18}H_{35}O_2$ solution.

16.17 The solubility in water of lanthanum molybdate, $La_2(MoO_4)_3$ (757.6 g/mol), is 1.8×10^{-3} g/100 mL at 25°C. (a) Convert this solubility to molarity. (b) Calculate the solubility product. (c) Calculate the solubility in g/100 mL of $La_2(MoO_4)_3$ in a 0.020 M solution of $LaCl_3$.

16.18 A 0.0050 M solution of $AgNO_3$ and a 0.0005 M solution of NaBr are mixed in equal volumes at 25°C. What is the reaction quotient? Is it possible for AgBr to precipitate?

16.19 A solution is prepared by mixing equal volumes of 0.010 M $MgCl_2$ and 0.020 M $Na_2C_2O_4$ solutions at 18°C. Is it possible for MgC_2O_4 to precipitate in the resulting solution?

16.20 30 mL of a 0.0020 M solution of $BaCl_2$ and 50 mL of a 0.050 M solution of NaF are mixed at 25.8°C. (a) Find $[Ba^{2+}]$ and $[F^-]$ in the mixed solution before any reaction occurs. (b) Is it possible for BaF_2 to precipitate?

16.21 These solutions are mixed: 50.0 mL 0.100 M $AgNO_3$ and 30.0 mL 0.100 M KBr. (a) What are the concentrations of Ag^+ and Br^- at equilibrium? (b) How many moles of AgBr precipitate?

16.22 A solution is 0.100 M with respect to Pb^{2+}. If 0.103 mole of Na_2SO_4 is added per liter of this solution (with no volume change), what percentage of the Pb^{2+} remains in solution?

16.23 A solution has $[Ag^+] = 0.020$ mol/L and $[Ba^{2+}] = 0.020$ mol/L. K_2CrO_4 is added slowly to this solution without increasing its volume appreciably. (a) What is $[CrO_4{}^{2-}]$ when Ag_2CrO_4 just begins to precipitate? (b) What is $[CrO_4{}^{2-}]$ when $BaCrO_4$ just begins to precipitate? (c) When enough K_2CrO_4 has been added so that $BaCrO_4$ begins to precipitate, what concentration of Ag^+ remains in solution? What percentage is this of the original Ag^+? Would adding $CrO_4{}^{2-}$ be a practical method of separating Ag^+ and Ba^{2+}?

16.24 (a) Calculate the concentration of $CrO_4{}^{2-}$ in saturated solutions of Ag_2CrO_4 and of $BaCrO_4$. (b) Which salt, Ag_2CrO_4 or $BaCrO_4$, has the greater solubility (expressed in mol/L) in water? Which salt has the larger solubility product? How do you account for any apparent disagreement?

16.25 A solution is 0.10 M with respect to K_2SO_4 and 0.10 M with respect to K_2CrO_4. A solution of $Pb(NO_3)_2$

PLATE 9
Redox titration using potassium permanganate. The flask contains an unknown quantity of Fe^{2+}. An excess of one drop of $KMnO_4$ turns the solution faintly purple ("pink"). *(Photos by E.R. Degginger and Grant Heilman)*

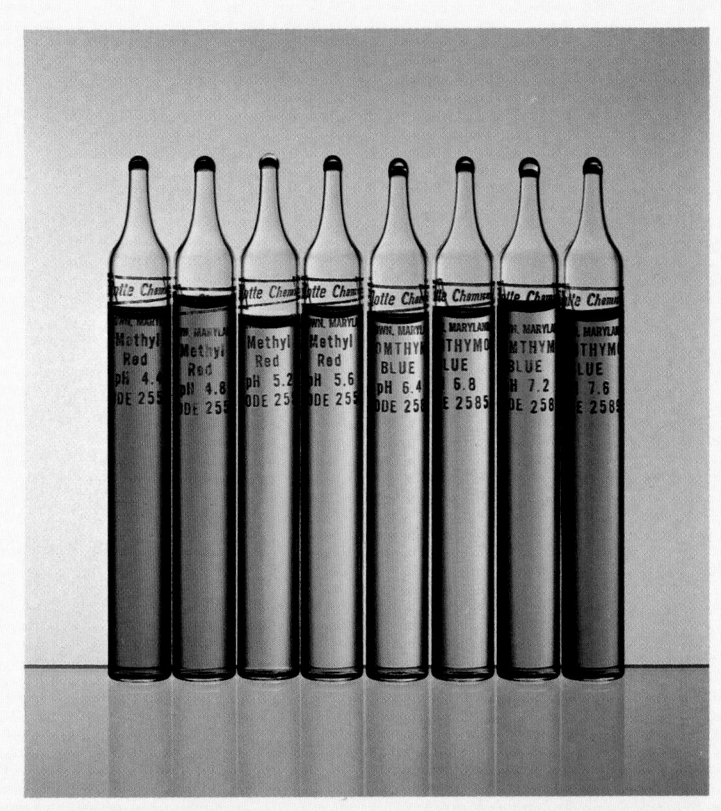

PLATE 10
Acid-base indicators in buffer solutions.
From left to right: methyl red at pH 4.4, 4.8, 5.2, and 5.6; bromthymol blue at pH 6.4, 6.8, 7.2, and 7.6.

PLATE 11
Formation of solid KBr from its elements; $K(c) + \frac{1}{2}BR_2(\ell) \rightarrow KBr(c)$.

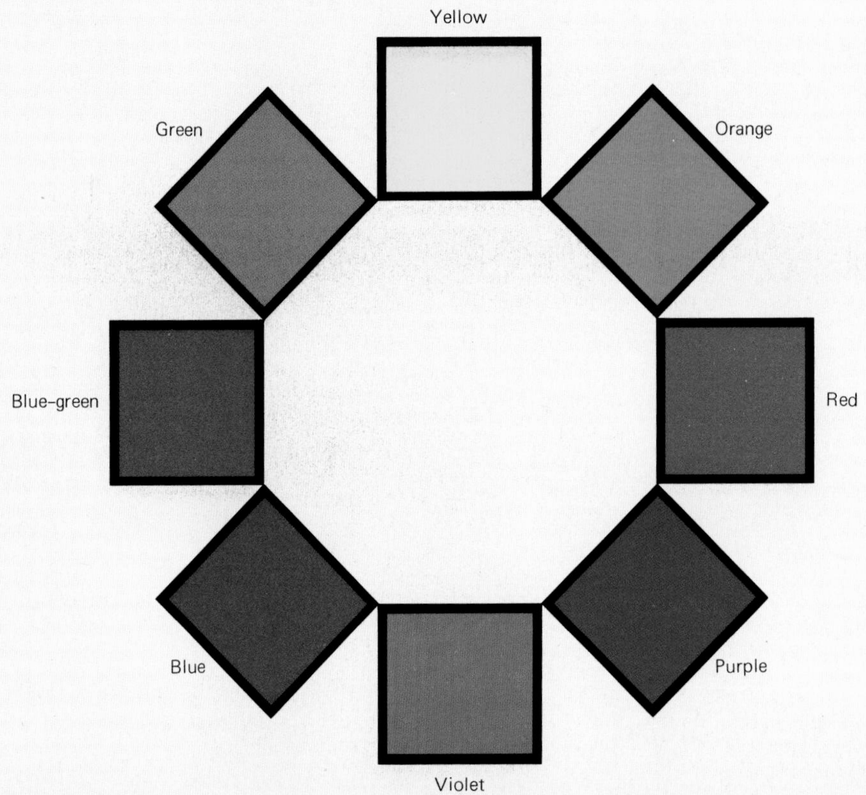

PLATE 12
Approximate color wheel.
Complementary colors are opposite each other.

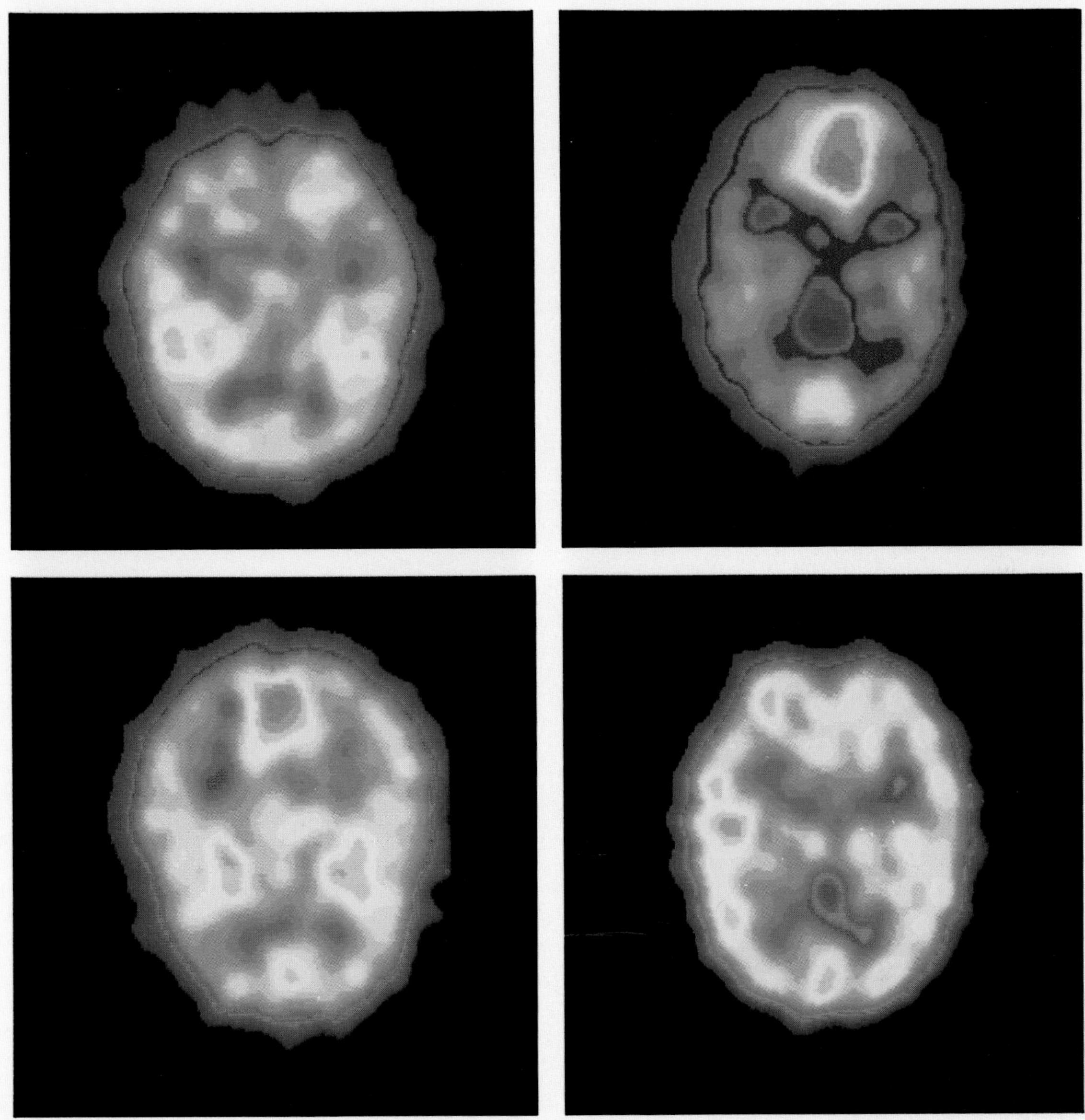

PLATE 13
PET cross-sectional images of glucose metabolism in the brain. Top photos: normal elderly person *(left)*; person with Alzheimer's disease *(right)*. Bottom photos: normal person *(left)*; person with schizophrenia *(right)*. *(Courtesy of Brookhaven National Laboratory and New York University Medical Center)*

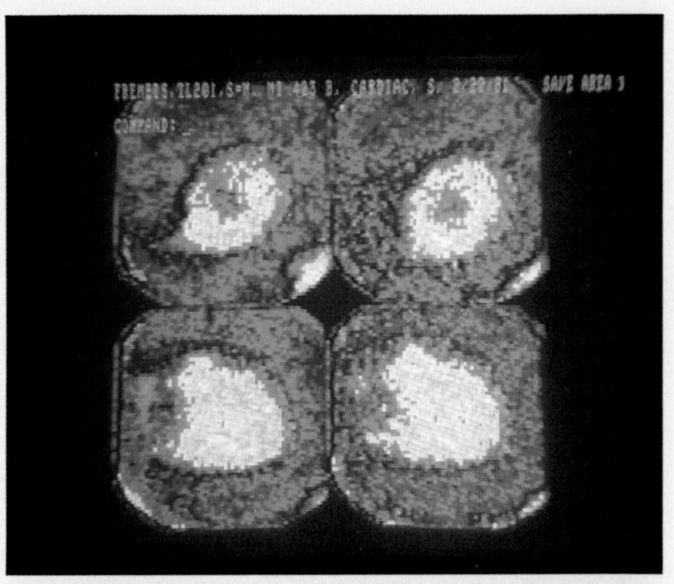

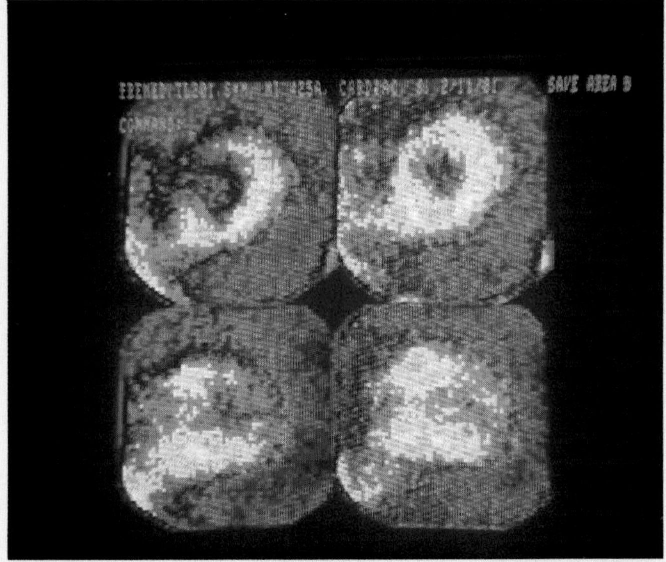

PLATE 14
Heart muscle of a normal person *(top)* and of a person with heart disease. $^{201}TlCl$ is the radioactive compound used. *(Courtesy of Dr. Sheldon B. Eisenberg, Pascack Valley Hospital, Westwood, N.J.)*

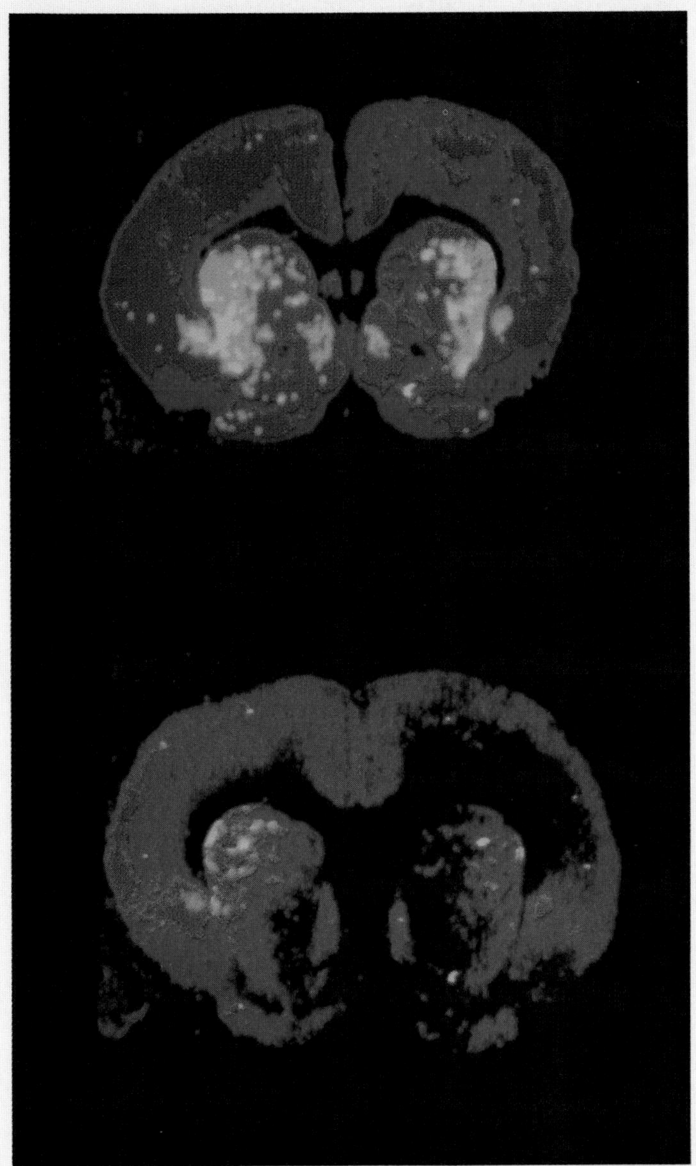

PLATE 15
$^3_1$H-autoradiographs of the effect of aging on opiate receptors in a coronal section of the brain of a young rat *(upper photo)* and an old rat. Quantitative analyses of neuroreceptors are readily made by automated autoradiogram systems. *(Autoradiographs by Luigi F. Agnati, University of Modena, Modena, Italy)*

PLATE 16
The Tokamak fusion test reactor.
(Courtesy of Princeton Plasma Physics Laboratory)

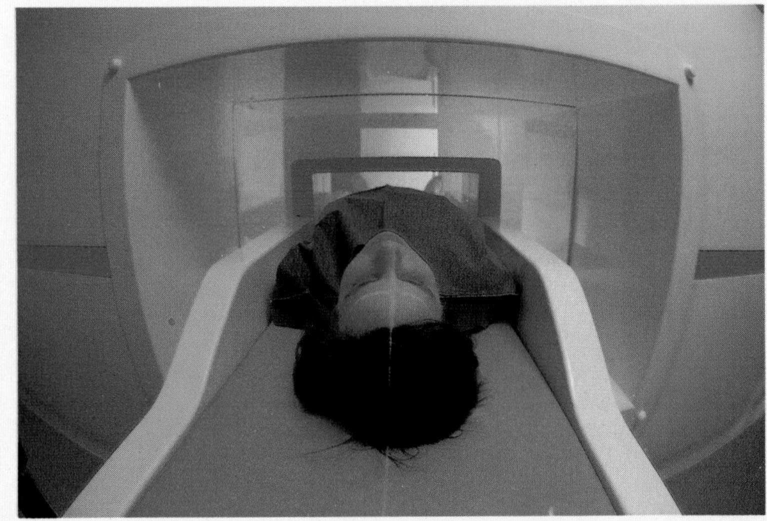

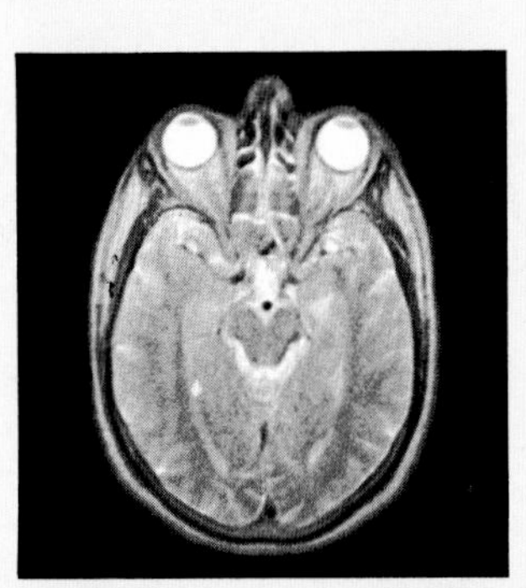

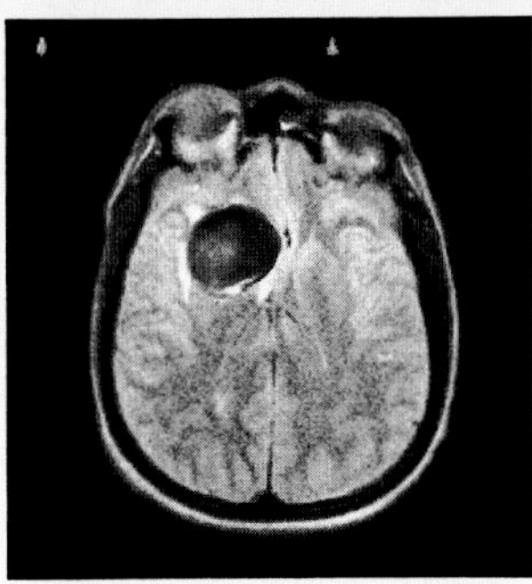

PLATE 17
A magnetic resonance imaging instrument. *(right, above) (Dan McCoy from Rainbow)*. An MRI normal brain image *(left, above)*, and a brain image clearly revealing an aneurysm *(right)*. *(Courtesy of Bruker Instruments, Inc., Billerica, Massachusetts)*

is added slowly without changing the volume appreciably. (a) Which salt, $PbSO_4$ or $PbCrO_4$, will precipitate first? (b) What is $[Pb^{2+}]$ when the salt in part a begins to precipitate? (c) What is $[Pb^{2+}]$ when the other lead salt begins to precipitate? (d) What are $[SO_4^{2-}]$ and $[CrO_4^{2-}]$ when the lead salt in part c begins to precipitate?

16.26 (a) What is the solubility of HgS in mol/L at 25°C? (b) Convert this solubility to number of ions/L. (c) Is there anything strange about your answer to part b? Can you explain what it means? (*Suggestion*: Ionic reactions usually occur very fast in both directions.)

SOLUBILITY AND pH

16.27 You have a 0.050 M solution of $FeCl_3$ and wish to precipitate nearly all the Fe^{3+} as $Fe(OH)_3$. You require that no more than 0.1% of the Fe^{3+} may remain in solution. What should the pH of the solution be after precipitation? To be on the safe side, should the pH be higher or lower than you calculated?

16.28 Find the solubility, in mol/L, of FeS in a solution in which $[H_2S] = 0.10$ mol/L and the pH is (a) 3.0 and (b) 9.0.

16.29 (a) Find the minimum pH at which ZnS will precipitate from a solution 0.10 M with respect to H_2S and 0.020 M with respect to Zn^{2+}. (b) Find the minimum pH at which NiS will precipitate from a solution 0.10 M with respect to H_2S and 0.020 M with respect to Ni^{2+}. (c) In what range should the pH be in order to separate Ni^{2+} and Zn^{2+} by precipitation of one sulfide and not the other?

16.30 Barium compounds are very toxic. However, a suspension of $BaSO_4$ is fed to patients with suspected disease of the digestive system because Ba^{2+} scatters X rays. (a) What is the solubility of $BaSO_4$ in water in g/100 mL at 25°C? (b) Does this solubility change much with pH? Why or why not? (c) Barium sulfite, $BaSO_3$, is also nearly insoluble in water. However, when $BaSO_3$ has been given to patients erroneously, fatalities have occurred. Why? (The pH of gastric juice is 1 to 3.)

16.31 Two common constituents of kidney stones are calcium phosphate, $Ca_3(PO_4)_2$, and calcium oxalate hydrate, $CaC_2O_4(H_2O)$. (a) Are stones containing these salts more likely to form when the urine is acidic or when it is basic? (b) If $[Ca^{2+}] = 5 \times 10^{-3}$ mol/L, at what minimum concentration of $C_2O_4^{2-}$ can $CaC_2O_4(H_2O)$ precipitate? Assume that the solubility is the same at body temperature (37°C) as at 25°C. (c) If the pH is 8.0, what are $[H_2C_2O_4]$ and $[HC_2O_4^-]$ when $[C_2O_4^{2-}]$ has the value calculated in part b? What is the total concentration of the three oxalate species? For $H_2C_2O_4$, $K_1 = 5.90 \times 10^{-2}$ and $K_2 = 6.40 \times 10^{-5}$. (d) Answer the question in part c for pH 5.0. Does the pH change make a great difference in the likelihood of forming stones? Why or why not?

16.32 The equilibrium constant of the reaction

$$Sn(OH)_2(c) + OH^- \rightleftharpoons Sn(OH)_3^-$$

is 1×10^{-2} at 25°C. What is the solubility (mol/L) of $Sn(OH)_2$ in a buffer solution of pH 12.5?

COORDINATION IONS

16.33 Find the concentration of Au^{3+} in a solution in which the other equilibrium concentrations are $[Cl^-] =$ 0.10 mol/L and $[AuCl_4^-] = 0.20$ mol/L.

16.34 Find the concentration of Cd^{2+} in a solution in which the other equilibrium concentrations are $[I^-] =$ 0.10 mol/L and $[CdI_4^{2-}] = 0.20$ mol/L.

16.35 A solution is prepared by dissolving 0.25 mol $Ni(NO_3)_2$ and 1.50 mol NH_3 in water to make 1.00 L of solution. What are the concentrations of $Ni(NH_3)_4^{2+}$, NH_3, and Ni^{2+} in this solution at equilibrium?

16.36 A solution is prepared by dissolving 0.010 mol $Cd(NO_3)_2$ and 0.052 mol KCN in water to make 200 mL of solution. Find the concentration of Cd^{2+} in the solution.

16.37 A solution is prepared by dissolving 0.010 mol $Co(NO_3)_3$ and 0.080 mol NH_3 in water to make 1.00 L. What are the concentrations of (a) NH_3, (b) $Co(NH_3)_6^{3+}$, and (c) Co^{3+} in the solution at equilibrium?

16.38 A buffer solution at 25°C has a constant pH of 11.00. To 1.00 L of this solution 0.0200 mol $Zn(NO_3)_2$ is added. (a) What is the ratio $[Zn^{2+}]/[Zn(OH)_4^{2-}]$ in the resulting solution? (b) Calculate the concentrations of Zn^{2+} and $Zn(OH)_4^{2-}$.

SELF-TEST

16.39 (20 points) Calculate the solubility of MnS in (a) pure water and (b) a 0.050 M solution of $Mn(NO_3)_2$.

16.40 (15 points) These solutions are mixed: 25.0 mL of 1.0×10^{-4} M $BaCl_2$ and 75.0 mL of 1.0×10^{-4} M Na_2SO_4. (a) Calculate $[Ba^{2+}]$ and $[SO_4^{2-}]$ in the mixed solution before any reaction occurs. (b) Is it possible for $BaSO_4$ to precipitate?

16.41 (30 points) A solution is 0.050 M with respect to KBr and 0.050 M with respect to KI. A solution of $AgNO_3$ is added slowly without changing the volume appreciably. (a) What is $[Ag^+]$ when AgI begins to precipitate? (b) What is $[Ag^+]$ when AgBr begins to precipitate? (c) What is $[I^-]$ when AgBr begins to precipitate? What percentage of the original I^- is still in solution?

16.42 (15 points) Find the minimum pH at which MnS will precipitate from a solution that is 0.10 M with respect to H_2S and 0.010 M with respect to $MnCl_2$.

16.43 (20 points) A solution is prepared by dissolving 0.020 mol $Hg(NO_3)_2$ and 0.100 mol NaBr in water to make 1.00 L. Find $[Br^-]$, $[HgBr_4^{2-}]$, and $[Hg^{2+}]$ in the solution.

17

ELECTROCHEMISTRY

All chemistry is electrical. Attraction between unlike charges is what holds atoms, molecules, liquids, and solids together. In a sense, then, all chemistry is "electrochemistry," but the term is reserved for processes in which an electric current passes through an ionic conductor, usually a solution. Electric current flows for one of two reasons: either it is pushed by an external force, as when a storage battery is charged, or it is generated by a chemical reaction, as when a battery is discharged.

The effects of electrochemistry abound in our everyday experience. Batteries start our cars, light our flashlights, power our portable radios, and make our calculators calculate. If present research on fuel cells and more compact storage batteries is successful, batteries may not only start our cars, but keep them running, with little noise and no pollution. Familiar products of electrochemical reactions include aluminum, copper, chromium plating on bumpers, silver plating on spoons, and chlorinated drinking water. On the unpleasant side, corrosion—a partially electrochemical process—eats up metals worth billions of dollars each year. This chapter will help you understand how all of these electrochemical processes work.

17.1 UNITS OF ELECTRIC CHARGE

A current is a flow of *electric charge*, measured in *coulombs* (C or coul; Appendix A.6). Electric current is measured in *amperes* (A or amp). When 1 A flows for 1 second, the charge passed through the circuit is 1 C:

$$\mathbf{1\ A = 1\ \frac{C}{sec}} \qquad \textbf{or} \qquad \mathbf{1\ C = 1\ A\ sec} \tag{1}$$

Originally, I stood for "intensity" and Q for "quantity."

Current is represented by I, charge by Q, and time by t. Thus, we can write

$$I = \frac{Q}{t} \qquad \text{or} \qquad Q = It$$

The coulomb is an arbitrary unit of charge. As section 2.5 showed—and some of the evidence will be reviewed in the next section—there is a natural unit of electric charge, the charge of 1 electron:

$$1 \text{ electron} = 1 \text{ electronic unit} = 1.602 \times 10^{-19}\text{ C} \tag{2}$$

An electron (e^-) has a *negative* charge, -1 electronic unit. Chemistry deals with quantities much larger than a single atom. That is why we usually talk about moles instead of molecules. In the same way, there is a unit of electric charge that is related to the mole, the **faraday**:

$$\begin{matrix}\textbf{1 mole} \\ \textbf{of electrons}\end{matrix} = \textbf{1 faraday} = \begin{matrix}\textbf{Avogadro's number} \\ \textbf{of electrons}\end{matrix} = N_A \textbf{ electrons} \quad (3)$$

The faraday is expressed in terms of the coulomb by multiplying the charge of one electron by Avogadro's number:

$$1 \text{ faraday} = 1.602 \times 10^{-19} \frac{\text{coulomb}}{\cancel{\text{electron}}} \times 6.022 \times 10^{23} \cancel{\text{electrons}}$$

$$= 9.65 \times 10^4 \text{ coulombs}$$

The conversion factor between coulombs and faradays is represented by $\mathscr{F}$ and is called **Faraday's constant.** The best value is $\mathscr{F} = 9.64853 \times 10^4$ C/faraday.

17.2 ELECTRODE REACTIONS

Graphite, though not a metal, conducts electricity like a metal and is a common electrode material.

We saw in Chapter 9 that an ionic solution conducts electricity. How does the current get into and out of the solution, and with what results?

Electricity is carried through metallic conductors by a flow of electrons. The metallic conductors in contact with the solution are called **electrodes**. When the electrons reach the solution at one electrode, they cannot cross over to the other electrode, because the solution is not an electronic conductor. The electrons must somehow be carried by the ions in the solution to the other electrode, where they resume their journey through the metal (Figure 17.1). There are many ways

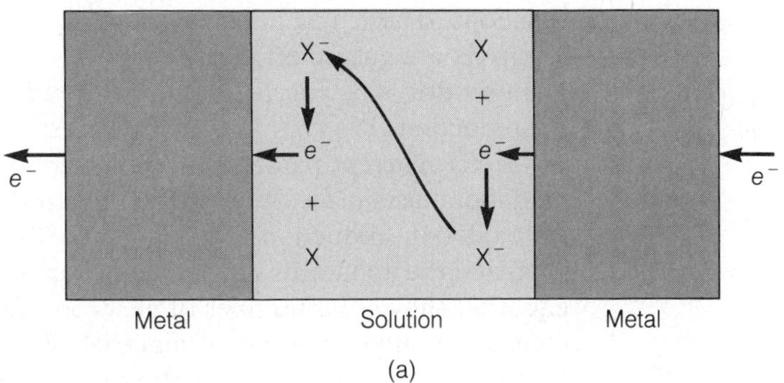

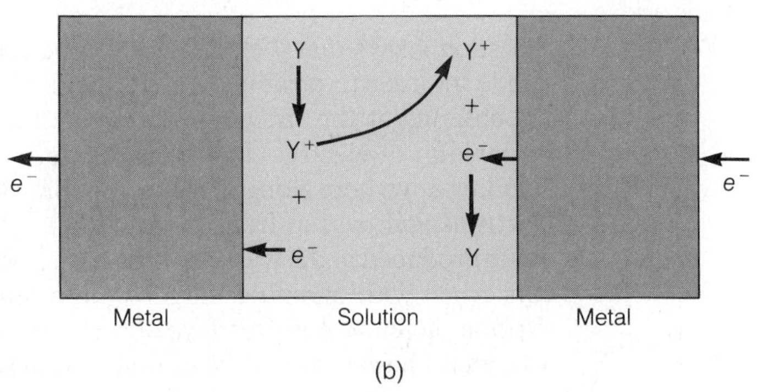

FIGURE 17.1
Two of the simpler ways for an electric current to flow through a circuit containing both electronic and ionic conductors. (a) Electrons from the metal on the right attach themselves to an atom or molecule X; $X + e^- \rightarrow X^-$. The X^- ion travels across the solution to the metal on the left, where it gives up its electron and becomes X again; $X^- \rightarrow X + e^-$. The net effect is that X is transported from right to left. (b) Y^+ ions are traveling from left to right in the solution. Electrons arriving at the right side of the solution attach themselves to Y^+; $Y^+ + e^- \rightarrow Y$. On the left, an atom or molecule Y gives up an electron to the metal and becomes Y^+; $Y \rightarrow Y^+ + e^-$. The net effect is that Y is transported from left to right.

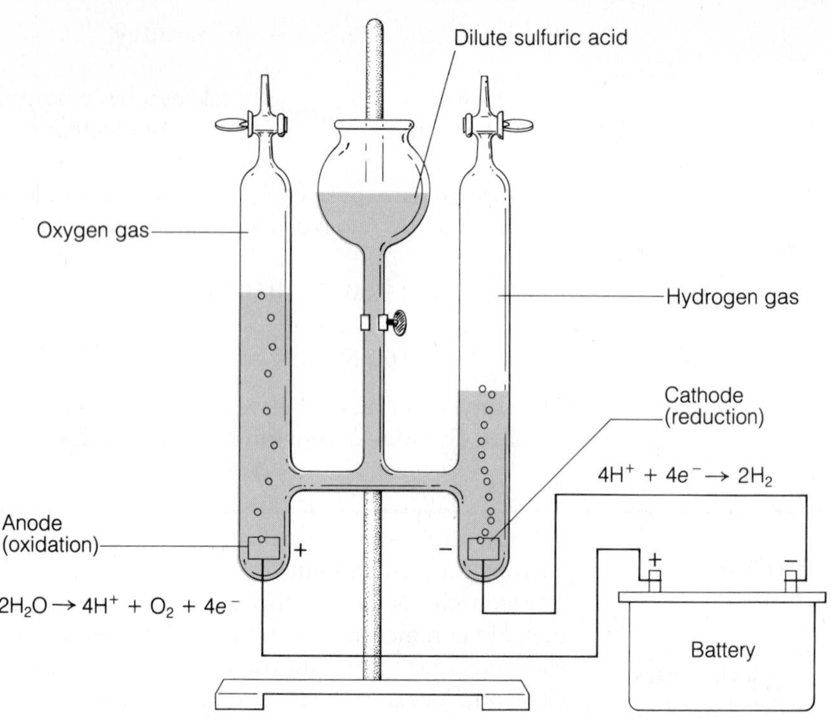

FIGURE 17.2
Laboratory apparatus for the electrolysis of water.

Alphabetical order, or vowels and consonants together:

oxidation	reduction
anode	cathode

of getting electrons into and out of the solution, but they all require chemical reactions. These reactions are necessarily oxidation–reduction reactions, since electrons are gained and lost.

In section 9.5, you learned how to write partial equations for oxidation and for reduction. We can now use these equations to describe what happens when an electric current passes through a solution. At one electrode, called the **anode**, oxidation takes place. At the other electrode, the **cathode**, reduction takes place. An oxidation–reduction reaction caused by an electric current is called **electrolysis**, and the apparatus in which it occurs is called an **electrolytic cell**. The partial equation shows the number of electrons that must be transferred to produce or consume a given amount of material. For example, suppose that water is electrolyzed by passing a current through it (Figure 17.2). The partial equations are

Or an "electrolysis cell."

Pure water is a very poor electrical conductor. To make it a better conductor, add a little sulfuric acid, $H_2SO_4(H^+ + HSO_4^-)$.

$$2H_2O \longrightarrow O_2(g) + 4H^+ + 4e^- \qquad \text{(oxidation, at anode)} \qquad (4)$$

$$4H^+ + 4e^- \longrightarrow 2H_2(g) \qquad \text{(reduction, at cathode)} \qquad (5)$$

$$2H_2O \longrightarrow 2H_2(g) + O_2(g)$$

Equation 4 says that 4 electrons must be removed from 2 water molecules at the anode in order to produce 1 molecule of O_2. At the same time, 4 other electrons are entering at the cathode to produce 2 molecules of H_2 (equation 5). There is no buildup of electric charge anywhere in the apparatus. Imagine that you are standing anywhere along the wire watching the electrons go by (Figure 17.3). Four electrons will pass in front of you while 1 O_2 molecule and 2 H_2 molecules are being produced and 2 water molecules are being decomposed. Since we are usually concerned with quantities larger than a few molecules, it is more convenient to express the charge in faradays and the amounts of matter in moles. Thus, when 4 faradays pass through the circuit, 2 moles of H_2O are decomposed, 2 moles of H_2 are produced, and 1 mole of O_2 is produced.

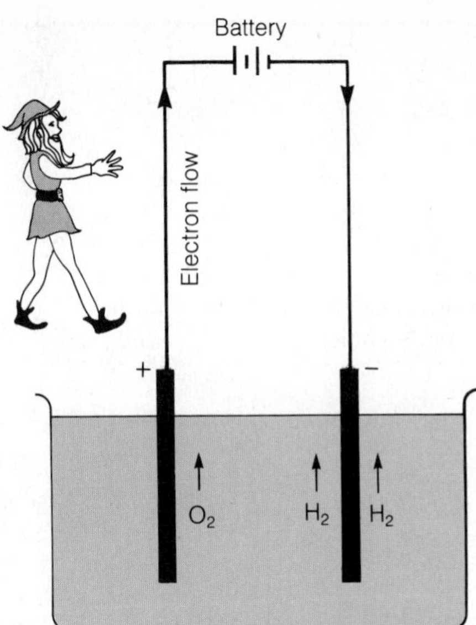

FIGURE 17.3
A molecule-sized observer counts 4 electrons when the cell produces 1 O_2 molecule and 2 H_2 molecules.

Michael Faraday, 1791–1867

The relationship between the charge and the quantity of matter in an electrode reaction was discovered by Michael Faraday in 1832–1833. Faraday did not know about partial reactions, electrons, or the unit that would be named after him. Rather, his work led to those concepts. All he had were chemicals, a balance, and electricity (from a battery—not from the electric company, of course). He did know something about atomic and molecular weights. His results are summarized in two laws, now famous as **Faraday's laws**. They are not mere approximations, like the ideal gas law and the law of chemical equilibrium; they are exact within the precision of the best experiments that have been performed:

1. **The mass of a substance produced or consumed is proportional to the quantity of charge (current × time) that has passed through the circuit.**
2. **The number of faradays that must pass through a circuit when one mole of a substance is produced or consumed is a whole number.**

The whole number is usually 1, 2, 3, or 4, hardly ever greater than 8.

The choice of the whole number depends on the substance and the reaction. For example,

$$2H_2O \longrightarrow O_2 + 4H^+ + 4e^- \quad \text{(4 faradays produce 1 mol } O_2\text{)}$$
$$2H^+ + 2e^- \longrightarrow H_2 \quad \text{(2 faradays produce 1 mol } H_2\text{)}$$
$$Ag^+ + e^- \longrightarrow Ag \quad \text{(1 faraday produces 1 mol Ag)}$$
$$Al^{3+} + 3e^- \longrightarrow Al \quad \text{(3 faradays produce 1 mol Al)}$$

PROBLEM 17.1 How many faradays must be transferred for each of the following? (First write the partial equations.) (a) To produce 1 mole of Au in the reaction

$$4HAuCl_4 + 6H_2O \longrightarrow 4Au + 3O_2 + 16H^+ + 16Cl^-$$

(b) To consume 1 mole of Cd in the reaction

$$Cd + 4CN^- + 2H_2O \longrightarrow Cd(CN)_4{}^{2-} + H_2 + 2OH^-$$ □

BOX 17.1

FARADAY'S LAWS AND THE QUANTIZATION OF ELECTRICITY

The fact that one mole of atoms or molecules always requires an integral number of faradays suggests the existence of a natural unit of charge (section 2.5). To see how this conclusion follows from Faraday's laws, think of silver. To produce one mole of Ag (N_A atoms) requires one faraday, or $\mathscr{F}$ coulombs:

$$1\ \cancel{\text{mol Ag}} \times \frac{1\ \cancel{\text{faraday}}}{1\ \cancel{\text{mol Ag}}} \times \frac{9.65 \times 10^4\ \text{C}}{1\ \cancel{\text{faraday}}} = 9.65 \times 10^4\ \text{C}$$

To produce one *atom* of Ag requires only $\mathscr{F}/N_A$ coulomb:

$$1\ \cancel{\text{atom Ag}} \times \frac{1\ \cancel{\text{mol Ag}}}{6.02 \times 10^{23}\ \cancel{\text{atoms Ag}}} \times \frac{9.65 \times 10^4\ \text{C}}{1\ \cancel{\text{mol Ag}}} = 1.60 \times 10^{-19}\ \text{C}$$

That does not tell us much until we see that the same unit, $\mathscr{F}/N_A = 1.60 \times 10^{-19}$ C, appears for *every* element:

$$\left.\begin{array}{r}\text{1 mole}\\ N_A \text{ atoms or molecules}\end{array}\right\} \text{requires } 1, 2, 3, \ldots \times \mathscr{F}\ \text{C}$$

1 atom or molecule requires $1, 2, 3, \ldots \times (\mathscr{F}/N_A)$ C
1 molecule O_2 requires $4\ \mathscr{F}/N_A$ C
1 molecule H_2 requires $2\ \mathscr{F}/N_A$ C
1 atom Ag requires $\mathscr{F}/N_A$ C
1 atom Al requires $3\ \mathscr{F}/N_A$ C

Thus, it appears that $\mathscr{F}/N_A$ is the natural unit of electricity—the electronic charge. If electrons could be subdivided, we would expect to find, sooner or later, a reaction in which one atom or molecule is produced by $\frac{1}{2}\mathscr{F}/N_A$ coulomb, $\frac{2}{3}\mathscr{F}/N_A$ coulomb, or some such fractional charge. No one has ever obtained such a result.

Faraday's experiments were thus the first indication that electricity, like matter, consists of indivisible particles. It was not Faraday who drew this conclusion, however. He was not even convinced that atoms were real. The concept and name of the electron were introduced by George Johnstone Stoney in 1874.

EXAMPLE 17.1 In the apparatus of Figure 17.2, 1000 C are passed through the circuit. How many (a) moles, (b) grams, and (c) milliliters measured at standard temperature and pressure (STP) of each gas are produced?

ANSWER It is best to start by converting the quantity of charge to faradays:

$$1000\ \cancel{\text{C}} \times \frac{1\ \text{faraday}}{9.65 \times 10^4\ \cancel{\text{C}}} = 1.036 \times 10^{-2}\ \text{faraday}$$

(a) From equation 5, 1 mole of H_2 requires 2 faradays:

$$2H^+ + 2e^- \longrightarrow H_2$$

$$1.036 \times 10^{-2}\ \cancel{\text{faraday}} \times \frac{1\ \text{mol}\ H_2}{2\ \cancel{\text{faradays}}} = 5.18 \times 10^{-3}\ \text{mol}\ H_2$$

From equation 4, 1 mole of O_2 requires 4 faradays:

$$2H_2O \longrightarrow O_2 + 4H^+ + 4e^-$$

$$1.036 \times 10^{-2}\ \cancel{\text{faraday}} \times \frac{1\ \text{mol}\ O_2}{4\ \cancel{\text{faradays}}} = 2.59 \times 10^{-3}\ \text{mol}\ O_2$$

(b) To find the mass of H_2 or O_2, use its molar mass:

$$5.18 \times 10^{-3}\ \cancel{\text{mol}\ H_2} \times 2.016\ \frac{\text{g}\ H_2}{\cancel{\text{mol}\ H_2}} = 1.04 \times 10^{-2}\ \text{g}\ H_2$$

$$2.59 \times 10^{-3}\ \cancel{\text{mol}\ O_2} \times 32.00\ \frac{\text{g}\ O_2}{\cancel{\text{mol}\ O_2}} = 8.29 \times 10^{-2}\ \text{g}\ O_2$$

(c) At STP, 1 mole of gas occupies 22.4 L = 2.24×10^4 mL (section 10.6):

$$5.18 \times 10^{-3}\,\cancel{\text{mol H}_2} \times 2.24 \times 10^4 \frac{\text{mL H}_2}{\cancel{\text{mol H}_2}} = 116 \text{ mL H}_2$$

$$2.59 \times 10^{-3}\,\cancel{\text{mol O}_2} \times 2.24 \times 10^4 \frac{\text{mL O}_2}{\cancel{\text{mol O}_2}} = 58.0 \text{ mL O}_2$$ ■

EXAMPLE 17.2 In a copper-refining plant, a current of 1.00×10^4 A passes through a solution. Copper is deposited on the cathode by the reaction

$$Cu^{2+} + 2e^- \longrightarrow Cu(c)$$

What mass of copper is deposited in 24 hours?

ANSWER To calculate the number of coulombs passed through, multiply the current in amperes by the time in seconds:

$$\underbrace{1.00 \times 10^4\,\cancel{A}}_{I} \times \underbrace{24\,\cancel{h} \times \frac{3600\,\cancel{\text{sec}}}{1\,\cancel{h}}}_{t} \times \frac{1\text{ C}}{\cancel{A}\,\cancel{\text{sec}}} = \underbrace{8.64 \times 10^8 \text{ C}}_{Q}$$

Then the number of faradays is

$$8.64 \times 10^8\,\cancel{C} \times \frac{1 \text{ faraday}}{9.65 \times 10^4 \text{ C}} = 8.95 \times 10^3 \text{ faradays}$$

From the given partial equation, 1 mole of Cu (63.55 g) requires 2 faradays:

$$8.95 \times 10^3\,\cancel{\text{faradays}} \times \frac{63.55 \text{ g Cu}}{2\,\cancel{\text{faradays}}} = 2.84 \times 10^5 \text{ g Cu} = 284 \text{ kg Cu}$$ ■

EXAMPLE 17.3 An aqueous solution of gold(III) nitrate, $Au(NO_3)_3$, is electrolyzed with a current of 0.500 A until 1.200 grams of Au has been deposited on the cathode. At the other electrode, the reaction is the oxidation of H_2O to O_2. NO_3^- is a spectator ion. Find (a) the number of moles and (b) the mass of O_2 that is liberated, (c) the number of coulombs passed through the circuit, and (d) the duration of the electrolysis.

ANSWER The atomic weight of Au is 197.0. The cathode reaction is

$$Au^{3+} + 3e^- \longrightarrow Au$$

Three faradays pass through the circuit for each mole of Au deposited. The number of faradays can be calculated from the mass of Au and its atomic weight:

$$1.200\,\cancel{\text{g Au}} \times \frac{1\,\cancel{\text{mol Au}}}{197.0\,\cancel{\text{g Au}}} \times \frac{3 \text{ faradays}}{\cancel{\text{mol Au}}} = 0.01827 \text{ faraday}$$

(a) As in Example 17.1, 1 mole of O_2 requires 4 faradays: $2H_2O \longrightarrow O_2 + 4H^+ + 4e^-$. Then

$$0.01827\,\cancel{\text{faraday}} \times \frac{1 \text{ mol O}_2}{4\,\cancel{\text{faradays}}} = 0.00457 \text{ mol O}_2$$

is liberated.

(b) $$0.004568\,\cancel{\text{mol O}_2} \times 32.00 \frac{\text{g O}_2}{\cancel{\text{mol O}_2}} = 0.1462 \text{ g O}_2$$

BOX 17.2

EQUIVALENTS AND NORMALITY

Faraday's second law says that 1 mole always requires a whole number of faradays: 1, 2, 3, ... Call this whole number n. *The quantity of a substance produced or consumed when 1 faraday passes through the circuit* deserves a special name: one **equivalent** (eq). An equivalent is $\frac{1}{n}$ mole, or 1 mol $= n$ eq. *The mass of 1 equivalent* is the **equivalent weight**. It is $\frac{1}{n}$ of the molecular weight (molar mass).

The number n, and thus the equivalent weight, depends on both the substance and the reaction. Thus, when Fe^{3+} is reduced to Fe,

$$Fe^{3+} + 3e^- \longrightarrow Fe$$

$n = 3$ eq/mol because producing 1 mole of Fe requires 3 faradays. The equivalent weight of Fe is

$$\frac{55.847\ \text{g}}{1\ \cancel{\text{mol}}} \times \frac{1\ \cancel{\text{mol}}}{3\ \text{eq}} = 18.616\ \text{g/eq}$$

However, Fe^{3+} can also be reduced to Fe^{2+}:

$$Fe^{3+} + e^- \longrightarrow Fe^{2+}$$

In this case, 1 mole requires 1 faraday and $n = 1$ eq/mol. The equivalent weight of Fe is

$$\frac{55.847\ \text{g}}{1\ \cancel{\text{mol}}} \times \frac{1\ \cancel{\text{mol}}}{1\ \text{eq}} = 55.847\ \text{g/eq}$$

An equivalent weight can also be defined for a compound. When MnO_4^- is reduced to Mn^{2+},

$$MnO_4^- + 8H^+ + 5e^- \longrightarrow Mn^{2+} + 4H_2O$$

each mole of MnO_4^- requires 5 faradays. If the compound is $KMnO_4$, each mole of $KMnO_4$ likewise requires 5 faradays: 1 mol MnO_4^- = 1 mol $KMnO_4$ = 5 eq, and the equivalent weight is

$$\frac{158.034\ \text{g}}{1\ \cancel{\text{mol}}} \times \frac{1\ \cancel{\text{mol}}}{5\ \text{eq}} = 31.6068\ \text{g/eq}$$

When MnO_4^- is reduced to MnO_2,

$$MnO_4^- + 2H_2O + 3e^- \longrightarrow MnO_2(c) + 4OH^-$$

$n = 3$ eq/mol; the equivalent weight of $KMnO_4$ is

$$\frac{158.034\ \text{g}}{1\ \cancel{\text{mol}}} \times \frac{1\ \cancel{\text{mol}}}{3\ \text{eq}} = 52.6780\ \text{g/eq}$$

It is ambiguous to speak of "the equivalent weight" of a substance, since it depends on the reaction. However, people often use the expression because some substances require the same number of faradays per mole in all of their reactions.

The concept of an equivalent goes beyond electrochemical reactions. Each partial equation in the preceding paragraph can represent part of any oxidation–reduction reaction, not just one that occurs in an electrolytic cell. Indeed, the reaction need not be an oxidation–reduction. A molecule or ion can gain or lose charge by accepting or giving up protons:

$$HCl + H_2O \longrightarrow H_3O^+ + Cl^-$$

This is an acid-base reaction (Chapter 14). Since 1 mole of HCl loses 1 mole of protons—with a charge of 1 faraday—we say that 1 mole of HCl is 1 equivalent. When C_2^{2-} accepts 2 moles of protons,

$$C_2^{2-} + 2H_2O \longrightarrow C_2H_2(g) + 2OH^-$$

1 mol C_2^{2-} = 2 eq. When 1 mole of H_3PO_4 gives up 1 mole of protons,

$$H_3PO_4 + OH^- \longrightarrow H_2PO_4^- + H_2O$$

1 mol H_3PO_4 = 1 eq, and its equivalent weight is

$$\frac{97.995\ \text{g}}{1\ \cancel{\text{mol}}} \times \frac{1\ \cancel{\text{mol}}}{1\ \text{eq}} = 97.995\ \text{g/eq}$$

However, H_3PO_4 can also give up 2 or 3 moles of protons:

$$H_3PO_4 + 2OH^- \longrightarrow HPO_4^{2-} + 2H_2O$$

$$H_3PO_4 + 3OH^- \rightleftharpoons PO_4^{3-} + 3H_2O$$

and its equivalent weight is then

$$\frac{97.995\ \text{g}}{1\ \cancel{\text{mol}}} \times \frac{1\ \cancel{\text{mol}}}{2\ \text{eq}} = 48.998\ \text{g/eq}$$

or

$$\frac{97.995\ \text{g}}{1\ \cancel{\text{mol}}} \times \frac{1\ \cancel{\text{mol}}}{3\ \text{eq}} = 32.665\ \text{g/eq}$$

Again, the equivalent weight depends not only on what the substance (or ion) is, but on what it does.

A third way for an ion to accept or give up charge is to combine with another ion. In the reaction

$$Pb^{2+} + 2I^- \longrightarrow PbI_2(c)$$

1 mole of Pb^{2+} accepts 2 faradays of negative charge in combining with 2 moles of I^-. One mole of Pb^{2+} thus counts as 2 equivalents, while 1 mole of I^- is 1 equivalent. For a precipitation reaction, the number (n) of equivalents per mole is the charge of the ion—sometimes, but not always, true in oxidation–reduction and acid-base reactions.

The **normality** (N) of a solution is *the number of equivalents per liter*. It is equal to the integer n times the molarity:

$$\frac{\text{N eq}}{1\ \text{L}} = \frac{n\ \text{eq}}{1\ \cancel{\text{mol}}} \times \frac{\text{M}\ \cancel{\text{mol}}}{1\ \text{L}}$$

Labeling a bottle of solution with normality is still a common practice, but an unfortunate one since it requires the user to read the labeler's mind. It survives because certain conventions are generally understood: HCl is 1 eq/mol; H_2SO_4 is 2 eq/mol; H_3PO_4 is 3 eq/mol; and—far from obviously—$KMnO_4$ is 5 eq/mol because it is usually used in acidic solutions, where MnO_4^- is reduced to Mn^{2+}.

(c) $0.01827 \cancel{\text{faraday}} \times \dfrac{9.649 \times 10^4\ \text{C}}{\cancel{\text{faraday}}} = 1.763 \times 10^3\ \text{C}$

(d) Since $Q = It$, the time is obtained by dividing the charge by the current. A current of 0.5 ampere means 0.5 coulomb per second:

$$t = \frac{Q}{I} = 1.76 \times 10^3\ \text{C} \times \frac{1\ \text{sec}}{0.500\ \cancel{\text{C}}} = 3.52 \times 10^3\ \text{sec}$$ ■

Luigi Galvani
1739–1804

PROBLEM 17.2 A current of 11.3 amperes is passed through a silver-plating bath for 2.00 hours. What mass of Ag is deposited at the cathode by the reaction $Ag(CN)_2^- + e^- \longrightarrow Ag(c) + 2CN^-$? □

PROBLEM 17.3 A current is passed through a solution of CaI_2. The following electrode reactions occur:

at the anode: $2I^- \longrightarrow I_2(c) + 2e^-$

at the cathode: $2H_2O + 2e^- \longrightarrow H_2(g) + 2OH^-$

Analysis of the solution shows that 53.5 mmol I_2 is formed. (a) How many faradays passed through the circuit? (b) How many coulombs passed through the circuit? (c) What volume of H_2 (measured at STP) is produced? □

17.3 GALVANIC CELLS

In 1786 Luigi Galvani, a professor of medicine at the university in Bologna, Italy, was experimenting with the effects of electricity on frog muscles. His source of electricity was the frictional machine common at the time. While doing his experiments, Galvani came across something unexpected: he could make the frog's legs twitch merely by bringing them into contact with two dissimilar metals (which also touched each other to make a complete circuit), such as copper and iron. Galvani thought he had discovered "animal electricity." However, Alessandro Volta, a physicist at Pavia (Italy), was skeptical. In 1796, Volta put together the famous "voltaic pile," which—dispensing with the frog—consisted of a stack of cardboards or pieces of cloth wet with salt water, each sandwiched between sheets of unlike metals (Figure 17.4). This was the first source of continuous electric current. It and others like it made possible the great electrochemical discoveries of the early nineteenth century. Today, the electrochemical cell, or battery of cells, has been generally superseded by the generator as a large-scale source of electricity. However, cells still have a place as portable sources of small amounts of power and they may be making a comeback in the form of fuel cells (section 17.12). There is another reason why cells are important for understanding chemistry: the electrical work that can be obtained from a chemical reaction is closely related to its equilibrium constant and to the intensity of its drive toward equilibrium.

Alessandro Volta
1745–1827

Also *voltaic* cell and *electrochemical* cell.

An apparatus in which chemical energy is converted to electrical energy is known as a **galvanic cell**. It is different from an electrolytic cell (described in the last section), which requires an outside power source to push the current through it and make the reactions occur. A *galvanic* cell is itself the power source. The reaction in a galvanic cell is *spontaneous*—it "wants" to occur—and therefore can do work. The opposite situation exists in an electrolytic cell: a reaction that would otherwise be impossible is forced to occur.

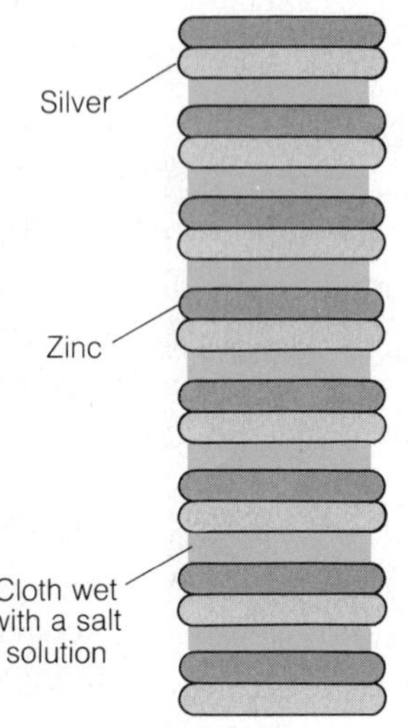

FIGURE 17.4
Volta's pile. (a) Archival photograph (courtesy of Museo Nazionale della Scienza e della Tecnica, Milan). (b) Diagram.

Sometimes the oxidation and reduction add up to give no *net* oxidation or reduction (section 17.10).

A reaction that can produce an electric current is a reaction in which electrons are transferred—that is, an oxidation–reduction reaction (section 9.4). A familiar example is the reaction between zinc and a soluble copper salt,

$$Zn(c) + Cu^{2+}(aq) \longrightarrow Cu(c) + Zn^{2+}(aq)$$

In this reaction, Zn is oxidized (loses electrons), while Cu^{2+} is reduced (gains electrons). The reaction is exothermic ($\Delta H = -217$ kJ). If a piece of zinc is added to a $CuSO_4$ solution, heat is given out but no work can be obtained. How can electrical work be obtained from this reaction? To see how this can be done, resolve the reaction into two partial reactions, one for oxidation and one for reduction:

$$Zn(c) \longrightarrow Zn^{2+}(aq) + 2e^- \qquad \text{(oxidation)} \qquad (6)$$

$$Cu^{2+}(aq) + 2e^- \longrightarrow Cu(c) \qquad \text{(reduction)} \qquad (7)$$

If these partial reactions occur in two physically separated places, then the electrons will have to flow through a wire from the place where the oxidation occurs (the anode) to the place where the reduction occurs (the cathode). This flow of electrons constitutes a current, which can be used to do work.

To separate the partial reactions, insert the Zn into a solution that does not contain Cu^{2+}. If Zn and Cu^{2+} were put together, they would simply react with each other, without sending electrons through the external circuit. This would defeat the purpose of a galvanic cell. It is not actually necessary to include Zn^{2+} or Cu, since they are products and will appear anyway. However, our principal interest in galvanic cells will be in using them to measure the work available from a reaction—a quantity of fundamental importance in chemistry. For this purpose, it is necessary to be able to make the reaction go either way (section 17.5); when the reaction is reversed, "products" become "reactants." We will therefore think of a cell as containing the products as well as the reactants, all at definite concentrations. Thus, we insert the Zn into a $ZnSO_4$ solution and, in a separate container, the Cu into a $CuSO_4$ solution. The Zn and Cu are connected by a wire (Figure 17.5a). The negative ion in each solution could be any ion that does not form an insoluble salt with the metal ion and that does not itself react; SO_4^{2-} is admirable in both respects.

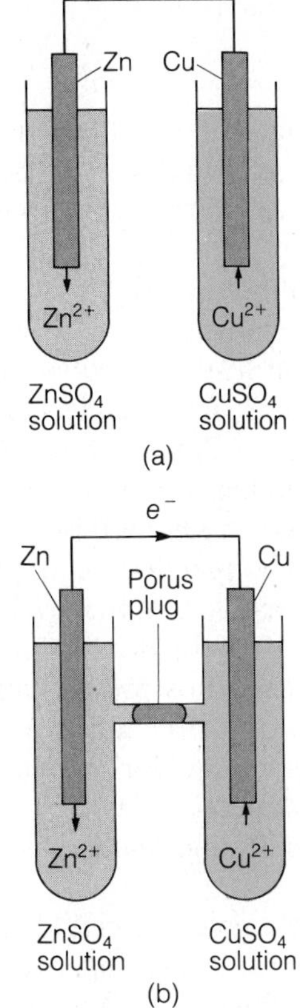

FIGURE 17.5
(a) An unsuccessful galvanic cell. The charges in the solutions cannot neutralize each other.
(b) A successful galvanic cell. The porous plug prevents the solutions from mixing.

The apparatus shown in Figure 17.5a is not yet a working galvanic cell. To see why, think about equation 6, $Zn(c) \longrightarrow Zn^{2+}(aq) + 2e^-$. The Zn^{2+} ions go into the solution. These extra positive ions are not balanced by negative ions, and they therefore give the solution a positive charge. Likewise, the partial reaction $Cu^{2+}(aq) + 2e^- \longrightarrow Cu(c)$ (equation 7) depletes the solution of positive ions, leaving an oversupply of SO_4^{2-} ions. Such a separation of positive and negative charges cannot occur without a very large input of energy. The oxidation of Zn to Zn^{2+} and the reduction of Cu^{2+} to Cu therefore stop before any detectable amount of Zn or Cu^{2+} has reacted.

How, then, can the cell be made to generate a current? As in any electrical apparatus, a complete circuit is needed. The buildup of positive charge (Zn^{2+}) in one solution and the depletion of positive charge (Cu^{2+}) in the other must be allowed to cancel out. The two solutions are connected by placing them in contact (directly or via a third solution) without allowing them to mix. Then SO_4^{2-} ions can flow into the $ZnSO_4$ solution and cancel the positive charge of the Zn^{2+} ions as they are produced; SO_4^{2-} ions can also flow out of the $CuSO_4$ solution to relieve its excess negative charge. Positive ions also move to help keep the solutions electrically neutral. Now there is a complete circuit, partly composed of metals (electronic conductors) and partly of solutions (ionic conductors). The positive charge produced in the anode solution by generating Zn^{2+} and the negative charge produced in the cathode solution by removing Cu^{2+} can then cancel each other by the migration of ions from one solution to the other. Figure 17.5b shows a successful cell. Electrons flow from the Zn to the Cu as shown.

In many galvanic cells, ions must be allowed to migrate between two solutions but the solutions should not be allowed to mix. Since solutions in contact mix spontaneously, these requirements present a problem. The problem is solved by various means, including porous plugs and felted fabrics (which make the mixing very slow without stopping it completely), semipermeable membranes (section 12.15), and solutions immobilized in pastes or gels (section 17.12). The method we will use most often in this chapter is that of connecting the solutions via a third solution, usually a concentrated solution of KCl or NH_4NO_3. Such a solution is called a **salt bridge**. The salt bridge and other techniques are not mutually exclusive; for example, the concentrated salt solution is often immobilized in a gel or fiber. When a cell is used for measurements rather than as a power source, a salt bridge provides the benefit of minimizing certain complications that result from a junction between two solutions. We will not try to explain why concentrated solutions of specific salts are used. A Zn-Cu cell modified by inclusion of a salt bridge is shown in Figure 17.6.

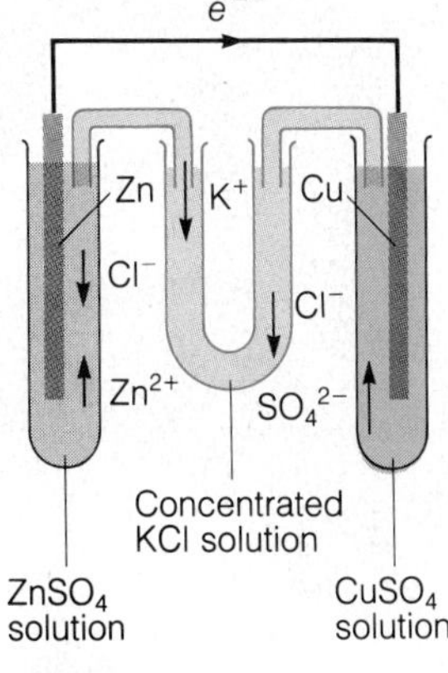

FIGURE 17.6
A galvanic cell with a salt bridge.

Salt bridges are not always needed. Sometimes a cell can be made with one solution. Figure 17.7 shows an example of such a cell. Two electrodes are immersed in a solution of $ZnCl_2$. One electrode is a piece of zinc; the other is a piece of silver with a porous coating of silver chloride. Because of the very low solubility of AgCl, the solution contains only a trace of Ag^+ ion. The partial reactions are

$$Zn(c) \longrightarrow Zn^{2+} + 2e^-$$

$$2AgCl(c) + 2e^- \longrightarrow 2Ag(c) + 2Cl^-$$

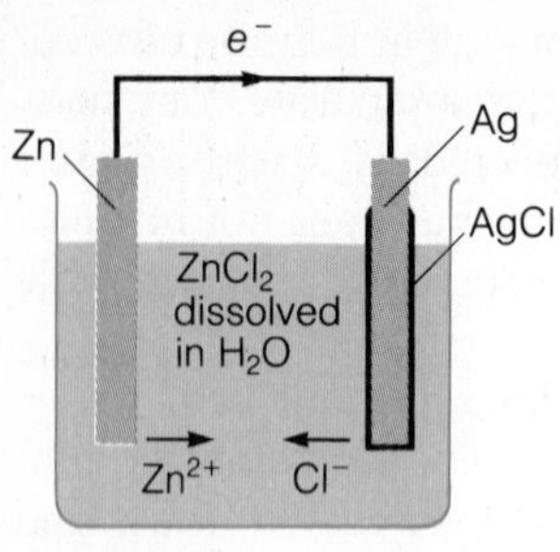

FIGURE 17.7
A galvanic cell with only one solution.

The net reaction is

$$Zn(c) + 2AgCl(c) \longrightarrow 2Ag(c) + Zn^{2+} + 2Cl^-$$

The two solids, Zn and AgCl, are separated simply by being on opposite sides of the cell. Since neither is in solution, they cannot react directly with each other, as Cu^{2+} and Zn could if we tried to make the cell of Figure 17.5b or 17.6 with only one solution.

In the cell in Figure 17.6, why do the electrons flow from the zinc to the copper instead of from the copper to the zinc? We could ask the same question about any chemical reaction—why does it go this way and not that way? An oxidation–reduction reaction is a competition for electrons. Either Cu^{2+} gets the electrons ($Cu^{2+} + 2e^- \longrightarrow Cu$), or Zn^{2+} gets them ($Zn^{2+} + 2e^- \longrightarrow Zn$). Each partial reaction has a certain tendency to occur as written. This tendency may be called the *driving force* of the partial reaction. The driving force for a reduction process is the tendency of the oxidized form to acquire electrons. The reaction occurs in the direction $Zn + Cu^{2+} \longrightarrow Zn^{2+} + Cu$. We therefore conclude that the driving force of the partial reaction $Cu^{2+} + 2e^- \longrightarrow Cu$ is greater than the driving force of the partial reaction $Zn^{2+} + 2e^- \longrightarrow Zn$. The first process thus prevails and drives the second process backward: $Zn \longrightarrow Zn^{2+} + 2e^-$. One direction of a reaction is "spontaneous"—the direction that brings it closer to equilibrium—and the other direction—away from equilibrium—is "nonspontaneous." We will explore the ideas of driving force and spontaneity in Chapter 18 and learn how to assign numbers to them.

The sport of arm wrestling provides a rough analogy of a galvanic cell. Each contestant tries to push the other's arm down to the table. If Joe is stronger than Bill and Bill is stronger than Mike, then Bill's arm is pushed down when he wrestles with Joe, but Bill is able to push Mike's arm down. We will see later how a partial reaction may go either way, like Bill's arm, depending on what the other partial reaction is. If Joe and Jim are equally strong (they are the fellows in the photograph), neither arm goes down; likewise, when the two reduction partial reactions have the same driving force, equilibrium is established and no current flows either way.

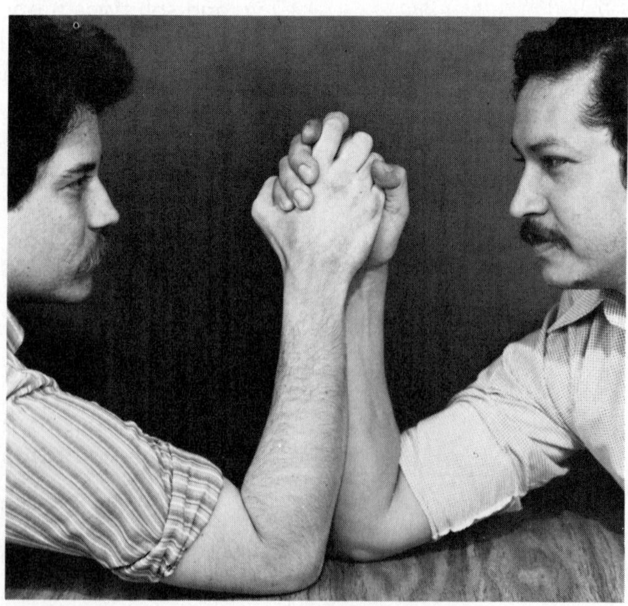

The idea that one partial equation has a greater driving force than another was already discussed in connection with the activity series (section 9.4). Zinc is a more active metal than copper and therefore displaces copper from solution: $Zn + Cu^{2+} \longrightarrow Zn^{2+} + Cu$. In turn, copper is a more active metal than silver and therefore displaces silver: $Cu + 2Ag^{+} \longrightarrow Cu^{2+} + 2Ag$. We will study this idea in a more quantitative way in section 17.7.

CELL NOTATION

We can describe a galvanic cell without going to the trouble of drawing a realistic picture of it. The notation used is a greatly simplified diagram of the cell. Boundaries between different phases (like solid and liquid) or between liquid solutions are indicated by |; a salt bridge, by ||. The cells previously discussed may be represented as follows (with spaces left for the concentrations):

We may also write, for example, $Zn^{2+} + SO_4{}^{2-}$ instead of $ZnSO_4$, or simply Zn^{2+} if the negative ion is not involved in the reaction.

Figure 17.5b $Zn|ZnSO_4(\_M)|CuSO_4(\_M)|Cu$

Figure 17.6 $Zn|ZnSO_4(\_M)||CuSO_4(\_M)|Cu$

Figure 17.7 $Zn|ZnCl_2(\_M)|AgCl|Ag$

In this kind of description, the *anode* is written on the *left* and the *cathode* on the *right*, a rule that may be remembered by observing the alphabetical order in each of the following pairs: *anode-cathode, oxidation-reduction, left-right.* How we write the cell description depends on how the equation for the reaction is written. If we write

$$Zn + Cu^{2+} \longrightarrow Zn^{2+} + Cu \qquad (8)$$

with Zn being oxidized and Cu^{2+} being reduced, we are assuming that Zn is the anode and Cu is the cathode:

$Zn|Zn^{2+}|Cu^{2+}|Cu$

anode cathode

If we write

$$Cu + Zn^{2+} \longrightarrow Cu^{2+} + Zn$$

with Cu being oxidized and Zn^{2+} being reduced, we are writing it as if Cu were the anode and Zn were the cathode; the cell in which this reaction occurs would be written

$Cu|Cu^{2+}|Zn^{2+}|Zn$

anode cathode

Ordinarily, the reaction occurs spontaneously as in equation 8, but it can be made to go either way by adjusting the concentrations of Cu^{2+} and Zn^{2+} (section 17.8).

To give a complete description of the cell, the notation should show the entire compounds used in preparing it, including spectator ions like $SO_4{}^{2-}$. However, if the spectator ions are irrelevant to the discussion, they can be omitted, as in "$Zn|Zn^{2+}|Cu^{2+}|Cu$." It is usually best to write the partial and net equations for the cell reaction as net ionic equations, unless you are specifically asked to do otherwise.

EXAMPLE 17.4 What net ionic reaction occurs in each of the following cells? Both the reactants and the products are given in the cell description.

(a) $Zn(c)|ZnCl_2(aq)||SnCl_2(aq)|Sn(c)$
(b) $Fe(c)|FeSO_4(aq)|PbSO_4(c)|Pb(c)$
(c) $Cu|CuBr_2(aq)|Br_2(\ell)|Pt$

ANSWER

(a) The reaction at each electrode must involve species that are present there. At the anode are Zn, Zn^{2+}, Cl^-, and H_2O. Look first for an element that is present in two different oxidation states. Zn is present in oxidation states 0 (Zn) and +2 (Zn^{2+}). A reasonable oxidation reaction is

$$Zn \longrightarrow Zn^{2+} + 2e^-$$

At the cathode are Sn, Sn^{2+}, Cl^-, and H_2O. Again, Sn is present in two oxidation states and the reduction

$$Sn^{2+} + 2e^- \longrightarrow Sn$$

is possible. The net reaction is obtained by combining the partial equations so that the electrons cancel out (section 9.5):

$$Zn + Sn^{2+} \longrightarrow Zn^{2+} + Sn$$

(b) The species present at the anode in this cell are Fe, Fe^{2+}, $SO_4{}^{2-}$, and H_2O. $SO_4{}^{2-}$ is very difficult to oxidize or reduce. Fe is present in two oxidation states (0 and +2); we therefore assume that the oxidation is

$$Fe \longrightarrow Fe^{2+} + 2e^-$$

At the cathode are Pb, $PbSO_4$ (nearly insoluble), Fe^{2+}, and $SO_4{}^{2-}$ (both from $FeSO_4$). Pb appears with two different oxidation numbers (0 and +2) and is thus a candidate for a reduction partial reaction:

$$PbSO_4 + 2e^- \longrightarrow Pb + SO_4{}^{2-}$$

This partial reaction illustrates a common kind: a slightly soluble salt on one side, the metal and the negative ion on the other. Adding these partial equations gives the net reaction:

$$Fe + PbSO_4 \longrightarrow Fe^{2+} + SO_4{}^{2-} + Pb$$

This cell needs only one solution—no salt bridge—because Fe and $PbSO_4$, being on opposite sides of the cell, cannot react directly. (The Pb electrode may be coated with solid $PbSO_4$, or powdered Pb and $PbSO_4$ may be pressed together to form the electrode.)

(c) At the anode are Cu, Cu^{2+}, and Br^-. ($CuBr_2$ is an ionic salt.) The most likely oxidation is

$$Cu \longrightarrow Cu^{2+} + 2e^-$$

Platinum (Pt) is commonly used as an inert (unreactive) electrode—one that gets the electrons to or from the solution without undergoing any change itself. It appears in the cell description but not in the partial or net equations, since it undergoes no change. It is also a good catalyst (section 19.10) for many reactions. The other species present at the cathode are Br_2, Cu^{2+}, and Br^-. The expected reduction reaction is

$$Br_2 + 2e^- \longrightarrow 2Br^-$$

The net reaction is then

$$Cu + Br_2 \longrightarrow Cu^{2+} + 2Br^-$$

which is the formation of $CuBr_2$(aq) from its elements. ■

When predicting what electrode reactions are likely to occur, it helps to recognize some that are *not* likely. Some common ions and electrode materials are very unreactive. It is usually a good guess that they are not oxidized or reduced in aqueous solutions at ordinary temperature.

unreactive ions: Li^+, Na^+, K^+, Mg^{2+}, Ca^{2+}, Ba^{2+}, F^-, $SO_4{}^{2-}$, $PO_4{}^{3-}$
unreactive electrode materials: Pt, C(graphite)

PROBLEM 17.4 What are the electrode (partial) reactions and the net reaction in each of these cells?

(a) $Sn(c)|Sn(NO_3)_2(aq)||Cu(NO_3)_2(aq)|Cu(c)$

(b) $Cd(c)|CdSO_4(aq)||KI(aq)|I_2(c)|Pt$ □

17.4 ELECTRICAL WORK

The preceding section showed how a cell can be constructed so as to obtain electrical work from a chemical reaction. The next question is, how much work can be obtained? According to Faraday's second law, when one mole of Zn is converted to Zn^{2+}, two faradays of charge, or $2 \times 9.65 \times 10^4$ C, pass through the external circuit. The quantity of work that can be obtained from a cell reaction is determined by the **difference of potential** between the electrodes and by the quantity of charge passing through the circuit from one electrode to the other. The difference of potential between two points is the work required or done when a unit charge is moved from one point to the other:

$$\text{difference of potential} = \frac{\text{work}}{\text{charge}}$$

Since work is in joules and charge is in coulombs, difference of potential is measured in joules per coulomb; 1 J/C = 1 **volt** (V):

$$\underset{\text{(in joules)}}{\textbf{work}} = \underset{\text{(in volts)}}{\textbf{difference of potential}} \times \underset{\text{(in coulombs)}}{\textbf{charge}} \qquad (9)$$

Work is done only by a spontaneous reaction—one that goes *toward* equilibrium—while work must be done on a system to take it *away* from equilibrium, in the nonspontaneous direction.

The potential difference between the two electrodes in a galvanic cell is called the **electromotive force**, or **emf** ($\mathscr{E}$), of the cell. It is the quantitative measure of the "driving force" we discussed in the preceding section. It is also the conversion factor that relates the charge passing through the circuit to the amount of work this charge can do. Let us assume that $ZnSO_4$ and $CuSO_4$ are both 1 M at 25°C; then $\mathscr{E} = 1.10$ V for the cell in Figure 17.6. With the usual understanding that each symbol (or formula) represents 1 mole, the cell reaction

$$Zn + Cu^{2+} \longrightarrow Zn^{2+} + Cu$$

describes a process in which 1 mole of zinc and 1 mole of Cu^{2+} are consumed. Resolving this reaction into the partial equations

$$Zn \longrightarrow Zn^{2+} + 2e^-$$

$$Cu^{2+} + 2e^- \longrightarrow Cu$$

tells us that when the reaction occurs as written, 2 faradays flow through the circuit. The work this electricity will do is

$$2\ \cancel{\text{faradays}} \times \frac{9.65 \times 10^4\ \cancel{\text{C}}}{\cancel{\text{faraday}}} \times 1.10\ \cancel{\text{V}} \times \frac{1\ \text{J}}{1\ \cancel{\text{V}} \times 1\ \cancel{\text{C}}} = 2.12 \times 10^5\ \text{J} = 212\ \text{kJ}$$

For any reaction, the work done by a cell when the reaction occurs as written is equal to $n\mathscr{F}\mathscr{E}$, where $\mathscr{F}$ is Faraday's constant and n is the number of faradays passing through the circuit:

$$\textbf{work done by cell} = n\mathscr{F}\mathscr{E} \tag{10}$$

n is also the number of electrons in each partial equation; for example, 1 in $Ag^+ + e^- \longrightarrow Ag$ and 5 in $MnO_4^- + 8H^+ + 5e^- \longrightarrow Mn^{2+} + 4H_2O$.

EXAMPLE 17.5 The net reaction in a galvanic cell is

$$2Al + 3Cd^{2+} \longrightarrow 2Al^{3+} + 3Cd$$

The emf of the cell is 1.20 V. (a) How many faradays pass through the circuit when the reaction occurs as written? (b) How much work can the cell do in that process? (c) How much work can the cell do when 1.00 g of Al reacts?

ANSWER

(a) Write the partial equations with the coefficients already given:

$$2Al \longrightarrow 2Al^{3+} + 6e^-$$

$$3Cd^{2+} + 6e^- \longrightarrow 3Cd$$

Each partial equation contains $6e^-$; therefore, 6 faradays pass through the circuit when 2 moles of Al and 3 moles of Cd^{2+} react.

(b) Charge is converted to work by using the emf as a conversion factor:

$$6\ \cancel{\text{faradays}} \times \frac{9.65 \times 10^4\ \cancel{\text{C}}}{1\ \cancel{\text{faraday}}} \times 1.20\ \cancel{\text{V}} \times \frac{1\ \text{J}}{1\ \cancel{\text{V}} \times 1\ \cancel{\text{C}}} = 6.95 \times 10^5\ \text{J}$$

The kilojoule is a more convenient unit here;

$$6.95 \times 10^5\ \cancel{\text{J}} \times \frac{1\ \text{kJ}}{10^3\ \cancel{\text{J}}} = 695\ \text{kJ}$$

(c) The answer to part b applies to 2 moles of Al. For 1.00 g,

$$1.00\ \cancel{\text{g Al}} \times \frac{1\ \cancel{\text{mol Al}}}{26.98\ \cancel{\text{g Al}}} \times \frac{695\ \text{kJ}}{2\ \cancel{\text{mol Al}}} = 12.9\ \text{kJ}$$

■

PROBLEM 17.5 (a) Calculate the work done when the reaction

$$2Fe^{3+} + Cu \longrightarrow 2Fe^{2+} + Cu^{2+}$$

occurs as written. The emf is 0.43 V. (b) How much work is done per gram of copper reacting?

□

A cell cannot read minds; its emf does not depend on how we write the chemical equation.

When all the coefficients in the chemical equation are multiplied by a constant, $\mathscr{E}$ is unaffected. Let us see why. An equation with doubled coefficients says that twice as many moles react when the reaction occurs as written; therefore, the cell does twice as much work. But there are also twice as many coulombs, or twice as many faradays, passing through the circuit. The work *per coulomb, $\mathscr{E}$*, is therefore unchanged. For example, compare the equations

$$\mathrm{Zn} + \mathrm{Cu}^{2+} \longrightarrow \mathrm{Zn}^{2+} + \mathrm{Cu} \qquad \text{2 faradays, work} = 212\ \mathrm{kJ},\ \mathscr{E} = 1.10\ \mathrm{V}$$

$$2\mathrm{Zn} + 2\mathrm{Cu}^{2+} \longrightarrow 2\mathrm{Zn}^{2+} + 2\mathrm{Cu} \qquad \text{4 faradays, work} = 424\ \mathrm{kJ},\ \mathscr{E} = 1.10\ \mathrm{V}$$

$$\tfrac{1}{2}\mathrm{Zn} + \tfrac{1}{2}\mathrm{Cu}^{2+} \longrightarrow \tfrac{1}{2}\mathrm{Zn}^{2+} + \tfrac{1}{2}\mathrm{Cu} \qquad \text{1 faraday, work} = 106\ \mathrm{kJ},\ \mathscr{E} = 1.10\ \mathrm{V}$$

The emf of a cell depends on the concentrations of the reactants and products. This dependence will be discussed in section 17.8. For the purposes of tabulating emf's and comparing different reactions, the **standard emf**, $\mathscr{E}^0$, of a cell is defined as the emf when each reactant or product is in its **standard state**. The standard state is the pure substance for solids (like insoluble salts and most metals) and for liquids like mercury. For solutes (molecules or ions), the standard state is the solute at a concentration of 1 mol/L. For a gas, the standard state is the gas at a pressure of 1 atm. The superscript ^{0} is not a zero exponent, nor is it a degree sign. It is simply a label—already used in ΔH^0 (section 4.6)—to indicate that each reactant or product is in its standard state. Do not confuse this "standard state" with the "standard temperature and pressure" or "standard conditions" used in gas problems (section 10.6). The standard state does not require any particular temperature, and indeed $\mathscr{E}^0$ changes with temperature. The one common feature of the two concepts is the standard gas pressure, 1 atm.

A more accurate definition of the standard state uses activities (Box 16.2, section 16.1). The solute is in its standard state when it is at an *activity* of 1 mol/L.

17.5 REVERSIBLE ELECTROMOTIVE FORCE

Section 17.4 does not tell the whole story of emf. The emf of a given cell can vary greatly, depending on how much current is drawn. The greater the current is, the more energy is lost as heat because of the resistance of the wires and the solution, as well as because of some specifically electrochemical effects (section 17.11). Thus, the emf—the available work per coulomb—drops off as more current is drawn. The quantity we were discussing in the last section, the quantity in which the chemist is most often interested, is the *maximum* emf of the cell. The maximum emf is what really measures the driving force of the cell reaction. The greater the maximum emf is, the greater is the driving force. To find this maximum emf, we must measure the potential difference between the electrodes when there is no loss by resistance. The way to have no loss is to *measure the emf when no current is flowing through the cell.* The **potentiometer** is the device used for this purpose.

An old-fashioned *voltmeter* draws some current and is used for less accurate measurements. Modern voltmeters, using solid-state technology, draw so little current that they are as good as potentiometers, or better.

The potentiometer provides a source of emf that can be varied continuously and read from a dial. The device is connected so that it will try to make a current flow through the cell in the direction opposite to the current generated by the cell reaction. The potentiometer is then adjusted until the net current is zero (Figure 17.8). At that point, the known emf of the potentiometer is equal to the emf of the cell, measured with no current flowing. The emf thus determined is a measure of the ideal maximum amount of work that can be obtained from the chemical reaction occurring in the cell.

Measuring the emf of a cell at zero current is the best illustration in chemistry of a profound and important concept—the **thermodynamically reversible process**. In such a process, everything happens in a delicately balanced way, so that the process can be made to go in either direction by an infinitesimal change in some condition. By a minute adjustment of the emf of the potentiometer, the cell reaction can be made to proceed very slowly in either direction. It is like weighing an object on a balance (ideally, a frictionless balance); a very slight adjustment of the weights causes the beam to swing one way or the other. When the beam

Infinitesimal means "as close to zero as you want it to be." Less precisely, it means "extremely small."

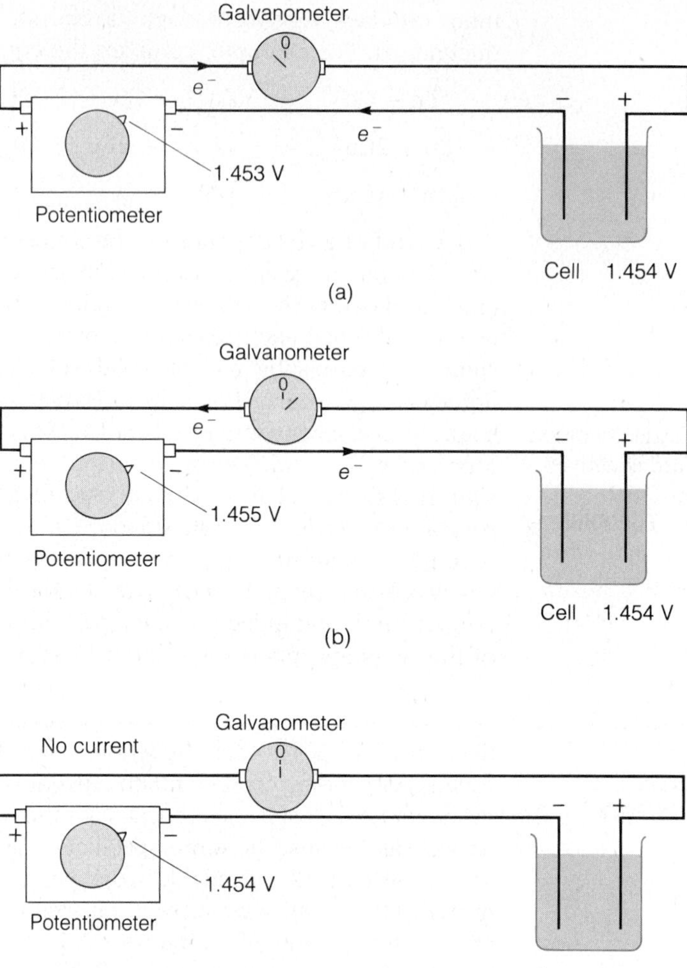

FIGURE 17.8
The use of a potentiometer to measure the emf of a cell. (a) The emf of the potentiometer is slightly less than that of the cell. A small current flows in the direction shown, causing the needle of the sensitive galvanometer to deflect. (b) The potentiometer has been adjusted until it overbalances the cell and the current flows the other way. (c) The correct setting has been found and no current flows.

does not swing either way, the balance is "balanced" and the mass of the object has been measured. The emf of the potentiometer balances the emf of the cell in the same way that the weights balance the object being weighed. A thermodynamically reversible process is an ideal process that can only be approached, not something that can actually be done. It is not enough that the process be merely very slow; it must be *infinitely* slow, and thus, carrying out a reaction reversibly would require an infinitely long time. From the human point of view, what takes an infinitely long time does not happen at all.

How is it possible even to think about—not to speak of making measurements on—a process that does not happen? "The reaction occurs reversibly" is a short way of saying something like, "I will not allow the reaction to occur; I will push back electrically just hard enough to keep it from occurring, without making it go the other way. By measuring how hard I have to push back, I can figure out the maximum work that could possibly be obtained if a certain number of moles were allowed to react."

The emf measured at zero current is known as the **reversible emf**. It may also be called the "ideal emf," since it is an ideal limit that can only be approached (like an ideal gas or an ideal solution). It is usually understood that "emf" means "reversible emf" unless otherwise specified. The reversible emf is the *maximum* emf that can possibly be obtained from the cell. When a cell actually delivers current, its emf is always less.

17.6 THE SIGN OF AN ELECTRODE

The electrodes in galvanic and electrolytic cells are commonly labeled "positive" and "negative," as well as "anode" and "cathode." It can be confusing that the relationship between these two sets of labels is reversed in the two types of cells.

In a *galvanic* cell, the signs are assigned from the perspective of the electrons in the external circuit. Electrons are left behind on the anode by the oxidation (loss of electrons) taking place there. The *anode* pushes these electrons into the external circuit and is therefore regarded as the *negative* electrode—the source of negative charge. The *cathode* pulls electrons in from the external circuit—they are in demand for the reduction process—and is therefore called the *positive* electrode. Dry cells (section 17.12) are marked "Center terminal is positive"; this means that the center terminal is the cathode.

Short notation used in circuit diagrams: —|⊢. The long electrode is positive.

In an *electrolysis* cell, the signs are assigned from the perspective of the ions in the solution. We forcibly feed electrons into the cathode by connecting it to the negative terminal (the electron source) of a generator or a galvanic cell. The cathode attracts positive ions (cations) in the solution and is therefore considered the *negative* electrode. Electrons are pulled out of the anode by the positive terminal of the power source, and the *anode* therefore attracts negative ions (anions). Like anything that attracts negative charges, it is considered *positive.* Figure 17.9 illustrates the usages described in this section.

Oxidation takes place at the anode and reduction at the cathode, in cells of both types.

When a cell changes from a galvanic to an electrolytic cell—as when a storage battery stops discharging and starts to be charged—the current is reversed; the anode becomes the cathode and the cathode becomes the anode. Nevertheless, the electrode that was positive remains positive, and the electrode that was negative remains negative. A storage battery is therefore permanently labeled with plus (+) and minus (−) signs, but "anode" and "cathode" labels would be useless.

EXAMPLE 17.6 (a) Example 17.4 refers to the cell

$$Zn(c)|ZnCl_2(aq)||SnCl_2(aq)|Sn(c)$$

in which the reaction is

$$Zn + Sn^{2+} \longrightarrow Zn^{2+} + Sn$$

Which electrode is positive? Which is negative? Which is the anode? Which is the cathode? (b) If the reaction is driven backward (in electrolysis),

$$Zn^{2+} + Sn \longrightarrow Zn + Sn^{2+}$$

which is the anode? Which is the cathode? What is the sign of each electrode?

ANSWER

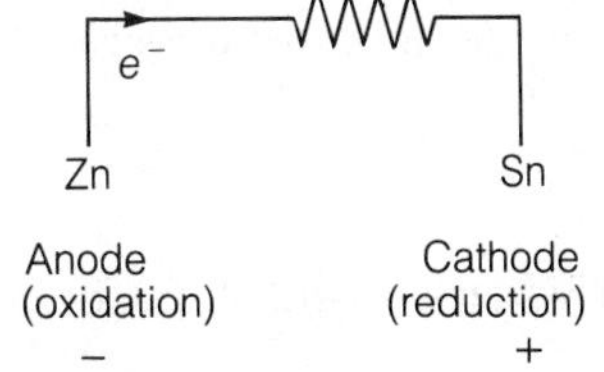

(a) Oxidation (loss of electrons) occurs at the zinc electrode: $Zn \longrightarrow Zn^{2+} + 2e^-$. Therefore, the Zn electrode, which gives out electrons, is negative and is the anode. Reduction (gain of electrons) occurs at the tin electrode: $Sn^{2+} + 2e^- \longrightarrow Sn$. The Sn electrode is the cathode. It takes electrons in and is therefore positive.

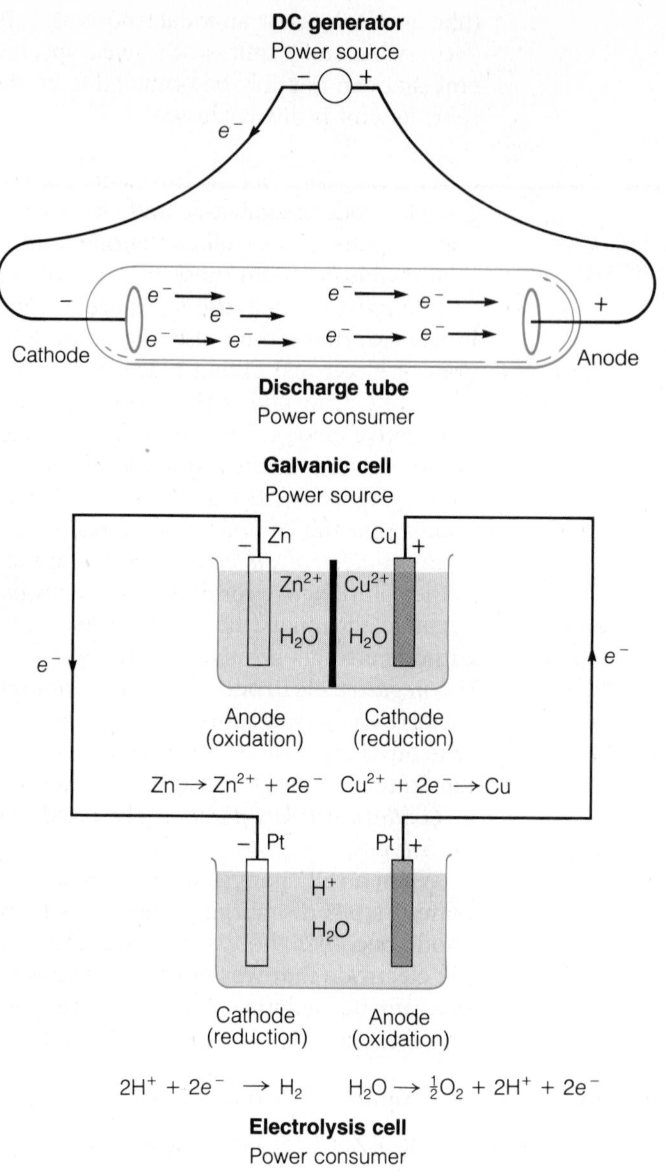

FIGURE 17.9
Examples of the meaning of "positive" and "negative" in electric circuits.

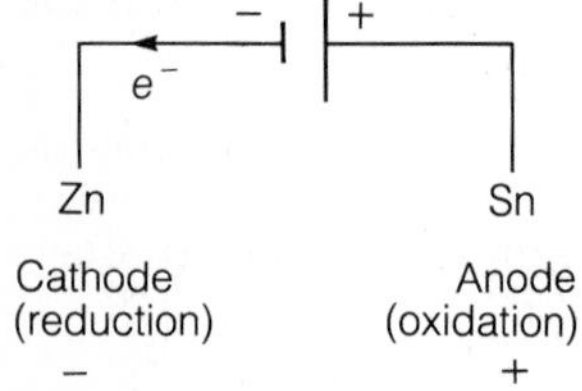

(b) Electrons are now being pushed into the Zn electrode. It is a cathode because reduction occurs there. However, it is still negative because it is connected to the negative side of a generator or battery and attracts positive ions. Sn is now the anode and is positive because electrons are being pulled out of it to the positive side of the power source. ■

PROBLEM 17.6 Identify the anode, cathode, positive electrode, and negative electrode in the cell

$$Sn(c)|Sn(NO_3)_2(aq)||Cu(NO_3)_2(aq)|Cu(c)$$

(a) when it is a galvanic cell, as in Problem 17.4a; (b) when current is pushed through it to drive the reaction backward. □

BOX 17.3

A discharge tube (section 2.5) is analogous to an electrolysis cell. The electrode into which electrons are fed is designated "negative" or "cathode," and it is this electrode from which the electron beam ("cathode rays") emanates inside the tube. Because discharge tubes are studied so widely, and so early in a scientific education, the misconception that "anode" always means "positive electrode" and "cathode" always means "negative electrode" can become deeply rooted. The equally familiar example of the dry cell shows how wrong this is.

17.7 HALF-CELL POTENTIALS

Any electrochemical cell—galvanic or electrolytic—requires two electrodes. Oxidation must occur at one electrode (the anode) and reduction at the other (the cathode). There is no way to measure the potential difference between an electrode and a solution without introducing another electrode. We always measure a potential difference between two electrodes. That difference is the emf of an entire cell. Also, we cannot measure the driving force of a *partial* reaction; a cell process is always a combination of two partial reactions, and it is the driving force of this combination that is measured as the emf of the cell.

As a bookkeeping device, however, it is useful to think of a cell as consisting of two **half-cells**, with each half-cell contributing its share to the emf of the cell. The contribution of a given half-cell is independent of the identity of the other half-cell. A half-cell consists of an electrode and the solution with which it is in contact. *It is possible to assign a potential to each half-cell* and then to combine the potentials to calculate the emf of the entire cell. Instead of tabulating the standard emf $\mathscr{E}^0$ for every cell, we tabulate it only for every half-cell, which makes for a much shorter table.

We can choose any half-cell, call its standard potential zero, and measure the standard potentials of all other half-cells by combining them with this reference half-cell. A uniform shift in all the half-cell potentials would make no difference because it would cancel out on subtraction. The **reference half-cell** chosen (by long-standing, worldwide custom) is shown in Figure 17.10; it consists of a platinum electrode in contact with both H_2 gas and a solution of H^+:

$$\mathrm{Pt}|\mathrm{H_2(g,\ 1.0\ atm)}|\mathrm{H^+(aq,\ 1.0\ M)}$$

The partial reaction in this half-cell is

$$\mathrm{H_2(g)} \longrightarrow \mathrm{2H^+} + 2e^-$$

For this half-cell or partial reaction, $\mathscr{E}^0 = 0$ by arbitrary convention. The negative ion (X^-) that goes with H^+ can be any ion for which HX is a strong acid; most commonly, it is Cl^-, Br^-, I^-, HSO_4^-, ClO_4^-, or NO_3^-.

Another arbitrary convention (Stockholm Convention, 1953) is more recent and still not universally accepted. The half-cell potential is the emf of a cell in which the reference half-cell is written on the *left* (as if it is the anode) and the other (variable) half-cell is written on the *right* (as if it is the cathode). Thus, a *positive* half-cell potential is associated with a half-cell that is actually a cathode when it is combined with the reference half-cell. The more positive the potential is, the more the half-cell "wants" to be a cathode and the more easily the reduction process occurs in it. A *negative* half-cell potential is a signal that this half-cell is really an *anode* when it is combined with the reference half-cell. To illustrate these ideas, some half-cell potentials are shown in Figure 17.11. The higher (more negative) the half-cell is on this scale, the more easily the oxidation occurs. The

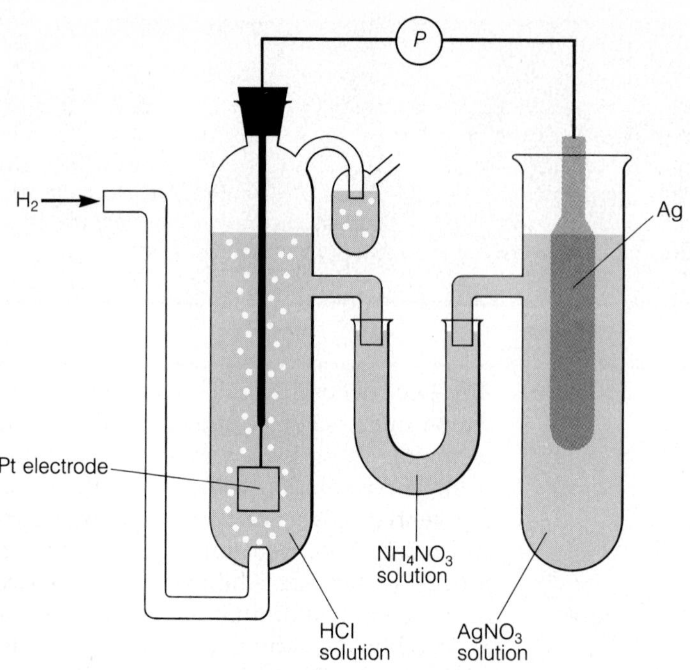

FIGURE 17.10 An example of the standard hydrogen half-cell ("hydrogen electrode") in the cell $Pt|H_2|HCl||Ag^+|Ag$. The potentiometer is indicated by *P*.

Potential (V)	Half-cell	Partial reaction
−3.040	$Li^+|Li$*	$Li^+ + e^- \rightleftharpoons Li$
−2.713	$Na^+|Na$*	$Na^+ + e^- \rightleftharpoons Na$
−0.7618	$Zn^{2+}|Zn$	$Zn^{2+} + 2e^- \rightleftharpoons Zn$
−0.1375	$Sn^{2+}|Sn$	$Sn^{2+} + 2e^- \rightleftharpoons Sn$
0	$H^+|H_2(g)|Pt$	$2H^+ + 2e^- \rightleftharpoons H_2$
+0.3419	$Cu^{2+}|Cu$	$Cu^{2+} + 2e^- \rightleftharpoons Cu$
+0.7996	$Ag^+|Ag$	$Ag^+ + e^- \rightleftharpoons Ag$
+1.3583	$Cl^-|Cl_2(g)|Pt$	$Cl_2 + 2e^- \rightleftharpoons 2Cl^-$
+2.866	$F^-|F_2(g)|Pt$*	$F_2 + 2e^- \rightleftharpoons 2F^-$

FIGURE 17.11 A scale of half-cell potentials. The higher a half-cell stands on this scale, the more it tends to be an anode. The numbers increase downward to conform to the order in Table 17.1. Half-cells marked with an asterisk cannot be prepared because the elements react with water; their potentials have been measured indirectly.

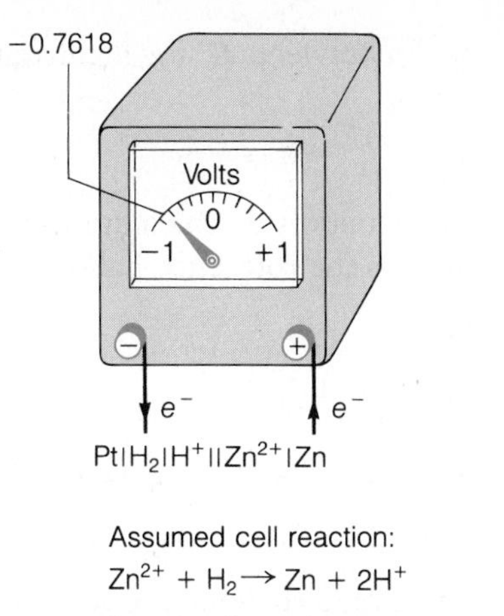

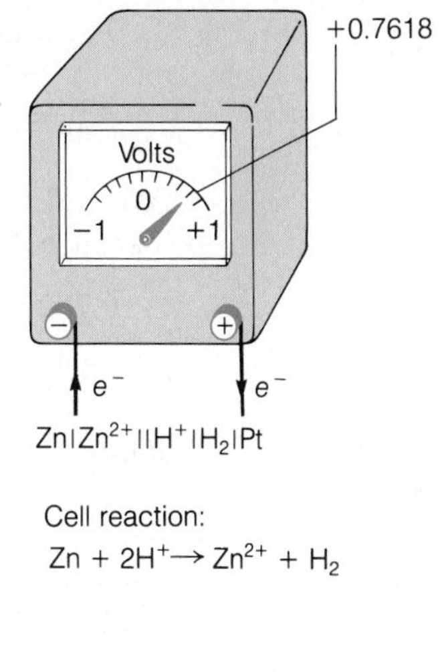

FIGURE 17.12
(a) A cell connected to a voltmeter according to the incorrect assumption that Zn is the cathode. Since electrons are being pushed into the "positive" terminal, the pointer indicates a negative voltage. (b) The same cell connected to the voltmeter correctly. The emf is positive.

lower (more positive) the half-cell is, the more easily the reduction occurs. The distance between two half-cells on this scale determines the emf of a cell made from them.

To determine experimentally which is the anode and which is the cathode, it is convenient to have a voltmeter in which the needle swings both ways (Figure 17.12). The needle shows the direction of electron flow. If the needle swings to the positive side, it means that electrons are flowing into the negative terminal and out of the positive terminal. If it swings to the negative side, it means that electrons are flowing into the positive terminal and out of the negative terminal. The anode is negative and the cathode positive (section 17.6). If the anode is connected to the negative terminal of the voltmeter and the cathode to the positive terminal, the measured emf is positive. If we guess wrong and connect the anode to the positive terminal, the measured emf is negative.

If the needle can swing only one way, you may have to reverse the wires.

Suppose we want to find the standard potential for the half-cell $Ag^+|Ag$. We must measure the standard emf of a complete cell in which this half-cell is combined with the reference half-cell. Prepare the cell

$$Pt|H_2(1\ atm)|H^+(aq,\ 1\ M)||Ag^+(aq,\ 1\ M)|Ag$$

corresponding to the partial reactions

$$\begin{array}{rcll} H_2(g) & \longrightarrow & 2H^+ + 2e^- & \text{(oxidation, at anode)} \\ 2Ag^+ + 2e^- & \longrightarrow & 2Ag(c) & \text{(reduction, at cathode)} \\ \hline H_2(g) + 2Ag^+ & \longrightarrow & 2Ag(c) + 2H^+ & \end{array}$$

When the cell and the reactions are written in this way, the assumption is that the reference half-cell is the anode (written on the left) and the unknown silver half-cell is the cathode (written on the right). Connect the Pt electrode to the negative side of the voltmeter and the Ag electrode to the positive side. The needle swings to the positive side, $\mathscr{E}^0_{cell} = 0.7996$ V at 25°C, which shows that the electrons are indeed coming out of the Pt and going into the Ag. This means that the left half-cell is the anode and the right half-cell is the cathode. The spontaneous cell reaction is, as assumed,

$$H_2(g) + 2Ag^+ \longrightarrow 2Ag(c) + 2H^+$$

The words *anode* and *cathode* apply strictly to the electrodes (the metallic conductors). However, it is common to refer to the half-cell (electrode + solution) in which oxidation occurs as the anode, the half-cell in which reduction occurs as the cathode, and either of them as an electrode.

The relationship between $\mathscr{E}^0$ for a galvanic cell and the $\mathscr{E}^0$'s of the half-cells is

$$\mathscr{E}^0_{\text{cell}} = \mathscr{E}^0_{\text{cathode}} - \mathscr{E}^0_{\text{anode}} \tag{11}$$

We will use this equation to calculate the desired half-cell potential. $\mathscr{E}^0_{\text{anode}} = 0$, because the anode is the reference half-cell; $\mathscr{E}^0_{\text{cell}} = 0.7996$ V by experiment. Then

$$0.7996 \text{ V} = \mathscr{E}^0_{\text{cathode}} - 0$$

$$\mathscr{E}^0_{\text{cathode}} = 0.7996 \text{ V}$$

Thus, $\mathscr{E}^0 = +0.7996$ V for the half-cell $Ag^+|Ag$, or for the partial reaction,

$$Ag^+ + e^- \longrightarrow Ag$$

Next, find the standard potential for the half-cell $Zn|Zn^{2+}(aq)$. This requires that we measure the $\mathscr{E}^0$ for the cell

$$Pt|H_2(g)|H^+(aq)||Zn^{2+}(aq)|Zn$$

which corresponds to the assumed reaction

$$Zn^{2+} + H_2(g) \longrightarrow Zn(c) + 2H^+ \tag{12}$$

Again, we connect the Pt electrode (the assumed anode) to the negative terminal of the voltmeter and the Zn electrode (the assumed cathode) to the positive terminal. When the cell is connected in this way, the needle swings to the *negative* side: $\mathscr{E}^0_{\text{cell}} = -0.7618$ V (Figure 17.12a). This means that the reference half-cell is really the *anode* and the unknown half-cell is really the *cathode*. However, equation 11 should be applied to the cell *as written*. Thus, we set $\mathscr{E}^0_{\text{anode}} = 0$ and $\mathscr{E}^0_{\text{cathode}} = \mathscr{E}^0_{\text{Zn}}$:

$$-0.7618 \text{ V} = \mathscr{E}^0_{\text{Zn}} - 0$$

$$\mathscr{E}^0_{\text{Zn}} = -0.7618 \text{ V}$$

The half-cell potential is negative—not surprising, because equation 12 represents a nonspontaneous reaction.

When the needle swings to the negative side, the natural reaction is "Oops! I connected it wrong!" Reversing the wires is the physical manifestation of changing your mind; now you treat the Pt electrode as the cathode—which it really is—and the Zn electrode as the anode:

$$Zn|Zn^{2+}(aq)||H^+(aq)|H_2(g)|Pt$$

This cell corresponds to the spontaneous reaction

$$Zn(c) + 2H^+ \longrightarrow Zn^{2+} + H_2(g)$$

The voltmeter now reads $\mathscr{E}^0_{\text{cell}} = +0.7618$ V (Figure 17.12b). This time, $\mathscr{E}^0_{\text{anode}} = \mathscr{E}^0_{\text{Zn}}$ and $\mathscr{E}^0_{\text{cathode}} = 0$, and the half-cell potential is the same:

$$\mathscr{E}^0_{\text{cell}} = \mathscr{E}^0_{\text{cathode}} - \mathscr{E}^0_{\text{anode}}$$

$$0.7618 \text{ V} = 0 - \mathscr{E}^0_{\text{Zn}}$$

$$\mathscr{E}^0_{\text{Zn}} = -0.7618 \text{ V}$$

If you want to avoid the embarrassment of writing a reaction or a cell and finding that it is backwards, assume that the half-cell with the more positive (or less negative) potential is the cathode and that the half-cell with the more negative (or less positive) potential is the anode. The standard emf of the cell will then automatically come out positive.

The half-cell potentials are called **electrode potentials** because they specify the potential of the electrode (the metal) relative to the solution, minus an unknown constant that is the same for all of them. Whatever the unknown constant is, it drops out when we calculate the difference of potential between the electrodes (Problem 17.36). There has been an effort for decades to arrive at a satisfactory definition of "absolute single electrode potentials" and—what must accompany the definition—a way to measure them. Since no agreement is in sight, everyone still uses the arbitrary convention that $\mathscr{E}^0 = 0$ for the standard hydrogen electrode.

The minus sign, indicating subtraction, goes with the reversal of the partial equation in Table 17.1 to give an oxidation.

Table 17.1 gives the standard potentials for a number of common half-cells. These potentials are for unit concentration (1 mol/L) of the ions and other solutes *appearing in the partial reactions.* Thus, a silver half-cell has $\mathscr{E}^0 = 0.7996$ V when $[Ag^+] = 1$ mol/L, 0.4470 V when solid Ag_2CrO_4 is present and $[CrO_4{}^{2-}] = 1$ mol/L, and 0.373 V when $[NH_3] = [Ag(NH_3)_2{}^+] = 1$ mol/L. Cell emf's are calculated from these potentials by subtracting the potential corresponding to oxidation (anode) from the potential corresponding to reduction (cathode).

Table 17.2 is an index for helping you to find potentials in Table 17.1. The emf's are listed alphabetically by the *symbol* of the element that is oxidized or reduced. When there is more than one entry for an element, formulas are given to distinguish among them.

In section 17.4 we noted that the emf of a complete cell is unchanged when the equation for the cell reaction is multiplied through by a constant. The same is true of an electrode potential. Thus, $\mathscr{E}^0 = +0.7996$ V for all of these:

$$Ag^+ + e^- \rightleftharpoons Ag$$

$$2Ag^+ + 2e^- \rightleftharpoons 2Ag$$

$$\tfrac{1}{2}Ag^+ + \tfrac{1}{2}e^- \rightleftharpoons \tfrac{1}{2}Ag$$

EXAMPLE 17.7 Find $\mathscr{E}^0$ at 25°C for the cell

$$Cu|CuCl_2(aq)|Cl_2(g)|Pt$$

in which the reaction is

$$Cu(c) + Cl_2(g) \longrightarrow Cu^{2+} + 2Cl^-$$

ANSWER First write the cathode and anode partial reactions. Beside each partial reaction, write the $\mathscr{E}^0$ from Table 17.1:

$Cl_2 + 2e^- \longrightarrow 2Cl^-$	(reduction)	$\mathscr{E}^0_{cathode} = +1.3583$
$Cu \longrightarrow Cu^{2+} + 2e^-$	(oxidation)	$-\mathscr{E}^0_{anode} = -(+0.3419)$

(A minus sign appears before $\mathscr{E}^0$ for the oxidation as a reminder that it is to be subtracted.) Then calculate $\mathscr{E}^0$ for the cell by using equation 11:

$$\begin{aligned}\mathscr{E}^0_{cell} &= \mathscr{E}^0_{cathode} - \mathscr{E}^0_{anode}\\ &= +1.3583 - 0.3419\\ &= +1.0164 \text{ V}\end{aligned}$$

■

EXAMPLE 17.8 Find $\mathscr{E}^0$ at 25°C for a cell in which the reaction is

$$2Ag^+ + Cd(c) \longrightarrow 2Ag(c) + Cd^{2+}$$

TABLE 17.1
STANDARD ELECTRODE (REDUCTION) POTENTIALS IN AQUEOUS SOLUTION AT 25°C

ELEMENT	PARTIAL REACTION	$\mathscr{E}^0$, volts
Li	$Li^+ + e^- \rightleftharpoons Li$	−3.040
K	$K^+ + e^- \rightleftharpoons K$	−2.931
Ba	$Ba^{2+} + 2e^- \rightleftharpoons Ba$	−2.912
Ca	$Ca^{2+} + 2e^- \rightleftharpoons Ca$	−2.868
Na	$Na^+ + e^- \rightleftharpoons Na$	−2.713
Mg	$Mg^{2+} + 2e^- \rightleftharpoons Mg$	−2.372
Al	$Al^{3+} + 3e^- \rightleftharpoons Al$	−1.662
H	$2H_2O + 2e^- \rightleftharpoons H_2(g) + 2OH^-$	−0.8277
Zn	$Zn^{2+} + 2e^- \rightleftharpoons Zn$	−0.7618
Cr	$Cr^{3+} + 3e^- \rightleftharpoons Cr$	−0.744
Au	$Au(CN)_2^- + e^- \rightleftharpoons Au + 2CN^-$	−0.60
Fe	$Fe^{2+} + 2e^- \rightleftharpoons Fe$	−0.447
Cd	$Cd^{2+} + 2e^- \rightleftharpoons Cd$	−0.4030
Ag	$Ag(CN)_2^- + e^- \rightleftharpoons Ag + 2CN^-$	−0.395
Hg	$Hg(CN)_4^{2-} + 2e^- \rightleftharpoons Hg + 4CN^-$	−0.37
Pb	$PbSO_4(c) + 2e^- \rightleftharpoons Pb + SO_4^{2-}$	−0.3588
Ni	$Ni^{2+} + 2e^- \rightleftharpoons Ni$	−0.257
V	$V^{3+} + e^- \rightleftharpoons V^{2+}$	−0.255
Ag	$AgI(c) + e^- \rightleftharpoons Ag + I^-$	−0.1522
Sn	$Sn^{2+} + 2e^- \rightleftharpoons Sn$	−0.1375
Pb	$Pb^{2+} + 2e^- \rightleftharpoons Pb$	−0.1262
Hg	$HgI_4^{2-} + 2e^- \rightleftharpoons Hg + 4I^-$	−0.04
Fe	$Fe^{3+} + 3e^- \rightleftharpoons Fe$	−0.037
H	$2H^+ + 2e^- \rightleftharpoons H_2(g)$	0.0000
Ag	$Ag(S_2O_3)_2^{3-} + e^- \rightleftharpoons Ag + 2S_2O_3^{2-}$	+0.015
Ag	$AgBr(c) + e^- \rightleftharpoons Ag + Br^-$	+0.0713
Hg	$HgBr_4^{2-} + 2e^- \rightleftharpoons Hg + 4Br^-$	+0.21
Ag	$AgCl(c) + e^- \rightleftharpoons Ag + Cl^-$	+0.2223
Hg	$Hg_2Cl_2(c) + 2e^- \rightleftharpoons 2Hg + 2Cl^-$	+0.2681
Cu	$Cu^{2+} + 2e^- \rightleftharpoons Cu$	+0.3419
Ag	$Ag(NH_3)_2^+ + e^- \rightleftharpoons Ag + 2NH_3(aq)$	+0.373
Hg	$HgCl_4^{2-} + 2e^- \rightleftharpoons Hg + 4Cl^-$	+0.38
O	$O_2(g) + 2H_2O + 4e^- \rightleftharpoons 4OH^-$	+0.401
Ag	$Ag_2CrO_4(c) + 2e^- \rightleftharpoons 2Ag + CrO_4^{2-}$	+0.4470
I	$I_2(c) + 2e^- \rightleftharpoons 2I^-$	+0.5355
Mn	$MnO_4^- + 2H_2O + 3e^- \rightleftharpoons MnO_2(c) + 4OH^-$	+0.595
Fe	$Fe^{3+} + e^- \rightleftharpoons Fe^{2+}$	+0.771
Hg	$Hg_2^{2+} + 2e^- \rightleftharpoons 2Hg$	+0.7973
Ag	$Ag^+ + e^- \rightleftharpoons Ag$	+0.7996
Hg	$Hg^{2+} + 2e^- \rightleftharpoons Hg$	+0.851
Au	$AuBr_4^- + 3e^- \rightleftharpoons Au + 4Br^-$	+0.854
Au	$AuCl_4^- + 3e^- \rightleftharpoons Au + 4Cl^-$	+1.002
Br	$Br_2(\ell) + 2e^- \rightleftharpoons 2Br^-$	+1.066
Pt	$Pt^{2+} + 2e^- \rightleftharpoons Pt$	+1.118
I	$2IO_3^- + 12H^+ + 10e^- \rightleftharpoons I_2(c) + 6H_2O$	+1.195
O	$O_2(g) + 4H^+ + 4e^- \rightleftharpoons 2H_2O$	+1.229
Cr	$Cr_2O_7^{2-} + 14H^+ + 6e^- \rightleftharpoons 2Cr^{3+} + 7H_2O$	+1.232
Cl	$Cl_2(g) + 2e^- \rightleftharpoons 2Cl^-$	+1.3583
Au	$Au^{3+} + 3e^- \rightleftharpoons Au$	+1.498
Mn	$MnO_4^- + 8H^+ + 5e^- \rightleftharpoons Mn^{2+} + 4H_2O$	+1.507
Pb	$PbO_2(c) + 3H^+ + HSO_4^- + 2e^- \rightleftharpoons PbSO_4(c) + 2H_2O$	+1.691
Au	$Au^+ + e^- \rightleftharpoons Au$	+1.692
F	$F_2(g) + 2e^- \rightleftharpoons 2F^-$	+2.866

TABLE 17.2
ALPHABETICAL INDEX OF STANDARD ELECTRODE (REDUCTION) POTENTIALS IN AQUEOUS SOLUTION AT 25°C†

SYMBOL OR FORMULA	$\mathscr{E}^0$, volts	SYMBOL OR FORMULA	$\mathscr{E}^0$, volts
Ag^+	+0.7996	H_2, H^+	0.0000
AgBr	+0.0713	H_2, OH^-	−0.8277
AgCl	+0.2223	Hg_2^{2+}	+0.7973
$Ag(CN)_2^-$	−0.395	Hg^{2+}	+0.851
Ag_2CrO_4	+0.4470	$HgBr_4^{2-}$	+0.21
AgI	−0.1522	Hg_2Cl_2	+0.2681
$Ag(NH_3)_2^+$	+0.373	$HgCl_4^{2-}$	+0.38
$Ag(S_2O_3)_2^{3-}$	+0.015	$Hg(CN)_4^{3-}$	−0.37
Al	−1.662	HgI_4^{2-}	−0.04
Au^+	+1.692	I^-	+0.5355
Au^{3+}	+1.498	IO_3^-	+1.195
$AuBr_4^-$	+0.854	K	−2.931
$AuCl_4^-$	+1.002	Li	−3.040
$Au(CN)_2^-$	−0.60	Mg	−2.372
Ba	−2.912	Mn^{2+}	+1.507
Br	+1.066	MnO_2	+0.595
Ca	−2.868	Na	−2.713
Cd	−0.4030	Ni	−0.257
Cl	+1.3583	O_2, H^+	+1.229
Cr^{3+}	−0.744	O_2, OH^-	+0.401
$Cr_2O_7^{2-}$	+1.232	Pb^{2+}	−0.1262
Cu	+0.3419	PbO_2	+1.691
F	+2.866	$PbSO_4$	−0.3588
Fe^{2+}	−0.447	Pt	+1.118
Fe^{2+}, Fe^{3+}	+0.771	Sn	−0.1375
Fe^{3+}	−0.037	V	−0.255
		Zn	−0.7618

† Arranged alphabetically by chemical symbols. See Table 17.1 for partial reactions.

ANSWER Write the partial equations, copy their $\mathscr{E}^0$'s from Table 17.1, and subtract:

Note that 0.7996 is *not* doubled when the coefficients in the partial equation are doubled.

$$\begin{array}{lll} 2Ag^+ + 2e^- \longrightarrow 2Ag & \text{(reduction)} & \mathscr{E}^0_{cathode} = +0.7996 \\ Cd \longrightarrow Cd^{2+} + 2e^- & \text{(oxidation)} & -\mathscr{E}^0_{anode} = -(-0.4030) \\ \hline 2Ag^+ + Cd \longrightarrow 2Ag + Cd^{2+} & & \mathscr{E}^0_{cell} = +0.7996 - (-0.4030) \\ & & = 0.7996 + 0.4030 \\ & & = 1.2026\ \text{V} \end{array}$$

■

When $\mathscr{E}^0$ for a cell comes out positive, it means that the reaction goes in the direction it is written; the anode and cathode are indeed what we assumed they would be. A negative $\mathscr{E}^0$ indicates that the reaction does not occur as it is written. It goes spontaneously in the opposite direction. Of course, we could force it to go in the nonspontaneous direction by applying a large enough potential (section 17.11), but then we would be dealing with an electrolytic cell, not a galvanic cell.

EXAMPLE 17.9 Find $\mathscr{E}^0$ for a cell in which the assumed reaction is

$$Hg(\ell) + 2Ag^+ \longrightarrow Hg^{2+} + 2Ag(c)$$

What does the answer tell you about this reaction?

ANSWER The partial reactions are

$$
\begin{array}{rcl}
2Ag^+ + 2e^- & \longrightarrow & 2Ag \\
Hg & \longrightarrow & Hg^{2+} + 2e^- \\
\hline
Hg + 2Ag^+ & \longrightarrow & Hg^{2+} + 2Ag
\end{array}
\qquad
\begin{array}{rl}
\ldots ode & -(+0.851) \\
\hline
\mathscr{E}^0_{cell} & = +0.7996 - 0.851 \\
 & = -0.051\ V
\end{array}
$$

The negative $\mathscr{E}^0_{cell}$ means that the reaction does not go in the direction assumed. The spontaneous direction is the opposite:

$$Hg^{2+} + 2Ag \longrightarrow Hg + 2Ag^+$$ ■

EXAMPLE 17.10 For the cell

KCl is the salt bridge.

$$Hg(\ell)|Hg_2Cl_2(c)|KCl(aq)|KRuO_4(aq), K_2RuO_4(aq)|Pt$$

$\mathscr{E}^0 = 0.32$ V at 25°C. Find $\mathscr{E}^0$ for the partial reaction

$$RuO_4^{\ -} + e^- \rightleftharpoons RuO_4^{\ 2-}$$

ANSWER The procedure is the same as in Examples 17.8 and 17.9, except that the unknown is in a different place. The partial equations are

$$
\begin{array}{rcll}
2RuO_4^{\ -} + 2e^- & \longrightarrow & 2RuO_4^{\ 2-} & \text{(reduction)} \quad \mathscr{E}^0_{cathode} = y \\
2Hg + 2Cl^- & \longrightarrow & Hg_2Cl_2 + 2e^- & \text{(oxidation)} \quad -\mathscr{E}^0_{anode} = -(+0.2681) \\
\hline
2Hg + 2Cl^- + 2RuO_4^{\ -} & \longrightarrow & Hg_2Cl_2 + 2RuO_4^{\ 2-} & \qquad\qquad \mathscr{E}^0_{cell} = 0.32\ V
\end{array}
$$

$$\mathscr{E}^0_{cell} = \mathscr{E}^0_{cathode} - \mathscr{E}^0_{anode}$$

$$0.32 = y - 0.2681$$

$$y = 0.59\ V$$ ■

PROBLEM 17.7 Find $\mathscr{E}^0$ at 25°C for a cell in which the reaction is

$$2Cr(c) + 3I_2(c) \longrightarrow 2Cr^{3+} + 6I^-$$ □

PROBLEM 17.8 The net reaction in the cell

$$Pt|Fe^{3+}, Fe^{2+}, SO_4^{\ 2-}||Hg^{2+}, NO_3^{\ -}|Hg(\ell)$$

is

$$2Fe^{2+} + Hg^{2+} \longrightarrow 2Fe^{3+} + Hg$$

(a) Write the partial reactions. (b) Calculate $\mathscr{E}^0_{cell}$ at 25°C. □

OXIDIZING AND REDUCING AGENTS

Table 17.1 is actually an expanded version of the activity series given in Table 9.5 (section 9.4). The "most active metal" (Li) is the strongest reducing agent (most easily oxidized), listed at the top; the "most active nonmetal" (F_2) is the strongest oxidizing agent (most easily reduced), listed at the bottom. The standard electrode potentials given in Table 17.1 are often called **reduction potentials** because they denote the tendency of a partial reaction to occur as a reduction. The more *positive* the potential, the greater is the tendency for the *reduction* to take place and for the electrode to behave as a *cathode.* A positive potential thus identifies a strong oxidizing agent—one that easily gains electrons. The more *negative* the potential, the greater is the tendency for the *oxidation* to take place and for the

Some older books give oxidation potentials, which are simply the negatives of ours, with the partial reactions written as oxidations.

electrode to be an *anode*. A negative potential thus identifies a strong reducing agent—one that easily loses electrons. The extreme cases are

$F_2 + 2e^- \rightleftharpoons 2F^-$, $\mathscr{E}^0 = +2.866$ V	$Li^+ + e^- \rightleftharpoons Li$, $\mathscr{E}^0 = -3.040$ V
Reaction has the most positive $\mathscr{E}^0$.	Reaction has the most negative $\mathscr{E}^0$.
Reaction has greatest tendency to proceed as a reduction.	Reaction has least tendency to proceed as a reduction.
F_2 is the strongest oxidizing agent in the table.	Li^+ is the weakest oxidizing agent in the table.
The reverse reaction (oxidation) has the least tendency to go.	The reverse reaction (oxidation) has the greatest tendency to go.
F^- is the weakest reducing agent in the table.	Li is the strongest reducing agent in the table.

EXAMPLE 17.11 Arrange the following species in the order of increasing strength as oxidizing agents: I_2(c), $AuCl_4^-$, $Au(CN)_2^-$, Fe^{2+}.

ANSWER Look in Table 17.1 for partial equations in which each species appears on the *left*. The following potentials are associated with the partial equations: I_2, +0.5355 V; $AuCl_4^-$, +1.002 V; $Au(CN)_2^-$, −0.60 V; Fe^{2+}, −0.447 V (*not* +0.771 V, which corresponds to Fe^{2+} behaving as a *reducing* agent). The more positive the potential is, the easier is the reduction and the stronger is the oxidizing agent. The order of increasing strength is $Au(CN)_2^-$, Fe^{2+}, I_2, $AuCl_4^-$. ■

PROBLEM 17.9 Arrange these species in the order of increasing strength as reducing agents: Na, Fe^{2+}, V^{2+}, Pb. □

The spontaneous direction of an acid-base reaction is the direction that produces the weaker acid and the weaker base (section 14.2).

An oxidation–reduction reaction tends to go in the direction that produces the weaker oxidizing agent and the weaker reducing agent. Thus, Example 17.11 concludes that $AuCl_4^-$ is a stronger oxidizing agent than Fe^{2+}. This means that a spontaneous reaction would be one in which $AuCl_4^-$ appears on the left as the oxidizing agent, while Fe^{2+} appears at the right as the product of the oxidation. (This prediction is reliable when concentrations are near 1 mol/L.) The net reaction is obtained by combining the two partial equations:

$$[AuCl_4^- + 3e^- \longrightarrow Au + 4Cl^-] \times 2$$
$$[Fe \longrightarrow Fe^{2+} + 2e^-] \times 3$$
$$\underset{\text{oxidizing agent 1}}{2AuCl_4^-} + \underset{\text{reducing agent 2}}{3Fe} \longrightarrow \underset{\text{reducing agent 1}}{2Au} + 8Cl^- + \underset{\text{oxidizing agent 2}}{3Fe^{2+}}$$

The *stronger* oxidizing agent ($AuCl_4^-$) is converted to the *weaker* reducing agent (Au); the stronger reducing agent (Fe) is converted to the weaker oxidizing agent (Fe^{2+}).

EXAMPLE 17.12 (a) Identify the two oxidizing agents and the two reducing agents in the reaction

$$Ni + I_2(c) \rightleftharpoons Ni^{2+} + 2I^-$$

(b) Which oxidizing agent is stronger? Which reducing agent is stronger? Which direction is spontaneous?

ANSWER

(a) In the reaction as written, Ni is being oxidized and is therefore the reducing agent; I_2 is being reduced and is therefore the oxidizing agent. In the reverse reaction, Ni^{2+} is the oxidizing agent and I^- is the reducing agent.

$$\underset{\text{reducing agent 1}}{Ni} + \underset{\text{oxidizing agent 2}}{I_2} \rightleftharpoons \underset{\text{oxidizing agent 1}}{Ni^{2+}} + \underset{\text{reducing agent 2}}{2I^-}$$

(b) The electrode potentials are

$$Ni^{2+} + 2e^- \longrightarrow Ni \qquad -0.257\text{ V}$$

$$I_2 + 2e^- \longrightarrow 2I^- \qquad +0.5355\text{ V}$$

The more positive potential corresponds to the easier reduction; therefore, I_2 is the stronger oxidizing agent. The more negative potential corresponds to the easier oxidation; therefore, Ni is the stronger reducing agent. The left-to-right direction is spontaneous. ■

PROBLEM 17.10 Identify the stronger and weaker oxidizing and reducing agents in the reaction

$$Cd(c) + 2Ag^+ \rightleftharpoons Cd^{2+} + 2Ag(c)$$ □

17.8 THE NERNST EQUATION

Thus far, we have concerned ourselves with the emf's of galvanic cells in which every dissolved reactant and product has a concentration of 1 mol/L and every gas has a pressure of 1 atm. The emf of a cell depends on the concentrations and pressures, which need not be anywhere near 1 mol/L and 1 atm. How the emf depends on concentration and pressure is the topic of this section.

The emf of a cell measures the net driving force of the cell reaction. For the cell of Figure 17.6 (section 17.3), the reaction is

$$Zn(c) + Cu^{2+} \rightleftharpoons Cu(c) + Zn^{2+}$$

Suppose that this reaction has come to equilibrium. According to Le Chatelier's principle (section 13.6), increasing the concentration of Cu^{2+} shifts the equilibrium to the right, and therefore must have increased the driving force of the left-to-right ("forward") reaction. On the other hand, increasing the concentration of Zn^{2+} increases the driving force of the reverse reaction, and thus decreases the net driving force of the reaction as it is written. *The greater the concentration of a reactant* (the left side of the equation), *the greater is the net driving force of the left-to-right reaction.* The greater the concentration of a product (right side of equation), the less is the net driving force.

The emf of a cell is a quantitative measure of the net driving force. When the emf is positive, the reaction tends to go from left to right; when the emf is negative, the reaction tends to go from right to left. When the emf is zero, the cell is dead and the reaction is in equilibrium. Therefore, increasing the concentration of Cu^{2+} (a reactant) increases the emf of the cell, and decreasing the concentration of Cu^{2+} decreases the emf of the cell. Increasing the concentration of Zn^{2+} (a product) increases the driving force of the *right-to-left* reaction, and therefore *decreases* the emf of the cell; decreasing the concentration of Zn^{2+} increases the emf of the cell. The emf does not depend on how much solid Cu or Zn is present, as long as *some* is present.

PROBLEM 17.11 This reaction occurs in a galvanic cell:

$$Cr + 3Fe^{3+} \longrightarrow Cr^{3+} + 3Fe^{2+}$$

How is the emf changed (increased or decreased) by increasing the concentration of (a) Fe^{3+}; (b) Cr^{3+}; (c) Fe^{2+}? □

A simple way to observe these effects in the laboratory is to set up a Zn-Cu cell and add a substance that reacts with Cu^{2+} or Zn^{2+}, thus decreasing its concentration. The equation $Zn + Cu^{2+} \rightleftharpoons Cu + Zn^{2+}$ predicts that decreasing $[Cu^{2+}]$ and decreasing $[Zn^{2+}]$ will have opposite effects. H_2S will do both jobs by forming the very insoluble salts CuS and ZnS. If H_2S is added to the $Cu|Cu^{2+}$ side of the cell, the emf drops sharply. The reason is that the reaction

$$Cu^{2+} + H_2S \longrightarrow CuS(c) + 2H^+$$

drastically decreases the concentration of Cu^{2+} from, say, 1 mol/L to something like 10^{-36} mol/L. On the other hand, adding H_2S to the $Zn|Zn^{2+}$ side of the cell increases the emf because $[Zn^{2+}]$ is decreased from 1 mol/L to, perhaps, 10^{-16} mol/L.

The quantitative relationship between the emf and the concentrations in this cell is

You should verify from the equation that changes in the concentrations have the effects described previously.

$$\mathscr{E} = \mathscr{E}^0 - \frac{RT}{n\mathscr{F}} \ln\left(\frac{[Zn^{2+}]}{[Cu^{2+}]}\right)$$

$\mathscr{E}$ is the emf of the cell in volts; $\mathscr{E}^0$ is the standard emf of the cell; R is the gas constant, 8.314 J/(mol K); T is the absolute temperature (K); $\mathscr{F}$ is Faraday's constant, 9.65×10^4 C/faraday; and n is the number of moles of electrons (faradays) transferred through the external circuit when the cell reaction takes place as written. For the reaction $Zn(c) + Cu^{2+} \rightleftharpoons Cu(c) + Zn^{2+}$, $n = 2$, since $2e^-$ appears in each balanced partial equation. As a check on the equation, note that when both concentrations are 1 mol/L, $\mathscr{E} = \mathscr{E}^0$, as it must be.

This famous equation is the **Nernst equation**. Its general form is

Walther Nernst, 1864–1941

$$\mathscr{E} = \mathscr{E}^0 - \frac{RT}{n\mathscr{F}} \ln Q \qquad (13)$$

Remember that $\ln Q$ is the natural logarithm of Q (base $e = 2.71828\ldots$). Since common logarithms (base 10) are more familiar, equation 13 is usually written in the form

$$\boldsymbol{\mathscr{E} = \mathscr{E}^0 - \frac{2.303RT}{n\mathscr{F}} \log Q} \qquad (14)$$

$\ln 10 = 2.302585092994\ldots$; usually rounded to 2.303.

where $\log Q$ is the common logarithm. The relationship (Appendix B.1) is $\ln Q = \ln 10 \times \log Q = 2.303 \log Q$. Q is the reaction quotient (section 13.2) for the reaction as written. For the Zn-Cu cell, the reaction quotient is

$$Q = \frac{[Zn^{2+}]}{[Cu^{2+}]}$$

The solids are omitted, as usual.

For the Zn-Ag-AgCl cell (Figure 17.7), the reaction is

$$Zn(c) + 2AgCl(c) \longrightarrow 2Ag(c) + Zn^{2+} + 2Cl^-$$

Since the only variable concentrations are $[Zn^{2+}]$ and $[Cl^-]$, the Nernst equation is

$$\mathscr{E} = \mathscr{E}^0 - \frac{2.303RT}{2\mathscr{F}} \log ([Zn^{2+}][Cl^-]^2)$$

When *gases* are involved in the cell reaction, write the partial pressure of each gas (in atmospheres), not its concentration. For the reaction

$$Zn(c) + 2H^+ \longrightarrow Zn^{2+} + H_2(g)$$

the reaction quotient is

$$Q = \frac{[Zn^{2+}]p_{H_2}}{[H^+]^2}$$

Since R and $\mathscr{F}$ are universal constants, the factor $2.303RT/\mathscr{F}$ can be calculated once and for all for 25°C or any other chosen temperature. When $T = 298.15$ K (25°C),

$$\frac{RT\ln 10}{\mathscr{F}} = \frac{8.31441 \dfrac{\text{J}}{\text{mol K}} \times 298.15 \text{ K} \times 2.30259}{9.64853 \times 10^4 \dfrac{\text{C}}{\text{faraday}}}$$

$$= 0.059159 \frac{\text{J faraday}}{\text{C mol}} = 0.059159 \frac{\text{V faraday}}{\text{mol}}$$

This number can usually be rounded to 0.0592 V faraday/mol. At 25°C, the Nernst equation is thus

0.0592 is used so often that many people remember it. Do not forget, however, that it is good only at 25°C.

$$\mathscr{E} = \mathscr{E}^0 - \frac{0.0592}{n} \log Q \qquad \text{(at 25°C)} \tag{15}$$

Most calculations will be at 25°C, and the Nernst equation will therefore be used in the form of equation 15.

EXAMPLE 17.13 For the cell

$$Zn(c)|ZnCl_2(aq, 0.100 \text{ M})||Cu(NO_3)_2(aq, 0.200 \text{ M})|Cu$$

$\mathscr{E}^0 = 1.1037$ V at 25°C. Calculate $\mathscr{E}$. (See Appendix B.1 for calculations with logarithms.)

ANSWER The equation is $Zn + Cu^{2+} \rightleftharpoons Cu + Zn^{2+}$, with $n = 2$ faradays. From the cell description, $[Zn^{2+}] = 0.100$ mol/L and $[Cu^{2+}] = 0.200$ mol/L. Then

$$\mathscr{E} = \mathscr{E}^0 - \frac{0.0592}{n} \log \frac{[Zn^{2+}]}{[Cu^{2+}]}$$

$$= 1.1037 - \frac{0.0592}{2} \log \frac{0.100}{0.200}$$

$$= 1.1037 - \frac{0.0592}{2} \log (0.500)$$

$$= 1.1037 - \frac{0.0592}{2} (-0.301)$$

$$= 1.1037 + 0.00891$$

$$= 1.1126 \text{ V}$$

■

Very large concentrations are impossible; even a pure substance has a finite concentration.

Note that $\mathscr{E}$ and $\mathscr{E}^0$ are not greatly different. This is the usual case with ordinary laboratory concentrations in the range 0.01 to 10 M. However, with very small concentrations, $\mathscr{E}$ can be very different from $\mathscr{E}^0$.

EXAMPLE 17.14 Calculate the emf of the cell in Example 17.13 when $[Zn^{2+}] = 0.100$ mol/L and (a) $[Cu^{2+}] = 1.0 \times 10^{-36}$ mol/L; (b) $[Cu^{2+}] = 1.0 \times 10^{-42}$ mol/L.

Problem 16.26 asks you to consider the meaning of an extremely low concentration.

ANSWER

(a)
$$\mathscr{E} = \mathscr{E}^0 - \frac{0.0592}{n}\log\frac{[Zn^{2+}]}{[Cu^{2+}]}$$
$$= 1.1037 - \frac{0.0592}{n}\log\frac{0.100}{1.0 \times 10^{-36}}$$
$$= 1.1037 - \frac{0.0592}{2}\log(1.0 \times 10^{35})$$
$$= 1.1037 - \frac{0.0592}{2} \times 35.00$$
$$= 1.1037 - 1.036 = 0.068 \text{ V}$$

(b)
$$\mathscr{E} = 1.1037 - \frac{0.0592}{2}\log\frac{0.100}{1.0 \times 10^{-42}}$$
$$= 1.1037 - \frac{0.0592}{2}\log(1.0 \times 10^{41})$$
$$= 1.1037 - \frac{0.0592}{2} \times 41.00$$
$$= -0.11 \text{ V}$$

■

In Example 17.14b, $\mathscr{E}$ is negative. The negative emf tells us that our assumption about the direction of the reaction is incorrect. The concentration of Cu^{2+} has become so low that the reaction goes spontaneously in the opposite direction, as expected from Le Chatelier's principle:

$$Cu + Zn^{2+} \longrightarrow Cu^{2+} + Zn$$

Cu has now become the anode and Zn, the cathode.

EXAMPLE 17.15 For the cell $Sn(c)|SnCl_2(0.10\ M)|AgCl(c)|Ag(c)$, the reaction is

$$Sn + 2AgCl \longrightarrow 2Ag + Sn^{2+} + 2Cl^-$$

Calculate $\mathscr{E}^0$ at 25°C, write the Nernst equation, and calculate $\mathscr{E}$.

ANSWER Write the partial equations and find the $\mathscr{E}^0$'s in Table 17.1:

$$\begin{array}{rll} 2AgCl + 2e^- \longrightarrow 2Ag + 2Cl^- & \text{(reduction)} & \mathscr{E}^0_{cathode} = +0.2223 \\ Sn \longrightarrow Sn^{2+} + 2e^- & \text{(oxidation)} & -\mathscr{E}^0_{anode} = -(-0.1375) \\ \hline Sn + 2AgCl \longrightarrow 2Ag + Sn^{2+} + 2Cl^- & & \mathscr{E}^0_{cell} = 0.2223 - (-0.1375) \\ & & = 0.3598 \text{ V} \end{array}$$

There are two electrons in each partial equation, so that $n = 2$. The reaction quotient is $Q = [Sn^{2+}][Cl^-]^2$. The Nernst equation is

$$\mathscr{E} = \mathscr{E}^0 - \frac{0.0592}{2}\log([Sn^{2+}][Cl^-]^2)$$

In a 0.10 M solution of $SnCl_2$, $[Sn^{2+}] = 0.10$ mol/L and $[Cl^-] = 0.20$ mol/L. Therefore,

$$\mathscr{E} = \mathscr{E}^0 - \frac{0.0592}{2} \log [(0.10)(0.20)^2]$$

$$= 0.3598 - \frac{0.0592}{2} \log (4.0 \times 10^{-3})$$

$$= 0.3598 - \frac{0.0592}{2} (-2.40)$$

$$= 0.3598 + 0.0710$$

$$= 0.4308 \text{ V}$$

■

The cell reaction in Example 17.15 could have been written, just as correctly,

$$\tfrac{1}{2}Sn + AgCl \longrightarrow Ag + \tfrac{1}{2}Sn^{2+} + Cl^-$$

In section 17.4 we noted that $\mathscr{E}^0$ and $\mathscr{E}$ must be unaffected when the coefficients in the equation are multiplied by a constant. $\mathscr{E}$ is work *per coulomb*, regardless of the number of moles appearing in the equation. However, Q depends on the coefficients in the chemical equation. It is therefore not immediately obvious that the Nernst equation gives the same answer for $\mathscr{E}$ when the coefficients are all multiplied by a constant. The next example illustrates how the answers are indeed the same.

EXAMPLE 17.16 Repeat the calculation in Example 17.15 for the reaction written

$$\tfrac{1}{2}Sn + AgCl \longrightarrow Ag + \tfrac{1}{2}Sn^{2+} + Cl^-$$

Is the answer the same as in Example 17.15?

ANSWER The partial equations now are

$$\begin{array}{rcll} AgCl + e^- & \longrightarrow & Ag + Cl^- & \quad +0.2223 \\ \tfrac{1}{2}Sn & \longrightarrow & \tfrac{1}{2}Sn^{2+} + e^- & \quad -(-0.1375) \\ \hline \tfrac{1}{2}Sn + AgCl & \longrightarrow & Ag + \tfrac{1}{2}Sn^{2+} + Cl^- & \quad \mathscr{E}^0_{cell} = 0.2223 - (-0.1375) \\ & & & \quad = +0.3598 \text{ V} \end{array}$$

The calculation of $\mathscr{E}^0$ is the same. This time, however, there is only one electron in each partial equation; $n = 1$. The reaction quotient is also different:

$$Q = [Sn^{2+}]^{1/2}[Cl^-]$$

The Nernst equation is

$$\mathscr{E} = \mathscr{E}^0 - \frac{0.0592}{1} \log ([Sn^{2+}]^{1/2}[Cl^-])$$

$$= 0.3598 - 0.0592 \log [(0.10)^{1/2}(0.20)]$$

$$= 0.3598 - 0.0592 \log (0.063)$$

$$= 0.3598 - 0.0592 \times (-1.20)$$

$$= 0.3598 + 0.0710$$

$$= 0.4308 \text{ V}$$

The answer is the same because $Q_{\text{Example 17.16}} = (Q_{\text{Example 17.15}})^{1/2}$. Therefore, $\log Q_{\text{Example 17.16}} = \frac{1}{2} \log Q_{\text{Example 17.15}}$, and the factor $\frac{1}{2}$ compensates for the change from $n = 2$ to $n = 1$. ■

EXAMPLE 17.17 Find $\mathscr{E}^0$ and $\mathscr{E}$ at 25°C for the cell

$$\text{Cu}|\text{CuCl}_2(0.50\text{ M})|\text{Cl}_2(1.50\text{ atm})|\text{Pt}$$

ANSWER The partial reactions, with their $\mathscr{E}^0$'s, are

$$\begin{array}{rll} \text{Cl}_2 + 2e^- & \longrightarrow 2\text{Cl}^- & \quad +1.3583 \\ \text{Cu} & \longrightarrow \text{Cu}^{2+} + 2e^- & \quad -(+0.3419) \\ \hline \text{Cu} + \text{Cl}_2 & \longrightarrow \text{Cu}^{2+} + 2\text{Cl}^- & \quad \mathscr{E}^0 = 1.3583 - 0.3419 \\ & & \quad\quad\;\; = +1.0164\text{ V} \end{array}$$

The Nernst equation is

$$\mathscr{E} = \mathscr{E}^0 - \frac{0.0592}{2} \log\left(\frac{[\text{Cu}^{2+}][\text{Cl}^-]^2}{P_{\text{Cl}_2}}\right)$$

Here, $[\text{Cu}^{2+}] = 0.50$ mol/L, $[\text{Cl}^-] = 1.00$ mol/L, and $P_{\text{Cl}_2} = 1.50$ atm. Thus

$$\mathscr{E} = \mathscr{E}^0 - \frac{0.0592}{2} \log\left(\frac{0.50 \times 1.00^2}{1.50}\right)$$

$$= 1.0164 - \frac{0.0592}{2} \log 0.3333$$

$$= 1.0164 + 0.014 = 1.030\text{ V}$$

■

PROBLEM 17.12 The reaction in a certain cell is

$$2\text{Al(c)} + 3\text{Fe}^{2+} \longrightarrow 2\text{Al}^{3+} + 3\text{Fe(c)}$$

(a) Calculate $\mathscr{E}^0$. (b) Calculate $\mathscr{E}$ when $[\text{Fe}^{2+}] = 0.0100$ mol/L and $[\text{Al}^{3+}] = 0.0500$ mol/L. □

PROBLEM 17.13 Calculate $\mathscr{E}$ for the cell

$$\text{Cd}|\text{Cd}^{2+}(0.100\text{ M}), \text{Cl}^-(0.200\text{ M})|\text{AgCl}|\text{Ag}$$

in which the net reaction is

$$\text{Cd} + 2\text{AgCl} \longrightarrow \text{Cd}^{2+} + 2\text{Cl}^- + 2\text{Ag}$$

□

Tabulations of data on aqueous solutions are so often given at 25°C that we may need to remind ourselves that there are other temperatures and that 0.0592 is peculiar to this one temperature. Problem 17.44 illustrates a calculation at another temperature.

17.9 MEASURING CONCENTRATIONS

The dependence of emf on concentration provides a convenient method for measuring concentrations. The method is especially useful with very small concentrations, which may not be measurable in any other way. The oldest and most important application is to the measurement of $[\text{H}^+]$ and, thus, that popular variable, pH (section 15.1). The pH range of most interest to physicians, biologists, farmers, fishermen, and others is about 6 to 8, with $[\text{H}^+]$ about 10^{-6} to

10^{-8} mol/L—too small to measure by ordinary analytical methods such as titration (section 3.11). Indicators (section 15.4) are useful for rough estimation, but not for distinguishing between, say, pH 7.40 and 7.42—and how are you going to see colors in blood or muddy water?

An example of an apparatus that can be used to measure pH is the cell

$$Pt(c)|H_2(g,\ P\ atm)|H^+||Cl^-|Hg_2Cl_2(c)|Hg(\ell)$$

The partial reactions are

$$\begin{array}{rcl} H_2 & \longrightarrow & 2H^+ + 2e^- \\ Hg_2Cl_2 + 2e^- & \longrightarrow & 2Hg + 2Cl^- \\ \hline H_2 + Hg_2Cl_2 & \longrightarrow & 2H^+ + 2Cl^- + 2Hg \end{array}$$

At 25°C, the Nernst equation is

Remember:
log $(xy) = \log x + \log y$,
and $\log x^2 = 2 \log x$.

$$\begin{aligned} \mathscr{E} &= \mathscr{E}^0 - \frac{0.0592}{2} \log\left(\frac{[H^+]^2[Cl^-]^2}{P_{H_2}}\right) \\ &= \underbrace{\mathscr{E}^0 - \frac{0.0592}{2} \log\left(\frac{[Cl^-]^2}{P_{H_2}}\right)}_{\mathscr{E}'} - 0.0592 \underbrace{\log [H^+]}_{-pH} \\ &= \mathscr{E}' + 0.0592 \text{ pH} \end{aligned}$$

We lump the first two terms (over the brace) together and call them $\mathscr{E}'$. The H_2 pressure, the concentration of Cl^- in the calomel (Hg_2Cl_2) half-cell, and the temperature are kept constant during a series of measurements. The apparatus is calibrated—that is, $\mathscr{E}'$ is determined—by measuring $\mathscr{E}$ with buffer solutions of known pH.

The hydrogen electrode requires a tank of hydrogen, a pressure regulator, and an expensive platinum electrode. A simpler device, the **glass electrode**, was introduced in the 1920s and has made hydrogen electrodes obsolete, except as the fundamental reference for the scale of half-cell potentials. The glass is made in the form of a thin membrane. It exchanges H^+ ions with the solution at each of its surfaces. A solution of constant pH is on the inside of the glass membrane, the unknown solution on the outside. A **pH meter** (Figure 17.13) consists of a glass electrode, a reference electrode, and a voltmeter that can respond to a very small (practically zero) current. Box 17.4 explains why the emf of the cell is given by the same equation as that for a cell containing a hydrogen electrode:

$$\begin{aligned} \mathscr{E} &= \mathscr{E}' - \frac{2.303RT}{\mathscr{F}} \log [H^+] \\ &= \mathscr{E}' + \frac{2.303RT}{\mathscr{F}} \text{pH} \end{aligned} \tag{16}$$

At 25°C,

$$\mathscr{E} = \mathscr{E}' + 0.0592 \text{ pH}$$

$$\text{pH} = \frac{\mathscr{E} - \mathscr{E}'}{0.0592}$$

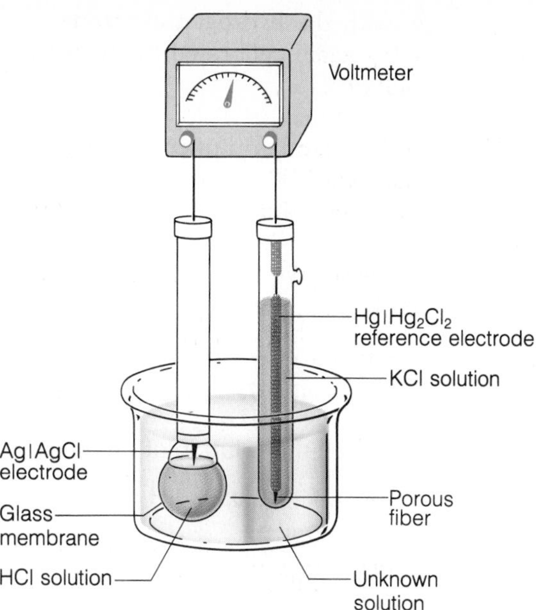

FIGURE 17.13
A pH meter using a glass electrode.

BOX 17.4
THE GLASS ELECTRODE

To understand how a glass electrode works, write the conventional representation of the cell illustrated in Figure 17.13.

$$\text{Ag(c)}|\text{AgCl(c)}|\text{HCl(aq, } c_1 \text{ mol/L)}|\text{glass}|\text{unknown solution } ([\text{H}^+] = c_2 \text{ mol/L})||\text{KCl(aq, } c_3 \text{ mol/L)}|\text{Hg}_2\text{Cl}_2\text{(c)}|\text{Hg}(\ell)$$

H^+ ions are present in the glass membrane near its surfaces and can go back and forth between the solution and the glass. In the interior of the glass, the ions that conduct electricity are mostly Na^+. The effect is that H^+ ions can enter the glass from the solution on one side while other H^+ ions leave the glass and enter the other solution on the other side. Actually, the glass is an electrolyte, not an electrode. The processes in this cell (if the salt bridge is ignored) are

$$\begin{array}{lrcl}
\text{anode:} & \text{Ag} + \text{Cl}^-(c_1) & \longrightarrow & \text{AgCl} + e^- \\
\text{glass:} & \text{H}^+(c_1) & \longrightarrow & \text{H}^+(c_2) \\
\text{cathode:} & \tfrac{1}{2}\text{Hg}_2\text{Cl}_2 + e^- & \longrightarrow & \text{Hg} + \text{Cl}^-(c_3) \\
\hline
\text{net:} & \text{Ag} + \tfrac{1}{2}\text{Hg}_2\text{Cl}_2 + \text{H}^+(c_1) + \text{Cl}^-(c_1) & \longrightarrow & \text{AgCl} + \text{Hg} + \text{H}^+(c_2) + \text{Cl}^-(c_3)
\end{array}$$

The Nernst equation, at 25°C, is

$$\begin{aligned}
\mathscr{E} &= \mathscr{E}^0 - 0.0592 \log (c_2 c_3/c_1^{\,2}) \\
&= \underbrace{\mathscr{E}^0 - 0.0592 \log (c_3/c_1^{\,2})}_{\mathscr{E}'} - 0.0592 \underbrace{\log c_2}_{-\text{pH}} \\
&= \mathscr{E}' + 0.0592 \text{ pH}
\end{aligned}$$

The concentrations c_1 and c_3 are kept constant during a series of measurements and are therefore included in $\mathscr{E}'$. The quantity to be measured is c_2, or $\text{pH} = -\log c_2$.

As with the hydrogen electrode, $\mathcal{E}'$ is a constant at any one temperature. The cell is again calibrated by measuring the emf with standard (known-pH) buffer solutions.

EXAMPLE 17.18 With a standard buffer solution of pH 6.50, the emf of a cell containing a glass electrode is $\mathcal{E} = 0.085$ V at 25°C. When the buffer solution is replaced with an unknown solution, $\mathcal{E} = 0.152$ V. What is the pH of the unknown solution?

ANSWER We first calibrate the apparatus by finding $\mathcal{E}'$. For the standard buffer of pH 6.50, equation 16 becomes

$$0.085 = \mathcal{E}' + 0.0592 \times 6.50$$

$$\mathcal{E}' = 0.085 - 0.0592 \times 6.50 = -0.300 \text{ V}$$

Then, for the solution of unknown pH,

$$\mathcal{E} = \mathcal{E}' + 0.0592 \text{ pH}$$

$$\text{pH} = \frac{\mathcal{E} - \mathcal{E}'}{0.0592} = \frac{0.152 - (-0.300)}{0.0592}$$

$$= 7.64$$

■

PROBLEM 17.14 A cell for measuring pH has an emf of 0.043 V with a buffer of pH 8.00 at 25°C. With an acidic solution, the emf is −0.072 V. (The needle swings the other way.) What is the pH of the acidic solution? □

In the 1960s, **ion-selective electrodes** for ions other than H^+ were developed. Electrodes for determining the concentrations of many ions, including Li^+, Na^+, K^+, NH_4^+, Ag^+, Pb^{2+}, Cu^{2+}, F^-, Cl^-, Br^-, I^-, CN^-, S^{2-}, and SO_4^{2-}, are available. The essential feature of an ion-selective electrode is a membrane through which (ideally) only one kind of ion can migrate. The membrane is usually a glass or an insoluble salt (Box 17.5).

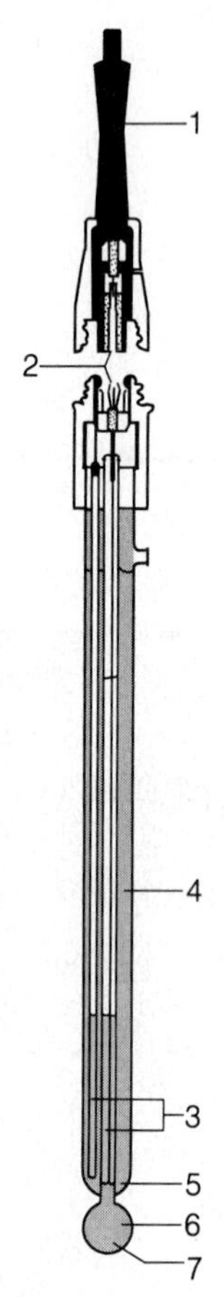

FIGURE 17.14
A combination electrode for use in a pH meter. (1) Connecting cable. (2) Plug-in contacts. (3) The reference half-cell (Ag|AgCl-type). (4) KCl electrolyte solution. (5) Platinum-in-glass liquid junction. (6) Internal buffer solution (pH 7). (7) pH-sensitive glass membrane. (Courtesy of Sargent–Welch Scientific Company.)

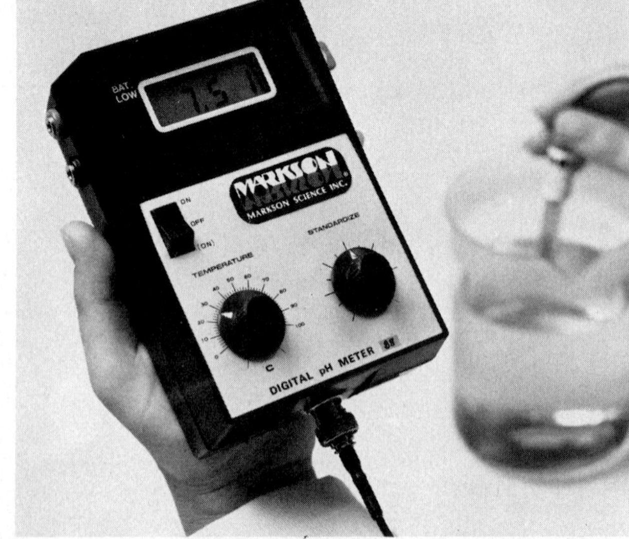

FIGURE 17.15
A battery-powered pH meter. (Courtesy of Markson Science, Inc.)

BOX 17.5

A silver sulfide (Ag_2S) membrane allows Ag^+ ions, but not the much larger S^{2-} ions, to migrate through it. The emf of a cell in contact with this membrane depends on the concentration of Ag^+ in the unknown solution:

$$\mathscr{E} = \mathscr{E}' - \frac{2.303RT}{\mathscr{F}} \log [Ag^+]$$

However, this solution is saturated with Ag_2S, and the concentrations of Ag^+ and S^{2-} must therefore satisfy the equilibrium condition (section 16.1)

$$[Ag^+]^2[S^{2-}] = K_{sp} = 6.6 \times 10^{-50} \text{ at } 25°C$$

$$[Ag^+] = (K_{sp}/[S^{2-}])^{1/2}$$

Thus, the emf depends on the concentration of S^{2-}. Substituting for $[Ag^+]$ in the equation and manipulating the logarithms leads to an equation similar to equation 16 for the pH meter:

$$\mathscr{E} = \mathscr{E}'' + \frac{2.303RT}{2\mathscr{F}} \log [S^{2-}]$$

The constant $\mathscr{E}''$ is determined by calibration with known solutions, as with a pH meter.

Biologists and physicians are finding uses for miniature ion-selective electrodes. Some are inserted directly into the bloodstream to measure pH or other ion concentrations. They have even been made small enough to be inserted into an individual living cell. They can be coated with enzymes and used to measure the product of an enzyme-catalyzed reaction. Thus, if the electrode is sensitive to NH_4^+ and the enzyme is urease (which catalyzes the decomposition of urea), the concentration of urea is determined by measuring the concentration of the ammonium ion to which it decomposes:

$$\underset{\text{urea}}{(H_2N)_2CO} + 2H^+ + H_2O \xrightarrow{\text{urease}} 2NH_4^+ + CO_2$$

Commercial pH meters—and their younger relatives, pNa meters, pF meters, and so on—are designed for convenience. Glass and reference electrodes and a salt bridge are often incorporated into a single probe (Figure 17.14), which needs only to be rinsed and inserted into the solution. The dial is read in pH units rather than volts, and the calculation is done internally. The meter is calibrated by setting a knob. Another knob is set according to the temperature to allow for the presence of T in the Nernst equation. A battery-powered meter (Figure 17.15) can be carried right down to the river when you want to see what the local polluters have been doing lately.

17.10 ELECTROMOTIVE FORCE AND EQUILIBRIUM

When a galvanic cell is allowed to deliver current, its emf gradually changes because the concentrations of reactants and products are changing. Eventually the emf drops to zero. The reason is that the cell reaction has come to equilibrium, and then there is no longer any current: no reaction, no emf. *The condition of equilibrium for a reaction is* $\mathscr{E}_{cell} = 0$.

When we first studied chemical equilibrium (section 13.2), the equilibrium constant K and the reaction quotient Q were distinguished from each other. Q, as an expression containing concentrations and/or partial pressures, may or may not be equal to K. At equilibrium Q and K are equal; away from equilibrium, they are unequal. Most of the reactions encountered before this chapter (especially ionic reactions) come rapidly to equilibrium. Therefore, Q has usually been equal to K, and the distinction between them may have faded from your memory. In a galvanic cell, however, equilibrium is *not* the normal condition. Equilibrium is established only in a "dead battery," with $\mathscr{E} = 0$. Because the reactants are physically separated in a galvanic cell, the reaction may take a long time (we always hope) to come to equilibrium. The distinction between Q and K is therefore essential.

The Nernst equation is

$$\mathscr{E} = \mathscr{E}^0 - \frac{RT}{n\mathscr{F}} \ln Q$$

At equilibrium, $\mathscr{E} = 0$ and $Q = K$. Therefore,

$$0 = \mathscr{E}^0 - \frac{RT}{n\mathscr{F}} \ln K$$

$$\mathscr{E}^0 = \frac{RT}{n\mathscr{F}} \ln K \qquad (17)$$

Solving for ln K gives

$$\ln K = \frac{n\mathscr{F}\mathscr{E}^0}{RT}$$

or

$$\mathbf{\log K = \frac{n\mathscr{F}\mathscr{E}^0}{2.303RT}} \qquad (18)$$

At 25°C,

$$\log K = \frac{n\mathscr{E}^0}{0.0592} \qquad (19)$$

For example, in the cell

$$\text{Pt}|\text{H}_2(\text{g})|\text{HCl(aq)}|\text{AgCl(c)}|\text{Ag(c)}$$

the partial reactions are

$$\begin{array}{rcl} \tfrac{1}{2}\text{H}_2 & \rightleftharpoons & \text{H}^+ + e^- \\ \text{AgCl} + e^- & \rightleftharpoons & \text{Ag} + \text{Cl}^- \\ \hline \tfrac{1}{2}\text{H}_2 + \text{AgCl} & \rightleftharpoons & \text{H}^+ + \text{Cl}^- + \text{Ag} \end{array}$$

For this cell, at 25°C, $\mathscr{E}^0 = 0.2223$ V and $n = 1$. Then

To preserve 4 significant figures, use 0.05916.

$$\log K = \frac{n\mathscr{E}^0}{0.05916} = \frac{0.2223}{0.05916} = 3.758$$

$$K = 10^{3.758} = 5.73 \times 10^3$$

The equilibrium condition is

$$\frac{[\text{H}^+][\text{Cl}^-]}{P_{\text{H}_2}{}^{1/2}} = 5.73 \times 10^3$$

Notice that when $\mathscr{E}^0$ is positive, K is greater than 1. Positive $\mathscr{E}^0$ and $K > 1$ both mean that when the reactants and products all start off at 1 mol/L or 1 atm ($Q = 1$), the forward reaction is spontaneous ($Q < K$, section 13.2). Conversely, negative $\mathscr{E}^0$ requires K less than 1; in this case, the reverse (left-to-right) reaction is spontaneous when the initial concentrations (or pressures) are 1 mol/L (or 1 atm). The emf of a cell is a measure of how far the reaction is from equilibrium, and in which direction. This interpretation of emf shows up especially well in the equation

$$\mathscr{E} = \frac{RT}{n\mathscr{F}} \ln (K/Q) = \frac{2.303RT}{n\mathscr{F}} \log (K/Q) \tag{20}$$

which is a form of the Nernst equation (Problem 17.62). $\mathscr{E}$ and the ratio K/Q both express the "distance" of the reacting mixture from equilibrium. At equilibrium, $Q = K$ and $\mathscr{E} = 0$.

The equilibrium constant of any reaction can be determined if the reaction can be resolved into two partial reactions for which the potentials are given in a table. From the resultant emf, $\mathscr{E}^0$, we calculate K as in equation 18. If gases are involved, the K obtained is K_p. Pressures must be in atmospheres.

EXAMPLE 17.19 Find the equilibrium constant at 25°C for the reaction (in aqueous solution)

$$Zn + Cu^{2+} \rightleftharpoons Zn^{2+} + Cu$$

ANSWER The partial reactions, with their standard potentials, are (Table 17.1)

$Cu^{2+} + 2e^- \rightleftharpoons Cu$	(reduction)	$+0.3419$
$Zn \rightleftharpoons Zn^{2+} + 2e^-$	(oxidation)	$-(-0.7618)$
$Zn + Cu^{2+} \rightleftharpoons Zn^{2+} + Cu$		$\mathscr{E}^0 = 0.3419 - (-0.7618)$

Thus, $\mathscr{E}^0 = +1.1037$ V and $n = 2$ (two electrons appear in each partial reaction). Then

$$\log K = \frac{n\mathscr{E}^0}{0.05916} = \frac{2 \times 1.1037}{0.05916} = 37.31$$

$$K = 10^{37.31} = 10^{0.31} \times 10^{37} = 2.1 \times 10^{37}$$

■

Even a modest voltage like 1 V—it would not give you a shock—corresponds to an enormous equilibrium constant. Example 17.19 shows that at equilibrium, $[Zn^{2+}]/[Cu^{2+}] = 2.1 \times 10^{37}$. The reaction thus goes almost completely to the right, in the sense that very little of the limiting reactant is present at equilibrium. If Zn is in excess, practically all of the Cu^{2+} will be consumed, leaving no $[Cu^{2+}]$ to be measured. Thus, K for this reaction could never be obtained by measuring the concentrations at equilibrium. Instead, it is calculated from measurements of emf. This is the best and most common method of finding very large and very small equilibrium constants. Even though oxidation must occur at one electrode and reduction at the other, the net reaction need not be an oxidation–reduction, as the next example will illustrate.

EXAMPLE 17.20 Find the equilibrium constant at 25°C for the reaction

$$Ag_2CrO_4(c) \rightleftharpoons 2Ag^+ + CrO_4^{2-}$$

ANSWER The only partial reaction in Table 17.1 involving Ag_2CrO_4 (with the corresponding $\mathscr{E}^0$) is

$$Ag_2CrO_4 + 2e^- \rightleftharpoons 2Ag + CrO_4^{2-} \qquad +0.4470$$

The other partial reaction should cancel out Ag and put in Ag^+:

$$Ag^+ + e^- \rightleftharpoons Ag \qquad +0.7996$$

Double and reverse this equation, and add the partial reactions:

$$\begin{array}{rcll} Ag_2CrO_4 + 2e^- & \rightleftharpoons & 2Ag + CrO_4^{2-} & \text{(reduction)} \quad +0.4470 \\ 2Ag & \rightleftharpoons & 2Ag^+ + 2e^- & \text{(oxidation)} \quad -(+0.7996) \\ \hline Ag_2CrO_4 & \rightleftharpoons & 2Ag^+ + CrO_4^{2-} & \mathscr{E}^0 = 0.4470 - 0.7996 \\ & & & \phantom{\mathscr{E}^0} = -0.3526\text{ V} \end{array}$$

Here $n = 2$ (two electrons appear in each partial reaction). Then

$$\log K = \frac{n\mathscr{E}^0}{0.05916} = \frac{2 \times (-0.3526)}{0.05916} = -11.92$$

$$K = 10^{-11.92} = 10^{0.08} \times 10^{-12} = 1.2 \times 10^{-12}$$

$$= [Ag^+]^2[CrO_4^{2-}] \text{ at equilibrium}$$

Ag appears in the partial equations, but it cancels out. The net reaction leaves each element in its original oxidation state; that is, +1 for Ag, +6 for Cr, and −2 for O. ■

You should recognize K in Example 17.20 as a solubility product (section 16.1). The negative $\mathscr{E}^0$ and the small K tell us the same thing: that this reaction tends to go to the left when $[Ag^+]$ and $[CrO_4^{2-}]$ are anywhere near 1 mol/L.

EXAMPLE 17.21 Find the equilibrium constant at 25°C for the reaction $H_2O \rightleftharpoons H^+ + OH^-$ by using Table 17.1.

ANSWER Partial reactions involving H_2O, H^+, and OH^- are needed. It is all right if something else appears, provided that we can make it cancel. Two suitable partial reactions are

$$\begin{array}{rcll} 2H_2O + 2e^- & \rightleftharpoons & H_2 + 2OH^- & \text{(reduction)} \quad -0.8277 \\ H_2 & \rightleftharpoons & 2H^+ + 2e^- & \text{(oxidation)} \quad 0.0000 \\ \hline 2H_2O & \rightleftharpoons & 2H^+ + 2OH^- & \mathscr{E}^0 = -0.8277\text{ V} \end{array}$$

The reaction is double the desired one. Multiply each partial reaction by $\frac{1}{2}$ to make the net reaction

$$H_2O \rightleftharpoons H^+ + OH^-$$

This does not affect the potentials, but makes $n = 1$ instead of 2, so that

$$\log K = \frac{-0.8277}{0.05916} = -13.99$$

$$K = 1.0 \times 10^{-14} = [H^+][OH^-]$$

We have thus calculated the ion product of water (section 15.1). ■

Most calculators cannot give antilogarithms of numbers > 100. *Remember*: $10^{123.4} = 10^{0.4} \times 10^{123}$.

PROBLEM 17.15 Calculate the equilibrium constant of each of these reactions at 25°C:

(a) $Mg + 2Fe^{3+} \rightleftharpoons Mg^{2+} + 2Fe^{2+}$
(b) $2Cr + 3HgBr_4^{2-} \rightleftharpoons 2Cr^{3+} + 3Hg + 12Br^-$
(c) $HgBr_4^{2-} \rightleftharpoons Hg^{2+} + 4Br^-$
(*Suggestion*: Hg also appears in the partial equations.) □

It is important to know whether a given reaction can occur when certain chemicals are present in certain amounts. To determine this, calculate the reac-

It is equally correct to calculate K from equation 18 and compare it with Q.

tion quotient, Q, for the reaction, using the initial concentrations or pressures of the species present. The next step depends on how the information is presented. If the equilibrium constant is available, compare Q and K, as in section 13.2. In this chapter, the usual datum is $\mathscr{E}^0$. Then you can use equation 15 to calculate $\mathscr{E}$. If $\mathscr{E}$ is positive, the reaction is spontaneous as written; if $\mathscr{E}$ is negative, the reverse reaction is spontaneous.

EXAMPLE 17.22 Will Fe reduce Fe^{3+} to Fe^{2+} when both ion concentrations are 0.1 M at 25°C?

ANSWER The reaction in question is $Fe + 2Fe^{3+} \longrightarrow 3Fe^{2+}$. For this reaction,

$$Q = \frac{[Fe^{2+}]^3}{[Fe^{3+}]^2} = \frac{0.1^3}{0.1^2} = 0.1$$

Now we must calculate $\mathscr{E}^0$. The reaction is resolved into the partial reactions

$$\begin{array}{lll} 2Fe^{3+} + 2e^- \longrightarrow 2Fe^{2+} & \text{(reduction)} & +0.771 \\ Fe \longrightarrow Fe^{2+} + 2e^- & \text{(oxidation)} & -(-0.447) \\ \hline Fe + 2Fe^{3+} \longrightarrow 3Fe^{2+} & & \mathscr{E}^0 = 0.771 - (-0.447) \\ & & = 1.218 \text{ V} \end{array}$$

The final step is to calculate $\mathscr{E}$ and see whether it is positive or negative. The Nernst equation gives

$$\begin{aligned} \mathscr{E} &= \mathscr{E}^0 - \frac{0.0592}{2} \log Q \\ &= 1.218 - \frac{0.0592}{2} \log (0.1) \\ &= 1.248 \text{ V} \end{aligned}$$

Since $\mathscr{E}$ is positive, the reaction is possible. A practical consequence of this result is that a solution of an iron(II) (Fe^{2+}) salt can be protected from atmospheric oxidation to iron(III) (Fe^{3+}) by keeping solid iron in the bottle; any Fe^{3+} formed will be reduced to Fe^{2+}. ■

A substantially positive $\mathscr{E}^0$, more than a few tenths of a volt, makes K very large. For the preceding example, $K = 1.5 \times 10^{41}$. With such a large equilibrium constant, it is safe to predict that the reaction will occur with any reasonable concentrations. Only an extremely small Fe^{3+} concentration would make Q greater than 1.5×10^{41}. Therefore, the sign of $\mathscr{E}^0$ is usually sufficient for prediction, unless $\mathscr{E}^0$ is very close to 0 or some concentrations are extremely small.

PROBLEM 17.16 Ascertain which way each of these reactions will proceed when all concentrations are near 1 mol/L:

(a) $Sn + HgI_4{}^{2-} \rightleftharpoons Sn^{2+} + 4I^- + Hg(\ell)$
(b) $Sn + Hg(CN)_4{}^{2-} \rightleftharpoons Sn^{2+} + 4CN^- + Hg(\ell)$ □

17.11 DECOMPOSITION POTENTIAL

When we discussed electrolysis earlier in the chapter, we took for granted that a potential difference had to be applied to the electrodes before anything would happen. We could not look into how big this potential difference should be because we first needed to study the emf's of galvanic cells. Now we are in a position to

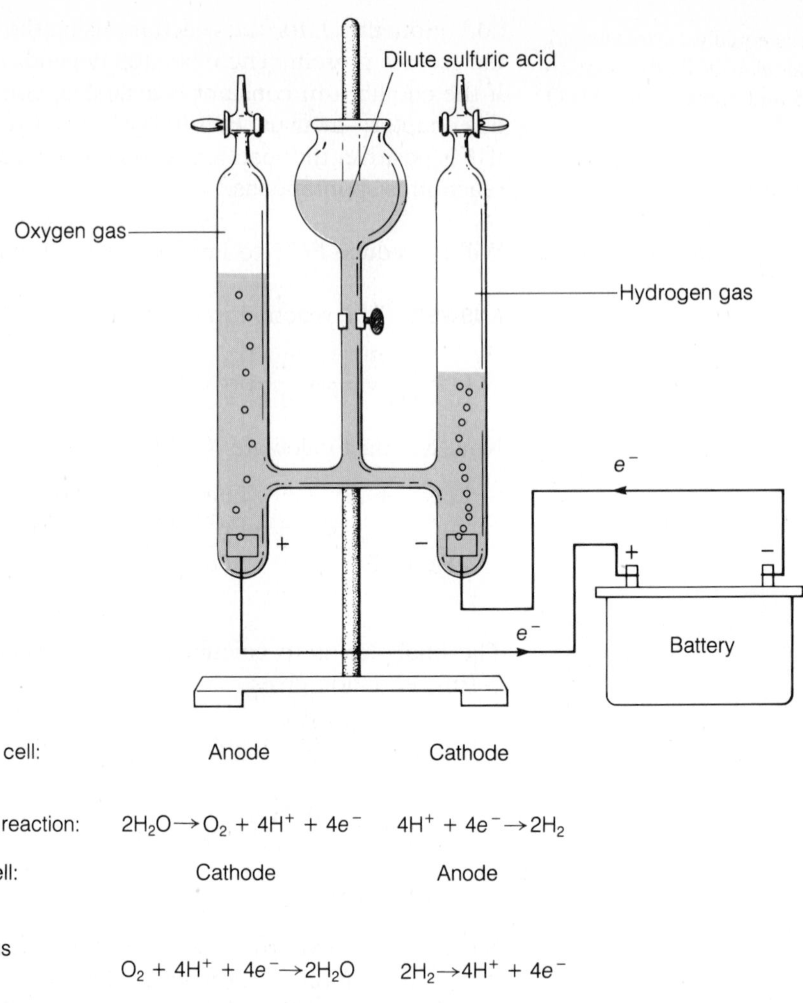

FIGURE 17.16
The cell $Pt|O_2|H^+|H_2|Pt$, necessarily created when an aqueous solution of H_2SO_4 is electrolyzed.

apply our knowledge of emf's to predicting what voltage is required to make a reaction occur.

Consider the electrolysis of water that contains some sulfuric acid to provide ions and thus make it a conductor (Figure 17.2, section 17.2). The partial reactions and $\mathscr{E}^0$ values are

$$\begin{array}{llr} 4H^+ + 4e^- \longrightarrow 2H_2(g) & \text{(reduction)} & 0 \\ 2H_2O \longrightarrow O_2(g) + 4H^+ + 4e^- & \text{(oxidation)} & -(+1.229) \\ \hline 2H_2O \longrightarrow 2H_2(g) + O_2(g) & & \mathscr{E}^0 = 0 - 1.229 \\ & & = -1.229 \text{ V} \end{array}$$

This process is of the kind called "impossible" or "nonspontaneous." However, it can be made to occur by an input of electrical work. The negative $\mathscr{E}^0$ means that the reaction tends to occur in the direction *opposite* to that written. The products of electrolysis, trying to undergo the reverse reaction, have made the apparatus into a galvanic cell. The spontaneous reaction in this cell would make a current flow in the wrong direction—that is, in the direction opposite to the current we are using to electrolyze the solution (Figure 17.16). In order to overcome this "back emf," the applied emf must be at least $-\mathscr{E}^0 = 1.229 \text{ V} \approx$

We round to 1.2 V because decomposition potentials are measured less precisely than reversible emf's.

1.2 V. If the applied emf is increased gradually from zero, practically no current will flow until 1.2 V is attained. Thereafter, the current rises with applied emf (Figure 17.17). The emf at which the rise begins is called the **decomposition potential** ($\mathscr{E}_d$) of the solution.

The decomposition potential must be at least equal to the reversible emf of the cell whose reaction opposes the desired electrolysis reaction. An additional potential proportional to the current must be applied to overcome ordinary resistance, as with any electrical apparatus. However, it is often necessary to increase the applied emf to considerably above the reversible emf before the reaction begins to occur at an appreciable rate. The additional emf required is the **overvoltage**. Then the decomposition potential is

$$\mathscr{E}_d = \mathscr{E}_{rev} + \mathscr{E}_{ov} \quad (21)$$

where $\mathscr{E}_{rev}$ is $\mathscr{E}$ calculated from Table 17.1 and the Nernst equation, and $\mathscr{E}_{ov}$ is the overvoltage. Overvoltages are especially large, up to about 2 V, when gases are produced. They are very much affected by the nature of the electrode surface. Another effect of a gas is to cover the electrode with an insulating layer, greatly increasing the resistance.

EXAMPLE 17.23 Calculate the emf required to bring about the reaction

$$Cu^{2+}(1\ M) + 2Cl^-(1\ M) \longrightarrow Cu(c) + Cl_2(g,\ 1\ atm)$$

with carbon electrodes. Assume that the overvoltage for the evolution of Cl_2 on a carbon surface is (a) 0; (b) 0.25 V.

ANSWER

(a) The desired reaction can be resolved into partial reactions, which have standard potentials listed in Table 17.1:

$$\begin{array}{lr} Cu^{2+} + 2e^- \longrightarrow Cu & +0.3419 \\ 2Cl^- \longrightarrow Cl_2 + 2e^- & -(+1.3583) \\ \hline Cu^{2+} + 2Cl^- \longrightarrow Cu + Cl_2 & \mathscr{E}^0 = 0.3419 - 1.3583 = -1.0164\ V \end{array}$$

The negative answer means that this reaction is not spontaneous; work must be done to make it occur. The reverse reaction,

$$Cu + Cl_2 \longrightarrow Cu^{2+} + 2Cl^-$$

is spontaneous, with $\mathscr{E}^0 = +1.0164$ V. The *minimum* required emf (the decomposition potential) is that needed to *prevent* the reverse reaction from occurring: $\mathscr{E}_d = 1.0164\ V \approx 1.02$ V.

(b) The overvoltage makes the desired reaction more difficult and thus increases the decomposition potential:

$$\mathscr{E}_d = \mathscr{E}_{rev} + \mathscr{E}_{ov} = 1.02\ V + 0.25\ V = 1.27\ V$$

■

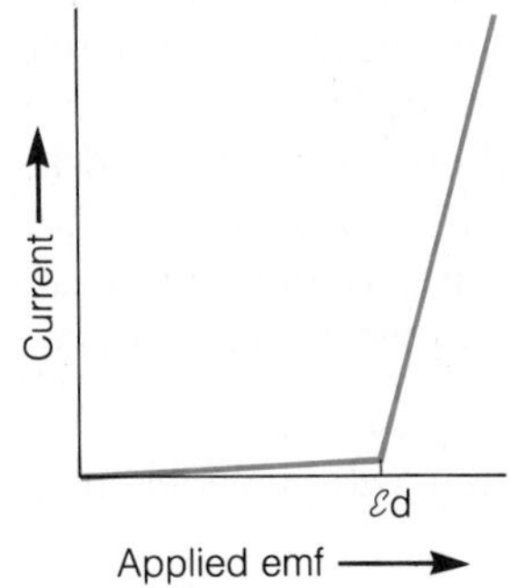

FIGURE 17.17
The dependence of current on emf in an electrolysis cell. $\mathscr{E}_d$ is the decomposition potential.

PROBLEM 17.17 Calculate the emf required for the reaction

$$2H_2O(\ell) \longrightarrow 2H_2(g) + O_2(g)$$

when the gases are produced at 1.0 atm, and (a) the electrodes are Pt, with overvoltage 0 for H_2 ($2H^+ + 2e^- \longrightarrow H_2$) and 0.60 V for O_2 ($2H_2O \longrightarrow O_2 + 4H^+ + 4e^-$); (b) the electrodes are Cu, with overvoltage 0.60 V for H_2 and 0.70 V for O_2.

□

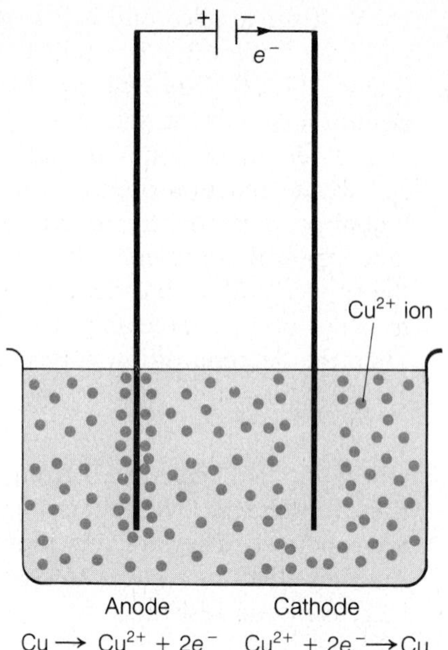

FIGURE 17.18
Concentration polarization in copper plating. If the solution is homogeneous, there is no net reaction and the required emf is zero. However, the buildup of Cu^{2+} near the anode and its depletion near the cathode make the process nonspontaneous, requiring an applied emf.

Even when there is no overvoltage, the potential necessary to bring about electrolysis at an appreciable rate may be considerably greater than we would calculate from the $\mathscr{E}^0$'s. Electrolysis causes nonuniformities in the concentration of the electrolyte—depletion of reactants, accumulation of products—always in such a way as to increase the required potential (Figure 17.18). This effect is called **concentration polarization**, and it can often be diminished simply by stirring the solution. A dry cell (section 17.12) contains paste instead of free liquid. Diffusion in the paste is especially slow, and concentration polarization is therefore especially large. That is why the emf of a dry cell drops so severely when a large current is drawn, and why a depleted cell makes a partial recovery when it is allowed to rest awhile.

Just as the positive reversible emf of a galvanic cell tells you the *maximum* work available from the cell, the negative reversible emf of an electrolysis cell tells you the *minimum* work required to make the reaction occur. In each case, the reversible emf is an ideal limit. When a process is carried out in the real world, less work will be obtained and more work must be put in. There is always a compromise between the infinitely slow reversible process, with its maximum efficiency, and the need to produce results for a customer who won't wait forever.

17.12 SOME PRACTICAL CELLS

Of the thousands of possible galvanic cells, only a relative handful are widely used as power sources. They fall into three categories:

1. **Primary cells** are used until the cell reaction comes close to equilibrium (or the emf declines for some other reason or the resistance of the cell becomes too large) and are then discarded.

Storage cells are also called *secondary cells* or *accumulators.*

2. **Storage cells** can be used repeatedly. When the reaction comes close to equilibrium, it is reversed (the cell is recharged) by passing current in the direction opposite to the discharge current. The anode during discharge becomes the cathode during charge, and vice versa.

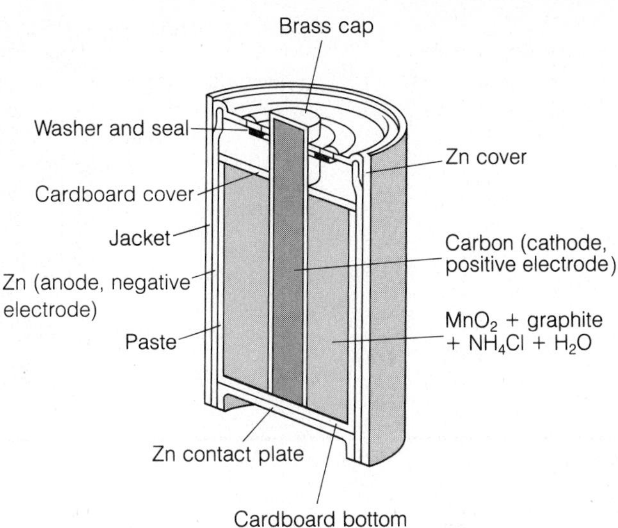

FIGURE 17.19
A Leclanché dry cell as used in flashlights.

A **battery** is a number of galvanic cells connected together. Commonly, however, a single cell is also called a "battery."

3. **Fuel cells** receive a continuous supply of reactants from the outside, and the products are continuously removed.

PRIMARY CELLS

Invented by Georges Leclanché in 1866.

The most familiar primary cell is the **dry cell** or **Leclanché cell** (Figure 17.19). The anode (on the outside) is zinc; the cathode (the rod in the center) is carbon, which is chemically inert under ordinary conditions and conducts electricity as if it were a metal. A paste consisting of graphite, MnO_2, NH_4Cl, and H_2O surrounds the cathode. Adjacent to the Zn is a layer of paste, usually made from wheat flour and corn starch, containing NH_4Cl, $ZnCl_2$, and H_2O, but not MnO_2. The cell is "dry" only in the sense that it contains paste rather than free liquid. It could not function if water were absent. The cathode reaction is complicated and subject to some controversy. One representation is

$$\text{anode:} \quad Zn \longrightarrow Zn^{2+} + 2e^-$$

$$\text{cathode:} \quad NH_4{}^+ + MnO_2 + e^- \longrightarrow MnO(OH) + NH_3(aq)$$

In the conventional notation, the cell is, when fresh,

$$Zn|NH_4{}^+, Zn^{2+}, Cl^-, H_2O|C, MnO_2, NH_4{}^+, Cl^-, H_2O|C$$

The carbon cathode in the center is the *positive* electrode (section 17.6). The maximum emf is 1.48 V. The emf drops sharply when a large current is drawn because of concentration polarization—more than in cells with liquid electrolytes. While a strictly dead cell is one that has come to equilibrium, cells of all kinds become useless long before equilibrium.

Trying to recharge a primary cell may result in an explosion (Problem 17.67).

Alkaline dry cells (Figure 17.20) use a paste that contains KOH as the electrolyte instead of NH_4Cl and $ZnCl_2$. The electrode processes again involve oxidation of Zn and reduction of MnO_2, but the cell reaction is different in a basic solution (Problem 17.66). They give current for a longer time than ordinary dry cells, and the emf drops less when heavy currents are drawn. However, they are more expensive, since a strong steel can and careful sealing are needed because KOH is harmful to skin and eyes.

The *mercury cell* is another type of dry cell. It can deliver a higher current than an ordinary dry cell of the same size; therefore, it can be made more compact and deliver the same current. The cell is

$$Zn(c)|ZnO(c)|KOH(aq)|HgO(c)|Hg(\ell)$$

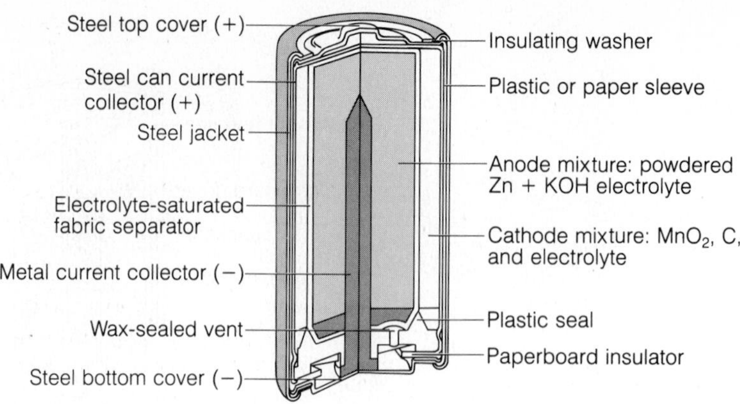

FIGURE 17.20
An alkaline dry cell.

The mercury does not flow around because it is mixed with HgO and graphite powders. The electrolyte is held by an absorbent cotton pad. The reactions are

$$Zn + 2OH^- \longrightarrow ZnO + H_2O + 2e^-$$

$$HgO + H_2O + 2e^- \longrightarrow Hg + 2OH^-$$

This cell is much more expensive than an ordinary dry cell because of the cost of mercury. It is used only where small size is important, especially in watches, hearing aids, and cameras. The emf is about 1.3 V.

Various cells containing *lithium* have recently come into use. Lithium has the most negative of all reduction potentials—conducive to a high-voltage cell. Its low atomic weight means that you get more atoms per gram—conducive to a low-mass, high-capacity cell. However, lithium can be used only in nonaqueous systems, since it reacts vigorously with water.

A small cell used in cardiac pacemakers is

$$Li(c)|LiI(c)|I_2(\text{in a polymer})|\text{stainless steel}$$

The reactions are

$$Li(c) \longrightarrow Li^+ \text{ (in LiI)} + e^-$$

$$\tfrac{1}{2}I_2 + e^- \longrightarrow I^- \text{ (in LiI)}$$

The polymer, a base, forms a weak bond with I_2, a Lewis acid. LiI is the electrolyte. It is an ionic conductor; the charge carriers are Li^+ ions, which can move into vacant sites in the imperfect crystal lattice. LiI has a high resistance, so the cell delivers a small current, which is desirable; it needs to last a long time—like ten years—since replacement requires major surgery. The emf is about 3 V, higher than most cells.

STORAGE CELLS

Invented in 1859 by Gaston Planté.

The *lead-acid storage cell* (Figure 17.21) is familiar to people who have looked under the hood of a car. In its charged state, it consists of an electrode of spongy lead and an electrode of finely divided solid PbO_2, each material supported by a grid of Pb-Sb or Pb-Ca alloy (stronger and more resistant to corrosion than pure Pb):

$$Pb|H^+, HSO_4^-, H_2O|PbO_2|Pb, Sb$$

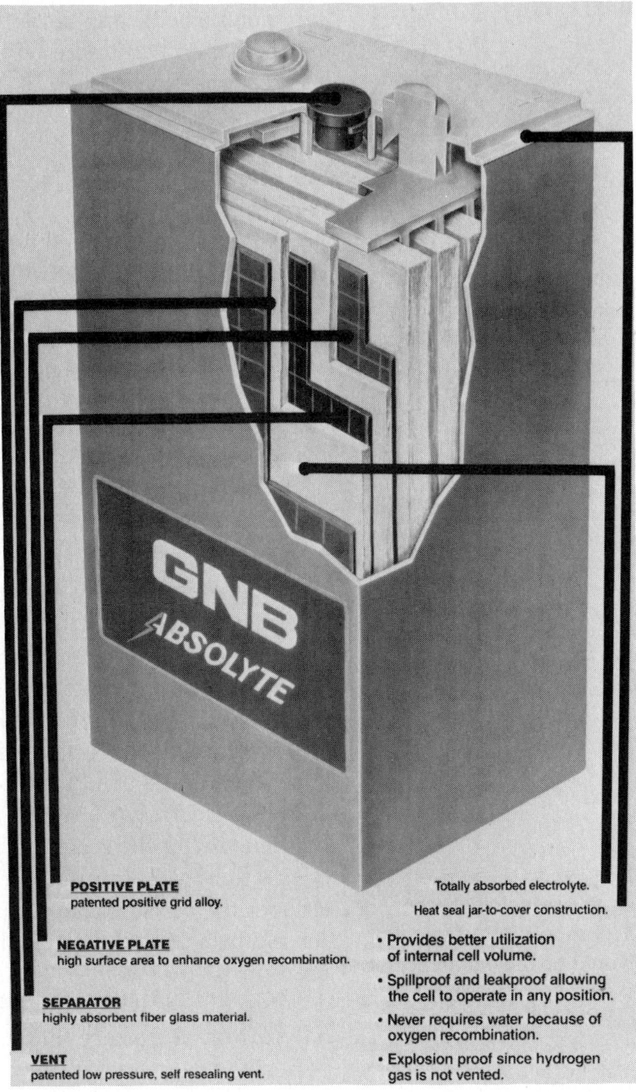
GNB
ABSOLYTE
POSITIVE PLATE
patented positive grid alloy.
NEGATIVE PLATE
high surface area to enhance oxygen recombination.
SEPARATOR
highly absorbent fiber glass material.
VENT
patented low pressure, self resealing vent.
Totally absorbed electrolyte.
Heat seal jar-to-cover construction.
• Provides better utilization of internal cell volume.
• Spillproof and leakproof allowing the cell to operate in any position.
• Never requires water because of oxygen recombination.
• Explosion proof since hydrogen gas is not vented.

FIGURE 17.21
A lead-acid storage battery. (Courtesy of GNB, Inc.—Industrial Battery Division)

The electrolyte is an aqueous solution of sulfuric acid. The reactions, on discharging, are

$$Pb(c) + HSO_4^- \longrightarrow PbSO_4(c) + H^+ + 2e^- \quad \text{(negative electrode)}$$
$$PbO_2(c) + 3H^+ + HSO_4^- + 2e^- \longrightarrow PbSO_4(c) + 2H_2O \quad \text{(positive electrode)}$$
$$Pb(c) + PbO_2(c) + 2H^+ + 2HSO_4^- \longrightarrow 2PbSO_4(c) + 2H_2O \quad (22)$$

The cell can be recharged by passing current in the reverse direction, with electrons entering at the Pb electrode and leaving at the PbO_2 electrode. On charging, both electrode reactions are reversed. The Pb electrode is the anode when the cell is discharging (acting as a galvanic cell), and is therefore negative. The same electrode is the *cathode* when the cell is *being charged* (acting as an electrolysis cell), and is therefore negative then, also. Thus, the Pb electrode can be permanently marked with a minus sign (black). By similar reasoning, the PbO_2 electrode is marked with a plus sign (red). "Anode" and "cathode" labels would make no sense, since each electrode is sometimes an anode and sometimes a cathode. When a car is "jump-started," the positive (red) terminal (cathode) of the good battery is connected to the positive (red) terminal (anode) of the dead (or weak) battery, and the negative (black) terminal (anode) of the good battery is connected to the negative (black) terminal (cathode) of the dead battery (or to the car frame if the negative side is grounded). Thus, the spontaneous reaction in the good battery provides the work that makes the reverse (nonspontaneous) reaction occur in the dead battery.

Sulfuric acid is consumed in the discharging process and regenerated on charging; the density of the solution provides an indication of the degree to which the cell is charged, for the density decreases as the acid content decreases. The solution in a charged cell usually contains 30% to 40% H_2SO_4, corresponding to a density of 1.2 to 1.3 g/mL. The emf of the cell is roughly 2 V. A 6- or 12-V battery thus requires 3 or 6 cells.

Lead-acid batteries have long been used in small electric trucks for short-distance use, but they are too heavy and bulky to power fast vehicles over long distances. There is active research on more compact storage batteries that may replace gasoline engines in cars of the future (Problems 17.71 and 17.72).

The obvious advantage of storage cells over primary cells is that they need not be replaced very often. The lead-acid battery is good for starting automobiles, its strongest point being its ability to provide a large current for a short time. It has a serious drawback, however, one that characterizes any cell containing an aqueous electrolyte. When the cell is overcharged—that is, when charging is continued after the reactants in the charging reaction are mostly consumed—water begins to be decomposed, producing H_2 and O_2. These gases must be allowed to escape; they may carry sulfuric acid spray and they may be ignited by a spark. An old-fashioned lead-acid battery must be kept in an approximately upright position so that acid is not spilled, and it loses water by both evaporation and decomposition, which must be replaced from time to time. It is very desirable to have a storage battery that can be sealed and used in any position, like a dry cell.

If H_2 and O_2 did not have large overvoltages, the decomposition of water would be the only reaction.

Sealed lead-acid batteries were introduced in the 1950s. They use sulfuric acid in a glass fabric made from SiO_2 and other oxides. The electrodes are designed so that any O_2 produced at the positive electrode (the anode when charging) can diffuse to the negative electrode (the cathode) and be reduced there. The reactant consumed at each electrode during normal charging is $PbSO_4$. The battery is made so that the $PbSO_4$ at the positive electrode is consumed before the $PbSO_4$ at the negative electrode is consumed. This means that O_2 production (at the positive electrode) begins before H_2 production (at the negative electrode) has a chance to begin. The overcharging reactions are

$$2H_2O \longrightarrow O_2 + 4H^+ + 4e^- \qquad \text{(positive electrode)}$$

$$O_2 + 4H^+ + 4e^- \longrightarrow 2H_2O \qquad \text{(negative electrode)}$$

The idea is to prevent the reaction $2H^+ + 2e^- \longrightarrow H_2$ from occurring at the negative electrode. H_2 is more difficult than O_2 to dispose of within the battery and must be vented, which introduces the danger of explosion. Nevertheless, the charging process in sealed lead-acid batteries must be controlled to prevent too much overcharging. These batteries are not completely gas-free, and are usually equipped with self-resealing vents.

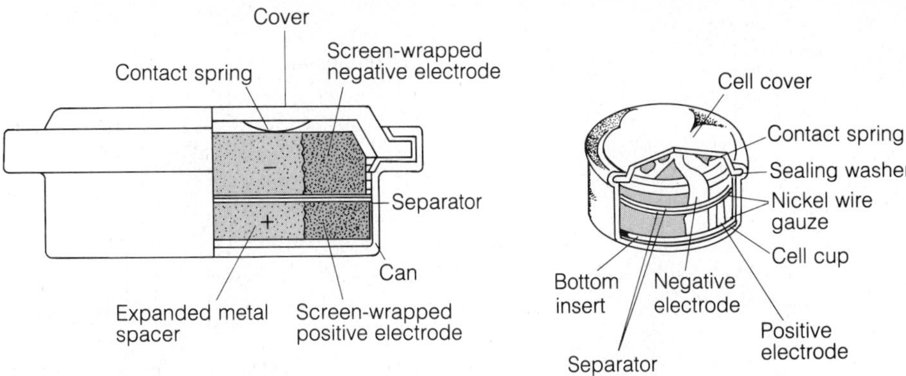

FIGURE 17.22
A nickel-cadmium button cell.

The first successful sealed storage cell (1953), still the most common, is the *alkaline nickel-cadmium cell* that is used in large numbers to power items such as calculators, electric shavers, and night-lights (Figure 17.22). The cell may be represented by

$$Cd(c)\|Cd(OH)_2(c)\|KOH(aq)|\text{``}Ni_2O_3\text{''}(c),\ Ni(OH)_2(c)|Ni$$

The approximate discharge reactions are

"Ni_2O_3" may be a solid solution of NiO and NiO_2.

$$\begin{array}{lll} Cd + 2OH^- \longrightarrow Cd(OH)_2 + 2e^- & & \text{(negative electrode)} \\ Ni_2O_3 + 3H_2O + 2e^- \longrightarrow 2Ni(OH)_2 + 2OH^- & & \text{(positive electrode)} \\ \hline Cd + Ni_2O_3 + 3H_2O \longrightarrow Cd(OH)_2 + 2Ni(OH)_2 & & \end{array}$$

The cell gives about 1.2 V. It is very resistant to overcharging and is usually connected to the power supply whenever it is not in use. As with lead-acid cells, the design permits O_2 evolution but not H_2 evolution. The O_2 produced at the positive electrode reacts with the Cd in the negative electrode, permitting a useless but harmless process to continue indefinitely:

$$\begin{array}{lll} 4OH^- \longrightarrow O_2 + 2H_2O + 4e^- & & \text{(positive electrode)} \\ O_2 + 2Cd + 2H_2O \longrightarrow 2Cd(OH)_2 & & \text{(negative electrode)} \\ 2Cd(OH)_2 + 4e^- \longrightarrow 2Cd + 4OH^- & & \text{(negative electrode)} \\ \hline \text{no net reaction} & & \end{array}$$

FUEL CELLS

The process in any galvanic cell involves the consumption of some materials and the production of others. The reactants—such as zinc, copper, mercury, and their compounds—are often rather expensive. This is one reason ordinary cells are unsuitable for generating large amounts of energy. Another reason is that the reactants are part of the cell; when a reactant is used up, the cell must be replaced or the user must take time out to recharge it from some other energy source. Also, since the products are not removed, their concentrations may increase as the cell is used, and this decreases the cell's emf according to the Nernst equation.

A better system would be a cell in which the reactants are relatively cheap fuels, plus oxygen from the air. Fuel and air would be fed to the cell, and the products removed, continuously. Such a device is a **fuel cell**. The only fully satisfactory fuel cells thus far developed require hydrogen as fuel, and they have been used only in special situations where cost is a secondary consideration, notably in space vehicles. H_2 is expensive because it does not occur in large natural

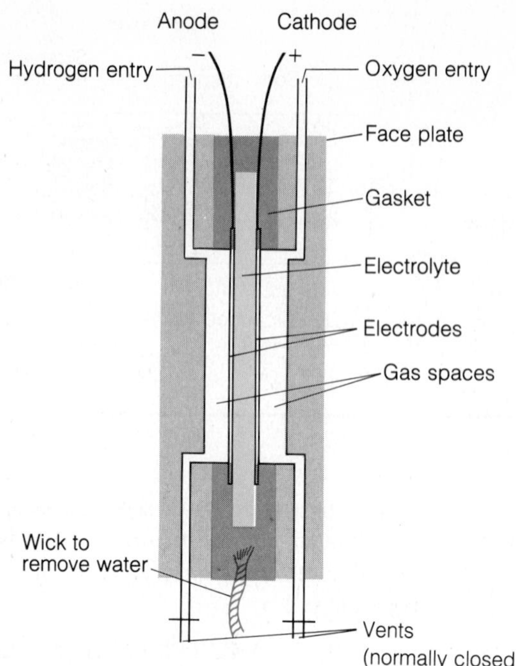

FIGURE 17.23
A hydrogen-oxygen fuel cell. The electrolyte is KOH solution absorbed in asbestos. The electrodes are alloys of Au, Pd, and Pt.

deposits (as do coal, petroleum, and natural gas), and its low density and low boiling point make it difficult to store and handle.

Figure 17.23 is a simplified diagram of a hydrogen-oxygen fuel cell similar to those used in spacecraft. If the pressures of the gases are 1 atm, the maximum emf of this cell at 25°C can be calculated from Table 17.1:

$$\begin{array}{rcll} \frac{1}{2}O_2 + 2H^+ + 2e^- & \longrightarrow & H_2O(\ell) & \text{(reduction)} \quad +1.229 \\ H_2 & \longrightarrow & 2H^+ + 2e^- & \text{(oxidation)} \quad 0 \\ \hline H_2 + \frac{1}{2}O_2 & \longrightarrow & H_2O(\ell) & \qquad \mathscr{E}^0 = 1.229\text{ V} \end{array}$$

The net reaction and emf are the same with a basic solution. In practice, the emf is 0.8 to 1.0 V. The only product should be H_2O, which the astronauts can use as drinking water, and testing its purity provides a diagnosis of the condition of the cell. A number of these cells, typically 30 or 40, are stacked together to make a battery that yields about 30 V. H_2 and O_2 are stored as liquids in insulated containers.

In theory, a fuel cell can be much more efficient than a steam engine or internal-combustion engine. It converts chemical energy directly to electrical work, instead of first converting it to heat by burning the fuel and then converting a part—only a part—of this heat to electrical work. Thus, a successful fuel cell uses less fuel for a given work output than an engine uses. Fuel cells that consume oil or natural gas could be of great value in making these fuels last longer. These fuel cells are still in the experimental stage. Problems with them have included high operating temperatures (with resulting corrosion and short lifetime), the large amounts of platinum used (expensive and available only in limited quantity), and the need to purify the fuel and the air to remove substances that would ruin the cell (such as CO_2, CO, and S).

The world supply of fossil fuels obviously cannot last forever. The time may not be far off when hydrogen, produced from water by nuclear or solar energy,

BOX 17.6

The most commercially advanced fuel cells use H_3PO_4 as the electrolyte. The electrodes are finely divided platinum supported on carbon. The fuel is a gaseous mixture of H_2, CH_4, C_2H_6, and other gases, which can be produced from oil, coal, or natural gas. The gas pressure is 1 to 8 atm; the temperature is 150°C to 250°C. Power plants using these cells have been tested in the United States and Japan. However, it is unlikely that portable batteries will soon replace internal-combustion engines in automobiles, trucks, and locomotives.

will be cheaper than oil or natural gas. The energy system of the future may be based on H_2, distributed by pipeline. It is much less expensive to pump a gaseous fuel than to transmit electricity over long distances—the ease of using electricity blinds us to the complexity and inefficiency of the transmission system that brings it to us. Several processes for producing hydrogen from water are being actively developed; for example,

1. Electrolysis seems like an obvious method, but there is plenty of room for ingenuity in improving it (Problem 17.74). The biggest need is for decreasing the energy required per unit of H_2 produced. However, hydrogen made by electrolysis must be more expensive (per unit of energy) than the electricity used to produce it. Therefore, electrolytic H_2 can compete with electricity as an energy source only to the extent that it can be transmitted more cheaply.

2. A very attractive idea is to use solar energy to bring about the reaction $H_2O \longrightarrow H_2 + \frac{1}{2}O_2$. There are major difficulties with this method, however. The most serious is that few photons with enough energy to break the H—O bond are present in sunlight at the surface of the Earth. Nevertheless, it may be possible to achieve hydrogen production this way by some roundabout sequence of reactions, if suitable light-absorbing catalysts can be found. This process is indeed used today, but with sunlight that arrived millions of years ago. The solar energy was used to produce plant tissue, which became coal, which is used in the reaction $C + H_2O(g) \longrightarrow CO(g) + H_2(g)$. Thus, the production of hydrogen by solar energy is possible; the problem is to speed it up.

3. *Photoassisted electrolysis* is a combination of processes 1 and 2. The electrons to be used in the electrolysis are raised to a higher energy level by the absorption of photons. The electrical energy required for electrolysis is thus decreased.

A long-established technology uses *heat* to operate a *refrigerator*.

If any of these processes becomes practical, we may look forward to a time when H_2, instead of natural gas (CH_4), is piped into our homes for heating, cooking, refrigeration, and air conditioning. Each building or each neighborhood would have a battery of fuel cells to generate electricity for lights and appliances. Whether H_2 will be practical as a fuel for automobiles and airplanes, which cannot be connected to pipelines, is more problematical. Storing H_2 as a compressed gas or a cold liquid is cumbersome and dangerous. Promising new methods use adsorption of H_2 on, or absorption in, a metal.

17.13 CORROSION

The oxidation (in the literal sense of "combination with oxygen") of most metals is spontaneous at ordinary temperatures. An undesired reaction—in most cases oxidation—of metals is called **corrosion**. In the particular case of iron, it is called "rusting." Corrosion is an economic loss, because metals are more useful than their oxides and other corrosion products, such as carbonates. Iron is structurally strong; its oxide is not. If a steel bridge is allowed to rust, it will ultimately collapse. Metals are lustrous; their oxides are dull. An object of copper or silver may be valued for its brilliant surface. If the object is not kept clean, however, the metal reacts with oxygen or compounds of sulfur, the surface becomes tarnished, and the beauty of the object is diminished.

By far the greatest environmental depletion resulting from corrosion is caused by the rusting of iron and steel. The annual costs in the United States are estimated to be in the billions of dollars, and the figures would be even higher if they included the cost of cleaning up the countryside by recovering scrap that is not worth gathering for its iron content. Our primary consideration here will therefore be the corrosion of iron.

For a metal to become oxidized it must lose electrons. The most direct pathway is provided by an electrochemical cell, in which the metal loses electrons at the anode. *Any* combination of parts that can function as a cell with iron as the anode, no matter how unlike a conventional cell it appears to be, will cause the iron to corrode. The necessary parts are

1. Some metal (or carbon) that can serve as the cathode to the iron anode. This need not be a different metallic object—it may simply be another area of the corroding iron. In other words, a given piece of iron may serve as both anode and cathode in different parts of itself (Figure 17.24). Small variations in composition or even in mechanical stress can make the difference.
2. A metallic conductor between the electrodes. The piece of iron itself will do.
3. A path between the electrodes for migrating ions. Any electrolyte, such as salt water, will serve. In fact, if *any* water is present, it will dissolve enough ionic matter from the environment, such as salts from the soil, to be a good enough electrolyte to promote the corrosion of iron. A bulk quantity of solution is not necessary; a moist, porous solid, such as soil or wood, will provide a path. Even the film of H_2O adsorbed by the iron from a moist atmosphere is sufficient.

The generally accepted theory is that Fe is oxidized to Fe^{2+} by H_2O or H^+:

$$Fe \longrightarrow Fe^{2+} + 2e^- \quad (23)$$

$$2H^+ + 2e^- \longrightarrow H_2(g) \quad (24)$$

If the H^+ concentration is large, bubbles of H_2 appear. In corrosion by natural waters, however, dissolved O_2 is usually reduced instead of H^+:

$$O_2(aq) + 4H^+ + 4e^- \longrightarrow 2H_2O \quad (25)$$

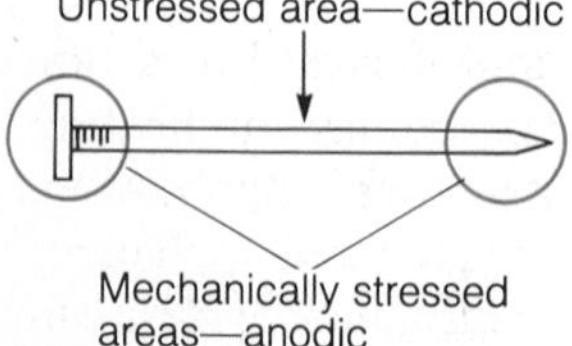

FIGURE 17.24 Anodic and cathodic areas of an iron nail.

The oxygen also oxidizes the Fe^{2+} ions liberated in reaction 23, producing hydrated iron(III) oxide (rust) (Figure 17.25):

$$4Fe^{2+} + O_2 + (4 + 2x)H_2O \longrightarrow 2Fe_2O_3(H_2O)_x + 8H^+ \quad (26)$$

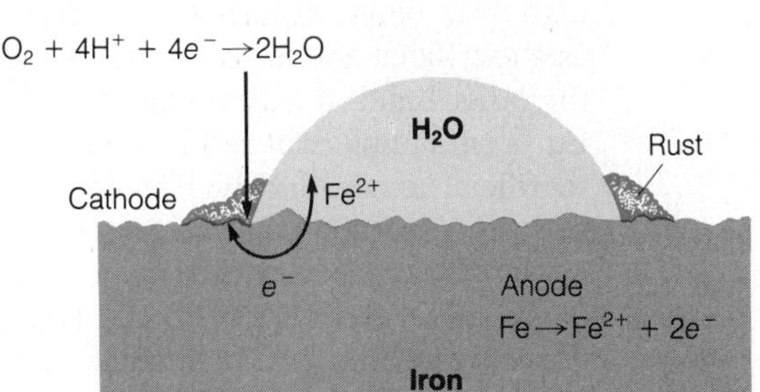

FIGURE 17.25
Corrosion of iron in the presence of water and air.

The H^+ consumed in reaction 24 or 25 is regenerated in reaction 26 so that H^+ is a catalyst; that is, it accelerates the reaction without being consumed. CO_2 is nearly always present in natural waters and it increases the H^+ concentration by the reaction $CO_2 + H_2O \rightleftharpoons H^+ + HCO_3^-$, thus contributing to corrosion. On the other hand, making the solution basic decreases $[H^+]$ and inhibits corrosion. Most salts accelerate corrosion by increasing the conductivity of the water, but salts that dissolve to give basic solutions (like Na_2CO_3) have the opposite effect.

The corrosion of an iron or steel pipe is greatly accelerated if it is attached to a copper pipe—reaction 23 occurs on the Fe; reaction 25 mostly on the Cu. Any metal less active (a poorer reducing agent) than iron will behave in the same way as copper. When only iron is present, reaction 23 takes place at the anodic sites, which will be corroded, and reaction 25 occurs at the cathodic sites, which will not be corroded.

If another metal more active (a better reducing agent) than iron—Zn or Mg, for example—is in contact with iron, the more active metal will be oxidized in preference to the iron, and the iron in the vicinity will be protected from corrosion (Figure 17.26). The iron is then said to be made **cathodic**. The active metal is called a "sacrificial anode," for it is allowed to corrode away in order to protect the iron. The same result can be achieved by imposing a negative electric potential on the iron and a positive potential on other electrodes placed in the water or the ground nearby. These methods, known as **cathodic protection**, are often used to protect water tanks, boilers, ship hulls, and submerged or buried pipes.

"Galvanized iron" is iron that has been given a protective coating of zinc. This coating is applied most often by dipping the iron into molten zinc, but it can also

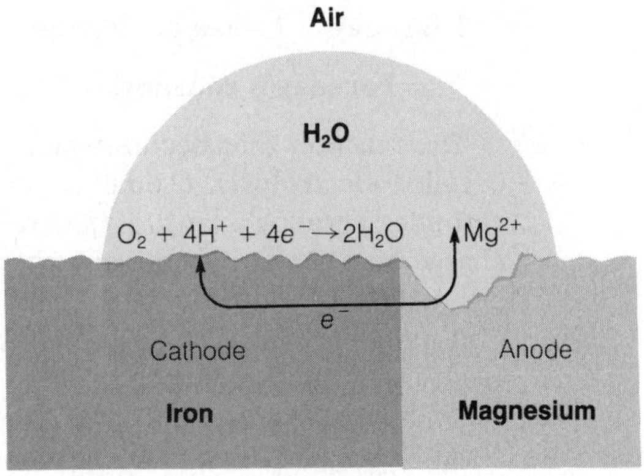

FIGURE 17.26
Cathodic protection of iron by magnesium.

be done by electroplating from an aqueous solution—possible only because of the high overvoltage required for hydrogen evolution (section 17.11). The zinc serves as a sacrificial anode and protects the iron even if the coating has a hole in it. On the other hand, if iron is plated with a metal less active than itself, such as Ag or Cu, a small hole will result in very rapid corrosion, for the iron will behave as a sacrificial anode and will protect the plating—hardly a desirable result. Tin is cathodic (less active) relative to iron in the presence of air, but in the absence of air (as in a can) and especially in the presence of acid (as in many foods), Sn becomes anodic relative to Fe. Thus, tin is useful for plating iron, even though the table of electrode potentials makes it appear that corrosion of iron might be accelerated by a hole in the tin. The reasons for this shift in potentials are not fully understood. Of course, an *unbroken* coating of an inactive metal provides protection.

These adherent oxides have crystal structures not very different from those of the metals.

Many metals—Al, Co, Cr, and Ni, for example—start to corrode by forming a thin, continuous, adherent layer of oxide that prevents further corrosion. These metals form oxides that adhere firmly to the underlying metal. Such metals are said to become **passive**. They are resistant to corrosion under ordinary conditions, even though Al and Cr, for example, are more active than Fe. FeO and Fe_2O_3, unfortunately, do not adhere to iron. The worst feature of rust is not that it forms, but that it falls off, continually exposing a fresh surface to be corroded. An especially impervious coating can be produced on many metals, even iron, by treating the metal with a strong oxidizing agent such as concentrated HNO_3 or Pb_3O_4 ("red lead"), or by using it as the anode in an electrolysis with high current density. The passivity of iron is easily destroyed merely by touching the passive metal with a piece of normal (active) iron, unless the oxidizing agent is kept in contact with the iron. Pb_3O_4 is used in paint for structural steel to protect it from corrosion. Basic zinc chromate, $ZnCrO_4(ZnO)_4$, is used to protect aluminum and magnesium aircraft parts. A metal can also be kept passive by applying a positive electric potential. These methods of maintaining passivity are examples of **anodic protection**. The resistance of platinum and related elements to electrolytic oxidation is much greater than their electrode potentials would lead us to expect; this is another example of passivity.

SUMMARY

When an electric current passes through an ionic solution or other ionic conductor, oxidation–reduction reactions occur at the **electrodes**.

current (in amperes, A) = charge (in coulombs, C) per second

1 **faraday** = 1 mole of electrons = 9.65×10^4 C

$\mathscr{F}$ = **Faraday's constant** = 9.65×10^4 C/faraday

The reaction that occurs when a current passes through an ionic conductor is called **electrolysis**. Oxidation takes place at the **anode**, and reduction at the **cathode**. Faraday's laws of electrolysis are (1) that mass is proportional to the charge passed through the circuit and (2) that the number of faradays required per mole is a whole number. This number appears in the partial equations.

A **galvanic cell** uses a spontaneous oxidation–reduction reaction to produce electrical work. It is described in a conventional notation with the anode on the left and the cathode on the right: anode|anode solution||cathode solution|cathode. A **salt bridge** (||) is a concentrated salt solution interposed between two other solutions.

The **difference of potential** between the electrodes in a galvanic cell is the **electromotive force (emf, $\mathscr{E}$)** of the cell, which specifies the driving force of the cell reaction. It is measured in **volts**. The work done by the cell when the reaction occurs as written is $n\mathscr{F}\mathscr{E}$, where n is the number of electrons appearing in each partial equation. $\mathscr{E} = \mathscr{E}^0$, the **standard emf**, when each reactant and product is in its standard state, defined as the pure substance, 1 mol/L for solutes, or 1 atm for gases.

The emf of a cell reaches a maximum when it is opposed by an equal emf from a **potentiometer**. The process is then **thermodynamically reversible**; it can be made to go either way by an infinitesimal change in the potentiometer setting. The emf thus measured is the **reversible emf**.

In a galvanic cell, the anode is negative and the cathode is positive. In an electrolytic cell, the anode is positive and the cathode is negative.

A cell can be divided into two **half-cells**. A **half-cell potential (reduction potential, electrode potential)** is the emf of a cell in which the **reference half-cell** is assumed to be the anode. A negative half-cell potential indicates that the reference half-cell actually functions as the cathode. The reference half-cell is arbitrarily chosen as $Pt|H_2(g)|H^+$. Standard half-cell potentials ($\mathscr{E}^0$) are tabulated. Then

$$\mathscr{E}^0_{cell} = \mathscr{E}^0_{cathode} - \mathscr{E}^0_{anode}$$

If $\mathscr{E}^0_{cell}$ is negative, the spontaneous reaction is opposite to what was assumed. The more positive the half-cell potential is, the greater is the tendency for reduction to occur, the stronger is the oxidizing agent, and the weaker is the reducing agent. In the same way, the more negative the half-cell potential is, the greater is the tendency for oxidation to occur, the stronger is the reducing agent, and the weaker is the oxidizing agent.

When the reactants and products are not in their standard states, the emf is given by the **Nernst equation**:

$$\mathscr{E} = \mathscr{E}^0 - (RT/n\mathscr{F}) \ln Q = \mathscr{E}^0 - (2.303RT/n\mathscr{F}) \log Q$$

where Q is the reaction quotient for the cell reaction. At 25°C, $\mathscr{E} = \mathscr{E}^0 - (0.0592/n) \log Q$.

The Nernst equation provides a method of measuring concentrations. A **pH meter** uses a **glass electrode**, which is sensitive to $[H^+]$. The emf therefore depends on pH: $\mathscr{E} = \mathscr{E}' + (2.303RT/\mathscr{F})$ pH. $\mathscr{E}'$ is determined by calibration with buffer solutions of known pH. Other **ion-selective electrodes** are used for measuring the concentrations of other ions.

The standard emf, $\mathscr{E}^0$, of a cell is related to the equilibrium constant, K, of the cell reaction by $\log K = n\mathscr{F}\mathscr{E}^0/2.303RT = n\mathscr{E}^0/0.0592$ at 25°C. If $\mathscr{E}^0$ is positive, K is greater than 1 and the reaction goes to the right when concentrations are near 1 mol/L.

The **decomposition potential** is the minimum emf needed to bring about an electrolysis reaction. It must be at least equal to the reversible emf of the galvanic cell created by the products of the reaction. It may be much greater because a large **overvoltage** may be needed, especially when gases are produced. **Concentration polarization**, resulting from depletion of reactants and accumulation of products, also increases the decomposition potential.

Primary cells are used and discarded. The **dry cell** is the most common example. **Storage cells** can be recharged. Lead-acid and nickel-cadmium storage batteries are widely used. **Fuel cells** use reactants that are supplied continuously, such as $H_2 + O_2$.

Corrosion is the undesired reaction of metals, usually combination with O_2. The corrosion of iron is called "rusting." It is an electrochemical process in which the metal is oxidized at an anodic site and O_2 is reduced at a cathodic site. In **cathodic protection**, a metal is made cathodic by connecting it to a more active metal or to the negative side of a power source. When a thin film of oxide adheres firmly, the metal becomes **passive**. Maintaining this film by coating the metal with an oxidizing agent or by connecting it to the positive side of a power source is referred to as **anodic protection**.

17.14 ADDITIONAL PROBLEMS

ELECTRODE REACTIONS

17.18 Gold is plated by the cathode reaction

$$Au(CN)_2^- + e^- \longrightarrow 2CN^- + Au(c)$$

A current of 1.50 A flows for 2.00 hours. What mass of Au is deposited?

17.19 A current of 0.300 A is used for the electrolytic decomposition of water. (a) How long (in hours) will it take to produce 1.00 L (STP) of H_2? (b) What mass of O_2 will be produced in that time?

17.20 Three electrolytic cells are connected in series. In the first, Cd is oxidized to Cd^{2+}. In the second, Ag^+ is reduced to Ag. In the third, Fe^{2+} is oxidized to Fe^{3+}. (a) 1.00 g of Cd is removed in the first cell. Find the number of faradays passed through the circuit. (b) What mass of Ag is deposited on the cathode in the second cell? What mass of $Fe(NO_3)_3$ can be recovered from the solution in the third cell?

17.21 (a) Assume that in the electrolysis of a $CuSO_4$ solution, the reaction at the anode is

$$Cu \longrightarrow Cu^{2+} + 2e^-$$

A current of 1.00 A passes for 3.00 hours. What mass of Cu reacts? (b) The loss in mass of the Cu anode was actually only 2.50 g. The reason is that the reaction

$$2H_2O \longrightarrow O_2 + 4H^+ + 4e^-$$

also occurred at the anode. What percentage of the current was used to produce Cu? What percentage of the current was used to produce O_2? (The *current efficiency* is the fraction of the current used for the desired reaction.)

17.22 Many metal carbides are very hard. A metal can be made much more resistant to abrasion by coating it with a layer of carbide. A new method of doing this is electrochemical deposition of the metal and carbon, which react to form the carbide. To deposit tantalum carbide (TaC, 192.96 g/mol), a molten mixture of LiF, NaF, and KF at 800°C is used as the solvent. K_2TaF_7 and K_2CO_3 are the solutes. Two cathode reactions occur,

$$TaF_7^{2-} + 5e^- \longrightarrow Ta + 7F^- \quad \text{and}$$

$$CO_3^{2-} + 4e^- \longrightarrow 3O^{2-} + C$$

followed by

$$Ta + C \longrightarrow TaC$$

The result of these three processes is the cathode partial reaction

$$TaF_7^{2-} + CO_3^{2-} + 9e^- \longrightarrow TaC + 7F^- + 3O^{2-}$$

(a) How many coulombs are needed to deposit 100 g of TaC? What are you assuming about the percentage of the current used for each of the two cathode reactions? How is your answer different if that assumption is incorrect? (b) The anode reaction is

$$2CO_3^{2-} \longrightarrow 2CO_2(g) + O_2(g) + 4e^-$$

When the process in part a occurs, what total volume of gas, measured at 800°C and 1.00 atm, is produced at the anode?

17.23 A new way to convert electrical energy to chemical energy has recently been developed. CO_2 is reduced to CH_4 at a ruthenium cathode in a 0.1 M aqueous solution of H_2SO_4. (a) Write the partial equation for this reduction. (b) About 30% of the current is used to produce CH_4. How many coulombs are needed to produce 100 L of CH_4, measured at standard temperature and pressure?

17.24 Aluminum can be given a more desirable color by *anodizing*. The aluminum is used as an anode, usually in an aqueous solution of H_2SO_4, to produce a porous layer of Al_2O_3, which can then be treated with a dye. (a) Write the partial equation for the anode reaction that produces Al_2O_3. (b) The density of Al_2O_3 is 3.97 g/cm^3. Assume that the oxide layer is 90% Al_2O_3 and 10% empty space, by volume. How many coulombs must be passed per cm^2 of surface to produce a layer 5.0×10^{-3} cm thick?

17.25 Most metals are plated from solutions that contain coordination ions (especially cyano complexes) because better deposits are obtained than from uncoordinated (actually hydrated) ions. (a) Write the equations for the anode and cathode partial reactions in the electroplating of Ag from a solution of $K^+[Ag(CN)_2]^-$ and K^+CN^- with an Ag anode. (*Hint*: The net process is Ag(anode) $\longrightarrow$ Ag(cathode).) (b) How many coulombs must pass through the circuit for each gram of Ag deposited? (c) Toward which electrode does the $[Ag(CN)_2]^-$ ion migrate to carry current through the solution? Is there anything strange about your answer? How do you reconcile your answer with the fact that the plating does occur?

17.26 The scale of atomic weights was changed in 1961. On the old chemical scale, the atomic weight of natural oxygen was exactly 16 by definition. On the new scale (based on ^{12}C), the atomic weight of natural oxygen is 15.9994. Did the number of coulombs per faraday change in 1961? If so, what is the ratio of the new faraday to the old faraday?

17.27 Electrochemical machining is a technique of shaping metal by controlled oxidation to ions. The metal to be shaped is the anode, and a complementary shape—with elevations where depressions are to be made, and vice versa—is the cathode. The anode and cathode are very close together but must not touch. High current densities are needed to make the process reasonably fast. (a) Why does an elevation in the cathode produce a depression in the anode? (*Hint*: What determines where the greatest current flows?) (b) Why must the electrodes be close together (besides minimizing resistance)? (c) Solution is continuously pumped through the gap betwen the electrodes. Why? What would happen if it were not pumped?

17.28 Adiponitrile, $N{\equiv}C(CH_2)_4C{\equiv}N$ (108.1 g/mol), is used in the manufacture of nylon. It is prepared electrochemically from acrylonitrile in basic solution:

$$2(H_2C{=}CH{-}C{\equiv}N) \longrightarrow NC(CH_2)_4CN \quad \text{(unbalanced)}$$

(a) Balance this partial equation. Is it an oxidation or a reduction? (b) How many coulombs are needed to produce 1.0 metric ton (10^6 g) of adiponitrile?

17.29 Malodorous materials can be removed from industrial stack gases by bubbling the gas stream through a solution that oxidizes the pollutants to harmless products. One process that accomplishes this is oxidation in acid solution by Ce^{4+} (which is reduced to Ce^{3+}), followed by electrolytic regeneration of Ce^{4+} with H_2 as a by-product. (a) If the pollutant is acrolein, C_3H_4O, the unbalanced partial equations are $C_3H_4O \longrightarrow CO_2 + H^+$, $Ce^{4+} \longrightarrow Ce^{3+}$, $Ce^{3+} \longrightarrow Ce^{4+}$, and $H^+ \longrightarrow H_2$. Balance these partial equations. (b) Write the equation for the net process. (c) How many coulombs are needed to dispose of 1.00 kg of acrolein? (Acrolein, named for its acrid odor, is produced when fats are decomposed at high temperature.)

17.30 Chromium is plated from an acidic aqueous solution containing the dichromate ion. The cathode reaction may be represented as

$$Cr_2O_7{}^{2-} + 14H^+ + 12e^- \longrightarrow 2Cr + 7H_2O$$

The current efficiency (see Problem 17.21) is only about 20%; that is, 20% of the current is used to produce Cr. (a) What cathode reaction accounts for the remaining 80% of the current? (b) How many grams of Cr can be deposited by a current of 8.0 A flowing for 3.00 h? Assume 20% current efficiency. (c) A student gave the following answer to part a of this problem: "The efficiency is low because the other 80% is dissipated as heat." Explain why this answer is incorrect.

17.31 A current of 0.0100 A is passed through a solution of rhodium sulfate. The only reaction at the cathode is the deposition of Rh. After 3.00 h, 0.038 g of Rh has been deposited. What is the charge on the ion Rh^{x+}?

17.32 A solution contains the unknown metal ion X^{z+}. A direct current is passed, depositing X on the anode: $X^{z+} + ze^- \longrightarrow X(c)$. When 5.00×10^3 C pass through the circuit, 1.983 g X is deposited. (a) Calculate the ratio of the atomic weight of X to the charge z. (b) Assume that z is 1, 2, or 3. Find the atomic weight for each of these possibilities. Select a likely metal, considering both its atomic weight and its position in the periodic table.

ELECTRICAL WORK

17.33 The emf of a galvanic cell is 1.20 V. (a) How much work is done by this cell when 10 C of charge pass from one electrode to the other? (b) How much power, in watts, is the cell producing when a current of 0.10 A flows? (1 watt = 1 joule/second) (c) How much work does the cell do when 1 faraday passes through the circuit? Assume that the emf of the cell is unaffected by the rate at which current is drawn.

GALVANIC CELLS, EMF, AND HALF-CELL POTENTIALS

17.34 For each of these cells, write the partial equations for oxidation and for reduction and the net equation for the cell reaction. Assume that the left electrode is the anode.

(a) $Cd|Cd(NO_3)_2(aq)||AgNO_3(aq)|Ag$

(b) $Cr|Cr_2(SO_4)_3(aq)||FeSO_4(aq)|Fe$

(c) $Ni|NiCl_2(aq)|Hg_2Cl_2(c)|Hg(\ell)$

17.35 For cells 1 to 5, the net reaction is given, in the direction consistent with the way the cell is written. (a) Write the partial equations for the anode and cathode processes. (b) Find the standard emf, $\mathscr{E}^0$, at 25°C. (c) Does the cell reaction actually occur as given or in the reverse direction when $\mathscr{E} = \mathscr{E}^0$?

(1) $Pb|Pb^{2+}||Ag^+|Ag$
$Pb + 2Ag^+ \longrightarrow Pb^{2+} + 2Ag$

(2) $Hg(\ell)|Hg^{2+}||Sn^{2+}|Sn$
$Hg + Sn^{2+} \longrightarrow Hg^{2+} + Sn$

(3) $Pt|Br_2(\ell)|SnBr_2(aq)|Sn$
$Sn^{2+} + 2Br^- \longrightarrow Sn + Br_2$

(4) $Fe|FeCl_2(aq)|Hg_2Cl_2(c)|Hg(\ell)$
$Fe + Hg_2Cl_2 \longrightarrow Fe^{2+} + 2Cl^- + 2Hg$

(5) $Pt|Cl_2(g)|KCl(aq)|AgCl(c)|Ag$
$2AgCl \longrightarrow 2Ag + Cl_2$

17.36 (a) Suppose we were to decide that $\mathscr{E}^0 = 0$ for the half-cell $Sn^{2+}|Sn$ instead of the hydrogen half-cell. On the basis of this choice, what would be $\mathscr{E}^0$ for $Cd^{2+}|Cd$ and for $Ag^+|Ag$? (b) Calculate $\mathscr{E}^0$ for the cell $Cd|Cd^{2+}||Ag^+|Ag$, based on the results in part a and directly from the data in Table 17.1. Compare the answers.

17.37 Select the stronger oxidizing agent in each pair: (a) Cr^{3+}, Cd^{2+}; (b) Br_2, $AuBr_4{}^-$; (c) $Ag(NH_3)_2{}^+$, I_2; (d) O_2 in acid, O_2 in base. Select the stronger reducing agent in each pair: (e) Al, Zn; (f) Hg in CN^- solution, Hg in I^- solution; (g) I^-, Ag; (h) H_2 in acid, H_2 in base.

THE NERNST EQUATION

17.38 The reaction in a cell is

$$3Be(c) + 2Ga^{3+} \longrightarrow 3Be^{2+} + 2Ga(c)$$

Is the emf of the cell increased, decreased, or unchanged when (a) more Be(c) is added; (b) $[Ga^{3+}]$ is increased; (c) $[Be^{2+}]$ is increased?

17.39 The reaction occurring in the cell

$$U|UO_2^{2+}, H^+||ReO_4^-, H^+|ReO_3(c)|Pt$$

is represented by these *unbalanced* partial equations:

$$U(c) \longrightarrow UO_2^{2+}$$

$$ReO_4^- \longrightarrow ReO_3(c)$$

(a) Write the balanced partial equations and the equation for the net reaction. (b) H^+ is at the same concentration throughout the cell. Is the emf increased or decreased when each of the following concentrations is decreased? $[H^+]$, $[UO_2^{2+}]$, $[ReO_4^-]$.

17.40 For each of the following cells, (a) write the equation for the cell process, assuming that the left electrode is the anode; (b) find $\mathscr{E}^0$ at 25°C; (c) find $\mathscr{E}$ at 25°C, using the Nernst equation.
(1) $Zn|Zn^{2+}(0.100\ M), I^-(0.200\ M)|I_2(c)|Pt$
(2) $Mg|Mg^{2+}(0.0500\ M),$
$NO_3^-(0.100\ M)||Cu^{2+}(0.200\ M), NO_3^-(0.400\ M)|Cu$

17.41 The reaction

$$H_2 + 2AgBr \longrightarrow 2Ag + 2H^+ + 2Br^-$$

occurs in the cell

$$Pt|H_2(1.00\ atm)|HBr(0.20\ M)|AgBr(c)|Ag$$

(a) Find $\mathscr{E}^0$ at 25°C. (b) Find $\mathscr{E}$ at 25°C, using the Nernst equation. (c) Calculate the pressure of hydrogen that will make the emf of the cell equal to 0 at the given HBr concentration.

17.42 (a) Calculate $\mathscr{E}^0$ at 25°C for a cell in which the reaction is $2H_2(g) + O_2(g) \longrightarrow 2H_2O(\ell)$. (b) Calculate $\mathscr{E}$ for this cell when the concentrations of $H_2(g)$ and $O_2(g)$ are each 1.00 mol/L. (First calculate the pressure corresponding to this concentration.) (c) What would $\mathscr{E}^0$ be for this cell if we decided to define $\mathscr{E}^0$ as the emf at unit concentration (1 mol/L) for gases as well as for solutes?

17.43 Suppose that you have a "silver" filling in a lower tooth and a gold filling in an upper tooth. The silver filling contains mercury and tin as well as silver. You have pain when you chew. A possible reason is that the reaction $Sn \longrightarrow Sn^{2+} + 2e^-$ occurs on the surface of the silver filling and the reaction $O_2 + 4H^+ + 4e^- \longrightarrow 2H_2O$ occurs on the surface of the gold filling. What is the potential difference between the two fillings? Assume that the pH in your mouth is 7, $[Sn^{2+}]$ is 0.01 mol/L on the filling surface, and the pressure of O_2 is 0.2 atm. Calculate as if the temperature were 25°C.

17.44 For the cell

$$Pt|H_2(g)|HBr(aq)|AgBr(c)|Ag$$

with the reaction

$$H_2 + 2AgBr \longrightarrow 2H^+ + 2Br^- + 2Ag$$

$\mathscr{E}^0 = 0.0567$ V at 50.0°C. (a) Calculate $RT \ln 10/\mathscr{F}$ at 50.00°C. (b) Calculate the emf of this cell at 50.0°C when $[H^+] = [Br^-] = 0.0100$ mol/L and $P_{H_2} = 1.00$ atm.

17.45 This galvanic cell is prepared, using the unknown metal X:

$$Zn(c)|Zn^{2+}(aq)||X^{y+}(aq)|X(c)$$

The cathode reaction is $X^{y+} + ye^- \longrightarrow X$. (a) Write the equation for the net reaction in the cell. (b) Write the Nernst equation, containing y as an unknown. (c) When $[X^{y+}]$ is changed from 0.0100 M to 0.100 M at 25°C, $\mathscr{E}$ changes by 0.059 V. Find y.

MEASURING CONCENTRATIONS

17.46 (a) A pH meter is calibrated with a buffer of pH 7.00. The emf is $\mathscr{E} = 0.493$ V at 25°C. Calculate $\mathscr{E}'$ (as in Example 17.18, section 17.9). (b) With an unknown solution, the emf is 0.346 V. Calculate the pH of the unknown solution.

17.47 The cell

$$Hg(\ell)|Hg_2Cl_2(c)|KCl(aq)|KNO_3(aq)|Ag^+|Ag(c)$$

can be used to measure the concentration of Ag^+. (a) Write the partial reactions, the net reaction, and the Nernst equation. (b) When $[Ag^+] = 0.0100$ mol/L at 25°C, the emf is 0.417 V. Calculate $[Ag^+]$ when the emf is 0.461 V.

17.48 In the cell

$$Ag(c)|AgI(c)|HI(aq)|H_2(g, 1.00\ atm)|Pt$$

the reaction is

$$2Ag + 2H^+ + 2I^- \longrightarrow H_2 + 2AgI$$

(a) Calculate $\mathscr{E}^0$ for the cell at 25°C. (b) The emf of the cell is 0.047 V at 25°C. Calculate the molarity of HI.

EMF AND EQUILIBRIUM

17.49 Ascertain, with as little calculation as possible, whether these reactions will go as written or in the reverse direction at 25°C, when all concentrations and pressures are close to 1:
(a) $2Cr^{3+} + 3Cu \longrightarrow 2Cr + 3Cu^{2+}$
(b) $Fe + Cl_2(g) \longrightarrow Fe^{2+} + 2Cl^-$
(c) $2Fe^{2+} + Cl_2(g) \longrightarrow 2Fe^{3+} + 2Cl^-$

17.50 For each of the following reactions in aqueous solution, (a) find $\mathscr{E}^0$ at 25°C for a cell in which the reaction occurs and (b) calculate the equilibrium constant.
(1) $Mg(c) + Ni^{2+} \longrightarrow Mg^{2+} + Ni(c)$
(2) $H_2(g) + Br_2(\ell) \longrightarrow 2H^+ + 2Br^-$

17.51 From the standard electrode potentials in Table 17.1, calculate (a) the solubility product of AgI and (b) the dissociation constant of $AuBr_4^-$. Compare your answers with Tables 16.1 and 16.2 (sections 16.1 and 16.3).

17.52 Find the equilibrium constants at 25°C for these reactions:
(a) $Ag(c) + H^+ + Br^- \rightleftharpoons AgBr(c) + \frac{1}{2}H_2(g)$
(b) $Ag(c) + H^+ \rightleftharpoons Ag^+ + \frac{1}{2}H_2(g)$
(c) $AgBr(c) \rightleftharpoons Ag^+ + Br^-$
Show how the equilibrium constants for reactions a and b may be used in determining the constant of reaction c (see section 13.5).

17.53 Use $\mathscr{E}^0$'s for two partial reactions involving O_2 (instead of H_2) to calculate the ion product of water at 25°C.

17.54 Will each of the following reactions proceed as written or in the reverse direction at 25°C when concentrations are about 1 *M*?
(a) $2Ag + Cu^{2+} \rightarrow 2Ag^+ + Cu$
(b) $2Ag + Cu^{2+} + 2Cl^- \rightarrow 2AgCl(c) + Cu$
(c) Would you use your best silver spoon to stir a solution of $Cu(NO_3)_2$; $CuCl_2$? Explain. (d) What is the molarity of the most concentrated solution of $CuCl_2$ that you would stir with a treasured silver spoon?

17.55 Will O_2 oxidize gold to $Au(CN)_2^-$ in a basic solution containing cyanide ion (CN^-)? The reaction would be

$$4Au + 8CN^- + O_2 + 2H_2O \longrightarrow 4Au(CN)_2^- + 4OH^-$$

Assume that the ions are 1 M and O_2 is at 1 atm.

17.56 Calculate the equilibrium constant of the reaction

$$Hg(\ell) + Hg^{2+} \rightleftharpoons Hg_2^{2+}$$

at 25°C. (*Suggestion*: Hg appears in both partial equations but does not cancel completely.) Why is a solution of a Hg(I) salt usually stored with a pool of mercury at the bottom of the bottle?

17.57 What value of $\mathscr{E}^0$ makes $K = 10^{10}$ at 25°C, when (a) $n = 1$ faraday and (b) $n = 2$ faradays?

17.58 For the reaction $V^{2+} + Md^{3+} \rightleftharpoons V^{3+} + Md^{2+}$, the equilibrium constant is approximately 15 at 25°C. Calculate $\mathscr{E}^0$ for $Md^{3+} + e^- \rightleftharpoons Md^{2+}$ at 25°C. (Mendelevium was the first actinide element found to exist as a stable 2+ ion in aqueous solution.)

17.59 The solubility product of CuS at 25°C is 1.3×10^{-36}. (a) Calculate $\mathscr{E}^0$ corresponding to the reaction $Cu^{2+} + S^{2-} \longrightarrow CuS(c)$, with $n = 2$ faradays. (b) What is $\mathscr{E}^0$ for the partial reaction $CuS(c) + 2e^- \longrightarrow Cu(c) + S^{2-}$?

17.60 The reaction

$$5I^- + IO_3^- + 6H^+ \rightleftharpoons 3I_2(c) + 3H_2O$$

can go either way, depending on the pH. Which direction is spontaneous when $[I^-] = [IO_3^-] = 1.0$ mol/L and the pH is (a) 2.0; (b) 11.0? (c) At what pH is this reaction at equilibrium when $[I^-] = [IO_3^-] = 1.0$ mol/L?

17.61 (a) Will the permanganate ion, MnO_4^-, liberate O_2 from water in the presence of acid (H^+)? The reaction (if any) is

$$4MnO_4^- + 12H^+ \longrightarrow 4Mn^{2+} + 5O_2 + 6H_2O$$

Assume that all three ions in the equation are 1 M and that O_2 is at 1 atm. (b) What is the maximum pH at which this reaction will occur? The other concentrations and pressure are the same as in part a.

17.62 Derive equation 20, $\mathscr{E} = (RT/n\mathscr{F}) \ln (K/Q)$, by substituting $\mathscr{E}^0$ from equation 17 into the Nernst equation.

DECOMPOSITION POTENTIAL

17.63 Find the minimum emf's required to bring about these reactions at 25°C. Assume the overvoltages are zero.
(a) $Ni^{2+}(1.0\ M) + 2Br^-\ (2.0\ M) \rightarrow Ni(c) + Br_2(\ell)$
(b) $Ni^{2+}(0.010\ M) + 2Br^-(0.020\ M) \rightarrow Ni(c) + Br_2(\ell)$

17.64 Find the minimum emf necessary to bring about the reaction

$$Cu(c) + 2H^+(1.0\ M) \longrightarrow Cu^{2+}(1.0\ M) + H_2(g,\ 1.0\ atm)$$

at 25°C, assuming that (a) the overvoltage is 0; (b) the overvoltage for H_2 evolution is 0.60 V.

PRACTICAL CELLS

17.65 Why is the dry cell so designed that Zn and MnO_2 do not come into contact? What reaction might occur if they were in contact? How would this reaction affect the usefulness of the cell?

17.66 In a basic solution, Zn^{2+} is converted to $Zn(OH)_4^{2-}$. Write partial and net equations for the reactions in an alkaline dry cell.

17.67 People sometimes try to recharge dry cells, with limited success. (a) What reaction would you expect at the cathode of a Leclanché cell (which was the anode during discharge)? (b) What problems would this reaction cause? (c) Would recharging an alkaline dry cell be more, or less, successful than recharging a Leclanché cell? Why?

17.68 The total charge that can be delivered by a 6-inch dry cell before its emf drops too low is usually about 35 amp-hours. (One amp-hour is the charge that passes through a circuit when 1 A flows for 1 h.) Find the mass of Zn consumed when 35 amp-hours of charge are drawn from the cell.

17.69 A galvanic cell must be so constructed that not all the reactants can come into contact and react directly. (a) How is this rule violated in the design of the lead-acid storage cell? (b) What reaction, resulting from this violation, helps to account for the fact that a charged storage cell slowly discharges itself when it stands idle?

17.70 (a) What is $\mathscr{E}^0$ for a lead-acid storage cell? Assume that the net reaction is $Pb + PbO_2 + 2H^+ + 2HSO_4^- \longrightarrow 2PbSO_4 + 2H_2O$. Make the approximation that $\mathscr{E}^0$ is the same for $PbSO_4 + 2e^- \longrightarrow Pb + SO_4^{2-}$ and for $PbSO_4 + H^+ + 2e^- \longrightarrow Pb + HSO_4^-$. (b) Assume that a fully charged cell contains a 36.0% H_2SO_4 solution with density 1.27 g/mL. Use the Nernst equation to calculate the emf of this cell at 25°C. (c) Calculate the emf when the solution is 5.0% H_2SO_4 with density 1.03 g/mL.

17.71 A proposed storage cell for electric automobiles is the sodium-sulfur cell:

$$Na(\ell)|Na_2Al_{22}O_{34}(c)|Na_2S_4(c)|S_8(\ell),\ C(graphite)$$

At 300 to 400°C, Na^+ ions can move through the solid electrolyte. "Na_2S_4" is actually a mixture of polysulfides

(Na_2S_y); they are semiconductors and thus do not obstruct the current. The liquid sulfur is dispersed in graphite felt. Write the partial equations and the net equation for the discharge process in this cell.

17.72 An experimental storage cell is based on the formation of an "intercalation compound," in which Li atoms are inserted between the layers of atoms in TiS_2:

$$y\text{Li(c)} + \text{TiS}_2\text{(c)} \underset{\text{discharge}}{\overset{\text{charge}}{\rightleftharpoons}} \text{Li}_y\text{TiS}_2\text{(c)}$$

where y is between 0 and 1. TiS_2 is an electronic conductor. The Li atoms are probably present in the compound as Li^+ ions and mobile electrons. The electrolyte is $LiClO_4$ dissolved in a mixture of ethers (section 24.12). Write a description of this cell, in conventional notation, in (a) its fully charged state and (b) its fully discharged state. (c) Na is much cheaper than Li, but cells using Na have not been promising. Suggest a reason for the difference, considering the structure of the intercalation compound.

17.73 Silver-zinc cells are commercially available. Their structure is $\text{Zn(c)}|\text{KOH(aq)}|\text{Ag}_2\text{O(c)}|\text{Ag(c)}$. Zn is oxidize to $Zn(OH)_4{}^{2-}$. Write the partial reactions in such a cell.

17.74 Solid polymer electrolytes have been developed for use in large-scale electrolysis of water. They are sheets of ion-exchange resin (section 12.18) about 0.3 mm thick, containing mobile H^+ ions and fixed $—SO_3{}^-$ groups. (a) Assume a current density of 5.0×10^3 A/m^2. How many kilograms of H_2 can be produced per hour per square meter of electrolyte? (b) What advantages does this method have over electrolysis in an ordinary liquid solution?

CORROSION

17.75 Which of these is the best way to store partly used steel wool? Which is most conducive to rusting? Explain. (a) Keep it immersed in plain tap water. (b) Keep it immersed in soapy water (a basic solution). (c) Just leave it in the sink.

17.76 Amines are related to ammonia and are therefore basic. Which would you recommend as an antifreeze additive for preventing corrosion: acetic acid (CH_3COOH), ethanolamine ($HOCH_2CH_2NH_2$), or NaCl?

17.77 Would you recommend plating Fe, to protect it from corrosion, with any of these metals: Au, Cr, Ba, Pb? Why or why not? (Do not consider the cost of the metals in making your decision.)

17.78 A magnesium bar weighing 5.0 kg is attached to a buried iron pipe to protect the pipe from corrosion. An average current of 0.030 A flows between the bar and the pipe. (a) What reaction occurs at the surface of the bar; of the pipe? In which direction do electrons flow? (b) How many years will it take for the Mg bar to be entirely consumed (one year = 3.16×10^7 sec)? (c) What reaction(s) will occur if the bar is not replaced after the time calculated in part b?

SELF-TEST

17.79 (20 points) An aqueous solution of $CuSO_4$ is electrolyzed between inert electrodes for 60.0 min. The cathode reaction is the deposition of 0.685 g Cu. The anode reaction is the formation of O_2. How many (a) moles, (b) grams, and (c) liters (STP) of O_2 are evolved at the anode? (d) How many coulombs pass through the solution? (e) What is the average current in amperes?

17.80 (21 points) For cells 1, 2, and 3, write (a) the partial equations for the anode and cathode processes, assuming that the left electrode is the anode, and (b) the equation for the cell reaction. (c) Find the standard emf $\mathscr{E}^0$ at 25°C. (d) Does the cell reaction actually occur as you wrote it in part b or in the reverse direction when $\mathscr{E} = \mathscr{E}^0$?

(1) $\text{Zn}|\text{Zn}^{2+}||\text{Ni}^{2+}|\text{Ni}$

(2) $\text{Al}|\text{AlCl}_3\text{(aq)}||\text{ZnSO}_4\text{(aq)}|\text{Zn}$

(3) $\text{Cd}|\text{CdSO}_4\text{(aq)}||\text{FeSO}_4\text{(aq)}, \text{Fe}_2(\text{SO}_4)_3\text{(aq)}|\text{Pt}$

17.81 (14 points) The reaction occurring in the cell

$$\text{Cr}|\text{Cr}^{3+}(0.010\text{ M})||\text{Sn}^{2+}(0.20\text{ M})|\text{Sn}$$

is

$$2\text{Cr} + 3\text{Sn}^{2+} \longrightarrow 2\text{Cr}^{3+} + 3\text{Sn}$$

Calculate (a) $\mathscr{E}^0$ at 25°C and (b) $\mathscr{E}$ at 25°C.

17.82 (15 points) The cell

$$\text{Ag}|\text{AgCl}|\text{K}^+(c_1), \text{Cl}^-(c_2)||\text{Cu}^{2+}(c_3), \text{Cl}^-(c_4)|\text{Cu}$$

can be used to measure the concentration of Cu^{2+}. The partial equation for the anode reaction is $\text{Ag} + \text{Cl}^- \longrightarrow \text{AgCl} + e^-$. (a) Write the partial equation for the cathode reaction and the net equation for the cell process. (b) Write the Nernst equation in terms of any or all of the concentrations c_1, c_2, c_3, and c_4. (c) At 25°C, the emf of the cell is 0.0593 V when the Cu^{2+} concentration (c_3) is 0.0100 mol/L. What is c_3 when the emf is 0.0719 V? The other concentrations are kept constant.

17.83 (16 points) Calculate the equilibrium constant K_p for each reaction at 25°C:

(a) $4\text{Ag} + \text{O}_2\text{(g)} + 4\text{H}^+ \rightleftharpoons 4\text{Ag}^+ + 2\text{H}_2\text{O}$

(b) $4\text{Ag} + \text{O}_2\text{(g)} + 4\text{H}^+ + 4\text{Cl}^- \rightleftharpoons 4\text{AgCl(c)} + 2\text{H}_2\text{O}$

17.84 (14 points) Will O_2 oxidize Hg to Hg^{2+} (a) at pH 0; (b) at pH 7? The reaction would be

$$\text{O}_2\text{(g)} + 2\text{Hg}(\ell) + 4\text{H}^+ \longrightarrow 2\text{Hg}^{2+} + 2\text{H}_2\text{O}$$

Assume that $[\text{Hg}^{2+}] = 1.0$ mol/L and $P_{\text{O}_2} = 1.0$ atm.

18 CHEMICAL THERMODYNAMICS

Backward, turn backward, O Time, in your flight,
Make me a child again just for tonight!
Elizabeth Akers Allen (1860)

Faust, Ponce de León, and Elizabeth Allen were three among many who have pursued an ancient impossible dream: the dream of reversing time, and thus restoring lost youth. Why is the dream impossible? And what does it have to do with chemistry? A reacting mixture, left to itself, always goes toward equilibrium. If, having reached equilibrium, it spontaneously departed from it and changed to a nonequilibrium state, the effect would be the same as making time turn backward. A dead battery that recharges itself, a run-down clock that winds itself up, and a dead organism that returns to life are outside our experience. Thermodynamics is the branch of physics and chemistry that makes possible a genuine understanding of equilibrium. We will see in this chapter that a system can do work only as it progresses toward equilibrium, whereas work must be done *on it* to move it away from equilibrium. The universal tendency toward disorder imposes these requirements, limits our ability to extract work from the surroundings, and predicts a dim but warm future for the universe.

18.1 WHAT IS THERMODYNAMICS?

Thermodynamics (Greek *therme*, "heat"; *dynamis*, "power") is the study of heat, work, and the conversion of one to the other. It began in the early nineteenth century as the application of science to that revolutionary new source of work, the steam engine. By the late nineteenth century, it had become one of the sturdiest pillars of physics, chemistry, and engineering, and so it remains 100 years later.

In thermodynamics, as in every branch of science, some words are used in special and precise ways. A **system** is a portion of the universe that we select for study. It is most often the contents of a vessel, such as a beaker or a flask. When no matter enters or leaves the system, it is a **closed system**, the kind that will concern us in this chapter. The rest of the universe, excluding the system, is the **surroundings**. In practice, of course, we do not have to stretch our minds to consider the entire universe. Only some local surroundings, like a tank in which the system is immersed, need to be considered. In this chapter we will discuss several special kinds of system. Table 18.1 is intended for reference (and to be learned by the end of the chapter—but not yet).

Even when a system is closed, it can interact with its surroundings. One kind of interaction is **work** (section 4.1). If the surroundings exert a force, F, on the system and this force is exerted along a path of length, ℓ, the surroundings do work, w, on the system:

$$w = F\ell \tag{1}$$

The older convention for the sign of w was opposite to ours: w was positive when work was done *by* the system *on* the surroundings. When consulting other books, be certain to note which convention is used.

When work is done *by the surroundings on the system*, w is *positive*. When the system does work on the surroundings, w is negative. If the system is a wind-up toy (such as a mechanical duck that waddles across the table) or a storage battery, w is positive when the duck is being wound up or the battery is being charged, because the surroundings (your hand, a generator) are doing work on the system. When the duck unwinds, it can do work by climbing a slope. When the battery discharges, it does work by starting an engine. In these processes, w is negative, because the system does work on the surroundings.

You may think the pressure must change as the gas is compressed. You are right, but we assume a constant pressure for simplicity.

Systems studied in chemistry transfer work to or from the surroundings in two common ways. The first is **pressure-volume work**. Suppose that a system consists of a gas in a cylinder. Now the gas is compressed by moving a piston with a constant applied pressure, P. Figure 18.1 shows how much work is done on the system. The force exerted by the piston on the gas is expressed by the formula $F = PA$, where A is the area of the piston (from the definition of pressure, Appendix A.4). This force is exerted along a distance ℓ. The difference between the initial volume, $V_{initial}$, and the final volume, V_{final}, is the area, A, times the distance, ℓ, through which the piston moves:

Recall: ΔX = change in $X = X_{final} - X_{initial}$. Volume of cylinder = area of base × height.

$$V_{initial} - V_{final} = A\ell$$

However, the volume is decreasing, and its change, ΔV, is therefore negative:

$$\Delta V = V_{final} - V_{initial} = -A\ell$$

Then

$$w = F\ell = PA\ell$$

and since $A\ell = -\Delta V$,

$$\boldsymbol{w = -P\Delta V} \quad \textbf{(pressure-volume work)} \tag{2}$$

Don't be fooled by the minus sign in equation 2: ΔV is *negative* and w is *positive*, since work is done *on* the system.

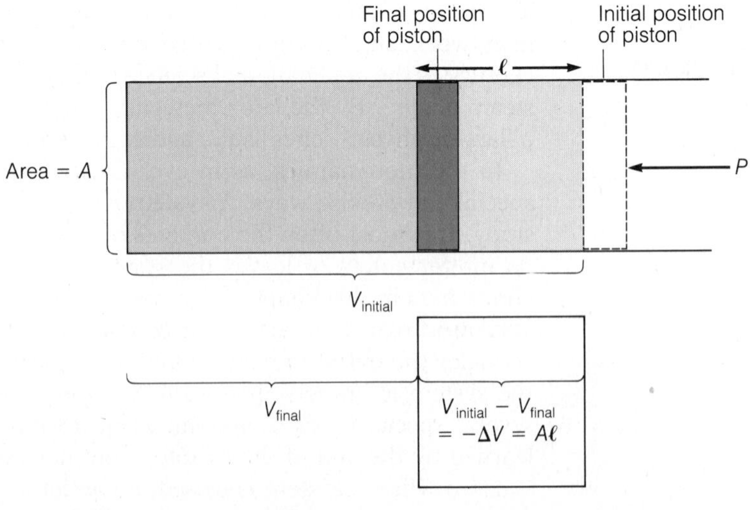

FIGURE 18.1
An example of pressure-volume work: a gas being compressed by an external pressure P.

VOLUME OF THE SYSTEM DECREASES.	VOLUME OF THE SYSTEM INCREASES.
ΔV is negative. Work is done *on* the system. w is positive.	ΔV is positive. Work is done *by* the system. w is negative.

Equation 2 gives the correct sign of w in either case.

EXAMPLE 18.1 A gas expands from 1.00 L to 1.50 L against a constant pressure of 2.00 atm. Calculate w in joules, with sign. What does work on what? See Appendix A.5 for unit conversions.

ANSWER

$$\Delta V = V_{\text{final}} - V_{\text{initial}}$$
$$= 1.50\ \text{L} - 1.00\ \text{L} = 0.50\ \text{L}$$
$$w = -P\Delta V$$

Since $P = 2.00$ atm,

$$w = -2.00\ \text{atm} \times 0.50\ \text{L} = -1.0\ \text{L atm}$$

The problem calls for an answer in joules. The conversion factor between J and L atm is needed:

$$w = -1.0\ \cancel{\text{L atm}} \times \frac{101.325\ \text{J}}{1\ \cancel{\text{L atm}}} = -1.0 \times 10^2\ \text{J}$$

The negative answer means that the system (the gas) does work on the surroundings. ■

The other common kind of work is electrical work (section 17.4). When a galvanic cell (the system) has emf $\mathscr{E}$ and it pushes a charge, Q, through the external circuit (the surroundings), the cell does work on the surroundings. The amount of work it does is the emf multiplied by the charge, $\mathscr{E}Q$. However, this work is done *by* the system *on* the surroundings and is therefore *negative*:

$$w = -\mathscr{E}Q \qquad (3)$$
$$\text{(J)} \qquad (\text{V} \times \text{C})$$

When the surroundings force a charge, Q, through an electrolytic cell (the system), the potential difference across the cell is *negative* (section 17.11), while the work is *positive* (surroundings do work on the system). Equation 3 still applies.

Another expression for property is "state function"; the property depends upon (is a function of) the state of the system.

The **state** of a system is specified by whatever information describes the system with enough detail for a particular purpose. For a typical chemical system, the state is adequately specified by the mass, chemical formulas, percentage composition, temperature, and pressure. Each of these is a **property** of the system. In thermodynamics the word *property* has a special flavor. A property of a system depends only on what the system is *now*, without our knowing anything about its history. A property is *any physical or chemical characteristic that can be measured or calculated from measurements.* Suppose that the system consists of 50.0 g H_2 at 25°C and 1.00 atm. Its mass (50.0 g), chemical composition (100% H_2), temperature (25°C), and pressure (1.00 atm) are properties. Its volume can also be measured, or it can be calculated by the ideal gas law (section 10.7):

We assume, as usual, that the gas is ideal.

$$V = \frac{nRT}{P} = \frac{50.0\ \cancel{\text{g}}}{2.016\ \cancel{\text{g}}/\cancel{\text{mol}}} \times \frac{(0.08206\ \text{L}\ \cancel{\text{atm}}/\cancel{\text{mol}}\ \cancel{\text{K}}) \times 298\ \cancel{\text{K}}}{1.00\ \cancel{\text{atm}}}$$
$$= 606\ \text{L}$$

The volume is therefore a property. However, if we ask "How was the gas prepared?" or "How much did the gas cost?", the answers are not properties because knowing the present state of the gas does not enable us to answer these questions.

Some properties have the peculiarity that only *changes* in them can be measured. Examples are energy and enthalpy (section 4.4). They are accepted as properties because the change is the same no matter how the system goes from one state to another—say, from state 1 to state 2. A **path** is a sequence of intermediate states, and it may be possible to go from state 1 to state 2 by many paths. Thus, a system can be heated from 40°C to 50°C, or it can be heated from 40°C to 100°C and then cooled to 50°C. The initial and final states, 40°C and 50°C, are the same for both of these paths; therefore, the change in any property of the system must be the same for both paths. Hess's law (section 4.5) says that the change in enthalpy (ΔH) does not depend on whether a reaction occurs in one step or several steps. That is another way of saying that *enthalpy is a property of the system.*

EXAMPLE 18.2 You are the system and your state is adequately defined by the number of the floor on which you stand. You go from the seventh floor to the eleventh floor. State whether each of the following quantities is a change in a property of the system: (a) the distance you walk, (b) the change in your altitude above the ground floor, (c) the amount of money you spend.

ANSWER

(a) Not a property. The distance walked depends on the path you choose; you might walk directly up from 7 to 11, or you might climb to 13, then down to 11, or . . . .

(b) A property. Whatever route you choose, the rises (+) and falls (−) add up to the same difference in altitude between the two floors.

(c) Not a property. If there is a soda machine on floor 13, you may buy a soda and thus spend more money when you go via that floor. ■

18.2 ENERGY AND HEAT

A process in an insulated system is *adiabatic* (Greek, "not going through").

Let us consider a kind of system called an **insulated system**. This system is closed (which means that no matter enters or leaves) and is heavily insulated so that no heat can flow in or out (Table 18.1). This system is, of course, an unattainable ideal—there is always some heat leakage—but that should not bother you; you are familiar by now with idealized systems, from the ideal gas to the reversible cell. An insulated system interacts with its surroundings only by doing work or having work done on it. All kinds of work are permitted: the volume of the system may be changed by moving a piston, the system may be penetrated by a shaft that winds a spring or turns a paddle wheel, and it may have wires through which the surroundings can do electrical work on it (or conversely). The fanciful contraption in Figure 18.2 is an insulated system equipped for all these kinds of work.

TABLE 18.1 KINDS OF SYSTEM

	OPEN	CLOSED[†]	INSULATED	ISOLATED
Can matter enter or leave?	yes	no	no	no
Can work be done on or by the system?	yes	yes	yes	no
Can heat flow into or out of the system?	yes	yes	no	no

[†] This column refers to a closed system that is *not* insulated or isolated. Insulated and isolated systems are necessarily closed.

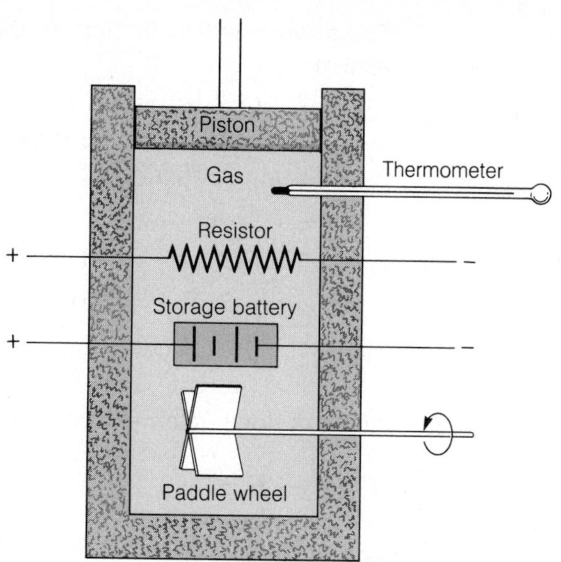

FIGURE 18.2
An insulated system equipped for exchanging work with the surroundings.

A refrigerator is an insulated system only as long as the door is not opened.

When we study an *insulated* system, we discover a remarkable fact When the system goes from state 1 to another state, 2, the *net* work done on it is always the same, regardless of the path. Various amounts of work of various kinds may be done along the way, but when the positive and negative *w*'s are added up, the total is always the same as long as the two states, initial and final, are the same.

EXAMPLE 18.3 Two states of the system in Figure 18.2 are defined as follows:

PROPERTY	STATE 1	STATE 2
Temperature	0°C	25°C
Volume	5.0 L	6.0 L
Density of battery acid	1.1 g/mL	1.3 g/mL

(a) The system goes from state 1 to state 2 by a three-step path: (1) The gas expands from 5.0 L to 6.0 L against a constant pressure of 1.5 atm, doing 152 J of work. (2) The storage battery is charged for 3.0 min; the work done on it is 450 J. (3) The battery is allowed to discharge for 3.0 min, doing 72 J of work. What is the net work done on the system?

(b) The system again goes from state 1 to state 2, but by a different path, also with three steps: (1) A current is passed through the resistor for 1.0 min, doing 36 J of work on the system. (2) The battery is charged for 10 min. (3) The gas expands, doing 101 J of work on the surroundings. How much work was done in charging the battery?

ANSWER

(a) The net work done on the system is the sum of the amounts of work done in the three steps. Watch the signs: for steps 1 and 3, w is negative because the system does work. In step 2, w is positive because work is done on the system. Then

$$\begin{aligned} w &= w_1 + w_2 + w_3 \\ &= -152\ \text{J} + 450\ \text{J} - 72\ \text{J} = +226\ \text{J} \end{aligned}$$

The positive w tells us that, in the process as a whole, work is done *on* the system.

(b) Since the initial and final states are the same as in (a) and the system is insulated, the net work must be the same as in (a). Let w_2 be the unknown work in step 2. Then

$$w = w_1 + w_2 + w_3$$
$$= +36\text{ J} + w_2 - 101\text{ J} = 226\text{ J, from (a)}$$
$$w_2 = 291\text{ J}$$

This is the work done in charging the battery. ■

The work done on an insulated system is the change in its *energy*, E:

$$\Delta E = E_2 - E_1 = w \qquad \textbf{(for an insulated system)} \tag{4}$$

There is no way to measure the absolute value (E) of the energy; only its change (ΔE) can be measured. The work, and thus the change in energy, depends only on the initial and final states—a necessary condition for energy to be called a "property."

Few systems are insulated; none is perfectly insulated. What is different about a noninsulated (but still closed) system? The answer is that it has another way, besides work, of gaining or losing energy. This way is the transfer of **heat** to or from the surroundings. As a result, for noninsulated systems, the net work depends on the path.

Originally, q stood for "quantity of heat."

The symbol for heat is q. The convention is that:

***q* is *positive* when the system *absorbs heat from the surroundings* (an endothermic process), and *q* is *negative* when the system *gives out heat* (an exothermic process).**

(As with work, whatever comes into the system is positive; whatever goes out is negative.) The energy change is thus the result of both heat and work:

$$\Delta E = q + w \tag{5}$$

Positive q (absorbed) and positive w (done on system) increase the energy. Negative q (given out) and negative w (done by system) decrease the energy. For an insulated system, $q = 0$ and equation 5 becomes $\Delta E = w$.

EXAMPLE 18.4 A system gains 35 J of heat, but it also does 65 J of work. Find ΔE.

ANSWER The addition of 35 J of heat *increases* the energy of the system by 35 J, but doing 65 J of work *decreases* the energy by 65 J. The net result is a loss of 30 J, and $\Delta E = -30$ J. Or we can write $q = +35$ J, $w = -65$ J. Substituting in equation 5,

$$\Delta E = q + w = +35\text{ J} + (-65\text{ J}) = -30\text{ J}$$ ■

EXAMPLE 18.5 A system evolves 50 J of heat, and 75 J of work is done on the system. Find ΔE.

ANSWER The loss of 50 J of heat *decreases* the energy of the system by 50 J, but 75 J of work done on the system *increases* the energy by 75 J. The net result is a gain of 25 J, and $\Delta E = +25$ J. Or we can write $q = -50$ J and $w = +75$ J. Substituting in equation 5,

$$\Delta E = q + w = -50 \text{ J} + 75 \text{ J} = +25 \text{ J}$$ ■

PROBLEM 18.1 (a) A system gives out 800 J of heat while 1200 J of work is done on it. What is ΔE? (b) The same system starts from the same initial state as in part a and ends at the same final state. However, it gives out 1000 J of heat. Calculate the work w, showing whether it is positive or negative. □

1 Btu = 1054 J

Joule measured work in foot-pounds, not in joules. 1 ft-lb = 1.356 J

The idea that heat and work are two ways of transferring the same thing—energy—was not generally accepted until the 1840s. Before then, heat was visualized as a fluid, called "caloric." It was therefore measured in units of its own, the *calorie* (section 4.3) and the *British thermal unit* (Btu), both of which are still in use. In a series of experiments extending from 1845 to 1878, James P. Joule studied the relationship between work and heat. His apparatus (Figure 18.3) consists of a paddle wheel that can be rotated in water in an insulated bucket. When operated in water, the moving blades do work on the water, increasing its temperature. Joule found that the rise in the water temperature is directly proportional to the amount of work done in turning the blades. The more the paddle wheel is rotated, the larger is the increase in the water temperature. He found that the same amount of work, 4.18 joule, *always* produces the same change in temperature as obtained from 1 calorie of heat. In other words, 4.18 joule of work has the same effect as 1 calorie of heat. Since Joule's time, heat and work have been recognized as energy, and we use the same unit, the joule (J), for all forms of energy. The calorie is now *defined* as exactly 4.184 joules.

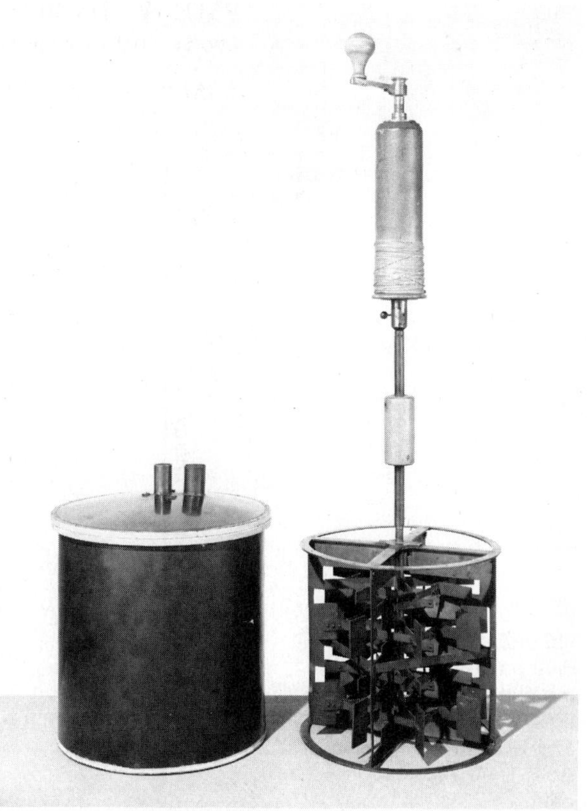

FIGURE 18.3
The insulated water bucket and the water-friction apparatus used by James Joule to raise the temperature of water by doing work on the water. Falling weights of known mass, connected to the handle by pulleys, were used to rotate the paddle wheel. (Science Museum, London)

You should now be convinced that ΔE is fixed when we fix the final and initial states of a system. However, this statement does not apply separately to the heat gained or lost by the system or to the work done on or by the system. These may have any value as long as ΔE is constant. For example, let us assign arbitrarily an energy value of 90 J to a particular quantity of a gas in an initial state, and suppose that its energy in the final state is 10 J more, or 100 J. This is the same as writing

$$\begin{aligned} \Delta E &= E_2 - E_1 \\ &= 100 \text{ J} - 90 \text{ J} = 10 \text{ J} \end{aligned}$$

This change may be accomplished by any of an infinite number of paths (methods), five of which are described here and illustrated in Figure 18.4:

PATH I Add 10 J of heat ($q = +10$ J) to the gas, *no work* being done on or by the gas ($w = 0$); then

$$\begin{aligned} \Delta E &= q + w \\ &= 10 \text{ J} + 0 \text{ J} = 10 \text{ J} \end{aligned}$$

PATH II Do 10 J of work *on* the gas ($w = +10$ J) *without permitting the evolution or absorption of heat* ($q = 0$); then

$$\Delta E = 0 \text{ J} + 10 \text{ J} = 10 \text{ J}$$

PATH III Add 35 J of heat ($q = +35$ J) and let the gas *do* 25 J of work ($w = -25$ J); then

$$\Delta E = 35 \text{ J} + (-25 \text{ J}) = 10 \text{ J}$$

PATH IV Do 40 J of work *on* the gas ($w = +40$ J) and let the gas *evolve* 30 J of heat ($q = -30$ J); then

$$\Delta E = -30 \text{ J} + 40 \text{ J} = 10 \text{ J}$$

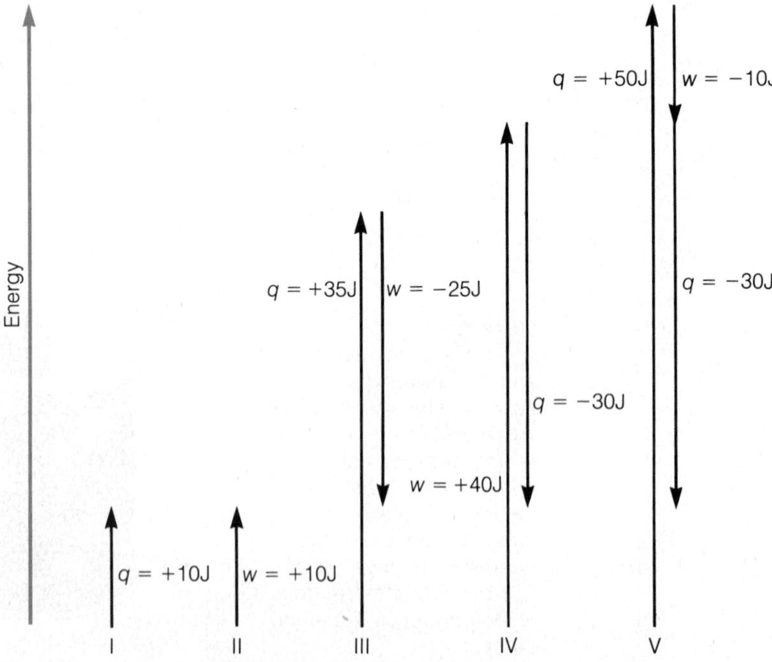

FIGURE 18.4
Five paths for changing the state of a system. Different amounts of heat and work are used, but $\Delta E = +10$ J for each path.

PATH V Add 50 J of heat ($q = +50$ J) and let the gas *do* 10 J of work ($w = -10$ J) and evolve 30 J of heat ($q = -30$ J); then

$$\Delta E = 50\text{ J} - 30\text{ J} - 10\text{ J} = 10\text{ J}$$

We can imagine many other paths by which this gas may pass from the initial to the final state. In words, we say that *ΔE is independent of the path between the initial and the final state, but q and w are path-dependent.* q and w are thus examples of quantities that are *not* changes in any property of the system. However, if either q or w is fixed at zero (like $q = 0$ for an insulated system), then the other becomes equal to a change in a property, ΔE—but only for those special systems in which q or w is always zero.

18.3 THE FIRST LAW OF THERMODYNAMICS

Thermodynamics starts with a few laws and derives from them a wealth of valuable conclusions. Its laws refer to the world as it is observed, without requiring any assumptions about its structure. If new evidence were to demolish the whole theory of atoms and molecules, the laws and results of thermodynamics would still stand.

The **first law of thermodynamics** is equation 5:

$$\Delta E = q + w$$

It says that any change in the energy of a closed system must be accounted for by heat, work, or some combination of them. Suppose that a system is *isolated* (Table 18.1): it has no interaction of any kind with its surroundings. Then $q = 0$, $w = 0$, and $\Delta E = 0$. *The energy of an isolated system is constant.* The first law, expressed in this form, is the **law of conservation of energy**.

An isolated system is an idealization because nothing can be perfectly insulated or perfectly rigid. However, there is one genuinely isolated system: *the entire universe.* If the universe is defined as all there is, it cannot interact with anything else because there is nothing else. Therefore, *the energy of the universe is constant.* Energy cannot be created or destroyed; it can only be transferred from one system to another, as from the Sun to the Earth or from a generating station to your light bulb. A persistent dream, along with the fountain of youth, is the *perpetual-motion machine.* In its most familiar form (the "first kind"), it would create energy out of nothing. Despite valiant efforts by generations of inventors, no one has ever made such a machine. This perennial failure is a fact of experience that has led to the acceptance of the first law.

A large proportion of chemical reactions occur at constant, usually atmospheric, pressure. For these reactions, the pressure-volume work is often considered unimportant and uninteresting: when w_{PV} is positive, the atmosphere does the work of compressing the system at no cost; when w_{PV} is negative, the system expands, pushing back the atmosphere and doing neither good nor harm. It is therefore often convenient to sweep pressure-volume work under the rug and forget it. Enthalpy (section 4.4) provides a way to do this.

The definition of enthalpy is

$$\boldsymbol{H = E + PV} \tag{6}$$

where P is the pressure and V is the volume of the system. As with energy, only changes in enthalpy (ΔH) can be measured. The enthalpy change in a process is

$$\Delta H = \Delta E + \Delta(PV) \tag{7}$$

$\Delta(PV)$ is the change in the *product*, PV, pressure multiplied by volume:

$$\Delta(PV) = P_2V_2 - P_1V_1$$

When P is constant ($P_1 = P_2 = P$),

$$\Delta(PV) = PV_2 - PV_1 = P(V_2 - V_1) = P\Delta V$$

Then, from equation 7,

$$\Delta H = \Delta E + P\Delta V$$

Equation 2 tells us that the pressure-volume work at constant pressure is

$$w_{PV} = -P\Delta V$$

and therefore

$$\Delta H = \Delta E - w_{PV}$$

Thus, for processes at constant pressure, ΔH may be described as the energy change, exclusive of pressure-volume work. Since $\Delta E = q + w$,

$$\Delta H = q + w - w_{PV} \tag{8}$$

When the only kinds of work are pressure-volume work and electrical work, $w = w_{el} + w_{PV}$, and equation 8 can be written

$$\Delta H = q + w_{el} \tag{9}$$

Equation 9 is a modified first law, with pressure-volume work subtracted from both sides. When the *only* work is pressure-volume work—no electrical work—equation 9 becomes the familiar equation from section 4.4:

$$\Delta H = q$$

Thus, in the case of constant presssure and no electrical work, ΔH is the heat absorbed by the system.

BOX 18.1

THE DIFFERENCE BETWEEN ΔE AND ΔH

From equation 7, $\Delta H - \Delta E = \Delta(PV)$. For a solid or liquid, $PV \approx 0$: the volume is so small that $\Delta(PV)$ can be neglected in comparison to ΔH and ΔE. Gases have much larger volumes. Suppose that there are n_1 moles of gaseous reactants and n_2 moles of gaseous products. The ideal gas law tells us that

$$(PV)_1 = n_1RT \quad \text{and} \quad (PV)_2 = n_2RT$$

Usually, we are concerned with a process whose initial and final temperatures are the same. Then

$$\begin{aligned}\Delta(PV) &= (PV)_2 - (PV)_1 \\ &= n_2RT - n_1RT \\ &= RT(n_2 - n_1) \\ &= RT\Delta n_{gas}\end{aligned}$$

where Δn_{gas} is the change in the number of moles of *gas*. Thus,

$$\Delta H - \Delta E = \Delta(PV) = RT\Delta n_{gas}$$

For the reaction

$$2H_2(g) + O_2(g) \longrightarrow 2H_2O(\ell)$$

$n_1 = 3$ mol, $n_2 = 0$ mol (the liquid does not count), and $\Delta n_{gas} = 0 - 3 = -3$ mol. Then, at 298 K,

$$\begin{aligned}\Delta H - \Delta E &= 8.314 \times 10^{-3}\ \text{kJ/(mol K)} \\ &\quad \times 298\ \text{K} \times (-3\ \text{mol}) = -7.43\ \text{kJ}\end{aligned}$$

ΔH for this reaction is -572 kJ; therefore,

$$\begin{aligned}\Delta E &= -572\ \text{kJ} + 7.43\ \text{kJ} \\ &= -565\ \text{kJ}\end{aligned}$$

ΔH and ΔE are not very different in this example. This is the usual result; $\Delta(PV)$ is a small correction. When no gases are involved, or the number of moles of gas does not change, $\Delta H = \Delta E$ within experimental error.

18.4 THE NATURE OF WORK AND HEAT

Thermodynamics does not depend on the atomic theory, but it does not contradict the theory. How can the distinction between work and heat be explained in terms of atoms and molecules? *Work is orderly, directed motion.* When you push an object, all its atoms move in the same direction. When an electric current flows through a metal, the net flow of electrons is in one direction. When a gas expands, its molecules move into or toward the newly available space. On the other hand, when *heat* is transferred, molecules (or atoms or electrons) in one system collide in a *random, disorderly* way with the molecules in another system.

A flow of heat occurs when there is a temperature difference between two systems in contact. The system at the higher temperature—hot gas produced in a flame, for example—loses energy, while the system at the lower temperature—your kettle of soup or beaker of water—gains energy. The amount of energy transferred is the amount of heat given out by the hotter system (negative q) and is also the amount of heat absorbed by the colder system (positive q). A system does not "have" heat any more than it "has" work. It has energy in various forms—the motion of its molecules (section 10.12), weak bonds that might be replaced by stronger bonds if a chemical reaction occurs (sections 4.7 and 7.8), a high position in a gravitational field (a shelf as opposed to the floor). Energy in any of these forms can be given out as heat or—with important limitations we will discuss later—as work.

The natural tendency of any system is toward disorder (section 11.14). An example of this tendency is that orderly motion—work—is easily changed into random, disorderly motion—heat. The reason is that there are many more ways for molecules to move randomly than there are for them to move in an orderly way like soldiers on parade. As a result, any effect that can be produced by heat can also be produced by work. All that is needed is a device for converting orderly motion to disorderly motion, such as an electric resistor or a paddle wheel. Suppose you want to raise the temperature of a beaker of water. The obvious way to do this is to introduce heat into it with a burner (Figure 18.5a). You could, in theory, also heat it by stirring it vigorously, perhaps in a blender (Figure 18.5b). The work done by the stirrer is quickly jumbled into random motion of water molecules. Thus, work is always a possible substitute for heat. The converse is *not true.* Some

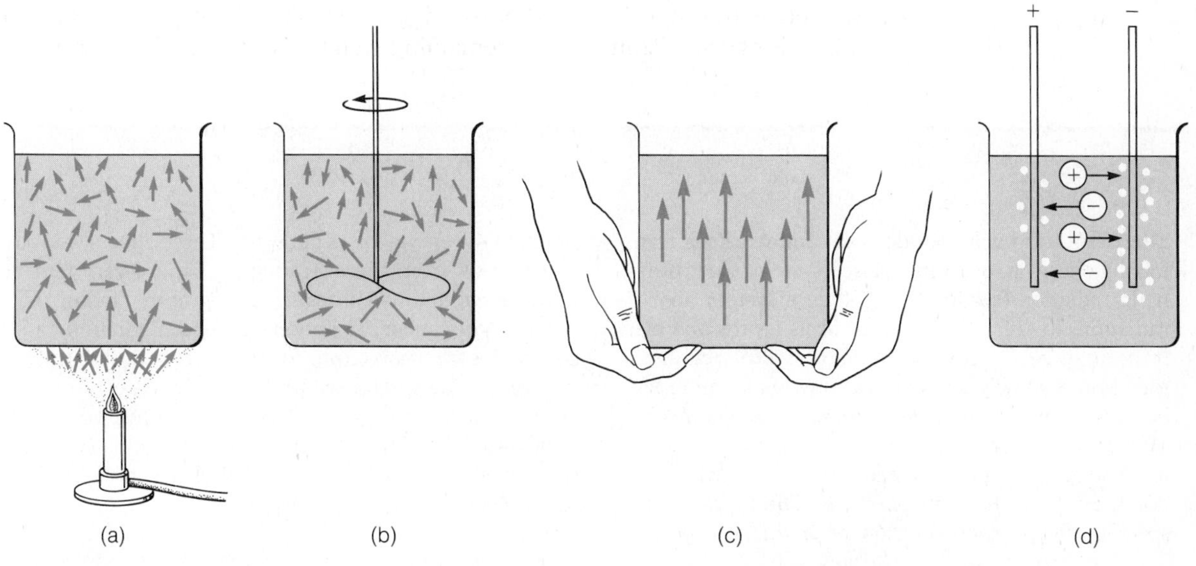

FIGURE 18.5
(a) A beaker of water absorbing heat. The energetic molecules in the flame collide randomly with the beaker, imparting kinetic energy to the atoms in the glass, which pass it on to the water molecules. (b) Work being done on water by stirring it. The orderly motion of the propeller produces random motion of the water molecules. (c) Another way to do work on a beaker of water. All the molecules are moving up, against the force of gravity. This motion is in addition to the random agitation that is always occurring. (d) Electrical work being done on a beaker of water that contains an ionic solute. All ions with the same charge move in the same direction.

effects can be produced only by work, not by heat, such as lifting an object (Figure 18.5c), decomposing water at room temperature (Figure 18.5d), compressing a spring, and charging a battery.

18.5 THE SECOND LAW OF THERMODYNAMICS

This is called a *perpetual-motion machine of the second kind.* The device that creates energy is of the *first* kind.

The first law of thermodynamics says that we cannot get something for nothing. There is no perpetual-motion machine that does work without an input of energy. That still leaves some intriguing possibilities, however. How about taking in air from the atmosphere—or water from the ocean—on one side of an engine, extracting heat from it, discharging it colder on the other side, and converting the extracted heat to work? Such a marvelous invention would make coal, oil, and nuclear power obsolete. It would be as good a perpetual-motion machine as the imaginary device that creates energy out of nothing. Unfortunately, this engine is impossible. The conclusion that it can never exist is the **second law of thermodynamics**.

A cyclic process :

To lead us to a formal statement of the second law, let us imagine a system that is kept at constant temperature. The system goes through a **cyclic process**: that is, a process in which the final state of the system is the same as the initial state. For this process, there is no change in energy: $\Delta E = q + w = 0$; $q = -w$. If the system were to absorb heat in the process (positive q), it would, by the first law, be obliged to do an equivalent amount of work (negative w). However, it would then be a perpetual-motion machine: when it returned to its initial state, it would be ready to repeat the cycle over and over. An **isothermal process** is a process in which the temperature of the system is constant. A statement of the second law is that:

> **A system cannot absorb heat and do work in a cyclic isothermal process.**

The second law will become clearer to you when we apply it to a specific system. For example, Figure 18.6a shows a cylinder of compressed gas fitted with a weighted piston. It is immersed in a tank of ice water, which remains at 0°C even as heat is added or removed. The gas is then allowed to expand when one of the weights is removed from the piston (Figure 18.6b). The expanding gas does work in raising the piston and the remaining weight. It obtains the energy to do

BOX 18.2

RADIATION AND HEAT

There is a third way, besides heat and work, of transferring energy to or from a closed system: electromagnetic radiation (section 5.1). When a system absorbs radiation, the resulting motions of its atoms and electrons may be quickly randomized. In this case, the radiation is merely a way of heating, as when food is heated in a microwave oven. Similarly, a hot body gives off radiation as it cools, and the effect may be no different from giving out heat in any other way. That is not the whole story, however. Radiation can cause specific effects, depending on its frequency (proportional to the energy of its photons). These effects would never result from mere heating. Some are very familiar: vision, photosynthesis, photography, fading of colors, skin cancer, radiation sickness. Radiation can sometimes be a substitute for work. This thought has inspired a widespread search for a way to decompose water by solar energy and thus produce hydrogen as a valuable fuel. This chapter does not emphasize the specific effects of radiation. It generally treats radiation as a form of heat.

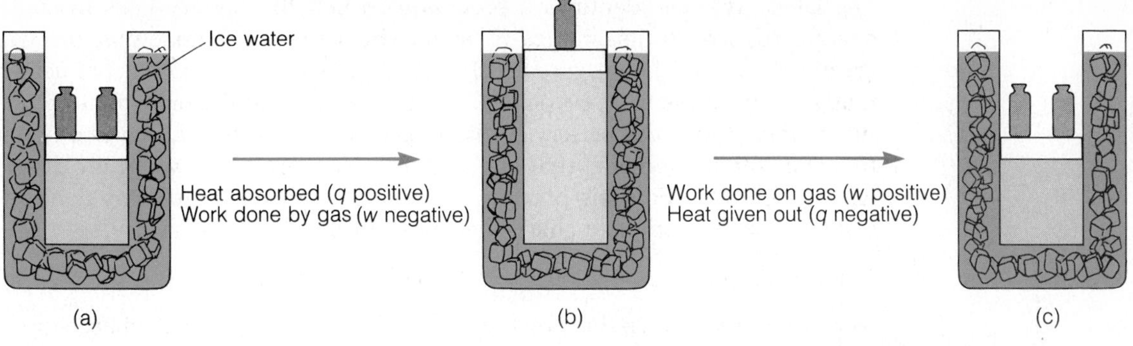

FIGURE 18.6
A gas undergoing a cyclic process. It is expanded at a constant temperature, 0°C, and then compressed again. (a) Before expansion. (b) After expansion. (c) After compression to its initial state.

this work by absorbing heat from the ice water, which causes some of the liquid water to freeze. So far, this looks good. Heat has been converted to work. This is just what a gasoline engine or a steam engine does. But a useful engine must do more than that; it must operate in repeating cycles. (After all, a gasoline "engine" whose piston moved only once would not do us much good.) Therefore, for the device of Figure 18.6 to operate as an engine, the piston must return from position b to position a, its original state (Figure 18.6c). The process must be *cyclic*—the initial and final states of the system must be the same. That's the catch; to put the piston and gas back into their initial state, we would have to do at least as much work in compressing the gas as the gas did for us in expanding. So we have failed to construct an engine that extracts heat from an ice bath by causing more water to freeze. This failure is so universal that it is accepted as a law of nature, the second law of thermodynamics. There is, however, no obstacle to doing work on the system and having it give out heat. Thus, the second law says that in a cyclic isothermal process, w must be zero or positive (work is done on the system), while q must be zero or negative (heat is given out by the system).

The second law is sometimes interpreted as the sheer cussedness of Mother Nature. A more scientific explanation is that it illustrates the universal tendency toward disorder (section 11.14). It is all too easy to convert orderly motion (work) to disorderly motion (heat). The reverse operation, converting disorder to order, is never easy and often impossible. The second law is a precise expression of this everyday experience.

The system that suffers from the second-law limitation need not, of course, be a cylinder of gas. Think of *any* system on which work can be done, which can then *do* work in return until it runs down, which can then have work done *on it* again, then *do* work, and so on, over and over, all at constant temperature. A good example is a storage battery (section 17.12). The battery has two states: fully charged (state c) and fully discharged (state d), as well as all the states in between. The voltage, and thus the amount of work, needed to charge the battery depends on the current; the larger the current, the more work is wasted because of resistance, overvoltage, concentration polarization, and so on (section 17.11). The least work is required when the charging process is thermodynamically reversible (or simply "reversible") (section 17.5). Then the current is zero and charging takes an infinitely long time. This process is, of course, an ideal limit that can be approached but never reached.

Suppose that, for a certain battery, the minimum work required to charge it is 4000 J:

$$w_{\mathrm{d}\rightarrow\mathrm{c},\,\mathrm{min}} = 4000\ \mathrm{J}$$

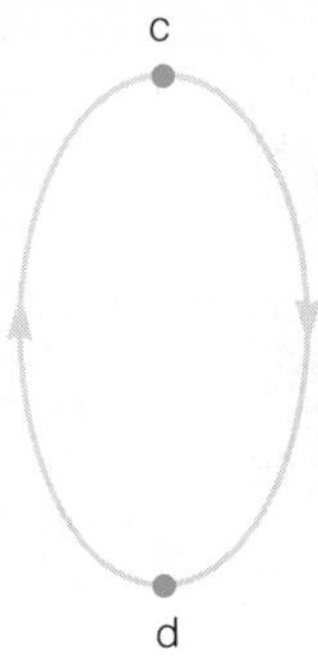

The label "d → c" identifies a process in which the battery goes from its discharged (d) state to its charged (c) state. The actual work done may be anything greater than 4000 J: $w_{d \to c}$ = 5000 J, 4200 J, 10,000 J, and so on. Let us assume that we come close to a reversible charging process and use only 4000 J to charge the battery. Next, the battery is discharged. How much work can be obtained? Imagine, for a moment, that the battery gives 4500 J of work on discharge: $w_{c \to d} = -4500$ J. For this process, w is *negative* because work is done *by* the system. For the complete charge-discharge cycle, the net work (w_n) is

$$w_n = w_{d \to c} + w_{c \to d} = 4000 \text{ J} + (-4500 \text{ J}) = -500 \text{ J}$$

Negative work is work done by the system. This battery, in each charge-discharge cycle, gives out 500 J of work. It can do this over and over. It is a perpetual-motion machine! It cannot exist. On awakening from our dream, we realize that the most work this battery can do on discharging is 4000 J. Since this is counted as negative work, the signs can be a bit troublesome:

$$w_{c \to d,\min} = -4000 \text{ J}$$

It is like two exothermic processes with $\Delta H_1 = -5$ kJ and $\Delta H_2 = -3$ kJ; more heat (5 kJ) is given out in the first process than in the second (3 kJ).

The actual value of $w_{c \to d}$ can be anything larger (less negative) than -4000 J. Thus, if only 3000 J is obtained on discharging, $w_{c \to d} = -3000$ J, which is larger than -4000 J. However, $w_{c \to d} = -4500$ J is impossible, for -4500 is *less* than -4000. *In a complete cycle at constant temperature, the net work cannot be negative*; it can only be zero or positive. This means that in a complete charge-discharge cycle, you must put in more work than you get back, or at best (in the ideal limit) break even. To charge a battery, you must put in at least as much work as you get back by discharging it. In the real world, "you can't break even"—a facetious, but not inaccurate, statement of the second law.

BOX 18.3

HEAT ENGINES

The second law says that a system cannot convert heat to work in a cyclic process *at constant temperature*. However, when heat reservoirs at two or more different temperatures are available, a system can absorb heat and do work in a cyclic process. Its temperature must change during the process. A system that goes through this kind of process is a **heat engine**. Assume, for simplicity, that the engine exchanges heat with its surroundings at only two temperatures. It accepts heat at a high temperature (T_h for hot) and discharges heat at a lower temperature (T_c for cold). What is the overall effect of this process? The engine absorbs heat from a hot reservoir (like a boiler), does work, and returns to its initial state (its state before it absorbed the heat). However, it also gives up heat at a lower temperature. Assume that in one cycle, it absorbs heat, q_h, at temperature T_h, gives up heat, q_c (a negative quantity), at T_c, and does work, w (also negative). The *efficiency* (e) of the engine is the ratio of useful output (work) to input (q_h):

$$e = \frac{\text{output}}{\text{input}} = \frac{-w}{q_h}$$

(Remember that $-w$ is positive; if 2000 J of work is done, $w = -2000$ J and $-w = 2000$ J.) It can be shown from the second law that the maximum possible efficiency of the engine is

$$e_{\max} = \frac{T_h - T_c}{T_h} \qquad (i)$$

Thus, the greater the temperature difference is, the more efficient the engine can be. When the temperature difference becomes 0, the efficiency becomes 0: no work can be obtained (the second law).

To visualize the meaning of equation *i*, imagine that there is a cold reservoir at absolute zero: $T_c = 0$ K. Then $e_{\max} = 1$. This imaginary engine could convert heat completely to work. Now suppose, more realistically, that $T_h = 800$ K and $T_c = 400$ K. In this case,

$$e_{\max} = (800 - 400)/800 = 0.50$$

When the heat "drops" halfway down to absolute zero, the maximum efficiency is $\frac{1}{2}$. Equation *i* gives the fraction of the "distance" from T_h to 0 K that the heat "falls"; this fraction is also the maximum fraction of the heat that can be converted to work.

PROBLEM 18.2 A system goes through a cyclic process (initial state = final state) at constant temperature. In the first half of the process, 500 J of work is done on the system. Which of the following are possible in the second half of the process, according to the second law? (a) $w = 400$ J; (b) $w = -600$ J; (c) $w = -300$ J. □

18.6 FREE ENERGY

When a quantity depends only on the initial and final states of a system, it can be identified as a change in some property of the system (section 18.1). The battery described in the preceding section points the way toward a new property. The work required to charge this battery by *any* reversible path at constant temperature is 4000 J. This is the *minimum* work required to charge the battery; $w_{min} = 4000$ J. Actual charging processes are not reversible and will require more work, like 4200 J or 5000 J. The work in excess of 4000 J, like 200 J or 1000 J, is wasted as heat. It is the 4000 J that does the battery some good and determines its final state—how fully charged it is.

The charged battery has 4000 J more of *something* than the discharged battery has. This "something" is a measure of the ideally available work—that is, the maximum work the charged battery could do in discharging. It has the *units* of energy, but it is not *equal to* the energy. It is another property, the **free energy**, represented by G. The word *free* is used in the sense of "free to do work." As with energy and enthalpy, only changes in free energy (ΔG) can be measured. *The change in free energy when a system passes from state* d (discharged) *to state* c (charged) *is equal to the minimum work that must be done on the system to take it from state* d *to state* c (all at constant temperature):

Also *Gibbs free energy, Gibbs energy, Gibbs function.* Josiah Willard Gibbs (1839–1903) was one of the founders of chemical thermodynamics.

$$\mathbf{\Delta G = G_c - G_d = \text{minimum work done on system}} \tag{10}$$

If work must be done *on* the system, then ΔG is *positive.* If the minimum work required to charge the battery is 4000 J, then

$$\Delta G = w_{min,\, d\rightarrow c} = +4000 \text{ J}$$

Battery is

G_c
↑ Minimum work
G_d required = 4000 J
$\Delta G = +4000$ J.

This makes sense. Since we are charging the system, its free energy increases and it becomes able to do more work. Conversely, if the system is being discharged, work is done *by* the system and this kind of work is *negative.* If the *maximum* work done by the system is 4000 J, then the *minimum* possible value

Plenty of free energy. Able to do work.

Plenty of energy, but only for warming bed. Not much free energy. No work done.

of w is

G_c Battery is being discharged. $\downarrow$ Maximum work done by battery = 4000 J G_d $\Delta G = -4000$ J.

$$w_{\text{min, c}\rightarrow\text{d}} = -4000 \text{ J}$$

ΔG is also negative, for the system is doing work, and its ability to do further work is decreasing:

$$\Delta G = G_d - G_c = -(\text{maximum work done by system}) = -4000 \text{ J}$$

EXAMPLE 18.6 In a reversible process from state 1 to state 2, a system does 30 kJ of work. The temperature is constant throughout. (a) What are w and ΔG for the process? (b) Give two other possible values of w for processes in which the system goes from state 1 to state 2. (c) If the system goes from state 2 to state 1 by a reversible process, what is w? (d) Give two other possible values of w when the system goes from state 2 to state 1.

Example of a real (irreversible) system doing work: $\Delta G = -4000$ J; G_c $\downarrow$ G_d; $w_d = -3500$ J. Less than maximum work obtained: only 3500 J.

ANSWER

(a) When the system does work, w is negative; $w = -30$ kJ. Since the process is reversible, this is also ΔG; $\Delta G = -30$ kJ.

(b) The work done *by* the system is a maximum in a reversible process. Therefore, w has its *most negative* (*minimum*) value for this process; $\Delta G = -30$ kJ. The actual w must be greater than -30 kJ; that is, less than 30 kJ of work is done. Possibilities are that 15 or 20 kJ of work is done by the system, corresponding to $w = -15$ kJ or -20 kJ (both greater than -30 kJ).

Example of work being done on a real system: $\Delta G = +4000$ J; G_c, G_d; $w_c = +4200$ J. More than minimum work needed: 4200 J.

(c) When the direction of the process is reversed, ΔG changes sign; $\Delta G = +30$ kJ. It is necessary to do at least 30 kJ of work on the system to take it from state 2 to state 1.

(d) Any amount of work greater than 30 kJ can be done on the system; $w = +35$ kJ, $+40$ kJ, and so on. ■

PROBLEM 18.3 For the process A $\longrightarrow$ B, $\Delta G = +5$ kJ. (a) Give two possible values of w for A $\longrightarrow$ B. (b) What is ΔG for B $\longrightarrow$ A? (c) Give two possible values of w for B $\longrightarrow$ A. □

When ΔG is negative, the *decrease* in free energy measures the *maximum work available* from the system when it passes "downhill" from one state to another at constant temperature. When ΔG is positive, the *increase* in free energy measures the *minimum work required* to take the system "uphill" from one state to the other. The work actually done by the system will always be *less* than the maximum; the work actually done *on* the system will always be *more* than the minimum.

We noted in section 18.3 that pressure-volume work is often brushed aside as unimportant. Actually, free energy is defined in accord with this preference. For a process at constant pressure,

$$\mathbf{\Delta G = w_{min} - w_{PV,min}} \tag{11}$$

ΔG is more closely related to ΔH (which also excludes w_{PV}) than to ΔE.

In the common situation where pressure-volume work and electrical work are the only kinds, ΔG is the minimum *electrical* work:

$$\mathbf{\Delta G = w_{el,min}} \tag{12}$$

If electrical work is done *by* the system, ΔG is negative.

The next section will show how free energy changes can be calculated from the reversible emf's of galvanic cells.

18.7 FREE ENERGY AND ELECTROMOTIVE FORCE

Recall the cell in Figure 17.6 (section 17.3). The partial reactions are

$$Zn \longrightarrow Zn^{2+} + 2e^-$$

$$Cu^{2+} + 2e^- \longrightarrow Cu$$

The superscript 0 (section 17.4) indicates 1 mol/L for solutes, 1 atm for gases, and the pure substance for solids and liquids.

so that 2 faradays (2 moles of electrons) pass through the circuit when 1 mole of Zn is consumed. For this cell, $\mathscr{E}^0_{cell} = 1.10$ V. When the cell reaction

$$Zn + Cu^{2+} \longrightarrow Zn^{2+} + Cu$$

occurs at unit concentrations (standard state), the maximum work available is

$$1.10\,\cancel{V} \times \frac{1\ J}{1\,\cancel{VC}} \times \frac{9.65 \times 10^4\,\cancel{C}}{1\ \cancel{faraday}} \times \frac{2\ \cancel{faradays}}{1\ mol\ Zn} = 2.12 \times 10^5 \frac{J}{mol\ Zn}$$

$$= 212 \frac{kJ}{mol\ Zn}$$

The positive emf tells us that the cell does work. Therefore, w is negative; it cannot be more negative than -212 kJ/mol Zn. Thus, $\Delta G = -212$ kJ. ΔG for a reaction is defined in a way similar to ΔH (section 4.4):

$$\begin{aligned}\Delta G &= G_{products} - G_{reactants}\\ &= G_{Zn^{2+}} + G_{Cu} - G_{Zn} - G_{Cu^{2+}}\\ &= -212\ kJ\end{aligned}$$

The negative ΔG signifies that the cell has done work, and its ability to do further work has declined. The decrease in free energy is not equal to the electrical work actually done when current is drawn from the cell, for the work done is always less than what *might* be done. It is the possibility, not the performance, that is expressed by the free energy change. The possibility is approached—never quite reached—in the unattainable limit of a thermodynamically reversible process.

As we have just seen, the reversible emf, $\mathscr{E}$, associated with a reaction is related to the free energy change. Let n be the number of faradays that pass through the circuit when the reaction occurs as written, and let $\mathscr{F}$ be the Faraday constant, 9.65×10^4 C/faraday. Then the free energy change is

$$\boldsymbol{\Delta G = -n\mathscr{F}\mathscr{E}} \qquad (13)$$

When the cell is producing current—doing work—$\mathscr{E}$ is positive, but ΔG is negative. That is why equation 13 contains a minus sign. $\mathscr{E}$ is in volts; ΔG is in joules (J) or kilojoules (kJ):

$$\begin{aligned}\Delta G = \quad & -n \times \mathscr{F} \times \mathscr{E}\\ & \cancel{faradays} \times \frac{\cancel{C}}{\cancel{faraday}} \times \cancel{V} \times \frac{1\ J}{1\,\cancel{CV}}\\ = \quad & -9.65 \times 10^4\ n\mathscr{E}\ J\end{aligned}$$

or

$$\Delta G = -9.65 \times 10^4\ n\mathscr{E}\,\cancel{J} \times \frac{1\ kJ}{10^3\,\cancel{J}} = -96.5\ n\mathscr{E}\ kJ$$

Many tables are still in kilocalories: 1 kcal = 4.184 kJ.

When all the coefficients in the chemical equation are multiplied by a constant, $\mathscr{E}$ is unaffected (section 17.4), but ΔG (like ΔH) is multiplied by the constant.

Compare the equations

$$Zn + Cu^{2+} \longrightarrow Zn^{2+} + Cu \qquad n = 2,\ \Delta G = -2\mathscr{F}\mathscr{E} = -212\ \text{kJ}$$

$$2Zn + 2Cu^{2+} \longrightarrow 2Zn^{2+} + 2Cu \qquad n = 4,\ \Delta G = -4\mathscr{F}\mathscr{E} = -424\ \text{kJ}$$

$$\tfrac{1}{2}Zn + \tfrac{1}{2}Cu^{2+} \longrightarrow \tfrac{1}{2}Zn^{2+} + \tfrac{1}{2}Cu \qquad n = 1,\ \Delta G = -\mathscr{F}\mathscr{E} = -106\ \text{kJ}$$

Think of two dry cells, a small one (size AA, 15 g) and a large one (size D, 135 g). They have the same emf, 1.5 V, when very little current is drawn. As long as the emf's are the same, the two cells will make a bulb glow equally brightly. However, the large cell will last longer because it contains 9 (135 divided by 15) times as much zinc and the other chemicals consumed in the reaction. In its longer lifetime, before the emf drops too low, it will do 9 times as much work: $w = -\mathscr{E}Q = -\mathscr{E}It$, and t is 9 times greater. If you want a higher voltage, what you need is not a larger cell, but several of them in series, making a battery. When galvanic cells are in series, their emf's are added. Nine-volt batteries (6 cells) are common.

Recall that Q = charge, I = current, t = time, and $Q = It$.

When all concentrations are 1 mol/L (pressures, 1 atm), $\mathscr{E} = \mathscr{E}^0$. The corresponding free energy change is known as the **standard free energy change**, ΔG^0:

$$\boldsymbol{\Delta G^0 = -n\mathscr{F}\mathscr{E}^0} \qquad (14)$$

The $\mathscr{E}^0$'s needed in equation 14 are calculated from the standard half-cell potentials in Table 17.1 (section 17.7).

EXAMPLE 18.7 Calculate ΔG^0 in kilojoules at 25°C for the reaction

$$Cu^{2+}(aq) + Ni(c) \longrightarrow Cu(c) + Ni^{2+}(aq)$$

Use the data in Table 17.1:

ANSWER We write the two partial reactions, look up $\mathscr{E}^0$ for each, and proceed as in section 17.7:

$$\begin{array}{rl} Cu^{2+} + 2e^- \longrightarrow Cu & \mathscr{E}^0_{cathode} = +0.3419 \\ Ni \longrightarrow Ni^{2+} + 2e^- & -\mathscr{E}^0_{anode} = -(-0.257) \\ \hline Cu^{2+} + Ni \longrightarrow Cu + Ni^{2+} & \mathscr{E}^0 = 0.3419 - (-0.257) = 0.599\ \text{V} \end{array}$$

When the reaction occurs as written, 2 faradays pass through the circuit; $n = 2$. Then

$$\begin{aligned} \Delta G^0 &= -96.5\ n\mathscr{E}^0\ \text{kJ} \\ &= -96.5 \times 2 \times 0.599\ \text{kJ} \\ &= -116\ \text{kJ} \end{aligned}$$

■

PROBLEM 18.4 Calculate ΔG^0 in kJ at 25°C for the reaction

$$Cr^{3+} + 3Ag(c) + 3Cl^- \longrightarrow Cr(c) + 3AgCl(c)$$

Use the data in Table 17.1. □

18.8 FREE ENERGY OF FORMATION

ΔG^0 for a reaction can be calculated from ΔG^0's for other reactions by Hess's law, just as we did with ΔH^0's in section 4.5. The data are commonly recorded as *standard free energies of formation* (ΔG^0_f), which are given in Appendix C

The superscript 0 does not refer to a specific temperature; ΔG^0 and ΔG_f^0 can be for a reaction at any constant temperature. The most common tabulations, like Appendix C, are for 25°C.

along with ΔH_f^0's at 25°C. ΔG_f^0 is ΔG^0 for the reaction in which *1 mole* of the substance is formed *from its elements* in their stable forms. For an element in the form that is stable at 1 atm and the given temperature, $\Delta G_f^0 = 0$. That simply says there is no free energy change in the reaction (or non-reaction) of forming the element from itself. However, ΔG_f^0 is not zero for an unstable form of an element. Thus, $\Delta G_f^0 = 163.2$ kJ/mol for $O_3(g)$, corresponding to the reaction $\frac{3}{2}O_2(g) \longrightarrow O_3(g)$. The most accurate ΔG_f^0's have been calculated, directly or indirectly, from the emf's of galvanic cells.

EXAMPLE 18.8 Calculate ΔG_f^0 in kilojoules per mole for $H_2O(\ell)$ at 25°C from the data in Table 17.1.

ANSWER The standard free energy of formation of $H_2O(\ell)$ is ΔG^0 for the reaction

$$H_2(g) + \tfrac{1}{2}O_2(g) \longrightarrow H_2O(\ell)$$

This is the reaction in which 1 mole of $H_2O(\ell)$ is formed from the elements in their stable forms, gaseous H_2 and O_2. The reaction can be resolved into the partial reactions

$$\begin{array}{rcl} 2H^+ + \tfrac{1}{2}O_2(g) + 2e^- & \longrightarrow & H_2O(\ell) \\ H_2(g) & \longrightarrow & 2H^+ + 2e^- \\ \hline H_2(g) + \tfrac{1}{2}O_2(g) & \longrightarrow & H_2O(\ell) \end{array} \qquad \begin{array}{rl} \mathscr{E}^0_{\text{cathode}} = & +1.229 \\ \mathscr{E}^0_{\text{anode}} = & 0 \\ \hline \mathscr{E}^0 = & +1.229\text{ V} \end{array}$$

$$\begin{aligned} \Delta G_f^0 &= -96.5\, n\mathscr{E}^0 \text{ kJ/mol} \\ &= -96.5 \times 2 \times 1.229 \text{ kJ/mol} \\ &= -237 \text{ kJ/mol} \end{aligned}$$

■

PROBLEM 18.5 Using data in Table 17.1, calculate ΔG_f^0 at 25°C for AgCl(c). The formation reaction is

$$Ag(c) + \tfrac{1}{2}Cl_2(g) \longrightarrow AgCl(c)$$

□

A table of ΔG_f^0's (Appendix C) can be used to calculate ΔG^0 for any reaction involving substances in the table. The procedure is the same as for ΔH^0 (section 4.6). Add up the ΔG_f^0's for the products, taking account of the coefficients in the equation. Then total the ΔG_f^0's for the reactants, and subtract the reactant sum from the product sum:

$$\Delta G^0 = \text{total } \Delta G^0_{\text{f,products}} - \text{total } \Delta G^0_{\text{f,reactants}}$$

Remember that $\Delta G_f^0 = 0$ for an element in its stable form.

EXAMPLE 18.9 Using data in Appendix C, calculate ΔG^0 at 25°C for the reaction

$$2Al(c) + 3BeF_2(c) \longrightarrow 3Be(c) + 2AlF_3(c)$$

ANSWER For the elements Al and Be, $\Delta G_f^0 = 0$. Appendix C gives $\Delta G_f^0 = -1425.0$ kJ/mol for AlF_3 and -979.4 kJ/mol for BeF_2. Then

$$\begin{aligned} \text{total } \Delta G^0_{\text{f,products}} &= 3\Delta G^0_{\text{f,Be}} + 2\Delta G^0_{\text{f,AlF}_3} \\ &= 3 \times 0 \quad + 2 \times (-1425.0) \\ &= -2850.0 \text{ kJ} \end{aligned}$$

$$\begin{aligned} \text{total } \Delta G^0_{\text{f,reactants}} &= 2\Delta G^0_{\text{f,Al}} + 3\Delta G^0_{\text{f,BeF}_2} \\ &= 2 \times 0 \quad + 3 \times (-979.4) \\ &= -2938.2 \text{ kJ} \end{aligned}$$

BOX 18.4

As mentioned in section 10.5, the standard pressure is now officially 1 bar = 0.98692 atm. Most changes in ΔG^0 are insignificant. If necessary, they can be calculated by the equation

$$\Delta G^{0(\text{bar})} = \Delta G^{0(\text{atm})} + RT(\ln 0.98692)\,\Delta n_{\text{gas}}$$

where Δn_{gas} is the change in the number of moles of gas when the reaction occurs as written. At 25°C,

$$\Delta G^{0(\text{bar})} = \Delta G^{0(\text{atm})} - 0.03263\,\Delta n_{\text{gas}}\ \text{kJ}$$

Then

$$\Delta G^0 = -2850.0\ \text{kJ} - (-2938.2\ \text{kJ}) = 88.2\ \text{kJ}$$ ■

PROBLEM 18.6 Calculate ΔG^0 at 25°C for the reaction

$$2CCl_4(\ell) + O_2(g) \longrightarrow 2CO(g) + 4Cl_2(g)$$

Refer to Appendix C. □

18.9 FREE ENERGY AND EQUILIBRIUM

Specific radiation effects like photosynthesis are excluded.

At the end of Chapter 13, we noted that a state of equilibrium is a kind of death. Less figuratively, this state is characterized by the inability to do work. Indeed, a system will move away from equilibrium only if work is done on it. To make this statement more precise, we will limit our discussion of equilibrium to states in which the system is closed (no matter enters it or leaves it) and is at constant temperature and pressure. *A closed system at constant temperature and pressure is in equilibrium when it cannot undergo a change unless work* (other than pressure-volume work) *is done on it.* For ordinary chemical systems, the work that must be done on it is *electrical* work. An equilibrium can be shifted by adding or removing matter—but then the system is not closed. It can be shifted by changing the temperature or pressure—but temperature and pressure are constant. When a process takes a system away from equilibrium, toward a less stable state, the electrical work required is positive and ΔG is positive. Therefore, a process with *positive* ΔG represents movement *away* from equilibrium, from stability to instability. Conversely, if ΔG is negative, then the process can occur with no input of work; indeed, electrical work can be done *by* the system if it is set up properly. Therefore, a process with *negative* ΔG can occur spontaneously and represents movement *toward* equilibrium, from instability to stability. The changes in free energy are as follows:

EQUILIBRIUM ⟶ DISEQUILIBRIUM	DISEQUILIBRIUM ⟶ EQUILIBRIUM
More stable state ⟶ less stable state	Less stable state ⟶ more stable state
Electrical work must be done on system.	System can do electrical work.
ΔG is positive.	ΔG is negative.
G increases.	G decreases.

The state of equilibrium at constant temperature and pressure is the state of minimum free energy.

A process with negative ΔG "should" happen, for it represents movement toward greater stability. Such a process is commonly described as "spontaneous," but this word can be misleading, for the process may not actually occur at a measurable rate. The reaction

$$\underset{\text{less stable}}{H_2(g) + \tfrac{1}{2}O_2(g)} \longrightarrow \underset{\text{more stable}}{H_2O(\ell)}$$

The system is kinetically stable (section 13.2).

has $\Delta G^0 = -237$ kJ at 25°C and is therefore "spontaneous," but the reaction does not occur at room temperature in the absence of a catalyst. The reverse reaction,

$$\underset{\text{more stable}}{H_2O(\ell)} \longrightarrow \underset{\text{less stable}}{H_2(g) + \tfrac{1}{2}O_2(g)}$$

has $\Delta G^0 = +237$ kJ, which means that there is no hope of decomposing water to its elements at 25°C and 1 atm unless work, in the amount of at least 237 kJ for every mole of H_2O, is done. This information about free energy does not guarantee that water can be decomposed at all. In fact, it can be, by electrolysis, and the electrical work required will be considerably more than 237 kJ/mol because of resistance, overvoltage, and the other inefficiencies inherent in any process.

The free energy change in a reaction is thus a measure of the driving force of the reaction. Actually, "drift" might be a better word than "drive," since molecules behave randomly, not purposefully. The only direction in which the net reaction can proceed spontaneously, at constant temperature and pressure, is the direction of lower free energy.

The ΔG^0 for a reaction is related to its equilibrium constant. Recall the equation (section 17.10)

$$\mathscr{E}^0 = \frac{RT}{n\mathscr{F}} \ln K$$

and the equation

$$\Delta G^0 = -n\mathscr{F}\mathscr{E}^0$$

Then

$$\Delta G^0 = -n\mathscr{F}\left(\frac{RT}{n\mathscr{F}}\right)\ln K$$

$$= -RT \ln K \qquad (15)$$

or

$$\mathbf{\Delta G^0 = -2.303RT \log K} \qquad (16)$$

Solving for log K gives

$$\log K = \frac{-\Delta G^0}{2.303RT} \qquad (17)$$

When ΔG^0 is in kJ, the value of R to use is 8.314×10^{-3} kJ/(mol K). At 25°C,

$$\log K = \frac{-\Delta G^0}{5.708 \text{ kJ}}$$

When gases are involved, K in these equations is K_p, not K_c.

BOX 18.5
HOW FREE ENERGY DEPENDS ON COMPOSITION

The graph shows how the free energy of a system reaches a minimum at equilibrium. The system is the same one used in section 13.1 to illustrate chemical equilibrium—a mixture of H_2, I_2, and HI in which the reaction

$$H_2(g) + I_2(g) \rightleftharpoons 2HI(g)$$

occurs at 445°C. Equal numbers of moles of H_2 and I_2 are put into the system, or pure HI is put in; in either case, the quantities are chosen so that the total pressure is 1 atm and the total number of moles is 1. The horizontal axis shows the mole percentage of HI in the mixture. The vertical axis gives the free energy of the mixture. The free energy reaches a minimum at 78% HI. This, then, is the composition of the equilibrium mixture, as shown also by the experimental data in Figure 13.1 (section 13.1).

The ΔG discussed in this chapter is actually the *slope* of a curve like that shown in the figure. This slope tells you which way the system has to change to approach equilibrium (down). When neither way is down—the slope is zero—equilibrium has been attained.

The free energy of a mixture of H_2, I_2, and HI at 445°C, plotted against the mole percent HI. The total pressure is 1 atm and the total number of moles is 1.

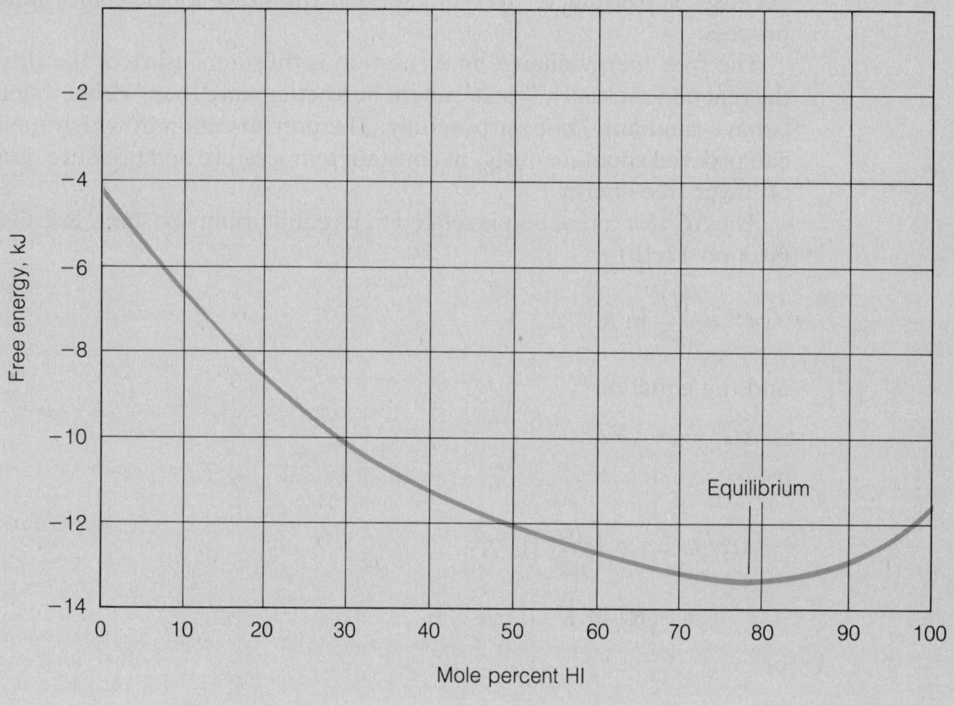

EXAMPLE 18.10 Use the data in Appendix C to calculate ΔG^0 and K_p at 25°C for the reaction

$$2HBr(g) + Cl_2(g) \rightleftharpoons 2HCl(g) + Br_2(\ell)$$

ANSWER

Note that $\Delta G_f^0 = 0$ for the elements $Br_2(\ell)$ and $Cl_2(g)$, which are the stable forms at 25°C and 1 atm.

$$\Delta G^0 = 2\Delta G^0_{f,HCl(g)} + \Delta G^0_{f,Br_2(\ell)} - (2\Delta G^0_{f,HBr(g)} + \Delta G^0_{f,Cl_2(g)})$$

$$= 2 \text{ mol} \times \left(-95.299 \frac{\text{kJ}}{\text{mol}}\right) + 0 - 2 \text{ mol} \times \left(-53.45 \frac{\text{kJ}}{\text{mol}}\right) - 0$$

$$= -83.70 \text{ kJ}$$

Then

$$\log K_p = \frac{-\Delta G^0}{2.303RT} = \frac{83.70 \text{ kJ}}{5.708 \text{ kJ}} = 14.66$$

$$K_p = 4.6 \times 10^{14}$$

As usual, the partial pressures are in atmospheres:

$$\frac{p_{HCl}{}^2}{p_{HBr}{}^2 p_{Cl_2}} = K_p = 4.6 \times 10^{14}$$ ■

PROBLEM 18.7 Using data in Appendix C, calculate ΔG^0 and K_p at 25°C for the reaction

$$NH_4Cl(c) \rightleftharpoons NH_3(g) + HCl(g)$$ □

When the pressures are not all 1 atm and the concentrations are not all 1 mol/L, ΔG is not equal to ΔG^0. The relationship between them is similar to that between $\mathscr{E}$ and $\mathscr{E}^0$ (section 17.8). The Nernst equation,

$$\mathscr{E} = \mathscr{E}^0 - \frac{RT}{n\mathscr{F}} \ln Q$$

can be combined with the equations $\Delta G = -n\mathscr{F}\mathscr{E}$ and $\Delta G^0 = -n\mathscr{F}\mathscr{E}^0$ (Problem 18.33) to give

$$\Delta G = \Delta G^0 + RT \ln Q \tag{18}$$

or, with common logarithms,

$$\Delta G = \Delta G^0 + 2.303RT \log Q \tag{19}$$

In gas reactions, Q is Q_p. At equilibrium, $\Delta G = 0$ and $Q = K$. These substitutions in equation 19 give equation 16.

EXAMPLE 18.11 For the reaction in Example 18.10,

$$2HBr(g) + Cl_2(g) \longrightarrow 2HCl(g) + Br_2(\ell)$$

calculate ΔG at 25°C when the partial pressures are 2.0×10^{-7} atm for HBr, 1.0×10^{-6} atm for Cl_2, and 0.10 atm for HCl.

ANSWER Q_p is calculated first:

$$Q_p = \frac{p_{HCl}{}^2}{p_{HBr}{}^2 p_{Cl_2}}$$

$$= \frac{0.10^2}{(2.0 \times 10^{-7})^2 \times 1.0 \times 10^{-6}}$$

$$= 2.5 \times 10^{17}$$

As we calculated in Example 18.10, $\Delta G^0 = -83.70$ kJ. Then equation 19 gives

$$\Delta G = \Delta G^0 + 2.303RT \log Q_p$$

$$= -83.70 \text{ kJ} + 2.303 \times 8.314 \times 10^{-3} \frac{\text{kJ}}{\text{mol K}} \times 298.15 \text{ K} \times \log(2.5 \times 10^{17})$$

$$= 15.62 \text{ kJ/mol } Cl_2 \text{ or } 15.62 \text{ kJ}$$ ■

BOX 18.6

In reading Example 18.11 you may have wondered how ΔG could be calculated "at" certain partial pressures. Wouldn't the pressures be expected to *change* when 2 moles of HBr and 1 mole of Cl_2 react? This reasonable, but troublesome, question can be answered "no" in several ways—most simply, by assuming that there is a very large tank of reacting mixture. Adding or subtracting a mere 1 or 2 moles, then, does not affect the partial pressures or concentrations.

The positive ΔG calculated in this example shows that the reverse (right-to-left) reaction is spontaneous. This reversal, in comparison with the negative ΔG^0, results from the very low partial pressures of the reactants.

PROBLEM 18.8 Calculate ΔG at 25°C for the reaction

$$2CCl_4(\ell) + O_2(g) \longrightarrow 2CO(g) + 4Cl_2(g)$$

when the partial pressures are 3.0×10^{-4} atm for O_2, 0.20 atm for CO, and 1.0×10^{-3} atm for Cl_2. Use the answer to Problem 18.6, $\Delta G^0 = -143.92$ kJ. □

Free energy changes are most accurately measured and most clearly understood through their relationship to emf's. However, the free energy concept is not dependent on galvanic cells. For processes that do not, or cannot, take place in a cell, free energy changes can still be calculated from other data. One way to get ΔG^0 is to analyze a mixture at equilibrium, calculate the equilibrium constant K from the measured concentrations or pressures, and calculate $\Delta G^0 = -2.303RT \log K$.

EXAMPLE 18.12 Calculate ΔG_f^0 for HI(g) at 350°C, given the following equilibrium pressures: $p_{H_2} = 0.132$ atm, $p_{I_2} = 0.295$ atm, and $p_{HI} = 1.61$ atm. At 350°C and 1 atm, the stable form of I_2 is the gas.

ANSWER The reaction

$$\tfrac{1}{2}H_2(g) + \tfrac{1}{2}I_2(g) \longrightarrow HI(g)$$

represents the formation of 1 mole of HI from the elements in their stable forms. For this reaction,

$$K_p = \frac{p_{HI}}{p_{H_2}{}^{1/2} p_{I_2}{}^{1/2}}$$

$$= \frac{1.61}{(0.132 \times 0.295)^{1/2}} = 8.16$$

$$\Delta G^0 = -2.303RT \log K_p$$

$$= -2.303 \times 8.314 \times 10^{-3} \frac{\text{kJ}}{\text{mol}\,\cancel{\text{K}}} \times (350 + 273)\cancel{\text{K}} \times \log 8.16$$

$$= -10.9 \text{ kJ/mol}$$

ΔG_f^0 for HI is ΔG^0 for the reaction. Therefore, $\Delta G_f^0 = -10.9$ kJ/mol. ■

PROBLEM 18.9 In $Br_2(g)$ at 1400 K and a total pressure of 1.00 atm, the partial pressure of Br atoms is 0.16 atm at equilibrium. Calculate K_p and ΔG^0 for the reaction

$$Br_2(g) \rightleftharpoons 2Br(g)$$

at 1400 K. □

18.10 ENTROPY AND HEAT

We introduced the concept of entropy in section 11.14 as a measure of the randomness, or disorder, of a system. The more disordered a system is, the greater is its entropy. Entropy is one of the most important, and one of the most subtle, concepts in physical science. The reason for its importance is that the universe tends to arrive at the most disorderly arrangement—the arrangement of greatest entropy—simply because there are many more ways to be disorderly than to be orderly.

How can we quantify such a nebulous idea as "randomness" or "disorder"? The concept of *heat* provides a clue. Heat is random, unorganized molecular motion (section 18.4). The way to increase the disorder of a system is to put energy into it by heating it. The symbol for entropy, for no good reason, is S. We expect the change in entropy, $\Delta S = S_{final} - S_{initial}$, to be related to heat; that is, when heat is absorbed (q is positive), S increases, and when heat is given out (q is negative), S decreases.

Something else in section 18.4 tells us to pause before concluding that we can calculate ΔS from q. There is always the option of putting in work instead of heat, or taking out heat instead of work. Entropy, like energy and enthalpy, is a *property* of a system. This means that ΔS depends on the initial state and the final state, not on the details of the process, such as whether energy is put in as work or as heat.

Let us consider specifically a system that absorbs heat and has work done on it. Two terms determine the change in the entropy of the system:

1. The heat actually put in (q).
2. Work that merely heats up the system, and thus could as well have been put in as heat. Rubbing, stirring, and passing a current through a resistor are typical examples. This kind of work may be called "frictional work" (w_{fric}).

From the viewpoint of the system, frictional work is equivalent to heat.

What is important in determining the final state of the system is the sum of the heat put in and the frictional work, which merely replaces heat:

$$q_{max} = q + w_{fric} \qquad (20)$$

In a reversible process, there is no frictional work; $w_{fric} = 0$. As much energy as possible is then put into the system in the form of heat rather than work; $q_{rev} = q_{max}$. Equation 20 also applies when heat is given out; q is then negative, and q_{max} is the *least negative* possible q.

The change in the disorder of the system depends not on q—any or all of which can be replaced by frictional work—but on q_{max}. For an isothermal (constant temperature) process (still the only kind considered), the relationship between the entropy change and the maximum heat is

$$\Delta S = \frac{q_{max}}{T}$$

or

$$\mathbf{q_{max} = T\Delta S} \qquad (21)$$

S is still a property when the temperature is not constant.

We will not attempt to explain why T (the absolute temperature) appears in equation 21. Just as with energy and enthalpy, we need to consider only *changes* in entropy. q_{max} is independent of the path at constant temperature, and therefore ΔS is also independent of the path, so that S is a property.

EXAMPLE 18.13 For the process of melting 1 mole of sodium,

$$Na(c) \longrightarrow Na(\ell)$$

at 98°C and 1.00 atm, $\Delta S = 7.11$ J/K. (a) Is the sign of ΔS consistent with the interpretation of entropy as randomness? (b) What is the maximum amount of heat absorbed in this process?

ANSWER

(a) The atoms in a crystal are arranged in an orderly pattern. This pattern is partly broken down in the liquid (section 11.3). It is therefore reasonable that the entropy increases (ΔS is positive) when a crystal melts.

(b) $q_{max} = T\Delta S = (98 + 273)\cancel{K} \times 7.11\ J/\cancel{K} = 2.64 \times 10^3\ J = 2.64\ kJ$ ■

EXAMPLE 18.14 (a) When one mole of mercury vapor (the system) condenses

$$Hg(g) \longrightarrow Hg(\ell)$$

at 336°C and 0.680 atm, the minimum heat given out is 57.3 kJ. What is q_{max}, with sign? Calculate ΔS. (b) Compare both the sign and the magnitude of ΔS with ΔS for the melting process in Example 18.13. How are they related to the disorder concept? (c) The condensation in part a occurs with 65 kJ of heat given out. Is this possible? (d) If the same change occurs with 50 kJ of heat given out, what is q? Is this process possible?

ANSWER

(a) $q_{max} = -57.3$ kJ. It is negative because heat is given out by the mercury.

$$\Delta S = \frac{q_{max}}{T} = \frac{-57.3\ kJ}{(336 + 273)K} = -0.0941\ kJ/K = -94.1\ J/K$$

(b) ΔS is negative, in keeping with the fact that a liquid is less disorderly than a gas. ΔS is much greater for vaporization of 1 mole of Hg than for melting of 1 mole of Na (94.1 versus 7.11 J/K). This suggests that only a little of the order in a crystal is lost when it melts; the liquid still has a high degree of order, which is lost when it vaporizes.

(c) Again, q is negative: $q = -65$ kJ. Since q is smaller (more negative) than q_{max}, the process is possible; -65 kJ is less than -57.3 kJ.

(d) $q = -50$ kJ, but -50 kJ is greater (less negative) than -57.3 kJ. Thus q is greater than q_{max}, which is impossible. ■

PROBLEM 18.10 For the process

$$HF(c) \longrightarrow HF(\ell)$$

at the melting point (−83.1°C), $\Delta S = 25.1$ J/(mol K). (a) What is q_{max}, with sign? (b) You hear a report that one mole of HF absorbs 5.5 kJ as it melts at −83.1°C. Is this process possible? (c) You are told that one mole of HF absorbs 4.0 kJ as it melts at −83.1°C. Is this process possible? □

Example 18.14 illustrates that there is a large entropy decrease when a gas condenses to a liquid or solid. Conversely, there is a large entropy increase when a liquid or solid becomes a gas. It is less obvious that the *sign* of ΔS in a *chemical* reaction can usually be predicted if the reaction produces or consumes gas. When

BOX 18.7
ABSOLUTE ENTROPY AND THE THIRD LAW

The phrases "absolute entropy" and "third-law entropy" are often used. Absolute entropy is based on the convention that $S = 0$ at absolute zero (0 K). At absolute zero, everything is in the state of lowest possible energy. For most substances, this state is a "perfect" crystal, with no disorder. (A few imperfections (section 11.8) do not make a significant difference.) It is natural to decide that the entropy is zero when there is no disorder.

The *third law of thermodynamics* says that, in any process involving perfect crystals at 0 K, $\Delta S = 0$. There is no disorder in the reactants, no disorder in the products, and therefore no increase or decrease in disorder. The choice $S = 0$ at 0 K is consistent with this law:

$$\Delta S = S_{\text{products}} - S_{\text{reactants}} = 0 - 0 = 0$$

Some substances, however, do not form perfect crystals, even at absolute zero. Carbon monoxide, CO, is an example. The reason is that the CO molecule has a very small dipole moment (section 7.12). It therefore makes little difference whether a molecule is oriented CO or OC. The molecules in the crystal are oriented more or less at random:

. . . CO CO OC CO OC CO CO . . .

As a result, if the reaction

$$\text{C(graphite)} + \tfrac{1}{2}O_2\text{(c)} \longrightarrow \text{CO(c)}$$

took place at absolute zero, there would be an increase in entropy; two perfect crystals, graphite and oxygen, would react to produce an imperfect crystal. Experimentally, ΔS for this reaction is 5.0 J/K at 0 K. (Actually, no reaction can occur at absolute zero, but ΔS can be calculated from measurements at higher temperatures.)

there is an increase in the number of moles of gas, the large increase in disorder is enough to make ΔS positive. When there is a decrease in the number of moles of gas, ΔS is negative. When there is no change in moles of gas, ΔS is small and can be positive or negative.

PROBLEM 18.11 Classify ΔS for each reaction as positive, negative, or indeterminable:

(a) $\text{Zn(c)} + Cu^{2+} \longrightarrow Zn^{2+} + \text{Cu(c)}$
(b) $2CO\text{(g)} + O_2\text{(g)} \longrightarrow 2CO_2\text{(g)}$
(c) $\text{C(graphite)} + O_2\text{(g)} \longrightarrow CO_2\text{(g)}$
(d) $\text{C(graphite)} + CO_2\text{(g)} \longrightarrow 2CO\text{(g)}$ □

18.11 ENTROPY, ENTHALPY, AND FREE ENERGY

Because of the universal tendency toward high entropy as well as toward equilibrium, we might expect to find a close relationship between entropy and equilibrium. Indeed, we shall see in this section that the entropy change is one of two changes—the other being the enthalpy change—that together determine the free energy change. In many cases these changes in entropy and enthalpy point in opposite directions, and the state of equilibrium is then a compromise.

Recall the modified first law (equation 9) in section 18.3:

$$\Delta H = q + w_{el}$$

According to the preceding section, the maximum heat absorbed is

$$q_{max} = T\Delta S$$

When q is a maximum, w_{el} is a minimum: $w_{el,min} = \Delta G$. Then

$$\begin{aligned}\Delta H &= q_{max} + w_{el,min}\\ &= T\Delta S + \Delta G\end{aligned}$$

or, as it is usually written,

$$\Delta H = \Delta G + T\Delta S \tag{22}$$

Since ΔH, ΔG, and ΔS depend only on the initial and final states, equation 22 is true for *any* isothermal process. Think of it as the first law written for minimum work and maximum heat, with pressure-volume work subtracted from both sides.

When everything is in its standard state (solute at 1 mol/L, gas at 1 atm, or pure solid or liquid), equation 22 becomes

$$\Delta H^0 = \Delta G^0 + T\Delta S^0$$

or

$$\Delta G^0 = \Delta H^0 - T\Delta S^0 \tag{23}$$

or

$$\Delta S^0 = \frac{\Delta H^0 - \Delta G^0}{T} \tag{24}$$

These equations make it possible to calculate any of the three quantities ΔH^0, ΔS^0, and ΔG^0 at a given temperature when the other two are known. We can use the data in Appendix C to calculate ΔH^0 and ΔG^0, and then ΔS^0 follows. In another book, you might find ΔH^0 and ΔS^0, which you could then use to calculate ΔG^0.

EXAMPLE 18.15 Use the data in Appendix C to calculate ΔH^0, ΔG^0, and ΔS^0 at 25°C for the reaction

$$2Ag(c) + Cl_2(g) \longrightarrow 2AgCl(c)$$

ANSWER For the elements Ag and Cl_2, $\Delta H_f^0 = 0$ and $\Delta G_f^0 = 0$. Then ΔH^0 and ΔG^0 are calculated as in section 18.8:

$$\Delta H^0 = 2\Delta H^0_{f,AgCl} - (2\Delta H^0_{f,Ag} + \Delta H^0_{f,Cl_2})$$

$$= 2 \text{ mol} \times \left(-127.068 \frac{\text{kJ}}{\text{mol}}\right) - (2 \times 0) - 0 = -254.136 \text{ kJ}$$

$$\Delta G^0 = 2\Delta G^0_{f,AgCl} - (2\Delta G^0_{f,Ag} + \Delta G^0_{f,Cl_2})$$

$$= 2 \text{ mol} \times \left(-109.789 \frac{\text{kJ}}{\text{mol}}\right) - (2 \times 0) - 0 = -219.578 \text{ kJ}$$

Watch the units. ΔH and ΔG are usually in kJ, but ΔS is in J/K.

Equation 24 for ΔS^0 gives

$$\Delta S^0 = \frac{\Delta H^0 - \Delta G^0}{T}$$

$$= \frac{-254.136 \text{ kJ} - (-219.578 \text{ kJ})}{298.15 \text{ K}}$$

$$= -0.11591 \text{ kJ/K} = -115.91 \text{ J/K}$$

■

The large decrease in entropy in Example 18.15 is typical of a reaction in which gas (Cl_2, disordered) is changed to solid or liquid (AgCl, relatively ordered). Since $T\Delta S$ is the *maximum* heat *absorbed*, $-T\Delta S$ is the *minimum* heat *given out*. In this reaction, $-T\Delta S = 34.6$ kJ/mol Cl_2. A galvanic cell using this reaction would inevitably give out *at least* 34.6 kJ of heat per mole of Cl_2 at 25°C. A large decrease in disorder (gas to crystal) corresponds to a need to remove much heat from the system. This 34.6 kJ is energy that is available only as heat, not as electrical work. There is sadness here, and waste, but not human error or negligence. On the contrary, whoever operated this cell and took out only 34.6 kJ as heat exercised infinite care and patience, for the cell was operated *reversibly*, therefore infinitely slowly. This is the way to get the most work possible out of it. Nonetheless, some heat was emitted. This irreducible minimum represents the waste imposed by the laws of nature.

In reality, the heat given out by a galvanic cell is more than $-T\Delta S$; correspondingly, the work obtained is less than $-\Delta G$. The total is always $-\Delta H$:

$$-\Delta H = -w_{el} - q$$

Figure 18.7 illustrates the possible values of work and heat for the reaction in Example 18.15. In part a, the reactants, Ag and Cl_2, are simply mixed and allowed to react. No work is done (excluding pressure-volume work); the entire 254 kJ $(= -\Delta H)$ is given out as heat. In part c, the reaction occurs in a reversible galvanic cell. Only 35 kJ is given out as heat; the remaining 219 kJ $(= -\Delta G)$ is available as work. Part b shows what happens in a real cell. The work done by the cell is less than 219 kJ (152 kJ in the figure); the heat given out is greater than 35 kJ (102 kJ in the figure).

The numbers are rounded so that they add up.

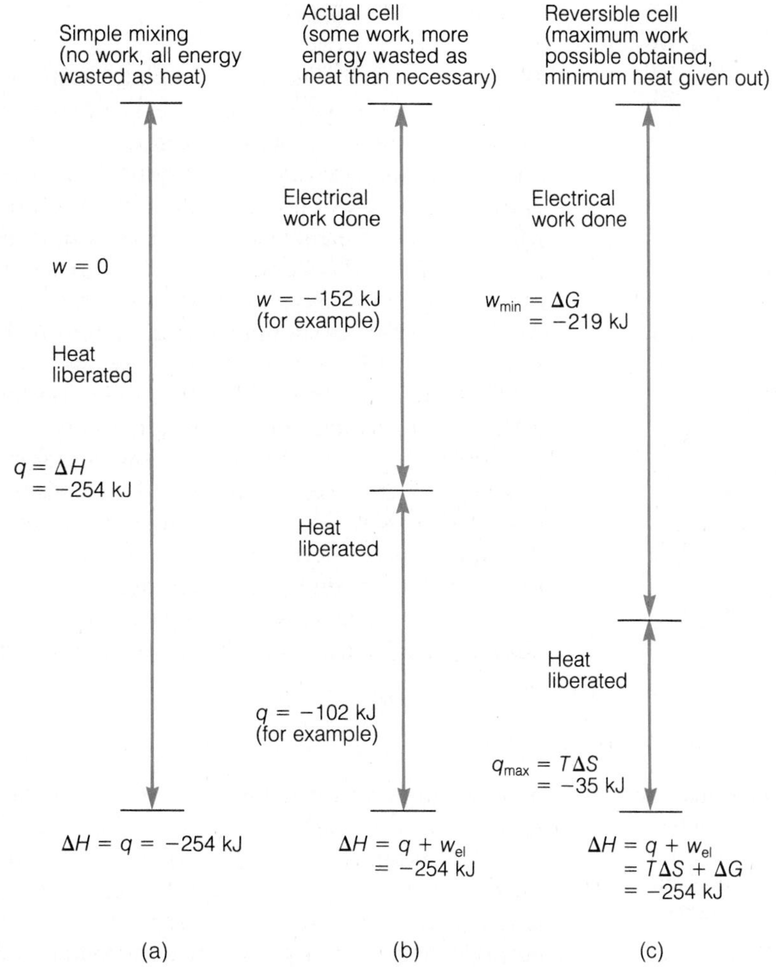

FIGURE 18.7 The division of the enthalpy change between work and heat for three ways of carrying out the reaction $2Ag + Cl_2 \longrightarrow 2AgCl$.

PROBLEM 18.12 Calculate ΔH^0, ΔG^0, and ΔS^0 at 25°C for the reaction

$$C_6H_6(g) \longrightarrow 3C_2H_2(g)$$ □

Sometimes, the maximum work that can be done is greater than the maximum heat that can be given out. In such cases, ΔS is positive and heat can be absorbed by the cell. An example is the reaction

$$Pb(c) + Hg_2Cl_2(c) \longrightarrow PbCl_2(c) + 2Hg(\ell)$$

for which $\Delta H^0 = -94.3$ kJ and $\Delta G^0 = -103.4$ kJ at 25°C. Here, the products have greater entropy than the reactants, even though they are lower in enthalpy.

18.12 TEMPERATURE AND EQUILIBRIUM

What determines the direction a system should go to reach equilibrium? Another way of asking the question is, what makes a process spontaneous? The spontaneous direction depends on the initial composition of the mixture. To make it definite, let us start with a mixture in which each reactant or product is in its standard state: pure solid or liquid, 1 atm for gases, 1 mol/L for solutes. For a fixed temperature, the spontaneous direction is the direction of negative ΔG^0. So what makes ΔG^0 negative? Equation 23 tells us the answer:

$$\boldsymbol{\Delta G^0 = \Delta H^0 - T\Delta S^0} \tag{23}$$

What makes ΔG^0 negative is negative ΔH^0 or positive ΔS^0. In the majority of cases, you can't have it both ways because states of high entropy (like gases) usually have high enthalpy; states of low enthalpy (like crystals) usually have low entropy. It appears that a competition exists between the tendency toward high entropy (the "law of universal sloppiness") and the tendency toward low enthalpy (the "law of universal laziness"). The balance between them depends on the temperature. When T is large, the $T\Delta S^0$ term is the important one. The direction corresponding to increasing entropy (positive ΔS^0) is then the direction that makes ΔG^0 negative—the spontaneous direction. At high temperature, the equilibrium state will have high entropy, like gaseous water above 100°C. When T is small, $T\Delta S^0$ is small and the ΔH^0 term is the important one. The spontaneous direction is then the direction of decreasing enthalpy (negative ΔH^0). At low temperature, the equilibrium state is a state of low enthalpy, like ice below 0°C.

Le Chatelier's principle (section 13.6) tells us in which direction an equilibrium is shifted by a temperature change. Now we can compare this prediction with the equations of thermodynamics, and even calculate the change in the equilibrium constant when the temperature changes.

Suppose a process is exothermic, which means that ΔH^0 is negative, and suppose that ΔS^0 is also negative. Is this process spontaneous in the standard state?

$$\begin{array}{cccc} \Delta G^0 & = \Delta H^0 & - & T\Delta S^0 \\ + \text{ or } - & - & - & (-) \end{array}$$

The process takes place now at one temperature, now at another, but each time the temperature is constant throughout the process.

At low temperature (small T), ΔG^0 is negative, and the process is therefore spontaneous. At high temperature (large T), $T\Delta S^0$ is the more important term. Since ΔS^0 is negative, $-T\Delta S^0$ is positive and ΔG^0 becomes positive, and therefore the process is nonspontaneous. The apparent competition between ΔH^0 and $T\Delta S^0$ is seen here. The negative ΔH^0 tends to make the process spontaneous, while the negative ΔS^0 tends to make it nonspontaneous. Raising the temperature helps ΔS^0 to win; lowering it helps ΔH^0 to win. Thus, when ΔH^0 and ΔS^0 are both negative, the process is favored by lowering the temperature, and disfavored by

BOX 18.8
WHAT REALLY DETERMINES STABILITY?

A state of high entropy is disordered; therefore it is probable, and therefore stable. But $T\Delta S$ is only one of two terms that tells us which way a system has to go to become stable. Why isn't stability determined by ΔS alone?

ΔG determines the spontaneous direction of a process for a system at constant temperature and pressure. The way to keep a system at constant temperature is to immerse it in a constant-temperature bath, such as a tank of ice water (Figure 18.6, section 18.5). For constant pressure, movable walls can be used to keep the system at atmospheric pressure. Now think of *the system and the bath together* as a supersystem, not interacting with anything else except through the constant pressure. The most stable state of this supersystem is the most probable, random, disorganized state—the state of highest entropy. (More will be said about this in section 18.13.) Thus, for a spontaneous process, the *total* entropy change of system and bath should be positive:

$$\Delta S_{total} = \Delta S_{sys} + \Delta S_{bath}$$

There is one requirement for the bath: it never does any work (excluding, as usual, pressure-volume work). Its state is determined entirely by how much heat it has absorbed or given out. For the bath, $q_{bath} = T\Delta S_{bath}$. No electrical work is done on or by the system; therefore, $q_{sys} = \Delta H_{sys}$. But the heat absorbed by the bath is the heat given out by the system, or conversely:

$$q_{bath} = -q_{sys} = -\Delta H_{sys}$$

$$\Delta S_{bath} = \frac{q_{bath}}{T} = \frac{-\Delta H_{sys}}{T}$$

Then

$$\Delta S_{total} = \Delta S_{sys} - \frac{\Delta H_{sys}}{T}$$

$$= \frac{-\Delta G_{sys}}{T}$$

So, when the total entropy of system and bath *increases*, the free energy of the system *decreases*. Using ΔG enables us to disregard the surroundings, which somehow maintain a constant temperature, and think only about the system. But the tendency of the *total* entropy to increase is what really determines the direction of greater stability.

---

raising the temperature. This prediction is the same one made by Le Chatelier's principle on the basis of ΔH^0 alone.

Another possibility is that ΔH^0 and ΔS^0 may both be positive. Then

$$\Delta G^0 = \Delta H^0 - T\Delta S^0$$

$$+ \text{ or } - \qquad + \qquad - \quad (+)$$

At low temperature, $T\Delta S^0$ is small, ΔH^0 is the more important term, and ΔG^0 is positive. At high temperature, T is large, $T\Delta S^0$ is the more important term, and ΔG^0 is negative. In this case, the process is favored by raising the temperature.

EXAMPLE 18.16 For the reaction

$$CaCO_3(c) \longrightarrow CaO(c) + CO_2(g, 1\text{ atm})$$

$\Delta H^0 = 179$ kJ and $\Delta S^0 = 160$ J/K. Is the reaction spontaneous at (a) low temperature; (b) high temperature?

ANSWER

(a) At low temperature, ΔH^0 (positive) is the important term. ΔG^0 is positive, and the reaction is nonspontaneous. The reverse reaction is spontaneous. (CaO combines readily with CO_2 at room temperature.)

(b) At high temperature, $T\Delta S^0$ (positive) is the important term. Positive $T\Delta S^0$ makes ΔG^0 negative. The reaction written is spontaneous at high temperature. (Preparation of quicklime (CaO) by "calcining" or "burning" limestone ($CaCO_3$) at about 1400°C is a step in the manufacture of cement.) ■

PROBLEM 18.13 For the reaction

$$4Fe(c) + 3O_2(g) \longrightarrow 2Fe_2O_3(c)$$

$\Delta H^0 = -1648$ kJ and $\Delta S^0 = -549$ J/K. (a) From these data, should iron rust (combine with oxygen) at low temperature? Is this conclusion supported by experience? (b) Do you think it might be possible to decompose Fe_2O_3 by heating it to a high enough temperature? □

This table summarizes all of the sign combinations, including two less common ones:

	ΔH^0	ΔS^0	$\Delta G^0 = \Delta H^0 - T\Delta S^0$	
			Low temperature	High temperature
More common	−	−	−	+
	+	+	+	−
Less common	−	+ or close to 0	−	−
	+	− or close to 0	+	+

The less common sign combinations result in processes that are spontaneous at all temperatures or nonspontaneous at all temperatures. A class of processes that show this behavior is the mixing of two completely miscible liquids, such as water and acetic acid (section 12.3). ΔS is positive, as we would expect for a mixing process. This means that heat would be absorbed, and net work done, in a thermodynamically reversible mixing. However, when the liquids are simply mixed, the process is highly irreversible. Usually ΔH is negative, which means that heat is given out in the mixing process when no work is done. Thus, with negative ΔH and positive ΔS, ΔG is always negative and the mixing is always spontaneous.

The ^{0}'s are omitted because the resulting solution may have any composition.

ΔH^0 and ΔS^0 change only slightly for moderate changes in temperature. Therefore, it is often a good approximation to treat ΔH^0 and ΔS^0 as constants, independent of temperature, and use them to calculate ΔG^0 at various temperatures. However, as equation 23 indicates, ΔG^0 is sensitive to temperature—that is, it changes considerably when T changes.

If ΔH^0 and ΔS^0 are assumed to be independent of temperature, the graph of ΔG^0 versus T is a straight line with a slope of $-\Delta S^0$. Figure 18.8 shows, in a simplified way, the variation of ΔG^0 with temperature for the four possible sign combinations. Where a line crosses the horizontal axis, the system is in equilibrium when everything is in its standard state ($\Delta G^0 = 0$). The only lines that cross the axis are those for which ΔH^0 and ΔS^0 have the same sign.

The equilibrium constant can be calculated from ΔG^0 by equation 17:

$$\log K_p = \frac{-\Delta G^0}{2.303RT}$$

EXAMPLE 18.17 (a) Estimate ΔG^0 and K_p at 500 K for the reaction

$$2Ag(c) + Cl_2(g) \longrightarrow 2AgCl(c)$$

using the results of Example 18.15: $\Delta H^0 = -254.136$ kJ and $\Delta S^0 = -115.91$ J/K. Assume that ΔH^0 and ΔS^0 are independent of temperature.

(b) More accurate values at 500 K are $\Delta H^0 = -250.0$ kJ and $\Delta S^0 = -105.2$ J/K. Use these data to calculate ΔG^0 at 500 K. Compare with the answer to part a.

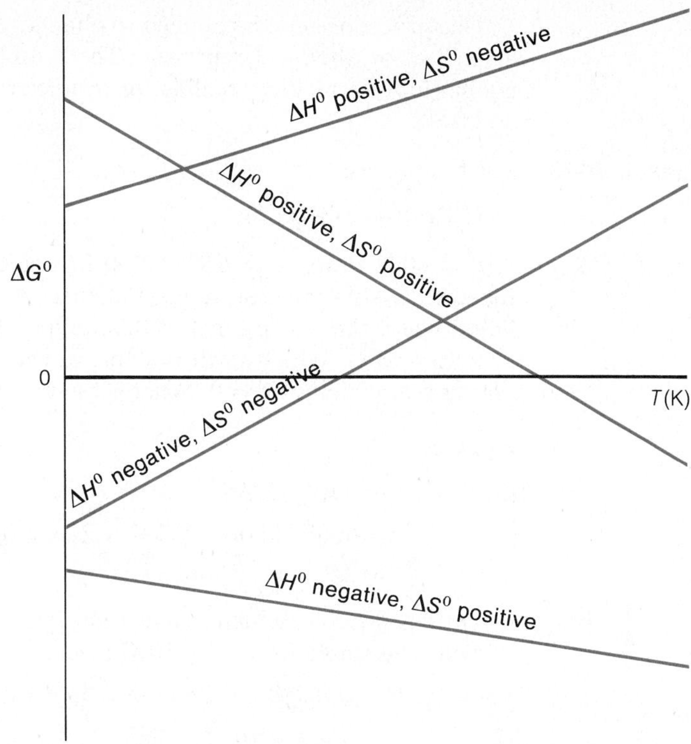

FIGURE 18.8
A graph showing the dependence of ΔG^0 on temperature for the four possible sign combinations of ΔH^0 and ΔS^0.

ANSWER

(a)
$$\begin{aligned}\Delta G^0 &= \Delta H^0 - T\Delta S^0 \\ &= -254.136 \text{ kJ} - 500 \text{ K} \times (-0.11591 \text{ kJ/K}) \\ &= -196.18 \text{ kJ}\end{aligned}$$

$$\log K_p = \frac{-\Delta G^0}{2.303RT}$$

$$= \frac{-(-196.18) \text{ kJ}}{2.303 \times 8.314 \times 10^{-3} \text{ kJ/(mol K)} \times 500 \text{ K}}$$

$$= 20.49$$

$$K_p = 3.1 \times 10^{20}$$

The process is still spontaneous, but less so than at 298 K. (ΔG^0 is less negative.)

(b)
$$\begin{aligned}\Delta G^0 &= \Delta H^0 - T\Delta S^0 \\ &= -250.0 \text{ kJ} - 500 \text{ K} \times (-0.1052 \text{ kJ/K}) \\ &= -197.4 \text{ kJ}\end{aligned}$$

The result is close to what was calculated in part a, partly because of cancellation of errors in subtracting. ■

PROBLEM 18.14 Estimate ΔG^0 at 1000 K for the reaction

$$C_6H_6(g) \longrightarrow 3C_2H_2(g)$$

Use the following results of Problem 18.12: $\Delta H^0 = 600.26$ kJ and $\Delta S^0 = 343.3$ J/K. Assume that ΔH^0 and ΔS^0 are constant. □

This procedure can be applied to changes of state, like melting or evaporation, as well as to chemical reactions. The temperature at which two phases are in equilibrium—a melting, boiling, or transition point—can be calculated from ΔH^0 and ΔS^0.

EXAMPLE 18.18 For the process

$$H_2O(c) \longrightarrow H_2O(\ell)$$

$\Delta H^0 = 6008$ J/mol and $\Delta S^0 = 22.0$ J/(mol K). Assume that ΔH^0 and ΔS^0 are independent of temperature. (a) Calculate ΔG^0 for the process at $-10°C$. Is ice or liquid water the stable form at this temperature? (b) Calculate ΔG^0 for the process at $+10°C$. Which form is stable at this temperature? (c) Calculate the temperature at which $\Delta G^0 = 0$. What is the physical significance of this temperature?

ANSWER

(a) $$\Delta G^0 = \Delta H^0 - T\Delta S^0$$
$$= 6008 \text{ J/mol} - 263 \cancel{\text{K}} \times 22.0 \text{ J/(mol}\cancel{\text{K}})$$
$$= 2.2 \times 10^2 \text{ J/mol}$$

ΔG^0 is positive, which means that liquid water has a higher free energy than ice. The stable form at $-10°C$ is ice.

(b) $$\Delta G^0 = 6008 \text{ J/mol} - 283 \cancel{\text{K}} \times 22.0 \text{ J/(mol}\cancel{\text{K}})$$
$$= -2.2 \times 10^2 \text{ J/mol}$$

At $+10°C$, liquid water has the lower free energy and is more stable.

(c) $$\Delta G^0 = \Delta H^0 - T\Delta S^0 = 0$$

$$T = \frac{\Delta H^0}{\Delta S^0} = \frac{6008 \cancel{\text{J}}/\cancel{\text{mol}}}{22.0 \cancel{\text{J}}/(\cancel{\text{mol}}\text{ K})}$$
$$= 273 \text{ K or } 0°C$$

At this temperature, the two forms of water are equally stable and can exist together in equilibrium under a pressure of 1 atm. ■

PROBLEM 18.15 For the process

$$SO_2(\ell) \longrightarrow SO_2(g)$$

at 30.0°C, $\Delta H^0 = 21.26$ kJ/mol and $\Delta S^0 = 82.68$ J/(mol K). (a) Calculate ΔG^0 at 30.0°C. (b) What is ΔG^0 at the normal boiling point of SO_2? (c) Estimate the boiling point. □

18.13 WHAT IS THE UNIVERSE COMING TO?

Section 18.10 showed that when a system goes from one state to another at constant temperature, the heat absorbed has an upper limit: q is always less than $T\Delta S$ or q/T is always less than ΔS. These statements are summarized by the inequality

$$\Delta S > \frac{q}{T} \tag{25}$$

This innocent-looking relationship turns out to have profound implications for the entire universe.

Suppose that a system is isolated from its surroundings. Then it can transfer no heat to or from its surroundings; q equals 0 for every process. Since ΔS is never

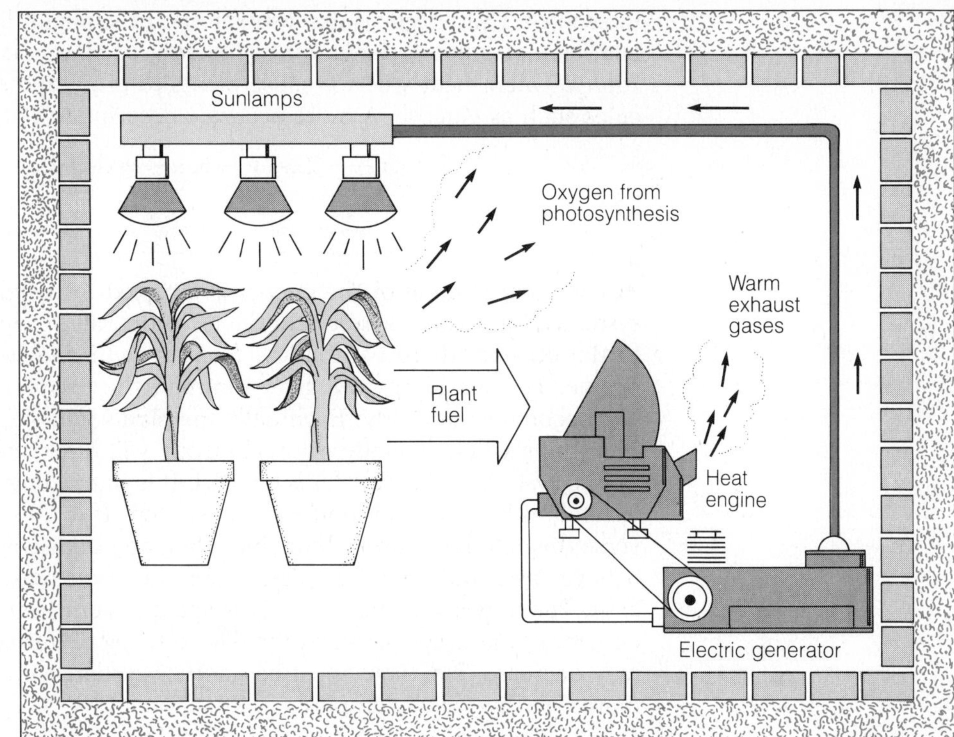

FIGURE 18.9
A futile attempt to recycle all energy in an isolated system. Plant matter is fuel for a heat engine that drives a generator to make electricity for sun lamps to promote photosynthesis. However, because the energy in the exhaust gases of the heat engine is in the form of random motion of the gas molecules, this energy cannot cause photosynthesis. The cycle dies down as entropy increases.

This conclusion is still true when the temperature is not constant.

less than q/T, and q equals 0, it follows that ΔS is never less than zero. The entropy of this isolated system can only increase. The most stable state—the state in which the system is most likely to be found—is the state of *highest entropy*. An isolated system always tends to go from a state of lower entropy to a state of higher entropy. *For any possible process in an isolated system, ΔS is positive.*

To understand how the entropy of an isolated system increases, think about the system in Figure 18.9. Plants are growing under sun lamps. By photosynthesis, they produce plant tissue and oxygen. The plants are burned as fuel in a heat engine, which powers an electric generator. The electricity is used to light the sun lamps. The system is completely isolated from its environment. The net process is

$$\text{light} \longrightarrow \text{plants} \longrightarrow \text{heat} \longrightarrow \text{electricity}$$

(with an arrow returning from electricity to light)

Could the system operate indefinitely, continuously recycling its energy and producing new fuel for the engine? To simplify the question, assume that the fuel is the sugar glucose, $C_6H_{12}O_6$. Glucose is a product of photosynthesis. When glucose burns it produces molecules of carbon dioxide and water:

$$C_6H_{12}O_6(c) + 6O_2(g) \longrightarrow 6CO_2(g) + 6H_2O(\ell)$$

Radiation has some of the quality of work: it can cause an increase in free energy, as in the photosynthesis of glucose.

A complex, orderly molecule of glucose is converted into a disordered set of more numerous, less complex molecules. Photosynthesis does just the reverse. If the photosynthesis produced as much glucose as was burned in the engine, the entropy of the entire system would remain constant. Such an event *never happens*. Why could it not happen in this case? The exhaust gases from the heat engine are hot, but their energy is in the form of the random motion of carbon dioxide and water molecules. This energy is not a substitute for sunlight; light can cause effects that heat cannot imitate. *A photon of the correct energy is needed to activate chlorophyll.* Mere heat will not cause plants to grow and synthesize complex molecules such as glucose. A more accurate description of the net process is

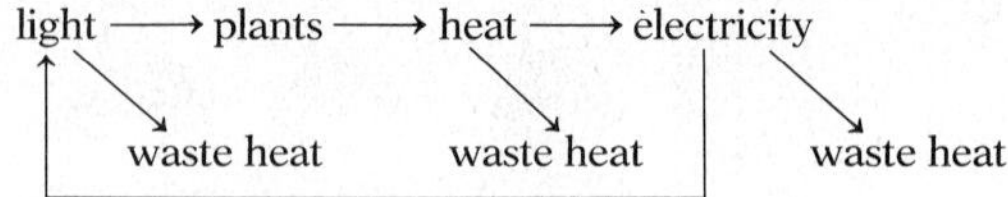

At each stage, some of the energy is diverted to merely warming up the entire system. Thus, as the system continues to operate, even though no energy is lost to the outside, there will be less plant growth and hence, less fuel to run the engine. The sun lamps will grow dimmer; the temperature will rise; the system will become less orderly. Eventually the plants will die, the engine will stop, and everything will be equally warm. Entropy will have reached its maximum.

The system in Figure 18.9 is fanciful. It is interesting, however, because, being isolated, it behaves as if nothing else existed. It is thus a small model of the universe (by definition, everything, including any undiscovered dark matter), which behaves as if nothing else existed—for the good reason that nothing else *does* exist. The universe is the one truly isolated system. *In every possible process, the entropy of the universe increases.* The entropy of one system within the universe may decrease, but there must be another system whose entropy increases even more. We give two examples:

1. Water can be frozen by placing it in contact with cooling coils in which liquid ammonia is boiling. The freezing of the water corresponds to a decrease in entropy—the solid is more orderly (has lower entropy) than the liquid—but this effect is more than compensated by the increase in the entropy of the ammonia as it changes from a liquid to a gas.

2. Of all the products that could possibly be formed from a fertilized egg plus an ample supply of food, only an unthinkably minute fraction would be adult organisms. The development of living organisms represents an enormous increase in order (and thus a decrease in entropy), but the developing embryo is not an isolated system. Its decrease in entropy is overbalanced by a greater increase in entropy somewhere else. Most of the increase in the entropy of the universe results from the dispersal of radiation from the Sun and other stars. This radiation provides green plants with a supply of free energy that they and their consumers (including ourselves) use to build and maintain their complex organizations. Remember that an increase in free energy is associated with either an increase in enthalpy or a decrease in entropy, or both.

This is why real processes are said to be "irreversible."

This inexorable increase in the entropy of the universe is one reason that time cannot go backward. "The moving finger writes; and having writ, moves on: nor all your piety nor wit shall lure it back to cancel half a line." Every real process increases the entropy of the universe, and all our piety and wit cannot lure the moving finger back to cancel the increase. From the viewpoint—perhaps limited—of thermodynamics, the universe, like the isolated greenhouse, should have as its eventual fate a uniformly warm "heat death." Rudolf Clausius (1822–1888), who invented and named the concept of entropy, summarized the first and second laws of thermodynamics in two famous sentences: "The energy of the universe is constant. The entropy of the universe tends toward a maximum."

A **system** is a well-defined portion of the universe; everything else is part of the **surroundings**. A system is **closed** when no matter can enter or leave it. **Work** (w) is positive when it is done *on* the system, negative when it is done *by* the system. $w = -P\Delta V$ (pressure-volume work); $w = -\mathscr{E}Q$ (electrical work, potential × charge). The **state** of a system is defined by its **properties**, such as mass, chemical composition, temperature, and pressure. A property depends on the system's present state, not on its history. A **path** is a sequence of states through which a system passes on the way from an initial state to a final state. A change in a property depends only on the initial and final states, not on the path.

An **insulated system** cannot exchange heat with its surroundings. For every process in an insulated system, w depends only on the initial and final states. The change in **energy** is $\Delta E = w$. A noninsulated system can gain or lose energy by **heat** as well as by work. $\Delta E = q + w$, where q is the heat absorbed by the system. q and w depend on the path.

The **first law of thermodynamics** (the **law of conservation of energy**) says that the energy of a system can change only by the amount of energy (heat and/or work) that enters or leaves it. An **isolated system** has no interaction with its surroundings. Its energy, in particular the energy of the universe, is constant. **Enthalpy** is related to energy by $H = E + PV$. At constant pressure, $\Delta H = \Delta E - w_{PV}$; when the only work is pressure-volume and electrical work, $\Delta H = q + w_{el}$.

Work is orderly motion; heat is random atomic, molecular, or electronic motion. Since orderly motion can easily be changed to disorderly motion, any effect that can be produced by heat can also be produced by work. However, effects of work cannot necessarily be produced by heat.

The **second law of thermodynamics** says that a system at constant temperature cannot absorb heat and do work in a cyclic process (final state = initial state). For any **isothermal process** (at constant temperature), the work done on a system (w) is never less than the work (w_{min}) that would be done on the system in a thermodynamically reversible process between the same two states.

The **free energy** (G) is a property defined by $\Delta G = w_{min} - w_{PV,min}$. If the only kinds of work are pressure-volume work and electrical work, $\Delta G = w_{el,min}$. When ΔG is positive, it is the minimum electrical work that must be done on the system. When ΔG is negative, $-\Delta G$ is the maximum electrical work available from the system.

Free energy is related to the emf, $\mathscr{E}$, of a reversible galvanic cell by $\Delta G = -n\mathscr{F}\mathscr{E}$; n faradays pass through the circuit for the reaction as written. For reactants and products in the standard state, the **standard free energy change** is $\Delta G^0 = -n\mathscr{F}\mathscr{E}^0$.

The **standard free energy of formation** (ΔG_f^0) of a substance is ΔG^0 for the formation of 1 mole from the elements in their stable forms. ΔG^0 for a reaction is

$$\Delta G^0 = \text{total } \Delta G^0_{f,\text{products}} - \text{total } \Delta G^0_{f,\text{reactants}}$$

ΔG is negative for a **spontaneous process**, positive for a nonspontaneous process. The equilibrium state (most stable state) of a system at constant temperature and pressure is therefore the state of minimum free energy. The standard free energy change is related to the equilibrium constant by $\Delta G^0 = -RT\ln K = -2.303RT\log K$. When gases are involved, $K = K_p$. For a nonstandard state, $\Delta G = \Delta G^0 + RT\ln Q$, where Q is the reaction quotient.

Entropy (S) is a quantitative measure of disorder. $\Delta S = q_{max}/T$, where q_{max} is the maximum possible heat absorbed as the system goes from its initial state to its final state at constant temperature.

In a reversible process, w_{el} has its minimum value and q has its maximum value: $\Delta H = w_{el,min} + q_{max} = \Delta G + T\Delta S$. When the reactants are merely mixed and allowed to react, $w_{el} = 0$ and $q = \Delta H$. When the reaction occurs in a reversible cell, $w_{el} = \Delta G$ and $q = T\Delta S$. In practice, w_{el} is between 0 and ΔG, and q is between ΔH and $T\Delta S$.

When the initial state is the standard state, spontaneity corresponds to negative ΔG^0. It is favored by negative ΔH^0 or positive ΔS^0. When ΔS^0 is positive, ΔG^0 decreases as the temperature increases, and the reaction is favored by high temperature. When ΔS^0 is negative, the reaction is favored by low temperature. If ΔH^0 and ΔS^0 have the same sign, $\Delta G^0 = 0$ at some temperature. ΔH^0 and ΔS^0 do not change sharply with temperature and can therefore be assumed to be approximately constant. ΔG^0 can therefore be estimated at various temperatures by $\Delta G^0 = \Delta H^0 - T\Delta S^0$. The equilibrium constant follows from the estimated ΔG^0.

For any process in an isolated system, ΔS is positive or zero. Since the universe is an isolated system, the entropy of the universe always increases.

18.14 ADDITIONAL PROBLEMS

Use the data in Appendix C and Table 17.1 (section 17.7) as needed.

ENERGY, WORK, AND HEAT

18.16 A gas expands from 2.00 L to 6.00 L against a constant pressure of 1.00 atm. How much work does the gas do, in joules?

18.17 When a gas expands suddenly, it may not have time to absorb a significant amount of heat: $q = 0$. Assume that 1.00 mol N_2 expands suddenly, doing 3000 J of work. (a) What is ΔE for the process? (b) The heat capacity (section 4.3) of N_2 is 20.9 J/(mol °C). How much does its temperature fall during this expansion? (This is the principle of most snow-making machines, which use compressed air mixed with water vapor.)

18.18 Calculate ΔE when a system (a) absorbs 65.0 kJ of heat and does 55.0 kJ of work; (b) evolves 75.0 kJ of heat and does 40.0 kJ of work; (c) evolves 35.0 kJ of heat and 35.0 kJ of work is done on the system.

18.19 For each of the following processes, calculate w and state whether work is done *on* or *by* the system. (a) A system absorbs 25.0 kJ of heat and its energy increases by 40.0 kJ. (b) A system gives out 17.0 J of heat and its energy decreases by 9.0 J.

18.20 (a) A system goes from state 1 to state 2. It gives out 150 J of heat while 200 J of work is done on it. Give q, w, and ΔE, with signs. (b) The system in part a is restored to state 1 and then goes again from state 1 to state 2 (the same states as in part a). This time, it gives out 220 J of heat. What is w, with sign? Is work done on or by the system?

SECOND LAW AND FREE ENERGY

18.21 State whether each process is possible at constant temperature. If it is not possible, which law of thermodynamics does it violate? (a) A gas expands, doing 100 J of work and absorbing 120 J of heat, and is then compressed to its initial state. This compression requires 110 J of work, and the gas gives out 120 J of heat. (b) A battery is charged by doing 10 watt-hours of work on it, while it gives out 1 watt-hour of heat. When the battery is discharged to its initial state, it yields 8 watt-hours of work and gives out 1 watt-hour of heat. (c) A gas is compressed by 15 liter-atm of work and it emits 15 liter-atm of heat. It then expands to its initial state, doing 16 liter-atm of work and absorbing 16 liter-atm of heat.

18.22 When the reaction

$$8Zn(c) + S_8(c) \longrightarrow 8ZnS(c)$$

occurs at 25°C, the maximum work that can be done by the system is 1610 kJ. (a) When the system does 1610 kJ of work, what is w? Is this the maximum or the minimum value of w? (b) Give two other possible values of w for the process. Express them as "$w = \ldots$" and as "work done on/by system $= \ldots$". (c) Since this process involves no gases, the volume change is negligible. What is ΔG?

FREE ENERGY, EMF, AND FREE ENERGY OF FORMATION

See also Problems 18.37–18.43, 18.45.

18.23 (a) Using Appendix C, calculate ΔH^0 and ΔG^0 at 25°C for the reaction

$$CH_4(g) + 2O_2(g) \longrightarrow CO_2(g) + 2H_2O(\ell)$$

(b) Calculate the reversible emf $\mathscr{E}^0$ at 25°C of a fuel cell in which this reaction occurs. (*Suggestion*: First write the two partial equations or figure out the oxidation numbers.) (c) $-\Delta H^0$ is the heat liberated when 1 mole of CH_4 burns. What is the maximum percentage of this heat available as work?

18.24 From data in Table 17.1, calculate ΔG_f^0 for AgI(c) at 25°C.

18.25 Combine the data in Table 17.1 and Appendix C to calculate $\mathscr{E}^0$ at 25°C for the partial reaction

$$PbCl_2(c) + 2e^- \longrightarrow Pb(c) + 2Cl^-$$

18.26 The figure shows three cells in series, all at the same temperature. When current flows through this circuit, a reaction occurs at electrode A in cell 1, and the reverse reaction occurs, to the same extent, at electrode A in cell 3. Similarly, the reactions at electrodes B and C in cell 2 are equal and opposite to the reactions at electrodes B and C in cells 1 and 3. (a) The net emf of this arrangement of cells must be zero ($\mathscr{E}_1 + \mathscr{E}_2 + \mathscr{E}_3 = 0$); otherwise, the second law of thermodynamics would be contradicted. Explain. (b) Choose A as the reference half-cell. This means that, by definition, $\mathscr{E}_B = \mathscr{E}_1$ and $\mathscr{E}_C = -\mathscr{E}_3$. Show that $\mathscr{E}_2 = \mathscr{E}_C - \mathscr{E}_B$. You have now proved the validity of expressing the emf of a cell as the difference of two electrode potentials, each of which is really the emf of a cell containing the reference half-cell.

18.27 Ancient organisms obtained energy by fermentation (as do some primitive ones in present times), illustrated by

$$C_6H_{12}O_6(c) \longrightarrow 2C_2H_5OH(\ell) + 2CO_2(g)$$

Since the atmosphere became rich in O_2, most organisms obtain energy by respiration, illustrated by

$$C_6H_{12}O_6(c) + 6O_2(g) \longrightarrow 6CO_2(g) + 6H_2O(\ell)$$

For glucose, $C_6H_{12}O_6$, $\Delta G_f^0 = -912$ kJ/mol at 25°C. Other data are in Appendix C. Calculate ΔG^0 at 25°C for each of these reactions. Why has fermentation gone out of fashion as an energy source in the last billion years?

FREE ENERGY AND EQUILIBRIUM

18.28 For the reaction

$$N_2(g) + 3H_2(g) \rightleftharpoons 2NH_3(g)$$

at 648 K, $\Delta G^0 = 43.2$ kJ. (a) Calculate K_p for the reaction. (b) In a mixture kept at 648 K, the three gases have these partial pressures: N_2, 1.0 atm; H_2, 2.0 atm; NH_3, 0.20 atm. In which direction can the reaction occur?

18.29 For this process of converting a single enantiomer to an equal mixture of enantiomers (section 24.11),

$$(+)\ CH_3CHNH_2COOH(aq) \longrightarrow CH_3CHNH_2COOH(aq)$$

alanine (enantiomer) — alanine (mixture)

$\Delta G^0 = -1.5$ kJ at 25°C. Calculate K.

18.30 Using the data in Appendix C, calculate (a) ΔG^0, (b) $\mathscr{E}^0$, and (c) K_p and K_c at 25°C for the reaction

$$Zn(c) + H_2S(g) \longrightarrow ZnS(c) + H_2(g)$$

18.31 For reactions 1–3 in aqueous solution, (a) find $\mathscr{E}^0$ at 25°C for a cell in which the reaction occurs; (b) calculate ΔG^0 in kJ; (c) calculate the equilibrium constant.
(1) $Cd(c) + Ni^{2+} \longrightarrow Cd^{2+} + Ni(c)$
(2) $H_2(g) + Br_2(\ell) \longrightarrow 2H^+ + 2Br^-$
(3) $5Sn(c) + 2MnO_4^- + 16H^+ \longrightarrow 5Sn^{2+} + 2Mn^{2+} + 8H_2O$

18.32 Compare the (a) ΔG, (b) K, and (c) $\mathscr{E}$ of reaction 1 with the corresponding quantities of reactions 2 and 3:
(1) $Zn + 2Ag^+ \rightleftharpoons Zn^{2+} + 2Ag$
(2) $\frac{1}{2}Zn + Ag^+ \rightleftharpoons \frac{1}{2}Zn^{2+} + Ag$
(3) $Zn^{2+} + 2Ag \rightleftharpoons Zn + 2Ag^+$
Explain the differing ways in which these three quantities are affected when the equation is written differently.

18.33 Show how equation 18 (section 18.9) follows logically from the equations that precede it.

ENTROPY

18.34 Is the entropy change in each process positive, negative, or too close to zero for you to decide? Make your predictions without looking up any data.
(a) $CO_2(c, -78°C) \longrightarrow CO_2(g, -78°C)$
(b) $O_2(g, 1\ atm, 25°C) \longrightarrow O_2(g, 10\ atm, 25°C)$
(c) $2O_3(g) \longrightarrow 3O_2(g)$ (constant temperature and pressure)
(d) $S(c) + O_2(g) \longrightarrow SO_2(g)$ (constant temp. and pressure)

18.35 Calculate ΔS^0 at 25°C for the reaction $H_2(g) + Cl_2(g) \longrightarrow 2HCl(g)$. Explain the sign of ΔS^0 in terms of the concept of entropy as disorder.

18.36 When a hydrogen-oxygen fuel cell operates reversibly at 25°C, is heat liberated or absorbed? How much heat is liberated or absorbed per coulomb? Will this amount of heat be increased or decreased in actual (irreversible) operation?

TEMPERATURE AND EQUILIBRIUM

18.37 (a) Calculate ΔH^0, ΔG^0, ΔS^0, and the equilibrium constant at 25°C for the reaction

$$H_2(g) + Br_2(g) \rightleftharpoons 2HBr(g)$$

Watch out for the physical states of Br_2. Do you have to distinguish between K_c and K_p? (b) Estimate ΔG^0 and K at 500 K. Make the usual assumption about ΔH^0 and ΔS^0.

18.38 (a) Calculate ΔH^0, ΔG^0, and ΔS^0 at 25°C for the reaction

$$C(graphite) + CO_2(g) \rightleftharpoons 2CO(g)$$

(b) Estimate ΔG^0 and K_p for the reaction at 1000 K.

18.39 (a) Calculate ΔH^0, ΔG^0, and ΔS^0 at 25°C for the reaction

$$N_2O_4(g) \rightleftharpoons 2NO_2(g)$$

(b) Calculate K_p at 25°C. (c) Estimate ΔG^0 and K_p at 600 K.

18.40 (a) Calculate ΔH^0, ΔG^0, ΔS^0, and K_p at 25°C for the reaction

$$4Ag(c) + O_2(g) \rightleftharpoons 2Ag_2O(c)$$

(b) What is the pressure of O_2 in equilibrium with Ag and Ag_2O at 25°C? (c) At what temperature would the pressure of O_2 in equilibrium with Ag and Ag_2O become equal to 1.00 atm? (*Hint*: What would K_p and ΔG^0 be at this temperature?)

18.41 (a) From the data at the end of section 18.11, calculate ΔS^0 for the reaction

$$Pb(c) + Hg_2Cl_2(c) \longrightarrow PbCl_2(c) + 2Hg(\ell)$$

at 25°C. (b) Estimate ΔG^0 at 500 K. (c) Is there any temperature at which $\Delta G^0 = 0$ for this reaction?

18.42 (a) Calculate ΔH^0, ΔG^0, and ΔS^0 at 25°C for the process

$$C(diamond) \longrightarrow C(graphite)$$

(b) Is there any temperature at which diamond is thermodynamically stable relative to graphite at 1 atm? How do you account for the formation of diamonds in the first place? (*Hint*: Diamond has a higher density than graphite.)

18.43 (a) Find ΔH^0, ΔG^0, K_p, and ΔS^0 for the reaction

$$3O_2(g) \longrightarrow 2O_3(g)$$

at 25°C. (b) Is there any temperature at which this reaction is spontaneous when both partial pressures are 1 atm? If there is, estimate it.

18.44 At ordinary temperature (about 300 K), the ΔH^0 of a reaction is usually much larger than $T\Delta S^0$ (ignoring the signs). (a) When this relationship is true, what can be said about the relationship between ΔG^0 and ΔH^0? (b) Are endothermic or exothermic reactions more likely to be observed in our everyday experience? Why?

18.45 (a) Calculate ΔH^0, ΔG^0, and ΔS^0 at 25°C for the reaction

$$2SO_2(g) + O_2(g) \rightleftharpoons 2SO_3(g)$$

(b) Estimate ΔG^0, K_p, and K_c for the reaction at 1000 K. (*Caution*: Watch units of R.) Compare your answers with the data in Table 13.1 (section 13.2). (c) Estimate the temperature at which $\Delta G^0 = 0$ for the reaction.

18.46 Sulfuric acid mists occur in the atmosphere as a result of the rapid reaction

$$SO_3(g) + H_2O(g) \longrightarrow H_2SO_4(\ell)$$

The SO_3 is derived from the sulfur in fossil fuels that are burned in air. Based upon your answers to Problem 18.45, do you think that most of the SO_3 is formed directly in the combustion flame, or by subsequent oxidation of SO_2 in the atmosphere at lower temperatures? Would you expect H_2SO_4 mists to be found only in close proximity to combustion sources, or to be widely dispersed in the atmospheres of industrial societies? Justify your answers.

18.47 When 1 mole of sulfuric acid, $H_2SO_4(\ell)$, is mixed with a large quantity of water, 96 kJ of heat is given out. (a) Is ΔH positive or negative for the process

$$H_2SO_4(\ell) + aq \longrightarrow H_2SO_4(aq)$$

(b) Mixing generally represents an increase in randomness. On the basis of this generalization, is ΔS positive or negative for this process? (c) Is ΔG positive or negative? (d) Assume that the signs of ΔH and ΔS remain the same when H_2SO_4 and water are mixed in any proportions at any temperature. Is the process in part a nonspontaneous at any temperature? At any concentration? What can you conclude about the solubility of H_2SO_4 in water?

18.48 Consider the equations

$$\Delta G^0 = \Delta H^0 - T\Delta S^0$$

$$R \ln K = \frac{-\Delta G^0}{T} = \Delta S^0 - \frac{\Delta H^0}{T}$$

When ΔH^0 is negative and ΔS^0 is positive, do (a) ΔG^0 and (b) K increase or decrease as the temperature increases? (c) Is there anything strange about your answers? Discuss this question with specific reference to the reaction $H_2(g) + Cl_2(g) \longrightarrow 2HCl(g)$, for which $\Delta H^0 = -184.6$ kJ and $\Delta S^0 = 0.02007$ kJ/K. Calculate ΔG^0 and $R \ln K$ for this reaction at 300 K and 350 K.

18.49 (a) Calculate ΔG^0 at 25°C for the reaction

$$Cd(c) + Br_2(\ell) \longrightarrow Cd^{2+}(aq) + 2Br^-(aq)$$

from Table 17.1. (b) For this reaction, $\Delta H^0 = -317.2$ kJ at 25°C. Calculate ΔS^0. (c) Estimate $\mathscr{E}^0$ at 100°C for a cell in which the reaction in part a occurs.

18.50 An experimental fuel cell uses carbon monoxide, CO(g), as fuel. The electrolyte is a solid solution of cerium(IV) oxide, CeO_2, and gadolinium oxide, Gd_2O_3. Oxide ions (O^{2-}) can move through this crystal at high temperatures (about 800°C). The cell can be described as

$$Ni, CeO_2|CO(g)|CeO_2, Gd_2O_3|O_2(g)|SnO_2, Sb$$

(a) For the reaction $CO(g) + \frac{1}{2}O_2(g) \longrightarrow CO_2(g)$, estimate ΔG^0 at 800°C. (b) Calculate the maximum possible emf of this cell at 800°C when each gas pressure is 1 atm.

18.51 For the process S_8(rhombic) $\longrightarrow$ S_8(monoclinic) around 80 to 110°C, $\Delta H^0 = 3213$ J/mol and $\Delta S^0 = 8.70$ J/(mol K). Estimate ΔG^0 for the process at (a) 80°C and (b) 110°C. What do these results tell about the stability of the two forms of sulfur at each of these temperatures? (c) Calculate the temperature at which $\Delta G^0 = 0$. What is the significance of this temperature?

18.52 (a) The equation $\Delta S = \Delta H/T$ sometimes appears. When this equation is correct, what can you say about ΔG? (b) The freezing point of benzene, C_6H_6, is 5.5°C. At each of the following temperatures, is ΔG for the process

$$C_6H_6(c) \longrightarrow C_6H_6(\ell)$$

positive, negative, or zero? 0°C, 5.5°C, 10°C. (c) At each

of these temperatures, is ΔS less than, greater than, or equal to $\Delta H/T$?

18.53 (a) For the vaporization of a liquid at its normal boiling point, what is ΔG^0? How are ΔH^0 and ΔS^0 for this process related? (b) *Trouton's rule* says that when 1 mole of a liquid is vaporized at its boiling point, $\Delta S^0 \approx$ 88 J/(mol K). If the boiling point of a liquid is 300 K, what is ΔH^0 of vaporization? (c) For water at 100°C, ΔH^0 of vaporization is 40.68 kJ/mol. What is ΔS^0 of vaporization? Does water conform to Trouton's rule? If it does not, can you think of a reason?

18.54 For the process $H_2O(\ell) \longrightarrow H_2O(g)$, $\Delta H = 40.68$ kJ/mol and

$$\Delta S = (109.0 - 19.14 \log P)\ \text{J/(mol K)}$$

where P is the pressure of $H_2O(g)$ in atm. (a) Write an expression for ΔG in terms of temperature and pressure. (b) Calculate ΔG at 110°C and 1.50 atm. Which state, liquid or gas, is more stable? (c) When $P = 1.20$ atm, at what temperature are the liquid and gas in equilibrium? (d) Calculate the vapor pressure of water at 90°C.

SELF-TEST

18.55 (12 points) When 1.00 mole of an ideal gas expands from 1.00 atm to 0.100 atm at 25°C, it does 5708 J of work. Its energy is unchanged. (a) What is w, with sign? (b) What is q, with sign? Is heat given out or absorbed?

18.56 (12 points) To bring about the reaction

$$2NaCl(c) \longrightarrow 2Na(c) + Cl_2(g)$$

at least 768 kJ of work (excluding pressure-volume work) must be done on the system. (a) What is ΔG? (b) Give two other possible values of w. State whether the work is done on or by the system.

18.57 (16 points) This cell has been proposed for use in electric automobiles:

$$Zn(c)|ZnO(c)(\text{dispersed in electrolyte})|Na_2SO_4(aq)|O_2(g)|C(\text{porous})$$

For the cell reaction

$$Zn(c) + \tfrac{1}{2}O_2(g) \longrightarrow ZnO(c)$$

at 25°C, $\Delta G^0 = -318.2$ kJ. (a) Calculate the standard emf of the cell at 25°C. (b) Assume that the actual emf is 70% of the value calculated in part a. How many kilowatt-hours of work could be obtained by the oxidation of 1 kg of Zn?

18.58 (15 points) From data in Table 17.1, calculate ΔG_f^0 for Hg_2Cl_2 at 25°C.

18.59 (18 points) (a) Calculate ΔH^0, ΔG^0, and ΔS^0 at 25°C for the reaction

$$H_2(g) + CO_2(g) \rightleftharpoons H_2O(g) + CO(g)$$

(b) Estimate ΔG^0 and K_p for this reaction at 700°C.

18.60 (12 points) (a) Calculate ΔH^0, ΔG^0, and ΔS^0 for the reaction

$$2H_2O_2(\ell) \longrightarrow 2H_2O(\ell) + O_2(g)$$

at 25°C. (b) Is there any temperature at which $H_2O_2(\ell)$ is stable at 1 atm?

18.61 (15 points) Given the following data at 25°C:

	ΔH_f^0, kJ/mol	ΔG_f^0, kJ/mol
$Hg(\ell)$	0	0
$Hg(g)$	61.317	31.85

(a) Calculate ΔS^0 at 25°C for the process $Hg(\ell) \longrightarrow Hg(g)$. (b) Estimate the temperature at which K_p for the process becomes equal to (1) 1/760 and (2) 1.00. What is the vapor pressure of Hg at each of these temperatures?

19
CHEMICAL KINETICS

Chemistry is concerned with changes in matter. The study of chemical equilibrium helps us to predict whether a reaction tends to go as it is written under specific conditions and to calculate the composition of the equilibrium mixture. However, **chemical kinetics** is concerned with the rate (speed) at which reactions occur. Some reactions are extremely rapid—coal dust (carbon) reacts explosively with air—and others are immeasurably slow—diamonds (carbon) "are forever." This chapter describes the conditions that affect the rates of reactions and explains the theory of reaction rates. Chemical kinetics is also concerned with finding the exact path by which reactants go to products. Unlike thermodynamic properties, the rate of a reaction is path-dependent, and finding the path is an intellectually interesting exercise with practical applications.

19.1 CHANGE OF CONCENTRATIONS WITH TIME

Kinetic stability refers to a given set of reactants and products at specified conditions of particle size, concentration (pressure), temperature, and presence/absence of a catalyst.

We have just learned that thermodynamics can predict the extent of a reaction from the change in free energy, ΔG, for the reaction. However, some reactions go very slowly so that even a thermodynamically "spontaneous" reaction (large negative ΔG) may not actually occur. ΔG tells us nothing about the rate of a chemical reaction or its *kinetic stability* (section 13.2). The time it takes for a reaction to occur is a measure of its kinetic stability; the *slower* the reaction, the *greater* is its kinetic stability. Thermodynamic instability and kinetic instability do not necessarily accompany one another. For example, the decomposition of dinitrogen oxide (nitrous oxide), $N_2O(g) \longrightarrow N_2(g) + \frac{1}{2}O_2(g)$, has a very strong thermodynamic tendency to go as written, $\Delta G^0 = -104.2$ kJ. Yet the oxide is used as the propellant in aerosol cans of whipped cream without fear that it may decompose. If any reaction occurs at room temperature, its rate is immeasurably slow. At the other extreme, the decomposition of pure hydrogen peroxide

$$H_2O_2(\ell) \longrightarrow H_2O(\ell) + \tfrac{1}{2}O_2(g) \qquad \Delta G^0 = -116.8 \text{ kJ}$$

may occur explosively. The decomposition is thermodynamically favored and also proceeds at a very fast rate.

In section 13.1, the rate of a chemical reaction was defined as the change in concentration per unit time. For a reactant, the concentration decreases, while for a product it increases.

We can determine the concentration of a substance undergoing reaction by a variety of methods. The concentration can be related to such measurable quantities as pressure (for a reaction accompanied by a change in the number of moles

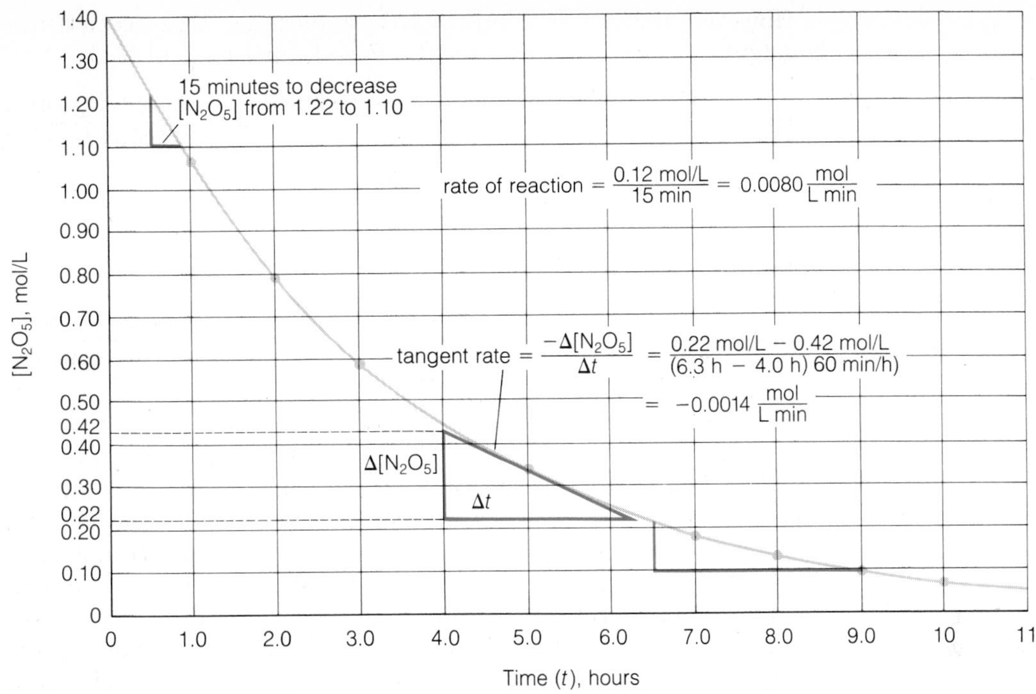

FIGURE 19.1
Plot of reactant concentration against time. The calculation of the slope at $[N_2O_5] = 0.34$ mol/L and 5.0 h is shown on a larger scale.

of gas), color, pH, and electrical conductivity. Changes in the concentration of a substance lead to a corresponding change in one or more of these quantities. Measurements are made at definite time intervals. Thus the rate of the reaction at any concentration can be obtained from the changes.

An example is the decomposition of dinitrogen pentoxide dissolved in liquid carbon tetrachloride:

$$N_2O_5(sol) \longrightarrow 2NO_2(sol) + \tfrac{1}{2}O_2(g)$$

The increase in the pressure of O_2 is related to the decrease in the concentration of N_2O_5. For every $\frac{1}{2}$ mole of O_2 that is formed, 1 mole of N_2O_5 disappears. Typical experimental results at 30.0°C are plotted in Figure 19.1. The rate of the reaction is expressed as the change in concentration of N_2O_5 divided by the change in time,

$$\text{rate of reaction} = \frac{\text{change in } [N_2O_5]}{\text{change in time}}$$

or, more briefly,

Δ has its usual meaning: Δ = final − initial.

$$rate = -\frac{\Delta[N_2O_5]}{\Delta t}$$

The minus sign indicates that the rate is equal to the decrease in $[N_2O_5]$ with time. The rate is always expressed as a positive quantity. For example, the rate between 40 min and 55 min (see Figure 19.1) is given by

$$-\frac{\Delta[N_2O_5]}{\Delta t} = -\frac{1.10\,\dfrac{\text{mol}}{\text{L}} - 1.22\,\dfrac{\text{mol}}{\text{L}}}{55\text{ min} - 40\text{ min}} = +\frac{0.12\,\dfrac{\text{mol}}{\text{L}}}{15\text{ min}} = 0.0080\,\frac{\text{mol}}{\text{L min}}$$

The rate calculated for a concentration-time interval is only an average rate for the chosen interval.

However, the rate in terms of NO_2 is twice as large because the chemical equation tells us that 2 moles of NO_2 form from 1 mole of N_2O_5:

$$rate = \frac{\Delta[NO_2]}{\Delta t} = \frac{0.0080 \cancel{\text{ mol } N_2O_5 \text{ consumed}}}{\text{L min}} \times \frac{2 \text{ mol } NO_2 \text{ formed}}{\cancel{1 \text{ mol } N_2O_5 \text{ consumed}}}$$

$$= 0.016 \frac{\text{mol } NO_2 \text{ formed}}{\text{L min}}$$

In terms of O_2, the rate is

$$rate = \frac{\Delta[O_2]}{\Delta t} = 0.0080 \frac{\cancel{\text{mol } N_2O_5 \text{ consumed}}}{\text{L min}} \times \frac{\frac{1}{2} \text{ mol } O_2 \text{ formed}}{\cancel{1 \text{ mol } N_2O_5 \text{ consumed}}}$$

$$= 0.0040 \frac{\text{mol } O_2 \text{ formed}}{\text{L min}}$$

Figure 19.1 shows that the rate of the reaction decreases as the reactant concentration decreases. The concentration of N_2O_5 drops sharply at the start of the reaction but decreases very slowly near the end of the reaction. A more accurate determination of the rate can be made at a *given concentration* from Figure 19.1. The rate is obtained from the slope (Appendix B.4) of the tangent to the curve. The tangent is drawn at a point on the curve that corresponds to a definite concentration and time. The slope then gives the rate at the specified concentration or time. Figure 19.1 shows how the rate corresponding to $[N_2O_5] = 0.34$ mol/L at 5.0 h is obtained:

The tangent rate is also known as the *instantaneous* rate because it gives the rate when the time interval is infinitely small. The slope is then the rate at a particular time instead of an average rate over a time interval.

$$rate \text{ at } 0.34 \frac{\text{mol}}{\text{L}} = -\frac{\Delta[N_2O_5]}{\Delta t} = -\frac{0.22 \frac{\text{mol}}{\text{L}} - 0.42 \frac{\text{mol}}{\text{L}}}{(6.3 \text{ h} - 4.0 \text{ h}) \times 60 \frac{\text{min}}{\text{h}}}$$

$$= +\frac{0.20 \frac{\text{mol}}{\text{L}}}{138 \text{ min}} = 1.4 \times 10^{-3} \frac{\text{mol}}{\text{L min}}$$

Modern techniques make it possible to study much faster reactions. Measurements in units of picoseconds, 10^{-12} sec, can now be made with lasers and other electronic devices.

Reactions are frequently classified as heterogeneous or homogeneous (section 13.7). The combustion of graphite is a typical **heterogeneous reaction**, a reaction that occurs only at the interface (boundary) between two phases:

$$C(\text{graphite}) + O_2(g) \longrightarrow CO_2(g)$$

The reactions

$$NO(g) \longrightarrow \tfrac{1}{2}N_2(g) + \tfrac{1}{2}O_2(g)$$

$$H_2O_2(aq) + 3I^-(aq) + 2H^+(aq) \longrightarrow I_3^-(aq) + 2H_2O$$

are typical **homogeneous reactions**, reactions occurring in only one phase.

19.2 CONDITIONS AFFECTING REACTION RATES

The rate of a reaction depends on the following factors: (1) the natures of the reactants, (2) the concentrations (or pressures) of the reactants, (3) the size of the particles of the solids or liquids in a heterogeneous reaction, (4) the temperature, and (5) the presence of a catalyst (section 13.6).

NATURE OF REACTANTS

Copper or silver reacts very slowly with oxygen, even in a flame, whereas magnesium reacts very rapidly. White phosphorus ignites spontaneously in air, but red phosphorus does not. Silver ions and chloride ions react rapidly,

$$Ag^+(aq) + Cl^-(aq) \longrightarrow AgCl(c)$$

whereas magnesium ions and oxalate ions react slowly:

$$Mg^{2+}(aq) + C_2O_4{}^{2-}(aq) \longrightarrow MgC_2O_4(c)$$

CONCENTRATION OF REACTANTS

At constant temperature the rate of a homogeneous reaction is usually proportional to some power of the concentration of each reactant. This generalization simply states that for the general net reaction $aA + bB \longrightarrow cC$ the rate of formation of C, or the rate of disappearance of A or B, is proportional to powers of the concentrations of A and B:

Recall that $\propto$ means "proportional to."

$$rate \propto [A]^x[B]^y$$

Then we may write a rate equation:

$$rate = k[A]^x[B]^y$$

Here k is a constant, called the **rate constant**, for the reaction. It is not a universal constant; it is a constant for a particular reaction at a specific temperature. It changes with temperature but is independent of reactant concentrations. The numerical values of the exponents x and y must be *determined experimentally. They cannot be deduced* from the coefficients in the net reaction or from any theory. Most important, they *are not necessarily the same as the coefficients a and b*. It is possible for some of these exponents to be fractions, zero, or even negative. For example, if $x = 0$, the rate equation is

$$rate = k[A]^0[B]^y = k \times 1 \times [B]^y = k[B]^y$$

The units of k depend on the concentration factors and the unit of time. The equation relating the rate of reaction to the concentrations of reactants is generally known as a **rate equation**.

If the rate depends on a product of concentration factors, such as

$$rate = k[A]^x[B]^y \ldots$$

then the *sum of the powers of the concentration factors in the rate equation* $(x + y \ldots)$ is called the **order of the reaction**. For example, the reaction $2NO + O_2 \longrightarrow 2NO_2$ follows the rate equation

$$rate = k[NO]^2[O_2]$$

The reaction is then described as *third-order* because the sum of the powers (2 + 1) is 3. The reaction is also said to be *second-order* with respect to NO and *first-order* with respect to O_2. The rate equation for the conversion of ammonium cyanate, NH_4CNO, to urea in water solution

$$NH_4{}^+ + CNO^- \longrightarrow NH_2CONH_2$$

is

$$rate = k[NH_4{}^+][CNO^-]$$

This reaction is second-order, since it is first-order with respect to both $NH_4{}^+$ and CNO^- (Box 19.1). However, the orders with respect to the reactants are often

BOX 19.1
KINETICS AND THE ARRHENIUS THEORY OF IONIC DISSOCIATION

The first kinetic study (1895) of the formation of urea from ammonium cyanate provided strong evidence for the highly controversial theory of ionic dissociation postulated by Svante Arrhenius in 1887. Chemists had struggled for most of the nineteenth century to verify the existence of atoms and molecules as the elementary particles of matter—a view that was further strengthened by the successes of the kinetic molecular theory of gases. The furor created by the assumption that large portions of solutes in electrolytic solutions exist not as molecules but as *ions* cannot be appreciated today. Arrhenius mailed a copy of his dissertation to Wilhelm Ostwald, generally recognized as the father of theoretical chemisty, and Ostwald wrote that he "got on the same day [the] dissertation, a toothache, and a nice daughter and [that] that was too much for one day. The poorest was the dissertation, for the others developed quite normally." The kinetic study (Problem 19.31) shows that the rate increases if a soluble cyanate salt, such as KCNO, or a soluble ammonium salt, such as $(NH_4)_2SO_4$, is added to an ammonium cyanate solution. Thus, the two particles involved in this second-order reaction are the ions NH_4^+ and CNO^-, not NH_4CNO molecules.

not identical to the coefficients of the reactants. An example of the lack of such a correlation is the reaction in acid solution between bromate and bromide ions forming bromine:

$$BrO_3^- + 5Br^- + 6H^+ \longrightarrow 3Br_2 + 3H_2O$$

The experimentally determined rate equation is

$$rate = k[BrO_3^-][Br^-][H^+]^2$$

This reaction is fourth-order—first-order with respect to BrO_3^-, first-order with respect to Br^-, and second-order with respect to H^+.

The decomposition of NH_3

$$2NH_3(g) \longrightarrow N_2(g) + 3H_2(g)$$

at 856°C on a platinum surface also illustrates that you cannot write the exponents by looking at the coefficients in the balanced chemical equation. The rate equation is

$$rate = k[NH_3]^0 = k$$

and the reaction is zero-order. This means that the reaction is independent of the NH_3 concentration. The order of zero is frequently found in reactions that occur on surfaces or in enzymatic reactions (Box 19.4). In this case, the surface is saturated with NH_3 molecules so that the surface concentration becomes independent of the measured NH_3 concentration in the gas phase. It is like a reaction occurring in a saturated solution in the presence of excess solute. As the dissolved solute reacts, it is replaced by excess undissolved solute.

PROBLEM 19.1 What is the order of each of these reactions?

(a) $2N_2O_5(sol) \longrightarrow 4NO_2(sol) + O_2(g)$; $rate = k[N_2O_5]$
(b) $CH_3CHO(g) \longrightarrow CH_4(g) + CO(g)$; $rate = k[CH_3CHO]^{3/2}$ □

The rate equation not only reveals the order of the reaction; as we will learn later (section 19.9), it also is the key to finding how reactants go to products.

BOX 19.2
PARTICLE SIZE AND EXPLOSIONS

Disastrous coal mine explosions do not always result from ignition of a gas. Bituminous (''soft'') coal dust suspended in air may be ignited by a match, producing a violent explosion. The rate of pressure rise reaches as much as 40 atm/sec. The pressure increase results from the rapid expansion of the gaseous products of the reaction and of air being heated by the reaction. This expansion is caused by the heat evolved from the combustion reaction. In general, combustible dusts, such as grain dust, suspended in air must be regarded as explosion hazards.

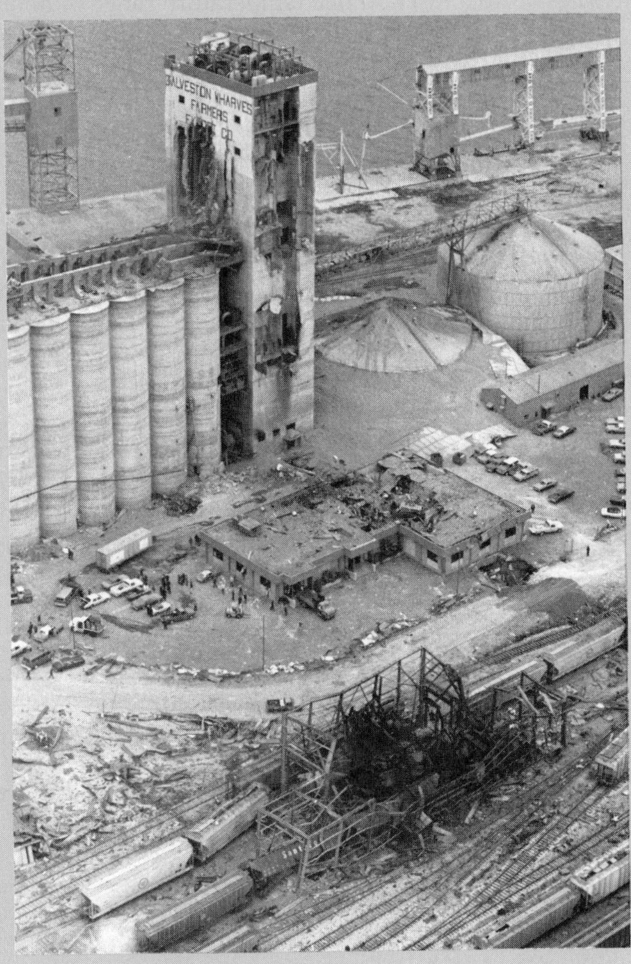

The explosion of this grain elevator in Galveston, Texas, in 1977 illustrated the hazard of combustible dusts suspended in air.

PARTICLE SIZE IN HETEROGENEOUS REACTIONS

Heterogeneous reactions occur at the surface boundary between reacting phases. Usually, one of the phases is a solid. Consequently, the rate of such a reaction is proportional to the surface area of the solid. When a given mass is subdivided into smaller particles, the surface area is increased and the rate of reaction increases. Lumps of coal or zinc, for example, are difficult to ignite in air, but when they are pulverized and dispersed in air, they react explosively (Box 19.2).

EFFECT OF TEMPERATURE

Temperature has a striking effect on the rate of chemical reactions. Our daily experiences provide many examples that show that the rate of a chemical reaction usually increases when the temperature is increased. For instance, we use a stove to cook food, a match to burn a piece of paper, and an ignition system to explode gasoline in an automobile engine. Conversely, storing food in a refrigerator slows down biochemical decomposition reactions and so retards food spoilage. The effect of temperature is the same for endothermic and exothermic reactions. The magnitude of the change in rate varies from one reaction to another and from one temperature range to another. A very rough but useful approximation is that an increase in temperature of 10°C doubles or triples the rate. Reaction rates that are negligibly slow at ordinary temperatures may become appreciably faster and even explosive at elevated temperatures. A rate "constant" is constant only as long as the temperature is constant.

EFFECT OF A CATALYST

Many reactions are speeded up by the presence of a substance that does not appear in the balanced equation for the net reaction (section 13.6). As we shall learn in section 19.10, a catalyst not only increases the value of the rate constant at constant temperature; it may also change the order of the reaction and the path by which reactants go to products.

19.3 DETERMINING REACTION ORDERS AND RATE CONSTANTS

Determining rate constants is not always as easy as illustrated by textbook examples. The measurement is especially difficult when dealing simultaneously with a large number of related reactions. For example, a minimum of 150 reactions are involved in polluted atmospheres. Atmospheric pollution problems cannot be analyzed (much less solved) until rate constants for these atmospheric reactions are established. The International Council of Scientific Unions has established a task group to study the problem.

Rate constants are calculated from rate equations. The following examples show how rate equations can be derived from the way in which the rate of reaction varies with reactant concentration.

EXAMPLE 19.1 The following data at 556 K are given for the decomposition of hydrogen iodide:

$$2HI(g) \longrightarrow H_2(g) + I_2(g)$$

[HI]	RATE OF DECOMPOSITION OF HI AT THE GIVEN CONCENTRATION
0.0100 mol/L	3.5×10^{-11} mol/(L sec)
0.0200	14×10^{-11}
0.0300	32×10^{-11}

(a) Write the rate equation for the reaction. What is the order of the reaction? (b) Calculate the rate constant for the reaction at 556 K. (c) Calculate the decomposition rate at 556 K at the instant when [HI] = 0.0325 mol/L.

ANSWER

(a) Note from the data that:

As the concentration is increased 2 times from 0.0100 to 0.0200,
the rate is increased 4 times (2^2) from 3.5×10^{-11} to 14×10^{-11}.

As the concentration is increased 3 times from 0.0100 to 0.0300, the rate is increased 9 times (3^2) from 3.5×10^{-11} to 32×10^{-11}.

In each case, the ratio of the rates is the square of the ratio of the concentrations. For example,

$$\frac{14 \times 10^{-11} \text{ mol/(L sec)}}{3.5 \times 10^{-11} \text{ mol/(L sec)}} = \left[\frac{0.0200 \text{ mol/L}}{0.0100 \text{ mol/L}}\right]^2 \quad \text{or } 4 = 2^2$$

The rate of this reaction is therefore proportional to the square of [HI], or

$$rate = k[\text{HI}]^2$$

The rate is proportional to the second power of the concentration of one reactant. Thus the reaction is second-order.

(b) Solve for k by substituting values for any one concentration:

$$k = \frac{rate}{[\text{HI}]^2} = \frac{3.5 \times 10^{-11} \text{ mol (L sec)}}{(0.0100)^2 \text{ mol}^2/\text{L}^2}$$

$$= 3.5 \times 10^{-7} \text{ L/(mol sec)}$$

k is 3.5×10^{-7} L/(mol sec). The other three concentration values give the same answer.

(c) From the previous calculation

$$rate = 3.5 \times 10^{-7} \frac{\text{L}}{\text{mol sec}} [\text{HI}]^2$$

so that when [HI] = 0.0325 mol/L

$$rate = 3.5 \times 10^{-7} \frac{\cancel{\text{L}}}{\cancel{\text{mol}} \text{ sec}} \times (0.0325)^2 \frac{\text{mol}^{\cancel{2}}}{\text{L}^{\cancel{2}}} = 37 \times 10^{-11} \frac{\text{mol}}{\text{L sec}}$$ ■

When two reactants participate in a reaction, the rate equation may be derived by keeping the concentration of one reactant constant while varying the concentration of the other.

EXAMPLE 19.2 The reaction $\text{NO(g)} + \frac{1}{2}\text{Cl}_2\text{(g)} \longrightarrow \text{NOCl(g)}$ has been studied at 323 K:

[NO]	[Cl_2]	RATE OF FORMATION OF NOCl AT THE GIVEN CONCENTRATIONS
0.250 mol/L	0.250 mol/L	1.43×10^{-6} mol/(L sec)
0.250	0.500	2.86×10^{-6}
0.500	0.500	11.4×10^{-6}

(a) Write the rate equation for the reaction. What is the order of the reaction? (b) Find the rate constant for the reaction at 323 K. (c) Calculate the rate of formation of NOCl at the instant when [NO] = [Cl_2] = 0.110 mol/L. (d) At the instant when Cl_2 is reacting at the rate of 2.21×10^{-7} mol/(L sec), what is the rate at which NO is reacting and NOCl is forming?

ANSWER

(a) Doubling the concentration of Cl_2 from 0.250 to 0.500 mol/L while holding [NO] constant at 0.250 M doubles the rate. This makes the rate of the reaction proportional to [Cl_2]. But doubling the concentration of NO from 0.250

to 0.500 mol/L while holding $[Cl_2]$ constant at 0.500 M quadruples the rate. This means that the rate is proportional to the square of [NO]. In summary,

	[NO]	$[Cl_2]$			RATE	
	0.250	0.250	× 2		1.43×10^{-6}	× 2
× 2	0.250	0.500		× 4	2.86×10^{-6}	
	0.500	0.500			11.4×10^{-6}	

$rate = k[NO]^2[Cl_2]$

The reaction is third-order: second-order with respect to NO and first-order with respect to Cl_2.

(b) k can be calculated from any one of the three sets of values:

$$k = \frac{rate}{[NO]^2[Cl_2]} = \frac{2.86 \times 10^{-6}\ \cancel{mol}/(\cancel{L}\ sec)}{(0.250\ mol/L)^2 \times 0.500\ \cancel{mol/L}} = 9.15 \times 10^{-5}\ \frac{L^2}{mol^2\ sec}$$

(c) $$rate = 9.15 \times 10^{-5}\ \frac{L^2}{mol^2\ sec}\ [NO]^2[Cl_2]$$

and when $[NO] = [Cl_2] = 0.110$,

$$rate = 9.15 \times 10^{-5}\ \frac{\cancel{L^2}}{\cancel{mol^2}\ sec} \times (0.110)^2\ \frac{\cancel{mol^2}}{\cancel{L^2}} \times 0.110\ \frac{mol}{L}$$
$$= 1.22 \times 10^{-7}\ mol/(L\ sec)$$

(d) From the chemical equation, 2 moles of NO must react for each mole of Cl_2 that reacts, and 2 moles of NOCl must form for each mole of Cl_2 that reacts. Therefore

$$\text{rate at which NO is reacting} = \frac{2.21 \times 10^{-7}\ \cancel{mol\ Cl_2}}{L\ sec} \times \frac{2\ mol\ NO}{1\ \cancel{mol\ Cl_2}}$$
$$= 4.42 \times 10^{-7}\ mol\ NO/(L\ sec)$$

$$\text{rate at which NOCl is forming} = \frac{2.21 \times 10^{-7}\ \cancel{mol\ Cl_2}}{L\ sec} \times \frac{2\ mol\ NOCl}{1\ \cancel{mol\ Cl_2}}$$
$$= \frac{4.42 \times 10^{-7}\ mol\ NOCl}{L\ sec}$$

■

PROBLEM 19.2 From the following data for the formation of HI

$$H_2(g) + I_2(g) \longrightarrow 2HI(g)$$

(a) find the rate equation, (b) calculate the rate constant, and
(c) calculate the rate when $[H_2] = 0.0400$ mol/L and $[I_2] = 0.00600$ mol/L.

$[H_2]$	$[I_2]$	RATE OF FORMATION OF HI AT THE GIVEN CONCENTRATIONS
0.0100 mol/L	0.0100 mol/L	8.40×10^{-9} mol/(L sec)
0.0100	0.0200	16.8×10^{-9}
0.0200	0.0200	33.6×10^{-9}

□

19.4 RELATIONSHIP BETWEEN CONCENTRATION AND TIME

Recall that rate constants are constant as long as the temperature is constant.

Rate equations do not show directly how concentrations depend on time. Concentrations change with time, whereas the rate constant remains constant. The rate equation must be further treated mathematically to obtain an equation that expresses concentration as a function of time. These equations allow us to find the concentration of a reactant remaining at any time after the reaction has started or how much time is required for a reactant concentration to reach a certain value. However, the form of the equation differs for different reaction orders. We therefore restrict this discussion to the simplest reaction orders. You will see that these equations can be written to fit the equation for a straight line.

FIRST-ORDER REACTIONS

Suppose the reaction

$$\text{R} \longrightarrow \text{products}$$

is first-order. Then the rate equation is

$$rate = -\frac{\Delta[\text{R}]}{\Delta t} = k[\text{R}]$$

This is solved by integral calculus.

Solving this equation gives an **integrated rate equation**:

$$\mathbf{2.303 \log \frac{[R]_0}{[R]_{\mathit{t}}} = \mathit{kt}} \qquad (1)$$

In equation 1, $[\text{R}]_0$ is the concentration of the reactant at the time $t = 0$. $t = 0$ need not correspond to the actual beginning of the experiment; it can be the time when the instrument readings are started. $[\text{R}]_t$ is the concentration at a later time, t.

$$\frac{[\text{R}]_0\ \cancel{\text{mol/L}}}{[\text{R}]_t\ \cancel{\text{mol/L}}}$$

Notice that for the first-order rate equation 1 the ratio appearing in the logarithm is *dimensionless*. The unit of k is therefore sec^{-1}. This means that k for first-order reactions is independent of the units chosen for concentration, and equation 1 may be rewritten as

$$\mathbf{2.303 \log \frac{\mathit{q}_0}{\mathit{q}_t} = \mathit{kt}} \qquad (1)$$

Now q_0, the quantity of reactant at $t = 0$, and q_t, the quantity at time t, are expressible in any convenient quantity unit—such as moles, grams, number of atoms, number of molecules, or pressure.

SECOND-ORDER REACTIONS

For a second-order reaction of the kind

$$\text{R} \longrightarrow \text{products}$$

the rate equation is

$$rate = -\frac{\Delta[\text{R}]}{\Delta t} = k[\text{R}]^2$$

With the aid of calculus, this equation leads to

$$\mathbf{\frac{1}{[R]_{\mathit{t}}} - \frac{1}{[R]_0} = \mathit{kt}} \qquad (2)$$

The units of k are L/(mol sec).

ZERO-ORDER REACTIONS

For a zero-order reaction of the kind

$$R \longrightarrow \text{products}$$

the rate equation is

$$rate = -\frac{\Delta[R]}{\Delta t} = k[R]^0 = k$$

This equation leads to the integrated rate equation

$$\mathbf{[R]_0 - [R]_t = kt} \qquad (3)$$

The units of k are mol/(L sec).

Clearly, the rate equations 1, 2, and 3 relating concentration, rate constant, and time are different for different reaction orders. These equations thus afford a method of finding the order of a reaction and its rate constant from a plot of the kinetic data.

DETERMINING THE REACTION ORDER AND RATE CONSTANT

Equations 1, 2, and 3 each take the form of a straight line as shown: the *zero-order* equation 3 is rearranged

$$[R]_t = [R]_0 - kt \qquad (3')$$

to the same form as the equation for a straight line:

$$y = b + mx$$

The rate constant, k, is given by the slope of the line when $[R]_t$ (y axis) is plotted against t (x axis). In Figure 19.2, the concentration of NH_3 in mmol/L present at time t is plotted against t in seconds for the decomposition of NH_3 on a tungsten wire at 856°C. The straight line obtained proves that this reaction can be only zero-order. The slope of the line is found by selecting any two points on the line

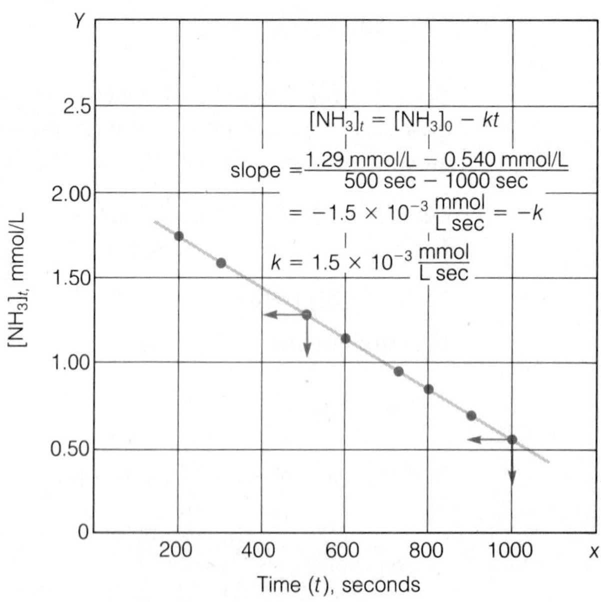

FIGURE 19.2
Plot of $[NH_3]_t$ against time for the decomposition of NH_3, $2NH_3(g) \rightarrow N_2(g) + 3H_2(g)$, on a tungsten wire at 856°C, a zero-order reaction.

and reading off the coordinates as indicated by the arrows in Figure 19.2:

$$\text{slope} = \frac{1.29\,\frac{\text{mmol}}{\text{L}} - 0.54\,\frac{\text{mmol}}{\text{L}}}{500\text{ sec} - 1000\text{ sec}} = -1.5 \times 10^{-3}\,\frac{\text{mmol}}{\text{L sec}}$$

A line slanting downward from the left has a negative slope.

The slope = $-k$, so that

$$-k = -1.5 \times 10^{-3}\,\frac{\text{mmol}}{\text{L sec}}$$

and

$$k = 1.5 \times 10^{-3}\,\frac{\text{mmol}}{\text{L sec}}$$

The $[R]_t$ intercept at $t = 0$ is equal to $[R]_0$.

PROBLEM 19.3 Data for the decomposition of N_2O_5 in a particular solution at 45.0°C are given:

$[N_2O_5]$, mol/L	t, minutes
2.08	3.07
1.67	8.77
1.36	14.45
0.72	31.28

Plot $[N_2O_5]_t$ against time, t. Is this reaction zero-order? □

Similarly, for a *first-order* reaction, equation 1

$$\log\,[R]_0 - \log\,[R]_t = \frac{kt}{2.303}$$

is rearranged to the form of the equation for a straight line:

$$\log\,[R]_t = \log\,[R]_0 - \frac{k}{2.303}\,t \qquad (1')$$

A plot of log $[R]_t$ against t is a straight line for a first-order reaction whose slope is $-k/2.303$ and whose intercept at $t = 0$ is log $[R]_0$.

In Figure 19.3, the log of the concentration in mmol/L of diethyl ether, log $[C_2H_5OC_2H_5]$, present at time t, is plotted against t in hours for the decomposition of the ether at 400°C:

$$2C_2H_5OC_2H_5(g) \longrightarrow 2CO(g) + 4CH_4(g) + C_2H_4(g)$$

The straight line obtained shows that this reaction is first-order.

PROBLEM 19.4 Use the kinetic data in Problem 19.3 to plot log $[N_2O_5]$ against t for the decomposition of N_2O_5. What is the order of the reaction? □

Finally, for a *second-order* reaction, rearrangement of equation 2 also yields an equation for a straight line:

$$\frac{1}{[R]_t} = \frac{1}{[R]_0} + kt \qquad (2')$$

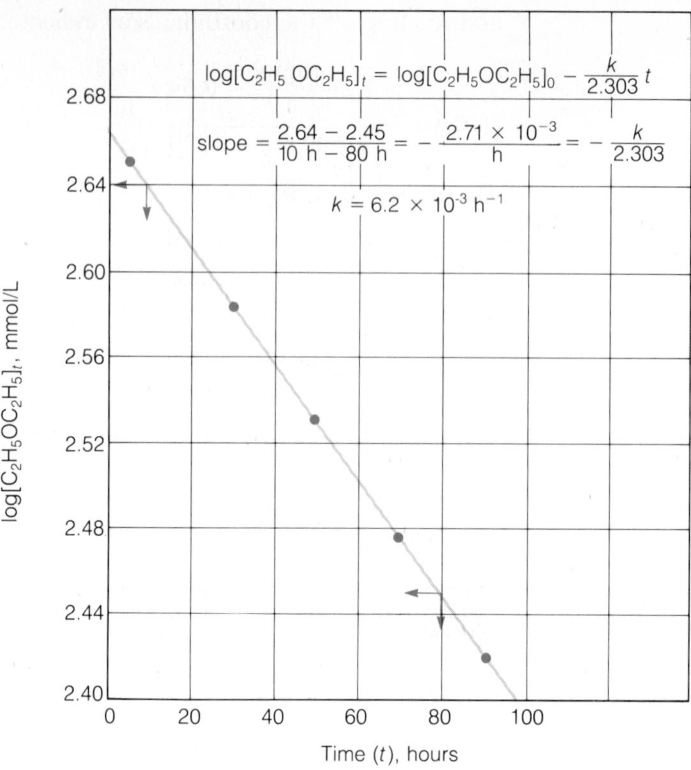

FIGURE 19.3
Plot of log $[C_2H_5OC_2H_5]$ against time for the decomposition of diethyl ether, $2C_2H_5OC_2H_5(g) \rightarrow 2CO(g) + 4CH_4(g) + C_2H_4(g)$, at 400°C, a first-order reaction.

Here, a plot of $\dfrac{1}{[R]_t}$ against t for a second-order reaction is a straight line whose slope is k and whose intercept at $t = 0$ is $\dfrac{1}{[R]_0}$. In Figure 19.4, kinetic data for the decomposition of hydrogen iodide at 507°C

$$HI(g) \longrightarrow \tfrac{1}{2}H_2(g) + \tfrac{1}{2}I_2(g)$$

is plotted as $[HI]_t$ against the time t, as log $[HI]_t$ against t, and as $\dfrac{1}{[HI]_t}$ against t. Obtaining a straight line only for the plot of $\dfrac{1}{[HI]_t}$ against t shows that this is a second-order reaction.

PROBLEM 19.5 The kinetic data for the reaction $R \longrightarrow$ products is given at 100°C:

[R], mol/L	t, min
0.0455	30.0
0.0218	90.0
0.0135	160
0.0102	220

Plot $[R]_t$, log $[R]_t$, and $\dfrac{1}{[R]_t}$ against t to find the order of the reaction. □

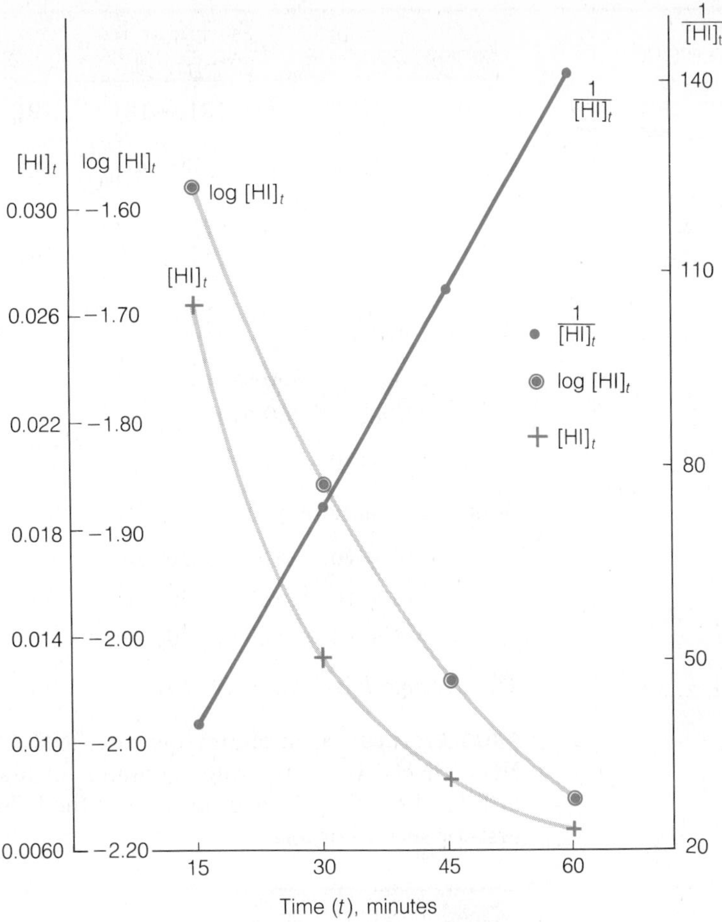

FIGURE 19.4 Plots of $[HI]_t$, $\log[HI]_t$, and $\frac{1}{[HI]_t}$ against time for the decomposition of hydrogen iodide, $HI(g) \rightarrow \frac{1}{2}H_2(g) + \frac{1}{2}I_2(g)$, at 507°C.

Equations 1, 2, and 3 may also be used to determine the value of k, as illustrated in Figures 19.2 and 19.3 and Example 19.3.

EXAMPLE 19.3 The decomposition of N_2O_5 in the gas phase at 55°C

$$N_2O_5 \longrightarrow NO_2 + NO_3$$

is a first-order reaction. For an initial N_2O_5 pressure of 300 torr, the following data were obtained:

TIME, sec	$p_{N_2O_5}$, torr
0	300
100	260
200	225
300	197
400	170

Find the average k at 55°C.

ANSWER The equation for a first-order reaction is

$$\frac{2.303}{t} \log \frac{q_0}{q_t} = k$$

TABLE 19.1
CHARACTERISTIC PROPERTIES OF REACTIONS OF THE TYPE R ⟶ PRODUCTS

ORDER	RATE EQUATION	INTEGRATED RATE EQUATION	PLOT	SLOPE	$t_{1/2}$	k UNIT
0	$k[\mathrm{R}]^0$	$kt = [\mathrm{R}]_0 - [\mathrm{R}]_t$	$[\mathrm{R}]_t$ vs. t	$-k$	$[\mathrm{R}]_0/2k$	mol/(L sec)
1	$k[\mathrm{R}]$	$kt = 2.303 \log \frac{[\mathrm{R}]_0}{[\mathrm{R}]_t}$	$\log [\mathrm{R}]_t$ vs. t	$\frac{-k}{2.303}$	$0.693/k$	sec^{-1}
2	$k[\mathrm{R}]^2$	$kt = \frac{1}{[\mathrm{R}]_t} - \frac{1}{[\mathrm{R}]_0}$	$\frac{1}{[\mathrm{R}]_t}$ vs. t	k	$\frac{1}{k\,[\mathrm{R}]_0}$	L/(mol sec)

Solving for k

$$k = \frac{2.303}{100\ \mathrm{sec}} \log \frac{300\ \cancel{\mathrm{torr}}}{260\ \cancel{\mathrm{torr}}}$$

$$= 1.43 \times 10^{-3}\ \mathrm{sec}^{-1}$$

Similar calculations yield

1.43 $k = 1.44 \times 10^{-3}\ \mathrm{sec}^{-1}$ at 200 sec

1.44 $= 1.40 \times 10^{-3}\ \mathrm{sec}^{-1}$ at 300 sec

1.40 $= 1.42 \times 10^{-3}\ \mathrm{sec}^{-1}$ at 400 sec

1.42

4) 5.69 The average k is $1.42 \times 10^{-3}\ \mathrm{sec}^{-1}$. ■

1.42

PROBLEM 19.6 Ammonium cyanate, NH_4CNO, rearranges in water to urea, NH_2CONH_2. Using the rate equations for first- and second-order reactions given in Table 19.1, find the average k from the following data at 50.1°C. Is the reaction first- or second-order?

t, min	$[NH_4CNO]$, mol/L
0	0.458
45.0	0.370
107	0.292
230	0.212
600	0.114

□

When the value of k is known, rate equations can be used to find how much of a given initial quantity is converted to products at some subsequent time. Or, conversely, we can use the value of k to calculate the length of time needed to convert a definite percentage of reactant to product, the subject of the next section.

19.5 HALF-LIFE AND REACTION ORDERS

The **half-life**, $t_{1/2}$, of a chemical reaction is the time it takes for a given concentration of a reactant, $[\mathrm{R}]_0$, to be decreased by one-half. It indicates the kinetic stability of a reactant; the longer the half-life, the greater is the reactant's stability. The form of the relationship between half-life and initial concentration is different for different orders of reactions.

FIRST-ORDER REACTIONS

The first-order rate equation (section 19.4) is

$$2.303 \log \frac{q_0}{q_t} = kt$$

This equation would also apply to [R]. The half-life is the time it takes to cut a given concentration or a given quantity in half. Therefore,

$$q_t = \frac{q_0}{2} \text{ at } t = t_{1/2}$$

and

$$2.303 \log \frac{q_0}{q_0/2} = 2.303 \log 2 = kt_{1/2}$$

Then,

$$t_{1/2} = \frac{2.303 \log 2}{k} = \frac{0.693}{k} \qquad (4)$$

Notice that both k and $t_{1/2}$ are independent of concentration for first-order reactions. We will see in Chapter 23 how the half-lives of radioactive isotopes are used in determining the ages of fossils, rocks, and ancient artifacts. Half-lives also determine the usefulness of radioisotopes for medical purposes; isotopes with short half-lives are desirable in order to avoid the risk of persistent radiation damage. A half-life can be determined from experimental data by plotting the quantity present at time t against t, as shown in Figure 19.5. The time required

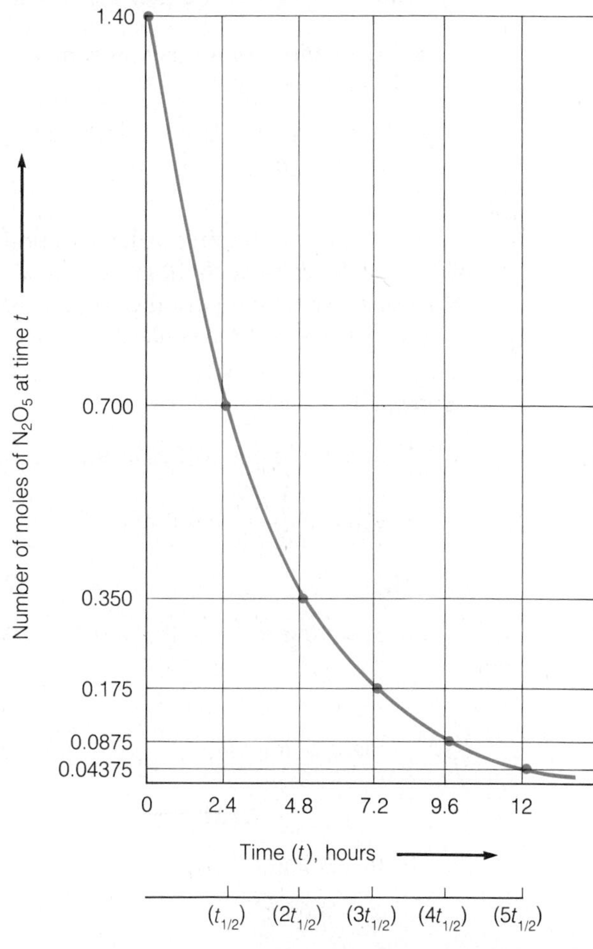

FIGURE 19.5
Determination of $t_{1/2}$ for a first-order reaction. The time required to reduce 1.40 moles of N_2O_5 to 0.70 mole is 2.4 hours, the same as the time required to reduce 0.70 mole to 0.35 mole. The time is also given in multiples of the half-life in parentheses.

for one-half of any chosen quantity to react is the half-life. For example, the half-life for the decomposition of N_2O_5 in carbon tetrachloride at 30°C is 2.4 hours. If we start with 10 grams of N_2O_5 at $t = 0$, then, after a period of 2.4 hours, 5.0 grams remain. After a second period of 2.4 hours (total of 4.8 hours), 2.5 grams remain. After a third period of 2.4 hours (total of 7.2 hours), 1.25 grams remain. For each half-life period, the quantity present is reduced by one-half; thus for 3 half-lives,

$$q_t = 10 \text{ g} \times 0.5 \times 0.5 \times 0.5 = 1.25 \text{ g}$$

of N_2O_5 remaining. This can be expressed as

$$q_t = 10 \text{ g} \times (0.5)^3 = 1.25 \text{ g}$$

in which 3 is the number of half-lives. Generalizing, we have

$$\boldsymbol{q_t = q_0(0.5)^{t/t_{1/2}}} \tag{5}$$

or

$$\log q_t = \log q_0 + \frac{t}{t_{1/2}} \log 0.5 \tag{5'}$$

in which $t/t_{1/2}$ gives the number of half-lives. This equation permits us to find

- the quantity of a reactant remaining (q_t) at any time (t) after the reaction has started, and
- the time (t) required to reduce a given reactant quantity (q_0) to a certain quantity (q_t).

EXAMPLE 19.4 The half-life for the first-order reaction $N_2O_5 \longrightarrow 2NO_2 + \frac{1}{2}O_2$ is 2.4 hours at 30°C. (a) Starting with 10 grams, how many grams will remain after a period of 9.6 hours? (b) What period of time is required to reduce 5.0×10^{10} molecules of N_2O_5 to 1.0×10^8 molecules?

ANSWER

(a) One half-life equals 2.4 h; 9.6 h corresponds to 4 half-lives:

$$9.6 \cancel{\text{h}} \times \frac{1\ t_{1/2}}{2.4 \cancel{\text{h}}} = 4.0\ t_{1/2}$$

Thus,

$$q_t = 10 \text{ g} \times 0.5 \times 0.5 \times 0.5 \times 0.5 = 0.63 \text{ g}$$

or

$$\log q_t = \log 10 + \frac{9.6 \cancel{\text{h}}}{2.4 \cancel{\text{h}}} \log 0.5 = 1.00 + 4.0 \times (-0.301)$$

$$= -0.204$$

from which

$$q_t = 0.63 \text{ g}$$

is the quantity remaining after 9.6 h.

(b) Using equation 5′

$$\log(1.0 \times 10^8 \text{ molecules}) = \log(5.0 \times 10^{10} \text{ molecules}) + \frac{t}{2.4\text{ h}} \log 0.5$$

and rearranging gives

$$\log \frac{1.0 \times 10^8 \text{ \sout{molecules}}}{5.0 \times 10^{10} \text{ \sout{molecules}}} = \frac{t}{2.4\text{ h}} \times \log 0.5$$

$$-2.70 = \frac{t}{2.4\text{ h}} \times (-0.301)$$

from which $t = 22$ h. ■

PROBLEM 19.7 The decomposition of gaseous dimethyl ether, $CH_3OCH_3 \longrightarrow CH_4 + CO + H_2$, at ordinary pressures is a first-order reaction. Its half-life is 25.0 min at 500°C. (a) Starting with 4.00 g, how many grams remain after (1) 100 min and (2) 135 min? (b) Calculate the time in minutes required to reduce 6.70 mg to 1.25 mg. □

The half-life may be evaluated from the rate constant or vice versa.

EXAMPLE 19.5 The rate constant for the first-order decomposition of N_2O_5 to NO_2 and NO_3 in the gas phase at 55.0°C is 1.43×10^{-3} sec^{-1}. Find $t_{1/2}$ for the reaction.

ANSWER

$$t_{1/2} = \frac{0.693}{k}$$

$$= \frac{0.693}{1.43 \times 10^{-3}\text{ sec}^{-1}} = 485 \text{ sec}$$ ■

PROBLEM 19.8 The half-life for the decomposition of N_2O_5 to NO_2 and NO_3 in liquid Br_2 solution is 333 sec at 55°C. Find k for this first-order reaction. □

ZERO-ORDER AND SECOND-ORDER REACTIONS

Similarly, we may find the relationship between k and $t_{1/2}$ for zero-order and second-order reactions. Substituting $t_{1/2}$ for t and $\frac{1}{2}[R]_0$ for $[R]_t$ in equation 3 for a zero-order reaction,

$$[R]_0 - \tfrac{1}{2}[R]_0 = kt_{1/2}$$

yields

$$\frac{[R]_0}{2k} = t_{1/2} \tag{6}$$

Making the same substitution in equation 2 for a second-order reaction

$$\frac{1}{\frac{1}{2}[R]_0} - \frac{1}{[R]_0} = kt_{1/2}$$

yields

$$\frac{1}{[R]_0} = kt_{1/2}$$

or

$$t_{1/2} = \frac{1}{k[\mathrm{R}]_0} \tag{7}$$

The half-life is independent of concentration only for first-order reactions. The characteristic kinetic properties of zero-, first-, and second-order reactions of the kind R $\longrightarrow$ products are summarized in Table 19.1.

19.6 COLLISION THEORY OF REACTION RATES

We now seek an explanation of the factors that govern reaction rates. We assume that molecules must collide in order to react. For example, if H_2(g) and I_2(g) molecules are to react, H_2 molecules must collide with I_2 molecules. The greater the number of collisions per unit time, the faster the reaction rate. The rate is proportional to the number of molecular collisions in one liter per second:

rate $\propto$ number of molecules (N) colliding per liter per second

Showing 4 possible collisions between 2 A and 2 B molecules and 16 possible collisions between 4 A and 4 B molecules.

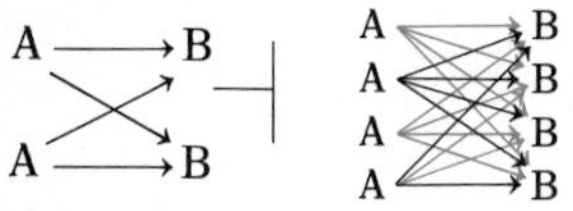

This assumption accounts for the dependence of the reaction rate on a product (and not a sum) of concentration terms. Let us assume that a molecule of A combines directly with a molecule of B to form AB; A + B $\longrightarrow$ AB. Visualize four molecules each of A and B in a box. In how many ways can the collision A + B occur? One A molecule can collide with any of four B molecules. It thus has four opportunities for collision. This is also true for each of the four A molecules. That makes 16 possible A + B collisions in all. Now if there were only two molecules each of A and B in the box, there would be only 2 × 2 possible collisions. Therefore, assuming the reaction rate is proportional to the number of collisions, the reaction in the box with four molecules each will be 16/4 or four times as rapid as in the box with two molecules each. Thus the rate is proportional to the *product* of the concentrations in molecules per liter, or converting to moles per liter,

rate $\propto$ [A][B]

The number of collisions between molecules calculated from the kinetic theory of gases is enormous—about 10^{32} per liter per second at standard temperature and pressure. This value does not vary much with different gases, so that if every collision led to product, all gaseous reactions should proceed at practically the same explosive rate. However, such is not the case. For example, for the same concentration of reactant at the same temperature, 300°C, the decomposition rate of gaseous hydrogen iodide is 4.4×10^{-3} mol/(L h), and the decomposition rate of gaseous nitrogen dioxide is 8.3×10^{2} mol/(L h). Clearly, *collisions between molecules cannot be the only factor involved in determining the rate of a reaction.*

About 10^9 mol/(L sec).

Chemical reactions involve redistribution of atoms, but the redistribution of atoms requires breaking bonds in the reactant molecules and the formation of bonds in the product molecules. For example, for the reaction 2CN(g) $\longrightarrow$ C_2(g) + N_2(g) to occur, the bond holding the C atom to the N atom must be ruptured, and bonds must form between C atoms and between N atoms. However, bond breaking is an endothermic process. We therefore postulate that molecules react only if in a collision they possess an energy equal to or greater than a certain critical value. *The critical energy needed for a reaction to occur is called the* **energy of activation**, E_a. *If the colliding molecules have an energy less than* E_a, *no reaction occurs.* If the colliding molecules have an energy equal to or greater than E_a, the reaction can occur. A portion of the kinetic energy is converted to

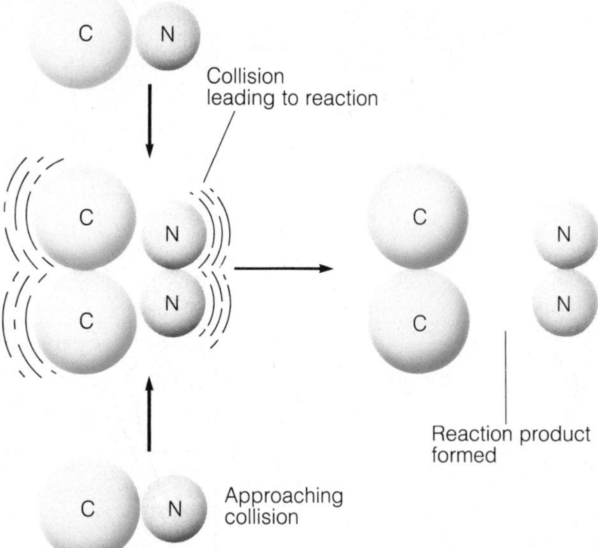

FIGURE 19.6
Chemical reaction pictured in terms of collisions. CN molecules acquire the energy necessary for the reaction (E_a) upon impact. This energy, converted to vibrational energy, weakens the C—N bonds.

vibrational energy (section 6.4) upon impact. The energy gets into the molecules so that vibrations along the chemical bonds will lead to bond rupture (Figure 19.6). The molecules are analogous to rapidly moving cars. Wreckage is produced when their kinetic energies are used to separate one part of a car from another.

For the reaction $2CN(g) \longrightarrow C_2(g) + N_2(g)$, the energy of activation is 180 kJ/mol. The colliding CN molecules must have a kinetic energy of at least 180 kJ/mol for the products to form. Figure 19.7 shows the energy relationships for a general exothermic reaction; E_r is the energy of the reactants and E_p is the energy of the products. The range of reaction rates from extremely slow to extremely fast is largely due to differences in energies of activation.

At any given temperature not all molecules possess the same kinetic energy (section 10.14). The speeds of molecules vary from almost no movement at all to very high speeds. Figure 19.8 illustrates an energy distribution curve. *The area*

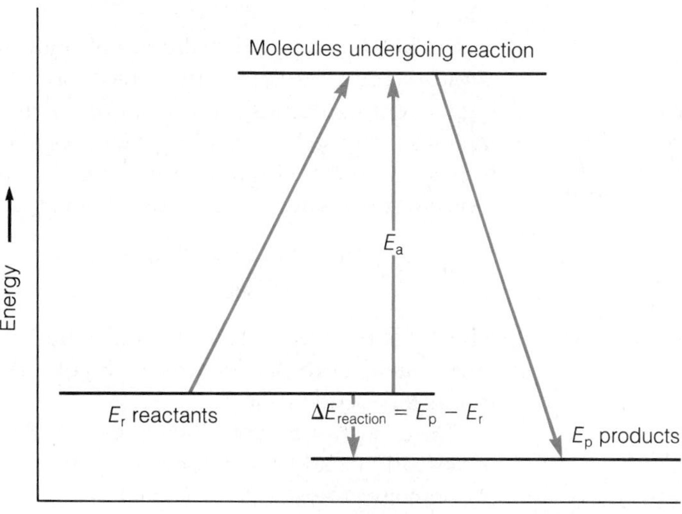

FIGURE 19.7
Energy relationships in an exothermic reaction.

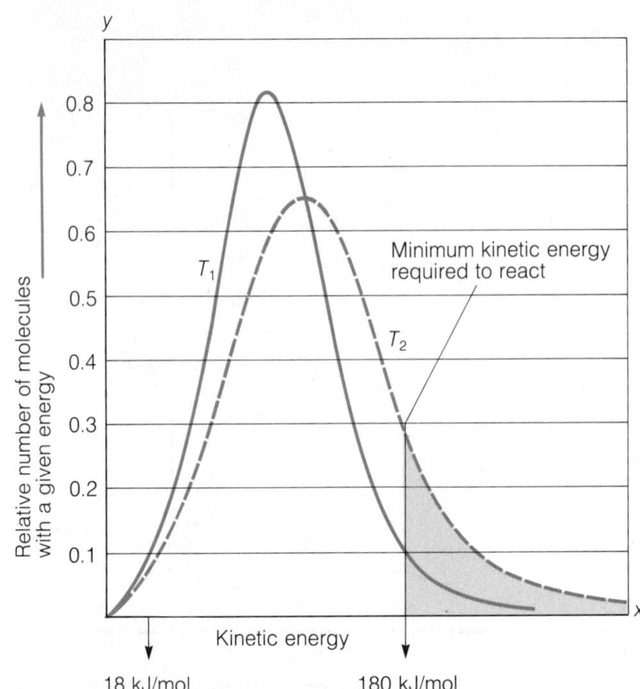

FIGURE 19.8
Kinetic-energy distribution curve. The *y* axis gives the relative number of molecules possessing the kinetic energy indicated in kilojoules per mole on the *x* axis. The shaded area represents the fraction of molecules having a minimum kinetic energy of 180 kJ/mol. Comparatively few molecules possess 180 kJ/mol. T_2 is a temperature higher than T_1. At T_2, a larger fraction of molecules have kinetic energies higher than 180 kJ/mol.

under the curve represents the total number of molecules. Note the kinetic energy of 180 kJ/mol on the figure; this is the kinetic energy that CN molecules must acquire before they can react. Then, the shaded area represents the number of CN molecules having a kinetic energy of at least 180 kJ/mol. Notice that comparatively few molecules have a kinetic energy of 180 kJ/mol or more at T_1, but that as the temperature increases to T_2 the number increases. Also notice that nearly all of the molecules have a kinetic energy of at least 18 kJ/mol. For the energy of activation equal to 180 kJ/mol, only a very small fraction of all the colliding molecules will react. Collisions between molecules with lower kinetic energies will not be effective in producing products. However, if the energy of activation requirement were only 18 kJ/mol, then practically all of the colliding molecules would react.

The increase in reaction rates with temperature cannot be attributed to an increase in the number of colliding molecules. An increase in temperature from 100°C to 110°C increases the number of colliding molecules only by 1%.

The kinetic energy distribution also affords an explanation of the effect of temperature on rates of reactions (section 19.2). Figure 19.8 illustrates that at the higher temperature T_2, the *fraction* of the molecules possessing at least 180 kJ/mol has increased sharply by a factor of about 2.5. This indicates that the increase in reaction rates with temperature is due largely to an increase in the fraction of molecules possessing the required energy of activation:

$$rate = N\frac{\text{molecules colliding}}{\text{L sec}} \times f$$

where f is the fraction of molecules having enough energy (E_a) to react. On the other hand, note particularly in Figure 19.7 that E_a and therefore the reaction rate is independent of $\Delta E_{reaction}$.

However, not all collisions between molecules possessing the required energy of activation lead to reaction. The manner in which they collide is also important. By analogy, the damage ("reaction") resulting from the collision of two automobiles depends not only on their speeds (energy) but also on their relative posi-

FIGURE 19.9 Influence of the orientation of molecules on the rate of the reaction $2AB \longrightarrow A_2 + B_2$. (a) Orientation that does not lead to reaction. (b) Orientation that leads to reaction.

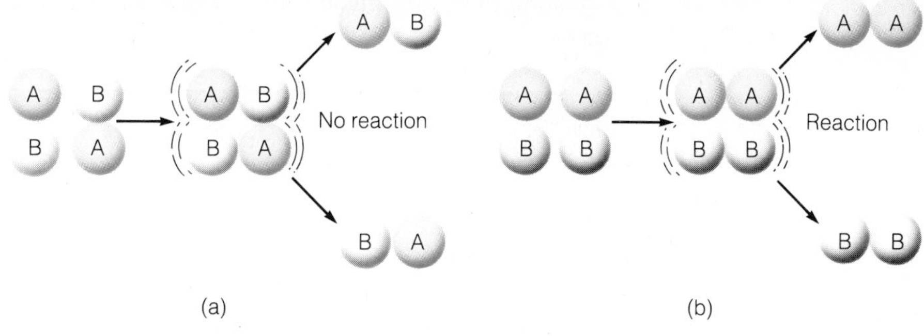

tions or orientation. For two cars traveling toward each other, a head-on collision poses more risk of damage than a side-swiping collision does. Some molecules must be oriented in a very specific manner for a reaction to occur (Figure 19.9). Other molecules may react when they collide in any way. For example, the combination of two H atoms to form an H_2 molecule requires no specific orientation.

The effect of orientation is dramatically illustrated with modern experimental techniques—laser technology, molecular beam methods, and mass spectroscopy. Experiments can be performed in which *individual gaseous molecules in precisely defined orientations with precisely known energies are selected and studied.* The reaction in the gas phase

$$K\cdot + CH_3I \longrightarrow KI + \cdot CH_3$$

is an example. The molecules K and CH_3I, methyl iodide, are aligned so that the K, C, and I atoms form a straight line. In one experiment, the I atom faces the K atom:

$$K\cdot + I{-}C{\lt}^{H}_{H}{-}H \xrightarrow[\text{approach"}]{\text{"head}} KI + \cdot CH_3$$

In a second experiment, the CH_3I molecule is turned 180° so that the $H_3C{-}$ group faces the K atom:

$$K\cdot + H{-}\underset{H}{\overset{H}{>}}C{-}I \xrightarrow[\text{approach"}]{\text{"tail}} \text{no products}$$

The rate constant for the "head approach"

$$H\cdot \longrightarrow H{-}Cl$$

in the reaction $H\cdot + HCl \rightarrow H_2 + Cl\cdot$ is roughly 10^{19} times greater than that for the approach at a 90° angle:

$$\underset{\underset{H\cdot}{\uparrow}}{H{-}Cl}$$

We see that while products are obtained in the "head approach," none forms in the "tail approach."

Since randomness is related to entropy (section 18.10), this orientation factor is related to the **entropy of activation**, $\Delta S^{\ddagger}$. *The more specific the orientation required for the reaction is, the smaller is the entropy of activation, and the slower will be its rate.* The entropy of activation may also be interpreted as a measure of the probability that the colliding molecules are favorably oriented for reaction. A higher entropy of activation thus means fewer restrictions on the kind of orientations needed for reaction. Thus the rate also depends on the fraction of colliding molecules having favorable orientations:

$$rate = N\frac{\text{molecules colliding}}{\text{L sec}} \times f \times p$$

where p is the probability that the energized colliding molecules are properly oriented for product formation.

In summary, at a given temperature, the reaction rate is determined by three factors:

FACTOR	CHANGE	EFFECT ON REACTION RATE
Frequency of collisions	higher	increase
	lower	decrease
Energy of activation	higher	decrease
	lower	increase
Entropy of activation	higher	increase
	lower	decrease

19.7 TRANSITION STATE THEORY OF REACTION RATES

More modern theories of reaction rates, developed by Henry Eyring and others, reexamine the idea that molecules must "collide" in a reaction. Transition state theory assumes that the arrangement of the atoms in molecules begins to change as molecules come close together. A chemical reaction is visualized as a continuous series of changes in bond distances as reactant molecules approach each other. Some bonds in reactant molecules may lengthen and finally break, while new bonds begin to form, finally forming product molecules. Energy changes accompany these continuous changes in the arrangement of atoms in molecules. Finally, *the reacting molecules must achieve a specific arrangement* before they can form the products of the reaction. This specific, transient arrangement possessing a definite energy is known as the **transition state.**

The transition state possesses more energy than either the reactant(s) or product(s) so that its energy is at a maximum as reactant(s) go to product(s). Thus the displacement of H by F in the reaction

$$F(g) + H_2(g) \longrightarrow HF(g) + H \qquad \Delta H = -133\ \text{kJ}$$

for which $E_a = 8.0$ kJ/mol may be visualized as shown in Sequences I and II below. (Distances between the atoms are given in nm units.) Dashed lines represent the breaking of bonds in the reactants and the formation of the new bonds that will eventually lead to products.

SEQUENCE I. From Reactants to Transition State

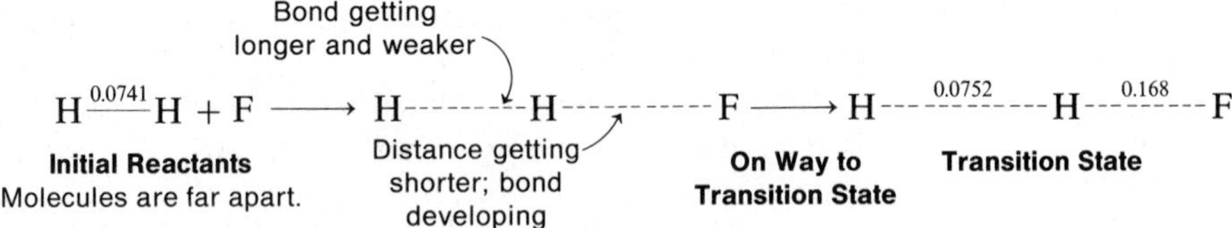

The calculations are based on Henry Eyring's mathematical ideas of rate processes (1934). He originally referred to a transition state as an *activated complex,* not to be confused with an activated *molecule,* which is a molecule in an excited electronic state.

The transition state is a transitory state based on mathematical theories. It is therefore impossible to isolate a transition state. It is a state the reacting molecules pass through, not a state in which they exist for a definite time interval. Nevertheless, the theory treats the transition state as a more or less real molecule with properties common to normal molecules—a molecular weight, interatomic distances, a definite enthalpy and entropy, and the ability to rotate and vibrate.

Once the transition state is formed, it does one of two things: it returns to the initial reactants or it proceeds to form products. In going from the transition state to products, the H---H bond distance increases further and the H---F bond distance decreases, as shown in Sequence II.

SEQUENCE II. From Transition State to Products

H---$\overset{0.0752}{}$---H---$\overset{0.168}{}$---F ⟶ H-----------H-------F ⟶ H + H$\overset{0.0917}{—}$F

Transition State — **On Way to Products** — Bond getting longer and weaker — Bond getting shorter and stronger — **Products** Molecules are completely separated.

The enthalpy changes accompanying the gradual breaking of the H---H bonds, and the gradual formation of the HF bonds, are illustrated in Figure 19.10a. H_r is the enthalpy of the initial reactants. As energy is absorbed, the interatomic distance increases in the H---H molecules. But simultaneously H---F bonds start to form, evolving energy. However, *in forming the transition state more energy is absorbed than evolved.*

In discussions of transition state, it has become customary to refer to the enthalpy, rather than the energy, of activation and to draw enthalpy diagrams like Figure 19.10. The transition state therefore always possesses a larger enthalpy than the initial reactants. The difference between the enthalpy of the transition state, H_{ts}, and the enthalpy of reactant(s), H_r, is called the **enthalpy of activation**, $\Delta H^‡$:

$$\Delta H^‡ = H_{ts} - H_r$$

It is usually assumed that $\Delta H^‡$ (the enthalpy of activation) and E_a (the energy of activation) are almost equal. Typical differences between $\Delta H^‡$ and E_a are about 2.5 kJ at 25°C and about 13 kJ at 500°C, depending on the kind of reaction. In this chapter, *energy* and *enthalpy* are used interchangeably. For the reaction $F + H_2 \longrightarrow H + HF$,

$$\text{enthalpy of activation} = H_{ts} - H_r = \Delta H^‡ \approx E_a = 8 \text{ kJ/mol}$$

As the transition state decomposes to products, the H---H atoms separate even farther. However, the energy required to separate them is now comparatively small. Simultaneously, the H---F atoms attract each other very strongly. Considerable energy is evolved. The product always possesses a smaller enthalpy, H_p, than the transition state. Note that in this reaction the products have a smaller enthalpy than the reactants:

The reaction is exothermic.

$$\Delta H = H_p - H_r = -133 \text{ kJ}$$

This is ΔH for the reaction. Note also that *ΔH for the reaction depends only on the difference between H_p and H_r* and not on H_{ts}. On the other hand, for an endothermic reaction H_p is larger than H_r (but smaller than H_{ts}) as shown in Figure 20.10b for the reaction $C_2H_6(g) \longrightarrow 2CH_3(g)$, for which $\Delta H^‡ = E_a =$ 384 kJ and $\Delta H = +377$ kJ.

The double dagger, ‡, is used to indicate differences between the transition state and the reactants:

$$\Delta H^‡ = \text{enthalpy of activation} = H_{\text{transition state}} - H_{\text{reactants}}$$
$$\approx E_{\text{a,energy of activation}}$$

$$\Delta S^‡ = \text{entropy of activation} = S_{\text{transition state}} - S_{\text{reactants}}$$

$$k = \frac{RT}{N_A h} e^{-\Delta H^‡/RT} e^{\Delta S^‡/R}$$

According to the transition state theory, the *f* term (section 19.6) in the collision theory is determined by $\Delta H^‡$, the *p* term (section 19.6) is determined by $\Delta S^‡$, and the number of colliding molecules per liter per second is related to a universal constant that is the same for *all* reactions in the gas phase. Interestingly, this

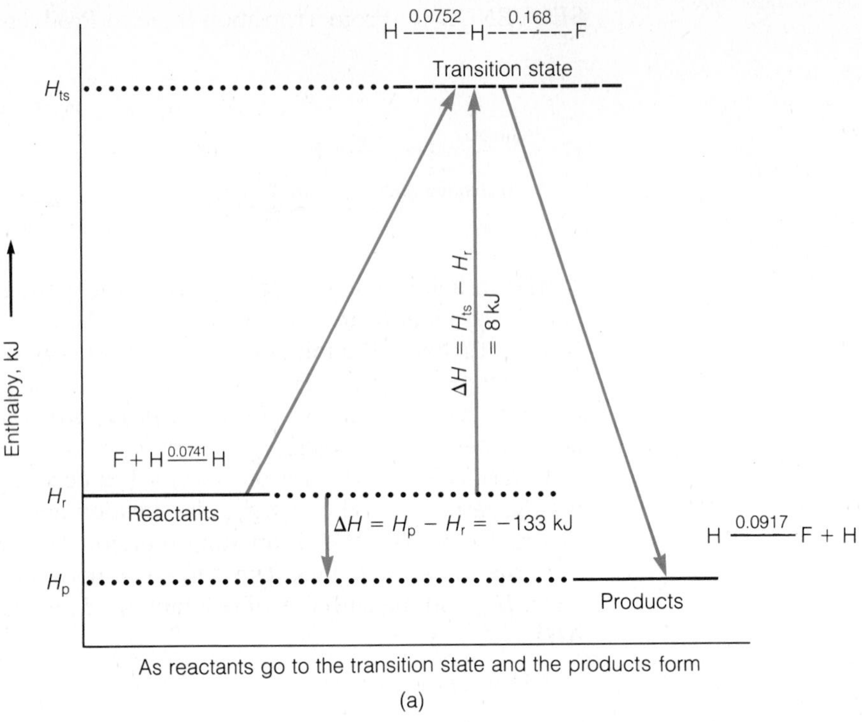

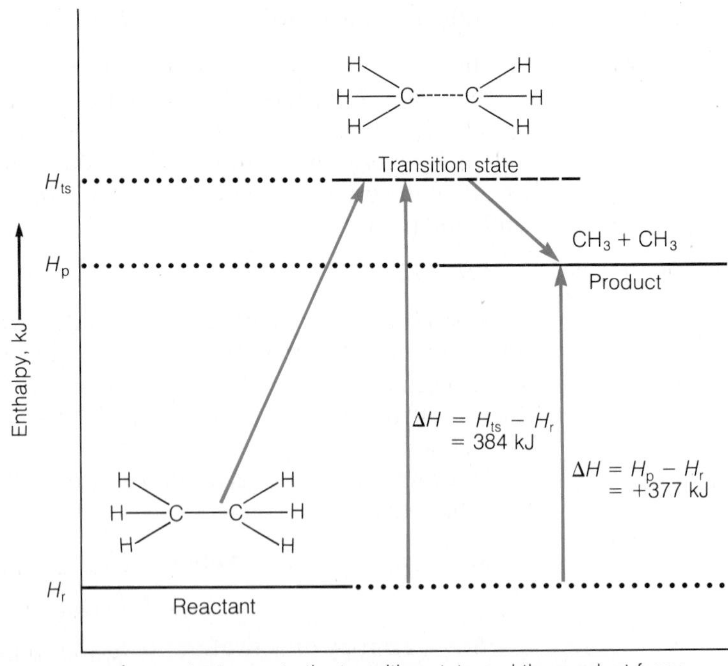

FIGURE 19.10 Enthalpy diagram. Bond distances are in nm units (not to scale). (a) $F + H_2 \longrightarrow HF + H$, $\Delta H^\ddagger = 8$ kJ, $\Delta H_{reaction} = -133$ kJ (exothermic). (b) $C_2H_6 \longrightarrow 2CH_3$, $\Delta H^\ddagger = 384$ kJ, $\Delta H_{reaction} = +377$ kJ (endothermic). (Adapted from *Journal of Chemical Education* 54, 288 (1977).)

universal constant is made up of the fundamental constants R (ideal gas), h (Planck's constant), and N_A (Avogadro's number).

Rate constants depend on both $\Delta H^{\ddagger}$ and $\Delta S^{\ddagger}$ at a given temperature. However, $\Delta H^{\ddagger}$ has a greater influence in most cases (but not all). Kinetic stability is thus controlled by the enthalpy of activation; the greater $\Delta H^{\ddagger}$ is, the smaller is the rate constant and the greater is the kinetic stability.

A knowledge of transition state theory permits the chemist to modify reaction conditions so as to affect the rate of reaction and thus alter production rates.

EXAMPLE 19.6 For the reaction

$$HC_2H_3O_2 + H_2O \longrightarrow H_3O^+ + C_2H_3O_2^-$$

a change of the solvent from water to nitromethane, CH_3NO_2, raises the enthalpy of the transition state more than it raises the enthalpy of the reactants. Do $\Delta H^{\ddagger}$ and the rate constant increase, decrease, or remain unchanged?

ANSWER The solvent change raises H_{ts} more than it raises H_r. Therefore H_{ts} is farther removed from H_r and $\Delta H^{\ddagger} = H_{ts} - H_r$ is now larger. The larger the enthalpy (energy) of activation, the slower is the reaction. The rate constant is therefore decreased. ■

19.8 THE ARRHENIUS EQUATION

Early in the development of chemical kinetics, Svante Arrhenius, famous for his contributions to our understanding of electrolytic solutions, also introduced (1889) the concept of an energy of activation. His empirical equation is

$$k = Ae^{-E_a/RT} \tag{8}$$

In logarithmic form, this is

$$\mathbf{\log k = \log A - \left(\frac{E_a}{2.303RT}\right)} \tag{9}$$

Notice in equation 9 that the term $E_a/2.303RT$ is *subtracted* from log A.

The factor A is approximately a constant, characteristic of a given reaction; it has the same units as k since the factor $e^{-E_a/RT}$ is dimensionless. R is the gas constant, 8.314 J/(mol K), and T is the absolute temperature. For most reactions, E_a is approximately independent of temperature.

The more accurate equation is

$$k = AT^n e^{-E_a/RT}$$

where n may be 2 to 2.5.

The Arrhenius equation concisely summarizes our concepts about reaction rates. In terms of the collision theory of reaction rates, the A factor is related to the number of molecules colliding per liter per second and the probability that the energized molecules are properly oriented. The $e^{-E_a/RT}$ factor gives the fraction (f) of colliding molecules that have kinetic energies at least equal to E_a.

In terms of collision theory, $A = N$(molecules colliding/L sec) × p. In terms of transition state theory

$$A = \frac{RT}{N_A h} e^{\Delta S^{\ddagger}/R}$$

Since E_a is positive, an increase in its value decreases the number of successful collisions. The fraction of molecules that will react also decreases, and thus, also the rate constant. A decrease in the value of E_a increases the number of successful collisions. This means the fraction of molecules that will react increases, which increases the rate constant. In summary,

LARGER E_a	SMALLER E_a
Fewer successful collisions	More successful collisions
Fewer reacting molecules	More reacting molecules
Smaller k value	Larger k value
Slower reaction	Faster reaction

BOX 19.3

THE e^x KEY AND THE ARRHENIUS EQUATION

If your calculator has an e^x key, you can solve for A directly from equation 8 as shown:

$$k = Ae^{-E_a/RT}$$

and

$$A = \frac{k}{e^{-E_a/RT}} = ke^{E_a/RT}$$

For the decomposition of HI,

$E_a = 187$ kJ/mol and $k = 3.16 \times 10^{-6}$ L/(mol sec) at 575 K

Then

$$A = 3.16 \times 10^{-6} \frac{\text{L}}{\text{mol sec}} e^{187/8.31 \times 10^{-3} \times 575}$$

$$= 3.16 \times 10^{-6} \times 9.9 \times 10^{16}$$

$$= 3.1 \times 10^{11} \frac{\text{L}}{\text{mol sec}}$$

and k in L/(mol sec) at other temperatures is given by

$$k = 3.1 \times 10^{11}\ e^{-187\ \text{kJ mol}^{-1}/RT}$$

and

$$\log k = 11.5 - \frac{187\ \text{kJ mol}^{-1}}{2.303\ RT}$$

See Problem 19.49.

On the other hand, the temperature factor is in the denominator. So, if the temperature is increased, the E_a/RT term in equation 9 decreases and a smaller number is subtracted from log A; thus, the value of k increases in agreement with the fact that rate constants almost always increase as the temperature increases.

A and E_a are treated as temperature-independent:

$$\begin{matrix} y & = & b & + & m & x \\ \log k & = & \log A & - & \dfrac{E_a}{2.303R} & \left(\dfrac{1}{T}\right) \end{matrix}$$

To see how E_a and A can be evaluated, write the Arrhenius equation in the form

$$\log k = \log A - \frac{E_a}{2.303R}\left(\frac{1}{T}\right)$$

This is an equation of a straight line when log k is plotted against $\frac{1}{T}$. E_a can be obtained from the slope of the straight line, which is the constant $\frac{-E_a}{2.303R}$. The intercept is log A. However, we can also substitute the value of E_a determined from the slope into equation 9 to find the value of A from a known k at a given temperature. (See Box 19.3.)

Rate constants are usually reported in the form of the Arrhenius equation because once the A and E_a terms are known, the rate constant at other temperatures can be found.

PROBLEM 19.9 The rate constant in sec^{-1} for the decomposition of dinitrogen pentoxide, $N_2O_5(g) \longrightarrow N_2O_4(g) + \frac{1}{2}O_2(g)$, is given by

$$\log k = 13.66 - \frac{103\ \text{kJ/mol}}{2.303RT}$$

Find the rate constant at 50.0°C. □

The energy of activation may also be evaluated by writing the Arrhenius equation at two different temperatures and subtracting. At T_1,

$$\log k_1 = \log A - \frac{E_a}{2.303RT_1}$$

At T_2,

$$\log k_2 = \log A - \frac{E_a}{2.303RT_2}$$

Subtracting,

$$\log k_1 - \log k_2 = \left(\log A - \frac{E_a}{2.303RT_1}\right) - \left(\log A - \frac{E_a}{2.303RT_2}\right)$$

$$= \frac{E_a}{2.303RT_2} - \frac{E_a}{2.303RT_1}$$

and rearranging yields

$$\log \frac{k_1}{k_2} = \frac{E_a}{2.303R}\left(\frac{1}{T_2} - \frac{1}{T_1}\right) \qquad (10)$$

Equation 10 may be used to calculate the energy of activation from known rate constants at different temperatures. It also provides a way of finding the rate constant at one temperature, T_1, when we know E_a and the rate constant, k_2, at another temperature, T_2.

PROBLEM 19.10 The energy of activation for the reaction

$$C_4H_8(g) \longrightarrow 2C_2H_4(g)$$

is 260 kJ/mol. At 800 K, $k_2 = 0.0315\ \text{sec}^{-1}$. Find k_1 at 850 K. □

19.9 MECHANISM OF A REACTION

A balanced chemical equation gives the identities and molar ratios of the reactants and products in a chemical conversion but it does not describe how the conversion occurred. Some reactions proceed in one step, but more frequently, reactions occur in a sequence of steps. The single step or the sequence of steps is called the **mechanism of the reaction**. Thermodynamic properties such as free energy describe only the initial and final states of a reaction. From the rate equation, however, we may learn how the reaction goes from the initial state to the final state. Therefore, chemical kinetics includes the study of the mechanisms of chemical reactions.

In this sense, a *step* is *a chemical equation* that describes an *assumed single molecular event*—such as the formation or rupture of a chemical bond or the displacement of atoms resulting from a single molecular collision. For example, the balanced equation,

Reactions are in the gas phase unless stated otherwise.

$$2NO(g) + Br_2(g) \longrightarrow 2BrNO(g) \qquad (11)$$

tells us nothing about the mechanism of the reaction. Do all three molecules combine together in one step?

$$NO + NO + Br_2 \longrightarrow 2BrNO$$

Or, does one NO molecule combine with one Br_2 molecule in one step

STEP 1. $NO + Br_2 \longrightarrow Br_2NO$

followed by a second step?

STEP 2. $Br_2NO + NO \longrightarrow 2BrNO$

Adding steps 1 and 2 yields the net reaction, equation 11. Steps 1 and 2 illustrate a typical proposed mechanism (Figure 19.11).

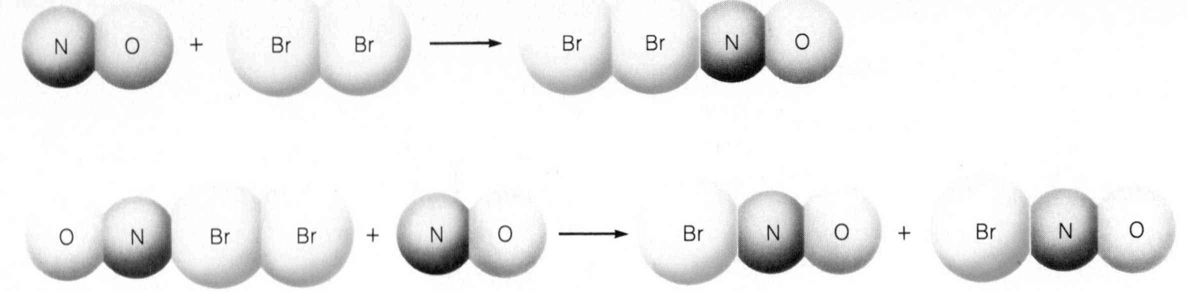

FIGURE 19.11
Schematic representation of the proposed two-step mechanism by which NO and Br_2 are converted to BrNO.

The set of steps that satisfactorily explains the kinetic properties of a chemical reaction constitutes a reaction mechanism.

MOLECULARITY OF STEPS

Each step can be described in terms of its reactant molecules. The *number of reactant molecules* (or atoms, ions, free radicals) coming together in a step is called its **molecularity**. Unlike the order of a reaction, the molecularity of a step must be a whole number. When *one molecule* is the reactant in a step, the step is said to be **unimolecular**. When two reactant molecules ($2A \longrightarrow$ products or $A + B \longrightarrow$ products) are written, the step is said to be **bimolecular**. For example, a two-step mechanism has been proposed for the decomposition of ozone, $2O_3 \longrightarrow 3O_2$:

STEP 1. $O_3 \longrightarrow O_2 + O$ (unimolecular reaction)

STEP 2. $O_3 + O \longrightarrow 2O_2$ (bimolecular reaction)

Step 1 is a unimolecular reaction. Step 2 is a bimolecular reaction. A step with three reactant molecules ($3A \longrightarrow$ products, $2A + B \longrightarrow$ products, or $A + B + C \longrightarrow$ products), is *termolecular*. In short, *molecularity* applies *only* to steps in a proposed mechanism and it is equal to the *sum of the coefficients* of the reactants written for the step. Most steps are uni- or bimolecular. Termolecular reactions are rare. The molecularity is also the number of molecules that must come together to form a transition state. The probability that four or more molecules may simultaneously combine to form a transition state is so small that molecularities higher than three are never postulated.

RATE EQUATIONS AND THE REACTION MECHANISM

The rate equation is an important clue for determining a reaction mechanism. Determining the mechanism of a reaction is an intellectual endeavor that should not be underestimated. Many seemingly simple reactions—for example, $2H_2 + O_2 \longrightarrow 2H_2O$—are very complicated, and after more than a half-century of labor by many scientists, the mechanism of this reaction is not yet completely understood. Therefore, we will be discussing only simple mechanisms here.

A study of the reaction $N_2 + 3H_2 \longrightarrow 2NH_3$, used for the industrial production of ammonia, reveals that the rate of the reaction is actually controlled by the rate at which N_2 dissociates into N atoms. Thus, *anything* that increases the rate of the reaction $N_2 \longrightarrow 2N$ will *automatically* increase the production rate of ammonia. This also reveals the most important point in the study of reaction mechanisms. When a reaction proceeds through a sequence of steps, **the slowest step determines the rate of the net reaction and therefore determines the observed rate equation**. In a sequence of steps, the bottleneck is the step with the lowest rate. This step is said to be **rate determining**. These ideas may be illustrated by an analogy. An assembly line for producing toasted corn puffs is set up to operate at these rates:

STEP 1. Explosion and toasting of corn kernels (maximum 125 lb/h)
STEP 2. Cooling puffed corn (maximum 150 lb/h)
STEP 3. Packaging (maximum 75 lb/h)
STEP 4. Sealing packages (maximum 150 lb/h)
STEP 5. Crating packages (maximum 125 lb/h)

Step 3 is the slow step; the measured production rate is only 75 lb/h. Acceleration in the other steps will not increase the production rate. On the other hand, any acceleration in step 3 will automatically increase the net production rate.

Let us now consider the reaction in which hypochlorite, ClO^-, undergoes self-oxidation–reduction to the chlorate, ClO_3^-, and the chloride, Cl^-, in aqueous solution:

$$3ClO^- \longrightarrow ClO_3^- + 2Cl^-$$

The experimental rate equation is

$$rate = k[ClO^-]^2$$

We can propose that the mechanism of this reaction involves two bimolecular steps:

STEP 1. $ClO^- + ClO^- \longrightarrow ClO_2^- + Cl^-$ (slow)

STEP 2. $ClO_2^- + ClO^- \longrightarrow ClO_3^- + Cl^-$ (fast)

Addition yields the net reaction. But there are other important features of a proposed mechanism:

- *A rate equation can be written for each step in a proposed mechanism.*
- In writing the rate equation for a step, the order for each kind of molecule in the step is equal to the coefficient of the molecule in that step. In short, *the molecularity and the order of a step are equal.* For example, a unimolecular reaction has an order of 1, and a bimolecular reaction has an order of 2.
- *Molecularity is never used to describe the net reaction.* Therefore, while a rate equation for a net reaction must be determined experimentally (section 19.2), a rate equation may be written for each step by inspection.
- Only the slowest step is important, because it determines the rate of the net reaction. The net reaction *cannot* be faster than the slowest step.

Slow and *fast* indicate only the relative rates of the steps.

Thus, for mechanism steps 1 and 2, we write for the slow step:

$$rate = k_1[ClO^-]^2$$ (rate equation derived from slow bimolecular step 1)

Note that the derived rate equation agrees with the experimental rate equation; $k = k_1$.

An interesting reaction is the oxidation of CO by NO_2,

$$CO + NO_2 \longrightarrow CO_2 + NO$$

At temperatures below 500 K the experimental rate equation is

$$rate = k[NO_2]^2$$

Notice that the experimental rate equation *does not* involve the concentration of carbon monoxide. It does involve, however, two NO_2 molecules. A proposed mechanism is

$$\text{STEP 1.}\quad NO_2 + NO_2 \longrightarrow NO_3 + NO \qquad \text{(slow)}$$

$$\text{STEP 2.}\quad NO_3 + CO \longrightarrow NO_2 + CO_2 \qquad \text{(fast)}$$

Remember that addition of the steps of a proposed mechanism must always yield the net reaction.

Addition yields the net reaction. Step 1 is the slow (rate-determining) step. Then, for the slow step, we write

$$rate = k_1[NO_2]^2 \qquad \text{(rate equation derived from slow bimolecular step 1)}$$

Thus, in agreement with the experimental rate equation, the rate is independent of the carbon monoxide concentration.

Let us try another example. From experiment, we know that for the reaction in the gas phase

$$2NO + O_2 \longrightarrow 2NO_2$$

the rate equation is

$$rate = k[NO]^2[O_2]$$

The simultaneous combination of the three molecules as shown in the net reaction is unlikely. A proposed mechanism A is

$$\text{(A) STEP 1.}\quad NO + NO \longrightarrow N_2O_2 \qquad \text{(slow)}$$

$$\text{STEP 2.}\quad N_2O_2 + O_2 \longrightarrow 2NO_2 \qquad \text{(fast)}$$

But we could propose other mechanisms, mechanisms B and C, that also agree with the net reaction:

Reversible here does not mean "thermodynamically reversible," but merely that the reaction occurs in both directions.

$$\text{(B) STEP 1.}\quad NO + O_2 \rightleftharpoons NO_3 \qquad \text{(fast reversible reaction)}$$

$$\text{STEP 2.}\quad NO + NO_3 \longrightarrow 2NO_2 \qquad \text{(slow)}$$

or

$$\text{(C) STEP 1.}\quad NO + NO \rightleftharpoons N_2O_2 \qquad \text{(fast reversible reaction)}$$

$$\text{STEP 2.}\quad N_2O_2 + O_2 \longrightarrow 2NO_2 \qquad \text{(slow)}$$

Each of the three proposed mechanisms yields the net reaction.

Next, we must derive a rate equation from each proposal. For mechanism A, we write

$$rate = k_1[NO]^2 \qquad \text{(rate equation derived from slow bimolecular step 1)}$$

Clearly, this derived rate equation does not agree with the experimental rate equation. We therefore rule out A as an acceptable mechanism.

For mechanism B, we write

$$rate = k_2[NO_3][NO] \qquad \text{(rate equation derived from slow bimolecular step 2)} \qquad (12)$$

Unlike the transition state, an intermediate is a real species.

This derived rate equation contains the concentration of NO_3, a molecule *not appearing* in the net reaction or in the experimental rate equation. NO_3 is a typical **intermediate**. Intermediates are usually very reactive species that are consumed in subsequent steps. They never appear in the net reaction because they are not among the substances originally mixed in measuring the rate of the reaction. Also, they are not found among the products. They do not appear in the experimental rate equation because the experimentalist is incapable of determining their minute concentrations in the reacting mixtures.

But notice that the NO_3 is formed in the fast reversible step 1. When an intermediate is formed in a previous fast reversible step, the equilibrium condition for that step can be used to find the concentration of the intermediate. Thus, the equilibrium condition for step 1 is used to find the NO_3 concentration in terms of the concentration of the reactants NO and O_2:

$$\frac{[NO_3]}{[NO][O_2]} = K_1$$

This gives us

$$[NO_3] = K_1[NO][O_2] \tag{13}$$

We can now substitute equation 13 into equation 12:

$$rate = k_2K_1[NO][O_2][NO] = k_2K_1[NO]^2[O_2]$$ (rate equation derived from mechanism B)

The derived rate equation is in agreement with the experimental rate equation. Thus, if mechanism B is correct, we conclude that the constant k in the experimental rate equation is equal to the product of two constants, k_2 and K_1.

For mechanism C, we write

$$rate = k_2[N_2O_2][O_2]$$ (rate equation derived from slow bimolecular step 2)

The equilibrium condition for step 1 gives us the N_2O_2 concentration in terms of the concentration of the reactant NO,

$$\frac{[N_2O_2]}{[NO]^2} = K'_1$$

This gives us

$$[N_2O_2] = K'_1[NO]^2$$

and

$$rate = k_2K'_1[NO]^2[O_2]$$ (rate equation derived from mechanism C)

This derived rate equation also agrees with the experimental rate equation. This assumed mechanism would make the constant k in the experimental rate equation equal to the product $k_2K'_1$.

Evidently, a mechanism that agrees with the net reaction and predicts a correct rate equation is not necessarily "the correct mechanism." However, the proposed mechanism must at least be consistent with experimental results. The detection of NO_3 but not of N_2O_2 at the conditions of the reaction strongly supports mechanism B.

NO_3 is thermodynamically and kinetically unstable.

This case emphasizes that the proposed mechanism of a chemical reaction is a theoretical pathway consistent with currently known data. Changes may be necessitated by new experimental or theoretical studies. Unraveling the mechanism of a reaction can be a very complicated game. Therefore you will only be asked to predict a rate equation from a given mechanism or to show whether or not a given proposed mechanism is consistent with given kinetic data.

EXAMPLE 19.7 Is the one-step mechanism involving the direct combination of three ClO^- ions proposed for the reaction $3ClO^- \longrightarrow ClO_3^- + 2Cl^-$

STEP 1. $ClO^- + ClO^- + ClO^- \longrightarrow ClO_3^- + 2Cl^-$

consistent with the rate equation give here?

$$rate = k[ClO^-]^2$$ (experimental rate equation)

ANSWER This one-step mechanism corresponds to the net reaction. To establish whether it is consistent with the kinetic data, we must write the rate equation for the given step. Step 1 involves three ClO^- ions:

$$rate = k_1[ClO^-]^3 \qquad \text{(rate equation derived from termolecular step 1)}$$

This derived rate equation clearly does not agree with the experimental rate equation. The proposed one-step mechanism therefore cannot be correct. ■

EXAMPLE 19.8 For the oxidation of I^- by H_2O_2 in acidic solutions,

$$H_2O_2 + 2H^+ + 2I^- \longrightarrow I_2 + 2H_2O$$

a proposed mechanism is

STEP 1. $H_2O_2 + I^- \longrightarrow H_2O + OI^-$ (slow)

followed by the rapid reactions

STEP 2. $H^+ + OI^- \longrightarrow HOI$ (fast)

STEP 3. $HOI + H^+ + I^- \longrightarrow I_2 + H_2O$ (fast)

For this mechanism to be consistent with kinetic data, what must be the experimental rate equation?

ANSWER The slowest step determines the rate of the net reaction. The slow step involves one molecule of H_2O_2 and one I^- ion. The derived rate equation is

$$rate = k_1[H_2O_2][I^-] \qquad \text{(derived from slow bimolecular step 1)}$$

The experimental rate equation should be the same as the derived; studies show that it is. Also, addition of the three steps gives the overall reaction. ■

EXAMPLE 19.9 A proposed mechanism for the decomposition of ozone, $2O_3 \longrightarrow 3O_2$, is

STEP 1. $O_3 \rightleftharpoons O_2 + O$ (fast reversible reaction)

STEP 2. $O + O_3 \longrightarrow 2O_2$ (slow)

Derive the rate equation for the net reaction.

ANSWER From the slow bimolecular step 2

$$rate = k_2[O][O_3] \tag{14}$$

But O does not appear in the net reaction. It is, however, formed in the fast and reversible step 1. The equilibrium condition for step 1 is used to find [O] in terms of $[O_3]$ and $[O_2]$:

$$\frac{[O_2][O]}{[O_3]} = K_1 \quad \text{and} \quad [O] = K_1\frac{[O_3]}{[O_2]} \tag{15}$$

Substituting equation 15 into equation 14 gives us the derived rate equation:

The rate equation can also be written as $k[O_3]^2[O_2]^{-1}$.

$$rate = k_2K_1\frac{[O_3]}{[O_2]}[O_3] = k\frac{[O_3]^2}{[O_2]} \qquad \text{(derived)}$$ ■

EXAMPLE 19.10 The experimental rate equation for the gas reaction $H_2 + I_2 \longrightarrow 2HI$ is

$$rate = k[H_2][I_2]$$

The formation of HI was long considered typical of many reactions in which the mechanism is simply the combination of two molecules. However, it is now known that this reaction occurs through a mechanism involving atoms. One such possibility is

STEP 1. $I_2 \rightleftharpoons 2I$ (fast reversible)

STEP 2. $2I + H_2 \longrightarrow 2HI$ (slow)

The mechanism is more complicated than given in step 2. A three-step mechanism is given in Problem 19.57.

Does this mechanism predict the correct rate equation?

ANSWER From the slow termolecular step 2:

$$rate = k_2[I]^2[H_2]$$

This does not agree with the experimental rate equation; there are no I atoms in the net reaction. The concentration of the intermediate [I] is obtained from reversible step 1:

$$\frac{[I]^2}{[I_2]} = K_1 \quad \text{and} \quad [I]^2 = K_1[I_2]$$

Make the appropriate substitution:

The mechanism of the $H_2 + I_2$ reaction is still an open question.

$$rate = k_2K_1[I_2][H_2] \qquad \text{(derived rate equation)}$$

The mechanism thus predicts the correct rate equation. ■

PROBLEM 19.11 The formation of nitrosyl bromide, $2NO + Br_2 \longrightarrow 2BrNO$, follows the rate equation, $rate = k[NO]^2[Br_2]$. Predict a rate equation based on each of these proposed mechanisms:

(a) $NO + Br_2 \rightleftharpoons Br_2NO$ (fast reversible)

$NOBr_2 + NO \longrightarrow 2BrNO$ (slow)

(b) $NO + NO \rightleftharpoons N_2O_2$ (fast reversible)

$N_2O_2 + Br_2 \longrightarrow 2BrNO$ (slow)

(c) $NO + Br_2 \longrightarrow BrNO + Br$ (slow)

$NO + Br \longrightarrow BrNO$ (fast)

(d) $NO + NO + Br_2 \longrightarrow 2BrNO$

Are all four mechanisms consistent with the given kinetic data? Explain. What intermediates are formed in mechanisms a and b? □

19.10 HOW CATALYSTS WORK

Many reactions are speeded up by the presence of a substance that does not appear in the balanced equation for the net reaction. For example, if hydrogen and oxygen are mixed in a flask at room temperature, no reaction occurs. However, if some platinum is dropped into the flask, the mixture explodes. Many industrial processes would be impossible without catalysts. The octane performance of a gasoline is improved by altering the structure of some of its molecules as shown:

```
                                   H
                                 H\ | /H
   H   H   H   H                 H  C  H
   |   |   |   |                 |  |  |
H—C—C—C—C—H  ——→  H—C—C—C—H
   |   |   |   |                 |  |  |
   H   H   H   H                 H  H  H
```

butane molecule — branched butane molecule

However, the rates of such reactions are negligible until aluminum chloride is added. In fact, the rate is then sufficiently increased so that the reaction is used on a commercial scale. Automobile exhaust systems incorporate a platinum catalyst to speed up the oxidation of unburned and partially burned gasoline to CO_2 and H_2O (Box 19.5).

Many instances of such effects are found in biochemical systems (Box 19.4). Life as we know it would be impossible without enzymes—the proteins responsible for catalysis in biochemical reactions. It is impossible to carry out the oxidation of sucrose

$$C_{12}H_{22}O_{11}(aq) + 12O_2(aq) \longrightarrow 12CO_2(g) + 11H_2O$$

in a test tube at room temperature without adding enzymes. Yet, reactions like this are the main energy sources of nearly all animal organisms.

These substances—platinum, $AlCl_3$, and enzymes—are typical catalysts. A *catalyst* is a substance that increases the rate of a reaction, but is not consumed in the reaction (section 13.6). The catalyst enters into the chemical reaction, but is subsequently regenerated. Here, we answer the question, "How does a catalyst increase the rate of a reaction?" The catalyst increases the rate by changing the mechanism of the reaction so as to decrease the energy of activation (Figure 19.12). Decreasing the energy of activation increases the fraction of molecules that reacts and thus increases the rate constant (section 19.8).

Typical is the catalytic effect of nitrogen oxide on the rate of the conversion of sulfur dioxide to sulfur trioxide. The reaction

$$2SO_2 + O_2 \xrightarrow{\text{slow}} 2SO_3$$

is slow. However, NO and O_2 react rapidly and NO_2, the product of this reaction, also reacts rapidly with SO_2:

$$2NO + O_2 \xrightarrow{\text{fast}} 2NO_2 \qquad (16)$$

$$NO_2 + SO_2 \xrightarrow{\text{fast}} NO + SO_3 \qquad (17)$$

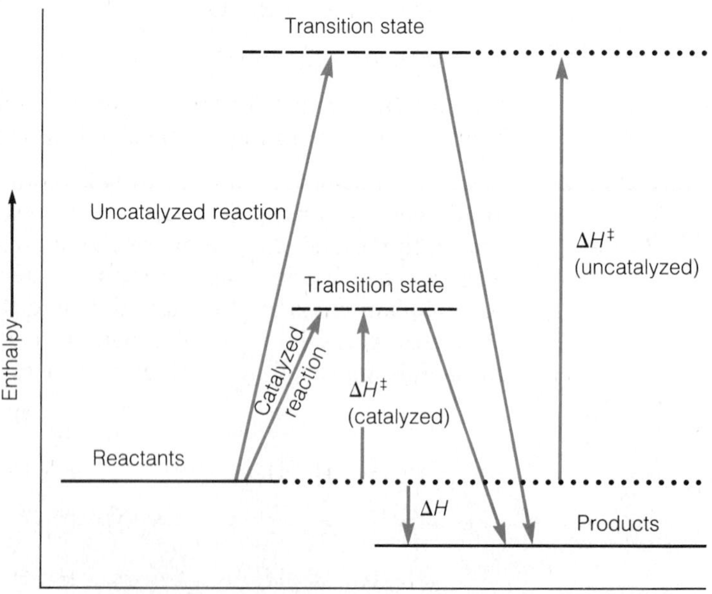

FIGURE 19.12
Enthalpy diagram for a catalyzed and uncatalyzed reaction. A catalyst speeds up a reaction by changing the mechanism so as to decrease the energy of activation. While the catalyst functions by lowering $\Delta H^{\ddagger}$, it has no effect on the ΔH of the reaction, a thermodynamic property independent of the pathway between products and reactants.

BOX 19.4

ZERO-ORDER DINING

Nearly all biological reactions are catalyzed by enzymes that accelerate reactions by a factor of at least 10^6. This mechanism accounts for the kinetic properties of enzymes:

$$\text{STEP 1.} \quad \underset{\text{enzyme}}{E} + \underset{\text{substrate}}{S} \rightleftharpoons \underset{\text{enzyme-substrate intermediate}}{ES}$$

$$\text{STEP 2.} \quad ES \xrightarrow{k} E + \underset{\text{product}}{P}$$

In step 1 the enzyme E combines reversibly with a substrate S (a reactant) to form an intermediate ES, which in step 2 forms the product P. For many enzymatic reactions, step 2 is rate-determining. So from the slow step,

$$rate = k[ES]$$

the equilibrium condition for step 1 gives [ES],

$$\frac{[ES]}{[E][S]} = K_c$$

$$[ES] = K_c[E][S]$$

and the rate equation is

$$rate = kK_c[E][S]$$

However, at comparatively high concentrations of S, step 1 is shifted almost completely to the right. Then very little free enzyme, E, is present; nearly all of it has been converted to ES so that [ES] equals the initial concentration of the enzyme, $[E]_0$. Under these conditions the rate reaches its maximum value and becomes independent of the concentration of S. The order is now zero with respect to S and the rate equation becomes

$$rate_{max} = kK_c[E]_0[S]^0 = kK_c[E]_0$$

This rate equation is true for most biological reactions because enzyme concentrations are very low (10^{-6}–10^{-5} M). This means that the enzyme molecules are saturated with S molecules; the enzymes are busy converting S molecules to product. The enzyme molecules remain saturated—and [ES] remains constant—as long as S is present to replenish the ES as it is converted to product. For example, in the degradation of glucose, an important energy source, at its normal concentrations in the blood the enzyme hexokinase is saturated and the rate becomes independent of the glucose concentration as shown in the figure. Initially, the rate rises rapidly, but at normal blood glucose concentrations (about 3×10^{-3} M), the rate becomes independent of the glucose concentration. At this point, addition of glucose does not increase the rate. Consequently, our food consumption rates have little effect on the rates of most of our metabolic reactions.

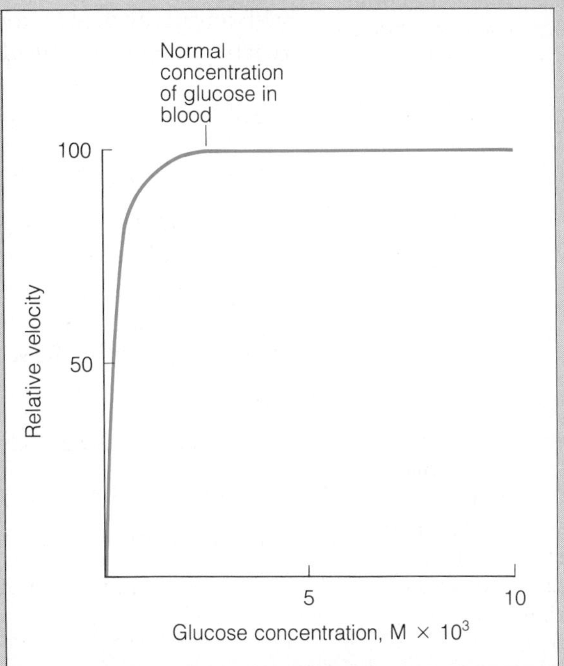

The plot of rate against glucose concentration in a glucose enzyme (hexokinase)–catalyzed reaction shows that the reaction is zero-order with respect to glucose. (Adapted from Albert Lehninger, *Biochemistry*, Worth Publishers, New York, N.Y., 2nd ed., 1975, p. 423.)

Addition of reaction 16 and reaction 17, after multiplying it by 2, yields a faster reaction than the uncatalyzed reaction:

$$2SO_2 + O_2 \xrightarrow[\text{faster}]{NO} 2SO_3$$

We have substituted two fast reactions for a slow one to yield the same net chemical reaction. The NO essentially functions as an oxygen carrier. This method, the "chamber process," was formerly used to make sulfuric acid, but it has been largely replaced by the contact process (section 20.2).

Catalysis provides a new mechanism. Although the uncatalyzed mechanism is still in operation, it is overshadowed by the catalytic mechanism (Figure 19.12).

The catalyzed conversion of SO_2 to SO_3 by NO is an example of homogeneous catalysis. In *homogeneous catalysis*, the reaction occurs in one phase; all reactants and the catalyst are in one phase. For example, the experimental rate equation for the homogeneous decomposition of acetaldehyde

$$CH_3CHO(g) \longrightarrow CH_4(g) + CO(g)$$

is

$$rate = k_1[CH_3CHO]^{3/2}$$

The energy of activation is 201 kJ/mol. k_1 is the rate constant for the uncatalyzed reaction. The mechanism of this reaction involves a complex chain reaction (section 19.11).

In the presence of iodine vapor, the experimental rate equation is

A catalyst appears in a rate equation but not in the net reaction (Problems 19.24 and 19.64).

$$rate = k_2[CH_3CHO][I_2]$$

with an energy of activation of 138 kJ/mol. k_2 is the rate constant for the catalyzed reaction. A possible mechanism is

STEP 1. $CH_3CHO + I_2 \xrightarrow{\text{slow}} CH_3I + HI + CO$

STEP 2. $CH_3I + HI \xrightarrow{\text{fast}} CH_4 + I_2$

The sum of these two steps gives the net reaction. At the same temperature (600 K), the catalyzed reaction is about 10^7 times faster than the uncatalyzed.

In *heterogeneous catalysis*, the catalyst, usually the surface of a solid, is not in the same phase as the reactants. The reaction occurs at the solid surface, and the rate depends on the surface area. The general mechanism involves the following steps:

In general, adsorbed molecules are bonded to surface atoms of the catalyst, stretching or breaking the bonds in these molecules and making them highly reactive.

1. Reactant molecules diffuse to the surface of the catalyst (Figure 19.13).
2. Reactant molecules are adsorbed on the surface.
3. Adsorbed reactant molecules react on the surface.
4. The products separate from the surface (desorption).
5. Products diffuse away from the surface.

For example, at about 800 K, the experimental rate equation for the homogeneous decomposition of hydrogen iodide

$$2HI(g) \longrightarrow H_2(g) + I_2(g)$$

is

$$rate = k_1[HI]^2$$

The energy of activation is 188 kJ/mol. But in the presence of platinum, Pt, the rate equation for the heterogeneous catalyzed reaction is

Industrial catalyst consumption is about 3×10^6 tonne, valued at over a billion dollars (1 tonne = 10^3 kg).

$$rate = k_2[HI] \times \text{surface area}$$

with an energy of activation of 59 kJ/mol.

During the lifetime of a heterogeneous catalyst, *one* active site (a surface atom or a group of atoms) may yield more than a *million* product molecules. If the

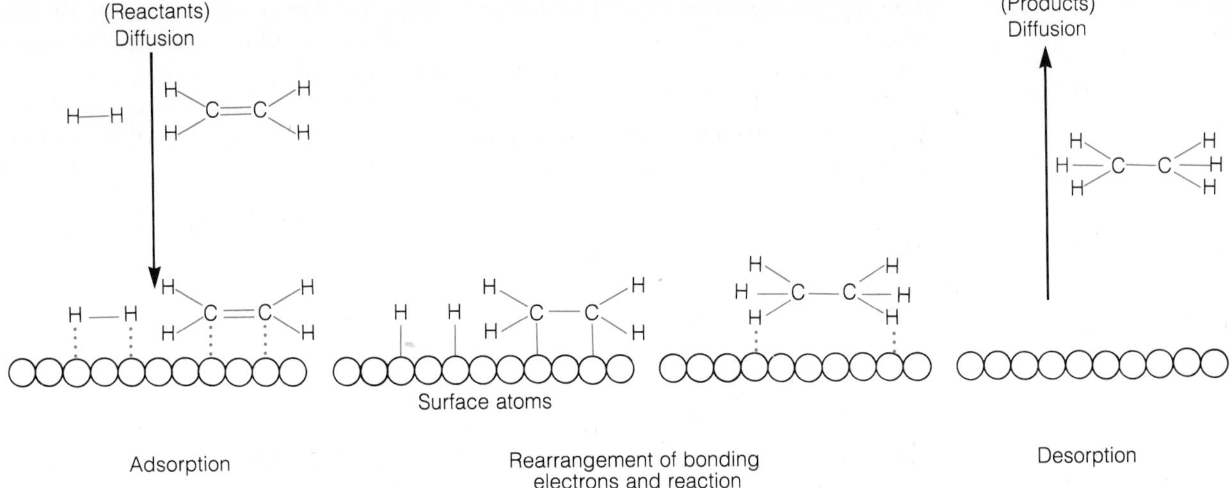

FIGURE 19.13
Schematic representation of the sequence of heterogeneous catalytic processes. The reaction of ethylene and hydrogen on a nickel surface: $C_2H_4(g) + H_2(g) \rightarrow C_2H_6(g)$.

reactant molecules form strong chemical bonds upon adsorption, catalysis ceases. Therefore, during the catalytic process, a reactant cannot be strongly bound; that would "poison" the catalyst. On the other hand, if the bonding is too weak, bond breaking does not occur. However, it is not easy to unscramble the mechanism of these heterogeneous reactions or to find the rate-determining step from experimental data.

A catalyst conserves energy by reducing activation energies; it conserves resources by promoting desired over undesired reactions.

The chemical industry favors the use of heterogeneous catalysts because it is relatively easy to separate them from the reaction products by filtration. However, a major disadvantage in this energy-conscious world is that these catalysts require high pressures and temperatures. On the other hand, homogeneous catalysts are soluble and thus difficult to separate from products, but reaction pathways at mild conditions are more easily deciphered and very high yields can be achieved.

19.11 CHAIN REACTIONS

This reaction is an efficient but dangerous alarm clock.

Suppose that hydrogen and chlorine are mixed in a flask one dark night near an east window. A reaction proceeds slowly until shortly after sunrise, when the mixture suddenly explodes. What would cause this to happen? What is the explanation for the explosion?

First, the chlorine absorbs photons, producing the free radicals (section 7.6), atomic chlorine:

Recall that free radicals are species with at least one unpaired electron.

$$Cl_2 + h\nu(\text{radiation}) \longrightarrow 2{:}\ddot{\underset{..}{Cl}}\cdot \qquad (18)$$

Then, *without* further absorption of photons, the $:\ddot{\underset{..}{Cl}}\cdot$ atom initiates a sequence of reactions. It reacts with H_2, forming HCl and an $H\cdot$ atom:

Bond energies:
Cl_2, 244 kJ/mol;
H_2, 436 kJ/mol.

$$:\ddot{\underset{..}{Cl}}\cdot + H_2 \longrightarrow HCl + H\cdot \qquad (19)$$

In turn, $H\cdot$ reacts with a Cl_2 molecule, producing another HCl molecule while regenerating the $:\ddot{\underset{..}{Cl}}\cdot$ atom:

$$H\cdot + Cl_2 \longrightarrow HCl + {:}\ddot{\underset{..}{Cl}}\cdot \qquad (20)$$

The $:\ddot{\underset{..}{Cl}}\cdot$ produced in step 20 then starts the cycle all over again via step 19. The sequence of steps 19 and 20 is referred to as a **chain reaction**, a reaction in which each step yields *one* species that is capable of continuing the cycle.

The initial step 18, in which free radicals are generated from non-free-radical species, is referred to as **chain initiation**. Steps 19 and 20 are referred to as **chain propagation**, because each reaction consumes a free radical and generates another.

However, the chain can be interrupted when two free radicals, the chain propagators, combine. For example, the reactions

$$:\ddot{\underset{..}{Cl}}\cdot + :\ddot{\underset{..}{Cl}}\cdot \longrightarrow Cl_2 \tag{21}$$

$$H\cdot + H\cdot \longrightarrow H_2 \tag{22}$$

$$:\ddot{\underset{..}{Cl}}\cdot + H\cdot \longrightarrow HCl \tag{23}$$

can stop the chain reaction. These steps, in which free radicals form a nonradical molecule, are referred to as **chain termination**:

$$\cdot R \;+\; \cdot R' \longrightarrow R{:}R'$$

radical + radical ⟶ molecule with no unpaired electrons

The propagators $:\ddot{\underset{..}{Cl}}\cdot$ and $H\cdot$ are present in very low concentrations. Therefore collisions between them, steps 21, 22, and 23, are rare compared to collision between propagators and reactants, steps 19 and 20. In summary, the chain reaction sequence is

Not all photon-induced reactions are chain reactions. In many photon-induced reactions, removal of the photon source stops the reaction.

$$Cl_2 + h\nu \xrightarrow[\text{chain starters}]{\text{supply of}} 2:\ddot{\underset{..}{Cl}}\cdot \xrightarrow{H_2} \left(\begin{array}{l} :\ddot{\underset{..}{Cl}}\cdot + H_2 \longrightarrow HCl + H\cdot \\ H\cdot + Cl_2 \longrightarrow HCl + :\ddot{\underset{..}{Cl}}\cdot \end{array} \right) \xrightarrow{\text{product(s)}} HCl$$

initiation — propagation — chain breaking

$$\left.\begin{array}{r} 2:\ddot{\underset{..}{Cl}}\cdot \longrightarrow Cl_2 \\ 2H\cdot \longrightarrow H_2 \\ H\cdot + :\ddot{\underset{..}{Cl}}\cdot \longrightarrow HCl \\ \cdot R + \cdot R' \longrightarrow R{:}R' \end{array}\right\} \text{termination}$$

Once the chain-propagating steps start, the overall reaction should proceed until reactants are consumed. *The net reaction is obtained from the sum of the chain-propagating steps* because they occur many times for each occurrence of the initiation and termination steps.

EXAMPLE 19.11 Given these steps

$$:\ddot{\underset{..}{Cl}}\cdot + CH_4 \longrightarrow \cdot CH_3 + HCl \tag{24}$$

$$Cl_2 + h\nu\,(\text{radiation}) \longrightarrow 2:\ddot{\underset{..}{Cl}}\cdot \tag{25}$$

$$:\ddot{\underset{..}{Cl}}\cdot + :\ddot{\underset{..}{Cl}}\cdot \longrightarrow Cl_2 \tag{26}$$

$$\cdot CH_3 + \cdot CH_3 \longrightarrow C_2H_6 \tag{27}$$

$$\cdot CH_3 + Cl_2 \longrightarrow CH_3Cl + :\ddot{\underset{..}{Cl}}\cdot \tag{28}$$

$$\cdot CH_3 + :\ddot{\underset{..}{Cl}}\cdot \longrightarrow CH_3Cl \tag{29}$$

Rearrange the steps in the order of chain initiation, chain propagation, and chain termination. Find the net reaction.

ANSWER A chain-initiation step produces free radicals from molecules. Step 25 is therefore the chain starter. A chain-propagation step produces a species capable of continuing the chain reaction. The chain propagator in free-radical chain reactions is always a free radical. Steps 24 and 28 are therefore chain propagation. Chain termination refers to the steps that break the chain reaction. In these steps, free radicals combine to form molecules in which all electrons are paired. Steps 26, 27, and 29 are therefore chain termination. In summary,

$$Cl_2 + h\nu \longrightarrow 2{:}\ddot{\underset{..}{Cl}}\cdot \qquad \text{(initiation)}$$

$$:\ddot{\underset{..}{Cl}}\cdot + CH_4 \longrightarrow HCl + \cdot CH_3 \qquad \text{(propagation)}$$

$$\cdot CH_3 + Cl_2 \longrightarrow CH_3Cl + :\ddot{\underset{..}{Cl}}\cdot$$

$$:\ddot{\underset{..}{Cl}}\cdot + :\ddot{\underset{..}{Cl}}\cdot \longrightarrow Cl_2 \qquad \text{(termination)}$$

$$\cdot CH_3 + \cdot CH_3 \longrightarrow C_2H_6$$

$$\cdot CH_3 + :\ddot{\underset{..}{Cl}}\cdot \longrightarrow CH_3Cl$$

Addition of the propagation steps yields the net reaction

$$CH_4 + Cl_2 \longrightarrow CH_3Cl + HCl$$

■

PROBLEM 19.12 (a) Rearrange and label these reaction steps in the order of chain initiation and chain propagation:

$$ClO\cdot + O\cdot \longrightarrow Cl\cdot + O_2$$

$$CF_2Cl_2 + h\nu \longrightarrow CF_2Cl\cdot + Cl\cdot$$

$$Cl\cdot + O_3 \longrightarrow ClO\cdot + O_2$$

(b) Suggest three possible chain-termination steps. (c) Find the net reaction. (d) Is it a catalyzed reaction? □

In general, activation energies are comparatively small for propagation reactions.

Such chain reactions are fast but not necessarily explosive. They become explosive when the heat evolved in the net reaction is so large that *constant temperature cannot be maintained.* As the reaction proceeds, the heat evolved increases the temperature. This in turn increases the rate of the reaction and the liberation of heat until, ultimately, the reaction becomes explosive (a thermal explosion).

On the other hand, branching-chain reactions are explosive without an increase in temperature. In a **branching-chain reaction**, *two* or more species capable of producing the reaction are regenerated for every *one* consumed. Consider these reactions:

$$O^* + H_2 \longrightarrow H_2O^* \qquad (30)$$

$$H_2O^* + O_2 \longrightarrow H_2O + 2O^* \qquad (31)$$

For every *one* excited oxygen atom (O^*) consumed in reaction 30, *two* are regenerated in reaction 31. The overall rate then increases without limit at constant temperature (Problem 19.68), and, if the reaction is exothermic, an explosion results. Branching-chain reactions typically produce more violent explosions than thermal explosions. At 550°C, for example, the rate of the reaction hydrogen + oxygen $\longrightarrow$ water increases slowly as the concentration of either reactant is increased. Then, *at the same temperature* (550°C), the mixture suddenly explodes, signaling the development of the branching-chain reaction. This is the kind of mechanism involved in nuclear (U, Pu) weapons (section 23.6).

BOX 19.5
CATALYTIC CONVERTERS AND AUTOMOBILE EXHAUST

The control of automobile exhaust emissions is an application of heterogeneous catalysis. The catalysts are essentially the heavy transition metals of Groups 9B and 10B, especially Pt, Pd, and Rh. The object of the catalyst is to drive the combustion of gasoline (typical component, C_8H_{18}) to completion, to produce only CO_2 and H_2O:

$$C_8H_{18}(g) + 12\tfrac{1}{2}O_2(g) \longrightarrow 8CO_2(g) + 9H_2O(g) \qquad \Delta G^0 = -4987 \text{ kJ}$$

The free-energy change at 25°C is

$$\Delta G^0 = 8\Delta G^0_{f,CO_2} + 9\Delta G^0_{f,H_2O} - \Delta G^0_{f,C_8H_{18}}$$
$$= 8(-394 \text{ kJ}) + 9(-229 \text{ kJ}) - (-226 \text{ kJ})$$
$$= -4987 \text{ kJ}$$

When hydrocarbons and other organic compounds are introduced into the atmosphere, smog forms.

The large negative value of ΔG^0 tells us that the reaction is thermodynamically very favorable. However, the oxidation involves a large number of separate steps, which necessarily include the breaking of all the C—C and C—H bonds in the octane, as well as the O—O bonds in O_2. Many other reactions, also with negative ΔG^0's, can occur along the way. Consequently, the gasoline is not completely burned. To make matters worse, these bond ruptures produce free radicals that initiate branching-chain reactions that, in turn, produce an explosion in the cylinder called an "engine knock." The knock makes the delivery of power to the piston inefficient and can also damage the cylinder wall, and is therefore undesirable. One way to prevent knock is to add a substance to the gasoline that is capable of terminating some of the reaction chains.

Research shortly after World War I culminated in the discovery of tetraethyllead, TEL: PB:$(CH_2—CH_3)_4$ During combustion TEL decomposes to ethyl radicals, $CH_3—CH_2\cdot$ and lead atoms, $:\dot{Pb}\cdot$, both of which serve as chain terminators and thereby reduce engine knock. However, the lead itself can produce harmful solid deposits in the engine. To get rid of such residues, dichloroethane, $C_2H_4Cl_2$, and dibromoethane, $C_2H_4Br_2$, are added to leaded gasoline. As a result, most of the lead is discharged from the exhaust in the form of the relatively volatile lead halides, $PbCl_2$, $PbBr_2$, and PbClBr.

Lead atoms are more strongly adsorbed on platinum and palladium than O_2 or hydrocarbon (gasoline) molecules. This adsorption of lead compounds poisons the automobile exhaust catalysts. TEL was therefore banished in order to protect the antipollution devices on automobiles. This law also has the desirable benefit of reducing the environmental dispersal of toxic lead compounds. Vehicles equipped with catalytic converters must use unleaded gasoline. However, the catalytic oxidation of the gasoline hydrocarbons generates heat within the catalytic converter. This heat promotes the undesired oxidation of N_2 to NO, which is also involved in the formation of smog.

The equilibrium constant for $N_2 + O_2 \rightleftharpoons 2NO$ decreases as the temperature decreases. We might therefore expect, optimistically, that the NO formed in a hot engine would decompose promptly to N_2 and O_2 when the exhaust gases are cooled. However, the rate of this decomposition, like most rates, decreases sharply as the temperature decreases. During the sudden cooling of the exhaust, there is no time for the NO to decompose, and once the exhaust is cooled, the decomposition is slow. The worst feature of NO is that at ordinary temperatures, it is easily oxidized to NO_2:

$$2NO + O_2 \longrightarrow NO_2$$

If a trace of ozone is present,

$$NO + O_3 \longrightarrow NO_2 + O_2$$

is faster. NO_2 is one of the worst air pollutants, in itself and because of its reactions with other pollutants derived from hydrocarbons.

Minimizing the production of oxides of nitrogen requires lowering the flame temperatures in automobiles or the boilers of electricity-generating plants. However, thermodynamics teaches us that lower temperatures lead to lower efficiency. The maximum efficiency of any engine—the original Ford Model T or an F-16 fighter plane jet engine—is determined solely by the ratio

$$\frac{T_2 - T_1}{T_2} \qquad \text{(a consequence of the second law; Box 18.3)}$$

where T_2 (K) is the flame temperature and T_1 (K) is the exhaust temperature.

So, some efficiency is sacrificed for pollution control. The engine's computerized control system maintains precise stoichiometric mixtures (molar ratios of the hydrocarbons, CO, NO, and O_2 required for complete reaction leaving no excess reactants) such that hydrocarbons and CO are oxidized,

$$C_2H_2(g)^{\S} + 2\tfrac{1}{2}O_2(g) \longrightarrow 2CO_2(g) + H_2O(g) \qquad \Delta G^0 = -1227 \text{ kJ}$$

$$CO(g) + \tfrac{1}{2}O_2(g) \longrightarrow CO_2(g) \qquad \Delta G^0 = -394 \text{ kJ}$$

while NO oxidizes hydrocarbons and is reduced to N_2:

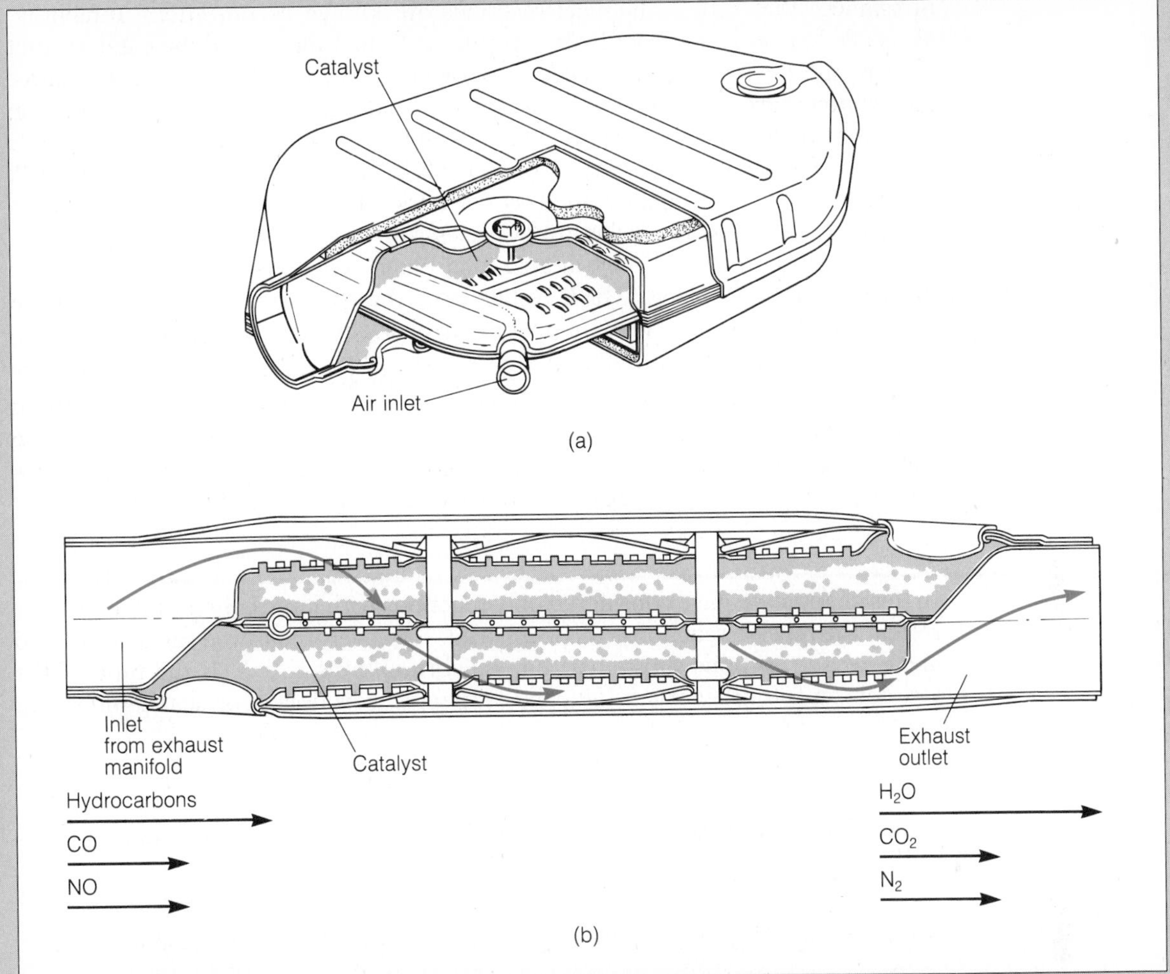

(a) Cross-sectional view of catalytic converter and (b) view showing flow of exhaust gases through the converter.

$$C_2H_2(g) + 5NO(g) \longrightarrow 2CO_2(g) + H_2O(g) + 2\tfrac{1}{2}N_2(g)$$
$$\Delta G^0 = -1707 \text{ kJ}$$

More recent catalytic studies over Rh/Al_2O_3 surfaces show that NO is also reduced to N_2 by CO,

$$2NO(g) + 2CO(g) \longrightarrow N_2(g) + 2CO_2(g)$$
$$\Delta G^0 = -687.4 \text{ kJ}$$

summarized in this mechanism (adsorbed is abbreviated ad):

$$2NO(g) \longrightarrow 2NO(ad)$$
$$NO(ad) \longrightarrow N(ad) + O(ad)$$
$$NO(ad) + N(ad) \longrightarrow N_2(g) + O(ad)$$
$$2CO(g) \longrightarrow 2CO(ad)$$
$$2CO(ad) + 2O(ad) \longrightarrow 2CO_2(g)$$

The rate constant and activation energy of each step have been measured, showing that the dissociation of NO(ad) is the slowest (rate-controlling) step. Thus any modification of the catalyst that increases the NO dissociation rate will improve its efficiency.

In this way, potential atmospheric pollutants are converted mainly to harmless substances. However, practical systems are seldom ideal; some undesirable gases such as SO_3 (from sulfur compounds in gasoline) and NH_3 are also produced.

Thus we see that the environmental aspects of automobile exhaust, characteristic of all atmospheric pollution problems, involve numerous chemical principles—kinetics and thermodynamics, catalysis, complex reaction pathways, free radicals—as well as old-fashioned trial-and-error chemistry.

§ C_2H_2 is a component of exhaust gases.

INHIBITORS

Substances that retard chemical reactions are known as *inhibitors*. Inhibitors, also called *poisons*, may act by interfering with the functions of the catalyst; that is, with the mechanism that leads to a lower energy of activation. Lead atoms, for example, are more strongly adsorbed on platinum than O_2 or hydrocarbon (gasoline) molecules (Box 19.5). This adsorption of lead "poisons" automobile exhaust catalysts when leaded gasoline is used; the Pb atoms block access to Pt atoms. Physiological poisons such as mercuric chloride and rattlesnake venom react with enzymes, rendering them useless for essential biochemical reactions.

Inhibitors may also act by interrupting chain reactions. Chain reactions are involved in food spoilage, which results from the oxidation of organic matter at room conditions. Free-radical reactions involving the superoxide radical anion, $:\dot{\underset{..}{O}}-\overset{..}{\underset{..}{O}}:^-$, formed by reduction of O_2 *in vivo*, have been implicated in a variety of diseases and biological aging. Inhibitors slow down such oxidation reactions by consuming free radicals, the chain starters. A small amount of inhibitor can stabilize materials for a long time. Inhibitors such as BHA—*b*utylated *h*ydroxy-*a*nisole, $C_6H_3(OH)(C_4H_9)(OCH_3)$—are used as food preservatives.

19.12 RELATIONSHIP BETWEEN *K* AND *k*

In this chapter we have considered only reactions with large equilibrium constants so that the reverse reactions can be ignored. However, measured rates of forward and reverse reactions show that the equilibrium constant, K, is the ratio of the rate constants for the forward, k_f, and reverse, k_r, reactions at or near equilibrium:

$$K_c = \frac{k_f}{k_r}$$

For example, the measured rate at 1100°C for the bimolecular reaction

$$\cdot CH_3(g) + HI(g) \longrightarrow :\overset{..}{\underset{..}{I}}\cdot(g) + CH_4(g)$$

is $rate_f = k_f[\cdot CH_3][HI]$, where $k_f = 10^9$ L/(mol sec). For the reverse reaction

$$:\overset{..}{\underset{..}{I}}\cdot + CH_4 \longrightarrow \cdot CH_3 + HI$$

the measured $rate_r = k_r\,[:\overset{..}{\underset{..}{I}}\cdot][CH_4]$, where $k_r = 10^{10}$ L/(mol sec). For

$$\cdot CH_3 + HI \rightleftharpoons :\overset{..}{\underset{..}{I}}\cdot + CH_4$$

the equilibrium condition is

$$K_c = \frac{[:\overset{..}{\underset{..}{I}}\cdot][CH_4]}{[\cdot CH_3][HI]} \tag{32}$$

But at equilibrium, $rate_f = rate_r$ and

$$k_f[\cdot CH_3][HI] = k_r[:\overset{..}{\underset{..}{I}}\cdot][CH_4]$$

from which

$$\frac{k_f}{k_r} = \frac{[:\overset{..}{\underset{..}{I}}\cdot][CH_4]}{[\cdot CH_3][HI]} \tag{33}$$

Comparing equations 32 and 33, we see that

$$K_c = \frac{k_f}{k_r}$$

Using the kinetic data, we obtain

$$K_c = \frac{10^9 \text{ L/(mol sec)}}{10^{10} \text{ L/(mol sec)}} = 10^{-1}$$

The measured value is $K_c = 10^{-1}$.

K is a measure of *thermodynamic stability*, whereas the two rate constants are measures of *kinetic stability*. Now notice that while the rate constants are high, the ratio K is low. Finally, satisfy yourself that the ratio can have the same value with high and low rate constants:

$$K = \frac{10^{10}}{10^{41}} = 10^{-31} \qquad K = \frac{10^{-41}}{10^{-10}} = 10^{-31}$$

thermodynamically stable; kinetically unstable — thermodynamically stable; kinetically stable

Similarly, thermodynamically unstable substances that are kinetically unstable (labile) or kinetically stable (inert) are also known. Thus, while equilibrium constants and rate constants do not parallel one another, they are nevertheless related.

SUMMARY

Chemical kinetics deals with the rates and the **mechanism of reactions**. The rate of a reaction depends upon the nature of the reactants, the reactant concentrations, the temperature, and the presence or absence of a **catalyst**. In **heterogeneous reactions**, the rate also depends on the surface area. A **homogeneous reaction** occurs in only one phase.

The relationship between the rate of a reaction and the concentration of reactants is expressed by an equation in which the concentrations are raised to powers that must be determined by experiment. For the general reaction $aA + bB \ldots \longrightarrow$ products, the rate equation is $rate = k[A]^x[B]^y \ldots$, where k is the **rate constant**. The sum of the exponents $(x + y \ldots)$ is the **order of the reaction**. The exponents need not be the same as the coefficients in the chemical equation. The rates of reaction are determined by following the decrease in the concentration of a reactant or the increase in the concentration of a product with time.

The relationship between the change in concentration and time is a linear equation.

- *zero-order reaction*: $[R]_0 - [R]_t = kt$
  for which a plot of $[R]_t$ versus t yields a straight line whose slope is $-k$ with the units of mol/(L sec). $[R]_0$ is the reactant concentration at $t = 0$ and $[R]_t$ is the reactant concentration at time t after the start of the experiment.
- *first-order reaction*: $2.303 \log \frac{[R]_0}{[R]_t} = kt$
  for which a plot of $\log [R]_t$ versus t yields a straight line whose slope is $-k$ with the unit of sec^{-1}.
- *second-order reaction*: $\frac{1}{[R]_t} - \frac{1}{[R]_0} = kt$
  for which a plot of $1/[R]_t$ versus t yields a straight line whose slope is k with the units of L/(mol sec).

The **half-life**, $t_{1/2}$, the time it takes for a given concentration of a reactant to be decreased by one-half, is related to the rate constant.

zero-order reaction: $t_{1/2} = \dfrac{[R]_0}{2k}$

first-order reaction: $t_{1/2} = \dfrac{0.693}{k}$ (independent of concentration or quantity)

second-order reaction: $t_{1/2} = \dfrac{1}{k[R]_0}$

The longer the half-life or the smaller the rate constant, the greater is the **kinetic stability** of the reactants.

According to the **collision theory**, molecules change to products after colliding only if they possess at least the minimum amount of energy required to form the products—the **energy of activation**. The lower the energy of activation, the larger is the number of successful collisions and the faster is the reaction; the converse is also true. The large temperature dependence of rate constants is explained by the sharp increase in the fraction of molecules having the necessary energy of activation.

For many reactions, specific orientations of the reactant molecules are necessary for successful collisions. These specific orientations are related to the **entropy of activation**. The *less specific* the orientation required for reaction, the *larger* is the entropy of activation and the *faster* is the rate.

Transition state theory modifies the collision theory by postulating that as reactant molecules approach, the arrangement of the atoms in molecules begins to change. In a continuous series of changes, some bonds weaken as they stretch (endothermic) while new bonds are formed (exothermic). At one point, the reactant molecules achieve a specific arrangement known as the **transition state**, a transitory state that can be described in terms of molecular properties. The transition state can either return to initial reactants or proceed to form products. In going to products, the bonds being broken are further stretched, while the lengths of the bonds being formed decrease. In forming the transition state, more energy is absorbed than evolved and, consequently, it always has a larger enthalpy than the initial reactants. This difference is the **enthalpy of activation**, $\Delta H^{\ddagger}$. As the transition state goes to product, more energy is evolved than absorbed and, consequently, the product always has a smaller enthalpy than the transition state.

The **Arrhenius equation** shows the dependence of the rate constant on temperature:

$$\log k = \log A - E_a/2.303RT$$

The energy of activation, E_a, is obtained from the slope, $-E_a/2.303R$, when $\log k$ is plotted against $1/T$.

A *step* is an *assumed* single molecular reaction whose **molecularity** is the number of reactant species that come together to form the transition state in that step. The molecularity is thus always a whole number. Unlike the net reaction, the molecularity and the order of a step are equal so that a rate equation may be written for each step. The experimental rate equation for the net reaction is an important clue to a proposed mechanism.

Most chemical transformations involve more than one step, producing and consuming one or more **intermediates** that do not appear as reactants or products in the net reaction. The slowest step, the **rate-determining step**, determines the rate of the net reaction. The set of steps that satisfies the net reaction and leads to the experimental rate equation is called a **reaction mechanism**.

Catalysts increase reaction rates by providing a new reaction mechanism with a lower energy of activation without affecting the equilibrium constant. The catalyst consumed in one step is regenerated in another. Catalysts are **homogeneous**, being in the same phase with the reactants, or **heterogeneous**, being in a different phase. **Inhibitors** (*poisons*) decrease reaction rates by interfering with the mechanism that lowers the energy of activation or by terminating a chain reaction.

Many fast and explosive reactions involve numerous steps in which an **initiation step**, the production of free radicals or excited species, is followed by a **chain reaction**—two **propagation steps** in which each step produces one species capable of repeating the other. Thus, a product of step 1 is a reactant in step 2, and a product of step 2 is a reactant in step 1. When free radicals are the chain propagators, they may combine and stop the chain reaction (**termination steps**). In a **branching-chain reaction**, two or more species capable of producing the reaction are regenerated for each species consumed. Although equilibrium constants and rate constants do not parallel one another, they are related by the equation $K_c = k_f/k_r$.

19.13 ADDITIONAL PROBLEMS

REACTION RATES

19.13 (a) What are the units of (1) a reaction rate and (2) a half-life for a first-order reaction? (b) Are the units of k the same for all reactions? If not, what are the units of k for a first-, second-, third-, and zero-order reaction?

19.14 A common danger confronts workers in grain elevators, coal mines, and flour mills but not in rock quarries. Identify and explain the danger.

19.15 (a) Can you write rate equations for these forward and reverse reactions in water?

$$H_3AsO_4 + 3I^- + 2H^+ \rightleftharpoons H_3AsO_3 + I_3^- + H_2O$$

If you cannot, briefly explain why. (b) Can you write the equilibrium condition for the reaction? If not, briefly explain why.

19.16 What is the experimental basis for the statement that not all collisions between reactant molecules lead to products?

19.17 Given the reaction $N_2 + 3H_2 \longrightarrow 2NH_3$. At the instant N_2 is reacting at a rate of 0.25 mol/(L min), what is the rate at which H_2 is disappearing and the rate at which NH_3 is forming?

RATE EQUATIONS, k

19.18 Give the order of each reaction: (a) 4Hb(hemoglobin) + 3CO $\longrightarrow$ $Hb_4(CO)_3$, *rate* = k[Hb][CO]; (b) $Ni(C^{18}O)_4 + C^{16}O \longrightarrow Ni(C^{18}O)_3(C^{16}O) + C^{18}O$, *rate* = $k[Ni(C^{18}O)_4][C^{16}O]^0$.

19.19 The following rate equations are suggested for the reaction $2NO + F_2 \longrightarrow 2NOF$: (1) first-order with respect to NO and first-order with respect to F_2; (2) second-order with respect to NO and first-order with respect to F_2; (3) second-order with respect to NO and zero-order with respect to F_2. (a) Write the rate equation and indicate the order of the reaction for each suggestion. (b) The experimental rate equation for the reaction is *rate* = $k[NO]^2[F_2]$. Under what conditions could the rate equation become *rate* = $k[NO]^2[F_2]^0$ or $k[NO]^0[F_2]$? Illustrate with one example.

19.20 Data are given at 660 K for the reaction

$$2NO + O_2 \longrightarrow 2NO_2$$

REACTANT CONCENTRATION, mol/L		RATE OF DISAPPEARANCE OF NO, mol/(L sec)
[NO]	$[O_2]$	
0.020	0.010	1.0×10^{-4}
0.040	0.010	4.0×10^{-4}
0.020	0.040	4.0×10^{-4}

(a) Write the rate equation for the reaction. (b) Calculate the rate constant. (c) Calculate the rate, mol/(L sec), at the instant when [NO] = 0.045 and $[O_2]$ = 0.025 mol/L. (d) At the instant when O_2 is reacting at the rate 5.0×10^{-4} mol/(L sec), what is the rate at which NO is reacting and NO_2 is forming?

19.21 The rate of the hemoglobin (Hb)–carbon monoxide reaction, $4Hb + 3CO \longrightarrow Hb_4(CO)_3$, has been studied at 20°C. Concentrations are expressed in micromoles per liter (μmol/L):

REACTANT CONCENTRATION, μmol/L		RATE OF DISAPPEARANCE OF Hb, μmol/(L sec)
[Hb]	[CO]	
3.36	1.00	0.941
6.72	1.00	1.88
6.72	3.00	5.64

(a) Write the rate equation for the reaction. (b) Calculate the rate constant for the reaction. (c) Calculate the rate, μmol/(L sec), at the instant when [Hb] = 1.50 and [CO] = 0.600 μmol/L.

19.22 Data are given for the decomposition of N_2O_5 in CCl_4 solution at 45°C,

$$2N_2O_5(sol) \longrightarrow 4NO_2(sol) + O_2(g)$$

The rate is determined by measuring the volume of O_2 produced.

$[N_2O_5]$, mol/L	RATE, mol O_2/(L sec)
0.600	3.7×10^{-4}
0.300	1.9×10^{-4}
0.100	6.2×10^{-5}

(a) What is the rate equation for the reaction?
(b) Calculate the rate constant.

19.23 The decomposition of NO_2,

$$2NO_2(g) \longrightarrow 2NO(g) + O_2(g)$$

has the rate 5.4×10^{-5} mol NO_2/(L sec) at 300°C when $[NO_2] = 0.0100$ mol/L. (a) Assume that the rate equation is $r = k[NO_2]$. What rate do you predict when $[NO_2] = 0.00500$ mol/L? (b) Assume that the rate equation is $r = k[NO_2]^2$. What rate do you predict when $[NO_2] = 0.00500$ mol/L? (c) The rate when $[NO_2] = 0.00500$ mol/L is 1.4×10^{-5} mol/(L sec). Which rate equation is correct? Calculate the rate constant.

19.24 The rate of the very fast reaction

$$N_2O_4(g) \longrightarrow 2NO_2(g)$$

was measured at 25°C in a mixture of N_2O_4 and N_2. It turns out that the rate depends on the concentrations of both N_2O_4 and N_2. Some typical data are

$[N_2O_4]$, mol/L	$[N_2]$, mol/L	RATE, mol N_2O_4/(L sec)
2.00×10^{-4}	0.0200	6.84
4.00×10^{-4}	0.0200	13.7
2.00×10^{-4}	0.0400	13.6

(a) What is the rate equation? (b) Calculate the rate constant.

19.25 (a) For a zero-order reaction, R ⟶ P, a plot of what concentration term versus time gives a straight line? (b) Answer the same question for a first-order reaction, R ⟶ P, and a second-order reaction, R ⟶ P. (c) In each case, what does the slope of the line determine?

19.26 The thermal decomposition of ammonia, $NH_3 \longrightarrow NH_2 + H$, at high temperatures was studied in the presence of inert gases. Data at 2000 K are given:

$[NH_3]$, mol/L	*t*, hours
8.000×10^{-7}	0
6.75×10^{-7}	25
5.84×10^{-7}	50
5.15×10^{-7}	75

Plot the appropriate concentration expression (Table 19.1) against time to find the order of the reaction. Find the rate constant of the reaction from the slope of the line. Use the given data and the appropriate integrated rate equation (Table 19.1) to check your answer.

19.27 The following data were obtained from a study of the decomposition of HI on the surface of a gold wire. (a) Plot the data to find the order of the reaction, the rate constant, and the rate equation. (b) Calculate the HI concentration in mmol/L at 900 sec.

t, sec	[HI], mmol/L
0	5.46
250	4.10
500	2.73
750	1.37

19.28 The decomposition of N_2O_5 in nitric acid is a first-order reaction. It takes 4.26 min at 55°C to decrease 2.56 mg N_2O_5 to 2.50 mg. Find *k* in min^{-1} and sec^{-1}.

19.29 The rate equation for the reaction of sucrose in water, $C_{12}H_{22}O_{11} + H_2O \longrightarrow 2C_6H_{12}O_6$, is $rate = k[C_{12}H_{22}O_{11}][H_2O]^0$. After 2.57 h at 25.0°C, 5.00 g/L of $C_{12}H_{22}O_{11}$ is decreased to 4.50 g/L. Find *k*.

19.30 The rate constant for the decomposition of nitrogen dioxide, $NO_2 \longrightarrow NO + \frac{1}{2}O_2$, with a laser beam is 3.40 L/(mol min). Find the time in seconds needed to decrease 2.00 mol/L of NO_2 to 1.50 mol/L.

19.31 Data at 25.0°C are given for the conversion of ammonium cyanate, NH_4CNO, to urea in water, $NH_4^+ + CNO^- \longrightarrow NH_2CONH_2$:

REACTANT CONCENTRATION, mol/L			RATE OF DISAPPEARANCE OF NH_4^+, mol/L min
$[NH_4CNO]$	$[NH_4Cl]$	[KCNO]	
0.200	0.000	0.000	4.48×10^{-4}
0.000	0.200	0.200	4.47×10^{-4}
0.100	0.100	0.000	2.24×10^{-4}
0.100	0.000	0.100	2.25×10^{-4}
0.100	0.200	0.100	6.74×10^{-4}
0.100	0.100	0.200	6.72×10^{-4}

(a) What is the order (1) with respect to NH_4^+ and to CNO^-; (2) of the reaction? (b) Find the rate constant and write the rate equation. (c) Calculate the rate, mol/(L min), when $[NH_4Cl] = 0.250$ and $[KCNO] = 0.120$ mol/L.

HALF-LIFE

19.32 The rate constant for the radioactive process $^{224}_{88}Ra \longrightarrow ^{4}_{2}\alpha + ^{220}_{86}Rn$, is $0.189\ day^{-1}$. Find the half-life of ^{224}Ra.

19.33 Calculate the remaining fraction of a reactant in a first-order reaction after ten half-lives.

19.34 The half-life for a first-order reaction is given by $t_{1/2} = 0.693/k$. Derive the corresponding expression for the time needed to decrease a given quantity to one-fourth of any given initial quantity.

19.35 The rate constant for the decomposition of gaseous azomethane, $CH_3N{=}NCH_3 \longrightarrow N_2 + C_2H_6$, is $40.8\ min^{-1}$ at 425°C. Find the number of moles of $CH_3N{=}NCH_3$ and N_2 in a flask 0.0500 min after 2.00 g $(CH_3N)_2$ is added.

19.36 The half-life of $Xe(CF_3)_2$ is 30 min. 7.50 mg $Xe(CF_3)_2$ is placed in a flask. Later, 0.25 mg $Xe(CF_3)_2$ remains. How long has the sample been in the flask?

19.37 Nitrogen oxides, NO_x (a mixture of NO and NO_2 collectively designated as NO_x), play an essential role in the production of pollutants manifested in photochemical smog. The average half-life for the removal of NO_x in the smokestack emissions of Boston during daylight is 3.9 h. Starting with 1.50 mg in an experiment, (a) what is the quantity of NO_x remaining after 5.25 hours? (b) What is the sun exposure time in hours when 2.50×10^{-6} mg remains?

THEORIES OF REACTION RATES

19.38 Use the collision theory to explain the increase in rate constants with temperature.

19.39 Draw and label an enthalpy diagram (such as Figure 19.10) for a one-step endothermic reaction.

19.40 Draw the enthalpy diagram for each given reaction in the gas phase:

(a) $Cl + H_2 \longrightarrow Cl\text{--}H\text{--}H \longrightarrow HCl + H$ $\quad \Delta H = +5.3$ kJ
(transition state) $\quad \Delta H^{\ddagger} = 20$ kJ

(b) $H + Br_2 \rightarrow H\text{--}Br\text{--}Br \rightarrow HBr + Br$ $\quad \Delta H = -173$ kJ
(transition state) $\quad \Delta H^{\ddagger} = 5.0$ kJ

(c) $C_2H_4 + C_4H_6$ $\quad$ transition state $\longrightarrow C_6H_{10}$
$\Delta H = -155$ kJ
$\Delta H^{\ddagger} = 112$ kJ

19.41 Find the ΔH of each reaction from the given data for the forward and reverse reactions:

(a) $C_2H_4(sol) + C_4H_6(sol) \rightleftharpoons C_6H_{10}(sol)$
(b) $H(g) + HF(g) \rightleftharpoons H_2(g) + F(g)$

	$\Delta H^{\ddagger}_{forward}$	$\Delta H^{\ddagger}_{reverse}$
(a)	469 kJ	1117 kJ
(b)	611 kJ	49.8 kJ

19.42 The enthalpy of activation ($\Delta H^{\ddagger}$) = 54.0 kJ and the entropy of activation ($\Delta S^{\ddagger}$) = 247 J/K for the reaction between $C_6H_3(NO_2)_3Cl$ and $C_5H_{11}N$ in one solvent. (a) In a second solvent, $\Delta H^{\ddagger}$ increases to 63.0 kJ while $\Delta S^{\ddagger}$ remains constant. Is the reaction rate increased or decreased in the second solvent? (b) In a third solvent, $\Delta H^{\ddagger}$ remains constant while $\Delta S^{\ddagger}$ increases to 795 J/K. Is the rate in this solvent greater than or less than the rate in the first solvent?

19.43 From each set of data for similar reactions of the same order of reaction, select the reaction with the larger rate constant.

REACTION	$\Delta H^{\ddagger}$, kJ/mol	$\Delta S^{\ddagger}$, J/(mol K)
Set 1. Solvent exchange in solvated Co^{2+} cations		
Solvent: ammonia	11.2	+10.2
Solvent: methanol	13 ± 2	+1.6
Set 2. Methanol exchange in +3 charge cations		
Cation: V^{3+}	26	−96
Cation: Ti^{3+}	14	−96
Set 3. Formation of $Pd(H_2O)_4{}^{2+}$ from Pd^{2+} salts		
Salt: $PdCl_2$	42	−25
Salt: $PdBr_2$	42	−13

19.44 Given this enthalpy diagram for the adsorption and dissociation of O_2 on a Pt surface:
(a) What is ΔH for (1) $O_2(g) \longrightarrow O_2(ads)$; (2) $O_2(ads) \longrightarrow 2O(ads)$; (3) $O_2(g) \longrightarrow 2O(ads)$? (b) What is $\Delta H^{\ddagger}$ for $O_2(ads) \longrightarrow 2O(ads)$?

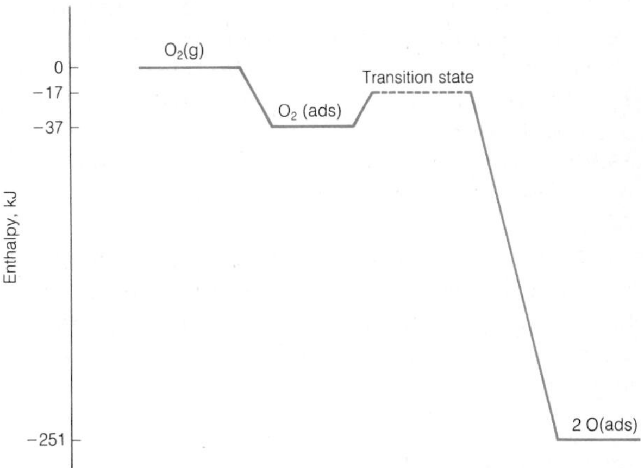

ARRHENIUS EQUATION, E_a

19.45 Nearly every collision between O atoms forms O_2 molecules. What can you say about E_a for this reaction?

19.46 At the same temperature, which reaction should have the larger rate constant? Explain. (Ignore $\Delta S^{\ddagger}$.)

$$O + O_3 \longrightarrow \text{transition state} \longrightarrow 2O_2 \qquad \Delta H = -392\ \text{kJ}, \quad E_a = +20.2\ \text{kJ}$$

$$H + H_2 \longrightarrow \text{transition state} \longrightarrow H_2 + H \qquad \Delta H = 0, \quad E_a = 33\ \text{kJ}$$

19.47 For the reaction $SiH_4 + \cdot CH_3 \longrightarrow \cdot SiH_3 + CH_4$,

$$\log k = \log 27.2 - \frac{16.7\ \text{kJ/mol}}{2.303RT}$$

Find k at 100°C. The unit of k is mL/(mol sec).

19.48 Given the rate constant as a function of temperature for the exchange reaction, $Mn(CO)_5(CH_3CN)^+ + NC_5H_5 \longrightarrow Mn(CO)_5(NC_5H_5)^+ + CH_3CN$:

k, min^{-1}	T(K)
0.0409	298
0.0818	308
0.157	318

Calculate E_a from a plot of log k versus $1/T$. Calculate log A and then find k at 311 K.

19.49 A small change in temperature causes a comparatively large change in k because k is exponentially related to $1/T$. (a) Find the ratio of k_2/k_1 for a 10-degree rise from 300 K to 310 K for a reaction whose E_a = (1) 50 kJ/mol; (2) 200 kJ/mol. (*Hint*: eliminate log A by subtraction.) Is the increase in k the same in both reactions? As E_a increases, does the effect on k for the same temperature range increase or decrease? (b) For the reaction for which E_a = 200 kJ/mol, find the k_2/k_1 ratio for a 10-degree rise from 500 K to 510 K. Is the increase in k from 500 K to 510 K the same as the increase for the rise from 300 K to 310 K? If it is not the same, for which temperature range is the ratio k_2/k_1 larger for a given E_a?

19.50 For the gas-phase reaction $2N_2O_5 \longrightarrow 2N_2O_4 + O_2$, E_a = 103 kJ and the rate constant is 0.0900 min^{-1} at 328.0 K. Find the rate constant at 338.0 K.

19.51 The rate constant of a reaction is doubled when the temperature is increased from 298.0 K to 308.0 K. Find E_a.

19.52 Egg-protein albumin is precipitated when an egg is cooked in boiling water. The E_a for this first-order reaction is 520 kJ/mol. Estimate the time required to prepare a three-minute egg at an altitude at which water boils at 90°C.

MECHANISM

19.53 Explain: the rate of the slow step is measured by measuring the rate of reaction.

19.54 (a) What is the molecularity of each step and the order of the reaction if the mechanism of the exchange reaction

$$[Co(CN)_5(H_2O)]^{2-} + I^- \longrightarrow [Co(CN)_5I]^{3-} + H_2O$$

is this two-step mechanism?

$$[Co(CN)_5H_2O]^{2-} \longrightarrow [Co(CN)_5]^{2-} + H_2O \qquad \text{(slow)}$$

$$[Co(CN)_5]^{2-} + I^- \longrightarrow [Co(CN)_5I]^{3-} \qquad \text{(fast)}$$

(b) What is the molecularity of the step and the order of the exchange reaction $CH_3Cl + Br^- \longrightarrow CH_3Br + Cl^-$ if it occurs in a one-step mechanism?

$$CH_3Cl + Br^- \longrightarrow CH_3Br + Cl^-$$

19.55 For the reaction between H_2 and CO at certain conditions, this mechanism is proposed:

$$H_2 \rightleftharpoons 2H \qquad \text{(fast reversible)}$$

$$H + CO \longrightarrow HCO \qquad \text{(slow)}$$

$$H + HCO \longrightarrow HCHO \qquad \text{(fast)}$$

Find the net reaction and the experimental rate equation.

19.56 For the gaseous decomposition of azomethane, $CH_3N{=}NCH_3 \longrightarrow N_2 + C_2H_6$, this mechanism is generally accepted:

$$CH_3N{=}NCH_3 + CH_3N{=}NCH_3 \rightleftharpoons CH_3N{=}NCH_3 + \text{activated } CH_3N{=}NCH_3 \qquad \text{(fast reversible)}$$

$$\text{activated } CH_3N{=}NCH_3 \longrightarrow N_2 + C_2H_6 \qquad \text{(slow)}$$

Derive the rate equation from the mechanism. What is the order of the reaction?

19.57 See Example 19.10 in section 19.9. Does this mechanism,

$$I_2 \rightleftharpoons 2I \qquad \text{(fast reversible)}$$

$$I + H_2 \rightleftharpoons IH_2 \qquad \text{(fast reversible)}$$

$$IH_2 + I \longrightarrow 2HI \qquad \text{(slow)}$$

also predict the correct rate equation, *rate* = $k[H_2][I_2]$?

19.58 At a distance of about 0.8 nm, a K atom transfers an electron (the "harpoon") to a Br_2 molecule, forming K^+ and Br_2^-. Then K^+ and Br_2^- strongly attract one another. The "harpoon" is drawn in forming an ionic molecule ($K^+ + Br^-$) and leaving Br behind. Write the equation for each step and for the net reaction.

19.59 A possible mechanism for the reaction $2NO + O_2 \longrightarrow 2NO_2$ is

$$NO + NO \rightleftharpoons N_2O_2 \qquad \text{(fast reversible)}$$

$$N_2O_2 + O_2 \longrightarrow 2NO_2 \qquad \text{(slow, rate constant} = k_2)$$

(a) Derive the experimental rate equation. (b) Step 1 is an exothermic reaction. In contrast to the usual behavior, the rate of this reaction decreases with increasing temperature. Offer an explanation based on this mechanism. Assume that k_2, as expected, increases with temperature.

CATALYSIS

19.60 What is the difference between a homogeneous and a heterogeneous catalyst?

19.61 What is the effect of a catalyst on (a) the enthalpy of the transition state and (b) the enthalpy of the reactants and products?

19.62 The economic benefits of catalysis in the production of ethanol, C_2H_5OH, are lost in the added expense of distilling the alcohol away from the catalyst. The following series of reactions has been proposed for directly producing anhydrous ethanol. (a) What is the net reaction? (b) What are the two original catalysts? (R represents a group of atoms that does not change.)

$$[NR_3CH_3]^+ + 2H_2 + CO \longrightarrow [NR_3H]^+ + C_2H_5OH$$

$$[NR_3H]^+ + HCO_2^- \longrightarrow H_2 + CO_2 + NR_3$$

$$NR_3 + HCOOCH_3 \longrightarrow HCO_2^- + [NR_3CH_3]^+$$

$$CH_3OH + CO \longrightarrow HCOOCH_3$$

19.63 This mechanism is suggested for the catalyzed combination of Cl atoms, $2Cl \longrightarrow Cl_2$:

$$N_2 + Cl \rightleftharpoons ClN_2 \qquad \text{(fast reversible)}$$

$$ClN_2 + Cl \longrightarrow Cl_2 + N_2 \qquad \text{(slow)}$$

Is this mechanism consistent with the experimental rate equation, $rate = k[N_2][Cl]^2$?

19.64 The enzyme carbonic anhydrase catalyzes the hydration of CO_2, $CO_2 + H_2O \longrightarrow H_2CO_3$, a critical reaction involved in the transfer of CO_2 from tissues to the lung via the bloodstream. One enzyme molecule hydrates 10^6 molecules of CO_2 per second. How many kg of CO_2 are hydrated in one hour in one liter by 5×10^{-6} M enzyme?

19.65 Biological reactions nearly always occur in the presence of enzymes, which are very powerful catalysts. For example, the enzyme catalase that acts on peroxides reduces the E_a from 72 kJ/mol (uncatalyzed) to 28 kJ/mol at 298 K. What is the increase in k? Assume A remains constant.

CHAIN REACTIONS

19.66 The chain mechanism suggested by Walther Nernst,

$$Cl_2 \rightleftharpoons 2Cl\cdot \qquad \text{(fast reversible)}$$

$$H_2 + Cl\cdot \longrightarrow HCl + H\cdot \qquad \text{(slow)} \quad E_a = 24 \text{ kJ}$$

$$H\cdot + Cl_2 \longrightarrow HCl + Cl\cdot \qquad \text{(fast)}$$

is generally accepted for the reaction $H_2 + Cl_2 \longrightarrow 2HCl$. Predict the rate equation for the reaction.

19.67 $CH_4 + Cl_2 \longrightarrow CH_3Cl + HCl$ proceeds as a chain reaction. The chain propagators are $Cl\cdot$ and $\cdot CH_3$. (a) Write equations showing (1) the chain-initiation step, (2) the chain-propagation steps, and (3) the possible chain-termination steps. (b) Does your mechanism agree with the net reaction?

19.68 (a) What is the essential difference between a chain reaction and a branching-chain reaction? (b) (1) At 550°C in a chain reaction, one molecule of HBr is produced in 10^{-9} second. Find the number of molecules and moles of HBr produced in a microsecond, 10^{-6} sec. (2) At 550°C in a branching-chain reaction, the number of H_2O molecules produced is doubled every 10^{-9} second. Find the number of molecules and moles of H_2O produced in 10^{-6} sec.

K AND *k*

19.69 Distinguish between thermodynamic stability and kinetic stability.

19.70 Select the correct words: The larger the ΔG^0, the (*greater, smaller*) is the tendency for reactants to go to products. The larger the $\Delta H^{\ddagger}$ (ignore $\Delta S^{\ddagger}$), the (*faster, slower*) is the rate of reaction. The larger the K, the (*greater, smaller*) is the tendency for reactants to go to products. The larger the k, the (*faster, slower*) is the reaction.

19.71 For the reaction $O_3 \rightleftharpoons O_2 + O$ at 300 K, $rate_f = 1.3 \times 10^{-2}[O_3]$ and $rate_r = 1.65 \times 10^{14}[O_2][O]$. Is K_c larger or small? Calculate K_c to confirm your answer.

19.72 Find the equilibrium constant at 25°C for the acidic ionization $Al(H_2O)_6{}^{3+}(aq) \rightleftharpoons [Al(H_2O)_5OH]^{2+}(aq) + H^+(aq)$ from $k_f = 1.1 \times 10^5$ and $k_r = 9.2 \times 10^9$. Is $Al^{3+}(aq)$ a strong, weak, or feeble acid?

19.73 For the rearrangement of NH_4CNO to $(NH_2)_2CO$, $K_c = 1.0 \times 10^4$ while $k_f = 5.76 \times 10^{-5}$. Find k_r.

SELF-TEST

19.74 (15 points) The reaction $2NO + 2H_2 \longrightarrow N_2 + 2H_2O$ has been studied at 904°C.

REACTANT CONCENTRATION, mol/L		RATE OF APPEARANCE
[NO]	$[H_2]$	OF N_2, mol/(L sec)
0.420	0.122	0.136
0.210	0.122	0.0339
0.210	0.244	0.0678
0.105	0.488	0.0339

(a) Write the rate equation for the reaction. (b) Calculate the rate constant at 904°C. (c) Find the rate of appearance of N_2 at the instant when [NO] = 0.550 and $[H_2] = 0.199$ mol/L.

19.75 (10 points) Data for the first-order decomposition of dinitrogen oxide, $2N_2O \longrightarrow 2N_2 + O_2$, on gold at 900°C are given. Find the rate constant from the slope of a graph. Write the rate equation and find the decomposition rate at 900°C when $[N_2O] = 35.0$ mmol/L.

t, min	$[N_2O]$, mmol/L
15.0	83.5
30.0	68.0
80.0	35.0
120	22.0

19.76 (10 points) The decomposition of gaseous dimethyl ether, $CH_3OCH_3 \longrightarrow CH_4 + CO + H_2$, at ordinary pressures is first-order. Its half-life is 25.0 min at 500°C. (a) Starting with 8.00 g, how many grams remain after (1) 125 min and (2) 145 min? (b) Calculate the time in minutes required to decrease 7.60 ng to 2.25 ng. (c) What fraction remains and what fraction has reacted after 150 min?

19.77 (10 points) The transition state for the reaction $Pb_2(g) + O_2(g) \longrightarrow 2PbO(g)$ is 4-centered,

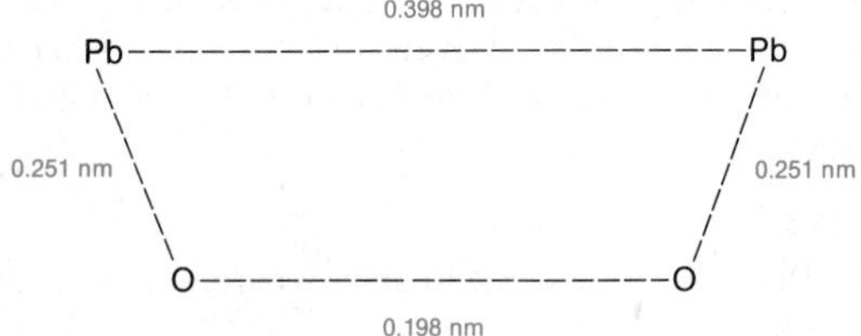

For this reaction, $\Delta H = -200$ kJ and $\Delta H^{\ddagger} = 267$ kJ. Draw the enthalpy diagram. Show bond breaking and bond formation. Identify which bond lengths are increasing and/or decreasing in going from reactants to transition state and from transition state to products. Bond lengths of initial reactants and final products: Pb—Pb, 0.308 nm; O—O, 0.121 nm; Pb—O, 0.203 nm

19.78 (10 points) For the second-order reaction $\cdot CF_3 + \cdot CF_3 \longrightarrow C_2F_6$, $E_a = 0$. What does $E_a = 0$ mean in terms of collision theory? $\log A = 13.50$. Find k at 20°C and 60°C.

19.79 (10 points) Three mechanisms are proposed for a reaction:

1. $NO + NO + Br_2 \longrightarrow 2BrNO$

2. $NO + Br_2 \rightleftharpoons Br_2NO$
   $Br_2NO + NO \longrightarrow 2BrNO$

3. $NO + NO \rightleftharpoons N_2O_2$
   $N_2O_2 + Br_2 \longrightarrow 2BrNO$

(a) Write the net reaction. (b) What is the molecularity of each step in each mechanism? (c) What are the intermediates formed in mechanisms 2 and 3?

19.80 (10 points) The reaction between ozone and nitrogen dioxide, $2NO_2 + O_3 \longrightarrow N_2O_5 + O_2$, has been studied at 231 K. The experimental rate equation is *rate* $= k[NO_2][O_3]$. (a) What is the order of the reaction? (b) Select the mechanism that is consistent with the given kinetic data. Show how you arrived at your answer.

1. $NO_2 + NO_2 \rightleftharpoons N_2O_4$ (fast reversible)
   $N_2O_4 + O_3 \longrightarrow N_2O_5 + O_2$ (slow)

2. $NO_2 + O_3 \longrightarrow NO_3 + O_2$ (slow)
   $NO_3 + NO_2 \longrightarrow N_2O_5$ (fast)

19.81 (10 points) The reaction of Br_2 with chloroform, $Br_2 + CHCl_3 \longrightarrow HBr + CBrCl_3$, in the gas phase at 175°C follows the rate equation, *rate* $= k[CHCl_3][Br_2]^{1/2}$. (a) What is the order of the reaction? (b) Is the following chain mechanism consistent with the kinetic data?

$Br_2 \rightleftharpoons 2{:}\ddot{Br}\cdot$ (fast reversible)

${:}\ddot{Br}\cdot + CHCl_3 \longrightarrow HBr + \cdot CCl_3$ (slow)

$\cdot CCl_3 + Br_2 \longrightarrow CCl_3Br + {:}\ddot{Br}\cdot$ (fast)

(c) Write the chain-initiation step, chain-propagation steps, and three possible chain-termination steps.

19.82 (10 points) Many biochemical reactions are acid-catalyzed. A typical mechanism consistent with the experimental rate equation, *rate* $= k[X][HA]^{1/2}$, in which HA is the acid and X is the reactant, is given:

$HA \rightleftharpoons H^+ + A^-$ (fast reversible)

$X + H^+ \rightleftharpoons XH^+$ (fast reversible)

$XH^+ \longrightarrow$ products (slow)

HA is the only source of H^+ and A^-. Is the measured rate constant independent of the acid strength? Explain.

19.83 (5 points) The reactants in reaction A,

$$SiCl_4(\ell) + 2H_2O(\ell) \longrightarrow SiO_2(c) + 4HCl(g)$$

$K_p = 4.7 \times 10^{54}$, exhibit both thermodynamic and kinetic instability. The reactants in reaction B,

$$CCl_4(\ell) + 2H_2O(\ell) \longrightarrow CO_2(g) + 4HCl(g)$$

$K_p = 8.5 \times 10^{72}$, are thermodynamically unstable but kinetically stable. $SiCl_4(\ell)$ and $CCl_4(\ell)$ are added to water. Which reaction(s) will occur?

20 THE CHEMISTRY OF THE REPRESENTATIVE ELEMENTS

In Chapter 6 we presented an overview of the physical and chemical properties of the elements (section 6.7). We showed how these properties are related to the electronic configurations of the atoms, which in turn underlie the arrangement of the elements in the periodic table. Now, fourteen chapters later, after discussions of chemical bonding, the rates and energetics of chemical reactions, and various other principles of chemistry, it is time to look more deeply into the chemical properties of the elements and their compounds.

This chapter is concerned with the representative elements: those in the sections of the periodic table that represent the stepwise addition of electrons to the *s* and *p* subshells of the atoms. These elements are designated as Groups 1 through 8. Chapter 21, which follows, is devoted to the transition elements.

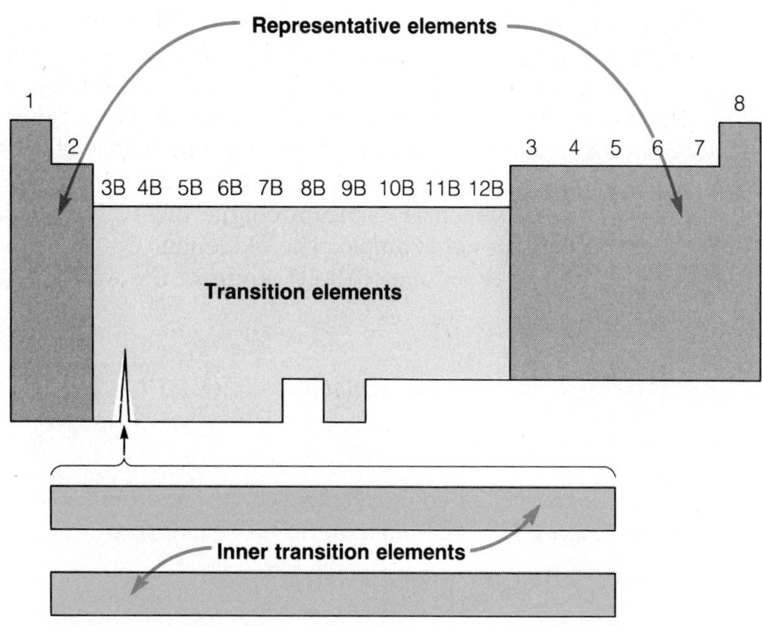

20.1 REPRESENTATIVE ELEMENTS

The chemistry of the 44 representative elements comprises an enormous body of literature. Whole books have been written about many of these elements, and some of them require very many books. The study of carbon and its compounds is an entire *discipline*—organic chemistry—to which Chapters 24 and 25 are devoted.

Zn, Cd, and Hg are sometimes counted as representative elements. When they are, there are 47.

Over 95% of the inorganic compounds that are common or important enough to be listed in one-volume handbooks of chemistry contain oxygen, hydrogen, sulfur, or a halogen—usually fluorine or chlorine. The fact that oxygen is part of very many compounds is not surprising: oxygen is the most abundant element in the Earth's crust and participates in both ionic and covalent bonding. Almost all of the elements form oxides, and many of the oxides, in turn, are chemically reactive to acids, bases, and water. Since we can hardly attempt to survey all of the compounds of the representative elements, an overview of the oxides will provide good descriptions of many important compounds and further illustrations of chemical principles.

The exceptions are 5 of the 6 noble gases, He, Ne, Ar, Kr, and Rn. Xenon forms unstable oxides.

What about the compounds of hydrogen? Although hydrogen is the most abundant element in the universe, it does not have that distinction on Earth, where it makes up only about 0.1% of the mass of the crust. Hydrogen, however, has unique chemical properties. Many elements form hydrides, and more important for our purposes here, the properties of the hydrides, like those of the oxides, serve to organize and illustrate important chemical relationships.

Most hydrogen on Earth is bound in the form of H_2O.

The chemistry of the halogens, too, is instructive. Fluorine is the most electronegative element, so it combines with almost everything that has electrons to spare—including even some of the noble gases. The properties of the halogens vary in interesting ways in the progression down Group 7 from fluorine to iodine, and by extrapolation to the all-but-unknown astatine.

Sulfur is another element we will use in this overview of descriptive chemistry, mainly because of the relationships between it and oxygen. Both their similarities and their differences, like those among the halogens, help to organize the chemical properties of the other elements that combine with them. Besides, as we shall see, sulfur is an interesting element in its own right.

20.2 BASIC AND ACIDIC PROPERTIES OF OXIDES

Since oxygen is the most electronegative of all elements except fluorine, the partial charge on oxygen in all oxides but OF_2 and O_2F_2 is negative. This partial charge reaches its most negative value of -2 in the ionic alkali metal oxides such as $(Na^+)_2O^{2-}$. As the partial negative charge on oxygen decreases, the oxide becomes less polar and more covalent, as in the oxides of nitrogen and chlorine.

Now consider the possible reactions between an oxide and water (section 8.6). Ionic oxides, in which O carries a charge of -2, are basic in water. Calcium oxide, CaO, is a good example. The oxide ion, O^{2-}, acts as the base, making a nucleophilic attack on one of the H atoms of the water molecule:

$$\underset{\text{base}}{:\ddot{\underset{..}{O}}:^{2-}} + \underset{\text{acid}}{H^{\delta+}{-}\ddot{\underset{..}{O}}^{\delta-}{-}H^{\delta+}} \longrightarrow (:\ddot{\underset{..}{O}}:H)^- + (:\ddot{\underset{..}{O}}:H)^- \qquad (1)$$

The products are the conjugate acid of O^{2-} and the conjugate base of H_2O, both of which are OH^-. The equation in molecular form is:

$$CaO(c) + H_2O(\ell) \longrightarrow Ca(OH)_2(c)$$

Anhydride, "lacking water."

Because CaO reacts with water to form OH^-, it is called a **basic anhydride**.

The O atoms in covalent oxides carry only a partial negative charge. These oxides, such as sulfur dioxide, SO_2, and sulfur trioxide, SO_3, are *acidic* in water. Sulfur trioxide has an electrophilic site because of unequal sharing of π electrons between S and O. Therefore H_2O acts as a Lewis base, its O atom making a nucleophilic attack on the S atom of the SO_3 molecule:

$$\underset{\text{base}}{H_2\ddot{O}:} + \underset{\text{acid}}{O{=}S^{\delta+}(O)=O^{\delta-}} \longrightarrow H{-}\overset{\oplus}{O}(H){-}\overset{2+}{S}(O^{\ominus})(O^{\ominus}){-}O^{\ominus} \tag{2}$$

The addition product then reacts with more water to produce H_3O^+ and HSO_4^- ions, which are the major components of an aqueous solution of sulfuric acid:

$$H{-}\ddot{O}(H): + H{-}\overset{\oplus}{O}(H){-}\overset{2+}{S}(O^{\ominus})(O^{\ominus}){-}O^{\ominus} \longrightarrow \underbrace{H_3O^+ + HSO_4^-}_{\text{sulfuric acid}} \tag{3}$$

and in molecular form:

$$H_2O(\ell) + SO_3(g) \longrightarrow H_2SO_4(\ell)$$

PROBLEM 20.1 Rewrite equation 2 using a resonance structure for SO_3 that has single bonds between S and each of the three O atoms. Show all formal charges. □

Because SO_3 reacts with water to form an acid, it is called an **acid anhydride**.

Recall from Chapter 14 that a Brönsted acid-base reaction involves a proton transfer, while a Lewis acid-base reaction involves an electron pair, but not a proton. Therefore in equation 1, which describes a proton transfer, O^{2-} is a Brönsted base and H_2O is a Brönsted acid. In equation 2, where no proton is transferred, H_2O is a Lewis base and SO_3 is a Lewis acid.

Other important reactions of acid anhydrides with water are

The *ligancy* of a given atom is the number of other atoms to which it is chemically bonded. Numerically, ligancy is the same as the coordination number; the only difference between the two is that coordination number (Chapter 11) denotes the number of "nearest neighbors" without any implication about chemical bonding.

$$P_4O_6(c) + 6H_2O(\ell) \longrightarrow \underset{\text{phosphorous acid}}{4H_2PHO_3(aq)} \tag{4}$$

$$N_2O_5(\ell) + H_2O(\ell) \longrightarrow \underset{\text{nitric acid}}{2HNO_3(aq)} \tag{5}$$

$$Cl_2O_7 + H_2O(\ell) \longrightarrow \underset{\text{perchloric acid}}{2HClO_4(aq)} \tag{6}$$

Some oxides are amphoteric. An example is Al_2O_3, which acts as a base toward a strong acid:

$$Al_2O_3(c) + 6H^+ \longrightarrow 2Al^{3+} + 3H_2O(\ell) \tag{7}$$

(see also section 14.8) and as an acid toward a strong base:

$$Al_2O_3(c) + 2OH^- + 3H_2O(\ell) \longrightarrow 2Al(OH)_4^-(aq) \tag{8}$$

The common basic and/or acidic oxides of the representative elements are displayed in Table 20.1.

TABLE 20.1
BASIC, ACIDIC, AND AMPHOTERIC OR NEUTRAL PROPERTIES OF THE OXIDES OF THE REPRESENTATIVE ELEMENTS

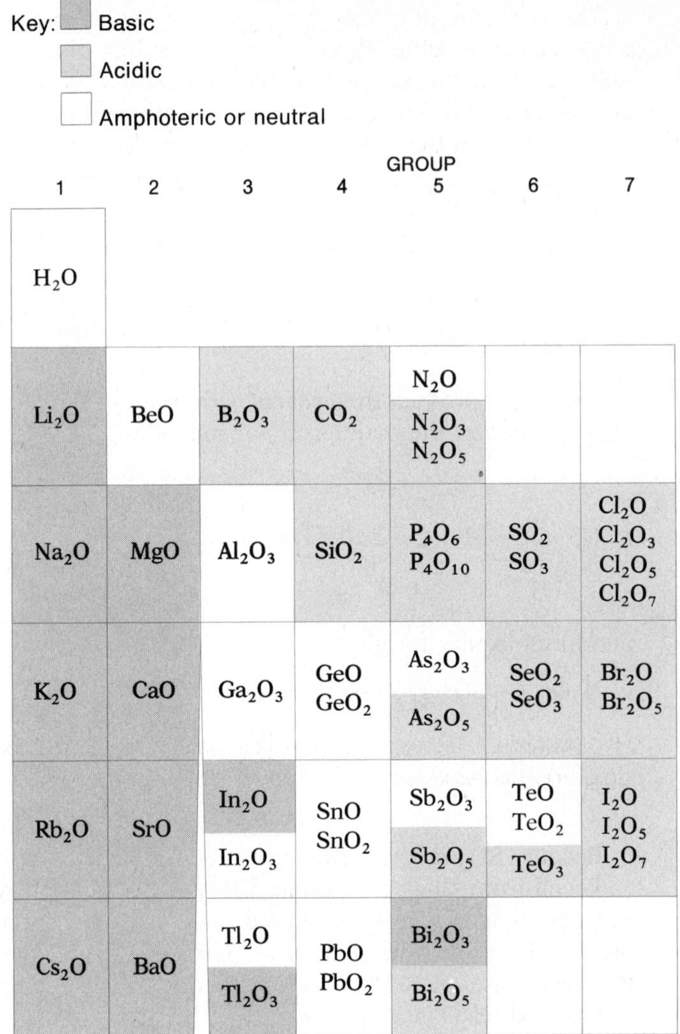

Key: Basic; Acidic; Amphoteric or neutral

GROUP 1	2	3	4	5	6	7
H_2O						
Li_2O	BeO	B_2O_3	CO_2	N_2O N_2O_3 N_2O_5		
Na_2O	MgO	Al_2O_3	SiO_2	P_4O_6 P_4O_{10}	SO_2 SO_3	Cl_2O Cl_2O_3 Cl_2O_5 Cl_2O_7
K_2O	CaO	Ga_2O_3	GeO GeO_2	As_2O_3 As_2O_5	SeO_2 SeO_3	Br_2O Br_2O_5
Rb_2O	SrO	In_2O In_2O_3	SnO SnO_2	Sb_2O_3 Sb_2O_5	TeO TeO_2 TeO_3	I_2O I_2O_5 I_2O_7
Cs_2O	BaO	Tl_2O Tl_2O_3	PbO PbO_2	Bi_2O_3 Bi_2O_5		

PROBLEM 20.2 Equations 7 and 8 are condensed because they do not show all of the H_2O molecules of hydration needed to satisfy the ligancy of 6 for Al in each of its ions in water. Rewrite and rebalance these equations by adding sufficient H_2O molecules so that Al will have a ligancy of 6 in each of its ions and H^+ will be changed to H_3O^+. □

The sequence from basic toward acidic anhydrides is shown for the oxides of the third period:

$$\underbrace{Na_2O,\ MgO,}_{\text{basic}}\quad \underbrace{Al_2O_3,}_{\text{amphoteric}}\quad \underbrace{SiO_2,}_{\text{reacts slowly with base}}\quad \underbrace{P_4O_{10},\ SO_3,\ Cl_2O_7}_{\text{acidic}}$$

PROBLEM 20.3 With the help of the periodic table, classify each of these oxides as acidic or basic, and identify the one oxide that is amphoteric: (a) I_2O_5; (b) Ga_2O_3; (c) BaO; (d) Rb_2O; (e) SeO_2. □

PREPARING ACIDS FROM NONMETALLIC OXIDES

The reaction of an acid anhydride with water offers an important method of preparing acids, as shown in equations 3 to 6. The manufacture of sulfuric acid, one of the most important of all industrial chemicals, starts from elemental sulfur. Its direct oxidation produces sulfur dioxide, SO_2, which can be subsequently oxidized in the presence of a vanadium pentoxide (V_2O_5) catalyst to SO_3. The hydration reaction of SO_3 is very exothermic. The heat liberated vaporizes the water, and the steam cloud carries away much of the SO_3 in the form of a sulfuric acid mist. For more effective results, the SO_3 is dissolved in concentrated H_2SO_4 to form pyrosulfuric acid, $H_2S_2O_7$, while water is added continuously to maintain a constant concentration of H_2SO_4:

Sulfur vapor is S_8. The molecule is an eight-membered, puckered ring

$$S(c) + O_2(g) \longrightarrow SO_2(g)$$

$$2SO_2(g) + O_2(g) \rightleftharpoons 2SO_3(g) \quad \text{(contact process, surface catalysis)}$$

$$SO_3(g) + H_2SO_4(\ell) \rightleftharpoons H_2S_2O_7(\ell), \text{ or } HO{-}SO_2{-}O{-}SO_2{-}OH$$

$$H_2S_2O_7(\ell) + H_2O(\ell) \longrightarrow 2H_2SO_4(\ell)$$

Phosphoric acid, too, is made by hydration of the anhydride (equation 4), which is obtained by burning phosphorus in air. The production of nitric acid, however, is not so direct. Nitrogen is not readily oxidized to N_2O_5, which is the anhydride of nitric acid, HNO_3. Instead, the first step is the production of ammonia, which is then oxidized in a series of three additional steps to nitric acid:

$$N_2(g) + 3H_2(g) \longrightarrow 2NH_3(g) \quad \text{(Haber process)}$$

$$4NH_3(g) + 5O_2(g) \longrightarrow 4NO(g) + 6H_2O(\ell) \quad \text{(Ostwald process)}$$

$$2NO(g) + O_2(g) \longrightarrow 2NO_2(g)$$

The final step is a self-redox reaction of NO_2 in water to give nitric acid and NO. **Self-redox** or **disproportionation** is an internal oxidation–reduction reaction in which one portion of a substance becomes oxidized while the other becomes reduced. The reaction here is:

$$3NO_2(g) + H_2O(\ell) \longrightarrow \underset{\text{nitric acid}}{2HNO_3(aq)} + NO(g)$$

The NO is recycled.

PROBLEM 20.4 Show that the $NO_2 + H_2O$ equation above describes a self-redox reaction by calculating the oxidation numbers of N in NO_2, in HNO_3, and in NO. □

Self-redox is typical of oxides of elements in Groups 5 and 7 of the periodic table when the element has an even-numbered oxidation state; the reaction results in higher and lower odd-numbered oxidation states. Self-redox in a basic medium results in formation of the oxoanions, as exemplified by the reactions of chlorine dioxide and dichlorine hexoxide:

$$\underset{\substack{+4\\ \text{chlorine dioxide}}}{2ClO_2(g)} + 2OH^- \longrightarrow \underset{+3}{ClO_2^-} + \underset{+5}{ClO_3^-} + H_2O(\ell)$$

$$\underset{\substack{+6\\ \text{dichlorine hexoxide}}}{Cl_2O_6(\ell)} + 2OH^- \longrightarrow \underset{+5}{ClO_3^-} + \underset{+7}{ClO_4^-} + H_2O(\ell)$$

Note that in the odd-numbered periodic groups a neutral atom has an odd number of electrons, which means that one electron must be unpaired. Examples are ${}_7N$, ${}_{15}P$ (both in Group 5), and ${}_9F$ and ${}_{17}Cl$ (both in Group 7). The addition of an odd number of electrons, however, gives an even total, as for N^{3-} or Cl^-. In the same way, the disproportionation to two *odd* oxidation states in oxides or oxyanions produces two species with an *even* number of electrons, so that all electrons can be paired and greater stability can be achieved.

PROBLEM 20.5 What is the total number of valence electrons in a molecule of ClO_2; in a ClO_2^- ion; in a ClO_3^- ion? □

20.3 THE CHEMISTRY OF SOME COMMON OXIDES

As mentioned earlier, practically all of the elements form oxides. Since oxygen is so abundant, many oxides are important compounds in our environment and familiar to us. Most elements form more than one oxide. The compounds with fewer O atoms per atom of the other element are called the lower oxides; those with more O atoms are the higher oxides. The lower oxide of an element can therefore serve as a reducing agent, and may even be able to burn.

OXIDES OF CARBON AND SILICON

Both carbon and oxygen readily form multiple bonds, which appear in all of the carbon oxides. The two major oxides are

:C≡O: (carbon monoxide)

:O=C=O: (carbon dioxide)

Two less common oxides are :O=C=C=C=O:, carbon suboxide (tricarbon dioxide), and :O=C=C=C=C=C=O:, pentacarbon dioxide.

Of these, CO_2 represents the most highly oxidized state of carbon and, therefore, it is the ultimate oxidation product. CO_2 is the major product obtained when carbonaceous matter is burned in an abundant supply of air or oxygen. When the supply of oxygen is insufficient to provide complete conversion to CO_2, the less highly oxidized product CO is formed, together with traces of other oxides that contain more carbon than oxygen.

PROBLEM 20.6 (a) What is the oxidation number of C in CO and in CO_2? (b) What is the average oxidation number of carbon in carbon suboxide, C_3O_2? (c) Is carbon suboxide combustible? □

Carbon dioxide is a colorless, odorless, noncombustible gas. When combined with water, it has a slightly acid taste. Together with water, it is the raw material for photosynthesis of glucose:

$$6CO_2 + 6H_2O \longrightarrow \underset{\text{glucose}}{C_6H_{12}O_6} + 6O_2$$

Humans lose consciousness when exposed for only a few minutes to air containing 10% CO_2.

Carbon dioxide is commercially available in steel cylinders as a liquid under the pressure of its own vapor, and also in the solid form ("dry ice") at its sublimation temperature, −78.5°C. The gas is not toxic in low concentrations, but *it is toxic in high concentrations.*

Carbon dioxide is soluble in water (90 mL CO_2 gas at 1 atm dissolve in 100 mL water at 20°C), and is the anhydride of carbonic acid, H_2CO_3:

H—O
    \
     C=O
    /
H—O

$$CO_2(g) + H_2O(\ell) \rightleftharpoons \underset{\text{carbonic acid}}{H_2CO_3(aq)}$$

Carbonic acid, however, is unstable; it does not exist in pure form. As pointed out in Chapter 15 (Table 15.2), less than 1% of the CO_2 that dissolves in water actually reacts to form H_2CO_3.

H_2CO_3 is a diprotic acid; its conjugate base is bicarbonate ion, HCO_3^-. The conjugate base of HCO_3^-, in turn, is carbonate ion, CO_3^{2-}. The first ionization constant, K_{a_1}, of carbonic acid is 4.30×10^{-7}. This value corresponds to the *overall* reaction of carbon dioxide and water as shown here:

$$CO_2 + H_2O \rightleftharpoons H_2CO_3 \qquad K \approx 1.4 \times 10^{-3}$$
$$H_2CO_3 + H_2O \rightleftharpoons H_3O^+ + HCO_3^- \qquad K \approx 3 \times 10^{-4}$$
$$CO_2 + 2H_2O \rightleftharpoons H_3O^+ + HCO_3^- \qquad K_{a_1} = 4.30 \times 10^{-7}$$

If all the CO_2 in a carbonated beverage were hydrated to H_2CO_3, soda water would be more acidic than vinegar (K_a for acetic acid $= 1.75 \times 10^{-5}$) and therefore would be too acidic to drink.

Carbonates, CO_3^{2-}, and bicarbonates, HCO_3^-, which are salts of carbonic acid, are stable and well known. The most abundant of these is calcium carbonate, $CaCO_3$, the major component of limestone, marble, coral, seashells, and pearls. In the household, the most familiar carbonate salt is sodium hydrogen carbonate, $NaHCO_3$, more commonly known as sodium bicarbonate, baking soda, bicarbonate of soda, or "bicarb." The pH of a freshly prepared aqueous solution of $NaHCO_3$ at 25°C is about 8.4. This is a very mildly basic solution, and therefore sodium bicarbonate can be taken internally as an antacid.

Another important carbonate salt, sodium carbonate, Na_2CO_3, known as soda ash, is used in large quantities for the manufacture of glass and other materials. The decahydrate, $Na_2CO_3(H_2O)_{10}$, known as sal soda or washing soda, is used in industry and in the home for washing and water softening. Its aqueous solutions are strongly basic (pH 11–12) and therefore toxic.

CO_2 gas may be prepared in the laboratory by the action of strong acids, such as H_2SO_4 or HCl, on carbonate or bicarbonate salts:

$$2H_3O^+ + \underset{\text{calcium carbonate}}{CaCO_3(c)} \longrightarrow Ca^{2+} + 3H_2O + CO_2(g)$$

$$H_3O^+ + \underset{\text{sodium bicarbonate}}{NaHCO_3(c)} \longrightarrow Na^+ + 2H_2O + CO_2(g)$$

CO_2 gas is produced commercially (a) by burning coke

$$\underset{\text{coke}}{C(amorph)} + O_2(g) \longrightarrow CO_2(g) \qquad \Delta H = -393 \text{ kJ/mol}$$

or petroleum, (b) as a by-product of fermentation, or (c) by the "calcining" of limestone (section 9.7):

$$\underset{\text{limestone}}{CaCO_3(c)} \longrightarrow \underset{\text{lime}}{CaO(c)} + CO_2(g)$$

Metals or carbon at incandescent temperatures reduce CO_2 to the element or to a lower oxide:

$$2Mg(c) + CO_2(g) \longrightarrow 2MgO(c) + C(amorph)$$

Since this reaction is exothermic, you should never attempt to extinguish a magnesium fire with carbon dioxide.

Carbon monoxide is an odorless, colorless, tasteless, flammable, highly toxic gas. Mixtures of CO in air in the concentration range between 12.5% and 74% CO by volume are explosive.

SAFETY NOTE: Carbon monoxide is insidious because it gives little sensory warning of its presence; a concentration of 0.1% by volume produces unconsciousness in one hour and death in four hours. Therefore, the combustion products of any carbonaceous fuel should be regarded as a possible source of CO and should not be allowed to accumulate without control in occupied spaces.

The gas is produced on an industrial scale by the reaction of steam with hot coal or coke, or with methane (water gas reactions):

$$C(\text{amorph}) + H_2O(g) \longrightarrow \underbrace{H_2(g) + CO(g)}_{\text{"water gas"}}$$

$$CH_4(g) + H_2O(g) \longrightarrow 3H_2(g) + CO(g)$$

Silicon forms two oxides, SiO and SiO_2, whose empirical formulas are analogous to CO and CO_2. Silicon monoxide, SiO, has been prepared under vacuum at high temperature, and its existence below 1200°C is in doubt. Silicon dioxide, SiO_2, also called silica, occurs widely in nature as quartz and other minerals. It is a hard, transparent network-covalent substance that melts around 1700°C.

OXIDES OF NITROGEN

Nitrogen, like carbon, forms double bonds with oxygen. Oxides corresponding to every oxidation state of N from 1 to 6 are known:

OXIDE		OXIDATION NUMBER OF N
N_2O	dinitrogen oxide; nitrous oxide	+1
NO	nitrogen oxide; nitric oxide	+2
N_2O_3	dinitrogen trioxide	+3
NO_2	nitrogen dioxide	+4
N_2O_4	dinitrogen tetroxide	+4
N_2O_5	dinitrogen pentoxide	+5
NO_3	nitrogen trioxide (unstable)	+6

As noted in section 9.7, ammonium nitrate is an explosive. Small quantities, however, are not particularly sensitive. In large masses, the buildup of internal pressures on heating may lead to detonation. Students should not heat this substance.

Dinitrogen oxide, N_2O, also known as nitrous oxide and "laughing gas," is prepared by heating ammonium nitrate above 200°C. This is a self-redox reaction:

$$NH_4{}^+NO_3{}^-(c) \longrightarrow 2H_2O + N_2O(g)$$

N_2O is a colorless gas (b.p. −88°C) with a slightly sweet odor and taste, and has anesthetic properties. Its molecular structure was described in section 8.5. It is stable at ordinary temperatures, but when heated above about 600°C it decomposes to nitrogen and oxygen, and therefore supports combustion:

$$2N_2O(g) \longrightarrow 2N_2(g) + O_2(g)$$

Nitrogen oxide (common name, nitric oxide) is a colorless gas. Its Lewis formula, $\cdot\ddot{N}{=}\ddot{O}{:}$, has an odd number of electrons. As expected, NO is a typical free radical and is paramagnetic. NO is produced by the high-temperature oxidation of N_2 or NH_3, and is therefore a by-product of all combustion processes that occur in air. As the temperature drops, NO is oxidized to NO_2:

$$2NO(g) + O_2(g) \longrightarrow 2NO_2(g)$$

As illustrated in section 8.7, NO has a relatively low ionization energy, because an electron is removed from an antibonding orbital. The resulting **nitrosonium ion,** NO^+, is isoelectronic with the stable molecules CO and N_2. Ionic nitrosonium compounds include the perchlorate, $NO^+ClO_4^-$, and the hydrogen sulfate, $NO^+HSO_4^-$.

Dinitrogen trioxide, $:\dot{O}=\ddot{N}-N(=\ddot{O}:)(-\ddot{O}:)$, can be formed by cooling an equimolar mixture of NO and NO_2 to about $-30°C$:

$$:\ddot{O}=\ddot{N}\cdot(g) + \cdot N(=\ddot{O}:)(-\ddot{O}:)\,(g) \longrightarrow N_2O_3 \qquad \text{(blue liquid)}$$

Note that this reaction is the combination of two molecules with unpaired electrons to form a nonradical product. At normal temperatures N_2O_3 is unstable and decomposes into NO and NO_2.

N_2O_3 is the anhydride of nitrous acid, HNO_2, which is also unstable:

$$N_2O_3(g) + H_2O \longrightarrow 2HNO_2(aq) \longrightarrow H_2O + NO(g) + NO_2(g)$$

Nitrites, salts of the NO_2^- ion, are well known and reasonably stable. Pure sodium nitrite, $NaNO_2$, can be prepared by the action of NaOH on an equimolar mixture of NO and NO_2.

PROBLEM 20.7 Write the balanced molecular equation for the preparation of sodium nitrite. □

Nitrogen dioxide, $:\ddot{O}-\dot{N}=\ddot{O}:$, like NO, has an unpaired electron, and its properties are consistent with its free-radical character. It is a reddish brown, paramagnetic gas. NO_2 molecules combine with each other by the pairing of their unpaired electrons to form the colorless dinitrogen tetroxide, N_2O_4:

$$\underset{\text{brown}}{2NO_2(g)} \underset{\text{high temperature}}{\overset{\text{low temperature}}{\rightleftharpoons}} \underset{\text{colorless}}{(:\ddot{O}=)(\ominus:\ddot{O}-)\overset{\oplus}{N}-\overset{\oplus}{N}(=\ddot{O}:)(-\ddot{O}:\ominus)\,(g)} \quad \text{or} \quad (O)_2\overset{\oplus}{N}-\overset{\oplus}{N}(O)_2\,(g)$$

At ordinary temperatures, the two gases exist in equilibrium with each other as a light brown mixture. Since the forward reaction involves only the formation of a bond, it is exothermic, and is therefore favored at lower temperatures. Thus the color of the equilibrium mixture fades as it is cooled, and deepens as it is warmed.

NO_2 is made industrially by the oxidation of NO,

$$2NO + O_2 \longrightarrow 2NO_2$$

and may be prepared in the laboratory by gently heating lead nitrate:

$$2Pb(NO_3)_2 \longrightarrow 2PbO + 4NO_2 + O_2$$

Victims have included workers who load silos with green fodder (silage), which emits oxides of nitrogen from biological oxidation of plant protein.

NO_2 is a deadly poison; its great danger is often unrecognized. Initial exposure may cause inflammation of the lungs with only slight pain, but death may occur a few days later as the lungs fill with fluid.

If NO_2 loses an electron, it forms the stable **nitronium ion,** NO_2^+, which is isoelectronic with CO_2. Its molecular structure is

$$:\ddot{O}=\overset{\oplus}{N}=\ddot{O}:$$

Nitronium salts include the perchlorate, $NO_2^+ClO_4^-$, and the hydrogen sulfate, $NO_2^+HSO_4^-$.

Dinitrogen pentoxide is represented in the vapor state by the structural formula

$$\overset{\oplus}{N}(=\ddot{O}:)(-\ddot{O}:^{\ominus})-\ddot{O}-\overset{\oplus}{N}(=\ddot{O}:)(-\ddot{O}:^{\ominus}) \quad \text{or} \quad {}^{\ominus}O_2\overset{\oplus}{N}-O-\overset{\oplus}{N}O_2{}^{\ominus}$$

In the solid form, however, it is an ionic salt, nitronium nitrate, $NO_2^+NO_3^-$. The solid melts at about 30°C and decomposes readily:

$$N_2O_5(c) \longrightarrow 2NO_2(g) + \tfrac{1}{2}O_2(g)$$

N_2O_5 is best prepared by reaction of NO_2 with ozone:

$$2NO_2(g) + O_3(g) \longrightarrow N_2O_5(c) + O_2(g)$$

N_2O_5 is the anhydride of nitric acid:

$$N_2O_5(c) + H_2O \longrightarrow \underset{\text{nitric acid}}{2HNO_3(aq)}$$

A nitrating agent is one that introduces NO_2 or NO_3 groups into a molecule.

The corresponding anion is nitrate ion, NO_3^-. Nitric acid is a strong acid, a strong oxidizing agent, and an effective nitrating agent. "Concentrated" nitric acid as used in the laboratory is about 68% HNO_3 by mass, or about 15 M. The common "dilute" nitric acid is about 17% nitric acid by mass, or about 3 M. Nitric acid is thermodynamically quite stable with respect to its elements (ΔG_f at 25°C is -80 kJ/mol), but the liquid decomposes spontaneously, especially in sunlight, and turns brown because of the NO_2 it produces:

$$2HNO_3 \rightleftharpoons 2NO_2 + H_2O + \tfrac{1}{2}O_2$$

When nitric acid acts as an oxidizing agent, the product to which it is reduced depends both on the nature of the reducing agent and on the concentration of the acid. The first two equations given below show that copper reduces dilute nitric acid to a greater extent than it reduces concentrated nitric acid, as shown by the oxidation numbers of the reduced products. The second and third equations show that Zn reduces dilute nitric acid to a greater extent than Cu does; this effect is related to the difference between their reduction potentials. The equations give only the major products; the actual circumstances are more complex. The nitric acid is usually reduced to a mixture of various nitrogen oxides, some N_2, and perhaps some NH_4^+.

$$Cu(c) + 4HNO_3(\text{conc aq}) \longrightarrow Cu(NO_3)_2(aq) + 2\overset{+4}{N}O_2(g) + 2H_2O(\ell)$$

$$3Cu(c) + 8HNO_3(\text{dilute aq}) \longrightarrow 3Cu(NO_3)_2(aq) + 2\overset{+2}{N}O(g) + 4H_2O(\ell)$$

Note that these are oxidations, so the electrode potentials are *minus* the values given in Table 17.1 (section 17.7) for the reduction potentials.

$$4Zn(c) + 10HNO_3(\text{dilute aq}) \longrightarrow 4Zn(NO_3)_2(aq) + \overset{-3}{N}H_4NO_3(aq) + 3H_2O(\ell)$$

$$Zn \longrightarrow Zn^{2+} + 2e^-, \quad \mathscr{E}^0 = +0.7628 \text{ V}$$

$$Cu \longrightarrow Cu^{2+} + 2e^-, \quad \mathscr{E}^0 = -0.3402 \text{ V}$$

PROBLEM 20.8 Rewrite the above molecular equations as net ionic equations. □

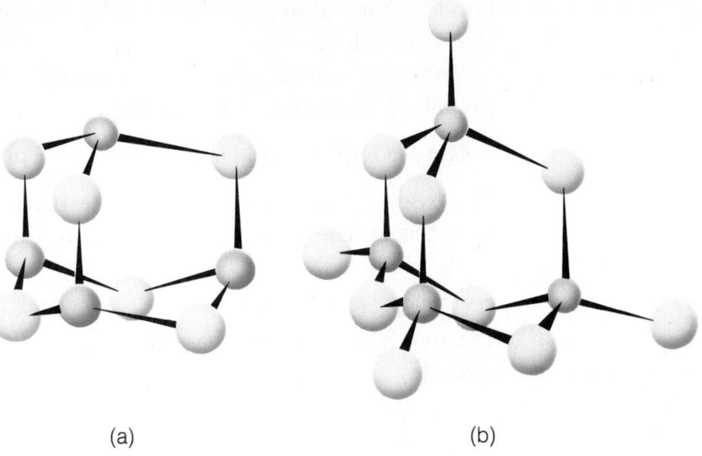

FIGURE 20.1 Structures of phosphorus oxides. (a) P_4O_6. (b) P_4O_{10}. Colored circles represent P atoms.

OXIDES OF PHOSPHORUS

The two common oxides are P_4O_6—tetraphosphorus hexaoxide or phosphorus(III) oxide, m.p. 22.5°C—and P_4O_{10}—tetraphosphorus decaoxide or phosphorus(V) oxide, a white solid that sublimes at 250°C. The tetraoxide, P_2O_4, is also known.

Older designations for the common P oxides used the empirical formulas and corresponding names: P_2O_3, phosphorus trioxide or phosphorus (III) oxide; and P_2O_5, phosphorus pentoxide or phosphorus (V) oxide.

The P_4O_6 structure is shown in Figure 20.1a. Note that the O and P atoms have their normal covalences of 2 and 3, respectively. If each P atom in P_4O_6 bonds to an additional O atom, P_4O_{10} is formed (Figure 20.1b).

Phosphorus(III) oxide is produced when white phosphorus is burned in a limited supply of air:

$$2P_4(c) + 6O_2(g) \longrightarrow 2P_4O_6(c)$$

The oxide is the anhydride of phosphorous acid,

$$P_4O_6(c) + 6H_2O \longrightarrow 4H_2PHO_3(aq), \quad (HO)_2P(=O)H$$

phosphorous acid

Recall that an H atom bonded to P is not acidic (section 14.2). Therefore, H_2PHO_3 is a diprotic acid, not triprotic. Its oxoanions are $HPHO_3^-$ and PHO_3^{2-}.

When phosphorus burns in an excess of oxygen, the principal product is P_4O_{10}. This is the anhydride of phosphoric acid, H_3PO_4, an acid of intermediate strength.

$$P_4O_{10}(c) + 6H_2O \longrightarrow 4H_3PO_4(aq), \quad (HO)_2P(=O)OH$$

phosphoric acid

The corresponding oxoanions include phosphate, PO_4^{3-}, and the mono- and diprotonated phosphate ions (section 14.3).

Because its hydration is rapid and goes practically to completion, and because it is a solid, P_4O_{10} is a convenient and effective *drying agent*, particularly for gases. It is also a *dehydrating agent*. You need to understand the difference between the two actions. A drying agent removes water from a wet substance or object. Thus, a towel and a wad of cotton, as well as P_4O_{10}, are drying agents. A dehydrating agent pulls H_2O molecules out of a substance in which it is chemically bonded. For example, sulfuric acid dehydrates sugar, $C_{12}H_{22}O_{11}$, leaving a residue that is mostly carbon.

The reduction of phosphate minerals by carbon is the industrial method for preparing elemental phosphorus:

$$2Ca_3(PO_4)_2(c) + 6SiO_2(c) + 10C(amorph) \longrightarrow 6CaSiO_3(c) + 10CO(g) + P_4(g)$$

The oxide P_4O_{10} is formed as an intermediate, and is then reduced by the carbon, with the overall process as depicted in this equation.

OXIDES OF SULFUR

The other known sulfur oxides, S_2O, SO, S_2O_3, S_2O_7, and SO_4, are unstable and very reactive.

The two common oxides are sulfur dioxide, SO_2, and sulfur trioxide, SO_3. **Sulfur dioxide,**

$$\overset{\oplus}{\ddot{S}}(=\ddot{O}:)(-\ddot{O}:^{\ominus}) \longleftrightarrow \overset{\oplus}{\ddot{S}}(-\ddot{O}:^{\ominus})(=\ddot{O}:) \quad \text{or} \quad O{\overset{\oplus}{\ddot{S}}}^{\ominus}O$$

is prepared industrially by burning sulfur or sulfides:

$$CuS(c) + O_2 \longrightarrow Cu(c) + SO_2(g)$$

$$2ZnS(c) + 3O_2 \longrightarrow 2ZnO(c) + 2SO_2(g)$$

SO_2 is a colorless, nonflammable gas that most people can detect in air by its bitter taste in concentrations around 1 ppm by volume. In concentrations of 3 ppm or more it has a pungent, irritating odor. It is probably the most significant single air pollutant, as well as the major source of acid rain (Box 20.1). Large amounts of SO_2 are discharged to the atmosphere by the burning of sulfur-containing fuels. SO_2 is the anhydride of sulfurous acid:

$$SO_2(g) + H_2O(\ell) \rightleftharpoons \underset{\text{sulfurous acid (unstable)}}{H_2SO_3(aq)} \quad \text{or} \quad O{=}S(OH)_2\ (aq)$$

The corresponding anions are hydrogen sulfite, HSO_3^-, and sulfite, SO_3^{2-}. Sulfurous acid is an unstable substance that has not been isolated in pure form. SO_2 in an aqueous solution could therefore be represented as hydrated sulfur dioxide, $SO_2(H_2O)_x$, but the formula H_2SO_3 is commonly used.

SO_2 reacts with hydrogen sulfide, H_2S, to form sulfur.

$$SO_2(g) + 2H_2S(g) \longrightarrow 2H_2O(\ell) + 3S(c)$$

In this redox reaction, the S in SO_2 is reduced from oxidation number 4 to 0 (elemental sulfur), while the S in H_2S is oxidized from oxidation number -2 to 0, so that sulfur is the product of both oxidation and reduction. This process is used to remove sulfur-containing substances from natural gas by converting them to useful sulfur.

Sulfur trioxide, SO_3, is prepared by oxidation of SO_2 (section 20.2). At room temperature or below, it is a solid that may exist in any of several different crystal modifications. It is the anhydride of sulfuric acid (section 20.2). The corresponding anions are hydrogen sulfate, HSO_4^-, which is itself an acid ($K_a = 1.2 \times 10^{-2}$), and sulfate, SO_4^{2-}.

Sulfuric acid is a moderately effective oxidizing agent:

$$H_2SO_4(aq) + 2H^+ + 2e^- \longrightarrow H_2SO_3(aq) + H_2O \qquad \mathscr{E}^0 = +0.20\ \text{V}$$

BOX 20.1 ACID RAIN

Rain that falls through an unpolluted atmosphere is slightly acidic (pH about 6), mainly because of its carbonic acid content. In addition, small concentrations of nitric acid are formed during lightning storms by the oxidation of nitrogen in the presence of water, as represented by the net equation:

$$2N_2 + 5O_2 + 2H_2O \longrightarrow 4HNO_3$$

In recent years, however, rain and snow in many parts of the world have become considerably more acidic. Much of this acid precipitation has been between pH 4 and 5, but more severely acidic episodes occur from time to time. For example, a rainstorm in Baltimore in 1981 had a pH of 2.7, which is about as acidic as vinegar, and a fog in Southern California in 1986 had a pH of 1.7, approaching the acidity of some solutions of hydrochloric acid used as toilet bowl cleaners. Acid rain is a worldwide problem that has been estimated to cause billions of dollars' worth of damage, including corrosion of metals, weathering of masonry, death of trees and fish, and reduction of certain agricultural crops.

Most of this excess acidity can be traced to a series of chemical reactions involving oxides of nitrogen and sulfur. The small concentrations of nitric acid formed by the action of lightning are augmented by those produced by the burning of fossil fuels. In the gasoline engine, for example, the electric spark generated by the spark plug serves as a substitute for lightning, and some nitrogen in the air in the cylinder is oxidized to NO. When the NO leaves the exhaust pipe and comes in contact with the air and moisture of the outside atmosphere, it is further oxidized to nitric acid:

$$4NO + 3O_2 + 2H_2O \longrightarrow 4HNO_3$$

All in all, the total atmospheric acidity that starts with sulfur is greater than that coming from nitrogen. The reason is that sulfur, being essential to life, existed in the organisms from which fossil fuels originated. As the remains of prehistoric plants and animals gradually became transformed to coal and oil, some of the hydrogen and most of the oxygen and nitrogen escaped. However, much of the sulfur stayed put. Therefore sulfur is still present in these carbonaceous fuels, especially in soft coal. As coal is burned, the sulfur is oxidized, mostly to sulfur dioxide. The ΔS^0 of the reaction

$$SO_2 + \tfrac{1}{2}O_2 \rightleftharpoons SO_3 \qquad \Delta G^0_{25^\circ C} = -371.1 \text{ kJ}$$

is negative because the more complex molecule, SO_3, is formed from less complex molecules. At high temperatures, this negative ΔS^0 makes the reaction *less* spontaneous, with the result that the formation of SO_2 is thermodynamically favored. (Recall that $\Delta G^0 = \Delta H^0 - T\Delta S^0$ and explain this in terms of the effect on ΔG.) At low temperatures, the $T\Delta S^0$ term is closer to zero and therefore has less effect on ΔG, and the oxidation of SO_2 to SO_3 is thermodynamically favored because ΔH is negative. However, the oxidation is slow in the absence of a catalyst. Once SO_3 is formed, it reacts rapidly with water to form sulfuric acid, which is washed down by rain.

The distance the acidic matter travels from the smokestack to where it is deposited depends on the wind and various other factors. The gases may be adsorbed on dust particles, which then fall at a rate determined by their density and size. The formation and size of acidic droplets are determined by the temperature, humidity, and the presence of small particles (''seeds'')—practically the same factors that influence the formation of clouds and rain.

Futhermore, the environmental damage is not a simple function of the total deposited acidity, for the following reasons:

- A small proportion of the acid precipitation that enters a body of water falls into it; a much greater proportion falls on the land and then runs off into the body of water. If the surrounding soil is rich in limestone, most of the acid may be neutralized before the rain runs off the land:

  $$CaCO_3(c) + H_2SO_4 \longrightarrow CaSO_4(c) + H_2O + CO_2$$

  In some instances the added sulfate even acts as an agricultural fertilizer.
- Observations in West Germany's Black Forest and elsewhere have indicated that trees on hilltops suffer much more damage from acid precipitation than trees at lower levels. This curious phenomenon is related to the prevalence of fogs at these summits. A fog or mist is not a gas but a collection of tiny liquid droplets. Sulfuric acid and nitric acid are highly soluble in water, and therefore they dissolve in fog droplets, where their concentration becomes about 10 times higher than their average concentration in rainwater. This difference is related in part to the fact that fog droplets are much smaller than raindrops. Therefore, the water in a fog droplet does not dilute the acid as much as the greater amounts of water in raindrops do. For this reason, fogs have been called the ''vacuum cleaners of the atmosphere.'' It is these high acid concentrations that cause the greatest damage to trees.
- Another unexpected effect is that nitric acid has been found to cause greater damage to pine trees than sulfuric acid, even though the atmospheric concentration of nitric acid is much less than that of sulfuric acid. The damage may be due to the nutrient effect of nitrogen, which fertilizes the pine trees and causes excess growth in the late fall, making them more susceptible to winter injury.

TABLE 20.2
THE HALOGEN OXIDES

	OXIDATION NUMBER OF THE HALOGEN								
	−1	+1	+3	+4	+5	+6	+7	+8	Mixed
F	OF_2								O_2F_2 O_5F_2 O_6F_2
Cl		Cl_2O	Cl_2O_3	ClO_2		Cl_2O_6	Cl_2O_7	Cl_2O_8	
Br		Br_2O		BrO_2 Br_2O_4					Br_3O_8
I				I_2O_4	I_2O_5				I_4O_9

It can oxidize bromides to bromine and iodides to iodine:

$$2NaBr(c) + 2H_2SO_4(\ell) \longrightarrow Br_2(g) + SO_2(g) + Na_2SO_4 + 2H_2O(\ell)$$

$$8KI(c) + 5H_2SO_4(\ell) \longrightarrow 4I_2(g) + H_2S(g) + 4K_2SO_4 + 4H_2O(\ell)$$

It will not oxidize fluorides or chlorides. Consequently, HCl and HF can be prepared by the action of concentrated H_2SO_4 on their respective salts, but HBr and HI, whose salts become oxidized, cannot.

The oxidizing action of hot, concentrated sulfuric acid sometimes produces unexpected and undesirable results. If even a small quantity of H_2SO_4 intended for some other function is reduced, it can form malodorous sulfur-containing products.

OXIDES OF THE HALOGENS

These generalizations may be made with regard to the properties of the halogen oxides (Table 20.2):

1. A halogen atom has an odd number of electrons. Therefore, any of the oxides with an odd number of halogens will also have an odd number of electrons and will be a free radical. In some such oxides two molecules tend to combine by pairing of the odd electrons (for example, $2BrO_2 \longrightarrow Br_2O_4$ and $2IO_2 \longrightarrow I_2O_4$).

2. The most stable compounds of the halogens are those, like HCl and KBr, in which the halogen has a negative oxidation number. In their oxides, however, all the halogens but F have positive oxidation numbers. Since all of the halogens in these compounds are electron-deficient, they can readily *gain* electrons. Therefore these oxides are all effective oxidizing agents. For example, ClO_2 has been used as a bleaching and water-purifying agent.

3. Bond dissociation energies of oxygen-to-halogen bonds are relatively low—for example, O—Br, 235 kJ/mol, compared with H—Br, 366 kJ/mol. As a result, the halogen oxides decompose readily to their elements, some exposively (such as ClO_2 and Cl_2O_7).

Recall from section 9.7 that chlorates and perchlorates, especially when mixed with any organic matter, are explosive.

The most important halogen oxoanions are those of chlorine. All of these oxoanions, from hypochlorite, OCl^-, to perchlorate, ClO_4^-, are strong oxidizing agents. Hypochlorites are used as disinfectants in swimming pools and laundry bleaching agents.

Perchloric acid, $HClO_4$, is a colorless, volatile, explosive liquid. It is marketed only in mixtures with water, usually containing 60% to 70% $HClO_4$. The acid is used as an oxidizing agent, and its magnesium salt, $Mg(ClO_4)_2$, is an effective drying agent.

20.4 PEROXIDES

In substances containing the *peroxide bond,* —O—O—, and in the peroxide ion, O_2^{2-}, the oxidation number of oxygen is -1. **Hydrogen peroxide**, H_2O_2, is a very pale blue, syrupy liquid that freezes at $-0.89°C$ and boils at 152°C, if it does not explode first. It is miscible with water in all proportions. The hydrogen peroxide solution used as an antiseptic is about 3% H_2O_2 in water; the more concentrated solutions sometimes used as bleaching agents are between 10% and 30% H_2O_2. Hydrogen peroxide decomposes exothermically in a self-redox reaction:

$$H_2O_2(\ell) \longrightarrow H_2O(\ell) + \tfrac{1}{2}O_2(g) \qquad \Delta H = -98.3\ \text{kJ}$$

Many substances, such as metals and metallic oxides, effectively catalyze this decomposition; therefore, small quantities of impurities may encourage pure H_2O_2 or concentrated solutions to decompose explosively. Commercial preparations contain an inhibitor to retard decomposition.

SAFETY NOTE: Concentrated H_2O_2 solutions should be kept clean, cold, and vented to avoid buildup of internal pressure. (Leave the stopper loose.) Also, since H_2O_2 is a strong oxidizing agent, it should be kept away from combustible materials.

The structure of the H_2O_2 molecule may be represented by two obtuse angles in different planes (Figure 20.2).

H_2O_2 can function as an oxidizing agent, by being reduced to water,

$$H_2O_2(aq) + 2H^+ + 2e^- \rightleftharpoons 2H_2O \qquad \mathscr{E}^0 = +1.776\ \text{V}$$

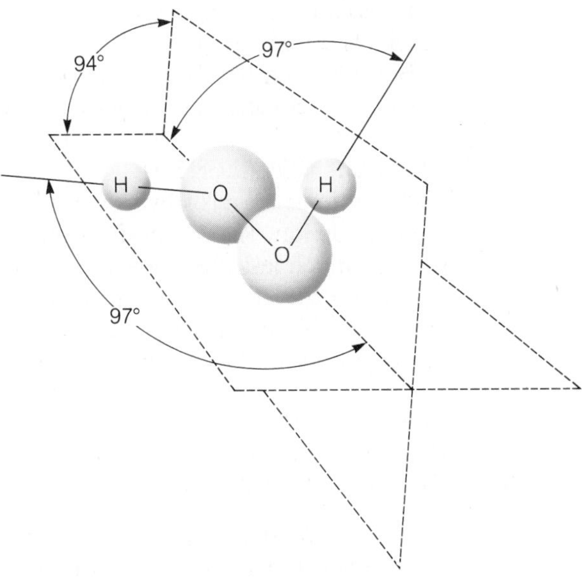

FIGURE 20.2
Structure of hydrogen peroxide.

or as a reducing agent, by being oxidized to O_2,

$$H_2O_2(aq) + 2OH^- \rightleftharpoons O_2 + 2H_2O + 2e^- \qquad \mathscr{E}^0 = +0.1146\ V$$

Metallic peroxides of the alkali metals $(M^+)_2O_2^{2-}$, and of most of the alkaline earth metals, $M^{2+}O_2^{2-}$, are well known. Examples are Na_2O_2, Rb_2O_2, and BaO_2. These ionic peroxides readily react with water or acids to form hydrogen peroxide:

$$O_2^{2-} + 2H_2O \longrightarrow 2OH^- + H_2O_2$$

PROBLEM 20.9 Write the molecular equation for the reaction of water with sodium peroxide, Na_2O_2; with barium peroxide, BaO_2. □

H_2O_2 is very weakly acidic in water; $K_a = 2.4 \times 10^{-12}$, weaker than HCN but stronger than HS^- or HPO_4^{2-}. The fact that it is acidic at all, however, results from an inductive effect (Chapter 14). The electron withdrawal by the two O atoms in H_2O_2 is stronger than that by the one O atom in H_2O. Therefore, in the competition for gaining a proton, the H_2O molecule, with its more available electrons, has a slight advantage over the H_2O_2 molecule, and the ionization reaction

$$H_2O_2 + H_2O \rightleftharpoons HO_2^- + H_3O^+$$

occurs to a slight extent.

Some peroxides, notably Na_2O_2 and BaO_2, are easily formed simply by heating the metal in air. The reaction of such peroxides affords an easy route for the laboratory preparation of H_2O_2. A particularly convenient procedure is the reaction of BaO_2 with sulfuric acid:

$$BaO_2(c) + H_2SO_4(aq) \longrightarrow BaSO_4(c) + H_2O_2(aq)$$

The insoluble $BaSO_4$ can be filtered off, and the remaining liquid phase is an aqueous solution of hydrogen peroxide.

20.5 HYDRIDES

The term **hydride** is used in two ways. It is the *generic* name for all binary hydrogen compounds, and the *specific* name for ionic hydrogen compounds such as Li^+H^- and $Ca^{2+}(H^-)_2$.

Hydrides can be classified according to two different schemes: whether they are ionic or covalent and whether they are basic or acidic.

IONIC OR COVALENT

The only ionic form of hydrogen that exists as an entity in chemical systems is the hydride ion, $H:^-$. Therefore the elements that form ionic hydrides are those that are much less electronegative than hydrogen. These are the alkali metals, Li to Fr, and the alkaline earth metals, Ca to Ra, as shown in Table 20.3. All the other hydrides are covalent, except possibly MgH_2, which is borderline. The blank spaces correspond to elements that do not form known, stable hydrides.

BASIC OR ACIDIC

Here the hydrides fall into three categories, related to the electronegativity of the other element (Table 20.3):

1. *Basic hydrides of elements that are much less electronegative than H.* This is the set of ionic hydrides referred to in the preceding scheme. The hydride ion is very basic and reacts with water to produce H_2

TABLE 20.3
THE MORE STABLE HYDRIDES OF THE REPRESENTATIVE ELEMENTS

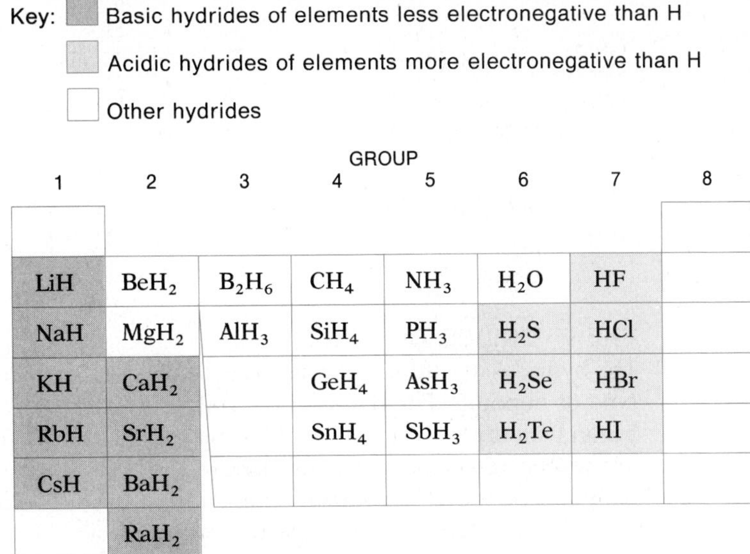
Key: (dark shading) Basic hydrides of elements less electronegative than H
(light shading) Acidic hydrides of elements more electronegative than H
(unshaded) Other hydrides

GROUP 1	2	3	4	5	6	7	8
LiH	BeH_2	B_2H_6	CH_4	NH_3	H_2O	HF	
NaH	MgH_2	AlH_3	SiH_4	PH_3	H_2S	HCl	
KH	CaH_2		GeH_4	AsH_3	H_2Se	HBr	
RbH	SrH_2		SnH_4	SbH_3	H_2Te	HI	
CsH	BaH_2						
	RaH_2						

and OH^-:

$$\underset{\substack{-1\\ \text{base}_1}}{\text{H:}^-} + \underset{\substack{+1\\ \text{acid}_2}}{\text{H}}\text{—}\ddot{\underset{\cdot\cdot}{\text{O}}}\text{—}\underset{+1}{\text{H}} \longrightarrow \underset{\substack{0\ \ 0\\ \text{acid}_1}}{\text{H—H}} + \underset{\substack{+1\\ \text{base}_2}}{(\text{:}\ddot{\underset{\cdot\cdot}{\text{O}}}\text{—H})^-}$$

oxidation number: −1 +1 +1 0 0 +1

It is noteworthy that this reaction is also an oxidation–reduction reaction, as deduced from the change in oxidation numbers shown below each H atom in the equation. In this respect, H^- is a reducing agent. The acceptance of a proton by hydride ion is both an acid-base and a redox reaction.

2. *Acidic hydrides of elements that are much more electronegative than H.* These are the hydrides of the Group-6 and -7 elements shown in Table 20.3. The acidic reactions in water, such as $HCl + H_2O \longrightarrow H_3O^+ + Cl^-$, were discussed in Chapter 14.

3. *Other hydrides.* The hydrides shown in the middle section of Table 20.3 exhibit various properties:

(i) H_2O undergoes intermolecular proton transfer but, of course, is neutral by definition (section 15.1).

(ii) The hydrides of the elements of Groups 4 and 5 (except for nitrogen) do not react with water. Included in this class of compounds are phosphine, PH_3, and the hydrocarbons, such as CH_4. Generally, in this group of compounds the central atom has an octet of electrons, and has an electronegativity close to that of hydrogen.

(iii) Ammonia is the only hydride that reacts to an appreciable extent as a base in water by virtue of a lone pair of electrons:

$$H_3N\text{:} + H_2O \rightleftharpoons NH_4^+ + OH^-$$

(iv) The known hydrides of Group 3 possess electron-deficient atoms, and so behave as Lewis acids. Of particular interest are their reactions with ionic hydrides:

$$AlH_3(c) + Li^+H^- \longrightarrow Li^+ \left[\begin{array}{c} H \\ H:\ddot{Al}:H \\ \ddot{H} \end{array} \right]^-$$

aluminum hydride $\qquad$ lithium aluminum hydride

$$B_2H_6(g) + 2Na^+H^- \longrightarrow 2Na^+ \left[\begin{array}{c} H \\ H:\ddot{B}:H \\ \ddot{H} \end{array} \right]^-$$

diborane $\qquad$ sodium borohydride

THE HYDRIDES OF BORON

The hydrides of boron are interesting compounds because they exhibit an unusual type of bonding. For the sake of comparison, we first consider the structure of boron trifluoride, BF_3, where B has only 6 electrons in its valence shell, as shown in the first resonance structure below. However, this electron deficiency is partially compensated by contributions from the unpaired electrons on F, as illustrated by the other three resonance structures:

$$:\ddot{F}{-}B(-\ddot{F}:){-}\ddot{F}: \longleftrightarrow :\ddot{F}{-}\overset{\ominus}{B}(=\overset{\oplus}{F}:){-}\ddot{F}: \longleftrightarrow \overset{\oplus}{F}:{=}\overset{\ominus}{B}(-\ddot{F}:){-}\ddot{F}: \longleftrightarrow :\ddot{F}{-}\overset{\ominus}{B}(-\ddot{F}:){=}\overset{\oplus}{F}:$$

In the compound BH_3, however, no such stabilization is possible because there are no unpaired electrons on the H atoms. BH_3 is therefore very unstable and spontaneously forms diborane, B_2H_6, which is the simplest stable member of the compounds of B and H called *boranes*. The structure of the diborane molecule is

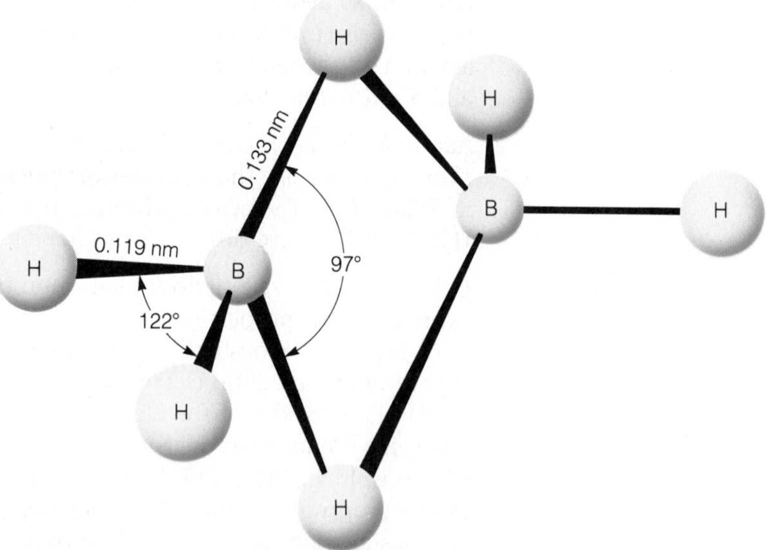

The total number of valence electrons is 12 (3 for each B and 1 for each H atom). The lengths of the four terminal B—H bonds (0.119 nm) are characteristic of covalent B—H bonds, and therefore require two electrons per bond. These account for 8 electrons, leaving only 4 valence electrons for the rest of the molecule. The distance between the boron atoms, about 0.235 nm, is too great for a B—B bond, which would normally be much shorter. Thus the 4 remaining valence electrons must account for the four 0.133 nm B—H linkages shown in the molecular structure. Therefore these cannot be ordinary 2-electron covalent bonds. Instead, the bonding is formed by overlap of the 1*s* orbital of the H atom with some sort of *s* and *p* hybrid orbital of each B atom. Such bonding of three nuclei with 2 electrons is called a **three-center bond** and the overall structure can be pictured as

```
H     H     H
 \  / \  /
  B     B
 /  \ /  \
H     H     H
```

Thus, the molecule is considered to have four B—H covalent bonds (two electrons each), and two three-center B—H—B bonds (also two electrons each).

20.6 HALIDES

We can predict some trends in the bonding types of binary halides on the basis of general principles. Let X represent the halogen and M represent the other element.

- For a given halogen, X, the halide is more ionic as the other element, M, is farther down in a group, and more covalent as M is farther to the right in a period:

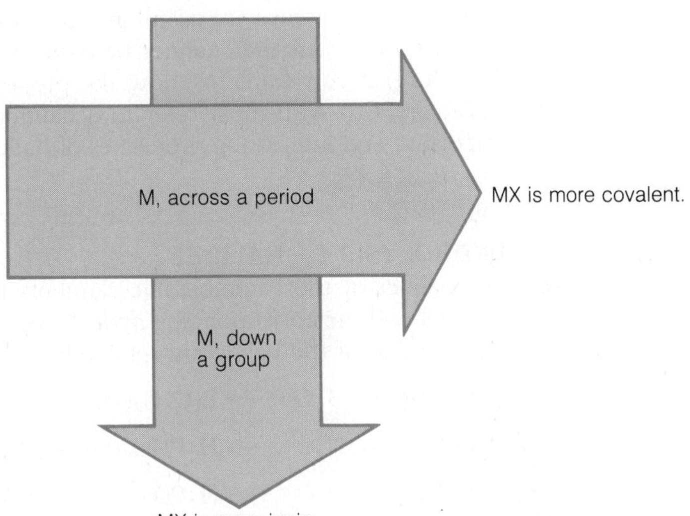

To illustrate the trend across the second period, LiCl is ionic (m.p. 614°C), $BeCl_2$ has considerable covalent character, and BCl_3 is definitely covalent (b.p. 13°C). Continuing to the right, the halides of the remaining nonmetals are also typically covalent.

Explanation (across a period): The halogens, in Group 7, have the highest electronegativities in their respective periods. The farther M is to the right, the closer is its electronegativity to that of the halogen, and the more covalent is the compound MX.

Explanation (down a group): The farther down M is in a group, the less electronegative it is, the greater the difference between its electronegativity and that of X is, and the more ionic the compound MX is.

- The halides of a given element, M, are more covalent as X is farther down in Group 7.

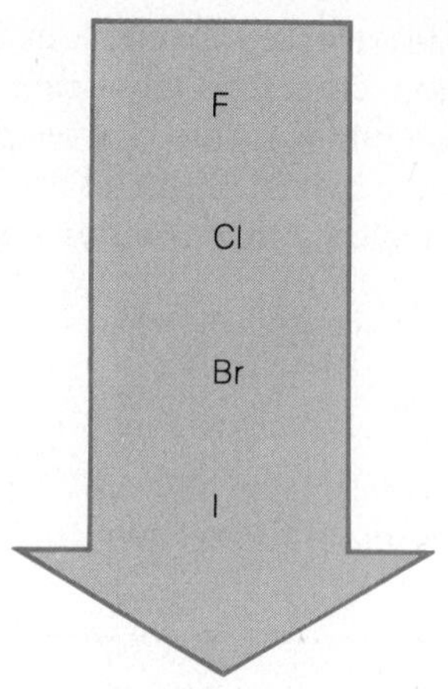

Explanation: The farther down in Group 7 X is, the lower is its electronegativity, the less difference in electronegativity there is between X and M, and the more covalent the compound MX is.

These relationships are useful in rationalizing differences among the properties of the halides, but they cannot be expressed as simple mathematical formulas. Therefore, if you want to know the properties of a particular halide—such as $BeCl_2$ or PCl_5—you must refer to a handbook or textbook of chemistry. Some important and interesting properties of halides are described in the remainder of this section.

HYDROLYSIS OF HALIDES

Hydrolysis means splitting by water.

The halides of most nonmetallic elements hydrolyze to give the oxoacid of the element and the corresponding hydrohalogen acid. The oxidation numbers of the element and of the halide are retained:

$$BCl_3(g) + 3H_2O \longrightarrow B(OH)_3(aq) + 3HCl(aq)$$

$$PBr_3(\ell) + 3H_2O \longrightarrow H_2PHO_3(aq) + 3HBr(aq)$$

$$PCl_5(c) + 4H_2O \longrightarrow H_3PO_4 + 5HCl(aq)$$

The reaction with PBr_3 is used to prepare hydrobromic acid. In the above examples, the halogen atom in the reactant molecule is more electronegative than the central atom. In such cases the reaction usually proceeds as shown. If the halogen is less electronegative, the hydrolysis proceeds in two steps as shown for nitrogen triiodide, in which N is more electronegative than I:

$$\overset{\delta-}{N}\overset{\delta+}{I_3}(c) + 3H_2O \longrightarrow NH_3(aq) + 3HOI(aq) \quad \text{(hydrolysis)}$$

$$NH_3(aq) + HOI(aq) \rightleftharpoons NH_4^+ + OI^- \quad \text{(neutralization)}$$

Both SbOCl and Sb_2O_5 are simplified formulas that omit waters of hydration. Hydrated Sb_2O_5 is sometimes called *antimonic acid,* but its exact composition is unknown.

As the metallic properties of the central atom increase—the trend on proceeding downward in a group—hydrolysis is incomplete and oxohalides are formed, as shown for antimony(III) chloride:

$$SbCl_3(c) + H_2O \rightleftharpoons \underset{\text{antimony oxychloride}}{SbOCl(c)} + 2HCl(aq)$$

Antimony(V) chloride reacts with water as a typical nonmetallic halide:

$$2SbCl_5(\ell) + 5H_2O \qquad Sb_2O_5(c) + 10HCl(aq)$$

Carbon tetrachloride, CCl_4, is one of the few nonmetallic halides that resist hydrolysis. It is immiscible with water and relatively inert.

In contrast to CCl_4, silicon tetrachloride, $SiCl_4$, reacts vigorously with water to form a compound called silicic acid, whose formula can be written as H_4SiO_4 or $Si(OH)_4$. This sharp difference between the behavior of CCl_4 and $SiCl_4$ results from a difference in kinetic, not thermodynamic, stability. Both halides are thermodynamically unstable in the presence of water, but only $SiCl_4$ is kinetically unstable (see Problem 19.83). CCl_4 does not react readily with water because (a) C does not have *d* orbitals available for bonding and (b) the four Cl atoms effectively block the C atom and prevent the approach of an H_2O molecule for a nucleophilic displacement. On the other hand, Si does have available *d* orbitals and its octet may be expanded by the bonding of additional H_2O molecules:

Silicic acid is not a stable compound. It undergoes partial loss of water to give a gel-like form of silica hydrated by an indefinite number of water molecules, $SiO_2(H_2O)_x$.

$$SiCl_4(\ell) + 2H_2O \longrightarrow SiCl_4(H_2O)_2 \qquad \text{(unstable)}$$

The hydrated form decomposes by the elimination of two HCl molecules and repetition of this cycle of addition of H_2O and elimination of HCl until all the Si—Cl bonds are broken. The overall reaction may be summarized as

$$SiCl_4(\ell) + 4H_2O \longrightarrow H_4SiO_4 + 4HCl(g)$$

PROBLEM 20.10 Write equations for the three intermediate steps (elimination, hydration, elimination) in the conversion of $SiCl_4(H_2O)_2$ to H_4SiO_4. □

The tetrachlorides of germanium and of tin, $GeCl_4$ and $SnCl_4$, both typically covalent liquid substances, also react vigorously with water. Thus, $SnCl_4$ vapor reacts with water vapor to give fumes consisting of stannic acid, $Sn(OH)_4(H_2O)_2$, and hydrogen chloride. In the anhydrous state, Sn(IV) combines readily with Cl^- to give the chlorostannate ion, $SnCl_6{}^{2-}$.

LOWER VERSUS HIGHER OXIDATION STATES

As we proceed down Groups 3, 4, and 5, we find that the stability of the higher oxidation state of the metal decreases and the stability of the lower oxidation state increases. Specifically, in each of these groups, the trend is for the $G-2$ oxidation state (where G is the group number) to become more stable at the expense of the G oxidation state as the atomic number of the element increases. This trend is illustrated by this sequence in Group 4:

$$\underset{\text{unknown}}{CCl_2} \longrightarrow \underset{\text{stable}}{CCl_4}$$

$$\underset{\text{known, but very unstable}}{2SiCl_2} \longrightarrow \underset{\text{stable}}{SiCl_4 + Si}$$

$$\underset{\text{unstable}}{2GeCl_2} \longrightarrow \underset{\text{stable}}{GeCl_4 + Ge}$$

$$Cl_2 + \underset{\text{stable}}{SnCl_2} \rightleftharpoons \underset{\text{stable}}{SnCl_4}$$

$$Cl_2 + \underset{\text{stable}}{PbCl_2} \longleftarrow \underset{\text{unstable}}{PbCl_4}$$

Among the Group-3 elements, the +3 oxidation state is the stable one for boron, aluminum, gallium, and indium. However, thallium, the heaviest element in the group, exists mainly in the +1 state.

20.7 SULFIDES

Oxygen and sulfur, both in Group 6, have the same number of electrons in their highest shells ($2s^2 2p^6$ and $3s^2 3p^6 3d^0$, respectively). Both typically exhibit a covalence of 2 and form divalent anions (O^{2-} and S^{2-}). However, there are three important differences between these two elements:

1. Sulfur, unlike oxygen, can use *d* orbitals in its highest shell and thus accommodate more than an octet of bonding electrons. As a result, sulfur forms stable compounds such as $SClF_5$ and SF_6, for which no oxygen analogs exist.

2. Sulfur atoms are larger than oxygen atoms and consequently have less tendency to form double bonds. Thus, sulfur exists as singly bonded S_8 cyclic molecules and as linear chains of S atoms in liquid sulfur. Sulfur also adds to S^{2-} ions to form linear polysulfides:

Sulfur is commonly represented by S for simplicity.

$$(Na^+)_2{:}\ddot{\underset{\cdot\cdot}{S}}{:}^{2-}(aq) + xS(c) \longrightarrow (Na^+)_2\left[{:}\ddot{S}{-}\ddot{\underset{\cdot\cdot}{S}}{-}\ddot{S}{-}\ddot{\underset{\cdot\cdot}{S}}{-}\ddot{S}{-}\text{etc.}\right]^{2-}(aq)$$

sodium polysulfide

However, sulfur can participate in p—p π-bonding with carbon, as in carbon disulfide, $:\ddot{S}{=}C{=}\ddot{S}:$, and with oxygen, as in SO_3.

3. Sulfur is less electronegative than oxygen. As a result, bonds to S are less polar than bonds to O. Thus, for example, the bond moments in CS_2 are smaller than those of CO_2. Since polarity contributes to the strength of a bond, the sulfides are less stable (with respect to thermal decomposition) and have lower melting points than the oxides. Oxygen can therefore replace sulfur in sulfides to form the more stable oxides. This is the transformation that generally occurs in the **roasting** of sulfide ores:

$$2ZnS + 3O_2 \longrightarrow 2ZnO + 2SO_2$$

Such replacement also occurs in aqueous solution, where soluble sulfides undergo hydrolysis:

$$Al_2S_3(c) + 6H_2O \longrightarrow 2Al(OH)_3(c) + 3H_2S(g)$$

Some sulfides, like As_2S_3, Sb_2S_3, and SnS_2, though less soluble than Al_2S_3, can be hydrolyzed in a strongly basic solution like KOH(aq). For example:

$$3SnS_2 + 6OH^- \rightleftharpoons Sn(OH)_6^{\ 2-} + 2SnS_3^{\ 2-}$$

Since the H—S bond is less polar than the H—O bond, the hydrogen bonding exhibited by water does not occur in H_2S. This circumstance is consistent with the fact that H_2S is a gas at ordinary temperatures (b.p. −60°C), while H_2O is a liquid.

H_2S may be prepared in the laboratory by the action of a strong acid on a metallic sulfide:

$$FeS(c) + H_2SO_4(aq) \longrightarrow FeSO_4(aq) + H_2S(g)$$

It can also be produced by the direct union of hydrogen and sulfur vapors at about 600°C. The gas has a characteristic "rotten egg" odor, detectable by most people in air in concentrations as low as about 5 parts per billion. Its ability to transform some biochemically essential metal cations, like Fe^{2+} and Cu^{2+}, into insoluble sulfides can result in physiological catastrophes.

Mixtures of H_2S and air are explosive in the range of 4.3% to 46% H_2S by moles ("by volume").

H_2S is flammable:

$$H_2S(g) + 1\tfrac{1}{2}O_2(g) \longrightarrow H_2O(g) + SO_2(g)$$

H_2S is fairly soluble in water, where it acts as a diprotic acid (section 14.3). The fact that the $[S^{2-}]$ is very sensitive to the pH of the solution is used to advantage in schemes of qualitative analysis of metal ions. In a saturated solution of H_2S,

$$[H_2S] = 0.10 \text{ mole/liter}$$

$$[S^{2-}] = \frac{6 \times 10^{-23}}{[H_3O^+]^2} \text{ mole/liter}$$

(See section 16.2.)

The sulfides of some metals—such as HgS, PbS, Bi_2S_3, CuS, CdS, and others—are so insoluble that they will precipitate even from acidic solutions in which $[H_3O^+]$ is large, making $[S^{2-}]$ very low. Separations of groups of ions may be made on the basis of the pH at which they will or will not precipitate as the sulfide. For example, when $[H^+]$ is 0.3 mole/liter (pH 0.5), and the resulting $[S^{2-}] = 7 \times 10^{-22}$ mole/liter, the following separation of sulfides will occur:

SOME CATIONS THAT WILL FORM INSOLUBLE SULFIDES IN 0.3 M HCl	SOME CATIONS THAT WILL NOT PRECIPITATE IN 0.3 M HCl
Ag^+, Cu^{2+}, Pb^{2+}, Hg^{2+}, Cd^{2+}, As^{3+}, Sn^{4+}, Sb^{3+}, Bi^{3+}	All of periodic Groups 1 and 2 Al^{3+}, Fe^{2+}, Fe^{3+}, Mn^{2+}, Ni^{2+}, Cr^{3+}, Zn^{2+}, CO^{2+}

Also helpful in qualitative cation analysis is the fact that some metallic sulfides have characteristic colors:

COLOR	SULFIDES
White	ZnS
Yellow	As_2S_3, CdS, As_2S_5, SnS_2
Orange	Sb_2S_3
Brown	Bi_2S_3
Black	CuS, HgS, PbS, CoS, NiS, FeS, Ag_2S
Pink	MnS

Some sulfides are so insoluble that they will not react with aqueous solutions of HCl or H_2SO_4, but they can be dissolved in nitric acid solution. The HNO_3 oxidizes the sulfide to free sulfur; the metallic cation remains in solution. Examples of such sulfides are CuS and PbS:

$$3PbS(c) + 2NO_3^- + 8H^+ \longrightarrow 3Pb^{2+} + 3S + 2NO(g) + 4H_2O$$

Mercuric sulfide, HgS, is the most insoluble of these sulfides ($K_{sp} = 6 \times 10^{-53}$) and will not dissolve even in HNO_3. It will, however, dissolve in a mixture of nitric and hydrochloric acids known as *aqua regia*, with the formation of $HgCl_4^{2-}$, a very stable ion:

Aqua regia, Latin for "royal water"; so-called because it dissolves gold.

$$3HgS(c) + 12Cl^- + 2NO_3^- + 8H^+ \longrightarrow 3HgCl_4^{2-} + 3S + 2NO + 4H_2O$$

20.8 THE NOBLE GASES

The noble gases:

helium, He
neon, Ne
argon, Ar
krypton, Kr
xenon, Xe
radon, Rn

(See section 6.8.)

The noble gases are the elements of Group 8, of which all but helium have s^2p^6 configurations in their valence shells. Helium's configuration is $1s^2$. As described in Chapter 7, the apparently inert chemical behavior of these elements gave the original and sustaining support to the significance of the stable electronic octet. The octet concept, however, reinforced the belief that these elements could not react with anything. The older designation of their periodic group as "Group 0" (implying zero valence) and their being named "inert" or "noble"—in the sense that they are aloof, and do not associate with other elements—supported this doctrine. Textbooks of chemistry stated, as fact, that these gases were completely unreactive. For all these reasons, chemists generally assumed that any attempt to study the reactions of these elements would be futile.

In 1962, however, Neil Bartlett noted that oxygen reacts with platinum hexafluoride, PtF_6, a dark red, unstable solid, to give an ionic product, dioxygenyl hexafluoroplatinate(V), $(O_2)^+(PtF_6)^-$. This reaction is unusual in that oxygen *loses* electrons. Since the first ionization energy of O_2 (1177 kJ/mol) is practically the same as that of xenon (1169 kJ/mol), it seemed reasonable to attempt the reaction with xenon. The PtF_6 vapor, when mixed with xenon, immediately formed a yellow solid, probably according to the reaction

$$Xe(g) + 2PtF_6(c) \longrightarrow (XeF^+)(PtF_6^-) + PtF_5(c)$$

Bartlett's finding shattered the previous assumptions about the "inert" gases and led quickly to further development of xenon chemistry, as well as to studies with other gases of this group. The only other noble gases that have been found to enter into chemical combination are krypton and radon. The only elements with which the noble gases are known to form stable bonds are F, O, Cl, and N.

These elements are still called the *noble gases,* partly because their ability to enter into chemical combination is small compared to that of other elements, partly out of habit, and partly because we have to call them something. We no longer say *inert,* however.

Fluorine is the only element with which xenon reacts directly, and three fluorides, XeF_2, XeF_4, and XeF_6, are known. These fluorides are thermally quite stable. They react readily with water or with glass or quartz (SiO_2); however, they can be safely stored in nickel containers. Xenon also forms two stable oxofluorides, $XeOF_4$ and XeO_2F_2, and two unstable oxides, XeO_3 and XeO_4. The oxides are dangerously explosive, white solids.

Since the valence shell of xenon is a complete octet, its chemical bonding necessarily requires octet expansion to involve the 5*d* orbitals. Some known geometries are:

$_{54}Xe$, $[Kr]4d^{10}5s^25p^6$

XeF_2, linear: F—Xe—F

XeF_4, square planar: F—Xe—F (with F above and F below Xe)

XeO_4, tetrahedral:

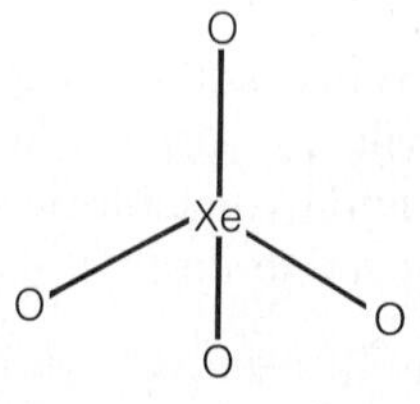

XeO_3F_2, trigonal-bipyramidal:

$$\begin{array}{c} F \\ | \\ O\!-\!Xe\!-\!O \\ O\diagup | \\ F \end{array}$$

The chemistry of krypton is much more limited. KrF_2 is known but is thermodynamically unstable.

The chemistry of radon is difficult to study because Rn is radioactive; the half-life of its most abundant and long-lived isotope is only 3.8 days. As a result, our knowledge of its chemistry is very limited.

What can be said about the significance of noble gas chemistry? On a practical level xenon fluorides are good fluorinating agents (reagents to introduce F atoms into another molecule), because the only by-product is xenon gas, which can be recycled. Similarly, the xenon oxides, XeO_3 and XeO_4, would be "clean" oxidizing agents if anyone used them. (The fact that the oxides are explosive doesn't enhance their popularity.) However, xenon compounds are expensive, and many other oxidizing and fluorinating agents are available. Seven years after his discovery Bartlett referred to noble gas compounds as "very useful, but not yet used." That description still holds today. It is for this reason that our discussion of noble gas chemistry has been limited.

On a theoretical level, the noble gas chemistry has extended our view of the periodic trends in the properties of the elements. For example, we can examine the trend in fluoride bond energies of the fifth-period elements Sb (439 kJ/mol), Te (368), I (259), and Xe (126). Interesting studies of the crystal forms of xenon compounds have been reported. No Nobel Prize has been awarded for noble gas chemistry, however, and the field is not at the forefront of chemical research activity. Nonetheless, the shattering of a widely held and, to some, precious assumption about an important group of elements was a major event in chemical science. Recall from Chapter 1 the problems that arise when experimental findings contradict firmly held theories. We must not forget that when facts stand firm, theories must bend.

SUMMARY

An overview of the chemistry of the representative elements can be obtained by a survey of the properties of their oxides, hydrides, halides, and sulfides.

Oxides of the alkali and alkaline earth metals are ionic and react with water to form basic solutions. Such oxides are therefore called **basic anhydrides**. Covalent oxides in which oxygen carries only a partial negative charge are acidic in water. Oxides that react with water to form an acid are called **acid anhydrides**. Some oxides in which the partial charge on O is closer to that of water are amphoteric. Oxides of Group-5 and -7 elements with even-numbered oxidation states typically undergo self-redox (disproportionation) reactions that result in higher and lower odd-numbered oxidation states in which all electrons are paired.

The familiar and important oxides are

- CO: colorless, odorless, toxic, combustible gas; formed by incomplete combustion of carbon or organic matter
- CO_2: colorless, odorless gas, anhydride of the weakly acidic, unstable carbonic acid; formed by complete oxidation of carbon; prepared in the laboratory by action of strong acids on carbonates or bicarbonates
- SiO_2: hard, transparent, insoluble solid; occurs in nature as quartz
- oxides of nitrogen: N_2O, NO, N_2O_3, NO_2, N_2O_4, and N_2O_5 are all well known and all are gases at ordinary temperatures except N_2O_5, which is a low-boiling liquid. NO is the primary oxide produced by the high-temperature oxidation of N_2 or of nitrogen-containing compounds. At lower temperatures in air or oxygen, NO oxidizes further to NO_2. N_2O_3 and N_2O_5 are acid anhydrides.
- oxides of phosphorus: The two important oxides are P_4O_6 and P_4O_{10}, both of which are acid anhydrides.
- SO_2: colorless, bitter-tasting, toxic gas produced by burning of sulfur or of sulfur compounds; it is the anhydride of sulfurous acid, H_2SO_3, and the major source of acid rain.
- SO_3: corrosive liquid or solid (depending on the crystal form) at room temperature; anhydride of sulfuric acid, H_2SO_4
- oxides of the halogens: various oxides with different oxidation states of the halogens; strong oxidizing agents
- Peroxides: substances containing the peroxide bond, —O—O—, in which the oxidation number of oxygen is -1; hydrogen peroxide, H_2O_2, is a pale blue liquid usually handled in the form of a solution in water.

The hydrides of the alkali and alkaline earth metals are ionic. All other hydrides are covalent. The ionic hydrides react with water to produce H_2 and OH^-. The hydrides of the Group-6 and -7 elements, which are much more electronegative than H, are acidic. Hydrides of the elements of Groups 4 and 5, except for nitrogen, do not react with water. Ammonia, NH_3, is basic in water by virtue of the lone pair of electrons on N. The hydrides of the Group-3 elements possess electron-deficient atoms, and so behave as Lewis acids. The boron hydrides, known as boranes, are characterized by three-centered B—H—B bonds.

For a given halogen, X, the halide is more ionic as the other element, M, is farther down in a group, and more covalent as M is farther to the right in a period. The halides of a given element are more covalent as X is farther down in Group 7. The trend down Groups 3, 4, and 5 is for the halide with the $G - 2$ oxidation state (where G is the group number) to become more stable while the halide with the G oxidation state becomes less stable.

Sulfur differs from oxygen in that it has *d* orbitals available for bonding, shows less tendency to form $p - p\,\pi$ bonds, and is less electronegative. Sulfides are typically less stable and have lower melting points than the corresponding oxides, to which they can be converted by reaction with oxygen. Sulfides of the heavy metals are generally very insoluble, even in acidic solutions.

The noble gases (Group 8) are more inert than any other group of elements, but xenon does form compounds with fluorine and oxygen. Krypton and radon also enter into some chemical reactions.

OXIDES

20.11 (a) Refer to Table 20.1. The higher the oxidation state of an element the greater the tendency of its oxide to be ________ (*acidic* or *basic*). (b) Defend your answer by examples and suggest a reason for the tendency.

20.12 (a) With the help of the information in Table 20.1, complete and balance these molecular equations:

(1) $SeO_2 + H_2O \longrightarrow$
(2) $K_2O + H_2O \longrightarrow$
(3) $BaO + H_2O \longrightarrow$
(4) $Cl_2O + H_2O \longrightarrow$
(5) $Ga_2O_3(c) + KOH(aq) \longrightarrow$
(6) $Ge_2O_3(c) + HBr(aq) \longrightarrow$

(b) Rewrite equations 5 and 6 in net ionic form.

20.13 Write the balanced equation for the complete oxidation of carbon suboxide (tricarbon dioxide).

20.14 Does the oxidation number of the central atom change when an anhydride combines with water to form an acid? Justify your answer for the anhydrides SO_2, SO_3, N_2O_3, N_2O_5, and P_4O_{10}.

20.15 Imagine that you wish to extinguish a magnesium fire in air. (a) Write the equation or equations for the combustion. (b) Give your recommendations for or against each of the following agents for extinguishing the fire. Support your answer, if necessary, with chemical equations: (1) water, (2) sand, (3) N_2, (4) Al_2O_3 powder, (5) Ar.

20.16 (a) Write an equation, using Lewis (electronic) formulas, for the dimerization of NO_2. Does every atom in your formula of N_2O_4 obey the octet rule? (b) Copy the Lewis formula for N_2O_3 from section 20.3. Identify the atoms that bear a positive or negative formal charge. Do any atoms violate the octet rule? The N—N bond in N_2O_3 is polar. Which N atom has the partial + charge and which has the partial − charge? Defend your choice.

20.17 Write the partial ionic equations for the disproportionation of NO_2 in water to yield nitric and nitrous acids. Specify which reaction is oxidation and which is reduction.

20.18 The disproportionation of NO_2 in water (see problem 20.17) may also yield a mixture of nitric acid and nitrogen oxide, NO, under some conditions. Write the partial ionic equations and the balanced net equation for the reaction that gives these products.

20.19 Write the molecular and the ionic equations for these reactions: (a) A snowblower is used to blow soda ash, Na_2CO_3, onto a massive spill of nitric acid from a tank car. (b) Marble architectural treasures such as the Taj Mahal and the Parthenon are being dulled by acid rain produced by the emissions from the burning of high-sulfur fuel.

20.20 A known cyclic silicate anion, $Si_3O_9^{6-}$, has this structure:

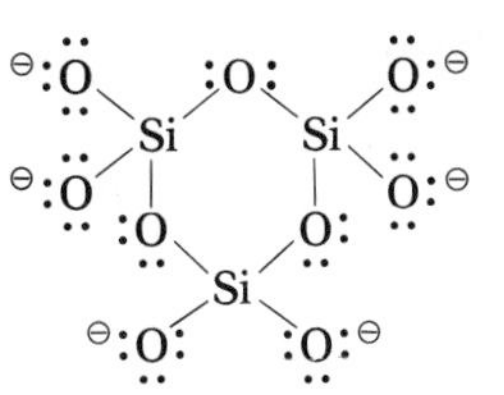

Show how this anion can be broken down into three SiO_4^{4-} anions by the action of the oxygen anions in lime, CaO. How many moles of CaO per mole of $Si_3O_9^{6-}$ are needed? Show the individual steps in the reactions, using Lewis formulas and curved arrows depicting electron transfers.

20.21 Write the balanced equation for the burning of sulfur to SO_2 in an atmosphere of pure N_2O.

20.22 Carbon suboxide, C_3O_2, can be produced by the dehydration of malonic acid, $H_2(C_3H_2O_4)$, with tetraphosphorus decoxide. Write the equation for the dehydration.

20.23 The decomposition of $(NH_4)^+(NO_3)^-$ to H_2O and N_2O is a self-redox reaction. Write the partial ionic equations and identify the oxidation and the reduction.

20.24 (a) Under some conditions, zinc reduces dilute nitric acid to nitrogen. Write the partial ionic equations and the overall ionic equation for this reaction. (b) Support or refute the statement, "Concentrated nitric acid is always a more powerful oxidizing agent than dilute nitric acid."

20.25 The molecular formulas of the (III) and (V) oxides of phosphorus differ numerically from those of the corresponding nitrogen oxides. Explain these differences in terms of the relative tendencies of these two elements to participate in multiple bonding.

PEROXIDES

20.26 The decomposition of H_2O_2 is a self-redox reaction. Write the partial ionic equations and state which is the oxidation and which is the reduction reaction.

20.27 Write the partial ionic equations and the balanced overall equation for the oxidation of Fe^{2+} to Fe^{3+} in acidic hydrogen peroxide solution.

HYDRIDES

20.28 Ionic hydrides such as CaH_2 give a basic reaction in water; covalent hydrides such as HCl are acidic in water. Does it follow that NH_3, which is basic in water, is more like an ionic hydride than is CH_4, which gives no basic reaction? Explain.

20.29 LiH melts at 689°C. Would you expect liquid LiH to conduct an electric current? Write the equations for the reaction at each electrode. Write the equations for the electrode reactions in the electrolysis of the solution formed by mixing water with LiH. (Assume that the electrodes are Pt in all cases.)

20.30 Arrange these sets in the order of increasing acidity in water. Account for your order. (a) H_2S, H_2Te, and H_2Se; (b) SiH_4, HCl, PH_3, and H_2S.

20.31 Arrange these sets in the order of increasing basicity in water. Account for your order. (a) HCl, PH_3, and H_2S; (b) NH_3, PH_3, and AsH_3.

20.32. Identify the four 3-centered bonds in the formula of tetraborane, B_4H_{10},

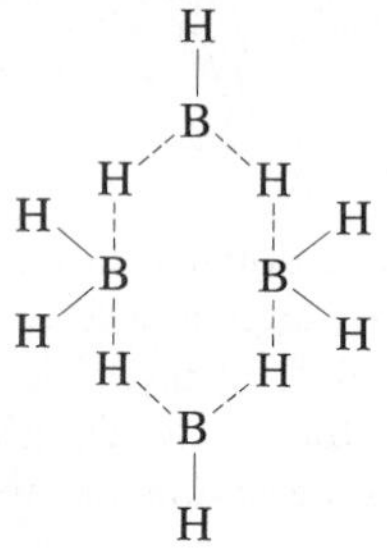

```
        H
        |
        B
      /   \
     H     H
 H   :     :   H
  \  :     :  /
   B         B
  /  :     :  \
 H   :     :   H
     H     H
      \   /
        B
        |
        H
```

HALIDES

20.33 Which compound, PCl_3 or PCl_5, would you expect to be the more effective chlorinating agent? Explain.

20.34 You are asked to select one of three chlorine compounds, Li^+Cl^-, CCl_4, or ClO_2, to be used as an agent for destroying fungus growths. Which compound would be the best choice? Justify your selection.

20.35 The decomposition of $PbCl_4$ and the disproportionation of $SiCl_2$ are both redox reactions. For each reaction, identify all changes in oxidation numbers and state what is being oxidized and what reduced.

20.36 List all the oxides shown in Table 20.2 that are likely to be paramagnetic. Is the number of such oxides large or small compared with the total number of known halogen oxides? Suggest a reason for your answer.

SULFIDES

20.37 SF_4 is a toxic gas that attacks glass and reacts violently with water. SF_6, however, is nontoxic, odorless, and very inert. Account for the difference in reactivity between these two compounds.

20.38 Write the partial ionic equations for the action of nitric acid on PbS.

20.39 Write the partial ionic equations for the oxidation by H_2SO_4 of (a) Br^- ion and (b) I^- ion.

20.40 Which compound, CS_2 or CO_2, do you think is more stable to heat? Defend your answer.

NOBLE GASES

20.41 No stable compounds of He, Ne, or Ar are known. If you wanted to try to prepare compounds of one or more of these elements, which ones, if any, do you think might be possible? Which, if any, are hopeless? Defend your answers.

SELF-TEST

20.42 (10 points) Classify each oxide as acidic, basic, or amphoteric: (a) SrO, (b) Cl_2O, (c) Cs_2O, (d) Al_2O_3, (e) N_2O_5.

20.43 (12 points) Complete and balance these equations. If no reaction occurs, write *no reaction*.

(a) $HI + H_2O \longrightarrow$

(b) $C_2H_6(g) + H_2O \longrightarrow$

(c) $LiH + H_2O \longrightarrow$

(d) $(CH_3)_3N\!: + H_2O \longrightarrow$

20.44 (12 points) Complete this table by inserting the corresponding acids and acid anhydrides.

ANHYDRIDE	ACID
SeO_2	______
______	H_3AsO_4
______	H_4SiO_4
I_2O_7	______

20.45 (12 points) Complete this table by inserting the corresponding bases and basic anhydrides.

BASIC ANHYDRIDE	BASE
MgO	______
______	RbOH
Li_2O	______
______	$Sr(OH)_2$

20.46 (40 points) Write the balanced molecular equation for each reaction: (a) burning of carbon monoxide in oxygen (b) reaction of steam with hot carbon ("water gas reaction") (c) decomposition of ammonium nitrate (d) reaction that occurs when NO_2 is converted to a colorless substance by cooling it (e) reaction of copper with dilute nitric acid (f) formation of lithium aluminum hydride (g) reaction between SO_2 and H_2S (h) hydrolysis of PCl_5 (i) formation of H_2S by the reaction of a strong acid with a metallic sulfide (j) burning of H_2S in oxygen.

20.47 (8 points) Arrange each set of halides in the order of increasing covalent character: (a) $AlCl_3$, NaCl, and PCl_5; (b) $AlCl_3$, AlF_3, and AlI_3.

20.48 (6 points) Three of these compounds are known and three have never been prepared. Identify the unknown compounds. (a) IF_7, (b) NCl_5, (c) HeF_2, (d) SiC, (e) $N^{3+}(H^-)_3$, (f) Co_2S_3.

21

TRANSITION ELEMENTS; COORDINATION COMPOUNDS

The compounds we have discussed in the previous chapters are mainly simple ionic and covalent substances, such as $CuSO_4$, NH_3, and C_6H_6. However, substances in which an ion combines with a covalent molecule to form an ionic compound, such as $Cu(NH_3)_4SO_4$, are also known. They are called coordination compounds. The first explanation (1893) of the nature of the bonds in these compounds led to the electronic theory of valence (Chapter 7).

This chapter is concerned with the transition elements—the metallic elements known for their strong tendency to form coordination compounds. We will describe the formation of these compounds, their properties—such as color, magnetism, geometry, and stability, their nomenclature, and the important roles they play in industrial processes and biological processes—roles that have earned them the designation "bioinorganic compounds." Our study of the shapes of coordination compounds will introduce you to another kind of isomerism. We will also discuss the relationship between oxidation number and the chemical behavior of some of the more prominent transition elements.

21.1 DEFINITION; GROUPS AND TRIADS

Ag(I) is the common oxidation state for Ag, but Ag(II) exists as in AgF_2.

In the broadest sense, the **transition elements** are those with partly filled *d* or *f* subshells in *any* of their oxidation states. Hence, the Group-11B elements copper, silver, and gold, which are $d^{10}s^1$ as atoms, are nevertheless treated as transition elements, because Cu^{2+} and Ag^{2+} have nine *d* electrons (d^9) and Au^{3+} has eight *d* electrons (d^8).

$$\underset{[Ar]3d^{10}4s^1}{Cu} - 2e^- \longrightarrow \underset{[Ar]3d^9}{Cu^{2+}}$$

This definition excludes the Group-12B elements zinc, cadmium, and mercury as transition elements, because their elementary states and all of their oxidation states have filled *d* orbitals. They were discussed in Chapter 6. The elements with partially filled *f* orbitals, the lanthanoids and actinoids (section 6.2), are not discussed in this chapter.

In this chapter we will stress the first-row (fourth-period) transition elements because they are representative of the subsequent rows. They also have great

TABLE 21.1
ELECTRON CONFIGURATIONS OF FIRST-ROW TRANSITION ELEMENTS†

	Sc	Ti	V	Cr	Mn	Fe	Co	Ni	Cu
GROUP	3B	4B	5B	6B	7B	8B	9B	10B	11B
ATOM	$3d^14s^2$	$3d^24s^2$	$3d^34s^2$	$3d^54s^1$	$3d^54s^2$	$3d^64s^2$	$3d^74s^2$	$3d^84s^2$	$3d^{10}4s^1$
ION§	$3d^0$	$3d^2$	$3d^3$	$3d^4$	$3d^5$	$3d^6$	$3d^7$	$3d^8$	$3d^9$

† Each atom has an argon inner-core electron configuration.
§ All ions are M^{2+} except for Sc^{3+}. All 4*s* orbitals in the ions are empty.

practical importance. However, all the transition elements are discussed when we describe periodic trends.

Table 21.1 shows the electron configurations. *The group number equals the number of the highest-energy d and s electrons* of the atom. To show the fate of the 4*s* electrons when ions are formed, the electron configuration of the M^{2+} ions are included. Since scandium has no M^{2+} cation, the electron configuration of its M^{3+} ion is given.

Similarity in chemical behavior is best discussed in terms of *vertical* periodic groups. For example, we refer to the 6B chromium group—chromium (Cr), molybdenum (Mo), and tungsten (W)—and to the 11B copper group—copper (Cu), silver (Ag), and gold (Au). However, for Groups 8B, 9B, and 10B, the chemistry is similar *across the period* rather than down the group. The elements in each period of these groups compose a **triad**. The name of the most prominent member is used to label the respective triad. Thus, the iron triad includes iron (Fe), cobalt (Co), and nickel (Ni); the palladium triad includes ruthenium (Ru), rhodium (Rh), and palladium (Pd); and the platinum triad includes osmium (Os), iridium (Ir), and platinum (Pt).

In general, the transition elements have several properties in common with each other:

1. They are metals.
2. They have multiple oxidation states.
3. When they have *partially* filled *d* orbitals their compounds are colored. The colors of the hydrated cations of the first transition series are given in Table 21.2. When no *d* electrons are present, as in Sc^{3+}, or when all *d* orbitals are filled, as in Cu^+, the ion is colorless.
4. Many of their compounds are paramagnetic. Metallic iron and, to a lesser extent, cobalt and nickel are much more magnetic than typical paramagnetic substances. They are *ferromagnetic*.

Approximately 10^6 times more magnetic.

TABLE 21.2
COLORS OF THE HYDRATED IONS OF THE FIRST TRANSITION SERIES

TRANSMITTED COLOR	IONS AND NUMBER OF 3*d* ELECTRONS
Colorless	Sc^{3+}(0), Cu^+(10), Ti(IV)†(0)
Red	Co^{2+}(7), Mn^{2+}§(5)
Green	Fe^{2+}(6), Ni^{2+}(8), V^{3+}(2), Cr^{3+}(3)
Purple	Ti^{3+}(1)
Violet	V^{2+}(3), Cr^{3+}(3), Mn^{3+}(4), Fe^{3+}§(5)
Blue	Cr^{2+}(4), Co^{3+}(6), Cu^{2+}(9)
Yellow	Fe^{3+}(5) (with halide ion)

† Does not exist as a Ti^{4+} ion.
§ Low intensity.

TABLE 21.3
SELECTED PHYSICAL PROPERTIES OF FIRST-ROW TRANSITION METALS

	Sc	Ti	V	Cr	Mn	Fe	Co	Ni	Cu
Atomic radius, nm	0.144	0.132	0.122	0.118	0.117	0.117	0.116	0.115	0.117
Ionization energies, kJ/mol									
First	644	657	649	653	715	761	757	736	745
Second	1235	1370	1414	1592	1509	1561	1646	1753	1958
Reduction potential†, V	−2.08	−1.63	−1.2	−0.91	−1.18	−0.44	−0.28	−0.25	+0.34
Melting point, °C	1539	1672	1890	1900	1244	1535	1492	1455	1083
Density, g/cm³	2.99	4.49	5.96	7.9	7.2	7.86	8.9	8.90	8.92
Electrical conductivity§	—	2	3	10	2	17	24	24	97

† For the reduction process $M^{2+} + 2e^- \rightarrow M$ (except for scandium, where the ion is Sc^{3+}).

§ Compared to an arbitrarily assigned value of 100 for silver.

5. Many of the metals and their compounds are industrial and biochemical catalysts.
6. They readily form ionic and molecular coordination compounds.
7. Many form metal clusters, which may be molecules, such as $Fe_3(CO)_2$, or ions possessing metal-to-metal bonds, such as $(Cl_4ReReCl_4)^{2-}$.

21.2 METALLIC BEHAVIOR

Tungsten, W, used for filaments in light bulbs, has the highest melting point of any metal (about 3400°C).

The transition elements are usually hard, high-melting, and good conductors of heat and electricity (Table 21.3). Many of these elements are used in forming alloys, especially steels. For example, molybdenum, tungsten, and chromium, even in small amounts, increase the hardness and strength of steel tremendously.

The transition metals are generally less reactive than the representative metals. Most transition elements are resistant to corrosion; iron, the most widely used transition element, unfortunately is an exception. Resistance to corrosion makes titanium, a relatively abundant element in the earth's crust (0.6%), useful in industry. Nickel and chromium plating are used to prevent the corrosion of iron products (section 17.13).

Many of the transition elements have negative standard reduction potentials for the reaction $M^{n+} + ne^- \longrightarrow M$, and therefore liberate hydrogen from strong acids such as HCl and H_2SO_4. This applies to the entire first row except Cu (see Table 21.3). A typical reaction is

$$Fe(c) + 2H^+ \longrightarrow Fe^{2+} + H_2(g)$$

(Titanium, Ti, is inert in HCl at room temperature, but it is attacked in the hot acid.) The platinum triad (called the "noble metals"), silver, gold, and the palladium triad are especially inert to acids—both the nonoxidizing acids, such as hydrochloric and hydrofluoric, and the oxidizing acids, such as nitric acid.† These

† An oxidizing acid is invariably an oxo acid whose central atom undergoes a decrease in oxidation number on reacting with a metal; for example, $NO_3^- + 2H^+ + e^- \longrightarrow NO_2 + H_2O$. With a "nonoxidizing" acid, it is the proton that is reduced, not the anion; $2H^+ + 2e^- \longrightarrow H_2$.

inert metals, however, dissolve in *aqua regia* ("royal water"), a 3 to 1 mixture by volume of concentrated hydrochloric and nitric acids, to form *chloro-*coordination ions, as typified by Pt:

HF may replace HCl to give fluoro-coordination ions such as WF_6^- and ZrF_6^{2-}.

$$3Pt(c) + 16H^+ + 18Cl^- + 4NO_3^- \longrightarrow 3[PtCl_6]^{2-} + 4NO(g) + 8H_2O \qquad \Delta G^0 = -212 \text{ kJ}$$

21.3 OXIDATION STATES AND BONDING

Most transition elements exhibit several oxidation states. (Exceptions are Zr, Hf, W, and the 3B metals; see Table 21.4.) Except for the triads and Group 11B, the maximum oxidation number is the group number which is the total number of highest-energy *s* and *d* electrons. Thus, chromium (6B) and manganese (7B) have maximum oxidation numbers of +6 and +7, respectively. When these high oxidation numbers are found, it is usually when the metal is combined with highly electronegative oxygen and fluorine—for example, CrO_4^{2-} and MnO_3F. On the other hand, iron (8B) and cobalt (9B) have maximum common oxidation numbers of only +3. The inability of the elements in the later groups—the triads and Group 11B—to attain maximum oxidation numbers is related to the fact that ionization energies increase to the right across the period. (See Table 21.3.) Table 21.4 lists the commonly observed oxidation states. The most prominent ones are in color.

Some general periodic trends are:

- The Group-3B elements have a single oxidation state of +3. They form cations with a +3 charge by losing three electrons, one from the highest-energy *d* orbital and two from the highest-energy *s* orbital: M − d^1s^2 electrons $\longrightarrow$ M^{3+}. The M^{3+} ion is stable because it has a noble-gas electron configuration.
- Within the other groups, as the atomic number of the element *increases*, the *higher oxidation states generally become more prevalent than the lower oxidation states.*
- Most elements in Groups 4B, 5B, 6B, and 7B prefer the higher oxidation states.
- The triads (Groups 8B to 10B) and the Group-11B elements prefer the lower oxidation states.

A few Cu(III), Ag(II), and Ag(III) compounds are known.

- Proceeding *across* a period to the right, the *lower oxidation states are increasingly preferred to the higher ones.* For example, across the first row, the +2 and +3 oxidation states increasingly dominate over the higher states, and the +2 state is increasingly more stable than the +3 state.

TABLE 21.4 COMMON OXIDATION NUMBERS OF TRANSITION ELEMENTS†

	3B	4B	5B	6B	7B	8B	9B	10B	11B
4	Sc **3**	Ti 3, **4**	V 2, 3, **4**, **5**	Cr 2, **3**. **6**	Mn **2**, 3, **4**, 6, **7**	Fe **2**, **3**	Co **2**, 3	Ni **2**	Cu 1, **2**
5	Y **3**	Zr **4**	Nb **4**, **5**	Mo **4**, 5, **6**	Tc 4, **7**	Ru 3, **4**, **6**, 8	Rh **3**	Pd **2**	Ag **1**
6	La **3**	Hf **4**	Ta **4**, **5**	W **6**	Re **4**, 5, **7**	Os 3, **4**, **6**, 8	Ir 3, **4**	Pt 2, **4**	Au 1, **3**

† The most commonly observed values are in color.

As the lower oxidation state becomes more stable than the higher state, the species in the higher state becomes a better oxidant because it is more readily reduced to the lower state. In summary:

Group:	3B	4B 5B 6B 7B	8B 9B 10B 11B
	only M^{3+}	higher oxidation states preferred	lower oxidation states preferred

Increased preferencce for lower oxidation states →

Increased preference for higher oxidation states ↑

EXAMPLE 21.1 Given the oxo anion $(VO_4)^n$. If vanadium, V, is in its highest oxidation state, find the charge on the ion.

ANSWER First locate V in the transition series of the periodic table and then determine its electron configuration. It is in Group 5B; V is a d^3s^2 element. Since this group is to the left of the triad, its maximum oxidation number is +5, its group number. Recall that oxygen almost always has an oxidation number of −2 (section 7.9). Therefore,

$$+5 + 4(-2) = n$$

The charge on the ion is −3; the ion is $(VO_4)^{3-}$. ■

PROBLEM 21.1 In $TcO_4{}^{n-}$, $ZrF_6{}^{n-}$, and $WO_4{}^{n-}$ the central atom has its highest oxidation state. Find the charge on each ion. □

EXAMPLE 21.2 Both Fe^{3+} and Ni^{3+} are reduced to the +2-charged cation ($M^{3+} + e^- \longrightarrow M^{2+}$). Which ion is more easily reduced (is a better oxidant)?

ANSWER Ni is to the right of Fe in the same period. Therefore, it is less stable in the +3 oxidation state and is more easily reduced to the preferred +2 state (a better oxidant). ■

PROBLEM 21.2 Which compound is the stronger oxidant, TiO_2 or MnO_2? □

The ions are probably coordinated with H_2O molecules, $M(H_2O)_x{}^{n+}$.

Elements in the +2, +3, and the rarely observed +1 oxidation states usually form ionic compounds. Cations with charges of +4 or more have very high ionization energies (section 7.4) and therefore do not form ionic bonds; they form covalent bonds. For example, titanium(IV) chloride is a liquid; vanadium(V) fluoride is a volatile white solid, m.p. 19.5°C; and osmium(VIII) oxide, OsO_4, is a very toxic liquid. Stable +1 oxidation states occur only in Group 1B—the Cu, Ag, and Au group.

In water, elements in unstable oxidation states often undergo self-oxidation–reduction to give two *more stable* oxidation states; one is higher and one is lower than the original. The behavior of Mn(III) and Mn(VI) is typical. (The oxidation numbers are shown.)

$$2Mn^{3+} + 2H_2O \longrightarrow Mn^{2+} + MnO_2(c) + 4H^+$$

Oxidation number: Mn^{3+} +3; Mn^{2+} +2; MnO_2 +4 (brown precipitate)

$$3MnO_4{}^{2-} + 4H^+ \longrightarrow 2MnO_4{}^- + MnO_2(c) + 2H_2O \quad (1)$$

Oxidation number: $MnO_4{}^{2-}$ +6 manganate (green); $MnO_4{}^-$ +7 permanganate (purple); MnO_2 +4

Not all self-redox reactions are pH-dependent; e.g., the reaction of Au(I)Cl,

$$3Au^+ \longrightarrow Au^{3+} + 2Au(c)$$

The equilibria in these reactions are dependent on pH. For example, in reaction 1, acid is needed for the self-redox of the manganate ion. Accordingly, stabilization of the manganate ion requires a very basic solution.

Elements in the less stable oxidation states can exist only in certain coordination ions or in insoluble solids. This behavior is illustrated by the chemistry of copper. The relative stabilities of the copper(I) and (II) states are indicated by the following reduction potentials:

$$Cu^+ + e^- \longrightarrow Cu \qquad \mathcal{E}^0 = 0.521$$

$$Cu^{2+} + e^- \longrightarrow Cu^+ \qquad \mathcal{E}^0 = 0.153$$

whence,

$$2Cu^+ \longrightarrow Cu + Cu^{2+} \qquad \mathcal{E}^0 = 0.368$$

This indicates that Cu^+ (less stable) should have a tendency to form Cu^{2+} (more stable) and Cu. Nevertheless, the more stable Cu(II) state can be reduced to the less stable Cu(I) state, for example, by I^- and CN^-:

$$\underset{\text{blue}}{2Cu^{2+}} + 5I^- \longrightarrow 2CuI(c) + \underset{\text{yellow-brown}}{I_3^-(aq)} \qquad \text{(precipitate formation)} \qquad (2)$$

$$2Cu^{2+} + 10CN^- \longrightarrow \underset{\text{colorless}}{2Cu(CN)_4^{3-}} + \underset{\text{cyanogen}}{(CN)_2(g)} \qquad \text{(coordination ion formation)}$$

Rather than undergoing self-redox, certain unstable cations undergo redox reactions with water. If the cations are oxidized, hydrogen is formed by the reduction of water:

Cr(OH)$_3$ is better represented as $Cr_2O_3(H_2O)_x$.

$$2Cr^{2+} + 4OH^- + 2H_2O \longrightarrow H_2(g) + 2Cr(OH)_3(c) \qquad \text{(reduction of } H_2O\text{)}$$

If the cations are reduced, water is oxidized with evolution of oxygen:

$$4Co^{3+} + 2H_2O \longrightarrow 4Co^{2+} + O_2(g) + 4H^+ \qquad \text{(oxidation of } H_2O\text{)}$$

Solutions of cations that are easily oxidized by oxygen must be protected from air or preserved with pieces of the corresponding metal. Thus, the Fe(II) state is maintained by placing iron nails in the solution of its salts. Any Fe^{3+} formed by oxidation would be reduced back to Fe^{2+} by the Fe, which in turn is oxidized to Fe^{2+}.

$$2Fe^{3+} + Fe \longrightarrow 3Fe^{2+}$$

21.4 HYDROXIDES AND OXO ANIONS

The hydroxides (and oxides) of the transition elements in different oxidation states are given in Table 21.5. Also shown are their relative basicities or acidities. Hydroxides of the elements in the lower oxidation states (I and II) tend to be basic. Hydroxy compounds of the elements in the higher oxidation states (IV to VIII) tend to be acidic; the intermediate ones (III) are either basic or amphoteric. This trend is illustrated by the hydroxy compounds of chromium:

COMPOUND	OXIDATION STATE	CHEMICAL BEHAVIOR
$Cr(OH)_2$	II	basic
$Cr(OH)_3$	III	amphoteric—dissolves in acid to form Cr^{3+}; dissolves in base to form the chromite ion $Cr(OH)_6^{3-}$
H_2CrO_4 or $H_2Cr_2O_7$	VI	very strong but unstable acids

TABLE 21.5
ACID–BASE PROPERTIES OF THE COMMON OXIDES AND HYDROXY COMPOUNDS OF THE FIRST-ROW TRANSITION ELEMENTS

ELEMENT								
3B	4B	5B	6B	7B	8B	9B	10B	11B
$_{21}$Sc	$_{22}$Ti	$_{23}$V	$_{24}$Cr	$_{25}$Mn	$_{26}$Fe	$_{27}$Co	$_{28}$Ni	$_{29}$Cu
3, B	4, WA	3, B 4, Am 5, A	2, B 3, Am 6, SA	2, B 3, B 7, SA	2, B 3, Am	2, B	2, B 3, B	1, B 2, B

Key: Numbers represent positive oxidation states.
B—base
A—acid
WA—weak acid
SA—strong acid
Am—amphoteric

Only the elements in Groups 3B through 7B have acidic hydroxy compounds and oxides, because only they can tolerate the necessary high oxidation states. These elements also form oxo anions. Recall that elements with higher atomic numbers (those at the lower end of the group) tend to prefer highly positive oxidation states. Hence, stable oxo anions comprise an important part of the chemistry of the higher-atomic-number elements such as Re, Mo, and W.

Oxo anions of elements with lower atomic numbers that are in their highest oxidation states are strong oxidizing agents. The central atom is easily reduced to one of its more stable lower oxidation states. Thus, oxo anions of the first-row transition elements, such as chromate, CrO_4^{2-}, dichromate, $Cr_2O_7^{2-}$, and permanganate, MnO_4^-, are good oxidants. Some typical examples are given in the half-equations below:

$$\underset{\text{deep purple}}{MnO_4^-} + 8H^+ + 5e^- \longrightarrow \underset{\text{pale pink}}{Mn^{2+}} + 4H_2O \qquad (3)$$

$$MnO_4^- + 2H_2O + 3e^- \longrightarrow MnO_2(c) + 4OH^- \qquad (4)$$

$$\underset{\text{orange}}{Cr_2O_7^{2-}} + 14H^+ + 6e^- \longrightarrow \underset{\text{green}}{2Cr^{3+}} + 7H_2O \qquad \text{(acid solution)}$$

$$CrO_4^{2-} + 4H_2O + 3e^- \longrightarrow Cr(OH)_3(c) + 5OH^- \qquad \text{(basic solution)}$$

The oxidizing capacity of MnO_4^- is pH-dependent. In acid (equation 3) MnO_4^- gains 5 electrons and is reduced to Mn^{2+}, whereas in basic or neutral solution (equation 4) it gains only 3 electrons to give MnO_2. Dichromate in acid and chromate in base are both reduced to the Cr(III) state.

Whether CrO_4^{2-} or $Cr_2O_7^{2-}$ exists in water solution depends on the pH. The equilibrium for their interconversion is

$$\underset{\text{yellow}}{2CrO_4^{2-}} + 2H^+ \rightleftharpoons \underset{\text{orange}}{Cr_2O_7^{2-}} + H_2O$$

Acid promotes the conversion of the yellow chromate ion to the orange dichromate ion by shifting the equilibrium to the right. Base favors the formation of CrO_4^{2-} by neutralizing the H^+, moving the equilibrium to the left. The reaction in acid can be conceived as occurring in two steps through the formation of the unstable $HCrO_4^-$ ion.

$$2CrO_4^{2-} + 2H^+ \longrightarrow 2HCrO_4^-$$

$$2HCrO_4^- \longrightarrow Cr_2O_7^{2-} + H_2O$$

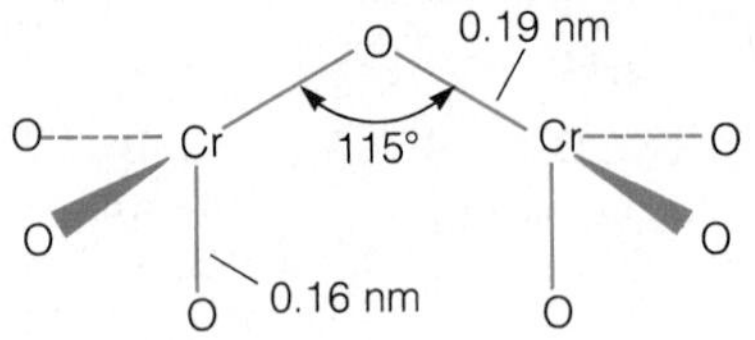

FIGURE 21.1
Structural formula for dichromate ion, $Cr_2O_7^{2-}$. A solid line indicates a bond in the plane of the page; a dashed line indicates a bond that projects behind the plane of the page; a solid wedge indicates a bond that projects toward you.

This sequence indicates that $Cr_2O_7^{2-}$ is the bimolecular dehydrated form of $HCrO_4^-$, and hence, the name *dichromate* is assigned to it. The structure of the $Cr_2O_7^{2-}$ ion, as observed in the ammonium salt, is shown in Figure 21.1. (The Cr—O—Cr bond angle of 115° indicates that the HO's used by the bridging O atom are intermediate between sp^2 and sp^3. (See Box 8.1, section 8.3.)

EXAMPLE 21.3 Which is the stronger oxidizing agent, permanganate, MnO_4^-, or perrhenate, ReO_4^-?

ANSWER Mn and Re are in the same periodic group and they both are in their highest oxidation states (+7) in these ions. Since Mn is above Re in the group, MnO_4^- is less stable in the higher oxidation state and is, therefore, a better oxidant than ReO_4^-. ■

PROBLEM 21.3 Give the formula of the oxo anion of the element in the chromium group (6B) that is the poorest oxidant. □

21.5 TRANSITION ELEMENTS AS SURFACE CATALYSTS

Many transition elements and their compounds, mainly the oxides, are used as heterogeneous surface catalysts (section 19.10) in industrial processes. Of particular importance are platinum, palladium, and nickel, especially when H_2 or O_2 is a reactant. The explosive conversion of H_2 and O_2 to H_2O in the presence of any of these metals is a typical example. Of interest also is the addition of H_2 to a C=C bond as shown for the reaction of ethene (ethylene):

Ni is the catalyst.

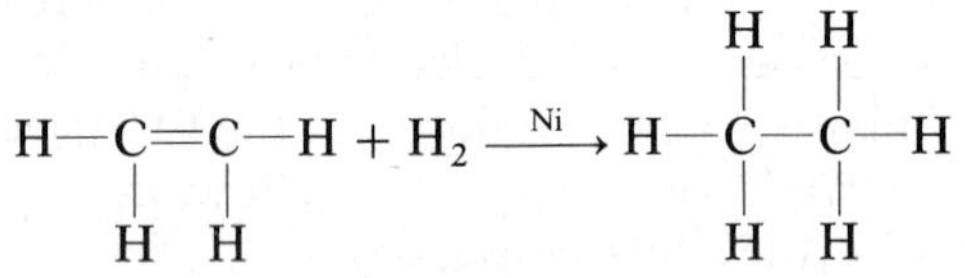

$$\text{H—}\underset{\text{H}}{\underset{|}{\text{C}}}\text{=}\underset{\text{H}}{\underset{|}{\text{C}}}\text{—H} + H_2 \xrightarrow{\text{Ni}} \text{H—}\overset{\text{H}}{\overset{|}{\underset{\text{H}}{\underset{|}{\text{C}}}}}\text{—}\overset{\text{H}}{\overset{|}{\underset{\text{H}}{\underset{|}{\text{C}}}}}\text{—H} \qquad \text{(a hydrogenation reaction)}$$

Industrially this type of reaction, called **hydrogenation**, is used to convert liquid fats, which have several C=C bonds, into margarine, whose solid fats have fewer C=C bonds.

We mentioned in section 19.10 that surface catalysts, such as palladium, function by adsorbing reactants, especially gases, on their surfaces. On being adsorbed on the surface of the metal, the H_2 molecules dissociate into H· atoms. The atoms react faster with the other reactant—which also may be adsorbed—than do the H_2 molecules. In this way Pd acts as a catalyst. This dissociation can be shown experimentally by adding H_2 to Pd in the absence of another reactant. Now, the adsorbed H· atoms diffuse into the solid and occupy the spaces (interstices) between the atoms of the metal. A kind of hydride is formed that does not follow the law of definite composition. The molecular formula is written as PdH_x to indicate the variable content of hydrogen; the value of x varies with the pressure of the H_2. Palladium metal has a remarkable ability to absorb large quantities of H_2 gas. Because of the nature of their compositions, such hydrides are classified as **interstitial compounds**. These interstitial hydrides are hard, nonvolatile solids.

The H_2 absorbed is as much as 1000 times the volume of the piece of Pd used.

Four of the more important industrial processes that use transition elements as catalysts are

Haber process: $3H_2(g) + N_2(g) \xrightarrow{Fe} 2NH_3(g)$

Ostwald process: $4NH_3(g) + 5O_2(g) \xrightarrow{\text{Pt or Rh}} 4NO(g) + 6H_2O(g)$ (5)

Contact process: $2SO_2(g) + O_2(g) \xrightarrow{\text{Pt or } V_2O_5} SO_3(g)$ (6)

Fischer–Tropsch process: $CO(g) + 2H_2(g) \xrightarrow{Fe} CH_3OH(g)$

In processes 5 and 6 the adsorbed O_2 molecules dissociate to O atoms, which react faster than the molecules do.

21.6 COORDINATION COMPOUNDS

Latin *ligare*, "to bind"

One of the most noteworthy chemical properties of the transition elements is their ability to form **coordination compounds** (**coordination complexes** or **complexes**). Recall that a coordination complex is an ion or molecule formed when a central atom coordinates (bonds) with anions and/or molecules called **ligands** (section 9.2). The number of atoms in the ligands bonded to the central atom is called the **coordination number**. Some common ligands are listed in Table 21.6.

TABLE 21.6
COMMON LIGANDS

LIGAND	NAME OF LIGAND IN COMPOUND
Anions	
bromide, Br^-	bromo
chloride, Cl^-	chloro
iodide, I^-	iodo
cyanide, CN^-	cyano
hydroxide, OH^-	hydroxo
oxide, O^{2-}	oxo
carbonate, CO_3^{2-}	carbonato
oxalate, $C_2O_4^{2-}$	oxalato
amide, NH_2^-	amido
nitrite†, $O\underline{N}O^-$	nitro
$\underline{O}NO^-$	nitrito
thiosulfate, $S_2O_3^{2-}$	thiosulfato
thiocyanate†, $\underline{S}CN^-$	thiocyanato
$SC\underline{N}^-$	isothiocyanato
ethylenediaminetetraacetate^{4-}	ethylenediaminetetraacetato
Molecules	
ammonia, NH_3	ammine
carbon monoxide, CO	carbonyl
water, H_2O	aqua
ethylene, C_2H_4§	ethylene
ethylenediamine, $H_2NCH_2CH_2NH_2$	ethylenediamine

† These are ambident ligands, they can be bonded at either of the underlined atoms.

§ Ethylene binds by using π electrons.

The central atom most often has a *positive oxidation number*, but it may also be an *uncharged atom*. Typical examples of ions and molecules are

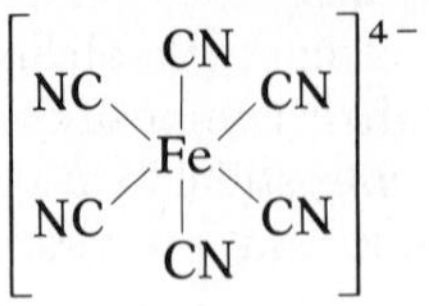

$[Fe(CN)_6]^{-4}$

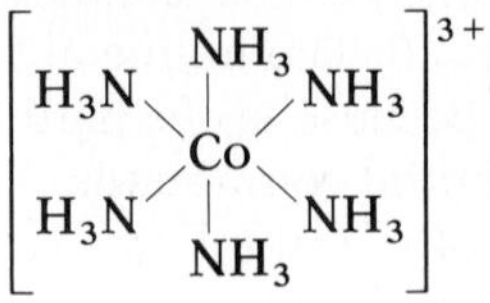

$[Co(NH_3)_6]^{3+}$

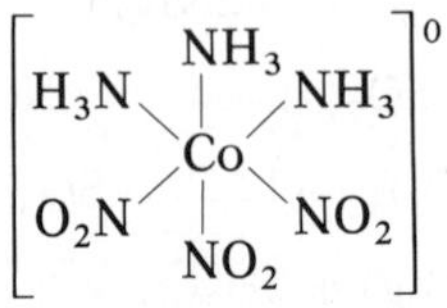

$[Co(NH_3)_3(NO_2)_3]$

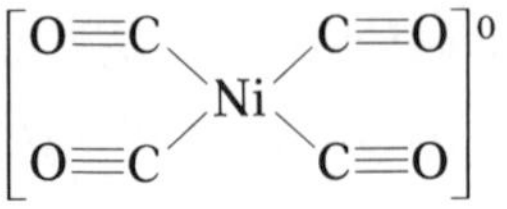

$[Ni(CO)_4]$

The coordinated complexes are written inside square brackets.

WERNER'S COORDINATION THEORY

In 1893, Alfred Werner proposed a structural theory to explain the unusual properties of coordination compounds. Werner's theory opened the way to the electronic theory of valence (Chapter 7) and to an understanding of the formation and structure of coordination compounds. He indeed laid the foundation for the modern development of coordination chemistry. We illustrate his theory with the several compounds having the general formula $PtCl_4(NH_3)_n$, where n can be 2, 3, 4, 5, or 6. Experimentally, 1 mole of $PtCl_4(NH_3)_6$ reacts instantaneously with 4 moles of silver nitrate to give 4 moles of insoluble AgCl:

$$PtCl_4(NH_3)_6 + 4Ag^+ \longrightarrow 4AgCl(c) + [Pt(NH_3)_6]^{4+}$$

The precipitation of AgCl is a test for the presence of Cl^- in the solution. To get 4 moles of AgCl, 1 mole of the platinum complex must furnish the solution with 4 moles of Cl^-. What, then, is the valence of Pt? Is it +4, as in $PtCl_4$? If so, what do we do with the six NH_3's? More puzzling yet, 1 mole of $PtCl_4(NH_3)_5$ reacts with only 3 moles of $AgNO_3$:

$$PtCl_4(NH_3)_5 + 3Ag^+ \longrightarrow 3AgCl(c) + [Pt(NH_3)_5Cl]^{3+}$$

Thus, one of the four chlorines remains with the Pt, while the other three must be in solution as Cl^- ions. Most startling is $PtCl_4(NH_3)_2$; it does not react with $AgNO_3$ at all:

$$PtCl_4(NH_3)_2 + Ag^+ \longrightarrow \text{no reaction}$$

Hence, no Cl^- ions are present in solution; the four chlorines stay with the Pt. In addition, none of these $PtCl_4(NH_3)_n$ compounds reacts with HCl. The NH_3 molecules stay with the Pt atom and are not in solution. Werner rationalized such behavior by suggesting that Pt has two kinds of valences. The first he called the "principal valence," such as +1 for Na in NaCl. The "principal valence" in all of the Pt compounds just cited is thus +4 which we now recognize as the oxidation number. The second he called an "auxiliary valence." For the platinum compounds the "auxiliary valence" is 6. We now recognize this as the coordination number. Thus, the reactions of the various complexes in aqueous solution may be formulated as:

$$[Pt(NH_3)_6]^{4+}(Cl^-)_4 + 4Ag^+ \longrightarrow 4AgCl(c) + [Pt(NH_3)_6]^{4+}$$

$$[Pt(NH_3)_5Cl]^{3+}(Cl^-)_3 + 3Ag^+ \longrightarrow 3AgCl(c) + [Pt(NH_3)_5Cl]^{3+}$$

$$[Pt(NH_3)_4Cl_2]^{2+}(Cl^-)_2 + 2Ag^+ \longrightarrow 2AgCl(c) + [Pt(NH_3)_4Cl_2]^{2+}$$

$$[Pt(NH_3)_3Cl_3]^+Cl^- + Ag^+ \longrightarrow AgCl(c) + [Pt(NH_3)_3Cl_3]^+$$

$$[Pt(NH_3)_2Cl_4] + Ag^+ \longrightarrow \text{no reaction}$$

The species containing the Pt are the complexes. The first four are ions; the last is a neutral molecule. In all cases the oxidation number (Werner's "principal valence") of Pt is +4, and the coordination number (Werner's "auxiliary valence") is 6. Only the molecules and/or anions inside the brackets are bonded to the central atom

The formulas so assigned on the basis of reaction with $AgNO_3$ agree with the number of ions in dilute solutions based on freezing-point depression and conductivity measurements:

COORDINATION COMPOUND	COLOR	MOLES OF IONS PER MOLE OF COMPOUND
$[Pt(NH_3)_6]Cl_4$	off-white	5
$[Pt(NH_3)_5Cl]Cl_3$	pale yellow	4
$[Pt(NH_3)_4Cl_2]Cl_2$	pale yellow	3
$[Pt(NH_3)_3Cl_3]Cl$	pale yellow	2
$[Pt(NH_3)_2Cl_4]$	orange-yellow or lemon-yellow	0

21.7 LIGANDS AND COORDINATION NUMBER

Some representative metals (such as Ca, Al, and Sn(IV)) and halogens (not F) also form complexes, such as AlF_6^{3-} and $[BrF_4]^-$.

Most ligands have at least one unshared pair of electrons; they are electron donors, Lewis bases. Examples are $:NH_3$, $:C{\equiv}N:^-$, $H_2\ddot{O}:$, $:\ddot{O}H^-$, and $:\ddot{F}:^-$. Cations and atoms of the transition elements act as electron acceptors—as Lewis acids. Examples are the cations Ag^+, Fe^{2+}, Co^{3+}, and Pt^{4+} and the atoms themselves, Fe, Co, and Ni. The formation of a coordination compound is therefore a Lewis acid-base reaction:

central atom	+	**ligands**	⟶	**coordination complex**
(metal atom or cation) electron acceptor Lewis acid		**(molecules or anions) electron donor Lewis base**		**(ion or molecule)**
Ag^+	+	$2:NH_3$	⟶	$[Ag(NH_3)_2]^+$
Fe^{2+}	+	$6:C{\equiv}N:^-$	⟶	$[Fe(CN_6)]^{4-}$
Ni(c)	+	$4:C{\equiv}O:(g)$	⟶	$[Ni(CO_4)](\ell)$

Note that the charge on the coordination complex is simply the sum of the charges (or oxidation numbers) on the central atom and the ligands:

$$[\overset{+1}{Ag}(\overset{0}{NH_3})_2]^+ \qquad 1 + 2(0) = +1$$

$$[\overset{+2}{Fe}(\overset{-1}{CN})_6]^{4-} \qquad 2 + 6(-1) = -4$$

$$[\overset{+2}{Cu}(\overset{0}{H_2O})_4]^{2+} \qquad 2 + 4(0) = +2$$

$$[\overset{+4}{Pt}(\overset{0}{NH_3})_2\overset{-1}{Cl_4}]^0 \qquad 4 + 2(0) + 4(-1) = 0$$

Reversing the process, we can obtain the oxidation number of the central atom from the charge on the complex and the sum of the ligand charges.

EXAMPLE 21.4 Find the oxidation state of Fe in (a) pentacarbonyl iron, $[Fe(CO)_5]$, a yellow liquid (b.p. 103°C); (b) $[Fe(CN)_6]^{3-}$ ion.

ANSWER

(a) The net charge of $Fe(CO)_5$ is 0. The charge of the ligand CO (carbon monoxide) is also 0. Thus, the oxidation state of Fe must be 0:

$$\text{oxidation number of Fe} + 5 \times \text{charge of CO} = \text{charge of complex}$$

$$Fe + 5(0) = 0$$

$$Fe = 0$$

(b) The charge of the CN^- ligand is -1. The oxidation number of Fe in $[Fe(CN)_6]^{3-}$ must be $+3$, as shown:

$$Fe + 6(-1) = -3$$

$$Fe = -3 + 6 = +3$$

■

EXAMPLE 21.5 Find the charge, x, of the Co(III) complex $[Co(NH_3)_3(NO_2)_3]^x$ if the NO_2 ligand is (a) nitrogen dioxide, NO_2; (b) nitrite ion, NO_2^-. (Both of these coordination complexes have been synthesized. They are isomers.)

ANSWER In each case the oxidation number of Co is $+3$ as given, and the charge on NH_3 is zero. The oxidation number of Co + (3 × the charge of NH_3) + (3 × the charge of NO_2) = x (the charge of the coordination compound.)
(a) The charge of NO_2 is zero. Therefore,

$$(+3) + 3(0) + 3(0) = +3$$

The charge, x, is $+3$.

(b) $(+3) + 3(0) + 3(-1) = 0$

The charge, x, is 0.

■

PROBLEM 21.4 Find the oxidation number of Co in (a) $[CoCl_4]^{2-}$ and (b) $[Co(NH_3)_5Cl]^{2+}$.

□

When no other ligand is present, most transition-metal cations in aqueous solutions are complexed with water molecules, $[M(H_2O)_x]^{n+}$, where x is usually 4 or 6 and n^+ is the charge on the cation, M^{n+}.

Latin *unidentatus*, "one-toothed"

Ligands like $:NH_3$ and $:\ddot{Cl}:^-$, with only one coordinating atom (bonding site), are described as **unidentate**. A unidentate ligand forms only one bond with the central atom. A single ligand may, however, have more than one bonding site that can bond to the same central atom. Such ligands are termed **multidentate**. 1,2-Diaminoethane (ethylenediamine), $H_2\ddot{N}CH_2CH_2\ddot{N}H_2$, is a typical **bidentate** ligand; it has two bonding sites. Cu^{2+} binds two ethylenediamine molecules, making its coordination number four, as shown:

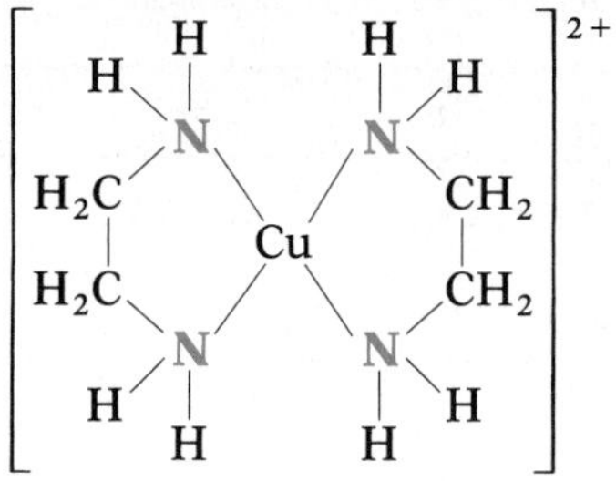

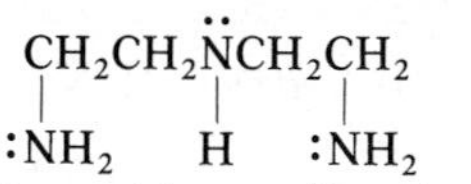

$$\left[\begin{array}{c} \text{H}_2\text{C}(\text{NH}_2)\text{-}\text{CH}_2\text{NH}_2 \text{ chelated to Cu (two ethylenediamine ligands)} \end{array}\right]^{2+}$$

$\underset{:NH_2}{CH_2}CH_2\underset{H}{\ddot{N}}CH_2\underset{:NH_2}{CH_2}$ is a *tridentate* ligand.

The coordination number is 4, the same as in $[Cu(NH_3)_4]^{2+}$.

A multidentate ligand is compared to an animal's claw; Greek *chele*, "claw."

Ligands with more than one binding site are also called **chelating agents**. An especially important, versatile, and powerful chelating agent is the hexadentate ligand ethylenediaminetetraacetate (EDTA) ion. Four oxygen atoms and two nitrogen atoms, indicated by color in the structure (Figure 21.2a), are the six binding sites. EDTA is used, for example, in the quantitative analysis of metal cations, as a food additive to prevent contamination by metals, and in the treatment of lead poisoning. The chelation of lead (a representative element) is shown in Figure 21.2b. EDTA is used in medicine because it is relatively nontoxic and forms very stable, excretable, water-soluble chelates. Unfortunately, it is nonspecific. It binds many transition- and representative-metal ions, especially the biochemically essential Ca^{2+} cation. Therefore, when EDTA is administered in large doses, it is supplemented with Ca^{2+} by using Ca_2EDTA. Note from the formula of these salts that the EDTA anion has a charge of -4. The ion to be chelated, such as Pb^{2+}, displaces the Ca^{2+} ion from Ca_2EDTA as shown:

$$2Pb^{2+} + Ca_2EDTA \rightleftharpoons Pb_2EDTA + 2Ca^{2+}$$

Since the lead complex is more stable than the calcium complex—despite the larger Ca^{2+} concentration in the body—the equilibrium of this reaction lies mainly to the right. In this way the harmful Pb^{2+} is complexed and excreted.

COORDINATION NUMBER

The value of the coordination number depends on the charge on the cation, the charge on the ligand, the relative sizes of the cation and the ligand, and the repulsion among ligands. The interplay of these factors cannot easily be evaluated

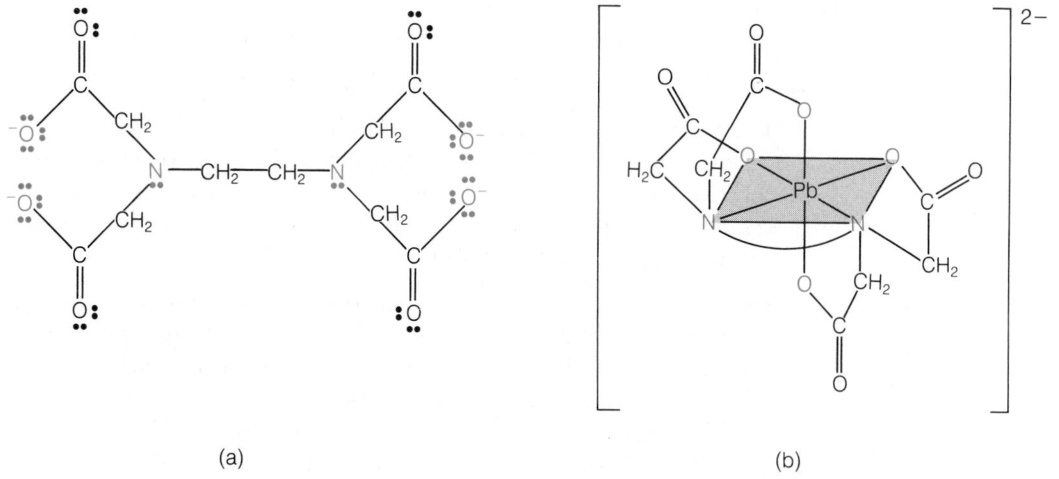

FIGURE 21.2
(a) Structure of ethylenediaminetetraacetate (EDTA) ion. (b) EDTA complexed with Pb^{2+}, a typical cation.

TABLE 21.7
MAJOR COORDINATION NUMBERS OF SOME COMMON TRANSITION ELEMENTS†

M^+	M^{2+}	M^{3+}	M(IV)
Cu^+ 2, 4	Fe^{2+}, Mn^{2+}, Cr^{2+} 6	V^{3+}, Mn^{3+}, Cr^{3+}, Ti^{3+}, Co^{3+} 6	Pt(IV), Ti(IV), V(IV), Pd(IV) 6
Ag^+ 2, 4	Co^{2+}, Ni^{2+}, Cu^{2+} 4, 6	Fe^{3+} 4, 6	
Au^+ 2, 4	Pt^{2+}, Pd^{2+} 4	Au^{3+} 4	

† Main coordination number is in color.

quantitatively. Hence, reliable predictions cannot be made. Furthermore, some cations display more than one coordination number, as does Fe^{3+} in $FeF_6{}^{3-}$ and $FeCl_4{}^-$.

The coordination number most commonly observed for transition-element cations is 6; the next most common one is 4. A coordination number of 2 is found principally in complexes of Cu^+, Ag^+, and Au^+, each of which has a d^{10} electron configuration. The number 3 is very rare and 5 occurs occasionally. Coordination numbers higher than 6 are found only among elements in the second and third transition periods; for example, $W(CN)_8{}^{3-}$. Table 21.7 gives the coordination numbers of some of the more common transition-element ions. The number usually found is in color.

21.8 NOMENCLATURE OF COORDINATION COMPOUNDS

The rules presented here apply only to the naming of coordination compounds with unidentate ligands and one central atom.

NAMING THE LIGANDS

Ethylene binds by using its π electrons.

The names for most *anionic ligands* end in *-o*. The names for *molecular* (*neutral*) *ligands* are typically unchanged. Thus, ethylene, C_2H_4, as a ligand is named ethylene. Some important exceptions to this are water (*aqua*), ammonia (*ammine*) (spelled with two *m*'s), and carbon monoxide (*carbonyl*). Table 21.6 (section 21.6) gives the names of some common ligands. For ligands such as $NO_2{}^-$ and SCN^-, where more than one kind of atom can bond to the metal, the bonding atom is underlined in the table.

NAMING THE COMPOUNDS

Some standard rules apply to the naming of coordination compounds:

1. As always, name the cation first, then the anion.
2. In naming the coordination compound, name the ligands first. The number of each kind of ligand is specified by the Greek prefixes *di-* for 2, *tri-* for 3, *tetra-* for 4, *penta-* for 5, and *hexa-* for 6. Then name the central atom followed by a Roman numeral (or 0) in parentheses to indicate its oxidation number.
3. For a cationic (positive) or neutral coordination complex, the regular name of the metal is used. See Table 21.8. Anions that are not themselves coordination complexes have their usual names; that is, chloride, bromide, sulfate. For example:

$[Ni(CO)_4]$	tetracarbonylnickel(0)
$[Co(NH_3)_6]Br_3$	hexaamminecobalt(III) bromide
$[Ag(NH_3)_2]Cl$	diamminesilver(I) chloride

TABLE 21.8
NAMING METALS IN COORDINATION COMPLEXES

CENTRAL ATOM	NAME IN CATIONIC COMPLEX	NAME IN ANIONIC COMPLEX
Ag	silver	argentate
Au	gold	aurate
Co	cobalt	cobaltate
Cr	chromium	chromate
Cu	copper	cuprate
Fe	iron	ferrate
Mn	manganese	manganate
Ni	nickel	nickelate
Pt	platinum	platinate

$[Cr(H_2O)_6]Br_3$ hexaaquachromium(III) bromide
$[Pt(NH_3)_2Cl_4]$ diamminetetrachloroplatinum(IV)
$[Cr(H_2O)_4Cl_2]Cl$ tetraaquadichlorochromium(III) chloride

4. When there are several kinds of ligands, they are named alphabetically. However, condensed formulas are usually written with neutral ligands first.

5. For an anionic (negative) coordination complex, the Latin stem name of the metal is used, followed by the suffix *-ate*. Table 21.8 lists the names for some common metals in anionic complexes. Cations that are not coordination complexes are given their usual names. For example:
   $K_3[Fe(CN)_6]$ potassium hexacyanoferrate(III)
   $Na_3[Ag(S_2O_3)_2]$ sodium dithiosulfatoargentate(I)
   $K[AgF_4]$ potassium tetrafluoroargentate(III)
   $K[PtCl_3(NH_3)]$ potassium amminetrichloroplatinate(II)
   $[Pt(NH_3)_4][PtBr_6]$ tetraammineplatinum(II) hexabromoplatinate(IV)

EXAMPLE 21.6 Name (a) $Na_2[PtCl_2Br_4]$ and (b) $[Cr(NH_3)_4Cl_2]Cl$.

ANSWER First determine the oxidation state of the central atom of the coordination compound.

(a) Since there are $2Na^+$ ions in the salt, the charge on the complex ion is -2. The ligands have a total charge of -6 $(2Cl^- + 4Br^-)$, and therefore the central atom is Pt(IV). The name is sodium tetrabromodichloroplatinate(IV).
(b) The single Cl outside the bracket means that there is one Cl^- ion in the salt. Therefore, the complex ion has a charge of $+1$. NH_3 has no charge, and two Cl^- ligand ions have a total charge of -2. The central atom is Cr(III). The name is tetraamminedichlorochromium(III) chloride. ■

PROBLEM 21.5 Name (a) $[Co(NH_3)_5Cl]Cl_2$ and (b) $Na_3[Co(NO_2)_6]$, a reagent used in the qualitative analysis for K^+. (The K^+ salt is insoluble and can be precipitated.) □

21.9 COORDINATION NUMBER AND SHAPE

Werner's ideas of the nature of the bonding within the complexes were rudimentary in comparison with our present concepts. However, he stated correctly that "principal valences" result from electron transfer, and his ideas about geometry

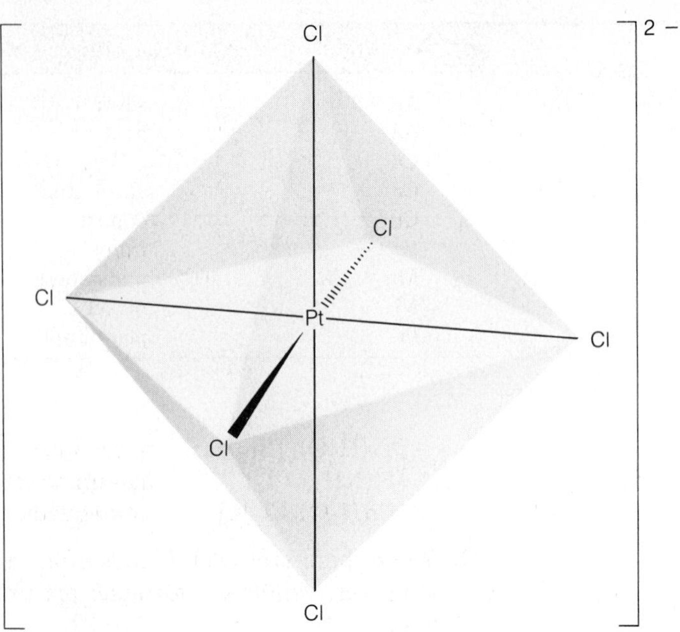

FIGURE 21.3
A typical octahedral six-coordinate complex.

were very good. He suggested that a definite spatial array is associated with each coordination number. His supposition agrees with the quantum-mechanical model for chemical bonding that is now accepted. The coordination numbers most frequently encountered are 6, 4, and 2.

The six-coordinate complexes are octahedral (Figure 21.3). Most four-coordinate complexes are tetrahedral; $[NiCl_4]^{2-}$ is typical (Figure 21.4). However, some are square-planar (Figure 21.5). They include the Pt(II) and Au(III) complexes, as in $[PtCl_4]^{2-}$ and $[AuCl_4]^{-}$, and many Ni(II) and Cu(II) complexes. (Refer to section 7.11 for a discussion of these shapes.)

A coordination number of two is observed especially for Cu, Ag, and Au in their +1 oxidation state; for example, $[CuCl_2]^{-}$, $[Ag(NH_3)_2]^{+}$, and $[AuCl_2]^{-}$. The shape of these complexes is linear.

$[:\ddot{\underset{..}{Cl}}:Au:\ddot{\underset{..}{Cl}}:]^{-}$

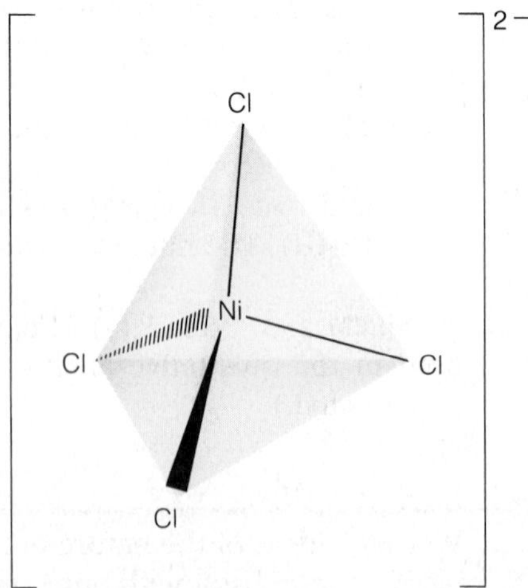

FIGURE 21.4
A typical tetrahedral complex, $[NiCl_4]^{2-}$.

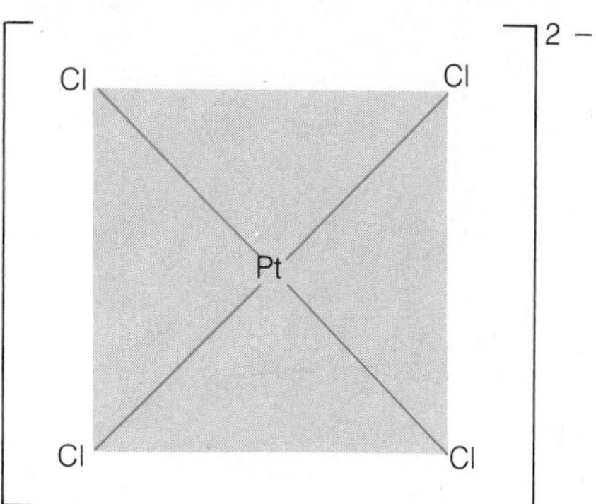

FIGURE 21.5
A typical square-planar complex, $[PtCl_4]^{2-}$

CIS-TRANS (GEOMETRIC) ISOMERISM

Alfred Werner deduced the geometry of the coordination complexes from his study of the observed number of isomers. Recall that isomers are two or more different compounds with the same molecular formula. For example, the reaction of ammonia with $K_2[PtCl_4]$ in water produces a neutral complex, $[Pt(NH_3)_2Cl_2]$,

$$[PtCl_4]^{2-} + 2NH_3 \longrightarrow \underset{\text{orange-yellow}}{[Pt(NH_3)_2Cl_2]} + 2Cl^-$$

A compound with the *same molecular formula* is also obtained from the reaction of HCl with $[Pt(NH_3)_4]Cl_2$ in water:

$$[Pt(NH_3)_4]^{2+} + 2H^+ + 2Cl^- \longrightarrow \underset{\text{lemon-yellow}}{[Pt(NH_3)_2Cl_2]} + 2NH_4{}^+$$

These two isomeric substances have distinctly different properties but identical molecular formulas. Werner proposed that the two compounds differ in the way the ligands are arranged about the central atom, Pt. He reasoned that if four-coordinate compounds are square-planar (Figure 21.5), a complex with two *different pairs of ligands* (general formula, MA_2Y_2) can have two different arrangements. The one with like groups *adjacent* to each other is called the *cis* isomer (Figure 21.6a). The one with like groups *diagonally opposite* from each

Latin *cis,* "on this side," and *trans,* "across"

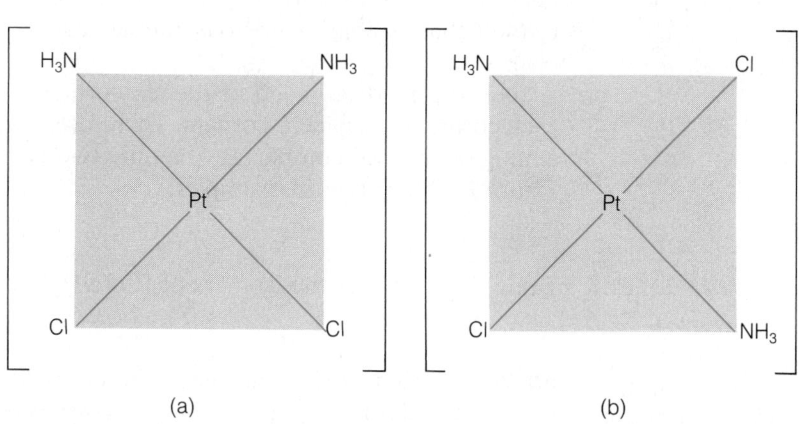

FIGURE 21.6
Stereoisomerism in $[Pt(NH_3)_2Cl_2]$, a square-planar complex, MA_2B_2. (a) Like groups are adjacent: *cis* isomer, orange-yellow, $\Delta H_f^0 = -469.9$ kJ, dipole moment = 10 D. (b) Like groups are diagonally opposite each other: *trans* isomer, lemon-yellow, $\Delta H_f^0 = -483.3$ kJ, dipole moment = 0

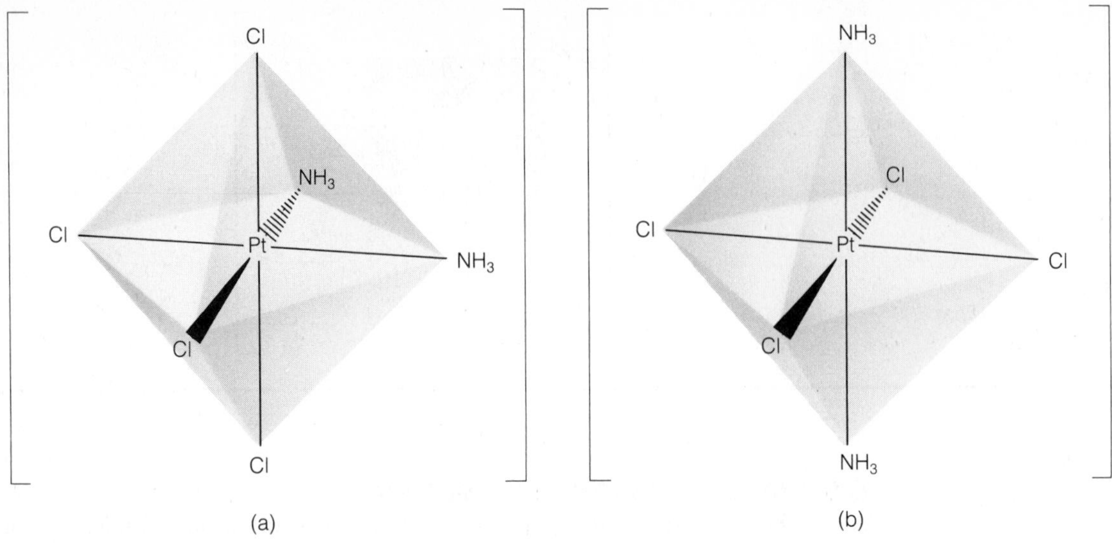

FIGURE 21.7
Stereoisomerism in $[Pt(NH_3)_2Cl_4]$, an octahedral complex, MA_2B_4-type. (a) Two NH_3 molecules adjacent to each other: *cis* isomer, orange-yellow. (b) Two NH_3 molecules opposite: *trans* isomer, lemon-yellow.

other is called the *trans* isomer (Figure 21.6b). The arrangement of the *trans* isomer is confirmed by the observation that the lemon-yellow compound has a zero dipole moment: the individual bond moments are on opposite sides of the molecule and cancel each other out. The orange-yellow compound, on the other hand, has a large dipole moment. Therefore, we conclude that the bond moments reinforce each other and so must be on the same side of the molecule. The orange-yellow compound must be the *cis* isomer. These ***cis-trans*** (also called **geometric**) **isomers**, differing in the spatial arrangement of atoms rather than in their linkages (Example 7.15, section 7.10), are types of **stereoisomers**. It is interesting that whereas *cis*-diamminedichloroplatinum(II) (Cisplatin), $[Pt(NH_3)_2Cl_2]$, is active against tumors and is widely used in the treatment of cancer, the *trans* isomer is inactive. Geometric isomers are also observed for the MX_2YZ- and MXWYZ-type square-planar coordination compounds.

Convince yourself that square-planar MXWYZ has three stereoisomers.

Cis-trans isomerism is impossible in tetrahedral complexes of the type MX_2Y_2. The four corners of a tetrahedron are all adjacent to one another, and none is opposite to another corner. Thus, any four-coordinate compound of this general formula that exhibits *cis-trans* isomerism cannot be tetrahedral; it must be square-planar.

In an octahedron, each of the six corners is opposite to another corner and each corner has adjacent corners. Therefore, *cis-trans* isomerism is also observed among octahedral complexes; the only two isomers of $[Pt(NH_3)_2Cl_4]$, shown in Figure 21.7, are typical examples.

EXAMPLE 21.7 Draw the shapes of the isomers of $[Co(NH_3)_4Cl_2]^+$.

The *cis* isomer is violet; the *trans* isomer is bright green.

ANSWER This is a six-coordinate ion and is, therefore, octahedral. Placing like pairs of ligands on adjacent or opposite corners of an octahedron leads to *cis-trans* isomerism (Figure 21.8). ■

2 Cl atoms adjacent
cis isomer
(a)

2 Cl atoms opposite
trans isomer
(b)

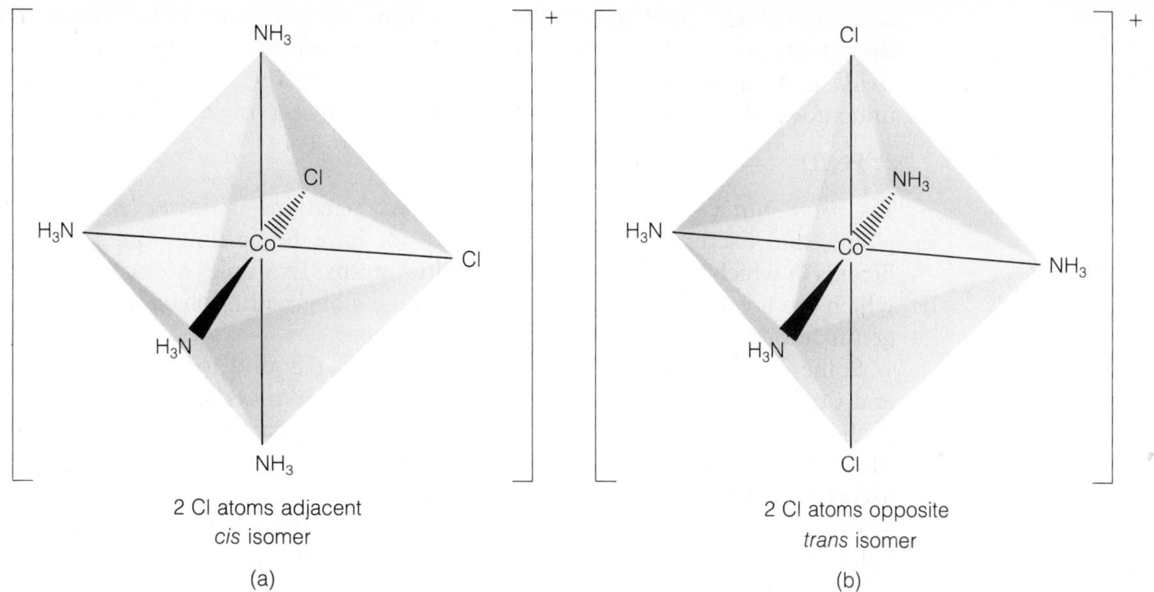

FIGURE 21.8
Stereoisomers of $[Co(NH_3)_4Cl_2]^+$: (a) *cis*; (b) *trans*.

PROBLEM 21.6 Draw the shapes of the two *cis-trans* isomers of (a) $[Pt(NH_3)_2Br_2]$ and of (b) $[Co(NH_3)_4(NO_2)_2]$. (The N atom of NO_2 is bonded to Co.) □

21.10 APPLICATIONS OF COORDINATION COMPOUNDS

APPLICATIONS IN BIOCHEMICAL PROCESSES

Coordination compounds, including several with transition elements, are important in many biochemical processes. Life would be impossible without them. These elements must be provided by the diet, often in trace amounts. The roles of several of these elements were revealed only recently when very accurate methods of analyzing trace amounts were developed.

Iron is the best-known trace transition element with regard to biochemical function. Iron complexes play two important roles: the transport of oxygen, and the transfer of electrons (redox). Two iron biochemicals transport O_2; hemoglobin brings O_2 to the cell, and myoglobin takes O_2 into the cell. Both have Fe(II) bound to a porphyrin molecule serving as a tetradentate ligand—this unit of structure is called a **heme** (Figure 21.9). Hemoglobin and myoglobin differ in the protein that

Chlorophyll, the catalyst of photosynthesis, is a similiar heme complex of Mg^{2+}.

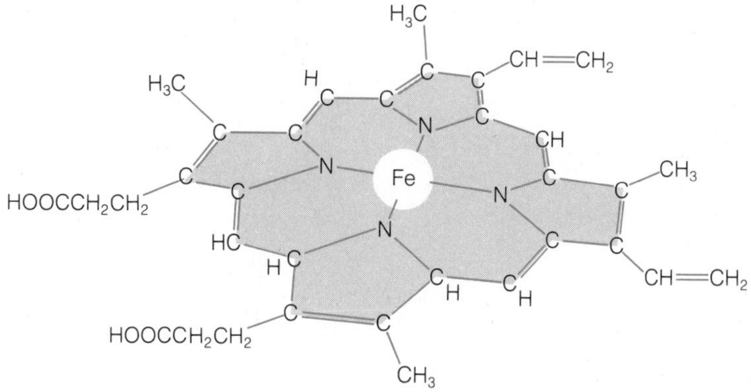

FIGURE 21.9
A heme, a square-planar complex, is composed of Fe(II) bound to the four N atoms of a porphyrin molecule. A hemoglobin molecule has four such hemes, each attached to a protein. One milliliter of blood contains about 5×10^9 red cells, each containing about 3×10^8 hemoglobin molecules (molecular weight 6.44×10^4).

is combined with the heme. Among the several electron-transfer biochemicals are the cytochromes, which also have the heme unit. A prime example is *cytochrome oxidase*; it catalyzes the reduction of O_2 to water. The iron atom in the heme undergoes an increase in oxidation number, thereby serving as the reducing agent:

$$\text{Fe(II)} \rightleftharpoons \text{Fe(III)} + e^-$$

The *iron-sulfur enzymes* function in redox reactions in lower animal forms, plants, and bacteria. These enzymes do not have heme units, but rather have ligands in which the binding sites are sulfur atoms. Examples are the *ferredoxins*, which are involved with photosynthesis and—along with molybdenum—in nitrogen fixation in plants.

Some cytochrome oxidases incorporate copper bound to a heme; the Cu(II) and Cu(I) oxidation states are involved in the redox reactions. In lower animals such as snails and insects, a Cu heme (hemocyanin) participates in O_2 transport. The Cu-containing enzyme ascorbic acid oxidase catalyzes the oxidation of ascorbic acid (vitamin C), with O_2 as the electron acceptor.

It's not necessary to memorize formulas and structures of these complex bioinorganic compounds.

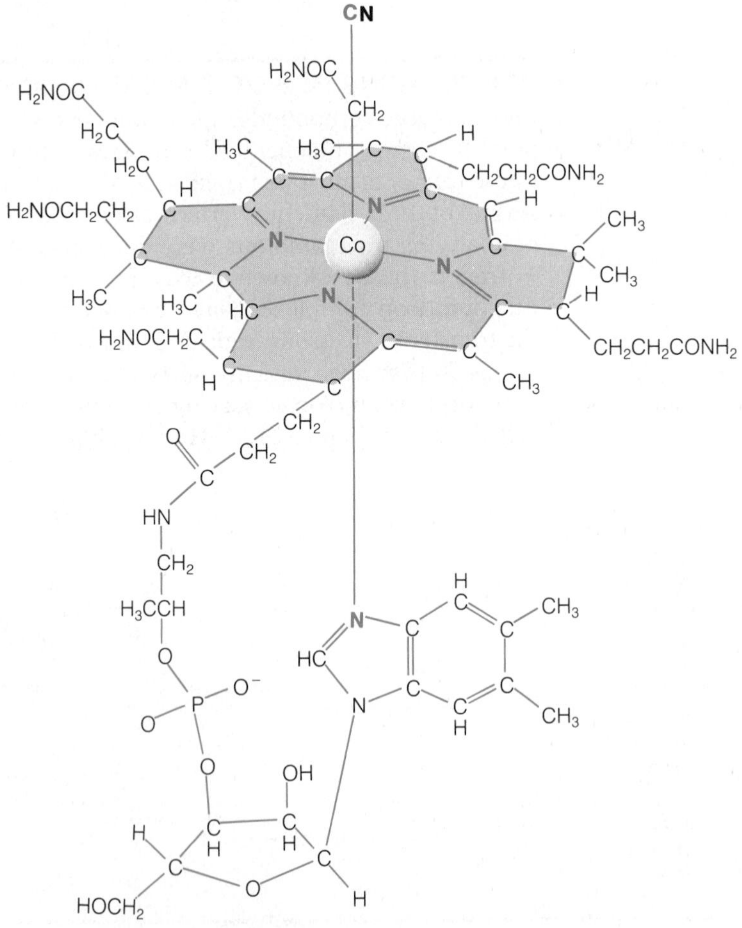

Cobalt is part of vitamin B_{12}, which prevents pernicious anemia because it helps to maintain a healthy level of hemoglobin in the blood. As shown in Figure 21.10, the Co atom is bonded to six sites (shown in color). First isolated in 1948, vitamin B_{12} is the largest and most complicated vitamin molecule known. (It was synthesized in 1969 by Robert Woodward and Albert Eschenmoser and a team of ninety-nine chemists from nineteen countries.)

FIGURE 21.10
Structure of vitamin B_{12} (cyanocobalamin, the most common commercial form of the vitamin). Notice that four N atoms form a square plane around Co, whereas the *trans* CN group above and the organic group below the plane give the molecule an octahedral arrangement around Co.

TABLE 21.9
TRACE TRANSITION ELEMENTS AND SOME OF THEIR BIOCHEMICAL ROLES

ELEMENT	CATALYTIC ROLE
Iron	O_2 transport; redox reactions; N_2 fixation
Copper	O_2 reduction; collagen formation; O_2 transport in insects
Cobalt	(in vitamin B_{12}) maintenance of hemoglobin level
Manganese	formation of urea; a component of urine
Vanadium	conversion of NO_3^- to NH_3
Molybdenum	formation of uric acid; N_2 fixation
Nickel	hydrolysis of urea
Chromium	glucose metabolism
Zinc†	CO_2 transport; hydrolysis of protein

† Also classified as a representative element.

Chromium is a component of the *glucose tolerance factor,* which may be essential for normal glucose metabolism in humans. In some cases, chromium supplements in the diet seem to alleviate a form of diabetes that afflicts some older people. Table 21.9 summarizes the biochemical roles of the transition elements.

These essential elements appear to function in one of three ways. (1) They may have an inherent activity that is enhanced by the protein portion of the enzyme. (2) The metal ion may function in redox reactions as seen for Fe and Cu. (3) They may coordinate with both the reactant and the active site of the enzyme, thus bringing them together in an active form. These essential trace elements are required in only milligram to microgram amounts per day; in more than trace amounts they are toxic. It is better to eat foods containing these elements as part of the diet than to take commercially prepared supplements.

INDUSTRIAL APPLICATIONS

FIXING PROCESS IN PHOTOGRAPHY The formation of coordination-complex ions is important in photography. A photographic film consists essentially of a silver compound, usually AgBr, uniformly suspended in a gelatin coating on a sheet of cellulose acetate or polyester. Exposure of the film to light excites the AgBr to a more chemically reactive state, leaving the unexposed AgBr in the less reactive ground state. When the film is treated with a reducing agent (the "developer solution"), the exposed reactive AgBr is reduced to silver, which appears black. The unexposed AgBr remains clear and must be removed. Sodium thiosulfate ("hypo," $Na_2S_2O_3$) is used in the "fixing" step to dissolve the AgBr,

$$AgBr(c) + 2S_2O_3^{2-} \longrightarrow [Ag(S_2O_3)_2]^{3-} + Br^-$$

ELECTROPLATING Coordination-complex ions play an important role in electroplating, a process for covering (plating) one metal with a thin layer of another metal. Many metals form a uniform plating that adheres more strongly and has a better appearance when the metal is electroplated from a solution in which it exists as a complex ion. Thus, in silver plating, $[Ag(CN)_2]^-$ is employed instead of Ag^+:

$$[Ag(CN)_2]^- + e^- \longrightarrow Ag(c) + 2CN^-$$

LEACHING OF METALS The formation of coordination-complex ions makes possible the isolation of pure silver and gold. These metals are thermodynamically stable in natural waters that contain oxygen and carbon dioxide (acidic), as shown for gold:

$$4Au(c) + O_2(aq) + 4H^+ \longrightarrow 4Au^+ + 2H_2O \qquad \Delta G^0 = +194 \text{ kJ}$$

In the presence of cyanide ion, however, gold becomes thermodynamically unstable in H_2O and O_2 and dissolves as cyanide complex:

$$4Au(c) + 8CN^- + O_2(aq) + 2H_2O \longrightarrow 4[Au(CN)_2]^- + 4OH^- \qquad \Delta G^0 = -407 \text{ kJ}$$

This reaction is involved in extracting (**leaching**) gold from ores. The pure Ag and Au are then recovered from their cyanide solutions by electrolysis.

HOMOGENEOUS CATALYSIS Heterogeneous catalysts (section 19.10) are widely used in industry because the solid catalysts used are easily removed from the reaction products by filtration. However, these catalytic reactions often require high temperatures and pressures. When oil was cheap, the cost of attaining these reaction conditions was of little consequence. As oil became more expensive, however, less expensive methods had to be developed. Recently, soluble catalysts that can be used in homogeneous systems at much lower temperatures have been developed. The use of soluble catalysts is called **homogeneous catalysis**.

An example of homogeneous catalysis is the small-scale laboratory addition of H_2 (hydrogenation) to the C=C group, which uses coordination complexes such as $[RhL_3Cl]$, L=$(C_6H_5)_3P$, as homogeneous catalysts:

$$H_2 + \;\rangle C{=}C\langle \;\xrightarrow{[RhL_3Cl]}\; -\overset{|}{\underset{H}{C}}-\overset{|}{\underset{H}{C}}-$$

CATION ANALYSIS

The separation and identification of the components of a mixture is called **qualitative analysis**. Coordination compounds and ions play an important role in the classical, "wet" qualitative analyses performed in the general chemistry laboratory. A few examples are presented here.

Wet analyses have been largely superseded by instrumental methods in industrial laboratories.

Silver chloride, AgCl, is dissolved and separated from mercurous chloride (Hg_2Cl_2) by adding a solution of NH_3:

$$AgCl(c) + 2NH_3 \longrightarrow [Ag(NH_3)_2]^+(aq) + Cl^-$$

When ammonia is added to a blue solution of Cu^{2+}, a pale blue precipitate of copper(II) hydroxide, $Cu(OH)_2$, forms. The hydroxide readily dissolves when more NH_3 is added to give a deep blue–colored solution. The hydroxide does not dissolve in excess OH^-; therefore, it is not amphoteric. Rather, it dissolves because of the formation of the tetraammine coordination complex. The pertinent ionic equations are

$$NH_3 + H_2O \longrightarrow NH_4^+ + OH^-$$

$$\underset{\text{(blue)}}{Cu^{2+}} + 2OH^- \longrightarrow \underset{\text{(pale blue)}}{Cu(OH)_2(c)}$$

$$Cu(OH)_2(c) + 4NH_3(aq) \longrightarrow \underset{\text{(deep blue)}}{[Cu(NH_3)_4]^{2+}(aq)} + 2OH^-(aq)$$

Several cations are detected by the formation of either insoluble coordination compounds or colored coordination ions as shown.

Probably $[Fe(H_2O)_5SCN]^{2+}$

- Fe^{3+} Add thiocyanate ion, SCN^-, to give the red-colored $[FeSCN]^{2+}$.

$$Fe^{3+} + SCN^- \longrightarrow [FeSCN]^{2+}$$

- Co^{2+} Add potassium nitrite in acetic acid solution to give the yellow potassium hexanitrocobaltate(III), one of the rare insoluble salts of K^+. Notice that this is a redox reaction.

$$Co^{2+} + 3K^+ + 7NO_2^- + 2HC_2H_3O_2 \longrightarrow K_3[Co(NO_2)_6](c) + NO + H_2O + 2C_2H_3O_2^-$$

(yellow)

HO—N=C(CH₃)—C(CH₃)=N—OH

dimethylglyoxime

The N atoms are the bonding sites.

- Ni^{2+} Add dimethylglyoxime, a bidentate ligand, in a properly buffered solution to give a red-gold coordination compound.

21.11 BONDING IN COORDINATION COMPOUNDS

Traditionally, a bond between atoms was thought to occur in one of two ways: either as an ionic bond involving electrostatic attraction between oppositely charged ions or as a covalent bond formed by the overlap of atomic orbitals in which a pair of electrons is shared. Both of these views are oversimplifications. We know that covalently bonded atoms may have partial charges, as in $\overset{\delta+}{H}—\overset{\delta-}{Cl}$. Likewise, there may be some orbital overlap even in ionic bonding. Thus many bonds have both ionic and covalent character. This same blending of bonding types is found in coordination compounds.

Any proposed bonding concept must pass the test of explaining the observed shapes, magnetic properties, and colors of the coordination compounds of transition elements. Three concepts have been proposed. The valence bond (VB) and molecular orbital (MO) concepts emphasize the covalent character of the bond. Only the VB concept is discussed here because it is simpler and conforms to the classical Lewis picture of a covalent bond. The **crystal field (CF) concept** emphasizes the electrostatic attraction. The ensuing discussion applies the VB and CF concepts to only the first-row transition elements.

Crystal field theory is applied here only to the octahedral shape.

We will find that the properties of coordination compounds depend on the number of d electrons remaining in the transition-metal cation. For this reason, such cations are classified by using the symbol d^n, where n is their number of d electrons. This value of n is obtained by subtracting the charge on the cation and 18 (the electron count in the Ar core) from the atomic number. Thus we know that $_{24}Cr^{3+}$ is a d^3 ion, since $24 - 3 - 18 = 3$.

21.12 VALENCE BOND CONCEPT

The ideas of using hybrid orbitals (HO's) developed for representative elements in section 8.1 are extended to the transition elements with one significant difference. For most representative elements each hybrid orbital used for bonding had one electron. However, since transition elements exclusively form coordinate covalent bonds in which they are the acceptors, they use hybrid orbitals with *no electrons.*

In coordination compounds with two ligands (2-coordinate compounds) and tetrahedral coordination compounds with four ligands (4-coordinate compounds) the metal atoms use sp and sp^3 HO's, respectively. Since the s and p AO's in the third ($n = 3$) principal energy shell are filled, the empty $4s$ and the empty $4p$

atomic orbitals are assumed to undergo hybridization. (Recall that for representative elements the partially filled $3s$ and $3p$ AO's are hybridized.) Cu^+ in $[CuCl_2]^-$ and Fe^{3+} in $[FeCl_4]^-$ illustrate these two kinds of hybridization.

$_{29}Cu^+$, a d^{10} atom ($29 - 1 - 18 = 10$), is formed when the $_{29}Cu$ atom, $[Ar]3d^{10}4s^1$, loses its $4s$ electron. In this section only the outermost electrons are shown; the [Ar] electron core is omitted.

***sp* hybridization of Cu^+ (d^{10}):**

ground state	*sp* hybrid state before bonding	*sp* hybrid state after bonding (Colored arrows indicate ligand electrons.)
$4p$ _ _ _	$4p$ _ _	$4p$ _ _
$4s$ _	$4sp$ _ _	$4sp$ ↑↓ ↑↓
$3d$ ↑↓ ↑↓ ↑↓ ↑↓ ↑↓	$3d$ ↑↓ ↑↓ ↑↓ ↑↓ ↑↓	$3d$ ↑↓ ↑↓ ↑↓ ↑↓ ↑↓

$_{26}Fe^{3+}$, a d^5 atom, is formed when the $_{26}Fe$ atom, [Ar] $3d$ ↑↓ ↑ ↑ ↑ ↑ $4s$ ↑↓, loses both the $4s$ and one of the $3d$ electrons.

***sp*³ hybridization of Fe^{3+} (d^5):**

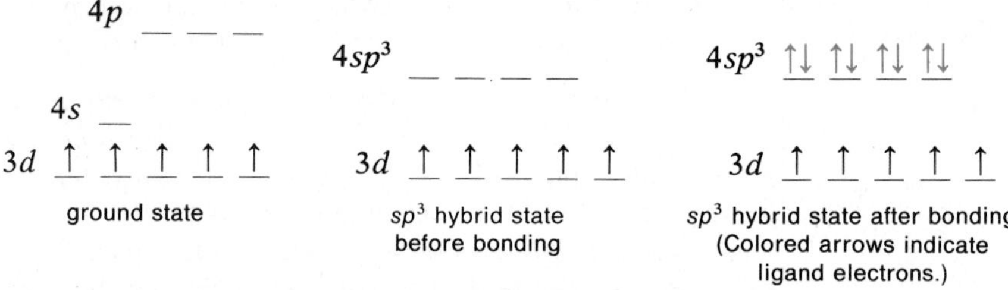

The most commonly observed 4-coordinate shape is square-planar. The four hybrid orbitals associated with this shape are one of the $3d$'s, the $4s$, and two of the $4p$'s. The HO's formed from these atomic orbitals are designated dsp^2. (Recall that this shape and HO type are not observed for representative elements.) Note that the orbitals are listed in the order of increasing n and l quantum numbers. $3d$ comes before and $4p$ comes after $4s$; hence the name dsp^2. We illustrate this type of hybridization with Ni^{2+} in the ion $[Ni(CN)_4]^{2-}$. Ni^{2+} (d^8) is formed when $_{28}Ni$, $[Ar]3d$ ↑↓ ↑↓ ↑↓ ↑ ↑ $4s$ ↑↓, loses its two $4s$ electrons ($28 - 2 - 18 = 8$).

***dsp*² hybridization of Ni^{2+} (d^8):**

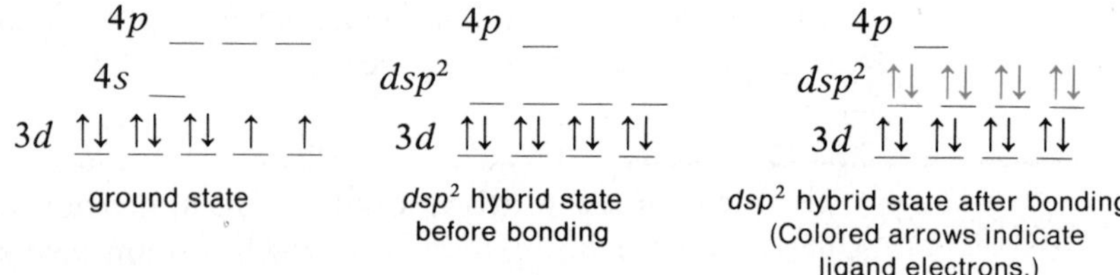

Notice that in order to make a vacant d orbital available, two d electrons, unpaired in the ground state, become paired.

The octahedral shape of the 6-coordinate compounds indicates that hybridization requires an s, three p, and two d orbitals. However, there are two ways the hybridization can be accomplished. In both ways the $4s$ and the three $4p$ orbitals are used. One way uses two $3d$'s to give d^2sp^3 HO's; the other way uses two $4d$'s to give sp^3d^2 HO's. These differences are illustrated

Convince yourself that Co^{3+} is a d^6 ion.

with coordination-complex ions of Co^{3+} (d^6), which are formed when $_{27}Co$, [Ar]$3d$ ↿⇂ ↿⇂ ↿ ↿ ↿ $4s$ ↿⇂ loses its two $4s$ electrons and one of its d electrons. Only the bonded hybrid states are shown, and the $4d$'s are included.

d^2sp^3 and sp^3d^2 hybridization of Co^{3+} (d^6):

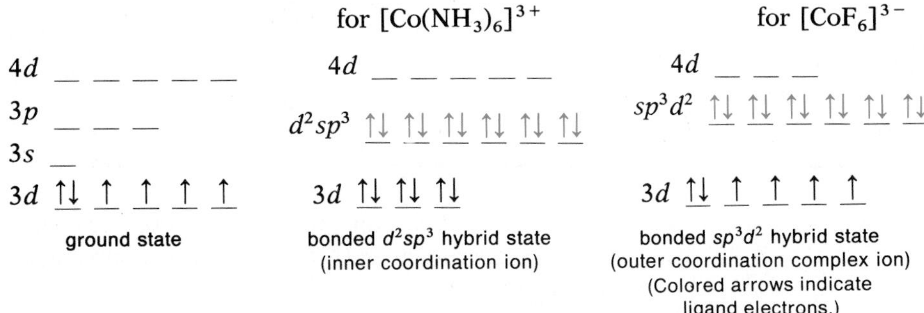

These two types of hybridization were suggested in order to explain the difference in the magnetic properties of the salts of these ions. $[Co(NH_3)_6]^{3+}$, called the **inner coordination complex** because Co uses $3d$ orbitals, is diamagnetic and therefore has no unpaired electrons. In order to get two vacant $3d$ orbitals for bonding purposes, the six d electrons must be paired in the remaining three $3d$ orbitals. Alternatively, $[CoF_6]^{3-}$, called the **outer coordination complex** because Co uses $4d$ orbitals, is paramagnetic; it has four unpaired electrons. The VB concept rationalizes this behavior by suggesting that two of the empty $4d$ orbitals are hybridized, which leaves the $3d$ *orbitals with their unpaired electrons unchanged.* The VB concept does not predict this difference in behavior of the Co^{3+} ion toward the different ligands. We will see in the next section that the crystal field concept addresses this question very nicely. Furthermore, the VB concept does not account for the colors of the coordination compounds, something else that the CF concept does very well.

Table 21.10 summarizes the relationship between the shapes of coordination compounds and the HO's used by the transition element.

EXAMPLE 21.8 $[NiCl_4]^{2-}$ has a tetrahedral shape. What is the d^n classification of Ni? Show the outermost electron configuration for the bonded Ni^{2+} ion. Can a difference in magnetic properties distinguish this ion from the square-planar $[Ni(CN)_4]^{2-}$?

ANSWER

$$n = \underset{\text{(atomic number)}}{28} - \underset{\text{(charge on Ni}^{2+}\text{)}}{2} - \underset{\text{(Ar core)}}{18} = 8$$

Ni^{2+} is a d^8 cation. The tetrahedral shape indicates the use of sp^3 HO's. The electron configuration for the bonded atom is $3d$ ↿⇂ ↿⇂ ↿⇂ ↿ ↿ sp^3

TABLE 21.10 SHAPES OF TRANSITION-METAL COORDINATION COMPOUNDS AND HO'S USED IN BONDING (VB concept)

SHAPE	HO'S USED
Linear	sp
Tetrahedral	sp^3
Square-planar	dsp^2
Octahedral	d^2sp^3 or sp^3d^2

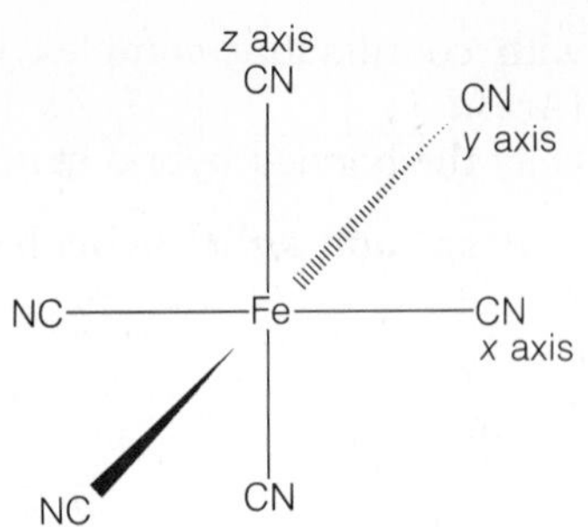

FIGURE 21.11
Octahedral shape of the $[Fe(CN)_6]^{4-}$ complex ion.

$\underline{\uparrow\downarrow}$ $\underline{\uparrow\downarrow}$ $\underline{\uparrow\downarrow}$ $\underline{\uparrow\downarrow}$. This ion has two unpaired electrons and is paramagnetic. The square-planar $[Ni(CN)_4]^{2-}/(dp^3)$ has only paired electrons (page 826) and is diamagnetic. ■

PROBLEM 21.7 Show the outermost electron configuration as predicted by the VB concept for $_{24}Cr$ in $[Cr(NH_3)_6]^{3+}$ assuming two kinds of hybridization if possible. □

21.13 CRYSTAL FIELD (CF) CONCEPT

We now consider the electrostatic nature of bonding between the metal cation and the ligands in coordination compounds. This viewpoint was first explored by physicists in the early 1930s in connection with the spectra of ionic crystals. The ideas about bonding derived from it are called the **crystal field concept**.

Its early proponents were Hans Bethe (1929) and John Van Vleck (1932).

CRYSTAL-FIELD SPLITTING

Consider, for example, the ion $[Fe(CN)_6]^{4-}$ (Figure 21.11). The six CN^- ligands are spread apart from each other as widely as is geometrically possible (the VSEPR concept, section 7.13). They are at the corners of an octahedron; for the closest CN^- ligands, the C—Fe—C angles are 90°. The six CN^- ligands may therefore be considered to be along the x, y, and z axes, two ligands on the end of each axis. Now recall the shapes and spatial orientations of the five d orbitals (section 5.6) that surround the nucleus of the Fe atom. Two of the orbitals lie along the

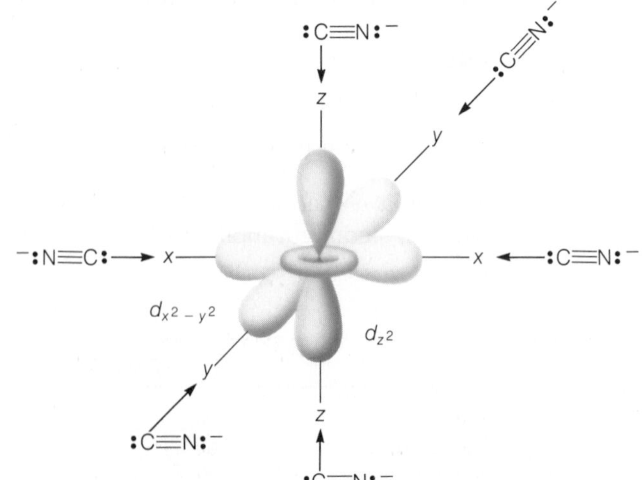

FIGURE 21.12
The d_{z^2} and $d_{x^2-y^2}$ orbitals lie along the x, y, and z axes. The other three d orbitals, which lie between the axes, are not shown.

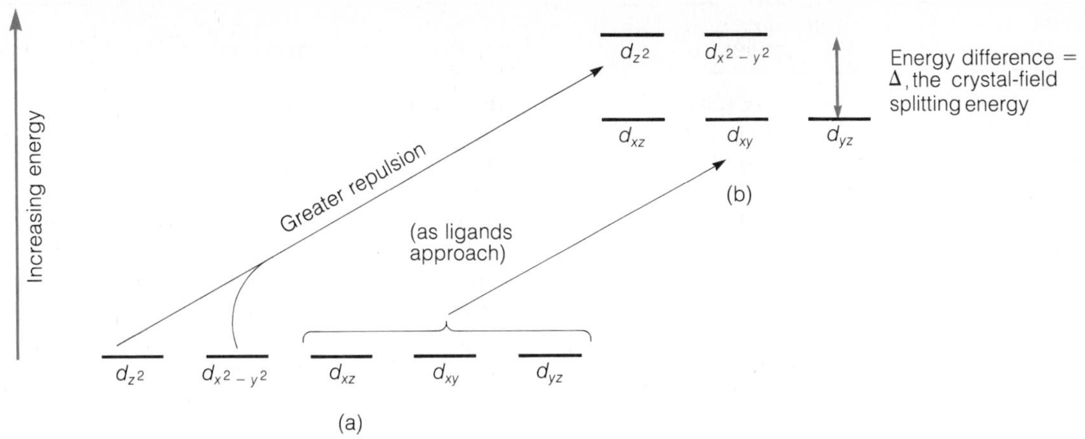

FIGURE 21.13
Crystal-field splitting energy of *d* orbitals in an octahedral field.

same axes as the ligands. One of these is the d_{z^2} orbital oriented along the *z* axis. The other is the $d_{x^2-y^2}$ orbital oriented along the *x* and *y* axes (Figure 21.12). The other three orbitals, d_{xy}, d_{yz}, and d_{xz}, lie *between* the axes. The d_{z^2} and $d_{x^2-y^2}$ orbitals are therefore *closer* to the ligands than are the other three orbitals. As a ligand, with its unshared electron pair, approaches the metal ion, the electron pair repels electrons in all of the *d* orbitals. However, since the electrons in the d_{z^2} and $d_{x^2-y^2}$ orbitals are directly in the path of approaching ligands (Figure 21.10), they are more strongly repelled than the electrons in the d_{xy}, d_{yz}, and d_{xz} orbitals. Consequently, when the ligands approach, the five *d* orbitals of the central atom no longer have the same energy. Instead, they are split into *two sets* that differ in energy. Electrons in the d_{z^2} and $d_{x^2-y^2}$ are more repelled and thus are at a higher energy than electrons in the other three. The energy difference between the two sets of orbitals is called the **crystal-field splitting energy**, symbolized by Δ, as shown in Figure 21.13. The separation of the five *d* orbitals into two sets of different energies is the main feature of this theory. The energy splitting pattern shown in Figure 21.13 is unique for the six-coordinate octahedral shape. Each of the other shapes has its own pattern as determined by the same reasoning just used for the octahedron. We will now discuss how these energy differences explain the color and paramagnetic properties of coordination compounds.

See Problem 21.50 for application to the tetrahedral shape.

COLOR AND THE CRYSTAL FIELD CONCEPT

A study of the color and, more specifically, the spectrum in the visible region of transition-metal coordination compounds helps us understand their structure. Before this discussion, however, it is necessary first to understand the origin of color.

Absorption of visible or ultraviolet radiation results in the excitation of an electron from a lower energy level (ground state) to a higher energy level (excited state). The difference in energy, ΔE, between the excited state, E_2, and the ground state, E_1, equals the energy of the absorbed radiation,

$$\underset{\text{energy of excited state}}{E_2} - \underset{\text{energy of ground state}}{E_1} = \underset{\text{energy of absorbed radiation}}{\Delta E}$$

TABLE 21.11
APPROXIMATE COMPLEMENTARY COLORS

ABSORBED WAVELENGTH REGION (λ), nm	COLOR ABSORBED	OBSERVED TRANSMITTED COLOR
<400	**ultraviolet**	**colorless**
400–420	violet	yellow-green
420–460	indigo	yellow
460–490	blue	orange
490–510	blue-green	red
510–540	green	purple
540–560	yellow-green	violet
560–590	yellow	indigo
590–650	orange	blue
650–750	red	blue-green
>750	**infrared**	**colorless**

The energy of the absorbed radiation is inversely proportional to the wavelength, λ; $\Delta E = \frac{hc}{\lambda}$ (section 5.3). The longer the wavelength of the absorbed light, the lower is the energy of the electronic excitation. Colorless substances usually absorb in the ultraviolet region. Colored substances absorb in the visible region, 400 to 750 nm. Hence, it takes less energy to electronically excite a colored substance than a colorless substance. Color appears when the energy level of the excited state is close to the ground-state energy level.

If all visible light is absorbed, the substance appears black. If all but one color is absorbed, the substance will have that color. More frequently, however, only one color is absorbed, in which case the substance has the complementary color (Table 21.11). In general, there are three complementary color pairs: red–green, orange–blue, and yellow–violet. The energy of excitation, ΔE, is smaller as the colors vary from yellow to orange to red to violet to blue to green.

Complementary colors are two colors that, when combined, produce white or nearly white light.

COLOR OF COORDINATION COMPOUNDS; STRONG- AND WEAK-FIELD LIGANDS

In general, compounds with transition elements having partially filled *d* orbitals are colored because visible light can excite electrons from lower- to higher-energy *d* orbitals. Specifically for octahedral coordination compounds, the electron excitation is from a lower-energy d_{xy}, d_{xz}, or d_{yz} orbital to a higher-energy d_{z^2} or $d_{x^2-y^2}$ orbital. For example, for a d^6 cation:

$$\underline{\uparrow\downarrow}\ \underline{\uparrow\downarrow}\ \underline{\uparrow\downarrow}\ \underline{\quad}\ \underline{\quad} \xrightarrow[\text{photon}]{\text{light}} \underline{\uparrow\downarrow}\ \underline{\uparrow\downarrow}\ \underline{\uparrow}\ \underline{\downarrow}\ \underline{\quad}$$

The observed color depends on the value of Δ, which determines the wavelength of the absorbed radiation. The larger the value of Δ is, the shorter is the wavelength of the absorbed radiation that causes the electron excitation. When the *d* orbitals are either all empty (d^0) or all filled (d^{10}), *electron excitation between d orbitals is impossible and the complex is colorless.*

We now examine the effect of different ligands on the relative Δ value and color of Co^{3+} coordination compounds. $[Co(H_2O)_6]^{3+}$ appears blue. We assume that H_2O induces a small Δ in Co^{3+} and that low-energy photons are absorbed. $[Co(NH_3)_6]^{3+}$ appears orange. We assume that ammonia induces a larger Δ and that higher-energy photons are absorbed. $[Co(CN)_6]^{3-}$ appears yellow. Thus, we assume that cyanide ion induces the largest Δ and that highest-energy photons are absorbed.

In summary:

$[Co(CN)_6]^{3-}$ yellow	$[Co(NH_3)_6]^{3+}$ orange	$[Co(H_2O)_6]^{3+}$ blue
Absorbs violet	Absorbs blue	Absorbs orange
↓	↓	↓
Wavelength of absorbed light is shortest.	Wavelength of absorbed light is intermediate.	Wavelength of absorbed light is longest.
↓ (~410 nm)	↓ (~475 nm)	↓ (~620 nm)
ΔE is largest.	ΔE is intermediate.	ΔE is smallest.
↓	↓	↓
Δ is largest.	Δ is intermediate.	Δ is smallest.
↓	↓	↓
CN^- has strongest splitting ability.	NH_3 has intermediate splitting ability.	H_2O has weakest splitting ability.

Studies of the absorption spectra of complex ions show that the value of Δ depends in large part on the nature of the ligand. The splitting ability for some of the more common ligands decreases in the order:

This order is called the spectrochemical series of ligands.

$$CN^- > NO_2^- > NH_3 > H_2O > OH^- > F^- > Cl^- > Br^- > I^-$$

strong-field ligands ⟶ weak-field ligands

The ligands at the left of the series have the greatest splitting ability; they are **strong-field ligands**. Those at the right of the series have the weakest splitting ability; they are **weak-field ligands**. Thus CN^-, at the head of the list and having the greatest splitting ability, is said to be a very strong field ligand. At the other end, I^- is a very weak field ligand.

EXAMPLE 21.9 Which of these compounds are most likely to be colorless: $ScCl_3$, $TiCl_2$, VF_4, VF_5, CuF?

ANSWER Find the number of electrons in the $3d$ orbitals of the metal ions; $_{21}$Sc(III) $3d^0$; $_{22}$Ti(II) $3d^2$; $_{23}$V(IV) $3d^1$; V(V) $3d^0$; and $_{29}$Cu(I) $3d^{10}$. The compounds with d^0 and d^{10} metals are colorless; they are $ScCl_3$, VF_5, and CuF. Those with partially filled d orbitals are colored. The colored compounds and their colors are $TiCl_2$, light brownish black, and VF_4, reddish brown. ■

PROBLEM 21.8 Select the colored and colorless compounds from among $TiCl_3$, $TiCl_4$, AgF, CuF_2, and $MnCl_2$. □

MAGNETIC PROPERTIES; HIGH- AND LOW-SPIN COMPLEXES

We saw in section 21.12 that the magnetic properties of $[Co(NH_3)_6]^{2+}$ and $[CoF_6]^{3-}$ are clues to the distribution of electrons in the d orbitals of the transition elements. The coordination complex in which all electrons are paired is diamagnetic; the one with unpaired electrons is paramagnetic. The larger the number of unpaired electrons, the greater the paramagnetism.

Recall that the distribution of a given number of electrons is determined by the tendency of electrons to occupy orbitals of lower energy before entering orbitals of higher energy, and by the tendency of electrons to resist pairing until each orbital of equal energy houses at least one electron (Hund's rule). Therefore, the electron distribution of an isolated d^6 atom is ↑↓ ↑ ↑ ↑ ↑ and not ↑↓ ↑↓ ↑↓ __ __. But as we have just learned, splitting of d-orbital energies

TABLE 21.12
DEPENDENCE OF ELECTRON DISTRIBUTION ON Δ[†]

d ELECTRONS	TYPICAL CENTRAL ION	WEAK FIELD SMALL Δ HIGH-SPIN COMPLEX	STRONG FIELD LARGE Δ LOW-SPIN COMPLEX
d^3	$_{24}Cr^{3+}$	↑ ↑ ↑ — —	— — ↑ ↑ ↑
d^4	$_{24}Cr^{2+}$	↑ ↑ ↑ ↑ —	— — ↑↓ ↑ ↑
d^5	$_{26}Fe^{3+}$, $_{25}Mn^{2+}$	↑ ↑ ↑ ↑ ↑	— — ↑↓ ↑↓ ↑
d^6	$_{26}Fe^{2+}$, $_{27}Co^{3+}$	↑↓ ↑ ↑ ↑ ↑	— — ↑↓ ↑↓ ↑↓
d^7	$_{27}Co^{2+}$[§]	↑↓ ↑↓ ↑ ↑ ↑	↑ — ↑↓ ↑↓ ↑↓

[†] Δ's are *not* to scale.

[§] Co^{2+} coordination compounds are easily oxidized to Co^{3+} compounds because the electron is lost from the high-energy *d* orbital.

You may verify that only one distribution is possible for d^1, d^2, d^3, d^8, d^9.

Chemical reactivity is also influenced by this dependence of electron distribution on Δ.

occurs when ligands approach the central atom or ion. When Δ is small, electrons ignore the slight difference in energy, and pairing does not occur until each of the five orbitals houses one electron. Thus, coordination complexes with a small Δ tend to have larger numbers of unpaired electrons and so tend to be highly paramagnetic (Table 21.12). They are called **high-spin complexes**.

When Δ is large, however, electrons respect the significant difference in energy of the split *d* orbitals. They fill, by pairing, the set of lower-energy orbitals before occupying the higher-energy set. Coordination complexes with large Δ's tend to have fewer unpaired electrons and so tend to be diamagnetic or less paramagnetic (Table 21.12). They are called **low-spin complexes**. When there are 4, 5, 6, or 7 electrons, different distributions are possible (Table 21.12). Evidently, the magnetic properties of coordination complexes depend strongly on the splitting energy. An Fe^{2+} complex may be a high-spin complex—highly paramagnetic—or a low-spin complex—diamagnetic—depending on Δ.

The *d*-electron distribution of the high-spin complex and the VB sp^3d^2 complex are identical. Similarly, the *d*-electron distribution of the low-spin complex and the VB d^2sp^3 complex are identical.

EXAMPLE 21.10 $[Fe(CN)_6]^{4-}$ is diamagnetic, whereas $[Fe(H_2O)_6]^{2+}$ has paramagnetic properties equivalent to four unpaired electrons. Explain this in terms of the crystal field concept.

ANSWER In both ions the metal cation is the d^6 Fe^{2+} cation. Differences in magnetic properties result from differences in the abilities of ligands to split the *d*-orbital energies. The distribution of the six *d* electrons is shown in Table 21.12 for a strong-field ligand (large Δ) and a weak-field ligand (small Δ). CN^- is a strong-field ligand and $[Fe(CN)_6]^{4-}$ is a low-spin complex with all electrons paired. H_2O is a weak-field ligand and $[Fe(H_2O)_6]^{2+}$ is a high-spin complex with four unpaired electrons. ■

EXAMPLE 21.11 (a) Draw the electron configuration for an octahedral six-coordination complex of Ni(III) having (1) a small Δ and (2) a large Δ. (b) Which typical ligand affords (1) a small Δ and (2) a large Δ value?

ANSWER

(a) First find the *d* classification for Ni(III)—it is d^7. (1) The seven *d* electrons are distributed in the five *d* orbitals ↑↓ ↑↓ ↑ ↑ ↑. (2) Six electrons are in the three lower-energy *d* orbitals, and the seventh electron is in one of the two higher-energy *d* orbitals, (lower energy) ↑↓ ↑↓ ↑↓ and (higher energy) ↑ __.

(b) (1) One with a weak splitting ability; for example, Br^-. (2) One with a strong splitting ability; for example, CN^-. ■

PROBLEM 21.9 Which ligand is more likely to cause an octahedral complex of Mn^{2+} to be paramagnetic, one with a small Δ or one with a large Δ? □

PROBLEM 21.10 Draw the electron configuration for an octahedral coordination complex of Mn(II) with (1) a small Δ and (2) a large Δ. □

21.14 STABILITY OF COORDINATION COMPOUNDS

Werner showed that a Cl^- ligand bonded to Pt(IV) cannot be removed by converting it to the white, insoluble AgCl by adding Ag^+ ions (section 21.6). Gold does not dissolve in aerated water, but it will dissolve if cyanide ion, CN^-, is added to form the complex $[Au(CN)_2{}^-]$. Evidence of this kind suggests that coordination compounds and ions may be thermodynamically very stable.

But also recall that kinetic stability and thermodynamic stability do not necessarily correlate with each other (section 19.12). For example, $[Co(NH_3)_6]^{3+}$ is thermodynamically unstable toward acid. The equilibrium constant for the decomposition reaction

$$[Co(NH_3)_6]^{3+} + 6H_3O^+ \rightleftharpoons [Co(H_2O)_6]^{3+} + 6NH_4{}^+ \quad \Delta G^0 = -367.9\text{ kJ}$$

is about 10^{64}, yet the coordination complex persists for weeks in an acid medium. The exchange of H_2O for NH_3 is *very slow*. This means that the ion is kinetically stable—the energy of activation is very large. $[Co(NH_3)_6]^{3+}$ is typical of **inert complexes**; they exchange ligands very slowly. $[Co(NH_3)_6]^{3+}$ is an example of a thermodynamically unstable coordination ion that is kinetically stable. On the other hand, $[Ni(CN)_4]^{2-}$ is thermodynamically stable:

$$[Ni(CN)_4]^{2-}(aq) + 4H_2O \rightleftharpoons [Ni(H_2O)_4]^{2+}(aq) + 4CN^-(aq)$$

$$\Delta G^0 = +172.0\text{ kJ},\ K = 7 \times 10^{-31}$$

Yet the exchange rate with radioactive $^{14}CN^-$ is extremely rapid. For example, when $^{14}CN^-$ is added to a solution of $[Ni(CN)_4]^{2-}$, the $^{14}CN^-$ reacts quickly with $[Ni(^{12}CN)_4]^{2+}$ to yield some $[Ni(^{14}CN^-)(^{12}CN^-)_3]^{2-}$ in solution. $[Ni(CN)_4]^{2-}$ is therefore kinetically unstable. It is a typical **labile complex**; it exchanges ligands very rapidly. It is also acceptable to describe ligands in complexes as labile or inert.

The octahedral coordination complexes of the first transition series are very labile *except* for those of Cr(III) (d^3) and Co(III) (d^6). The inert coordination complexes of these two cations were among those studied by Werner. It was indeed fortunate that he chose these inert coordination complexes. Once formed, inert coordination complexes remain unaltered long enough to be studied. Had he chosen the labile coordination complexes, he could not have made his significant discoveries. The octahedral coordination complexes of these cations are inert when they are coordinated to a strong-field ligand (large Δ). Under these conditions, the lower-energy d orbitals are half-filled (d^3) or filled (d^6). Therefore it appears that when the three lower-energy d orbitals of a strong-field complex are half-filled or filled, a special kinetic stability is imparted to the complex. The electron arrangements are shown in Table 21.12.

Some interesting relationships exist between kinetic stability and the chemistry of coordination compounds. For example, a correlation has been found between the antitumor activity of platinum complexes and the lability of their ligands. The ligands of the most active and least toxic complexes are moderately labile, like the Cl^- in $[Pt(NH_3)_2Cl_2]$. Ligands with high lability are associated with active but highly toxic complexes, whereas inert ligands are associated with nontoxic, inactive complexes.

SUMMARY

Transition elements are those with partially filled d orbitals in at least one oxidation state. They are metals with multiple oxidation states (except Group-3B elements). In the lower oxidation states, the elements form ionic bonds; in the higher oxidation states the bonding is covalent. Higher oxidation states become more stable on proceeding down a group, but become less stable on proceeding across a period. As the oxidation number of an atom increases, its hydroxy compounds change from weak ionic bases to amphoteric hydroxides to oxo acids. Many of their compounds are colored and paramagnetic.

Transition elements and their cations readily act as Lewis acids to form ionic and molecular **coordination compounds** with ligands that act as Lewis bases. The charge on the coordination molecule or ion is the sum of the charge on the central metal atom and the charges on the ligands. The **coordination number** is the number of atoms bonding to the central atom. **Multidentate** ligands (**chelating agents**) bind simultaneously at more than one binding site to produce **chelates**. The coordination number determines the shape of the coordination compound:

COORDINATION NUMBER	SHAPE
2	linear
4	tetrahedral or square-planar
6	octahedral

Certain square-planar and octahedral coordination compounds (for example, MX_2Y_2 and MX_4Y_2, respectively) have isomers (**stereoisomers**) that differ in the spatial arrangement of the ligands. The kind of stereoisomerism observed in these examples is called ***cis-trans* (geometric) isomerism**. In the ***cis*-isomer** the Y groups are adjacent; in the ***trans*-isomer** the Y groups are diagonally opposite.

Transition-metal complexes play important roles in the life processes of plants and animals. As a result, transition elements such as Fe, Cu, Co, Cr, Mn, Mo, Ni, and V are essential dietary components for humans. Coordination compounds are commercially useful in photography, electroplating, and leaching (dissolving of metals or ores). Many of the metals and their compounds are used as heterogeneous and homogeneous catalysts. Coordination compounds are used in cation **qualitative analysis**, the separation and identification of cations in a mixture.

The valence bond concept stresses the covalent nature of the metal–ligand bond by assuming that a hybrid orbital (HO) of the metal atom overlaps with an orbital of the binding-site atom of the ligand. The shape of the coordination compound relates to the HO's used by the transition element: linear, *sp*; tetrahedral, sp^3; square-planar, dsp^2; and octahedral, d^2sp^3 (*inner coordination complex*) or sp^3d^2 (*outer coordination complex*).

The **crystal field concept** explains the bonding of coordination compounds by emphasizing the electrostatic nature of the bond between the central atom and the ligand. The concept assumes that the unshared electron pair in the approaching ligand repels the *d*-orbital electrons of the central atom. However, electrons in *d* orbitals lying in the direct path of the incoming ligand are repelled more than those that are out of the direct path. Consequently, not all of the *d* orbitals in the coordination compound have the same energy; those in the direct path have a higher energy. This splitting of the energies of the five *d* orbitals, the **crystal-field splitting energy** (Δ), depends on the shape of the coordination compound. For example, in the octahedron, the d_{z^3} and $d_{x^2-y^2}$ have the same higher energy, and the d_{xy}, d_{xz}, and d_{yz} have the same lower energy.

Transition-metal compounds are colored because they absorb visible light, which causes the excitation of electrons from lower-to higher-energy *d* orbitals. The value of Δ can be found from the wavelength of the light absorbed by the coordination compound. The color seen is the complement of the color absorbed.

Ligands that are strong Lewis bases (**strong-field ligands**), such as CN^- and CO, engender relatively large Δ values to give **strong-field coordination complexes**. In the strong-field case, the *d* electrons belonging to the metal atom fill the lower-energy *d* orbitals before entering the higher-energy *d* orbitals. Ligands that are weak Lewis bases (**weak-field ligands**), such as the halide anions, give **weak-field coordination complexes** in which there is little splitting of the energies of the five *d* orbitals. In the weak-field case, the *d* electrons are placed in the five *d* orbitals so that each orbital is half-filled before pairing occurs. When there are 4, 5, 6, or 7 *d* electrons in the transition element of an octahedron, there are fewer unpaired *d* electrons in the strong-field than in the weak-field coordination complexes, as deduced from the different paramagnetisms. The strong-field ("**low-spin**") coordination compound is less paramagnetic than the corresponding weak-field ("**high-spin**") coordination compound.

Coordination compounds may exchange their ligands for water or other ligands in solution. Exchange is slow for kinetically stable (**inert**) coordination compounds and rapid for kinetically unstable (**labile**) compounds. Kinetic and thermodynamic stability do not necessarily correlate with each other.

21.15 ADDITIONAL PROBLEMS

TRANSITION METALS

21.11 Write the electron configuration and give the d^n classification for each of these transition metals: (a) Y, (b) Zr, (c) Tc, (d) Ag.

21.12 Write the electron configuration and give the d^n classification for each of these cations: (a) Y^{3+}, (b) Zr^{3+} and Zr^{4+}, (c) Mo^{4+}, (d) Nb^{2+}, (e) Ru^{2+} and Ru^{3+}, (f) Pd^{2+}.

21.13 Give the oxidation states of these elements when they are d^0 or d^{10}: (a) Ta, (b) Hf, (c) Ag, (d) Sc, (e) Tc.

21.14 Which is the better oxidizing agent of each pair? (a) CrO_4^{2-} or WO_4^{2-}, (b) Cr^{3+} or Ni^{3+}.

21.15 Write a balanced ionic equation for the reaction that results in each observation: (a) SO_2 gas turns the orange solution of acidified $Cr_2O_7^{2-}$ to the green Cr^{3+}

(actually hydrated). (b) H_2S gas decolorizes the purple solution of MnO_4^-, leaving the brown-black precipitate of MnO_2 mixed with yellow flecks. (c) Solid MnO_2 is dissolved by acidified H_2O_2, leaving a pale pink solution of Mn^{2+} while a gas is evolved. (d) Fe dissolves in dilute hydrochloric acid to give a gas.

21.16 What is an interstitial hydride? How does it differ from a hydride such as AlH_3?

21.17 Explain how platinum catalyzes the hydrogenation of the C=C group to the —CH—CH— grouping.

21.18 In terms of the principle of charge dispersal (section 14.9), account for the fact that permanganate, MnO_4^-, is a weaker base than perrhenate, ReO_4^-. Which is the stronger acid, $HMnO_4$ (unstable) or $HReO_4$?

21.19 Why do most transition-metal ions have several oxidation states, whereas many representative elements have only one?

COORDINATION COMPOUNDS

21.20 Define (a) coordination compound, (b) ligand, (c) chelate, and (d) bidentate.

21.21 What structural feature is common to all ligands? What structural feature do all transition elements possess that enables them to form coordination compounds?

21.22 What are the coordination number and oxidation number of the central atom in (a) $[PdBr_6]^{2-}$, (b) $[FeF_4]^{2-}$, (c) $[CuCl_2]^-$, (d) $(NH_4)_2[OsBr_6]$, (e) $K_2[VO_2F_3]$, and (f) $[Ti(H_2O)_6]Cl_3$?

21.23 What are the coordination number and the oxidation number of the central atom in (a) $Na[Co(CO)_4]$, (b) $K_4[Ni(CN)_4]$, (c) $K_2[Ni(CN)_4]$, (d) $K_2[NiF_6]$, and (e) $[FeBr(H_2O)_5]^{2+}$?

21.24 What is the coordination number of Ni in each complex? (a) $[Ni(NH_3)_6]^{2+}$; (b) the complex ion formed in aqueous solution when 3 moles of ethylenediamine, $\ddot{N}H_2—CH_2—CH_2—\ddot{N}H_2$, combine with 1 mole of Ni^{2+}.

21.25 Which of the following ligands are capable of forming one bond to give structural isomers (are ambident)? N_3^-, OCN^-, Br^-, H_2NOH, $NH_2CH_2CH_2NH_2$, $\left[CH_3C\begin{smallmatrix}{\approx}O\\{\approx}O\end{smallmatrix}\right]^-$, $NH_2CH_2CH_2OH$.

NOMENCLATURE

21.26 Name each compound: (a) $Rb_2[CrF_6]$, (b) $Na_2[MnCl_6]$, (c) $[Mn(H_2O)_2]SO_4$, (d) $KCu(II)[Fe(CN)_6]$, (e) $Cu_3[Fe(CN)_6]_2$, (f) $Na[AuCl_4(H_2O)_2]$.

21.27 Name each compound: (a) $K_3[CoBr_6]$, (b) the 2 structural isomers (not stereoisomers) of $[Co(NH_3)_3(NO_2)_3]$, (c) $[Cr(CO)_6]$, (d) $NH_4[Ag(CN)_2]$.

21.28 Write the formula for each compound: (a) sodium dibromotetracyanocobaltate(III), (b) hexacarbonyl tungsten(0), (c) hexaamminechromium(III) hexachlorochromate(III), (d) tetraamminedichloroplatinum(IV) chloride.

21.29 Write the formula for each compound: (a) pentacarbonyl osmium(0), (b) sodium diaquatetrafluorovanadate(III), (c) potassium diamminediaquadisulfatochromate(III), (d) copper(II) hexacyanoferrate(II), (e) tetraamminecopper(II) sulfate.

WERNER THEORY

21.30 An aqueous solution of the compound $CoN_6H_9O_6$ gives a negative test for NH_3 and NO_2^-. The solution also does not conduct electricity. Write the formula of this coordination compound.

21.31 An aqueous solution of the compound $CrN_6H_{18}Cl_3$ does not react with HCl. The solution has an electrical conductivity corresponding to 4 moles of ions per mole of compound. It yields 3 moles of AgCl per mole of compound. Write the formula of this coordination compound.

21.32 A compound contains 21.35% Cr, 28.76% N, 6.209% H, and 32.68% Cl by mass. It does not react with HCl. On reaction with $AgNO_3$, it gives 2 moles of AgCl per mole of compound. It has an electrical conductivity corresponding to 3 moles of ions per mole of compound. Write the formula of the compound.

SHAPE

21.33 What is the shape of a 2-, 4- (two possibilities), and 6-coordinate complex? Give an example of each.

21.34 Draw the *cis-trans* isomers of $[Co(NH_3)_2Cl_4]^-$ and identify each one.

21.35 (a) Draw the geometric isomers for $[Co(NH_3)_3Cl_3]$. (b) Draw three geometric isomers for $[Co(NH_3)_3Cl_2Br]$.

21.36 (a) $[CdBr_2Cl_2]^{2-}$ has no isomers. What is its shape? (b) There are two isomers of $[NiBr_2Cl_2]^{2-}$. What is its shape?

21.37 Explain why there are more than two isomers of square-planar $[Pt(NH_3)_2(SCN)_2]$. How many are possible?

BONDING, COLOR, MAGNETIC PROPERTIES

21.38 Which of these compounds are colorless? TiO_2, Cr_2O_3, TaF_5, $Fe(NO_3)_2$, YCl_3, $AuCl_3$, AuCl.

21.39 Does a violet or yellow compound absorb the higher-energy photon?

21.40 What is the difference between an inner and outer octahedral coordination complex?

21.41 (a) Use the VB concept to predict the electron configuration of the bonded atom in (1) $[Cu(CN)_2]^-$, (2) $[Cu(NH_3)_4]^{2+}$, (3) $[CuF_6]^{3-}$, (4) $[Ni(NH_3)_6]^{2+}$, and (5) $[VCl_6]^{3-}$. (b) Could $[Cu(NH_3)_4]^{2+}$ be square-planar? Explain. (c) Could $[CuF_6]^{3-}$ be an inner complex? Explain.

21.42 (a) Use the VB concept to determine the electron distribution for the bonded Cr in $[Cr(NH_3)_6]^{2+}$, assuming it uses (1) d^2sp^3 HO's; (2) sp^3d^2 HO's. (b) What could you experimentally test for in order to distinguish between the two hybrid conditions?

21.43 For which *d*-state (d^n) cations would the d^2sp^3 and sp^3d^2 octahedral coordination complexes have the same magnetic properties?

21.44 $[V(CN)_6]^{3-}$, $[VF_6]^{3-}$, and $[V(H_2O)_6]^{3+}$ are octahedral coordination-complex ions. (a) Predict the order of increasing energy of the photons they absorb. (b) Indicate the general direction of the change in color in going from the complex ion that absorbs the lowest-energy photon to the one that absorbs the highest-energy photon. See Table 21.9, section 21.11.

21.45 (a) Arrange these ions in the order of increasing splitting-energy Δ: $[Cu(NH_3)_4(H_2O)_2]^{2+}$, $[Cu(H_2O)_6]^{2+}$, $[Cu(NH_3)_2(H_2O)_4]^{2+}$, $[Cu(NH_3)(H_2O)_5]^{2+}$, and $[Cu(NH_3)_3(H_2O)_3]^{2+}$. (b) Arrange them in the order of increasing wavelength of absorbed light.

21.46 $[Ti(H_2O)_6]Cl_3$ is violet. What is a likely color of $Na_3[TiF_6]$? See Table 21.2 (section 21.1).

21.47 Give the d electron configurations for a small-Δ and a large-Δ octahedral complex of (a) V(IV) (d^1) and (b) Ni(III) (d^7).

21.48 Which central atom in an octahedral complex is more likely to be paramagnetic: (a) Cr(IV) (d^2) (small Δ) or Cr(IV) (large Δ); (b) Ni(IV) (d^6) (small Δ) or Ni(IV) (large Δ)?

21.49 Give the distribution of the d^6 electrons of the central atom according to the crystal field concept for (a) $[Co(NH_3)_6]^{3+}$, diamagnetic; (b) $[CoF_6]^{3-}$, paramagnetic.

21.50 In the tetrahedral 4-coordinate complex the four ligands approach the cation between, rather than along, the x, y, and z axes. Consequently, the CF concept suggests that the five d orbitals are split into two groups. The higher-energy group has three d orbitals and the lower-energy group has two d orbitals. (a) Place the five d orbitals in the two groups. (b) Give the distribution of the d electrons of Co in the tetrahedral $[CoCl_4]^{2-}$ ion.

21.51 In the square-planar 4-coordinate complex, the four ligands approach along the x and y axes. Which d orbital has the highest energy as a result? Which d orbital has the next-highest energy?

STABILITY

21.52 With the aid of an enthalpy diagram, explain why reactants that are thermodynamically stable need not be kinetically stable. (See section 19.7.)

21.53 Draw an enthalpy diagram for a reaction in which the reactants are both thermodynamically and kinetically (a) unstable; (b) stable. (See section 19.7.)

21.54 $[Fe(CN)_6]^{3-}$ and $[Fe(CN)_6]^{4-}$ are both thermodynamically stable with respect to the central atom and cyanide ion. $[Fe(CN)_6]^{3-}$ is very poisonous, but $[Fe(CN)_6]^{4-}$ is not. Offer an explanation.

21.55 Given this information for aqueous solutions:

$$Ni^{2+} + 2NH_3 \rightleftharpoons [Ni(NH_3)_2]^{2+} \qquad K = 1.0 \times 10^5$$

$$Ni^{2+} + 2en \rightleftharpoons [Ni(en)_2]^{2+} \qquad K = 7.2 \times 10^{13}$$

("en" stands for ethylenediamine.) Which ligand should be more effective in dissolving insoluble Ni^{2+} compounds? Explain.

APPLICATIONS

Consult Table 21.6 when necessary.

21.56 How many binding sites are there in (a) heme of hemoglobin and (b) vitamin B_{12}?

21.57 Write a balanced ionic equation for the solution of zirconium in a mixture of HF and HNO_3. ($NO_2(g)$ is one of the products.) Why doesn't Zr dissolve in either acid individually?

21.58 Oxalic acid, HO—C(=O)—C(=O)—OH, ($H_2C_2O_4$), a bidentate ligand, dissolves iron rust, Fe_2O_3. Write an ionic equation for the reaction.

21.59 Write a balanced ionic equation for the leaching of Ag.

21.60 Write a balanced ionic equation for the solution of gold in *aqua regia*. (NO_2 is one of the products.)

21.61 Write balanced ionic equations for the following reactions. (Assume all transition-metal complexes are 4-coordinate.) (a) NiS(c) dissolves in excess KCN. (b) Nickel(II) sulfate and NH_3 in water gives a precipitate that dissolves in excess NH_3. (c) Addition of concentrated hydrochloric acid to a solution of copper(II) sulfate (blue) gives a bright green solution.

SELF-TEST

21.62 (10 points) Select the most appropriate lettered item for each numbered item.

__________ 1. weak field
__________ 2. *cis-trans* (geometric) isomerism
__________ 3. labile complex
__________ 4. *trans* isomer
__________ 5. inert complex
__________ 6. triad
__________ 7. crystal-field splitting energy
__________ 8. strong ligand
__________ 9. coordination number
__________ 10. transition metals

(a) class of elements whose common oxidation states are characterized by atoms having partly filled d subshells; (b) number of atoms attached to the central atom of a species; (c) three periodic groups whose chemistry is similar across the period; (d) complex that reacts very slowly; (e) square-planar complex containing two different pairs of ligands in which the like groups are diagonally opposite; (f) complex characterized by a small crystal-field splitting energy so that electron pairing in the d orbitals is discouraged; (g) Lewis base whose strong interaction with a transition-metal ion causes a splitting of the d orbitals of the ion into two groups of significantly different energies; (h) energy difference between the two groups of d orbitals of a complexed transition-metal ion; (i) existence of two or more compounds with the same molecular formula but different physical and chemical properties because of dissimilar spatial arrangements of the ligands about the central metal ion; (j) complex that undergoes rapid reactions.

21.63 (24 points) Fill in the blanks. Consult the periodic table when necessary. (a) The maximum oxidation state expected for tantalum, Ta, is ________. (b) The most stable oxidation state for tungsten, W, is ________. (c) The number of unpaired electrons expected in an octahedral weak-field coordination complex of the d^7 ion Co^{2+} is ________. (d) *Cis-trans* (geometric) isomerism is *not* observed for the ________-shaped 4-coordination complex. (e) The charge on the cobalt(II) coordination complex $[CoCl_2Br_2]^n$ is ________. (f) The coordination number of Os in $(NH_4)_2[OsBr_6]$ is ________. (g) The oxidation state of V in $K_2[VO_2F_3]$ is ________. (h) The molecular formula for the colorless oxide of titanium, Ti, is ________. (i) The number of geometric isomers for the square-planar coordination complex $[PtCl_2BrI]^{2-}$ is ________. (j) The number of unpaired electrons in a d^2sp^3-coordination complex of a d^5-transition metal is ________. (k) The oxide of the Group-6B elements giving the most acidic oxo acid is ________. (l) Among the Group-6B elements in their highest oxidation state, the oxo anion that is the weakest oxidizing agent is ________.

21.64 (12 points) Two ionic isomers have the molecular formula $K_2PtCl_2Br_2$. No precipitate with Ag^+ is formed. (a) Give the formula of the anion and show its charge. (b) Give the oxidation state of Pt. (c) Name the compound. (d) Write the structural formulas for the two isomers and identify each one.

21.65 (6 points) Write a net ionic equation for the solution of $Cu(OH)_2$ in an ammonia solution.

21.66 (4 points) Predict whether each oxide is predominantly an acidic or basic anhydride: NiO, Ag_2O, CrO_3, Re_2O_7.

21.67 (12 points) $[Cu(H_2O)_4]^{2+}$ is blue-green, $[Cu(NH_3)_4]^{2+}$ is violet, and $[Cu(NO_2)_4]^{2-}$ is yellow. (a) Which absorbs the highest-energy photons and which absorbs the lowest-energy photons? (b) Explain the change in color in terms of crystal field theory.

21.68 (14 points) Draw the energy-level diagram for an octahedral complex of (a) $Ni^{3+}(d^7)$ with a large Δ; with a small Δ; (b) $Cu^{2+}(d^9)$ with a small Δ; with a large Δ. How many unpaired electrons are in each ion?

21.69 (12 points) Explain in terms of thermodynamic and kinetic stability: (a) $[HgI_4]^{2-}$ has a dissociation constant of 5.3×10^{-31}. When radioactive iodide, $^{131}I^-$, is added to a solution of this coordination-complex ion, $[Hg^{131}I_4]^{2-}$ can be isolated within a short period of time. (b) $[Fe(CN)_6]^{4-}$ has $K_{diss} = 10^{-35}$. After 100 hours in an aqueous solution of $^{14}CN^-$, less than 2% of the cyanide in the complex has exchanged.

21.70 (6 points) Show the outermost electron configuration as predicted by the VB concept for the transition element in (a) $[VCl_6]^{3-}$ and (b) square-planar $[Ni(NH_3)_4]^{2+}$.

22

METALLIC BONDING AND METALLURGY

From the discussion in Chapter 7, we can generalize about bonding between types of elements:

$$\text{metal} + \text{nonmetal} \longrightarrow \text{ionic bond} \quad \text{(by transfer of electrons)}$$

$$\text{nonmetal} + \text{nonmetal} \longrightarrow \text{covalent bond} \quad \text{(by sharing of electrons)}$$

What about the third combination?

$$\text{metal} + \text{metal} \longrightarrow \text{metallic bond} \qquad \text{(but how?)}$$

Since metallic elements do not accept electrons, they cannot be negative ions with ionic bonds. Furthermore, since metallic elements have a small number of valence electrons (usually three or fewer), they cannot acquire an octet of electrons by sharing electrons with each other.

In this chapter we discuss the electronic nature of metallic bonds in terms of the molecular orbital (MO) concept. We will see how this concept led to the development of semiconductors—the raw materials of the present age of computers. We will also present an overview of the metallurgical processes for obtaining pure metals from their naturally occurring compounds (minerals) and a discussion of alloys.

22.1 METALLIC PROPERTIES

Before we can discuss the models of metallic bonding, we will first summarize the typical properties of metals. These properties must be explained by any concept that is proposed.

1. *Metals have lustrous, shiny surfaces because they reflect light.* Exceptions are spongy or finely divided forms of metals, such as platinum, which appear gray to black. Metals in these forms have irregular surfaces that trap light.
2. *Metals have high electrical and thermal conductivities.* The electrical conductivity decreases with increase in temperature—a behavior peculiar to metals. Furthermore, the passage of current does not produce any chemical change in a metal.
3. *Metals are malleable* (can be hammered) *and ductile* (can be drawn into a wire). Especially significant is the fact that the crystalline structure of metals is preserved under moderate deformations.

4. *Metals display photoelectric and thermionic effects.* Both effects involve the emission of electrons. Photoelectric emissions (section 5.1) occur when metals are irradiated. Thermionic emissions occur when metals are heated.

5. *Many metals have high densities and high boiling points* (Table 7.1, section 7.1). The densest of all substances are the metals.

6. *Metals react chemically to acquire a positive oxidation number—they lose electrons.*

22.2 ELECTRON SEA MODEL

These properties are consistent with the fact that metallic atoms readily lose electrons. This led to the early "electron sea" model of metallic bonding, which pictures the metal as being a fixed network of metallic cations submerged in a "sea" of delocalized electrons (Figure 22.1). The bonding electrons, freed from their atoms, move freely throughout the entire piece of metal, forming the "electron sea."

The electron sea model explains electrical conductivity as follows: An electron enters the "sea" at one end of the piece of metal. It then starts a sequence of electron repulsions through the sea along the length of the metal. Finally an electron is ejected from the metal at the other end. The entire conduction process is practically instantaneous. However, the electron sea model is too simple. For example, it cannot account for the behavior of semiconductors. Therefore, this model has been superseded by a more sophisticated one.

22.3 MO CONCEPT OF METALLIC BONDING—BAND THEORY

The usual symbol N_A is here simplified to N.

Lithium ($1s^2$, $2s^1$), the first metallic element in the periodic table, is a convenient example for this discussion. We will start with the covalent molecule Li_2 and then see the consequences of adding more Li atoms in stages until we get a mole of lithium metal, Li_N. We have already analyzed the molecular orbitals of Li_2 (section 8.7). The ground-state representation showing all the molecular orbitals, not just the valence MO's, is

$$2Li(1s^2 2s^1) \longrightarrow Li_2(\sigma_{1s})^2(\sigma^*_{1s})^2(\sigma_{2s})^2(\sigma^*_{2s})^0$$

6 e⁻'s in 4 AO's — 6 e⁻'s in 4 MO's

Although the σ^*_{2s} molecular orbital is empty, it nevertheless has the capacity to house two electrons. Empty MO*'s are significant in explaining metallic bonding. The relative energy levels of these four molecular orbitals are shown in Figure 22.2a.

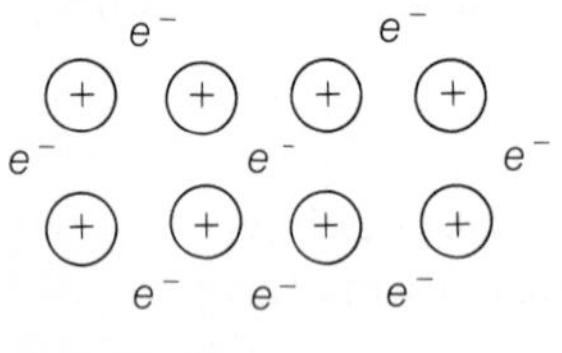

We next consider Li_4. Four Li atoms, each with two interacting AO's, give rise to eight molecular orbitals—two each of σ_{1s}, σ^*_{1s}, σ_{2s}, and σ^*_{2s}:

$$4Li(1s^2 2s^1) \longrightarrow Li_4[\underbrace{2(\sigma_{1s})^2 2(\sigma^*_{1s})^2 2(\sigma_{2s})^2}_{\text{filled}}\underbrace{2(\sigma^*_{2s})}_{\text{empty}}]$$

12 e⁻'s in 8 AO's — 12 e⁻'s in 8 MO's

e⁻ e⁻
(+) (+) (+) (+)
e⁻ e⁻ e⁻
(+) (+) (+) (+)
e⁻ e⁻ e⁻

FIGURE 22.1
"Electron sea" model of metallic bonding. A plus sign represents a metallic cation. In the actual metal, the cations are practically touching—they are closely packed.

With Li_4 two important new effects emerge:

- The molecular orbitals of each kind do *not* have the same energy (Figure 22.2b). For example, one of the σ_{2s} MO's has a slightly higher energy than the other σ_{2s} MO.

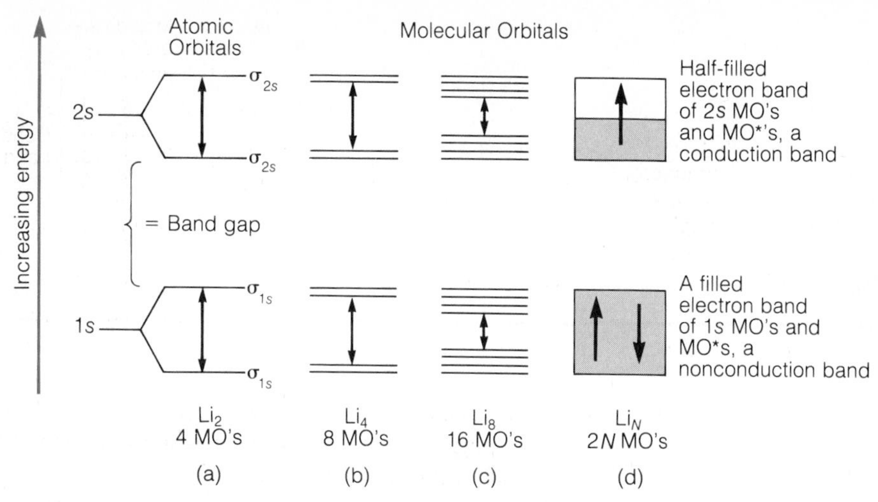

FIGURE 22.2 Formation of molecular orbitals from overlap of the 1*s* and 2*s* orbitals of lithium. ↕ represents the energy gap between bonding and antibonding molecular orbitals within the same shell. (a) Li_2. (b) Li_4. (c) Li_8. (d) Li_N (1 mole). Note that ↕ decreases as the number of Li atoms increases.

- The energy difference between the groups of bonding and antibonding molecular orbitals within a given shell, indicated by ↕ in Figure 22.2, diminishes as the number of atoms involved in bonding increases.

Any sample of Li large enough to be studied is metallic; it could be much less than a mole.

These effects are further shown by Li_8 (Figure 22.2c). For Li_8, the energies of the MO's and MO*'s within the same shell are closer than they are for Li_4. In a large quantity, such as a mole, Li_N, these bonding and antibonding molecular orbitals become so close in energy that they form a continuous sequence called a **band**. Lithium metal actually forms two bands. A lower-energy 1*s* band is composed of 1*s* inner-shell molecular orbitals. A higher-energy 2*s* band is composed of 2*s* valence-shell molecular orbitals. The energy difference, or gap, between these two bands is called a **band gap** (Figure 22.2d). The 1*s* band is completely filled because it is composed of filled molecular orbitals. Such a completely filled band is called a **nonconduction band**. Electrons in filled shells of an atom generally are found in nonconduction bands. These electrons remain where they are and are not involved in conduction. The 2*s* band is made up of closely packed 2σ-bonding MO's and $2\sigma^*$-antibonding MO*'s. Since there is only one electron in the 2*s* atomic orbital of each Li atom, only half of the total number of molecular orbitals generated from these AO's are filled. An incompletely filled band is called a **conduction band**. In summary,

The valence electrons are in the conduction band.

$$N\ 1s \text{ AO's with } 2N\ e^-\text{'s} \longrightarrow N \text{ molecular orbitals with } 2N\ e^-\text{'s}$$

(All molecular orbitals are filled, composing a nonconduction band.)

$$N\ 2s \text{ AO's with } N\ e^-\text{'s} \longrightarrow N \text{ molecular orbitals with } N\ e^-\text{'s}$$

(Half of the molecular orbitals are filled, composing a conduction band.)

Heavy arrows are used to indicate the electron population of a band (Figure 22.2d). No arrow means the band is empty; one arrow means it is half-filled, and two arrows—pointing in opposite directions—mean it is filled. Only electrons in partially filled shells are found in conduction bands. *Electrons in conduction bands are responsible for metallic properties.* The difference in energy between the conduction band and the nonconduction band is called the **band gap**. No electron has an energy that would place it within the band gap. Furthermore,

For this reason the band gap is also called the **forbidden zone**.

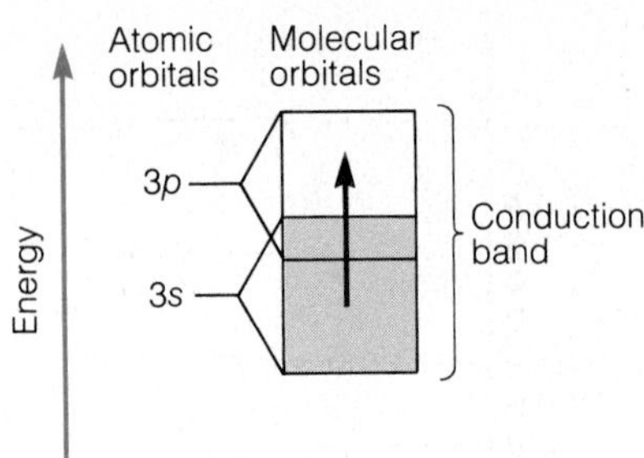

FIGURE 22.3 Formation of a conduction band from overlap of the filled 3*s* and empty 3*p* atomic orbitals of a mole of Mg atoms.

under ordinary conditions, electrons in a nonconduction band *cannot* absorb enough energy to enter the conduction band. This MO model of metallic bonding is also called the **band theory**.

EXAMPLE 22.1 Which AO's of Na form the molecular orbitals that compose the conduction band? To simplify matters, do not use the 3*p* AO's.

ANSWER The electron configuration of Na is $1s^22s^22p^63s^1$. Since the AO's in the first two shells are filled, the MO's they generate are filled. The σ_{3s} and σ^*_{3s} molecular orbitals formed from the half-filled 3*s* AO are half-filled and make up the conduction band. ■

PROBLEM 22.1 Which AO's of K form the molecular orbitals that compose the conduction band? Do not use the 3*d* or 4*p* orbitals. □

For the Group-2 elements such as Mg ($1s^22s^22p^23s^2$), all bands derived from inner shells 1 and 2 are filled. The band comprising the molecular orbitals from the filled 3*s* AO's is also filled. From these filled bands we might predict that Mg would not be metallic, but Mg is a typical metal. Such a prediction overlooks the fact that the molecular orbitals from the 3*s* and 3*p* AO's are very close in energy and blend into a single, partially filled conduction band (Figure 22.3). This electronic structure is consistent with the behavior of Mg as a metal. Except for boron, the elements of Group 3 demonstrate typical metallic bonding. The conduction band is composed of molecular orbitals arising from combination of *s* and *p* AO's. (Elementary boron exists as covalent molecules with the molecular formula B_{12}.) In Groups 4 and 5, only the heaviest members, Sn and Pb in 4 and Bi in 5, demonstrate metallic bonding.

METALLIC PROPERTIES IN TERMS OF MO BAND THEORY

How does an incompletely filled band account for metallic properties? In an incompletely filled band small amounts of energy, such as electric potentials, excite electrons to slightly higher energy, unoccupied molecular orbitals *within* the conduction band. These excited electrons, unlike electrons in filled bands, are free to move in the direction of the applied potential. As a result, the metal conducts electricity. A rise in temperature increases vibration of the positive ions in the crystal lattice; such vibrations interfere with the movement of electrons. Consequently, electrical conductivity of metals *decreases* as the temperature *increases*—a characteristic property of metals.

Electrons in conduction bands can also absorb heat. The mobility of these thermally excited electrons accounts for the high thermal conductivity of metals.

Salts are not ductile. The strongly attracting, oppositely charged ions can't easily move away from each other.

The absorption and reemission of photons of visible light by conduction electrons account for the high reflectivity of metals. Metals are ductile and malleable because under mechanical stress the cations of the crystal can move past each other with very little resistance. This is so because there is no change in the bonding environment. Before and after the stress, a cation is still surrounded by electrons.

22.4 INSULATORS AND SEMICONDUCTORS

Substances that offer great resistance to the passage of electric current are called **insulators**. Typical examples are glass and rubber. The attempt to force a current through an insulator may destroy the insulator before conduction is noticed. Thus, if we tried to push an electric current through, say, a rubber stopper by turning up the voltage, the stopper would go up in smoke before it would allow much current to pass. Recall that covalently bonded materials are insulators. Salts at room temperature are also insulators, but they become conductors in the molten state (section 7.1).

Some solid salts are conductors at high temperatures.

In terms of the MO band theory, what is the difference between a conductor and an insulator? Whereas in a metal the conduction band is partially filled with electrons (Figure 22.4a), *in an insulator the corresponding band has no electrons.* Also, the band gap is very large, which prohibits the excitation of electrons into the conduction band when a voltage is applied (Figure 22.4b). With no electrons in the molecular orbitals of the conduction band, there is no conduction.

Also called *semimetals* or *metalloids.*

"Semiconductor" has become a household word. Semiconductors are used in numerous "solid-state" (no vacuum tubes) electronic devices such as transistors, lasers, solar-energy cells, and photovoltaic cells. We would not have the convenient

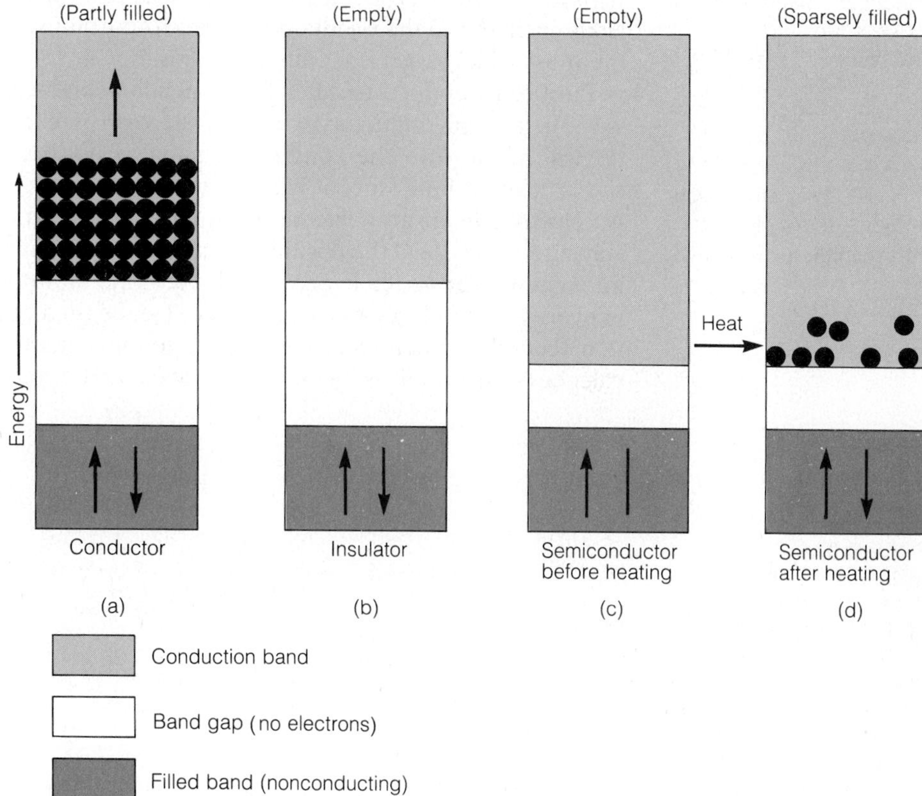

FIGURE 22.4 Band theory of solids. (a) Conductor. (b) Insulator. (c) Semiconductor before being heated and (d) after being heated (nonconduction band is no longer filled).

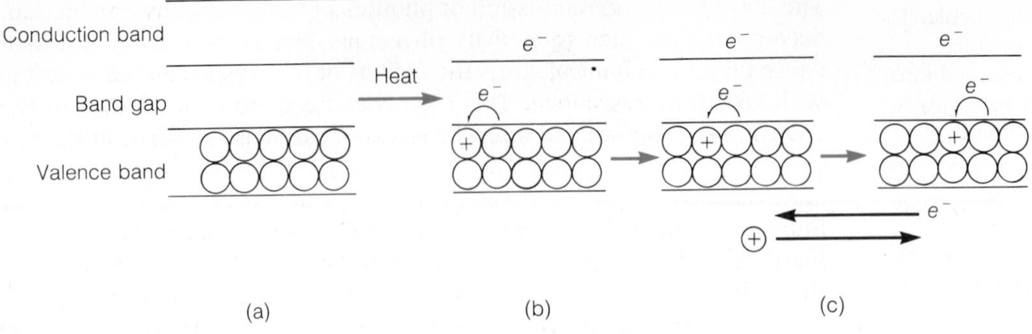

FIGURE 22.5 MO band concept for pure semiconductors. (a) Ground state. (b) Excitation of an e^- from valence band to conduction band with formation of a hole, shown as ⊕, in the valence band. (c) Conduction in valence band; electron movement from right to left is equivalent to movement of hole, ⊕, from left to right. Electrons in the conduction band also move freely during conduction of electricity. Electrons moving in valence band do not enter band gap.

In some cases absorption of light excites electrons into the conduction band.

transistor radios, hand calculators, computers, video cassette recorders, and modern television sets were it not for semiconductors. Nor would we have our modern dry photocopying machines or self-winding watches.

How do semiconductors differ from conductors (metals) and insulators (nonconductors)? Like insulators (Figure 22.4b), semiconductors have empty conduction bands. But unlike insulators, which have a large band gap, *semiconductors have a narrow band gap* between the filled nonconduction band and the "empty" conduction band (Figure 22.4c). The nonconduction band of semiconductors is often called the **valence band** because it is composed of valence AO's. Very small amounts of heat energy can excite some electrons into the conduction band (Figure 22.4d), where they can move throughout the solid and thus cause conduction. For this reason, *conductivity of semiconductors,* unlike that of metals, *increases with temperature.* The higher the temperature, the greater is the number of electrons excited into the conduction band. Important semiconductors are silicon, Si, the most widely used one, and germanium, Ge.

Pure semiconductors differ from metallic conductors in another fundamental way. In a semiconductor, when an electron is excited from the valence band (Figure 22.5a) into the conduction band, a positive site remains. This positive site is called a **hole** since it is a site where an electron is missing (Figure 22.5b). An electron from an adjacent atom can then be attracted to the positive hole, leaving a new hole at the position it previously occupied. The shifting of an electron into a newly formed hole occurs from atom to atom (Figure 22.5c). The hole acts in every respect as a carrier of charge. Hence, in a pure semiconductor, conduction occurs by simultaneous and opposite movement of holes and electrons in the valence band as well as by movement of excited electrons in the conduction band.

Missing e^- leaves a hole.

Extra e^-

Pure Si (a)

p-type (b)

n-type (c)

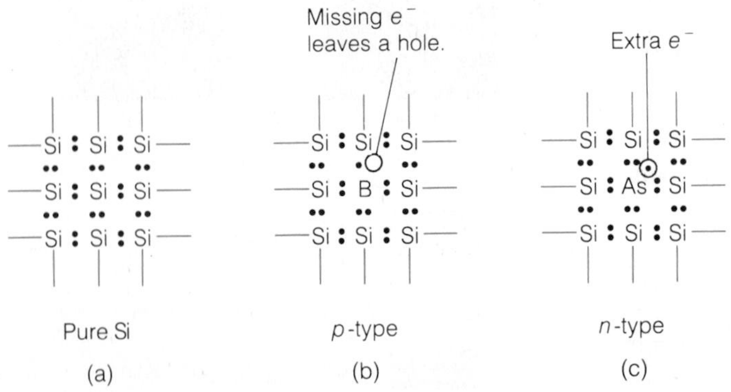

FIGURE 22.6 MO band concept for doped semiconduction. (a) Pure substance; Si. (b) *p*-Type; B in Si. ○ is the electron-deficient site. (c) *n*-Type; As in Si. ⊙ is an extra e^-. Dashes represent bonds to other Si or doping atoms.

FIGURE 22.7
Conduction in *p*-type semiconductor (Si doped with B). The curved arrow, ↼, shows arbitrarily chosen movement of an e^- from right to left, which causes the ⊕ to move simultaneously from left to right.

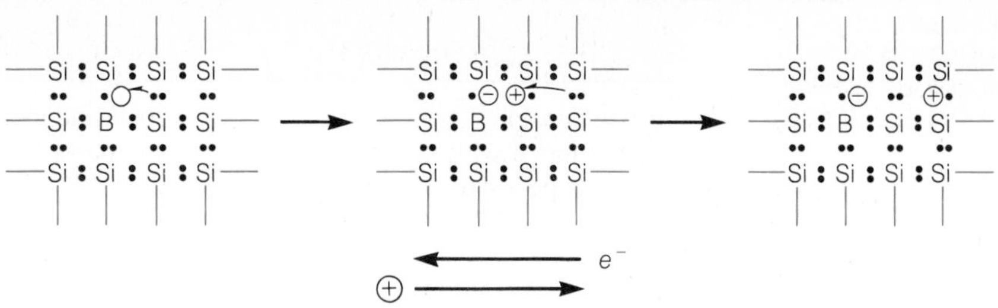

Dope is derived from the Dutch word for "dip," and refers generally to a material that is added to something else.

A few B atoms go a long way.

Semiconduction can be enhanced by adding appropriate impurities ("doping") to an ultrapure form of a semiconducting material such as silicon (Figure 22.6a). For example, when one boron (B) atom is added per million Si atoms, the conductivity of the silicon is multiplied about 100,000 times. A boron atom differs from an Si atom in having one less valence electron. When ·Ḃ· is inserted into the ·Si· covalent network, a hole is created (Figure 22.6b). The presence of this hole initiates conduction in the valence band (Figure 22.7a). The presence of B atoms obviates the need to excite electrons into the empty conduction band. Nearly all conduction then occurs in the valence band. Adding a Group-3 atom (B) changes the valence band of pure Si or Ge, a Group-4 element, into a conduction band. This type of semiconductor is designated a ***p*-type**, *p* for "positive," because the charge carriers are positively charged holes.

Traces of Group-5 elements, such as As, are also used to dope Si and Ge. Each added atom of As has one more electron than is needed to bond to Si (Figure 22.6c). These extra electrons go into a narrow energy region just below the energy level of the empty conduction band. At room temperature many of these electrons are easily excited into the conduction band, enabling conduction to occur. These are called ***n*-type** semiconductors, because the dominant charge carrier is the electron (negative charge) (Figure 22.8).

Conduction in *n*-type semiconductors occurs only in the conduction band—none occurs in the valence band because there are no holes present.

High-purity GaAs is best made on space shuttles.

A widely studied semiconductor is gallium arsenide, GaAs. Circuits using chips made from GaAs are five times faster and operate over a wider temperature range than circuits using Si chips. Compared to Si, Ga, a Group-3 element, has a hole and As, a Group-5 element, has an extra electron. The extra electrons of As fill the holes of Ga. The bonding situation is then the same as in Si. Unfortunately, GaAs is hazardous to make because of the toxicity of arsenic compounds.

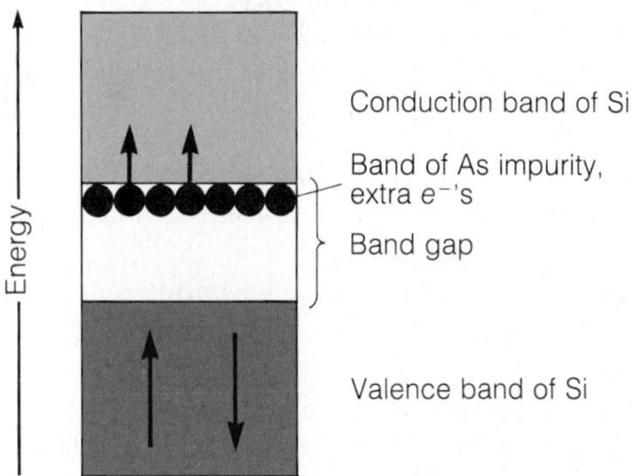

FIGURE 22.8
n-Type semiconductor. Si is doped with As. (Arrows show excitation of excess e^-'s of As into conduction band of Si.)

TABLE 22.1
CONDUCTION IN SEMICONDUCTORS (GROUP-4 ELEMENTS)

TYPE OF SEMICONDUCTOR	CHARGE CARRIER	BANDS WHERE CONDUCTION OCCURS
Pure (intrinsic)	holes and electrons	valence and conduction
p-Type (doped with Group-3 element)	holes and electrons	valence
n-Type (doped with Group-5 element)	electrons	conduction

Table 22.1 summarizes the three types of semiconductors. The transistor, which initiated the recent electronic revolution, is made of *p*-type and *n*-type semiconductors. (Box 22.1 is a discussion of an application of combined *p*-type and *n*-type semiconductors.)

EXAMPLE 22.2 What type of semiconductor is formed when gallium, Ga, is added to germanium, Ge?

ANSWER Ga is a Group-3 element and has one electron less than Ge, a Group-4 element. A *p*-type semiconductor is formed. ■

BOX 22.1
SOLAR (PHOTOVOLTAIC) CELLS

Much research is underway to create solar cells for harnessing the Sun's energy. A solar cell directly converts sunlight into electricity. One kind of solar cell utilizes *n*- and *p*-type semiconductors. A thin layer of a *p*-type semiconductor is pressed together with a layer of an *n*-type semiconductor. In the dark, at the junction between the two layers (the *p–n* junction), some electrons from the *n*-type move into some of the holes of the *p*-type. As the excess of electrons appear in the holes of the *p*-type layer, positive holes appear in the *n*-type layer. The further flow of electrons into the *p*-type layer stops. Thus an equilibrium is reached in which some extra electrons are trapped in the *p*-type layer.

When sunlight hits the upper *p*-type surface, the following sequence of electron movements (⟶) occurs:

> Excess electrons in *p*-type layer above the junction are excited ⟶ *n*-type layer ⟶ wire connected to the *n*-type layer ⟶ external circuit doing useful work ⟶ to wire connected to *p*-type layer ⟶ *p*-type layer.

In this way sunlight falling on the cell induces an electric current to flow through the external circuit. This is why the cell is also called a **photovoltaic cell**.

Individual cells are combined to make solar batteries. Solar batteries are used to supply electric current on a small scale, especially in spacecrafts and hand calculators. The major drawback to their large-scale, commercial use is their high cost. Research goals are to improve the efficiency (now at about 27%) and thereby lower the cost. Solar batteries have been developed to run automobiles, but they are too expensive. A few photovoltaic plants have been built to produce electricity. The world's largest facility, in San Luis Obispo County, California, has a generating capacity of 4.5 megawatts of electricity. We await the time when solar batteries or other solar conversion devices can become sufficiently economical to decrease our dependence on oil, coal, and nuclear fuels.

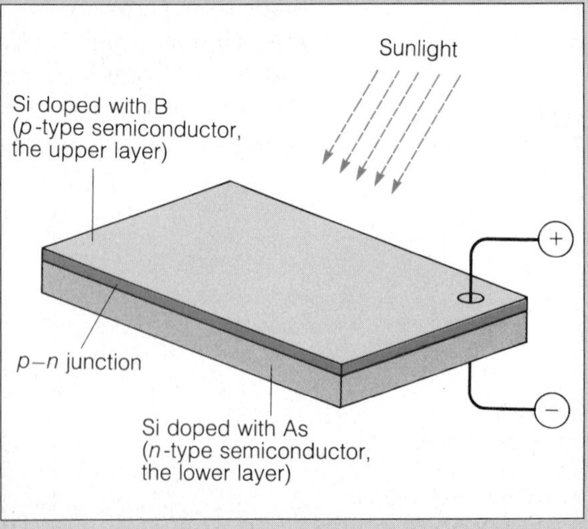

A solar cell.

PROBLEM 22.2 What type of semiconductor is formed when phosphorus, P, is added to Si? □

22.5 EPILOGUE ON METALLIC BONDING

About 10^6 atm is required.

Metallic Xe and S have also been made.

With confidence in the MO band concept, we can take an imaginative step forward. A metallic substance is characterized by a partially filled conduction band formed by the interaction of the AO's of all the atoms in the substance. The main idea is that an extremely large number of AO's produce the same large number of molecular orbitals. The energy levels of some of these molecular orbitals are so closely spaced that they form a continuous, partially filled band (conduction band). Now we come to the bold idea. Theoretically, insulators may be changed to metal-like conductors by pushing together their AO's to give partially filled, very closely packed molecular orbitals. This idea can be extended even to hydrogen. If, by applying high pressures, all the *s* AO's in hydrogen could be squeezed together and made to interact as in metallic Li, metallic hydrogen should form. Attempts to prepare metallic hydrogen have been successful. Astronomers believe that the deep interiors of the largest planets, Jupiter and Saturn, consist of metallic hydrogen.

22.6 METALLURGICAL PROCESSES

Bronze is an alloy of copper and tin.

Table 22.3 is on page 856.

Occasionally steps b and c merge into one step.

Metals are a touchstone of civilization: civilization progressed as the Stone Age evolved into the Bronze Age and then into the Iron Age. Now we are feeling the impact of the Plastics Age. The reasons for the usefulness of metals to humankind can be inferred from their properties (section 22.1).

Most metallic elements do not occur in their elementary (free or native) state in nature. Rather they must be obtained from ores by industrial processes that require energy. **Metallurgy** is the technology of all of these processes. An **ore** contains a metal in an economically exploitable quantity. Many metals occur in compounds that are not useful, such as Al in clay (silicates). The compound that contains the metal is called the **mineral**. Minerals are most commonly oxides, sulfides, or carbonates. These are shown in Table 22.3. Notice that some minerals, such as $CuFeS_2$, Pb_2VO_4Cl, and $FeWO_4$, have more than one metal. The very active metals such as Na and K are found extensively as their chloride salts in seawater and in land deposits that were formerly ancient seas. (The largest land deposit of KCl, 10 billion tons, is in Saskatchewan, Canada.)

Metallurgical processes include (a) purification of the mineral, (b) changing of the mineral to a compound from which the metal can be practically obtained, (c) obtaining the metal, and (d) purification of the metal. An optional fifth step is the addition of other elements to provide the metal with specific desired properties. A metal that is a mixture or compound of elements (metallic or nonmetallic) is called an **alloy**. This step, alloy formation, is exemplified by the conversion of pure iron into alloy steels.

22.7 MINERAL PURIFICATION

An ore is essentially a mixture of minerals and **gangue**—impurities such as sand, clay, and rocks (mostly silicates). After mining, the gangue must be removed from the ore before the metal can be obtained. One process, using a **cyclone separator** (Figure 22.9), is useful when the mineral is more dense than the gangue. The crude ore is pulverized and air-blown into the separator, where it is subjected to a centrifugal force. The denser minerals settle to the bottom of the separator where they are drawn off. At the same time the less dense gangue is removed at the top with the stream of air.

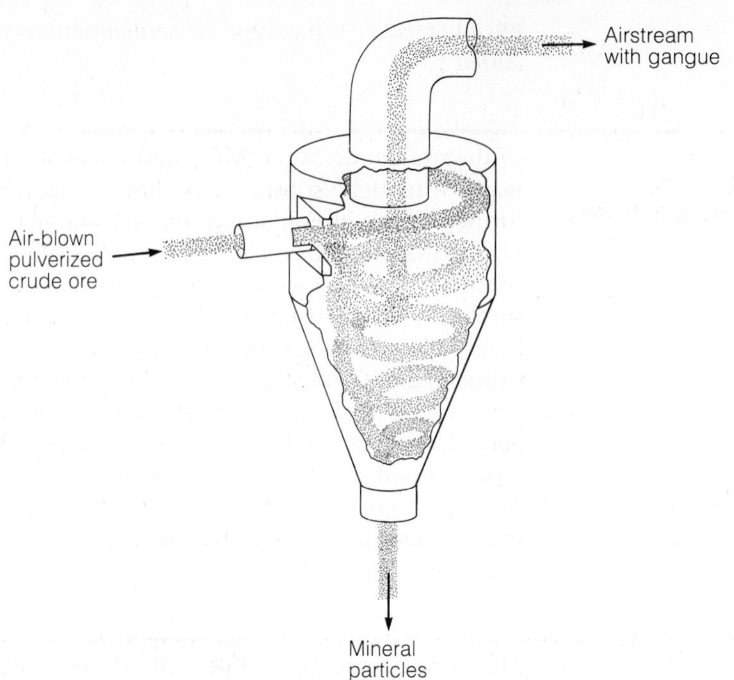

FIGURE 22.9
A cyclone separator.

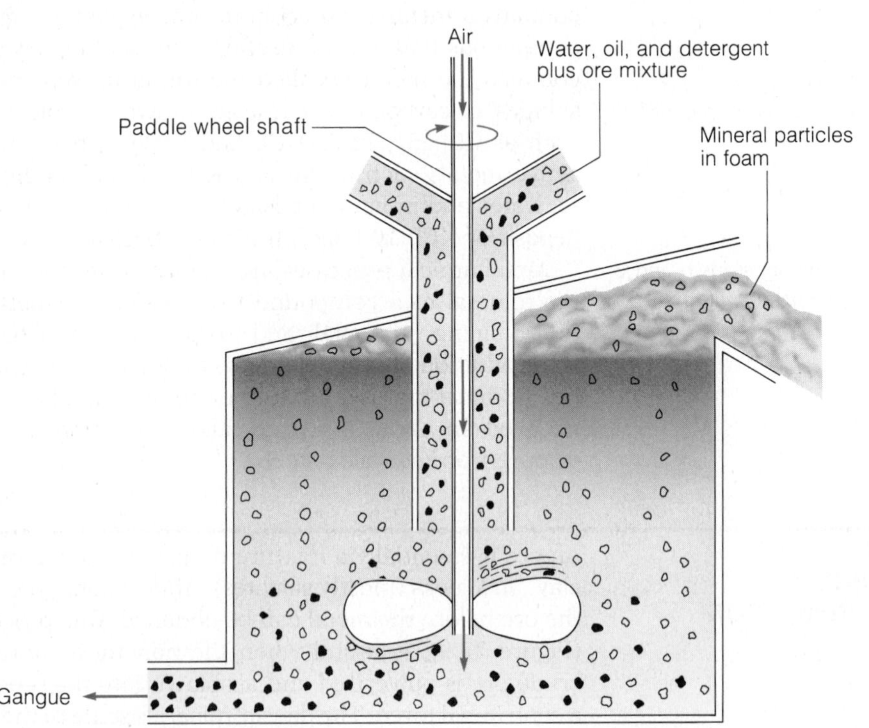

FIGURE 22.10
The flotation process for purifying ore.

The **flotation process** (Figure 22.10) can be used when oil clings to the mineral particles but not to the gangue particles. The pulverized crude ore is fed into a tank containing a water-oil-detergent mixture. A stream of air blown through this mixture causes tiny bubbles to form on the oil-covered mineral particles, which then float to the surface where they are skimmed off. The nonoily particles of gangue settle to the bottom of the tank. The oil is then removed from the mineral particles.

Chemical methods are also used to purify minerals. Leaching of gold was discussed in section 21.10. Another purification method, used for aluminum, will be discussed later in this chapter.

22.8 ISOLATING METALS; THE NEED FOR REDUCTION

The metals used in early times were those found in the elementary state; they included copper, gold, and silver. We now know that metals that occur as elements are those with *positive reduction potentials* (Table 17.1, section 17.7), which means that they do not displace hydrogen from water or from acidic solutions. Using Cu as an example, we assume the following cell reaction:

reduction:	$2H^+ + 2e^- \longrightarrow H_2$	$\mathscr{E}^0 = 0$ V	(cathode)
oxidation:	$Cu \longrightarrow Cu^{2+} + 2e^-$	$\mathscr{E}^0 = 0.3402$ V	(anode)
net reaction:	$Cu + 2H^+ \longrightarrow Cu^{2+} + H_2$	$\mathscr{E}^0_{cathode} - \mathscr{E}^0_{anode} = -0.3402$ V	

The negative value of $\mathscr{E}^0_{cell}$ for the net reaction indicates that the reaction will *not* go as it is written; it does not occur in nature, and therefore elementary Cu survives in an acidic, watery environment. Such metals appear below hydrogen in the activity series (sections 9.4 and 9.5). In addition to Cu, they include bismuth, mercury, silver, gold, and the platinum triad. Although bismuth, copper, mercury, and silver are found as free metals, they are mainly mined as their sulfides. In general, only the metals below silver in the activity series, such as gold and platinum, are mined in the elementary state.

The metals in all of their naturally occurring minerals are in positive oxidation states. The essential chemical process in metallurgy is therefore *reduction*. Refer to Table 17.1 (section 17.7), which lists chemical reductions and their standard electrode potentials, and recall that the reducing agents are the *products* of the reactions shown in the table. The most powerful chemical reducing agents are those at the top of the table (such as Li, K, and Na), and the least powerful are those at the bottom (such as F^- and Au). There is, however, a reducing agent that can be more powerful than anything appearing in the table. That agent is not another chemical, but rather the cathode of an electrochemical cell, to which any desired potential can be applied, using an electric generator as a source. Table 22.2 lists some common metals and relates the $\mathscr{E}^0$ range to the methods of reduction discussed in later sections.

22.9 THERMODYNAMICS OF METALLURGY; ROASTING

The choice of a practical reducing agent cannot be made simply from a consideration of electrode potentials such as are shown in Table 17.1, or from the series of Tables 9.4 and 9.5 (sections 9.2 and 9.4). For one thing, standard state conditions rarely prevail. For example, the $\mathscr{E}^0$ values given in such tables refer to ions in aqueous solutions at 25°C, not a common condition in metallurgy. Cost, too, is a consideration that the $\mathscr{E}^0$ values do not take into account. The physical natures of the reducing agent and its oxidation products and the consequent steps needed to obtain the metal in pure form must also be considered.

TABLE 22.2
STANDARD ELECTRODE POTENTIALS OF COMMON METALS AND METHODS OF REDUCTION

STANDARD ELECTRODE POTENTIAL, volts	METAL	REDUCING AGENT	EXAMPLES
-2.925	potassium, K[†]		
	barium, Ba	electrolysis	$2NaCl \longrightarrow 2Na + Cl_2$
	calcium, Ca		
	sodium, Na		
	magnesium, Mg		
-1.66	aluminum, Al		
	manganese, Mn	aluminum (hot)	$3MnO_2 + 4Al \longrightarrow 3Mn + 2Al_2O_3$
-0.7628	zinc, Zn	carbon	$2ZnO + C \longrightarrow 2Zn + CO_2$
	chromium, Cr	hydrogen	$TiO_2 + 2H_2 \longrightarrow Ti + 2H_2O$
	iron, Fe	carbon monoxide	$PbO + CO \longrightarrow Pb + CO_2$
	cadmium, Cd		
	nickel, Ni		
	tin, Sn		
	lead, Pb		
0.00	hydrogen, H_2[§]		
	copper, Cu		
	antimony, Sb		
	bismuth, Bi		
$+0.851$	mercury, Hg	heat alone	$2HgO \longrightarrow 2Hg + O_2$
	silver, Ag		
	gold, Au	reduction not needed; found in elementary state	
	platinum, Pt		

[†] K^+ is reduced with Na vapor.
[§] H_2 is included as a reference point.

For a general overview of metallurgical processes it will be helpful to consider the free energy relationships involved in obtaining a metal from its oxide. The oxide is chosen because usually this is the compound that is reduced to the metal. Other minerals, such as sulfides and carbonates, are usually first changed to oxides, as will be discussed in more detail later in this section.

These relationships are best revealed by the free energies of formation of the oxides shown by the general form:

$$m\text{M(c)} + \frac{n}{2}\text{O}_2\text{(g)} \longrightarrow \text{M}_m\text{O}_n\text{(c)}(+ \text{ or } -\Delta G_f^0)$$

Thus,

- The more negative is the ΔG_f^0 of the oxide, the more thermodynamically stable is the oxide, and the more difficult it is to reduce it to the metal.
- The more positive is the ΔG_f^0 of the oxide, the *less* thermodynamically stable is the oxide, and the easier it is to reduce it to the metal.

Remember that ΔG^0 is for pure solids and liquids and for gases at 1 atm, but it may refer to *any* temperature.

The role of temperature in metallurgical reductions becomes evident by plotting ΔG^0 versus temperature, T, as in Figure 22.11. The equations for the formation of H_2O, CO_2, and CO are included for reference purposes to solve problems. ΔG^0 values, rather than ΔG_f^0 values, are plotted, because several of the equations given have data for more than one mole of the oxide. Notice these important points:

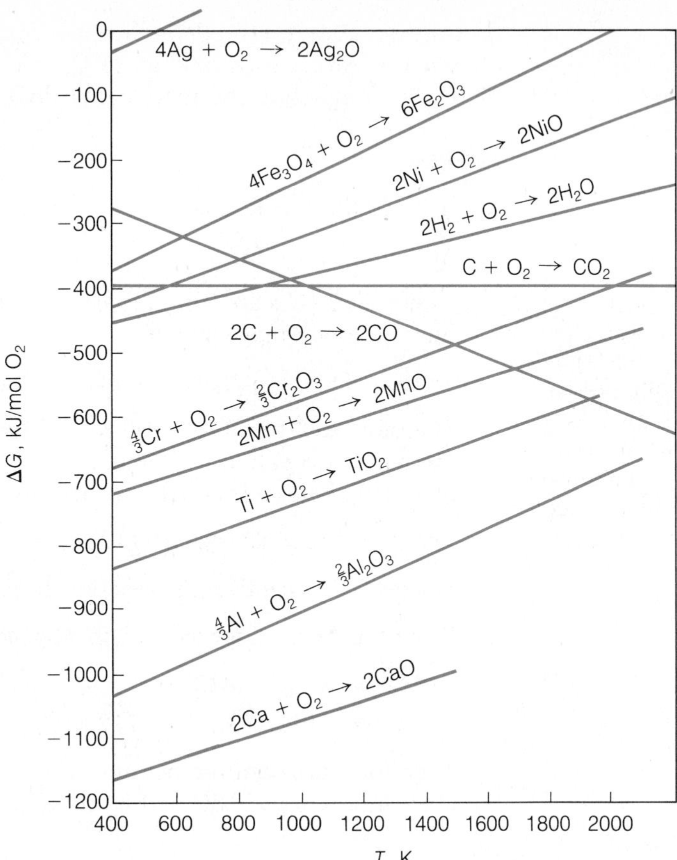

FIGURE 22.11
Free energies of formation of oxides plotted against temperature.

- All the ΔG^0 values for the metallic oxides are negative. This means that the plotted oxides are stable relative to the free metals.
- The ΔG^0 values for the metallic oxides become less negative as the temperature increases. This observation shows why metallurgical reductions of oxides to metal are often carried out at elevated temperatures.
- With most oxides, the ΔG^0 values remain negative at the high temperatures. Hence, at high temperatures the conversion of the oxide to the metal still does not become spontaneous (thermodynamically favorable), so that simple heating does not suffice to decompose the oxide. It is for this reason that reducing agents such as carbon are usually used. Silver oxide, Ag_2O, is an exception because its ΔG^0 value becomes *positive* above approximately 600 K. Above this crossover temperature the oxide decomposes spontaneously since the metal is then thermodynamically more stable than the oxide.

EXAMPLE 22.3 (a) Referring to the appropriate lines in Figure 22.11, estimate ΔG^0 at 1600 K for the reaction

$$NiO(c) + C(graphite) \longrightarrow Ni(c) + CO(g)$$

(b) Is the reaction thermodynamically favorable at 1600 K? (c) Compare the value obtained in part a with ΔG^0 at 25°C, calculated from the data given in Appendix D. Is the reaction thermodynamically favorable at room temperature?

ANSWER

(a) ΔG^0 for the reaction is

$$\Delta G^0 = \Delta G^0_{f,CO} - \Delta G^0_{f,NiO}$$

To make reasonable estimates from a graph (such as Figure 22.11), you need straight lines at right angles to the axes. A draftsman's triangle or any other right-angle guide should give estimates good to 1 or 2 significant figures.

where the ΔG^0_f's are free energies of formation (section 18.8) per mole. From the figure, we may estimate

$$2Ni + O_2 \longrightarrow 2NiO \qquad \Delta G^0 \text{ at } 1600 \text{ K} = -200 \text{ kJ}$$

Therefore ΔG^0_f of NiO at 1600 K = -200 kJ/2 mol NiO or -100 kJ/mol. From the figure we can also estimate ΔG^0_f of CO at 1600 K = -500 kJ/2 mol CO or -250 kJ/mol. Then ΔG^0 for the given reduction reaction at 1600 K is

$$\Delta G^0 = -250 \text{ kJ} - (-100 \text{ kJ}) = -150 \text{ kJ}$$

(b) Since ΔG^0 is negative, the reaction at 1600 K is thermodynamically favorable.

(c) From the ΔG^0_f values at 25°C in Appendix D, we have

$$\begin{aligned}\Delta G^0 &= \Delta G^0_{f,CO} - \Delta G^0_{f,NiO} \\ &= -137.15 \text{ kJ} - (-212 \text{ kJ}) = +75 \text{ kJ}\end{aligned}$$

Therefore, at room temperature the reaction is thermodynamically unfavorable. A mixture of NiO and C would just sit there; nothing would happen. ■

PROBLEM 22.3 Aluminum (m.p. 660°C) is prepared by electrolysis of Al_2O_3 at a high temperature (section 22.9). Is it possible to replace this expensive process by reduction with H_2? To find out, use Figure 22.11 to estimate ΔG^0 for the reaction

$$Al_2O_3(c) + 3H_2(g) \longrightarrow 2Al(\ell) + 3H_2O(g)$$

at (a) 800 K and (b) 1600 K. □

ROASTING

Ores used in ancient metallurgy were, for the most part, surface deposits of oxides, carbonates, or free metals. The oxides and carbonates were reduced to metal by heating with wood charcoal, and the gaseous products, CO and CO_2, escaped readily to the atmosphere. Heating converted the carbonate to the oxide

$$MCO_3 \longrightarrow MO + CO_2$$

The heat-induced conversion of the carbonate to the oxide is called *calcining.*

and it was the oxide that was reduced by the carbon of the charcoal. As mining developed in the pursuit of ores at greater depths, sulfides were increasingly encountered. It was found advantageous to convert these sulfides to oxides prior to reduction. **Roasting** is the process whereby such conversion is accomplished by heating in air. Representative equations are

$$2ZnS(c) + 3O_2(g) \longrightarrow 2ZnO(c) + 2SO_2(g)$$

$$4FeS_2(c) + 11O_2(g) \longrightarrow 2Fe_2O_3(c) + 8SO_2(g)$$

Elements such as molybdenum that are more stable in higher oxidation states may also undergo oxidation during roasting:

$$2MoS_2(c) + 7O_2(g) \longrightarrow 2MoO_3(c) + 4SO_2(g)$$

The sulfide of a less active metal, such as copper, can be roasted to the free metal:

$$Cu_2S(c) + O_2(g) \longrightarrow 2Cu(\ell) + SO_2(g) \qquad (1)$$

Sometimes roasting may form a sulfate rather than an oxide:

$$CoS(c) + 2O_2(g) \longrightarrow CoSO_4(c)$$

The water-soluble sulfates may be leached away from insoluble impurities. (Hydrometallurgy is discussed in section 22.11.)

PROBLEM 22.4 Write an equation for the roasting of PbS (galena). □

PROBLEM 22.5 When HgS (cinnabar) is roasted, it reacts as Cu_2S does. Write an equation for the roasting of HgS. □

Why is roasting necessary? To answer the question we look to thermodynamics. The reason is that direct reduction of sulfides with carbon is thermodynamically less satisfactory than the reduction of the oxides. In fact, the reduction of sulfides is usually impossible. Unfortunately, considerable damage to the environment, such as acid rain (Box 20.1), can result from the release of SO_2 formed in the roasting process.

22.10 SOME ELECTROCHEMICAL METALLURGICAL PROCESSES

As mentioned earlier, electricity is used to reduce compounds of metals to the free metal when chemical reductants are not practical. We will now describe industrial electrochemical reduction processes for the production of aluminum and sodium.

ALUMINUM

The Earth's crust contains about 8.8% Al by weight.

Aluminum is the most abundant of all metals in the Earth's crust, ranking just ahead of iron. Furthermore, it is less dense than iron and more resistant to corrosion. Most of the aluminum on earth occurs as complex aluminum silicates, found widely dispersed in clays and common rocks. Unfortunately, silicates are not a practical source of aluminum. Bauxite, which contains Al_2O_3, is the only concentrated ore from which the extraction of aluminum is economically feasible. The impurities in bauxite include silicates and iron(III) oxide, Fe_2O_3. Aluminum must be free even of traces of silicon and iron if it is to be commercially useful.

The U.S. imports almost all of its bauxite—nearly 50% from Jamaica (its major export).

The process for separating Al_2O_3 from Fe_2O_3 takes advantage of the fact that Al_2O_3 dissolves in both base and acid; it is amphoteric (section 14.8). Fe_2O_3 dissolves in acid but not in base. The Al_2O_3 is dissolved in hot aqueous NaOH, and the solution is poured off from the insoluble Fe_2O_3 and insoluble silicates. When the basic solution is acidified with the weak acid carbon dioxide, hydrated aluminum hydroxide is precipitated and then recovered. If a strong acid were used, it would dissolve the aluminum hydroxide. The hydroxide is heated and thoroughly dehydrated to give pure Al_2O_3.

The Al_2O_3 must be thoroughly dried before being reduced. Water is more easily reduced than Al_2O_3.

$$Al_2O_3(c) + 2OH^-(aq) + 7H_2O \longrightarrow 2[Al(OH)_4(H_2O)_2]^-(aq)$$

$$[Al(OH)_4(H_2O)_2]^-(aq) + CO_2(g) + H_2O \longrightarrow Al(OH)_3(H_2O)_3(c) + HCO_3^-(aq)$$

$$2Al(OH)_3(H_2O)_3(c) \longrightarrow \underset{\text{anhydrous aluminum oxide}}{Al_2O_3} + 9H_2O$$

Aluminum oxide, Al_2O_3, is very stable even at high temperatures. Carbon, the reducing agent used for metals such as iron, is unsuitable for the production of

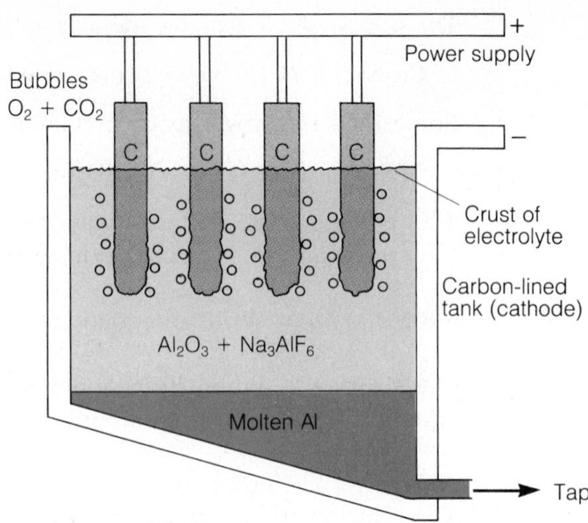

FIGURE 22.12
Schematic diagram of apparatus for electrolytic production of aluminum by the Hall–Héroult process, using carbon anodes.

aluminum at any feasible temperature (see Table 22.2). The method of choice (the Hall–Héroult process) is electrolytic reduction, which requires a solvent that does not undergo electrolysis itself but permits ionic dissociation of the Al_2O_3. The solvent now used is sodium hexafluoroaluminate (**cryolite**), Na_3AlF_6. The cryolite itself undergoes very little electrolysis because $AlF_6{}^{3-}$ is resistant to oxidation and Na^+ is resistant to reduction at the voltages used. Molten aluminum accumulates at the cathode and is tapped off. Carbon is used as the anode and the molten aluminum itself is the cathode. A simple representation of the reaction is

The process was invented independently in 1886 by both Charles M. Hall and Paul Héroult.

$$\text{cathode:} \quad 4Al^{3+} + 12e^- \longrightarrow 4Al(\ell)$$
$$\text{anode:} \quad 6O^{2-} \longrightarrow 3O_2 + 12e^-$$

A schematic diagram of the cell appears in Figure 22.12.

One interesting aspect of the Hall–Héroult process is that the carbon anodes react with some of the oxygen produced at their surfaces and are thus slowly oxidized away. When this reaction is taken into account, the net process is

$$\begin{array}{r} 2Al_2O_3 \longrightarrow 2Al + 3O_2 \\ 3C + 3O_2 \longrightarrow 3CO_2 \\ \hline 2Al_2O_3 + 3C \longrightarrow 2Al + 3CO_2 \end{array}$$

The production of aluminum requires enormous amounts of energy. Energy is required to purify the bauxite, dry the hydrated alumina, and carry out the reduction. By comparison, the energy required to recover aluminum by recycling of aluminum scrap (mainly Al cans) is only about 4% of what is needed for primary production.

In the U.S., Al production uses between 4% and 5% of all the electric power.

Recently a more economical, energy-efficient process has been developed. Anhydrous aluminum oxide is chlorinated with Cl_2 to give aluminum chloride, $AlCl_3$, which is melted and electrolyzed to Al and Cl_2.

SODIUM AND CHLORINE PRODUCTION

We would expect that a metal more active than hydrogen could not be produced in an aqueous solution. It should be easier to liberate H_2 by reducing water than to liberate, for example, Na and Mg by reducing their ions. This expectation is usually correct. Sodium is an extreme case; even if we managed to produce some of it, it would react immediately with water:

$$2Na + 2H_2O \longrightarrow 2Na^+ + 2OH^- + H_2(g)$$

There is, surprisingly, a way to produce Na in the presence of water, and we will describe it later, but the obvious thing is to exclude water entirely.

Fused (melted) salts conduct electricity for the same reason aqueous solutions conduct: ions can move through the liquid. Sodium chloride, by far the most abundant sodium compound, melts at 801°C, which is inconveniently high—especially since Na boils at only 892°C. The melting point can be lowered to about 600°C by adding $CaCl_2$. When an electric current is passed through this molten mixture, the reactions are

Ca^{2+} from $CaCl_2$ is not reduced under the conditions used.

$$\text{cathode:} \quad 2Na^+ + 2e^- \longrightarrow 2Na(\ell)$$
$$\text{anode:} \quad 2Cl^- \longrightarrow Cl_2(g) + 2e^-$$

A typical apparatus is shown in Figure 22.13. Cl_2 is also a valuable product.

The Downs process.

There is another way to produce Na and Cl_2 from NaCl. As noted earlier, we would not expect that Na could be produced in the presence of water. However, a mercury cathode permits the reduction of Na^+ to occur in water. The Hg cathode has two things going for it: it is difficult to produce H_2 on a Hg surface (the overvoltage is high, see section 17.11), and Hg dissolves Na to form an amalgam. Thus, about 95% of the current is used to make Na rather than H_2, and the Na, by being dissolved in the Hg, is protected from the water long enough to reach the next step in the process. Since mercury boils at 357°C and sodium at 892°C, it is possible to separate the sodium and the mercury by distillation. Because of the

An amalgam is a solution of another metal in mercury.

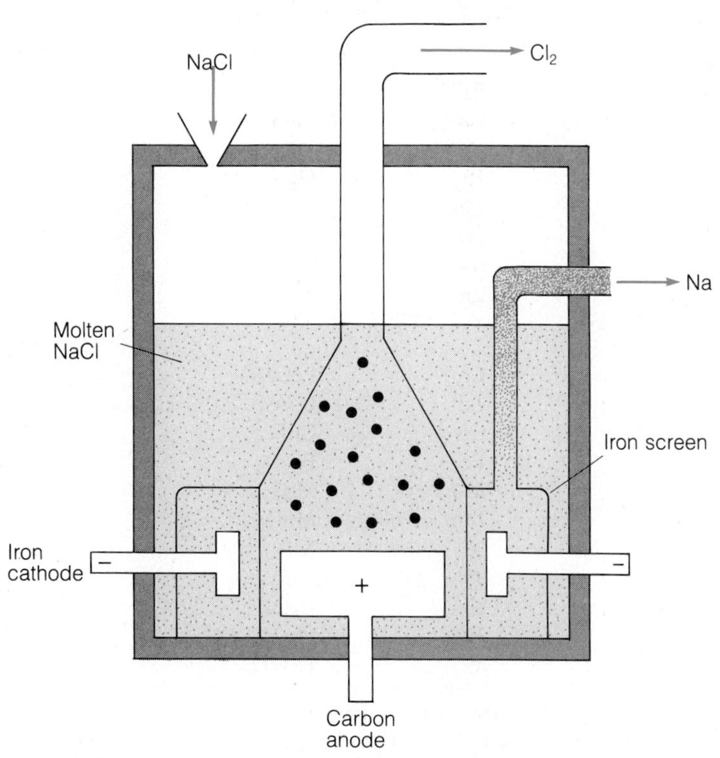

FIGURE 22.13
Schematic diagram of the apparatus used to produce sodium and chlorine by the electrolysis of molten sodium chloride. Voltage is 7 V.

high energy requirements of this process, it has not been used industrially to produce sodium. Instead, the mercury cell is used to produce NaOH and Cl_2. The sodium amalgam flows from the electrolytic cell to another compartment, where it reacts with water:

$$2Na(in\ Hg) + 2H_2O \longrightarrow 2Na^+ + 2OH^- + H_2(g)$$

This process seems roundabout; any cathode in the NaCl solution would yield NaOH directly:

$$2Na^+ + 2H_2O \longrightarrow 2Na^+ + 2OH^- + H_2 + 2e^-$$

But this NaOH is in the same solution with NaCl, and separating them is another big job.

22.11 CHEMICAL REDUCTANTS IN METALLURGY

Tungsten carbide is one of the hardest substances known.

The chemical reducing agent of choice is carbon. See Table 22.2 for reduction of ZnO. Carbon, in the form of coke obtained by heating coal in absence of air (pyrolysis), is cheap and easy to handle. However, some metals such as Cr, Mn, V, and Co form very stable compounds with carbon, called **carbides**. For this reason, carbon is not a practical reducing agent for these metals. In the metallurgy of such elements, active metals such as Al and Mg are used. (The Goldschmidt process is discussed in section 9.4.) Table 22.3 summarizes the metallurgy of some common metals, showing the minerals, the compounds reduced, and the reducing agents.

TABLE 22.3 METALLURGY OF SOME COMMON METALS USING CHEMICAL REDUCTANTS

METAL	MINERAL	COMPOUND REDUCED	REDUCTANT
		Representative Elements	
Bi†	Bi_2S_3	Bi_2O_3	C
Ca	$CaCO_3$	CaO	Al
Cd	CdS	$CdSO_4$	Zn, H_3O^+
Hg†	HgS	HgS	O_2§, heat
Mg	MgO, $MgCO_3$	MgO	C
Pb	PbS	PbO	C, CO
Sn	SnO_2	SnO_2	C
Zn	ZnS	ZnO	C
		Transition Elements	
Cr	$FeCr_2O_7$	Cr_2O_3	Al
Co	CoAsS	Co_3O_4	Al
Cu	$CuFeS_2$	Cu_2S	O_2§, heat
Fe	Fe_2O_3	Fe_2O_3	CO
Mo	MoS_2	MoO_3	Al or H_2
Mn	MnO_2	MnO_2	Al
Ni	NiS	NiO	CO or H_2
Ti	TiO_2	$TiCl_4$	Na or Mg
U	U_3O_8	UCl_4	Ca
V	Pb_2VO_4Cl	V_2O_5	Al
W	$FeWO_4$	H_2WO_4	C

† Also found in the free state.

§ Oxidizing agent; the reducing agent is the mineral. The free metal is obtained on roasting; $MS + O_2 \longrightarrow M + SO_2$.

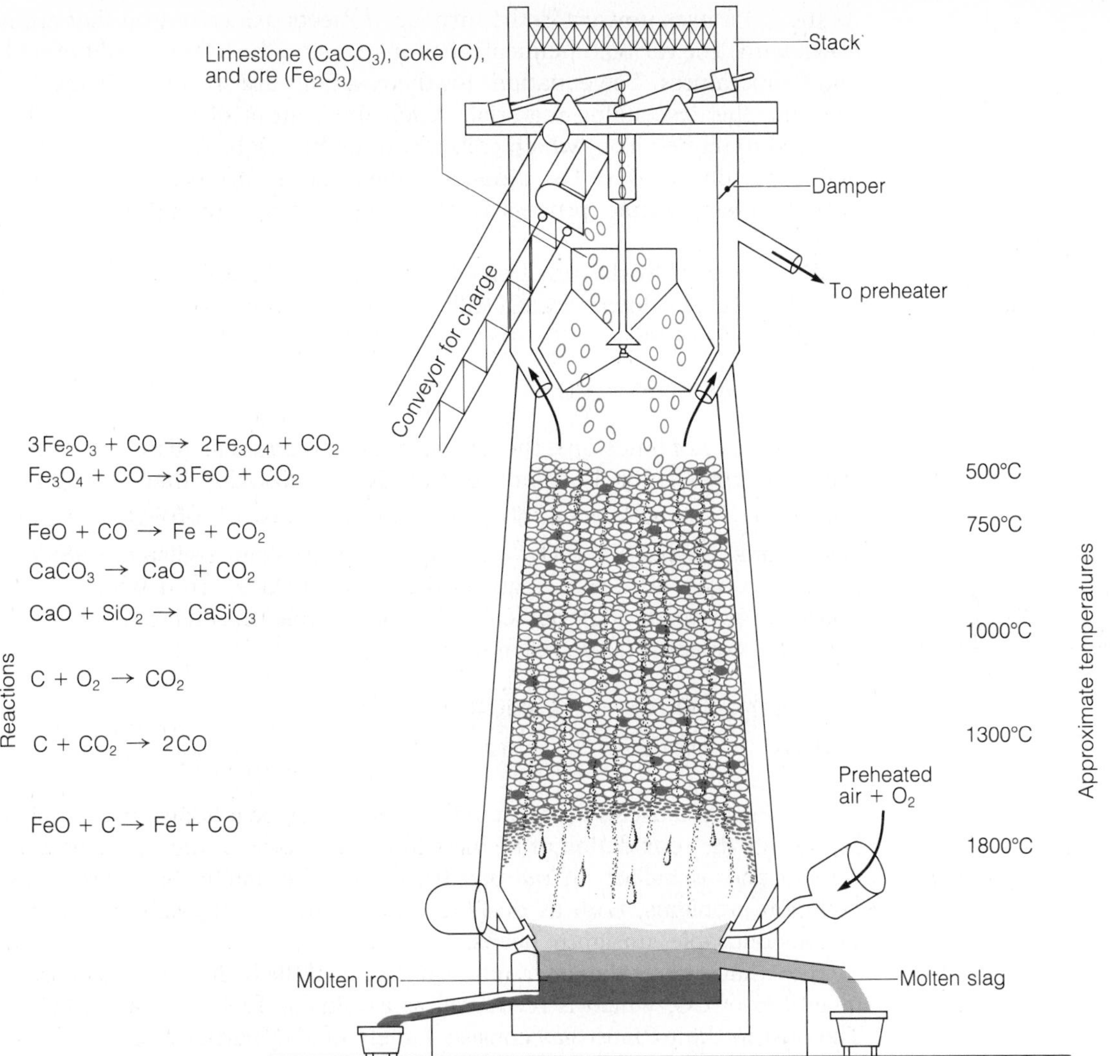

FIGURE 22.14
A blast furnace.

METALLURGY OF IRON

The transition from bronze to iron—the metallurgical event that ushered in the Iron Age—occurred some 3000 years ago. Iron is still the most important metallic element of our civilization, for reasons that have not changed in all the intervening centuries: it is abundant in the Earth's crust (ranking fourth after O, Si, and Al); it has many useful applications; and it is relatively easy to recover from its ores. The important sources are its two naturally occurring oxides, Fe_2O_3 (hematite) and Fe_3O_4 (magnetite). Although carbon is added initially, it is carbon monoxide that is the actual reducing agent.

Iron is made in a blast furnace (Figure 22.14), which operates continuously around the clock. A mixture of iron ore, coke (C), and limestone ($CaCO_3$) enters at the top, while preheated air (~ 1000 K), enriched with O_2, is introduced at the bottom. The oxygen reacts with some of the carbon in the coke to produce CO, thereby liberating considerable heat.

$$C(c) + \tfrac{1}{2}O_2(g) \longrightarrow CO(g) \qquad \Delta H^0 = -110.5 \text{ kJ}$$

As the CO moves upward in the furnace, it meets oxides of iron that are moving downward. The various chemical reactions occur in the furnace at different heights and temperatures. The equations for the reactions are shown in Figure 22.14. By the time the descending materials reach the bottom of the furnace, the ore is reduced to molten iron, which collects on the hearth below. Here it is tapped off about four times a day. The product of the blast furnace is called **pig (cast) iron**. This high-temperature process is an example of **pyrometallurgy**—*pyro* means "fire."

The CaO is called a flux because it is the agent that causes the SiO_2 to melt.

We now turn to the function of the limestone, $CaCO_3$, in the process. This ingredient breaks down to CaO, called a **flux**, which helps remove impurities consisting mostly of SiO_2.

$$CaCO_3 \longrightarrow CaO + CO_2$$

In the absence of limestone, these impurities would remain dissolved in the molten metal after reduction is complete. CaO is a basic oxide that contributes oxide anions, $:\ddot{\underset{..}{O}}:^{2-}$, which break down the covalent network of SiO_2 (section 7.1) to silicate anions, $SiO_3{}^{2-}$. $SiO_3{}^{2-}$ and Ca^{2+} form calcium silicate, $CaSiO_3$, which is called a **slag**. The slag floats to the metal surface, from which it is readily poured or skimmed off periodically. The slag formation can be summarized by the simplified equation (also shown in Figure 22.14)

The anion in "$CaSiO_3$" is not $SiO_3{}^{2-}$, but rather a covalent network structure, $Si_nO_{3n}{}^{2n-}$.

$$\underset{\text{flux}}{CaO(c)} + \underset{\text{impurity}}{SiO_2(c)} \longrightarrow \underset{\text{slag}}{CaSiO_3(\ell)}$$

The slag has a useful function in this process. Since it floats on the molten iron, it prevents the reoxidation of Fe back to Fe_2O_3. Some of the slag is used to make cement and as ballast on railroad tracks. Unfortunately, however, slags cause pollution problems, both as dusts discharged to the atmosphere and as mountainous unusable, unsightly heaps.

The exhaust gases that are emitted at the top of the furnace contain considerable quantities of CO, which is recovered as a valuable fuel for heating the furnace. The dust in the exhaust gas consists largely of slag particles, most of which are trapped by pollution-control devices. These devices are now mandated by the federal government.

Blast furnaces are operated at 1800°C in order to melt the iron, and thus, energy is a major cost item in Fe production. To reduce this cost, the **direct reduction process** was developed in which high-quality iron ore is reduced with a mixture of H_2 and CO. The process operates at about 900°C so that the iron ore remains solid, and after reduction occurs, a spongy form of iron is recovered. This "sponge iron," often along with scrap iron, is refined by being melted with a flux in an electric arc furnace. The liquid slag is removed, and then the purified molten iron is alloyed with different metals to form a variety of steels. The process is feasible in small steel plants and is amenable to a readily automated continuous operation. Thus, not only is there a savings in energy and labor costs, but there is also a reduction in capital cost for equipment needed to produce a given tonnage of Fe.

The H_2–CO reducing mixture is produced from natural gas, which is mostly methane, CH_4. Thus, the impressive growth in the number of plants producing sponge iron has occurred in regions of the world where natural gas is cheap, such as the Middle East. The sponge iron made in these regions is then shipped to steel producers all over the world.

HYDROMETALLURGY

We have already seen that the metallurgy of sulfide ores may require a high-temperature roasting step. Roasting is less than desirable because it consumes much energy and produces SO_2, a major pollutant responsible for acid rain. Therefore, a technology called **hydrometallurgy** (*hydro* means "water") has been developed to eliminate the roasting step.

Essentially, hydrometallurgy involves two steps:

STEP 1. Conversion of the insoluble sulfide to a saturated solution of a water-soluble salt of the metal cation. This is achieved by bubbling O_2, under high pressure and moderate temperature, into a water slurry of the sulfide. This step, called **leaching**, often involves formation of a coordination ion of the metal.

STEP 2. Reduction of the cation to pure metal electrolytically or with hydrogen at high pressures.

Steps 1 and 2 are illustrated with CuS and NiS.

- CuS
  1. Leaching:

  $$CuS(c) + 2H^+(aq) + \tfrac{1}{2}O_2 \longrightarrow Cu^{2+}(aq) + S(c) + H_2O$$

  2. Reduction by electrolysis:

  $$2Cu^{2+}(aq) + 2H_2O \longrightarrow 2Cu(c) + O_2(g) + 4H^+(aq)$$

  The Cu needs no further purification. The S, recovered in the first step, is converted into saleable by-products such as $(NH_4)_2SO_4$ (fertilizer) and pure sulfur.

- NiS
  1. Leaching:

  $$NiS(c) + 2O_2(g) + 4NH_3(aq) \longrightarrow \underbrace{[Ni(NH_3)_4]^{2+}(aq) + SO_4{}^{2-}(aq)}_{\text{tetraamminenickel(II) sulfate}}$$

  2. Reduction with H_2:

  $$[Ni(NH_3)_4]^{2+}(aq) + H_2(g) \longrightarrow Ni(c) + 2NH_4{}^+(aq) + 2NH_3(aq)$$

  Ammonia facilitates the leaching (80°C) by forming tetraamminenickel(II) sulfate. The reduction at 180°C with H_2 at 35 atm yields 99.9% Ni powder.

22.12 REFINING THE METAL

Most metals obtained in the reduction step are too impure to be used directly and must be **refined**—purified. Refining of metals is achieved by physical and chemical (including electrochemical) means.

PHYSICAL METHODS; ZONE REFINING

Zone refining is also used to purify solid covalent compounds.

Low-boiling metals such as mercury (b.p. = 356°C) and zinc (b.p. = 907°C) are refined by simple distillation. When highly purified metals are needed, **zone refining** may be used (Figure 22.15). This process relies on the fact that many impurities do not fit into the crystal lattices of the pure metal. An electric heater surrounding the impure metal bar is slowly moved from one end of the bar to the other. As the heater moves, it melts portions of the bar, which slowly re-crystallize as the heater moves on. The impurities do not easily fit into the newly

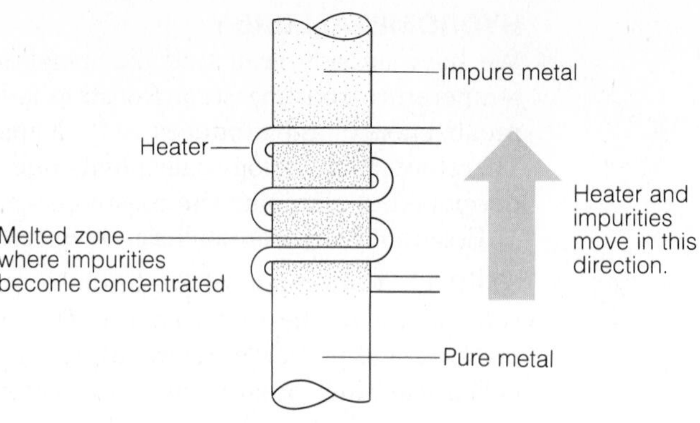

FIGURE 22.15
Zone refining.

Silicon used for semiconductor chips is purified this way.

formed metal lattice. Instead, they are pushed along in the moving molten portion until they reach the end of the bar. Repeated passes of the heater produce a bar of high purity. The impure end is cut off from the bar and recycled.

ELECTROREFINING OF COPPER

Much copper is obtained from natural deposits of copper(I) sulfide, Cu_2S. The ore is first purified and then roasted to give Cu directly. (See equation 1 in section 22.8.) The product is about 99% Cu, but that is not pure enough. Much copper is used as a conductor of electricity; even traces of impurities impair the conductivity substantially. Besides, the impurities include gold and silver, which are worth recovering. Copper is refined by oxidizing it electrically to Cu^{2+}, which travels a few centimeters through a solution and is reduced back to Cu:

$$\begin{array}{ll} \text{anode:} & \text{Cu(impure)} \longrightarrow \text{Cu}^{2+} + 2e^- \\ \text{cathode:} & \text{Cu}^{2+} + 2e^- \longrightarrow \text{Cu(pure)} \\ \hline \text{net reaction:} & \text{Cu(impure)} \longrightarrow \text{Cu(pure)} \end{array}$$

Since there is no net chemical change, the decomposition potential is zero, but about 0.3 V is needed to push the ions through the solution. The electrolyte is an aqueous solution of $CuSO_4$ and H_2SO_4. The process is illustrated in Figure 22.16.

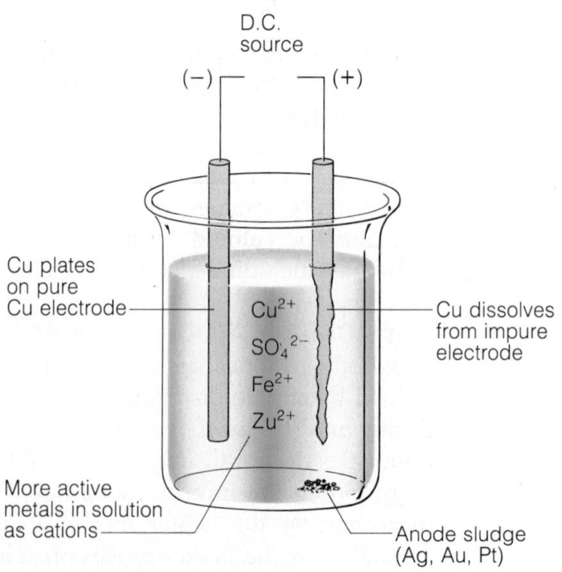

FIGURE 22.16
Electrorefining of copper.

What happens to the impurities? They can be divided into two classes: those that are more active than copper (more easily oxidized, stronger reducing agents), and those that are less active than copper (less easily oxidized, weaker reducing agents). The principal impurities more active than copper are Zn, Ni, Pb, Sb, and As. They are oxidized to ions along with the Cu: $Ni \longrightarrow Ni^{2+} + 2e^-$. At the low voltage used, Ni^{2+} remains in solution because it cannot be reduced back to Ni.

The impurities less active than Cu—mostly Ag, Au, and SiO_2—are not oxidized at all. As the anode is oxidized, the impurities are deprived of the support of their copper-atom neighbors and settle to the bottom as an "anode sludge." Despite its unexciting name, this sludge pays for the whole process.

CHEMICAL REFINING

The impurity elements are often removed by being oxidized to either gaseous oxides or solid oxides that are removed as slag. The refining of iron illustrates this method.

REFINING IRON; MAKING STEEL Pig iron still contains impurities, mostly C (about 4%), in addition to some P, Si, Mn, and S. Iron in this form is brittle and suitable only for casting, not for bending or rolling into structural shapes. The impurities are removed by bubbling O_2 through the molten pig iron containing some added powdered $CaCO_3$, which furnishes the flux, CaO. The impurities are oxidized to their oxides and either are evolved as gases, such as CO_2 and SO_2, or form a removable slag with the CaO:

$$Si \longrightarrow SiO_2 \longrightarrow CaSiO_4$$

$$P \longrightarrow P_4O_{10} \longrightarrow Ca_3(PO_4)_2$$

impurities — oxides — slags

The product of this controlled **basic oxygen process**, which may contain up to 1.5% carbon, is **carbon steel**. The amount of carbon affects the properties of the steel. As the carbon content increases, the steel becomes less ductile and harder. Stainless steels have a low carbon content but must contain chromium (14%–18%) and nickel (7%–9%).

22.13 ALLOYS

Recall that we have defined an alloy as a metallic substance that contains at least two elements. Alloys can be classified as homogeneous solid solutions, heterogeneous mixtures, or compounds.

HOMOGENEOUS SOLID SOLUTIONS

Solid-solution alloys, like any solution, are single-phase, homogeneous mixtures. They have the less abundant elements (the solutes) randomly dispersed throughout the host metallic elements (the solvents). The solute elements are usually, but not always, metallic. There are two types of solid solution alloys:

- In **substitutional alloys**, some of the host solvent atoms are replaced by other *metal* solute atoms of approximately the *same size* (Figure 22.17a). For example, copper (diameter = 0.234 nm) and zinc (diameter = 0.25 nm) form solid solutions in all proportions. The brasses are the alloys of Cu and Zn.
- In **interstitial alloys**, the solute atoms occupy the interstitial spaces (section 21.5) in the crystal lattice of the solvent atoms (Figure 22.17b).

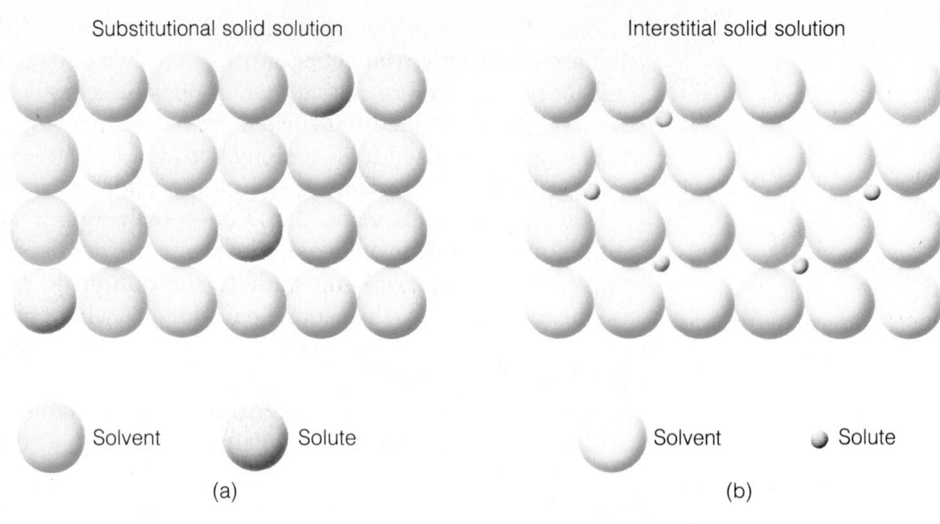

FIGURE 22.17
Types of homogeneous solid solutions. (a) Substitutional solid solution. (b) Interstitial solid solution.

The formation of interstitial solid-solution alloys requires that the dispersed solute atoms be small enough to fit into the small interstices. The most commonly used elements that meet this requirement are carbon and hydrogen. Ordinary steel is an interstitial solution alloy of carbon in iron. The interstitial hydrides discussed in section 21.5 are examples of interstitial alloys.

HETEROGENEOUS MIXTURES

In **heterogeneous-mixture alloys**, the host metal and the alloying substance are each distinct phases—the components of the alloy are not dispersed uniformly. For example, in one form of steel (pearlite), two distinct phases of Fe and the carbide Fe_3C (cementite) are present in alternating layers. In some homogeneous alloys, clumps of alloying substance may be dispersed throughout the host phase. The physical properties of the heterogeneous-mixture alloy depend not only on the composition but also on the mode of dispersal. The mode of dispersal is influenced by the manner in which the solid is formed from the molten mass. Rapid or slow cooling produces alloys with distinctly different properties.

COMPOUNDS

Compound alloys are homogeneous alloys that follow the law of definite composition. They have definite molecular formulas. For example, aluminum and copper form a compound alloy, $CuAl_2$, known as Duralumin. It is not uncommon for two metals to form more than one compound. For example, Cu and Al also form Cu_3Al and Cu_9Al_4. The usual concept of valence does not apply to these compounds. Theories for rationalizing compositions have been developed, but they are beyond the scope of this book. (See Problem 22.46.)

Compound alloys are used extensively in our technology because of their special properties. The strong yet light-weight alloy, Ni_3Al, is a major component of jet engines. The very hard Cr_3Pt coats the new, longer-lasting razor blades. Strong permanent magnets are made from alloys with the general formula Co_5M, where M is a lanthanoid. The most prominent one is Co_5Sm, which is used in light-weight headsets.

Lanthanoids have atomic numbers between 58 and 71.

TABLE 22.4
SOME COMMERCIALLY IMPORTANT ALLOYS

COMMON NAME	HOST ELEMENT	MASS PERCENT COMPOSITION	PROPERTIES	USES
Alnico	Fe	Fe(50), Al(20), Ni(20), Co(10)	magnetic	magnets
Aluminum bronze	Cu	Cu(90), Al(10)	tough	crankcases, connecting rods
Brass	Cu	Cu(67–90), Zn(10–33)	ductile, high polish	plumbing, hardware, decorative uses
Bronze	Cu	Cu(70–95), Zn(1–25), Sn(1–18)	durable	bearings, bells, medals
Carbon steel	Fe	Fe(98–99.5), C(0.5–2)	variable hardness	structural metal
Cast iron	Fe	Fe(96–97), C(3–4)	supports weight	castings
Coinage (U.S.)	Cu	Cu(75), Ni(25)	durable	5¢, 10¢, 25¢, 50¢ coins
Dental amalgam	Hg	Hg(50), Ag(35), Sn(15)	workable	dental fillings
Duriron	Fe	Fe(84), Si(14), C(1), Mn(1)	hard	pipes, kettles, condensers
German silver	Cu	Cu(60), Zn(25), Ni(15)		teapots, jugs, faucets
Gold, 18-carat	Au	Au(75), Ag(10–20), Cu(5–15)	durable	jewelry
Gold, 10-carat	Au	Au(42), Ag(12–20), Cu(38–46)	durable	jewelry
Gunmetal	Cu	Cu(88), Sn(10), Zn(2)	shockproof	gun barrels, machine parts
Lead battery plate	Pb	Pb(94), Sb(6)		storage batteries
Lead shot	Pb	Pb(99.8), As(0.2)		shotgun shells
Magnalium	Al	Al(70–90), Mg(10–30)	low density	aircraft bodies
Manganese steel	Fe	Fe(86), Mn(13), C(1)	hard	safes, armor plate, rails
Monel	Ni	Ni(60–70), Cu(25–35), Fe, Mn	resists corrosion	instruments, machine parts
Nichrome	Ni	Ni(60), Fe(25), Cr(15)	high resistance	electrical resistance wire
Pewter	Sn	Sn(70–95), Sb(5–15), Pb(0–15)		tableware
Silver solder	Ag	Ag(63), Cu(30), Zn(7)		high-melting solder
Solder	Pb	Pb(67), Sn(33)	low melting point	solder joints
Stainless steel	Fe	Fe(73–79), Cr(14–18), Ni(7–9)	resists corrosion	instruments, sinks, tableware
Sterling silver	Ag	Ag(92.5), Cu(7.5)	shiny	tableware, jewelry
Wood's metal	Bi	Bi(50), Pb(25), Sn(13), Cd(12)	low melting point	automatic sprinkler systems

A remarkable property of some compound alloys is **superconductivity**, which refers to conduction of an electric current without resistance. Superconductivity occurs now at the very low temperature provided by liquid helium (b.p. = −268.6°C). The most prominent superconductor is Nb_3Sn. Passage of a very large current through the superconducting compound generates a very strong magnetic field. Such supercooled strong magnets are smaller in size than permanent magnets with the same field strength. This property has led to their use in modern instruments requiring magnets, one of which is the nuclear magnetic resonance spectrometer (Box 23.5). Since it takes considerable energy to liquefy He, which is not an inexhaustible substance, search is underway to synthesize compounds that will be superconducting at higher temperatures.

In 1986, in one of the most dramatic episodes in recent science, materials were discovered that are superconducting at temperatures as high as 100 K. If these materials—compounds of La, Ba, Cu, and O—can be put to practical use, superconduction could occur at the temperature of liquid nitrogen (b.p. 77 K). Liquid nitrogen is an inexhaustible resource and is far cheaper and easier to handle than liquid helium. The dream of superconduction at room temperature no longer seems like a hollow fantasy.

Table 22.4 lists some commercially important alloys.

22.14 ARE METALS HERE TO STAY?

The Iron Age, augmented by the influx of many other metals, has continued for more than 3000 years. Today, however, we are faced with the question, Will metals decline in importance? It appears that the answer may be in the affirmative, the reason being the increased use of plastics. For example, plastics have been steadily displacing steel in the manufacture of automobiles. There is now an oversupply of certain metals—such as Cu, Zn, and Al—because of declining consumption.

Readily accessible sources of ores are becoming depleted. If it should become necessary to develop more inaccessible sources, such as the ocean floor or the moon, metals will become much more expensive. To delay the exhaustion of readily available ores, the reclamation of scrap metals must be expanded.

Plastics in general have some advantages over metals: they are lighter-weight, more resistant to corrosion, and less expensive. Furthermore, the invention of plastics that conduct electricity may negate one of the major attractions of metals. Unfortunately, there may be a limit to the potential of plastics. The raw material of most plastics is petroleum, which, of course, is also a major fuel. Petroleum may not be inexhaustible. Therefore, in the years ahead a crucial decision may have to be made—is petroleum to be used for energy or for the manufacture of plastics? The dilemma could be resolved if another, relatively inexpensive source of energy that does not threaten the environment could be developed.

Clearly, in the future, supplies of both metals and plastics may dwindle coincidentally with a burgeoning demand due to the growing world population and increased demand by developed and developing nations. However, do not despair. Research is in progress to develop materials to replace both metals and plastics. Graphite and specially treated wood could be such materials. However, ceramics appear to be the most promising—sand and silicates, from which most ceramics are made, are inexpensive and inexhaustible. Ceramics are being developed that have the specialized properties needed for our present technology.

Actually, we are not living in a *metal* (iron) Age; we live in an Age of *mixed materials*—mainly of metals and plastics. How the world progresses from here depends on research, supplies of raw materials, energy sources, energy consumption, protection of the environment, and population growth. The problems that arise are subject to political solutions, of course, and we would trust that those decisions will be based on the results of good, successful, worldwide scientific research, especially in chemistry, chemical engineering, and material science.

SUMMARY

Since metals conduct heat and electricity, they must have electrons that move easily. This led to the early, oversimplified "**electron sea**" **model** which suggests that the metal (M) has electrons moving through a crystal lattice made up of cations (M^{n+}). A better model, called the **band theory**, is based on the MO concept. According to the band theory, when the AO's of a very large number (say a mole) of metallic atoms are combined in a piece of metal, the MO's and MO*'s with the same principal quantum number have very closely spaced energy levels that form energy bands. The band formed from inner-shell molecular orbitals is filled with electrons and is called the **nonconduction band**. The band formed from the valence-shell molecular orbitals, called the **conduction band**, is partially filled with electrons. The easy movement of electrons in the conduction band accounts for the typical metallic properties such as conduction of heat and electricity.

Insulators have empty conduction bands and it takes too much energy (large **band gap**) to promote electrons from the lower-energy nonconduction band to the empty higher-energy band where conduction could occur. The pure Group-4 **semiconductor** elements, such as silicon, are similar to insulators, except that they have a small band gap. Heat and electricity can promote electrons from the nonconduction band, now also called the **valence band** because it has the valence electrons, into the empty higher-energy band, which then becomes a conduction band. The excitation of electrons from the former nonconduction band leaves electron-deficient sites, called **holes**. Then, conduction occurs also in the valence band because of the simultaneous movement of electrons and holes in opposite directions.

The semiconduction of pure Si is greatly enhanced by adding trace amounts (doping) of a Group-3 element (boron) to give a ***p*-type semiconductor**. Since B has one less electron than Si, a hole is created for each B atom, permitting conduction in the valence band without necessitating excitation of electrons into the empty higher-energy band. Doping pure Si with a Group-5 element such as arsenic gives ***n*-type semiconductors**. Arsenic has one more electron than Si. The band gap is greatly narrowed so that excitation of these extra electrons into the empty higher-energy band can then occur at room temperature.

Metallurgy is the technology of obtaining a pure metal from an economically accessible source called an **ore**. The ore is first purified to remove the **minerals**, which contain the metal, from the **gangue**, which contains the impurities. The minerals are mainly oxides, sulfides, and carbonates. **Cyclone separation** can be used when the minerals are more dense than the gangue. The **flotation process** can be used when only the mineral is "wetted" by oil and made to float away in air bubbles from the gangue. Chemical methods are also used to purify the mineral in the ore. For example, Al_2O_3 is purified by taking advantage of its amphoteric nature.

Only a few inactive metals—such as Au, Ag, and Pt—are mined in the free state. Most metals are obtained by reducing their oxides, or chloride salts in the case of elements in Groups 1 and 2. Since it is thermodynamically more efficient to reduce the oxide, the carbonates are heated and sulfides are **roasted** to give the oxide. High-temperature roasting and reduction, called **pyrometallurgy**, is energy intensive and produces the air pollutant, SO_2. To bypass these defects, **hydrometallurgical** low-temperature processes have been developed to convert sulfides to water-soluble salts (**leaching**) that are then reduced. Active metals such as Na and Al are reduced electrochemically under nonaqueous conditions. Chemical reductants include carbon (the cheapest), CO (reduction of Fe_2O_3 to Fe), Al, and H_2. During the pyrometallurgical reduction step, when SiO_2 is the major impurity, it is removed by adding a **flux** (CaO from $CaCO_3$) to give an easily removable product called a **slag** ($CaSiO_3$).

The metal obtained after the reduction step usually has to be **refined** (purified). Low-boiling metals (Zn and Hg) are distilled. **Zone refining** relies on the fact that impurities do not fit into the crystal lattices formed as the pure molten metals cool. **Electrorefining** is used, for example, for copper. Chemical refining, as used for Fe, is achieved by oxidizing the impurity elements to gaseous oxides, which are emitted, or solid oxides, which are removed as slag.

Alloys are metals composed of more than one element. Homogeneous **compound alloys** have definite molecular formulas. Some are **superconducting**—they conduct an electric current without resistance when cooled to the temperature of liquid helium. Some alloys are **heterogeneous mixtures**, in which the components form distinct phases. The third type comprises two kinds of homogeneous **solid solutions**. In the **substitutional** type, some of the host atoms (the solvent) in the crystal lattice are replaced by other metal atoms (solute) of approximately the same size. In the **interstitial** type, small solute atoms, such as C or H, fit into the interstices of the crystal lattice of the host atoms. In carbon steels the interstitial atoms are carbon and the host atoms are Fe.

22.15 ADDITIONAL PROBLEMS

METALLIC BONDING

22.6 (a) Relate the preference for metallic bonding to the electron configuration of the element. (b) Why, under ordinary conditions, do metals *not* form covalently bonded structures? (c) What type of elements have both covalent and metallic allotropes?

22.7 What feature must a substance possess in order for it to be a metallic conductor?

22.8 Which atomic orbitals make up the conduction band in Na and in Al?

22.9 Explain (a) the high electrical conductivity and (b) the very good heat conductivity of metals.

22.10 Why are metals malleable, whereas network-covalent structures (for example, diamond and quartz) and salt crystals are not?

22.11 All of the transition elements are metallic. Which AO's form the molecular orbitals of the conduction band?

22.12 What fraction of the conduction band is filled with electrons in Li, Be, Al, Pb, and Bi?

22.13 Should it be easier to convert H_2 or F_2 to a metallic state? Justify your answer in terms of the band theory.

SEMICONDUCTORS

22.14 Use the band theory to differentiate among conductors, semiconductors, and insulators.

22.15 Explain why semiconductors, unlike metals, show an increase in electrical conductivity with temperature increases.

22.16 Differentiate between *n*-type and *p*-type semiconductors.

22.17 Differentiate between a pure semiconductor and a *p*-type semiconductor.

22.18 Name two elements that would make Si (a) a *p*-type semiconductor; (b) an *n*-type semiconductor.

22.19 State whether each of the following is likely to be an insulator, a metallic conductor, a pure semiconductor, an *n*-type semiconductor, or a *p*-type semiconductor: (a) germanium, (b) germanium doped with Ga, (c) germanium doped with As, (d) pure CaO, (e) pure Ta.

22.20 Tin exists in two allotropic forms; gray tin has a diamondlike network covalent structure, and white tin has a closely packed structure. Which allotrope would you expect to be more metallic in character?

METALLURGY

22.21 Differentiate between an ore, a mineral, and gangue.

22.22 What kind of metals are most likely to be found in the elementary (free) state?

22.23 List the four steps involved in the pyrometallurgical method of getting a pure metal from a mined sulfide ore.

22.24 Describe the flotation process for ore purification.

22.25 (a) Where in the periodic table would you be most likely to find metals that are obtained by electrolysis? (b) At which electrode is the metal found?

22.26 Describe the metallurgy of (a) Na and (b) Al.

22.27 CaO is manufactured from the very abundant $CaCO_3$. Why is the carbonate rather than the oxide added as the flux in the reduction of Fe?

22.28 Write chemical equations for the purification of Al_2O_3.

22.29 How is Fe_2O_3 reduced (a) in a blast furnace; (b) by the direct reduction process?

22.30 Draw and label an electrochemical cell for electrorefining Sn. What would be the likely fate of the Zn, Fe, Pb, and Cu impurities?

22.31 Write chemical equations for these metallurgical steps: (a) reduction of Co_3O_4 with Al, (b) reduction of MoO_3 with H_2, (c) roasting NiS, (d) reducing UCl_4 with Ca, (e) reducing SnO_2 with C, (f) reducing PbO with CO.

22.32 In the hydrometallurgy of Zn, ZnS(c) reacts with aqueous H_2SO_4 at 150°C under oxygen to form water-soluble zinc sulfate and S(c). Write a chemical equation for the reaction.

22.33 The ore $CuCo_2S_4$ is leached with aqueous H_2SO_4 to give a solution of Cu^{2+} and Co^{3+} ions. What is the fate of the sulfur in the ore? Write a chemical equation for the reaction.

22.34 Write a balanced chemical equation to illustrate the role played by each of these materials in a blast furnace: (a) air, (b) coke, (c) limestone.

22.35 In the obsolete Castner process, Na was made by electrolysis of molten NaOH (m.p. 318°C): $2Na^+ + 2OH^- \longrightarrow 2Na(\ell) + H_2 + O_2$. (a) This reaction may be broken down mentally into 3 steps: oxidation of OH^- to O_2 and H_2O, reduction of Na^+ to Na, and reaction of the H_2O produced in the first step with Na. Write the equation for each step. Verify that they add up to the net equation. (b) From the partial reactions, how many faradays are needed to produce 1 mol Na? (c) How many faradays are needed to produce 1 mol Na in the Downs process? (d) Does your answer help explain why the Castner process is obsolete?

22.36 In a chemical plant, molten NaCl is electrolyzed with a current of 5.00×10^4 A. (a) How many kilograms of Na are produced every 24 h? (b) How many liters of Cl_2 at standard conditions are produced?

THERMODYNAMICS AND METALLURGY

22.37 Joseph Priestley discovered O_2 when he heated HgO. What thermodynamic property should an oxide have so that it could be practically used for a Priestley-type discovery? Would he have been more apt to make the same discovery by heating MnO or Ag_2O? (See Figure 22.11.)

22.38 (a) Write separate balanced chemical equations for the reduction of ZnO to Zn with (1) C and (2) CO. (b) Calculate the ΔG^0 for each reaction at 25°C. ΔG_f^0 of ZnO is −318 kJ/mol. In terms of these values, forgetting about cost, which would be the better reductant?

22.39 Use Figure 22.11 to estimate the free-energy change of the following reactions at 1200 K, and state whether the reductions are feasible.

(a) $NiO(c) + CO(g) \longrightarrow Ni(c) + CO_2(g)$

(b) $TiO_2(c) + 4H_2(g) \longrightarrow Ti(c) + 4H_2O(g)$

(c) $TiO_2(c) + 2CO \longrightarrow Ti(c) + 2CO_2$

22.40 In terms of the entropy change, explain why the ΔG_f^0 of (a) metallic oxides become more positive as T increases; (b) CO_2 is nearly independent of T; (c) CO becomes less positive as T increases.

22.41 (a) Referring to the appropriate lines in Figure 22.11 estimate ΔG at 1600 K for the reaction

$$TiO_2(c) + 2C(\text{graphite}) \longrightarrow Ti(c) + 2CO(g)$$

Would the reaction be spontaneous at this temperature? (b) At what temperature does the CO line intersect the TiO_2 line? Is this also the temperature at which ΔG^0 for the reaction in part a equals zero? Explain. (c) Does the procedure in part b apply to the intersection with the NiO line? Explain.

ALLOYS

22.42 Give the three broad classifications of alloys.

22.43 How do steels differ from pig iron?

22.44 Account for the fact that (a) carbon steels are alloys but tungsten carbide is not; (b) Ni_3Al is an alloy but aluminum carbide, Al_4C_3, is not.

22.45 Is (a) brass and is (b) an alloy of a heavy metal with Li an interstitial or substitutional solid solution? Explain.

22.46 Hume-Rothery (1926) suggested that stable intermetallic compounds form when the ratio of the number of valence electrons to the number of atoms in the compound is 21:12, 21:13, or 21:14. In terms of this suggestion, (a) show why Cu and Zn form CuZn, Cu_5Zn_8, and $CuZn_3$; (b) find the number of Al atoms in Cu_5Al_x; (c) show whether $Cu_{31}Sn_8$ is a reasonable intermetallic compound.

22.47 (10 points) Matching

___ 1. band gap	___ 6. conduction band
___ 2. semiconductor	___ 7. valence band
___ 3. *p*-type semiconductor	___ 8. nonconduction band
___ 4. insulator	___ 9. positive hole
___ 5. band	___10. *n*-type semiconductor

(a) partially filled band that permits conduction of heat and electricity; (b) energy difference between a filled band and an empty or partially filled band; (c) very large group of closely spaced bonding and antibonding molecular orbitals within a material; (d) band filled with electrons; (e) material that resists conduction of heat or electricity; (f) material that becomes a conductor of electricity as the temperature rises; (g) nonconduction band in semiconductors that becomes conducting because of doping or electron excitation; (h) positive site in a valence band created because of a missing electron; (i) semiconductor in which conduction occurs in the valence band because of the movement of positive holes; (j) semiconductor in which conduction occurs because of easy excitation of excess electrons from the valence band into what was formerly an empty conduction band.

22.48 (18 points) Matching

___ 1. ore	___10. interstitial alloy
___ 2. mineral	___11. solid-solution alloy
___ 3. gangue	___12. flux
___ 4. roasting	___13. substitutional alloy
___ 5. flotation process	___14. zone refining
___ 6. hydrometallurgy	___15. slag
___ 7. alloy	___16. heterogeneous alloy
___ 8. pyrometallurgy	___17. refining
___ 9. compound alloy	___18. superconductivity

(a) purifying of metals; (b) economically accessible source of a metal; (c) metal composed of more than one element; (d) homogeneous alloy having the alloying element uniformly dispersed in the host metallic element; (e) homogeneous alloy of more than one metal with a definite molecular formula; (f) alloy in which components form distinct phases; (g) solid-solution alloy in which some host atoms are replaced by other atoms of similar size; (h) solid alloy in which the solute atoms fit into some of the interstices of the lattice of the host atoms; (i) impurities in an ore; (j) substance added to chemically aid in removal of impurities; (k) easily removable substance formed from an impurity; (l) metallurgical processes requiring high temperatures; (m) metallurgical processes occurring at relatively low temperatures in water solutions; (n) method of purifying ores; (o) physical method of purifying metals; (p) conduction without resistance; (q) metallic compound in an ore; (r) conversion of a mineral to an oxide by heating.

22.49 (16 points) Multiple-Choice

(A) A semiconductor conducts electricity, when heated but not before, because of (a) a partially filled conduction band before heating, (b) a filled conduction band, (c) a very large band gap, (d) a very small band gap.

(B) A possible reason for the decrease in conductivity of metals as the temperature increases is that (a) increased vibrations of the metal cations interfere with electron flow, (b) the conduction band acquires too many electrons, (c) electrons cannot be excited from the nonconduction band to the conduction band, (d) there is an increase in collisions between metal cations.

(C) Which of the following is *not* a reason that insulators do not conduct electricity? (a) absence of electrons in conduction band, (b) large band gap, (c) temperature insufficient to excite electrons from conduction to nonconduction bands, (d) positive holes in the valence band.

(D) Which of the following is *not* a reason for conduction of electricity? (a) flow of metal cations, (b) movement of positive holes in valence band, (c) movement of electrons in conduction band, (d) flow of electrons in valence band.

22.50 (20 points) Write balanced chemical equations for these metallurgical reactions: (a) roasting of Bi_2S_3, (b) reduction of Fe_3O_4 with CO, (c) reduction of V_2O_5 with Al, (d) reduction of $CdSO_4(aq)$ with Zn in acid, (e) reduction of $TiCl_4$ (ℓ) with Mg.

22.51 (16 points) Fill in the blanks.

(a) On electrofining copper, any lead impurity would be removed as ________. (b) In the pyrolytic refining of Fe the sulfur impurity is removed as ________. (c) The purification of Al_2O_3 relies on the ________ nature of the oxide, in contrast to Fe_2O_3. (d) A gaseous impurity that forms with Cl_2 at the anode during the electrolysis of NaCl(aq) is ________. (e) Stainless steel contains substantial amounts of the element ________ in addition to Fe. (f) Ni_3Al is an example of a(n) ________ alloy. (g) A(n) ________ is formed when palladium adsorbs hydrogen. (h) If CaO were the gangue, a flux that could be added is ________.

22.52 (10 points) A small particle of sodium contains 1×10^{-6} mole. (a) How many atoms, (b) how many filled or half-filled atomic orbitals, and (c) how many filled molecular orbitals does this particle contain? (d) How many molecular orbitals are there in the conduction band? (e) Give the distribution of electrons in subshells for this particle of sodium.

22.53 (10 points) (a) Referring to Figure 22.11, estimate ΔG^0 at 1400 K for the reaction

$$\tfrac{1}{2}Cr_2O_3(c) + \tfrac{3}{2}C(\text{graphite}) \longrightarrow Cr(c) + \tfrac{3}{2}CO(g)$$

Would the reaction be spontaneous at this temperature? (b) Estimate ΔG^0 at 1000 K for the reaction

$$\tfrac{1}{2}Cr_2O_3(c) + Al(\ell) \longrightarrow \tfrac{1}{2}Al_2O_3(c) + Cr(c)$$

Compare Al with C as a reducing agent for metallic oxides.

23

NUCLEAR CHEMISTRY

In previous chapters, we have discussed the electronic structures of atoms, molecules, and ions and their relationship to the chemistry of elements and compounds. We have shown that these particles are the principal building blocks of matter and the principal emitters of radiation. A chemical reaction is the result of interactions between electrons and the absorption and emission of radiation results from transitions of electrons between two energy levels. Yet it is the composition of the nucleus of an atom that fixes the number of electrons and so determines its atomic properties.

In this chapter, we probe beneath the electron clouds to study nuclear structure and the nature of the bond (forces) that holds nucleons together in nuclei. We will also study nuclear reactions such as radioactivity, artificial transmutations, fission, and fusion. Whenever possible, we will discuss these topics in terms of the principles we have previously applied to the properties of atoms and molecules. We will also discuss the medical applications of radioactive elements and the issues surrounding the use of nuclear power plants.

23.1 RADIOACTIVITY

By the end of the nineteenth century, the concept of a chemical element as a fundamental substance that could not be decomposed into other substances was firmly established. The indivisibility of the atom appeared to stand on firm ground scientifically. However, this complacency was upset by the discovery of radioactivity, which revealed that the nuclei of certain atoms are unstable and spontaneously change to other kinds of nuclei.

Soon after the discovery of X rays by Röntgen (section 2.5), Henri Becquerel searched for a possible relationship between the glow induced in minerals by X rays and the light emitted by minerals upon exposure to sunlight. In the course of this research, in 1896 he accidentally discovered that uranium compounds emit a radiation similar to X rays. *Elements such as uranium that spontaneously emit energy without absorbing energy are said to be naturally* **radioactive**.

Experiments performed by E. Rutherford and P. Villard (1898).

They were simply labeled with the first 3 letters of the Greek alphabet, α, β, and γ, because their natures were unknown.

Experiments using electric or magnetic fields showed that there are three distinct types of emission. As shown in Figure 23.1, the emission from a naturally radioactive source is split by an electric field into three beams labeled **alpha** (α), **beta** (β), and **gamma** (γ). The α beam, deflected to the negative plate, must be composed of positively charged particles. The β beam, more sharply deflected to the positive plate, must be composed of lighter, negatively charged particles. The γ beam, undeflected, must be electrically neutral. The α particle carries two unit positive charges and has the mass of a helium nucleus. Furthermore, when a stream of α particles picks up electrons, $\alpha + 2e^- \longrightarrow \text{He}$, it becomes helium gas, which can be detected spectroscopically. Therefore, an α particle is, in fact,

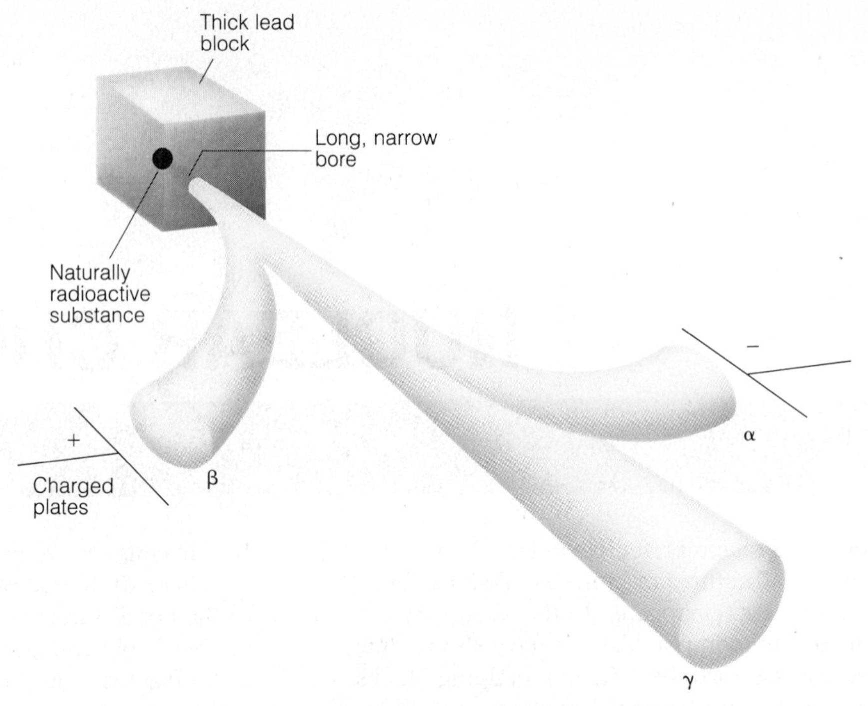

FIGURE 23.1
The behavior of α, β, and γ emissions in an electric field. Apparatus is enclosed in a vacuum.

a helium nucleus, designated as $^{4}_{2}He^{2+}$. The electric charge and mass of a β particle show it to be an electron, represented by $_{-1}^{0}\beta$. γ rays do not consist of charged particles but they are a highly energetic radiation. They are therefore similar to X rays, except that their wavelengths are shorter. Gamma rays are photons several million times higher in frequency than visible light (Color plate 4 and Table 23.1).

Reminder: The superscript is the mass number; the subscript is the atomic number. $_{-1}^{0}\beta$ means that the atomic weight of an electron is closer to zero than it is to 1; a mass number of zero does *not* mean massless; it means that no nucleons are present.

It was also discovered that radioactive atoms (radioisotopes) nearly always produce other elements. Rutherford found that radium, for example, spontaneously emits α particles and a radioactive gas—later named *radon*. To account for these observations, Rutherford and Frederick Soddy proposed (1902) that radioactivity is the result of a change in the atoms of one element into the atoms of another element. (Box 23.1)

This sensible, though revolutionary, idea posed a new problem: radioactive changes seemed to generate an utter confusion of new elements. Mendeleev, who was still alive, rejected Rutherford's explanation of radioactivity out of fear that

TABLE 23.1
PROPERTIES OF α, β, AND γ EMISSIONS

	ALPHA	BETA	GAMMA
Symbol	α, $^{4}_{2}He^{2+}$	β, $_{-1}^{0}\beta$, $_{-1}^{0}e$	γ
Charge	+2	−1	0
Mass	6.65×10^{-24} g	9.11×10^{-28} g	0
Kind of Particle	High-speed $^{4}_{2}He$ nuclei	High-speed electrons	High-energy photons
Relative Ability to Ionize Air†	10^4	10^2	1
Relative Penetrating Power	1	10^2	10^4

† The number of ions formed per millimeter of beam path for a fixed area.

BOX 23.1
RADON POLLUTION

Marie and Pierre Curie were the first to note that air in contact with radium compounds becomes radioactive. Later it was shown that the radioactivity came from the radium itself; and the product was therefore called "radium emanation." Rutherford and Soddy succeeded in condensing this gas to a liquid at low temperature and thus confirmed that the emanation is a real substance, the noble gas radon, Rn.

It is now known that the air in uranium mines is radioactive because of the presence of radon. The mines must therefore be ventilated well to protect the miners. Nearly all rocks contain very small quantities of radium, as do various soils, and the resulting emission of radon makes the air in improperly ventilated buildings more radioactive than outdoor air.

Uranium miners are exposed to less radiation than many people are in their homes. The solution for those people appears to be to lower the air pressure in the soil beneath the basement so as to minimize the quantity of radon that is drawn into the home.

Radon itself is not really the significant hazard; it is chemically inert, and inhaled radon is promptly exhaled. Rather, the major dose of radiation to lung tissue comes from the decay products of radon, mainly the isotopes of polonium (Figure 23.7) that can stick to bronchial or lung tissue.

there would be no place in his periodic table for the additional elements. Fortunately, the concept of isotopes (Greek, "same place") saved the day and preserved the periodic table. (See Box 23.2.)

Radioactive nuclei are unstable and emit particles spontaneously. These **nuclear reactions** (transmutations) involve a change in the atomic number or the mass number (or both) of the radioisotope. The transmutation of radium may be represented by the equation

$$ {}^{226}_{88}\text{Ra} \longrightarrow \underset{\alpha \text{ particle}}{{}^{4}_{2}\text{He}} + {}^{222}_{86}\text{Rn} \tag{1} $$

Recall (section 2.6) that the atomic number is also known as the *proton number* and the mass number as the *nucleon number*.

The subscript is the atomic number—the number of protons in the nucleus—and the superscript is the mass number—the number of nucleons (protons and neutrons) in the nucleus. This is a typical **radioactive decay** reaction or, simply, a **decay** reaction.

In chemical reactions, atoms are rearranged; they are not created or destroyed. There are two analogous conservation laws for nuclear reactions. In nuclear reactions, nucleons are rearranged; they are not created or destroyed. Hence, *mass*

BOX 23.2
ISOTOPES AND ORDER

OLD NAMES AND SYMBOLS OF SOME "CONFUSING" ELEMENTS	MODERN ISOTOPE DESIGNATION	ACCEPTED NAME
Uranium X_1, UX_1	${}^{234}_{90}\text{Th}$	thorium-234
Ionium, Io	${}^{230}_{90}\text{Th}$	thorium-230
Actinouranium, AcU	${}^{235}_{92}\text{U}$	uranium-235
Radium A, RaA	${}^{218}_{84}\text{Po}$	polonium-218
Radioactinium, RdAc	${}^{227}_{90}\text{Th}$	thorium-227
Radium emanation, niton, Nt	${}^{222}_{86}\text{Rn}$	radon-222
Actinon, An	${}^{219}_{86}\text{Rn}$	radon-219

(nucleon) numbers are conserved: The sum of the mass (nucleon) numbers of reacting nuclei and particles must equal the sum of the mass (nucleon) numbers of the nuclei and particles that are produced. *Charge is also conserved*: The sum of the atomic (proton) numbers of reacting nuclei and particles must equal the sum of the atomic (proton) numbers of the products. Equation 1 and Example 23.1 illustrate these laws.

EXAMPLE 23.1 Write the balanced equations for the reactions (a) ${}^{27}_{13}\text{Al} + {}^{4}_{2}\text{He} \longrightarrow {}^{1}_{0}\text{n} + ?$, (b) ${}^{240}_{92}\text{U} \longrightarrow {}^{0}_{-1}\beta + ?$, and (c) ${}^{240}_{93}\text{Np} \longrightarrow {}^{0}_{0}\gamma + ?$. Replace the question marks with the isotopes that are formed. ${}^{1}_{0}\text{n}$ is the symbol for the neutron.

ANSWER

(a) The sum of the atomic numbers of ${}_{13}\text{Al}$ and ${}_{2}\text{He}$ is 15 and the atomic number of the neutron is zero. To balance the charge (conservation of charge), the isotope must have an atomic number of 15. Inspection of the periodic table tells us it is an isotope of phosphorus, ${}_{15}\text{P}$. The sum of the mass numbers of ${}^{27}_{13}\text{Al} + {}^{4}_{2}\text{He}$ is 31 and the mass number of the neutron is 1. To balance the mass numbers (conservation of mass (nucleon) numbers), the isotope must have a mass number of 30. The balanced equation is

$$ {}^{27}_{13}\text{Al} + {}^{4}_{2}\text{He} \longrightarrow {}^{1}_{0}\text{n} + {}^{30}_{15}\text{P} $$

(b) The charge of an electron is -1 and its mass number is zero. Therefore, in order to balance the charge and the mass numbers, the isotope formed must be an isotope of element 93, neptunium, with a mass number of 240. The balanced equation is

$$ {}^{240}_{92}\text{U} \longrightarrow {}^{0}_{-1}\beta + {}^{240}_{93}\text{Np} $$

(c) Since the atomic and mass numbers of a gamma photon are zero, the atomic and mass numbers of the atom from which it is emitted are not changed:

$$ {}^{240}_{93}\text{Np} \longrightarrow {}^{0}_{0}\gamma + {}^{240}_{93}\text{Np} $$

■

PROBLEM 23.1 Balance the nuclear reaction

$$ {}^{25}_{12}\text{Mg} + {}^{4}_{2}\text{He} \longrightarrow {}^{1}_{0}\text{n} + ? $$

by replacing the question mark with the symbol for the isotope formed. □

Neutrons, *outside the nucleus*, disintegrate into protons and electrons. However, they are *not composed of* protons and electrons. Robert Hofstadter (1958) bombarded protons and neutrons with high-energy electrons and studied the deflection patterns of the electrons. These experiments are analogous to the Rutherford experiments on the deflection of α particles by nuclei (section 2.6). The distribution of the deflected electrons indicated the presence of electrically charged particles within the protons and neutrons. Hofstadter suggested that the proton consists of a neutral core surrounded by two positively charged clouds, and that the charges of the two clouds add up to one unit positive charge. He also proposed that the structure of a neutron is similar except that one of the clouds is negative, and that the charges of the clouds cancel each other electrically. It is now accepted that protons and neutrons are not fundamental particles. Rather, they are each made up of three particles called **quarks** (Box 23.3). Of the neutron-proton-electron triad, only the electron remains as a fundamental particle—a structureless particle that is not composed of smaller units.

BOX 23.3

MOLECULES, ATOMS, NUCLEONS, QUARKS, AND GLUONS

Hofstadter's proposal is consistent with the quark theory of particles, which says that nucleons are composed of more fundamental particles called *quarks*. Quarks are subparticles, with various properties specified by quantum numbers, but which cannot exist separately as independent particles. Their most astonishing property is that they have electric charges smaller than the electronic charge. A proton is composed of two quarks, each carrying a charge of $+\frac{2}{3}$, and one quark carrying a charge of $-\frac{1}{3}$ so that the proton electric charge equals $(\frac{2}{3} + \frac{2}{3} - \frac{1}{3}) = +1$. A neutron is composed of two quarks, each carrying a charge of $-\frac{1}{3}$, and one quark carrying a charge of $+\frac{2}{3}$ so that the neutron electric charge equals $(-\frac{1}{3} - \frac{1}{3} + \frac{2}{3}) = 0$. An electron, however, carries a single charge of -1. Fractional charges are not observed, because quarks always form combinations in which the sum of their charges is either 0, $+1$, or -1. The reality of quarks is supported by the success of the quark theory in predicting the experimental results of collisions between particles traveling at nearly the speed of light. The famous theoretical physicist Paul Dirac, one of the founders of quantum mechanics, wrote in 1947, ". . . One can never arrive at the ultimate structure of matter [because] it becomes necessary to postulate that each part is itself made up of smaller parts." Matter is made up of atoms and molecules; molecules are made up of atoms; atoms are made up of protons, neutrons, and electrons; protons and neutrons are made up of quarks; quarks are held together by massless particles called *gluons*, and gluons may be made up of . . . . The predictions of the quark theory have been so well verified by experiment that the existence of quarks is no longer doubted.

23.2 INTERCONVERTIBILITY OF MATTER AND ENERGY: NUCLEAR ENERGY

The law of conservation of matter (section 2.2) and the law of conservation of energy (section 18.3) were originally conceived as independent relationships, but we now know that they are in fact identical. From a very small loss in mass of uranium, we get a very large amount of heat; matter is converted into energy. We can illustrate the meaning of this statement with an example. Suppose we extracted 0.01 g of energy from a 10.00 g mass of matter; we would then have 0.01 g energy + 9.99 g matter = 10.00 g. Adding 0.01 g of energy to the 10.00 g mass of matter would give us a total of 10.01 g of matter. This is the meaning of the statement that matter and energy are interconvertible. It is the principle on which nuclear reactors and nuclear explosives operate. The equivalence of mass and energy escaped everyone's attention until the twentieth century for only one reason: the change in mass accompanying the emission or absorption of energy in non-nuclear reactions is so small that it is undetectable. For instance, when water is boiled, the H_2O molecules in the steam have a greater mass than they had in the water, but the difference is too small to be detected by present instrumentation. The mass of CO_2 produced at constant temperature in the combination of C and O_2 is less than the mass of the consumed reactants. But, again, the difference is not detectable. Now to the details.

In any exothermic reaction, matter is converted to energy; conversely, in any endothermic reaction, the energy absorbed is converted into matter. This means that matter and energy are not conserved separately; rather, the principle of conservation is broadened to include matter as another form of energy or energy as another form of matter.

An ideal calorimeter is a perfectly insulated, indestructible box.

The quantity of energy liberated or absorbed, expressed in mass units, *is exactly equal to the quantity of matter destroyed or created*, expressed in mass units. Imagine a reaction with a large heat output taking place in an *ideal calorimeter* (Figure 23.2). No change in mass occurs until the heat is allowed to escape. The calorimeter registers a decrease in mass as the heat escapes. For example, a decrease of 0.1 mg means that 0.1 mg of matter is destroyed, but it appears as 0.1 mg

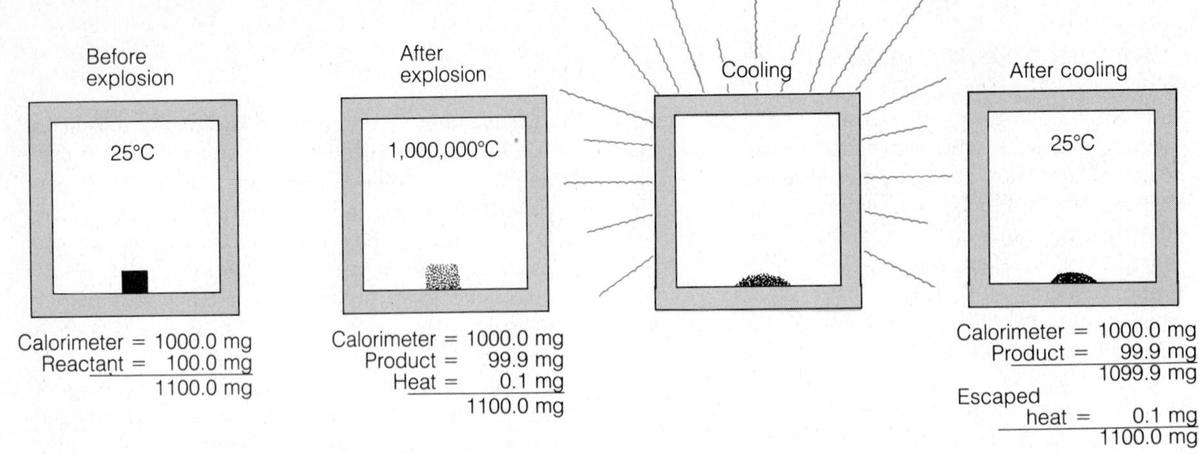

FIGURE 23.2
An imaginary nuclear reaction occurring in an ideal calorimeter.

Nuclear physicists express mass in energy units.

of heat. However, most of us have not been educated to express energy in mass units. We therefore use a conversion factor to change the mass unit (kg) to the energy unit (J).

The interconvertibility of matter and energy was predicted by Albert Einstein in 1905 (Box 23.4). It is described by the equation $\mathbf{\Delta E = \Delta m \times c^2}$. ΔE is the energy change in joules when Δm, the quantity of matter converted, is in kilograms, and the constant c, the speed of light, is in m/sec (3.00×10^8 m/sec).

EXAMPLE 23.2 In the combustion of 400 g of gasoline, 2.1×10^7 J are evolved:

$$\text{reactants} \longrightarrow \text{products} \qquad \Delta E = -2.1 \times 10^7 \text{ J}$$

Calculate the decrease in mass in grams that accompanies this reaction.

ANSWER The mass of the products will be less by the amount calculated from the Einstein equation ($\Delta E = \Delta m \times c^2$):

$1 \text{ J} = 1 \text{ kg} \dfrac{\text{m}^2}{\text{sec}^2}$

$$\Delta m = \frac{\Delta E}{c^2}$$

$$= \frac{-2.1 \times 10^7 \cancel{\text{J}}}{\left(3.0 \times 10^8 \dfrac{\cancel{\text{m}}}{\cancel{\text{sec}}}\right)^2} \times \frac{1 \text{ kg} \dfrac{\cancel{\text{m}^2}}{\cancel{\text{sec}^2}}}{1 \cancel{\text{J}}} = -2.3 \times 10^{-10} \text{ kg}$$

BOX 23.4
ALBERT EINSTEIN AND $E = mc^2$

The formula $E = mc^2$ has become as common as toothpaste advertisements. Yet, when Einstein submitted the article (1905) in which he derived the equation from concepts he had previously published, he wrote to one of his friends, "This line of thought is amusing and fascinating, but I cannot know whether the good Lord does not laugh at it and has been pulling my leg." His doubt, however, did not prevent him from suggesting that the formula could be checked experimentally by comparing the mass loss in nuclear reactions with the energy evolved by the reaction. Some thirty years later, experimentation finally caught up with theory and the laws of conservation of matter and conservation of energy became one law.

Now change the mass in kilograms to the mass in grams:

$$\Delta m = -2.3 \times 10^{-10}\ \cancel{kg} \times \frac{10^3\ g}{1\ \cancel{kg}} = -2.3 \times 10^{-7}\ g$$

$$\frac{2.3 \times 10^{-7}\ g}{400\ g} \times 10^6 = 5.8 \times 10^{-4}\ \text{ppm}$$

2.3×10^{-7} g or 5.8×10^{-4} parts per million is the quantity of matter liberated as heat to the surroundings. This quantity is so small that the total mass of matter undergoing a chemical change or a phase change is practically constant. ■

EXAMPLE 23.3 Calculate ΔE in kJ/mol for the transmutation of radium from the given molar masses:

$$^{226}_{88}Ra \longrightarrow {}^{4}_{2}He + {}^{222}_{86}Rn \qquad \Delta E = ?$$

molar masses, g/mol: 226.0254 4.0026 222.0176

ANSWER The energy, ΔE, of a nuclear reaction is calculated from the difference between the masses of products and reactants and the Einstein equation. First, find Δm from the given molar masses and then use $\Delta E = \Delta m \times c^2$:

$$\begin{aligned}\Delta m &= m_{products} - m_{reactants} \\ &= 4.0026\ g/mol + 222.0176\ g/mol - 226.0254\ g/mol \\ &= -0.0052\ g/mol\end{aligned}$$

$$\frac{5.2 \times 10^{-3}\ g}{226\ g} \times 10^6 = 23\ \text{ppm}$$

and

$$\Delta E = -5.2 \times 10^{-3}\ \frac{\cancel{g}}{mol} \times \frac{1\ \cancel{kg}}{10^3\ \cancel{g}} \times \left(3.0 \times 10^8\ \frac{\cancel{m}}{\cancel{sec}}\right)^2 \times \frac{1\ \cancel{J}}{1\ \cancel{kg}\ \frac{\cancel{m^2}}{\cancel{sec^2}}} \times \frac{1\ kJ}{10^3\ \cancel{J}}$$

$$= -4.7 \times 10^8\ \frac{kJ}{mol}$$ ■

The calculated value agrees precisely with the measured value; $\Delta E = -4.7 \times 10^8$ kJ/mol. Notice that this value is many orders of magnitude larger than ΔE for typical chemical reactions.

PROBLEM 23.2 Calculate ΔE in kJ/mol for the reaction

$$^{26}_{13}Al \longrightarrow {}^{1}_{1}H + {}^{25}_{12}Mg$$

Isotopic weights: $^{26}_{13}Al$, 25.99410; $^{1}_{1}H$, 1.00783; $^{25}_{12}Mg$, 24.98584. (An isotopic "weight," like an atomic "weight," is really an isotopic relative mass.) □

Often, nucleons, but not electrons, participate in nuclear reactions. However, tables of atomic (isotopic) constants record atomic (isotopic) weights rather than nuclear weights. Notice in Example 23.3 and Problem 23.2 that the masses of the extranuclear electrons of the products (86 + 2 and 1 + 12) are canceled by the masses of the extranuclear electrons of the reactants (88 and 13). It is therefore correct to use isotopic weights in these problems.

In an actual experiment, shown in Figure 23.3, a high-speed electron ("matter") slows down, losing kinetic energy. This energy appears as a γ-ray photon. In turn, this photon ("energy") is materialized, forming matter—an electron and a positron. The **positron** ($^{0}_{+1}\beta$) is a particle with the same mass as an electron ($^{0}_{-1}\beta$) but the opposite charge.

The positron is a typical *antiparticle*. Antiparticles have the same mass as particles and differ only in the sign of the charge. If sufficient energy is available, the electron and its antiparticle can be created (Figure 23.3). However, when a particle and its antiparticle come into contact, the particles annihilate one another, forming two high-energy photons:

Paul Dirac predicted (1928) the existence of the positron, as well as its antimatter properties, from a solution of the relativistic Schrödinger equation for electrons.

$$^{0}_{+1}\beta + {}^{0}_{-1}\beta \longrightarrow 2\ \gamma\text{-ray photons}$$

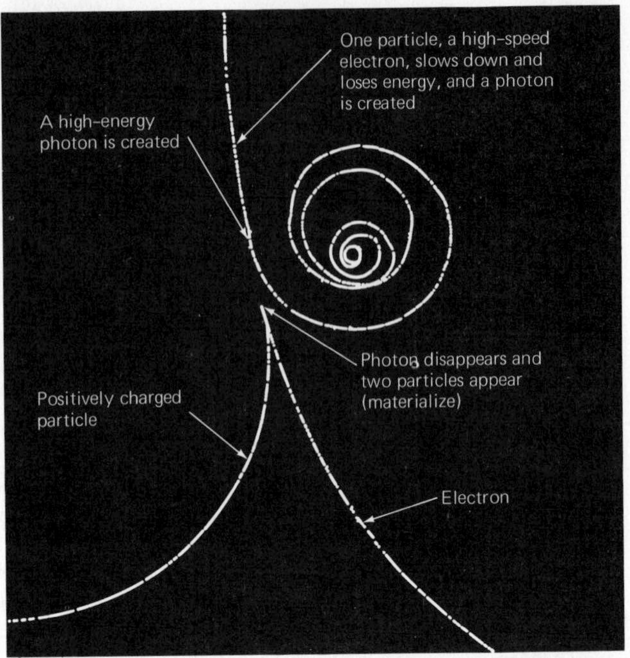

FIGURE 23.3
An experiment performed by Luis Alvarez that illustrates the meaning of the matter-energy conservation principle. Kinetic energy of an electron is converted to a photon, which in turn is converted to two oppositely charged particles. The photon does not show because it is uncharged.

A system composed of equal numbers of particles and antiparticles cannot exist; mutual annihilation is a characteristic property of these two kinds of matter (Figure 23.3).

The energy of radioactivity is consumed as the emissions pass through matter. Some of the energy excites or ionizes atoms and molecules; some is absorbed in breaking chemical bonds. Alpha particles usually cannot penetrate more than a few sheets of paper. The walls of an ordinary glass beaker generally stop β particles. X and γ rays are highly penetrating and are stopped only by thick layers of lead or concrete (Table 23.1). Distance from the source of the rays is also important in minimizing exposure; the radiation's intensity varies inversely with the square of the distance from the source.

The order of penetrating ability is $\gamma > \beta > \alpha$.

23.3 NUCLEAR STRUCTURE

The magnetic moment of a proton is about 2×10^3 times weaker than the magnetic moment of an electron.

The $^2_1$D nucleus has 1 proton and 1 neutron. Hund's rule is applied separately to neutrons and protons.

In certain respects the structure of the nucleus is analogous to the electronic structure of the atom. Quantum mechanics shows that the energy states (levels) of a nucleon in the nucleus are quantized and characterized by four quantum numbers. The Pauli principle also applies to nucleons; that is, no two protons and no two neutrons in a nucleus can possess the same four quantum numbers. One of these numbers corresponds to m_s, the magnetic-spin quantum number. Protons and neutrons may thus have two spin orientations, corresponding to quantum numbers of $+\frac{1}{2}$ and $-\frac{1}{2}$ (section 5.6). Therefore protons and neutrons, like electrons, possess a characteristic (but much smaller) permanent magnetic moment. However, spin pairing, $\underline{\uparrow\downarrow}$, occurs. A nuclear orbital can accommodate two protons or two neutrons with opposite spin. Measurements show that nuclei with *an even number of protons* and *an even number of neutrons*—for example, $^4_2$He, $^{12}_6$C, $^{16}_8$O, $^{32}_{14}$Si—do not have a permanent magnetic moment. In $^2_1$D, however, the spins of the proton and neutron are parallel (not paired) and deuterium nuclei possess

BOX 23.5
NUCLEAR MAGNETIC RESONANCE IN MEDICAL DIAGNOSIS

When a sample of sodium vapor, $^{23}_{11}Na$, is placed in a magnetic field, the unpaired electrons give rise to closely spaced electronic levels. Transitions between these energy levels occur in the microwave region, giving rise to a characteristic spectrum. Similarly, when a sample containing hydrogen atoms, ^1_1H, is placed in a magnetic field, the unpaired protons give rise to closely spaced nuclear energy levels. Transitions between these energy levels occur in the radiowave region (radio frequencies), giving rise to a characteristic spectrum. However, the frequency at which a nucleus absorbs radiation is strongly dependent upon its chemical environment. The ^1_1H atoms in H_2O, H_2S, and HF, in fat, protein, and carbohydrate, in normal and diseased organs, absorb at different frequencies. The technique for detecting ^1_1H atoms in different molecular environments is the basis of **nuclear** (nuclei with unpaired protons or unpaired neutrons) **magnetic** (in an external magnetic field) **resonance** (absorb photons with characteristic frequencies), **NMR**.[†] NMR is a powerful tool, a workhorse, for determining the molecular structure of compounds. An observed absorption peak at a definite radio frequency can be correlated with a specific group of H atoms in a molecule. For example, the three different H environments,

```
     H   H
     |   |
 H—C—C—OH
     |   |
     H   H
```

CH_3, CH_2, OH, in ethanol give three different absorption peaks, while the two CH_3 groups in ethane

```
     H   H
     |   |
 H—C—C—H
     |   |
     H   H
```

being equivalent, give only one peak. Furthermore, the areas under the peaks are proportional to the number of H atoms of each kind; for ethanol, the ratios are $3(CH_3):2(CH_2):1(OH)$.

The application of NMR to generate images of human organs or tissues, called **MRI**[§] (magnetic resonance imaging), has resulted in one of the latest, most exciting, most expensive techniques for medical diagnosis. For example, malignant tumors show a distinctly higher absorption intensity than the surrounding tissues. The atherosclerotic thickening of the aortic wall is clearly detected. Color plate 16 shows a magnetic resonance imaging instrument, an MRI normal brain image, and a brain image revealing an aneurysm.

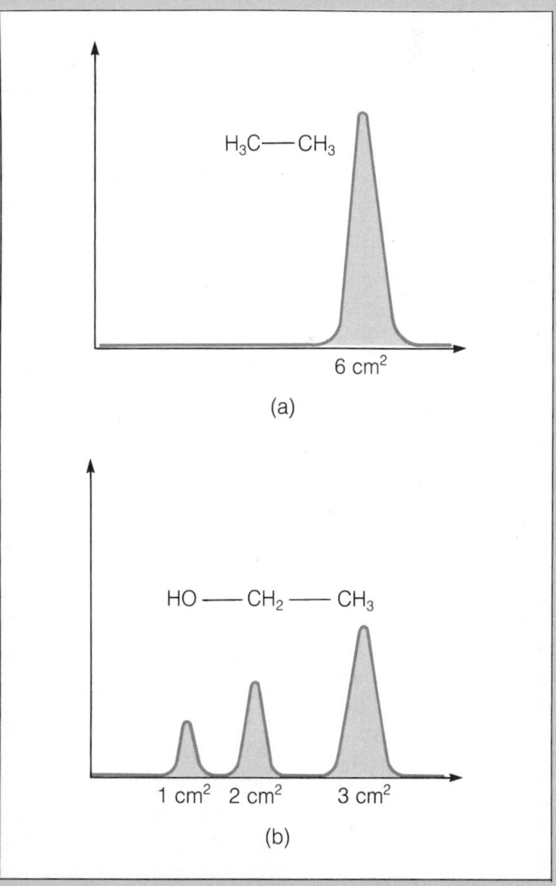

Simplified representation of NMR spectrum of (a) ethane, $H_3C{-}CH_3$, and (b) ethanol, CH_3CH_2OH.

† Developed by Edward Purcell, Henry Torrey, and Robert Pound (1946) and, independently, by Felix Bloch, William Hansen, and Martin Packard (1946). The external magnetic field strength may be varied at a fixed radio frequency or the radio frequency may be varied at a fixed magnetic field strength.

§ MRI instruments use huge superconducting magnets, which are electromagnets cooled by liquid helium. It is expected that they will be displaced by smaller supermagnets that require no cooling.

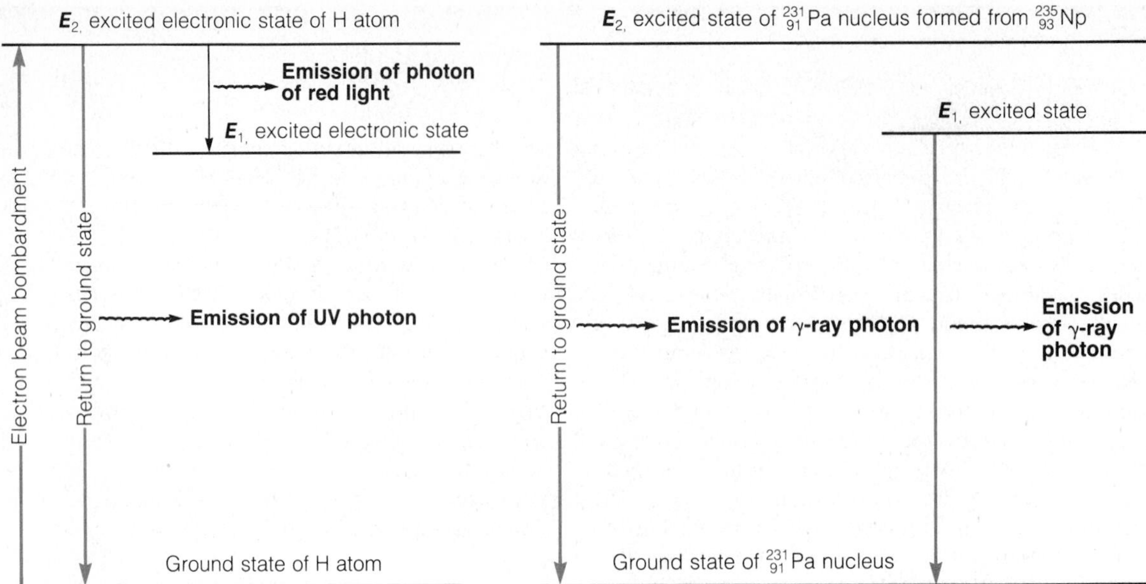

FIGURE 23.4
(Not to scale) (a) The excited electronic states of the H atom may be produced by electron-beam bombardment. (b) The excited nuclear states of Pa are produced when Np emits an α particle. A photon is produced when an electron or nucleon returns to a lower energy state (arrow). The energy of the photon is determined by the difference in the energies of the two states.

These are referred to as "magic numbers." They differ from what could be called the "chemical magic numbers," $2(_2He)$, $10(_{10}Ne)$, 18, 36, 54 . . . , because the number of subshells per shell is different. (See Problem 23.55.)

a magnetic moment. $^{14}_{7}N$, with 7 protons and 7 neutrons, also has one unpaired neutron. These nuclei, like electrons, may be made to orient with respect to an applied external magnetic field (Box 23.5).

Nucleons are assigned to nuclear orbitals in the same way that electrons are assigned to atomic orbitals (section 5.7). Just as filled subshells—for instance, He $1s^2$ and Ne $1s^2 2s^2 2p^6$—possess a special stability, closed sets of nuclear energy levels ("shells") correspond to great nuclear stability. Nuclei containing 2, 8, 20, 28, 50, 82, or 126 nucleons of each kind are particularly stable. This means that an especially large amount of energy is required to separate the nucleons. For example, the ^4_2He nucleus contains two protons and two neutrons, corresponding to a closed shell. This combination accounts for the extreme stability of the ^4_2He nucleus and for the fact that many naturally radioactive nuclei spontaneously emit α particles. The less stable nuclei, ^1_1H (proton), ^2_1H (deuteron), ^3_1H (triton), ^3_2He, and ^6_3Li, however, are never seen as products of natural radioactive decompositions. Of the isotopes of oxygen, ^{14}O, ^{15}O, ^{16}O, ^{17}O, ^{18}O, ^{19}O, ^{20}O, the most stable is ^{16}O; its nucleus contains 8 protons and 8 neutrons, corresponding to a closed shell. $^{40}_{20}Ca$ forms the next closed shell. $^{208}_{82}Pb$ is also a very stable nucleus. It contains a "magic number" of both protons (82) and neutrons (126). This model (nuclear shell model) does not always yield correct predictions. It is nevertheless very useful in correlating much information about nuclei.

A nucleus remaining after the emission of an α or β particle often emits a γ ray (photon). In terms of the energy-level model of nuclei, the nucleus is in an excited state. It thus emits a photon (the γ ray) in a manner analogous to the emission of a photon by an excited hydrogen atom (Figure 23.4).

23.4 THE STABILITY OF NUCLEI

Recall that the energy required to separate a molecule in the gaseous state into its component atoms equals the energy evolved when these atoms recombine at the same temperature (section 4.7). Similarly, the **binding** (bond) **energy of a**

nucleus is the energy required to separate the nucleus into its component nucleons. It is also the energy evolved when the nucleons recombine to form the nucleus. However, the heats of nuclear reactions are about a million times greater than the heats of chemical reactions; more matter per mole of reactant is converted to energy.

EXAMPLE 23.4 Calculate the binding energy in kJ/mol for the formation of (a) $^{30}_{15}P$ and (b) $^{31}_{15}P$. Find the energy released per nucleon. Isotopic weights: $^{30}_{15}P$, 29.9880; $^{31}_{15}P$, 30.97376; $^{1}_{1}H$, 1.00783; $^{1}_{0}n$, 1.00867.

ANSWER

(a) The binding energy for the formation of the nucleus of $^{30}_{15}P$ is the energy evolved or absorbed when 15 protons (frequently designated as $^{1}_{1}p$), and 15 neutrons, $^{1}_{0}n$, combine: $15^{1}_{1}p + 15^{1}_{0}n \longrightarrow {}^{30}_{15}P$. The energy change can be calculated from the change in mass, Δm, and the Einstein equation, $\Delta E = \Delta m \times c^2$. However, isotopic weights and not nuclear weights are listed. We therefore form the $^{30}_{15}P$ atom from 15 $^{1}_{1}H$ atoms and 15 neutrons; $15^{1}_{1}H + 15^{1}_{0}n \longrightarrow {}^{30}_{15}P$. The mass of the extranuclear electrons of the product $^{30}_{15}P$ atom is canceled by the mass of the 15 extranuclear electrons of the 15 $^{1}_{1}H$ atoms. Then, Δm for the formation of the $^{30}_{15}P$ nucleus from 15 protons is the same as the formation of the $^{30}_{15}P$ atom from 15 $^{1}_{1}H$ atoms:

$$15^{1}_{1}H \quad + \quad 15^{1}_{0}n \quad \longrightarrow \quad {}^{30}_{15}P$$

molar masses, g/mol: 15 × 1.00783   15 × 1.00867   29.9880

(30.2475)

$$\Delta m = 29.9880 \text{ g/mol} - 30.2475 \text{ g/mol} = -0.2595 \text{ g/mol}$$

$$\Delta E = \Delta m \times c^2$$

$$= -0.2595 \frac{\text{g}}{\text{mol}} \times \frac{1 \text{ kg}}{10^3 \text{ g}} \times \left(2.9979 \times 10^8 \frac{\text{m}}{\text{sec}}\right)^2 \times \frac{1 \text{ J}}{1 \text{ kg} \frac{\text{m}^2}{\text{sec}^2}} \times \frac{1 \text{ kJ}}{10^3 \text{ J}}$$

$$= -2.332 \times 10^{10} \text{ kJ/mol}$$

(b) For

$$15^{1}_{1}H \quad + \quad 16^{1}_{0}n \quad \longrightarrow \quad {}^{31}_{15}P$$

15 × 1.00783   16 × 1.00867   30.97376

(31.25617)

$$\Delta m = 30.97376 \text{ g/mol} - 31.25617 \text{ g/mol} = -0.28241 \text{ g/mol}$$

$$\Delta E = \Delta m \times c^2$$

$$= -0.28241 \frac{\text{g}}{\text{mol}} \times \frac{1 \text{ kg}}{10^3 \text{ g}} \times \left(2.99793 \times 10^8 \frac{\text{m}}{\text{sec}}\right)^2 \times \frac{1 \text{ J}}{1 \text{ kg} \frac{\text{m}^2}{\text{sec}^2}} \times \frac{1 \text{ kJ}}{10^3 \text{ J}}$$

$$= -2.5381 \times 10^{10} \text{ kJ/mol}$$

2.54×10^{10} kJ are needed to separate 1 mole of $^{31}_{15}P$ into its nucleons, whereas 2.33×10^{10} kJ are needed to separate 1 mole of $^{30}_{15}P$.

The average energy released per nucleon is then given by ΔE divided by the number of nucleons. First calculate the number of nucleons per mole of nuclei:

$$\text{For } ^{30}\text{P:} \quad \frac{6.02 \times 10^{23}\ \cancel{\text{nuclei}}}{1\ \text{mol}\ ^{30}\text{P}} \times \frac{30\ \text{nucleons}}{1\ \cancel{\text{nucleus}}} = \frac{180.6 \times 10^{23}\ \text{nucleons}}{\text{mol}\ ^{30}\text{P}}$$

$$\text{For } ^{31}\text{P:} \quad \frac{6.02 \times 10^{23}\ \cancel{\text{nuclei}}}{1\ \text{mol}\ ^{31}\text{P}} \times \frac{31\ \text{nucleons}}{1\ \cancel{\text{nucleus}}} = \frac{186.6 \times 10^{23}\ \text{nucleons}}{\text{mol}\ ^{31}\text{P}}$$

The average energy released per nucleon is then given by:

$$\text{For } ^{30}\text{P:} \quad \frac{-2.33 \times 10^{10}\ \text{kJ}}{1\ \cancel{\text{mol}\ ^{30}\text{P}}} \times \frac{1\ \cancel{\text{mol}\ ^{30}\text{P}}}{1.806 \times 10^{25}\ \text{nucleons}} = -1.29 \times 10^{-15}\ \frac{\text{kJ}}{\text{nucleon}}$$

$$\text{For } ^{31}\text{P:} \quad \frac{-2.54 \times 10^{10}\ \text{kJ}}{1\ \cancel{\text{mol}\ ^{31}\text{P}}} \times \frac{1\ \cancel{\text{mol}\ ^{31}\text{P}}}{1.866 \times 10^{25}\ \text{nucleons}} = -1.36 \times 10^{-15}\ \frac{\text{kJ}}{\text{nucleon}}$$

■

PROBLEM 23.3 Calculate the binding energy in kJ/mol for the formation of $^{10}_{5}\text{B}$. Isotopic weights: $^{10}_{5}\text{B}$, 10.01294; $^{1}_{1}\text{H}$, 1.00783; $^{1}_{0}\text{n}$, 1.00867. □

The greater the energy released per nucleon in the formation of a nucleus, the greater is the stability of the nucleus. We can therefore conclude that ^{31}P is more stable than ^{30}P. The nature of the force holding nucleons together must be fundamentally different from the electrical force involved in chemical bonding. Just as with chemical bonding, gravitational forces are too weak to be significant. Furthermore, the force *cannot* be electrical. For example, the deuteron, composed of one charged particle (the proton) and a neutral particle (the neutron), cannot be held together by electric forces. An even more striking illustration is the great stability of the nucleus of $^{3}_{2}\text{He}$ with two protons and one neutron; the electrostatic interaction of the two protons is repulsive.

Greek *mesos*, "intermediate"; the mass of the meson is between that of an electron and a nucleon.

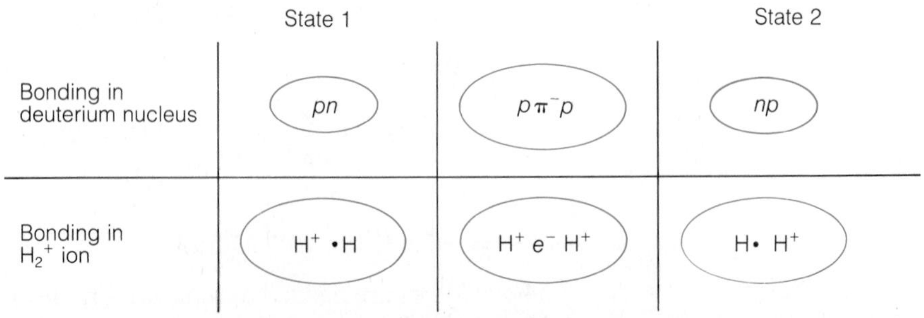

In 1935 Hideki Yukawa suggested that a *particle oscillating between nucleons* with practically the speed of light *is the force* that holds the nucleons together. The attractive force between nucleons is the consequence of the exchange of the particle back and forth between the nucleons. He further predicted that this exchange particle has a mass of about 275 ± 25 times the electron mass and may be electrically neutral, positive, or negative. These particles, called "pi-mesons" or "pions," were discovered later, and they are represented as π^0, π^+, π^-. As a result of the continuous charged pion transfer, protons are changed to neutrons and neutrons are changed to protons.

	State 1		State 2
Bonding in deuterium nucleus	pn	$p\pi^-p$	np
Bonding in H_2^+ ion	$H^+ \cdot H$	$H^+ e^- H^+$	$H \cdot H^+$

FIGURE 23.5
The deuterium nucleus and the H_2^+ ion illustrate the analogy between pion sharing in nuclei and electron sharing in molecules.

At any given instant, the nucleus has a fixed number of nucleons with a definite positive charge equal to the atomic number. At any given instant, two nucleons are sharing the same exchange particle. The sharing of pions by nucleons is analogous to the sharing of electrons by bonded atoms in molecules (Figure 23.5). The two attractive forces are alike in that they drop off sharply beyond a certain critical distance—about 10^{-8} cm between bonded atoms and about 10^{-13} cm between adjacent nucleons. As with chemical bonding, the closer the nucleons are, the stronger the nuclear bond is. However, a small separation of nucleons beyond 10^{-13} cm leads to fragmentation of the nucleus.

23.5 NUCLEAR REACTIONS

RADIOACTIVITY

Radioactivity is associated with unstable nuclei. However, it is not easy to predict whether a given isotope is radioactive or how it transforms (decays). We have already seen that greater nuclear stability is associated with a "magic number" of nucleons. Thus, while $^{208}_{82}Pb$ is stable, $^{210}_{82}Pb$ is radioactive.

Although $^{2}_{1}D$ and $^{3}_{1}T$ (tritium) exist, $^{2}_{2}He$, which lacks a neutron, does not exist. The influence of one neutron in stabilizing the nucleus of $^{3}_{2}He$ prompts us to look for a general relationship between nuclear stability and the neutron/proton ratio of nuclei. A plot of the number of neutrons against the number of protons in the nuclei of nonradioactive (stable) isotopes is shown in Figure 23.6. We observe that the stable nuclei fall within a narrow band, called the *stability band* or *belt.*

With the light elements, the ratio of N (number of neutrons) to Z (number of protons) is close to 1. With the heavier, stable elements, on the other hand, the

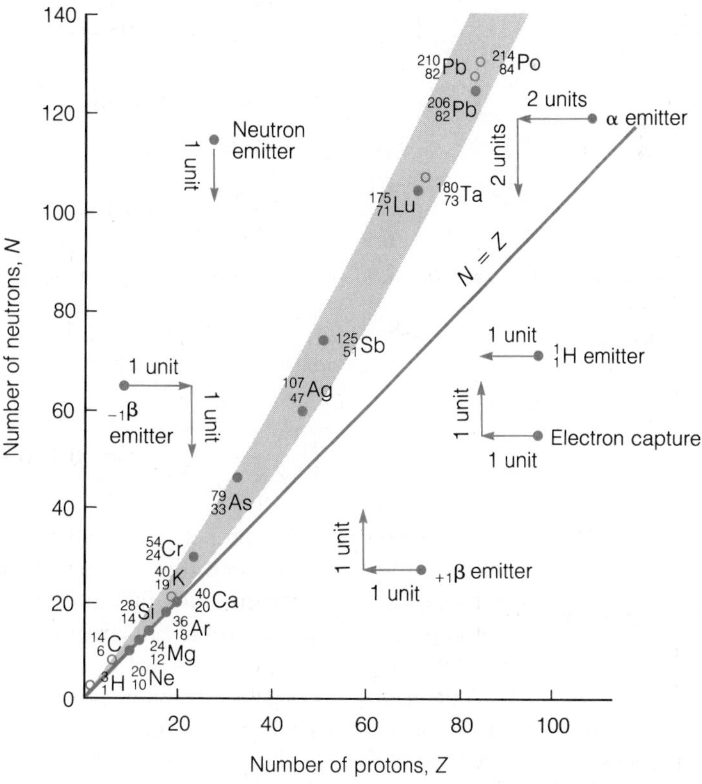

FIGURE 23.6
The stability band, approximately represented by the shaded area. The solid line locates nuclei with equal numbers of protons and neutrons. Several stable nuclei are included in the illustration to serve as reference points. A few naturally occurring radioactive nuclei are indicated by open circles.

ratio of neutrons to protons increases to about 1.5. Apparently, relatively more neutrons are needed in the heavier nuclei to dilute the electrostatic repulsion between the protons. If some of these neutrons were converted to protons, the electrostatic repulsion between the protons would make the nucleus unstable. All nuclei with atomic numbers greater than 83 are radioactive. Of these, only $^{235}_{92}U$, $^{238}_{92}U$, and $^{232}_{90}Th$ occur on Earth in relatively large amounts. The only stable nuclei that contain fewer neutrons than protons are $^{1}_{1}H$ and $^{3}_{2}He$. It is noteworthy that about 86 mole percent of the nuclei in the Earth's crust (excluding the hydrogen in water) have *even mass (nucleon) numbers.*

Their half-lives are comparable to the age of the Earth, 5×10^9 years.

Nuclei outside the stability band spontaneously transform (decay) to nuclei closer to or within the stability band. Nuclei *below* the stability band have a relative excess of protons and transform by emission of $_{+1}^{0}\beta$, $^{1}_{1}H$, or $^{4}_{2}He$ or by electron capture. These reactions decrease the number of protons, whereas the number of neutrons in a reactant nucleus is increased, decreased, or unchanged (Figure 23.6), depending upon the reaction. In short, they decay so that the product has a *greater* N/Z ratio:

$43N$ represents 43 neutrons, and $35Z$ represents 35 protons.

- *Positron emission* results in one less proton and one more neutron.

$$^{78}_{35}Br \longrightarrow ^{78}_{34}Se + _{+1}^{0}\beta$$

$$\left(\frac{43N}{35Z} \longrightarrow \frac{44N}{34Z} \quad \text{or} \quad 1.23 \longrightarrow 1.29\right)$$

- *Proton emission* results in one less proton.

$$^{43}_{21}Sc \longrightarrow ^{42}_{20}Ca + ^{1}_{1}H$$

$$\left(\frac{22N}{21Z} \longrightarrow \frac{22N}{20Z} \quad \text{or} \quad 1.05 \longrightarrow 1.10\right)$$

- *Alpha emission* results in two less protons and two less neutrons.

$$^{168}_{78}Pt \longrightarrow ^{164}_{76}Os + ^{4}_{2}He$$

$$\left(\frac{90N}{78Z} \longrightarrow \frac{88N}{76Z} \quad \text{or} \quad 1.15 \longrightarrow 1.16\right)$$

The captured electron is usually from the lowest-energy atomic orbital ($1s$ shell).

- *Electron capture* results in one less proton and one more neutron.

$$^{7}_{4}Be + \underset{\text{orbital electron}}{e^-} \longrightarrow ^{7}_{3}Li$$

$$\left(\frac{3N}{4Z} \longrightarrow \frac{4N}{3Z} \quad \text{or} \quad 0.750 \longrightarrow 1.33\right)$$

When you read Chapter 2 you may have wondered, "Why doesn't the electron fall into the nucleus, since they attract each other?" Now you have the answer. Sometimes the electron *does* fall into the nucleus. However, stable nuclei do not capture orbital electrons. For example, the highly endothermic process $^{1}_{1}H$ + orbital electron $\longrightarrow$ $^{1}_{0}n$ does not occur.

Thermodynamic principles are applicable to nuclear reactions. However, entropy changes are relatively so small that for nuclear reactions, $\Delta G = \Delta E$.

Nuclei *above* the stability band have a relative excess of neutrons and decay by $_{-1}^{0}\beta$ or $^{1}_{0}n$ emission. These reactions may increase the number of protons while decreasing the number of neutrons in a reactant nucleus. In short, they decay so that the product has a *smaller* N/Z ratio:

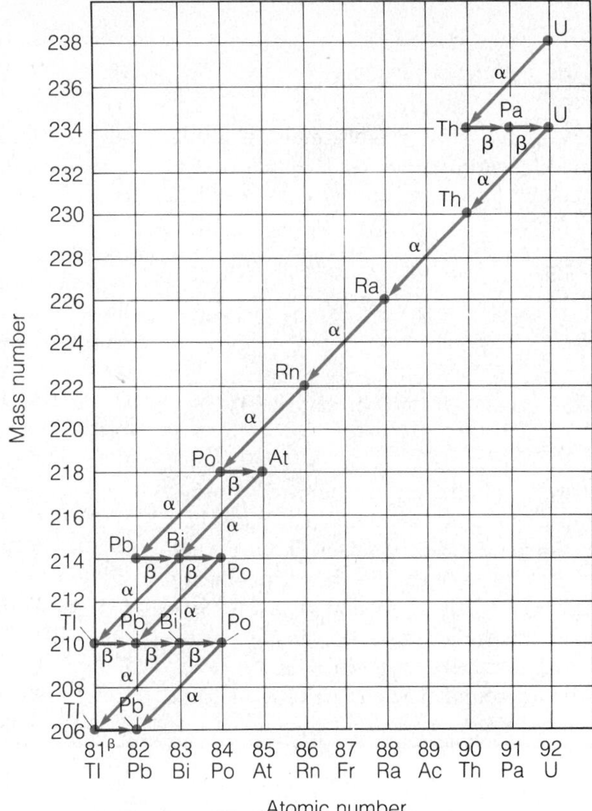

FIGURE 23.7
Uranium-238 radioactive series. $^{238}_{92}U$ is the parent; $^{206}_{82}Pb$, nonradioactive, is the end product. The radioisotopes emit either α or $_{-1}\beta$ particles. Occasionally, a fraction of the nuclei of an element—for example, $^{218}_{84}Po$—emits α particles, while the remainder emit $_{-1}\beta$ particles.

- *Electron emission* (beta decay) results in one more proton and one less neutron (Box 23.6):

$$^{87}_{36}Kr \longrightarrow {}^{87}_{37}Rb + {}^{0}_{-1}\beta$$

$$\left(\frac{51\ N}{36\ Z} \longrightarrow \frac{50\ N}{37\ Z} \quad \text{or} \quad 1.42 \longrightarrow 1.35\right)$$

- *Neutron emission* results in one less neutron.

$$^{87}_{36}Kr \longrightarrow {}^{86}_{36}Kr + {}^{1}_{0}n$$

$$\left(\frac{51\ N}{36\ Z} \longrightarrow \frac{50\ N}{36\ Z} \quad \text{or} \quad 1.42 \longrightarrow 1.39\right)$$

Like the reactant nucleus, the product nucleus of a nuclear reaction may not be stable. Several nuclear reactions may occur before a stable nucleus is finally formed. A *radioactive series* is a sequence of nuclear reactions beginning with an unstable nucleus (the *parent*) and terminating with a stable nucleus (the *end product*). $^{235}_{92}U$, $^{238}_{92}U$, and $^{232}_{90}Th$, the parents, respectively, of three natural radioactive series, have an excess of protons and decay mostly by α-particle emission. The successive transformations of the ^{238}U series are shown in Figure 23.7. A stable isotope of lead is the end product of each of these series: $^{235}_{92}U \longrightarrow {}^{207}_{82}Pb$, $^{238}_{92}U \longrightarrow {}^{206}_{82}Pb$, and $^{232}_{90}Th \longrightarrow {}^{204}_{82}Pb$. Thus, lead is an element whose isotopic composition is highly variable.

BOX 23.6
THE NEUTRINO AND THE LIFE OF THE UNIVERSE

A particle called the **neutrino**, ${}^{0}_{0}\nu$, is also emitted during β decay. This neutral particle, with no detectable mass, is needed to account for energy changes. β decay is a three-step process:

$$ {}^{78}_{35}\mathrm{Br} \longrightarrow {}^{78}_{34}\mathrm{Se} + \pi^{+}\text{(pion)} $$
$$ \pi^{+} \longrightarrow \mu^{+}\text{(meson)} + {}^{0}_{0}\nu $$
$$ \mu^{+} \longrightarrow {}^{0}_{+1}\beta + 2{}^{0}_{0}\nu $$

The net reaction, usually written without the $3{}^{0}_{0}\nu$, is

$$ {}^{78}_{35}\mathrm{Br} \longrightarrow {}^{78}_{34}\mathrm{Se} + {}^{0}_{+1}\beta + 3{}^{0}_{0}\nu $$

Neutrinos travel at almost exactly the speed of light. Having no electric charge, they are unaffected by electric forces. The interaction with nuclei is so weak that neutrinos can pass through the Earth without being captured. Consequently, the detection of a few neutrinos requires an intense beam passed through large, specially designed chambers filled with carbon tetrachloride deep underground. The carbon tetrachloride serves as a source of chlorine atoms, which capture neutrinos; ${}^{37}_{17}\mathrm{Cl} + {}^{0}_{0}\nu \longrightarrow {}^{37}_{18}\mathrm{Ar} + {}^{0}_{-1}\beta$. The detection of ${}_{-1}\beta$ particles announces the presence of neutrinos.

The neutrino is often referred to in debates about the ultimate fate of the Universe. There is no accepted evidence that it has a mass. However, if the neutrino has an atomic weight even as small as 10^{-9} (atomic weight of the electron is 5.5×10^{-4}), there are enough of them in the Universe to supply sufficient mass to cause eventual reversal of the expansion of the Universe. The ultimate consequence would be a gravitational collapse, perhaps followed by another big bang and. . . . The Moscow Institute for Theoretical and Experimental Physics has assigned an unconfirmed atomic weight of about $(1\text{–}5) \times 10^{-9}$ to the neutrino. Studies of the decay reaction ${}^{3}_{1}\mathrm{H} \longrightarrow {}^{3}_{2}\mathrm{He} + {}^{0}_{-1}\beta + {}^{0}_{0}\nu$, may settle the question of the neutrino mass and the life of the Universe. Meanwhile, we may rest comfortably with the thought that theoreticians have estimated that its atomic weight is 8.6×10^{-12}.

ARTIFICIAL TRANSMUTATIONS

Everyone has heard the stories of the alchemists who attempted to transform cheap metals into gold. Their dreams of artificial transmutation were shattered by Dalton's atomic theory, but were once again revived with the postulation of the nuclear transmutation theory of the atom.† The first induced transmutation was demonstrated in 1919 by Rutherford. He exposed nitrogen to α particles from radium and detected the production of protons:

$$ {}^{14}_{7}\mathrm{N} + {}^{4}_{2}\mathrm{He} \longrightarrow {}^{17}_{8}\mathrm{O} + {}^{1}_{1}\mathrm{H} $$

With the removal of the particle source, the reaction stops. In 1934, while treating light elements such as boron or aluminum with α particles, Irène and Frédéric Joliot-Curie detected the emission of positrons and neutrons. More important, they observed that the emission of positrons does not stop when the source of the α particles is removed. The reactions are

$$ {}^{10}_{5}\mathrm{B} + {}^{4}_{2}\mathrm{He} \longrightarrow {}^{13}_{7}\mathrm{N} + {}^{1}_{0}\mathrm{n} $$

followed by

$$ {}^{13}_{7}\mathrm{N} \longrightarrow {}^{13}_{6}\mathrm{C} + {}^{0}_{+1}\beta $$

${}^{13}_{7}\mathrm{N}$ was the first radioisotope to be produced artificially.

† The theory had to contend with a century of work that established the idea of unalterable atoms that could *not* be transmuted into other atoms. "Soddy turned to his colleague and blurted: 'Rutherford, this is transmutation!' Rutherford rejoined: 'For Mike's sake, Soddy, don't call it *transmutation*. They'll have our heads off as alchemists.' Rutherford and Soddy were careful to use the term 'transformation' rather than 'transmutation.'" (*Scientific American* (August 1966), 91.)

Before a positively charged particle can be captured by a nucleus, it must possess sufficient kinetic energy to overcome the repulsive force that develops as the positive particle approaches the positive nucleus. Particle accelerators, such as cyclotrons, were invented to increase the kinetic energy of charged particles to the levels required for capture by nuclei with high atomic numbers. With the accelerators available in 1934, it was not possible to induce radioactivity in the elements beyond potassium. However, a neutron is electrically neutral. Enrico Fermi reasoned that a neutron's entry into a nucleus would not be opposed by repulsive forces. It should be possible to transform all known elements by exposure to neutron sources. He was correct. Practically all known elements have been transformed, and many transuranium elements—elements with atomic numbers greater than 92—have been synthesized. For example,

The half-life of free neutrons is 12 minutes. This time is ample for using them in nuclear reactions.

$$^{238}_{92}U + ^{1}_{0}n \longrightarrow ^{239}_{92}U \longrightarrow ^{239}_{93}Np + ^{0}_{-1}\beta$$

$$^{239}_{93}Np \longrightarrow ^{239}_{94}Pu + ^{0}_{-1}\beta$$

Nuclear reactors, in which controlled fission reactions (next section) are carried out, are excellent sources of neutrons. Neutrons produced this way are used to bombard atomic nuclei to synthesize useful isotopes, such as ^{14}C:

$$^{14}_{7}N + ^{1}_{0}n \longrightarrow ^{14}_{6}C + ^{1}_{1}H$$

A solution of a nitrate salt is run through the reactor. Exposure to neutrons converts the nitrogen in the NO_3^- to $^{14}_{6}C$, creating ^{14}C-labeled carbonate ion, $^{14}CO_3^{2-}$. It is recovered and sold as solid $Ba^{14}CO_3$. Other isotope syntheses are the preparation of radioactive $^{60}_{27}Co$—a source of γ rays for cancer therapy—and tritium:

$$^{59}_{27}Co + ^{1}_{0}n \longrightarrow ^{60}_{27}Co \qquad ^{6}_{3}Li + ^{1}_{0}n \longrightarrow ^{3}_{1}H + ^{4}_{2}He$$

See Figure 23.8.

SYNTHESIS OF TRANSURANIUM ELEMENTS

0, *nil*; 1, *un*; 2, *bi*; 3, *tri*; 4, *quad*; 5, *pent*; 6, *hex*; 7, *sept*; 8, *oct*; 9, *enn*.

1 0 9	(atomic number)
un–nil–enn–ium	(name)
Une	(symbol)

The IUPAC-recommended names for elements 104, unnilquadium, Unq, and higher are derived from the numerical roots shown in the margin. All elements beyond $_{92}U$ have such short lives that they are not found in nature. $_{93}Np$ and $_{94}Pu$ (nuclear fuel and explosive) are easily synthesized by exposure of $^{238}_{92}U$ to neutrons as shown above. Heavier elements up to $_{101}Md$ may be synthesized by intense neutron bombardment of $_{94}Pu$ in which the addition of neutrons is followed by $_{-1}\beta$ decay or by alpha-particle bombardment, for example:

$$^{239}_{94}Pu + 4^{1}_{0}n \longrightarrow ^{243}_{94}Pu \longrightarrow ^{243}_{95}Am + ^{0}_{-1}\beta$$

$$^{243}_{95}Am + ^{1}_{0}n \longrightarrow ^{244}_{95}Am \longrightarrow ^{244}_{96}Cm + ^{0}_{-1}\beta$$

$$^{242}_{96}Cm + ^{4}_{2}\alpha \longrightarrow ^{245}_{98}Cf + ^{1}_{0}n$$

The elements past $_{101}Md$ have been synthesized by bombardment of nuclei with heavier nuclei, for example:

$$^{252}_{98}Cf + ^{10}_{5}B \longrightarrow ^{257}_{103}Lw + 5^{1}_{0}n$$

$$^{58}_{26}Fe + ^{209}_{83}Bi \longrightarrow ^{267}_{109}Une \longrightarrow ^{266}_{109}Une + ^{1}_{0}n$$

$$^{266}_{109}Une \longrightarrow ^{4}_{2}\alpha + ^{262}_{107}Uns \longrightarrow ^{4}_{2}\alpha + ^{258}_{105}Unp \longrightarrow ^{258}_{104}Unq + ^{0}_{-1}\beta$$

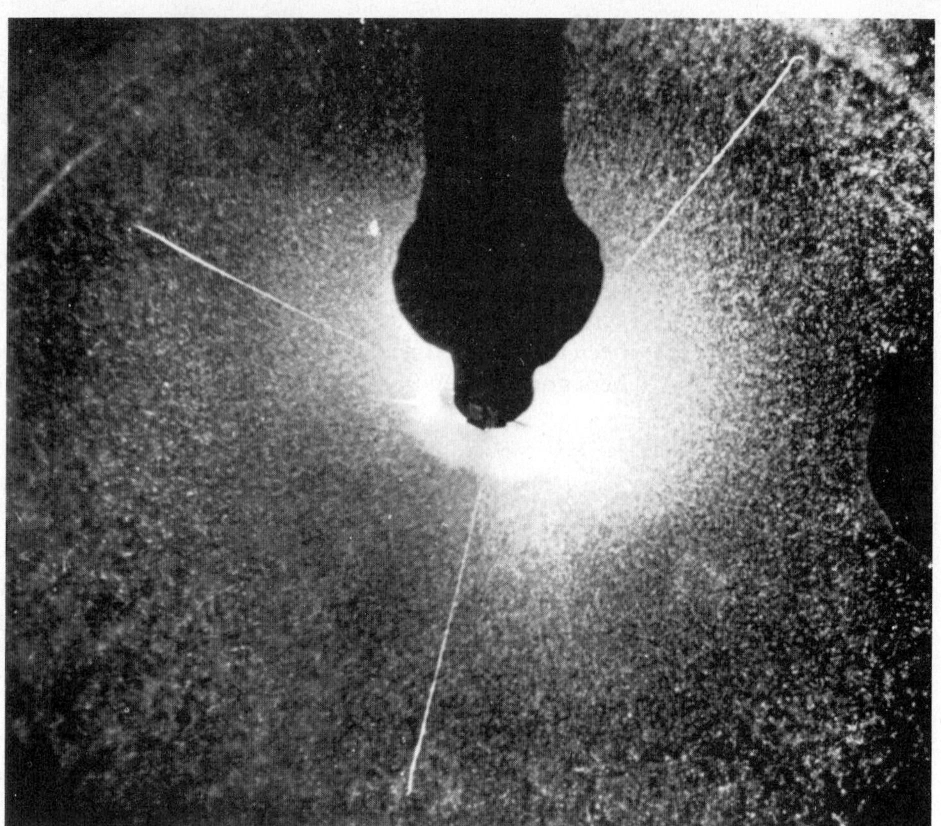

FIGURE 23.8
The transmutation of boron into helium by proton bombardment, $^{11}_{5}B + ^{1}_{1}H \longrightarrow 3^{4}_{2}He$. The three lines are the tracks of the α particles. (Photographed by P. I. Dee and C. W. Gilbert, *Proceedings of the Royal Society* A 154, 296 (1936).)

23.6 NUCLEAR FISSION

Figure 23.9 is a plot of the average energy released per nucleon when a nucleus is formed from nucleons. The more energy released per nucleon, the greater is the stability of the nucleus. As you can see in the graph, the very light and very heavy nuclei are less stable than the nuclei in the vicinity of $^{56}_{26}Fe$. We might expect that spontaneous changes should include the fusion of very light nuclei into heavier nuclei, and the **fission** (splitting) of a very heavy nucleus into two nuclei. These reactions should be rich sources of energy compared with chemical changes.

Spontaneous fission has been detected in $^{232}_{90}Th$ and heavier elements. However, the rate is too slow for any practical purpose because of the high energy of activation required to produce the deformation of the nucleus (a slight separation of nucleons) necessary for fission. The energy of activation (roughly 5.9×10^8 kJ/mol) is supplied by a neutron. The binding energy of the added neutron suffices to produce an unstable deformation of the nucleus (Figure 23.10). This unstable intermediate undergoes fission immediately. For example,

$$^{235}_{92}U + ^{1}_{0}n \longrightarrow ^{236}_{92}U \longrightarrow ^{137}_{53}I + ^{97}_{39}Y + 2^{1}_{0}n \qquad (2)$$

This situation, illustrated in Figure 23.10, is comparable to a two-step chemical reaction (section 19.9).

Section of the periodic table accepted in the thirties:

$_{73}Ta$ $_{74}W$ $_{75}Re$ $_{76}Os$

$_{91}Pa$ $_{92}U$ $_{93}?$ $_{94}?$

In 1934 Fermi claimed that the synthesis of transuranium elements occurs by neutron bombardment of uranium. However, many investigators doubted this interpretation of the chemical analytical results. In particular, Ida Noddack suggested (1934) that the uranium nuclei were split into smaller nuclei. Practically

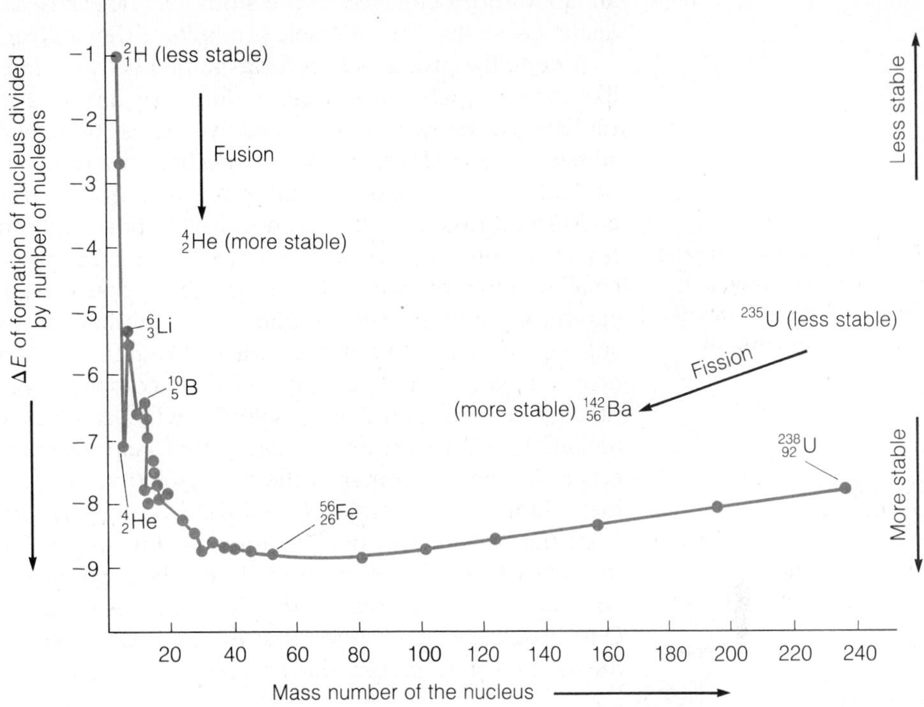

FIGURE 23.9
The average energy (in MeV) evolved per nucleon in the formation of nuclei from the separate nucleons (1 MeV/nucleon = 1.60×10^{-16} kJ/nucleon).

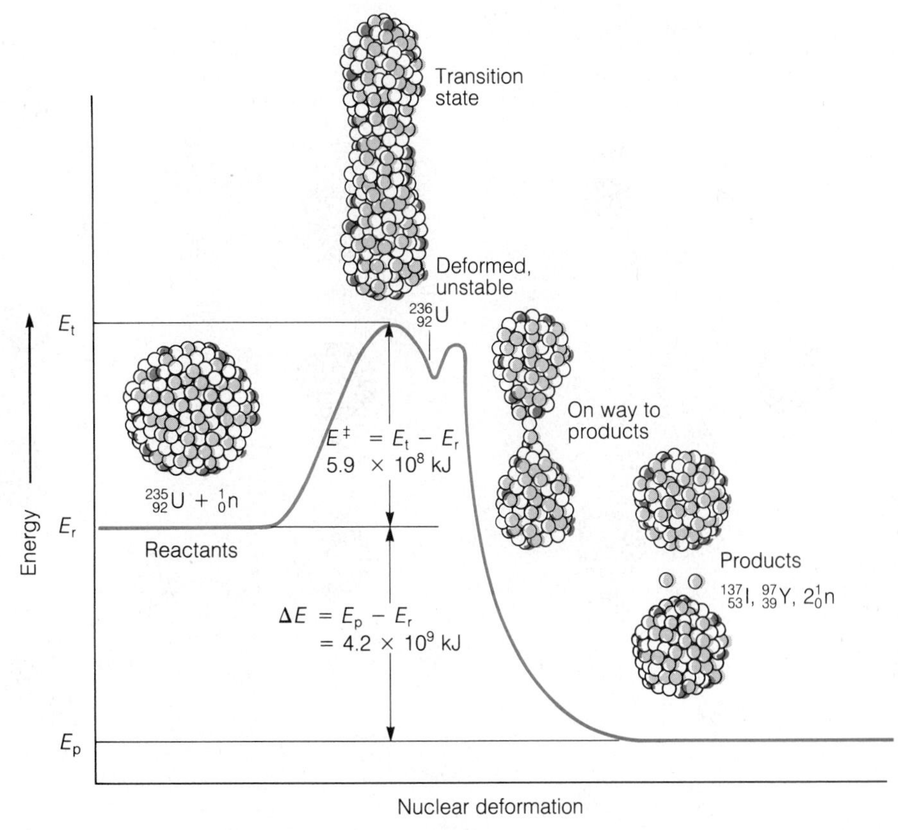

FIGURE 23.10
The change in energy with deformation of a nucleus. For a chemical reaction, *molecular deformation* would be plotted on the horizontal axis. (Not drawn to scale.)

all laboratories engaged in the study of nuclear reactions at that time began an analytical study of the possible products of the neutron bombardment of uranium.

Among the products found was a radioactive element that behaved chemically like the Group-2 elements (strontium, barium, and radium). Among these, only radium was known to be radioactive. Thus the new radioactive element was assumed to be radium. However, because its radioactivity differed from that of natural radium, it was thought to be some isotope of radium. In 1939 Otto Hahn and Fritz Strassman attempted to isolate this "radium" by adding a larger quantity of barium to serve as a "carrier." The carrier was used to help separate the small quantity of radium from the other impurities. This was followed by a final careful separation of the radium from the barium by crystallization. Barium is still a good carrier to remove a trace of radium out of a mixture because the properties of barium and radium and their compounds are similar. For example, a precipitate of $BaCO_3$ from a solution will also carry with it any trace of radium as $RaCO_3$. However, this similarity that helps the barium carry the radium from other elements also makes the final isolation of the desired radium difficult. In fact, Hahn and Strassman found that the final separation of the new "radium" from the barium was not just difficult, but *impossible*. They ultimately showed that the radioactive element could not be separated from barium for the simple and yet unexpected reason that it was an isotope of *barium*! Lise Meitner and Otto Frisch then concluded that the uranium nucleus had divided into two fragments, and they named the process *fission*. It was immediately realized that a

This was the method used by Marie Sklodowska Curie and Pierre Curie when they discovered radium.

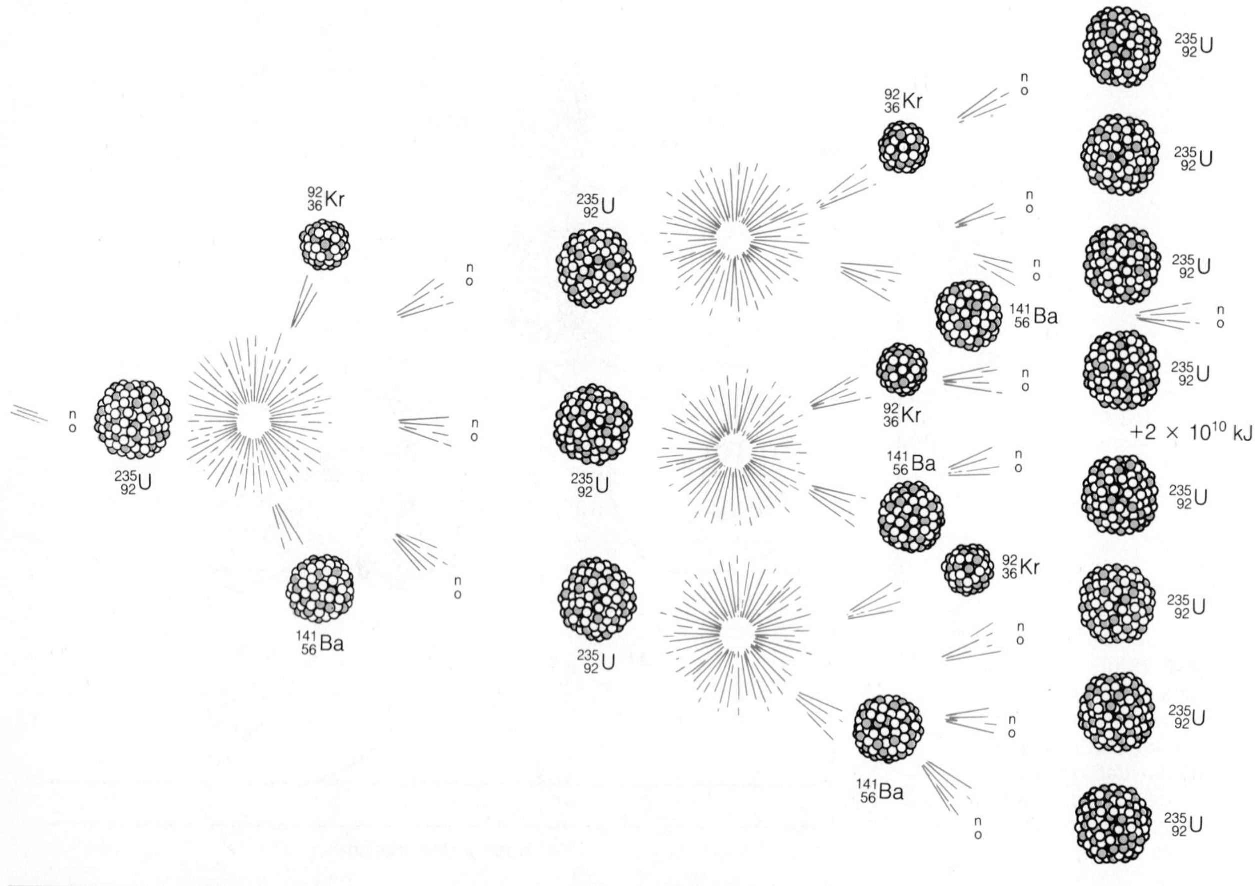

FIGURE 23.11 Illustration of a branching-chain reaction initiated by the capture of a stray neutron. (This reaction produces many pairs of different isotopes, but only one kind of pair is shown.)

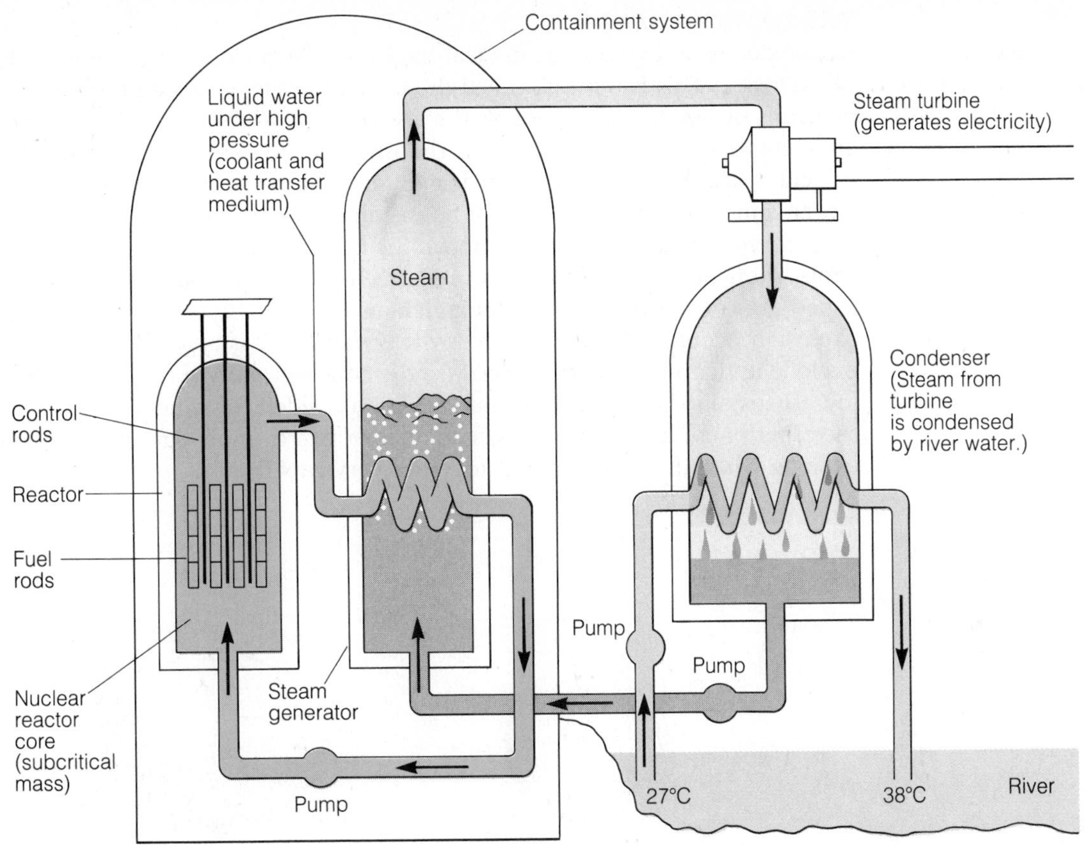

FIGURE 23.12
Essential parts of a conventional (USA) nuclear power plant.

tremendous amount of energy would accompany this division. This was verified by the large readings produced by the energy in an ionization chamber. Similar readings obtained by previous investigators had been dismissed as "electronic noises." Nuclear fission was thus discovered by chemists.

From the principle of conservation of charge, the other nucleus formed with $_{56}$Ba must be $_{36}$Kr. A typical fission reaction is

See equation 2 (section 23.6); fission may occur in many ways. For example, $^{236}_{92}U$ may yield $^{144}_{58}Ce + ^{88}_{34}Se + 4^{1}_{0}n$, $^{90}_{38}Sr + ^{143}_{54}Xe + 3^{1}_{0}n$, or $^{99}_{43}Tc + ^{131}_{49}In + 6^{1}_{0}n$.

$$^{235}_{92}U + ^{1}_{0}n \longrightarrow ^{236}_{92}U \longrightarrow ^{142}_{56}Ba + ^{91}_{36}Kr + 3^{1}_{0}n \qquad \Delta E = -1.9 \times 10^{10}\ \text{kJ/mol}$$

In nuclear fission, more than one neutron is produced for each one consumed, and a ^{236}U nucleus undergoes fission in about 10^{-9} sec. Thus it is possible to develop a branching-chain reaction (section 19.11) that leads to a nuclear explosion (incorrectly called an "atomic explosion"). Unprecedented quantities of energy are released in extremely short periods of time (Figure 23.11). However, through the use of "chain breakers"—materials such as cadmium that absorb neutrons—the number of neutrons available for the fission of $^{235}_{92}U$ may be decreased by about two-thirds. The rate of the fission reaction is thereby decreased and an explosion is prevented. In this way, the reaction is controlled, and the heat of the reaction can be utilized as a practical energy source (Figure 23.12).

A typical capture reaction is

$$^{113}_{48}Cd + ^{1}_{0}n \longrightarrow ^{4}_{2}\alpha + ^{110}_{46}Pd$$

Chain breakers play the role of the termination step in a free-radical mechanism (section 19.11).

The first controlled and sustained fission reaction (nuclear reactor) was operated in 1942 at the University of Chicago. The first branching-chain fission reaction was successfully carried out in a test explosion with ^{239}Pu in 1945 near Alamogordo, New Mexico.

THE CRITICAL MASS

1 mg of gunpowder explodes as well as, if not better than, a 1-kg sample.

Unlike chemical explosions, nuclear explosions require a certain minimum mass of explosive. This is a mass, called the **critical mass**, at which the production of neutrons by fission exactly equals the loss of neutrons by nonfission capture (such as by impurities) and by escape. If the mass of the explosive is smaller than the critical mass, the loss of neutrons is greater than the gain by fission reactions and a branching-chain reaction cannot be sustained. Consequently, the fissionable isotope *cannot explode*. However, above the critical mass, the escape path for neutrons is longer, and more of them react before they get out. As a result, there is a net *gain* of neutrons, a branching-chain reaction is sustained, and a nuclear explosion occurs. *Nuclear reactors are subcritical*; it is impossible for them to explode under any conditions because the concentration of the nuclear fuel is too low to sustain the branching-chain reaction needed for a nuclear explosion. On the other hand, a nuclear weapon consists of two or more subcritical masses that explode when they are brought together very rapidly.

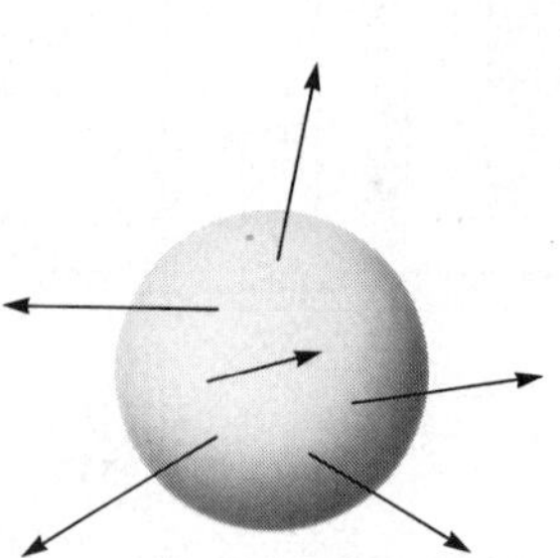

Subcritical mass
Too many neutrons escape to sustain a branching–chain reaction, so there are not enough for a nuclear explosion.

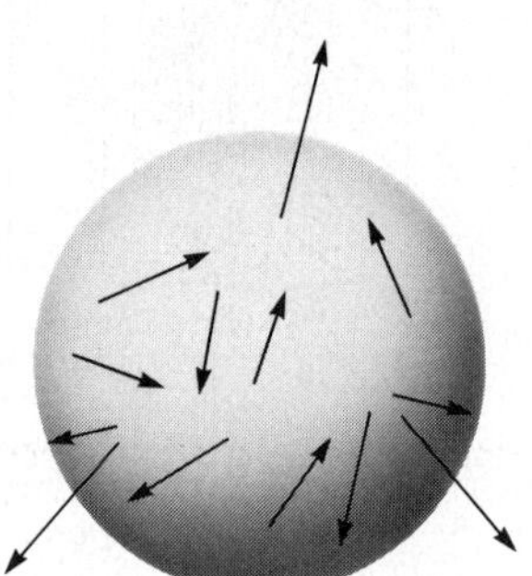

Supercritical mass
Most neutrons are captured, sustaining a branching–chain reaction. There are sufficient neutrons for a nuclear explosion.

23.7 NUCLEAR FISSION REACTORS

Recall that the function of any power plant is to drive a turbine to generate electricity. The only difference among power plants is the source of the energy that drives the turbines. In a nuclear plant, the source of energy is a nuclear fission reaction. ^{235}U (fissionable but not abundant) and ^{239}Pu (does not occur in nature but may be manufactured from abundant ^{238}U) are the fissionable isotopes on which the current nuclear fission energy program in the United States is based.

The essential parts of a conventional (USA) nuclear power plant are shown in Figure 23.12. Liquid water at a pressure of 130 atm is heated to about 325°C by being circulated through the reactor. The hot water converts other water to steam in the steam generator. The steam so produced is used to drive an electric generator. The steam is condensed back to liquid water, which then returns to the steam generator, where the water is again converted to steam, and so forth.

The main components of a nuclear reactor are

- A nuclear fuel (^{235}U, ^{239}Pu). Typically the fuel is a ceramic form of uranium dioxide, UO_2. This form is much superior to metals in its ability to retain most fission products, even when overheated. The uranium fuel consists of natural uranium (99.3% ^{238}U and 0.7% ^{235}U), enriched if necessary to about 3% ^{235}U. The fuel is inserted in the form of pellets into long, thin tubes, "fuel rods," made of a zirconium alloy.

These fuel rods are then bundled into assemblies that are inserted into the reactor core (Figure 23.12).

Ordinary water is used as the moderator in a "light"-water reactor.

- A moderator (H_2O, D_2O, graphite). Nuclear fission produces fast neutrons that must be slowed down so they can be captured by fissionable nuclei. They are slowed down by collisions with nuclei of substances that do not capture neutrons. Such a substance is called a *moderator*. Almost all reactors in the United States use ordinary "light" water (H_2O) as the moderator, which is very convenient because it can also serve as the coolant.

  Although water is cheap and can serve two purposes, H nuclei do capture some neutrons; ${}^1_1H + {}^1_0n \longrightarrow {}^2_1H$. Consequently, light-water reactors cannot use natural uranium. Instead, the ${}^{235}U$ concentration must be enriched to make up for the neutron loss to the light water.

  On the other hand, D_2O is nearly the ideal moderator. However, it is an extremely expensive substance, concentrated by the electrolysis of ordinary water (0.015% D). The advantage of D_2O is that it allows the use of natural rather than enriched uranium. D_2O is the basis of the Canadian CANDU design (Canadian, D_2O, natural U).

  Pure graphite was the moderator in the first successfully operated nuclear reactor.

- A coolant (H_2O, $Na(\ell)$, $He(g)$) for transferring heat from the reactor core to an external boiler. Without a cooling fluid that circulates through the reactor core, the heat generated would *melt down* the metal-clad fuel rods in the reactor core.

- A system of control rods (containing Cd or B) to control the rate of the fission reaction by nonfission capture of neutrons; ${}^{113}_{48}Cd + {}^1_0n \longrightarrow {}^4_2\alpha + {}^{110}_{46}Pd$, ${}^{10}_5B + {}^1_0n \longrightarrow {}^4_2\alpha + {}^7_3Li$. When all control rods are fully set between the fuel rods, the fission reaction is completely stopped. As the control rods are withdrawn, fewer neutrons are absorbed, and the fission reaction occurs at a faster rate. The control rods are set at the elevation where the reaction produces the desired temperature.

  However, the control rods also serve other important functions. Fission products (impurities) capture neutrons. If there is no control of neutrons, the neutron flow would gradually slow down and the fission reaction would die out before much fuel is used up. For this reason, the neutron flow must be gradually increased during the life of the fuel to make up for the loss caused by the impurities. The way to do this is to design the system for an extra large neutron flow at the outset and then limit the actual flow by the control rods. As fuel is consumed, the control rods are gradually withdrawn to compensate for the accumulation of neutron-absorbing impurities. This amounts to a neat balancing of impurities to maintain steady power production. The other purpose of the control rods is to serve as an emergency shut-off system. When the fission reaction must be quenched, the rods are pushed rapidly all the way into the core.

- A containment system to prevent the escape of radioactive nuclides. The entire reactor is enclosed in thick steel walls and housed in thick, high-density steel–reinforced concrete. This containment system imposes lower limits on reactor size—there will never be pocket-sized generators nor even fission engines for motorcycles or motor vehicles.

56 nuclear power plants are under construction; 80 plants have been canceled since 1970, and several outworn reactors are closed.

When fissionable components have been consumed and the fission reaction stops, the heat released by decay of fission products can melt the rods. Spent rods are therefore stored in water pools on site until they can be transported to a radioactive waste disposal facility.

At present, about eighty nuclear power plants supply approximately 15% of the electricity generated in the United States.

BREEDER REACTOR

"Breeding" is a nuclear process in which more nuclear fuel is produced than is consumed. ^{235}U captures slow neutrons and undergoes fission. The more abundant ^{238}U has a higher energy of activation, and slow neutrons do not split it. However, the capture of fast neutrons by ^{238}U produces fissionable ^{239}Pu. Thus, some ^{239}Pu is produced in ordinary nuclear reactors. However, in a breeder reactor, conditions are established so as to *manufacture* ^{239}Pu from ^{238}U, obtaining more than one fissionable ^{239}Pu atom for each fissionable ^{239}Pu consumed. Thus, the fission of ^{239}Pu yields 3 neutrons:

Breeder power stations are in operation in the U.S.S.R., France, Great Britain, and West Germany.

$$^{239}_{94}Pu + ^{1}_{0}n(\text{slow}) \longrightarrow ^{137}_{55}Cs + ^{100}_{39}Y + 3^{1}_{0}n(\text{fast}) + \text{heat of fission}$$

One neutron is used to fission ^{239}Pu, and the other two are used to convert two atoms of ^{238}U to ^{239}Pu,

$$2^{238}_{92}U + 2^{1}_{0}n(\text{fast}) \longrightarrow 2^{239}_{94}Pu + 4^{\ 0}_{-1}\beta$$

leaving a gain of one atom of fissionable ^{239}Pu. In this way, a breeder reactor provides a way of using nonfissionable ^{238}U to manufacture more fuel than it consumes—an attractive means of utilizing uranium resources more completely. ^{239}Pu, however, is the stuff of nuclear weapons.

However, the use of ^{232}Th offers a breeding effect *without* producing plutonium. When ^{232}Th is mixed with ^{235}U, the ^{232}Th (which is abundant but nonfissionable) converts to fissionable ^{233}U (which does not occur in nature), but plutonium is not produced.

No moderator is used in a breeder reactor. This allows ^{239}Pu to produce enough fast neutrons for breeding. The heat output is too rapid for water to be an effective coolant. (Furthermore, water is a moderator, which is unwanted in a breeder reactor.) Liquid sodium is used as the coolant. It becomes highly radioactive, but its saving virtue is that it remains in the liquid state over a very wide temperature range—98°C to 883°C at atmosphere pressure—and it is an excellent heat conductor, capable of transferring heat away from the reactor core rapidly. The steam generator is shielded from the radioactive sodium by a second, nonradioactive, sodium coolant system.

$Na(\ell)$ reacts with water to produce H_2; if air is also present, the heat of the reaction can spark an explosion of the hydrogen.

OTHER REACTORS

Some proponents of a nuclear future believe that the public's confidence in nuclear energy—shattered by the Three Mile Island (Pennsylvania) and the Chernobyl (USSR) nuclear incidents—can be regained only by designing an inherently safe reactor. Such a reactor would need to operate without the intervention of human operators and without the use of electronic and mechanical devices for insuring safety. It should depend on the physical and chemical behavior of all parts and afford effective protection against accidents caused by equipment or operator failure, earthquake, sabotage, or attack with conventional (non-nuclear) explosives. Such an inherently safe reactor may not be an impossible dream. In fact, such reactors are being developed.

One such innovation is the so-called pebble-bed reactor under construction in West Germany and the United States. This design does away with the metal-clad fuel rods now used. Tiny uranium-235 oxycarbide particles are embedded in graphite and silicon carbide and then clad in "pebbles," small graphite spheres, about 6 cm in diameter. A number of advantages are realized: (1) Graphite, which serves both as the cladding (covering for the fuel) and the moderator, remains solid up to 3925 K, above which temperature it vaporizes rather than melts. (2) The graphite pebbles have a larger heat capacity than metals, making the temperature rise much slower than in metal-clad fuel rods. (3) The pebbles serve to retain fission products. (4) The coolant is helium, so no water comes in contact with the reactor core. (5) The design permits the continuous examination of the fuel spheres during operation. (6) As an added feature, the plant output is about half that of the conventional nuclear power plant, and the addition of ^{232}Th has a desirable breeding effect. The small size and low power density of the reactor (only 3 kW per liter) and other schemes eliminate the need for control rods: if the coolant were lost, the nuclear chain reaction would be *self-terminated* after a modest rise in temperature. The reactor is small and has a large surface-to-volume ratio so that thermal radiation and heat loss to the surroundings would stabilize the reactor temperature at a safe level. Finally, all coatings and graphite blocks, core, and moderator are chemically and physically stable above temperatures that could possibly be reached even in the most severe scenario. A meltdown is therefore practically impossible.

23.8 NUCLEAR FUSION

Tremendous amounts of energy can be released by the **fusion** of small nuclei into larger nuclei. The largest energy release occurs with the lightest elements. Two of the more important reactions are

$$^{2}_{1}H + ^{3}_{1}H \longrightarrow ^{4}_{2}He + ^{1}_{0}n \qquad \Delta E = -1.8 \times 10^{9} \text{ kJ}$$

$$^{2}_{1}H + ^{2}_{1}H \longrightarrow ^{3}_{2}He + ^{1}_{0}n \qquad \Delta E = -2.9 \times 10^{8} \text{ kJ}$$

Energy from the fusion of one pair of nuclei induces the fusion of other pairs. Such a sustained reaction results in an explosion, specifically described as **thermonuclear**.

Recall that the $^{1}_{1}H$ nucleus is a proton.

A tank of hydrogen does not explode to form deuterium (D or $^{2}_{1}H$) (see equations below), because the rate of the reaction is infinitesimally small. The rate is proportional to the number of collisions between the H nuclei (not the H atoms or H_2 molecules). However, the number of collisions between nuclei is essentially zero because of repulsive electrostatic forces. To increase the number of collisions between nuclei, the temperature of the reactants must be increased to about 10^7 K. At these temperatures, atoms of light elements are completely stripped of electrons, forming a **plasma**, a gas containing equivalent numbers of positive ions and electrons. At these extremely high temperatures, the kinetic energy of the nuclei can overcome repulsive forces. Therefore, the number of collisions between H nuclei increases greatly. For the hydrogen isotopes, the probability of reaction per collision appears to be highest for the reaction

$$^{3}_{1}H + ^{3}_{1}H \longrightarrow ^{4}_{2}He + 2^{1}_{0}n$$

and least for the reaction

$$^{1}_{1}H + ^{1}_{1}H \longrightarrow ^{2}_{1}H + ^{0}_{+1}\beta$$

The hydrogen bomb is now out of military style; improvements in the accuracy of delivery systems make gigantic fusion explosions unnecessary.

The hydrogen bomb makes use of nuclear fusion. The high temperature needed to activate and fuse nuclei is supplied by a fission bomb. To date, ways of controlling nuclear fusion for the generation of electricity have not been devised. One of the major stumbling blocks is the inability to handle the plasma. It must be kept away from the walls of its container. Contact with the wall cools the plasma and stops the reaction. The wall may even melt or vaporize. Also, the very reactive ions attack the wall, forming metal hydrides (deuterides and tritides). Several schemes have been tried, but experiments in which ions are trapped by magnetic fields show the greatest potential for success. The fusion reaction is

$$^{2}_{1}H + ^{3}_{1}H \longrightarrow ^{4}_{2}He + ^{1}_{0}n$$

$^{2}_{1}H + ^{2}_{1}H \longrightarrow ^{3}_{1}H + ^{1}_{1}H$

but the reactor must also breed ^{3}H, which is radioactive and extremely rare in nature. The ^{2}H–^{3}H fuel must be heated to about 10^{8}°C, six times hotter than the interior of the sun—estimated to be 1.5×10^{7}°C.

In this design, extremely strong magnetic fields squeeze the plasma into a doughnut shape and heat it while preventing it from reaching the walls. The largest experimental fusion reactors are based on a Soviet design, the *Tokamak*—from the Russian acronym *to* (*toroidal*, "doughnut-shaped") *ka* ("chamber") *mak* ("magnetic"). The European Economic Community, the Japanese, and the French are also experimenting with tokamaks. The break-even point (energy output equals energy input) may be reached before 1990. In fact, a plasma temperature of 2×10^{8} K has been reached in experiments in the Tokamak Fusion Test Reactor at Princeton, New Jersey (Color plate 15). However, no one can yet predict when a self-sustaining commercial device for generating electricity economically will be achieved. Nevertheless, because of unpleasant experiences with fission reactors, utility companies may prefer to stay with gas, oil, and coal. There also appears to be serious concern that fusion as it is now being developed may be too expensive and unreliable for commercial use. While the helium product ($^{2}H + {}^{3}H \longrightarrow {}^{4}He + {}^{1}n$) is benign, the neutrons produced can damage and induce radioactivity (section 23.5) in the structural components of the reactor. Instead, the reaction recommended for study, $^{1}_{1}H + {}^{11}_{5}B \longrightarrow 3\,^{4}_{2}He$, is superior to the D-T reaction; neither the fuel nor the products are radioactive and no neutrons that can induce radioactivity are produced.

Natural boron is 81% ^{11}B.

SOLAR ENERGY

The mean temperature of the sun's interior, about 1.5×10^{7}°C, suffices for fusion among ^{1}H, ^{2}H, and ^{3}He,

$$^{1}_{1}H + ^{1}_{1}H \longrightarrow ^{2}_{1}H + ^{0}_{+1}\beta \qquad \text{(slow step)}$$

$$^{2}_{1}H + ^{1}_{1}H \longrightarrow ^{3}_{2}He$$

$$^{3}_{2}He + ^{3}_{2}He \longrightarrow ^{4}_{2}He + 2\,^{1}_{1}H$$

Once the sun has synthesized ^{3}He and ^{4}He, further nuclear reactions lead to the production of ^{12}C. The ^{12}C then starts a carbon-nitrogen-oxygen cycle:

$$^{12}_{6}C + ^{1}_{1}H \longrightarrow ^{13}_{7}N \longrightarrow ^{13}_{6}C + ^{0}_{+1}\beta$$

$$^{13}_{6}C + ^{1}_{1}H \longrightarrow ^{14}_{7}N$$

$$^{14}_{7}N + ^{1}_{1}H \longrightarrow ^{15}_{8}O \longrightarrow ^{15}_{7}N + ^{0}_{+1}\beta$$

$$^{15}_{7}N + ^{1}_{1}H \longrightarrow ^{12}_{6}C + ^{4}_{2}He$$

Other reactions, still unknown, also contribute to the energy output of the Sun.

The last reaction replenishes ^{12}C and closes the cycle; ^{12}C serves as a catalyst. The overall reaction is the fusion of four protons to form an He atom:

$$4\,^{1}_{1}H \longrightarrow {}^{4}_{2}He + 2\,{}^{\;\,0}_{+1}\beta \qquad \Delta E = -2.38 \times 10^9 \text{ kJ}$$

23.9 RATE OF RADIOACTIVE DECAY

It is impossible to predict when, if ever, a particular atom or molecule will react. Fortunately, the chemist is rarely concerned with the behavior of a single particle. Even in a small mass of radioactive material, there are many trillions of atoms. For example, 1 mg of ^{238}U contains 2.5×10^{18} atoms. With such large numbers of atoms or molecules, it is possible to make precise predictions about reaction (decay) rates. Radioactive decay rates are directly proportional to the number of atoms present: if 10^{20} atoms of a radioisotope have a decay rate of 1000 atoms per minute, then 4×10^{20} atoms of the same isotope will have a decay rate of 4000 atoms per minute. Radioactive decay is a typical first-order reaction,

$$rate = k \times N$$

where N is the number of atoms. The decay rates of radioisotopes are nearly always given in terms of their half-lives (Table 23.2). These values are direct measures of the stabilities of the isotopes: *the shorter the half-life, the faster is the rate of decay*. The half-life is a characteristic of each radioisotope and it is independent of the number of atoms.

The half-life of $^{15}_{8}O$ is 2.0 minutes. Given 8 mg of this isotope, 4 mg would remain after 2.0 minutes, 2 mg would remain after 4.0 minutes, and 1 mg would remain after 6.0 minutes. The relationship between q_t, the quantity at time t, and q_0, the initial quantity, is

$$rate_t = k \times N_t$$

$$rate_0 = k \times N_0$$

and

$$\frac{rate_t}{rate_0} = \frac{N_t}{N_0} = \frac{q_t}{q_0}$$

$$q_t = q_0 \times (0.5)^{t/t_{1/2}} \tag{3}$$

(section 19.4). $t_{1/2}$ is the half-life and $t/t_{1/2}$ gives the number of half-lives. Rates in atoms per unit time may be substituted for q_t and q_0 in equation 3, since the ratio of the rates is the same as the ratio of the number of atoms.

EXAMPLE 23.5 Radon samples collected from radium bromide and sealed in capillary tubing were formerly used for cancer treatment. The half-life for radon, $^{222}_{86}Rn \longrightarrow {}^{218}_{84}Po + {}^{4}_{2}He$, is 3.82 days. (a) Starting with 0.0200 mg, how many milligrams will remain after 14.50 days? (b) How many milligrams will have decayed (reacted)? (c) A radon sample initially emitted 7.0×10^4 α particles per second; it now emits 2.1×10^4 α particles per second. What is the age of the sample?

TABLE 23.2
HALF-LIVES OF SOME RADIOISOTOPES

ISOTOPE	DECAY REACTION	$t_{1/2}$
$^{14}_{6}C$	$^{14}_{6}C \longrightarrow {}^{14}_{7}N + {}_{-1}\beta$	5.73×10^3 years
$^{60}_{27}Co$	$^{60}_{27}Co \longrightarrow {}^{60}_{28}Ni + {}_{-1}\beta$	5.272 years
$^{59}_{26}Fe$	$^{59}_{26}Fe \longrightarrow {}^{59}_{27}Co + {}_{-1}\beta$	44.51 days
$^{3}_{1}H$	$^{3}_{1}H \longrightarrow {}^{3}_{2}He + {}_{-1}\beta$	12.3 years
$^{131}_{53}I$	$^{131}_{53}I \longrightarrow {}^{131}_{54}Xe + {}_{-1}\beta$	8.040 days
$^{32}_{15}P$	$^{32}_{15}P \longrightarrow {}^{32}_{16}S + {}_{-1}\beta$	14.28 days
$^{222}_{86}Rn$	$^{222}_{86}Rn \longrightarrow {}^{218}_{84}Po + {}^{4}_{2}He$	3.8234 days
$^{90}_{38}Sr$	$^{90}_{38}Sr \longrightarrow {}^{90}_{39}Y + {}_{-1}\beta$	29 years
$^{238}_{92}U$	$^{238}_{92}U \longrightarrow {}^{206}_{82}Pb + 8\,{}^{4}_{2}\alpha + 6\,{}^{\;\,0}_{-1}\beta$	4.468×10^9 years

ANSWER

(a) Substitute the given data into equation 3:

$$q_t = 0.0200 \text{ mg} \times (0.5)^{14.50 \text{ days}/3.82 \text{ days}}$$

$$\log q_t = \log 0.0200 + \frac{14.50 \text{ days}}{3.82 \text{ days}} \log 0.5$$

$$= -1.699 - 1.143 = -2.842$$

$$q_t = 1.44 \times 10^{-3} \text{ mg}$$

The quantity left is 1.44×10^{-3} mg.

(b) The quantity decayed is 0.0200 mg − 0.00144 mg = 0.0186 mg.

(c) By using the rates in place of q_t and q_0, equation 3 gives

The first-order equation, $2.303 \log \frac{q_0}{q_t} = kt$, may also be used after calculating k from $t_{1/2}$ (sections 19.4 and 19.5).

$$2.1 \times 10^4 \frac{\cancel{\text{particles}}}{\cancel{\text{second}}} = 7.0 \times 10^4 \frac{\cancel{\text{particles}}}{\cancel{\text{second}}} (0.5)^{t/3.8 \text{ days}}$$

and solving

$$\log (2.1 \times 10^4) = \log (7.0 \times 10^4) + \frac{t}{3.8 \text{ days}} \times \log (5 \times 10^{-1})$$

$$4.322 = 4.845 + \frac{t}{3.8 \text{ days}} (-0.301)$$

$$0.523 = \frac{0.301\, t}{3.8 \text{ days}}$$

from which $t = 6.6$ days. The age of the sample is 6.6 days. ■

23.10 RADIOCHEMISTRY

RADIOCHEMICAL DATING

Hominid refers to a very remote ancestor of the human species.

Radiochemical dating has made important contributions to paleoanthropology and geology. The full significance of the discoveries of hominid fossils and tools at Lake Turkana (Kenya), Laetoli (Tanzania), and Hadar (Ethiopia), for example, could not be realized until the ages of the rock beds at Lake Turkana (2 to 3×10^6 years) and at Hadar and Laetoli (3 to 4×10^6 years) were accurately determined. Other radiochemical measurements show that human beings lived in South America many centuries before 11,000 B.C. In radiochemical dating, ages are *not* measured directly, as in the counting of tree rings. Rather, the activity is measured. The **activity** is the number of disintegrations per unit time per gram of object or material (such as a sedimentary bed, the rock formed from sediment deposited by a river, lake, or sea) associated with the object. Ages are then calculated on the bases of certain assumptions.

Dating with ${}^{14}_{6}C$, conceived by Willard Libby in 1948, is based on the following reasoning. Natural carbon contains a small fraction of ${}^{14}_{6}C$. Its activity, as found in *living* plants and animals and in the air, is approximately constant at 14 disintegrations per minute per gram (d/min g) of natural carbon. The concentration of ${}^{14}C$ remains constant because it is replenished by the bombardment of ${}^{14}N$ in the atmosphere by cosmic-ray neutrons; ${}^{14}_{7}N + {}^{1}_{0}n \longrightarrow {}^{15}_{7}N \longrightarrow {}^{14}_{6}C + {}^{1}_{1}H$. Calculations and measurements show that the rate of formation of ${}^{14}C$ equals the decay rate of ${}^{14}C$. The simplest assumption is then made: the ${}^{14}_{6}C$ activity in the

BOX 23.7

THE AGE OF THE EARTH AND THE MOON

The oldest rock discovered on Earth has an age of 3.7×10^9 y. Moon rocks are older by about a billion years. However, this does not indicate that the Moon is older than the Earth. After volcanic action had ceased on the Moon, it continued on Earth. Old rocks continued to melt, and new ones recrystallized. The age of the Earth, 5×10^9 y, is estimated from the age of its oldest rocks, 4×10^9 y, plus a cooling allowance of roughly 1 billion years for the solidification of its crust.

air was the same in ancient times.† This $^{14}_{6}C$ is then incorporated into living tissues through photosynthesis and consumption of plants by animals. Then, as long as a plant or an animal is alive, the activity of $^{14}_{6}C$ per gram of carbon remains constant. Upon the death of the plant or animal, however, the $^{14}_{6}C$ continues to decay without replenishment, and the activity decreases. From the known half-life of 5.73×10^3 years for $^{14}_{6}C \longrightarrow {}^{\ 0}_{-1}\beta + {}^{14}_{7}N$, the age of archeological samples can then be determined.

EXAMPLE 23.6 The activity of the hair of a mummified Egyptian woman is 7.50 d/(min g) of carbon. What is the age of the hair?

ANSWER From the statement of the problem,

$$q_0 = 14 \frac{\text{d}}{\text{min g}} \quad \text{and} \quad q_t = 7.50 \frac{\text{d}}{\text{min g}}$$

Using equation 3,

$$7.50 \frac{\cancel{\text{d}}}{\cancel{\text{min g}}} = 14 \frac{\cancel{\text{d}}}{\cancel{\text{min g}}} \times (0.5)^{t/5730\text{ yr}}$$

from which $t = 5.2 \times 10^3$ years. The age of the hair is therefore 5.2×10^3 years. ■

PROBLEM 23.4 The activity of ^{14}C in the linen wrappings of the Book of Isaiah of the Dead Sea Scrolls—Hebrew manuscripts found near the Dead Sea in 1947—is about 11 d/(min g). Calculate the approximate age of the scrolls. □

Since uranium is present in virtually all rocks and soils, it was the first radioclock used.

The age of rocks (Box 23.7) can be determined from the relative amounts of ^{238}U and ^{206}Pb. The ^{238}U isotope, originally present when the rock crystallized, slowly decays through a series of disintegrations to ^{206}Pb (Figure 23.7, section 23.5). A rock sample thus contains ^{238}U and ^{206}Pb isotopes in amounts depending on its age. The older the rock, the higher is the percentage of ^{206}Pb. One mole of ^{238}U decays to 1 mole of ^{206}Pb, with a half-life of 4.47×10^9 yr. Rocks containing rubidium are dated by determining the relative amounts of ^{87}Sr and ^{87}Rb. The half-life for the reaction $^{87}_{37}Rb \longrightarrow {}^{87}_{38}Sr + {}^{\ 0}_{-1}\beta$ is 4.89×10^{10} yr.

While ^{14}C-dating still depends largely on the determination of the ^{14}C activity per gram of carbon, other dating methods now use mass spectrometry to find

† The number of cosmic-ray neutrons striking the earth per second and the concentration of nonradioactive CO_2 in the atmosphere resulting from the combustion of fossil fuels vary with time. However, ^{14}C dates have been calibrated against tree rings of known ages so that ^{14}C-dating may be used with confidence.

See Problem 23.39 for the Ar-K dating method.

See Problem 23.41.

the relative numbers of a pair of atoms (product/reactant), such as $^{206}Pb/^{238}U$, $^{87}Sr/^{87}Rb$, or $^{40}Ar/^{40}K$. However, recent experiments have made mass spectroscopic methods available for ^{14}C dating. This means that much smaller masses of historic artifacts will need to be destroyed and age measurements will be extended to at least 10^5 years.

EXAMPLE 23.7 A cluster of fossil cells has been found in a rock in Swaziland, Africa. Analysis of the rock by mass spectrometry shows the relative number of ^{87}Sr and ^{87}Rb atoms is $^{87}Sr/^{87}Rb = 0.0520$. Find the age of the fossil cells.

ANSWER The age of the fossil cells is taken to be the same as the age of the rock in which they were found. Let ^{87}Rb represent the number of rubidium-87 atoms now present in the rock; then $q_t = {}^{87}Rb$. Next, find q_0, the initial quantity of ^{87}Rb present when the rock last crystallized. We assume that no ^{87}Sr atoms were present when the rock crystallized. From the equation, ${}^{87}_{37}Rb \longrightarrow {}^{87}_{38}Sr + {}^{0}_{-1}\beta$, we see that one atom of strontium-87 appears for every atom of rubidium-87 that decays. Let ^{87}Sr represent the number of strontium-87 atoms now present in the rock. Then the original number of rubidium-87 atoms in the rock must have been $^{87}Rb + {}^{87}Sr$; hence $q_0 = {}^{87}Rb + {}^{87}Sr$. Using equation 3,

$$q_t = q_0 \times (0.5)^{t/t_{1/2}}$$

$$^{87}Rb = (^{87}Rb + {}^{87}Sr) \times (0.5)^{t/4.89 \times 10^{10}\ yr}$$

and dividing by ^{87}Rb,

$$1 = \left(1 + \frac{^{87}Sr}{^{87}Rb}\right) \times (0.5)^{t/4.89 \times 10^{10}\ yr}$$

Substituting the given value for $^{87}Sr/^{87}Rb$

$$1 = (1 + 0.0520) \times (0.5)^{t/4.89 \times 10^{10}\ yr}$$

and solving,

$$t = 3.6 \times 10^9\ yr$$

■

Recall that ^{206}Pb is the end product of ^{238}U decay (section 23.5).

PROBLEM 23.5 The analysis of a rock from the Grand Canyon of Arizona shows that the relative number of ^{206}Pb and ^{238}U atoms is $^{206}Pb/^{238}U = 0.01733$. Calculate the age of this rock. $t_{1/2}$ for $^{238}U \longrightarrow {}^{206}Pb$ is 4.47×10^9 y. □

TRACER ELEMENTS

With very few exceptions, the disintegration rate of a given radioisotope is, unlike chemical changes, independent of the chemical state and physical conditions. Radioisotopes are almost chemically identical with the stable isotopes of the same element. They are therefore used to "label" specific atoms in a given molecule. In this way, **radioactive tracers** (Figure 23.13) can be "traced" through chemical and physical changes to yield valuable information about the mechanism of such changes. For example, it was once impossible to study the rate of the reaction

$$Fe^{3+}(aq) + Fe^{2+}(aq) \longrightarrow Fe^{2+}(aq) + Fe^{3+}(aq)$$

because the products are the same as the reactants. For every 1 mole of Fe^{3+} that disappears as a reactant, 1 mole appears as a product; no change in $[Fe^{3+}]$ occurs during the course of the reaction. However, the rate of the reaction can now be studied by using radioactive ${}^{55}_{26}Fe^{2+}$; ${}^{55}_{26}Fe \longrightarrow {}^{55}_{25}Mn + {}^{0}_{+1}\beta$, $t_{1/2} =$ 2.68 yr. Separation of Fe^{2+} from Fe^{3+} in a portion of the solution shows that the activity of Fe^{2+} decreases with time, whereas the activity of Fe^{3+} increases.

FIGURE 23.13
Rosalyn S. Yalow, recipient of the Nobel Prize in Medicine and Physiology for her pioneering applications of radiotracers in detecting extremely small amounts (radioimmunoassay) of complex biochemicals for early detection of pregnancy and the detection of diseases.

Radioisotopes can also be used to determine the fraction of ingested cholesterol that terminates in the bloodstream of a subject. This is done by feeding the subject cholesterol enriched with ${}^{14}_{6}C$ of known activity, and then measuring the ${}^{14}_{6}C$ activity of cholesterol per unit volume of blood.

In the 1950s Melvin Calvin unraveled the sequence of the complex steps in the photosynthesis of CO_2 and H_2O to sugars, starches, cellulose, and oxygen by labeling CO_2 with radioactive ${}^{14}C$. The biosynthesis of many other biochemical compounds, such as cholesterol and fats, have been determined.

The uptake of plant nutrients (fertilizers) from water and soil and the mechanism of plant growth can be studied with radioactive tracers such as ${}^{32}P$.

In drug design research, how the body handles a new drug must be studied. It is practically impossible to do this without the use of a radioisotope labeled drug.

MEDICAL APPLICATIONS

TREATMENT Radiation is used to treat cancer, because cancerous cells are more sensitive than normal cells to radiation. The idea is to select a dosage that destroys the malignant cells but not the normal cells. Gamma- or X-radiation is applied to a cancerous growth in one of three ways: (1) by directing an external source of radiation, usually ${}^{60}Co$ (cobalt-60), into the body so that it passes through the growth; (2) by inserting a radioisotope directly into the cancerous tissue; or (3) by inserting the radiation source into an accessible body cavity near the cancerous growth. One difficulty is that cells at the center of a tumor may be deprived of oxygen. Unfortunately, the central cells become more resistant to γ-ray photons, thus requiring a dosage that also damages healthy cells. A small but significant percentage of patients develop radiation-caused cancers five or more years later. However, the oxygen-deprived cancerous cells remain sensitive to neutrons, pions,

BOX 23.8
DETECTING AND MEASURING NUCLEAR RADIATIONS

The detection and measurement of nuclear radiations (alpha particles, electrons, positrons, and γ rays) from radioisotopes is based on the radiations' abilities to ionize matter (Table 23.1). A radiation detector tube consists of a wire (anode) inserted through the center of one end of the tube. The other end of the tube is closed by a mica window sufficiently thin to permit passage of the particles to be counted. The tube contains argon gas under reduced pressure. A voltage insufficient to cause a current flow is maintained between the wire and the metal inner side (cathode) of the tube. When a particle enters the tube, the gas is ionized and a pulse of current flows from the cathode to the anode. This current is amplified and used to operate a mechanical counter. Such an apparatus is known as a *Geiger counter.*

The stronger the radiation, the greater is the number of counts per second. Translated into a computer image, the color intensity of an area reflects the number of counts per second. The half-lives of radioisotopes used for diagnostic purposes need to be as short as possible to minimize radiation dosages received by patients, yet long enough to obtain the necessary diagnostic information.

A Geiger tube, named after its inventor, Hans Geiger. Geiger is also known for his work on alpha-particle deflection experiments with Lord Rutherford (Chapter 2).

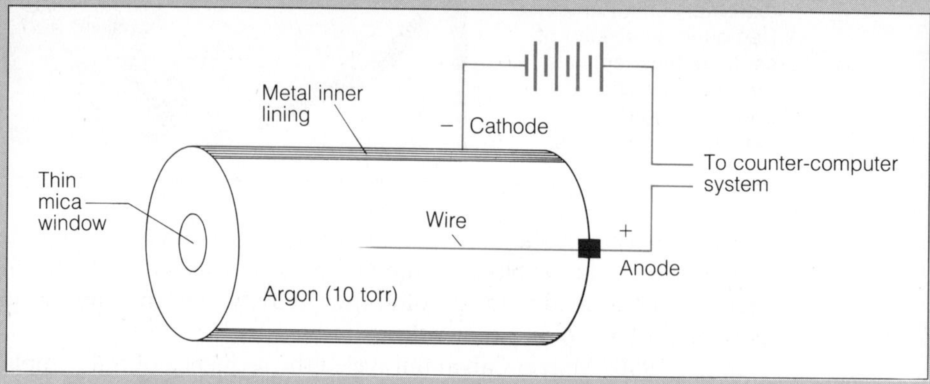

and heavy ions. Alpha particles, for example, are destructive to cancer cells. Carboranes (carbon-boron compounds) bind to substances in cancer cells. Then, neutron irradiation of these cells also produces alpha particles; $^{10}B + {}^{1}n \longrightarrow \alpha + {}^{7}Li$. These kinds of radiation may prove useful in selectively destroying cancer tissue.

An experimental cancer treatment uses the beta radiation emitted by yttrium-90 whose half-life is only 64 h ($^{90}_{39}Y \longrightarrow {}_{-1}\beta + {}^{90}_{40}Zr$). The radiation penetrates body tissue only 0.2 cm so that insertion into a cancerous tumor minimizes damage to healthy tissue.

The heat generated by the decay of $^{238}_{94}Pu \longrightarrow {}^{234}_{92}U + {}^{4}_{2}He$ is converted to electrical energy in heart pacemakers. When two wires composed of dissimilar metals, such as Pt and Rh, are connected at their ends and one of the junctions is heated, a voltage is generated. The current is proportional to the temperature difference of the junctions.

George Hevesy is regarded as the father of radiotracer and radioanalytical techniques and nuclear medicine (1913).

DIAGNOSIS Radioisotopes used for diagnostic purposes localize in specific organs or tissues within a body. This enables a specialist with a radiation detector–computer system to generate a picture of the area being studied. The pictures afford insights into the pathology of organs without surgery (Box 23.8).

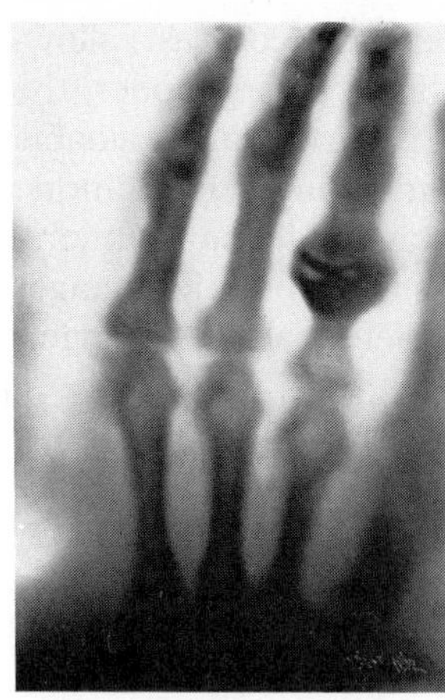

FIGURE 23.14
One of the first images Röntgen obtained with X rays, an X ray of a hand taken 22 December 1895. Notice that the person was wearing a ring. (Deutsches Museum, Munich.)

- Hemoglobin contains iron. The rate at which injected radioactive ^{59}Fe appears in the blood provides a measurement of the red blood cell production.
- The volume of blood in the body is determined by measuring the dilution of ^{14}C activity after injecting a ^{14}C-labeled compound into the bloodstream.
- Xenon-133 or krypton-85 is used for evaluating pulmonary functions, imaging the lungs, and assessing cerebral blood flow.
- $Na^{131}I$ is used in the diagnosis of thyroid gland disorders. Secretions from the thyroid gland, located in the neck, affect the rate of growth and metabolism. Disorders in the gland can be diagnosed by measuring the rate at which ^{131}I appears in the gland. Thus, a patient is given a solution of labeled NaI, and a detector is placed near the thyroid gland to measure the uptake of iodine in the gland.
- Obstruction of a blood vessel by a clot or a bubble (embolism) is detected with fibrinogen, a soluble protein, labeled with ^{125}I. The labeled fibrinogen piles up at the obstruction and signals an above-average blood activity.
- Brain tumors can be located very accurately because of their greater tendency to absorb certain radioisotopes—such as ^{111}In and ^{64}Cu—from the bloodstream.
- Gallium-67 is 95% accurate in locating deeply hidden lesions resulting from Hodgkin's disease.
- Bone-density measurements are made in patients with bone-demineralization conditions by comparing the intensities of the two X-ray beams emitted by gadolinium-153 after passage through the bone.

X-ray photons penetrate the lower atomic–numbered atoms of teeth (H, C, N, O, P, Ca) better than they penetrate the higher atomic–numbered atoms of fillings (Ag, Hg).

NUCLEAR IMAGING *X-ray technology* has been with us since the close of the nineteenth century. Traditional X-ray technology is efficient for imaging bones and lungs—areas of high contrast—since the higher the atomic number of an atom, such as the Ca in bone, the greater is its ability to absorb photons (Figure 23.14). Soft-tissue organs, however, are not distinguished by X rays because their absorption of photons is almost the same as that of water. Patients ingest a suspension of barium sulfate ($BaSO_4$) to sharpen the visibility of the digestive system.

Computerized instrumentation now produces images of blood flow, the structure of tissues, and the size, shape, and position of organs and tumors. It is also capable of locating and defining the shape of a diseased area. With these methods, particles or photons emitted by radioactive atoms are converted to a visual figure that can be stored in the computer bank as part of the patient's permanent medical record. In nuclear imaging, a compound labeled with a gamma-emitting isotope is injected into the patient. Since these isotopes are selectively absorbed by particular organs, nuclear-imaging technology has become prominent in the detection of organ pathology.

A. M. Cormack and G. N. Hounsfield developed the mathematical process for converting signals to an image.

In a *CAT* (*computed axial tomography*) scan, a source of a very narrow X-ray beam and an X-ray detector slowly rotate around the body. Extremely brief pulses of X rays from all angles thus pass through a cross-sectional area of the patient. The recordings are then combined mathematically to form an image on a screen.

An iodine compound injected into the bloodstream increases the contrast, allowing visualization of soft tissues such as blood vessels, the heart, lymph nodes, the kidney, and the white and gray matter in the brain. CAT is particularly valuable for visualizing various parts of the brain and diagnosing hemorrhages and tumors. However, it sometimes fails to distinguish between benign and malignant tumors.

A modified CAT scanner (*Imatron*) produces three-dimensional motion images of a beating heart. This eliminates the need for cardiac catheterization, a dangerous surgical procedure.

isotope	$t_{1/2}$, min
^{11}C	20.4
^{18}F	110
^{15}O	2.05
^{13}N	9.98

These isotopes are manufactured in a nuclear reactor on-site. For this reason, the PET is not used for routine diagnosis.

Positron-emission tomography (*PET*) utilizes the emission of positrons by short-lived isotopes such as $^{11}_{6}C$, $^{13}_{7}N$, $^{15}_{8}O$, and $^{18}_{9}F$. A biologically active molecule like glucose ($C_6H_{12}O_6$) or a related compound is labeled with either $^{11}C^{12}C_5H_{12}O_6$ or $C_6H_{11}{}^{18}FO_5$. When the emitted positron approaches a neighboring electron, the two particles are mutually annihilated with the production of two γ photons:

$$_{+1}\beta + {}_{-1}\beta \longrightarrow 2\gamma$$

Thus, unlike the CAT, the gamma-radiation is generated *within* the body. The two gamma photons travel in opposite directions. Detectors placed 180° apart record the positron-electron annihilations. The distribution of such nuclear events is then mapped by a computer, producing an image corresponding to the distribution of the labeled molecule. Defects in the biochemistry of the brain can thus be detected. Glucose, for example, is the sole energy source for the brain. Therefore, changes in the mechanism of glucose metabolism indicate some kind of brain disorder. The glucose metabolism patterns in the brains of normal persons, persons with Alzheimer's disease, and diagnosed schizophrenics differ sharply (Color plate 12). The PET is therefore a powerful tool for studying metabolic disorders in the vital organs of the body. Disorders such as gallbladder obstructions are readily diagnosed by radioactive labeling of bile acids. By labeling blood platelets, the development of atherosclerosis in blood vessels can be studied.

Heart disease ranks first among the causes of death in humans. Therefore, the assessment of the flow of blood to the heart muscle is critical in the decision to choose between medication or by-pass surgery. Heart damage is very reliably appraised by the *thallium stress test*. While exercising on a treadmill, the patient is injected with thallium-201.

$$^{201}_{81}Tl + e^- \longrightarrow {}^{201}_{80}Hg + \gamma,\ t_{1/2} = 73.1\text{ h}$$

Tl^+ mimics K^+, which is found in nearly every living cell in the body. Healthy heart tissue therefore has a strong affinity for Tl. The γ photons picked up by a detector placed over the heart are converted into an image on a screen (Color plate 13). Scars and areas deficient in blood supply are easily seen. In a normal heart, the blood flow is uniform throughout the heart muscle. In a damaged heart, blood fails to reach some parts of the heart. Heart conditions commonly overlooked in electrocardiograms are quickly diagnosed from the image.

$$^{98}_{42}Mo + {}^{1}_{0}n \longrightarrow {}^{99}_{42}Mo \longrightarrow {}^{99}_{43}Tc + {}_{-1}\beta$$

excited state

$$^{99}_{43}Tc \longrightarrow {}^{99}_{43}Tc + \gamma$$

excited state → ground state

$t_{1/2} = 6.02$ h

$^{99}_{43}Tc$ imaging duplicates the ^{201}Tl information. $^{99}_{42}Mo$ decays to an excited state of $^{99}_{43}Tc$, which decays to the ground state with the emission of a γ photon. ^{99}Tc localizes in normal heart tissue, yielding sharp images of the heart.

Technetium-99, the workhorse of nuclear medicine, is also used in scanning the brain for the location and size of tumors and for imaging internal organs—bone, liver, kidney, lungs, thyroid, spleen—in the detection of disease.

It is necessary to dispel the notion that medical imaging may displace the physician. In all cases, computers convert signals to images. Are these images equivalent to the object seen with the human eye? Not always; it is still the diagnostician

Technetium, the first artificially produced element (1937); from Greek *technetos*, "artificial."

that must make the decisions on the basis of what is known about the disease and the medical history of the patient.

Biochemical effects of radiation also require a word of caution. The high-energy particles used in nuclear medicine (γ rays and X rays) lose their energy mainly by ionizing atoms and molecules. In animals, the energy is lost largely by ionizing water, a process that yields free radicals as the final product. A plausible mechanism is

$$
\begin{array}{lll}
 & \mathrm{H{:}\underset{\cdot\cdot}{\ddot{O}}{:}H \longrightarrow H{:}\underset{\cdot}{\ddot{O}}{:}H^{+} + e^{-}} & \\
 & \mathrm{H{:}\underset{\cdot}{\ddot{O}}{:}H^{+} \longrightarrow H^{+} + \cdot\underset{\cdot\cdot}{\ddot{O}}{:}H} & \text{(free radical)} \\
 & \mathrm{H^{+} + e^{-} \longrightarrow H\cdot} & \text{(free radical)} \\
\hline
\text{net reaction:} & \mathrm{H_2O \longrightarrow H\cdot + \cdot OH} &
\end{array}
$$

Hydrogen peroxide may also be formed ($\mathrm{2\cdot\underset{\cdot\cdot}{\ddot{O}}{:}H \longrightarrow H{:}\underset{\cdot\cdot}{\ddot{O}}{:}\underset{\cdot\cdot}{\ddot{O}}{:}H}$).

The dangers involved in low radiation exposures are surrounded by controversy. Averaged over the whole (U.S.) population, the exposure for diagnostic purposes is essentially equal to the background sources, 0.1 rem/g (Appendix A.7). As in all life activities, the risk must be balanced against the benefits. Radiation should be used for medical purposes with discretion and with good reason. Both the patient and the technician should be protected from unnecessary exposure.

INDUSTRIAL APPLICATIONS

The decrease in the intensity of radiation depends on the quantity of matter through which it passes. This makes it possible to use radiation to gauge and control the thickness of industrial products ranging from metal plates to plastic films. The wear-resistant properties of the moving parts of an engine are easily measured by using radioactive parts and determining the transfer of radioactivity to the lubricating oil.

The technology of *food irradiation* as an alternative to food preservatives and pesticides has been available since the 1950s. Gamma rays from cobalt-60 or cesium-137 or electrons from an electron accelerator are used to retard the sprouting of potatoes and onions and to sterilize fruit, vegetables, and meat. The dosage is sufficient to kill adult and larval insects (such as the medfly), viruses, and bacteria (including salmonella and botulinum) in various foods. For example, irradiation preserves prepared meats without the use of nitrites. Just as tumor irradiation does not induce radioactivity in patients, sterilization does not make foods radioactive. The shelf lives of foods can be extended to about seven years, and unlike canning and freezing, irradiation does not make food soft and mushy.

Spoilage ruins an estimated 25% of the world's food supply.

Astronauts have consumed foods sterilized by irradiation. Such foods are available in Europe and Japan, and several independent international organizations including the World Health Organization have approved the technique. However, consumers' negative image of "radiation" has delayed acceptance in the United States. Even though experiments have amply demonstrated that the process is safe, it is still surrounded by controversy.

The FDA has approved the use of radiation in some fruits and vegetables and in pork to destroy the pork parasite *Trichinella spiralis*.

Industrial applications of radiation chemistry now include sterilization of medical supplies, modification of various plastics and paints, decomposition of industrial effluents, and treatment of sewage sludge to control the population of pathogens.

NEUTRON-ACTIVATION ANALYSIS

Methods of quantitative analysis based on the formation of radioactive nuclei by neutron capture have been developed. First, the sample is exposed to neutrons.

Each constituent element forms a specific radioisotope with a characteristic half-life. Then the mass of the element is calculated from the activity of the radioisotope that it produces. The sensitivity of the method makes it very useful for the *nondestructive* analysis of trace elements in complex biological systems, archeological samples, meteorites, and lunar rocks. For example, the claim has been made that Sir Francis Drake landed on the California coast in 1579 and installed an inscribed brass plate claiming the land for Queen Elizabeth I. Neutron-activation analysis for about 30 elements in what was claimed to be the original brass plate indicates that it is a fake. The composition does not match the composition of brasses made in Drake's time.

About one part or less per billion.

The method has also found application in investigations of crimes. It is possible to determine whether a suspect has recently fired a gun. The detonation of the primer leaves extremely small quantities of antimony, barium, and copper on the hand. A wax cast made of the hand picks up these elements. The cast is then subjected to neutron-activation analysis.

Nondestructive techniques are important when analyzing art objects, which can be damaged by sampling. For example, paintings, covered with photographic film, are placed in a nuclear reactor, and various radioactive isotopes are produced by neutron absorption. As the isotopes in the paint decay, a series of pictures is produced by radiation from the painting. Such a photograph is called an *autoradiograph*. Details about the nature and composition of the paints and canvas material can be discerned from the autoradiograph. Color plate 14 shows autoradiograms obtained in studies on the aging process.

SUMMARY

Naturally **radioactive** (unstable) nuclei decay by emission of **beta particles** ($_{-1}^{0}\beta$ electrons), **positrons** ($_{+}^{0}\beta$), **alpha particles** ($_{2}^{4}He^{2+}$), and **gamma-radiation** (γ rays). Nuclear transformations also occur by capture of $1s$ electrons from the $1s$ shell and by spontaneous fission. Stable nuclei are made radioactive by bombarding them with neutrons or high-energy charged particles. Isotopes above atomic number 92 that do not occur naturally have thus been synthesized.

Like an atomic orbital, a nuclear orbital can accommodate two protons and two neutrons. Nuclei containing unpaired protons or unpaired neutrons have a magnetic moment. **Magic numbers** are the numbers of protons or neutrons in completed shells of protons and neutrons. Nuclei with magic numbers of protons and/or neutrons are particularly stable. An even number of protons or neutrons generally increases nuclear stability.

Nuclear reactions are represented by nuclear equations in which the charge (atomic numbers) and the number of nucleons (mass numbers) are conserved. Radioactive decay reactions are first-order, making the half-life independent of quantity or concentration. Half-lives of radioisotopes can therefore be used to determine the ages of rocks, fossils, and ancient organic matter that still contain radioisotopes.

Nuclear reactions, unlike chemical reactions, are accompanied by comparatively large changes in mass of matter and energy, related by the **Einstein equation**, $\Delta E = \Delta m \times c^2$.

The **binding energy of a nucleus**, the energy per nucleon required to separate a nucleus into its component nucleons (protons and neutrons) or the energy per nucleon evolved in forming the nucleus from nucleons, is calculated from the change in mass, Δm, when nucleons combine to form a nucleus. The most stable nuclei are in the region of $_{26}^{56}Fe$. Therefore very light nuclei tend to fuse into heavier nuclei (**fusion**), and very heavy nuclei tend to split into lighter nuclei (**fission**). Fusion is the source of solar energy, and fission is the source of heat in **nuclear reactors**. In **breeder reactors** more nuclear fuel is produced than is consumed.

Nuclei outside the **stability belt**, based on the neutron–proton ratio, decay so as to move closer to the stability belt. Nuclei above the stability belt have a high n/p ratio and move down toward the belt by $_{-1}\beta$ emission ($^1_0n \longrightarrow {}^1_1p + {}_{-1}\beta$), which increases the number of protons while decreasing the number of neutrons. Nuclei below the stability belt have a low n/p ratio and move up toward the belt by $_{+1}\beta$ emission ($^1_1p \longrightarrow {}^1_0n + {}_{+1}\beta$), proton emission, or electron capture, all of which increase the ratio. Nuclei with atomic numbers above 83 usually decay by alpha emission, and several such nuclei initiate a sequence of nuclear reactions (a **radioactive series**) that finally terminates with the formation of a stable nucleus.

The ease with which radioisotopes can be detected accounts for their use as **radioactive tracers** in chemical analysis and mechanistic studies and for their wide use in biochemical research and *nuclear medicine,* especially in **nuclear imaging**. Gamma-radiation from certain radioisotopes is especially useful in cancer therapy. Radioisotopes also have many commercial applications.

23.11 ADDITIONAL PROBLEMS

NUCLEAR CHEMISTRY

23.6 Which statements are true? (a) The internal structures of nuclei affect the chemical properties of atoms and molecules. (b) In the atom, electrons are distributed in atomic orbitals; in the nucleus, nucleons are distributed in nuclear orbitals. (c) Electrostatic force is the same between all electron pairs; nuclear force is the same between all nucleon pairs. (d) Unlike the electronic charge, a nuclear charge may be altered. (e) The total number of nucleons in a nuclear reaction can be changed. (f) Hydrogen can be "burned" to helium. (g) Nucleons are structureless (have no internal structure). (h) Electrons are structureless. (i) The total charge of a nucleus always equals the sum of the charges of Z protons and N neutrons. (j) The sum of protons and neutrons is changed during fission. (k) Confined electrons and nucleons can only have quantized energies. (l) The π-meson (pion) is the "glue" that keeps nucleons together. (m) The proton is a structureless point charge.

23.7 *Scientific American,* April 1903: "Professor Curie has announced to the French Academy of Sciences that radium possesses the extraordinary property of continuously emitting heat without combustion, without chemical change of any kind, and without change in its molecular structure. Radium, he states, maintains its own temperature at a point 1.5°C above the surrounding atmosphere. Despite this constant activity, the salt apparently remains just as potent as it was at the beginning." Would you, at the present time, accept this statement as correct? Justify your position.

23.8 High-speed protons, α particles, and $_{-1}^{0}\beta$ particles pass out of nuclei through surrounding electron clouds *without* picking up electrons. What are these particles called when they come nearly to rest?

23.9 How is the number of protons in a nucleus changed when (a) a γ-ray photon, (b) a $_{-1}^{0}\beta$ particle, (c) an α particle, is emitted from the nucleus?

23.10 Balance these nuclear reactions. (Indicate the symbol where possible, the mass number, and the atomic number for **?**)

(a) $^{47}_{20}Ca \longrightarrow {}_{-1}^{0}\beta + ?$

(b) $^{48}_{23}V \longrightarrow {}_{+1}^{0}\beta + ?$

(c) $^{240}_{93}Np \longrightarrow {}^{238}_{93}Np + ?$

(d) $^{14}_{7}N + {}^1_0n \longrightarrow 3\,{}^4_2He + ?$

(e) $^{249}_{98}Cf + {}^{15}_{7}N \longrightarrow 4\,{}^1_0n + ?$

(f) $^{188}_{78}Pt \longrightarrow {}^4_2\alpha + ?$

(g) $^{248}_{96}Cm + {}^{40}_{20}Ca \longrightarrow 4\,{}^1_0n + ?$

(h) $^{37}_{17}Cl + {}^0_0\nu$ (neutrino) $\longrightarrow {}_{-1}^{0}\beta + ?$

(i) $^{212}_{86}Rn \longrightarrow 2\alpha + \gamma + ?$

(j) $^{232}_{90}Th \longrightarrow 6\alpha + 4\,{}_{-1}^{0}\beta + ?$

23.11 $^{235}_{92}U$ decays through a series of steps, ending with a stable isotope. The particles successively emitted in the series are α, β, α, α, β, α, α, α, β, α, β. Write the nuclear reaction for each step. What is the end product?

23.12 Write balanced equations for the preparation of positron emitters for use in PET studies. Also write the equation for the positron emission:

$^{14}N + {}^1H \longrightarrow \alpha + ?$ $\qquad$ $^{20}Ne + {}^2H \longrightarrow \alpha + ?$

$^{14}N + {}^2H \longrightarrow n + ?$ $\qquad$ $^{16}O + {}^1H \longrightarrow \alpha + ?$

23.13 (a) A novel mode of natural decay has been discovered in ^{223}Ra. For every billion ^{223}Ra atoms ejecting an alpha particle, one ^{14}C nucleus is emitted. Write a balanced equation for the alpha emission and for the ^{14}C emission. (b) Another new type of radioactivity is the ejection of a pair of protons by ^{22}Al and ^{26}P. Write a balanced equation for each reaction.
23.14 Give examples of the possible modes of decay for isotopes whose neutron/proton ratio place them (a) above and (b) below the band of stability.
23.15 List the following nuclei in the order of increasing stability.

$$^{16}_{8}O,\ ^{4}_{2}He,\ ^{206}_{82}Pb,\ ^{55}_{24}Cr$$

23.16 Assume that the neutrons produced by ^{235}U fission travel an average distance of 5.0 cm before being captured by ^{235}U. The density of ^{235}U is 19 g/mL. What sphere mass must not be exceeded if an explosion is to be prevented on exposure to a stray neutron? Show how you arrive at your answer.
23.17 What is the probability of developing nuclear energy sources with isotopes in the region of $_{26}Fe$? Explain.

ΔE

23.18 Calculate ΔE in kJ/mol for the reaction

$$^{204}_{82}Pb + ^{4}_{2}\alpha \longrightarrow ^{208}_{84}Po$$

Isotopic weights: ^{204}Pb, 203.9735; ^{4}He, 4.002603; ^{208}Po, 207.9824.
23.19 Calculate the mass loss for the combustion of 1.00 mole of fuel oil (142 g/mol) whose heat of combustion is 44.4 kJ/g. Is the law of conservation of matter applicable to chemical reactions?
23.20 For the process,

$$\underset{\text{excited state}}{^{141}Cs} \longrightarrow \underset{\text{ground state}}{^{141}Cs} + h\nu \qquad \textbf{(1.194 MeV/atom)}$$

calculate Δm, the difference in mass between the final and initial atoms, in g/mol. See Appendix A.5. 1.602×10^{-19} J = 1 eV.
23.21 When a positron is slowed down in matter, annihilation of matter occurs, producing two γ-ray photons,

$$^{0}_{+1}\beta + ^{0}_{-1}\beta = 2\gamma$$

The electron mass is 9.11×10^{-28} g. Calculate the energy in MeV and the frequency of each photon. (Assume the energy is evenly divided between the two photons.) The measured frequency of this radiation is 1.24×10^{20} sec^{-1}. The reverse process is also observed experimentally (see Figure 23.3, section 23.2). 1 eV = 1.60×10^{-19} J.
23.22 Find the mass loss for the formation of 1 mole of $^{204}_{80}Hg$, isotopic weight 203.9735, from neutrons, protons, and electrons. If no mass loss occurred, would the principle of conservation of mass numbers be valid? (Calculate mass numbers by rounding isotopic weights.)
23.23 Calculate the binding energy in kJ/mol for the formation of $^{209}_{83}Bi$ and find the energy released per nucleon. Isotopic weights: ^{1}H, 1.00783; ^{1}n, 1.00867; ^{209}Bi, 208.9804.
23.24 Calculate ΔH and ΔG per mole for the nuclear reaction

$$^{239}_{94}Pu + ^{1}_{0}n \longrightarrow ^{144}_{56}Ba + ^{85}_{38}Sr + 11^{1}_{0}n$$

ΔS is negligible and $\Delta H = \Delta E$. Isotopic weights: ^{239}Pu, 239.0522; ^{1}n, 1.00867; ^{144}Ba, 143.9051; ^{85}Sr, 84.9089.
23.25 Find ΔE in kJ/mole of $^{7}Li^{1}H$ for the fusion reaction $^{7}Li^{1}H \rightarrow 2^{4}_{2}He$. Isotopic weights: ^{7}Li, 7.016005; ^{1}H, 1.00783; ^{4}He, 4.002603.
23.26 A suggestion for harnessing fusion reactions involves a two-step reaction:

STEP 1. $^{2}_{1}H + ^{3}_{1}H \longrightarrow ^{4}_{2}He + ^{1}_{0}n$

STEP 2. $^{6}_{3}Li + ^{1}_{0}n \longrightarrow ^{4}_{2}He + ^{3}_{1}H$

(a) What is the net reaction? (b) This reaction is referred to as "catalytic" burning of Li. What substance acts as the catalyst? (c) Calculate ΔE in kJ/mole of ^{6}Li for the net reaction. (d) Calculate the heat evolved in kJ when 70.00 g ^{6}Li reacts. Isotopic weights: ^{6}Li, 6.0151; ^{2}H, 2.0141; ^{4}He, 4.0026.
23.27 (a) Show that the energy of a $_{-1}^{0}\beta$-decay nuclear reaction, for example,

$$^{35}_{16}S \longrightarrow ^{35}_{17}Cl + _{-1}\beta$$

expressed in grams per mole, is equal to the difference between the molar masses of $^{35}_{17}Cl$ and $^{35}_{16}S$. (b) Show that the energy of a $_{+1}^{0}\beta$-decay nuclear reaction, for example,

$$^{64}_{29}Cu \longrightarrow ^{64}_{28}Ni + _{+1}\beta$$

expressed in grams per mole, is equal to the difference between the molar masses of $^{64}_{28}Ni$ and $^{64}_{29}Cu$ plus the molar mass of 2 electrons.

HALF-LIFE

23.28 The decay rate constant for ^{199}At is 356 h^{-1}. Find its half-life in seconds.
23.29 ^{131}I has a half-life of 8.04 days. Find its rate constant in sec^{-1}.
23.30 ^{126}Ce has a half-life of 50.0 sec. Starting with 10.0 g ^{126}Ce, how much remains after (a) 50.0 sec; (b) 100 sec; (c) 200 sec?
23.31 ^{60}Co, $t_{1/2} = 5.27$ yr, has largely displaced the X-ray tube as a radiation source for the treatment of cancer. Starting with a supply of 100 mg, how many milligrams remain after (a) 5.000 half-lives; (b) 26.35 yr; (c) 28.00 yr?

23.32 Sodium technetate(VII), $Na^{99}TcO_4$, is used in scanning the brain for the location and the size of tumors. Plot the following data and find the ^{99}Tc half-life:

d/min	TIME, h
100	0
72.0	2.50
58.0	5.00
43.0	7.50
32.5	10.0
25.5	12.5
19.0	15.0
13.5	17.5

23.33 The rate constant for the decay of ^{49}Ca is 0.0795/min. A sample containing ^{49}Ca has an activity of 2.5×10^5 d/sec. (a) How many atoms are disintegrating per second? (b) Find the number of atoms, moles, and picograms of ^{49}Ca in the sample at the time the measurement is made. Do the data in this problem justify looking up the accurate isotopic weight of ^{49}Ca?

23.34 An artificial isotopic mixture of lead with an average atomic weight of 207.2 is assembled in a laboratory from ^{207}Pb (isotopic weight = 206.98) and radioactive ^{210}Pb (isotopic weight = 210, $t_{1/2} = 22.3$ yr). Find the activity of this artificially prepared lead sample in d/(g yr).

DATING

23.35 Organic matter associated with several clay tablets on which is inscribed some writing of the ancient Sumerians has a ^{14}C-activity of 8.50 d/(min g). Calculate the age of the organic matter and the clay tablets.

23.36

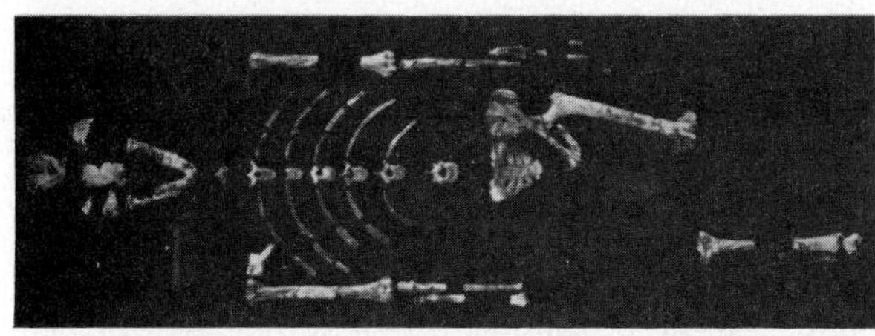

"Lucy," an erect, walking hominid. Height, 1.2 m. (Courtesy of The Institute of Human Origins.)

The sedimentary rock bed at Hadar, Ethiopia, where the famous "Lucy" skeleton was discovered, has been radio-dated by several methods. The Rb-Sr method ($t_{1/2} = 4.89 \times 10^{10}$ yr) yields a $^{87}Sr/^{87}Rb$ ratio of 4.96×10^{-5}. How old is "Lucy"?

23.37 The supply of tritium, ^{3}H ($t_{1/2} = 12.3$ yr), in the atmosphere is maintained nearly constant (in the absence of nuclear bomb tests) by cosmic-ray bombardment. The wine in a bottle reportedly recovered from a wrecked ancient ship contains 3.72×10^5 ^{3}H atoms per mole of ^{1}H. Newly fallen rainwater contains 4.8×10^5 ^{3}H atoms per mole of ^{1}H. Find the age of the wine.

23.38 Natural fission reactors, operating billions of years ago, were discovered by the French Atomic Energy Laboratories in Oklo, Gabon (Africa). Present natural U contains 0.720% ^{235}U. How many years ago did natural uranium contain 3.00% ^{235}U, an amount sufficient to sustain a natural reactor? $t_{1/2}$ for $^{235}U = 7.04 \times 10^8$ yr.

23.39 The K-Ar method of dating finds many significant applications. $^{40}_{19}K$ decays to $^{40}_{18}Ar$ by electron capture and decays about 8 times faster to $^{40}_{20}Ca$ by $_{-1}^{0}\beta$ emission. The half-life of ^{40}K is 1.31×10^9 yr. Thus

$$q_t = {}^{40}K \text{ and } q_0 = {}^{40}K + {}^{40}Ca + {}^{40}Ar$$

(a) Show that

$$1 = \left(1 + 9 \times \frac{^{40}Ar}{^{40}K}\right) \times (0.5)^{t/1.31 \times 10^9 \text{ yr}}$$

(b) The marine beds in Patagonia, Argentina, have a $^{40}Ar/^{40}K$ ratio of 5.81×10^{-4}. Calculate the age of the beds.

23.40 A sample of the cellulose (dead part of each tree ring) from bristlecone pinewood of age 7000 $\pm$ 100 years, known by counting growth rings, possesses an activity of 6.53 d/(min g) of carbon. Another cellulose sample, with a growth-ring age of 2300 yr, has an activity of 10.99 d/(min g) of carbon. (a) Calculate the ages of these samples from radiochemical evidence. (b) What does this tell us about the ^{14}C content per gram of carbon over the ages? (c) How should the expanded use of fossil fuels in modern times affect the ^{14}C activity per gram of carbon in the atmosphere—increase, decrease, no change? Explain. (d) The present ^{14}C activity is 16 d/(min g) of carbon, not 14. Explain. (Sudden bursts of natural cosmic-ray activity do not account for the increase.)

23.41 Counting atoms instead of disintegrations extends the range of ^{14}C dating. A cyclotron technique detects one ^{14}C atom in a 150-mg C sample with a low efficency of 3×10^{-4} (which means that $(0.5)^{t/t_{1/2}}$ is multiplied by 3×10^{-4}). There are 99 ^{12}C atoms to 100 C atoms and 10^{-12} carbon-14 atoms to one ^{12}C atom. Calculate the maximum age of samples that can be dated by this method.

23.42 Charcoals and charred remains of edible plants found in a buried hearth at Wadi Kubbaniya, Egypt, were analyzed in a tandem accelerator–mass spectrometer to give the ratio $(^{14}C/^{13}C)$ sample/$(^{14}C/^{13}C)$ modern tree = 0.101. Calculate the ages of the sample and the occupation site.

TRACER CHEMISTRY

23.43 Exchange of Cr between Cr^{3+} and $CrO_4{}^{2-}$ does not occur. Using radioactive $^{51}Cr^{3+}$ and nonradioactive

$^{52}CrO_4^{2-}$ in a chromic acid plating bath, the Cr deposit is not radioactive. Using $^{51}CrO_4^{2-}$ in a similar bath, the Cr deposit is radioactive. What is the source of Cr in chromic acid plating baths?

23.44 Balance the equation $SO_3^{2-}(aq) + ClO_3^-(aq) \longrightarrow SO_4^{2-}(aq) + ClO_2^-(aq)$ by the ion-electron method. ^{18}O-labeled chlorate is prepared, $[Cl^{18}O_3]^-$, and the reaction is allowed to occur in $H_2{}^{16}O$. (a) If the partial equations written to balance the equation represent the actual mechanism for the reaction, what should be the isotopic composition of H_2O and SO_4^{2-} at the end of the reaction? (b) Experimentally, it is found that the fourth O atom needed to convert SO_3^{2-} to SO_4^{2-} does *not* come from H_2O.

$$\left[{}^{16}O{-}\ddot{S}({}^{16}O)({}^{16}O)\right]^{2-} + \left[{}^{18}O{-}\ddot{Cl}({}^{18}O){-}{}^{18}O\right]^{-} \longrightarrow \left[{}^{16}O{-}S({}^{16}O)({}^{16}O){-}{}^{18}O\right]^{2-} + \left[{}^{18}O{-}\ddot{\underset{\cdot\cdot}{Cl}}{-}{}^{18}O\right]^{-}$$

Suggest a mechanism for this oxidation–reduction reaction.

23.45 A 71.0-kg person is injected with a few drops of ^3HHO having an activity of 9.0×10^9 dpm. After allowing some time for uniform distribution, 1.00 mL of plasma water is withdrawn and measured at 1.80×10^6 dpm. Calculate the volume of plasma water.

23.46 ^{44}Ti has a γ activity of 180 d/(min g), and ^{18}F has a $_{+1}^{0}\beta$ activity of 444 d/(min g). A sample of titanium-44 fluoride-18 is synthesized; it has a γ activity of 7.90 dpm and a $_{+1}^{0}\beta$ activity of 24.0 dpm. (a) How many grams of Ti and F are in the sample? (b) Find the empirical formula of the fluoride.

23.47 Evidence shows that a radioisotope in solution (an amount too small to be measured by separation) is an isotope of either Al^{3+} or Cu^{2+}. What chemical experiments (write equations) would you perform to determine whether the radioisotope is Al^{3+} or Cu^{2+}?

RADIATION EFFECT

23.48 Alpha particles can be stopped by a piece of paper or by the outer skin layers. Does it follow that an α emitter, internally located, presents no hazard to an animal?

23.49 (a) Plutonium emits alpha particles that are incapable of penetrating the outer layer of dead skin, but Pu is highly carcinogenic when it is in contact with living tissues. However, Pu usually exists as PuO_2, a solid that is highly insoluble in body fluids and readily absorbed by indigestible food (fibrous materials). Would you classify Pu or PuO_2 as nonhazardous to human health? (b) Measurable levels of radioactivity are found in foods. For example: dried apricots, 1.07 dpm/g; shelled Brazil nuts, 0.17 dpm/g; dehydrated bananas, 0.034 dpm/g. For comparison, natural KCl has an activity of 2.36 dpm/g. Would you classify these foods as nonhazardous to human health?

23.50 A sample has an activity of 5.55×10^6 dps per cm^2 at a distance of 1.00 m. What is the activity at 3.50 m? (The activity is inversely proportional to the square of the distance.)

23.51 Probable *early* effects of exposure of the *whole body* to a single dose of radiation are given (Appendix A.7):

DOSE, rems	PROBABLE EFFECTS
0–10	No obvious effect
10–25	Small temporary changes in blood of some exposed persons
25–50	Low white and red blood cell counts in nearly all exposed persons
50–100	Nausea within 12 hours for 5% of exposed persons
100–200	Vomiting in 5 to 50% of those exposed; hair loss in up to 10% of exposed population in 5 to 10 days
225	Death within 60 days for 5% of exposed persons without medical care
400	Death within 60 days for 50% of exposed persons without medical care

(a) A technician unknowingly was exposed to a neutron source; $t_{1/2} = 6.75$ days. The condition was discovered after 30 days. At this time, the activity of the neutron source was 0.414 rad/day. (1) What was the activity at the start of exposure? (2) What are the probable effects? (b) Exposure at 0.50 m from an alpha source is 4.0 rads/min. (1) How far away from the source should the workers remain to reduce their exposure to 0.030 rem (average dose from a chest X ray) per min. (2) Nuclear workers may be exposed to 5 rems per year. Estimate the time that a worker may work in a year at the distance found in part 1. See Problem 23.50 for the relationship between rems and distance.

23.52 Biological half-lives (the time required for one-half of a substance to leave the body) are determined by both radioactive decay rates and excretion rates. The biological half-life of ^{239}Pu for the whole body is 43×10^2 days. Find the time required to reduce an internal dosage of 160 d/(min g) of ^{239}Pu to the *present* background count of ^{14}C, namely, 16 d/(min g).

23.53 Biological radiation damage starts with the production of highly reactive species like $\cdot H$, $\cdot OH$, $\cdot OOH$, H_2O_2, and $e^-(H_2O)_x$. Write balanced equations showing how these could be formed from water.

ANSWER

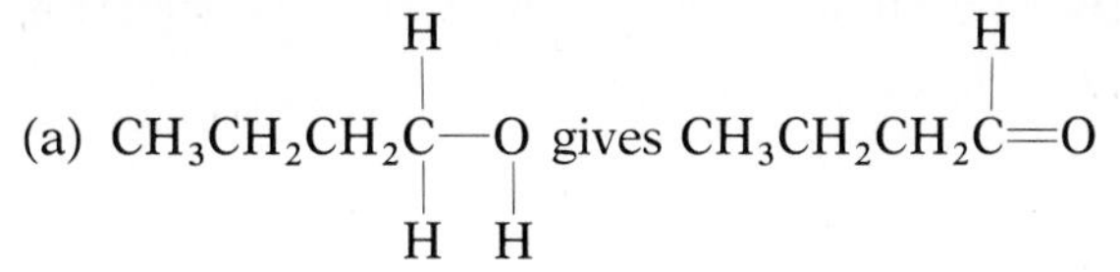

(a) $CH_3CH_2CH_2C(H)(H)-O-H$ gives $CH_3CH_2CH_2C(H)=O$ (butanal, an aldehyde)

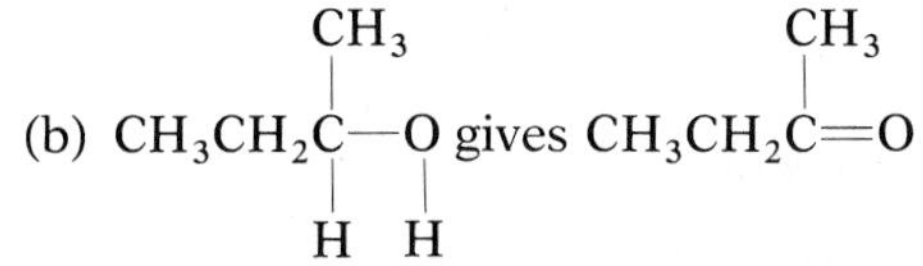

(b) $CH_3CH_2C(CH_3)(H)-O-H$ gives $CH_3CH_2C(CH_3)=O$ (2-butanone, a ketone)

(c) $(CH_3)_3C-O-H \longrightarrow$ no reaction

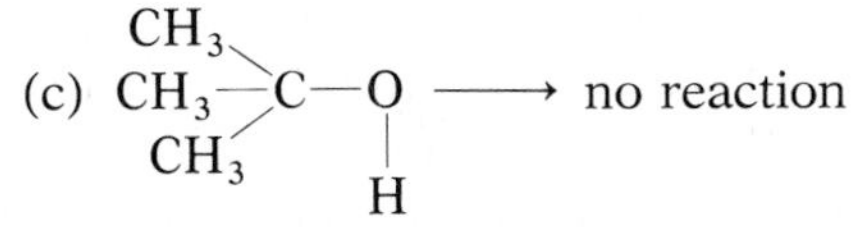

There is no H atom on the C atom bonded to the OH group. ■

PROBLEM 24.12 Give the structures of the alcohols needed to make the carbonyl compounds (a) $CH_3CH_2CH_2C(H)=O$, (b) $CH_3C(=O)CH_2CH_3$, and (c) 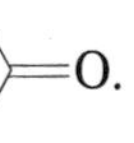 cyclohexanone (ring)=O. □

Some other important alcohols are shown in Figure 24.13. Notice that the names of most alcohols end in the suffix *-ol*. Compounds with two OH groups are called **diols**.

FIGURE 24.13
Some important alcohols.

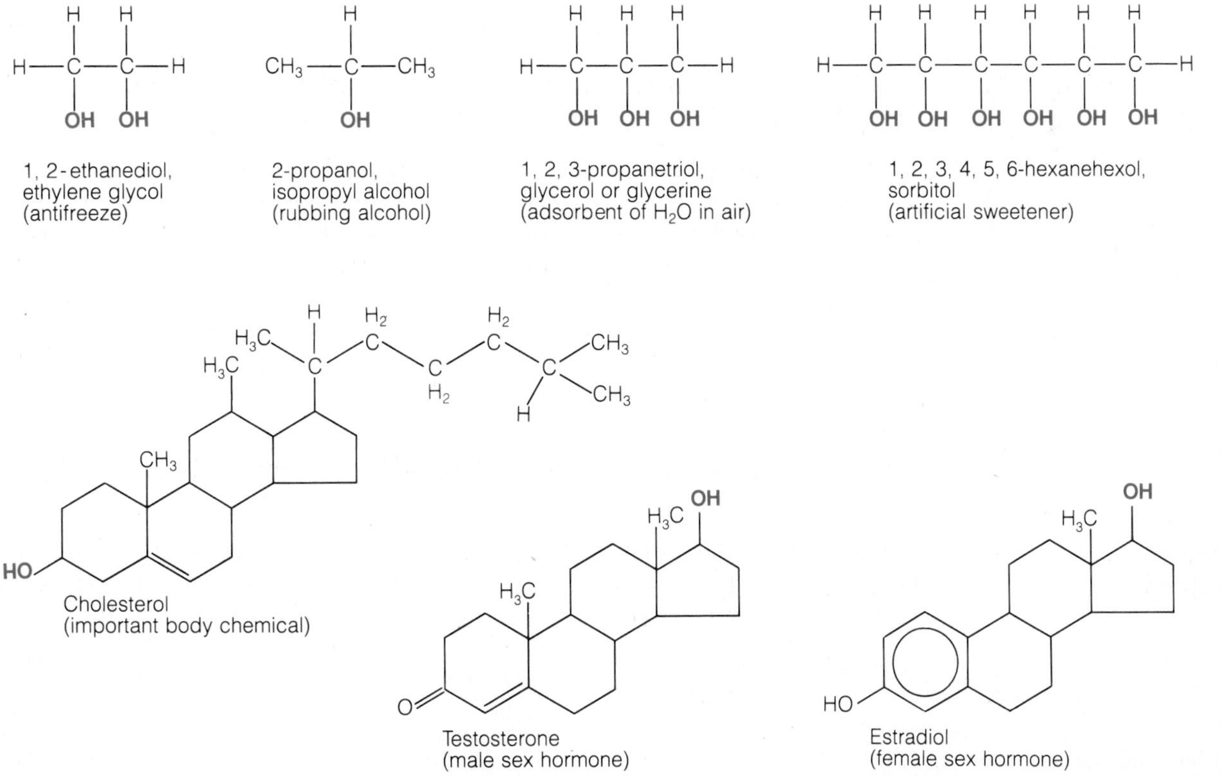

$HOCH_2-CH_2OH$

1, 2-ethanediol,
ethylene glycol
(antifreeze)

$CH_3-CH(OH)-CH_3$

2-propanol,
isopropyl alcohol
(rubbing alcohol)

$HOCH_2-CH(OH)-CH_2OH$

1, 2, 3-propanetriol,
glycerol or glycerine
(adsorbent of H_2O in air)

$HOCH_2-CH(OH)-CH(OH)-CH(OH)-CH(OH)-CH_2OH$

1, 2, 3, 4, 5, 6-hexanehexol,
sorbitol
(artificial sweetener)

Cholesterol
(important body chemical)

Testosterone
(male sex hormone)

Estradiol
(female sex hormone)

Phenols, $K_a \approx 10^{-10}$, are more acidic than alcohols, $K_a \approx 10^{-18}$.

Compounds with an OH group bonded directly to a benzene ring are called **phenols** rather than *alcohols*. The class is named for its simplest member, phenol (carbolic acid),

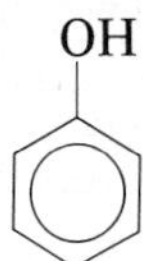

It was the first germicide used in medicine, by Joseph Lister in 1865.

ETHERS, ROR OR ROR (ArOR, ArOAr)

Replacement of the H atom of the OH group of an alcohol by an alkyl (or aryl) group gives an **ether**. Since ethers have no OH group, they do not react with Na. Nor do they have H atoms for hydrogen bonding. Hence, an ether (CH_3OCH_3, b.p. −24°) has a much lower boiling point than the isomeric alcohol (CH_3CH_2OH, b.p. 78.5°). Ethers have unshared electron pairs on the $-\ddot{\underset{\cdot\cdot}{O}}-$ atom and behave as bases toward strong Brönsted acids and Lewis acids:

$$\underset{\substack{\text{base}_1\\\text{(solvent)}}}{C_2H_5\ddot{\underset{\cdot\cdot}{O}}C_2H_5} + \underset{\text{acid}_2}{H\!:\!\ddot{\underset{\cdot\cdot}{I}}\!:} \longrightarrow \underset{\substack{\text{acid}_1\\\text{diethyloxonium ion}}}{\left[\begin{array}{c}H\\|\\C_2H_5\underset{\cdot\cdot}{O}C_2H_5\end{array}\right]^+} + \underset{\text{base}_2}{:\!\ddot{\underset{\cdot\cdot}{I}}\!:^-}$$

$$\underset{\substack{\text{Lewis base}\\\text{(solvent)}}}{C_2H_5\ddot{\underset{\cdot\cdot}{O}}C_2H_5} + \underset{\text{Lewis acid}}{BF_3} \longrightarrow C_2H_5\overset{\oplus}{\ddot{O}}C_2H_5\text{ with }O\text{–}\overset{\ominus}{B}F_3$$

EXAMPLE 24.12 Explain why ethers such as CH_3OCH_3 are more soluble in water than are alkanes of comparable molar mass such as $CH_3CH_2CH_3$.

ANSWER Although ethers have no H atoms able to form a hydrogen bond with the O atom of H_2O, they do have an O atom for H-bonding with the H of H_2O. This H-bonding increases the attractive force between CH_3OCH_3 and H_2O. Since alkanes have no way to H-bond with H_2O, they are much less soluble.

$$(CH_3)_2O\cdots H{-}O{-}H \quad (\text{H bond})$$

■

PROBLEM 24.13 Are alcohols more soluble in ethers than in alkanes? Explain.

This is an *inter*molecular dehydration.

Some ethers of the ROR type (the two alkyl groups are identical) may be made by removal of H_2O from two molecules of alcohol. Diethyl ether, called simply "ether," is made from ethanol in this way:

$$\underset{\text{ethanol}}{C_2H_5OH + HOC_2H_5} \xrightarrow[\text{heat}]{H_2SO_4(\ell)} \underset{\substack{\text{diethyl ether}\\\text{(an anesthetic)}}}{C_2H_5OC_2H_5} + H_2O$$

High temperatures favor alkene formation.

This reaction is competitive with alkene formation (equation 4).

To make an ether of the ROR′ as well as the ROR type, a nucleophilic displacement (equation 2) is usually used, as shown:

$ROH + Na \longrightarrow RO^-Na^+ + \frac{1}{2}H_2$

$$CH_3CH_2CH_2Br + CH_3\ddot{\underset{..}{O}}:^- \xrightarrow{CH_3OH(solvent)} CH_3CH_2CH_2OCH_3 + :\ddot{\underset{..}{Br}}:^-$$

methoxide ion
(a nucleophile)

or

$$CH_3Br + CH_3CH_2CH_2\ddot{\underset{..}{O}}:^- \xrightarrow{CH_3CH_2CH_2OH(solvent)} CH_3CH_2CH_2OCH_3 + :\ddot{\underset{..}{Br}}:^-$$

EXAMPLE 24.13 Give two reactions for the preparation of di-*n*-propyl ether, $CH_3CH_2CH_2OCH_2CH_2CH_3$.

ANSWER

$$CH_3CH_2CH_2OH + HOCH_2CH_2CH_3 \xrightarrow{H_2SO_4(\ell)} (CH_3CH_2CH_2)_2O + H_2O$$

$$CH_3CH_2CH_2\ddot{\underset{..}{O}}:^- + BrCH_2CH_2CH_3 \longrightarrow (CH_3CH_2CH_2)_2O + :\ddot{\underset{..}{Br}}:^-$$ ■

PROBLEM 24.14 Give two ways to synthesize isopropyl *n*-propyl ether, $CH_3CH_2CH_2O\underset{\displaystyle CH_3}{\underset{|}{C}}HCH_3$. □

Cyclic compounds that have at least one atom in the ring other than C are classified as **heterocyclic**.

The ether oxygen, $-\ddot{\underset{..}{O}}-$, can be part of a ring. Three examples of cyclic ethers are

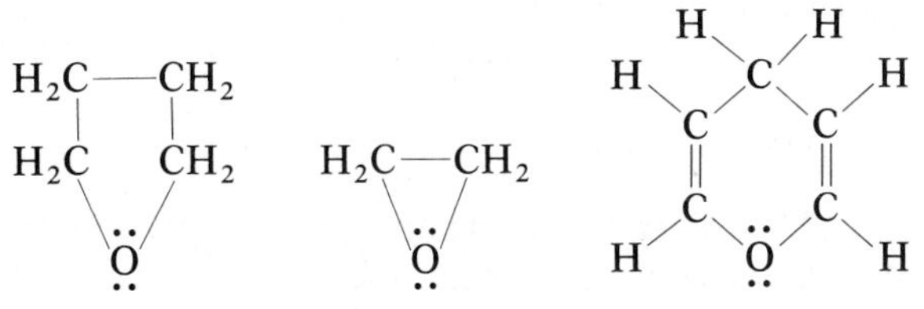

tetrahydrofuran (H_2C-CH_2, $H_2C\;\;CH_2$ joined through $\ddot{O}$)

ethylene oxide (an epoxide) (H_2C-CH_2 bridged by $\ddot{O}$)

pyran (CH_2 with two $HC=CH$ units joined through $\ddot{O}$)

$CH_2{=}CH_2 \xrightarrow[Ag]{O_2} H_2C-CH_2$ (bridged by O)

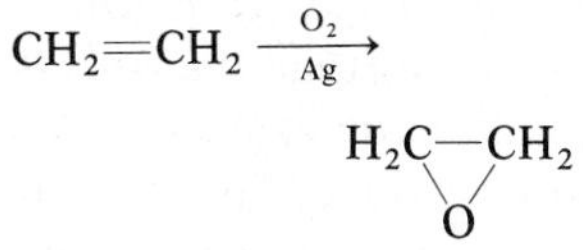

Epoxides, the three-membered ring compounds, are of particular interest. They are synthesized from alkenes, and because of the angle strain, they readily undergo ring-opening reactions. For example, ethylene oxide reacts with H_2O in the presence of H_3O^+ or OH^- to form ethylene glycol (an antifreeze):

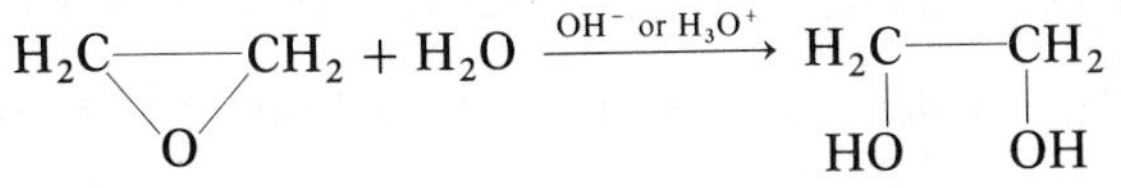

$$\underset{\diagdown O \diagup}{H_2C-CH_2} + H_2O \xrightarrow{OH^-\text{ or }H_3O^+} \underset{HO\quad\;\; OH}{\overset{}{H_2C-CH_2}}$$

(a diol)

ALDEHYDES AND KETONES

Aldehydes and **ketones** are two closely associated classes of compounds called **carbonyl compounds**. They are called this because they possess the carbonyl, $-C{=}O$, functional group. An aldehyde has at least one H atom bonded to the carbonyl carbon; a ketone has two R groups (no H's) attached to the carbonyl

BOX 24.3
EPOXIDES AND CANCER

Unwanted, extraneous compounds that enter the body, mainly by ingestion or inhalation, are chemically changed—usually becoming water-soluble—so that they can be excreted in the urine. The biochemical modification of several alkenes and aromatic hydrocarbons, especially fused benzene rings, involves the enzyme-induced reaction sequence, alkene $\longrightarrow$ epoxide $\longrightarrow$ diol. The diol is further changed to make it more water-soluble. Unfortunately, this process has a built in "time bomb." Epoxides are cancer-inducing agents. Therefore, compounds that are converted in the body to epoxides may be carcinogens. Ethylene oxide, itself, is now classified by the Environmental Protection Agency (EPA) as a carcinogen.

A common food additive, BHA, butylated hydroxyanisole, $(HO)C_6H_3(OCH_3)(C(CH_3)_3)$, appears to be a scavenger for cancer-inducing epoxides; it is an **anticarcinogen**. The use of BHA is believed to be partly responsible for the decline in stomach cancers in the United States. There are also compounds, called **cocarcinogens**, that make carcinogens more potent. For example, something in cigarette smoke is believed to inhibit the enzyme that catalyzes the conversion of the carcinogenic epoxides to the much less harmful diols. When this enzyme is inhibited, the concentration of epoxides in the body increases, which increases the chances for cancer cells to be formed.

carbon:

R could also be Ar.

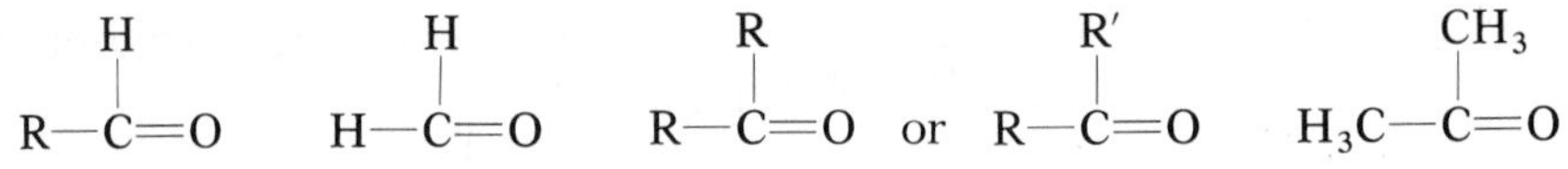

$$\underset{\text{any aldehyde}}{R-\overset{\overset{H}{|}}{C}=O} \qquad \underset{\text{formaldehyde}}{H-\overset{\overset{H}{|}}{C}=O} \qquad \underset{\text{any ketone}}{R-\overset{\overset{R}{|}}{C}=O \text{ or } R-\overset{\overset{R'}{|}}{C}=O} \qquad \underset{\text{acetone}}{H_3C-\overset{\overset{CH_3}{|}}{C}=O}$$

Oxidation of alcohols is one way carbonyl compounds are synthesized (equations 5 and 6).

Formaldehyde is toxic and a suspected carcinogen. Methanol is toxic because it is enzymatically oxidized in the liver to the toxic formaldehyde.

The enzyme is alcohol dehydrogenase.

$$CH_3OH \xrightarrow{\text{enzyme}} H_2C=O$$

In small doses, $H_2C=O$ irritates the mucous membranes and the skin. You should definitely avoid inhaling the vapors when you use formalin. Formalin, a preservative familiar to biology students, is a 37% water solution of formaldehyde.

Ingested ethanol is enzymatically oxidized to acetaldehyde (equation 5) in the liver. The enzyme has been discovered to be more plentiful in diagnosed alcoholics. The enzyme also seems to activate carcinogens and make normally harmless agents, such as analgesics, highly toxic to the liver. Also, unlike the livers of most normal drinkers, the livers of alcoholics have been found to produce insufficient amounts of a different enzyme that eliminates the acetaldehyde. The accumulation of acetaldehyde in the liver is believed to lead to the addiction.

Unlike C=C, C=O does not add Cl_2, Br_2, HCl, or HBr.

The multiple bond in the $\rangle C=O$ group is an unsaturated site. It undergoes addition reactions, as for example, in catalytic hydrogenation:

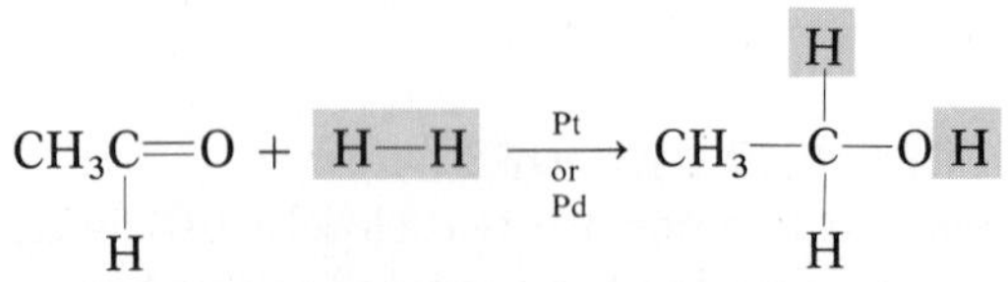

$$CH_3\underset{\underset{H}{|}}{C}=O + H-H \xrightarrow[\text{or Pd}]{\text{Pt}} CH_3-\overset{\overset{H}{|}}{\underset{\underset{H}{|}}{C}}-O\,H$$

As a result of this addition of H_2, the carbonyl compound is reduced to an alcohol.

At room temperature, aldehydes are oxidized to carboxylic acids, $RC\begin{matrix} \diagup\!\!\!\!=O \\ \diagdown OH \end{matrix}$

$$H{-}\overset{\overset{\displaystyle H}{|}}{\underset{\underset{\displaystyle H}{|}}{C}}{-}\overset{\overset{\displaystyle H}{|}}{C}{=}O \xrightarrow{[O]} H{-}\overset{\overset{\displaystyle H}{|}}{\underset{\underset{\displaystyle H}{|}}{C}}{-}\overset{\overset{\displaystyle OH}{|}}{C}{=}O$$

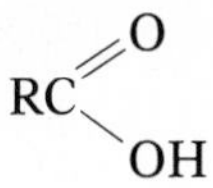

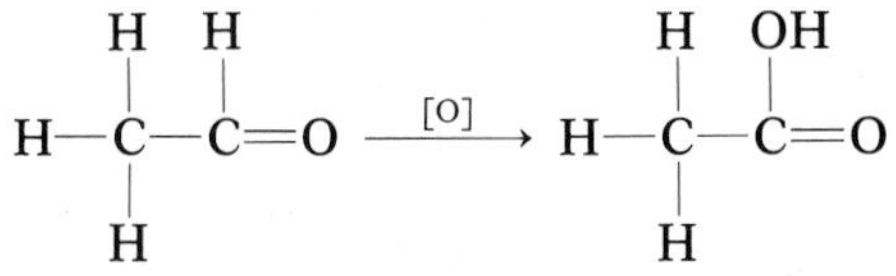

acetaldehyde — acetic acid

Under the same conditions, ketones are unreactive.

CARBOXYLIC ACIDS, $R{-}C\begin{matrix} \diagup\!\!\!\!=O \\ \diagdown OH \end{matrix}$

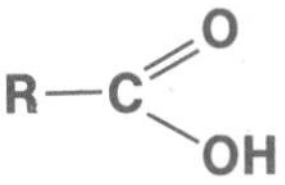

Carboxylic acids, $K_a \approx 10^{-5}$, react with hydroxides to form salts, as shown for acetic acid:

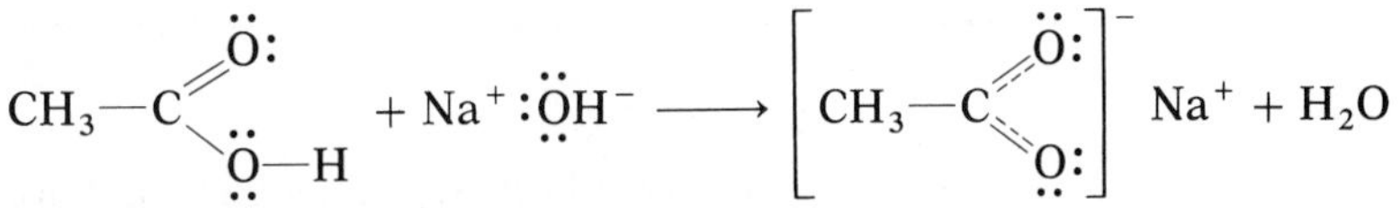

$$CH_3{-}C\begin{matrix} \diagup\!\!\!\!= \ddot{\underset{..}{O}}: \\ \diagdown \ddot{\underset{..}{O}}{-}H \end{matrix} + Na^+ :\ddot{\underset{..}{O}}H^- \longrightarrow \left[CH_3{-}C\begin{matrix} \diagup\!\!\!\!= \ddot{O}: \\ \diagdown\!\!\!\!= \underset{..}{O}: \end{matrix} \right]^- Na^+ + H_2O$$

acid$_1$ acetic acid; base$_2$; base$_1$ sodium acetate; acid$_2$

Note the delocalized π bond in the acetate ion. This results in greater stability of CH_3COO^- than $CH_3CH_2O^-$, where delocalization does not occur (section 14.9). Therefore carboxylic acids are more acidic than alcohols.

Carboxylic acids are often referred to as *fatty acids.* This term is mainly used for the naturally occurring long-chain saturated and unsaturated carboxylic acids. The most abundant of these naturally occurring "fatty" acids are

1. Saturated fatty acids (no C=C bonds, solids at room temperature):

   $CH_3(CH_2)_{14}COOH$, palmitic acid (m.p. 63°C)

   $CH_3(CH_2)_{16}COOH$, stearic acid (m.p. 70°C)

The configuration of groups around these C=C bonds is *cis.*

2. Monounsaturated fatty acids (one C=C bond, liquids except when refrigerated):

   $CH_3(CH_2)_7CH{=}CH(CH_2)_7COOH$, oleic acid (m.p. 16°C)

3. Polyunsaturated fatty acids (two or more C=C bonds, liquids):

   $CH_3(CH_2)_4CH{=}CHCH_2CH{=}CH(CH_2)_7COOH$, linoleic acid (m.p. −5°C)

   $CH_3CH_2CH{=}CHCH_2CH{=}CHCH_2CH{=}CH(CH_2)_7COOH$, linolenic acid (m.p. −11°C)

The salt of a long-chain carboxylic acid (fatty acid) is called a **soap**. Hard soaps are Na^+ salts; liquid soaps are K^+ salts.

Palmitic, stearic, and oleic acids are synthesized in the bodies of animals from enzyme-catalyzed combinations of acetate ions, CH_3COO^-. Since each acetate ion used has two carbon atoms, it should not be surprising that naturally occurring fatty acids have an even number of carbon atoms.

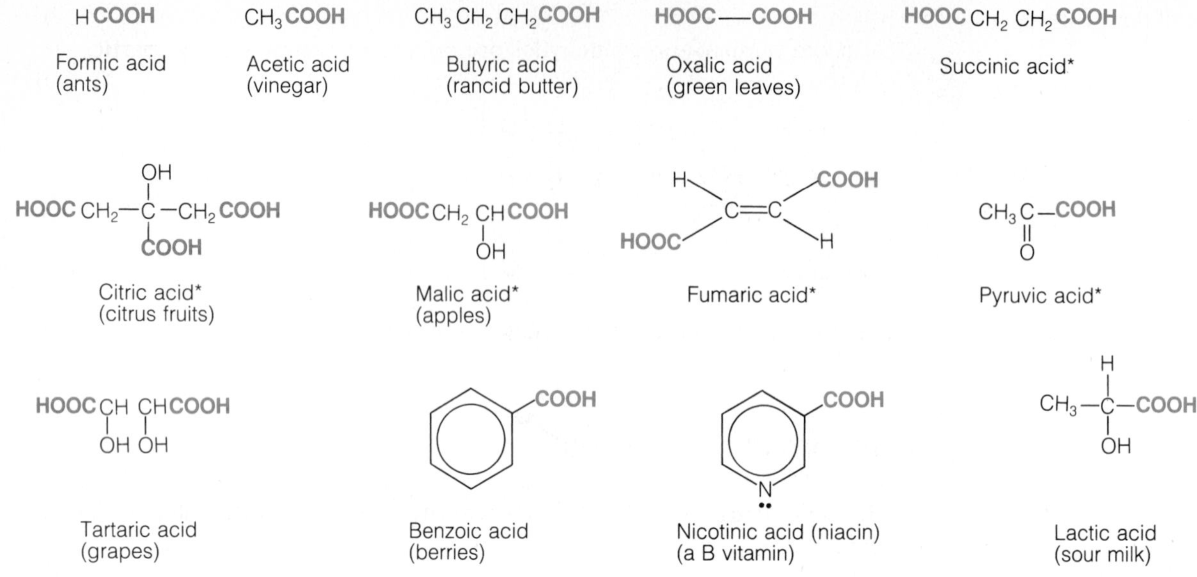

FIGURE 24.14
Some common carboxylic acids and their natural sources. Asterisks indicate acids that are involved in the citric acid cycle.

The polyunsaturated fatty acids, linoleic and linolenic, are made only by plants. The polyunsaturated fatty acids are called **essential fatty acids** because they must be part of the human diet. A diet deficient in sources of these acids results in poor growth, skin lesions, kidney damage, and impaired fertility. The acids serve as precursors for the biosynthesis of **prostaglandins**, a group of compounds discovered in 1960, that have a very wide variety of important physiological effects. Some common naturally occurring carboxylic acids are shown in Figure 24.14. Citric, succinic, fumaric, and malic acids are part of the citric acid cycle, the end of the pathway for the oxidation of fatty acids, carbohydrates, and amino acids. Pyruvic and lactic acids are products of the metabolism of sugar and fats.

Sir Hans Krebs postulated this cycle in 1937.

DETERGENCY

The cleansing action of soaps, their **detergency**, results from the opposing ways their two structural parts interact with water. The *polar head*, which in the case of a soap is the carboxylate group, COO^-, has a strong affinity for water; it is **hydrophilic**. The long-chain-hydrocarbon portion, the *nonpolar hydrocarbon tail group*, has no affinity for water; it is **hydrophobic**. The colloidal dispersion of soap in water consists of **micelles**, approximately spherical aggregates of 50–150 soap molecules (Figure 24.15). We picture the micelle as a ball with the polar head groups on the surface exposed to the water molecules and the nonpolar tails buried on the inside. The micelle satisfies the needs of both groups. The polar heads are close to water, and the nonpolar tails are close to each other on the inside of the micelle. The small micelles do not coalesce because of the repulsions between their like-charged ionic surfaces.

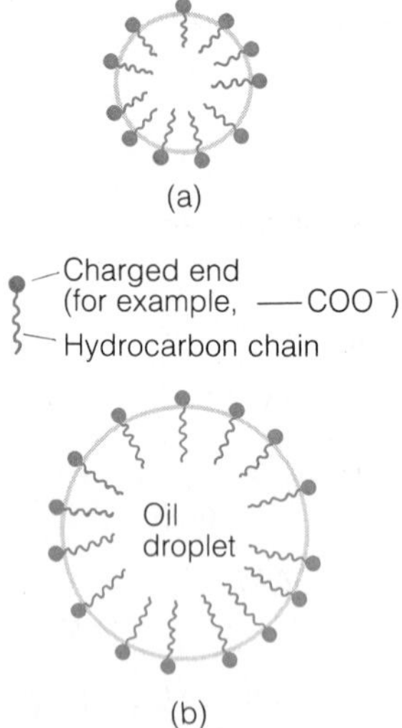

FIGURE 24.15
Micelles (a) in soapy water and (b) in soapy water with tiny oil droplets.

Micelles play a key role in detergency. Most dirt is held to surfaces, such as clothes or skin, by a thin film of oil or grease. Oil and grease are nonpolar, hydrocarbonlike substances that are insoluble in water. When the dirty object is agitated by rubbing or scrubbing in soapy water, the grease or oil is dispersed into very thin droplets. The droplets dissolve in the hydrophobic interiors of the micelles (Figure 24.15b). The micelles with the droplets of oil or grease still have an affinity

Recall that "like dissolves like" (section 12.6).

for water because of their ionic surfaces. Therefore, these micelles are carried into colloidal dispersion (an emulsion) and down the drain. Dirt particles, freed from the grease or oil, also wash away.

When added to water containing such cations as Ca^{2+}, Mg^{2+}, and Fe^{3+}, commercial soaps form insoluble deposits that hamper their detergency.

$$2RCOO^-Na^+(aq) + Ca^{2+}(aq) \longrightarrow (RCOO^-)_2Ca(c) + 2Na^+(aq)$$

These deposits cause the dull films on clothes and "rings" on bathtubs. Water that contains these ions is called **hard water**. For this reason synthetic detergents that do not have this drawback have been developed. The most common synthetic detergents have either sulfate, $—OSO_3^-$, or sulfonate, $—SO_3^-$, as the ionic polar head group along with the nonpolar hydrocarbon tail group. Two examples are

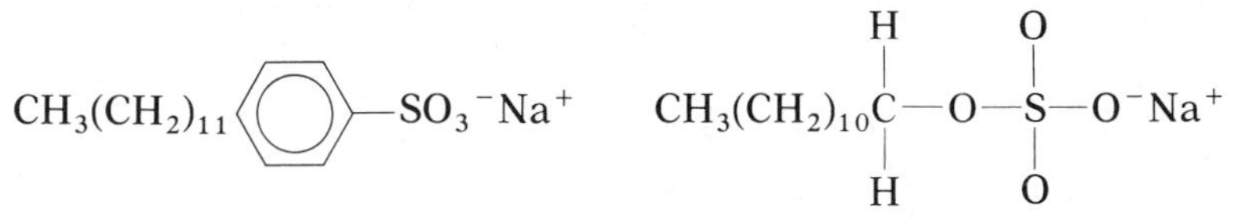

$$CH_3(CH_2)_{11}\text{—}C_6H_4\text{—}SO_3^-Na^+$$

sodium *p*-dodecylbenzenesulfonate
(A sulfonate has a C—S bond.)

$$CH_3(CH_2)_{10}CH_2\text{—}O\text{—}SO_2\text{—}O^-Na^+$$

sodium lauryl sulfate
(A sulfate has a C—O bond.)

These are salts of a sulfonic acid, RSO_3H, and an alkyl sulfuric acid, $ROSO_3H$. (Note that not all organic acids are carboxylic acids.)

Soaps, sulfates, and sulfonates are **anionic** detergents because the polar head group has a negative charge. There are also **cationic** detergents (see amines) and **nonionic** detergents. The example of a nonionic detergent shown in Figure 24.16 has a nonpolar tail with 12 C atoms and a polar group with several ether linkages and an OH group. Nonionic detergents are used in window-cleaning preparations, since incomplete rinsing does not leave a visible residue and the water drains more smoothly when a trace of detergent is still present.

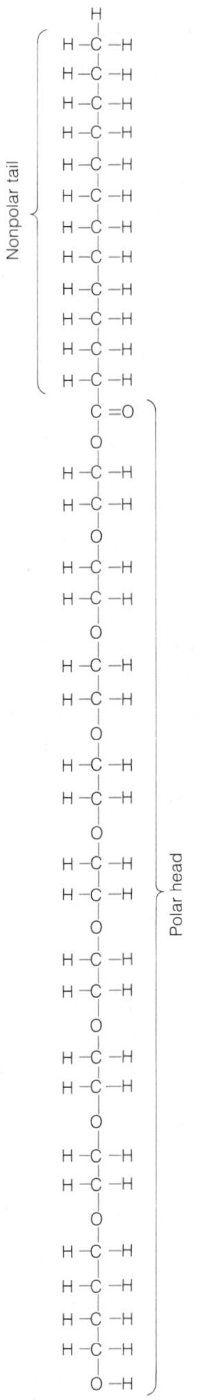

$CH_3(CH_2)_{11}C(=O)O(CH_2CH_2O)_{9}CH_2CH_2CH_2CH_2OH$ (structure shown vertically; labels: Nonpolar tail, Polar head)

FIGURE 24.16
A nonionic detergent. The O atoms make the lower part of the molecule, the "head," hydrophilic.

ESTERS, $R—C(=O)—O—R$ OR $R—C(=O)—O—R'$

Carboxylic acids react with alcohols to form **esters** by loss of water. The reaction is catalyzed by strong acids such as HCl.

$$CH_3C(=O)\text{—}OH + H\text{—}OC_2H_5 \underset{\text{HCl(aq) hydrolysis}}{\overset{\text{dry HCl esterification}}{\rightleftharpoons}} CH_3C(=O)\text{—}OC_2H_5 + H_2O$$

ethyl acetate
(an ester)

Esterification is not analogous to the neutralization reaction between an acid and a base, such as

$$CH_3C(=O)OH + HO^-Na^+ \longrightarrow [CH_3CO_2]^- Na^+ + H_2O$$

The H of H_2O comes from the carboxylic acid; the OH of H_2O comes from OH^-. In esterification, the OH group in the product H_2O comes from the carboxylic acid, not from the alcohol.

The reverse reaction, **hydrolysis**, occurs in aqueous acid. An ester is also hydrolyzed in the presence of hydroxide ion. This process is called **saponification** because the carboxylate salt is a soap when R′ is a long hydrocarbon chain:

$$\underset{\text{ester}}{R'COOR} + Na^+OH^- \longrightarrow \underset{\text{soap}}{R'COO^-Na^+} + \underset{\text{alcohol}}{ROH}$$

Triglycerides are examples of lipids, which are water-insoluble nonpolar biochemicals.

Natural fats are *solid* esters, and fatty oils are *liquid* esters of glycerol, a trihydroxy alcohol. Saponification of these **triglycerides** yields a mixture of soaps:

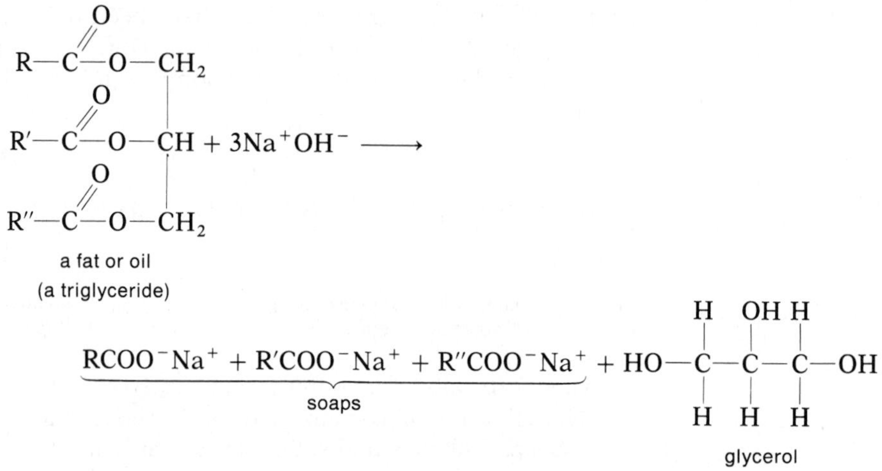

$$\underset{\substack{\text{a fat or oil}\\\text{(a triglyceride)}}}{\begin{array}{l} R-C(=O)-O-CH_2 \\ R'-C(=O)-O-CH \\ R''-C(=O)-O-CH_2 \end{array}} + 3Na^+OH^- \longrightarrow$$

$$\underbrace{RCOO^-Na^+ + R'COO^-Na^+ + R''COO^-Na^+}_{\text{soaps}} + \underset{\text{glycerol}}{HO-CH_2-CH(OH)-CH_2-OH}$$

Fats such as lard, butter, and chicken fat are esters of saturated fatty acids. Oils such as those of olives, sunflowers, safflower seeds, corn, and peanuts are esters of monounsaturated and polyunsaturated fatty acids. Oils are the source of the essential fatty acids in our diets. (See carboxylic acids earlier in this section.)

Palm and coconut oils are saturated.

EXAMPLE 24.14 (a) Give the ester formed from CH_3OH and CH_3CH_2COOH (propanoic acid). (b) Give the products formed on hydrolysis of $CH_3C(=O)OCH_2CH_2CH_3$ in (1) HCl, an acid, and (2) NaOH, a base.

ANSWER

(a) $$CH_3CH_2\overset{O}{\overset{\|}{C}}OH + HOCH_3 \xrightarrow{\text{acid}} CH_3CH_2\overset{O}{\overset{\|}{C}}OCH_3 + H_2O$$

(b) (1) $$CH_3\overset{O}{\overset{\|}{C}}OCH_2CH_2CH_3 + HOH \xrightarrow{\text{acid}} \underset{\text{(a carboxylic acid)}}{CH_3\overset{O}{\overset{\|}{C}}OH} + \underset{\text{(an alcohol)}}{HOCH_2CH_2CH_3}$$

(2) $$CH_3\overset{O}{\overset{\|}{C}}OCH_2CH_2CH_3 + NaOH \xrightarrow{\text{base}}$$

$$\underset{\text{(a carboxylate salt)}}{\left[CH_3\overset{O}{\overset{\vdots}{C}}{=\!=}O\right]^- Na^+} + \underset{\text{(an alcohol)}}{HOCH_2CH_2CH_3}$$

■

PROBLEM 24.15 Which two esters could be formed when ethylene glycol, $HOCH_2CH_2OH$ (1,2-ethanediol), reacts with different amounts of butanoic (butyric) acid, $CH_3CH_2CH_2COOH$? □

Phospholipids, which have an essential role in the chemistry of cell membranes, are closely related to the fats since they are also esters of glycerol. Two oxygens of glycerol are esterified with fatty acids, but one of the terminal oxygens is esterified with a special type of organic phosphate derivative as shown:

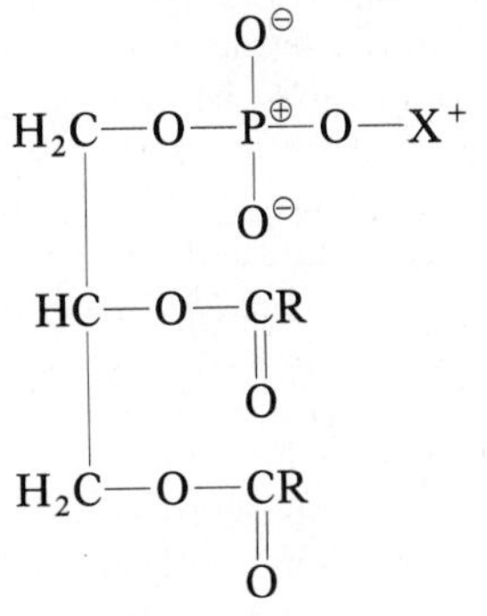

$$\begin{array}{l} H_2C-O-\overset{\overset{\displaystyle O^{\ominus}}{|}}{\underset{\underset{\displaystyle O^{\ominus}}{|}}{P^{\oplus}}}-O-X^+ \\ \;| \\ HC-O-\underset{\underset{\displaystyle O}{\|}}{C}R \\ \;| \\ H_2C-O-\underset{\underset{\displaystyle O}{\|}}{C}R \end{array}$$

(X^+ can be, for example, $-CH_2CH_2NH_3^+$ or $-CH_2CH_2N(CH_3)_3^+$.)

a phospholipid

The phospholipid has a net negative charge on the O atoms and a positive charge in the X group. It is an example of a *dipolar* ion.

Also called a *zwitterion*.

EXAMPLE 24.15 Is a phospholipid related to a detergent? Explain.

ANSWER Yes. The end with the phosphate group is ionic and polar. The rest of the ester, the R and R′ groups, is hydrocarbonlike and nonpolar. ■

AMINES, RNH_2, R_2NH, AND R_3N

Amines are derived from NH_3 when at least one H atom is replaced by a hydrocarbon group. Some examples are

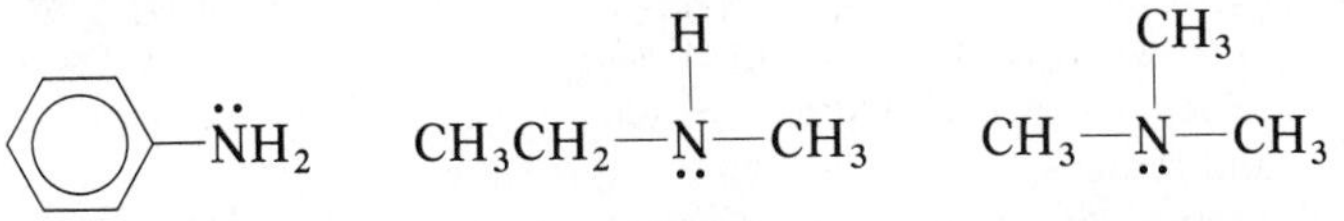

$$C_6H_5-\ddot{N}H_2 \qquad CH_3CH_2-\underset{\cdot\cdot}{\overset{\overset{\displaystyle H}{|}}{N}}-CH_3 \qquad CH_3-\underset{\cdot\cdot}{\overset{\overset{\displaystyle CH_3}{|}}{N}}-CH_3$$

aniline	ethylmethylamine	trimethylamine
a primary amine	a secondary amine	a tertiary amine
(1 H is replaced.)	(2 H's are replaced.)	(3 H's are replaced.)

Aniline is used to manufacture certain dyes.

Like NH_3, amines are weak bases. They react with acids to form substituted ammonium salts:

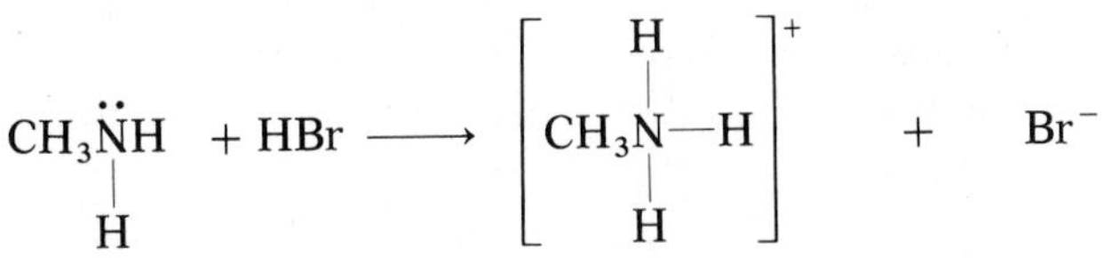

$$\underset{\underset{\displaystyle H}{|}}{CH_3\ddot{N}H} + HBr \longrightarrow \left[CH_3\overset{\overset{\displaystyle H}{|}}{\underset{\underset{\displaystyle H}{|}}{N}}-H \right]^+ + Br^- \qquad (7)$$

base$_1$ methylamine; acid$_2$; acid$_1$ methylammonium ion (a substituted ammonium salt); base$_2$ bromide

The methylammonium ion in equation 7 is related to the ammonium ion, NH_4^+, except that an alkyl group such as CH_3 replaces an H atom. There are

BOX 24.4
THE CHOLESTEROL STORY

Cholesterol, $C_{27}H_{46}O$, is a biochemically important alcohol found in all animal tissues but few plant tissues. It is the essential biochemical source of sex hormones, adrenal cortex hormones (corticosteroids), and bile acids. Found in eggs, whole milk, animal fats, and shellfish, cholesterol is part of the human diet. It is also synthesized by the body from the acetate ion. Unfortunately, cholesterol can also be harmful. Gallstones are almost pure cholesterol. More damaging, cholesterol is also found in the plaque that thickens the inner walls of arteries. This thickening of the artery walls, called **atherosclerosis**, contributes to high blood pressure, strokes, angina pectoris, and other cardiovascular diseases.

Because cholesterol is insoluble in water, it is transported through the blood plasma (an aqueous medium) bound to water-soluble lipoproteins, which are phospholipids that are bound to proteins. Lipoproteins are classified according to their density. Of major interest in the cholesterol story are the low-density lipoproteins (LDL) and the high-density lipoproteins (HDL). One of the most fascinating discoveries in recent years is that cholesterol bound to HDL reduces the risk of coronary disease, whereas cholesterol bound to LDL increases the risk of coronary disease. Consequently, the level of LDL in blood plasma is a better predictor of risk than the *total* cholesterol serum level. Even if the total cholesterol level is high, the risk for heart disease may be moderate or low if the HDL level is high and the LDL level is low.

LDL's are believed to transport and deposit cholesterol in all parts of the body, including the coronary and brain arteries. In this way there is a buildup of the plaque that leads to blockage of the arteries, which causes heart attacks and strokes.

HDL behaves differently. It appears to transport cholesterol exclusively to the liver, where it is converted to the useful bile acids. In fact, the HDL can take cholesterol deposited in the plaque, where it is harmful, and transport it to the liver where it is beneficial. This property of HDL is believed to help prevent heart attacks.

How can these high levels of HDL and low levels of LDL be achieved? An answer may lie in the structural differences of HDL and LDL. The fatty acids comprising LDL are saturated, whereas those in HDL are polyunsaturated. Hence, we are advised that our total intake of lipids should be moderate but high in polyunsaturated oils (fish is a good source) and low in saturated fats. Although olive oil is monounsaturated, recent research indicates that it nevertheless helps to lower the risk of coronary heart disease. Further recommendations are to use moderation in eating foods rich in cholesterol (such as eggs), cut down on caloric intake (don't be overweight), have a sensible exercise program (don't be sedentary), and abstain from smoking.

A space-filling model of cholesterol, $C_{27}H_{46}O$. The black balls represent C atoms; the white balls represent H atoms. The arrow points to the O atom.

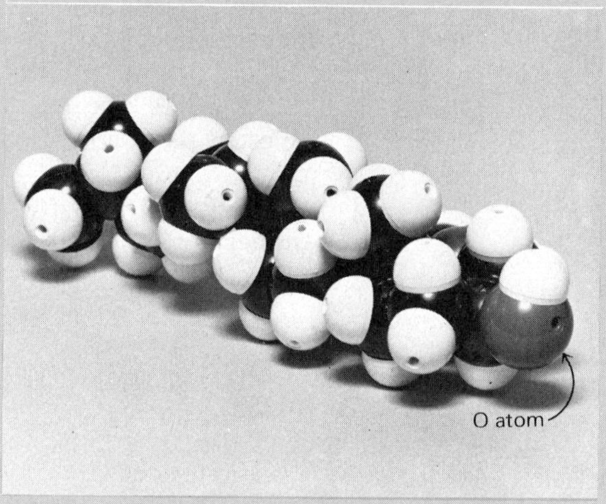

ammonium cations in which all four H atoms of NH_4^+ are replaced by alkyl groups, NR_4^+. When at least one R group in NR_4^+ is a long-chain alkyl group the compound is a cationic detergent. Such compounds are also used as disinfectants, especially in mouthwashes.

Hexadecyl is also called *cetyl*.

$$\left[\begin{array}{c} \quad\quad\quad\quad\quad CH_3 \\ \quad\quad\quad\quad\quad | \\ CH_3(CH_2)_{15}-N-CH_3 \\ \quad\quad\quad\quad\quad | \\ \quad\quad\quad\quad\quad CH_3 \end{array}\right]^+ Cl^- \quad \text{(a cationic detergent)}$$

hexadecyltrimethylammonium chloride

BOX 24.5

NITROSAMINES AND CANCER

Amines react with nitrous acid, HNO_2. Of particular interest here is the reaction of secondary amines, R_2NH, to form a class of compounds called **nitrosamines**. Ethylmethylamine is an example:

$$CH_3CH_2\ddot{N}(CH_3)H + HO\ddot{N}{=}\ddot{O}: \longrightarrow CH_3CH_2\ddot{N}(CH_3){-}\ddot{N}{=}\ddot{O}: + H_2O$$

a secondary amine (1 H on N) ($H^+ + NO_2^-$ from $NaNO_2$)

a nitrosamine

Nitrosamines are potent carcinogens in laboratory animals and presumably in humans. Secondary amines occur naturally in the body and they can react with the nitrous acid formed in the body from ingested nitrite salts ($NaNO_2$).

Since nitrite salts are the precursors for the very harmful nitrosamines, they are classified as carcinogens. Nitrite salts are added to bacon, frankfurters, and other cured meats as preservatives for preventing the growth of the bacteria causing botulism. Nitrites are also used to give meats such as corned beef an attractive red color. When making decisions about food additives, we always weigh the risks (in this case cancer) and the benefits (preventing botulism). Thus we might favor using minimal amounts of nitrites as preservatives but ban their use as coloring agents. The National Academy of Sciences has recommended limiting the consumption of cured and smoked meat, fish, and poultry in favor of fresh or fresh-frozen varieties. The expected benefit is a reduction in the incidence of cancers of the stomach and esophagus. Efforts should also be made to find other meat preservatives that are not harmful.

Nitrates in food are also suspect because they can be converted in the body to nitrites. However, nitrates are naturally present in most vegetables and therefore cannot be excluded from the diet.

AMIDES, $R{-}\overset{O}{\overset{\|}{C}}{-}NH_2$

Ammonia and amines react with carboxylic acids to form salts. When the solid salts are heated, **amides** are formed:

$$RCOOH + NH_3 \longrightarrow RCOO^-NH_4^+ \xrightarrow{\text{heat}} R{-}C({=}O){-}NH_2 + H_2O$$

$$RCOOH + R'NH_2 \longrightarrow RCOO^-[NH_3R']^+ \xrightarrow{\text{heat}} R{-}C({=}O){-}NHR' + H_2O$$

ammonium carboxylate salts; amides

SUMMARY

Organic chemistry is the study of the very large number of carbon compounds. The carbon atoms bond to each other in **continuous** or **branched** chains or in **cyclic** molecules. **Hydrocarbons**, which contain only C and H atoms, are classified into **homologous series** described by general formulas. **Alkanes** have only chains of single bonds between the carbon atoms (C_nH_{2n+2}), **alkenes** have a C=C bond (C_nH_{2n}), **alkynes** have a C≡C bond, (C_nH_{2n-2}), and **cycloalkanes** have a ring of

C's with only single bonds (C_nH_{2n}). **Aromatic** hydrocarbons normally possess at least one benzene ring.

Structures that differ as the result of rotation about a C—C bond are called **conformations**—they are not isomers. Any molecule with a sequence of at least three single bonds can have different conformations. These are exemplified by the **staggered** and **eclipsed** conformations of ethane. Cyclic compounds such as cyclohexane can be nonplanar (puckered). Their different conformations are due to C—C bonds of the ring bending up or down.

The number of isomers increases as the number of C atoms increases because there are more varieties of branching, more locations of multiple bonds, and more differences in the size of rings. ***Cis-trans* isomerism** is observed in alkenes when each doubly bonded C atom has two different atoms or groups attached to it. The *cis*-alkene has the like groups on the same side of the doubly bonded C's. The *trans*-alkene has the like groups on opposite sides.

Alkanes are chemically unreactive under ordinary conditions. However, at high temperatures they react with O_2 to give CO_2 and H_2O—combustions that provide most of the world's energy. When excited by ultraviolet light or high temperatures, alkanes undergo **substitution reactions**, in which an H is replaced by some other atom or group, typically a Cl or Br atom. The multiple bonds in alkenes and alkynes are sites of chemical reactivity. They undergo **addition reactions** with reactants such as H_2, HCl, Br_2, and H_2O. For this reason hydrocarbons with multiple bonds are said to be **unsaturated**; those with only single bonds are **saturated**. Although the aromatic compounds have multiple bonds, they undergo substitution rather than addition reactions—they behave in this respect like saturated hydrocarbons. In this way the stable benzene ring is preserved. The benzene ring has less energy and is more stable and less reactive toward additions than a typical alkene with three double bonds. This is so because its three pairs of π electrons are not in fixed positions, but rather are delocalized over all the C atoms of the ring.

A carbon atom tetrahedrally bonded to four different atoms or groups is **chiral (asymmetric)** because its mirror images, called **enantiomers**, are nonsuperimposable. Enantiomers are examples of stereoisomers often called optical isomers. An equal mixture of enantiomers is called a **racemate**. Enantiomers have the same physical and chemical properties in nonchiral environments. However, they react at different rates with chiral reagents, catalysts (such as enzymes), and solvents.

Sites of reactivity in organic molecules, such as multiple bonds and atoms other than carbon, are called **functional groups**. **Alkyl halides**, RX, where X is F, Cl, Br, or I and R is the **alkyl** group composed of C and H atoms, react with bases in two competing ways. They undergo **elimination** of HX from adjacent atoms to give alkenes. They also undergo nucleophilic **displacement** reactions, in which the very weakly basic leaving group X^- is displaced by a stronger base. When OH^- is the nucleophile, displacement gives an **alcohol**, ROH. Alcohols are also formed by adding H_2O to alkenes in the presence of $(H_3O)^+$. Alcohols undergo acid-catalyzed elimination of H_2O to give alkenes. Alcohols form hydrogen bonds. In **phenols** the OH group is bonded to a benzene ring. One molecule of H_2O can be lost from two molecules of alcohol to give an **ether**, ROR. Ethers with different R groups result when RX undergoes nucleophilic displacement with an alkoxide ion, $R'O^-$, from $R'O^-Na^+$, a reagent formed by adding Na to an alcohol.

Controlled oxidation of an alcohol with the general formula RCH_2OH gives an **aldehyde**, RHC=O; RR'CHOH gives a **ketone**, RR'C=O. The functional group in aldehydes and ketones is the **carbonyl** group, >C=O. Aldehydes are further oxidized to the weakly acidic **carboxylic acids**, RCOOH, featuring the **carboxyl** group, $-C\begin{matrix}\nearrow OH \\ \searrow O\end{matrix}$. A **soap** is the salt of a long-chain carboxylic acid (**fatty acid**).

Carboxylic acids react with alcohols to give **esters**, $\overset{O}{\overset{\|}{RC}}OR'$, and under the proper conditions with NH_3 to give **amides**, $\overset{O}{\overset{\|}{RC}}NH_2$. In each case a molecule of H_2O is lost from the two reactants. **Fats** (solids) and **oils** (liquids), both triglycerides, are esters of glycerol, $C_3H_5(OH)_3$, with saturated and unsaturated fatty acids, respectively. When heated with NaOH (**saponification**), they produce glycerol and soaps. The cleansing action (**detergency**) of soaps results from **micelle** formation; the nonpolar alkyl (R) end dissolves in a very small oil particle, and the polar COO^- end dissolves in water. In this way an emulsion forms that washes away with the loosened dirt. **Phospholipids** are glycerol esters in which one of the fatty acids is replaced by a substituted phosphoric acid; they have essential roles in life processes.

The **amines**, RNH_2 (primary), R_2NH (secondary), and R_3N (tertiary), are derived from NH_3 by replacing at least one H with an R group. Like NH_3, they are weak bases; they form substituted ammonium salts with acids, $-\overset{|}{\underset{|}{N}}H^+X^-$.

24.13 ADDITIONAL PROBLEMS

GENERAL

24.16 Discuss with the aid of illustrations the validity of each statement. (a) The ring atoms of an aromatic compound must be in a plane. (b) Alkynes do not have *cis-trans* isomers. (c) Replacement of an H by a functional group increases the number of isomers. (d) The compound whose formula is C_3H_6 must be an alkene. (e) In order to have different conformations a molecule must have at least four σ-bonded atoms in sequence, —A—B—C—D—. (f) All molecules with carbon-to-carbon multiple bonds readily undergo addition reactions. (g) All bonds to carbon are covalent. (h) The compound whose molecular formula is C_4H_8O must have a C=C bond or be a cyclic compound. (i) Benzene normally undergoes substitution reactions rather than addition reactions.

24.17 Use the ion-electron method to write a balanced ionic equation for the oxidation of CH_3CHO to CH_3COOH by acidified MnO_4^- that is reduced to Mn^{2+}.

24.18 Select pairs of the following reactions that are (a) competitive; (b) reversible.

(1) $CH_3CH(OH)CH_3 \xrightarrow{H_2SO_4} CH_3CH{=}CH_2 + H_2O$

(2) $CH_3CHBrCH_3 + {}^-OCH_3 \longrightarrow CH_3CH(OCH_3)CH_3 + Br^-$

(3) $CH_3\overset{O}{\overset{\|}{C}}{-}OCH_2CH_3 + H_2O \xrightarrow{H^+} CH_3\overset{O}{\overset{\|}{C}}{-}OH + CH_3CH_2OH$

(4) $2CH_3CH(OH)CH_3 \xrightarrow{H_2SO_4} (CH_3)_2\underset{H}{\underset{|}{C}}O\underset{H}{\underset{|}{C}}(CH_3)_2 + H_2O$

(5) $CH_3\overset{O}{\overset{\|}{C}}{-}OH + CH_3CH_2OH \xrightarrow{H^+} CH_3\overset{O}{\overset{\|}{C}}{-}OCH_2CH_3 + H_2O$

(6) $CH_3CH{=}CH_2 + H_2O \xrightarrow{H_2SO_4} CH_3CHOHCH_3$

(7) $CH_3CHBrCH_3 + {}^-OCH_3 \longrightarrow CH_3CH{=}CH_2 + HOCH_3 + Br^-$

24.19 Give the IUPAC name for (a) $CH_3CHBrCH_2CH_3$, (b) $CH_3CHBrCHBrCH_3$, (c) $CH_3CBr_2CH_2CH_3$, (d) $BrCH_2CH_2CHBrCH_2Br$.

24.20 (a) In what kind of hybrid orbital (*sp*, *sp*^{2}, or *sp*^{3}) do we find the unshared pair of electrons of nitrogen in each of the following compounds, given in order of decreasing basicity?

$$CH_3CH_2\ddot{N}H_2 > CH_3\underset{H}{\underset{|}{C}}{=}\underset{\cdot\cdot}{N}H > CH_3C{\equiv}N{:}$$

(1) ethylamine (2) ethylimine (3) acetonitrile

(b) In view of your answer to part a, suggest a relationship between basicity and the *s* character of the orbital holding an unshared pair.

24.21 Given this stereochemical observation:

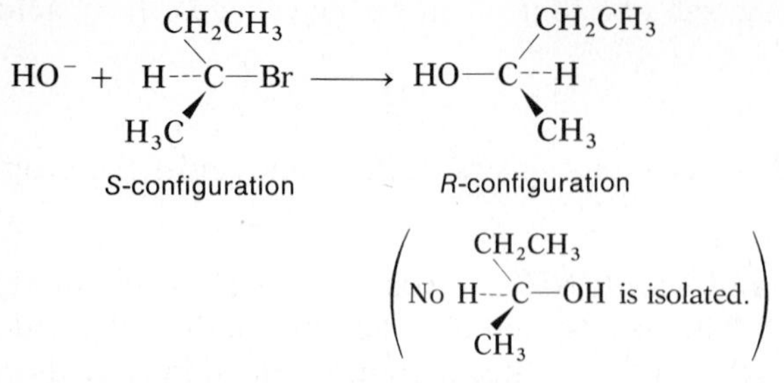

$$HO^- + H\cdots\overset{CH_2CH_3}{\underset{H_3C}{C}}—Br \longrightarrow HO—\overset{CH_2CH_3}{\underset{CH_3}{C}}\cdots H$$

S-configuration *R*-configuration

(No $H\cdots\overset{CH_2CH_3}{\underset{CH_3}{C}}—OH$ is isolated.)

(a) In this reaction, does the OH^- approach the carbon atom from the same side as the C—Br bond or from the opposite side? (b) Offer a reason for the kind of approach in terms of electrical repulsion.

24.22 Account for the fact that borazine and pyridine are aromatic compounds.

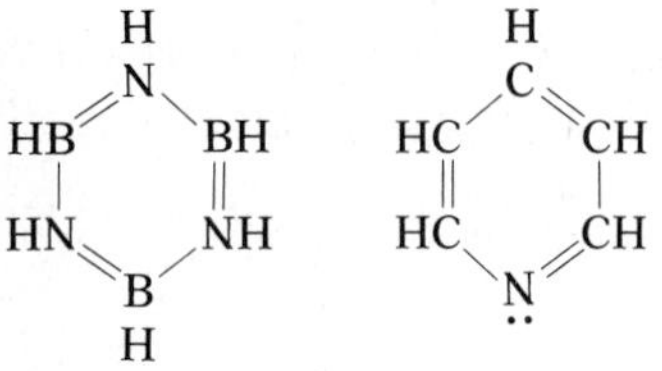

borazine pyridine

HYDROCARBONS

24.23 (a) Give the structural formulas for all isomers of (1) C_4H_{10}, (2) C_5H_{12}, and (3) C_6H_{14}. (b) Give the IUPAC name for each.

24.24 Find the molecular weight of (a) an alkane with 10 carbon atoms, (b) an alkyne with 6 carbon atoms, and (c) a benzene compound with 9 C's.

24.25 Give the structural formulas for the four alkyl groups with the formula C_4H_9.

24.26 Group together the classes of hydrocarbons with the same general formula: alkene, alkyne, cycloalkane, alkane, cycloalkene, alkadiene, cycloalkadiene.

24.27 Give the IUPAC names for

(a) 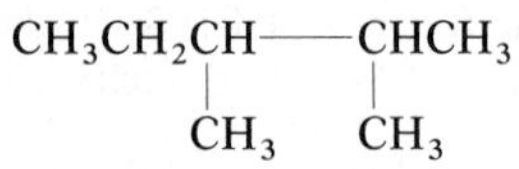 $CH_3CH_2\underset{CH_3}{CH}—\underset{CH_3}{CH}CH_3$

(b) 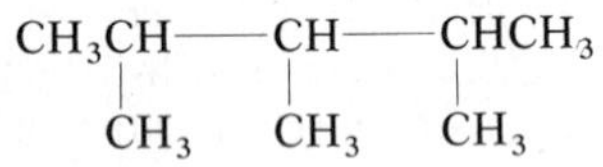 $CH_3\underset{CH_3}{CH}—\underset{CH_3}{CH}—\underset{CH_3}{CH}CH_3$

(c) 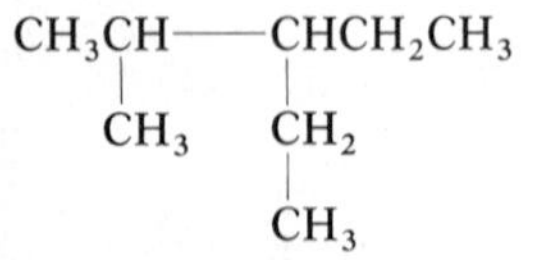 $CH_3\underset{CH_3}{CH}—\underset{CH_2CH_3}{CH}CH_2CH_3$

(d) 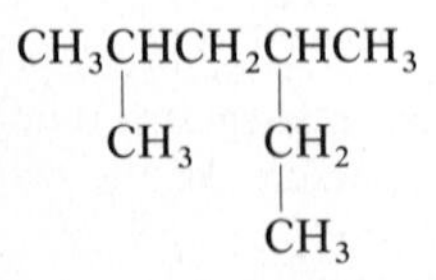 $CH_3\underset{CH_3}{CH}CH_2\underset{CH_2CH_3}{CH}CH_3$

24.28 Identify the error in each name and give the correct IUPAC name and the structural formula: (a) 1-methylpropane, (b) 3,3-dimethylbutane, (c) 2,2-methylpentane, (d) 2-ethylbutane.

24.29 Why can you dispense with the numbers in naming (a) $CH_2{=}CHCH_3$ and (b) $CH_3\underset{CH_3}{CH}CH_3$?

24.30 (a) Write the structural formulas for the isomers of trimethylbenzene, $C_6H_3(CH_3)_3$. Name each one. (b) Write the structural formulas for the isomers of methylnaphthalene.

24.31 What is the general formula for the naphthalene homologous series?

24.32 Why are these compounds not aromatic?

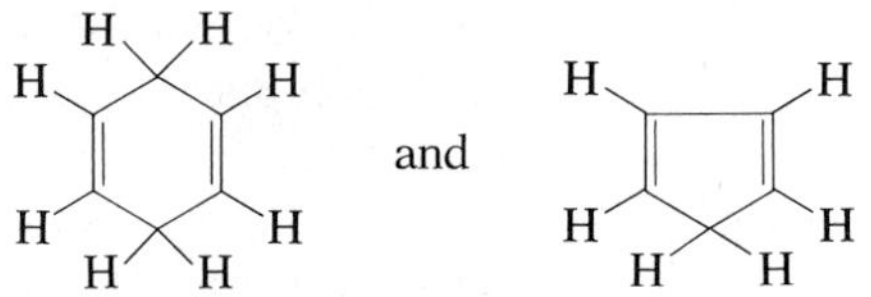

and

24.33 Acetylene is produced when calcium carbide, CaC_2, reacts with water. The other product is $Ca(OH)_2$. Write an equation for the reaction. What category of reaction is this? How is CaC_2 related chemically to acetylene?

CONFORMATION AND STEREOCHEMISTRY

24.34 Which of the following compounds have stereoisomers (*cis-trans* isomers or enantiomers)? State which in each case, (a) $CH_3CH{=}CHCH_2CH_3$, (b) $(CH_3)_2C{=}CHCH_3$, (c) $CH_3CH{=}C(CH_3)CH_2CH_3$, (d) $H_2C{=}CH—CH{=}CHCH_3$, (e) $H_3C\ddot{N}{=}\ddot{N}CH_3$, (f) $CH_3SiHClBr$, (g) H_2NCH_2COOH.

24.35 (a) Give a structural formula for the simplest chiral (1) alkane, (2) alkene, (3) alkyne, and (4) benzene hydrocarbon. (b) Name each one by the IUPAC method. (Recall that the C_6H_5 group is called the phenyl group.)

24.36 Show structural formulas (like those in Figure 24.3) for (a) the two eclipsed conformations of butane, $CH_3CH_3CH_2CH_3$, and (b) the two staggered-type conformations of butane. (Rotate around the $C_2—C_3$ bond.)

24.37 Account for the stereochemical results in each step of this sequence:

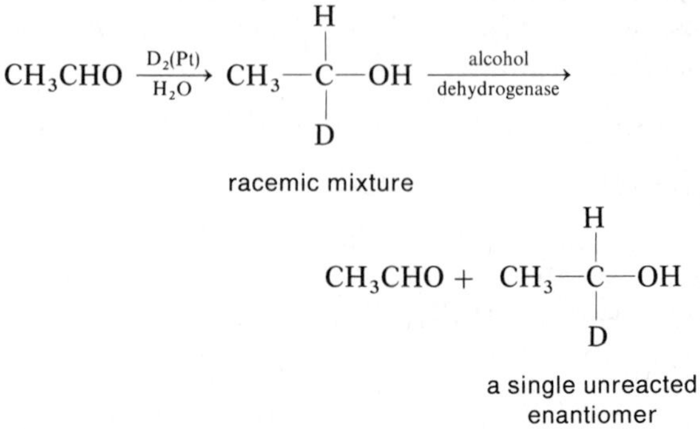

$$CH_3CHO \xrightarrow[H_2O]{D_2(Pt)} CH_3—\overset{H}{\underset{D}{C}}—OH \xrightarrow[\text{dehydrogenase}]{\text{alcohol}}$$

racemic mixture

$$CH_3CHO + CH_3—\overset{H}{\underset{D}{C}}—OH$$

a single unreacted enantiomer

REACTIONS OF HYDROCARBONS

24.38 Give the structural formulas for all the possible products formed on the chlorination of ethane. Do not restrict the number of Cl atoms introduced into the hydrocarbon.

24.39 Give the structural formulas for (a) the five monochlorination products of C_4H_{10} (include stereoisomers), (b) the four monochlorination products of toluene, and (c) the two monochlorination products of naphthalene.

24.40 Write a balanced equation for the combustion of an octane.

24.41 Complete these reactions:

(a) $CH_3CH_2CH{=}CH_2 + Br_2 \longrightarrow$

(b) $CH_3CH{=}CHCH_3 + HBr \longrightarrow$

(c) $HC{\equiv}CCH_3 + H_2 \xrightarrow{Pt}$

(d) $HC{\equiv}CCH_3 + 2H_2 \xrightarrow{Pt}$

(e) $CH_3CH{=}CHCH_3 + H_2O \xrightarrow{H_2SO_4}$

(f) (benzene) $+ Cl_2 \xrightarrow{Fe}$

24.42 Which hydrocarbon is needed to make CH_2BrCH_2Br? Write a balanced equation. (Bromination of an alkane would give a complex mixture, and individual products would not be formed in good yields.)

24.43 (a) Which alkenes with the molecular formula C_4H_8 give more than one alcohol when reacted with acidified water? (b) Does cyclohexene give more than one alcohol on the addition of H_2O? Explain.

24.44 Give a simple test to distinguish between $CH_2{=}CHCH_2CH_2CH_3$ and its isomer cyclopentane.

24.45 When 1 mole of benzene and 1 mole of H_2 react under severe conditions $\frac{1}{3}$ mole of cyclohexane and $\frac{2}{3}$ mole of unreacted benzene are isolated. No cyclohexene or cyclohexadiene is isolated. Explain.

```
        H
        C
      /   \\
  H2C       CH
   |         |
  H2C       CH
      \   //
        C
        H
```

1,3-cyclohexadiene

24.46 (a) Write the equation for the chlorination of ethane, C_2H_6, to give ethyl chloride, C_2H_5Cl. (b) Ethyl chloride can be dichlorinated to give two isomers, A and B, of formula $C_2H_4Cl_2$. A can be chlorinated to give two isomers of $C_2H_3Cl_3$, whereas B gives only one isomer of $C_2H_3Cl_3$ on chlorination. Write the equations for the chlorination of ethyl chloride. (c) Give the structural formulas of A and B.

24.47 (a) What two monobromo substitution products arise from the reaction of $(CH_3)_3CH$ with Br_2 in sunlight? (b) The major product is $(CH_3)_3CBr$. What conclusion can you draw about the reactivity of the two different sets of H's in $(CH_3)_3CH$ toward bromination? (c) Give an equation for the preparation of $(CH_3)_3CBr$ by an addition reaction.

FUNCTIONAL GROUPS

24.48 Name the type of compound: (a) CH_3CH_2F, (b) $CH_3OCH_2CH_3$, (c) $CH_3CH_2CH_2CH_2OH$, (d) CH_3NH_2, (e) $CH_3CH_2CH_2\overset{H}{\overset{|}{C}}{=}O$, (f) $CH_3\overset{O}{\overset{\|}{C}}CH_2CH_3$, (g) $H\overset{O}{\overset{\|}{C}}{-}OH$, (h) $CH_3\overset{O}{\overset{\|}{C}}{-}NH_2$, and (i) $CH_3\overset{O}{\overset{\|}{C}}{-}OCH_3$.

24.49 Use no more than three carbon atoms to give a structural formula for a typical (a) alkyl bromide, (b) alcohol, (c) ether, (d) aldehyde, (e) ketone, (f) three kinds of amines, (g) carboxylic acid, (h) ester, (i) amide.

24.50 (a) Write the structural formulas for all the isomers of (1) C_2H_4ClBr and (2) $C_4H_{10}O$. (b) Which isomers are chiral?

24.51 Write the structural formulas for the isomers of $C_2H_4O_2$ that do *not* have a C=C bond or a ring structure.

24.52 Give structural formulas for the possible isomeric products formed from monobromination of the rings of (a) *o*-xylene, (b) *m*-xylene, and (c) *p*-xylene.

REACTIONS OF FUNCTIONAL GROUPS

24.53 Which functional groups make a water solution (a) acidic; (b) basic?

24.54 Why do alcohols and amines have higher boiling points than alkanes of similar molecular weight?

24.55 Complete these reactions.

(a) $CH_3Br + OH^- \longrightarrow$

(b) $CH_3CH_2CH_2OH + H_2SO_4 \longrightarrow$
(two different reactions)

(c) $CH_3OCH_3 + BF_3 \longrightarrow$

(d) $CH_3CH_2CH_2OH \xrightarrow{[O]} ? \xrightarrow{[O]} ?$

(e) $H\overset{O}{\overset{\|}{C}}{-}O^-NH_4^+ \xrightarrow{heat}$

(f) $CH_3\overset{O}{\overset{\|}{C}}CH_3 + H_2 \xrightarrow{Pt}$

24.56 (a) Write balanced ionic equations for the reactions of aqueous NaOH with (1) CH_3COOH, (2) CH_3CH_2Cl, (3) $CH_3\overset{O}{\overset{\|}{C}}{-}OCH_2CH_2CH_3$, and (4) $[CH_3CH_2NH_3]^+Cl^-$. (b) Write a balanced ionic equation for the reaction of NH_3 with $CH_3CH_2CH_2I$. (c) Write balanced ionic equations for the reactions of hydrochloric acid with (1) $CH_3CH_2NH_2$, (2) $(CH_3)_2NH$, and (3) $CH_3CH_2CH_2\overset{O}{\overset{\|}{C}}{-}O^-Na^+$.

24.57 Use a chemical equation to help show why a solution of diethyl ether in concentrated H_2SO_4 conducts an electric current although neither of the pure substances does.

24.58 Use structural formulas to complete these reactions in H_2SO_4:
(a) $CH_3COOH + CH_3CH_2CH_2OH \longrightarrow$
(b) $HCOOH + CH_3CH_2OH \longrightarrow$

24.59 Write balanced equations for (a) the formation of potassium oleate, (b) the formation of calcium palmitate, (c) the hydrolysis of methyl acetate, and (d) the saponification of glyceryl trilinoleate.

24.60 Write structural formulas for the three isomeric fats that give, on saponification, a mixture of these soaps: sodium oleate, $C_{17}H_{33}COO^-Na^+$; sodium stearate, $C_{17}H_{35}COO^-Na^+$; and sodium palmitate, $C_{15}H_{31}COO^-Na^+$.

24.61 Give a simple chemical reaction to distinguish between

(a) $CH_3CH_2C(H){=}O$ and $CH_3C({=}O)CH_3$,

(b) $CH_3CH_2OCH_2CH_3$ and $CH_3CH_2CH_2CH_2OH$,

(c) $CH_3CH_2C({=}O){-}OH$ and $CH_3C({=}O){-}OCH_3$.

24.62 Write an equation for the reaction of ethyl bromide with each of the following: (a) $H\ddot{S}\!:^-$; (b) $CH_3\ddot{O}\!:^-$; (c) $:H^-$; (d) $:CH_3{}^-$; (e) $CH_3\ddot{N}H_2$.

24.63 Account for the fact that bases such as $:NH_3$ do not typically add to a C=C group but do add to a C=O group as shown:

$$CH_3C(H){=}O + NH_3 \longrightarrow \left[CH_3C(H)(NH_2){-}OH\right] \quad \text{(not stable)}$$

(See section 14.10 for a discussion of the electronic structure of the C=O group.)

24.64 Show steps in the synthesis of the following compounds from CH_3CH_2OH and any inorganic compound: (a) CH_2BrCH_2Br, (b) CH_3COOH, (c) $CH_3C({=}O){-}OCH_2CH_3$, (d) CH_3CH_2Br, (e) $CH_3CH_2NH_2$.

SELF-TEST

24.65 (15 points) Match each lettered item with the most appropriate numbered item.

________ 1. aromatic compound
________ 2. conformations
________ 3. elimination reaction
________ 4. chiral molecules
________ 5. soap
________ 6. hydrocarbon
________ 7. racemate
________ 8. functional group
________ 9. homologous series
________ 10. addition reaction
________ 11. nucleophilic displacement
________ 12. polyunsaturated oil
________ 13. saponification
________ 14. fat
________ 15. enantiomers

(a) reaction whereby one nucleophile, a strong base, displaces another nucleophile, a very weak base, from some atom; (b) a compound whose molecules contain at least one ring that has considerable stabilization because of extensive delocalization of *p* electrons; (c) reaction in which the removal of two atoms or groups from adjacent atoms introduces a multiple bond into a molecule; (d) reaction of an ester with OH^-; (e) solid ester of glycerol; (f) 50/50 mixture of enantiomers; (g) different structures of a molecule arising from rotation around σ bonds in chain molecules and ring flipping in cyclic compounds; (h) liquid ester of glycerol with carboxylic esters having two or three double bonds; (i) molecules with nonsuperimposable mirror images; (j) salt of a carboxylic acid, typically one with a long chain of carbon atoms; (k) series of compounds characterized by a general formula; (l) reaction in which the net effect is the breaking up of a reactant into two parts that bond to multiply bonded atoms; (m) site of reactivity, other than an H atom, in a substituted hydrocarbon; (n) compound of carbon and hydrogen; (o) nonsuperimposable mirror images.

24.66 (10 points) (a) How many different substances do these structural formulas represent?

(1) $CH_3{-}CH(CH_3){-}CH_2{-}CH(CH_3){-}CH_2{-}CH_3$

(2) $CH_2(CH_3){-}CH(CH_3){-}CH_2{-}CH(CH_3){-}CH_3$

(3) $CH_3{-}CH(CH_3){-}CH(CH_3){-}CH_2{-}CH_2(CH_3)$

(4) $CH_2(CH_3){-}CH(CH_3){-}CH_2{-}CH(CH_3)(CH_3)$

(b) Which of these compounds can exist in different conformations? (1) CH_2ClBr, (2) H_2NNH_2, (3) $HC{\equiv}CH$, (4) $CH_3CH_2CH_3$, (5) CH_3NH_2, (6) HOCl.

24.67 (10 points) Given the compounds (a) C_4H_{10}, (b) C_4H_8, and (c) C_4H_6. (1) Name and write a general formula for all the homologous series (when more than one is possible) to which each of these hydrocarbons belongs. (See Table 24.3.) (2) Write a structural formula consistent with each series.

24.68 (15 points) Write structural formulas and give the names of the homologous series for (a) five isomers of C_6H_{14} and (b) six isomers of C_4H_8.

24.69 (9 points) With the aid of equations, show that at least two isomers (include stereoisomers) are possible from each of these reactions and indicate the type of isomerism: (a) 1 mol $CH_3C{\equiv}CCH_3$ + 1 mol H_2(Pt); (b) $CH_3CH_2CH_3$ + 1 mol Cl_2 in light; (c) $CH_3\overset{\overset{\displaystyle O}{\|}}{C}CH_2CH_3$ + H_2(Pt).

24.70 (12 points) (a) Draw the structural formulas for the four alcohols with the molecular formula $C_4H_{10}O$. (b) Which alcohol has enantiomers? (c) Write the structural formulas for the three ethers with the molecular formula $C_4H_{10}O$. (d) Give a simple chemical test to distinguish alcohols from ethers.

24.71 (10 points) Which of these compounds have stereoisomers (*cis-trans* isomers or enantiomers)? State which in each case. (a) $H_2C{=}CClCH_3$, (b) $ClFC{=}CHCl$, (c) $CH_3CH_2CH{=}CHCH(CH_3)_2$, (d) $CH_3CHClCOOH$, (e) $CH_2{=}CHCHClCH_3$.

24.72 (6 points) Give the IUPAC names for

(a) $CH_3\underset{\underset{\displaystyle CH_3}{|}}{C}HCH_2CH_3$ and (b) $CH_3\underset{\underset{\displaystyle CH_3}{|}}{C}HCH_2\underset{\underset{\displaystyle CH_3}{|}}{C}HCH_3$

24.73 (7 points) Classify each reaction as (a) substitution or displacement, (b) addition, or (c) elimination. Consider only the organic reactant.

(1) $CH_3CH_3 + Cl_2 \xrightarrow{\text{light}} CH_3CH_2Cl + HCl$

(2) $CH_3CH_2OH + [O] \longrightarrow CH_3CH{=}O + H_2O$

(3) $CH_3CH{=}O + HCN \longrightarrow CH_3\underset{\underset{\displaystyle CN}{|}}{\overset{\overset{\displaystyle H}{|}}{C}}{-}OH$

(4) $CH_3CH{=}CH_2 + HOCl \longrightarrow CH_3\underset{\underset{\displaystyle OH}{|}}{C}HCH_2Cl$

(5) $CH_3CH_2Br + CN^- \longrightarrow CH_3CH_2CN + Br^-$

(6) $CH_3\underset{\underset{\displaystyle Br}{|}}{C}HCH_2Br + Zn \longrightarrow CH_3CH{=}CH_2 + ZnBr_2$

(7) $C_6H_6 + HO\underset{\underset{\displaystyle O}{|}}{\overset{\overset{\displaystyle O}{|}}{S}}OH \longrightarrow C_6H_5\underset{\underset{\displaystyle O}{|}}{\overset{\overset{\displaystyle O}{|}}{S}}OH + H_2O$

24.74 (6 points) Complete the equations:

(a) $CH_3CH_2OH + HCOOH \xrightarrow{H_2SO_4}$

(b) $(CH_3)_3N + HBr \longrightarrow$

(c) $CH_3CH_2COO^-NH_4^+ \xrightarrow{\text{heat}} H_2O + ?$

25 POLYMERS AND BIOCHEMICALS

The previous chapters of this book have dealt largely with substances of low molecular weight, substances whose complete structural formulas can be written on a fraction of a page. However, many chemists in commerce and industry deal with substances whose molecular weights are so large that one of their structural formulas, if written out in full, would occupy an entire book. These are the substances most familiar to us in our daily lives. They include natural materials such as wood, starch, rock, and the tissues of our bodies. They also include most of the manufactured materials—synthetic plastics, rubbers, and textiles.

Is the chemistry of substances with large molecular weights, which are known as **polymers**, different from what you have learned in previous chapters? Yes and no. The same kinds of chemical bonds exist in both large and small molecules. The kinds of chemical structures that produce large molecules, however, are different from those that produce small molecules. Furthermore, polymers have properties not seen in substances whose molecules are small.

In this chapter we will study the chemical makeup of polymers and learn how they are synthesized. We will also consider how the properties of polymers are related to their molecular structures. Finally, we will describe three important biochemical polymers—carbohydrates, proteins, and nucleic acids.

25.1 SMALL MOLECULES AND LARGE MOLECULES

A molecule is a group of atoms linked by chemical bonds. This concept that matter is composed of molecules was quite useful in the study of gases during the eighteenth and nineteenth centuries. The molecular weights of the common gases range from 2 (for H_2) to more than 100. Even the densest vapors produced by the evaporation of liquids or solids, however, rarely exceed about 300 in molecular weight, and their molecules are rarely larger than about 10 Å in diameter. Substances whose molecules do not exceed these limits can generally be identified by their vapor pressure or their melting, boiling, or sublimation temperatures. The molecular weights of these substances can readily be determined from their vapor densities or the colligative properties of their solutions (section 12.16). Moreover, such measurements can be made with laboratory equipment that has been available at least since the advent of modern chemistry. As a result of this happy combination of concept and method, the first 150 years of modern chemistry were largely directed to the study of materials of low molecular weight.

During their investigations, however, chemists sometimes encountered materials that did not respond to such studies—materials that decomposed instead of melting or vaporizing, materials that were gummy, gluey, waxy, or resinous and insoluble in most solvents. Even when these materials did dissolve, they yielded solutions whose boiling points, freezing points, and vapor pressures were hardly distinguishable from those of the pure solvents. Chemists found these materials so difficult to work with that they usually threw them into the trash can. Gradually, however, it was recognized that some valuable substances like cellulose, rubber, and protein resembled those sticky residues that plagued experimenters. During the latter part of the nineteenth century, chemists began to suspect that what these classes of substances had in common was molecules of very large size. It was evident that entirely new methods of investigation would have to be developed for studying such materials.

The concept of large molecules was in fact vigorously opposed by many chemists even as late as the 1930s. They believed substances such as protein and rubber really consisted of small molecules that had a great tendency to stick together.

Many substances of high molecular weight can be characterized by their decomposition products. When natural rubber is heated, the hydrocarbon isoprene, C_5H_8, distills off. If starch is chewed it breaks down to form maltose, a sugar. If egg albumin, a protein, is boiled in dilute sulfuric acid, the amino acids leucine, alanine, serine, glutamic acid, methionine, and some thirteen others are produced. It is reasonable to assume that these fragments detected in the decomposition of large molecules are indeed their units of structure. Thus, starch is called a **polysaccharide**, derived from the Greek words for "many" and "sugar."

When two molecules of the same substance combine with each other, the resulting doubled molecule is called a **dimer**. For example,

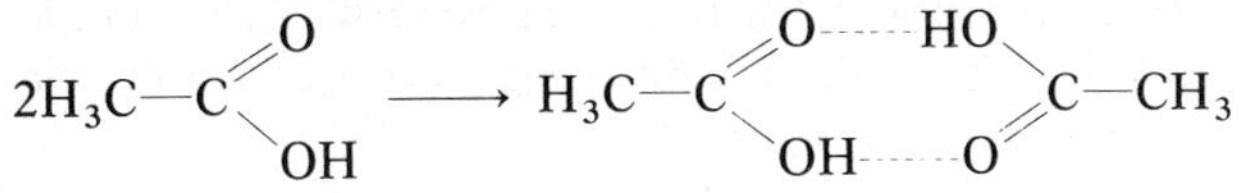

$$2H_3C{-}C{\overset{\displaystyle /\!\!/ O}{\underset{\displaystyle \backslash OH}{}}} \longrightarrow H_3C{-}C{\overset{\displaystyle /\!\!/ O\cdots HO \backslash}{\underset{\displaystyle \backslash OH\cdots O /\!\!/}{}}}C{-}CH_3$$

acetic acid — dimer of acetic acid formed by H-bonding

Similarly, a tripled molecule is called a **trimer**. For example,

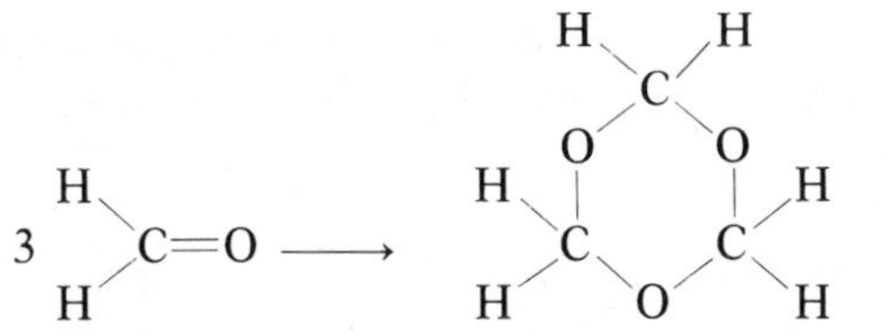

$$3\ H_2C{=}O \longrightarrow (CH_2O)_3$$

formaldehyde — "trioxymethylene," a trimer of formaldehyde

Greek, "many parts."

When the number of repeating units is large, the substance is called a **polymer**, and it is said to consist of **macromolecules**. Molecular weights of some polymers reach the range of 10^4 to 10^9. Such large molecules take the form of very long chains, usually branched, and often with linkages between chains. The small molecule that is repeated in a polymer is called a **monomer**. Thus, formaldehyde is the monomer of trioxymethylene, and ethylene, $H_2C{=}CH_2$, is the monomer of polyethylene, $\left(\begin{array}{cc} H & H \\ | & | \\ -C- & C- \\ | & | \\ H & H \end{array}\right)_n$, where n is a large number.

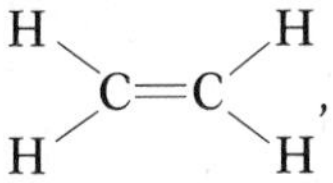

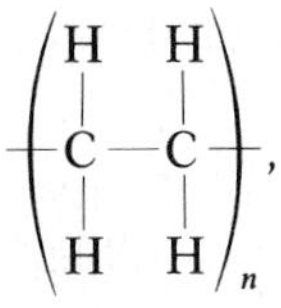

When two different monomers combine in a repeating pattern, the product is called a **copolymer**. Examples are the nylons described in section 25.3.

25.2 NATURAL POLYMERS

The structure of isoprene (2-methyl-1,3-butadiene) produced by the destructive distillation of natural rubber is

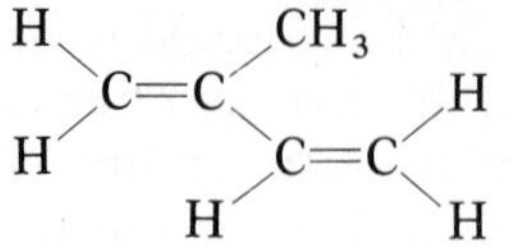

$$\begin{matrix} H & & CH_3 & & \\ & C{=}C & & & H \\ H & & & C{=}C & \\ & & H & & H \end{matrix}$$

Natural rubber itself is a long, chainlike structure of the type

$$\underbrace{-CH_2-\overset{CH_3}{\overset{|}{C}}=CH-CH_2}_{\text{repeating unit}}\underbrace{-CH_2-\overset{CH_3}{\overset{|}{C}}=CH-CH_2}_{\text{repeating unit}}\underbrace{-CH_2-\overset{CH_3}{\overset{|}{C}}=CH-CH_2-}_{\text{repeating unit}}$$

or

$$\left(-CH_2-\overset{CH_3}{\overset{|}{C}}=CH-CH_2-\right)_n$$

(For a more accurate rendition of bond angles, see Table 25.1.) The formula of this macromolecule may thus be written as an indefinite series of repeating units. Thus, information about the structure of a polymer can be obtained by identifying its decomposition products.

EXAMPLE 25.1 Show how the structure of the isoprene monomer can be deduced from the structure of the repeating unit of its polymer.

ANSWER Rewrite the repeating unit, using dots for the unbonded and π-bonded electrons. Then redistribute the dots to pair up the electrons into π bonds:

A half-arrow is used to represent the shift of a single electron. Full arrows are used for pairs of electrons.

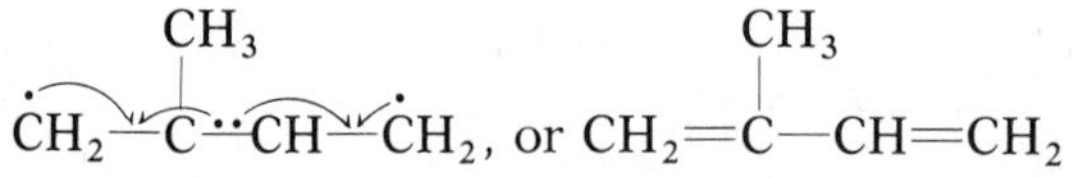

$$\dot{C}H_2-\overset{CH_3}{\overset{|}{C}}{\cdot\cdot}CH-\dot{C}H_2, \text{ or } CH_2=\overset{CH_3}{\overset{|}{C}}-CH=CH_2$$ ■

Geometrical (*cis-trans*) isomerism, which is a consequence of restricted rotation around double bonds, can account for differences between polymers.

PROBLEM 25.1 (a) Is isoprene capable of exhibiting *cis-trans* isomerism? (b) Is the repeating unit in Example 25.1 capable of exhibiting *cis-trans* isomerism? Can the stereochemical structure of polyisoprene be deduced from the fact that its decomposition product is isoprene? □

Table 25.1 gives the formulas of some natural organic polymers and their decomposition products. Keep in mind, however, that the identification of the decomposition product is far from a full description of the polymer. This is so for the following reasons:

- The mode of linkage of the monomer influences the properties of the macromolecule. Note in Table 25.1 that when isoprene units are

TABLE 25.1
SOME NATURAL POLYMERS AND THEIR DECOMPOSITION PRODUCTS

POLYMER	FORMULA[†]		DECOMPOSITION PRODUCT(S)
Cellulose (primary structural material of plants)		$\xrightarrow[(+\text{acid})]{H_2O}$	glucose
Chitin (protective shell of insects, crustaceans, etc.)		$\xrightarrow[(+\text{acid})]{H_2O}$	aminoglucose
Natural rubber	*cis* form	heat	$H_2C{=}C(CH_3){-}CH{=}CH_2$ isoprene
Gutta-percha	*trans* form	heat	isoprene
Silk[§] (a protein)	$-CO-CH(R)-NH-CO-CH(R)-NH-CO-CH(R)-$	$\xrightarrow[(+\text{acid})]{H_2O}$	$HO-C(=O)-CH(R)-NH_2$ amino acid

[†] The colored portions are the repeating units.
[§] $R = H$, CH_3, CH_2OH, or any of about 12 other groupings.

linked in the *cis* form, the polymer is natural rubber, which is soft and flexible. The same isoprene units linked in the *trans* form, however, are gutta-percha, a hard, somewhat brittle polymer formerly used as cores for golf balls.

- The sequence and extent of monomeric linkages are important determinants of polymer structure. For example, cellulose and starch are both glucose polymers, yet they are different substances. The rigid cellulose consists of linear macromolecules of about 3500 glucose units each. The more pliable starch, on the other hand, has only about 500 glucose units per molecule, with structural linkages different from those in cellulose and with extensive chain-branching.

- The shape of the macromolecules and their orientation in space with respect to each other are critical determinants of polymer structure. An illustration of this is provided by raw and cooked egg white (albumin). The shapes of their constituent protein molecules are different.

Two kinds of relationships between monomers and polymers are evident in Table 25.1. In rubber and gutta-percha, the repeating unit of the macromolecule is identical to the decomposition product in atomic composition and skeletal structure. If the reaction could be reversed, simple *addition* of the isoprene units would form the polymer. Such polymers are therefore called **addition polymers.**

Cellulose, chitin, and silk are decomposed by hydrolysis, which requires the uptake of water. Since the formation of the polymer is the reversal of its decomposition, polymerization must involve the elimination of water. Such a reaction is called *condensation*, and the substances produced are called **condensation polymers**. Condensation reactions are not defined only by the loss of water. In general, condensation is a union of two or more molecules in which a smaller molecule, such as H_2O, NH_3, or HCl, is eliminated.

Addition polymerization is also called *chain-reaction polymerization*, and condensation polymerization is sometimes called *step-reaction polymerization.*

Note that all the polymers shown in Table 25.1 contain C atoms as parts of the chain. Such polymers are said to be *organic*. Polymers that do not have C atoms in the main chain are said to be *inorganic*, even if they contain carbon atoms in branches off the main chain.

25.3 CONDENSATION POLYMERS

The controlled synthesis of polymers by condensation methods became commercially successful as early as the 1930s, largely as a result of the pioneer work of Wallace H. Carothers. A condensation polymer must be made from monomers that contain at least two functional groups so as to enable intermolecular reactions to proceed continuously.

To illustrate this point consider the reaction of an amine with a carboxylic acid:

$$\underset{\text{an amine}}{CH_3{-}CH_2{-}N\begin{matrix}H\\ \\ H\end{matrix}} + \underset{\text{a carboxylic acid}}{\begin{matrix}HO\\ \\ O\end{matrix}C{-}CH_2{-}CH_3} \xrightarrow{\text{water is eliminated}}$$

$$\underset{\substack{\text{an amide}\\ \text{(no free } NH_2 \text{ or COOH)}}}{CH_3{-}CH_2{-}\underset{\displaystyle H}{\underset{|}{N}}{-}\underset{\displaystyle O}{\underset{\|}{C}}{-}CH_2{-}CH_3} + H_2O$$

This condensation does not lead to a polymer because the product no longer has an NH_2 or COOH group with which to continue the reaction. However, the reaction of a diamine with a diacid leaves two reactive groups, and the condensation can be repeated many more times to form a polymer:

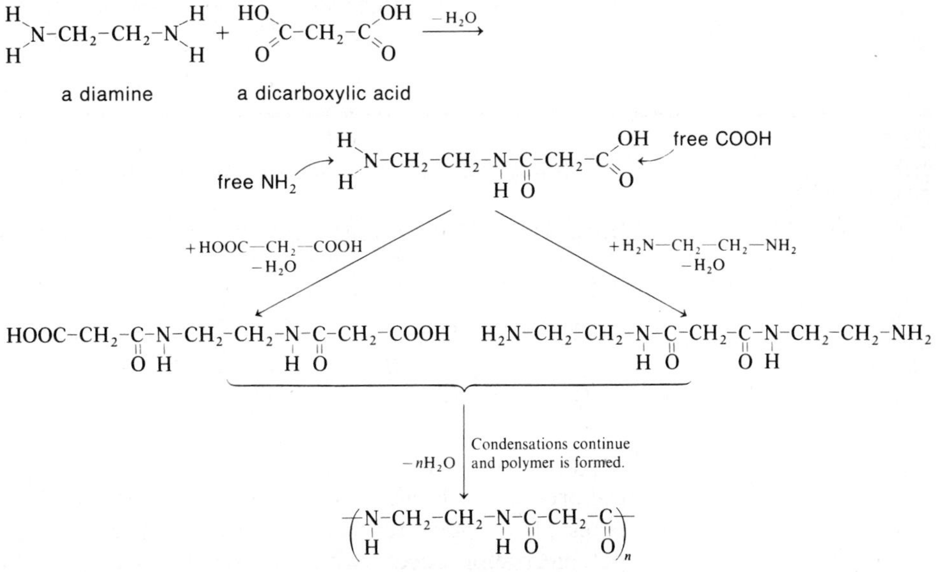

$$\underset{\text{a diamine}}{H_2N{-}CH_2{-}CH_2{-}NH_2} + \underset{\text{a dicarboxylic acid}}{HO{-}\overset{O}{\overset{\|}{C}}{-}CH_2{-}\overset{O}{\overset{\|}{C}}{-}OH} \xrightarrow{-H_2O}$$

$$\overset{\text{free } NH_2}{H_2N}{-}CH_2{-}CH_2{-}\underset{H}{N}{-}\overset{O}{\overset{\|}{C}}{-}CH_2{-}\overset{\text{free COOH}}{COOH}$$

$+HOOC{-}CH_2{-}COOH$, $-H_2O$ (left branch); $+H_2N{-}CH_2{-}CH_2{-}NH_2$, $-H_2O$ (right branch)

$$HOOC{-}CH_2{-}\overset{O}{\overset{\|}{C}}{-}\underset{H}{N}{-}CH_2{-}CH_2{-}\underset{H}{N}{-}\overset{O}{\overset{\|}{C}}{-}CH_2{-}COOH \qquad H_2N{-}CH_2{-}CH_2{-}\underset{H}{N}{-}\overset{O}{\overset{\|}{C}}{-}CH_2{-}\overset{O}{\overset{\|}{C}}{-}\underset{H}{N}{-}CH_2{-}CH_2{-}NH_2$$

$-nH_2O$ — Condensations continue and polymer is formed.

$$\left(\underset{H}{N}{-}CH_2{-}CH_2{-}\underset{H}{N}{-}\overset{O}{\overset{\|}{C}}{-}CH_2{-}\overset{O}{\overset{\|}{C}}\right)_n$$

repeating unit of the polymer

Carothers succeeded in producing condensation polymers by heating the salts of diamines with dicarboxylic acids at 200° to 250°C, with elimination of water. These substances are called **nylons**. A common hosiery polymer is Nylon-66, the copolymer of a diamine and a dicarboxylic acid, each of which has six carbon atoms per molecule. The 66 indicates that each monomer molecule has six carbon atoms.

$$H_2N{-}(CH_2)_6{-}NH_2 + HOOC{-}(CH_2)_4{-}COOH \xrightarrow{-H_2O} \text{dimer} \ldots \longrightarrow$$

$$\ldots{-}CO{-}\boxed{NH{-}(CH_2)_6{-}NH{-}CO{-}(CH_2)_4{-}CO}{-}NH{-}\ldots$$

polymer, Nylon-66
(The colored portion is the repeating unit.)

A nylon can also be made by the polymerization of a single amino acid, such as 6-aminohexanoic acid, which is polymerized to the material called Nylon-6:

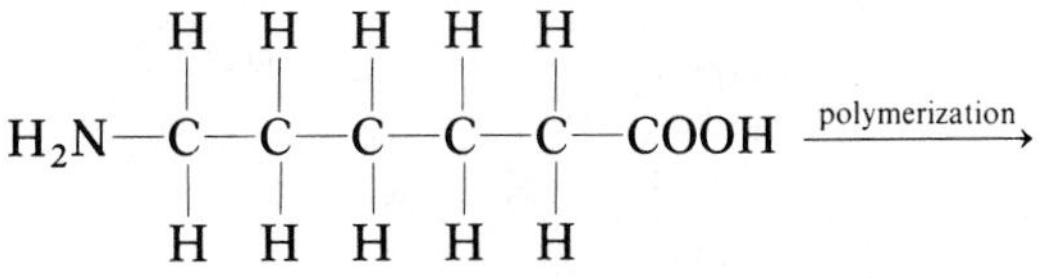

$$H_2N{-}CH_2{-}CH_2{-}CH_2{-}CH_2{-}CH_2{-}COOH \xrightarrow{\text{polymerization}}$$

6-aminohexanoic acid

$$\ldots{-}CO{-}\boxed{NH{-}(CH_2)_5{-}CO}{-}NH{-}\ldots$$

polymer, Nylon-6
(Colored portion is the repeating unit.)

The single 6 indicates that this nylon consists of repetitions of a single kind of six-carbon monomer.

PROBLEM 25.2 Can a condensation polymer be formed by the reaction of (a) a diamine with a monocarboxylic acid; (b) a dicarboxylic acid with a monamine? □

25.4 ADDITION POLYMERS

The functional group of a molecule of ethylene, $H_2C{=}CH_2$, is its double bond. In an addition reaction between ethylene and, say, bromine,

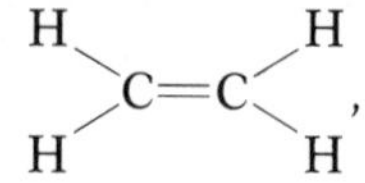

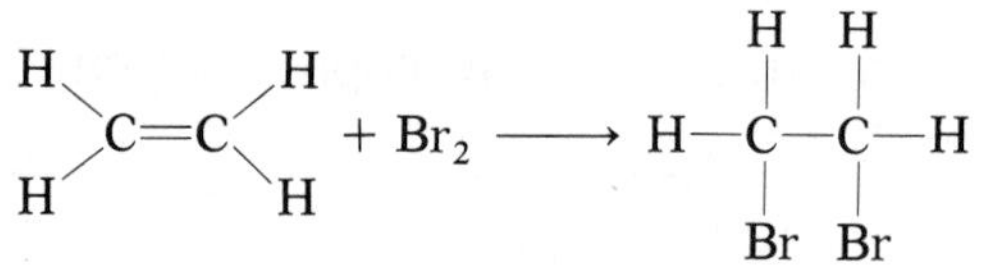

$$H_2C{=}CH_2 + Br_2 \longrightarrow H{-}CHBr{-}CHBr{-}H$$

the product no longer contains a double bond, and therefore no further addition takes place. Such termination prevents the formation of a polymer. Addition polymerization occurs when the reactive feature of the molecule is *regenerated* after each addition; thus termination does not occur. In this way the addition reactions may proceed continuously. Pioneer studies in this field were carried out by Hermann Staudinger in the 1920s. An example of free-radical addition polymerization is shown below. (The curved arrows represent electron shifts that occur during the reaction.) The hydroxyl radical reacts at the carbon-to-carbon double bond to produce a new free radical. The unshared electron in the dimer is associated with the carbon bearing the CN group:

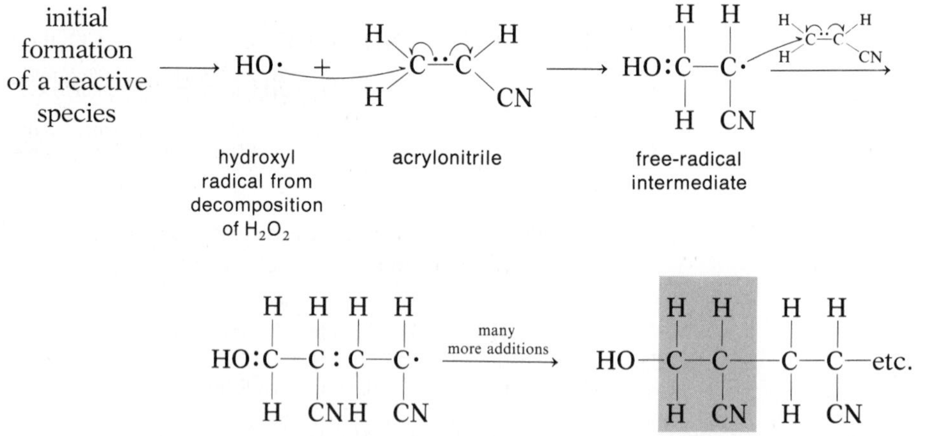

initial formation of a reactive species $\longrightarrow$ HO· + $H_2C{::}CH(CN)$ $\longrightarrow$ HO:CH_2—$\dot{C}H$(CN) $\xrightarrow{H_2C::CH(CN)}$

hydroxyl radical from decomposition of H_2O_2 — acrylonitrile — free-radical intermediate

HO:CH_2—CH(CN):CH_2—$\dot{C}H$(CN) $\xrightarrow{\text{many more additions}}$ HO—CH_2—CH(CN)—CH_2—CH(CN)—etc.

dimeric free-radical addition product

macromolecule of polyacrylonitrile (Acrilan, Orlon) (Colored portion is the repeating unit.)

The preservation of the reactive feature (the unshared electron) is facilitated by its delocalization with the π bond of the adjacent C≡N group:

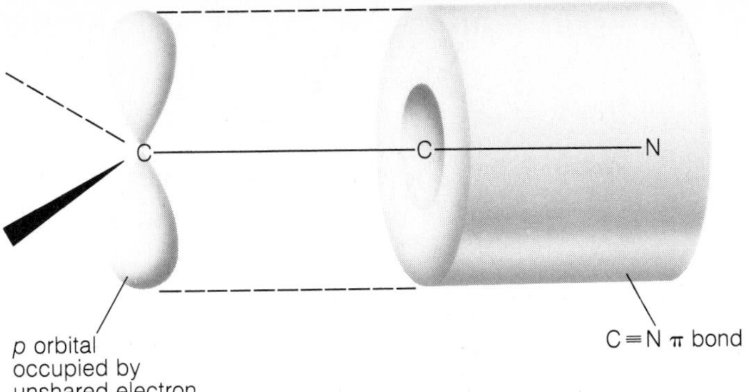

An example of ionic polymerization is the use of sodium amide in the production of polyvinylidene nitrile:

$Na^+(\ddot{N}H_2)^- + H_2C{=}C(CN)_2 \longrightarrow [H_2\ddot{N}{-}CH_2{-}\ddot{C}(CN)_2]^- \xrightarrow[\text{another molecule of vinylidene nitrile}]{H_2C=C(CN)_2}$

sodium amide; vinylidene nitrile; anionic intermediate

$[H_2\ddot{N}{-}CH_2{-}C(CN)_2{-}CH_2{-}\ddot{C}(CN)_2]^- \xrightarrow{\text{many more additions}} H_2\ddot{N}{-}(CH_2{-}C(CN)_2)_n{-}$

dimeric anion; polyvinylidene nitrile (Colored portion is the repeating unit.)

The unshared pair of electrons on the carbon is the reactive site.

An example of cationic polymerization is the action of a strong acid on an alkene:

$H^+ + H_2C{=}CHR \longrightarrow H{-}CH_2{-}C^+HR \xrightarrow[\text{another molecule of alkene}]{H_2C=CHR}$

(from H_2SO_4); alkene; cationic intermediate; a carbocation

$H{-}CH_2{-}CHR{-}CH_2{-}C^+HR \xrightarrow{\text{more additions}} CH_3{-}(CH_2{-}CHR)_n{-}$

dimeric cation; polymer (Colored portion is the repeating unit.)

25.5 LINKAGE PATTERNS IN POLYMERS

Addition polymers formed by a free-radical process do not necessarily produce linear chains. Instead, the products often have highly branched structures. Modern synthetic methods, however, utilize catalysts that control the steps in the addition polymerization of ethylene, for example, to produce polyethylene with

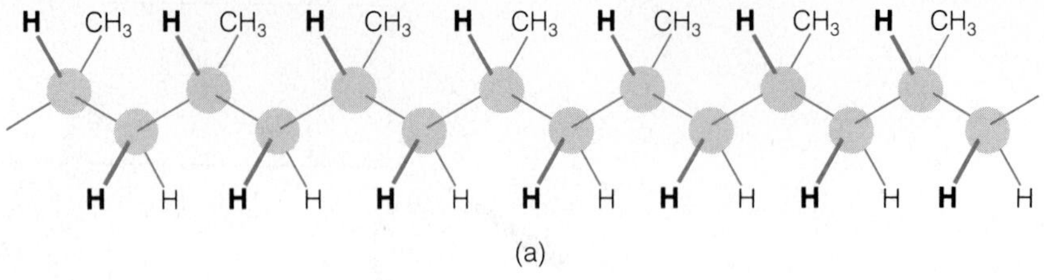

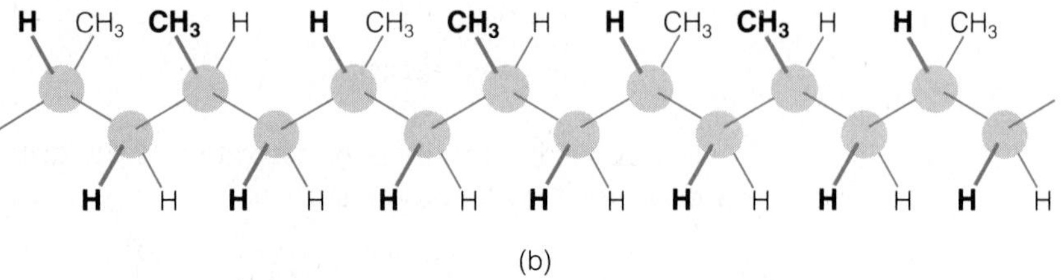

H CH3 CH3 H CH3 H CH3 H H CH3 H CH3 CH3 H

H H H H H H H H H H H H H H

(c)

FIGURE 25.1
Polypropylene chain arrangements. (a) Isotactic arrangement; all CH_3 groups are on one side of the chain. (b) Syndiotactic arrangement; CH_3 groups are on alternate sides. (c) Atactic arrangement; CH_3 groups occur randomly on both sides.

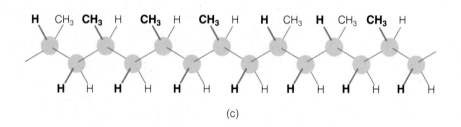

(a)

(b)

FIGURE 25.2
(a) Straight-chain amylose starch. (b) Branched-chain amylopectin starch. (These are just structural skeletons.)

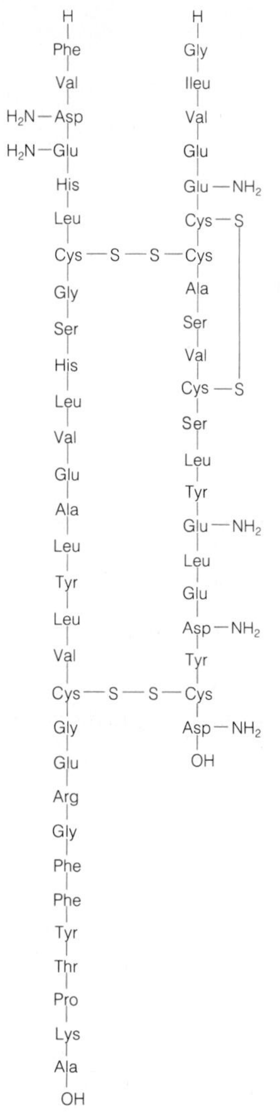

unbranched chains. The resulting polymer is denser, higher-melting, and stronger than a polymer whose chains contain many branches.

In addition polymerization of propylene,

$$n\mathrm{CH_2{=}CH{-}CH_3} \longrightarrow \left(\begin{array}{cc} \mathrm{H} & \mathrm{H} \\ | & | \\ -\mathrm{C} & -\mathrm{C}- \\ | & | \\ \mathrm{H} & \mathrm{CH_3} \end{array}\right)_n$$

catalytic methods can even direct the positioning of the methyl side chains. First note the chirality of the carbon atoms within the chain to which the methyl groups are attached. The four different bondings that make these atoms chiral are (1) the side-chain methyl group, (2) the H atom, and (3 and 4) the segments of chain on either side, which are of different lengths and terminate in different groupings. Because of this, the configurations of these carbons cannot be changed by rotation around the single bonds. Consequently, three different patterns are possible: (a) all the methyl groups lie on the same side of the chain (**isotactic**); (b) the methyl groups alternate regularly from one side to the other (**syndiotactic**); and (c) the methyl groups are randomly arranged on either side (**atactic**) (Figure 25.1). The isotactic and syndiotactic polypropylenes are highly crystalline, dense polymers that are used in the fabrication of many products. The atactic variety is soft and rubbery.

We have seen that molecules must provide two reaction sites to produce a condensation polymer. If more than two sites per molecule are available, then the chain may branch and a two- or three-dimensional polymeric network structure may be formed. Chains connected to each other by occasional bridges are called **cross-linked polymers**.

An example of a branched modification of a linear polymer is amylopectin, a type of starch. The difference due to branching is shown in Figure 25.2.

A typical cross-linking bridge in protein is that of the diamino acid, cystine,

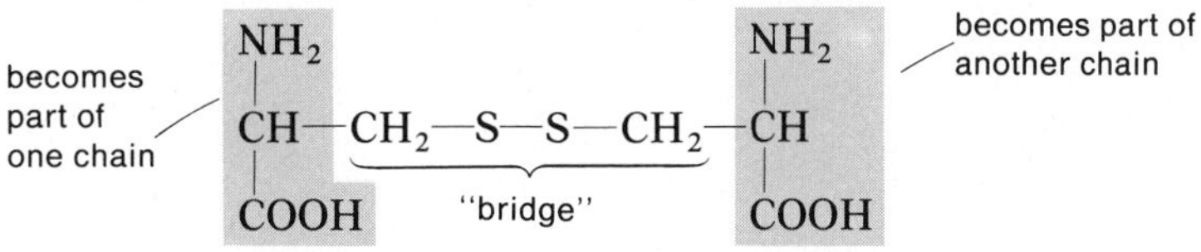

$$\begin{array}{ccccccc} \mathrm{NH_2} & & & & & & \mathrm{NH_2} \\ | & & & & & & | \\ \mathrm{CH} & -\mathrm{CH_2} & -\mathrm{S} & -\mathrm{S} & -\mathrm{CH_2} & - & \mathrm{CH} \\ | & & & & & & | \\ \mathrm{COOH} & & & & & & \mathrm{COOH} \end{array}$$

becomes part of one chain; "bridge"; becomes part of another chain

which has four sites for reaction. (The functional groups are in the colored portions.) Thus, cystine can serve as a ladder rung between amino acid chains. (See also section 25.10.) Such cross-links are part of the structure of the insulin molecule (Figure 25.3).

H	H
Phe	Gly
Val	Ileu
H_2N—Asp	Val
H_2N—Glu	Glu
His	Glu—NH_2
Leu	Cys—S
Cys—S—S—Cys	
Gly	Ala
Ser	Ser
His	Val
	Cys—S
Leu	Ser
Val	Leu
Glu	Tyr
Ala	Glu—NH_2
Leu	Leu
Tyr	Glu
Leu	Asp—NH_2
Val	Tyr
Cys—S—S—Cys	
Gly	Asp—NH_2
Glu	OH
Arg	
Gly	
Phe	
Phe	
Tyr	
Thr	
Pro	
Lys	
Ala	
OH	

FIGURE 25.3
Amino acid sequence in beef insulin, determined in 1955 by Frederick Sanger. Note the —S—S— bridges between the two chains. (See Table 25.4 for explanation of the abbreviations.)

A good example of the difference between linear and network polymers is the formation of polyesters from dihydroxy or trihydroxy alcohols. As was shown for diamines in section 25.3, reaction of a dihydroxy alcohol with a diacid gives a linear copolymer. If a diacid reacts with a trihydroxy alcohol, a network copolymer is formed.

EXAMPLE 25.2 Alkyd resins, produced from glycerol (a trihydroxy alcohol) and phthalic acid (a dicarboxylic acid), are used as ingredients of paints and varnishes. Using the letter *G* for glycerol and *P* for phthalic acid, draw a possible portion of the formula of the copolymer showing the cross-linking. See Figure 25.4.

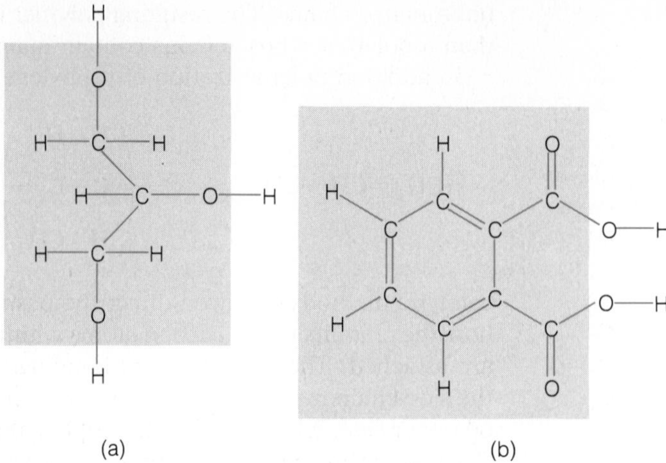

FIGURE 25.4
(a) Glycerol. (b) Phthalic acid. The colored portions are those that remain in the condensation copolymer.

Glycerol is also known by its common name, *glycerine.*

ANSWER Since glycerol is a trihydroxy alcohol, it may be represented with three bonds as —$\overset{|}{\text{G}}$—. Similarly, the dicarboxylic phthalic acid is represented with two bonds, —P—. The extra bond to glycerol provides the cross-linking, perhaps with cycles:

```
              |
              P
              |        /
  —P—G—P—G—P—G
     |                 \
     P
     |        P
   —G—P—G  /    \  G /
          |        |
          P        P
            \    /
              G
              |
```

■

25.6 CRYSTALLINITY OF POLYMERS

The degree to which molecules of a substance are arranged in an orderly pattern is, of course, a measure of crystallinity. So it is, also, with macromolecules. Longer molecules have more geometrical opportunities for partially crystalline arrangements than do shorter ones, however. Consider as examples a pile of neatly stacked, long, flexible rods, and an adjacent neatly stacked pile of short rods of the same material. Now, if both piles are disturbed at the same time by a mild earthquake, the pile of long rods will become disordered (lose "crystallinity") much more slowly than the pile of short ones. This example illustrates that the distinction between "crystalline" and "noncrystalline" is less sharp for macromolecular material than it is for substances of low molecular weights.

A small region of a macromolecular material in which portions of large molecules are arranged in some regular way is called a **crystallite**. It is the orderly orientation of crystallites with respect to each other that builds up crystallinity in a polymer (Figure 25.5).

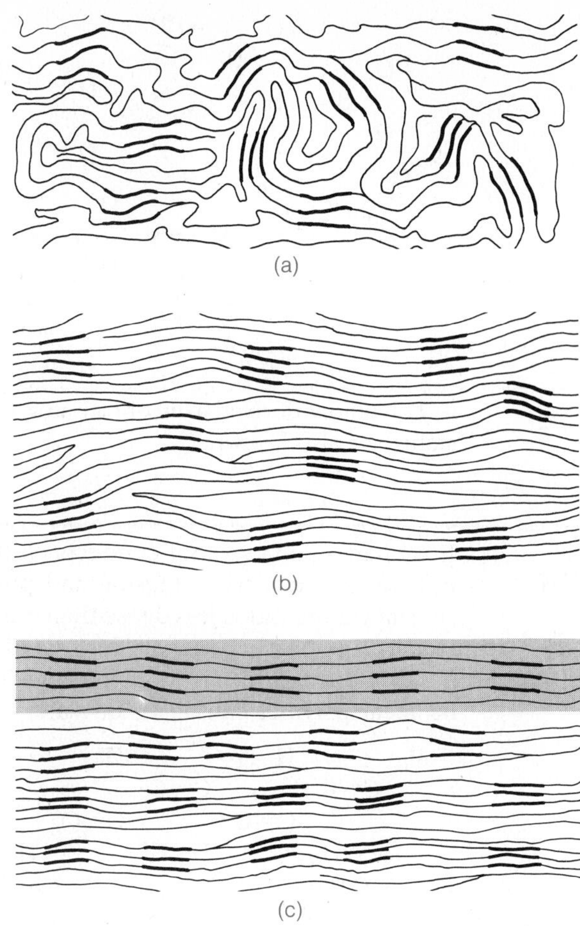

FIGURE 25.5
Crystallinity in chain polymers. (a) Unoriented crystallities in a polymer that is otherwise amorphous. (b) Oriented crystallites. (c) Parallel crystalline areas separated by amorphous regions.

25.7 IONIC POLYMERS

A number of inorganic oxoanions and oxocations form polymers in which oxygen atoms alternate with atoms of another element. The covalence of oxygen is generally 2; that of the other element may be 2 or more. Cross-linking may lead to two- or three-dimensional network structures:

```
                 |
                 O
         |       |
—O—M—O—M—O—
         |       |
                 O
                 |
               —M—
```

An example of a group of oxoanion polymers is the class of complex silicates. A cyclic silicate polymer derived from six monomeric units is the $Si_6O_{18}{}^{12-}$ ion, which occurs in the mineral beryl, $Be_3Al_2Si_6O_{18}$. Beryl crystals are precious gemstones; emerald and aquamarine are varieties of beryl to which traces of impurities impart characteristic tints.

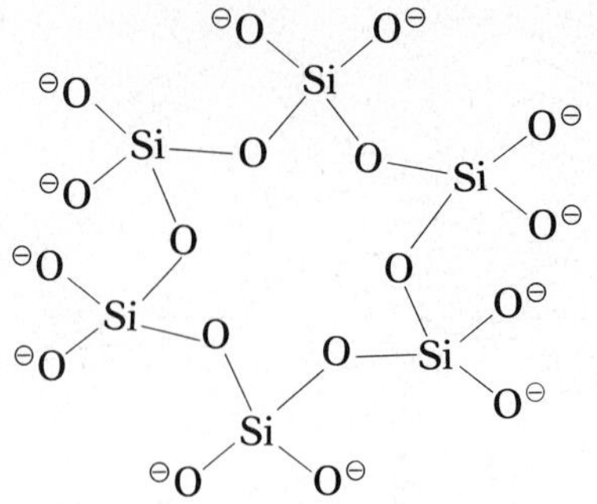

$Si_6O_{18}^{12-}$

Polyanions derived from chromates, phosphates, molybdates, sulfates, and tungstates are also known.

25.8 POLYMERS THAT CONDUCT ELECTRICITY

If we had a polymer that conducts electricity we could produce electrochemical cells or batteries with unusual and potentially useful properties. Since organic substances are much less dense than metals, a very useful product could be a lightweight automobile battery, which would make all-electric cars more economical.

One such polymer is polyacetylene. The formula of acetylene is HC≡CH, and its linear polymer can be *cis* or *trans*:

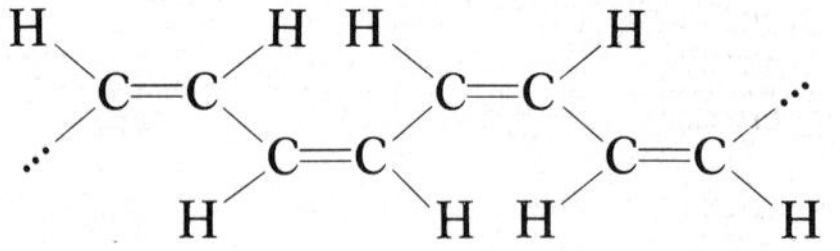

cis-polyacetylene

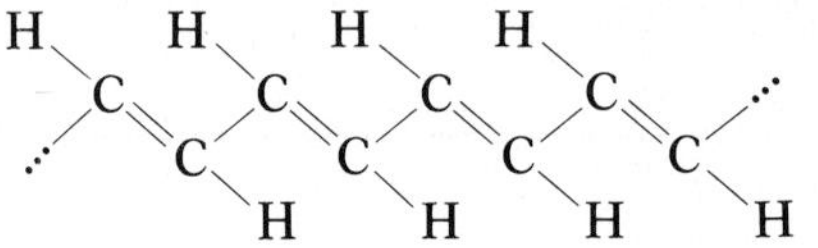

trans-polyacetylene

These chains can contain segments with electronic "defects," in which there are unpaired electrons. *Trans*-polyacetylene illustrates this:

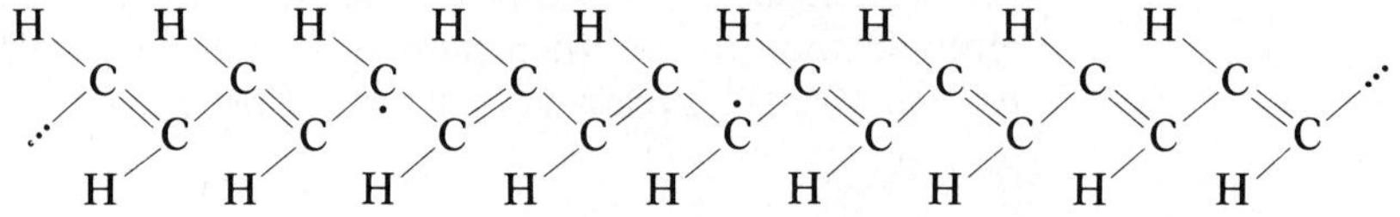

Note that this structure, too, is a polymer of acetylene; all the repeating portions are CHCH units. The special feature about this structure is the presence of unpaired electrons in *p* orbitals that can overlap with the adjacent π bonds. The π-bonding over the entire polymer chain results in extensive delocalization of the unpaired electrons. This description would fit that of delocalization in a metal, except for one thing: the unpaired electrons are in a band consisting of π-molecular orbitals. This band is practically filled and is therefore nonconducting—like the filled bands in a metal (section 22.3). Electrical conductivity could be attained if electrons were removed from the valence band, leaving holes that would make the polymer a *p*-type semiconductor. This removal can be effected by "doping" the polymer with a substance capable of pulling away the lone electrons, namely a Lewis acid. Arsenic pentafluoride, AsF_5, has been used. The chemistry of these polymers is complex, and much research is currently directed to it.

Another conducting polymer is polyphenylene,

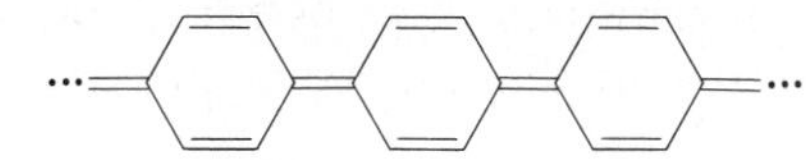

$$\cdots\!=\!\langle\;\rangle\!=\!\langle\;\rangle\!=\!\langle\;\rangle\!=\!\cdots$$

which can be viewed as a polyacetylene chain with —CH=CH— bridges spanning successive 4-carbon portions:

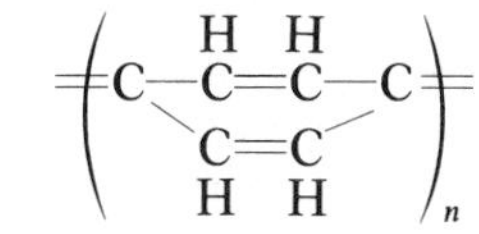

$$\left(\!=\!C-\overset{H}{C}=\overset{H}{C}-C\!=\!\right)_n \quad \text{bridged by } \underset{H}{C}=\underset{H}{C}$$

PROBLEM 25.3 Still another conducting polymer is polypyrrole:

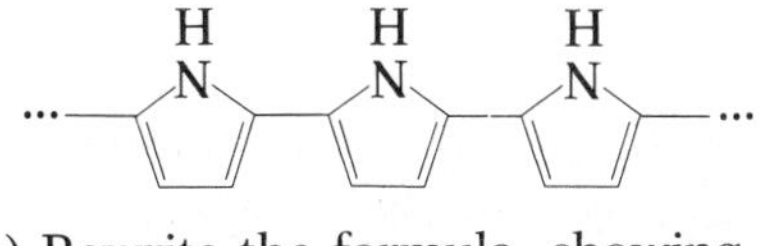

$$\cdots-\text{(pyrrole, N–H)}-\text{(pyrrole, N–H)}-\text{(pyrrole, N–H)}-\cdots$$

(a) Rewrite the formula, showing all C, H, and N atoms, in a form that can be viewed as a polyacetylene chain with —NH— bridges. (b) Is the polyacetylene chain *cis* or *trans* in polyphenylene; in polypyrrole? □

25.9 PROPERTIES OF POLYMERS

Polymers display wide varieties of physical and chemical properties. This should not be surprising when we consider the many possible compositions and arrangements. The broad range of fibrous, adhesive, plastic, filmy, foamy, and rubbery materials attest to this versatility. What generalizations can we make regarding the relationships between the properties of polymers and structure?

In large measure, the chemical reactivity of a polymer is the reactivity of its functional groups. Natural rubber, for example, undergoes deterioration when ozone attacks the double bonds of the polymer chain. A saturated hydrocarbon chain like polyethylene has no functional group and is more resistant to such attack. Celluloses (see Table 25.1) offer their hydroxyl groups to a variety of reagents, and thus we are able to modify its properties. Reaction with nitric acid produces **nitrocellulose**, from which explosive (guncotton) and plastic (celluloid) products are formed. Reaction with acetic acid produces **cellulose acetate**, which can be fabricated into films, sheets, and other useful forms:

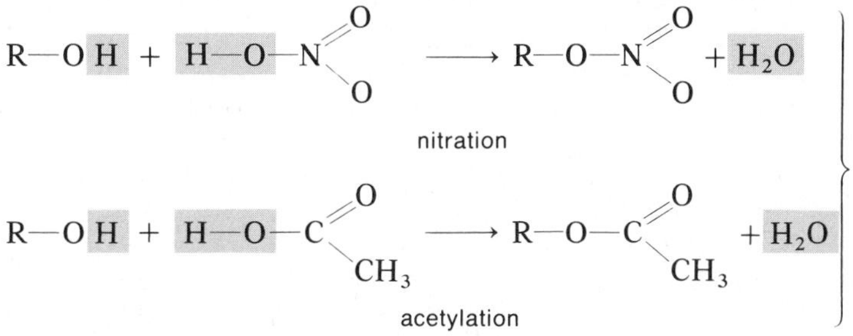

$$R-OH + H-O-NO_2 \longrightarrow R-O-NO_2 + H_2O$$

nitration

$$R-OH + H-O-C(=O)CH_3 \longrightarrow R-O-C(=O)CH_3 + H_2O$$

acetylation

These reactions are ester formations (section 24.11).

More drastic chemical differences yield wider variations in chemical properties. Especially notable is the family of silicone polymers, in which the macromolecular chains contain —Si—O— linkages,

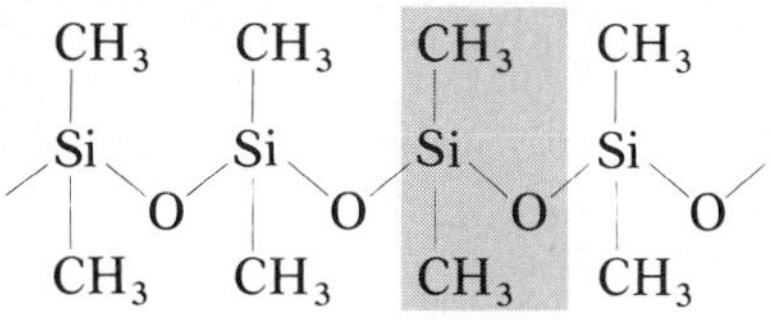

$$-\left[Si(CH_3)_2-O\right]-\left[Si(CH_3)_2-O\right]-\left[Si(CH_3)_2-O\right]-\left[Si(CH_3)_2-O\right]-$$

methylsilicone polymer
(The colored portion is the repeating unit.)

The great thermal stability of the O—Si bond makes it possible to use silicone products at high temperatures. The hydrocarbon side chains contribute oily or lubricating properties. If the silicone is cross-linked, a hard solid polymer is obtained:

```
 CH3    CH3
  |      |
 /Si\  /Si\
  |   O  |   O/
  O      O
  |      |
 /Si\  /Si\
  |   O  |   O/
 CH3    CH3
```

Many attempts have been made to extend the range of "inorganic" polymers by using other linkages for the chain backbones. Potential candidates include boron-carbon, boron-oxygen, arsenic-oxygen, and beryllium-oxygen bonds. The problems are difficult, in part because our understanding of inorganic linkages lags behind our knowledge of the simpler σ and σ-π bonds typical of common organic molecules. Another difficulty is the annoying—to the polymer chemist, anyway—tendency of inorganic systems to cyclize in units of relatively low molecular weight. What's more, third-period elements such as Si and P do not readily form multiple bonds; as a result it becomes more difficult to use the addition-polymerization techniques that work so well with organic monomers like ethylene. Despite such difficulties, however, many inorganic polymers have been prepared (Figure 25.6).

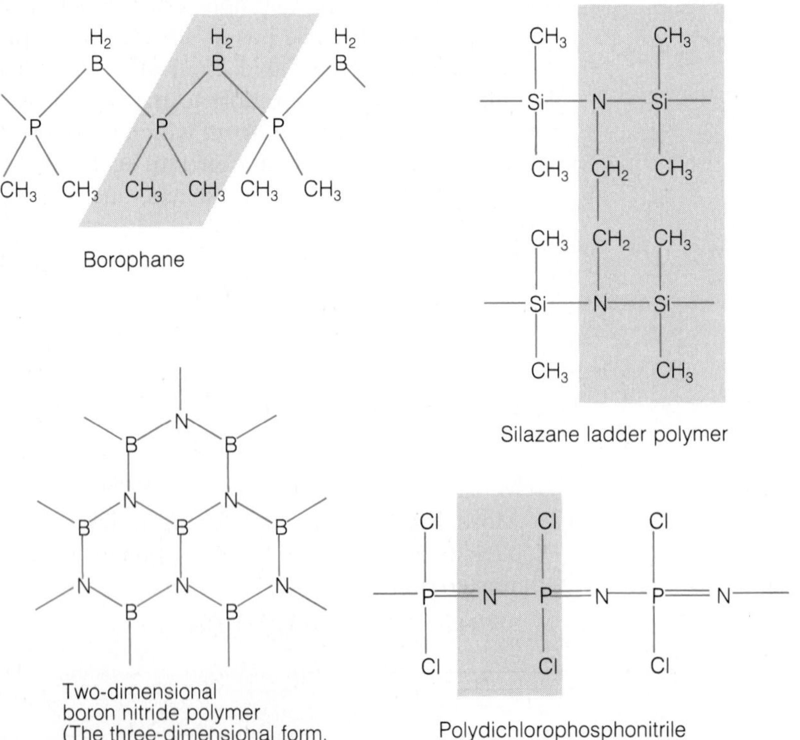

FIGURE 25.6
Some inorganic polymers. The colored portions are the repeating units.

TABLE 25.2
PROPERTIES AND MOLECULAR MAKEUP OF POLYMERS

PHYSICAL NATURE OF POLYMER	MOLECULAR REQUIREMENTS
Broad Categories	
Thermoplastic (able to soften and assume new shapes by application of heat and pressure)	linear macromolecules little cross-linking relatively low molecular weight **Example:** cellulose acetate
Thermosetting (not able to melt or flow without decomposition)	high molecular weight highly cross-linked or network structure **Example:** urea-formaldehyde resin
Specific Properties	
Hard (difficult to scratch)	high molecular weight high crystallinity cross-linking or network structure **Example:** phenol-formaldehyde copolymer (Bakelite)
Strong (cannot easily be pulled apart)	high molecular weight cross-linking or network structure partly crystalline—regions of high crystallinity embedded in an amorphous matrix that acts like a cement **Example:** polymethylmethacrylate (Lucite, Plexiglas) reinforced with glass fiber
Fibrous	high molecular weight linear macromolecules long parallel arrangements of crystalline and amorphous regions **Example:** polyacrylonitrile (Acrilan, Orlon)
Leathery	high molecular weight linear macromolecules with slight degree of cross-linking; fragments of the chains are free to move under stress low crystallinity **Example:** vinyl chloride-vinyl acetate copolymer (vinyl floor covering)
Rubbery	linear macromolecules with little cross-linking; entire chains are free to move under stress, but rotation is restricted by C=C double bonds high molecular weight low crystallinity (but crystallinity increases with elongation) **Example:** polybutadiene (Buna rubber)
Soft, waxy	low molecular weight ($<10{,}000$) low crystallinity **Example:** polyvinyl acetate chewing gum

The properties of a polymer are also determined by the form and arrangement of its macromolecules. The critical factors are the molecular weight (which depends on the degree of polymerization); the extent of branching, cross-linking, or network structuring; the spatial distribution of the monomeric units; and the degree and kind of crystallinity of the macromolecules. Certainly there is room enough for variation even without altering chemical functional groups. Some of these relationships are shown in Table 25.2. The molecular structures of some synthetic polymers are given in Table 25.3.

TABLE 25.3

MOLECULAR STRUCTURES OF SOME COMMON SYNTHETIC POLYMERS

MONOMER ⟶ POLYMER	COMMON OR PROPRIETARY NAME	INDUSTRIAL APPLICATIONS
$H_2C{=}CH_2 \longrightarrow (-CH_2-CH_2-)_n$ ethylene	polyethylene	coatings, containers, toys
$H_2C{=}CHCl \longrightarrow (-CH_2-CHCl-)_n$ chloroethene; vinyl chloride	polyvinyl chloride (PVC)	plastic tubing and film
$H_2C{=}CCl_2 \longrightarrow (-CH_2-CCl_2-)_n$ 1,1-dichloroethene; vinylidene chloride	saran	plastic tubing and film
$F_2C{=}CF_2 \longrightarrow (-CF_2-CF_2-)_n$ tetrafluoroethene	Teflon	plastic parts resistant to high temperatur and solvents
$H_2C{=}C(CH_3)C({=}O)OCH_3 \longrightarrow (-CH_2-C(CH_3)(COOCH_3)-)_n$ methyl methacrylate	Lucite; Plexiglas	transparent molded objects
$H_2N-C(=O)-NH_2 + O{=}CH_2 \longrightarrow H_2O + (-NH-C(=O)-NH-CH_2-)_n$ urea formaldehyde	Melmac (a urea-formaldehyde network copolymer)	plastic dinnerware
$H_2C{=}CH(C_6H_5) \longrightarrow (-CH(C_6H_5)-CH_2-)_n$ styrene	polystyrene	clear cast objects; light, insulating foa
$C_6H_5OH + O{=}CH_2 \longrightarrow H_2O + -CH_2-C_6H_2(OH)(CH_2-)-CH_2-C_6H_3(OH)-CH_2-$ (with $-CH_2-C_6H_3(OH)-$ cross-link) phenol formaldehyde linear and cross-linked	Bakelite (a linear or hard, cross-linked polymer)	resin binder for fibrou glass insulation (line polymer); hard obje (cross-linked polym

Chain-branching decreases the ability of molecules to come into close contact with each other, becuase the branches get in the way. The result is that the London forces (section 11.2) are weaker. Therefore, such polymers are softer than those whose molecules consist of unbranched chains.

The strength and thermal stability of three-dimensional network polymers are not attributable to stronger chemical bonds, nor even to a greater number of bonds. Instead, these properties result from the fact that the network polymer is better able than a linear polymer to preserve its structural integrity even after some bonds are broken, as shown:

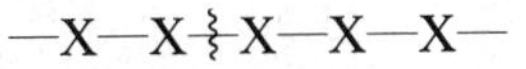

```
—X—X⌇X—X—X—
```

Linear polymer: Breaking a bond severs the chain.

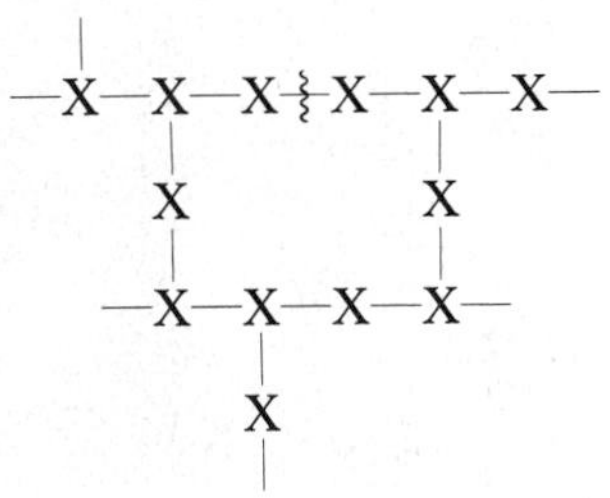

```
 |
—X—X—X⌇X—X—X—
     |       |
     X       X
     |       |
    —X—X—X—X—
       |
       X
       |
```

Network polymer: Breaking a bond does not sever the chain.

Shifts in properties also accompany the changes in molecular arrangements produced by a rise in temperature. On heating, a polymer may change from a rigid glassy state, through a partly flexible leathery condition, to a rubbery condition, and finally to a flowing viscous liquid. Mechanical deformation also changes the properties of a polymer. Perhaps the most striking instance is the stretching of rubber. In the unstretched form, rubber molecules are oriented randomly. On stretching, random motion is restricted, entropy is reduced, and the molecules assume an orderly, linear arrangement. Stretching is thus analogous to freezing, an exothermic process. To demonstrate this release of energy, stretch a wide rubber band, touch it immediately to your lips, and feel its sudden warmth.

25.10 BIOCHEMICAL POLYMERS

Biochemistry is the chemistry of organisms. An **organism** is anything that is alive (a "living organism") or that was once alive (a "dead organism"). No organism on Earth is made up entirely of low-molecular-weight units; as far as we know, polymeric material is essential to life. Defined most broadly, living things are characterized by their ability to use a source of energy to reproduce themselves at the expense of materials from the environment, and to retain any accidental changes (**mutations**) that may occur in the pattern of reproduction.

Some organisms are considered to be on the very borderline of life—viruses, for example (Figure 25.7). A **virus** consists of particles several hundred angstrom units in length or diameter; these particles can reproduce themselves in a suitable environment, but they do not ingest food, grow, or carry on any other metabolic processes. The reproductive mechanisms of these and all other organisms, however, involve processes that can occur only with macromolecular materials.

The next three sections are devoted to the three types of biochemical polymers that most universally characterize living things: carbohydrate polymers, proteins, and nucleic acids.

FIGURE 25.7
Electron micrograph of crystalline tobacco necrosis virus; magnification × 16,000.

25.11 CARBOHYDRATES

Common table sugar is sucrose, $C_{12}H_{22}O_{11}$, obtained from sugar cane and from sugar beets. Sucrose is broken down by hydrolysis to two simpler sugars: glucose and fructose, which are isomers with the formula $C_6H_{12}O_6$. Chemists noted that the ratio of H to O in these and other sugars is 2:1, as in H_2O. Thus, the early formulas were written $C_{12}(H_2O)_{11}$ and $C_6(H_2O)_6$, as if they were hydrates of C. Even though these formulas are misleading, because they imply incorrect bonding in the molecules, they are responsible for the group name **carbohydrate**.

Glucose and fructose *cannot* be broken down to simpler sugars by hydrolysis. They are therefore regarded as monomer sugars (**monosaccharides**), and sucrose is regarded as a dimer, or co-dimer, of glucose and fructose. Carbohydrates—particularly cellulose, starch, and glycogen—also exist in the form of polymers. For a brief survey of this important class of substances, we will first examine the structures of simple sugar molecules, and then go on to their dimers and polymers.

Simple sugars are polyhydroxy aldehydes or ketones with at least one chiral center. The simplest sugar of all is glyceraldehyde, $C_3H_6O_3$. Its structure is shown here; the asterisk identifies the chiral center.

```
H—C=O
   |
   C*
H ◄ | ► OH
  CH₂OH
```

glyceraldehyde

The most abundant sugar is **glucose**, $C_6H_{12}O_6$. Its cyclic molecular structure is shown on the left here. (See also Table 25.1.) In aqueous solution, this cyclic form is in equilibrium with a small proportion of the open-chain pentahydroxy aldehyde, on the right:

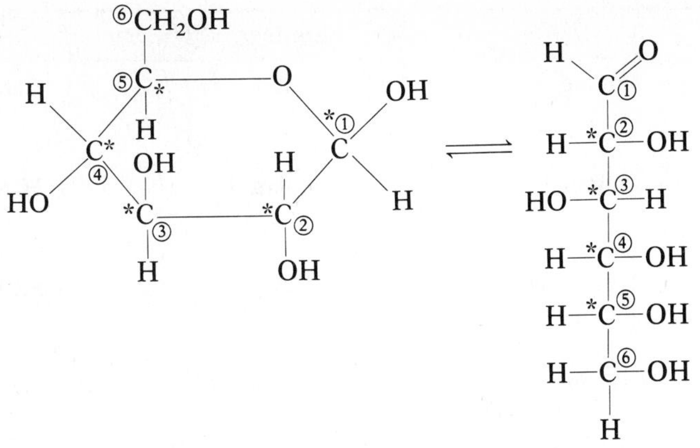

cyclic form of glucose

open-chain form of glucose

The carbon atoms are correspondingly numbered in the two structural formulas. Note that carbons 2, 3, 4, and 5 (asterisks) are chiral in both the cyclic and open-chain structures. This chirality accounts for the existence of 2^4, or 16 isomers of the six-carbon aldehyde sugars, known as **aldohexoses**.

Three important sugar dimers, known as **disaccharides**, are sucrose, lactose (milk sugar), and maltose (malt sugar). The structure and hydrolysis of sucrose are shown here:

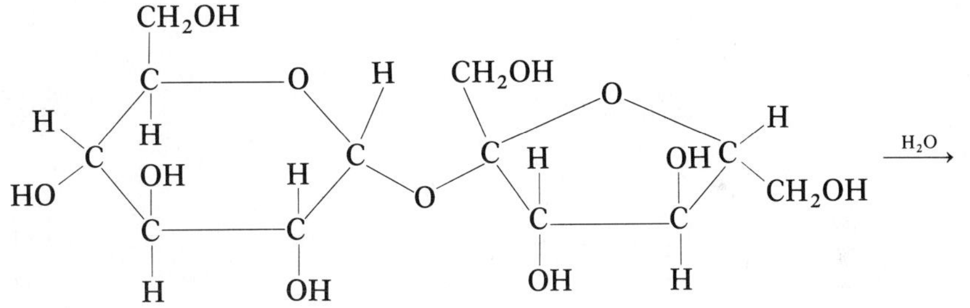

sucrose $\xrightarrow{H_2O}$ glucose + fructose

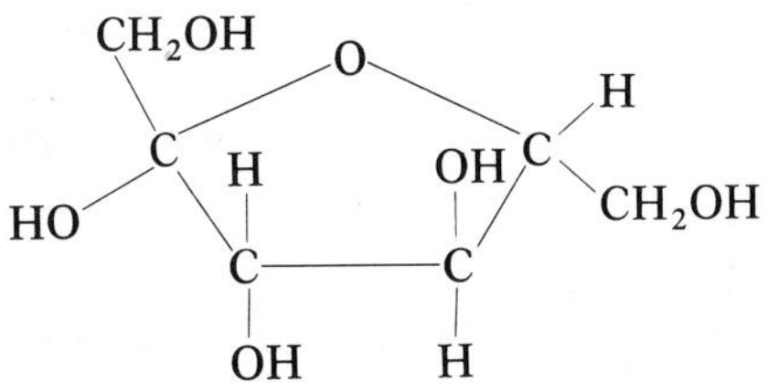

The hydrolysis of 1 mole of maltose yields 2 moles of glucose, which shows that maltose is a glucose dimer. A mole of lactose, on the other hand, hydrolyzes to produce 1 mole of glucose and 1 mole of galactose, which is another aldohexose.

The most important carbohydrate polymers are cellulose and starch.

- **Cellulose** (see Table 25.1) is a polymer made up of some 10,000 glucose units. It is the main structural material of plants and is probably the most widely occurring organic material on Earth, if you don't count fossil fuels. Humans cannot digest cellulose, but many other organisms, including cows, deer, and many species of insects, can.

TABLE 25.4
THE AMINO ACIDS

NAME	ABBREVIATION	FORMULA
Alanine	Ala	$CH_3—CHCOO^-$ with $^+NH_3$ on the CH
Arginine†	Arg	$H_2NCNHCH_2CH_2CH_2CH_2CHCOO^-$ with $=^+NH_2$ on the first C and NH_2 on the CH
Aspartic acid	Asp	$HOOCCH_2CHCOO^-$ with $^+NH_3$ on the CH
Cysteine	Cys	$HSCH_2CHCOO^-$ with $^+NH_3$ on the CH
Cystine	$(Cys)_2$	$^-OOCCHCH_2SSCHCHCOO^-$ with $^+NH_3$ on each CH
Glutamic acid	Glu	$HOOCCH_2CH_2CHCOO^-$ with $^+NH_3$ on the CH
Glutamine	Gln	$H_2NCCH_2CH_2CHCOO^-$ with $=O$ on the first C and $^+NH_3$ on the CH
Glycine	Gly	$H_3\overset{+}{N}CH_2COO^-$
Histidine†	His	HC=N—CH=C—CH_2CHCOO^- (ring: HC, N, CH, C, NH) with $^+NH_3$ on the CH
Isoleucine†	Ileu	$CH_3CH_2CH—CHCOO^-$ with CH_3 on the first CH and $^+NH_3$ on the second CH
Leucine†	Leu	$(CH_3)_2CHCH_2CHCOO^-$ with $^+NH_3$ on the CH

† An essential α-amino acid. It cannot be synthesized in the body.

■ **Starch** is also a glucose polymer, but the linkages between the glucose portions are sterically different from those in cellulose. It is this steric difference that makes it possible for humans to digest starch. Starch is hydrolyzed to maltose by the enzyme amylase, which occurs in saliva. Maltose is further hydrolyzed to glucose in the small intestine. Straight- and branched-chain skeletons of starch molecules are shown in Figure 25.2.

25.12 AMINO ACIDS, PEPTIDES, AND PROTEINS

AMINO ACIDS

Amino acids are examples of compounds that have more than one type of functional group, in this case $—NH_2$ and $—COOH$. Of particular importance are the **α-amino acids**.

NAME	ABBREVIATION	FORMULA
Lysine†	Lys	$H_3\overset{+}{N}CH_2CH_2CH_2CH_2CH(NH_2)COO^-$
Methionine†	Met	$CH_3SCH_2CH(^+NH_3)COO^-$
Phenylalanine†	Phe	$C_6H_5-CH_2CH(^+NH_3)COO^-$
Proline	Pro	H_2C-CH_2 / $H_2C-\overset{+}{N}H_2-CHCOO^-$ (ring)
Serine	Ser	$HOCH_2CH(^+NH_3)COO^-$
Threonine†	Thr	$CH_3CH(OH)-CH(^+NH_3)COO^-$
Tryptophan†	Try	$CH_2CH(^+NH_3)COO^-$ attached to indole ring (N–H)
Tyrosine	Tyr	$HO-C_6H_4-CH_2CH(^+NH_3)COO^-$
Valine†	Val	$(CH_3)_2CHCH(^+NH_3)COO^-$

$$R-\overset{NH_2}{\underset{H}{C^{\alpha}}}-\overset{O}{\overset{\|}{C}}-OH$$

an α-amino acid

The Greek letter alpha, α, designates the C to which the COOH group is attached. In an α-amino acid, the NH_2 group is also attached to the α carbon. Table 25.4 lists the biologically important α-amino acids. Those with daggers cannot be synthesized in the human body and are therefore essential in the diet. Except for glycine, $CH_2(NH_2)COOH$, where R is H, the α carbon of α-amino acids is attached to four different groups and is therefore asymmetric. With very rare exceptions, the naturally occurring α-amino acids have the same configuration, designated L, as shown:

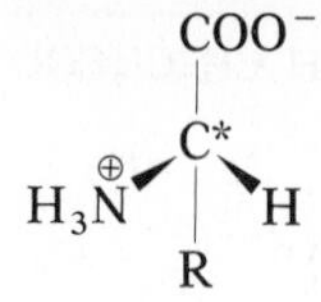

$$\begin{array}{c} COO^- \\ | \\ \overset{\oplus}{H_3N} - C^* - H \\ | \\ R \end{array}$$

a typical L α-amino acid

Low-molecular-weight carboxylic acids such as formic acid (HCOOH, b.p. 100°), acetic acid (CH_3COOH, b.p. 118°), and propionic acid (CH_3CH_2COOH, b.p. 141°), are liquids at room temperature. Low-molecular-weight amines, such as aminomethane (CH_3NH_2, b.p. −7.5°), aminoethane ($CH_3CH_2NH_2$, b.p. 17°), and 1-aminopropane ($CH_3CH_2CH_2NH_2$, b.p. 49°), are gases or liquids. Yet all α-amino acids, such as glycine (m.p. 262°) and alanine (m.p. 314°), are solids with fairly high melting points. How do we explain these high melting points? There must be a much greater intermolecular attraction between molecules of amino acids than between amine molecules or between carboxylic acid molecules. First, $R—\ddot{N}H_2$ is basic ($K_b \approx 10^{-5}$) and R—COOH is acidic ($K_a \approx 10^{-5}$), so they react to form a dipolar ion, as shown for glycine:

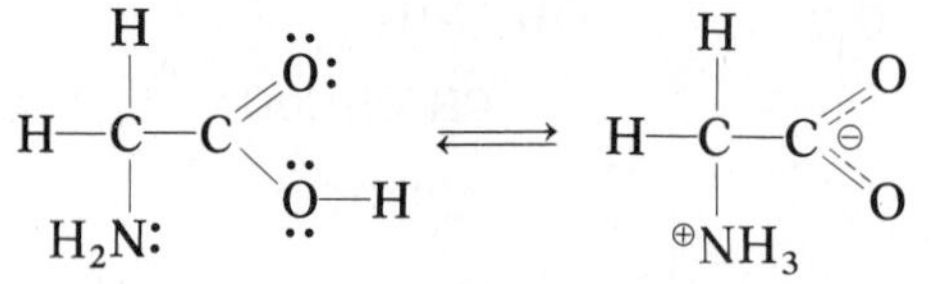

$$\begin{array}{c} H \\ | \\ H—C—C(=\ddot{O}:)—\ddot{O}—H \\ | \\ H_2N: \end{array} \rightleftharpoons \begin{array}{c} H \\ | \\ H—C—C(O)_2^{\ominus} \\ | \\ {}^{\oplus}NH_3 \end{array}$$

dipolar ion of glycine
a typical amino acid

The dipolar ion is also called a **zwitterion**.

These charges can then exert strong intermolecular electrical attractions. An amino acid crystal is like an ionic crystal, even though the individual units have zero net charge.

PEPTIDES AND PROTEINS

Peptides are made up from two to about fifty amino acids. When the number of amino acids is small, the appropriate prefix is used; *di*peptide, *tri*peptide, and so on. With larger numbers of amino acids, up to about fifty, the term *poly*peptide is used. When a peptide contains more amino acids than that, the molecule is called a **protein**. The individual amino acids are joined when a water molecule is eliminated, forming an amide linkage, called a **peptide bond**. This is the same linkage that exists in nylons (section 25.3). Two different amino acids can form two different dipeptides, depending on which amino acid furnishes the —COOH group and which one furnishes the $—NH_2$ group to the peptide bond. The two structures are shown in the non-ionized form for the reaction of glycine with alanine:

1. Glycine furnishes the —COOH group

$$\underset{\text{glycine (Gly)}}{H—\overset{H}{\underset{NH_2}{C}}—\overset{O}{\overset{\|}{C}}—OH} + \underset{\text{alanine (Ala)}}{H—N(H)—\overset{H}{\underset{CH_3}{C}}—\overset{O}{\overset{\|}{C}}—OH} \longrightarrow \underset{\text{glycylalanine (Gly-Ala)}}{H_2\underset{NH_2}{C}—\overset{O}{\overset{\|}{C}}—\overset{H}{N}—\underset{CH_3}{C}(=O)OH} + H_2O$$

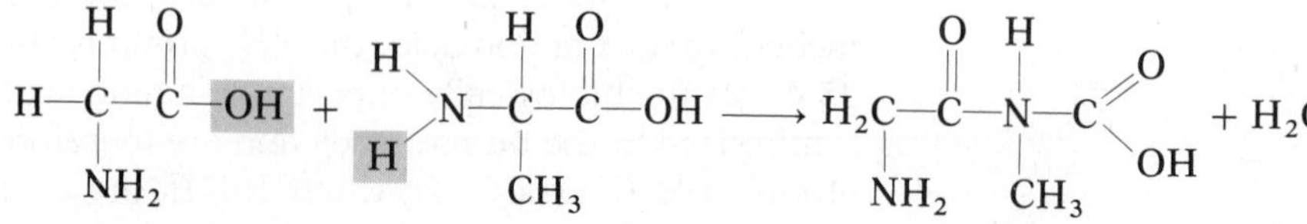

glycine (Gly) | alanine (Ala) | glycylalanine (Gly-Ala)

2. Alanine furnishes the —COOH group

$$\underset{\text{alanine}}{CH_3{-}\overset{\displaystyle H}{\underset{\displaystyle NH_2}{C}}{-}C\!\!\begin{array}{l}{}^{\textstyle //O}\\ {}_{\textstyle \backslash}\boxed{OH}\end{array}} \;+\; \underset{\text{glycine}}{\begin{array}{l}H\backslash\\ \boxed{H}/\end{array}\!N{-}CH_2{-}C\!\!\begin{array}{l}{}^{\textstyle //O}\\ {}_{\textstyle \backslash}OH\end{array}} \longrightarrow$$

$$\underset{\substack{\text{alanylglycine}\\ \text{(Ala-Gly)}}}{CH_3{-}\overset{\displaystyle H}{\underset{\displaystyle NH_2}{C}}{-}\overset{\displaystyle O}{\overset{\|}{C}}{-}\overset{\displaystyle H}{\underset{\displaystyle H}{N}}{-}C\!\!\begin{array}{l}{}^{\textstyle //O}\\ {}_{\textstyle \backslash}OH\end{array}} + H_2O$$

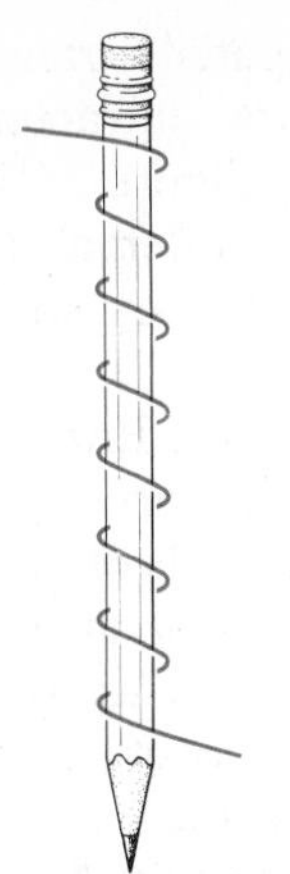

FIGURE 25.8
A string coiled around a cylinder forms a helix.

PROBLEM 25.4 Write the structural formula for two dipeptides formed from valine and serine (see Table 25.4). □

Protein is derived from the Greek *proteos,* "primary."

Proteins have molecular weights in the range of 6000 to 50,000. A protein may be characterized according to its **primary** and **secondary structures**. The primary structure of a protein shows the sequence of linkages of amino acids in the protein molecule without regard to conformation or hydrogen bonding. The first formulation of this kind was reported in 1955 by Frederick Sanger for the protein **insulin**. Figure 25.3 (page 967) depicts the primary structure of insulin. Note the bridging between amino-acid strands by the —S—S— bonds.

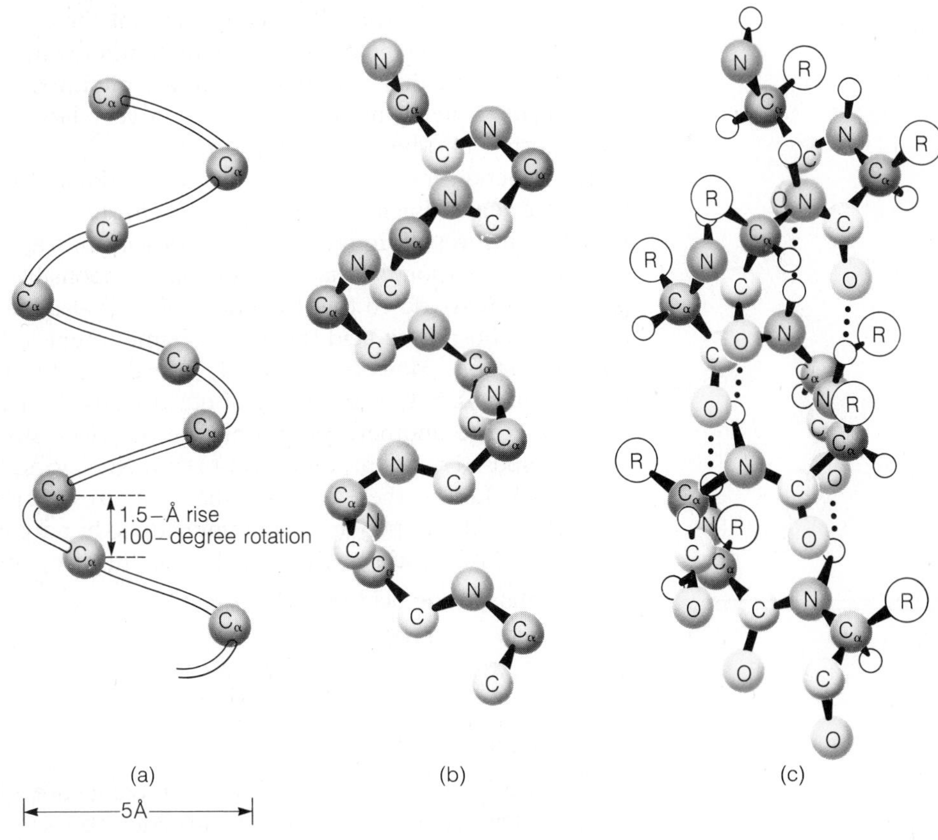

FIGURE 25.9
Models of a right-handed α helix. (a) The α carbon atoms on the helical thread. (b) The backbone nitrogen atoms (N), α carbon atoms (C_α), and carbonyl carbon atoms (C) on the helical thread. (c) The entire helix. Hydrogen bonds (· · ·) between NH and CO groups stabilize the helix. (From L. Stryer, *Biochemistry*, 2nd ed., W. H. Freeman, San Francisco.)

The secondary structure of a protein is determined by the spatial arrangement of the polypeptide chain. Evidence obtained mainly from X-ray diffraction patterns has shown that the chain is typically wound into a helix (Figure 25.8). The helical form is maintained by hydrogen bonds located at spaced intervals (Figure 25.9). The entire structure is called the **alpha helix**. Other secondary structures of proteins include pleated sheets and random coils.

Proteins that catalyze biochemical reactions are called **enzymes**. Enzymes are very highly specific, each one being capable of catalyzing only a particular reaction of one type of substance. Thus a cell needs to contain hundreds of enzymes, nearly as many as there are biochemical reactions in the cell. Much of an enzyme's specificity lies in the particular configurations of its chiral centers, as well as in its preferred conformations. Figure 25.10 shows a schematic model of an enzyme-catalyzed decomposition of a molecule. Under the influence of the approaching molecule, the enzyme bends so as to grip the molecule in a perfect fit. A complex is formed in which chemical bonds within the molecule are weakened so that decomposition occurs more rapidly than it would without the enzyme.

25.13 NUCLEIC ACIDS

Proteins are continuously being synthesized and destroyed in living organisms. The chemical natures of the proteins can differ from cell to cell, from tissue to tissue, and from species to species. How are the "right" proteins made? What is the chemical mechanism of reproduction?

Biologists have long known that reproduction occurs as a series of cell divisions that start within the nuclei of the cells. It is also recognized that structures called **chromosomes**, which can be seen with the aid of a microscope, are involved in the process. It is reasonable to conclude that the chromosomes contain the "set of instructions" that guides the reproduction. Chromosomes consist of molecules, and any molecule that can carry such a vast amount of information as that required to duplicate a living organism must be a very large molecule, one made up of many parts—in other words, a polymer.

The genetic "instructions" are actual portions of the chromosome molecules known as **genes**. Each gene is responsible for the production of a specific protein, which in turn results in the expression of a given trait in the individual organism.

It was once thought that the polymers responsible for heredity were nuclear proteins. It was known that the nuclei of cells also contained organic substances related to phosphoric acid (hence the name "nucleic" acids), but their role was not understood. However, in 1944, Oswald T. Avery showed that **deoxyribonucleic acid (DNA)** could bring about a transfer of hereditary traits from one bacterial cell to another. Avery's finding therefore strongly suggested that genes were composed not of protein, but of DNA. Nucleic acids also include **ribonucleic acid, RNA**. Let us, then, examine the structures of these remarkable molecules.

Nucleic acids are polymers containing up to hundreds of millions of monomeric units, called **nucleotides**. Each nucleotide, in turn, consists of three portions:

1. A phosphoric acid portion

```
 O     OH
   \  /
    P
   /  \
HO     OH
```

phosphoric acid

The two hydroxyl groups (shown in color) are eliminated by condensation with two sugar molecules to form two molecules of H_2O. Also, some ionization takes place in the aqueous medium of the cell, which leaves a negative charge on the remaining phosphate ion.

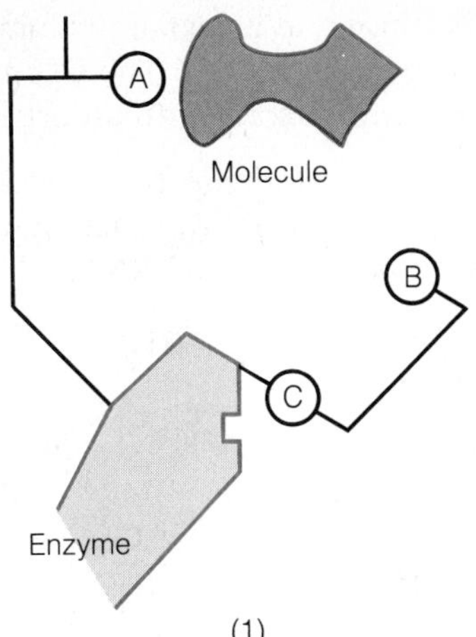

(1)

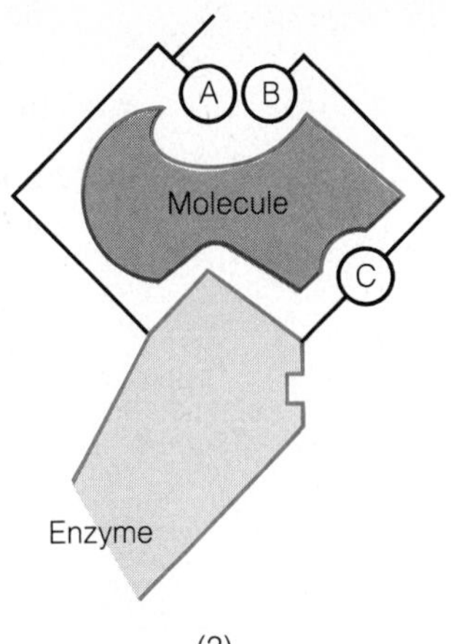

(2)

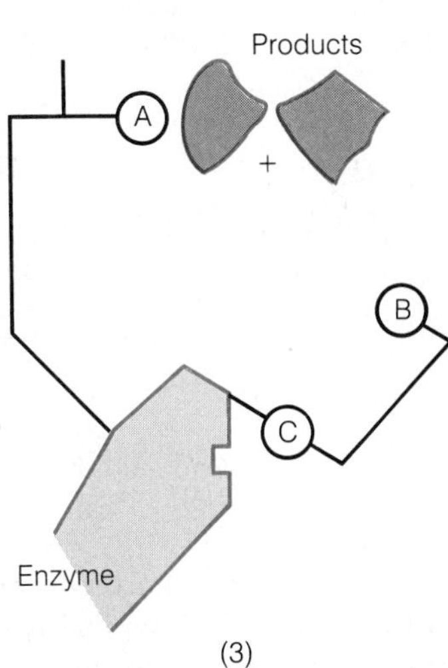

(3)

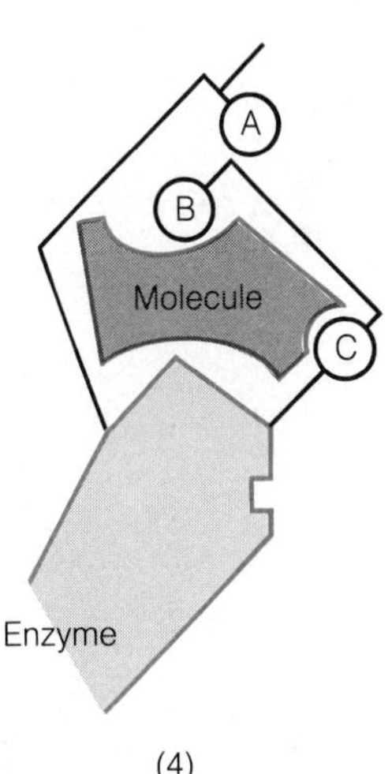

(4)

FIGURE 25.10
Schematic representation of the steps in enzyme action. (1) Molecule and enzyme before the reaction. (2) Molecule binds to enzyme via group C, forming a complex. The binding induces proper alignment of catalytic groups A and B. (3) Reaction ensues, yielding products and the original enzyme. (4) Molecules that are either too large or too small may be bound, but they fail to react, because the alignment of the catalytic groups is improper.

---

2. A five-carbon sugar

ribose (part of RNA)

deoxyribose (part of DNA)

The H atoms (shown in color) are eliminated by condensation with phosphoric acid to form H_2O. The OH group (also shown in color) is eliminated by condensation with an organic base to form H_2O.

3. An organic base that contains N. There are mainly five bases involved. Thymine occurs only in DNA, uracil only in RNA, and the other three are common to both DNA and RNA.

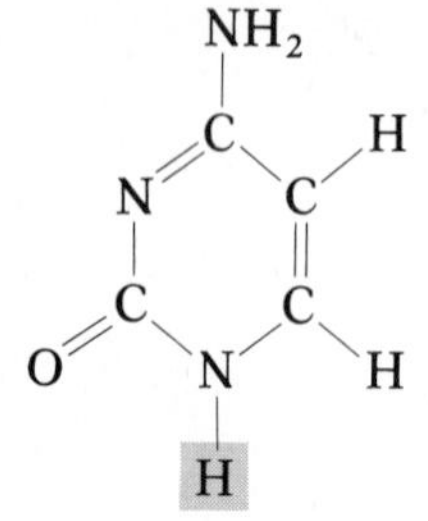

cytosine, C
(in DNA and RNA)

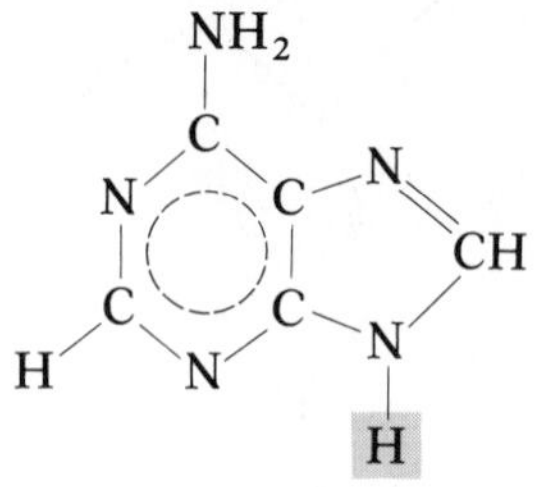

adenine, A
(in DNA and RNA)

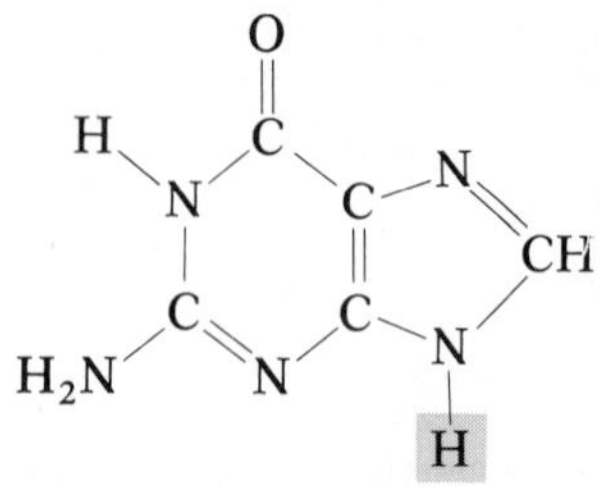

guanine, G
(in DNA and RNA)

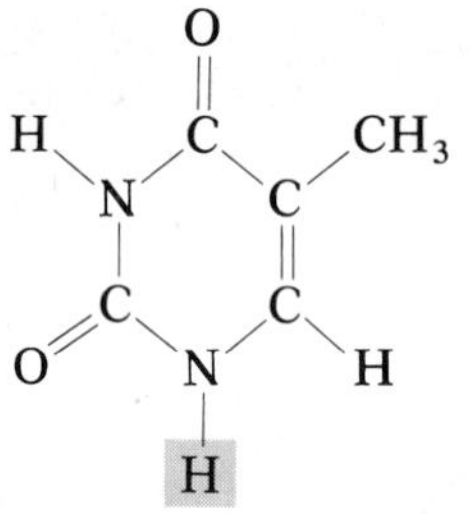

thymine, T
(in DNA)

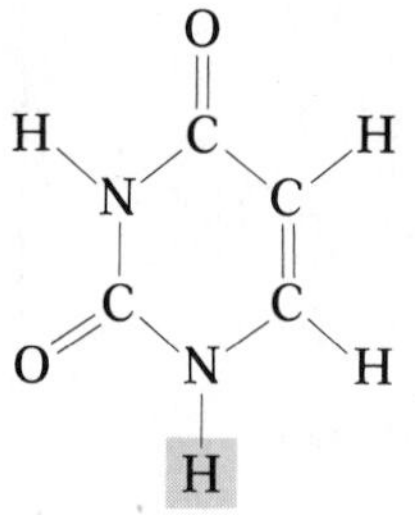

uracil, U
(in RNA)

The H atoms (shown in color) are eliminated by condensation with the sugar portion to form H_2O. A typical nucleotide of DNA, then, may consist of phosphoric acid, deoxyribose, and, say, cytosine:

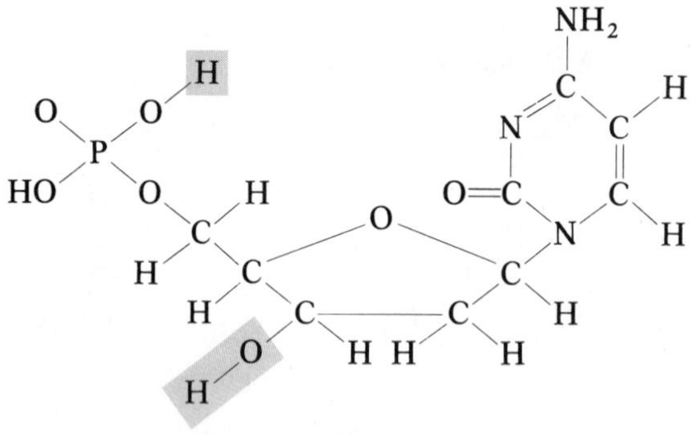

a typical nucleotide of DNA

The colored portions are eliminated by condensation. A nucleic acid chain may be pictured as a large number of nucleotides linked together, with the elimination of H_2O molecules (Figure 25.11).

Thus far, we have shown only the primary structure of a nucleic acid chain. Of course, if DNA determines the reproduction of cells, there must be much more

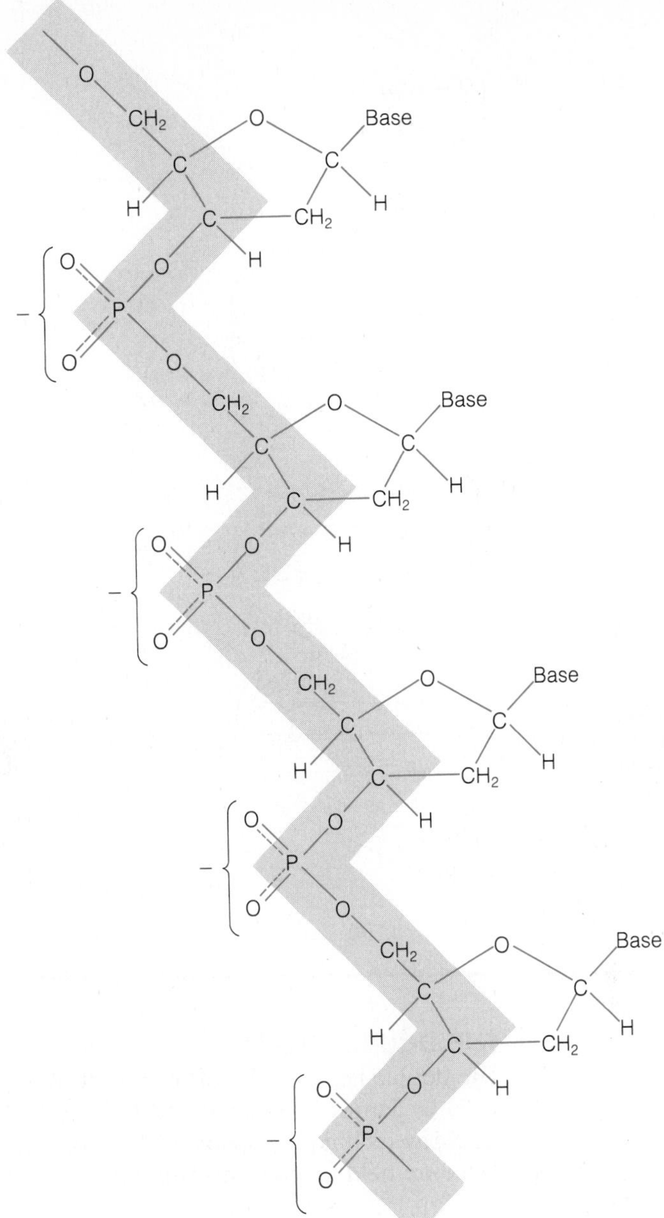

FIGURE 25.11
DNA chain. The colored area represents the chain formed by elimination of H_2O molecules. The phosphoric acid portions are shown in their ionized state.

to tell. We must now consider three important aspects of the DNA molecule:

1. DNA consists of a very long, double chain, not a single chain. The two chains wind around themselves in the form of a double helix.
2. The DNA molecule can reproduce itself.
3. The sequence of bases in the two chains embodies a pattern, or code, for the synthesis of proteins.

Let us examine each of these aspects.

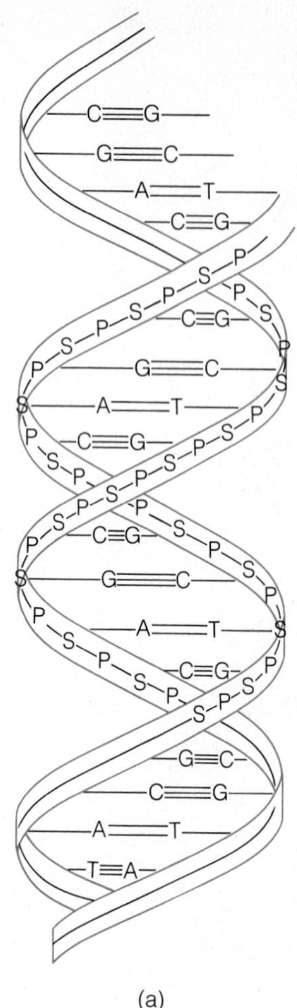

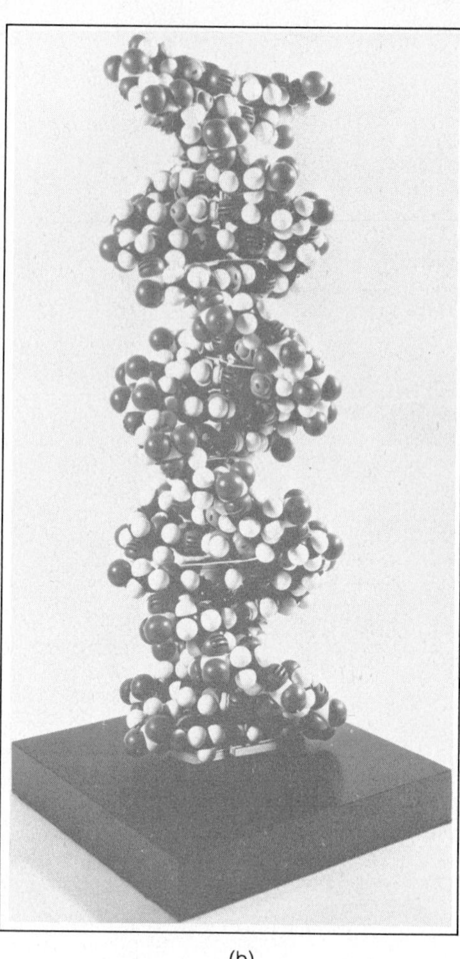

FIGURE 25.12
(a) Schematic illustration of DNA double helix. S represents sugar; P represents phosphate. G, C, A, and T are the paired bases. (b) Molecular model of the DNA helix. (Courtesy of Professor M. H. F. Wilkins, Medical Research Council, Biophysics Unit, Kings College, London.)

THE DOUBLE HELIX

The double-helix model of the structure of DNA (Figure 25.12) was proposed in 1953 by James D. Watson and Francis H. C. Crick. In 1973 it was confirmed by an X-ray technique capable of "seeing" individual atoms. The two chains of the helix are held together by H-bonding between pairs of bases. The only pairs that

FIGURE 25.13
H-bonding holding together base pairs in DNA.

Thymine ====== Adenine

Cytosine ======= Guanine

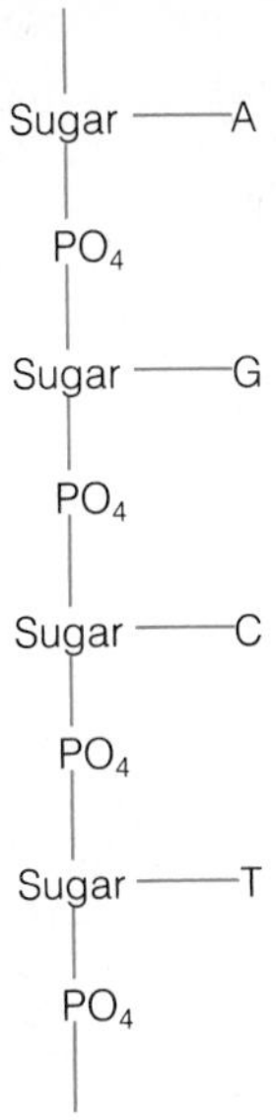

Sugar — A
PO_4
Sugar — G
PO_4
Sugar — C
PO_4
Sugar — T
PO_4

FIGURE 25.14
Section of a single DNA chain (schematic).

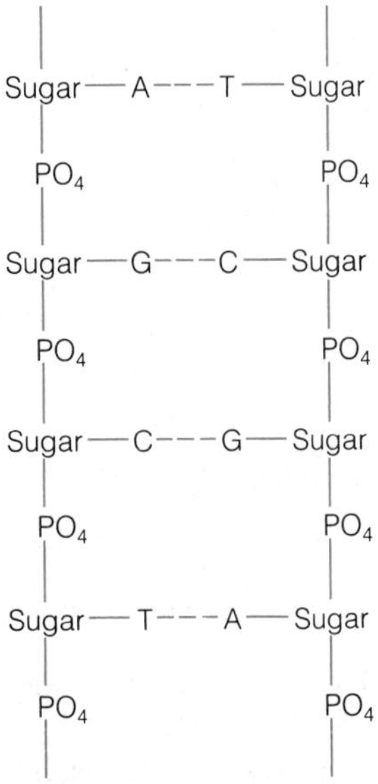

bond to each other in DNA are thymine and adenine (T–A) and cytosine and guanine (C–G) (Figure 25.13).

THE PAIRING OF DNA MOLECULES

As we have stated, the genetic code is embodied in the sequence of bases in the DNA chain. Now, imagine that a single strand of DNA could be isolated and stretched out along a straight line. We would have a structure like that shown in Figure 25.14. Reading from top to bottom, we see that this particular stretch of DNA gives the base sequence —AGCT—. But as we have seen, the DNA molecule consists of two chains, and the chains are arranged spatially so that the bases from the two separate chains are brought into very close proximity. Furthermore, since only A–T and G–C combinations can match, we can extend our schematic diagram one step further, as shown in Figure 25.15. Thus, the two chains of DNA are not identical, but they complement each other in a *fixed relationship of base-pairing.* A DNA molecule reproduces itself when its two chains separate and each strand serves as the pattern for the formation of a new strand with the complementary bases. This pairing gives rise to two new DNA molecules, each identical to its parent (Figure 25.16).

THE SYNTHESIS OF PROTEINS

The sequence of bases in DNA ultimately determines the sequence of amino acids in proteins. To help describe some of the intermediate steps in this very complex process, we list the essential molecular structures involved.

- DNA The molecule that contains all of the information needed to transfer hereditary traits. The information is comprised in the sequence of bases in the polymer chains.
- messenger RNA (mRNA) A single-stranded RNA molecule that carries the information from the DNA molecule to structures outside the cell nucleus where protein synthesis takes place. The mRNA determines the particular protein to be synthesized.
- transfer RNA (tRNA) A much smaller single-stranded RNA molecule that selects specific amino acids and "escorts" them to the growing protein chain so that they join at the proper position.
- ribosome A structure outside the cell nucleus where protein synthesis occurs.
- ATP (adenosine triphosphate) A molecule that "stores" energy and releases it under specific conditions.
- α-amino acids $R{-}\overset{\displaystyle NH_2}{\underset{\displaystyle H}{C}}{-}COOH$
- activated amino acids Amino acids complexed with ATP.

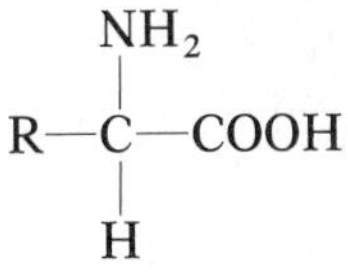

Sugar — A - - - T — Sugar
PO_4 PO_4
Sugar — G - - - C — Sugar
PO_4 PO_4
Sugar — C - - - G — Sugar
PO_4 PO_4
Sugar — T - - - A — Sugar
PO_4 PO_4

FIGURE 25.15
Base-pairing in double-strand DNA (schematic).

How can the "base language" of DNA be translated to the "amino acid language" of proteins? Remember that there are only four kinds of bases in DNA and 20 amino acids in proteins. Think of it this way: If we have an alphabet of four letters (A, C, G, and T), from which we must compose 20 different "words" (representing amino acids), how many letters must be used for each word? The answer cannot be one, because that would yield only 4 one-letter words—A, C, G, and T. Nor can the answer be two, because that can yield only 4 × 4, or 16, two-letter words, such as AC, CA, GA, CT, and so on—not enough to represent 20

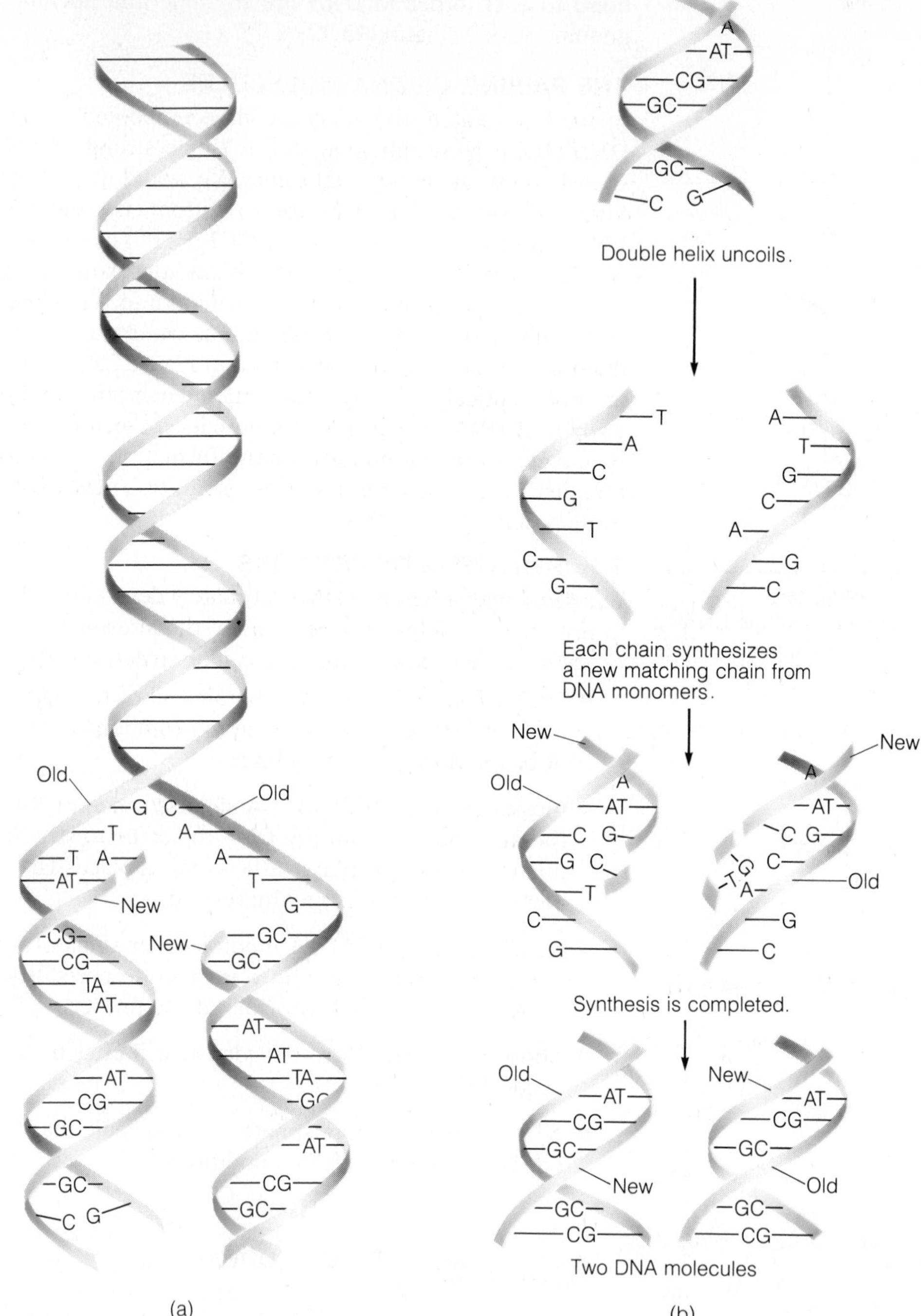

FIGURE 25.16
Duplication of a DNA molecule. (a) The double helix gradually uncoils. Specific base pairing results in the formation of two new molecules that are identical to the old one. (b) The several steps in the process are shown in sequence. (Adapted from James Watson, *Molecular Biology of the Gene*, W. A. Benjamin, New York, 2nd ed., 1970 (267).)

amino acids. However, three-letter words such as ACG, TAC, and so on, would do the job, since there are $4 \times 4 \times 4$, or 64, of them—more than enough.

The intermediate steps in the "translation" may now be represented schematically:

- mRNA is synthesized in the cell nucleus by pairing with an untwisted portion of a DNA chain.

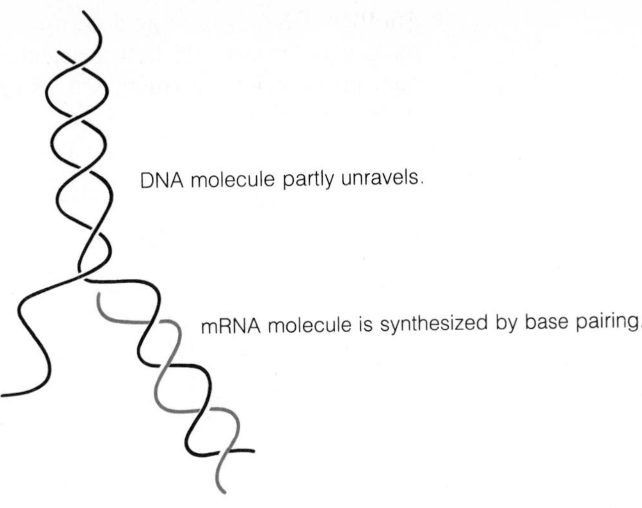
DNA molecule partly unravels.
mRNA molecule is synthesized by base pairing.

- The mRNA then migrates outside the nucleus and becomes complexed with a ribosome, Ri.

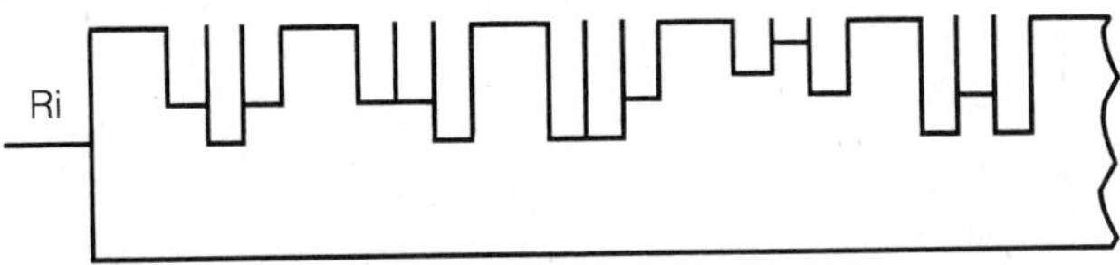
Ri

- An amino acid interacts with a molecule of ATP and becomes activated so that it can combine with a molecule of tRNA.

$$a_1 + ATP \longrightarrow a_1^* \text{ (activated amino acid)}$$

- The activated amino acid combines with a molecule of tRNA that has the proper structure to accept this particular amino acid.

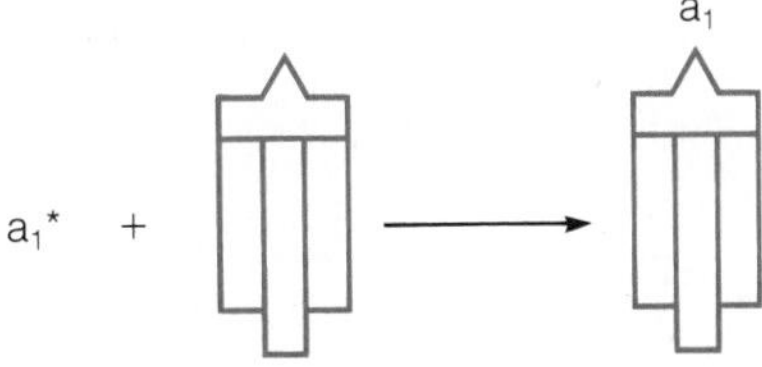
a₁* +
a₁

- The tRNA–amino acid complex then undergoes base-pairing with an appropriate section of the mRNA. Such a section comprises three nucleotides, and the set of three bases included in this section is called a **codon**. One codon corresponds to a particular amino acid.

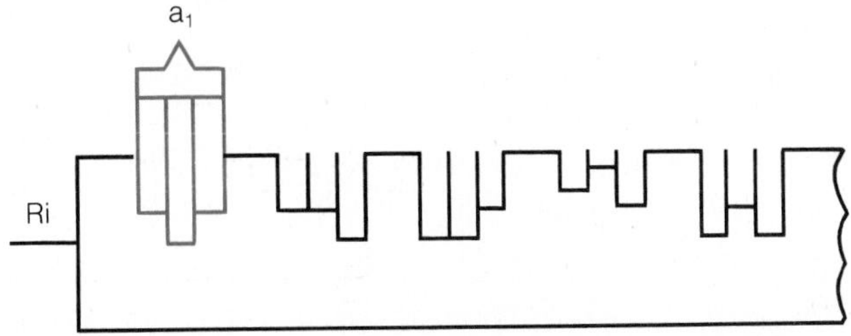
a₁
Ri

- Another tRNA–amino acid complex undergoes a similar process. The particular amino acid that is selected is determined by which codon is next in the mRNA sequence. This amino acid then forms a peptide bond with its neighbor.

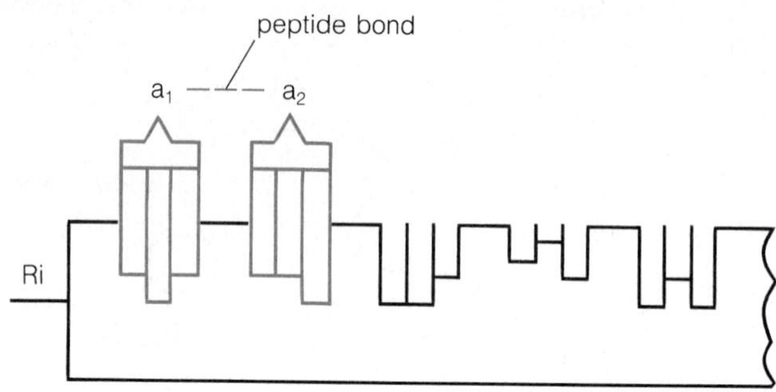

- As the process continues, the polypeptide chain continues to grow, forming a protein molecule. As the protein chain breaks loose, the tRNA molecules are released and are reused. (We can say they are "recycled.")

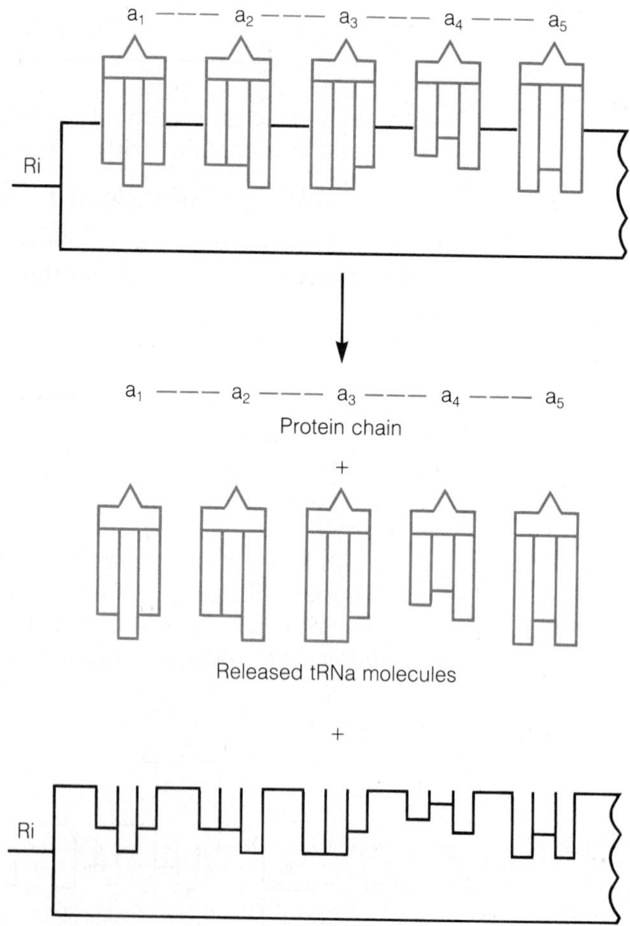

BOX 25.1
CANCER, DNA, AND NUCLEOPHILIC DISPLACEMENT

Many *carcinogens* exist in our environment. The carcinogenicity of nitrosamines was discussed in Box 24.5. In addition, most combustions produce carcinogenic products. When organic matter burns in air—whether in an open flame, industrial boiler, automobile cylinder, or cigarette—the evolving gases escape from the flame and cool off before the reaction is complete. The H atoms of the organic matter bond to oxygen more strongly than the C atoms do. As a result, the incompletely oxidized products are low in hydrogen and relatively rich in carbon. These products typically contain polycyclic aromatic hydrocarbons (section 24.7) such as benzo[*a*]pyrene, $C_{20}H_{12}$, many of which are known carcinogens. These compounds may be viewed as condensation polymers of benzene. Tar contains many such compounds.

An important question is how a carcinogen causes cancer. It is believed that one pathway involves reaction with DNA in a way that alters the genetic code that the DNA provides. The altered DNA may then trigger the formation of cancerous tumors.

DNA participates in the reaction by serving as a nucleophile, and the displacement occurs on the carcinogen molecule. One of the N or O atoms in the organic bases of DNA, with its unshared electron pair, can act as a nucleophile and attack a carbon atom of the carcinogen molecule, displacing a negatively charged atom or group. The consequent alteration of the DNA is pictured here schematically:

$$\text{N(H):} + \text{R:X} \longrightarrow \text{N}^{\oplus}\text{(H)(R)} + \text{:X}^-$$

schematic representation of DNA, showing nucleophilic site | carcinogen | altered DNA | displaced group

Fortunately, the body has some mechanisms for repairing altered DNA molecules. In some individuals, however, especially older people, the repair mechanisms function poorly. This impairment allows the altered DNA to do its damage. There is some evidence that compounds in tobacco smoke interfere with the repair mechanisms.

25.14 EPILOGUE

Your DNA molecules are in the chromosomes located in the nuclei of your cells; each of your chromosomes contains one DNA molecule. Reproduction of DNA occurs when your cells divide into two cells. All forms of life on Earth, from the smallest organisms to the largest, are composed of the same amino acids and DNA monomers. Evidence strongly supports the conclusion that the mechanisms of DNA duplication, RNA formation, and protein synthesis are essentially the same in all organisms.

The complete biochemical synthesis of a biologically active DNA of a virus was accomplished by Arthur Kornberg in 1967 with the use of two enzymes. The first complete determination of the sequence of repeating units of a single-strand DNA molecule was reported by Frederick Sanger in 1977. This molecule contained the 5375 repeating units making up the known nine genes of a virus that infects a bacterium. It is now possible to transplant the genetic code, in the form of DNA, from viruses and animal cells into bacterial cells, as well as from one bacterium to another. Great human benefits could result from planned changes in the gene stock of living organisms, a practice that is called **genetic engineering**. The hope is that hereditary diseases such as hemophilia and diabetes might be ''engineered'' out of the human burden.

SUMMARY

A molecule consisting of many repeating units is a **polymer**. When the number of repeating units is small and specified, the appropriate numerical prefix is used, as in *di*mer, *tri*mer, and so on. The molecule that forms the repeating unit is called the **monomer**. When two different monomers combine in a repeating pattern, the product is called a **copolymer**.

Information about the structure of a polymer is obtained by identifying its decomposition products. Polymers in which the repeating units (such as $—CH_2—CH_2—$) have the same composition as the monomer ($CH_2{=}CH_2$) are called **addition polymers**. Addition polymerization occurs when a reactive feature of the monomer is regenerated after each addition. The reaction is typically initiated by a free radical or an ion.

Polymers made up of monomers from which a smaller molecule such as H_2O, NH_3, or HCl has been eliminated are called **condensation polymers**. An example of the latter category is nylon, the generic term for one of a group of synthetic polyamides.

Modern synthetic methods utilize catalysts that control the linkage patterns in addition polymerization. Thus, polyethylene can be synthesized to have mostly unbranched chains, imparting greater density and strength to the product. Polypropylene synthesis can be controlled to establish specific positionings of the methyl side chains. Other structural features that influence the properties of polymers include geometrical isomerism (as in polyisoprene or polyacetylene), cross-linking between chains, and crystallinity. Polymers can also be derived from inorganic sources.

Polymers whose chains have a sequence of conjugated double bonds, such as polyacetylene, may contain some sections with unpaired electrons in *p* orbitals. Extensive delocalization of these electrons, as in the delocalization of electrons in metals, results in the formation of a filled MO bonding band and an empty MO* antibonding band. The removal of electrons from the valence band by appropriate doping results in electrical conductivity.

Polymers display a wide variety of physical properties. The controlling factors are chemical composition, molecular weight, crystallinity, extent of cross-linking, conformation, and stereoisomerism.

Carbohydrates include monosaccharides and their dimers and polymers. Monosaccharides are polyhydroxy aldehydes or ketones with at least one chiral center. The most abundant sugar is **glucose**. Three important sugar dimers (**disaccharides**) are sucrose (cane or beet sugar), lactose (milk sugar), and maltose (malt sugar). Cellulose is a polymer of glucose, indigestible by humans. Starch is also a glucose polymer, but the linkages between glucose units differ from those in cellulose in a way that makes starch digestible by humans.

Amino acids have two functional groups: $—NH_2$ and $—COOH$. The α-amino acids, in which the NH_2 group is attached to the α carbon, are biologically important. Except for glycine, $CH_2(NH_2)COOH$, the α carbon of α-amino acids is attached to four different groups and is therefore chiral. Amino acids exist largely in a dipolar ionic form, represented as $RCH(NH_3^+)COO^-$. **Peptides** are condensation dimers or polymers consisting of up to about 50 amino-acid units.

Proteins are higher-condensation polymers of amino acids with molecular weights in the range of 6000 to 50,000. They are characterized by the sequence of amino acids and by the conformations of the polypeptide chains, which are established by intramolecular H-bonding. Proteins that catalyze biochemical reactions are called **enzymes**.

Deoxyribonucleic acid (DNA) and **ribonucleic acid (RNA)** are the polymeric repositories for the storage and transfer of genetic material. Their monomers, called **nucleotides**, consist of a phosphoric acid portion, a sugar portion (deoxyribose for DNA and ribose for RNA), and a nitrogen-containing organic base. DNA consists of two chains wound around themselves in the form of a double helix. The chains are held together by H-bonding between pairs of bases, and the genetic code for the synthesis of proteins is embodied in the sequence of the bases. The process involves the synthesis of an RNA polymer by the pairing of bases with an untwisted portion of a DNA chain. The RNA then migrates outside the cell nucleus where particular sections of the chain accept specific amino acids. The linking of amino acids to form peptide bonds starts a polypeptide chain that grows into a protein. Alteration of DNA by reaction with extraneous compounds may cause cancer.

25.15 ADDITIONAL PROBLEMS

POLYMERS

25.5 Paraldehyde is a trimer. Draw the structural formula of the monomer.

(paraldehyde)

Hint: Write the molecular formula of the trimer, divide it by three to get the molecular formula of the monomer, then write the corresponding structural formula.

25.6 State which of the following properties or behaviors are typical of substances of low molecular weight, which are typical of polymers, and which can belong to either category: (a) forms a concentrated solution with no detectable freezing-point depression; (b) decomposes when heated; (c) is a solid substance that, when heated, decomposes without melting or vaporizing to form a single product of low molecular weight; (d) forms a concentrated solution that exhibits an extremely low osmotic pressure; (e) vaporizes rapidly at room temperature; (f) is a solid substance produced from a pure liquid substance in the absence of any other reactant and without heating; when the solid is then heated, it does not melt.

25.7 Explain why the addition of a Lewis acid imparts electrical conductivity to polyacetylene.

25.8 Neoprene, a synthetic rubber introduced commercially in 1931, is able to withstand ozone and oils more effectively than natural rubber. It is made from "chloroprene," $CH_2{=}CCl{-}CH{=}CH_2$, and is known to have this structural formula:

$$\cdots-\underset{H}{\overset{H}{C}}-\overset{Cl}{C}=\overset{H}{C}-\underset{H}{\overset{H}{C}}-\underset{H}{\overset{H}{C}}-\underset{CH=CH_2}{\overset{Cl}{C}}-\underset{H}{\overset{H}{C}}-\overset{Cl}{C}=\overset{H}{C}-\underset{H}{\overset{H}{C}}-\underset{H}{\overset{H}{C}}-\underset{CH=CH_2}{\overset{Cl}{C}}-\cdots$$

(a) Circle each monomeric unit in the formula. How many such units are shown? (b) Write the molecular formula of the monomeric units you have circled. Are they the same or different? Is neoprene an addition polymer or a condensation polymer?

25.9 Dimethylbutadiene

$$H_2C{=}\underset{CH_3}{C}-\underset{CH_3}{C}{=}CH_2$$

can undergo free-radical polymerization as follows:

$$R\cdot + \; H_2C{=}\underset{CH_3}{C}-\underset{CH_3}{C}{=}CH_2 \longrightarrow$$

$$\text{radical} \xrightarrow[\text{of dimethylbutadiene}]{\text{many more molecules}} \text{polymer}$$

(a) Draw the formula of the radical produced by the first addition. (b) Draw the formula of polydimethylbutadiene showing three repeating units. Identify one of the repeating units. (c) Draw the skeleton formulas (carbon atoms, single and double bond, but no hydrogen atoms) of *cis*-polydimethylbutadiene and of *trans*-polydimethylbutadiene.

25.10 The structural formula of butadiene is

$$H_2C{=}CH{-}CH{=}CH_2$$

(a) Draw the structural formulas of *cis*-polybutadiene and *trans*-polybutadiene, assuming in each case that the carbon skeleton is unbranched. Show which portion of each formula is the repeating unit. (b) Under the influence of ultraviolet radiation, an added "sensitizer" such as diphenyl disulfide,

$$C_6H_5{-}S{-}S{-}C_6H_5$$

is converted into radicals that can add to the double bonds in the polymer to form transitory radical structures that rotate freely. When the sensitizer radicals subsequently become detached, the double bonds reform. Under these conditions, irradiation of *cis*-polybutadiene converts it to a mixture of *cis* and *trans* forms. Show these transformations with the aid of structural formulas. Explain why the transitory radical derivative of polybutadiene is freely rotating.

25.11 Early attempts at polymerization of isoprene failed to produce a product with the properties of natural rubber. Write the structural formula of an isoprene polymer that might be produced by random polymerization.

25.12 Write the structural formula for a condensation polymer formed from each of these monomeric units, indicating which portion of each formula is the repeating unit.

(a) Copolymer of

$$HO-\overset{O}{\overset{\|}{C}}-C_6H_4-\overset{O}{\overset{\|}{C}}-OH \quad \text{(terephthalic acid)}$$

with $HO-CH_2-CH_2-OH$ (ethylene glycol).

(b) Copolymer of

$$H-O-C_6H_4-\underset{CH_3}{\overset{CH_3}{\overset{|}{\underset{|}{C}}}}-C_6H_4-O-H \quad \text{and} \quad Cl-\underset{O}{\underset{\|}{C}}-Cl$$

a phenol derivative — phosgene

with elimination of HCl as shown, to form **polycarbonate plastic.**

25.13 Using Lewis formulas, show the products of each of these addition reactions and identify the repeating unit of each polymer.

(a) $R\cdot + F_2C{=}S \longrightarrow \ ? \xrightarrow{\text{many more additions of } F_2C=S} \ ?$

(b) $R:^- + H_2C{=}\underset{H}{\underset{|}{C}}-C(=O)OC_2H_5 \longrightarrow \ ?$

ethyl acrylate

$\xrightarrow{\text{many more additions of ethyl acrylate}} \ ?$

acrylic plastic

25.14 (a) Write the structural formula for the addition product of the vinyl chloride radical to a molecule of vinyl chloride:

$$H-\underset{H}{\overset{H}{C}}-\underset{Cl}{\overset{H}{C}}\cdot \ + \ \underset{H}{\overset{H}{}}C{=}C\underset{Cl}{\overset{H}{}} \longrightarrow \ ?$$

vinyl chloride radical

(b) Write the structural formula for a portion of a molecule of polyvinyl chloride. Indicate the repeating unit.

25.15 Write the structural formula for the repeating unit of (a) Nylon-88 and (b) Nylon-8.

25.16 In the reactions of cellulose with HNO_3 and with CH_3COOH, the free hydroxyl groups are converted to $-ONO_2$ and $-OCOCH_3$ groups, respectively. Draw the structures of the monomeric units of cellulose nitrate and cellulose acetate.

25.17 When polyethylene is subjected to ionizing radiation from a nuclear reactor, it becomes much stronger, more difficult to melt, and less soluble. Account for these effects. (Refer to Table 25.2.)

25.18 Which of these condensation reactions can proceed to form a polyester?

(a) $HO-C(=O)-C_6H_4-C(=O)-OH + HO-CH_2-CH_2-CH_2-CH_3 \longrightarrow$

(b) $CH_3-CH_2-CH_2-C(=O)-OH + HO-C_6H_4-OH \longrightarrow$

(c) $HO-C(=O)-C_6H_4-C(=O)-OH + HO-CH_2-CH_2-CH_2-OH \longrightarrow$

(d) $HO-C(=O)-CH_2-CH_2-C(=O)-OH + HO-C_6H_4-C_6H_4-OH \longrightarrow$

(e) $C_{17}H_{35}COOH + \underset{OH}{\underset{|}{CH_2}}-\underset{OH}{\underset{|}{CH}}-\underset{OH}{\underset{|}{CH_2}} \longrightarrow$

25.19 Write the structural formula for a condensation polymer formed from each reaction. Enclose the repeating unit in brackets.

(a) Copolymer of

$$HO-C_6H_4-OH \ \text{with} \ HO-C(=O)-CH_2-C(=O)-OH \xrightarrow[\text{formation}]{\text{polyester}} \ ?$$

(b) Copolymer of

$$H_2N-C_6H_4-C_6H_4-NH_2 \ \text{with} \ Cl-\underset{O}{\underset{\|}{C}}-Cl \xrightarrow[\text{of HCl}]{\text{elimination}} \ ?$$

(c) Copolymer of

$$H_2N-C_6H_4-NH_2 \text{ with } HO-C(=O)-C_6H_4-C(=O)-OH \xrightarrow[\text{formation}]{\text{polyamide}} ?$$

25.20 Draw a section of the glycerol–phthalic acid network copolymer pictured in section 25.6, placing the respective atoms in each monomeric unit in place of the symbols P and G.

BIOCHEMICAL POLYMERS

25.21 Starch (amylose), $(C_6H_{10}O_5)_x$, is a polymer of the sugar glucose, $C_6H_{12}O_6$, linked by —O— (ether) bonds; each starch molecule contains 200 to 1000 glucose units. A mixture of amino acids added to a solution of amylose produces no detectable reaction. One molecule of the enzyme amylase, however, which contains the same amino acids, hydrolyzes 4000 —O— bonds per second. Estimate the mass in picograms of glucose produced by an amylase molecule in 1 day and the heat in joules liberated to a cell by the oxidation of the glucose:

$$C_6H_{12}O_6(aq) + 6O_2(g) \longrightarrow 6CO_2(g) + 6H_2O(\ell)$$
$$\Delta H^0 = -2870 \text{ kJ}$$

25.22 (a) Write the structural formulas for the two dipeptides formed from alanine and phenylalanine. (b) Write the structural formula of the tripeptide met-thr-tyr (refer to Table 25.4).

25.23 (a) Nylon-66 is characterized by amide linkages. Argue for *or* against the proposition that Nylon-66 is a synthetic protein. (b) Nylon-6 is an amino-acid polymer. Argue for *or* against the proposition that Nylon-6 is a synthetic protein.

25.24 Using the information in Figure 25.3, write the portion of the structural formula for insulin that includes the first four amino acids. The insulin molecule starts with glycine. Do not use abbreviations for the amino acids; write your formula completely, using a symbol for each atom.

25.25 Match each function with its structure in the synthesis of proteins:

STRUCTURE	FUNCTION
(a) DNA	(1) serves as "factory" outside the cell where proteins are synthesized
(b) tRNA	(2) stores energy and releases it under specific conditions during protein synthesis
(c) mRNA	(3) unravels to offer base sequence to synthesize mRNA
(d) ATP	(4) condenses to form polypeptide chain
(e) ribosome	(5) carries base sequence to site of protein synthesis
(f) amino acid	(6) transfers amino acids to the growing polypeptide

25.26 Look at Figure 25.9. Is this object chiral? Hold the page in front of a mirror to see if the object is identical with its mirror image.

25.27 Experiments have been carried out in which bacteria are nourished and reproduce in a nutrient medium in which all of the nitrogen is the ^{15}N isotope. It is found that 50% of the nitrogen in the DNA of the first new generation of bacteria is ^{15}N. (a) How does this evidence support the double-helix structure of DNA? (b) If the first new generation of bacteria were transferred to a nutrient medium containing only ^{14}N, what isotopic makeups would you expect to find in the DNA of the second new generation?

25.28 Draw the structural formula of (a) a DNA nucleotide containing adenine and (b) an RNA nucleotide containing uracil.

25.29 The structural formulas of ribose and deoxyribose appear in section 25.12. What is the difference between them?

25.30 Consider the sequence DNA → mRNA → tRNA → polypeptide → protein. At what juncture does the "translation" from "base language" to "amino acid language" take place?

SELF-TEST

25.31 (16 points) State whether each substance described is or is not a polymer, or that the evidence is insufficient for a decision. Defend your answers. (a) cortisone—crystalline material melting at about 222°C; used for the medical treatment of arthritis and many other diseases; slightly soluble in benzene, giving a solution whose freezing-point depression can be measured with a good thermometer. (b) Dylene (trade name)—clear, colorless substance that does not dissolve in water or alcohol but does dissolve in benzene; freezing-point depression, if any, of the benzene solution cannot be detected with thermometer; when heated, decomposes to yield an odorous product that is liquid at room temperature; analysis of the decomposition product shows it to be styrene, $C_6H_5-CH=CH_2$. (c) Plyophen (trade name)—hard, opaque substance that does not dissolve in any common solvent; can survive indefinitely in boiling water without even softening; can be synthesized by a chemical reaction between phenol, C_6H_5OH, and formaldehyde, H_2CO, in which water also is produced. (d) bergamot oil—liquid extracted from fresh rind of the fruit *Citrus bergamia*; has pleasant odor and is used in perfumes and hair oils; evaporates slowly when warmed; can be obtained from the citrus rind by distillation at low pressure in the absence of air.

25.32 (15 points) A steel tank is filled with a pure gas under pressure *P* (above 1 atm) at temperature *T*. The mass of the tank and its contents is *m*. Match the symptoms and diagnoses.

SYMPTOMS	DIAGNOSIS
(a) decreasing *T*, *m*, and *P*	(1) The tank is provided with a relief valve that opens at the pressure now showing.
(b) constant *m*; increasing *P* and *T*	(2) Someone read the gauges incorrectly.
(c) constant *m* and *T*; decreasing *P*	(3) As the temperature in the tank rises, the pressure also rises but the tank does not leak.
(d) constant *P*; increasing *T*; decreasing *m*	(4) The tank is leaking.
(e) constant *T* and *P*; increasing *m*	(5) The gas is polymerizing.

25.33 (15 points) Using Lewis formulas, show the first product of each addition reaction, and give the repeating unit of the polymer.

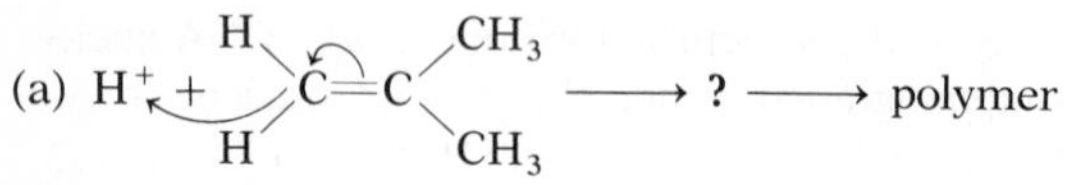

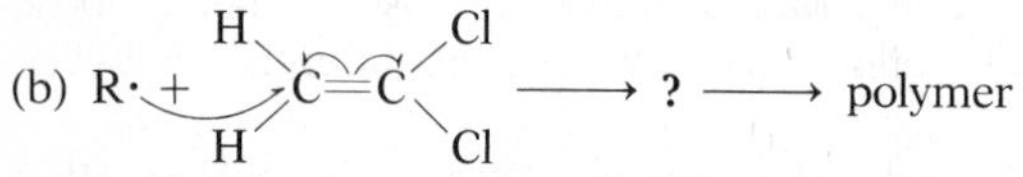

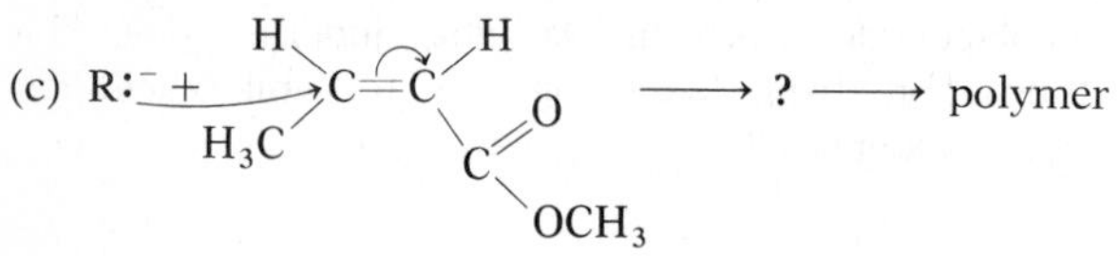

(a) $H^+ + H_2C{=}C(CH_3)_2 \longrightarrow ? \longrightarrow$ polymer

(b) $R\cdot + H_2C{=}CCl_2 \longrightarrow ? \longrightarrow$ polymer

(c) $R{:}^- + (H)(H_3C)C{=}C(H)C(=O)OCH_3 \longrightarrow ? \longrightarrow$ polymer

25.34 (15 points) For each pair of substances, state whether the indicated reaction could or could not produce a polymer. Explain your answers.

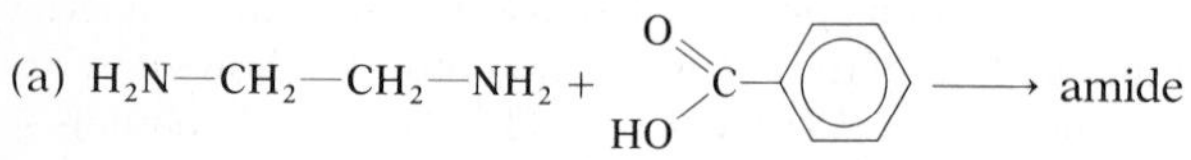

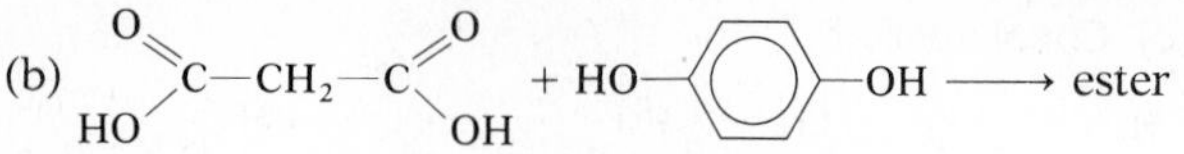

(a) $H_2N{-}CH_2{-}CH_2{-}NH_2 + C_6H_5{-}C(=O)OH \longrightarrow$ amide

(b) $HO(O=)C{-}CH_2{-}C(=O)OH + HO{-}C_6H_4{-}OH \longrightarrow$ ester

(c) $CH_3{-}CH_2{-}O^-Na^+ + Cl{-}CH_2{-}CH_2{-}Cl \longrightarrow$
ether formation by elimination of NaCl

25.35 (15 points) The structural formula of 1,3-pentadiene is $CH_2{=}CH{-}CH{=}CH{-}CH_3$. A portion of its polymer is represented as:

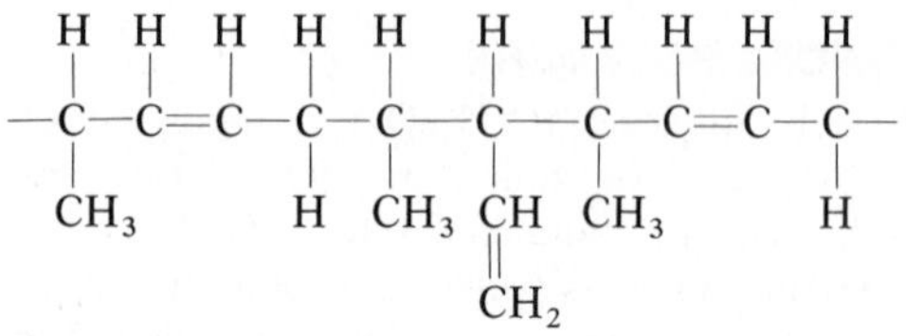

```
H  H  H  H  H   H   H  H  H  H
|  |  |  |  |   |   |  |  |  |
—C—C=C—C—C——C——C—C=C—C—
 |        |  |   |   |        |
CH3       H CH3  CH  CH3      H
                 ||
                 CH2
```

(a) Circle each monomeric (repeating) unit. (b) Write the molecular formula of the monomeric units you have circled. Write the molecular formula of 1,3-pentadiene. (c) Is the polymer an addition polymer or a condensation polymer?

25.36 (12 points) Choose the correct answer or answers (there may be more than one): (a) The primary structure of a protein is ________ (1) the alpha helix, (2) the sequence of amino acids in the chain, (3) determined by the Sanger method, (4) deduced from X-ray diffraction patterns, (5) established largely by hydrogen bonding, (6) the spatial arrangement of the polypeptide chain. (b) The secondary structure of a protein is ________ (Select the correct answer(s) from 1–6 above.)

25.37 (12 points) Complete the sentences. (a) DNA is a polymer of ________ (*phosphoric acid, amino acids, nucleotides, five-carbon sugars, organic bases*). (b) The organic bases in nucleic acids all contain C, H, and ________ (*O, N, S, P, Cl*). (c) The difference between ribose and deoxyribose is ________ (*water, C, O, H, the size of the ring*). (d) The sequence of organic bases in a DNA strand complementary to –GCTA– is ________ (*–CGAT–, –ACGT–, –GCTA–, –ACTG–, –TCGA–*). (e) The molecules eliminated in the condensations that form nucleic acids are ________ (*NH_3, O_2, H_2, HCl, H_2O, H_3PO_4, none of these*). (f) The two strands of the DNA double helix are held together by ________ (*ionic bonds involving phosphoric acid, hydrogen bonding between organic bases inside the strands, hydrogen bonding between organic bases outside the strands, London forces, covalent bonds formed in the condensation steps*).

APPENDIXES

APPENDIX A
Review of Physical Concepts

A.1 MATTER

The tendency to maintain a constant velocity is called **inertia**. Thus, unless acted on by an unbalanced force, a body at rest will remain at rest, and a body in motion will remain in motion with uniform velocity. **Matter** is anything that exhibits inertia; the quantity of matter is its **mass**.

A.2 MOTION

Motion is the change of position or location in space. Motions of objects may be classified as:

- **Translation** occurs when the center of mass of an object changes its location. *Example*: an arrow in flight.
- **Rotation** occurs when each point of a moving object moves in a circle about an axis through the center of mass. *Examples*: a spinning top, a rotating molecule.
- **Distortion** a change of shape, or the motion of the points of an object relative to one another. *Example*: a sagging metal bar.
- **Vibration** periodic distortion and recovery of original shape. *Examples*: a struck tuning fork, a vibrating molecule.

A.3 FORCE AND WEIGHT

Force is that which changes the velocity of a body; it is defined as

force = mass × acceleration

The SI unit of force is the **newton**, N, whose dimensions are kg m/sec^2. A newton is therefore the force needed to change the velocity of a mass of 1 kg by 1 m/sec in a time of 1 sec. However, an older unit, the **dyne** (defined in the table) is still in use.

The **weight** of a body is the force exerted on it by gravity, and is therefore equal to the mass of the body times the acceleration due to gravity:

$$\text{weight} = \text{mass} \times \text{acceleration due to gravity}$$

Since the Earth's gravity is not the same everywhere, the weight corresponding to a given mass is not a constant. However, at any given spot on Earth gravity is constant, and therefore weight is proportional to mass. When a balance tells us that a given sample (the "unknown") has the same weight as another sample (the "weights," as given by a scale reading or by a total of counterweights), it also tells us that the two masses are equal. The balance is therefore a valid instrument for measuring the mass of an object independently of slight variations in the force of gravity.

We can summarize force conversions as:

FROM	TO	MULTIPLY BY
dyne	newton, N	10^{-5} newton/dyne (exactly)
lb (force)	newton, N	4.4482 N/lb

A.4 PRESSURE

Pressure is force per unit area. The SI unit, called the **pascal**, Pa, is

$$1 \text{ pascal} = \frac{1 \text{ newton}}{\text{m}^2} = \frac{1 \text{ kg m/sec}^2}{\text{m}^2} = \frac{1 \text{ kg}}{\text{sec}^2 \text{ m}}$$

The International System of Units also recognizes the **bar**, which is 10^5 Pa and, as we shall see, is close to standard atmospheric pressure.

Chemists also express pressure in terms of the heights of liquid columns, especially water and mercury. This usage is not completely satisfactory, because the pressure exerted by a given column of a given liquid is not a constant, but depends on the temperature (which influences the density of the liquid) and the location (which influences gravity). Such units are therefore not part of the SI, and their use is now discouraged. However, the older units have not been removed from books and journals, and therefore chemists must still be familiar with them.

The pressure of a liquid or a gas depends only on the depth (or height), and is exerted equally in all directions. At sea level, the pressure exerted by the Earth's atmosphere will support a column of mercury about 0.76 m (76 cm, or 760 mm) high.

One **standard atmosphere** (atm) is the pressure exerted by exactly 76 cm of mercury at 0°C (density 13.5951 g/cm^3) and at standard gravity, 9.80665 m/sec^2.

One **torr** is the pressure exerted by exactly 1 mm of mercury at 0°C and standard gravity.

Let us now calculate the pressure in bars exerted by a 0.76 m column of mercury (in other words, the number of bars per standard atmosphere). Assume our column has an area of 1 m^2. Then its volume is $0.76 \text{ m} \times 1 \text{ m}^2 = 0.76 \text{ m}^3$. The density of mercury is 13.5951 g/mL, or 13.5951 kg/L, or 13.5951×10^3 kg/m^3. Therefore the mass of our column of mercury is $0.76 \text{ m}^3 \times 13.5951 \times 10^3 \text{ kg/m}^3 = 1.03323 \times 10^4$ kg. The pressure is given by

$$\frac{\text{force}}{\text{area}} = \frac{\text{mass} \times \text{acceleration of gravity}}{\text{area}}$$

TABLE A.1
PRESSURE CONVERSIONS

FROM	TO	MULTIPLY BY
atmosphere	torr	760 torr/atm (exactly)
atmosphere	lb/in^2	14.6960 $lb/(in^2\ atm)$
atmosphere	kilopascal	101.325 kPa/atm
atmosphere	bar	1.01325 bar/atm
bar	$dyne/cm^2$	10^6 $dynes/(cm^2\ bar)$ (exactly)
bar	pascal	10^5 Pa/bar (exactly)
bar	lb/in^2	14.5038 $lb/(in^2\ bar)$
mm of mercury	torr	1 torr/mm mercury (exactly)
pound (force)/in^2	pascal	6894.73 Pa in^2/lb

Since the pressure exerted by this column of mercury is, by definition, 1 atm, we have

Of course, we can choose *any* area for the column of mercury, not just 1 m^2, and get the same result. You may wish to verify this.

$$1 \text{ atm} = \frac{1.03323 \times 10^4 \text{ kg} \times 9.80665 \text{ m/sec}^2}{1 \text{ m}^2} = 1.01325 \times 10^5 \text{ Pa}$$

$$= 1.01325 \text{ bar}$$

and

$$1 \text{ bar} = 1/1.01325 \text{ atm} = 0.986923 \text{ atm}$$

Thus, the bar is the SI unit that is closest (almost 99%) to the standard atmosphere. For this reason, the International Union of Pure and Applied Chemistry (IUPAC) has recommended that the bar be adopted as the SI unit for ''standard-state pressure.''

A convenient way to calculate the pressure exerted by any uniform column of liquid is derived as follows:

$$\text{pressure} = \frac{\text{force}}{\text{area}} = \frac{\text{mass} \times \text{acceleration}}{\text{area}}$$

But the mass per unit area of a column of liquid equals the height of the column times the density of the liquid:

$$\frac{\text{mass (kg)}}{\text{area (m}^2)} = \text{height (m)} \times \text{density} \left(\frac{\text{kg}}{\text{m}^3}\right)$$

Then, substituting this product in the equation for pressure,

$$\text{pressure (Pa)} = \text{height (m)} \times \text{density} \left(\frac{\text{kg}}{\text{m}^3}\right) \times \text{acceleration} \left(\frac{\text{m}}{\text{sec}^2}\right)$$

The acceleration due to gravity at sea level is 9.81 m/sec^2. Therefore the pressure exerted by a column of liquid of a given density and height at standard gravity is:

$$\text{pressure (Pa)} = 9.81 \frac{\text{m}}{\text{sec}^2} \times \text{height (m)} \times \text{density} \frac{\text{kg}}{\text{m}^3}$$

For more convenient units, let

h = height in cm

d = density in g/cm^3 or g/mL

Then,

$$\text{pressure (Pa)} = 98.1\ hd \qquad (h \text{ in cm; } d \text{ in g/cm}^3)$$

When two columns of different liquids balance each other, the acceleration due to gravity is canceled out, and

$$h_1 d_1 = h_2 d_2$$

This relationship is the basis of devices that measure the density of an unknown liquid by noting the height of a column of that liquid needed to balance a given height of a standard liquid of known density.

A.5 ENERGY, WORK, HEAT, AND POWER

The SI unit of energy is the product of the units of force and distance, or (kg m/sec^2) × m, which is kg m^2/sec^2; this unit is called the **joule**, J. The joule is thus the work done when a force of 1 newton acts through a distance of 1 meter.

Work may also be expressed in terms of pressure and volume:

$$\text{work} = \text{pressure} \times \text{volume change}$$

Work may also be done by moving an electric charge in an electric field. When the charge being moved is 1 coulomb, and the potential difference between its initial and final positions is 1 volt, the work is 1 joule. Thus

$$1 \text{ joule} = 1 \text{ coulomb volt (CV)}$$

Another unit of electrical work that is not part of the International System of Units but still in use is the **electron volt**, eV, which is the work required to move an electron against a potential difference of 1 volt. (It is also the kinetic energy acquired by an electron when it is accelerated by a potential difference of 1 volt.) Since the charge on an electron is 1.602×10^{-19} coulomb, we have

$$1 \text{ eV} = 1.602 \times 10^{-19} \text{ CV} \times \frac{1 \text{ J}}{1 \text{ CV}} = 1.602 \times 10^{-19} \text{ J}$$

and

$$1.602 \times 10^{-19} \text{ J} \times \frac{1 \text{ kcal}}{4184 \text{ J}} = 3.829 \times 10^{-23} \text{ kcal}$$

Thus 1 eV equals 1.602×10^{-19} J, or 3.829×10^{-23} kcal. If these values are multiplied by the Avogadro number, we obtain the energy involved in moving 1 mole of electronic charges (1 faraday) in a field produced by a potential difference of 1 volt:

$$1 \frac{\text{eV}}{\text{particle}} = 1.6022 \times 10^{-19} \frac{\text{J}}{\text{particle}} \times 6.0221 \times 10^{23} \frac{\text{particles}}{\text{mol}} \times \frac{1 \text{ kJ}}{1000 \text{ J}}$$

$$= 96.49 \frac{\text{kJ}}{\text{mol}}$$

$$= 96.49 \frac{\text{kJ}}{\text{mol}} \times \frac{1 \text{ kcal}}{4.184 \text{ kJ}}$$

$$= 23.06 \frac{\text{kcal}}{\text{mol}}$$

TABLE A.2
ENERGY CONVERSIONS

FROM	TO	MULTIPLY BY
erg	joule	10^{-7} J/erg (exactly)
calorie (thermochemical)†	joule	4.184 J/cal (exactly)
kilocalorie, kcal, also called kilogram-calorie, or Calorie (capital C, used in expressing food energies for nutrition)	calorie	10^3 cal/kcal (exactly)
kilocalorie	joule	4.184×10^3 J/kcal (exactly)
liter atmosphere	joule	101.325 J/L atm
liter atmosphere	calorie	24.2173 cal/(L atm)
liter atmosphere	liter torr	760 torr/atm (exactly)
electron volt, eV	joule	1.60218×10^{-19} J/eV
electron volt	calorie	3.8293×10^{-20} cal/eV
electron volt	erg	1.60218×10^{-12} erg/eV
British thermal unit, Btu	calorie	252 cal/Btu
electron volt per particle	kJ/mole	$96.485 \frac{\text{kJ particle}}{\text{eV mol}}$
electron volt per particle	kcal/mole	$23.060 \frac{\text{kcal particle}}{\text{eV mol}}$
coulomb volt, CV	joule	1 CV/J (exactly)
kilowatt hour	kcal	860.4 kcal/(kWh)
kilowatt hour	joule	3.6×10^6 J/(kWh) (exactly)

† The "small calorie" or "gram-calorie" is the quantity of heat required to warm 1 g of water from 3.5 to 4.5°C. The "normal calorie" involves the temperature change from 14.5 to 15.5°C, and the "mean calorie" is 1/100 the heat needed to warm 1 g of water from 0 to 100°C. All of these units are nearly the same.

We may do work on an object and yet fail to convert such work into equivalent energy of motion, electricity, magnetism, radiation, or chemical or physical change. For example, we may bend an iron bar back and forth several times, or stir a liquid or a gas in a confined space, or force an electric current through a copper wire and, after all such expenditures of energy, observe that the object retains substantially its original form and position in space. We assume that this energy has been conserved, and exists as an internal energy in the form of random motion of all the elementary particles of a body. When this energy is transferred from one body to another without work being done, **heat** is transferred. The traditional unit of heat is the **calorie**, cal, or **kilocalorie**, kcal, but neither of these is part of the International System of Units.

Power is the amount of energy delivered per unit time. The SI unit is the watt, W, which is a joule per second. One kilowatt, kW, is 1000 watts. Watt hours and kilowatt hours are therefore units of energy (see the preceding table). For example,

$$10^3 \text{ W h} \times \frac{1 \text{ J}}{1 \text{ W} \times 1 \text{ sec}} \times \frac{3.6 \times 10^3 \text{ sec}}{1 \text{ h}} = 3.6 \times 10^6 \text{ J}$$

A.6 ELECTRIC CURRENT AND CHARGE

When electric current flows along parallel conductors, a magnetic force is produced between them. The unit of electric current, the **ampere**, A, is the constant current which, if maintained in each of two parallel conductors of infinite length and 1 meter apart in a vacuum, would produce between them a force of 2×10^{-7}

newton per meter of conductor length (exactly). An ampere is also a current of 1 coulomb per second.

The magnitude of the current that can pass from one point of matter to another point of matter depends on the potential difference between the two points and on the resistance imposed by the matter. The unit of resistance is the ohm, Ω. Current, resistance, and potential difference are related by **Ohm's law**:

$$\text{electric current (amperes)} = \frac{\text{potential difference (volts)}}{\text{resistance (ohms)}} \quad \text{or} \quad \text{A} = \frac{\text{V}}{\Omega}$$

Therefore, an ohm is equivalent to a volt per ampere, or V/A. The reciprocal of the resistance is called the **conductance** (unit: reciprocal ohm, or mho, Ω^{-1}).

To account for observed electrostatic interactions, it is postulated that two kinds of electric charge exist, called **positive** and **negative**. Unlike charges attract and like charges repel each other. The magnitude of electrostatic force in a vacuum is given by Coulomb's law. Using older units, the equation is:

$$F = \frac{q_1 q_2}{r^2}$$

(F is in dynes, q_1 and q_2 are in electrostatic units, and r is in cm.) Using SI units, the relationship becomes

$$F = 9.0 \times 10^9 \frac{q_1 q_2}{r^2}$$

In media other than a vacuum, the force is reduced by the factor $1/D$, where D is the dielectric constant of the medium. For gases at standard atmospheric pressure, D is very close to 1 (it is 1.0006 for air), so the electrostatic force is about the same as it is in a vacuum.

A.7 RADIOACTIVITY

Radioactivity units are summarized in this table:

UNIT	ABBREVIATION	DEFINITION AND EXPLANATION
becquerel or disintegrations per second	Bq	the number of a given isotope disintegrating every second
curie	Ci	37 billion (3.7×10^{10}) Bq
microcurie	μCi	one-millionth of a curie; 37,000 Bq
roentgen	R	unit of the energy received from a radioactive dose (One R delivers 8.4×10^{-3} J of energy to 1 kg of air.)
rad		another measure of radiation dosage, equivalent to the absorption of 0.01 J/kg of biological tissue (*rad* is acronym for radiation absorbed dose.)
gray	Gy	100 rad

Table Continued

UNIT	ABBREVIATION	DEFINITION AND EXPLANATION
rem		a measure of the effect of radiation exposure on a human being that incorporates these biological damage factors: X rays, gamma rays, and electrons (factor = 1); neutrons, protons, and alpha particles (factor = 10); and high-speed, heavy nuclei (factor = 20). The rem is then defined by the relationship: rems = rads × biological damage factor. For example, 0.01 J/kg (X rays) = 1 rad × 1 = 1 rem; 0.01 J/kg (neutrons) = 1 rad × 10 = 10 rems. *rem* is acronym for roentgen equivalent man.
sievert	Sv	100 rem

APPENDIX B Mathematical Review

B.1 EXPONENTS AND LOGARITHMS

Most computations involving exponents and logarithms are now carried out with the aid of pocket calculators. However, you still need to understand the operations involved. They are explained in this section.

In the expression x^n, x is called the base and n the exponent. The expression x^{-n} is the reciprocal of x^n:

$$x^{-n} = \frac{1}{x^n}$$

Thus,

$$4^{-1} = 1/4 \text{ and } 2^{-3/2} = \frac{1}{2^{3/2}} = \frac{1}{\sqrt{2^3}} = \frac{1}{\sqrt{8}} = \frac{1}{2\sqrt{2}} = \frac{\sqrt{2}}{4}$$

Any number (except 0) to the 0th power is 1:

$$(\tfrac{1}{2})^0 = 1^0 = 8^0 = 1000^0 = (10^{23})^0 = e^0 = x^0 = 1$$

When exponential expressions having the same base are multiplied, the exponents are added; when such expressions are divided, the exponents are subtracted:

$$10^2 \times 10^3 = 10^5 \qquad 10^6 \times 10^{-4} = 10^2 \qquad \frac{10^5}{10^2} = 10^3 \qquad \frac{10^5}{10^{11}} = 10^{-6}$$

When a power is raised to a power, the exponents are multiplied:

$(10^8)^3 = 10^{24}$ and $(10^{-1})^2 = 10^{-2}$

To add and subtract numbers in exponential notation, follow these steps:

1. Convert all numbers to the same exponent of 10.
2. Add or subtract as required.
3. Adjust to the proper number of significant figures.
4. Convert to an exponent of 10 such that the decimal point in the coefficient follows the first digit.

EXAMPLE Add the following numbers in exponential notation: 3.48×10^5, 1.23×10^6, -0.78×10^4.

ANSWER

Numbers to be added	Convert to same exponent of 10	
3.48×10^5	3.48×10^5	
1.23×10^6	12.3×10^5	step 1
-0.78×10^4	-0.078×10^5	
	15.702×10^5	step 2
	15.7×10^5	step 3
	1.57×10^6	step 4

■

A logarithm is an exponent:

$N = a^x$ (a is the base; x is the exponent.)

$\log_a N = x$ (a is the base; x is the logarithm.)

The base of logarithms used as an aid to ordinary computations is 10. Then

$10{,}000 = 10^4$ and $\log_{10} 10{,}000 = \log_{10} 10^4 = 4$

10 is generally the base, except in purely mathematical writings, including tables of integrals, which use natural logarithms exclusively.

When the base is not specified, 10 is generally understood:

$\log 0.001 = \log 10^{-3} = \log_{10} 10^{-3} = -3$

Because logarithms are exponents, logarithms of products are added and logarithms of quotients are subtracted:

$\log (a \times b) = \log a + \log b$

and

$\log \frac{a}{b} = \log a - \log b$

Logarithms that cannot be expressed as integral exponents are found in tables or by using pocket calculators. Logarithmic tables present logarithms of numbers between 1 and 10; a decimal point is assumed to follow the first digit of the number and to precede the first digit of the logarithm. Thus the table says that the logarithm of 191 is 281; this means that $\log 1.91 = 0.281$.

A number that is not between 1 and 10 may be written in proper exponential form and the logarithm obtained as follows:

$$\begin{aligned}\log 7040 &= \log (7.040 \times 10^3)\\ &= \log 7.040 + \log 10^3\\ &= 0.8476 + 3 = 3.8476\end{aligned}$$

$$\begin{aligned}\log 0.0006250 &= \log (6.250 \times 10^{-4})\\ &= \log 6.250 + \log 10^{-4}\\ &= 0.7959 + (-4) = -3.2041\end{aligned}$$

The procedure can be reversed. The number N, whose logarithm has a given value x, is the *antilogarithm* of the value:

$$\left.\begin{aligned}\log N &= x\\ 10^x &= N\\ \text{antilog } x &= N\end{aligned}\right\} \quad (x \text{ is the log; } N \text{ is the antilog.})$$

It is possible to work with the exponential forms in antilog calculations, as shown below:

$$\begin{aligned}\text{antilog } (-11.05) &= 10^{-11.05} = 10^{(0.95-12)} = 10^{0.95} \times 10^{-12}\\ &= (\text{antilog } 0.95) \times 10^{-12}\\ &= 8.9 \times 10^{-12}\end{aligned}$$

Logarithms to the base e (where $e = 2.71828\ldots$) are **natural logarithms** and are given the symbol ln. Thus:

$$\ln e^x = x$$

$$\ln e^{0.2} = 0.2$$

$$\ln 10 = 2.303$$

Therefore, the relationship of the logarithms to the two bases, e and 10, is $\ln x = 2.303 \log x$.

To solve for the value of e^x, for example, $e^{-3.0}$, we recommend this procedure:

1. Take the natural logarithm of $e^{-3.0}$; $\ln e^{-3.0} = -3.0$.

2. Convert to the base 10 by dividing by 2.303.

$$\frac{\ln (e^{-3.0})}{2.303} = \log (e^{-3.0})$$

$$\frac{-3.0}{2.3} = -1.3$$

3. Take the antilogarithm.

$$-1.3 = 0.7 - 2$$

$$\begin{aligned}\text{antilog } (-1.3) &= \text{antilog } 0.7 \times \text{antilog } (-2)\\ &= 5 \times 10^{-2}\end{aligned}$$

Therefore,

$$e^{-3.0} = 5 \times 10^{-2}$$

The notation $\exp(x)$ is often used to represent e^x.

Values of exponential functions and natural logarithms are given in tables in standard reference books. However, these tables are usually less adequate than the tables of logarithms to the base 10.

RULES FOR SIGNIFICANT FIGURES IN LOGARITHMS AND ANTILOGARITHMS

The number of *decimal places* (digits after the decimal point, including zeros) in the logarithm (x) should be equal to the number of significant figures in the number (N).

$$\log 1.21 = 0.083$$

Note that 1.21 has three significant figures, and its logarithm has 3 decimal places.

The number of significant figures in the antilogarithm is equal to the number of significant decimal places in the logarithm. For example,

$$\log 100 = \log 10^2 = 2$$

$$\text{antilog } 2 = 10^2 = 100$$

We assume that 2 and 100 are exact numbers.

$$\begin{aligned}\text{antilog } 6.3909 &= \text{antilog } (6 + 0.3909)\\ &= \text{antilog } 6 \times \text{antilog } 0.3909\\ &= 10^6 \times 10^{0.3909}\\ &= 10^6 \times 2.460 \text{ (from log table)}\\ &= 2.460 \times 10^6\end{aligned}$$

Note that only four significant figures are permitted.

$$\begin{aligned}\text{antilog } (-0.0079) &= \text{antilog } (0.9921 - 1)\\ &= \text{antilog } 0.9921 \times \text{antilog } (-1)\\ &= 9.820 \times 10^{-1}\\ &= 0.9820\end{aligned}$$

Note that four significant figures are permitted because there are four figures after the decimal point in -0.0079.

$$\begin{aligned}\text{antilog } (-9.42) &= \text{antilog } 0.58 \times \text{antilog } (-10)\\ &= 3.8 \times 10^{-10}\end{aligned}$$

Note that only two significant figures are permitted.

B.2 APPROXIMATE SOLUTIONS TO QUADRATIC EQUATIONS

A quadratic equation is one in which the highest exponent to which a variable is raised is 2. Any quadratic equation can be written as

$$ax^2 + bx + c = 0$$

The equation has two solutions, given by

$$x = \frac{-b \pm \sqrt{b^2 - 4ac}}{2a}$$

If the term in x or the term in x^2 (not both) is very small compared with the constant c, it can be dropped:

The symbol $\approx$ means "equals approximately." The symbol $\ll$ means "very much smaller than."

If $bx \ll c$, then:

$$ax^2 + c \approx 0$$

$$x^2 \approx -\frac{c}{a}$$

$$x \approx \pm\sqrt{-\frac{c}{a}}$$

If $ax^2 \ll c$, then:

$$bx + c \approx 0$$

$$x \approx -\frac{c}{b}$$

The case that frequently arises in problems of ionic equilibrium (Chapter 15) is that in which the x term is dropped. For example,

$$\frac{x^2}{0.10 - x} = 2.6 \times 10^{-4}$$

Solve by using the quadratic formula:

$$x = 5.0 \times 10^{-3}$$

Drop the x term. Then,

$$\frac{x^2}{0.10} = 2.6 \times 10^{-4}$$

$$x = 5.1 \times 10^{-3}$$

We see that the two answers nearly agree. But how can we know this without solving the equation both ways (which does not save any work)? The idea is this: Make the simplifying assumption and solve the equation. Then note whether or not the assumption is true (in this case, whether 5.1×10^{-3} is much less than 0.10; it is). If it is, then the equation $x^2/(0.10 - x) = 2.6 \times 10^{-4}$ and the equation $x^2/0.10 = 2.6 \times 10^{-4}$ are essentially the same, because $0.10 - x \approx 0.10$. Therefore, they have essentially the same solution. The answer can also be checked by the usual method of substituting it for x in the original equation to see whether an identity is obtained.

Let us now try this method in a case where it does *not* work:

$$\frac{x^2}{0.10 - x} = 2.6 \times 10^{-2}$$

We drop the x term. Then,

$$\frac{x^2}{0.10} = 2.6 \times 10^{-2}$$

$$x = 5.1 \times 10^{-2}$$

We see that 5.1×10^{-2} (=0.051) is *not* small compared to 0.10. Therefore the x term may *not* be dropped and the quadratic formula must be used. The correct solution thus obtained is

$$x = 4.0 \times 10^{-2}$$

B.3 PREVENTING ERRORS, OR HOW TO KEEP YOUR CALCULATOR HONEST

If you were to purchase a box of cereal, a quart of milk, and a dozen oranges and your cash register total came to \$3756.72, you would know that something was wrong. You would know this because you have a *sense of expected magnitude* of the total cost of your purchases. You have developed this sense through your experience with previous purchases, and by applying very simple logic—such as knowing that if you bought fewer of the same items today than you bought yesterday, your bill should be lower today.

You can develop the same sense for chemical calculations. It is easy, it is logical, and it is good insurance against computational disasters.

USING CHEMICAL COMMON SENSE

Most of the computational problems in this textbook ask you to calculate some desired quantity from one or more given quantities. Your success with these problems will be greater if you can predict whether the desired quantity will be greater or less than a given quantity, since that allows you to eliminate an entire set of possible wrong answers.

For example, let us calculate the mass of silver produced by the complete decomposition of 7.0 g of silver oxide according to the equation,

$$2Ag_2O \longrightarrow 4Ag + O_2$$

Since the silver oxide loses oxygen and does not gain anything, the final mass of the silver must be less than the original mass of the oxide. If you get an answer such as 65 g, you will know something must be wrong. In that case, you should look for your mistake and try the calculation again. The correct answer is 6.5 g.

As another example, calculate the mass of ammonia that can be produced when 13.5 g of hydrogen reacts with nitrogen according to the equation,

$$N_2 + 3H_2 \longrightarrow 2NH_3$$

Note that these "greater than" and "less than" predictions cannot be made on the basis of the differences in the coefficients of the balanced equations. Why not?

In this reaction all of the hydrogen has become NH_3, and some nitrogen has been added to it, so the answer must be greater than 13.5 g.

Another caution: If your calculations ever show a molecular weight less than 1, you have made an error. The lowest value is 1, for the H atom. The highest values you will typically meet (except in the study of polymers) will be several hundred, such as 342 for sucrose or 572 for uranium sulfide, U_2S_3. If you get a value in the thousands, check your work.

You can also make predictions for other types of calculations. Consider this problem based on the gas laws: A 20-L sample of helium gas at 1.0 atm pressure and 400 K is compressed to 2.0 atm and cooled to 389 K; calculate the final vol-

BOX B.1
A COMMON, DISASTROUS CALCULATOR ERROR

Computations that involve terms in both numerator and denominator, such as

$$\frac{A \times B \times \ldots}{C \times D \times \ldots}$$

are done rapidly in a series of calculator steps. As an example, consider this correct calculation:

$$\frac{58.2 \times 760 \times 300}{780 \times 273} = 62.3$$

How could you go wrong?

The last term in the denominator, 273, is a divisor, just as 780 is, so the last operation should be a division. A careless student, however, noting the × sign before 273, might enter a multiplication command and get 4,644,400 as the answer. Don't ever do that.

ume of the gas. Since compression and cooling both cause a decrease in volume, the answer must be a value less than 10 L; any other answer is wrong.

The preceding examples are rather simple, but the types of errors they illustrate are common. Here is an example that requires a bit more chemical common sense: Calculate the mass of carbon dioxide that is produced by the complete oxidation of 10 g of glucose. The equation is

$$C_6H_{12}O_6 + 6O_2 \longrightarrow 6CO_2 + 6H_2O$$

Notice first that all of the carbon in glucose becomes CO_2, and thus there is no difference in the mass of carbon in the two compounds. Next, whereas there is a gain of 6 oxygen atoms from the O_6 in glucose to the 12 oxygen atoms in $6CO_2$, the 12 H atoms are lost. Which represents the greater change of mass, the gain of 6 oxygen atoms or the loss of 12 H atoms? Surely you remember that the atomic weight of hydrogen is 1, and you must also know that the atomic weight of oxygen is more than 2 (it is 16). Therefore 6 oxygen atoms gained is more than 12 H atoms lost, and the answer must be a quantity greater than 10 g. (It is 15 g.)

ESTIMATING YOUR ANSWER

The previous examples showed you how to spot answers that are in error because they differ from a given quantity in the wrong direction. Now we consider how to spot errors that are too far from the correct answer, even though they may be in the right direction.

Return to the problem of the 20 L of helium gas at 1.0 atm and 400 K that was compressed to 2.0 atm and cooled to 389 K. Suppose your calculated answer is 0.9725 L. Can that be right? Consider the following (in your head, now, no calculator or pencil and paper): The change from 1.0 to 2.0 atm doubled the pressure, which means that the volume was halved, from 20 L to 10 L. Now consider the change from 400 K to 389 K, and compare it with, say, a discount in the price of a pair of shoes from $40.00 to $38.90, which you would probably regard as only a modest reduction. The change from 10 L to 0.9725 L is a much more drastic change, so 0.9725 L must be wrong. Your recheck on your calculator gives an answer of 9.725 L, which is more like it.

One final suggestion: You may not believe this, but with a little practice, and without even being a mental wizard, you can teach yourself how to solve many of these problems in your head—without a calculator, pencil and paper, or even a memorized equation. To illustrate, we return again to the gas law example: To what volume is 10 L of helium at 400 K reduced by cooling it to 389 K at constant pressure? The drop in temperature is 11 K, and 11 out of 400 is almost 3 out of 100, or 3%. If 10 L is reduced by 3%, it comes down to 9.7 L, which is the correct answer, since only 2 significant figures are allowed. Mental calculation of this sort is just a series of easy steps, requiring no great fund of memory. It can be fun to practice, and you may amaze your instructor and classmates.

B.4 GRAPHS AND PROPORTIONALITY

A straight line is represented algebraically by the equation

$$y = mx + b$$

where x and y are variables, b is the value of y when $x = 0$ ("the y intercept"), and m is the slope of the line—the change in y per unit change in x. The plot is as shown in Figure B.1.

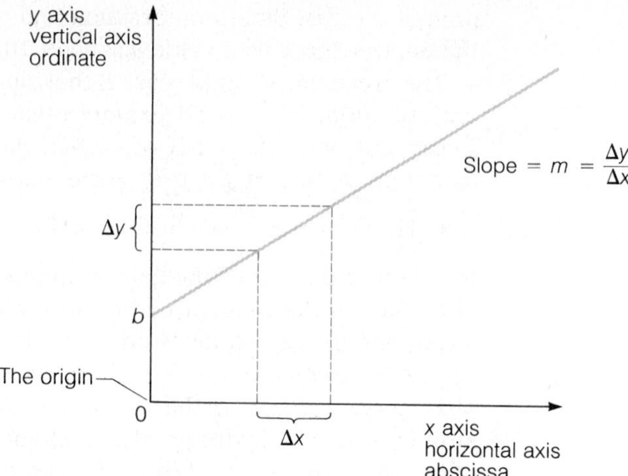

FIGURE B.1
The straight line.

A nonlinear equation can be represented by a straight line if the variables are suitably expressed. Boyle's law (Chapter 10) will serve as an illustration:

$$V = k\left(\frac{1}{P}\right)$$

If the variables are considered to be V and P, the plot is not linear. However, a plot of V versus $(1/P)$ corresponds to the equation for a straight line and will pass through the origin:

$$\begin{array}{c} y = mx + b \\ \downarrow \quad \downarrow \searrow \quad \searrow \\ V = k\left(\frac{1}{P}\right) + \text{zero} \end{array}$$

A **direct proportionality** is a linear relationship that passes through the origin, such as $x = ky$. Any change in one of the variables produces a proportionate (of equal portion, or equal percentage) change in the other. The symbol $\propto$ means "is proportional to." Then $x \propto y$ is the same as $x = ky$.

In an inverse proportionality (see Boyle's law),

$$x \propto \frac{1}{y} \quad \text{or} \quad x = \frac{k}{y}$$

When any one variable changes by a given factor, the other variable changes by the reciprocal of that factor. Thus when x is doubled, y is halved; when x is quadrupled, y is quartered.

Of course, other proportionalities are possible. According to Graham's law (Chapter 10), the rate (u) of effusion of a gas is inversely proportional to the square root of its density (d),

$$u \propto \frac{1}{\sqrt{d}} \quad \text{or} \quad u = \frac{k}{\sqrt{d}} = kd^{-1/2}$$

A physical scale that has a *natural* beginning at zero—that is, the real value of the measurement is zero when the number 0 is assigned to it—is called an

absolute scale. Therefore, measurements may have to be converted to absolute scales before proportionalities can be established. An example is the conversion of the Celsius scale to the Kelvin scale (Chapter 1).

B.5 VECTORS

A **vector** is a quantity that has both magnitude and direction; examples are velocity, displacement, and force. By contrast, a **scalar** quantity, such as mass or volume, has magnitude but no direction.

Consider a vector represented by a line from point O to point A, designated $\overrightarrow{OA}$, and another from the same origin to point B, designated $\overrightarrow{OB}$:

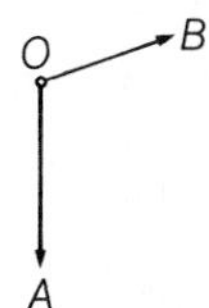

The lengths and directions of the two vectors could represent, for example, the magnitudes and directions of two different forces exerted on a body at point O, such as a force of gravity, $\overrightarrow{OA}$, and a force exerted by a magnetic field, $\overrightarrow{OB}$.

To add the two vectors, we construct a parallelogram:

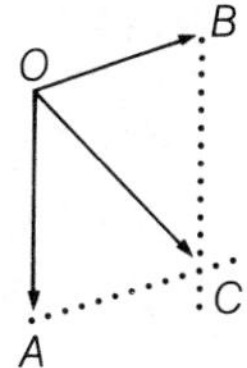

Then the vector $\overrightarrow{OC}$ drawn from O to the opposite vertex of the parallelogram is the **vector sum** (or **resultant**) of $\overrightarrow{OA} + \overrightarrow{OB}$. This sum has physical meaning because the actual object will move just as if a force were exerted in the direction $\overrightarrow{OC}$ with a magnitude represented by the length of the $\overrightarrow{OC}$ line vector.

Note that a dipole moment is a vector quantity. It is the vector sum of the individual bond moments and unshared electron pair moments of the molecule.

APPENDIX C Standard Enthalpies and Free Energies of Formation at 25°C (298.15 K)

This table gives standard enthalpy changes and standard free energy changes for reactions in which 1 mole of a substance is formed from its elements in their stable form at 25°C and at 1 atmosphere pressure.

SUBSTANCE	ΔH_f^0, kJ/mol	ΔG_f^0, kJ/mol
Ag(c)	0	0
AgCl(c)	−127.068	−109.789
AgBr(c)	−100.37	−96.90
AgI(c)	−61.84	−66.19
Ag_2O(c)	−31.05	−11.20
Al(g)	+326.4	+285.7
AlF_3(c)	−1504.1	−1425.0
Al_2O_3(c) (corundum)	−1675.7	−1582.3
As_2S_3(c)	−169.0	−168.6
BCl_3(ℓ)	−427.2	−387.4
$BaCl_2$(c)	−858.6	−810.4
$BaSO_4$(c)	−1473.2	−1362.2
BeF_2(c)	−1026.8	−979.4
$BiCl_3$(c)	−379.1	−315.0
Br_2(ℓ)	0	0
Br_2(g)	+30.907	+3.110
C(graphite)	0	0
C(diamond)	+1.895	+2.900
C(g)	+716.682	+671.257
CCl_4(ℓ)	−135.4	−65.21
C_2Cl_4(ℓ)	−52.3	+4.7
CH_4(g)	−74.81	−50.72
C_2H_2(g)	+227.73	+209.20
C_2H_4(g)	+52.26	+68.15
C_2H_6(g)	−84.68	−32.82
C_3H_8(g)	−104.9	−24.6
C_6H_6(ℓ)	+49.04	+124.5

Source: The NBS [National Bureau of Standards] Tables of Chemical Thermodynamic Properties, *Journal of Physical and Chemical Reference Data*, Volume II, Supplement 2 (1982).

SUBSTANCE	ΔH_f^0, kJ/mol	ΔG_f^0, kJ/mol	SUBSTANCE	ΔH_f^0, kJ/mol	ΔG_f^0, kJ/mol
$C_6H_6(g)$	+82.93	+129.7	$KClO_4(c)$	−432.75	−303.09
$C_2H_5OH(\ell)$	−277.69	−174.78	$KF(c)$	−567.27	−537.75
$CO(g)$	−110.525	−137.168	$MgCl_2(c)$	−641.32	−591.79
$CO_2(g)$	−393.509	−394.359	$NH_3(g)$	−46.11	−16.45
$Ca(g)$	+178.2	+144.3	$NH_4Cl(c)$	−314.43	−202.87
$CaH_2(c)$	−186.2	−147.2	$NH_4NO_3(c)$	−365.56	−183.87
$Ca(NO_3)_2(c)$	−938.39	−743.07	$NO_2(g)$	+33.18	+51.31
$CdS(c)$	−161.9	−156.5	$N_2O_4(g)$	+9.16	+97.89
$Cl_2(g)$	0	0	$Na(g)$	+107.32	+76.761
$CoCl_2(H_2O)_6(c)$	−2115.4	−1725.2	$NaCl(c)$	−411.153	−384.138
$CuSO_4(H_2O)_5(c)$	−2279.65	−1879.745	$NaClO_3(c)$	−365.774	−262.259
$Fe(CO)_5(\ell)$	−774.0	−705.3	$NiO(c)$	−239.7	−211.7
$Fe_2O_3(c)$ (hematite)	−824.2	−742.2	$O_2(g)$	0	0
$H(g)$	+217.965	+203.247	$O_3(g)$	+142.7	+163.12
$H_2(g)$	0	0	$PCl_5(g)$	−374.9	−305.0
$HBr(g)$	−36.40	−53.45	$PbCl_2(c)$	−359.41	−314.10
$HCl(g)$	−92.307	−95.299	$PbO(c)$ (yellow)	−217.32	−187.89
$HI(g)$	+26.48	+1.70	$S_8(c)$ (rhombic)	0	0
$H_2O(\ell)$	−285.830	−237.129	$SO(g)$	+6.259	−19.853
$H_2O(g)$	−241.818	−228.572	$SO_2(g)$	−296.830	−300.194
$H_2O_2(\ell)$	−187.78	−120.35	$SO_3(g)$	−395.72	−371.06
$H_2S(g)$	−20.63	−33.56	$Si(g)$	+455.6	+411.3
$H_2SO_4(\ell)$	−814.989	−690.003	$SiCl_4(\ell)$	−687.0	−619.84
$HgO(c)$ (red)	−90.83	−58.539	$SiO_2(c)$ (quartz)	−910.94	−856.64
$HgS(c)$ (black)	−53.6	−47.7	$SnCl_4(\ell)$	−511.3	−440.1
$I_2(c)$	0	0	$ZnS(c)$ (sphalerite)	−205.98	−201.29
$I_2(g)$	+62.438	+19.327	$ZnSO_4(c)$	−982.8	−871.5

APPENDIX D
Vapor Pressure of Water

TEMPERATURE, °C	PRESSURE, torr	TEMPERATURE, °C	PRESSURE, torr
0	4.6	23	21.1
5	6.5	24	22.4
10	9.2	25	23.8
11	9.8	26	25.2
12	10.5	27	26.7
13	11.2	28	28.3
14	12.0	29	30.0
15	12.8	30	31.8
16	13.6	35	42.2
17	14.5	40	55.3
18	15.5	60	149.4
19	16.5	80	355.1
20	17.5	100	760.0
21	18.7	110	1075
22	19.8		

APPENDIX E
Handy Conversion Factors

TO CONVERT FROM	TO	MULTIPLY BY
Centimeters	angstroms	10^8 Å/cm (exactly)
	nanometers	10^7 nm/cm (exactly)
	feet	0.0328 ft/cm
	inches	0.394 in/cm
	meters	0.01 m/cm (exactly)
	micrometers	1000 μm/cm (exactly)
	millimeters	10 mm/cm (exactly)
Feet	centimeters	30.48 cm/ft (exactly)
	meters	0.3048 m/ft (exactly)
	micrometers	304800 μm/ft (exactly)
Gallons (U.S., liquid)	cubic centimeters	3785 cm^3/gal
	cubic feet	0.133 ft^3/gal
	cubic inches	231 in^3/gal
	cubic meters	0.003785 m^3/gal
	cubic yards	0.004951 yd^3/gal
	liters	3.785 L/gal
Grams	kilograms	0.001 kg/g (exactly)
	micrograms	1×10^6 μg/g (exactly)
	ounces (avdp)	0.03527 oz/g
	pounds (avdp)	0.002205 lb/g
Inches	centimeters	2.54 cm/in (exactly)
	meters	0.0254 m/in
Kilograms	ounces (avdp)	35.27 oz/kg
	pounds (avdp)	2.205 lb/kg
Liters	cubic centimeters	1000 cm^3/L (exactly)
	cubic feet	0.0353 ft^3/L
	cubic inches	61.03 in^3/L
	cubic meters	0.001 m^3/L (exactly)
	gallons (U.S., liquid)	0.264 gal/L
	quarts (U.S., liquid)	1.0567 qt/L
Meters	feet	3.2808 ft/m
	inches	39.37 in/m
	miles (statute)	0.0006214 mi/m
	yards	1.0936 yd/m
Miles (statute)	centimeters	160934 cm/mi
	feet	5280 ft/mi (exactly)
	inches	63360 in/mi (exactly)
	kilometers	1.609 km/mi

(continued)

TO CONVERT FROM	TO	MULTIPLY BY
Miles (statute)	meters yards	1609 m/mi 1760 yd/mi (exactly)
Ounces (avdp)	grams pounds (avdp)	28.35 g/oz (avdp) 0.0625 lb/oz (avdp)
Pounds (avdp)	grams kilograms ounces (avdp)	453.6 g/lb 0.454 kg/lb 16 oz/lb (exactly)
Ounces (troy)†	grams ounces (avdp)	31.1 g/oz (troy) 1.097 oz (troy)/oz (avdp)
Pounds (troy)	grams ounces (troy)	373 g/lb (troy) 12 oz (troy)/lb (troy) (exactly)
Tonnes (metric tons)	kilograms pounds (avdp) long tons (2240 lb) short tons (2000 lb)	10^3 kg/tonne (exactly) 2204.6 lb/tonne 0.9842 long ton/tonne 1.1025 short ton/tonne

† The price of gold and other precious metals is commonly quoted in troy weight. Note that whereas a troy ounce (31.1 g) is heavier than an avdp ounce (28.35 g), a troy pound (12 troy oz) is lighter than an avdp pound (16 avdp oz).

APPENDIX F Abbreviations

ampere, A
angstrom, Å
atmosphere, atm
calorie, cal
coulomb, C
concentration, *c*
cubic centimeter, cm^3
cubic foot, ft^3
degree Celsius (centigrade), °C
degree Fahrenheit, °F
disintegrations per second, dps
electron volt, eV
energy, *E*
enthalpy, *H*
entropy, *S*
foot, ft
free energy, *G*
gram, g
hertz, Hz
hour, h
inch, in
joule, J
kelvin, K
kilocalorie, kcal or C
kilogram, kg
liter, L
minute, min
mole, mol
molal (mol/kg solvent), *m*
molar (mol/L), M
newton, N
ohm, Ω
parts per million, ppm
pascal, Pa
pound, lb
second, s or sec
volt, V
watt, W
year, yr

GLOSSARY

absolute zero 0 K or $-273.15°C$. No substance can be cooled below this temperature.

accidental property A property that depends on the size or shape of a specific sample. See also *extensive property, intensive property.*

accuracy The agreement of a measurement with the accepted value of the quantity.

acid (1) (Arrhenius) A substance that ionizes in water, forming H^+ ions. (2) (Brönsted) A molecule or ion that donates a proton. (3) (Lewis) A molecule or ion that accepts a pair of electrons in forming a coordinate covalent bond.

acid anhydride A nonmetallic oxide, such as SO_3, that can react with water to form an acid.

acid-base indicator *Indicator* (definition 1).

acid(ic) solution A solution in which $[H^+] > [OH^-]$.

acid rain Rain rich in H_2SO_4 with some HNO_3.

actinoids The inner transition elements of period 7, from actinium (89) to nobelium (102). Also, "actinides."

activation (1) Preparation of an adsorbent by heating and other processing to remove volatile components and leave a network of pores. (2) Any process in which a chemical species becomes more reactive.

activity (1) A property of a solute that becomes equal to its concentration in a very dilute solution and that, when substituted for concentration, makes the law of chemical equilibrium exact. (2) The number of disintegrations of a radioisotope per unit time per unit mass.

activity coefficient The ratio of *activity* (definition 1) to concentration.

activity series A list of metals in the order of decreasing tendency to be oxidized.

addition reaction A reaction in which the net effect is breaking up a reactant into two parts that bond to multiply bonded atoms.

adsorbent A solid capable of collecting on its surface a large quantity of another substance.

alcohol A compound with an OH (hydroxyl) group attached to an alkyl group; general formula is ROH.

aldehyde A compound with an H atom attached to a carbonyl group; general formula is

$$\begin{array}{c} RC{=}O \\ | \\ H \end{array}$$

alkali A strong base.

alkali metals The elements of Group 1 from lithium to francium.

alkaline earth metals The elements of Group 2 from magnesium to radium.

alkane A hydrocarbon with singly bonded C atoms in chains; the general formula is C_nH_{2n+2}.

alkene A hydrocarbon with a C=C bond; the general formula is C_nH_{2n}.

alkyl group A group of C and H atoms, represented by R, remaining after the removal of an H atom from a nonaromatic hydrocarbon.

alkyl halide A compound, RX, having a halogen atom (F, Cl, Br, or I) bonded to an alkyl group.

alkyne A hydrocarbon with a C≡C bond; the general formula is C_nH_{2n-2}.

allotropes Forms of an element with the same physical state but different properties.

alloy A metallic substance or mixture composed of more than one element.

alpha (α) particle The nucleus of a helium atom, He^{2+}, emitted at high speed from the nuclei of certain radioactive atoms.

amide The product from the loss of a molecule of H_2O between an acid and an amine; the general formula is

$$\begin{array}{c} O \\ \| \\ RCNH_2 \end{array}$$

. (The H's can be replaced by R's.)

amine An organic base having the general formula RNH_2 (primary), R_2NH (secondary), or R_3N (tertiary). (The R groups may be different.)

ampere A unit of current: the current that, when flowing in each of two straight parallel conductors of infinite length 1 meter apart in a vacuum, produces between the conductors a magnetic force of 2×10^{-7} newton per meter of length, 1 coulomb/sec. Symbol, A.

amphoteric Able to function both as an acid and as a base.

amu Atomic mass unit.

anode (1) The electrode at which oxidation occurs. (2) The positive electrode in a vacuum tube.

anodic protection Maintenance of passivity by coating with an oxidizing agent or applying a positive potential.

antibonding molecular orbital A molecular orbital, higher in energy than the atomic orbitals from which it is derived, whose electrons weaken bonds and induce instability in molecules and ions. Symbol, MO*. (The asterisk denotes antibonding.)

AO Atomic orbital.

Arrhenius equation See *energy of activation.*

asymmetric carbon atom Chiral carbon atom.

asymmetry Chirality.

atmosphere A unit of pressure: 1 atm = 101,325 pascal; the pressure exerted by a column of mercury 760 mm in height at 0°C and standard gravity, 9.80665 m/sec^2. Also, "standard atmosphere."

atom The smallest unit of an element.

atomic mass unit A mass unit defined as exactly 1/12 of the mass of a ^{12}C atom. Also, "amu," "dalton."

atomic number The number of protons in the nucleus of an atom; also, "proton number."

atomic orbital An orbital in a specific atom.

atomic theory Elements are composed of characteristic atoms and a chemical change involves the combination, separation, or rearrangement of atoms.

atomic weight The mass of an atom relative to the mass of a ^{12}C atom.

aufbau principle The rules for writing the electron configuration of the ground state of an atom by filling orbitals in order of increasing energy.

autoradiograph A photograph produced by radiation from an object itself.

average bond energy ΔH for breaking all bonds of a given kind in a molecule in the gaseous state, divided by the number of bonds broken; data are given in ΔH per mole of bonds. Also, "bond energy." See also *bond dissociation energy.*

average deviation The average of the absolute differences between the measurements in a series and the average of the measurements.

Avogadro's law Equal volumes of all gases at the same temperature and pressure have the same number of molecules.

Avogadro's number The number of atoms in exactly 12 g of the pure isotope ^{12}C: $N_A = 6.02214 \times 10^{23}$ particles/mol.

band A sequence of MO's and MO*'s of closely spaced energy levels resulting from the combination of AO's (with the same principal quantum number) of very large numbers of metallic atoms.

band gap The difference in energy between an empty or partially filled higher-energy band and a filled (nonconducting) lower-energy band.

band theory An application of the MO method to account for the bonding in metals.

barometer An instrument for measuring atmospheric pressure.

base (1) (Arrhenius) A substance that ionizes in water, forming OH^- ions. (2) (Brönsted) A molecule or ion that accepts a proton. (3) (Lewis) A molecule or ion that donates a pair of electrons in forming a coordinate covalent bond.

basic anhydride A metallic oxide, such as CaO, that can react with water to form a basic hydroxide.

basic-site atom The donor atom with the unshared pair of electrons in a base.

basic solution A solution in which $[OH^-] > [H^+]$.

battery (1) A set of galvanic cells connected in series (usually) or parallel. (2) A galvanic cell.

beta particle An electron, ${}_{-1}^{0}\beta$, emitted from an unstable nucleus.

bimolecular Having a molecularity of 2.

binary acid An acid consisting of hydrogen and one other element.

binary compound A compound consisting of two elements.

body-centered cubic lattice A cubic crystal lattice in which atoms are at the corners and body centers of unit cells.

boiling point The temperature at which the liquid and gaseous phases of a substance are in equilibrium at a given total pressure. The *normal boiling point* is the boiling point at standard atmospheric pressure.

boiling-point elevation The increase in the boiling point of a solvent when a solute is added.

bond angle The angle between two covalent bonds with an atom in common.

bond axis A straight line joining the nuclei of bonded atoms.

bond dissociation energy ΔH for breaking one bond in a molecule in the gaseous state into products in the gaseous state; data are given in ΔH per mole of bonds. Also, "bond energy." See also *average bond energy.*

bond molecular orbital A molecular orbital, lower in energy than the atomic orbitals from which it is derived, whose electrons strengthen bonds and induce stability in molecules and ions. Symbol, MO.

bond order One-half the difference of the number of electrons in MO's and MO*'s.

bond stability A concept quantified by the energy needed to break a bond. The larger the energy, the greater is the bond stability.

Boyle's law The pressure times the volume of a fixed mass of gas at constant temperature is constant.

branching-chain reaction A reaction in which two or more chain carriers are produced for each one consumed.

breeder reactor A nuclear fission reactor that produces more nuclear fuel than is consumed.

Brönsted acid, Brönsted base See *acid, base.*

Brownian motion The quivering motion of very small particles in a liquid caused by their collisions with molecules of the liquid.

buffer solution A solution containing a conjugate acid-base pair, having a pH that changes little on addition of moderate amounts of strong acids or bases.

buret(te) A long tube with volume markings and a stopcock, used in titration.

calorie A unit of energy: 1 cal = 4.184 J exactly. When spelled with a capital C in tables of food energies, 1 Cal = 1 kcal = 4.184 kJ.

carbohydrate An aldehyde or ketone containing a number of OH groups, with H and O atoms in the ratio 2:1.

carbonyl group $-\overset{|}{C}=O$

carboxyl group $-\overset{\overset{O}{\|}}{C}-OH$; compact form, —COOH.

carboxylic acid An organic acid containing the carboxyl group; the general formula is RCOOH.

catalysis The effect of a catalyst in increasing the rate of a reaction.

catalyst A substance that increases the rate of a reaction without being consumed.

cathode (1) The electrode at which reduction occurs in an electrochemical cell. (2) The negative electrode in a vacuum tube.

cathode ray An electron beam ejected from the cathode of a high-voltage vacuum tube.

cathodic protection Prevention of corrosion by connecting a metal to a more active metal or to the negative side of a power source.

Celsius scale The temperature scale defined as 273.15 degrees below the Kelvin scale. The common practical reference points are 0°C for the freezing point of water and 100°C for the normal boiling point of water. Older name: centigrade scale.

centigrade scale Celsius scale.

central atom An atom that is covalently bonded to more than one other atom.

chain carrier A product of a step in a chain reaction that reacts in another step.

chain reaction A self-sustaining reaction. In a two-step sequence, a product of step 1 is a reactant in step 2, and a product of step 2 is a reactant in step 1.

chelating agent A ligand that forms more than one bond with the same metal atom or ion.

chemical equation A description of a chemical reaction showing the formulas of reactants and products with coefficients representing relative numbers of moles.

chemical equilibrium, law of A mixture in equilibrium has a composition that satisfies the equilibrium condition(s) for the reaction(s) that have come to equilibrium.

chiral carbon atom A carbon atom singly bonded to four different atoms or groups.

chirality The property of not being identical with its mirror image.

***cis* isomer** A geometric isomer in which like atoms, groups, or ligands occupy the same side of an approximately rectangular array. When bonding is to C=C, the like groups are on different C atoms.

colligative property A property of a solution that is proportional to the mole fraction, molality, or molarity of the solute, with a proportionality constant characteristic of the solvent.

collision theory A theory that assumes that molecules must collide and have a definite minimum amount of energy for reaction to occur.

colloid (1) Colloidal dispersion. (2) The dispersed phase in a colloidal dispersion.

colloidal dispersion A mixture containing particles larger than ordinary molecules, but small enough to remain dispersed for a long time and too small to be seen clearly with a microscope using visible light.

combination reaction A reaction in which two substances combine to form another substance.

common ion effect Increasing the concentration of an ion decreases the ionization or solubility of an electrolyte that produces this ion.

complex A coordination ion or a molecule with a similar structure.

complex ion Coordination ion.

concentrated solution A solution containing a large quantity of solute(s) relative to solvent.

concentration (1) Molarity. (2) The number of moles of a gas per liter. (3) Quantity (in any units) per unit volume (in any units).

condensation Transformation of gas to liquid.

condensation temperature The temperature at which a vapor at a specified partial pressure is in equilibrium with the liquid.

conduction band A partially filled band in a metal or, in a semiconductor, a band with electrons excited from a filled lower-energy band or from impurity levels.

conformations Different nonisomeric spatial arrangements resulting from rotation about C—C or other single bonds, as the eclipsed and staggered structures of chain compounds, or from puckering of cyclic compounds such as cyclohexane.

conjugate acid The substance or ion formed when a base accepts a proton.

conjugate base The substance or ion formed when an acid loses a proton.

conservation of energy, law of The energy of an isolated system is constant.

conservation of matter, law of The mass remains constant during a chemical change.

conservation of orbitals, principle of The number of molecular orbitals formed must equal the number of AO's combined.

continuous spectrum See *spectrum.*

contributing structures See *resonance.*

conversion factor (1) A multiplier that converts a value expressed in one unit or set of units to the same value expressed in another unit or set of units. (2) Any multiplier that converts one quantity to another.

cooling curve The time vs. temperature graph plotted when a substance is cooled and passes through one or more changes of state.

coordination compound A compound containing a coordination ion or a molecule with similar structure.

coordination ion An ion consisting of a metal atom or ion covalently bonded to anions or molecules (ligands).

coordination number (1) The number of nearest neighboring atoms to an atom in a given crystalline, molecular, or ionic structure. (2) The number of bonds between the central atom and the ligands in a complex.

corrosion An undesired process in which a metal is converted to a compound.

coulomb A unit of electric charge: the charge passed when 1 ampere flows for 1 second. Symbol, C.

covalence (1) The number of covalent bonds that an atom usually forms (when it has no charge). (2) The number of covalent bonds that an atom forms in a given compound.

covalent bond A bond formed when atoms share electrons.

covalent molecule A molecule with covalent bonds.

critical mass The minimum mass of fissionable isotope required to maintain a branching-chain reaction.

cross-linking Bonding between the chains of atoms in a polymer.

crystal field concept A concept that stresses the electrostatic nature of bonding in coordination compounds by assuming that the approaching ligands split the energies of the *d* orbitals in the same way that they would be split by the neighboring ions in an ionic crystal.

crystal lattice The repeating pattern that represents the spatial arrangement of atoms, ions, or molecules in a crystalline solid.

crystalline solid A solid whose component atoms, molecules, or ions are arranged in an orderly pattern.

crystallization The process of becoming crystalline.

cubic crystal lattice A crystal lattice in which the unit cell is a cube.

cyclic compound A compound in which bonded atoms form at least one ring.

cyclic process A process in which the initial and final states of the system are the same.

dalton Atomic mass unit.

Dalton's law of partial pressures The total pressure of a mixture of gases is the sum of the partial pressures.

decomposition potential The applied difference of potential needed to cause a reaction in an electrolytic cell.

decomposition reaction The breakdown of a substance into two or more substances.

definite composition, law of The mass composition of a pure substance is always the same.

deionization Removal of ions from water by passage through both a cation exchanger containing mobile H^+ ions and an anion exchanger containing mobile OH^- ions.

delocalized π system An electronic configuration with π bonding extending over more than two atoms; an alternative description of bonding in resonance hybrids.

delta (δ) bond A bond formed by side-to-side overlap of all four lobes of each of two *d* orbitals.

density The mass per unit volume of a material.

detergent A substance used for cleansing, with a long-chain hydrocarbon group and a terminal ionic or polar group.

diamagnetic Repelled from a magnetic field.

dielectric constant The ratio of the force (F_0) between charged bodies in a vacuum to the force (F) between them when immersed in the substance: $D = F_0/F$.

diffusion The dispersal of a gas in space or of one component in a solution by the random motion of its molecules.

dilute solution A solution containing a small quantity of solute(s) relative to solvent.

dipole A separation of partial positive and negative charges in an individual bond or in a molecule.

dipole moment The product of either charge in a dipole and the distance between them; a measure of polarity.

dispersion medium The phase in which colloidal particles are dispersed.

displacement reaction A reaction in which one element displaces another from a compound.

disproportionation Self-oxidation-reduction.

dissociation constant (1) The equilibrium constant for the process in which a coordination ion dissociates into the metal atom or ion and the ligands. (2) Ionization constant.

distillation The process of vaporizing a liquid and condensing the vapor back to liquid, commonly used for purification.

double bond The sharing of two pairs of electrons.

double displacement A reaction in which cations and anions exchange partners.

dry cell A galvanic cell in which the electrolyte is not a free-flowing liquid.

effusion The flow of gas through a narrow opening.

electrode (1) A metallic or other electronic conductor that gives electrons to or receives electrons from an ionic conductor or a low-pressure gas. (2) Half-cell.

electrode potential Half-cell potential.

electrolysis A reaction caused by passage of an electric current.

electrolyte A substance that conducts an electric current by movement of ions.

electrolytic cell An apparatus in which the passage of an electric current causes a nonspontaneous reaction to occur.

electromagnetic radiation *Radiation* (definition 1).

electromotive force (1) The difference of potential

between the electrodes in a galvanic or electrolytic cell. Symbol, $\mathscr{E}$. (2) Reversible electromotive force.

electron The unit particle of electricity carrying a negative charge of 1.602177×10^{-19} C.

electron capture (1) Radioactive decay by capture of an electron by a nucleus. (2) Addition of an electron to an atom, molecule, or ion.

electron cloud density diagram A diagram showing the probability of finding an electron in a volume of space at a point in an orbital about the nucleus of an atom or the nuclei of a molecule.

electron configuration The arrangement of electrons in the orbitals of an atom or a molecule.

electronegativity The net tendency of a covalently bonded atom to attract bonded pairs of electrons.

electronic unit The charge of 1 electron, without sign.

electron-sea model A model that assumes a metal has a crystal lattice of metal cations with free-moving electrons.

electrophile Lewis acid. See *acid.*

element A substance consisting of atoms, all of which have the same atomic number.

elimination reaction The removal of atoms or groups from adjacent atoms to form a small molecule (e.g., H_2O), leaving a multiple bond.

emf Electromotive force.

emission spectrum See *spectrum.*

empirical formula The formula that gives the smallest whole-number ratios of atoms in a substance. Also, "simplest formula."

enantiomers Isomers that are nonsuperimposable mirror images.

endothermic reaction A reaction that absorbs heat from the surroundings. ΔH for the reaction is positive.

end point The stage of a titration at which the signal that the reaction is complete appears.

energy A property whose change is equal to the work done on the system when it is perfectly insulated; a measure of the ability to do work or transfer heat. Symbol, E.

energy levels The allowed energy states of an atom or molecule.

energy of activation The minimum energy E_a that molecules must have for reaction to occur, calculated from the Arrhenius equation which relates the rate constant to temperature: $\log k = \log A - E_a/2.303RT$.

enthalpy A property equal to the sum of the energy and the product of pressure times volume: $H = E + PV$. Its change (ΔH) is the heat absorbed when the pressure is constant and only pressure-volume work is done.

enthalpy of activation The difference between the enthalpies of the transition state and the reactants, $\Delta H^{\ddagger} = H^0_{ts} - H^0_r$, the value of which does not differ greatly from the energy of activation.

entropy A property (symbol, S) whose change, in a process at constant temperature, is the maximum possible heat absorbed by the system divided by the absolute temperature: $\Delta S = q_{max}/T$. It is a quantitative measure of randomness, disorder, or probability.

entropy of activation The difference between the entropies of the transition state and the reactants, $\Delta S^{\ddagger} = S^0_{ts} - S^0_r$. A greater $\Delta S^{\ddagger}$ is associated with a greater number of orientations of reacting molecules that lead to products.

equilibrium (1) A condition in which two opposing processes occur at the same rate, resulting in no net change. (2) A mixture that is in equilibrium. (3) The composition of a mixture in equilibrium.

equilibrium condition An equation that is satisfied by the concentrations or partial pressures of the reactants and products when a reaction is at equilibrium. For the reaction $aA + bB + \cdots \rightleftharpoons cC + dD + \cdots$, the equilibrium condition is $[C]^c[D]^d \cdots /[A]^a[B]^b \cdots = K_c$ or $p^c_C p^d_D \cdots /p^a_A p^b_B \cdots = K_p$.

equilibrium constant The constant K_c or K_p appearing in an equilibrium condition.

equivalence point The stage of a titration at which the reactants have been mixed in exactly the molar ratio shown in the chemical equation.

equivalent The quantity of a substance or ion that accepts, gives up, or combines with 1 faraday of charge in the form of protons, electrons, or other ions.

equivalent weight The mass in grams of 1 equivalent.

error The determined value of a quantity minus the accepted value.

ester The product derived from the loss of a molecule of H_2O between an acid (usually a carboxylic acid) and an alcohol; the general formula is $\mathrm{R\overset{\displaystyle O}{\underset{}{C}}OR'}$ (RCOR′ with =O on the C).

evaporation Vaporization.

excited state See *ground state.*

exothermic reaction A reaction that emits heat to the surroundings. ΔH for the reaction is negative.

extensive property A property that is proportional to the mass of a sample of matter. See also *accidental property.*

face-centered cubic lattice A cubic crystal lattice in which atoms are at the corners and faces of the unit cells.

Fahrenheit scale The temperature scale in the British system defined by the practical reference points of 32°F for the freezing point of water and 212°F for the normal boiling point of water.

faraday The charge (without sign) of 1 mole of electrons: 9.64853×10^4 C.

Faraday's constant The conversion factor between coulombs and faradays: 9.64853×10^4 C/faraday. Symbol, $\mathscr{F}$.

Faraday's laws (1) The mass of a substance produced or consumed in electrolysis is proportional to the charge passed through the circuit. (2) The number of faradays that pass through a circuit when 1 mole of a substance is produced or consumed is a whole number.

fat A solid ester of glycerol and three molecules of fatty acids.

fatty acid A long-chain carboxylic acid.

ferromagnetic substance A substance that is strongly attracted into a magnetic field and may remain magnetized when removed from the field.

first law of thermodynamics The change in the energy of a closed system is the sum of the heat absorbed by the system and the work done on the system: $\Delta E = q + w$.

fission The splitting of a heavy nucleus into two lighter nuclei of comparable mass.

flocculation Removal of impurities from polluted water by collection on hydrophilic colloids.

flotation process A method for purifying ores that relies on the fact that the mineral is "wetted" by the oil in an oil-water system, while the impurities are not.

flux A compound (e.g., CaO) added during a metallurgical process to combine with impurities, forming low-melting products.

formal charge A charge assigned to an atom by a convention that assumes equal sharing of bonding electrons.

formula A representation of the composition of a substance with element symbols and with subscripts that give the numbers of atoms of the constituent elements.

formula weight The sum of the atomic weights of all atoms in a formula.

free energy A property (symbol, G) whose change is the minimum work, excluding pressure-volume work, done on the system in a process between specified states at constant temperature and pressure: $\Delta G = w_{\min} - w_{PV,\min} = \Delta H - T\Delta S$. ΔG is negative for a spontaneous process at constant temperature and pressure.

free-energy change, standard The change in free energy when each reactant or product is in its standard state. Symbol, ΔG^0.

free energy of formation The standard free-energy change when 1 mole of a substance is formed from the elements in their stable forms. Symbol, ΔG_f^0.

free radical A species with at least one unpaired electron. Also called "radical."

freezing point The same temperature as the melting point, but approached experimentally by cooling the liquid.

freezing-point depression The decrease in the freezing point of a solvent when a solute is added.

frequency The number of complete waves that pass a given point per second.

fuel cell A galvanic cell in which the reactants are supplied and the products are removed continuously.

functional group A site of reactivity, other than an H atom, in a substituted or unsaturated hydrocarbon (e.g., Cl, OH, a multiple bond).

fusion (1) Melting; sometimes used in the sense of uniting by being melted together. (2) Combination of light nuclei to form a heavier nucleus.

galvanic cell An apparatus in which a spontaneous reaction produces electrical work.

gamma radiation High-energy radiation, ${}_0^0\gamma$, emitted from nuclei.

gangue The impurities in an ore, mostly SiO_2.

Gay-Lussac's law of combining volumes When gases react, the volumes consumed and produced, measured at the same temperature and pressure, are in ratios of small whole numbers.

geometric isomers Isomers in which the same atoms and bonds exist but in which there are different arrangements in an approximately rectangular array, as

```
   W  X          X
   |  |          |
in C==C  or  W---M---Y,  where the latter represents a
   |  |          |
   Y  Z          Z
```

square-planar complex or part of an octahedral complex.

Gibbs energy, Gibbs free energy, Gibbs function Free energy.

glass A state of matter characterized by rigidity but whose component atoms or molecules are disordered.

glass electrode A half-cell including a glass membrane, with a potential dependent on the concentration of a specific ion, usually H^+.

Graham's law The rates of effusion of gases are inversely proportional to the square roots of their densities.

gravimetric analysis Quantitative analysis in which a substance is isolated from a sample and weighed in a pure form.

ground state The lowest energy level or the most stable state of an atom or a molecule. All higher energy levels are *excited states*.

group One of the vertical columns in the periodic table, showing elements with similar electronic configurations in their valence shells.

half-cell One of two parts into which a galvanic cell is conceptually divided.

half-cell potential A potential associated with a half-cell such that the difference between two of them is the emf of a cell consisting of those half-cells.

half-life The time required to decrease an initial reactant concentration by one-half.

half-reaction Partial equation.

halogen One of the elements in Group 7 of the periodic table, from fluorine to astatine.

hard water Water containing salts that form precipitates with soap.

heat Energy transferred to or from a system by random motion of atoms, molecules, or electrons, when two systems at different temperatures are in contact, considered positive for the system whose energy is increased. Symbol, q.

heat capacity The quantity of heat needed to raise the temperature of an object by 1°C, expressed in J/°C.

heat of combustion The heat emitted when a unit quantity of a substance burns at constant pressure with O_2 to form CO_2, H_2O, SO_2, N_2, Cl_2, etc., and the initial temperature is restored.

heat of fusion The amount of heat required to melt a unit quantity of solid at a given temperature (usually its melting point).

heat of sublimation The amount of heat required to sublime a unit quantity of solid at a given temperature.

heat of vaporization The amount of heat required to vaporize a unit quantity of liquid at a given temperature (usually its normal boiling point).

Henderson–Hasselbalch equation A form of the law of chemical equilibrium for the ionization of a weak acid or base: $pH = pK_a + \log([base]/[acid])$.

Henry's law The molarity of a gaseous solute in a saturated solution is approximately proportional to the pressure of the gas: $[B] \approx k_B P_B$.

hertz The SI unit of frequency, sec^{-1}. Symbol, Hz.

Hess's law The enthalpy change of a reaction is the same whether carried out in one step or in several steps at constant pressure and temperature.

heterogeneous Having non-uniform properties within a sample.

heterogeneous equilibrium A chemical equilibrium in which the reactants and products are in two or more phases.

heterogeneous reaction A reaction involving more than one phase.

high-spin complex A complex (weak-field) that may be more paramagnetic than the corresponding strong-field complex because it has more unpaired electrons.

HO Hybrid orbital.

HOMO The highest (in energy) molecular orbital occupied by at least one electron.

homogeneous Having uniform properties throughout the sample.

homogeneous equilibrium A chemical equilibrium in which all the reactants and products are in one phase.

homogeneous reaction A reaction that occurs in one phase.

homologous series Sets of compounds with the same general formula (e.g., alkanes, C_nH_{2n+2}).

Hund's rule Electrons are added singly to a subshell until each orbital has one electron before pairing them.

hybridization A mathematical blending of a number of different-energy AO's in a single atom to give a new set of the same number of equal-energy orbitals.

hybrid orbital An atomic orbital formed as the result of hybridization. Symbol, HO.

hybrid orbital number The number of HO's needed by an atom, which is the sum of the number of σ bonds and unshared pairs of electrons on the atom. Symbol, HON.

hydrate A crystal containing water molecules in definite proportion.

hydration Solvation when the solvent is water.

hydration energy The enthalpy change when 1 mole of gaseous solute is dissolved in water. For an ionic solute, the initial state is gaseous ions.

hydrocarbon A compound of carbon and hydrogen only.

hydrogen bond The relatively weak chemical bond formed between an H atom covalently bonded to a highly electronegative atom and an atom of another molecule that offers an unshared electron pair.

hydrogen ion Hydronium ion, especially when written in the short form H^+.

hydrometallurgy A metallurgical procedure that converts sulfide ores to soluble salts that are then reduced to pure metal.

hydronium ion The conjugate acid of water, H_3O^+. See also *hydrogen ion*.

hydrophilic colloid A colloidal dispersion in which water is the dispersion medium and there are strong attractive forces between water and the surface of the colloidal particles.

hydrophobic colloid A colloidal dispersion in which water is the dispersion medium and there are only weak attractive forces between water and the surface of the colloidal particles.

hypothesis An assumption useful as a basis for designing future experiments.

ideal solution A solution described accurately by Raoult's law.

indicator (1) A weak acid (or base) differing in color from its conjugate base (or acid), showing gradations of color over a certain range of pH. (2) Any solute that shows by its color some condition in the solution.

inductive effect The shift of electron density from one atom to another through σ bonds.

inhibitor A substance that decreases the rate of a reaction. Also, "poison."

inner transition elements The set of elements characterized by the progressive addition of electrons to the *f* subshells of their atoms. They include the lanthanoids and actinoids.

intensive property A property of matter that is independent of the size of the sample. Also, "specific property."

intermediate A species produced in a chemical reaction that does not appear in the equation for the net reaction.

International System of Units (SI) The international system of measurement, comprising base units, derived units, and prefixes.

interstitial alloy A solid solution in which small atoms (usually C and H) (the solute) fit into the interstices of the host element (the solvent).

ion An atom (monatomic ion) or group of atoms (polyatomic ion) carrying a positive or negative charge.

ion exchanger A solid bearing fixed charges that attract mobile ions of the opposite sign. It is a *cation* or *anion exchanger* according to the sign of mobile ions.

ionic bond A bond resulting from electrostatic attraction between a cation and an anion in an ionic compound.

ionization The formation of ions from electrically neutral atoms or molecules.

ionization constant The equilibrium constant for an ionization reaction, such as $HA \rightleftharpoons H^+ + A^-$ (K_a for HA) or $B + H_2O \rightleftharpoons BH^+ + OH^-$ (K_b for B).

ion product of water The equilibrium constant (K_w) for the self-ionization of water, $H_2O \rightleftharpoons H^+ + OH^-$.

ion-selective electrode A half-cell whose potential depends on the concentration of a specific ion.

isoelectronic Having the same number of electrons in similar orbitals.

isomers Compounds with the same molecular formula but different arrangements of atoms.

isotonic solutions Solutions with the same osmotic pressure.

isotopes Atoms of the same element having the same number of protons but different numbers of neutrons.

joule The SI unit of energy: $1\ J = 1\ kg\ m^2/sec^2$.

kelvin The SI unit of temperature, defined as 1/273.16 of the triple point of water. Symbol, K.

ketone A compound with two alkyl or other hydrocarbon groups bonded to a carbonyl group; general formula is $R_2C{=}O$.

kinetic energy The energy of a body in motion, $\frac{1}{2}mu^2$, where m is the mass and u is the speed.

kinetics, chemical The study of the rates and mechanisms of chemical reactions.

kinetic stability The property of changing only at a very low or immeasurable rate.

lanthanoids The inner transition elements of period 6, from lanthanum (57) to ytterbium (70). Also, "lanthanides," "rare earth elements."

lattice energy The enthalpy change when 1 mole of an ionic crystal is converted to gaseous ions.

law A reliable predictor of a natural phenomenon, usually expressed in the form of a mathematical equation.

Le Chatelier's principle When a system in equilibrium is disturbed, the equilibrium shifts in the direction that decreases the effect of the disturbance.

Lewis acid, Lewis base See *acid, base.*

Lewis structure A structural formula showing all valence electrons as dots. (A shared pair may be shown with a dash.) Also, "electron-dot structure," "Lewis formula," "electron-dot formula."

Lewis symbol The symbol for an atom with a dot for each valence electron.

ligand A base bonded to the metal atom or ion in a complex.

limiting reactant The substance that is completely consumed in a reaction and determines the quantity of product formed. Also, "limiting reagent."

line spectrum See *spectrum.*

liquefaction The process of becoming liquid.

liquid A state of matter that maintains a definite volume but which flows readily and takes the shape of its container under the stress of gravity.

liquid crystal A state of matter characterized by a degree of order of its component atoms or molecules that is intermediate between that of a liquid and that of a crystalline solid. Liquid crystals can retain their shape if not stressed, but can easily be made to flow.

liter 1 cubic decimeter, or 1000 cm^3; a very close approximation to the volume of 1 kilogram of ice-cold water. Symbol, L.

London forces Intermolecular attractions between induced dipoles. Also, "dispersion forces," "van der Waals forces."

lone pair Unshared pair.

low-spin complex A complex (strong-field) that is typically less paramagnetic than the corresponding weak-field coordination compound because it has fewer unpaired electrons.

main-group elements Representative elements.

mass number The sum of the numbers of protons and neutrons in an atomic nucleus. Also, "nucleon number."

mass percentage of A The mass of A divided by the total mass, times 100%.

mass spectrometer An instrument that measures or compares the mass/charge ratios of ions by acceleration in an electric field and deflection in a magnetic field.

matter wave A particle in motion considered with reference to its wave properties.

mechanism of reaction A sequence of steps whose sum is the net reaction and which is consistent with the experimental rate equation for the reaction.

melting Transformation of solid to liquid; fusion.

melting point The temperature at which the solid and liquid phases of a substance are in equilibrium. The temperature is approached experimentally by warming the solid.

metal An element that, when solid or liquid, conducts electricity by movement of electrons in a partially filled band.

metal cluster (1) A transition metal compound with single or multiple bonds between metal atoms. (2) Any small group of metal atoms bonded together.

metallurgy Processes for purifying, changing the composition of, and reducing ores, and purifying the metal isolated from them.

metathesis Double displacement.

micelle (1) A spherical colloidal particle formed by several soap or detergent molecules in water, having the hydrocarbon ends pointing to the center of the sphere and the terminal polar ends on the surface jutting out into the water. (2) Any colloidal particle consisting of atoms, molecules, or ions loosely bound together.

mineral The compound of a metal found in its ore.

MO (1) Molecular orbital. (2) Bonding molecular orbital.

MO* Antibonding molecular orbital.

molality The number of moles of solute per kilogram of solvent. Symbol, m.

molarity The number of moles of solute per liter of solution. Symbols: M, c, or [B], where B is the formula of the solute. See also *concentration.*

molar mass The mass of 1 mole of a substance, numerically equal to the molecular weight but in the unit g/mol.

mole The amount of substance that contains as many particles (atoms, molecules, or sets of atoms specified by the formula) as there are atoms in exactly 12 g of ^{12}C.

molecular formula (1) A formula showing the actual number of atoms of each kind in a molecule. (2) A formula for an entire compound rather than an ion.

molecularity The number of molecules coming together in a step in a reaction mechanism.

molecular orbital An orbital encompassing the entire molecule that results from the combination of AO's of the individual bonding atoms. It is a "localized MO" when it can be represented by a dash or two dots in a Lewis structure.

molecular orbital (MO) method A method of describing chemical bonding in terms of molecular orbitals.

molecular shape (1) The arrangement of atoms around a central atom. (2) The geometrical arrangement of the atoms in a molecule.

molecular weight The sum of the atomic weights of the atoms in one molecule or the atoms shown in a formula; the molar mass with units omitted.

molecule A chemical combination of atoms in fixed whole numbers.

mole fraction The number of moles of a specified component in a solution divided by the total number of moles of all components: $x_A = n_A(n_A + n_B + \cdots)$.

mole percentage Mole fraction times 100%.

multidentate ligand Chelating agent.

multiple bond The sharing of more than one pair of electrons.

multiple proportions, law of When two elements form more than one compound, the masses of one element that combine with a fixed mass of the other are in ratios of whole numbers.

Nernst equation An equation relating the reversible emf $\mathscr{E}$ of a cell to the standard reversible emf $\mathscr{E}^0$: $\mathscr{E} = \mathscr{E}^0 - (2.303RT/n\mathscr{F}) \log Q$, where R is the gas constant, T is the absolute temperature, n is the number of faradays passing when the cell process occurs as written, $\mathscr{F}$ is Faraday's constant, and Q is the reaction quotient for the cell process as written.

net ionic equation An equation showing only the atoms, molecules, or ions that undergo a chemical change.

network covalent substance A substance that is a large aggregate of covalently bonded atoms with no individual molecules.

neutral (1) Having zero electric charge; neither positive nor negative. Also, "electrically neutral." (2) Neither acidic nor basic.

neutralization (1) A reaction between the conjugate acid and the conjugate base of the same substance. (2) Any reaction between an acid and a base.

neutral solution A solution with $[H^+] = [OH^-]$.

neutron A neutral particle present in the nucleus of an atom, having nearly the same mass as a proton.

neutron activation analysis A method of chemical analysis that depends on the conversion of stable nuclei to radioactive nuclei by neutron bombardment.

noble gas An element of Group 8, from helium to radon. Formerly, "inert gas."

node A point of zero amplitude of a wave; applied to electrons in atoms or molecules, it is the plane or other surface on which the electron density is zero.

nonbonding molecular orbitals Molecular orbitals whose electrons do not participate in bonding and are not antibonding.

nonconducting band A band filled with electrons.

nonelectrolyte A compound that is not ionized in water.

nonpolar molecule A molecule with no net dipole moment.

normality The number of equivalents per liter of solution.

***n*-type semiconductor** A semiconductor formed by adding a trace of a Group-5 element to a pure Group-4 semiconductor, thereby introducing extra electrons, which are easily excited to the empty high-energy band enabling conduction to occur.

nuclear binding energy The energy per nucleon required to separate a nucleus into its nucleons (endothermic) or the energy evolved per nucleon when a nucleus is formed from nucleons (exothermic); the energy is calculated from $\Delta E = \Delta m \times c^2$.

nuclear imaging The production of computerized images of internal organs by use of gamma rays and radioactive isotopes.

nuclear reaction The conversion of one isotope into another.

nuclear reactor A nuclear-fission reactor in which fissionable isotopes produce heat for the generation of electricity and/or neutrons for the manufacture of radioactive isotopes.

nucleon number Mass number.

nucleons The protons and neutrons found in atomic nuclei.

nucleophile Lewis base. See *base*.

nucleus (1) The small, dense, positively charged core of an atom. (2) A small particle or bubble on which a new phase grows.

nuclide A specific isotope of a specific element.

octahedron The shape of XY_6-type species with the central atom (X) in the center of a square and a Y atom above and below X and at each corner of the square.

octet rule Atoms form bonds to acquire eight electrons in their valence shell.

oil (1) A liquid ester of glycerol with fatty acids. (2) Any compound (usually organic) or mixture that is liquid at ordinary conditions, nearly immiscible with water, more or less viscous, and with low vapor pressure.

optical isomerism The type of stereoisomerism that results from chirality.

orbital A mathematical function associated with an electron having a definite energy, related to the probability of finding the electron at each point in the vicinity of a nucleus or nuclei.

orbital overlap A condition in which the same region of space is occupied by parts of orbitals belonging to two atoms.

order The exponent of a specified concentration in an experimentally determined rate equation.

order of reaction The sum of the exponents of all concentrations in an experimentally determined rate equation.

ore An economically accessible source of a metal that contains the mineral and gangue.

organic chemistry The study of carbon compounds.

osmosis The flow of solvent through a semipermeable membrane.

osmotic pressure The excess pressure that must be applied to a solution to prevent osmosis and keep it in equilibrium with the pure solvent.

outer-shell electrons Valence electrons.

overvoltage The difference between the decomposition potential and the reversible emf associated with the reverse (spontaneous) reaction.

oxidant Oxidizing agent.

oxidation A loss of electrons or an increase in oxidation number.

oxidation number A charge on an atom obtained by arbitrarily assigning each bonding pair to the more electronegative atom. Also, "oxidation state."

oxidation–reduction reaction A reaction in which there is a transfer of electrons or changes in oxidation numbers. Also, "redox reaction."

oxidizing agent A substance that oxidizes another substance and is reduced at the same time. Also, "oxidant."

oxo acid An acid with at least one OH group bonded to a central atom, with or without lone O atoms.

oxoanion A polyatomic anion containing oxygen and another element.

p The negative of the common logarithm of a specified concentration (molarity) or equilibrium constant, as in $pH = -\log [H^+]$; $pOH = -\log [OH^-]$; $pK = -\log K$.

paramagnetic Drawn into a magnetic field.

partial charge A charge less than 1 electronic unit, resulting from an unequal sharing of electrons in a covalent bond.

partial equation or **reaction** An equation showing an oxidation or reduction process, with electrons as products or reactants.

partial pressure The pressure that a gas in a mixture would exert in the same space if no other gases were present.

pascal A unit (symbol, Pa) of pressure: $1\ Pa = 1\ newton/m^2$.

passivity An unreactive state of a metal caused by oxidation of its surface.

path The sequence of states by which a system passes from one state to another.

Pauli exclusion principle An orbital (atomic or molecular) can hold no more than two electrons.

percentage (1) Mass percentage. (2) Mole percentage. (3) The quantity of a part divided by the quantity of the whole (the two quantities in the same units), times 100%.

percentage by volume (1) Mole percentage (for gases). (2) The volume of a pure substance (before being dissolved) divided by the volume of the solution containing it, times 100%.

period One of the horizontal rows in the periodic table, showing elements with the same number of electron shells.

periodic table An arrangement of the symbols of the elements in the order of increasing atomic numbers that forms a pattern of rows and columns that highlights recurrences of chemical and physical properties and of some aspects of the electronic configurations of their atoms.

peroxide A compound whose molecules contain O—O bonds such as hydrogen peroxide, H—O—O—H, or sodium peroxide, Na_2O_2.

pH The negative common logarithm of the concentration (molarity) of hydrogen (hydronium) ion: $pH = -\log [H^+]$.

phase A homogeneous portion of matter that is separated from other parts of the sample by a definite surface or boundary.

phase diagram A graph of temperature vs. pressure showing the conditions under which each phase or combination of phases is stable.

phenol (1) A compound with an OH group attached to a benzene ring. (2) The simplest phenol, C_6H_5OH.

pH meter An instrument consisting of a potentiometer or voltmeter and a galvanic cell whose emf depends on the pH of a solution.

phospholipid An ester of glycerol with two molecules of fatty acids and one molecule of a substituted phosphoric acid.

photoelectric effect The ejection of electrons from the surface of a metal by light (photons).

photon A particle (quantum) of radiation moving with the speed of light and having an energy equal to $h\nu$.

pi (π) bond A bond resulting from the side-to-side overlap of *p* AO's, in which the electron density lies mainly in the plane of the overlapping AO's.

pion One of the particles that are exchanged between nucleons, producing the strong force that binds nucleons together in nuclei.

Planck's constant The constant that relates the energy of a photon to its frequency: $h = 6.62608 \times 10^{-34}$ J sec/particle.

plasma An electrically neutral gas of positive ions and electrons at very high temperature.

polar covalent bond A bond in which the more electronegative bonded atom has a partial negative charge and the less electronegative atom has a partial positive charge.

polar molecule A molecule with a net dipole moment.

polyatomic ion An ion composed of at least two covalently bonded atoms.

polyprotic Capable of giving up or accepting more than one proton.

positron A particle with the same mass as an electron but carrying a positive charge, ${}_{+1}^{0}\beta$.

potential, difference of The work required or done when a unit charge moves from one point to another.

potential energy The energy a body has because of its position, configuration, or composition.

potentiometer An instrument for measuring the reversible emf of a galvanic cell by opposing it with an adjustable emf so that no current flows.

precipitate A nearly insoluble solid formed in a reaction, which may settle to the bottom of the container.

precision The degree of mutual agreement of repeated determinations; hence, a measure of the reproducibility of an experiment.

primary cell A galvanic cell used as a practical source of energy and discarded when it is depleted.

principal energy level Shell.

principal quantum number The whole-number values of 1, 2, 3, . . . that define the principal energy levels (shells). See also *quantum numbers.*

product A substance appearing to the right of the arrow in a chemical equation.

property (1) A quantity or description determined by the state of a system, independently of its history. (2) A quantity whose change is determined by the initial and final states of a system, independently of the path joining them. Also, in either sense, "state function."

proton A particle carrying a unit positive charge equal to the negative charge of an electron but with a mass almost equal to the mass of an H atom.

proton number Atomic number.

***p*-type semiconductor** A semiconductor formed by adding a trace of a Group-3 element to a pure Group-4 semiconductor, thereby introducing holes (electron deficiencies), enabling conduction to occur in the originally filled band.

pure substance A substance consisting of only one kind of atom (a pure element) or one kind of molecule or other combination of atoms (a pure compound).

qualitative analysis The identification of the substances in a sample.

quantitative analysis The determination of the quantity(ies) of one or more substances in a sample.

quantum numbers Dimensionless numbers, designated as n (principal), ℓ (shape), m_ℓ (orientation), and m_s (electron spin), that determine the energy, size, shape, and spatial orientation of an orbital and the spin of an electron. See also *principal quantum number.*

quarks The particles that make up nucleons and pions.

racemate An equal mixture of enantiomers of the same chiral compound. Also, "racemic mixture."

radiation (1) Energy emitted, absorbed, or transmitted in the form of photons traveling at the speed of light. Also, "electromagnetic radiation," "radiant energy." (2) High-energy particles emitted in nuclear processes.

radical (1) A group of atoms that appears in many compounds. (2) Free radical.

radio- Radioactive.

radioactive isotope An isotope that undergoes a change in nuclear composition or a transition between nuclear energy levels with the emission of high-energy particles or photons or the capture of an orbital electron.

radioactive tracer A radioactive isotope that can be followed through a number of chemical or physical changes.

Raoult's law The vapor pressure of a component in a solution is approximately equal to its mole fraction times the vapor pressure of the pure component: $p_A \approx x_A p_A^0$.

rare earth elements Lanthanoids and actinoids.

rate (1) The change in a quantity per unit time. (2) Reaction rate.

rate constant The proportionality constant (k) appearing in a rate equation.

rate equation An equation giving the rate of a reaction in terms of concentrations, usually of the form $rate = k[A]^a[B]^b \ldots$, where k and the powers are experimentally determined.

reactant A substance appearing to the left of the arrow in a chemical equation.

reaction quotient The expression containing concentrations (Q_c) or partial pressures (Q_p) that appears in an equilibrium condition.

reaction rate The decrease in the concentration of a reactant per unit time or the increase in the concentration of a product per unit time.

reagent (1) Reactant. (2) An especially pure chemical suitable for laboratory use.

redox reaction Oxidation–reduction reaction.

reducing agent A substance that reduces another substance and is oxidized at the same time. Also, "reductant."

reduction A gain of electrons or a decrease in oxidation number.

reduction potential Half-cell potential with the sign convention that the more positive is the potential, the more the electrode tends to be a cathode.

reference half-cell A half-cell to which the potential 0 is arbitrarily assigned.

representative elements The set of elements characterized by the stepwise addition of electrons into the *s* and *p* subshells. Also, "main-group elements."

resonance The use of two or more plausible but fictitious Lewis structures (contributing structures) to represent the actual structure (resonance hybrid).

reverse osmosis Flow of solvent through a semipermeable membrane from a more concentrated to a less concentrated solution, resulting from a pressure exceeding the osmotic pressure.

reversible electromotive force The emf of a cell when the current is 0 and the cell process is thermodynamically reversible.

reversible process or **reaction** (1) Thermodynamically reversible process. (2) A process that can occur in either direction under appropriate conditions.

roasting The conversion of (mainly) carbonate and sulfide ores to oxides by heating in air.

rusting Corrosion of iron by formation of its oxide.

salt An ionic compound.

salt bridge A concentrated solution of a salt interposed between two other solutions in a galvanic or electrolytic cell.

saturated hydrocarbons Hydrocarbons with only single bonds between the C atoms. They do not undergo addition reactions.

saturated solution A solution that is or can be in equilibrium with undissolved solute.

secondary cell Storage cell.

second law of thermodynamics A law stated in various ways, including: (1) A system cannot absorb heat and do work in a cyclic process at constant temperature. (2) The entropy of the universe increases in every possible process.

self-ionization A reaction in which one molecule of an amphoteric substance, acting as an acid, transfers a proton to another molecule of the same substance, acting as a base.

self-oxidation-reduction A reaction in which a single substance undergoes both oxidation and reduction to form two products, one more highly oxidized and one less highly oxidized than the original substance. Also, "disproportionation," "self-redox."

semiconductor A material, often a Group-4 element, with a moderate conductivity that increases with temperature.

semipermeable membrane A membrane that allows solvent molecules, but not solute molecules, to pass through.

shell The set of orbitals having the same principal quantum number, n. Also, "principal energy level."

SI International System of Units.

sigma (σ) bond A bond, formed by head-to-head overlap of AO's, that has a symmetrically distributed electron density around the bond axis.

significant figure A digit in a measurement or in a computation from measurements that is believed to be correct or nearly so.

simple cubic lattice A cubic crystal lattice in which atoms are only at the corners of the unit cells.

simplest formula Empirical formula.

slag A product formed from impurities and flux in a metallurgical process that is easily separated from the metal.

soap A salt of a fatty acid, $RCOO^-M^+$, typically with a long carbon chain, often having cleansing action.

solid A state of matter characterized by rigidity. See also *crystalline solid.*

solubility The extent to which a solute dissolves in a solvent, specified by giving the composition of the saturated solution.

solubility product The equilibrium constant for the process of dissolving a slightly soluble salt. Also, "solubility product constant."

solute (1) A component of a solution present in smaller quantity than the solvent. (2) An originally solid or gaseous component of a liquid solution.

solution A mixture in which the components are dispersed as molecules, atoms, or ions.

solvation The association of solute molecules or ions with solvent molecules in a solution.

solvent (1) The component of a solution present in largest quantity. (2) The liquid component of a liquid solution in which the other components were solids or gases before being dissolved.

specific heat The quantity of heat needed to raise the temperature of 1 gram of a substance by 1°C; unit, J/(g °C).

specific property Intensive property.

spectator ions Ions that remain unchanged during a reaction and appear on both sides of an ionic equation.

spectrum The distribution of various frequencies of electromagnetic radiation emitted (emission spectrum) or absorbed (absorption spectrum) by a substance. A *continuous spectrum* consists of radiation that is continuously distributed over a broad range of frequencies. A *line spectrum* consists of an array of separated spectral lines of specific frequencies.

splitting energy The energy difference, caused by the approaching ligands, between the sets of orbitals on the central atom of a complex. Symbol, Δ.

spontaneous process A process that can occur at constant temperature when no work is done on the system, excluding pressure-volume work.

stainless steel A low-carbon steel that contains nickel usually and at least 10% chromium.

standard electromotive force The emf of a cell when each reactant and product is in its standard state. Symbol, $\mathscr{E}^0$.

standard heat of formation The enthalpy change in the formation of one mole of a substance in its standard state from its elements in their standard states. Symbol, ΔH_f^0.

standard state A state chosen, at any temperature, as follows: a pure substance at 1 atm, a gas in a mixture at a partial pressure of 1 atm, or a solute at 1 mol/L.

standard temperature and pressure (STP) 0°C and 1 atm.

state The condition of a system, commonly specified by mass, chemical composition, temperature, and pressure.

state function Property.

step A single reaction in a proposed reaction mechanism. See also *mechanism of reaction, chain reaction.*

stereoisomerism A kind of isomerism resulting from differences in the spatial arrangement of bonded atoms or groups rather than in the sequence of bonds.

stereoisomers Isomers, such as *cis-trans* and optical isomers, that differ only in the spatial arrangement of the atoms.

stoichiometry Calculation of quantities involved in a chemical reaction.

storage cell A galvanic cell used as a practical source of energy that can be recharged by reversing the cell process.

strong acid An acid that reacts almost completely with water, forming H_3O^+.

strong base (1) A soluble ionic compound that exists almost completely as cations and OH^- ions in water. (2) A base that reacts almost completely with water, forming OH^-.

strong electrolyte A substance that exists almost completely as ions in a dilute aqueous solution.

strong-field coordination compound A coordination compound with extensive splitting of the *d* orbitals caused by strong-field ligands.

strong-field ligand A ligand that is a strong Lewis base and causes extensive splitting of the energy levels of *d* orbitals.

structural formula A formula showing the bonding arrangement of the atoms in the molecule.

sublimation (1) Transformation of a solid to a gas without passing through the liquid state. (2) Transformation of a solid to a gas and back to solid, often used as a method of purification.

subshell A set of orbitals having the same principal quantum number, n, and the same ℓ quantum number. The number of subshells per shell equals n. Also, "sublevel."

substituted hydrocarbon A hydrocarbon with a carbon atom bonded to an atom or group of atoms other than an H atom.

substitutional alloy A solid solution alloy in which a number of atoms of the crystal lattice are replaced by other atoms of approximately the same size.

substitution reaction The replacement of an H atom, usually of a hydrocarbon, by another atom or group.

superconductivity The conduction of a current without resistance at very low temperature.

supercooling The cooling of a liquid below its freezing point or of a vapor below its condensation temperature (a nonequilibrium condition). Also, "undercooling."

superheating The heating of a liquid above its boiling point (a nonequilibrium condition).

supersaturated solution A solution more concentrated than a saturated solution (a nonequilibrium condition).

surroundings (1) Everything except the system. (2) Whatever is in the vicinity of a system and interacts with it.

suspension A mixture containing coarse particles that settle out rapidly.

system A portion of the universe selected for study.

system, closed A system that matter cannot enter or leave.

system, insulated A closed system that interacts with its surroundings only by work because no heat can enter or leave.

system, isolated A system that does not interact with its surroundings.

system, open A system that matter can enter or leave.

systematic error A error that repeatedly occurs in the same direction and has about the same magnitude.

tetrahedron The shape of most XY_4-type molecules having four triangular faces with X at the center and a Y at each of the four corners.

theoretical yield The quantity of product that should be obtained as calculated from the balanced equation.

theory A strongly based concept that serves to organize much scientific knowledge.

thermochemical equation A chemical equation that includes the ΔH or ΔE of the reaction as written.

thermodynamically reversible process A process that can be made to occur in either direction by an infinitesimal change in one condition.

thermodynamic instability The condition of not being in equilibrium.

thermodynamics The branch of physics, chemistry, and engineering concerned with heat and work and their interconversion.

thermodynamics, laws of See *first law, second law, third law.*

thermodynamic stability The condition of being in equilibrium.

third law of thermodynamics In any process involving perfect crystals at 0 K, $\Delta S = 0$.

three-centered bond A chemical bond in which two electrons serve to link a series of three atoms.

tincture A solution in ethanol.

titration The process of adding a solution to a sample until a reaction is complete and measuring the volume added.

titration curve A graph showing the pH of a solution being titrated as a function of the volume of the titrating solution added.

torr A unit of pressure: 1 torr = (1/760) atm

***trans* isomer** A geometric isomer in which like atoms, groups, or ligands occupy diagonally opposite corners of an approximately rectangular array.

transition element One of the elements, in the fourth and later periods of the periodic table, not having counterparts in the earlier periods and typically having partially filled *d* or *f* subshells in some of their common oxidation states, but usually including the zinc subgroup.

transition state A specific geometrical structure that is the condition of highest enthalpy in the progress of a reaction.

transition state theory A theory that assumes that the path from reactants to products is a continuous change in a few bond distances and angles in the reactant species that come together to form a transitory unstable configuration.

trigonal bipyramid The three-dimensional structure of XY_5-type molecules having X at the center of a triangle with a Y above and below X and at each corner of the triangle.

trigonal-planar Having the flat structure of XY_3-type molecules with X at the center of a triangle and a Y at each corner.

triple point A point on a phase diagram where three lines meet, corresponding to equilibrium among three phases.

unit cell The smallest representative portion of a crystal lattice. The lattice can be generated by repetitions of the unit cell, all in contact with each other and similarly oriented in space.

unshared pair A pair of electrons assigned only to one atom, not shared with any other atom. Also, "lone pair."

valence bond (VB) method A concept of bonding that assumes the AO's of the bonding atoms overlap to form covalent bonds which consist of pairs of localized electrons.

valence electrons The electrons in the outer shell of an atom.

valence shell electron pair repulsion (VSEPR) theory A theory, based on the repulsion between pairs of valence electrons, that predicts the shape of molecules.

van't Hoff factor The ratio of the measured freezing-point depression (Δt_f) of an ionic solution to the freezing-point depression ($K_f m$) calculated without considering ionization: $i = \Delta t_f / K_f m$.

vaporization The process of becoming a vapor.

vapor pressure The partial pressure of a vapor in equilibrium with its liquid phase.

vapor-pressure depression The decrease in the vapor pressure of a solvent (A) when a solute (B) is added: $\Delta p = p_A - p_A^0 \approx x_B p_A^0$.

volt A unit of electric potential: 1 joule/coulomb. Symbol, V.

voltmeter (1) An instrument that measures difference of potential approximately by measuring the current that flows through a high resistance. (2) An instrument that applies a difference of potential to a transistor or vacuum tube, thus controlling a current that can be measured.

volumetric analysis Quantitative analysis in which quantities are determined by measuring the liquid volumes used in titrations.

warming curve The time vs. temperature graph plotted when a substance is warmed and passes through one or more changes of state.

wavelength The distance between successive crests or troughs of a wave. Symbol, λ.

weak acid An acid whose reaction with water is limited, yielding only a weakly acidic solution.

weak base A base whose reaction with water is limited, yielding only a weakly basic solution.

weak electrolyte A substance that is only partially ionized in water.

weak-field complex A complex in which there is little splitting of the energy levels of the *d* orbitals because its ligands are weak-field.

weak-field ligand A ligand that causes little splitting of the energy levels of *d* orbitals.

weight The force of gravity exerted on a body.

work The product of force times the distance through which it is exerted, considered positive when it increases the energy of the system. Symbol, *w*.

work, pressure-volume The work done on or by a system when the change in its volume is ΔV; when the pressure is constant, $w_{PV} = -P\Delta V$.

x ray A radiation emitted by some radioactive atoms or by an object hit with high-speed electrons.

yield, percent The ratio of the actual quantity of product obtained to the theoretical yield, multiplied by 100%.

zinc subgroup The elements zinc, cadmium, and mercury.

zone refining A way of refining a substance that relies on the fact that impurities do not fit into the crystal lattice formed as the impure liquid freezes.

ANSWERS†

CHAPTER 1

1.1 (a) 10^4 μm/cm (b) 10^{12} ng/kg
1.2 (a) −37.97°F, 234.28 K (b) 260°C, 533 K
1.3 (c) 5 significant figures
1.4 (a) 0.2 (b) 0.206
1.5 88 km/h, 2.5×10^3 cm/sec, 1.5×10^3 m/min
1.6 8.9 g/mL, 8.9×10^3 mg/mL, 8.9×10^3 kg/m^3
1.7 (a) Student A: ave. −0.1°C, ave. dev. 0.2°C; Student B: ave. 273.16 K, ave. dev. 0.02 K (b) Student A error −0.1°C, Student B error +0.01 K; Student B is more accurate and more precise.
1.9 (a) 10^6 mg/kg (b) 10^3 μL/mL (c) 10^3 μL/cm^3 (d) 10^7 nm/cm (e) 10^9 psec/msec
1.10 (a) 10^8 Å/cm (b) 10^7 Å/mm (c) 10 Å/nm
1.11 (a) 0.040 cm^3, 40 μL (b) 4.2 mm
1.13 3.51 mg/mm^3 = 3.51 g/cm^3
1.14 19.3 g/mL
1.15 529 cm^3
1.16 10.2 kg
1.18 (a) 646 K (b) −263.00°C (c) −25.6°F (d) −35.6°C
1.19 −40°C = −40°F
1.21 32.5°N
1.22 (a) 4 (b) 6 (c) 3
1.23 (a) 24.1 km/h, 670 cm/sec
1.24 100 ¢/$ and 1 gal/3.785 L
1.25 $4096
1.26 41.7 cm/sec, 1.50 km/h
1.28 (a) Method I: ave. 2.4 g/mL, ave. dev. 0.15 g/mL; Method II: ave. 2.703 g/mL, ave. dev. 0.001 g/mL (b) Method I error −0.30 g/mL; Method II error +0.001 g/mL; Method II is more precise and more accurate.
1.32 (a) 10^{15} pg/kg (b) 10^3 nsec/μsec (c) 10^5 cm/km (d) 10^6 mL/m^3
1.33 (a) 38°C (b) 68°F (c) 233 K
1.34 (a) 5 (b) 4 (c) 2 (d) 5 (e) 1
1.35 (a) 1.2×10^3 L, 1.2×10^6 mL (b) 3.2 g/cm^3, 3.2 g/mL
1.36 (a) Student X: ave. 135°C, ave. dev. 2°C, error 0°C; Student Y: ave. 138°C, ave. dev. 0°C, error +3°C (b) Student Y's average is more precise; Student X's average is more accurate.
1.37 (a) pure homogeneous (b) pure heterogeneous (c) impure heterogeneous (d) impure homogeneous (e) pure homogeneous

CHAPTER 2

2.2 $(m_N/m_O)_1 = 2 \times (m_N/m_O)_1$
2.3 5.67×10^{18} electrons/sec
2.4 92p, 141n; 92p, 142n; 92p, 143n
2.5 19p, 21n, 19e^-; 19p, 21n, 18e^-
2.13 5.39 g
2.14 12.0 g C, 2.99 g H, 8.00 g O
2.15 (a) 8.00, 4.00, 16.0, 12.0, 20 g O
2.17 (a) 2.3×10^{14} g/cm^3 (b) 3.34×10^{-3} g/cm^3, 1.0×10^{13}
2.18 78p 114n, 78p 116n, 78p 117n, 78p 118n, 78p 120n; 78e^-
2.19 83p 126n 80e^-, 77p 116n 77e^-, 23p 28n 18e^-, 35p 46n 36e^-, 42p 56n 38e^-, 16p 16n 18e^-
2.20 (a) 3p 3n, 3p 4n (b) 3e^- (c) 3p 3n 2e^-, 3p 4n 1e^-, 3p 3n 0e^-, 3p 4n 0e^-
2.21 46p 59n 46e^-, 34p 46n 36e^-, 48p 66n 47e^-
2.22 $^{197}_{79}Au$, 118n, 79e^-; $^{59}_{27}Co^{2+}$, 25e^-; $^{66}_{30}Zn$, 30e^-; $^{80}_{34}Se^{2-}$, 36e^-; $^{84}_{36}Kr$, 48n, 36e^-; $^{19}_{9}F^-$, 10e^-; $^{18}_{8}O^{2-}$, 10n, 10e^-; $^{64}_{30}Zn$, 30e^-
2.23 (a) (2) (b) (3) and (4)
2.26 (a) 1.6×10^{-19} C (b) 10^{-20} C/e
2.27 (a) 1.59×10^{-19} C (b) 1, 7, 6, 10, 4 (d) ave. dev. 0.00, error -0.01×10^{-19} C
2.28 (a) 9.11×10^{-28} g (b) $9.1093894 \times 10^{-28}$ g

† It is not uncommon for two calculations of the same problem to yield answers that differ only in the last significant figure, such as 103 g and 104 g. The value of the last digit may depend on the number of digits used in the conversion factors, and therefore the difference does not mean that one of the calculations must have been "wrong."

2.44 2.50 mg Mg
2.00 mg O_2
4.50 mg initial and final mass
x mg O_2 + 4.15 mg MgO = 4.50 mg
$x = 0.35$ mg O_2

2.45 $\frac{1.60}{2.43} = 0.658 = \frac{0.658}{1.00} = \frac{2.29}{3.48}$

2.46 (a) 23 electrons (b) 14 electrons

2.47

233	235		69	
−92p	−92p	92p	−31p	31p
141n	143n	$-90e^-$	38n	$-28e^-$
$92e^-$		2+		3+

12		239		14	
−6p	6p	−94p	94p	−7p	7p
6n	$-10e^-$	145n	$-93e^-$	7n	$-10e^-$
	4−		1+		3−

2.48

${}^{19}_{9}F^-$	9p	10n	1−	$10e^-$
${}^{24}_{12}Mg$				12
${}^{90}_{38}Sr^{2+}$				36
${}^{32}_{16}S^{2-}$	16	16	2−	

2.49 (a) Fe^{2+} (b) SO_2^{3+} (c) P^{3-}

2.50 $Mg_3(PO_4)_2$, $Ga_2(SO_4)_3$, $Co_2(CO_3)_3$, $(NH_4)_2S$, $Cr_2(CO_3)_3$

2.51 lead(II) chloride, lead(IV) chloride, ammonium sulfite, barium nitrate, copper(II) acetate, copper(I) bromide, sodium nitrite, chlorate, peroxide

2.52 $SiO_2 + 2C \longrightarrow Si + 2CO$
$PBr_3 + 3H_2O \longrightarrow H_3PO_3 + 3HBr$
$C_{12}H_{22}O_{11} + 12O_2 \longrightarrow 11H_2O + 12CO_2$
$3NO_2 + H_2O \longrightarrow 2HNO_3 + NO$
$(NH_4)_2Cr_2O_7 \longrightarrow Cr_2O_3 + 4H_2O + N_2$

2.53 $2Al + Fe_2O_3 \longrightarrow Al_2O_3 + 2Fe$
$2KOH + Zn(ClO_3)_2 \longrightarrow Zn(OH)_2 + 2KClO_3$
$2AgNO_3 + H_2S \longrightarrow Ag_2S + 2HNO_3$
$Na_2CO_3 + 2HCl \longrightarrow 2NaCl + CO_2 + H_2O$
$(NH_4)_2SO_4 + 2NaOH \longrightarrow 2NH_3 + Na_2SO_4 + 2H_2O$

CHAPTER 3

3.1 2.7×10^{23} H_2O molecules
3.2 14.007, 14.007 g/mol
3.3 68.5
3.4 7.44 g
3.5 (a) 0.899 mol SO_3 (b) 5.41×10^{23} SO_3 molecules (c) 1.62×10^{24} O atoms
3.6 6.43×10^3 g
3.8 75.0% C, 25.0% H
3.9 $C_6H_{10}O_5$
3.10 N_2O_4
3.11 46.069
3.12 342 g/mol
3.13 (a) 18.7 mg (b) 47.6 mg
3.14 3.11 g
3.15 52.7 mL
3.16 0.672 mol
3.17 28.1 g
3.18 7.06 g
3.19 49.0 mL
3.20 90 cars, 10 car bodies excess
3.21 (a) 3×10^4 mol H_2S excess (b) 7.06 g S, 1.90 g SO_2 excess
3.22 30.0%
3.23 0.1 mol SO_2, 1.8 mol O_2, 2.4 mol SO_3
3.24 1.630
3.26 24.305
3.27 52.00
3.28 (a) 11.01 (b) ${}^{63}Cu$ (69.2%), ${}^{65}Cu$ (30.8%)
3.29 1.28/1
3.32 (a) 10.73 (b) 66.68
3.35 (a) 17.0 (b) 120 (c) 127 (d) 298
3.36 (1) 70.906 (2) 18.0153 (3) 183.18 (4) 291.992 (5) 172.20 (6) 397.28 (7) 1000.0258
3.37 184
3.39 44.5 g, 534 g, 2.23×10^{21} kg
3.41 (a) 11.0 g (b) 3.07 g (c) 3.41 g
3.42 17.8 g
3.43 (a) 2.34 mol (b) 2.12 mol (c) 1.06 mol (d) 0.939 mol (e) 1.29 mol (f) 0.245 mol (g) 2.06 mol (h) 0.177 mol (i) 1.28 mol (j) 0.381 mol
3.44 (a), (b) 6.02×10^{23} molecules (c) 1.36×10^{23} molecules (d) 2.72×10^{23} molecules
3.45 (a) 0.900 mol P, 27.9 g P; 1.80 mol Na, 41.4 g Na (b) 5.6 g N
3.46 (a) 1.50×10^{20} C and O atoms, 3.0×10^{20} H atoms (b) 6.00×10^{21} C and O atoms, 1.20×10^{22} H atoms
3.47 (a) 13.3 mol (b) 0.500 mol (c) 6.63 mol (d) 40.0 mol
3.48 (a) 26.5 mol (b) 1.00 mol (c) 13.3 mol (d) 160 mol
3.49 (a), (b) 0.526 mol (c) 2.12×10^{-3} mol (d) 8.73×10^{-4} mol
3.50 (a) 3.00 mol (b) 2.00 mol (c) 4.00 mol
3.54 183 mL
3.56 1.007276, −0.05443%; 2.01355, −0.02724%; 6.01457, −0.009120%; 24.98529, −0.002196%
3.58 $NaHCO_3$
3.59 (a) NaOCl (b) CH_4N (c) $C_{17}H_{21}O_4N$
3.60 CH_3N
3.61 Fe_3O_4
3.62 $MgSiHO_4$, 117.396 g/mol
3.63 C_4H_9
3.64 $C_2H_3NCl_2Br_2$
3.67 NaCl, Na_2Cl_2, 58.443 g/mol
3.68 $C_{17}H_{28}N_4O_7S$

3.69 C_2H_6
3.70 (a), (b) 10.1% (c) 13.1% (d) 4.79% (e) 79.3% (f) 96.1%
3.71 52.4%, 20.8%
3.72 (a) 8.00 g O, 3.01×10^{23} O atoms (b) 9.12 g O, 3.43×10^{23} O atoms
3.73 0.775 M
3.75 (a) 0.175 M (b) 1.75×10^{-4} mol
3.76 (a) 7.3×10^{-4} M (b) 124 mg
3.77 8.2 g
3.78 8.38 kg
3.79 160 mL
3.80 361 g
3.81 (d) 56.2 g
3.82 24.8 g
3.83 7.24 g
3.84 (a) 266 g (b) 204 g
3.85 (a) 36.2 lb (b) 1.98×10^5 L
3.86 34.4 g
3.87 287 g
3.89 50.0 mL
3.90 0.114 M
3.91 154 mL
3.92 14.9 g, 0.96 g
3.93 0.74 g, 6.99 g
3.94 94.9%
3.95 21.5%
3.96 (a) 200 g (b) 1.33×10^3 g
3.97 (a) 107.8682 (b) Li_3
3.98 0.372 mol, 77.0 g, 2.24×10^{23} atoms
3.99 (a) NH_2Cl (b) NH_4Cl, $Cl_2 + 2NH_3 \longrightarrow NH_2Cl + NH_4Cl$
3.100 $C_{15}H_{24}O_2$, $C_{15}H_{24}O_2$, 236.354
3.101 (a) 6.76 g Ni, 4.94 g S (b) Ni_3S_4
3.102 260 g
3.103 1.96 M
3.104 1.57 g B_2H_6 (formed), 1.77 g $LiAlH_4$ (excess)
3.105 0.80 mol CH_4, 0.60 mol H_2O, 4.80 mol H_2, 1.20 mol CO_2

CHAPTER 4

4.2 (a) 0.808 kJ (b) 27, 26, 26, 29, 27, 28 J
4.3 $0.18 \frac{J}{g\,°C}$
4.4 1.69×10^3 kJ/mol (b) 101 kg
4.5 1065 kJ
4.6 123.8 kJ
4.7 −473.8 kJ
4.8 −1940 kJ
4.11 (a) 10.5 kJ (b) 236 J
4.12 22°C
4.15 198 J
4.16 199 J
4.17 98.2°C
4.18 (a) 0.439 J/(g °C) (b) 25.8 J/(mol °C)
4.19 5.18°C
4.20 2.2 J
4.22 (a) 5.09×10^3 kJ/mol (b) 2.35 g
4.23 (a) 71.4 kJ (b) 598 g
4.24 (a) 346 kJ (b) (1) 176 kJ (2) 40.6 kJ
4.25 (a) 559 kJ (b) 1.12×10^3 kJ
4.26 221 g
4.27 (a) −393.5 kJ, −568.8 kJ
4.28 1.7×10^3 kcal, 2.0×10^3 kcal
4.30 (a) $272 \frac{kJ}{g}$ (b) $2124 \frac{kJ}{day}$
4.31 −988.4 kJ
4.32 −125.0 kJ
4.36 (a) −257.9 kJ (b) −90.8 kJ (c) −24.5 kJ (d) −518.0 kJ (e) −794.4 kJ (f) +15.8 kJ (g) +714.8 kJ (h) −521.4 kJ (i) −429.8 kJ
4.37 −187.8 kJ
4.38 −391.8 kJ
4.39 (a) 3.03 kJ (b) 4.45 kJ (c) 109 g
4.40 463 kJ
4.41 (a) +109 kJ (b) −899 kJ (c) +2292 kJ (d) −133 kJ (e) −535 kJ
4.44 (a) 5.590 kJ/°C (b) 1.336 kJ
4.45 0.7469°C
4.46 (a) 7.3×10^8 kJ (b) 1.2×10^2 °C
4.47 (a) 34 m^2/day (b) 78 × 10 L/day
4.48 (a) 8 kJ (b) 95 kJ gain/mol C
4.49 130 kJ/mol
4.50 7.0×10^{16} kJ, 1.5×10^{17} kJ
4.51 $H_2O(g) \longrightarrow H_2O(\ell)$ is an exothermic reaction, giving out heat to the surrounding air; increase. $H_2O(\ell) \longrightarrow H_2O(g)$ is an endothermic reaction, absorbing heat from the surrounding air; cool.
4.52 3.96 J
4.53 (a) 0.461 J/(g °C) (b) 5.53 J, 66.4 J, 25.7 J
4.54 (a) 327 kJ/mol (b) 17.5 kJ, 128 kJ (c) 3.90 g
4.55 0.235 g
4.56 −336.7 kJ
4.57 (a) −852.4 kJ (b) +176.0 kJ (c) −6602.6 kJ
4.58 −66 kJ
4.59 (a) 2.88 kJ/°C (b) 22.514°C

CHAPTER 5

5.1 3.00×10^2 m
5.2 3×10^{-8} g
5.3 (a) 2.21×10^{-25} nm (b) 1.2×10^{-21} m/sec
5.4 (a) 6.539×10^{-18} J/photon (b) 10^7
5.5 1.362×10^{-19} J/atom, 82.01 kJ/mol
5.6 4.576×10^{-19} J/photon, 2.756×10^2 kJ/mol
5.7 4.341×10^2 nm
5.16 3.0×10^9 sec^{-1}
5.17 2.60077×10^{-3} m
5.18 (a) 1.00000000 m, 2.99792458×10^8 sec^{-1} (b) 1.1963×10^{-4} kJ/mol
5.19 5.92×10^4 photons
5.20 57.24 nm

5.21 52.49 kJ/mol
5.22 3.022×10^{-19} J/photon
5.23 (d) 3.197×10^{15} sec^{-1}
5.24 91.16 nm
5.25 7.830×10^{-19} J/electron
5.26 0.037 sec
5.27 (a) 4.443×10^{-19} J/photon
(b) 2.122×10^{-19} J/photon
5.28 (b) 2.961×10^{-19} J/atom (c) 2.90×10^{3} m/sec
5.30 0.118 nm
5.31 (a) -2.615×10^{-17} J, 7.596 nm;
-1.634×10^{-18} J, 121.6 nm
(b) 7.844×10^{17} J/atom, 2.179×10^{-18} J/atom
5.34 (b) 1, 3, 5
5.56 (a) 1.32×10^{-5} nm (b) 3.98×10^{-23} nm
5.57 0.192 nm
5.61 (a) 5.3×10^{-27} m/sec (b) 5.8×10^{5} m/sec
5.62 7.2×10^{25} photons
5.63 (a) no
5.64 821.0 nm
5.65 3
5.66 (a and b) 3, (c and h) 1, (d) 25, (e and f) 5,
(g) 7 orbitals
5.67 3, 3, 2, 1, 0 unpaired electrons
5.68 (a) ${}_{23}V^{3+}$ (b) ${}_{23}V^{3+}[Ne]3s^2 3p^6 4s^0 3d^2$
5.69 (a) 1*p*, 2*d*, 3*f*, 4*g* (b) ten *s*, eighteen *p*, twelve *d*
5.70 ${}_{16}S$, ${}_{25}Mn^{2+}$, ${}_{27}Co$, ${}_{27}Co^{3+}$, ${}_{18}Ar^{-}$, ${}_{18}Ar^{+}$
5.71 (a) 1.0×10^{-4}, greater (b) zero

CHAPTER 6

6.1 (a) $5s^2 5p^5$ (b) $2s^2 2p^5$
6.2 (a) $4d^2 5s^2$ (b) $3d^{10} 4s^2$
6.4 (a) 2,6 (b) 2,5
6.5 0.214 nm
6.16 (a) 0.194 nm (b) 0.186 nm and 0.214 nm
6.17 0.328 nm and 0.205 nm
6.21 (a) 3− (b) 2+ (c) 2− (d) 0 (e) 1+
6.23 K^{9+}
6.46 See periodic table.
6.47 KH, MgH_2, SiH_4, H_2Te
6.48 (a) isoelectronic (b) allotropes (c) transition elements (d) noble gases
6.49 (a) Ca > Si > Cl (b) $P^{3-} > Mg^{2+} > Al^{3+} > Si^{4+}$
6.50 (a) F > C > Li (b) Na > K > Rb
(c) Ne > N > Be (d) $Ag^{2+} > Ag^{+} > Ag$
6.51 (a) F (b) Cl
6.52 See section 6.8.
6.53

Hydrogen:	colorless, less dense than air
Alkali metals:	lustrous, soft, rapidly reactive with water
Group 2:	reactive with oxygen when heated, divalent
Group 3:	all metallic except boron; all shiny except boron; moderately reactive
Group 4:	range from nonmetallic to metallic; include some semimetals
Group 5:	range from nonmetallic to metallic; include elements that exhibit allotropy
Group 6:	range from nonmetallic to metallic; exhibit allotropy; include some elements whose hydrides stink
Group 7:	include elements in three different states of matter; is the only such group; are all colored
Group 8:	colorless, unreactive
Transition:	all metals; include the densest substances known
Zinc subgroup:	form some toxic compounds
Lanthanoids:	all metals; similar to each other in many properties
Actinoids:	all metals; all radioactive

CHAPTER 7

7.28 −404 kJ/mol
7.29 −348 kJ/mol
7.30 647.5 kJ/mol

7.68 (a) Sr:, :C̤̈l·, and ·Ṡ̤i·
(b) (1) Sr: + 2:C̤̈l· → Sr^{2+} + 2:C̤̈l:⁻
(2) 4:C̤̈l· + ·Ṡ̤i· → :C̤̈l:S̤̈i:C̤̈l: (with :C̤̈l: above and below Si)
(c) $\overset{\delta+}{Si}$—$\overset{\delta-}{Cl}$

7.69 (a) (1) H:C̤̈:B̤̈r: (with H above and below C) (2) [:B̤̈r:F̤̈: (with :F: above Br)]⁺ (3) :C̤̈l:C:::N:
(b) (1) tetrahedral (2) bent (3) linear

7.70 (a) H—N̈=$\overset{\oplus}{N}$=$\overset{\ominus}{N}$: or H—$\overset{\ominus}{N}$—$\overset{\oplus}{N}$≡N:
(b) :$\overset{\ominus}{C}$≡$\overset{\oplus}{O}$:

7.71 (a) (1) S(+6) (2) V(+4)
(b) (1) yes, there are changes in the oxidation state of Cl:

$$\underset{0}{Cl_2} + H_2O \longrightarrow \underset{-1}{HCl} + \underset{+1}{HOCl}$$

(2) no, there are no changes in any of the oxidation states:

$$\underset{+3\,-1}{PBr_3} + \underset{+1\,-2}{3H_2O} \longrightarrow \underset{+1\,+3\,-2}{H_3PO_3} + \underset{+1\,-1}{3HBr}$$

7.72 (a) (1) Sr (2) F (b) Greater atomic number means lower IE.

7.73 (a) (1) Cl—P (with F above, F and Cl to the right, Cl below) (2) Cl—P (with F above, Cl and Cl to the right, F below)

(b) The P—F bond is very polar; the polarity shortens the bond.
(c) BN is a network covalent substance:

(d) H:C̈l: H H
H:C:C:C̈l: :C̈l:C:C:C̈l:
H H H H

(e) HBr has a polar bond, $\overset{\delta+}{H}—\overset{\delta-}{Br}$, whereas Br—Br does not. (f) Electronegativity decreases on going from F to Cl to Br to I, and the bonds to C become less polar and weaker in the same order. (g) CN^-(:C≡N:$^-$), CO(:C≡O:), and C_2^{2-}(:C≡C:$^{2-}$); and a second group, CO_2(:Ö=C=Ö:) and N_3^-(:N̈=N=N̈:$^-$)

CHAPTER 8

8.47 (a) F (b) F (c) T (d) F (e) T

8.48 (a) *sp*, linear; sp^2, trigonal-planar; sp^3, tetrahedral (b) order of decreasing energy: $p > sp^3 > sp^2 > sp > s$ (c) better overlap, get larger bond angles (d) sp^3d, triangular bipyramid; sp^3d^2, octahedral

8.49 (a) (1) sp^3 (2) sp^3 (3) sp^2 (4) *sp* (b) (1) There is an unshared pair in an sp^3 orbital; each of the other three sp^3 orbitals overlaps with the *s* orbital of an H atom. (2) Each of the four sp^3 orbitals of N overlaps with the *s* orbital of an H atom. (3) Each N atom has two σ bonds formed from overlap of sp^2 orbitals. The remaining sp^2 orbital has the unshared pair. The *p* orbitals left, one on each N atom, overlap to form the π bond. (4) The N uses an *sp* orbital to form a sigma bond to C, which also uses an *sp* orbital. The second *sp* orbital of N has the unshared pair. The two remaining *p* orbitals overlap laterally with corresponding *p* orbitals of the C atom to form two π bonds.

8.50 (a) :C̈l—Hg—C̈l:, Hg uses *sp* HO's, linear

(b) BF_3 (:F̈: groups on B), B uses sp^2 HO's, trigonal-planar

(c) $[BF_4]^-$ (:F̈: groups on B), B uses sp^3 HO's, tetrahedral

(d) :F̈—S̈e—F̈:, Se uses sp^3 HO's, bent

(e) $AsCl_5$ (:C̈l: groups on As), As uses sp^3d HO's, trigonal bipyramidal

(f) SbF_6 (:F̈: groups on Sb), Sb uses sp^3d^2 HO's, octahedral

8.51 H—Ö—S^{2+}(—Ö:$^\ominus$)(—Ö:$^\ominus$)—ÖH ↔ H—Ö—S$^{\oplus}$(—Ö:$^\ominus$)(=Ö:)—ÖH ↔

H—Ö—S$^{\oplus}$(=Ö:)(—Ö:$^\ominus$)—ÖH ↔ H—O—S(=Ö:)(=Ö:)—O—H

8.52 (a) (1) sp^2 (2) *sp* (3) sp^3 (4) $H_2\overset{1}{C}=\overset{2}{C}=O$; C^1 uses sp^2 AO's, C^2 uses *sp* AO's (b) (1) trigonal-planar (2) linear (3) tetrahedral at each C atom

(4) $H_2\overset{1}{C}=\overset{2}{C}=O$; C^2 has linear bonds, C^1 has trigonal-planar bonds.

8.53 (a) (1) $(\sigma_{1s})^2$ (2) $(\sigma_{1s})^2(\sigma^*_{1s})^1$ (3) $(\sigma_{1s})^2(\sigma^*_{1s})^2$ (4) $(\sigma_{1s})^2(\sigma^*_{1s})^2(\sigma_{2s})^1$ (b) (1) 1 (2) 0.5 (3) 0 (4) 0.5 (c) $H_2 > He_2^+ = He_2^- > He_2$

8.54 (a) $(\sigma_{1s})^2(\sigma^*_{1s})^2(\sigma_{2s})^2(\sigma^*_{2s})^2(\pi_{2y})^2(\pi_{2z})^2$ (b) no

8.55 A *p* orbital on the central atom—C in (a) and S in (b)—overlaps laterally with a *p* orbital on each of the attached oxygen atoms.

(a) $[O{=}C(O)(O)]^{2-}$ (b) :Ö—S̈—Ö: (delocalized)

CHAPTER 9

9.14 10.74%

9.15 14.02%

9.16 8.412%

9.43 23.6 g Zn

9.47 0.07969 M

9.48 (a) (b) 1.21 M

9.49 1.39 g

9.50 (a) 0.4700 g (b) 8.283%

9.51 38.98%

9.52 (a) 16.7 mL (b) 197 mL (c) 51.8 mL

9.53 3.536%
9.54 16.08%
9.63 AgCl, PbS, $PbCl_2$, $CaSO_4$, $BaCrO_4$, Hg_2Cl_2, $BiPO_4$
9.64 (a) $Ag^+ + Cl^- \longrightarrow AgCl(c)$
(b) $Ba^{2+} + SO_4^{2-} \longrightarrow BaSO_4(c)$
(c) $Ag^+ + Cl^- \longrightarrow AgCl(c)$ or $2Ag^+ + S^{2-} \longrightarrow Ag_2S(c)$
(d) $Pb^{2+} + 2Cl^- \longrightarrow PbCl_2(c)$
9.65 (b) $Hg_2^{2+} + 2Cl^- \longrightarrow Hg_2Cl_2(c)$
(e) $H^+ + CN^- \longrightarrow HCN(g)$ (The reaction $8HCN + HClO_4 \longrightarrow 4C_2N_2 + 4H_2O + HCl$ may also occur.)
9.66 (a) $2Al + Cr_2O_3 \longrightarrow 2Cr + Al_2O_3$; Cr_2O_3
(b) no reaction
(c) $2Al + 6H^+ \longrightarrow 2Al^{3+} + 3H_2$; H^+ or H_2SO_4
(d) $3Zn + 2Cr^{3+} \longrightarrow 3Zn^{2+} + 2Cr$; Cr^{3+}
(e) no reaction
9.67 (a) Ca (b) K (c) Zn (d) Zn (e) H_2
9.68 (a) $ZnCO_3 \longrightarrow ZnO + CO_2$
(b) $TlClO_3 \longrightarrow TlCl + 1\frac{1}{2}O_2$
(c) $Cr_2(SO_3)_3 \longrightarrow Cr_2O_3 + 3SO_2$
9.69 (a) CdO (b) ZnO (c) CuO (d) PtO
9.70 $Ca + Cl_2 \longrightarrow CaCl_2$; $N_2 + 3H_2 \longrightarrow 2NH_3$;
$2Cr + 3S \longrightarrow Cr_2S_3$; $S + O_2 \longrightarrow SO_2$;
$4P + 3O_2 \longrightarrow P_4O_6$; $2As + 2\frac{1}{2}O_2 \longrightarrow As_2O_5$;
$CaO + SiO_2 \longrightarrow CaSiO_3$; $BaO + CO_2 \longrightarrow BaCO_3$;
$MgO + H_2O \longrightarrow Mg(OH)_2$;
$P_2O_3 + 3H_2O \longrightarrow 2H_3PO_3$ or
$P_4O_6 + 6H_2O \longrightarrow 4H_3PO_3$; $Li_2O + H_2O \longrightarrow 2LiOH$;
$As_2O_5 + 3H_2O \longrightarrow 2H_3AsO_4$;
$SO_3 + H_2O \longrightarrow H_2SO_4$
9.71 (a) $Cr_2O_7^{2-} + 3H_2S + 8H^+ \longrightarrow 2Cr^{3+} + 3S + 7H_2O$
(b) $7ClO_3^- + 3N_2H_4 + 6OH^- \longrightarrow 7Cl^- + 6NO_3^- + 9H_2O$
9.72 (a) 0.06900 M (b) 0.1728 g
9.73 16.42%

CHAPTER 10

10.1 0.276 atm, 0.280 bar, 28.0 kPa
10.2 0.932 atm
10.3 33°C
10.4 8.5×10^{22} molecules
10.5 3.9×10^{-3} kg
10.6 321 L
10.7 184 L
10.8 137 g/mol
10.9 0.654 g/L
10.10 28.9 g/mol
10.11 52.8 g
10.12 36.5 L
10.13 O_{2n}, where n is an integer; O_2
10.14 2.1 atm H_2S, 1.4 atm CO_2, 42.5 atm N_2
10.15 47.0 mL
10.16 742.5 torr, 97.7 mL
10.17 He, 3.2 times as fast
10.22 **1 atm** = (exactly) 760 torr = 1.01325×10^5 Pa = 101.325 kPa = 1.01325 bar
593 torr = 0.780 atm = 7.91×10^4 Pa = 79.1 kPa = 0.791 bar
133 bar = 131 atm = 9.98×10^4 torr = 1.33×10^7 Pa = 1.33×10^4 kPa
33.7 kPa = 0.333 atm = 253 torr = 3.37×10^4 Pa = 0.337 bar
10.23 (a) 858 torr, 1.13 atm (b) 808 torr, 1.06 atm
10.24 11 m
10.25 (a) 4.00 mL (b) 4.00×10^3 mL (c) 15.9 mL (d) 3.04×10^3 mL (e) 3.47×10^5 mL
10.26 (a) 2.9 atm (b) 43 mL
10.27 3.4×10^3 balloons
10.29 373 K, 25.00°C, 194.7 K, 1.5×10^7 °C
10.30 (b) 32.1 L (c) 41.8 L (d) 25.0 L
10.31 4 m^3, 4×10^3 L
10.32 185°C
10.33 (a) 1.431 L (b) 31 cm
10.35 6.0×10^{23} molecules/mol
10.36 (a) 1.5 L of hydrogen and 0.5 L of nitrogen
10.38 (c) 10.0 times (d) 7.25 times
10.39 318 L
10.40 5.18 mol
10.41 0.0125 mol
10.42 1.0×10^2 atm
10.43 4.65×10^7 L
10.44 46°C
10.45 99.6 g/mol
10.46 2.42 g/L
10.47 48.0 g/L
10.48 58 g
10.49 390 L
10.50 1.95 kg
10.51 26.4 mL
10.52 385 mL
10.53 163 torr
10.54 p_{O_2} = 30.80 torr; p_{CO_2} = 26.95 torr; p_{H_2O} = 5.85 torr; $p_{\text{inert gases}}$ = 706.4 torr
10.56 (a) air: $116/x$ g/L, neon: $80.7/x$ g/L, helium: $16.0/x$ g/L, unknown gas: $160/x$ g/L
(b) neon/air: 1.20, helium/air: 2.69, unknown gas/air: 0.852
10.57 5.2 km/min
10.59 See text.
10.61 (a) 2.04 L (b) 128 mL
10.62 (a) 6.79 atm, 6.88 bar
10.63 53.9 mL
10.64 44.1 g/mol
10.65 307 mL
10.66 27.5 g
10.67 Ar, 1.81 times as fast
10.68 (a) Intermolecular forces are in the order $CO_2 > N_2 > H_2$. (b) No, because the effect of molecular volume predominates at this pressure and therefore the PV values do not indicate relative molecular attractions.

CHAPTER 11

11.2 8
11.3 2 atoms/unit cell
11.4 4 each
11.5 35 kJ
11.6 309 J/g
11.7 718 torr
11.24 (a) 1/4 (b) 1/2 (c) 1
11.26 (a) 8, 8 (b) 1, 1
11.27 19.0 g/cm^3
11.29 Pb
11.31 1.16×10^3 J
11.32 9.7×10^2 J
11.33 34.0×10^3 J
11.34 1.29×10^9 J
11.36 2.18×10^{-6} g/L; 2.352×10^{-3} g/L
11.44 1.45 Å
11.45 1.37 Å
11.46 1.25 Å
11.47 (a) crystalline solid (description fits iodine) (b) gas or vapor (c) liquid (d) glass (e) crystalline solid
11.48 (a) gas (b) condense or liquefy, liquid, freeze or crystallize, crystalline solid (c) glass (d) sublimation, gas or vapor
11.49 a and d
11.50 (a) $2 < 3 < 4 < 1 < 5$ (b) all but 2
11.51 2 atoms/unit cell, body-centered cubic
11.52 (a) F (b) F (c) T (d) F (e) T (f) T (g) F (h) T (i) F (j) T
11.53 (a) solid (b) solid (c) slightly lower (d) The temperature is below the triple point.
11.54 (b) yes (c) no
11.55 $20.0 \text{ g } \{11.3 \text{ J/g} + [357 - (-39)]°\text{C} \times (0.138 \text{ J/g °C}) + 276 \text{ J/g}\} = 6.84 \times 10^3$ J

CHAPTER 12

12.1 (a) 16.7% (b) 6.00 kg
12.2 1.25 M
12.3 (a) 0.07494 M (b) 0.966%
12.4 1.13 M
12.5 (a) 200 mL (b) 0.120 M
12.6 10.85%
12.7 0.45 g
12.10 4242 kJ/mol
12.11 0.719, 0.281
12.12 (a) 322.7 torr
12.13 1.30 m
12.14 0.261 M
12.15 4.95°C, 173.4°C
12.16 1.70, 2
12.17 0.319 m
12.18 7.7 atm
12.19 (a) 338 g/mol ($C_{12}H_{22}O_{11}$, 342 g/mol) (b) 2.69 atm
12.20 (a) 25.0 g (b) 245 g
12.21 (a) 1.03 M (b) 3.26 M (c) 0.0681 M
12.22 1.10 g/mL
12.23 (a) 63.3 mL (b) 50.1 mL (c) 109.5 mL (d) 3.9 mL less
12.24 1.207 M
12.25 0.0999 M
12.26 2.79%
12.27 (a) 5.99% (b) 16.2%
12.28 (a) 16.7% (b) 1.10 g/mL
12.29 0.250 M
12.30 16.7 mL
12.31 12.5 mL
12.32 400 mL
12.33 (a) 28% (b) 15 M (c) 13 mL
12.34 1.018 M
12.35 (a) 19.2% (b) 2.92 M
12.36 (a) 1.04×10^{-6} mol/L torr (b) 7.88×10^{-3} M
12.37 (a) 4.04×10^{-2} g/100 g (b) 3.09×10^{-2} mL O_2/mL H_2O
12.38 1×10^{-5} M
12.39 (a) 1.35×10^{-2} M (b) 0.022%
12.48 (a) 101 g (b) 60 g, 41 g (c) 50.2%, 37.5%
12.49 888 g
12.50 (b) 34.7 g
12.51 −2339 kJ/mol
12.52 (a) 11.0% (b) 11.6 mol%
12.53 59.6 g $C_2H_6O_2$, 40.4 g H_2O
12.54 (a) 6.5×10^{-2} (b) 2.1%
12.55 (a) 96.2 torr
12.56 71 torr
12.57 256 torr
12.58 81%
12.60 (a) 1.80×10^{-3} (b) 4.28×10^{-2} torr
12.61 (a) 0.178 m (b) 0.183 m (c) 1.87 m
12.62 (a) 3.33 m (b) 0.0566
12.63 0.864%
12.64 (a) 0.26 (b) 20 m
12.65 1.05 m
12.66 (a) 8.27% (b) 0.481 m
12.67 (a) 11.9% (b) 18.7% (c) 0.830 M
12.68 (a) 3.6×10^{-5} M (b) 4.6×10^{-5} m
12.69 (a) 18 M (b) 2.4×10^2 m
12.71 (a) 0.561 m, 1.133 m (b) 0.544 M, 1.069 M (c) 2.02 (d) 1.96
12.72 (a) 17.9 m (b) −33°C (experimental, −40°)
12.73 (a) 172 g, 334 g
12.74 7.0°C kg/mol
12.75 (a) −0.056°C (b) −0.074°C
12.76 3
12.77 1.76, 2
12.78 (a) 2.06
12.79 (a) 8.00×10^{-2} torr (b) −0.348°C (c) 100.0959°C (d) 4.44 atm
12.81 (a) 0.040 mol/L (b) 1.4%
12.82 (a) −2.1°C (b) 27 atm
12.83 256 g/mol, S_8
12.84 1.7×10^2 g/mol

12.85 (a) 4.9×10^4 g/mol (b) -3.8×10^{-4} °C
12.86 (a) 6.5×10^5 g/mol (b) 2.9×10^{-5} °C
12.87 115, 62 g/mol
12.91 (a) 5.24×10^5 nm^3, 3.14×10^4 nm^2
(b) 1.91×10^{15}, 60 m^2
12.95 (a) 18.0 mmol/L (b) 21 kg/month
12.96 (a) 0.2060 M (b) 0.04120 M
12.97 14.3 M
12.98 (a) 0.29 mol/L (b) 1.3 L N_2O/L solution
12.99 (a) $(C_2H_5)_2O$ (b) $(C_2H_5)_2CO$
12.100 (a) 0.394, 0.606 (b) 3.12 torr
12.101 (a) 0.144 *m* (b) -0.268°C
12.102 (a) 0.0153 *m* (b) 3.3 (c) 4; $3K^+ + Fe(CN)_6^{3-}$
12.103 (a) 125 g/mol; actual value, 113 g/mol
($C_4H_7N_3O$) (b) 1.81 atm
12.104 (a) hydrocarbon, starch (b) hydrocarbon

CHAPTER 13

13.1 (a) 5.0×10^{-4}
13.2 53
13.3 9.1×10^{-4}
13.4 53
13.5 (a) 2.3×10^{21} (b) 4.6×10^{-6}
13.6 2.7×10^{33}
13.10 3.2×10^{20}
13.12 (a) 7.9×10^{-5} mol H_2, 1.81×10^{-3} mol HI
(b) 88
13.13 (a) 0.348 atm, 0.152 atm (b) 2.95
13.14 (a) 0.035 mol C_6H_{10}/L, 0.015 mol $C_6H_{10}I_2$/L
(b) 0.082
13.15 0.95 atm
13.16 1.55×10^{-3} mol/L, 1.55×10^{-3} mol/L,
8.9×10^{-3} mol/L
13.17 1.0 atm, 1.3×10^{-4} atm
13.18 (a) 0.58 (b) 1.33 atm CO, 1.03 atm CO_2
(c) 2.20 atm CO, 1.40 atm CO_2, 3.60 atm total
(d) 1.69 atm CO, 1.66 atm CO_2, 3.35 atm total
13.19 $[C_5H_{10}] = 0.034$ mol/L,
$[CH_3COOH] = 0.004$ mol/L,
$[CH_3COOC_5H_{11}] = 0.016$ mol/L
13.21 (a) 54.1, 51.6, 48.1; ave. 51.3, ave. dev. 2.1
(4.1%)
13.22 (b) 3.1×10^{-4} mol
13.23 (a) 254, 364, 367, 481, 530; ave. dev. 85 (21%)
13.27 4.32×10^{-3}
13.28 (a) 7.9×10^{11} (b) 8.9×10^5 (c) 1.3×10^{-12}
(d) 1.6×10^{-24}
13.29 2.4
13.30 (a) 6.88×10^{-6} (b) 4.63×10^{-2}
13.31 1.3×10^{-7}
13.32 13.7
13.34 (a) 24.2 (b) 0.346
13.36 (a) 4.4×10^{-22} (b) 0.20
13.37 (a) 0.254 (b) 1.36×10^{-7}
13.39 (a) 1.6×10^{19}
13.40 ave. 1.82, ave. dev. 0.08 (4%)
13.42 10^{-3} atm
13.43 (a) 0.0178 mol/L (b) 0.0289 mol/L (c) 0.53
13.44 (a) 7.7×10^{-5} mol H_2, 1.8×10^{-3} mol HI (b) 89
13.45 0.54
13.46 (a) 0.011 (b) 0.54
13.47 1.6×10^{-9}
13.48 3.5
13.49 19 atm
13.50 0.028 atm
13.51 1.4×10^{-7} atm
13.52 (a) 7.28×10^{-2} mol (b) 16.7%
13.53 (b) 4.0×10^2 (c) 0.50%
13.54 (a) 2.36×10^{-3} mol (b) 8.68×10^{-8}
13.55 (a) 5.0×10^{-3} atm
13.56 (a) 8.28×10^{-4} mol/L (c) 1.46×10^{-4} mol/L,
6.82×10^{-4} mol/L, 82%
13.57 (a) 0.011 mol/L (b) 2.9×10^{-3} mol PCl_3/L,
1.2×10^{-2} mol Cl_2/L, 1.6×10^{-3} mol PCl_5/L
13.58 (a) 0.862 atm (b) 0.806 atm CO, 1.306 atm Br_2
(c) 1.55 atm
13.59 6.0×10^{-4} atm
13.60 1.3×10^{-10} atm
13.61 (a) 33.2%
13.62 (a) 0.022 mol/L (b) 7.0×10^{-3} mol/L
13.63 (a) 0.062 (b) 11%
13.64 (a) 0.293 mol/L (b) 4.28×10^{-3} mol NO_2/L,
0.291 mol N_2O_4/L
13.65 95%
13.66 (a) 0.019 (b) 43%
13.67 1.1×10^{15}, 1.4×10^{17}
13.68 3.8×10^{33}
13.69 (a) decrease (b) increase (c) increase (d) decrease
(e) increase (f) increase (g) no change (h) increase
(i) increase (j) no change (k) no change
13.70 (a) 1.4×10^{-5} (b) 5.6×10^{-2}
13.71 2.0%
13.72 1.26×10^{-5} mol/L
13.73 (a) 8.5×10^{-6} (b) 0.22%

CHAPTER 14

14.70 (a) (1) $\underset{\text{base}_1}{HCO_3^-} + \underset{\text{acid}_2}{H_3O^+} \longrightarrow \underset{\text{acid}_1}{H_2CO_3} + \underset{\text{base}_2}{H_2O}$

(2) $\underset{\text{acid}_1}{HCO_3^-} + \underset{\text{base}_2}{OH^-} \longrightarrow \underset{\text{base}_1}{CO_3^{2-}} + \underset{\text{acid}_2}{H_2O}$

(b) amphoteric
14.71 (a) (1) $H_2PO_4^-$ (2) NH_3 (3) O^{2-} (4) H_2SO_4
(5) PH_4^+ (6) HPO_4^{2-} (b) HNO_3
14.72 (a) $2HCN \longrightarrow H_2CN^+ + CN^-$
(b) $H_2CN^+ + CN^- \longrightarrow 2HCN$
(c) One is the reverse of the other.
14.73 (a) weaker than (b) stronger than
14.74 (a) $Sn(OH)_2(c) + 2H_3O^+ \longrightarrow Sn^{2+}(aq) + 4H_2O$
(b) $Sn(OH)_2(c) + 2OH^- \longrightarrow SnO_2^{2-}(aq) + 2H_2O$

14.75 (a) $F^- < OH^- < NH_2^- < CH_3^-$
(b) $H_2S < HS^- < S^{2-}$
(c) $HSO_4^- < HSeO_4^- < HTeO_4^-$
(d) $ClO_4^- < HSO_4^- < H_2PO_4^- < H_3SiO_4^-$

14.76 HA, HA, No

14.77 periodic acid, periodate; hypoiodous acid, hypoiodite; iodous acid, iodite; iodic acid, iodate

14.78 Lewis acids: a, b, e, i, j, m; Lewis bases: d, k; both: f, g; neither: c, h, l

14.79 (a) 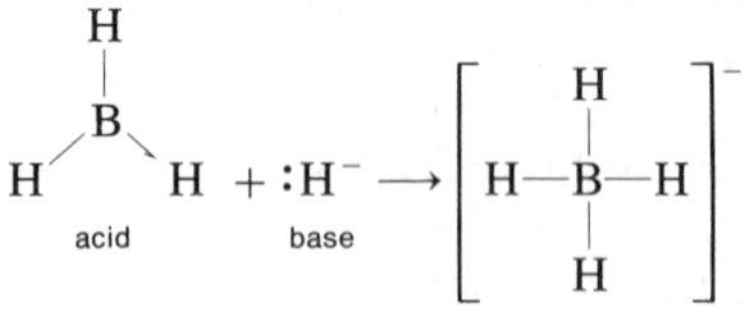 BH_3 (H—B(H)—H) + $:H^-$ ⟶ $[H\text{—}B(H)_2\text{—}H]^-$
acid base

(B atom has only $6e^-$.)

(b) 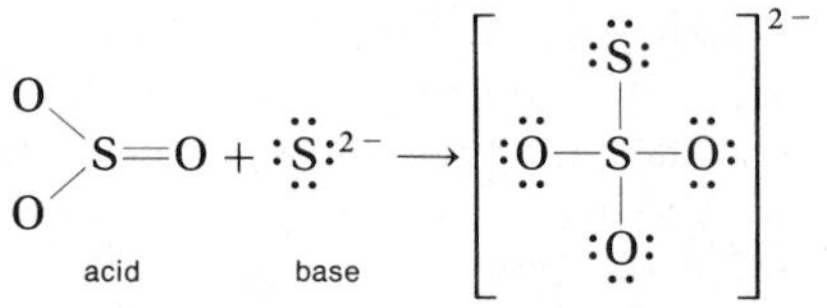 SO_2 (O=S=O) + $:\ddot{S}:^{2-}$ ⟶ $[:\ddot{O}\text{—}S(\text{—}\ddot{S}:)(\text{—}\ddot{O}:)\text{—}\ddot{O}:]^{2-}$
acid base

(multiple bond between atoms of dissimilar electronegativities)

(c) $Au^+ + 2:C{\equiv}N:^- \longrightarrow [:N{\equiv}C\text{—}Au\text{—}C{\equiv}N:]^+$
acid base

(Au^+ is a transition element, does not have a noble gas configuration

(d) 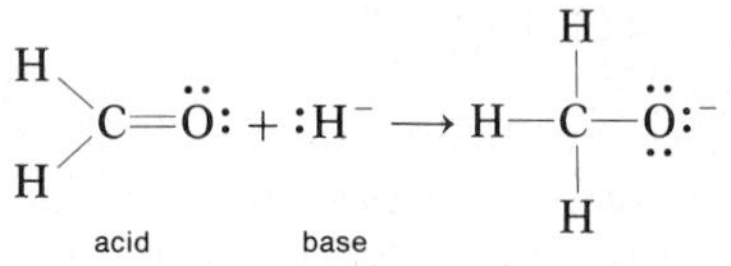 $H_2C{=}\ddot{O}:$ + $:H^-$ ⟶ $H\text{—}C(H)(H)\text{—}\ddot{O}:^-$
acid base

(same reason as for b)

(e) 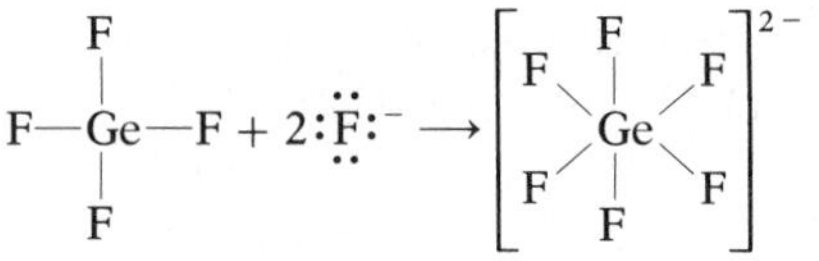 $F\text{—}Ge(F)(F)\text{—}F + 2:\ddot{F}:^- \longrightarrow [GeF_6]^{2-}$

(Ge can acquire more than $8e^-$.)

14.80 (a) HSe^-; Se is larger. (b) Br^-; it has more unshared pairs of electrons. (c) HCO_2^-; it has more extended π bonding. (d) PO_4^{3-}; the P—O bond is shorter and there is more effective delocalized π bonding. (e) BrO_4^-; more lone O atoms provide more delocalized π bonding.

CHAPTER 15

15.1 (a) 9.47 (b) 2.4×10^{-4} mol/L
15.2 5.9×10^{-7}, 1.7×10^{-8} mol/L
15.3 1.0×10^{-7}, 1.1×10^{-8} mol/L
15.4 9.307
15.5 (a) 13.686 (b) 9.69
15.6 (a) 3.30 (b) 11.48
15.7 0.032 M
15.8 1.2×10^{-2} mol/L, 1.93
15.9 1.02×10^{-3} mol/L, 2.99
15.10 3.20, 2.1%
15.11 8.683
15.12 10.30, 5.0×10^{-11}
15.13 2.12×10^{-11}, 10.673
15.14 3.17
15.15 11.43
15.16 1.1×10^{-8}
15.17 1.16, 8.6×10^{-7} mol/L
15.18 (a) 1, 1.6×10^{-7} (b) 13, 1.6×10^{-7} mol/L
15.19 4.65
15.20 6.7×10^{-6} mol/L
15.21 2.1×10^{-8} mol/L, 3×10^{13}
15.22 (a) 1.60×10^{-7}, 5.9×10^{-13}
(b) 3.1×10^{-19} mol/L
15.23 5.28
15.24 (b) 2.86
15.25 5.6
15.26 (a) 9.03 (b) 9.47
15.27 9.55
15.28 (a) 1.85×10^{-11} (b) 5.92
15.29 9
15.30 6.35
15.31 (a) 6.77 (b) 5.57 (c) 4.36
15.32 (a) 1.44×10^{-7} mol/L (b) 6.843
15.33 (a) 1.00×10^{-8}, 6.000
(b) 2.0×10^{-6}, 5.6×10^{-10}
(c) 1.0×10^{-7}, 5.4×10^{-7}
(d) 4.00×10^{-6}, 5.398
15.34 7.435
15.35 (a) 3.2×10^{-17} mol/L (b) 1.0×10^{-30} mol/L
15.36 4.00
15.37 10.30
15.38 7.00
15.39 6.3×10^{-4} M
15.40 5.29
15.41 8.9×10^{-4} mol/L
15.42 4.92
15.43 1.85
15.44 12.18
15.45 (a) 1.88×10^{-3} mol/L (b) 1.71×10^{-3} mol/L
15.46 (c) less than 2.5×10^{-3}
15.47 (a) 3.29×10^{-2} mol/L (b) 1.1×10^{-5} mol/L
(c) 5.66
15.48 1.5×10^{-5}, 4.84
15.49 0.058 mol/L
15.50 (a) 1.6×10^{-3} M (b) 16 mL
15.52 (a) 3 (b) 1 (c) 0.3
15.53 (a) between 4.4 and 5.2
15.54 8.97
15.55 (a) 2.688 (b) 4.88×10^{-12}, 11.312
15.56 (a) 1.9×10^{-6} (b) 5.2×10^{-9}
15.57 (a) 2.34×10^{-5} (b) 2.82

15.58 8.62
15.59 9.00
15.60 (a) 10^{-8} (b) 1.00×10^{-7}
15.62 2.6×10^{-9}
15.63 (a) 1.30 (b) 6.24×10^{-8} mol/L
15.64 3.04
15.65 (a and b) 2.39×10^{-2} mol/L
(c) 6.23×10^{-8} mol/L (d) 1.3×10^{-18} mol/L
15.66 3.00
15.67 1.7×10^{-2}
15.68 4.4×10^{-4} mol/L
15.69 (a) 6.9×10^{-6} mol/L (b) 3.4×10^{-11} mol/L
15.70 7.9×10^{-20} mol/L
15.71 (a) 1.0×10^{-4} mol/L (b) 4.7×10^{-8} mol/L
(c) 1.1×10^{-4} mol/L (d) 10
15.72 (a) 3.66 (b) 3.56
15.73 (a) 9.35 (b) 9.25 (c) 1.0
15.74 (b) 4.42 (c) 4.10
15.75 0.19 mol/L
15.76 1.8
15.77 6.44×10^{-2}, 3.56×10^{-2} mol/L
15.78 0.066, 0.13 mol/L
15.79 (a) 0.30 mol/L
15.80 (a) chloroacetic acid (b) 0.90
15.81 (a) 2.79 (b) 8.94 (c) 4.57
15.82 0.9
15.83 (a) 1.00 (b) 3.2×10^{-5}
15.84 5.00
15.85 (a) -0.30 (b) $+0.30$
15.87 737 mL, 263 mL
15.88 2.81
15.90 8.94
15.91 6.23
15.92 (a) 2.58 (b) 3.45 (c) 5.75 (d) 7.84
(e) 9.92
15.93 5.14, 7.2×10^{-6}
15.94 5.0×10^{-6} mol/L, 5.30
15.95 (a) 12.00 (b) 1.80 (c) 8.93 (d) 11.53
(e) 5.43
15.96 methyl red, bromcresol purple
15.97 (a) 2.43 (b) 2.47×10^{-6} mol/L
15.98 4.95
15.99 3.2×10^{-5} mol/L
15.100 (a) 1.41 (b) 1.83
15.101 1.6
15.102 7.93, cresol red

CHAPTER 16

16.1 (a) 3.43×10^{-4} mol/L (b) 4.0×10^{-9} mol/L
16.2 2.1×10^{-14} mol/L
16.3 1.6×10^{-6}
16.4 (a) 1.25×10^{-3}, 1.5×10^{-13} mol/L
(b) 1.25×10^{-4} mol
16.5 4.1×10^{-4} mol/L
16.6 (a) 1.3×10^{-4} mol/L (b) 1.3×10^{-2} mol/L
(c) 9.3×10^{-7} mol/L, 9.3×10^{-3}%.
16.7 2.1
16.9 1.8
16.10 0.13 mol/L
16.11 0.500, 0.10 mol/L (b) 2.5×10^{-15} mol/L
16.12 7.56×10^{-3} mol/L
16.13 4.3×10^{-5} mol/L
16.14 (a) 1.3×10^{-4} mol/L (b) 1.8×10^{-7} mol/L
16.15 1.5×10^{-5}
16.16 (a) 1×10^{-12} (b) 2×10^{-8} g/100 mL
16.17 (a) 2.4×10^{-5} mol/L (b) 8.2×10^{-22}
(c) 3.2×10^{-5} g/100 mL
16.18 6×10^{-7}
16.20 (a) 7.5×10^{-4}, 3.1×10^{-2} mol/L
16.21 (a) 0.0250, 2.13×10^{-11} mol/L
(b) 3.00×10^{-3} mol
16.22 6×10^{-3}%
16.23 (a) 3.0×10^{-9} mol/L (b) 1.1×10^{-8} mol/L
(c) 0.010 mol/L, 52%
16.24 (a) 6.7×10^{-5}, 1.5×10^{-5} mol/L
16.25 (b) 2.0×10^{-15} mol/L (c) 1.8×10^{-7} mol/L
(d) 0.10, 1.1×10^{-9} mol/L
16.26 (a) 8×10^{-27} mol/L (b) 10^{-2} ion/L
16.27 2.6
16.28 (a) 3×10^{-3} mol/L (b) 3×10^{-15} mol/L
16.29 (a) 0.3 (b) 1.5 (c) 0.3 to 1.5
16.30 (a) 2.4×10^{-4} g/100 mL
16.31 (b) 4×10^{-6} mol/L
(c) 1×10^{-16}, 6×10^{-10}, 4×10^{-6} mol/L
(d) 1×10^{-10}, 6×10^{-7}, 5×10^{-6} mol/L
16.32 3×10^{-4} mol/L
16.33 1.4×10^{-22} mol/L
16.34 6×10^{-4} mol/L
16.35 0.25, 0.50, 4.5×10^{-8} mol/L
16.36 1.9×10^{-15} mol/L
16.37 (a) 0.020 mol/L (b) 0.010 mol/L
(c) 2×10^{-26} mol/L
16.38 (a) 4.5×10^{-4} (b) 9.0×10^{-6}, 0.0200 mol/L
16.39 (a) 2.1×10^{-7} mol/L (b) 9.2×10^{-13} mol/L
16.40 (a) 2.5×10^{-5}, 7.5×10^{-5} mol/L (b) yes
16.41 (a) 1.7×10^{-15} mol/L (b) 1.1×10^{-11} mol/L
(c) 7.7×10^{-6} mol/L, 1.5×10^{-2}%
16.42 5.44
16.43 0.020, 0.020, 2.5×10^{-17} mol/L

CHAPTER 17

17.1 (a) 3 (b) 2
17.2 91.0 g
17.3 (a) 0.107 faraday (b) 1.03×10^4 C (c) 1.20 L
17.5 (a) 83 kJ (b) 1.3 kJ/g
17.7 1.280 V
17.8 (b) 0.026 V
17.12 (a) 1.215 V (b) 1.181 V
17.13 0.6963 V
17.14 6.06
17.15 (a) 2×10^{106} (b) 10^{97} (c) 10^{-22}
17.17 (a) 1.83 V (b) 2.53 V

17.18 22.0 g
17.19 (a) 7.98 h (b) 0.714 g
17.20 (a) 1.78×10^{-2} faraday (b) 1.92 g, 4.31 g
17.21 (a) 3.56 g (b) 70.3%, 29.7%
17.22 (a) 4.50×10^5 C (b) 308 L
17.23 (b) 1.1×10^7 C
17.24 (b) 1.0×10^2 C/cm^2
17.25 (b) 894 C/g
17.26 0.99996
17.28 (b) 1.8×10^9 C
17.29 (c) 2.41×10^7 C
17.30 (b) 1.6 g
17.31 Rh^{3+}
17.32 (a) 38.26 g/faraday (b) 38.26, 76.52, 114.8 g/mol
17.33 (a) 12 J (b) 0.12 W (c) 116 kJ
17.35 (b) (1) 0.9258 V (2) −0.989 V (3) −1.204 V (4) 0.715 V (5) −1.1360 V
17.36 (a) −0.2655 V, +0.9371 V (b) 1.2026 V, 1.2026 V
17.40 (1) (b) 1.2973 V (c) 1.368 V (2) (b) 2.714 V (c) 2.732 V
17.41 (a) 0.0713 V (b) 0.154 V (c) 6.2×10^{-6} atm
17.42 (a) 1.229 V (b), (c) 1.291 V
17.43 1.0 V
17.44 (a) 0.06412 V faraday/mol (b) 0.3132 V
17.45 (c) 1
17.46 (a) 0.079 V (b) 4.51
17.47 (b) 0.056 mol/L
17.48 (a) 0.1522 V (b) 0.13 M
17.50 (1) (a) 2.115 V (b) 3.2×10^{71} (2) (a) 1.066 V (b) 1.1×10^{36}
17.51 (a) 8.2×10^{-17} (b) 2×10^{-33}
17.52 (a) 6.2×10^{-2} (b) 3.0×10^{-14} (c) 4.9×10^{-13}
17.54 (d) 0.03 M
17.56 7 × 10
17.57 (a) 0.592 V (b) 0.296 V
17.58 −0.185 V
17.59 (a) 1.062 V (b) −0.720 V
17.60 (c) 9.29
17.61 (b) 7.83
17.63 (a) 1.31 V (b) 1.48 V
18.64 (a) 0.34 V (b) 0.94 V
17.68 43 g
17.70 (a) 2.050 V (b) 2.129 V (c) 2.017 V
17.74 (a) 0.19 kg/(m^{2}h)
17.78 (b) 42 years
17.79 (a) 5.39×10^{-3} mol (b) 0.172 g (c) 0.121 L (d) 2.08×10^3 C (e) 0.578 A
17.80 (1) (a) $Zn \longrightarrow Zn^{2+} + 2e^-$, $Ni^{2+} + 2e^- \longrightarrow Ni$
(b) $Zn + Ni^{2+} \longrightarrow Zn^{2+} + Ni$
(c) 0.505 V (d) as written
(2) (a) $Al \longrightarrow Al^{3+} + 3e^-$, $Zn^{2+} + 2e^- \longrightarrow Zn$
(b) $2Al + 3Zn^{2+} \longrightarrow 2Al^{3+} + 3Zn$
(c) 0.900 V (d) as written
(3) (a) $Cd \longrightarrow Cd^{2+} + 2e^-$, $Fe^{3+} + e^- \longrightarrow Fe^{2+}$
(b) $Cd + 2Fe^{3+} \longrightarrow Cd^{2+} + 2Fe^{2+}$
(c) 1.174 V (d) as written
17.81 (a) 0.607 V (b) 0.625 V
17.82 (a) $Cu^{2+} + 2e^- \longrightarrow Cu$,
$Cu^{2+} + 2Ag + 2Cl^- \longrightarrow Cu + 2AgCl$

(b) $\mathscr{E} = \mathscr{E}^0 - \frac{0.0592}{2}\log\frac{1}{[Cu^{2+}][Cl^-]^2}$

$= \mathscr{E}^0 + 0.0296 \log (c_3 c_2{}^2)$

(c) 0.0266 mol/L
17.83 (a) 1×10^{29} (b) 1.2×10^{68}
17.84 (a) $\mathscr{E} = 0.38$ V, yes (b) $\mathscr{E} = -0.04$ V, no

CHAPTER 18

18.1 (a) +400 J (b) +1400 J
18.3 (a) any number ⩾5 kJ (6 kJ, 7 kJ, etc.)
(b) −5 kJ
(c) any number ⩾ −5 kJ (−4 kJ, −3 kJ, etc.)
18.4 280 kJ
18.5 −109.61 kJ
18.6 −143.92 kJ
18.7 91.12 kJ, 1.1×10^{-16}
18.8 −200.3 kJ
18.9 0.030, 40.6 kJ
18.10 (a) 4.77 kJ/mol
18.12 600.26 kJ, 497.9 kJ, 343.3 J/K
18.14 257 kJ (experimental: 249 kJ)
18.15 (a) −3.80 kJ/mol (b) 0 (c) −16°C (experimental: −10°C)
18.16 405 J
18.17 (a) −3000 J (b) −144°C
18.18 (a) 10.0 kJ (b) −115.0 kJ (c) 0
18.19 (a) 15.0 kJ (b) 8.0 J
18.20 (a) −150 J, +200 J, +50 J (b) +270 J
18.22 (a) $w_{min} = -1610$ kJ
(b) $w = -1600$ kJ, −1000 kJ, . . . (any number ≥ −1610 kJ); work done by system = 1600 kJ, 1000 kJ, . . . (any number ⩽ 1610 kJ)
(c) −1610 kJ
18.23 (a) −890.36 kJ, −817.90 kJ (b) 1.0596 V (c) 91.9%
18.24 −66.35 kJ/mol
18.25 −0.2694 V
18.27 −226 kJ, −2877 kJ
18.28 (a) 3.3×10^{-4}
18.29 1.8
18.30 (a) −167.73 kJ (b) 0.86921 V (c) 2.4×10^{29} (both)
18.31 (1) (a) 0.146 V (b) −28.2 kJ (c) 8.6×10^4
(2) (a) 1.066 V (b) −205.7 kJ (c) 1.1×10^{36}
(3) (a) 1.645 V (b) −1587 kJ (c) 1×10^{278}
18.35 20.07 J/K
18.36 0.252 J/C liberated
18.37 (a) −103.71 kJ, −110.01 kJ, 21.1 J/K, 1.9×10^{19}
(b) −114 kJ, 10^{12}
18.38 (a) 172.459 kJ, 120.023 kJ, 175.87 J/K
(b) −3.4 kJ, 1.5 (experimental: 3.7)

18.39 (a) 57.20 kJ, 4.75 kJ, 175.9 J/K (b) 0.147 (c) −48.3 kJ, 1.6×10^4

18.40 (a) −62.10 kJ, −22.40 kJ, −133.2 J/K, 8.40×10^3 (b) 1.19×10^{-4} atm (c) 466 K

18.41 (a) 31 J/K (b) −110 kJ

18.42 (a) −1.895 kJ, − 2.900 kJ, 3.371 J/K

18.43 (a) 285.4 kJ, 326.4 kJ, 6.5×10^{-58}, −138 J/K

18.45 (a) −197.78 kJ, −141.73 kJ, −188.0 J/K (b) −9.8 kJ, 3.2, 2.7×10^2 (c) 1052 K

18.48 (c) −190.6 kJ, 0.6354 kJ/K; −191.6 kJ, 0.5475 kJ/K

18.49 (a) −283.5 kJ (b) −113 J/K (c) 1.43 V

18.50 (a) −190 kJ (b) 0.99 V

18.51 (a) 141 J (b) −120 J (c) 96°C

18.53 (a) 0 (b) 26 kJ/mol (c) 109.0 J/(mol K)

18.54 (b) 0.21 kJ/mol (c) 105.3°C (d) 0.70 atm (experimental: 0.692 atm)

18.55 (a) −5708 J (b) +5708 J, absorbed

18.56 (a) 768 kJ (b) $w = 800$ kJ, 1000 kJ, etc. (any number $\geqslant$768 kJ); *on* the system

18.57 (a) 1.649 V (b) 0.95 kWh

18.58 −210.38 kJ/mol

18.59 (a) 41.166 kJ, 28.619 kJ, 42.083 J/K (b) 0.21 kJ, 0.97 (experimental: 0.62)

18.60 (a) −196.10 kJ, −233.56 kJ, 125.6 J/K (b) no

18.61 (a) 98.83 J/K (b) (1) 125°C, 1 torr (2) 347°C, 1 atm (experimental: (1) 126.2°C (2) 356.6°C)

CHAPTER 19

19.2 (b) 8.40×10^{-5} L/(mol sec) (c) 2.02×10^{-8} mol/(L sec)

19.6 0.0113 L/(mol min)

19.7 (a) 0.250 g, 0.0947 g (b) 60.6 min

19.8 2.08×10^{-3} sec^{-1}

19.9 1×10^{-3} sec^{-1}

19.10 0.314 sec^{-1}

19.20 (b) 25 L^2/(mol^2 sec) (c) 1.3×10^{-3} mol/(L sec) (d) 1.0×10^{-3} mol/(L sec)

19.21 (b) 0.280 L/(μmol sec) (c) 0.252 μmol/(L sec)

19.22 (b) 6.2×10^{-4} sec^{-1}

19.23 (a) 2.7×10^{-5} mol/(L sec) (b) 1.4×10^{-5} mol/(L sec)

19.24 (b) 1.71×10^6 L/(mol sec)

19.26 9.2×10^3 L/(mol h)

19.27 (a) $k = 5.46 \times 10^{-3}$ mmol/(L sec) (b) 0.55 mmol/L

19.28 5.57×10^{-3} min^{-1}, 9.28×10^{-5} sec^{-1}

19.29 4.10×10^{-2} h^{-1}

19.30 2.94 sec

19.31 (b) 0.0112 L/(mol min) (c) 3.36×10^{-4} mol/(L min)

19.32 3.67 days

19.33 9.77×10^{-4}

19.34 1.39/k

19.35 4.48×10^{-3} mol

19.36 147 min

19.37 (a) 0.59 mg (b) 75 h

19.41 (a) −648 kJ (b) +561 kJ

19.44 (a) (1) −37 kJ (2) −214 kJ (3) −251 kJ (b) +20 kJ

19.47 0.125 mL/(mol sec)

19.48 52.0 kJ/mol, 7.725, 0.0983 min^{-1}

19.49 (a) 1.9, 13 (b) 2.6

19.50 0.275 min^{-1}

19.51 52.9 kJ/mol

19.52 303 min

19.64 8×10^2 kg/(L h)

19.65 5.2×10^7

19.68 (b) 2×10^{-21} mol HBr, 10^{277} mol H_2O

19.71 7.9×10^{-17}

19.72 1.2×10^{-5}

19.73 5.8×10^{-9}

19.74 (a) *rate* = $k[NO]^2[H_2]$ (b) 6.32 L^2/(mol^2 sec) (c) 0.380 mol/(L sec)

19.75 0.0127 min^{-1}, 0.44 mmol/(L min)

19.76 (a) (1) 0.250 g (2) 0.144 g (b) 43.9 min (c) 0.0156 remains, 0.984 reacted

19.77 (Lengths are in nm and not to scale.)

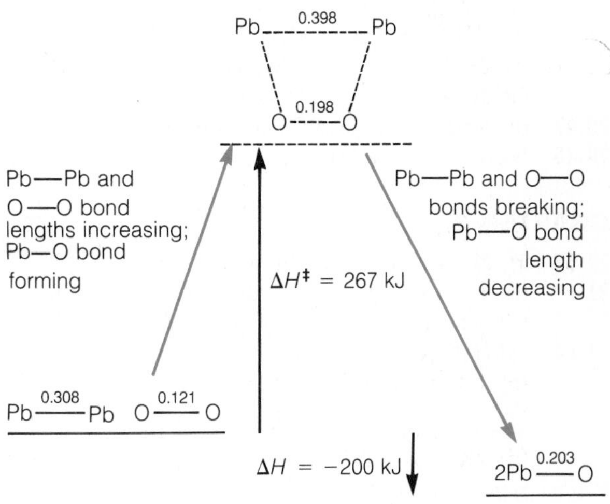

19.78 $E_a = 0$ means that all collisions lead to products. $k = 3.2 \times 10^{13}$ at 20°C and 60°C.

19.79 (a) $2NO + Br_2 \longrightarrow 2BrNO$ (b) (1) termolecular (2 and 3) bimolecular, bimolecular (c) (2) Br_2NO (3) N_2O_2

19.80 (a) second (b) (1) *rate* = $k_2K_1[NO_2]^2[O_3]$ (incorrect) (2) *rate* = $k_1[NO_2][O_3]$ (correct)

19.81 (a) $1\frac{1}{2}$ (b) yes, *rate* = $k_2K_1^{1/2}[CHCl_3][Br_2]^{1/2}$ (c) step 1, steps 2 and 3, and

$$Br\cdot + Br\cdot \longrightarrow Br_2$$

$$\cdot CCl_3 + \cdot CCl_3 \longrightarrow C_2Cl_6$$

$$\cdot CCl_3 + \cdot Br \longrightarrow CCl_3Br$$

19.82 ($rate = k_3K_1^{1/2}K_2[X][HA]^{1/2}$) No, a stronger acid means a larger K_1; a larger K_1 increases the measured rate constant ($= k_3K_1^{1/2}K_2$).
19.83 reaction a

CHAPTER 20

20.6 (b) 4/3
20.42 acidic: b, e; basic: a, c; amphoteric: d
20.43 (a) $HI + H_2O \longrightarrow H_3O^+ + I^-$
(b) no reaction
(c) $LiH + H_2O \longrightarrow LiOH + H_2$
(d) $(CH_3)_3N + H_2O \longrightarrow (CH_3)_3NH^+ + OH^-$
20.44 H_2SeO_3, As_2O_5, SiO_2, HIO_4
20.45 $Mg(OH)_2$, Rb_2O, LiOH, SrO
20.46 (a) $2CO + O_2 \longrightarrow 2CO_2$
(b) $C + H_2O \longrightarrow CO + H_2$
(c) $NH_4NO_3 \longrightarrow 2H_2O + N_2O$
(d) $2NO_2 \longrightarrow N_2O_4$
(e) $3Cu + 8HNO_3 \longrightarrow 3Cu(NO_3)_2 + 2NO + 4H_2O$
(f) $AlH_3 + LiH \longrightarrow LiAlH_4$
(g) $SO_2 + 2H_2S \longrightarrow 2H_2O + 3S$
(h) $PCl_5 + 4H_2O \longrightarrow H_3PO_4 + 5HCl$
(i) $2HCl + CuS \longrightarrow H_2S + CuCl_2$ (for example)
(j) $2H_2S + 3O_2 \longrightarrow 2H_2O + 2SO_2$
20.47 (a) $NaCl < AlCl_3 < PCl_5$ (b) $AlF_3 < AlCl_3 < AlI_3$
20.48 NCl_5, HeF_2, $N^{3+}(H^-)_3$

CHAPTER 21

21.62 1f, 2i, 3j, 4e, 5d, 6c, 7h, 8g, 9b, 10a
21.63 (a) V (b) VI (c) 3 (d) tetrahedral (e) 2− (f) 6
(g) +5 (h) TiO_2 (i) 2 (j) 1 (k) CrO_3 (l) WO_4^{2-}
21.64 (a) $[PtCl_2Br_2]^{2-}$ (b) II
(c) potassium dibromodichloroplatinate(II)

(d) $2K^+\left[\begin{array}{ccc} Br & & Cl \\ & Pt & \\ Cl & & Br \end{array}\right]^{2-}$ $2K^+\left[\begin{array}{ccc} Br & & Cl \\ & Pt & \\ Br & & Cl \end{array}\right]^{2-}$

trans *cis*

21.65 $Cu(OH)_2(c) + 4NH_3(aq) \longrightarrow Cu(NH_3)_4^{2+} + 2OH^-$
21.66 basic: NiO, Ag_2O; acidic: CrO_3, Re_2O_7
21.67 (a and b) $[Cu(CN)_3^{2-}]$ absorbs the highest-energy photons and has the largest Δ; $[Cu(H_2O)_4]^{2+}$ absorbs the lowest-energy photons and has the smallest Δ.
21.68 (a) $\underline{\uparrow\downarrow}\ \underline{\uparrow\downarrow}\ \underline{\uparrow\downarrow}\ \underline{\uparrow}\ \underline{\ }$, $\underline{\uparrow\downarrow}\ \underline{\uparrow\downarrow}\ \underline{\uparrow}\ \underline{\uparrow}\ \underline{\uparrow}$; unpaired electrons: 1, 3
(b) $\underline{\uparrow\downarrow}\ \underline{\uparrow\downarrow}\ \underline{\uparrow\downarrow}\ \underline{\uparrow\downarrow}\ \underline{\uparrow}$, $\underline{\uparrow\downarrow}\ \underline{\uparrow\downarrow}\ \underline{\uparrow\downarrow}\ \underline{\uparrow\downarrow}\ \underline{\uparrow}$; unpaired electrons 1, 1
21.69 (a) High thermodynamic stability but low kinetic stability (a labile complex) (b) high thermodynamic stability and high kinetic stability (an inert complex)
21.70 (a) $3d\ \underline{\uparrow}\ \underline{\uparrow}\ \underline{\ }$ (b) $3d\ \underline{\uparrow}\ \underline{\uparrow}\ \underline{\ }\ \underline{\ }\ \underline{\ }$

CHAPTER 22

22.47 1b, 2f, 3i, 4e, 5c, 7g, 8d, 9h, 10j
22.48 1b, 2q, 3i, 4r, 5n, 6m, 7c, 8l, 9e, 10h, 11d, 12j, 13g, 14o, 15k, 16f, 17a, 18p
22.49 Ad, Ba, Cd, Da
22.50 (a) $2Bi_2S_3 + 9O_2 \longrightarrow 2Bi_2O_3 + 6SO_2$
(b) $Fe_3O_4 + 4CO \longrightarrow 3Fe + 4CO_2$
(c) $3V_2O_5 + 10Al \longrightarrow 6V + 5Al_2O_3$
(d) $Cd^{2+} + Zn \longrightarrow Cd + Zn^{2+}$
(e) $TiCl_4 + 2Mg \longrightarrow Ti + 2MgCl_2$
22.51 (a) Pb^{2+} (b) $SO_2(g)$ (c) amphoteric (d) O_2 (e) Cr (f) compound alloy (g) interstitial (h) SiO_2
22.52 (a) 6×10^{17} atoms (b) 36×10^{17} AO's
(c) 33×10^{17} molecular orbitals (d) 6×10^{17}
(e) $1s^{12\times10^{17}}\ 2s^{12\times10^{17}}\ 2p^{36\times10^{17}}\ 3s^{6\times10^{17}}$
22.53 (a) +40 kJ, not spontaneous (b) −250 kJ, spontaneous; Al is better.

CHAPTER 23

23.2 -3.9×10^7 kJ/mol
23.3 -6.252×10^9 kJ/mol
23.4 2.0×10^3 yr
23.5 1.11×10^8 yr
23.16 9.9 kg
23.18 5.7×10^8 kJ/mol
23.19 -6.9×10^{-8} g/mol
23.20 -1.282×10^{-3} g/mol
23.21 0.512 Mev/photon, 1.24×10^{20} sec^{-1}
23.22 −1.7280 g/mol
23.23 -1.5835×10^{11} kJ/mol, -1.26×10^{-15} kJ/nucleon
23.24 -1.362×10^{10} kJ/mol
23.25 -1.67×10^9 kJ/mol
23.26 (c) -2.16×10^9 kJ/mol (d) 2.51×10^{10} kJ
23.28 7.01 sec
23.29 9.98×10^{-7} sec^{-1}
23.30 (a) 5.0 g (b) 2.5 g (c) 0.63 g
23.31 (a)(b) 3.13 mg (c) 2.52 mg
23.32 6.1 h
23.33 (b) 1.9×10^8 atoms, 3.1×10^{-16} mol, 1.5×10^{-2} pg
23.34 7×10^{18} d/(g yr)
23.35 4.1×10^3 yr
23.36 3.50×10^6 yr
23.37 4.5 yr
23.38 1.4×10^9 yr
23.39 (b) 1×10^7 yr
23.40 (a) 6.3×10^3 yr, 2.0×10^3 yr
23.41 1×10^5 yr
23.42 1.90×10^4 yr
23.45 5.0 L
23.46 (b) TiF_3
23.50 4.53×10^5 dps/cm^2
23.51 (a) (1) 9.01 rads (b) (1) 18 m (2) 3 h/yr
23.52 39 yr
23.55 (a) (1) 2.7 (2) 1.4 (3) 1 (4) 0.71 (5) 0.38

23.56 a, b, c, e
23.57 (a) $^{239}_{92}U$, $^{137}_{49}In$ (b) $^{127}_{52}TeO_4{}^{2-}$ (c) $^{29}_{15}P$ (d) $^{6}_{2}He$ (e) $^{1}_{0}n$
23.58 7 α and 4 $_{-}\beta$
23.59 (a) 7.1×10^8 kJ/mol (b) 1.26×10^9 kJ/mol; reverse
23.60 1.56 days
23.61 1.10×10^7 yr
23.62 5.81×10^{-6} g/100 mL

CHAPTER 24

24.65 1b, 2g, 3c, 4i, 5j, 6n, 7f, 8m, 9k, 10l, 11a, 12h, 13d, 14e, 15o
24.66 (a) 1 and 4 are identical, 2 and 3 are identical (b) 2, 4, and 5
24.67 (a) (1) alkane, C_nH_{2n+2} (2) $CH_3CH_2CH_2CH_3$ (b) (1) alkene or cycloalkane, C_nH_{2n} (2) alkene, $H_2C{=}CHCH_2CH_3$; cycloalkane, $H_2C{-}CH_2$ / $H_2C{-}CH_2$ (ring) (c) (1) alkyne, alkadiene, or cycloalkene, C_nH_{2n-2} (2) alkyne, $HC{\equiv}CCH_2CH_3$; alkadiene, $H_2C{=}CHCH{=}CH_2$; cycloalkene, $H_2C{-}CH$ / $H_2C{-}CH$ (ring, with C=C between the CH groups)
24.68 (a) alkanes $CH_3(CH_2)_4CH_3$, $(CH_3)_2CHCH_2CH_2CH_3$, $CH_3CH_2CH(CH_3)CH_2CH_3$, $(CH_3)_3CCH_2CH_3$, $(CH_3)_2CHCH(CH_3)_2$ (b) alkenes $CH_2{=}CHCH_2CH_3$, *cis- trans*-$CH_3CH{=}CHCH_3$, $(CH_3)_2C{=}CH_2$ and cycloalkanes $H_2C{-}CH_2$ / $H_2C{-}CH_2$ (ring) and cyclopropane ring CH_2, H_2C, CH bearing CH_3
24.69 (a) *cis-* and *trans*-$CH_3CH{=}CHCH_3$ (b) $CH_3CH_2CH_2Cl$ and $CH_3CHClCH_3$ (structural isomers) (d) enantiomers of $CH_3CH(OH)CH_2CH_3$
24.70 (a) $CH_3CH_2CH_2CH_2OH$, $CH_3CH(OH)CH_2CH_3$, $(CH_3)_2CHCH_2OH$, $(CH_3)_3COH$ (b) $CH_3CH(OH)CH_2CH_3$ (c) $CH_3OCH_2CH_2CH_3$, $CH_3OCH(CH_3)_2$, $CH_3CH_2OCH_2CH_3$ (d) With alcohols, but not with ethers, potassium causes evolution of H_2.
24.71 (a) none (b) *cis-trans* isomers (c) *cis-trans* isomers (d) enantiomers (e) enantiomers
24.72 (a) 2-methylbutane (b) 2,4-dimethylpentane
24.73 (1) substitution (2) elimination (3) addition (4) addition (5) displacement (6) elimination (7) substitution
24.74 (a) $HC({=}O){-}OCH_2CH_3$ (b) $(CH_3)_3NH^+Br^-$ (c) $CH_3CH_2C({=}O){-}NH_2$

CHAPTER 25

25.21 0.10 pg, 1.6 10^{-9} J
25.31 (a) Not a polymer because measurable freezing-point depression indicates low molecular weight. (b) Polymer, because it is a material of high molecular weight (freezing-point depression too low to measure) that yields a single decomposition product. The polymer is polystyrene. (c) Polymer, because it is a hard, insoluble substance synthesized from compounds of low molecular weight. It is a phenol-formaldehyde copolymer. (d) Not a polymer, because its volatility indicates low molecular weight.
25.32 a4, b3, c5, d1, e2
25.33 (a) $H{-}CH_2{-}C^+(CH_3)_2$ [H–C(H)(H)–C⁺(CH₃)(CH₃)], $\left({-}CH_2{-}C(CH_3)_2{-}\right)_n$
(b) $R{-}CH_2{-}\dot{C}Cl_2$ [R–C(H)(H)–C·(Cl)(Cl)], $\left({-}CH_2{-}CCl_2{-}\right)_n$
(c) $R{-}CH_2{-}C{:}^-(CH_3)(C({=}O)OCH_3)$ [R–C(H)(H)–C:⁻(CH₃)(C=O, OCH₃)], $\left({-}CH_2{-}C(CH_3)(COOCH_3){-}\right)_n$
25.34 Reactions a and c could not produce a polymer, because in each instance one of the starting materials has only 1 functional group. Reaction b could, because both starting materials are bifunctional.
25.35 (a) $-CH_2{-}C(CH_3){=}CH{-}CH_2{-}CH(CH_3){-}CH(CH{=}CH_2){-}CH_2{-}C(CH_3){=}CH{-}CH_2-$ (isoprene units outlined)
(b) C_5H_8, C_5H_8 (c) addition polymer
25.36 (a) 2, 3 (b) 1, 4, 5, 6
25.37 (a) nucleotides (b) N (c) O (d) —CGAT— (e) H_2O (f) hydrogen bonding between the bases inside the strands

INDEX

Symbols used: A, Appendix; *f*, figure; *p*, problem; *t*, table.

C

D

E

I

J

K

L

M

N

O

S

T

A 8
B 9
C 0
D 1
E 2
F 3
G 4
H 5
I 6
J 7